PHYSICS FOR THE INQUIRING MIND

The Methods, Nature, and

Philosophy of Physical Science

PHYSICS FOR THE INQUIRING MIND

THE METHODS, NATURE, AND PHILOSOPHY OF PHYSICAL SCIENCE

BY

ERIC M. ROGERS

1960

PRINCETON, NEW JERSEY

PRINCETON UNIVERSITY PRESS

London: Oxford University Press

L. C. Card 59-5603

SBN 691-08016-X

Publication of this book has been aided by grants
from The Rockefeller Foundation,
The Alfred P. Sloan Foundation, and The
Whitney Darrow Publication Reserve Fund
of Princeton University Press

Printed in the United States of America

Second Printing 1961

Third Printing 1961

Fourth Printing 1962

Fifth Printing 1963

Sixth Printing 1964

Seventh Printing 1965

Eighth Printing 1966

PUBLISHER'S NOTE TO THE EIGHTH PRINTING:
In this printing there has been a significant rearrangement and revision of the material in Chapter 30, beginning on page 450 and continuing to the end of the chapter. Various other minor corrections have been made running throughout the text.

Ninth Printing 1968

Tenth Printing 1970

TO JANET TRAN ROGERS

PREFACE

THIS BOOK offers a course in physics to non-physicists who wish to know physics and understand it. Here are the general reading, problems, and laboratory instructions of a one-year course given at Princeton to undergraduates whose chief field of study lies outside technical physics: economists, students in humanities and in life sciences, and many premedicals.* That course is open alike to those who have studied physics before and those who have not. Like that course, this book neither requires a previous physics course nor repeats in material or treatment the normal content of high-school physics—so it welcomes all readers.

This book treats a series of topics intensively: topics chosen to form a coordinated structure of knowledge. Although mathematics provides the essential tools of physics, only the simpler parts of high-school algebra and plane geometry are used here. On the other hand, critical reading, good reasoning and clear thinking are asked for again and again. The problems, which are of primary importance, are not plug-in slots for formulas, but ask for reasoning and critical thinking. In this way, both text and problems ask readers to learn by their own thinking.

A NOTE TO ALL READERS: THE PROBLEMS

The problems are an important part of the book's teaching, because they ask you to discuss and reason and polish up your own knowledge. There is much discussion and reasoning in physics. To *understand* how experimental knowledge is fitted with theory and new results extracted, you need to do your own reasoning and thinking. Of course it would be quicker and easier, for both teacher and student, if the text stated all the results and outlined all the reasoning; but it is hard to remember such teaching for long, and harder still to extract a fine understanding of science from it. So, in this book, many of the problems ask you to do your own thinking; and for that reason they form a very important part of the book's teaching.

Some problems are dissected into a series of steps, not as spoon-feeding but to take the place of worked examples—you should work through those as you read the text.** Some problems give necessary preparation for later chapters. Some problems raise general questions whose discussion can do much to advance your understanding. Such general questions ask for opinions as well as reasoning; and they obviously do not have a single, completely right answer. Yet, thinking your way through them and making your own choice of opinion, and discussing other choices, is part of a good education in science.

A NOTE TO INSTRUCTORS

The Origin of The Course

A dozen years ago, our concern for the good name of science moved some of us to design new courses for non-scientists: courses in physical science for general education in college. In this age of science, educated non-scientists need an understanding knowledge of physics; and they deserve to enjoy that knowledge as part of their intellectual outlook throughout their lives. Bankers, lawyers, business men, and administrators of all kinds have to deal with scientists and their work; and educated people everywhere find scientific knowledge offering to influence their interests, outlook, and philosophy. What kind of physics courses could answer such needs? Not the routine training courses in facts and formulas and principles that were designed for future physicists and engineers and that are still offered as standard fare. To many non-scientists those courses fail to give an appreciative understanding of science—and we even hear doubts whether they give professional scientists the best start. Nor does a smörgåsbord acquaintance-course of items of information meet the need—a course that gives students a temporary sense of satisfaction but cannot convey much lasting understanding.

So we designed a "block-and-gap" course that

* In describing their physics requirements, Deans of medical schools now stress thoroughness of understanding more than completeness of coverage. They ask for a course that teaches its material thoroughly and encourages constructive thinking and careful experimenting. So we now welcome many pre-medical students in this course. To cover some extra ground that is important for them, we add classes in acoustics and a series of laboratory sessions with optical instruments, ranging from the eye to the microscope. Those pre-medicals who prefer a more mathematical or "technical" treatment join physicists and engineers in another course.

** Some of the most important problems, which are intended to teach by a series of steps, have been printed in this book as reduced photographs of typewritten sheets. Instructors or Physics Departments can obtain single sample copies of the full-size original sheets from the author, c/o Palmer Physical Laboratory, Princeton University, Princeton, N.J. These specimens may be reproduced by photo-offset, or they may serve as copy for typewriting. Sets of lithographed copies are also available.

builds important blocks of material into a connected framework. We taught those blocks carefully to give a sense of genuine understanding; we discussed the connections between one block and another; and we tried to show the building of science as a whole. We were teaching both science and philosophy of science—without using that forbidding phrase. There was plenty of solid physics (more than half the content of an orthodox one-year course); our treatment was thorough, (within the limitations of mathematical tools); and we aimed for knowledge and a sense of understanding rather than a wealth of information.

The gaps, where topics were omitted, gave time for careful teaching and for students to learn by reading and thinking things out for themselves; and they gave space and time for a developing perspective of science. We considered the loss of omitted topics unimportant. If such a course succeeds, its students will be well prepared, in both background and attitude, to read more science on their own, to fill any gaps they wish. But if it is to succeed, the course must encourage that depth of learning which comes with each student's own reasoning and creative thinking: the course must ask questions rather than hand out results. That is the kind of course for which this book was made.

The Essential Plan of This Book

To enable students and other readers to understand physics as scientists know it, we must show its connected framework of knowledge and thought. This book tries to do that by linking one chapter with another so that information here leads to commentary there and a fuller explanation later still; so that tools developed in one place are used in others; and, in general, so that knowledge grows as an organized system.

Since relevance to the structure of the course and its teaching is more important than coverage, there is no single ideal choice of topics for a course like this. Some basic elements are universal (such as Newton's Laws of Motion), and some are demanded by most teachers or many students (such as planetary astronomy for a discussion of Theory; nuclear physics to keep up with modern knowledge). Yet there is much room for individual choice, depending on interests of teachers and students, and on equipment available, and on method of teaching. Originally this book contained one man's selection of topics—a workable course but a prejudiced choice. However, to allow for wider choice by other teachers, some chapters have been expanded and others added. These additions raise the danger of overcrowding, if users try to cover all the chapters. On the other hand, they bring the course of the book into line with recent recommendations, such as those of the Carleton Conference.

As assurance that the material is less crowded than the traditional full menu, here is a list of the topics *omitted or treated trivially* in the book: *hydrostatics, statics, calorimetry, ray optics, sound,* and *parts of electricity and magnetism.* Thus, the course is mainly concerned with DYNAMICS, PLANETARY ASTRONOMY, MOLECULAR THEORY, and PARTS OF ELECTRICITY AND MAGNETISM, AND "ATOMIC PHYSICS"—interwoven with general discussion.

In the Princeton course for which this book is used, we treat the chapters as shown opposite. (The broken lines indicate roughly—and optimistically—the quarter-stages in a year's course.)

A Word to Critics

In its method of presenting physical science, this book also makes first moves towards studies of history and philosophy of science. I hope that experts in those fields will pause before condemning ignorance or mistaken judgments in my treatment, and will remember that this is an attempt to teach science itself at first hand.

Historian, philosopher, and scientist: *each* feels that the others are rich in vision but lack some knowledge of *his* field. To the historian, the scientist lacks perspective and accurate knowledge of history; to the philosopher, the scientist lacks critical skill and accurate knowledge of philosophy. To the scientist, the works of philosopher and historian are a great delight; but he finds that they pre-suppose (rather than lack) a full knowledge of scientific material and a *first-hand understanding of the nature of scientific work.* To convey the latter to non-scientists seems to me the essential first move in giving them an understanding of science for use in later life and work—an understanding comparable with the knowledge of music that a good music course conveys to non-musicians.

So I believe the non-scientist needs a course that he himself considers a course *in* science, not a course *about* science. Although the history and philosophy of science are of the essence in understanding science at a sophisticated level, a first study of them in expert hands is apt to strike the stranger as difficult commentary rather than science itself. To *understand* science, each student must be, in his own mind, a "scientist for the day."

I fear it may be as difficult for a philosopher or

historian to immerse the beginner in science itself as it is for a scientist to embrace the historian's perspective or do justice to the philosopher's knowledge. Therefore, I have rashly written as a physicist, without shame but in all humility; and I hope that those others will forgive narrowness and errors and will regard this course as a base on which some students will build the next story in their special fields.

Thanks

A physics book, like physics itself, is a cooperative effort. It cannot be produced in an ivory tower. This book owes much to many people: to colleagues who have offered experiments, problems, and valuable suggestions; and to many scientists of past and present generations on whose teaching I have drawn and whose problems I have borrowed, often unconsciously. To all who have thus contributed, whether knowingly or not, to this work in physics teaching, I give the most heartfelt thanks. I am grateful to them all, from my own first teachers many years ago to the newest generation of physicists who have taught this course with me, and the thousand students who have read and argued their way through preliminary forms of this book.

In particular, I am specially grateful to Professor Frederick A. Saunders of Harvard University, who welcomed me to physics teaching in America. He provided the original inspiration and showed me that a physics text can be humane.

And I am very grateful to many others who have helped with technical skills to produce the book.

Some who have helped deserve special mention for devoted skill beyond the measure of general thanks. In naming some of these below, I know that I am equally grateful to many others.

I want to thank one colleague for his part in starting the whole project of course and book. A dozen years ago, when new science courses for nonscientists were being planned, I spent many visits discussing aims, methods, and progress with Professor Edwin C. Kemble at Harvard. We hammered out the idea of a "block-and-gap" course and planned content and treatment that would favor our aims. We discussed genuine laboratory work, decided on planetary astronomy (following Sir Oliver Lodge) as our example of theory; decided to treat energy-conservation with more discussion and less assertion. We looked for inquiring problems and examination questions that asked for thought; and we planned to require long essays or reading papers as a "pay-off" for the course. Many of these ideas are now commonplace characteristics of science courses in many institutions; but we thought we were making our own discoveries in education; and I shall always be grateful. At this distance in time I do not remember which of us made each suggestion, but I do know that those early discussions cultivated the ground from which this book and its course have grown.

And I am grateful to those who continued such discussions as the course was given. Among those teaching, I owe special thanks for criticism and suggestions to Mr. Gordon Likely and Drs. Robert Dicke, John Fletcher, Wayland Griffith, Claude Kacser, Aaron Lemonick, Robert Naumann, John Wheeler, and many other colleagues, past and present.

Recently, I spent long, equally stimulating visits at M.I.T. with the Physical Science Study Committee, where I found congenial views on physics flourishing. I only wish that the P.S.S.C. program had started some years earlier, so that this book could

Use of Chapters of This Book in an Actual One-year Course

USED FOR THOROUGH STUDY		TREATED LIGHTLY	OMITTED, OR USED FOR REFERENCE ONLY
In class, lecture, & reading	In lab only		
1, 7, 8	4	2, 3, 5, 6*, 9*, 10	6*, 9*, 11
16, 17, 18, 19, 21, 22, 25, 26	(27), 28	12, 13, 14, 15, 23, (27)	20, 24
29, 30 (parts), 36	32, (34)	30 (rest), 33, 34	31, 35
37, 38, 39, 40, 43, 44 (parts)	41 (parts)	42	41 (rest), 44 (rest)

* *Surface Tension* (Ch. 6) and *Fluid Flow* (Ch. 9) are examples of an individual teacher's favorites that other teachers can safely exclude. Yet, most of us would like to include a measurement of oil-molecule length; and some would like to eliminate a few demons by a brief showing of Bernoulli paradoxes as simple examples of Newton's Second Law.

have fitted with it more closely. As it is, I am especially grateful to Professor Francis Friedman for many a wise word of good physics in time to save an error or add a note in proof.

As a special critic, Professor Frederic Keffer of the University of Pittsburgh read most of the chapters in manuscript. I am exceedingly grateful to him for skillful criticism and many suggestions of good changes.

I owe special thanks to Professor Henry D. Smyth and Professor Allen G. Shenstone who, as Chairman, encouraged the course; and to Dean J. Douglas Brown and Dean Donald R. Hamilton who did much to encourage the publication of the book.

I and this book owe great thanks to Rudolph N. Schullinger, M.D., of Presbyterian Hospital, for great skill and kindness.

And I and this book owe lifelong thanks to my wife for affection and encouragement.

In making the book itself, many have given technical help. I am especially grateful to:

Mrs. J. E. Reef and Mrs. W. Parfian have carried out tasks of technical typewriting with great skill and have safeguarded the making of the book in many ways. Mrs. A. E. Sorenson has made typewritten problems for photographing.

The proofs have been read by Mrs. P. H. E. Aul, Mrs. M. R. Willison, and Mr. J. L. Snider, and informally by many teachers and students.

Mrs. T. H. Logan has prepared the index and read proofs.

Reducing my drawings to forms suitable for the Press proved to be a large task. Many people helped, some of them students, others professional draftsmen, all under the direction of Mr. Tracy Logan. In the later stages, Mrs. G. L. Carlson took charge of finishing and processing diagrams.

Throughout, Mr. Howard Schrader did marvelous photography to produce special effects and changes of scale.

The italic lettering on the diagrams all through the book has been done by Mrs. C. C. Pratt, with skill and devotion that can only be rewarded by an artist's delight in creative work.

Above all, I wish to give special thanks to my editorial associate, Mr. Tracy H. Logan, now of the Physics Department, Hobart College. In the course of years, he has taken charge of many preparations for me. He has drawn some diagrams, processed other diagrams for the Press, supervised draftsmen; he has searched out data, given valuable criticism, kept account of changes, organized records; in general, he has managed the production of material. He has taken a great load of responsibility, with authority and judgment; and I am very grateful. Without his skill and devotion, this book would not be published now.

This book owes its present form to the warm encouragement and admirable skill of the Princeton University Press. I wish to thank the Director, Mr. Herbert Bailey, and the two Editors who worked with me, Mr. Benjamin F. Houston and Mr. John Boles: all gave skillful help and welcome encouragement. I would also like to thank the unseen experts of a great Press: the two typesetters who transformed manuscript into linotype; the proofreaders who saved errors with consummate skill; and the hand-compositors who assembled pages.

And I wish to thank Mr. Jay F. Wilson, now Science Editor of the Princeton University Press, who took charge in the final stages. I am grateful for his good advice and his able, enthusiastic help.

I wish to thank the Carnegie Corporation of New York for contributing, long ago, to the writing of the book by providing part of a sabbatical year of leave. The University Press and I join in thanking the Sloan Foundation and the Rockefeller Foundation for financial help in connection with the book as an experiment in educational material, and the Eugene Higgins Fund for Education in Natural and Physical Sciences for financial help in production of diagrams, etc. I am grateful to these foundations not only for their financial help but also for this clear sign that they consider such aspects of education important.

CONTENTS

PART ONE

MATTER, MOTION, AND FORCE

"Give me matter and motion, and I will construct the universe."

—René Descartes (1640)

". . . from the phenomena of motions to investigate the forces of nature, and then from these forces to demonstrate the other phenomena; . . . the motions of the planets, the comets, the moon and the sea. . . ."

—Isaac Newton (1686)

"No one must think that Newton's great creation can be overthrown by [Relativity] or any other theory. His clear and wide ideas will forever retain their significance as the foundation on which our modern conceptions of physics have been built."

—Albert Einstein (1948)

PRELIMINARY PROBLEMS LEADING TO CHAPTER 1

A wise explorer reviews his maps before he starts on the expedition. You would be wise to review your present knowledge and prejudices before this chapter offers you new knowledge. The problems below are not intended to discomfort you by asking for answers before you are prepared. They are only intended to clear the ground for discussion. Some ask you to check minor matters of vocabulary. Others raise major questions that will appear again and again through the course.

1. (a) "I shall try an experiment. . . ." Suppose you have just made such a remark with your present knowledge and views. Write a note of a few lines to show what you would mean by it.
   (b) Write a similar note for, "I have a theory that. . . ."
   (c) Write a similar note for, "I shall treat this scientifically."

(At this stage, before you begin the course, we do not expect you to know all the answers to questions like this. Here you are asked to describe your present views. Later you may change them.)

2. Look up the word "logical" in a good dictionary; then write short answers to the following:
   (a) State in your own words the proper meaning of the word "logical."
   (b) State in your own words the colloquial or slang use of the word.
   (c) What word(s) could be used aptly for the meaning in (b), leaving "logical" for its important use in science, philosophy, etc.?
   (d) Do you consider algebra logical? Give reason(s) for your answer.

3. Look up the word "data." Then write short answers to the following:
   (a) What is its origin?
   (b) Which of the following statements do you consider correct language and which incorrect? (Where incorrect, mention reason.)
      (i) These data were obtained by my partner.
      (ii) This data was obtained by my partner.
      (iii) This set of data was obtained by my partner.

4. (a) What is the plural of the word "apparatus"?
   (b) What does "phenomenon" mean?
   (c) What is the plural of phenomenon?

5. (The following questions ask for written answers. Try to make them short. Some may require considerable thought. Consult dictionaries if you like. It is hoped that you will enjoy finding answers to these questions. If you have fun puzzling these out, your education will gain; if you do them with a feeling of headache, it will lose. So you are advised to treat these rather lightly, yet fairly seriously.)

   "The sun rose in the east this morning, yesterday morning, the morning before that, and for many mornings before that." This is a statement of observations. Scientists and others make statements like the following: "*I expect* the sun will rise in the east tomorrow morning."

   A number of other statements with some differences in wording may be made, and eight of these are shown below. With each of the statements:

   (a) Write a short explanation of its meaning, paying special attention to the part played by the words *in italics*. (For example, in the version above, where the word "*expect*" is used, your answer might run thus: "Because the sun has risen in the east so regularly in the past, I *look for* the same thing tomorrow with some confidence, and I shall make my everyday plans on this basis, because I have noticed that most such natural events continue to repeat themselves uniformly." Note, however, the dangers of this last view for an insect, hatched in the summer, anticipating an endless series of warm mornings, and completely ignorant of the snowstorm which will end its life.)
   (b) For each case say whether or not you consider the statement a wise or safe one for a scientist to use. (In other words, do you consider the statement scientific or superstitious, safe or risky, "right" or "wrong"?)
   (c) Give a brief reason for each answer in (b).

Statements:

A. I *predict* the sun will rise in the east tomorrow morning.
B. I *deduce* the sun will rise in the east tomorrow morning.
C. I conclude from *inductive* reasoning that the sun will rise in the east tomorrow morning.
D. I *believe* the sun will rise in the east tomorrow morning.
E. I *know* that the sun will rise in the east tomorrow morning.
F. I consider it *highly probable* that the sun will rise in the east tomorrow morning.
G. These observations lead to a law which *proves* that the sun will rise in the east tomorrow morning.
H. Investigations show that the world is a solid spinning body and the Principle of Conservation of Angular Momentum proves that it will continue to spin thus, and therefore proves that sunrise, which is due to this spin, will continue to occur in this same way each day.

6. Look up the verb "infer."
   (a) What is the proper meaning? (This is the one to use in science.)
   (b) What is the colloquial use? Give a better verb to replace it.
   (c) The word "infer" is used correctly in one of the statements below and incorrectly, or poorly, in the other. Explain what you think "infer" is intended to mean in each of the passages. Say which passage has the correct use and suggest a substitute for "infer" in the other one.
      (i) "Are you trying to infer, by your remarks, that my uncle is a fool?"
      (ii) "From his behavior, I infer that your uncle is a fool."

CHAPTER 1 · GRAVITY, A FIELD OF PHYSICS

"What distinguishes the language of science from language as we ordinarily understand the word? How is it that scientific language is international? . . . The super-national character of scientific concepts and scientific language is due to the fact that they have been set up by the best brains of all countries and all times. In solitude and yet in cooperative effort as regards the final effect, they created the spiritual tools for the technical revolutions which have transformed the life of mankind in the last centuries. Their system of concepts has served as a guide in the bewildering chaos of perceptions, so that we learned to grasp general truths from particular observations."

—A. Einstein, *Out of My Later Years*

Introduction

In this book, and the course that goes with it, we shall study the nature and methods of physical science. We shall do that by studying some parts of physics thoroughly and leaving out other parts to gain time for discussion. In the samples we study, you will learn many scientific facts and principles, some useful for life in general, others important groundwork for discussions in the course. To gain much from the course, you need to learn this "subject matter" thoroughly. In itself it may seem unimportant—such factual knowledge is easily forgotten,[1] and we are concerned with a more general understanding which will be of lasting value to you as an educated person—but we shall use the factual knowledge as a means to more important ends. The better your grasp of that factual knowledge, the greater your insight into the science behind it. And this course is concerned with the ways and work of science and scientists.

To begin by discussing scientific methods or the structure of science would be like arguing about a foreign country before you have visited it. So we shall plunge at once into a sample of physics—gravity and falling bodies—and later discuss the general ideas involved.

What to Do about Footnotes

You are advised to read a chapter straight through first, omitting the footnotes. Then reread carefully, studying both text and footnotes. Some of the footnotes are trivial, but many contain important comments relevant to the work of the course. They are not minor details put there with a twinge of conscience to avoid their being omitted altogether. They are moved out to make the text more continuous for a first reading. Often the footnotes wander off on a side issue and would distract attention if placed in the main text. Yet this developing of new threads itself shows the complex texture of scientific work; so at a second reading you should include the footnotes.

Falling Bodies

Watch a falling stone and reflect on man's knowledge of falling objects. What knowledge have we? How did we obtain it? How is it codified into laws that are clearly remembered and easily used? What use is it? Why do we value scientific knowledge in the form of laws? Try the following experiment before you read further. Take two stones (or books or coins) of different sizes. Feel how much heavier the larger one is. Imagine how much faster it will fall if the two are released together. You might well expect them to fall with speeds proportional to their weights: a two-ounce stone twice as fast as a one-ounce stone. Now hold them high and release them together. . . . Which are you going to believe: what you saw, what you expected, or "what the book says"?

People must have noticed thousands of years ago that most things fall faster and faster—and that some do not. Yet they did not bother to find out carefully just *how* things fall. Why should primitive people want to find out how or why? If they speculated at all about causes or explanations, they were easily led by superstitious fear to ideas of good and evil spirits. We can imagine how such people living a dangerous life would classify most normal occurrences as "good" and many unusual ones as "bad"—today we use "natural" as a term of praise and "unnatural" with a flavor of dislike.

This liking for the usual seems wise: a haphazard unregulated world would be an insecure one to live

[1] Once learned, it is easily relearned if needed later. Much of the difficulty of learning a piece of physics lies in understanding its background. When you understand what physics is driving at, the rules or calculations will seem sensible and easy.

in. Children emerge from the sheltered life of a baby into a hard unrelenting world where brick walls make bruises, hot stoves make blisters. They want a secure well-ordered world, bound by definite rules, so they are glad to have its quirky behavior "explained" by reassuring statements. The pattern of seeking security in order, which we find in growing children today, probably applied to the slower growing-up of primitive savages into civilized men. As civilization developed, the great thinkers codified the world—inanimate nature and living things and even the thoughts of man—into sets of rules and reasons. Why they did this is a difficult question. Perhaps some were acting as priests and teachers for their simpler brethren. Perhaps others were driven by childish curiosity—again a need to know definitely, born of a sense of insecurity. Still others may have been inspired by some deeper senses of curiosity and enjoyment of thinking—senses rooted in intellectual delight rather than fear—and these men might be called true philosophers and scientists.

You yourself in growing up run through many stages of knowledge, from superstitious nonsense to scientific sense. What stage have you reached in the simple matter of knowledge of falling objects? Check your present knowledge by actually watching some things fall. Take two different stones (or coins) and let them fall, starting together. Then start them again together, this time throwing both outward horizontally (Fig. 1-1). Then throw one outward

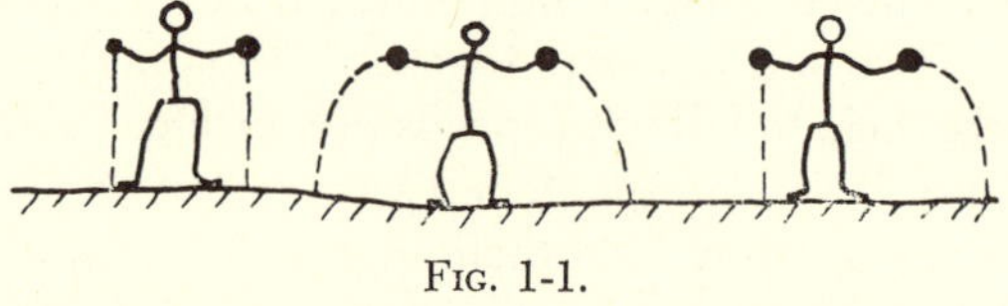

Fig. 1-1.

and at the same instant release the other to fall vertically. Watch these motions again and again. See how much information about nature you can extract from such trials. If this seems a childish waste of time, consider the following comments:

(i) This *is* experimenting. All science is built with information from direct experiments like yours.

(ii) To physicists the experiment of dropping light and heavy stones together is not just a fable of history; it shows an amazing simple fact that is a delight to see again and again. The physicist who does not enjoy watching a dime and a quarter drop together has no heart.

(iii) In the observed behavior of falling objects and projectiles lies the germ of a great scientific notion: the idea of *fields of force*, which plays an essential part in the development of modern mechanics in the theory of relativity.

(iv) And here is the practical taunt: if you use all your ingenuity and only household apparatus to try every relevant experiment you can think of, you will still miss some of the possible discoveries; this field of investigation is so wide and so rich that a neighbor with similar apparatus will find out something you have missed.

Mankind, of course, did not gather knowledge this way. Men did not say, "We will go into the laboratory and do experiments." The experimenting was done in daily life as they learned trades or developed new machines. You have been doing experiments of a sort all your life. When you were a baby, your bathtub and toys were the apparatus of your first physics laboratory. You made good use of them in learning about the real world; but rather poor use in extracting organized scientific knowledge. For instance, did your toys teach you what you have now learned by experimenting on falling objects?

Out of man's growing-up came some knowledge and some prejudices. Out of the secret traditions of craftsmen came organized knowledge of nature, taught with authority and preserved in prized books. That was the beginning of reliable science. If you experimented on falling objects you should have extracted some scientific knowledge. You found that the small stone and the big one, released together, fall together.[2] So do lumps of lead, gold, iron, glass, etc., of many sizes. From such experiments we infer a simple general rule: *the motion of free fall is universally the same, independent of size and material.* This is a remarkable, simple fact which people find surprising—in fact, some will not believe it when they are told it,[3] but yet are reluctant to try a simple experiment.[4]

[2] Yes; if you did not try the experiment, you now know the result of at least part of it. This is true of a book like this: by reading ahead you can find the answers to questions you are asked to solve. When you work on a crossword puzzle you would feel foolish to solve it by looking at the answers. In reading a detective story, is it much fun to turn to the end at once? Here you lose more still if you skip: you not only spoil the puzzle, but you lose a sense of the reality of science; you damage your own education. It is still not too late. If you have not tried the experiments, try them now. Drop a dime and a quarter together, and watch them fall. You are watching a great piece of simplicity in the structure of nature.

[3] Notice your own reaction to this statement: "A heavy boy and a light boy start coasting down a hill together on equal bicycles. In a short run they will reach the bottom together." The statement is based on the same general behavior of nature. See a demonstration. In a *long* run they gain high speeds and air resistance makes a difference.

The result *is* surprising. Would you expect a 2-pound stone to fall just as fast as a 10-pound one? Wouldn't it seem more reasonable for the 10-pound one to fall five times as fast? Yet direct trial shows that 1-pound, 2-pound, and 10-pound lumps of metal, stone, etc., all fall with the same motion.

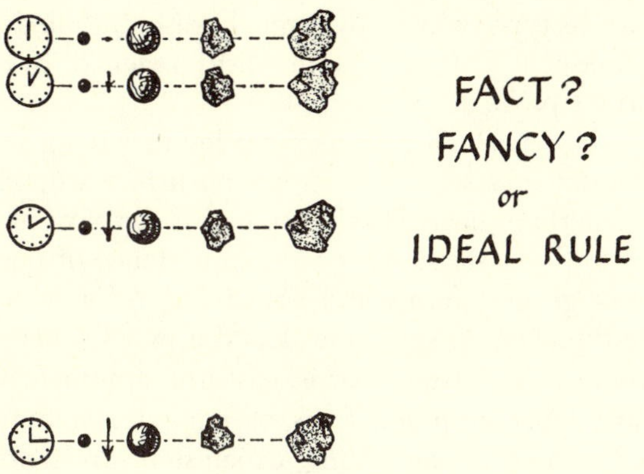

FACT?
FANCY?
or
IDEAL RULE

Fig. 1-2. Free Fall

Early Science of Falling Bodies

What is the history of this piece of scientific knowledge? There may have been a long gap between casual observation and careful experimenting. Interest in falling objects and projectiles grew with the development of weapons. Spears, arrows, catapults, and more ambitious "engines of war" favored a simple vague knowledge of ballistics. But that took the form of craftsmen's working rules rather than scientists' understanding—unspoken familiarity rather than extracted simplicity. Two thousand years ago the Greeks thought and wrote about nature with genuine scientific interest, possibly inspired by similar activity earlier still in Egypt and Babylon. They gave rules and reasons for falling, but not very reliable ones. Though some of the ancient scientists must have experimented sensibly with falling objects, medieval use of the Greek written tradition set down by Aristotle ($\sim$ 340 B.C.) clouded the matter rather than clarified it, and led to a muddle which lasted for many centuries. Gunpowder greatly increased the interest in projectiles, but early cannon were still chiefly used to frighten the enemy when Galileo ($\sim$ 1600) rewrote the science of ballistics in clear rules that agreed with experiment. Those were rules for heavy slow cannonballs that ignored air resistance. Since then, speeding up of projectiles has made air resistance more and more important, requiring modification of Galileo's simple treatment.

[4] We all rely in many matters on authority embedded in home teaching or common sense; we are reluctant to risk disturbing our sense of security. If you do not believe this accusation, wait and you will presently catch yourself.

Aristotle and Philosophy

The great Greek philosopher and scientist Aristotle appears to have supported the popular idea that heavy things fall faster than light ones. Aristotle was a pupil of Plato and for a time the tutor of Alexander the Great. He founded a great school of philosophy and wrote many books. His writings were the authoritative sources of learning for centuries—through the dark ages when there were still no printed books but only handwritten ones copied and handed down by devout scholars in a rough and troubled world.

Why should philosophers be concerned with science? How is science related to philosophy? What *is* philosophy? Philosophy is not a weird high-brow scheme of impractical argument; it is *man's thinking about his own thoughts and knowledge.* Professional philosophy consists of criticizing knowledge, evolving systems of knowledge with rules of logic for critical argument. Philosophers are interested in questions of truth and nonsense, right and wrong, and in judgments of values. Just as professional physicians advise us on health, eating, sleeping, etc., so professional philosophers offer us advice on thinking and understanding, on all our intellectual activities. You and I indulge in amateur philosophy when we think intelligently about our life and its relation to the world around us, whenever we ask questions such as: "Is this really true?" "Does that really exist?" "What does it mean when I say something is right?" "Why is arithmetic right?" "Is happiness real or imaginary?" "Does a pin cause pain in the same sense that it causes a puncture?" Thinking about our place in the world is closely tied with our scientific knowledge of the world, so it is not surprising that the great philosophers studied science and influenced its progress. You cannot embark on science without a first step in philosophy. You need a philosophical assumption and a philosophical interest: you practically have to assume that there is an external world; and you have to wish to find out about it and "understand" it. And when you collect facts, formulate scientific laws, or invent theories, the philosopher in you will ask, "Are these *true*?" As you brood on that question you may change your view of science. When you have studied this course, you may not have settled a general philosophy, but you will have done some philosophical thinking and you will have started making your own Philosophy of Science.

Aristotle inherited a general philosophical viewpoint from Plato. In trying to answer the question of ultimate truth and reality, Plato sheared away individual differences among the things we observe and extracted simple ideal forms. From dogs he extracted an ideal class, DOG; from all varieties of stones, an ideal STONE, and so on. Then he set forth the view that only these primary types or ideal forms really exist. These forms or essences remain universal and unchangeable, and individual examples of them are just shadows of the ideal. Aristotle used this insistence on classes of things as a basis for logical argument (If . . . , then . . . .). Yet, as a great observer and classifier of nature, he had to credit individual stones and dogs with some real existence; so his outlook was a compromise. Later students of his work gave increasing reality to ordinary objects, and came to treat the underlying classes as mental concepts or even mere names. This later view, that individual things are important and real, is a comfortable one for a scientist experimenting on objects and events in nature—he would like to feel that he is working with real things. Sir William Dampier seems to call some such view "nominalism" in the passage below, though modern philosophers use this name somewhat differently.

"Whatever be the truth of Plato's doctrine of ideas from a metaphysical point of view, the mental attitude which gave it birth is not adapted to further the cause of experimental science. It seems clear that, while philosophy still exerted a predominating influence on science, nominalism, whether conscious or unconscious, was more favourable to the growth of scientific methods. But Plato's search for the 'forms of intelligible things' may perhaps be regarded as a guess about the causes of visible phenomena. Science, we have now come to understand, cannot deal with ultimate reality; it can only draw a picture of nature as seen by the human mind. Our ideas are in a sense real in that ideal picture world, but the individual things represented are pictures and not realities. Hence it may prove that a modern form of [Plato's view of] ideas may be nearer the truth than is a crude nominalism. Nevertheless, the rough-and-ready suppositions which underlie most experiments assume that individual things are real, and most men of science talk nominalism without knowing it . . .

"The characteristic weakness of the inductive sciences among the Greeks is explicable when we examine their procedure. Aristotle, while dealing skilfully with the theory of the passage from particular instances to general propositions, in practice often failed lamentably. Taking the few available facts, he would rush at once to the widest generalizations. Naturally he failed. Enough facts were not available, and there was no adequate scientific background into which they could be fitted. Moreover, Aristotle regarded this work of induction as merely a necessary preliminary to true science of the deductive type, which, by logical reasoning, deduces consequences from the premises reached by the former process."[5]

While Aristotle may be regarded as having given *experimental* science a strong push forward, Plato was perhaps nearer to the modern *theoretical* physicist with his insistence on the importance of underlying general forms and principles. As a tool for his thinking, Aristotle developed a magnificent system of formal logic, that is, cast-iron argument that starts from admitted facts or agreed assumptions and draws a compelling conclusion. In treating science he first tried to extract some general scientific principles from observations, a process we call *induction*. Then he reasoned logically from these principles to deduce new scientific knowledge. His system of logic was itself a magnificent discovery, but it cramped the development of early experimental science by directing too much attention to argument. It has influenced the growth of our civilization profoundly. Most of us never realize how much our pattern of thinking has been influenced by the age-long tradition of Aristotelian logic, though many thinkers today question its rigid simplicity. It argued from one absolute yes or no to another absolute yes or no; argued with good logic to a valid conclusion, provided the starting-point was valid. "Is every man mortal?" "Do 4 times 3 make 14?" "Do 2 plus 2 make 4?" "Do all dogs have 7 legs?" We answer any of these with an absolute "yes" or "no" and then deduce answers to questions such as, "Is Jones mortal?" "Does my terrier have 7 legs?" But try the following:

"Is self-sacrifice good?"

"Was Lincoln a success?"

"Is my Boyle's law experiment right?"

These are important questions, but we can make fools of ourselves by insisting on a yes or no answer. If instead we spread our judgments over a wider scale of values, we may lose some "logic" but gain greatly in intellectual stature. It is well to beware of people who try to dissect every problem or dis-

[5] Sir William Dampier in *A History of Science* (4th edn., Cambridge University Press, Cambridge, Eng., 1949), pp. 34-35, from which some of this discussion is drawn.

cussion into components that have an absolute yes or no.

Aristotle's logic was safe as far as it went; modern logicians regard it as restricted and unfruitful but "true."[6] The damage to your thinking and mine comes from centuries of medieval scholarship drawing blindly and insistently on his writings—"the ingrown, argumentative, book-learned, world-ignorant atmosphere of medieval university learning." That medieval Aristotelian tradition is built into today's language and thought, and people often mistakenly require an absolute yes or no. For example, people trained to think they must choose between complete success and complete failure are heartbroken when they find they cannot attain the impossible goal of complete success. We are all in danger: students in college, athletes in contests, men and women in their careers, older people reviewing their life—all face terrible discouragement or worse if they demand absolute success as the only alternative to failure. Fortunately, many of us achieve a wiser balance; we stop judging ourselves by an absolute yes or no and enjoy our own measure of success. We then find the conflicting mixture of our record easier to live with.[7]

In science, where simple logic once seemed so safe, we are now more careful. Asked whether a beam of light is a wave, we no longer assume there is an absolute yes or no. We have to say that in some respects it is a wave and in others it is not. We are more cautious about our wording. Remembering that our modern scientific theory is more a way of regarding and understanding nature than a true portrait of it, we change our question, "*Is* it a wave?" to "Does it *behave as* a wave?" And then we can answer, "In some circumstances it does, in other circumstances it does not." Where an Aristotelian would say an electron must be either inside a certain box or not inside it, we have to say we would rather regard it as both! If you find such cautiousness irritating and paradoxical, remember two things: first, you have been brought up in the Aristotelian tradition (and perhaps you would be wise to question its strong authority); second, physicists themselves shared your dismay when experiments first forced some changes of view on them, but they would rather be true to experiment than loyal to a formality of logic.

Aristotle and Authority

Aristotle's chief interests lay in philosophy and logic, but he also wrote scientific treatises, summing up the knowledge available in his day, some 2000 years ago. His works on Biology were good because they were primarily descriptive. In his works on Physics he was too much concerned with laying down the law and then arguing "logically" from it. He and his followers wanted to explain *why* things happen and they did not always bother to observe *what* happens or *how* things happen. Aristotle explained why things fall quite simply: they seek their *natural place*, on the ground. In describing how things fall, he made statements such as these: ". . . just as the downward movement of a mass of lead or gold or of any other body endowed with weight is quicker in proportion to its size. . . ." ". . . a body is heavier than another which in equal bulk moves down more quickly. . . ." He was a very able man, discussing as a philosopher the *why* of falling, and he probably had in mind a more general survey of falling bodies, knowing that stones do fall faster than feathers, blocks of wood faster than sawdust. In the course of a long fall air-friction brings a falling body to a steady speed, and he probably referred to that.[8] But later generations of thinkers and teachers who used his books took his statements baldly and taught that "bodies fall with speeds proportional to their weights."

The philosophers of the Middle Ages grew more and more concerned with argument and disdained experimental tests. Most of the earlier writings on geometry and algebra had been lost and experimental physics had to wait until they were found and translated. For centuries, right on through the "dark ages," the authority of Aristotle's writings

[6] Roughly speaking, Aristotelian logic deals with classes of things, and its arguments can be carried out by machines, e.g., "electronic brains" in which "yes" or "no" is signified by an electron stream being switched "on" or "off." Modern logic deals with *relations* (such as ". . . larger than . . . ," ". . . better than . . .") as well as classes (such as "dogs," "mammals") and, nowadays, with implicational relations between complete propositions. Its arguments, too, can probably be carried out by machines, though that may be more difficult to arrange. But a machine cannot criticize the system of logic that it is asked to administer. Only man still thinks he can do that, making judgments of value.

For descriptions of machines see the following numbers of *Scientific American*:

Vol. 183, No. 5, "Simple Simon" (a small mechanical brain); Vol. 180, No. 4, "Mathematical Machines" (a detailed account of electronic calculators); Vol. 182, No. 5, "An Imitation of Life" (mechanical animals that learn); Vol. 185, No. 3, "Logic Machines"; Vol. 192, No. 4, "Man Viewed as a Machine" (excellent article by a philosopher); Vol. 197, No. 3, a complete issue on self-regulating machines.

[7] For a fuller discussion, see Ch. I of *People in Quandaries* by Wendell Johnson (Harper and Bros., New York, 1946).

[8] A denser body (or a bigger one) has to fall farther before approaching its limiting speed; and then that speed is much greater.

remained supreme, in a misinterpreted form at that. Simple people, like children, love security more than freedom; they will worship authority blindly, and swallow its teaching whole. You may smile at this and say, "We are civilized. We don't behave like that." But you may presently ask, "Why doesn't this book give us the facts and tell us the right laws, so that we can learn reliable science quickly?"; and that would be *your* demand for simple authority and easy security! We now condemn "Aristotelian dogmatism" as unscientific, yet there are still people who would rather argue from a book than go out and find what really does happen. The modern scientist is realistic; he tries experiments and abides by what he gets, even if it is not what he expected.

Logic and Modern Science

Wholesale appeal to Aristotle's logic may restrict our intellectual outlook, and medieval wrangling with it certainly hampered science; but logic itself is an essential tool of all good science. We have to reason inductively, as Aristotle did, from experiments to simple rules. Then we often assume such rules hold generally and reason deductively from them to predictions and explanations. Some of our reasoning is done in the shorthand logic of algebra, some of it follows the rules of formal logic, and some of it is argued more loosely.

In extracting scientific rules from old laws we trust the "Uniformity of Nature": we trust that what happens on Friday and Saturday will also happen on Sunday; or that a simple rule which holds for several different spiral springs will hold for other springs.[9] Above all, we rely on the agreement of other observers. That is what makes the difference between dreams and hallucinations on one hand and science on the other. Dreams belong to each of us alone, but scientific observations are common to many observers. In fact, scientists often refuse to accept a discovery until other experimenters have confirmed it.

Scientists do more than assume that nature is simple, that there are rules to be found; they also assume that they can apply logic to nature's ways. There lies the essential distinction that enabled science to emerge from superstition: a growing belief that *nature is reasonable.* As science grows, mathematics and simple logic play an essential part as faithful servants. The modern scientist puts them to more use than ever, but he goes back to nature for experimental checks. In a sense, the ideal scientist has his head in the clouds of speculation, his arms wielding the tools of mathematics and logic, and his feet on the ground of experimental fact.

Greeks to Galileo

> "In studying the science of the past, students very easily make the mistake of thinking that people who lived in earlier times were rather more stupid than they are now."—I. BERNARD COHEN

Aristotle's authority grew and lasted until the 17th century when the Italian scientist Galileo attacked it with open ridicule. Meanwhile, many people must have privately doubted the Aristotelian views on gravity and motion. In the 14th century a group of philosophers in Paris revolted against traditional mechanics and devised a much more sensible scheme which was handed down and spread to Italy and influenced Galileo two centuries later. They talked of *accelerated motion*, and even of *uniform acceleration* (under archaic names), and they endowed moving objects with "impetus," meaning motion or momentum of their own, to carry them along without needing a force.

Galileo (~ 1600) was a great scientist. He started science advancing to a new level where critical thinking and imagination join with an experimental attitude—a partnership of theory and experiment. He gathered the available knowledge and ideas, subjected them to ruthless examination by thinking, experimenting, and arguing; and then taught and wrote what he believed to be true. He lost his temper with the Aristotelians when they disliked his teaching and disdained his telescope; and he wrote a scathing attack on their whole system of science, setting forth his own realistic mechanics instead. He cleared away cobwebs of muddled thinking and built his scheme on real experiment—not always his own experiments, more often those of earlier workers whose results he collected.

Thought-Experiments

In his books and lectures, Galileo often reasoned by drawing on common sense, quoting "thought-experiments." For example, he discussed the breaking-strength of ropes in this manner: suppose a rope 1 inch in diameter can just support 3 tons. Then a rope of double diameter, 2 inches, has four times the cross-section area (πr^2) and therefore four times

[9] The obvious condition, "all other circumstances remain the same," is often difficult to maintain, and we blame many an exception to the Uniformity of Nature on some failure in that respect. Magnetic experiments in towns that have streetcars may give different results on Sundays, when fewer cars are running.

as many fibers. Therefore, the rope of double diameter has four times the strength—it should support 12 tons. In general, STRENGTH must increase as DIAMETER^2. Galileo gave this argument and extended it to wooden beams, pillars, and bones of animals.[10] Some thought-experiments deal with simplified or idealized conditions, such as an object falling *in a vacuum.*[11]

Ideal Rules for Free Fall

Galileo realized that air resistance had entangled the Aristotelians. He pointed out that dense objects for which air resistance is relatively unimportant fall almost together. He wrote: ". . . the variation of speed in air between balls of gold, lead, copper, porphyry, and other heavy materials is so slight that in a fall of one hundred cubits a ball of gold would surely not outstrip one of copper by as much as four fingers. Having observed this, I came to the conclusion that, in a medium totally devoid of all resistance, all bodies would fall with the same speed."[12] By guessing what would happen in the imaginary case of objects falling freely in a vacuum, Galileo extracted ideal rules:

(1) All falling bodies fall with the same motion; started together, they fall together.

(2) The motion is one "with constant acceleration": the body gains speed at a steady rate; it gains the same addition of speed in each successive second.

Having guessed the rules for the ideal case, he could test them in real experiments by making allowances for the complications of friction.

Galileo's ? Experiment: Myth and Symbol

There is a fable that Galileo gave a great demonstration of dropping a light object and a heavy one from the top of the leaning tower of Pisa.[13] (Some say he dropped a steel ball and a wooden one, others say a 1-pound iron ball and a 100-pound iron ball.) There is no record of such a public performance, and Galileo certainly would not have used it to show his ideal rule. He knew that the wooden ball would be left far behind the iron one, but he said that a taller tower would be needed to show a difference for two unequal iron balls. He certainly tried rough experiments as a youth and knew as you do what does happen, but he did not suddenly turn the course of science with one fabulous experiment. He did accelerate the growth of real physics by refuting the Aristotelians' silly dogmatic statements. And he did start science on a new kind of growth by applying his simplifying imagination to experimental knowledge. These, and not the leaning tower, made him a landmark in scientific history. Many a myth is attached to great figures in history—stories about cherry trees, burning cakes, etc. Though scholars delight in debunking these anecdotes, they also use some of them to show how the people of the great man's day thought of him. The leaning tower story is not even credited with that advantage. Yet we might use it, quite apart from Galileo and the growth of science, as a symbol of a simple experiment. In your own experiment with unequal stones, they fell almost together, and not, as some people expect, the heavy one much faster. We shall use this Myth & Symbol in our course as a reminder of two things: the need for direct experiment, and a surprising, simple, important fact about gravity.

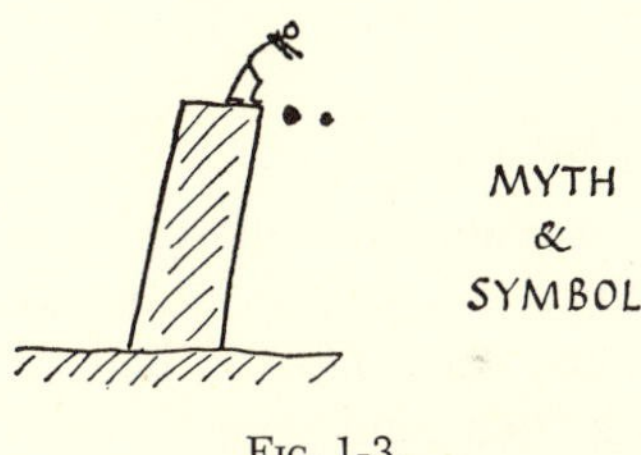

FIG. 1-3.

Honest Experimenting vs. Authority

Your own experiments did not show that all things fall together; they did not even show that large and small stones fall *exactly* together; and if in obedience to book or teacher you said, "They fall exactly together," you were cheating yourself of honest science. Small stones lag slightly behind big ones—the difference growing more noticeable the farther they fall. Nor is it simply a matter of different sizes: a wooden ball and a steel ball of the same size do not fall exactly together.

Once you accept Galileo's view that air resistance obscures a simple story, you can interpret your own observations easily—though that still leaves air re-

[10] See problems in Chapter 5.

[11] The Aristotelians had argued themselves into believing a vacuum to be impossible, so they cut themselves off from Galileo's satisfying simplification.

[12] From *Dialogues Concerning Two New Sciences*, by Galileo Galilei, English translation by H. Crew and H. de Salvio, Northwestern University Press, p. 72.

[13] Pisa. The leaning tower is a charming little building in a friendly Italian town. It is a round tower of white marble, built beside the cathedral. It began to lean as it was built, and it now has a remarkable tilt, about 5° from the vertical. The visitor who climbs its winding stair or walks around one of its open slanting balconies has strange sensations of shifting gravity. The tower was built long before Galileo's day, and he must have tried using it for some experiments. In his lifetime a pro-Aristotelian used the tower, to demonstrate *unequal* fall.

sistance to be investigated. Or you might pretend you had never heard Galileo's view, and proceed towards it yourself through a series of experiments with denser and denser objects. Finding the motion more and more nearly the same for larger or denser bodies, you might guess the rule for the ideal case. To examine the blame against air resistance, you might try streamlining, or reducing, using some object such as a sheet of paper.

Galileo's Guess: Newton's Crucial Experiment

Galileo could only *decrease* air resistance. He could not remove it completely, so he had to argue from real observations with less and less resistance to an ideal case with none. This intellectual jump, from real observations to an ideal case, was his great contribution. Then, looking back, he could "explain" the differences in real experiments by blaming air resistance. He could even study air resistance, codify its behavior, and learn how to make allowances for it. Not long after his time, air pumps were developed which enabled people to try free-fall experiments in a vacuum. Newton pumped the air out of a tall glass pipe and released a feather and a gold coin at the top. Even this extreme pair fell together. *There* was a crucial test of Galileo's guess.

Scientific Explanations

When we "explain" the differences of fall by air resistance, the term "explain" means, as so often in science, to point out a likeness between the thing under investigation and something else already known. We are saying essentially, "You know about wind resistance, when you move a thing along in the air. Well, the falling bodies experience wind resistance which depends in some way on their bulkiness. A wooden ball and a lead ball of the same size moving at the same speed would suffer the same air resistance—how could the air know or mind what is inside?—but the lead ball weighs more, is pulled harder by gravity, *so the air resistance matters less to it in comparison with the pull of the Earth.*"[14]

Further Investigation

The explanation leads to a whole new line of enquiry: wind resistance, fluid friction, streamlining—with applications to ballistics and airplane design—new science from more accurate study of some simple rule of behavior, from a study of its failures.

You could extend your series of experiments in the other direction, making more and more resistance, first with air, then with water, and find things of importance in the design of ships and planes. For simple experiments with fluid friction, try dropping small balls in water instead of in air. Balls of different sizes do not fall together. Moreover they fail to move any faster after a while in a long fall. Each ball seems to reach a fixed speed and then move steadily down at that speed. What is happening then? Investigations might lead you to Stokes' Law for fluid friction on a moving ball, a law which plays a vital part in measuring the electric charge of a single electron. If you investigated still smaller falling bodies, specks of dust or drops of mist, you would discover surprising *irregularities* in their fall, and these in turn could lead to useful information in atomic physics.

Galileo's experimenting and thinking, which you have been repeating, led to a simple rule that applies accurately to objects falling in a vacuum. For things falling in air, it applies with limited precision. In other words, the simple statement "all freely falling bodies fall together" is an artificial extract distilled by scientists out of the real happenings of nature. This is a good scientific procedure: first to extract a general rule, under simplified or restricted conditions, secondly to look for modifications or exceptions and then to use them to polish up the rule and to extend our knowledge to new things. In the case of falling bodies we can now test the extracted rule by dropping things in a vacuum. Ask to see Newton's "guinea and feather" experiment. In many cases in physics, however, we have to be content with knowing that our rule is an extracted simplification, believing in it as a sort of ideal statement, with only indirect evidence to justify our full belief.

Restricting the Number of Variables

Apart from ignoring air resistance, we have restricted our study of falling bodies in another way: we have concentrated attention on just one aspect of them, their comparative rate of fall. We have not observed what noise they make as they fall, or watched how they spin, or looked for temperature changes, etc. By narrowing our interests, temporarily, we have better hopes of finding a simple guiding rule. Again this is good scientific procedure. In many investigations we not only concentrate on a few aspects but even arrange to hold other aspects

[14] At this stage, the explanation ends in unsupported dogmatism that might be "straight out of Aristotle." Wait for studies of mass, force, and motion to make it good science.

constant so that they do not muddle the investigation. In physics we nearly always try to limit our investigation to one pair of variables at a time. For example, we compress a sample of air and measure its VOLUME at various PRESSURES, while we *keep the temperature constant.* Or we warm up the gas and measure the PRESSURE at various TEMPERATURES, while we *keep the volume constant.* From these experiments we can extract two useful "gas laws" that can be combined into one grand law. If we did not make restrictions but let TEMPERATURE and PRESSURE and VOLUME change during our experiments, we could still discover the grand law but our measurements would seem mixed and complicated—it would be harder to see the simple relationship connecting them. But other sciences such as biology and psychology, following the successful example of physics, have found this method very dangerous. While restricting attention to one aspect of growth or behavior, the investigator may lose sight of the body or mind as a whole. In attempts to apply the methods of natural sciences to social sciences such as economics this danger is even more severe.

Why Do Things Fall?

Aristotle was concerned with the answer to "Why?" Why *do* things fall? What is *your* answer to the question? If you say, "because of gravitation or gravity," are you not just taking refuge behind a long word? These words come from the Latin for heavy or weighty. You are saying, "things fall because they are heavy." Why, then, are things heavy? If you reply "because of gravity," you are talking in a circle. If you answer, "because the Earth pulls them," the next question is, "How do you know the Earth goes on pulling them *when they are falling*?" Any attempt to demonstrate this with a weighing machine during fall leads to disaster. You may have to answer, "I know the Earth pulls them because they fall"—and there you are back at the beginning. Argument like this can reduce a young physicist to tears. In fact, physics does not explain gravitation; it cannot state a cause for it, though it can tell you some useful things about it. The Theory of General Relativity offers to let you look at gravitation in a new light but still states no ultimate cause. We may say that things fall because the Earth pulls them, but when we wish to explain why the Earth pulls things all we can really say is, "Well it just *does.* Nature *is* like that."[15] This is disappointing to people who hope that science will explain everything, but we now consider such questions of ultimate cause outside the scope of science. They are in the province of philosophy and religion. Modern science asks *what?* and *how?* not the primary *why?* Scientists often explain why an event occurs, and you will be asked "why . . . ?" in this course; but that does not mean giving a first cause or ultimate explanation. It only means relating the event to other behavior already agreed to in our scientific knowledge. Science can give considerable reassurance and understanding by linking together seemingly different things. For example, while science can never tell us what electricity *is,* it can tell us that the boom of thunder and the crack of a man-made electric spark are much the same, thus removing one piece of fearful superstition.

Aristotle's explanation of falling was: "The natural place of things is on the ground, therefore they try to seek that place." People today call that a silly explanation. Yet it is in a way similar to our present attitude. He was just saying, "Things *do* fall. That's *natural.*" He carried his scheme too far however. He explained why clouds float upward by saying that *their* natural place is up in the sky and thus he missed some simple discoveries of buoyancy.[16] Aristotle was much concerned with stating the "natural place" and "natural path" for things, and he distinguished between "natural motion" (of falling bodies) and "violent motion" (of projectiles). He might have produced good science of force and motion except for a mistake of applying common knowledge of horses pulling carts to all motion. If the horse exerts a constant pull, the cart keeps going with constant speed. This probably suggested Aristotle's general view that a constant force is needed to keep a body moving steadily; a larger force maintains larger speed in proportion. This is a sensible explanation for pulling things against an adjustable resisting force, but it is misleading for falling bodies and projectiles. In all cases it forgets the resisting force is there and prevents our seeing what happens when there is no resistance.

To explain the motion of projectiles, the Greeks

[15] Parents often give answers such as, "Well it just *is*" or "because it *is*" to children's questions. Such answers are not so foolish as they sound. For a child they provide the reply that is really needed at that stage, an assurance that everything is normal, that the matter asked about is a part of a consistent world. When a child asks "Why is the grass green?" he does not want to have a lecture on chlorophyll. He merely wants to be reassured that it is o.k. for the grass to be green.

[16] Buoyancy affects falling objects. When a thing falls in water its effective weight is lessened by buoyancy, and this makes falling in water quite different for different objects. Even air buoyancy has some effect, trivial for cannon balls, overwhelming for balloons.

imagined a "rush of air" to keep them going; and even more mysterious agents were required to keep the stars and planets moving. On their view—shove is needed to maintain motion—an arrow was kept moving by the push of the bowstring until it left the bow. After that another pushing agent had to be invoked to keep the arrow moving. Aristotelian philosophers imagined a rush of air pushing the arrow, not just a gust of wind travelling with it, but a circulation of air, with the air ahead of the arrow being pushed aside and running around to shove the arrow from behind. This rush of air satisfactorily prevented the unthinkable vacuum forming behind the arrow.[17] So firmly established was the idea of a rush of air, with embellishments of initial commotions, that it was used as an argument to show that projectiles could not move in a vacuum. "In a vacuum with zero resistance any force would maintain infinite velocity," argued the Greeks, "therefore a vacuum is impossible." God could never make a vacuum. Aristotle himself understood that all things would fall equally in a vacuum, but he considered that too a proof that a vacuum could not exist.

Mass

Whatever gravity really is, falling bodies do fall together except for effects of air resistance. This hints at a useful idea which we shall meet again and again: the idea of mass. Suppose we have a 2-pound lump of lead and a 1-pound lump. When we hold them, we feel the big lump being pulled more; we feel its greater weight. That is why we expect it to fall faster. Yet it does not, so there must be some other factor involved, something that the doubled weight-pull has to contend with. The reason is: *there is twice as much lead to be moved.* The double chunk needs twice the pull to give the same motion to its double quantity of lead. Galileo felt his way towards this idea of quantity of matter, which we shall call mass, but it had to wait for Newton to state it clearly. Mass is not an easy idea to grasp, but we shall return to it many times, because it plays a very important part in physics. At this stage, the amazing thing about gravitation is this: whatever the materials, gravitational pulls are exactly proportional to the amounts of stuff being pulled. Gravity, the mysterious pulling agent, seems to be ready to pull indiscriminately on any body, whatever it is made of, ready to pull just twice as hard on two bricks as on one, four times as hard on 4 cubic feet of lead as on 1 cubic foot, so that the object with more stuff in it to be pulled on has just the bigger pull needed to make it fall with the same motion as a smaller object.

Gravitational Field

We give a name to this state of affairs all around us of gravity-prepared-to-pull. We say there is a *gravitational field.* We are not really explaining anything by inventing a new term,[18] but we shall find it useful later. For the moment you should think of a gravitational field as waiting to clutch (proportionally) on any piece of matter put there, to pull it down towards the Earth, to make it fall. Near a magnet there is a similar state of affairs for bits of iron, a *magnetic field* waiting to clutch them. In your television tube, *electric fields* and *magnetic fields* clutch the whizzing electrons, speed them up, and swing the beam to and fro to sketch the picture.

Here we have been letting our thoughts run away with new words and ideas, such as MASS and FIELD, arising from simple experimenting. If we just worship such new ideas and phrases we are liable to fall back to a state of witchcraft. But if we use them to develop our knowledge, and if we put our suggestions to experimental tests, they may help the progress of science.

Galileo's Argument

Galileo was a great arguer. The Aristotelians had woven a web of "scientific" arguments based on Aristotle's statements, but Galileo beat them with their own weapons. An argument would upset them more than an experimental demonstration. So he revived a thought-experiment which ran thus: Take three equal bricks, A, B, C. Release them together, to fall freely. Now chain A and B together (by an invisible chain which is not really there) so that they form one object A + B which is twice as heavy as C.

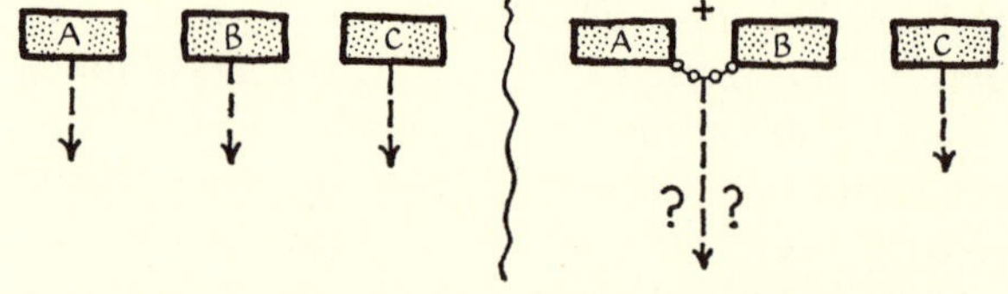

FIG. 1-4. GALILEO'S THOUGHT-EXPERIMENT

Again release them. The Aristotelians would now expect A + B to fall twice as fast as C; yet since it

[17] For a fuller and very interesting account of these views on motion, see H. Butterfield, *The Origins of Modern Science* (New York, 1952), Ch. 1.

[18] Cf. the great use in psychology or biology of special words such as "repression," "complex," "heliotropism," etc. Such new terms, coined or adopted for scientific use, cannot explain things, but they can aid clear thinking and discussion.

is really only two separate bricks it will fall just as before, at the same rate as C. Therefore, the double brick A + B and the single one C must fall together. "Ah no," says the Aristotelian in the argument, "there is the chain that joins A and B. One of the bricks will somehow get a little ahead of the other and then it will drag the other downward, making the combine fall faster." "I see," says the Galilean spokesman, "then the other, being a little behind, drags the first one back, making the combine fall slower!" Can you see in the comparison of A + B and C, the germ of the idea of mass?

The Motion of Free Fall

If all freely falling bodies have the same motion, that motion itself is worth detailed investigation. It might tell us something about nature in general, something common to all falling things. We can see that falling bodies move faster and faster; they accelerate. (This is merely a word meaning "move faster"—using it does not make our statement more scientific.) Just what kind of accelerated motion do they have?

(1) Does the SPEED increase by sudden jumps? Experiment says no.
(2) Does the SPEED increase in direct proportion to the DISTANCE TRAVELLED? Galileo devised an ingenious argument to show that this is very unlikely.[19]
(3) Does the SPEED increase in direct proportion to the TIME?
(4) or to the $(\text{TIME})^2$,
(5) or in some other, more complicated manner?

Since we are asking a question about real nature, only experiments on real nature can answer it. (If you want to know how tall Abraham Lincoln was you must find out from someone who actually measured his height. Information from books is useless unless it came originally from real measurements. Algebra alone cannot possibly tell you.) We might go straight to the laboratory and experiment wildly and boldly, hoping to extract the essential story from a host of measurements. Or we might do some thinking first, guess cunningly at some simple types of motion, calculate the consequences of each, and then go into the laboratory and experiment on the consequences. Both methods have contributed to the growth of science.

[19] Galileo's argument was ingenious but not quite sound. It ran thus: "Compare two trips, each starting from rest, trip A of a certain distance, trip B of twice that distance. Then if speed increases in proportion to distance travelled, the speeds at corresponding stages (half way, ¾ way, etc.) of trip B are twice those of A. Then the double trip, B, is travelled with doubled speeds. Therefore B would take the same total time as A; which is absurd." But this argument supposes that the motion could start from rest. A sound version of the argument requires calculus to show that such a motion could never start from rest. Given a start, however, such a motion would continue in an ever increasing rush, its speed growing with compound interest.

Inductive and Deductive Methods

The first method is named the *inductive* method. We gather information either in a laboratory or from the accumulated lore of some trade; then we extract from it some simple rule, or story about nature. We call this extracting process "inductive inference" or simply "induction." We first gather experimental data and then infer some general rule or scientific law from the data. For example, after watching the Moon for some years an observer might extract the general rule that the Moon travels around the Earth regularly, about 13 times a year, and it might seem a safe inference by induction that this will continue. Again, from an extensive record of eclipses, we might *infer inductively* a rule that eclipses of the Moon run in several regular series, with a fixed time-interval of about 18 years between successive eclipses in any one series.

The second method is named the *deductive* method. We start with some general rules or ideas then derive particular consequences or predictions from them by logical argument. If we are scientists, we then test the predictions experimentally. If experiment confirms the predictions, we continue our scheme. If it disagrees with our predictions, we throw doubt on our original assumptions and try to modify them. For example, we might assume that eclipses of the Moon are due to the Earth getting in the Sun's light, casting a shadow on it; assume simple orbital motions for Sun and Moon, and then *infer deductively* (or *deduce*) that an eclipse must occur again after an interval of time sufficient for Sun *and* Moon to return to the same positions relative to the Earth. This interval must be the "lowest common multiple" of one Moon-month and one Sun-year. So, by combining simple observations with sensible assumptions we could make a striking deductive prediction of the 18-year repeat cycle of eclipses. (For a successful calculation, the "Sun-year" must be a special, short year geared to the Moon's changing orbit.)

As Lancelot Hogben points out:

"Readers of crime fiction will be familiar with two types of detectives. One adopts the card index method of Francis Bacon, collecting all relevant information piece by piece. The other follows a hunch, like Newton, and, like Newton, abandons it at once when it comes into conflict with observed facts. From time to time the philosophers of science emphasize the merits of one or the other, and write as if one or the other were the true method of science. There is no one method of science. The unity of science resides in the nature of the result, the unity of theory with practice. Each type of detection has its use, and the best detective is one who combines both methods, letting his hunch lead him to test hypotheses and keeping alert to new facts while doing so."[20]

And here is an overall view, from a leading American physicist, P. W. Bridgman:

"I like to say that there is no scientific method as such, but that the most vital feature of the scientist's procedure has been merely to do his utmost with his mind, *no holds barred.*"[21]

Accelerated Motion: Inductive and Deductive Treatment

Much of the *early* growth of science was made by induction; general laws were inferred from the knowledge gained in crafts and trades. In a simple way we have treated falling bodies inductively, inferring from many observations a general statement that all bodies falling freely in a vacuum fall together. When Galileo studied the details of this falling motion, he probably used a mixture of two approaches. He was good at making guesses, and he used geometry and reasoning powerfully.

We shall now follow the second method, deduction, in our study. We shall start by *assuming* a likely rule, and then we shall make a test comparing its consequences with real falling motion.

We choose guess (3) above and *assume that a falling body gains speed steadily, gaining equal amounts of speed in equal stretches of time.* We can express this more conveniently if we give a definite meaning to the word *acceleration*, so that we can say "the acceleration is constant." Therefore, we give the name ACCELERATION to

$$\frac{\text{GAIN OF SPEED}}{\text{TIME TAKEN}} \text{ or RATE-OF-CHANGE OF SPEED}$$

In making this definition of acceleration, we are really *choosing* the thing (GAIN OF SPEED)/(TIME TAKEN) to work with, and then giving it a name. We are not discovering some true meaning which the word acceleration possessed all along! We make this choice and assign it a name because it turns out to be useful in describing nature easily.

We shall start using the grander word *velocity* instead of *speed*, and presently we shall make a distinction between their meanings. Since we shall often deal with changing things, we want a short way of writing "change of . . ." or "gain of . . . ." We choose the symbol Δ, a capital Greek letter D, pronounced "delta." It was originally used to stand for the *d* of "difference." Then our definition[22] of acceleration states that:

$$\text{ACCELERATION} = \frac{\text{GAIN OF VELOCITY}}{\text{TIME TAKEN}}$$

$$= \frac{\text{CHANGE OF VELOCITY}}{\text{CHANGE OF TIME-OF-DAY}}$$

$$a = \frac{\Delta v}{\Delta t}$$

where a, v, and t are obvious shorthand.

Deductive Treatment of Motion with Constant Acceleration

Now we express our assumption about falling bodies in this new terminology. We are *assuming* that:

$\frac{\Delta v}{\Delta t}$ is *constant*, for bodies freely falling (in vacuum). This states a huge assumption regarding real nature. Is it true? Is $\Delta v/\Delta t$ constant? To test this directly we should need an accelerometer to measure the acceleration of a body, $\Delta v/\Delta t$, at each stage of its fall. Such instruments are manufactured, but they are complicated gadgets which would not provide convincing proofs at this stage. Instead, we follow Galileo's example and ask mathematics, the logical machine, to grind out a consequence of our assumption, and then we test the consequence by experiment. The machine tells us that:

IF the acceleration a ($= \Delta v/\Delta t$) is constant, and s is the distance travelled in time t with this constant acceleration, THEN

$s = \frac{1}{2} at^2$, if the motion starts from rest

$s = v_0 t + \frac{1}{2} at^2$, if the motion starts with velocity v_0 at the instant $t = 0$, when the clock is started.

(The logical argument of this *IF . . . THEN . . .*

[20] *Science for the Citizen* (Allen and Unwin, London, 1938), p. 747.

[21] "New Vistas for Intelligence" in *Physical Science and Human Values*, ed. E. P. Wigner (Princeton, 1947), p. 144.

[22] In calculus, VELOCITY, v, at an instant, is defined by $v = \frac{ds}{dt}$, and ACCELERATION, a, at an instant, is $\frac{dv}{dt}$ or $\frac{d^2s}{dt^2}$.

is given in Appendix A of this chapter.) In these relations, ½ a is a constant number, since we are assuming a is constant; so, for motion starting from rest,

DISTANCE = (constant number) (TIME)^2
or DISTANCE increases in direct proportion to TIME^2
or DISTANCE varies directly as TIME^2
or DISTANCE $\propto$ TIME^2, this being shorthand for any of the versions above.

For example, if a body moving with fixed acceleration falls so far in one second from rest, then it will fall four times as far in two seconds from rest, nine times in three seconds, and so on.

★ PROBLEM 7. A CHART OF ACCELERATED MOTION

(a) Suppose a beetle crawls home with a motion for which it is true that *distance* $\propto$ *time*^{2}. Starting from rest he travels ¼ of an inch in the first second. How far will he travel in 2 seconds from his start? in 3 secs? in 4, 5, 6 secs?

(b) Draw a line across a sheet of paper; mark a starting-point near one end, and mark a rough scale of inches on it. Make marks to show the beetle's position at the end of each second.

★ PROBLEM 8. A SIMPLER RULE

Galileo announced the relation $s \propto t^2$ for uniformly accelerated motion, (where s is the *total* distance travelled in total time t from rest); but he stated another simple rule for such motion, relating the *distances* $d_1, d_2, \ldots$ *covered during 1 second,* in successive one-second intervals: (that is, the distance travelled in the first second, the distance travelled during the next period of one second, &c.) *Look for such a rule in the example of Problem 7, and state it. (Hint*: Calculate $d_1 = s_1 - 0$, $d_2 = s_2 - s_1, \ldots$ and look for some rule relating these one-second distances.)

★ PROBLEM 9. SCIENTIFIC THINKING

(a) You might have foreseen the rule of Problem 8, by common-sense thinking about accelerated motion, without using special algebra or studying an actual example. Why? (*Hint*: the distance travelled in any period of one second is a measure of . . ? . . in that period.)

(b) Is the rule of Problem 8 restricted (like $s \propto t^2$) to motion starting from rest when $t = 0$, or does it apply to any motion with constant acceleration?

★ PROBLEM 10. ANALYZING MOTIONS

Here are the records of four cyclists, moving with different motions. They all passed a post P at the instant the clock was started. Their distances from P after 1 second, 2, 3, 4, 5 seconds were as follows:

Time from Start	*1 sec*	*2 sec*	*3 sec*	*4 sec*	*5 sec*	
Cyclist A	1.8	7.2	16.2	28.8	45.0	feet
Cyclist B	1.8	3.6	5.4	7.2	9.0	feet
Cyclist C	1.8	5.2	10.2	16.8	25.0	feet
Cyclist D	1.8	14.4	48.6	115.2*	225.0*	feet

* These are distances Cyclist D would have travelled if he could have continued his motion.

(a) Try analyzing each of these motions, *looking for constant acceleration,* not by asking if $s \propto t^2$ but in the light of answers to Problems 8 and 9 above.

(b) Where the motion does not have constant acceleration describe its general nature if you can.

Experimental Investigations

The converse can be shown to be true. IF the distance s varies directly as t^2, THEN the acceleration is constant.[23] That gives us a relation to test in investigating real motions. We can arrange a clock to beat equal intervals of time, and measure the distances travelled from rest by a falling body, in total times with proportions 1:2:3: . . . If the total distances run in the proportions 1:4:9: . . . and so on, we may infer a fixed acceleration. Or, as in one form of laboratory experiment, we can measure the time t for various total distances s, and test the relation $s =$ (constant number) (t^2) by arithmetic, or by graph-plotting.

Over three centuries ago Galileo used this method, though he had neither a modern clock nor graph-plotting analysis. Galileo was one of the first to suggest an accurate pendulum clock, but he probably never made one. All he used to measure time was a large tank of water with a spout from which water ran into a cup. He estimated times by weighing the water that ran out—a crude method yet accurate enough to test his law. However, free fall from reasonable heights takes very little time—the experiment was too difficult with Galileo's apparatus.[24] So he "diluted" gravity by using a ball rolling down a sloping plank. He measured the times taken to roll distances such as 1 foot, 2 feet, etc., from rest.

On the basis of rough experimenting and sturdy guessing, Galileo decided that a ball rolls down a sloping plank with *constant acceleration.* Believing that this would be true for *any* slope, and arguing from one slope to greater slopes and greater still, he expected it to hold for a vertical plank, that is for free fall.[25] The idea of constant acceleration had

23 By calculus: if $s = kt^2$, then velocity $\frac{ds}{dt} = 2\,kt$; and acceleration $\frac{dv}{dt} = \frac{d}{dt}\left(\frac{ds}{dt}\right) = 2k$, which is constant.

24 Galileo's apparatus was rough. He used it to illustrate his argument rather than to measure acceleration.

25 He convinced himself that the speed acquired by a body sliding down a frictionless incline depends only on the height, h, not on the length of slope, L. If so, a body falling freely through a vertical height, h, would acquire the same speed, since this would be like a vertical incline. Then he could argue safely from his experiments to vertical fall.

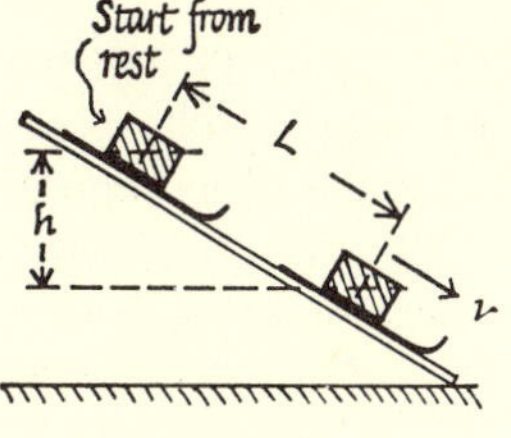

FIG. 1-5.

been suggested by earlier scientists—who were scorned for it. Galileo did his best to minimize friction, which threatened to complicate matters—though we now know that constant friction would not spoil the simple relationship. His results were rough, but seemed to convince him that his guess was right. It was the simplest kind of accelerated motion he could imagine, and he was probably influenced by the general faith, which has inspired scientists from the Greeks to Einstein, that nature is simple.

Later experiments, with improved apparatus, confirmed Galileo's conclusion: the motion *is* one with constant acceleration, i.e., with $\Delta v/\Delta t =$ constant, in all the following cases:

for a ball or wheel rolling down a straight inclined plank;
for a body sliding down a smooth inclined plank, or a truck with wheels running down it;
for free fall.

Yet each such test has only shown that the acceleration is constant for that one set of apparatus, on that one occasion and within the limits of accuracy of that experiment. If as scientists we want to believe in a general rule inferred from these experiments, if we want to codify nature's behavior in a simple "law" as a starting point for new deductions, then we need a great body of consistent testimony as a foundation for our inference. The more the better, in quantity and variety, and no witness is unwelcome. If any experiment contradicts this general story—and some do—it thereby offers a searching test. "The exception proves the rule" is a fine scientific proverb—though often misunderstood—if "proves" means "tests" (as in "proving-grounds" for artillery, the "proving" of bank accounts). If "proves" had the modern common meaning of "shows it to be right" the proverb would be nonsense.[26] Exceptions do *not* show that the rule is correct. Exceptions do put a rule to fine tests and show its limitations. They raise the question "What is to blame?" and they lead either to limitations of the rule or to greater care in experimenting. Either way, the rule emerges more clearly established.

Experimental Tests in Lecture and Laboratory

Therefore you should see and make some tests of accelerated motion yourself. Not only will these make you feel that the experimental basis of science is more real, but they will enable you to add your assurance to the accumulated body of testimony. Galileo made little more than wise guesses; others have added careful measurements, and you should add your measurements and judgment.

Demonstration Experiment

We let a small truck run down a long sloping track, and make measurements to estimate its acceleration. It is not easy to measure speeds in a lecture experiment but rough estimates will suffice to show how the acceleration is derived.

We measure the truck's speed at some station, A, early in the run, and again at B farther down the track. The difference between these speeds gives us the gain of speed Δv. The time taken for this gain, Δt, is the time taken by the truck to travel from A to B. Then the acceleration is $\Delta v/\Delta t$. To measure Δt we equip the truck with a thin mast, M, and measure with a stopclock the time taken for M to travel from A to B.

To estimate the truck's speed at A we have to time it over a short run in the region of A. We might install a short billboard there, with its mid-point at

FIG. 1-6.

A, as in Fig. 1-6, and measure the time taken by the mast to run the length of the billboard. But human errors are inconveniently big for such short timings, so it is better to install the billboard on the moving truck and time its transit past A with the help of an electric eye (photocell). Fig. 1-7 shows a good arrangement. A lamp sends a beam of light across the track into the electric eye, where it produces a tiny electric current. The current is amplified and used to run an electromagnet. The electromagnet keeps an electric clock switched off. When the light is obstructed, the electromagnet releases the clock and lets it run. The truck carries a long strip of cardboard which obstructs the light while the truck carries it past. Thus the clock runs while the truck is passing the electric eye at A, and records the time

[26] The original legal meaning is amusing, but irrelevant here: "the quoting of an exception makes it clear that the rule exists."

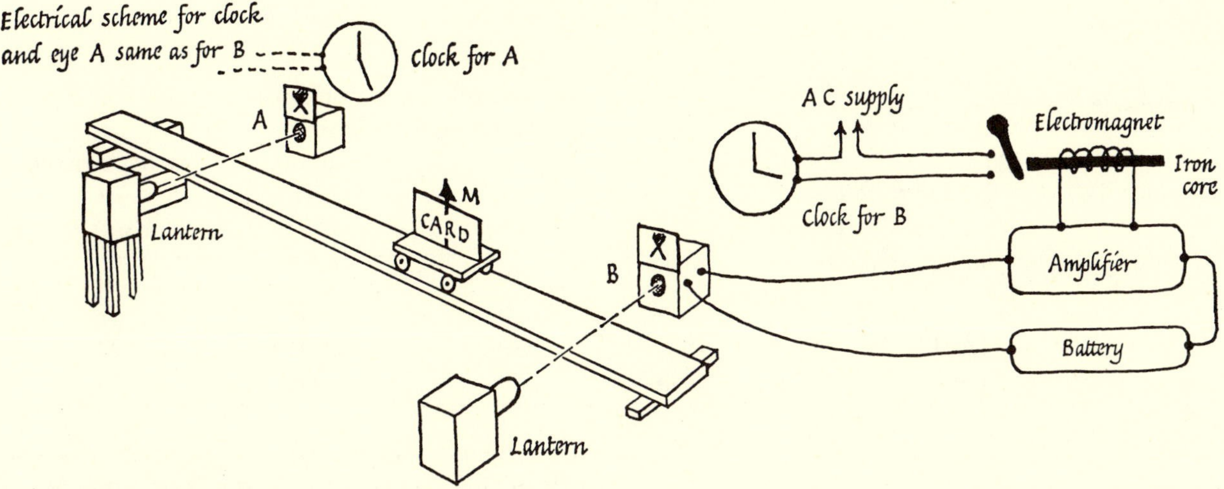

FIG. 1-7. EXPERIMENTAL ARRANGEMENT: measuring acceleration of model truck running down hill, with two electric eyes and three electric clocks. (Clock C for total travel-time not shown, operated by hand.)

taken for the card-length to travel past. In this time, the truck must travel the card-length. Then

CARD LENGTH/OBSTRUCTION TIME

gives the truck's speed.

We need three clocks, one to measure the total time between the two stations A and B where the speeds are estimated, one to measure obstruction time for the card passing A, and another for similar measurements at B. The following problem illustrates the calculation of the acceleration.

PROBLEM 11

Suppose the card on the truck is 2 feet long, and the obstruction at A takes 0.30 seconds. What is the truck's speed when running past A? If the obstruction at B takes 0.10 seconds, what is the truck's speed there? What is the gain of speed, Δv? If the truck takes 2.0 seconds to travel from A to B, what is its acceleration?

Nothing is said in Problem 11 about starting from rest. The truck is already moving when it passes A, and we can start it with any shove we like before that. So we can repeat the experiment with a variety of starting speeds. We can even start the car with an uphill shove so that it is moving backwards when it first passes A; but then we must be careful about + and − signs. The measurements tell us the acceleration whatever the starting speed. Whether the acceleration is the *same* for different starting speeds is a question about nature. To answer that you must see the real experiment.

In laboratory, you may experiment with a wheel rolling down sloping rails. You cannot easily measure the acceleration directly, or the (increasing) velocity. Instead you should measure DISTANCE TRAVELLED and TIME TAKEN, from rest, and then test whether they fit with the relation

DISTANCE varies directly as $(\text{TIME})^2$.

When you have collected reliable measurements, you should make the test both by arithmetic and by graph plotting.

Example of an Accelerated Motion Experiment

Meanwhile we shall proceed with a fictitious case of constant acceleration. Suppose measurements on a moving body gave the following results:

TABLE 1

DISTANCE TRAVELLED *from starting point* *(feet)*	TIME TAKEN FOR TRAVEL *(seconds)*			
0	0			
2	5.1	5.4	5.0	5.3
8	10.1	10.3	9.6	10.4
18	15.6	15.0	15.9	15.5

These measurements are too few and too poorly spaced for a good test, but will suffice for illustration. The four measurements 5.1, 5.4, 5.0, 5.3 are the result of four attempts to time the motion for 2 ft from start. Averaging is likely to remove some chance errors—though some errors may remain, such as the effect of impatient stopping of the watch too early. So we average these; we add them, and divide by 4:

$$\text{average time} = \frac{(5.1 + 5.4 + 5.0 + 5.3)}{4} = \frac{20.8}{4} = 5.2 \text{ secs}$$

Treating the other timings similarly we can make this table.[27]

TABLE 2

DISTANCE TRAVELLED *from starting point* *(feet)*	AVERAGE OF TIMINGS TIME OF TRAVEL *(secs)*
0	0
2	5.2
8	10.1
18	15.5

A glance at these numbers tells us that the times do not increase in proportion to the distances. Plotting the values on Graph (a) tells us the same thing. The graph shows clearly that the body is covering ground faster and faster, i.e., accelerating. It does not tell us whether the acceleration is constant.[28] To test that we plot a different graph, which will give a straight line *if* the acceleration is constant. We get a hint of what to plot by *assuming* constant acceleration and deducing DISTANCE $\propto$ TIME^2, which suggests we should plot DISTANCE against (TIME)^2. We make Table (3).

[27] A sensible experimenter in a real laboratory would save trouble by combining the two tables. He would leave a spare column for "average time" in his first table. If he foresaw the need for Table 3 he would leave another column for TIME^2. Even if he foresaw no need, an experienced experimenter would leave some blank columns, and blank lines below 18, for possible later use.

[28] We can make an indirect test by drawing tangents to the graph. See next section.

TABLE 3

DISTANCE TRAVELLED *from starting point,* s *(feet)*	AVERAGE OF TIMINGS, TIME OF TRAVEL t *(secs)*	(TIME OF TRAVEL)^2 t^2 *(secs)*^{2}
0	0	0
2	5.2	27
8	10.1	102
18	15.5	240

Then we plot Graph (b). To see whether the acceleration is constant, we draw a "best" straight line through the origin. We deliberately draw it straight, as a test, but we try to make it pass "as near as possible to as many as possible" of the plotted points. In this example, the points lie close to a straight line. If we think their displacements from the line are genuinely accountable by the incompetence of our apparatus, then we say that *so far as we can tell* from our measurements, the motion may well have constant acceleration.

Very Honest Graphing: Showing Likely Experimental Errors

If we wish to be more outspoken about our experimental uncertainties, we may spread each plotted point out into a patch to exhibit uncertainties of timing and distance measurement. Graph (c) in Fig. 1-10 shows this, with black points given by measurements surrounded by grey uncertainty patches. The timing is more risky than the distance measurement, so each patch is wider than it is tall.

Since we do not know how big our errors *are* but only how big they are *likely* to be, each patch should extend an indefinite distance out from its point; but we should show that the outer regions represent very unlikely errors. This might be done by shading

FIG. 1-8. GRAPH (a)

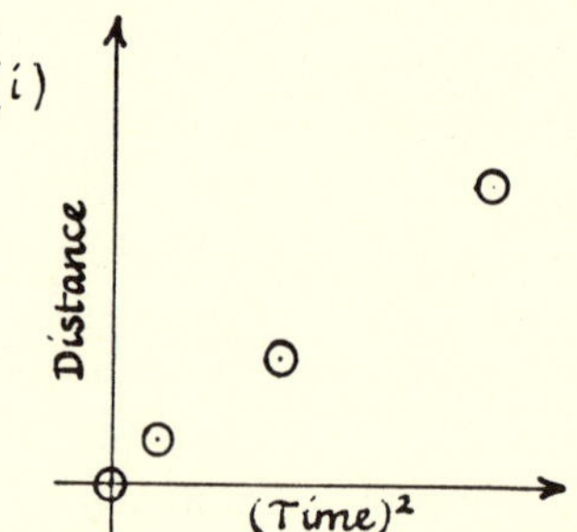

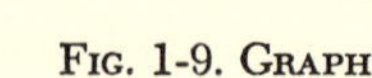

FIG. 1-9. GRAPH (b)

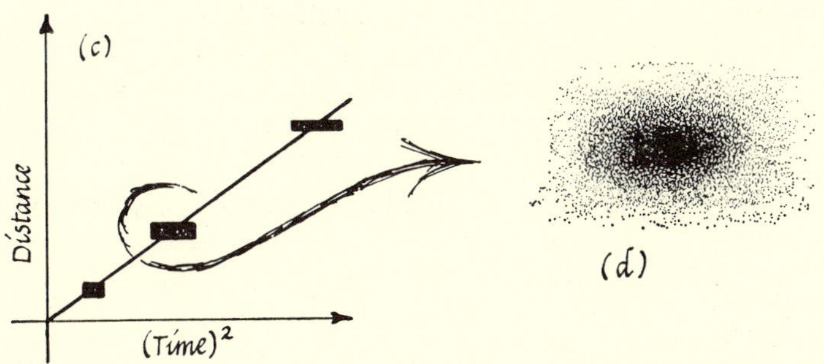

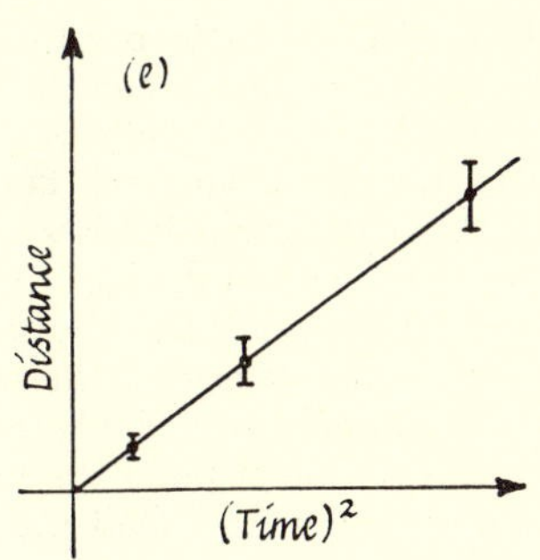

FIG. 1-10. ERROR BOXES

the patch as in (d). Shading is too tedious, so the general custom is to mark a box of definite size such that the betting that the true value lies inside the box has some standard value—say 50-50 for and against. The dimensions of the box show the errors the experimenter is willing to admit to.

Professional scientists often show errors or uncertainties on graphs, but they lump them together and express horizontal and vertical uncertainties combined as an uncertainty of the thing plotted upward. The experimenter estimates his likely error, Δy, in the measurement plotted upward. He estimates his likely error in the horizontal plot, Δx, and then asks, "If I did make that error, Δx, how big an error in y at the same time would just compensate for it?" That tells him Δy^*, the equivalent of his Δx. He draws a vertical error-line running $(\Delta y + \Delta y^*)$ above and below his experimental point. Then each plotted point carries an uncertainty line as in (e).

Finding Velocity by Tangents

We could make a direct study of acceleration if we could plot a graph showing how *velocity* changes with time. For that we need estimates of velocities at various instants.

We can estimate velocities by drawing tangents to the curved graph of distance against time. If a tangent is drawn touching the curve at some point, the slope of the tangent gives the speed of the moving object at that time and place. To see why this is so, choose a point P on this curve, then move to a point Q a little farther up the curve corresponding to a time which is a little later. At P on the graph, the moving object has travelled a certain total distance in a certain total time. From P to Q it travels a small further distance, Δs, in a small further time Δt. Then the AVERAGE VELOCITY in the interval between stage P and stage Q is

$$\frac{\text{DISTANCE TRAVELED BETWEEN STAGE P AND STAGE Q}}{\text{TIME TAKEN BETWEEN STAGE P AND STAGE Q}}$$

$\therefore$ AVERAGE VELOCITY $= \dfrac{\Delta s_{PQ}}{\Delta t_{PQ}}$ see sketch (f)

= HEIGHT/BASE for the small triangle PQM

= HEIGHT/BASE for any similar big triangle

$= h/b$ in sketch (f)

= SLOPE, or HEIGHT/BASE, of chord joining PQ

If P and Q are very close, the line joining them is *very nearly* a tangent to the curve at a "point" PQ, and the velocity is still given by the slope of this "tangent." In the mathematical limit, when P moves up to coincide

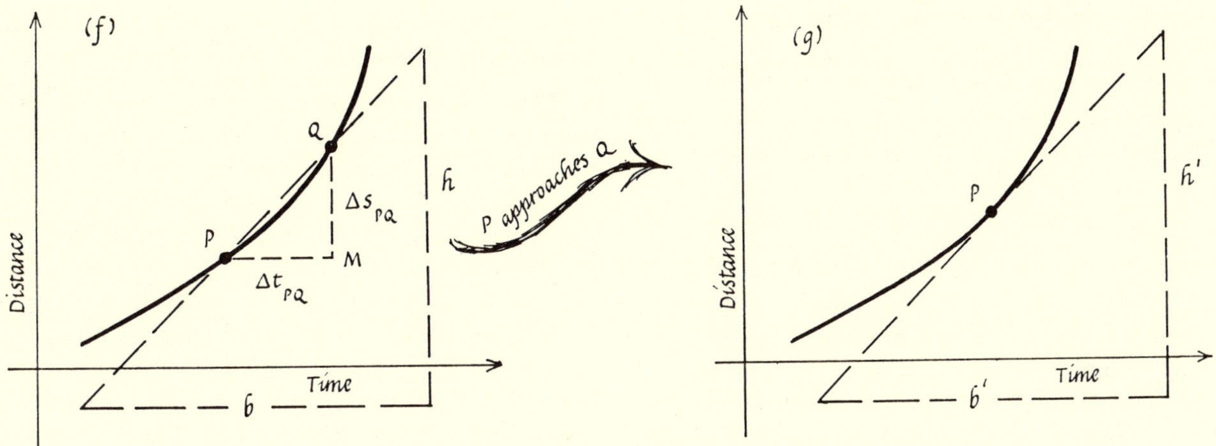

FIG. 1-11. VELOCITY = SLOPE OF TANGENT

with Q making a single point, the chord becomes a tangent at that point; Δs and Δt become zero, but the *ratio* $\Delta s/\Delta t$ still has a definite value, given by h'/b' for any large triangle having the tangent as sloping side, as in sketch (g). When PQ is a chord, the slope gives the *average* velocity in the motion from stage P to stage Q. In the limit, when P and Q coincide, the slope of the tangent gives the velocity at the instant of time represented by the single point P where the tangent is drawn. This is because the slope of the tangent is the same as the slope of an infinitely short piece of the curve which represents the motion at that point. By drawing tangents at a number of points on the curve and measuring their slopes we could obtain several velocities which could be plotted on a new graph showing VELOCITY *vs.* TIME. The shape of this new graph would tell us whether the acceleration is constant, but tangent-drawing is not easy and the original graph would have to be drawn very carefully, with many more plotted points to give a reliable set of tangents. So in practice we test for constant acceleration by plotting the second graph, DISTANCE against (TIME)^2.

However, we can use this tangent-property to help us in drawing the original graph. Though our graph (a) passes through the origin, it is hard to see how the curve runs *near the origin*, since measurements of very short travels are difficult. We are not sure which of the possible curves in Fig. 1-12 is the true one. We can gain a hint by arguing thus: the body started from rest, according to the record. Therefore, at the start its *speed* was zero. Therefore, at the origin of the graph the tangent-slope must be zero, the tangent horizontal. Of the three possible curves in Fig. 1-12, this argument tells us that the middle one is probably the true one.

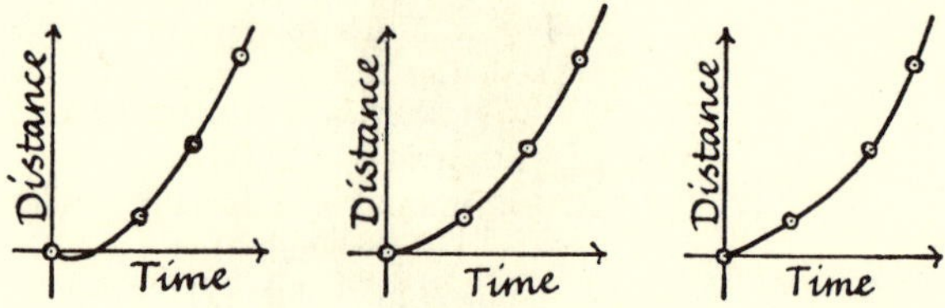

FIG. 1-12. VERSIONS OF GRAPH (a). Distance against Time

Arithmetical Test of Constant Acceleration

We can also make a test in our fictitious experiment by arithmetic. If the acceleration is constant, then

$$\text{DISTANCE} = (\text{constant})\ (\text{TIME})^2.$$

Therefore, DISTANCE/TIME^2 = constant. And conversely, if DISTANCE/TIME^2 *is* constant, the acceleration is constant. So we enlarge our table with an extra column to investigate this.

To draw any fair conclusion from the numbers in the last column, we should need to know something about the accuracy of the measurements. Otherwise, all we can say is "the motion seems to have a fairly constant acceleration."

TABLE 4

DISTANCE TRAVELLED s (*feet*)	TIME OF TRAVEL t (*secs*)	(TIME OF TRAVEL)^2 t^2 (*secs*)^2	DISTANCE/TIME^2	s/t^2 *feet*/(*secs*)^2
0	0	0	$\frac{0}{0}$	?
2	5.2	27	$\frac{2}{27}$	0.074
8	10.1	102	$\frac{8}{102}$	0.078
18	15.5	240	$\frac{18}{240}$	0.075

Both the graph and the arithmetical test above suffer badly from having too few data. *However, this is only a fictitious example. The real test should be made in your own experiments.*

The work of many scientists, professional researchers and amateurs such as yourselves, has built up great faith in Galileo's discovery: *bodies falling freely under gravity, and bodies sliding or rolling down a straight slope, under diluted gravity, move with constant acceleration.*

Further experiments show that the acceleration has the same constant value even if the body does not start from rest but is given a push to start it. If it already has speed v_0 when the clock starts, then the simple relation $s = \frac{1}{2}at^2$ no longer holds; we must use $s = v_0t + \frac{1}{2}at^2$. (See discussion in Appendix A.) But the acceleration, a, is the same. It could hardly be different: how could the ball know that it had started with a shove instead of rolling down an earlier piece of the same incline?

The Actual Acceleration

Experiments do more than just assure us the acceleration is constant: they tell us its actual value. If a is constant, then DISTANCE = $(\frac{1}{2}a)$(TIME)^2, so DISTANCE/TIME^2 = $\frac{1}{2}$ ACCELERATION. Thus, in the fictitious example, 0.076, etc., are estimates of $\frac{1}{2}a$. These give a =0.152, or 2/13. But 2/13 is incomplete—two-thirteenths of what? Such a number is useless unless it carries a tag to show its units. We calculated this number by dividing distance, in feet, by time^2. Since the time is in seconds the answer must be feet/sec^2. (This is read "feet per second squared" or "feet per second per second.")

Units for Acceleration

Return to the definition of acceleration to look for its units directly;

$$a = \frac{\Delta v\text{, measured in velocity-units, e.g., feet/second}}{\Delta t\text{, measured in time-units, e.g., seconds}}$$

$$= \text{acceleration measured in acceleration-units, e.g., } \frac{\text{ft/sec}}{\text{sec}}.$$

Thus we expect to measure acceleration in units such as $\frac{\text{ft/sec}}{\text{sec}}$, which we write ft/sec/sec or ft/sec^2.

The Use of "per" in Science

The word "per" is of great use in science. We started using it above to mean "divided by" or "for each . . . ," as it does in ordinary arithmetic. Later we shall concentrate on a different aspect of its meaning, when it is used for ratio or proportion.

In arithmetic we divide 10 cents by 5 and get 2 cents. Or we divide 10 sheep by 5 sheep, and get 2 flocks. We feel doubtful about dividing 10 sheep by 5 cents—we object that they are different kinds of thing. But sometimes we do divide one kind of thing by another; such as 10 cents divided by 5 boys, which gives a pocket-money proportion of 2 cents per boy. Again, 60 cents divided by a dozen oranges gives a PRICE of 5 cents per orange. In science we often make divisions like these, and we preserve the truth by preserving the units as well as the number in the answer. If a beetle crawls steadily 10 feet in 2 hours, we can say "10 feet divided by 2 hours, or 10 feet/2 hours gives *5 feet per hour.*" The answer shows the *distance* it crawls in *each hour*, but the statement does not restrict the beetle to one hour's travel. It applies to ¼ hour, ½ hour, 1½ hours, perhaps 2½ hours. But it also applies to very short time intervals; the beetle can still have a speed of *5 feet per hour* during a few seconds. We can, in imagination, shorten the time interval more and more, and still picture the beetle moving *5 feet per hour.* In the limit, we speak of the beetle having a *speed of 5 feet per hour at some particular instant.* This is a new idea, speed *at an instant* of time, at a certain mark on the clock. We can no longer divide a distance by a time—zero divided by zero is meaningless—yet a speedometer can register 5 feet per hour at an instant. True, a real beetle moves unevenly, but we can easily imagine an ideal one moving smoothly. Then the unit "one foot per hour" is no longer the result of division, but a thing-of-itself, a unit of rate; and the speed, 5 feet per hour, is a rate, *a limiting value*, glimpsed at an instant.

Mathematical limits appear in physics as well as calculus—which *is* the algebra of calculating limits. To understand the essential idea of a *limit* look at the sum to many terms of the series: 1, ½, ¼, ⅛, 1/16, ... The sum of the first two terms is 1½; of three terms 1¾; of ten terms 1 511/512; &c. However far you go, the sum is never quite 2, but you can get as near as you like to 2 by taking enough terms. (Notice that the sum always falls short of 2 by just the last term that is included; so you can make that failure as small as you like.) So we say 2 is the *limit* of the sum of many terms. You met a limit in tangent-slope, the limit of the slope of a chord through two points on a graph.

Until this century, physics dealt with many smooth ratios, such as speed, density, illumination. But now, much as we find a real beetle's speed uneven, we find many physical quantities jumpy or chunky; we cannot reduce them smoothly to limiting values. As an obvious example consider the ratio mass/volume, which we call density. We can divide the mass of a large chunk of aluminum by its volume, or the mass of a small chunk by *its* volume, obtaining the same density. But if we try to push our determination of density to the limit of smaller and smaller samples we are stopped when we meet a single atom. What ratios in physics can be pushed to the mathematical limit? What things are not "atomic"? This is a question worth watching, to which we shall return at the very end of the course.

At present, you should take "per," or the sign / used for it, to mean "divided by" or "for each," but you should think about letting it take its place in the idea of a ratio.

Scientific Units

In ordinary life, we measure speeds in *feet per second* or in *miles per hour*, and engineers often use these units. We express accelerations in *feet/second per second*, or sometimes in stranger units such as *miles/hour per second.* But scientists all over the world have agreed to use the metric system of units in their measurements, and we shall use one version of this, the Meter-Kilogram-Second system. In this "MKS" system, lengths and distances are measured in meters instead of feet, masses of stuff in kilograms instead of pounds, and times in seconds. A meter is almost 10% longer than a yard, its exact length being defined by a bar of fireproof metal which is carefully preserved, with copies in standardizing laboratories throughout the world. A kilogram is roughly 2.2

Table of Units and Abbreviations

	Ordinary system used by householders and engineers (FPS system)	Metric system used by scientists: MKS system used in this course	Metric system used by scientists: CGS system (in common scientific use; not used in this course)
Length	foot (ft)	meter (m.)	centimeter (cm)
Mass	pound (lb)	kilogram (kg)	gram (gm)
Time	second (sec)	second (sec)	second (sec)

Conversion Factors

1 foot = 12 inches

1 meter = 100 centimeters
= 1000 millimeters

1 inch = 2.540 centimeters = 0.02540 meters

1 foot = 0.3048 meter

1 meter = 39.37 inches
≈ 1.1 yards

1 pound = 454 grams = 0.454 kilogram

1 kilogram ≈ 2.2 pounds

pounds, 10% more than 2 pounds. It is defined by a standard lump of fireproof metal. A meter is subdivided into 100 centimeters, each about a fingerbreadth, and a kilogram is subdivided into 1000 grams, each about 1/28 ounce. Though many science courses use centimeters and grams, we shall follow the new fashion and use meters and kilograms, to make it easier to understand electrical units such as amps and volts. Scientists write m. as an abbreviation for meter or meters, but as this is easily confused with an algebra symbol m for mass, it is better to write it in full as meter(s). We write kg as an abbreviation for kilograms.

The gram was originally made of such size that one cubic centimeter of water weighs one gram. This gives the density of water, mass/volume, the useful value 1.00 gram per cubic centimeter (useful, but misleading because it can be left out so harmlessly). The density of water is *not* 1.00 kilogram per cubic meter. Nor is it 1 pound per cubic foot. If you make a hollow box with internal dimensions 1 ft × 1 ft × 1 ft, you will find it holds 62.4 pounds of water. The density of water is therefore:

62.4 pounds per cubic foot,
or 1.00 grams per cubic centimeter,
or . . ? . . kilograms per cubic meter.

In our scientific MKS system we measure speeds in *meters/second*, accelerations in *meters/second per second*. The acceleration in the example above, 0.076 *feet/sec per sec*, is the same as about

0.076 × 0.3 *meters/sec per sec*

since each foot is about 0.3 meter.

Acceleration of Free Fall

For free fall, the acceleration can be measured. To show that the acceleration is constant as a body falls faster and faster is difficult, though of course it can be done with modern timing apparatus, some of which can measure to one-millionth of a second. If we *assume* the acceleration is constant, then it is fairly easy to measure its value by timing free fall for one known distance from rest and using the relation $s = \frac{1}{2}at^2$. This leads to $a = 2s/t^2$. As a reminder that we are dealing with a characteristic constant acceleration "due to gravity," we label this particular acceleration "g" and write $g = 2s/t^2$. Using experimental values of s and t we can compute g. However, air friction limits the accuracy; it is difficult to make sure that we start the timing just when the falling body starts from rest, and the time of fall itself is a very short one; so such measurements do not give an accurate value of g. Yet we need to know g accurately for a number of uses in physics. Could we possibly eliminate the effects of friction? And could we lump together many falls, say several thousand, and measure the total time for the whole bunch to obtain the time for one fall with greater accuracy? These look like hopeless ambitions. Yet they can be achieved in a simple, easy experiment which Galileo foreshadowed, and which you will meet.

Measurements give a value about 9.8 meters/sec^2 for g, or 32.2 ft/sec^2. For ordinary calculations, 32 ft/sec^2 will suffice: accurate within 1%.

At the Equator, g is slightly smaller; and at the North Pole g is slightly greater.

Force and Acceleration

We think of a falling body as being pulled down by a force which we call its weight. To hold a body suspended we must support its full weight. If we cut the suspending cord we imagine the weight still acting, now unopposed by our supporting pull. If we suppose the body's weight remains constant while the body is falling, we may picture this constant force "causing" the constant acceleration of free fall. Trucks running down a slope have a smaller acceleration, a fraction of g; but only a fraction of their weight is available to pull them down along the slope. Later you will find what this fraction is. It depends on the slope of the hill. If you knew this fraction, you could follow Galileo in comparing downhill FORCE and downhill ACCELERATION. What kind of relation would you expect[29] to find between the force and the acceleration? You can see how early experimenters like Galileo could guess at it by studying falling and rolling bodies. That relation, to be discussed soon, is a very important piece of physics, a basic relation governing the motion of stars and the action of atoms, one of obvious importance in engineering.

While looking forward to discussing force and acceleration, we will end on a note of doubt. How do you know the weight of a body pulls it while it is falling freely? When you sit on a chair you feel the supporting force of the chair, and you believe you feel your own weight. But if you jump out of a window, do you feel your weight while you are falling? Suppose you jump out of a window with a lump of metal in your hand and try to weigh the lump as you fall. To make the temporary laboratory more comfortable, for a time, suppose you and the lump and the weighing apparatus are enclosed in a vast box which has been dropped from a tower and is falling freely. Suppose the box has no windows. When you release the lead lump inside the box, will it fall to the floor? If you think about this, you will see that gravity will seem to have disappeared. Can you possibly tell whether gravity has really disappeared or whether your laboratory is accelerating downwards? If you cannot tell the difference, *is* there any difference? Discussion of these questions would lead you towards the Theory of Relativity.

[29] Do we mean "expect" or "hope"? If *expect*, on what basis? If *hope*, is this scientific or not?

PROBLEMS FOR CHAPTER 1

1-6. These are at the beginning of the chapter.
7-11. These are in the text of Chapter 1.

★ 12. METHODS

Write a short note distinguishing between inductive and deductive methods.

★ 13. YOUR PRESENT VIEWS

Write a short note ($\frac{1}{4}$ page to 2 pages) saying what *you* think are (or should be) the parts played by *experiment* and *theory* in a science like Physics. (*Note*: At this stage of the course we do not expect you to know all the answers to questions like this. Later you should know more of them. So we ask you now just to write some general comments stating *your present views*. Please do not extract some complicated statements from a book.

14. EXPLANATIONS

(a) How would Aristotelians explain the rising of a helium balloon?
(b) How would modern scientists explain it?

15. DENSITIES

(a) Look up the relative densities of gold, silver, aluminum, brass, stone, iron, and wood in reference tables (often at the end of physics books).
(b) Why did Newton use a gold guinea?

16. SCIENTIFIC WRITING

(a) Write a short essay, (half a page at most), giving your answers to the following questions:
  (i) Do you consider it good scientific writing to use long words wherever possible?
  (ii) Why do you suppose people who are trying to imitate a scientist tend to use long words?
  (iii) Do you consider it good scientific writing to avoid long technical words?
(b) Rewrite the following passage, replacing long words by suitable shorter ones wherever you can: "Henderson conducted considerable experimentation concerning the relationship between superficial area and electrical charge of aqueous solutions atomized into numerous spherical particles of microscopic dimensions. He theorized that the phenomenon of electrification was attributable to friction."
(c) Rewrite the following passage replacing a word by its technical equivalent wherever you feel that the change would make the passage more scientific: "When the r.p.m. of the fan is pepped up, the atoms of air whiz down the tube at a great rate of speed; and when they hit the thermometer its mercury rises and registers more degrees of heat."

17. Do you agree with Bernard Cohen's remark on page 8? Is it a mistake? Discuss briefly.

18. In the discussion of mass it is stated that ". . . gravitational pulls are exactly proportional to the amounts of stuff being pulled." On what piece of experimental knowledge is this statement based?

19. Suppose on a certain (fictitious) island it is a custom for each member of a family to give a small present to every other member of his family, and a present to him-

self as well, on New Year's Day. Suppose this custom is followed in every family, and that each present costs the same amount of money.

(a) How will the total expenditure of any one family be related to the number of members in it? (Find this empirically, that is, by trial and error, if you like.)
(b) Sketch a graph showing *total cost* plotted upward against *number of members in the family*.
(c) What graph do you suggest plotting, using these things, to obtain a straight line?

★ 20. IMPORTANT PUZZLE

When a ball is thrown vertically upward, it continues up until it reaches a certain point, then falls down again. At that highest point it stops momentarily and is not moving up or down.

(a) Is it accelerating at that point?
(b) Give reasons for your answer to (a). (*Hint*: See Problem 21.)
(c) Devise an experiment (given any apparatus you need) to find out whether it *is* accelerating at that point.

★ 21. PROBLEM TO HELP SOLVE PUZZLE

A man leans out of a window high above the ground, and throws a ball vertically up. The ball rises till it is about 30 feet above the man, then falls. (See Fig. 1-13.)

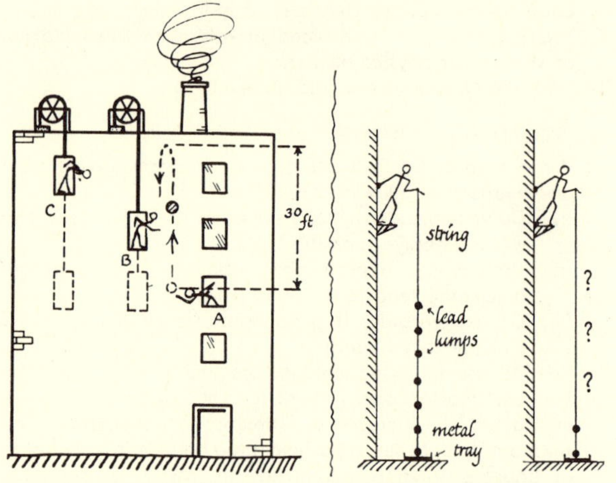

Fig. 1-13. Problem 21 Fig. 1-14. Problem 22

(a) Give a short description of the motion of the ball, as seen by the man, A.
(b) At the instant that the man throws the ball, an elevator running up the outside of the building is passing the window with the same upward speed that the man gives the ball. The elevator continues upward with constant speed, carrying an observer, B, who watches the ball. Describe the motion of the ball as seen by B (who forgets that he is moving, and thinks that all the motion he observes belongs to the ball).
(c) Another elevator runs beside the first carrying an observer, C, steadily upward, with smaller constant speed. C is just passing the window when the man throws the ball. Describe C's observations of the motion of the ball.
(d) In the light of your answers to (a), (b), (c), comment on the puzzle of Problem 20. (Note that A, B, C all agree that the ball has the usual decelerated and accelerated motions, but they disagree in one respect.)

★ 22. DEMONSTRATING CONSTANT ACCELERATION

A lecturer wishing to demonstrate the constant acceleration of free fall drops a chain of lead lumps down a stairwell and asks his audience to listen to the sounds of them hitting a metal tray at the bottom. He makes one such chain by tying quarter-pound lumps of lead to a light string every foot along the string. Then holding the string so that the lowest lump is just on the ground, he has lumps 1 ft, 2, 3, 4 ft, and so on, from the ground. When he releases the string the lumps hit the ground with a tattoo of *increasing frequency*.

(a) What does this tell the audience about the motion of falling bodies?
(b) The lecturer wishes to test for constant acceleration by arranging the lumps *unevenly* on the string in such a way that if the acceleration is constant the audience will hear an *evenly spaced tattoo*. He ties one lump to the bottom of the string on the ground, the next 1 ft above the ground. Where should he tie the next five lumps? (See Fig. 1-14.)

23. HISTORY

Read Galileo's description of his own experiment on accelerated motion (available in Magie's *Source Book in Physics*, New York, 1935) and write a short account of it. Indicate the apparatus he used and the results he got.

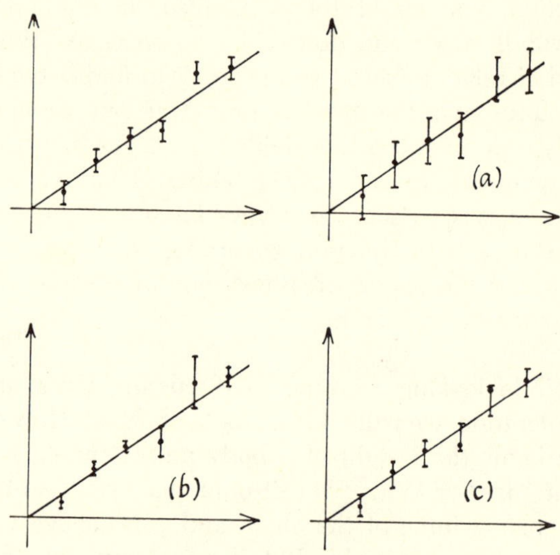

Fig. 1-15. Problem 24

★ 24. ERROR-BOXES ON GRAPH

A student, S, making an experimental investigation finds when he plots a graph of his measurements that his points do not lie on a straight line as he hoped. However, they are fairly near a "best straight line," and he knows his measurements are liable to some small errors. So he draws lines as "error-boxes" through each point on his graph. He finds that even the error-boxes miss his best line in several cases. He is, therefore, tempted to change his error-boxes (a) by making all of them taller or (b) by making *some* of them taller or (c) by sliding them up or down through each point until they hit the line.

Is any of the changes (a) or (b) or (c) a wise one for a scientist to make? Discuss with student S what he is really saying about his experiment in each case, (a), (b), (c).

★ 25. MKS UNITS

(a) Make a rough estimate of your height in meters.
(b) What do you weigh in kilograms?
(c) What is the width of this page in meters?
(d) What is the thickness of a page of this book in meters?
(e) Explain how you made the estimate asked for in (d) without using any special micrometer gauge.
(f) Atomic scientists find molecules and atoms so small that they like to use a much smaller unit than a meter for measuring them. They use an "Ångstrom Unit," which is one ten-billionth of a meter or 10^{-10} meter. What is the thickness of this page, in Å.U.?
(g) Most atoms are a few Å.U. in diameter, say 3 Å.U. How many atoms thick (roughly) is this page?

★ 26. DENSITIES IN MKS UNITS (Learn these values)

(a) What is the density of water in *kilograms/cubic meter*?
(b) Show the reasoning by which the answer to (a) can be obtained from the data in Chapter 1.
(c) Lead has a "specific gravity" of about 10. This means its density is 10 times that of water. What is the density of lead on the MKS system of units?
(d) Olive oil has a specific gravity of about 0.8. What does this mean?
(e) What is the density of olive oil in the MKS system?
(f) The specific gravity of mercury is 13.6. The atmosphere presses on each square inch of table, chair, our bodies, walls, . . . etc. with a force that can balance a column of mercury of cross-section 1 sq. inch and height about 30 inches. That is the "height of the barometer" in which mercury with a vacuum above it inside balances atmospheric pressure outside. What would be the height of a water barometer?

★ 27. A USEFUL CONVERSION FACTOR

Show that 60 miles/hour = 88 ft/sec.

28. A SPECIMEN ACCELERATED MOTION

The motor of a certain elevator gives it an upward acceleration of 150 ft/min/sec. The elevator starts from rest, accelerates thus for 2 secs, then continues steadily with constant speed.

(a) Explain what this statement of acceleration means.
(b) What is the final speed after 2 secs?
(c) Calculate the speed after 0 sec, 0.5 sec, 1 sec, 1.5 secs, 2, 3, 4, 5 secs. Sketch a rough graph showing speed (upward) against time from start (along), for the first 5 seconds.
(d) How far has the elevator risen 1 second from the start? How far has it risen 2 secs, 3 secs, 4 secs, from the start? Sketch a rough graph of distance against time.

★ 29. CALCULUS STATEMENTS

In this question, v is a symbol for speed or velocity; *a* is a symbol for acceleration, *t* for time.

(a) What does Δv mean?
(b) What does the statement "$\Delta v/\Delta t$ = constant" mean?
(c) What does the statement "$\Delta a/\Delta t$ = constant" mean? (Make an intelligent guess.)

30. FORMAL LOGIC*

Here is an example of a syllogism, a type of perfect deduction—too restricted to be much use in science but an important part of classical logic.

(1) All dogs have 4 legs. (the "major premise," a generalization)
(2) Fido is a dog. (the "minor premise")
(3) ∴ Fido has 4 legs. (the conclusion)

These three steps involve three "terms":

(a) 4-legged creatures (the major term, a large class)
(b) dogs (the middle term, a smaller class)
(c) Fido (the minor term, a member of a class)

The argument holds true if:
(c) falls wholly within the class (b) and (b) falls wholly within the class (a). Then, (c) must fall within the class (a). A corresponding argument can be carried out if (c) falls wholly within (b) but (b) falls wholly *outside* (a). Then (c) must fall *outside* (a).

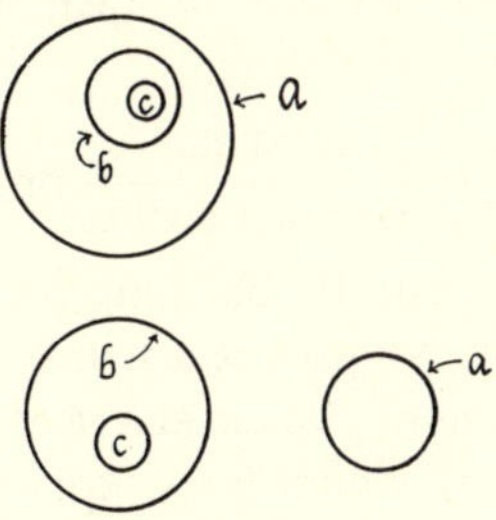

FIG. 1-16.

The inference or conclusion drawn in a "syllogism" may be untrue

(i) because the major or minor premise is not true,
(ii) because the reasoning is untrue, e.g., (b) falls partly within (a),
(iii) because there is confusion of language, e.g., ambiguous terms.

In each of the following examples there is something that makes the conclusion untrue or at least unjustified. Point out the defect in each case.

(A) All plums are vegetables.
This is a plum.
∴ This is a vegetable.
(B) All poisons are harmful.
Sugar is a poison.
∴ Sugar is harmful.
(C) All dogs are animals.
An antelope is an animal.
∴ An antelope is a dog.
(D) All salts dissolve in water.
George is an old salt.
∴ George dissolves in water.
(E) Caesar Augustus was a Roman Emperor.
Julius Caesar was a Roman general.
∴ Julius Caesar was the uncle of Augustus.

* Drawn from "Clear Thinking" by R. W. Jepson (Longmans, Green and Co., London, 1936).

APPENDIX A · THE ALGEBRA MACHINE

Grinding Out Useful Formulas for "Constant Acceleration"

In this appendix we shall neither discover new physics nor review old physics, but only do some intellectual machine-shop work. We shall start with a clear assumption, namely *motion with constant acceleration*, and make the algebra-machine grind out some logical consequences. The results, old information rehashed in new form, will be useful in examining the real world; but while we are deriving them we can sit in our ivory tower, confident that our work is necessarily true—true to its own assumptions—the work of perfect logic.

Definition. We choose to deal with (rate of change of speed), which is

$$\frac{\text{CHANGE OF SPEED}}{\text{TIME TAKEN FOR CHANGE}} \text{ or } \frac{\Delta v}{\Delta t}.$$

Since this turns out to be a useful thing-to-deal-with, or *concept*, we give it a name, ACCELERATION. Then the statement "ACCELERATION $= \Delta v/\Delta t$" is merely dictionary definition, explaining what we give that name to.

Assumption. We assume that acceleration is constant. (That is, we investigate the kind of motion which has constant $\Delta v/\Delta t$. There are many other types of motion common in nature; but this one is both simple and important, so we investigate it in detail.)

$$\frac{\Delta v}{\Delta t} = \text{constant, whose value we will call } a$$

In our elementary algebraic method, we shall assume that the *average* speed of an object moving with constant acceleration is just the average of the speeds at the beginning and end of the trip. We assume that

$$\text{AVERAGE SPEED} = \frac{\text{INITIAL SPEED} + \text{FINAL SPEED}}{2}$$

We also say

$$\text{DISTANCE TRAVELED} = \text{AVERAGE SPEED} \cdot \text{TIME}$$
$$\text{or } s = \bar{v} \cdot t$$

Note that we use · as a sign for multiplication, modern practice that we shall extend to *units* that are multiplied, such as man · hours, and we use v with a bar over it for "average v."

Terminology. Let:

(i) acceleration be a meters/second per second

(ii) moving object's speed be v_0 meters/second at the instant when the clock is started, that is, when $t = 0$. We shorten this description to:
initial speed $= v_0$ meters/second, at $t = 0$

(iii) moving object's speed be v meters/second when the clock reads t seconds; or:
final speed $= v$ meters/second.

(iv) distance traveled, in time t secs, be s meters.

These are merely descriptions of the letters we use. We can make a more connected statement, thus: the moving object starts with speed v_0, travels a distance s in time t, with acceleration a, reaching a final speed v.

Relationships. Now let the algebra machine grind out relationships.

(1) $v = v_0 + at$

$$\text{ACCELERATION, } a = \frac{\Delta v}{\Delta t} \quad \text{(dictionary definition)}$$
$$= \frac{\text{GAIN OF SPEED}}{\text{TIME TAKEN}}$$
$$= \frac{\text{FINAL SPEED} - \text{INITIAL SPEED}}{\text{TIME TAKEN}}$$
(if acceleration is constant)
$$= \frac{(v - v_0)}{t}$$

The last line gives only an average acceleration, *unless* the acceleration is constant, as we assume here. To obtain a useful expression for final speed, v, we must now rearrange the algebra. We start with the equality-truth

$$a = \frac{v - v_0}{t}$$

and arrive at an equally true statement by multiplying both sides by t.

Then: $\quad a \cdot t = v - v_0$

We arrive at another equally true statement by adding v_0 to both sides.

Then: $\quad v_0 + at = v - v_0 + v_0 = v$

$\therefore \; v_0 + at = v$

And, reversing this,

$$\underline{v = v_0 + at.}$$

All the changes we have made from $a = \dfrac{v - v_0}{t}$ are merely changes allowed by the rules of logic. The result, $v = v_0 + at$, is just as true or untrue as the starting-point, $a = (v - v_0)/t$. In this case we can see that the new "formula" is just a new version of the old starting-point, because it says

FINAL SPEED = INITIAL SPEED + RATE OF GAIN · TIME

(and this must be the gain of speed — RATE OF GAIN · TIME)

which says,

FINAL SPEED = INITIAL SPEED + GAIN OF SPEED = FINAL SPEED.

This discussion must have seemed unnecessarily long if you are familiar with algebra and trust it. You could just say,

$$a = \frac{v - v_0}{t} \qquad \therefore\ at = v - v_0 \qquad \therefore\ v = v_0 + at$$

the essence of the logic being coded in the ∴ signs. But if you regard formula-making as a mystery, you should read the detailed discussion carefully. The novice might take our word for the algebra, but he needs to grow out of mistaken ideas about the "truth" of formulas or about the mystery of their derivation.

(2) $s = \frac{1}{2}(v + v_0)\,t$

Our experimental tests will deal with DISTANCE, not SPEED. To see what our assumption of constant acceleration predicts for the relation between DISTANCE TRAVELED and TIME TAKEN, we need some way of computing distance when the speed is changing. We make a common-sense guess that we should use the AVERAGE SPEED, $\bar{v}$, got by adding the initial and final speeds and dividing by 2. Thus,

$$\text{AVERAGE SPEED, } \bar{v} = \frac{v_0 + v}{2}.$$

We use this AVERAGE SPEED as a steady speed to replace the real changing speed and we find the DISTANCE TRAVELED by multiplying AVERAGE SPEED by TIME.

Then, DISTANCE $s = \bar{v} \cdot t$

or $s = \frac{1}{2}(v + v_0)\,t$

In this relation, acceleration, a, and its definition do not appear. Yet the relation is not true unless the motion is one with constant acceleration (see Problem A-1 at the end of this appendix). This statement is not just a rearrangement of the earlier one. It contains the assumption about average speed. That assumption, so far only a guess from "common sense," can be checked by calculus (special algebra) or by Galileo's clever geometrical treatment (see Problem A-1). These show that this use of average speed is correct for motion with fixed acceleration. For other types of motion, a different kind of average, not the arithmetic mean, is needed.[30] So our assumption is a lucky one, correct for constant acceleration—we use it in teaching only because we know from calculus that it is a safe one. Thus the elementary presentation has been tailored to give the right results. Though this is sometimes unavoidable it gives a regrettable impression of glib plausible assumptions and fails to show the careful feeling-of-the-way and honest testing which are characteristic of science. Therefore, you should study Problem A-1.

(3) $s = v_0 t + \frac{1}{2}at^2$

We still want to express DISTANCE TRAVELED in terms of TIME and ACCELERATION, without using the FINAL SPEED. We obtain this from the other two relations, (1) and (2), by using one of them to provide an expression for v which can then be inserted instead of v in the other.

Thus, $s = \dfrac{(v_0 + v)}{2}t$ and $v = v_0 + at$

$$\therefore\ s = \frac{(v_0 + v_0 + at)}{2}t$$

$$= \frac{(2v_0 + at)}{2}t = \frac{2v_0 \cdot t}{2} + \frac{at \cdot t}{2}$$

$$\therefore\ s = v_0 t + \tfrac{1}{2}at^2$$

This gives a relation, belonging with constant acceleration, which is useful in experimental tests.

If the timing starts from the instant when the moving thing is at rest, the initial speed is zero, $v_0 = 0$, and the relation becomes

$$s = \tfrac{1}{2}at^2$$

Since a is constant, $\frac{1}{2}a$ is constant, so we can say

$$s = (\text{constant}) \cdot t^2 \text{ or } s \propto t^2.$$

Thus we can say, "theory predicts that $s \propto t^2$ for constant acceleration from rest." When we say "theory predicts," we mean that starting from some assumptions and using reasoning-machinery (which includes mathematics) we have recast those assumptions in what looks like a different form. *If* experiment agrees with this new form, we may decide our assumptions (and our machinery) are "true" or "justified." Yet often we cannot be sure that our chosen assumptions give the only possible true underlying story. We should be safer to say that they fit the facts so far. If you found in experi-

[30] For example, if the acceleration is not constant but starts with a large value and soon dwindles to zero, the moving object makes most of its gain of speed quite early in its trip, and then the proper average speed is not $(v_0 + v)/2$ but greater than that.

ments on falling bodies that distances and timings agreed closely with the relation $s \propto t^2$, then you could say that they agree with the relation predicted for constant acceleration. You could say that falling bodies seem to move with constant acceleration. In experiments on balls rolling down a plank, Galileo found that distances and timings fitted fairly well with the relation $s \propto t^2$. So they agreed with his prediction for constant acceleration.

Notice that the experiments do not prove the formula is the right one for constant acceleration. The formula itself is necessarily, logically, true for any motion which does have fixed acceleration. Experiments only show that the rolling motion, in agreeing with the formula (probably) has constant acceleration. When we compare experimental data with the formula we can discover something about nature.

Arriving at the formula involved the following stages:

Definition of acceleration: We invented it, chose a name, then used it.

Decision to think about motion with constant acceleration. This is one of the many choices we might have to try for real falling bodies. But, once made, the decision enables us to proceed with algebra. In making this decision we are not discovering anything about nature.

Algebra: A logical sausage-making machine. Mathematics cannot manufacture scientific facts, though it may help us to discover them.

Common-sense assumption that the proper $\bar{v}$ to use is $(v_0 + v)/2$. The risk in this can be avoided by Galileo's geometry (Problem 1), or by a calculus investigation, which would justify it for fixed acceleration.

Algebra again

Result: A useful relationship, deduced from our assumptions, useful in experimental tests.

(4) $v^2 = v_0{}^2 + 2\,as$ [This is a form which we shall not need for a long time yet. This section may be postponed till it is needed.]

We can use further algebra, a few more turns of the sausage machine, to change the formulas to other forms. We already have three relations:

(1) involving v, v_0, a, t, but not distance, s;
(2) involving s, v, v_0, t, but not acceleration, a;
(3) involving s, v_0, a, t, but not final speed, v.

Later we shall want a relation expressing v in terms of v_0, a, s, but not involving the time t explicitly. Since we want it without t, we obtain it from any two of the earlier relations by eliminating t. For example, we can use (1) and (3). Then $v = v_0 + at$ gives $t = \dfrac{(v - v_0)}{a}$ and we substitute this in $s = v_0 t + \tfrac{1}{2} at^2$.

Then: $s = v_0 \left[\dfrac{(v - v_0)}{a} \right] + \tfrac{1}{2} a \left[\dfrac{(v - v_0)}{a} \right]^2$

Will this lead to the formula (4) quoted above? Yes, if you use courage and algebra. You will have to square and cross-multiply and rearrange and simplify. The work will be clumsy and messy, but the final expression for v^2 will be $v_0{}^2 + 2\,as$. Try it, if you like.

The professional mathematician has a strong poetic sense of form in his own language of mathematics and he would consider the method above horribly clumsy. He would say, "Here is a more elegant derivation . . ." and would produce the answer quickly and neatly. Non-mathematicians who see him do this are mystified by his superior knowledge, and may be annoyed by the magical atmosphere. The real story is a sordid one. The mathematician is quite human, and feels his way in several trials, like any other explorer—though in simple problems his exploring may have all been done before and stored in his mind as "mathematical common sense." When he has found the answer by *any* method, clumsy or not, he may try working *backwards* from it to find a neat method of deriving it, like a mountaineer seeking a better path. There is no sin in this, but then he often forgets to tell the layman about the previous work, and startles him by producing the elegant method out of his hat. Let us try such an analytical search, thinking aloud as we go. The answer we want is $v^2 = v_0{}^2 + 2\,as$, so far obtained by algebraic drudgery. Try to undo it. Does it look as if it could be twisted or changed easily by algebra? Does it simplify or split up in any obvious way? No. Then we must push it around. Try shifting something across the $=$. Then we can have $v^2 - v_0{}^2 = 2\,as$. Is *this* easily attacked by algebra? Yes, the left hand side is an old friend, with factors $(v + v_0)(v - v_0)$. We could manufacture it from those factors if we could obtain them separately from somewhere. Where have we seen $(v + v_0)$ before? In the relation (2), $s = \tfrac{1}{2}(v + v_0)t$. Then $v + v_0 = 2s/t$. Where have we seen $(v - v_0)$? In the definition of acceleration, which we wrote $a = (v - v_0)/t$. Therefore, $(v - v_0) = at$. Now we want $v^2 - v_0{}^2$, which we can get by multiplying $(v + v_0)$ and $(v - v_0)$. We do this, using $(v + v_0) = 2\,s/t$ and $(v - v_0) = at$.

$$(v + v_0)\,(v - v_0) = (2\,s/t)\,(at)$$

$\therefore\ v^2 - v_0^2 = 2\,as$, which leads to the form we want. Now, having found the method by analysis, we erase the details of our search and start afresh, thus:

To derive $v^2 = v_0^2 + 2\,as$ by an elegant method, start with the definition of acceleration,

$$a = (v - v_0)/t,$$

and with the formula for distance travelled in terms of average speed, $s = \tfrac{1}{2}\,(v + v_0)t$, and just multiply these two equations together, obtaining $a \cdot s = \tfrac{1}{2}\,(v^2 - v_0^2)$ which reduces to

$$v^2 = v_0^2 + 2\,as$$

Here, then, are four relations between v, v_0, a, s, and t.

$$v = v_0 + at \qquad s = \tfrac{1}{2}\,(v + v_0)\,t \qquad s = v_0t + \tfrac{1}{2}\,at^2$$

$$v^2 = v_0^2 + 2\,as$$

They provide a quick way of calculating the value of any one of these quantities, given the values of three others.

Algebra Yields Net *Distance*

The numerical values must be given appropriate + and − signs. For example, if the initial velocity is 6 ft/sec eastward and the acceleration 2 ft/sec/sec eastward, we can say $v_0 = +6$ and $a = +2$. However, if v_0 is 6 ft/sec eastward but the acceleration is in the opposite direction, 2 ft/sec/sec westward, then one of them must have a *minus* value. If we say $v_0 = +6$ we must say $a = -2$, using + signs for eastward velocities, accelerations and travel-distances, and − signs for westward ones. Then s is the *net* distance travelled in time t, not the arithmetic sum of westward and eastward travels. This is because in calculating each part of the trip the algebra will give + sign to eastward travels and − sign to westward ones and in adding up these + and − parts to find s the algebra will give the net difference. With $v_0 = +6$ and $a = -2$ the motion is decelerated: slower and slower forward for 3 secs, then at rest, then faster and faster backward. In 5 seconds it will show a path like Fig. 1-17, with 9 ft forward travel, then 4 ft backward, giving a net travel 5 ft.

Algebra gives:

$$s = v_0t + \tfrac{1}{2}\,at^2 = (+6)(5) + \tfrac{1}{2}\,(-2)(5)^2$$
$$= 30 - 25 = 5\text{ ft.}$$

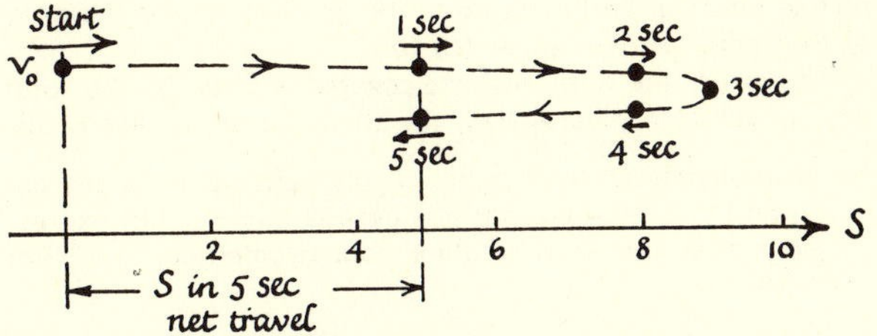

FIG. 1-17. S IS NET DISTANCE

Thus s always gives the *net* distance from start to finish.

These useful relations are tools, not vital pieces of science. They are absolutely true for motion with constant acceleration, and they are not reliable for other motions. Only experiment can tell us where they apply in the real world.

PROBLEMS FOR APPENDIX A

★ A-1. NON CALCULUS PROOF

Galileo, lacking the help of calculus and preferring geometry to algebra, dealt with uniformly accelerated motion as follows: Imagine a graph with time plotted along and velocity of a moving body plotted upwards. *If* the body has constant acceleration, its velocity must increase *steadily* as time goes on. The graph must be a straight line. It will not necessarily pass through the origin, but will start at the initial velocity, v_0 when time is zero, and run up to some value v at time t.

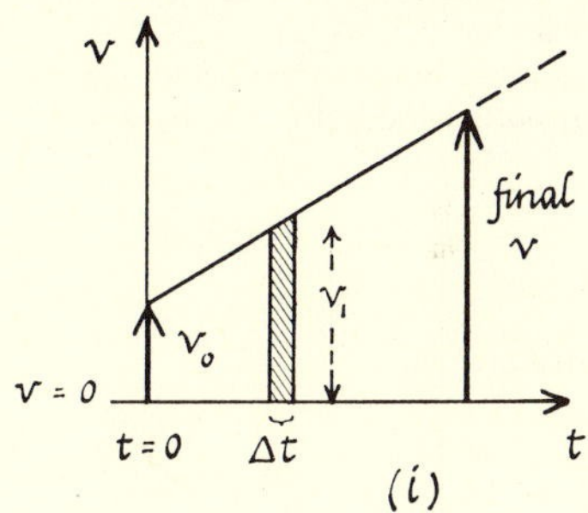

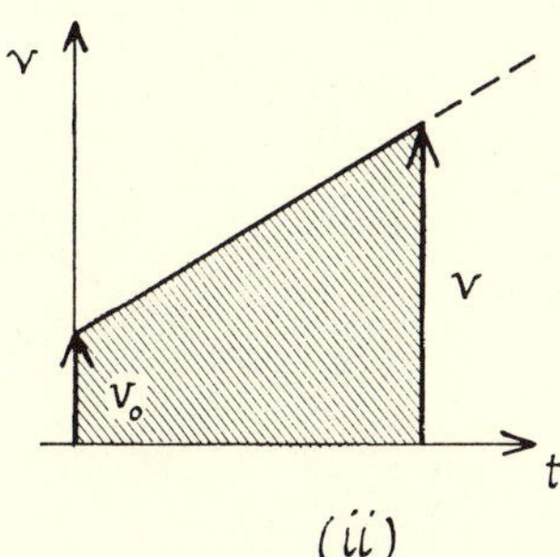

FIG. 1-18. GALILEO'S PROOF

Now consider what happens in some very short interval of time Δt, when the velocity is, say, v_1. (Of course v is increasing, but we can take v_1 as the average during short Δt.) Then the body moves a distance $[(v_1) \cdot (\Delta t)]$ in that time. But on the graph $[(v_1) \cdot (\Delta t)]$ is the [*height* · *width*] of the small pillar resting on Δt and running up to the graph-line. It is the *area* of that pillar, shaded in sketch (i).

Therefore, the *total* distance covered is given by the total area of all such pillars—i.e., the shaded area in sketch (ii).

(a) If in sketch (ii) the heights of this patch at its edges are v_0 and v as marked, and the base is time t, what expression gives the *area*? (Outline your geometrical argument briefly.)

(b) If the heights at the edges are v_0 and $v_0 + at$ (which follows from the definition of acceleration), what expression gives the area? (Outline your argument briefly.)

(c) Write the results of (a) and (b) as expressions for s the distance covered by the body in time t.

(d) Now suppose the acceleration is *not* constant but starts with a smaller value, rising to a greater one, so that the velocity still changes from v_0 to v in time t, but *not* steadily. (i) Sketch the new graph picture. (ii) Will the expressions from (a) and (b) apply now? (iii) What weakness in the earlier algebraic discussion in Appendix A has now been removed?

★ A-2. CALCULUS PROOF

In the limit, velocity, v, is rate-of-change of distance, ds/dt, and acceleration, a, is rate-of-change of velocity $\frac{dv}{dt}$ or $\frac{d}{dt}\left(\frac{ds}{dt}\right)$ or $\frac{d^2s}{dt^2}$ · Show that if a is constant, each of the following is true:

(i) $dv/dt = a$ integrates to $v = v_0 + at$
(where v_0 is a constant, the value of v at time $t = 0$)

(ii) $v = v_0 + at$ integrates to $s = v_0t + \frac{1}{2}at^2$
(*Hint*: remember $v = ds/dt$.)

(iii) $dv/dt = a$ integrates to $v^2 = v_0^2 + 2as$
(*Hint*: try multiplying both sides by v.)

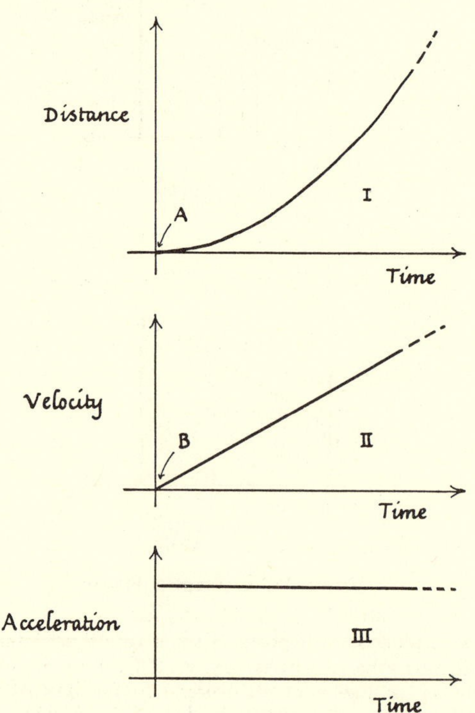

Fig. 1-19. Problem A-3, parts a, b, and c

A-3. GRAPHS OF MOTION

Fig. 1-19 shows an arrangement of three time-graphs for the motion of an object along a straight track. Graph I shows *distance* plotted against *time*; graph II *velocity* against *time*; graph III *acceleration* against *time*. They are drawn with matching time-scales. The graphs sketched relate to an object moving with constant acceleration, starting at $s = 0$ (shown by A) and velocity $v = 0$ (shown by B) at $t = 0$. In graphs for more complicated motions, all three lines may be curved.

(a) In the general case of any motion, one or more of the graphs can be derived from another of the three by tangent slopes. Which one(s)? Explain why.

(b) In the general case, one or more of the graphs can be derived from another of the three by measuring areas under the curve. Which one(s)? Explain why.

(c) A motorcycle policeman starts from rest, accelerates 15 ft/sec² for 6 secs; runs at constant velocity for 10 secs; then skids to a stop in 4 secs, with constant deceleration. Sketch a trio of graphs I, II, III, for his motion.

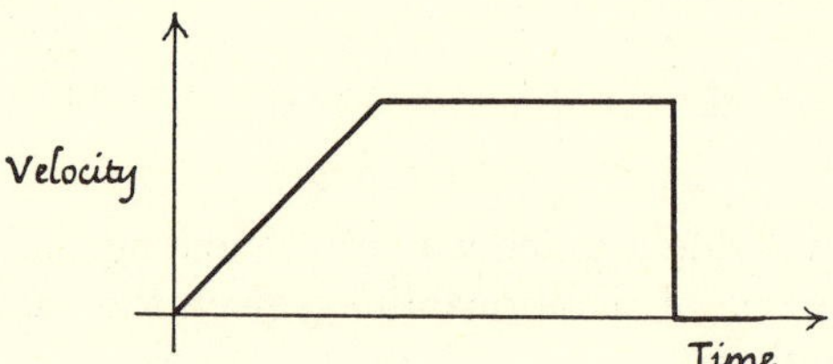

Fig. 1-20. Problem A-3, part d

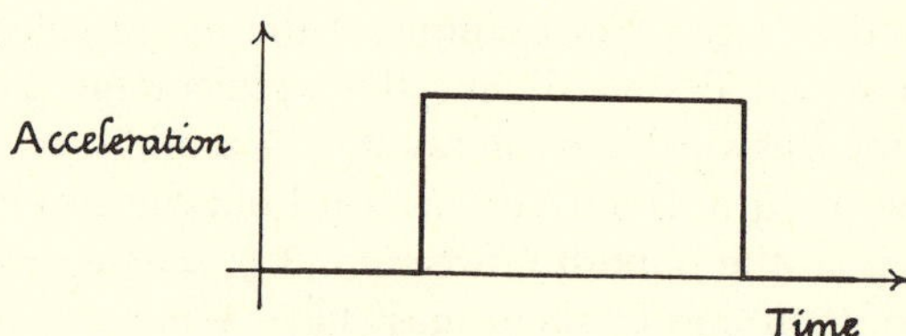

Fig. 1-21. Problem A-3, part e

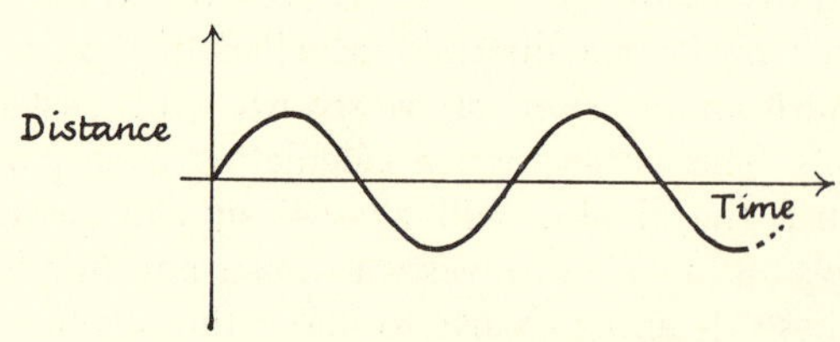

Fig. 1-22. Problem A-3, part f

(d) Fig. 1-20 shows graph II for the motion of a car. Copy it and add sketches of graphs I and III.

(e) Fig. 1-21 shows graph III for the motion of a truck. Copy it and add sketches of graphs I and II.

(f) Fig. 1-22 shows graph I for the motion of the bob of a long pendulum along its almost-straight path. Copy it and add sketches of graphs II and III. (*Difficult*: Deserves careful guessing.)

APPENDIX B · "g"

Measurement of "g"

We have glibly announced the value of "g" as 9.8 meters/sec^2 (or 32 ft/sec^2), but this came from laboratory measurements. You will use it for simple calculations concerning falling bodies, and for important calculations of forces when you treat "g" as gravitational field-strength. "g" is such a useful quantity that you should see its value measured before you use it. You could make a very rough estimate with a stone and a stopwatch and a meter-stick.

PROBLEM B-1. ROUGH MEASUREMENT OF "g"

An experimenter drops a big stone from a 14th-story window and finds it takes "just over" 3 seconds to reach the ground. If the window is 150 ft from the ground,

(a) Make an estimate of "g."

(b) Taking 150 ft to be about 46 meters, estimate "g" in meters/sec^2.

A better measurement can be made with an electric clock, as illustrated in Fig. 1-23, and you should see some such demonstration. For very accurate measurements you must wait for the promised scheme which avoids friction and takes a group of falls.

PROBLEM B-2. MORE ACCURATE MEASUREMENT OF "g"

A metal ball is allowed to fall from ceiling to floor. At the ceiling it is held against two metal pins so that it makes an electrical connection which *prevents* the electric clock from starting. The ball is released abruptly, and the clock starts.

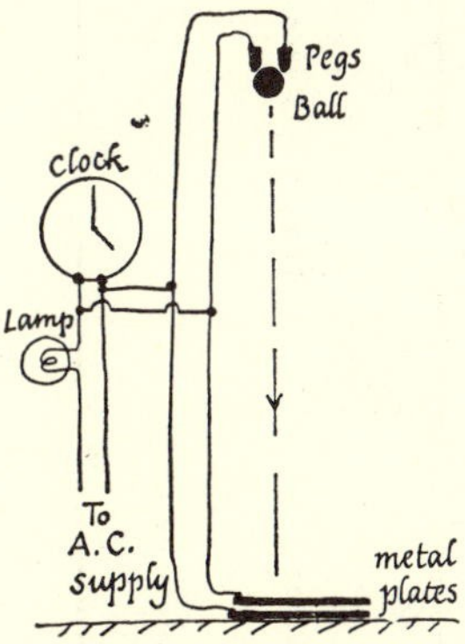

FIG. 1-23. MEASURING "g"

As it reaches the floor, the ball pushes two light metal plates together, making another electrical connection which stops the clock. In an actual experiment, the height of the fall was 7.00 meters from ceiling pegs to floor contacts, and the clock recorded a time of 1.20 secs.

(a) Estimate the value of "g," using these data.

(b) Say what assumptions you made in (a) concerning the type of motion; the apparatus; the conduct of the experiment. (Give details; avoid prim generalities such as "apparatus accurate" or "avoided personal error.")

Values of "g" in various localities

"g" has been measured very precisely at a few standard laboratories. Comparative measurements have then provided accurate values of "g" at many places all over the world.

	New York	Equator	Pole
Value in meters/sec/sec	9.80267	9.780	9.832
Value in feet/sec/sec	32.16	32.09	32.26

For ordinary calculations, in problems or in experiment-design, you should use the rough values $\underline{g = 9.8 \text{ meters/sec/sec}}$, $\underline{g = 32 \text{ ft/sec/sec}}$.

Arithmetical Problems on Free Fall: Dissected Problems

When you know the value of "g," you can make simple calculations about dropping stones, arrows shot at monkeys, etc. Such calculations are occasionally used by physicists in designing apparatus or in dealing with some experiment, but they are not important physics. Elementary textbooks and examinations make much of them "because they make accelerated motion clearer." Students trained to solve them mechanically may gain little but a damaging prejudice that "physics consists of putting numbers in the formulas." We wish to avoid that foolish picture of science, and we would not give you such problems in this course except for two reasons: (1) You may meet similar calculations, that are important, in atomic physics; (2) They will show you something important about the place of mathematics in physics. For these two reasons you should work through Problems B-3, 4, 5, and 6. Even so, if earlier studies have made you a convinced formula-monger you had better omit these problems unless you are prepared to start with an open mind.

Problems B-3 to B-6 have been dissected. You should answer them step by step, on question sheets reproduced from the small ones printed here. This scheme—which you will meet several times in the course—is intended to give you preliminary help and teaching towards later problems to be done on your own. Note that this insulting simplicity is meant to help you with the mathematics but not to save you from thinking out the physics for yourself. As you work such problems you should stop to notice that you are learning a method of solving them, but you should then concentrate on the physical results that emerge.

The problems here are intended to be answered on typewritten copies of these sheets. Work through the problems on the enlarged copies, filling in the blanks, (______), that are left for answers.

PROBLEMS ON ACCELERATED MOTION NAME________________

Work through these problems, filling in the blanks which are left for answers. Doing that will show you how to solve such problems and will, it is hoped, give you a pleasant understanding of the use of mathematics in physics. You will see that mathematics is a faithful servant, but sometimes a slightly dumb one, carrying out instructions relentlessly.

PROBLEM B-3. A rock falls from rest, with constant acceleration 32 ft/sec per sec, downward. (GIVEN: The acceleration is constant, and air friction is negligible, in this case.)

(a) What will be its velocity after 3 seconds of fall?

(b) How far will it fall in 3 seconds?

A. ARITHMETICAL METHOD.

(a) Acceleration 32 ft/sec per sec means that the rock gains in velocity

by ________ . ________ (units) in each second.

∴ In 3 secs of fall it gains in velocity. . . . . ________ ft/sec

∴ Since it starts from rest, its final velocity is. . . . ________ ft/sec

(b) The velocity grows from ________ ft/sec (at start) to. . . ________ ft/sec

∴ Average velocity is $\frac{1}{2}$(________ + ________) or. . . ________ ft/sec

∴ Distance travelled, with this average velocity,

in 3 sec is (________) · (________) or. . ________ ft

B. ALGEBRAIC METHOD.

(a) The acceleration, a = 32 ft/sec/sec Time, t = 3 secs

Initial velocity, $v_o = 0$

Substituting these values in $v = v_o + at$

we have: final v = ________ + ________ = ________ ft/sec

(b) Substituting the values above in $s = v_o t + \frac{1}{2}at^2$

we have: s = ________ + $\frac{1}{2}$(________) = ________ ft

(NOTE: In using algebra, always state the "formula" clearly first, as above. Also state the values you are going to substitute, attaching units to them. For example, "t = 3 secs", not "t = 3".)

PROBLEMS ON ACCELERATED MOTION SHEET 2

PROBLEM B-4. A ball is thrown downward, and released with velocity 10 ft/sec, to fall freely at the instant when the clock is started.

(a) What will be its velocity after 3 seconds of fall?

(b) How far will it fall in 3 seconds?

A. ARITHMETICAL METHOD.

(a) Acceleration 32 ft/sec per sec means that

the ball gains in velocity by ________ ________ (units) in each second.

∴ in 3 secs of fall it gains in velocity . . . . ________ ft/sec

∴ since it starts with velocity 10 ft/sec downward,

its final velocity will be ________ ft/sec

(b) The velocity grows from ________ ft/sec at start to

the final value of ________ ft/sec

∴ average velocity is $\frac{(\_\_\_\_)+(\_\_\_\_)}{2}$ or . . ________ ft/sec

Distance ball would travel in 3 sec with this

average velocity is. . . . ________ ft

B. ALGEBRAIC METHOD.

(a) The acceleration, a = 32 ft/sec/sec, downward

Initial velocity, v_o = 10 ft/sec, downward Time of travel, t = 3 secs

Substituting these values in $v = v_o + at$, we have

Final velocity, v = ________ + ________ = ________ ft/sec

(b) Substituting the values above in $s = v_o t + \frac{1}{2}at^2$, we have

Distance s = ________ + $\frac{1}{2}$(________) = ________ ft

PROBLEM B-5. From a place on the top of a tower, a ball is thrown upward, released with velocity 10 ft/sec upward at the instant the clock is started.

(a) What will its velocity be after 3 seconds?

(b) How far below its starting point will it have fallen after 3 secs?

PRELIMINARY DISCUSSION

In this case, the ball moves upward at first, but more and more slowly, having a downward acceleration 32 ft/sec per sec (equivalent to an upward deceleration) all the while. It reaches a highest point or vertex, (still with the same downward acceleration); and then it falls, (still with the same downward acceleration). Questions (a) and (b) do not inquire about the highest point; and, if acceleration and velocity are given + and - signs for down and up, the algebra should carry right through the highest point and yield the net distance, s. (So, although you may have been shown methods which make you calculate first the trip up to the vertex, then the trip down, avoid such methods -- see the discussion below.)

A. ARITHMETICAL METHOD.

(a) Acceleration 32 ft/sec per sec down means that

in every second ball gains in downward velocity by ________ ________

∴ In 3 secs, its gain in downward velocity is ________ ft/sec

But it started with 10 ft/sec upwards; so, with the

gain of downward velocity, the final velocity must be ________ ft/sec down

(b) To calculate the net distance fallen we need the average downward velocity. The velocity is upwards at first and downwards at the finish. To find the proper average velocity, we cannot just add the two numbers for the velocities and then halve the total, because that would be the procedure for a ball thrown downwards, as in the previous problem.

PROBLEM B-5 CONTINUES ON NEXT SHEET

PROBLEM B-5 (ARITHMETICAL METHOD continued)

This is like the problem of averaging a bank balance which changes steadily from $10 in the red to $40 in the black.

The average is not ½($10 + $40) or $25; it is ½(-$10 + $40) or $15.

To see that this is right, try making the change from $10 in the red to $40 in the black by a series of daily jumps, each of $5. The daily balance runs:

red	red		black	black . . .						. . . black
$10	$5	zero	$5	$10	$15	$20	$25	$30	$35	$40

The balance on the middle day is $15. The series is better shown thus:

-$10 -$5 0 +$5 +$10 +$15 +$20 +$25 +$30 +$35 +$40

So, in this problem, we attach + signs to all downward velocities and accelerations, but call the upward velocity at the start -10 ft/sec.

Then average velocity is given by $\bar{v} = \frac{-10 + (\quad\quad)}{2}$ = ________ ft/sec

Ball falling for 3 secs with this steady average velocity would travel

a downward distance given by (average velocity)·(time) = ________ ft

THIS GIVES THE NET DOWNWARD TRAVEL, NOT THE TOTAL LENGTH OF PATH UP AND DOWN. It gives the direct distance down from the start to the finish point. This is because we have used the kind of average that will yield this net distance.

B. ALGEBRAIC METHOD.

The acceleration, $a = +32$ ft/sec/sec (plus meaning "downward")

Initial velocity, $v_o = -10$ ft/sec (minus because upward)

Time of travel, $t = 3$ seconds

(a) Substituting in $v = v_o + at$, we have

final v = ________ + ________ = ________ ft/sec

(b) Substituting in $s = v_o t + \frac{1}{2}at^2$, we have

distance s = ________ + ½(____________) = ________ ft (up?/down?)

PROBLEMS ON ACCELERATED MOTION SHEET 5

PROBLEM B-6. (THE MOST IMPORTANT PROBLEM) A bird sits in a tree, 48 feet above the ground. A man on the ground vertically below shoots a projectile vertically up at the bird with initial velocity 64 ft/sec upward. How long will the projectile take to reach the bird?

A. ARITHMETICAL METHOD. Here the methods of arithmetic and/or common-sense become almost impossibly clumsy. You could find out where and when the "vertex" is, and work the problem from there, but the algebraic method is neater and more interesting. The difficulty is you do not know the velocity of the projectile when it reaches the bird.

B. ALGEBRAIC METHOD. Here we must distinguish between up and down. It does not matter which you label + so long as you keep to your choice. (Try choosing each way, and you will reach the same equations and the same answers in both cases.) It feels more comfortable to call upward distances and velocities and accelerations all + here. We will work the problem with that choice.

In that case, the (downward) acceleration must be called -32 ft/sec/sec.

Then v_o = +64 ft/sec; s = +48 feet; a = -32 ft/sec/sec

We wish to find the time, t, taken to reach the bird, 48 ft above the ground.

Substituting in the relation $s = v_ot + \frac{1}{2}at^2$, we have

$$\_\_\_\_\_\_ = (\_\_\_\_\_\_)t + \tfrac{1}{2}(\_\_\_\_\_\_)t^2$$

This is an ordinary quadratic equation. Like any quadratic, it has two answers. Simplify it and solve it by any method, in the space below.

Answers: t = ______ seconds <u>or</u> t = ______ seconds

One answer gives the time taken to fly up and hit the bird. Suggest below a meaning for the other answer. ________________________________

____

With his limited instructions, how could the faithful servant, mathematics, do otherwise than yield both answers?

PROBLEMS FOR APPENDIX B

Problems B-1 and B-2 are in the text.

B-3, -4, -5, -6. Dissected problems, to be worked on special sheets.

B-7. DOUBLE ANSWERS. (Another problem, like Problem B-6. Try this problem, using the methods shown in dissected problems, B-3 to B-6, and the hints given here. Leave it if you find it too hard, but try it anyway.)

A man standing on the top of a tower throws a stone up into the air with initial velocity 32 ft/sec upward. The man's hand is 48 ft above the ground.

(a) How long will the stone take to reach the ground? [*Hint*: Once again you must use + and — signs. If you choose + for upward, the *acceleration must have a negative value and the distance s from hand to ground being downward must have a negative value; but the initial* velocity will be +. If, disliking negative signs, you choose to use + for downwards, then you will still get the same equations and answers. Try both if you like, *but do not mix the two in one set of calculations.*

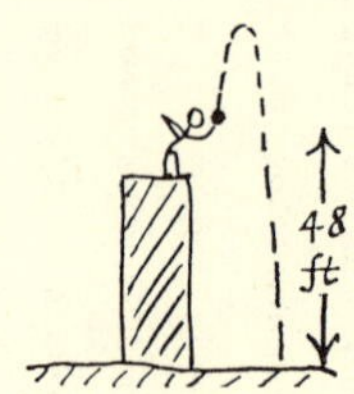

FIG. 1-24. PROBLEM B-7

(b) Once again, you will have a quadratic with two answers. Try to state a meaning for "the other answer." In doing this, ask yourself, "Was the mathematical machine ever told that the man actually threw the stone?"

EASY PROBLEMS ON FREE FALL MOTION* (Neglect air resistance)

v	?
v_0	−5 ft/sec
a	+32 ft/sec/sec
s	X
t	2 sec
$\bar{v}$	X

* In working problems on accelerated motion, you will find it pays to organize your information clearly, like a good engineer. A table like this is worth making. Write your data in the table, with ? where you seek information, and X where you neither have it nor want it. (This specimen shows the data and question in Problem B-10(a).) Then you can see which algebraic relation will be useful. (In this example it must be the one that does not contain s.)

B-8. A helicopter, remaining still above the ground, drops a small mailbag. When the bag has fallen for 2 seconds:
(a) What is its speed?
(b) How far has it fallen?

★ B-9. FREE FALL FROM MOVING OBJECT

A helicopter, falling steadily 5 ft/sec without acceleration, releases a small mailbag. After 2 seconds:
(a) What is the speed of the bag?
(b) How far has it fallen?
(c) How far is it below the helicopter?

★ B-10. A helicopter, rising steadily 5 ft/sec, releases a small mailbag. After 2 seconds:
(a) What is the speed of the bag?
(b) How far has it fallen?
(c) How far is it below the helicopter?

★ B-11. FREE FALL FROM MOVING OBJECT

What common property is shown by the answers to Problems 8, 9, 10?

★ B-12. IMPORTANT PROBLEM (Answer needed for later problems)

A man standing on a shelf 4 ft above the floor steps off and falls to the floor.
(a) How long does he take to fall?
(b) What is his speed just before landing?

★ B-13. CAR BRAKES

A certain car with smooth tires on a wet road can have an acceleration of 1/5 of "g" but not more. (To accelerate, the car must be pushed by some real, external, agent. The agent is the road, pushing the car by friction. With these tires, friction can provide up to $g/5$ acceleration but, if asked to provide more, the wheels begin to slip and friction falls to an even lower value, giving a smaller acceleration.)
(a) What speed will the car gain in 4 secs, with this maximum acceleration?
(b) How far can it travel from rest in 4 secs?

★ B-14. CAR BRAKES AND SAFETY

A car with good brakes but smooth tires on a wet road can have a deceleration of 1/5 of "g" but not more (see Problem B-13). Discuss the stopping of this car by answering the following questions:
(a) Driving at 30 miles/hour (= 44 ft/sec) the driver takes 1 sec to react to danger, decide to stop and get the brakes working; then he makes the brakes give maximum deceleration.
   (i) How far does he travel in the 1 second before braking?
   (ii) How much time do the brakes then take to reduce the speed from 30 miles/hour to zero?
   (iii) How far does the car travel in the braking time?
   (iv) How far does the car travel in the *total* time from seeing the danger until stopped?
(b) If the car is travelling twice as fast, 60 miles/hour, how far does it travel in the *total* time, as in (iv) above?
(c) The car is travelling 30 miles/hour and the driver (after 1 second of thought, etc.) jams the brakes on so that the tires skid, commanding less friction, giving a deceleration of only $g/8$. How far does the car travel in coming to rest? (Sliding friction in a skid is unable to provide such a large maximum force as non-slip friction.)
(d) With new tires on dry concrete, the car has maximum deceleration $g/2$. (Friction of rubber on concrete can do much better than that, but many a brake mechanism can not.) Again calculate the total distance of stopping, from 30 mi/hr.

B-15. "g" IN A MOVING LABORATORY

A portable timing apparatus can now be made to time free fall of a few feet from rest accurately enough to give a value of "g" reliable within 1% or better. Suppose such an apparatus gave g = 32 feet/sec^2. What would you expect it to give:
(a) If used in a railroad train running smoothly at fixed speed along a level track? (Think what happens when you drop something, say an orange, in a moving train.)
(b) In an elevator moving downward at constant speed? (*Hint*: Think . . .)
(c) In an elevator falling freely after its cable has broken?
(d) In elevator accelerating downward 16 ft/sec^2? (Make a bold guess.)
(e) In an elevator accelerating up 16 ft/sec^2?

MORE SIMPLE PROBLEMS ON FREE FALL

B-16. How long would it take a freely falling body to fall 400 feet from rest?

B-17. A ball is thrown upward with speed 80 ft/sec. How high will it rise?

B-18. An explorer discovers a deep crevasse in a rocky mountain. He drops a stone into it and 4 seconds later he hears the sound of the stone hitting the bottom of the crevasse.
(a) Estimate the depth.
(b) Comment on the accuracy of this method.

B-19. A stone thrown vertically upward with initial velocity 40 ft/sec takes 1 second to reach a bird.
(a) What is the vertical height of the bird above the thrower?
(b) A time of 1.5 seconds gives the same answer for the bird's height. Give a physical reason for this duplicity.

PROBLEMS ON APPARATUS OF PROBLEM B-2

B-20. Why is the lamp (or some other resistance) necessary in the arrangement sketched in Problem B-2?

B-21. In the experiment of Problem B-2, the following troubles may occur:
(a) The clock may lag a few tenths of a second in starting.
(b) The clock may lag a few tenths of a second in stopping.
(c) The pegs at the top, being compressed when the ball is held there, may give the ball a small downward shove when it is released.
(d) Air friction may have an appreciable effect.
   (i) For each of the troubles (a)-(d), say whether it, operating alone, would make the estimated value of "g" too big or too small; and give a brief reason for your answer.
   (ii) What would happen if (a) and (b) operated together, about equally?
   (iii) Suggest experiments to test for each trouble, (a)-(d). Describe them with sketches where possible.

CHAPTER 2 · PROJECTILES: GEOMETRICAL ADDITION: VECTORS

"What hopes and fears does the scientific method imply for mankind? I do not think that this is the right way to put that question. Whatever this tool in the hand of man will produce depends entirely on the nature of the goals alive in this mankind. Once these goals exist, the scientific method furnishes means to realize them. Yet it cannot furnish the very goals. The scientific method itself would not have led anywhere, it would not even have been born without a passionate striving for clear understanding."

—A. Einstein, *Out of My Later Years*

Experiments

This chapter might start with crisp statements of simple rules of projectile motion. Or you could consult a modern textbook on "Ballistics, the science of projectiles," which would give you profuse information and more abstruse rules. The text would mention ancient prejudices only to sneer at them, and tell you that Galileo's simple rules are of little use in modern gunnery. But with such a start you would miss a share of the delight of the great experimenters. Instead, please start with your own experiments.

Throw stones or coins outwards and watch their motion. Try this with a variety of objects ranging from a heavy stone to a crumpled sheet of paper. Try releasing two stones simultaneously, dropping one to fall freely downward, projecting the other horizontally. Make any other investigations and comparisons that occur to you; and try to extract simple rules or generalizations.

Watch a stone or baseball follow a curved path. Labelling this curve a "parabola" is neither true nor helpful at this stage. But it is good science to note that the curve is almost symmetrical, like (a) in Fig. 2-1, and unlike (b) or (c). This suggests that

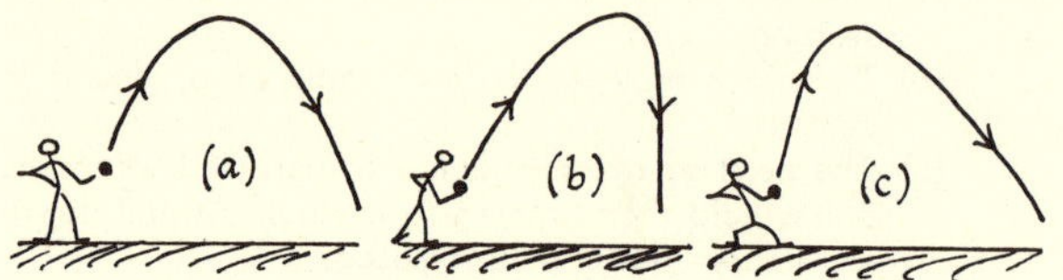

Fig. 2-1. Projectile Paths?

the motion in the downward half somehow matches the motion in the upward half. Perhaps the upward motion and downward motion take equal times—a suggestion to be investigated directly.

A careful experimenter trying a series of materials such as lead, stone, wood, cork, paper finds that for the later members of that series (b) is nearer the truth than (a).

As late as the 16th century, people believed the traditional statement that heavier things fall faster in proportion to their weight. And their beliefs about the path of a projectile were stranger still. It was said to be made up of three parts (see Fig. 2-2): (A) the violent motion (straight out unaffected by gravity)[1]; (B) the "mixed motion"; (C) the "natural motion" (where the bullet falls splosh on the victim below). You may see from your

Fig. 2-2. Medieval Idea of Projectile Path

own experiments with a ball of crumpled paper how such an idea arose, and you can see why it was foolish to apply it to dense and slow-moving cannon balls. Air resistance and gravity were making a confusing mixture. Galileo got rid of air resistance by thinking out what would happen if it were negligible. Cannon balls of his day moved so slowly that air resistance did matter very little, and thus his rules might have helped artillery men to hit their mark. As usual, practical men took little notice of scientists' suggestions for a long time; and by the

[1] If you feel inclined to jeer at this ancient picture, you should, as Lloyd Taylor suggested, ask your friends what path the bullet from a modern rifle takes when it first leaves the muzzle. Does it travel straight ahead, or does it begin to fall at once?

time Galileo's theory was taken up by gunners it had long been rendered useless by higher speeds. Meanwhile Newton and others had produced more useful theory which included air resistance. By now, three centuries later, projectiles move so fast that air resistance modifies their path tremendously. Fig. 2-3 shows paths for a large high speed projectile, (a) the "ideal" path without air resistance, as Galileo would have sketched it, (b) the actual path in air for the same elevation and muzzle velocity. Modern ballistics involves much more mathematics and even requires electronic-brain calculators to cope with the details of real problems. These are matters of engineering or applied mathematics which do not help our study of the growth of mechanics. Here we shall keep to the simple case of negligible air resistance.

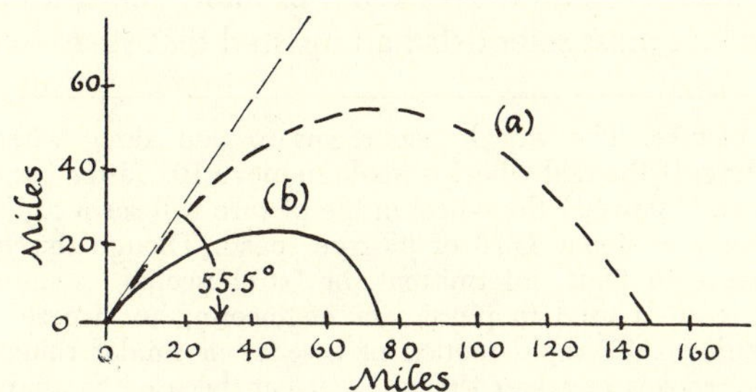

Fig. 2-3. Path of a Projectile
Curves (a) and (b) show paths of a projectile shot with initial velocity 1 mile/second in a direction making 55.5° with horizontal. (From *Science for the Citizen* by Lancelot Hogben; Allen and Unwin, London.)

Galileo tried to separate the up and down (vertical) motion of a projectile from its horizontal motion. Experiment vouches for this treatment by showing that these two motions are independent. Try this yourself. Throw one stone out horizontally and at the same moment release another to fall vertically. They both hit the floor at the same instant. Stone B moving in a curve has to fall the same *vertical* distance to reach the floor as stone A falling vertically. They take the same time. Do A and B keep abreast at intermediate stages of their fall? You need not place special observers to sight them at various levels. Instead, you can move the floor up to catch them earlier and repeat the experiment. Or, more easily, you can move the starting point down nearer to the floor. If A and B arrive at the same instant whatever height they start from, you can say fairly that they keep abreast all the way down. Notice how a series of experiments can be used to replace a difficult complex of simultaneous observations. In trusting our inference from such a set of experiments, we assume the "Uniformity of Nature."

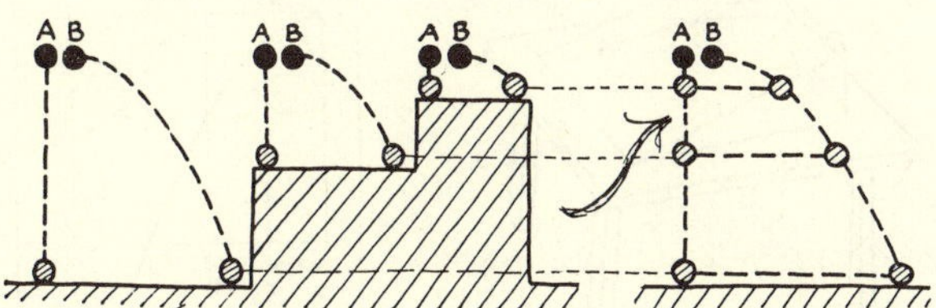

Fig. 2-4. Experimental Comparison of Motions: inferring the general result that falling stone and projected stone keep level all the way.

Demonstration Experiments

(1) Vertical and horizontal motions independent. Fig. 2-5 shows a simple demonstration experiment in which two metal balls are released by a small spring gun to fall like A and B. You should watch this experiment carefully, and ask to see it repeated with a different height.

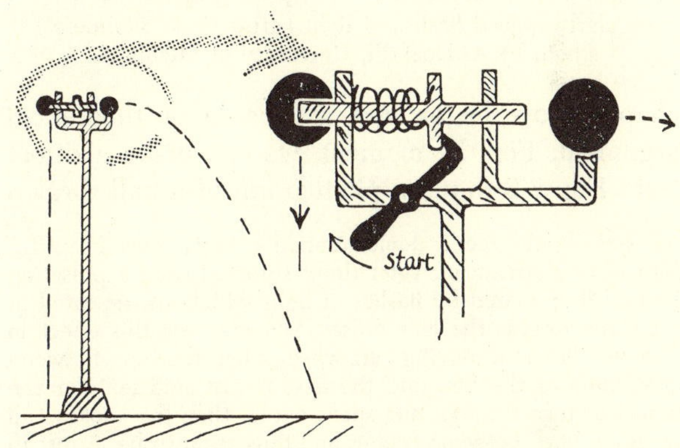

Fig. 2-5. Demonstration Experiment.
The spring gun releases one ball to fall freely at the instant that it projects another ball horizontally. A latch releases the gun's piston. The piston, driven by a compressed spring, hits the second ball, which is resting loosely on a support. The first ball has a hole in it which accommodates the other end of the piston until the piston is released; then the first ball is left behind to fall freely.

(2) Horizontal motion unchanging. A projectile moves *vertically* with the acceleration of gravity quite independently of its horizontal motion. How does its *horizontal* motion behave? The symmetrical path of a stone or ball suggests that it does not move slower and slower horizontally, or the path would be more like Fig. 2-1b. Galileo, revolting against the medieval view that any motion needs a force to keep it going—gravity, or demon or rush of air—suggested that the horizontal motion just continues unchanged, since there is no pull like gravity to increase or decrease it. You will see in a later

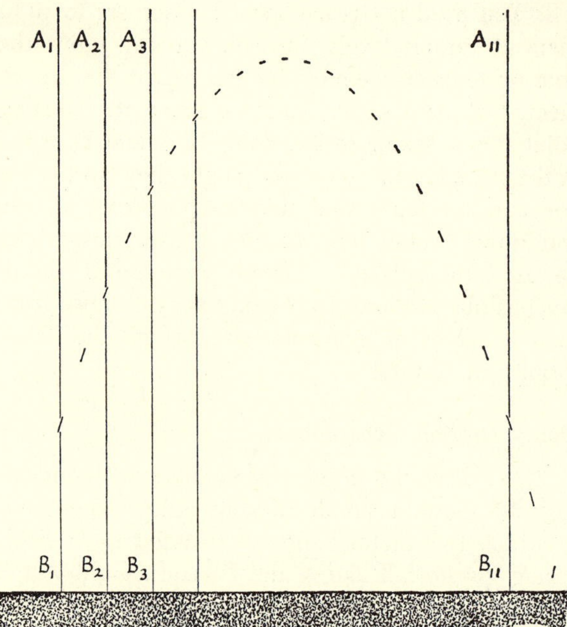

FIG. 2-6. PROJECTILE PATH: photographed with regularly spaced flashes of light (after F. A. Saunders.) Photo by A. Dockrill, University of Michigan.

chapter how he arrived at this by a theoretical argument. For the moment, we can make a direct test.[2] Fig. 2-6 shows a photograph of a ball thrown into the air and illuminated by a series of short flashes of light, evenly spaced in time. Measure the picture for yourself drawing in lines like A_1B_1, A_2B_2, A_3B_3. You will find that the lines are evenly spaced: $A_1A_2 = A_2A_3 = \ldots$ etc. Therefore the ball moved across steadily, moving neither faster nor slower horizontally while it rose vertically slower and slower then fell faster and faster. Once thrown, it kept its horizontal motion unaltered.

Galileo recognized this property of moving things and handed it on to Newton. For many centuries before him most scientists had insisted that steady mo-

[2] You should see a demonstration of this. One beautiful form uses a stream of water drops squirted from a pulsating jet and illuminated by flashes of light which are repeated at the same rate as the jet's pulses. You may see this effect in a movie film of a moving cart-wheel when the time between one frame of the film and the next is just sufficient for the wheel to turn through one spoke-angle; then the spokes "all move on one" between frames and thus seem to be at rest in the picture. The wheel then seems to skid along without rotating. If the real wheel is made to move 10% faster (or the camera is slowed) the wheel in the picture will seem to turn, but only at about 1/10 of its true speed. Though this is a nuisance in films, intermittent, or "stroboscopic," illumination is often used in physics or engineering to "freeze" or slow down the rapid motion of a series of similar things—wheel spokes or water drops. Or it can be used to study a single vibrating object which is repeating a motion rapidly (e.g., a bell, or a violin string). Fig. 2-7 shows the arrangement for water drops. Water is fed from a tank to a small glass nozzle through a rubber pipe which is squeezed by an electromagnet. The magnet, run by an alternating current, squeezes the pipe 120 times a second (twice per cycle of the A.C.) making the stream emerge in drops at a steady rate, 120 a second. The stream is shadowed on a screen by light from a small lantern. With steady illumination, the stream looks continuous. But when a spinning shutter is interposed, the flashes getting through show up individual drops. The shutter, a disc with a slit in it, can be spun by a synchronous motor run on the same A.C. supply. Then the flashes are synchronized with the drops and the pattern stays still. A rectangular grid of wires can be shadowed as well, for measurements.

As a simpler demonstration, balls or water drops can be projected in front of a blackboard and the curve of their path sketched and analyzed. Or you can try your own experiment, rolling a ball with diluted gravity on a slanting table. See Fig. 2-8.

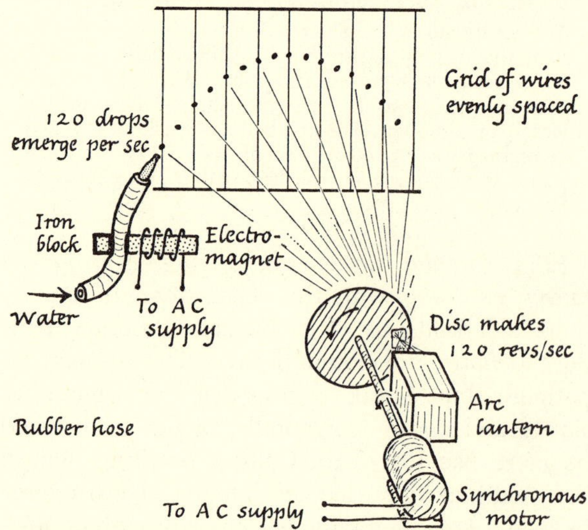

FIG. 2-7. STROBOSCOPIC ILLUMINATION OF A STREAM OF WATER DROPS. The drops emerge regularly, 120 a second, from the pulsed water jet and are illuminated by flashes of light, 120 a second.

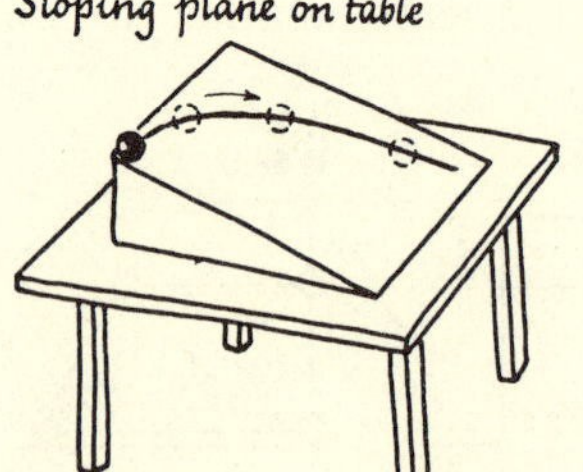

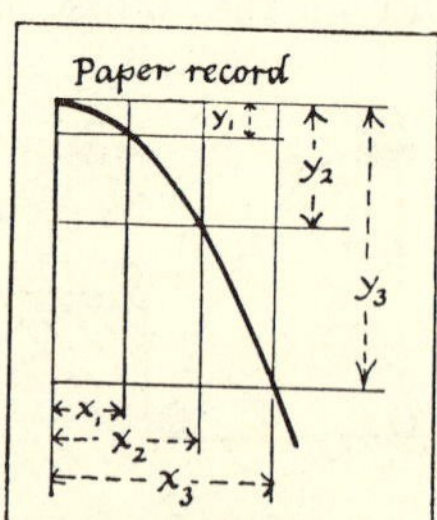

FIG. 2-8. DEMONSTRATION AND ANALYSIS of projectile motion with diluted gravity. A ball rolls across and down a sloping plane, on carbon paper to mark its path. Analysis: On the paper record, draw lines with $x_2 = 2x_1$, $x_3 = 3x_1$, and so on. Measure y_1, y_2, etc., and find out whether $y_2 = 2^2y_1$, $y_3 = 3^2y_1$, and so on.

MEDIEVAL VIEW OF MOTION

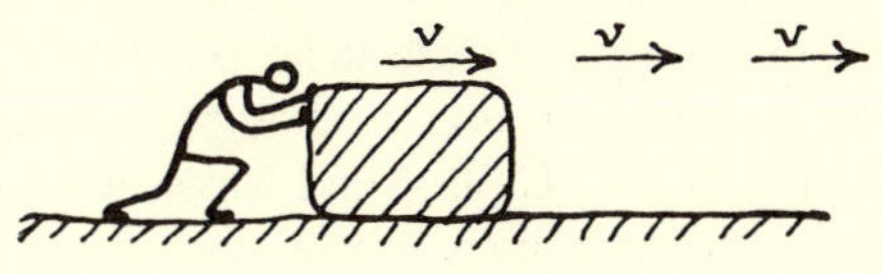

Fig. 2-9.

tion requires a force to keep it going. That ancient idea appeals to common sense today. To keep a box moving along the floor you have to shove; a car running along the level uses gasoline for its engine and the engine somehow provides a steady shoving force. "If you leave a moving thing alone," said the ancients, "it will come to a stop." But to Galileo and Newton, a rough floor and blowing wind do not leave a moving body alone; they exert forces opposing the motion (what we call friction-forces or air resistance). A massive cannon ball moving slowly experiences only trivial resistance; it *is* almost being left alone, so far as horizontal motion is concerned, and it keeps that motion. Hence the new view of a moving body, that it possesses something intrinsic in its motion that keeps it going, unless it is opposed. This something was called "impetus" by some 14th-century thinkers in Paris and Oxford. Their writings reached and influenced Leonardo da Vinci by 1500 and Galileo by 1600—if printing had been available, modern views on motion might have spread three centuries before Galileo. Impetus is a useful name for this quality of a moving body, with a comfortable feeling of "driving ahead" in our modern vocabulary. Later we shall change the name to "momentum," with a more precise meaning. Note that neither word explains anything; at best they are suggestive labels, reminders that a moving body carries its own motion with it and needs no maintaining push. Both are Latin words meaning motion; a Latin dictionary gives them different flavors.

Watching the motion of a cannon ball, Galileo said the gun gives it impetus *which it retains.* The horizontal part of this impetus remains unaltered. The vertical part is changed by the pull of gravity

GALILEAN AND NEWTONIAN VIEW OF MOTION

F v v v −F

Steady motion requires no RESULTANT force

v v v

A truck running along a level track has "impetus"

When resultant force acts, impetus increases

v v V V

Downhill force is a share of gravity

F v F V F V

Fig. 2-10.

Fig. 2-11. Attempted Demonstration. One observer drags a box along a rough "floor," measuring his pull with a spring balance. The "floor" is supported on frictionless rollers. Another observer uses a spring balance to restrain the "floor."

as in the motion of any other falling body. When a box needs a steady push to keep it moving along the floor, the floor exerts a dragging force opposing the motion, and Galileo and Newton would say our steady forward push is just enough to counteract that drag. Then in this case, too, the total force is zero while the box moves steadily. Fig. 2-11 shows an experiment intended to demonstrate this, but there is a serious flaw concealed in its argument. It is difficult or impossible to make an honest experimental test of the assertion that our forward push and the floor's friction-drag are just equal and opposite. At this stage you should accept it as a description of our viewpoint. We shall return to this idea. Meanwhile remember that direct observation shows that for all projectiles in cases where air resistance is negligible:

(I) *the motion is independent of the size or mass of the object,*
(II) *the vertical and horizontal motions are independent of each other,*
(III) *the vertical motion has a constant downward acceleration, the same as that of any falling body,*
(IV) *the horizontal motion continues unchanged.*

★ PROBLEM 1. PROJECTILE MOTION

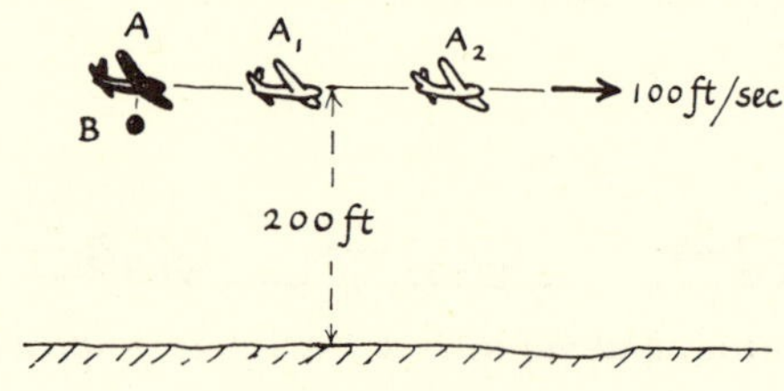

Fig. 2-12. Problem 1

Make your own sketch illustrating projectile motion by applying the simple discussion above to a plane dropping a bomb, as follows: An airplane, A, flying horizontally 200 feet above the ground at a speed of 100 feet/second, drops a bomb, B.

(i) Draw a sketch showing the positions of A and B, roughly to scale, at times 1, 2, 3, 4, 5 seconds after the release of B. Mark these positions A_1 and B_1, A_2 and B_2, etc. Pretend that air resistance is negligible. (Experiments show that it has only a small effect on a compact dense bomb at such speeds.)

(ii) Suppose the plane drops a larger bomb, twice as heavy. How will its path compare with that of the first bomb, if air resistance is again negligible?

(iii) Suppose the plane drops a wooden bomb, W, on which air resistance has a noticeable effect. Making sensible guesses, sketch the probable path of W, marking several positions of W and A.

PROBLEM 2

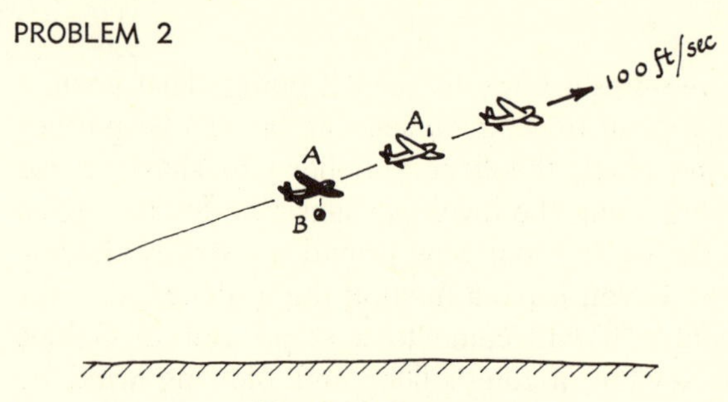

Fig. 2-13. Problem 2

Suppose the airplane of Problem 1 is not flying horizontally but is climbing steadily, so that it flies up a slanting straight line at constant speed.

(i) With what motion does the bomb now *start* when released?

(ii) Forgetting the plane for the moment, describe in words and with a sketch the path of *any* massive projectile when started thus. (If uncertain, try an experiment.)

(iii) Make a sketch showing several positions of B. Add to your sketch the corresponding positions of A. [*Hint*: Since the plane flies steadily, *it* has a constant horizontal component of velocity. The bomb. . . .]

PROBLEM 3

If you succeeded in answering Problem 2, you made an extra assumption concerning the vertical part of projectile motion, not needed in Problem 1. What was that assumption?

You can see how the "Rules" extracted from experiment can be turned to practical use. In Galileo's hands they might have aided medieval gunnery. Nowadays they form the starting point of modern ballistics, which adds the complexities of air resistance, Earth's motion, and even variable gravity.

Projectiles and Relative Motion

Galileo imagined experiments on board ship, to show that motions could be separated and that a steady motion of one's "laboratory" could be ignored. The following problem explores some properties of relative motion.

★ PROBLEM 4 (Hard, but important). A BEGINNING OF GALILEAN RELATIVITY

A passenger in a railroad coach lets an orange drop. It falls on his feet. In the following discussion assume air resistance is negligible.

(a) Suppose the coach is stationary. What is the path of the orange?
(b) Suppose the coach is moving steadily forward, say 10 miles/hour. Then before the orange is released it is moving with the man and coach, steadily forward 10 miles/hour. Thus when it is released the orange has a forward motion of 10 miles/hour, and begins to fall.
    (i) What happens to its downward motion, as time progresses?
    (ii) What happens to its forward motion, as time progresses?
    (iii) Suppose a stationary observer, standing on the roadside, watches through a window. Sketch the path of the orange, *as he sees it.* Mark three or four positions of the orange in its fall, O_1, O_2, . . . etc. and mark the corresponding positions of the passenger's feet, F_1, F_2, . . . etc.
    (iv) Sketch the path observed by the passenger in the coach, marking several stages.
    (v) Suppose the window shades of the coach are drawn down so that the passenger inside cannot see the outside world. Pretend the train moves smoothly with no bumps. Could experiments with the orange inside the coach tell him the coach is moving? If so, what observations would tell him that? If not, how *real* is the coach's motion? Does it make any difference to the passenger, so far as orange-throwing experiments are concerned, whether the coach is moving forward or, instead, the whole countryside is moving backward? (Such questions mark the beginning of discussions of Relativity; first the relativity of slow steady motion which Galileo knew all about and which is being discussed in this question; then the Relativity that Einstein discussed. Following Einstein, modern scientific philosophers take the view that if experiment cannot answer a question, then the question itself is an improper one, an unscientific attempt to imply knowledge where we cannot know.)

A projectile does not really have separate horizontal and vertical motions. As it moves along its curved path, its motion at any instant is directed along the tangent. While it rises from A to B to C

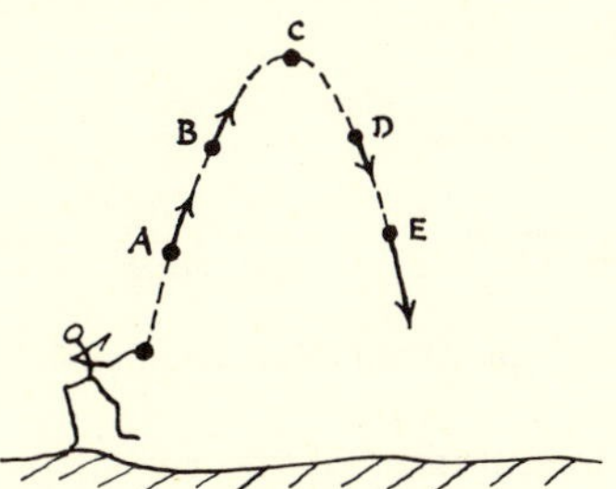

FIG. 2-14. PROJECTILE MOTION

it moves slower and slower, then falling from C to D to E it moves faster and faster, the speed changing as the vertical part of the motion is changed by "gravity."

PROBLEM 5

In the photo of Fig. 2-6, the ball made a small mark during each short, standard exposure.

(a) What information can you gather from the *lengths* of the marks?
(b) What information can you gather from the *directions* of the marks?
(c) How could the marks themselves (and not their spacing) tell you whether the horizontal motion is constant?
(d) How could the marks themselves tell you about the vertical acceleration?
(e) The highest mark looks almost like a point. Should it be a point or a streak? Why?
(f) What modification of the experiment would you suggest to prove your answer to (e)?
(g) In the photographing, the ball was not just thrown once and photographed, but many pictures had to be taken and one selected. Suggest a reason for this, other than the photographer's incompetence.

The splitting up of the actual motion along the path into horizontal and vertical motions, called *components,* is an artificial trick which we have been taking for granted. What rules govern resolving, or splitting into components, and the reverse process of compounding components? The process of compounding separate motions into a single motion which we call the *resultant,* is important in navigation where motions of ship and ocean currents, or plane and wind, are to be combined. In the next section we shall study such adding of motions.

Geometrical Addition

No one watching the curved flight of a stone flung in the air would automatically separate it into vertical motion and unchanging horizontal motion; yet as scientists we are encouraged to make this separation or analysis when we discover that the two motions are of different types and are independent of each other. Attempting such analysis at once raises the questions: (i) How is a single slanting motion split up into two ingredients or "components"? (ii) How are two separate motions to be compounded together into one single motion? We can guess the answer to the second question, and use it to answer the first. If we try to add two or more *motions,* we have to keep track of simultaneous movements in different directions. Instead, let us allow the motions to proceed for some specified time, say one hour, and then deal with the *distances travelled* in that time. Then the problem of adding motions becomes a simple one of adding travelled distances, or journeys or trips.[3] Is the addition rule the same as in arithmetic, as in adding 2 and 3 to make 5?

[3] The technical term for such directed distances is "displacements."

Experiment soon shows us this will not work unless the separate journeys to be added are straight ahead in the same direction. Then we see 4 ft due North and 3 ft due North do make a total trip of 7 ft due North as in Fig. 2-15. (And, therefore a

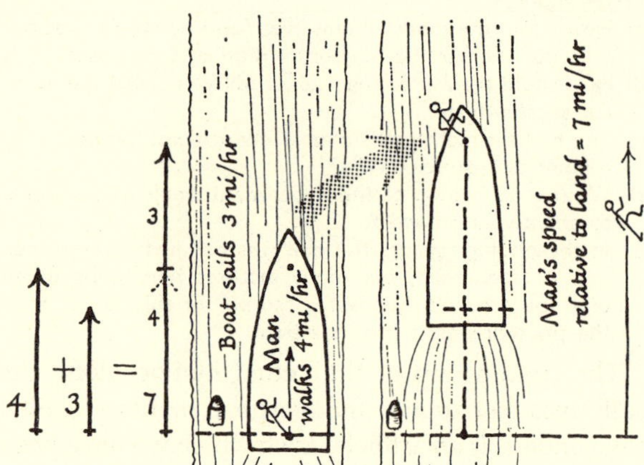

Fig. 2-15. Adding Motion in Same Direction

speed of 4 ft/sec and a speed of 3 ft/sec both due North do make a total speed of 7 ft/sec due North. And 4 miles/hr plus 3 miles/hr both in the same direction do make a total speed of 7 miles/hr.)

However, if the directions are different simple arithmetic does not work. A trip of 4 ft due East added to 3 ft due North does not make a trip of 7 ft. Nor does a speed 4 miles/hr due East plus a speed 3 miles/hr due North make a speed of 7 miles/hr in any direction. To fit the facts of the world, we have to use another kind of addition, which we call *geometrical addition.* Common sense—in this case simple knowledge accumulated in crawling, walking, driving, sailing, etc.—suggests how geometrical adding should be done. Suppose you wish to add trips of 4 ft to the East and 3 ft Northward, to find the *single trip that would carry you from the starting point to the destination.* Though it seems childish, try this for yourself. Stand facing North with your feet together. Then try to make both these trips, i.e., step four paces to the right and three paces forward at the same time. You could try this by doing one trip with each foot; sideways with your right foot and forward with your left foot, simultaneously; but

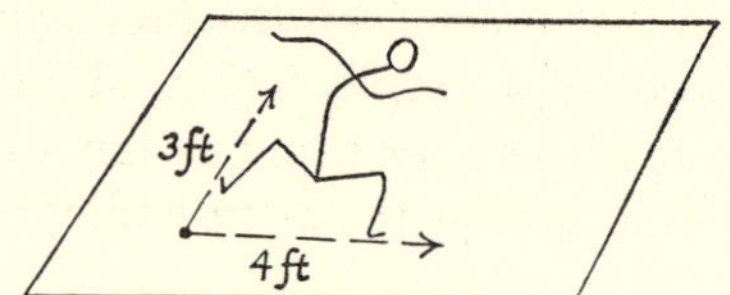

Fig. 2-16. Trying to Add Two Motions in Different Directions

the result is uncomfortable (Fig. 2-16). Instead you had better take one trip first, then the other, thus: move 4 paces to the right *then* 3 paces forward (Fig. 2-17). Or you can take them in the other

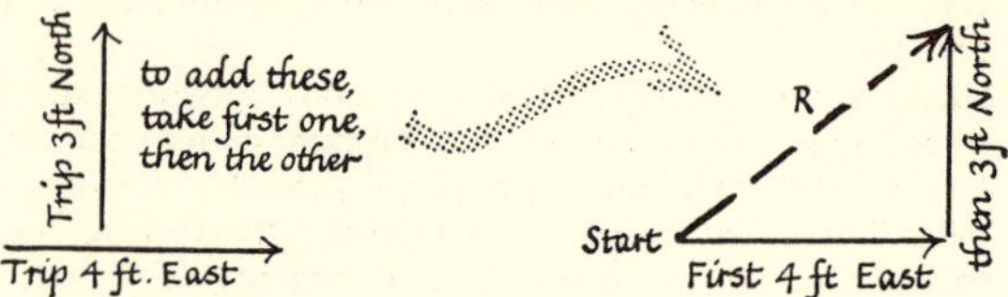

Fig. 2-17. Adding Motions

order, and arrive at the same destination. If you could somehow make the two trips simultaneously you should reach the same end-point. In fact this can be done if you have a rug which can be drawn across the floor by an electric motor. Then have the motor drag the rug with you on it (or a toy, as in Fig. 2-18) 4 paces to the right while you move

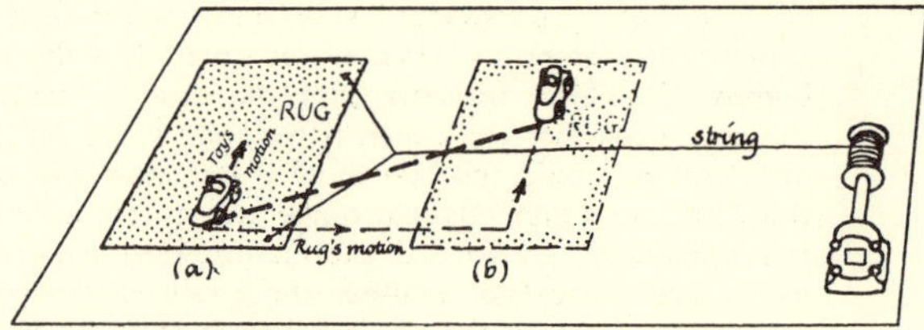

Fig. 2-18. Adding Motions. The toy crawls along the rug while an electric motor pulls the rug across the floor. The toy has a diagonal motion over the floor.

3 paces forward at the same time. On the rug—relative to the rug—you only move 3 paces forward. From a bird's eye view you make both journeys simultaneously and reach the same destination as if you made first one journey then the other. What single trip could replace these two, whether they are taken simultaneously or separately, and get you to the same destination? The simple single trip is along the straight line from starting point to finish. This

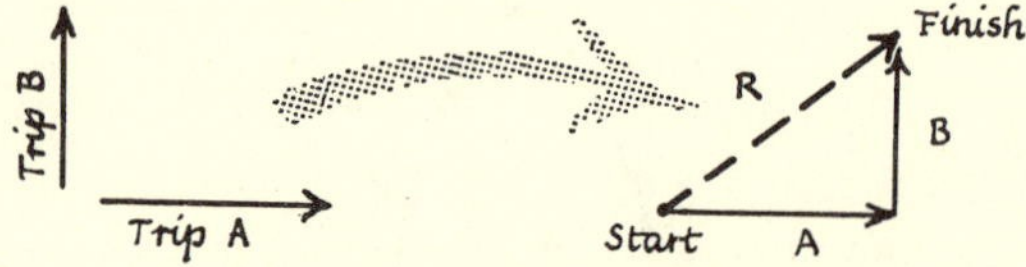

Fig. 2-19. Adding Perpendicular Trips

is called the *resultant* of the two trips. If the trips are drawn to scale on paper, as in Fig. 2-19, then the single trip which would replace them (if they are taken separately) is trip **R**. If the trips are not at right angles, a similar scale drawing will work, as in Fig. 2-20. If the trips are taken simultaneously—as when a plane flies in a wind—we can

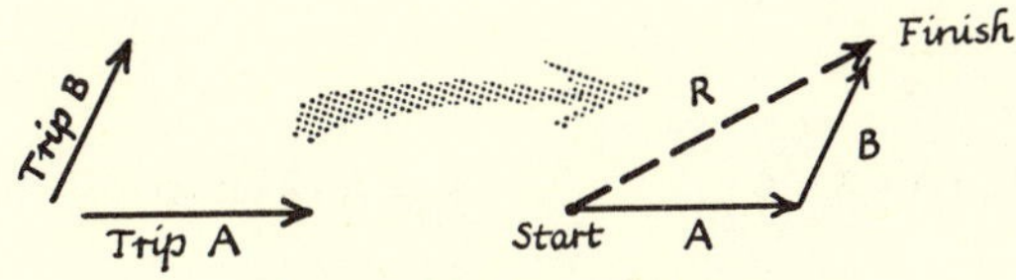

FIG. 2-20. ADDING TRIPS

still pretend to take first one then the other, and arrive at the resultant **R**, as in Fig. 2-21.

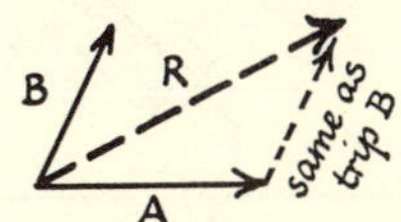

FIG. 2-21. ADDING TRIPS

We find the resultant by taking first one trip then the other, as in Fig. 2-22a or Fig. 2-22b. Combining these figures in Fig. 2-22c, we see that the resultant is given by the diagonal of the parallelogram whose sides are the original trips.

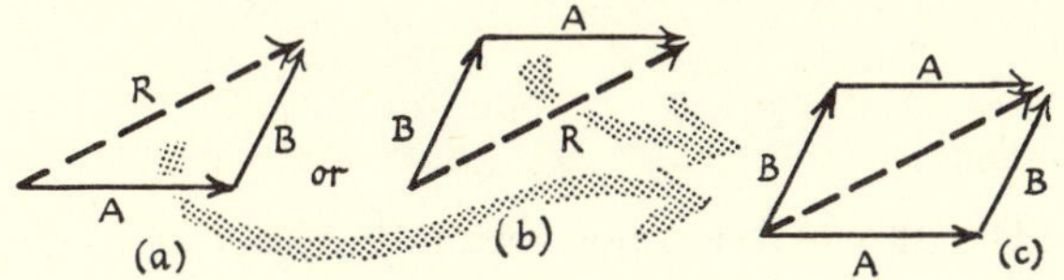

FIG. 2-22. ADDING TRIPS

This system is obviously right for adding trips: we are assured by common sense, drawing on experience ranging from nursery exploration to complex navigation.

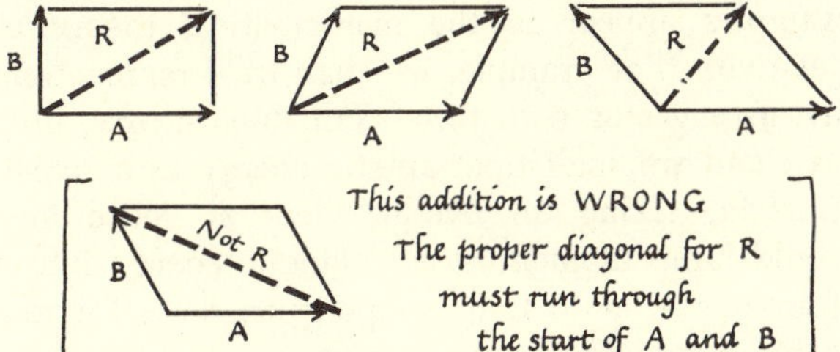

FIG. 2-23. EXAMPLES OF PARALLELOGRAM ADDITION

The system can be reversed, and the trip **R** split into components **A** and **B**. They are one possible pair that would combine to make **R**. There are an infinite number of such pairs, each adding to the same **R**.

PROBLEM 6

(i) Sketch (a) in Fig. 2-24 shows a trip **R** split into two components, $\mathbf{A}_1$ and $\mathbf{B}_1$; and (b) shows the same **R** split into a different pair, $\mathbf{A}_2$ and $\mathbf{B}_2$. Copy these sketches, and add several more, all showing the same **R** split into different components, $\mathbf{A}_3$, $\mathbf{B}_3$; $\mathbf{A}_4$, $\mathbf{B}_4$; etc.

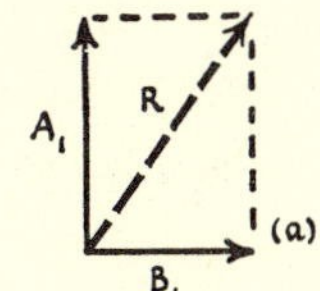

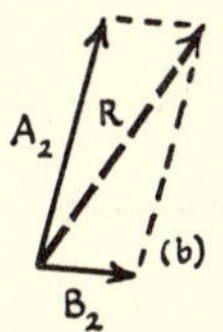

FIG. 2-24. PROBLEM 6. The vector R may be split up into components A_1 and B_1, or into components A_2 and B_2, or into other pairs of components. The components need not make 90° with each other.

(ii) Show that we can assign component **A** *any direction and any size*, and still find a **B** to fit, so that **A** and **B** add up to **R**. (This is equivalent to subtracting vectors, **R—A**, useful in later physics.)

Velocity and Speed

The *direction* of a motion is just as important as its size. We now need a name for the idea of a *definite speed* associated with a *definite direction.* We call this *velocity.*[4] Velocity then has two qualities: size (= speed) and direction. Do velocities add by the geometrical system? Or, as a scientist would say, are velocities "vectors"?

Vectors: Definition

Vectors are those things which are added by the geometrical system. They are called "vectors," because we can draw[5] a line to represent them, showing both their size (to some scale) and their direction.

RULE FOR ADDING TWO VECTORS

The following rule describes geometrical addition. Our definition of vectors makes it automatically true for vectors.

Geometrical addition: *To add two vectors, choose a suitable scale, and draw them to scale starting from the same point. Complete the parallelogram. Then, on the same scale, their resultant is represented by the diagonal from the starting-point to the opposite corner.*

In this, the *resultant* of a set of vectors is defined as *that single vector which can replace, or has the same physical effect as, the original vectors taken together.*

[4] In ordinary language, speed and velocity mean the same thing: how fast an object is moving. In physics, it is useful to reserve the name velocity for speed-in-a-particular-direction, which is a vector. From now on, we shall use speed to mean just rate of covering distance along some path whether straight or crooked—a worm's measure of progress. A speed is specified by a number with a unit, such as 15 miles/hour. A velocity needs a number with a unit *and* a direction to specify it, e.g., 15 miles/hour Northward.

[5] Vector and vehicle come from the Latin verb meaning to carry or convey.

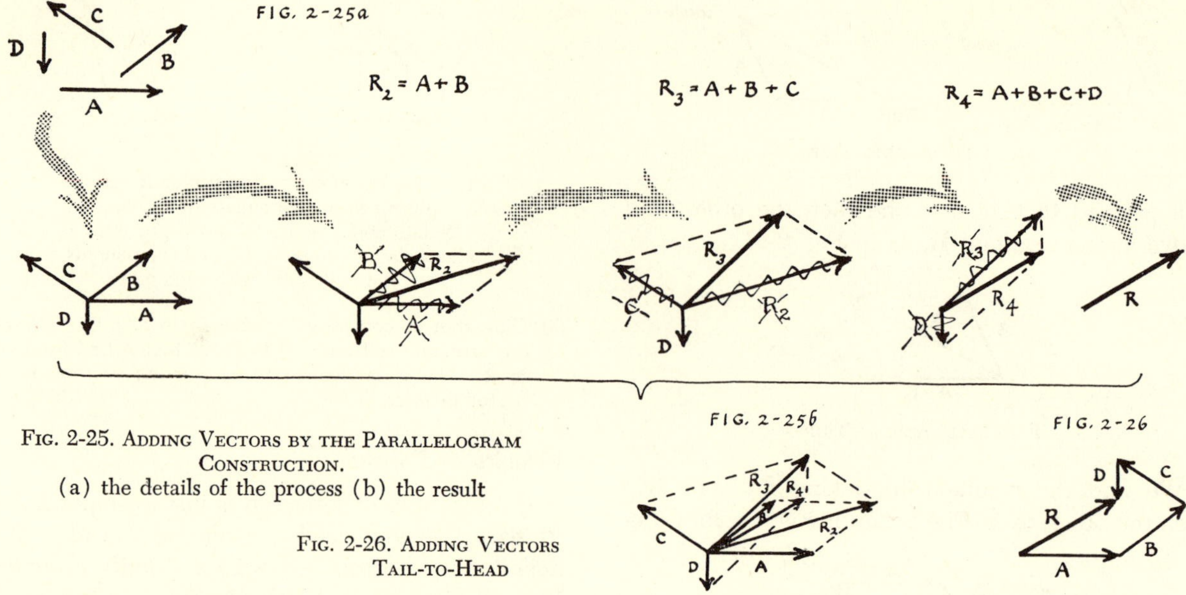

FIG. 2-25. ADDING VECTORS BY THE PARALLELOGRAM CONSTRUCTION.
(a) the details of the process (b) the result

FIG. 2-26. ADDING VECTORS TAIL-TO-HEAD

Just as vectors **A** and **B** add to give resultant $\mathbf{R}_2$ in Fig. 2-25 so we can add vectors **A** and **B** and **C** by adding **C** to $\mathbf{R}_2$ to get $\mathbf{R}_3$. Further addition of vector **D** would give $\mathbf{R}_4$ and so on. Or, more simply, any set of vectors can be added tail-to-head as in Fig. 2-26 (which is only a simplification of Fig. 2-25b), and their resultant is shown by the single vector joining start to finish.

What things are vectors? That is, which things in science do add geometrically by the parallelogram construction? Trips, or to give them a more official name "directed distances" or "displacements," are vectors. If trips are vectors, we need only divide by the time taken to travel them to see that velocities are vectors too. If we use as vectors the length travelled in unit time, then these vectors, which add geometrically as trips, themselves represent velocities. As an extension of this, we see that accelerations are vectors too.[6] We shall find other vectors, other things that can be measured with instruments and which obey geometrical addition. At the moment an important question arises: are forces vectors, i.e., do they obey geometrical addition? This cannot be answered by thinking about it.[7] It is not obvious. It needs experimental investigation. See Chapter 3.

[6] *Trips* are vectors. *Velocities* are *trips per hour*, say. Therefore velocities are vectors. Therefore *changes of velocity* (which are themselves each a velocity gained or lost) are vectors. *Accelerations* are *changes of velocity per hour*, say. Therefore *accelerations* are vectors.

[7] Unless we are prepared to *define* forces as things which add geometrically and then take the consequences of our definition in our later development of mechanics!

Scalars

Things which are not vectors but have only size, without any direction attached, are called *scalars*; for example, volume, speed, temperature. There are other things which are neither vectors nor scalars: vague things such as kindness, and some definite ones, some of them "super-vectors" called tensors. The stresses in a strained solid provide an example of tensors: pressure perpendicular to any sample face and shearing forces along it. More complicated examples appear in the mathematical theory of Relativity. For example, we shall treat momentum, **mv**, as a vector with three components, $\mathbf{mv}_x$, $\mathbf{mv}_y$, $\mathbf{mv}_z$; and we shall treat kinetic energy as a scalar. Einstein, taking an overall view of space-time, would lump momentum and kinetic energy into a "four-vector" with four components, three for momentum, one for kinetic energy.

Addition of Many Vectors

Two vectors are added by the parallelogram method. At the top of Fig. 2-27, **A + B = R** (the heavy **+** and **=** referring to geometrical addition). We can work back from this definition to the crude "first one trip then the other" method of adding, as in Fig. 2-27. This tail-to-head method is the easiest way of adding several vectors. If we wish to add vectors **A**, **B**, **C**, **D**, we could add them by applying the parallelogram construction again and again—getting the resultant of **A + B**, adding the latter to **C**, adding the new resultant to **D**. But the drawing is tedious, and if we perform all the

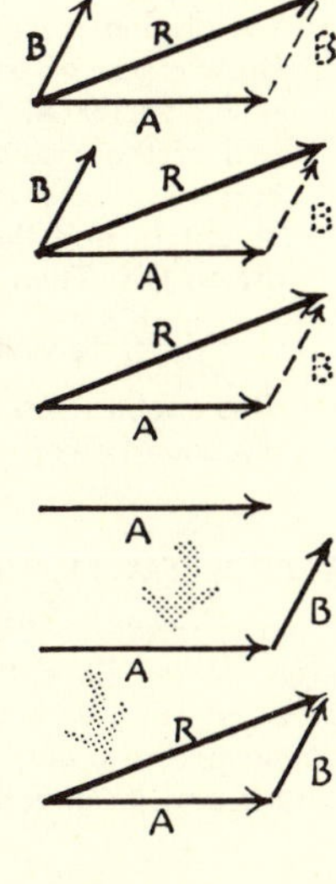

FIG. 2-27. "TAIL-TO-HEAD" ADDITION. Adding two vectors by parallelogram method is equivalent to "tail-to-head" addition.
Starting with parallelogram addition, we can omit part of the drawing and still obtain *R*.
We can economize still further and draw only a triangle, and we are back to our first discussion of trips, where we added them by taking first one trip and then the other. This leads to an easy *rule for adding vectors*:
DRAW ONE OF THEM FIRST.
THEN DRAW THE SECOND, STARTING IT WHERE THE FIRST ONE ENDED—that is, draw them one after the other, "tail-to-head."
THEN DRAW THE LINE JOINING START TO FINISH, AND THAT REPRESENTS THE RESULTANT, *R*.

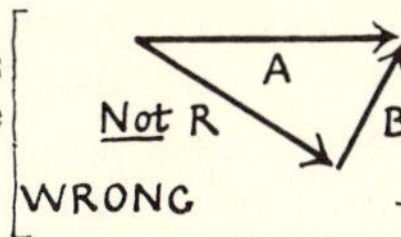

FIG. 2-28. BEWARE of adding vectors "*head*-to-head." That gives quite the wrong answer, not their resultant.

stages on one diagram it is a gorgeous mess (Fig. 2-29, I). Instead, we add **A** and **B** by the tail-to-head method, then add **C** to their resultant, tail to head, then add **D**. We can omit the intermediate resultants, and *find the main resultant* **R** *by joining the start of the first vector to the end of the last* (Fig. 2-29, II).

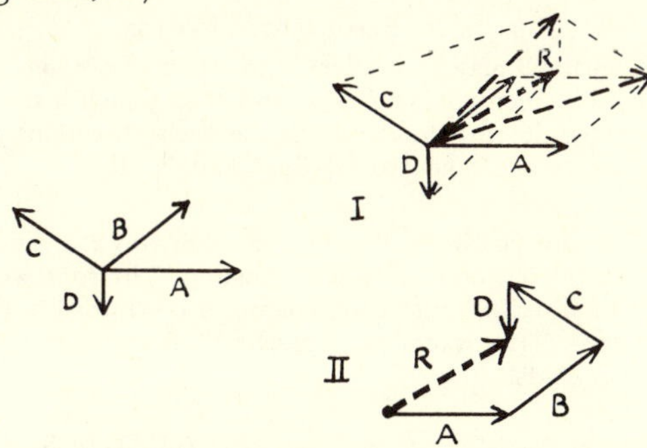

FIG. 2-29. ADDING A SET OF VECTORS:
(I) by the piecemeal parallelogram method.
(II) by the tail-to-head polygon method.

Drawing Parallel Lines

To transfer a vector from one place on a sheet of paper to another, we must draw a line in the new place with the same length and the same direction as the old line; so it must be parallel to the old line. There are geometrical methods and machines for drawing a line parallel to another line. If you are not familiar with at least one good method, ask for instructions. Difficult drawing of angles is quite unnecessary.

Fig. 2-30 shows one easy method using a ruler and a book cover (or any rectangle or triangle). To

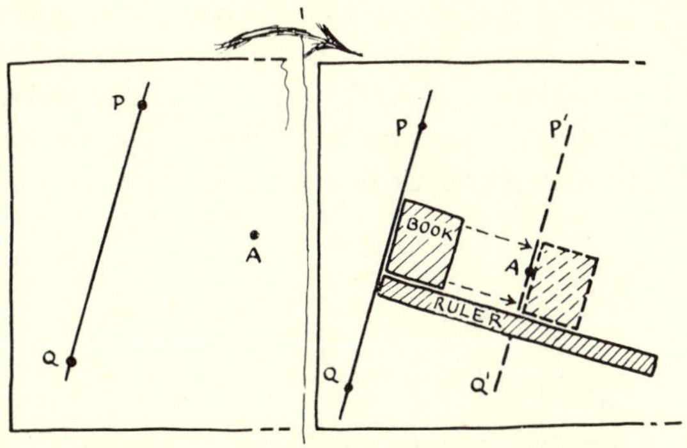

FIG. 2-30. AN EASY WAY TO DRAW PARALLELS. To draw a line P′Q′ through a given point A and parallel to a given line PQ, set the edge of a book along PQ; then slide the book along the ruler until its edge passes through A; then draw P′Q′ along the edge through A.

transfer from line PQ to a parallel line through point A, place one edge of the book on PQ. Place the ruler along the other edge of the book. Hold the ruler fixed, and slide the book along it till its first edge passes through A. Draw along that edge the required line through A.

★ PROBLEM 7. COMPOUNDING VELOCITIES

A ship surrounded by fog is pointed due North and sailing, as the navigator thinks, 4 ft/sec due North in still water. Actually it is in a current moving 4 ft/sec due East. If the fog disperses and the navigator can observe nearby islands, in what direction will he find he is really moving? How fast?

★ PROBLEM 8. CALCULATING A RESULTANT

A ship sails "northward 4 ft/sec" in a fog, as in Problem 7. It is really moving in water that is flowing eastward 3 ft/sec. What is its speed relative to land?

★ PROBLEM 9. NAVIGATION

A navigator trying to follow a narrow channel through reefs, has to sail in a fog.

(i) He knows the channel runs North-east, and that the ocean current carries him eastward 10 ft/sec. His propeller carries him ahead 10 ft/sec. In what direction should he steer, by his compass? (*Hint*: Sketch the known current-vector. From its starting point sketch the direction of the resultant; and at its end point fit in the engine-vector. Complete the parallelogram.)

(ii) Suppose the channel runs due North, the current is 10 ft/sec eastward and his propeller speed 20 ft/sec. Draw a diagram to show the direction in which he should steer by the compass.

(iii) Suppose the channel runs due North and the current is 10 ft/sec eastward. Prove that he cannot follow the channel unless his propeller speed is greater than 10 ft/sec.

Does the Order in Which the Vectors Are Added Affect the Resultant?

In adding vectors tail-to-head one after another, we might choose them in a different order—**A**, **D**, **C**, **B**, . . . instead of **A**, **B**, **C**, **D**, . . . , say—making quite a different pattern. *Will this give the same resultant?* The problems below explore this question.

★ PROBLEM 10

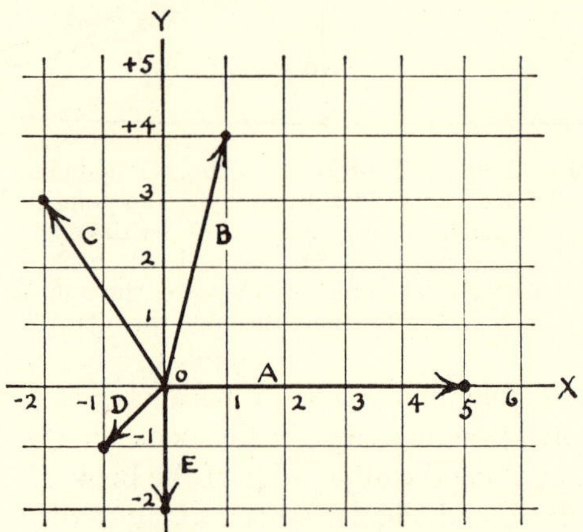

Fig. 2-31. Vectors for Problem 10.
(Copy this diagram on a larger scale, making each square 1 inch by 1 inch.)

Fig. 2-31 shows a set of vectors, **A, B, C, D, E,** all starting from one point, O. Add vectors by the "polygon method" of drawing them tail-to-head following the instructions below. The sketch shown here is too small for accurate drawing and measurement, so you should first reproduce the sketch on a larger scale on a sheet of graph paper. Expand the squares ruled lightly on the sketch to one-inch squares. Then, starting with **A** already drawn, add **B**, then **C**, then **D**, then **E**, tail-to-head. For this you will have to transfer **B, C, D, E,** by some parallel ruling method. (Either use the information given by the graph grid of Fig. 2-31, or use the method of Fig. 2-30.) Mark the result. Measure and record its size. To record the direction of the resultant you could either measure some angle or find its slope. Try both, as follows:

(a) Measure and record the angle between the resultant and the original vector, **A**.

(b) Draw a pair of perpendicular axes, OX and OY, with OX along vector **A**. Then drop a perpendicular *h* from the end of the resultant on to OX (you need not draw this carefully. Just measure *h* without drawing). Measure height *h*, and the base *b* which the perpendicular cuts off on OX. Then calculate the fraction

$$\frac{\text{height, } h}{\text{base, } b}\text{, which is called the slope of the line } \mathbf{R}.$$

This enables you to specify **R** as a vector of size . ? . and direction having slope . ? .

★ PROBLEM 11. UNIQUE RESULTANT

Is the resultant different if the vectors are added in a different order? Repeat Problem 10 on a new sheet of graph paper, starting with vector **A** as before but then adding the rest in a new order, **B, E, D, C**. Record the size of the resultant and its direction.

PROBLEM 12: ARGUMENT CONCERNING VECTOR ADDITION

Think of the vectors, **A, B, C, D, E** in Problem 10 as navigated trips to be taken one after the other. Think of the axes OX, OY, as compass directions, East, North. Then one trip, say **B**, carries us a certain amount Northward and a

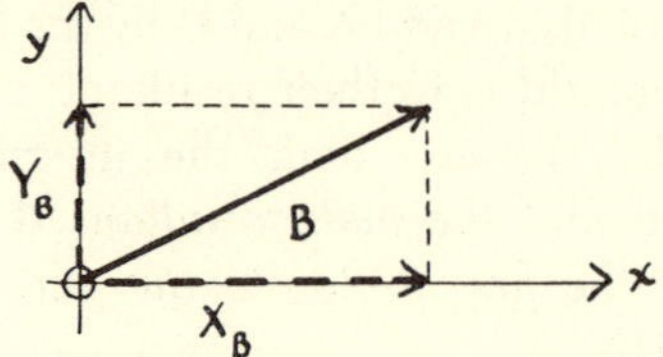

Fig. 2-32a. Resolving a Vector.
A vector B may be resolved into a pair of perpendicular "components," X_B and Y_B, which can replace it. In Problem 12 the two directions x and y are taken to be East and North.

certain amount Eastward. We may say that trip **B** gives us so much "northing" and so much "easting." In fact we are thinking of **B** as split into components, a northward one and an eastward. This is called "resolving" **B** into North and East components.

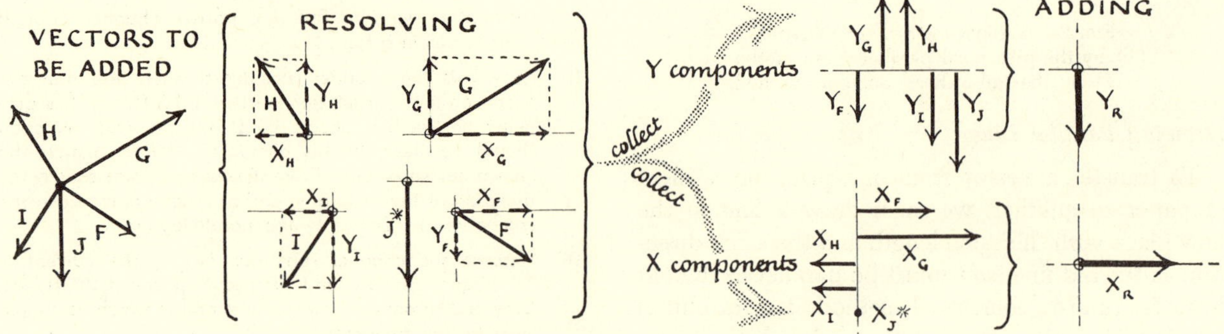

Fig. 2-32b. Illustration of Problem 12.
A set of vectors, F, G, H, I, J, resolved into components in directions x (East) and y (North).
These components are then added to give the components of the resultant R.
* Note that the vertical (southward) vector J has zero X-component.
Its Y-component is, of course, the full vector J itself.

We can resolve all the vectors like this. Some of the eastings or northings may be negative. The resultant, **R**, could also be imagined resolved into an Eastward and a Northward component. When we add the vectors, **A**, **B**, . . . , each will contribute its own share of easting and its own share of northing.

(a) How would you expect the easting of the resultant to be related to the eastings of the vectors, **A**, **B**, . . . ?
(b) How about its northing?
(c) Are your answers to (a) and (b) altered by changing the order in which the separate vectors are to be added?

PROBLEM 13: ANOTHER WAY OF ADDING VECTORS

The last problem shows another way of adding vectors which is very useful when there are many of them, particularly if the angles are given and we can use trigonometry. We shall hardly use it in this course, but it is given here to show what a clean brutal method is available.

(i) Suppose you knew the easting, $\mathbf{X_R}$, of the resultant **R** and its northing, $\mathbf{Y_R}$. The sketch shows **R**, $\mathbf{X_R}$, $\mathbf{Y_R}$. How

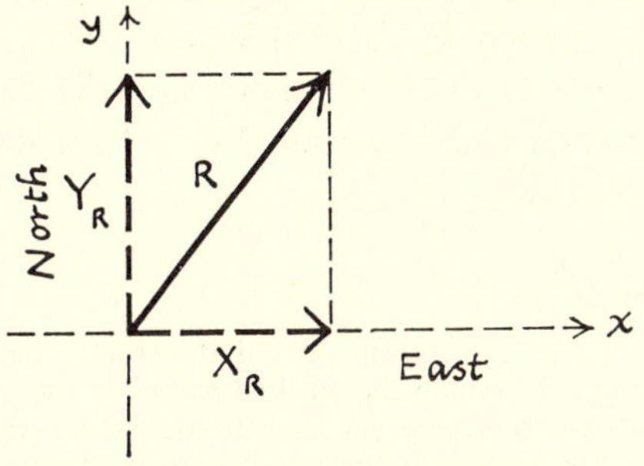

Fig. 2-33. Problem 13

could you *calculate* the size of **R** from $\mathbf{X_R}$ and $\mathbf{Y_R}$? *Write an equation.*

(ii) Suppose you knew the easting and northing of each of the vectors, **A**, **B**, . . . etc. (Suppose vector **A** resolves into $\mathbf{X_A}$, $\mathbf{Y_A}$, and similarly for the rest.) How would you calculate the Eastward and Northward components of **R**? Write equations for $\mathbf{X_R}$ and $\mathbf{Y_R}$. [*Note*: Mathematicians often use Σ, a capital Greek S, to mean "the Sum of all things like this." For example: if the members of a gang A, B, C, et al., have cash M_A dollars, M_B dollars, etc., the gang's total cash is $M_A + M_B + M_C + \ldots$ etc., and this is written ΣM. Use Σ here.]

(iii) Write instructions for calculating the size of the resultant **R** of a set of vectors, given the eastward and northward components of all the separate vectors.

(iv) Carry out your instructions for the vectors of Problem 10. Copy the vectors on a fresh sheet of squared paper. Draw suitable perpendiculars and measure the E. and N. components (giving minus signs to any components pointing West or South). Calculate the size of the resultant. Calculate the slope of the resultant, using vector **A** as horizontal base line.

PROBLEM 14 (for students familiar with trigonometry)

In Problem 10, the vectors, **A**, **B**, **C**, **D**, **E**, make the following angles with the vector **A**, which is taken as due East, **A**: 0°; **B**: 76.0°; **C**: 123.7°; **D**: 225.0°; **E**: 270.0°; and their lengths, when drawn on inch squared paper are about: **A**: 5.00; **B**: 4.12; **C**: 3.61; **D**: 1.41; **E**: 2.00 inches. Use trigonometry to find the North and East components of each vector, following the instructions (a, b, c) below.

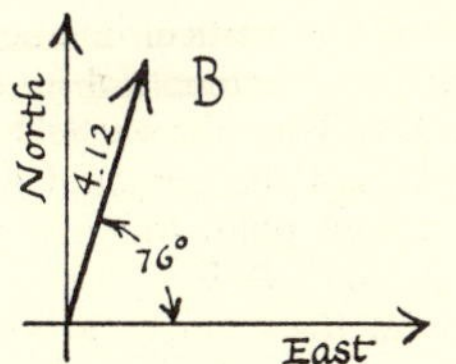

Fig. 2-34. Problem 14

(a) Which of the following is the easting of **B**?

$4.12 \cos 76°$ $\quad$ $4.12 \sin 76°$ $\quad$ $4.12 \tan 76°$

$\frac{4.12}{\cos 76°}$ $\quad$ $\frac{4.12}{\sin 76°}$ $\quad$ $\frac{4.12}{\tan 76°}$

(b) Calculate the numerical value of the easting of **B**, using trig. tables. (4-figure tables will do; 3-figure ones would be better. Do not waste time with bigger tables.) Also calculate the northing of **B**. Call these two components $\mathbf{X_B}$ and $\mathbf{Y_B}$.

(c) Do the same for each vector. Calculate the resultant's size and slope. Note that this method does not involve drawing to scale. Of course, it cannot be an ivory-tower method and avoid drawing completely. The equivalent of the real drawing is concealed in the trig. tables. Like π, sines and cosines can be computed arithmetically from infinite series; but the series are based on geometrical assumptions which have been tested against the real world.

PROBLEM 15

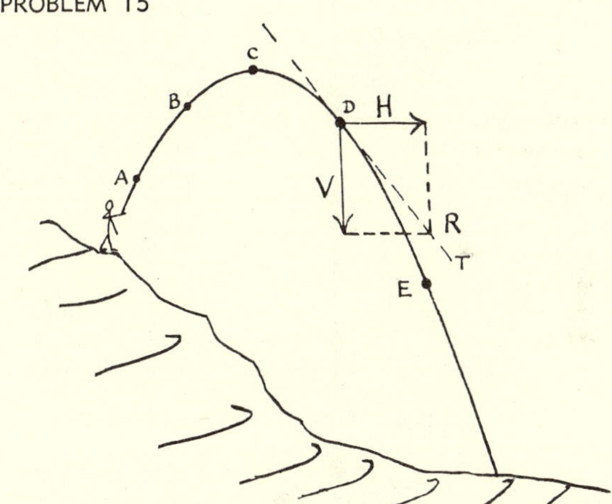

Fig. 2-35. Problem 15. The motion at D is along the tangent to the curve at D. A vector to represent this motion splits into components *H* and *V* as shown.

Use your knowledge of vectors to show the horizontal and vertical motions of a projectile. Fig. 2-35 shows the path of a stone thrown in the air. Copy it roughly on a larger scale. Choose a number of points on the path such as A, B, C, D, E, and for each point sketch the horizontal velocity, the vertical velocity and the actual (resultant) velocity of the motion, following the example given for point D.

Analysis for point D. At D draw the tangent DT. Then the actual velocity is along DT. From D draw a horizontal line **H** to represent the horizontal motion. (Since you do not know how fast the stone is moving, pretend its horizontal velocity is suitably represented by a line, **H**, shown 1.1 cm long in Fig. 2-35.) We are treating the actual motion along DT as made up of horizontal and vertical components, so we draw a parallelogram which is a rectangle, with **H** as one side, and with diagonal along DT. Then the vertical side, **V**, shows the vertical velocity at D and the diagonal **R** shows the actual velocity along the curved path. Make a similar analysis at each of the points, A, B, C, D, E, . . . on your sketch, just by rough drawing, to show the changes in the motion. In doing this, remember the important property of the horizontal motion of a projectile.

Projectiles and Parabolas

We can analyze the shape of a projectile's path with the help of geometry or algebra.

Geometrical analysis: Suppose a stone is thrown out horizontally. Then its horizontal motion carries it the same distance horizontally every second, while it accelerates vertically. It falls 16 feet vertically in the first second from its start, 64 feet in the first two seconds, 144 in the first three seconds, etc. Make a scale map of its positions at several instants of time. Choose total times from the start which run in the proportions 1:2:3:4. . . . In these times it travels steadily sideways, covering distances in the same proportions 1:2:3:4 . . . ; but it falls vertical distances proportional to the squares of these numbers, to 1, 4, 9, 16 . . . because

$$\text{VERTICAL FALL} = \tfrac{1}{2}g(\text{TIME})^2$$

and $(\text{TIME})^2$ has values in proportions 1:4:9. . . . Map its position at these equally spaced instants of time by drawing vertical lines evenly spaced, say at intervals of 2 inches across the map; and horizontal lines 1 inch down from starting level, 4″ down, 9″ down, and so on to show vertical falls. Then the predicted path is marked by the crossings of these lines, as shown in Fig. 2-36. This can be demonstrated by shooting balls or water-drops in front of a blackboard on which such lines have been ruled.

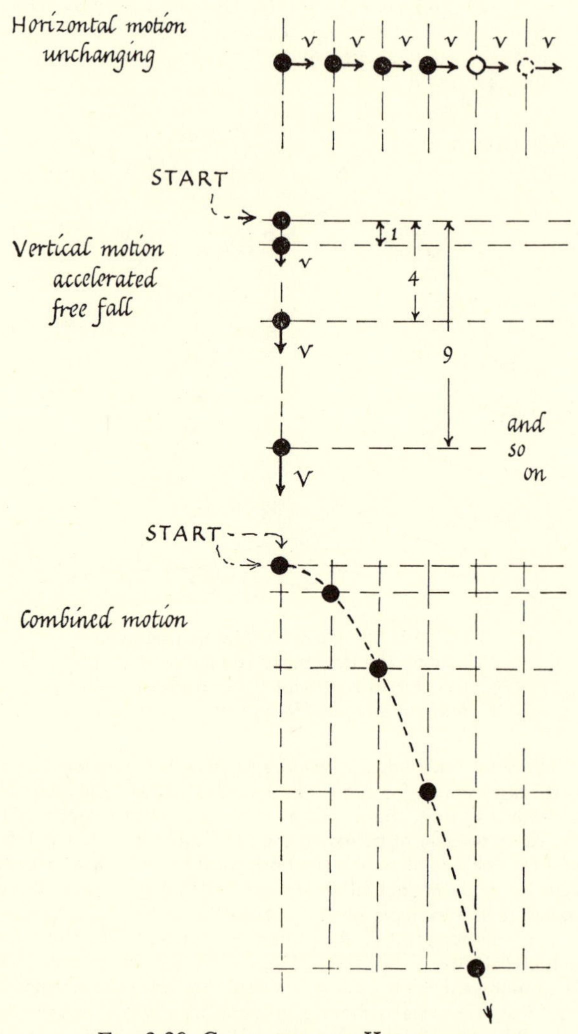

FIG. 2-36. COMBINING THE HORIZONTAL AND VERTICAL MOTIONS OF A PROJECTILE.

PROBLEM 16

If you see in a demonstration that actual projectiles do follow the curve through the marked points on the grid, what assurance does this give you concerning motion in nature? In such a demonstration the initial horizontal speed of the projectile must be chosen to fit the markings (Fig. 2-37a). Suppose you then decrease the speed and mark the new path on the board; how could you test whether it follows the same type of pattern? (See Fig. 2-37b.)

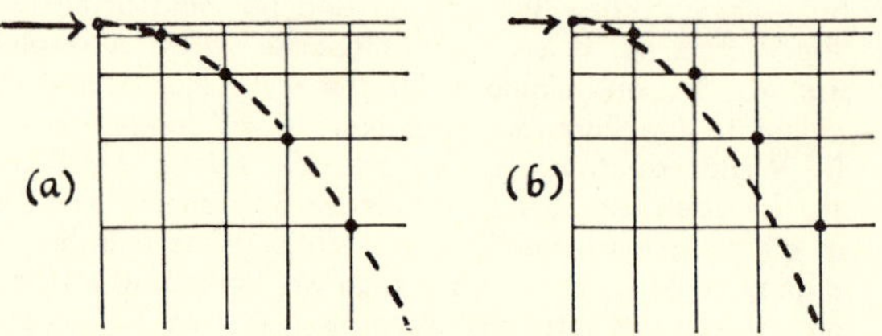

FIG. 2-37. PROBLEM 16

Algebraic analysis: In imagination, plot the stone's path on an x-y graph and find its equation. Suppose the stone is thrown out horizontally from the origin (0, 0) with velocity 10 feet/second. Then in each second it will move 10 feet horizontally. After t seconds from the start it will have moved $10t$ feet horizontally; so we say

$$\text{DISTANCE MOVED HORIZONTALLY, } x, = 10t \text{ feet.}$$

In t seconds, falling from rest, it would drop a vertical distance, y, given by:

$$\begin{aligned}\text{VERTICAL FALL, } y, &= \tfrac{1}{2}\text{ ACCELERATION } t^2\\ &= \tfrac{1}{2}(32)\, t^2 \text{ feet}\\ &= 16\, t^2 \text{ feet.}\end{aligned}$$

As the stone moves along its path, both these statements are true at each stage; so we say

$$x = 10t$$
$$y = 16t^2$$

To find a single equation that describes the path, we ask, "What relation between x and y makes both requirements above true at each stage in the path?" If we choose any point on the path, its x and y values must satisfy the two equations above, for the appropriate value of t. That value of t must be the same in both the equations—it is the time when the stone reaches that chosen point. Therefore we can get rid of t by making one equation yield an expression for t which can be substituted in the other equation; thus:

$x = 10t$ gives $t = x/10$, and we can use $x/10$ for t in $y = 16t^2$, which then becomes
$$y = 16\ (x/10)^2 \text{ or } y = (16/100)x^2.$$

The equation of the path is then $y = (0.16)x^2$.

More generally, if the stone is thrown horizontally with initial velocity v_H feet/second, and falls with vertical acceleration g feet/second per second,

$$x = v_H t \text{ and } y = \tfrac{1}{2}gt^2$$

$$\therefore\quad y = \frac{1}{2}g\left[\frac{x}{v_H}\right]^2 = \frac{1}{2}\left[\frac{g}{v_H{}^2}\right]x^2$$

$\therefore\quad y = (constant)\ x^2$ since $\tfrac{1}{2}\ g/v_H{}^2$ is constant.

This is the equation of a parabola.[8]

You can plot beautiful parabolas on graph-paper by starting with an equation like this. Try plotting the graph given by $y = (\tfrac{1}{2})x^2$ on paper marked in inch squares, taking $x = -4$ inches, $x = -3, -2, -1, 0, 1, 2$, etc. Try to match this curve with a real projectile. Put the paper with the sketched curve on a sloping table and experiment with a rolling ball. Or hold the paper upright and throw a small object up in front of it.

Projectile from Tilted Gun

If the projectile is not thrown out horizontally but starts upward in some slanting direction, its path is still a parabola. Algebraically this can be shown by starting with $s = v_0t + \tfrac{1}{2}gt^2$ instead of $s = \tfrac{1}{2}gt^2$. Or, we appeal to the visible symmetry

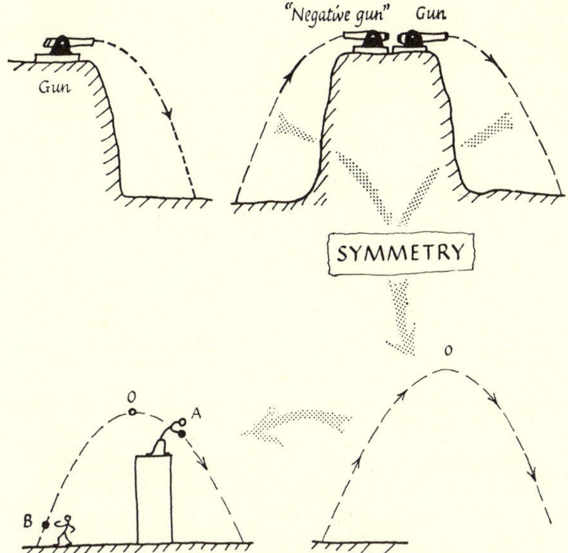

FIG. 2-38. PROJECTILE MOTION UP AND DOWN. Symmetry of path suggests that motion up to a "negative gun" on a hill, is similar to motion down from a gun on the hill. These two motions join to give a full parabola. Then a projectile *started* along this parabola, at any point *A* should follow the same path as if it started (earlier) at the vertex, O. Symmetry extends this argument to the whole parabola.

of the curved path and say that the decreasing upward motion to the top must match the increasing downward motion from the top, and so we draw the complete path from the horizontally projected case. This is only good guessing but experiment confirms it. Or we may argue thus: on the downward part of the path from the vertex, O, the stone cannot know whether it started at O or earlier or later. So a stone started on this part of the path, say at A by being thrown outward and downward must therefore follow the same path as one thrown horizontally from some earlier vertex, O. (See Fig. 2-38.) Similarly for one thrown upward at B.

This implies an extension of the idea of independence of motions. The *vertical component* of the starting motion also continues *unchanged*, while the accelerated motion of falling is added to it. This constant vertical motion is responsible for the distance v_0t in the relation $s = v_0t + \tfrac{1}{2}gt^2$. Then we might lump together the two constant motions, the vertical and horizontal parts of the initial throw, and say that *the initial slanting motion given by the thrower remains unchanging in flight*, while the gain of vertical falling motion responsible for $\tfrac{1}{2}gt^2$ is added to it. Thus a stone thrown as in Fig. 2-39 may be regarded as having two motions, its initial motion continuing along the line AB, and free fall measured from successive positions on AB.

[8] Originally described as one of the shapes made by slicing a cone, a parabola is often defined now as a curve whose graph-equation is of the form
$$y = (\text{constant})\ x^2, \text{ or } y \propto x^2.$$
A piece of algebraic geometry shows that the algebraic and geometrical definitions are equivalent.

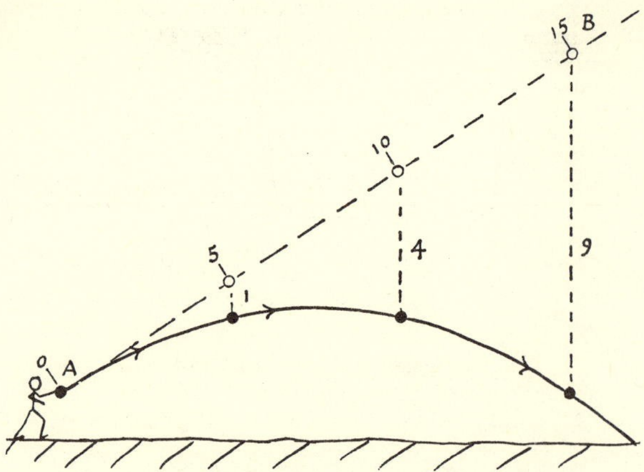

Fig. 2-39. Analysis. The motion of any projectile may be regarded as being made up of an unchanging motion along the initial direction and a free fall from that line.

We can demonstrate this by the "monkey and gun" experiment. Suppose a hunter, ignoring gravity, sights through his rifle barrel on a monkey hanging by one arm from a tree. If the hunter fires, the bullet will miss the monkey because of its falling motion, as in Fig. 2-40. Now suppose the monkey

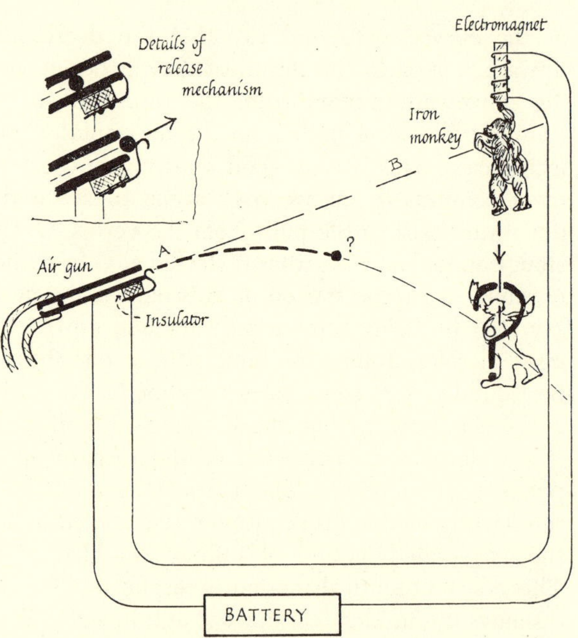

Fig. 2-40. The Monkey and Gun Experiment. At the instant the bullet emerges from the gun, it breaks a contact and allows the electromagnet to release the "monkey." The electrical connection is maintained by a small spring touching the metal gun barrel until the emerging bullet moves it.

watches the gun and lets go at the instant the bullet leaves the gun, when he sees the flash of the gun. From this instant, both monkey and bullet are accelerated downwards by gravity; the monkey falls from rest, the bullet—according to our recent view—falls from the line AB of its "undisturbed path." What will happen? This can be demonstrated by using an iron monkey released by an electromagnet which is switched off when the bullet trips a switch at the muzzle of an air-gun aimed at the monkey.

Such experiments confirm our guess that vertical fall is quite independent of the initial motion, which continues unchanged. Any projectile drops freely from its starting line, from the very beginning. It falls 1, 4, 9, 16, . . . feet in 1, 2, 3, 4, . . . quarter-seconds from start. If the starting line slants upwards, the projectile's actual path rises at first and then falls when the accelerated rate of free fall has beaten the steady rate of rise due to the initial motion. (See Fig. 2-41. Note that this path is a parabola.)

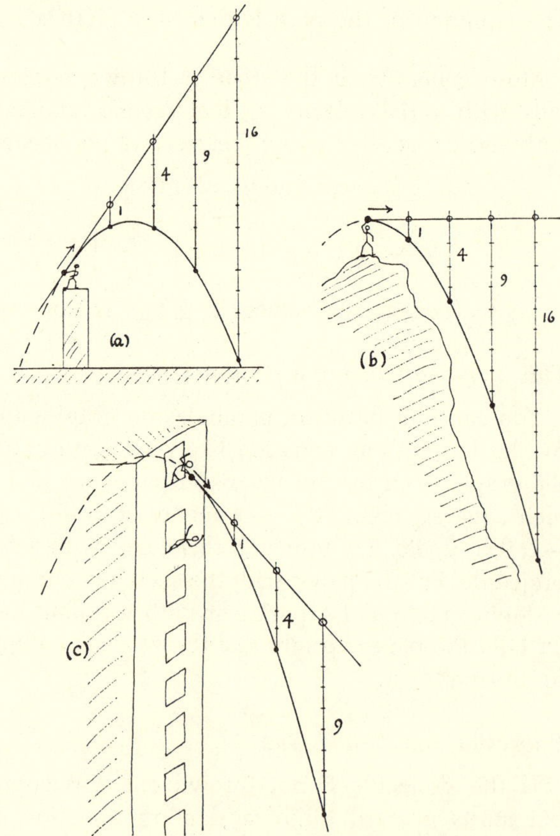

Fig. 2-41. Free Fall of a Projectile. However it is started, a projectile falls with the same "free fall" from its original starting-line as an object released from rest. The accelerated motion of fall is independent of both the vertical and horizontal components of the initial motion.

Notice how our discussion has torn the problem of projectile motion to pieces, leaving it easier to deal with, ready for further studies by experts in ballistics. We have not so much set forth new information as made existing knowledge easier to use.

When the projectiles are speedy electrons (or, in other cases, charged atoms) pulled by electric and magnetic fields instead of gravity, we assume that similar "rules" apply and use our measurements of the curved path to obtain information about electric charge and mass and speed. We use that information in turn—still assuming the same rules for projectile behavior—to predict the effects of fields on particles moving at other speeds. Then we are becoming literally electronic engineers, designing television tubes and other radio devices; and we are becoming atomic scientists, bending streams of electrons or atoms to our bombarding uses or sorting light atoms from heavy ones by differences of their projectile paths.

PROBLEMS FOR CHAPTER 2

1-16. These are in the text of Chapter 2.

★ 17. POLICE WORK

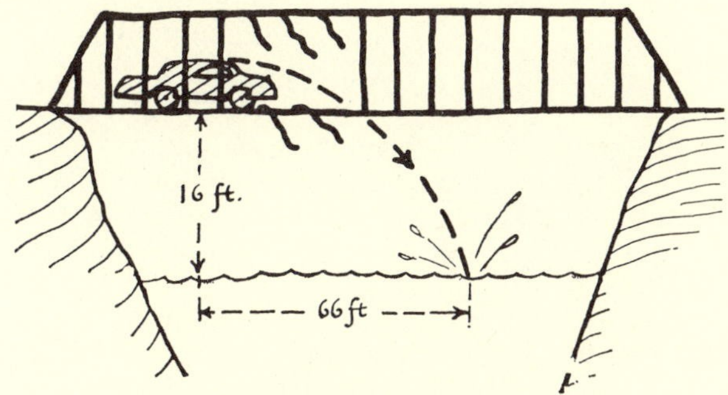

FIG. 2-42. PROBLEM 17

A lunatic, driving too fast on a bridge, skids, crashes through the railings along the side of the bridge, and lands (man + car) in the river 16 feet below the level of the roadway of the bridge. (*Note*: 16 feet is the vertical distance between bridge and river.) The police find that the car is not vertically below the break in the railings, but is 66 ft beyond it horizontally.

(a) Estimate the speed before the crash.
(b) Say whether this is probably an overestimate or an underestimate; and say why.
(c) State clearly the properties of falling bodies that you assumed in making the calculation of (a).

★ 18. ELECTRON STREAM

An electron moving 6 million meters/sec (which is quite slow, as electrons go) along a horizontal path, runs into a region where a vertical electric field gives it a downward acceleration of

40,000,000,000,000 m./sec/sec or 4×10^{13} m./sec/sec.

This region extends for 0.30 meter, in the direction of the original path; so the electron travels along in a region of no field, then for 0.30 meter (horizontally) of vertical field, then out into a region of no field again.

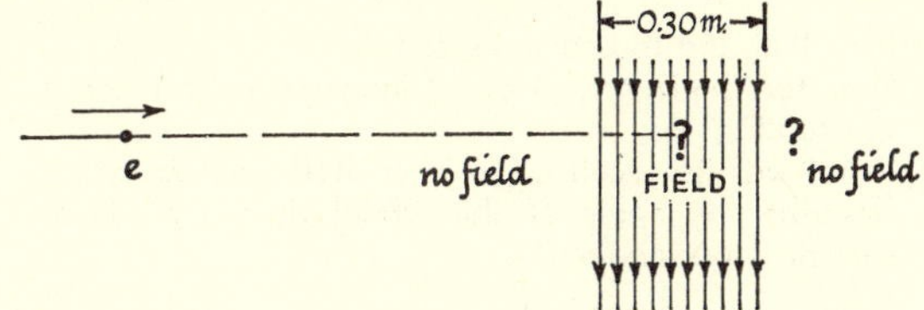

FIG. 2-43. PROBLEM 18

(a) Do you expect the vertical acceleration to affect the horizontal motion?
(b) Calculate the time taken by the electron to travel across the field-region.
(c) Calculate the distance it will fall in the field. (This is the distance an experimenter would measure to investigate the electron's behavior.)
(d) Calculate the electron's vertical velocity at the instant it emerges from the field.
(e) Predict the path of the electron, and draw a rough sketch showing the path before, through and after the field.
(f) Why is it unnecessary to take gravity into account in this question? (It *does* act on the electron.)

19. RANGE OF PROJECTILE (A problem using algebra and trigonometry.)

(a) An ancient gun projects a cannon ball at an elevation of 45° with speed 141.4 feet/sec.
   (i) Split this velocity into horizontal and vertical components.
   (ii) Calculate the time from the start of the ball till it reaches the ground again.
   (iii) Calculate the range.
(b) An ancient gun projects a cannon ball with velocity v_0 in a direction making angle A with the horizontal.
   (i) Resolve v_0 into horizontal and vertical components.
   (ii) Calculate the time taken by the vertical motion, from the start of the ball till it reaches the ground again.
   (iii) Calculate the horizontal distance the ball travels (i.e., its range).
(c) Show, by trig. or calculus, that range is maximum for $A = 45°$, for a given v_0.

(Remember that $2 \sin x \cos x = \sin 2x$.)

★ 20. MEASURE HOW FAST YOU CAN THROW A BALL

A scientist wants to find out how fast he can throw a baseball. He throws it out horizontally at shoulder height, 4 ft above the ground. It lands on the ground 20 ft away from his feet.

(a) What was the ball's original speed? (See Problem 17.)
(b) Apart from any formula for accelerated motion, an important general principle concerning projectiles (formulated by Galileo) has to be used in calculating the answer to (a). What is it?
(c) Instead of throwing the ball, the scientist runs along at the speed calculated in (a) above, carrying the ball at shoulder height. While running, he releases the ball so that it can fall. Describe carefully the path of the falling ball:
   (i) as seen by a stationary observer,
   (ii) as experienced by the running scientist.

21. An automobile travelling 96 ft/sec (over 65 miles/hour) along a horizontal mountain road failed to make a corner and crashed into a snowdrift 144 ft (vertically) below.

(a) How long did the car take to fall?
(b) How far (horizontally) did it land from the place it left the road?
(c) What was its acceleration when half way down?
(d) Describe the angle of the tunnel that it made in the snowdrift on landing.

22. A man holds a rifle 9 ft above the level ground and aims it horizontally.

(a) How long is it from the instant of firing until the bullet hits the ground?
(b) If the cartridge is ejected horizontally to the side, just as the bullet leaves the barrel, when will the cartridge hit the ground?
(c) Would the man be able to shoot farther (with this aim) on the Moon?
(d) Give a clear reason for your answer to (c).

★ 23. A man inside a large elevator throws a ball straight out from him horizontally with speed about 10 ft/sec. In each of the cases (a), (b), etc., sketch the path of the ball as observed by the man in the elevator.

(a) The elevator is moving downward with constant velocity 10 ft/sec.
(b) The elevator is accelerating steadily with downward acceleration 32 ft/sec/sec.
(c) The elevator is accelerating steadily with downward acceleration 10 ft/sec/sec.
(d) The elevator is accelerating with downward acceleration 64 ft/sec/sec (suitable machinery being used to achieve this).

CHAPTER 3 · FORCES AS VECTORS

Brute force, unsupported by wisdom, falls of its own weight.

—Horace, *Odes*, III, 4

Forces are pushes and pulls: things you feel when they act on you, things that stretch springs, things that make moving bodies accelerate. We shall measure forces with spring balances. As these instruments are commonly graduated in pounds or in kilograms, we shall use those units for force at present. Later we shall change to more proper units.

Engineers are much concerned with adding forces in bridges—cranes, buildings, machinery—or with subtracting them to find the remaining force needed to hold some system balanced. We can show that forces are vectors, i.e., that they obey geometrical addition. The vector treatment of balanced forces is called "Statics." It is a bulky but dull part of physics, and most texts spend a lot of space teaching tricks for solving engineering statics problems. We shall give only a few examples, and even they might be better omitted to give time for more study of force and motion.

First we must have some assurance that forces *are* vectors. To say that they *must* be vectors because they have size and direction is risky. That does not make sure they add geometrically. Though it looks plausible—especially to people who deal with ropes on ships or tents—we ought to test it directly. You should see the demonstration described below.

Demonstration Experiment

Fig. 3-1 shows a large contraption set up in front of a blackboard. O is a metal ring pulled by two ropes, OA, OB, with spring balances A and B to measure the pulls. The ropes must exert considerable pulls to hold O in the position shown because it is pulled the opposite way by a large spring S, which is anchored to the wall at its other end. The rope-pulls are together sufficient to stretch the spring and hold O in its present position. The position of the ring O is marked, and the lines of the ropes, OA, OB, are marked. The balances A and B are read to give the pulls $\mathbf{F}_A$ and $\mathbf{F}_B$.

The resultant of these two pulls is found by drawing, assuming geometrical addition of forces. For

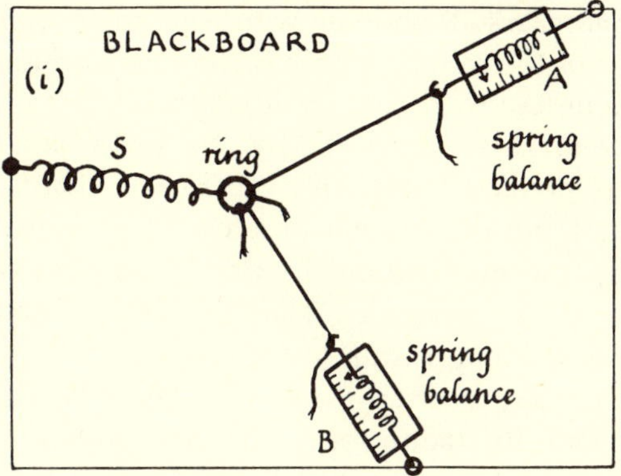

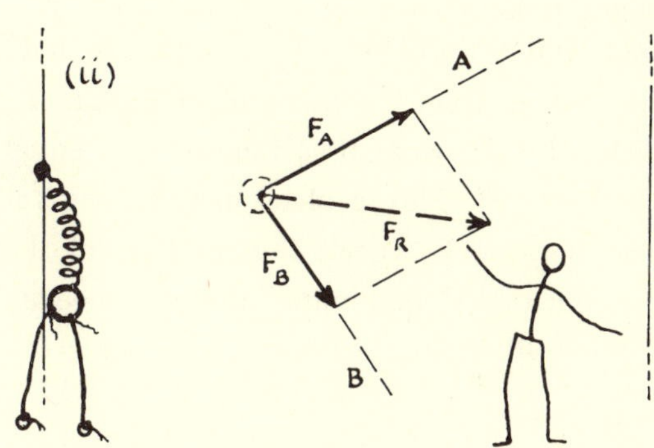

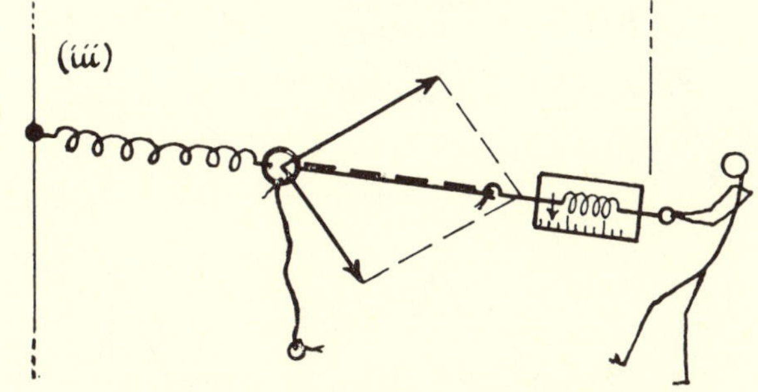

Fig. 3-1. Demonstration Test.
Are forces vectors? (i.e. do forces add by geometrical addition?)

(i) Two ropes exert measured pulls on a ring O, pulling it out to a marked position against the pull of spring S.

(ii) The ropes and spring are removed, and the predicted resultant of the pulls F_A and F_B is obtained by geometrical addition.

(iii) Then the prediction is tested by measuring the *actual* force needed to pull the ring out to its marked position with a single rope.

this, a suitable scale is chosen; forces $\mathbf{F}_A$ and $\mathbf{F}_B$ are marked off along OA and OB; the parallelogram is completed; and the diagonal, $\mathbf{F}_R$, is marked, measured, and interpreted to scale. We then know the *predicted* resultant, $\mathbf{F}_R$, that single force which can replace the two pulls, *if geometrical addition applies.*

We then measure the *real* resultant directly, by undoing the two ropes and pulling the ring out to the same marked position with one rope alone. The size of the resultant force is shown by a spring balance on the rope, and its direction is shown by the rope itself. We then compare the real resultant with the predicted one. This experiment affords a single test, but the accumulated testimony of many such experiments confirms the view that forces behave properly as vectors. The wealth of indirect evidence is stronger still.

Another test is often used. It is easier but less direct, and its indirectness is often dishonestly ignored. Two forces $\mathbf{F}_A$ and $\mathbf{F}_B$ are arranged (by weights and pulleys, or by spring balances) to pull on a knot, and a third force $\mathbf{F}_C$ is added to hold the knot at rest. Then the resultant of $\mathbf{F}_A$ and $\mathbf{F}_B$ is predicted by drawing. It is found to be equal and opposite to $\mathbf{F}_C$. This involves an extra stage of argument, because $\mathbf{F}_C$ is not the resultant of the other two but their "equilibrant," the force needed to hold them in check.

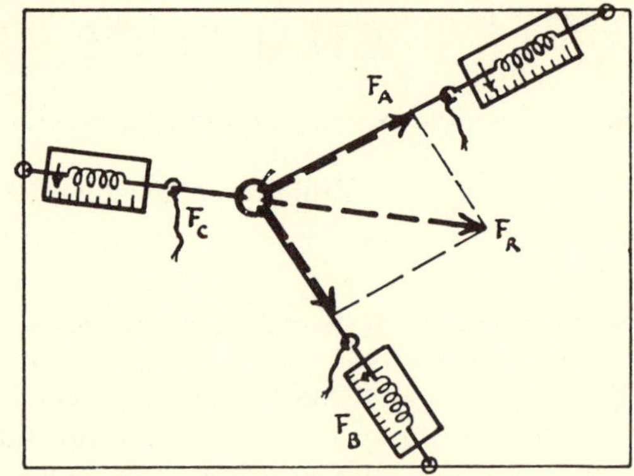

Fig. 3-2. Indirect Test of Vector Addition of Forces. The resultant of F_A and F_B is predicted by drawing. It is compared with the equilibrant F_C, that must be applied to make equilibrium.

Forces in Equilibrium

When part of a crane or bridge is acted on by several forces, the engineer's ambition is to have this part be at rest and remain at rest, and this requires the resultant force on it to be zero. Then, according to Galileo's view, that part will either move steadily or stay at rest steadily.[1] We then say that the forces are "in equilibrium" (= balance). If a set of forces has zero resultant, their vector addition diagram must show this by having the line joining starting-point to finish a line of zero length. That means the diagram must be a *closed figure.* Thus, if the resultant is zero, the end of the tail-to-

[1] In reality it moves steadily (?) with the Earth.

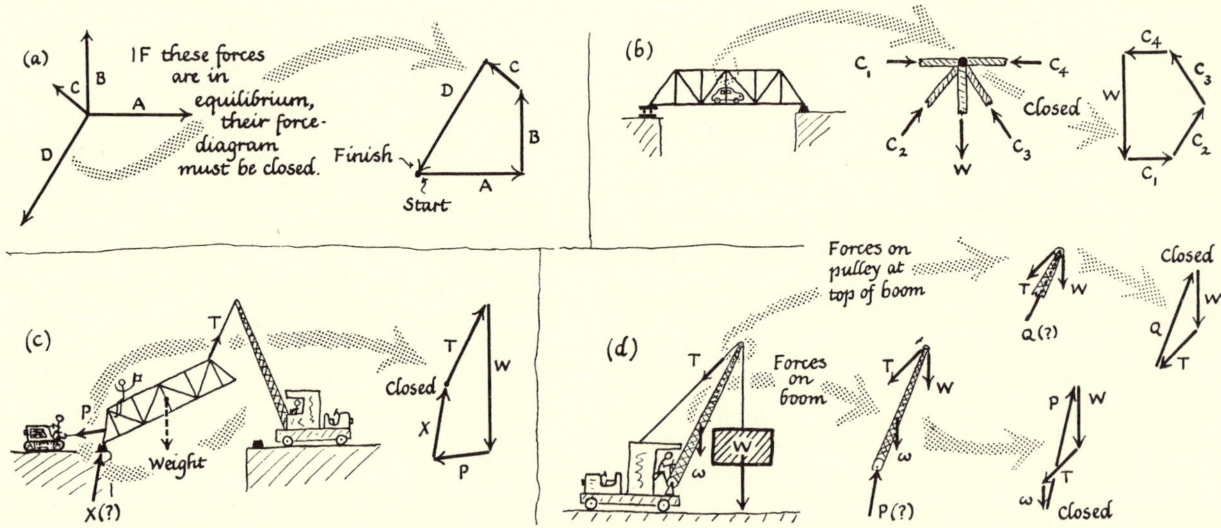

Fig. 3-3. Forces in Equilibrium. (Note that the examples above are not discussed in the text. They are sketched here to illustrate the engineering problems mentioned.)
(a) If these forces are in equilibrium, their force-diagram must be closed.
(b) Closed force-diagram for the forces acting on a joint in a bridge.
(c) Closed force-diagram for the forces on a bridge being installed.
(d) Closed force-diagram for crane lifting a load.

head pattern must be back at the start. The sketches of Fig. 3-3 illustrate this. This zero-resultant-for-equilibrium condition must apply to the whole arrangement—a complete crane or bridge, for example—but it must also apply to each separate part of the structure in equilibrium. We often obtain information by applying it to some particular part, such as the boom of a crane, or one strut of a bridge, or a peg tying several parts of a bridge together, or the bob of a pendulum. In such cases we are careful to include every force that acts *on that part*, and then we can insist on the whole set of forces making a closed diagram—if the part is in equilibrium.

In solving problems yourself, be careful not to include forces that apply to other victims. First choose, and then label clearly, the victim you are claiming to be in equilibrium.

Three Forces in Equilibrium: Force Triangle

If there are three forces in equilibrium their *vector diagram must be a closed triangle.* We can calculate the size and direction of one force if the other two are known, and this is useful in solving engineering problems. In many simple structures there are just three forces acting on each important part. For a steady structure each part must remain at rest: the resultant force on it must be zero. So if there are three forces on any part we draw a *closed triangle* for them. We shall now work through some

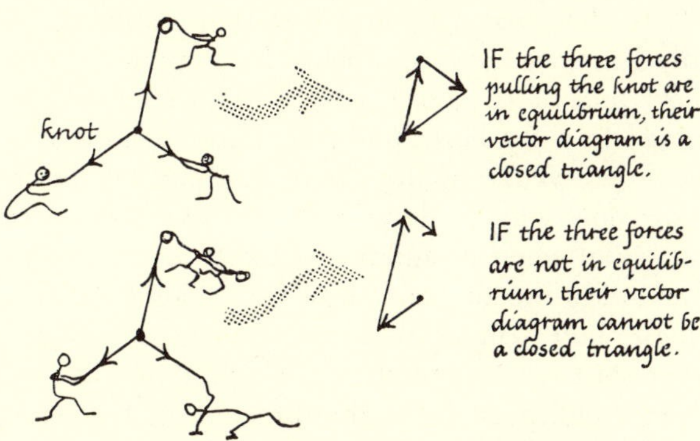

Fig. 3-4. Three Forces in Equilibrium
can be represented by the sides of a closed triangle.

examples of engineering force problems (statics problems). Follow them through; then try some of the problems at the end of the chapter.

PROBLEM 1

Three boys pull horizontally on a large iron ring in different directions. Suppose there are no other pulls on the ring, not even gravity. *Each boy pulls with a pull of 20 pounds* and the ring stays at rest.

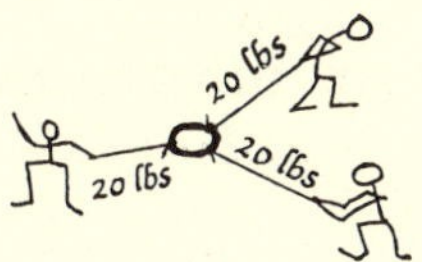

Fig. 3-7. Problem 1.

(a) What is the size of the resultant of the pulls?

(b) Sketch a force diagram adding up these three pulls.

(c) Sketch a view of the ring from above showing the directions of the pulls.

(d) Suppose one boy lets go suddenly, but the others pull as before. What is the size and direction of the resultant pull due to the other two boys?

Example A

A heavy pendulum consists of an 8-pound bob hung on a string 5 feet long. The bob is pulled aside by a horizontal cord, which pulls with a *horizontal* force of 6 pounds.

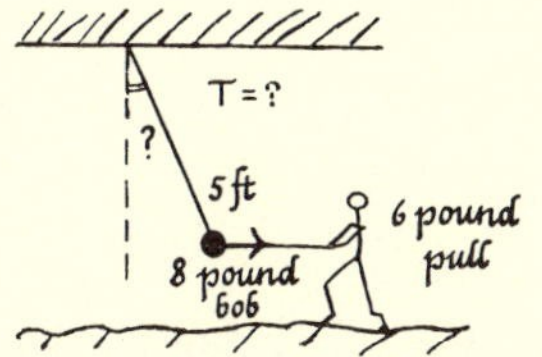

Fig. 3-5a. General Sketch to Illustrate Example A.

(a) Calculate the tension in the string.

(b) What angle will the pendulum make with the vertical?

There are three forces pulling on the *bob*:

(i) its weight, 8 pounds vertically down

(ii) the horizontal pull, 6 pounds

(iii) the string tension, of unknown size, pulling up along the line of the string. (A string must pull along its length.)

To calculate the tension, draw two diagrams, which should be quite separate since they deal with quite different things. The *fact picture* is a sketch showing the structure involved. It may be drawn to scale, or it may be just a sketch with dimensions marked. The force diagram is a vector diagram in which forces are represented by drawn lines. *It should not be drawn on top of the fact picture,* although they may bear a close similarity. In this problem, we choose to draw a force diagram for the three forces acting on the bob. When the bob settles down and remains at rest, the resultant of these forces must be zero. Therefore their vector lines, drawn to scale, must make a closed triangle.

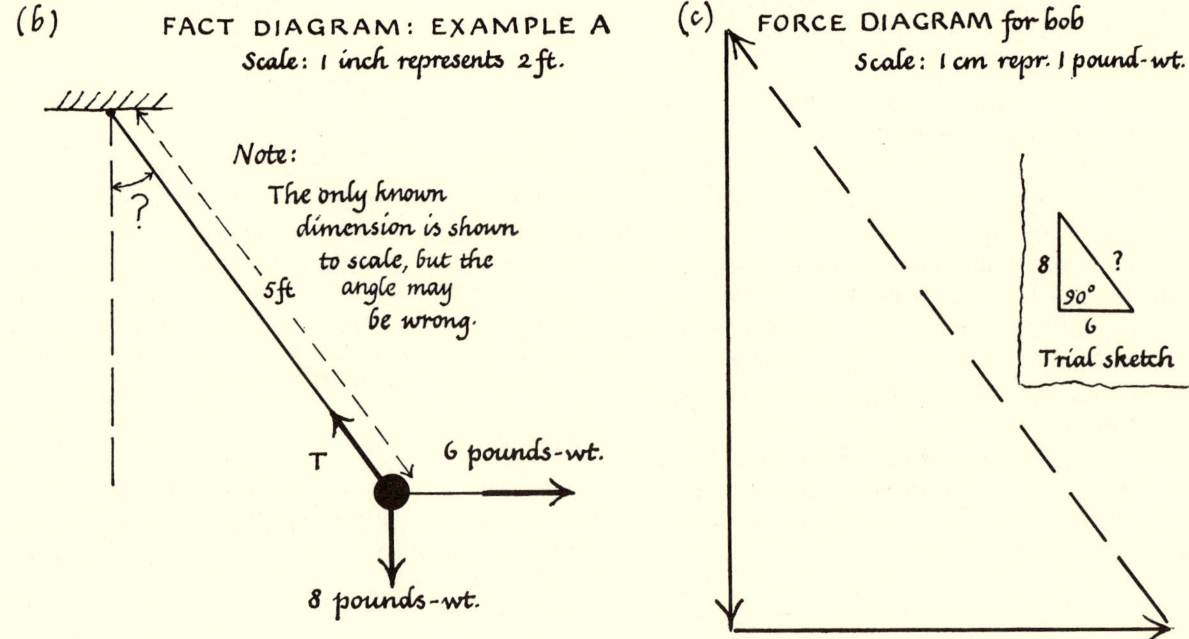

FIG. 3-5b,c. DIAGRAMS FOR EXAMPLE A.

We start by drawing a vector we know all about: the vertical pull of 8 pounds. We represent this by an 8-centimeter line **AB** drawn vertically, with an arrow head to show it is downwards.* We then add the other vector we know all about, the 6-pound horizontal pull, representing it by a line **BC**, 6 centimeters long. The third line for force must close the triangle since the resultant is zero. Therefore the third force must be in the line **CA**. Measuring this on a carefully drawn diagram we find it is 10 centimeters long, representing 10 pounds tension in the slanting string.

Or we can in this case look at a rough sketch and use Pythagoras and say the length is $\sqrt{8^2 + 6^2}$ or $\sqrt{100}$ or 10 centimeters. The direction makes an angle with the vertical whose slope (tangent) is ¾. Therefore from trig. tables, or by measurement, the angle is about 37°. Transferring to the actual pendulum we then say: The string tension must be 10 pounds and the string must make an angle 37° with the vertical.

* The points A, B, C are not labelled on Fig. 3-5c. Mark them.

Example B

A pendulum consisting of a 10-pound bob on a 5-foot string has its bob pulled aside by a horizontal pull **F**. If the bob is thus displaced 3 feet sideways, what is the size of the force? The fact diagram and the stages of the force diagram are shown in Fig. 3-6. In the force diagram we start by drawing the only force we know all about, the 10-pound downward pull of the bob's weight, **AC**. Then we try to add to it the horizontal pull, tail-to-head, but as we do not yet know the size of that pull we do not know how long to draw its line. However, we do know that when we add the string's pull to the other two forces the diagram must be a closed triangle (if the bob is in equilibrium). So the string's pull must start where **F** ends and finish at A. Also the string's pull must be in the direction of the string itself. (Can you visualize a string pulling in any direction but along itself?) So we transfer the string direction from the fact-diagram to the force-diagram and draw a line through A parallel to the string. This slanting line gives the third side of the force triangle, **BA**, for string tension. The corner B lies on the slanting line and on the horizontal

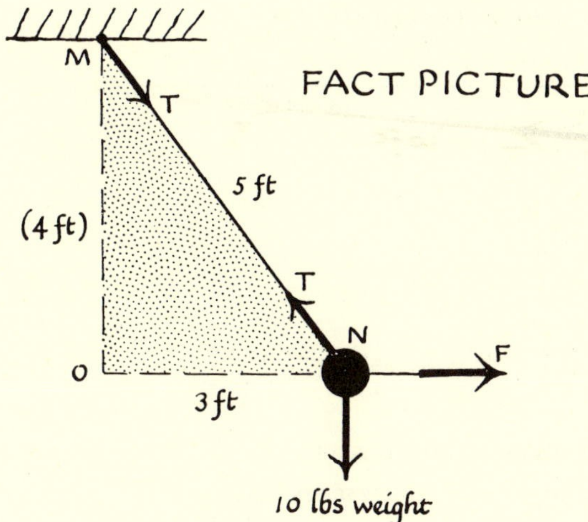

FIG. 3-6. EXAMPLE B.
These show how a force-diagram is built up from information shown in the fact picture. Since a triangle can be specified completely by two angles and a side, the force-diagram can be constructed in this case.

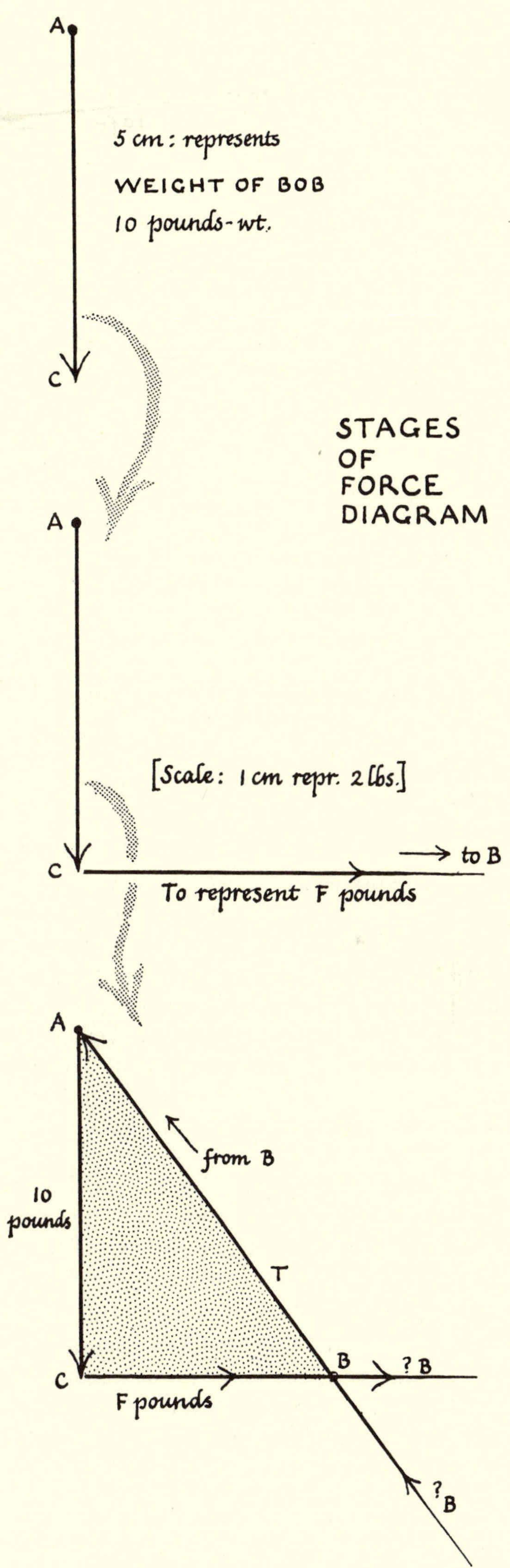

line; so it must be the place where these two lines intersect. Now that we know where B is, we know the size of **F**, and, incidentally, the string tension also. We can find the size by careful drawing and measurement.

Or, in this case where the data provide easy geometry, we can calculate **F** from rough sketches, arguing thus: The sides of the force triangle ABC are parallel to the sides of the triangle MNO in the fact picture. Therefore[2] these triangles are similar.

(*Note*: Pythagoras' Theorem tells us that OM = 4 feet.)

$\therefore \quad \dfrac{\text{F pounds}}{\text{10 pounds}}$ in the force triangle

$= \dfrac{\text{3 feet}}{\text{4 feet}}$ in the fact picture.

$\therefore \quad F = (10 \text{ pounds})(3/4)$

$\therefore$ Horizontal Pull, $F = 7.5$ pounds

Similarly, $\dfrac{\text{T pounds}}{\text{10 pounds}} = \dfrac{\text{5 feet}}{\text{4 feet}}$

$\therefore$ String tension, $T = 12.5$ pounds

[2] If you are not familiar with the properties of similar triangles, review them in a geometry book, or ask for instruction. You will need to use them confidently.

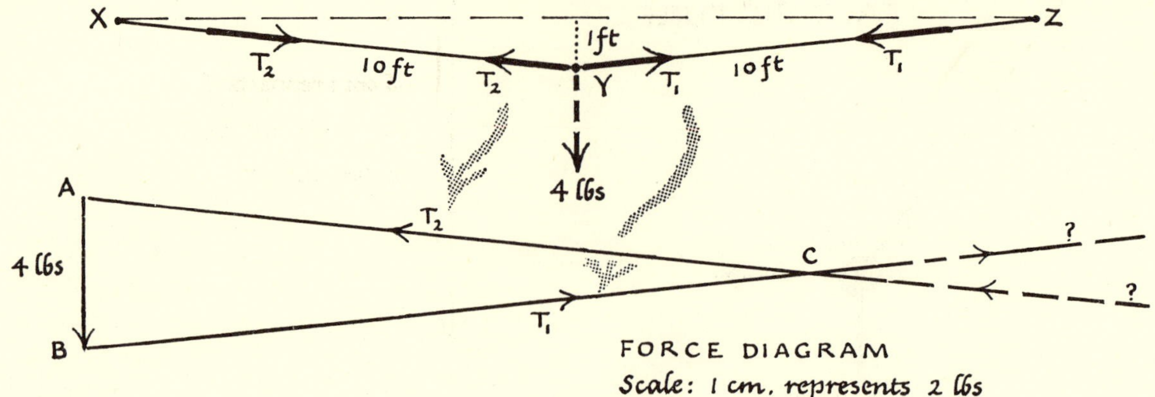

FIG. 3-9. DIAGRAMS DRAWN TO SCALE FOR EXAMPLE C

Example C

A 20-foot telephone wire is strung loosely between two supports. A 4-pound bird perches on the

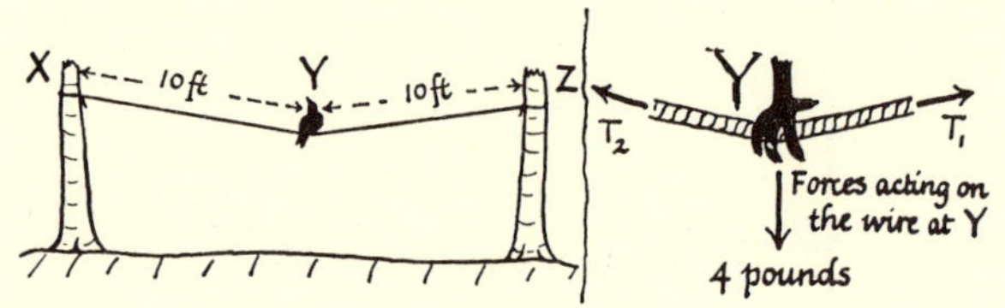

FIG. 3-8. EXAMPLE C

mid-point of the wire. The wire is thus pulled into a shallow V. Its mid-point is then 1 foot below the level of the end-supports. *Calculate the tension in the wire.* (This may seem a trivial artificial problem, like so many mechanics problems in statics, but it is a very serious matter for telephone and power companies. As the answer to this problem suggests, birds and ice may produce huge tensions in wires and overstretch or break them.)

We draw a force diagram for the small central bit of wire, Y, where the bird perches.[3] Three forces act on it, the weight of the bird downward and the slanting pulls of the wire tensions $\mathbf{T}_1$ and $\mathbf{T}_2$. Call the angle between wire and horizontal E.

To draw the force diagram for Y (shown in Fig. 3-10) we start with the fully known weight of the bird, drawn as a downward vertical vector, **AB**, 2 centimeters long to represent 4 pounds weight.

[3] How do we know that it is wise to choose that bit of wire as the victim to draw the diagram for, rather than half the wire, XY, or the whole wire, XYZ? To know which victim to choose is one of the "tricks" of solving mechanics problems, soon learned, not of much value in serious science.

From B we draw BC parallel to the right-hand section of wire to represent its tension, and then another vector parallel to the other section of wire. This must close the triangle, since the resultant force on Y must be zero. But we do not know how long to make the vectors for the tensions. So we draw unlimited lengths, one slanting at angle E upwards *from* B and the other slanting at angle E upwards *to* A, and thus fix C by their intersection. We now have a triangle of forces, which we could measure and interpret by the scale we use.

We could avoid measuring if we could link up the fact-picture and the force-diagram by similar triangles. The force-triangle ABC is *not* similar to the fact-triangle XYZ, but, as in most of these problems, we can find similar triangles by playing with the diagrams and adding simple construction lines. In this case we can add the broken lines in the diagrams and use the argument shown below.

The triangles WYZ and DBC are similar.
In triangle DBC, **DB** represents *half* the bird's weight, or ½(4 pounds).
In triangle WYZ, **WY** is the vertical sag of wire, given as 1 foot.

$$\therefore \quad \frac{10 \text{ feet}}{1 \text{ foot}} \text{ in triangle WYZ}$$

$$= \frac{T_1 \text{ pounds}}{2 \text{ pounds}} \text{ in triangle DBC}$$

$\therefore$ Tension, $T_1 = (2 \text{ pounds})(10/1) = 20$ pounds
Similarly, $T_2 = 20$ pounds

A 4-pound bird can produce a 20-pound tension. If the wires were less slack, sagging only one inch instead of one foot, what would the tension be?

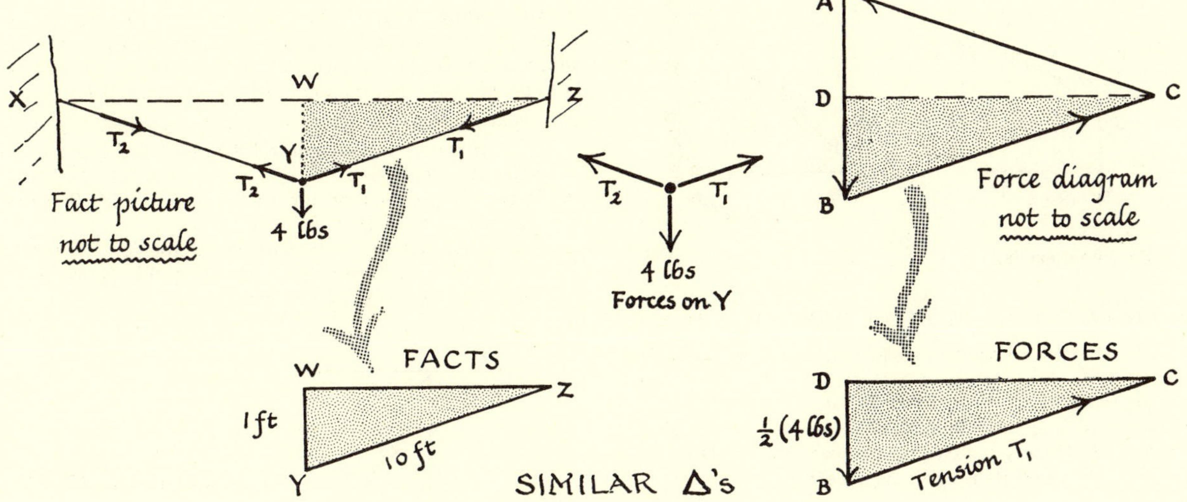

Fig. 3-10. Example C. These sketches, not to scale, illustrate the treatment of Example C by geometrical argument with similar triangles.

PROBLEMS FOR CHAPTER 3

1. In text.

2. A pendulum consisting of a 12-pound bob on a 10-foot cord is pulled aside by a horizontal pull on the bob. The tension in the slanting cord is then 20 pounds.
(i) How big is the horizontal pull on the bob?
(ii) Describe the slant of the cord.

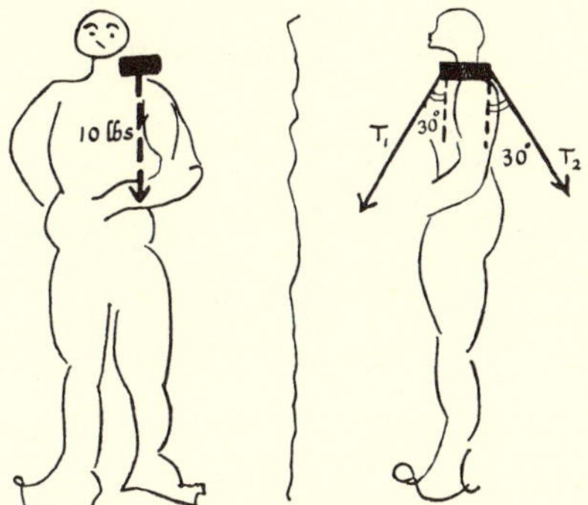

Fig. 3-11. Problem 3

3. A surgeon wishes to apply a vertical force of ten pounds down on a special splint on the shoulder of his patient. He proposes to do this by pulling the splint down with a string. But the patient's shoulders and ribs get in the way, so the surgeon decides to use two strings, one slanting forward and downward, the other backward and downward, from the patient's shoulder, each string making an angle of 30° with the vertical. (See Fig. 3-11.)
(i) Calculate the tension that each string should have. (Give a large clear force diagram.)
(ii) Explain your calculation.

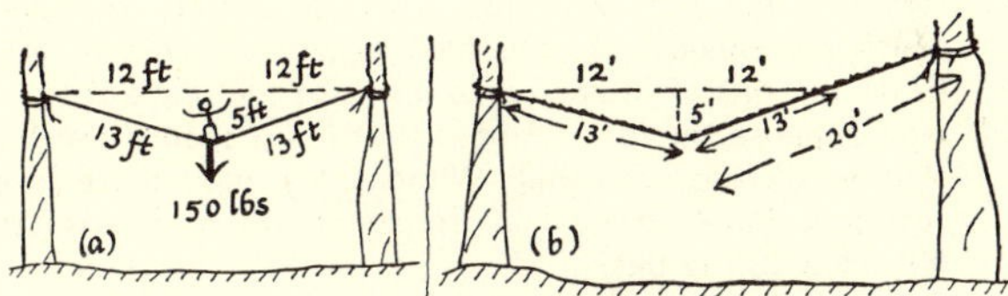

Fig. 3-12. Problem 4.

4. (a) A 150-pound tight-rope walker stands on the mid-point of a 26 ft wire strung between two posts 24 ft apart. Find the tension in the wire, giving diagrams and clear explanation. (*Note*: With these dimensions, sag at middle is 5 ft. See sketch, Fig. 3-12.)

(b) Suppose extra wire is added so that the wire at one end extends farther and rises to a higher support as in sketch (b), but the angles made by the two sections of wire remain unaltered. How will the tension(s) be affected?

5. (a) Which of the following do you consider must be vectors? (A vector is a quantity which obeys the geometrical addition rule.) Force, volume, acceleration, velocity, temperature, density, kindness, humility, humidity, electric field.

(b) Write, in two lines at most, a definition of the *resultant* of a set of vectors. (Do not give a rule for finding the resultant. Give a clear *description*, showing what it *is*.)

(c) Show by sketches and a little description how the parallelogram method of adding vectors (i.e., geometrical addition) leads to the polygon method of adding vectors tail-to-head.

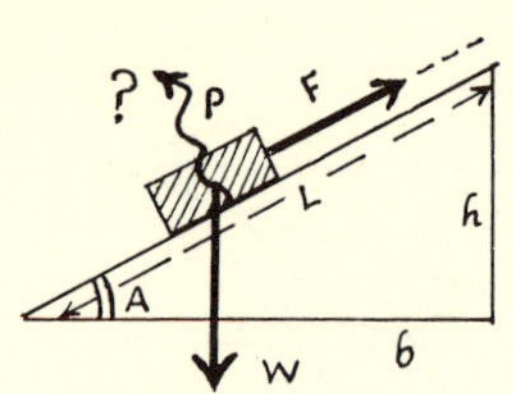

Fig. 3-13. Problem 6

Fig. 3-14. Problem 8.

★ 6. AN IMPORTANT RELATION: LOAD ON A HILL

A sled rests on a frictionless incline which makes an angle A with the horizontal (or rises h feet vertically for a distance L feet up the slope so that $\sin A = h/L$). The sled is prevented from sliding down hill by a rope which pulls uphill with tension F. Gravity pulls vertically down on the sled with a force which we call its weight, W. (See Fig. 3-13.)

(a) Draw a sketch of the sled on the hill and add arrows to show the directions of W and F. Add another arrow to show the direction of P, the push of the hill. Assume that, since the hill is frictionless, P must be perpendicular to the slope. Make all these arrows sprout out from the sled.

(b) Now draw another sketch showing the vectors W, P, and F all adding up to a resultant zero.

(c) If you agree that your two sketches contain triangles which are similar, use this idea to *write down the ratio F/W in terms of h, etc.,* or in terms of the angle A.

(d) Now suppose the rope is cut so that F disappears and the sled accelerates downhill. Without the rope there is a *resultant force* downhill of the same size as F was uphill. How big is this?

7. Treat the sled problem (6) by a different method. Resolve the weight W into components F downhill and P across the hill (i.e., perpendicular to it). Express F in terms of W and h, etc. This gives the tension the rope must have; or, if there is no rope, the downhill resultant accelerating force.

8. Shortly before Galileo's work, Stevin published an ingenious "thought experiment," arguing thus: Imagine a necklace of smooth beads hung on a triangular wedge or prism, as in Fig. 3-14. The necklace must be in equilibrium—we do not expect it to slide round and round, faster and faster, just because there are more beads on the slope. Cut off the loop that hangs freely underneath. Since that loop is symmetrical, its removal cannot spoil the equilibrium. From this, Stevin predicted that F/W must $= h/L$ for a load on an incline. Try to continue and complete his argument and make that prediction. (*Hint*: Condense all the beads on the slope into one lump, all the beads in the vertical portion into another lump. Connect the lumps by a thread over a pulley.)

9. A designer wishes to incorporate a pendulum in his apparatus, with a string to pull it aside *with a pull perpendicular to the pendulum-cord; i.e., along the tangent to the arc along which the bob moves.* (See Fig. 3-15.) The pendulum cord is to be 10 feet long, its bob a 20-pound lump of iron.

(a) What pull, P, is needed to pull the pendulum aside by 1 ft *horizontally?* Give careful diagrams and explanations of your calculations.

(b) Repeat the calculation for the bob pulled aside 2 ft horizontally, 3 ft, 4 ft, 5 ft. . . .

(c) What can you say in general about the force P needed for such deflections? (This contains the germ of the theory of swinging pendulums.)

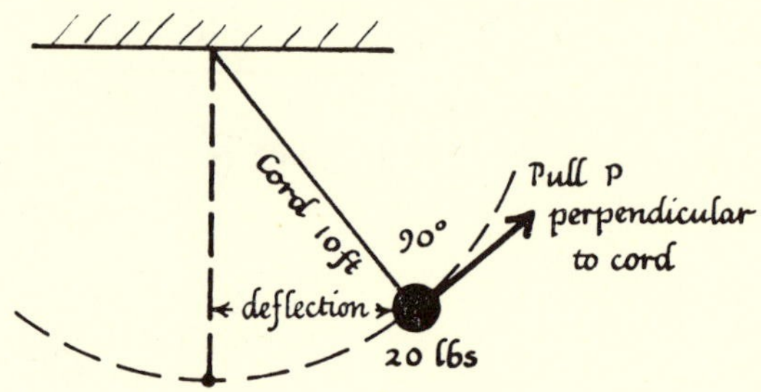

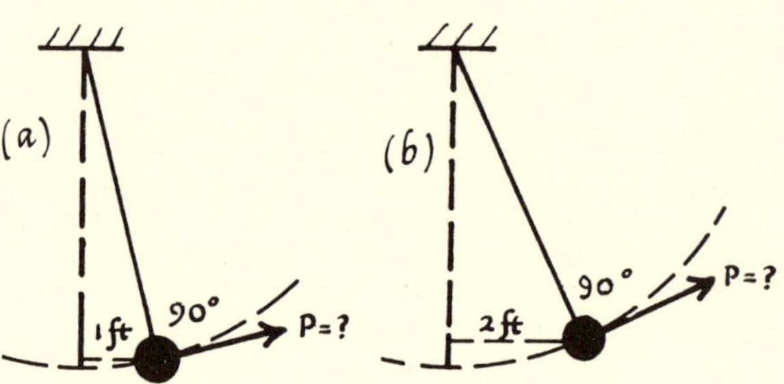

Fig. 3-15. Problem 9.

CHAPTER 4 · "IT'S YOUR EXPERIMENT"

LABORATORY WORK

". . . *'For I neither conclude from one single Experiment, nor are the Experiments I make use of, all made upon one Subject: Nor wrest I any Experiment to make it quadrare* [square] *with any pre-conceiv'd Notion. But on the contrary, I endeavour to be conversant in all kinds of Experiments, and all and every one of those Trials, I make the standards (as I may say) or Touchstones by which I try my former Notions. . . .'* Thus the young man of twenty-six set down his scientific creed in an age when the experimental method was still making its way."

—E. N. da C. Andrade, quoting Robert Hooke (1661)[1]

[How much you work in a laboratory yourself, and what experiments you do, will depend on the arrangements for your course.

If you do no laboratory work, you should omit some chapters that ask for experiments and see demonstrations for others, particularly for accelerated motion, for pressure and Boyle's Law (this chapter) and electric circuits (Ch. 32). (Often, a single demonstration can replace several laboratory periods, so far as factual information is concerned.)

If you work in a laboratory yourself your experiments will depend on the equipment available; so this chapter offers only general suggestions and comments. Almost any laboratory experiment, old or new, with simple apparatus or complex, can be used to great benefit; or, if spoiled by childish instructions, it may do more harm than good to your understanding of science.]

Welcome

It may seem strange for someone who teaches in one lab to write a speech of welcome to students in another lab far away. But, if this course is to make a lasting contribution to your education, your work in laboratory will play an important part for good or ill. It will do harm if you find it a dull following of cookbook instructions. It can do essential good if the laboratory welcomes you as a participating scientist, to plan, discuss, and do experiments, to criticize and use them, so that you experience the delights and sorrows of experimental scientists. To be profitable, your experiments should show you how scientists work: how they measure and experiment, how they trust and doubt their results, how they draw conclusions and balance them with speculation—in short, you should gain an insight into the working relationship between experiment and theory.

Why does the value of your lab work to you depend so greatly on your attitude; why don't we trust it to train you scientifically anyway? Modern medical men discuss the technical details of their treatment with educated adult patients. For much the same reason, we shall discuss the benefits of your lab work before you begin.

Earnest educators—deans, parents, and others—often expect science courses to teach students accuracy and skill with instruments and scientific honesty and logic and "scientific method"—in fact, *to make students scientific.* They trust that the training acquired in a science course will provide universal benefit by spreading to other studies, and to life in general. You yourself probably hope for such benefits, and would be glad if the course could quickly endow you with them—though you may also have lighthearted hopes of fun in lab. Curiously enough, the lighthearted fun is likely to be valuable, while more serious benefits may prove almost unattainable. However, the most important value is concealed behind both: *learning to understand the work of scientists*, and that is helped by doing some of their work yourself in lab.

Why are we so modest, offering only a general experience of scientific work as an outcome of lab? Because investigations by psychologists in the last half century have thrown great doubts on the grander hopes of "transfer of training" from a course to life-in-general. Direct training is easy: soldering for the radio repair man, calculus for the future mathematician, weighing techniques for the pharmacist; but most of such training seems to remain in its own narrow channel. (Look at the scientists themselves. Are they noticeably better in general ways for their training in method: tidy, systematic, scientifically critical and unbiased in their thinking?

[1] In *Proceedings of the Royal Society*, vol. 201 A., March 1950, p. 442.

Some show these virtues, but only the same proportion of the population as in other professions.)

This raises a key question in planning education. Can our minds transfer training in some skill or in the use of some idea or method, from, say, a science course to life in general? If the answer is No, this course is of little use in your general education. The only useful thing it could then promise would be *information*; but facts are easily forgotten—and better looked up in reference books when needed. If the answer is Yes, our hopes should be grand indeed: courses can make people scientific, far-seeing, well-balanced, . . . , a generation of supermen. Since your own hopes and interests are seriously involved, we shall discuss the question further and review the psychologists' findings.

"Transfer of Training"

Over the last fifty years and more, careful experiments have been made, with tests before and after training, with control groups, etc. One of the earliest was a personal trial by the psychologist William James. He measured his speed in learning French verse. Then he switched to learning English verse and practiced techniques for several weeks. Then, back to French verse: no improvement, no gain from the English practice. Other early results were equally discouraging: almost no transfer.[2]

Take as an example training in accurate weighing with a chemical balance. We can teach students good techniques and train them to weigh quickly and accurately in a good chemistry lab. But when they go out to a factory or home the training is often lost. Such results would suggest that higher education can provide only technical training; and they would spell disaster for your hopes and mine in this course. Fortunately, fuller investigations have shown that transfer *does* occur, though not nearly so easily as had been hoped, and only in certain special circumstances.[3] To get the most out of your laboratory work—where transfer of training could offer great benefits—you should know what these favorable circumstances are. They seem to be:

(1) *Common ground.* Transfer is likely and easy if there is common ground between the field of training and the other field. For example, if you are trained to weigh accurately in a chemistry lab, it is certain that the training will transfer to another chemistry lab, so you will weigh more accurately there; it is fairly certain you will carry your good training to any weighing you do in a physics lab (particularly if the apparatus looks similar!); it is much less likely that you will carry it to weighing done at home or in your business; and it is *very unlikely* that the training in accurate weighing will produce a general habit of being accurate in other activities. Another example: training in logical argument learned in geometry is likely to transfer to further geometrical studies; fairly likely to help your arguments in physics; unlikely to make you think critically about arguments in newspaper advertisements; and *very unlikely* to make you a more logical economist. Yet, the end of each tale of examples is what we most want from Education; and fortunately we can lessen the gloomy doubts of *very unlikely* by attending to conditions (2) and (3) below.

(2) *Generalizing, with hope of transfer should be encouraged*: If you are aware of your gains in one field and see that they could be applied to other fields, transfer is more likely. It pays to want transfer—thus establishing common ground. That is why all this is being discussed here: so that you can set your own aims and thereby increase your profit.

(3) Training does transfer over a long bridge to a different field *when carried by strong intellectual feeling.* If you *enjoy* the power of a piece of study, or *take a strong delight* in some method and its skills, or if you feel inspired by an idea or intrigued by philosophical questions raised in your learning, then you are more likely to retain and generalize your gains.

Thus, returning to our examples: a student who develops a delight in accurate weighing, making accuracy almost a minor ideal, may carry the techniques and attitude of seeking accuracy far and wide in all his activities. A student who learns skill in geometrical argument *and* feels inspired by its sure method may become the cleverer economist or lawyer[4] by the transfer of some of that training.

Again, the economist, business man, or administrator who enjoys attacking problems like a scientist may bring to them an imaginative, critical experimental attitude which makes his work not only very valuable to others but a creative delight to him.

We should all of us be wise to think about wider uses of what we learn, and we should give great value to our enjoyment of some experiment, argument, theory, speculation. There is a strange threat, then, in our invitation to laboratory: "Come and work in the lab—and you had better enjoy it too!" We who offer the lab cannot command your enjoyment, or ensure your right feelings. We can only tell you how much your present attitude matters and then give you a genuine laboratory for your experimenting.[5] For that, almost any experi-

[2] True, students who had done brilliantly at Latin were very clever in other studies. But Classics had not necessarily trained their minds or polished their wits—they had always been exceptionally able. Their teachers, who claimed great benefits, should have been more careful to distinguish between *propter hoc* and *post hoc*, between ability *because of this* (study of classics) and ability observed there *after this.*

[3] The experts still disagree on amounts of transfer—it is difficult to make fair comparisons and even more difficult to interpret the results fully and correctly—but the account here summarizes generally agreed opinion.

[4] A lawyer who argues on the model of Euclid's clinching proofs is a terrifying opponent. Like Euclid, he starts by setting forth his axioms and assumptions and securing agreement on them—". . . self evident. We all know that. . . ." Then, he knows his case is proved, when it seems barely started. It is only a matter of logic (IF . . . , THEN . . . SINCE WE ARE AGREED THAT . . . , AND SINCE . . . , THEREFORE . . .), before his opponent is pinned inescapably in a corner.

[5] Experimenting is just a longer word, from the Latin, for trying things. As you work in a lab you will probably come to

ment offers a fruitful field, and almost any equipment gives you good opportunities. But you need time to bring each experiment to a close—or sometimes to extend an experiment along a new line. Thus it is good if the laboratory organization lets you extend the time for some experiments instead of requiring each to be run through in a single period. Given time, you can work on your own, needing only occasional advice from instructors—if you ask them for cookbook instructions, they will remind you, *"It's your experiment."*

The Objectives of Laboratory Work

Now, with doubts and conditions of "transfer" in mind, we can review the purposes of laboratory work realistically.

You can learn "facts" of physics, from a small detail to a general law, very comfortably in lab—but very slowly. Tests have shown that information is acquired quicker and just as well in classes or by reading. It seems a waste of time and expensive equipment to use labs to teach you facts. Yet you will find yourself learning what you learn in lab with a fuller sense of understanding. In that way laboratory provides a valuable sense of understanding the information of science.

However, laboratory can provide much more important gains if it can teach you scientific ways and give you a more general understanding of science. For that you must make your experiments your own work, work that you like and regard as part of your present life and future hopes. If you work as a scientist yourself, you are on common ground with scientists, gaining an understanding of science.

So we offer laboratory work with a variety of aims: learning some physics with the thoroughness that comes from doing it yourself; gaining an appreciation of scientific techniques by using them; and, above all, gaining an understanding of science and scientists by experiencing for yourself the delights of scientific work and its sorrows, its honesty and its risks, its successes and failures, its uncertainties as well as sure results. When you come to the lab, make yourself "a scientist for a day," and you will gain understanding that will outlast all information.

"Open" Experiments

On some days in lab you will be free to find out all you can (within a region of physics) by experimenting and making inferences—an "open" lab. Though some apparatus is provided, you may get other equipment by asking for it. (Modern laboratory store-rooms have vast resources, and you should receive extra equipment if it is reasonably possible. You may have to explain what you want it for, but your plans will not be laughed at.) A good lab encourages independent experimenting and should offer good apparatus. In return you will be expected to treat apparatus with care. Scientific instruments are expensive,[6] often made with great care by skilled craftsmen; and, as agents to extend human senses and skills, they deserve your respect.

You will also meet simple harmless instruments like rulers and watches which you take for granted as good. Do not be too sure: there are crooked rulers, and shrunk ones of modern plastic, and there are rough watches in brightly polished cases. Make sure they are as good as you need, or ask for better.

Those who provide for "open" labs foresee many of your investigations and needs; but someone, you or a neighbor, will branch out in unexpected directions, sometimes very fruitfully, and such work will be welcomed and provided for. It is not necessary for you to do everything your neighbors do—methods and results can be pooled in conferences.

Discoveries?

You may think of new experiments or devise new methods, but you are not likely to discover entirely new physics unknown to professional scientists: we shall not deceive you into thinking that, nor should you deceive yourself—yet you can enjoy making what is to you a new discovery, uncovering a delightful simplicity in nature, or exploring some surprising phenomenon.

Classical Experiments

On other days of laboratory, you may find the field narrowed to a definite requirement, perhaps trying for yourself some famous experiment. In

restrict the word to *systematic trial and planned investigation*, in contrast with casual playing around with apparatus that gives entertainment with little promise of new knowledge. The distinction cannot be made a hard and fast one without a loss of good science, and yet you will find that the word "experimenting" develops a clear flavor as you proceed. The longer word "experimentation" has a respectable history, but it now seems to many professional scientists a childish use of a longer word to make science sound grander.

[6] e.g.: A single set of electrical apparatus for one of your later experiments will cost \$50 to \$150. The jewelled pivots of its ammeters are so fine that the small weight they bear exerts a *pressure* of several tons/sq. inch at the points. Putting a meter down on the table abruptly applies many times that pressure; and that may blunt the pivots and reduce the meter to a sticky unreliable one which you do not deserve. The repair is costly, chiefly in skilled instrument-maker's time.

such cases, you will gain more if you do not regard it as routine repetition but take it as an opportunity to share the original scientist's delight.

Reports

As a scientist for a day, you are entitled to ask for good equipment, to work hard, extract inferences, argue with all who come, and trust your own results. In all this activity you should keep some record of your work, a diary of *what you did* and *what you observed*, with a discussion of *inferences and conclusions.* Such a record is usually called a Laboratory Report. It can be drawn out by formal requirements into a long work of futile drudgery, and that is not genuine science. Yet without some record you would miss the scientist's systematic care. Every professional scientist treasures the pencil-written notebook that is his companion as he experiments. He copies it and expands it for formal publication, and then he dreads the errors that creep in when data are copied. So you should make simple records. We hope you will value them as a diary of your work, and, like professional scientists, will keep them long after the experiment is over.

A good report need not be long. Write it at the time of the lab, and do not copy it out later in a mistaken worship of neatness. As a written workshop for data and calculations of your experiment, your report is more efficient if fairly tidy, but we should not expect it to be as clean as a workshop that is not in use. What would you think of a carpenter who was afraid to use his shop for fear of making it untidy with sawdust and shavings? Mistakes crossed out, poor measurements put in [ ], circuits redrawn with improvements, calculations tried roughly then repeated in detail, . . . are all welcome as signs of honest work. (Never erase—that looks like suppressing evidence—but cross out when you change your mind.)

As a guide: your report should serve to tell you clearly what you did, if you consult it a year later. Perhaps it should enable another student to carry the experiment through quickly, if he comes to the lab alone with no instructor but with the same apparatus and takes your record as a specimen, with your warnings of useful techniques.

A good report should contain:

(i) a full record: measurements and other observations—the actual pointer reading you saw on each meter or scale, not some number calculated from it.

(ii) analysis: calculations, graphs, etc.

(iii) comments, inferences and conclusions. These are essential, the fruitful outcome of your work.

(iv) a critical survey of the report's accuracy—something a good professional scientist would add. If other scientists want to use his results they would hardly welcome it alone like this:

In fission of Li nucleus,
energy released = 17.4 Mev

They want the scientist's own estimate of his likely error—or they may prefer to study his discussion of errors and make their own assessment too. They would want at least this:

energy released = 17.4 ± 0.2 Mev (see note 7)

Estimating errors and discussing them is a fine thing to do, but it takes time and skill; do not spend time on it unless you feel the need and enjoy it.

(i) *Description.* This is a timewaster and should be reduced to notes jotted while you work. Omit the obvious. (e.g., if your record says

Temperature of water . . . . . . 16.2°C

it is quite unnecessary to say in your description, "We took a thermometer and measured the temperature of the water.") It is a waste of time to list the apparatus used (unless some was of special quality or a numbered instrument that you wish to use again). On the other hand, if you took special precautions, say so (e.g., "We stirred the water carefully before taking maximum temperature"). A few lines will usually suffice for (i)—an account of what *you did,* not a copying of some official instructions that you did or did not follow.

(ii) *Record of measurements.* This is the central part of your report. Record everything that might be useful. To avoid re-copying, plan your original record to be easy to read and use. There are two standard forms used by scientists:

(a) sets of measurements in labelled columns and rows;

(b) a collection of separate measurements like this:

```
MEASUREMENT OF THERMAL CONDUCTIVITY    --   RECORD

Temperature of water . . . . . . . . . . . .  16.2 °C

Diameter of rod    2.48  2.46 }
                   2.46  2.44 } . . . average  2.46 cm

  ∴ radius of rod  =  2.46/2 . . . . .  =  1.23 cm
    (see note 8)

  ∴ area     =  πr²  =  3.14 x 1.23²     =  4.75 cm²

Length of rod ... (and so on)
```

(iii) + (iv) *Analysis and conclusions* make half the main value of your experiment—the other half is in doing it. Be critical and ingenious, bold and imaginative. Squeeze all the inferences you can from your observations, and outline your reasoning. Where you can draw a definite conclusion, state it clearly—but avoid pious statements such as "I verified the law," "The difference is due to experimental error." (This last is a catch-all, an alibi used by amateurs.)

Partners

You may be asked to work in pairs, or larger groups. Most laboratories require this as a necessary economy in equipment and staff. A partner is a doubtful advantage. If you rely on him, a partner will take both apparatus and thinking away from you, to your short-term

[7] That means "17.4 is the best guess, from my measurements. I do not think the correct value *is* 17.4 + .2 or 17.4 − .2, but from my work I consider the true value may be anywhere in the region between."

[8] The *diameter* of a rod is the easy thing to measure, not the *radius.* Record the diameter. Then calculate the radius, setting that line in your record a short distance in, to show it is derived.

gain and ultimate loss. But you can gain from your partner if you treat him as a fellow scientist and critic: plan your experiment with him; criticize his techniques; watch his measurements and have him watch yours; and compare your results with his independent ones. (Use his measurements in your report as well as yours, but make their origin clear in case they prove doubtful.)

SUGGESTED EXPERIMENTS

(This list is tentative and incomplete, to be supplemented by suggestions from other chapters and from your Instructors in laboratory.)

EXPERIMENT A. Falling Bodies and Projectiles (An informal, OPEN lab)

This was mentioned in Ch. 1. If you have an opportunity for laboratory work at the very beginning of the course, try any physical experiments you like relating to falling bodies and projectiles (in air or other media). Start with old, simple, obvious ones such as dropping unequal coins: try them quickly and record, in a few lines, what you did, what you observed, and any inferences. Then proceed quickly to any more complicated investigations you can devise. Try for ingenuity and variety. This is a lighthearted stage, at which careful planning and long systematic investigations are not called for.

EXPERIMENT B. Investigation of Springs (OPEN lab)

Find out anything you can about the physics of springs. (You will be provided with a ready-made spiral spring of steel wire, but you can make other springs for yourself, e.g., by winding copper wire on a rod. Other materials, and other shapes than spirals, are easily available for investigations of springy behavior.) This is a wide field, usually called "Elasticity," that allows you to push your investigations in many directions: stretching, twisting, effects of various treatments, . . . You would be wise to start with the simple "official experiment" below and then branch out on investigations of your own devising—and it is the latter that carry most "credit" in this course, and offer you most lasting benefit.

"Official Experiment"

Find out how the STRETCH of your spring depends on the LOAD hung on it. STRETCH is defined as INCREASE-OF-LENGTH, from the spring's length when unloaded to its length when carrying a given load. (Thus, STRETCH is reckoned from the original, unloaded position each time.)

Record the position of some pointer on a scale of centimeters or meters (and then compute the stretch in a later column), and the load in grams or kg. Plot a graph to exhibit your measurements. (*After* this experiment you should, at some time, consider the meaning of a Law in Science.)

EXPERIMENT C. A Precise Study of Accelerated Motion

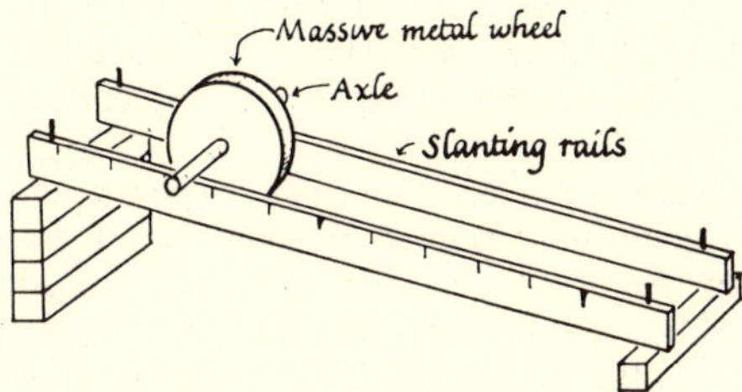

FIG. 4-1. EXPERIMENT C.

This is the "rolling wheel" investigation, mentioned in Ch. 1. Try some form of that experiment (see Fig. 4-1). Make very careful measurements, and exhibit your results by plotting graphs. Use a graph of s against t^2 to see how closely your wheel's motion fits the simple motion described by

ACCELERATION = constant.

If possible, make a general review of accuracy, after your experiment, in conference with instructors and other experimenters. If you can decide on likely errors for your average values of times and distances, mark error-boxes around your graph points.

Analysis Questions for Rolling Wheel Experiment

When you have finished Experiment C, give your results an analytical "work-out" along the lines of the questions below. These questions—which must be modified to fit the arrangements of your apparatus—were framed for a wheel that rolled down slanting rails a meter long, taking about 25 seconds from start. The experimenter is supposed to have measured times for travels of 0.2, . . . , 0.6, 0.8, . . . meter from rest and used them for his graphs. (He also recorded the time for an intermediate distance, 0.7 meter, and he also noted the distance travelled in 15 secs from rest, but he did not plot these data on his graph; he kept them secret for use as checks in this analysis.)

(1) Did your Graph I (of s against t) suggest that s varies directly as TIME?

(2) Did your Graph II (of s against t^2) suggest that s varies directly as TIME^2?

(3) *Interpolation.* Assume that your graph-lines are "true" and interpolate to answer the following questions:

(a) What is the value of t for $s = 0.7$ meter,

(i) by reading from Graph I?

(ii) by reading from Graph II?

(iii) by your direct measurement (stored up for this check)?

(b) What is the value of s for $t = 15$ secs,

(i) by reading from Graph I?

(ii) by reading from Graph II?

(iii) by your direct measurement (stored up for this check)?

(4) Why is it more accurate to use Graph II than Graph I for the interpolation asked for in (3)?

(5) (a) If $s = \frac{1}{2}at^2$ (the acceleration a being constant) how would you expect t for 0.8 meter to compare with t for 0.2 meter?

(b) Write down the ratio of your average measurements for these two times, and calculate its value.

(6) *Calculating Acceleration from Graph-Slope.* Choose two convenient points on your second graph, far apart, read off their values of s and t^2, record them. Calculate a from them, assuming $s = \frac{1}{2}at^2$ applies.

(7) Why is the method of (6), using a graph-slope, a more accurate way of estimating a than using measured values of s and t for one distance alone, say 0.8 meter?

(8) *Estimating Velocity by Several Methods.* Choose a suitable place in the motion, say at 0.6 meter, and find the velocity there by three methods:

(a) Use $v = at$, with your measured value of t for that instance and the value of a from (6).

(b) Use $v^2 = 2as$, with your chosen s and the value of a from (6).

(c) Use the slope of Graph I. Draw a tangent at $s = 0.6$ meter. Extend the tangent right across the paper; choose two points on the tangent, far apart; read off their coordinates and record them. Thence calculate the slope. (See the discussion in Ch. 1. Also see notes on graphs in Ch. 11.)

(d) Compare the results of methods (b) and (c). Express the resulting difference between VELOCITY given by (c) and VELOCITY given by (b) as a % of that velocity (see discussion of % differences in Ch. 11).

(9) *If the wheel were to slide without friction instead of rolling, would you expect the same acceleration, or greater or less? Why?* (This is a very important question, which deserves a guess at an answer now. The full answer involves more complicated physics. You would be wise to leave your guess without asking now whether it is wholly "right." A later chapter will supply a clear answer.)

EXPERIMENT D. PENDULUMS

Narrow your field of investigation down to a *simple pendulum*, a small bob swinging to and fro on a long thread; and narrow it still more to the question, "How does the time-of-swing or *period* of a pendulum depend on each of the physical factors that might affect it?" Thus restricted, this is still a complex investigation unless you follow the good scientific practice of holding all the other factors constant while you change one chosen factor at a time.

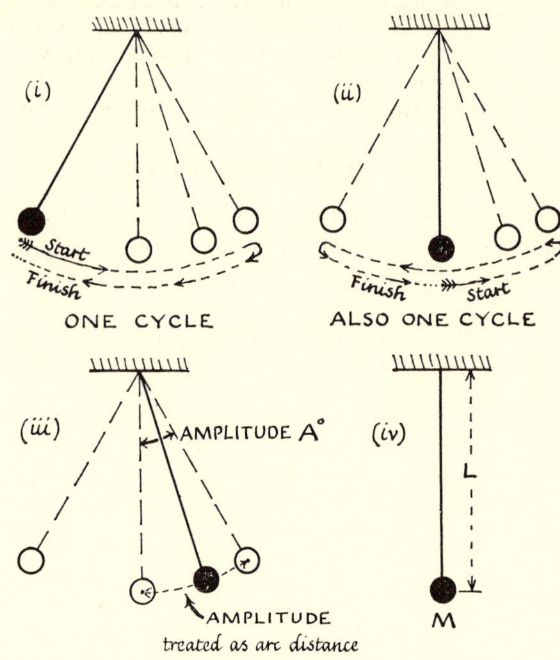

FIG. 4-2. SIMPLE PENDULUM

We define the PERIOD as the time of one "swing-swang," one complete cycle. What factors might affect the period? Obviously the LENGTH of the pendulum—most easily defined as distance from support to center of bob, but not easily, or wisely, measured *directly* in that form. We all know a longer pendulum takes more time to swing to-and-fro. But just how is period T related to length L? Is there a simple mathematical rule? (There is. We can deduce it from other experimental knowledge, e.g., from force vectors and Newton's Laws of Motion; but here you should make an "empirical investigation"—ask a straight question of nature in your own experiments.)

What other factors might affect T? Mass or weight of bob; *amplitude*, or size of swing; and possibly other factors?

Start by investigating how T depends on
length of pendulum, L
amplitude of swing, $A°$ each side of vertical
mass of bob, M

To avoid confusing several effects, keep two of the three, L, A, M, constant while you change the third, and measure T. Does it matter which of the three you choose to vary first? In this case it does: there is one logically correct choice. While you are considering this, make some preliminary measurements to try out techniques.

EXPERIMENT D (0). Preliminary Measurements

A good scientist does not expect a stream of accurate measurements to flow at once from his apparatus. He experiments on his experiment, trying out techniques, gaining skill by practice. Choose a long pendulum, 2 or 3 ft long, and make accurate timings of its period with a good stop-watch (or use a magnifying glass over the seconds hand of an ordinary watch while a partner gives you signals.) Record your measurements. Compare them with your partner's. Look at the methods being used by neighbors, and get ready to criticize.

Discussion Group

When you have practical experience of your apparatus, meet with other students and the instructor, as a "research council" to discuss difficulties and techniques. This is the time to suggest good tricks you have discovered, to criticize mistakes you saw neighbors making, to discuss the reliability of the equipment, and to decide on good techniques and plan the order of experiments. There is little use in such a discussion before you have done preliminary experimenting: that would lead to childish guessing or else your instructor would have to step in with cookbook directions. As in a professional research group, an adult discussion needs practical experience of the apparatus.

At this time you will find there is a good reason for investigating the effect of AMPLITUDE first, before MASS or LENGTH.

EXPERIMENT D (1). Measurements: T vs. A

Make careful measurements of *T* for various amplitudes such as 80°, 60°, 40°, 30°, 20°, 10°, . . . Plot a *rough* graph of *T* against *A* as you go, to guide your further measurements. If you find the graph-points use only a narrow region of your paper, you should plot another graph with one coordinate expanded, so that the blown-up graph reveals the shape better. (A blown-up graph need not have the origin on the paper—in fact its origin may be many inches off the paper.) If you run into difficulties, discuss them with your instructor—treat him as a source of good advice from another scientist, not as a short cut to "right answers."

When you have enough good measurements, plot a careful graph of *T* against *A*—with a blown-up version, too, if that seems called for.

At this stage, another conference is likely to be useful. Comparing your graph with the graphs of others, you will probably decide there is a definite relationship, but many graphs may show such large accidental errors that their form is obscured. An accurate answer to "How does *T* depend on *A*?" is essential: *the other parts of this investigation will be impossible without it.* You will need very careful measurements of *T* for a certain range of amplitudes. It will be obvious from your present graphs what that range is, and why measurements outside that range need not be very accurate.

More accurate measurements

With increased skill from practice and better knowledge of techniques, make the measurements needed to settle the essential question about *T* and *A* and plot them in the graph of *T* against *A*. (This sounds like a long piece of drudgery, fussing over better precision. It will take time and trouble, but the outcome is rewarding.) When you have settled the question write your answer or conclusion clearly, and use it in D(2).

EXPERIMENT D (2). Measurements: T vs. M

How much faster would you expect a heavy bob to pull the pendulum to and fro than a light one? Measure *T* for a fairly long pendulum, using the conclusion of your *T* vs. *A* experiments to guide your arrangements. (Of course you should not repeat your *T* vs. *A* investigation all over again for each bob in this investigation. Once that is settled it is settled, and its results can be used without repeating the investigation.)[9] Change to another bob much heavier or much lighter and repeat your measurements. Make sure that the *L*, from support to center of bob, is the same for both. (That is why we offered the cookbook suggestion of a *long* pendulum: the longer the thread the smaller the % change in *L* if you make a small mistake in changing bobs.)

EXPERIMENT D (3). Measurements of T vs. L

The period *T* changes greatly as *L* changes, and a good graph (or other investigation) of the relationship between them needs many sets of measurements. Here is where cooperation among the whole laboratory group is welcome. With the answers to questions D(1) and D(2) known, it is easy to arrange for a pooling of comparable measurements from every member of the group.

Each student (or pair) should measure *T* for one length, and then the group should pool their meas-

[9] We assume that the factors *A* and *M*, *L*, etc. affect *T* independently, so that changing *M* here does not affect the *T vs. A* story. That is a safe assumption in most places in physics. In some other fields of study—such as biology, psychology, economics—it would be a very dangerous assumption.

urements for graphs. One student should choose a length between 1 meter and 1.1; another student between 0.9 and 1 meter; and so on. (Very short pendulums are more difficult, but auxiliary apparatus can help those who decide to try them.) Then everyone in the group will need a record of all results, to plot a graph of T against L and to look for the relationship.

"The Pendulum Formula." Measuring g

There *is* a definite relationship between T and L, a well-known one that you are likely to meet in physics books or as an example in algebra books. If you do not already know it, try guessing at it from the numbers in the communal record, or from the shape of your graph. Then plot a second graph of something else ($1/T$?, $\sqrt{T}$?, . . . ?) against L that you expect to give a straight line. Or, if you prefer, ask an instructor to tell you the relationship—as a piece of well-known hearsay—and use it to plot a straight-line graph. Draw the "best straight line" that runs "as near as possible to as many points as possible."

The pendulum formula relating T and L also contains "g," and pendulum timings offer the most precise methods of measuring g today. If you do not know it, get the full formula from your instructor and calculate "g" from your straight-line graph. For that you need first the *slope* of that line—by using the slope instead of single measurement, you are taking a weighted average of many measurements. Also calculate "g" from your own single measurement which you contributed to the pool.

EXPERIMENT D (4). *Other Systems with "Pendulum Motion"*

When you have completed the earlier parts of Experiment D, you should study the "Theoretical Discussion" of pendulum motion, in the problem at the beginning of Ch. 10. There you will find a suggestion of quite different mechanical systems that should have the essential properties of pendulum motion. Take that hint, set up suitable apparatus, and investigate its motion. (This experiment is quick, and its results are surprising.)

EXPERIMENT E. Pressure Measurement and Boyle's Law

Introduction: The Importance of These Experiments

Scientific investigation grew to a new and flourishing life in the 17th century. Genuine skeptical experimenting began to replace the "argument about experiments," which Galileo used to combat the dogmatic science of the dark ages. Science was encouraged by the awakening intellectual life but also by the growing needs of engineering. Slave labor, or its near equivalent, had given place to a system where the interests and health of workers had to be considered. Machinery was needed to make heavy work lighter, to lessen the dangers of such work as mining. There were no factories like ours today, but mining was important and dangerous. Water had to be pumped out to avoid floods; and ventilation needed improvement. So scientific investigation had its industrial aspect; and when Galileo and Boyle investigated pumps and air pressure, they were working for business as well as starting science on a new era of experimenting.

Hooke made his discovery of the simple behavior of springs about 1676. Boyle investigated the "spring of the air" a little earlier. In experiments on a spring we can enclose it in a glass tube without much harm, and without any benefit so long as we *stretch* the spring. However, if we wish to *compress* the spring[10] an enclosing glass tube is useful. Boyle put some air in a glass tube and treated it like an enclosed spring. It was easy to measure the stretch or compression of the air, by measuring the length of the enclosed sample. He used a frictionless piston of mercury,[11] driven up into a tube of air closed at the top. He then took the PRESSURE of the mercury as the "force" on the springy air. He had to allow for the help provided—whether he wanted it or not—by the atmosphere.

Boyle made a remarkable discovery—a clear rule relating PRESSURE and VOLUME for air—which delighted him greatly. The relationship is of practical use in dealing with gases (in chemistry, physics, biology), and it links up in a satisfying way with molecular theory.

In laboratory, you will be able to follow some of Boyle's work, and perhaps test his discovery more rigorously than he did. But first you need to understand the use of PRESSURE; so you should measure some common pressures as a preliminary experiment. Make the measurements and calculations of E(1) before proceeding to Boyle's Law, E(2) or E(3). If you are not familiar with the laws of pressure and the use of U-tubes as gauges, read the preliminary notes below first. Also try the pressure problems.

Notes on Pressure and its Measurement

When we put water in open U-tubes, it takes the same level on both sides, whatever the shape of the tube[12] (see demonstrations). This suggests that it is not just the weight of liquid in each arm that determines the balancing. In an uneven tube there is a much greater *weight* on the wide side, but it is supported by the shoulders of the tube. There is something driving liquid equally each way through the bottom bend of the tube,

[10] Will compression shorten a steel spring as much for each kilogram push as each kilogram pull stretches it? Only experiment can give a trustworthy answer, though we may make some guesses by looking at our stretching graph and thinking about the structure of the spring.

[11] Mercury is a metal which can be roasted out of certain ores. It happens to be liquid at ordinary temperatures. At the North Pole, mercury is as solid as lead.

[12] Except for the effects of surface tension, which become noticeable with very small tubes.

and we find this is not the *weight* of liquid above, but the *force per unit area,* or *push on each square inch* which we call *pressure.*

A cat may look at a king; and a scientist may freely give the name PRESSURE to FORCE PER UNIT AREA without thereby implying any scientific facts. But we give special names only to things that are specially useful. For example, MASS · VELOCITY is useful because it is conserved, and useful in estimating forces, so we name it MOMENTUM. Again the ratio MASS/VOLUME, which tells us the mass of each unit volume, is useful,[13] so we call it DENSITY. We give a name to FORCE/AREA (= the force pushing on each unit area) because it is so useful in dealing with liquids and gases.

PRESSURE = FORCE/AREA

Working through Problems 1-5 will help you to deal with pressure.

PROBLEM 1

A pound-weight is a "bad" unit of force, a poundal a "good" unit.

(a) Name one other "bad" unit of force, and one other "good" unit of force.
(b) Name the unit of *pressure* that corresponds to each of these four units of force.

PROBLEM 2

(a) A 150-pound man stands on a rectangular brick of soft clay. The top of the brick measures 5 inches × 3 inches, and is wholly covered by his shoe, which is larger than the brick.
 (i) What *force* and (ii) what *pressure* does he exert on the clay?
(b) Now suppose he steps off, places a small wooden cube, 1″ × 1″ × 1″, on the clay brick, then steps on to the top of the cube, balancing himself there on one foot.
 (i) What *force* and (ii) what *pressure* is now exerted on the clay?
(c) How would you expect the effect on the clay to differ between cases (a) and (b)?
(d) Which gives the more useful measure for dealing with the distortion of clay, etc., *force* or *pressure?*

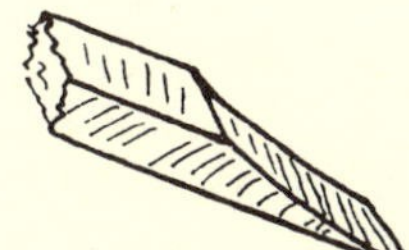

FIG. 4-3. PROBLEM 2e

(e) Chisels for carving wood, etc., have a wedge blade that thins to a sharp edge. Discuss the reasons for this shape. (There are at least two.)

[13] In designing buildings, machinery, apparatus, etc., we can calculate the volume of some part from our drawings, but we often need to know its mass (or its weight)—how many pounds or kilograms to order. We can find the MASS by multiplying VOLUME by the MASS/VOLUME ratio which we call DENSITY. This is like finding the COST of a lot of trucks by multiplying the NUMBER of trucks by the COST/NUMBER ratio, which we call the PRICE of a truck. So, in a way, DENSITY is the "price" of the material, in mass per unit volume.

PROBLEM 3

(These questions explore the ground for barometers.)

(a) The density of water is 62.4 pounds/cubic foot. What does this mean?
(b) The specific gravity of mercury is 13.6. What does this mean?
(c) State the density of mercury in pounds/cu. ft, in factors.
(d) State the density of mercury in kilograms/cu. meter. (Density of water in MKS units is 1000 kg/cu. m.)

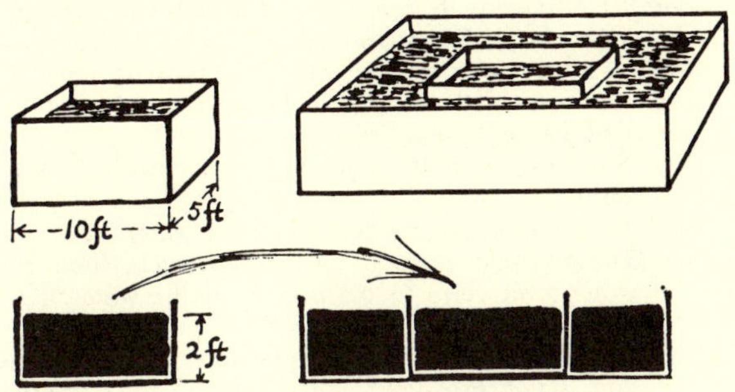

FIG. 4-4. TANKS WITH MERCURY. (PROBLEM 3)

(e) Suppose a rectangular tank of mercury has base 10 ft × 5 ft and is filled with mercury to a depth of 2 ft. (See Fig. 4-4.)
 (i) What is the *weight* of the mercury in the tank? (in *pounds-weight*)
 (ii) What is the *force* on the bottom of the tank? (in *pounds-weight*)
 (iii) What is the *pressure* at the bottom? (expressed in *pounds-wt./sq. ft*)
 (iv) How many square inches are there in 1 sq. foot?
 (v) What is the *pressure* at the bottom? (expressed in *pounds-wt./sq. inch*)
(f) Now suppose that the tank is placed inside a much larger tank, and the larger tank is also filled with mercury outside the original tank to a depth 2 ft so that there is mercury 2 ft deep inside the original tank and outside it. (See Fig. 4-4.)
 (i) Will this change make any difference to the answers to (e) for the inner tank?
 (ii) Now suppose the original tank dissolves away, the mercury remaining, with the base of the original tank still there. Will the pressure at the base remain the same? (Mercury depth remains 2 ft.)
(g) In calculating pressure at the bottom of a tank of mercury, what factors are needed? What factor is *not* needed, and why not?
(h) Pressure can be calculated by the crude, simple, but long method of dividing total weight of liquid by the area of base on which it sits.
 (i) What base-area makes this calculation easiest?
 (ii) Is this method true if the sides are sloping?

PROBLEM 4

FIG. 4-5. PROBLEM 4 FIG. 4-6. PROBLEM 5

(See Fig. 4-5.) Fresh water is held back by a dam which is 100 ft wide and 40 ft high. The water level is 10 ft below the top of the dam. The water extends for a distance of 2.93 miles behind the dam.

(a) The total weight of water held back by the dam can be found. To calculate it we must use the 2.93 miles. Why does the 2.93 miles not enter into the calculation of water *pressure* on the dam? (In other words, how can the pressure be the same as if the water extended only 1.93 miles?)

(b) The atmosphere presses on the open *outer* face of the dam. It also adds its pressure to the water pressure on the *inner* face. These two contributions subtract out when we are trying to find the forces pushing the dam over. So in the following calculations, *atmospheric pressure may be neglected.* Calculate:
   - (i) The pressure at the water's open surface. (*Answer*: zero)
   - (ii) The pressure at the bottom of the water.
   - (iii) The average pressure over the region from the water-level down to the bottom of the water. (Use common sense.)
   - (iv) The total *force* with which the water pushes on the dam. (*Hint*: *Pressure = force/area.* ∴ *force = pressure · area.* Now use the *average pressure* to calculate *force.*)

PROBLEM 5 (HARD)

A dam has been built incompetently low, so that the water-level behind it is higher than the dam, and water gushes over the top of the dam to a height 2 ft above the dam top (Fig. 4-6). The dam is 100 ft wide, 40 ft high, and the water-depth behind it is 42 ft. Repeat the calculations of Problem 4 to find the total force pushing the concrete dam. (Ignore any strange pressure-changes due to rapid motion of water, such as "Bernoulli effects.")

Laws of Pressure (Due to Pascal)

We find that pressure has the following useful characteristics, in fluids at rest:[14]

(I) The PRESSURE is the same all over the bottom of a rectangular tank of liquid. More generally, the pressure is the same at all points which are at the same level in one liquid (or gas).

(II) Fluid PRESSURE on any surface is perpendicular to it. (A diver carrying a coin finds the pressure perpendicular to its surface whatever direction it faces.)

(III) At any place in a fluid, PRESSURE pushes equally in all directions. (A diver carrying a coin finds the same pressure on the coin whatever direction it faces.)

(IV) PRESSURE is transmitted without loss from one place to another throughout a fluid. (Push a piston in at one place in a hydraulic system and the pressure you exert is carried to every wall and any other pistons in the system.)

(V) The DIFFERENCE IN PRESSURE between any two places in a single fluid is given by $h \cdot d$ where h is the *vertical* difference of level and d is the density of fluid. This leads to an easy way of measuring pressures. It is derived below.

[14] When there is motion, there are complications, such as fluid friction and "Bernoulli effects." (See Ch. 9.)

Algebra and Pressure Laws I and V

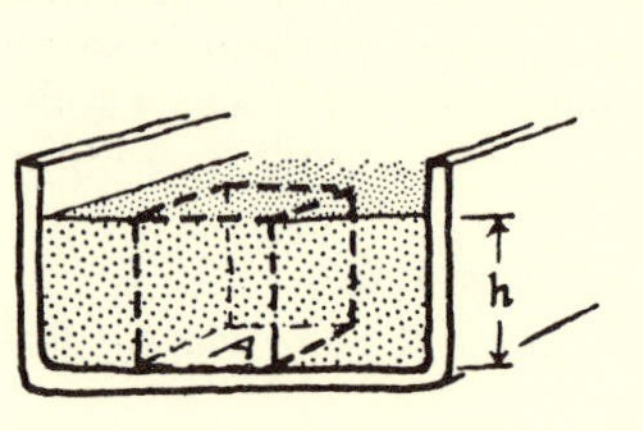

Fig. 4-7. Law I.

Fig. 4-7. Law I. The pressure is the same all over the bottom of a rectangular tank of liquid.

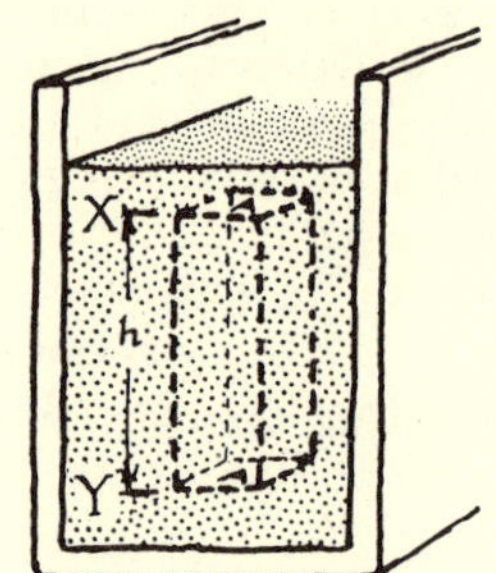

Fig. 4-8. Law V.

Fig. 4-8. Law V. Pressure Difference between two places in a fluid is Δ(Height) · Density.

(I) *The pressure is the same all over the bottom of a rectangular tank of liquid.* We can calculate the pressure on any area of bottom thus:

Choose an area A sq. inches.

Find the weight of the vertical pillar of liquid which sits on A (= the pull-of-the-Earth on that liquid). Then divide this WEIGHT by the AREA A, to find the pressure.

VOLUME of pillar = HEIGHT · AREA = $h \cdot A$

MASS of liquid in this pillar = VOLUME · (MASS/VOLUME) = VOLUME · DENSITY = $hA \cdot d$

In "bad" units (such as kg-wt.) the MASS of the pillar of liquid, in kg, tells us the WEIGHT of liquid, in kg-weight.

∴ PRESSURE, p
= FORCE/AREA
= (WEIGHT OF PILLAR)/(AREA OF BASE)
= $hAd/A = hd$

Thus the pressure on *any* base area is

DEPTH OF LIQUID · DENSITY

and is *independent of the area chosen.*

If we want the weight in "good" units, such as newtons, we must multiply MASS by gravitational FIELD-STRENGTH, g (9.8 newtons/kilogram). Then,

PRESSURE = $hd \cdot$ (FIELD-STRENGTH, g)

PRESSURE on *any* base area = DEPTH OF LIQUID · DENSITY · g

(V) *Pressure difference between two places in a fluid is* Δ (HEIGHT) · DENSITY. To find the difference of pressure, $p_Y - p_X$, between Y and X, we imagine a rectangular box, or vertical pillar, drawn in the liquid, with base of area A and height h from Y to X. The fluid in this block is in equilibrium, so the resultant of all vertical forces on it must be zero. These forces are:

WEIGHT of fluid in block, $h \cdot A \cdot d$
PUSH DOWN of neighboring fluid on top, $p_X \cdot A$
PUSH UP of neighboring fluid on bottom, $p_Y \cdot A$

$$\therefore \quad p_Y \cdot A = p_X \cdot A + h \cdot A \cdot d$$
$$\therefore \quad p_Y - p_X = h \cdot d$$

In "good" (*absolute*) units

$$p_Y - p_X = hdg$$

U-tubes for Measuring Pressure-Differences

To measure pressures, we often use liquid in a U-tube, which need not be uniform in bore. We apply the last result, PRESSURE-DIFFERENCE = *hd*.

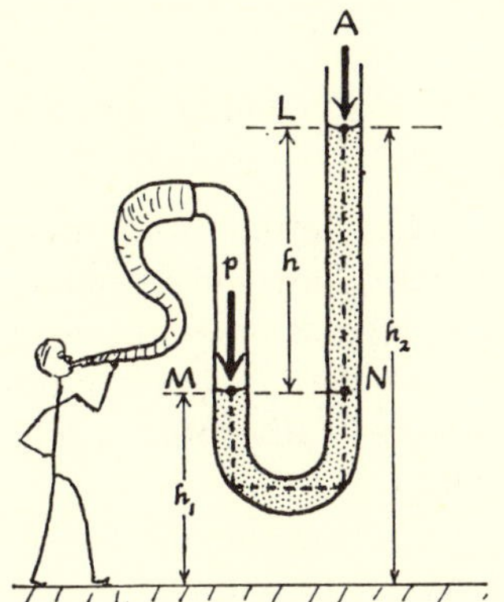

FIG. 4-9. MEASURING PRESSURE

For example, in Fig. 4-9, the man's breath exerts a pressure p which we wish to measure. So the pressure at M is p. The pressure at N, opposite M, is also p. (We may argue our way down from M to the bend, then across, then up to N, getting back to the same pressure p at the same level.) The pressure at L is the atmospheric pressure A.

But (PRESSURE at N) = (PRESSURE at L) + (*hd*)
$\therefore$ pressure, $p = A + hd$

Units for Pressure

When we use *hd*, we obtain pressure differences in "engineering" units such as pounds/sq. inch or kilograms/sq. meter. (This is how they are written and spoken. Strictly, their force-unit should be pounds-weight or kg-wt.)

If we then multiply by g, the Earth's gravitational field-strength (9.8 newtons/kg), we obtain pressure in "absolute" units such as newtons/sq. meter.

Sometimes pressures are expressed in liquid heights such as "inches-of-water," just as a mountain distance may be expressed in "hours (of climbing)."

Sometimes pressures are expressed in "atmospheres" using a standard average value for atmospheric pressure.

EXPERIMENT E(1). Simple Pressure Measurements

Use U-tubes with liquid to make the measurements listed below. It is difficult to measure accurately from one level to the other. It is much wiser to make two measurements, each *from the table* to the liquid level. Surface tension makes the liquid surface in each tube curve into a meniscus. Since you want a level difference, you should measure to the same part of the meniscus on both sides. Professional observers consider the bottom of the meniscus bowl the best for this—*viewed with eyes level with it.* (Do you also need the original levels, before the pressure is applied? Why?)

(i) Measure your lung-pressure in *inches-of-water*, excess over atmospheric pressure. Then calculate your lung pressure in lbs/sq. inch. Call the atmospheric pressure, which you do not yet know, A, simply writing $+ A$ where necessary.

(ii) If you like, also measure your *minimum* lung pressure, using suction.

(iii) Measure your lung pressure in *meters-of-mercury*, excess over atmospheric pressure. Thence calculate it in (a) kilograms-wt./sq. meter; (b) newtons/sq. meter. Call the atmospheric pressure, A, adding it as $+ A$.

(iv) Measure the excess pressure of the illuminating gas in *inches-of-water.*

FIG. 4-10. BAROMETER

(v) Demonstration Experiment. A barometer will be set up to measure the pressure of the atmosphere at the time and place of your experiments. Record the "barometer height" in inches-of-mercury and meters-of-mercury. Calculate the atmospheric pressure in (a) pounds/sq. inch; (b) kilogram-wt./sq. meter; (c) newtons/sq. meter. (It is likely to be near an easily remembered round-number value in these units. You will need this in Ch. 25.)

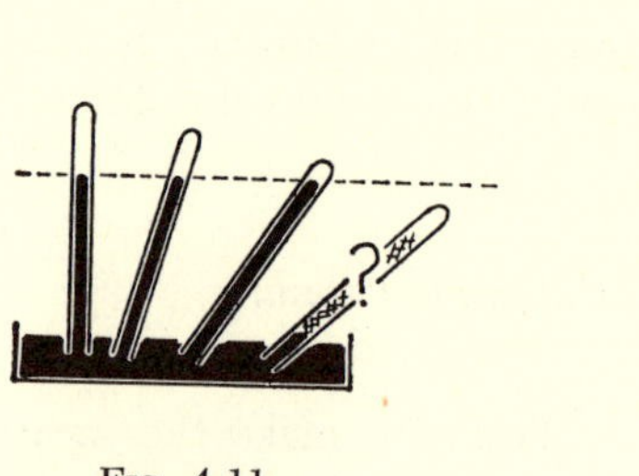

FIG. 4-11. TESTING FOR A GOOD VACUUM

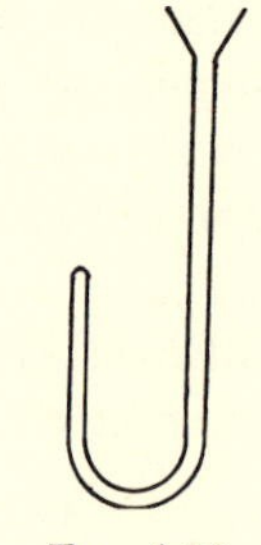

FIG. 4-12. BOYLE'S LAW APPARATUS

PROBLEM 6

In calculating air pressure from barometer height, we assume there is a vacuum at the top of the tube.

(a) Why do we expect a vacuum? Give detail of experimental procedure that makes us expect it.

(b) What practical test makes us believe there is a vacuum?

EXPERIMENT E(2). Boyle's Law Test (Original Form)

(This is a simple, single test using Boyle's own arrangement.)

Robert Boyle gave an account of his experiments on the "Spring of the Air" in a paper communicated to the Royal Society of London in 1661. The quotation below is an extract from his account. With a supply of mercury, and a J-shaped glass tube, like the one in Fig. 4-12, carry out the test described by Boyle. (Record the two stages of the experiment by two sketches, with measurements marked on them.)

"We took then a long glass tube, which by a dexterous hand and the help of a lamp was in such a manner crooked at the bottom, that the part turned up was almost parallel to the rest of the tube, and the orifice of this shorter leg . . . being hermetically sealed, the length of it was divided into inches (each of which was subdivided into eight parts) by a straight list of paper, which, containing those divisions, was carefully pasted along it. Then putting in as much quicksilver as served to fill the arch or bended part of the siphon that the mercury standing in a level might reach in the one leg to the bottom of the divided paper and just to the same height or horizontal line in the other, we took care, by frequently inclining the tube, so that the air might freely pass from one leg into the other by the sides of the mercury (we took, I say, care), that the air at last included in the shorter cylinder should be of the same laxity with the rest of the air about it. [The same density and pressure as the atmosphere.] This done, we began to pour quicksilver into the longer leg of the siphon, which by its weight pressing up that in the shorter leg did by degrees strengthen the included air, and continuing this pouring in of quicksilver till the air in the shorter leg was by condensation reduced to take up but half the space it possessed . . . before, we cast our eyes upon the longer leg of the glass, on which was likewise pasted a list of paper carefully divided into inches and parts, and we observed not without delight and satisfaction that the quicksilver in that longer part of the tube was 29 inches higher than the other . . . the same air being brought to a degree of density about twice as great as that it had before, obtains a spring twice as strong as formerly."

Boyle made more extensive measurements and obtained close agreement between the pressure observed and "what that pressure should be according to the hypothesis that supposes the pressures and expansions [= volumes] to be in reciprocal proportions."

EXPERIMENT E(3). Boyle's Law Test with Modern Apparatus

Use some modern form of apparatus for an accurate test of Boyle's Law for a sample of some gas over as wide a range of pressures as possible. (You would be wise to regard this as a test of your skill—you against Nature—rather than a routine verifying of a well-known law.)

The tube containing the sample of dry air (or other gas for the test) must have a uniform bore: otherwise you will be testing the taper of the tube as well as a gas law.

If the pressures are measured by a mercury column open to the atmosphere at the top, there is a useful trick for calculating the pressure of the sample. Since the atmosphere presses on the open mercury, replace it by an extra column of "imaginary mercury," thus: (1) read the open mercury level; (2) add the barometer height, to obtain a new "open level" with atmosphere allowed for; (3) then continue with any subtraction, etc. . . .

Make Boyle's test with your measurements by multiplying PRESSURE of gas, p, by its VOLUME, V. Also plot two graphs:

Graph I PRESSURE against VOLUME

Graph II (taking a hint from Boyle of what to plot for a straight line)

$$\text{PRESSURE against } \frac{1}{\text{VOLUME}}$$

PROBLEM 7

(a) If the points on Graph II fit a straight line through the origin, show that $pV =$ constant expresses the behavior of the gas.
(b) Since the gas is enclosed by a leak-proof piston, its mass, m, is constant. If Graph II is a straight line through the origin, what does that tell you about the density of the gas?

PROBLEM 8

To see the shape of a "Boyle's Law Graph" more clearly sketch an extended $p:V$ graph. *Assume* that Boyle's Law gives the behavior of air quite accurately over a much wider range than that of your laboratory experiment, and obtain more "data" by extrapolation as follows. Suppose your experiment ran from $\frac{1}{2}$ atm. to 2 atm. Calculate the (average) value of pV for your measurements, and then calculate V for, say, $\frac{1}{4}$ atm., $\frac{1}{8}$ atm., and for 4 atm., 8 atm. Plot these "data" and your measurements on a graph with a suitable scale—distinguishing carefully between *true points* from your experiment and *guesses by extrapolation.*

EXPERIMENT F. General Investigation of Heat-Transfer

[These experiments, with closely defined field and some cook-book instructions, are intended to offer an OPEN field for making inferences. They can be done early in the course, as they need only a simple idea of heat (supplemented by the notes given here). They need not wait for the experiments on the measurement of heat suggested in Ch. 27.]

Introduction

Work in a scientist's laboratory ranges from specific measurements or tests through carefully planned study of some new phenomenon to general, freehanded investigation of some field. This last extreme was the way in which much early science developed, and it is still useful today when a new field opens up. Scientists carry out such investigations with flexible hand-to-mouth planning as the work proceeds, with open eyes and ears for unexpected possibilities of future experimenting or hints of new knowledge. In science, "chance favors only the prepared mind,"[15] and science itself favors the alert, flexible mind.

In Experiments *F(1)-(10)* you are asked to find out all you can, from your own observations, concerning the *transfer of heat.* Apparatus is provided, some of it already set up with definite instructions. However, you should apply for any extra apparatus you need, and if time permits you should devise further experiments of your own. First, read the notes below.

[15] From the writings of Louis Pasteur, whose preparation in physics, chemistry, and good scientific thinking served him well in his brilliant researches in biology.

Descriptive Notes on Forms of Heat-transfer

Heat can be transferred from one thing to another; and, besides being a nuisance in experiments, such transfer can be of great importance, e.g. in heating houses and in chemical manufacture. There are three distinct methods of transferring heat, rather like the three ways of transferring a message: by handing a note from person to person in a crowd; by a runner carrying it; by sound waves.

Conduction. When heat is handed on from one piece of material to the next without the material moving visibly, we call the process *conduction.* Heat is *conducted* along a poker from red-hot end to colder end; or up a silver spoon dipping in coffee.

In terms of atoms or molecules—discussed fully later—we imagine the hotter particles of material jostling their less agitated neighbors so that the molecular motion which we call heat is handed on. In liquids and gases, the process is just a progressive sharing of energy, in collisions between richer (hotter) molecules and poorer ones. In solids we picture molecular vibrations being handed on by elastic binding forces. (Sometimes modern theory treats this slow diffusion of heat through solids as a case of waves ganging together into a group that travels slowly under some quantum restriction.)

Convection. When a piece of hot material moves as a whole, thus carrying heat with it to another region, we call the process *convection.* Chunks of the hot material *convey* heat elsewhere. A red-hot poker carried across the room is a case of convection, if we must give it a name, but the word is usually applied to warm currents carrying heat through a fluid while colder reverse currents complete the flow. In this sense, convection occurs in liquids and gases but not in solids. Winds are convection currents on a vast scale.

When some hot water or air moves upward in such a current, people say, "hot water rises" or "hot air rises." These statements are poor science. They merely repeat the observation, in a dogmatic voice. Taken literally they are obviously untrue, but they can be expanded to become sensible. Hot coffee does not shoot up out of a cup. Hot air does not rise when all alone any more than a cork rises when all alone. On the other hand, corks do rise when released under water . . . , and there is the hint of the proper explanation. A chunk of hot water in cold water is pushed up by the presence of the denser water around it, a case of buoyancy. Hot gases are pushed up the chimney by the denser cold air outside the chimney. One current moves up and another down—often in a pattern of circulation. (Usually the hotter material moves up, but not always. Water expands, growing less dense, when heated from 4° to 10°C and on up to boiling point. But it shows an unusual behavior below 4°. As it warms from 0°C, melted ice, to 4°C it *contracts*—though very little, only 0.013%. How does that peculiarity affect a lake when its surface water is cooled by freezing winds or warmed by bright sunshine?)

Radiation. There is another way in which heat travels, or rather disappears in one place and reappears in another. This form of transfer occurs extremely fast, along straight lines. We call this "radiation" after the Latin "radius" for the spoke of a wheel. Though scientists use the word for anything spreading out along straight spokes, we use it here for some process that transfers warming from a glowing fire to us. This includes warming by the Sun—carried through millions of miles of vacuum—and warming by light, visible and invisible, through the vacuum of some electric lamps. So we are dealing with transfer which can take place through a vacuum, and also through glass, ice water, . . .[16] This can hardly be conduction or convection, as we have pictured them. It is not actual heat travelling, because the material it travels through remains unwarmed. As an extreme example, a lens made of ice can be used as a burning glass to focus sunshine without the ice melting. Later experimenting shows all such radiation to be electromagnetic waves, which include light. We then picture the hot source producing waves, at the expense of some of its heat, and the waves travelling till they reach a receiver, where they are stopped and heat is generated again.

Record and Inferences

Write your record as you work, making very short notes of what you did, writing clear statements of what you observed. Then add conclusions, or inferences. These conclusions should be the facts that you extract from your observations, or the guesses you make by inference from them, or even generalizations.

If you use some other knowledge, from books or previous science courses, to *explain* your observations, you are reversing the logic asked for and are missing the whole point of this experimenting.

For example, suppose in some very simple experiment on another aspect of heat, your record ran: "Plunged thermometer in hot water. Saw mercury run up the tube." You might proceed to the *inference*: "*I conclude* that (or *I infer* that . . ." or "*Therefore* . . .) mercury expands when heated."[17] Or, on the other hand, you might try to explain your observation, saying, instead of any inference: "This is *because* mercury expands when heated." The two look almost the same, but the second form spoils the logic of this investigation. Please avoid such "explanations" here, even where you are sure of them, and pretend you are restricted to what you can draw from your experiment.

In general, there are great possibilities of drawing a rich variety of conclusions from some of these experiments.

EXPERIMENT F(1). General Experimenting

You are provided with Bunsen burner, glass beaker, test tubes, samples of metal wire (iron and copper), glass rod, "dye" (crystals of potassium permanganate), and you may ask for other apparatus relevant to your researches. Find out all you can about the travelling or transfer of heat.

Pre-arranged Experiments. When you have tried informal experiments with the materials of F(1), try Experiments F(2), F(3), etc. Though these are given with some cookbook instructions you are hardly told what to look for and, still less, what kind of conclusions to hope for. In your record, make notes of what you do and what you observe. Then add clear conclusions, squeezing out all the inferences you can, even at the risk of guessing.

EXPERIMENT F(2). Experiments with Water

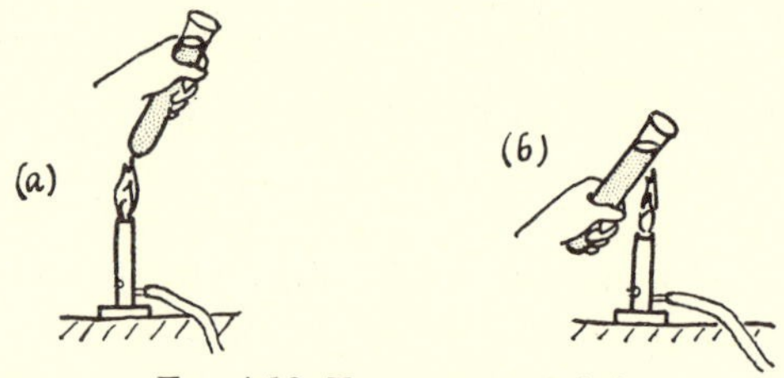

Fig. 4-13. Experiment F(2)

Heat some cold water in a pyrex test tube. To mark any currents in the water, drop a crystal of "dye" (potassium permanganate) into the water and let it fall to the bottom without stirring. It will leave little color; but if there is any circulation it will color the stream and show it. Perform two experiments, in each case *holding the tube with bare fingers at one end* and heating it with a Bunsen flame at the other end.

(a) Hold the tube near the top of the water, but *not above the water level.* Heat with flame at the bottom of the tube as long as you can hold it. Watch the "dye."

(b) *Cool the tube carefully* and refill with cold water. When the water is still, add a crystal of "dye" without stirring. Hold the tube at the bottom, and heat with flame near the top just below the water surface. Continue as long as you can, and watch the "dye."

Record your observations. Infer.

[16] And some forms of it through other materials; e.g., infra-red radiation through hard-rubber, X-rays through cardboard or flesh, radio waves through brick walls.

[17] In fact this is not the logically safe inference! What *can* you infer? This is a good scientific puzzle. When you have guessed the right answer, you will know you are right; and you can suggest an experiment to settle the matter. And in fact the observation is not quite correct. When the thermometer is first plunged in hot water its mercury dips down momentarily, then rises. With some common-sense knowledge, *this* provides a discriminating inference.

EXPERIMENT F(3). Comparison of Thermal Conductivities (Demonstration)

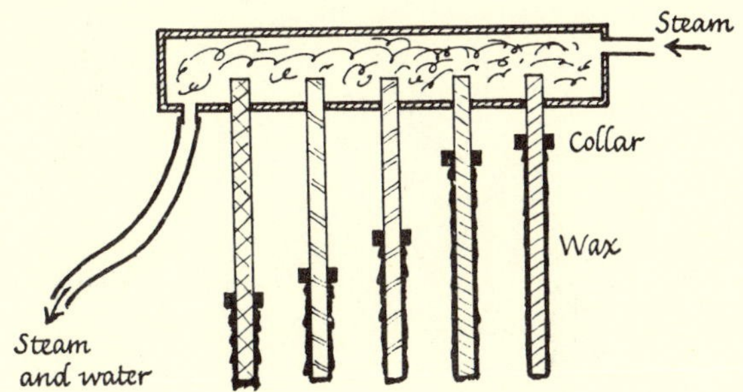

Fig. 4-14. Experiment F(3)

Watch a demonstration ("Ingen-Hausz" apparatus) that gives a rough comparison of conductivities. Long rods of the same size and different materials hang from a steam-heated box. Each rod carries a loose metal collar, but is coated with paraffin wax (candle wax) to prevent the collar sliding down. Make a note of the final positions of the collars which mark the melting point of the wax. The upper ends of the bars are at 100°C.

This is a rough experiment. Conductivities are not proportional to melted lengths. For rough estimates you may take the conductivities to be proportional to the squares of the melted lengths.[18]

(The *speed* at which a collar travels depends on the specific heat as well as the conductivity of the metal of the bar, so the early stages of the test should not be used to compare conductivities.)

EXPERIMENT F(4). Hot Gases in Flames (Demonstration)

A small source of light such as an arc is used to cast a shadow of a Bunsen flame. Observe what happens. Try shadows of other flames.

EXPERIMENT F(5). Radiation Transmitted through Air

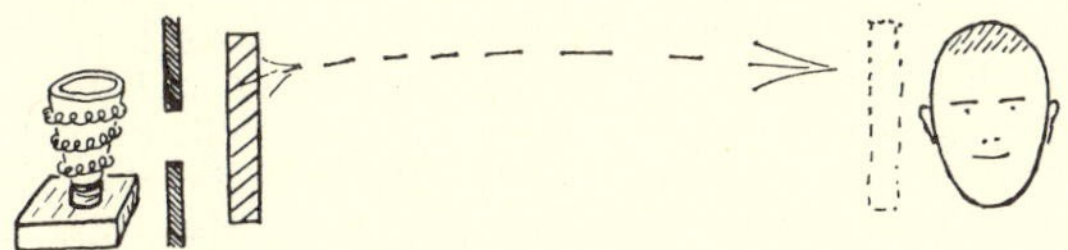

Fig. 4-15. Experiment F(5)

Use a glowing electric heater as a source, shielded by an asbestos sheet with a hole to let through a stream of radiation (chiefly infra-red). Use your cheek as a detector. Place your face 10 inches or so from the hole in the shield. Try inserting a solid obstruction such as a book or a block of wood between the source and your face. How quickly does the warming stop? What difference does it make if the obstruction is near your face instead of near the source? (See Fig. 4-15.)

[18] This relationship can be derived by ingenious argument from some general experimental properties of heat-conduction. If you are interested, ask an instructor to carry this "thought experiment" through with you.

EXPERIMENT F(6). Glass and Infra-red Radiation

Let the electric heater radiate (mainly infra-red) to your cheek, but hold a sheet of glass between the shield and your cheek. Move close to the radiator if that is comfortable. Then remove the glass. Record and interpret your observations. If a small slab of rock salt is available, try that like the glass.

EXPERIMENT F(7). Emission of Radiation from Black and Bright Surfaces

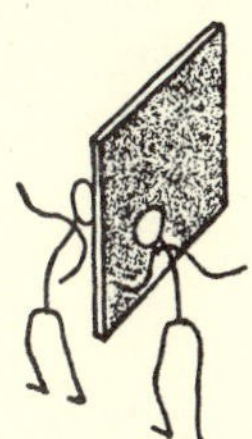

Fig. 4-16. Experiment F(7)

A large sheet of copper, polished on one side and painted black on the other, is heated with several Bunsen flames. The flames are removed and the sheet used as a radiator for a few minutes. (The sheet can be replaced by a steam-heated copper box, but a single sheet kept much hotter by frequent flame heating is much more impressive.) When the copper sheet has been heated, hold your cheek (or the back of your hand) first near the black side, then near the bright side. Record your sensations of warming by radiation and state your conclusions concerning the comparison between black and bright surfaces for radiation-emission. (The high conductivity of copper brings both faces to practically the same temperature. In heating the sheet beforehand, we apply the flame to the bright side, to "lean over backwards" in case of doubts about equality.)

EXPERIMENT F(8). Absorption of Radiation by Black and Bright Surfaces

Use your hand as a detector and a glowing electric heater as a source of radiation (chiefly infra-red).

(a) Hold the back of your hand close to the hole in the screen near the source, as long as you can stand it. Record a rough estimate of time.

(b) Have an instructor cover the back of your hand with very thin aluminum leaf.[19] Again hold it near the source, and estimate the time.

(c) Keep the leaf on your hand, but apply black paint (soot + alcohol) over it. *Wait until the black paint is quite dry,* then hold your hand near source. Record your observations. What conclusion do you draw concerning black and bright surfaces as absorbers of radiation? For this infra-red radiation, is your bare skin "black," "grey," or "bright"?

EXPERIMENT F(9). Reflection of Radiation

A pair of similar parabolic mirrors (of metal) are placed at the ends of a table with a glowing electric heater at the focus of one. Place your hand at the focus of the other mirror.

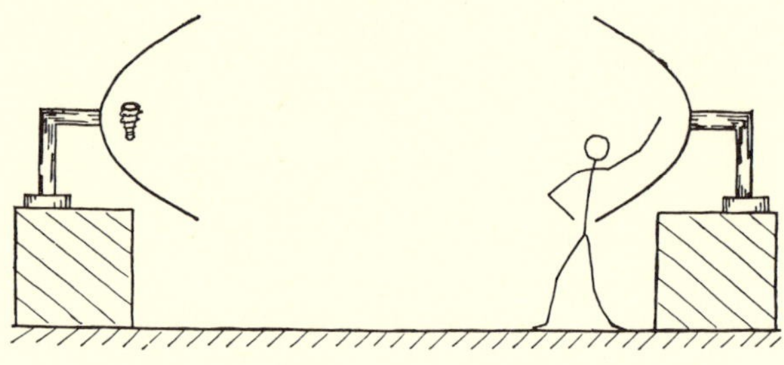

Fig. 4-17. Experiment F(9)

If good mirrors are available they can be used to demonstrate the great speed of radiation. With the arrangement described above, have someone hold a large sheet of cardboard just in front of *one* mirror to stop all radiation from reaching your detecting hand or cheek. When the cardboard is removed abruptly, try to note how soon the radiation seems to arrive. Then repeat the experiment with the cardboard just in front of the other mirror.

[19] To make the leaf stick, proceed thus: clench your fist so that the back of your hand is smooth; lick it until it is quite wet with saliva, which is a good temporary glue for metal leaf; have some one place the metal leaf on the wet skin; unclinch your fist a little, to prevent the leaf being pulled and cracked; use your hand like that for the test. Very thin aluminum leaf, like the gold leaf used for gilding shop windows, can now be obtained in booklets, interleaved with tissue paper. This is much better than thin kitchen foil which has to be held against your skin.

To blacken the leaf—and the copper sheet used in Experiment F(7)—paint on a soupy mixture of lamp-black and alcohol. To remove the black paint (and leaf) hold your hand under a running faucet. Do not rub it—that pushes the soot into the skin.

EXPERIMENT F(10). Spectrum (Lecture Demonstration)

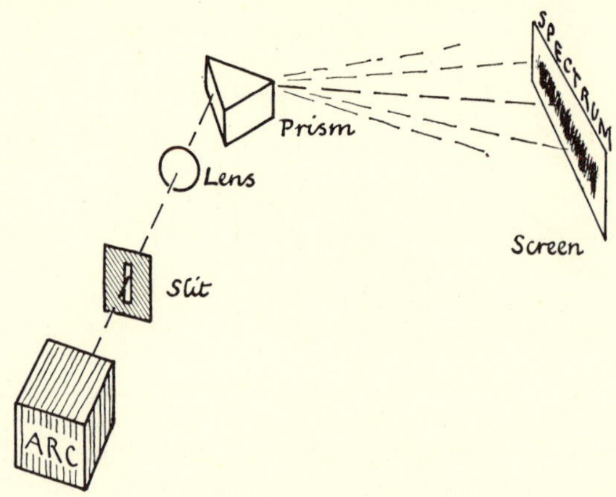

Fig. 4-18. Experiment F(10)

A large carbon arc provides radiation (ultra-violet, visible light, and infra-red radiation). Lenses concentrate some of this radiation on a slit and another lens then forms an "image" of the slit on a distant screen. When a prism is interposed this image is spread into a band of overlapping colored images which we call a spectrum. Human eyes are sensitive to only a narrow portion of this band. The radiation in the invisible regions each side of the visible spectrum carry energy in just the same way as visible light and this energy takes the form of heat when the radiation is absorbed.[20]

We measure the power stream in various regions of the spectrum by catching it on a "thermopile" connected to a micro-voltmeter. The thermopile is a pile of pairs of wires of two different metals joined in series. When alternate "junctions" are warmed, small voltages are developed which can be read (proportionally) on the micro-voltmeter. The radiation enters a polished parabolic horn which focuses it on the alternate "junctions" of the thermopile. These are blackened so that the radiation

[20] There are no special "heat rays" in the infra-red region or elsewhere. It merely happens that a glowing arc emits a greater stream of power in the infra-red region than in the visible spectrum—that is why the infra-red radiation delivers heat fastest on absorption. 420 watts of infra-red radiation deliver 1/10 Calorie per second to an absorber; 420 watts of green light deliver 1/10 Calorie per second, and so would 420 watts of ultra-violet light. But while it is easy to make an arc which emits 420 watts of infra-red radiation, such an arc would emit far less than 420 watts of green light and still less in the ultra-violet region.

There is a children's riddle which asks, "Why do white sheep give more wool than black sheep?" Answer: "Because there are more of them." The question, "Why does infra-red light radiation (from most sources) give more heat than visible light?" deserves a similar answer. You should avoid the unscientific name "heat radiation." *All* radiation turns its energy into heat when absorbed.

reaching them is absorbed and gives them a small temperature-rise leading to a corresponding voltage. (The absorbing metal quickly warms up until it is losing heat by convection, etc., as fast as it is gaining it from the radiation; and its temperature-rise gives a measure of the rate-of-receiving radiation.)

Ordinary glass happens to be transparent for the visible spectrum but only a little way beyond it at each end. In the far ultra-violet and in most of the infra-red glass is BLACK. Since glass is used in the spectrum apparatus we find a sharp "cut-off" when we reach the limit of glass-transmission in the infra-red. This is a defect due to our choice of apparatus, not a real cut-off in the energy spectrum.

Make a note of the micro-voltmeter readings for various portions of the spectrum. (Remember it may have a "zero reading" due to other radiation reaching it.) Sketch a rough graph.

If time and equipment are available, experiment with various color filters (which *subtract* some colors from the spectrum) and colored spotlights (which *add* colors on a screen).

CONNECTIONS OF EXPERIMENTS IN THIS CHAPTER WITH OTHER CHAPTERS

Ch. 1: Mentions open-lab investigation of falling bodies and projectiles; also precise investigation of motion down a hill. These are *Experiments A* and *C* of this chapter.

Ch. 5: Gives a discussion of Hooke's Law, which should *follow Experiment B* (Springs) of this chapter.

Ch. 10: Gives pendulum analysis and discussion of S.H.M., which should *follow Experiment D* (Pendulums) of this chapter.

Ch. 25: Assumes a knowledge of Boyle's Law from *Experiment E* of this chapter.

SUGGESTIONS FOR EXPERIMENTS IN OTHER CHAPTERS

Ch. 10: Young's fringes, rough estimate of wavelength of light.

Ch. 21: Test of $F = Mv^2/R$ for motion around a circle.

Ch. 27: Calorimetry, *Experiments A-E*: simple measurements of heat.

Ch. 28: Measurements of power.

Ch. 29: "J": measuring the mechanical equivalent of heat.

Ch. 32: Electric circuits, *Experiments A-W*. Intended to give a knowledge of "Current Electricity"—from simple circuits to a diode radio tube—by laboratory work and reading, without other teaching.

Ch. 34: Mapping fields of magnets and currents.

Ch. 39: Measuring radioactive decay (depends on facilities).

Ch. 41: "Laboratory work with electrons," *Experiments A-J*. These are:

A-G "Magnets and Coils" (electromagnetic induction, generators, transformers, capacitors).

H Triode tube, used for amplifying.

I Electrons: measurement of e/m and velocity.

J Oscilloscope: working and use.

CHAPTER 5 · LAW AND ORDER AMONG STRESS AND STRAIN[1]

At this moment the King, who had been for some time busily writing in his note-book, called out "Silence!" and read out from his book, "Rule Forty-two. All persons more than a mile high to leave the court."

Everybody looked at Alice.

"I'm not a mile high," said Alice.

"You are," said the King.

"Nearly two miles high," added the Queen.

"Well, I sha'n't go, at any rate," said Alice: "besides, that's not a regular rule: you invented it just now."

"It's the oldest rule in the book," said the King.

"Then it ought to be Number One," said Alice.

The King turned pale, and shut his note-book hastily. "Consider your verdict," he said to the jury, in a low trembling voice.

—LEWIS CARROLL, *Alice in Wonderland*

What is a scientific Law? Who makes it, who obeys? Who uses it, the great thinker or the practising engineer? In this chapter we select one aspect of your work on springs, their proportional stretching, to discuss it as an example of a scientific law; and to show how engineers put it to use.

Hooke's Discovery

In 1676 Robert Hooke announced that he had made a discovery concerning springs. It was a simple law, accurate over a wide range, destined to play an important part in physics and engineering. Hooke was delighted with his discovery but jealous of his colleagues and anxious lest someone should steal the credit for it. The publishing of discoveries in regular scientific journals was only beginning to replace personal books and private letters; and there was still danger that when one revealed a discovery others would jump up and say, "Oh, we found that long ago." So Hooke gave his law of springs as an anagram:

c e i i i n o s s s t t u v

This was like patenting his discovery. He gave his rivals two years in which to claim *their* discoveries about springs; then he translated his puzzle: "ut tensio, sic vis," or "as the stretch, so the force."[2]

[1] You are advised to postpone the reading of this chapter until you have finished your laboratory investigation of springs.

[2] The Latin word "tensio" means stretch (extension) not tension.

He had discovered that when a spring is stretched by an increasing force the stretch varies directly as the force.

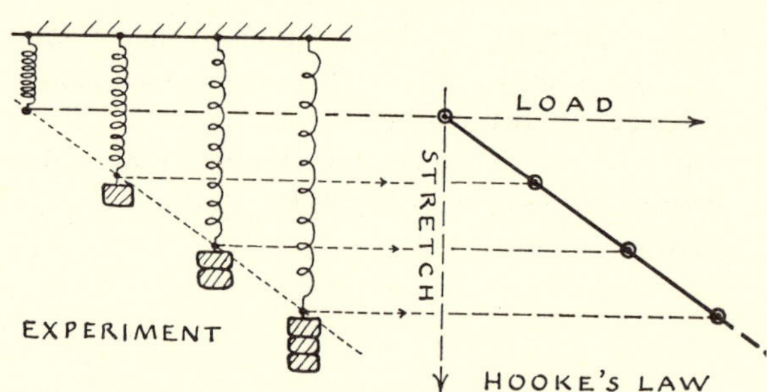

FIG. 5-1. EXPERIMENT AND GRAPH FOR SPRING (Note: In this case the graph is plotted *downward* to match the experiment.)

As you know from your own work, this simple relation holds for a steel spring with remarkable accuracy over a wide range of stretches. It holds for springs of other materials, perhaps best of all for a spiral of quartz (pure melted sand). It would not be so surprising, or so useful, if it only applied over a narrow range of small stretches—almost any curve can be treated appropriately as a straight line for short distances. But this relation continues until the spring's stretch is several times its original length. It gives many of us, as it did Hooke, a delightful feeling of success to discover something so clear and simple about nature.

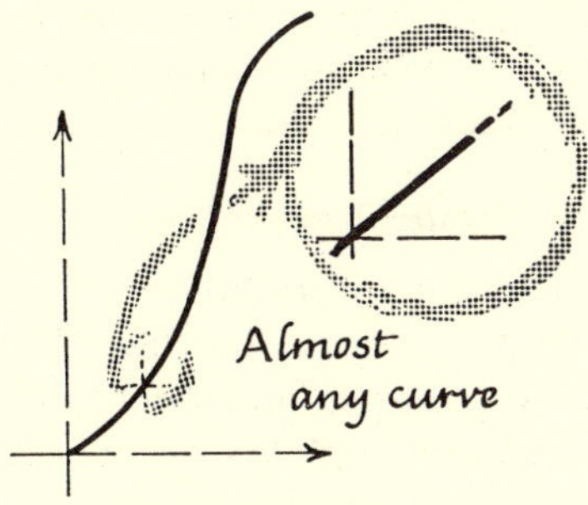

Fig. 5-2.

We meet similar Hooke's-Law behavior in many cases of stretching, compression, twisting, bending, all varieties of elastic deformation. Here are some examples:

(i) for a wire being pulled: STRETCH ∝ TENSION.

(ii) for a rod being stretched or compressed: Δ LENGTH ∝ FORCE.

(iii) for a rod being twisted: ANGLE OF TWIST ∝ TWISTING FORCE.

(iv) for a beam being bent: SAG OF BEAM ∝ LOAD.

(v) for a sample of solid or liquid being compressed: CHANGE OF VOLUME ∝ APPLIED PRESSURE.

and, in general,

DEFORMATION ∝ DEFORMING FORCE.

This general rule is called "Hooke's Law" in honor of Hooke's discovery. The sketches in Fig. 5-3 show devices for investigating some examples of Hooke's Law. See them demonstrated.

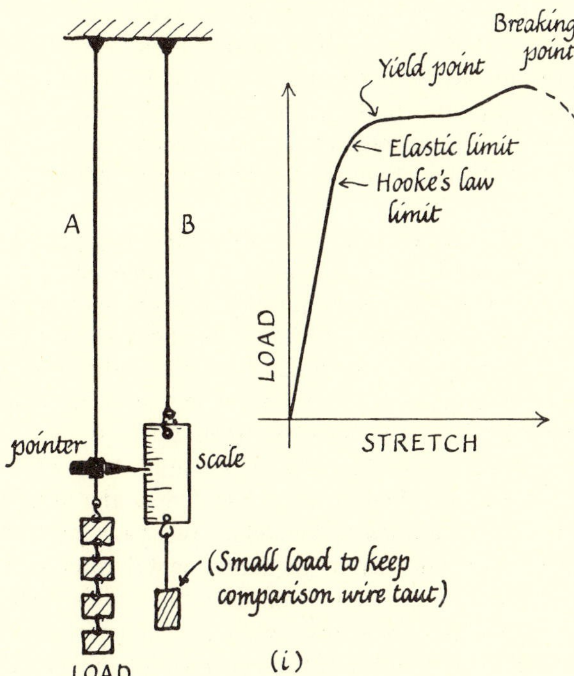

Fig. 5-3. Demonstrating Elastic Changes
Fig. 5-3 (i) Stretching a wire.

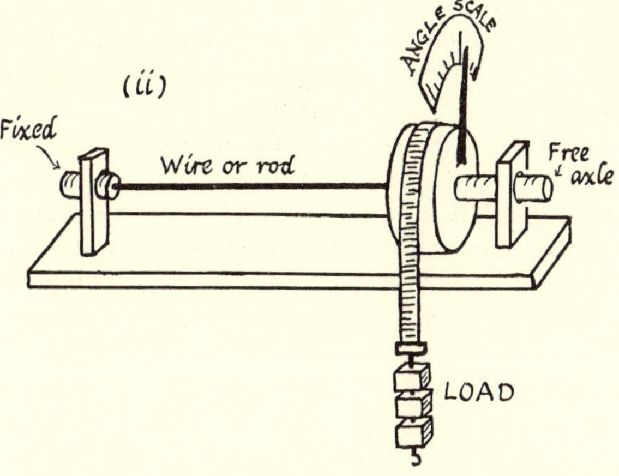

Fig. 5-3 (ii) Twisting a metal rod or wire. The left-hand end of the specimen is clamped and cannot turn. The right-hand end is attached to the large wheel which is free to turn. Loads are hung on a tape which is wrapped around the wheel's circumference. A pointer on the wheel shows the angle of twist.

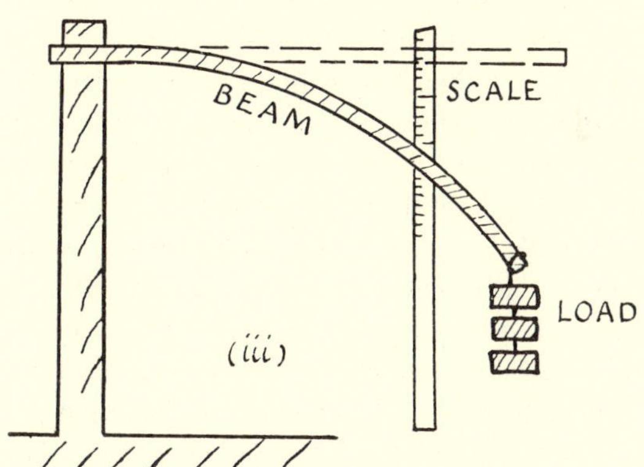

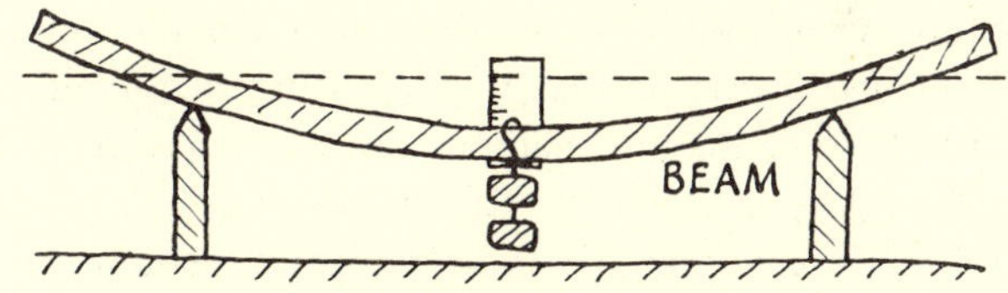

Fig. 5-3 (iii) Bending. A wooden beam, anchored at one end, is loaded at the other end. The vertical deflection is measured near the loaded end. Or the beam may be supported near its ends and loaded in the middle. This avoids doubts about the choice of deflection for measurement.

Scientific Laws

When we say that wires "obey" Hooke's Law, for small stretching loads, we do not mean that either Hooke or his Law compels them to behave in this simple way. We mean they just do behave thus—as shown by experiment—and that this is an example of the general behavior which is summed up by Hooke's Law. The word "Law" is misleading. It is used in science for a relationship, or description of behavior which has been discovered, and which seems very general and appeals to us as simple and important.

Most scientific laws are first derived inductively from experiment, as in the case of Hooke's Law. Others are first deduced from some theoretical scheme: chemistry's Law of Multiple Proportions developed from atomic theory; the Law of Equipartition of energy among molecules was deduced from mathematical mechanics (and turned out to be partly inapplicable). Sometimes a different title is awarded: "principle" or "rule" or even the honest word "relation"; for example: the Principle of Conservation of Energy, the Quantum Rules, the Mass-Energy Relation $E = mc^2$. Law, principle, rule[3]—at present you may regard these as all much the same; all are summaries of what we find, or think, does happen in nature. So it is unfortunate that scientists say, ". . . obey . . . Law." Scientific laws do not command nature like policemen. Nor should we use them to "explain" the observations that suggested them—though they can throw light on *other* experiments; they come *from* experiments themselves, and can hardly be taken *to* experiments as heaven-sent causes. Laws are, rather, simple guiding threads which we have drawn from the tangled web we study, the main threads of experimental knowledge which we weave into the fabric of science. Science gets nowhere if knowledge is just a vast tangle of facts or random observations.

We take it for granted that there *are* simple laws to be found, and that they are true descriptions of nature when we find them. But modern philosophy of science warns us that we are being over-confident. It reminds us that our whole behavior in seeking law is artificial. The nature we codify is just our idea of nature. Our laws are man-made because we make assumptions to suit our hopes. Even in deriving Hooke's Law we assumed we could simply add up the weights we put on the spring to find the load. We have no possible way of proving that load 200 + load 300 makes load 500; we simply assume that as a definition of "total load." So some of the simplicity is of our own making; we do not force nature into a simple mould, but we do force our description of nature to be simple. If this cynicism irritates you, you are in good company with many physicists.

[3] There is a tendency to use "law" for great simple outcomes of experiment, "principle" for general beliefs which are built into theory, and "rule" for more earthy working statements. Perhaps there are signs of the times, reflections of changes in scientific philosophy, in the shift in popularity from proud laws in the 17th and 18th centuries to great principles in the 19th, and now hard-working unassuming rules.

Another View of Laws

Once extracted, must scientific laws live a precarious existence, ready to be discredited by the discovery of exceptions or limitations? Some modern philosophers quarrel with such poor courtesy to laws and award them much more permanent privilege. They take the view that the law is there, a clear statement of possible simple behavior, with no question of its being wrong or untrue. It just states what it states, a peg on which to hang our information. The important science, they say, is the knowledge that goes with the law concerning its limitations.

In discussing Hooke's Law, "stretch varies as load," we should not ask, "Is that statement true?" but rather, "How closely do the facts fit the statement? Do many substances in many shapes 'obey' it? Does it apply over a small range of stretch or a large one?" When we find that most springs and wires obey it over a large range of stretching, we consider it a useful law, worth naming. We may picture the law itself as going on for ever, right out towards infinite stretches and back into compressions, but we have no illusion that real materials obey it over such a range. Instead we pride ourselves on a cunning knowledge (drawn of course from experience) of its limitations. We consider we know within what range of stretches it is likely to apply to, say, a steel wire, and in that range how closely experimental measurements are likely to fit it. And we keep track of special substances, such as glass and clay, that we suspect of serious deviations.

On this view, a law is rather like the permanent timetable of a railroad. A timetable just says what it says; there is no question of its being wrong (apart from foolish misprints). But how accurately real trains follow it each day is quite another question, the important one for travellers and one that may take an experienced railroad scientist to answer. Notice that this view of scientific law is not as different as it looks from our first view, that a law sums up experimental behavior. We have merely pushed the knowledge of experimental tests and limitations of the law itself into the guide-book knowledge that now goes with it. Then we must think of every good scientist as carrying an invisible "little black pocket-book" of detailed knowledge. That is what makes him an expert.

The body of knowledge that we call Science will stay much the same however you regard laws, but thinking about these views may help you to see how real nature, which seems complex indeed, can be interpreted with the help of a framework of simple laws.

We take it for granted that there are simple laws to be found—whichever of the two views above we choose to favor. Extracting laws is one of the great activities of physical science, but there is imaginative thinking, too, and above all much scheming to combine laws together, hoping to find a common key or reveal new predictions. We shall return in a later chapter to a discussion of laws and concepts and theories. Meanwhile, in following this course, you should watch for laws, give each a critical welcome, and look forward to seeing for yourself the growth of science when laws are combined.[4]

Stretching beyond the Hooke's Law Range

In the apparatus of Fig. 5-3(i), the stretch of a copper wire, A, several meters long is shown by a pointer which slides on a scale held by another wire B suspended from the same ceiling. Up to a certain load (many kilograms for a copper wire a millimeter in diameter), the wire "obeys" Hooke's Law, stretching a few millimeters. As the load is increased still more, the stretches are slightly greater than Hooke's Law would predict, then suddenly far greater, the wire running visibly. Finally after stretching hundreds of millimeters, perhaps 40% of its length, the wire breaks. Try this for yourself with a short copper wire, well anchored. The broken end is worth examining.

Physicists are interested in these changes and try to interpret them in terms of metal crystals distorting or slipping. There are surprises; the irregularity introduced by jamming many small crystals together makes a much stronger wire than a single crystal whose atom-layers slide easily on each other. The forces between atoms which bind crystals together are still being explored. We know that these forces change rapidly with distance, so it is surprising to find that they can produce anything as simple as Hooke's Law, even for the smallest stretches.

[4] For example: combining Hooke's Law with Newton's Second Law of Motion can produce surprising and useful predictions about the bouncing of a load hung on a spring, the vibrations of a tuning-fork and the motion of the balance-wheel in a watch, and even certain vibrations of atoms in molecules. These diverse motions and many others are shown to be linked together by a common characteristic, which you will meet later. Without the help of combined laws, the common behavior might have remained unknown, and some of the motions never put to use or fully understood.

Engineers and Elasticity

To the engineer, Hooke's Law offers an easy way of allowing for elastic changes in his structures when loaded. He can calculate the sag of a bridge before he builds it, or estimate the twisting force in a ship's propeller shaft by measuring its slight twisting-distortion. For such uses, he needs to know the exact behavior of a measured sample of the material; he can argue from that to his large structure. He is also interested in properties beyond the Hooke's Law range, such as the breaking load; and these too he calculates from measurements on a sample. How do experimenters who compile engineers' reference tables remove the unwanted details of the sample? How do they reduce their measurements to a number that belongs to the material itself, not just the sample?

The descriptions and questions in the special Problems on Elasticity will show you some of the engineer's treatment, and answering the questions will enable you to think out the "rules of the game." Some of the questions are "thought experiments" drawing on common sense. Others are merely illustrations of the useful terms introduced by engineers.

"The problems on pages 82, 83, and 84 are intended to be answered on typewritten copies of these pages. Work through the problems on the enlarged copies, filling in the blanks, (———), that are left for answers."

Name ______________________

(Use the symbol $\propto$ to mean "varies directly as" or "is proportional to.")

Preliminary discussion

Like springs, a wire or rod of a solid, such as steel, stretches when loaded. Up to a certain limit the stretch varies directly as the load. This is an example of the general rule called Hooke's Law to commemorate his discovery of it with springs. The limit beyond which this simple relation fails is appropriately called the "Hooke's-Law-limit."

The "elastic limit" is the stage beyond which a sample acquires permanent changes.

Some substances suddenly show great yielding when loaded to a point somewhat beyond the elastic limit. This is called the "yield point".

With still greater loads, a loaded sample breaks at the "breaking point", which comes soon after the yield point, if there is one.

Engineers and physicists are interested in knowledge of breaking point, yield point, elastic limit, Hooke's-Law-limit, and the relationship between load and deformation in the Hooke's Law region. By using some common sense, you can guess how to predict some of these things for wires or rods of one size when you are given experimental information for some other size. The questions below show how to do this.

BREAKING FORCES.

1. Suppose the breaking force for a certain wire is 100 pounds. The breaking force for a bundle of 4 such wires together would be . . . . __________ pounds.
If all 4 wires were fused into one thick wire (without change of length), we should expect this thick wire also to break with a force of __________ pounds.
The area of cross-section of the thick wire would be __________ times the area of the single original wire.
This reasoning from common sense suggests that the relationship between BREAKING FORCE, F_B, and CROSS-SECTION AREA, A, of a wire or rod is likely to be . . . . ______________________________
write an algebraic statement using $\propto$

2.(a) Suppose we have rods of square cross-section; a small one 1" x 1", and a fat one 2" x 2". The fat one would need a breaking force __________ times as big.
In general, the relationship between BREAKING FORCE, F_B, and WIDTH, w, of a square rod should be. . . . . ______________________________
write an algebraic statement using $\propto$

2.(b) Suppose we have round rods or wires. (Remember that the area of a circle is πr^2 or $\pi d^2/4$ where d is the diameter.) If we make the diameter of a circle twice as big, we double its radius also and make its area _______ times as big.
If we make the diameter of a circle ten times as big, we make its area . . . . _______ times as big.
In general, the relationship between DIAMETER, d, and AREA, A, for a circle is ______________________________
∴ in general, the relationship between BREAKING FORCE, F_B, and DIAMETER, d, for rods and wires should be . . . ______________________________
write algebraic statement using $\propto$

3. Try an interesting application of this to the possible size of elephants. The mammoths died out, perhaps because they were too heavy for their legs. An animal of the same shape but built on double scale so that it has twice the height, twice the length, and twice the width has a volume ______ times as big, and therefore weighs ______ times as much, if made of similar flesh and bones. However, its legs, having doubled diameter, are only ______ times as strong. So there is a limit to the safe size of an animal. Does the same limitation apply to whales? ______ Why? ______________________________

4. Suppose we have a wire which is twice as long (as some test wire) and hang on it just the breaking load of the test wire. This pull is transmitted all along the double length and the breaking is just as likely to occur as before.

(We certainly do not expect to have to double the breaking force, nor should we expect to break the long wire with half the force. If, as is often the case, the break occurs at some place of slight extra weakness, then using a longer wire may perhaps give greater chances of finding weak places. In the latter case a longer wire might be easier to break, but you should ignore this argument in answering the question below.)

∴, in general, the relationship between LENGTH OF WIRE, ℓ, and BREAKING FORCE, F_B, should be ______________________________

STRESS.

5. Looking back on your answers above, you will see that when we deal with wires and rods of different sizes, but of the same material, the thing that settles whether a wire will break is not just the size of the force (load) applied: the cross-section area is involved as well.

For wires of different sizes, the breaking force is different; but the ratio or fraction BREAKING-FORCE/CROSS-SECTION AREA should be the same for all samples.

On the basis of your answers above, do you agree with this? ________

This fraction is therefore the best way of specifying how hard we must load such material (rather than any particular rod of it) to break it.

This fraction $\frac{\text{BREAKING FORCE}}{\text{CROSS SECTION AREA}}$ is called breaking stress.

By using stress we can make statements that are independent of the shape and size of the sample. Given the breaking stress for a material, we can calculate the breaking force for any particular rod or wire.

6. Stress, calculated as FORCE/AREA, gives a general measure of the treatment applied to the material. The loads which give the Hooke's-Law-limit, the elastic limit, and the yield point, all follow much the same relations as the load for breaking -- though they are different in size. Thus, we may have "yielding stress" and "Hooke's-Law-limit stress", etc.

If all the loads are measured in pounds and all the diameters in inches, each of these STRESSES must be measured in the following units: ________ units

If the FORCES are in newtons, DIAMETERS in meters,

all the STRESSES will be in . . . . . ________ units

Incidentally, these are also units for . . . . ___?___

HOOKE'S LAW REGION

7. For the Hooke's Law stretches we may again imagine a bundle of wires turned into one fat wire; and from this we argue out the way in which we should expect to find the FORCE NEEDED TO PRODUCE SOME DEFINITE STRETCH related to the DIAMETER OF THE WIRE.

For the same stretch, a bundle of 4 wires would need ______ times the force.

Such a bundle fused into a single wire would have ______ times the area of cross-section.

∴ in general, relationship between FORCE, F, needed for some standard stretch and AREA of cross-section, A, should be ___?___ ___?___

For round wire the relation between FORCE, F, (for standard stretch) and DIAMETER, d, should be. . . . . ___?___ $\propto$ ___?___

8. Here again, the ratio STRETCHING FORCE/CROSS-SECTION AREA is the thing which really determines the stretch for a given material. We call this ratio STRESS. Then if the same stress is applied to wires of several sizes but of the same lengths and material, the stretch should be the same for all those wires.

Explain briefly why this should be so. ______________________________

____

____

9. Within the Hooke's Law region, changing to a wire of double length gives two lengths of wire each of which will stretch with the original stretch.

So the total stretch will be ________ times as big, for the same load.

In general the relationship between STRETCH, $\Delta\ell$, and LENGTH, ℓ, of wire, for several different wires of same material carrying the same load will be ______________________________

STRAIN.

10. In dealing with wires of different lengths, we see that the ratio STRETCH/LENGTH should be the same for all wires of the same material with the same stress, though they have different lengths. Do you consider this statement risky, reasonable, probably right, right? ____________________

This useful ratio is called STRAIN. By using it we get rid of the length of the sample, and specify the state of the material itself. If we measure STRETCH in inches and LENGTH in inches, STRAIN must be measured in __________ units

MODULI.

11. Engineers and physicists often want to record the elastic properties of a material in some form which will hold for a variety of shapes and sizes of sample, and for a variety of applied forces. To do this, we use:

STRESS, which is $\frac{\text{FORCE}}{\text{AREA to which it is applied}}$ instead of just FORCE (or load);

STRAIN, which is $\frac{\text{CHANGE OF LENGTH (or size)}}{\text{ORIGINAL LENGTH (or size)}}$ instead of just CHANGE OF LENGTH.

Then in the Hooke's Law region, where the simplest statement says STRETCH $\propto$ LOAD, (or LOAD/STRETCH = constant), we manufacture a much grander fraction, which, like LOAD/STRETCH, is constant. But this grander fraction does not depend on the shape or size of sample used. It is the same for all samples of a given material. To make this grander fraction we use stress and strain instead of load and stretch. And we now write Hooke's Law in the general, grand form, $\frac{?}{?}$ is constant.

We call such a constant a "modulus." The easier a substance is to stretch (or compress), the __________________ (larger? / smaller?) its modulus must be.

Elastic Moduli

Using stress and strain, we can state Hooke's Law in general form, $\frac{\text{stress}}{\text{strain}}$ is constant; that is, the fraction

$$\frac{\text{force/area}}{\Delta \text{ length (or size)/(original length or size)}}$$

is constant.

We call such a $\frac{\text{stress}}{\text{strain}}$ fraction a "modulus." Within the Hooke's Law range, moduli are constants characteristic of the material, different for different types of distortion but independent of the shape and size of the sample and the force applied. The bigger the force a material needs for a given distortion the bigger the modulus. So the size of the modulus indicates the *stiffness* of the material, not its ease of stretching, etc.

For *simple stretching* of a rod or wire by tension—which is what we have been discussing—the modulus given by *stress/strain* is called *Young's modulus.* It applies to compression too. Engineers use it to predict changes in bridge girders when pulled or compressed.

Bending an elastic beam stretches some fibers and compresses others, therefore Young's modulus is involved in bending. Mark a rubber tube or block of eraser with ink and try stretching it and bending it.

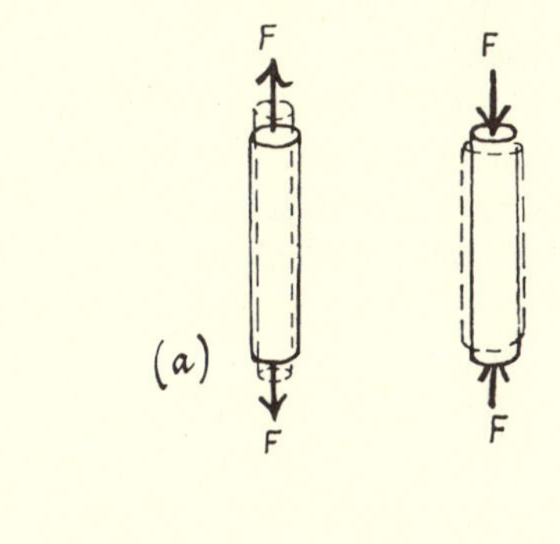

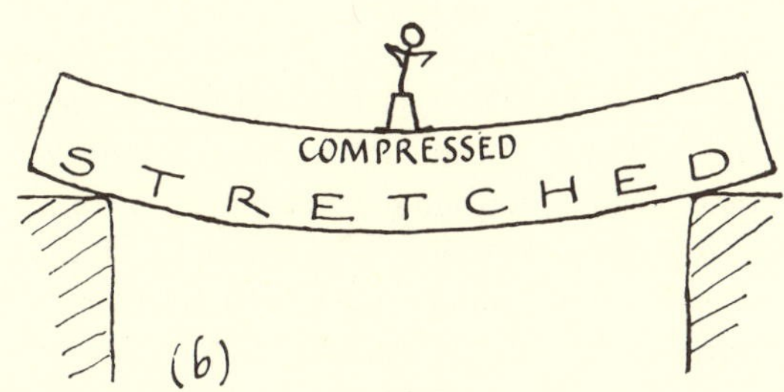

FIG. 5-4
(a) Stretching or compressing a rod or a wire
(b) Bending a beam

It is the outermost fibers of a bent beam that are greatly compressed and stretched and therefore exert great pressures and tensions to oppose the

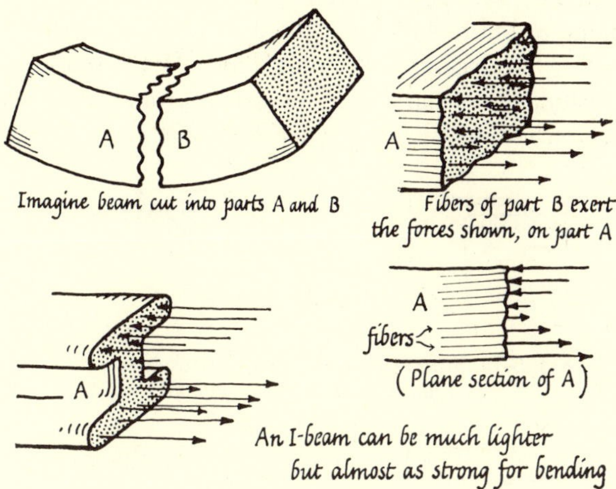

FIG. 5-5. BENT BEAM.
Stretching and compressing oppose bending.

bending. The inner fibers suffer little strain, so they exert only small forces; they can be removed with little loss of strength but valuable saving of weight. That is why solid beams are scooped out into I-girders and solid rods hollowed into tubes in bicycle frames.

Other types of distortion lead to other moduli. For *pure change of size,* without change of shape (i.e., pure compression) we have the "bulk modulus." The compression stress is easily applied by fluid pressure.

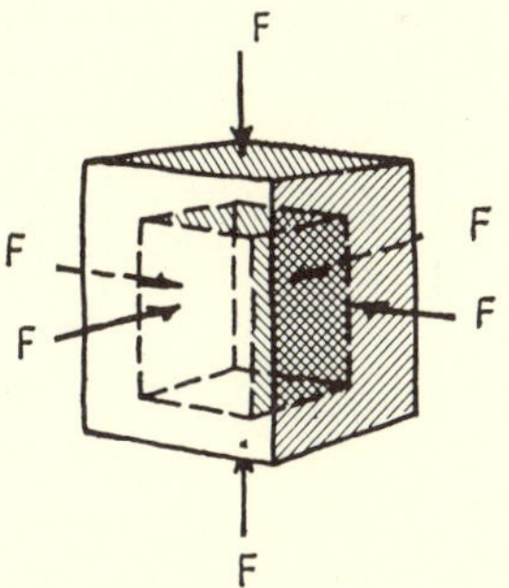

FIG. 5-6. PURE CHANGE OF SIZE

For *pure change of shape,* without change of size (shearing) we have the "rigidity modulus." Torsion or twisting of a rod involves shearing, and therefore involves the rigidity modulus. Try twisting rubber blocks or tubes marked with ink. Place a fat book on the table and push the cover, so that the pages slide. A square, pencilled on the end edges of the pages, distorts into a diamond. The book is being sheared; its shape changes while its volume stays the same. You might imagine each layer of atoms or molecules—represented by a page of the book—being urged to slide over the next layer and experiencing increasing restraining force. When a rod is twisted, fibers parallel to its axis are made to lean over: they are sheared. (See Fig. 5-7c.)

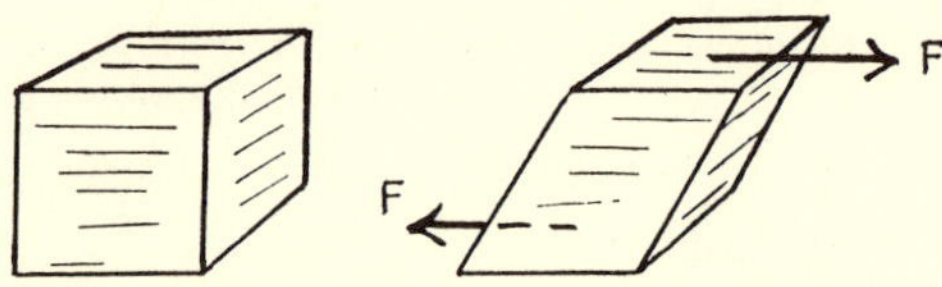

Fig. 5-7a. Shearing. As a cubical block is sheared, its square sides become diamonds.

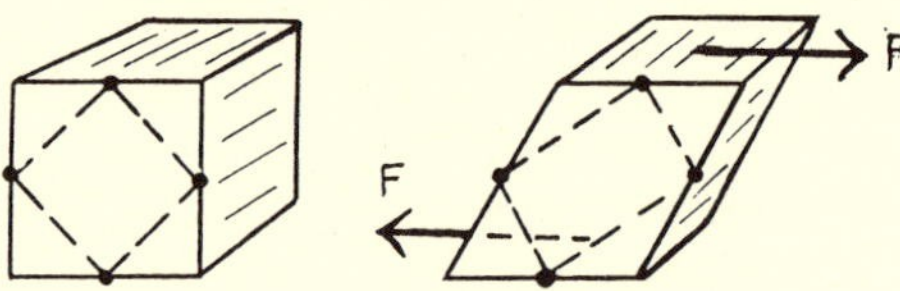

Fig. 5-7b. Shearing.
Alternative view of the same shear-distortion; slanting fibers are stretched and compressed so that a "45° square" becomes a rectangle. Try this on the top edge of a large book.

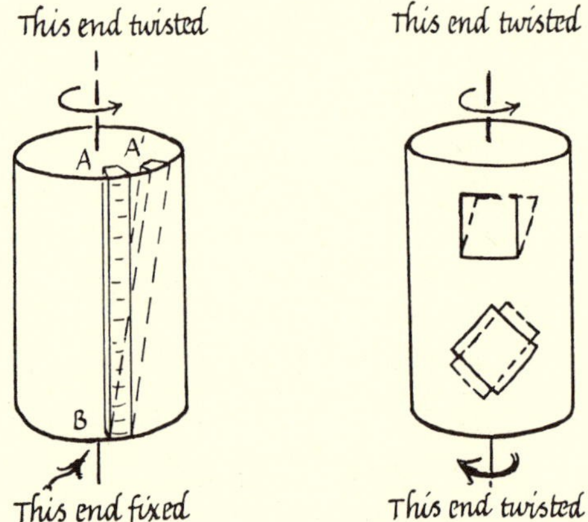

Fig. 5-7c. Torsion.
When a cylinder is twisted, a fiber AB is sheared to a slanting position A'B and squares drawn on its surface show shearing distortions.

Here again the *inner* layers of a twisted rod suffer relatively small strains, produce small opposing stresses, and so contribute little to its strength against twisting. A tube is almost as strong as a solid rod and far lighter.

Various Strains in Various Materials

Liquids and gases offer no permanent opposition to change of shape, so they do not have a rigidity modulus for shearing. But they are elastic for changes of volume, and have a bulk modulus for compression. Liquids obey Hooke's Law, with their small changes of volume, over a large range of pressures; but gases soon deviate from it, and another law must be sought for them. For solids, the simple changes of shearing and compression can be combined into more complicated distortions, e.g., in coiled springs or loaded machinery, and in all cases the simple Hooke's behavior holds over a wide range:

$$\frac{\text{stress (due to applied forces)}}{\text{strain (distortion)}}$$

is a constant for the material; or stress $\propto$ strain.

Hooke's Law

The general form of Hooke's Law, "STRESS/STRAIN is constant," applies to all materials (within limits) and to many types of distortion. The law is remarkable, and useful, not just for its simplicity but for its wide range. A closely coiled steel spring may stretch to five or ten times its original length before reaching its Hooke's Law limit. A wooden beam may be bent, or a hair-spring coiled up, through a large angle and still fit with Hooke's Law. Even a simple metal wire being stretched fits Hooke's Law over a surprising range of stretches—far beyond the tiny expansion caused by heating. Its atoms, being dragged apart against electrical attractions, seem to experience individual Hooke's Law forces.

If we plot a graph of y, to represent STRAIN, against x, to represent STRESS, Hooke's Law would be shown by a straight line through the origin. That line would represent a relation $y = kx$. The accurate statement for real materials might be a much more complicated mathematical relation, but in many cases where $y =$ (a complicated function of x) we can express it as a series:

$$y = A + Bx + Cx^2 + Dx^3 + \ldots$$

where $A, B, C, \ldots$ are constants. In this case, $y = 0$ when $x = 0$ (no stress applied, no strain). So A must $= 0$. Now the experimental fact that Hooke's Law fits well suggests that $C, D, \ldots$ are very small. Then $y \approx Bx$, for Hooke's Law. However, when x increases x^2 and x^3, etc. grow even larger in importance (since doubling x makes x^2 4 times as big, x^3 8 times as big, etc.). So unless $C, D, \ldots$ are exactly zero, we should expect their terms to make themselves felt with big stresses. The wide range of Hooke's Law tells us that those constants are remarkably small. Yet they are there, so we must regard our great simple Hooke's Law as only a very close guess at nature. Have we discovered that simplicity, or have we manufactured it?

CHAPTER 6 · SURFACE TENSION: DROPS AND MOLECULES

"We especially need imagination in science. It is not all mathematics, nor all logic, but it is somewhat beauty and poetry." —MARIA MITCHELL (*American Astronomer*, ~ 1860)

[This chapter is not an essential part of the course. It is inserted because:

(1) It provides many pretty experiments, some of them simple household ones. See the experiments first and enjoy them; then read the chapter.

(2) It shows, on a local scale, how scientists investigate a piece of nature: observing, interpreting, guessing, testing, thus increasing both useful knowledge and scientific understanding.

(3) In addition to comments on practical matters ranging from shampooing to gold-mining, it provides a measurement of the size of a single molecule, useful in our studies of atoms.]

Demonstration Experiments

Start, like a scientist yourself, collecting information by watching the behavior of liquid surfaces.

I. *General Observation.* Look at the shapes of small drops:

(a) Drops forming on a dripping faucet.

(b) Pools of liquid on a table: (i) water on clean glass, (ii) water on waxed glass, (iii) mercury on glass. The sketches of Fig. 6-1(b) show the shapes roughly, but as a wise scientist you should observe the real shapes and pay little attention to pictures in books except as reminders. Does a real teardrop on a heroine's cheek have the shape shown in the story books?

(c) Raindrops are perfect spheres. Accurate direct observation is difficult, but we get indirect assurance from two sources: first, the shape and position of rainbows. These appear exactly where

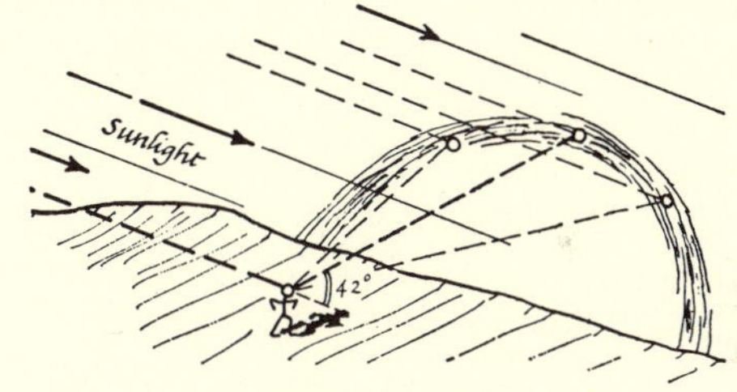

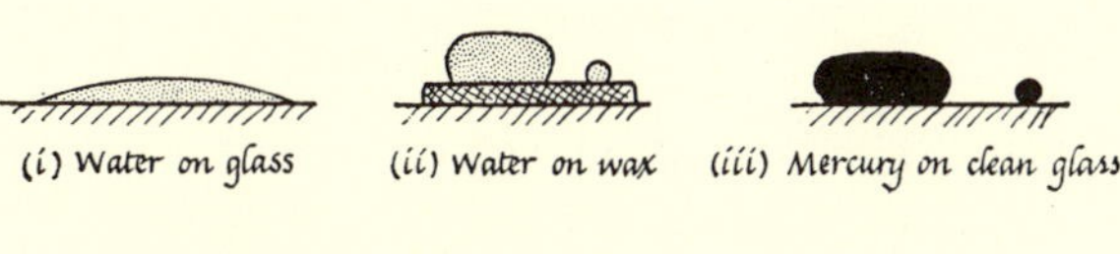

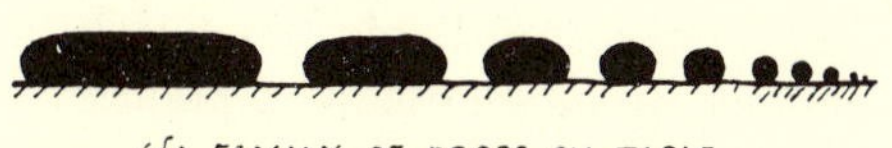

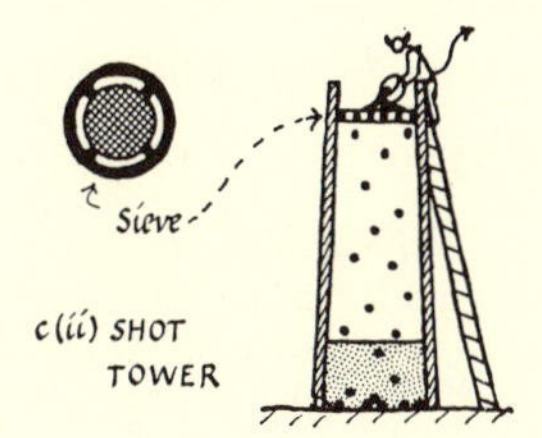

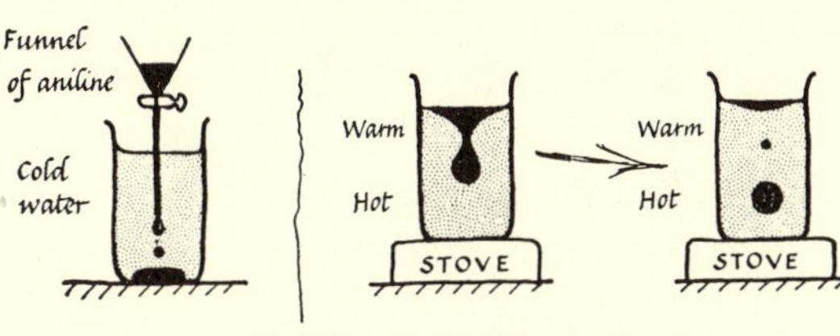

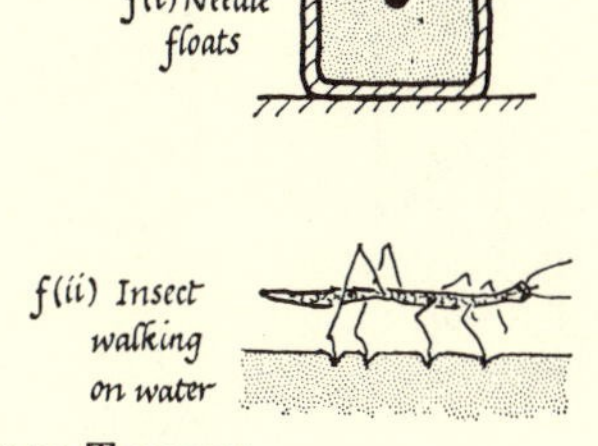

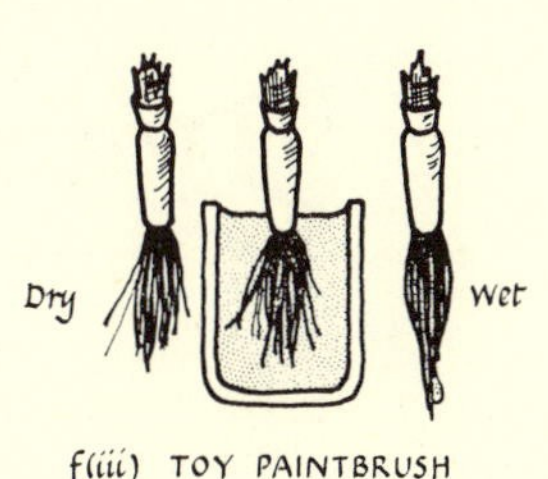

FIG. 6-1. SURFACE TENSION

round raindrops would place them. Distorted drops would shift the bows. Second, the shape of lead shot made in an old fashioned shot tower—molten lead poured through a sieve fell in a rain of drops through air into a deep tank of water and arrived as round balls.

PROBLEM 1

A tiny raindrop resting on a woolly sleeve is spherical, but a large drop of water on a waxed floor takes a flatter shape. Why?

II. *Special Apparatus.* The next scientific move is to use instruments or apparatus to help us. With a projection lantern, observe the drops of (a) and (b) in Fig. 6-1 on an enlarged scale. (If water now seems to move too fast, try dropping viscous oil from a medicine dropper.)

(d) If you observe a whole "family" of pools and drops of different sizes, as in Fig. 6-1(d), you should be able to infer (by induction) several interesting rules. Look for properties common to most of them.

(e) Remove most of the effects of gravity by using a supporting liquid. Crude aniline, a brown poisonous liquid, is slightly denser than water. When it drips from a funnel submerged in water the drops form very slowly; a bag of aniline forms and grows deeper, then a narrow waist develops and the drop quickly breaks off, the waist turning into a smaller drop which follows.[1] If the apparatus is jogged, the hanging drop vibrates.

(f) Sometimes small objects which we expect to sink will float on the surface of water; for example, a needle, or a razor blade if it is a little greasy, and some kinds of water-beetle. Strange supporting forces seem to be available.

III. *Soap Films.* Soap bubbles show liquid surface effects on a large scale. They are "all surface and no bulk," with little weight to compete against surface forces. See the following, sketched in Fig. 6-2.

(g) A soap bubble on a funnel contracts, blowing air out against a candle flame.

(h) The "window shade." A flat soap film is formed on a wire frame whose lower edge is a movable slider. The film can be stretched by pulling the slider down by a thread which is then released.

(i) A flat soap film is formed on a square wire frame. A silk thread knotted into a loose loop is thrown on to the film. Then the film inside the loop is broken.

(j) The window-shade experiment is repeated with a movable bar at the top as well as the bottom. The upper bar is held by a small spring. A soap film is formed between the two bars. The lower bar is then pulled up and down with a thread.

(k) Two soap bubbles, unequal in size, are blown on a T-shaped tube. The blowing inlet is then closed, leaving the bubbles connected.

★ PROBLEM 2. INFERENCES

For each of the experiments (g) to (k) described above, first record your observations and then say what you can infer from them regarding soap films and their "surface tension." (*Note*: The plane figure with *maximum area* for a given perimeter is a *circle*.) Warning: An important inference from (h) will rule out the simplest interpretation of (k).

Extracting General Comments

What do these experiments show about liquid surfaces? The drops forming on a faucet look as if they were supported by a rubber bag. We can make a giant artificial "drop" with a real skin of rubber, which goes through similar shapes as more and more water is poured into it; but the increasing tension of the rubber spoils the strict analogy.

Raindrops and pools of liquid on a table seem to be pulled towards round shapes, again suggesting a skin holding them together against the pull of gravity. Thinking about our observations, we may extract two general comments, vague and risky, but worth further study:

(A) The surfaces of liquids seem to behave as if held together by an elastic skin, pulling them into round shapes.

[1] Another method: Some aniline is poured into a tall glass beaker of hot water which is heated steadily at the bottom. Hot aniline is less dense than hot water, so the aniline starts at the top as a huge drop hanging in the water. When it cools it drips down through the water to the bottom where it is warmed again and rises to repeat the giant dripping motion.

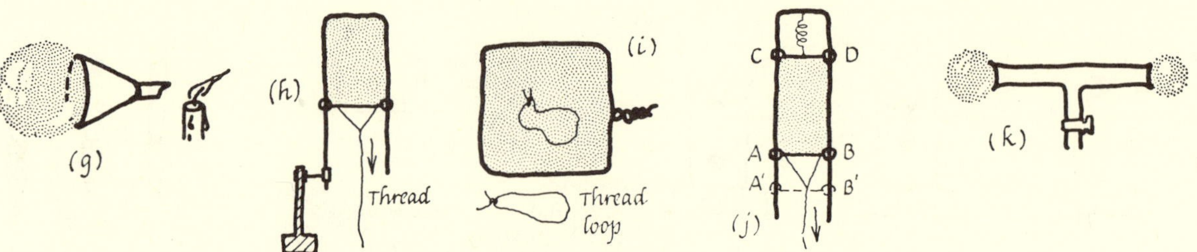

FIG. 6-2. SOAP BUBBLES

(B) These skin-effects are more noticeable on a small scale (tiny drops) than on a large one (pool of water), but when gravity is balanced out they take charge even on a huge scale.

Classifying and Naming

Surface Tension. We call these "surface tension effects," and say the liquid has a surface tension, like the tension of a stretched rubber skin. This is merely useful naming and cannot in itself prove or explain anything. At best it promotes easy reference and discussion. At worst it misleads people into thinking there is a real tensile skin that can be flayed off a drop like skinning a rabbit.

Angle of Contact. Pools of liquid resting on a table show two distinct shapes:

(i) The liquid clings to the table and spreads, as in Fig. 6-3a. We say the liquid "wets" the table surface.

(ii) The liquid humps itself up into a roundish drop, opposing the effects of gravity more visibly, as in Fig. 6-3b. It does not wet the surface. If the table is tilted such drops may roll off. We distinguish the two cases by the angle A, which we call the *angle of contact*. This is the angle inside the liquid, between table and liquid surface, where they meet. We find the same angle at other boundaries, e.g., where the surface of water in a drinking-glass meets the walls. If A is small the liquid wets the solid surface. Again this is just naming. By picking out this angle and naming it we have neither

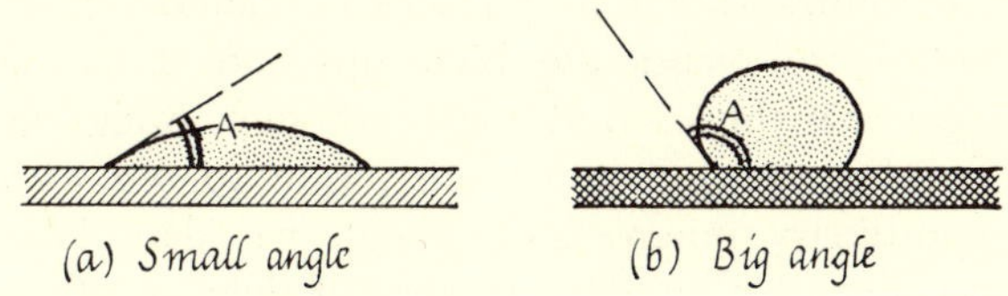

FIG. 6-3. ANGLE OF CONTACT.
When a small pool of liquid rests on a solid table, the angle marked *A* is called angle of contact.

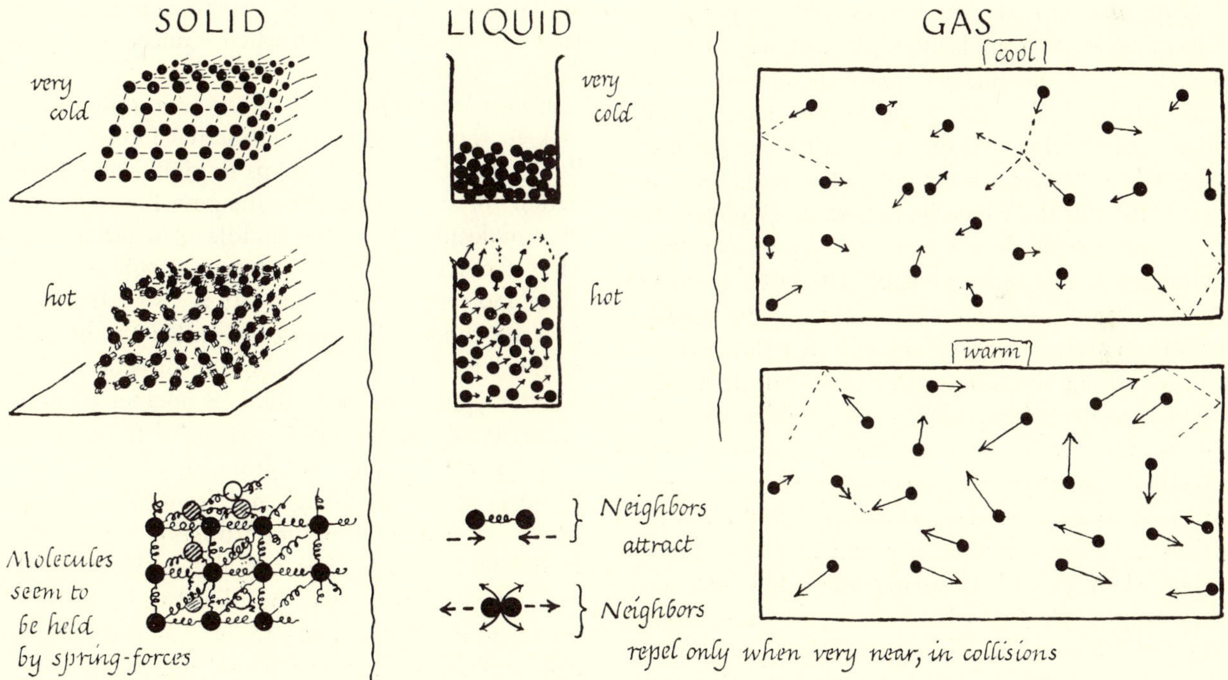

FIG. 6-4. MOLECULES IN SOLID, LIQUID, AND GAS.

Molecules in solids are arranged in a regular pattern—all true solids are crystalline. They keep more or less the same positions but vibrate more and more as the solid is made hotter.

Molecules in liquids are close together, as in solids, but move about freely among neighbors. The higher the temperature, the faster the motion and the more violent the frequent collisions.

Molecules in gases are far apart, moving fast—the higher the temperature, the faster they move—colliding occasionally. During (brief) collisions they must repel; otherwise they exert negligible forces on each other.

Analogy: If gas molecules correspond to football players moving fast, with occasional collisions in an open field, molecules of liquid are like people in a fluid crowd jostling their way in to see a game, but still loosely held in family parties; and molecules of solids are like the same people well-seated in orderly rows—but still vibrating with enthusiasm.

learned nor explained anything but we have made discussion easier.[2]

Attempt to Form a Theory

Molecules. Let us take a hint from chemical talk of molecules—the tiniest pieces of material from which large pieces are built up—and do some imaginative thinking. If there are such things as molecules, they are too small for us to see with any ordinary microscope—though we shall later find convincing indirect evidence of them—so they must be very small and numerous. From the way liquids pour, their molecules must be able to slide over each other easily. The difficulty of compressing liquids suggests that their molecules are crowded close to each other. Other evidence which you will meet later suggests they are in rapid motion, jostling like people in an excited crowd—and with rising temperature the motion increases. In fact, liquid behavior can be imitated with steel balls or grains of sand if these large "molecules" are kept in rapid agitation.

Molecular forces: attractions and repulsions. Assuming such a molecular picture, we look at liquids and at once make a guess about their molecules: that they resist being pulled apart, they attract each other. Water in a half-filled jar just stays there; it does not expand or fly apart. But gas in a bottle fills the whole bottle and quickly spreads or diffuses if the bottle is open. Liquid stays together, so its molecules probably attract each other. At this stage, the attraction is only a vague guess. It is from surface tension, among other things, that we get strong support for this idea. The fact that liquids resist compression strongly tells us that liquid molecules also resist being pushed closer together; thus they must repel each other. So must gas molecules, when *very* close,[3] and molecules of solids, e.g., the molecules of your finger and thumb repel when shoved close together—how else do they stop sinking in when you try to press them together? But solids also oppose attempts to pull them apart; their molecules must attract each other. So we picture molecules of solids as having two sets of mutual forces: repulsions which we find experimentally act only at very small distances, *short range forces*; and attractions which are important at greater distances, *long range forces.* In a normal unrestrained solid, each molecule takes up a neutral position so that the resultant of these forces is zero. If we compress the solid, the repulsions rise and oppose our efforts. If we stretch it, the repulsions decrease more than the attractions, and we are left with tension. We know experimentally that even the attraction does not have a great range—only one or two molecular diameters.[4] For liquid molecules we picture similar forces, repulsions when extra close (as in collisions) and attractions spreading farther out. (This view raises one puzzle: Liquids should stretch slightly under tension, but in real life they just give way and develop bubbles of vapor when we try to stretch them. However, by taking great care to get rid of air bubbles we can make liquids bear tension, with freak behavior: water or mercury sticks at the top of a barometer far above the "atmospheric height," and a siphon *can* work in a vacuum! It is sabotage by some small air bubble when liquids show themselves "as weak as water.")

Molecular explanation of surface tension. We look to these long range attractive forces to hold liquids together, and perhaps to yield some "explanation" of surface tension. Imagine the experience of a molecule, A, in the middle of a jar of water (Fig. 6-5). It is jostled by frequent collisions with other molecules on all sides. It is also attracted to near neighbors all around it—pulled to the right, to the left, up, down, in slanting directions—with resultant pull zero. Now think of another specimen molecule, B, on the open surface of the water. It too is jostled, though not from all directions; and it too is attracted, though not from all directions. There are neighbors within attracting range below it and to each side, but not above. The resultant pull of attraction will be downward—into the liquid from the surface—to be balanced by the effects of collisions from below. Thus B experiences a peculiar downward attraction, like an extra weight. In a large round drop, molecules in the inner regions will be like A, attracted from all sides,

[2] For all that, some naming is silly and some sensible. If we took care to name the length of the whole table on which the pool rests, that would be silly. If we made a special name for the width of the drop, we should not find it very useful. But the angle A turns out to be worth naming. It is characteristic of the materials—if you look at a "family" of pools like the one sketched in Fig. 6-1(e), you will see that all the drops have the same angle A.

[3] When colliding with each other or with their container. What else makes the pressure of gas on its container?

[4] How can we possibly find out about such a tiny range by experiment? The actual estimate is made for liquids by combining surface tension measurements and a simple measurement of heat of evaporation. Very simple experiments will show you the range must be minute. Try to make pieces of metal join by pressing them together. Try with freshly-broken glass. Yet very gentle heating *can* remove a crack which has started in a piece of glass.

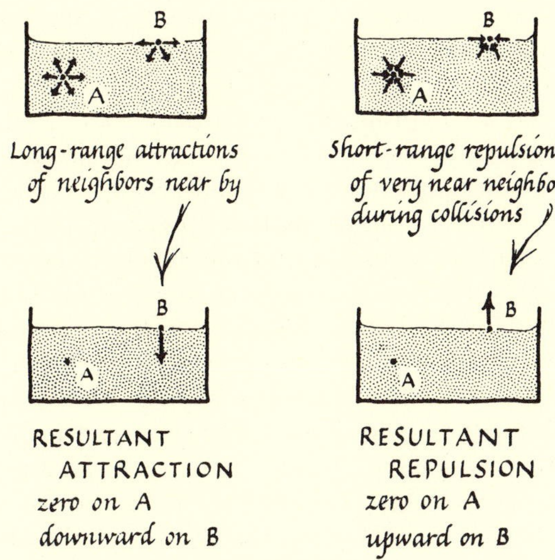

Fig. 6-5. Forces on Molecules in a Liquid

pulled in no particular direction; molecules on the surface will be like B, pulled *inward.* With all such "B" molecules trying to get in towards the middle of the drop, the surface will try to shrink: in fact it will *seem* to have a stretched skin. Obviously, if

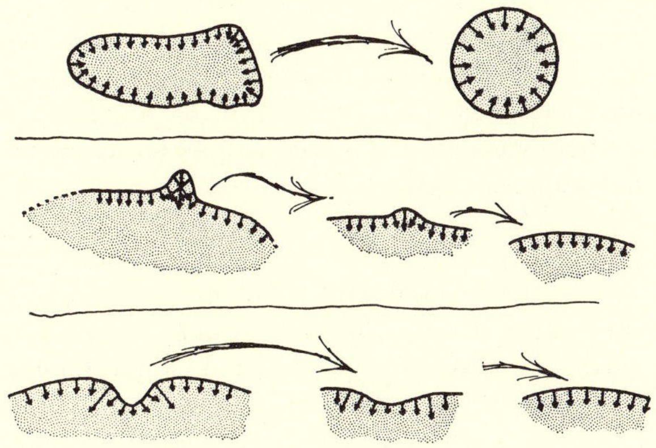

Fig. 6-6. Surface Forces on Small Drop of Liquid. The attractions of neighbors on "B" type molecules tend to pull the liquid mass into spherical shape. (Note that a sphere is the shape with minimum surface for a given volume.) If small irregularities develop on surface, surface forces tend to remove them—the full discussion of such effects is very complicated.

a pimple developed on the surface, molecular attractions would flatten it out, against jostling opposition. (A small depression, a dimple, would also be removed, though this is less obvious; molecular attractions would flatten the convex corners of its rim; Fig. 6-6.) To picture the general effect, compare a drop full of molecules with a crowd of people attracted by a street fight. Fig. 6-7b shows a bird's eye view of a crowd collecting. More and

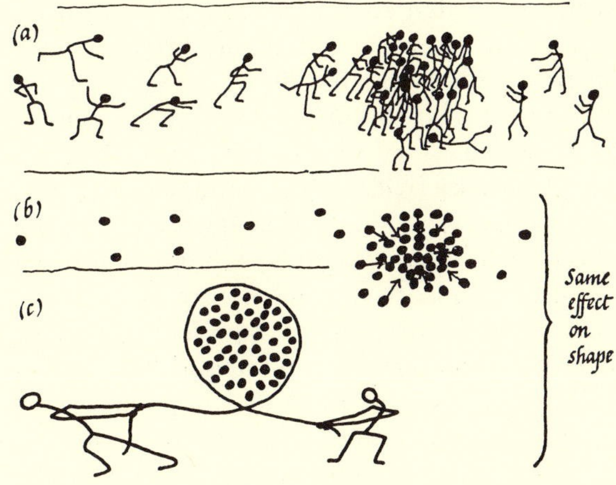

Fig. 6-7. A Crowd Collects

more interested spectators arrive. Later arrivals, finding it hard to see, press in towards the middle—they are attracted by the fight, but they would push just the same if attracted by neighbors farther in. What is the effect of this inward attraction on the crowd as a whole? A fluid crowd is shoved into a round shape with minimum outside perimeter. (A circular patch has less rim than any other shape with the same total area.) A person, A, well inside the crowd gets squeezed; and if he is tall enough he sees that his unpleasant sensations are caused by the outer members, B, pushing inward. But we could make him suffer in the same way by running a great belt of rope around the crowd and pulling it tight. A belt in tension would have much the same effect on the external shape of the crowd, and its internal discomfort, as inward attractions acting on the outermost members. Using this analogy,[5] can you see how molecular attractions might produce the effects of a stretched elastic skin in tension all over a liquid surface? On this view, there is a privileged layer on liquid surfaces, a layer of outer "B" molecules, not a real skin like a rabbit's.

Surface-Effects versus Volume-Effects. Tragedy in a Bug's Life

Why does this "skin" pull tiny drops into a perfect ball, in defiance of gravity, while larger pools compromise? On our molecular view—on our theory, if you like—the skin effect is due to the peculiar experience of surface molecules, B; thus its forces

[5] *Analogy,* often a help in learning, can never prove anything. Some theories are really analogies; for example, the older mechanical models of atom-structure. While we should welcome their help to our thinking and give them credit for fruitful suggestions, we should not make the mistake of thinking they must tell us "the real truth" and we should not cling to them when their usefulness is over.

should be related to the surface, and should not involve the main bulk of liquid inside. Gravity, however, pulls on all the liquid, outer layers and inner ones alike. Surface tension is a "surface-effect," weight is a "volume-effect"; and their *relative* importance will change with actual size of drop or pool. To study this contrast, pretend that surface forces increase in direct proportion to surface area, while weight, of course, increases in direct proportion to volume. Consider the change from a small drop to one ten times as big. For geometrical simplicity, pretend the drops are shaped as *cubes*:[6] a small cube, C_1 (Fig. 6-8), with each edge of length a, and a big cube, C_2 with edge $10a$.[7]

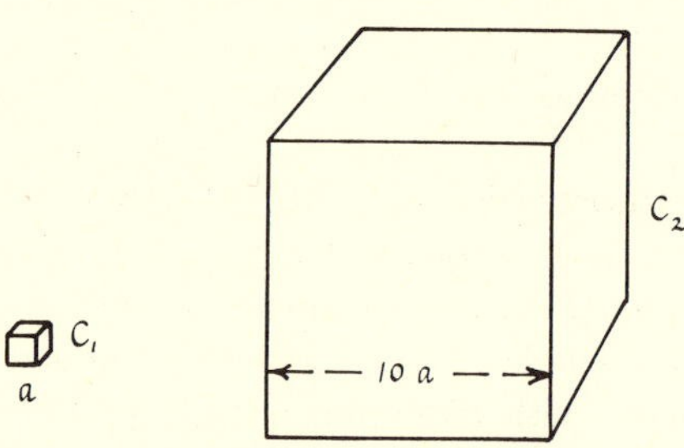

FIG. 6-8. CUBICAL "DROPS."
Illustrating surface and volume comparisons.

How do their surfaces compare? Each cube has six sides. C_1 has area $6a^2$; and C_2 has area $6(10a)^2$ or $600a^2$. The larger cube, with 10 times the linear size, has 10^2 or 100 times the area. How do the volumes compare? They are a^3 and $(10a)^3$ which is $1000a^3$. The larger cube has 10^3 or 1000 times the volume and therefore would contain 1000 times the weight of water. When we change from small cube to large, surface-effects increase only one hundredfold; but gravity-effects increase a thousandfold and thus grow tenfold in relative importance.

Actually, surface tension forces appear to tug at any *boundary* or *rim* in the surface. So they increase in direct proportion to *linear* dimensions, edge or radius, and their comparison with volume-forces is even more extreme.

For a very large pool, gravity literally outweighs surface tension effects by a huge factor: ponds are flat and a pailfull of water poured on the floor spreads out under gravity's control. For small drops surface tension has an important effect on shape. For very small drops it becomes paramount. A man diving into a lake contends with gravity-pressures. A tiny bug interviewing a raindrop finds surface-tension-forces insurmountable. Now can you see why *small* water-spiders can run about on the surface of a pond without falling in? They are safe enough: most of them are waterproof and *cannot* fall in. Even if pushed under water they bob out through the surface with skin helping them. Other small creatures with a wettable body find a drop of water a clutching prison. Some partially waterproof ones can keep above water if they are small enough, but once in, once through the terrifyingly tough "skin" which they encounter, they can never get out. To still smaller creatures, bacilli for example, surface forces are everything and weight hardly matters. Their surface is their channel of life, through which all food must come and which they must change if they wish to move. No wonder their life can be ruined by surface poisons, which cover their surface as a dye covers cloth fibers.

Imaginative thinking has carried us far beyond the experimental facts. Some of the ideas set forth are justified by further experimenting; others remain little more than wild picturing, to be suspected of romantic lying and to be used only so far as fruitful suggestions emerge.

Molecular View of Angle of Contact

Yet we might carry molecular pictures one stage further and discuss the way liquids hang on to solids: questions of wetting and waterproofing. Reverting to small pools on a table, and our classification by angle of contact, we picture the pool being humped together by surface forces on "B" molecules. However, at the edges where the pool meets the table the corner molecules, C, must be attracted by the table as well. How do the combined attractions tilt the surface and determine the angle of contact? Adding the attractions as vectors, we could obtain the resultant attraction, R, due to neighboring molecules of both liquid and table. The liquid surface will treat this resultant as a local "vertical" and will set itself perpendicular to it, just as the surface of a big pool takes a horizontal position perpendicular to gravity's vertical. The direction of the resultant attraction, R, determines angle of contact; but before discussing that we should give a more detailed account of the forces that mold the surface.

[6] Cubical drops are unreal, but lead to the same result as spheres—or any other pair of similar shapes. If you know the formulas for a sphere, surface $4\pi r^2$ and volume $(4/3)\pi r^3$, argue with them instead. The result must be general, since we have to measure surface in units like ft^2 and volume in ft^3.

[7] One of the soap film demonstrations supports this view that surface tension is independent of the main bulk of liquid. So does another simple experiment, sketched in one part of Fig. 6-1.

Fig. 6-9. Molecular View of Surface Tension and Angle of Contact

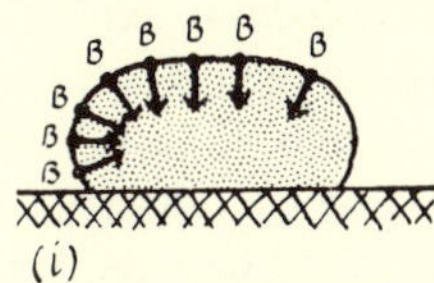

(i) Attractions of neighbors on "B" molecules in the surface make skin effects.

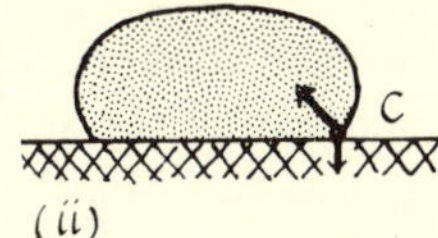

(ii) and (iii) Corner molecule C is attracted by neighbors in liquid, and by molecules of table: resultant attraction R.

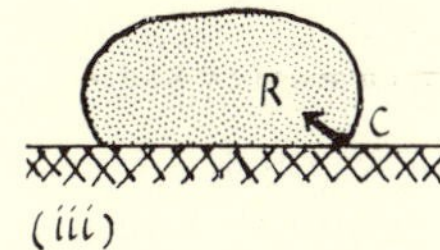

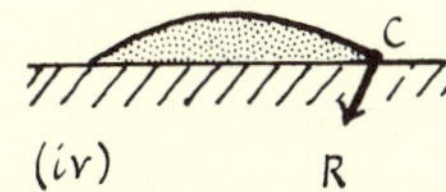

(iv) In this case C is attracted more strongly by table than by liquid neighbors.

Molecular Forces and Liquid Surface

To see why the liquid surface sets itself perpendicular to the resultant attraction, R, return to the general discussion of forces on a molecule. Molecules are acted on by:

"long-range forces"
- (a) gravity
- (b) attraction of neighbors (range only a few molecular diameters)

"short-range forces"
- (c) violent repulsions during "collisions" with neighbors (range, a small fraction of molecular diameter)

"Equilibrium" is a doubtful term in a molecule's detailed life, but we can say that each molecule, in a liquid at rest, is *on the average* in equilibrium. For any surface molecule, "B", the *short-range* forces come from neighbors at each side and below; and, by symmetry, their average resultant will be perpendicular to the surface. Because it balances that short-range force, the resultant *long-range* force must have the opposite direction: it too must therefore be perpendicular to the surface. Putting the last comment the other way round, the surface must be perpendicular to the resultant long-range attraction—all the forces involved will push the surface about until it is so. (Of course, looked at in molecular detail, the surface itself would vanish into a hubbub of irregular motion, like the edge of any crowd. It is only when we view it from afar with gross human eyes that it seems so smooth; we are then taking a molecular average.) Two of these forces on a surface molecule belong with the surface and change their direction when the surface tilts. These are the short-range repulsion and the long-range attraction of neighbors. The third force, Earth's gravity, always acts vertically down. In a large pond, vertical gravity gives the defining direction, and pulls the whole surface into a horizontal plane, and that makes the other two forces then vertical also. In the case of molecules very near a solid wall or in the surface of a small curved drop, the effect of the neighbors' attractions is much greater than that of gravity. So we neglect gravity in a first attempt to explain a curved meniscus or an angle of contact. We simply say: "the surface will set itself perpendicular to the resultant long-range attraction on a surface molecule."

Angle of Contact and Molecular Forces

To interpret angle of contact in terms of molecular forces, consider the attractions acting on a corner molecule, C, where a liquid pool meets a solid table (see Fig. 6-11). A wedge of liquid neighbors pulls with resultant attraction, F_1, along the bisector of the wedge-angle—the direction suggested by symmetry. The molecules of the solid table within range of C exert a resultant pull, F_2, perpendicular to the table—symmetry again.

Vector addition gives the resultant of these pulls, R, and we expect the liquid surface near C to set itself perpendicular to R. This is sketched in Fig. 6-11 with F_1 drawn much smaller than F_2, showing

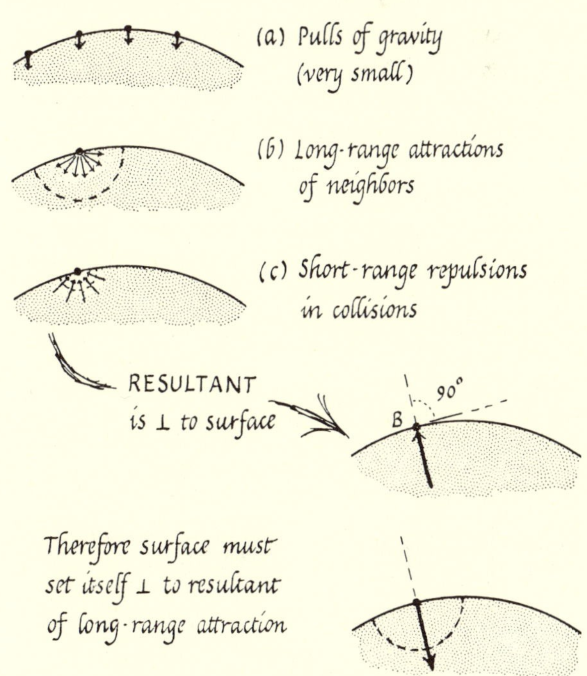

Fig. 6-10. Long- and Short-Range Forces

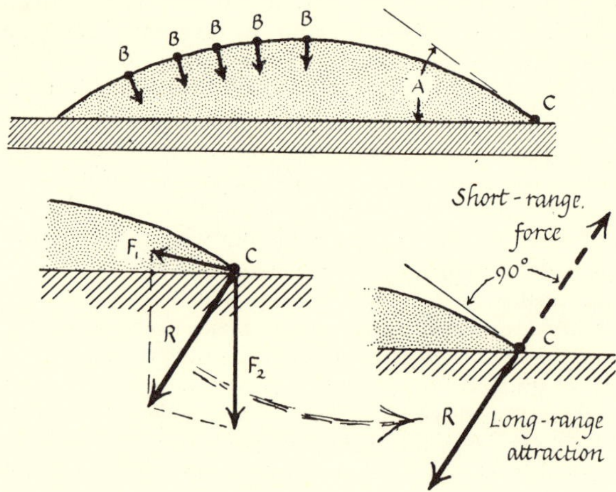

FIG. 6-11. FORCES ON A MOLECULE AT THE EDGE OF A SMALL LIQUID POOL.
The pool is on a table that attracts liquid molecules strongly.

the molecule C attracted less by its brethren than by the table. This leads to a small angle of contact, with the liquid wetting the table. We can picture the highly attractive table encouraging the liquid to spread. Thus wetting appears to be a matter of relative molecular attractions. If liquid molecules are pulled harder by neighboring solid molecules than by neighboring molecules of the liquid itself, the liquid will wet the table and spread.

On the other hand, if the liquid molecule C likes its brethren better than the table molecules, the force F_1 must be drawn larger than F_2, and the pattern swings over to Fig. 6-12 showing a large angle of contact. "Waterproofing" (non-wetting) seems to require the table molecules to exert relatively small pulls on a liquid molecule nearby compared with the pulls of liquid neighbors.

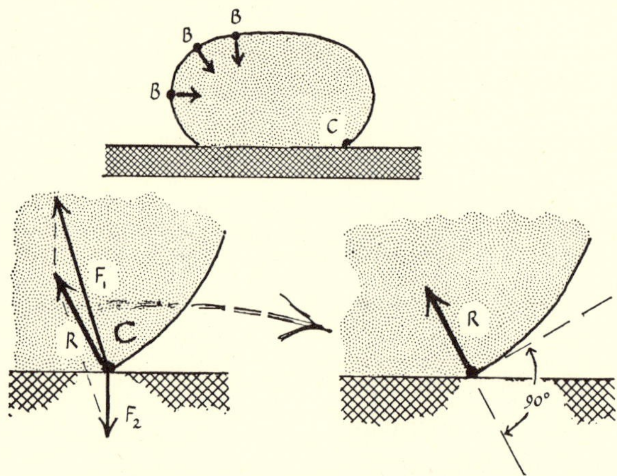

FIG. 6-12. FORCES ON A MOLECULE AT THE EDGE OF A SMALL LIQUID POOL. The pool is on a table that does not attract liquid molecules strongly.

Waterproofing and Wetting

Now we have a molecular explanation of wetting and angles of contact. Explanation? Is it anything more than an interpretation in terms of a fairy story woven to fit the facts? It is not as bad as that because it uses molecular ideas which fit in well elsewhere in physics and chemistry. And it makes useful suggestions such as the following:

(1) To promote wetting (the laundryman's dream), make F_2 larger than F_1: have the liquid molecules attracted more by the solid than by their own brethren. This can be done by using middleman-molecules, which are in fact molecules of soap. This is the secret of soap and has pointed the way to new synthetic soaps.

(2) To make a large angle of contact (the waterproofer's hope), coat the textile fibres with something that has F_2 small compared with F_1. In answer to the question, "how thick a coating?" (the waterproofer's worry), footnote 4 says a very thin layer will suffice, just a few molecules thick. (When the waterproofer's purchasing department says, "how thick *is* a molecule?" our own studies later in this chapter will provide an answer.)

(3) Offered a situation where surface forces are important, a liquid with small angle of contact (F_2 greater than F_1) will try to crawl over the solid surface, even climbing upward. This is specially noticeable when liquids climb up very narrow tubes: "capillarity," a useful behavior, which we shall now discuss.

Capillarity

Demonstration experiment. Melt a piece of glass tubing, draw it out into a very thin tube, and dip one end into ink (Fig. 6-13). The dyed water runs up, in defiance of gravity, refuting the adage "water finds its own level." Yet the apparatus used to

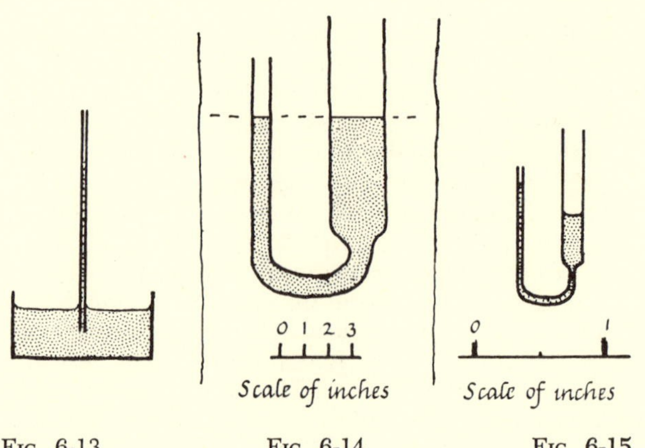

FIG. 6-13. FIG. 6-14. FIG. 6-15.

demonstrate that adage, a U-tube with unequal arms, shows the liquid at the same level on both sides (Fig. 6-14). Remembering the discussion of surface-effects *vs.* volume-effects, we guess that a *small-scale* apparatus would show surface tension effects more clearly. A tiny U-tube shows this (Fig. 6-15). Of course, this was already shown, slightly disguised, by dipping the thin tube in ink. The sketches of Fig. 6-16 show the gradation from

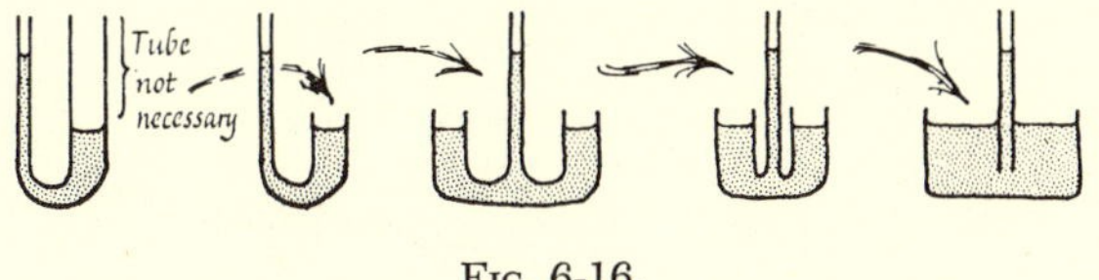

FIG. 6-16.

one experiment to the other. If liquid runs up fine tubes, it should run farther up finer tubes. Test this. (See Fig. 6-17.)

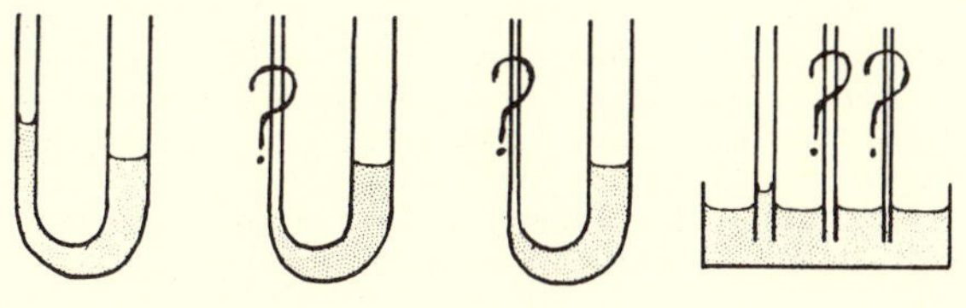

FIG. 6-17.

Since this effect of surface tension shows up in tubes "as fine as a hair" it is named after the Latin word for hair, *capilla.* Capillarity, then, is an old name for surface tension, still used, particularly for the action of liquids climbing up fine tubes. It is a nice name, but it does not *explain* liquid rise. If we say water runs up a small tube "because of *capillarity*" we are saying "because of small-tube behavior." Magnifying the meniscus (liquid surface)

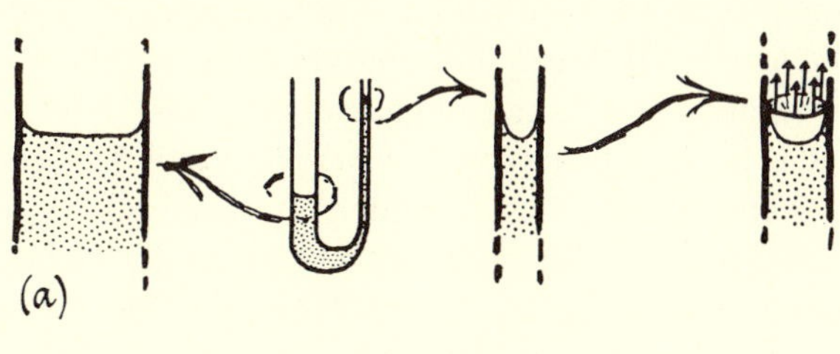

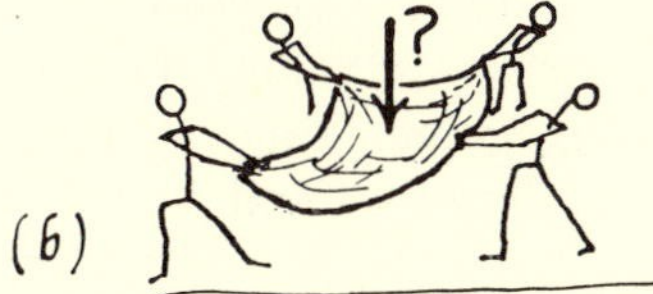

FIG. 6-18.

in each tube we see it hanging like a curved bag hitched up on the glass rather like a fireman's blanket receiving a heavy citizen. We are back at the rubber skin idea. If we measure the hitch-up forces involved, we find they are the same as those holding small drops together. We may even talk of the sagging skin supporting the weight of liquid which rises up the tube,[8] but it is much more realistic to talk of molecules climbing up the tube's inner surface to make the slanting meniscus.

The liquid does not have to have a round glass capillary tube to run up. Any narrow spaces will show capillarity. When water runs among the bristles of a paintbrush or seeps up from a bathtub into your hair, it is not filling hollow hairs but running into narrow spaces between hairs. This behavior has many uses: pulling oil up lamp wicks, water into bath towels, etc.

PROBLEM 3. (Difficult) CAPILLARITY FORMULA

Suppose you accept the view that capillary rise is determined by a pressure-difference across the meniscus. Look back at the demonstration of two soap-bubbles connected together (Fig. 6-2k). What can you infer, *simply from that demonstration,* concerning the relation between capillary rise and diameter of capillary tube?

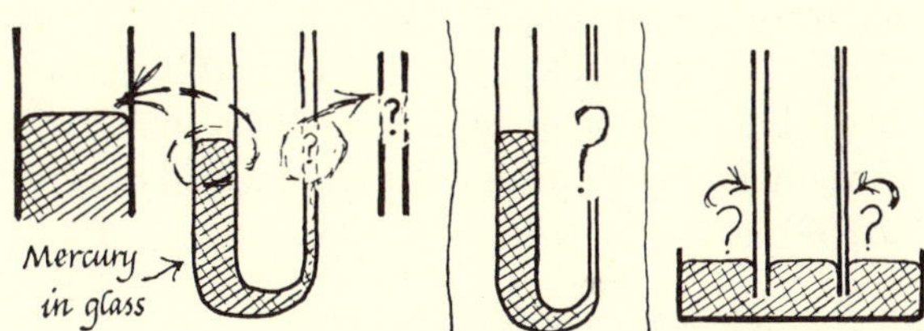

FIG. 6-19. PROBLEM 4

★ PROBLEM 4. CAPILLARITY IN A "WATERPROOF" TUBE

Suppose the liquid makes a *large* angle of contact on the tube. Fig. 6-19 shows mercury, for example, in a glass tube. The mercury meniscus in the large tube is shown but the diagrams are unfinished. Sketch all the diagrams and complete them.

---

[8] We may use this idea to derive a formula, much beloved of the problemsetter in old-fashioned examinations, for measuring the surface-tension, T, pulling on each inch or meter of rim: Pull up of skin = weight of liquid supported in tube.

$$T \cdot [\text{rim, } 2\pi r] = [\text{VOLUME, } (\pi r^2) \cdot (\text{RISE HEIGHT})] \cdot [\text{DENSITY}] \cdot [\text{FIELD STRENGTH, } g]$$

Therefore, $T = \frac{1}{2}(g) \cdot \text{DENSITY} \cdot \text{RISE HEIGHT} \cdot \text{TUBE RADIUS}$. This formula is more or less right and is used in rough measurements of T; but this derivation is almost a swindle. There *is* no rubber skin hitched onto the glass; and the T in the real formula relates to the liquid/air surface and is not a hitch-on-to-glass force. There is a curved surface, however (the meniscus); and, as in any balloon, the pressure is greater "inside" (above this meniscus) than outside. Using this pressure-difference we can both explain capillary rise and derive the formula honestly.

Using Capillarity

To promote capillary rise—not necessarily vertical rise, but any running of liquid into pores—a small angle of contact between liquid and walls is essential. A large angle of contact will have the opposite effect, tend to keep things dry. Here are some examples:

(1) *Require small angle of contact; preferably with large surface-tension*:

Water, on fibers of dish-driers, bath towels, etc.
Ink on pen-point. (Slit in pen-point carries ink to paper by capillarity. On old-fashioned steel pens the angle of contact was large when the pen was new; saliva was used as a wetting agent.)
Ink on writing paper (but the paper's pores must be blocked up).
Blood on bandages.
Nose-drops on nose's lining and hairs.
Solder on metal (flux is used to make small angle of contact).
? Saliva on food.
Paint-liquid on dry paint-powder.
Liquid paint on surfaces to be painted (raises problems in art).
Laundry-water on dirty clothes.
? Water on spectacle glasses (no narrow spaces here: but with small angle of contact, water condensing on glass makes a flat film instead of misty drops).

(2) *Require large angle of contact*:

Water on duck's back, on fibers of umbrella and tent fabric.
? Pancake on griddle.
Water on good bathroom floor.
? Water on spectacle glasses (misty drops evaporate fast).

In gardens, capillarity plays an important part. Water runs up the fine spaces between grains of soil. Hoeing and digging change the size of these spaces and alter the water supply from deep soil to surface, avoiding waste to the air.

Bricks are porous. Brick houses should have a "damp-course" of slate or other non-porous substance, a foot or more above the ground.

Molecular View of Capillarity

A very thin layer of liquid, perhaps one molecule thick, probably spreads all the way up the tube, and then the main body of liquid crawls after it to make a curved meniscus. The sketches of Fig. 6-20 show the forces F_1 and F_2 in the cases of small angle of contact and large. They are rather misleading because gravity has been omitted.

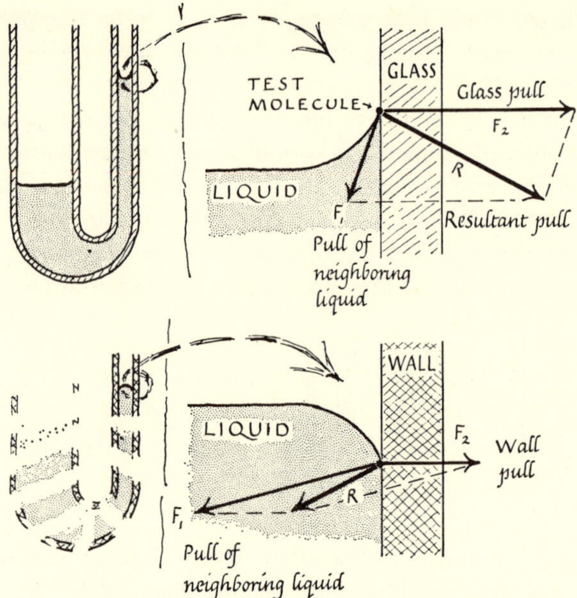

FIG. 6-20. MOLECULAR FORCES, ANGLE OF CONTACT, AND CAPILLARITY.
The liquid surface sets itself perpendicular to the resultant R of the long-range attractive forces on the molecules in it. The short-range forces involved in collisions with other molecules make it do this. (When the angle of contact is zero, the glass wall is probably covered by a thin layer of liquid all the way up, a few molecules thick. The meniscus climbs up this liquid layer.)

Wetting Agents: Soaps and Miracle Cleaners

In many cases where we want a small angle of contact, Nature provides a large one. Sheep have waterproof wool which offers no welcome to sheep dip (a delousing shampoo); dinner dishes offer a duck's back to water, and even drinking glasses have waterproof fingerprints; and new dishtowels come from the store with a maddening waxy finish. We need go-between molecules to act as middlemen and make a small angle of contact between the water and the greasy dishes, waxy cloth fibers, etc. This can now be done by "wetting agents" of which soap was the forerunner. Soap itself attacks grease by surface tension, helping water to crawl in under the grease and detach small particles of it, which are washed away in an "emulsion" (a crowd of tiny particles of grease suspended in water). One end of the soap molecule[9] likes water—a chemical or electrical attraction—and the other end is inert to water but

[9] "He is a tall fellow with a greasy head and feet that like paddling in water."

attaches easily to grease. While the grease-liking ends cluster round particles of grease, the water-liking ends face outward and welcome water. Modern synthetic soaps or washing powders are mostly "wetting agents" whose molecules act like soap as go-betweens and reduce angles of contact. They crawl into any crack between grease and dish to make invading water welcome.

We can picture ourselves as dishwashing physicists, going to a group of chemists and saying, "please design and manufacture a suitable go-between substance to act as a wetting agent. It must have the following properties:

(i) Its molecules must be attracted to grease (or, for other purposes, to textile fibers).

(ii) Its molecules must also be attracted to water.

(iii) It must dissolve fairly easily in water so that its molecules can swim around and reach the water/grease boundary where their services are needed.

(iv) P.S. It must be cheap to manufacture."

Modern organic chemists would reply, "That's easy." To hitch on to wax or grease the molecules should have a long carbon-hydrogen chain like this:[10]

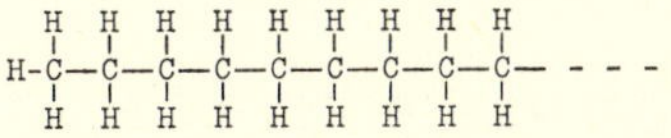

```
  H H H H H H H H H
  | | | | | | | | |
H-C-C-C-C-C-C-C-C-C- - - -
  | | | | | | | | |
  H H H H H H H H H
```

but not too long or it would not dissolve in water. Waxes and greases have similar chain structures, and they should welcome a chain like this. Then it should have something which water will like, such as a sodium atom at one end. Anything will do that will detach in water and leave an electric charge, since water molecules carry + and − charges at their ends and will cluster around another charge.

Such molecules have been designed and made, and we now buy them in vast quantities in grocery stores. Here are examples of an ancient soap and a modern synthetic detergent with similar structure.

SOAP (SODIUM LAURATE)

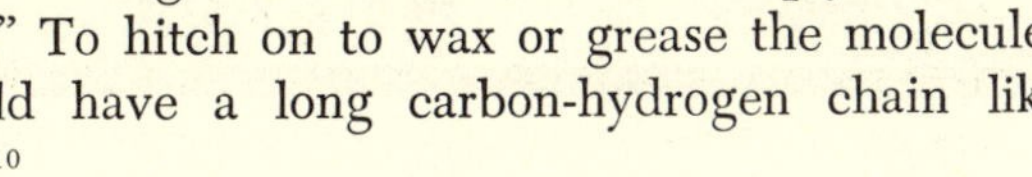

```
  H H H H H H H H H H H O
  | | | | | | | | | | | ‖
H-C-C-C-C-C-C-C-C-C-C-C-C-O⁻ · · · Na⁺
  | | | | | | | | | | |
  H H H H H H H H H H H
```

SYNTHETIC DETERGENT (SODIUM DODECYL SULFATE)

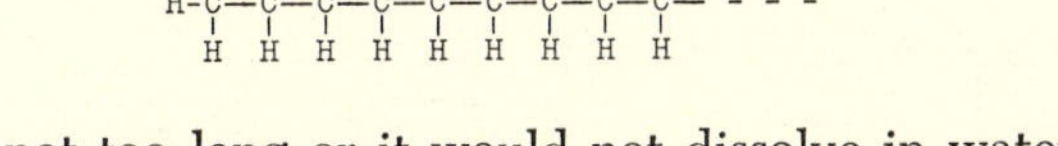

```
  H H H H H H H H H H H H   O
  | | | | | | | | | | | |   ‖
H-C-C-C-C-C-C-C-C-C-C-C-C-O-S-O⁻ · · · Na⁺
  | | | | | | | | | | | |   ‖
  H H H H H H H H H H H H   O
```

(In each case, the sodium atom, Na, detaches as an ion, Na^+, in water. This leaves the rest of the soap molecule with a negative charge, a long negative ion.)

One which is used in photography and other scientific work is Aerosol O.T.[11] Its molecule is a long chain with waxy ends and a water-seeking sodium atom in the middle, thus:

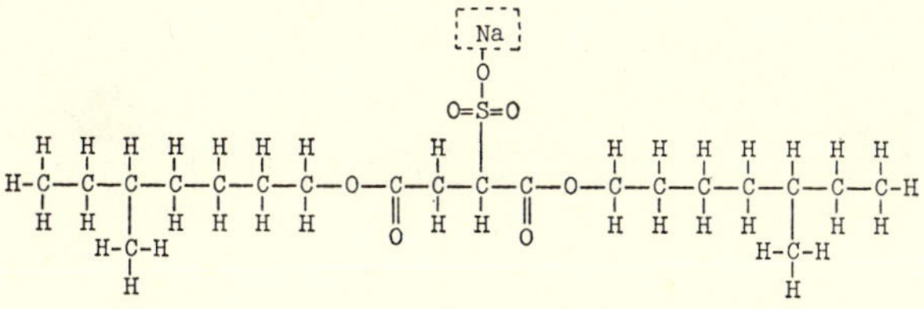

```
                          [Na]
                            |
                            O
                            |
                          O=S=O
H H H H H H H         H     |     H H H H H H H
| | | | | | |         |     |     | | | | | | |
H-C-C-C-C-C-C-C-O-C-C-C-C-O-C-C-C-C-C-C-C-H
| | | | | | |     ‖ | | ‖ | | | | | | |
H H | H H H H     O H H O H H H H | H H
  H-C-H                             H-C-H
    |                                 |
    H                                 H
```

Demonstration Experiments

(1) A drop of clean water is placed on waxed glass. A trace of wetting-agent solution is added on the end of a match. Observe the change of angle of contact. (2) A dishtowel, with the waxy behavior of a cheap new one, is cut in two pieces and spread on a sloping table. A strong solution of dye is poured on one piece. The dye hardly takes; most of it runs off. Then the rest of the dye, with a little wetting agent added, is poured on the other piece.

The Action of Soaps and Wetting-Agents. When a solution of wetting-agent meets a waxy surface its molecules cluster on the wax, their waxy ends towards it, and their water-liking ends outward. These outer ends then present a water-attracting coating to the surrounding water and thus encourage wetting. (Aerosol O.T., with its double-length molecule, hitches both ends on to wax or grease or cellulose and humps its water-liking middle, like a looper caterpillar. The humped backs make a water-attracting coating.)

Dishwashing. Most wetting-agents or soaps have the water-attracting part at one end. Their action in dishwashing is sketched in Fig. 6-22, drawn from an excellent account in the *Scientific American*, October 1951, pages 26-30.

Soap Bubbles. Soap bubbles seem strong; they can bounce, and they last well, if evaporation is prevented. This is because (i) soap molecules collect at each surface of the film, with their water-liking ends facing inward and inert ends outward, making a neutral unattractive surface coating.[12]

[10] Chemists write H for a hydrogen atom, C for a carbon atom, etc., and join them with "bonds" to show the make-up of a molecule. Such "pictures" are based on expert chemical experimenting and reasoning, and we find them trustworthy and useful.

[11] Solutions are sold in photographic stores to promote wetting of films.

[12] This coat of crowded soap molecules fences the water in and impedes evaporation. A bubble kept in very wet air, to discourage all evaporation, lasts even longer—the record runs to months.

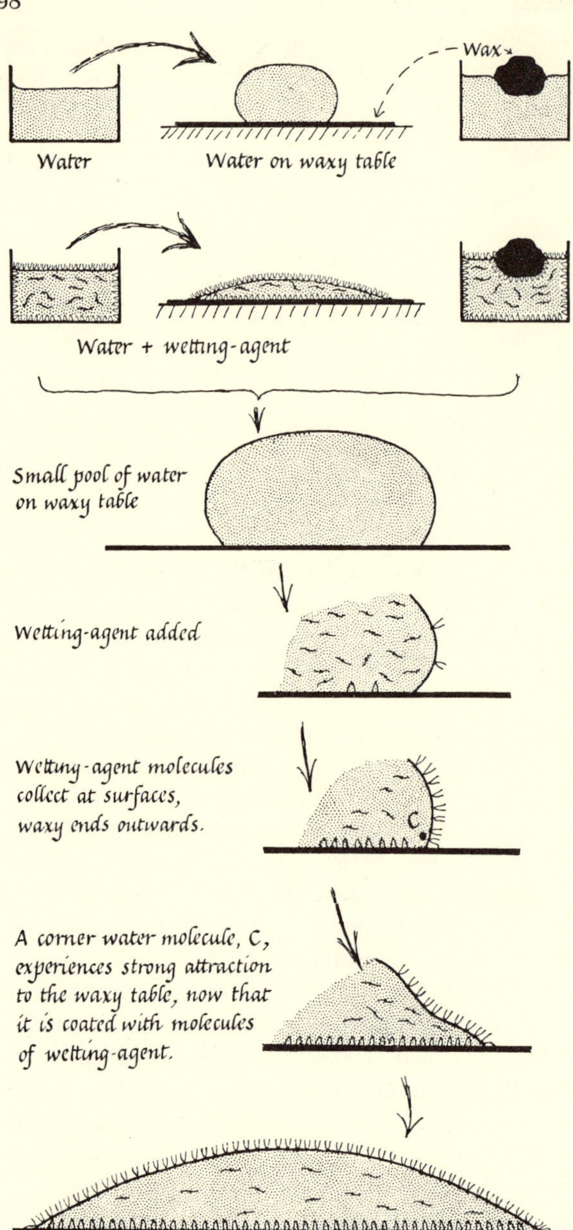

Fig. 6-21. Action of Wetting-Agent (Aerosol O. T.)
Note: Molecules of Aerosol O. T. wetting-agent are shown utterly out of scale—far too large.
The long molecule, whose structure is sketched in the text is shown by a line with a dot at the water-liking center. At a waxy surface this molecule hitches both ends to the wax and humps its water-liking middle inward. At a free surface it humps itself with its inert ends outward.

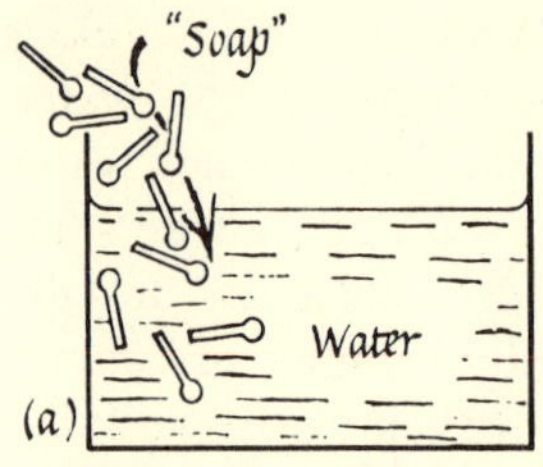

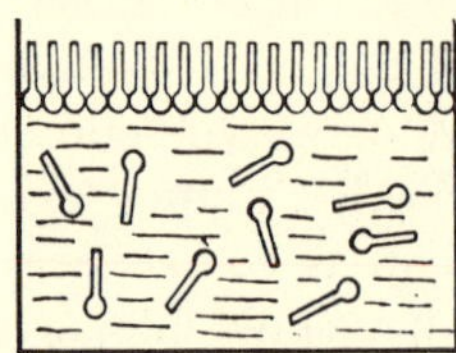

(b)

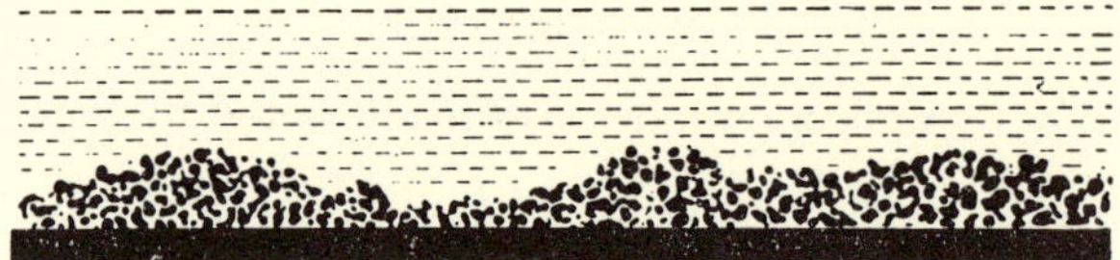

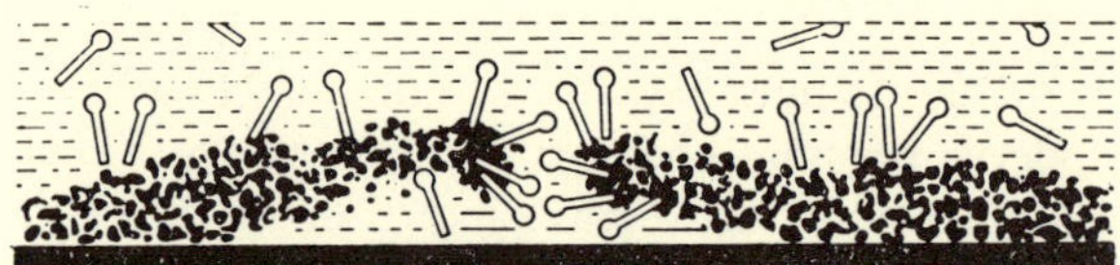

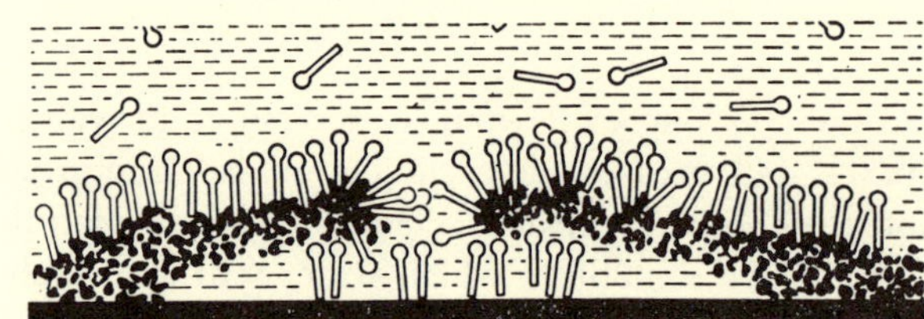

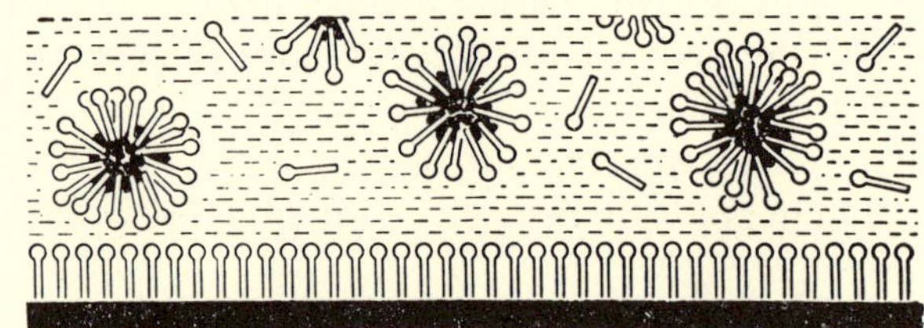

Fig. 6-22.* Action of Detergent (Soap or Synthetic). (a) Molecules of "soap" being added to water. The molecules are shown out of scale—much too big. These are molecules with one water-liking end and the other end waxy so that it can attach itself to grease or wax. The water-liking end is shown as a round blob. Many of the dissolved detergent molecules crowd onto the surface of the water and line up there with their water-liking ends in water and their inert ends out. They also crowd on the walls of the container and on any wax or grease. (b) The detergent (= cleaning) action of soap or modern synthetic detergent.

* From *Scientific American*, October 1951, pages 26-27.

And (ii) soap solution, an uneven mixture, provides the film with a slightly variable surface-tension which enables it to carry extra weight near the top and to pull any irregularities back to normal. A pure liquid, however, seldom forms stable bubbles or froth—beware of drinking from ponds with froth.

Waterproofing. To waterproof a raincoat, we make surface tension discourage water from running through the pores. This is done, without blocking the pores, by coating the fibers with wax, to give a large angle of contact with water. Then, if the pores are small, the water does not run right through but is restrained by bulgy skin surfaces. The sketches of Fig. 6-23 show water being poured on coated fibers—a magnified version of rain meeting umbrella fabric or tent canvas. The stages can be demonstrated by a model in a lantern, or the general effect can be shown by a small sieve of metal screen netting. When its wires have been waterproofed with molten paraffin wax, the sieve will hold water poured in gently. A wet finger touching the sieve underneath will break the bulging water droplets and start a deluge, like the unbelieving Boy Scout's wet head in the tent.

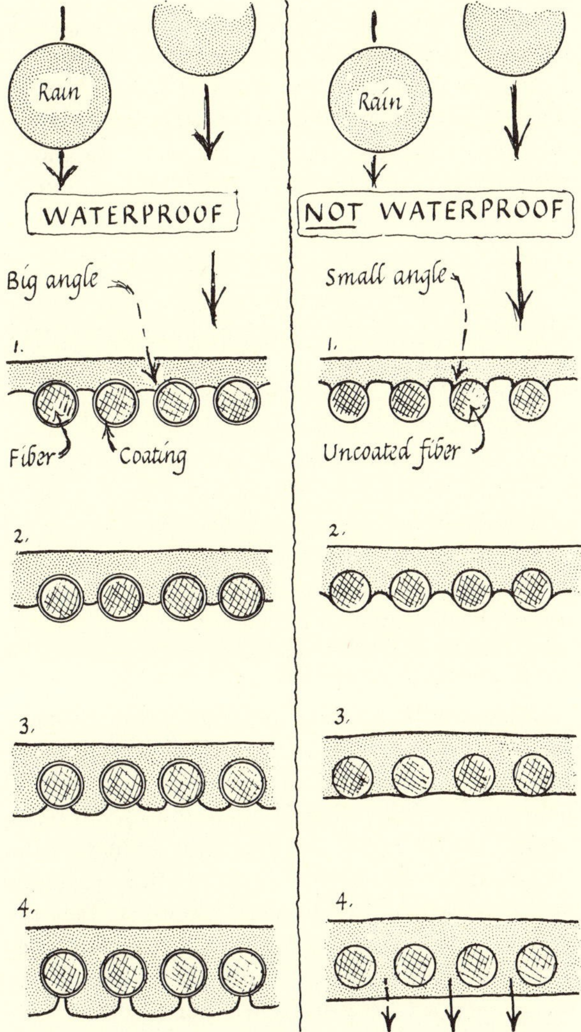

FIG. 6-23. WATERPROOFING AND WETTING.
These diagrams show the fibers of a woven fabric, in section, greatly enlarged, with water descending on them. The fabric might be umbrella fabric or tent canvas. The pores are not blocked up, but where the fibers are coated to make a large angle of contact (between water and coated fiber), the water bulges through with surface tension in opposition.

Surface Chemistry and Mining Miracles

The chemistry of angle-of-contact-changers opens fields of technical miracles: wetting-agents to help launderers, sheep-dippers, window-cleaners; trifling additions to nose-drops to help them wriggle past the barrage of hairs in the patient's nose; waterproofing agents for raincoats and industrial filters; and differential waterproofing/wetting agents to separate valuable mineral from useless rock. In this last use, rock containing metal ore is pounded to dust in a mill, then stirred in a vat of water. A differential agent added to the water coats the particles of ore and makes them "float"[13] easily; but it lets the useless sand be wet and sink to the bottom as a sludge which is thrown away. The open air surface of the water is insufficient to collect all the waterproofed ore particles; so a froth of air bubbles is blown in the muddy mixture to carry the ore up to the top, where it is skimmed off. This "froth flotation" scheme is no impractical toy. It is a successful process, used in mining to separate a million tons of rock a day. There is much clever chemistry in finding agents that will grip the ore with a proofing coat and refuse to protect sandy rock. Some agents go further, selecting one metal in mixed ores and refusing another—cleverer chemistry still. Strange new uses of froth-flotation include separating ergot fungus from rye grain, sorting peas for canning, recovering particles of waste rubber; but the main business of separating lead, zinc, silver, etc., has grown to a vast industry in which surface-tension is the essential worker.

Amoebas and Surface-Tension

How do small simple creatures living in water travel and find their food? You can get hints from crude chemical models like the wriggling camphor boat and the following demonstration of a fake mercury "amoeba" (Fig. 6-24). A small pool of

[13] To float a needle or razor blade on water, first waterproof it with a little wax or grease. A large angle of contact enables surface-tension forces to give much greater support.

mercury, resting in a watch-glass in a dish, is covered with dilute nitric acid. A crystal of potassium bichromate is dropped in near the mercury. The mercury's amoeba-like movements are due to changes of surface tension caused by chemical or electrical effects. A real amoeba pushes and pulls its irregular shape in a similar way, possibly using changes of surface-tension.

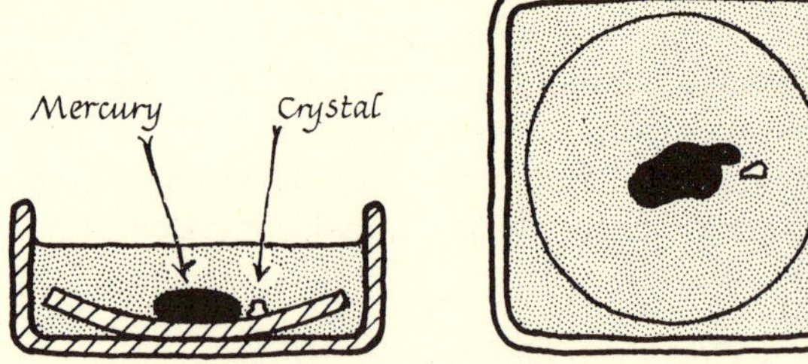

Fig. 6-24. A Mercury "Amoeba"

More Demonstrations

Changing the surface skin of water. Here are some pretty experiments which show changes in surface-tension.

(i) A sewing needle or a tiny leaf of metal can be made to float in a dish of water. If surface-tension is lessened, the boat sinks. Try adding alcohol or soap to the water.

(ii) Sprinkle the surface of clean water with waterproof dust (soot, talc, or lycopodium seeds). The weakening of the surface skin can be shown by motion of dust. When alcohol is dropped on the surface the dust rushes away. The usual explanation is this: the alcohol provides a weak skin and the dust is pulled away by the strong skin of plain water farther out. But some people prefer to say that the alcohol molecules push the powder as they spread, with a "surface pressure." Though the two views differ, either is useful in interpreting experiments.

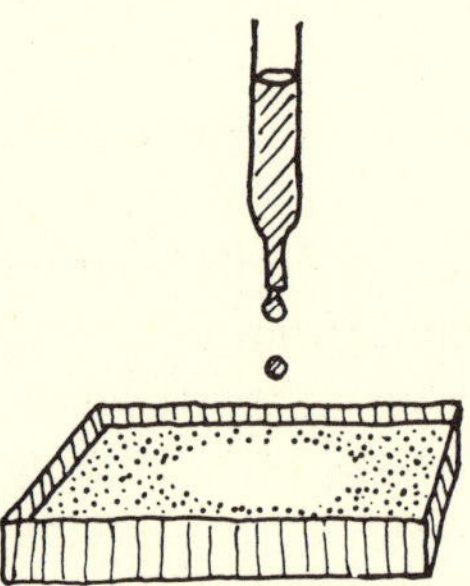

Fig. 6-25. Drops of Alcohol Fall on Water, which has been dusted to mark surface motions

(iiia) Add olive oil to a clean water surface sprinkled with dust. So little is needed that a match stick dipped in oil and then wiped dry will suffice. Even your finger rubbed in your hair will collect enough natural oil. In experiment (ii), the surface recovers from alcohol, but the effect of oil remains; hence the need for very clean grease-free apparatus in these experiments. Soap and saliva have effects like that of alcohol.

Mosquito larvae live in ponds and protrude tail-whisker breathing tubes through the surface. Oil placed on the surface gets into these tubes and kills the larva. The old explanation, that oil so weakened the skin that the larva could not hang on to breathe, is discredited.

(iiib) A tiny drop of oil placed on a huge dish of lightly powdered clean water, spreads very fast to a big round patch, then seems to have no further interest in spreading. This is observed with vegetable oils, which are "fatty acids" with one end of their molecule attracted to water, the acid end

$$-\mathrm{C}\begin{matrix}\nearrow\!\!\!\!/\ \mathrm{O} \\ \searrow\ \mathrm{O-H}\end{matrix}$$

(Mineral-oil molecules, which have both ends inert, seem to lie flat on the water surface and move around like a two-dimensional gas, spreading more casually—waxy unemployables.) In all cases, the oil film seems to exert an outward "pressure" on the surface boundaries—this seems a more realistic explanation than "weakening the skin tension of water." Nowadays we measure this outward push with delicate balances which weigh the sideways push of the oil film on a movable boom.

Uses of Long Oil Molecules

Lubrication. In modern lubrication of high-speed bearings molecules of vegetable oil attach themselves to the metal—the metal ousting hydrogen at the acid end of the oil molecule—and the oil forms monomolecular velvet carpets whose inert outer surfaces slide comfortably on each other. (Mineral oils are added to provide inert oily rollers between these velvets.) Under extreme ill-treatment even the velvet monolayers are torn off the metal; the moving metals then grip each other ("seize") with great force at close quarters, and there is serious damage.

In a similar way lanolin grease will grip your skin and soak in, to bring it drugs or general comfort, while inert mineral oil wanders about on the surface in a greasy mess—beware the druggist who prefers mineral oil to lanolin in ointments. Again, molecules of good shoe polish grip the leather; but paraffin wax (a longer version of mineral oil) merely makes messy smears.[14] Shoe-polishing promotes the grip and aligns the chains.

Calming of storms at sea. The calming of rough seas by oil is no mere fable. Quite small amounts of suitable oils poured overboard will spread over a big area. As the wind tries to whip up high waves by pushing the shallow humps which start them, the oil is blown into patches whose different surface-tensions hamper the wind's effect, with a sort of surface friction. Then some of the big waves may never be formed. And big ones arriving from far away are at least discouraged from breaking into damaging whitecaps. Surface-tension plays an important part in the breaking of a crest into spray, and oil can weaken the surface and stop the breaking.

More Experiments

(iv) How would you expect surface-tension to be affected by temperature rise? Try warming a dusted water surface by bringing a red hot poker near it.

(v) Sprinkle crumbs of camphor on *clean* water. Each crumb swims about irregularly. Camphor dissolves slowly in water, making a weaker skin. Each crumb is pulled forward by clean water ahead and less strongly backward by weaker camphorated water, so it sails ahead like a little boat, steered by its own irregular shape. Then try adding a little oil. This "kills" the camphor movements permanently. A pretty little experiment, but rather a childish one? By no means a useless toy. It played an important part in one of the great simple experiments of atomic physics: measurement of a molecule's size.

The Size of a Molecule

Sixty years ago, Lord Rayleigh watched oil spread on water. At a time when scientists were speculating about molecular sizes, he made a very clever guess. He guessed that the thinnest sheet of oil that could cover a water surface completely would be just one molecule thick, and he set out to estimate this thickness. He pictured a spreading drop of oil as a huddle of molecules tumbling and crawling over each other till each reached the water surface and could hitch one end to water—for these oils have long chain molecules with a water-liking chemical group at one end. Once all the oil molecules are thus attached they should keep together in a monomolecular carpet, showing little tendency to spread more. With just enough oil for a given

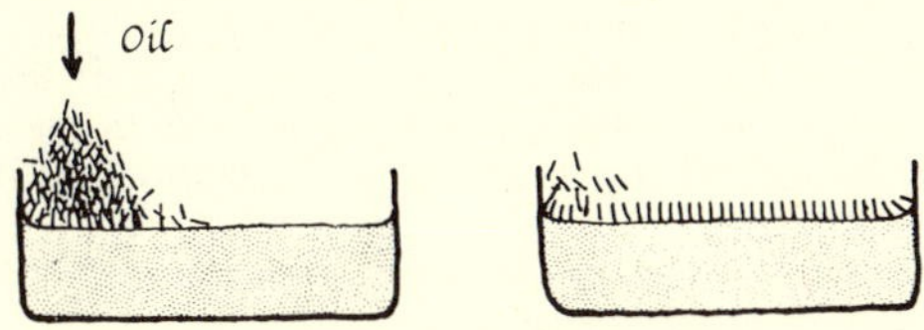

Fig. 6-26. Oil on Water.
A drop of oil placed on a clean water surface spreads to cover it with a layer one molecule thick. The oil molecules probably stand on end like a velvet carpet.

water surface, the layer would be one molecule thick, with the molecules packed close and upright, like the pile of velvet. With less oil, patches of open water should be revealed. With more oil, there should be excess puddles on the water[15] (as on greasy soup).

Lord Rayleigh cleaned a large dish and filled it with water—the biggest washtub he could buy, 33 inches across. He placed a weighed drop of oil on the surface and watched it spread to cover the surface completely. Then he started again with clean water and used a smaller drop, then smaller still, until he found a size that failed to cover the water completely. How did he know it failed? Dusting the surface beforehand might spoil it. He used camphor crumbs afterward, the childish toy. As long as the water surface was completely covered with oil the crumbs could find no clean water to spoil and dance around on; but when the oil drop was too small it left patches of clean water. In Problem 5 below, follow Rayleigh's calculation, using data based on his actual measurements, and find out how tall an oil molecule is.

PROBLEM 5. MEASURING A MOLECULE

Rayleigh placed a drop of olive oil on clean water in a large tub. For simplicity, pretend the tub was rectangular so that the water surface measured 0.55 meter by 1.00 meter. (This would have the same area as Rayleigh's round tub.)

[14] For a pleasant discussion, on which this section has drawn, see W. H. White, *A Complete Physics: Written for London Medical Students* (London, 1935), Chs. xxii and xxiii (pp. 250-269). Also see *Science News*, No. 20. May 1951.

[15] As a poor analogy, picture a herd of pigs released near a long food-trough. Just as one end of a vegetable oil molecule likes water, one end of a pig likes food. They scramble and push till every pig reaches the trough. If the herd is too big an unsatisfied crowd is left waiting (like thick drops of excess oil on water). If the numbers are just right, they form a mono-porcine line, all crowded perpendicular to the trough. If too few, they are unevenly oriented and there are vacant patches.

Camphor movements showed that the oil just sufficient to cover that water was a drop weighing 8/10 of a milligram, or 0.00000080 kilogram.

The density of the oil was 900 kg per cubic meter (0.90 times the density of water). Assume the density remained the same, even in a very thin film. (Remember that since the oil is *less* dense than water, its volume must be *greater* than the volume of the same mass of water.)

(a) Estimate the thickness of the oil film made by the drop when it spread over the water.

(b) Assume Rayleigh was correct in guessing that the film just sufficient to stop camphor movements is one molecule thick, and use the chemists' suggestion that this oil has "long" molecules one end of which is strongly attracted to water; say what you infer from (a) concerning the molecule's dimensions.

The length is minute: millions of molecules to make a line one inch long. At the time when Rayleigh made his estimate, scientists had been making rough, rash guesses at the size and mass of molecules; roundabout guesses based on gas friction, on scattering of sunlight by molecules in the sky, and on some risky electrical arguments. Here was an amazingly simple measurement, probably reliable.

The method has since been refined and extended by many, especially by Langmuir in the United States. Rayleigh's olive oil was a vague mixture of oily substances. Later workers used pure chemical compounds, often measuring several members of a "homologous series" (= chemical family). For example, Langmuir used the "fatty acids." These are derived from natural oils and fats, and they combine with sodium or potassium to make soap. They have long molecules with one end inert and the other an "acid" end which likes water. There is a whole series of such compounds, each member's molecule being larger than its predecessor's by one carbon atom and two hydrogen atoms. Long ago, chemists sketched the molecules of different members of this series with structural formulas like the three shown below.

PALMITIC ACID

```
                                                    acid end
  H H H H H H H H H H H H H H H  O
  | | | | | | | | | | | | | | |  ‖
H-C-C-C-C-C-C-C-C-C-C-C-C-C-C-C--C-O-H
  | | | | | | | | | | | | | | |
  H H H H H H H H H H H H H H H
```

MARGARIC ACID

```
  H H H H H H H H H H H H H H H H  O
  | | | | | | | | | | | | | | | |  ‖
H-C-C-C-C-C-C-C-C-C-C-C-C-C-C-C-C--C-O-H
  | | | | | | | | | | | | | | | |
  H H H H H H H H H H H H H H H H
```

STEARIC ACID

```
  H H H H H H H H H H H H H H H H H  O
  | | | | | | | | | | | | | | | | |  ‖
H-C-C-C-C-C-C-C-C-C-C-C-C-C-C-C-C-C--C-O-H
  | | | | | | | | | | | | | | | | |
  H H H H H H H H H H H H H H H H H
```

These patterns were only guesses based on chemical evidence but they suggested long chain molecules, longer for every $-\underset{\text{H}}{\overset{\text{H}}{\text{C}}}-$ link added to the chain in changing from one family member to the next. Problem 6 is based on improvements of Rayleigh's method made by Langmuir, Adam and others.

PROBLEM 6. MEASURING MOLECULES PRECISELY

N. K. Adam used a rectangular bath 0.14 meter wide and half a meter long. The bath was filled brim-full of water and the experimental region was limited by two booms A and B across the top, about 0.4 meter apart. (See Fig. 6-27.)

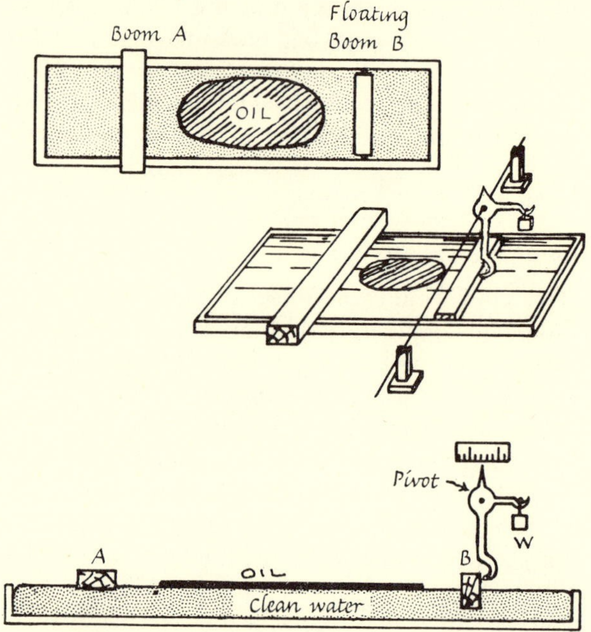

FIG. 6-27. SIMPLIFIED SKETCH OF N. K. ADAM'S APPARATUS. Oil film is confined between Booms A and B.

Boom B was free to move; it floated freely, but was prevented from drifting by a weighing-device with spring or weight to indicate any horizontal push on the boom. Boom A was placed across the bath, resting on the sides, and could be moved along by hand. The bath and the booms were waxed, so that the water level could be slightly above the top of the bath—thus booms A and B cut off a central section of the surface.

Starting with boom A far from B, Adam placed a minute measured quantity of palmitic acid on the water surface between the booms. Boom B showed no push. He then moved boom A along, crowding the oil film into smaller and smaller area, until B suddenly showed a considerable push, this suggesting that the oil molecules had been pushed together into a close-packed velvet. (In the actual experiments, the push did not suddenly rise from zero to its full value absolutely abruptly. It started at a certain area, and rose rapidly with further compression, reaching a constant value, after which still further compression probably made the "velvet" buckle. From a graph it was easy to find the point at which a considerable push developed.)

Adam added the fatty acid by dissolving it in benzene and placing a few drops of the solution on the water surface. The benzene quickly evaporated.

Here are typical measurements (not Adam's actual ones, but these data are based on his record):

Benzene solution: proportions, 4 grams of palmitic acid dissolved in 996 grams of benzene. So every kilogram of solution contained 0.004 kg of palmitic acid.

Size of drops: 100 drops of solution dropped into a bottle and weighed. Mass of 100 drops of solution: 0.33 grams or 0.00033 kg.

Main experiment: 5 drops of solution placed on water. When benzene had evaporated (leaving the palmitic acid as an invisible, insoluble, surface film), the boom A was moved towards the boom B. Boom B showed a strong push on it when the distance between A and B was 0.23 meter. Then the water surface between the booms was 0.23 meter long by 0.14 wide.

Density of palmitic acid (when liquid) is 850 kg/cubic meter; (0.85 times as dense as water).

The problem: assuming the density of the palmitic acid in the film is the same, estimate the dimensions of its molecule, using the instructions below.

(a) Calculate *length* of molecule, using Rayleigh's assumption as before.
(*Note*: One mistake of arithmetic can make nonsense of this problem. Calculating the volume of oil (palmitic acid) used is a simple fraction-problem, like the ingredients of a cake or the dilution of a blood-count. It needs elementary arithmetic and confidence. To avoid mistakes, carry it out in stages, thus: From the amount of solution (5 drops) placed on water, calculate

(i) mass of *solution* placed on water

(ii) mass of palmitic acid contained in that mass of solution

(iii) the volume which that mass of palmitic acid would occupy by itself. (850 kg occupy 1 cubic meter, therefore. . . .)

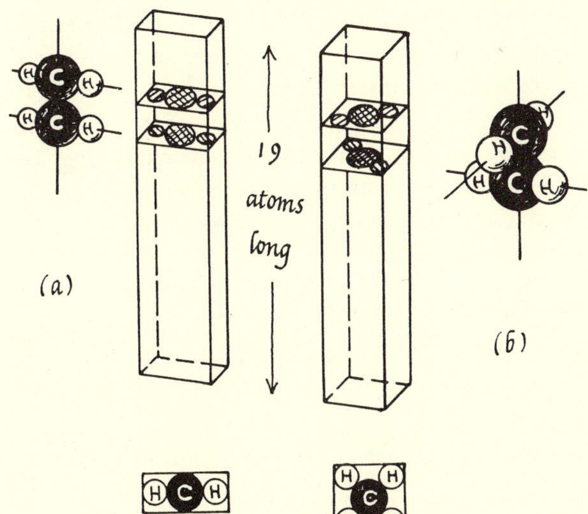

Fig. 6-28. Fanciful Thinking about the Shape of a Palmitic Acid Molecule.
Modern chemists, stacking carbon and hydrogen atoms into molecules, assign sharp sizes to them—carbon much larger than hydrogen. Here we are showing how the earliest guesses at such sizes might be made, so it would be unfair to use modern values and the C atoms are drawn only a little larger than the H. Is the cross-section "oblong" (a), or "square" (b)?

(b) Guess the *width* of the molecule, on the basis of the following discussion. The chain pattern gives a picture of the molecule 19 atoms long and only a few atoms wide. It is hard to guess the shape of the molecule's cross-section, and the H atoms might be thinner than the C atoms of the spine. It might be 3 atoms wide by one atom thick; or alternate links might swivel round, making the cross-section squarish, say 3 atoms by 3 atoms. Guessing wildly,* let us pretend the cross-section is a square, 1.5 to 3 atoms wide. Forgetting that different atoms may have different sizes, we guess the molecule's *width* may be somewhere between 1/10 and 1/5 of its 19-atom *length*. It would be foolish to try to narrow these limits. (See Fig. 6-28.)
Calculate the *volume* of a palmitic acid molecule, using the length you obtained in (a) above, and taking its cross-section to be a square, of dimensions (1/10) length by (1/10) length. Repeat the calculation with the other limit, 1/5.

(c) Taking the density as 850 kilograms per cubic meter, calculate the mass of one molecule, on each basis ("1/10" and "1/5").
(850 kg occupy 1 cubic meter, therefore. . . .)

(d) Simple chemical measurements (analysis by roasting and weighing, etc.) tell us that the palmitic acid molecule has mass 256 times the mass of a hydrogen atom. (Chemical experiments cannot tell us the actual masses of individual atoms and molecules, but *relative* measurements are easy.) From your result above, *calculate the mass of a single hydrogen atom* on each basis, ("1/10" and "1/5").

(e) From (d) say roughly how many hydrogen atoms would you therefore expect to find in a kilogram of hydrogen (about 11 cubic meters at atmospheric pressure).

(f) We now have quite other means of estimating the mass of a hydrogen atom (roundabout but reliable ones, using the electric charge of an electron, or using certain measurements in radioactivity). They yield the result:

mass of hydrogen atom = 1.66×10^{-27} kilograms.

Assuming *this* value is correct and working backwards, what can you now say about the shape of a molecule of palmitic acid? Was either "1/10" or "1/5" near the mark?

(*Note*: To work all the way backwards in detail may be tedious. Short cuts to rough estimates would be suitable here.)

* For a glimpse of the mass of a single molecule, even wild guessing seems worthwhile.

PROBLEM 7. CHAIN MOLECULES

Measurements with boom and balance, as in Problem 6, gave the following estimates for the lengths of molecules of several members of the fatty acid series. The lengths are given in special units, obviously *not* meters. (The units are Ångstrom units, each worth 10^{-10} meter, often used in atomic physics.)

Name	*Structure suggested by chemists* (number of atoms guaranteed; arrangement of atoms suggested)	*Molecule length from film method* (special units)
myristic acid	H—C(H)(H)— . . altogether 13 lots of —C(H)(H)— . . —C(=O)—O—H	21.1
pentadecanoic acid, as above but 14 lots . .		22.4
stearic acid, as above but 17 lots . .		26.2
behenic acid, as above but 21 lots . .		31.4
palmitic acid, as above but 15 lots		*

(Note: the number of lots listed includes the first carbon atom with three hydrogen atoms)

Do these measurements support the chain-molecule idea? Analyze them by making a graph. (* Insert your value for palmitic acid, if you like, taking 1 special unit = 10^{-10} meter).

Physical Tests of Chemical Patterns

It would be poor science teaching to make you feel familiar with these oil molecules by mere talk of "chain of links" or "velvet pile" for a thin film; but if calculations like those above give you a feeling of gaining knowledge, you are making genuine progress in understanding science. These structural-formula pictures were clever imaginings, vouched for indirectly by chemical arguments, but quite untested till Rayleigh's method brought most satisfying confirmation—a long thin molecule with a standard increase of length for each added group of —C(H)(H)—. Still, Rayleigh took several risks in his argument: an independent measurement would be welcome. In this century, X-rays have provided an even finer measurement of molecular sizes. Freezing the oils into waxes, we can make the molecule layers of their crystals reflect X-rays and we can read off layer-spacing (or molecule size) from the X-ray reflections, much as physicists of Rayleigh's day could read off the rib-spacing on a butterfly's wing from the colors of light it reflected.[16] Some account of these "diffraction grating effects" will be given in a later chapter. X-ray measurements confirmed Rayleigh's guess with pleasing closeness, then added refining comments.

Looking back on matters of wetting and waterproofing, you can now guess about the amounts of material needed. A layer one molecule thick will probably suffice, so the quantities required are minute. We have come to think of monomolecular layers as definite, familiar things. They are too thin to see with ordinary light, though X-rays or electron diffraction could detect them. But Blodgett has developed a scheme for adding layer upon layer on a glass plate, folding 2 more layers on each time the plate is dipped into water with a floating monomolecular layer. The dipping is repeated until the thickness can be measured by common instruments that measure paper thickness, giving at last a direct measurement. (Such films provided an early way of coating glass with the right thickness to make it "non-reflecting." Coating is applied to modern camera lenses by another method, vacuum evaporation.)

In this chapter, we started with simple observation and some naming, then borrowed ideas of molecules, did some rash imagining and some more experimenting, and emerged with a variety of results ranging from practical things like shampoos and boot polish to a measurement of the length of a molecule. The "physics and chemistry of surfaces"[17] now forms a science of its own.

[16] We can use the colors of soap bubbles to estimate the thickness between inner and outer faces.

[17] If you would like to see scientific writing that is authoritative and thorough and yet in fine English, you should look at N. K. Adam's huge treatise, *The Physics and Chemistry of Surfaces* (3rd edn., London, 1941). Some of the mathematical treatment is given in texts on "Properties of Matter." An old book, *Soap Bubbles and the Forces which Mold Them* (New York, 1890, now reprinted in paperback, Doubleday and Company, Inc., 1959) by C. V. Boys, gives some delightful experiments.

CHAPTER 7 · FORCE AND MOTION: $F =\!\!| M \cdot a$

"Pooh," said the Elephant's Child . . . "I'll show you."

Then he uncurled his trunk and knocked two of his dear brothers head over heels. "O Bananas!" said they, "where did you learn that trick, and what have you done to your nose?" "I got a new one from the Crocodile . . ." said the Elephant's Child.

"It looks very ugly," said his hairy uncle, the Baboon.

"It does," said the Elephant's Child. "But *it's very useful*," and he picked up his hairy uncle, the Baboon, by one hairy leg, and hove him into a hornet's nest.

—RUDYARD KIPLING, *Just So Stories*

(This is a long, hard, important chapter: ugly but very useful. It may need several readings and much careful thought; but without it you would make little of astronomy or atomic physics, which play important parts in the course.

If you find this study of motion difficult, reflect that it took mankind a long time to master it. Greek scientists had a good knowledge of the easy things in physics, levers and simple machines and floating bodies, etc., but they were muddled and foggy about motion. Much of the fog remained until three or four centuries ago. It took mankind over sixteen hundred years to reach a clear understanding of motion; you should hardly be impatient if it takes you several weeks.)

Force and Changing Motion

How can a rocket propel itself in a vacuum? How do we know the charge of a single electron? How can we predict the behavior of gases by theory? How can we explore the structure of atoms with alpha-particles shot from radium? How can we predict the energy-release in nuclear fission? All these things can be done; but, to understand what happens or how scientists make the measurements, you must know the relation between *force* and *motion*. This chapter will explore that relation in detail, not for the sake of dull problems on speeding bicyclists but as a necessary foundation for almost all the most important physics, ancient as well as modern.

To the modern scientist, motion is not very interesting unless it changes. He expects steady motion to continue of its own accord; but if he sees a moving thing speed up or travel in a curve with changing direction, he considers he can gain useful information. He thinks there is force at work. Changing motions enable him to study the play of "force" in the physical world, perhaps even to make rash surmises about cause and effect. The world is full of *changing* motions: cars accelerate, cannonballs rise and fall, baseballs "curve," pendulums swing, the Moon sweeps around its orbit, planets wander across the sky in looped patterns, gas molecules reverse their motion violently when they bounce on the walls of their container, a beam of charged atoms squirted through an electric field is tugged into a parabola, a fine stream of electrons is wiggled up and down by magnetic fields in a television tube; and it would be a final marvel if even rays of light fell in a curve under gravity.[1] In this book you will study all these examples of changing motion. Each involves force, and if we are to go beyond mere cataloguing description we must know the relationship between "force"—whatever "force" may be—and changing motion. We shall call any push or pull a force, and we shall measure such forces by simple spring-balances (without assuming Hooke's Law).

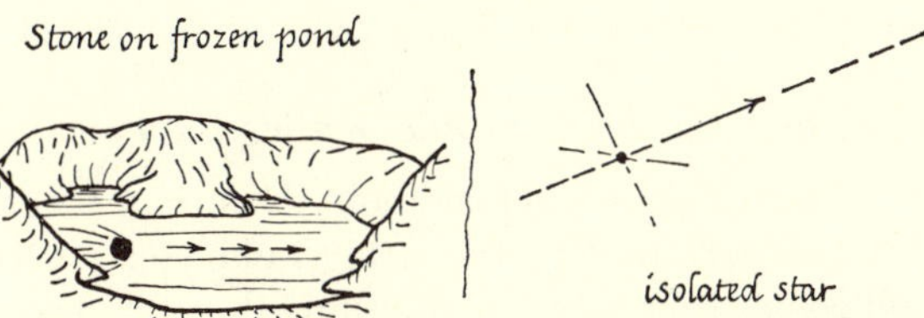

FIG. 7-1a. STEADY MOTION WITH CONSTANT VELOCITY
Fixed speed in fixed direction.

This is the time for more experiments, mostly demonstrations.

[1] Perhaps they do. How could you tell whether a beam of light is curved? How do you test whether a ruler is straight?

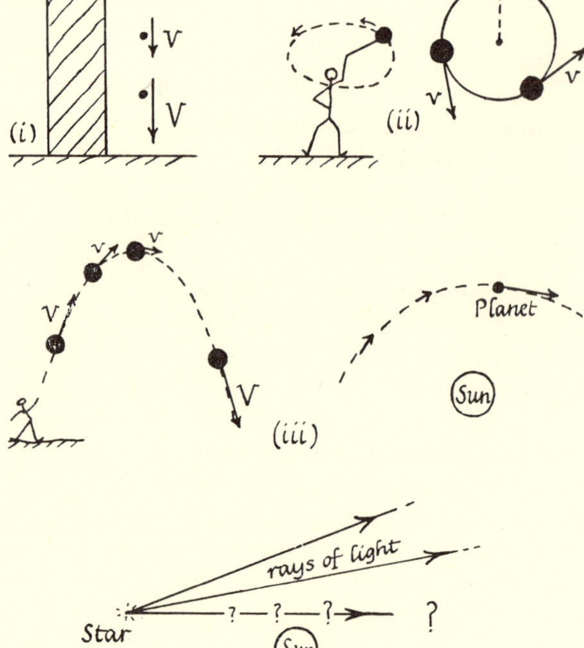

Fig. 7-1b. Changing Motion
Velocity changes in size (i), direction (ii), or both (iii).

Force and Acceleration: Anticipating the Laws

Tie a rock to a string and hold the string. You can feel the rock pulling the string, and you say this pull comes from something pulling the rock down—the Earth or "gravity" or simply the rock's weight. That downward pull on the rock is balanced by

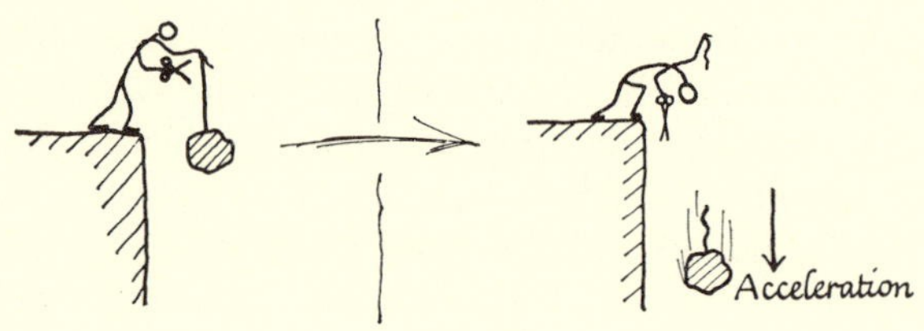

Fig. 7-2.

your upward pull. Now cut the string: the rock falls with constant acceleration. You have stopped pulling the rock *upward*, but you may assume that the same *downward* pull still acts on it, and is the only force, a constant downward force. In that case, *a constant force produces a constant acceleration.* This is the beginning of good knowledge of force and motion. Put with it Galileo's teaching that *when there is no force there is no acceleration: an object with zero RESULTANT force acting on it stays at rest or moves with constant velocity.*

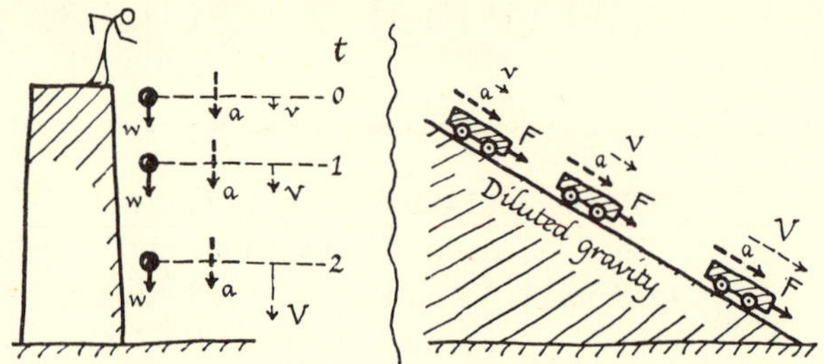

Fig. 7-3. Acceleration with Constant Force

To investigate force and motion further, we must apply forces of different sizes, and we must try various moving objects. In this course, we shall use a primitive spring balance to measure FORCE, in

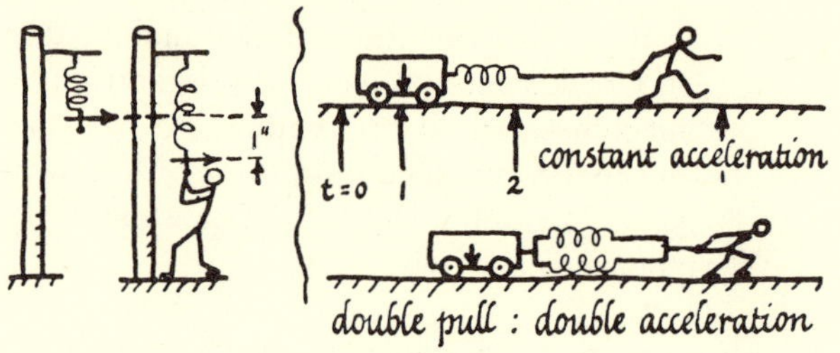

Fig. 7-4.

arbitrary units. Take a good steel spring and pull it to stretch some standard amount, say 1 inch. Call that pull "unit force," one "strang." (*Strang* is a new name for a new unit, coined for use here. Presently it will be replaced by an official unit.) Then we can apply one strang to accelerate some "victim"—a small truck or a block of ice on a level table—by pulling with the spring kept at standard stretch. It is no easy job to pull a cart along with a constant pull as it runs faster and faster. For the moment we shall pretend we can do that and look ahead to the results of such experiments. Measuring times and distances would show that *the acceleration is constant.* The victim's travel-distances in 1 second, 2, 3, . . . secs from rest would show the proportions 1:4:9: . . . . (Or from measurements of s and t we could calculate $2s/t^2$ and we should find it constant.) Then apply double force, 2 strangs, by a pair of identical springs side by side "in parallel," each at standard stretch. We should get double acceleration. *The acceleration increases in the same proportion as the force.*

To provide a whole range of forces, 1, 2, 3, 4, . . . strangs, make several equal[2] springs. Then accelerate the victim with 1 strang, 2, 3, . . . and we should

[2] For a discussion of the philosophy of making forces "equal" and adding them, see later in this chapter.

find accelerations in proportions 1:2:3: . . . . Then ACCELERATION increases in the same proportion as the accelerating FORCE, or $a \propto F$, for a given victim.

So far we have always pulled the same victim. Now change to different victims, different quantities

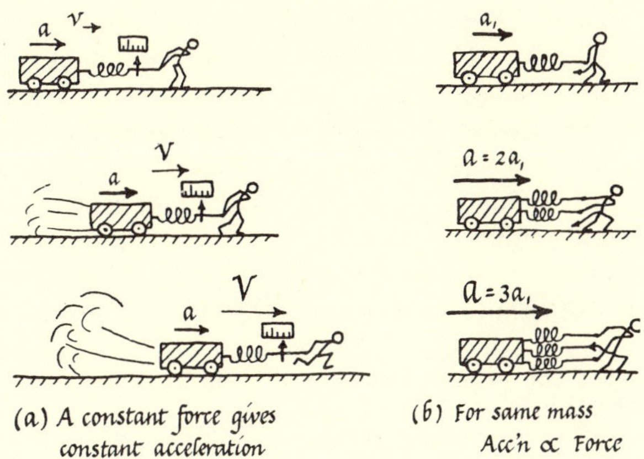

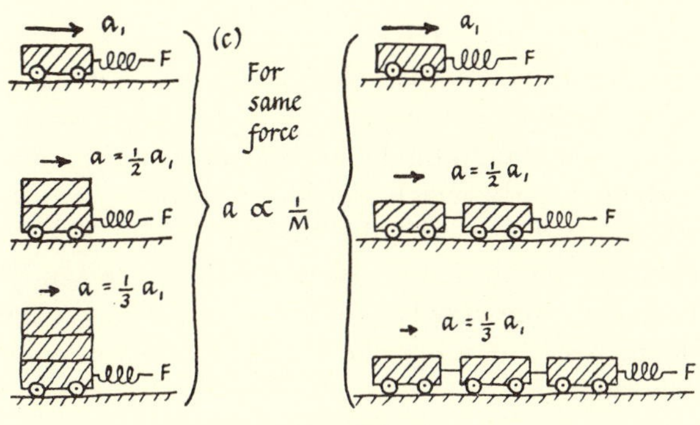

FIG. 7-5.

of moving matter, to twice and three times the MASS. Make several identical victims[3] (trucks or blocks of ice). For double mass, tie two together—or pile one on top of the other—and pull with 1 strang. Then put three victims together and pull them. With double mass, we should find half the acceleration; with triple mass, one third. The *acceleration* decreases in the proportion that the *mass* increases, or $a \propto 1/M$, where M is measured by counting the number of trucks. This is a harder relationship to picture, so we ask our question another way: how must the force change if we want the *same acceleration* for different masses? With double mass, 1 strang gives half acceleration, so 2 strangs would give the original acceleration. Then, *for the same acceleration*, single mass, double mass, and triple mass need forces in proportion 1:2:3. The FORCES needed are proportional to the MASSES, $F \propto M$.

[3] For a discussion of the philosophy of making masses "equal" and adding them, see later in this chapter.

Here, by mass we mean how much stuff is to be accelerated, how many equal trucks (or blocks of ice).

Summary

Then we have two important relationships:

(i) ACCELERATION ∝ FORCE, for a constant mass, or FORCE ∝ ACCELERATION

(ii) FORCE ∝ MASS for a constant acceleration.

These can be combined[4] into

FORCE ∝ MASS · ACCELERATION
or FORCE = (constant) · MASS · ACCELERATION

Newton's Second Law of Motion

We have pretended glibly that we can apply constant forces to moving masses and measure the accelerations accurately. And we have assumed that the pull of our springs is the *only* horizontal force acting on the victim, so that it is the resultant pull. The relation we have announced,

RESULTANT FORCE ∝ MASS · ACCELERATION,

is true. It is Newton's great SECOND LAW OF MOTION (which includes his FIRST LAW, and assumes his THIRD LAW in any experimental tests). This law relating FORCE, ACCELERATION and MASS is essential in later physics. It fits experimentally the motion of all large bodies, from toy trucks and tennis balls to jet planes and planets; and we extend it, *by assumption*, to atoms, electrons, and nuclei. To understand this law clearly and use it well, you must understand its basis of experiment and definition. So it is very important for you to see good experimental tests or demonstrations. Before describing some demonstrations, we shall discuss the special case $F = 0$.

No Force: Unchanging Motion—Newton's First Law of Motion

If $F \propto M \cdot a$, then in the special case $F = 0$, the acceleration must be zero; the motion must continue without change. You could infer this from projectile motion: you see in the vertical acceleration the effect of Earth-pull, and you should also see in the horizontal motion the effect of any horizontal force. Apart from air friction (which is not involved in this ideal case) there is no horizontal

[4] See note 16 for the algebra of this combining. For the moment, compare this with the cost of labor for some job:
COST ∝ NUMBER OF MEN working
COST ∝ NUMBER OF HOURS worked
combine into
COST ∝ (NUMBER OF MEN) · (NUMBER OF HOURS)

force. Yet a cannon-ball continues to move forward with constant horizontal velocity. Then you can make the guess that with no force the velocity continues unchanged. In such cases the horizontal *acceleration* is zero, but the *velocity* does not have to be zero: it can maintain any constant value. So physicists say, "it takes no force to keep steady motion going." This seems absurd at first sight. It takes a large and continuous shove to keep a box

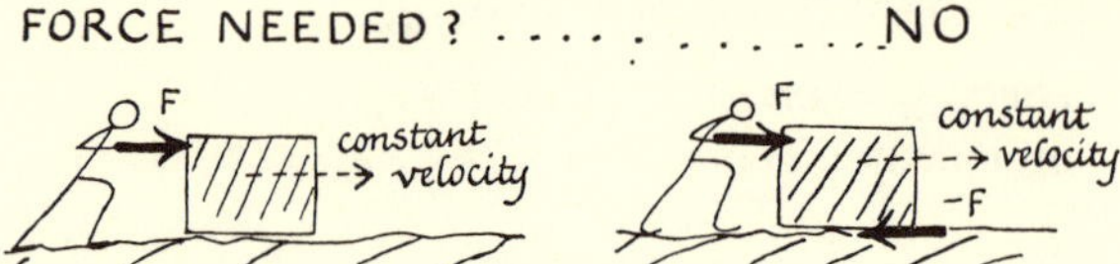

Fig. 7-6. Ancient Question and Modern Answer

moving along a rough floor or to keep a car moving steadily on a level road. But then we are taking a limited view; we are forgetting the backward shove of the floor's friction or of air resistance. If we include that, the resultant force may well be zero. And it is zero *resultant* force that we say goes with constant velocity (see Fig. 7-6). Even in the case of the flying cannon-ball's horizontal motion we could add a contraption of springs with resultant

Fig. 7-7. Forces Do Not Affect Motion, If Their Resultant Is Zero!

force zero, one pulling forward, the other backward, and the horizontal motion would still continue unchanging (see Fig. 7-7).

DEMONSTRATION EXPERIMENTS

(Here are descriptions of some experiments. The forms you see must depend on the equipment available.)

I. *Motion of a body with "no resultant force"—the skater's dream.* It is impossible to demonstrate this honestly. We cannot provide a moving body which *obviously* has no force acting on it. We run into trouble from gravity, friction, or logic. All we can show are experiments that illustrate our line of thought, running towards the ideal case, itself an imaginary experiment extrapolated from all real ones.[5] The rule "no force: constant motion" applies whether there is friction or not. We only look for *frictionless* experiments to demonstrate it because friction is difficult to measure and allow for.

Demonstration I(a). *Watch a ball roll along a level table.* Unfortunately it slows and stops; we blame friction. (And, unfortunately, a rolling ball is also used as a test to find out whether the table is level; so there seems to be a danger of arguing in a circle. But you can avoid that if you experiment intelligently.)

★ PROBLEM 1. SCIENTIFIC EXPLANATION vs. DEMONS*

How do you know it is friction, not demons, that brings a rolling ball to rest? Suggest experiments to test or support your view.

* This problem, which looks like a joke at first sight, raises the whole question of the nature of scientific explanations and laws. Try to make a logical defense—but remember that an opponent defending demons could claim a variety of properties for them.

Demonstration I(b). *Watch a large block of "dry ice"* (solid carbon dioxide) *slide* along a level table of aluminum or plate glass. The block is kept from contact with the table by a cushion of gaseous carbon dioxide which is constantly being roasted off its bottom surface by the table. The block is cold, far colder than ordinary melting ice; and it finds the table very hot, so it evaporates to gas and skates over the table like a block of ice on a hot sidewalk.

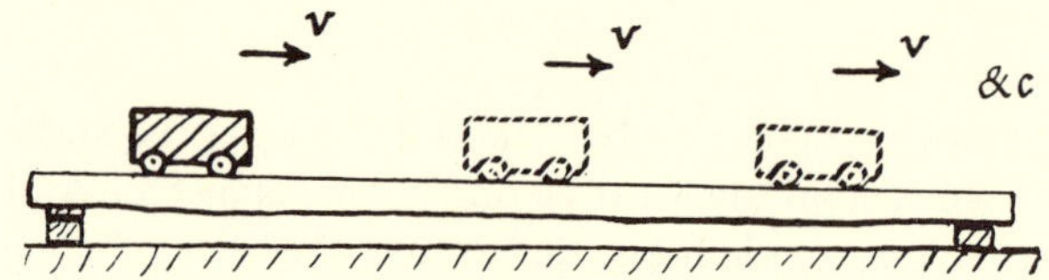

Fig. 7-8. "Newton's Law I"
Truck on friction-compensated track.

Demonstration I(c). *"Model railroad"* (see Fig. 7-8). Honestly admitting defeat by friction, we can make a friction-compensated railroad by tilting the track. A truck on a tilted track is pulled downhill by a fraction of the Earth-pull on it, and we can adjust the slight slope to make that small downhill

[5] Perhaps the ideal experiment is *unthinkably* difficult, requiring a single moving body infinitely isolated from all others which might disturb it. Then how could we observe its steady motion? Where would we be, and where would our mile posts be? Since it would be impossible to observe such motion if it existed, are we wise to talk about it as part of scientific knowledge? We are safer to put up with minor disturbances from friction or perturbations by gravity.

force just counteract friction. Then we start the truck with a momentary push and watch it move. This is a fake demonstration—how did we find out how much to tilt the track? Yet it is interesting to watch the truck creep along, the slope of the track being almost invisible. In fact, we believe the resultant force *is* zero. Pull of the Earth and push of the track and drag of friction combine by vectors to make zero. If the truck is started with a bigger push, it maintains its new speed all the way along. Loaded with sand or metal and given a start, it again moves steadily. Without measurements this is an unconvincing demonstration, really telling things about friction rather than about motion with no force, but we shall find the friction-compensated track useful in later experiments.

Demonstration I(d). We meet illustrations of straight-line travel in the paths of very fast projectiles; rifle bullets move so fast that their gravity-fall is unnoticeable in a short travel. This only shows us that the path is nearly straight; it gives no assurance about unchanging speed. Or streams of electrons (and other atomic particles) moving faster still can be shot through pinholes in a series of barriers in a long pipe (Fig. 7-9). If the pinholes are not lined

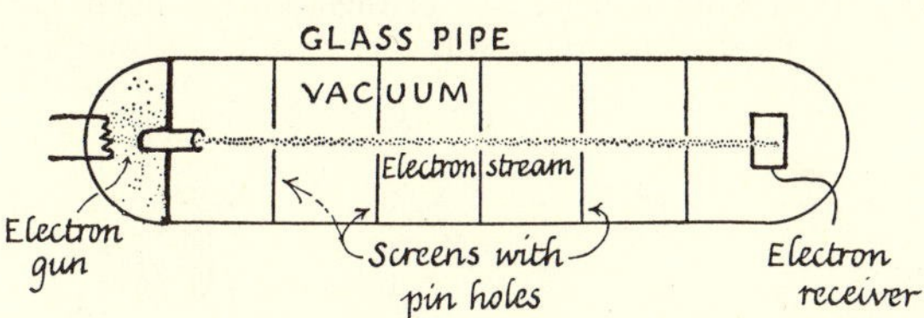

Fig. 7-9. Electron Stream Travels Along a Straight Line, if Left Alone

up in a straight line, the stream does not get through.[6]

Some fast atomic particles make a black track when they pass through sensitive photographic gelatine. Shot at glancing angles through photographic films they make very straight streaks. (See emulsion photographs of the tracks of electrons, protons, etc. from cosmic rays.)

Demonstration II. *Force and Acceleration.* The relationships suggested above (with optimistic stories about measurements with springs) form a great basic law of physics, so you should see real demonstrations. To test whether accelerations are proportional to forces, as Galileo's writings suggested to Newton, we measure the acceleration of a small truck pulled along a railroad by various

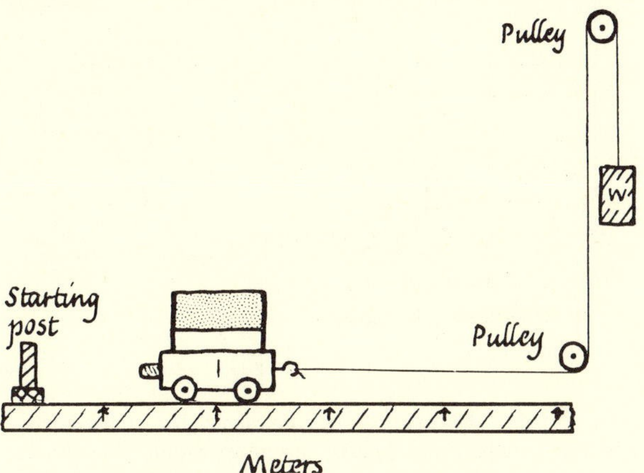

Fig. 7-10a. Demonstration Experiment illustrating the relations of force, mass, and acceleration. A small load W pulls a massive truck by means of a thread running over very good pulleys. The track is tilted slightly to compensate for friction. This is the apparatus used in the demonstrations II a,b,c. (For details of timing system, see Fig. 7-10b.)

forces. We tilt the track slightly to compensate for friction. That is not dishonest, if we publish our precaution. It is good to compensate for friction, or keep it small: the laws of motion do not fail when there is friction, but friction adds another force that must be measured separately if we want to know the resultant force. And it is the *resultant* force that appears in the simple laws.

Detailed arrangements of track and timing system depend on the equipment available. The track should be long, with steel rails as straight as possible and carefully supported. The truck should have the very best ball-bearing wheels. Electrical timing with a large clock as recorder is more convenient than schemes of ink spots or wavy traces. The clock can be started by an electric contact and stopped by an electric eye (= *photocell*) as in Fig. 7-10b. This may seem complex and mysterious at this stage. You will meet this machinery later—photocells, amplifiers, etc. All you need to know now can be got by watching the actual working of the railroad system and clock. You will see that the clock starts when the truck leaves the starting-post and stops when it reaches the electric eye. If you use the clock simply on this basis of direct observation you are doing nothing worse than when you use any clock—you assume its behavior is reasonable, but you keep an eye open for unwanted errors.

If you see this fundamental demonstration done with elaborate apparatus, just check it also with your own watch.

[6] How would you make sure the pinholes were properly lined up? An experimental physicist would probably use a flashlight. If he found it embarrassing to rely on the straightness of rays of light, he could use a taut thread, allowing for its sag, like a surveyor.

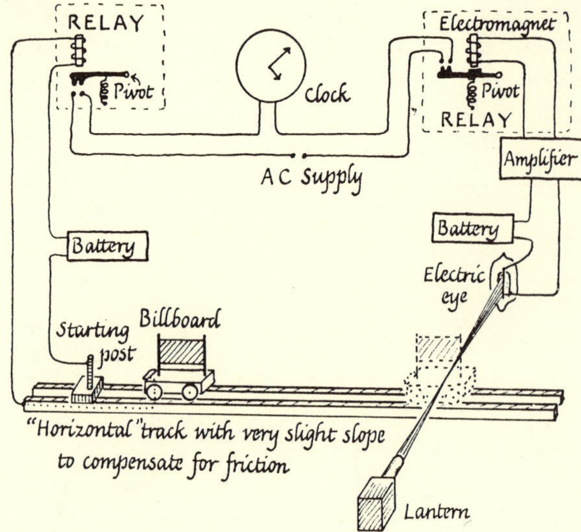

FIG. 7-10b. DEMONSTRATION APPARATUS for showing relations between force, mass, and acceleration. The clock starts when the truck leaves the starting-post, because an electrical connection (running through the rails and truck to insulated starting-post) is broken then. That stops the current through the electromagnet (= relay), which then releases a lever to switch the clock on. When the truck reaches the finishing point, its billboard eclipses the beam of light directed at the electric eye. As long as the electric eye receives light, it produces an electron stream which is amplified by a radio tube and made to run an electromagnet. When the light is cut off, the electromagnet releases the clock switch. So the clock records the time of travel over a measured distance, starting from rest.

Measuring Forces: "Strang-meter"

Free fall is a common example of an accelerated motion; and the pull of the Earth that makes things fall is a common example of a force. Yet it is unwise to use Earth-pull in a first investigation of forces and accelerations, because it leads to serious confusion over the very important concept *mass*. Instead, use a "strang-meter": a good steel spring with pointer and scale, marked to measure the pulling force in "strangs." As above, one strang is an arbitrary unit, the pull that stretches the spring a standard amount, say one inch. To mark the scale for other forces, such as 2, 3, 4, . . . strangs, there is no need to rely on Hooke's Law. Make several more equal springs—tested for equality by hanging the same load on each, or by pulling them all in line.[7] Then use them to pull the strang-meter—one for 1 strang, two pulling the strangmeter in parallel for 2 strangs—and mark the scale. Then three, then four, . . .[8] The scale can be subdivided by interpolation. Then we have a "strang-meter" to measure forces by spring-stretch.

Use the strang-meter to measure the pull that accelerates a small truck along a horizontal track; and measure

[7] The ideal scheme—simple in theory, troublesome in practice—is to adjust the springs to equal acceleration-producing-ability. Try them in turn on the same truck and clip or twist them until they all give the same acceleration at some standard stretch.

[8] Or if you prefer you can make several equal lumps of metal, each pulled by the Earth enough to pull the spring 1 strang; then hang 1, 2, 3, . . . of them on the main spring and mark its stretches on a scale: 1, 2, 3, . . . strangs.

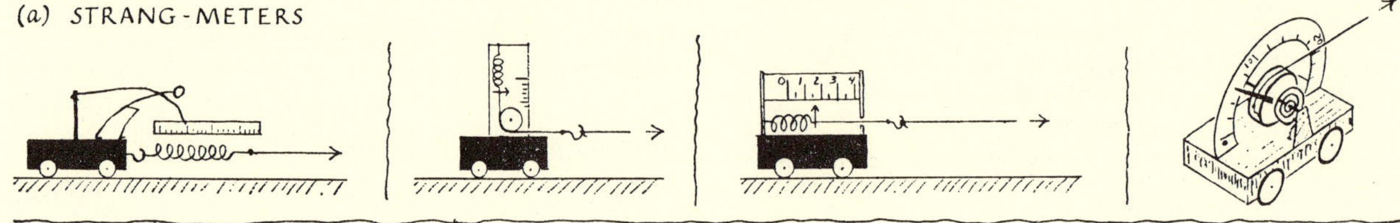

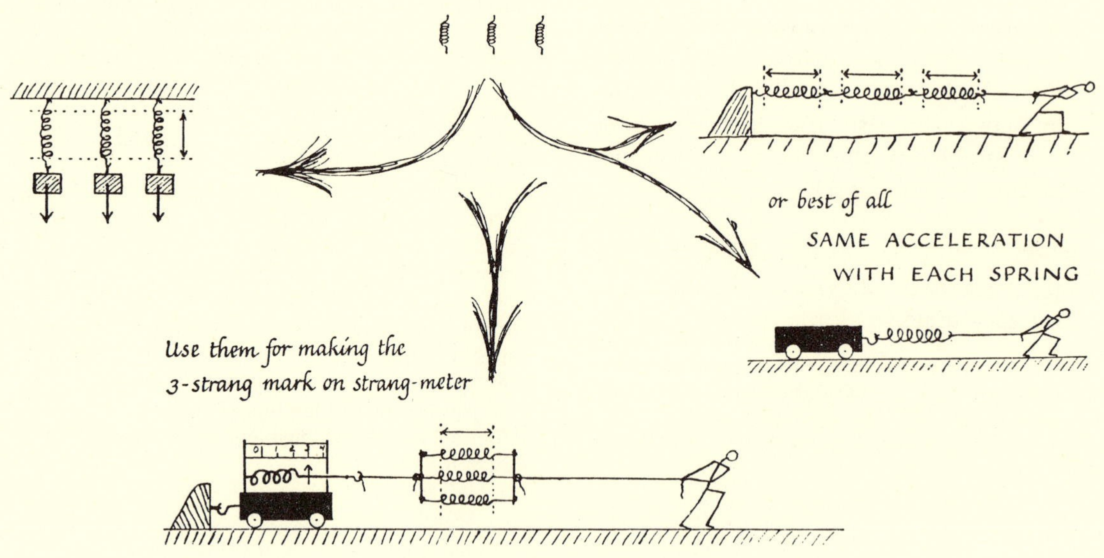

FIG. 7-11a, b. STRANG-METERS

the accelerations. Does double force give double acceleration, &c? It is best to install the strang-meter on the truck itself. Then the thread used to pull the truck is attached to the strangmeter's spring and the pull is measured *while the truck is accelerating.*[9]

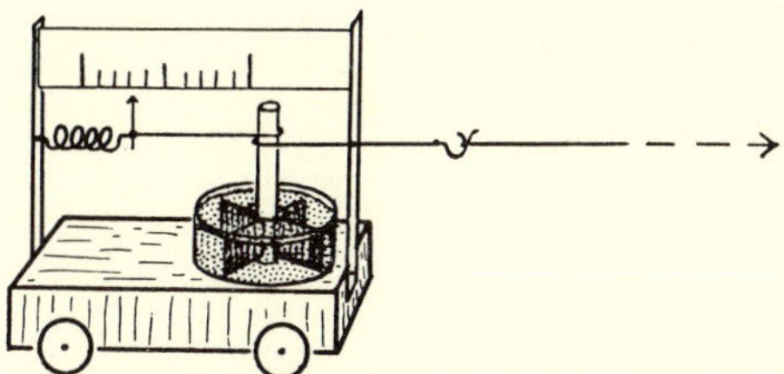

FIG. 7-11c. A GOOD STRANG-METER with paddle in oil to damp out oscillations.

If we measure the accelerating force with a strang-meter on the truck, it does not matter how that pull is applied to the other end of the pulling thread. The strangmeter will tell us how big the pull of the thread is, and whether it remains constant. For convenience, we pull the thread by running it over a pulley wheel and hanging a small load on it. The Earth pulls that load, and that pulls the thread, and that pulls the truck with a force measured by the strang-meter. The load can be a small sandbag, either of arbitrary size or adjusted to give some desired pull. For simple demonstrations we may want pulls that increase in simple proportion, 1 to 2 to 3, . . . Then it is easy and honest to pre-arrange several sandbags to give such pulls in the conditions of the experiment, say 1.2 strangs, 2.4 strangs, 3.6, . . . The fact that the pull of the Earth on the sandbag produces the pull on the thread is irrelevant here. We should get the same results if we employed a trained rabbit to pull the thread in such a way that the strang-meter maintained the same reading.

(In most apparatus for demonstrating FORCE-MASS-ACCELERATION relationships, the truck has no strang-meter—though it is easy to install one. Instead, the force is measured by the Earth-pull on the little load hung on the thread. That Earth-pull *is* the accelerating force, but it accelerates two victims together: the truck *and the hanging load itself.* That makes the reasoning harder: if we want to keep the total moving mass constant, we cannot just add more pulling load when we want to increase the pull. We must take some mass off the truck and add it to the pulling load—"promote a passenger to be the extra locomotive engineer.")

[9] The spring will show unwelcome bouncing vibrations unless some friction is added. Sticky friction of rough surfaces would lead to inaccuracy, but liquid offers the ideal friction to "damp" the vibrations of a measuring machine: its friction-forces increase with velocity but are zero when the liquid is at rest. (Hang a pendulum in liquid and watch the swings die down. The more viscous the liquid the bigger the forces opposing *motion*; yet in every case the pendulum settles down to a vertical position.) Fluid friction never interferes with the final equilibrium. In the strang-meter, the pulling thread should run around an axle that carries a paddle in thick oil.

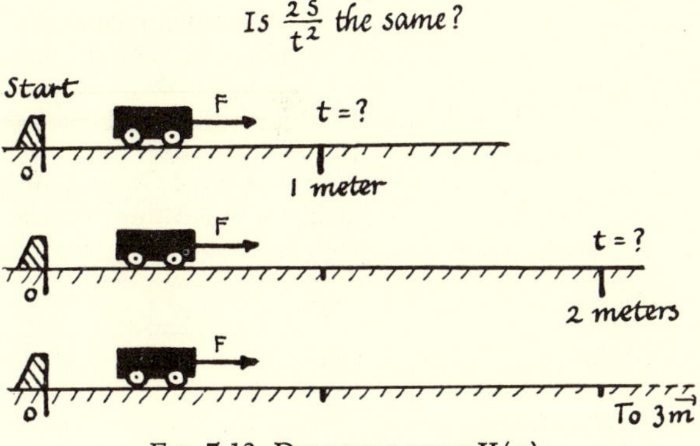

FIG. 7-12. DEMONSTRATION II(a)

Demonstration II (a). *Effect of constant force.* First we want to make sure *that a constant force produces a constant acceleration.* (Remember the laboratory experiment on a rolling wheel.) If not, there is little point in proceeding to experiments with various forces! We use a small load hung on a thread to apply a constant pull, as shown by the strang-meter, throughout the motion. We test for constancy of acceleration by the method used for the rolling wheel: that is, we time the travel from rest for various distances and see whether $2s/t^2$ has a constant value. Table A refers to a particular arrangement of apparatus on a long bench. Only a few specimen measurements are inserted to show how the table would be used. Supplying complete data would be insulting: this experiment is something to see for yourself.

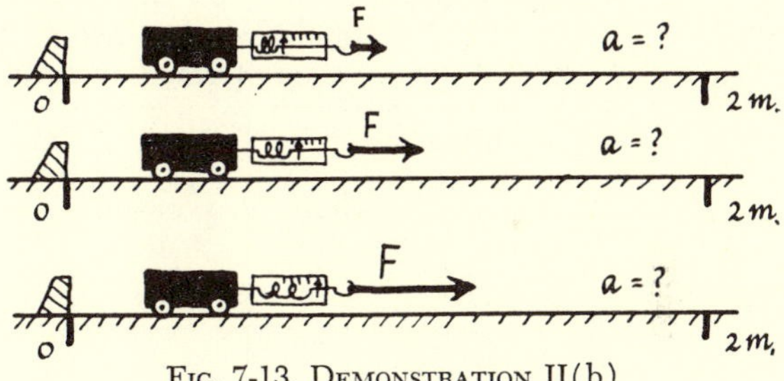

FIG. 7-13. DEMONSTRATION II(b)

Demonstration II(b). *Force and acceleration.* If we are convinced that the acceleration is constant—that is, if the values of $2s/t^2$ agree with each other within the likely errors of the apparatus (including the observer's foibles)—we can thereafter use a single travel-distance, such as 2 meters. We then try several different forces in turn, each measured by the strang-meter, and measure the acceleration by timing as before. Then we ask whether measured ACCELERATIONS are proportional to measured FORCES. (See Table B.)

Demonstration II(c). *Mass.* So far the total amount of moving stuff, the mass of the truck, has been kept constant. According to the usual custom in physical investigations, we have kept all the variables fixed except the two, FORCE and ACCELERATION, whose relation is under investigation. Now we change to different quantities of moving matter, to twice and three times the "MASS." If we wish to think of *mass* as a *quantity of matter to be moved*, we should be able to double the

TABLE A Specimen record of experiment to test whether a constant force (acting on a fixed mass) gives a constant acceleration.

Track tilted slightly to compensate for friction.
Moving mass (truck): 2:00 kilograms Pulling force (constant): 1 strang
Electric eye set up near end of track. Starting post moved to positions which provide total run of 1 meter, 2m., 3m. Each run timed three times.
If acceleration is constant, $2s/t^2$ should be same for all distances; so, $2s/t^2$ is calculated (see note * below).

Measurements				Calculations and Test			
DISTANCE TRAVELLED FROM REST s (meters)	TIME TAKEN (secs)			AVERAGE TIME t (secs)	(TIME)^2 t^2 (sec)^2	ACCELERATION* $\frac{2s}{t^2}$ (meters/sec/sec)	SPARE COLUMN FOR CONCLUSION
3.00	3.50	3.54	3.53	3.52	12.39	$\frac{2 \times 3.00}{(3.52)^2} = \frac{6.00}{12.39} = 0.485$	
2.00							
1.00							

* We are trying to find out whether there is a constant acceleration. The relation $a = 2s/t^2$ assumes there is a constant acceleration, and does not hold true for a changeable acceleration. However, in calculating $2s/t^2$ we are estimating an "average acceleration" for the region of motion concerned, and if we obtain the same value for several different values of travel-distance, s, we infer acceleration is constant.

TABLE B Specimen record of experiment to investigate relation between force and acceleration, for a constant mass.

Track tilted to compensate for friction. Moving mass (truck): 2.00 kilograms
Electric eye arranged to time travel distance, s: 2.00 meters from rest

Measurements		Calculations			Test
PULLING FORCE "strangs"	TIME TAKEN secs	AVERAGE TIME, t secs	(TIME)^2 (sec)^2	$\frac{2s}{t^2}$ m./sec/sec	Spare space for test: acceleration/force
1.20					
2.40					
3.60					

TABLE C Specimen record of test of "guessed" relation between moving mass and force required, for fixed acceleration.

Track tilted to compensate for friction.
Electric eye arranged to time travel of . . . 2.00 meters from rest
Masses chosen in proportions 1 : 2 : 3
Pulling loads adjusted to give pulling forces in same proportions, 1 : 2 : 3 .

Measurements				Test—conclusions
MASS OF TRUCK kg	PULLING FORCE "strangs"	TIME TAKEN secs	AVERAGE TIME secs	
2.00	2.10			
4.00	4.20			
6.00	6.30			

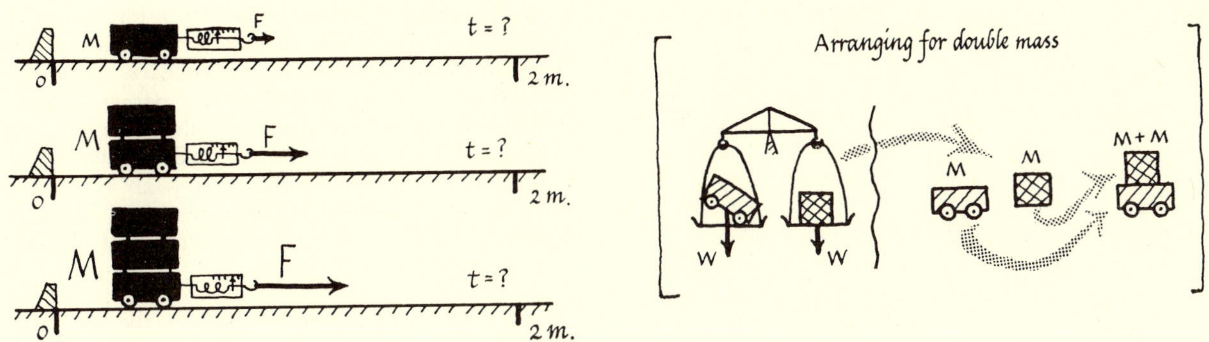

FIG. 7-14. DEMONSTRATION II(c)

mass by putting two identical trucks together and making the puller accelerate them both.

How can we make sure the victims are all identical in mass? For a first attempt, we might just make them all the same size of the same materials. We might imagine a demon assistant checking them by counting atoms—actually a human experimenter can now do that with radioactive tracers and a Geiger counter. However, we really want victims that are the same for *pulling-and-acceleration experiments.* So, having manufactured several that we think are equal, we should test their equality by trying each in turn with the same pull. If they move with the same acceleration we believe they are equal, that is, have equal "masses." Then we *assume* we can make double mass, triple mass, etc., by piling them on top of each other or hooking them together in series.[10]

If we used the same force for masses M, $2M$, and $3M$, we should expect smaller and smaller accelerations. Hoping for a simple rule we might try whether the same force gives accelerations with proportions 1:½:⅓. However, we save trouble by anticipating the result and devising a simpler test. Using different masses, we try to arrange the pulling force to produce the *same acceleration for each mass.* We guess that double and treble masses will need double and treble forces. (Heaven knows, the Myth-and-Symbol experiment shouts that at us.) Then we can word our final test as a crucial question: If we change the moving mass from M to $2M$ to $3M$ and change the pulling force in the same proportions, F to $2F$ to $3F$, *will the acceleration remain the same*? But if the acceleration does remain the same, the times of travel will be the same; so our test is simpler still: *examine the times of travel.* (See Table C.)

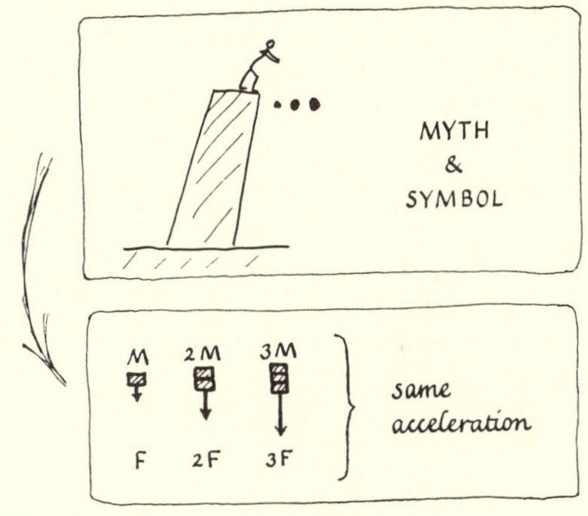

FIG. 7-15

[10] *Notes on Masses in II(c).*

Arranging for double and treble mass. Making several equal trucks and piling them together is inconvenient in a lecture demonstration. Instead, we can make a truck of double mass by loading the original truck with an equal mass of metal, found by "weighing." We find the load of metal that will balance the empty truck on weighing scales. Then we know that gravity pulls with equal forces on load and truck. We also know that, falling freely with gravity pulling alone, both would fall with the same acceleration. Therefore the *same force* gives the *same acceleration* to both. Therefore the masses of load and truck are equal—that is our definition of equality of mass (see Fig. 7-14). However this takes it for granted that gravitational mass and inertial mass are equal, or at least proportional.

Allowing for the inertia of the truck's wheels. As the truck rolls along, its wheels rotate and add some *extra* motion of their rims to the motion of general progress. This demands a little *extra* accelerating force, as if the truck had a little extra mass. You will meet the effects of this "rotational inertia" elsewhere in the course. If you accept the idea on trust we can use it to improve the test of II(c). We shave a small mass off the truck, to allow for the extra-mass-behavior of the wheels. For a simple test with truck masses M, $2M$, $3M$, we take the same "wheel-allowance" off the truck each time—and in subsequent discussion pretend we have not lost it, since it resides in the wheels. This "wheel allowance" can be calculated from measurements of the wheels: or it can be estimated by trial and error. In the latter method, we use up two measurements of the main experiment for trial and error, leaving only a single test of the main question, in the third measurement.

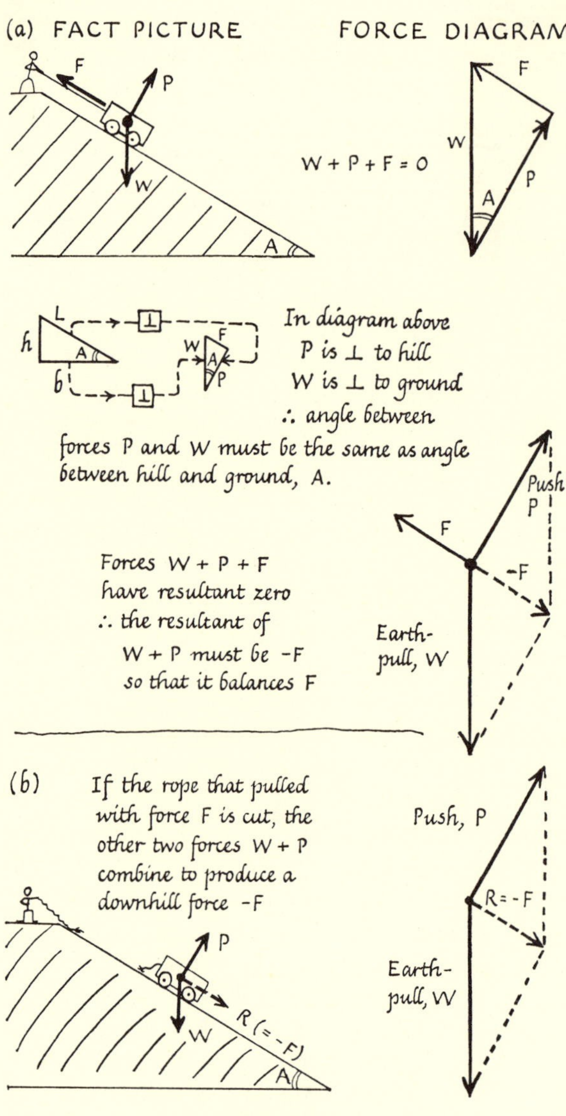

Fig. 7-16
(a) Object held at rest on a hill.
(b) Object on hill is released.

Galileo's original approach to Newton's Law II. Motion down a Hill

If you measured the motion of a wheel rolling down rails in laboratory, you have already seen that *diluted gravity* gives constant acceleration. Galileo, in his teaching of motion, made great use of sloping hills ("inclined planes") to dilute gravity. If you are interested in this early approach to modern mechanics, see the account below, or consult Galileo's book "Two New Sciences."[11]

Shortly before Galileo, Stevin showed that if a body remains at rest on a frictionless hill, held by an uphill rope exerting a pull F, the pull is given by:

$$\frac{\text{UPHILL FORCE, } F}{\text{EARTH-PULL } (= \text{WEIGHT}),\ W} = \frac{\text{HEIGHT OF HILL, } h}{\text{SLANT LENGTH OF HILL, } L}$$

The hill pushes out on the body with a force P (given by $P/W = [\text{BASE}, b]/L$). If there is no friction, P must be perpendicular to the hill, as we have assumed in drawing our triangle. If P were *not* perpendicular to the hill it would have a component along the hill, dragging the body uphill or downhill, and this would be a friction drag.[12] With real hills, there is always friction, ready to oppose motion; but here we have chosen the ideal case of a frictionless hill, which must therefore push straight out.

Push P and pull F together balance the Earth-pull, W. If the rope is then cut and the body accelerates downhill, we may imagine that the other forces, Earth-pull W, and the hill's out-push, P, are unchanged. Then, if W and P balanced F before, they must themselves have resultant $-F$, a *downhill* force of size F. Thus we may think of the freely sliding body being accelerated by a downhill pull F such that $F/W = h/L$ or $F = W \cdot (h/L)$. The fraction h/L is constant, the same all the way down the hill. For any given hill, F is therefore the same all the way down; and experiments such as the rolling wheel investigation show that this *constant force produces a constant downhill acceleration.*[13] If the slope is changed, the downhill force is changed, and so is the acceleration.

Galileo considered motion down hills of different slopes, and decided that the acceleration of a rolling ball[14] varies directly as the fraction h/L.[15] Then:

DOWNHILL ACCELERATION = (constant) $\cdot h/L$
AND, DOWNHILL FORCE F = (EARTH-PULL W) $\cdot h/L$

So in this case the acceleration increases in the same proportion as the resultant force. Thus Galileo provided a clear basis for the general rule

$$\text{ACCELERATION} \propto \text{RESULTANT FORCE}$$

that Newton incorporated in his Law II. This was a discovery of great importance, grasped at by people before Galileo, but not formulated clearly. It is the key relation between *force* and *motion*, needed in dealing with projectiles, planets, electrons, rockets, trains, machinery . . .

---

[11] Translated into English by H. Crew and A. de Salvio, Northwestern University Press. (Now in Dover paperback edition.)

[12] If this seems strange, use your own curved forehead as a hill, and press your finger on it. Your forehead pushes the finger straight out, except for friction which you can feel. Try imagining this with no friction.

[13] If you wish to be a very cautious natural philosopher you should avoid the prejudicial word "produces." All we really know is that force and acceleration go together. In many cases, we have no separate evidence that there are forces acting; we merely guess there is a force because we see the acceleration.

[14] The *rolling* motion of a ball introduces a complication which we have slurred over, so we investigate this important matter of force and acceleration with a "frictionless" slider or a railroad truck and track. There the truck moves straight along, and only its wheels roll.

General Relationship

From many demonstrations, ranging from rough timings quoted by Galileo to indirect evidence from astronomy and modern ballistics, we can produce a general relationship as in the preliminary summary.

With a constant resultant force acting on it, a body moves with constant acceleration. If the force is doubled or tripled the acceleration is increased in the same proportion:

ACCELERATION $\propto$ FORCE, if the mass is kept the same or FORCE $\propto$ ACCELERATION.

Again, for the same acceleration, double and triple masses require double and triple forces:

FORCE $\propto$ MASS, if the acceleration is kept the same.

These can be combined[16] in

FORCE $\propto$ MASS · ACCELERATION
or $F = K \cdot M \cdot a$

The relation $F = K \cdot M \cdot a$, which says that:

RESULTANT PULLING FORCE
= (constant) · MASS · ACCELERATION

is a useful summary of the behavior of accelerating objects. Our demonstrations do not prove that it is right; but they illustrate it and add their mite of testimony. $F = K \cdot M \cdot a$ is our version of Newton's Second Law of Motion, which we shall state formally later. We use that relation for calculating real forces: the push of the floor when we land from a jump, the force on a car in a smash, the pressure of gas on its container, the pull of the Earth on the Moon. We shall first comment on mass and force (and weight) and then show how we can reduce $F = K \cdot M \cdot a$ to a simpler form for useful calculations.

Mass and Force

We have talked glibly of mass, but did we ever define it in a clear scientific manner? Suppose you consider you know what force is—a familiar push or pull—and agree to assume that two equal springs side by side will pull with twice the force of one; then you can say that you know about force, F, and acceleration, a, in $F = K \cdot M \cdot a$. You can then describe *mass* as something proportional to F/a, since $F/a = K \cdot M$. The bigger the mass the bigger the force needed to give it some chosen acceleration. And the bigger the mass the smaller the acceleration a standard force will give it.

We have put masses together on the assumption that separate chunks add up to the total mass, that mass can be measured by the number of equal trucks lumped together; and this fits with the behavior summed up in $F \propto M \cdot a$. Mass, then is additive, and shows the difficulty-of-acceleration of a body. Mass is the price, in force, of unit acceleration, much as ordinary price is the price, in money, of some kind of goods. Newton said that mass means quantity of matter, and in further explanation he made use of density and volume without really

[15] Suppose we have several frictionless hills, *all of the same height h*, but with different slopes, and start a body sliding from rest at the top of each. Then *if* the accelerations are proportional to the values of fraction h/L we can predict that *all the sliders arrive at the bottom with the same velocity, v.* For uniformly accelerated motion from rest $v^2 = 2as$ (see Ch. 1, App. A): and if $a = C \cdot (h/L)$, where C is a constant, $v^2 = 2as = 2 \cdot C \cdot (h/L) \cdot$ (DISTANCE L) $= 2Ch$, which is the same for all the hills. Conversely, if v is the same for all hills of the same height, h, accelerations must be proportional to h/L. Galileo felt convinced from his general knowledge of mechanical behavior, and from a clever experimental test, that the "same velocity" property is true; and he used that as a starting-point for many discussions of accelerated motion.

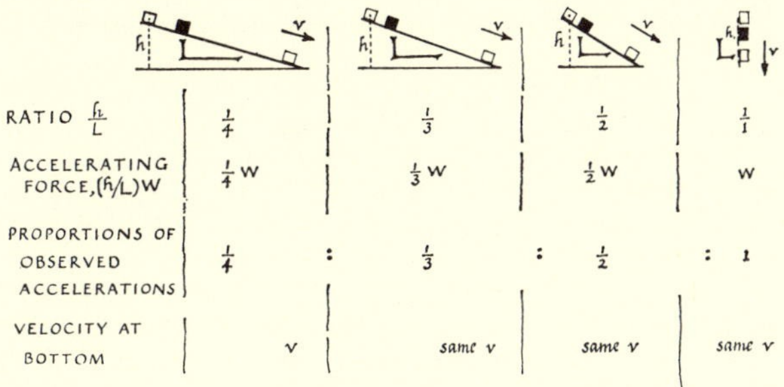

FIG. 7-17.

[16] It is not easy to see straight away that $F \propto a$ and $F \propto M$ combine into the statement $F = K \cdot M \cdot a$. Remember that the first two statements contain some unwritten conditions. The first really says: "*$F \propto a$, when the mass, M, is kept constant.*" But when M is constant we can rewrite the more general statement $F = K \cdot M \cdot a$. Thus:

$$F = (K \cdot M) \cdot a = (\text{constant}) \cdot a \text{ or } F \propto a$$

So the statement $F = K \cdot M \cdot a$ contains "$F \propto a$, if M is constant."

Again, the second statement says $F \propto M$ when acceleration a is kept the same. But when a remains constant we can rewrite the combined statement $F = K \cdot M \cdot a$ like this:

$$F = (K \cdot a) \cdot M = (\text{constant}) \cdot M \text{ or } F \propto M.$$

So $F = K \cdot M \cdot a$ does contain $F \propto a$ and $F \propto M$, subject to the conditions implied.

$F = K \cdot M \cdot a$

$F = K \cdot M \cdot a$ → $F = K_1 \cdot a$ If M is constant, K_1 is constant. → $F \propto a$

$F = K \cdot a \cdot M$ → $F = K_2 \cdot M$ If a is constant, K_2 is constant. → $F \propto M$

Fig. 7-18. Ideas of Force, Mass, and Acceleration

Fig. 7-18a. Force is Familiar.
We feel we know about forces—in fact, we *feel* forces, though we should find it difficult to measure them by muscular sensations or by pain.

Fig. 7-18b. Acceleration Seems Obvious Enough
We can measure it with a speedometer + a clock.

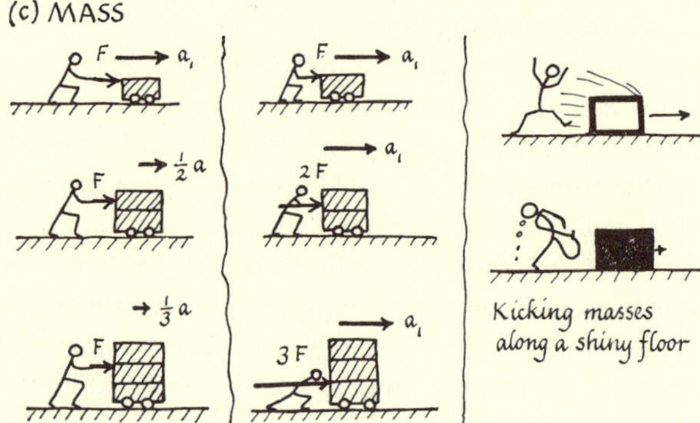

Fig. 7-18c. Mass is a Less Familiar Idea. To understand it we had better watch how it is *used* in physics—in fact, we might say the idea was invented for such uses.

helping things further. If you want a feeling for the nature of mass, some slang descriptions may help, such as "unaccelerability," "difficultness-of-getting-goingishness"; but these must not be offered as scientific definitions! Scientifically, we can say definitely: MASSES are proportional to values of the fraction

$$\frac{\text{RESULTANT FORCE PRODUCING ACCELERATION}}{\text{ACCELERATION PRODUCED}}$$

(And we shall soon choose our force unit to make MASS *equal* to F/a.) We have described force as something that stretches springs or something that can be obtained by hanging loads on strings. And yet in many moving systems we cannot see a spring stretching or feel gravity pulling, and we only guess there is a force because we observe an acceleration. In more advanced physics, we even describe force as something producing acceleration, and we measure force by the acceleration it gives to a standard mass. Then we are perilously near to arguing in a circle.[17] How much of $F = \text{K} \cdot M \cdot a$ is definition of force and mass, and how much is experimental? This is a difficult question. At least $F = \text{K} \cdot M \cdot a$ fits with experiment, and we can use it to predict real events.

Units of Mass: the Kilogram and the Pound

The *pound* was chosen long ago as a standard unit of quantity of matter, and it is our unit of mass on the FOOT-POUND-SECOND system. The *kilogram* was devised for similar use and it is our unit of mass on the METER-KILOGRAM-SECOND system. The standard pound and standard kilogram are carefully preserved in the form of lumps of untarnishable metal.

Weight, a Local Force

Instead of pulling a truck with a spring to accelerate it, we can use a cord that runs over a pulley and has a load hung on the other end. Then the accelerating force is due to the *weight* of that load, instead of being due to changes of forces between the atoms of a spring or our muscles. Again, when a body falls freely, the accelerating force is its whole *weight.* Unfortunately, "weight" is used with several meanings, and that confuses this part of physics. Therefore we shall develop the scientific meaning of "weight" carefully, and meanwhile we shall often use the name "pull-of-the-Earth" or 'Earth-pull" instead.

In science, *weight* is the official name for the force that seems to pull things towards the ground—the "pull of gravity," whatever that may mean. We may "explain" weight by saying that it is the pull of the Earth, if we like—and the fact that weight always pulls towards the center of the Earth makes this seem sensible—but we have no assurance that this explanation is right until we study universal gravitation[18] (Ch. 23).

Whatever it is called, WEIGHT *is a force.* It is just like any other force, except for two peculiarities: it is *vertical* and *unavoidable.*

[17] For our use, saying that MASS is FORCE/ACCELERATION makes the unfamiliar thing, mass, easier to think about; but professional physicists usually reverse the logic. They say that mass is an obvious (!) stodginess of stuff—calling it "inertia" is mere naming—and they define FORCE as MASS · ACCELERATION. Of course this does not remove logical doubts like those in our scheme; it only shifts them. Either viewpoint is satisfactory for working purposes; but a mixture of them would be horribly illogical.

[18] A fish might say that bubbles rise because of buoyant forces due to Earth-repulsions.

To measure a body's weight directly, we should use a spring-balance marked in force units.[19] Since that is often inconvenient, we compare one weight with another by a lever balance; that is, we find the ratio

$$\frac{\text{EARTH-PULL on X}}{\text{EARTH-PULL on standard mass}}$$

For example, suppose a body X is pulled three times as much as a standard kilogram. Then we say the EARTH-PULL on X is 3 *"kilograms-weight,"* meaning "*3 times the Earth-pull on a kilogram of stuff.*" Unfortunately, this leads to confusing units for *weight* (and other *forces*) because we shorten them to "kilograms," which are proper units for *masses.* It would indeed be unwise to use the same name for units of such different things as force and mass. We shall return to this question of units for forces.

If we measure the weight of any object very precisely, with a spring balance, and carry the apparatus to different places, we find slight variations from one region on the Earth to another; and we now know we should find the weight much smaller far away from the Earth, or far down inside it. Thus, a "kilogram-weight" is not only a confusing unit for weight (and other forces) but a variable one. We try to avoid variable units in science; so we have devised a better, standard, unit for *all* forces (including weights). Before discussing that, we shall continue our study of mass.

"Mass NEVER Changes"

We may imagine ourselves repeating the demonstration experiments on $F \propto Ma$ with trucks and strangmeters on the Moon. We suspect that gravity is much weaker there, so a given sandbag would pull with less weight. But, with the strangmeter spring stretched *to the same mark* (by a bigger sandbag), the FORCE would be the same as ever. Would the MASS of the truck be the same on the Moon? Scientists brooding on this question have long believed that MASS would stay unchanged. Even at the center of the Earth—where gravity, pulling in every direction, would have zero resultant pull—an object would still have the same mass. Nowadays we have evidence in the light from stars that, if atomic forces are the same in those remote fields, atomic masses are the same as on Earth.

[19] Or we may do an acceleration experiment: we let the Earth pull the object unopposed. It falls freely with acceleration g. Then we apply $F = Ma$, as in the discussion later in this chapter.

We do have a feeling of something solid and definite about matter, something that remains constant whatever we do to an object: heat it, melt it, compress it, . . . , even carry it to the Moon. A lump of

FIG. 7-19. MASS AND WEIGHT IN VARIOUS PLACES
Mass, estimated by the difficulty of accelerating a small truck, will be the same everywhere, on Earth, in Earth, on Moon. Weight, estimated by the stretch of a spring-balance (and by muscular sense of force in the arm of the man holding the balance) will seem much less on the Moon, practically zero at center of Earth.

lead on roller skates would be just as difficult to accelerate on the Moon, or at the Earth's center. On the other hand, its *weight,* the downward pull on it, would be quite different (Fig. 7-19).

A massive wheel balanced on frictionless bearings is not pulled around either way by its weight, but when we push its rim to make it spin we become aware of its mass; and we expect it to be just as difficult to spin if taken to the Moon or anywhere else. A pound of chocolate if eaten provides other things besides a sense of heaviness due to its Earth-pull. It provides bulk and nourishment; and, if our health were unchanged, we should expect it to provide these equally on the Moon. Even if we set up a laboratory in a freely falling box we should find masses unchanged—but should we notice things being pulled by the Earth as usual?

In formulating the idea of mass by vague descriptions such as quantity of matter, "unaccelerability," "stodginess of stuff," etc., or by the clear-looking definition

$$\text{MASS} = \text{FORCE}/\text{ACCELERATION},$$

we think we are describing something universal, *an unchangeable property of all kinds of matter, something that lasts as long as matter itself.*

Mass and Weight

How big are the Earth-pulls on different masses—how do the weights of things compare? Take two equal lumps of lead, say 1 kilogram each. The Earth pulls each with the same force—the weight of 1 kg. If we put them together to make 2 kg, the vertical pulls simply add: the Earth pulls 2 kg twice as much as 1 kg. We get the same double pull if we melt them into one lump or place one on top of the other. *For any one material*, the pulls of gravity simply add, and there is no absorption, or shielding of one piece of matter by another.[20]

For any one material, WEIGHT $\propto$ MASS. So we think of the Earth as having a "gravitational field" radiating out along vertical lines, ready to pull on every piece of matter. The field exerts the same pull on every kilogram of, say, lead. But how about its pulls on equal masses of *different materials*—such as 1 kg of lead and 1 kg of aluminum? The answer—in fact the meaning of the question—depends on what we mean by equal masses. Comparing masses of two objects by acceleration experiments (e.g., with truck and track) is a clumsy business; but it can be done, and we could then compare the weights of those masses with a spring balance. However you know quite well that the easy way to compare masses, in science and in commerce, is simply to use a seesaw balance. That method uses the *forces* pulling the two loads down, and is quite rightly called "weighing." But if we obtain equal masses of, say, lead and aluminum by weighing we are already *assuming* that equal masses have equal weights. No further force-measuring experiment can answer our question about mass and weight—there seems to be a threat of circular argument. In fact, wc are talking of two utterly different kinds of mass: inertial mass and gravitational mass. The distinction carries the essential hint of General Relativity; yet for most of the time from Newton to Einstein it seemed unimportant—went almost unrecognized—and that made the physics of mass, motion, force, weight, and gravity more confusing in elementary studies. We shall discuss the two kinds of mass briefly, giving them distinctive labels M^* and $M^\dagger$.

The Two Kinds of Mass

Inertial mass. The constant M in $F = \mathrm{K} \cdot M \cdot a$ is *inertial mass.* Experiments with trucks being accelerated by springs show M as the "stodginess of stuff," the unaccelerability of the body concerned, measurable by its F/a. This mass is a measure of inertia, the tendency of all mechanical systems to oppose change. We call it "inertial mass," and label it M^*. If we restrict ourselves to a single chemical element we can compare one M^* with another or with the standard 1-kg*, by counting atoms. (We *can* count atoms nowadays, but even the fastest Geiger counter working day and night would take billions of years to count one kilogram directly in atoms.) More realistically, we can compare masses* by the way M^* was defined, by acceleration-and-force experiments. (E.g. we use some standard force such as the pull of a standard spring to pull, on a "frictionless horizontal" track, as in Fig. 7-23:

(a) an empty truck, unknown mass $[M_0^*]$
(b) truck + standard 1-kg*, $[M_0^* + 1^*]$
(c) truck + mass M^* to be measured, $[M_0^* + M^*]$

We measure the acceleration in each case, (a), (b), and (c); apply $F = \mathrm{K} \cdot M^* \cdot a$ to each case; and use algebra to obtain $M^*/1$ kg*, which is M^* in kilograms.[21] This is a tedious method, seldom carried out, except in imagination to clear up the meaning of mass*. We shall describe a more practical device, the "wig-wag machine," but that too is only good for demonstrations of principle. So far in our discussion inertial mass has shown no direct connection with gravity. Mass* is something that we believe would be the same anywhere, near the Earth, at the Moon, far out in space, at the Earth's center. What is its connection with gravity; and what does weighing really do?

Gravitational mass. Quite separately from inertial mass we can develop the idea of a gravitational mass,

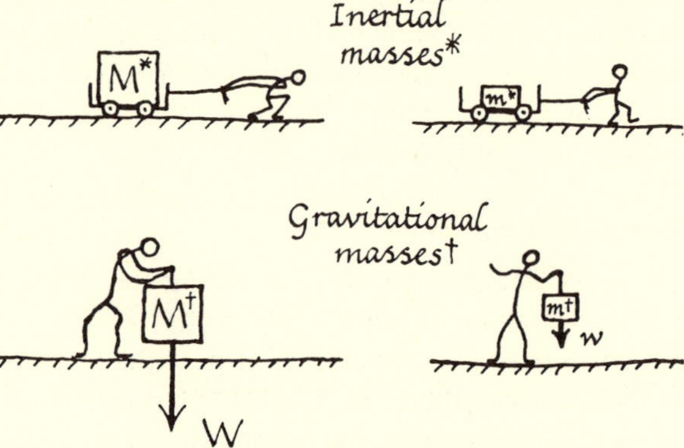

FIG. 7-20. THE TWO KINDS OF MASS

[20] No method of shielding or cutting off gravity has ever been found, and we do not expect to find any—gravitational pulls stretch right through any wall. In this, gravitational fields are unlike the other two common fields of physics. Magnetic fields are partially shielded by iron; and electric fields completely stopped by a closed metal box.

[21] Note that this is not really measuring mass* absolutely—we cannot set some machine going and have an absolute number emerge for M as in counting rabbits, or atoms, or electrons. We simply choose our standard kilogram and somehow count "how many kilograms" which is measuring a *ratio*,

[MASS OF UNKNOWN, X]/[MASS OF UNIT, 1 kg] or X/1

Yet we call that an absolute mass since we say M is in kilograms, in contrast with a comparison between any two masses, such as

MASS OF X/MASS OF Y = 2

Compare this with the statements, "A's age is 40 years"; "B's age is twice C's age." The first says "A's age/1 year = 40"; the second says "B's age/C's age = 2." We may call the first an absolute measurement because it uses a standard unit, while the second is called a relative measurement. In a sense, both measurements are comparisons—*every* measurement is a process of deciding how many steps of one item fit into the other.

the quantity of matter pulled on by a gravitational field. We assume that the Earth offers the same field to all comers, but we assign different masses to different victims in proportion to the pulls on them. This kind of mass is gravitational mass, $M^{\dagger}$. We say that different objects have different weights because they have different masses† to be pulled on by any gravitational field. Thus gravitational masses† are by definition proportional to weights. The gravitational mass of an object determines how hard the Earth will pull on it. Later, we shall find (as Newton's Law III suggests) that gravitation is symmetrical: if the Earth pulls a stone, the stone pulls the Earth equally. Then a body's gravitational mass† also determines how strongly it will attract another body, the Earth. Thus gravitational mass measures the amount of stuff to be pulled by gravity, and also the amount of stuff for pulling gravitationally. We might say a body's $M^{\dagger}$ measures its "size" for gravitational *interaction* with other bodies. (As you will see in Ch. 23, every piece of matter pulls, and is pulled by, every other piece; but only the Earth has a big enough $M^{\dagger}$ to make a noticeable pull on small objects around us.)

When we compare bodies by weighing we are comparing their gravitational masses. (When two objects balance on an equal-armed weighing scale, we know their *gravitational* masses are equal, but that observation alone does not tell us whether their *inertial* masses are equal.)

The connection between gravitational mass and inertial mass. Gravity pulls twice as hard on two equal lead lumps as on one. For lumps of lead, gravitational masses must be proportional to inertial masses, since both are clearly proportional to the numbers of atoms of lead. The same applies to lumps of any other material we choose, say wax. But how do a lead lump and a wax lump compare with each other in this respect?

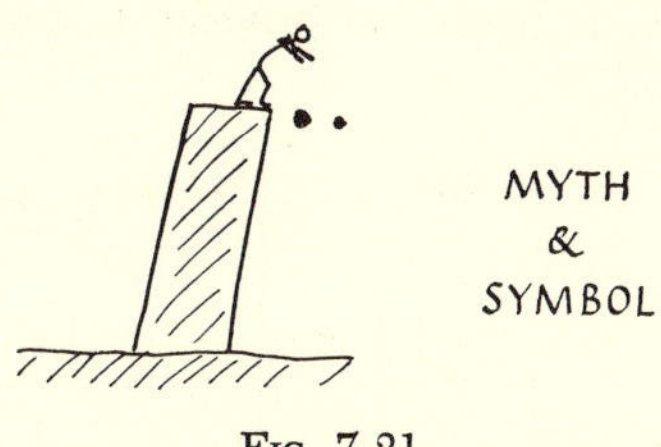

FIG. 7-21.

This question cannot be answered by common sense or reasoning alone. It is answered by the Myth-and-Symbol experiment, which covers the whole gamut of sizes and materials. Drop two lumps of any sizes and any materials, and they fall with the same acceleration, g. The accelerating force on each is its WEIGHT, the pull of the Earth on it. We know that WEIGHTS are proportional to GRAVITATIONAL MASSES—that is the definition of gravitational mass†. But since WEIGHTS which are Earth-pulls give all objects the same acceleration, g, WEIGHTS must be proportional to INERTIAL MASSES. Therefore the two kinds of mass keep the same proportions for any variety of bodies. If we start with a standard kilogram as a unit for both, GRAVITATIONAL MASS and INERTIAL MASS must be equal, for any body of any size, any material, in any place.

Here is the argument with algebra: release two bodies A and B to fall freely. Each falls with acceleration g. The accelerating force on each is its WEIGHT, W. Apply $F = K \cdot M \cdot a$ to each falling body. Then F is the Earth-pull, W; M is the body's INERTIAL MASS M^*; and a is the acceleration of free fall, g. Then $F = K \cdot M \cdot a$ gives:

for body A, $\quad W_A = K \cdot M_A^* \cdot g$

for body B, $\quad W_B = K \cdot M_B^* \cdot g$

Dividing, we cancel the general constant K and use the Myth-and-Symbol experiment's assurance that g is the same for different bodies

$$\therefore \quad \frac{W_A}{W_B} = \frac{M_A^*}{M_B^*}$$

But $\quad \dfrac{W_A}{W_B} = \dfrac{M_A^{\dagger}}{M_B^{\dagger}} \quad$ by the definition of $M^{\dagger}$

$$\therefore \quad \frac{M_A^{\dagger}}{M_B^{\dagger}} = \frac{M_A^*}{M_B^*}$$

Gravitational masses have the same proportion, for A and B, as inertial masses.

Or we can say $\quad \dfrac{M_A^{\dagger}}{M_A^*} = \dfrac{M_B^{\dagger}}{M_B^*}$

Therefore the ratio GRAVITATIONAL MASS/INERTIAL MASS is the same for A and B, and all other bodies. If we choose 1 kilogram as unit for both kinds of mass, the ratio becomes 1, and GRAVITATIONAL MASS = INERTIAL MASS for all bodies.[22]

The Surprising Identity

This is the surprising property of nature shown by the Myth-and-Symbol experiment, that the two kinds of mass are the same for all pieces of matter. It is surprising because we describe the two masses so differently: one measures the body's stodginess, its inertia for velocity-changes; and the other measures the body as both receiver and giver of gravitational pulls. If you take a casual view, you may say, "Oh that's obvious: both properties simply go by the amount of matter there." If you take the inquiring view of an Einstein you may say "If they are equal, if no experiment shows a difference, must not our framework of nature be such that we *cannot* tell them apart? And in that case are we wise even to talk about two kinds of mass as if they could be distinguished?" That would lead to Einstein's treatment of a gravitational field (involving $M^{\dagger}$) as *equivalent* to an accelerating observer (involving M^*);

[22] If you find this a long hair-splitting argument, look at a short concrete form: compare a standard kilogram of platinum with a rock of unknown mass. We compare their inertial masses by an acceleration experiment—dragging each in turn along a horizontal track. Suppose we find the rock has a mass of 5.31 kg. Gravity is not involved in that comparison. Then we compare their gravitational masses by measuring the gravitational attraction between each of them and some standard object—the Earth is easiest. We can do that by just weighing them. We shall find the rock's gravitational mass too is 5.31 kg.

thus General Relativity describes space-&-time in a way that makes M^* and $M^\dagger$ necessarily the same.

Since the Myth-and-Symbol experiment's result is important, we need a far more precise test than simple observation of falling bodies in air. We want to lump together thousands of falls in a single comparison. And we would like to remove the trouble of air resistance. This is the problem mentioned in Ch. 1, with the promise that you would meet a solution. There is a method that is simple and very precise. You will recognize it when you meet it. (It does not involve vacuum pumps or electronic clocks—though we now expect such methods to supersede the simple test within the next few years.) Newton knew it, and used it as a Myth-and-Symbol test for such diverse things as lead, gold, sand, salt, wood, water, and even wheat. Early in this century, J. J. Thomson and others used it for a further test, on the $M^\dagger$ *and* M^* of what we now call nuclear energy. There was even then a suspicion that energy as well as matter has inertial mass. Does it have equal gravitational mass? Radioactive atoms were known to release a great deal of energy when they break up; so they must contain a store of releasable energy with, probably, an appreciable inertial mass. The experimenters repeated Newton's test, comparing radioactive samples with ordinary materials: result, same *g*.

A Simpler Treatment of Weight and Mass

Since the upshot of our discussion seems to be that the two kinds of mass have the same value, we shall omit the distinguishing marks, forget the difference, and call mass M.

Now we can start afresh and give a quick, careless discussion of weight and mass. The Myth-and-Symbol experiment tells us that g is the same for any freely falling bodies A and B. Their WEIGHTS,

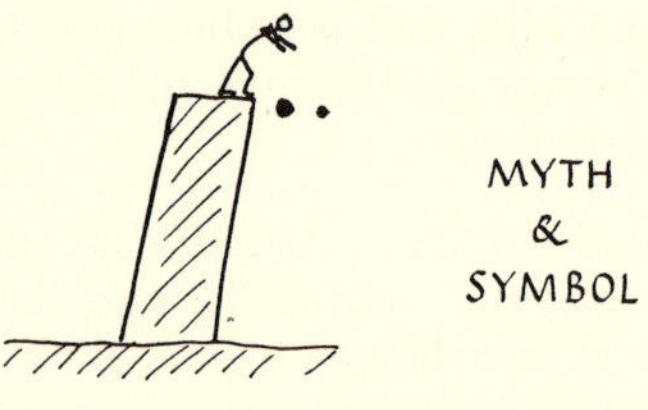

FIG. 7-22.

W_A and W_B, pull on their MASSES, M_A and M_B, and give each an acceleration g.

Apply $F = K \cdot M \cdot a$ to each

$$\therefore \quad W_A = K \cdot M_A \cdot g$$

and

$$W_B = K \cdot M_A \cdot g$$

$$\therefore \quad \frac{W_A}{W_B} = \frac{M_A}{M_B}$$

Therefore we can compare masses by weighing. That is what we do in practice; we compare or balance the FORCES W_A and W_B, and say we are comparing the MASSES M_A and M_B. (We already offered carelessly to do that in arranging masses M, $2M$, $3M$ by weighing for the lecture demonstration.)

Measuring Masses by Weighing

Therefore we can compare masses by weighing. Spring balances and weighing scales, which deal with forces, are much easier to use than trucks on tracks. So all precise mass measurements are made by weighing—and our assurance of the "Conservation of Mass" is based on precise weighing.

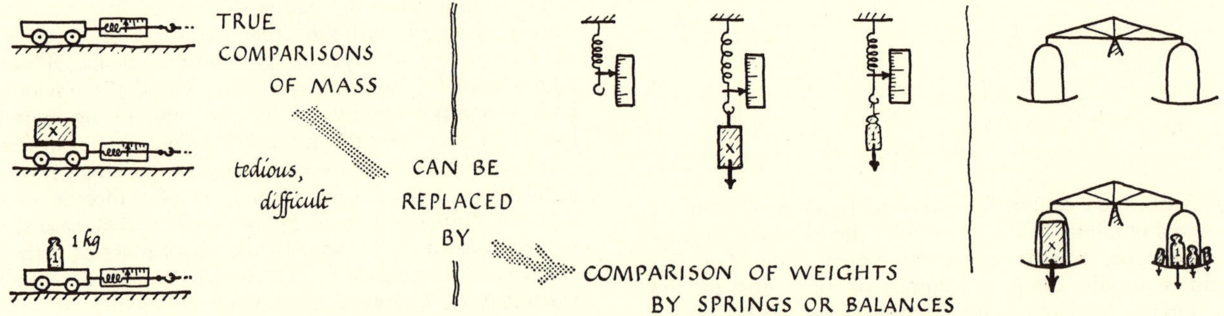

FIG. 7-23. COMPARING INERTIAL MASSES: THE TRUE WAY AND THE EASY WAY

Truck and track acceleration measurements give a true comparison of mass X with standard mass. (A third measurement is necessary to eliminate the unknown mass of the truck, etc.)
Spring-balances compare the Earth-pulls, weight of X with weight of standard kilogram. Since the two measurements are done in the same locality, with same "g," the Myth-&-Symbol experiment vouches for this as an indirect comparison of masses.
A common balance can be used to weigh the Earth-pull on X against Earth-pull on standard kilograms and fractions. This too is a direct comparison of weights; and the Myth-&-Symbol experiment vouches for it as an indirect comparison of masses.
(NOTE. FIG. 7-25 shows how masses can be compared by acceleration experiments using a pulling load without a strang-meter. The argument is then more difficult, because the pulling load's mass must be included in the total moving mass.)

However the fact that $W_A/W_B = M_A/M_B$ by no means makes MASS and WEIGHT similar things, any more than, in dealing with milk, COST and VOLUME are the same things just because $C_A/C_B = V_A/V_B$.

Conservation of Mass

The development of Chemistry, which came surprisingly late after the development of Newtonian mechanics, was helped by the idea of unchanging total mass. Chemical changes swap ingredient atoms to-and-fro, but there is no change of total mass. This was tested by increasingly skilful weighings, the later ones in miniature chemical laboratories sealed in glass. Not even the most accurate of these experiments in the past century could possibly detect the minute mass carried away, as we now believe, by the heat-energy evolved in some of those chemical reactions. Thus, we have long trusted Conservation of Mass, the idea that something measuring the total amount of matter stays constant in all changes of motion and through all chemical transformations. It was not until this century that this view proved to be too narrow. However difficult to define, MASS grew to seem simple and real to those who worked with it. Physicists built the mechanics of motion on the assumption that mass is a constant property of matter, that mass is conserved. Chemists tested its conservation with increasing rigor and then trusted it for further development of chemical knowledge. In civil life as well as in science and engineering, we still take Conservation of Mass for granted.

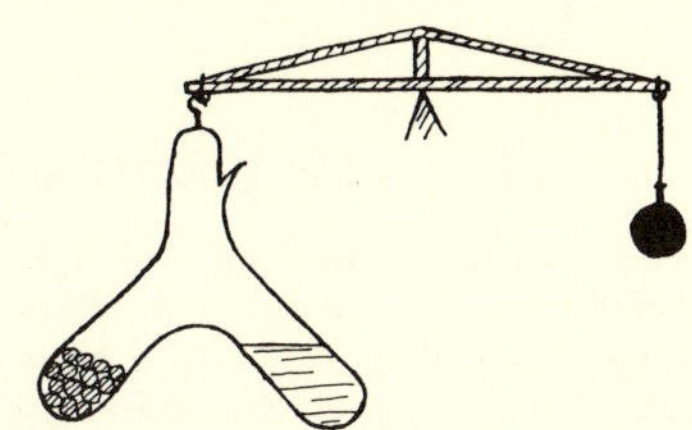

FIG. 7-24. MINIATURE CHEMICAL LABORATORY
The apparatus is counterpoised in a sensitive balance. The tube is tilted and the ingredients react chemically. When the apparatus has returned to room temperature the counterpoise is tried again.

In the last century, Conservation of Energy (Ch. 26 and Ch. 29) emerged with similar growing assurance: first the idea of energy, then guesses, then rigorous tests; finally a blazing into complete assurance as the evidence converged. It is only in this century that we have realised fully that energy itself has mass, so that the two great conservation laws can be combined into one law of enormous importance and universal scope.

The Lunatic Fringe of Scientific Vocabulary

We say we are "weighing" things when we are really aiming at comparing masses. This is a true term since most mass comparisons are done by weighing; but it is confusing to people who are learning the nature of mass. More unfortunately, standard masses (lumps of metal) are always called "weights" in science, commerce, and common life. This is a wrong naming that aids confusion, but we must follow established custom and use it. Worst of all, we say, "The man weighs 100 kilograms," when we mean his MASS is 100 kg.

Constant Mass, Changing Weight

The kilogram in our "box of weights" is a standard mass which is universal. A kilogram is the same everywhere, as a *mass*, though its weight would be far less on the Moon, and would dwindle to nothing if carried to the center of the Earth.

"Mass NEVER changes" is our working rule. Relativity mechanics makes us whisper, "Well, hardly ever," but the changes are only noticeable when matter gains stupendous speeds or makes stupendous changes of energy. At all ordinary speeds, from snails' to rockets', mass keeps its same value everywhere. Hence scientists' liking for MASS as a property to deal with.

Problem 6, on a later page, may seem difficult, but working through it should clarify your understanding of mass and weight.

Comparing Masses Directly: The Wig-Wag Machine

Suppose we wish to measure or compare inertial masses directly, instead of using the indirect, but precise, method of weighing. Then we must do acceleration experiments such as those sketched in Fig. 7-23 or those in Fig. 7-25. These measure mass

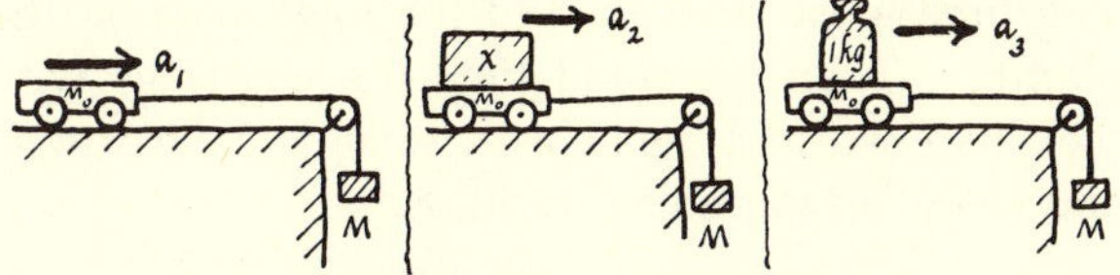

FIG. 7-25. COMPARING MASSES DIRECTLY

The *same* force is applied to:

EMPTY TRUCK $F = K \cdot (M_0 + M) \cdot a_1$

TRUCK + UNKNOWN $F = K \cdot (M_0 + M + X) \cdot a_2$

TRUCK + 1 KILOGRAM $F = K \cdot (M_0 + M + 1) \cdot a_3$

Measure the acceleration in the three cases and use $F = K \cdot M \cdot a$ and algebra to find the ratio (X kilograms)/(1 kilogram).

(Note: the force F, the masses M_0 and M of truck and puller, need not be known; nor need the value of constant K.)

properly in terms of the definition MASS $= F/a$. The *same* force is applied to

[empty truck],
[truck + unknown],
[truck + 1 kg].

A gadget that compares masses truly is shown in Fig. 7-26. It is far less accurate than simple weighing and serves only to teach the idea of true mass-measurement. The shove-meter (or better, the unshovability-meter) has a platform P supported by two stout strips of heavy clockspring, S, S. When pushed sideways and released, P oscillates to-and-fro with a wig-wag motion. When masses are placed on P, they too must be given acceleration by the same springs. The bigger the mass, the smaller the acceleration, the slower the wig-wags. To compare an unknown mass with a set of standards, we must find the collection of standard masses that will replace the unknown on P and give the same wig-wag time. Or we might use interpolation. Or we might hire a mathematician to analyze the motion and work out (with $F = \text{K} \cdot M \cdot a$) how the wig-wag time must be related to the total mass on the platform. Then we could time the wig-wags with the unknown and again with some standard mass and calculate the unknown mass.

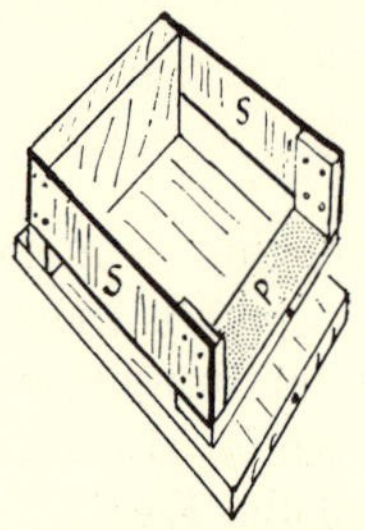

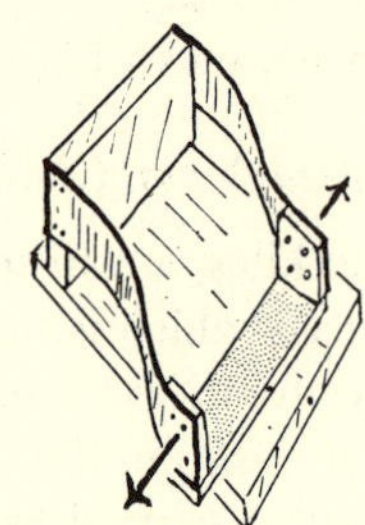

FIG. 7-26. THE WIG-WAG MACHINE
The platform P is attached to the main frame by two strips of steel spring, S, S. When the platform is moved to one side and released it moves to-and-fro with a wig-wag motion, whose cycles are easily timed. With extra masses placed on the platform, the wig-wag cycle takes more time.

★ PROBLEM 2. MATHEMATICAL ATTACK ON THE WIG-WAG MACHINE. (HARD; but worth trying for the sake of Problem 3.)

(i) By staring hard at the general relation $F = \text{K}\cdot M \cdot a$ and the special relation $s = \frac{1}{2} at^2$ for uniformly accelerated motion (which this is not), try to guess the kind of relationship between T, the time of one wig-wag, and M, the total mass of platform and load (i.e., guess, with some reasoning, whether $T \propto M$ or $T \propto M^2$, or what). Assume some form of Hooke's Law relating F and s.

(*Hint*: Though neither force nor acceleration is constant, the same kind of relationships may apply. For a given distance, s, moved to one side, the force, F, exerted by the bent springs is the same whether the platform is loaded or not. In fact, F/s is some kind of Hooke's-Law constant for the springs. What do the springs care? If they are bent they push, whatever there is on the platform.)

(ii) If you guess the relation, say how you would use it to compare an unknown mass of lead with a standard 1 kilogram mass. Remember that the platform, P, itself has some mass (unknown) which you must allow for somehow.

Notice that gravity plays no part in measurements with this apparatus; they are true mass-comparisons. This is only a crude demonstration experiment, but the corresponding measurement with atoms vibrating in a molecule can be very revealing.

★ PROBLEM 3. ATOMIC MASS COMPARISONS

Spectroscopists, examining the glow of light from electrically excited molecules, can measure the time of vibration (wig-wag) of a hydrogen atom tied into a massive chemical molecule. If you solved the problem above, guess how the vibration time would be changed by substituting an atom of heavy-hydrogen (twice the mass) for the hydrogen atom, if the spring-forces holding it remain the same. Essentially, when heavy hydrogen, now so important in nuclear research, was discovered this gave an early test of its mass. Fig. 7-27 shows the results of one such experiment. *Do they agree with your guess above?* The experiment was performed on methane gas, CH_4.

A methane molecule has 4 hydrogen atoms at the 4 corners of a symmetrical pyramid. They are arranged symmetrically in space around a carbon atom.

In one of the molecule's modes of vibration all 4 hydrogen atoms move OUT... IN... OUT... &c The data refer to this motion.

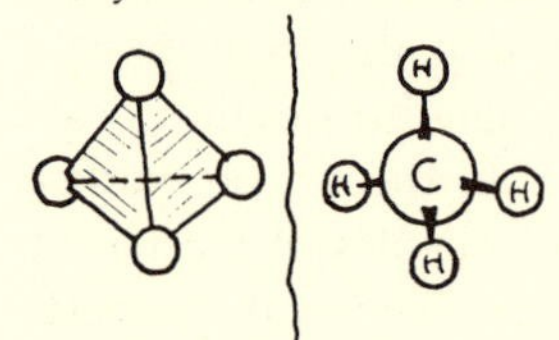

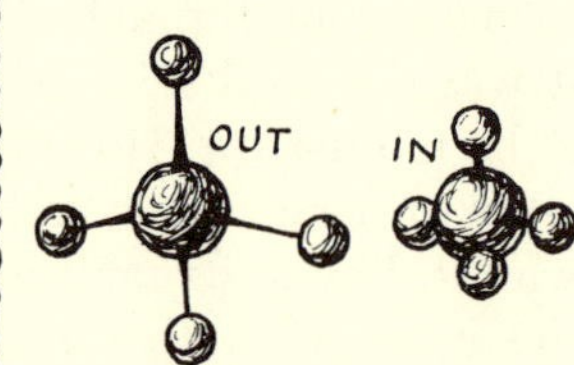

Time of vibration of hydrogen atom in a molecule of methane, CH_4 or H-C-H (with H above and below C) is 0.0000000000000114 sec, or 1.14×10^{-14} second.

Time of vibration of heavy-hydrogen * atom in the corresponding methane molecule, CH_4 or **H-C-H** (with **H** above and below C) is 0.0000000000000160 sec, or 1.60×10^{-14} second.

* Here **H** is used for heavy-hydrogen atom, instead of the usual D for "deuterium."

FIG. 7-27. A METHANE MOLECULE VIBRATING

PROBLEM 4. MASS AND WEIGHT

A physicist moving house packs some of his belongings in empty grocery cartons, all alike. After packing several such boxes with books and several with pillows, he discovers he has forgotten to label them and wishes to find out which is which. He can distinguish them by two methods:

(i) he bends over and tries lifting each box;
(ii) he kicks each box, to shove it along a very smooth floor.
  (a) What is he comparing, in (i), values of masses or weights?
  (b) What is he comparing in (ii)?
  (c) Give a short reason for your answers to (a) and (b).
  (d) If he repeats (ii) on a very rough floor, what comparison is he making?

Simpler Version of $F = K \cdot M \cdot a$; Absolute Units for Force

We can make $F = K \cdot M \cdot a$ look simpler by forcing the constant K to take the value 1.0000 so that $F = M \cdot a$. We do this by choosing a special unit for force.[23] We need a unit for force anyway, since we regard the standard kilogram and pound as unchanging units of the unchanging thing mass. We need a good unit for force, a universal one that will not pull with different strengths in different localities, as a "weight" unit would. In our demonstrations we used a home-made (arbitrary) unit, a strang, but we must now define a standard unit.

As long as we write $F = K \cdot M \cdot a$, we can choose any units we like for F and M and a, and make the constant K take the value needed to keep the relation true to real life.[24] However, if we fix the value of K, choosing to make $K = 1$, we cannot take just any units for F and M and a. We can choose the units for two of these, and our choice of $K = 1$ will settle the unit for the third. We choose kilograms for M, meters/second per second for a, and then find that $F = M \cdot a$ defines a unit of force for us. Pretend we have done this and find what size of unit we get. In $F = K \cdot M \cdot a$, put $K = 1$ (our choice) and suppose we are giving 1 kilogram an acceleration of 1 meter/sec^2. Then $M = 1$ and $a = 1$ and the FORCE, F is given by $K \cdot M \cdot a = 1 \cdot 1 \cdot 1 = 1$. Here is FORCE of value 1, unit force. We call this unit one *newton* (a mere dictionary matter, but a very suitable name). We see that 1 newton is the FORCE that will give a MASS of 1 kilogram an ACCELERATION of 1 meter/sec^2. One *poundal* is the similar unit for 1 pound mass accelerating 1 foot/sec^2. These force units are universal. Wherever we take our apparatus, the one-kilogram mass is the same, and wherever we give it a 1 meter/sec^2 acceleration the force will be just the same standard size, as shown by any spring-balance or strangmeter. The units newton and poundal are called *absolute units* of force.

"Good" and "Bad" Units for Force

We call these absolute units, the newton and the poundal, "good" units because they are universally constant; and we shall use them for all forces, including weights, in $F = M \cdot a$ calculations. We shall call kilogram-wt. and pound-wt. "bad" units because their size depends on locality.

LOCATION	SIZE OF FORCE UNIT: *Value of 1 newton*	SIZE OF FORCE UNIT: *Value of 1 kilogram-weight*
Surface of Earth, at equator	1 newton (= 1 kg m./sec^2)	Earth-pull 9.78 newtons
Surface of Earth, at North Pole	same	Earth-pull 9.83 newtons
4000 miles above surface of Earth	same	Earth-pull 2.45 newtons
Surface of Moon	same	Moon-pull 1.6 newtons (see note)
Center of Earth	same	Earth-pull 0 (see note)

Note: There are also *very* small pulls, due to Earth and Sun at the Moon, and due to Moon and Sun at the Earth; but the effect of these is not noticeable. See Ch. 31.

The kg-wt. unit of force depends on what there is to pull on the kilogram and give it weight. The variations over the surface of the Earth are small enough to be neglected in engineering. So we find engineers using kg-wt. (and likewise pounds-wt.) in their planning.[25] In this course we shall refer to

[23] We can make $K = 1$ by a choice of units, just as scientists of Napoleon's time made the density of water = 1 by choosing the size of the gram to make it so. They decided to define the gram as the mass of one cubic centimeter of water. So they tried, not quite successfully, to make the standard kilogram out of enough metal to balance 1000 cubic centimeters of water. If successful, this would make the density of water exactly 1.000 gram/cubic centimeter. Notice that the density is not plain 1 but 1 gram/cu. cm. Our K is not really plain 1, but 1 (newton)/(kg · meter/sec^2). However, that is seldom mentioned.

[24] For example, experiment shows that a force equal to the pull of the Earth on 2 tons when pushing 1000 kilograms gives an acceleration about 730 inches/sec^2. If we wish to use these uncouth units, we must adjust K to the proper value to keep $F = K \cdot M \cdot a$ true. Then 2 tons-weight = K · (1000 kg) (730 inches/sec^2). Therefore K must be given the value 2/730000. The relation $F = (2/730000) \cdot M \cdot a$ is true for the data above, and we expect it to hold for any other set of measurements in the same crazy units.

[25] Until recently, many engineers found that their major problems involved the *weights* of objects near the Earth's surface; so the units they chose—pounds or kilograms for forces—were good ones for the work in hand. Meanwhile,

these "engineering" units, kg-wt. and pounds-wt., as "bad" units of force, because:

(i) they are not universally constant, as newtons and poundals are.
(ii) they do not fit in $F = Ma$, but require a modification, $F = (W/g) \cdot a$, which leads to confusion.
(iii) they make it very hard to form a clear concept of mass.

How Big is a Newton?
"Absolute" Units and "Bad" Units

To see how the new "absolute" unit compares with the "bad" unit, a kilogram-weight, do an imaginary $F = M \cdot a$ experiment. Let one kilogram of matter be pulled by its own weight (1 kilogram-weight) unopposed: that is, let it fall. Measure the acceleration. It is about 9.80 meters/sec². Then the force on it, given by $F = M \cdot a$, is (1 kilogram) (9.80 meters/sec²) or 9.80 newtons.[26] Therefore, a force of 1 kilogram-weight is about 9.80 newtons, in New York. (At the North Pole 1 kg-wt. is about 9.83 newtons. It is the kilogram-weight that is different, because the Earth pulls harder there; the standard newton is unchanged.)

The same conversion-rate applies to any weight. A 10-kilogram rock falls with acceleration 9.8 meters/sec², so the force pulling it (its weight) must be (10 kg)(9.8 meters/sec²) or 98 newtons. The weight of a mass M kilograms of matter is $9.80 \cdot M$ newtons.

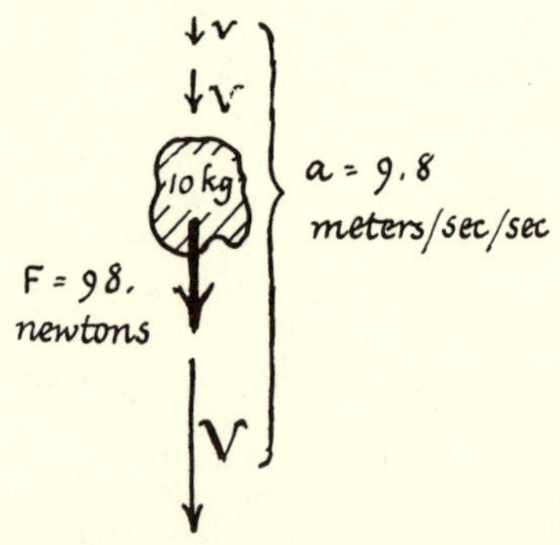

Fig. 7-28. How Big is a Newton?

Conversely, a force (9.80 M) newtons is M kilograms-weight. Therefore

$$1 \text{ newton} = \frac{1}{9.80} \text{ local kg-wt.} \approx \frac{1}{10} \text{kg-wt.}$$

$$\approx \frac{1}{4} \text{ local pound-wt.}$$

So if you put a ¼-pound block of butter on your hand you will feel a force of about one newton pressing down on it.

We often have to change between "bad" units, such as kilograms-weight, and "good" ones such as newtons. This is because we often apply forces by hanging loads on things; so what we know is the mass of the load in kilograms. Then the force it applies is the pull of the Earth on it (its weight). To use that force in $F = M \cdot a$, we need to change so many kilograms-weight into newtons. Again, we are used to judging forces by weights; so in common life, as in engineering, we speak of a force of so many kilograms or so many pounds. Weighing-scales are primarily force-measurers, but are gradu-

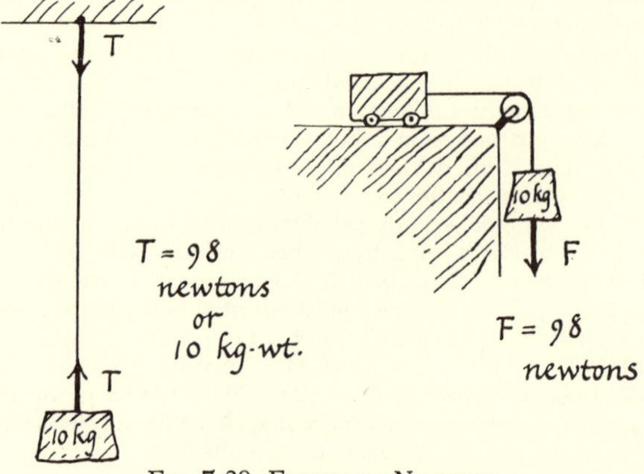

Fig. 7-29. Forces in Newtons

physicists, prying into atoms, found *masses* all-important in their problems, while weights were trivial for atoms and misleading for astronomy; so they chose $F = Ma$, with a pound and a kilogram as units of mass. In the present age, engineering problems have taken a new turn: space-travel runs into regions where g has different values, and nuclear engineering deals with atomic particles whose masses, and even changes of mass, are important. The new engineering is merging with the new physics in a treatment that uses absolute units for force and makes an understanding of mass an important aim.

[26] Change of units: It looks as if we had made a sudden switch of units here. Actually we have only made a change of name, from kg · meters/sec² to the shorter name newtons. When we use $F = M \cdot a$, we express F in newtons, and therefore have some newtons on the left-hand side of the equation; but on the right we have kilograms multiplied by meters/sec². Therefore we must regard *newtons* as the same as *kilogram · meters/sec²*. Or, in particular, if $F = 1$ and $M = 1$ and $a = 1$, we have

$$1 \text{ newton} = 1 \text{ kilogram} \cdot 1 \text{ meter/sec}^2$$

and this tells us the relation between the units, 1 newton = 1 kg · meter/sec². This relation comes from the simple fact that one newton will give one kilogram an acceleration of 1 meter/sec²—it comes from the definition of a newton, in fact. (Notice that in all this we have forgotten the constant K = 1 concealed in $F = M \cdot a$. If we assign units to K as suggested in an earlier note, the relation among units becomes more self-evident but clumsier, and we have to remember to insert units for K every time. In this course we shall forget K and treat a newton as a name for a kilogram · meter/sec².)

ated in kg or pounds. As long as we are dealing with forces in equilibrium (e.g., in problems on levers, cranes, pulleys, etc.), we can keep them in "bad" units, since we are only concerned with ratios. Even so, as a reminder that they are *force* units, we should write them kg-wt. (= kilograms-weight) to distinguish them from plain kg properly used for masses. But *in any use of* $F = M \cdot a$ *we MUST use force units which fit with it; that is, ABSOLUTE UNITS, such as newtons* or *poundals which make* $K = 1$ *in* "the expression" $F = K \cdot M \cdot a$.

Engineering Units

What we have called "bad" units, e.g., kg-wt., are usually called "engineering" units. Many an engineer uses them with gusto and faith, finding simple problems easy to solve with them. But the confusion from using the same units for force and mass makes it all the harder for him to understand mass. Even in practical calculations he is therefore liable to lose a factor 9.8 and end in disaster with an answer nearly 10 times too big or too small. (More often, using pounds, he dreads errors of 32.)

Advice to Students who have used W/g Systems

The engineering system disguises its defects by using W/g (=WEIGHT/ACCELERATION of gravity) instead of M. This makes very elementary problems easy to work, but in more advanced physics it looks uncouth and aids confusion of thought. The factor $1/g$ in $F = (W/g)(a)$ is the proper value of K for forces in engineering units; and the use of W instead of M is an attempt to make an honest equation of it. It is honest, but confusing. MASS, so important in dealing with motion, energy, and atoms, fails to emerge as a clear concept.

If you have used this "W/g" system before, *you are strongly advised to make a fresh start with absolute units.*

> ". . . Peace, you mumbling fool!
> Utter your gravity o'er a gossip's bowl;
> For here we need it not."
>
> —*Romeo and Juliet*

Gravitational Field Strength

Without really saying what we think gravity is, we may express the experimental connection between weight and (gravitational) mass by imagining that there is a "field of force" extending out from the Earth like a set of clutching tentacles waiting to pull down on any mass placed there. The field is not itself a force but a state of waiting-to-pull-on-mass.

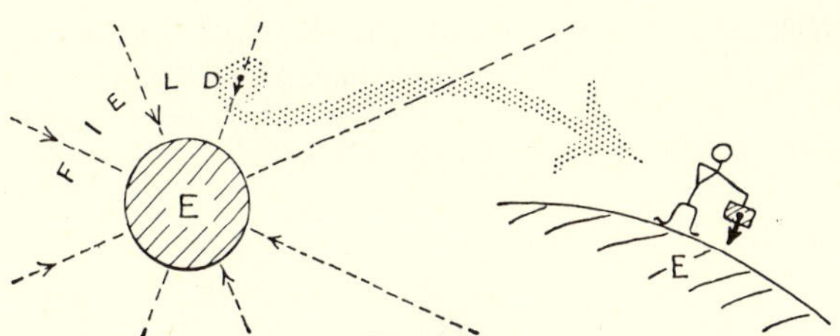

FIG. 7-30. THE EARTH'S GRAVITATIONAL FIELD

If we place a mass of 1 kg near the Earth the pull is 9.80 newtons. The pull on 10 kg is 98 newtons; and on a mass of M kilograms, (M) (9.8) newtons.

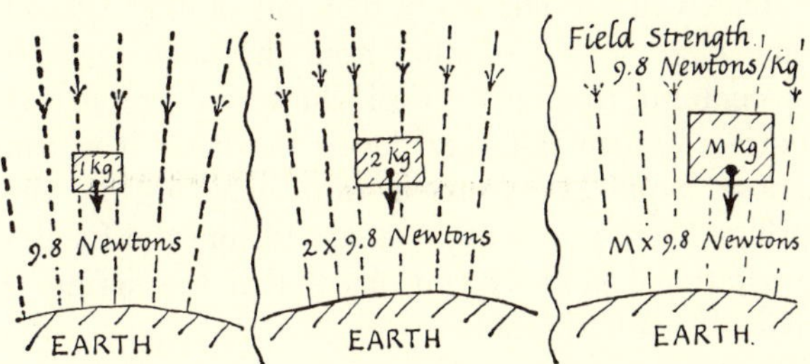

FIG. 7-31. THE EARTH'S GRAVITATIONAL FIELD pulls with forces proportional to masses. Its "strength" is the pull it is ready to exert on each kilogram.

So we say the *"strength" of the field is 9.8 newtons per kilogram.* We picture it waiting to pull on any lump of matter to the tune of 9.8 newtons on each kilogram of the lump. This field strength,

9.8 newtons/kg,

provides a very useful way of dealing with problems which involve weights. The actual number 9.8 comes from treating the measurement of free-fall acceleration as an $F = M \cdot a$ experiment; but in using the factor you should think of it as a field strength of 9.8 newtons/kg, not as an acceleration of 9.80 m./sec/sec. To find the *force* with which gravity pulls on any mass, multiply mass (in kg) by field strength (9.8 newtons/kg near the Earth).

In any problem involving $F = M \cdot a$ *(or other relations derived from it), the forces must be in absolute units, newtons* or *poundals,* and if any of them are provided in the form of weights (e.g., in kg-wt.), *you must first use the field strength (9.8 newtons/kg) to find the value of that weight in absolute units.*

Conversely, if you use $F = M \cdot a$, or any of its family, to calculate a force, the answer will emerge in absolute units; and, if you want to know how big a mass would be pulled with that force by the Earth's gravity, you must make use of the field strength.

PROBLEM 5. EARTH'S FIELD STRENGTH IN FPS SYSTEM

The Earth's gravitational field strength is 9.8 newtons/kg in the system based on meters, kilograms, seconds. Give its value *and units* in the system using feet, pounds and seconds.

Personal Experiment

Try to "feel" the pull of the Earth's gravitational field by raising and lowering a heavy book. With your eyes open you know you are just pulling against the book's weight. With your eyes shut, pretend you are not doing that but pulling against some huge spring running from the book towards the center of the Earth. "Feel" the spring stretch and contract as you raise and lower the book. You can almost hypnotize yourself into believing in the spring. Then, in a sense, you are feeling the Earth's gravitational field. If you object that the spring is invisible, so is the field.

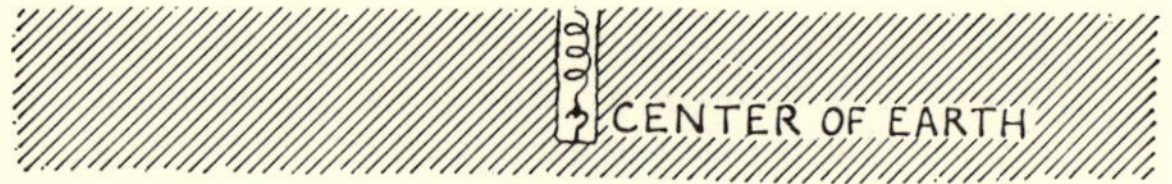

Fig. 7-32

PROBLEM 6. MASS AND WEIGHT, ON MOON AND EARTH

Suppose that in some future age "Lunar Research Corporation" sets up laboratories on the Moon and on the Earth, maintaining a rocket service between them. They use standard kilograms, which are universally the same, sending some to the Moon and using some on the Earth. They acquire a block of ice, trim it till its mass is just 10.0 kg and use it as a truck on a level frictionless table, for acceleration experiments, on the Earth and on the Moon. The block of ice does not chip or melt, when sent to and fro by rocket—it remains a constant 10-kg block. See Fig. 7-33.

(a) When the block is on the Moon, is its MASS the same as on the Earth?

The experimenters have a spring balance, A, marked to read *newtons* (by preliminary experiments with a track and truck system). They use this to pull the block along the table with a horizontal pull of *4 newtons*.

(b) In the Earth laboratory, with 4 newtons pull, what acceleration will the block have? (Explain briefly.)

(c) In the Moon laboratory, with 4 newtons pull, what acceleration will the same block have? (Explain briefly.)

The experimenters obtain an unmarked spring balance, B, and graduate it in "kilograms-weight" of Earth-pull, by hanging universal kilograms on it, in the Earth laboratory.

They obtain another unmarked spring balance, C, send it to the Moon, and graduate it there in "kilograms-weight" of Moon-pull, by hanging universal kilograms on it, in the laboratory on the Moon, where the gravitational field is much weaker.

(d) In the Earth laboratory they pull the same block of ice with spring balance B, (calibrated by kilograms in Earth laboratory). If the spring balance reads 2.0, what is the acceleration of the block? Explain briefly.

CALIBRATION OF BALANCES

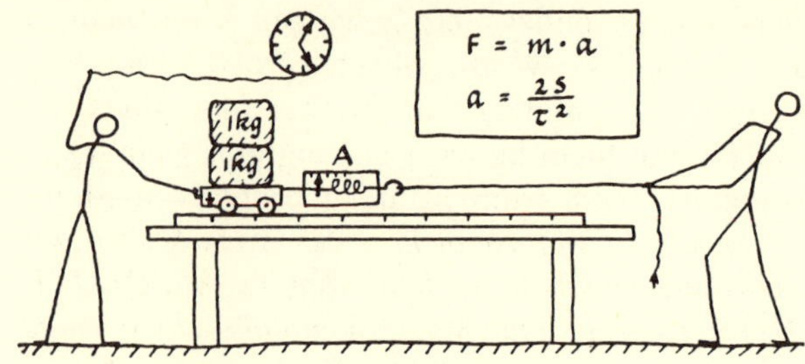

Balance A is calibrated in newtons

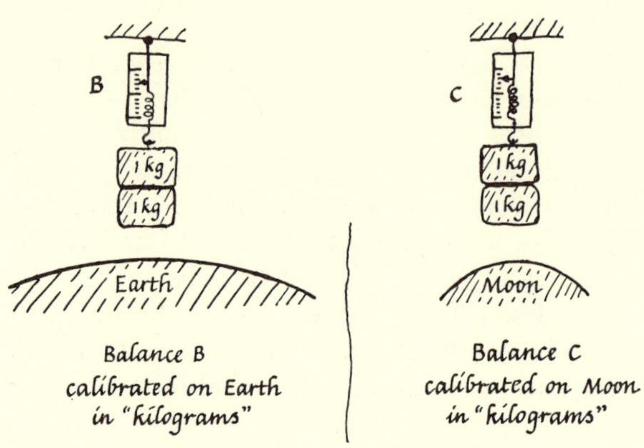

MAIN EXPERIMENTS (a) to (e)

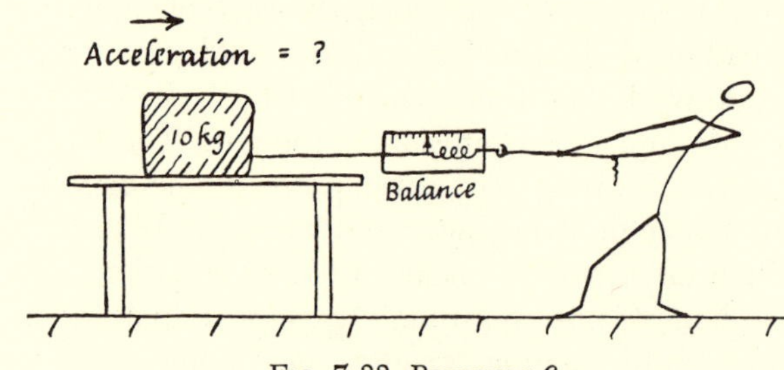

Fig. 7-33. Problem 6

(e) In the Moon laboratory they pull the same block of ice with the same spring balance B (calibrated by kilograms in the Earth laboratory, and then sent by rocket to the Moon). If the spring balance B reads 2.0, is the acceleration of the block greater, the same or less than in (d)? Give a clear reason for your answer.

(f) In the Moon laboratory, they pull the same block of ice with spring balance C (calibrated by hanging kilograms on it in Moon laboratory). If the spring balance C reads 2.0, how does the acceleration of the block compare with that in (e)—is it the same, or greater, or less? Give a clear reason for your answer.

Problems on Force and Motion

$F = M \cdot a$ will lead us into discussions of momentum and energy which we shall need in dealing with planets and atoms. Problems on $F = M \cdot a$ are apt to seem artificial and dull, but they provide useful practice in thinking about forces. Try Problems 7, 8, 9, 10.

"PROBLEM SHEETS"

The problems here are intended to be answered on typewritten copies of these sheets. Work through the problems on the enlarged copies, filling in the blanks, (———), that are left for answers.

PROBLEMS USING $F = M \cdot a$ NAME ____________

Remember that whenever you use $F = M \cdot a$, M should be in pounds or kilograms but force, F, must be in poundals or in newtons.

If some of the forces are weights (= Earth-pulls on objects), these are forces just like the rest and they too must be expressed in poundals or in newtons.

If you use $F = M \cdot a$ to calculate any force, remember it will automatically deliver the answer in poundals or in newtons.

PROBLEM 7. A car moving 40 ft/sec (about 36 miles/hr) crashes into a wall and comes to rest with final velocity zero. The whole collision takes 0.10 sec.

Calculate the collision-force involved, by answering the following:

(i) The final velocity is . . . . ____________ ft/sec

the initial velocity is . . . . ____________ ft/sec

∴ the change of velocity is . . . . ____________ ft/sec

the time taken for this change is . . . . . . ____________ seconds

∴ the acceleration is . . . . ____________ ft/sec/sec

(This acceleration is just the average value during the time taken by the stopping process. We have no guarantee there is constant (negative) acceleration; but this calculation gives the effective average value which we need for calculating the effective average force.)

NOTE: We get this result more quickly if, treating acceleration as constant, we substitute in the standard relation $v = v_o + at$. We were really doing this above.

(ii) To calculate the force involved we must know the mass of the car.

Suppose we are given the mass of the car as 3200 pounds.

Using the acceleration above in $F = M \cdot a$ we find the

force acting on the car during stopping must be ____________

Since we have used $F = M \cdot a$, the units of F must be ____________

To show what this force would feel like, express it in more familiar units.

Suppose you wish to apply this force to a steel cable hanging vertically, by hanging a load on it in the form of a huge iron block. The quantity of iron needed would be ____________ pounds or roughly ____________ tons.

PROBLEMS USING $F = M \cdot a$ SHEET 2

PROBLEM 8. A man jumps from a window ledge about 4 feet high to a hard floor.

Estimate the force exerted on him by the floor while he is stopping, by answering the questions below. Suppose he is a 200-pound man and that he foolishly forgets to bend his knees while landing, so that the total "give" of his feet, etc., is only 1 inch (in compression of floor, soles, ankles, shoes, spine, etc.) during the stopping process.

To calculate the force, we need the man's (negative) acceleration during stopping. For that, we need to know his velocity just before stopping, his velocity just after stopping, and the time taken by the stopping process.

(i) Calculate his speed at the end of his fall, just before landing, by using

$v = v_o + at$. For this, the time-of-fall, t, is needed, so calculate it first.

Since he starts from rest and falls 4 ft with acceleration 32 ft/sec^2,

$s = v_o t + \frac{1}{2}at^2$ gives ____________ = ____________ + ____________

∴ time-of-fall, t, is . . . . ____________ sec

Then $v = v_o + at$ gives final $v =$ ________ + ________ = ________ ft/sec

(ii) His final velocity, when he has just finished landing, is ________ ft/sec

(iii) To calculate the time taken by the landing process we must find his average speed during the landing process. Using the results above, we see his average velocity during the process of landing is ________ ft/sec

With this average speed he travels the "give" distance of 1 inch, or 1/12 foot, in a time . . . . ________ sec

(iv) In the time calculated above his velocity changes from ________ ft/sec just before landing to ________ ft/sec just after landing.

∴ his acceleration during landing is . . . . ________ ft/sec^2

(v) ∴ force exerted on him by floor during landing is ________ ________ units

(vi) This force, expressed in more familiar "bad" units is ________ pounds-weight

(vii) MORAL: WHEN JUMPING LIKE THIS ____________________________

(viii) Suppose the man does bend his knees, making the total "give" 10 inches; then the force on him during landing is about ____________ pounds-weight

PROBLEM 9. (See Problem 7 for method.) A player kicks a $\frac{1}{2}$-kilogram football and gives it a velocity 14 meters/second, starting from rest. If the contact between foot and ball lasts 1/50 second, estimate the force.

The average acceleration during contact must be. . __________ · __________ units

The average force must be. . . . . __________ · __________ units

This force, in "bad" units, is . . . . __________ kg-weight

This force, in other "bad" units, is approximately __________ pounds-weight

(NOTE: 1 kilogram = approximately 2.2 pounds.)

PROBLEM 10. (An important problem for later Relativity discussion.) A 160-pound man stands on a spring weighing-scale in an elevator. Show, by answering the questions below, what the scale may be expected to read in each of the following cases, if Newton's Laws are safe summaries of what does happen.

(a) Elevator permanently at rest,

(b) Elevator moving up steadily 1 ft/sec, with no acceleration,

(c) Elevator accelerating up 8 feet/second/second,

(d) Elevator accelerating down 10 feet/second/second.

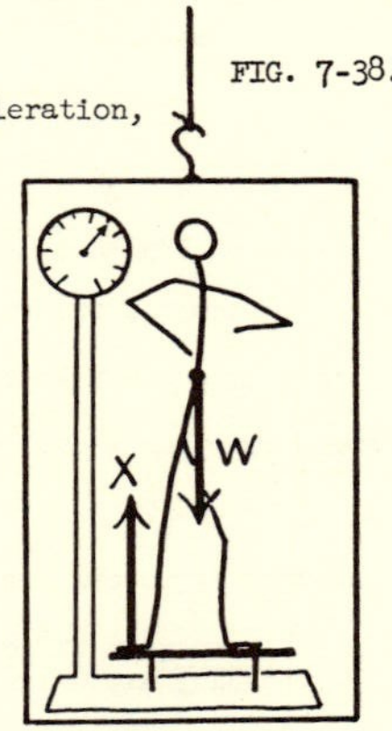

FIG. 7-38.

General discussion.

The spring scale does not tell us the weight of (= pull of the Earth on) the man. It only tells how hard his feet press down on the platform of the scale.

The only external forces on the man are:

(i) the Earth-pull on him, (his weight), W.

(ii) the push-up of the scale platform on his feet, X.

The resultant of these external forces, X - W, is the force that accelerates him.

Assume Newton's Law III (Action = Reaction) gives a true account in all cases. That tells us that the push-up on his feet is always equal and opposite to the push-down of his feet on the platform; and that push-down is what the scale indicates. Note that Law III does not say that X and W are always equal.

PROBLEM 10 CONTINUES ON NEXT SHEET

PROBLEM 10. (continued)

(a) When elevator is at rest, not accelerating, resultant force on man is __________

and ∴ the push-up, X, on him must be . . . . __________ pounds-weight

and ∴ the scale will read. . . . . __________ pounds-weight

(b) When elevator is moving up steadily 1 foot/second,

not accelerating, the resultant F on the man is __________

∴ the scale will read. . . . . __________ pounds-weight

Can the man know whether the elevator is moving or not, if it is windowless and moves smoothly and silently? . . __________

(c) When elevator is accelerating up 8 feet/second/second, the resultant force on man must be just the force needed to give him this upward acceleration; and the relation F = M·a tells this needed force, F.

F = M·a = (   )·(   ) = __________

Since we have used F = M·a, F must be in units called __________

This F must be the resultant of scale's push-up and man's weight. But man's weight, in same units as F, is __________ __________ units

Resultant F = (push-up, X) - (weight, down, W)

∴ push-up of scale, X, must be __________ __________ units

This is equal and opposite to the push-down of man on scale, and the scale reads that force.

∴ during the acceleration scale will read . . . . . . __________ pounds-weight

(d) On a spare sheet calculate, with a brief explanation, the scale-reading when elevator is accelerating downward with acceleration 10 feet/second/second.

Scale will read (in pounds-weight) . . . . __________ pounds-weight

(e) Can the man be sure that the elevator is accelerating, if it is windowless and moves smoothly and silently? Instead of thinking the elevator acquired an acceleration, he might think another change had occurred in his locality.

What change? __________

Action and Reaction

A thread transmits a pull or tension unchanged. If you attach a long thread to a truck and pull the loose

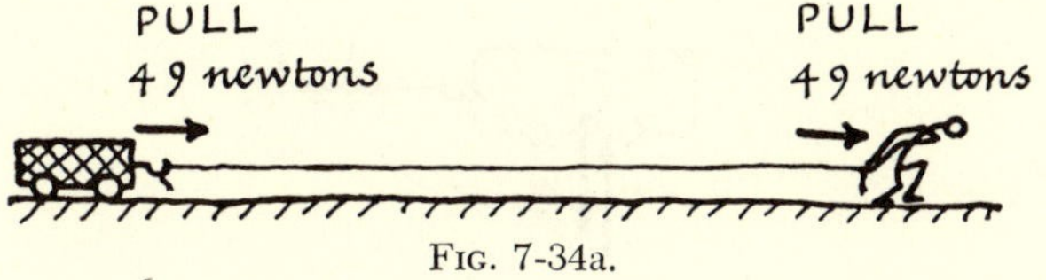

FIG. 7-34a.

end with 49 newtons, then the thread carries the 49-newton pull along and applies it to the truck. *You* pull the *thread* 49 newtons forward and at the same time the *thread* pulls *you* 49 newtons backward—try it, and feel it. These two forces where your hand meets the thread are called "action and reaction," and we shall

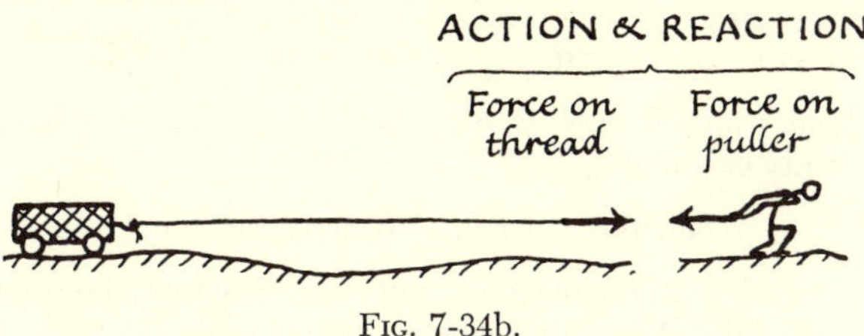

FIG. 7-34b.

take it for granted they are equal and opposite, in all circumstances. You cannot pull the thread without its pulling you. (You cannot push against a wall or floor without its pushing back against you. You cannot punch a man's head without his head hitting back on your fist.) Where the thread joins the truck, there is another pair of action and reaction. There the thread pulls the truck forward and the truck pulls the thread backwards.

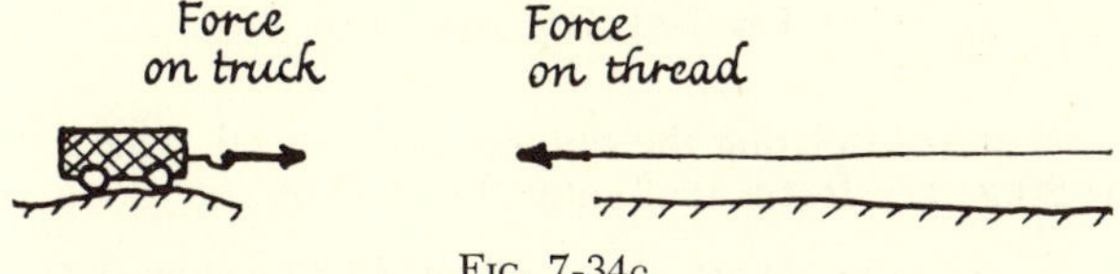

FIG. 7-34c.

An action-and-reaction pair do not cancel out to zero force because they act on different victims. At your hand, your forward pull acts on the *thread*; and the reaction pull of the thread acts only on your *hand*—feel it. At the truck, only the backward pull of the truck acts on the thread. So the thread is pulled outward at both ends, by hand and truck, trying to stretch it. These are the only pulls on the thread, and if they do not balance out to zero their resultant will accelerate the thread. Apply $F = M \cdot a$ to the negligible mass of a fine

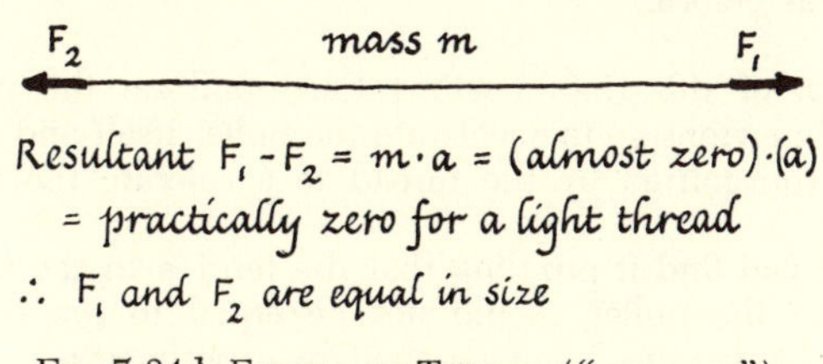

FIG. 7-34d. FORCES ON THREAD ("ISOLATED")

thread, and you see that the resultant F must be negligible. Therefore the pulls at the two ends of the thread must balance even if the thread is accelerating. (Of course, a massive rope that is accelerating requires pulls that differ enough to provide the accelerating force.)

Assume that action and reaction are equal and opposite, at your end of the thread, and again at the truck end. Then, just as a light thread is pulled equally by its supports, it must pull them inward equally. So it

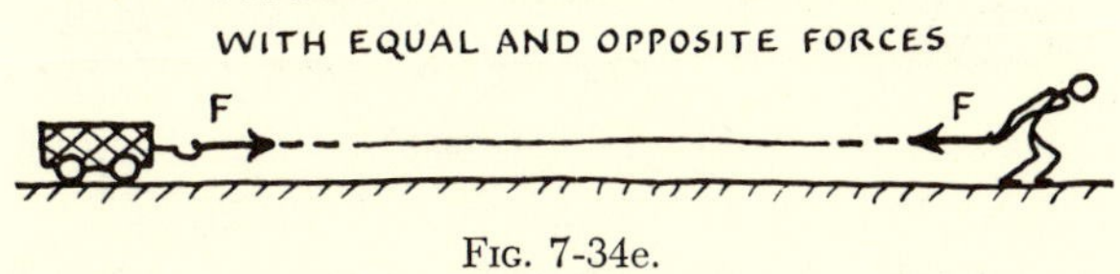

FIG. 7-34e.

pulls the truck forward just as much as it pulls your hand backward. This brings us back to our starting point, pull transmitted all along the thread.

Tension

We call this pull in a thread the *tension*. Thus the tension is the pull at either end, or the pull the thread exerts if it is cut at any point and tied to a wall. To make a tension of 49 newtons in a thread, hang 5 kg on it; or, for a horizontal thread, hang 5 kg on it over a

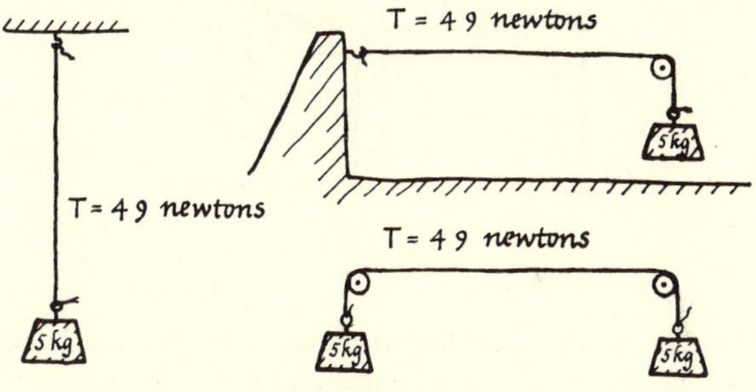

FIG. 7-35. TENSION

pulley at one end. The other end must be anchored, or it too must carry a 5-kg load. The tension is still 5 kg-weight (49 newtons), and not twice that although there are two 5-kg loads, one at each end.[27] That is the

---

[27] If you find this puzzling, reflect that there are three statements of force concerning the thread:

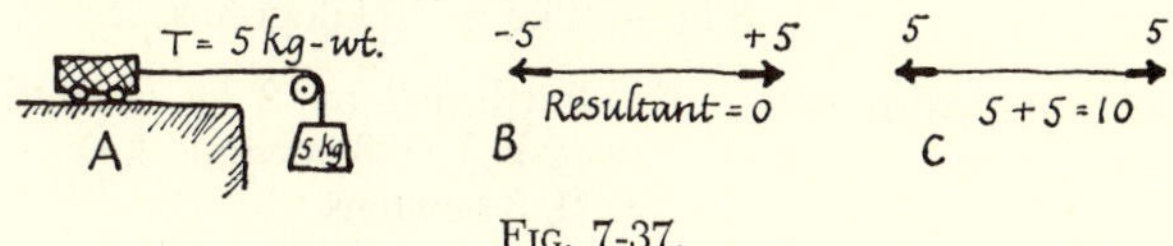

FIG. 7-37.

A. Force with which thread pulls (or is pulled) at any end, 5 kg-wt.

B. Resultant force on thread, (5 kg-wt.) + (−5 kg-wt.) = zero

C. "Total" force (5 kg-wt.) + (5 kg-wt.) = 10 kg-wt.

The force in A is very useful—it tells us what the thread can do to things it pulls—so it is worth naming, "tension." B is true and dull. We already have the name, "resultant-force." The force in C is of no practical use, so we do not name it.

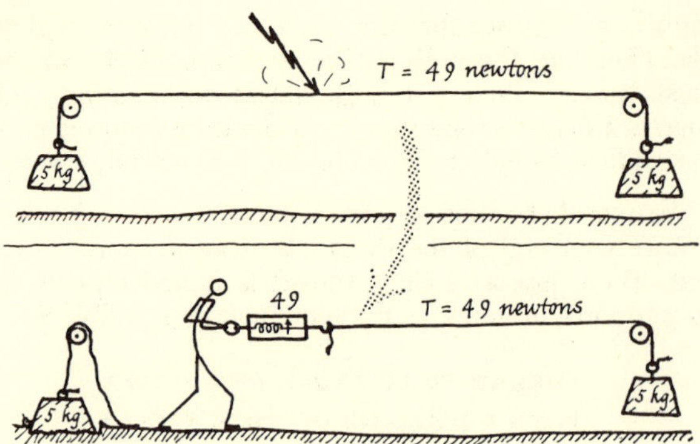

FIG. 7-36. TENSION

force the thread pulls with wherever you cut it; that is what a spring balance would read if you cut the thread and inserted it.

A Problem in Detail: Mysterious Loss of Tension

Suppose we use a hanging load to accelerate a truck on a level table. Then the Earth-pull on the load (its weight) is the resultant force accelerating the combine.

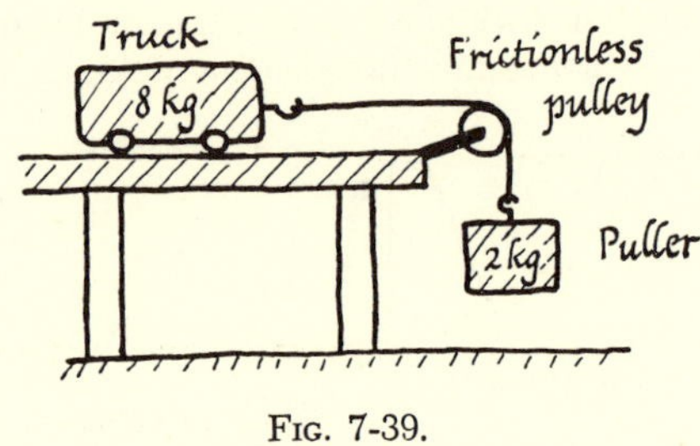

FIG. 7-39.

(The truck also is pulled downward by the Earth but the level table applies an equal and opposite supporting force.) The *mass* being accelerated by this force is the total mass of truck + pulling load, because the load accelerates just as much (downward) as the truck (along). Before, we used a strangmeter to tell us what force pulled the truck, and we could apply $F = M \cdot a$ to the truck alone. Now, with no strangmeter, we only know the resultant accelerating force on the combine. So we say, in the example sketched:

MOVING MASS, $M =$ (MASS OF TRUCK + MASS OF PULLER)
$= 8 \text{ kg} + 2 \text{ kg} = 10$ kilograms.

ACCELERATING FORCE, $F =$ Earth-pull on 2 kg of stuff
$= (2 \text{ kg})\ (9.8 \text{ newtons/kg})$
$= 19.6$ newtons

$$\therefore \quad \text{ACCELERATION} = \frac{\text{RESULTANT FORCE}}{\text{MASS BEING ACCELERATED}} = 19.6 \text{ newtons}/10 \text{ kg} = 1.96 \text{ meters/sec}^2$$

We can predict an interesting measurement: the tension in the thread from puller to truck. This is *not* 19.6 newtons (2 kg-weight), because some of that force is used to accelerate the puller, and only some of it is carried by the thread to accelerate the truck. To find the tension, apply $F = M \cdot a$ to the truck alone, thus:

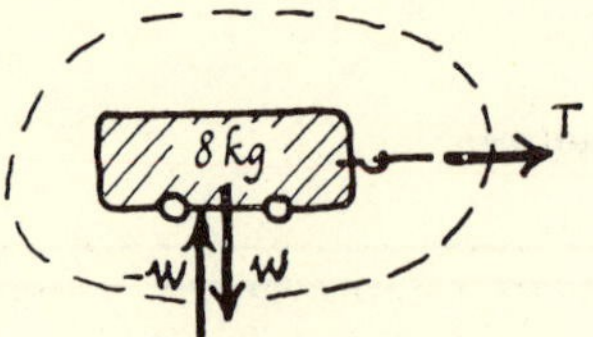

FIG. 7-40. ISOLATED TRUCK

"isolate" the truck by drawing a ring round it. Then take the mass in the ring, and the resultant of all forces that enter the ring from outside. The MASS is 8 kg of truck. The forces that act on it are:

ITS WEIGHT DOWN } these
THE PUSH UP OF THE TABLE } cancel } resultant force T newtons
THE PULL OF THE THREAD, T newtons

Then, using $F = M \cdot a$ with the acceleration we have already calculated,

$$T \text{ newtons} = (8 \text{ kg})\ (1.96 \text{ meters/sec}^2)$$

$\therefore$ $T = 15.68$ newtons, not 19.6 newtons, because some of that Earth-pull is used to accelerate the pulling load itself.

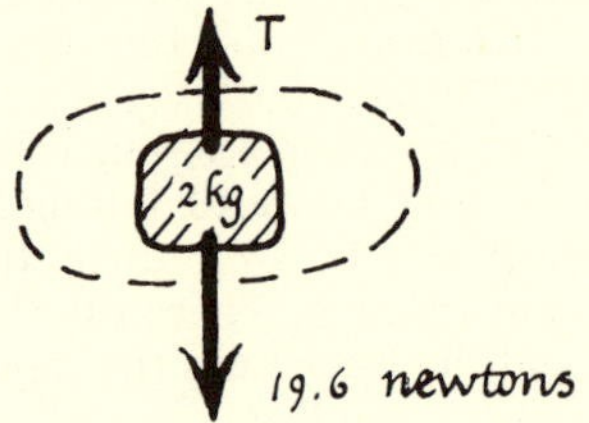

FIG. 7-41. ISOLATED LOAD

Now try isolating the pulling load instead. The MASS is 2 kg. The forces that enter the ring are:

EARTH-PULL ON PULLER $= (2 \text{ kg})\ (9.8 \text{ newtons/kg})$
$= 19.6$ newtons, *downward*

TENSION OF THREAD, T newtons, pulling puller upward.

Then the RESULTANT ACCELERATING FORCE, F, is $(19.6 - T)$ newtons, *downward*

Therefore, applying $F = M \cdot a$ to the puller alone, with the same acceleration, 1.96 meters/sec²,

$$19.6 - T = (2 \text{ kg})\ (1.96 \text{ meters/sec}^2)$$
$$T = 19.6 - 3.92 = 15.68 \text{ newtons},$$

just as before.

Thus, of the 19.6 newtons Earth-pull on the puller, 3.92 newtons go to accelerate the puller itself and 15.68 are transmitted by the thread to accelerate the truck.

If you find it puzzling that the tension in the thread above the puller should not be equal to the puller's weight, try the following thought experiment:

Fig. 7-42. Feeling a Fraction of Weight

Fig. 7-43. Problem 11(a)

(i) Hold a 2 kg load at rest on the palm of your hand. What force do you feel?

(ii) Lower your hand with downward *acceleration* just a little more than g (Fig. 7-42). What force do you feel? What force do you exert on the load? What is the pull of the Earth on the load doing to it?

(iii) Lower your hand with downward *acceleration* about ½ g. What is the pull of the Earth on the load doing to it now? What force do you exert on the load? What force do you feel now? (This is, of course, repeating the discussion above in a simpler form, but starting with the extreme case of acceleration g makes the result easier to accept.)

(iv) Repeat the discussion of (i)-(iii) with the 2 kg hung from your hand by a string.

PROBLEM 11. "LOSS OF TENSION"

(a) Loads of 2 kg and 1 kg are hung on a string over a light frictionless pulley as in Fig. 7-43. When an experimenter clamps the string and pulley with his hand, to prevent any motion the string on the left has tension 9.8 newtons (= 1 kg-wt.) and on the right 19.6 newtons. When the experimenter releases the string and pulley, there is acceleration.

(i) Which way does the 2 kg load on the right accelerate? Is the tension of its string the same as before or greater or less? Why?

(ii) Is the tension of the string above the 1 kg load the same as before or greater or less? Why?

(iii) If the pulley wheel is frictionless and massless, what relation would you *expect* between the tensions of string on the two sides, when the arrangement is free to move?

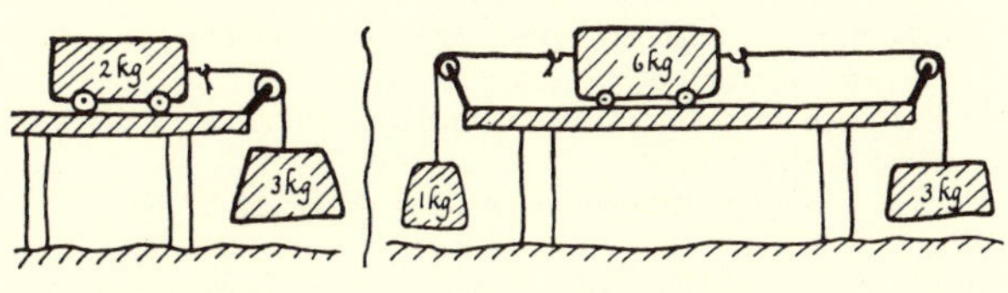

Fig. 7-44.

Fig. 7-45.

(b) Calculate the acceleration and thread tensions in the arrangement sketched in Fig. 7-44. Assume friction is negligible.

(c) Calculate the acceleration and thread tensions in the arrangement sketched in Fig. 7-45. Assume friction is negligible.

Newton's Laws of Motion

The ideas we have been discussing were codified by Newton in his Laws of Motion. Here they are in modern wording:

Law I: *Every body remains at rest or moves with constant velocity (in a straight line) unless acted on by an external force.*

Law II: *When an external force acts on a body, the product* MASS · ACCELERATION *varies directly as the force, and the acceleration is in the direction of the force.* (Later we shall find that Newton's original version in terms of momentum is better.)

Law III: *Action = Reaction.* (This will be discussed in Ch. 8.)

Even after demonstrations and all this discussion, Newton's statements of Laws I and II may seem odd and unreal. The mistake is in the omission of the word "resultant." *External* force means *resultant* force. Laws I and II make good sense when the word "resultant" is inserted. They say:

"When there is no *RESULTANT* force acting on a body, its motion continues"; and

"*RESULTANT* FORCE = MASS · ACCELERATION."

Newton stated these beliefs about force and motion when he wrote his great treatise on mechanics and astronomy. He tested them, in a way, on the Moon and planets; and we have taken a risk and extended them to molecules, atoms, and now even parts of atoms.

In most elementary textbooks, Newton's Laws are stated formally at the beginning of the chapter instead of here at the end where they properly belong as summaries. And they are announced so firmly that students think Newton got them from Heaven. He did not—he was merely re-wording the views of Galileo and others who had experimented and thought about motion. He gave them as working rules, partly based on experiment, partly definition and clarification of terms. Right down to this day scientists have disagreed over the status of Newton's Laws. His immediate followers may have thought them simply experimental, drawn from knowledge of the real world—rather like Hooke's Law. Nowadays we are more cautious and see Law I as chiefly a description of force, and Law II as a definition of force-measurement—it says that FORCE is MASS · ACCELERATION, assuming that mass is intuitively obvious. But a few enthusiastic arguers go further still and claim that the laws are wholly definitions or conventions, and contain no experimental ties to the natural world. That is misleading, if not silly. We could certainly conceive of a universe in which the behavior of moving things could not be summed up in Newton's Laws. Perhaps the best comment is that of Poincaré,[28] one of the most distinguished mathematical physicists. He says in his book *La Science et l'Hypothèse* (Flammarion, Paris):

"We shall see that there are several kinds of hypotheses, that some are verifiable and when once confirmed by experiment become truths of great fertility; that others may be useful to us in fixing our ideas; and finally that others are hypotheses only in appearance, and reduce to definitions or conventions in disguise. These last are to be met with especially in mathematics and in the sciences to which mathematics is applied. From them, indeed, those sciences derive their rigor; such conventions are the result of the unrestricted activity of the mind, which in this domain admits no obstacle. For here the mind may decree because it lays down its own laws; but let us clearly understand that while these laws are imposed on *our* science, which otherwise could not exist, they are not imposed on Nature. Are they then arbitrary? No; for if they were, they would not be fertile. Experience gives us freedom of choice, but it guides us by helping us to discern the most convenient path to follow. Our laws are therefore like those of an absolute monarch who is wise and consults his council of state. . . . Are the laws of acceleration and of the composition of forces only arbitrary conventions? Conventions, yes; arbitrary, no—they would be so if we lost sight of the experiments which led the founders of science to adopt them, and which, however imperfect, were sufficient to justify their adoption."

PROBLEMS FOR CHAPTER 7

Problems 1-11 are in the text.

12. In the third investigation with truck and track (Table C) we changed both the pulling force and the total mass. What relationship would you look for if distance and force were kept constant, and the time measured for total mass 2 kilograms, 4 kg, 6 kg?

★ 13. A man pushes a box along a rough horizontal floor, exerting a push of 40 newtons. Friction exerts an opposite drag of 10 newtons on the box.
(a) What is the actual accelerating force?
(b) How hard should the man push, if friction stays the same, to double the acceleration of the box?

★ 14. (a) Copy the following statement and complete it: "A newton is defined as that force which will. . . ."
(*Note*: The proper completion should contain the words "mass" and "acceleration.")
(b) Write the corresponding definition for a poundal.

★ 15. A force of 5 newtons acts on 20 kg of ice on a frictionless table. The force is horizontal, due North. Calculate the acceleration of the ice.

★ 16. (a) What force is needed to give an acceleration of 4 meters/sec² to a mass of 4.90 kilograms? Give your answer in absolute units and name the units.
(b) Suppose you wish to apply the force calculated in (a) above to a thread by hanging a lump of iron on it. What mass of iron would you need?

★ 17. One kilogram-weight is a force of approximately 9.80 newtons (for people living on the Earth's surface).
(a) What easy, simple experiment demonstrates this?
(b) Give the argument which leads from that experiment to this result.

★ 18. Suppose you wish to experience a force of one poundal on your hand, acting vertically down. How much metal should you place on your hand? Give the reasoning by which you reached your answer.

★ 19. A 20-ton boxcar (40,000 pounds) is at rest on a slightly inclined railroad which runs east-west. The road is tilted just enough to compensate for friction for a boxcar moving eastward. A child pushes the boxcar steadily eastward with a force of 2 pounds-weight. Having nothing else to do, the child continues to push for 5 minutes (300 seconds).
(a) What speed will the car acquire in those 5 minutes?
(b) How far will the child walk in the 5 minutes?

20. MASS AND WEIGHT

An engineer has two huge iron castings, each hanging from a crane by a 50-ft steel rope. They look alike outside, but one is solid and the other largely hollow. To find out which is which, he makes several tests:
(i) He tries to lift each. Each is far too heavy to lift.
(ii) He tries pulling each out sideways, about 1 ft, with a cord, and estimates his pull for that.
(iii) He pulls each out about 1 ft sideways and lets it go, and estimates the time it takes for its first swing back to the original position with the rope vertical.
(iv) He pounds on the side of each with his fists and estimates the ensuing motion.
(v) He pushes each for some time in such a way as to make it spin slowly around the vertical axis of its rope. They do not continue to spin when released, but slow down, stop, spin the opposite way, . . . &c., making twisting oscillations. He estimates the time of one complete oscillation for each.
(a) For each test, (i)-(v), say whether it compares *masses* or *weights* or neither.
(b) Give a brief reason for each decision in (a).

21. A 150-pound track man starts from rest and 2 seconds later is running 25 ft/sec.
(a) What horizontal force must be acting on him during his start?
(b) How does it compare with his weight?*

* In each case where a "comparison" with the object's weight is asked for, it is best to give the answer as a fraction: as in "the force is $\frac{3}{4}$ of the truck's weight," "the elephant's pull is 10% of his weight," etc. To make these fractions you must express the force in question in the same units as the weight.

[28] "Not the Prime Minister, but his cousin the mathematician, who was a great man."—Bertrand Russell.

22. An 80-kilogram track man starts from rest and two seconds later is running 8 meters/sec.

(a) What horizontal force must act on him?

(b) How does this force compare with his weight?*

★ 23. A 2000-pound car accelerates 8 ft/sec/sec.

(a) What is the resultant force acting on the car?

(b) How does this force compare with the car's weight?*

(c) Who or what exerts this force on the car?

(d) Car brakes are assessed by the % of "g" that they can produce when decelerating the car at their best. If this car's brakes are just able to prevent it accelerating as above (when the stupid owner drives with the brakes on) what is their rating, as a %?

★ 24. An electrically charged hydrogen atom ("proton," = hydrogen nucleus = hydrogen atom that has lost its one and only electron), moving along a horizontal path in a vacuum with speed a million meters/second ($v = 10^6$ meters/sec), passes through a region 0.20 meters long where there is a vertical electric field which pulls on the atom's charge. The field produces a force 0.0000000000000032 newtons (3.2×10^{-15} newtons) on the atom vertically down. The mass of the atom = 0.00000000000000000000000000166 kilograms (1.66×10^{-27} kg).

(a) What acceleration will the atom have, while in the field?

(b) Why can we ignore gravity in this problem? (*Note*: The data are quite usual ones for experiments with protons.)

(c) What shape of path will the atom pursue in the field?

(d) How long will the atom take to cover the horizontal distance of 0.20 meter in the field?

(e) How far will it move vertically in the field region?

(f) What vertical velocity will it acquire in the field region?

(g) If after the 0.20 meter travel across the field it emerges into a region where there is no field, what path will the atom then pursue? (Be as specific as you can. Give a sketch.)

★ 25. If an electron is fired *with the same speed* as the proton of problem 24, through the *same electric field*, what will its path look like, compared with the proton's? Just describe the difference generally. (*Note*: An electron has an equal but opposite electric charge; so the same electric field will exert the same force on it but in reverse direction. But it has only about 1/2000 of the mass of a proton.)

★ 26. In this course "A 200-pound man" in a problem means the man's *mass* is 200 pounds (and this would be the same anywhere: Earth, Moon, etc.). Where you need his mass M for use in $F = M \cdot a$, use 200 pounds. Near the Earth's surface, the Earth pulls on the man with a *force* called his "weight."

(a) What value, with units, should you use for this force when you wish to use it for F in $F = M \cdot a$?

(b) Explain how you arrived at the value in (a).

(c) Describe the experimental evidence and reasoning that led to the special factor that you used in calculating the value of F in (a) above.

(d) Is that special factor different on the Moon?

* In each case where a "comparison" with the object's weight is asked for, it is best to give the answer as a fraction: as in "the force is $\frac{3}{4}$ of the truck's weight," "the elephant's pull is 10% of his weight," etc. To make these fractions you must express the force in question in the same units as the weight.

★ 27. There are two essentially different ways of comparing the masses of two bodies. What are these two ways and how do they differ essentially?

★ 28. PREPARATION FOR RELATIVITY DISCUSSION

Suppose we do the following experiments to test Newton's Laws I and II.

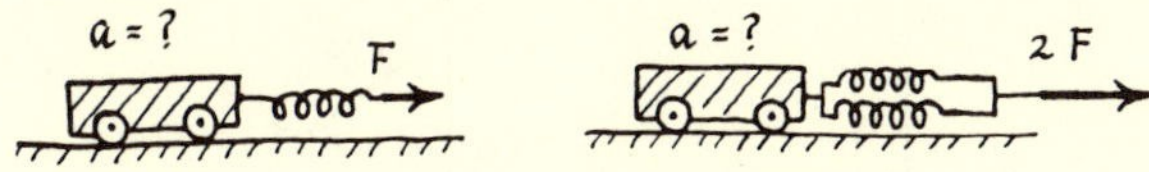

FIG. 7-46. PROBLEM 28

FORCE/ACCELERATION is the same for both these experiments in Case 1. Is it the same in Case 2? Case 3?

(a) We observe the motion of a block of "dry ice" (solid carbon dioxide) on a level table.

(b) We pull a small truck along a level track by measured forces (with friction "removed" by compensating tilt).

These are demonstration experiments in which we show that:

(i) *A constant force gives constant acceleration.*

(ii) *For the same total moving mass, acceleration ∝ force.*

(iii) *For a constant acceleration, force ∝ mass.*

And thus we conclude that $F \propto M \cdot a$.

Now suppose these experiments are done inside a railroad coach in each of the following cases, and suppose that we use the coach for our framework of reference (= our coordinate system). The track of experiment (b) is parallel to the railroad. For each case, say what you think would be observed and say whether the observations would seem to agree with Newton's Laws I and II.

Case 1: Coach at rest.

Case 2: Coach moving with constant velocity, without acceleration.

Case 3: Coach moving with *constant acceleration* forward along level railroad. The motion in the experiments inside is *along* the direction of the railroad.

★ 29. In Case 3 of Problem 28 certain peculiarities of behavior would be observed. Suppose that an investigator trying to interpret them does the following additional experiments:

(c) He installs a pendulum to serve as a plumb line so that he can observe the "vertical." And he installs a bucket of water (or preferably oil or molasses) so that he can observe the "horizontal." What will he observe?

(d) In a frantic attempt to preserve his belief in the laws of nature he learned earlier, he hires a carpenter to build a laboratory inside the coach, tilted at such a slope that the plumb line and water of (c) are parallel to the walls and floor of the new lab—the "vertical" and "horizontal" now seem normal. What will he then observe if he repeats the experiments of (a) and (b) in this laboratory, referring them to his new "horizontal" and "vertical" floor and walls?

(e) Suppose the new tilted lab has no windows; and no message from outside tells the investigator that he is accelerating. Then, instead of realizing that he is accelerating, he might decide that a certain common physical quantity had changed to a new value in his lab. What quantity? What kind of change would it show?

(f) Could such an investigator ever distinguish, by experiments inside the coach, between a real acceleration and the change mentioned in (e)?

30. A man on the ground hauls a bucket of water up to the top of a building by a rope that runs over a pulley at the top of the building and down to him. The bucket+water weighs 20 pounds. The man starts pulling with a force of 40 pounds-weight on his end of the rope. What acceleration will the bucket have? (*Note*: There is a catch in this question.)

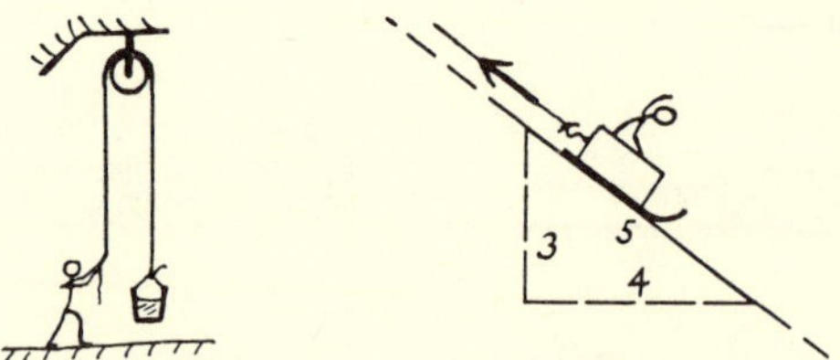

Fig. 7-47. Problem 30 Fig. 7-48. Problem 31

★ 31. A 200-pound sled (including man) slides down a frictionless hill which falls vertically 3 feet for every 5 ft down the slope (and every 4 ft along horizontal base).

(a) If the sled were held by an uphill rope parallel to the slope, what force would that rope have to exert to hold the sled at rest?
(b) With the rope cut, what acceleration would the sled have?
(c) With an extra 150-pound man *added* (making 350 pounds of men + sled in all) what acceleration would the loaded sled have?
(d) Why should (c) require no extra calculation?

32. A squirrel with an armful of nuts is given a push along a frictionless horizontal table top, and then left to slide. He finds himself getting dangerously near the edge. He understands Newton's laws of motion and saves himself from falling over. How?

33. A child pulls a 30-pound sled along a frictionless level road by a slanting rope. The rope is 5 ft long. Its upper end is 3 ft vertically above its lower end, and 4 ft ahead horizontally, as shown in the sketch. The child pulls the rope with a force of 10 pounds.

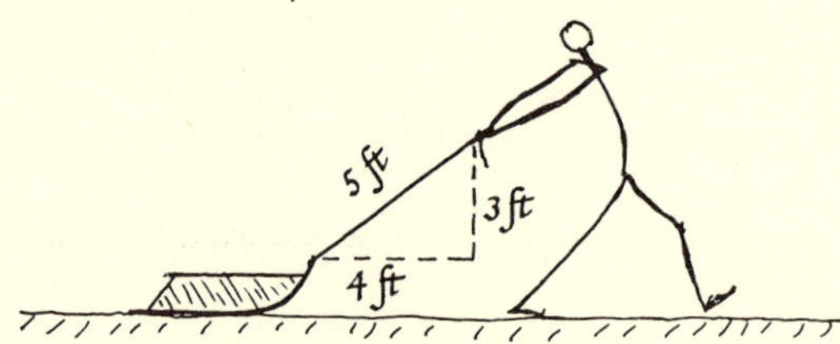

Fig. 7-49. Problem 33

(a) What is the horizontal force that accelerates the sled? (Sketch a diagram to illustrate your calculation.)
(b) What is the acceleration of the sled?

34. Stakes are being driven into a hard river-bed by a "pile-driver" which is a 200-pound block of iron hauled up until it is some distance above the top of the stake and then allowed to fall freely until it hits the top of the stake. It then drives the stake 1/20 of a foot (just over $\frac{1}{2}$ inch) deeper into the mud. The iron block stops dead on the top of the stake and does not rebound. The block, falling freely, has acquired a velocity of 24 ft/sec by the time it reaches the stake.

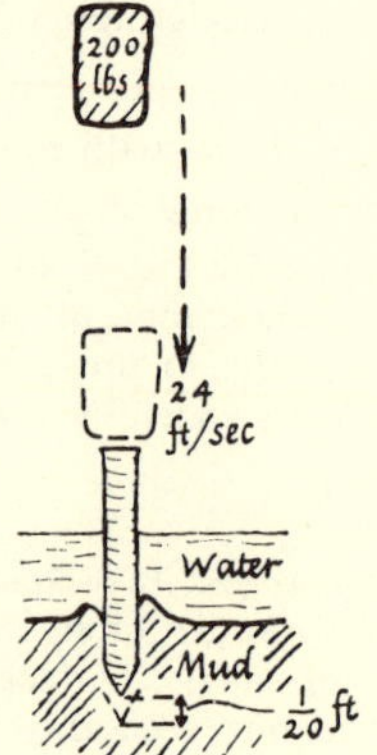

Fig. 7-50. Problem 34

(a) Calculate the time taken by the collision (during which the block comes to rest from this high speed).
(b) Calculate the force which acts on the block during collision.
(c) Express this force in "bad" or engineering units.

35. A 3000-pound automobile moving 40 ft/sec crashes into a massive wall and comes to a stop. During the crash the center of the car moves forward 1 foot (from the time when the car starts to hit the wall to the time when it is at rest). Calculate the average force involved during the collision.

36. A certain thread pulling a block of ice along a frictionless table can just accelerate an 8-pound block 40 ft/sec². Attempts to make the thread exert a bigger pull simply break the thread.

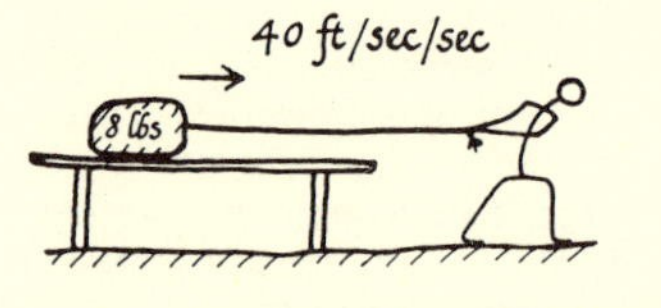

Fig. 7-51. Problem 36

(a) If a piece of the same thread is hung vertically with a lump of iron on it, how many pounds of iron could it support without breaking?
(b) On the Moon, where "g" is about one-sixth as great, would the maximum iron lump on a piece of the same thread be much greater, much smaller, or the same?
(c) On the Moon, with the same thread, would the maximum acceleration the man could give to the same block of ice on a level table be much greater, much smaller, or the same?

CHAPTER 8 · CRASHES AND COLLISIONS. MOMENTUM

"Action is equal to reaction."—NEWTON

"If he won't dim his lights, I won't dim mine."†

"PROBLEMS" A. A 2-pound projectile is moving horizontally 2000 ft/sec. What is its force?

B. A 10-ton truck moving 30 miles/hour crashes into a wall and stops. What is the force of the collision?

These look like sensible questions. The answers might be important. Yet in fact the question is meaningless in each problem, as it stands. The moving projectile does not *have* a force along its direction of motion nor does it need one to keep it going—the problem is trying to lead you into the mistake the Greeks and medieval Aristotelians made. No force is involved until the motion changes, and then the acceleration requires a force exerted by an outside agent. Even when we know the change of speed, as we do in "Problem" B the question is still unanswerable, because we do not know the time taken for that change, and therefore we cannot calculate the acceleration.[1] The belief that these problems are reasonable comes from a mistaken feeling about force and motion. When we say mistaken, we are not just condemning one viewpoint and asserting another; we are referring back to experimental possibility as a test of reality. No engineer or physicist can produce instrument or apparatus to measure the "force" of a projectile in flight. Spring balances attached to it would show no force at all as long as it moves freely. And instantaneous photographs of the projectile itself would show it to be neither stretched nor compressed—thus indicating no stress. So, with attempts to measure the "force" yielding no useful answer, we do not think the idea of "force" in this case a useful one. But when the moving projectile (or the moving truck) hits something and changes its speed a real force is involved, the kind of push or pull that you can feel, the kind of thing that can be measured by spring balances, or indicated by its elastic effects. Photographs of a projectile taken during its collision with a steel wall show marked compression and have been used to estimate the force involved.[2]

Calculating Force from Changes of Motion

If we accept $F = Ma$ as a true summary of the behavior of nature, we can use it to calculate the force involved, *if* we know the victim's mass, M, and its acceleration, a. This is done in problems that arise in engineering and almost every branch of physics, ranging from planetary orbits to the interior of atoms.

In "Problem" B above we know the mass but cannot find the acceleration until we are told the time taken for the truck's change of velocity from 30 miles/hour to zero. We need to know the duration of the collision. Suppose we are told that the crash lasts 0.1 second. Then we can proceed to calculate the force thus:

$$\text{ACCELERATION} = \frac{\Delta v}{\Delta t}$$

$$= \frac{(0, \text{for truck at rest}) - (30 \text{ miles/hr})}{0.1 \text{ sec}}$$

$$= \frac{-30 \text{ miles/hour}}{0.1 \text{ sec}} = \frac{-44 \text{ ft/sec}}{0.1 \text{ sec}}$$

$$= -440 \text{ ft/sec}^2$$

The minus sign here shows that the motion is *decelerated*. The minus sign which then appears in the force answer shows that the force is in the opposite direction to the motion, removing the momentum the truck had. Then

[1] On the other hand the following question, which looks similar to "Problem" B, has a definite answer:

C: A firehose delivering 50 gallons per sec shoots a horizontal stream of water with speed 40 ft/sec at a vertical wall. The water is stopped by the wall and trickles down its face. What is the force of the water on the wall?

[2] Of course we see a projectile changing its *vertical* motion and can estimate the effect of gravity from those changes. Here we are discussing other, more violent, changes of *horizontal* motion.

FORCE $F = Ma$

$$= (20{,}000 \text{ pounds})(-440 \text{ ft/sec/sec})$$
$$= -8{,}800{,}000 \text{ pounds} \cdot \text{ft/sec}^2 \text{ or } poundals$$
$$\approx 275{,}000 \text{ pounds-weight}$$
or over 130 tons-weight

This is the backward push that the wall exerts, during 0.1 sec, on the truck to stop it. Direct use of $F = Ma$ produced the answer, but this is a roundabout method. We are given MASS, CHANGE OF VELOCITY, and TIME, and we want the FORCE. Can $F = Ma$ be changed into another form which uses F, t, m, and *change of* v? Yes, this is easily done, and we find:

$$Ft = \Delta(Mv)$$

which we shall show below is another form of $F = Ma$. Try using it:

FORCE · TIME = change of MASS · VELOCITY

$$Ft = \Delta(Mv)$$

Then $F\ (0.1 \text{ sec}) = \Delta(Mv)$

$$= \underbrace{(20{,}000 \text{ lbs} \times 0)}_{\text{FINAL } Mv} - \underbrace{(20{,}000 \text{ lbs} \times 44 \text{ ft/sec})}_{\text{INITIAL } Mv}$$

$$F = (0 - 880{,}000 \text{ pounds} \cdot \text{ft/sec})/(0.1 \text{ sec})$$
$$= -8{,}800{,}000 \text{ pounds} \cdot \text{ft/secs}^2 \text{ or } poundals$$

Here is how we shift $F = M \cdot a$ into the other form, which was actually Newton's original form. (We assume mass, M, remains constant during the change).

SIMPLE VERSION

$$F = M \cdot a$$
$$= M\frac{v - v_0}{t}, \text{ using definition of acceleration}$$

Multiply both sides by t.

$$F \cdot t = M(v - v_0) = Mv - Mv_0$$
$$= (\text{new } Mv) - (\text{old } Mv), \text{ since } M \text{ remains the same}$$
$$F \cdot t = change\ of\ Mv \text{ or } \Delta(Mv)$$

CONDENSED VERSION

(Here we use Δt instead of t, for the time the force acts.)

$$F = M \cdot a = M\frac{\Delta v}{\Delta t}$$
$$F \cdot \Delta t = M \cdot \Delta v$$
$$= \Delta(Mv) \text{ since } M \text{ is constant}$$
$$F \cdot \Delta t = \Delta(Mv) \text{ or } change\ of\ (Mv)$$

CALCULUS VERSION of the algebra above.

$$F = M \cdot a = M\frac{dv}{dt}$$
$$\int F\,dt = \int M\,dv = M\int dv, \text{ since } M \text{ is constant},$$
$$= M \cdot \Delta v = \Delta(Mv)$$

If F is constant, the left side becomes $F\int dt$, which is $F \cdot \Delta t$.

Then $F \cdot \Delta t = \Delta(Mv)$.

If F is not constant, $\int F\,dt$, the "impulse," gives us AVERAGE FORCE · Δt. Then we can say

$$(\text{AVERAGE } F) \cdot \Delta t = \Delta(Mv).$$

When M is not constant (e.g. in a belching rocket), $F = M \cdot a$ proves unsuitable, but CHANGE OF MOMENTUM, $\Delta(Mv)$, is still equal to $\int F\,dt$ or (AVERAGE F) · Δt. That leads back to a general definition of force,

$$F = \frac{d(Mv)}{dt},$$

or FORCE = RATE OF CHANGE OF MOMENTUM. That is Newton's original statement, which holds true always, even in Relativity mechanics.

Momentum

We call Mv "momentum" (plural: "momenta"). This Latin word for "motion" should hereafter be reserved for technical use, as a name for MASS · VELOCITY. Calling Mv momentum is mere dictionary-work; but the product Mv is very useful in science, so we give it a name, choosing one that reminds us of motion. Then the relation $F \cdot t = \Delta(Mv)$ reads "FORCE multiplied by the TIME-DURING-WHICH-IT-ACTS = CHANGE OF MOMENTUM."[3] In a sudden change, the time t is often small, so we write it Δt, meaning "change of time-of-day." This hints that Δt may be a small time, like the interval between 3 hours 42 minutes 4.60 seconds and 3 hours 42 minutes 4.72 seconds. Then we say:

$$F \cdot \Delta t = \Delta(\text{MOMENTUM}) \text{ or } \Delta(Mv).$$

Units

Since this relation $F \cdot \Delta t = \Delta(Mv)$ comes from $F = Ma$ it requires the same absolute units for force, newtons or poundals. If M is in kg, v in

[3] We give FORCE · TIME = the name "IMPULSE." So we can say, IMPULSE = CHANGE OF MOMENTUM

The name "impulse" carries a useful reminder that the same change of momentum can be produced by a small force acting for a long time or a huge force acting for a short time. In many violent blows and crashes we do not know the size of the huge force, F, or its very short duration, t; we only know their product, Ft, the impulse, measured by the change of momentum.

"PROBLEM SHEET" This problem is intended to be answered on a typewritten copy of this sheet. Work through the problem on the enlarged copy, filling in the blanks, (——), that are left for answers.

PROBLEMS USING $F\cdot\Delta t = \Delta(Mv)$ NAME__________

$F\cdot\Delta t = \Delta(Mv)$ is really another form of $F = M\cdot a$, and for many purposes a quicker one. Forces must be in poundals or newtons. If we use this to calculate a force, the answer will automatically be in poundals or newtons.

PROBLEM 1(a). A man applies a force of 200 newtons to a moving football, which contains 0.500 kg of total material, for a time of 1/50 second. How much faster will the ball be moving after this kick?

Force applied is. .__________newtons

Time, Δt, during which this force acts is . . . . .__________seconds

∴ gain of momentum must be. . . . . .__________newton·seconds

(NOTE: newton·seconds must be the same as kilogram·meters/seconds)

∴ since M is 0.500 kg, gain of velocity must be . . . .__________meters/second.

PROBLEM 1(b). A player kicks a $\frac{1}{2}$-kg football and gives it a velocity 14 meters/sec, starting from rest. The contact between foot and ball lasts 1/50 sec. Calculate the force involved in this impact.

(Here, use $F\cdot\Delta t = \Delta(Mv)$ instead of the clumsier method of Problem 9 in Chapter 7.)

The change of momentum is. . . . . .__________kg·m./sec (or newton·sec)

∴ the force involved must be . . . . . . .__________newtons

This force in "bad" units is about . . . . .__________kg-wt.

And this is about. . . . . .__________pounds-wt.

PROBLEM 1(c). A player kicks a $\frac{1}{2}$-kg football that is flying towards him at 10 meters/sec. It bounces straight back at 14 meters/sec. The impact lasts 1/50 sec. Calculate the average force involved.

(Velocity and momentum are vectors. Note the use of + and - signs.)

v_o = -10 meters/sec v = +14 meters/sec t = 1/50 sec

In 1/50 sec, momentum changes from __________ to __________kg·m./sec

The change of momentum is . . . . .__________kg·m./sec

∴ force involved . . . . = __________.__________ units

≈ __________pounds-weight.

meters/sec, the MOMENTUM Mv is in (kilograms) multiplied by (meters/sec), and this is written[4] kg · m./sec. If Mv is in kg · m./sec and t in sec, then F must be in newtons.[5]

Jumps and Crashes

Try using $F \cdot \Delta t = \Delta(Mv)$ in problems on jumping men and colliding cars. We shall make use of it in later problems, and later still in developing a molecular theory of gases, where it will enable us to make powerful predictions.

F in $F \cdot \Delta t = \Delta(Mv)$ is a real force: it is the actual force needed to make the momentum change that much in that time Δt. If the force is not provided by the ground or a wall or something, the moving thing's momentum will not change.

A big change of Mv may occur in a short time—for example, when a jumper lands on the ground or a car smashes into a wall. Then $\Delta(Mv)$ is large and Δt is small. $F \cdot (\text{small } \Delta t) = \text{large } \Delta(Mv)$. So F must be VERY LARGE. The forces in collisions are huge, and though they act for only a short time they can do great damage. To lessen F and save damage to bones, etc., we need to increase Δt. Bending knees, soft shoes, springy fenders, may do this. The ball catcher wears padded gloves, and lets his arm move back, to lengthen Δt during which the ball is stopping. In the problem in Ch. 7 of the jumper landing on the floor, his speed changed from 16 ft/sec to zero in about 1/100 sec.

Then $F \cdot \Delta t = \Delta(Mv)$ gives

$$F \cdot (1/100 \text{ sec}) = (200 \text{ pounds} \times 0) - (200 \text{ pounds} \times 16 \text{ ft/sec}) = -3200 \text{ pounds} \cdot \text{ft/sec or poundal} \cdot \text{secs}$$

$F = -320{,}000$ poundals (F is in poundals since we used pounds for mass, ft/sec for v, and F must be in absolute units)

$\approx 10{,}000$ pounds-wt. or 5 tons-weight

Five tons is a terrible force to allow the floor to use on his feet and drive up through his spine, even for 1/100 second. Do not try landing so abruptly—the penalty is pain and serious damage. Yet you *can* land safely after a 4-ft jump: simply bend your knees, making Δt ten or twenty times 1/100 sec, and thus reducing F by a factor of 10 or 20.

The footballer who gives a 1-pound football a speed of 60 ft/sec by kicking it for 1/100 sec exerts a force given by

$$F \cdot (1/100) = (1 \text{ pound} \times 60 \text{ ft/sec}) - (1 \text{ pound} \times 0)$$

$$F = 60/0.01 = 6000 \text{ poundals} \approx 180 \text{ pounds-wt.}$$

He needs tough toes.

The wrestler thrown to the ground tries to make his time-of-arriving-on-the-ground as long as possible, by relaxing his muscles and spreading the crash into a series of blows, as ankle, knee, hip, ribs, and shoulder fold onto the floor in turn.

Momentum Sharing

When things collide, they exchange or share momentum. Watch the following experiment to see whether momentum is gained or lost in collisions.

DEMONSTRATION EXPERIMENT. A truck running on a friction-compensated track hits a stationary truck and the two lock buffers and proceed together. The first truck carries a billboard so that an electric eye and clock can give its speed. A second electric eye and clock show the speed of the combine after the crash. (See Fig. 8-1.) Here is a fictitious example of this demonstration, necessarily unconvincing, given merely to show how a test is made. A 2.00 kg truck A hit a stationary 4.00 kg truck B head-on; the two interlocked and the 6.00 kg

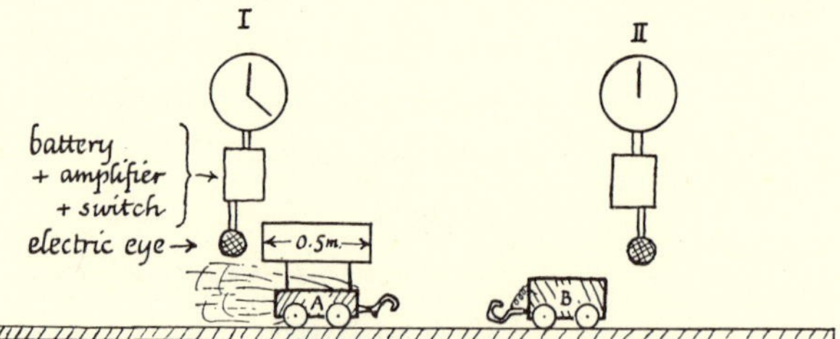

FIG. 8-1. COLLISIONS

Clock I shows transit-time of billboard on truck A, moving alone before collision. Clock II shows transit-time of billboard after collision, when trucks are locked together. Truck A is given a shove and released to move before it passes the electric eye. In this sketch it has just passed the first electric eye, is moving steadily, and will soon smash into the second truck. Truck B is at rest until truck A hits it. When the trucks collide, the hook-couplings interlock and are held locked by a spring. After collision, the combine moves past the second electric eye.

[4] The dot between the units means the units are multiplied together. You met this multiplication of units long ago in arithmetic problems with "man-hours." There, a hyphen was used to show multiplication, offering a silly confusion with a subtraction sign. A hyphen is often used when scientific units are multiplied. Here, we adopt a more modern symbol, the multiplying dot, as in newton · meters, man · hours, etc.

[5] These units are consistent. Remember that, with $F = Ma$, 1 newton gives a mass of 1 kg an acceleration 1 meter/sec².

$$F = M \cdot a$$
$$1 \text{ newton} = (1 \text{ kg}) \cdot (1 \text{ meter/sec}^2)$$
$$1 \text{ newton} = 1 \text{ kg} \cdot \text{meter/sec}^2$$
$$\underset{F}{(1 \text{ newton})} \cdot \underset{t}{(1 \text{ sec})} = \underset{M}{1 \text{ kg}} \cdot \underset{v}{\text{meter/sec}}$$

combine proceeded more slowly. Before the collision, the 0.5 meter billboard on truck A (which was already moving steadily) took 0.40 seconds to pass the first electric eye. (And it would have taken 0.40 seconds to pass the second electric eye, if it had reached it without collision.) After the collision, the billboard took 1.20 seconds to pass the second electric eye. (It was carried on truck A, but as they were locked together it represented the combine.) Then, before collision,

SPEED OF TRUCK A $= 0.50$ meters/0.40 sec
$= 1.25$ m./sec

SPEED OF TRUCK B $= 0$ meters/sec

After collision,

SPEED OF COMBINE A + B $= 0.50$ meters/1.20 secs
$= 0.417$ m./sec

Now calculate the total momentum before and after collision to see whether momentum is gained or lost in the collision.

Before collision:

MOMENTUM OF TRUCKS

$$= \underbrace{(2.00\text{ kg})(1.25\text{ m./sec})}_{\textit{Truck A}} + \underbrace{(4.00\text{ kg})(0)}_{\textit{Truck B (at rest)}}$$

$$= 2.500\text{ kg}\cdot\text{m./sec}$$

After collision:

MOMENTUM OF COMBINE

$$= (6.00\text{ kg})(0.417\text{ m./sec})$$
$$= 2.502\text{ kg}\cdot\text{m./sec}$$

The agreement in this case is magnificent, but then we invented the numbers. You should see as many real demonstrations as possible, and you should know that all real experiments have shown, within the accuracy which their apparatus can provide, that momentum is neither gained nor lost, but only exchanged or shared. *This is independent of the kind of collision.* It can be a delicate springy bounce, a sticky amalgamation, or a fearsome smash with lots of kinetic energy wasted as heat, and yet momentum is conserved. This gives us a very valuable guiding principle for dealing with collisions:

MOMENTUM GAINED = MOMENTUM LOST

and in another form:

TOTAL MOMENTUM NEVER CHANGES

Collisions and Conservation of Momentum

Collisions are important: gas molecules exerting pressure on the walls they bombard; helium atoms shying away from a gold atom's nucleus as they hurtle through gold leaf, and emerging to tell us about atomic structure; neutrons showing their mass by knocking hydrogen atoms straight ahead; fast electrons flinging other electrons out of atoms; and even quantum-packets of light bouncing like bullets on electrons—all these are collisions to which we may profitably apply our new general rule and obtain hints of new knowledge or better understanding. We believe the same rule applies to remote "collisions" such as the gravitational effects of the Sun on the Earth, or of one planet on another, and the slow silent "collisions" of the Moon and our ocean that we call tides. Details of forces may differ, but one rule seems to guide all collisions and interactions, a rule summed up by Newton in a form that has been extended into atomic physics and incorporated in the re-thinking of Einstein's Relativity. The rule is this:

> IN ANY INTERACTION IN A CLOSED SYSTEM (ON WHICH NO RESULTANT FORCE ACTS FROM OUTSIDE) MOMENTUM IS CONSERVED, AS A VECTOR.

Momentum is a Vector

In $F \cdot \Delta t = \Delta(Mv)$, there is a vector on each side. FORCE is a vector, but TIME does not have a direction in space: it is just a number (like "number of ticks of the clock") to be used as a multiplying factor. VELOCITY is a vector, but MASS has no direction. Mass is a "scalar" a simple number (like "number of trucks") to be used as a multiplying factor. (Multiplying 3 ft/sec *due East* by 5 pounds produces 15 pounds · ft/sec *due East.*) So we expect to find IMPULSE $\boldsymbol{F} \cdot \Delta t$ a vector, and MOMENTUM $\boldsymbol{Mv}$ a vector; and experiment confirms this. The full statements of Newton's Law II contain a reminder of this: the acceleration, and therefore the change of momentum, takes place *in the direction of the applied force.* This may not seem very important in head-on collisions where all the motion is along one line; but in collisions at other angles we must treat momentum as a vector. When cars coming from different directions collide and interchange momentum, we find that the *Mv*'s obey vector addition. Fig. 8-2 shows a bird's-eye view of a collision in which car A moving eastward crashes into car B moving northward on an icy level road. If they cling together, they move in a slanting direction with momentum which is the vector sum of the two original momenta. Fig. 8-3 shows a bomb sliding on an icy pond. When it explodes into two fragments, their momenta combine by vector addition

FACT PICTURES

VECTOR DIAGRAM FOR MOMENTUM

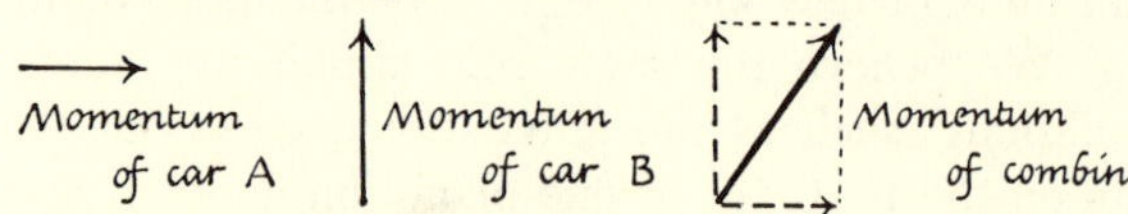

FIG. 8-2. MOMENTUM AS A VECTOR

CARS COLLIDING ON AN ICY ROAD. Car A has eastward momentum. Car B has northward momentum. When the cars collide there is no change of northward momentum, no change of eastward momentum. The combined wreck continues with these two momenta, which add as vectors to give wreck's momentum.

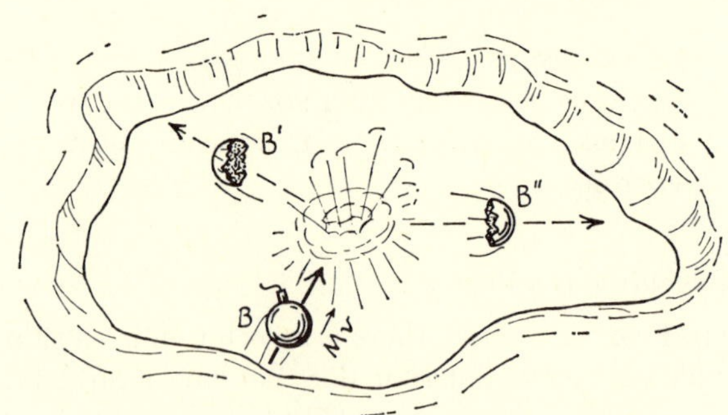

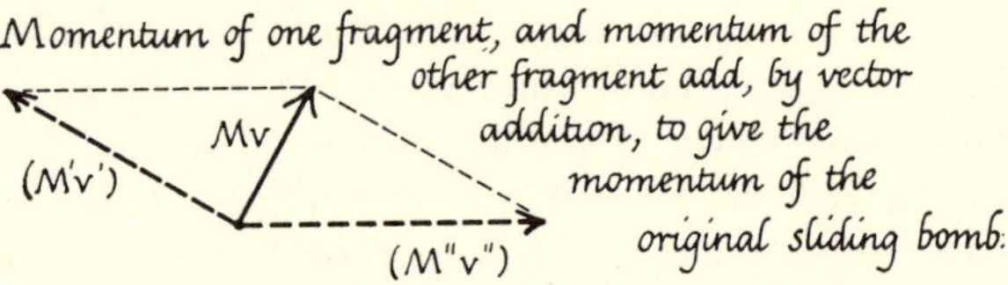

FIG. 8-3. BOMB SLIDING ALONG AN ICY POND

When the bomb bursts, the momenta of its fragments add up (by vector addition) to the original momentum.

to equal the original momentum of the sliding bomb.[6]

To test the vector conservation of momentum, we must give up the model railroad track, and observe blocks of dry ice colliding on an aluminum table. Or we can use pendulums: steel balls hung on long threads.[7] Then we find, in every case, that the momenta after collision add by vectors to the same total as the momenta before. Or we can analyze our measurements by splitting each Mv into components along two perpendicular directions. If just one body is moving originally, we choose our x-axis along the motion and y-axis across it; and we split each momentum into x- and y- components. Then we find the total of the x- components after collision = the original momentum. And the two y-components after collision are equal and opposite.

It may seem artificial and useless to sketch and analyze the tracks of colliding bodies like this. But we can photograph the tracks of single atoms and parts of atoms making collisions; and the analysis of those tracks is of tremendous importance in atomic physics. Electrons and charged helium atoms and other atomic particles make clear tracks as they fly through a "cloud chamber" (discussed in Ch. 39). When there is a collision the track shows a sudden bend and a new track branches out, the track of the recoiling victim, usually a gas atom that has been hit. If we know the masses of the colliding atomic bodies, we can gain useful information about their velocities by drawing a vector diagram for momentum. Or if we know the velocities a vector diagram will tell us the ratio of the masses.

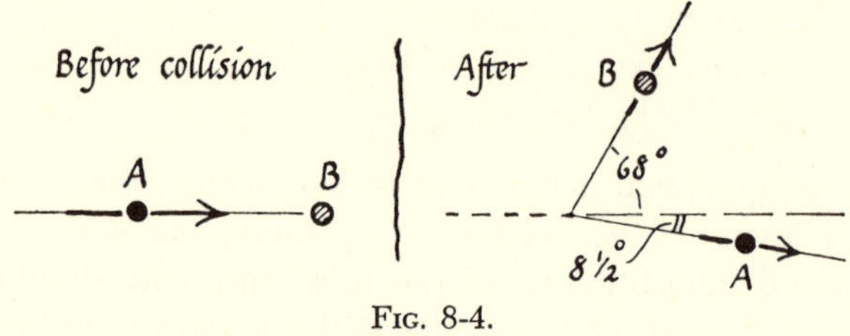

FIG. 8-4.

PROBLEM 2. A NUCLEAR COLLISION

Measurements of an actual cloud-chamber photograph yielded the following data,* for a fast alpha particle (= helium nucleus), A, hitting a stationary particle, B. (The velocities are in arbitrary units.)

Before collision, A was moving with speed 2.00 units/sec.

After collision, A moved with speed 1.90 units/sec, along a path making 8.5° with its original path.

After the collision, B moved with speed 1.25 units/sec along a path making 68° with A's original path (so that the Y-shaped fork had an angle 76.5°).

* Professional physicists analyzing such photographs assume conservation of momentum and conservation of kinetic energy, and use algebra and simple trig to find M_B/M_A in terms of the angles alone. That saves the trouble of roundabout estimating of velocities. However, in the rare cases where the collision is not elastic, the velocities are estimated from the lengths of tracks; and the velocities given in the data here are the kind of values such estimates would give. The arbitrary unit used for velocity is worth about 10,000,000 meters/sec.

[6] Further geometry extends this to a remarkable result: when any projectile—bullet, rocket, or atomic nucleus—explodes, the center-of-gravity of its fragments pursues the same path after the explosion as before. For example, suppose a rocket, travelling in an ellipse in the Earth's gravitational field, explodes or shoots out a sub-rocket. The center-of-gravity (= "centroid" or weighted mass-center) of the pieces continues along the ellipse as if nothing had happened—until one fragment hits the Moon or returns to Earth, or until air-friction spoils the isolation of the system. No wonder nuclear physicists like to measure collisions from the center-of-mass of the participants.

[7] Velocity measurements are more difficult without a track, but they can be made by taking a series of flash photographs on one film, with equal times between exposures.

Identify the particle B, by finding how its mass compares with the mass of A, following the stages (a), (b) below. For convenience, adopt the relative scale of atomic masses used in chemistry, on which the helium nucleus A has mass 4.0 "atomic mass units." Then if B were an oxygen nucleus its mass would be 16.0 a.m.u.; if nitrogen, 14.0; if helium, 4.0; if heavy hydrogen, 2.0; if ordinary hydrogen, 1.0. Use this list for your identification. (If you get some fractional answer, like 0.2 or 5.3, you have discovered a new atomic particle, which should be named in your honor.)

(a) On a large sheet of paper, draw a vector diagram of momenta, to a suitable scale, thus: draw A's momentum before and after collision, and mark the momentum that B must gain to keep momentum conserved.
(b) Measure B's momentum-vector and use the velocity from the data to calculate B's mass.
(c) You probably used the 8.5° angle in (a), but not the 68°. In that case, measure the appropriate angle on your diagram and compare it with 68°. (The agreement you get is a partial test of the addition and conservation of momentum—which you assumed in (a) above).
(d) When you are familiar with *kinetic energy* $= \frac{1}{2}mv^2$, (Ch. 26), return to this problem. Use 4.00 for A's mass and your value for B's and find how nearly kinetic energy is conserved. If it is conserved, the collision is a simple *elastic* one with no nuclear changes. If kinetic energy is not conserved, nuclear energy must have been stored or released.

Conservation Laws

Whatever happens, momentum lost by one thing is gained by some other thing(s): the vector resultant never changes. To keep this rule universally true we now know that we must include the momentum carried by electromagnetic fields, e.g. in light waves; and though we still calculate momentum as MASS · VELOCITY we give MASS its relativistic property of increasing with motion. The relativistic change is imperceptible at ordinary speeds, even at most astronomical ones, but it raises mass and momentum up and up towards infinite values as we observe atomic particles with speeds approaching that of light.

This simple account-keeping rule—total Mv after collision is the same as total Mv before—makes momentum enormously important as well as useful. Studying the mechanics of moving bodies, whether planets or atoms, is like tracking a gang of criminals who frequently change their disguise. The detective looks for recognizable marks which his shadowing agents can follow through all changes of disguise—the jagged ear, the golden tooth, limping foot, or twitching finger. In mechanics, scientists found mass to be one such continuing characteristic: mass is conserved, they said. For centuries matter has been regarded as indestructible, chemical changes being only exchanges of matter. Careful weighing of chemicals in a bottle before and after chemical reactions showed no measurable change in total mass; so scientists claimed conservation of mass—which they took to imply conservation of matter—as a universal rule. But this rule alone was not sufficient for full account-keeping of collisions (and in recent times we have found the rule itself unreliable in its baldest form). Then momentum, Mv, emerged as a useful measure of a body's motion, as it was found to be conserved. Clinging to Mv as the best clue of all to motion and its changes, we still make conservation of momentum a great basis of mechanics.

We now have two rules for any *closed*[8] system:

> IN ANY INTERACTION, TOTAL M REMAINS UNCHANGED

(MASS M is a scalar) This is called the Law of Conservation of Mass.

> IN ANY INTERACTION, TOTAL Mv REMAINS UNCHANGED

(MOMENTUM is a vector) This is called the Law of Conservation of Momentum.

Such constant-total rules enable us to make predictions or extract useful information from measurements. In a way, such rules are the heart and soul of our physical science, the codification of nature. (They are like the rules of balance-sheets that accountants use, such as "The debits column and the credits column must add up to the same totals"—a rule based on a belief in conservation of cash.) Are there any more such rules? Are Mv^2, Mv^3, &c., also conserved? Measurements of M and v in collisions tell us certainly not for Mv^3 and Mv^4—so we lose interest in them and do not waste a name on them. But Mv^2 proves interesting: in *some* cases it is conserved; and in others it disappears into different forms of the useful thing that we call energy. So we do deal with Mv^2, giving it, or rather $\frac{1}{2}Mv^2$, the name "kinetic energy." We shall return to this in a later chapter. (CHANGE OF MOMENTUM is given by FORCE · TIME. We shall find that CHANGE OF KINETIC ENERGY is given by FORCE · DISTANCE, another simple product containing force and a measurement—no wonder $\frac{1}{2}Mv^2$ is simple and useful.)

Other Demonstrations of Momentum Changes

(i) Rough experiment with tables on wheels. Students A and B each sit on a table with good ball-bearing wheels. Starting at rest, far apart, they pull them together with a rope till they collide. (Fig.

[8] If two trucks collide and you keep track of the momentum of only *one*, you will not find momentum conserved. For conservation, you must keep track of the momentum of both trucks; in general, of *all* participants. That is what the closed-system qualification means: include all the interacting bodies.

8-5) If the masses [student + table] are unequal the taut rope gives them unequal accelerations. They gain unequal velocities as they approach. But when they collide they destroy each other's momentum and there is no motion left. This illustrates momentum changes but it provides no proof of conservation unless we measure the motions acquired.

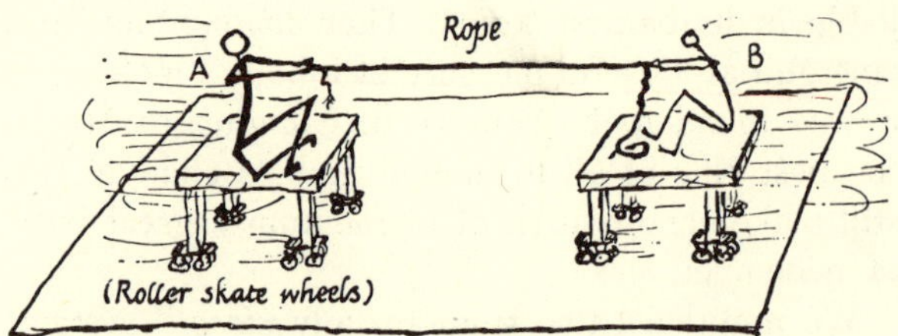

Fig. 8-5. Rough Experiment:
Students on tables with good wheels. Students A and B pull steadily on rope till tables collide.

★ PROBLEM 3

While the tables in the demonstration above are being pulled together, the rope is kept taut by one or both of the riders pulling it in. Suppose that:

(a) Rider A holds his end firmly while B pulls the rope in, keeping the tension constant at 100 newtons; or
(b) Rider B holds his end firmly while A pulls the rope in, keeping the tension constant at 100 newtons; or
(c) Both A and B pull the rope in, keeping the tension constant at 100 newtons.

What differences would you expect to notice between schemes (a), (b), and (c) as regards:

(i) the relative motion with which the two tables approach?
(ii) the motion of A and his table? (*Hint*: Rider A + his table is a definite mass, pulled by a 100-newton force in each case.)
(iii) the piling up of rope on each table?

(ii) *Model trucks pushed apart by spring.* See Fig. 8-6(a). Two trucks are placed at the mid-point of a track which is friction-compensated for trucks moving outward. A compressed spring tries to push the trucks apart, but they are tied together by a thread. When the thread is burned the spring pushes the trucks apart, giving each some momentum; then the spring falls out of action, and the trucks run steadily.

If the trucks have equal masses, experiments show that they run away with equal and opposite velocities—for example, while truck A travels 1 meter to some finishing post, truck B travels 1 meter in the opposite direction. Then, with equal and opposite velocities, v and $-v$, and equal masses, B's gain of momentum, Mv, is equal and opposite to A's gain, $-Mv$.

If the truck B has twice the mass of A, it is found

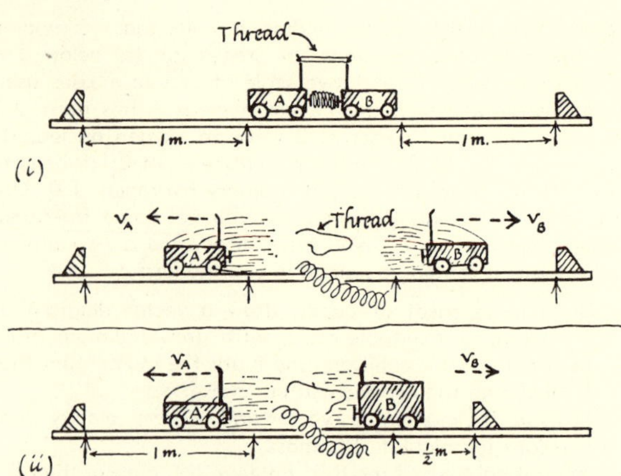

Fig. 8-6a. Model Trucks Pushed Apart by a Spring
The spring is compressed and the trucks are tied together with a thread. When the thread is burned, the spring pushes the trucks apart, then drops out of action, leaving each truck with a steady motion. Do they have equal and opposite momenta?

to gain only half as much velocity as A: while A travels 1 meter B travels ½ meter. Then, while A gains momentum $-Mv$, B gains opposite momentum $2M(\frac{1}{2}v)$.

In each of these cases, the changes of momentum are equal and opposite. Other masses, such as M hitting $3M$, or $2M$ hitting M, give similar results.

(iii) *Model trucks pulled together by spring.* See Fig. 8-6(b). This is a small version of (i). Two

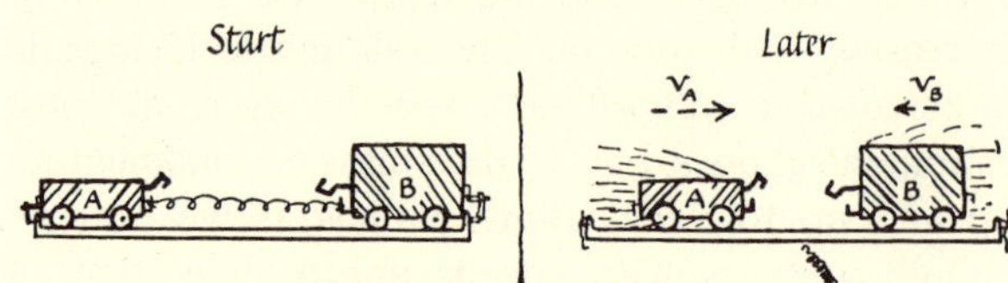

Fig. 8-6b. Model Trucks Pulled Together by a Stretched Spring
Before release, the trucks are held apart by hand. After release, they approach, collide, and lock. What motion then remains?

trucks far apart on a level track are pulled towards each other by a long stretched spring, which then falls out of use, leaving the trucks to approach, hit, and lock together with a hook coupling. If they are released at rest originally, the combine *finishes at rest.* We say that the two masses gained equal and opposite momenta, which cancelled out on collision. This happens whether the masses are equal or unequal.

(iv) *Measurement of rifle bullet speed.* When a rifle bullet bores its way into a block of wood and comes to rest the collision is very inelastic—the bullet's *energy* of motion largely disappears, being converted into heat, but its momentum is just handed on to the wood. We use a calculation of momentum to measure the bullet's speed, as shown in the problem below.

★ PROBLEM 4. RIFLE BULLET SPEED MEASUREMENT

A rifle bullet is fired horizontally into a block of wood on a truck which runs on a friction-compensated track, sketched in Fig. 8-7. The truck, originally at rest, then proceeds

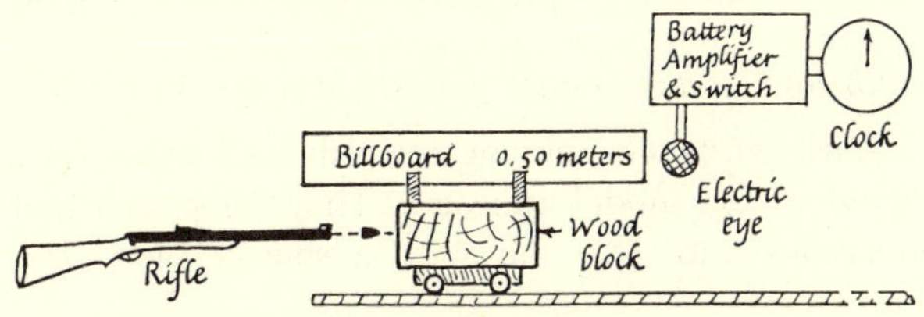

Fig. 8-7. Measuring the Speed of a Rifle Bullet, Assuming Momentum is Conserved

steadily along the track, with the bullet embedded in it. A billboard 0.50 meter long carried on the truck takes 0.80 second to pass an electric eye. The bullet, extracted from the wood, is found to have mass 0.00300 kg (3 grams or about 1/10 ounce). The [wood + truck] has mass 2.997 kg. To find the original speed of the bullet, call it V meters/second, and assume momentum-conservation.

(i) What is the speed of the [wood + truck + bullet] after collision?
(ii) Call the bullet's speed before collision V meters/second. What was its momentum before collision?
(iii) What was the momentum of the [wood + truck] at rest before collision?
(iv) What was the total momentum before collision?
(v) What is the *mass* of the combine after collision?
(vi) What is the *momentum* of the combine after collision?
(vii) *Assume* momentum-conservation, and write your assumption as an equation, using the results above. Find the value of V.

This provides a practical method used in ballistics. (The wood block is usually suspended as a pendulum instead of running on rails. This alters the geometry of the experiment but not its essential principle.) The method assumes that momentum is conserved; but we can measure bullet speeds by other methods, and the results lend support to our assumption. Fig. 8-8 shows an alternative method that we can use as a check.

Newton's Law III

If we are sure that momentum, Mv, is conserved—never lost or created but only exchanged—we can infer that two things which collide or interact must exert equal and opposite forces on each other. This is Newton's Third Law of Motion,

"Action = Reaction."

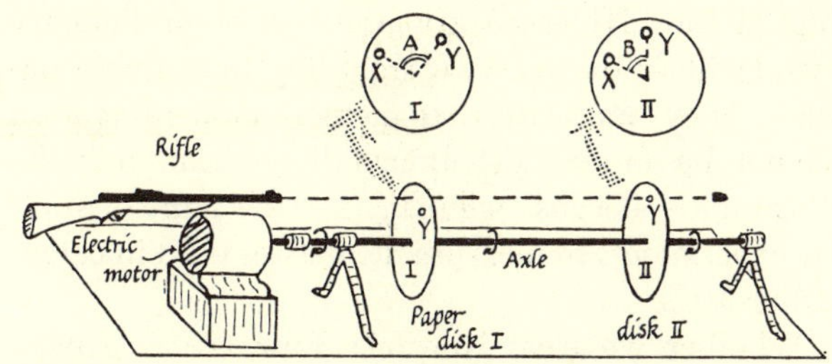

Fig. 8-8. Measuring Speed of a Rifle Bullet: Alternative Method. The bullet is fired through two paper discs a measured distance apart, making holes Y. The discs are carried on an axle which spins at a measured rate. Preliminary shot, fired with axle at rest, makes "standardizing holes" X in each disc. Angular displacement of holes ($\hat{A} - \hat{B}$) measures the time of flight of the bullet.

Here is the argument:

Suppose two bodies A and B collide (or exchange momentum in any other way). Call A's change of momentum $\Delta(Mv)_A$, and B's change of momentum

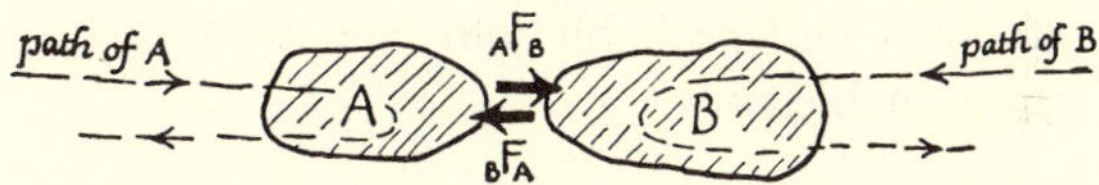

Fig. 8-9. Forces in a Collision. When they collide, *each body* pushes the other with *one force.*

$\Delta(Mv)_B$. Then, if momentum is conserved, $\Delta(Mv)_A$ and $\Delta(Mv)_B$ must be equal and opposite.

So $\Delta(Mv)_B = -\Delta(Mv)_A$.

(Or, putting the same thing another way, the total change $\Delta(Mv)_A + \Delta(Mv)_B$ must be zero).

But, for A, the change of momentum,

$$\Delta(Mv)_A = (\text{FORCE on A}) \cdot \Delta t$$

And, for B, the change of momentum,

$$\Delta(Mv)_B = (\text{FORCE on B}) \cdot \Delta t$$

The time Δt is the same for both, since A cannot collide with B for a longer time than B spends colliding with A: Δt is simply the duration of the collision.

∴ if momentum is conserved,

$$(\text{FORCE on A}) \cdot \Delta t = -(\text{FORCE on B}) \cdot \Delta t$$

$$\therefore\ (\text{FORCE on A}) = -(\text{FORCE on B})$$

∴ (FORCE on A) and (FORCE on B) are equal and opposite;

ACTION = REACTION

Some scientists claim that momentum-conservation is well established by experiment and therefore consider Newton's Law III well vouched for. Others regard Law III as an axiom, a sort of preliminary statement of the way we are going to examine nature. They warn us that momentum-conservation cannot be proved experimentally—it can only be illustrated—because we really use the same or similar experiments to measure the masses used in calculating Mv's.

Whether we treat Newton's Law III as experimental fact or as basic unprovable axiom, we now use it all through science; and it moulds our thinking without leading to inconsistency. Newton himself did not announce Law III from nowhere and try to use it like a dictator. He stated it as a working opinion which he was going to use and take for granted in developing mechanics; but he also gave it a thorough test by measurements on colliding pendulums. (See Newton's account of his experiments,[9] showing his ingenious way of coping with air friction.)

If you understand the meaning of Law III—it is often misunderstood, even misrepresented in textbooks—you will probably join Newton in using it as a reasonable rule.

A Powerful Tool: Independent of Details

Now you can see how powerful the law of Conservation of Momentum can be in solving problems. When an event occurs within a system, many internal forces may arise between one part of the system and another, but they appear as equal and opposite pairs (Newton Law III) so they cannot make any net change of momentum. We can make our overall calculations without knowing, or caring about, the internal strains and motions. And when we divide our system into two parts—as in treating a collision—and say that the momentum gained by one must be lost by the other, *we need not know anything about the forces involved in that interchange.* They make an equal-and-opposite pair of Action and Reaction. We know they make equal and opposite changes of momentum whether they are constant forces, or forces that increase rapidly and then decrease again; whether they are sudden forces of abrupt contact or gentle gravitational pulls; whether they drive molecules into vibration (heat) or wind up springs (potential energy) or return the original energy of motion in full. For example, when we fire a fast bullet into a block of wood on a frictionless table, the wood (+ bullet) coasts away with a velocity that we can calculate from the masses and original speed, assuming momentum is conserved. The calculation needs no knowledge of the detailed fate of the bullet. Normally the bullet plows its way through the wood fibers, rubbing and scraping and leaving them hotter, until it has spent all its motion in heat. Or if it hits a block of metal inside the wood the bullet itself heats up, and is likely to melt. However we could conceal a device inside the wood to catch the bullet and let it spend its motion compressing a spring, or in setting a small wheel spinning. In every case the final speed of the target is the same—provided the bullet stays in it.

[9] Given in *A Source Book in Physics* by W. F. Magie (New York, 1935).

Collisions and "Contact"—a Misleading Word

Start two trucks moving towards each other on a "frictionless" model railroad. They keep constant velocities until they hit with a smack; then after

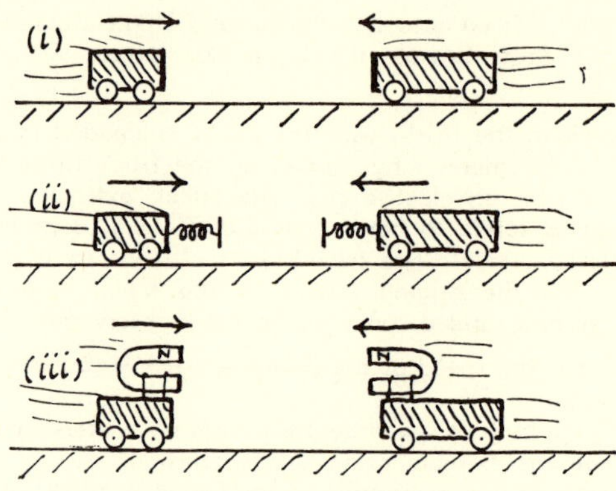

FIG. 8-10a. COLLISIONS
(i) Contact (ii) Spring-buffers (iii) Magnets repel

very brief contact they rebound with different velocities but the same total momentum. Now equip the trucks with buffers of good steel springs. The collision takes longer, and we can examine its stages in detail. The final rebounding velocities may be greater than those after an abrupt smack, but momentum is conserved. At every intermediate stage during the collision, momentum is conserved: the total momentum of the two trucks is equal to their total momentum before the collision started. In the middle of the collision, when they are closest and have compressed the springs most, the two trucks and springs are all moving together with one velocity—and we could calculate its value since we know the total mass and the total momentum. We could install a latch to hold the trucks together at that stage, and then we could measure the velocity of the combine and check our calculation, as a test of momentum conservation. That is what we did with the colliding trucks in the first Demonstration Experiment of this chapter. Now make the "collision" even more gentle by arranging two large magnets, one on each truck, to repel each other. Then, as the trucks approach, the magnets push harder and harder like a spring being compressed. The trucks rebound—again with momentum conserved—without having touched each other. At first sight, this seems an unreal collision, yet it is really a typical collision, a large-scale, slow-motion, model of a sharp contact collision. In each type of collision (magnets repelling, springs pushing, smack of contact) equal and opposite forces develop, at some stage of the approach, that push the colliding bodies outward and go on pushing until they are apart again.

The magnetic repulsions begin to make themselves felt quite far away and increase greatly at closer approach. Just how close the trucks have to approach to be turned back by these repulsions depends on the original speeds. Over a wide range of speeds the magnetic repulsions suffice.

The smack of contact produces similar forces, but at much smaller distances. They are "short-range" atomic forces that remain practically zero until the front atoms of one body are *very* close to the front atoms of the other body—far less than a molecular diameter apart. Then these strong repulsions appear; and they grow much stronger still, at slightly closer approaches.[10] And that is all contact is, a sudden set of repulsions at very close range. There is no such thing as one atom "touching" another. On the "micro" scale of atomic behavior there are only fields of force that push and pull on atoms or parts of atoms, fields that change rapidly with distance. Press your finger on the table, and you feel the table's atoms repelling your finger's atoms when they are too close. However firmly or lightly you touch the table, you are only suffering small repulsions that are transmitted to the flesh of your finger and affect your nerves.[11] Except in this sense of developing big forces at very short distances, there is no "touching" or "contact" between finger and table—those words have a misleading suggestiveness. Your finger with its sense of touch is only what research engineers would call a "strain gauge."

To illustrate collisions in a different way, replace the trucks by rolling balls, and instead of providing repulsive forces let the balls roll up a hill of suitable profile. Fig. 8-10b shows hill models for the three cases we have just considered. Note that for the

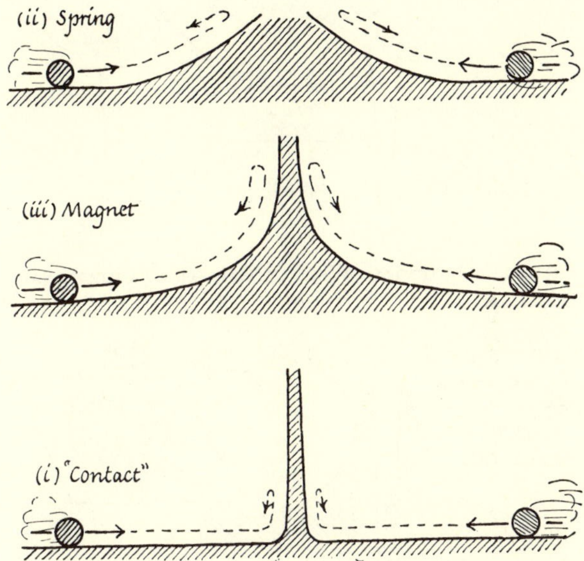

Fig. 8-10b. Potential Hills. These hills should be pictured with two equal balls approaching with equal speeds, or one ball can be fixed at the center of the hill while the other ball is rolled towards it.

contact smack the hill is very sudden and steep; but it is *not* a vertical wall. (These are called "potential diagrams" because the heights on the hill represent the potential energy stored in spring—or magnetic

[10] At very close approach, these atomic repulsions outgrow the "long-range" attractions that hold solids together, produce "surface tension," and provide some chemical bonds. The attractions extend out a few molecule diameters and increase at closer approach, but they do not increase as rapidly as the repulsions. The repulsions catch them up, and their effects must just balance at the equilibrium position of atoms on the surfaces of solids, etc. When we try to push atoms closer (in any collision) the repulsions exceed the attractions and oppose our push. The repulsions must be there or, with attractions alone, matter would not just hold together strongly: it would collapse inward. We consider both sets of forces electrical in origin. The short-range repulsions arise when one atom tries to interpenetrate the structure of another. Then electrons resist trespass by other electrons into their orbits; and nuclei—no longer fully shielded by electrons—repel. The mechanisms are not easy to picture or explain. For some further comments, see Ch. 44.

[11] When you slide your finger along a rough table, you feel sideways rubbing forces. Again, these are only strong forces, and there is no "contact" between solid atoms. They are repulsions between your finger and irregular humps (on the micro scale) on the table. When you feel friction between very smooth surfaces—e.g. metal sliding on metal—atomic *attractions* play a part (at slightly greater distances) in dragging tiny pieces of metal off one face on to the other. When iron rubs on copper, chemical analysis shows that a little copper is dragged off by the iron; and sometimes twigs of uprooted copper can be seen with a microscope. We even know for certain, now, that copper rubbing on copper exchanges minute quantities between the faces, even when no twigs are visible. Here is a puzzle: how can this be known, and the exchange measured, when no chemical analysis can possibly distinguish the two lots of copper? In reading later chapters, you will be able to guess the remarkable method used.

field, or atomic fields—during the collision; see Ch. 26. Such potential hills are very useful in discussing collisions in nuclear physics. When there are *attractive* forces the hill becomes a potential *well.* Fig. 8-11 shows the potential well of the Earth's gravitational field; and the potential well of an atomic nucleus with its protective potential hill around it.)

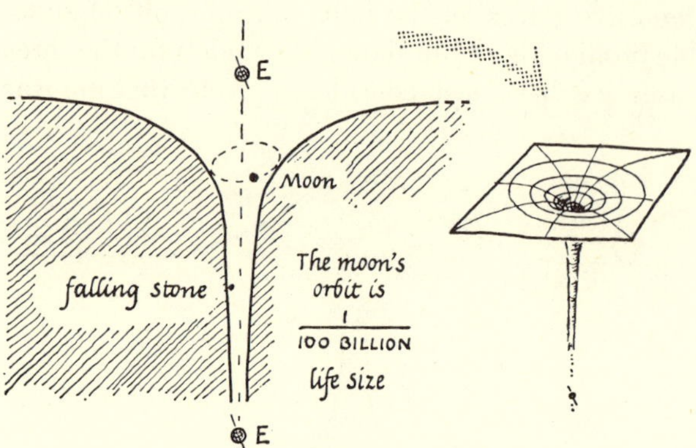

FIG. 8-11a. POTENTIAL WELL OF EARTH'S GRAVITATIONAL FIELD

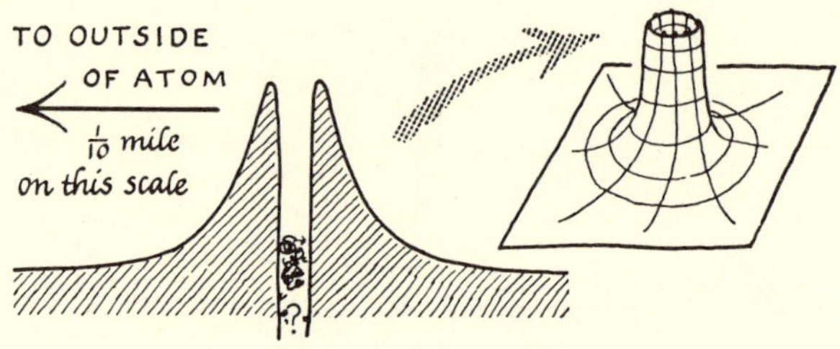

FIG. 8-11b. POTENTIAL WELL AND HILL OF NUCLEUS

Thus we consider all collisions as essentially alike. The differences lie in the form of the force-field involved, and those do not affect our overall treatment by momentum conservation. All the force fields we deal with in large-scale physics seem to exert equal and opposite forces: gravitational attractions, electrical repulsions and attractions, magnetic forces (which we consider are due to electricity in motion), and the molecular and atomic forces which we consider are electrical in origin. As yet we know only part of the story about nuclear forces.

Since "Action = Reaction" underlies our whole treatment of forces and motion, it is important to understand its meaning; so we shall now return to a further discussion.

Meaning of "Action = Reaction"

You cannot push me without finding that I push you back. Suppose we join hands and you push me with a force of 100 newtons eastward (see Fig. 8-12). Automatically I must push you with a force of 100 newtons westward. We cannot have one force without the other. Any attempt to push leads to both forces or none.

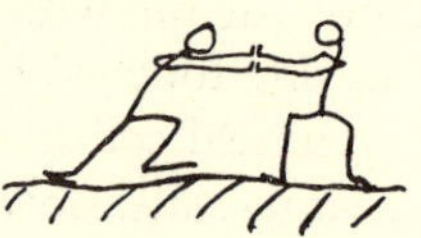

FIG. 8-12a.

The forces are still equal and opposite if we move steadily or accelerate. If you are on roller-skates, my push will accelerate you. I must run after you faster and faster to keep up with you as I push. But you still push me just as hard as I push you, whatever our motion. These two forces are equal and opposite, but they do not just cancel out to no force at all. *My push* is a force *on you,* and you

FIG. 8-12b.

feel it. *The fact that you are also pushing me back does not constitute a force on you.* Of the two forces only *my* push acts on you. Unless it is balanced by other outside forces also acting on you, it will make you accelerate.[12]

The Horse and Cart Paradox

Suppose a horse is pulling a cart along. The cart pulls the horse back as much as the horse pulls the cart forward. How do they ever progress? If you are worried

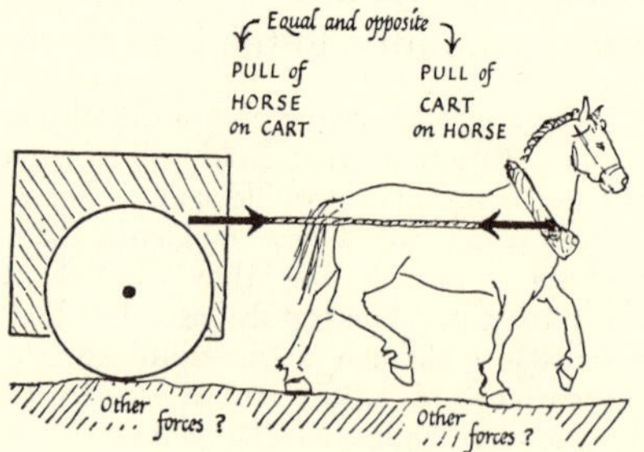

FIG. 8-13. THE HORSE AND CART PROBLEM

[12] Wear your roller-skates and let me shove my fist against your stomach with a force of 100 newtons. It will still hurt while you are accelerating. At the same time your reaction, or counterpush, acts on *me, but not on you,* and my fist feels a 100-newton push back on it. If *I* am on roller-skates, I, too, shall accelerate, backwards.

by this question, read the discussion below—otherwise you can omit it without serious loss.

The confusion of this "paradox" arises from failure to watch which force acts on what. Suppose the horse pulls the cart forward with 100 newtons eastward. This acts *on the cart only*, trying to accelerate it. The fact that the horse exerts this pull on the cart does not constitute a force on the horse. There *is* a force on the horse, the backward pull of the cart, 100 newtons westward, and that is a pull on the horse alone. Each of these two forces *acts on* one thing only, the thing that it pulls, the thing that it tries to accelerate; but the force is *provided* by the other thing.

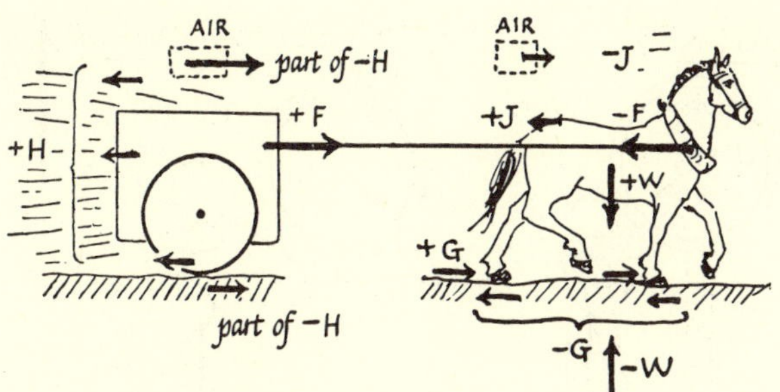

FIG. 8-14. THE HORSE AND CART PROBLEM: SETS OF FORCES

Pull of horse on cart +F
Pull of cart on horse −F } are equal and opposite

and, quite separately, for horizontal forces

Horizontal shove of ground on horse +G
Horizontal shove of horse on ground −G } are equal and opposite

Drag of friction on cart +H
Drag of cart on ground and air −H } are equal and opposite

Drag of air on horse +J
Drag of horse on air −J } are equal and opposite

and, for vertical forces,

Pull of Earth on horse +W
Pull of horse on Earth −W } are equal and opposite

with a similar pair for cart and Earth.

The *horse provides* 100 newtons eastward,
which *acts on* the *cart*.
The *cart provides* 100 newtons westward,
which *acts on* the *horse*.

Cart. There are other forces acting on the cart, friction drags of ground and air. If the friction on the cart is just enough to balance the horse's pull, the resultant force on the cart is zero and the cart will stay at rest or move along steadily. Then

PULL OF HORSE − FRICTION DRAG = *zero*.

Therefore, no acceleration. If the horse's pull is more than the friction drag, then

LARGER PULL OF HORSE − FRICTION
= FORWARD RESULTANT FORCE,
which accelerates the cart.

Horse. Meanwhile, the horse is pulled back by the cart; and if he is to progress, he must kick the ground backwards and thus make the ground kick him forwards (another pair of equal and opposite forces). The horse, spurning the ground, is pushed forward by it. If

FORWARD PUSH OF GROUND − DRAG OF CART
− DRAG OF AIR FRICTION ON HORSE = *zero*

then the horse proceeds steadily. But if he kicks the ground harder, so that

LARGER FORWARD PUSH OF GROUND
− DRAG OF CART − AIR FRICTION
= FORWARD RESULTANT FORCE
(which acts on horse)

then he will accelerate.

"Well, then," you may object, "if I take both horse and cart together, why don't the two pulls, +100 newtons and −100 newtons, cancel?" Of course they do now. They cancel and have no more effect on the progress of the combined [horse + cart] than you have if you shake hands with yourself and tug one hand with the other when trying to run. The [horse + cart] is pushed forward by the ground acting on the horse and it is dragged back by friction drags. Its motion depends on which is greater.

Each object, [horse], [cart], [horse + cart], is acted on by several forces. Newton's Law III does not say whether the two main forces acting on any one of these are equal and opposite. It does require the forces in each pair listed in Fig. 8-14 to be equal and opposite. In each pair we have equal and opposite forces such as F and $-F$, G and $-G$, etc., but how F and G, or F and H, compare with each other is quite a different matter, no concern of Newton's Law III. (But adding up *all the forces on one body*, such as F and H *on the cart*, will enable Newton's Second Law to predict the body's acceleration.)

"Action = Reaction" Almost an Axiom

In building a great system of astronomical and earthly mechanics Newton had to deal with the pull of the Earth on the Moon and the pull of the Moon on the Earth. Unless he could say such pairs are equal and opposite, the development of mechanics would be complicated if not impossible—perhaps it would even be nonsensical, for in a way this equal-and-opposite property is a definition of how we handle forces in mechanics. When you weigh yourself on bathroom scales you really measure the *push-down of your feet on the scale*. But you are trying to measure the *pull of the Earth on your body*, and if you are in equilibrium that pull is balanced by the *push-up of the scale on your body*. In the sketch of Fig. 8-15 we want to measure the Earth-pull, W. We assume (Newton I) that *in equilibrium* $W = -F_1$, where F_1 is the push-up of scale. We assume (Newton III) that F_1 is equal and opposite to F_2, the push-down exerted by the body on the scale; and we let the machine measure F_2. Newton's Law III does not say anything about the relation between W and either F. It only says F_1 and F_2 are equal and opposite. (Of course W itself has an equal and opposite reaction upward, a pull exerted *by* your body *on* the great Earth.) If this whole arrangement is accelerated upward (as in an elevator starting), F_1 must be greater than W, so that the resultant force $[F_1 - W]$ accelerates your body upward too, according to $F = Ma$; but F_2 is still equal and opposite to the new, larger, F_1. Then the scale tells you F_2 (or F_1) but not W.

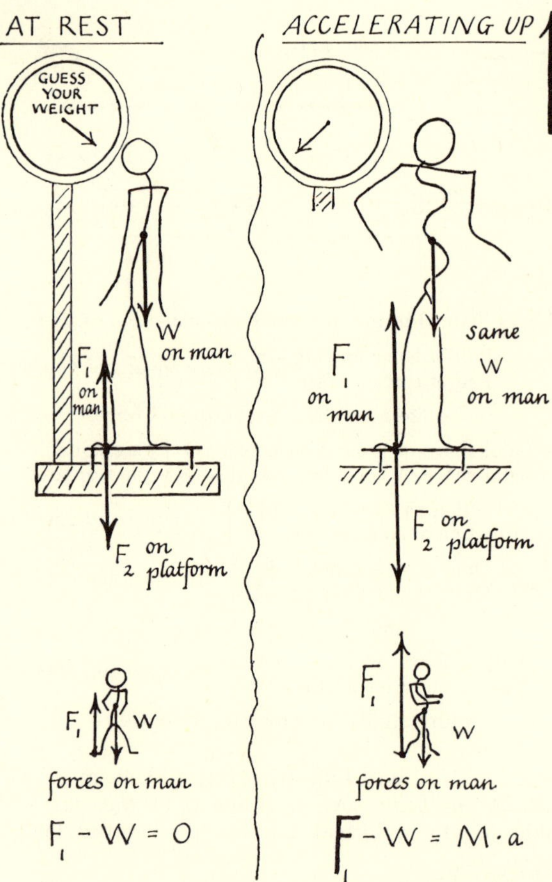

FIG. 8-15. EXPERIMENT IN ELEVATOR
Man weighs himself in accelerating elevator.

Demonstrations of Action and Reaction

If "Action = Reaction" seems to you an obvious[13] piece of symmetry, you might accept it as a self-evident "fact" of nature like $2 + 2 = 4$ and deduce conservation of momentum. But most scientists would call that too trusting and would defy you to prove the equality of action and reaction other than by measuring momentum changes.

Encouraging Experiments

We can show experiments that illustrate the idea of ACTION = REACTION even if they do not prove it. Those sketched in Figs. 8-17 and 8-18 look helpful, but they can be argued into being tests of the springs themselves, tests that prove nothing unless you assume what you are trying to prove.

FIG. 8-17. ATTEMPTS TO DEMONSTRATE "ACTION = REACTION"
Spring balances a and b read the same, even if A (and/or B) is on roller-skates and accelerating. Allegedly, balance a reads the pull of man A, and balance b the pull of man B. But how do the balances know which is measuring whose pull?

Perhaps the best of these is the one which contains least confusing detail. A hoop of clockspring is used to show forces by its deformation. (See Fig. 8-18.) Belief in symmetry prevents our ascribing the deformation to the agent pulling at one end rather than to the agent pulling at the other end, and encourages us to believe the pulls are equal and opposite. The hoop distorts into the same symmetrical oval whether the agents are walls or men, at rest or in any kind of motion. (At best these experiments give us comfort. At worst they are a simple swindle to make us think. It is a tempting thought that this experiment *can* provide some proof of Newton's Law III. Imagine the hoop of spring pared down till it has practically no mass. Then, even when accelerating, it would require practically no resultant force on it—Newton's Law I. Therefore the pulls of the two agents on it must be equal and opposite. Newton's Law III proved? No. Those are not the forces we want to prove equal, but are forces exerted by two different agents on a common victim! We still want to know whether the reaction of the loop on one agent is equal and opposite to the pull of that agent on the loop. We have a Newton's Law III question *at each end of the loop*. Worse instead of better! This is a warning against the dangers of glib argument.)

---

[13] "Fallacious proofs." See Fig. 8-16. Beware of the following argument, sometimes offered. "A book rests on my hand. *Then, since the book remains at rest*, I must push it up just as much as it pushes me down." The final phrase states that ACTION = REACTION; but the phrase in italics offers no proof at all. There are two fallacies:

(A) Only one of the two forces mentioned acts on the book, my hand's push-up, P. Another quite separate force acts on it too, the Earth's pull, W. If the book remains at rest, we believe $P = -W$ (Newton I). But this does not assure us that $P = -Q$ where Q is the push-down of the book on my hand (Newton III).

(B) The book need not *remain* at rest. It may accelerate. If I move my hand with a downward acceleration, I decrease P; then the forces P and W on it do not balance. Yet we believe action and reaction between book and hand, P and Q, are still equal and opposite—though our belief does not rest on this false argument.

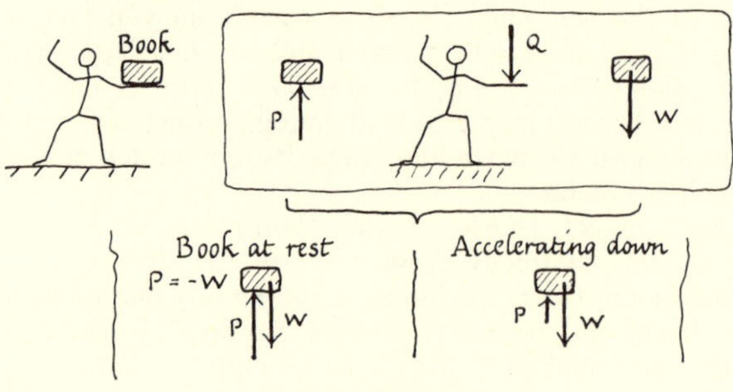

FIG. 8-16.

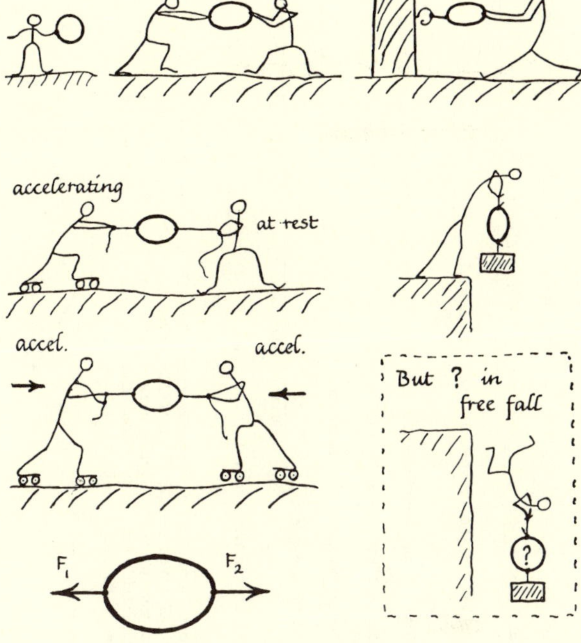

The symmetrical spring hoop suggests that F_1 and F_2 are equal & opposite — this is true but not really Newton's Law III.

FIG. 8-18. ACTION AND REACTION?
A simple hoop of clockspring pulls into an oval when pulled at opposite ends of a diameter.

Universal Momentum Conservation

From now, we shall take Action = Reaction as a working rule and we shall therefore assume momentum is *always* conserved (unless new experimental evidence persuades us to change our viewpoint). We are not dismayed by apparent contradictions such as the mysterious "creation" of momentum when a man at rest starts walking fast—we believe his feet give the great Earth an equal and opposite backward momentum. Experimenting with the man on a movable "ground" (a small platform on rollers) reassures us. The "ground" starts moving backwards when he starts forwards. A similar demonstration is given by a toy electric train on a circular track that is free to rotate. The track is supported by a good bicycle wheel with its axle vertical. When the train starts forwards the track starts moving backwards.

FIG. 8-19. MOMENTUM CONSERVATION
(a) Man starts running on movable "ground"—a platform on rollers.
(b) Toy electric train starts running around; track runs around backwards.

We take the view that momentum conservation holds true among molecules and atoms, even parts of atoms. We believe it applies to elastic collisions between bouncing molecules and to inelastic collisions when one atom hits another and tears pieces off it. So useful is this general principle that if we did find a case of momentum "disappearing to nowhere"—say in some atomic change—we might be tempted to invent a tiny unseen particle to carry off the missing momentum. We should be inventing a special demon, a dangerous path to tread. Unscientific? Yes, if we find no other evidence for the demon or other uses for him.

We do find momentum disappearing, unaccountably, when a radioactive atom shoots out an electron as a beta-ray.

So we have resorted to such a demon in nuclear physics, a tiny "neutrino" with practically no mass and no electric charge with which to make his presence felt. For years, he was unseen; in fact, many thought it impossible to detect him—and in that case was it honest to claim he existed? Perhaps we were unwise to treat him as real; yet, as a way of specifying some minute but important failures of momentum-conservation, he seemed no worse than any other way of stating the experimental truth. He aided clear thinking, and we called on him to do several jobs: as well as carrying away momentum he carried off energy and spin-momentum, to keep those balance-sheets true too. And that made him more respectable—a one-armed demon has no place in science. Recently a great experimental search succeeded: neutrinos have been detected. Our demon has become a respectable device in our system of remembering science.

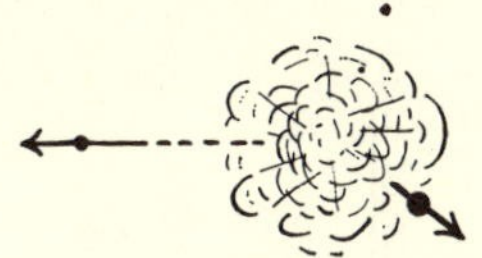

"PROBLEM SHEETS" The problems here are intended to be answered on typewritten copies of these sheets. Work through the problems on enlarged copies, filling in the blanks, (............), that are left for answers.

PROBLEMS ON CONSERVATION OF MOMENTUM NAME____________________

In the problems below calculate the total momentum, in a given direction, before the "collision" and then calculate the total after the "collision". Use X for any velocity or force that you do not know. Then write these two totals equal to each other (i.e., assume conservation) and solve for X.

PROBLEM 5. A 3000-pound car driving 20 feet/second overtakes and smashes into a 4000-pound truck driving 10 feet/second in the same direction. Find the speed with which the combined wreck proceeds.

MOMENTUM IS ALWAYS CONSERVED (i.e., its total is unchanged in ANY collision)

Initial momentum of car . . . . = (________)·(________)·________ units

Initial momentum of truck . . . = (________)·(________)·________ units

After impact total moving mass is . . . . ________ pounds

Let velocity of wreck, after impact, be X feet/second.

∴ in terms of X, momentum of wreck is ____________·________ units

Since total momentum in the forward direction is same before and after,

__________ + __________ must = ____________________

∴ Solving for X, the final velocity: X = ______________ feet/second

PROBLEM 6. A 3000-lb car driving 20 feet/second hits a 4000-lb car driving 10 feet/second, head-on. Find the speed of the combined wreck, if they interlock.

Total momentum before impact = momentum of car I + momentum of car II

= ____________ + ____________·________ units

MOMENTUM IS A VECTOR; BE CAREFUL ABOUT + AND - SIGNS.

If final velocity is X, momentum of wreck = ____________·________ units

ASSUMING MOMENTUM IS CONSERVED, WE OBTAIN THE EQUATION

____________________ = ____________________

∴ Final velocity X is. . . . . .__________ feet/second

PROBLEMS ON CONSERVATION OF MOMENTUM SHEET 2

PROBLEM 7. A 4000-pound truck crawling due North 20 miles/hour along an icy street is hit by a 2000-pound car driven by a lunatic due East 30 miles/hour, out of a side street. The two interlock. Calculate the velocity of the combined wreck, indicating its size and its direction.

NOTE: Do not change miles to feet, but use queer units containing miles. This will not matter in a CONSERVATION OF MOMENTUM equation because____

____

The momentum of the truck = __________·______________ due __________
units direction

The momentum of the car = __________·______________ due __________
units direction

*The resultant of these is momentum of size ______________·________ units

in a direction given by tan A = __________ where angle A is the angle between the resultant and the due North direction.**

If the final velocity is V miles/hour, final momentum = ________·________ units

If momentum (VECTOR) is conserved, then V must be ____________ miles/hour

PROBLEM 8. A 200-pound gun resting on a frictionless frozen pond shoots a 2-pound cannon ball horizontally 1000 feet/second. Find recoil speed of gun.

The cannon ball emerges with momentum . . . . . . ____________·________ units

The momentum of the gun must be ______________________________
same? equal and opposite? larger? smaller?

∴ velocity of gun must be . . . . __________ ft/sec

*NOTE: Momentum is a vector, having direction as well as size. Since FORCE · TIME gives CHANGE OF MOMENTUM, momentum must obey vector addition just as force does. Draw a sketch of these two momenta (which are perpendicular to each other) and use this to find the resultant. Triangles with sides proportional to 3, 4, 5 are right-angled triangles.

** In the sketch, tan A is $\frac{h}{b}$

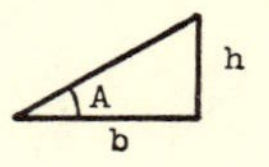

A h b

FIG. 8-21.

PROBLEMS FOR CHAPTER 8

★1a, b, c. Special problems, to be answered on a printed sheet.

2, ★ 3, ★ 4. In text of the chapter.

5, 6, 7, 8. Special problems, to be answered on printed sheets.

PROBLEMS ON CHANGE OF MOMENTUM

9. A man wishes to break a 1 ft length off a 4 ft length of stout string. He bruises it at the place where he wants the break. Then he hangs a 2-pound lump of iron on the lower end and ties the top end to a nail in the wall. (Fig. 8-22) He raises the lump till it is level with the nail—so that the string hangs in a slack loop—and releases it. The lump falls (4 ft, freely) and pulls the string taut with a jerk. From the sound it makes, the jerk seems to last about 1/100 second. The string just breaks. Estimate its breaking force.

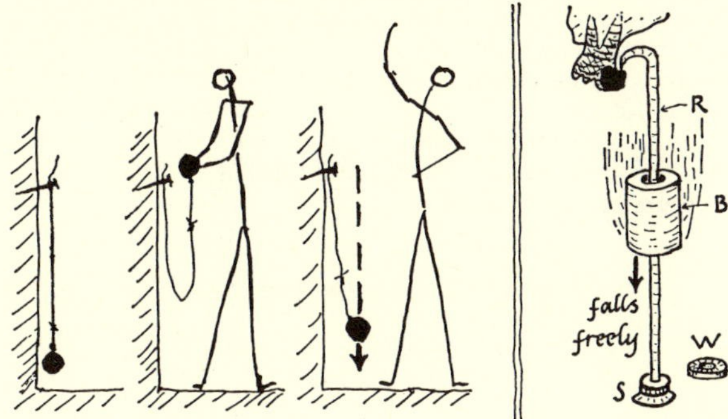

Fig. 8-22. Problem 9

Fig. 8-23. Problem 12

10. HAMMER BLOWS

(a) A 2-pound hammer head, moving 10 ft/sec, hits a nail. The nail is already embedded in hard wood and moves very little. The hammer rebounds with negligible velocity. An electronic timing device shows the time of contact to be about 2/1000 of a second. Estimate the average force acting on the nail during contact.

(b) Suppose the hammer in (a) hits a springy nail, remains in contact 2/1000 of a second but rebounds almost as fast as it approached, about 10 ft/sec. Estimate the average force on the nail during contact. (*Note*: If you arrive at the answer zero you have obviously done something wrong. Try inserting your finger as a test object between hammer and nail.)

★ 11. JET ENGINE

A jet engine on test is held stationary in a wind-tunnel where air is flowing with velocity 300 ft/sec. The engine takes in 35 pounds of air per second, which enters with velocity 300 ft/sec. After compression and heating, that air is discharged with velocity 1500 ft/sec.

(a) What is the thrust of the engine in poundals?

(b) What is the thrust in pounds-weight?

(c) Suppose this engine is in a plane flying along at 300 ft/sec ($\approx$ 200 miles/hour), scooping up the air at that speed. The thrust will be about the same and it is now pushing the plane forward at 300 ft/sec (together with help from other engines). As you will find in a later chapter, the *power* of such an engine can be calculated by multiplying thrust by speed.

(i) Estimate the power of this engine in "bad" units, ft · pounds/sec.

(ii) Estimate its power in "horsepower" given that 1 HP = 550 ft · pounds/sec.

12. Modern dentists use the device shown in Fig. 8-23 to remove recalcitrant fillings. A metal bob, B, slides loosely on a smooth metal rod, R. The top of the rod has a hook that hitches into the filling. The bottom of the rod has a metal stop, S, to prevent the bob falling off. The dentist places the hook in the filling, raises the bob to the top of its slide and lets it fall. Suppose in such a device the bob is a .1-kg lump of metal; and its maximum speed, at the end of its fall, is 0.5 meter/sec.

(a) Suppose the patient's jaw is firmly clamped so that the sliding bob comes to rest in 1/1000 sec. Estimate the tug.

(b) Suppose the patient lets his head sink softly, prolonging the time-of-stopping to 1/100 sec. Estimate the tug.

(c) Suppose the makers offer the dentist a rubber washer, W, that can be placed on the rod just above the bottom stop. What would be the effect of inserting this ring?

13. SWIMMER'S START

A 200-pound racing swimmer can shove with his feet against the end of the pool with a force of 500 pounds-weight.

(a) How much speed can be gained by such shoving, in 1/10 sec?

(b) Why not encourage him to push like that for 3/10 sec?

14. A 250-pound thief (including booty) runs over a sleeping dog. This encounter lasts about 0.1 second, and takes 5 ft/sec off his speed. What force did his feet experience?

15. A party of invaders storming a medieval castle attacks the main gate with a battering-ram. The ram is an 800-pound log carried on the shoulders of a dozen men who rush the log (end-on) towards the gate with speed 10 ft/sec. Just as the ram reaches the gate the men release their grip on it (as they stop) and let it slide forward on their smooth leathern shoulders. It breaks the gate, pushing it in 6 inches before it comes to a stop.

(a) Estimate the force exerted by the ram on the gate during the smash.

(b) The chief stalwart of the castle's defenders says, "If you had let my men push on the door from within they could have saved it. Each of them can push with a force of 100 pounds." How many stalwarts were needed to save the gate?

(c) One of the storming party forgets to let go of the ram and clings to it. What happens to him? Make a rough calculation.

PROBLEMS ON CONSERVATION OF MOMENTUM

★ 16. A 200-pound man, skating 30 ft/sec on frictionless ice hits another 200-pound man at rest, head-on. The two cling together and proceed as one unit.

(a) With what speed will they proceed?

(b) Suppose the two men carry big magnets, so that they attract each other strongly while the moving one is approaching the other. How will this affect their speeds just before the collision?

(c) If they attract as above, how will the final speed of the pair be affected? Why?

17. A 200-pound man stands at one end of a long flat boat, at rest on a lake. He suddenly starts running along the deck to the other end. The boat (+ any water that moves with it) has mass 100 pounds. An observer standing on the bank sees the man running past him at 10 ft/sec.
(a) What speed will the observer note for the boat?
(b) How fast will the man find he is running along the boat?

18. A medieval gunner sets up his 400-pound cannon at the edge of the flat roof of a high tower. It shoots a 10-pound cannon ball horizontally. The ball lands on the ground 800 feet from the base of the tower. The gun also moves, *on frictionless wheels*, and reaches the ground. (See Fig. 8-24.)

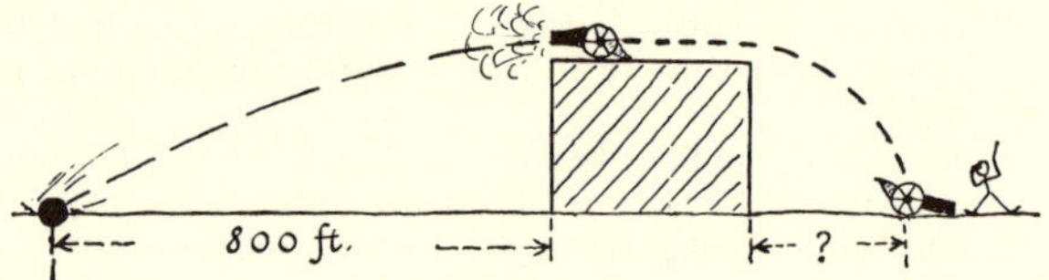

Fig. 8-24. Problem 18

(a) What is the horizontal distance of the gun's landing-place from the place where it left the roof? (*Note*: No further data are supplied.)
(b) Why does the width of the top of the tower not enter into the calculation for (a)?

★ 19. (*Note*: This problem looks rather silly, but is useful in discussing collisions of molecules.)

A $\frac{1}{2}$-kg squirrel sits on a frictionless, icy, horizontal, flat roof top. A man throws a 1/10-kg stone at him horizontally with speed 6 meters/sec.

(a) The squirrel catches the stone and holds it. Calculate:
   (i) the moving mass before and after the catching.
   (ii) the momentum before.
   (iii) the speed with which the squirrel (and his burden) will recoil (assuming momentum is conserved).
(b) The squirrel catches the stone, immediately sees it is not a nut and throws it back at the man in disgust, with horizontal speed 2 meters/sec, relative to the ground. Calculate the speed with which the squirrel recoils.
(c) Explain why it makes no difference to the answer in (b) if the squirrel holds the stone for a few seconds before throwing it back.

20. A 200-pound man leaps from a dock with a horizontal velocity of 12 ft/sec and lands in a 100-pound rowboat.

Fig. 8-25. Problem 20

(This 100 pounds includes any water that moves with the boat.) The boat is at rest before the man jumps.

(a) Calculate the velocity with which boat and man start away from the dock.
(b) If water friction and air resistance are negligible, how far will the boat move in 3 seconds from the instant the man lands on it?

21. MAKING A FORMULA

(a) A 50-kilogram boy is standing on a 500-kilogram raft floating on a lake. The raft is at rest. It can move on the surface of the lake with negligible friction. Starting from rest, the boy begins to walk with constant speed 1 meter/sec (relative to the ground), and continues to walk for 20 seconds. How far does the raft move in this time?
(b) Suppose the boy walks twice as fast for only half as long a time. How far does the raft move?
(c) Suppose the boy walks with velocity v meters/sec for a time t secs. Suppose the boy weighs m kg and the raft M kg. Find the distance the raft moves, in terms of the letters v, t, m, M.
(d) Discuss the answers to parts (a) and (b) in terms of your analysis in part (c).

22. MEASURING BULLET'S SPEED

In Ch. 8 a method of measuring the speed of a rifle bullet is described. You could make a similar measurement in the laboratory. You would be provided with the same truck containing a large chunk of wood, running on rails, with almost frictionless wheels. Instead of the electric eye, you would have a stopwatch. Describe the measurements you would make; show how you would calculate the bullet's velocity from your measurements.

PROBLEMS ON MOMENTUM AND FORCE

23. What is a bazooka gun? Explain why the man holding it feels no "recoil" when it is fired.

24. Two cyclists, Albert and Bertram, coast side-by-side on a level road at a speed of 10 ft/sec. Albert is a 160-pound man. Bertram is a 100-pound boy. In the following calculations, neglect the effects of friction. Albert gives Bertram a shove in the forward direction, after which Bertram's speed is 18 ft/sec.

(a) What is Albert's new speed?
(b) Bertram notices that the shove lasts for 2 seconds. What is the average force with which Albert pushes him?

25. VARIOUS COLLISIONS

A massive block of metal is at rest on a frictionless horizontal table.

(a) A 2-pound ivory ball is thrown horizontally at the block, hits it head-on and rebounds.
(b) The experiment is repeated with a 2-pound lump of clay thrown with the same speed. It hits the block head-on, drops dead and does not move.
(c) The experiment is repeated with a 2-pound lump of sticky clay thrown with the same speed. It hits the block head-on, clings to it and moves with it.
   (i) In which case(s) will the block of metal gain most speed, (a), (b), or (c)?
   (ii) In which case(s) will the block of metal gain least speed?
   (iii) Explain how you argued your answers to (i) and (ii) above.

26. A machine-gun manufacturer boasts "Our gun is so powerful that it can keep itself floating in the air by firing a steady stream of bullets downward." (See Fig. 8-26.) Find out how rapidly the gun would have to fire, using the data and suggestions below.

The gun weighs 50 pounds. Each bullet is 1/20 of a pound of steel and emerges with muzzle velocity 4000 ft/sec.

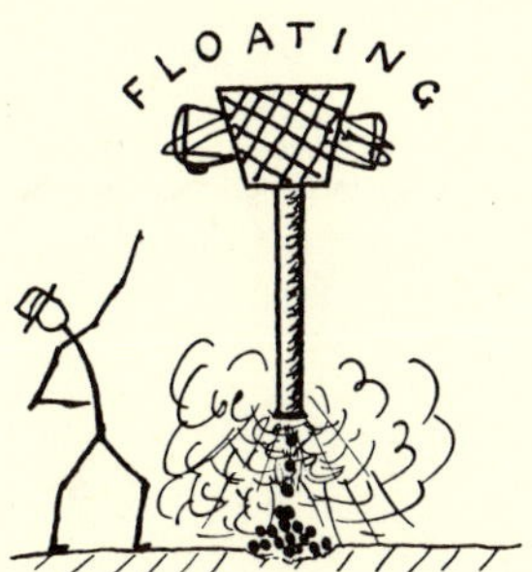

FIG. 8-26. PROBLEM 26

The explosions push the bullets down the barrel and exert an equal and opposite recoil force on the gun. (The force is not steady but a bumpy one, one bump each time a bullet is fired. However, if the bumps follow each other rapidly, we can imagine them smoothed out into a steady force. It is this smoothed-out force that is being used here.)

Suppose the rate of firing is X bullets per second.

(a) Calculate the momentum carried away by bullets in a period of 10 sec.

(b) Calculate the recoil force.

(c) Assuming this recoil force is just sufficient to keep the gun floating in the air, calculate the rate of firing.

27. NEWTON'S LAW II WITH CHANGING MASS

A freight train travels along a level track. For each of the four cases below, answer the following questions:

(a) Does the momentum of the train change? (Give a clear reason for your answer.)

(b) If so, what external object or agent provides the force for this change?

Case (i) The train is travelling steadily with constant velocity.
Case (ii) The train is accelerating steadily along a straight horizontal track.
Case (iii) The train, originally at rest, is suddenly started.
Case (iv) The train, passing at constant speed under a stationary coal chute, gains mass by having coal (falling vertically) poured into open cars, and *the engineer takes any steps necessary to keep the train's velocity constant, during this loading.*

(c) Describe the forces that act on the coal in case (iv) as it arrives in the truck and settles there.

In the light of this discussion, $F = \Delta(Mv)/\Delta t$ appears to be a better statement of Newton's Law II than $F = Ma$. Why?

28. ROTATIONAL MOMENTUM

As in Problem 21, suppose a boy is standing on a raft floating, at rest, on a lake. The boy starts walking around a large circle on the raft, and continues walking around it at constant speed. (Suppose water-friction is negligible.) Predict the behavior of the raft. (Make an intelligent guess. An answer will be suggested in Chapter 22.)

CHAPTER 9 · FLUID FLOW

"How quaint the ways of Paradox.
At common sense she gaily mocks."

—W. S. GILBERT

How can a baseball "curve" sideways? Why does the air-blast in a sprayer suck liquid up instead of pushing it down? A number of freakish effects of wind and water-flow are really examples of accelerated motion "obeying" Newton's Law II. Accelerating a car by pushing it seems obvious enough, and we should expect a speeding fluid to show equally obvious effects; but in fact it shows some unexpected ones, studied by the mathematician Bernoulli and named after him. Some of these are useful elsewhere in physics, and some show important patterns. We shall describe a few and show that they are part of ordinary mechanical behavior.

DEMONSTRATION EXPERIMENTS AND GENERAL DISCUSSION OF FLUID FLOW

Demonstrations (i) and (ii) show two "Bernoulli paradoxes."

(i) *A blast of air through a glass funnel attracts a light ball.* (Fig. 9-1) Air flowing down lifts the ball into the funnel and holds it there, despite gravity.

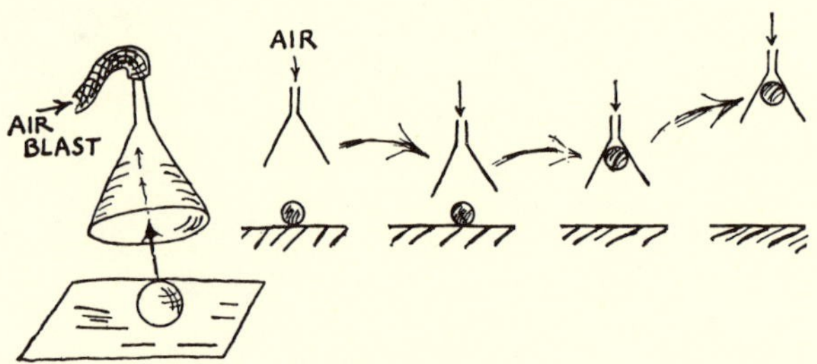

FIG. 9-1. PARADOX
Air blast through funnel picks up ball and holds it.

How does this paradoxical lifting work? Where the ball is closest to the funnel, the air-flow, squeezed into a narrow passage, must be faster, and we might expect it to push the ball away—yet the ball seems to be attracted towards it.

(ii) *A jet of air can support a light ball.* (Fig. 9-2) Even if the jet is tilted over, the ball keeps near it

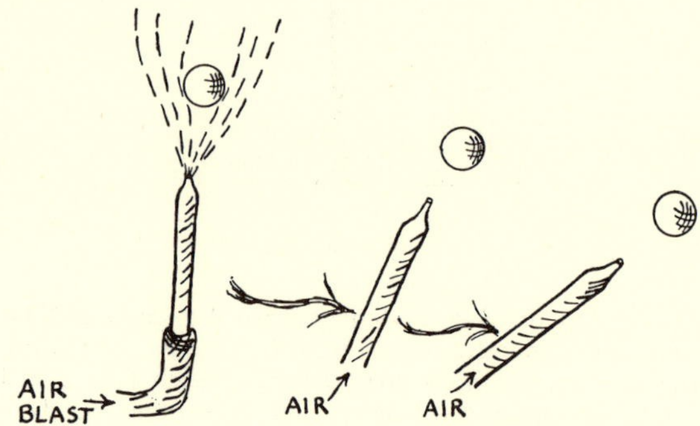

FIG. 9-2. PARADOX. Air jet supports ball.

and does not fall. The air-stream squirts over the *top* of the ball; and again we would expect its faster flow to push the ball—but it does not.

Streamline Flow and Eddy Motion

To explain these paradoxes, we must study streamline flow. When a fluid (= liquid or gas) flows steadily through a pipe, it moves in *streamlines*, smooth paths of steady flow shaped by the solid boundaries. (Figs. 9-3 and 9-4) With faster flow,

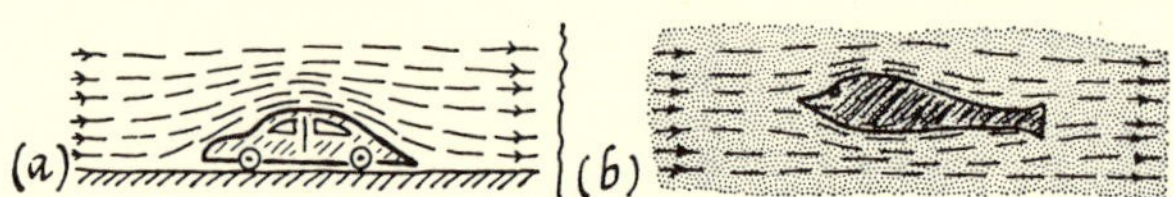

FIG. 9-3. STREAMLINES OF FLUID FLOWING AROUND AN OBSTACLE
(a) Wind blowing over stationary car.
(b) River flowing past stationary fish.

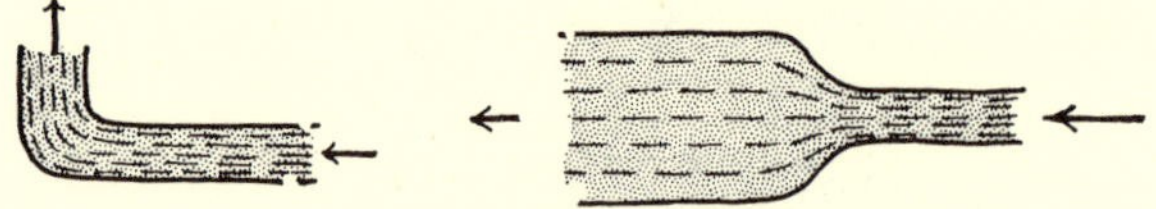

FIG. 9-4. STREAMLINES OF FLUID FLOWING THROUGH PIPES

the streamlines passing any obstruction in the pipe may curl into *eddies* or *vortices*; and with still faster flow the streamlines, even in a straight pipe, disappear into a mess of irregular *turbulence*.

Demonstration (iii). Specimen streamlines are shown in slow-flowing water by streamers of ink as in Fig. 9-5; or water is colored as it flows past crystals of dye (potassium permanganate), as in Fig. 9-6.

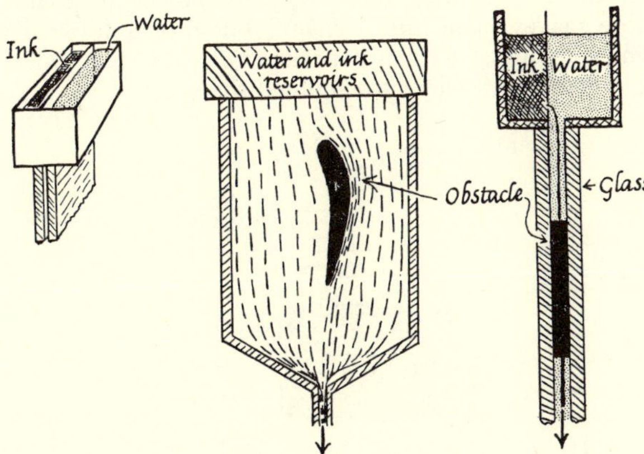

FIG. 9-5. DEMONSTRATION OF STREAMLINES

Water flows down between two glass plates from a narrow slit in a water tank. Ink, fed in at a series of points along the slit, marks streamlines. In the central sketch, an obstacle shaped like the section of an airplane wing modifies the streamlines.

(NOTE. The "model" experiments of FIGS. 5 and 6 show streamlines with very slow flow, controlled by fluid-friction [viscosity]. Faster flow, with pressures determined by momentum-changes rather than friction, follows much the same streamline pattern. With much faster flow still, the streamlines change to eddies and finally disappear into muddled turbulence.)

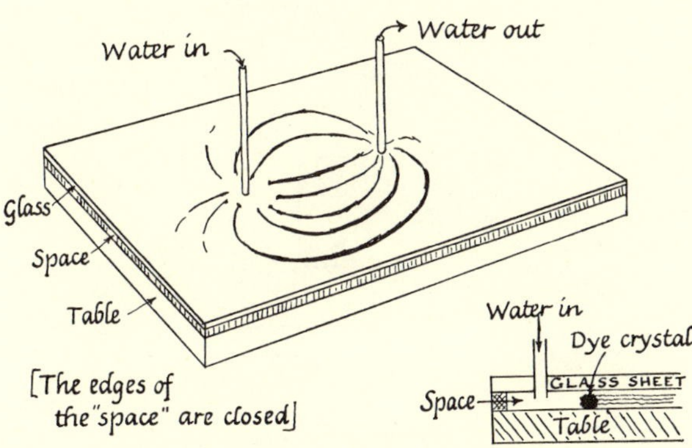

FIG. 9-6. STREAMLINE DEMONSTRATION: "SOURCE AND DRAIN" IN A LAKE

Water flows in a narrow horizontal space, sandwiched between a glass sheet and a table top. A small steady flow is fed in by one pipe and out by another. Near the inlet, crystals of dye sprinkled on the table make colored streamlines.

Demonstration (iv). A spoon moved through soup, or a finger through a dish of dusty water, leaves visible eddies in its wake. A streamer of dye inserted in water flowing along a pipe shows a streamline for slow flow; but above a certain critical speed it wobbles and spreads into eddies and general turbulence so that all the water is colored. (Figs. 9-7a, b, c)

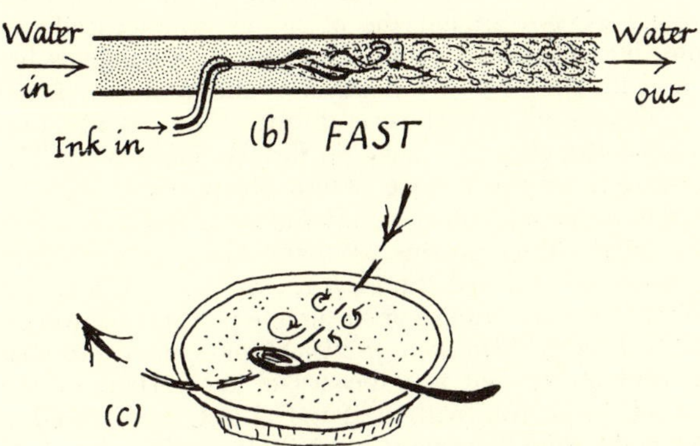

FIG. 9-7. STREAMLINE FLOW AND TURBULENT FLOW

(a) With slow flow, an ink streamer follows the streamline.
(b) With fast flow, there is turbulence.
(c) A spoon moved fast through soup leaves turbulent whirlpools behind.

Instead of fluid flowing in pipes or rivers, consider the motion of a solid object, such as a fish or an airplane, through stationary fluid. The fluid has to move out of the way as the object moves through it. Those temporary motions are difficult to picture, so instead we make the fluid move in a steady stream and hold the object at rest, like a model in a wind-tunnel. Then the fluid moves in streamlines which swerve around the object. The fluid sandwiched between two marked streamlines must continue between them. When the streamlines twist and turn, crowd closer or splay out, that fluid must run between them as boundaries, like a river between its banks. (Since these streamlines *are* the lines of motion, there can be no flow across them.) Where a pipe narrows and crowds the streamlines closer, the fluid must move faster, because the same mass of fluid has to flow, each second, through a narrower space. (Fig. 9-8) In general, crowded streamlines go with fast flow.

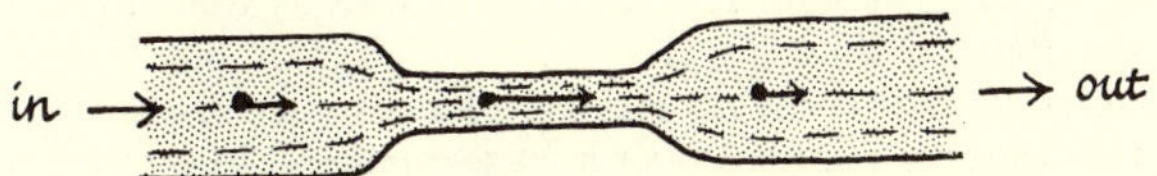

FIG. 9-8. CROWDING OF STREAMLINES INDICATES SPEED

Where the pipe narrows, crowding the streamlines, there must also be increase of speed.
(The arrows show speed in one streamline.)

Types of Flow

When fluid flows past a stationary object, the streamline pattern and the forces on the object take different forms according to the speed of flow. We shall now discuss some types of flow for fluid passing a stationary object.

(*a*) *Flow of ideal frictionless fluid.* If the fluid were frictionless (an unfortunate fiction), the streamlines would swerve around the object as symmetrically as possible and would continue as smooth streamlines beyond it. The fluid would travel at the same rate in all of them—full speed—except for some speeding-up around the object to make up for the obstruction. The pressures on the surface of the object would yield *a resultant force of zero*—no lift and no drag in a frictionless fluid! Although this predicted behavior seems unnatural, we still find the ideal frictionless fluid a useful fiction for some studies of streamline patterns. However, *all real fluids show some friction.* Fluid cannot slip past a solid object, but is stationary at the surface of the object (or moves with it if the object moves). Even when the fluid flows very fast it finds a polished surface quite rough *on a molecular scale* and it is dragged to a standstill at the surface. So we shall never observe the odd behavior—no lift and no drag—predicted for an ideal fluid.

Fluid friction modifies the pattern of streamlines and the distribution of flow-speeds. For very slow creeping flow the streamlines bend smoothly around the object; and for very fast flow they are whirled into a complex wake of eddies behind the object. We shall now describe these extreme forms, and the intermediate stage, for a real fluid flowing past a solid object.

(*b*) *Very slow streamline flow.* This is wholly controlled by fluid friction. The streamlines have much the same shape as for an ideal fluid, but the speeds are quite different. Far out from the object, where the flow is undisturbed, the fluid runs at full speed. At the object's surface, the fluid is stationary. And there is a regular increase of speed from streamline to streamline, out from the object. (See Fig. 9-9b.) In this creeping motion, fluid friction ("viscosity") controls the pattern and speeds, and it produces a drag on the object which varies directly as the general flow-speed ($F \propto v$).

(*c*) *Fast flow past a blunt object: eddy flow.* When we increase the flow-speed, fluid friction fails to hold complete control, and large-scale momentum-changes grow more important. The streamlines diverge as before where they meet the object, but do not close in again completely. Instead they curl up behind the object into a roaring street of eddies (= vortices). The forming of these eddies produces a drag that far outweighs the small drag of fluid friction. It increases as the square of the general flow-speed ($F \propto v^2$). Thus, a blunt object moving fast through air experiences an air-resistance that varies as SPEED^2 over a large range of speeds. (And then the *power* needed to maintain the motion varies as SPEED^3; so that double speed costs 8 times the power—a very serious matter for ships.)

The width and strength of the avenue of eddies—the wake of the object—depends on the object's shape. A blunt object, even a round ball—any shape that is not "streamlined"—presents a big eddy-forming face to the stream and suffers a large drag. A rounded or pointed nose makes some improvement; but profitable streamlining requires a long tapering tail (see Fig. 9-28). Fishes have excellent streamlined shapes.

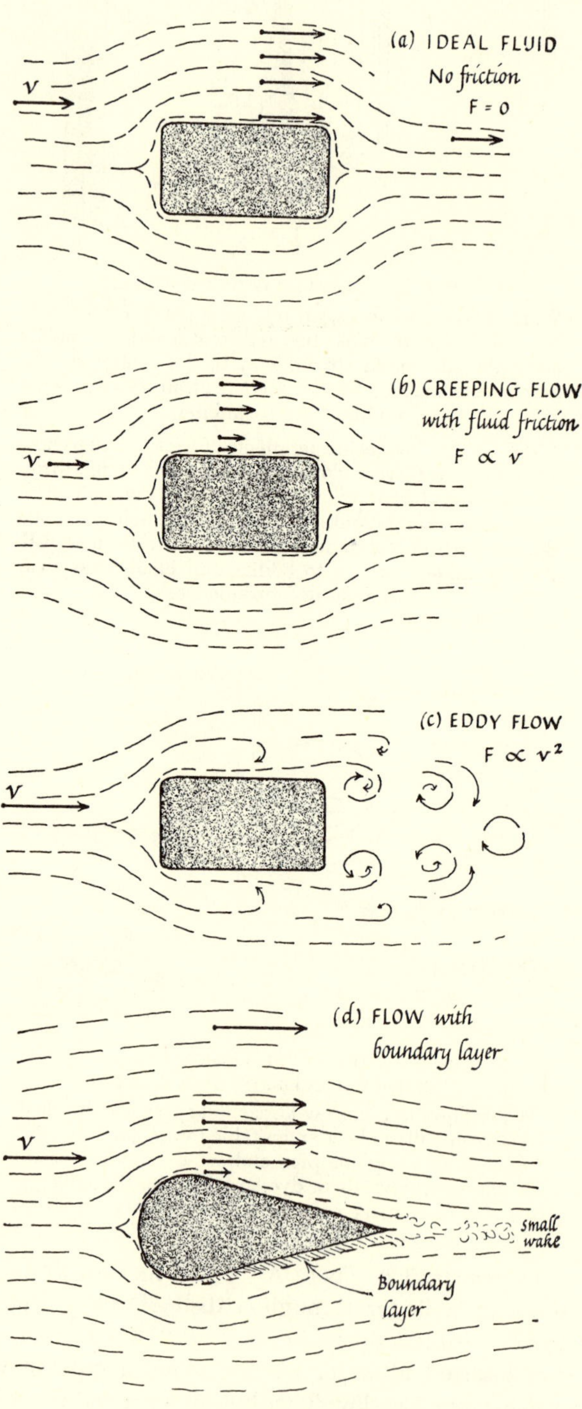

Fig. 9-9. Streamline Flow

(*d*) *Fast flow past a streamlined object: boundary-layer flow.* The streamlines keep somewhat the same shape as for creeping flow, though the pattern may become unsymmetrical; the velocities are quite different; and some eddying wake will form. With an object of good streamlined shape, the wake is small and the main pattern is a streamline one. This is the important case for airplanes and ships. There is fluid-friction flow, with its characteristic gradation of speeds close to the object; but now, with fast flow, the drag of the obstruction does not have time to make itself felt far out from the object. (In a sense, the fluid has got past the object before a message of dragging can travel out to it.) So the region of graded velocities crowds in very close to the object in a thin "boundary layer," and all the streamlines farther out run at full speed, almost like the ideal case. Across the boundary layer, the streamline speeds range from zero to full in a very thin space, and their fluid friction exerts a drag on the object. (Fig. 9-9d) The faster the flow the closer the region of varying speeds crowds in, the thinner the boundary layer. That makes the drag increase more rapidly with speed. (Detailed analysis gives $F \propto v^{3/2}$ or $F \propto \sqrt{v^3}$.)

Fluid friction often has another effect at these speeds. It can promote a circulating motion—e.g. of air around an airplane wing—and that circulation leads to unsymmetrical streamlines and useful forces. (Fig. 9-10)

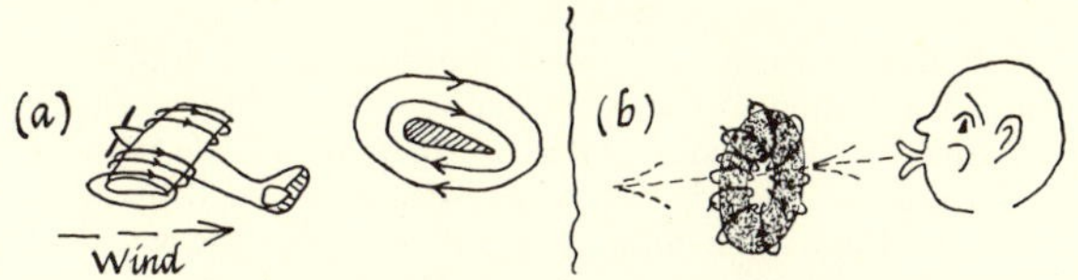

FIG. 9-10. FRICTION STARTS CIRCULATION
(a) Fluid friction creates a circulation around an airplane wing, when wind meets the wing.
(b) Fluid friction of smoky air against lips produces a circulation, which is a smoke ring.

The boundary layer grows thicker towards the rear of the object, and there, where it seems to cling less closely, it may peel off and start eddies. The art of the airplane designer is concentrated on preventing that peeling off from happening at too early a place on the supporting wings—if it does, lift is decreased and drag is increased, and the plane cannot fly.

Even with good design, there is some eddy-flow with its serious drag ($F \propto v^2$) as well as the boundary-layer friction. However, aerodynamic experts cannot use a simple formula for the combined drags (or a corresponding one for lifts), such as:

$$F = kv + k'v^{3/2} + k''v^2$$

where each k is a constant for a given body, because the whole pattern of flow changes from one range of speed to the next, and therefore the k's change. So their mathematical procedure is a more involved one; and in practical flight there is the further complication that the controls move flaps that change the shape of the moving body itself. Beware of trusting simple formulas with "constants" in them for flight or fluid flow, in this chapter or elsewhere. (And in reading books on these matters, look first at the publication date and avoid those more than a dozen years old.)

The Paradoxes

The Bernoulli paradoxes described below belong to the intermediate flow-speeds where the motion is still in streamlines but so fast that the friction forces are small compared with the pressure-differences involved in momentum-changes where the flow in a streamline changes speed or direction.[1] In our first paradox, the attracting funnel that picks up the ball, there is faster flow, and crowding of streamlines, in regions above the ball where it is close to the funnel. We might expect such faster flow to push harder and help the ball to fall, yet the ball seems to be pulled up towards the funnel. (Fig. 9-11) Taking the hint that a region of faster flow seems to have a peculiar effect, we shall investigate pressure and flow, both experimentally and by thinking about them.

[1] The forces due to friction vary as v for creeping flow or as $v^{3/2}$ for boundary layer flow; but the forces due to momentum-changes where the flow changes its velocity are found to vary as v^2. That is why the latter grow most important at higher speeds—before eddies start and take charge. (In FORCE = (CHANGE OF Mv)/TIME, Mv contains a factor v; but TIME, being the time taken by that M to pass by, varies as $1/v$. Therefore FORCE involves changes of v^2.)

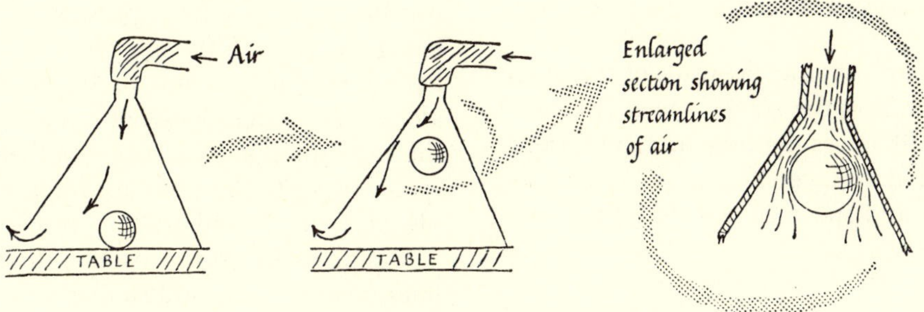

FIG. 9-11. FUNNEL AND BALL PARADOX.

We shall start with flow in a pipe. In a uniform pipe the streamlines are all parallel. With an ideal fluid they would all have the same speed, from center to walls of the pipe (Fig. 9-12)—and of

FIG. 9-12. IDEAL (FRICTIONLESS) FLUID FLOWING WITH STREAMLINES IN A PIPE
The speed is the same in all regions.

course there would be no drag on the pipe, and no end-pressure needed to maintain the flow once it was started (Newton Law I). With a real fluid, the flow is fastest in the central streamline on the axis, slower in its neighbors, slower still farther out; and at the pipe walls the fluid remains at rest. Fig. 9-13 shows the speeds for creeping flow. (For faster

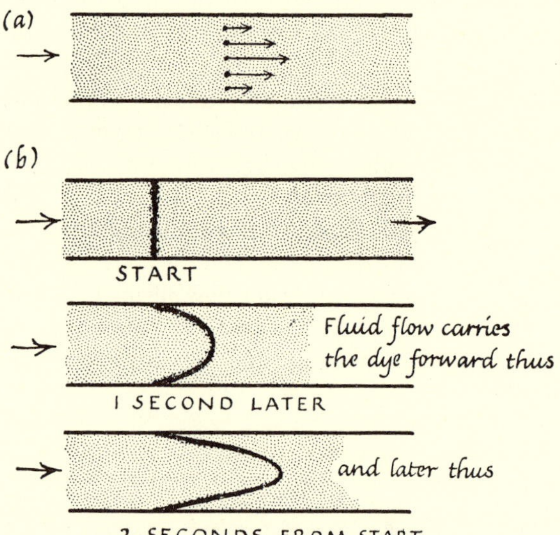

FIG. 9-13. STREAMLINE FLOW
(a) Streamline flow of real fluid in a pipe. The arrows indicate flow-speeds in various regions.
(b) If liquid flowing slowly in a pipe is suddenly "labelled" with a line of dye drawn across it, the dye then shows velocities in various regions.

flow, with a boundary layer at the walls of the pipe, the flow is still fastest at the center, but it is nearer to uniform speed all across the tube and only drops greatly in the boundary layer.)

Watch pressure and speed in a water pipe.

Demonstration

(v) *Water flow through uniform narrow pipes.* (Fig. 9-14) Water flowing through pipes always suffers some fluid friction drag. To avoid letting

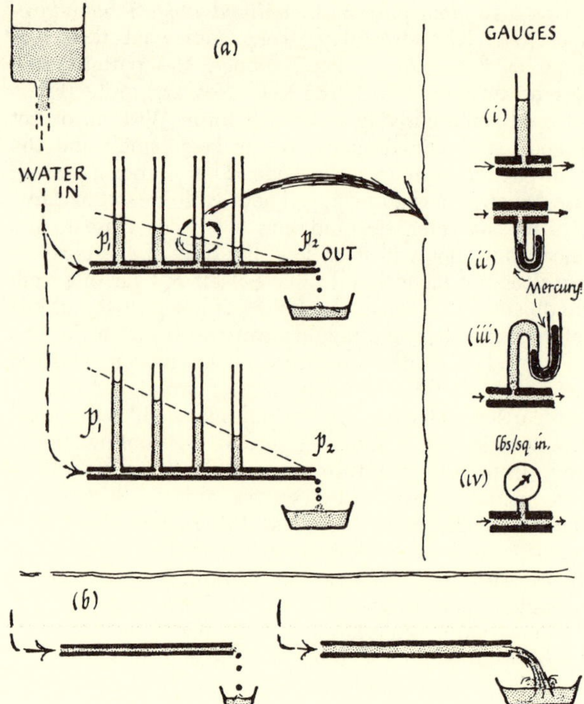

FIG. 9-14. SLOW WATER-FLOW THROUGH UNIFORM NARROW PIPE
(a) If the pressure is doubled, the flow-rate is doubled. Gauges attached to the side-openings show the pressure exerted by the flowing water on the walls of the pipe. Any sensitive pressure-gauge will serve, such as: (i) Vertical stand-pipe; (ii) and (iii) U-tube with mercury in it; (iv) dial gauge, containing an uncurling metal tube connected to a pointer (Bourdon gauge).
(b) If the same pressure is applied to a pipe of double the diameter, the flow rate is 16 times as great.

this muddle our paradoxes, look at friction effects first. We shall use them later in illustrating electric current flow (Ch. 32), and before that we shall suggest a molecular explanation for the friction of gases (Ch. 30).

In the experiment shown in Fig. 9-14, the tube is *very* narrow, a "capillary" with liquid creeping very slowly in streamlines through the pipe, driven by a pressure-difference between the ends. The pressure gauges show a progressive fall of pressure along the pipe. Although there is a pressure-difference between the ends of the pipe driving liquid along, the water does not accelerate—the flow is the same all along the pipe—so there must be other forces to make the resultant force zero on any sample of water. Fluid friction provides those forces. The pipe walls exert a friction-drag on the streamlines nearest them, and similar drags are handed on from one streamline to its neighbor all through the flowing liquid from pipe-wall to the pipe-axis, where flow is fastest.

To change the speed of this steady flow to a faster rate all along the pipe we must change the applied pressure. Faster flow needs larger pressure to maintain it. In fact, experiment shows that, for a given pipe,

FLOW-RATE varies directly as PRESSURE-DIFFERENCE

between the two ends of the pipe (until turbulence sets in at very fast flow). This is a general characteristic of fluid streamline friction:

$$v \propto (p_1 - p_2).$$

When we change to a wider pipe, there is a similar streamline pattern, and fluid friction is still there but it is less noticeable. The slow fluid near the wall of the pipe is a smaller fraction of the total moving fluid. Then a much smaller pressure-difference is needed for the same speed of central streamline. And for the same pressure-difference we get much greater flow in a wider pipe. Experiment gives the following relationships *for slow streamline flow* driven through various long thin pipes by a difference of pressure between the ends:

SPEED, averaged over the whole set of streamlines in the pipe

$$\propto \frac{\text{PRESSURE-DIFFERENCE BETWEEN ENDS OF PIPE}}{\text{LENGTH OF PIPE}}$$

SPEED, averaged over the whole set of streamlines in the pipe

$$\propto (\text{DIAMETER OF PIPE})^2$$

Multiplying this *average* speed by the cross-section area of the pipe gives the volume flowing past any section in unit time. This is because

$$\text{SPEED} = \text{LENGTH TRAVELLED IN UNIT TIME, SO}$$
$$\text{SPEED} \cdot \text{AREA} = (\text{LENGTH} \cdot \text{AREA})/\text{TIME}$$
$$= \text{VOLUME}/\text{TIME}$$

Then we have for slow streamline flow through various pipes:

VOLUME flowing through per second

$$\propto \frac{\text{PRESSURE-DIFFERENCE BETWEEN ENDS OF PIPE}}{\text{LENGTH OF PIPE}}$$

And VOLUME flowing through per second

$$\propto (\text{DIAMETER OF PIPE})^4$$

Notice the enormous effect of changing diameter of pipe. Think of the differences of blood flow through fine veins compared with its flow through arteries. In very fine capillaries, blood corpuscles may actually jam up the pipe, decreasing the flow even more than we predict by the simple engineers' formula above.

PROBLEM 1

There is an easy "graphical" description for the fraction (pressure difference)/(length), used above, in terms of the main diagram of Fig. 9-14. Can you suggest a technical phrase for that fraction?

★ PROBLEM 2

Taking the flow of oil in a pipeline as slow streamline flow, and assuming that the experimental relations mentioned above apply, say how you would expect *doubling the pipe diameter* to affect:

(a) The volume flowing per day past any section of pipe for same pressure of pumping system.
(b) The cost of metal for the pipe, *if the pipe wall-thickness is the same*, and the total length the same.

Demonstration

(vi) *Changes of flow-speed and pressure: Bernoulli effect.* (Fig. 9-15) Now make fast water

FIG. 9-15. FAST FLOW THROUGH WIDE PIPES: BERNOULLI EFFECT

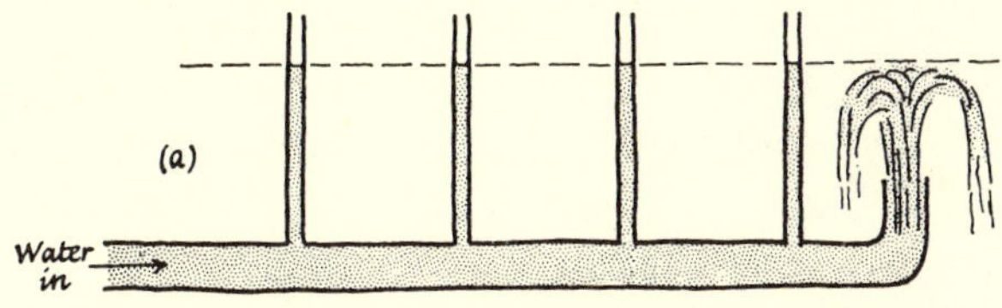

FIG. 9-15a. When there is streamline flow in a wide pipe, fluid friction is much less important; so the pressure is almost the same all along a wide uniform pipe.

change its speed in the course of flowing along a pipe. This can be done by inserting a section of narrower pipe. Make the whole apparatus of very wide tubing to make fluid friction practically negligible. Then in addition to the small effects of fluid friction we see a sudden drop of pressure where the water meets the narrower tube (Fig. 9-15b).

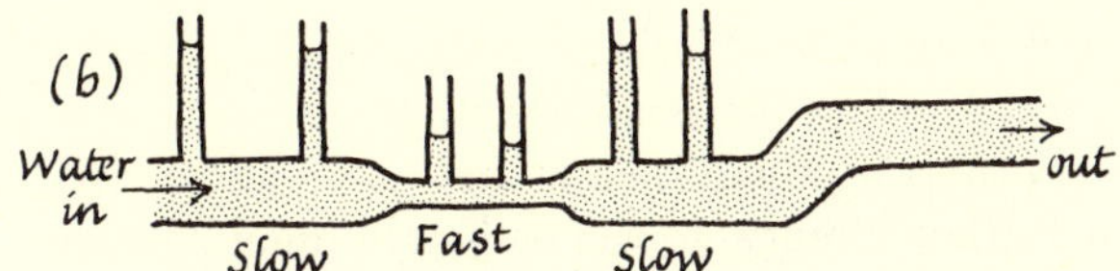

FIG. 9-15b. When the pipe has a narrow section, there is a surprising change of pressure from wide pipe to narrow pipe. (Again, fluid friction is unimportant, barely noticeable in these big pipes.)
NOTE the crook in the exit-pipe, to make the water pile up to visible heights in the stand-pipes.

When the supply pressure is raised to make all the flow faster, friction is greater, but the new effect is *much* greater. So with wider tubes and fast flow we can neglect friction and observe new pressure effects: changes of pressure wherever the flow changes its speed as the pipe narrows or widens. Or we can lose sight (rather dishonestly) of friction effects by using a single pressure gauge for each section (Fig. 9-15c,d). At faster flow still, the pressure in the

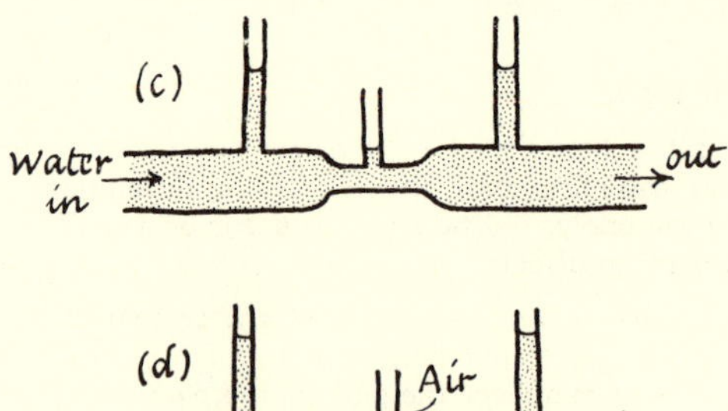

Fig. 9-15c and d. The faster the flow the bigger this change of pressure.

narrower section falls below atmospheric and the tube takes in air bubbles.

PROBLEM 3

Use the effect mentioned above to design a simple spraying device.

PROBLEM 4

Fig. 9-16 shows the apparatus of Fig. 9-15c inverted with its side pipe dipping in ink. What will happen? Give reasons.

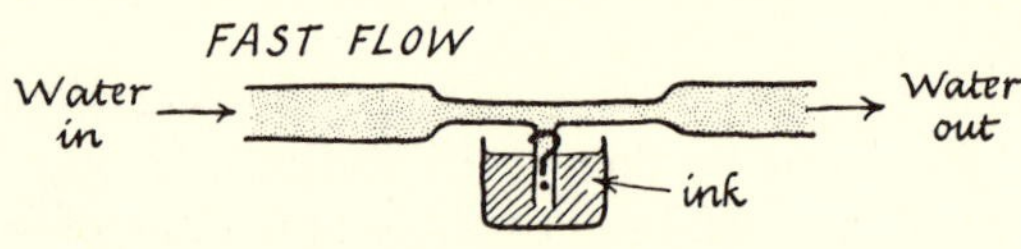

Fig. 9-16. Problem 4

Bernoulli's Principle: The Key to the Paradoxes

In the demonstration of Fig. 9-15, the pressure is less where the flow is faster. This is true of streamline flow in general, and expressed in more definite form it is called *Bernoulli's Principle.*

Transferring this experimental observation, without further inquiry, to the ball-and-funnel paradox, look at its streamline pattern. The streamlines are sketched in Fig. 9-17a. At D, where the air-flow spreads into the open air, the pressure is atmospheric. In the narrow gap, C, flow is faster because the same amount of air must flow through a narrower gap. Will the pressure be greater or smaller? How is the ball held up?

Bernoulli's Principle and Its Explanation

"Less pressure where the flow is faster," holds for streamline flow in general, in gas or liquid. It is peculiar, yet not so unreasonable as it looks; in fact

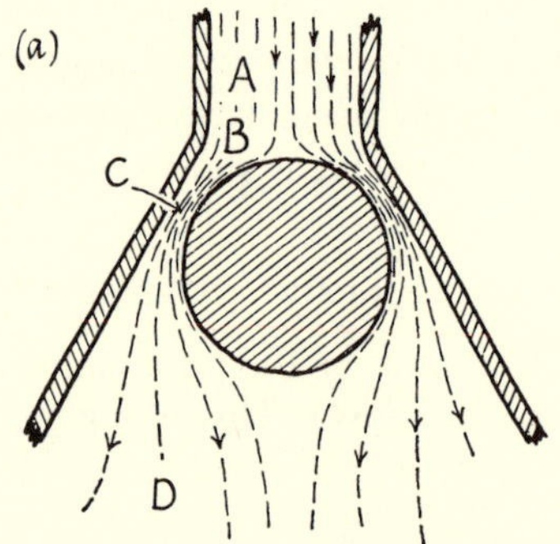

Fig. 9-17a. Streamlines of Air Around Ball in Funnel
The pressure is smaller at C where the flow is faster.

you might have predicted it from knowledge you already have, Newton's Law II, by the following argument.

Consider a small chunk of fluid flowing along in streamline at A in Fig. 9-17b. When it reaches B

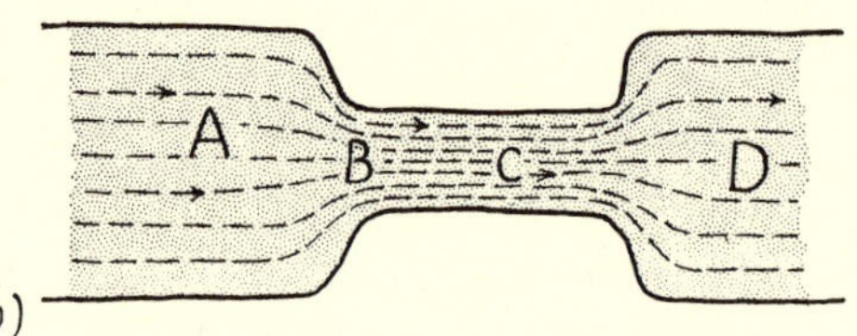

Fig. 9-17b. Streamlines of Fluid in a Pipe

it is flowing faster, and therefore has more momentum. Somehow between A and C it must have accelerated obviously at the narrowing neck, B. But this acceleration requires a forward force; and, for the sample of moving fluid, such a force must be provided by the fluid pressure of neighboring fluid on it. This suggests vaguely that the pressure must be greater at A than at B. If the pressure were the same all along A, B, C, how could the fluid experience any accelerating force—what else does it know of the outside world and its forces but the pressure of neighboring fluid? So the paradoxical Bernoulli effect appears to be an illustration of Newton's Law II: *the pressure difference must be there to produce the acceleration.*

To see this more clearly, imagine a tiny cube-shaped submarine which sails in the streamline flow, carried along by the fluid. Where the flow is faster the boat moves faster too; it too must accelerate in progressing from the wide tube A to the narrow tube C. From C to D it must decelerate. (See Fig. 9-18.) Its acceleration must be caused by a difference of pressure. Pressures on its sides have no effect on its forward motion—their effects must cancel out. But pressures on its front and hind ends

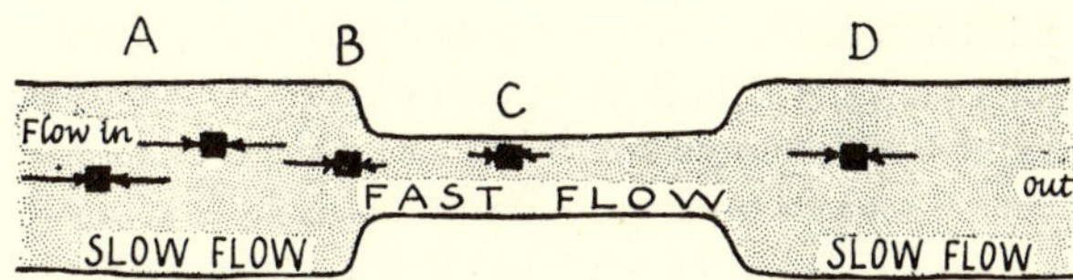

FIG. 9-18. BERNOULLI'S PRINCIPLE EXPLAINED
A small cubical submarine sailing along a streamline would experience large equal pressures on its bow and stern at A and D. In region B the pressures must be unequal, making a resultant force on the submarine to accelerate it. At C the pressures must therefore be smaller, but again equal.

must provide a resultant force when it accelerates or decelerates. Therefore when the boat is accelerating at B, in the transition between A and C, the forward push on its stern (hind end) must be greater than the backward push on its bow (front end). The pressure must be greater on its stern than the pressure on its bow. But its stern is then in region A of slower flow and its bow in the region C of faster flow; so the pressure must be smaller where flow is faster. When the boat passes from C to D the fast-flow pressure on its stern is smaller than the slow-flow pressure on its bow and it is decelerated.

This vague argument is sound as far as it goes—the pressure-difference does accelerate the fluid—but to carry it further we should have to discuss energy-changes in detail. Here we shall use Bernoulli's Principle in the vague form we have predicted: *in streamline flow the pressure is less where flow is faster.* It does not apply to eddy flow or turbulent flow. Even with streamline flow we cannot stride across from one streamline to another with the principle—since no specimen chunk of water does that. However, since there is no flow *across* streamlines, there is in general not much pressure difference in crossing from one streamline to a neighboring one. Bernoulli's Principle is interesting, but not a fundamental piece of physics that all should know. It is given here as an example of special behavior—and peculiar behavior at that—which can be "explained" in terms of general knowledge, without needing special laws of physics, assumed just for this purpose.[2]

[2] These would be called "*ad hoc* assumptions" or a whole framework of them might be called an "*ad hoc* theory." "*Ad hoc*" means "to this (purpose)." Primitive magicians' explanations are full of *ad hoc* assumptions of special spirits or influences. Modern science sinks to them sometimes—e.g., when biologists "explain" plants growing towards the light by saying they have a tendency to face the sun. However, we regard such explanations as poor, if not downright dishonest, except when they help to link together several different behaviors.

Examples of Bernoulli Effects

In Fig. 9-19a, an air jet blows fast across the upper end of an open tube dipping in liquid. The air stream moves faster at A than at B where it has spread into the wide open air. Therefore the pressure at A is smaller than atmospheric and atmospheric pressure at D can drive the liquid up the tube where it is blown into a spray.

In Fig. 9-19b two ping-pong balls are hung on flexible wires a small distance apart. An air jet blown through the space between them pulls them together.

In Fig. 9-19c the tube AB feeds air through a hole in the center of a fixed disc C. A loose disc D rests a short distance below C. The air flowing through

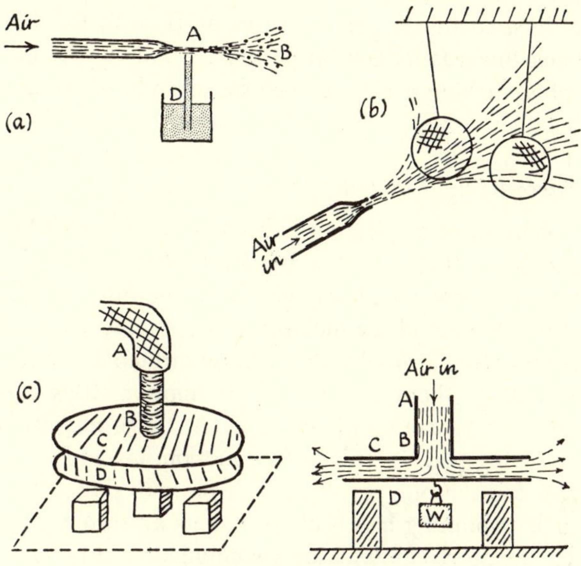

FIG. 9-19. DEMONSTRATIONS
(a) Sprayer.
(b) Air jet directed between two light balls suspended close together.
(c) The loose plate D is pulled up to C when the air-flow is turned on.

AB turns and flows out horizontally through the narrow space between C and D before escaping into the open air. The loose disc D is pulled up towards C, even if it is loaded with a weight W. If D is very light and shackled loosely to prevent its sliding away sideways, it will rattle against C with a piercing shriek. The small boy's blade-of-grass squeaker works on this scheme. The action of our vocal cords has something in common with it.

In Fig. 9-19d a ball is supported on a jet of air or water. The wonder here is not that the jet can push the ball up—that is just a matter of letting the ball kill a stream of upward momentum—but

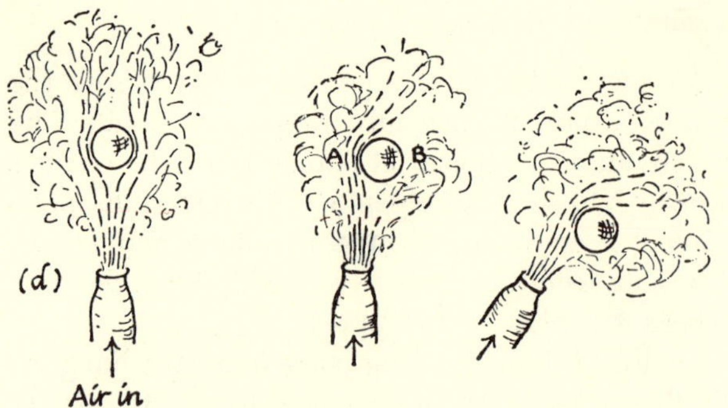

FIG. 9-19d. A light ball is supported on a jet of air.

that the ball does not topple over sideways. The equilibrium looks unstable but is not. If the ball moves to one side B, most of the jet flares up past the other side, A. With faster flow at A, the pressure is less, so the larger pressures at B push the ball back towards its central position. (Usually the ball spins, making a further, helpful, change of streamlines.)

Curved Flight of Ball

Why does a spinning ball move in a curve? We can show that this is a Bernoulli effect. Any ball, however smooth, is rough from the microscopic point of view of air molecules. A spinning ball hits neighboring air molecules with its surface roughness and drags them into sharing its motion. It is surrounded by spinning layers of air, those nearest it revolving with its full surface motion, layers farther and farther out revolving slower and slower.* If such a spinning ball is also flying along through the air, the air streamlines are a combination of two motions, the circulation of air around the ball and the steady stream of air past it. Imagine an observer running along keeping level with the ball to observe the streamlines and make Bernoulli-effect predictions. The runner will see the ball stationary beside him and both he and the ball will encounter a steady wind blowing past. The "wind" blows with the speed of the ball's flight, in the opposite direction.

Another scheme which has the same virtue is to imagine a great, general wind blowing backward, with velocity just equal and opposite to that of the ball. Then the observer can stand still on the ground and watch the ball, fixed opposite him.[8] Such a wind would have straight, parallel streamlines, as in Fig. 9-20a. To see why a spinning ball can fly in a curve, it is worth while trying to sketch the two patterns of streamlines and then combine them by intelligent guessing. Fig. 9-20b shows the spinning ball, with layers of air spinning with it. To show that the air motion is slower farther from the ball, the outer streamlines have been spaced farther apart and marked with shorter arrows. To combine the two motions, place one pattern on top of the other (Fig. 9-20c) and add the two vector velocities at each point. At specimen point, P, draw two little velocity-vectors, v_1 for the uniform flow, v_2 for the motion due to spin, and sketch the parallelogram to find their resultant (Fig. 9-20d) which is the velocity of the combined flow at that specimen point. Repeat this sketching at specimen points all over the pattern, using the same horizontal velocity v_1 everywhere, and making v_2 tangent to the circles. Make v_2 small far out, large close to the ball. Then, with enough resultant vectors to guide the sketching of flow-lines, erase unwanted scaffolding, leaving just a short arrow through each point to show the direction of resultant flow (Fig. 9-20e, f). (These arrows need not show the speed by their length.) Then with considerable guessing continuous streamlines which everywhere have the directions of the arrows can be sketched. As a further guide use the following hints from common sense: (i) very far from the ball, the spin motion is negligible, so the flow is uniform v_1—the streamlines are horizontal and evenly spaced; (ii) very near the ball, the spin motion predominates—the streamlines are practically circles; (iii) at some point, N, below the ball v_2 and v_1 just cancel, making a "neutral point" of no motion. With more tedious work and more imagination—or a cheating glimpse of the real pattern, obtained in some other way—we can fill in the picture. Such sketching can give only a rough indication of the resultant pattern. For a reliable map we must drag in algebra to do the geometry for us, first investigating the gradation of the spin velocity v_2 in detail. More careful mapping gives the diagram

* This applies reasonably to a rough ball like a baseball, spinning fast. The full story is more complicated: see *Amer. Journal of Physics*, 27 (1959), p. 589. A very smooth ball, with moderate spin, carries only a thin "boundary layer" of air around, and often curves the "wrong way"!

[8] This "adverse wind" scheme is a useful one in physics. It can be used, for example, in thinking about sound waves; and, with some mathematical application of Newton Law II, enables us to predict that the speed of sound in air will be $\sqrt{(7/5)(\text{pressure of air})/(\text{density of air})}$.

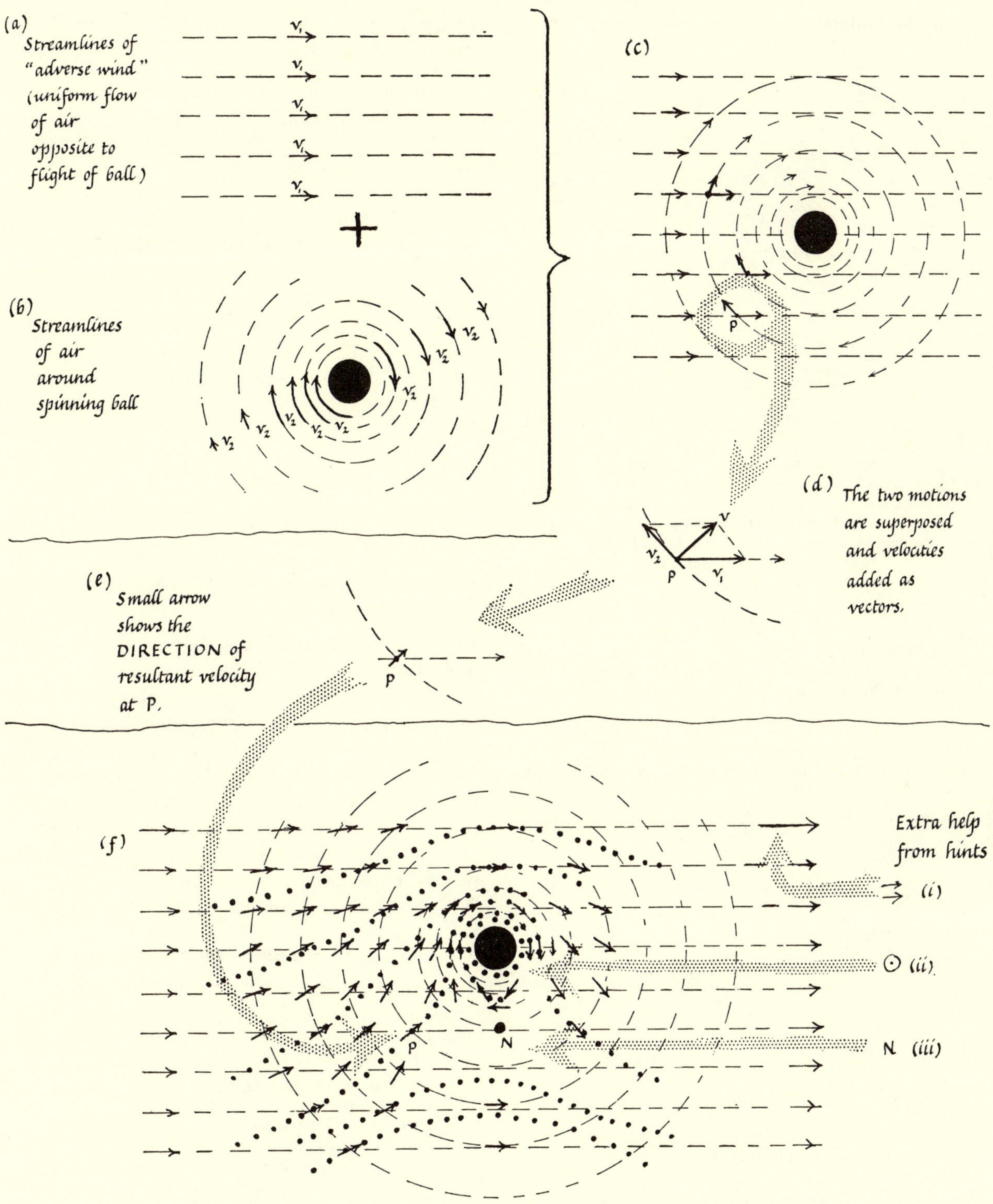

FIG. 9-20. STREAMLINES FOR SPINNING BALL MOVING THROUGH AIR

below for a rod (instead of ball) spinning in a uniform current of air. The pattern for a ball is somewhat similar.

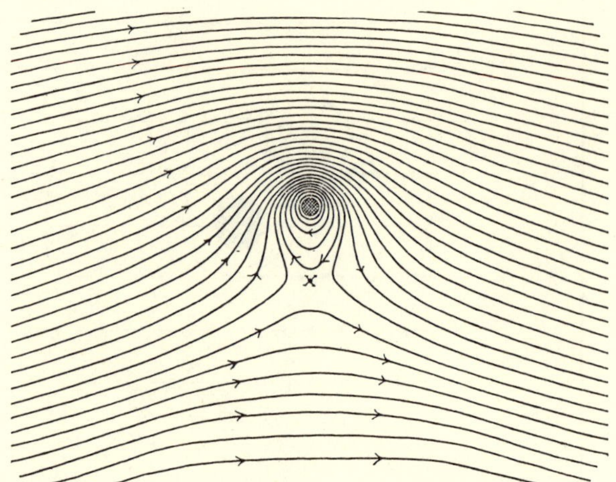

FIG. 9-21. STREAMLINES FOR A SPINNING CYLINDER IN UNIFORM WIND. These are sketched fairly accurately from the streamline pattern predicted geometrically from $\nabla^2 V = 0$, the basic mathematical rule governing streamline patterns of this kind and all other "inverse-square law" patterns.

PROBLEM 5

If you have studied physics before, you may have met this pattern in an entirely different part of physics. If so, where? Is the likeness purely coincidental? Can the likeness be of any practical use?

PROBLEM 6. (This is well worth while, if done roughly and quickly.)

Using the method of Fig. 9-20, sketch the streamlines for the flow described below. (See Fig. 9-22.)

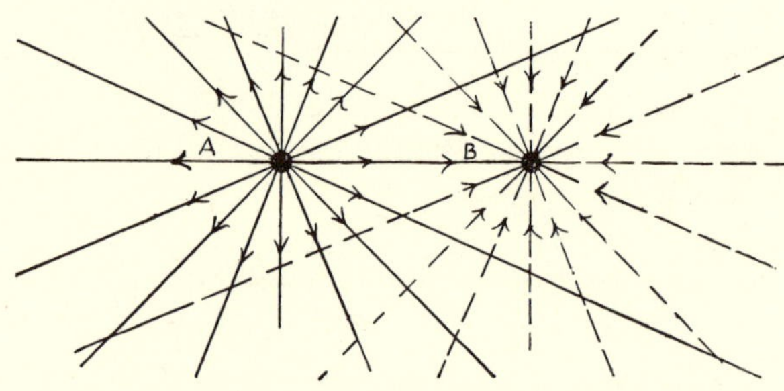

FIG. 9-22. PROBLEM 6. Streamlines for a source and an equal drain in an infinite lake of uniform depth.

In a huge shallow lake of still water, a spring at A delivers a steady inflow of water, and a drain at B carries away an equal outflow. Sketch the streamlines in the lake using the following help. The spring *by itself* would produce streamlines which are straight spokes radiating from A. Near A, where the streamlines are crowded close, the speed of this outflow along radii would be great. Farther out from A the speed would be smaller.* The drain *by itself* would produce a similar pattern of radial inflow to B.

Mark A and B a few inches apart on a full sheet of paper, sketch both sets of streamlines right across the paper and find the resultant pattern, by drawing and guessing. (What corresponds to hints (i, ii) above?)

Where have you met a similar pattern?

* In a "shallow" lake the flow speed would vary as $1/r$. But if A and B were immersed in a "deep" ocean, the flow speed would vary as $1/r^2$. The patterns of resultant streamlines for the two cases look somewhat similar.

Now we can return to the flying baseball. Viewed by an *observer running along beside it*, the ball is surrounded by a streamline pattern like Fig. 9-23.

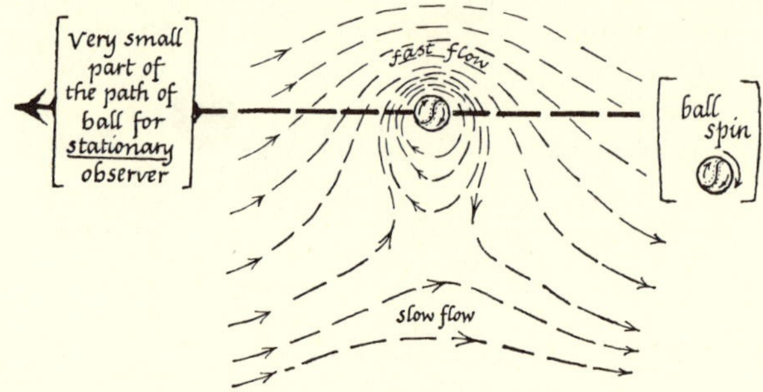

FIG. 9-23. STREAMLINES OF AIR FLOWING PAST A SPINNING BALL

Air is flowing past it faster above it than below it, so there is a region of lower pressure above it and high pressure below. Air pressure therefore pushes the ball upward, distorting its path. Similarly a ball with a spin around a vertical axis is pushed to one side and will swerve. There have been many arguments about this matter, but the "curving" of a spinning baseball has been measured. However, prejudice from a pitcher's reputation may make player or spectator see more curves than are there. With a lighter ball spinning fast—e.g. a cut tennis ball—real curves are easily seen.

★ PROBLEM 7. THROWING CURVES

Suppose a massive baseball and a much lighter ball of the same size are projected horizontally, side by side with the same speed and with *equal spins* around a vertical axis.

(a) *Apart from* spin effects and air friction, how would their paths compare?
(b) How would the sideways forces (produced by the Bernoulli effect just discussed) on the two balls compare?
(c) How would the sideways (curve) motions compare?
(d) Give a clear reason for your answer to (c).

Demonstrations: Curved Flight

(vii) A cork ball is thrown across the room from a rough cardboard tube. The thrower holds the tube behind his shoulder, and slings the ball out by sweeping the tube forward and downward. The ball, "left behind" by the tube's motion, rolls along the inside upper surface of the tube and emerges spinning fast round a horizontal axis. The upward curve in its flight is visible. (Fig. 9-24)

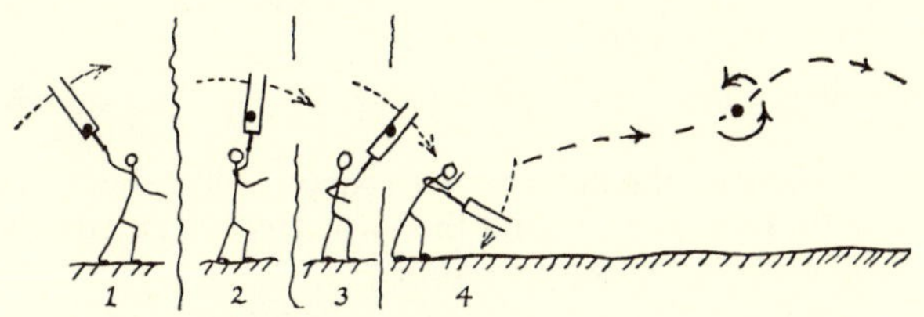

Fig. 9-24. Throwing a Spinning Ball. The cardboard tube is rough inside and has room for the ball to roll.

(viii) A cardboard cylinder is thrown forward by a catapult which sets it spinning at the same time. The Bernoulli forces are so big that it may even loop the loop. (Fig. 9-25)

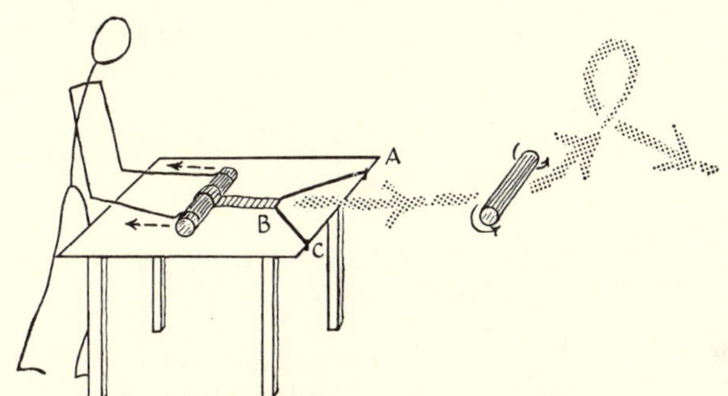

Fig. 9-25. Throwing a Spinning Tube
ABC is a piece of rubber cord fixed to the table at A and C. From its center B a piece of cloth tape runs back to the cardboard cylinder, and is wound around the central part of the cylinder several times. The operator pulls the cylinder back across the table, stretching the rubber, then lets it go.

Airplane Flight

We can look at streamline flow past a model airplane wing by marking water flow with ink streams, or air flow with smoke, and we see a clear crowding of the lines *above* the wing. Then Bernoulli effects provide the lifting force, by making the air pressure less above the wing than below. But how does the wing secure this fortunate pattern of streamlines? Geometry and mechanics show that a frictionless fluid would give a more symmetrical pattern without crowding above, and therefore no lifting force and no drag. But with real air or water, a "vortex" like a smoke ring of air-circulation around the wing develops as the plane starts, and moves along with the plane. (Fig. 9-26) This motion combines with

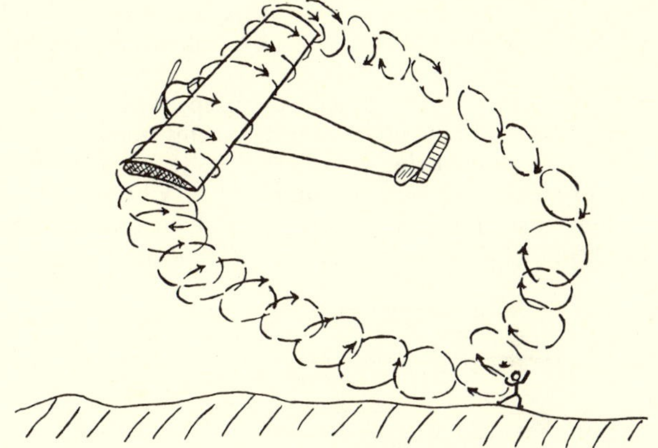

Fig. 9-26a. Circulation Around Airplane Wing
When a plane starts flying, air friction starts a "smoke-ring" of circulation. This motion is combined with the steady flow of air past the plane.

the steady motion of the air past the plane to give a resultant pattern rather like that for a spinning cylinder in flight—the wing does not spin but its shape produces an equivalent air circulation. This vortex cannot end at the wing tips like a cut-off sausage but continues out and around behind the plane. As the plane flies away it carries the front of the vortex with it, leaving the rest behind, with lengthening vortex streamers from the wings. (It is the portion left behind that removes your hat if you stand too near a starting plane.)

Fig. 9-26b. Streamlines of Air Past Airplane Wing
This gives an idealized picture of streamline flow. In real flight some eddy motion develops behind the plane. If too much develops at an early stage, the plane stalls.

Wind Resistance ("drag")

(This discussion of air friction is difficult and may be omitted.) A flying plane wing leaves a streamer of circulating air behind its wing tips; and other eddy motions trail away from the wings and body. The air behind the wing is thus given considerable eddy motions (with considerable kinetic energy) and some forward drift. Thus the wing is continually giving away forward momentum—therefore it experiences a backward force, the "drag" of the air on it—and the rest of the plane must pull forward to provide momentum to make up. The complete plane neither gains nor loses momentum, when flying steadily. Its propeller drives a roaring stream of air backwards, giving that air backward momentum and the wings + body leave a string of eddies with forward momentum. Thus the plane leaves behind a mixture of air motions with no resultant momentum forward or backward.[4]

How does wind-drag depend on speed, for an airplane wing or any other object leaving eddies in its trail? The wing, flying with speed v, leaves behind a slab of air drifting forward after it. Let A be the cross-section area of this slab, the "vertical face-section" of the wing. (Fig. 9-27) Let F be the drag on the wing,

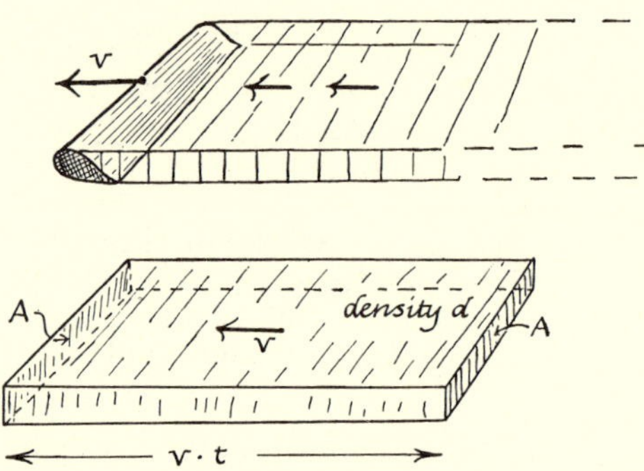

Fig. 9-27. Moving wing leaves behind a slab of forward-moving air, moving with velocity which is really only a fraction of the plane's velocity, v (but it is taken as v for this simple discussion). In real flight the air does not form a simple moving slab—the motion spreads and is mixed by eddies.

due to the continual losing of forward momentum. To predict the size of F, pretend, to start with, that the slab of air is left with the full velocity of the wing, v.

Then, from $F \cdot \Delta t = \Delta(mv)$,

(FORCE, F) (TIME, t secs) = MOMENTUM lost by wing in t secs
= MOMENTUM given in t secs to the chunk of air set moving behind wing.

In t seconds the wing moves ahead a distance vt, leaving behind a chunk of moving air of length vt, area A, therefore volume $A \cdot vt$. This new moving air has

MASS = (DENSITY) (VOLUME) or $(d)(A \cdot v \cdot t)$.

If its speed is v, its momentum is

(MASS) (VELOCITY acquired)
or $(d \cdot A \cdot v \cdot t)(v)$ or $d \cdot A \cdot v^2 t$.

$\therefore\ F \cdot t = d \cdot A \cdot v^2 \cdot t$

$\therefore\ F \quad = d \cdot A \cdot v^2$

FORCE = (DENSITY) (AREA) (SPEED)^2 [see note 5]

In real cases, the air does not acquire full v, but some definite fraction of v; and the area A is not exactly the wing section, but we still have

$$F = (\text{constant})(\text{some area})(\text{air density})(v^2).$$

The constant depends on the geometrical shape of the wing *and* on the speed-range. This shape-factor is large for unstreamlined objects such as a flat plate meeting the air head-on, or even for a round ball. For a "streamlined" body, presenting the same area to the wind but having a properly designed teardrop shape, the factor is 20 to 100 times smaller because far less eddy motion is left behind. This drag due to eddies-left-behind is quite different from the friction-drag of creeping streamline motion.

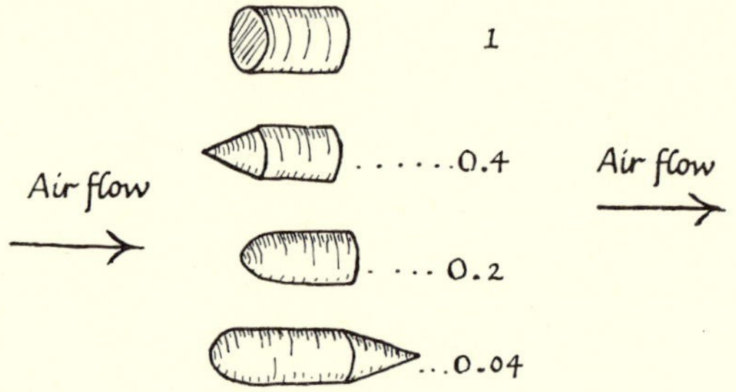

Fig. 9-28. Relative Shape-Factor for Air Resistance with Fast Flow

The Mechanism of Friction-Drag

Creeping friction-drag is not produced by wholesale carrying away of motion but by individual momentum-nibbling and exchange by colliding molecules. Here is a rough discussion of it: fluid molecules nearest the moving object catch some of its forward momentum when they collide with it, and hand on some of their catch to neighbors when they hit them. Such molecules, buzzing to and fro with their own speedy random motion, act like mice, nibbling a little forward momentum from the creeping object. This taking-away of forward

[4] The energy of this turbulent motion must be paid for—by the man who buys the gasoline. Ultimately the eddy motions are socialized into individual molecular motions, heat, making the air behind the plane a little warmer, just as much warmer as it would have become if the gasoline had been burned directly to heat it!

[5] This is (exactly) equivalent to keeping the wing still in an adverse wind approaching with speed v, and having the wing stop dead all the air that hits it. Then again in t seconds the wing stops a chunk of air of length $v \cdot t$ and MASS = (DENSITY) $A \cdot v \cdot t$. This mass changes velocity from v to zero, and loses MOMENTUM = (DENSITY) $(A \cdot v \cdot t)(v)$. $\therefore F \cdot t = d \cdot A \cdot v^2 \cdot t$. $\therefore F = d \cdot A \cdot v^2$. Notice that this calculation is just like that of a problem where a stream of water from a firehose hits a wall. In fact this is a case of a firehose-stream of air hitting the wing. As in that problem, the FORCE $\propto v^2$.

momentum makes the moving object experience a dragging force. ($F \cdot t =$ loss of momentum in time t).

How will this friction-drag depend on the speed of the moving object? Suppose we make the object move twice as fast: then it has twice as much momentum. At each collision, the fluid molecules probably take the same *fraction* of the object's new doubled momentum as before.[6] Thus they carry away twice as much momentum at each collision. And collisions are as frequent as ever, because the object's creeping is trivial compared with molecular speeds. So, with doubled speed, the object loses twice as much momentum in the same time. Therefore, it should experience a doubled friction-drag, so we should expect the friction-drag to vary directly as the object's velocity. $F \propto v$. Experiment confirms this for slow streamline motion in gas or liquid.

At higher speeds, there is wholesale robbery by chunks of fluid in eddies, organized gangs of molecules taking momentum. Then, as suggested before, we expect the drag-force to vary as v^2.

Then CREEPING DRAG-FORCE $\propto v$, for very slow motion (e.g., tiny raindrops in a cloud, or sediment settling in a pond) and EDDY DRAG-FORCE $\propto v^2$, for faster motions. A modern airliner flies so fast that even the best streamline design produces drag forces that vary nearly as v^2. In applying physics to real flying you must remember that the controls are changed for different speeds, so the shape factor changes. Then the drag will appear to be more complicated still, showing a minimum at some optimum speed.

★ PROBLEM 8. TERMINAL VELOCITY (*This problem prepares for an important use in atomic physics*)

A small streamlined body is allowed to fall in air. Released from rest it accelerates at first but soon reaches a constant speed of fall (called "terminal velocity"). Test this statement for yourself with a small chip of paper or a toy parachute.

(a) Why does it not continue to accelerate?
(b) When it is falling with constant speed what is the *resultant* force acting on it? What can you therefore say about the drag-force on it?
(c) Can you tell just by watching the body fall whether the force is the friction-drag kind ($F \propto v$), or the eddy-loss kind, ($F \propto v^2$)?
(d) Suppose by some chance collision with a mosquito the falling object is slowed slightly, or speeded slightly. Explain why it will return to its original speed provided the friction-drag increases with speed (as it does for either $F \propto v$ or $F \propto v^2$).
(e) Suppose the falling body is hollow, and that it can be filled up till its total mass is 4 times as large. How would you expect this change to affect its "terminal velocity," v, (i) if $F \propto v$? (ii) if $F \propto v^2$?

Personal Experiments

(A) Hold a small flexible sheet of paper with two hands at one end, so that it is horizontal at that end and the other end sags under its own weight. Place your mouth just above the horizontal end and blow a steady blast of air out just above the paper. (Fig. 9-29) Watch what it does and explain it. Here indeed is the airplane reduced to simplest form.

FIG. 9-29.

(B) *The motion of a streamlined sheet of paper.* (Remember the taunt in Ch. 1 that in trying falling-body experiments you were likely to leave some simple ones undiscovered.)

(i) Drop a small sheet of paper, and watch it fall. (If you like, compare this with the fall of a similar sheet crumpled.)

(ii) Give the sheet a little streamlining by bending up a small wall along each edge, to make a tray, as in Fig. 9-30. Watch it fall.

FIG. 9-30. PAPER TRAY

(iii) Try obvious variations on (ii). This offers great opportunities for inventiveness and critical thinking.

(iv) From your experiments decide whether this is slow streamline motion with friction-drag predominant, ($F \propto v$), or faster motion with eddy formation, ($F \propto v^2$). Make sure you can give a clear reason for your decision. (It is true that you might get a hint by marking air currents around the falling tray with smoke; but you should try to arrive at a much more precise proof.)

Bernoulli Effects: Demons or Science?

Though technical scientists use Bernoulli's Principle in developing flying machines, and engineers enlist its help in the design of gadgets, it is not a vital part of the fabric of physical science. Yet here we have spread its chapter, which could easily be omitted, over many pages, not to show practical uses but to show scientific explanation at work. Starting with paradoxes of attracting funnels and curved flight, each of which seems to need a special, private demon to explain it, we end with a single

[6] A problem at the end of Ch. 26 will show you that a ball hit elastically by a massive moving object gains twice the object's velocity. Then the ball takes the same fraction, $2m/M$, of the object's momentum whatever the velocity. Here the random motion of the molecules is superimposed on such encounters, without altering the net effect.

principle to explain those paradoxes and predict new ones. At first our knowledge is "empirical" (= straight from experiment), a mere summary: "where streamlines are crowded and flow is *faster*, pressure is *lower*." Then, thinking about this, we take a suggestion from common sense: "if the flow changes from slow to fast, the fluid must accelerate." Next we bring in "theory" in the form of our general belief in Newton's Law II, $F = M \cdot a$: "where there is acceleration, there must be a corresponding resultant force." Applying that belief to a simple case, such as liquid flowing in an uneven tube, we *predict* "low pressure with fast flow." So, if we trust $F = M \cdot a$ as a universal rule, we must expect Bernoulli effects as examples of $F = M \cdot a$ behavior. (Then if they did not occur, we should have to doubt the general nature of $F = M \cdot a$.) More theory, using the principle of energy-conservation and some algebra, will predict an algebraic relation between flow-speed and pressure—a relation confirmed by experiment. The prediction is, essentially:

½ · (density of fluid) · (speed of flow)² + (pressure in fluid) = constant. That is $(\frac{1}{2}d \cdot v^2 + p)$ should have the same value at all points along a streamline.

Then

$$\underset{\text{in region 1}}{[\tfrac{1}{2}d_1v_1^2 + p_1]} = \underset{\text{in region 2}}{[\tfrac{1}{2}d_2v_2^2 + p_2]} = \underset{\text{in region 3}}{[\tfrac{1}{2}d_3v_3^2 + p_3]}, \text{\&c.}$$

(If the level changes, we must allow for changes of potential energy too.)

Thus we have reduced several separate whimsical demons to one single mechanism, a combination of $F = M \cdot a$ and geometry—both of the latter are "unexplained" but are common to many parts of science. We have lessened the number of mysteries, by tying all our examples to one mystery, $F = M \cdot a$. As Conant would put it, we have lessened the "degree of empiricism" in our knowledge of fluid behavior, thus forwarding good science. The Principle, which we give to engineers to help build pumps and flowmeters and airplanes, and the further studies of fluid friction which we shall need in dealing with molecules and atoms, now appear as reasonable parts of the mechanics that we have been building without any reference to paradoxes.

This chapter on Bernoulli effects will be worth while if it increases your feeling that science makes sense; that its essence of progress is simplification rather than increasing complexity.

or $\frac{1}{2}d_1v_1^2 + p_1 = \frac{1}{2}d_2v_2^2 + p_2$

DEMONS OR FORMULA ?

PROBLEMS FOR CHAPTER 9

Problems 1-8 are in the text of the chapter.

9. Which would you advise for the water used in washing drinking glasses, etc., streamline or eddy motion? What would happen if the water had the other motion?

10. Two small boats are anchored side by side in the middle of a swift river, by ropes that run upstream from their bows to two anchors. When first anchored they are a few feet apart. Would you expect the stream to drive them farther apart or pull them together? Give a reason for your answer, with a sketch.

11. The Rotor Ship was an odd craft that sailed across the Atlantic successfully by using a Bernoulli effect. Instead of masts it had huge vertical cylinders that were kept spinning by motors. Suppose there is a steady wind blowing northward and such a ship wants to sail eastward. Should the cylinders spin clockwise or counter-clockwise, as seen from above? Illustrate your answer with a sketch.

★ 12. Devise a simple illustration of Bernoulli effects using two small sheets of paper.

Fig. 9-32. Problem 13

★ 13. Suppose a steady wind blows across a level plain and up and over a range of mountains at the edge of the plain. An airplane, using a pressure gauge (barometer) to measure its altitude flies over the plain. A pilot flying by night, tries to maintain a course at constant altitude sufficient to carry him over the mountains. Explain how the wind might lead to disaster.

14. The vocal cords of our larynx are strips of muscle with a slit between them through which we breathe. Suggest a way in which the vocal cords may be kept vibrating when we talk.

15. A steady wind blows over an ocean in which low humps and troughs of waves have been formed. Describe ways in which the wind may increase the humps and troughs.

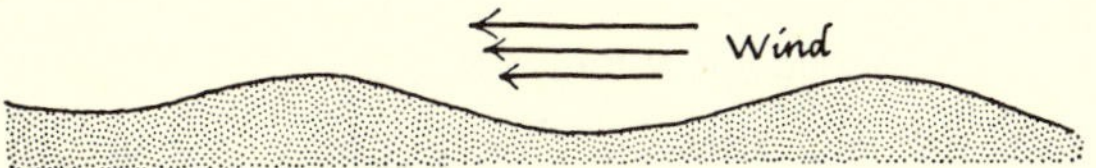

Fig. 9-33. Problem 15

16. Fig. 9-34 shows the arrangement of a flowmeter to measure the rate of flow of a liquid in a chemical factory (not its pressure, but its speed, in, say, gallons per minute). The liquid flow through a pipe ABC is measured by attaching a pressure-difference gauge between a hole at A on the pipe and a hole at B where the pipe is narrower. In the sketch the gauge is just a U-tube of mercury.

(a) Explain why the gauge can tell the flow speed.

(b) Explain why the gauge tells nothing about the pressure of the fluid in the main pipe.

(c) By arguing (e.g., "Suppose the flow speed is doubled, the pattern of streamlines remaining the same . . .") find out how the gauge reading of pressure should be related to the flow rate. (Would you expect the gauge reading to vary as (flow rate), or as $(\text{flow rate})^2$, or what?)

Note: This does not require the algebra of Bernoulli's Principle. It can be reasoned out from the simple explanation, but the reasoning requires care and courage.

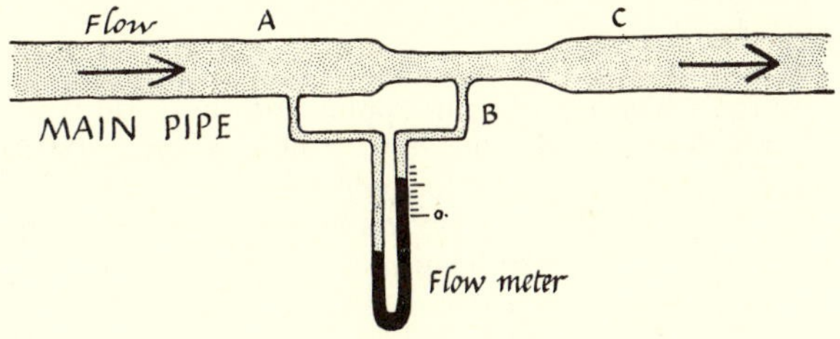

Fig. 9-34. Problem 16

17. Re-draw Fig. 9-14(a) with the liquid flowing *half* as fast as in Fig. 9-14(a). Write a note saying what change you expect to observe.

18. Re-draw Fig. 9-15(c) with the liquid flowing *half* as fast. Write a note saying what change you expect to observe.

CHAPTER 10 · VIBRATIONS AND WAVES

"An ocean traveler has even more vividly the impression that the ocean is made of waves than that it is made of water."

—A. S. EDDINGTON, *The Nature of the Physical World* (Cambridge, 1929)

TWO MATHEMATICAL REVOLUTIONS IN PHYSICS

1822 . . .

FOURIER'S THEOREM (first given by Fourier, still being investigated and elaborated and put to many uses in science):

ANY *(repeating) motion can be treated as the sum of a series of Simple Harmonic Motions.* ANY *wave, whatever its form, can be treated as the sum of a set of Simple Harmonic Waves.*

1924 . . .

DE BROGLIE'S DUALITY (suggested by de Broglie, now elaborated and used as a foundation of the physics of atoms today):

ANY *moving particle (electron, atom, neutron, . . . baseball, . . . even a quantum of light) is an extensive* WAVE *in some of its behavior, and a compact* PARTICLE *in some of its behavior.*

Simple Harmonic Motion, the standard component of all vibrations, is very common and very useful and it plays an essential part in the physics of sound and music and in our present "atomic physics" of waves and particles.

The study of wave motion is a large branch of physics, which has obvious applications to ocean waves, earthquakes, sound, etc., but it achieved even greater importance for the development of science when light was "proved" to be waves, and again when de Broglie's suggestion started the new revolution in physics.

In the choice of material for this course, both these topics are largely neglected, and most of this chapter can be omitted or postponed without missing connections with neighboring chapters. However, in the final chapter, Ch. 44, it is *essential* to have some knowledge of light-waves, spectra, and interference. The latter part of this chapter provides for that. Readers interested in fuller treatment should consult other texts on General Physics, Mechanics, Optics, Mathematical Physics—the choice depending on their mathematical equipment.

Pendulum Motion and Timekeeping

A pendulum has a surprising property—surprising to Galileo timing with his pulse, surprising to a modern student with his stop watch. *It takes practically the same time for a narrow swing as for a wide one.* Starting with a huge swing, say 80° each side of the vertical, a pendulum's PERIOD (= time for one cycle) lessens by a few % as its amplitude dies down to 60° . . . 40° . . . 20°; then from 20° to the smallest noticeable swing it changes by less than 1%. For all amplitudes below 5° the periods are the same within 0.05%.

This is useful as well as surprising. Galileo suggested using a pendulum as a regulator for a clock. In his day, the escapement of clocks was driven by a weight and controlled by a crude windmill device using air resistance. A pendulum would compel even timekeeping, since it would take the same time for small swings and for large ones when driven farther by some chance gust of wind. A century after Galileo, pendulum clocks were in good use, but navigators still needed accurate clocks to measure their longitude at sea. A prize was offered for a sea-going clock that could keep true enough time, and Harrison won it with his chronometer controlled by a balance wheel and special hairspring.

This property, equal times for all amplitudes, has an elaborate name, *isochronous*,[1] a Greek word for "equal in time." We say that pendulum motion is (approximately) isochronous, for small amplitudes. The property is worth naming, because it proves so useful.

Problem 1 leads from pendulums to other systems with isochronous motion. The problem is not easy but it is worth trying as an example of theoretical physics. It will show you how to argue from a simple experimental fact to a prediction of a new field. If you have carried through the "pendulum analysis" of Problem 1 successfully, you have predicted other systems with isochronous motion, some of them even better suited to regulating clocks. In fact, the revolution in timekeeping which started with Galileo's suggestion has continued, from clocks with pendulums to watches with balance wheels and

[1] Pronounced eye-sok'-r'n-us.

"PROBLEM SHEETS" The problems here are intended to be answered on typewritten copies of these sheets. Work through the problems on enlarged copies, filling in the blanks, (..........), that are left for answers.

DISCUSSION OF PENDULUM MOTION NAME____________

Complete the "theoretical discussion" of pendulums indicated below.

(i)Experiment shows that for small amplitudes the period, T, is practically independent of amplitude. In all this discussion we shall limit ourselves to such small amplitudes. When the amplitude is doubled, T is the same though the bob travels twice as far. ∴ bob must travel__________ as fast for double amplitude.

The motion is not one with fixed speed, nor even one with constant acceleration. Yet the way in which the bob changes its speed looks similar for different amplitudes, so we may guess that: though the speed is different at different stages of the swing, the speed at corresponding stages of the doubled-amplitude motion will be . . . . __________ as great as speed of the single-amplitude motion, otherwise T would not be the same.

(ii)Taking a risk and generalizing the argument of (i), we should expect to find for all (small) amplitudes, the speeds at corresponding stages of the swing are related to the amplitude of swing; thus:____________________

(iii)Return to (i), which compared amplitude and double amplitude. Since the effect of doubling the amplitude is to make the corresponding speeds __________ as great and since the bob acquires these speeds in the same time*, its acceleration, $\Delta V/\Delta t$, must be . . . . . __________ as great for the double amplitude as for the original one. (Again, the acceleration is not constant, but we are comparing values of acceleration at corresponding stages of the swing.)

(iv)Generalizing the argument of (iii), we can say that the relation between acceleration (at any chosen stage of swing) and amplitude must be: ____________________

DISCUSSION OF PENDULUM MOTION SHEET 2

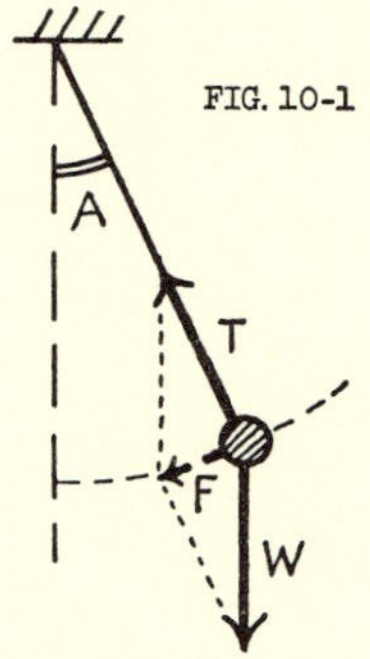

FIG. 10-1

(v)Out at the end of the swing, although the bob is not moving, it has its largest (inward) acceleration. This acceleration is due to the combined pulls of gravity and string. Gravity and string combine to give a resultant pull F along the motion.

From (iv) it seems likely that this resultant pull, F, at the end of the amplitude must be related to the amplitude, A, as follows:** ____________________

(vi)This relationship between force and distance-out must apply at each stage of swing. It seems similar to: . . . . . ____________________

(vii)From (vi) we might expect the same type of motion, in which period T is independent of amplitude, for such things as . . . . __________ and for these things the independence-of-amplitude is likely to be:

____
(restricted to small amplitude? / unrestricted? / or ???)

*NOTE. The changes of velocity have to occur in the same interval of time because the period (total time for all changes) is the same for all small amplitudes. Thus the period-independent-of-amplitude property is used twice over in this argument.

**The result is obtained here for the forces at the ends of various amplitudes; but the same relationships must hold between force, F, and distance-out, x, at various stages of one swing. Gravity can hardly know or care whether the pendulum is at the end of a small swing or at the same angle part way out in a large swing.

hairsprings, to oscillating quartz crystals, and now to the vibrations and spins of atoms themselves.

Simple Harmonic Motion

All these repeating isochronous vibrations have the same type of motion, with the same shape of time-graph, a "sine curve." We call such motion Simple Harmonic Motion—"harmonic" because of its importance in music. A pendulum moves with

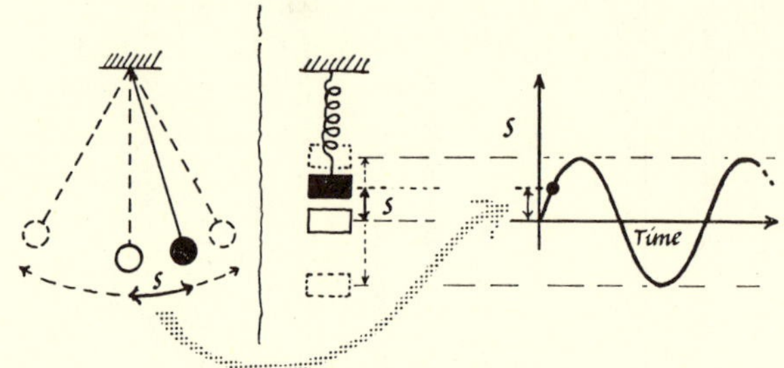

Fig. 10-2. Isochronous Oscillations and Time-Graph

(approximately) S.H.M. for small amplitudes. A load hung on a spring bounces up and down with S.H.M., for any amplitude over a wide range. (Try a quick experiment in the lab. You will find it very satisfying.) A spring with a load, a flexible beam, a stretchable wire, a twistable rod—*any elastic system obeying Hooke's Law*—will oscillate with S.H.M.

We use the name S.H.M. for a particular type of repeating motion—the pendulum motion and the spring motion that matches it—and not just for any motion with constant period. (A tribe of creatures who crawl out of the ground each morning to feed and back each night would have "isochronous" motion in a sense—period 24 hours, however deep their burrow—but they certainly would not have S.H.M.) If we analyze the pendulum's motion with a little geometry, we find the essential characteristics:

> The motion has a *changing acceleration*, which is always *directed inward*, towards the mid-point, and which *varies directly as the distance out from the mid-point.*

If s is the distance along the path (say of the pendulum bob) and a the acceleration we shall find that:

$a \propto s$ or $a = -k^2 s$, where k is a positive constant.

The minus sign is there to show the acceleration is inward. (When the bob is out to the right, where we reckon positive, the acceleration is backward to the left, so we give it a minus value.)

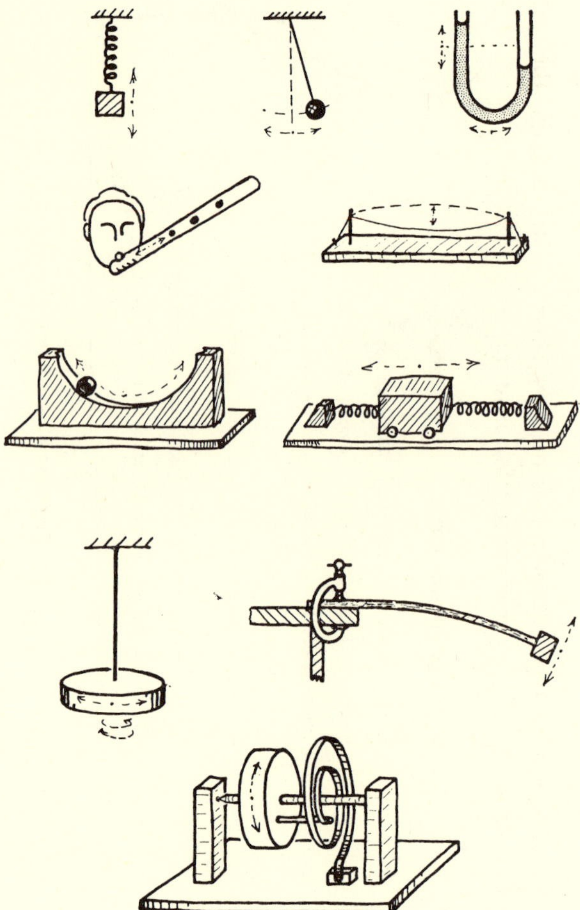

Fig. 10-3. Many Systems Vibrate with S.H.M.

Mechanics of Pendulum Motion

To show that $a \propto s$ for a pendulum bob (for small amplitudes) consider the forces on it. The thread pulls with a tension along a radius, and cannot change the bob's speed. The only other force is the pull-of-the-Earth, the bob's WEIGHT vertically down. Split that vector into components

F_1, along the arc, which will accelerate the bob,
F_2, along the radius, opposing the string tension.

By similar triangles (Fig. 10-4):

$$\frac{\text{"DOWNHILL FORCE," } F_1}{\text{WEIGHT, } Mg} = \frac{\text{HORIZONTAL DISTANCE, } x}{\text{LENGTH, } L}$$

$$\frac{F_1}{Mg} = \frac{x}{L} \qquad \therefore\ F_1 = \frac{Mgx}{L}$$

$$\therefore \text{ bob's ACCELERATION} = \frac{\text{FORCE}}{\text{MASS}} = \frac{-F_1}{M}$$

$$= \frac{-Mgx/L}{M} = \frac{-gx}{L}$$

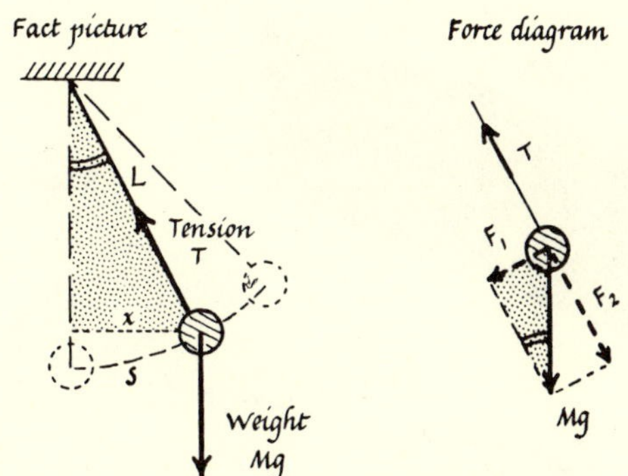

FIG. 10-4. FORCES ON PENDULUM BOB

Thus a is *inward* and $a \propto x$, but we do not have $a \propto s$ along the bob's track. However, for large amplitudes the pendulum's motion is not S.H.M. For small amplitudes, it is almost exactly S.H.M.; and, for small amplitudes,

x, the bob's horizontal displacement,

is almost exactly the same as

the curved arc s, the bob's distance out along its path.

Then we can change from $a = -(g/L)x$ to

$a = -(g/L)s$, (approximately for pendulum),

and this *is* our description of S.H.M.:

a is *inward* and $a \propto$ DISPLACEMENT, s

which we describe by the general expression

$a = -k^2 s$ where k^2 is a constant

From this it can be proved that the PERIOD, T, is given by:

$$T = 2\pi/k$$

(This is most easily proved by calculus—see below. There are proofs that avoid calculus, but they are roundabout and clumsy. See standard texts on General Physics.) So *whenever we find a system whose forces lead to $a = -k^2 s$, we know it can oscillate with S.H.M., with period $2\pi/k$.*

S.H.M. and Hooke's Law

Now we can follow up the hint of Problem 1. For a load hung on a Hooke's-Law spring, the spring's TENSION just balances the load's WEIGHT when the load remains at rest, or when it passes through its central rest-position while oscillating. At all other positions, there is a small EXTRA TENSION (+ or −) proportional to the EXTRA STRETCH (from Hooke's Law); and that gives the load an acceleration. The ACCELERATION is always towards the central position and varies directly as the displacement, s, from the central position (from Hooke's Law). So we have $a = -k^2 s$, and this is what we name S.H.M. The period $2\pi/k$ can be calculated from the mass of the load, M, and the spring's "spring-strength," K, which is the FORCE/STRETCH constant, the slope of a Hooke's Law graph. For extra stretch s the extra force is Ks, and the acceleration is $-Ks/M$. Therefore k^2 is K/M, and T is given by

$$T = 2\pi/\sqrt{\text{SPRING-STRENGTH, } K/\text{MASS, } M}$$

This enables us to calculate (predict) the period of a S.H.M. It also provides an excellent way of estimating a spring-strength from a measured period. We do that when we measure g with a pendulum—there, gravity provides an equivalent "spring-strength" of Mg/L. In Cavendish's measurement of the gravitation constant, G (Ch. 23), the fiber is far too weak for direct measurements of its spring-strength for twisting, so we time its period of twisting S.H.M. and calculate the strength from that.

S.H.M. Is Common

On this basis, any system that provides a restoring-force proportional to displacement will oscillate with S.H.M.:

any pendulum (for small amplitudes)
any Hooke's Law system (e.g. loaded spring, bending beam, twisting hairspring—which is really a curled-up bending beam—etc.)
atoms attached to a molecule by elastic electric forces. (The "wig-wag" problems in Ch. 7 were an informal approach to S.H.M.)
air vibrating as sound waves go through it, or the air in a flute. (The p:V graph of Boyle's Law does not look like a Hooke's Law straight line, but only a short piece of it is involved in these very small changes of pressure—so short that it is practically straight.)
liquid in an open U-tube[2]
strings of musical instruments

We call this simple *harmonic* motion because musical instruments make such oscillations, and give out corresponding waves, when they make a pure musical note. As the vibration dies down it keeps the same period, the waves have the same frequency, and we hear the same musical note.

[2] In this case, we can calculate how the pressure due to level-difference provides the "restoring force" that makes the liquid accelerate. Then we find the period is the same as that of a simple pendulum whose length is half the total length of liquid in the tube. Check this in lab if you like. The period is the same whatever the liquid—you could have predicted that by simple argument.

Personal Experiment

If you watch a very long pendulum swinging it will clarify your knowledge of S.H.M. Or you can walk to-and-fro along a line with an attempt at S.H.M. As

FIG. 10-5. S.H.M.

you walk, lean over towards the center to indicate your *acceleration*—lean strongly inward at the end of the path; run through the center upright, with maximum speed. At the center where you are moving fast*est* you cannot be moving fast*er*, so your acceleration is zero. At the ends your velocity is momentarily zero, yet it is changing most rapidly, from outward, through zero, to inward: you have a large inward acceleration there, as your feet will tell you. (Compare this with the early question about a projectile's acceleration at the "vertex.")

S.H.M. has a Sine Curve as Time-Graph

Further mathematics shows that S.H.M., defined by $a = -k^2s$, has a time-graph of form $s = A \sin kt$ where A is the amplitude. Fig. 10-6 shows a demonstration: a pendulum bob draws a time-graph of its motion. The

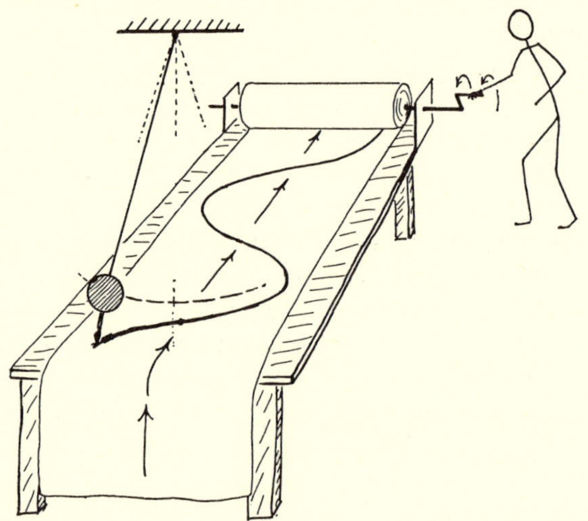

FIG. 10-6. TIME-GRAPH OF PENDULUM MOTION

bob carries an inked brush that writes on a long sheet of paper. Sliding the paper steadily along under the brush draws the graph.

Fig. 10-7 shows a similar demonstration for a vibrating tuning-fork. A small mirror is attached to one arm of the fork (by a flimsy strip of celluloid or mica to magnify the motion). When the fork vibrates a small beam of light reflected by the mirror swings up and down and makes a vertical streak on a remote wall. That beam is intercepted by a large mirror which rotates steadily and sweeps the spot across the wall horizontally, thus plotting a time-graph of any up-and-down motion. The tuning fork is really a bending beam with its ends curled up to form a U. It is elastic and obeys Hooke's Law so we should expect S.H.M. for its vibrations. The demonstration shows the sine curve characteristic of S.H.M. As the fork gives out sound, the motion dies down in *amplitude*, as shown by the curve on the wall; but it keeps the same rate of vibration, as shown by the spacing of the bumps of the curve. A careless blow with a metal hammer gives the fork two modes of vibration at once, and you can see complex harmonic motion.

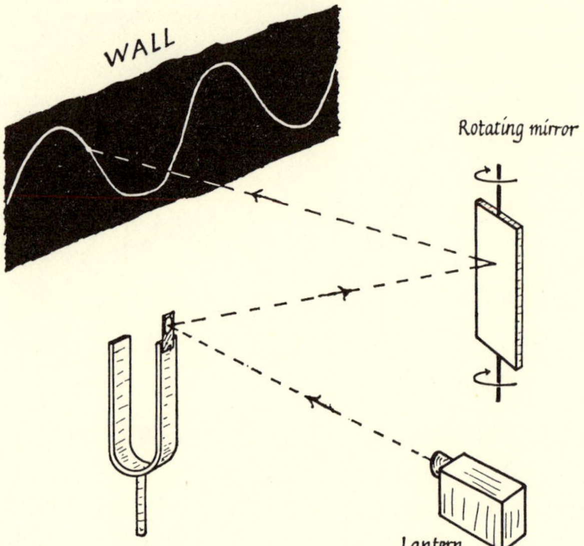

FIG. 10-7. TIME-GRAPH OF TUNING-FORK MOTION

S.H.M. as Projection of Circular Motion

In beginning trigonometry you find *sines* described with the help of a circle; and that tells us that the sine graph also describes the *projection of motion around a circle*, thus: imagine a point P traveling around a vertical circle at constant speed; view the circle edge-on; or shadow it (edge-on) by horizontal sunlight on to a vertical wall (see Fig. 10-8). Then, Q, the shadow of

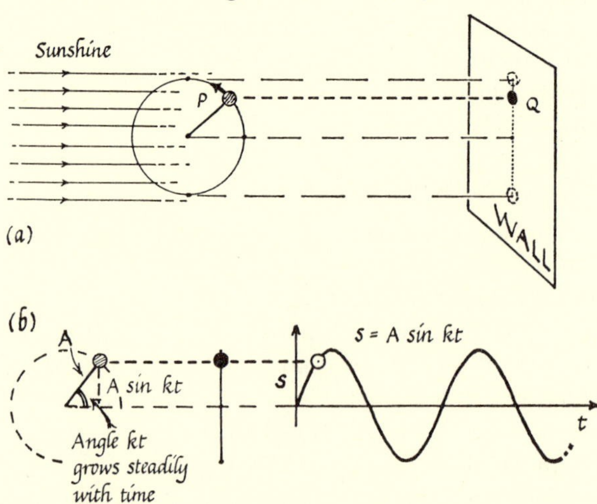

FIG. 10-8. THE PROJECTION OF CIRCULAR MOTION

P, moves up and down. Q's time-graph can be shown to be a sine curve (with equation $s = A \sin kt$, where A is the radius of the circle), and that is also the time-graph of S.H.M. Therefore the projection of motion around a circle *is* S.H.M. Figs. 10-9 and 10-10 show demonstrations to compare the motion of a pendulum, or an oscillating spring, with the projection of motion around a circle. (Electrical engineers have much to do with

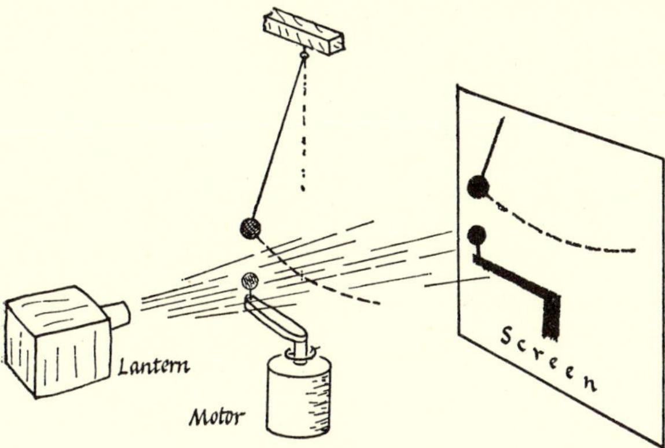

FIG. 10-9. COMPARISON OF PENDULUM MOTION WITH PROJECTED CIRCULAR MOTION

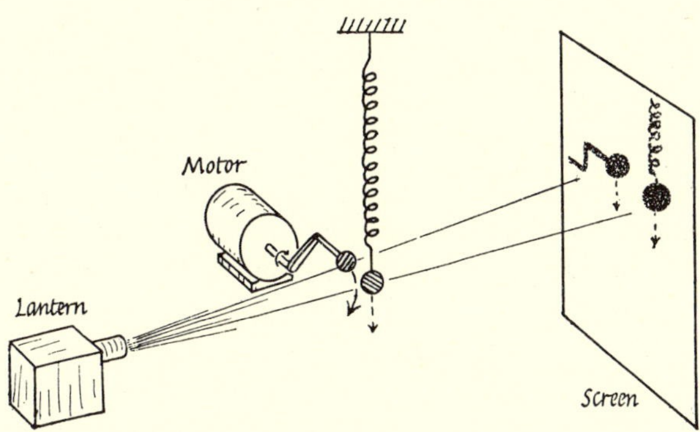

FIG. 10-10. COMPARISON OF S.H.M. OF LOAD ON STRING WITH PROJECTED CIRCULAR MOTION

alternating currents, which are simple harmonic and have sine curve time-graphs. To condense their calculations, they often represent such a current, or voltage, by a circle with a rotating radius for its amplitude. The projecting of these circles is taken for granted—and forgotten.)

An electric motor (e.g. a small clock motor) carries a ball B on the end of an arm steadily around a circle. The motion of B seen from the side is compared with that of a small pendulum, by shadowing. If the pendulum is chosen with the right length and started at the right epoch the two shadows keep exactly in step. Similarly, shadowing B and a load bouncing on a spring shows motions that agree at every stage.

Many Descriptions of S.H.M.

There are several alternative descriptions of S.H.M.:

(i) It is the *to-and-fro motion of a pendulum bob* (with small amplitude); or the *up-and-down motion of a load on a spring* (or any other Hooke's Law system).

(ii) It is motion, to-and-fro along a track, such that *acceleration* (along the track, always towards the center) *varies directly as distance out from center.*

(iii) It is the *projection of steady motion around a circle* (e.g. circular motion seen edge-on; or motion around a vertical circle shadowed on the ground by vertical sunlight).

(iv) It is motion whose *time-graph is a sine curve.* Algebra—with some simple mechanics for version (i)—will show that all these define the same motion. All we shall do here is give some of the algebra and mention some demonstrations, to link these versions together.

The Importance of S.H.M.

For great value in describing nature, S.H.M. ranks with motions of constant velocity and constant acceleration, because:

(1) It is common. (e.g. pendulums, musical instruments, vibrating machinery, ocean tides, alternating currents, light of a spectrum line, . . .)

(2) Its period is independent of amplitude. (Timekeeping.)

(3) It can be described by a simple mathematical form:

$$s = A \sin kt, \text{ which leads to } T = 2\pi/k$$
$$\text{where } k^2 = \text{SPRING-STRENGTH}/\text{MASS}$$

So T can be predicted. In other cases, T is measured and used to estimate a spring-strength.

(4) According to Fourier's Theorem, any repeating motion can be analyzed into Simple Harmonic components (see below). The analysis is easily done by algebra (where the original form has a formula) or by machine (where it has only a graph) so we can use the simple mathematics of S.H.M. to treat much more complicated motions: e.g. the tides in a harbor, music from a clarinet, voice wave-forms, earthquake waves, . . . the motions of electrons in an atom. In the case of sound, our ear-mechanism probably performs a "Fourier analysis" and sorts out a complex sound into pure notes when we listen.

Fourier Analysis

Fourier's Theorem applies so generally that it is difficult to specify its limits. It is not even restricted to periodic motions or repeating wave-forms. Here are a few examples:

(a) The wave-form of sound from a flute is shown in Fig. 10-11. The analysis of wave-form (iii) is

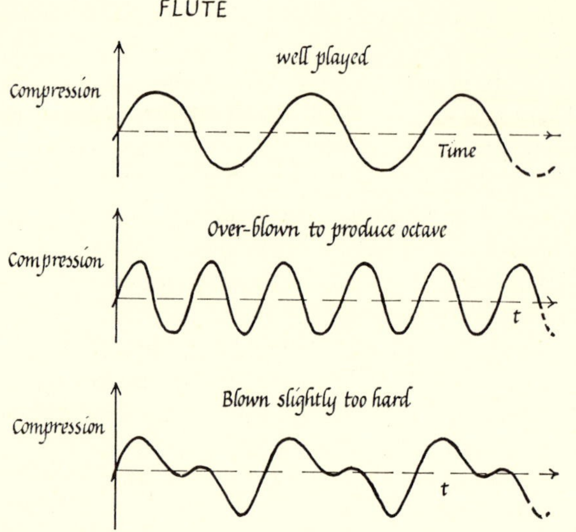

FIG. 10-11. WAVE FORMS OF SOUND FROM FLUTE

obvious: a lot of wave-form (i) with some of (ii) added. (Slight overblowing produces a combination of the original note and some of its octave, a pleasant musical note, though not a welcome one from a flute.)

(b) We can easily produce a "square wave" electronically, and show its form on a cathode ray oscilloscope. (Sound from a squeaker with metal jaws that snap open and shut abruptly might have a square-wave form.) Fig. 10-12 shows an attempt at its Fourier analysis. The main component has the same "wavelength" as the square wave. That leaves some bumps of excess and defect to be smoothed out. The next component needs a "wavelength" ⅓ of the main one, or 3 times the frequency. That leaves residual troubles which are largely cured by a small component of 5 times the original frequency, and so on. An exact match requires an infinite series of components with frequencies, 1, 3, 5, 7, . . . times the original. But even the first few produce a fair match (except for an unwelcome spike that extends the vertical part). This provides a useful test of loudspeakers, microphones, etc.: feed a square wave to the instrument and see how closely it reproduces the square shape. If it does well, that shows it can handle very high frequencies as fairly as low.

(c) Voice wave-forms often have a complicated shape. Fig. 10-13 shows a fairly simple one, a light "u" sound being sung. You can guess the Fourier analysis: the basic singing note + a much higher frequency, which we find is characteristic of that vowel. Such analysis is valuable in designing telephone systems to carry speech, economical artificial-voice-coders for cable telephony, hi-fi sets to do

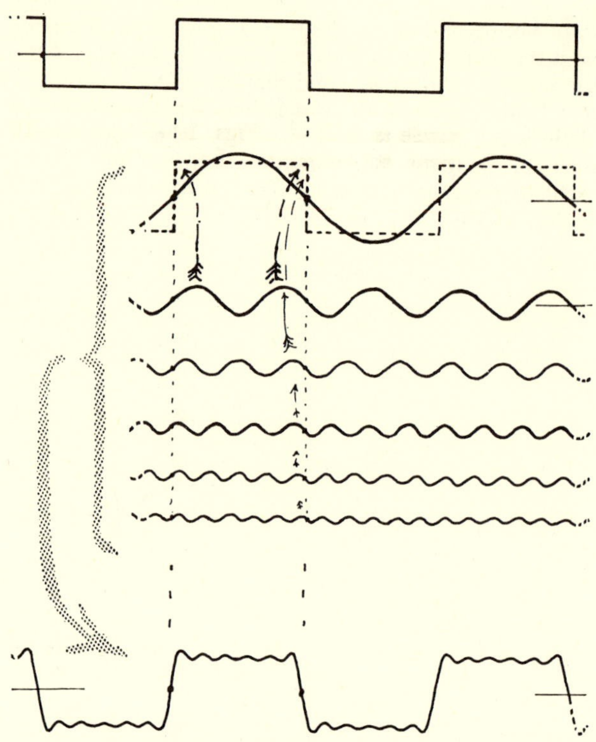

FIG. 10-12. FOURIER ANALYSIS OF SQUARE WAVE FORM (Note: even with a great number of components, the sum shows sharp unwanted peaks at the corners of the pattern.)

justice to speech. Other vowels, and less skillful singers, show wave-forms that look much more complex but are easily analyzed into a few essential components.

(d) A "wave packet." We can apply Fourier analysis to a single pulse (such as the sound of a slap, the radio wave from a lightning stroke), or to a short wave-train like the wavy hump that marks the position of a moving electron in modern theory. There we must add an infinite range of component frequencies for a perfect match, but the components of appreciable amplitude are spread smoothly over a band of frequencies around the original wave frequency. We have to add up a basic wave of the same wavelength as the original wave-train + one of slightly longer wavelength + . . . slightly longer still . . . + etc., and a similar range of shorter wavelengths. All these agree for one or more humps at the center; but they get out of step farther away

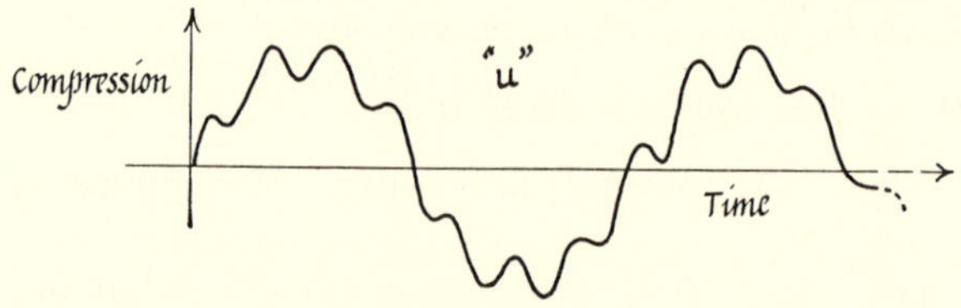

FIG. 10-13. WAVE FORM FOR VOWEL "u" BEING SUNG

and neutralize each other. If the original wave-train is many wavelengths long the important components are spread over a narrow range of frequencies and wavelengths—the longer the train the narrower the analysis-band's range. In contrast, a very short wave-train—and in the extreme case a single bump or pulse—needs a wide band of frequencies for a good match.

(This is *not* obvious—without the mathematics you can at best say that you see it might work.) This becomes a very important matter in modern atomic theory.

Fig. 10-14. Fourier Analysis

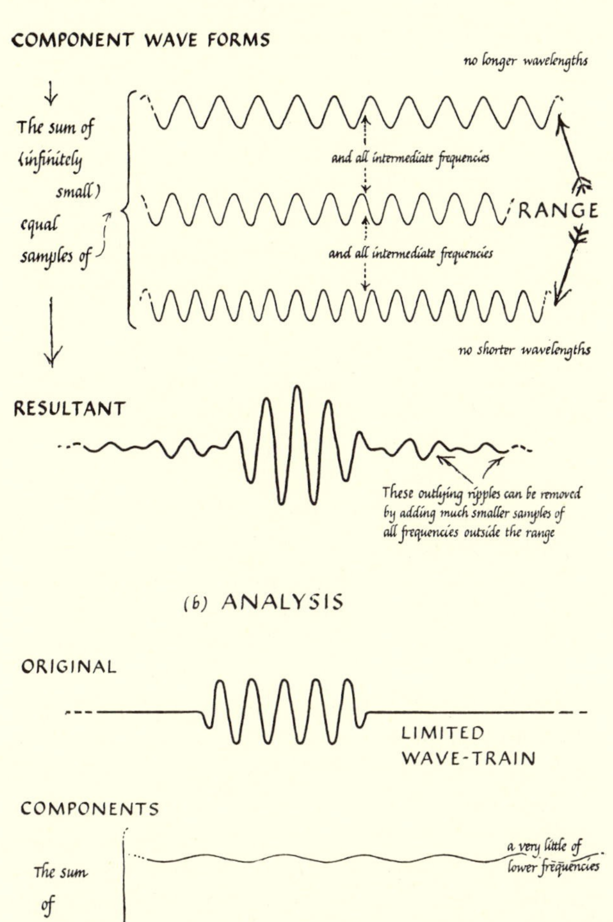

(c) PARTICLE & WAVE BEHAVIOR

of electrons etc.

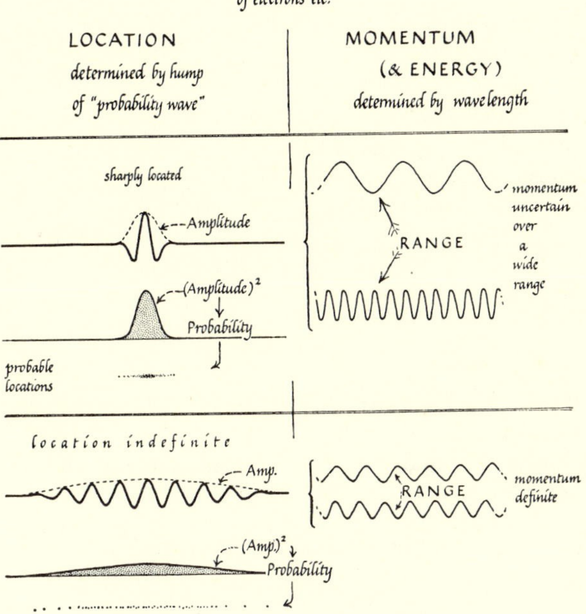

Fig. 10-14. Fourier Analysis

(a) Making a "Wave-Packet" by Adding Simple Harmonic Components. This synthesis requires samples of *all* frequencies (or all wavelengths) from zero to infinity, to make a short wave-packet with no disturbance before or after it. Therefore we must add up an infinite number of infinitely small samples—and that requires the calculus-process of integration. The most important samples fall in a central "range" of frequencies (or wavelengths) —outside that, the samples needed are relatively much smaller still. The narrower that range, the longer the wave-packet, the more wavelengths in its pattern.

(b) Analysis of a Limited Wave-Train. If we direct a continuous stream of waves at a gateway and open the gate for a short time only, we expect to chop the train and have a limited wave-train pass through. This chopped sample can be analyzed into an infinite number of infinitely small component samples, the most important of which cover a central "range." The shorter the original section of wave-train, the wider that range of components in the analysis.

(c) Particles and Waves. (This illustrates the basis of modern atomic physics, discussed in Chapter 44.) Nowadays, we think of all moving particles—electrons, nuclei, etc.—as guided by waves: in some respects they *are* waves. A particle can be regarded as a wave-packet. The wave inside the wave-packet represents the particle's position and motion in the following way: the *square of the amplitude* of the wave motion at each place in the packet shows the *probability* of the particle being located there; and the *wavelength* gives its *momentum*, by $mv = h/\lambda$.
Thus, if we want to locate a moving particle precisely, we must chop its wave to a short train or a sharp wave-packet, but that will have a wide range of component waves, a wide range of possible momentum values. Then we cannot tell its momentum precisely. On the other hand, if we insist on precise momentum we require a narrow range of wave components and we must therefore show the location by a long wave-packet that leaves the particle's position very indefinite.

The essential value of Fourier analysis—which Fourier's Theorem says always applies—is that it enables us to treat complicated motions by the simple mathematics of S.H.M. It forms a welcome tool in engineering and in physics, for use by telephone designers, radio engineers, compilers of ocean tide-tables, . . . and nowadays by the mathematical physicists who describe the behavior of atoms and electrons with Fourier components.

Calculus Treatment and the Pendulum Formula

Start with a motion defined by

$$s = A \sin kt,$$

where A is the amplitude and k is a constant. Differentiate the displacement, s, with respect to time, t, to find the velocity; and a second time to find the acceleration:

$$v = \frac{ds}{dt} = k A \cos kt$$

$$a = \frac{dv}{dt} = -k^2 A \sin kt$$

$$= -k^2 s$$

This shows how to calculate the period, T, of this motion:

$$\begin{aligned} \mathrm{T} &= \text{time from } t = 0 \text{ to } t = T \\ &= \text{time for one cycle of } s \\ &= \text{time for } (kt) \text{ to run from 0 to } 2\pi \end{aligned}$$

$$\therefore\ kT = 2\pi$$

$$\therefore\ \text{PERIOD } T = 2\pi/k$$

Therefore, wherever we find a system whose forces give it an acceleration $-k^2s$, we can say "this can oscillate with S.H.M., and its period will be $2\pi/k$."

The "Pendulum Formula"

(This is derived below from the calculus treatment. There are methods that avoid calculus, but they are more roundabout and clumsy. See texts on General Physics.) For a pendulum with small amplitude, we showed earlier that

$$\text{bob's ACCELERATION} = -(g/L)\, s$$

Compare this with the calculus result above

$$\text{ACCELERATION} = -k^2 s$$

Then $[k^2]$ in the general form is $[g/L]$ for a pendulum

$$\text{Then, PERIOD } T = \frac{2\pi}{k} = \frac{2\pi}{\sqrt{g/L}} = 2\pi\sqrt{\frac{L}{g}}$$

This is the "pendulum formula" used in measuring g accurately with a simple pendulum.

WAVES

Any moving form—some shape or pattern that travels along without carrying all the medium with it—is called a wave. Water waves move fast, making the water itself move up-and-down or even in circles as they pass, but they do not carry the water itself far with them—watch a floating cork or boat bob up and down as waves go by. Think of waves along a rope, ripples in a pool, sound waves in air. A gust of wind sweeping across a field of wheat shows a wave of bending stalks. We can even say a rumor travels as a wave in a crowd.

Velocity, Wavelength, Frequency

The velocity of a wave, V, is the speed at which its wave-form travels along, the speed of any labeled piece of disturbance, such as a crest or a trough or a region of compression.

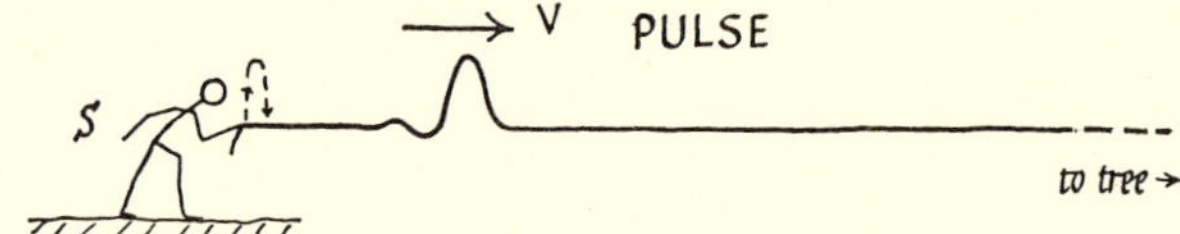

SIMPLE HARMONIC WAVE

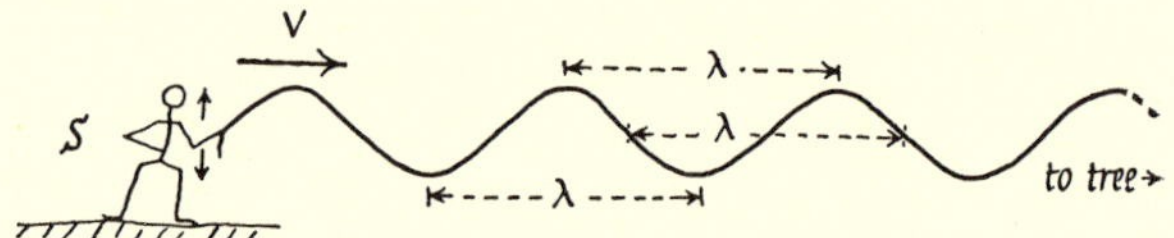

FIG. 10-15. WAVES

Transverse waves travel along a taut rope with a definite speed, and if we waggle the rope with S.H.M. we have a stream of *simple harmonic waves*, with definite *wavelength*, which we label λ (a Greek letter ell, pronounced "lambda"). The wavelength is the shortest distance from crest to crest, or from trough to trough, or between any such pair of points where the medium is at the same stage (*phase*) of its cycle. It is the repeat-distance of the wave-form.

If the source, S, moving with S.H.M., performs f complete cycles per second, we say its *frequency* is f. The waves leave S at a rate f cycles per sec and they must pass any observer O at the same rate f cycles/sec, otherwise waves must be getting lost or created somewhere between S and O.

$$\text{FREQUENCY } f = \text{number of cycles/sec}$$

$$= \frac{1 \text{ second}}{\text{TIME FOR ONE CYCLE}}$$

$$= \frac{1 \text{ second}}{\text{PERIOD } T \text{ secs}} = \frac{1}{T}$$

Therefore for any simple harmonic wave (also for any S.H.M.) $f = 1/T$.

Wave velocity V meters/sec means that a given hump travels V meters (along the rope or other medium) in one second. Therefore in one second a

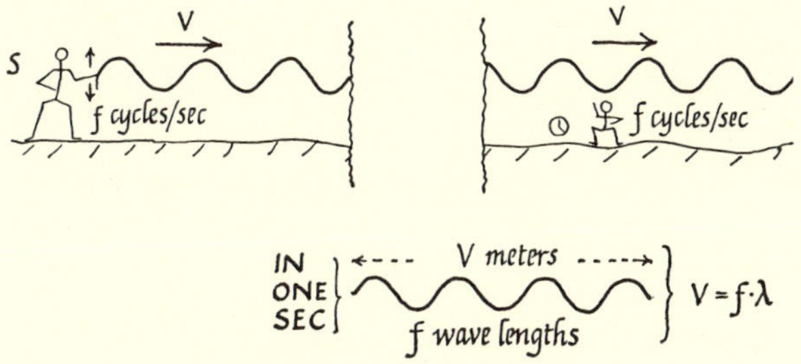

Fig. 10-16. Waves

wave-train V meters long travels out from the source. But in one second the source performs f cycles, each producing one wavelength. Therefore a wave-train, of length V meters contains f wavelengths, each of length λ.

$$\text{VELOCITY} = \text{FREQUENCY} \cdot \text{WAVELENGTH}$$
$$V = f \cdot \lambda$$

This applies to any periodic waves.

Symbols for Light-Waves

When we deal with light-waves we shall follow tradition and use special symbols:

- c for the velocity (in air or vacuum)
- ν for frequency (this is a Greek "n" pronounced "new")
- λ as above for wavelength

How Waves Travel

Essentially one piece of medium disturbed by a wave disturbs the next piece of medium ahead, and gives up the motion to it. Look at Fig. 10-17 which shows successive stages of a wave on a rope. At stage (i) the bit of rope at B has moved upward. At stage (ii), shortly after, the wave pattern has moved along, and bit B is still higher; so B must have been moving upward in stage (i)—and a glance at the sketch shows you it is still moving upward at stage (ii), though not so fast. However the bit of rope at A in (i) has reached maximum "displacement" and is not moving. The bit at C is not displaced, but it is moving upward fast. (These "particle velocities" of bits of medium are quite distinct from the wave speed, V.) The wave progresses because each particle of medium is:

> moving (at most times), *and*
> pulled by neighbors ahead, *and*
> pulled by neighbors behind,
> *usually with a different force.*

(In Fig. 10-17 stage (i), look at the pulls of neighbors on the bit of rope at B. They are not quite parallel, and have a downward resultant. That must be decelerating B, which *was* moving upward as fast as C now is, and *will be* at rest as A now is.) Most of the waves we deal with in physics carry MOMENTUM along the medium; and they carry ENERGY too. (See Ch. 26.) The details of their progress can be worked out, and their velocity calculated, from a knowledge of forces and masses in the medium. The easiest methods use calculus. We shall not give details here.

Sound waves are *longitudinal*: the displacements are *along* the wave-travel instead of across. (This can be tested by watching smoky air under a microscope. We shall not discuss it here beyond giving Fig. 10-18, which shows a longitudinal wave, together with the easy transverse representation used by physicists. Longitudinal displacements plotted in a transverse graph turn the pattern into a "wave on a rope").

Properties of Waves

Waves are reflected (sound from a wall, water waves from a breakwater); and they are "refracted"—their line of travel bends if they meet a region where they have different velocity. These

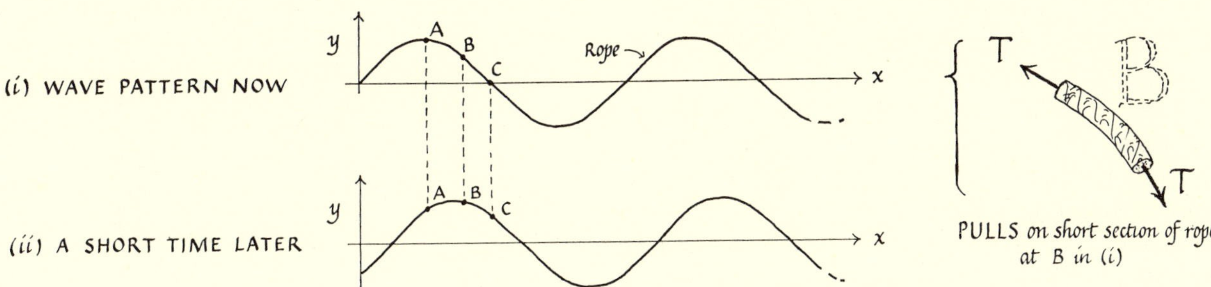

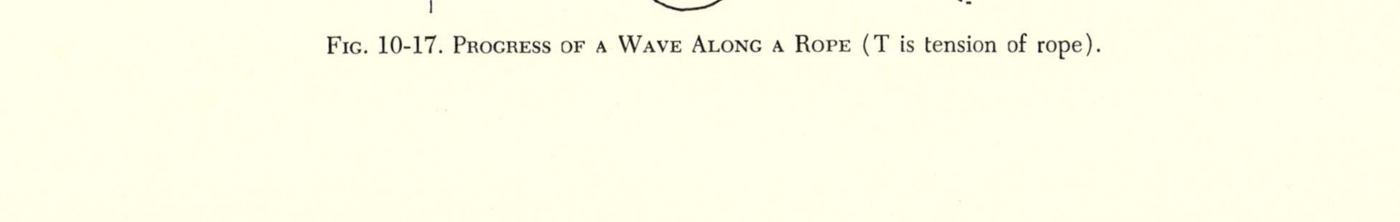

Fig. 10-17. Progress of a Wave Along a Rope (T is tension of rope).

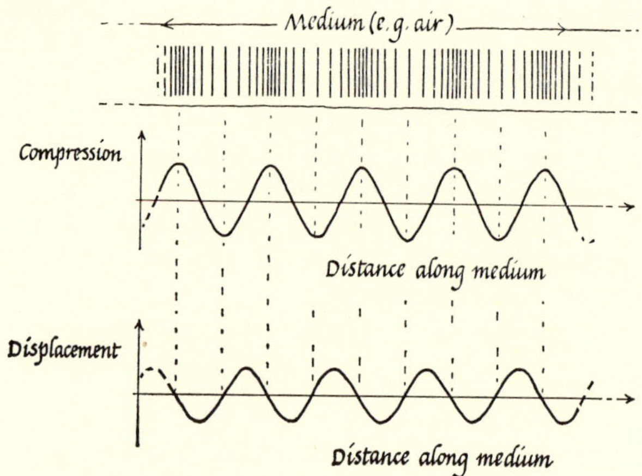

FIG. 10-18. LONGITUDINAL WAVES and their representation by a transverse pattern.

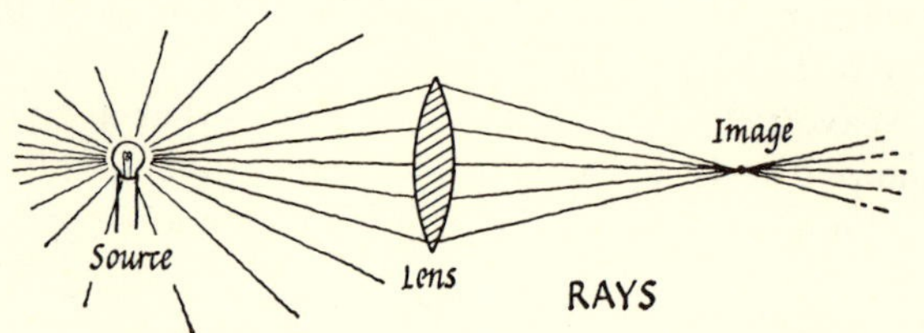

FIG. 10-19. LIGHT RAYS FORM AN IMAGE

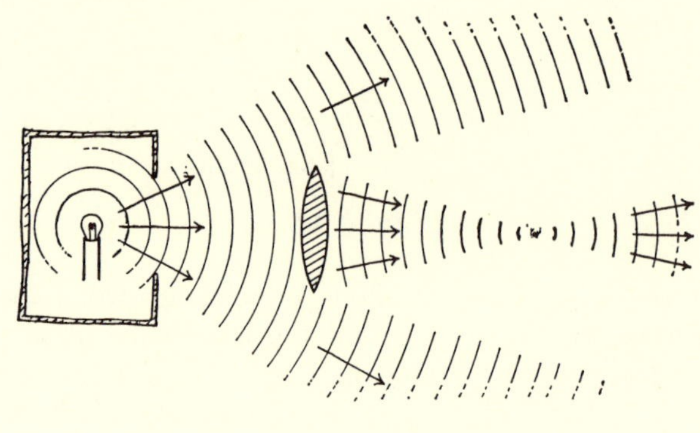

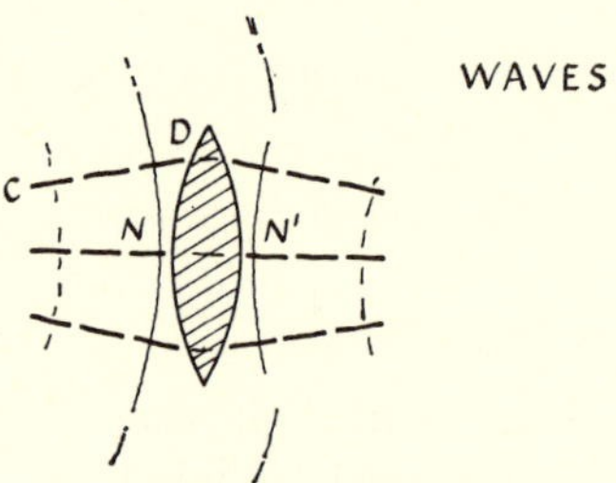

FIG. 10-20. LIGHT-WAVES

properties are discussed in other books, particularly texts on Optics. There it is shown that reflection and refraction of waves follow the laws already known experimentally for reflection and refraction of beams of light. Huygens, at the time of Newton, discussed these properties in detail and suggested that light consists of waves. Newton himself demurred because he doubted if waves would cast such sharp shadows. He suggested that light is a stream of bullets which on simple mechanical theory would also be reflected and refracted in the same way.

Here is one example of a wave treatment in optics. Fig. 10-19 shows light being focussed by a lens—rays from a small white-hot source being brought to a burning spot, an *image* of the source. In terms of waves we imagine the source giving out spherical waves (as in Fig. 10-20) which grow until they reach the lens. Beyond the lens the waves must shrink as they converge to the image, passing practically through a point there. (The image is this region of thickest energy-flow.) How did the wave get changed from convex to concave by the lens? Obviously the thick center of the lens must have delayed the wave there so that the *forward* bulge *N* of the wave (which hits the center of the lens) was delayed most and left behind as a backward bulge *N'*. Therefore the *wave must travel slower in glass than in air.*

However, bullets, to follow the same bent paths on meeting the lens must travel *faster* in glass than air. Fig. 10-21 shows a bullet-path along a ray of light. A bullet traveling along ray CDE must be attracted by the glass at D (like a vapor molecule returning to liquid), and therefore must move faster in the glass. Here was a chance of a "crucial experiment" to test between wave theory and bullet theory of light: compare the speeds of light in air and glass (or some other dense medium such as water). This was not tried successfully until 1850 and then measurements showed that *light travels slower in water than in air.* Before this conclusive result was obtained, other evidence pointed clearly to light-waves: diffraction and interference.

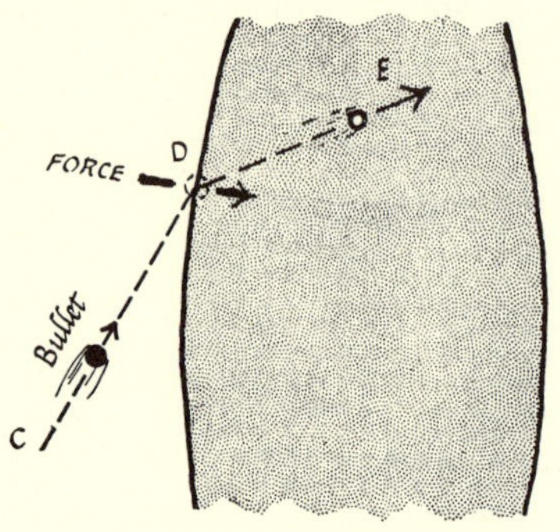

FIG. 10-21. A BULLET OF LIGHT

Diffraction: Bending of Waves

Watch water ripples passing through a gap in a barrier. A wide gap (many λ's wide) lets through a sample beam, so that the waves seem to travel straight on through the gap and leave a calm shadow out at the sides. But a narrower gap seems to encourage spreading; the waves splay out after passing through. A very narrow gap leads to complete splaying; waves travel out in all forward directions. (As Huygens pointed out, we should expect that. Waves arriving at the barrier make the water bounce up and down in the narrow gap and that generates circular ripples—how does the water beyond know that there is not, say, a human finger bouncing up and down in the gap and starting ripples? We should, then, expect ripples to spread in all directions from a narrow gap that is only a fraction of λ wide.) This spreading is called *diffraction*—bending away, or bending around corners by waves.

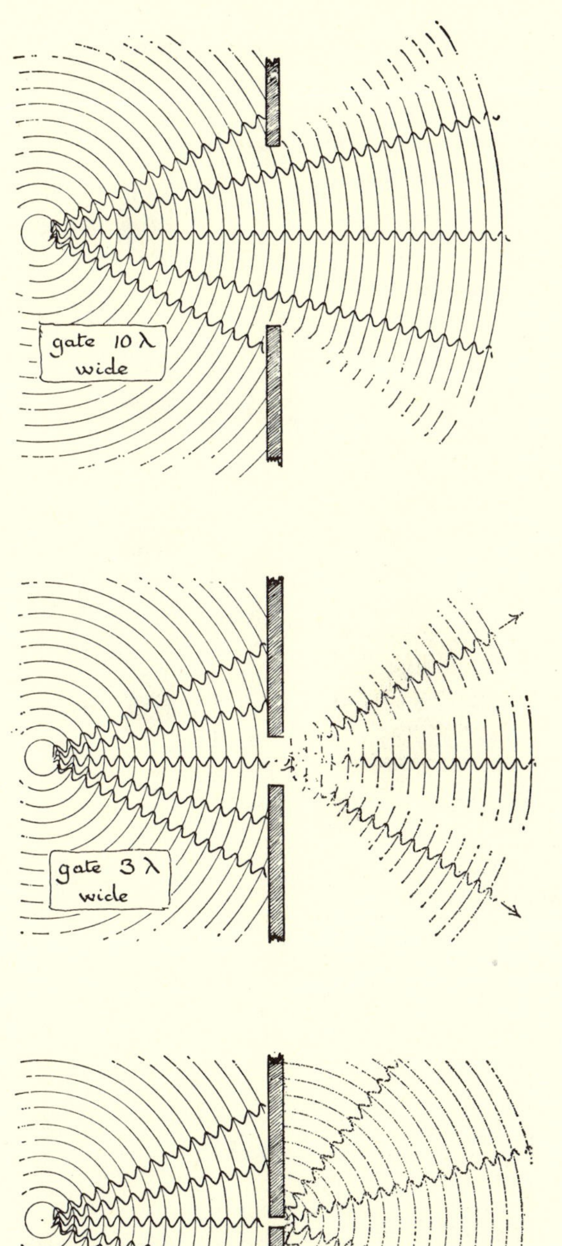

Fig. 10-22. Diffraction of Waves Passing Through an Opening

If light is waves, why does sunlight pass through a pinhole as a sharp beam, why does it not spread out? Because an ordinary pinhole is a wide gap—a thousand λ's wide as we now know! If light meets a very small pinhole in a barrier it does spread. Try this experiment. Look at a distant street lamp at night through a pinhole in a card, or a chink made by two fingers and thumb. You will see the lamp sharp—no noticeable spreading or diffraction. Then try smaller pinholes. As well as getting less light through, you will see the small street lamp spread out when viewed through a very small hole: diffraction. Or use a grid of very small holes: a fine sheet of cloth, such as a taut umbrella or a silk handkerchief. Now you will see the small street lamp repeated in a pattern of bright spots. From measurements of that pattern physicists can estimate λ. Waves can do that—they must do that—when the holes are a few λ apart, but bullets could not. Pour a stream of sand (to represent bullets) through a sieve of fine wire screen. The sand will make a mountain on the table, but it will not arrange itself in a pattern of separate hills.

See demonstrations of diffraction of light: the effect of light passing through a narrow slit and spreading out; light passing the edge of a solid wall and spreading into fine bands in the shadow; also the strange story of the "hole in every coin" mentioned in footnote 1 in Ch. 31.

Interference

The strongest evidence for waves, and perhaps their most important property, is interference. When two trains of waves arrive in some region and overlap, their effects add up. Suppose we have two neighboring sources, S_1 and S_2 connected together to give out waves in step. (For sound waves, this is easily done by two loudspeakers connected in series, carrying the same current. For light, we illuminate two fine slits or pinholes, side by side, then light spreads out—diffraction—from each, with similar waves in step.) Watch those waves arrive at an observer far away. Straight ahead the two wave trains have traveled equal distances and arrive in step. While one drives the medium at P flip-flap-flip- . . . , the other does the same, flip-flap-flip. . . . The resultant is bright, FLIP-FLAP-FLIP-. . . . But now move the observer to Q where one train travels half a λ more than the other. There while one train drives the medium flip-flap-flip . . . , the other drives it flap-flip-flap . . . , and the resultant is *zero.* (This is the Principle of Interference: that waves do not upset each other but simply add up algebraically: flip + flip = FLIP, but flip + flap = zero.) In 1803 Thomas Young produced convincing evidence that light is waves. He let light from a single source fall on a pair of slits close together and examined the

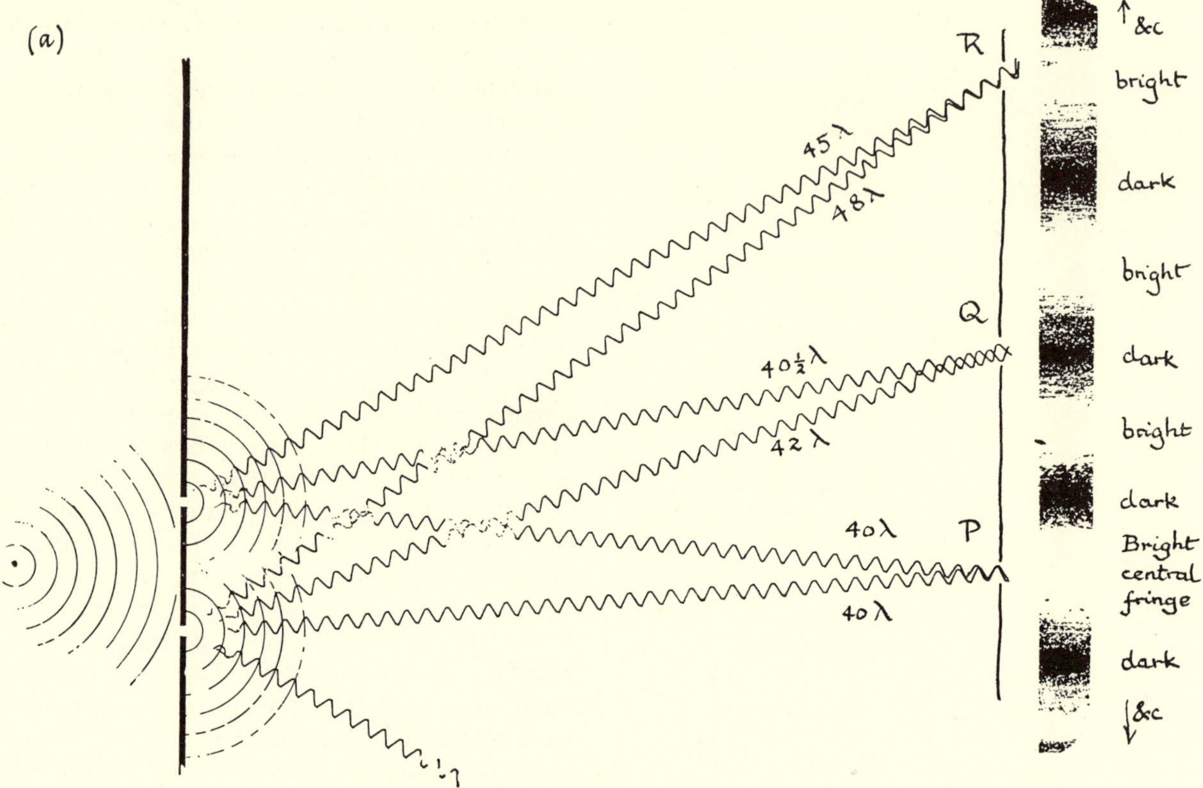

Fig. 10-23. Interference of Waves
"Young's Fringes," formed when waves that pass through two openings meet on a distant wall.

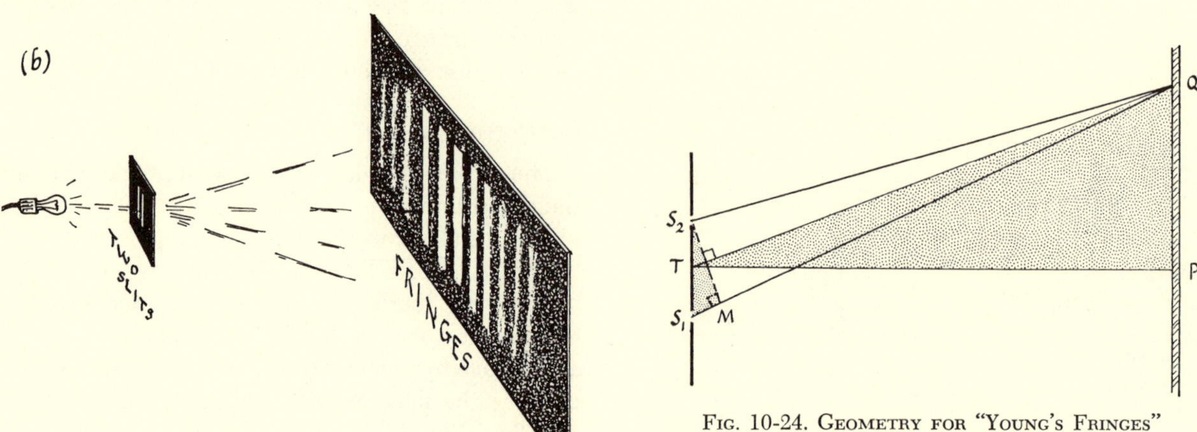

Fig. 10-24. Geometry for "Young's Fringes"

light reaching a remote wall. There he found bands of dark and light, *interference fringes*, that are characteristic of waves. There is a bright white band at the center, a black band each side, then bright and dark bands farther out; but there the bands are colored.

With light of a single color[3] all of one wavelength, many light and dark bands are clearly visible. The paths of waves from the two slits to the central bright band are equal; but the paths to other bands differ. Wherever the path-difference is λ or 2λ, etc., a whole number of wavelengths, there is a bright band—LIGHT + LIGHT = MORE LIGHT, there. There is a *dark* band wherever the path-difference is $\frac{1}{2}\lambda$ or $1\frac{1}{2}\lambda$, etc., an odd half wavelength more than a whole number—and there LIGHT + LIGHT = NO LIGHT. That is called "interference," although it is really no more than two waves with opposite motions adding up to zero.

If different colors are tried, they give different spacings of bands: red wider than green, and green wider than blue, showing differences of wavelength. So if white light is used the bands farther out are blurred by overlapping of colors.

You should see these "Young's Fringes" for yourself, as evidence that light is waves, and as an indication of their very small wavelength. (Then you should see the "photoelectric effect" as evidence that light is *not* spreading waves, but a stream of compact bullets of energy! That paradox will be discussed in the final chapter.)

EXPERIMENT. Measure the wavelength of light, roughly.

Use an electric lamp with a straight filament as source. A few yards away place a pair of slits parallel to the filament. Several yards beyond that, look at the Young's fringes through a sheet of ground glass or frosted celluloid. (On white paper viewed from the front the fringes may be too faint. A translucent screen makes them much easier to see.) The double slit is made by scratching two lines on a blackened photographic plate (or on the silver backing of an old mirror). The lines should be about ½ millimeter apart, or even closer.

Measure the spacing of the bright bands roughly and calculate λ. (With white light this must be a *very* rough estimate, for an average wavelength.)

With a green filter interposed, more bands are visible and a better estimate can be made, λ for green light. However this is an experiment of Principle rather than of Precision.

Use the geometry sketched in Fig. 10-24. If the central band is at P and the next bright band at Q, the path-difference, $S_1Q - S_2Q$ must be λ. Draw S_2M perpendicular to TQ. Then S_1M *is* the path-difference λ. With the big distances and small angles involved, the triangle S_1S_2M is practically similar to PQT. Then, by similar triangles:

$$\frac{\lambda}{S_1S_2} = \frac{PQ}{TQ}$$

$$\therefore \quad \lambda = (S_1S_2)(PQ)/(TQ)$$

$$\lambda = \frac{(\text{DIST. BETWEEN SLITS})(\text{DIST. BETWEEN FRINGES})}{\text{DISTANCE FROM SLITS TO FRINGES}}$$

Interference of Water-Ripples

Look at ripples excited in a shallow pool by a vibrating fork with two prongs that act as a pair of sources. You will see streams of strong ripples traveling out—these make the "bright fringes"—with regions of little disturbance between them. Each strong stream is a hyperbola of points like X such that

$S_1X - S_2X = \lambda$ for one hyperbola,
2λ for the next
and so on

Diffraction Gratings: Spectra

Instead of a pair of slits we now use many slits, all parallel and equally spaced apart. In that way we get more light through to form the pattern (and in fact the pattern is sharper). To make the pattern wider we make the spacing between slits much smaller (say 1/10,000 inch instead of 1/50 inch in demonstrations of Young's fringes). This set of slits, a *diffraction grating*, is made by ruling furrows on a sheet of glass, using a fine diamond and a very precise ruling-engine to space the furrows evenly. The spaces between the furrows act as transparent slits.

Offered a beam of white light, such a glass diffraction grating splays out the "Young's fringes" so much that we see widespread bands of color (*spectra*) out at each side of the narrow central white band. A lens is used to gather up contributions in each splay-out direction and focus them to an image of the original source-slit. The image is a sharp narrow band for each color; but with white light the multitude of these images overlap making a wide spectrum of colors.

[3] "Monochromatic" light, such as yellow light from a salted flame or a sodium street-lamp. White light filtered by green glass is not monochromatic, but consists of a range of greens, whose wavelengths may differ by 10% between extremes.

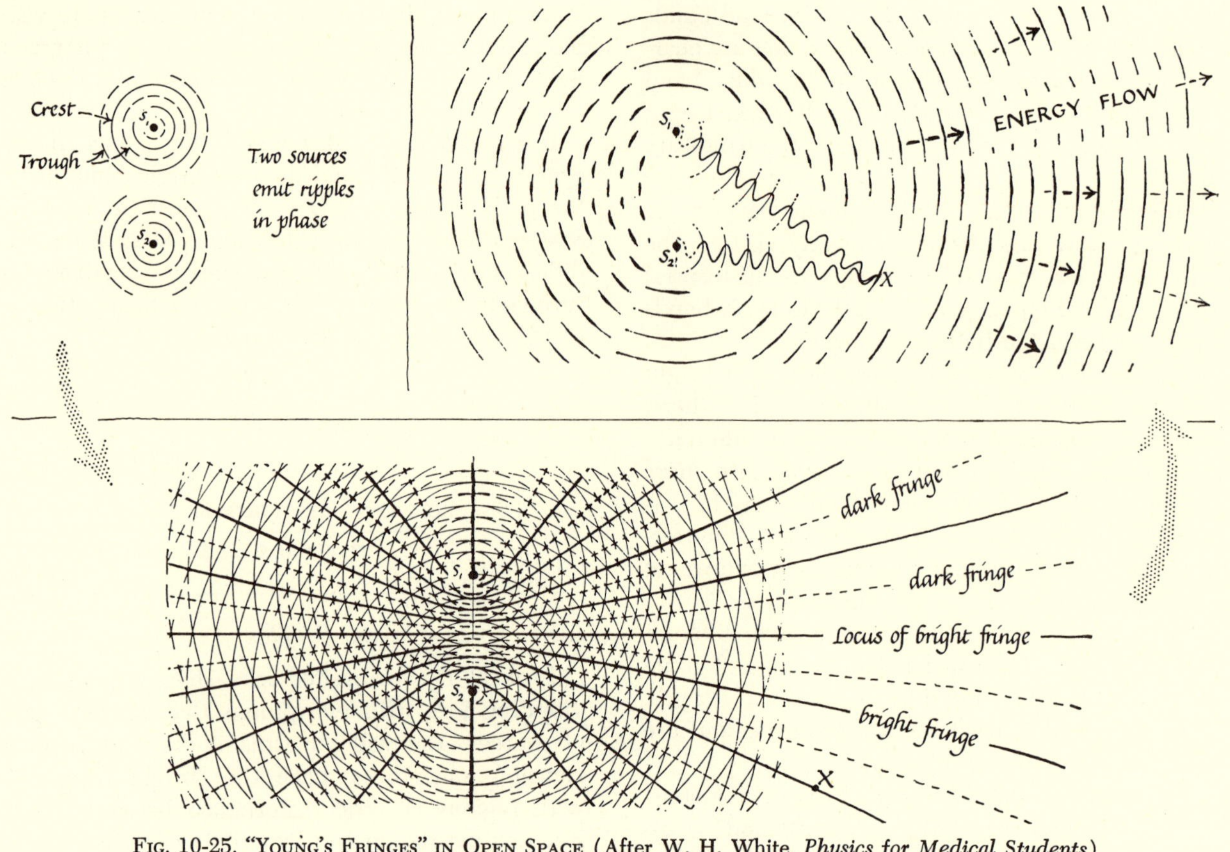

FIG. 10-25. "YOUNG'S FRINGES" IN OPEN SPACE (After W. H. White, *Physics for Medical Students*)

FIG. 10-26. DIFFRACTION GRATING

FIG. 10-27. DIFFRACTION-GRATING GEOMETRY

In the first spectrum on each side ("first order") the contribution from each transparent slit travels one λ more (or less) than the contribution from the next slit. In the next spectrum band ("second order"), contributions from adjacent slits differ by 2λ, and again they all arrive in step for that color.

Send pure yellow light from a salted flame through such a grating and you will see a central yellow "line" (the image of the source-slit in front of the flame), and equally sharp yellow lines in each first order, second order, &c. Geometry—see Fig. 10-27—gives

$$\text{WAVELENGTH} = d \sin A$$

for the first order swung out by an angle A, where d is the grating space between furrows, known from the original ruling engine. Thus a good grating enables us to measure wavelengths accurately. (You can make a rough measurement with a long-playing record as a reflecting grating. To measure its d, measure the record, play it, and count the turns.)

With white light a diffraction grating gives a wide spectrum in the first order, a wider one in the second order, &c.

Red light is swung out most (therefore its λ is greatest), then orange, yellow, green, blue, violet. Measurements of angles show that the wavelengths run roughly:

COLOR	WAVELENGTH	
	in meters	*in Ångström Units*
Red	7×10^{-7}	7000
Yellow	6×10^{-7}	6000
Green	5×10^{-7}	5000
Violet	4×10^{-7}	4000

Beyond the Visible Spectrum

Outside the visible range, there is infra-red radiation which has greater wavelength, easily measured with coarse diffraction gratings. Beyond infra-red, radio waves continue the spectrum out through microwaves, short waves to ordinary radio waves with λ measured in hundreds of meters. On the other side, ultraviolet light has shorter wavelengths than visible, measured by fine gratings operating in a vacuum to eliminate absorption by air.

SOME SOURCES OF ELECTROMAGNETIC WAVES

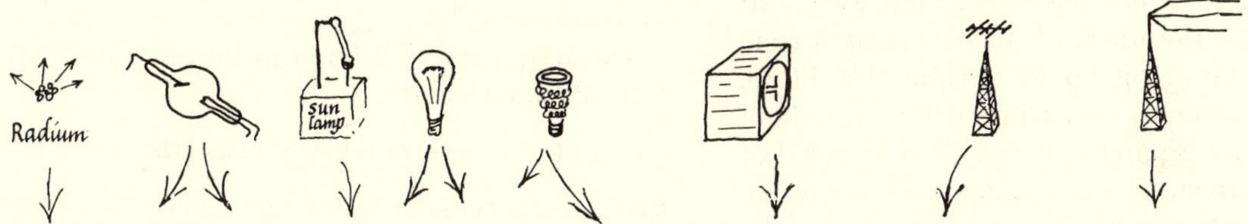

SPECTRUM OF ELECTROMAGNETIC WAVES

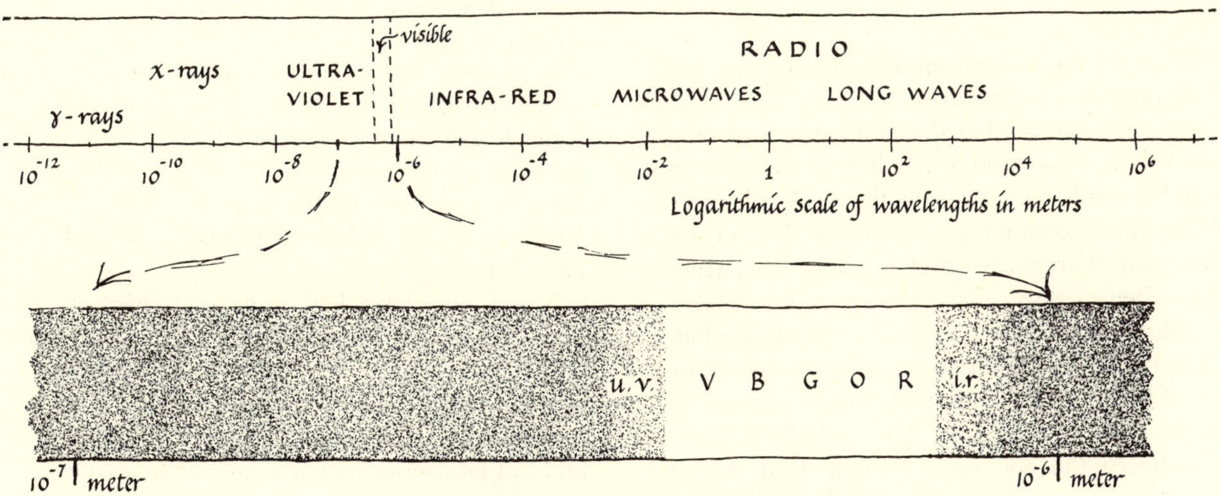

Fig. 10-28. The Electromagnetic Wave Spectrum

X-ray Spectra

X-rays have far shorter wavelength, one Å.U. or so, in contrast with many thousands for visible light.

We could hardly rule a fine enough grating, with furrows say 10 Å.U. apart (though ordinary gratings have been used obliquely, so that the X-rays see them foreshortened). Instead, we use the layers of atoms in crystals. The electrons of the atoms in each layer scatter X-rays in a faint "reflected wave." At a suitable angle the waves of one wavelength from successive layers gang up to a noticeable beam, just like an ordinary spectrum where the light gangs up from a grating's rulings. Thus a crystal of known structure enables us to measure X-ray wavelengths. Then we can use X-rays to investigate the arrangement of atoms in other crystals. We find that all solids are really crystalline; and even liquids have some local ordering among their molecules.

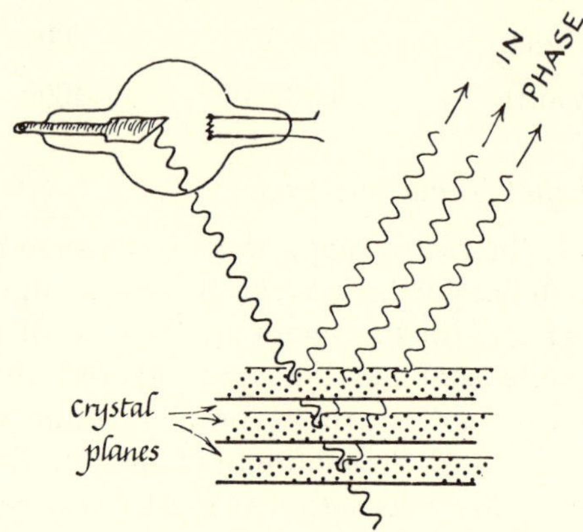

Fig. 10-29. X-Ray Diffraction by Crystal
X-rays (*very* short wavelength "light") are reflected by layers of atoms and the waves reflected from many layers add up to a strong wave in certain directions.

Line Spectra

When the light comes from a glowing gas—such as sodium in a salted flame, or neon in an advertising sign—the grating shows it consists of only a few colors, *very* narrow patches of the spectrum with the rest black, so narrow that each color makes a thin "line." Sodium gives a yellow line—really a close pair of twins. Neon gives many lines. Hydrogen, when made to glow, gives a series of lines—a red, a green-blue, a blue, a violet, obviously spaced along the spectrum according to some simple law. Mercury gives a pair of yellow lines, a very bright green line, a violet line, and others, but no red—hence the odd color of light from mercury street lamps.

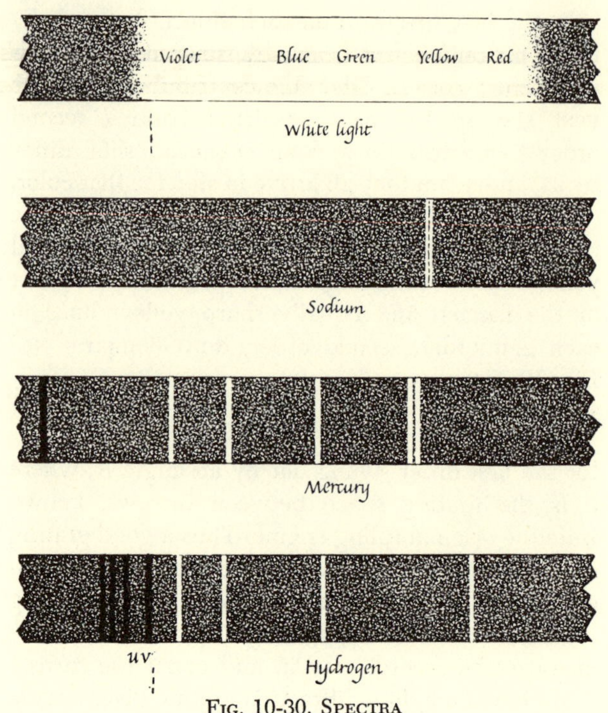

Fig. 10-30. Spectra

Such line spectra provide sensitive, unique tests for analysis. Each element gives its own characteristic lines when its atoms are excited. The lines of an element fall into series when classified by measured wavelengths.

The frequency of a line can be calculated easily from its wavelength:

FREQUENCY = VELOCITY/WAVELENGTH $\quad \nu = c/\lambda$

Frequencies came to be used instead of wavelengths in sorting out series of lines, and nowadays we are glad that tradition grew up. Not only do frequencies fit an even simpler mathematical form for each series, but modern theory makes FREQUENCY the essential measure of each *quantum* of ENERGY in a stream of light.

A century ago line spectra were catalogued in series; useful rules relating the frequencies in a series began to be discovered. Though some of the rules (e.g. for hydrogen) had simple mathematical form they did not fit the early pictures of atomic structure so the "origin of spectra" remained puzzling for many years.

X-rays, too, have both a general spectrum, like white light on a thousand times smaller scale of λ; and some sharp "lines," added to that general spectrum. The frequencies of the lines are characteristic of the atoms of the target of the X-ray tube, and fall into series with a simple code.

Look at various spectra yourself. For a qualitative

survey, a prism of glass can replace the grating. It spreads the colors by a different mechanism; and its spectrum-spreading geometry is too complex to serve for direct wavelength measurement; but it gives us a cheap and easy way of looking at spectra.

Absorption Spectra

Hot solids and liquids emit "white light" which a grating spreads into a full spectrum. Sometimes white light passes through a gas or vapor that is glowing but cooler than the white hot source. That happens when sunlight from the Sun's white-hot core passes through its cooler atmosphere. Then we get a "negative line spectrum," an *absorption spectrum* in which the characteristic lines are "black," that is, missing.[4] The cooler gases seem to absorb just the colors they emit if glowing alone.[5] This is obviously some kind of resonance or "sympathetic response" by the gas atoms to light of their natural frequency, but its mechanism was not fully understood until Bohr developed his theory.

Spectroscopy

Spectroscopy, the science of spectra and their measurements, has become a science of great precision. We can measure wavelengths today to one part in ten million—and small shifts of a line to even finer detail. The standard *meter* hitherto preserved as a metal bar with a fine mark near each end is now to be defined in wavelengths of light. The new standard will probably be:

1 meter = 1,650,763.73 wavelengths of a line from krypton gas.

Spectra and Atomic Physics

The obvious sharpness of spectral lines, their orderly arrangement of frequencies, their shifts when the source is placed in magnetic or electric fields, all these seemed to offer a great deal of information about atoms when they were discovered. Yet most of the information remained puzzling until the first quarter of this century. Then the story of spectra "all came true" when Bohr developed his theory to give a very satisfying common explanation of line-spectra, absorption spectra, even X-ray spectra—all linked to other electronic properties of atoms. (A partial account is given in Ch. 44.)

As atomic theory changes and grows today, spectroscopy still plays a prime part in providing essential measurements of superb precision for our study of atomic structure.

Stationary Waves

In modern atom models, we often picture the behavior of electrons and nuclear particles in terms of *stationary waves* (or *standing waves*). These are not in themselves proper waves, but they are wavy patterns of vibration that do not travel along. Before showing why they are called waves at all, we shall describe them simply as modes of vibration.

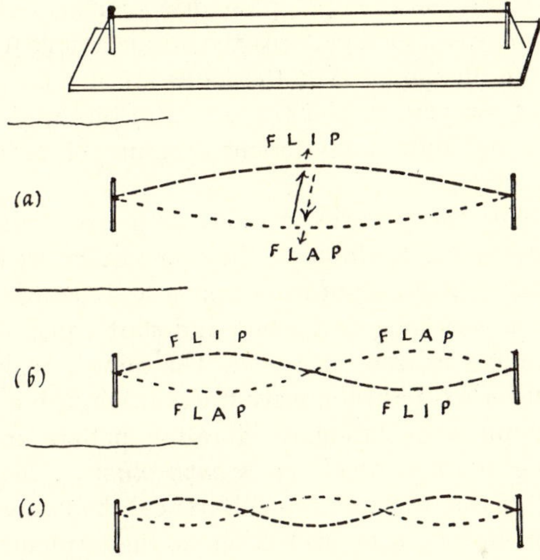

Fig. 10-31. Modes of Vibration of Taut String

A violin string can vibrate in many simple modes: a single loop that sways to and fro, flip-flap-flip- . . . &c; two loops that move in opposite phase, one moving flip-flap- . . . and the other flap-flip- . . . ; or 3 loops; or 4, 5, . . . any number of loops. When bowed (or when plucked carelessly), the string moves with a mixture of many modes; but any simple mode is easily excited alone by plucking with one finger (or bowing gently) at a suitable place while another finger touches the string gently to discourage unwanted modes. The touching finger is placed at a *node*, a place of no-motion for the desired pattern of loops. In each simple mode the flip-flap motion is simple harmonic, and the violin gives out simple harmonic sound waves of the same frequency. Pythagoras described the harmony of musical notes in terms of proportions of string-lengths; and Galileo gave rules for the frequency of vibration of a string. The same string vibrating in

[4] Hold a bright sewing needle in sunlight as source and look through a glass prism held close to your eye. You should just be able to see the dark lines in the Sun's spectrum. These can tell you there are sodium and hydrogen, etc., in the Sun's atmosphere. They even led to the discovery of helium when it was unknown on Earth.

[5] As they absorb these colors they re-emit them, but *in all directions*—so the forward contribution is too faint to show in contrast with all the other colors from the hotter core.

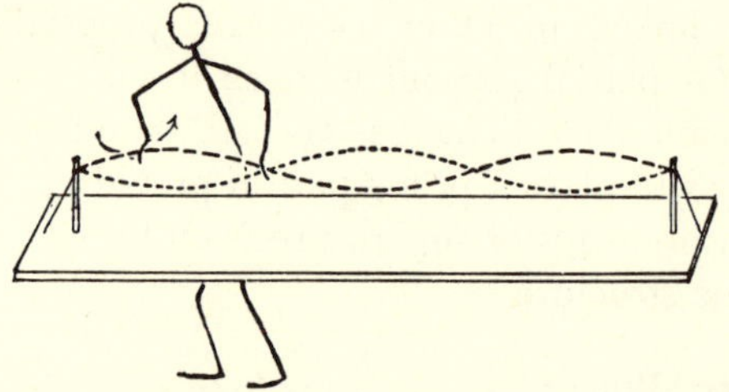

FIG. 10-32. EXCITING A SINGLE MODE by touching string gently at a node while plucking to make a loop.

1, 2, 3, . . . &c. loops has frequencies in proportions 1:2:3: &c. We now think of similar series of modes with characteristic frequencies when we picture wave patterns in the essential structure of atoms. The simple orbits of electrons in earlier atom models have become closed rings of stationary waves with more and more "loops" in the orbits farther out. We picture some such patterns, too, inside atomic nuclei. But in all those cases, the loops are not pieces of string swaying to and fro, not even pieces of electron, but only a mysterious measure of probable position.

While the patterns of loops on a string are just modes of stable vibration, they can also be regarded as being made up of travelling waves. Suppose we take a very long taut string and start equal waves travelling in from each end. The middle region is undisturbed until the wave-trains reach it, but then they make a stationary vibrating pattern in the string *as they travel across each other.* (This is a case of the principle of interference: the two wave-trains do not upset each other, so the resultant pattern is just the sum of the two.) At an instant when the two crossing wave-trains are in opposite phase, (a) in Fig. 10-33, the resultant is zero; the string is straight and undisturbed, but pieces of it are moving fast, sideways, through their "zero position." At an instant ¼ cycle later, one wave has travelled ¼λ forward the other ¼λ the opposite way and they are in phase; so the resultant pattern has double height. Then ¼ cycle after that they cancel again; and ¼ cycle later still they form a wave pattern of double height the other way up. The sketches (a)-(d) of Fig. 10-33 show the stages at ¼-cycle intervals. Detailed drawing, or algebra + trig, will show that the intermediate stages have a wavy resultant pattern, just like the maximum pattern sketched but lower. The crests and troughs always occur at the same places in the string, the *nodes.* The whole motion can be summed up in (e) of Fig. 10-33. In fact, the motion is just a set of vibrating loops, like a long violin string vibrating in

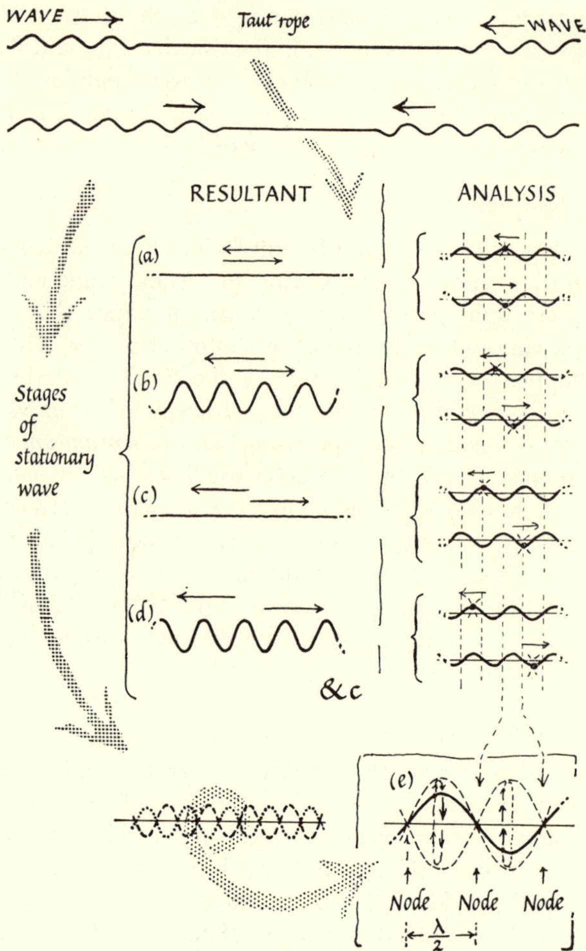

FIG. 10-33. GENERATING STATIONARY WAVES by adding two travelling wave trains.
Note that the nodes, regions of no motion, are ½λ apart where λ is the wavelength of either travelling wave.

many loops. Then we can regard the pattern of loops of, say, a vibrating string as made up of two equal wave-trains with opposite velocities adding up to make a pattern of stationary waves. Look at the sketches and you will see that the nodes of the stationary wave pattern are ½λ apart, where λ is the wavelength of the component waves. The advantage of this artificial[6] description of loop vibrations as a stationary wave is that it tells us the wavelength of ordinary travelling waves of the same

[6] Not completely artificial: we can manufacture the loop pattern by sending waves along a rope whose other end is fixed to a wall and letting them mix with their own reflection, the waves that come back from the other end. The child who excites standing waves in a bathtub of water is using a mixture of direct waves from his splashing hand and reflected waves from the wall of the tub. We use such a scheme to measure short radio wavelengths. And the effects of stationary light waves have been seen in photographic emulsion exposed with a mirror behind.

frequency. That wavelength, λ, is twice the length of one loop.

We treat a vibrating string as a piece of stationary wave pattern, the ends of the string being necessarily nodes where there is no motion. The length of the string, L meters, is ½ wavelength when it is vibrating in one loop: $L = \frac{1}{2}\lambda_1$. When it is vibrating in 2 loops, the component wavelength, λ_2, is shorter, and L is two half wavelengths: so $L = 2(\frac{1}{2}\lambda_2)$. With 3 loops $L = 3(\frac{1}{2}\lambda_3)$; &c. Then the wavelengths run thus:

$$\lambda_1 = 2L \qquad \lambda_2 = (2L)/2 \qquad \lambda_3 = (2L)/3 \qquad \text{\&c.}$$

But, for any *travelling* waves, velocity $v = f\lambda$. So the frequencies of the string's loop vibrations are:

$$f_1 = \frac{v}{\lambda_1} = \frac{v}{2L}$$

$$f_2 = \frac{v}{\lambda_2} = 2\frac{v}{2L}$$

$$f_3 = \frac{v}{\lambda_3} = 3\frac{v}{2L} \qquad \text{\&c.}$$

Thus we predict (with simple wave geometry, and the assumption that waves add geometrically) that a string has natural frequencies that run in proportions 1:2:3: . . . (So does the air in a pipe such as a flute or organ pipe. But many musical instruments, e.g. bells, have modes whose frequencies do not form a simple series of whole numbers—and that is why they sound less harmonious when hit.) If we knew the speed, v, of travelling waves on the string, we could calculate the actual frequencies. (See the next section for a derivation of v for waves on a string or rope.) Conversely, if we measure λ by stationary waves of measured frequency, we know v without needing any measurement of waves travelling. We use that to find the speeds of sound waves and short radio waves. And in designing an antenna for radio we try to adjust its length so that the incoming radio waves will excite stationary waves of voltage and current in the antenna system.

Speed of Waves along a Rope: Simple Theory

If you wish to see a wave-speed derived, look at this scheme;[7] otherwise omit it. We shall offer a somewhat similar derivation of the speed of electromagnetic waves—light and radio—in Ch. 37.

[7] I am grateful to Professors Uno Ingard and Francis Friedman for suggesting this very ingenious derivation. Other methods either use calculus or try to avoid it by more complex schemes.

Run a taut horizontal rope, with tension T newtons, from a "source," S, to a tree very far away. Suppose S suddenly starts raising the rope with vertical speed u and continues to raise it indefinitely. That will make a "knee" start out along the rope. (See Fig. 10-34.) The knee is a wave disturbance

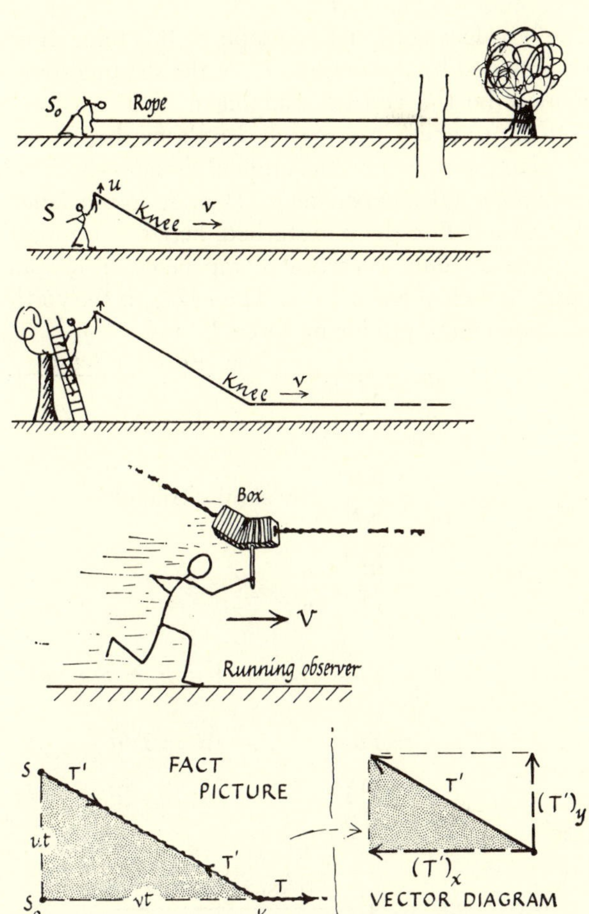

Fig. 10-34. Derivation of Wave Speed Along a Taut Rope

which will travel along the rope with wave-speed v. Any other shape of wave could be imagined to be made up of a lot of knees fitted together; so if we can calculate the speed, v, at which the knee travels along the rope we shall know the speed of any wave along the rope. After time t secs (Fig. 10-34), the knee has travelled vt along the rope and S has raised his end ut. Suppose an observer runs along beside the knee with speed v, carrying a box which just conceals the knee without touching the rope. Then in a short interval of time Δt the observer sees a length of rope $v \cdot \Delta t$ run horizontally into the box and come out moving in a slanting direction, where it has a *vertical* component of velocity u. The mass of this sample of rope is $d \cdot v \cdot \Delta t$, where d is its "line density" or mass per unit length, in kg per meter. The sample gains vertical momentum $(d \cdot v \cdot \Delta t)(u)$ in time Δt. Therefore there must be a vertical force on it given by:

$$F = \text{GAIN OF MOMENTUM/TIME}$$
$$= (d \cdot v \cdot \Delta t)(u)/(\Delta t) = d \cdot v \cdot u$$

The box does not touch the rope, so this force must be provided by the tension, T', of the slanting rope: F must be the vertical component of T'. (Notice that the source S has to pull a little harder, T', on the slanting rope than the original tension—because the horizontal component of T' must just balance the tension T of the undisturbed, horizontal, rope.) Resolve T' into a vertical component $(T')_y$ and horizontal component $(T')_x$. Then $(T')_y$ *is* the vertical momentum-producing force F; and $(T')_x = T$.

$$\therefore \quad \frac{F}{T} = \frac{(T')_y}{(T')_x}$$

$$= \frac{S_0S}{S_0K} \quad \text{by similar triangles}$$

$$= \frac{ut}{vt} = \frac{u}{v}$$

$$\therefore \quad F = T\frac{u}{v} \quad \text{But we had } F = d \cdot v \cdot u$$

$$\therefore \quad d \cdot v \cdot u = Tu/v \qquad \therefore \quad v^2 = T/d$$

$$\therefore \quad v = \sqrt{\frac{T}{d}}$$

$$\text{WAVE SPEED} = \sqrt{\frac{\text{TENSION OF ROPE}}{\text{MASS PER UNIT LENGTH}}}$$

As a simple example, use this result to predict the frequency of vibration of a piano wire 2 meters long, hanging vertically with a 10-kg load on its end, vibrating in *five* loops. Suppose that 900 meters of the wire weigh 2 kg; and take any information you need from the previous section. Answer: 262.5 cycles per second, which would give a musical note close to middle C.

Resonance

A system that can vibrate has natural modes of motion with definite frequencies—e.g., for a taut string, motion with one loop, 2 loops, &c.—and if we set it in motion and leave it alone, it will vibrate in one of those natural modes, or in a mixture of several of them. Any natural motion, however complex its pattern looks, is just a mixture of the system's natural modes. However, if we drive the system with a simple harmonic force it will respond with a small vibration with the frequency of the driving force. Putting that another way, incoming waves can generate a small stationary wave of their own frequency. But if the applied force, or incoming wave, has the same frequency as a natural mode of the system, it builds up an enormous response. This is called *resonance.* (A few generations ago the name was sympathetic vibrations.) In tuning your radio set you trade on resonance to obtain a large response. So does the child who builds up an oversplashing vibration in a bathtub. And the atomic particle that tunnels unexpectedly into a nucleus as it sails by seems to be trading on resonance between its own wavy nature—see Ch. 44—and some natural mode in the nucleus.

Thus waves and stationary waves are very important in physics. Watch for wave effects, sometimes in concealed form, as the course proceeds. You will meet them in fullest importance at the end, in our present view of atomic structure.

INTERLUDE

APPENDIX ON ARITHMETIC

Swift, capable calculations with very big numbers and with very small ones; sure, sensible use of percentages, to describe accuracy in experiments; clear graphs, accurately plotted and quickly read, to analyze results and exhibit relationships; clear-headed use of proportional arguments, to discuss connections; and above all, a wise understanding of the place of rough answers in good science—all these skills of hand and eye and brain are daily working tools for the professional physicist, and for all who wish to try his work or share his outlook. His major "power tools"—algebra, geometry, calculus—are provided by a separate course of mathematics; but these essential "hand tools" associated with arithmetic are treated briefly here.

> "All things began in order, so shall they end, and so shall they begin again; according to the ordainer of order and the mystical Mathematics of the City of Heaven."
>
> —Sir Thomas Browne, *The Garden of Cyrus*, (1658)

CHAPTER 11 · APPENDIX ON ARITHMETIC

> "Round numbers are always false."
>
> —DR. JOHNSON

Standard Form

Atoms and electrons are tiny and numerous. When we measure their masses in ordinary units, the numbers are extremely small, and when we measure electric currents by electron flow the numbers are huge. The arithmetic of atomic physics and of astronomy contains huge numbers and tiny fractions, always expressed as decimals. Both kinds of extreme numbers are unmanageably clumsy to handle if written out in full as in elementary arithmetic. Instead, scientists usually reduce them to "standard form," that is to: [a-number-with-just-one-figure-in-front-of-the-decimal-point] multiplied by [a-power-of 10], e.g., 2.3×10^6. Numbers written in standard form are easy to multiply (or divide) since the powers of 10 simply add (or subtract). They are ready for use on a slide rule. (And they are ready for use with log tables. The power of 10 shows the whole-number-part of the log directly, and tables give the decimal part of the log directly from the first part of the standard form number.) And *we can indicate the accuracy of the data or result* by the number of figures we give after the decimal point.

E.g., the breadth of a hair is about 0.00015 meters.
If you multiply this by 10 it becomes 0.0015
by 10 again it becomes 0.015
by 10 again it becomes 0.15
by 10 again it becomes 1.5 and it is ready for standard form.

Then $0.00015 \times 10 \times 10 \times 10 \times 10$ is 1.5; or 0.00015×10^4 is 1.5.

$$\therefore\ 0.00015 \text{ is } \frac{1.5}{10^4} \text{ or } \underline{1.5 \times 10^{-4}}$$

This is standard form for 0.00015.

If we want to split hairs about the breadth of a hair and measure with great accuracy, we might find a particular hair with diameter 0.0001502 meters. We should call this 1.502×10^{-4} meters. If we then repented or distrusted such precision, we might discard the final 2 and say 1.50×10^{-4} meters.

Note: 1.50×10^{-4} is *not* the same as 1.5×10^{-4}. The number 1.50 says, "I guarantee the 1, I guarantee the 5, and I think the next number is 0; it is nearer to 0 than to 1 or to the preceding 9"; whereas, 1.5 means "anywhere between 1.45 and 1.55."

The mass of a hydrogen atom is

0.00000000000000000000000000166 kg.

To get rid of each 0 you must multiply by 10. To get rid of all twenty-six zeros, you must multiply by 10^{26}. That would give you 0.166. Then one more multiplying by 10 gives you 1.66. Move the decimal point 27 places or multiply by 10^{27} to get 1.66. To pay for multiplying by 10^{27} and keep the actual mass unaltered, you must therefore divide 1.66 by 10^{27}. "Divided by 10^{27}" is written "multiplied by 10^{-27}." $\therefore$ the mass is $\underline{1.66 \times 10^{-27} \text{ kg}}$.

One kilogram of helium contains

150000000000000000000000000 atoms of helium.

In standard form this is $\underline{1.5 \times 10^{27}}$.

To multiply numbers in standard form: Multiply their "main numbers" by slide rule, and ADD their powers of 10 to get the new power.

To divide: Divide the main numbers, and SUBTRACT the powers of 10.

e.g., (a) $(3.1 \times 10^4) \times (2.0 \times 10^3) = 6.2 \times 10^{4+3} = 6.2 \times 10^7$

(b) $(3.1 \times 10^{-4}) \times (2.0 \times 10^{+1}) = 6.2 \times 10^{-4+1} = 6.2 \times 10^{-3}$

(c) $\dfrac{3.1 \times 10^4}{2.0 \times 10^3} = 1.55 \times 10^{4-3} = 1.55 \times 10^1$

(d) $\dfrac{3.1 \times 10^{-4}}{2.0 \times 10^{+1}} = 1.55 \times 10^{-4-1} = 1.55 \times 10^{-5}$

(e) $\dfrac{3.1 \times 10^{-4}}{2.0 \times 10^{-7}} = 1.55 \times 10^{-4-(-7)} = 1.55 \times 10^{+3}$

(f) $\dfrac{3.1 \times 10^{-4} \times 6.0 \times 10^7}{2.0 \times 10^{-3} \times 1.55 \times 10^2}$
$= 6.0 \times 10^{-4+7-(-3)-2}$
$= 6.0 \times 10^{-4+7+3-2} = 6.0 \times 10^4$

Slide Rules

Slide rules will do multiplication and division for you quickly and easily, if you learn to estimate tenths of their small spaces. But they will not tell you "where the decimal point is." To locate the decimal point, either use common sense or reduce all the numbers involved to "standard form" for a rough calculation. Example: for $\dfrac{126 \times 79.2 \times 0.074}{0.00521 \times 876}$ a slide rule gives 1618. *Common sense* says this is roughly

$$\frac{120 \times 80 \times 7/100}{5/1000 \times 800} \text{ or } \frac{120 \times 7}{5} \text{ or roughly } 160$$

so the decimal point should be placed thus: 161.8

Standard form gives

$$\frac{(1.26 \times 10^{2}) \times (7.92 \times 10^{1}) \times (7.41 \times 10^{-2})}{(5.21 \times 10^{-3}) \times (8.76 \times 10^{2})}$$

$$\approx \frac{1.2 \times 8 \times 7}{5 \times 9} \times 10^{2+1-2+3-2} \quad \approx 1.6 \times 10^{2}$$

$\therefore$ Answer is 161.8

Percentages

The sign % merely means /100 so that 2 % means 2/100. 6.21 % means 6.21/100 and 0.03 % means 0.03/100. If you want to express 3/20 as a % you must turn it into an equal fraction which has 100 at the bottom. In this case it is easy. 3/20 is the same as 15/100. So 3/20 is 15 %, and 15 % is simply another way of writing the fraction. To turn 3.2/23 into % we must change it into an equal fraction with 100 at the bottom. To do that, we write it as a grand fraction with 1 at the bottom, then multiply top and bottom by 100, then clear up the top by dividing, thus:

$$3.2/23 = \frac{3.2/23}{1} = \frac{3.2/23 \times 100}{1 \times 100} = \frac{320/23}{100} = \frac{\text{about } 14}{100}$$

Then 3.2/23 is the same as 14/100 which is written 14 %. "Express 3 as a percentage of 20" merely means "Write the fraction 3/20, then turn it into an equal fraction with 100 at the bottom, and write the new fraction with the help of a % sign." We simply say 3/20 is 15/100; therefore, the answer is 15%. To express 0.032 as a percentage of 7.91 we write the fraction 0.032/7.91, then reduce its top and bottom to whole numbers, 32/7910, then change it to a fraction with 100 at the bottom, and we have the answer:

$$\frac{0.032}{7.91} = \frac{32}{7910} = \frac{32/7910 \times 100}{1 \times 100} = \frac{3200/7910}{100}$$

$$= \frac{.4...}{100} \approx .4\,\%$$

Percentage Errors and Agreements in Experiments

When two measurements of a thing disagree slightly, we exhibit their disagreement as a percentage of the whole measurement, as in the examples below:

(i) Experimenters A and B time a race. A records 506 secs and B records 504 secs. They differ by 2 secs in a total of just over 500 secs. To show how closely they agree, we express their difference as a fraction of the total time, $\dfrac{2\text{ secs}}{500\text{ secs}}$. The difference, 2 secs, is 2/500 of the time. Reducing this to a fraction with 100 at the bottom, we have: 2/500 = 0.4/100 = 0.4 %. We say the measurements differ by 0.4 %.

(ii) Two weighings of the same object give 2.130 kg and 2.132 kg. These differ by 0.002 kg in a total of 2 kg. So the interesting fraction is 0.002/2 or 0.001 or 0.1 %. We say the weighings disagree by 0.1 %.

If the two measurements are regarded as equally reliable—e.g., those made by two good students—we can exhibit their disagreement by a *% difference,* but we should not call this a *% error.* If an experimenter is trying a new apparatus to measure a well-known quantity, he can exhibit the disagreement between his measurement and the standard value as a % and call this a *% error,* to be ascribed to the instrument. Sometimes a research group makes many measurements of a certain quantity and takes the average of their results, hoping thereby to eliminate the effect of chance errors. Then they may find the % differences between their separate results and their average and call these their % errors.

When a percentage "error" expresses some carelessness in experimenting or defect in the apparatus, it is necessarily a statement of *vagueness* somewhere in the machinery or argument. There is, therefore, no point in being *very accurate* when stating how *vague or inaccurate* you have been. That would be illogical. (If you do chop up the dining table into firewood, why sandpaper the pieces after chopping?) Thus if an error is calculated out to be 0.4219365 %, the statement is silly and unscientific, as it stands; but to say 0.4 % may well be sensible and useful.

Therefore, *it does not matter whether we divide by one main measurement or by the other, or by an average, or by a round number near them.* As we are using the percentage to express a lack of precision, it is unscientific to putter with precise arithmetic in calculating it.[1] In example (ii) above, we may divide by 2.130 or 2.132 or just rough 2. Here are the answers:

$$0.002/2.130 = 0.0939\,\% \qquad 0.002/2.132 = 0.0938\,\%$$
$$0.002/2 = 0.1000\,\%$$

All these round off to the same 0.1 %, and the last one, with easy arithmetic, is the one any sane scientist would choose.

Calculations with Errors

When several measurements are *multiplied together* to calculate some required result, the % errors (or uncertainties) of these factors must be *added together* to give the % error of the result. For example, suppose a man measuring a rectangular field is careless and overestimates both the length and the width. Suppose his length is 2 % too big and his width 3 % too big. Then if he calculates the area his estimate will be 2 % + 3 % or 5 % too big—not 2 % × 3 % which would be 0.06 %. Study the following problems.

PROBLEM 1. ERRORS IN FACTORS

(a) (Arithmetical example.) A rectangular field is 400 ft long by 300 ft wide. A man measures the field roughly. His estimate is 408 ft by 309 ft.
- (i) Calculate the true area of the field.
- (ii) Calculate the man's estimate of the area, from his measurements.
- (iii) Express the man's error in the LENGTH as a % of the LENGTH. Also find his % error in the WIDTH.
- (iv) Express the man's error in AREA as a % of the AREA.
- (v) To obtain AREA we *multiply* LENGTH by WIDTH. Which rule applies to the % error of area in this example, "multiply % errors of length and width" or "add % errors"?

(b) (More formal version.) Treat the problem above as follows:
- (i) The man's estimate of length is
  408 ft or [400] + [2% of 400].
  We can write this as
  400 + (2/100) · 400
  and factor it thus: 400 · (1 + 2/100).
  Treat the estimate of breadth similarly. Calculate the man's estimate of area by multiplying the two factored forms, thus:
  [400 · (1 + 2/100)] [300 · (   )]
  which becomes
  400 · 300 · (   ) (   ) or
  120,000 · (   ) (   )
  Since 120,000 sq. ft is the true area, the (   ) (   ) will reveal the area's % error, when multiplied out. *Multiply it out* as you would with (   ) (   ) in algebra. Then, just as 400 · (1 + 2/100) shows the error in the 400 ft length to be 2 %, this result shows the error in the area to be . . . . %.

(c) (Algebra version.) A rectangular field measures X ft by Y ft. A man overestimates the length by x %, saying it is X + (x/100) · X ft; and he overestimates the width by y %. As in method (b), express the man's estimates in factored form. Multiply these together to find his estimate of area. Interpret part of the result as a % error in his estimate of area. (*Note* that the error does not confirm the simple rule exactly. The multiplied out (   ) (   ) contains an extra fraction which is very small, with denominator 10,000. This fraction contributes a very tiny addition to the % error, so it may be neglected. Satisfy yourself by trying actual numbers, such as 2 for x and 3 for y, that this fraction is negligible.)

(d) (Geometrical version) Sketch a rectangular field. Then mark the position of one end moved outward to make the rectangle x % longer; and that of one side moved upward to make the rectangle y % wider. What fractions of the original area do the extra strips make?

PROBLEM 2. + AND − ERRORS IN FACTORS

Suppose the man in Problem 1 overestimates the length but underestimates the breadth of the field. Show, by algebra, or by an arithmetical example, that his % error in calculated area will be the *difference* between his % errors in length and width, or their *algebraic sum* if the underestimate error is counted negative.

PROBLEM 3. ERRORS IN SQUARED FACTORS, ETC.

If the rectangle in the problems above is a square, and the man measuring it knows it is, he just measures one side, X, with an error of x % and squares it to find the area.
- (i) What is the % error in his estimate of area?
- (ii) In general, if a product contains the *square* of a factor X, then an error of x % in X produces an error in the product of . . . . %.
- (iii) If a product contains X^3, an error of x % in X makes an error in the product of . . . . %.
- (iv) If a product contains X^n, an error x % in X makes an error in the product of . . . . %.

PROBLEM 4. ERRORS IN SQUARE ROOTS

Suppose a product contains $\sqrt{X}$ as a factor, how will an error of x % in X affect the product? Try to guess an answer by either of the following methods:
- (i) Writing $\sqrt{X}$ in the form $X^{\frac{1}{2}}$, assume that the answer to Problem 3 (iv) applies even when *n* is a fraction.
- (ii) If the factor $\sqrt{X}$ occurred *twice*, we should have $\sqrt{X} \cdot \sqrt{X}$ or $(\sqrt{X})^2$ which would be plain X. Then an error of x % in X would make an error of x % in the product. Therefore, when the factor $\sqrt{X}$ occurs only *once*, we expect an error of . . ? . . %. (*Note*: This has an important use in discussing the separation of uranium^{235} for atomic energy release. See Problem in Ch. 30.)

PROBLEM 5. ERRORS IN DIVISORS

Suppose we have to calculate a quotient, X/Y. If Y suffers from an overestimate of y %, how will the quotient suffer? Suppose we increased X as well as Y, by the *same* percentage. Then the quotient would be $\dfrac{X(1 + y/100)}{Y(1 + y/100)}$ or $\dfrac{X}{Y}$, unchanged. A y % overestimate in the bottom of the fraction just undoes a y % overestimate in the top. These are equal and opposite in effect. Therefore, a y % overestimate in the bottom of the fraction must have the same effect as a y % *underestimate* in the top. Then an error of + y % in the bottom of a quotient X/Y makes an error of − y % in the quotient. (*Note*: This also follows from Problem 3 (iv).)

---

[1] It is "anti-scientific" because it gives science a bad name for unrealistic precision.

PROBLEM 6. RESULT WITH MANY FACTORS

Suppose an experiment leads to the result:

$R = \frac{126 \times (9.25)^3 \times 0.0740}{0.00521 \times (29.62)^2}$ and the experimenters state that their measurements have likely percentage errors as follows: 126 may be wrong by ± 1 %

9.25 may be wrong by ± 0.2 %

0.0740 by ± 0.1% 0.00521 by ± 0.1%

29.62 by ± 0.2 %

If *all* these measurements were too *small* by the full likely margin, then

(i) the top of the fraction for *R*, as written above, would be too small by . . ? . . %.

(ii) the bottom of the fraction for *R* would be too small by . . ? . . %.

(iii) That would make the final result ($R = 1530$) too . . ? . . by . . ? . . %.

At very worst, all the measurements in the top might be too small by the full likely margin, and all those in the bottom too big by the full likely margin.

(iv) Then the result would be too . . ? . . by . . ? . . %. (In practice, we hope there will not be such a grand conspiracy. Nevertheless the last result gives us a wise warning.)

Desperate Measures: "Judging"

We often need to make a rough guess at an answer where we have not the data for an accurate calculation, or where we have not enough time or energy to make use of all the data fully. For example, there is sudden snow in a big city, and the authorities want to know how many men to hire to clear the snow. They do not mind whether the answer is 3219 men or 3456 men: all they need to know is that 3000 to 4000 men are needed. And that estimate is itself needed quickly—to delay and putter with questions of whether exactly 3219 or 100 men more or 50 less are needed will cost time and money, and may even lead to serious danger. Snow-clearing is an old problem where the estimate can be based on experience of earlier storms. But there are new problems needing a quick answer where we must even guess at the data we need. The General asks, "How many men, Colonel, could that district on the map feed for a month?" He welcomes an immediate answer, "About 7000," though it is unreliable. A thorough survey by scouts followed by painstaking adding up of supplies and needs, with detailed consideration of transport, might yield a more reliable answer, say 9250, but he cannot get the data until he has made the invasion! Again, a tax-reformer needs a rough estimate of tobacco imports, quickly. Even a 40% error will not hurt his deliberations. A careful study leading to a result with only 0.1% error would be wasteful and unscientific—the result will play only a small part in a large scheme, and it will be combined with other information that *cannot* be accurate.

At the frontiers of new knowledge, even the main result of an experiment may be only a rough guess, and yet we may be very glad to have it. For example,[2] in the early days of atomic physics, experiments yielded a guess that "carbon atoms have about 6 electrons each." Nowadays we know that every neutral carbon atom has exactly 6 electrons, but fifty years ago physicists were glad to know that it is about 6 and not 2 or 20. They took a risky bet on 6 being right and built a theory of atoms that was very successful in guiding further experimenting and thinking. The experimental tests suggested by the theory supported it and finally vouched for the choice of 6 in retrospect.

We meet many such problems, where finding a precise answer is either wasteful or impossible, and yet a rough answer is valuable. Then we resort to desperate intelligent guessing: a process that we shall call "judging."

Judging, in business or in government or in science, is not an easy job to be done by a careless, stupid guesser. It needs cunning and skill as well as a wide range of experience and educated knowledge, and it needs a ruthless spirit. Watch for opportunities for judging in your own work, now in your studies and later in your profession. If you rise high in your profession you will certainly do much judging—skill therein is a prime quality of good administrators. Rightly used, judging, with its rough answers, is good science. In fact it may yet become a science in itself—as John W. Tukey suggests, the specialist in judging must be a "generalist" in knowledge.

Here are two examples: (A) "How long will it take to mow that lawn?" The mower looks less than two feet wide. Allowing for overlaps it should cut a swath at most 1.5 feet wide or at least one foot. The lawn looks about 100 feet long by 30 feet wide. The 30 foot width contains 30/1.5 to 30/1, or 20 to 30 swaths; decide on 30 swaths. The mower must take 30 trips each about 100 feet long, say 3000 feet of travel, or 3/5 of a mile. The man pushing it is unlikely to walk briskly at 4 miles/hour but should be able to make 2 miles/hour, in which case he needs (3/5 mile)/(2 miles/hour) or 3/10 hour. Our answer, 20 minutes, is a rough guess but useful as a check on wages.

(B) "How does the Sun compare with the Earth, as regards mass?" A rough answer is very useful

[2] For examples in current physics look at some of the estimates and guesses being made in the field of Cosmic Rays. Though some experiments with Cosmic Rays reach high precision, others are yielding rough guesses that are, nevertheless, of vital importance to new theory.

because it tells us whether the Earth is massive enough compared with the Sun to disturb the orbits of comets and other planets seriously. Astronomers can "weigh" the Sun against the Earth using Newton's universal gravitation law; and the accurate data that are available give

$$\frac{\text{Mass of Sun}}{\text{Mass of Earth}} = \frac{93600000 \times 93600000 \times 93600000 \times 27.3 \times 27.3}{240000 \times 240000 \times 240000 \times 365.2 \times 365.2}$$

Ruthless rounding-off to give a quick rough answer yields

$$\frac{\text{Mass of Sun}}{\text{Mass of Earth}} \approx \frac{200000}{1} \text{ or } \frac{300000}{1} \text{ or } \frac{400000}{1}$$

according to the choice of rounding-off. This is certainly wrong—or rather "inaccurate" compared with the accuracy of the original information—and all it really tells us is that the answer is somewhere between 100,000 and 500,000. But this is quite enough to tell us that the Earth is only a tiny fraction of the Sun and will have hardly any effect on the orbits of other planets such as Mars around the Sun.

Quick Arithmetic in Judging

From its very nature, judging deserves only *rough* arithmetic. The data should be rounded off to hasten the calculations. Rounding everything to the nearest power of 10 may be too ruthless. That would make 8 become 10, 67 become 100, 1453 become 1000. Where numbers are near to 10, or a power of 10, that will suffice. You should call 8, 9, 10, 11, or 12, any of them, plain 10; and 1193 would become 1000. Rounding 73 off to 100 looks too rough; yet rounding it off to 70 will give you many small numbers like 7 to multiply and divide—still too much trouble; bolder judging is needed. So you should use two rules together for rounding off in judging.

RULE I. ROUND OFF TO A POWER OF 10, where the number is near a power of 10.

RULE II. Where the number is not near a power of 10, ROUND IT OFF TO THE NEAREST POWER OF 2, OR TO A POWER OF 2 TIMES A POWER OF 10. On this scheme, 7 becomes 2^3, 4200 becomes $2^2 \times 10^3$
67 becomes $2^3 \times 10^1$, or, better, $2^6 (= 64)$
3 becomes $2^2 (= 4)$, or, better, $2^5/10$ $(= 32/10)$

In some cases you have a choice: 8 can become 2^3 or 10. And 27.3 can become 2 times 10, or 2^2 times 10 (actually it is about midway between) or, better, 2^5; or a cunning calculator calls it 25 which is 100/4 or $10^2/2^2$. A very cunning calculator having to use the factor 27.3 squared pushes it *upward* to $2^2 \times 10$ one time and *downward* to 2×10 the other time—but that is neurotic unless it is done happily from sure habit.

A mixture of Rule I and Rule II looks queer but it gives safe rough answers. It yields a result consisting of a power of 10 and a power of 2, which can then be converted to a simple final answer. You need to remember the powers of 2 in the table. Note that $2^{10} \approx 1000$ or 10^3. This gives an easy way of getting rid of huge powers of 2 at the end. If you reduce all numbers to "standard form," you can round the first figure off to 1, 2, 4, 8, or 10, and the rest *is* a power of 10. This is very rough. To be less rough, take the first two figures and round them off to one of the following, each of which is a case of Rule I or Rule II: 10 16 20 32 40 64 80 100. Example:

TABLE OF POWERS OF 2
$2^1 = 2$
$2^2 = 4$
$2^3 = 8$
$2^4 = 16$
$2^5 = 32$
$2^6 = 64$
. . . . . .
$2^{10} = 1024$
$\approx 10^3$

$$\frac{\text{Mass of Sun}}{\text{Mass of Earth}} = \frac{93600000 \times 93600000 \times 93600000 \times 27.3 \times 27.3}{240000 \times 240000 \times 240000 \times 365.2 \times 365.2}$$

$$\approx \frac{100000000 \times 100000000 \times 100000000 \times 32 \times 32}{320000 \times 320000 \times 320000 \times 400 \times 400}$$

(*Note*: For fairness, we have rounded *all* factors *upward*, in top *and* in bottom.)

$$\approx \frac{10^8 \times 10^8 \times 10^8 \times 2^5 \times 2^5}{2^5 \times 10^4 \times 2^5 \times 10^4 \times 2^5 \times 10^4 \times 2^2 \times 10^2 \times 2^2 \times 10^2}$$

$$\approx \frac{2^{10} \times 10^{24}}{2^{19} \times 10^{16}} \approx 2^{10-19} \times 10^{24-16} \approx 2^{-9} \times 10^{+8}$$

$$\approx 2^1 \times [2^{-10}] \times 10^8 \approx 2 \times [10^{-3}] \times 10^8 \approx 2 \times 10^5$$

$$\approx \underline{200{,}000}$$

This is not a bad guess. More tricky judging will give answers nearer to the accurate value 330,000.

Rough Answers and Sure Knowledge

For some scientific work, measurements that are reliable within 1% will suffice: specific heats for apparatus-design, X-ray measurements of atom-spacing in crystals for theory of chemical structure, the half-life of a radioactive element to identify it, the period of revolution of an Earth satellite to estimate its average distance. In many cases measurements reliable to 0.1%, of data known to 1 in 1000, are needed to decide between one theoretical viewpoint and another. And in some cases the experimental data must be known to 1 in a million or even 1 in a billion to settle an essential question in the growth of scientific knowledge. For example, atomic masses must be known to great precision (mass spectrograph) before their small differences can provide reliable predictions of nuclear energy releases; wavelengths of light must be known to 1 in a million to resolve modern questions of atomic structure; measurements of gravitation need an accuracy of 1 in a billion to make a further critical test of the basis of general relativity.

At the other end of the scale, a very rough measurement can settle some important questions—we should not mind a 20% uncertainty in a chemical valency (which *has* to be a small whole number), or in the temperature of a nuclear fusion reaction, or in the age of the universe. It is not always very scientific to be very precise. So scientists do not just wear out their lives seeking greater and greater precision in every measurement, from sheer love of accuracy. They only take great pains where they see a great use ahead, though sometimes precision is promoted by a general sense of duty or a delight in designing better apparatus and has to wait for an important use to appear. When a scientist makes a measurement he also states its reliability according to his best judgment. He does not just say "I have measured g and find $g = 9.8$ meters/sec^2." He says "with probable error $\pm$ 0.1 meter/sec^2." In fact other scientists wishing to make use of the result are often as much interested in this estimate of possible error as in the result itself. Without the $\pm$ 0.1 his result is hardly a safe piece of data for a stranger to use. And there is much skill in estimating that error: looking at the scatter of measurements, avoiding known sources of error, guessing at concealed one-way errors, and an overall wise judgment that comes from the experimenter's intimate knowledge of his apparatus and his use of it. (In your own laboratory work, look back after you have worked with a piece of apparatus for a time, notice your increase in skill, and reflect on your growing feeling for the reliability of your measurements.)

Approximations: Signs $=$, $\approx$, $\sim$

Here are three ways of specifying a measurement that is considered by the experimenter to be reliable to about 1 in 1000:

$$x = 3.1642 \pm 0.003$$
$$x = 3.1642 \pm 0.1\%$$
$$x = 3.164$$

In the third form, the last figure, the 4, is understood to be unreliable. It is not clear there that it is uncertain by $\pm$ 3. All that form suggests is an uncertainty of a few in the last figure—which is usually precise enough as a statement of imprecision. The final 2 in the other forms is clearly unjustified; it makes a claim that is denied by the error just following it, and it makes a fool of the experimenter who keeps it.

When an experiment yields a rough measurement, or when a rough answer emerges from judging, we should not write the result in the form $x = 800$, since that conflicts with the precise meaning of "$=$." Instead we should write

$$x \approx 800$$

meaning "x is approximately 800." The symbol $\approx$ is usually read, rather illogically, "equals approximately" or "equals roughly."

For still rougher judgments, we may write

$$y \sim 1000$$

meaning y is nearer to 1000 than to 100 or 10,000. Such rough "order of magnitude" guesses are often of enormous importance at intermediate stages of knowledge—the size of an atom a century ago, the curvature (if any) of the universe now. Also there are many cases where an order of magnitude measurement is sufficient—a temperature-rise known to be small enough to be neglected, a weight known to be large enough to make surface tension trivial; a rough date in history where exact dates would add irrelevant distraction. If you express results in standard form, as $z = 2.34 \times 10^6$ and $w = 7.8 \times 10^3$, their order of magnitude is given by $z \sim 10^6$ and $w \sim 10^4$.

In your own work, whatever its field, you are likely to find good uses for $=$, $\approx$, and $\sim$ and the distinctions between them. Note that the symbols are not completely standardized. Some writers and printers replace $\approx$ by other signs.

Proportionality, the Key to Many Laws

In codifying our knowledge of nature in simple laws, we look first for constancy: the mass of a body remains constant; total electric charge remains constant; momentum is conserved; all electrons are the same. Almost as simple and equally fruitful is direct proportionality, when two measured quantities increase together in the same proportion: stretch of a spring with its load; force and acceleration; gas pressure and gas density.

We say that for a good spring (within the Hooke's Law range):

STRETCH is proportional to LOAD
or STRETCH varies[3] as LOAD
or we write STRETCH $\propto$ LOAD

Like percentages, proportion and variation are often given special treatment in elementary teaching that makes them seem mysterious and hard. Without this conditioning people would find them obvious pieces of common sense. So we shall discuss some simple examples.

Suppose that in providing potatoes to feed a camp, we find the weekly needs are

for a camp of 100 men, 400 pounds of potatoes

200	800
300	1200
500	2000

The mass of potatoes increases proportionally with the size of the household to be fed. Here is the simple type of relationship that we meet so often[4] in physics. We shall give several wordings to describe it:

(i) MASS OF POTATOES *is proportional to* NUMBER OF MEN
(ii) MASS OF POTATOES *varies (directly) as* NUMBER OF MEN
(iii) MASS OF POTATOES $\propto$ NUMBER OF MEN
(This is mathematical shorthand for (i) or (ii)
(iv) MASS OF POTATOES = (constant) · NUMBER OF MEN

[3] Some mathematicians and physicists advocate *varies as* or *varies directly as* for the continuous relationship and keep "proportional" for sets of isolated items, when they can say *are proportional to*.

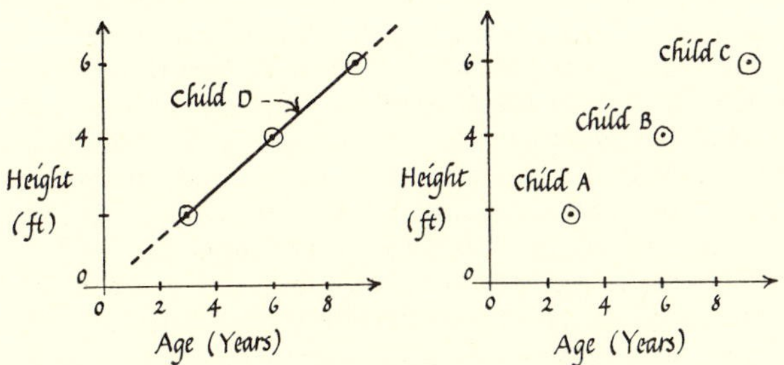

FIG. 11-1.

Example: Suppose a child D grows steadily from height 2 ft at age 3 years through 4 ft at 6 years to 6 ft at 9 years. Then, over that age range, we could say "D's HEIGHT *varies as* D's AGE."

Suppose three different children, A,B,C, aged 3 years, 6 years, 9 years have heights (today) of 2 ft, 4 ft, 6 ft: we then could say "the HEIGHTS of A,B,C *are proportional to* their AGES." These two stories are shown graphically in Fig. 11-1.

[4] Because we hunt for it.

Versions (i) and (ii)—and their mathematical form, (iii)—are simply attempts to say the simple obvious things, *"The two go up in the same proportion.* If we double one, the other doubles; if we triple one, the other triples; and so on." With that view in mind, you can easily solve problems, without any calculating or using the "constant" of version (iv). For example, given that 100 men need 400 lbs of potatoes, how much do 600 men need? Six times the number need six times as much food, 2400 lbs.

Take a different example. Given that: the VOLUME of a sphere varies as RADIUS^3. If a sphere is blown up to 5 times its original radius what happens to its volume? If RADIUS grows to 5 times original size, RADIUS^3 grows to 5^3 times the original value (since $R^3 = R \cdot R \cdot R$ and $5R \cdot 5R \cdot 5R = 5^3R^3$). Therefore the volume grows to 125 times the original volume. This should seem a clear matter of common sense without any use of $(4/3)\pi R^3$. In both examples the best attack is just to use your sense of the two things increasing together in the same proportion. There is no need to "set up a proportion"—whatever that may mean—or to calculate a constant K.

The "Proportionality Constant"

Version (iv) above,

MASS OF POTATOES = (constant) · NUMBER OF MEN

is much the same as version (iii), but to the scientific eye it does not emphasize the idea of relationship so brightly; therefore, you should avoid it where you can use common sense, as in the two examples above.

Clearly in each of the data given for potatoes

MASS OF POTATOES, $P = 4 \cdot$ NUMBER OF MEN, N

so we can include all the examples in the statement $P = 4N$. The essence of this last statement is the relationship: not so much the actual value 4 as the fact that the 4 stays the same, remains 4, remains *constant.* (Actually, it is the individual consumption, 4 lbs per man.) Since the 4 is constant, we can say

$$P = (\text{constant}) \cdot N.$$

This general statement should also apply to camps of men from a different district where all are great potato-eaters, consuming 10 lbs of potatoes a week. Then the statement would take the numerical form $P = 10 \cdot N$. (Of course, if we mix the two types of potato-eater arbitrarily, the whole story fails—and we must beware of the corresponding danger in scientific laws.)

Tests of Proportionality

How can we recognize simple proportionality in analyzing measurements? In the potato example, the numbers given make the relationship[5] obvious, but we need easy tests for more obscure cases. Here they are:

TEST A. *Divide one measurement by the other and look for a constant result.* In our example:

NUMBER OF MEN, N	MASS OF POTATOES, P	MASS OF POTATOES / NUMBER OF MEN
100	400 lbs	4 lbs/man
200	800 lbs	4 lbs/man
300	1200 lbs	4 lbs/man
and so on,		
always giving the result . . .		4 lbs/man

Here is an unfailing test of direct variation or proportionality. It works either way of course: if we divide NUMBER OF MEN by MASS OF POTATOES, we get another constant answer, ¼ man per pound.

TEST B. *Graphs.* Soon after Galileo's time the French mathematician and philosopher Descartes invented the method of plotting graphs with x and y coordinates. Nowadays we take graphs for granted and read them as easily as printed words—in fact there may even be a danger of making a generation of statistical illiterates by letting newspapers present all our statistics in graphical form, thus avoiding the clear use of words and figures. Yet a few generations ago, graphs were regarded by many as puzzling things involving difficult techniques. In your generation, all you need, at most, is practice in quick accurate plotting and reading, and for this you can gain much by using certain standard scales and maintaining a standard accuracy by estimating tenths of the smallest division.

As a way of exhibiting a relationship, graphs are magnificent. If you have a set of observations of two things (e.g., men and potatoes), you can represent them by a set of points, using vertical distances for one measured thing, horizontal for the other, each on a convenient scale. The arrangement of the points may tell you the relationship between the measured things. Fig. 11-2 shows the graph (A), of the data above of potatoes and men. So far as our data go, we have no right to fill in the intervening points, as if we knew the needs of every possible number of men (including fractions). However, we may guess that many an intermediate

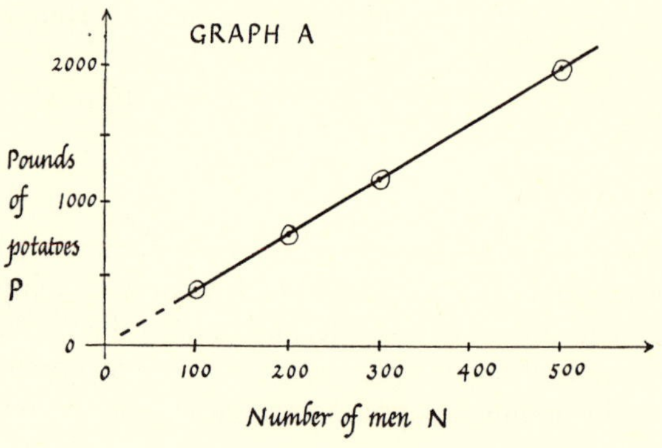

FIG. 11-2. GRAPH A

point is just as true as the ones we have plotted, and such guessing would help us budget for other numbers of men. To find or show the relationship between our data we jump ahead to the simple relationship of direct variation or proportionality and work back from it. Suppose we know that the MASS OF POTATOES, P, *does* vary directly as NUMBER OF MEN, N, and want to predict what the graph of P against N will look like. We know that P/N is a constant. But, for any graph-point, P/N is the slope of the line joining that point to the origin. So the lines from the origin to all our plotted points must have the same slope. They must all be the *same* sloping line. Therefore all the plotted points lie on the same sloping line through the origin. Conversely, if all the plotted points do lie on a straight line through the origin when P is plotted against N, we can say that P/N is a constant. Here the graph is a straight line through the origin, showing that MASS OF POTATOES does go up in direct proportion to NUMBER OF MEN.

Linear Relationships

The graphs B and C of Fig. 11-3 both show a "linear relationship" between y and x. In graph B the plotted points x_1, y_1, etc., all lie on a straight line through the origin, 0, 0; and the shaded triangles are similar, each having the same y/x ratio for height/base which gives the slope of the slanting side. In graph C the points x_5, y_5, etc., lie on a straight line that does not pass through the origin, 0, 0. We can no longer say we have direct variation or proportionality between y and x. The slanting sides of the shaded triangles have different slopes. It is *not* true that y_5/x_5 and y_6/x_6 and y_7/x_7, etc., are all the same. We must be careful over this in seeking relationships. Yet there is obviously *some* simple relationship in graph C. We should get back to the simple story if we could start afresh with an origin on the line itself. We can do that if we deal with *increases* (or changes) of x and y from some chosen value on the line, instead of dealing with the full values. Thus in graph D (Fig. 11-4) we have sketched new axes (broken lines) and reckoning from this new starting point 0′ we can say (increase of y, from value at 0′)

[5] Note that neither *relation* nor *relationship* means a ratio or fraction in mathematics, but only a definite connection.

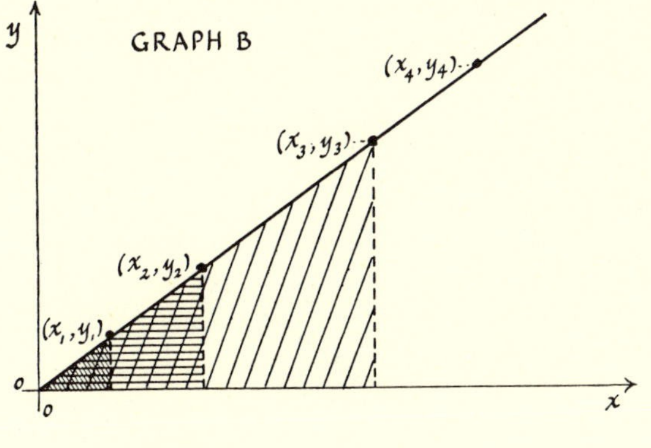

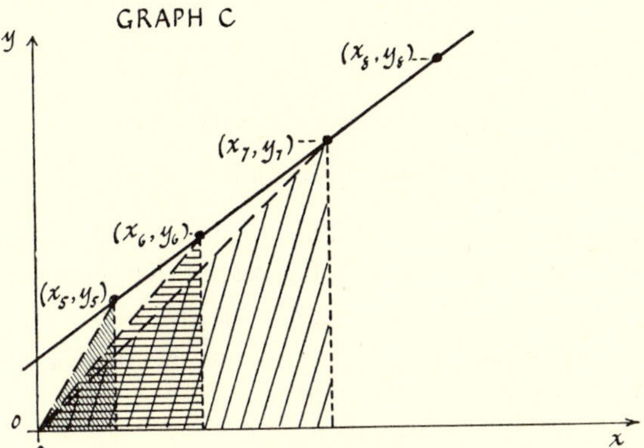

FIG. 11-3. LINEAR RELATIONSHIPS: GRAPHS B AND C

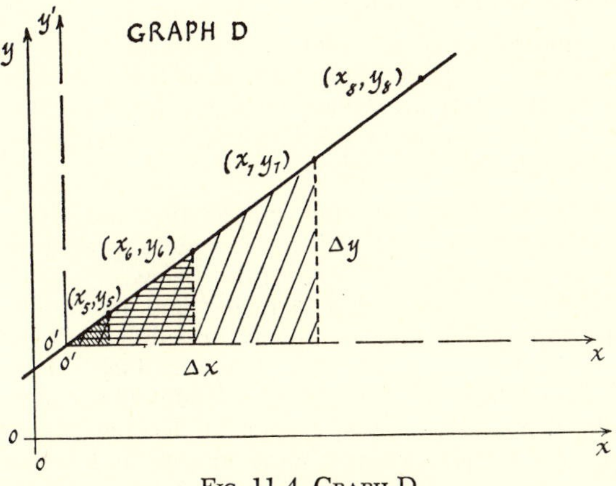

FIG. 11-4. GRAPH D

does vary directly as (increase of x), or $\Delta y \propto \Delta x$. We use Δ as mathematical shorthand for "increase of," or "change of," or "difference of." Then we can say for graph D $\Delta y \propto \Delta x$, or $\Delta y / \Delta x$ is constant (for all points on the line). Or we can deal with the matter in another way, as in graph E. We can jump from point to point, always using the changes in y and x. Then whatever points we choose on the line we can again say $\Delta y \propto \Delta x$, or $\Delta y / \Delta x$ is constant, since again we have similar triangles, though in this case Δy and Δx have slightly different meanings. In both graphs C and D the ratio $\Delta y / \Delta x$ gives the slope of the line, just as y/x gives it in graph A. These slopes are the constant value involved in the statements of proportionality.

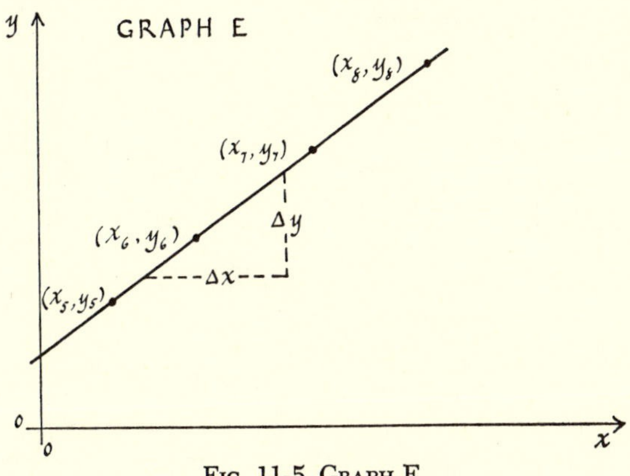

FIG. 11-5. GRAPH E

This is the scientific procedure. We plot a graph of our observations. Then, looking for a simple relationship, we draw a straight line, to see how closely our plotted points fit with a simple relationship which the line implies. For this test we try to draw the "best" straight line—that is, one which passes "as near as possible to as many of the points as possible." (These fine-sounding instructions for a best straight line will not bear logical scrutiny, yet their intention is clear—interpret them as common law.) If we want to test for direct variation or proportionality between the things plotted, ($y \propto x$), we *make* the line pass through the origin. If this fits poorly, we may want to draw another straight line that misses the origin, testing only whether $\Delta y \propto \Delta x$. In any case our "best straight line" is a "question-asking" line. It is neither a statement of the right answer nor an attempt to link up the plotted points with some allowance for a mysterious experimental error, but only a statement of a simple relation which we have guessed at, drawn beside our plotted facts for comparison. The plotted points are the real *facts*; they represent what we did observe.

If we can make the line run very near our plotted points, we may say our observations very nearly fit the simple relationship. We may perhaps guess that they fit exactly (whatever that may mean) with the simple relationship, the small misfits on the graph being due to mistakes on our part in observing. This alibi of "experimental error" is a convenient one and a comforting one until we examine it carefully. Then we find that we are really saying—almost boasting—that we are careless experimenters, or that our instruments are very poor. If we ask, "how careless?", "how poor?", we can state reasonable limits of the errors we will admit to. Then if the errors of our plotted points lie within such limits, we may safely say that, for all we can tell, the simple story fits the facts.

Graph F (Fig. 11-6) shows plotted data for camps with two kinds of potato eaters, with best straight lines drawn. These data are fictitious, but they have been made up with minor variations—such as data from real camps would certainly show—and they are not easy round numbers fitting smoothly on the lines as in the original example. If we decide the straight lines represent the real relationship underlying the data, we can

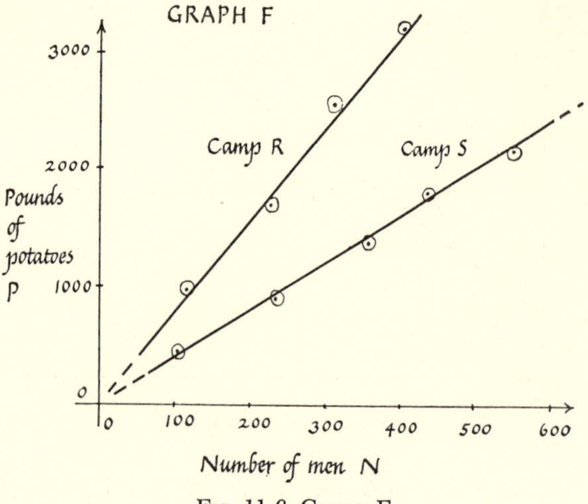

FIG. 11-6. GRAPH F

reduce the story of each line to a statement such as $P \propto N$ and we can even state

$$P = 4.1\ N \text{ for one line}$$
$$P = 8.0\ N \text{ for the other}$$

In each case the constant, (4.1, or 8.0), is best obtained from the slope of the *line* rather than from any individual point or piece of data. Drawing the line takes an automatically *weighted* average.

Weighted Averages

A weighted average is one in which extra weight is given to the data considered most reliable, and very little weight to any which are suspected of grave errors. When we make such an average arithmetically we weight a good piece of data heavily by adding it in several times—like stuffing a ballot—and adding a poor piece only once. Then we divide by the total number of ingredients, counting the repeats of course. This kind of averaging may be fair and wise, but it is risky because it may encourage us to get the answer we hope for! In drawing a line among points on a graph, we may find we can draw a line which runs very close to nearly all the points, leaving one or two points far out in the cold. If we settle on this line, its slope will give a weighted average, with little weight to one or two remote points. The odd positions of the latter may be due to some carelessness, in which case we are wise to take the hint and largely ignore them. On the other hand the agreement of the majority with the line may be due to chance errors; or the few odd points may even be the key to an important story. So there is danger of wishful prejudice in drawing the line, yet if we are careful we may hope to obtain a very cleverly weighted average that is more helpful than risky.

Direct Variation or Proportion

If in drawing our question-asking line we are asking, "Is this a case of direct variation?", we must make the line pass through the origin. The origin may or may not be a plotted point—a piece of experimental data—but in any case we compel our line to pass through it. We may decide this restriction is unwise. Graph G (Fig. 11-7) is plotted for a camp where the permanent kitchen staff eat potatoes but are not counted in the population. Here the broken straight line through the

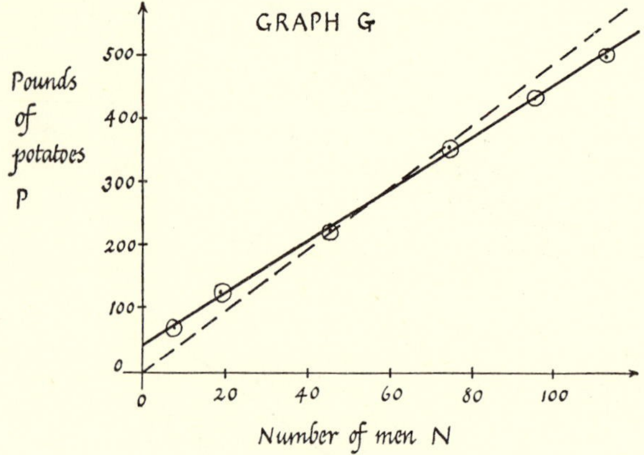

FIG. 11-7. GRAPH G

origin runs far away from several points, while the full line runs very close to all. In this case we had better say

$$\Delta P \propto \Delta N \quad \text{or} \quad \Delta P = 4.2\ \Delta N$$

If we read the "intercept" made by the line on the vertical axis at 42.0 we can say

$$P = 42 + 4.2N$$

and we may guess that the kitchen staff eats 42 pounds per week and that there are probably ten of them.

Hints for Graph Plotting

ROUGH GRAPHS. Plot a rough graph as you experiment, to guide your work. It will enable you to see whether you are taking enough measurements. A blunt pencil and paper ruled with coarse rulings (e.g., ¼ inch squares) will suffice.

PRECISE GRAPHS. For graphs that are easy to plot and easy to read and yet provide a precise test, you are advised to follow the rules below:

Paper: use paper ruled in inches and tenths of an inch. These small squares are easy to subdivide by eye in tenths (that is, in hundredths of an inch). Larger ones or smaller ones are harder to subdivide accurately.

Use scales that are easy to plot by decimal guessing. Suppose you are plotting some masses in kilograms. The easiest scale is 1 inch represents 1 kg; and 1 inch represents 10 kg, or 100 kg, . . . , or 0.1 kg, . . . , etc., are all easy. The next easiest family of scales is 1 inch represents 2 kg, or 20, . . . , or 0.2, . . . , etc. These scales are easy to use if you halve the measurements in your head and then plot directly. Another easy family is 1 inch represents 5 kg, or 50, . . . , or 0.5, . . . , etc. These involve doubling the measurements in your head before plotting them. All other families such as 4 kg to the inch, etc., are more troublesome and usually lead to mistakes in plotting. Choose among the three families: 1's, 2's and 5's to the inch.

Choose scales that will make your graph extend several inches sideways and several inches upwards—there is no gain in spreading out your graph one way if you crowd it in the other direction. If possible, its general slope should be between, say, 30° and 60° with the horizontal. To obtain that, you may have to choose quite different scales for the two axes.

Line: After plotting your experimental points, use a taut thread to decide on the "best straight line," if you wish to draw one. Then draw the line. Then draw a small circle (or, if you have the necessary information, an error box) around each plotted point. Draw the line first, or the circles will distract your eye. If a straight line seems unsuitable, draw a smooth curve.

Interpolation and Extrapolation

Even if a straight line seems unsuitable and you just draw a smooth curve through (or close to) your plotted points, you can use your graph to read off further information. Assuming that the curve represents the behavior of your apparatus, you can mark intermediate points on it and confidently read off values for which you made no measurements. This is *interpolation.* Or you can sketch a continuation of the curve and read off values outside the range of your data. This is *extrapolation.* As an example: if you knew that a train leaves Boston at 2 P.M. and arrives at New York at 6 P.M. you could estimate its time of passing through New Haven by interpolation. You could also estimate its time of arrival at Washington, D.C., by extrapolation—but with much greater risk, since the train might terminate at New York.

Obviously both interpolation and extrapolation are easier if the graph is a straight line. Even then they are not equally safe as sources of information. Which of the two, interpolation and extrapolation, would you as a scientist value more? In the final chapter you will find interpolation and extrapolation playing important roles in a discussion of the progress of science.

PROBLEMS FOR CHAPTER 11

Problems 1-6 are in the text.

7. The value of $\dfrac{3.14 \times 75200 \times 373 \times (0.00162)^2}{8 \times 9.8 \times (0.0282)^2}$
is given as 375 by a slide rule, with the decimal point not placed. Find where the decimal point should be, *showing your method clearly.*

8. STANDARD FORM. Express the numbers of the following data in standard form.
Density of mercury, d_{Hg} = 13,600 kg/cu. meter.
Distance of Moon from Earth, R_M = 240,000 miles.
Charge on electron = —0.000,000,000,000,000,000,16 coulomb.

9. The "kinetic energy" (= energy of motion: see Ch. 26) of a mass M kilograms moving with speed v meters/sec is $(\frac{1}{2})Mv^2$ joules. A nitrogen molecule N_2 in air at room temperature has
mass 0.000,000,000,000,000,000,000,000,0465 kilogram, and speed 520 meters/second.

Energy is also measured in "electron • volts" or ev. One ev is equal to 0.000,000,000,000,000,000,16 joules.

(a) Express each of these data in standard form.
(b) Calculate the kinetic energy of a nitrogen molecule at room temperature and state it in standard form,
 (i) in joules,
 (ii) in electron • volts.
(c) When a mass M kilograms of matter disappears and turns into radiation, the energy of that radiation, E joules, is given by $E = Mc^2$, where c is the speed of light, 300,000,000 meters/sec. Calculate what would be released if one complete nitrogen molecule could be annihilated thus. Express in standard form in electron • volts.
(d) The calculation (c) refers to a wild surmise, very unlikely ever to be observed. The fission of uranium atoms is observed, but that is a mere splitting, with only a small fraction of the total matter of the atom disappearing. The fission of a uranium^{235} atom releases about 200,000,000 electron • volts. Compare this energy with the answers to (b) and (c).

10. A student measures a pendulum thread's length with a meter stick graduated in meters, centimeters, millimeters (but the millimeter marks are thick and he cannot easily judge tenths of a millimeter): STRING LENGTH 1.186 meter. He measures the diameter of the bob with calipers and halves it, obtaining: RADIUS = 0.01425 meter.

(a) Why is he unscientific if he then says the length of his pendulum to the center of bob is 1.20025 meters?
(b) What statement is he making about the length to center of bob, if he says it is 1.2 meters?

11. A student obtains 9.83 m./sec^2 for "g" where the standard value is 9.80.

(a) What is his % error?
(b) If he converts his result into ft/sec^2 and compares it with the standard value for "g" in ft/sec^2, what will he find his % error is?
(c) Give reason for (b).

12. Two weighings give 0.040593 kg and 0.040591 kg. What is their % difference? (Note warning against "sandpapering" early in this chapter.)

13. In calculating an acceleration by using $a = 2s/t^2$ with measurements of s and t, a student uses an estimate of s that is 4% too big, and an estimate of t that is 3% too big. What will be the % error in the value of a that he calculates?

PART TWO

ASTRONOMY: A HISTORY OF THEORY

ASTRONOMY is part of Physical Science, but we shall not use it here as just one more bunch of topics. It is included for a very important purpose: *to provide a clear example of the growth and use of THEORY in science.* With that aim, this section deals with the history of our knowledge of the solar system—Sun, Moon, Earth, and planets—from early watching and simple fables to the magnificent success of Newton's gravitational theory. By giving a simple historical treatment we can make the nature of theory clearer than by describing it ready-made. A genuine understanding of theory and its relationship with experiment is essential if you wish to know science; and gravitational planetary theory gives the best example here because you need not just learn its results but can see why it is needed and how it is formed.

It would be a great further advantage if this history could also show the interplay between scientific discovery and the social environment, and the reactions between scientific theory and other branches of philosophy. However, that would require far more discussion of historical background. So this section is not a fair setting-forth of history. If accounts of people show a one-sided tendency to moralize, or if a few great scientists seem to stand like isolated lighthouses in a formless ocean, remember that this is not fair history, but only a restricted summary for a special purpose.

"The relative individualism of the history of science, as opposed to general history, is also due to the fact that if it is not altogether easy to analyze and to estimate a man's contributions in the field of science, at least it is a good deal easier than in any other field, except art. The cleverest general cannot win a battle without an army, and how much of the victory must be credited to him and how much to the brave soldiers who executed his orders? The scientists are not alone in the world, yet they win their battles without armies behind them; they win them to a large extent by their own unassisted efforts. . . . And yet the history of science is not simply the history of great scientists. When one investigates carefully the genesis of any discovery, one finds that it was gradually prepared by a number of smaller ones, and the deeper one's investigation, the more intermediary stages are found. Our first impression of scientific progress is like that of gigantic stairs, each enormous step representing one of those essential discoveries which brought mankind almost suddenly up to a higher level, but that impression is imperceptibly obliterated as we pursue our analysis. The big steps are broken into smaller ones, and these into others still smaller, until finally the steps seem to vanish altogether,—yet they never vanish. . . . No scientific victory was ever won by sheer numbers or by the mass of projectiles. Each was won by a series of efforts the humblest of which was deliberate to a degree."

—GEORGE SARTON, *Science and the New Humanism*

[Henry Holt & Co., N.Y.] 1931

CHAPTER 12 · MANKIND AND THE HEAVENS

"An undevout astronomer is mad."

—EDWARD YOUNG (~1700)

The Beginnings of Man[1]

ASTRONOMY is almost as old as man himself. How old *is* man, and why did he bother about astronomy?

Man began to emerge as a distinct creature several thousand centuries ago. The records of the caves and rocks—the only records stretching so far back—are far from completely explored. Anthropologists warn us not to guess too confidently, and not to guess at all unless we first decide what we mean by Man. What essential characteristics distinguish man from animals? Solving problems? Rats can solve a maze: ants organize war. Using tools? Apes use sticks and stones to solve an immediate problem, and some build simple tree huts. Planning for the future? Perhaps a clear difference begins there. Man makes tools for *future* use. Such planning involves a simple form of reasoning: if . . . then. Man makes arrows for game that *may* come; and he builds graves to comfort the dead *if* they have some future life. Planning for food supply and shelter could lead, with the help of speech, to larger plans . . . community life . . . traditions . . . laws. . . . Thus, man emerged as a planner and reasoner, a worrier. He used his toolmaking hands and worrying brain in adapting himself to changes of environment. Unlike animals, man changed his living-equipment quickly to meet new conditions: shifts of climate, invasions, floods, famines. Instead of depending on heredity—the chance survival of an animal's mutation in several generations—man controlled his own adaptation. He changed tools, clothing, housing, feeding and defense to meet each new environment. That gave him far greater chances of survival—and hopes of progress.

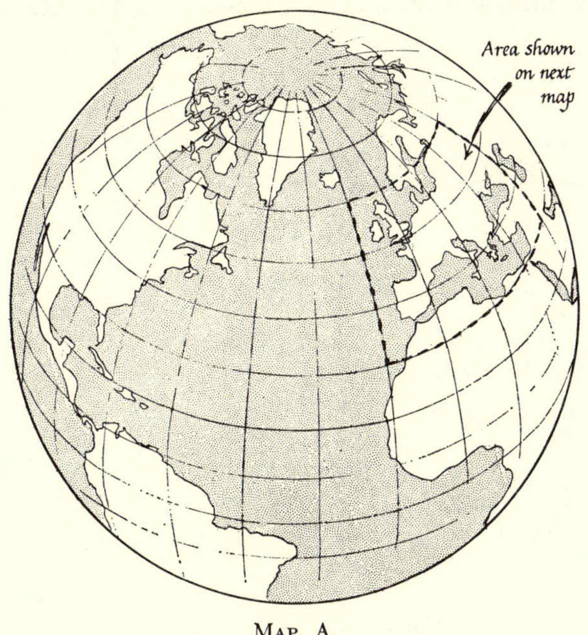

MAP. A.

MAP B. Copernicus lived and worked near Thorn. Kepler was born in Weil and first taught at Gratz.

How old is man? Perhaps as long as 1,000,000 years ago, very primitive tool-making man was showing divergencies from ape-like "cousins." Some 200,000 years ago simple "Neanderthal man" developed along a side-line of our family tree, and was later displaced by our more capable ancestors. He was a simple hunter with rough tools, but he made some use of fire, and he buried his dead with care. There are few signs of our direct ancestors until, say, 100,000 years ago,[2] when men using carefully

[1] Some of the comments here are drawn from an excellent popular book on primitive man, *Man Makes Himself*, by V. Gordon Childe, published in paperback as a Mentor Book (New York, 1951).

[2] These spans of time may be wrong by a factor of 2. Even if they are true of man's development in one region of the world, they are untrue in others.

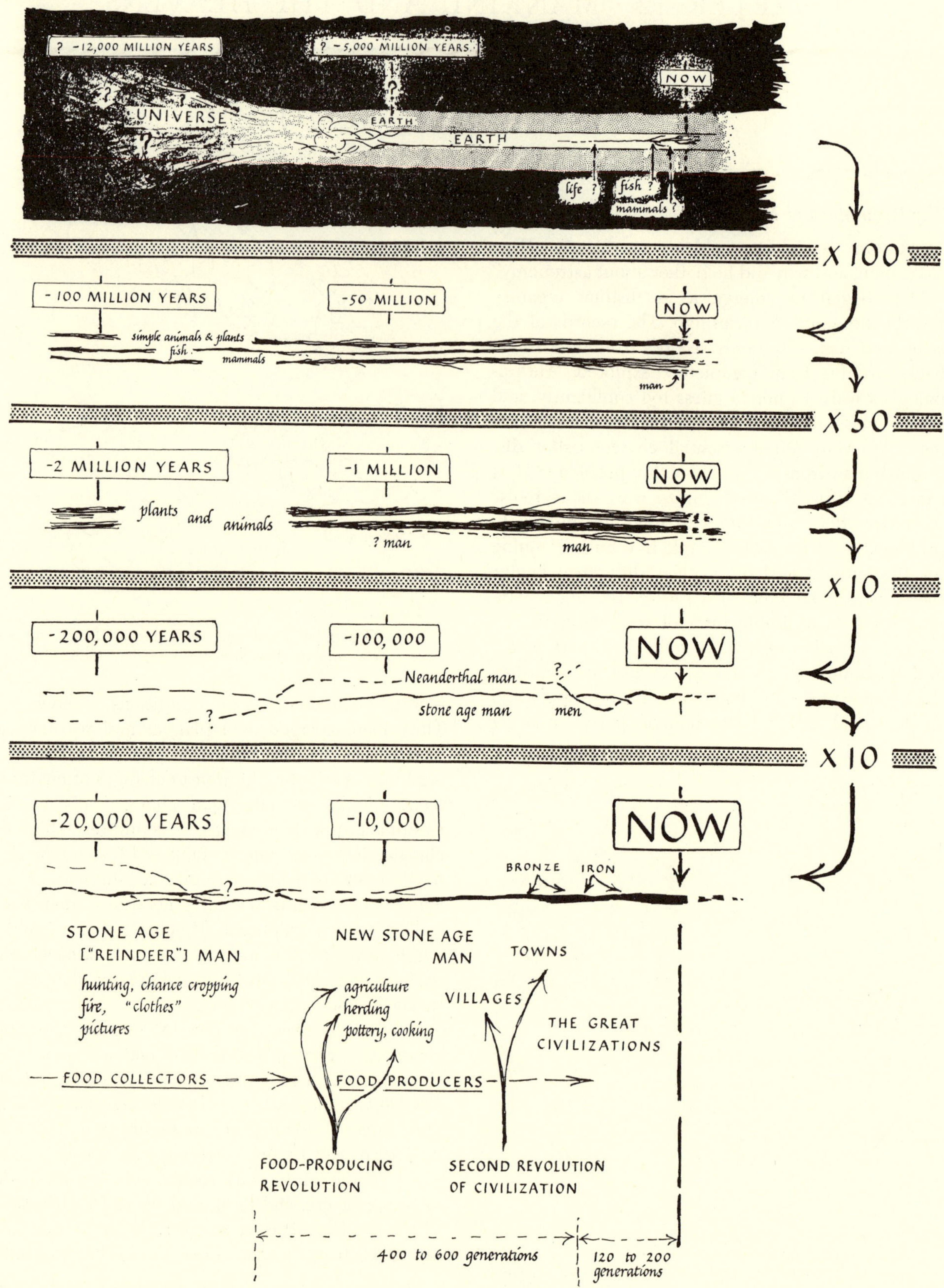

Fig. 12-1. A Rough Table of Ages

The times for the stages of man differ greatly with locality. "The age of the Universe" not only seems a fantastic guess but is also entirely dependent on our choice of time-scale: yet astrophysicists are making scientific speculations.

chipped stones pushed their way across Europe. In the next 80,000 years these stone tools and weapons improved, bone needles were made and used, carvings and pictures were added; but man remained a *food-collecting* savage, living in small groups, with leisure when game was plentiful. Stone-age artists made little statues to symbolize fertility and drew animals on cave walls, for simple magic. Some are masterpieces of artistic skill and sympathetic insight.

It was not till some 12,000 years ago that an *age of cultivation* began. Then a new level of human life developed—perhaps even new races of man—in which better tools were used, agriculture began to supplement chance cropping, herding began to replace hunting of wild animals, and pottery and cooking came into use. Village life developed in this *food-producing* culture, and simple trade was carried on.

Then, five or six thousand years ago, a new revolution of man's living started: groups of villages were gathered into states, with farm and town developing their distinct activities—the beginning of the great civilizations. Large cities were built to maintain secondary industries, fed by organized farming outside and enriched by far reaching trade. Craftsmen's knowledge brought metals to replace stone for tools: copper, bronze, then iron. In cities, problems of building and trade and government required arithmetic; geometry; measures of weight, length, area, volume; and timekeeping. Organized agriculture to feed the cities required a good calendar to arrange for planting, animal breeding, and irrigation by river floods. Long trade routes by sea and land required signposts for navigation. Compass, clock, and calendar were essential in the early civilizations, as they are now: *astronomy provided all three.*

Man's Growth

Three thousand years ago, there were flourishing civilizations with good practical knowledge of astronomy. Observations of the Sun, Moon, and stars, recorded, codified, and extended to predictions, provided a daily clock and an accurate calendar of months and years and a compass for steering journeys. *Thirteen thousand* years ago, man had been a simple food-gatherer, using Sun, Moon, and stars as rough guides at most. A great science of astronomy grew up in the ten thousand years between. If that seems long, write it in generations: from savage with simple magic to civilization with working astronomy in 400 generations, another 120 to our knowledge of science today. That is rapid progress, both in power over environment and in intellectual grasp.

The Beginning of Astronomy

Yet, earliest knowledge of the heavens grew slowly. For centuries very primitive man must have watched the stars, perhaps wondering a little, taking the Sun for granted, yet using it as a guide, relying on moonlight for hunting, and even reckoning simple time by moons. Then slowly gathered knowledge was built into tradition with the help of speech. The Sun offered a rough clock by day and the stars by night.[3] For simple geography, sunrise marked a general easterly region, and sunset a westerly one; and the highest Sun (noon) marked an unchanging South all the year round. At night the pole-star marked a constant North.

As the year runs through its seasons the Sun's daily path changes from a low arc in winter to a high one in summer; and the exact direction of sunrise shifts round the horizon. Thus the Sun's path provided a calendar of seasons; and so did the midnight star-pattern, which shifts from night to night through the year.

With the age of cultivation, herding and agriculture made a calendar essential. It was necessary to foretell the seasons so that the ground could be prepared and wheat planted at the right time. Sheep, among the first animals domesticated, are seasonal breeders, so the early herdsmen also needed a calendar. Crude calendar-making seems easy enough to us today, but to simple men with no written records it was a difficult art to be practised by skilled priests. The priest calendar-makers were practical astronomers. They were so important that they were exempt from work with herds or crops and were paid a good share of food, just as in many savage tribes today.

When urban civilizations developed, clear knowledge of the apparent motions of Sun, Moon, and stars was gathered with growing accuracy and recorded. These regular changes were worth codifying for the purpose of making predictions. In the great Nile valley, where one of the early civilizations grew, the river floods at definite seasons, and it was important for agriculture and for safety to predict these floods. And fishermen and other sailors at the mercy of ocean tides took careful note of tide regularities: a shifting schedule of two tides a day, with a cycle of big and little tides that follows the

[3] An experienced camper can tell the time by the stars within ¼ hour.

Moon's month. In cities, too, time was important: clocks and calendars were needed for commerce and travel.[4]

Timekeeping promoted an intellectual development: "In counting the shadow hours and learning to use the star clock, man had begun to use geometry. He had begun to find his bearings in cosmic and terrestrial space."[5]

Astronomy and Religion

Apart from practical uses, why did early man attach such importance to astronomy and build myths and superstitions round the Sun, Moon, and planets?

The blazing Sun assumed an obvious importance as soon as man began to think about his surroundings. It gave light and warmth for man and crops. The Moon, too, gave light for hunters, lovers, travellers, warriors. These great lamps in the sky seemed closely tied to the life of man, so it is not surprising to find them watched and worshipped. The stars were a myriad lesser lights, also a source of wonder. Men imagined that gods or demons moved these lamps, and endowed them with powers of good and evil. We should not condemn such magic as stupid superstition. The Sun *does* bring welcome summer and the Moon *does* give useful light. The simple mind might well reason that Sun and Moon could be persuaded to bring other benefits. The very bright star, Sirius, rose just at dawn at the season of the Nile floods. If the Egyptians reasoned that Sirius caused the floods, it was a forgivable mistake—the confusion of *post hoc* and *propter hoc* that is often made today.

When a few bright stars were found to wander strangely among the rest, these "planets" (literally "wanderers") were watched with anxious interest. At a later time, early civilized man evolved a great neurotic superstition that man's fate and character are controlled by the Sun, Moon, and wandering planets. This superstition of astrology, built on earlier belief in magic, added drive—and profit—to astronomical observation.

Thus the growth of astronomy was interwoven with that of religion—and they still lie very close, since modern astronomy is bounded by the ultimate questions of the beginning of the world in the past and its fate in the future. The next two pages contain speculations on the early stages of that development.[6]

Science, Magic, and Religion

Science began in magic. Early man lived at the mercy of uncontrolled nature. Simple reasoning made him try to persuade and control nature as one would a powerful human neighbor. He tried simple imitative magic, such as jumping and croaking like a frog in the rain to encourage rainfall, or drawing animals on cave walls to promote success in hunting. He buried his dead near the hearth, with logical hope of restoring their warmth; and he gave them tools and provisions for future use. In a way, he was carrying out scientific experiments, with simple reasoning behind them. It was not his fault that he guessed wrong. The modern scientist disowns magicians because they refuse to learn from their results—that is the essential defect of superstition, a continuing blind faith. Primitive man, however, had neither the information nor the clear scientific reasoning to judge his magic.

As he carried out jumping ceremonies, or squatted before magic pictures, man could form the idea of presiding spirits: kindly deities who could help, malicious demons who brought disaster, powerful gods who controlled destiny. Like a child, man tried to please these gods so that they would grant good weather, health, plentiful game, fertility. The original reasons and purposes were then forgotten and the ceremonies continued by habit.

Speech was the essential vehicle of this development. Earliest man, just beginning speech, was forging his own foundations of thought, slowly and uncertainly. Other creatures communicate—bees dance well-coded news of honey, dogs bark with meaning—but man's speech opened up greater intellectual advances. In the course of a long de-

[4] The ancient civilizations had no reliable mechanical clocks, but only sand-glasses, simple lamps, and water-clocks. For accurate timekeeping they used the Sun and stars. Pendulum clocks and good portable watches are recent medieval inventions to meet the demands of ocean trade.

[5] Lancelot Hogben, *Science for the Citizen* (Allen and Unwin Ltd., London, 1938). In the early chapters, the author discusses how and why astronomy developed; and he gives a fine account of astronomical measurements and their use in navigation. Some of the material of the present section is drawn from that stimulating book on the social background of scientific discovery.

[6] Such speculative guessing is very risky. It is the misfortune of the young Science of Prehistory that laymen, and even scientists from other fields, think they can guess correctly how man grew and even what he thought. (History suffers similarly from amateur speculation; Education is almost built on it.) Yet here we need some picture of the background of the beginning of science. Make your own speculations, if you prefer; or look at the early chapters of H. G. Wells' *Outline of History*, on which some of the comments here are based. In that book, often criticized for inaccuracy of fact and view, the general reader finds what the experts fail to give him, a connected story—though a risky one.

velopment, it not only gave him a rich vocabulary for communicating information but it enabled him to store information in tales for later generations; and then it blossomed into a higher intellectual level with *words for abstract ideas.* Thus speech opened up a new field of *ideas and reasoning.* Of course, this development did not happen suddenly.[7] Early verbal thinking must have been crude and confused, with reasoning left unfinished, and words mistaken for the things they represent—as they are in children's thinking today and even in simple people's attitude to science.

With speech could come the beginnings of religion and science: rules of conduct for the individual and the community; and, in a different sense, rules for nature. Even before speech developed, family life involved obedience; but speech could hand on the tradition of "the strong father, the old man whose word was law, whose possessions must not be touched," and of a kindlier mother-figure. As families gathered into groups . . . villages . . . tribes . . . , these traditions crystallized into law and custom that restricted the individual's freedom for the general good. Out of such feelings and traditions, out of hopes of success and fears of disaster, grew a sense of being bound together in a community, a sense of *religion.*

Primitive religion wove together myths of gods, magic ceremonies, and tales of nature, in attempts to codify both the natural world and man's growing social system. Astronomy played an important part in this ceremonial religion. The priests—wise men of the village or tribe—were the calendar-makers, the first professional astronomers. Their successors were the powerful priesthood of the first urban civilizations. In early Babylon, for example, the priests were bankers, physicians, scientists, and rulers—they *were* the government. In their knowledge, and that of their craftsmen, there were the foundations of many sciences. The practical *information* was there, at first unrecognized, then held secret, then published in texts. Was it *science*?

[7] Anthropologists warn us: do not assume that primitive savage communities today give a reliable picture of our equally primitive ancestors long ago. Contemporary savages may be primitive in technology and yet maintain a complex of customs or religion developed over thousands of years. It may be safer to base surmises on the behavior of civilized children.

Science, the Art of Understanding Nature

Curiosity and collecting knowledge go back before earliest man. Primitive man collected knowledge and used it: a beginning of applied science. Then, as a reasoner, he began to organize knowledge for use and thought. That is a difficult step, from individual examples to generalization—watch a child trying to do it. It is difficult to grasp the idea of a common behavior or a general law or an abstract quality. Yet that is the essential step in turning a "stamp collection" of facts into a piece of science. Science, as we think of it and use it now, never was just a pile of information. Scientists themselves, beginning perhaps with the early priesthood, are not just collectors. They dig under the facts for a more general understanding: they extract general ideas from observed events.

Scientists feel driven to *know,* know what happens, know how things happen; and, for ages, they have speculated why things happen. That drive to know was essential to the survival of man—a generation of children that did not want to find out, did not want to understand, would barely survive. That drive may have begun with necessity and fear; it may have been fostered by anxiety to replace capricious demons by a trustworthy rule. Yet there was also an element of wonder: an *intellectual delight* in nature, a delight in one's own sense of understanding, a delight in creating science. These delights may go back to primitive man's tales to his children, tales about the world and its nature, tales of the gods. We can read wonder and delight in stone-age man's drawings; he watched animals with intense appreciation and delighted in his art. And we meet wonder and delight in scientists of every age who make their science *an art of understanding nature.*

As scientists, we have travelled a long way from capricious gods to orderly rules; but all the way we have been driven by strong forces: a sense of urgent curiosity and a sense of delight.

Fear and anxiety, wonder and delight—these are two aspects of *awe,* mainspring of both science and religion. Lucretius held, 2000 years ago, that "Science liberates man from the terror of the gods." Walt Whitman grieved for man's anxiety while he rejoiced in the scientist's delight:

I believe a leaf of grass is no less than the journeywork of the stars,
And the pismire is equally perfect, and a grain of sand, and the egg of the wren,
And the tree-toad is a chef-d'œuvre for the highest,
And the running blackberry would adorn the parlors of heaven,
And the narrowest hinge in my hand puts to scorn all machinery,
And the cow crunching with depressed head surpasses any statue,
And a mouse is miracle enough to stagger sextillions of infidels,
And I could come every afternoon of my life to look at the farmer's girl boiling
her iron tea-kettle and baking shortcake.

. . . .

I think I could turn and live awhile with the animals . . . they are so placid
and self-contained,
I stand and look at them sometimes half the day long.
They do not sweat and whine about their condition,
They do not lie awake in the dark and weep for their sins,
They do not make me sick discussing their duty to God,
Not one is dissatisfied . . . not one is demented with the mania of owning things,
Not one kneels to another nor to his kind that lived thousands of years ago,
Not one is respectable or industrious over the whole earth.

Leaves of Grass
Doubleday, N.Y.

CHAPTER 13 · FACTS AND EARLY PROGRESS

> "To speculate without facts is to attempt to enter a house of which one has not the key, by wandering aimlessly round and round, searching the walls and now and then peeping through the windows. Facts are the key." —JULIAN HUXLEY, *Essays in Popular Science*

Facts

Before showing how astronomy was organized into great schemes of thought, we shall review the knowledge that you—or primitive man—could gain by watching the heavens. If you have lived in the country you will be familiar with most of this. If you have grown up in a town this will seem a confusing pile of information unless you go out and watch the sky—now is the time to observe.

The Sun as a Marker

Each day the Sun rises from the eastern horizon, sweeps up in an arc to a highest point at noon, due South,[1] and down to set on the western horizon. It is too bright for accurate watching, but it casts a clear shadow of a vertical post. At noon, mid-day between sunrise and sunset, the shadow is shortest and points in the same direction, due North, every day of the year. The positions of the noon Sun in the sky from day to day mark a vertical "meridian" (= mid-day) plane running N-S.

In winter the shadows are longer because the Sun sweeps in a low arc, rising south of East and setting south of West.[1] In summer the arc is much higher, shadows are shorter, daylight lasts longer. Half-way between these extremes are the "equinox" seasons, when day and night are equally long and the Sun rises due East, sets due West.

On the horizon—simple man's extension of the flat land he lived on—sunrise marked a general eastern region for him and the exact place of sunrise showed the season. The length of noonday shadows also provided a calendar. The shadow of a post made a rough clock. Though it told noon correctly, its other hours changed with the seasons—until some genius thought of tilting the post at the latitude-angle (parallel to the Earth's axis) to make a true sundial.

Stars

The stars at night present a constant pattern, in which early civilizations gave fanciful names to prominent groups (constellations). The whole pat-

[1] This description applies to the northern latitudes of the early civilizations.

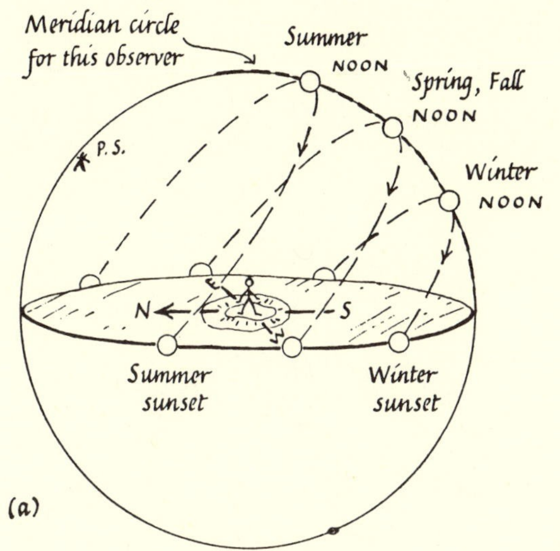

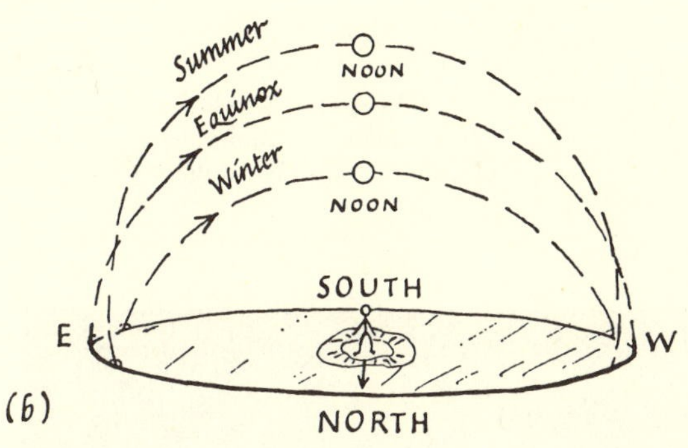

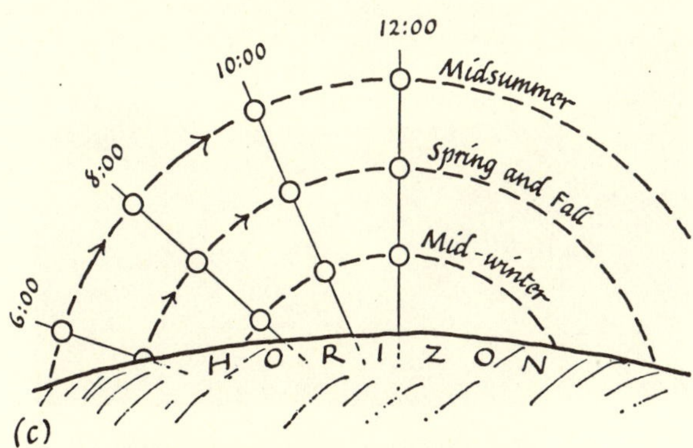

FIG. 13-1. THE PATH OF THE SUN IN THE SKY CHANGES WITH THE SEASONS

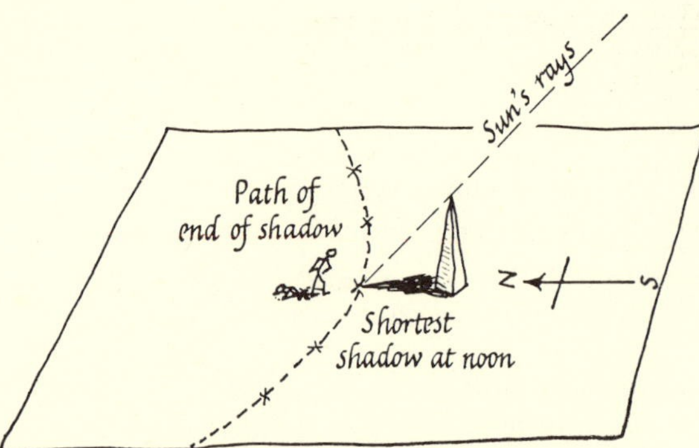

Fig. 13-2. Noon. Sunlight casts the shadow of a pillar on horizontal ground. The shadow is shortest at noon.

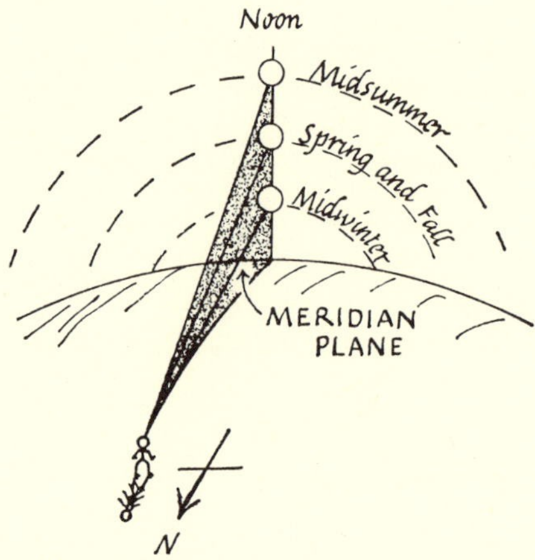

Fig. 13-3. Meridian. Noon-day Sun is due South (or due North). The meridian (= mid-day) plane is a vertical plane passing through the Sun's noon positions.

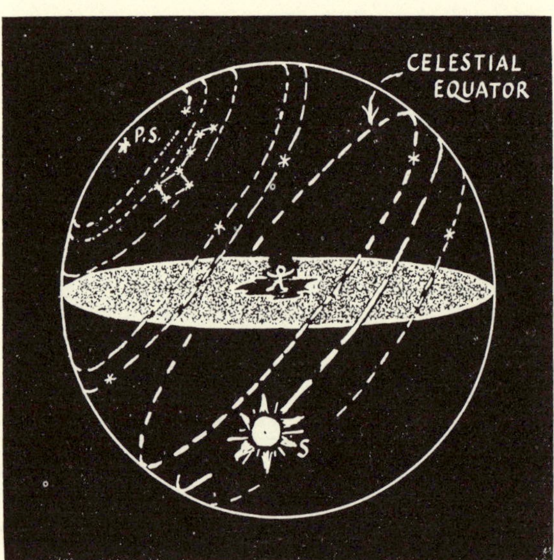

Fig. 13-4. The Star Pattern Revolves

Fig. 13-5. A Photo of the Sky Near Pole Star
Taken with an eight-hour exposure. The Pole Star itself made the very heavy trail near the center. Photo from Lick Observatory.

tern whirls steadily across the sky each night, as if carried by a rigid frame. One star, the pole-star, stays practically still while the rest of the pattern swings round it. Watch the stars for a few hours and you will see the pole-star, due North, staying still while the others move in circles round it. Or, point a camera at the sky with open lens, and let it take a picture of those circles. Night after night, year after year, the pattern rotates without noticeable change. These are the "fixed stars."[2] The pole-star is due North, in the N-S meridian plane of the noonday Sun. The star pattern revolves round it at an absolutely uniform rate. This motion of the stars gave early man a clock, and the pole-star was a clear North-pointing guide.[3]

The simplest "explanation" or descriptive scheme for the stars is that they are shining lights embedded in a great spinning bowl, and we are inside at the center. That occurred to man long ago, and you would feel it true if you watched the sky for many nights. It was a clever thinker who extended the bowl to a complete sphere, of which we see only half at a time. This is the celestial sphere, with its axis running through the pole-star and a celestial equator that is an extension of the Earth's equator.

[2] However, if you could live for many centuries you would notice changes in the shape of some constellations. Stars are moving.

[3] The procession of the equinoxes carries the Earth's spin-axis-direction slowly round in a cone among the stars, so only in some ages has it pointed to a bright star as pole-star. (See sketches in later chapters.) It is near to one now, our present pole-star, and was near to another when the pyramids were built. In A.D. 1000 there was no bright pole-star—perhaps that lack delayed the growth of navigation.

The celestial sphere revolves steadily, carrying all the stars, once in 24 hours. The Sun is too dazzling for us to see the stars by day, so we only see the stars that are in the celestial hemisphere above us at night, when the Sun is in the other hemisphere below. The Sun's daily path across the sky is near the celestial equator; but it wobbles above and below in the course of the year, 23½° N in summer, 23½° S in winter.

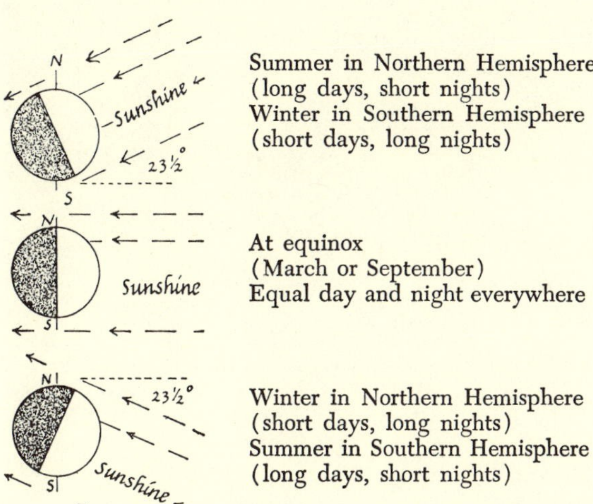

FIG. 13-6a. EARTH AND SUNSHINE
Day and night at various seasons

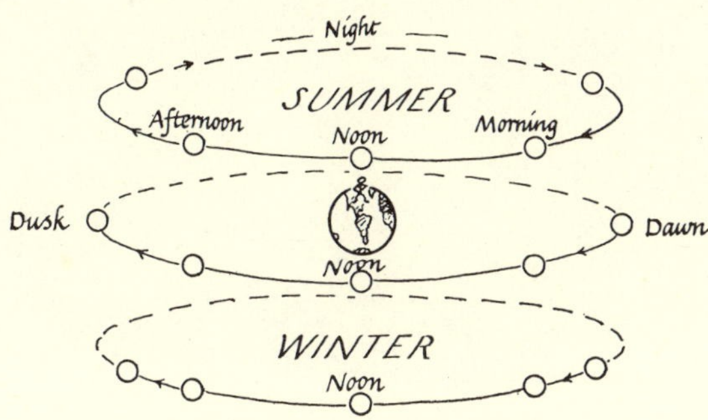

FIG. 13-6b. PATH OF SUN.
Viewed from stationary Earth, at various seasons. The Sun-positions are labeled noon, afternoon, etc., for observers in the longitude of New York. If such an observer could watch continuously, *unobstructed by the Earth*, he would see the Sun perform the "spiral of circles" sketched in FIG. 13-6c, during the course of 6 months from summer to winter; then he would see the Sun spiral upwards, revolving the same way, from winter to summer.

Though the pattern of stars has unchanging shape, we do not see it in the same position night after night. As the seasons run, the part of the pattern overhead at midnight shifts westward and a new part takes its place, a whole cycle taking a year. Stars that set an hour after sunset are 1° lower in

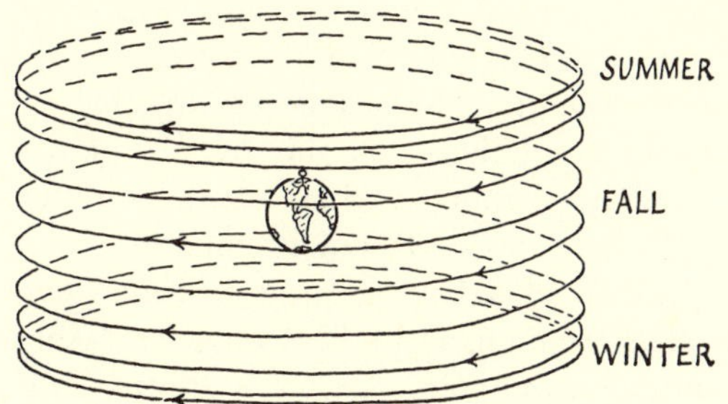

FIG. 13-6c. THE SUN'S SPIRAL OF CIRCLES,
in course of half a year of seasons.

the West next night and set a few minutes earlier; and two weeks later they are level with the Sun and set at sunset. Thus, in 24 hours the celestial sphere makes a little more than one revolution: 360° + about 1°. It is moving a little faster than the Sun, which makes one revolution, from due South to due South in the 24 hours from noon to noon. The celestial sphere of stars makes one extra complete revolution in a year.

Sun and Stars

This difference between the Sun's daily motion and that of the stars (really due to the Earth's mo-

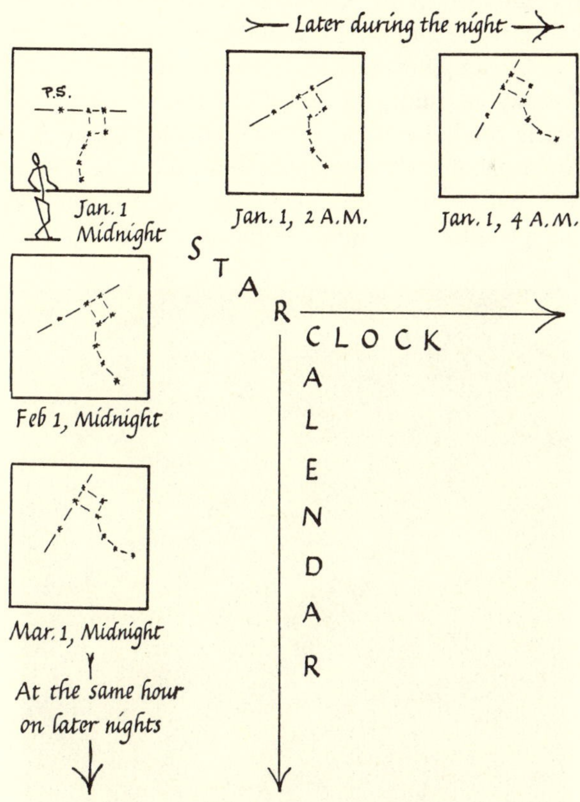

FIG. 13-7. THE STAR-PATTERNS
keep constant shape but revolve nightly and advance 30° per month, compared with Sun's noon and midnight.

tion in its orbit round the Sun) was obvious, and suggested that the Sun is moved by a separate agent. The Sun-god became a central figure in many primitive religions, and his travels were carefully traced by shadows and recorded by alignments of great stones in ceremonial temples.

Instead of saying the star pattern "gains" (like a clock running fast) 1° a day, we take the very constant star motion as our standard and say the Sun lags behind it 1° a day. We may stick the Sun as well as the stars on the inside of the celestial sphere; but since the Sun lags behind the stars it does not stay in a fixed place in the starry sphere; it crawls slowly *backwards* round the inside of the sphere, making a complete circuit in a year. Thus we can picture the Sun's motion compounded of a *daily motion shared with the stars* of the celestial sphere and a *yearly motion backward through the star pattern.*

Ecliptic and Zodiac

This is a sophisticated idea, a piece of scientific analysis: to separate out the Sun's yearly motion from its daily motion across the sky with the star-pattern. Once this idea was clear it was easy to map the Sun's yearly track among the stars—not directly, because the Sun outglares the nearby stars by day, but by simple reference to the pattern of stars in the sky at midnight instead of noon. The Sun's yearly track is not the celestial equator but a tilted circle making 23½° with the equator. It is this tilt that makes the Sun's *daily* path across the sky change with the seasons. At the equinoxes the Sun's yearly track crosses the equator. In summer, the tilted track has carried the daily path 23½° higher in the sky and in winter 23½° lower. This tilted yearly track is called the *ecliptic.*

As the Sun travels round the ecliptic in the course of the year, it passes through the same constellations of stars at the same season year after year. This broad belt of constellations containing the ecliptic is called the *zodiac.* The constellations in it were given special names long ago by the astrologer priests, a named group for each month in the year.

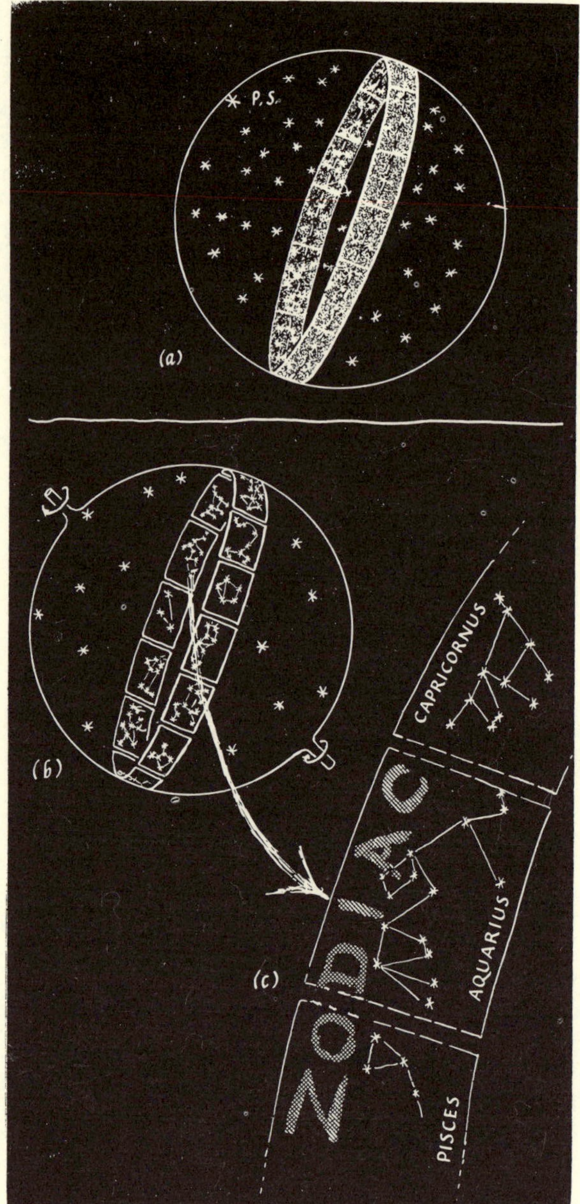

FIG. 13-9. THE ZODIAC, a belt of the celestial sphere, tilted 23½° from the equator. The Sun's yearly path (the ecliptic) runs along the middle line of this belt. The paths of Moon and planets lie within this belt. The Zodiac was divided into twelve sections named after prominent star-groups or constellations. (Zodiac patterns after H. A. Rey, *The Stars.*)

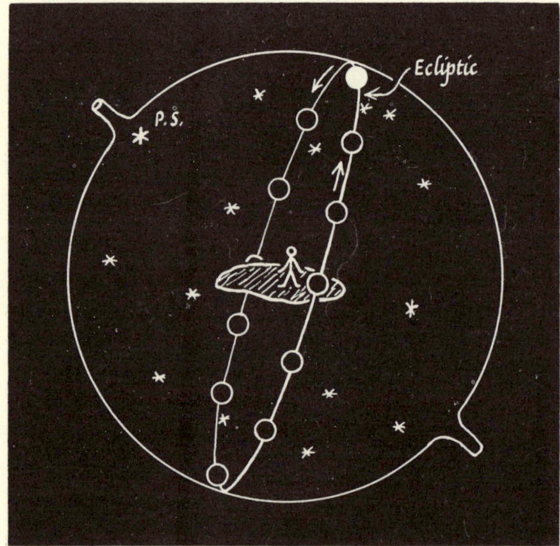

FIG. 13-8. THE ECLIPTIC, the Sun's track through the star pattern in the course of a year. Here the daily motion is imagined "frozen."

Moon

The Moon is obviously moving round the Earth and illuminated by sunlight. Watch it for a week or two; think where the Sun is each time and see if it accounts for the Moon's lighting. The Moon swings across the sky with the neighboring stars, but even in one night it lags noticeably behind the stars. It lags much quicker than the Sun, 90° in a week; right round the star-pattern in a month. Its monthly track is tilted about 5° from the ecliptic, but it is still within the broad band of the zodiac.

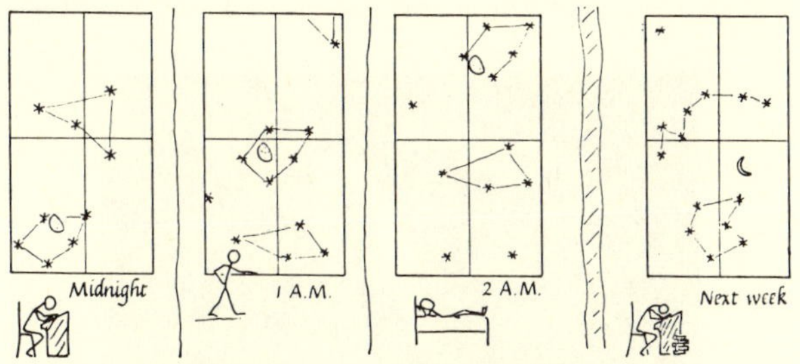

FIG. 13-10. THE MOON'S MOTION
The Moon, while moving across the sky with the stars each night, slips rapidly backward through the star-pattern, a whole circle in a month.

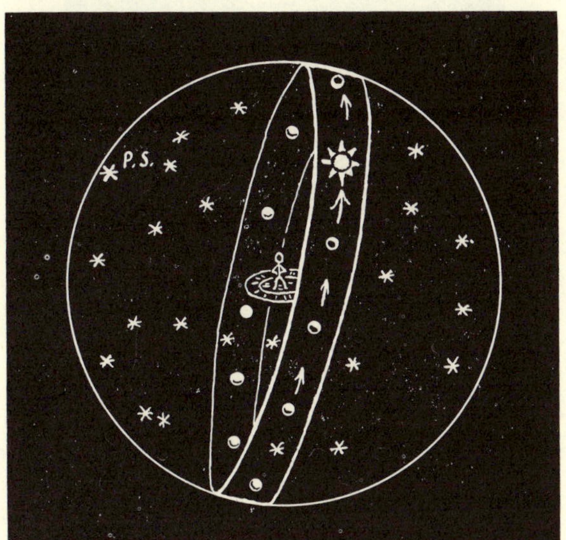

FIG. 13-11. ZODIAC BELT WITH POSITIONS OF MOON, IN VARIOUS PHASES, IN THE COURSE OF A MONTH
The daily motion of the celestial sphere is "frozen" here.

Eclipses

Eclipses are impressive events. Something seems to take a bite out of the Sun or Moon. A total eclipse of the Sun is awe-inspiring even to educated people—daylight disappears and it grows strangely cold.

"In the priestly calendar lore, magic and genuine science were inextricably entangled. . . . As liaison officers to the celestial beings, the priests found it paid to encourage the belief that nature can be bought off with bribes like a big chief. One of their most powerful weapons was their ability to forecast eclipses. Eclipses were indisputable signs of divine disapproval, and divine disapproval provided a cogent justification for raising the divine income-tax. No practical utility other than the advancement of the priestly prestige and the wealth of the priesthood can account for the astonishingly painstaking attention paid to these phenomena."[4]

Later, men realized that eclipses are just shadows. When the Moon is eclipsed the Sun throws the Earth's shadow on the Moon. The Sun is eclipsed when the Moon *gets in the light* between us and

FIG. 13-12. ECLIPSES
NOT TO SCALE. See FIG. 14-19b, c.

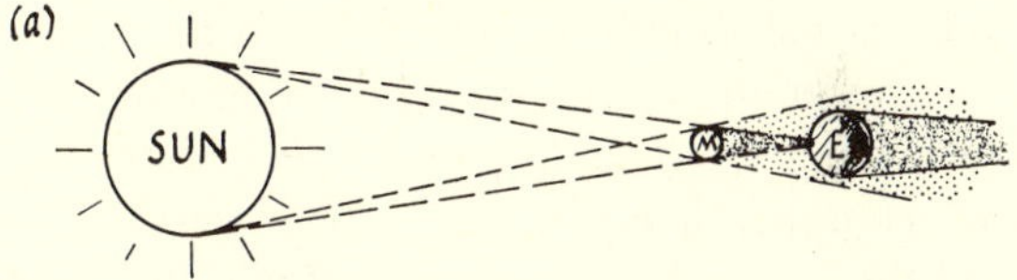

FIG. 13-12a. Eclipses of the Sun occur when the Moon gets in the light between Sun and Earth. The Moon's size and distance are such that total eclipses are just possible.

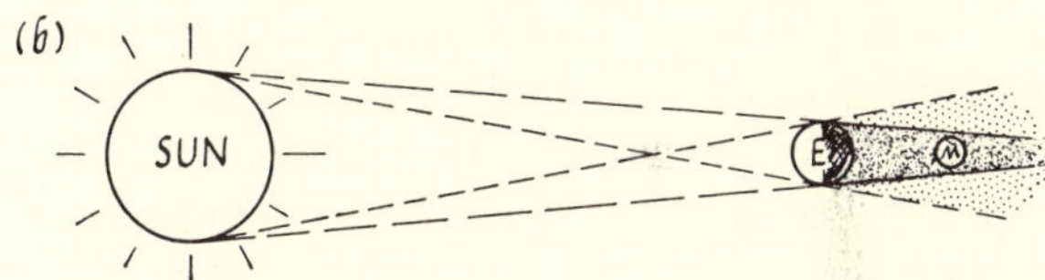

FIG. 13-12b. Eclipses of the Moon occur when the Moon passes through the Earth's shadow.

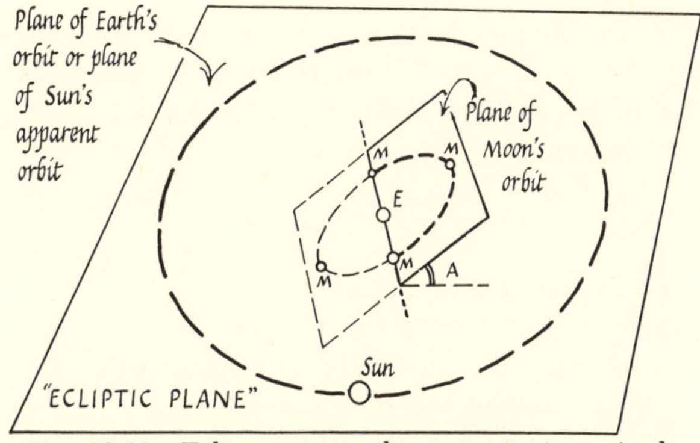

FIG. 13-12c. Eclipses occur only at certain times. Angle A is about 5°. However, as the line where the Moon's orbit-plane cuts the ecliptic plane slowly moves around —due to perturbing attractions—eclipses do not always occur at the same season of our year.

[4] Lancelot Hogben, *Science for the Citizen*, Allen and Unwin, Ltd., pp. 43-44.

the Sun, and we are in the Moon's shadow, which sweeps rapidly across (part of) the Earth.

For an eclipse, Sun, Moon, and Earth must be in line. There is a chance of this only when the Moon in its tilted orbit crosses the ecliptic plane, which by definition contains Sun and Earth—hence the name "ecliptic." Even then, the necessary alignment is rare. An eclipse of the Moon is a shadow *on the Moon*, and looks the same from everywhere on the Earth. So an eclipse of the Moon observed from different parts of the Earth occurs at different times *by the local clocks*—a proof that the Earth is round, not flat.

Calendar Periods

Day. The Sun's motion from noon to noon marks an almost constant day. However, it varies slightly; sundial noon runs ahead of a constant clock at some seasons and behind at others, sometimes by many minutes. The Sun's apparent motion along the ecliptic is not exactly regular through the year—it moves faster in winter—so its daily motion shows slight changes. (The Moon's motion shows even more complex irregularities.) The stars' motion round the celestial axis through the pole-star defines a constant, slightly shorter, day—man's ultimate standard of timekeeping until the perfection of electronic clocks.

Month. The Moon was probably the first calendar-marker for men. The month from full moon to full moon is about 29½ days. The full moon is exactly opposite the Sun, so that is the month judged relative to the Sun. In 29½ days the Sun travels almost 29° along the ecliptic, so the Moon makes more than one revolution, 360° + 29°, relative to the stars to catch up with the Sun. Relative to the stars as fixed markers, the Moon takes 27.3 days for one revolution. We use a 29½-day month, like the early-calendar makers, to predict full moon, new moon, etc.; but we shall use the 27.3-day period when we calculate the Moon's motion under gravity.

Year. The idea of a year grew up as:

(a) the repeat-time of the seasons
(b) the time the Sun takes to return to the same place among the stars (or the stars to return to the same midnight position in the sky). This differs slightly from (a).
(c) a period of 12 (or 13) moon-months. Easy to observe, such a year soon gets out of step with the solar year of seasons.

Planets

A few bright "stars" do change their positions, and move so unevenly compared with Sun, Moon, and the rest that they are called *planets*, meaning wanderers. These planets, which look like very bright stars, with less twinkling, wander across the sky in tracks of their own *near the ecliptic.* They

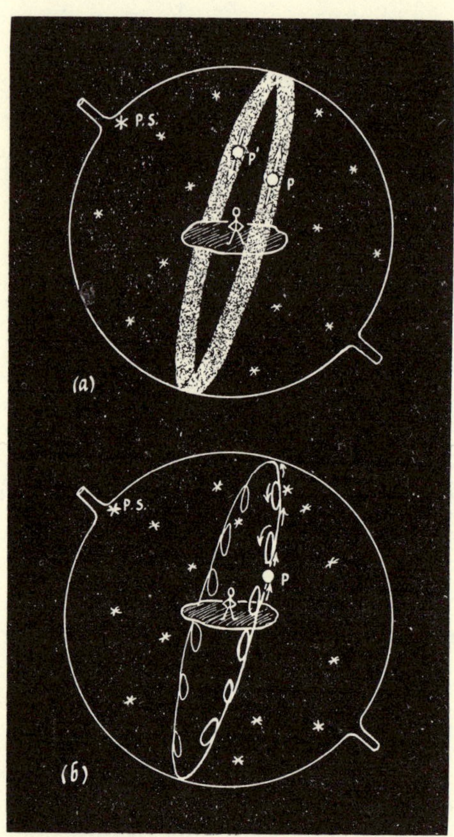

FIG. 13-13. THE PATH OF A PLANET
All the planets wander through the star pattern in a belt near the ecliptic—the zodiac belt.
(a) General region of a planet's path—the zodiac belt.
(b) In detail, a planet's path has loops—an epicycloid seen almost edge-on.

follow the general backward movement of the Sun and Moon through the constellations of the zodiac, but at different speeds and with occasional reverse motions. Primitive man must have observed the brighter planets but cannot have got any good use from his observations, unless like eclipses they were used to impress people.

Zodiac

Thus the zodiac belt includes the Sun's yearly path, the Moon's monthly path, and the wandering paths of all the planets. In modern terms, the orbits of Earth, Moon, and planets all lie near to the same plane. Astrology assigned fate and character by the

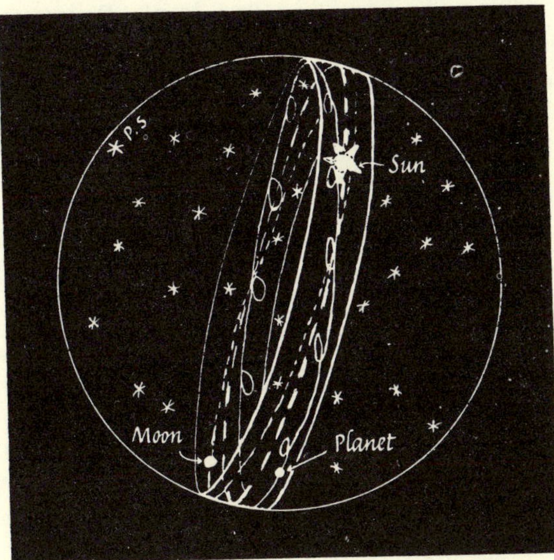

FIG. 13-14. ZODIAC BELT WITH PATHS OF SUN (in one year), MOON (in one month), and a specimen PLANET (in planet's "year"). The daily motion of the celestial sphere is "frozen" here.

places in the zodiac occupied by Sun, Moon, and planets at the time of a man's birth.

The Planets and Their Motions

Five wandering planets were known to the early civilizations, in addition to the Sun and Moon which were counted with them. These were:

Mercury and Venus, bright "stars" which never wander far from the Sun, but move to-and-fro in front of it or behind, so that they are seen only near dawn or sunset. Mercury is small and keeps very close to the Sun, so it is hard to locate. Venus is a great bright lamp in the evening or morning sky. It is called both the "evening star" and the "morning star"—the earliest astronomers did not realize that the two are the same.

Mars, a reddish "star" wandering in a looped track round the zodiac, taking about two years for a complete trip.

Jupiter, a very bright "star" wandering slowly round the ecliptic in a dozen years.

Saturn, a bright "star" wandering slowly round the ecliptic, in about thirty years.

Jupiter and Saturn make many loops in their track, almost one loop in each of our years.

When one of the *outer* planets, Mars, Jupiter or Saturn, makes a loop in its path it crawls slower and slower eastward among the stars, comes to a stop, crawls in reverse direction westward for a while, comes to a stop, then crawls eastward again, like the Sun and Moon.[5]

The sketches show the looped tracks of planets through the star patterns. Once noticed, planets presented an exciting problem to early scientists: what gives them this extraordinary motion? From now our main concern in this section on astronomy will be: how can we explain (produce, predict) the strange motions of the planets, which excited so much wonder and superstition? We study this to show how scientific theory is made.

Epicycloid

Nowadays we call the looped pattern of a planet's track an epicycloid (from the Greek for outer-circle) because we can imitate it by rolling a little circle round the circumference of a big one. Fig. 13-17 shows a scheme to manufacture an epicycloid which imitates a planet's track. A large wheel W spins steadily round a fixed axle. At some point A on its rim, there is an axle carrying a small wheel,

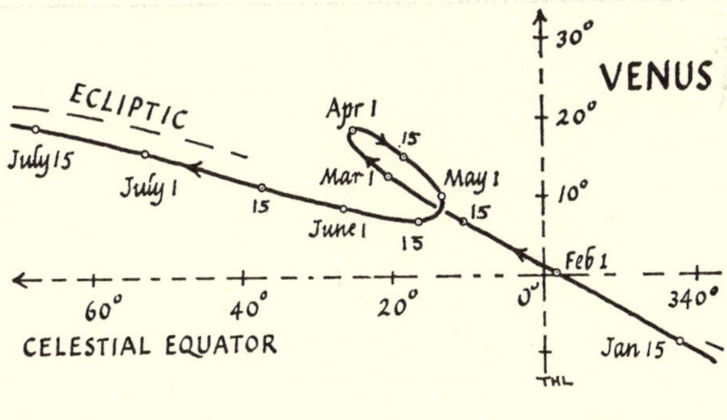

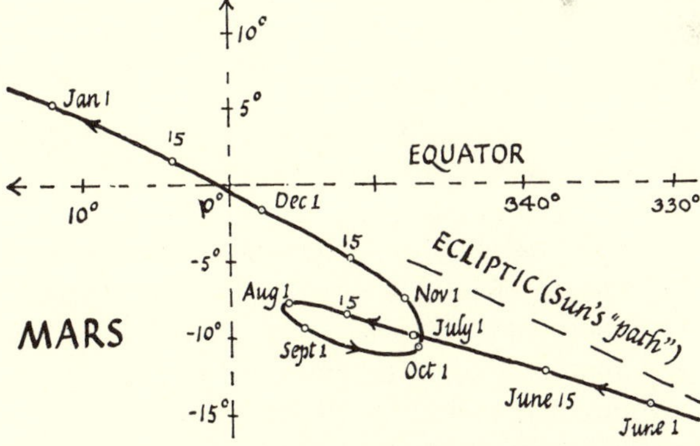

FIG. 13-15. PATHS OF PLANETS THROUGH THE STAR PATTERN
(a) Venus (January-July 1953)
(b) Mars (June-December 1956)
(The ecliptic is the Sun's apparent path. The planets' orbits run close to the ecliptic, because the planes of those orbits are close to the plane of the Earth's orbit (or the Sun's apparent orbit, the ecliptic).

[5] Remember that though the Sun sweeps from East to West in its daily motion with the stars, it crawls backward, or West to East, in its yearly motion round the ecliptic. So do planets.

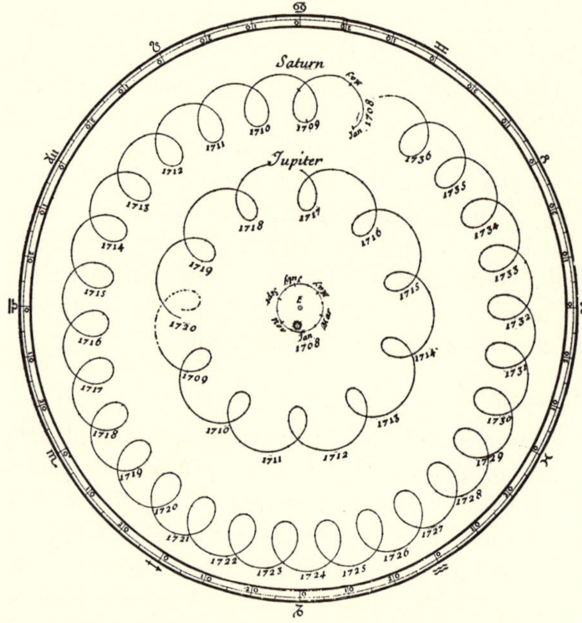

Fig. 13-16. Paths of Planets in the Sky.
This sketch shows the apparent paths of Jupiter and Saturn, plotted for many years, as they would appear to an observer attached to the Earth but viewing them from *far* out from the Earth, so that the epicycles are seen face-on, without the foreshortening really observed. The apparent orbit of the Sun is also shown. The Earth is at the center. When the astronomer Cassini constructed this diagram in 1709 he used Copernican measurements of orbit sizes.

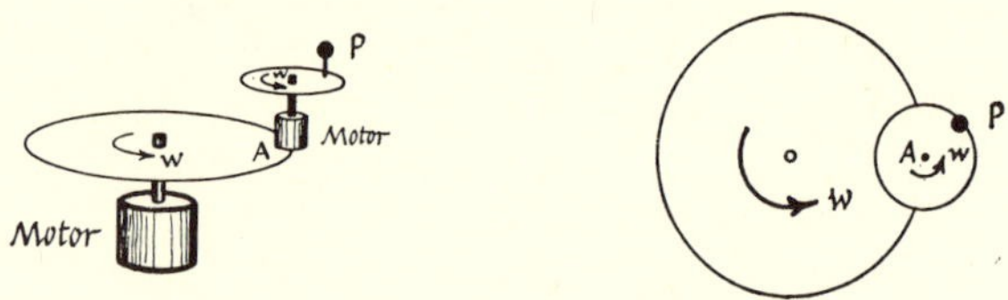

Fig. 13-17. Machine for Drawing Epicycloids

w, which spins steadily, much faster than W. Then a point P on the rim of the small wheel traces out an epicycloid. The planetary path we observe is like this epicycloid seen obliquely, as if the whole contraption of wheels were at eye level. (There is a strong hint in this model that a planet's apparent motion is compounded of two circular motions. The hint grows stronger when we find one of those motions is a yearly one for every planet, taking almost one of our years. The ancient astronomers did not take this hint.)

Observation

Many city-dwellers today take little notice of the sky, but to anyone living out of doors at night the planets are obvious strange bright things. Once you have seen them, you are not likely to miss them again. With even the simplest telescope or field glasses you can see surprising details: the crescent-shaped phases of Venus, Jupiter's moons, perhaps Saturn's rings. With a telescope the planets look bigger, while the fixed stars do not. This is because the planets are much nearer. The fixed stars are farther away and bigger, but so much farther than bigger that they look like points.[6]

Planets and Stars

We now know that these nearby wanderers, the planets, are things of much the same size as our Earth, and, like the Earth and the Moon, shine only by reflected sunlight. (As a test of this we now examine their light with a spectroscope and find it carries all the characteristic absorption-lines of sunlight.) The fixed stars, however, are lights in their own right. They are white-hot furnaces like our Sun (and the spectroscope tells us how they differ in composition and temperature).

Parallax

You and I know the Earth moves round the Sun, swinging 186,000,000 miles across its orbit in six months. Then we ought to see some changes in the pattern of stars when we make that huge move—parallax shifts as they are called. Try this *personal experiment on parallax.* Look at a group of people, or chairs or other objects at different distances. Walk to-and-fro sideways, or walk round in a circular orbit, and notice how the relative positions change in the group. Those in the remote background, seem to stay still, while nearer ones move in little orbits against the background, with a motion the reverse of yours. Such *parallax* shifts are used unconsciously by people in judging distances by a wagging head; and modern astronomers use them to judge the distance of the Moon, planets, and stars.

Even if the stars were all embedded in a single spherical bowl, the move across the Earth's orbit would bring us nearer some of the bowl and distort the pattern by foreshortening. The ancient astronomers saw no such changes and concluded that the Earth must be at rest at the center of the universe. The only alternative explanation was that the stars are infinitely far away compared with the diameter of the Earth's orbit. Nowadays very delicate telescopic measurements show that there *are* small parallax shifts, which place even the nearest stars

[6] Telescopes fail to show the real *size* of stars. Even the most powerful telescope would still show stars as points but for the spreading of light waves, which affects the images formed by all optical instruments and makes a small disc pattern for each star—the *bigger* the telescope, the *smaller* the disc.

at vast distances. Much easier measurements place the planets a million times closer. If we could measure distances by clocking a flash of light from each body to us, we should find that light takes 8 minutes from the Sun, a few minutes from the nearer planets and a few hours from the farthest, while from the nearest fixed star it takes several years.

Early Progress

Early astronomy, then, had three driving motives:
(a) practical uses: for compass, clock, and calendar
(b) magic to impress people; and astrology to predict fates and good and bad luck—such beliefs made many a king in later ages support a good astronomer
(c) pure scientific interest: as men grew up, there were scientists like those today whose delight in nature and interest in understanding are the driving forces.

Astronomy in the Early Civilizations

[It is difficult to tell who made the great discoveries because these were probably made in stages and then spread slowly, being re-discovered and claimed many times. Therefore, the notes which follow are not reliable and are offered only as general indications. They cover the development of astronomy from the age when it mattered for corn and animals to the stage when it had taken its place as a science. Here and later, we give short notes instead of a developed account.]

Urban civilizations developed in several great river valleys 5000 or more years ago. Much "applied science" had already been discovered—a few thousand years before then—such as: artificial irrigation of crops by canals and ditches; plow, sailboat, wheeled vehicles; use of animals for motive power; production and use of copper, bricks, glazes; and, finally, a solar calendar, writing, a number system, and the use of bronze.[7]

Sumerians, Babylonians, and Chaldeans (Three separate peoples in Mesopotamia)

By 4000 years ago (2000 B.C.) there were flourishing towns with extensive trade. They had excellent commercial arithmetic that was almost algebra: they could solve problems leading to quadratics, even cubics; they knew value of $\sqrt{2}$ accurately, but took π as roughly $= 3$; and they used similar triangles and knew Pythagoras' rule. They had good weights and measures, sundials, and water-clocks. Records and teaching texts were stamped in clay tablets, which we can now decipher.

Their astronomical observations were not the wonders often claimed, but made a good working basis for calendars. Near the equator the Sun's daily path does not provide an obvious calendar, as it does farther North. The Moon is much easier. So the Babylonians based their calendar on new Moons; but they had to reduce that to a solar calendar of seasons for agriculture and seasonal ceremonies. This required careful observations of Sun and Moon. Positions were mapped with reference to the zodiac, which was divided into 12 sections. Stars were catalogued, eclipses recorded, planets observed, and the returns of the planet Venus specially studied.

A thousand years later, the Babylonians developed a marvelous mathematical system for predicting the motions of Sun and Moon accurately. This consisted, essentially, of rules for calculating, like zigzag time-graphs of the uneven motions. These were empirical, with no theoretical basis, but they maintained an accurate calendar and could even predict probable eclipses. A similar scheme, with rougher interpolation, gave positions of planets. Belief in omens (prophetic signs) flourished, and astrology took a strong hold.

Egyptians

More than 4000 years ago, the Egyptians were living a comfortable life—with more friendly gods—as the Nile floods renewed the soil's fertility every year. Their mathematicians served magic and commerce. Their papyrus texts dealt with measuring corn stores, dividing property, building an exact pyramid. Their magnificent building projects necessitated good mathematics for the organization of the work and the management of armies of workmen. They had good weights and measures and ingenious water-clocks.

Their astronomy was simpler than the Babylonians'. They had an efficient solar year of 12 months of 30 days + 5 extra days; so they paid less attention to eclipses, the Moon, and planets. The Sun-god was supreme in the state religion. Two thousand years later they recorded accurate planet observations, probably for astrology.

Greeks

Some 3000 years ago, Greek civilization began to evolve. It produced mathematicians, scientists, philosophers, who made such important advances that we shall spend a chapter on them—even so,

[7] See V. Gordon Childe, *Man Makes Himself* (Mentor Books, New York, 1951).

the choice of the few names to be mentioned is unfair.

PROBLEMS FOR CHAPTER 13

1. Sketch the relative positions of Sun, Earth, Moon at the stages (a-e) asked for below. (In your sketches you cannot do justice to the real proportions of sizes and distances; but you should not show the Earth as big as the Sun or as small as the Moon.) The real measurements are:

Sun:	distance from Earth	≈	93,000,000 miles
	personal diameter	≈	860,000 miles
Earth:	diameter	≈	8,000 miles
Moon:	distance from Earth	≈	240,000 miles
	personal diameter	≈	2,000 miles

(a) at full moon; (b) at "new moon"; (c) at half moon; (d) at a total eclipse of the Sun; (e) at an eclipse of the Moon.

2. During the winter, in the Northern Hemisphere, the Earth is actually a little *nearer* the Sun than in the summer. Why then is the winter colder?

★ 3. APPARENT MOTIONS

Contrast prevents us seeing the stars in daytime. Suppose we could see them in daytime: then we should see some pattern of stars near the Sun.

(a) Suppose we could note that pattern at noon in June. When should we see the same star pattern in the same position in the *midnight* sky?
(b) What path should we see the Sun take relative to the unchanging patterns of stars, from month to month (disregarding *daily* motion of stars, etc.)?
(c) Describe the path of an "outer" planet such as Jupiter or Mars relative to the unchanging patterns of stars (disregarding *daily* motion of stars, etc.)?
(d) Describe the path of an "inner" planet such as Venus, relative to the stars.

★ 4. What are equinoxes? When do they occur?

5. FINDING LATITUDE AND LONGITUDE

(a) State rough values for the latitude of: New York, San Francisco, London, North Pole, Arctic Circle, equator.
(b) State rough values for the longitude of New York, San Francisco, London, Tokyo.
(c) Suppose you are making an exploration in a small boat and are wrecked on an unknown desert island, far off your course. You wish to find your position, but have no radio or other modern electronic equipment, and no special instrument such as a sextant. You do have a simple stick with markers for sighting stars etc. and a plumb line and a protractor for measuring angles. Explain how you would estimate the following: (Say what measurements you would make and how you would treat them. Give a practical explanation that an untrained sailor could use—avoid technical phrases such as "obtain a fix").
  (i) your latitude by observing star(s) on a clear night
  (ii) your latitude by observing the Sun
  (iii) your longitude by observing Sun or stars. (For this a certain auxiliary instrument is *essential*. What instrument?)
(d) (i) How could accurate *predictions* of eclipses of the Moon help in a rough determination of longitude in remote places?
  (ii) Why was this use of eclipses seriously considered in ancient times?

How great love is, presence best tryall makes,
But absence tryes how long this love will bee;
To take a latitude,
Sun, or starres, are fitliest view'd
At their brightest, but to conclude
Of longitudes, what other way have wee,
But to marke when, and where the darke
eclipses bee?

—John Donne (~ 1600)

CHAPTER 14 · GREEK ASTRONOMY: GREAT THEORIES AND GREAT OBSERVATIONS

"If science is more than an accumulation of facts; if it is not simply positive knowledge, but systematized positive knowledge; if it is not simply unguided analysis and haphazard empiricism, but synthesis; if it is not simply a passive recording, but constructive activity; then, undoubtedly [ancient Greece] was its cradle."

—GEORGE SARTON*

Theory, a House for the Facts, "To Save the Phenomena"

Astronomical knowledge grew up with the early civilizations from simple noticing to systematic observing that provided an official priesthood of astronomers with material for calendar-making on one hand, and on the other a growing tangle of superstitious astrology. With this knowledge came stories to teach children or reassure simple folk. Describing the Sun as a god, worshipping the planet Venus, telling of the "abode of the blessed" above a crystal globe of stars: these were not merely superstitious myths, they were the forerunners, too, of theoretical science. They were not real science: their relationship with fact was thin and fanciful; but they set the pattern of a speculative scheme to "explain" the facts. When Greek civilization formed out of neighboring groups, the wisest thinkers brought a new attitude to science: they sought *general* schemes of explanation that would appeal to the inquiring mind, not simple myths to satisfy public curiosity. Their aim, as they put it, was to "save the phenomena," or save the appearances, meaning to make a scheme that would *account for the facts.* This was a grander business than either collecting facts or telling a new tale for each fact. This was an intellectual advance, the beginning of great scientific theory.

The earliest of the Greek "natural philosophers" gave simple pictures of the structure of the Universe, but as more information was gathered and intellectual tradition grew they evolved schemes to save the phenomena in detail: first simple tales to tell about the Earth, then fuller ones to explain the motion of the sky as a whole and the detailed motions of the Sun, Moon and planets.

At each stage, these philosophers tried to start with a few simple assumptions or general principles and draw from them as logically as possible a complete "explanation"—or setting forth—of the observed behavior. This explanation would serve to coordinate the information and to make future predictions, but above all to give a feeling that there is a *pattern* that holds diverse behavior together, that *nature makes sense.* Although some of the search for a good scheme was prompted by practical needs such as calendar-making, this delight in a unified clear explanation went far beyond that. Driven by an urge to ask WHY, the Greek philosophers were seeking and making scientific theory. Though our modern tradition of experimenting and our modern wealth of scientific tools have made great changes, we still hold the Greek delight in a theory that will save the phenomena.

This chapter gives an account of some of the Greek scientists. Watch how they built their theory.

Early Greek Astronomy

As Greek civilization developed, some 3000 years ago, the poets (Homer) told the history of earlier neighbors and tried to answer some of the great WHY questions that intelligent people were asking about mankind and about the world. The Earth was pictured as an island surrounded by a great river and covered by a huge bell of the heavens. The home of the gods was at the "ends of the Earth." Hell or the land of the dead was also at the ends of the Earth, or perhaps underneath. A daily Sun rose from the surrounding river and swept over the vault above.

By about 2500 years ago, we hear of great "natural philosophers" telling fuller stories with clearer thinking.

Thales (~ 600 B.C.) was a founder of Greek science and philosophy. In later centuries his reputation as one of the "seven wise men" grew so

* *Introduction to the History of Science* (1927), Vol. 1, page 8, Carnegie Institute of Washington.

fabulous that impossible feats were ascribed to him, such as predicting a solar eclipse. He collected geometrical knowledge, perhaps from Egypt, and began to reduce geometry to a system of principles and deduction—the beginning of a science that Euclid was to bring to full flower. He thought the Earth was a flat disc floating on water; yet he knew the Moon shines by reflected sunlight, so he had applied reasoning to common observations. He is said to have known that lodestone, native rock magnet, will attract pieces of iron; and he is rumored to have discovered electricity by rubbing amber (= "electron" in Greek). Moreover, he went beyond these bits of knowledge and set forth a general explanation of the Universe: that water is

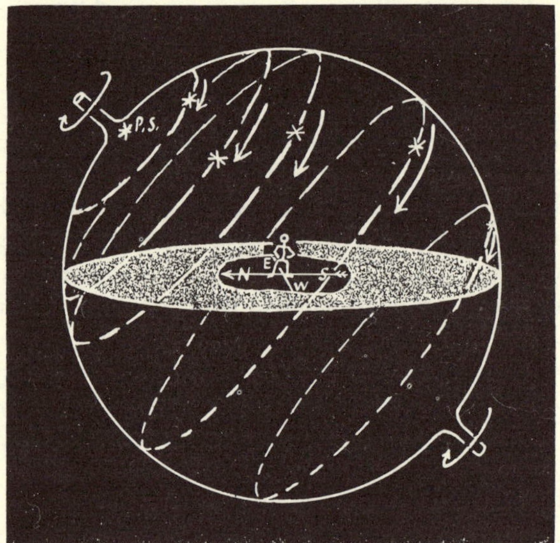

Fig. 14-1. The Universe According to Thales

the "first principle," the basic material of which all else is made. This was a bold beginning in "natural philosophy." He was a man of science, who *assumed that the whole universe could be explained by ordinary knowledge and reasoning.*

Others described the stars as set in a rotating sphere and discovered the obliquity of the ecliptic, the slant of the Sun's yearly path among the stars. This sorting out of the Sun's yearly motion from its daily motion was a useful step. The belt of star patterns along the Sun's yearly path came to be divided into twelve equal sections, the "signs of the zodiac," each named for a constellation. The Moon's path and the planets' paths are very near the Sun's, so they too travel through the signs of the zodiac.

Pythagoras ($\sim$ 530 B.C.). By the time Pythagoras established his school of philosophy—religion, sci-

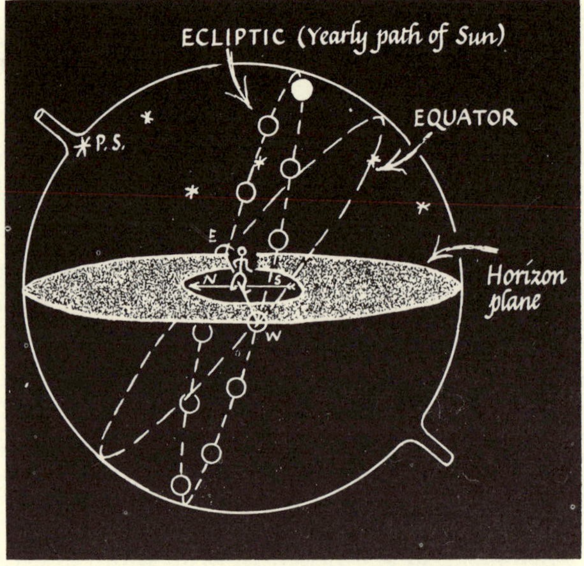

Fig. 14-2. Early Greek View
The Sun's yearly path through the star patterns was mapped. This is the tilted band called the ecliptic. The Sun is shown in one position (near mid-summer) and other positions are sketched. Here the celestial sphere is not spinning, but "frozen" with one star pattern overhead.

ence, politics . . . —the time was ripe for the idea of a round Earth. Travellers' tales of ships and stars would suggest a curved Earth to an inquiring mind. Yet the picture of the Earth as a round ball is hard to believe. You accept it easily because you were indoctrinated when very young—but watch a child first learning about the antipodes where the people

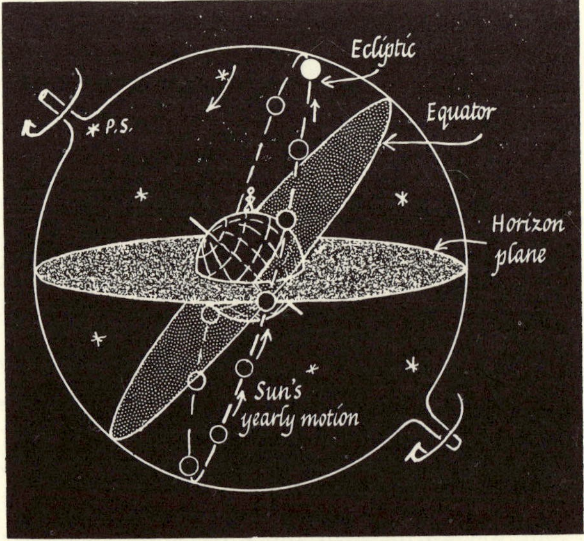

Fig. 14-3. Pythagorean View
The Pythagorean school adopted spherical Earth; and separated the general daily motion of stars, Sun, Moon, and planets, from the slow, backward motion of Sun, etc., through the star pattern.

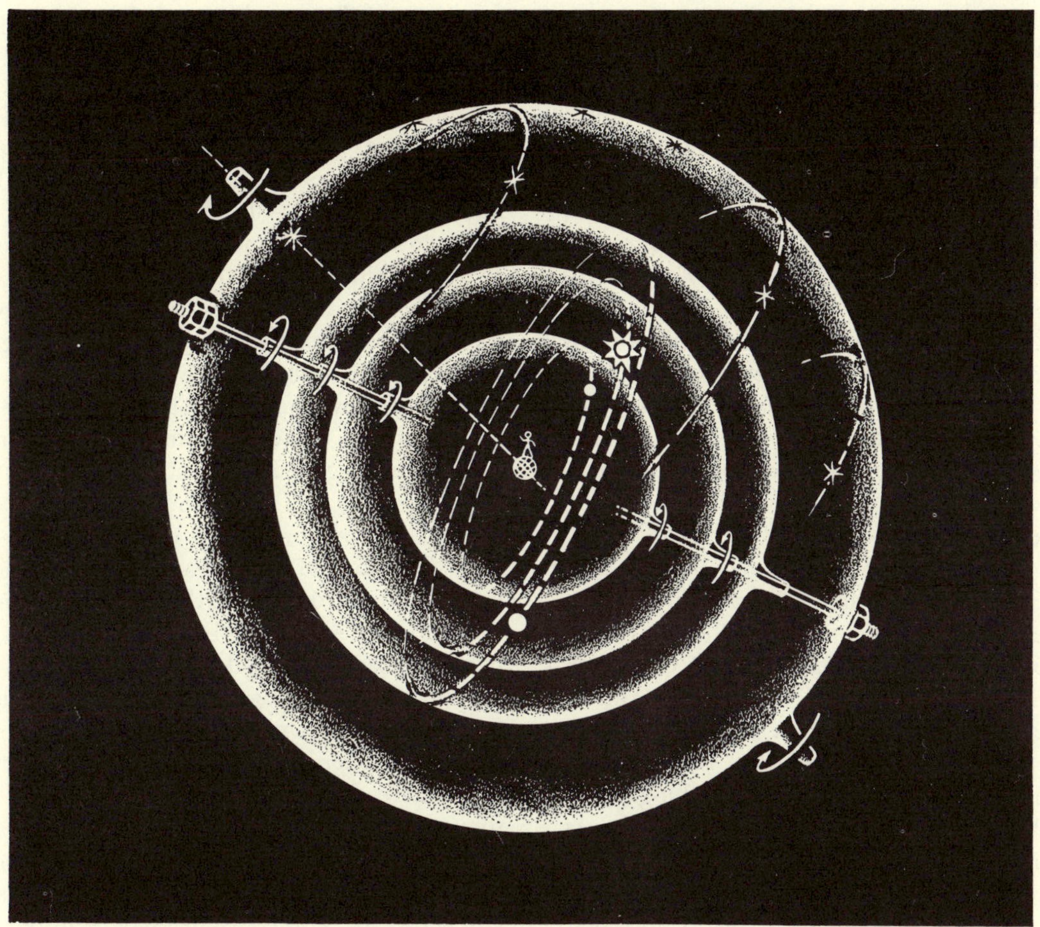

Early Greek System of Crystal Spheres. (Pythagoras)

Fig. 14-4a. Part of the system, showing the rotating spheres of the Sun and two planets, carried around by the outer sphere of stars which spins daily.

are "upside down"! Pythagoras himself probably taught that the Earth is round; but we do not know whether most of the Pythagorean discoveries and views were his or those of later members of his school, which flourished for some 200 years. For the heavenly system, they pictured a round Earth inhabited all over, surrounded by concentric transparent spheres each carrying a heavenly body. The innermost sphere carried the Moon, obviously closer to the Earth than the rest. The outermost sphere carried the stars, and the intermediate ones carried Mercury, Venus, the Sun, Mars, Jupiter, and Saturn. The outer sphere of stars rotated once in a day and a night; the others ran slightly slower to show the lagging course of Sun, Moon, and planets. Here was a simple scientific theory, a conceptual scheme of rotating spheres that was simple (plain spheres, steady spins) and which could claim to be based on a simple general principle (spheres are the "perfect" shape and uniform rotation is the "perfect"

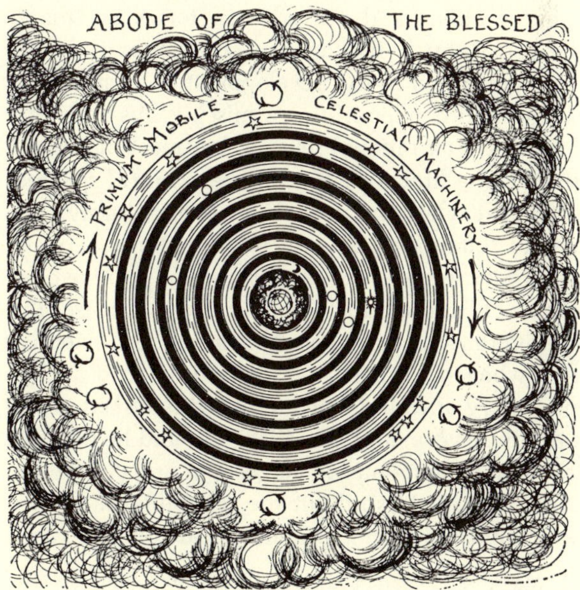

Fig. 14-4b. Early Greek System of Crystal Spheres
A "section" of the whole system in the ecliptic plane.

motion for a sphere). The spheres carrying planets were arranged in order of their spinning speeds: Saturn, moving almost as fast as the stars—lagging only one revolution in thirty years—was placed just inside the stars; then Jupiter, Mars, Sun; with Venus and Mercury just inside or outside the Sun's sphere. This arranging by speeds was a lucky guess. We now know Saturn, Jupiter, and Mars are the "outer" planets, farther from the Sun than the Earth, with Saturn outermost, Mars nearest.

Some members of the school realized that a common 24-hour rotation could be separated, so they made the outer sphere of stars carry all the inner spheres with it. Then the inner ones had to spin slowly *backwards* within the outermost, thus carrying Sun, Moon, and planets backwards through the zodiac band of stars. Each inner sphere had its appropriate rate, once round in a year for the Sun, once in a month for the Moon, . . . once in 12 years for Jupiter. . . .

Pythagoras made discoveries in geometry. Though his "square on the hypotenuse" theorem was known long before him, he showed how to derive it. And he developed a theory of numbers. He preached that "numbers are the very essence of things," the basis of all natural knowledge; and his school was much concerned with the arithmetical properties of numbers and their use in science. He thus attached to the study of numbers mystical values that have appealed to thinking men from long before him till long after. Among primitive men, superstition gave lucky and unlucky numbers magic powers, and to this day reputable scientists discuss the structures of atoms and universes in terms of "magic numbers."[1] Such Pythagorean mysticism turns up again and again in the development of science. The hardheaded condemn it as a mischievous rock that can shipwreck rational science, but most of us welcome it as a lifebuoy that can keep fruitful speculation afloat when the way ahead seems stormy. The layman today finds it hard to distinguish between useful mysticism—such as dreams of a positive electron or of "anti-matter"— and cranky nonsense. But the difference is sharp enough: the modern scientist, even when he is being most mystical, uses a clear vocabulary of well-defined terms with agreed meaning between him and his colleagues; and he not only draws on experiment for suggestions and checks but insists on critical study of the reliability of the experimental evidence. The crank can quote experiments to suit his purpose but fails to win confidence by his biased choice. There is, in fact, a corporate sanity among scientists that guides thinking in wise channels without restricting fruitful imagination.

Pythagoras was a sane scientist. In developing the science of music—a fine field in which to look for number properties—he assigned simple number ratios to musical harmonies. We keep these today: to be in perfect tune, two notes an octave apart must have vibration-frequencies that are as 2:1; two notes a musical fifth apart as 3:2. If different lengths of a harp string are chosen to give these harmonic intervals, they show the same proportions: the lengths are 2:1 for an octave, 3:2 for a fifth. Other simple ratios like 4:3 give a pleasant chord, but outlandish ones like 4.32 : 3.17 make an ugly dissonance in our ears trained by generations of the classical musical scale. This idea of ruling harmonic proportions was extended to astronomy. Pythagoras' pupils imagined the planetary spheres to be arranged by musical intervals: sizes and speeds had to fit simple number proportions. In rolling round with appropriate motion, each sphere made a musical note. The whole system of spheres made a harmony, "the music of the spheres," unheard by ordinary men, though some held that the master Pythagoras was privileged to hear it. Even this fanciful scheme was hardly unscientific, for its time. There was an utter absence of data; the distances of Sun and planets were unknown, and there was no prospect of measuring them; so the celestial harmonies merely added zest to thinking. A historian eight centuries later, wrote romantically: "Pythagoras maintained that the universe sings and is constructed in accordance with harmony; and he was the first to reduce the motions of the seven heavenly bodies to rhythm and song."[2]

Philolaus. The Sun, Moon, Venus, Mercury, Mars, Jupiter, Saturn—the seven planets as the Greeks listed them—all travel slowly *across the stars* from West to East. The star pattern carries the whole lot daily from East to West. This unfortunate reversal that spoiled the simplicity could be removed if the central Earth revolved instead of the stars; then all would revolve the same way. Philolaus, a pupil of Pythagoras, recorded such a view: instead of the Earth being the center of the Universe, there is a central fire—"the watchtower of the gods"—and the Earth swings round this fire daily in a small orbit, its inhabited part always facing outward away from the fire. This daily motion of the Earth would account for the daily motion of

[1] "Magic numbers" in nuclear physics are useful, so they are respectable. For an elementary modern example of almost meaningless number-witchcraft see the account of the Bohr atom in some beginning Chemistry texts.

[2] Hippolytos, quoted by G. Sarton in *A History of Science* (Harvard University Press, 1952), p. 214.

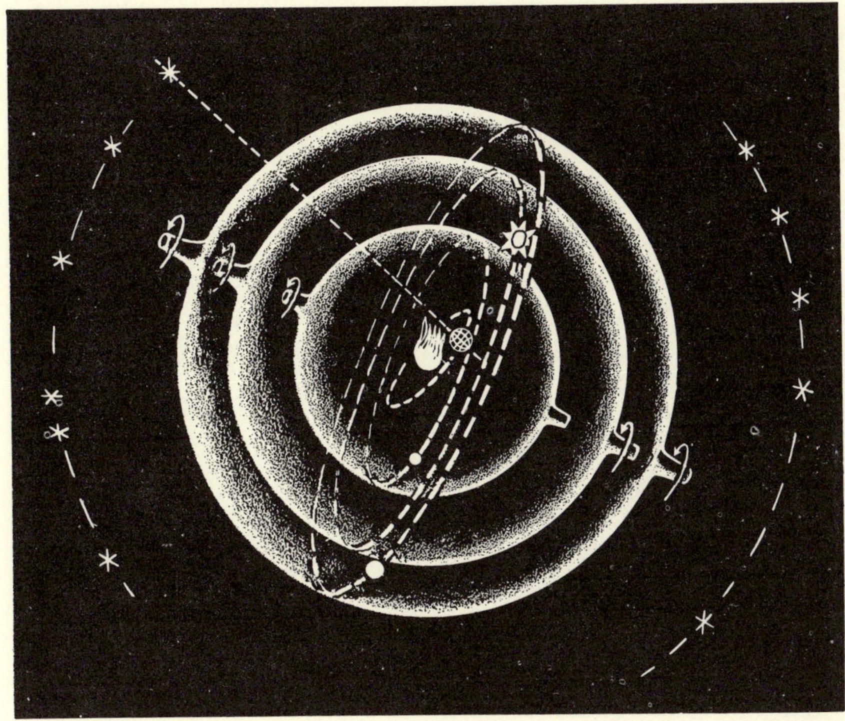

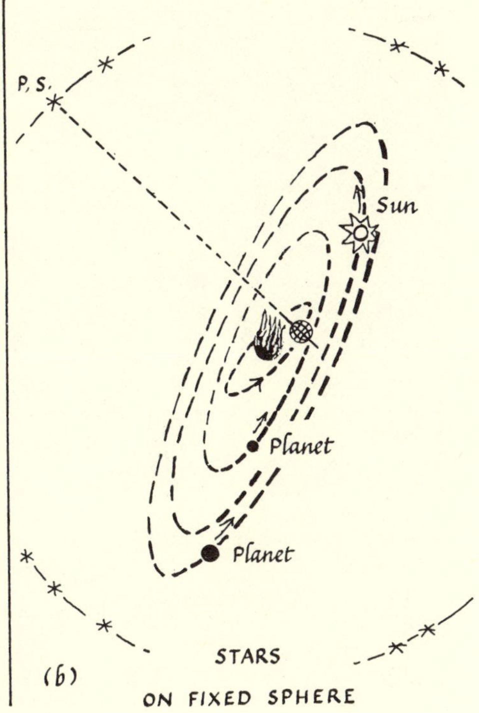

Fig. 14-5. Scheme of the Pythagorean Philolaus,
who pictured the Earth swinging around a central fire once in twenty-four hours.
This accounted for the *daily* motion of stars, Sun, Moon and planets.
Then spheres spinning slowly in the *same* direction carried the Moon, Sun, planets.
(a) View of spheres. (b) Skeleton scheme of orbits.

the stars across the sky: the outermost crystal sphere could be at rest. (Some went further and imagined an extra planet interposed between the Earth and the central fire. That counter-earth would protect the antipodes from being scorched—or perhaps it just *was* the antipodes—and it brought the total number of heavenly bodies up to the sacred Pythagorean number of ten.)

This fantastic scheme was revolutionary: it treated the Earth as a planet instead of making it the divine center, and it pointed out that the rotation of the starry sphere could be transferred to a daily revolution of the Earth. It may have paved the way for later theories of a moving Earth, but it did not last long, and it never suggested that the *Sun* is at the center nor even that the Earth is simply spinning. That last simplifying idea was suggested soon after, but it too did not find favor.

The Pythagoreans knew the Earth is round. They based their belief on a simple principle (spheres are perfect) and on practical facts. And they described the motions of the heavenly bodies by a rough but simple scheme that could be called a *theory*, in contrast with the more accurate workaday *rules* that were developed in Babylon. As a machine for making predictions, this first Greek system of uniform spins was hopelessly inaccurate; but as a frame of knowledge it was indeed superior: it gave a feeling that the heavenly scheme of things makes sense.

Socrates (~ 430 B.C.). The great philosopher championed clear thinking with careful definitions, and condemned the astronomers for their wild conjectures. Thus he probably helped astronomy towards becoming an inductive science that extracts its picture from experimental observations.

At much the same time, two philosophers, *Democritus* and *Leucippus*, were constructing a theory of atoms to explain the properties of matter and even the structure of the world. It was unthinkable, they held, that matter could be chopped up into smaller and smaller pieces without limit. There must be tiny hard unsplittable atoms. Though they had no experimental evidence but only fanciful speculations, they managed to sketch a theory of fiery particles that looks sensible today.[3] They provided the idea

[3] It would be silly historical mysticism to call this Greek atomic theory, for all its modern flavor, a foreseeing of Dalton's atomic chemistry of 1800. It was not a scientific discovery 2000 years before its time. It was a great idea that had to wait 2000 years to direct the course of scientific thinking.

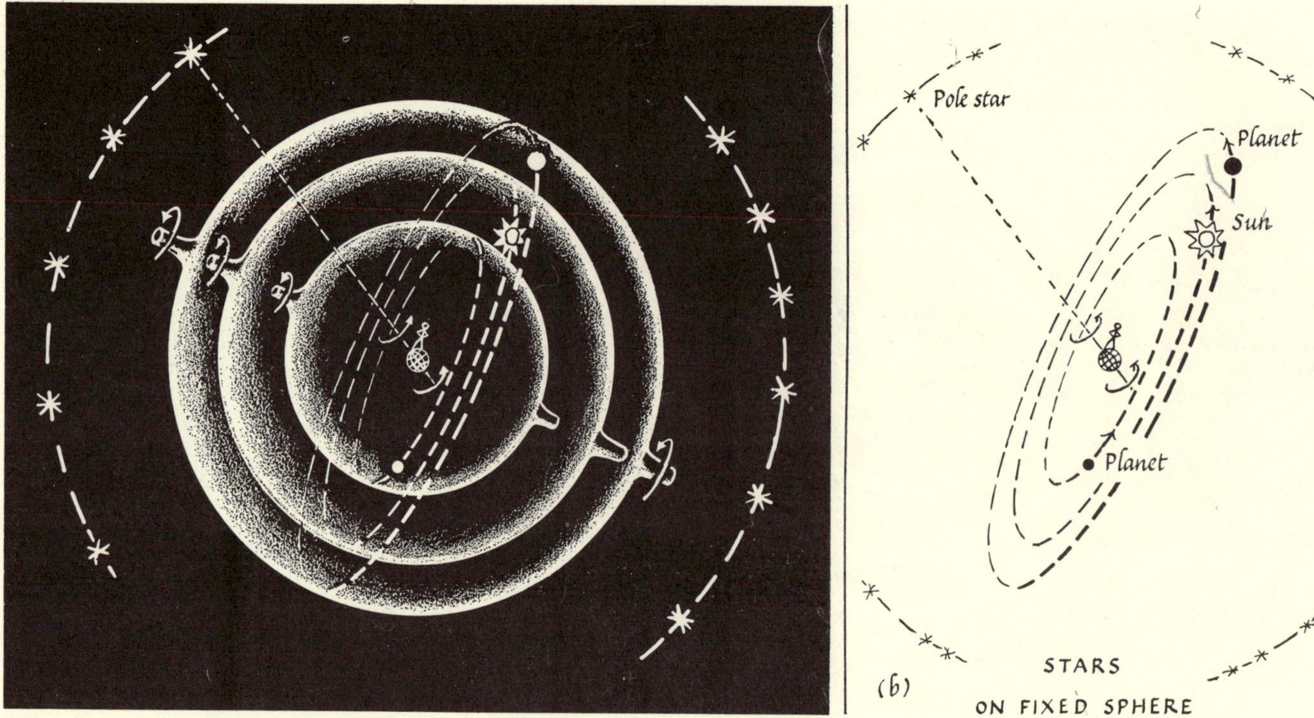

FIG. 14-6. LATER PYTHAGOREAN VIEW
The round Earth *spinning* accounted for the daily motion of stars, Sun, Moon and planets. Then spheres rotating slowly in the same direction carried the Moon around once in a month, the Sun around once in a year, and each planet around once in that planet's "year."
(a) View of spheres carrying Sun and two planets. (b) Skeleton scheme of orbits.

of atoms, to be mulled over, and occasionally used, through the centuries until chemical knowledge finally allowed an experimental atomic theory, in the last two hundred years. Their writings are lost, but the Latin poet Lucretius recorded their ideas in a magnificent poem two centuries later. He held that "reason liberates man from the terror of the gods"—a poetic version of the modern view "science cures superstition."

Though atomic theory did not concern astronomy directly, its insistence on a vacuum between atoms made it easier to think of empty space between heavenly bodies, or beyond them—contrary to the usual Greek view that space is limited and filled with an invisible æther.

Plato ($\sim$ 390 B.C.) was not much of an astronomer. He favored a simple scheme of spheres, and placed the planets in order of their speeds of revolution: Moon; Sun; Mercury and Venus travelling with the Sun; Mars; Jupiter; Saturn. The first great scheme that seemed to give a successful account of planetary motions came from Eudoxus, possibly at Plato's suggestion.

Eudoxus ($\sim$ 370 B.C.) studied geometry and philosophy under Plato, then travelled in Egypt, and returned to Greece to become a great mathematician and the founder of scientific astronomy. Gathering Greek and Egyptian knowledge of astronomy and adding better observations from contemporary Babylon, he devised a scheme that would save the phenomena.

The system of a few spheres, one for each moving body, was obviously inadequate. A planet does not move steadily along a circle among the stars. It moves faster and slower, and even stops and moves backwards at intervals. The Sun and Moon move with varying speeds along their yearly and monthly paths.[4] Eudoxus elaborated that scheme into a vast family of concentric spheres, like the shells of an onion. Each *planet* was given several adjacent spheres spinning about different axes, one within the next: 3 each for Sun and Moon, 4 each for the planets; and the usual outermost sphere of all for the stars. Each sphere was carried on an axle that ran in a hole in the next sphere outside it, and the axes of spin had different directions from one sphere to the next. The combined motions, with suitably chosen spins, imitated the observed facts. Here was a system that was simple in form (spheres) with a

[4] For example, the four seasons, from spring equinox to midsummer to autumn equinox, &c., are unequal. The Babylonians in their schemes for regulating the calendar by the new Moon had, essentially, time-graphs of the uneven motions of Moon and Sun.

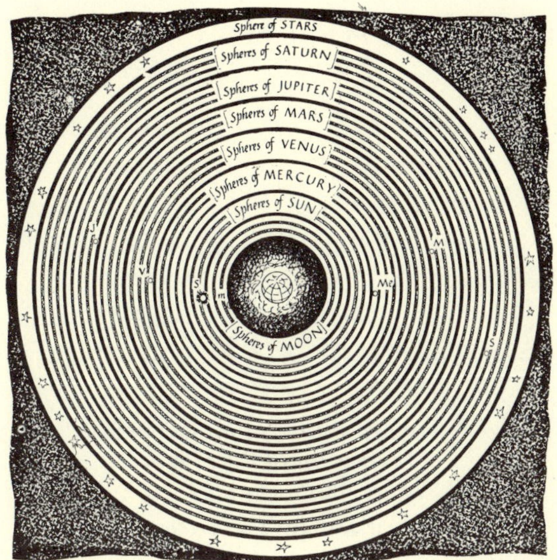

Fig. 14-7. Eudoxus' Scheme of Many Concentric Spheres
Each body, Sun, Moon or planet, had several spheres spinning steadily around different axes. The combination of these motions succeeded in imitating the actual motions of Sun, Moon and even planets across the star pattern.

simple principle (uniform spins), adjustable to fit the facts—by introducing more spheres if necessary. In fact, this was a good theory.

To make a good theory, we must have basic principles or assumptions that are simple; and we must be able to derive from them a scheme that fits the facts reasonably closely. Both the usefulness of a theory and our aesthetic delight in it depend on the simplicity of the principles as well as on the close fitting to facts. We also expect fruitfulness in making predictions, but that often comes with these two virtues of simplicity and accuracy. To the Greek mind, and to many a scientific mind today, a good theory is a simple one that can save all the phenomena with precision. Questions to ask, in judging a good theory, are, "Is it as simple as possible?" and "Does it save the phenomena as closely as possible?" If we also ask, "Is it *true*?", that is not quite the right requirement. We could give a remarkably *true* story of a planet's motion by just reciting its locations from day to day through the last 100 years; our account would be true, but so far from simple, so spineless, that we should call it just a list, not a theory.[5] The earlier Greek pictures with real crystal spheres had been like myths or tales for children—simple teaching from wise men for simple people. But Eudoxus tried to devise a successful machine that would express the actual motions and predict their future. He probably considered his spheres geometrical constructions, not real globes, so he had no difficulty in imagining several dozen of them spinning smoothly within each other. He gave no mechanism for maintaining the spins—one might picture them as driven by gods or merely imagined by mathematicians.

Here is how Eudoxus accounted for the motion of a planet, with four spheres. The planet itself is carried by the innermost, embedded at some place on the equator. The outermost of the four spins round a North-South axle once in 24 hours, to account for the planet's daily motion in common with the stars. The next inner sphere spins with its axle pivoted in the outermost sphere and tilted 23½° from the N-S direction, so that *its* equator is the ecliptic path of the Sun and planets. This sphere revolves in the planet's own "year" (the time the planet takes to travel round the zodiac), so its motion accounts for the planet's general motion through the star pattern.[6] These two spheres are equivalent to two spheres of the simple system, the outermost sphere of stars that carried all the inner ones with it, and the planet's own sphere. The third and fourth spheres have equal and opposite spins about axes inclined at a small angle to each other. The third sphere has its axle pivoted in the zodiac of the second, and the fourth carries the planet itself em-

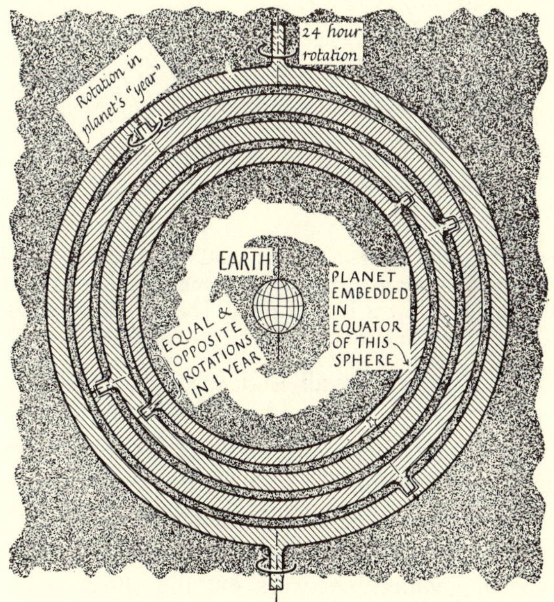

Fig. 14-8. Part of Eudoxus' Scheme:
Four Spheres to Imitate the Motion of a Planet
The sketch shows machinery for one planet.
The outermost sphere spins once in twenty-four hours.
The next inner sphere rotates once in the planet's "year." The two innermost spheres spin with equal and opposite motions, once in our year, to produce the planet's epicycloid loops.

[5] Young scientists are urged, nowadays, not to be satisfied with just collecting specimens, or facts or formulas, lest they get stuck at the pre-Greek stage.

[6] In terms of our view today, the spin of the outermost sphere corresponds to the Earth's daily rotation; the spin of the next sphere corresponds to the planet's own motion along its orbit round the Sun; the spins of the other two spheres combine to show the effect of viewing from the Earth which moves yearly around the Sun.

EVIDENCES FOR ROUND EARTH

ANCIENT

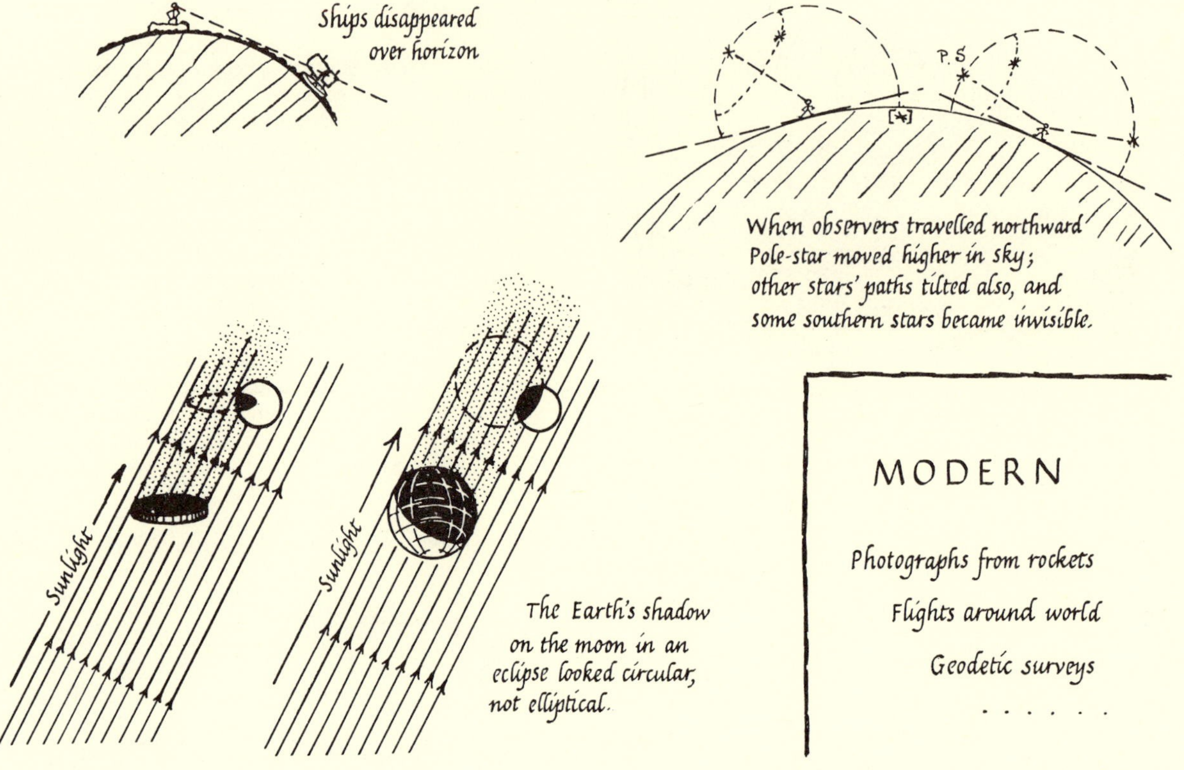

Fig. 14-9.

bedded in the equator. Their motions combine to add the irregular motion of stopping and backing to make the planet follow a looped path. The complete picture of this three dimensional motion is difficult to visualize.

With 27 spheres in all, Eudoxus had a system that imitated the observed motions quite well: he could save the phenomena. The basis of his scheme was simple: perfect spheres, all with the same center at the Earth, spinning with unchanging speeds. The mathematical work was far from simple: a masterpiece of geometry to work out the effect of four spinning motions for each planet and choose the speeds and axes so that the resultant motion fitted the facts. In a sense, Eudoxus used harmonic analysis—in a three-dimensional form!—two thousand years before Fourier. It was a good theory.

But not very good: Eudoxus knew there were discrepancies, and more accurate observations revealed further troubles. The obvious cure—add more spheres—was applied by his successors. One of his pupils, after consulting Aristotle, added 7 more spheres, greatly improving the agreement. For example, the changes in the Sun's motion that make the four seasons unequal were now predicted properly. Aristotle himself was worried because the complex motion made by one planet's quartet of spheres would be handed on, unwanted, to the next planet's quartet. So he inserted extra spheres to "unroll" the motion between one planet and the next, making 55 spheres in all. The system seems to have stayed in use for a century or more till a simpler geometrical scheme was devised. (An Italian enthusiast attempted to revive it 2000 years later, with 77 spheres.)

Aristotle (340 B.C.), the great teacher, philosopher, and encyclopedic scientist was the "last great speculative philosopher in ancient astronomy." He had a strong sense of religion and placed much of his belief in the existence of God on the glorious sight of the starry heavens. He delighted in astronomy and gave much thought to it. In supporting the scheme of concentric spinning spheres, he gave a dogmatic reason: *the sphere is the perfect solid shape*; and this prejudiced astronomical thinking about orbits for centuries. By the same token, the Sun, Moon, planets, stars must be spherical in form. The heavens, then, are the region of perfection, of unchangeable order and circular motions. The space

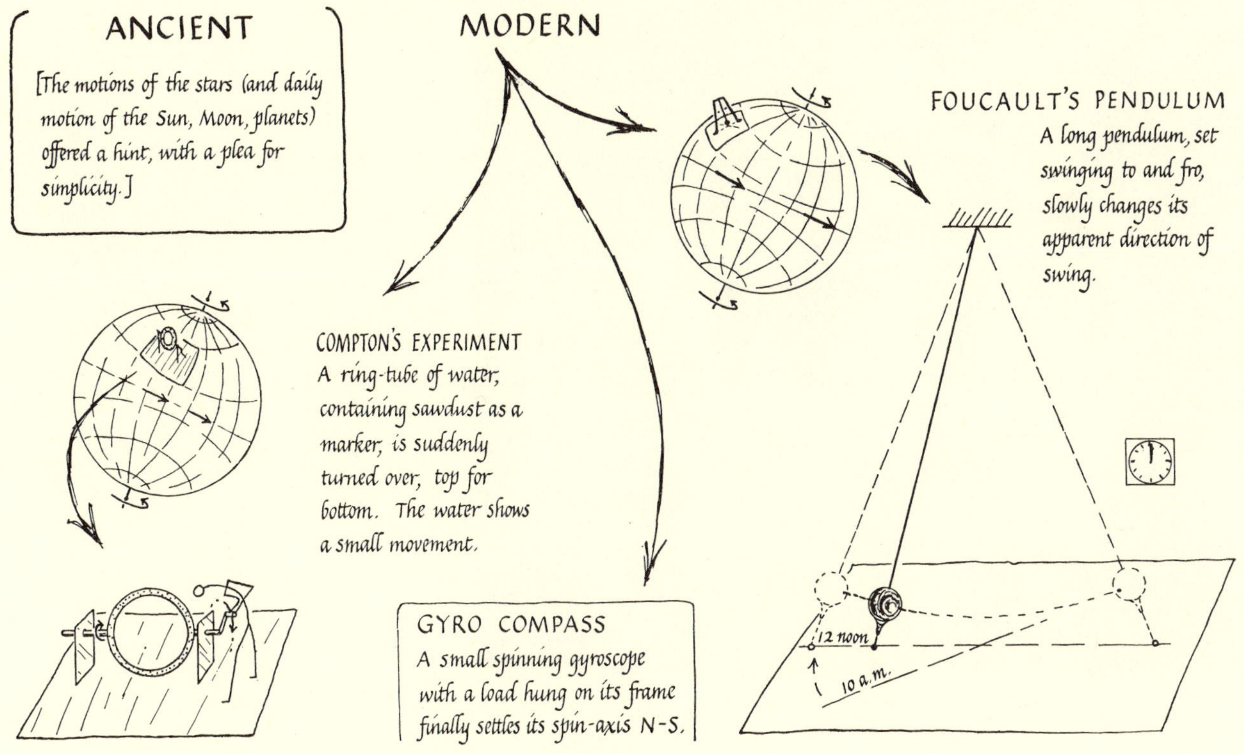

Fig. 14-10.

between Earth and Moon is unsettled and changeable, with vertical fall the natural motion.

For ages Aristotle's writings were the only attempt to systematize the whole of nature. They were translated from language to language, carried from Greece to Rome to Arabia, and back to Europe centuries later, to be copied and printed and studied and quoted as the authority. Long after the crystal spheres were discredited and replaced by eccentrics, those circles were spoken of as spheres; and the medieval schoolmen returned to crystal spheres in their short-sighted arguments, and believed them real. The distinction between the perfect heavens and the corruptible Earth remained so strong that Galileo, 2000 years later, caused great annoyance by showing mountains on the Moon and claiming the Moon was earthy; and even he, with his understanding of motion, found it hard to extend the mechanics of downward earthly fall to the circular motion of the heavenly bodies.

Aristotle made a strong case for the Earth itself being round. He gave theoretical reasons:

(i) *Symmetry*: a sphere is symmetrical, perfect

(ii) *Pressure*: the Earth's component pieces, falling *naturally* towards the center, would press into a round form

and experimental reasons:

(iii) *Shadow*: in an eclipse of the Moon, the Earth's shadow is always circular: a flat disc could cast an oval shadow.

(iv) *Star heights*: even in short travels Northward or Southward, one sees a change in the position of the star pattern.

This mixture of dogmatic "reasons" and experimental common sense is typical of him, and he did much to set science on its feet. His whole teaching was a remarkable life work of vast range and enormous influence. At one extreme he catalogued scientific information and listed stimulating questions; at the other extreme he emphasized the basic problems of scientific philosophy, distinguishing between the *true physical causes* of things and *imaginary schemes to save the phenomena.*

Euclid, soon after Aristotle, collected earlier discoveries in geometry, added his own, and produced his magnificent science: geometry developed by deductive logic. Such mathematics is automatically true to its own assumptions and definitions. Whether it also fits the natural world is a matter for experiment. Therefore, we should neither question the truth of a piece of mathematics, nor call it a *natural* science.

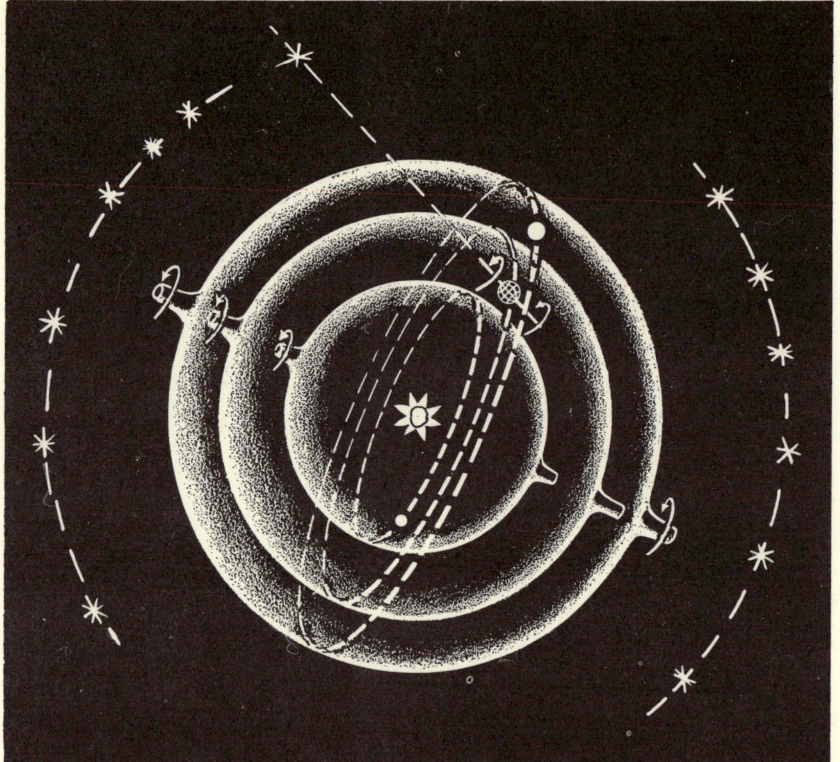

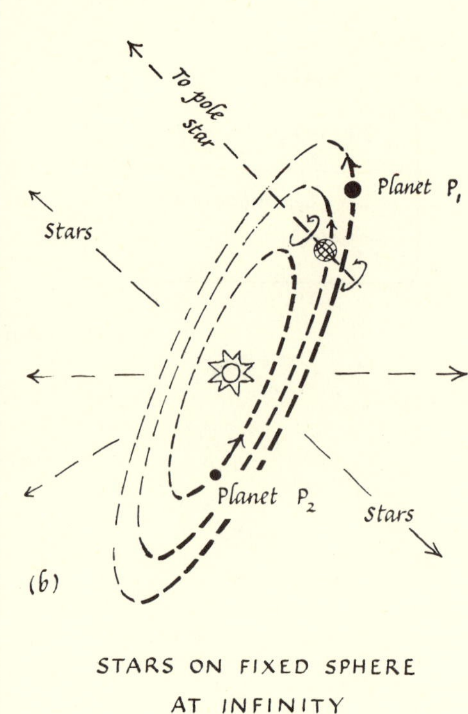

FIG. 14-11. ARISTARCHUS' SCHEME
Only two specimen planets are shown. P_1 might be Mars, Jupiter or Saturn.
P_2 might be Mercury or Venus.
(a) View of spheres. (b) Skeleton scheme showing planetary orbits.

The Scientific School at Alexandria

Alexander the Great built a huge empire, sweeping in a dozen years from Greece through Asia Minor, Egypt, Persia, to the borders of India and back to Babylon. Early in his campaign, he founded the great city of Alexandria at the mouth of the Nile. Greek scholars collected there, and the Museum or University of Alexandria grew to be a great center of learning. A school of astronomers started there about 330 B.C. and flourished for centuries. Some made new and more accurate observations, devising new instruments; some made a new kind of advance by trying to measure the actual sizes and distances of the Sun and Moon; and some produced new and better theories.

Before the new school changed the spinning spheres into eccentric circles, one Greek astronomer, *Aristarchus* (~ 240 B.C.), made two simplifying suggestions:

(i) *The Earth spins*—and *that* accounts for the daily motion of the stars. (Others had made this suggestion.)

(ii) *The Earth moves round the Sun in a yearly orbit*; and the other planets do likewise—that accounts for the apparent motions of the Sun and planets across the star patterns.

This simple scheme failed to catch on: tradition was against it; and it was merely an idea, not backed by a reasoned set of measurements, such as Copernicus gave many centuries later. Earth moving around an orbit raised mechanical objections that seemed even more serious in later ages; and it raised a great astronomical difficulty immediately. If the Earth moves in a vast orbit, the pattern of fixed stars should show parallax changes during the year. None were observed, and Aristarchus could only reply that the stars must be almost infinitely far off compared with the diameter of the Earth's orbit. Thus he pushed the stars away to far greater distances. He also released them from being all on one great sphere. As long as they are far enough away, they may be scattered through a great range of space, at rest while the Earth spins.

Measurements of Size and Distance

Astronomers were now trying to gauge the actual sizes of Sun, Moon, and Earth, and their distances apart. Earlier, there had been vague guesses: the Sun and Moon are very far away, or they are only just beyond the clouds; the Sun is the size of Greece, the Moon smaller. . . . Definite measurements would turn astronomy into a much more real science, but they were difficult to make.

Man judges ordinary distances by his eye-muscles, estimating the angle of squint when both eyes are directed on the object. For remote objects, our eyes are too close, and we use a longer base line and actually measure angles. Then we draw a diagram to scale, or use trigonometry. We now know that, for the Moon, a base line of 1000 miles gives an angle-of-squint of only ¼°. And for the Sun it would be only 1/600°, a very difficult difference to measure even today, with observers so far apart.

The *size* of the Sun (or Moon) is connected with its *distance* from us by an easy measurement: its *angular diameter*. Hold a coin at arm's length and move it until it just blots out the Sun's disc. Measure the coin's diameter and distance from your eye: their ratio, coin diameter/coin distance, gives the ratio, Sun's diameter/Sun's distance. The ratio is about 1/110. Or use some instrument to measure the angle the Sun's diameter subtends—almost exactly ½°. Draw a triangle with vertex angle ½° on a big sheet of paper and measure its sides. Or use simple trigonometry. You will find the distance from base to vertex is about 110 times the base. *The Sun's distance from us is 110 times its diameter.* Almost the same proportion holds for the Moon—Moon and Sun *look* about the same size, and this is confirmed in total eclipses of the Sun, when the Moon only just manages to cover it. A measurement of *one* of the two quantities DIAMETER and DISTANCE then combines with 110 to give the other. The usual measurement is DISTANCE, estimated by squint.

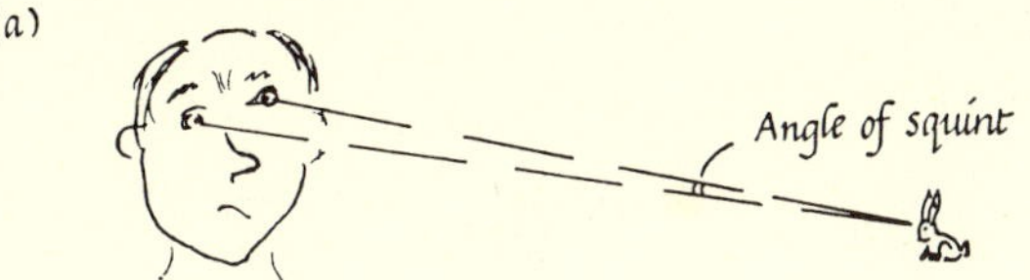

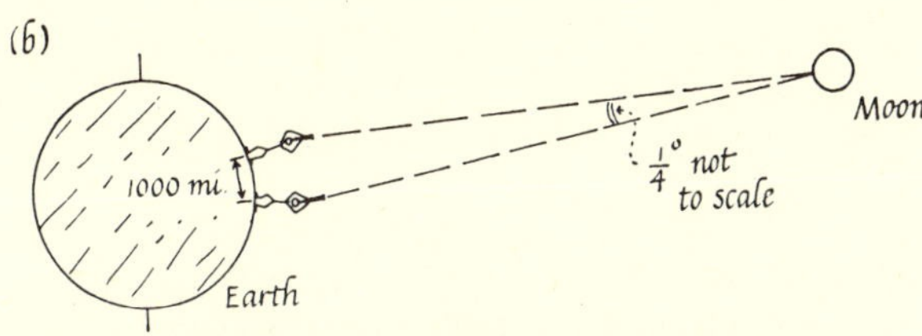

FIG. 14-13. ESTIMATING DISTANCES
(a) Judging distance by squint.
(b) Judging Moon's distance by squint would require observers 1000 miles apart to notice a difference of ¼°.

Size of the Earth

The first measurement to be made was the size of the Earth itself, and the other measurements emerged in terms of the Earth's radius.

Eratosthenes (~ 235 B.C.) made one of the early estimates. He compared the direction of the local vertical with parallel beams of sunlight at two stations a measured distance apart. He assumed that the Sun is so remote that all sunbeams reaching the Earth at any instant are practically parallel.

He needed *simultaneous* observations at two stations far apart. Good clocks that could be compared and transported were not available. So he obtained

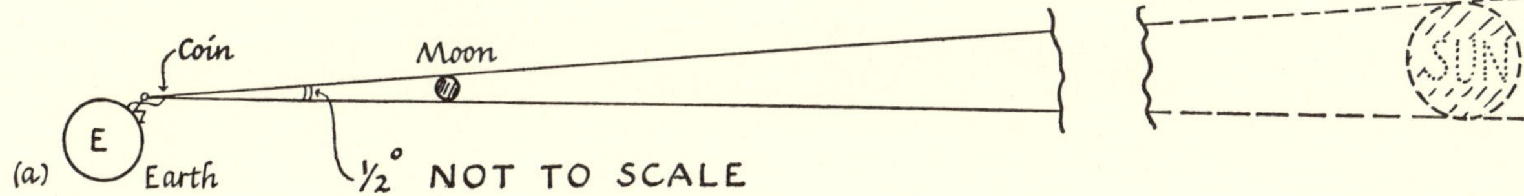

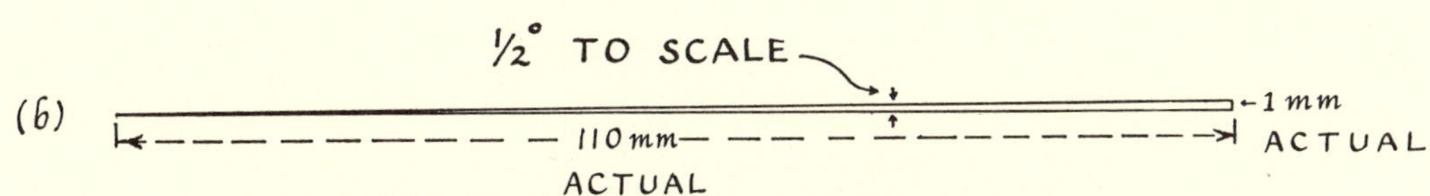

FIG. 14-12. RELATION BETWEEN SIZE AND DISTANCE can be found by holding measured coin at measured distance. This does not tell us absolute size or distance.
Sketch (a) is *not* to scale. Sketch (b) shows the "angular size" of Sun and Moon *drawn to scale*. Measurements show the Sun and Moon each subtend about ½° at Earth. Measurements or trigonometry tables give a proportion of about 1:110 for base:height.

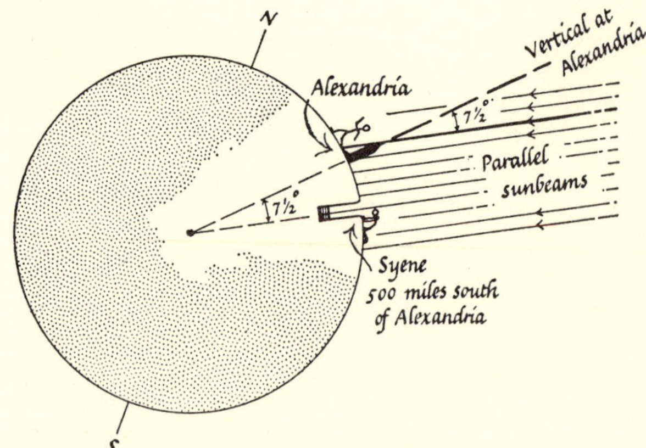

FIG. 14-14. HOW ERATOSTHENES ESTIMATED THE SIZE OF THE EARTH

simultaneity by choosing noon (highest Sun) on the same day at stations in the same longitude. He used observations at Alexandria, where he worked, and at Syene,[7] 500 miles farther south. The essential observation at Syene was this: at noon on midsummer day, June 22, sunbeams falling on a deep well there reach the water and are reflected up again. Eratosthenes knew this from library information. Therefore the noonday Sun must be vertically overhead at Syene on that day. At noon on the same day of the year, he measured the shadow of a tall obelisk at Alexandria and found that the Sun's rays made 7½° with the vertical. He assumed that all sunbeams reaching the Earth are parallel. So it was the vertical (the Earth's radius) that had different directions. Therefore the Earth's radii to Alexandria and Syene make 7½° at the center. Then if 7½° carries 500 miles of Earth's circumference, 360° carries how much? The rest was arithmetic. Measuring the 500 miles separation was hard—probably a military measurement done by professional pacers. There is doubt about the units he used, but some say his error was less than 5%—a remarkable success for this early simple attempt. He also guessed at the distances of Sun and Moon.

[7] Modern name: Aswân, where the great dam has been built on the Nile.

Moon's Size and Distance

The size of the Moon was compared with the size of the Earth by watching eclipses of the Moon. Timing the Moon as it crossed the Earth's shadow, Aristarchus found that the diameter of the round patch of Earth's shadow out at the Moon was 2½ Moon diameters. If the Sun were a point-source of light at infinity, it would shadow the Earth in a parallel beam as wide as the Earth itself. In that case we should have:

EARTH'S DIAMETER = 2½ MOON DIAMETERS;
or MOON'S DIAMETER = ⅖ EARTH'S DIAMETER,
∴ MOON'S DISTANCE, which is 110 MOON DIAMETERS
= (⅖) 110 EARTH DIAMETERS
= 44 EARTH DIAMETERS or 88 EARTH RADII.

That, with Eratosthenes' value of about 4000 miles for the Earth's radius, would place the Moon 350,000 miles away. Taking the Sun at infinity is reasonable. Treating it as a point source is not, and of course

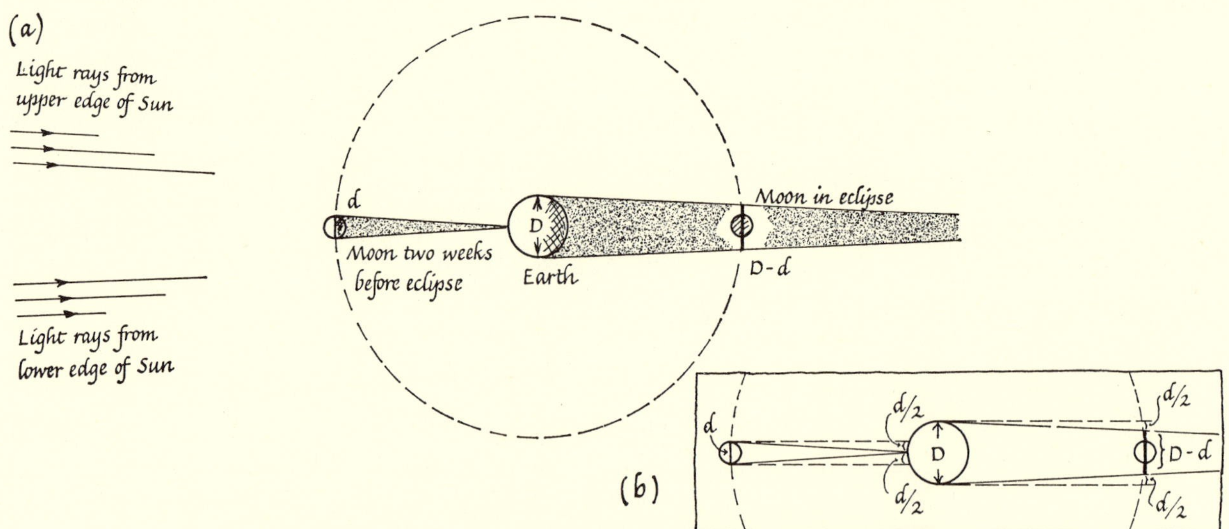

FIG. 14-15. EARLY GREEK MEASUREMENT OF SIZE OF THE MOON (AND THEREFORE ITS DISTANCE). Observations of eclipses showed that the width of the Earth's shadow at the Moon is 2.5 Moon-diameters. However, the Earth's shadow narrows as its distance from Earth increases because the Sun is not a point-source. Since the Moon's shadow almost dies out in the Moon-Earth distance, the Earth's shadow must narrow by the same amount—one Moon-diameter—in the same distance. Then Earth-diameter must be 3.5 Moon-diameters.

Aristarchus knew that. The Sun is a great flaming globe that casts tapering shadows (of angle about ½°). When there is a total eclipse of the *Sun* the Moon can only just shut the whole Sun off from our eyes; so the Moon's shadow cone only just reaches us—it ends practically at the Earth. Therefore, in running the Moon-to-Earth distance, the Moon's shadow narrows by the whole Moon diameter. In an eclipse of the *Moon*, the *Earth's* shadow is thrown the same distance, Earth-to-Moon, so it too must lose the same amount of width, one Moon diameter. Thus, Aristarchus argued:

EARTH'S DIAMETER — one MOON DIAMETER
= 2½ MOON DIAMETERS,
∴ EARTH'S DIAMETER = (1 + 2½) MOON DIAMETERS
= 7/2 MOON DIAMETERS,
∴ MOON'S DISTANCE = 110 MOON DIAMETERS
= 2/7 (110) EARTH DIAMETERS
= 31.4 EARTH DIAMETERS or 63 EARTH RADII.

More accurate measurements, used by Aristarchus and his successors, gave 60 Earth's radii (within 1% of the modern measurement), about 240,000 miles.

Later, the Moon's distance was measured by the squint method, thus: Observers at two stations far apart, in the same longitude, sight the Moon at the same time on an agreed night. They measure the angle between the *Moon's direction* and a *vertical plumb-line* (which gives the local zenith direction). These angles u, v suffice to locate the Moon if the base-line distance between the stations is known. With stations far apart, that distance would be a difficult measurement for early astronomers; but the angle between the Earth's radii to the two stations will do instead. So the observer at each station measures the angle between his local vertical

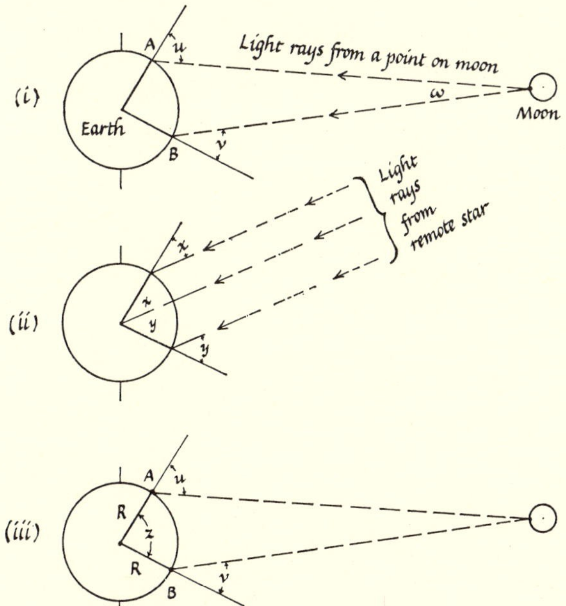

FIG. 14-16. MEASURING MOON'S DISTANCE BY SQUINT METHOD

and the light from a standard star. Any star will do: a perfect pole star, or any other star observed at its highest point. Then, as in sketch (ii) of Fig. 14-16, the sum of these two measured angles (x + y) gives the angle z at the center. Now in diagram (iii) the three angles u, v, z are known, and the two radii R and R are known

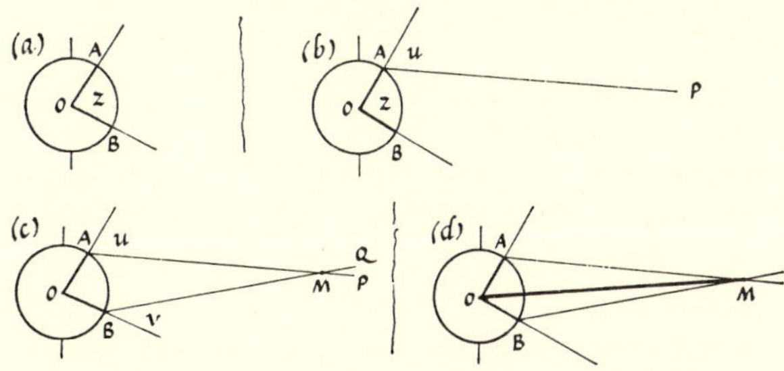

FIG. 14-17. SCALE DIAGRAM Method for Calculating (MOON'S DISTANCE)/(EARTH'S RADIUS) from measurements

to be equal. To find the Moon's distance, either use trigonometry or make a simple drawing to scale, thus: on a big sheet of paper (sand on the floor for early astronomers), draw a circle and mark radii OA and OB making the angle z (known by adding measurements x + y). Continue these radii out to represent the zenith verticals at stations A and B. From A draw a "Moon-line" AP making measured angle u with the radius there. Draw BQ from B. Where they cross is the Moon's position M on the scale diagram. Measure OM and divide it by the radius OA. The result gives the Moon's distance as a multiple of Earth's radius.

Accurate measurements give

MOON'S DISTANCE = just over 60 EARTH'S RADII
≈ 240,000 miles

Sun's Size and Distance

The Sun's distance is much harder to estimate even today because the Sun is in fact so far away, so bright, and so big. The squint method angle at the Sun is too small to measure without telescopic accuracy. However Aristarchus used an ingenious scheme to get a rough estimate. He watched the Moon for the stage at which it showed *exactly* half moon. Then sunlight must be falling on the Moon at right angles to the observer's line of sight, EM. At that instant, he measured the angle between the

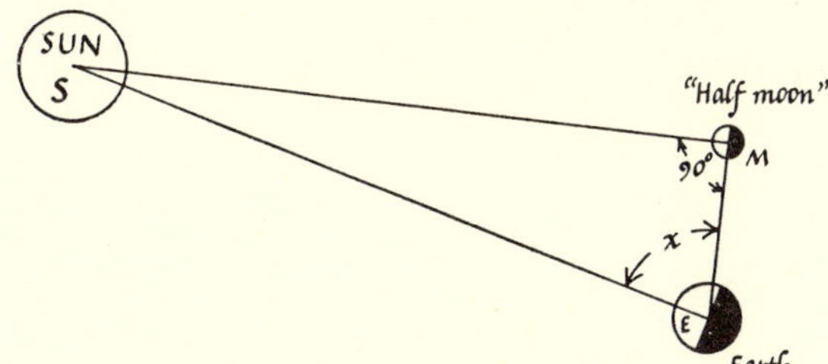

FIG. 14-18. SUN'S DISTANCE
Early Greek estimate of the Sun's distance from the Earth, in terms of the Moon's known distance. Greek astronomers tried to measure the angle x (or SEM), which is itself nearly 90°.

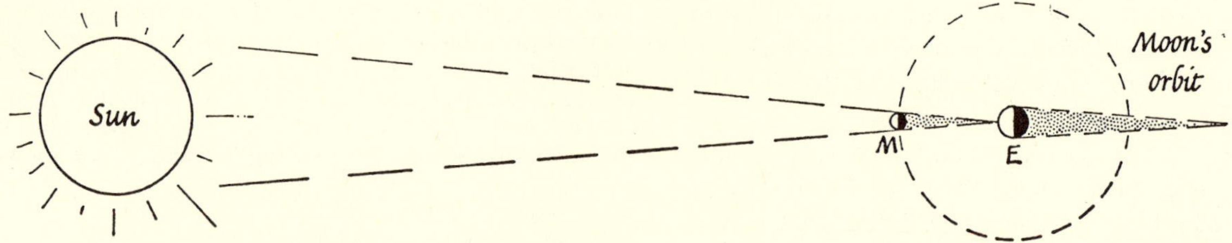

Fig. 14-19a. Sun, Moon, Earth. Sketch *not* to scale. The Sun is shown much too near, and the Moon is too big and too near.

Moon's direction and the Sun's direction. This angle, SEM, was nearly a right angle but not quite. Then in the great triangle, SEM, two angles were known. The third angle, ESM, is the small one that gives the essential measure of the Sun's distance. It is got by subtracting from 180°. It is small: 3° as Aristarchus estimated, but really only ⅙°. So Aristarchus' conclusion that the Sun's distance is about 20 times the Moon's was an underestimate, about 20 times too small. This proportion, SUN'S DISTANCE/MOON'S DISTANCE, is got from the angles by a scale drawing or by very simple trigonometry. (EM/ES is the cosine of the measured angle SEM. Therefore $\frac{ES}{EM} = \frac{1}{\cos SEM}$, easily taken from trig. tables.)

Thus, the astronomers at Alexandria had estimates of the dimensions of the heavenly system, and these were used by their successors with little change for many centuries:

Earth:	radius 4000 miles
Moon:	distance from Earth, 60 Earth's radii or 240,000 miles personal radius 1100 miles
Sun:	distance from Earth ?? 1200 Earth's radii. (Thought to be inaccurate: it was.) personal radius ?? 44,000 miles
Planets:	distances quite unknown, but presumed farther than Moon
Stars:	distances quite unknown, presumed beyond Sun and planets

Looking at these estimates, you can see that the usual pictures in books to illustrate eclipses are badly out of scale. Fig. 14-19 shows more realistic pictures based on modern measurements. No wonder eclipses are rare. It is easy enough to miss the slender cones of shadow. The Moon's orbit is tilted about 5° from the Sun's apparent path, and that makes eclipses still rarer.

New Theories: Eccentrics; Epicycles

In the school at Alexandria, the bold suggestion of making the Earth spin and move round a central Sun did not find favor. A stationary central Earth remained the popular basis, but spinning concentric

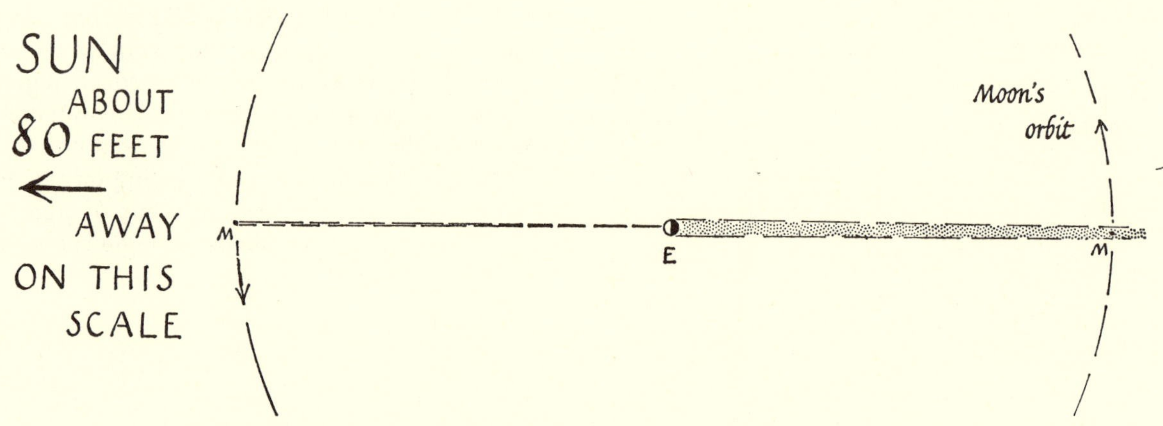

Fig. 14-19b. Sketch to scale showing the shadow cones of Moon and Earth.

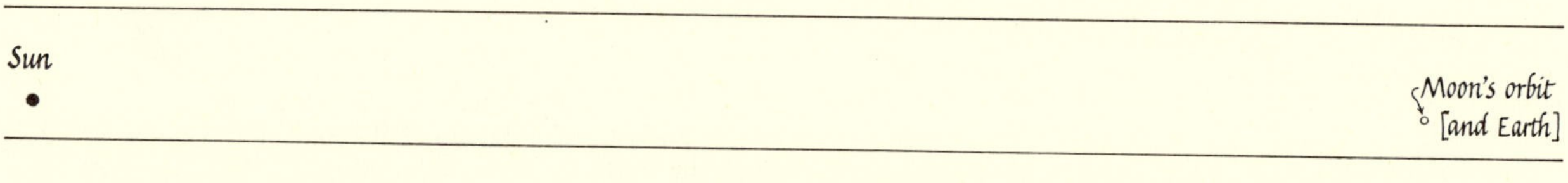

Sun

Moon's orbit [and Earth]

Fig. 14-19c. Sketch to scale. Here the scale has been reduced so that Sun, Moon and Earth are in the picture. The small circle is the Moon's orbit. The Earth, at the center of that circle, is too small to show. On this scale it is a dot 1/1000 inch in diameter. The Moon is much too small to show.

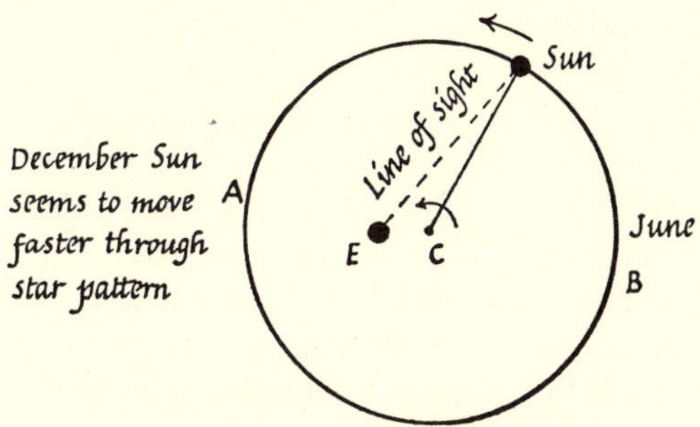

FIG. 14-20. THE ECCENTRIC SCHEME FOR THE SUN
The Sun is carried around a circular path by a radius that rotates at constant speed, as in the simplest system of spheres. The observer, on the Earth, is off-center, so that he sees the Sun move unevenly—as it does—faster in December, slower in June.

spheres made the model too difficult. Instead, the slightly uneven motion of the Sun around its "orbit" could be accounted for by a single eccentric circle: the Sun was made to move steadily round a circle, with the Earth fixed a short distance off center. Then, as seen from the Earth, the Sun would seem to move faster at some seasons (about December, at A) and slower 6 months later (at B). This was still good theory. For good theory, the scheme should have an appealing simplicity and be based on simple assumptions.[8] These needs were met: a perfect circle of constant radius, and motion with *constant* speed round it. Such constancies were necessary to the Greek mind—in fact to any orderly scientific mind. Without them theory would degenerate into a pack of demons. Placing the Earth

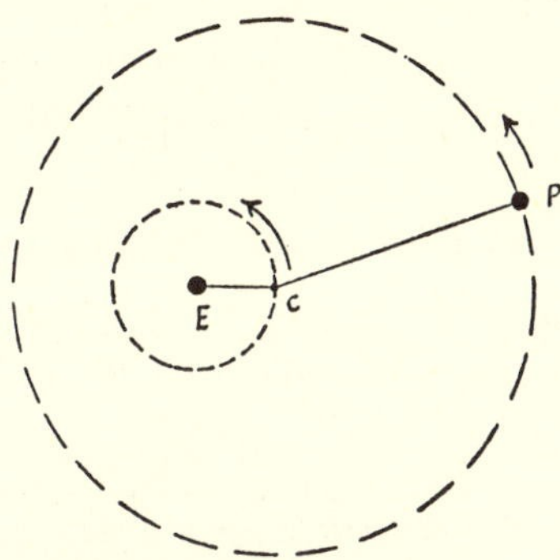

FIG. 14-21. THE ECCENTRIC SCHEME, FOR A PLANET
Each planet is carried at the end of a radius that rotates at constant speed; but this whole circle—center, radius and planet—revolves once a year round the eccentric Earth. (To picture the motion, imagine the radius CP continued out to be the handle of a frying pan. Then imagine the frying pan given a circular motion (of small radius, EC) around E as center, by a housewife who wants to melt a piece of butter in it quickly. Then make the handle CP revolve too—very slowly, for an outer planet like Jupiter.)

off-center was a regrettable lapse from symmetry, but then the Sun's speed *does* behave unsymmetrically—our summer is longer than our winter. A similar scheme served for the Moon, but the planets needed more machinery. Each planet was made to move steadily around a circle, once around in its own "year," with the Earth fixed off center; but then the whole circle, *planet's orbit and center* was made to revolve around the Earth once in 365 days. This added a small circular motion (radius EC) to the large main one, producing the planet's epicycloid track. The daily motion with the whole star pattern was superimposed on this.

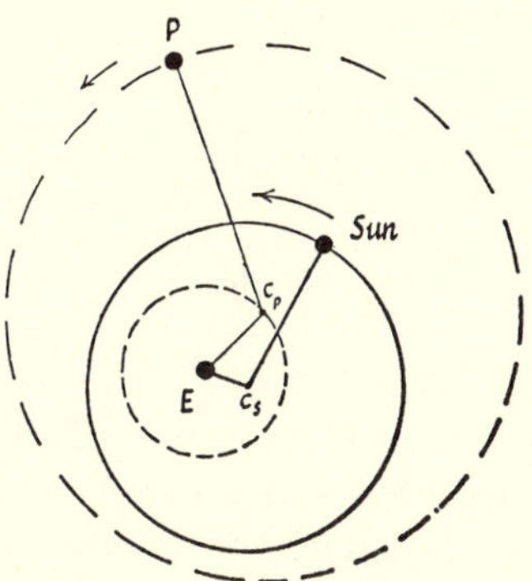

FIG. 14-22. THE ECCENTRIC SCHEME
Sketch showing machinery for motions of Sun and one planet.

Another scheme to produce the same effect used a fixed main circle (the deferent) with a radius arm rotating at constant rate. The end of that arm carried a small sub-circle (the epicycle). A radius of that small circle carried the planet round at a steady rate, once in 365 days. Though these schemes operate with circles, they were described more grandly in terms of spinning spheres and sub-spheres. For many centuries, astronomers thought in terms of such "motions of the heavenly spheres"—the spheres themselves growing more and more real as Greek delight in pure theory gave place to childish insistence on authoritarian truth.

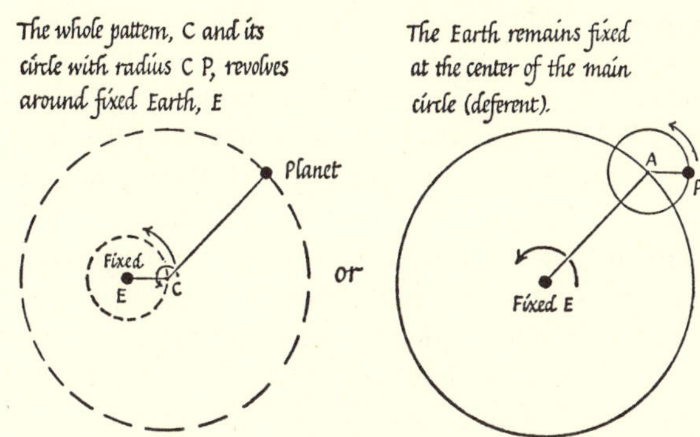

FIG. 14-23.
THE ECCENTRIC SCHEME AND THE EPICYCLE SCHEME

[8] The assumptions should be *logical* in the schoolboy slang sense—a sense that would have pained the Greek thinkers. In *their* use, it was the argument from the assumptions that had to be logical.

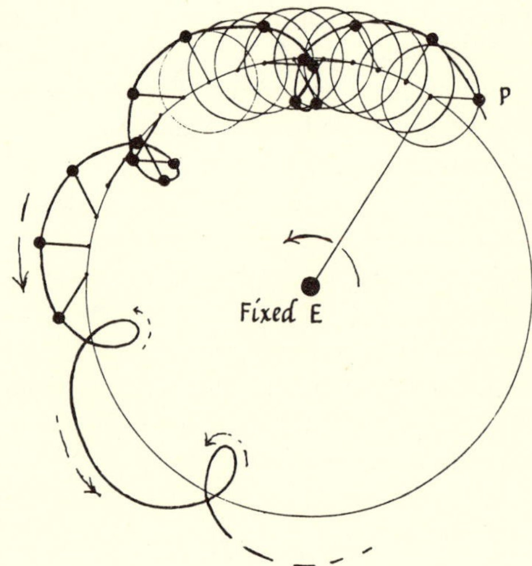

Fig. 14-24.
Making the Path of a Planet by the Epicycle Scheme.
This sketch shows how the two circular motions combine to produce the epicycloid pattern that is observed for a planet.

Hipparchus (~ 140 B.C.), "one of the greatest mathematicians and astronomers of all time,"[9] made great advances. He was a careful observer, made new instruments and used them to measure star positions. A new star that blazed into prominence for a short time is said to have inspired him to make his great star catalogue, classifying stars by brightness and recording the positions of nearly a thousand by celestial latitude and longitude. He made the first recorded celestial globe. There were no telescopes,[10] nothing but human eyes looking through sighting holes or along wooden sticks. Simple instruments like dividers measured angles between one star (or planet) and another, or between star and plumb line. Yet Hipparchus tried to measure within ⅙°.

He practically invented spherical trigonometry for use in his studies of the Sun and Moon. He showed that eccentrics and epicycles are equivalent for representing heavenly motions. Adding his own observations to earlier Greek ones and Babylonian records, he worked out epicyclic systems for the Sun and Moon. The planets proved difficult for want of accurate data, so he embarked on new measurements.

From Greek observations dating back 150 years Hipparchus discovered a very small but very important astronomical creep: the "precession of the equinoxes." At the spring equinox, midway between winter and summer, the Sun is at a definite place in the zodiac, and it returns there each year. But Hipparchus discovered that the Sun is not quite in the same patch of stars at the next equinox. In fact it gets around to the old patch of stars about 20 minutes too late, so that at the exact instant of equinox it is a little earlier on its zodiac path, about 1/70° after one year, nearly 1½° after a century. Hipparchus found this from changes in star longitudes between old records and new. Longitudes are reckoned along the zodiac, from the spring equinox where the equator cuts the ecliptic. Since all longitudes seemed to be changing by one or two degrees a century, Hipparchus saw that the zodiac girdle must be slipping round the celestial sphere at this rate, carrying all the stars with it, leaving the celestial equator fixed with the fixed Earth.[11] This motion seems small—a whole cycle takes 26,000 years—yet it matters in astronomical measurements, and has always been allowed for since Hipparchus discovered it. The discovery itself was a masterpiece of careful observing and clever thinking.

Precession remained difficult to visualize until Copernicus, sixteen centuries later, simplified the story with a complete change of view (see Ch. 16). Even then it remained unexplained—unconnected with other celestial phenomena—until Newton gave a simple explanation. Discovered as a mysterious creep, precession is now a magnificent witness to gravitation.

Hipparchus left a fine catalogue of stars, epicycle schemes, and good planetary observations—a magnificent memorial to a great astronomer. These had to wait two and a half centuries for the great mathematician Ptolemy to organize them into a successful theory.

[9] Sarton, *History of Science*.

[10] That invention was seventeen centuries in the future. By magnifying a patch of sky it enables much finer measurements to be made.

[11] The Sun's ecliptic path cuts the celestial equator in two points. When the Sun reaches either of these it is symmetrical with respect to the Earth's axis. Day and night are equal for all parts of the Earth: that is an equinox. Precession is a slow rotation of the whole celestial sphere, including the zodiac and the Sun, around the *axis of the ecliptic*, perpendicular to the ecliptic plane. From one century to the next, the creeping of precession brings a slightly different part of the zodiac belt to the equinox-points (*where ecliptic cuts equator*)—hence the name. The whole celestial sphere joins in this slow rotation round the ecliptic axis. This applies, for example, to the stars near the N-S axis, which is fixed with the Earth and 23½° from the ecliptic axis; so the motion carries the current pole-star away from the N-S axis and brings a new one in the course of time. Thus, in some centuries there is a bright star in the right position for pole-star, and in others there is no real star, only a blank in the pattern. In the 40-odd centuries between the building of the pyramids and the present, this motion has accumulated a considerable effect. In fact, by examining the pyramid tunnels that were built to face the dog-star Sirius at midnight at the Spring equinox, we can guess roughly how long ago they were built.

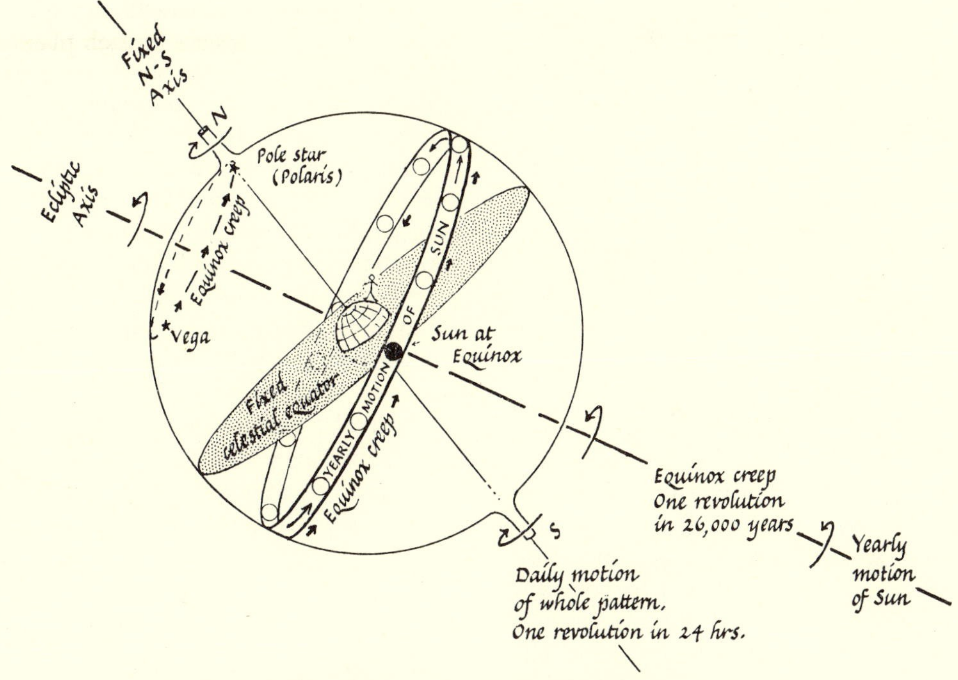

Fig. 14-25. Precession of the Equinoxes
In addition to (a) the daily motion of the whole heavens around the N-S axis fixed in the fixed Earth and (b) the yearly motion of the Sun around its ecliptic path in the zodiac band of stars, Hipparchus discovered (c) a slow rotation of the whole pattern of stars around a different axis, *the ecliptic axis* (perpendicular to the zodiac belt).

Ptolemy (~A.D. 120) made a "critical reappraisal of the planetary records." He collected the work of Hipparchus and his predecessors, adding his own observations, evolved a first-class theory and left a masterly exposition that dominated astronomy for the next fourteen centuries. The positions of the Sun, Moon, and planets, relative to the fixed stars, had been mapped with angles measured to a fraction of a degree. He could therefore elaborate the system of eccentric crystal spheres and epicycles and refine its machinery, so that it carried out past motions accurately and could grind out future predictions with success. He devised a brilliant *mathematical machine*, with simple rules but complex details, that could "save the phenomena" with century-long accuracy. In this, he neglected the crystal spheres as moving agents; he concentrated on the rotating spoke or radius that carried the planet around, and he provided sub-spokes and arranged eccentric distances. He expounded his whole system for Sun, Moon, and planets in a great book, the *Almagest*.

Ptolemy set forth this general picture: the heaven of the stars is a sphere turning steadily round a fixed axis in 24 hours; the Earth must remain at the center of the heavens—otherwise the star pattern would show parallax changes; the Earth is a sphere, and it must be at rest, for various reasons—e.g., objects thrown into the air would be left behind a moving Earth. The Sun moves round the Earth with the simple epicycle arrangement of Hipparchus, and the Moon has a more complicated epicyclic scheme.

In his study of the "five wandering stars," as he called the planets, Ptolemy found he could not "save the phenomena" with a simple epicyclic scheme. There were residual inequalities or discrepancies between theory and observation. He tried an epicycle scheme with the Earth eccentric, moved out from the center of the main circle. That was not sufficient, so he not only moved the Earth off center but also moved the center of uniform rotation out on the other side. He evolved the successful scheme shown in Fig. 14-26. C is the center of the main circle; E is the eccentric Earth; Q is a point called the "equant," an equal distance the other side of C (QC = CE). An arm QA rotates with constant speed around Q, swinging through equal angles in equal times, carrying the center of the small epicycle, A, round the main circle. Then a radius, AP, of the epicycle rotates steadily, carrying the planet P. It was a desperate and successful attempt to maintain a scheme of circles, with constant rotations. The arm of the little circle carrying the planet rotated at constant rate. Ptolemy felt compelled to have an arm of the main circle also rotating at constant rate. To fit the facts, that arm could not be a

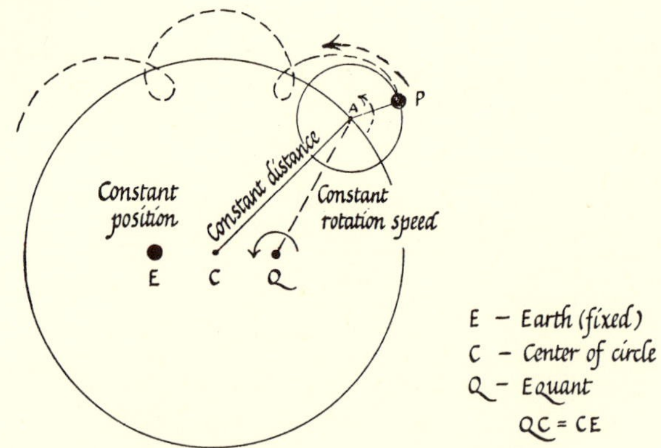

Fig. 14-26. The Ptolemaic Scheme
This system imitated the motions of Sun, Moon, and planets very accurately.

radius from the center of the main circle, as in the simplest epicycle scheme. Nor could it be the arm from an eccentric Earth E. But he could save the phenomena with an arm from the equant point Q that did rotate at constant speed. Thus for each planet's main circle, there were three points, all quite close together, each with a characteristic constancy.

E	C	Q
the Earth in *fixed position*	the center of the main circle with arm CA of *constant length*	the equant point with arm QA rotating at *constant speed*

By choosing suitable radii, speeds of rotation, and eccentric distance EC (= CQ), Ptolemy could save the phenomena for all planets (though Mercury required a small additional circle). The main circle was given a different tilt for each planet, and the epicycle itself was tilted from the main circle.

Here was a gorgeously complicated system of main circles and sub-circles, with different radii, speeds, tilts, and different amounts and directions of eccentricity. The system worked: like a set of mechanical gears, it ground out accurate predictions of planetary positions for year after year into the future, or back into the past. And, like a good set of gears, it was based on essentially simple principles: circles with *constant radii*, rotations with *constant speeds*, *symmetry* of equant (QC = CE), *constant* tilts of circles, and the Earth fixed in a constant position.[12]

In the *Almagest*, Ptolemy described a detailed scheme for each planet and gave tables from which the motion of each heavenly body could be read off. The book was copied (by hand, of course), translated from Greek to Latin to Arabic and back to Latin as culture moved eastward and then back to Europe. There are modern printed versions with translations. It served for centuries as a guide to astronomers, and a handbook for navigators. It also provided basic information for that extraordinary elaboration of man's fears, hopes, and greed—astrology—which needed detailed records of planetary positions.

The Ptolemaic scheme was efficient and intellectually satisfying. We can say the same of our modern atomic and nuclear theory. If you asked whether it is true, both Greeks and moderns would question

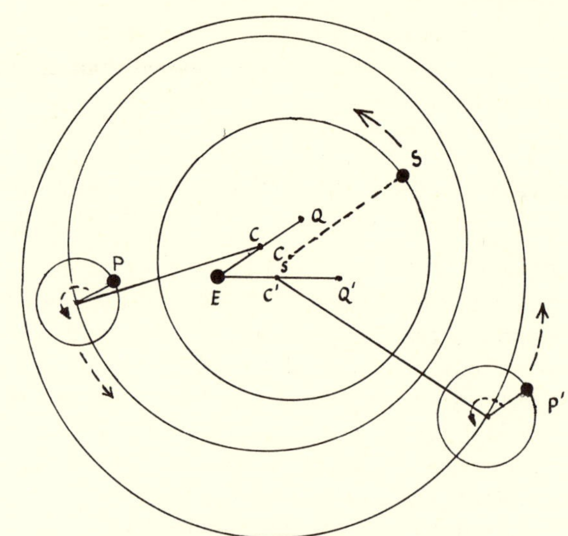

Fig. 14-27. The Ptolemaic Scheme
for the Sun, S, and two planets, P, and P′.

your word "true"; but if you offered an alternative that is simpler and more fruitful, they would welcome it.

PROBLEM FOR CHAPTER 14

★ 1. EARLY MEASUREMENTS

Describe in your own words, with large clear diagrams, a method used by the Greeks to estimate (a) the Earth's radius; (b) the Moon's distance from the Earth; (c) the Sun's distance from the Earth.

---

[12] If this insistence on circles seems artificial—a silly way of dealing with planetary orbits—remember:

(i) that you have modern knowledge built into your own folklore,

(ii) that though this now seems to you an unreal model, it is still fashionable as a method of analysis. Adding circle on circle in Greek astronomy corresponds to our use of a series of sines (projected circular motions) to analyze complex motions. Physicists today use such "Fourier analysis" in studying any repeating motion: analyzing musical sounds, predicting tides in a port, expressing atomic behavior. *Any* repeating motion however complex can be expressed as the resultant of simple harmonic components. Each circular motion in a Ptolemaic scheme provides two such components, one up-and-down, one to-and-fro. The concentric spheres of Eudoxus could be regarded as a similar analysis, but in a more complex form. Either scheme can succeed in expressing planetary motion to any accuracy desired, if it is allowed to use enough components.

CHAPTER 15 · AWAKENING QUESTIONS

[*The Angel Raphael discusses the alternative views with Adam:*]

Hereafter, when they come to model Heav'n
And calculate the Stars, how they will wield
The mighty frame, how build, unbuild, contrive
To save appearances, how gird the Sphere
With Centric and Eccentric scribbl'd o'er
Cycle and Epicycle, Orb in Orb:

. . . .

. . . What if the Sun
Be Center to the World, and other Stars
By his attractive virtue and their own
Incited, dance about him various rounds?
Their wandering course now high, now low, then hid,
Progressive, retrograde, or standing still,
In six thou seest, and what if seventh to these
The planet Earth, so stedfast though she seem,
Insensibly three different Motions move? . . .

—John Milton, *Paradise Lost*, Book VIII (1667)

Such was the Ptolemaic picture of the heavens—a complicated, clumsy system but workable and successful. To your mind and mine, it may seem unreal or unthinkable, but to Ptolemy and to many after him it was the alternative that seemed unthinkable, a spinning Earth whizzing round the Sun in a vast orbit. *Would not objects be thrown off a spinning Earth or left behind by a moving Earth?* And would not the nearer of the fixed stars change their apparent positions when the Earth moved right across the diameter of its orbit in half a year? Mankind's muddled ideas about motion had to wait for Galileo's teaching and Newton's clear thinking before the fog of the first objection cleared away. The second objection would still be serious if the stars did show absolutely no parallax shifts. We now know there are shifts, but they are too small to be observed without *very* delicate instruments. The first successful observation was made in 1832. These scientific objections to the Sun-in-center alternative were overshadowed by one that seemed to arise naturally from man's sense of his own importance. The Earth on which We live must be the center of the Universe—other things must revolve round Us.[1] This view, suggested by observation, gained easy support from simple self-centered thinking and from humanist teaching. It was firmly placed in people's beliefs. So we must not be surprised to find that Ptolemy's picture, with Man's world at the center, held the field in man's beliefs right on through the Dark Ages until the Renaissance brought questionings with awakening flexibility of mind.

The other view, with the Sun as center and a spinning Earth travelling round it, was suggested by some Greek astronomers and discussed by an occasional philosopher or churchman in the 12th to 15th centuries, but the suggestions were put

[1] Hesitate before you condemn this arrogance unsympathetically. You, too, have passed through just such attitudes. Healthy children today start with much this kind of view: "I am the important person, almost the only person, and the world is arranged around Me." The process of growing up socially involves the broadening of this view, perhaps in stages such as "ME"; "Me and Mother"; "Me and family"; "Me and you"; "Me and other people"; "Me and my country"; and, finally, for a few, "me and the world." Most people fail to advance through all these stages. A man's success in social adjustment and wisdom would seem to depend on how far he progresses along this series. Many fail to advance beyond the first few stages, and though they may be very successful in material matters we are unlikely to find much to admire in them spiritually or intellectually. Some of the world's dictators got stuck at the "ME" stage, though others got as far as "Me and Mother" and regressed. At the other extreme, the rare souls who think and feel in terms of "me and the world" are the great philosophers and prophets.

apologetically as unreal theories, and gained little acceptance. For a thousand years the Ptolemaic picture was believed and hardly questioned. People in Europe had little interest in science, except as a basis for wordy discussions, arguing from authority instead of inferring principles from experiment. The growing Church was responsible for what teaching there was, and it treated science with the same dogmatic authority that maintained its religious structure. Any dispute with its teachings, even a simple appeal to experiment and observation, would bring disturbing questions to threaten its authority. Such disturbances were not welcome in an age when simple men were directed and taught in all matters by the Church and nobles and kings ruled by the authority of the Church.

In the thousand years between the Greek astronomers and the awakening of scientific experimenting there must have been questioning scientists, but their work remains forgotten. After centuries of "dark ages," came glimmerings of the new light. Roger Bacon, an English monk (~1250), almost cried aloud, "Experiment, experiment." He was an honest but intemperate critic, attacking churchmen and other thinkers in his insistence on the need to experiment and to collect real knowledge instead of poring over bad Latin translations. In his books he attacked ignorance and prejudice and made wise suggestions for gathering more knowledge. One writer pictures him shouting to mankind, "Cease to be ruled by dogmas and authorities; *look at the world*!"[2] His tactless manner brought him into conflict with his brother friars and with the Church itself even while he was writing his books. His teaching and books, and those of others like him, were probably suppressed and certainly forgotten for a long time—he was centuries ahead of his time. (The later Bacon, Francis, is credited with formulating the new scientific attitude, but some doubt whether his contribution was very helpful.)

Two hundred years later Leonardo da Vinci (~1480) thought and experimented and wrote and drew as a scientist as well as an artist. In mechanics he began to sort out ideas of force and mass and motion; and he formulated scientific ideas and sketched skillful models. His famous notebooks are a storehouse of mechanical inventions and include some of the finest drawings in the history of art. In making these notebooks, he was both historian and prophet, recording interesting ideas from others and ingenious schemes he had thought of. A new approach was starting which would have warmed Roger Bacon's heart.

[2] H. G. Wells, *The Outline of History* (London, 1923).

Meanwhile, astronomical records had grown, with observations from Arab astronomers and others. The needs of medicine and navigation kept scientific teaching alive and ultimately impressed scientific growth on the Renaissance. Alphonso X of Castille (~1260), ordered his school of navigators to construct new tables to predict the motions of the heavenly bodies. These tables were collected, printed some 200 years later, and used for a further 100 years. Alphonso is rumored to have said, when he first had the complicated Ptolemaic system explained to him, that if he had been consulted at the Creation, he would have made the world on a simpler and better plan.

Others added measurements and in turn made mathematical refinements of the Ptolemaic machinery, but even in the intellectual awakening of the early Renaissance the alternative idea, Sun-at-center, was not put forward seriously until Copernicus wrote his great book. Then, running right through the Renaissance and up to the present day, came the great series of scientists who developed the science of mechanics from the foggy views of the dark ages to its present precise and powerful form, using the solar system (and, later, atomic systems) as a vast laboratory with frictionless apparatus. We are concerned not only with the physics they developed but also with the interaction between their work and the life and thought of other men. Therefore, we shall give some account of their lives as well as their work. First we shall give a set of short notes showing how their main contributions were related.

SHORT NOTES

[In these notes, as in earlier ones, the dates given are not the date the man was born or died but "average" dates usually showing when he was about 40.]

Nicolaus Copernicus, a Polish monk (~1510), suggested that the Sun-at-center (heliocentric) picture of the planetary system would be simpler. He wrote a great book setting forth the details of this system, showing calculations of its size, etc., and predicting tests. After his death this view spread, though it was not universally accepted for a long time.

Tycho Brahe, a Danish nobleman (~ 1580) who, fired with curiosity about the planets, became a brilliant observer, a genius at devising and using precise instruments. He built the first great observatory. He knew of Copernicus' suggestion but did not wholly accept it, and was not much concerned with theory. He constructed far more accurate planetary tables than any before him, and left his pupil Kepler to complete their publication.

Johannes Kepler, a German (~1610), was a powerful mathematician with a gift for subtle speculation and a great belief that there are simple underlying rules in nature. Using Tycho's observations, he extracted three general rules for the motion of the planets. He could not find any underlying explanation of these rules.

Meanwhile, *Galileo Galilei* (~ 1610), was experimenting and thinking and teaching new scientific knowledge of mechanics and astronomy. To the dismay of classical philosophers, and to his own danger, he preached the need to abide by experiment. With the newly-invented telescope he gained evidence supporting Copernicus' picture which he advocated violently until the Church stopped him.

René Descartes, French philosopher (~1640), described a world system deduced from general principles which he felt were implanted by God. He opposed the idea of a vacuum and filled space with whirling vortices to carry the planets. His greatest contribution to science was the invention of co-ordinate geometry: the use of x-y graphs to link algebra and geometry. This enabled calculus to develop.

GREAT SCIENTIFIC SOCIETIES were formed for the exchange of knowledge, and experimental science throve publicly. (1600- . . . , 1700- . . . now.)

Isaac Newton (~1680), gathered up the results of Galileo's experimenting and the work and thinking of others into clearly-worded "laws" summing up the experimental facts concerning mass, motion, and force, with the help of clear ideas and definitions. He extended the force of gravity to universal inverse-square-law gravitation, showing that this would account for the moon's motion, for Kepler's three planetary laws, for the tides, etc., thus building a great deductive theory. In the course of this he needed calculus as a mathematical tool, and invented it. He experimented and speculated in other branches of physics, too, particularly optics.

In terms of his summary of mechanics, Newton showed that universal gravitation would "explain" the whole of the behavior of the Moon and planets as described by the Copernicus-Kepler system. In the next two centuries further consequences were worked out by other mathematicians and physicists, including the French mathematicians, *Joseph Louis Lagrange* and *Pierre Simon de Laplace*; and a new planet was discovered by its minute gravitational effects on the known ones.

Early in this century, *Albert Einstein* suggested modifications and reinterpretation of the laws of mechanics. These do not destroy Newton's work, but enable us to account for such things as a small unexplained motion of the planet Mercury, and to deal successfully with very rapid atomic motion. In addition to such modification of the "working rules" of mechanics, the great value of Relativity lies in the light it throws on the relation between experiment and theory, ruling out unobservable things from even the speculation of wise scientists.

CHAPTER 16 · NICOLAUS COPERNICUS (1473-1543)

> Beauty is truth, truth beauty,—that is all
> Ye know on Earth, and all ye need to know.
>
> —John Keats

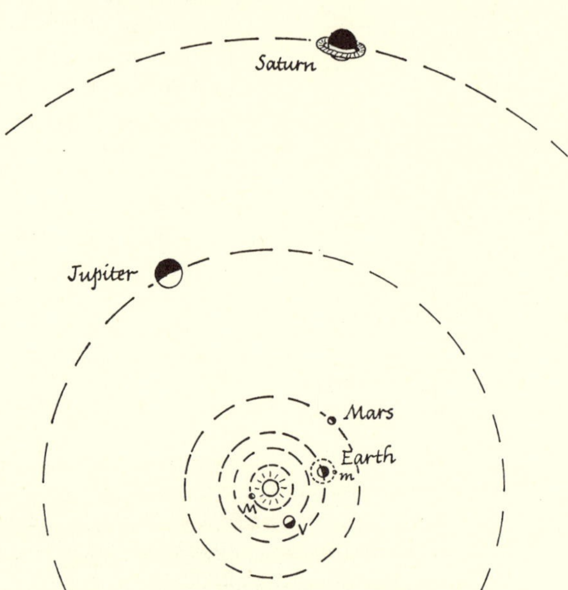

FIG. 16-1A. COPERNICUS' PLANETARY SYSTEM

Nicolaus Copernicus was born in Polish Prussia. He seems to have lived a quiet, uneventful life. He was pious, capable, not brilliant, but inspired by a love of truth. He had the clear vision and the courage to challenge traditional authority, but he had no delight in entering into conflict with it.

Copernicus was brought up by his uncle, who was Bishop and ruler, practically prince, of the district. His uncle intended him to serve the Church and sent him to school and university near home to study the great classics. He also studied some astronomy and learned to use the clumsy astronomical instruments of the time. Then he travelled to Italy, learned Greek, and studied Church Law, in which he received the degree of Doctor. He also continued to study astronomy, and was now able to read the original Greek texts. After a few years, when he was 26, he visited Rome, and while there gave a course of lectures in mathematics (?=astronomy). Meanwhile his uncle had named him canon in the cathedral but allowed him to spend two more years in Italy to study medicine.

At last, in his early thirties, Copernicus returned to his uncle's cathedral near home. There he spent the rest of his life, as a scholarly monk, dividing his time between church duties, account keeping, occasional medical consultations, and meditation on the System of the World.

For his meditations he liked solitude, and he seems to have made few friends, though his reputation as a scholar drew several students to him. He had no use for the long wrangling arguments that were fashionable; yet, when asked to help a government committee which was trying to simplify the coinage, he accepted willingly and presented a clear capable report, which the senate adopted.

Copernicus was impressed by the variety and disagreement of opinions on planetary motions. The Ptolemaic system, with its artificial equants, seemed to him too clumsy to be God's best choice. He believed that the planetary system, spheres and all, was a divine creation; but he believed God's arrangement would be a simple one, all the more splendid for great simplicity. He collected together observations of the planets in more reliable tables than had so far been available; and in thinking about the planetary motions he was struck by the simplicity that would come from changing to a Sun-in-center picture. He made an intuitive guess that the Earth is a planet like the rest—an extraordinary shift of view. Then he pictured all the planets moving in circular orbits around a fixed Sun. He made the Earth travel once around the Sun in a year, spinning once in 24 hours as it goes. The "fixed stars" and the Sun could then remain at rest in the sky.

This scheme replaced Ptolemy's epicycles and equants with simpler circular motions. The *daily* motion of the stars, carrying Sun, Moon, and planets as well, could obviously be replaced by a daily spinning Earth. That alternative had often been discussed, but had been turned down because the critics did not understand the mechanics of motion. (They claimed that there would be a howling wind of air left behind, and that the ground would outstrip a stone dropped from a high tower. On the other hand, the stars etc. could well be carried around by Ptolemy's spheres because spheres and rotations were natural in the heavenly region.)

The slower, irregular motions of Sun and planets through the star pattern could be simplified by circular motions around the Sun. This was Copernicus' main contribution: to stop the Sun and place it at the center of the planetary system. Then the Sun's yearly motion around the ecliptic was only an apparent one due to the Earth's yearly motion around the Sun. And the complex epicycloid of a planet was simply a compound of the planet's own motion around a circle and the Earth's yearly motion. (On this view, the epicycloid picture is making us pay for ignoring the Earth's motion.) This tempting idea of a Sun-in-center scheme had been thought of before, but not strongly supported. Copernicus, searching early records for such ideas, had both the clear mind and the store of data to develop it.

His detailed explanation of a planet's epicycloid ran like this. Suppose the Earth travels around a circular orbit and Jupiter more slowly around a bigger orbit, both with the Sun at the center, as in Fig. 16-2. The fixed stars must be *much* farther away, because no parallax-shifts are observed. Then in marking the position of Jupiter among the fixed stars we look along a sightline running from Earth to Jupiter and on, far beyond, to the pattern of the

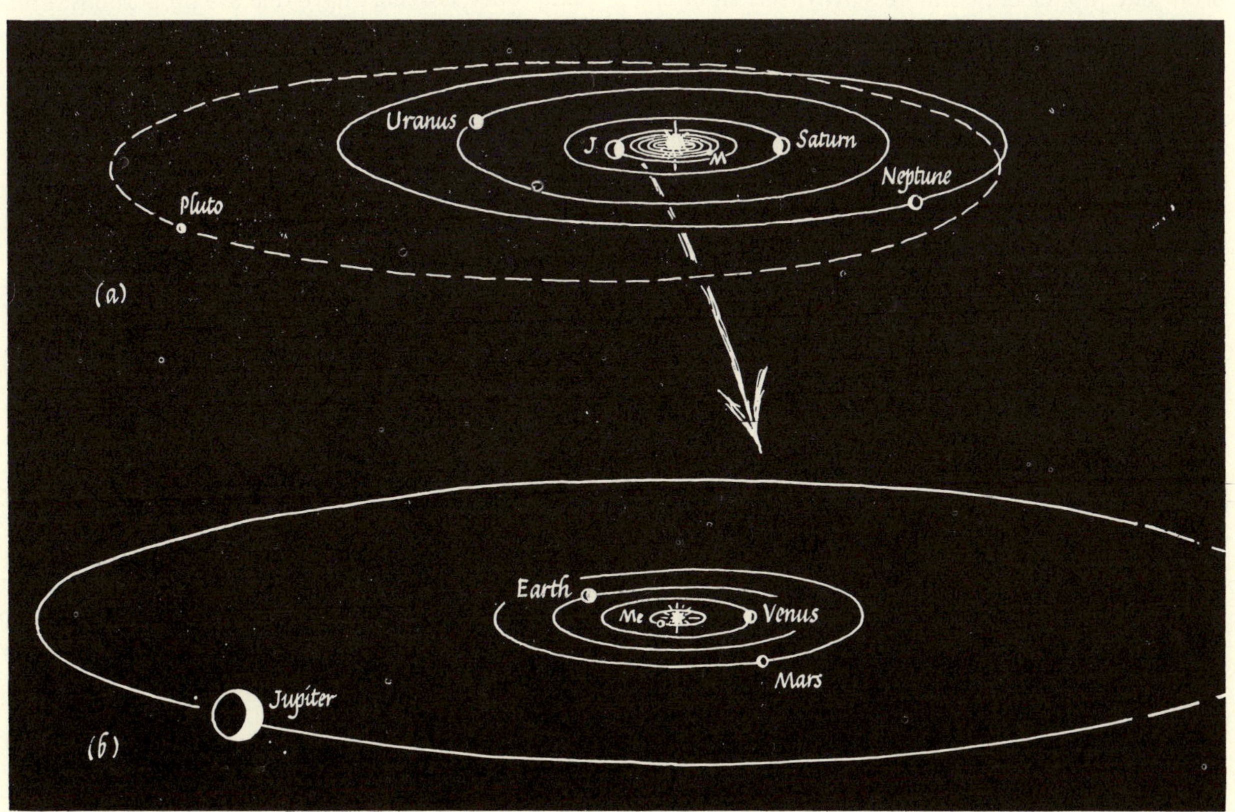

Fig. 16-1B. The Copernican System (with later additions) seen in perspective
(a) The whole system. (b) Inner region of (a) magnified several times.
The orbits are almost circular, with the Sun only a little off center.
The arrangement of orbits is shown here, but the sizes are *not to scale.*
The planets themselves are *much* smaller in proportion than these sketches show—on the scale used here for orbits the planets would be invisible dots.

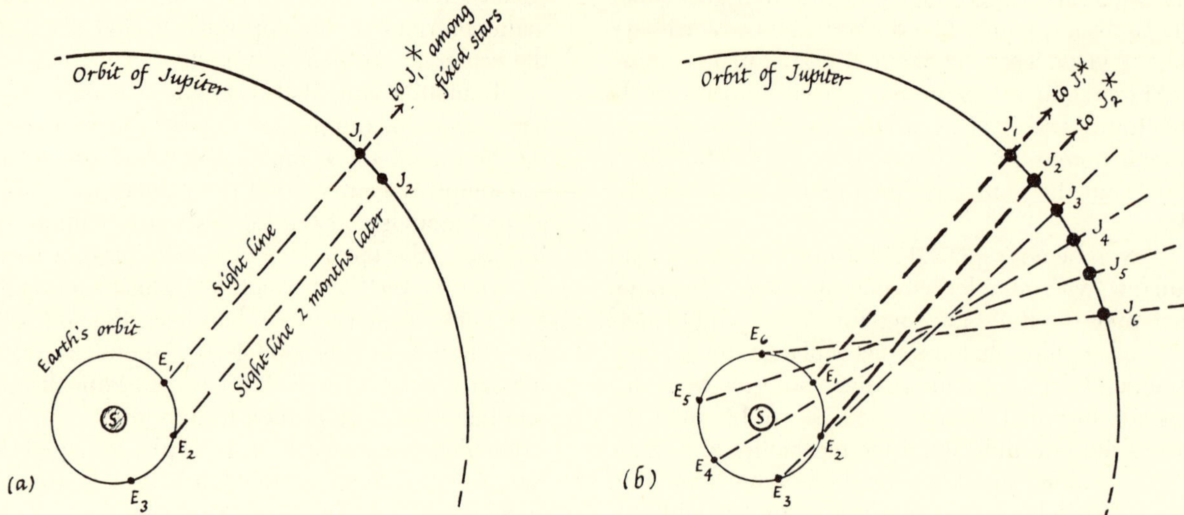

Fig. 16-2. Copernicus' Explanation of Planetary Epicycloids
The lines E_1J_1, E_2J_2, etc., are sight-lines from positions of the Earth every two months through Jupiter's position towards the stars.
(a) Two stages sketched (b) More stages sketched.

stars. As the Earth sweeps round and round its orbit and Jupiter crawls more slowly, this sightline wags to and fro as it goes around, marking an epicycloid among the stars. When the Earth is at E_1, Jupiter is at J_1, and an observer looking along the sightline E_1J_1 sees Jupiter among the stars at J^*_1. As the Earth travels from E_1 to E_2 to E_3, E_4, E_5, E_6, &c., Jupiter travels steadily but slowly forward from J_1 to J_2 to J_3, J_4, J_5, J_6, &c. Then the observer on E sees J^* in directions that swing mostly forwards but sometimes backwards. To see this, look at Fig. 16-3, which shows Fig. 16-2 condensed to a small scale with the sightlines continued out to a remote background of stars.

Copernicus accounted for the epicycloids of Mars, Jupiter, and Saturn by making them move around large circular orbits outside the Earth's orbit. He made Venus and Mercury move around smaller orbits, nearer the Sun than the Earth's. This accounted for their observed behavior—keeping close to the Sun and swinging to and fro each side of it. Thus the same scheme served for both the "inner" planets and the "outer" ones.

Copernicus did not just offer an alternative that looked simpler; he extracted new information from his scheme: the order and sizes of the planetary orbits, a remarkable advance contributed by theory. In the Ptolemaic scheme the main circles could be chosen with any sizes—it did not even matter which planet was put outermost. In fact, Ptolemy was just drawing patterns with a mathematical machine, to fit the observations, on the celestial sphere. In the Sun-in-center scheme, the orbits must be in a definite order and must have definite proportions. From the planets' apparent motions in the sky it was obvious to Copernicus whose orbits were largest and whose least.[1] The order must be (as in Fig. 16-1):

SUN, stationary at the center
Mercury, nearest the Sun
Venus
Earth, with the Moon travelling around it
Mars
Jupiter
Saturn, farthest of the planets then known.

Treating the orbits as simple circles, Copernicus calculated their relative radii from available observations; he could thus plot a fairly accurate scale map of the system. To obtain the actual radii from these relative values, he needed an absolute measurement of any one of them, say the distance from Sun to Earth. This was known only roughly,[2] so the *scale* of his complete picture was unreliable.

[1] Try this theoretical discussion, with Figs. 16-2 and 16-3. Suppose that Jupiter and the Earth each keep their present times of revolution (our year, and Jupiter's "planetary year"), but that the radius of the Earth's orbit is changed. What would happen to the shape of the loops of Jupiter's apparent path, among the fixed stars,

(a) if the Earth's orbit changed to a *very small* radius?

(b) if the Earth's orbit changed to be nearly as big as Jupiter's?

[2] In fact he used a Greek estimate that was 20 times too small.

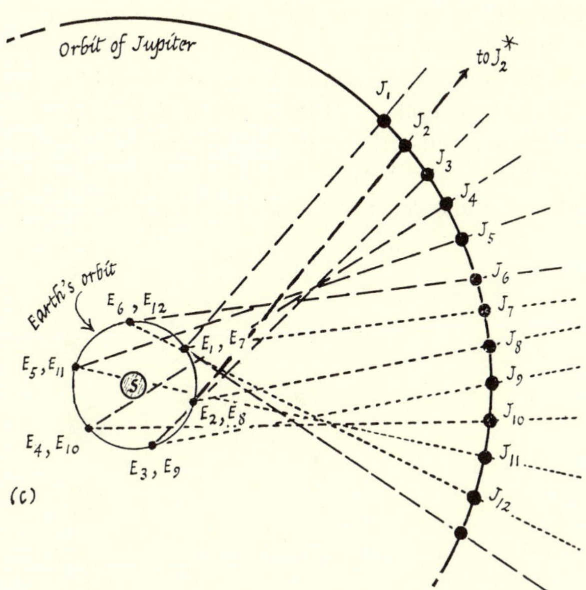

(c) Many stages sketched. The sight-line EJ wags up and down in a complicated way.

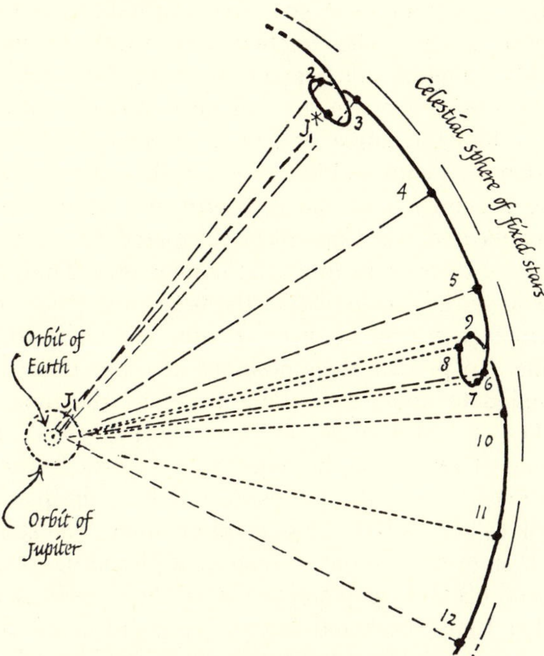

Fig. 16-3. Copernicus' Explanation
The apparent positions of Jupiter in the background of fixed stars. This shows Fig. 16-2c redrawn on a much more condensed scale with the sight-lines from Earth to Jupiter continued on out to the stars. (E.g. the line to J_1* here is continuation of EJ_1.) The specimen sight-lines are drawn parallel to the corresponding ones in Fig. 16-2c.

Estimating Orbits

To see how he calculated relative radii, suppose you are attacking the problem for an inner planet, say Venus. Venus, nearer the Sun than the Earth, travels in a small orbit round the Sun. This circle is seen practically edge-on from the Earth; so Venus seems to swing to-and-fro in front of the Sun or behind it, travelling only a small way each side of the Sun before it turns back. Thus it is seen only near the Sun as a morning or evening "star." When Venus seems farthest to one side of the Sun, just about to turn back, it must be at a point such as C lying on a tangent from the Earth to its orbit (Fig. 16-4). In positions A, B, D, . . . etc., it would seem nearer the Sun. This tangent is perpendicular to the radius, SC, of the orbit (by geometry of circle). So the triangle ECS has a right angle at C and an angle at E that can be measured by sighting from the Earth. Given these, you can draw a scale model of this triangle (i.e., a similar triangle), and by measuring that you can find the proportion between SC and SE which are the radii of the orbits of Venus and the Earth. To measure the required angle at E, you must find the angular distance between Venus and the Sun at the moment when Venus seems farthest from the Sun. In trying to make a direct measurement you might be prevented by the glare of the Sun, but you can wait till the Sun has set, calculating where it will have got to, and then observe Venus day after day till the separation is greatest. So you may imagine yourself measuring the angle with two jointed sticks with peep-

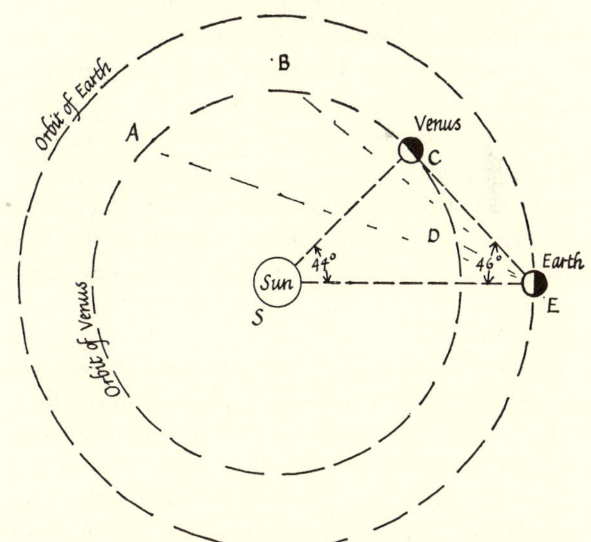

Fig. 16-4. Estimating Relative Radii of Orbits
Venus is shown farthest from the sun.

holes, though the real method must be slightly less direct. Actual observations show that this angle is about 46°. Drawing and measuring a triangle with angles 46°, 90°, and 44° will show you that the fraction (medium side) / (longest side) is about 72/100. This tells you that the orbit-radii for Venus and Earth are in the proportion 72:100. You need not draw the triangle, if you have trig. tables, as Copernicus had. The fraction you want (medium side)/(longest side) is sin 46°, and, from tables, this is 0.72. Copernicus had measurements which gave him this angle, and he performed this calculation for Venus and Mercury. For the outer planets the argument and the geometry are rather more complicated, but Copernicus calculated the relative sizes of their orbits in much the same way. Thus, he could draw a scale map of the orbits and place the planets correctly in them at some chosen starting time. To predict their positions at other times he needed to know each planet's "year," the time it takes to travel round its orbit. These "years," or times of revolution, he found from recorded observations. Essentially, he found how long the planet took to get back to the same place among the stars.

Using recorded data, Copernicus placed the planets on his scale map and predicted their positions at other times, past and future. He could check the past ones, and thus test his "picture," or "theory" as we should now call it. These tests were encouraging, but there were some disagreements which led, through long careful calculations, to modifications of the simple picture.

Copernicus gave other points in support of his theory:

(i) Mars is much brighter (seeming larger) in some seasons, obviously because it is nearer the Earth then. On the Ptolemaic system its slightly eccentric orbit round the Earth could not possibly provide big enough changes of distance. But on the Copernican scheme the distance ranges from the sum of the orbit radii to the difference. In fact Mars *is* brightest when that distance is least, at the seasons when Mars and the Earth are on the same side of the Sun—Mars in "opposition" to the Sun, overhead at midnight.

(ii) Just when an outer planet makes the *reverse* part of a loop it is exactly in opposition to the Sun. Ptolemy could give no reason for this. It is obvious from the Copernican geometry (study Fig. 16-2).

(iii) If Venus and Mercury are nearer the Sun than the Earth is and travel round the Sun in small orbits, then when we look at them we should see only part of them brightly lit, the side facing the Sun, not the whole planet (see Fig. 16-5). Thus these two planets should show stages or "phases," like the Moon as it changes from new Moon to half Moon to full Moon and so on.[3] With the large planet Venus, these stages should, the critics thought, be visible. As they were not observed, the critics claimed the Sun-in-center idea was wrong. There is a story, almost certainly mythical, that Copernicus replied: "If the sense of sight could ever be made sufficiently powerful, we should see phases in Mercury and Venus." Within a century, Galileo's telescope showed the phases of Venus.

[3] On a pure Ptolemaic scheme, Venus would also show phases but not the whole range from crescent to "half moon" to full.

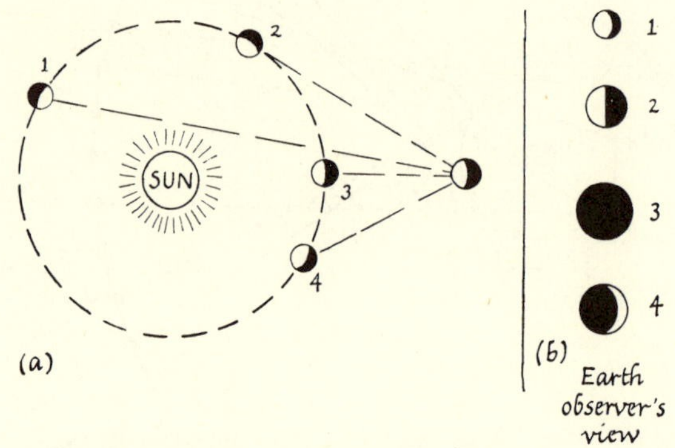

Fig. 16-5. Phases of Venus, as Seen from the Earth

As a crowning virtue of simplicity, Copernicus gave a new interpretation of the precession of the equinoxes. Precession, as discovered by the Greeks, was described as the whole star system (and the Sun) crawling slowly around the axis of the *ecliptic*, while the Earth and its equator plane and N-S axis stayed still. Copernicus reversed the description, saying the Sun and its ecliptic plane stay fixed; that is, the plane of the Earth's orbit stays fixed. And the Earth's equator-plane (and celestial equator) swings slowly around, always tilted 23½° to the ecliptic. Then Copernicus could describe precession simply: the Earth's spin-axis has a slow conical movement; carrying the equator-plane, it gyrates around a cone of angle 23½° in 26,000 years. Though Copernicus gave this clear picture of what happens in the precession of the equinoxes, he had no idea what "caused" it. That problem had to wait for Newton, who showed that, like so many astronomical phenomena, it is a result of universal gravitation.

Though the simplicity delighted him, Copernicus presently found that steady motion in simple circular orbits would not fit the facts. He had to make the orbits eccentric, and even add little epicycles. In doing this he spoiled the simplicity somewhat, and perhaps made it harder to guess the underlying simple rules which Kepler found later; but, like Ptolemy, he was insistent on making his machinery fit observations accurately. In this respect both sys-

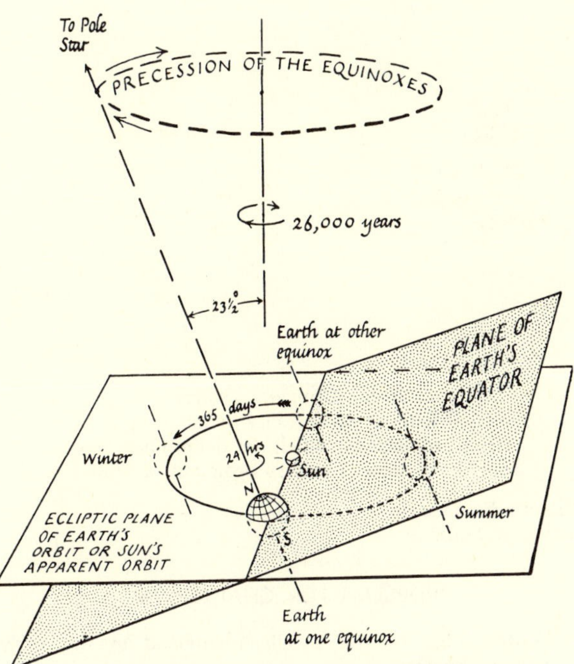

Fig. 16-6. Sketch Showing Motion Called the Precession of the Equinoxes

tems were good descriptions of the observed motions, and we ought not to call either "wrong."

Copernicus spent twenty years or more constructing and perfecting his scheme. During this time he became well known among mathematicians and astronomers, and some came to talk and study with him and carry away his powerful idea. He sent a small outline of his scheme to friends. Yet he had no wish for fame. Many urged him to publish his work: his star maps, his tables of observations, and his great scheme of the Solar System with its full defense and all the details he had worked out. Even after a friend had published a preliminary review of his scheme, he delayed for many years because he wanted to correct and improve it before he offered such a revolutionary change of view to a conservative public. He had no fear of conflict with the Church of which he was a well-accepted member; and the friend who offered to pay for the publishing was himself a Cardinal. But he knew that such a book would arouse opposition, and he feared ridicule. Great clearness, critical reasoning, and organized data would be needed to convince a prejudiced world. It was a tremendous thing to try to overthrow the Ptolemaic system, founded and perfected by the great men of the past and made almost sacred by tradition and practical use. For a long time the Ptolemaic scheme had been failing in accuracy—even the calendar had accumulated a big error. Yet no one doubted its essential rightness. Astronomers merely tried to correct its radii and shift its equants for better agreement. Copernicus wanted to be very sure of his ground. He could afford to wait, for he knew that truth is very strong.

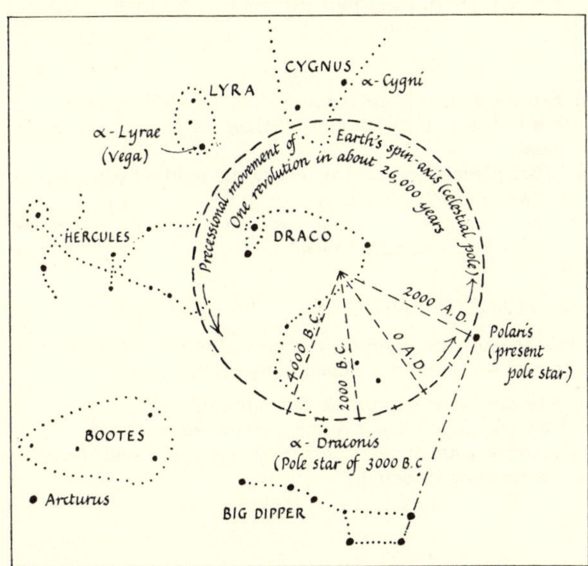

Fig. 16-7. The Precession of the Equinoxes
Sketch of a large patch of Northern sky (about 90° by 90°), showing the slow movement of the celestial North Pole among the stars. The point where the Earth's spin-axis cuts the pattern of the stars moves slowly around a roughly circular path making one revolution in about 26,000 years. (After Sir Robert Ball.)

At last he was persuaded, and he wrote a great book which was set up in type and published at the very end of his life. Its title was *De Revolutionibus Orbium Cœlestium*, "On the Revolutions of the Heavenly Spheres." He dedicated it to the Pope, saying in his dedication,

"I can easily conceive, most Holy Father, that as soon as some people learn that in this book . . . I ascribe certain motions to the Earth, they will cry out at once that I and my theory should be rejected. For I am not so much in love with my conclusions as not to weigh what others will think about them, and although I know that the meditations of a philosopher are far removed from the judgment of the laity, because his endeavor is to seek out the truth in all things, so far as this is permitted by God to the human reason, I still believe that one must avoid theories altogether foreign to orthodoxy. Accordingly, when I consider in my own mind how absurd a performance it must seem to those who know that the judgment of many centuries has approved the view that the Earth remains fixed as center in the midst of the heavens, if I should, on the contrary, assert that the Earth moves; I was for a long time at a loss to know whether I should publish the commentaries which I have written in proof of its motion."

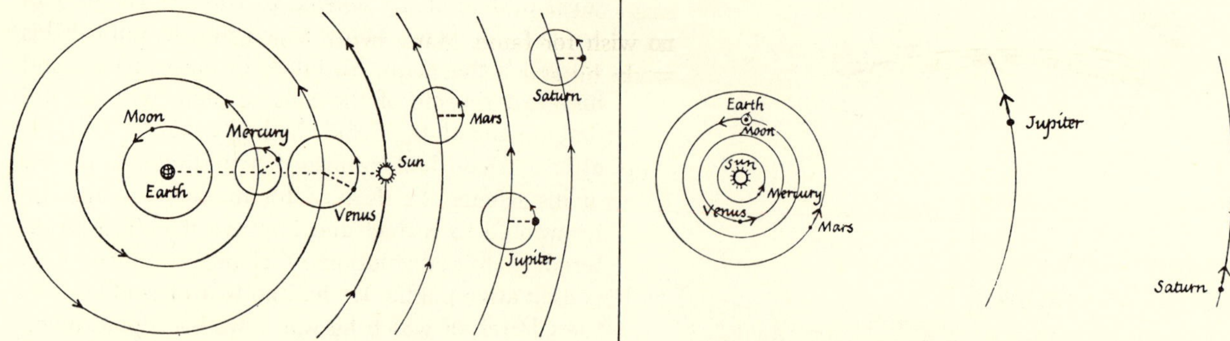

Fig. 16-8a. Ptolemaic System, sketched without eccentricity or equants. Order and proportions of orbits not determinate. Epicycle radii not "to scale."

Fig. 16-8b. Copernican System, sketched without eccentricity or minor epicycles. Orbit proportions, which *are* determinate, are roughly to scale. (Moon's orbit out of scale.)

Fig. 16-8. Comparison of Simple Ptolemaic Scheme and Simple Copernican Scheme

He further says,

"If there be some babblers, who, though ignorant of all mathematics, take upon them to judge of these things, and dare to blame and cavil at my work, because of some passage of Scripture which they have wrested to their own purpose, I regard them not, and will not scruple to hold their judgment in contempt."

In the book he gave his star tables and planetary observations as well as his great exposition of the new view of the Solar System.

The book was set in type and printed far away, in Nuremberg. It starts with a general description of the scheme and its advantages.[4] Then chapters expound the necessary trigonometry, and a section develops the rules of spherical astronomy. Then the Sun's "motion"—or rather the Earth's—is discussed in full, with the explanation of precession of the equinoxes. The Moon's motion is then discussed, and the last two sections deal with the motions of the planets in full detail. In the latter, he included discussions of measurements of the distances of Sun and Moon and the sizes of planetary orbits.

It was a great book indeed, destined to have far-reaching effects. Copernicus never read it in its final printed form. While he was waiting for its publication, an old man of 70, he was taken very ill, partly paralyzed. On May 23, 1543 the first printed copy was sent him, so he saw it and touched it. That night he died peacefully.

[4] A friend added a timid preface to say this was only a theory—far from that, Copernicus believed his scheme was *true*.

PROBLEMS FOR CHAPTER 16

1. Write a list of the people mentioned in the early Astronomy chapters,

(a) who clearly held the Copernican view (give dates where you can),

(b) who clearly held the Earth-in-center view (give dates where you can).

2. TIME CHARTS

(a) On a full sheet of graph paper or ordinary paper, make a "time chart" showing: primitive man, Babylonians, and other groups who gathered astronomical knowledge, to and including people in the Greek School at Alexandria.

(b) Draw another chart running from Early Greeks to Copernicus, and mark important astronomers on that.

★ 3. ORBIT SIZES

(a) Explain with a clear diagram how Copernicus estimated the radius of a planet's orbit. Give the geometrical argument.

(b) What planets can be treated by the method you gave in answering (a)?

(c) What other astronomical measurement strongly resembles the method you gave in (a)?

★ 4. PLANETS' PATHS

Planets such as Jupiter are observed to move through the pattern of the stars in a looped path, an "epicycloid."

(a) How did Ptolemy produce this pattern?

(b) How did Copernicus say the pattern was caused?

(Illustrate your answers with clear diagrams and describe the mechanisms in detail.)

★ 5.

(a) Why would you expect Venus to show phases like the Moon?

(b) Why would you not expect Jupiter to show phases? (Give sketches.)

CHAPTER 17 · TYCHO BRAHE (1546-1601)

"The fault, dear Brutus, lies not in our stars. . . ."

The Copernican Revolution

Copernicus struggled to escape from the atmosphere of Aristotelian dogmatism and argumentation in which he had been brought up. He loved truth, and he succeeded in producing a clearer, better picture of the heavens. Though insistent, he was calm and peaceful and considerate, and he persuaded many to his view. After his death, knowledge of his system spread. His tables were checked, corrected, and printed. His estimate of the length of the year (365 days, 5 hours, 55 minutes, 58 seconds) was used by a later Pope to reform the calendar, which was by then days out of gear with the seasons. When Copernicus himself had been consulted on calendar reform some eighty years earlier, he had demurred because he wanted first to clear up his own ideas about the system of the world. Here is a tribute to his final work: the reformed calendar, which we still use, has its scheme of leap years so well arranged that it will remain true to the Sun's seasonal calendar within *one day* for the next 3,000 years!

But as time went on others taught the new system less quietly and less tactfully, and its real effect began to be felt. It upset philosophers, disturbed ordinary people, started scientists towards new thinking, and brought the official Church into open opposition. Within a century of his death, Copernicus' work, originally dedicated to the Pope, became the center of one of the most violent intellectual controversies that the world has known. How could it do this? Because it upset knowledge that was taken for granted and seemed obviously right; and because it upset the intellectual outlook of the times. Copernicus was attacking—though he did not realize he was—a great interwoven structure of thought and belief. In those days scholarly knowledge was not divided into separate fields of study such as physical science, biology, physiology, psychology, sociology, languages, arts, and philosophy. We who are taught in organized classes and specialized courses can hardly picture the confused but strongly knit *general* intellectual outlook of the medieval intellectual world. The medieval scholars who met and discussed and taught were masters of all fields, anxious to maintain a scheme that gave all knowledge the same assured standing. They tied all studies together in a unified system.

The natural world was regarded as made up of four basic elements: earth, air, fire, and water. Liquids flowed because they were mostly water; solids were dense and strong because they were mostly earth; and so on. Four corresponding "humours" ruled men's behavior—according to which of the four is dominant in him, a man is angry, sad, calm, or strong in his temperament. The planets, marked out among the stars by their wandering, were given qualities that linked them both to man and to the four physical elements. They were linked to human temperament and fate, and they were linked to the metals whose properties depend on their proportions of earth and fire. (For example, Mars, the reddish planet, might represent the god of war, angry temperament, fiery metal.) The planets and stars in turn linked man to divine providence. The celestial system of spheres formed a vast ideal model for man himself. And beyond the outermost sphere of stars was heaven—placed on top of the astronomical world in the proudest place of the celestial scheme. Thus an attack on the planetary system threatened to change all teaching about nature and man, and even man's relation to God.

But did Copernicus attack the planetary system with its moving spheres? He only moved the center from the Earth to the Sun. To see why that upset the whole system of thought, first see why the spheres were taken for granted.

Professor Herbert Dingle gives one historian's view as follows:

"They were not seen or directly observed in any way; why, then, were they believed to be there? If you imagine the Earth to be at rest and watch the sky for a few hours, you will have no difficulty in answering this question. You will see a host of stars, all moving in circles round a single axis at precisely the same rate of about one revolution a day. You cannot then believe that each one moves independently of the others, and that their motions just happen to have this relation to one another. No one but a lunatic would doubt that he was looking at the revolution of a single sphere with all the stars attached to it. And if the stars had a sphere, then the Sun, Moon and planets, whose movements were almost the same, would also be moved by the same kind of mechanism. The existence of the spheres having thus been established and accepted implicitly for century after century, men no longer thought of the reasons which demanded them, but took them for granted as facts of experience; and Copernicus himself, despite his long years of meditation on the fundamental problems of astronomy, never dreamed of doubting their existence."[1]

But presently the full meaning of Copernicus' work was realized. A spinning Earth would remove any need for a spinning bowl to carry the stars. The stars needed no bowl: they could just hang in the sky at *any* great distance, and if at *any* perhaps at *many different* distances. That pushed the stars back into the remoter region of Heaven. That was a terrifying change—where was the Heaven of theology now to be located, the Home of God and the abode of departed souls? A century after Copernicus, Giordano Bruno suggested such a picture of infinite space, peopled with stars which were distant suns. He was burnt at the stake as a heretic for his unfortunate views.

Again, Copernicus pushed the Earth out of its ruling position as center of the universe, to join Mars and Jupiter, as a planet ruled by the Sun. That made the machinery of spheres seem less complex and less necessary; and it completely upset the picture of planets holding controlling influences over a central Earth and Man.

"We can hardly imagine today the effect of such a change on the minds of thoughtful men in the sixteenth century. . . . With the spheres went a localised heaven for the souls of the blessed, the distinction between celestial and earthly matter and motion became meaningless, the whole place of man in the cosmic scheme became uncertain.

". . . Copernicus himself . . . was thoroughly medieval in outlook, and had he been able to foresee what his work was to do we may well believe that he would have shrunk in the utmost horror from the responsibility which he would have felt. But what he did was to make it possible for the new scientific philosophy to emerge."[1]

[1] From *Copernicus and the Planets*, a talk by Herbert Dingle, Professor of the History and Philosophy of Science in the University of London. The rest of the discussion on this page and the previous one also draws on Professor Dingle's talk, published as Chapter III of a symposium of talks on the *History of Science* (Cohen and West, Ltd., London).

If this was an upheaval for scholars, what might it not do to the vast uneducated crowd once they understood it? Not only was the machinery of the spheres wrecked, but the general outlook on knowledge was threatened with a change, and so was men's faith in God's provision for their souls.

Copernicus' work reached the common people through its effect on astrology. For many centuries, superstitious people—almost everyone, from king to beggar—believed that the Sun, Moon, and planets influenced their fate. Casting a man's horoscope was a system of fortune-telling based on the positions of the planets at his birth. A horoscope by an accredited astrologer was trusted to help kings rule wisely (or safely) and to foretell any man's character and future. As long as the Earth was considered the center of the heavenly system, with planets wandering strangely around our important selves, it was natural to think that planetary motions occurred for our benefit and might even control our fate. When Copernicus moved the Sun into the center of the planetary system *and described the Earth as merely one of the planets*, he started the downfall of astrology from its position as counsellor of kings and comforter of common men to its present rank as amusing fiction. Yet civilized man is still gullible enough—insecure enough, or perhaps romantic enough—to keep the publishing of astrology books a thriving business.

Opposition to the Copernican system arose from the Church, probably for two reasons; first because the system contradicted the Church's teaching of astronomy and philosophy; and second because preaching the new system meant a questioning of authority—questioning, even defying, the authority and tradition of the Ptolemaic system. Rebellious people who like new ideas, and enjoy proving others wrong, seldom find themselves popular. In those days the Church was anxious to maintain its tremendous power and insisted that all people obey its authority strictly and believe its teachings completely. Anyone who questioned the Church's authority or disagreed with its teaching was, to his own knowledge, risking his life and, in the Church's view, risking his soul. There were a few such men at that time. They were the martyrs. Though a scientist preaching a new *theory* might seem harmless, one insisting that it was *true* would seem an annoying rebel. He who rebels in one field may rebel in other fields, too, so the provocative arguer in any field is apt to strike the Church (or the State—authority wherever it is) as a dangerous man politically. Within less than a century Copernicus' book,

De Revolutionibus Orbium Cœlestium, had been placed on the list of forbidden books by the Church. (It remained there for 200 years, and was quietly dropped about 1830.)

In this stormy time for Astronomy, came Tycho Brahe, the great observer whose amazingly accurate measurements formed the basis for Kepler's discoveries and in turn Newton's explanations. Galileo lived at the same time as Kepler, and bore some of the worst of the storm. A few centuries later, Copernican advocates seemed harmless. As Lodge[2] points out, this happens in age after age. In the early 1800's geologists met with violent condemnation as impious critics of the Bible story of Creation. Later in the century geology was safe, but theories of evolution were condemned and the teaching of them forbidden—perhaps this continues. *Every* age has one or more groups of intellectual rebels who are persecuted, condemned, or suppressed at the time but, to a later age, seem harmless. Some are cranks and others are wise prophets—pacing the growth of man and man's knowledge—men ahead of their time.

Tycho Brahe

Tycho Brahe was the eldest son of a noble Danish family—"as noble and ignorant as sixteen undisputed quarterings [of their coat of arms] could make them." Hunting and warfare were regarded as the natural aristocratic occupations: though books were fashionable, study was fit only for monks, and science was useless and savoured of witchcraft. Tycho would probably have been brought up to be a soldier had not his uncle, a man of more education, adopted him, his parents agreeing unwillingly when their second son was born. His uncle gave him a good education. Tycho began Latin at the age of 7; his parents objected, but his uncle urged this would help him with law. At 13 he went to the university to study philosophy and law, and there his interest was suddenly turned toward astronomy by a special event. An eclipse of the Sun took place. The astronomers had predicted it, and people turned out to watch for it with great excitement. Tycho watched and was amazed and delighted. When it happened at the predicted time it awoke in him a great love for the marvelous that made him long to understand a science that could do such wonders. He continued to study law, but his heart was set on astronomy. He spent his pocket-money on astronomical tables and a Latin translation of Ptolemy's *Almagest* for serious study. After three years, his uncle sent him abroad with a tutor, to travel and to read law in German universities. There he continued to work at astronomy secretly. He sat up much of each night, observing stars with the help of a celestial globe as small as his fist. He bought instruments and books with all the money he could get from his tutor without revealing his purpose. He found the tables of planetary positions inaccurate. Ptolemaic tables and Copernican tables disagreed, and both differed from the facts. As a boy of 16, he realized—what the professional astronomers of Europe had missed—that a long series of precise observations was needed to establish astronomical theory. A few planetary observations made at random could not decide between one system for the heavens and another. Here was the start of his life work.

When he was 17 he observed another special event, a conjunction of Jupiter and Saturn. Two planets are in "conjunction" when they cross the same celestial longitude together, very close to each other. This strange "clashing" of two planets could be predicted with the help of tables such as Ptolemy's or Copernicus'. Such events were regarded by superstitious people as bringing good or bad luck. The enthusiastic young Tycho observed the conjunction and compared his observed time with the predictions of the planetary tables. He found Alphonso's revision of the Ptolemaic tables wrong by a month,[3] and Copernicus' tables wrong by several days. He decided then to devote his life to the making of better tables—and he succeeded with a vengeance. He became one of the most skillful observers the world has known. Neither his aristocratic birth nor his education had saved him, however, from being superstitious, full of belief in occult influences; and he believed that this conjunction foretold, and was responsible for, the great plague which soon after swept across Europe.

Tycho started observing with a simple instrument: a pair of jointed sticks like compasses, one leg pointed at a planet and the other at a fixed star. Then he measured the angular separation by plac-

[2] I owe a debt of inspiration and delight to Sir Oliver Lodge, who first showed me how the history of astronomy illuminates physics, in his *Pioneers of Science* (Macmillan, London, 1893). In the next few chapters, I have drawn on Sir Oliver's book for general ideas and in many places for words and phrases. I am grateful to him and to those earlier writers on whom he drew in turn. Modern historians of science have weeded out some mistaken information, and they quite rightly plead for a wider view with less hero worship; yet the book was the work of a physicist and a great man with vision.

[3] A month's error may seem large in predicting a meeting of planets; but those tables dated back, essentially, to Ptolemy fourteen centuries before. One month in 1400 years seems little enough, a great credit to the Ptolemaic system as representing the facts accurately, however clumsily.

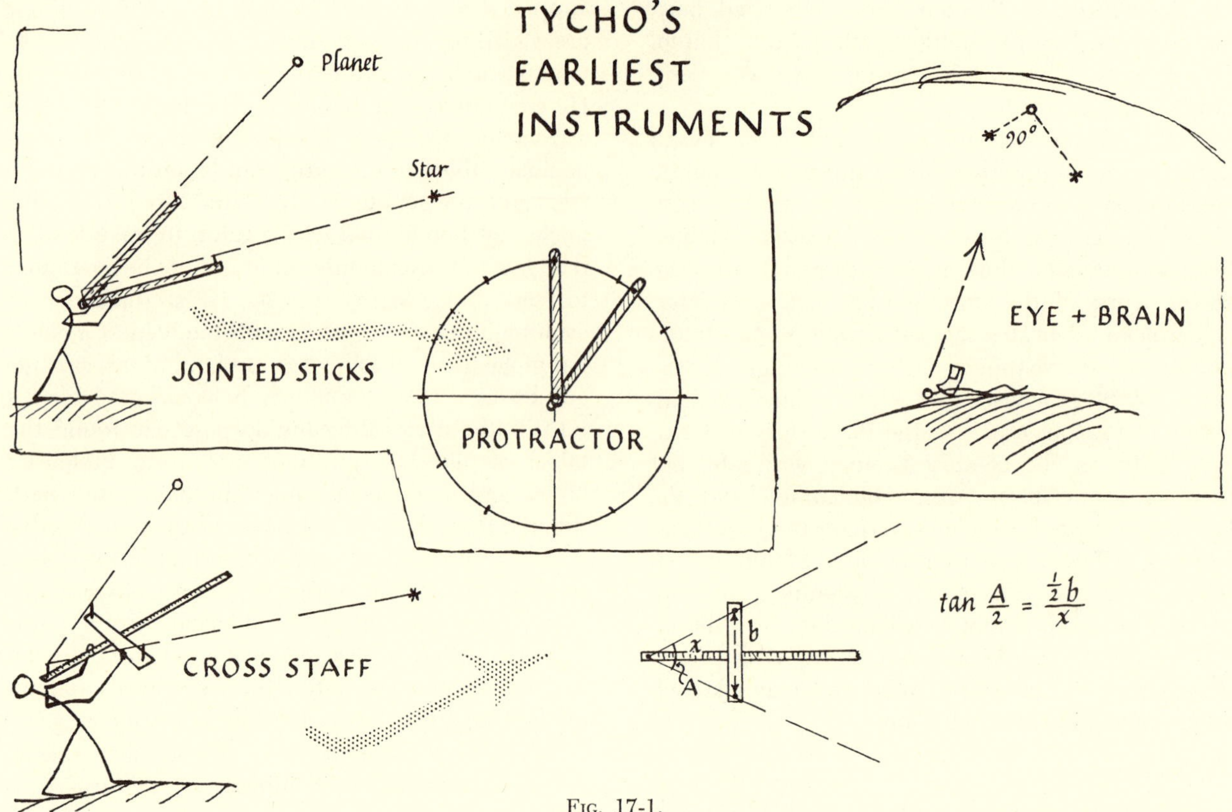

FIG. 17-1.

ing the compasses on a graduated circle drawn on paper. He often used eyes + brain alone to mark a planet's position when it formed a 90° triangle with two stars that he knew. Soon he obtained a "cross staff," a graduated stick with a slider at right angles that carried sights at its ends. An observer looking through a peephole at the near end of the main stick could set the sights on two stars, and thus measure the angle between them. He found his instrument was not accurately graduated, so he made a careful table of corrections, showing the error at each part of the scale—a method of precision that he used all his life. This is the way of the best experimenters: for great precision they do not try to make an instrument "perfectly accurate," but they make it robust and sensitive, and then they calibrate it and record a trustworthy table of errors.

Presently he was called home by threats of war. His uncle died and the rest of his family did not welcome him. They blamed him for neglecting law, and they despised his interest in star-gazing. Disappointed, Tycho left Denmark for Germany to continue his studies. In his travels, he made friends with some rich amateur astronomers in Augsburg. He persuaded them that very precise measurements were needed, and they joined in constructing an enormous quadrant for observing altitude-angles by means of a plumb line and sighting holes fixed on the arm of a huge graduated circle. This wooden instrument was so big that it took twenty men to carry it to its place in a garden and set it up. Its circle had a radius of 19 feet. It had to be big for accurate measurements—there were no telescopes in those days, merely peep-holes for sighting. The quadrant was graduated in sixtieths of a degree. Tycho and his enthusiastic friends also had a huge sextant of radius about 7 feet. This was the beginning of his accurate planetary observations.

In his stay in Germany he met with a strange accident. His violent temper led him into a quarrel over mathematics, and that led to a duel which was fought with swords at seven o'clock one December night. In the poorly lit fight, part of Tycho's nose was cut off. However, he made himself a false nose (of metal or putty, probably painted metal). He is said to have carried around with him a small box of cement to stick the nose on again when it came off.

After four years in Germany, Tycho went home again, this time to be received well as an astronomer of growing fame. His aristocratic relatives thought more kindly of science and received him with admiration. When Tycho's father died, another uncle welcomed Tycho to his estate and gave him an

QUADRANS MAXIMUS QUALEM OLIM
PROPE AUGUSTAM VINDELICORUM EXSTRUXIMUS

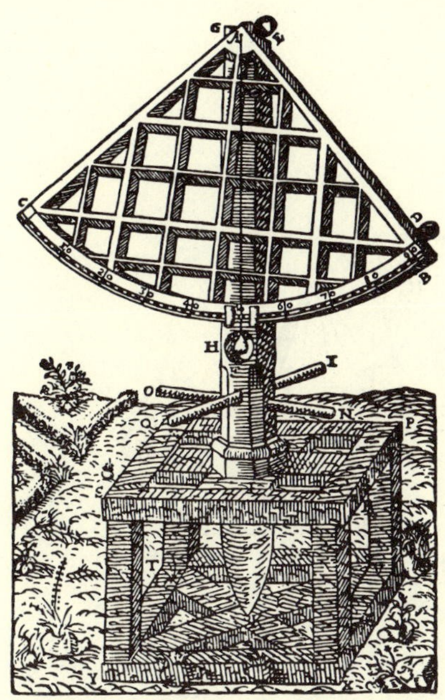

FIG. 17-2. TYCHO'S EARLY QUADRANT, built with his friends when he was a young man, travelling in Germany. (Radius of circle: about 19 feet.) Sighting the Sun or a planet through peepholes D, E, the observer could read its "altitude" by the plumb line AH on the scale graduated in sixtieths of a degree. This picture and the following ones are reprinted from *Tycho Brahe's Description of His Instruments and Scientific Work* by permission of The Royal Danish Academy of Sciences and Letters.

extra house as a laboratory for alchemy. Tycho's fascination with the marvelous had drawn him to alchemy. This was not a complete break with astronomy because the astrology of the time linked the planets closely with various metals and their properties. Alchemy had a useful side too: it gave Tycho knowledge of metals for instruments. Thereafter he often combined a little alchemy with his astronomical work and even concocted a universal medicine.

The New Star

The year after his return, a new star blazed up in the sky and was visible for many months. At first it was as bright as Venus, and could even be seen in daylight.[4] Tycho, amazed and delighted, observed it carefully with a large sextant, and found it was very far away, one of the fixed stars, "in the 8th sphere, previously thought unchangeable." After much careful watching and recording, Tycho published a report on it.

Tycho's fame was growing, and a group of young nobles asked him to give a course of lectures on astronomy. He refused at first, thinking this below the dignity of one of noble birth; but he was persuaded when he received a request from the King.

At this time, Tycho married a peasant girl—to the horror of most of his family—and thereafter seems to have modified some of his aristocratic prejudices.

The Great Observatory, Uraniborg

Finding his life as a noble interfered with astronomy, he embarked on another move to Germany; but King Frederik II of Denmark, understanding that Tycho's work would bring the country great honor, made him a magnificent offer. If Tycho would work in Denmark, he should have an island for his observatory, estates to provide for him, a good pension, and money to build the observatory. Tycho accepted with enthusiasm. Here at last was a chance to carry out his ambitions.

Tycho built and equipped the finest observatory ever made—at enormous cost.[5] He called it Uraniborg, The Castle of the Heavens. It was built on a hill on the isle of Hveen, surrounded by a square wall 250 feet long on each side, facing North, East, South, and West. In the main building there were magnificent living quarters, a laboratory, library, and four large observatories, with attic quarters for students and observers. There were shops for making instruments, a printing press, paper mill, even a prison for recalcitrant servants. Tycho made and installed a dozen huge instruments and as many smaller ones. These instruments were the best that Tycho could devise and get made—all constructed, graduated, and tested with superb skill and fanatical attention to accuracy. Some of them were graduated at intervals of 1/60 degree, and could be read to a fraction of that.

In the library, Tycho installed the great celestial

[4] We now know that a new star, or *nova*, appears in the sky fairly frequently—some sudden condensation, or other change, heats a star to higher temperature. Much more rarely—averaging once in several centuries in our galaxy—there is a far brighter outburst, a *supernova*. Hipparchus probably saw one, and Tycho's new star was one. A recent speculative theory suggests that the appearance of a supernova involves the radioactive element californium. Tycho's careful comparisons of the brightness of his star with standard stars as it died down fit well with the "half life" of californium—a fantastic modern use of his careful work.

[5] Some years later, Tycho stated the total cost. His estimate was equivalent to about 17,000 English pounds at that time. Translated in terms of cost-of-living this would be at least $200,000 today. Translated in terms of luxury and equipment, Uraniborg would be a multimillion dollar observatory.

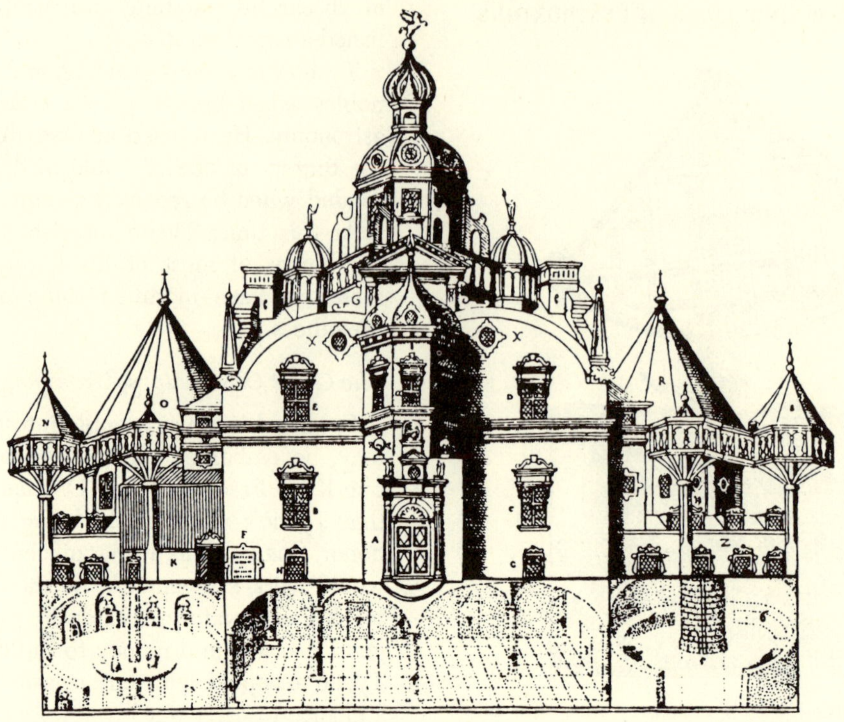

FIG. 17-3. URANIBORG. Design of the main building, built about 1580.

globe he had ordered in Augsburg some years before. It was covered with polished brass, an accurate sphere as high as a man. As the work of the observatory proceeded, star positions were engraved on it. Making and marking it took 25 years in all.

In Tycho's study, a quadrant was built on the wall itself, a huge arc with movable sights for observing stars as they passed a peg in a hole in the wall opposite. This was one of his most important instruments, and Tycho had the empty wall space inside the arc decorated with a picture showing himself and his laboratory, library, and observatories. Fig. 17-5 is an engraving of this mural, with observers in front using the quadrant and the primitive unreliable clocks of Tycho's day. (Tycho said the portrait was a good likeness.)

It was a gorgeous temple of science, and Tycho worked in it for twenty years, measuring and recording with astounding precision. Students came from far and wide to work as observers, recorders, and computers. This was Tycho's great work, to make continuous accurate records of the positions of Sun, Moon, and planets. Then he proposed to make a theory for them. At first he did not concern himself much with theory—though he insisted that without *some* theory an astronomer could not proceed with his work. Later in life he put forward a useful compromise that acted as a stepping stone for thinkers

GLOBUS MAGNUS ORICHALCICUS

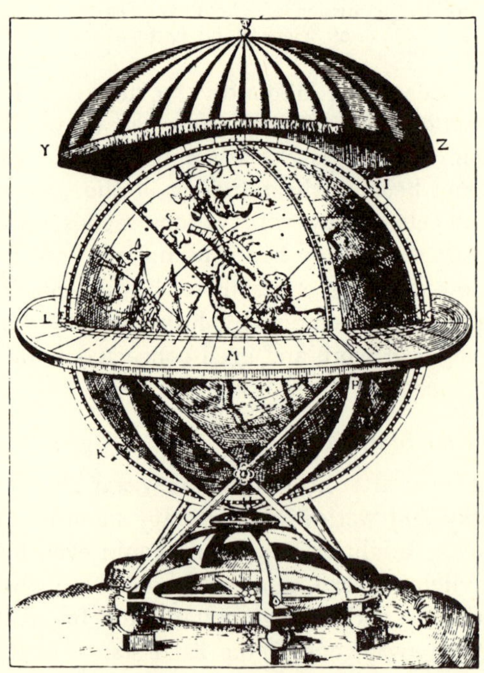

FIG. 17-4. TYCHO'S GREAT GLOBE
Tycho had this globe made very carefully, at great expense, so that he could mark his measurements of star positions on its polished brass surface. He ordered it before he started Uraniborg, had it brought and installed there, and took it with him when he moved to Prague.

QVADRANS MVRALIS
SIVE TICHONICUS

Fig. 17-5. Tycho's Mural Quadrant
The huge brass arc was firmly fixed in a western wall, with its center at an open window in a southern wall. The empty wall above the arc was decorated by a huge painting showing: Tycho observing; students calculating; Tycho's globe, books, dog; and some of Uraniborg's main instruments. An observer sighted the star (or Sun) by pinholes at F and a marker in the window. The brass arc (radius over 6 ft.) could be read to ⅓₆₀ degree. This sketch, from Tycho's own book, shows an observer at F, a recorder, and a timekeeper with several clocks. Good clocks had not been invented, but these were the best Tycho could make.

who found the jump from Ptolemy to Copernicus too big. He pictured the five planets (without the Earth) all moving in circles around the Sun. Then the whole group, Sun and planets, moved around the Earth; and so did the Moon. Geometrically, this is equivalent to the Copernican scheme, but it avoids the uncomfortable feeling of a moving Earth.

Tycho became the foremost man of science in Europe. Philosophers, statesmen, even kings, and many scientists, came to visit him. They were received in grand style and shown the wonders of the castle and its instruments. Yet Tycho could be hot tempered and haughty to people he thought stupid or visiting only for fashionable gossip. To such he appeared to be a rude contradictory little man with a violent temper, but to the wise he was a great experimental scientist with a passion for accuracy and a delight in marvels.

For all his scientific fervor, Tycho was vain and superstitious. He kept a half-witted dwarf in his household; and at banquets, with his peasant-born wife presiding, Tycho insisted on listening to the dwarf's remarks as prophetic. "It must have been an odd dinner party, with this strange, wild, terribly clever man, with his red hair and brazen nose, sometimes flashing with wit and knowledge, sometimes making the whole company, princes and servants alike, hold their peace and listen humbly to the ravings of a poor imbecile."[6]

Troubles

While Tycho's grand observatory attracted visitors from far and wide, his impetuous ways brought him troubles. He made jealous enemies at court, and he had serious troubles with his tenants. When the King had given Tycho the island for life, the peasants who had small farms on it were bound to do some work for him as his tenants. They did much of the work of building Uraniborg, and after that

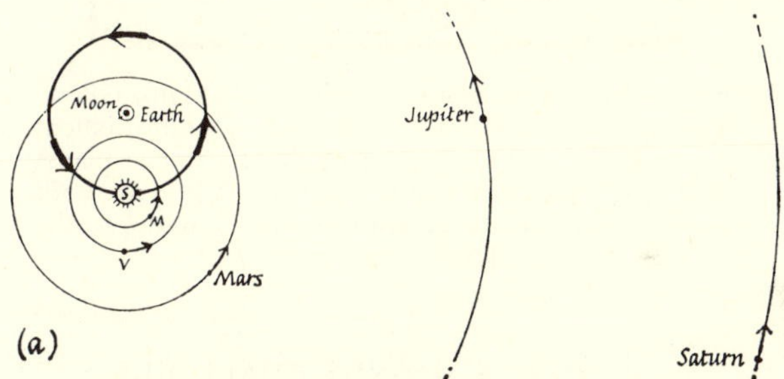

Fig. 17-10a. Tycho Brahe's Theory of Planetary Motion
The Sun moves around a fixed Earth and carries all the rest of the Copernican system with it.

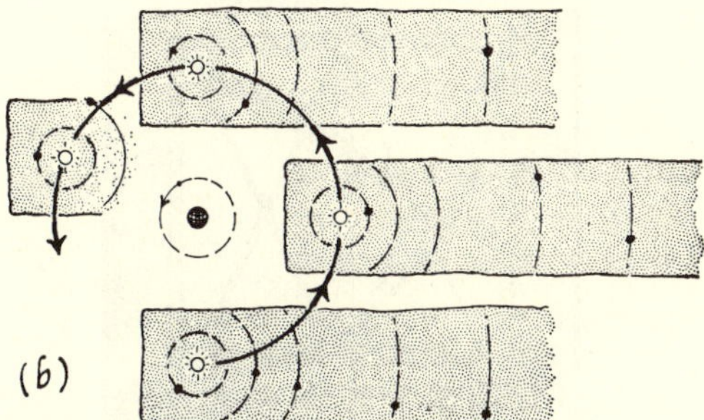

Fig. 17-10b. Tycho Brahe's Theory of Planetary Motion
A sketch, *not to scale*, showing successive positions of the system in January, April, July, September.
(The planetary system moves like a frying-pan given a circular motion by a housewife who wants to melt a piece of butter in it quickly.)

---

[6] Professor Stuart quoted by Sir Oliver Lodge.

Figs. 17-6, 7, 8, 9. Some of Tycho's Instruments in Uraniborg.
Reprinted from *Tycho Brahe's Description of His Instruments and Scientific Work* by permission of The Royal Danish Academy of Sciences and Letters.

ARMILLÆ ZODIACALES

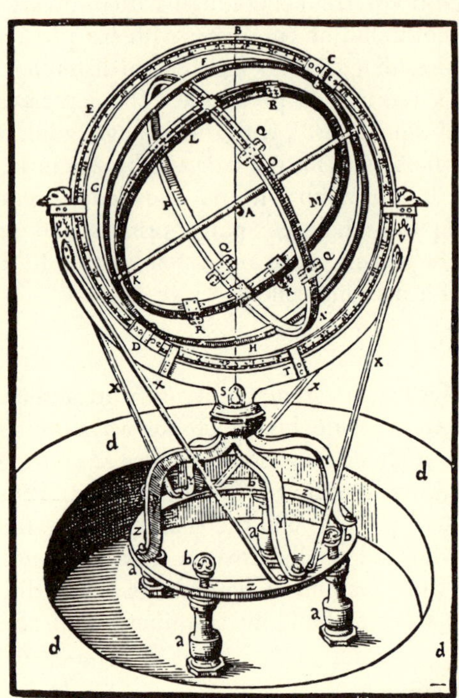

Fig. 17-6. An "Astrolabe" Built by Tycho, following the design used by Hipparchus. This instrument measures the latitude and longitude of a star or planet directly. (Diameter of circles: about 4 ft.) Tycho built several improved forms, with one axis parallel to the Earth's polar axis.

SEXTANS ASTRONOMICUS TRIGONICUS PRO DISTANTIIS RIMANDIS

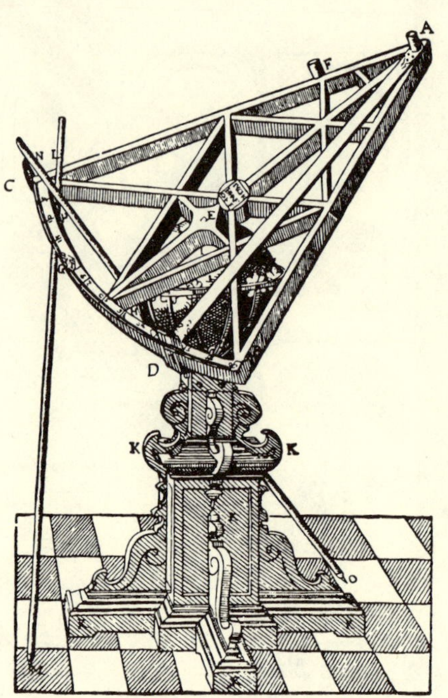

Fig. 17-7. One of Tycho's Sextants
This instrument, with brass scales and wooden frame, was used to measure the angle between the directions of two stars, by two observers sighting simultaneously along arms AD, AC. It was carried on a globe which could twist in firm supports, so that it could be tilted in any direction. (Length of arms: about 5 ft. Angles estimated to $\frac{1}{240}$°.)

QUADRANS MAGNUS CHALIBEUS.
IN QUADRATO ETIAM CHALIBEO COMPREHENSUS.
UNAQUE AZIMUTHALIS

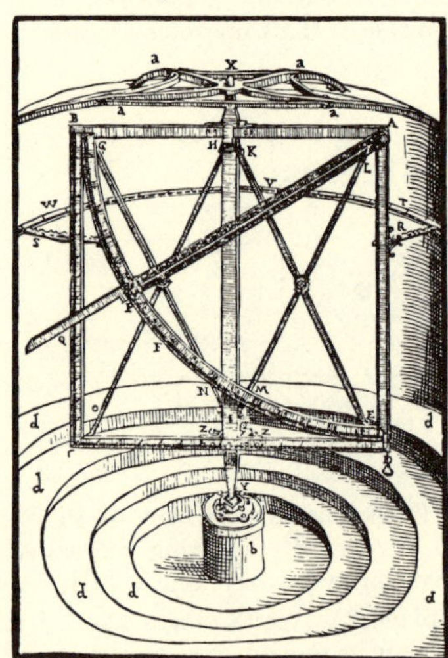

Fig. 17-8. Tycho's Great Quadrant
Radius about 6 ft.

PARALLATICUM ALIUD SIVE REGULÆ
TAM ALTITUDINES QUAM AZIMUTHA
EXPEDIENTES

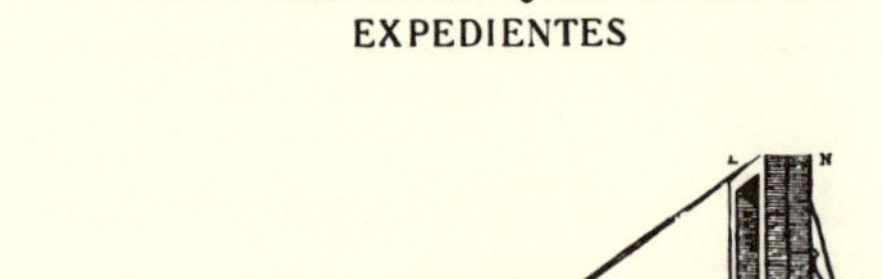

Fig. 17-9. A Ruler Built by Tycho, to measure altitudes and azimuths.

Tycho made them do chores for his household. He was unreasonable in his demands and haughty in his treatment—he may even have put unwilling workers in his prison. Several times complaints reached the King, who had to intervene. The King had given Tycho other estates as well, whose rents were to provide him with money for living and running the observatory. In return, Tycho was expected to keep the estates in reasonable repair; and, again, complaints reached the King that he failed to do so.

King Frederik's intense interest in Tycho's work protected him, but when the King died, eleven years after the observatory was built, Tycho's troubles began to grow. Young King Christian IV came to the throne and was surrounded by nobles who were less favorable to Tycho. Some of Tycho's estates were withdrawn, and he began to worry about the future. He wrote to a friend saying he might have to leave the island, comforting himself that "every land is home to a great man," and that wherever he went the same heavens would be over his head. The young King was sympathetic but needed to economize. More of Tycho's estates were taken. Tycho, seeing that he would not have enough money to maintain Uraniborg, moved to the mainland. Feeling unwelcome there, he decided to leave his ungrateful country and look for a new patron and place to work. He took his smaller instruments with him and set them up in temporary quarters in Germany while he negotiated with the Emperor Rudolph of Bohemia, an enlightened ruler with a great interest in science. He wrote a long haughty letter to young King Christian of Denmark offering to return, and received only a chilly reply. Meanwhile, he printed a large illustrated catalogue of his instruments and sent elegantly bound copies to possible patrons, including Rudolph.

The New Observatory in Prague

Finally, after two years of travelling and visiting, he arrived in Prague and was welcomed by Rudolph, who gave him a castle for observatory and promised him a huge salary. The Emperor was genuinely interested in astronomy (and probably astrology too), but he was careless as a ruler and could not always pay Tycho in full. Yet he did reestablish Tycho, and he deserves great credit for thus saving Tycho's work—credit which he received when Tycho's records were published with the name "Rudolphine Tables."

In his new castle, Tycho lived the same weird, earnest life he had lived in Denmark. He fetched his big instruments across Germany from Denmark and gathered round him a small school of astronomers and mathematicians. But his spirit was broken; he was a stranger in a strange land. He continued his observations and began setting up the Rudolphine Tables, but he became more and more despondent. After less than three years in Prague, he was seized with a painful disease and died. In the delirium of his illness he often cried, "Ne frustra vixisse videar," "Oh that I may not appear to have lived in vain." His life had been given him not just to enjoy but to achieve a great work and, still yearning for this, his life's ideal, he died. This doubt was undeserved by an astronomer who had catalogued a thousand stars so accurately that his observations are still used, a man who recorded the planets' positions for twenty years with accuracy calculated to 1/60 of a degree, a man who gave Kepler and Newton the essential basis for their work in turn. He had succeeded in his original intention, and his work was not in vain.

Just before he died, free from delirium, Tycho gathered his household around him, asked them to preserve his work, and entrusted one of his students, Johannes Kepler, with the editing and correcting and publishing of his planetary tables. The great instruments were preserved for a time but were smashed in the course of later warfare: only Tycho's great celestial globe now remains. The island observatory was broken up and there is little sign of it today. Denmark lost its great name as a center of science; and it was not until this century that its fame grew again as such a world center, this time around the name of Niels Bohr.

PROBLEMS FOR CHAPTER 17

★ 1. TYCHO'S PRECISION

Tycho made some of his observations with plumb lines and peepholes like rifle sights. His final estimates were usually to be trusted to 1 minute of angle (= 1/60 of 1 degree). To see how careful he must have been, answer the following questions:

(a) Suppose he had pointed his peephole sights at a planet, and that they carried a graduated angle-scale with them on which he could read the position of a vertical plumb line. Suppose his angle-scale was part of a circle of radius 7 feet. (Your protractor in an ordinary box of geometrical instruments has a radius about 3 inches.) How thin must the thread of his plumb line have been so that a mistake of 1 thread-thickness on the scale made an error of 1 minute of angle? (Give your answer as a fraction of an inch.) (*Hint*: With $r = 7$ ft, length of whole scale of 360° round the circumference would be . . . . Then 1° must take a length of scale about . . . . Then 1 minute must take . . . .)

(b) Does your estimate call for a cord, a string, a thread, or a spider filament?

2. Why did Kings often support Astronomers?

★ 3. PARALLAX AND STARS

(a) Suppose all the stars in some group or constellation were *infinitely* far away except one star and that single star was only a few billion miles away. What would its parallax motion look like? How long would the motion take for one cycle? Describe the pattern of this (apparent) motion for a close star (i) near the ecliptic; (ii) near the pole of the ecliptic (90° from ecliptic).

(b) To see whether Tycho Brahe had a hope of detecting the tiny parallax motion of the nearest fixed stars try the following calculation. In 6 months the Earth swings around half its orbit from one end of a diameter to the other, 186,000,000 miles, *straight across* from A to B. (Fig. 17-11) Suppose Tycho looked at a very near star, S, against a background of other stars which are very much farther away. In watching the position of S against the background, Tycho would swing his sight-line through an angle ASB which he would measure as an angular displacement of star S, among the background stars. Suppose this angle to be 1/360 of a degree. It seems doubtful that he could have detected a smaller shift than that. Use rough arithmetic ("judging") to answer the following:

(i) Taking the angle ASB to be 1/360 of a degree, estimate the distance, AS, of the star from the Earth. Use a method like that of Problem 1 above—do not try to use trigonometry for these extreme angles. (Take the shallow arc, AB, with center at S, as 186,000,000 ≈ 200,000,000 miles.)

(ii) Compare the result of (i) above with modern measurements. These are usually expressed by giving the time light takes to travel the distance considered. Light travels the *diameter* of the Earth's orbit, AB, in about 16 minutes (8 minutes from the Sun to us). Light from the nearest star takes about 4 years (≈ 2,000,000 minutes) to reach us. What value does *this* give for the angle ASB? (Avoid trig. Argue simply by proportions.)

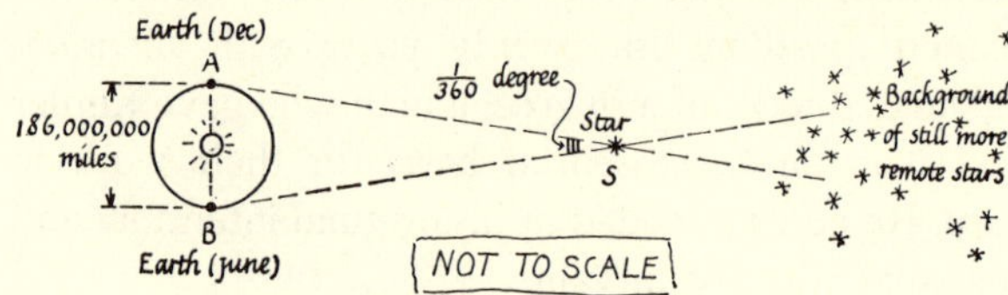

Fig. 17-11. Problem 2.
(Diagram *not to scale*—the real angle is only 1/360°)

4. Describe Tycho's big outdoor quadrant. Give a simple sketch, and explain what angle it measured and how it was used.

CHAPTER 18 · JOHANNES KEPLER (1571-1630)

Out of the night that covers me,
Black as the pit from pole to pole
I thank whatever gods there be
For my unconquerable soul.

—W. E. Henley (in hospital, 1875)

"For there is a musick where ever there is a harmony, order, or proportion: and thus far we may maintain the musick of the sphears; for those well-ordered motions and regular paces, though they give no sound unto the ear, yet to the understanding they strike a note most full of harmony."

From Religio Medici
Sir Thomas Browne (1642)

Kepler, the young German to whom Tycho Brahe left his tables, was well worthy of this trust. He grew into one of the greatest scientists of the age—perhaps equalled in his own time only by Galileo and later outshone only by Newton. As Sir Oliver Lodge points out, Tycho and Kepler form a strange contrast: Tycho "rich, noble, vigorous, passionate, strong in mechanical ingenuity and experimental skill, but not above the average in theoretical power and mathematical skill"; and Kepler "poor, sickly, devoid of experimental gifts, and unfitted by nature for accurate observation, but strong almost beyond competition in speculative subtlety and innate mathematical perception."[1] Tycho's work was well supported by royalty, at one time magnificently endowed; Kepler's material life was largely one of poverty and misfortune. They had in common a profound interest in astronomy and a consuming determination in pursuing that interest.

Kepler was born in Germany, the eldest son of an army officer. He was a sickly child, delicate and subject to violent illnesses, and his life was often despaired of. The parents lost their income and were reduced to keeping a country tavern. Young Johannes was taken from school when he was nine and continued as a servant till he was twelve. Ultimately he returned to school and went on to the University where he graduated second in his class. Meanwhile, his father abandoned his home and returned to the army; and his mother quarreled with her relations, including her son, who was therefore glad to get away. At first he had no special interest in astronomy. At the University he heard the Copernican system expounded. He adopted it, defended it in a college debate, and even wrote an essay on one aspect of it. Yet his major interests at that time seem to have been in philosophy and religion, and he did not think much of astronomy. But then an astronomical lecturership fell vacant and Kepler, who was looking for work, was offered it. He accepted reluctantly, protesting, he said, that he was not thereby abandoning his claim "to be provided for in some more brilliant profession." In those days astronomy had little of the dignity which Kepler himself later helped to give it. However, he set to work to master the science he was to teach; and soon his learning and thinking led to more thinking and enjoyment. "He was a born speculator just as Mozart was a born musician"[1]; and he *had* to find the mathematical scheme underlying the planetary system. He had a restless inquisitive mind and was fascinated by puzzles concerning numbers and size.[2] Like Pythagoras, he "was convinced that God created the world in accordance with the principle of per-

[1] Sir Oliver Lodge, *Pioneers of Science.*

[2] Most of us have similar delights, though less intense. You have probably enjoyed working on series of numbers, given as a puzzle or an "intelligence test," trying to continue the series. Try to continue each of the following. If you enjoy puzzling over them (as well as succeeding) you are tasting something of Kepler's happiness.

(a) 1, 3, 5, 7, 9, 11, . . . How does this series probably go on?
(b) 1, 4, 9, 16, 25, . . . ?
(c) 5, 6, 7, 10, 11, 12, 15, 16, . . . ?
(d) 2, 3, 4, 6, 8, 12, 14, 18, . . . ?
(e) 4, 7, 12, 19, 28, . . . ?
(f) 1 7 3 6 5 5 7 4 9 . . . ?
(g) 0 1 8 8 1 1 0 2 4 1 5 6 2 5 . . . ?

[Note that in (f) and (g) you must also find where to put the commas.]

FIG. 18-1. ? LAW RELATING SIZES OF PLANETARY ORBITS ?

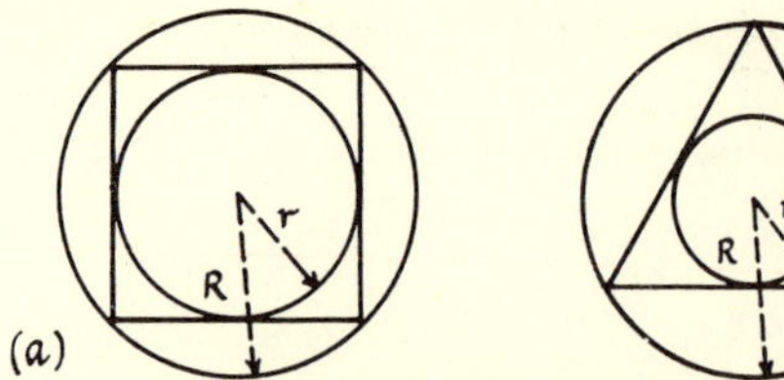

FIG. 18-1a. KEPLER'S FIRST GUESS
A regular plane figure (such as a square) can have a circle inscribed, to touch its sides. It can also have an outside circle, through its corners. Then that outside circle can be the inner circle for another, larger plane figure. The ratio of radii, R/r, is the same for all squares; and it has a different fixed value for all triangles. Geometrical puzzle: what is the fixed value of R/r for the inner and outer circles of a square? What is the value for a triangle?

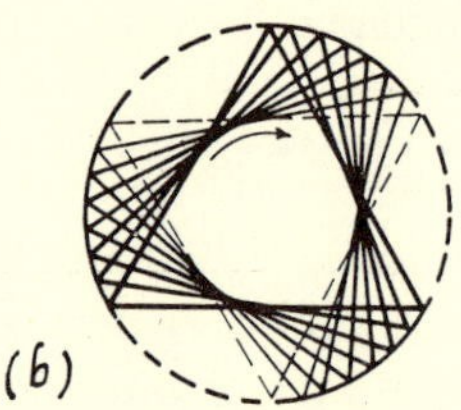

FIG. 18-1b. The same two circles can be generated by letting the figure (here a triangle) spin around its own center, in its own plane. Its corners will touch the outer circle, and its sides envelop the inner one.

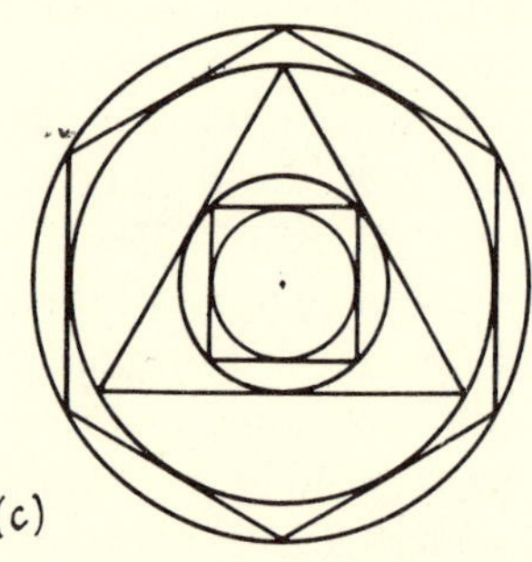

FIG. 18-1c. A series of regular plane figures, separated by inner and outer circles, provides a series of circles which might show the proportions of the planetary orbits. Even the best choice of figures failed to fit the solar system.

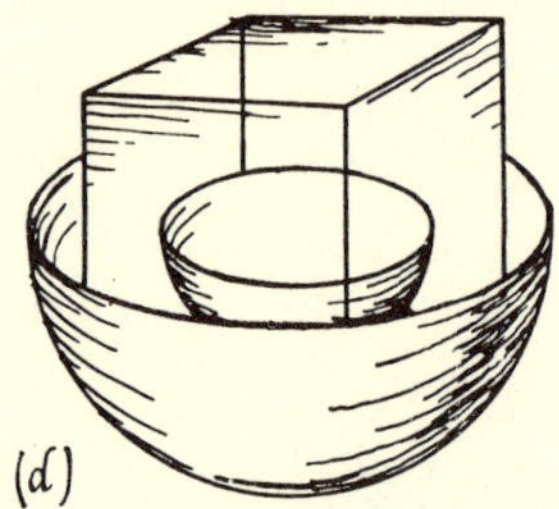

FIG. 18-1d. KEPLER'S SECOND GUESS
This shows the basis of Kepler's final scheme. He chose the order of regular solids that gave the best agreement with the known proportions of planetary orbits.

fect numbers, so that the underlying mathematical harmony . . . is the real and discoverable cause of the planetary motions."[3] Kepler himself said, "I brooded with the whole energy of my mind on this subject."

His mind burned with questions: Why are there only *six* planets? Why do their orbits have just the proportions and sizes they do? Are the times of the planets' "years" related to their orbit-sizes? The first question, "Why just six?" is characteristic of Kepler's times—nowadays we should just hunt for a seventh. But then there was a finality in facts and a magic in numbers. The Ptolemaic system counted seven planets (including Sun and Moon, excluding the Earth) and even had arguments to prove seven must be right.

Kepler tried again and again to find some simple relation connecting the radius of one orbit with the next. Here are *rough relative* radii from Tycho's observations, calculated for the Copernican scheme: 8 : 15 : 20 : 30 : 115 : 195. He tried to guess the secret in these proportions. Each guess meant a good deal of work, and each time he found it did not fit the facts he rejected that guess honestly. His mystical mind clung to the Greek tradition that circles are perfect; and at one time he thought he could construct a model of the orbits thus: draw a circle, inscribe an equilateral triangle in it, inscribe a circle in that triangle, then another triangle inside the inner circle, and so on. This scheme gives successive circles a definite ratio of radii, 2:1. He hoped the circles would fit the proportions of the planetary orbits if he used squares, hexagons, etc., instead of some of the triangles. No such arrangement fitted. Suddenly he cried out, "What have flat patterns to do with orbits in *space*? Use solid figures." He knew there are only five completely regular crystalline solid shapes (see Fig. 18-3). Greek mathematicians had proved there cannot be more than five. If he used these five solids to make the separating spaces between six spherical bowls, the bowls would define six orbits. Here was a wonderful reason for the number six. So he started with a sphere for the Earth's orbit, fitted a dodecahedron outside it with its faces touching the sphere, and another sphere outside the dodecahedron passing through its corners to give the orbit of Mars; outside that sphere he put a tetrahedron, then a sphere for Jupiter, then a cube, then a sphere for Saturn. Inside the Earth's sphere he placed two more solids separated by spheres, to give the orbits of Venus and Mercury.

[3] Sir William Dampier, *History of Science* (4th edn., Cambridge University Press, 1949).

THE REGULAR SOLIDS. A geometrical intelligence test

How many different shapes of regular solid are possible? To find out, follow argument (a); then try (b).

A regular solid is a geometrical solid with identical regular plane faces; that is, a solid that has:

all its edges the same length
all its face angles the same
all its corners the same
and all its faces the same shape.
(See opposite for shapes that do not meet the requirements.)

Fig. 18-2.

For example, a cube is a regular solid.

The faces of a regular solid might be:

all equilateral triangles
or all squares
or all regular pentagons
or . . . and so on . . .

&c.

(a) Here is the argument for square faces. Try to make a corner of a regular solid by having several corners of squares meeting there.

We already know that in a cube each corner has three square faces meeting there. Take three squares of cardboard and place them on the table like this, then try to pick up the place where three corners of squares meet. The squares will fold to make a cube corner.

Therefore we can make a regular solid with three square faces meeting at each of the solid's corners. (We need three more squares to make the rest of the faces and complete the cube.)

Could we make another regular solid, with only one, or two, or four square faces meeting at a corner?

With *one square*, we cannot make a solid corner.

With *two squares*, we can only make a flat sandwich.

With *three squares*, we make a cubical corner, leading to a cube.

With *four squares* meeting at a corner, they make a flat sheet there, and cannot fold to make a corner for a closed solid.

Thus, SQUARES CAN MAKE ONLY *ONE* KIND OF REGULAR SOLID, A CUBE.

(b) Now try for yourself with regular pentagons, and ask how many regular solids can be made with such faces.

Then try hexagons, and other polygons.

Then return to triangles and carry out similar arguments with triangular faces.

THE RESULT: Only FIVE varieties are possible in our 3-dimensional world. (Fig. 18-3)

(NOTE that these arguments need pencil sketches but can be carried out in your head without cardboard models.)

THE SOLIDS BELOW ARE
NOT REGULAR SOLIDS

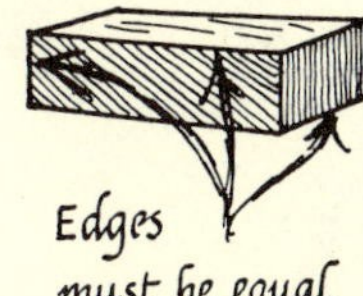

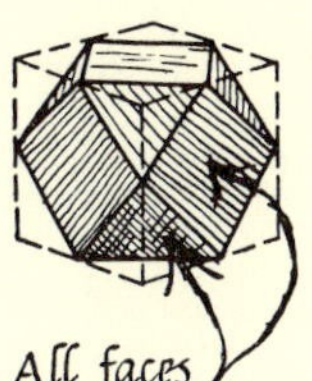

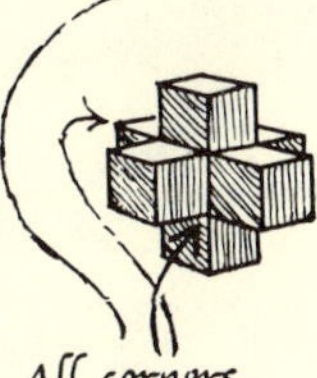

THE REGULAR SOLIDS

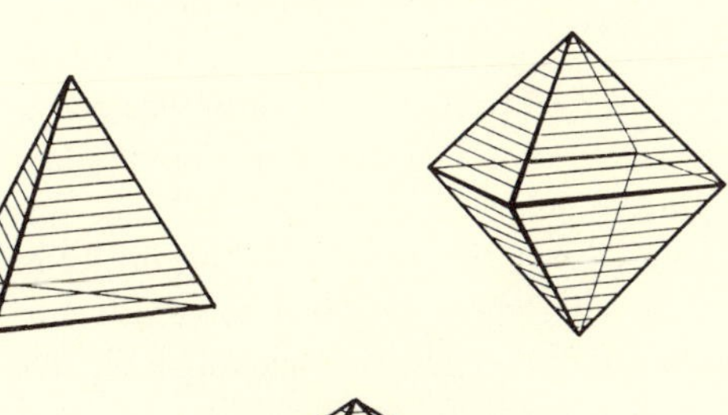

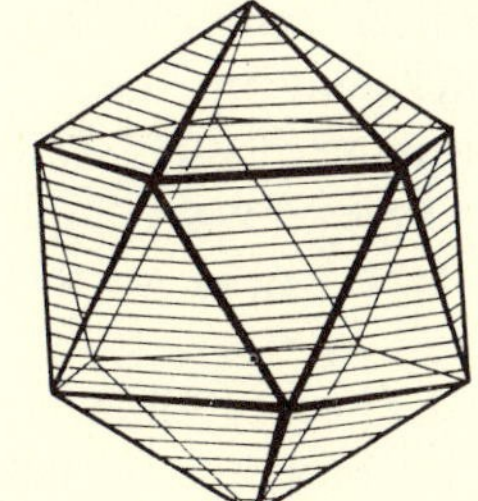

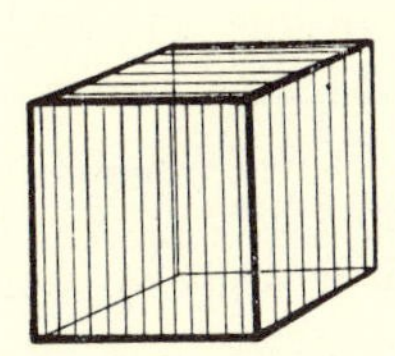

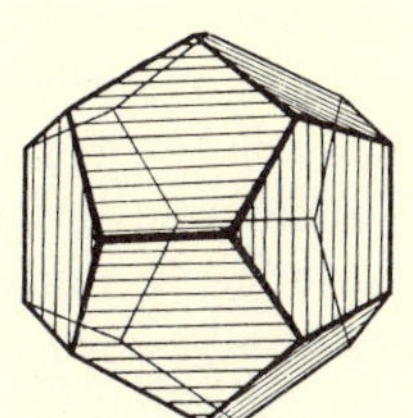

Fig. 18-3.
The five regular solids are drawn after D. Hilbert and S. Cohn-Vossen in *Anschauliche Geometrie* (Berlin: Julius Springer, 1932).

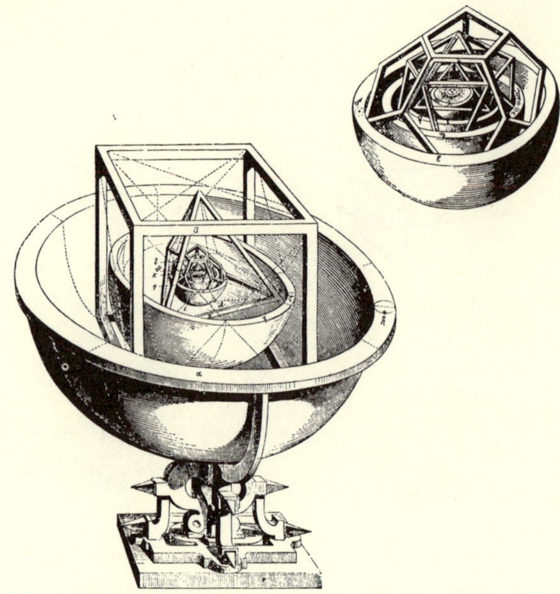

FIG. 18-4. KEPLER'S SCHEME OF REGULAR SOLIDS, FROM HIS BOOK
The relative sizes of planetary orbits were shown by bowls separating one solid from the next. The bowls were not thin shells but were just thick enough to accommodate the *eccentric* orbits of the planets.

The relative radii of the spheres, calculated by geometry, agreed fairly well with the proportions then known for planetary orbits, and Kepler was overjoyed. He said: "The intense pleasure I have received from this discovery can never be told in words. I regretted no more the time wasted; I tired of no labor; I shunned no toil of reckoning, days and nights spent in calculation, until I could see whether my hypothesis would agree with the orbits of Copernicus, or whether my joy was to vanish into air."

We now know the scheme was only a chance success. In later years, Kepler himself had to juggle the proportions by thickening up the bowls to fit the facts; and, when more planets were discovered centuries after, the scheme was completely broken.[4] Yet this "success" sent Kepler on to further, great discoveries.

He published his discovery in a book, including an account of all his unsuccessful trials as well as the successful one. This unusual characteristic appeared in many of his writings. He showed *how* his discoveries were made. He had no fear of damaging his reputation but only wanted to increase human knowledge, so instead of concealing his mistakes he gave a full account of them. "For it is my opinion," he said, "that the occasions by which men have acquired a knowledge of celestial phenomena are not less admirable than the discoveries themselves. . . . If Christopher Columbus, if Magellan, if the Portuguese when they narrate their *wanderings*, are not only excused, but if we do not wish these passages omitted, and should lose much pleasure if they were, let no one blame me for doing the same."

The book also contained an admirable defense of the Copernican system, with good solid reasons in its favor. Young Kepler sent copies of his book to Tycho Brahe and Galileo, who praised it as a courageous beginning. This started Kepler's lifelong friendship with them.[5] In the same book, he made the suggestion that each planet may be pushed along in its orbit by a spoke carrying some influence from the Sun—a vague and improbable idea that later helped him discover his second Law.

Kepler was a Protestant, and he found himself being turned out of his job by Roman Catholic pressure on the administration. Worrying about his future, and anxious to consult Tycho on planetary observations, he travelled across Germany to Prague. Tycho, busy observing Mars, "the difficult planet," wrote to him: "Come not as a stranger but as a friend; come and share in my observations with such instruments as I have with me." While the work of the observatory proceeded, Tycho was turning to detailed "theory," schemes to fit his long series of observations. Kepler was soon set to work on Mars, working with Tycho to find a circular orbit that fitted the facts. Sensitive, and sick, Kepler complained that Tycho treated him as a student and did not share his records freely. Once, driven half crazy by worry, he wrote Tycho a violent letter full of quite unjust reproaches, but Tycho merely argued gently with him. Kepler, repenting, wrote:

"Most Noble Tycho,

How shall I enumerate or rightly estimate your benefits conferred on me? For two months you have liberally and gratuitously maintained me, and my whole family . . . you have done me every possible kindness; you have communicated to me everything you hold most dear. . . . I cannot reflect without consternation that I should have been so given up by God to my own intemperance as to shut my eyes

[4] There *is* a rough empirical rule relating orbit-radii to each other, called Bode's Law; but until recently no reason for it could be found. However, see G. Gamow, *1, 2, 3, . . . Infinity* (New York, Mentor Books, 1953) for a suggested reason.

[5] In a later edition, Kepler took special trouble to avoid any appearance of stealing credit from Galileo. In one of his rejected theories he assumed a planet between Mars and Jupiter. Fearing a careless reader might take this to be a claim anticipating Galileo's discovery of Jupiter's moons, he added a note, saying of his extra planet, "Not circulating round Jupiter like the Medicaean stars. Be not deceived. I never had them in my thoughts."

on all these benefits; that, instead of modest and respectful gratitude, I should indulge for three weeks in continual moroseness towards all your family, in headlong passion and the utmost insolence towards yourself. . . . Whatever I have said or written . . . against your excellency . . . I . . . honestly declare and confess to be groundless, false, and incapable of proof."

When Kepler ended his visit and returned to Germany, Tycho again invited him to join him permanently. Kepler accepted but was delayed by poverty and sickness, and when he reached Prague with no money he was entirely dependent on Tycho. Tycho secured him the position of Imperial Mathematician to assist in the work on the planets.

Tycho died soon after, leaving Kepler to publish the tables. Though he still held the imperial appointment, Kepler had difficulty getting his salary paid and he remained poor, often very poor. At one time he resorted to publishing a prophesying almanac. The idea was abhorrent to him, but he needed the money, and he knew that astrology was the form of astronomy that would pay. For the rest of his life, over a quarter of a century, he worked on the planetary motions, determined to extract the simple secrets he was sure must be there.

The Great Investigation of Mars

When Tycho died, Kepler had already embarked on his planetary investigations, chiefly studying the motion of Mars. What scheme would predict Mars' orbit? Still thinking in terms of circles, Kepler made the planet's orbit a circle round the Sun, with the Sun a short distance off center (like Ptolemy's eccentric Earth). Then he placed an equant point Q off center on the other side, with a spoke from Q to swing the planet around at constant speed. He did not insist, like Ptolemy, on making the eccentric distances CS and CQ equal, but calculated the best proportions for them from some of Tycho's observations. Then he could imagine the planet moving around such an orbit and compare other predicted positions with Tycho's record. He did not know the direction of the line SCQ in space, so he had to make a guess and then try to place a circular orbit on it to fit the facts. Each trial involved long tedious calculations, and Kepler went through 70 such trials before he found a direction and proportions that fitted a dozen observed longitudes of Mars closely. He rejoiced at the results, but then to his dismay the scheme failed badly with Mars' latitudes. He shifted his eccentric distances to a compromise value to fit the latitudes; but, in some parts of the orbit, Mars' position as calculated from his theory disagreed with observation by 8′ (8 sixtieths of one degree). Might not the observations be wrong by this small amount? Would not "experimental error" take the blame? No. Kepler knew Tycho, and he was sure Tycho was never wrong by this amount. Tycho was dead, but Kepler trusted his record. This was a great tribute to his friend and a just one. Faithful to Tycho's memory, and knowing Tycho's methods, Kepler set his belief in Tycho against his own hopeful theory. He bravely set to work to go the whole weary way again, saying that upon these eight minutes he would yet build up a theory of the universe.

It was now clear that a circular orbit would not do. Yet to recognize any other shape of orbit he must obtain an accurate picture of Mars' real orbit from the observations—not so easy, since we only observe the apparent path of Mars from a moving Earth. The true distances were unknown; only angles were measured and those gave a foreshortened compound of Mars' orbital motion and the Earth's. So Kepler attacked the Earth's orbit first, by a method that had all the marks of genius.

Mapping the Earth's Orbit in Space and Time

To map the Earth's orbit around the Sun on a scale diagram, we need many sets of measurements, each set giving the Earth's bearings from *two* fixed points. Kepler took the fixed Sun for one of these, and for the other he took Mars *at a series of times when it was in the same position in its orbit.* He proceeded thus: he marked the "position" of Mars in the star pattern at one opposition (opposite the Sun, overhead at midnight). That gave him the direction of a base line Sun-(Earth)-Mars, SE_1M. Then he turned the pages of Tycho's records to a time *exactly one Martian year later.* (That time of Mars' motion around its orbit was known accurately, from records over centuries.) Then he knew that Mars was in the same position, M, so that SM had the same direction. By now, the Earth had moved on to E_2 in its orbit. Tycho's record of the position of the Mars in the star-pattern gave him the new apparent direction of Mars, E_2M; and the Sun's position gave him the direction E_2S. Then he could calculate the angles of the triangle SE_2M from the record, thus: since he knew the directions E_1M and E_2M (marked on the celestial sphere of stars) he could calculate the angle A between them. Since he knew the directions E_1S and E_2S, he could calculate the angle B, between them. Then on a scale diagram he could choose two points to represent S and M and

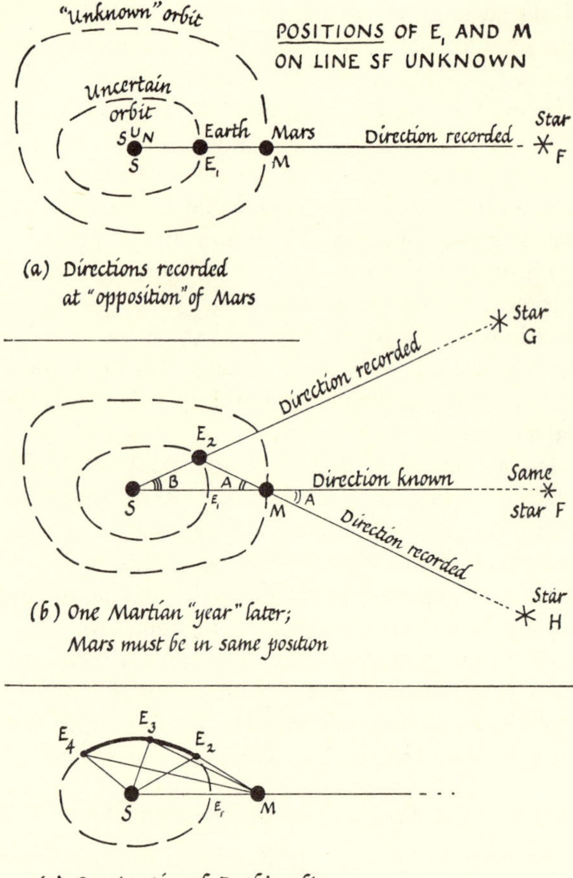

FIG. 18-5. KEPLER'S SCHEME TO PLOT THE EARTH'S ORBIT

locate the Earth's position, E_2, as follows: at the ends of the fixed baseline SM, draw lines making angles A and B and mark their intersection E_2. One Martian year later still, he could find the directions E_3M and E_3S from the records, and mark E_3 on his diagram. Thus Kepler could start with the points S and M and locate E_2, E_3, E_4, . . . enough points to show the orbit's shape.

Then, knowing the Earth's true orbit, he could invert the investigation and plot the shape of Mars' orbit. He found he could treat the Earth's orbit either as an eccentric circle or as slightly oval; but Mars' orbit was far from circular: it was definitely oval or, as he thought, egg-shaped, but he still could not find its mathematical form.

Variable Speed of Planets: Law II

Meanwhile his plot of the Earth's motion in space showed him just how the Earth moves unevenly along its orbit, faster in our winter than in summer. He sought for a law of uneven speed, to replace the use of the equant. His early picture of some pushing influence from the Sun suggested a law to try. He believed that motion needed a force to maintain it, so he pictured a "spoke" from the Sun pushing each planet *along* its orbit, a weaker push at greater distance. He tried (with a confused geometrical scheme) to add up the effects of such pushes from an eccentric Sun; and he discovered a simple law: *the spoke from Sun to planet sweeps out equal areas in equal times.* It does not swing around the Sun with constant speed (as Ptolemy would have liked), but it does have a constancy in its motion: constant rate of sweeping out area (which Ptolemy would probably have accepted). Look at the areas for equal periods, say a month each. When the planet is far from the Sun the spoke sweeps out a long thin triangle in a month; and as the planet approaches the Sun the triangles grow shorter and fatter—the planet moves faster. Later on, when Kepler knew the shape of Mars' orbit he tried the same rule and found it true for Mars too. Here he had a simple law for planetary speeds: each planet moves around the Sun with such speeds that the radius from Sun to planet sweeps out equal areas in equal times. Kepler had only a vague "reason" for it, in terms of solar influences, perhaps magnetic; but he treasured it as a true, simple statement, and used it in later investigations. We treasure it too, and assign a first-class reason to it. We call it Kepler's Second Law. His First Law, discovered soon after, gave the true shape of planetary orbits.

The Orbit of Mars: Law I

When he had plotted Mars' orbit (forty laboriously computed points), Kepler tried to describe its oval shape mathematically. He had endless difficulties—at one time he says he was driven nearly out of his mind by the frustrating complexity. He wrote to the Emperor (to encourage finances), in

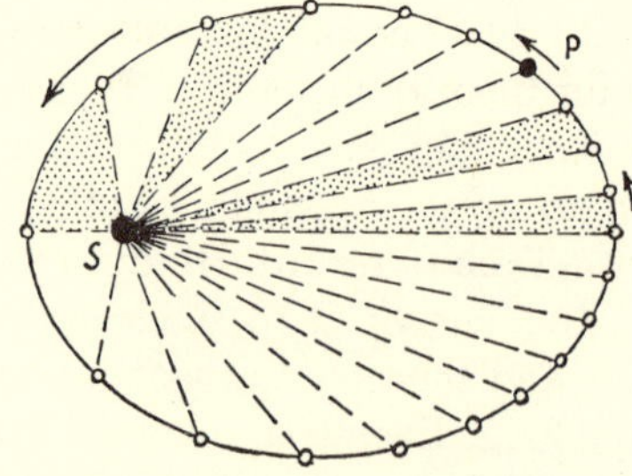

FIG. 18-6. KEPLER'S DISCOVERIES FOR MARS
An ellipse with the Sun in one focus fits the orbit of Mars. The spoke from Sun to Planet sweeps out equal areas in equal times. The positions marked here show planet's positions at equal intervals of time, 1/20 of its "year" apart. The planet moves with such speeds that all the sectors marked here—a few of them shaded—have equal areas.

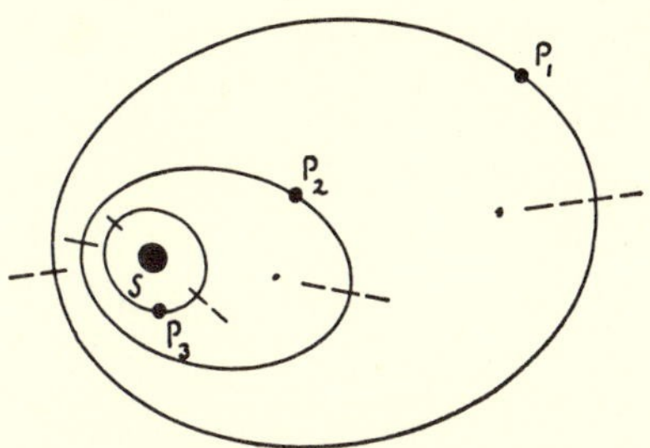

Fig. 18-7. A Solar System with Elliptical Orbits Around a Common Sun
(The planets' orbits in our own Solar System have much smaller eccentricities. But some comets move in elliptical orbits with great eccentricity.)

his grandiose style: "While triumphing over Mars, and preparing for him, as for one already vanquished, tabular prisons and equated excentric fetters, it is buzzed here and there that my victory is vain, and that the war is raging anew. For the enemy left at home a despised captive has burst all the chains of the equations, and broken forth from the prisons of the tables."

Finally, he found the true orbit sandwiched between an eccentric circle that was too wide and an inscribed ellipse that was too narrow. Both disagreed with observation, the circle by $+8'$ at some places, the inner ellipse by $-8'$. He suddenly saw how to compromise half way between the two, and found that gave him an orbit that is *an ellipse with the Sun in one focus.* He was so delighted with his final proof that this would work that he decorated his diagram with a sketch of victorious Astronomy (Fig. 18-8). At last he knew the true orbit of Mars.[6] A similar rule holds for the Earth and other planets. This is his First Law.

[6] It may seem strange that he did not think of an ellipse earlier. It was a well-known oval, studied by the Greeks as one of the sections of a cone. But then *we* know the answer. Besides, ellipses were not so important then. It was Kepler who added greatly to their fame. (An ellipse is easy to draw with a loop of string and two thumb-tacks. If you have never tried making one for yourself you should do so. This is an amusing experiment which will show you a property of ellipses that is valuable in optics.)

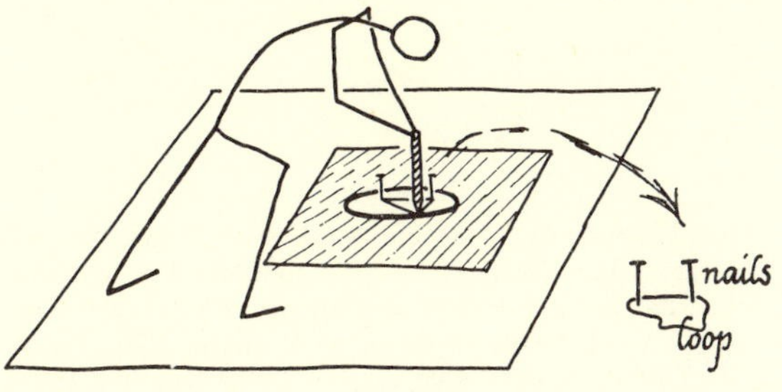

Fig. 18-9.
Drawing an Ellipse, with a loop of thread and two nails

Law III

Kepler had then extracted two great "laws" from Tycho's tables, by his fearless thinking and untiring work. He continued to brood on one of his early questions: what connection is there between the sizes of the planets' orbits and the times of their "years"? He now knew the average radii[7] of the orbits; the times of revolution ("years") had long been known. (As the Greeks surmised, the planets with the longest "years" have the largest orbits.) He felt sure there was some relation between radius and time. He must have made and tried many a

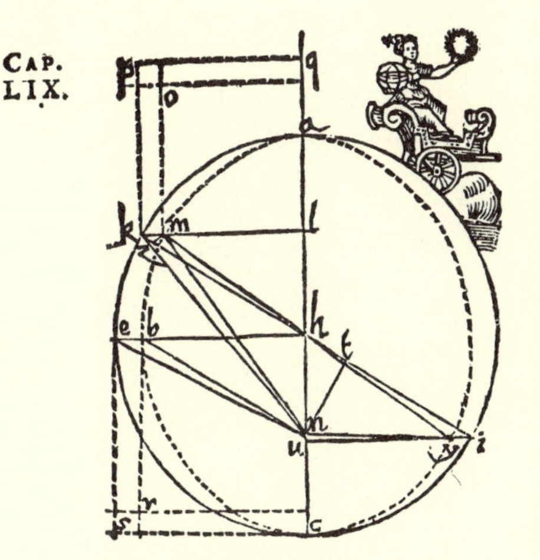

Fig. 18-8.
Kepler's Triumphant Diagram, From his Book on Mars
When he succeeded in proving that an ellipse with the Sun in one focus could replace an oscillating circular orbit and maintain an "equal area" law, Kepler added a sketch of Victorious Astronomy, to show his delight and to emphasize the importance of the proof.

guess, some of them sterile ones like his early scheme of the five regular solids or wild mystical ones like his speculation of musical chords for the planets. Fortunately there *is* a connection between radii and times, and Kepler lived to experience the joy of finding it. He found that the fraction R^3/T^2 is the same for all the planets, where R is the planet's average orbit-radius, and T is the planet's "year," measured in our days. See the table.

[7] Assuming circular orbits, Copernicus made rough estimates, and Tycho made better ones. Kepler knew these when he tried his strange scheme of regular solids, and he traded on their roughness to let his test of that theory seem "successful."

PLANETARY DATA—TEST OF KEPLER'S THIRD LAW
(These are modern data, more accurate than Kepler's)

Planet	Radius of planet's orbit R (miles)	Time of revolution (planet's "year") T (days)	R^3 (miles)^3	T^2 (days)^2	$\frac{R^3}{T^2}$ $\frac{(miles)^3}{(days)^2}$
Mercury	3.596×10^7	87.97	46.50×10^{21}	7739.	6.009×10^{18}
Venus	6.720×10^7	224.7	303.5×10^{21}	50490.	6.011×10^{18}
Earth	9.290×10^7	365.3	801.8×10^{21}	133400.	6.010×10^{18}
Mars	14.16×10^7	687.0	$2839. \times 10^{21}$	472100.	6.015×10^{18}
Jupiter	48.33×10^7	4332.	$112900. \times 10^{21}$	18770000.	6.015×10^{18}
Saturn	88.61×10^7	10760.	$695700. \times 10^{21}$	115800000.	6.008×10^{18}

The test of Kepler's guess is shown in the last column.

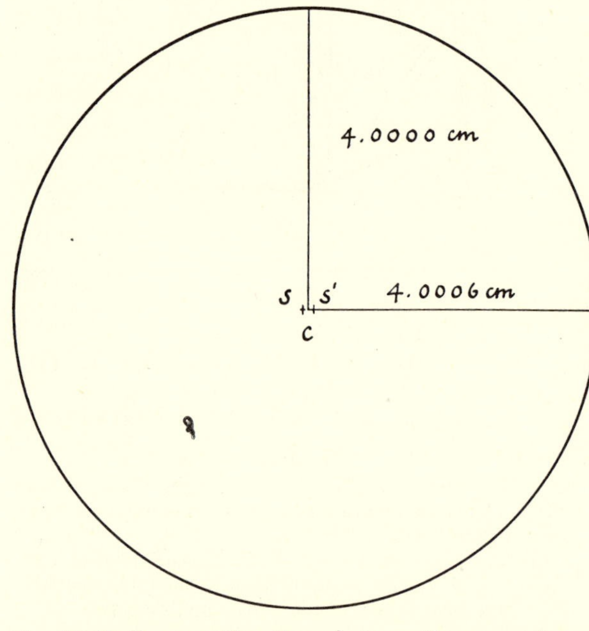

FIG. 18-10. ELLIPSE: THE EARTH'S ORBIT DRAWN TO SCALE
The actual eccentricity of planetary orbits is very small. The orbits are almost circles, yet Tycho's observations enabled Kepler to show that they are not circles but ellipses. The sketch above shows the Earth's orbit drawn to scale. If a 4.0000 centimeter line is used, as here, to represent the minimum radius, which is really some 93,000,000 miles, the maximum radius needs a line 4.0006 centimeters long. The eccentricity of Mars' orbit is over thirty times as big, but even then the ratio of radii is only 1.0043 to 1.0000. Mercury is the only planet with a much greater eccentricity of orbit, with radii in proportion 1.022 to 1.000. Even this eccentricity of orbit seems small, but it is sufficient to involve Mercury in such speed changes around the orbit that Relativity mechanics predicts a very slow slewing around of the orbit—a precession of only 1/80 of a degree per *century*, discovered and measured long before the Relativity prediction!

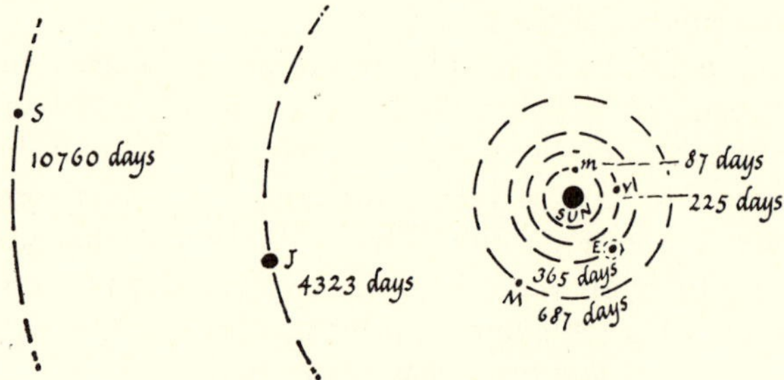

FIG. 18-11a. ? RELATIONSHIP BETWEEN *RADIUS* AND *"YEAR"* FOR PLANETARY ORBITS ?
(Planetary orbits roughly to scale.)

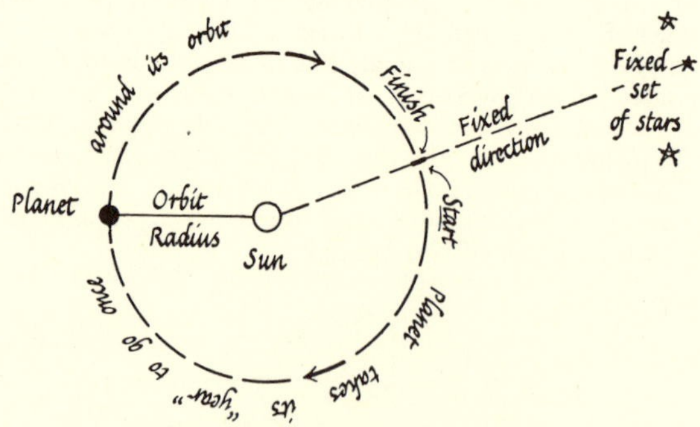

FIG. 18-11b. PLANET'S "YEAR"
The planet's year is the time it takes to go once around its orbit. This is the time-interval from the moment when its direction hits some standard mark in the star-pattern until it returns to the same mark. (The Earth moves too. An allowance for the Earth's motion must be made when extracting the planet's true year from observations.)

Again he was overjoyed at wresting a divine secret from Nature by brilliant guessing and patient trial. He said:

"What I prophesied two-and-twenty years ago, as soon as I discovered the five solids among the heavenly orbits—what I firmly believed long before I had seen Ptolemy's "Harmonies"—what I had promised my friends in the title of this book, which I named before I was sure of my discovery—what sixteen years ago, I urged as a thing to be sought—that for which I joined Tycho Brahe, for which I settled in Prague, for which I have devoted the best part of my life to astronomical contemplations, at length I have brought to light, and recognized its truth beyond my most sanguine expectations. It is not eighteen months since I got the first glimpse of light, three months since the dawn, very few days since the unveiled sun, most admirable to gaze upon, burst upon me. Nothing holds me . . . the die is cast, the book is written, to be read either now or by posterity, I care not which; it may well wait a century for a reader, as God has waited six thousand years for an observer."

Kepler's Laws

These investigations took years of calculating, changing, speculating, calculating. . . . Kepler discovered—among other "harmonies" that he valued—three great laws that are clear and true. Here they are:

LAW I — EACH PLANET MOVES IN AN ELLIPSE WITH THE SUN IN ONE FOCUS.

LAW II — THE RADIUS VECTOR (LINE JOINING SUN TO PLANET) SWEEPS OUT EQUAL AREAS IN EQUAL TIMES.

LAW III — THE SQUARES OF THE TIMES OF REVOLUTION (OR YEARS) OF THE PLANETS ARE PROPORTIONAL TO THE CUBES OF THEIR AVERAGE DISTANCES FROM THE SUN. (Or R^3/T^2 is the same for all the planets)

Once guessed, the first two laws could be tested with precision with available data; so Kepler could make sure he had guessed right. Law III was tested in its discovery. Only relative values of orbit-radii were needed.

Kepler had done a great piece of work. He had discovered the laws that Newton linked with universal gravitation. Of course that was not what Kepler thought he was doing. "He was not tediously searching for empirical rules to be rationalised by a coming Newton. He was searching for ultimate causes, the mathematical harmonies in the mind of the Creator."[8] He emerged with no general reason for his ellipses and mathematical relationships; but he delighted in their truth.

Guessing the Right Law

Guessing the third law was a matter of finding a numerical relationship which would hold for several pairs of numbers. An infinite variety of "wrong" guesses can be made to fit a limited supply of data, in this case values of T and R for only six planets. Many such guesses that succeed with six planets fail when applied to a seventh planet (Uranus, discovered later). Of those that still succeed, many would fail if tried on an eighth planet (Neptune). So trials with more and more sets of data can help to remove "wrong" guesses, leaving the "right" one. But in what sense is the "right" one right? Some of us believe there is a really true story behind the things we see in Nature. Kepler, Galileo, and Newton probably thought like that. Others now say that the right rule is merely (*a*) *the rule that applies most generally* (for example, to the greatest variety of planets). In this sense Kepler's R^3/T^2 guess was right because it applies to later-discovered planets and to other systems such as Jupiter's moons. His five-regular-solids rule was wrong, because it did not agree well with data for the original six planets and failed completely when required to deal with more than six. And, they say, the right rule is (*b*) *the rule that fits best into a theoretical framework which ties together a variety of knowledge of Nature.* If that theory has been manufactured just to deal with the problem in hand, then (*b*) is nonsense—it would merely say that the rule is right because it agrees with its own theory constructed to agree with it. We call that an *ad hoc* theory. If, however, the theory connects the problem in hand to other natural knowledge, then (*b*) is a cogent recommendation. Newton, guessing at universal gravitation, made a theory that connects falling bodies and the Moon's motion and planetary motion and tides, etc. He showed that Kepler's Law III (as well as the other two) was a necessary deduction from this theory. Thus Kepler's R^3/T^2 rule seems "right" on both scores, (*a*) and (*b*), *generalness* and *agreement with wide theory.* It might have been a "wrong" guess, waiting like the early "five-regular-solids" law for more data to refute it and for theory to fail to "predict"[9] it.

[8] Sir William Dampier, *op.cit.*

[9] Scientists use "predict" in this way, but it is an unfortunate choice of word. Here it means "coordinate with other knowledge."

A Fictitious "Kepler Problem"

To see something of the hazards involved in an investigation like Kepler's let us trace through a specimen problem using imaginary data, with a fictitious relationship. Suppose you have invented a planetary puzzle and know the scheme you have used, but ask me to try to find the scheme. You present me with the following data.

Data			*Problem*
"Planet"	R	T	What is the "law" connecting R and T?
A	1	3	
B	2	6	
C	4	18	

You know the scheme, since you have invented it. (It is not an inverse square law system: the "planets" are not real ones!) In fact, you got T by squaring R and adding 2. That is, you *chose* the relation $T = R^2 + 2$ and used it. (Make sure our data fit this formula.) So if a new planet D is discovered with $R = 5$ it will have $T = 5^2 + 2$, or 27. Suppose you give me the data for A, B, C (holding D up your sleeve). In looking for a rule, I try to find some algebraic combination of T and R which will be the *same for each of these planets.* Starting with planets A and B, I notice that T/R is 3/1 for A, 6/2 for B, the same for both. Hoping I have found the right rule (T/R the same for all), I try this on planet C. For C, T/R is 18/4 and this is not the same as 3/1. I must therefore reject this simple guess. In trying other schemes which give the same answer for planets A and B, I find several more which fail for C. But presently I find that I get the same answer for planets A and B if I proceed thus: I divide R into 8 and add 7 times R and subtract T; that is, I find the value of $8/R + 7R - T$.

For planet A, $8/1 + 7 \times 1 - 3 = 12$;
and for planet B, $8/2 + 7 \times 2 - 6 = 12$.

So the answer is the same, 12, for both A and B. Trying the same rule on planet C,

I have $8/4 + 7 \times 4 - 18 = 12$ again.

So I am delighted to find the rule works for C and A and B. Confident that I have got the right rule, I plan to publish it, but you then divulge the data for planet D: $R = 5$ and $T = 27$. Trying my rule on planet D,

I obtain $8/5 + 7 \times 5 - 27 = 9.6$.

After asking you whether your data might be wrong enough to excuse the difference between 9.6 and 12.0, I start all over again. If I am lucky as well as patient, I may hit upon a scheme such as this: add 2 to the square of R and divide by T. This yields an answer 1.000 for all four planets, A, B, C, D.[10] Therefore it has a better chance of being the right rule than the others. Tests on more data would improve its reputation further and if some general theory could endorse it I might feel sure I had the right rule. Summing up this investigation in a table, we have

"PLANET"	DATA		ATTEMPTS TO OBTAIN CONSTANT NUMBERS		
			1st Trial	Nth Trial	Qth Trial
	R	T	$\frac{T}{R}$	$\frac{8}{R} + 7R - T$	$\frac{R^2+2}{T}$
A	1	3	3	12	1
B	2	6	3	12	1
C	4	18	4.5	12	1
D	5	27	5.4	9.6	1
e	3	11	3.667	12.67	1

Note that at the last moment another "planet" has been discovered, e, which is so small that it was not noticed before. It too fits with the final rule (of course it does, in this game, since *you* manufactured its data by using your private knowledge of that rule), and it fails to fit with the earlier rules. Notice, however, that it nearly fits with the second rule, giving 12.67 instead of 12.00. If the data for planet e had been available when I was working on my second rule, should I not have been tempted to say "12.67 is near enough; the difference is due to experimental error"?

Kepler's Writing

Kepler wrote many books and letters setting forth his discoveries in detail, describing failures as well as successes. His account of his Laws is immersed in much mystical writing about other discoveries and ideas: planetary harmonies, schemes of magnetic influence, hints about gravitation, and a continuing delight in his earliest scheme of the five regular solids. Remember Kepler did not know the "right answers." He had no idea which of his theories would be validated by later discoveries and thought. He finally managed to get the Rudolphine tables printed—paying some of the cost himself, which he could hardly afford—so that at last really good astronomical data were available. Among his own books, he wrote a careful fairly popular book on general astronomy in which he explained the Copernican theory and described his own discoveries. The book was at once suppressed by the Church authorities, leaving him all the poorer by making it hard to get any of his books published and sold.

Comments on Kepler

"When Kepler directed his mind to the discovery of a general principle, he . . . never once lost sight of the explicit object of his search. His imagination, now unreined, indulged itself in the creation and invention of various hypotheses. The most plausible,

[10] There is no special virtue in the answer being 1.000. If I divide by $5T$ instead of by T the answers would all be 0.200, but the essential story is unchanged.

or, perhaps, the most fascinating of these, was then submitted to a rigorous scrutiny; and the moment it was found to be incompatible with the results of observation and experiment, it was willingly abandoned, and another hypothesis submitted to the same severe ordeal. . . . By pursuing this method he succeeded in his most difficult researches, and discovered those beautiful and profound laws which have been the admiration of succeeding ages."[11]

Sir Arthur Eddington says:[12]

"I think it is not too fanciful to regard Kepler as in a particular degree the forerunner of the modern theoretical physicist, who is now trying to reduce the atom to order as Kepler reduced the solar system to order. It is not merely similarity of subject matter but a similarity of outlook. We are apt to forget that in the discovery of the laws of the solar system, as well as of the laws of the atom, an essential step was the emancipation from mechanical models. Kepler did not proceed by thinking out possible devices by which the planets might be moved across the sky—the wheels upon wheels of Ptolemy, or the whirling vortices of later speculation. I think that is how most of us would have attacked the problem; we should have hunted for some concrete mechanism to yield the observed motion, and have approached the laws of motion through an explanation of the motion. But Kepler was guided by a sense of mathematical form, an aesthetic instinct for the fitness of things. In these later days it seems to us less incongruous that a planet should be guided by the condition of keeping the Action stationary than that it should be pulled and pushed by concrete agencies. In like manner Kepler was attracted by the thought of a planet moving so as to keep the growth of area steady—a suggestion which more orthodox minds would have rejected as too fanciful. I wonder how this abandonment of mechanical conceptions struck his contemporaries. Were there some who frowned on these rash adventures of scientific thought, and felt unable to accept the new kind of law without any explanation or model to show how it could possibly be worked? After Kepler came Newton, and gradually mechanism came into predominance again. It is only in the latest years that we have gone back to something like Kepler's outlook, so that the music of the spheres is no longer drowned by the roar of machinery."

[11] Sir David Brewster, *Martyrs of Science*, 1848.

[12] In his Introduction to the Tercentenary Commemoration book on Kepler's life and work, *Johann Kepler*, Williams and Wilkins Company, for The History of Science Society, 1931.

Kepler carried astronomy through a great stage of development. His laws are landmarks in knowledge, rules true today for planetary systems and perhaps even for atomic models.

PROBLEMS FOR CHAPTER 18

1. (a) What did Kepler's discovery of the "five regular solids" relation refer to?
   (b) Give a brief account of this scheme.

★ 2. KEPLER'S LAWS

(a) Give short descriptions, with sketches, of Kepler's Laws I, II, III (his main laws, not his "regular solids" scheme, "Law O").

(b) Kepler himself believed something was needed to keep the planets moving and thought perhaps some kind of spoke running from Sun to planet did this. Would such a spoke turn at constant rate? If not, at what stage of the planet's year would it turn fastest?

★ 3. PUZZLE FOR A MODERN KEPLER

Radioactive atoms shoot out small atomic "projectiles," which are pieces of their own innermost core or nucleus. Many such atoms, including radium itself, shoot out a projectile that is itself an electrically charged helium atom (= a helium nucleus, = a helium atom stripped of its two electrons). These are called alpha-particles, or α-particles. The atomic "explosion" in which a radioactive atom shoots out an alpha-particle occurs spontaneously, the parent atom then changing into an entirely different kind of atom with different chemical properties. This is the characteristic of radioactivity. Such radioactive changes give us information about atomic structure. But they also provide "projectiles" which can be used to investigate the structure of other atoms, somewhat in the way in which a boxer investigates the structure of other boxers' faces. In particular a stream of alpha-particles was used to investigate the structure of atoms of gold in a very famous experiment which led to a revolutionary change in atomic theory. The problem below refers to that experiment.

A stream of α-particles was shot at a very thin leaf of gold in a vacuum. Most of them passed straight through, missing severe collisions with any gold atoms in the very thin leaf. But a few of the α-particles bounced out in new directions, having suffered severe collisions. A very few even bounced back. These observations suggested a new theory which then predicted just how many should bounce back in some chosen direction, out of every million fired. The theory predicted a definite relationship between the number of α-particles bouncing back (per million) and the speed with which they were travelling when they hit the gold leaf. The theory was tested by a crucial experiment, reported by Geiger and Marsden in *Philosophical Magazine*, Vol. 25, page 620, 1913. Some of the measurements are given below:

v Velocity of helium atoms (In "arbitrary units"*)	N Number of helium atoms bouncing back per minute in a standard chosen direction
2.00	25
1.91	29
1.70	44
1.53	81
1.39	101
1.13	255

* These velocities are in arbitrary units. One such unit was probably worth about 10,000,000 meters/second.

These data provide a problem somewhat like the one that faced Kepler when he had planetary orbit data but had not guessed his third law. There is a fairly simple relationship between N and v.

Can you find this relationship? Try this, as Kepler would, with courage and care, without any help from a theory or a book. If you find the relationship, show how closely the data fit it. Of course, the original experimenters had an advantage over you; they knew what relation to try first—but then they had to do a difficult experiment. In these difficult experiments of counting *single atoms* as they bounce away from the gold, you must not expect great accuracy; so, unlike Kepler's, your constant may wobble by 10% but not in any particular direction.

CHAPTER 19 · GALILEO GALILEI (1564-1642)

"Science came down from Heaven to Earth on the inclined plane of Galileo."

Galileo's life overlapped that of Kepler. When Tycho Brahe had moved to Prague with those instruments he had saved, and Kepler was starting his attack on Mars, Galileo, in his thirties, was growing famous as a mathematician and natural philosopher. In his life, Galileo did many great things for science; perhaps the greatest was establishing mathematical argument, tied to experiment, as the basis of scientific knowledge. He experimented, and he drew on the experiments of others, until he had an instinct for sound science; but he was above all a thinker and a teacher, and so good an arguer that he could out-argue the traditional philosophers on their own ground. He liked to use what we call his "thought experiments": hypothetical experiments devised for use in argument.[1] In these he appealed to common-sense knowledge of nature, or sometimes to specific experiments, and then argued out predictions of behavior or relationships. Thus, rather than call him the father of experimental science—as used to be the fashion—we might look on him as the first modern theoretical physicist.

Galileo gathered and taught the facts and ideas from which Newton formed his Laws of Motion. He drew on many contributions from earlier experimenters and thinkers—we even know, from the phrasing, which edition of certain earlier writers he copied in his books. He did not pull the new mechanics out of his hat, all his own discovery; but he did begin to build it into a comprehensive picture and he did make it *public* and *convincing*. He constructed one of the earliest telescopes and with it gathered new evidence to support Copernicus' theory and even Kepler's Third Law. He expounded the Copernican Theory with such compelling clearness that he upset traditional authorities. And he preached honest experiment and clear thinking with such exasperating fervour that he started physics on a new life.

Galileo and the New Science

Galileo's greatest contribution to the new physical science was a change of treatment. He brought back the scientific attitude of Pythagoras and Archimedes: *experimental knowledge should be codified by abstract mathematical ideas.* For example, he stated clearly that for an object falling freely the distances fallen in times 1, 2, 3, 4, from rest run in the proportions 1:4:9:16: . . . (which we now express compactly by algebra: $s \propto t^2$). In stating this he cleared away modifications made by air-resistance, spinning, horizontal motion, etc., and described an ideal case for a particle falling in a vacuum. He derived, by simple mathematical reasoning, an alternative form: the distances fallen in successive equal intervals of time increase steadily, 1:2:3:4: . . . , or, as we now say, Δs (for $\Delta t = 1$) $\propto t$ (see note 2). Galileo, and his successors down to the present day, do not spoil science when they "think away" real conditions such as air-resistance. The modern scientist can formulate ideal mechanical laws for frictionless materials, weightless carts, unstretchable strings, . . . and then add the real conditions to modify the ideal laws.

Galileo also promoted a complete change of thought in astronomy: he broke the sharp distinction between heavenly affairs and earthly science. Copernicus had maintained the mystical ideal of perfect spheres, but Galileo tried to treat the planets and Sun and Moon as ordinary earthy bodies. He started applying the same treatment to a ball rolling downhill and a planet in the sky. He did not carry this democratic treatment through—he still endowed planets with a natural circular motion—yet he drew man's understanding of the whole universe towards a scheme of general mathematical laws.

For his mathematical treatment, Galileo had to deal with things that could be measured definitely. So he gave importance to "primary" qualities of matter, such as length, volume, velocity, force; and he disowned, as outside proper science, such subjective things as color, taste, smell, and musical sound, which he said, just disappear when the observer is not present.[3] Shakespeare hinted at this

[1] Examples: His argument about three bricks falling (Ch. 1); the arguments in Ch. 5 (due to Galileo) on breaking-force of wires, etc.

[2] This is Galileo's simpler rule, asked for in Problem 8 of Ch. 1 (page 15). If $\Delta t = 1$, Δs is the velocity; and this progression of Δs-values is a statement of constant acceleration—velocity increasing steadily with time.

[3] Such exclusions could be revoked if a scheme of measurement appeared. For example, the invention of a thermometer with a definite scale may bring our sense of warmth into good science.

(in *The Merchant of Venice*):

> The crow doth sing as sweetly as the lark,
> When neither is attended, . . .

Thus Galileo moved science towards the hardheaded mathematical treatment that followed Newton; and he carried philosophy towards the complete separation of matter and mind that followed Descartes. His teaching helped to make matter and motion seem true and real, while taste and color, etc., seem unreal, mere sensations in the observer's mind produced by shapes or motions of atoms—though those atoms themselves are well disciplined by mathematical laws. A century later Berkeley suggested that even the primary qualities of matter are unreal; they too come to our minds only as "sense-perceptions." On that view the whole organization of scientific laws and knowledge which Galileo helped to build is a framework of abstractions, a picture that we extract from the sense impressions the world sends us. It is a good picture, comforting, useful, interesting: but it is not the world itself. The world itself—whatever real or concrete world there is outside our senses—may well be far more complex than we can "know" in our scientific way. If we believe our scientific picture is completely real and true, we may find the laws of mechanics offering to trace the course of every atom from now into the future, and thereby threatening to predict all events, including our own decisions and actions. That would take away all our choice of action, all free will—a very distressing prospect. But that offer applies only to the abstract world of Newtonian science, not to the complex concrete world beyond. We should not let ourselves be frightened by the "fallacy of the misplaced concrete."[4]

Galileo's Life and Work—Pisa

Galileo was the son of an Italian nobleman who was himself a philosopher and musician. His home was in Pisa, near Florence. Though young Galileo wanted to be a painter, his father sent him to the University to study medicine, a field that was much respected and well paid. There, at the University of Pisa, he seized on a chance to learn geometry. (There is a story that he overheard a lecture on Euclid, was thrilled with it, and implored the lecturer to teach him.) His father opposed this new interest—mathematicians were poorly paid—but Galileo's enthusiasm could not be stopped. He devoured the works of Euclid, then read Archimedes, and soon started his own investigations of the properties of centers of gravity.

[4] A. N. Whitehead's phrase.

When he was 25, Galileo was appointed by the Duke, one of the ruling family of Medici, to the post of lecturer in mathematics—at a miserable salary. With great energy and zeal, but little tact, he set to work on the mechanics of moving bodies: reading the earlier books, sorting sense from nonsense, and putting statements and ideas to the test of experiment. He enjoyed annoying the Aristotelian philosophers around him by showing up the mistakes in their teaching. Though he was right and they were wrong, his tactless manner was not wise.

"The detection of long-established errors is apt to inspire the young philosopher with an exultation which reason condemns. The feeling of triumph is apt to clothe itself in the language of asperity, and the abettor of erroneous opinions is treated as a species of enemy to science. Like the soldier who fleshes his first spear in battle, the philosopher is apt to leave the stain of cruelty on his early achievements. . . . Galileo seems to have waged this stern warfare against the followers of Aristotle; and such was the exasperation which was excited by his reiterated and successful attacks, that he was assailed, during the rest of his life, with a degree of rancour which seldom originates in a mere difference of opinion."[5]

Galileo's realistic discussions of falling bodies and accelerated motion upset traditional teaching and were not welcome; nor were his arguments exposing the fallacies of old doctrines. While he gathered enthusiastic followers he also made enemies. Malice and jealousy made his position at Pisa so uncomfortable that he accepted an invitation to move to the University of Padua in the neighboring republic of Venice. At Padua, he found philosophers already talking of free fall as due to a force, and doubting whether it was wise science to rely on "natural places" or to look for "first causes." The time was ripe for Galileo's teaching. He taught with vigor and amazing skill, and he wrote on motion, mechanics, astronomy. Even then he was poorly paid. He had to run a lodging house for his students, and he set up an instrument-making shop.[6]

Padua

In his new post at Padua, his reputation grew. He loved to expound and argue. He was formidable in argument because he started by expounding his

[5] Sir David Brewster, *Martyrs of Science* (1848).

[6] He manufactured a "military compass" that combined the uses of a protractor and a slide rule, and received orders for it from many parts of Europe. He also sent some as presents to important people, to show how he could aid "the military art."

opponents' case more clearly than they could and then he demolished it—he was an intellectual truck-driver. He stayed at Padua twenty years, during which he gathered much knowledge of mechanics and developed his defense of Copernican astronomy. He lectured to large audiences; and princes and nobles came to study under him.

When a bright new star suddenly appeared in the sky, he gave three lectures on it. Crowds came to hear him, but he rebuked them for paying attention to a temporary phenomenon while they overlooked the wonders of everyday nature. His lectures became so popular that even the grêat hall of the school of medicine was sometimes too small and he lectured in the open air. He taught honest science with compelling force.

Copernican Astronomy

Early in his career Galileo was converted to the Copernican system, and he taught it quietly at first, incautiously later. In a dialogue he describes what was probably his own conversion:

"I must upon this occasion relate some accidents that befell me when I first began to hear of this new doctrine [the Copernican system]. Being very young, and having scarcely finished my course of philosophy, . . . there chanced to come into these parts . . . a follower of Copernicus, who in an Academy made two or three lectures upon this point, to whom many flocked as auditors, but I, thinking they went more for the novelty of the subject than otherwise, did not go to hear him, for I had concluded with myself that that opinion could be no other than a solemn madness. Questioning some of those who had been there, I perceived they all made a jest of it, except one. He told me that the business was not altogether to be laughed at, and, because this man was reputed to be very intelligent and wary, I repented that I was not there. From that time forward, as often as I met with anyone of the Copernican persuasion, I demanded of them if they had been always of the same judgment; and, of as many as I examined, I found not so much as one who did not tell me that he had been a long time of the contrary opinion but had changed it for this, as convinced by the strength of the reasons in its favour. Afterwards, questioning them one by one, to see whether they were well possessed of the reasons of the other side, I found them all to be very ready and perfect in them; so that I could not truly say that they had taken up this opinion out of ignorance, or vanity, or to show the acuteness of their wits.

"On the contrary, of as many of the [Aristotelians] and Ptolemeans as I have asked (and out of curiosity I have talked with many) what pains they had taken in the book of Copernicus, I found very few that had so much as superficially perused it. But of those whom I thought had understood it, not one; and, moreover, I have enquired amongst the followers of the [Aristotelian] doctrine if ever any of them had held the contrary opinion and likewise found none that had. In other words, there was no man who followed the opinion of Copernicus who had not been first on the contrary side and who was not very well acquainted with the reasons of Aristotle and Ptolemy; and, on the contrary, there is not one of the followers of Ptolemy who had ever been of the judgment of Copernicus and had left that to embrace this of Aristotle. Considering these things, I began to think that a man who leaves the opinion imbued with his milk and followed by very many to take up another owned by very few and denied by all the schools, one that really seems a very great paradox, must needs have been moved, not to say forced, by more powerful reasons. For this cause I am become very curious to dive into the bottom of this business."[7]

Galileo's Mechanics of Motion

At Pisa and Padua, Galileo collected his knowledge of mechanics that he set forth much later in *Two New Sciences.* One of his early discoveries was the remarkable property of pendulums: that (for small amplitude) the time of swing is independent of amplitude. There is a fable that he discovered this in his student days in Pisa by timing the decreasing swings of a long lamp hanging in the cathedral. He had no accurate clock—in fact he was discovering the basis for good clocks—so he used his own pulse for the timing.[8] He then turned his discovery to use in medicine by constructing a short adjustable pendulum for timing pulses.

At some early stage, Galileo studied falling bodies, and he knew it was nonsense to say that heavier bodies fall faster *in proportion to their weight.* This had come from Aristotle—who probably gave it as a sensible description of final velocity in a very long fall, when air friction has increased

[7] Reprinted from the Salusbury translation of Galileo's *Dialogue,* as revised by Giorgio de Santillana, by permission of the University of Chicago Press. Copyright, 1953 by The University of Chicago Press. All rights reserved. Pages 142-144.

[8] The lamp quoted in the fable was not installed till some years later, so the story is doubted. Yet Galileo wrote in a dialogue ". . . Thousands of times I have observed vibrations, especially in churches where lamps, suspended by long cords, had been inadvertently set in motion."

till it balances weight. However, this was being taught as a rule for simple fall from rest—a piece of nonsense made almost sacred by ages of dogmatic teaching. Galileo saw that unequal bodies fall together, with motion *independent* of their weight, except for relatively slight differences. He satisfied himself by experiment and argument that those differences are due to air resistance. He pointed out that though a piece of gold falls fast, the same piece when beaten out to a thin leaf flutters down slowly. And he suggested a clinching proof: drop a scrap of lead and a wisp of wool in a vacuum—impossible in his day, but later carried out by Newton. He complained bitterly at the Aristotelians who claimed that in the time taken by a 100-pound cannon ball to fall 100 feet a 1-pound ball would fall only 1 foot. Actual experiment, he said, showed only a difference of finger-breadths. "How can you hide Aristotle's 99-foot difference behind a couple of finger-breadths?" he asked, to ridicule his opponents.

Galileo confirmed his belief of equal fall by comparing pendulums with light and heavy bobs on equal threads. The time of swing is the same, whatever the bob. Here was gravity-fall diluted, in a form that was easily timed accurately (a bunch of swings), and practically free from friction-troubles. (Since the time-of-swing is independent of amplitude, *air friction should not affect the time of swing*. Friction does reduce the amplitude from one swing to the next, but that does not matter!) This result contributed to the ideas of mass and gravitation that Newton later extracted and used. A heavy body has more weight than a light one—the Earth pulls it more. On that score we should expect it to fall faster. However, it also has more stuff-in-it-to-be-moved than a light one, a greater quantity of matter or mass, as Newton later called it. It has more "inertia," needs more force for its acceleration.[9] Therefore, when experiment shows that heavy and light bodies fall with the same acceleration (or swing with the same motion as pendulums) it suggests that the heavier body has greater mass in just the same proportion as it has greater weight. This is a remarkable property of gravitation, that Earth-pulls are proportional to the *inertial* masses of matter pulled. Galileo seems to have accepted it without seeking its cause. He did not formulate the concept of mass clearly in his studies of force and motion. It was Newton who put it to use. In this century mass gained new importance when we came to think of it as specially related to energy.

To investigate the motion of a falling body in detail, Galileo diluted gravity by using an inclined plane. He describes an experiment of rolling a ball

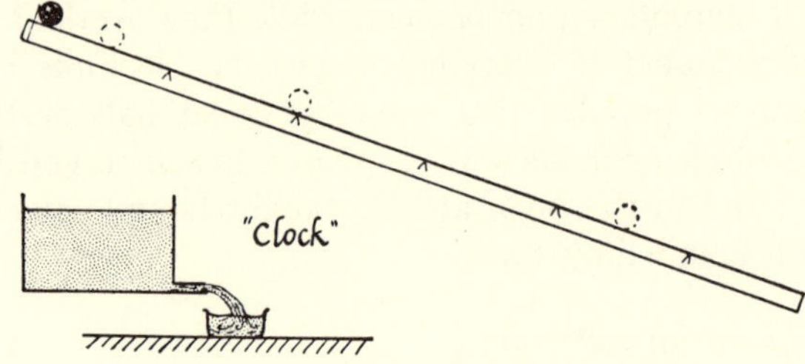

Fig. 19-1. Galileo's Experiment

down a long shallow incline, a grooved plank with a smooth parchment lining. Time-of-travel was measured with a simple water clock: the experimenter had a tank of water with a small spout and weighed the water that ran out. Measurements of time and distance agreed with the simple relation:

$$\text{DISTANCE TRAVELLED FROM REST} \propto \text{TIME}^2$$

It is not clear whether Galileo actually did the experiment or just quoted it from earlier scientists. Anyway, the measurements were rough, but Galileo was confident he knew the correct "law." By an ingenious geometrical argument, he proved that this is the necessary rule for motion with constant $\Delta v/\Delta t$. Therefore the rolling ball moved with constant acceleration. By confident extrapolation from his shallow incline to a steeper one to a still steeper one and finally to vertical fall, he argued that freely falling bodies have constant acceleration; hence he knew their law of fall.

On any chosen incline, the force producing the acceleration must be the same all the way down. (It is a constant fraction of the ball's *weight.*) Already part of Newton's Law II had emerged: a constant force leads to constant acceleration.

Continuing with hills of different slopes, Galileo was on the verge of finding the main relation of Newton's Law II: ACCELERATION varies as FORCE; but he kept this in geometrical forms which obscured the part played by force. He was preparing the way for an experimental science of motion which could be applied to a variety of problems: projectiles, pendulums, the planets themselves; and, later, moving machinery and even the moving parts of atoms.

[9] Though mass is an idea partly invented and partly extracted from properties of nature, it is not something that can be described successfully in a few words. One has to gain a feeling of its nature by working with it—calling it "inertia" is mere naming. When Newton defined it as "quantity of matter" he simply moved the doubt on to the definition of matter. Yet this description of Newton's was not as worthless as some critics claim. As a descriptive phrase it helped the scientists of Newton's day to understand what he meant. Perhaps it is useful in the same way for students today.

Speed at Bottom of Hill

Galileo guessed that if a ball rolls down one incline, A, and up another, B, *it will roll up to the original level*, whatever the slopes; and that led him to a very important general assumption from which he made many predictions. Imagine several downhill slopes A_1, A_2, A_3, *all of the same height*, all leading to the same uphill slope B. Then if his guess were true, the ball would rise to the same height on B *whichever of the A slopes it descended.* At the bottom, just about to run up B, the ball has the momentum needed to carry it up to the final point on B. That momentum must therefore be the same at the bottom of hill A_1, hill A_2, . . . ; the same for all slopes. Therefore the ball must have the same speed at the bottom of A, whatever the slope. So Galileo made his general assumption: *The speeds acquired by any body moving down planes of different slopes are equal if the heights are equal.* This is the property sketched in Ch. 7, where we showed it belongs with Newton's Law II. Galileo generalized this to curved hills. He deduced many geometrical predictions for motion down inclines from this assumption, combined with his proven knowledge of constant acceleration.

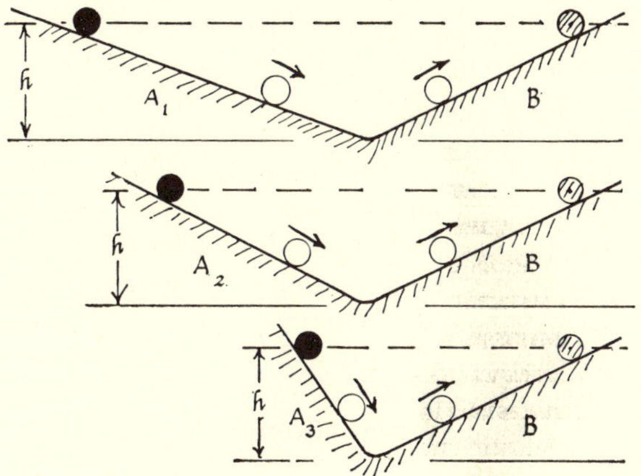

FIG. 19-2a. GALILEO'S ESSENTIAL FACT
Ideal downhill-and-uphill motion.

The Downhill = Uphill Guess

Friction would prevent a satisfactory demonstration of a ball rolling down one hill and up another to the same height. Galileo probably based his guess on a mixture of experiment and thinking—he had a genius for making the right intuitive guesses with the help of rough experiments. It seemed plausible. For his colleagues he made it more plausible by a careful argument about compounding motions downhill and uphill. Look at the following irritating Galilean "thought experiment" (due to a later author): Suppose the ball finished *higher* on the further slope B. We could insert an extra plank, C, and let it roll back to its starting-place on A, and then start again with the velocity it had gained. This could continue, with the ball gaining more and more motion, cycle after cycle, which seems absurd. If the ball finished *lower* on the further slope B, then we might start it on B instead, at that lower point. If it rolled down B we should expect it to retrace its path to the original point on A, thus ending up higher in this backward journey. Again with an extra plank, D, we could arrange for cycle after cycle of an absurd increasing motion. Both cases seem absurd and therefore the ball must rise to the same height on the opposite hill. The weak point in the argument is the claim in the second case that the motion must be exactly reversible. Even apart from this, the method of solving problems by argument does not seem to us very scientific. But Galileo lived in an age of arguers and knew that such attacks would carry considerable weight. Besides, here, as elsewhere in physics, arguments *can* help to clarify the problem, to suggest what to think out and what to investigate.

FIG. 19-3. A GALILEAN "THOUGHT EXPERIMENT"

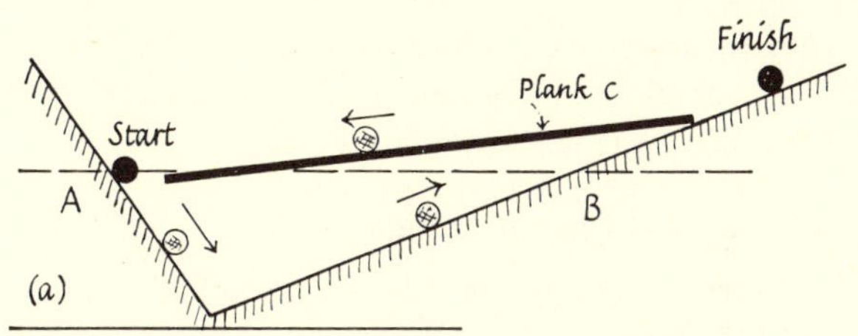

FIG. 19-3a. If the ball finishes higher, let it run back down a temporary extra plank, C. Absurd increasing motion.

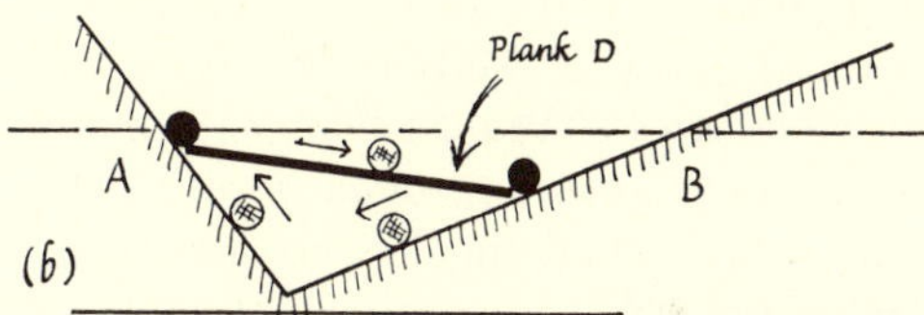

FIG. 19-3b. If the ball finishes lower, let it make the return trip, presumably (?) retracing its path and ending higher on A than on B. Then let it roll back, via a temporary plank D, to the same place on B. Absurd, perpetually increasing motion.

Galileo himself clinched his contention by producing what seems impossible, a frictionless version of the downhill = uphill experiment, an amazingly simple but convincing demonstration, his "pin and pendulum" experiment. In this a peg catches the thread of a pendulum as it swings through its lowest point, thus converting the pendulum abruptly from a long one to a short one. In all cases, the bob after

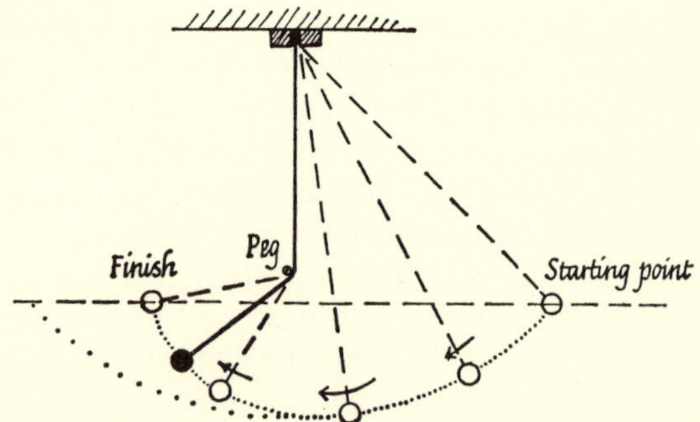

Fig. 19-4. The Pin and Pendulum Experiment
The pendulum bob rises to the same vertical level, in spite of the change of motion caused by the peg.

falling down a long shallow arc rises up a steep arc to the same height. Try this experiment of genius yourself.

In this property of nature, Galileo had the key to one aspect of conservation of energy, though that general idea was not formulated till later.

Newton's First Law of Motion

The downhill=uphill rule suggested one thing more. Galileo argued to the extreme case when the second hill is horizontal, with no slope. Then a ball that rolls down the first hill must continue along the horizontal forever. Thus he found the essence of Newton's First Law of Motion: *Every body continues to move with constant velocity (in a straight line), unless acted on by a resultant force.* Here, had he but known, he had the key to one puzzle of planetary motion: What forces maintain the motion of the planets, Moon, etc.? What pushes them *along* their orbits? The new answer was to be: no force; none is needed, because motion continues of its own accord.

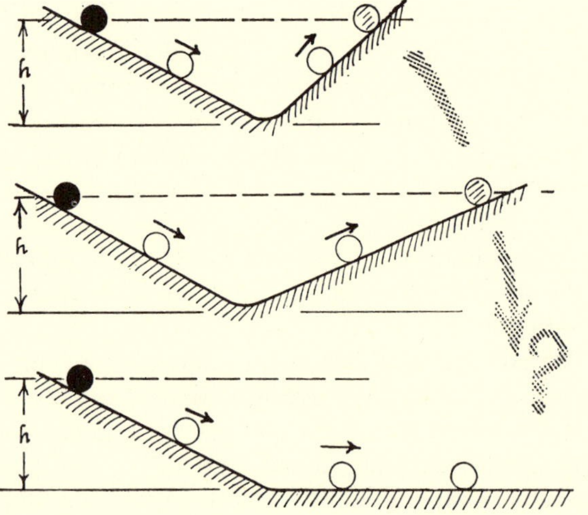

Fig. 19-2b. Galileo's Argument

Independence of Motion

The parallelogram addition of velocities, forces, etc., was just being recognized or discovered—essentially this implies that one vector does not disturb another: they act independently and just add geometrically. All through his experiments Galileo insisted that motions (and forces) are independent of each other. For example, a vertical accelerated motion and a constant horizontal motion simply add by vectors—one motion does not modify the other, but each has its full effect. He applied this to ideal projectiles and showed that their paths are parabolas.

Galileo preached this independence of vectors again and again in his dialogues, as an essential reply to critics of Copernicus. When they claimed that a moving Earth would leave falling bodies far behind, Galileo asked about things dropped from the mast of a ship that was sailing steadily. If they then murmured about wind, he repeated the "thought experiment" in the ship's cabin. He said that clouds and air, which already have the motion of the Earth's surface, simply continue to move with it. He carried his readers through problems like those at the end of Chapters 1 and 2 and showed that a steady motion of the laboratory does not affect experiments on statics, free fall, or projectiles. A laboratory's steady motion cannot be detected by any mechanical experiments inside. That is *Galilean Relativity.*

Leaving Padua

After 20 years at Padua, Galileo was tempted to return to his home university at Pisa. He had kept in touch with the Medici family there, and now he negotiated with the Duke for an appointment in Florence with better pay and more leisure. His public duties at Padua took only an hour a week, but to supplement his salary—which was still small, though an admiring university increased it several times—he had to do private teaching. "He was weary of universities, of lecturing, tutoring, and boarding students; he had had enough of the stuffed robes against which he had written satirical poems; . . . of the closed and petty atmosphere of Padua. . . . He wanted to be in his own land, in his own native light and air, free, and among friends of his own choosing."[10] He needed leisure to study and write, and the support of noble patronage. In return

[10] G. de Santillana in his "Historical Introduction" to Galileo's *Dialogue*, *op.cit.*, p. xi.

for a better salary he promised the Duke he would write a series of books: ". . . principally two books on the System of the Universe, an immense design full of Philosophy, Astronomy, and Geometry; then three books on Motion, three on Statics, two on the Demonstration of Principles, one of Problems; also treatises on Sound and Speech, Light and Colors, Tides, the Composition of Continuous Quantity,[11] the Motion of Animals, and the Military Art." This gives some indication of his eagerness and wide interests.

The Telescope

While he was considering the move to Pisa and Florence, Galileo happened to hear of the invention of the telescope. It is said that a Dutch spectacle-maker[12] had found an arrangement of two lenses that made distant things seem large and close. Hearing about this, Galileo made a simple telescope, magnifying only three times, by fixing two lenses in a pipe. He used a weak convex lens for the first lens, and a concave lens for the eyepiece. This may have differed from the invention he heard of, in which case he was the first to make an "opera glass" which we now sometimes call a Galilean telescope. He was delighted with his new instrument and the fame of this marvel soon spread. The telescope was the talk of society, and crowds came to look through it. The Venetian Senate hinted they would like a copy. Galileo presented them with one. His salary was doubled soon after!

[11] The *composition of continuous quantity* sounds like an attempt at the calculus problem of integration. The need for calculus as a mathematical tool was growing. Galileo himself needed it, and made preliminary attempts. By the next generation the time was ripe for its development, and it is not surprising that Newton and Leibniz invented it independently.

[12] The spread of printing in the century before had increased the use of spectacles, so the time was ripe for the discovery of telescopes by the chance putting together of lenses.

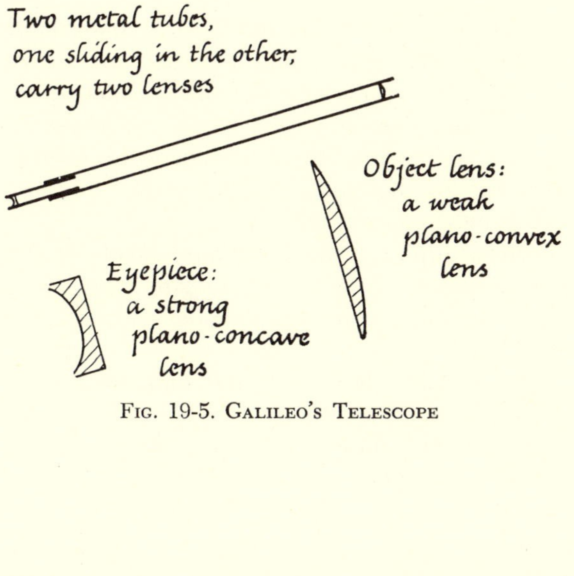

Two metal tubes, one sliding in the other, carry two lenses

Object lens: a weak plano-convex lens

Eyepiece: a strong plano-concave lens

Fig. 19-5. Galileo's Telescope

Galileo looked at the Moon and then the planets and stars with, he says, "incredible delight." If you have never seen the Moon through a telescope, borrow any small instrument, field-glasses or a toy, and try it.

On observing the Moon, Galileo saw mountains and craters. He even estimated the heights of Moon-mountains from their shadows. What he saw was unwelcome to many who had been taught that the Moon is a smooth round ball. Mountains and craters made the Moon earthy and broke the Aristotelians' sharp distinction between the rough, corruptible Earth and the polished unchangeable heavens. The telescope dealt a smashing blow to the old astronomy of perfect spheres and globes. Human beings are conservative and do not like to have their settled opinions changed by a newcomer who proves he is right. Far from pleased at being shown something new, they are angry to find their beliefs upset, particularly if those beliefs have been firmly established in childhood—their sense of security is assailed. So Galileo found some people angry over his discovery. When he offered a convincing look through his telescope, many were delighted, but some refused, and others looked and then said they didn't believe it. One Aristotelian admitted the mountains were there but explained away the damage by saying that the valleys between them are filled with invisible crystal material to bring the surface back to a perfect sphere. Sure, said Galileo, and there are mountains of invisible crystal there as well, that stick out ten times as far!

Good lenses were hard to obtain and Galileo had to grind and polish his own in his instrument shop. He made better ones than most, so that his telescopes succeeded where others failed. Even so, his telescope (preserved in Florence) gives a poor image compared with modern instruments. He made a second instrument which magnified eight times, then another which magnified 30 times and which involved great labor: the grinding of a block of glass into the right shape is a tedious, difficult business, and much of the final performance depends on care in polishing. Through the new telescope the planets appeared as bright discs. The stars looked brighter and farther apart, but were still just points. Galileo was delighted to find how many more stars he could see. The luminous haze of the "milky way" was resolved into a myriad of stars.

FIG. 19-6a.
A MODERN PHOTOGRAPH OF THE MOON NEAR LAST QUARTER (Photograph by J. H. Moore and J. F. Chappell with 36-inch refractor, at Lick Observatory.) The sunshine catching the mountain tops near the edge shows the rough landscape that Galileo's telescope revealed, to the dismay of many people. [This shows the Moon inverted, as we see it with a modern telescope.]

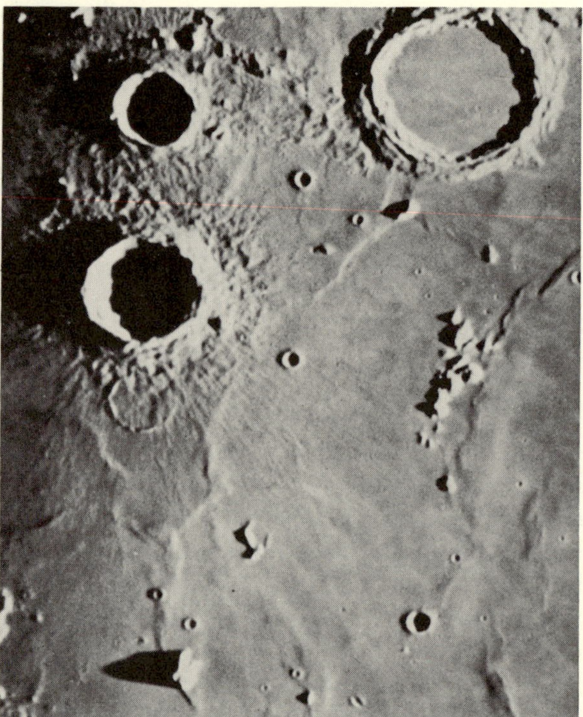

FIG. 19-6b. A MODERN PHOTOGRAPH OF MOON MOUNTAINS (Lick Observatory). This is a section of the picture (a), enlarged about 6 times. The peak with the long shadow near the bottom is the mountain "Piton."

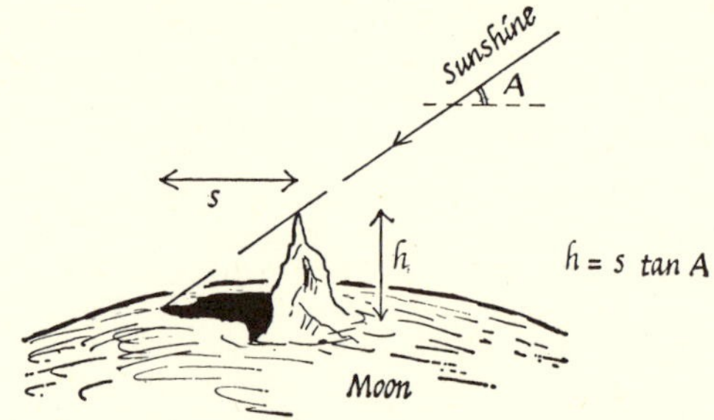

FIG. 19-6c. SKETCH OF PITON AND ITS SHADOW, to show how mountain heights are estimated. (After Whipple, in *Earth, Moon and Planets.* Harvard University Press, 1941.)

Jupiter's Moons

With this new powerful telescope Galileo made a still more important discovery. On the night of January 7, 1610, he observed three small stars in a line (Fig. 19-7) near Jupiter, two to the East of Jupiter, and one to the West. He thought they were fixed stars and paid little attention to them. The next night he happened to look at Jupiter again and found all three of the stars were to the West of Jupiter and nearer one another than before. He ignored the latter peculiarity and thought the shift was due to Jupiter's motion; but then he realized that this would require Jupiter to have moved in the *wrong direction,* for Jupiter was on a backward loop. This was mysterious indeed. He waited anxiously to observe them again the next night but the sky was cloudy. The night after that, on January 10, only two of the stars appeared, both to the East of Jupiter. Jupiter could hardly have moved from West to East in one day and then from East to West in two days by such amounts. Galileo decided the "stars" themselves must be moving and he set himself to watch them. The sketches show his record of what he saw. He had really discovered four small moons moving around Jupiter.

Look at Jupiter yourself, with any small telescope, even field glasses. You cannot miss the moons, which you will see better than Galileo did with his simple lenses.

Kepler, when Galileo wrote to him about his discovery, shared Galileo's delight, although the extra moons seemed rather contrary to his limit of six

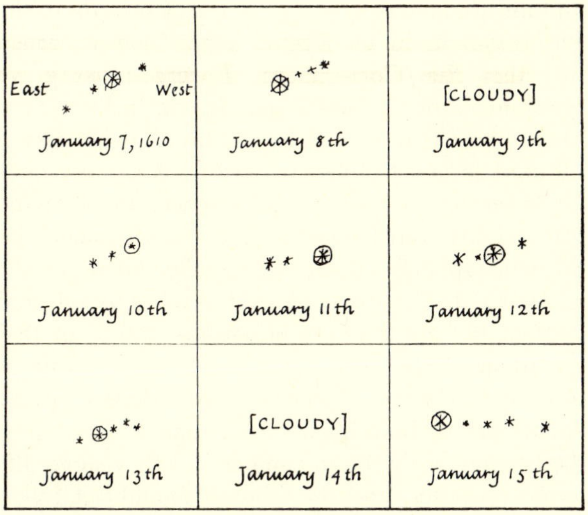

East West January 7, 1610	January 8th	[CLOUDY] January 9th
January 10th	January 11th	January 12th
January 13th	[CLOUDY] January 14th	January 15th

FIG. 19-7. GALILEO'S OBSERVATIONS OF JUPITER'S MOONS These sketches are copied from Galileo's handwritten record. (The orbits of the moons are nearly in planes containing our line of sight from Earth to Jupiter; so the moons are often in front of Jupiter or behind, and they are often eclipsed by moving into Jupiter's shadow. They move quickly around their orbits. That is why the pattern changes so quickly and why, often, less than four moons are visible.) (For a copy of Galileo's written record, see *Galileo* by J. J. Fahie.)

planets. The Aristotelian philosophers did not welcome the discovery because it made Jupiter an important center, spoiling the Earth's unique position, supporting, in fact, the Copernican theory. One argued,

"There are seven windows in the head, two nostrils, two eyes, two ears, and a mouth; so in the heavens there are two favorable stars, two unpropitious, two luminaries, and Mercury alone undecided and indifferent. From which and many other similar phenomena of nature, such as the seven metals, etc., which it were tedious to enumerate, we gather that the number of planets is necessarily seven. Moreover the satellites of Jupiter are invisible to the naked eye, and therefore can have no influence on the earth, and therefore would be useless, and therefore cannot exist. . . ."

Galileo wrote to Kepler:

"Oh, my dear Kepler, how I wish that we could have one hearty laugh together! Here, at Padua, is the principal professor of philosophy whom I have repeatedly and urgently requested to look at the moon and planets through my glass, which he pertinaciously refuses to do. Why are you not here? What shouts of laughter we should have at this glorious folly! And to hear the professor of philosophy at Pisa labouring before the grand duke with logical arguments, as if with magical incantations, to charm the new planets out of the sky."

Jupiter and his moons provided a small scale model of the Copernican system of Sun and planets —a compelling argument for the Copernican view. Kepler used Galileo's measurements in a rough test to see whether his Law III applied to Jupiter's moons. He found it does apply, though the constant obtained for R^3/T^2 is different from the value given by the Sun's planets. The table below shows modern, more accurate, data. More moons, smaller and farther out, have been discovered since. We now know a dozen.

Return to Pisa and Florence

When Galileo accepted the new post at Florence he had to resign his professorship at Padua. This unexpected resignation was most unwelcome there; it seemed ungrateful and even unfair, but the new post offered better opportunities. Galileo made the move and lost some friends. Though he now enjoyed the leisure he needed for his work, the move proved unwise in the long run, because it was a return to enemies as well as friends. (Back in his student days there he had been known as "the

JUPITER'S SATELLITES AND KEPLER'S THIRD LAW

NAME OF SATELLITE	DISTANCE FROM JUPITER in Jovian diameters	DISTANCE FROM JUPITER in miles (R)	TIME OF REVOLUTION in hours (T)	CALCULATIONS FOR TEST OF LAW III: R^3 (miles)^3	T^2 (hours)^2	$\frac{R^3}{T^2}$
Io	3.02	262,220	42.36	1.803×10^{16}	1802.8	TRY
Europa	4.80	417,190	85.23	7.261×10^{16}	7264.	THIS
Ganymede	7.66	665,490	171.71	29.473×10^{16}	29,484.	(See
Callisto	13.48	1,170,700	400.54	160.440×10^{16}	160,430.	note 14)
	(see note 13)					

[13] It is simplest to measure the moons' orbits in terms of Jupiter's diameter. The radii could remain in those units for a test of Kepler's Law III; but, if these data are to be used in gravitational theory (e.g., to compare Jupiter's mass with the Sun's), then the same units, e.g., miles, must be used on both sides of the comparison.

[14] The test is made easy by a lucky chance arising from the choice of units, miles and hours. Look at the numbers.

wrangler," and he had been violent in his attacks on those he called the "paper philosophers.") He was a sincere man but not tactful, and the opposition which his discoveries and arguments excited was more a subject of triumph to him than sorrow. "The Aristotelian professors, the temporizing Jesuits, the political churchmen, and that timid but respectable body who at all times dread innovation, whether it be in religion or science, entered into an alliance against the philosophical tyrant who threatened them with the penalties of knowledge."[15]

At Florence he continued to study the planets with his new telescope and soon made more discoveries. He found that Saturn seemed to have a small knob on each side, almost fixed to it. Modern telescopes show Saturn as a bright ball with a flat ring round it like a hat-brim; and we now know that the ring is made up of small pebbles, possibly ice, all circulating independently around Saturn—an army of Kepler III examples. In Galileo's telescope the ring was not clear—seen almost edge-on, it looked like side-planets. Galileo next found that Venus showed phases like the Moon. Here was direct confirmation of the Copernican picture. On the Sun, Galileo found sunspots, black blotches that moved and changed—another blow at the purity of the heavens.

He took his telescope on a tremendously successful visit to Rome, where he was welcomed with enthusiasm. His telescope was the wonder of the day, and a Church committee approved his discoveries.

Growing Troubles

Galileo returned to Pisa full of plans for a great treatise on the Constitution of the Universe. The Copernican system seemed right in its simplicity and compellingly vouched for by his telescope. An Earth spinning with its own momentum would remove the improbable daily motion of an outermost sphere of stars, driven by "an immense transmission belt from nowhere,"[16] with uncouth gears to run the inner spheres. A central Sun, with Earth a planet, simplified many heavenly motions and predicted what he saw in his telescope. He even saw a model of the Solar System in Jupiter and his moons. But that Copernican view conflicted with the simple poetic astronomy of the Bible, taught with full authority of the Churches, Roman Catholic and Protestant alike. Just as Galileo felt confident he could prove the Copernican case, Church disapproval suddenly hardened. He was attacked in sermons, and his arguments in reply were sent secretly to the Inquisition in Rome. From Rome, friends warned him the Copernican doctrines were under grave question. Galileo's pupils and friends, including the Duke himself, defended him nobly, but he started his most serious troubles by writing dangerous public letters on scripture and science. In these he claimed that the language of the Bible should be taken metaphorically and not literally, where science is discussed. The Bible, he said, teaches us spiritual matters, but not the facts of nature—and he quoted a cardinal: "The Holy Spirit teaches us how to go to Heaven, not how the heavens go." Since scripture and nature are both works of the same divine author, they cannot really be in conflict, but they serve different purposes; and the Church should not try to make astronomers disbelieve what they see. Nor should people condemn Copernicus' book without first reading and understanding it. Such was his open defense.

Authorities in Rome were still more disturbed. Church astronomers withdrew their support of Galileo in the light of this subversive attack. Galileo, alarmed, went to Rome to investigate his own safety. There he argued persistently with friend and foe alike; but "to such minds, Galileo could not communicate what he and Kepler were alone to see: the three forces of mathematics, physics, and astronomy converging toward a junction which would make them irresistible and creating a physical science of the heavens."[16] Meanwhile the Church had appointed a group of theological experts to examine the Copernican teaching and they reported on two key propositions:

THAT THE SUN DOES NOT MOVE: ". . . false and absurd in philosophy, and formally heretical."

THAT THE EARTH BOTH MOVES AND SPINS: ". . . false and absurd, and at least erroneous in faith."

Galileo stayed on in Rome to help the discussion, as he thought. He was summoned and told that the Copernican doctrine was condemned as "erroneous." Copernicus' book was suspended—no devout Catholic could read it until it had been "corrected." And Galileo himself was not to hold or defend the doctrine as true. He waited a short while, to show a brave face, then returned home, in good standing as a devout Catholic, but reproved and bitterly disappointed.

He remained in Florence for half a dozen years. A new Pope was elected to the throne who was friendlier to science and was in fact a friend of

[15] Sir David Brewster, *Martyrs of Science* (1848).
[16] G. de Santillana, *op.cit.*

Galileo. Delighted, but in poor health, Galileo made an uncomfortable journey to Rome to congratulate the new Pope, and he had a marvelous visit there. He had several audiences with the Pope, who gave him rich presents and honors. He even argued delicately on the Copernican system, emphasizing its simplicity. The cardinals were reserved, but the Pope himself commented, "The Church has not condemned this system. It should not be condemned as heretical but only as rash." However, when Galileo pressed his views, the Pope replied sharply that the earlier prohibition must stand. He told Galileo not to limit the wisdom of God to a scientific scheme: *God could devise any scheme he pleased*—a very able argument that can stop all science. However, the Pope finally agreed that Galileo might write a *non-committal* book explaining the arguments on both sides between Copernicus and Ptolemy. That would be a mere theoretical discussion, leaving any question of fact and truth to be decided by the higher wisdom of the Church.

The Great Dialogue

Galileo returned home, disappointed yet honored and confident. He was confident that he had permission to write his long-planned book on the System of the Universe. But he was overconfident, and perhaps undergrateful to the Church. He continued surreptitious teaching of Copernican ideas and developed his book. He wrote it in the form of a dialogue—a very acceptable form of teaching in those days. After some difficulty with Church censors, one of them a personal friend, Galileo got the book published. Its title runs

THE DIALOGUE
OF
GALILEO GALILEI, Member of the Academy of Lincei
PROFESSOR OF MATHEMATICS IN THE UNIVERSITY OF PISA
And Philosopher and Principal Mathematician to
THE MOST SERENE
GRAND DUKE OF TUSCANY
Where he discusses, in four days of discourse, the two
GREAT SYSTEMS OF THE WORLD
THE PTOLEMAIC AND THE COPERNICAN
Propounding impartially and indefinitely the Philosophical and Physical arguments, equally for one side and for the other side

It begins with a preface addressed "to the prudent reader" which looks like a most imprudent attack on the Inquisition. The dialogue is conducted by Salviati, a philosopher, who sets forth with able arguments Galileo's Copernican views; Sagredo, who, as a sort of attorney to Salviati, asks questions and raises difficulties and cheers up the dialogue with his wit; and Simplicio, a dogged follower of Aristotle and Ptolemy, who is beaten in argument by Salviati every time and made a fool of by Sagredo.

The *Dialogue* was not written in Latin for scholars but in Italian, for the general reader, in rolling prose with long discussions and cunning arguments. One critic says it "meanders at ease across the whole cultural landscape of the time";[17] but essentially it was a great setting forth of the nature of motion, terrestrial and celestial, with the fullest arguments in favor of the Copernican scheme. It had much good teaching and argument, and some serious shortcomings. Galileo was a great man, but the new science that he taught was still unfinished and sometimes unclear. He never realized that circular orbits need an inward pull. He maintained *vertical fall* as natural for *earthly* bodies and *circular motion* as natural for *celestial* bodies—an Aristotelian prejudice —despite his understanding of inertia in earthly mechanics. He never taught Kepler's elliptical orbits, perhaps because of that principle of circles, or perhaps because as a hard-headed teacher he realized how *very* close to circles the actual ellipses are. His explanation of tides—caused by a breathing Earth —seems even more pig-headed.

The book proved popular and carried conviction. By contrast, Copernicus' book was difficult—few had understood its full import, and now it was prohibited. There had been talk and covert discussion, but most educated people could not "piece together the great puzzle that stayed disassembled by superior orders. The *Dialogue* did exactly that: it assembled the puzzle and for the first time showed the picture. It did not go into technical developments; it left all sorts of loose ends and hazardous suggestions showing to the technical critic. But it was exactly on the level of educated public opinion, and it was able to carry it irresistibly. It was a charge of dynamite planted by an expert engineer."[18]

News of this tremendous attack soon reached Rome and the Pope, good friend though he was, ordered the Inquisition to forbid the book and re-examine Galileo. Galileo, aged and sick, was summoned to

[17] G. de Santillana, "Historical Introduction" to Galileo's *Dialogue*, *op.cit.*, p. xxx.
[18] G. de Santillana, *op.cit.*, p. xxxi.

Rome. There he was well treated on the whole and and comfortably housed—they knew he was a great man—but the Inquisition proceeded to a strict examination, formulating his offenses then asking him to defend his actions. Galileo knew he had written dangerously, but with permission. The original prohibition had only instructed him not to teach Copernican astronomy as *true*, and he had obeyed the dictum, thinly and insincerely. So he felt fairly safe until a document, probably forged, was produced that showed he had promised never to teach or discuss the Copernican system at all. In that case, the indictment was very serious: heretical teaching and writing in the face of a pledge to desist. Failing repentance and abjuration, a heretic faced terrifying threats of torture, and then torture. Galileo was in a very grave position. He had disobeyed instructions of the Church; he had set forth the Copernican picture *in print* (with a thin pretense that it was only a piece of theorizing); and he had even criticized the interpretation of the Scriptures. The powerful, ruthless Church, which crushed questioners with severe punishment and condemned defiant martyrs to be burned at the stake, would stand no such behavior. Outside the Inquisition's court, he was well treated; inside, they were lenient at first, reasoning with him and asking him to defend his position. Yet he was being questioned by a court that held both the physical powers of torture and the spiritual powers of a great Church. His health grew worse; he was questioned and questioned again. Still he held on to his beliefs, holding on to his real life. A friendly examiner suggested he should confess to false pride as the cause of his writing and be let off lightly. Galileo, giving up hope of arguing his position, at last agreed. However, the highest court of the Church overruled this lenient compromise and insisted on unconditional surrender. Galileo was summoned to a "rigorous inquiry." He did not emerge from the court till three days later. We do not know how far he was taken in the steps towards torture. He was not tortured physically—that was ruled out by his great age—yet to his intellect much of the proceedings must have had the horrors of mental torture. In the course of this inquiry he agreed to recant completely, to withdraw his unorthodox statements and deny his own earlier beliefs. He accepted the judgment of the Inquisition as a penitent—remember he was a pious if argumentative member of the Church—and he knelt and read the abjuration required of him, swearing never again to believe in or teach the Copernican system. It was a long grim document of abject apology, confession of errors, complete recantation of views, and absolute promises for the future under severest penalties. Kneeling, he signed.

There is a tale that as he rose from his knees he muttered "E pur si muove"—"and yet it [the Earth] does move"—but that is hardly likely. There was no friend there to hear, it was far too dangerous, and Galileo was a broken old man. As Bertrand Russell has said, "It was the world that said this—not Galileo."

Galileo was imprisoned for a time, then allowed to return home, under some restrictions. His health was poor, but his head, he complained, was "too busy for his body." He composed his great discourse on *Two New Sciences*. It contained the account of his work on accelerated motion which formed the basis for Newton's laws, his rules for elasticity of beams, and his foundations for calculus. This was no popular account, but a great technical text. While he was working on it he became blind in one eye and soon the other one became blind too. He says of this calamity: "Alas! your dear friend and servant has become totally and irreparably blind. These heavens, this earth, this universe, which by wonderful observation I had enlarged a thousand times beyond the belief of past ages, are henceforth shrunk into the narrow space which I myself occupy. So it pleases God; it shall therefore please me also." He was now allowed more freedom, and in spite of much sickness, he continued his writing with the help of friends. His health grew worse, and he died at the age of 78.

The Contest between Science and Church

Galileo brought into the open the differences between authoritarian churchmen and independent scientists. By his tactless manner and powerful arguments he brought great troubles on himself, and on science too. His biographers differ in their views of his conflict with the Roman Catholic Church, according to their own feeling about authority. Some paint him as almost a martyr, threatened with torture by a bigoted Inquisition, suspected, persecuted, imprisoned, and forbidden to teach the great Truths he had helped to discover—with the Church as the villain of the piece taking the side of superstition and prejudice, trying to suppress, in the interest of dogmatic authority, the simple things of nature that should be to the glory of a worldwide religion. Others show Galileo bringing his troubles on himself by his hotheaded arguments and exasperating manner of setting people right; they paint him as ungrateful towards the Church which listened to his teaching and honored him with pensions; and they point out that his conflict with the Church arose directly from his attack on scriptural science—in which he was meddling in the Church's rightful province. Others again regret his subservient behavior—in not making himself a martyr for science—but this seems a cruel criticism of one who was so far from subservient most of his life.

Bertrand Russell says,

"The conflict between Galileo and the Inquisition is not merely the conflict between free thought and bigotry or between science and religion; it is a conflict between the spirit of induction and the spirit of deduction. Those who believe in deduction as the method of arriving at knowledge are compelled to find their premises somewhere, usually in a sacred book. Deduction from inspired books is the method of arriving at truth employed by jurists, Christians, Mohammedans, and Communists. Since deduction as a means of obtaining knowledge collapses when doubt is thrown upon its premises, those who believe in deduction must necessarily be bitter against men who question the authority of the sacred books. Galileo questioned both Aristotle and the Scriptures, and thereby destroyed the whole edifice of mediaeval knowledge. His predecessors had known how the world was created, what was man's destiny, the deepest mysteries of metaphysics, and the hidden principles governing the behavior of bodies. Throughout the moral and material universe nothing was mysterious to them, nothing hidden, nothing incapable of exposition in orderly syllogisms. Compared with all this wealth, what was left to the followers of Galileo?—a law of falling bodies, the theory of the pendulum, and Kepler's ellipses. Can it be wondered at that the learned cried out at such a destruction of their hard-won wealth? As the rising sun scatters the multitude of stars, so Galileo's few proved truths banished the scintillating firmament of mediaeval certainties. . . . Knowledge, as opposed to fantasies of wish-fulfilment, is difficult to come by. A little contact with real knowledge makes fantasies less acceptable. As a matter of fact, knowledge is even harder to come by than Galileo supposed, and much that he believed was only approximate; but in the process of acquiring knowledge at once secure and general, Galileo took the first great step. He is, therefore, the father of modern times. Whatever we may like or dislike about the age in which we live, its increase of population, its improvement in health, its trains, motor-cars, radio, politics, and advertisements of soap—all emanate from Galileo. If the Inquisition could have caught him young, we might not now be enjoying the blessings of air-warfare and poisoned gas, nor on the other hand, the diminution of poverty and disease which is characteristic of our age."[19]

The principal mistake was the same one on both sides: failure to regard the Copernican scheme as a useful picture, an hypothesis or piece of "theory." Galileo and the Church argued whether it was *true.*

". . . it is entirely false to assert that the Church stopped the scientific experiments of Galileo. Galileo's difficulties with the Church had nothing to do with his experiments. They developed, apart from purely personal causes, out of his refusal to yield to the request that he treat the Copernican hypothesis as an hypothesis, which in the light of modern relativity was not an unreasonable request. There seem to be as many myths about Galileo as about any of the saints."[20]

Galileo himself, with his love of truth, might have decided more calmly to treat the Copernican view as an hypothesis—though in mechanics he did believe in absolute fixed space. He might well have defended the wisdom of such an open mind.

The new scientific temper was unwelcome to the academic philosophers, busy with cultured discourse—though their spirited Renaissance predecessors might have enjoyed it. The Church, busy maintaining political power and spiritual good, feared it. (Earlier in Galileo's lifetime, Bruno had been burned at the stake for heretical views, among them his use of Copernican astronomy. He had pointed out that, with the outermost "sphere" of stars at rest, the stars could be spread into infinite space beyond—myriad suns occupying Heaven. That dissolving of our universe's neat outer shell was a shocking novelty to the medieval mind.) Galileo's *Dialogue* was placed on the forbidden list, and his abjuration was read in churches and universities, "as a warning to others." The new Protestant Church was no less intolerant. With no papal authority, its leaders placed even greater importance on the literal truth of the Bible. Martin Luther had said of Copernicus, "the fool will overturn astronomy." In Italy and elsewhere, the new science was

[19] Bertrand Russell, *The Scientific Outlook* (W. W. Norton and Company, Inc., New York, 1931), pp. 33-34.

[20] Morris R. Cohen, *The Faith of a Liberal: Selected Essays*, in which there is a delightful essay, reviewing Galileo's book *Two New Sciences.* These three pages (417-419), are well worth reading. (Henry Holt and Company, New York, 1946.)

discouraged by authority and restricted by fear for another half century.

This struggle is not unique, an isolated battle of the past. There is some such struggle in every age. Conditions of living, interacting with men's knowledge and powers and beliefs, produce struggles between conservative elements—not always the same kind of people—and rebellious ones—not always with the same kind of causes. In each age, from Galileo's to now, the struggles of the day have usually seemed just as serious, making it just as dangerous to take, or even to discuss, the rebel's side. Often, a generation or two later, both sides regret the quarrels; yet mankind still learns sadly little from those regrets. Age after age, current quarrels seem as vital and as dangerous as their forgotten or regretted ancestors did in their own time.

Here is a translation of a note in Galileo's handwriting in the margin of his own copy of the *Dialogue*:

"In the matter of introducing novelties. And who can doubt that it will lead to the worst disorders when minds created free by God are compelled to submit slavishly to an outside will? When we are told to deny our senses and subject them to the whim of others? When people devoid of whatsoever competence are made judges over experts and are granted authority to treat them as they please? These are the novelties which are apt to bring about the ruin of commonwealths and the subversion of the state."[21]

Whatever we make of Galileo's life, his work remains, as a foundation of physics, a monument of method for all who came after, and a basis of knowledge on which the science of mechanics was already being built.

[21] From the English translation edited by G. de Santillana, preceding p. xi. Chicago University Press, 1953.

CHAPTER 20 · THE SEVENTEENTH CENTURY

"Great ideas emerge from the common cauldron of intellectual activity, and are rarely cooked up in private kettles from original recipes."

—James R. Newman
in *Scientific American*

A Hundred Centuries of Astronomy

In the hundred or more centuries between the earliest civilized men and the time of Galileo, Astronomy grew from simple beginnings rooted in primitive man's curiosity and wonder and fear to a well-ordered science ready to furnish a great laboratory in which mechanics, the basis of all physics, could be developed and tested. For many centuries, astronomy remained in the hands of the calendar-makers and priests, with only a rare scientist to observe more carefully than the rest or to extract some new thread of law from the tangle of observation. Astrologers throve on men's superstitious fears; yet, just as alchemy did much for chemistry, astrology helped early astronomy. Then came philosophers and scientists with greater interest in knowledge for its own sake. They gathered careful observations and extracted rules for the motions of planets, Moon, and Sun—working rules, though clumsy ones. They invented reasons or causes which now seem fanciful and complex. Astronomy waited centuries more, almost asleep but for a few keen observers, while civilization reformed itself toward a new awakening. In those dark ages, the teaching of the Church and the method of deductive argument based on the authority of books were supreme in intellectual matters. Tradition replaced experiment, prejudice overrode science.

Yet there were practical needs, in medicine and navigation, to keep good science alive. Then came the cry with growing fervor, "Watch what *does* happen; stop arguing about what *ought* to happen." Prejudice was being pushed aside by careful thinking in terms of experimental observations.

The Renaissance

In the three centuries preceding 1600, the Renaissance grew and spread across Europe, a great awakening of new interests in art and literature and a new outlook in religion. The tight hold of the traditional scholastics loosened, and the study of Greek writers in the original opened out after centuries of obscurity; paper and printing came to spread knowledge, and the renewed interest in knowledge, far and wide; great navigations brought new markets, new wealth, new outlooks, and new leisure for intellectual growth; the arts flourished more freely and helped to release man's restless spirit of inquiry.

The Renaissance brought a new attitude of mind: the idea of *man as an individual* developed, in contrast with man as a servant to his group and its traditions. The enthusiasm for study and the general spirit of free inquiry that came with the renewed "humanities" prepared the ground for the development of science in the 17th century: ". . . the humanists . . . played the chief part in that widening of the mental horizon which alone made science possible. Without them, men with scientific minds would never have thrown off the intellectual fetters of theological preconception; without them, external obstacles might have proved insurmountable."[1]

In Renaissance science, one man towers above the rest, Leonardo da Vinci, a new scientist a century or two before his time. A genius at whatever he turned his hand and mind to—painting, sculpture, architecture, engineering, . . . physics, biology, . . . philosophy, . . . —he regarded observation and experiment as the only true approach to science. "He dismissed scornfully the follies of alchemy, astrology . . . ; to him nature is orderly, non-magical, subject to immutable necessity."[2] He trusted the logical conclusions of arithmetic and geometry because they are based on concepts of universal truth; but science must be based on experiment. He said, "those sciences are vain and full of errors which are not born from experiment, the mother of all certainty, and which do not end with one clear experiment."[3] His own experimenting with techniques for art and architecture and engineering led him to much scientific knowledge: machines and forces;

[1] Sir William Dampier, *A History of Science*, 4th edn. (Cambridge University Press, 1949), p. 98.

[2] *op.cit.*, p. 107.

[3] Quoted by Dampier, *op.cit.*, p. 105.

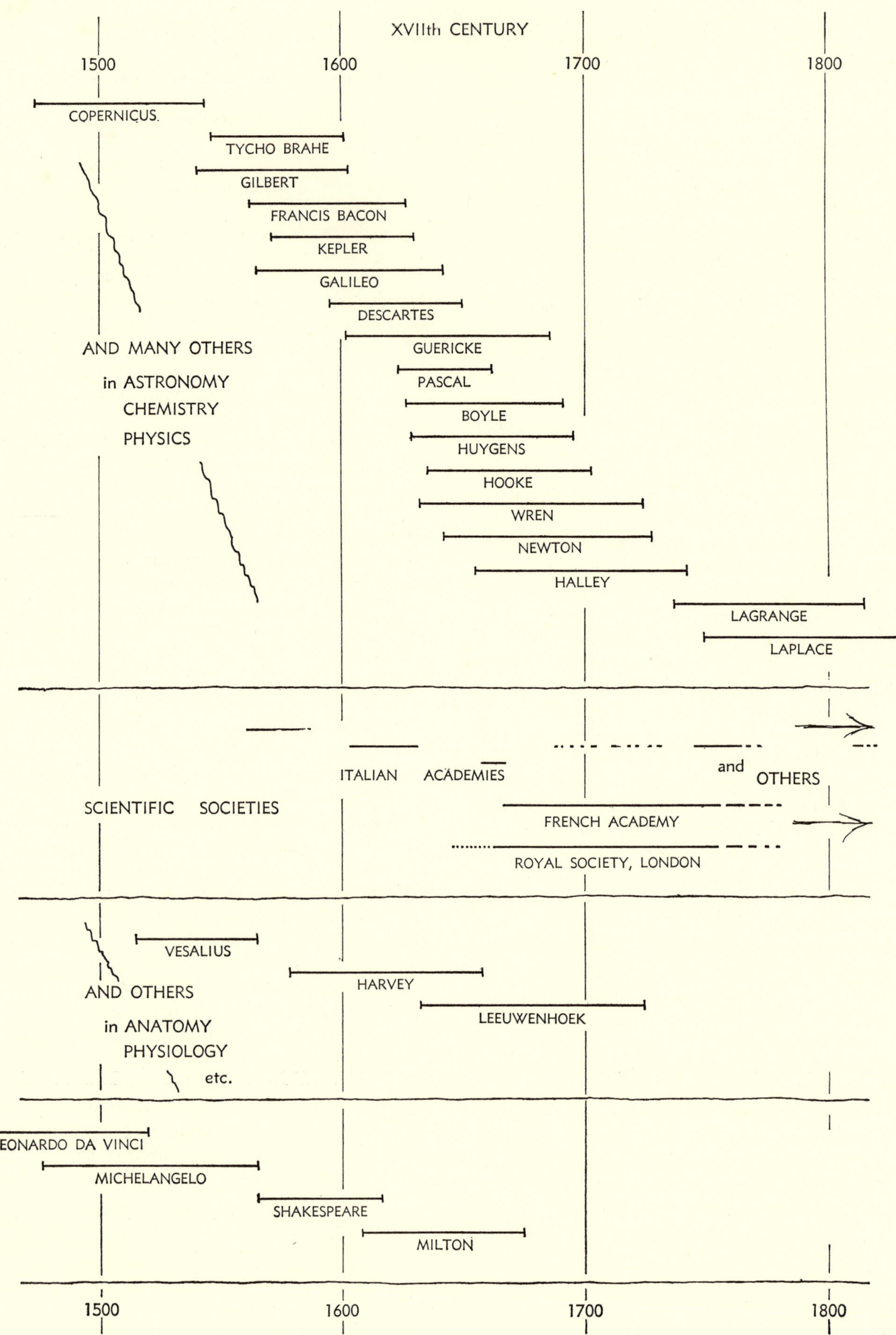

Fig. 20-1. A Time Chart

properties of motion; Newton's Law I in simple form, long before Galileo taught it; properties of liquid pressure and fluid flow; waves in water and sound waves in air; the impossibility of perpetual motion. He advanced optics with studies of the eye and perspective; and he may have constructed a pendulum clock 200 years before Huygens. He saw in fossils the evidence of geological history; he may have discovered the circulation of the blood; and he set forth human anatomy in brilliant drawings that used the knowledge from many dissections. We learn all this from his drawings and from scrappy notes in his private notebooks. "Had he published his work, science must at one step have advanced to the place it reached a century later."[4]

"Progress"

Sometime then the idea of *progress* appeared, as a new outlook. We now take progress as an obvious aim—progress toward a better state, a higher education, a finer man, etc.—and we might think our ancestors always aspired to progress. Yet, for many centuries, as men looked back on a golden age in the past and tried to model their conduct on tradition, the idea of progress in earthly life had been almost unknown. When a forward outlook emerged from the Renaissance, it offered science new encouragement.

The Seventeenth Century

The simpler Sun-in-center idea suggested by Copernicus gained ground faster and faster, as new experimental evidence came to its support and freedom of speech and teaching increased. Tycho and Kepler untangled the actual motions of the planets, extracting simple general "laws." Galileo explored and expounded mechanics; he argued and taught and preached to establish realistic science. It was a tremendous advance from earliest watching of the planets and awed marking of eclipses to Galileo's telescope and Kepler's Laws. If ten thousand years seems a long time even for that, remember it is only some four hundred human generations. Are four hundred generations too many for man's understanding to grow from simple superstitions to mathematical certainty? Many of us think this was rapid advance. Yet in the three or four generations of the following century the awakened experimental science made still greater strides.

By 1600 the new growth of science was well under way. Kepler and Galileo were at work. Astronomy was nearly ready to provide a vast frictionless laboratory for mechanics, and experimenting was becoming fashionable. By 1700, planetary laws extracted from observations had already been used to test general laws and theories of mechanics. Newton was developing a new scientific method, in which general theory, devised by shrewd guessing, was made to yield a variety of results by deduction—an ancient method in itself, but now the results were tested by experiment. The deductive method, so dangerous and unscientific when used with wordy arguments alone, was taking its proper place in science, fostering a real alliance between theory and experiment. There were changes in the political, social, and religious structure of Western civilization which gave science greater opportunities. Throughout the century science was blossoming and becoming fashionable; experimenting and realistic argument were all the rage.

Many people helped to build the science of mechanics and to put the backbone of theory into astronomy. Some invented or improved the mathematical tools needed by the physicists—though the latter, like the true blacksmiths, mostly learned to make their own tools to suit their needs. Some carried over the new experimental attitude and the zest for clear thinking into other fields of science; with growing interchange of knowledge, many sciences developed together. Scientists brought fame to their country, and royal favors were bestowed on them from pride instead of superstition. Besides, there was a suspicion that scientists could be useful in commerce and manufacture and war—an early glimpse of their contributions to industry today[5]—so perhaps it paid to favor them!

This was also the time of starting of scientific societies. An Academy of Science was founded in Florence and another in Paris; and the Royal Society was founded in London. These helped science to emerge from the secrecies of the dark ages. They supported some experimenting and encouraged much discussion and interchange of problems and knowledge, but their greatest contribution was publication. No longer were scientific discoveries transmitted by letter to a few friends. They were tested and extended by experiment and argument, then published in print for others to teach and use. With so much keen discussion among able, leisured people, scientific questions were in the air, and the time was ripe for rapid progress.

The names of Copernicus, Tycho Brahe, Kepler,

[4] Dampier, *op.cit.*, p. 108.

[5] Seventeenth-century science is said by some to have laid the foundations of the Industrial Revolution. Others see that later change growing quite independently of scientific attitudes, and only drawing on some factual knowledge.

Galileo, and Newton serve as landmarks, but there were many others who helped to build science in the seventeenth century. Here are short notes on a few of those concerned with physics and astronomy. Others made tremendous advances in biology and medical science (circulation of the blood, mechanism of breathing, embryology . . .).

William Gilbert (1540-1603). Physician. Experimented with magnetism and wrote a very good book on it. Also experimented on static electricity.

Francis Bacon (1561-1626). Brilliant writer who laid down rules for making discoveries by experimenting and induction. These rules are not very practical and did not contribute much to the development of science. He sneered at the work of Gilbert and Galileo and rejected the Copernican theory. However, he did help to spread the idea that nature should be investigated by experiment and not just described by argument.

René Descartes (1596-1650). Philosopher and mathematician. Contributed greatly to Newton's work. Born in France of a rich family, he took life easily yet accomplished much. Did a great deal for philosophy and mathematics, and contributed to anatomy. In physics, he studied optics and motion—going some way toward Newton's Laws—and the nature of matter. He produced an ingenious vortex theory to account for gravity, cohesion, and the motion of the planets. His greatest contribution to physics was his *invention of graph-plotting* with x and y coordinates, leading to the use of algebraic equations for curves, tangents, etc. This prepared the ground for the invention of calculus, which is concerned with drawing tangents and finding areas on such graphs by calculation from their equations instead of by measurement. Our x-y graphs are named "cartesian" after him.

Otto von Guericke (1602-1686). Devised a workable vacuum pump and used it to give demonstrations of atmospheric pressure with the "Magdeburg hemispheres," great hollow cups that teams of horses could not pull apart when the air had been pumped out.

Evangelista Torricelli (1608-1647). Physicist. Made first barometer.

Blaise Pascal (1623-1662). Theologian and scientist. Started the mathematics of probability. Stated laws of pressure for fluids at rest.

Robert Boyle (1626-1691). A great experimenter: vacuum, gas law, chemistry. One of the original members of the Royal Society. Wrote "The Skeptical Chymist."

Christian Huygens (1629-1695). Mathematician and physicist. Developed a wave theory of light. Built a very good clock (probably the first working pendulum clock); and even arranged to correct for slight increase of pendulum period with large amplitude. Studied mechanics, and derived v^2/R for centripetal acceleration earlier than Newton.

Robert Hooke (1632-1702). Began scientific work as Boyle's assistant, but soon rose to be a great scientific experimenter and thinker. His rivalry with Newton obscured his fame and hurt him greatly. But for Newton's overshadowing genius, Hooke would have been classed as one of the great scientists of the age. He bitterly claimed some of Newton's mechanics as his own discovery. Original member of the Royal Society.

Edmund Halley (1656-1742). Astronomer. Friend of Newton. Did much to help the publication of Newton's *Principia.* Important member of the Royal Society.

Science.

There were five important growths in science: (i) the growth in *respectability* of experimental science with increasing freedom of speech; (ii) the growth of *factual knowledge and theory* used in describing it; (iii) the growth of *mathematical tools*; (iv) the invention of *new instruments* for experimenting; and (v) the change in *scientific method and attitudes.*

(i) *Respectability.* We see the growth of science's respectability in Galileo's own life. His father regarded mathematics and science as a poor academic occupation; yet Galileo, rebel though he was, was respected as one of the world's great men in his later years. Newton, Boyle, and Hooke did not have to defend their interest in science; they argued about their discoveries, but not about the spirit of discovery itself. They wrote with little fear of condemnation or ridicule, only with anxiety lest they miss priority or fame. The discussions and publications sponsored by the new societies started scientific knowledge on its way to becoming public and universal—thus the idea of scientific truthfulness began to react on the thinking of mankind.

(ii) *Knowledge.* The growth of actual knowledge was great and varied: e.g. Kepler's Laws, Halley's orbit for his repeating comet, Hooke's Law for springs, Harvey's discovery of the circulation of the blood, Boyle's chemical discoveries and his gas Law.

(iii) *Mathematics.* Coordinate geometry (x-y graphs) was invented, calculus followed, and each helped the other.

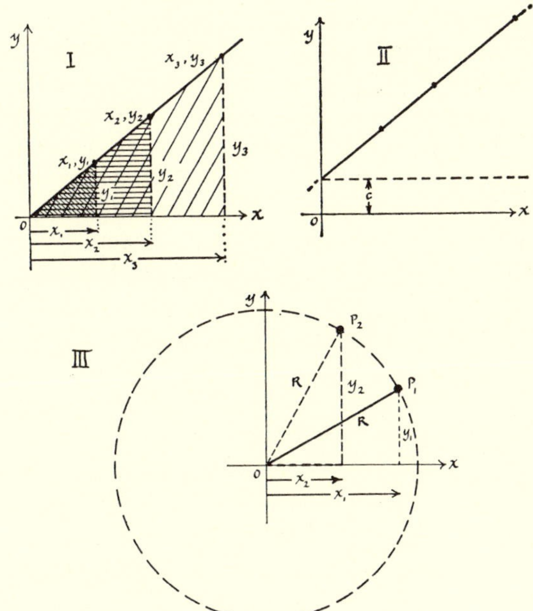

FIG. 20-2. CARTESIAN GRAPHS

Graphs link algebra with geometry by converting geometrical forms and operations to compact algebra, and by exhibiting algebra for easy survey.

Graph I shows a straight line through the origin, with several points (x_1,y_1), (x_2,y_2), . . . marked. By similar triangles, the fractions y_1/x_1, y_2/x_2, . . . are all equal—the same for every point on the line. Call the constant value of this fraction k. Then every point on the line represents a pair of values, (e.g. x_1,y_1), which fit the relationship $y/x = \mathrm{k}$ or $y = \mathrm{k} \cdot x$. This is the algebraic description of the graph, and the line is the geometric picture of the relationship. If y and x are a pair of physical measurements (e.g. s and t^2 for a falling body) the straight line expresses the relationship $y =$ (constant) x, or $y \propto x$, and the slope of the line gives the (constant).

In Graph II every y has a constant bonus, c; so its equation is $y = \mathrm{k}x + c$. In this case we can *not* say that $y \propto x$, but only that $\Delta y \propto \Delta x$.

Graph III is a circle, and

for a point P_1, $x_1{}^2 + y_1{}^2 = R^2$;
for P_2, $x_2{}^2 + y_2{}^2 = R^2$,

so the equation of this circle is:

$$x^2 + y^2 = R^2.$$

Or we can write this: $\dfrac{x^2}{R^2} + \dfrac{y^2}{R^2} = 1.$

An ellipse can be made by distorting a circle with uniform strain. Draw a circle on a sheet of rubber and stretch it. Radii R and R become semi-axes a, b.

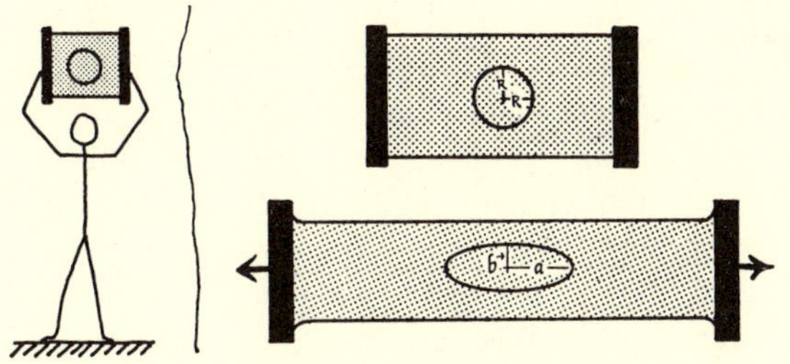

FIG. 20-3. CIRCLE STRAINS INTO ELLIPSE

Then, at a guess, a circle with

equation $\dfrac{x^2}{R^2} + \dfrac{y^2}{R^2} = 1$ and AREA $= \pi R^2 = \pi R \cdot R$

becomes an ellipse with

equation . . ? . $= 1$ and AREA $=$?

Thus, coordinate geometry could reduce orbit ellipses to equations that are easier to handle.

Science had developed two serious mathematical needs, both filled by calculus: to *calculate* tangent-slopes of curved graphs, and to *calculate* areas inside curves, *by algebra. Differential calculus* does the first, and the reverse process, *integration*, does the second. A tangent-slope gives a rate-of-change. Calculus is just an algebraic process to find a rate-of-change *at an instant.* It enables us to calculate ACCELERATIONS from an equation that specifies VELOCITIES, or VELOCITIES from an equation relating DISTANCE and TIME. (For example: given $s = 16t^2$, calculus tells us that $v = 32t$; and it then tells us that $a = 32$, a constant value.) Integration—again just a refined algebra-logic-machine—adds up an infinite number of infinitely small contributions: tiny patches to find an area (e.g. for Kepler's Law II), or tiny attractions to find a total gravitational pull.

You used graphs and calculus for your early investigation of a wheel rolling downhill:

Stage 1. EXPERIMENT → EMPIRICAL GRAPH. You plotted s against t^2, each point coming from an experimental timing. *The points are the facts.* A line sketched to follow them is a summary of facts—in the grammar of graphs, an "indicative" graph.

Stage 2. THINKING → THEORY. Guess at constant acceleration, as a possible simple rule for nature. Use calculus to predict the necessary relation between s and t. Integration adds all the little distances travelled with growing speed, and shows that *if* acceleration is constant, *then* s must vary as t^2.

Stage 3. TEST. Draw a straight line through origin to represent the relationship $s \propto t^2$ from theory.

IF your points lie close to this line, THEN *the wheel's motion in the experiment is close to constant acceleration.* That straight line on your graph is an "interrogative" line: drawing it is asking a question, "do the facts fit constant acceleration?" You would be wise to call such a graph line a "question-asking line." Drawing the "best straight line" on your graph of an experiment exhibits your *hypothesis*—a temporary rule, guessed at and to be trusted for the moment (e.g. "suppose $\Delta v/\Delta t$ is constant . . .").

(iv) *Instruments.* A new instrument, as well as a new mathematical tool, can promote great developments of science. The 17th century was a century of great inventions of instruments: the telescope, the microscope, the vacuum pump, the barometer, the pendulum-controlled clock, and the first crude thermometers were the tools that promoted a new range of experimental science.

(v) *Attitude and Method.* From the Greeks to Galileo, science was being built by collectors, accurate observers, makers of schemes, authoritarian philosophers.[6] The collectors gathered a lot of knowledge which by itself would have been too diverse to be called science. The scheme-makers organized this knowledge and extracted rules that were good working prescriptions, able to summarize the facts and often to make predictions. Rules and knowledge, together with techniques for gaining more knowledge, made the beginning of the new science.

Meanwhile the thinkers were busy devising explanations—statements that would make knowledge fit together better and become easier to "understand" or easier to accept. Many explanations or reasons were drawn from their own thinking with only remote experimental background—e.g., for epicycles, "circles are perfect"; for the barometer, "invisible threads hold the mercury up." Some explanations seem to be little more than a statement that nature "is like that," put with authority—e.g., for falling bodies, "the lowest place on the ground is *natural.*" Man needed such reassurance that the external world of nature has a simple organization in it; otherwise his fears of a profuse unknown would have driven him to more superstition and even madness. As general working rules emerged—e.g., the epicycle scheme, Hooke's Law, Kepler's Laws—the sense of security and comfort increased, and the early belief that nature is definite and reasonable gained ground as a basic belief in science. The Greeks *deduced* their explanations and schemes for nature from a few general ideas which they just assumed—e.g., from "circles are perfect" they deduced epicycloids. In the course of the 17th century that kind of deductive reasoning fell into disfavor; it was really philosophical speculation flourishing with authority, rather than science. By the middle of that century experiment was regarded as the real source and test of science. Men were occupied with extracting rules or laws by inductive reasoning from experiment. In doing this, they too were making assumptions: that nature is simple, and that nature is uniform—that is, that in the same conditions the same behavior will occur again and again. They still assumed that there are *causes* for things, but the meaning of causality remained as difficult a problem as ever.

Though this inductive method was an honest one leading to good rules, it lacked the general tying-together and mental satisfaction that a grand theory can give. Newton, with greater insight, looked at experiment then jumped to theory and worked back deductively from theory, predicting results that could be tested. This brought theory back into science as a framework of thought, but in a more respectable and responsible form. Theory was again considered valuable—e.g., the theory of universal gravitation—but *as a servant to science rather than as master.*

Later still, say in the last century, theory was subjected more and more to the test of productiveness. Scientists asked, "Can this theory make (further) predictions?" If not, it was shelved or modified. That now seems too harsh a treatment for theory. Its use may lie not only in its ability to make predictions but also in the scheme of thinking that it offers us.

Descartes' New Philosophy

While the viewpoint of working scientists was thus changing with the temper of the times, René Descartes in France advanced a new philosophy that had a lasting effect on scientific thinking. And with it he announced a new model of the universe that remained popular for a century. Seeing faults in classical philosophy, Descartes turned his consideration of the world inward to his own thoughts and feelings and set himself to doubt every stage of his knowledge. From this examination he evolved a

[6] We may trace some survivals from childhood in these activities: a child's delight in collecting may turn to scientific data-collecting instead of adult stamp-collecting or even money-hoarding; the grim determination of many adolescents not to be beaten by a hard problem may turn to a Tycho's drive for accuracy, instead of to some more cruel behavior; the insecure child's craving for a definite framework of rules may turn to hunting for simple laws instead of to some more neurotic form of worry.

dualism, a picture of two sharply-divided worlds existing together, *each as real as the other*: a world of matter, with size and shape and motion, and a world of soul and mind. Just as two clocks side by side can keep the same time, so the two worlds, although entirely separate, keep in tune, because "God made them so."

In this scheme, matter is entirely dead, without spirit, and it can only exchange motion with other matter by contact. The motion of matter must have been started originally by God. Thus, to Descartes, God was not a presiding power who controlled the world for man's life in it; but God was the First Cause, who started the Universe with motion, laid down rules for its running, and then left it to run. Thereafter, motion can only be carried from one piece of matter to another by some matter. Therefore, the open spaces in the solar system cannot be empty. They must be full of an invisible material "aether" that carries the motion. Since a moving region of aether cannot continue out indefinitely, it must be arranged in a closed circuit, a whirlpool or *vortex*. All space in fact is full of vortices of aether, large and small all geared together, conveying the motions of visible bodies. The planets are carried around their orbits by a huge vortex belonging to the Sun. The Earth, carried with the other planets in that vortex, has a smaller vortex of its own to draw objects inward. Thus, inward fall under gravity corresponds to the sweeping of floating straws towards the center of a whirlpool in water. On a smaller scale the picture accounted for cohesive forces that hold small pieces of matter together. This scheme of whirlpools within whirlpools, all invisible, sounds fantastic today; yet it proved very popular at the time, because it explained the whole system of the universe by a vast machine started by God but then kept running by constant mechanical rules. In fact, Descartes' picture of a "full" universe with no vacuum, running as a machine, became a serious rival when Newton published his gravitational theory. Newton favored a vacuum and gave no ultimate cause for gravitation. Descartes' theory offered more explanation, but it rejoiced in unsupported speculation—the vortices were undetectable except by the motions they were to explain. Newton attacked them mathematically, showing they could not fit with Kepler's Law III, and he attacked them on principle when he claimed, "I will not feign hypotheses."

Out of his systematic doubting, Descartes seems to have emerged with a certainty of God. God started the universe and provided laws to govern it. Therefore, the laws of nature must be completely right: God would neither make a careless mistake nor accept a rough average. This view of the laws of nature influenced the next generation of scientists—Newton and his contemporaries felt they were looking for great laws established by God's command and waiting to be found.

If you think this a strange digression in a discussion of hardheaded physics, reflect that the same problems remain today at the borders of science and philosophy: What is the nature of space (that carries electromagnetic and gravitational fields, and obeys Relativity geometry)?; What do the laws of nature mean?; How did the world begin?; How old is time?; Will time go on?

Francis Bacon

While Descartes tried to explain the universe with a grand *deductive* theory operated by mathematics, Francis Bacon in England advocated a grand *inductive* treatment of systematic experimenting. He sought universal knowledge by great organized schemes of research, to produce stores of data from which scientific knowledge could be extracted. He claimed that science could not advance by pure deduction and argument, nor could it advance rapidly by haphazard data collecting—children's play. Rather, scientists should plan their experimenting carefully and treat it by a formal system of inductive reasoning and testing.

Thus, Bacon saw clearly the difference between "good experimenting" and "just playing around with apparatus." If you have enjoyed your own laboratory work, you now understand that difference; though you would find it difficult to describe that essential criterion of science to a child busy with haphazard "experimentation."

In your laboratory investigation of downhill motion and its analysis with graphs, you are following Galileo and Newton in one scheme of scientific method: collect information, extract rules, frame hypothesis, deduce consequences, test deductions (straight line on graph), &c.

Francis Bacon set forth ideal schemes like this for science, but if you watch scientists at work you will see there is no one scientific method. Physical science does not develop as a simple rigid chess-game of alternating moves; there is far more variable and complex interplay of pieces, players, and the board itself. Nor is the progress just a series of forward leaps. A first round of thinking and experimenting may even lead back to the starting point—"this is where we came in"—but, as with seeing a movie over again, we have a richer knowledge with which to pursue the second round. As J. R. Zacharias puts it, "science bootstraps itself."

Bacon wrote with great eloquence, advocating a vast organization of professional experimenters and reasoners. His grandiose scheme was too artificial for successful science, and it aimed more at practical values than ultimate understanding of nature. (It had some of the misplaced zeal that we might expect today from a non-scientist research-director placed in charge of a large commercial research laboratory.) His proposals were only schemes on paper, yet they had great influence on the starting of the Royal Society and the experimental work of its members, in particular Boyle. By mid-century, "art was giving way to science under the pressure of Bacon's influence, . . ."[7] Nowadays, two centuries later, we see that in a deeper sense science *is* an art.

[7] Dorothy Stimson, *Scientists and Amateurs, A History of the Royal Society* (Abelard-Schuman Ltd., New York, 1948), p. 36.

The Growth of Theory: A Necessity for Modern Science

1600 to 1700 was an exciting century for astronomy—as for all the life of the intellect. At its beginning, there was a growing stock of facts and rules, pressing for explanation. Speculations on general causes were in the air: the time was ripe for a comprehensive view. By the end of the century, knowledge and interest had grown and spread; but, above all else in importance, Newton had built and published a great theoretical scheme that pulled astronomy into a single "explanation" and projected it into a future of rich promise.

If you wish to understand modern physical science, you need to know what theory does; you need to have a feeling for "good theory." You can hardly learn that from sermons *about* theory. Instead, study a good example of it. The next four chapters describe and discuss Newton's great theory of gravitation.

CHAPTER 21 · CIRCULAR ORBITS AND ACCELERATION

"... all science as it grows towards perfection becomes mathematical in its ideas."
—A. N. Whitehead (1911)

The Problem of Orbital Motion

Why do the planets move in Kepler orbits? Why do they keep moving, and why are their orbits ellipses? These questions naturally followed Kepler's discoveries, in the long tradition of asking WHY, dating from the Greek philosophers. Astronomers had measured and recorded WHAT the planets do. Copernicus and Kepler had shown HOW the planets' motions can be expressed in a simple scheme; but they had only the traditional answers to WHY. Copernicus thought of spheres rolling around, though with simpler motion; Kepler imagined a spoke of influence from the Sun carrying each planet and pushing it *along its orbit*; and he talked mystically of magnetism shaping the orbits. His spokes were real in one sense; they were needed to express his

"PROBLEM SHEET"

The problem here is intended to be answered on a typewritten copy of this sheet. Work through the problem on the enlarged copy, filling in the blanks, (———), that are left for answers.

PROBLEM ON SUBTRACTING VECTORS (Problem 1.) NAME____________________

We shall need to subtract vectors in studying planetary motion. This problem is to give you practice.

I. Ordinary (arithmetical) subtraction. Suppose we want to subtract 2 from 5.

This can be regarded in several ways:

(a) We can say, 2 subtracted from 5 makes . . . . . . __________

or, the same thing in other words, 5 - 2 is . . . __________

(b) Or we can change the sign of the 2 to -2 and ask an "addition" question: 5 + (-2) makes ? . . . . __________

(c) Or we can put this more childishly and ask: What must we add to 2 to make 5 ? . . . . __________

This last form gives the key to subtracting vectors (or finding the difference between two vectors, or finding the change of vector from one vector to another.)

II. Vectors. Suppose we have an "old vector" and a "new vector" and want to find the change (or gain, or difference). We ask, "What vector must be added to the old one to make the new one?" (This is like form I(c) above; but it requires geometrical addition.)

(a) If the vectors both point due East as below, old vector 2 and new vector 5, what is the change or difference? "What must be added to the vector 2 to make the vector 5 ?" Show it by drawing on the sketch below.

FIG. 21-1.

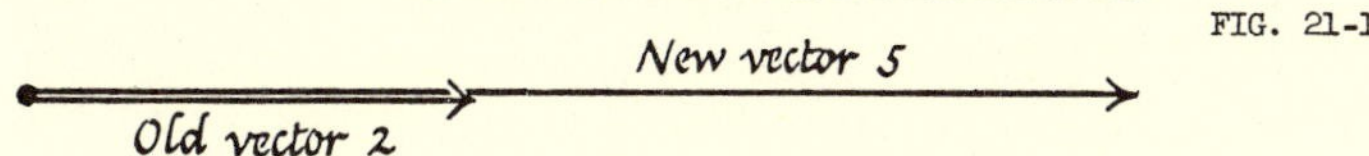

(b) If the vectors (old vector A and new vector B) have different directions, as in the various cases shown below, what is the change or difference? "What must be added to the vector A to make vector B?" Show it for each case by drawing an arrow, on this sheet. In each case, you want B - A.

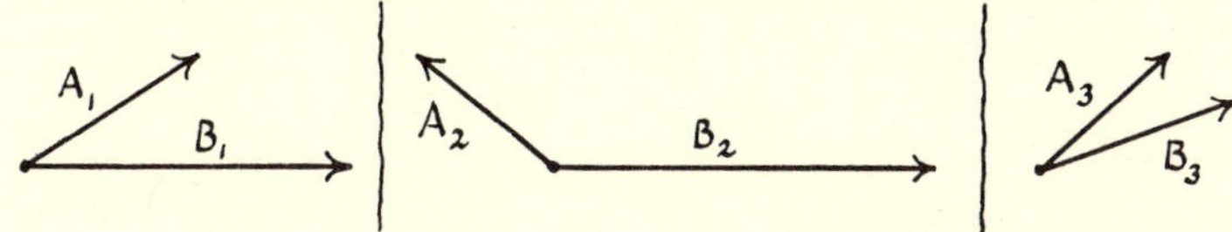

(c) If the vectors do not sprout from a common starting point, you must first transfer one or both till they do. Then find difference, B - A, in each case below, again using an arrow to show your answer.

FIG. 21-2.

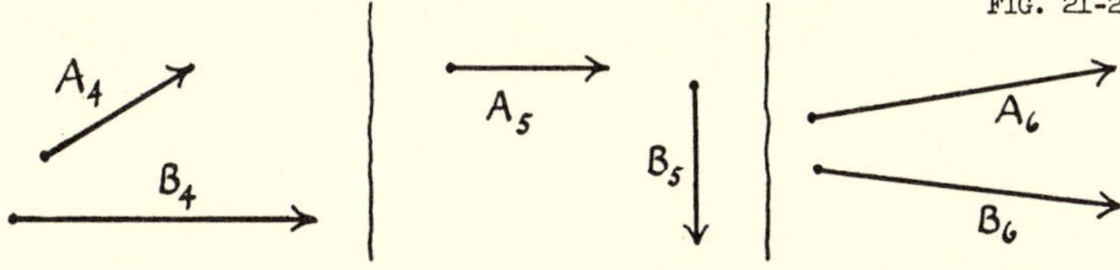

Law II; and they are still with us, as geometrical lines that sweep out areas. As solid arms to propel the planets Kepler's spokes soon seemed unnecessary: Galileo's new teaching put things in a different light. A moving thing, said Galileo, will continue if it is left alone, and he gave a clever thought-experiment to justify his view. A generation later, Newton expressed this view as a working rule: Law I, "Every body will remain at rest or continue to move with constant speed in a straight line unless acted on by a resultant force." Then Newton crystallized the vague idea of motion into definite momentum, to be calculated by multiplying mass (which he tried to define) by velocity, and said: Law II, "When there is a resultant force, there are changes of momentum in the *direction of the force.* RATE OF CHANGE OF MOMENTUM varies directly as the RESULTANT FORCE." This was equivalent to "MASS · ACCELERATION varies as RESULTANT FORCE."

From Galileo to Newton these new views of motion—groped for by philosophers far earlier, half stated by Leonardo long before Galileo and by Descartes after him—were getting ready to play a profound part in astronomy. The members of the newly-formed Royal Society, which soon welcomed Newton as a younger Fellow, discussed Kepler's laws eagerly, asking quite a different kind of WHY question. They no longer worried about an agent to push the planets *along.* Galileo told them no push—and therefore no pusher—is needed for that; the planet will continue to move of its own accord if left alone, like a block of ice on a frozen pond or a bullet in space. Scientists had dissolved away Kepler's spoke. What they sought instead was an *inward* force, along the spoke perhaps, to pull the planet into a curved orbit instead of a straight line. Such a force pulling sideways, "across the motion" of the planet, would give it momentum in a new direction. What kind of forces would do this? This new question was in the air. Hooke, Huygens, and Newton all attacked it. Taking the planetary orbits as roughly circular, they argued back from Kepler's Law III and suggested that there is an inward attraction, pulling planets to the Sun, a pull that decreases with increasing distance according to an "inverse square law"—a scheme we shall discuss in the next chapter. But would such a force, half suspected and quite unexplained, produce elliptical orbits fitting with Kepler's Laws I and II? This was too difficult for all but Newton. It required clear formulation of laws of motion, and then clever mathematics. Newton not only solved it but extended his solution into a magnificent framework of good theory. Before you study his work, you should extend your discussion of force and motion to this new case of "sideways forces" that pull a moving body's path into a curve. You have already met this with projectiles, where gravity adds vertical momentum to horizontal motion, making a curved path.

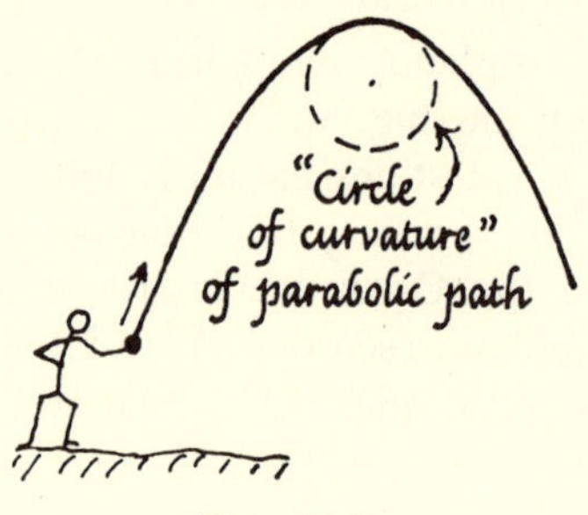

FIG. 21-3.

But acceleration in that motion with its increasing speed seems easier to understand—dare we say it "seems more natural"?—than in steady motion around a circular orbit, with unchanging speed.

Acceleration of Body Moving Around a Circle

Suppose we have a planet moving around a circle (or a stone on a string, or an airplane, or an atom). Does it have any acceleration? If not, we hardly expect to find any resultant force acting on it; and in that case why does it not continue straight ahead? Does it have any acceleration? Certainly not any acceleration *along* its path—we have chosen a case of constant speed, with no speeding-up along the path. Is there any acceleration *across* the path, perpendicular to it? Try drawing vectors to look for changes of (vector) velocity. The moving body P

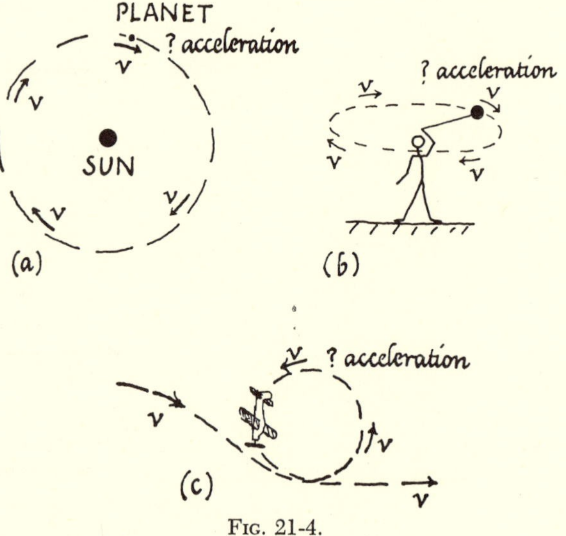

FIG. 21-4.

travels around a circle of radius R with fixed *speed* v.[1] Then v is also the size of P's *velocity*, but that velocity, as a vector, has direction as well as size and the direction changes from instant to instant. When the moving body, P, is at A it has velocity v in the direction shown, along the tangent. If it moves with unchanging speed, the *size* of v is the same at B as at A, but its *direction* is different; the two vectors are not identical. There *is* a change of velocity between A and B. (And therefore an acceleration, and therefore . . . on to planetary astronomy with a wonderful tale, if this is so.) Calculate the change of velocity and divide it by the

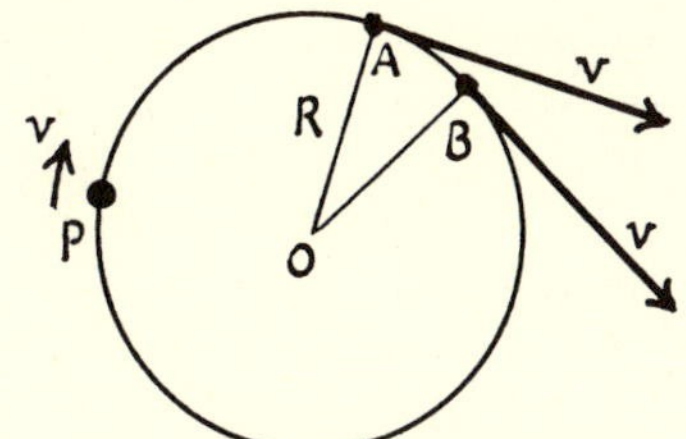

FIG. 21-5. VELOCITY VECTORS

time taken, to find the "acceleration." This involves subtracting vectors to find a change as in the preliminary problem at the beginning of this chapter.

Derivation of Expression $a = v^2/R$

While P moves from A to B it changes its velocity from (**v** along AT) to (**v** along BT′). Make a vector diagram to find the change of velocity. Transfer the two velocity vectors to a common starting point, X, and draw XY to represent VELOCITY **v** at A and XZ to represent VELOCITY **v** at B. Then XY is the "old velocity" and XZ the "new velocity." What is the change? What velocity must be added to the old to get the new? The change is shown by YZ, the vector marked Δ**v** in the sketch. Then

(old **v**) + Δ**v** makes (new **v**), *by vector addition*

To see the direction of Δ**v**, redraw the original picture, and slide the **v**'s along their tangent lines till they both sprout out from the common point C. Then we can treat C like X and draw old **v** and new **v** from it and mark Δ**v**. Look at the direction of Δ**v**. It is parallel to CO, from C to the circle's center. If we took B very close to A, the Δ**v** would have to be along the radius from the region AB to the center. Δ**v** is an inward velocity vector, towards the center of the circle.

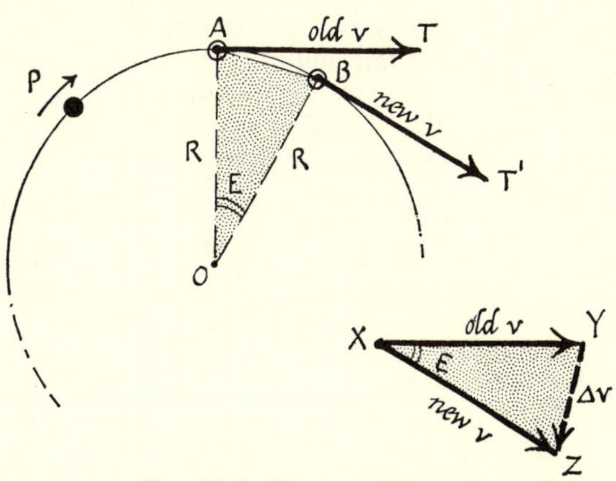

FIG. 21-6. VELOCITY CHANGE
Since velocities are along tangents, which are perpendicular to radii, the fact-picture triangle OAB and the vector-diagram triangle XYZ are similar.

If there are velocity changes, there is acceleration.[2] Calculate the acceleration by dividing velocity change, Δv, by the time it takes, Δt. Time Δt is the time taken by P to move from A to B with speed v along the orbit. In fact SPEED v is $\frac{\text{arc } \widehat{AB}}{\Delta t}$. To calculate $\Delta v/\Delta t$ in terms of v and R, etc., we need some geometry discovered by Newton's contemporaries. Here it is. Join A and B by the chord $\overline{AB}$. As often in solving geometrical problems, the trick is to add one construction line, here the chord $\overline{AB}$. Now look for similar triangles between the fact-picture and the vector diagram of velocities. Radii OA, OB in the fact-picture make a small angle E. The velocity vectors are along tangents, perpendicular to the

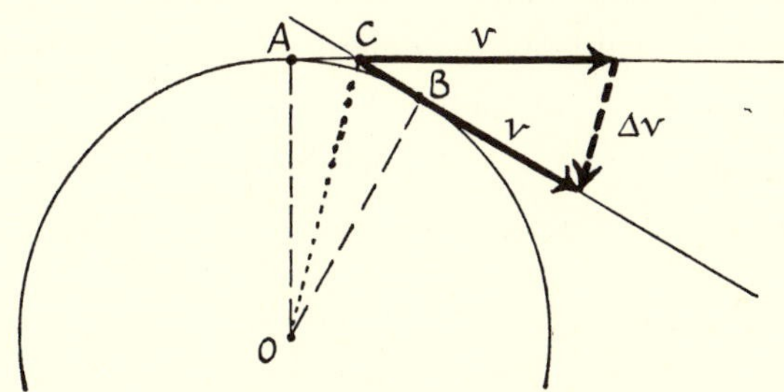

FIG. 21-7. DIRECTION OF VELOCITY CHANGE
Here the vectors for old v and new v have been slid along their lines in the fact picture so that they both start from C. The change of velocity, Δv, is parallel to CO. The change of velocity during travel from A to B is inward towards the center.

[1] Remember we use "speed" for rate of travel along any path, straight or curved—the drunkard's speedometer-reading. Speed is a scalar. Velocity is a vector, with direction as well as size.

[2] From now on we shall stop showing **v** and Δ**v** as vectors by boldface type because we are going to calculate the *size* of the acceleration, using the *speed* v, which is the *size* of the velocity, and Δv, the *size* of the change of velocity. Remember, however, that the acceleration has the direction of the vector Δ**v**.

radii, so the velocity vectors (old v) and (new v) make the *same* small angle E.[3] Then in the fact-picture we have triangle OAB with two equal sides, R and R, enclosing angle E; and in the vector-diagram we have triangle XYZ with two equal sides v and v, enclosing the same angle E. Therefore the two triangles OAB, XYZ are similar.

$$\text{Then } \frac{\text{“short side,” } \Delta v}{\text{“equal side,” } v} \text{ must} = \frac{\text{“short side,” } \overline{AB}}{\text{“equal side,” } R}$$

in vector triangle — in fact triangle

$$\therefore \frac{\Delta v}{v} = \frac{\overline{AB}}{R} \qquad \therefore \Delta v = \frac{v \cdot \overline{AB}}{R}$$

Now we can calculate the "acceleration."

$$\text{ACCELERATION} = \frac{\Delta v}{\Delta t} = \frac{v \cdot \overline{AB}}{R} \Big/ \Delta t = \frac{v}{R} \cdot \frac{\overline{AB}}{\Delta t}$$

To go further we need to know what $\overline{AB}/\Delta t$ is. What *is* $\overline{AB}/\Delta t$? What is the fraction [(CHORD $\overline{AB}$) divided by (TIME OF TRAVEL from A to B)]? We know what (arc $\widehat{AB}$)/Δt is. That is distance/time, around the orbit from A to B, so it is the SPEED v. But, for a very short arc, with B very close to A, the curved arc $\widehat{AB}$ is very nearly the same as the chord straight across, $\overline{AB}$. Look at the series shown in Fig. 21-9. As we move A and B closer and closer together the arc $\widehat{AB}$ and the chord $\overline{AB}$ get smaller but they also show *much* less difference from each other. Like mathematicians inventing calculus, we crawl towards the "limit" when B coincides with A. We never get to that limit, but we can crawl as close as we like and make the difference between arc and chord as trivial as we like. We do not merely make the *difference*, $\widehat{AB} - \overline{AB}$, trivially small: we make the *fraction* (difference/chord), or $(\widehat{AB} - \overline{AB})/\overline{AB}$, trivial. This makes the ratio $\frac{\widehat{AB}}{\overline{AB}}$ very nearly 1. So, with a big separation between A and B, we can say *arc is somewhat greater than chord*; with small separation we can say *arc = chord approximately*; with still smaller separation, *arc = chord very nearly*; and we can get as near as we like to the limit *arc = chord*. Mathematicians prefer to describe this limit thus: Limit $\left(\frac{\text{arc}}{\text{chord}}\right) = 1$. Now we want the acceleration *at an instant of time*, when B and A do practically coincide—we do not want a vague average over a long separation. We want the *limit* when B coincides with A. So we say: in the limit, arc = chord, $\widehat{AB} = \overline{AB}$.

[3] If you take two lines making an angle X, and turn each line through 90°, you turn the whole pattern through 90° and the two lines in their new position will still make angle X.

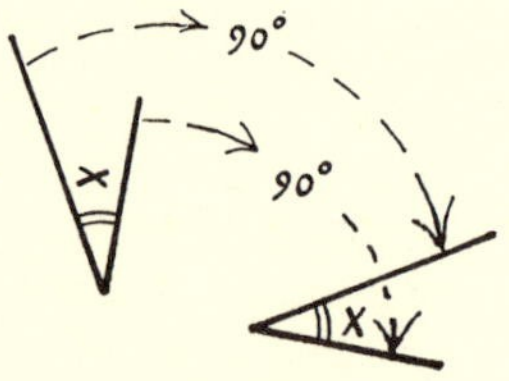

FIG. 21-8.

$$\text{Then } \frac{\text{arc}}{\Delta t} = \frac{\text{chord}}{\Delta t}$$

$$\text{or } \frac{\widehat{AB}}{\Delta t} = \frac{\overline{AB}}{\Delta t}, \text{ in the limit.}$$

$$\text{Then acceleration} = \frac{\Delta v}{\Delta t}$$

$$= \frac{v}{R} \cdot \frac{\overline{AB}}{\Delta t} =, \text{ in the limit, } \frac{v}{R} \cdot \frac{\widehat{AB}}{\Delta t}$$

$$= \frac{v}{R} \cdot (v) \text{ since } \frac{\widehat{AB}}{\Delta t} \text{ is } v.$$

$$\text{Then the acceleration } \frac{\Delta v}{\Delta t} = \frac{v}{R} \cdot (v)$$

$$= \frac{v^2}{R} \text{ or } \frac{(\text{SPEED AROUND ORBIT})^2}{\text{RADIUS OF ORBIT}}$$

This relation, ACCELERATION $= v^2/R$, is of great importance. We shall use it in planetary theory, in studying electron streams, in making a mass spectrograph for atoms, in designing a cyclotron—in fact wherever we meet motion around orbits. It is so important that you should retrace its derivation for yourself and make sure it is sensible. Once you understand the derivation you will see that you can reduce it to a short explanation + two sketches + a few lines of algebra.

Two Important Questions

The result, ACCELERATION $= v^2/R$, brings two questions:

I. How can a moving thing have an acceleration and yet neither go any faster nor get any nearer the center?

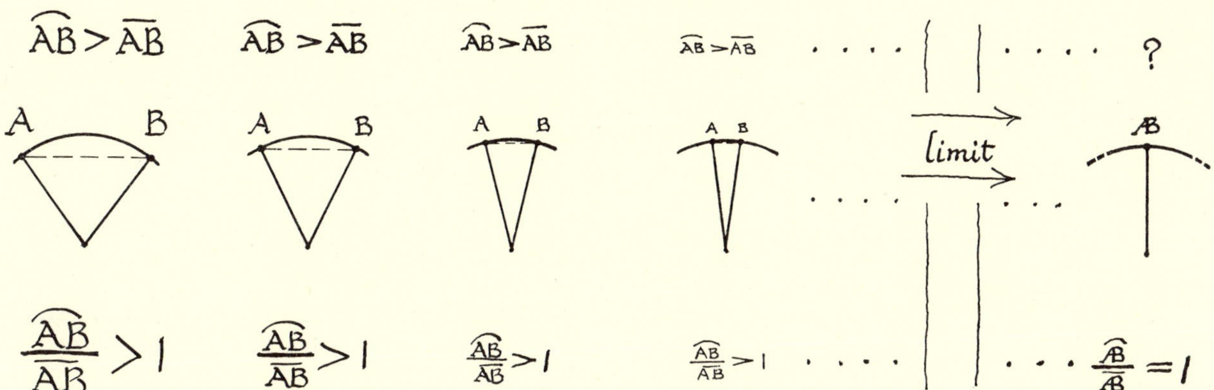

FIG. 21-9. As we proceed from larger arc to smaller and smaller arcs of the same circle, we proceed from large chord to smaller and smaller chords, but the chord grows more and more nearly equal to the arc.

IN the limit

$$\frac{\text{arc}}{\text{chord}} = 1$$

(If you disbelieve this, and claim that the disagreement between arc and chord remains unchanged and is only *disguised* by moving A and B closer, examine the following case (see Fig. 21-10): Choose some size of AB, then change to chord ab half as long, but blow up the new picture to double scale, so that the new *chord* a′b′ returns to the full length that you chose originally for AB. Now look at the new chord a′b′. Is it nearer to its arc? Note that the blowing-up does not itself alter the relative proportions of a chord and its arc—it does not change the angles, but merely acts like a magnifying-glass.)

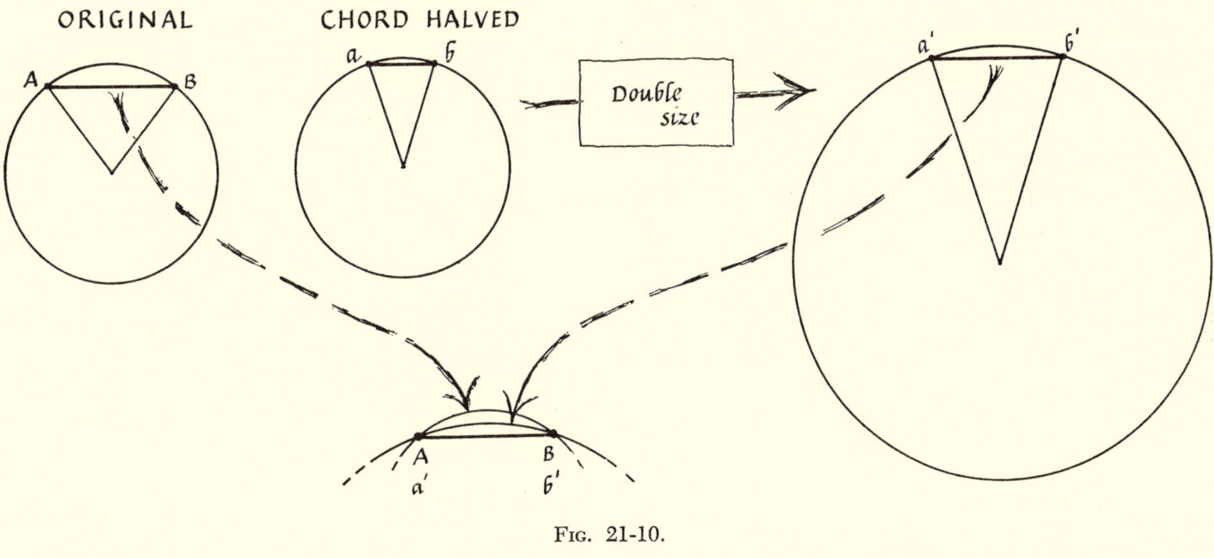

FIG. 21-10.

II. Does this kind of acceleration need a force, according to $F = M \cdot a$, just like a speeding-up acceleration along the path? Does a force $M \cdot v^2/R$ act on a mass M moving around a circle?

Real Forces Needed?

Both questions express real difficulties which kept mankind from jumping to the explanation of planetary orbits at once. Question II is answered by experiment, "Yes every real motion around a circle *does* need a real force, inwards; and Mv^2/R *does* predict the size of that force correctly." To make it move in a circle a body must be pulled or pushed inward by real external agents, such as string or spring or gravity.[4]

[4] People sometimes think that motion in a circle *manufactures* the inward force, provides the inward force needed to maintain itself. A child who *wants* candy does not find his need provides the money to buy it. Some real outside agent such as a rich uncle or an employer must provide the money to buy with—otherwise, no candy. The condition for circular motion is similar. A real outside agent must provide the inward force—otherwise, no curved orbit.

Look for real forces in the following examples:

A. Whirl a stone around with a string. You pull on the string, and the string pulls the stone inward.

FIG. 21-11. WHIRLING STONE ON A STRING
The force on the whirling stone is an *inward* pull, exerted by the string. (The string also pulls the man's hand *outward*—but he is not moving in a circle. He is held in equilibrium by extra forces on his feet.)

The string tugs the stone, gives it some momentum in a new direction, changing the direction of its velocity. Think of the string as giving a series of small tugs: tug to change velocity direction, tug to change it again, tug to change it again, . . . all around the circle. If you release the string, the tugging stops, the velocity no longer changes, so the stone *continues steadily along a tangent.* (To say that it "flies off on a tangent," is a misleading description.)

Swing a stone on a string in a horizontal circle, with a spring or a weight providing a measurable inward force. See Fig. 21-12. Any of the experiments sketched can be used as a test of the prediction $F = Mv^2/R$.

B. Watch a "conical pendulum." The bob, which moves in a horizontal circle, is pulled by two real forces, its weight and the string tension.

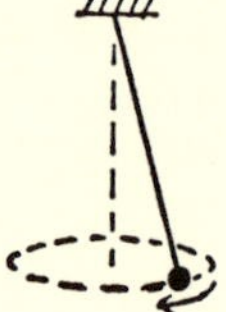

FIG. 21-13.

(If you measure these forces and add them by vectors you will find they produce a resultant horizontal force inward, towards the center of the bob's orbit. With measurements of dimensions and time of revolution you can test the prediction $a = v^2/R$.)

C. A smooth ball rolls around inside a glass funnel.

D. A steel ball rolls on a horizontal sheet of glass in the field of a magnet pole.

E. The motions of Moon and planets.

FIG. 21-12. TESTS OF $F = Mv^2/R$

FIG. 21-12a. A metal ball, tied to a steel spring by a cord, is whirled steadily in a circle. The spring stretches till it applies a suitable pull, and the length, R, of cord + spring remains constant during whirling. The motion is timed and the predicted value of the inward force, Mv^2/R, is calculated. The force actually exerted by the spring is found by hanging loads on it in a separate experiment. Some indicating device is needed to show how much the spring is stretched during whirling. This arrangement is shown in detail later.

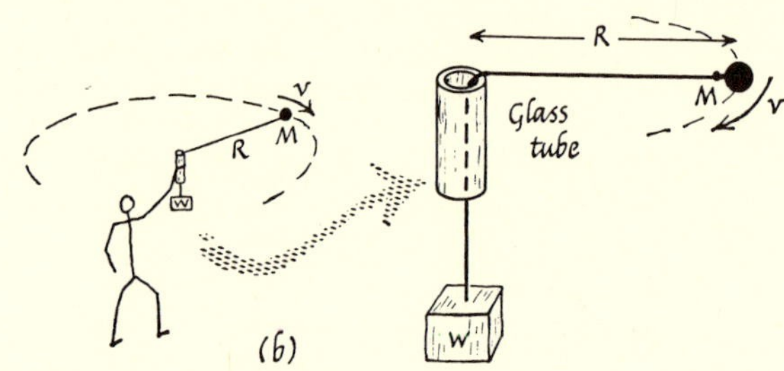

FIG. 21-12b. A metal ball, tied to a cord, is whirled steadily in a circle. The cord runs down a glass tube with smooth open ends and carries a pulling load W at its lower end. By moving the tube around in a tiny circle the experimenter keeps the ball moving around a constant horizontal circle. The motion is timed, and the predicted value of the inward force Mv^2/R is calculated. The force actually exerted by the cord on the ball is its tension, and that is practically equal, but for slight friction, to the pull of the weight of the load W.

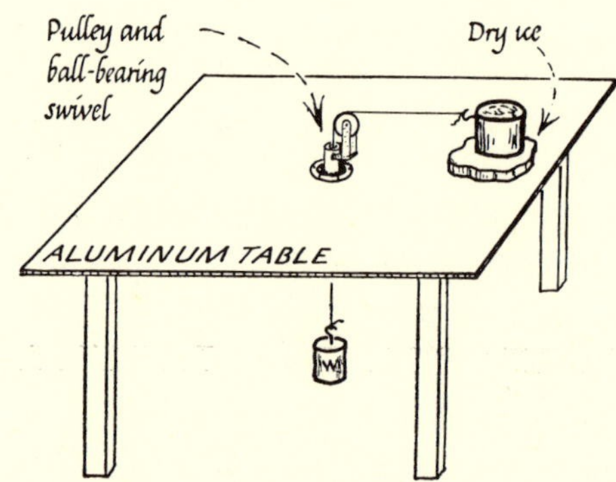

FIG. 21-12c. A practically frictionless version of (b) replaces the ball by a massive block seated on dry ice, coasting on an aluminum table. The cord runs through a hole in the center of the table, and the glass tube is replaced by a small pulley that swivels around the hole on very good bearings.

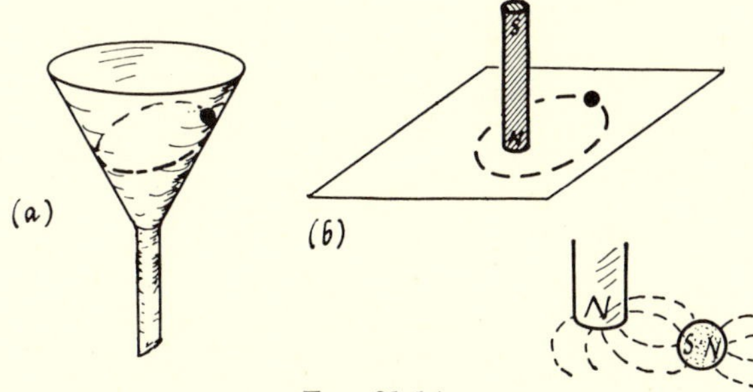

Fig. 21-14.
Examples of Orbital Motion with Centripetal Force
(a) Steel ball rolling in elliptical orbit in a glass funnel.
(b) Steel ball rolling on a smooth table in the field of a magnet. (In practice an electromagnet is used in showing this. It is often placed under the table.) The magnet's magnetic field magnetizes the ball (? temporarily) in such a way that it is attracted by the magnet's pole. Given a suitable push, the ball will then pursue an orbit around the magnet pole.

PROBLEM 2

Accept for the moment the idea that a body moving in a circle *must* be pulled inward by a real force (a force supplied by real agents such as strings, springs, roads). In each of the following cases, say what agency produces the needed inward force. (The first answer is given as a specimen.)

(a) Stone whirled on a string, in horizontal circle. (Answer: "String tension," or "string pulls it.")

(b) (i) Roller-coaster on rails going around sharp corner on level section of track.

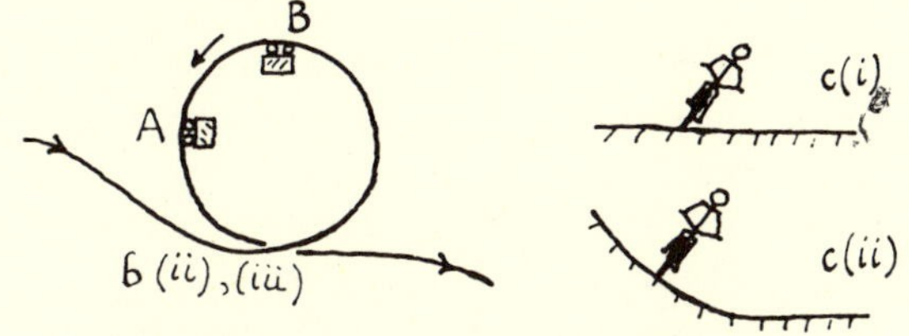

Fig. 21-15 and Fig. 21-16. Problem 2.

(ii) Roller-coaster looping-the-loop at A.
(iii) Roller-coaster looping-the-loop at B.

(c) (i) Bicyclist rounding sharp corner on level road.
(ii) Bicyclist rounding sharp corner on properly banked track.

(d) Plane flying around a curve.

(e) Negative electron pursuing circular orbit around positive nucleus of atom.

Acceleration with No Change of Speed

Experiments answer question II above: motion around a circle *does* need an inward force; and Mv^2/R *does* predict the size of the needed force. Now return to question I: how can a thing accelerate towards the center of a circle and yet neither move faster nor get nearer the center? This is still puzzling, but it now seems to be more a matter of wording than physics. The facts are clear; there are circular motions, and inward forces are needed to maintain them. These inward forces produce inward momentum changes which swing the moving object's velocity around, changing its direction without adding to its size. If we like to include such changes of velocity in our definition of acceleration, $a = \Delta v/\Delta t$, then there *are* accelerations in circular motion. But *if* we restrict "acceleration" to its earlier meaning, "going faster and faster," then there can be *no* "acceleration" in steady motion around a circle. If we take that restricted view we must then announce a new set of forces, in addition to those given by:

"OLD" FORCE = MASS · RATE OF GOING FASTER AND FASTER

The new forces would be given by:

"NEW" FORCE = MASS · (SPEED)^2/RADIUS

These new forces would have to be real forces, experiment assures us, forces that must be provided to make a body move in a circle. However, to save trouble, we avoid the restricted view and use the name acceleration for *all* kinds of $\Delta v/\Delta t$, because we find *by experiment*[5] that $F = M \cdot a$ then predicts the resultant force involved in all cases. On this view we must undertake two sets of tests and illustrations of $F = M \cdot a$, one set with little carts pulled along a track, the other set with objects whirled around in a circle.

As it goes around a circle, the moving body does fall *in*, towards the center, in from the tangent it would otherwise pursue and in again, from the new tangent line, and in again, and so on, continually falling in without ever getting nearer the center. If this seems paradoxical, you may get some comfort by watching a skater making a small circle on ice—he leans inward and is falling, yet never falls over.

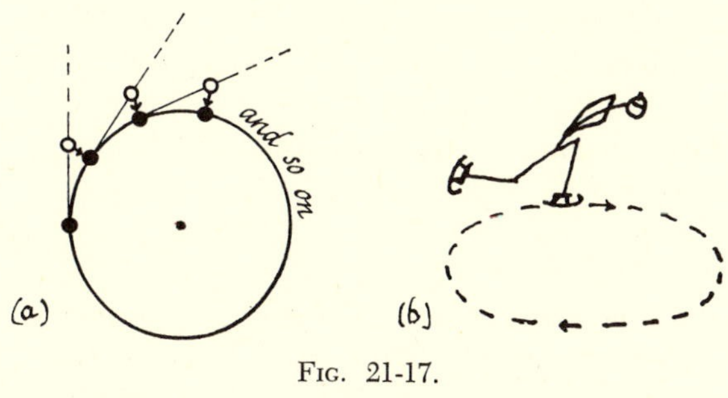

Fig. 21-17.

[5] Galileo really told us this when he said that a projectile's horizontal and vertical motions are independent of each other. At the "nose" of its parabola, the vertical acceleration is perpendicular to the motion and does need a real vertical force.

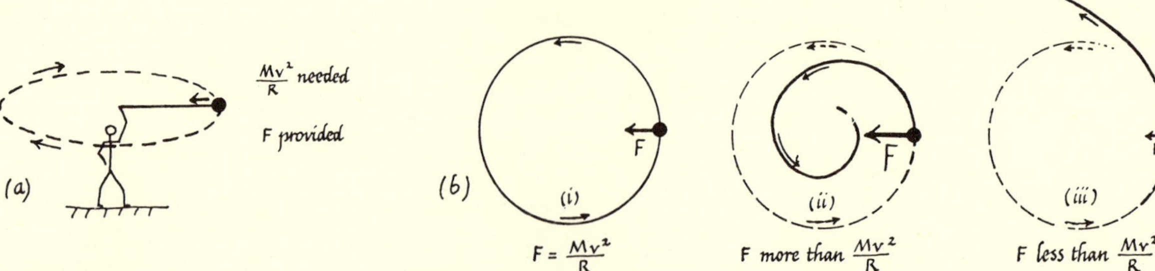

FIG. 21-18. MOTION IN A CIRCLE

(a) Motion in a circle requires changes of *direction of velocity*; and, in that sense, involves a definite, inward $\Delta v/\Delta t$. Experiment shows that this, like "speeding-up acceleration," *needs* a force $M \cdot (\Delta v/\Delta t)$. Geometry shows that the inward $\Delta v/\Delta t$ is given by v^2/R. Then some real agent must provide the needed force $M \cdot v^2/R$.

(b) Of course the pulling agent may fail to provide just the required force Mv^2/R—although strings stretch and rails compress, and friction changes its drag, to adjust the force provided to the need in many cases. The sketches above show what happens when the agent provides (i) the right force, (ii) too big a force, (iii) too small a force. You can arrange these yourself with a stone on a string. In some atomic orbits there may be sudden changes. In planetary motion there are changes which take place smoothly, modifying a circular orbit to an ellipse.

The essential answer to question (I) is this: *the acceleration is perpendicular to the motion*, so it does not increase the speed along the motion. And, added to zero velocity across the motion, the acceleration's contribution just pulls the object in to the standard radius-distance.

Centripetal or Centrifugal Force?

When the force pulls towards the center of a round orbit, producing changes of velocity-*direction* only, we call it a *centripetal* force (from the Latin, meaning a center-seeking force. The opposite of this, a flying-out-from-center force, is called *centrifugal* force).[6] You have often heard that name, but unfortunately it is a misleading term when applied to the moving object. Of course there *is* an outward, centrifugal, pull on the "other fellow" at the center; e.g., on the man who holds the string that whirls a stone. But this gets confused with the force on the moving body, so you will learn good physics most easily if you avoid the phrase "centrifugal force." However, since the idea and phrase are in common use, and are much trusted by many formula-supported engineers, we will discuss it briefly in a later section.

Centripetal Force. Mv^2/R

Real agents must provide the needed force which is equal to Mv^2/R, if the mass M is to follow a circular orbit with radius R at fixed speed v. If the real forces on M produce a greater resultant than this needed force, the body will speed up inward; it will spiral in. If the actual forces are too small it will again fail to follow the circle; it will spiral outward. In many cases the mechanical system adjusts the force to the needed amount, as in the examples below.

[6] These two words have opposite meanings, but their proper pronunciations sound somewhat alike: cen*trip*'t'l and cen*trif*'g'l. To avoid dangerous confusion in discussions, it would be wise to mispronounce them: centri-*pet*'l and centri-*fewg*'l.

EXAMPLES OF CENTRIPETAL FORCE

We shall now discuss in detail some important examples of circular motion ranging from a stone on a string to the modern centrifuge.

Stone whirled on string: the string stretches till its tension is Mv^2/R pulling inward

Roller-coaster looping the loop: the car is pulled by gravity, and it is pushed by the rails. Apart from friction, the rails push perpendicular to their own surface, out from the track. Continue this discussion by answering Problem 3 below.

PROBLEM 3

Suppose a roller-coaster is looping-the-loop as shown in sketch. At A, a force is needed to make the truck move in a circle.

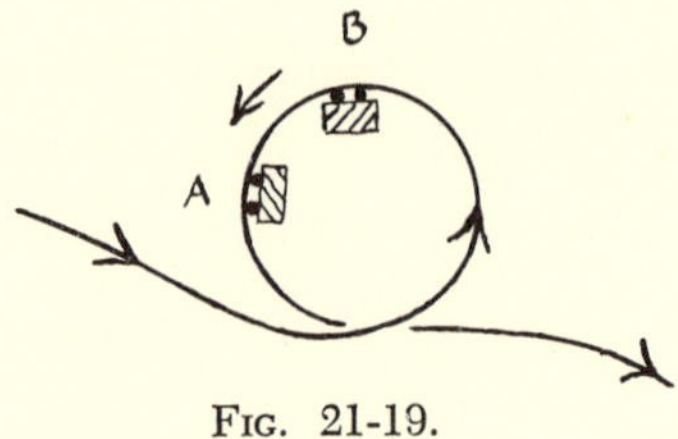

FIG. 21-19.

(a) What direction must that force have?
(b) What provides that force?
(c) What other real force(s) must act on truck at A?

(d) What effect must the other force(s) have on the truck's motion?

(*Note*: In answering (d), forget the force discussed in (a) and (b). It must be there, a real force, but it does a special job. Combining by vectors into a resultant force would not be helpful here.)

At B, a certain force is needed to make the truck move in a circle.

(e) What direction must that force have at B?

(f) How is that force provided?

(g) If the truck is moving much slower, a much smaller force is needed. *Why?*

(h) Why may the force *provided* at B be too big? If it is too big, what happens?

Bicycling. When a bicyclist rounds a corner, travelling in a horizontal circle, some real agent must provide an inward, centripetal force. On a rough road, friction does this; it provides a horizontal force, pushing the tires sideways. (On an icy road friction is not available, and the cyclist does not make the corner—he skids straight ahead, thinking, by contrast to his wishes, that he is skidding round a reverse corner.) He leans sideways as he goes round. Leaning does not in itself help friction, but it is necessary, because otherwise the road's push topples him over. If he rides on a tilted cycling track instead of a flat road, he may not need help from sideways friction; the banked track pushes him straight out from its surface with a force which has a vertical component to balance his weight and a horizontal component that provides the needed centripetal force.

Airplanes. A pilot flying round a corner must bank so that air-pressures on his plane push it towards the center of its circular orbit. Instead of just balancing the weight, as they do in steady, straight flight, much greater "lift forces" must be provided through a change of control flaps.[7] The pilot himself must also move in the curve, and thus needs an inward push to make him do it. In the banked plane, his seat pushes him with the required extra force. But what about his blood, circulating to his head (which is pointed towards the center of the banked turn)? The blood must not only make

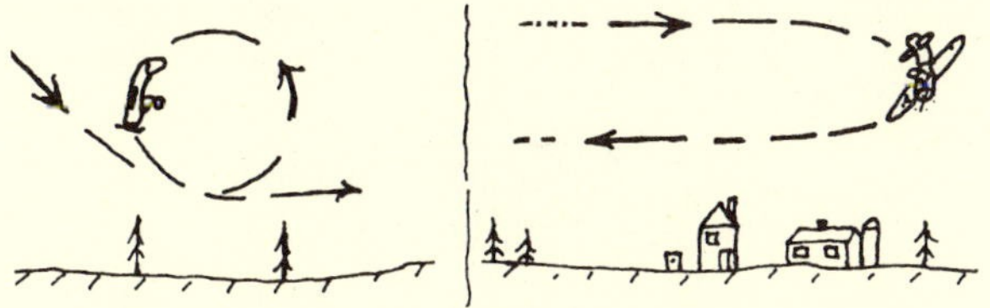

FIG. 21-20a. When a plane loops-the-loop, the pilot's head points towards center. When a plane makes sharp turn, it banks, tilting the pilot over with his head towards center of turning circle.

[7] Gravity, often trivial in a turn, is neglected in this discussion.

FIG. 21-20b. Air exerts forces on wings and body, pushing plane towards center of circle, in loop or turn. (In this sketch, gravity is ignored. In sharp turns it is relatively unimportant.)

the turn, it must move inward ("upward") from his heart to his head. His heart must pump extra hard to feed his brain with an adequate supply. If it cannot pump strongly enough, the blood fails to reach his brain and he faints or blacks out. In ordinary life when you are standing your heart must push a mass m kilograms of blood upward with force $(m \cdot 9.8)$ newtons to keep it moving steadily up against gravity. If the plane flies fast (large v)

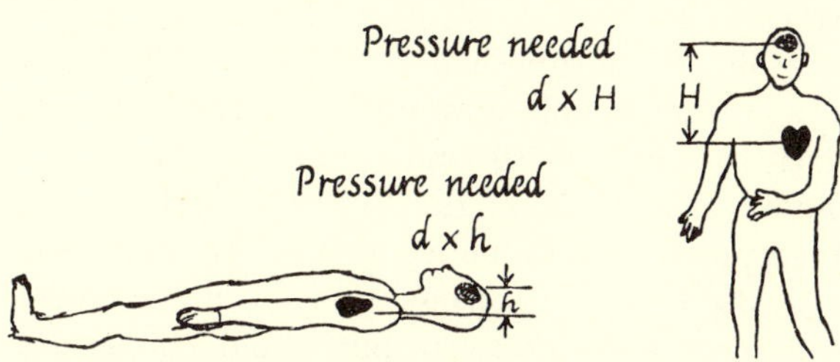

FIG. 21-21. In a man at rest, standing or lying, the heart must provide pressure to pump blood up to brain. In a plane making a turn or loop, the pilot's heart must provide extra pressure for centripetal force, or blood will fail to reach his brain.

around a sharp curve (small R), the needed force from heart to head, $m(v^2/R)$, may be many times bigger than $m \cdot 9.8$. The pilot's heart may be unable to provide the force, mv^2/R, so he may soon black out. If you lie down instead of standing upright, you make much smaller demands on your heart. If, instead of sitting "upright" with his head pointed towards the center of his flying curve the pilot lies down with his body along the curve, he can remain conscious much longer. Fig. 21-22 shows the results of some experiments. The centripetal acceleration of the plane, v^2/R, is expressed as a multiple of "g."

Electrons in atoms. In a later chapter we shall discuss atomic structure and picture electrons whirling around tiny orbits. We assume Newton's Laws apply, and invoke electric attractions to provide mv^2/R.

Cream separators and centrifuges. If we whirl a bottle of liquid around on a string, every chunk of liquid must be given the needed centripetal force mv^2/R, or it cannot make the orbit. This extra inward force must be provided by real pressure-dif-

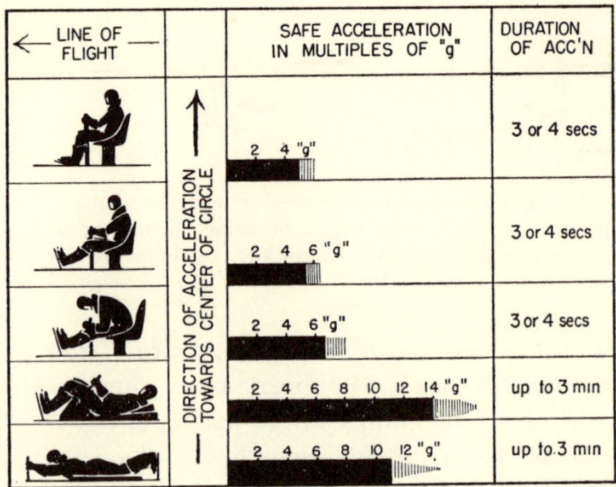

FIG. 21-22. EFFECT OF POSTURE ON TOLERANCE OF ACCELERATION
The safe accelerations are those for which the pilot does not black out. (Data from Ruff) (From *Nature*, June 10, 1944, vol. 153.)

ferences. The outer end of a chunk must experience bigger pressure than the inner end. Thus there is a gradation of pressure increasing outwards much like the vertical gradation of pressure due to gravity in a liquid at rest, but with fast whirling these pressures may be far greater. An air bubble, or anything else less dense than the liquid, would experience these pressures but would not need so big an inward force—m being extra small, its needed mv^2/R would be extra small. It would get more than it needs and be driven inward. In moving inward it would accelerate for a short time until it reached that speed at which fluid friction just balances the available extra force. Then:

INWARD FORCE	—	OUTWARD FORCE	—	FRICTION DRAG
due to large pressure on outer end		due to smaller pressure on inner end		due to motion inward through liquid
	=	RESULTANT FORCE which is the needed inward force mv^2/R		

That is what happens to cream in a whirling cream separator. Anything extra dense—a piece of sand in muddy water, for example—receives the same inward force from surrounding pressure, but needs *more*, so fails to make the orbit, spirals out and lands on the outer edge of the bottle. Take a bottle of muddy water containing air bubbles. If you stand it on the table, mud will settle and air bubbles move to the top with slow steady motions against fluid friction. If you whirl it on a string, mud settles and bubbles rise much faster. Machines called centrifuges

FIG. 21-23. WHIRLING A BOTTLE OF LIQUID ON A CORD

In each case the actual inward force provided is the resultant of (END AREA) · (PRESSURE AT OUTER END) *minus* (END AREA) · (PRESSURE AT INNER END) and these pressures depend on radii and speeds, but not on contents of sample. Therefore, the *force provided is the same, for a given volume, in all cases; but the force needed differs according to density of sample*—hence the separating action.

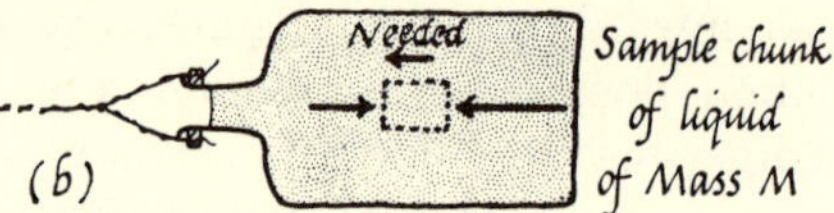

(b) Any specimen chunk of liquid, such as the one marked in the sketch, is moving in a circle and therefore requires an inward resultant force to keep it in its orbit. That force must be provided by difference of fluid-pressure on its ends. The pressure on the outer end of the chunk must be greater than the pressure on its inner end.

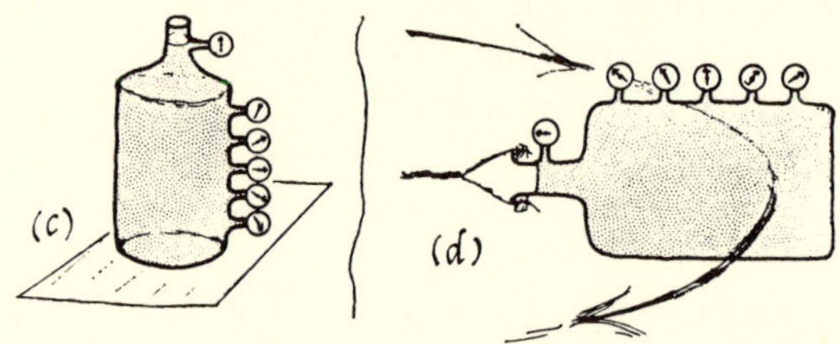

(c) A bottle of liquid *at rest* standing on a table, with pressure gauges to indicate the gradation of pressure. (d) A bottle of liquid being whirled in a horizontal circle, with pressure gauges to indicate the gradation of pressure. (The vertical effects due to gravity, relatively unimportant, are ignored here.)

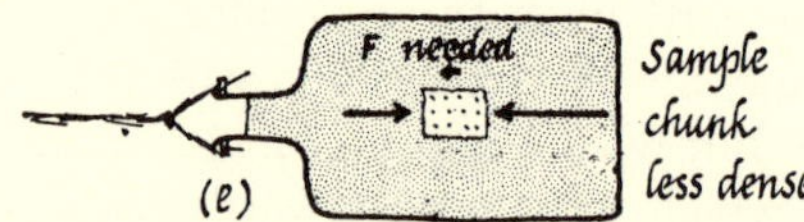

(e) If the immersed chunk is not just a sample of liquid but has smaller density, it has smaller mass, therefore needs smaller Mv^2/R. But pressures supply the *same* resultant inward push, so sample accelerates IN.

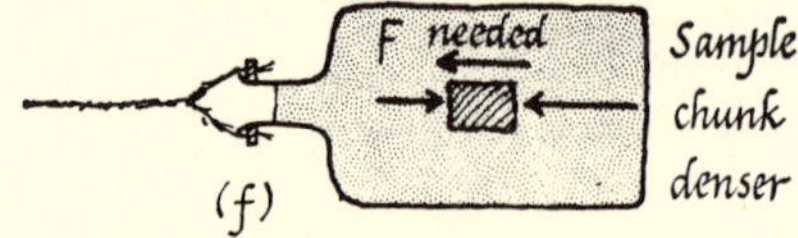

(f) If the sample has greater density than the liquid, it has greater mass and needs greater Mv^2/R than it gets. Therefore it accelerates OUT, and settles to what will be the "bottom" of the bottle when it is placed upright on the table after whirling.

are used to separate cream, to promote settling of fine sediment, and even to sort out oversized molecules of proteins by their rate of settling, against streamline friction, in water.

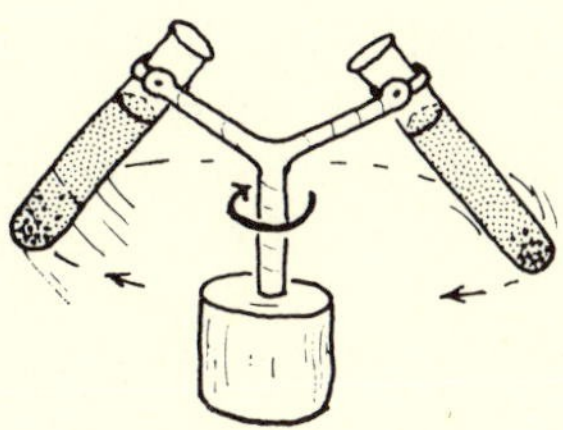

FIG. 21-24. CENTRIFUGE

Square-dancing. When partners "swing" in a square dance, they interlock hands or arms and rotate around a common axis (possibly one or more heels). The partners A and B pull each other inward, the pulls of their arms providing the needed centripetal forces. Even if A has large mass and B small mass, the pulls are equal and opposite (Newton Law III), but the system adjusts its orbit-radii to make these forces just suffice. As a result the dancers rotate around an axis that passes through their common center of gravity.

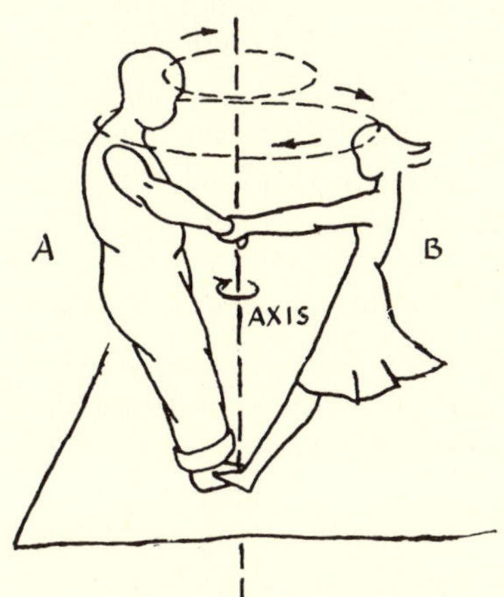

FIG. 21-25. SQUARE-DANCERS

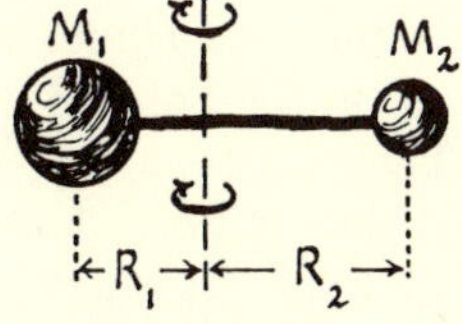

FIG. 21-26.

Here is the algebra of square-dance swinging: Suppose A and B have masses M_1 and M_2 and move around their common axis in circles of radii R_1 and R_2, with speeds v_1 and v_2. One revolution takes time T, the same for A and B. Then $v_1 = 2\pi R_1/T$ and $v_2 = 2\pi R_2/T$ (and $\therefore \dfrac{v_1}{v_2} = \dfrac{R_1}{R_2}$ though this is not needed). The two centripetal forces must be equal and opposite, since they are provided by the action and reaction of the dancers.

$$\therefore \frac{M_1 v_1^2}{R_1} = \frac{M_2 v_2^2}{R_2} \quad \therefore \frac{M_1(2\pi R_1/T)^2}{R_1} = \frac{M_2(2\pi R_2/T)^2}{R_2}$$

$\therefore M_1R_1 = M_2R_2$ (cancelling $(2\pi)^2$, and T^2, etc.).

Then MASS · DISTANCE FROM AXIS OF ROTATION is the same for both partners. The massive partner must therefore take a correspondingly smaller distance from the axis. If you consult the rules in physics books for finding the position of the center of gravity of two bodies, you will find that if R_1 and R_2 are measured from an axis through the center of gravity of M_1 and M_2, then M_1R_1 must $= M_2R_2$, as a property of centers of gravity. And here we find $M_1R_1 = M_2R_2$ *for* the rotating motion. Therefore the rotating pair *must* revolve around their common center of gravity.

This piece of physics is true of square dancers, but it is not important to them. It is also true of the motion of the Moon and Earth, and it is important for an understanding of tides. Astronomers use it for double stars. You will meet other examples of centripetal force in astronomy and in atomic physics.

CENTRIFUGAL FORCE AND THE ENGINEER'S HEADACHE-CURE

Motion in a circle needs a real inward force, provided by real external agents. This view of centri*petal* force will help you to deal with all real problems of circular motion. Then what is centri*fugal* force? You often hear of it, may find yourself speaking of it when you whirl something around, and will find books using it to explain things in physics. Here are a variety of opinions on it. You may choose according to your taste.

OPINION I: *"Centrifugal force is a phony force, imagined through a misinterpretation of evidence confusing agent and victim."*

If you whirl a stone on a string, the string-tension pulls your hand outwards (just as it pulls the stone inwards). This is a real centrifugal force on your stationary hand, not on the whirling stone. You feel your hand being pulled outwards, so you say, "I feel the stone and string pulling my hand outwards. That tells me the stone is being pulled outwards, by some centri*fugal* force, and the string is just

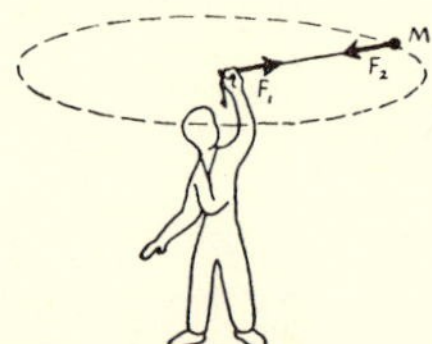

FIG. 21-27. ? CENTRIFUGAL FORCE ? OPINIONS I AND II
Some people confuse the inward pull, F_2, on the stone with the string's outward pull, F_1, on the hand.

transmitting that force." That is where you are mistaken. There is no outward force on the stone. Really the string, in a state of tension, pulls at both its ends. While it pulls your hand outwards it pulls the stone inwards. The only real force *on the stone* is inward, centri*petal.*

Again, suppose you visit one of those amusements at a fair in which people sit on a floor that rotates.

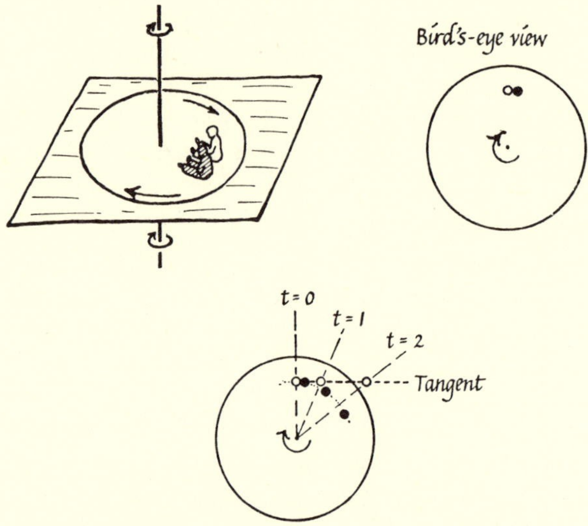

FIG. 21-28. AMUSEMENT AT FAIR OR CIRCUS
The polished floor rotates around a central vertical axis. If one visitor is allowed to slide, he appears, to the other visitor to move radially outward, with acceleration. But an outsider, bird or man, sees him move along tangent.

You and a friend enter the room while the floor is at rest, and sit on the polished floor. Knowing the trick of the performance, you glue yourself to the floor. When the floor begins to spin you note that a mysterious force seems to pull you outward; and, but for the glue, it would make you slide out to the wall. Your friend will slide out to the wall if you do not hold on to him, exerting an inward pull on him. Both of you feel you are struggling against "centrifugal force." But now let a stationary observer take a bird's-eye view from above. Seen from outside the spinning room, you are both moving in a circular orbit, and you both need real *inward* forces to keep you in your orbit. For your friend, the force is the inward pull you provide; for you it is the pull of the sticky floor on you. Once again, you merely imagined an outward force on your friend because you had to apply a real inward force to him. As the outsider sees, these inward forces are not neutralizing a mysterious outward force, they are *making an inward acceleration*; they are making you move in a curve. The outside observer offers a further comment. When you let go of your friend he then continues along a tangent (if there is no friction). His successive positions along that tangent are farther and farther out from the center of the circle; so, as seen by you spinning with the floor he *seems* to be sliding out along a radius. But really he is just *continuing a straight (tangent) path, a simple example of Newton Law I.*

OPINION II: *"Centrifugal force is a delusion arising from living in the rotating system and trying to forget it."*

The rotating-floor discussion leads straight to this view. To people sitting on the table in a concealing fog—and ignoring its motion—there is an outward field of force, endowing every mass M with an outward force Mv^2/R. Unless some real agent applies an inward force to balance this, any object left alone will seem to slide outward with acceleration v^2/R. Preferring to take a sober view from outside, we say that both the outward field of force and the outward sliding are delusions due to living in a rotating framework and not allowing for its motion.

OPINION III: *The Engineer's Headache-Cure*

Here is a good use for centrifugal force. Let us be rude and say, with some truth, that a weak engineering student prefers "Statics," the physics of things at rest (in equilibrium), to the physics of motion. Problems involving acceleration and rotation make his head ache; and he wishes they could be reduced to simple statics problems that he is so good at—forces in bridges and cranes. And they can. Consider, for example, the problem of a pendulum whirling around in a conical motion. The two real forces acting on the bob are its weight and the string tension. These two real forces must add up to a resultant force Mv^2/R inward—otherwise the bob could not continue around the orbit. Here then are two forces W and T which have horizontal resultant Mv^2/R inward. Let us turn this into a statics problem with equilibrium (resultant zero) by adding an extra fictitious force. *What fictitious force must we add to W and T to make zero?* The third force would have to be $-Mv^2/R$, or Mv^2/R *outward.* So

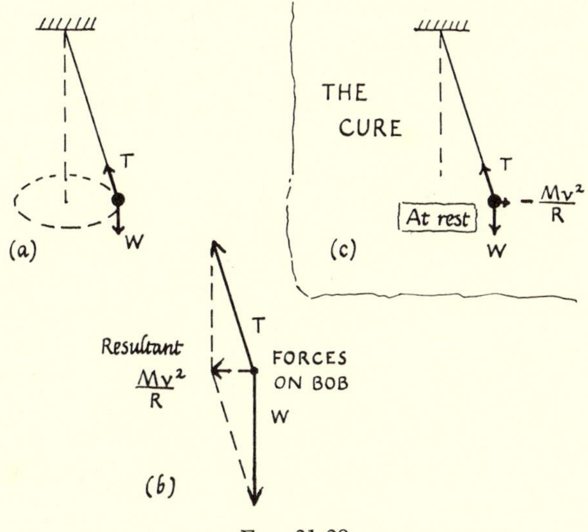

Fig. 21-29.

some teachers say this to the faint-hearted engineer: "Yes, you can turn any problem with circular motion into a statics problem if you *take all the real forces acting on the moving body and ADD a fictitious centrifugal force, Mv^2/R outward, and then write an equation stating that these forces (including the fictitious one) have resultant zero.* Solving the equation will give you the same information as the method of making the real forces combine to produce inward acceleration v^2/R."

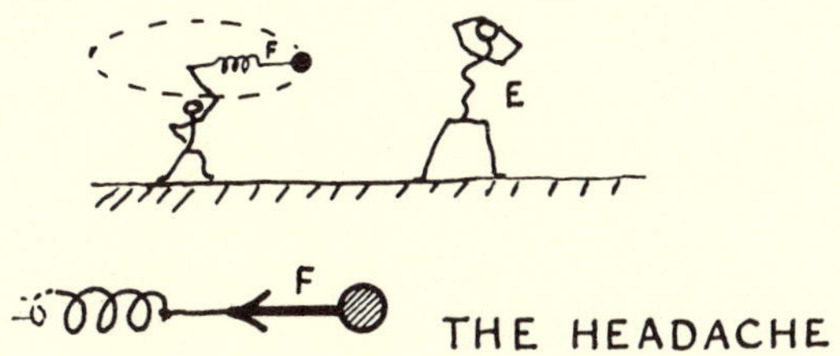

The spring (= agent) provides the real force, F, to make acceleration v^2/R

NO MOTION

E

F $\frac{-Mv^2}{R}$

THE CURE

Imaginary force $-\frac{Mv^2}{R}$ *+ real force F make equilibrium*

Fig. 21-30.

On this view, centrifugal force is a fictitious force, but a useful one, to cure the engineer's headache. It is also used thus in advanced physics, to save trouble—but then it is a sophisticated trick in the hands of skilled craftsmen. As used by most students, it gives the right answer but makes some of the theory harder to understand—how can it help that when it reduces obvious motion to fictitious rest? The trustful user, with his right answer, is confused about the forces: he is not sure which are real or which way they pull. *If you value your understanding of physics, avoid this headache-cure at all costs.* Of course, a *mixture* of this centrifugal headache-cure with centripetal forces will produce utter confusion!

Opinion IV: *Relativity.*

(This opinion sketches some comments from very sophisticated relativity theory. Read it for amusement or for a good moral warning, but do not let it convert you to the engineer's headache-cure for problems. This relativity-view is true, but only within the framework of definitions constructed for it.) Can nothing better be said of centrifugal force? Returning to Opinion II, some scientists ask, "Why is it so wicked to view things from a rotating framework? After all, we live on a spinning Earth. Are the 'centrifugal forces' that arise from our rotating-framework viewpoint really different from other forces, and less real? Who are we to say which is really rotating, ourselves or everything else?" (We are back to Copernicus vs. Ptolemy.) This last question is like the problem of testing Newton's laws in an accelerating railroad coach. By building a tilted room in the coach we could still find the same laws, though we should find "gravity" changed in size and direction. We suspect that we *cannot* distinguish between the effect of acceleration and a change of gravity—Einstein built General Relativity theory on an elaboration of that "cannot."

Relativity theory starts with an axiomatic statement, that we cannot tell which is moving, ourselves or "the other fellow," that there is no such thing as *absolute* motion. If that is so, "absolute space" is meaningless; it should not be used, and cannot be needed, in science. In that case, the working geometry of "space" must be such that we discover the same physics whether we think we are moving or "the other fellow" is. And that makes us modify the simple geometry of space and motion that Euclid assumed and Galileo and Newton used. For constant velocities, we have many experimental failures to distinguish absolute motion even with the help of light-signals, so we feel justified in accepting the Relativity principle and its modified geometry. In practical life, the modifications are not noticeable, and they only affect experiments noticeably when very high speeds are involved, as they are in astronomy and in atomic physics. Extending the Relativity attitude to accelerated motion we assume that a local observer will find the effects of acceleration indistinguishable from a local change of gravity; and thus we decide that gravitational fields can be treated as local changes of geometry in space-&-time. This is Einstein's Principle of Equivalence. Though the viewpoint is entirely new, its practical form shows only small deviations from Newton's law of gravitation.

Extending this idea to rotation, we suggest that a local observer cannot distinguish between the effects of a rotating framework and a local change of gravity, if he is moving with that frame. In that case centrifugal force tugging outward would be just as real to him on his spinning floor as an extra, horizontal pull of gravity. Then, to a bug in a centrifuge, centrifugal force-fields should appear just like real gravitational fields, only some thousands of times as strong as ordinary gravity. To the bug, gravity would take on a new direction—he would quite forget about its old direction—and it would be enormously stronger. This General Relativity view has proved useful in coordinating thinking and successful in making predictions; and so far we have not observed anything inconsistent with it. In this way, centrifugal force has grown to be respectable. When we want to test the effects of large gravitational fields, unattainable on Earth, we think we may use a centrifuge instead.

The general principle of equivalence forbids us to call the motions of the Earth absolute. It therefore leads to a new mechanics and geometry that will predict the same effects whether the Earth spins and moves around the Sun, or the stars and Sun move around us. On General Relativity theory, a rotating universe would produce "centrifugal forces" at a stationary Earth; so tests of a spinning Earth, with a Foucault pendulum[8] or equatorial changes of "g," could not distinguish between the two causes: Earth spinning or everything-else-spinning. Faced with the old question, "Is Copernicus right and Ptolemy wrong?" we must demur at Galileo's cocksure insistence and say, "Both views *may* well be equally true, though one is a simpler description for practical thinking and working." Here is Hegel's development: thesis . . . antithesis . . . synthesis.

[8] See Fig. 14-10 in Ch. 14.

Opinion on the Four Opinions?

Make your own choice. However, for problems and experiments in this course, you are advised to use only centri*petal* force.

LABORATORY EXPERIMENT:

A Simple Test of "Needed Force = Mv^2/R"

Instead of a complicated arrangement sold by "Apparatus Suppliers," with cookbook instructions to enable students to "verify the law," this is a crude gadget intended to offer a rough but obvious test. Try it with a partner. If the measurements and calculations are not clear to you, use common sense and think them out, rather than ask for help. First, listen to a fictitious account of the apparatus being designed by a research group consisting of: *Sagredo* the chairman, *Salviati* the chief experimenter, and *Simplicio* his assistant and critic. (This apparatus was developed in a series of such discussions, with intermediate experiments to try out suggested designs.)

SAGREDO We want to test the prediction that motion around a circle requires an inward force given by $F = Mv^2/R$. We propose to whirl a stone on a string and measure the actual force that we use to hold it in its orbit. Then we shall compare that with the force predicted by assuming $a = v^2/R$ and $F = Ma$.

SIMPLICIO We shall need nice complicated apparatus: a speedometer for v. . . .

SALVIATI Nonsense: simply measure the time for a complete revolution. Then v is the circumference $2\pi R$ divided by that time.

SIMPLICIO . . . and a complicated machine to measure the force.

SAGREDO Foolish; we want a clear simple test.

SALVIATI Use the stretch of a spring to measure the pull directly. We can insert the spring instead of some of the whirling string. Then we can find the force by hanging the spring up afterwards and loading it to the same stretch. We know a good steel spring is trustworthy.

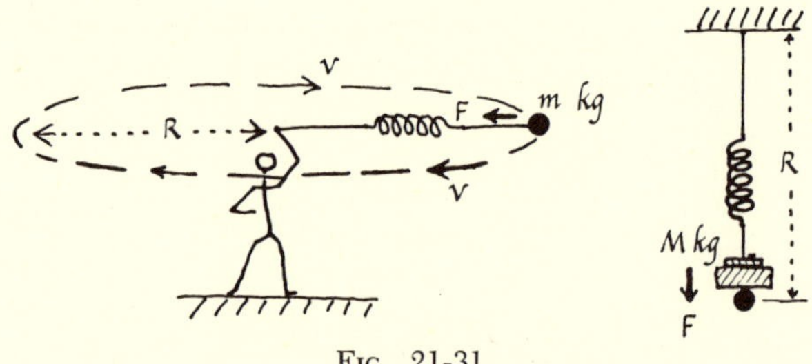

Fig. 21-31.

SIMPLICIO Yes. I know springs obey Hooke's Law; and laws are absolutely true.

SAGREDO Silly! Even if you must trust laws in that fatuous way, you should remember they are limited. Don't you know that springs can stretch beyond Hooke's Law behavior?

SIMPLICIO Well, we can stop that by tethering the ends of a spring together with a long loose thread to stop it stretching too much.

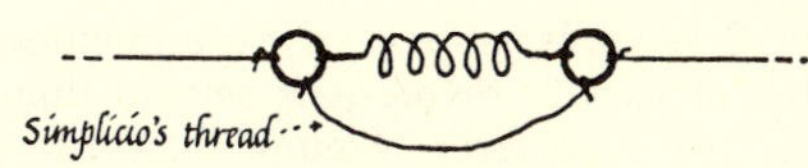

FIG. 21-32.

SAGREDO Oh, fine! Then when the thread is loose it isn't needed; and when it is taut it will spoil the experiment by carrying some of the pull; then we shall not know the total pull.

SALVIATI I think we can put Simplicio's suggestion to some use. We shall want to know just how far this spring is stretched during the whirling, so that we can load it to the same stretch afterwards. If we tether its ends with Simplicio's loose thread, we can whirl faster and faster until the thread is *just* pulled taut. Then we can call that "standard stretch."

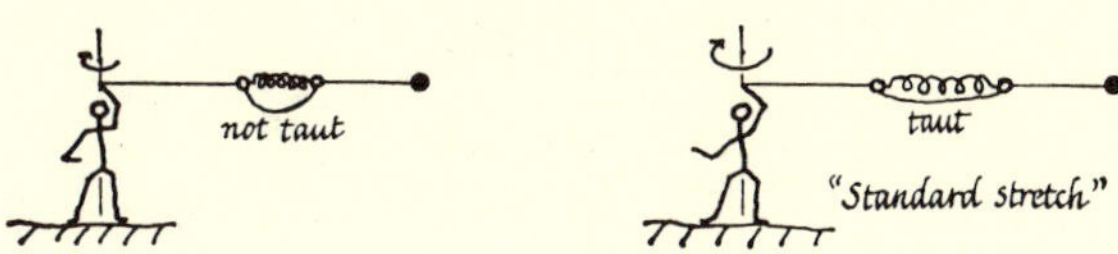

FIG. 21-33.

SIMPLICIO I see how we find the mass and the speed; but how do we measure *R*? Shall we have to hold the meter stick under the whirling string?

SALVIATI We can measure *R* later when the spring is hung up and loaded to standard stretch.

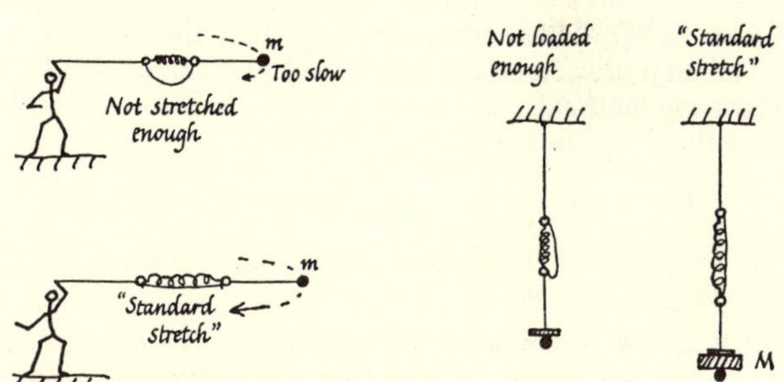

FIG. 21-34 and FIG. 21-35.

SIMPLICIO I see the point of a standard stretch; but if that makes the string taut we are back at your objection to my string—it will add an unknown pull of its own.

SALVIATI Well, *nearly* taut, then.

SIMPLICIO I don't see how we can tell if it is nearly taut.

SALVIATI Let's use another spring for that, a tiny sub-spring attached to the string. The sub-spring will remain coiled up until the string begins to pull taut.

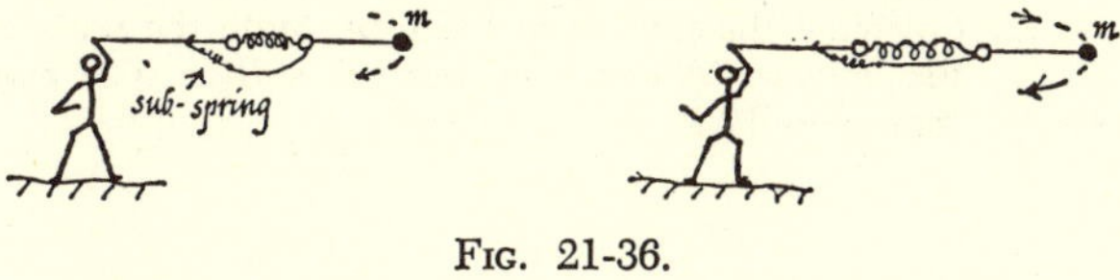

FIG. 21-36.

SIMPLICIO Shall we have to measure the tension of the sub-spring too?

SALVIATI No, the sub-spring is just a signal, an indicator. It will not pull at all until it begins to stretch; and then it can be so weak that its pull is trivial.

SIMPLICIO Even I can see that this will get into a dreadful tangle. The string will rub on the main spring.

SALVIATI We had better run the string through inside the main spring instead of outside.

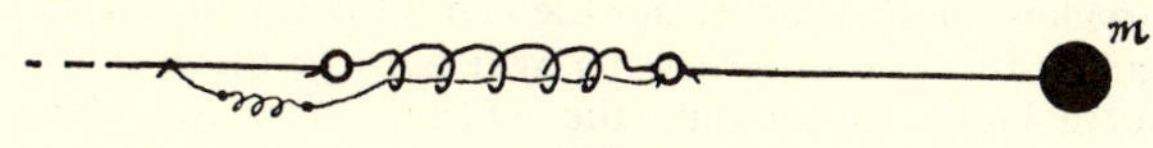

FIG. 21-37.

SIMPLICIO I am afraid we shall spoil the sub-spring before we can make measurement. We shall swing too fast and stretch it hopelessly.

SAGREDO Well, *you* had a cure for that before.

SIMPLICIO Use my suggestion again. Tether the sub-spring by a loose loop of thread.

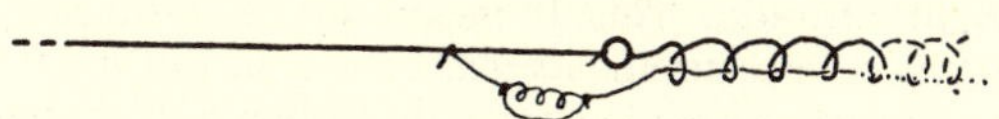

FIG. 21-38.

SAGREDO Good. Then we could whirl faster and faster until we see the sub-spring just about half stretched; and that will be our standard stretch of main spring.

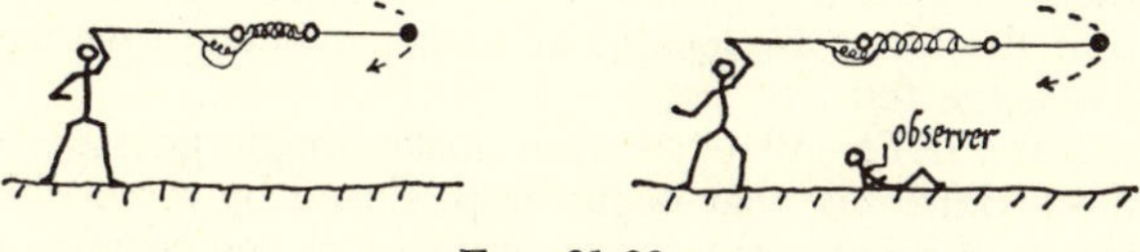

FIG. 21-39.

SALVIATI It will be easier to whirl if we tie the main cord to a big ring and slip the ring over a vertical peg in the experimenter's hand. We can prevent it slipping off the peg by adding two washers and nails.

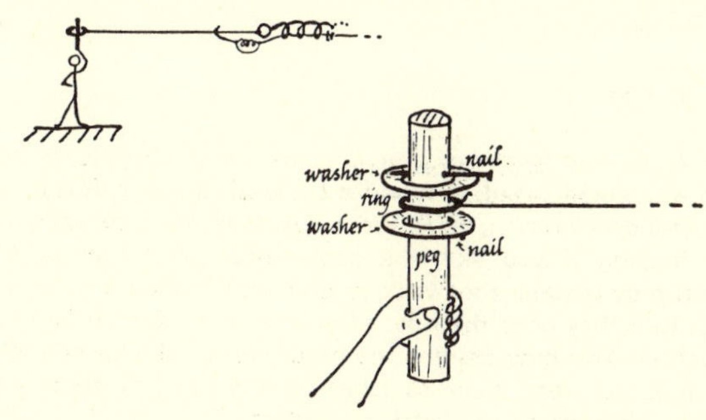

FIG. 21-40.

SAGREDO Now we have a contraption worth trying.

SIMPLICIO But we haven't used Hooke's Law yet.

SAGREDO No, we do not need it. We simply hang the spring up and measure the load that will stretch it to the same length as in whirling. The experimenters will need some training till they can whirl at just the right rate to keep the sub-spring half-stretched; then they can time the motion. After that they can hang the spring up vertically and measure its tension at standard stretch by hanging loads on it. And that solves the problem of measuring the orbit-radius too; they can measure R when they have got the spring hung up and loaded.

SIMPLICIO Oh, so I shall not have to run around and around with a meter stick.

SALVIATI No, but you can still help with the measurement of R. I am afraid the central peg will not stay quite still: the experimenter will have to move it in a small circle to maintain the whirling. You can watch and guess how much to add or subtract to get the true radius for whirling from the measurement when the spring is hung up.

SIMPLICIO That adds one more difficulty. This is now a tricky contraption. I don't believe we can secure accurate measurements and make a test.

SALVIATI Oh yes I can, if I practice the techniques and take great care.

SAGREDO I do not think either of you is taking the right attitude. This is meant to be a simple apparatus. The measurements *will* be rough. But if Mv^2/R is the correct expression then the differences between directly measured force and calculated force Mv^2/R will be due to chance errors in measurement. Those differences will be as often + as −. They should cluster around zero if the experiment is done many times, preferably by many different observers. I suggest we ask a large team of experimenters to try our apparatus, working in pairs. Let each pair express the difference between their measured force and their "calculated" force as a % of the force. By looking at their percentages we can judge the test.

SALVIATI We can even make rough guesses of the % difference that might be produced by reasonable errors of measurement. For example, if one revolution takes about 2 seconds, a pair of experimenters might time a batch of twenty revolutions several times. They could hardly make an error of more than a few tenths of a second in the total time of 40 secs, say 0.2 sec in 40 sec, or 2 in 400, or .5 in 100 or .5%. The time of revolution is used in v, and v occurs *twice* in Mv^2/R. So, the error in timing might contribute .5% + .5% or 1% possible error in Mv^2/R. We can estimate possible errors due to other errors in measurement. . . .

SAGREDO Let the experimenters do that for themselves. They will know best about their own reliability.

Note: A professional physicist looking at this experiment at once objects that the whirling string + spring do *not* move in a horizontal plane, but describe a wide cone.

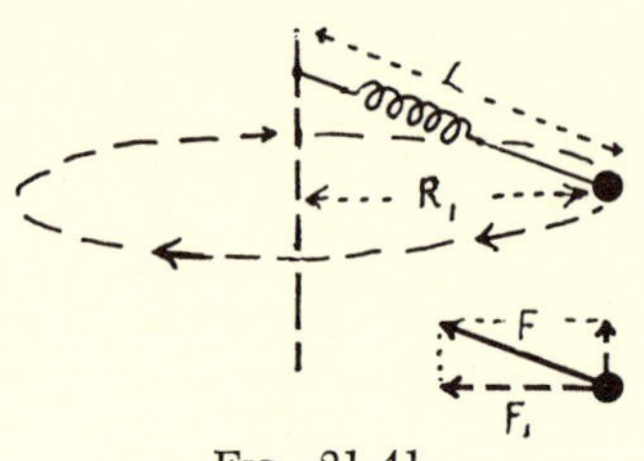

Fig. 21-41.

The radius, R_1, of the bob's circle is *not* the whole measured length of string + spring but only a certain fraction of it. However, the horizontal force making the bob move in a circle is *not* the whole pull of the string + spring but only a certain fraction of it. A look at the actual experiment reassures the critic—he sees that the slant is so small that the fractions concerned are almost 1.00. A short geometrical calculation confounds him—he finds that the fractions compensate exactly!

PROBLEMS FOR CHAPTER 21

1. Special sheet of problems on subtracting vectors (at beginning of chapter).

2. In text.

3. In text.

★ 4. Geometry shows that a point (or small object) moving with fixed speed, v, round a circle of radius R, has a centripetal acceleration, v^2/R. Write out, from memory, refreshed by reading if you like, the geometrical derivation of this. (You may assume that velocity and acceleration are vectors; i.e., that they obey the rules of geometrical addition and subtraction. You may assume the properties of similar triangles. Give large, very clear sketches: one sketch of *facts*—the circle, etc.—and one sketch of *vectors*.)

★ 5. In the following problems assume: (i) that the centripetal acceleration is v^2/R, and (ii) that $F = Ma$ applies to this motion. (*Remember that whenever $F = Ma$ is involved the force must be in newtons or poundals, if the mass is in kilograms or pounds.*)

(a) A 2.00-kilogram stone is whirled in a horizontal circle on a level frictionless table by means of a string. The string is 4.0 meters long, so the circle has radius 4.0 meters. The stone moves with speed 7.0 meters/sec around its orbit. Calculate, with a word or two of explanation:
  (i) the stone's acceleration. (Leave it in factors.)
  (ii) the tension in the string (state the units of your answer).

(b) Suppose the string just breaks under the tension calculated in (a). What is its breaking force in *kilograms-weight?*

(c) As in (a), a 2.00-kilogram stone is whirled in a circle by a rope 4.0 meters long but with such a speed that it makes 5 revolutions in 2 seconds.
  (i) Calculate the orbital speed. (Leave answer in factors, keeping π as π.)
  (ii) Calculate the tension in the rope. State the units of the answer. (A rough answer will suffice. You may take $\pi^2 \approx 10$.)

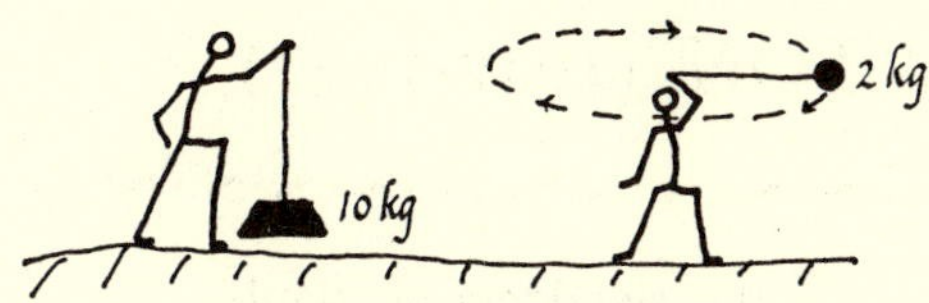

FIG. 21-42. PROBLEM 6

6. A certain kind of string can just carry 10.0 kilograms hung on it vertically, but breaks with the smallest increase of load.

(a) What is its breaking force in kilograms-weight?
(b) What is its breaking force in newtons?

A piece of this string 1.00 meter long is used to whirl a 2.0-kilogram stone in a horizontal circle faster and faster till the string breaks.

(c) Calculate the stone's maximum *speed* around its orbit, giving a short explanation of your calculation.

7. A lacrosse player running with the ball weaves his stick to and fro in front of him as he runs. He does not move it in a straight line from side to side, but swings it in a curve that is concave towards him. Explain how this motion prevents the ball from falling out of the net of his stick.

8. A plane flying 600 ft/sec (410 miles/hr) is pursuing a small slow plane flying 300 ft/sec. The slow plane turns around and runs away by flying in a horizontal semicircle; and the fast plane tries to follow. The pilot in each plane can just stand an acceleration of 5*g*.

(a) Calculate the radius of the smallest semicircle the pilot of the slow plane can safely make at 300 ft/sec.
(b) How long (roughly) will the slow plane take to turn around its *semi*circle?
(c) Calculate the radius of the smallest circle the pursuer can safely make.
(d) Where will the pursuer be when the slow plane has finished its semicircle? (Mark its path on a sketch.)

9. ALTERNATIVE DERIVATION OF $a = v^2/R$ (Newton's method).

As the body moves around the circle from A to B, *treat it as a falling body with constant acceleration downward.* Then in time *t* it falls distance *h* with acceleration *a*, from rest.

(a) Write an equation for *a* in terms of *h* and *t* assuming *a* is constant.
(b) Using a geometrical property of chords of a circle, write an equation expressing *h* in terms of other measurements in the diagram.
(c) Substitute the expression for *h* in the equation of (a).

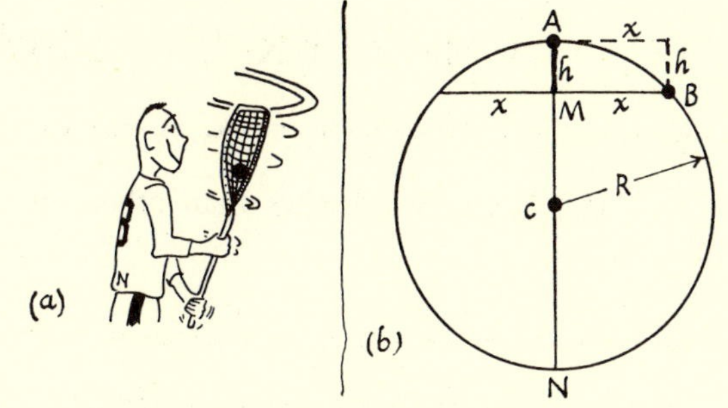

FIG. 21-43.
(a) PROBLEM 7 (b) PROBLEM 9

(d) Now imagine B is moved closer and closer to A. As B → A, the horizontal distance x → arc $\widehat{AB}$. And, as B → A, the chord MN → full diameter, 2*R*. Make these changes in your expression for acceleration.

10. CENTRIFUGING (Estimate roughly throughout this problem).

(a) A centrifuge whirls its test-tube in a circle of average radius 1 ft, at 5000 revolutions/*minute*, thus placing the contents in a force-field of strength many times "*g*." How many times "*g*"?
(b) A sample of muddy water contains particles about the size of blood-corpuscles (diameter 10^{-5} meter). If it is standing in a vertical test-tube, the particles fall at constant speed about $\frac{1}{4}$ inch/minute. So a sample about 4 inches high clears in about $\frac{1}{4}$ hour. (The particles do not settle completely to the bottom. Brownian-motion diffusion keeps some in suspension.) How long would the same sample take to clear in the centrifuge of (a)?
(c) Protein molecules (several hundred times smaller in diameter than the mud of (b), but large compared with the simpler molecules of, say, salt or air) fall about 300,000 times slower in water than the mud particles of (b). How long would a 4-inch tube of a suspension of such protein molecules in water take to clear in the centrifuge of (a)?
(d) Without the use of the centrifuge, the suspension of protein would never clear. Why?
(e) From the rates of clearing, the diameters of the particles involved can be compared, if their densities are known. (Friction drag on a small sphere ∝ radius and ∝ velocity.) The mud specks of (b) can be measured with a microscope. The protein molecules are invisible. "Chemical" measurements (e.g., osmotic pressure) show the protein molecules are about 10^6 times as massive as a hydrogen atom. What important *atomic* information could the centrifuging yield?

CHAPTER 22 · ISAAC NEWTON (1642-1727)

"If I have seen farther than others, it is by standing on the shoulders of giants."
—An old saying, quoted by Newton.

PRELIMINARY PROBLEM FOR CHAPTER 22

★ PROBLEM 1. NEWTON'S FIRST TEST OF GRAVITATION.

Newton did not explain why apples fall. (Calling the cause of weight "gravitation," from the Latin and French for "heavy," does not explain. Saying "the Earth attracts the apple" does put the blame on the Earth and not the sky, but beyond that gives no further information about the attraction.) However, faced with the problem "what keeps the Moon and planets moving in their orbits?" Newton was able to suggest an "explanation" in the sense that the same property of Nature is involved in these motions as in the already-well-known apple fall. "Explaining" therefore just meant linking together under a common cause—but this linking is very useful for further predicting and for simplifying our understanding of Nature.

Operating first on the Moon, Newton calculated its actual acceleration, v^2/R. This is much smaller than the ordinary value of "g," 32.2 ft/sec². Therefore, the Moon cannot merely be "falling under gravity" unless Earth's gravity out there is much smaller, diluted by distance. Newton tried using a simple form of distance-dilution, an inverse square law: he assumed that at twice the distance "g" would be ¼ as big, at ten times the distance 1/100 as big, &c.

Using the data below, repeat Newton's test, calculating (accurately—see note below*):

(a) The Moon's *actual* acceleration, in ft/sec², assuming $a = v^2/R$.

(b) The *expected* value of "g" at the Moon, in ft/sec², assuming the ordinary value of "g" is diluted according to an inverse square law. (Assume that the Earth pulls an apple as if all the material of the Earth is concentrated at the center, one Earth's radius from the apple.)

Since the answers are asked for in FEET/sec², it is important to express *distances* in *feet* and *times* in *seconds before using them in any calculation.* However, you may write in the conversion factors and postpone the multiplying-out if you like, till you find you have to do it. A mixture of miles, hours, feet, seconds, and hope will lead to frustration.

Data: Radius of Earth = 3957 miles.
Radius of Moon's orbit is 60.3 times radius of Earth
1 month is 27.3 days. (This is the Moon's absolute period, relative to the stars.)
1 mile is 5280 feet.
"g" for the apple is 32.2 feet/sec².

* This was a crucial calculation, a great test of a theoretical idea. Therefore, in trying it for yourself you need to do the arithmetic correctly, or it is a complete waste of time. The two answers to (a) and (b) which are to be compared should each be calculated to three significant figures (which will require five decimal places in ft/sec²). First show your answers to (a) and (b) in factors without any cancelling, and then reduce to final decimals.

Newton's Life and Work

Newton was born in the year Galileo died. Even as a boy he enjoyed experimenting—like Galileo and Tycho he made ingenious toys such as water mills,[1] and he even measured the "force" of the wind in a great storm by noting how far he could jump *with* the wind and *against* it. He went to grammar school, doing poorly at first in its principal subject, Latin, then showing unusual promise in mathematics. His uncle, who acted as informal guardian, sent him to the university when he was 19. There, at Cambridge, he devoured a book on logic and Kepler's treatise on optics (so fast that he found no use in attending the lectures on that subject). He read Euclid's geometry, finding it childishly easy; then started on Descartes' geometry. He had to work hard to master this, but he threw himself heart and soul into the study of mathematics. Soon he was doing original work. While still a student he discovered the binomial theorem,[2] and by the time he was 21 he had started to develop his work on infinite series and "fluxions"—the beginning of calculus. He was too absorbed in his work, or too shy, to publish his discoveries—this curious absent-mindedness or dislike of public argument lasted

[1] Early play with bricks, toys, stoves, and bathtubs provides a store of experimental knowledge which we call common sense. When we say "common sense tells us that" we are often appealing to such knowledge—though sometimes to prejudice or tradition instead.

[2] Binomial theorem:

$$(1+x)^n = 1 + \frac{n}{1}x + \frac{n(n-1)}{1\cdot 2}x^2 + \frac{n(n-1)(n-2)}{1\cdot 2\cdot 3}x^3$$

$+ \ldots$ etc. If n is a positive whole number, the series ends after $(n+1)$ terms. If not, the series is infinite and x must not exceed 1 for the statement to be true. When x is much smaller than 1, we can say $(1+x)^n \approx 1 + nx$, because the later terms are so much smaller. This provides some useful approximations such as $(1+x)^3 \approx 1 + 3x$, if x is small

$\sqrt{1+x} \approx 1 + \frac{1}{2}x$, if x is small.

Note how this shows that an error of $y\%$ in a factor Y makes an error of $3y\%$ in Y^3, or of $\frac{1}{2}y\%$ in $\sqrt{Y}$.

Examples: When a solid is heated and its linear dimensions expand by 0.02%, its volume expands by 0.06%. When a clock's pendulum expands in length from winter to summer by 0.02%, its time of swing increases only 0.01%.

through his life. He was also interested in astronomy, observing Moon halos and a comet. Later he was to design and make his own telescopes. He took his bachelor's degree, continuing to work on mathematics and optics, helping the professor of mathematics with original suggestions. In the next two years he turned his attention to the solar system. He began to think of common gravity extending to the Moon and holding the Moon in its orbit, like a string holding a whirling stone. He arrived at the formula for centripetal acceleration, $a = v^2/R$, which he needed to test his idea for the Moon—he found this before Huygens published his version of it. Then he extended the idea to the planets and imagined them held in their orbits by gravitational pulls from the Sun. Thus he guessed at universal gravitation: attractions between Earth and apple, Earth and Moon, Sun and Mars, Sun and Earth, . . . Kepler's Law III told him that these attractive forces must decrease at greater distances, and that the attraction must vary as $1/\text{DISTANCE}^2$. He was already making his great discoveries. When asked how he made his discoveries, Newton replied that he did it by thinking about them.[3] This seems to have been his way: quiet, steady thinking, uninterrupted brooding. This is probably the way in which much of the world's greatest thinking is done. Genius is not solely patience or "an infinite capacity for taking trouble"; yet patience and perseverance must go with great ability and insight for the latter to bear their fullest fruit.

In his use of the Moon's motion to test his new theory Newton met a serious difficulty, so he put that work on one side, shelving it for some years, and threw himself heart and soul into optics, buying prisms, grinding lenses, delighting in his experiments on the spectrum.

By the time he was 24 he had laid the foundations of his greatest discoveries: differential and integral calculus, gravitation, theory of light and color; but he had revealed little of his results. Then his mathematics teacher consulted him about a new discovery in mathematics that was being discussed. Newton made the surprising reply that he had worked this out himself, among some other things, some years earlier. The papers he fetched showed he had gone further and solved a more general form of the problem. This made such an impression that, when the professor retired shortly after, Newton was elected, at 26, to one of the most distinguished mathematical professorships in Europe. In his new post, he lectured on optics but he still did not publish his work on calculus. He was invited to give a discourse to the newly-formed Royal Society of London on his reflecting telescope. The members were delighted with his talk and proceeded to elect him a Fellow. In later lectures he expounded to them his discoveries concerning color.

It was then, after six years, that he returned to his work on astronomy. He could now carry through his test on the Moon's motion with delightful success. Yet for a dozen years more he worked on in silence. Meanwhile Kepler's Laws were begging for explanation. The idea of gravitation was in the air. The members of the new and flourishing Royal Society were arguing about it. They could prove that some inverse-square-law force would account for circular orbits with Kepler's Law III, but they found elliptical orbits too difficult. One of them appealed to Newton for help, and he calmly explained that he had already solved the problem, that he knew, and could prove, that *an inverse-square-law-force would require planets to obey all three of Kepler's Laws*!

Then came a time of writing and publishing (not always willingly) and extending his work on mechanics and astronomy and the mathematics that went with it. His friends in the Royal Society persuaded him to publish his theory of the solar system. The book he then produced was far more than that; it was the world's greatest treatise on mechanics: definitions, laws, theorems, all beautifully set forth and then applied to a general gravitational theory of the solar system, with explanations, examples, and far reaching predictions—a magnificent structure of knowledge. This was the *Principia*, "The Mathematical Principles of Natural Philosophy."

An appreciative people made him a Member of Parliament and later Master of the Mint. This was a way of rewarding him financially as well as honoring him. Newton took the work seriously and did some work on metallurgy, extending his earlier interest in chemistry.[4] Through most of his life he seems to have wanted to be a man of importance and property. This appointment, as well as his election to Parliament, fulfilled the wish in some measure. When he was 61 he was made President of the Royal Society and held that very distinguished office for twenty-four years, the rest of his life. When

[3] His remark was "by always thinking unto them. I keep the subject constantly before me and wait till the first dawnings open slowly by little and little into a full and clear light."

[4] At Cambridge he had carried out long chemical investigations, recording a wealth of detail, but even he was hampered by the chaotic state of chemical knowledge and thinking at that time.

he was 65, he was knighted, becoming Sir Isaac Newton. The people of his country, and their neighbors far and wide, realized that they had more than a great man, a very great man, and they did their best to honor him and provide for him in a time when scientists were only just coming to their own.

When he died at 85, Newton left books on laws of motion, gravitation, astronomy and mathematics, and optics, in addition to many writings on religion. (He was a religious man with devout, though rather unorthodox views—perhaps something like those of Unitarians today. Having unravelled astronomy with magnificent success, he hoped to do the same for religion.) He had raised astronomy to an entirely new plane in science, bringing it to order by a general explanation in terms of laws which he laid down and tested. Since we are concerned here with the growth of astronomy, we now return to an account of Newton's work in that field.

Laws of Motion

To bring all the heavens into one explanatory scheme, Newton needed rules for motion. He found in Galileo's writing clear statements about force and motion, with a less clear understanding of the nature of mass. He re-stated Galileo's findings in clear usable form, carefully defining his terms. In his *Principia* he published his statements in the form of two laws, after he had been using them in his work for some time, and he added a third law supported by his own experiments.[5]

In this he was a great law-maker, a codifier, the Moses of Physics. Of course, Moses stated laws with an entirely different attitude, codifying heavenly commands to be obeyed by man, while Newton was codifying Nature's ways and man's interpretation of them. Yet both were extracting, codifying, and teaching. Moses did not invent all the laws and and rules he set forth; he gathered them together, edited them so to speak, and put them clearly so that the people could understand them. As a great law-maker he was a great teacher. Newton, like Moses, was a great teacher, though shy and modest, almost teaching himself rather than others. Much of his writing was done originally to make things clear for himself, but it served when published to clarify, for the scientists of his day and of ages to come, what had been difficult and unclear.

> Nature and Nature's Laws lay hid in Night.
> God said, "Let Newton be"; and all was Light.
> —Alexander Pope.

Newton wrote his laws to be clear, not pompous; though, since he disliked argumentative criticism by amateurs, he kept his mathematics tough and elegant. He used Latin for his *Principia* because it was the universal language among scholars. When he wrote in English (e.g., his book on Optics) he used the English of his day, and where that now seems pompous or obscure it is because vocabulary and usage have changed with time. If he were writing his laws now he would want them well worded and would write them in English, avoiding the involved phrases that lawyers love. Here are the three laws in their original Latin from the *Principia*, followed by versions in ordinary English.

LEX I

Corpus omne perseverare in statu suo quiescendi vel movendi uniformiter in directum, nisi quatenus a viribus impressis cogitur statum illum mutare.

LEX II

Mutationem motus proportionalem esse vi motrici impressae et fieri secundum lineam rectam qua vis illa imprimitur.

LEX III

Actioni contrariam semper et aequalem esse reactionem: sive corporum duorum actiones in se mutuo semper esse aequales et in partes contrarias dirigi.

Newton added comments and explanations, in Latin, after each law. Using modern terminology we translate the laws thus:

LAW I

Every body remains at rest or moves with constant velocity (in a straight line) unless compelled to change its velocity by a resultant force acting on it.

[5] He probably formulated his Law III, when he found the need for it in developing mechanics systematically—others had already put forward the general idea. He tested it by experiment thus: he allowed moving pendulum bobs to collide, and measured their velocities before and after collision. He calculated the *changes* of momentum and found these were equal and opposite. Arguing from his own Law II, he concluded the forces involved must be equal and opposite. This is an honest test of Law III. (Attempts to "prove" Law III by pulling two spring balances against each other, or by discussing the forces on a book at rest on a table are fallacious. They conceal a circular argument, or else they demonstrate Law I instead.) Newton experimented with great care, allowing ingeniously for the effects of air-friction on his pendulums. He gives a detailed account in his *Principia*, available in good translations, and quoted in Magie's *Source Book in Physics.*

LAW II

When a force acts on a body, RATE OF CHANGE OF MOMENTUM, (Mv), varies directly as the RESULTANT FORCE; *and the change takes place in the direction of that force.*

OR

The product MASS · ACCELERATION varies directly as the RESULTANT FORCE, and *the acceleration is in the direction of that force.*

LAW III

To every action there is an equal and opposite reaction.

Or, when any two bodies interact, the force exerted *by the first body on the second* is equal and opposite to the force exerted *by the second body on the first.*

Note that in the English version of Law II, Newton's form which uses momentum is given first. This is still the best, most general, form today; but the second form, using acceleration, is often used in elementary teaching because it seems simpler. In Chapter 8 we showed the two forms are equivalent. Here is the reverse change from the momentum version to $F \propto Ma$:

RATE OF CHANGE OF MOMENTUM varies as FORCE

$$\therefore \frac{\Delta(Mv)}{\Delta t} \propto F \quad \therefore \frac{(Mv_2 - Mv_1)}{\Delta t} \propto F$$

$$\therefore \frac{M(v_2 - v_1)}{\Delta t} \propto F \text{ if } M \text{ remains constant.}$$

$$\therefore M(\Delta v/\Delta t) \propto F \quad \therefore Ma \propto F \quad \text{or, } F \propto Ma.$$

Or, in one stride with Newton's own invention, calculus:

$$\frac{d(Mv)}{dt} = M\frac{dv}{dt} = Ma \text{ if } M \text{ is constant.}$$

We assumed above that the moving mass remains constant. When the mass is changing the first version, $\Delta(Mv)/\Delta t \propto F$, gives correct predictions, and this is the version that Newton chose. He must have seen that it would apply to a moving object gaining mass (e.g., a truck with rain falling into it). He could not have foreseen its extension to modern Relativity where we still use it to define force in terms of mass which increases with velocity—an increase that is undetectable *except at very high speeds.*

Earlier views

Motion had been worrying scientists for some time.

Leonardo da Vinci (150 years before Newton), had stated, probably just as guesses copied from still earlier writers:

(1) If a force moves a body in a given time over a certain distance, the same force will move half the body in the same time through twice the distance.

(2) Or: the same force will move half the body through the same distance in half the time.

(3) Or: half the force will move half the body through the same distance in the same time.

Descartes (some 40 years before Newton), stated:

(1) All bodies strive with all their might to stay as they are.

(2) A moving body tends to keep the same speed and direction. (He gave a theological reason.)

(3) The measure of a body's force is its mass (not clearly defined) and its speed.

Query: How many of these early views on motion seem true to Nature? (At least one of them seems quite wrong.)

Out of such earlier statements, together with Galileo's books and his own thinking, Newton produced his three Laws of Motion. Today we trust them to predict many kinds of motion: a ball rolling downhill, a rocket starting, planets in their orbits, and even streams of electrons deflected by fields. Einstein has added a modification, but the original summaries are very close to the real behavior of Nature.

Newton's Laws: Natural Truths or Definitions?

Like any modern scientist, Newton tried to give clear definitions of velocity, momentum, and force. In science, a definition is not an experimental fact, or a risky assumption or a speculative idea. It is a piece of dictionary-work explaining as precisely as possible how we are going to use a word or phrase or even an idea. For example, we define "acceleration" as "$\Delta v/\Delta t$" and thereafter whenever we say "acceleration" we mean, definitely, GAIN OF VELOCITY/TIME TAKEN, OR RATE-OF-GAIN-OF-VELOCITY, and we do not mean something else such as $\Delta v/\Delta s$, or something vague such as "going faster." We define "gravitational field-strength at a point" as "FORCE, due to gravity, on UNIT MASS, placed at that point"; and that is both a description of what we mean by field-strength and a definite statement showing how it is measured.

Newton's Laws were clear, powerful rules, based on observation of mechanical behavior, meant to be used to predict other cases of mechanical behavior. However, they were not merely statements extracted from experiment. They incorporated definitions and descriptions of words and ideas such as mass and momentum; and they provided a consistent scheme of prediction *in terms of those definitions.* Definitions often take part in theory like that. For example, two centuries after Newton the science of thermodynamics was developed to produce amazing predictions of heat-properties, in terms of a temperature scale. But this scale had to be a particular scale *defined in the scheme of thermodynamics itself.* We find disagreement when we compare this scale (the Kelvin scale) with other scales such as that given by the mercury thermometer, or that given by the gas thermometer. Yet we cannot say one scale is "right" and another "wrong." Each scale is defined clearly and unambiguously, and all three are equally right as ways of giving a precise measure to man's vague sense of hotness. But when we have a particular use in mind one scale may be best, and when we have a particular theory we may be restricted to the scale whose definition is incorporated in the theory. In using thermodynamic theory and its predictions we *must* use the Kelvin scale of temperature. Fortunately the Kelvin scale is almost the same as the common mercury thermometer's scale, so we can put the predictions to practical uses.

This close interweaving of experiment and definition to make the structure of a theory is characteristic of modern science. So if you examine Newton's Laws critically you may come to the conclusion that Law I merely explains what is meant by force—in fact, defines the nature of force—and that Law II defines the measurement of either force or mass. Then perhaps the laws are just our own inventions: a put-up job? That is going too far. Both laws refer to real nature as revealed by experiment and there is a solid core of fact in them—though that may be hard to disentangle logically from the definitions involved.

Two centuries after Newton stated his laws, further doubts and difficulties began to appear. Newton had adopted "Galilean relativity." In his mechanics it did not matter whether the observer was at rest or moving with constant velocity. Yet Newton thought absolute space must be identifiable by rotational effects. (If the Earth were at rest with all the heavens spinning around it once a day, should we observe the Earth's bulge, differences of gravity, Foucault's pendulum slewing around?) So Newton wrote of absolute motion, forces producing absolute accelerations—not just accelerations relative to some moving frame. Yet where is the fixed, unmoving frame in space? Earth, Sun, stars, may all be moving. *Is* there a real fixed frame? If we cannot find one, we may be foolish to talk in terms of it. Out of these doubts grew Relativity theory.[6] In your present studies of mechanics you would be wise to forget these doubts and take Newton's Laws as good simple working rules. In using them for solving problems, remember that they are at best carefully worded summaries of agreed definitions and experimental knowledge. They are not sacred laws to be cited to make things true! All they do is remind us how to interpret past experiments, and how to predict what will happen in some future ones. At the same time they teach us useful ideas or concepts, such as mass and momentum.

Newton and Planetary Motion

Newton formulated his laws so that he could use them. Turning then to astronomical problems, he at once had the answer to the problem which had misled the Greeks and had puzzled Kepler and even Galileo: "What keeps the Moon and the planets moving along their orbits?" Crystal spheres, natural circular motions, rotating spokes of magnetic influence, vortices—had all been suggested. Newton saw that these explanations provided an agency where none is needed. To keep a planet moving, no force is needed (Law I). Left alone it would just go straight on forever. But a force *is* needed to pull its path into a curved orbit, away from the no-force straight line. How big an inward force is needed? And what agency provides *that* force? These were Newton's new questions. If Law II holds for this motion, the required force F must be MASS · ACCELERATION. But what *is* the acceleration for this motion in an orbit? Newton attacked uniform motion in circular orbits first. (The orbits of the Moon and most planets are nearly circular.) He arrived at the same result as others who were working on the same problem around that time: ACCELERATION, directed inward along the radius, $= v^2/R$, where v is the orbital speed and R the orbit radius. (See Ch. 21 for the derivation of this from the definition of acceleration. Geometry is involved, but no knowledge of force or mass. Newton's proof was an unusual one, treating the moving body as a projectile and each short piece of the circle's circumference as a "nose" of the projectile's parabola.) Then the force must be

[6] Simple forms of Einstein's Relativity deny any fixed framework of "space" but more thoroughgoing discussions of General Relativity still refer their predictions (such as the slow slewing of Mercury's orbit) to a framework made by the most distant stars.

Mv^2/R, inward along the radius. So the Moon, moving steadily around a circular orbit, is always accelerating inward toward the Earth, yet never getting any nearer. You may think of it as falling in to the circle from a straight tangent path, reaching the orbit just in time to start falling in enough to reach the next part of the orbit, still without getting any nearer. If this seems paradoxical, remember that any projectile thrown in a parabola is *accelerating* down ("g") just as much as ever at the nose or vertex of its parabola—yet at that point it is not *moving* upward or downward, not getting any nearer the ground. Thus, it is possible to have acceleration without any velocity at the moment in its direction. The Moon's orbit is a series of "noses," so to speak.

Now came Newton's guess to explain what provides this force. He suggested that the agency which makes bodies near the Earth fall may also pull the Moon in to follow its orbit. There is a story, probably true, that he was thinking about this problem while sitting in an orchard, and an apple falling on his head suggested the answer. We call such pulls "gravity," a word which just means heaviness or implies some connection with weight, and this name in itself is no explanation. The commoner

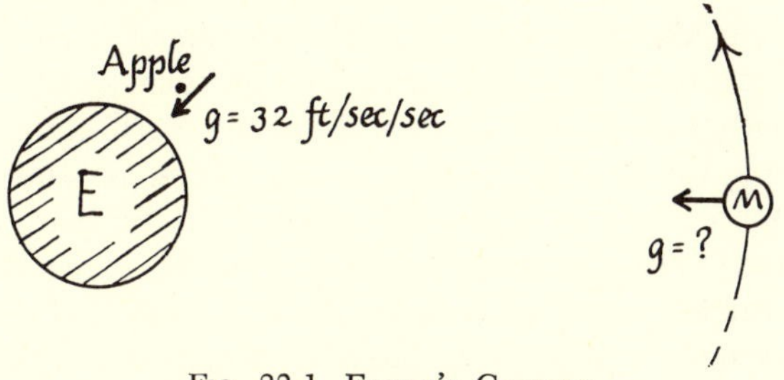

FIG. 22-1. EARTH'S GRAVITY

word, weight, is better for most purposes. Newton suggested that it is the Moon's *weight* that keeps it in its orbit. If the Moon were very near the Earth's surface its weight would give it an acceleration, "g," about 32 ft/sec²—the same for the Moon as for an apple, except for the bulkiness of the Moon spoiling such an experiment. Does the Moon out in its real orbit have this acceleration? Is v^2/R for the Moon about 32 ft/sec²? The Moon takes a month of 27.3 days to travel once around its orbit relative to the fixed stars. Newton knew the Moon's orbit-radius, R, was 60 Earth's radii, $60r$. He had a rough value for the Earth's radius, r. So he could calculate the Moon's speed, v, in the form (circumference $2\pi R$)/(time T, one month) and thence the Moon's actual acceleration v^2/R. The answer is *far* smaller than 32 ft/sec². If "gravity" is responsible, it must be much weaker out at the Moon's orbit. Newton guessed at a simple dilution-rule, the inverse-square law. This is the rule governing the thinning out of light (and radio) and sound, and the force due to a magnetic pole or electric charge—the rule for anything that spreads out in straight lines from some "source" without getting absorbed.[7] He had got a hint by working backwards from Kepler's third law! He tried this inverse-square-law rule. For a Moon at 60 Earth's radii, instead of an apple at only one Earth's radius from the center of the Earth, the pull should be reduced in the proportion of $1:\frac{1}{60^2}$ or $1:\frac{1}{3600}$. Then the Moon's acceleration should be not 32 ft/sec² but $\frac{32}{3600}$ ft/sec². You can easily compute v^2/R for the Moon and will find that it comes out very close to that "predicted value." Think what a joy it would have been if you had discovered this agreement for the first time. It is a successful test of a combination of $F = Ma$ and $a = v^2/R$ and the inverse-square law of gravitation. You would have made a first check of a tremendous theory; a great discovery would have been yours.

Yet Newton himself, eager but far-seeing, was not entirely happy about this test. He mysteriously put the whole calculation aside for some years. He was probably worried about calculating the attraction of a bulky sphere like the Earth. He diluted "g" by a factor 60^2, but this dilution, 1 to $\frac{1}{60^2}$, assumes that the body near the Earth, with "g" = 32 ft/sec², is *one Earth's radius* from the attracting Earth. Does the great, round Earth attract an apple as if the whole attracting mass were 4000 miles below the surface, at its center? Some of the Earth's mass is very near the apple and must pull very hard (according to an inverse-square law, which for the moment is being assumed). Some of the Earth is 8000 miles away and must pull very little.

[7] Suppose a small sprayer acts as a butter-gun and projects a fine spray of specks of butter from its muzzle, in straight lines in a wide cone. If the cone just covers one slice of bread

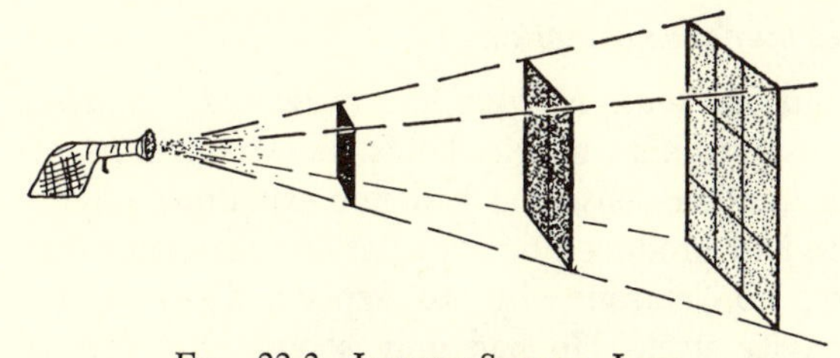

FIG. 22-2. INVERSE-SQUARE LAW

held one foot from the muzzle, it will cover four slices at 2 feet or nine slices at 3 feet, &c.; therefore, for specimen slices of bread placed 1 foot, 2 feet, 3 feet, . . . from the muzzle, the *thickness of buttering* will be in the proportion $1:\frac{1}{4}:\frac{1}{9}$. . . . This is the "inverse-square law of buttering."

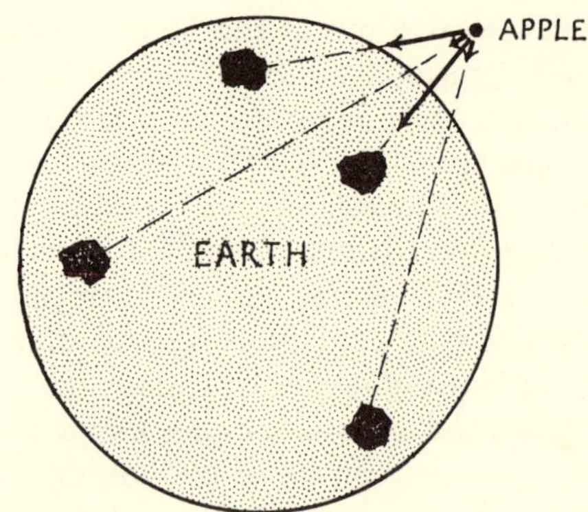

FIG. 22-3. NEWTON'S PROBLEM: ATTRACTION OF SPHERE
Apple being attracted by various parts of the Earth. (Four specimen blocks of Earth are shown.) What will be the resultant, for whole Earth?

Other parts of it pull in slanting directions. What is the resultant of all these pulls? Here was a very difficult mathematical problem; the adding up of an infinite number of small, different pulls. It is easily done by integral calculus, but that fine mathematical tool was just being constructed. Newton himself invented it, for this and other purposes in his work, simultaneously with the German mathematician Leibniz. He shelved his Moon calculation till he knew from calculus that *a solid sphere attracting with inverse-square-law forces attracts as if its mass were all at its center.* "No sooner had Newton proved this superb theorem—and we know from his own words that he had no expectation of so beautiful a result till it emerged from his mathematical investigation—than all the mechanism of the universe at once lay spread before him."[8] Then he could attack the Moon's motion again and test, in one single calculation, his laws of motion, his formula, v^2/R, and his great guess of inverse-square-law gravity as the cause of the Moon's round orbit. This time he was overjoyed with the calculation. The agreement was excellent; the needed force *is* provided by diluted gravity. He had explained the mystery of the Moon's motion.

Newton's Explanation

In one sense Newton had *explained* the mystery, by saying that gravity holds the Moon in its orbit. In another sense, he had not explained anything. He had produced no explanation whatever of gravity, no "reason why" to explain the mystery of gravity itself. He had only shown that the same agency causes, or is concerned in, both an apple's fall and the Moon's motion. This finding of general causes common to several things is called "explaining" in science. If you are disappointed, reflect that this process does simplify our picture of Nature and note that in common speech "to explain" means to make things clearer. It also means to give reasons for, and in Newton's work, as in most science, the *basic* reasons or *first causes* do not emerge; but things that seemed to have different causes are shown to be related. Thus, while we learn more about Nature by finding these common explanations, the basic questions of how the Universe started, and why things in it behave just so, remain unanswered.

Universal Gravitation

So, plain gravity—or rather diluted gravity—is the tether that holds the Moon in its orbit. How about the planets? Does a similar force hold them in their orbits? Since they move round the Sun rather than round the Earth, the force must be a Sun-pull not an Earth-pull. To deal with this, Newton guessed at universal gravitation: a universal set of mutual attractions, with an inverse-square law. He said he reasoned it must be that, by working backwards from Kepler's Law III.

Every piece of matter in the Universe, he guessed, attracts every other piece of matter. He knew, from the Myth-and-Symbol experiment, that the weights of bodies (Earth-pulls on them) are proportional to their masses. Thus the Earth's attraction varies as the mass of the victim. So the attraction exerted by Earth of mass M_1 on mass M_2 varies as M_2. If the attraction is mutual (Law III), symmetry vouches for a factor M_1 to match M_2. For changes with distance, the Moon's motion had given a single successful test of inverse-square attraction.

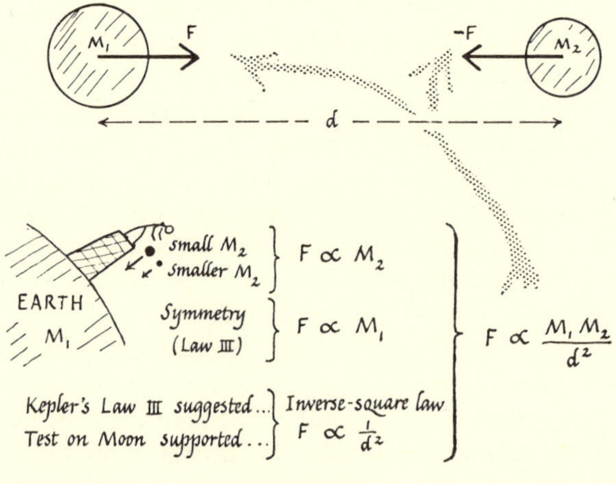

FIG. 22-4.

[8] J. W. L. Glaisher, on the bicentenary of the *Principia*, 1887, quoted in Dampier, *A History of Science* (Cambridge University Press, 1952), p. 153.

So Newton tried a factor $1/d^2$ in his general law. Thus he formulated his Law of Gravitation:

$$F \propto \frac{M_1M_2}{d^2} \qquad \text{or } F = (\text{constant}) \cdot \frac{M_1M_2}{d^2}$$

$$\text{or } F = G\frac{M_1M_2}{d^2}$$

where G is a *universal* constant, M_1 and M_2 are the two masses and d is the distance between them. F is the force with which each pulls the other. Note that G, the universal constant, is quite a different physical quantity from g, the local value of "gravity."[9]

Would this general law account for planetary motion? Newton showed it would. He proved that such an attraction would make a planet move in an ellipse with the Sun in one focus. It was easy (for him) to show that Kepler's other two laws also followed from his guess of universal gravitation. (These laws hold for a sun attracting alone. Then we must add the effects of other planets attracting the moving planet. In the solar system those are trivial compared with the pull of the massive Sun—trivial but far from negligible in the precise accounting of modern astronomy.)

Thus Newton carried his simple idea for the Moon far out to the whole planetary system. He assumed that every piece of matter attracts every other piece, with a force that varies directly as each of the two masses and inversely as the square of the distance between them; and from that he deduced the whole detailed motion of the solar system, already codified in laws that had been tested by precise measurements over two centuries. Satellites of planets formed a similar scheme. Even comets followed the same discipline. All these motions were linked with the gravity that was well known on Earth. Newton explained the heavenly system in a single rational scheme.

This is so great an achievement that you should see for yourself how Newton derived Kepler's three laws—and then put them to further use. The first proof, the ellipse, requires either calculus, which Newton devised, or cumbersome geometry (which Newton had to provide as well, to convince those who distrusted his new calculus). So, with great regret, we shall not give it here. We shall now derive Kepler's Law III, and then Law II—equal areas in equal times. Law II follows from *any* attractive force-law, provided the force always pulls *directly from planet to Sun*; but Kepler's first and third laws fit only with inverse-square attraction.

Kepler's Law III

To deduce Kepler's third law, Newton had merely to combine his laws of motion with his law of universal gravitation. For elliptical orbits, calculus is needed to average the radius and to deal with the planet's varying speed, but the same law then follows.

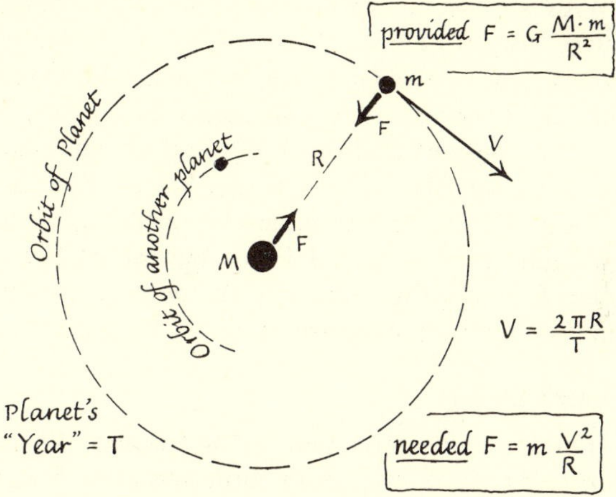

FIG. 22-5. PLANETARY MOTION

For circular orbits, suppose a planet of mass m moves with speed v in a circle of radius R around a Sun of mass M. This motion requires an inward resultant force on the planet, mv^2/R to produce its centripetal acceleration v^2/R (see Ch. 21). Assume that gravitational attraction between sun and planet just provides this needed force. Then

$$G\frac{Mm}{d^2} \text{ must} = \frac{mv^2}{R}$$

and distance d between m and M = orbit-radius, R.

$$\text{But } v = \frac{\text{circumference}}{\text{time of revolution}} = \frac{2\pi R}{T}$$

where T is the time of one revolution

$$\therefore\ G\frac{Mm}{R^2} = m\frac{(2\pi R/T)^2}{R} \qquad \therefore\ G\frac{Mm}{R^2} = \frac{4\pi^2mR^2}{T^2R}$$

To look for Kepler's Law III, collect all R's and T's on one side; move everything else to the other.

$$\therefore\ \frac{R^3}{T^2} = \frac{GM}{4\pi^2}$$

Now change to another planet, with different orbit radius R' and time of revolution T', then the new

[9] For the case of the Earth and the apple, M_1 is the Earth's mass and M_2 the apple's; and their distance apart is the Earth's radius, r. So the apple's *weight*, M_2g, must be $G\,M_1M_2/r^2$. This shows the relation between g and G:

$$g = G\,M_1/r^2$$

ACCELERATION (OR FIELD-STRENGTH), g,

$$= \frac{\text{GRAVITATION CONSTANT} \cdot \text{MASS OF EARTH}}{(\text{RADIUS OF EARTH})^2}$$

value of $(R')^3/(T')^2$ will again be $GM/4\pi^2$; and *this has the same value* for all such planets, since G is a universal constant and M is the mass of the sun, which is the same whatever the planet. Thus R^3/T^2 should be the *same for all planets* owned by the sun, in agreement with Kepler's third law. For another system, such as Jupiter's moons, M will be different (this time the mass of Jupiter) and R^3/T^2 will have a different value, the same for all the moons.

The planet's mass, m, cancels out. Several planets of different masses could all pursue the same orbit with the same motion. You might have foreseen that—it is the Myth and Symbol story on a celestial scale.

With any other law of force than the inverse-square law, R^3/T^2 would *not* be the same for all planets. An inverse-cube law, for example, would make R^4/T^2 the same for all; then values of R^3/T^2 would be proportional to $(1/R)$, and *not* the same for all planets. In fact, as Kepler found, they are all the same. The inverse-square law is the right one.

Calculus predicts Law III for elliptical orbits too, where R is now the average of the planet's greatest and least distances from the Sun.

Kepler's Law II

Here is a crude derivation, due to Newton. We use Newton's Law II, CHANGE OF MOMENTUM $= F \cdot \Delta t$. Then changes of mv are vectors, along the direction of F, proportional to F.

First suppose we have a planet moving under *zero* force. We can still draw a radius to it from a non-attracting sun, S ! The planet, P, moves with fixed speed in a straight line, AF (Newton Law I). Mark the distances travelled by the planet in equal intervals of time: AB, BC, CD, . . . etc. Since the speed is constant, AB = BC = . . . , etc. Consider the areas swept out by the radius SP. How do the following triangles compare, SAB, SBC, SCD? All these triangles have the same height, SM, and equal bases, AB, BC, CD. Therefore all their areas are equal: the spoke from S sweeps out equal areas in equal times. This simple motion obeys Kepler's second law.

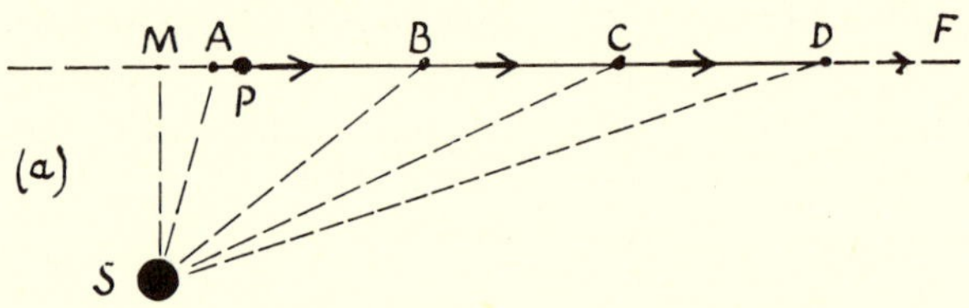

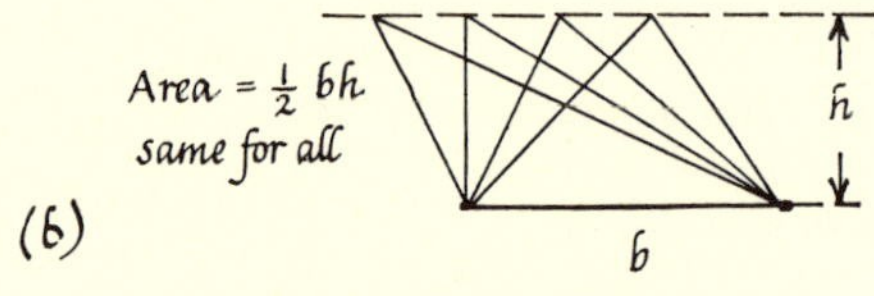

FIG. 22-6.
(a) MOTION OF A PLANET WITH NO ATTRACTION. Planet P moves in a straight line with constant speed. SP sweeps out equal areas in equal times.
(b) THE PROPERTY OF TRIANGLES USED HERE. All triangles on the same base and with the same height have the same area. Another version: If triangles have the same base and their vertices lie on a line parallel to the base, their areas are equal.

Now suppose the planet P moves in an orbit because the Sun pulls it inward along the radius PS. But, to simplify the geometry, suppose the attraction only acts in sudden big tugs, for very short times, leaving the planet free to travel in a straight line betweenwhiles. Then it will follow a path such as that shown in Fig. 22-7. Suppose it travels AB, BC, CD, etc., *in equal times*, the inward tugs occurring abruptly at B, at C, at D, etc. The planet moves steadily along AB; then, acted on by a brief tug at B, along BS, it changes its velocity abruptly and moves (with new speed) along BC. Except for the tug at B the planet would have continued straight on, as in the simple case discussed above. On this continuation, mark the point X an equal distance ahead, making AB and BX equal. But for the tug at B, the planet would have travelled AB and BX in equal times, and the radius from S would have swept out equal triangles, SAB and SBX. But in fact the planet reaches C instead of X. Does this change spoil the equality of areas? If the planet travels to C, the two areas are SAB and SBC. Are these equal? To change the motion from along AB to along BC, the tug at B pulls straight towards the Sun, along BS. This tug gives the planet some inward momentum along BS, which must combine with the planet's previous momentum to make the planet move along BC. The planet's previous momentum was along AB.

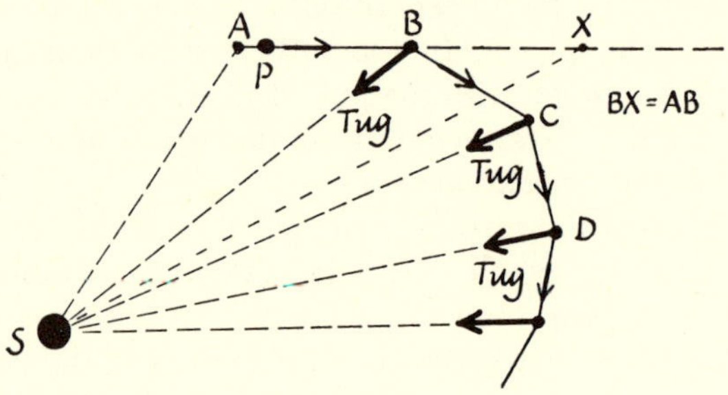

FIG. 22-7. MOTION OF A PLANET WITH TUGS OF ATTRACTION
Without tug at B, P would move on to X.

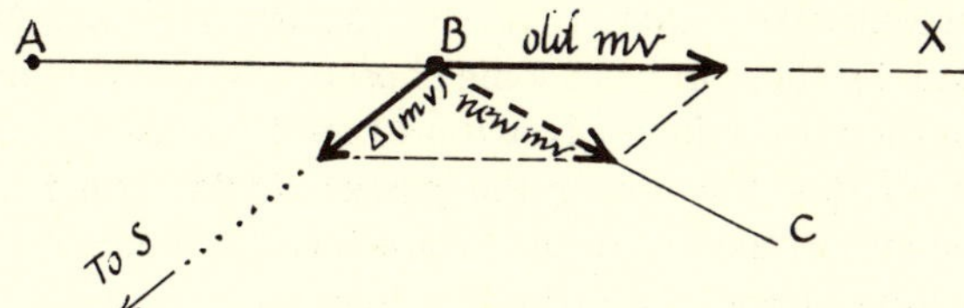

FIG. 22-8. ENLARGED SKETCH OF MOMENTUM-CHANGE AT B

Therefore,

original momentum along AB	+	gain of momentum inward along BS	must =	new momentum along BC.

Newton's Law II reminds us that momentum is a vector. So the adding must be done *by vector addition* (see Fig. 22-8). As the planet's mass is constant, we may cancel it all through and use velocities thus:

$$\frac{\text{velocity}}{\text{along AB}} + \frac{\text{gain of velocity}}{\text{along BS}} \text{ must } = \frac{\text{velocity}}{\text{along BC}}$$

Let us use the *actual distance* AB to represent the planet's velocity along AB. Then, BX must also represent this velocity and BC must represent the planet's new velocity along BC (since all these are distances travelled in equal times). Using this scale, we make a vector diagram (see Fig. 22-9) expressing the equation above. Use BX (= AB) for the original velocity before the tug. Use BC for the velocity after. The change of velocity must be shown by some vector BY along BS straight towards S. Complete the parallelogram, with BC the diagonal giving the resultant. Because this is a parallelogram *the side XC is parallel to BY*, so C lies on a line parallel to BS.

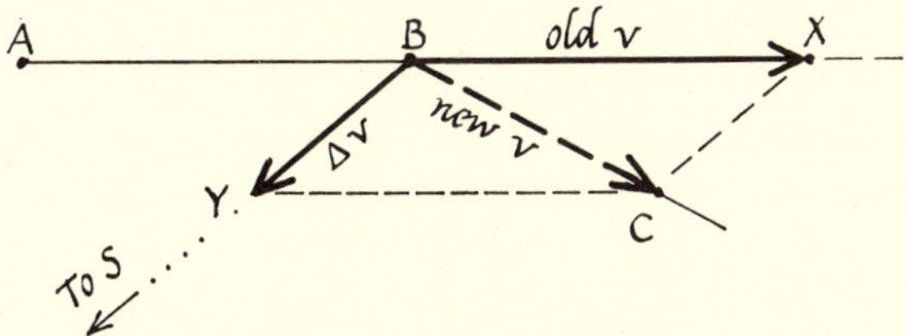

FIG. 22-9. COPY OF FIG. 22-8, WITH VECTORS TO SHOW VELOCITIES AT B
Scale has been chosen so that AB or BX represents original velocity along AB, before tug acts at B.

Now look at the triangles SBC and SBX, in Fig. 22-10. They have the same base, BS, and lie between the same parallels, BS and XC, so they have equal areas. Therefore, area of SBC = area SBX, which = area SBA. Therefore, the triangles SBA and SBC have equal areas. By a similar argument, the triangles SBC and SCD have equal areas, so all the triangle areas are equal, and Kepler's second law does hold for *this* motion. This argument only holds if all the tugs come from the same point S. If we now make the tugs more frequent (but correspondingly smaller) we have an orbit like Fig. 22-11, nearer to a smooth curve, and Kepler's law still holds, *provided the tugs are directed straight from planet to Sun.* If we make the tugs still more frequent, we approach the limit of a continual force, with an orbit that is a smooth curve. The argument extends to this limit, so Kepler's Law II holds for a smooth curved orbit.

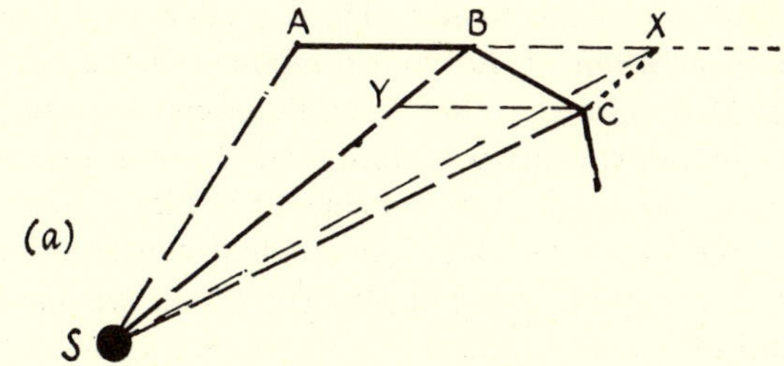

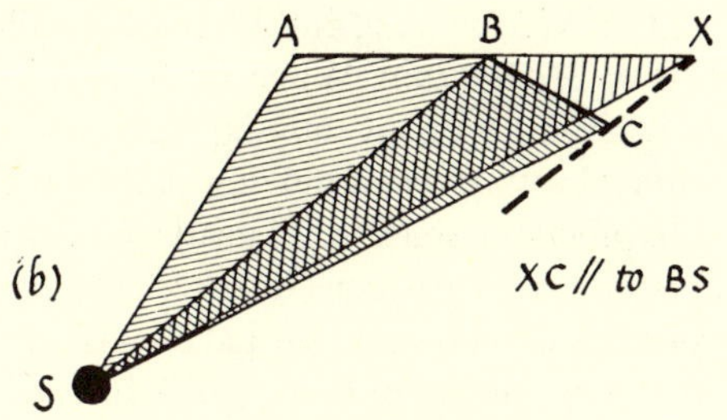

FIG. 22-10.
(a) FIG. 22-7, redrawn with C in its proper place on XC parallel to BY or BS.
(b) FIG. 22-10a, redrawn with equal-area triangles shaded.

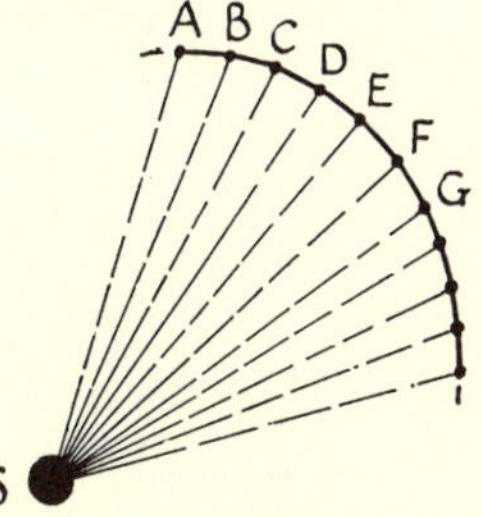

FIG. 22-11.
The equal time-intervals from A to B to C . . . are much shorter. Orbit is nearer to a smooth curve. With a smooth-curve orbit, the segments swept out in equal times may each be regarded as a bunch of small triangles like those here. So the segments must have equal areas.

Kepler's Law II and Rotational-momentum.

Newton deduced Kepler's second law from his mechanical assumptions. The inverse-square law is not necessary; any *central* attraction, directed straight to Sun as center, will require this law of equal areas.

In advanced mechanics, this is treated as a case of *Conservation of Rotational-momentum* (*or Angular Momentum*). What is rotational-momentum,[10] and why are we sure it is conserved? Here is a short account, too rough to be convincing but intended to give you an idea of this fundamental conservation law. For motion along a straight track, we have quantities such as: DISTANCE (s), VELOCITY (v), ACCELERATING FORCE (F), . . . , and laws or relations such as $F \cdot \Delta t = \Delta(Mv)$, . . . , and Principles such as *Conservation of Momentum.* When we have a body that is only spinning without moving along, we can apply Newton's Laws to each moving part of it and produce an equivalent scheme. Instead of DISTANCE MOVED we have ANGLE TURNED (in revolutions or in radians). Instead of VELOCITY we have RATE-OF-ROTATION (in r.p.m. or in radians/sec). Instead of FORCE we have the MOMENT OF FORCE or TORQUE, which is FORCE · ARM—the common-sense agent to make a thing spin faster and faster. Then instead of

FORCE · TIME = GAIN OF MOMENTUM,

Newton's Law II gives

TORQUE · TIME = GAIN OF ROTATIONAL-MOMENTUM.

Guess at rotational-momentum and you will probably guess right: just as TORQUE is FORCE · ARM ($F \cdot r$), so ROTATIONAL-MOMENTUM is MOMENTUM · ARM, ($Mv \cdot r$). Multiply F and Mv each by the arm from some chosen axis, and the new rotational version of Newton's Law II is just as true. In all this, the arm is the *perpendicular arm* from the axis to the line of the vector force or momentum.

Now suppose two bodies that are not spinning collide and exert forces on each other so that one body is left rotating. The FORCES are equal and opposite (Newton Law III); and the ARM from any chosen axis is the same arm for both the forces. Therefore, the TORQUES around that axis on the two bodies are equal and opposite during the collision. Therefore, the ROTATIONAL-MOMENTUM gained by one body must be just equal and opposite to the ROTATIONAL-MOMENTUM gained by the other. Therefore, the total ROTATIONAL-MOMENTUM generated must be zero. If one body develops rotation, the other body must also develop a counter rotation around the same axis. In *any* collision or other interaction: *rotational-momentum is conserved*—it can only be exchanged without loss, or created in equal and opposite amounts.

Therefore, an isolated spinning body (e.g., a skater whirling on one toe) cannot change its rotational-momentum. It cannot change the total of all the $Mv \cdot r$ products of its parts. Suppose it shrinks (skater draws in his outstretched arms). The "r" decreases for the parts drawn in, and if the total rotational-momentum stays constant, Mv must increase: *the body must spin faster.* Watch a skater. Whether he wishes to or not, he spins faster when he draws in his arms or curls up an extended leg.

[10] For a spinning body we might call this "spin-momentum" or simply "spin"—which we do for an electron—but, for a planet swinging around a remote Sun, we use the general name "rotational-momentum." This name applies to spinning objects and objects revolving in an orbit.

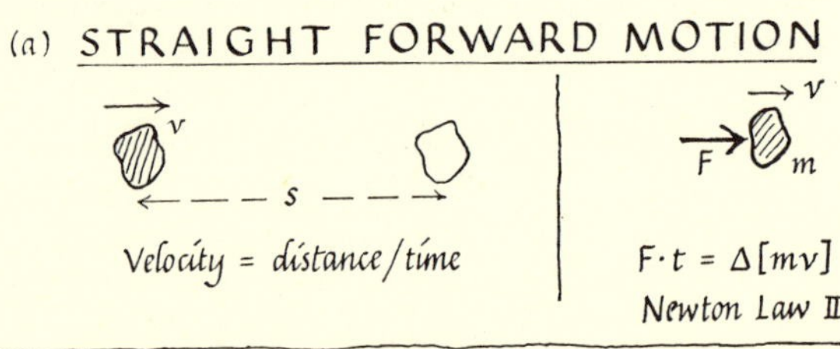

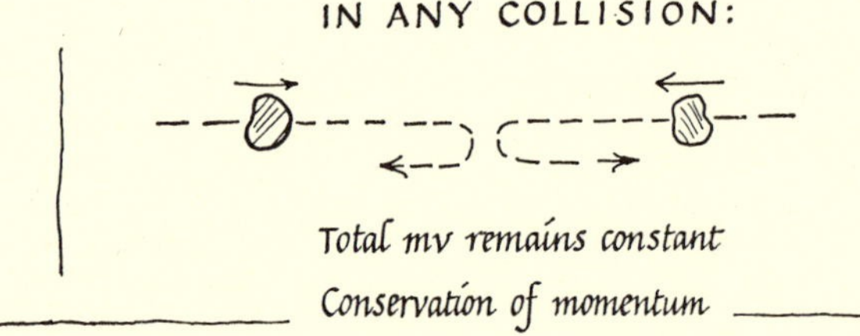

(b) ROTATIONAL MOTION

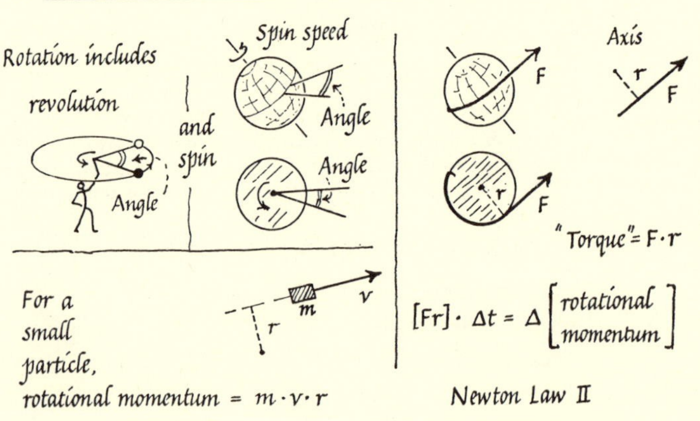

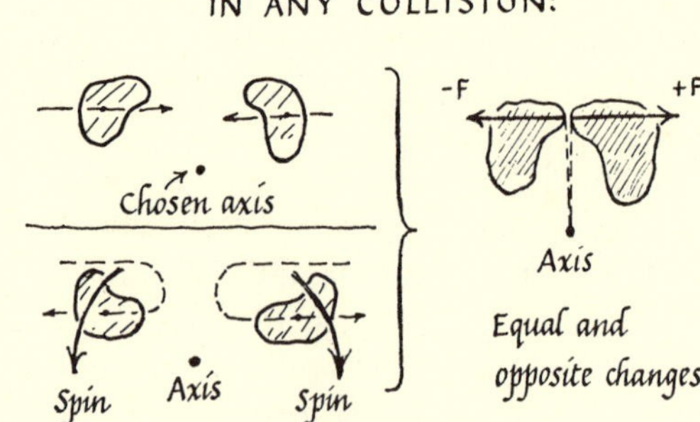

FIG. 22-12. ROTATION

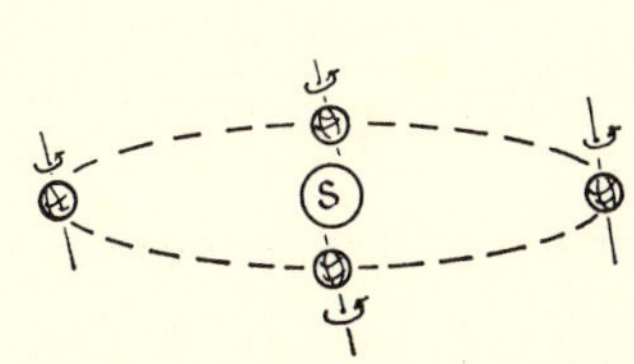

FIG. 22-13. A SPINNING BALL, with no torque acting on it from outside, keeps its rotational-momentum constant in size *and* direction. (e.g. spinning Earth).

"An *isolated* spinning body cannot change its rotational-momentum." Apply that to the spinning Earth. Apply that to a man spinning on a *frictionless* piano stool. Turn yourself into an "isolated spinning body" as follows: pivot yourself by standing on one heel so that you can spin around several times before friction stops you. (Better still, stand or sit on a

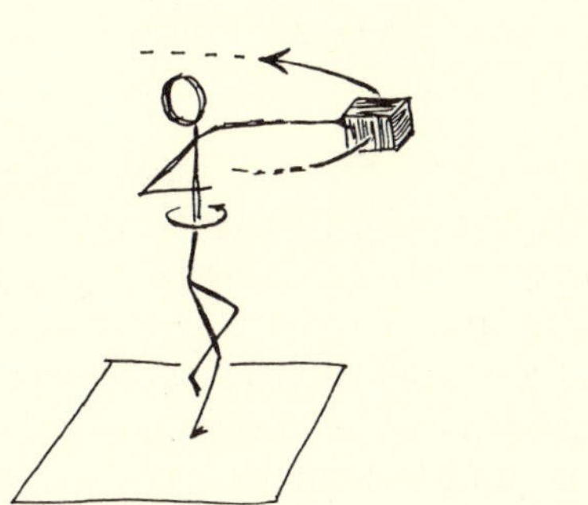

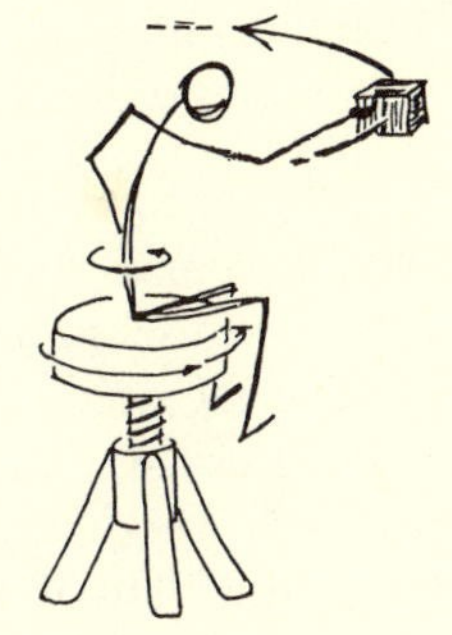

FIG. 22-14. MAN ON "FRICTIONLESS" PIVOT changes *speed* of spin when he pulls load in nearer to axis.

turntable that rotates freely.) Hold a massive book at arm's length. Now, as you spin, bend your arm and pull the book suddenly toward you. Notice what happens to your speed. Here is rotational-momentum being conserved. But here is Kepler's Law II: the book is a "planet" drawn nearer by you as "Sun" as it revolves in its orbit. (Your own great mass is hopelessly involved in this rough test, so you will not observe Kepler's Law accurately.)

For a real planet, if the Sun pulls along the direct spoke, *that pull has no torque around an axis through the Sun; so that pull cannot change the planet's rotational-momentum around the Sun.* A real planet has rotational-momentum $Mv \cdot r$ around the Sun, where r is *not* Kepler's spoke but the *perpendicular* arm from the Sun to the line of velocity v (the orbit tangent). As the planet moves nearer the Sun, r decreases and, to keep Mvr constant, v must increase in the same proportion. Suppose that in a very short time t the planet moves along a short bit of arc s with velocity $v = s/t$. Then the planet's rotational-momentum around the Sun is $M(s/t) \cdot r$, or Msr/t. But $s \cdot r$ is BASE · PERPENDICULAR HEIGHT for the thin triangle swept out by the

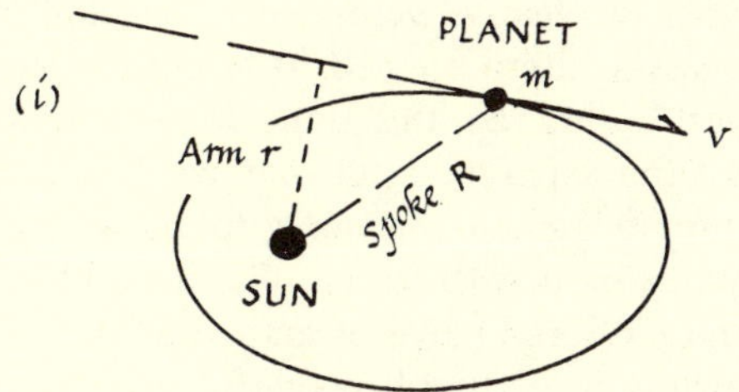

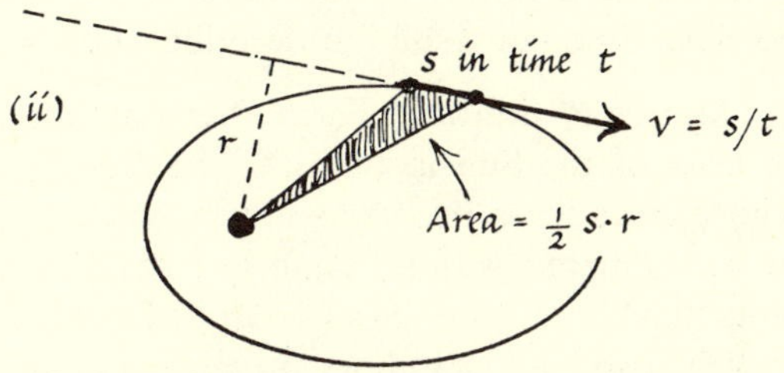

FIG. 22-15. ROTATIONAL-MOMENTUM OF PLANET Equals $mvr = m \cdot (s/t) \cdot r = m$ (TWICE AREA SWEPT OUT)/TIME.

spoke in this time. That is twice the triangle's area. Therefore,

$$\text{planet's ROTATIONAL-MOMENTUM} = \frac{(\text{MASS},M)\,(\text{twice the AREA SWEPT OUT by spoke})}{\text{TIME},\ t}$$

For direct pull by Sun, $\frac{\text{AREA SWEPT OUT}}{\text{TIME}}$ cannot change: the RATE-OF-SWEEPING-OUT-AREA is constant; Kepler's Law II must hold. Conversely, when Kepler discovered his equal-areas-in-equal-times law, he was showing that the only force on the planet is directed straight towards the Sun, and that there is no other kind of force such as a friction-drag due to viscous æther.

Of all the great conservation-rules of mechanics—holding the total constant for mass, momentum, etc.—the Conservation of Rotational-momentum is as universal as any. In atomic physics we shorten it misleadingly to Conservation of "Spin" and expect it to hold through thick and thin, even in violent interchanges between atomic particles and radiation.

Fruitful Theory

Newton formed his theory: he framed his laws as starting points by intelligent guessing helped by hints from experimental knowledge; then he deduced consequences, such as Kepler's laws, then tested those deductions against experiment. In the case of Kepler's laws, the experiments were already done. Tycho's observations had provided rigorous tests; so when Newton deduced them his experimental tests of theory were ready for him. With these successes, there seemed little doubt that the theory was "right." By this stage it seemed worth more than the separate facts that went into it. It gave a simple general meaning to planetary behavior by linking it with the familiar facts of falling bodies, and it offered hopes of many further predictions. Newton, armed with powerful mathematical methods and guided by an uncanny insight, applied his theory to a variety of problems in his book the *Principia.* Some of these are described below.

1. *Masses of Sun and Earth.* Newton calculated the mass of the Sun in terms of the Earth's mass. (The Earth's mass itself was not known and could not be estimated without some terrestrial measurements like Cavendish's. See Ch. 23.) His calculation can be carried out as follows. Subscripts $_s$ and $_e$ and $_m$ refer to Sun and Earth and Moon.

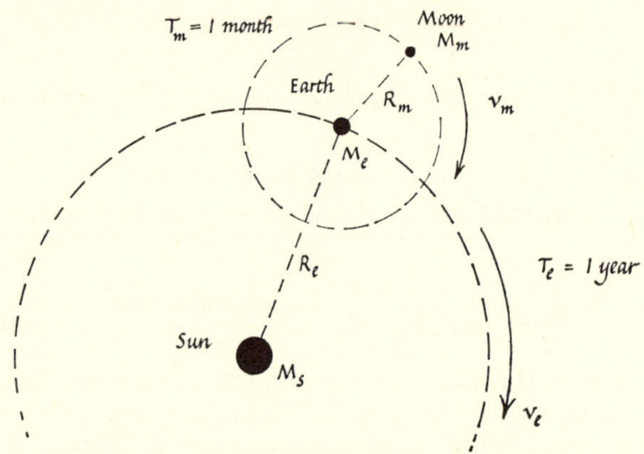

FIG. 22-16.
CALCULATING THE RATIO OF SUN'S MASS TO EARTH'S M_s/M_e, by using the motion of the Moon. The Moon's mass M_m cancels out. For each motion, Earth-around-Sun and Moon-around-Earth, just write an equation stating that the needed force Mv^2/R is provided by gravitation.

For the Earth's motion around the Sun in its yearly orbit,

$$G\frac{M_s M_e}{R_e^2} = M_e\frac{V_e^2}{R_e} = M_e\frac{4\pi^2 R_e^2}{R_e T_e^2}$$

$$\therefore\ M_s = \frac{4\pi^2}{G}\left[\frac{R_e^3}{T_e^2}\right]$$ Note that the Earth's mass, M_e, cancels

For the Moon's motion around the Earth in its monthly orbit,

$$G\frac{M_e M_m}{R_m^2} = M_m\frac{v_m^2}{R_m} = M_m\frac{4\pi^2 R_m^2}{R_m T_m^2}$$

$$\therefore\ M_e = \frac{4\pi^2}{G}\left[\frac{R_m^3}{T_m^2}\right]$$ Again, the Moon's mass, M_m, cancels

Therefore, dividing one equation by the other

$$\frac{M_s}{M_e} = \left[\frac{R_e^3/T_e^2}{R_m^3/T_m^2}\right] = \frac{R_e^3}{R_m^3}\,\frac{T_m^2}{T_e^2}$$

$$= \left[\frac{\text{DISTANCE OF SUN}}{\text{DISTANCE OF MOON}}\right]^3\left[\frac{1\text{ month}}{1\text{ year}}\right]^2$$

With the known values of these times and orbit radii, the ratio of the Sun's mass M_s to the Earth's mass M_e can be calculated.

2. *Masses of planets.* Newton could make similar estimates for the mass of Jupiter or any other planet with a satellite, in terms of the Earth's or Sun's mass. (Our own Moon has no obvious satellite yet; so its mass, which cancels out in the first equation applied to it, seems difficult to find.)

3. *"g" at equator.* Since the Earth spins, an object should seem to weigh less at the equator than at the pole, because some of its weight must be used to provide the needed centripetal force to keep it moving in a circle with the Earth's surface. An object at rest on a weighing scale must be *pushed up* by the scale less than it is pulled down by gravity (its weight). Therefore, the object's *push down* on the scale (which is what the scale indicates), must be less than its weight by the small centripetal force mv^2/R. The Earth's gravitational field-strength must *seem* less. Newton calculated this small modification of "g," which is now observed, together with the effects of the Earth's spheroidal shape.

4. *The Earth's bulge.* Newton calculated the bulging shape of the Earth, arguing as follows. Suppose the Earth was spinning with its present motion when it was a pasty half-liquid mass. What shape would it take? To answer this, consider a pipe of water running through a *spherical* Earth from the North Pole to the center and out to the Equator. If this were filled with water, just to the Earth's surface at the North Pole, where would the water-level be in the equatorial branch of the pipe? The water pressure at the bottom of the polar pipe is due to the weight of the water in that pipe; and this pressure pushes around the elbow at the bottom and out along the other branch, trying to push the column of water up that branch. The weight of water in that branch

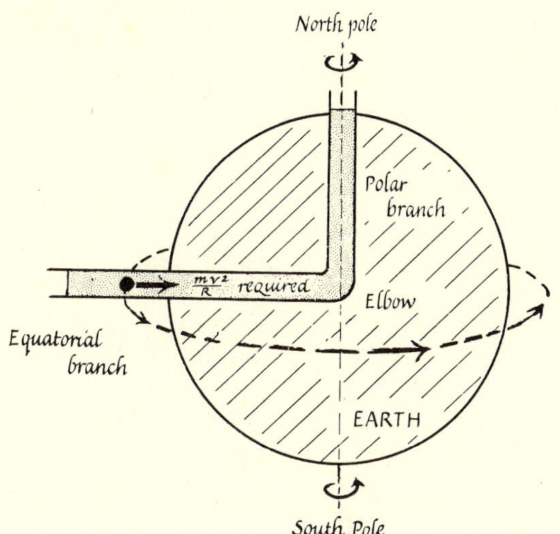

FIG. 22-17. TO ESTIMATE THE BULGE OF A SPINNING EARTH, imagine a pipe of water running from North Pole to center and out to equator. Calculate the extra height of water in equatorial branch needed to provide mv^2/R forces for spin. This gives extra radius of bulge for a pasty Earth congealing while spinning.

pulls it down. But these two forces on water in the *equatorial branch* must be unequal. They must differ by enough to provide an inward centripetal force to act on the water in that pipe, which is being carried around with the spinning Earth. The weight of the water in that branch must *exceed* the upward push from the water at the elbow by the amount needed for mv^2/R forces. Therefore, the water column in this pipe must be taller than that in the polar pipe. The equatorial pipe must extend out beyond the Earth's surface to carry the extra height of water. Newton calculated the extra height and found that 14 miles would be required. He argued that the Earth at an early pasty stage would bulge out about this distance. This bulge had not yet been observed. A short time later, measurements of the Earth confirmed the prediction. Jupiter shows a more marked elliptical shape.

5. *Precession.* Newton explained the precession of the equinoxes thus: the axis of spinning Earth is made to slew around in a cone by the pulls of Sun and Moon on its equatorial bulge. The Earth's axis is tilted and not perpendicular to the ecliptic plane of the Earth's orbit; so the equatorial bulge is subject to unsymmetrical gravitational pulls by Sun and Moon. Here we shall describe the effect of the Sun's pulls. The Sun would pull a *spherical* Earth evenly, as if all the Earth's mass were at its center. The resultant pull would run along the line joining centers of Earth and Sun, whether the Earth spins or not (Fig. 22-19a). A spheroid with an equatorial bulge is subject to small extra pulls on the bulge (Fig. 22-19b). These pulls are uneven, the largest pull being on the portion of bulge nearest the Sun (Fig. 22-19c). These small extra pulls are equivalent to an average pull on the whole bulge, along the line of centers, *plus* a small residual force, f, which tries to rock the spin-axis (Fig. 22-19c). Since the Earth's axis is tilted, this force f is a slanting one, off center, with greatest slants at midsummer and midwinter. When such a slanting pull acts on any *spinning* body it does *not* succeed in rocking the body over in the expected way. Instead, it produces a very curious motion, called precession, which you have seen when a spinning top leans over while spinning fast. The pull of gravity on the leaning top does not

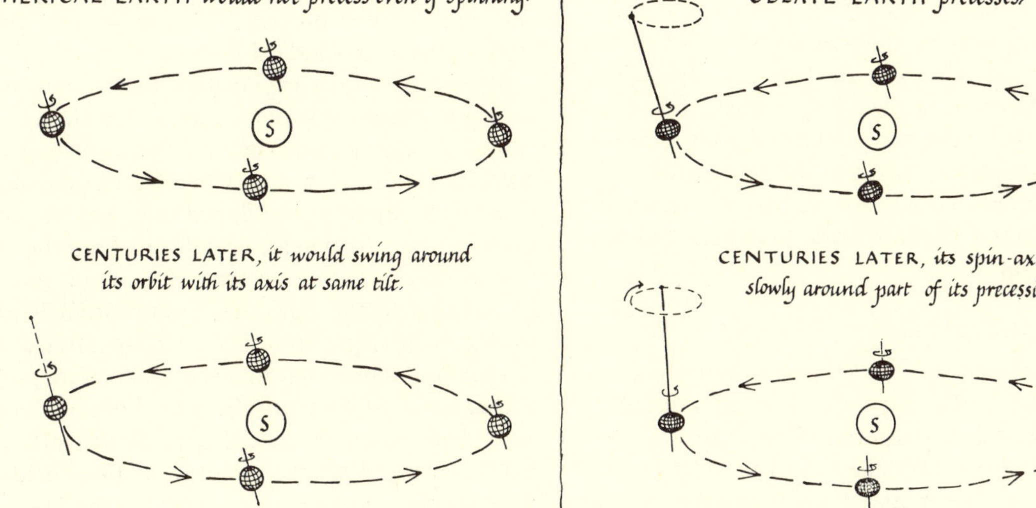

FIG. 22-18. PRECESSION OF THE EQUINOXES

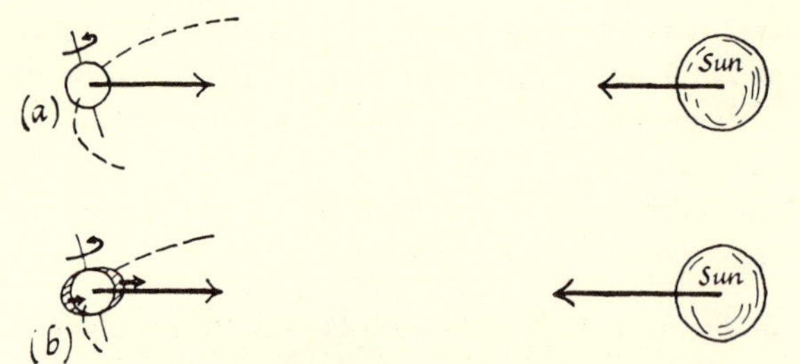

Fig. 22-19. Precession
(a) The Sun would pull a spherical Earth with a central pull along line joining centers, whether the Earth spins or not.
(b) The Sun exerts extra unequal pulls on the bulge of oblate Earth.

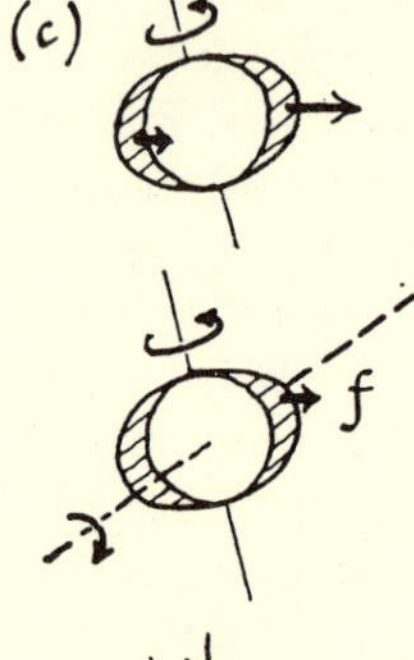

(c) The Sun pulls the nearer part of the bulge harder than it pulls the remoter part.

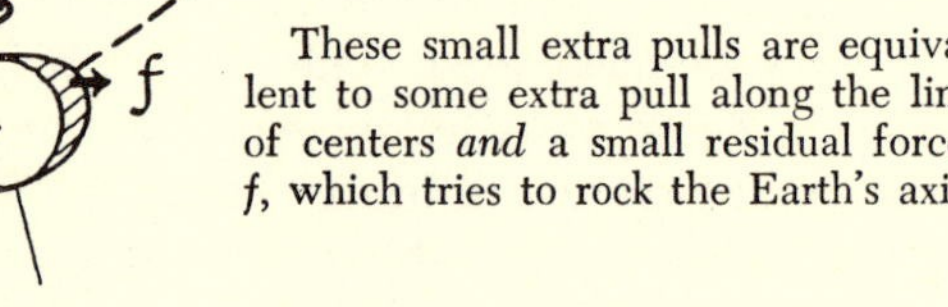

These small extra pulls are equivalent to some extra pull along the line of centers *and* a small residual force, *f*, which tries to rock the Earth's axis.

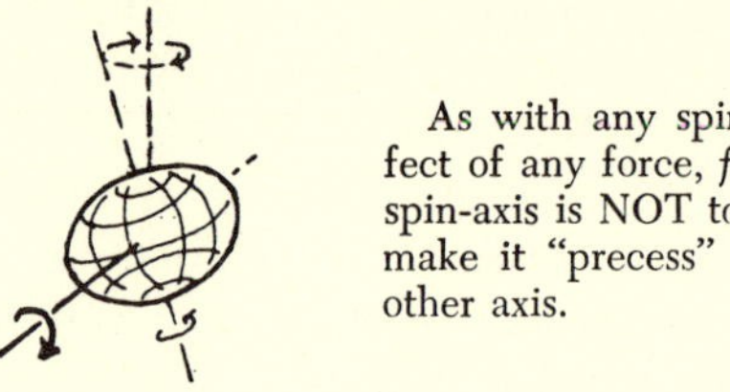

As with any spinning body, the effect of any force, *f*, tending to tilt the spin-axis is NOT to tilt the axis but to make it "precess" instead around another axis.

make it fall over, but makes the spin-axis slew slowly round a cone. Newton showed that the attractions of the Sun, and even more the Moon, would make the Earth's axis precess round a cone of angle 23½° taking some 26,000 years to complete a cycle (Fig. 22-19c). Here was a deduction of the precession that was observed by the Greeks and expressed more simply by Copernicus, but entirely unexplained until Newton's day. It had seemed such a strange motion that men must have had little hope of finding a simple explanation. Yet Newton showed that it is just one more result of universal gravitation: the spinning Earth is made to precess like an unbalanced top.

Demonstration Experiment

Fig. 22-20 shows an experiment to illustrate precession of the Earth. A frame carrying a rapidly spinning flywheel is hung on a long thread. The thread and frame enable the wheel to twist freely about a vertical axis or about a horizontal axis (in the frame, perpendicular to the vertical axis). Left

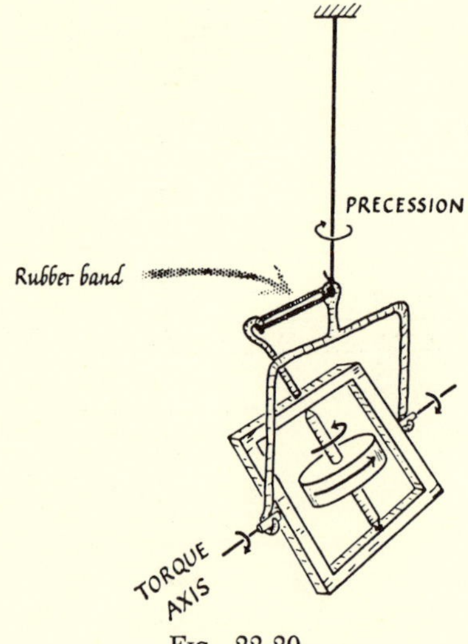

Fig. 22-20.
Experiment to Illustrate the Precession of the Earth

alone, the tilted spinning wheel continues to spin without other motion. Then a rubber band is attached so that it pulls the frame with a force that tries to rock the spinning wheel about the horizontal axis; the wheel does not obey the rocking force in the obvious way, but instead precesses around the vertical thread as axis.

Explaining precession of a gyroscope

The Earth, and a spinning top, and a "mysterious gyroscope" all precess in the same way, for the same reason (Fig. 22-21). Precession looks mysterious, but it is just a complicated example of Newton's Laws applied to the parts of a spinning body. With no off-center force, the body retains its spin, unchanging in size or direction. With an off-center force making a rocking torque, we can compound rotational-momentum changes as vectors to show that the axis will precess. The geometry of precession is given in standard texts.

Here is a simpler explanation that shows precession as a straight case of Newton's Law II. Fig. 22-22 shows a large wheel with massive rim hung by a cord, PQ, and precessing. Consider the momentum of a small chunk of rim at the top, A. It is moving forward; but the wheel's weight, rocking the wheel over a little, moves A to the right, giving A a little momentum to the right. This momentum is added to A's main momentum forward: so after a short time A has momentum in a skew direction, *forward and to the right.* Similarly B, at the bottom, develops skew momentum *backwards and to the left.* Then, for A and B to have these momenta, *the wheel must have twisted round the vertical axis* (i.e., precessed). Here is a hint of the mechanism of preces-

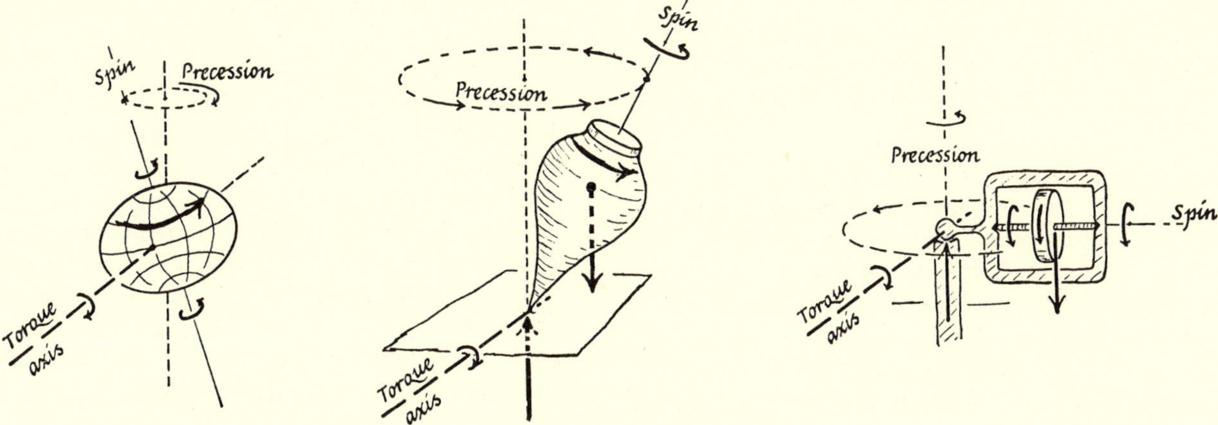

Fig. 22-21. Precession
The Earth, a spinning top, and a "mysterious gyroscope" all precess in the same way, for the same reason. In the sketches above "torque axis" means the axis around which the tilting force tries to rock the spinning object.

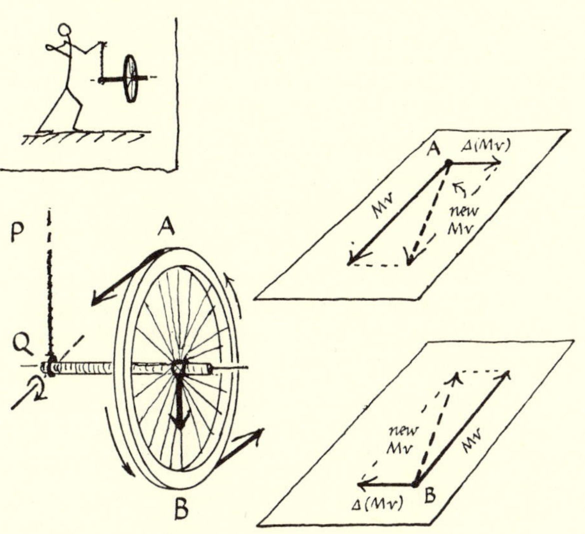

Fig. 22-22. Precession, treated as a case of Newton's Law II. The sketch shows a large bicycle wheel with massive rim, precessing.

sion but it is hard to visualize the extension of this discussion to the rest of the moving rim.

6. *Moon's motion.* The Moon suffers many disturbances from a uniform circular motion. For one thing, it moves in a Kepler ellipse, as any satellite may, with the Earth in one focus. But that orbit is upset by small variations[11] in the *Sun's* attraction. The Moon is nearer the Sun at new moon than two weeks later at full Moon, and that makes changes of attraction that hurry and slow it in the course of a month. This effect is exaggerated when the Sun is nearer in our winter, so there is also a yearly variation of speed. In addition, the changes in Sun-pull make the Moon's orbit change its ellipticity; they tilt the Moon's orbit up and down a little; they slew the orbit-plane slowly around; and they make the oval orbit revolve in its own plane. Newton predicted these effects on the Moon's motion, making estimates of their size where he could. Some effects had long ago been observed; some were actually being sorted out—and Newton begged the Astronomer Royal for the measurements; some were not observed till afterwards. For the revolution of the elliptical orbit in its own plane, measured as 3° a month, Newton's first calculations predicted only 1½°. For years after Newton, mathematicians wrestled with the problem, trying to explain the disagreement—they even tried modifying the inverse-square law with an inverse 3rd power term. Then one of them found that some terms of Newton's algebra had been neglected unjustifiably and that with them theory agreed with experiment. Still later, it was found from Newton's papers that Newton himself discovered the mistake and obtained the correct result.

Thus Newton showed that the marked irregularities of the Moon's motion fit a system controlled by universal gravitation. He did not work out the effects of the Sun's pull in complete detail; and the Moon's motion has remained a complex problem, solved with increasing detail from then till now. The ideal method would be the general one: attack the disease, not each of its many symptoms, and simply calculate the path of the Moon in the combined gravitational field of Earth and Sun. This is the "Three Body Problem": given three large masses, heaved into space with any given initial velocities, work out their motions for all time there-

[11] The Sun exerts a strong, almost constant, attraction on the Moon, and the average of this is just sufficient to make the Moon accompany the Earth in its yearly orbit: a solar Myth-and-Symbol.

after. Though this problem looks simple in its ingredients, it has remained a challenge for centuries; its possibilities have proved too complex for a complete explicit solution. Methods are still being explored.

7. *Tides.* Ocean tides had long been a puzzle, crying for some explanation connected with the Moon. Yet men found it hard to imagine a real connection and even Galileo laughed at the idea. Newton showed that tides are due to *differences* of the Moon's attraction on the water of the ocean. The Moon does not make its orbit around the Earth's center; but Moon and Earth together swing around their common center of gravity like two unequal square-dance partners. That center of gravity is 3000 miles from the center of the Earth, only 1000 miles under ground. While the Earth pulls the Moon, the Moon pulls the Earth with an equal and opposite pull (Law III), which just provides Mv^2/r to keep the whole Earth moving around the common center-of-gravity, once around in a month. The portion of ocean nearest the Moon is pulled extra hard (because it is nearer) so it rises in a lump which is a high tide. The ocean farthest from the Moon *needs* the same pull as all the rest, but is pulled less than average; so, local gravity, "g", must provide the remainder of the needed pull.[12] That makes the ocean farthest from the Moon slightly lighter; so it is pushed up into a hump away from the Earth—another high tide.

Therefore, there are two high tides in twenty-four hours. As the Earth spins, its surface travels around, while the tidal humps, held by Moon and Sun, stay still, so the tides surge up and down the shores that are carried under them. The ocean water moves around with the Earth; but the tidal humps move like a wave from shore to shore. Friction and inertia delay those surgings in a complicated manner, so high tide is not just "under the Moon" but often lags as much as ¼ of a day. The Sun also produces tides—not so big, because its greater distance makes the *differences* of pull smaller. Twice a month, when the Sun's and Moon's tides coincide, we have large "spring" tides. When the two sets of tides are out of step we have small "neap" tides.

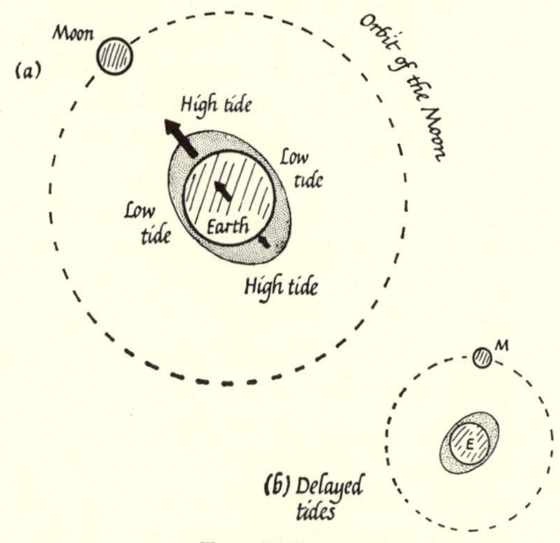

Fig. 22-23.
Ocean Tides are Caused by Differences of Moon's Attraction
(a) The extra-large pull on ocean nearest the Moon raises one high tide. The extra-small pull on ocean farthest from the Moon lets it flow away into another high tide.
(b) Delayed tides. Actually the high-tide humps are delayed by inertia, tidal friction, and effects of rotation. As the Earth spins, they are not opposite the Moon. In most places they arrive about 1/4 cycle (6 hours) late.

We can estimate the "tide-generating-forces" that act on some standard chunk of material in various parts of the Earth. Take a sample chunk that weighs 30,000,000 newtons† at all places on the Earth's *surface.** Then at Earth's *center*, E, Earth's pull on the chunk is zero; Moon's pull just provides the needed force $mv^2/$(radius EG) for the chunk's monthly motion—and calculation shows that to be 100 newtons. At all other places, A,B,C, . . . , the *needed* force is the same, 100 newtons towards the Moon.[12] But Moon's pull is only 97 at A, and 103 at C. So the *radial* pulls on a sample chunk are:

at A, 30,000,000 + 97, which will provide the needed 100 and leave 29,999,997 for effective "g"

at B, 30,000,000 + the vertical component of Moon's pull, which is now slanting. That component is $^{4000}/_{240000}$ of 100, or about 1½. This makes 30,000,001½ for effective "g".

at C, 30,000,000 (inward) and 103 (outward), which will provide the needed 100 (outward) and leave 29,999,997 (inward) for effective "g".

Thus, at A and C the chunk is "lighter" than at B: it feels an outward tide-generating-force of 4½ newtons. That is the force that piles up the two humps, only 4½ newtons on each 3300 tons of ocean.

[12] The Moon travels in its orbit around the common center-of-gravity, G, in a month. And, since it always keeps the same face to the Earth, it also spins, one complete turn per month—as seen by a stationary observer, at rest among the stars. But the Earth does not keep the same face always to the Moon. The whole Earth moves around G, one revolution in a month; but it does not also have a monthly spin. Instead, it keeps a fixed orientation as it moves around G (apart from daily spin, neglected here*). Thus, all parts of the Earth move in circles of the same size—like the circular motion of a man's hand cleaning a window, or the frying-pan motion of Fig. 17-10b on p. 257. This motion requires every part to have the same acceleration v^2/r, where $r =$ EG, towards the Moon.

† Mass of sample $= 3 \times 10^7/9.8 \approx 3 \times 10^6$ kg ≈ 3300 tons. So it is a 3300-ton chunk of rock (the size of a house) or of water, or even of air.

* We neglect the differences of "g" from equator to poles. These differences are real, but they do not produce a noticeable effect on the oceans. That is because the oceans suffer the same effects of the Earth's spin as the "pasty" Earth of Newton's calculation (4 above); so the values of "g" are just those that will spread the oceans evenly over the bulgy spinning Earth. (I.e., we assume that daily spin has given the Earth an "equilibrium shape," and expect it to do the same for the oceans).

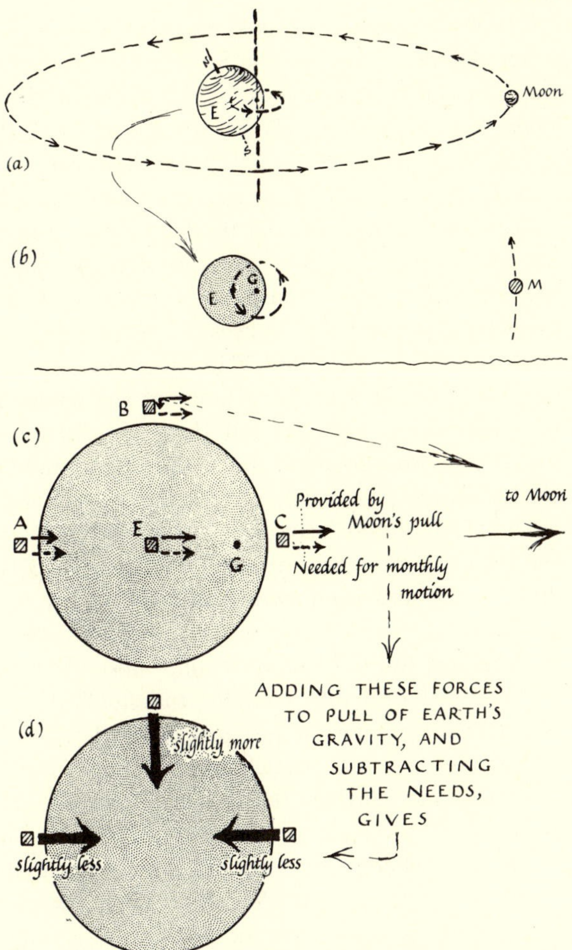

Fig. 22-24. The Tide-Producing Force

8. *The Moon's Mass.* By comparing neap tides and spring tides we can separate out and compare the effects of the Sun and the Moon.[13] Newton did this and was able to make an estimate of the Moon's mass from the size of the tide it causes. In other words the Moon has always had a satellite after all, the hump of ocean which we call high tide. No independent check of the Moon's mass could be made for two centuries, until man could send up "Moon-probe" satellites.

9. *Comets.* Newton explained the nature of comets, those visitors in the solar system which had always aroused interest and even fear. (It is strange to note how comets are still regarded as mysterious things by the popular press. A tabloid newspaper hesitates to call an eclipse a mystery because it will be laughed at; but when a visible comet appears, or even rumors of one, most newspapers make a fuss about the "mysterious event in the heavens." This superstition survives with some of the feelings that made astrology powerful for centuries—a sad reminder of the way in which the pressures of civilized life curdled man's simple wonder into fear.)

Tycho and Kepler had shown that comets are not just "vapor in the clouds" but that they travel across the planetary orbits, making a single trip, as was thought, through the solar system. They seemed to be illuminated by sunlight, and therefore could only be seen when fairly near. Newton showed how comets move in a long elliptical orbit, with the Sun in one focus. They are controlled by gravity just like the planets; but they are small and they have far more eccentric orbits so they are only visible when they are near the Sun. Such comets travel out beyond the farthest planets, slower and slower (Kepler Law II); finally turn around the remote

[13] At shores where there are bays and river mouths, tides may pile up to great heights; but at islands in mid-ocean the spring tides rise only about 4 ft, and the neap tides only 2 ft. Therefore: tide due to [Moon + Sun] is 4 ft, and tide due to [Moon − Sun] is 2 ft. This makes the Moon's tide 3 ft and the Sun's tide 1 ft. Thus we can estimate the ratio of Moon's to Sun's mass. However, momentum and friction make the problem very complicated.

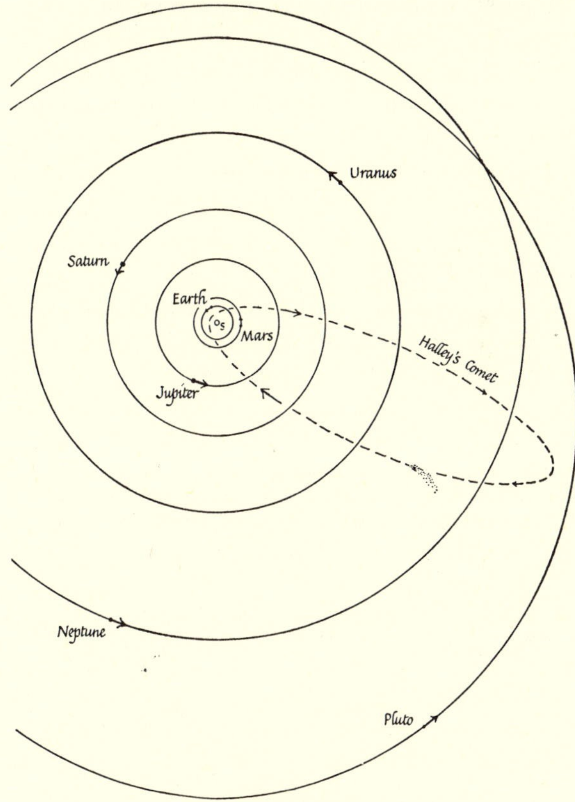

Fig. 22-25.
Sketch of the Solar System, with Halley's Comet Shown
The most recently discovered planet, Pluto, is very small and pursues an elliptical orbit extending from within Neptune's to a much greater distance. (Mercury and Venus are not shown)

nose of their ellipse (Kepler Law I); and after a long time (Kepler Law III), come hurtling back into our region; then they swing, at maximum speed, around the Sun and away again. The elliptical orbit can be measured and the comet's return predicted. One of the most famous, named after its discoverer Halley (who saw the *Principia* through the press) was the first case of successful prediction, 76 years from one visit to the next. Newton pointed to an earlier record by Kepler at just the right time, and predicted future returns. When these occurred on time, comets should have lost all mystery, though not their glory. Their regular return on predicted dates gives a test of our orbit-observations, and a further confirmation of the law of gravitation. We can carry the calculations back and identify some of the comets of history. For example, Newton's comet, observed by him in 1680 and expected to return in 2255, may have been the comet that was thought to herald the death of Julius Caesar.

Occasionally a comet suffers a severe gravitational perturbation in passing near a big planet, and changes to a new orbit with a different cycle of returns. That is how we know comets have small mass: *they* are affected, but the planet, pulled with an equal and opposite force, fails to show a noticeable effect.

If a comet arrives from outer space *very fast*, it swings around the Sun and away in its new direction in a hyperbola instead of an ellipse, and then it never comes back.[14]

Comets move so fast and visit so seldom that we are still waiting to apply the most modern methods of investigation to a big one. We believe they are collections of rock, dust, gas, etc. all travelling together. As they approach the Sun, they reflect more and more sunlight and look brighter and brighter. When very near they may be heated so much by sunlight that they emit some light of their own. The Sun's radiation roasts vapor out of some, adding to the light-scattering material, making these comets look bulkier and brighter. Many comets develop a "tail" of bright (but transparent) material which streams out behind and curves sideways from the orbit, *away from the Sun.* Why does this tail not keep up with the rest? The body of the comet is made up of many pieces, but these will all move around the orbit together since the Sun's gravitational pulls are proportional to the masses—remember the Myth-and-Symbol experiment. The tail however seems to be an exception. It fails to keep up with the rest and even swings sideways. This suggests some repulsive force between Sun and comet which affects the *tail proportionally more than the rest.* The tail probably contains the smallest particles—dust, perhaps, or just gas molecules. What forces affect a small particle proportionally more than a big one? Surface tension, fluid friction, and any forces that vary directly as the surface *area* of the particle; whereas masses, and therefore gravity forces, vary as the *volume.* The most likely "surface forces" on comets' dust are the pressure of sunlight and the pressure of ions streaming from the Sun. Making a particle ten times smaller in linear dimensions reduces its mass by a factor of 1000 but reduces its surface area by a factor of only 100, making surface-forces 10 times more important than they were, compared with gravitational pulls in their

---

[14] If we ask, as Newton's fellow scientists did, "Given inverse-square law attraction, what shape must a planet's —or a comet's—orbit have?", the mathematical machine replies, as it did in Newton's hands, "The orbit will be a

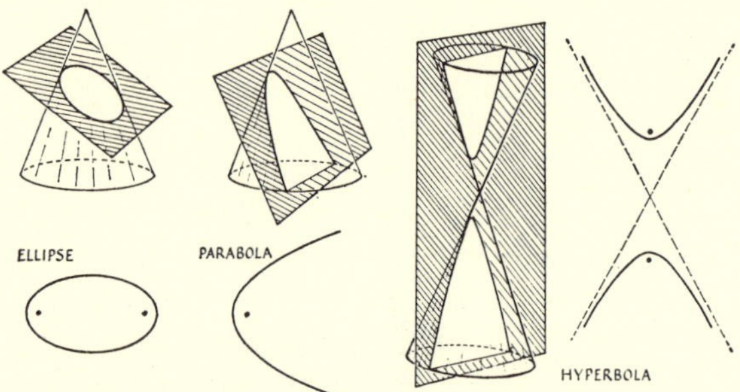

Fig. 22-26. Conic Sections

conic section with the Sun in one focus." Conic sections are curves got by taking plane slices of a solid circular cone. A cone sliced "straight across" gives a circle. If the slice is slanted, the section is an ellipse. With greater slant, just "parallel to the cone's edge," the section is a parabola. With still greater slant, the section is a hyperbola. These curves all belong to one geometrical family. Their algebraic equations are similar:

$$x^2 + y^2 = 9, \text{ a circle}$$

$$\frac{x^2}{9} + \frac{y^2}{4} = 1, \text{ an ellipse}$$

$$\frac{x^2}{9} - \frac{y^2}{4} = 1, \text{ a hyperbola}$$

The equations for parabolas look different (e.g., $x^2 = 9y$) but are closely related to the others. In physics, we meet ellipses in planets' orbits, all these curves in comets' orbits; and hyperbolas when alpha-particles are shot at other atomic nuclei. From measurements of such rebounding of alpha-particles, we can calculate the arrangement of forces that must cause these rebounds; and we find the forces must be inverse-square ones between the alpha-particle and some very tiny core of the atom it "hits." We guess that these forces are inverse square repulsions between electric charges. From further measurements we can even estimate the electric charges of different atom cores. That is how, in this century, Newtonian mechanics established the nuclear atom.

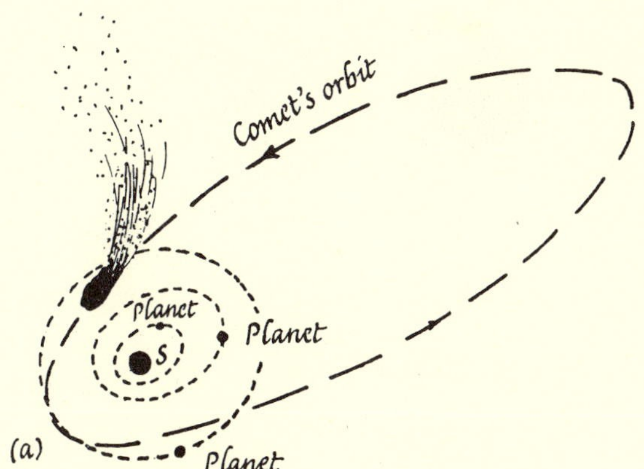

FIG. 22-27a. COMET, MOVING IN AN ELLIPTICAL ORBIT WITH THE SUN IN ONE FOCUS, PASSES THROUGH SOLAR SYSTEM

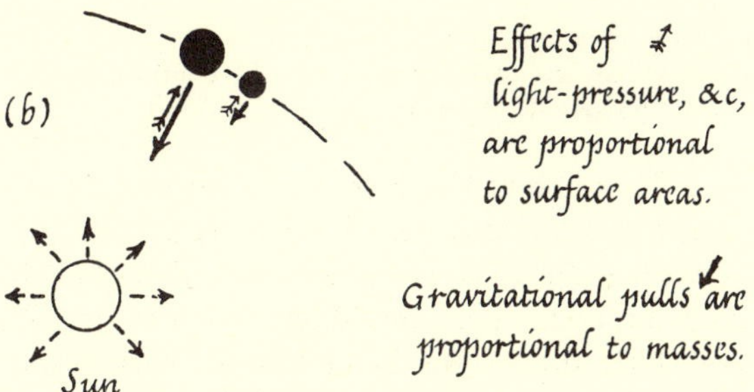

FIG. 22-27b. FORCES ON PARTICLES IN A COMET'S TAIL
If one particle of the comet has 10 times the diameter of another, its mass, for the same density, is 10 x 10 x 10, or 1000, times as great as the other's. Gravitational forces on it will be 1000 times as big. But surface-forces, e.g., light-pressure from sunlight, will be only 10 x 10, or 100, times as great. Therefore, light-pressure matters proportionally more for small particles; less for large ones. It can push the very tiny particles in the comet's tail away.

effect on the mass. Near the Sun, the radiation is intense, also streams of protons gush out, and the pressure on the smallest particles becomes important. It is thought that such pressures drive the comet's tail away from the Sun.

10. *Gravity inside the Earth.* Newton showed by calculus that a hollow spherical shell of matter attracts a small mass *outside* it as if the shell's mass were concentrated at its center. By imagining the Earth built of concentric shells (even of different densities), Newton was able to proceed with the "apple and Moon" comparison, knowing that the Earth would attract as if its whole mass were at its center. He also showed that a hollow spherical shell would exert *no force at all* on a small mass *inside* it. This result is not much use in treating the Earth's gravity, though it is very important in the corresponding theory of electric fields. There it provides a first-class test of the inverse-square law of forces between charges. We shall derive it in the chapter on electric fields.

These two results for a spherical shell give an interesting picture of the gravitational field of a solid uniform sphere. Outside, the field decreases with an inverse-square law: "g" varies as $1/R^2$, where R is the distance from the center. If we are at an inside point, we are *inside* some of its concentric spherical shells, and we lose their attraction completely. We are *outside* the remaining central

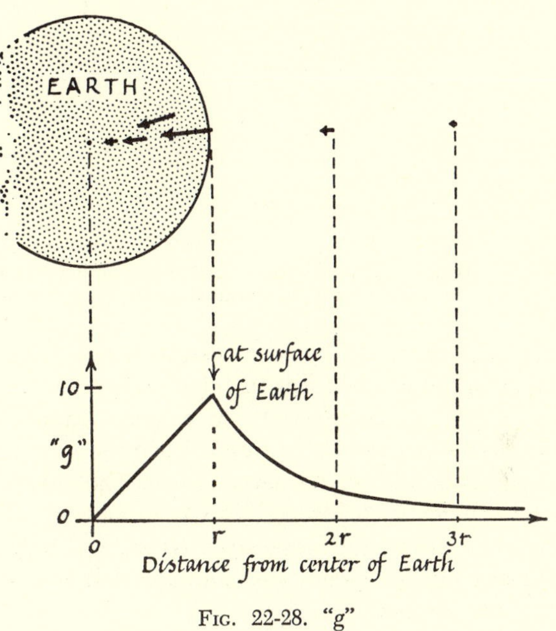

FIG. 22-28. "g"

batch of shells. These make a smaller attracting mass, but we are closer to the center. The resultant attraction makes "g" vary directly as R, inside the sphere.

11. *Artificial satellites.* Newton pointed out that any projectile is an Earth-satellite. Suppose a man on a mountain top fires a bullet horizontally. A slow bullet falls to the ground in a "parabola," which has its focus just below its nose. The path is really a Kepler ellipse, with the lower focus at the Earth's center. Parabola and ellipse are indistinguishable in the small part of the orbit observed before the ball hits the ground. (To obtain a true parabola we would need a great flat Earth, not a round one with radial directions of "g"). A faster bullet still makes an ellipse, but not so eccentric—still faster, an even rounder ellipse. One fired fast enough would go on around the Earth, like a little moon, travelling round its circular orbit again and again (provided the man got out of the way one "little-moon-month" after firing the bullet). Here was Newton's picture of an artificial satellite. It and the Moon would form a Kepler-Law-III group, with Earth as owner. We now have satellites that do this.

Fire the bullet still faster than for a circular orbit,

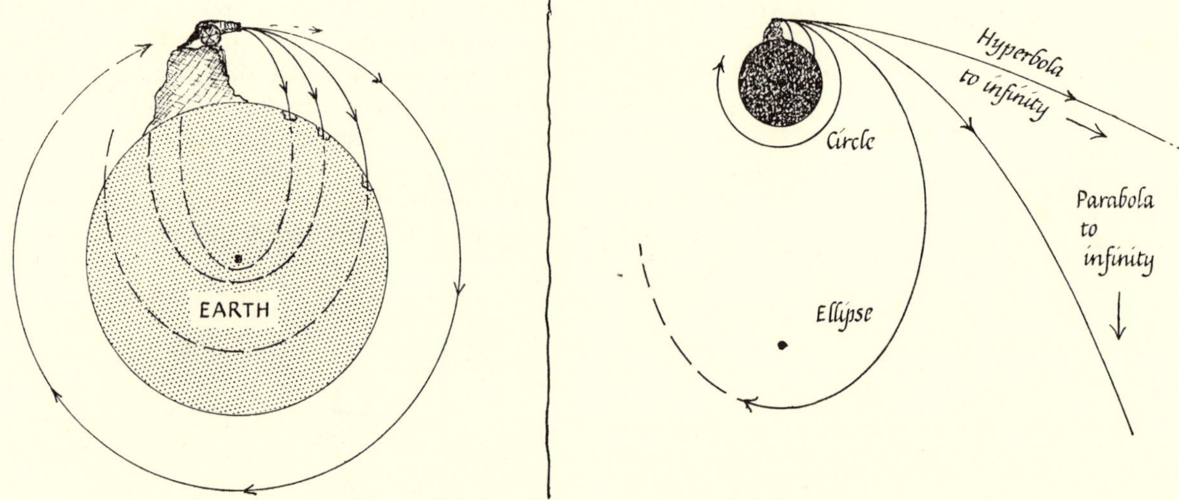

FIG. 22-29. EARTH-SATELLITE ORBITS (from Newton's sketch)
Where elliptical orbits are drawn through the Earth, they show how the original path would continue outside a tiny concentrated mass. They do not allow for the decrease of attraction inside the Earth.

and its path would be an ellipse with the Earth's center in the nearer focus this time. Still faster and the ellipse would elongate to a great parabola. Faster still and the bullet would leave the Earth in a hyperbola and would never return. The velocity necessary for such "escape" can be calculated—an important matter for future space travel, and important long ago for speedy gas molecules that escaped from the atmosphere.

12. *Planetary perturbations. The great discovery.* The major influence on a planet's motion comes from the Sun; but the other planets, acting with the same universal gravitation, also apply small forces which "perturb" the simple motion. Newton began the study of these perturbations. For example, the great planet Jupiter attracts neighboring Saturn enough to make noticeable changes in Saturn's orbit. The attraction changes in direction, since Jupiter and Saturn are both moving in their orbits; and it changes greatly in amount, as the planets move from greatest separation to closest approach.[15] This small, tilting, changing pull builds up changes of motion that accumulate in tiny perceptible changes of orbit. Newton estimated this effect and showed that it fitted with observed peculiarities of Saturn's motion. However, the general problem is very difficult, and Newton only made a start on it.

A study of planetary perturbations looks like fiddling with trivial details; yet over a century later it led to a great triumph, the discovery of a new planet. Before that the first planet beyond the five known to Copernicus was discovered by telescopic observation. In 1781 Herschel noticed a star that looked larger than its neighbors and was found to move. This proved to be a planet, soon named Uranus. The new planet was found to be twice as far away as Saturn, and its orbit radius and "year" fitted into Kepler's Law III.

The continued observations of Uranus showed small deviations from the Kepler orbit. Some of these could be explained as perturbations due to Saturn and Jupiter. However an unexplained "error" remained—a mere 1/100 of a degree in 1820. Some astronomers questioned whether the inverse-square law was exactly true of nature; others speculated about another, unknown planet perturbing Uranus. That was ingenious but posed almost impossible problems. However, two young mathematicians, Adams in England and Leverrier in France, set out to locate the planet. It is hard enough to compute the effect of one known planet on another. Here was the reverse problem, with one of the participants quite unknown: its mass, distance, direction and motion all had to be guessed and tried from the tiny residual deviations of Uranus from its Kepler orbit.

Adams started on the problem as soon as he finished his undergraduate career. Two years later he wrote to the Astronomer Royal telling him where to look for a new planet. Adams was right within 2°; but the Astronomer Royal took no notice, beyond asking Adams for more information. Then, as now, professional scientists were besieged with letters from cranky enthusiasts and had to ignore them.

Meanwhile, Leverrier was working on the problem quite independently. He examined several

[15] Their distance apart changes from the sum of orbit radii to the difference: 880 million miles + 480 million to (880 − 480) million, a change in proportion 3½ to 1. This makes the perturbing attraction increase in proportion 1 to 12.

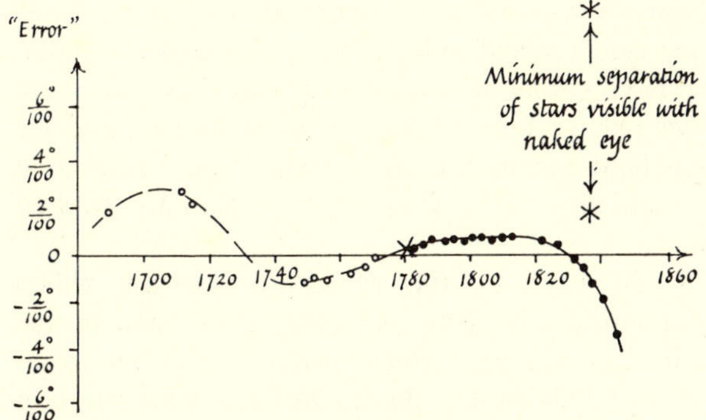

Fig. 22-30.
Residual "Unexplained" Perturbations of Uranus (a.d. 1650-1850)
The "error" is the difference between the observed position of Uranus and the expected position (for a Kepler orbit) after known perturbations had been subtracted. The point X marks the discovery of Uranus by Herschel. Working back to its orbit in earlier times, astronomers found that Uranus had been observed and recorded as a star in several instances. These earlier records are marked by ° on the graph.
(After O. Lodge, *Pioneers of Science*)

hypotheses, decided on an unknown planet, and finally managed to predict its position, near to Adams'. He too wrote to the Astronomer Royal, who then arranged for a careful but leisurely search. By this time other astronomers began to believe in the possibility—"We see it as Columbus saw America from the shores of Spain." Leverrier wrote also to the head of the Berlin observatory, who looked as directed, compared his observation with his new star-map, and saw the planet! The discovery raced around the world and was soon confirmed in every observatory. This new planet, discovered by pure theory, was named Neptune.

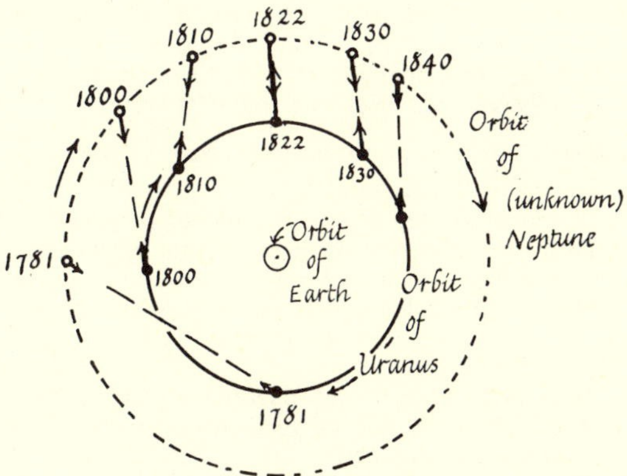

Fig. 22-31. Perturbing Forces on Uranus, Due to Neptune
The sketch shows positions of the planets in the years marked. Before 1822 Neptune's pull made Uranus move faster along its orbit so that it reached positions ahead of expectation. After 1822 Neptune's pull retarded Uranus. (After O. Lodge, *Pioneers of Science*)

Newton's Method

Newton set forth his treatment of astronomy in the *Principia.* He was using *deduction* to derive many things from a few laws, but his treatment was essentially different from the deductive methods of the Greeks and their followers. Newton devised his theory with the help of guesses from experiment; then drew from that theory many deductions; and *then* tested as many of these deductions as he could by experiment. Thus his theory was a framework of thought and knowledge, tied to reality by experiment and clear definitions, able to make new predictions which in turn were tested by experiment. A theory, as Newton used it, "explained" a variety of mysteries by referring them to a few familiar things.

Newton's successors mistook Newton's view of gravitation. They thought he treated it as "action at a distance," a mysterious force that arrives instantaneously through a vacuum, in contrast with Descartes' picture of space filled with whirlpool vortices that transmit force and motion. Newton himself merely said that an inverse-square field of force will account for Kepler's laws and many things besides. For this "explanation" he did not need to know how the force got there. He said clearly that he did not know the *cause* of gravitation. He suggested it must be some kind of influence that travels out from every piece of matter, and penetrates every other piece, but that was only description of observed properties. He insisted he did not know its ultimate cause. *"Hypotheses non fingo"*—"I will not feign hypotheses"—he wrote fiercely at one time. He meant he would not invent unnecessary details in his description of Nature, or pretend to explanations *that could never be tested.* Yet in later writing he offered many a keen guess, at the nature of light, at the properties of atoms, and even at the mechanism of gravitation.

Newton is usually described as a genius of cold, unemotional logic and clear insight who set the style for modern science. But one of his biographers, Lord Keynes, who studied many of Newton's writings, found him a difficult recluse who treated Nature as a mystical field of magic.

"Newton was not the first of the age of reason. He was the last of the magicians, the last of the Babylonians and Sumerians, the last great mind which looked out on the visible and intellectual world with the same eyes as those who began to build our intellectual inheritance rather less than 10,000 years ago. Isaac Newton, a posthumous child born with no father on Christmas Day 1642, was the last wonder-child to whom the Magi could do sincere and appropriate homage."[16]

He felt himself the magician who had unlocked God's riddle of the solar system by reasoning from clues strewn around by God in recorded measurements, in experiments waiting to be done, in useful folklore, and even in some inspired hints in ancient writings. He succeeded in this by his extraordinary gift of concentrating continuously on a problem in reasoning, "his muscles of intuition being the strongest and most enduring with which a man has ever been gifted."[17] Could he not in the same way unlock the great riddle of Religion: discover the nature of God, explain the behavior of matter and man's mind, and reveal the whole progress of time from the original creation to the ultimate heaven? This tremendous mind aspired, as Keynes sees him, to be "Copernicus and Faustus in one."[18] Be that as it may, all the biographers, from Newton's contemporaries to Keynes and Einstein, regard Newton as the greatest mathematical scientist of a thousand years.

Newton's "Guesses"

As an architect of science Newton set a new building style, much of which is still in great use. As a thinker about science he seems to have been a remarkably lucky guesser—he guessed right more often than mere chance would make likely. He formulated laws of motion which we still use and consider very nearly "right." He guessed correctly at universal gravitation. He even made a guess, on scanty evidence, at the mass of the Earth—a guess which could not be tested at the time but had to wait for Cavendish's experiment. He argued, speculatively, that none of the solid ground can be less dense than water, or it would float up into mountains. On the contrary, the central regions must be much denser than the outer rocks. So, he guessed, the average density of the whole Earth must be between 5 and 6 times the density of water. We now know it is 5½ times! The Earth's mass is 5.5 times that of an equal globe of water. Again, Newton devised a theory of light waves that would explain both the properties of rays and the interference colors of thin films (which he discovered and measured). It was a curious scheme in which light consists of bullets accompanied by waves to arrange where they shall travel. A century later, the wave theory of light displaced and discredited the bullet theory. For many years scientists laughed at Newton's queer mixed scheme. Now, two centuries later, we have gathered clear evidence to show that light *does* behave both as waves and as bullets, and we now hold a composite theory which has a surprising resemblance to Newton's! Once again he guessed right.

I do not think this successful guessing which characterizes Newton and other great men is due to luck or to a mysterious intuition, or divine inspiration. I believe it is due to Newton using more of his knowledge; gathering in and using chance impressions and other things barely perceived and soon quite forgotten by ordinary men; and thinking with great flexibility of mind. He had unusual intuition just because he drew upon a greater fund of detailed knowledge—he was sensitive and could remember, where most of us are insensitive and forgetful; and he was willing to turn his thinking quickly in different directions. Just as the great actor is aware of his audience and can draw upon a rich knowledge of other people's emotions and behavior, so Newton was aware of nature and could draw on a rich fund of observation. Perhaps in some respects that is where greatness lies, in sensitive awareness of the world around—the world of people or the world of things.

[16] "Newton the Man" by J. M. Keynes, in *Newton Tercentenary Celebrations of The Royal Society of London* (Cambridge University Press, 1947), p. 27.
[17] *Ibid.*, p. 28. [18] *Ibid.*, p. 34.

PROBLEMS FOR CHAPTER 22

1. In text.

★ 2. KEPLER'S LAW III

Newton guessed at universal, inverse-square-law gravitation. We express his guess in the form $F = GM_1M_2/d^2$. From his guess (a "principle") he deduced (predicted) the behavior of the Moon, planetary systems, tides, etc.

Predict Kepler's Law III by following the instructions (A) and (B) below. Suppose a Sun of mass M holds a planet of mass m in a circular orbit of radius R by gravitational attraction. Suppose the planet moves with fixed speed v, taking time T (its "year") to travel once round its orbit.

(A) *Algebra.* State in algebraic form, each of the following:

- (a) The planet's acceleration.
- (b) The force *needed* to give this planet this acceleration.
- (c) The force *provided* by gravitational attraction, if it follows Newton's law of gravitation.
- (d) The planet's velocity v in terms of R and T.

(B) *Argument*:

- (i) Now write down, as an algebraic equation, Newton's guess, that the needed force (b) is just provided by gravitational pull (c).
- (ii) Then get rid of v in this equation by using relation (d).
- (iii) Now move all the R's and T's in the equation over to the left hand side of the equation and everything else to the right hand side, thus obtaining a new (version of the old) equation.
- (iv) In the new equation, do you find R^3/T^2 on the left? (If not, check your algebra.) Do you now find the

right hand side is the same for all planets; that is, a constant which does not contain m, or R, or T?

(v) Would this new equation hold, *with the same right hand side* for other planets with different masses, orbits, and years, but with same Sun? Then does Newton's guess predict Kepler's Law III?

★ 3. KEPLER'S LAW II

Kepler's Law II is the "Equal Area" law.

(a) What did the law state? (Include a clear sketch.)
(b) Newton showed that this law must hold for any planetary motion provided that . . . ?
(c) Review Newton's geometrical proof of (b), then write out your own version of the proof with your own diagrams. (Give several clear sketches rather than one crowded one.)

4. RELATIVE MASSES OF PLANETS

(a) Starting with Newton's laws of motion, and $a = v^2/R$, and universal gravitation, $F = G\,M_1M_2/d^2$, show how the ratio (JUPITER'S MASS)/(SUN'S MASS) can be obtained from astronomical measurements. Show your full working; do not just quote algebraic results.
(b) Make a *rough estimate** of the ratio. (See below for data)
(c) Make a similar *rough estimate** for the ratio (EARTH'S MASS)/(SUN'S MASS).
(d) From terrestrial experiments, such as Cavendish's, we can estimate the mass of the Earth. It is about 6.6×10^{21} tons. *Calculate roughly*,* from (c), the Sun's actual mass, in tons.

DATA (some of which may not be needed):
Radii of planetary orbits: see table in Ch. 18.
Lengths of planetary "years": see table in Ch. 18.
Data for satellites of Jupiter: see Ch. 19. (Do not use the values of orbit-radii given in multiples of Jupiter's personal radius; but use the values in miles. The "times" are given in hours. Convert them to the units you use for all other "times" in the calculation.)
Data for Earth: personal radius $\approx$ 4,000 miles
time of revolution about axis = 24 hours
radius of orbit $\approx$ 93,000,000 miles $\approx 1.5 \times 10^{11}$ meters
1 year $\approx$ 365 days $\approx 3 \times 10^7$ seconds
Data for Moon: radius of orbit $\approx$ 240,000 miles ($\approx$ 60 Earth-radii).
Personal radius $\approx$ 1,000 miles
1 month = 27.3 days. (This is the Moon's absolute period, relative to stars.)

5. ARTIFICIAL SATELLITES

(a) Suppose an Earth-satellite pursues a circular orbit 4000 miles above the Earth's surface—that is, at radius 8000 miles from the Earth's center. From your knowledge of planetary motion, estimate the time required for the satellite to make one trip around.
Give your answer (i) in factors with no cancelling
(ii) reduced to a *rough numerical estimate** in minutes or hours, or days or years.
(Obtain any data you want from earlier problems above. You do not need the value of G).
(b) Some television engineers propose to establish a satellite that would re-broadcast short-wave radio, enabling people on the West Coast to receive programs from New York. They wish the satellite to stay in place, hovering permanently over Chicago (for example), without using any motive power once it is there.
  (i) Describe the behavior of such a satellite, as seen by an observer far away from the Earth.
  (ii) Calculate the height at which the satellite would have to hover. (Give your answer first in factors, then worked out *roughly** in miles.)
(c) A satellite is clocked at 90 minutes per revolution around the Earth (relative to the stars). Assuming that its orbit is circular, *estimate** its height above the surface of the Earth.
(d) Suppose a projectile could be shot out of a gun horizontally so fast that it never hits the ground but continues around the world just above the ground.
  (i) How long* would it take to return to its starting point (if air resistance were negligible)?
  (ii) *Estimate** its speed.
  (iii) The speed asked for in (ii) is the speed that a point on the Earth's equator would have if the Earth were spinning so fast that . . . . . ?
(e) [A quick question to try on your neighbor or rival in the course: time limit 15 secs.] How long would a 1-ton satellite take to go around a circular orbit around the Earth, with radius a quarter of a million miles?

6. BOHR ATOM-MODEL

Bohr constructed his simplest model of a hydrogen atom with an electron pursuing a circular orbit around a massive nucleus, which exerted an electrical *inverse-square-law attraction* on it. (This form of atom-picture is now considered misleading. Yet it is still used in advertising, and even by physicists when they need a crude picture to aid rapid thinking.) The "quantum theory restriction" formulated by Bohr stated, essentially, that only those circular orbits are allowable for which
(momentum of electron) · (circumference of orbit) = $n \cdot h$
where h is a universal "quantum-constant" and n is a *whole number*: 1, or 2, or 3, 4, 5, 6, etc. . . .
Thus, $(mv) \cdot (2\pi r) = nh$ allows the atom to have its electron only in orbits, with $n = 1, 2, 3$, etc.

(a) With help from Kepler and Newton, show that the radii of allowed orbits must be proportional to n^2, so that they run in proportions 1:4:9:. . . . (Thus if an "unexcited" atom had radius x, the atom in higher excited states would have radius $4x$, $9x$, etc.
(b) For simple hydrogen atoms with $n = 1$, $r \approx 0.5$ Ångstrom Unit (0.5×10^{-10} meter), and this seems to be the "size" of the atoms. Excited hydrogen atoms in stars have been observed with n as large as 30. What would be the "size" of such atoms?

---

* In these questions where a VERY ROUGH ANSWER is asked for—to give a general idea of relative masses, or the size of some force—precise arithmetic would miss the point and give no advantage for this purpose. Therefore, you are strongly advised to proceed as follows:

(i) USE ALGEBRA UNTIL AS LATE A STAGE AS POSSIBLE.
(ii) Then insert arithmetical values, with no cancelling, and SHOW THE RESULT IN FULL FACTORS. (DO NOT CANCEL THAT FIRST "RESULT" BUT LEAVE IT UNTOUCHED, IN CASE YOU NEED TO RETURN TO IT OR READERS WISH TO CHECK IT.)
(iii) Re-copy the "result" of (ii) and make ruthless approximations to find a rough answer. Here are three good schemes for approximation:
  (a) Use rough arithmetic; follow the advice on "judging" in Ch. 11. That will produce satisfactory results faster than any other schemes except those that require special skill or experience. (Reduce all data to powers of 2 and powers of 10. Then cancel. Remember: $2^{10} \approx 10^3$.)
  (b) Use log tables—good if you are very quick. Logs to TWO decimal places would suffice.
  (c) Use slide-rule. Also, do very rough calculation to find "where the decimal point is." This is slow because it uses two processes: (a) is better.

If your result is wrong by a factor of 1000 through careless cancelling, it is worthless; but if it only suffers from a 40% error due to rough arithmetic, it still carries a useful message in such problems.

CHAPTER 23 · UNIVERSAL GRAVITATION

"... You must be satisfied beyond all reasonable doubt. . . ."
(part of Judge's charge to Jury in criminal trials)

The idea of inverse-square-law gravitation was "in the air" when Newton made his calculations. Other scientists were speculating on a cause for Kepler's laws and asking whether planetary motions could be explained by an attraction spreading from the Sun, thinning out as it spreads. Newton rescued the question from mere speculation and extended the guess of some-kind-of-pull-from-the-Sun to universal gravitation. He tested his guess of inverse-square-law gravity by treating the Moon's motion, and by showing it led to Kepler's Laws. Further tests on Jupiter's satellites showed that the same kind of force acts between planets and satellites as between Sun and planets. So the $1/d^2$ factor in the relation $F = G\,M_1M_2/d^2$ seemed well established by experimental evidence from the solar system.

The "Myth-and-Symbol" experiment guaranteed the factor M_2, the mass of the attract*ed* body. Since all bodies fall freely with the same acceleration, *g*, the Earth must pull them with gravitational attractions that are proportional to their masses, M_2 M_2'.

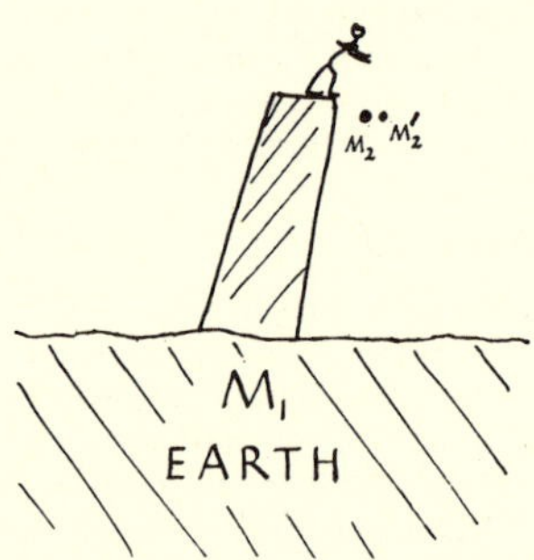

Fig. 23-1. Myth-&-Symbol

Newton trusted his Law III, action equals reaction, which he considered he had tested somewhat by his pendulum experiments on momentum-conservation. Then the gravitational pull of M_1 on M_2 must be equal and opposite to the gravitational pull of M_2 on M_1. That is, ${}_1F_2 = {}_2F_1$. Therefore, G must be the same in the two forces below

$${}_1F_2 = G(M_1M_2)/d^2 \qquad {}_2F_1 = G(M_2M_1)/d^2$$

Thus the attracted and attracting bodies are interchangeable in this story, and gravitational attraction must also vary directly as the mass of attract*ing* body. This seems very likely, even certain, to anyone believing in symmetry; but it can not be proved experimentally by astronomical measurements since we have no other way of estimating *astronomical* masses—until rocket explorers can bring back surveys and samples.

Trusting his general theory, Newton could estimate the ratios of celestial masses, $\dfrac{\text{Mass of Jupiter}}{\text{Mass of Sun}}$, $\dfrac{\text{Mass of Earth}}{\text{Mass of Sun}}$, and even, by guesses from tides, $\dfrac{\text{Mass of Moon}}{\text{Mass of Sun}}$; but he could not calculate actual masses separately because he did not know the value of the universal gravitation constant, G. To find the value of G required terrestrial experiments to measure the very small attraction between two *known* masses.

Measuring G

The gravitation constant, G, remained unknown for over half a century after Newton. A rough estimate of G from guesses like Newton's of the average density of the Earth showed that the attractions between small objects in a laboratory must be almost hopelessly small. The common forces of gravity seem strong; but they are due to an Earth of huge mass. And the Sun, with enormously greater mass still, controls the whole planetary system with gravitational pulls. But the pulls between man-sized objects are so small that we never notice them compared with Earth-pulls and the short-range forces between objects in "contact." It was clear that to measure G very delicate and difficult experiments would be needed.

As a desperate attempt, several scientists at the end of the 18th century tried to use a measured mountain as the attracting body. They estimated G by the pull of the mountain on a pendulum hung near it. They had to measure, *astronomically*, the tiny deflection of the pendulum from the vertical, caused by the sideways attraction of the mountain. They had to estimate, *geologically*, the mass of the mountain and its "average distance" from the pendulum. Substituting these measurements in $F = G\,M_1M_2/d^2$ gave an estimate of G.

About the same time Cavendish, and later many others, measured the attraction between a large lump of metal and a small metal ball by a form of direct "weighing." Cavendish placed a pair of small metal balls on a light trapeze suspended by a long thin fiber. He brought large lead balls near the small ones in such positions that their attractions on the small balls pulled the trapeze round the fiber as axis, twisting the fiber until its Hooke's-Law forces balanced the effects of the tiny attractions.

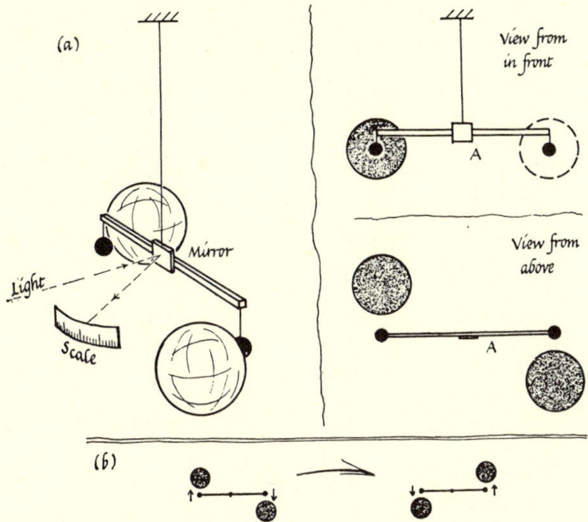

FIG. 23-2. THE CAVENDISH APPARATUS
(a) The trapeze carrying the small lead ball was hung on a very fine twistable fiber. When the big balls were brought into position, their attraction made the trapeze twist the fiber slightly. This minute twist was shown by rays of light reflected by a small mirror, A, on the trapeze.
(b) To double the measured twist, the big balls were then moved across to the "opposite" positions, so that they pulled the trapeze around the other way.

He measured the masses and the distances between small balls and large attracting balls; but, to calculate the value of G, he also needed to know the attraction forces, and to find these he needed to know the twisting spring-strength of the fiber. The fiber was far too thin and delicate for any direct measurements. So Cavendish let the trapeze and its balls twist to and fro freely with simple harmonic motion (Ch. 10) and timed the period of that isochronous motion. From that, with measurements of mass and dimensions of the trapeze, he could calculate the twisting strength of the fiber. Then, he obtained a good estimate of G, confirmed by similar measurements made more carefully by Boys, Heyl, and others. In all cases the apparatus is so delicate that the slightest air currents will spoil the measurements. To avoid convection currents, Cavendish placed his apparatus in a box, then placed the box in a closed room and observed the apparatus with a telescope from outside the room.

Results of Measurements of G

The table shows some details of many measurements of G made over the past 220 years. It not only shows increasingly reliable values for the important quantity G, but it also gives great support to the relationships

$$\left.\begin{array}{l} F \propto M_1 \\ F \propto M_2 \\ F \propto 1/d^2 \end{array}\right\} \text{which are combined in } F = G\,M_1M_2/d^2.$$

It gives this support because it shows that a *great variety* of masses, materials, and distances all yield the same value of G, within forgivable experimental errors. If we wished to show how we know the value of G accurately (agreed for some reason or other to be universally constant) we should describe just one very good experiment; but since we want to give evidence for the validity of Newton's great theory, we give many experiments with great variety.

Modern Uses of the Cavendish Experiment

Early rough estimates of G gave a good idea of the general size of gravitational forces. The attraction between two people seated side by side is almost immeasurably small; the attraction between the Sun and the Earth is unbelievably big—a steel cable as wide as the Earth could just about replace it. And the *electric* attraction between electron and nucleus in a hydrogen atom exceeds their *gravitational* attraction by a stupendous factor of about

2000000000000000000000000000000000000000.

Later measurements of G gave us the value fairly accurately, with a likely error of less than 0.2%. As recently as 1942, Heyl, at the National Bureau of Standards in Washington, made one of our most trusted measurements of this fundamental constant. Unless some new theory asked for much more precise measurements, the Cavendish Experiment would hardly be repeated. However, the apparatus has been refined into a differential gravity meter which can estimate tiny differences of gravitational field due to local deposits of rock of unusual density. This instrument is used by geologists for surveying the Earth's crust, and by oil companies to look for geological peculiarities that might yield oil. (The commercial explorers treat the instrument and its technicians with an attitude of naïve empirical hopefulness.) In one form, the two balls of a small Cavendish Apparatus are hung at *different levels.* The balls would be pulled unevenly by a shallow deposit of

MEASUREMENTS OF *G*

Date (approx.)	*Experimenter*	*Attracting Mass*: Description	*Attracting Mass*: Mass kg	*Attracted Mass*: Mass kg	*Attracted Mass*: Description	*Distance apart* meters	*Result* G newton · m.^{2}/kg^2
A							
1740	Bouger	Mountain	many millions of millions of kg	pendulum mass: a few tenths of a kg to a few kg	pendulum	several thousand meters	12 $\times 10^{-11}$
1774	Maskelyne	Mountain			pendulum		7 to 8 " "
1821	Carlini	Mountain			pendulum		8 " "
1854	Airy	Outer shell of Earth	3×10^{20}		pendulum	6,000,000 meters	5.7 " "
1854	James	Mountain	many millions of millions of kg		pendulum	several thousand meters	7 " "
1880	Mendenhall	Mountain			pendulum		6.4 " "
1887	Preston	Mountain			pendulum		6.6 " "
B							
1798	Cavendish	lead ball	167	0.8	lead ball	0.2	6.75 $\times 10^{-11}$
1842	Baily	lead ball	175	0.1 to 1.5	balls of: lead, zinc, platinum, glass, brass	0.3	6.5 to 6.6 " "
1881	von Jolly	lead ball	45,000	5	metal ball	0.5	6.46 " "
1891	Poynting	lead ball	160	23	lead ball	0.3	6.70 " "
1895	Boys	lead ball	7	0.0012	gold ball	0.08	6.658 " "
1896	Braun	brass ball; iron ball	5; 9	0.05	brass ball	0.08	6.66 " "
1898	Richarz and Krigar-Menzel	lead cube	100,000	1	copper ball	1.1	6.68 " "
1930–1942	Heyl and Chrzanowski	steel cylinder	66	0.05	platinum, glass, gold	0.1	6.673 " "

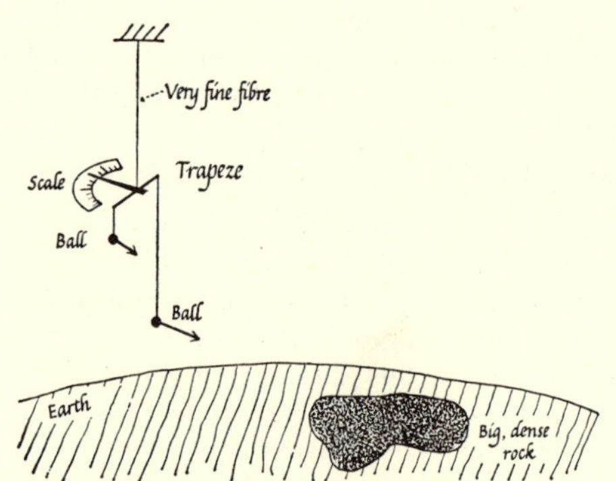

FIG. 23-3. DIFFERENTIAL GRAVITY-METER
A compact, highly sensitive form of Cavendish apparatus, with the two small balls hung at different levels. An extra dense rock nearby has a slightly greater pulling-around-effect on the lower ball than on the upper one. This is because: (i) The rock is slightly nearer the lower ball and inverse-square-law force is bigger; (ii) the pull is slightly nearer to horizontal, so that the horizontal component, which does the pulling around, is a slightly greater fraction of it. The instrument is set up, brought to an even temperature, and observed (with a telescope) in several different orientations.

dense rock nearby, and the trapeze would show a small twist when oriented suitably. This is the physicist's form of the "divining rod." To make such an instrument rugged enough to be portable and sensitive enough to be useful is a triumph of skill. Oil explorers are now replacing these differential gravity meters by more direct instruments that measure small differences of vertical g itself.

Modifications of Cavendish's experiment have been done to see whether gravitational attractions are influenced by temperature changes, or shielded by intervening slabs of matter, or dependent on crystalline form, etc. No changes have been observed—so far. *G* seems to be universally the same even when M_1 or M_2 includes the mass of some *releasable* nuclear energy in radioactive material: the relationship $F = G M_1 M_2/d^2$ still holds, with *G* the same.

Speculations

At present, most physicists regard the gravitation constant as a true constant, to be ranked with the speed of light, the charge of the electron, and a few others that seem to be the same for all matter in all circumstances, fundamental constants of the universe. However, others, bold speculators but wise ones, have suggested that G may be slowly changing as time goes on (see below). A G that was much larger in earlier times could bring gravitational forces nearer to matching electrical ones in the distant past.

Good theoretical physicists are trying to coordinate gravitational fields with electric and magnetic fields in a single "unified field theory."

Some enterprising thinkers hope to show a connection between G and other basic physical constants—possibly in connection with magnetism or perhaps in terms of the total population of atomic particles in the universe.

Time is of the Essence

Some physicists and astronomers speculating on the properties of space, time, and matter have suggested a slowly changing G, *as measured by* apparatus using *atomic clocks*. (If we used a pendulum clock to time the period of the trapeze, or if we held a pendulum near a mountain, we should find no change, because we should be comparing G with Earth's g which contains G as a factor.) This raises the whole question of TIME. What do we mean by time; and how do we know one second is the same length as the second before? We have several kinds of "clocks." Some use pendulum-swings for their constant unit of time; others use the S.H.M. of a loaded spring (Ch. 10); others use the spin of the Earth (*sidereal day*); others the Earth's revolution around the Sun (*solar year*); and others the motions of atoms (spectral lines, atomic spins, . . .); and the decay of radioactive material (Ch. 39) has been suggested. Many of these depend chiefly on *atomic* properties (e.g. radioactive decay; and even the Earth's spin, which stays practically the same even if gravity changes). But some involve gravitation directly (e.g. pendulums, the solar year). So we may have to deal with two different scales of time.

Geologists and astronomers are making good guesses at the age of the Universe from measurements of radioactivity, star temperatures, speeds and distances of nebulae. Estimates run to some 10 billion years but these are on an *atomic* scale of time. There might be such a different gravitational scale of time (e.g. by pendulum clock or by solar years) that *its* date for the beginning of the Universe stretches much farther back, perhaps even to "minus infinity."[1] That would put questions of the "beginning of the world" in a different light.

Much of this is fanciful speculation in the boundary layer of metaphysics between philosophy and science. Yet hardheaded experiments on g and G are proceeding and the next ten years may see surprising developments in our views of gravitation, with implications ranging from standards of timekeeping to theoretical cosmology.

---

[1] Here is one scheme that might be imagined. Suppose that G is slowly decreasing, as measured by "atomic" clocks. Then a pendulum clock, keeping gravitational time, must have ticked faster (judged by the spinning Earth, an atomic clock) in earlier days. So the Earth must have been spinning slower (judged by the pendulum clock) in earlier days. We shall take the pendulum clock's gravitational time as our reference standard and watch the changes in atomic time-scale. (Neither time-scale is more true than the other—only modern prejudice makes us think that atomic time is the *right* one that flows at constant rate.)

Imagine, for example, that the relationship between the two time-scales is an "exponential" one, such that, compared with pendulum clocks, atomic clocks double their rate in a trillion pendulum days. This is just one out of countless possible relationships—we choose it for a simple illustration. To make the illustration still simpler, imagine that the sliding scale connecting the two clock-schemes does not change smoothly but in sudden jumps, thus: after each trillion days by the pendulum clock, atomic clocks suddenly double their rate—twice as many atomic ticks to the pendulum day as there were before. (Such a scheme of jumps is hopelessly unlikely. Calculus offers to deal just as easily with a smooth, exponential version, which might be true of nature.) Then for each trillion days that we travel back into the past by pendulum clock, we should find atomic clocks running only half as fast—half as many atomic ticks to the pendulum day. Counting back from the present, the first trillion days would be the same on both time-scales; for the next trillion pendulum-clock days the atomic clocks would run at half rate and register only ½ trillion days; and so on. As we travel back into the past in trillion-day periods by the pendulum clock, the tally runs:

PENDULUM CLOCK: 1 trillion days + 1 trillion + 1 trillion + . . .

ATOMIC CLOCK: 1 trillion days + ½ trillion + ¼ trillion + . . .

The second series never adds to more than 2 trillion, but the first mounts to infinity. So while we count back to a definite beginning of time by atomic clocks (2 trillion days ago in this example), the pendulum clock's tally runs back from now to minus infinity.

In this example, if we measure time by radioactive changes or by the Earth's spin (which defines the sidereal day), we should find the Universe beginning some 2 trillion days ago, but we could never get back to that beginning by counting pendulum days or solar years.

PROBLEMS FOR CHAPTER 23

★ 1. Newton guessed at universal, inverse-square-law gravitation. We express his guess in the form $F = G\,M_1M_2/d^2$. From his guess he deduced (predicted) the behavior of the Moon, planetary systems, tides, etc.

(a) Say what each letter in the relation above stands for and give its proper units in the meter, kilogram, second system. Copy the example, then proceed similarly for the rest of the letters. (*Example*: "G is a universal *constant*, the same for all attracting bodies. It is measured in newtons · meters²/kg².")

(b) G is a universal constant measured by Cavendish and others. If we use newtons for the force, and kg for the masses, and meters for the distance, G has the value 6.66×10^{-11} (or 0.0000000000666) newtons · meters²/kg². From this, make a rough estimate of the mass of the Earth as follows:

(i) Using the relation above, with the value of G above, calculate the attraction of the Earth on a 0.40-kg apple near the surface. (Assume the attraction is the same as if Earth's mass were all at its center, 4000 miles from apple.) The radius of the Earth is about 4000 miles or about 6,400,000 meters. Call the mass of the Earth M kg.*

(ii) Using your ordinary knowledge of physics, calculate the *weight* of (= pull of the Earth on) the 0.40-kg apple in *newtons.*

(iii) Assuming the answers to (i) and (ii) are the same, write an equation and solve it for the mass of the Earth. This will be in kilograms. Convert it to pounds, then tons. Use "judging."** (1 kg ≈ 2.2 pounds)

* Since in stage (i) you do not know the mass of the Earth, you must call it M in your answer to (i).

** See Chapter 11 for explanation of "judging.". Also see footnote at end of Ch. 22, on rough answers.

2. HOW BIG IS GRAVITATIONAL ATTRACTION?

As an indication of the size of the Sun's pull on the Earth, carry out the following calculations roughly. Suppose the Sun's gravitational attraction could be replaced by a steel wire running from the Sun to the Earth, the wire's tension holding the Earth in its orbit. Good steel has a breaking stress of 100 tons-weight per square inch.

(a) Calculate *very roughly* the cross-section area of the wire which could just hold the Earth in its orbit.

(b) Calculate *very roughly* the wire's diameter.

Data: $G = 6.7 \times 10^{-11}$ newtons · meters²/kg²
$\approx 1.0 \times 10^{-9}$ poundals · feet²/pounds²
Distance of the Sun from Earth is 93 million miles.
Mass of the Sun: about 2×10^{27} tons
Mass of the Earth: 6.6×10^{21} tons.

3. HOW SMALL IS GRAVITATIONAL ATTRACTION?

Calculate roughly the gravitational attraction between two boys assumed spherical, one a 70-kilogram boy, the other a 90-kilogram boy, seated with their centers 0.80 meter apart.

4. OTHER FORCES: COMETS

Comets are probably collections of separate solid lumps, dust, and gas.

(a) Explain why, if the motion is controlled by gravity, you would expect to find the comet travelling as a whole without changing shape (big lumps and little lumps keeping pace together), and not having a tail that lags behind or swings away.

(b) What type of forces would be needed to "explain" the behavior of comets' tails? (State the *essential characteristic* of these forces by describing their mathematical form rather than their physical nature.)

(c) Give clear reason for your answer to (b).

CHAPTER 24 · SCIENTIFIC THEORIES AND SCIENTIFIC METHODS

"A time to look back on the way we have come, and forward to the summit whither our way lies."

—J. H. Badley

The Fable of the Plogglies

"The prescientific picture is represented by a little story about the 'plogglies.'

"According to this story, there were once two very perplexing mysteries, over which the wisest men in the land had beat their heads and stroked their beards for years and years. . . . Whenever anyone wanted to find a lead pencil he couldn't, and whenever anyone wanted to sharpen a lead pencil the sharpener was sure to be filled with pencil shavings.

"It was a most annoying state of affairs, and after sufficient public agitation a committee of distinguished philosophers was appointed by the government to carry out a searching investigation and, above all, to concoct a suitable explanation of the outrage. . . . Their deliberations were carried out under very trying conditions, for the public, impatient and distraught, was clamoring ever more loudly for results. Finally, after what seemed to everyone to be a very long time, the committee appeared before the Chief of State to deliver a truly brilliant explanation of the twin mysteries.

"It was quite simple, after all. Beneath the ground, so the theory went, live a great number of little people. They are called plogglies. At night, when people are asleep, the plogglies come into their houses. They scurry around and gather up all the lead pencils, and then they scamper over to the pencil sharpener and grind them all up. And then they go back into the ground.

"The great national unrest subsided. Obviously, this was a brilliant theory. With one stroke it accounted for both mysteries."

—Wendell Johnson[1]

By the eighteenth century, science was being moulded into a body of experimental knowledge connected by logical thought. Before going on to study further developments of physics, look at the ingredients of this system that we call science.

Why do we call Newton's work good theory and the plogglies a bad theory? What makes good theory good; and why do we put such trust in theory today? We shall spend this chapter discussing such questions.

We shall not give a compact definition of scientific theory or scientific knowledge—that would make a mockery of their varied nature and great importance. You need to develop an educated taste for them as you do for good cooking; in a sense, scientific theory is a form of intellectual cookery. All we can do here is provide some general background and vocabulary for your own thinking. This chapter offers comments on the kind of theory developed by Newton's day. You will find further discussion and other views of theory in later chapters, particularly the final chapter, Chapter 44. You are advised to read the present chapter quickly, to watch its to-and-fro discussion rather than to extract any final answers. Your view of science must be of your own making.

[1] From *People in Quandaries* (Harper and Brothers, New York, 1946).

VOCABULARY

Here is a dictionary of terms for use in your thinking:

Facts

Most physical scientists believe they are dealing with a real external world—or at least they act as if they believe that in building up their first scheme of knowledge. Even if they have philosophic doubts, they start with "sense impressions" or "pointer-readings on instruments" as their *facts* of nature.

We trust such facts because they are agreed on by different, independent observers. In common life, our facts may be vague—e.g., "Uncle George is bad tempered." In physical science they are usually definite measurements, the results of experiment—e.g.:

the crystal has 8 faces,
this sheet of paper is 8.5 inches wide,
aluminum is 2.7 times as dense as water,
a freely falling stone gains 32 ft/sec in each second,
orbit of Mars is twice as wide as orbit of Venus,
the gravitation constant is 6.6×10^{-11} MKS units,
an atom is a few Ångström Units wide.

To be completely clear and true, each of these needs some commentary: definitions of terms, explanation of accuracy, limitations of applicability; but among scientific colleagues we usually leave these unsaid—just as a family may agree that Uncle George is bad tempered without worrying over accurate definitions of temper. As the list progresses we get farther and farther from direct sense-impressions, and our "facts" are more and more dependent on our choice of theory. When we get to the statement "the diameter of a hydrogen atom is 10^{-10} meter," the "fact" has little meaning unless we say what behavior of atoms we are dealing with and even what theory of atoms we are using to express the behavior.[2]

Nevertheless, we do have a vast supply of facts that we trust as coming more or less directly from experiment. They have the essential quality of the *Uniformity of Nature*; they are the same on different days of the week and in different laboratories, and they are the same for different observers. "Are your results repeatable?" is one of the first questions the research director asks the young enthusiast.

Laws

We try to organize facts into groups and extract common pieces of behavior (e.g., all metals carry electric currents easily; stretch of a spring varies as load). We call the extracted statement or relation a rule, a *law*, occasionally a principle. Thus a law is a generalized record of nature, not a command that compels nature. Some scientists go further and idealize laws—along the lines of the second view of Law in Chapter 5. They take each law as simple and exact; but then they gather a guide-book full of real knowledge telling them just how nature fits the ideal law and within what range, etc. This invisible guide-book is what distinguishes the experienced scientist from the amateur who only knows the formal wording of the laws. It is no handbook of densities and log tables, but a very valuable pocket book of understanding—theory and experimental knowledge combined.

When we are trying to extract a law, we usually restrict our attention to particular aspects of nature. When we are finding Hooke's Law, our spring may be twisting, the loads may be painted different colors, the loads may even be evaporating; but we ignore those distractions. Or our spring may be growing hotter in an overheated laboratory; and then we find the stretch changing less simply. Discovering that temperature affects our measurements we arrange to keep the temperature constant. (This is an important precaution when we investigate gas expansion. Rough experiments suggest it is negligible for investigations with steel springs, but careful measurements show that good spring balances should be "compensated for temperature.")

Most laws in physics state the relationship between measurements of two quantities. For example:

STRESS/STRAIN = constant
PRESSURE $\propto$ 1/VOLUME
FORCE $\propto M_1M_2/d^2$
RADIATION RATE $\propto T^4$

Almost all laws can be reworded with the word "constant" as their essential characteristic:

TOTAL MOMENTUM remains constant in any collision

PRESSURE · VOLUME = constant
$F \cdot d^2/M_1M_2$ = constant

We look for laws because we enjoy codifying these regularities in the behavior of nature.

Concepts

In ordinary discussion, "concept" is a highbrow word for an idea or a general notion. In discussing science we shall give it several meanings.[3]

A. Minor Concepts

(i) *Mathematical Concepts* are useful tool-ideas, such as: the idea of direct proportionality or variation (e.g. STRETCH $\propto$ LOAD); the idea of a limit (e.g. pressure *at a point*; speed as *limit* of $\Delta s/\Delta t$).

[2] Measurements of size by collisions will give different results if the collisions are made much more violent, because the colliding atoms squash to smaller size. And we have to make some theoretical assumptions in extracting a size from our indirect measurements.

[3] This follows an excellent discussion by James B. Conant in a report on "The Growth of the Experimental Sciences" (Harvard, 1949). For more detailed discussion of the "tactics and strategy of science," see the same author's *On Understanding Science*.

(ii) *Name Concepts*: the ideas in some descriptive names that help us to classify and discuss. We may name a group of materials (e.g. *metals*) or a common property (e.g. *elasticity*).

(iii) *Definition Concepts*: the ideas that we invent and define for our own laboratory use. These may be manufactured from simple measurements (e.g. pressure from force and area; resultant of a set of forces; acceleration $= \Delta v/\Delta t$). Or they may describe some arrangement (e.g. constant temperature; equilibrium of a set of forces).

B. *Major Concepts*

(iv) *Scientific Concepts*: useful ideas developed from experiment such as:
- the resultant of a set of vectors treated as things that add geometrically
- heat as something that makes things hotter
- momentum as a useful quantity to keep track of in collisions
- a molecule as a basic particle.

(v) *Conceptual Schemes*: more general scientific ideas that act as cores of thinking, such as:
- heat as a form of molecular motion
- heat as a form of energy
- the Copernican picture of the solar system
- Newton's Laws of Motion
- the picture of the atmosphere as an ocean of air surrounding the Earth

(vi) *Grand Conceptual Schemes* such as:
- The whole Greek system for planetary motions etc.
- Newtonian gravitational theory
- Conservation of Energy
- Conservation of Momentum
- Kinetic Theory of Gases

Speculative Ideas

Most scientific concepts arise from experiment, or are vouched for by experiment to some extent. Other parts of scientific thinking are pure speculation, yet they may be helpful, and they are safe as long as we remember their status. We might label these *speculative ideas.* The crystal spheres were a speculative idea—invisible and quite uncheckable. In fact, the Ptolemaic scheme was not ruined when a comet was found to pass through the spheres: only the spheres were smashed. In examining any conceptual scheme, be careful to sort out its necessary concepts from the speculative ideas that accompanied its birth.

"Theory" and "Hypothesis"

Many scientists would call a grand conceptual scheme a *theory* and they would say that a speculative idea is much the same as a *hypothesis.* Both these words have become vague in general use, and almost confused with each other; so it might be better to avoid using them. However, we shall use them here; and you may profitably use them, distinguishing them as follows:

Hypotheses are single tentative guesses—good hunches—assumed for use in devising theory or planning experiment, intended to be given a direct experimental test when possible.

Theories are schemes of thought with assumptions chosen to fit experimental knowledge, containing the speculative ideas and general treatment that make them grand conceptual schemes.

THE BUILDING OF SCIENTIFIC KNOWLEDGE

Our knowledge of nature is first gained by *induction*, by extracting general rules from experimental data (see Ch. 1). Then, when we trust our rule we assume that nature will do the same thing next time—we bet on the Uniformity of Nature. If you look back on early astronomy (and on your own early laboratory work) you will see that although inductive knowledge is reasonably sure—e.g. planetary paths, Hooke's Law—it is not very fruitful in explanations or predictions. For greater, more fruitful knowledge we turn to *deductive theory.* There we start with assumptions and rules—guessed at, snatched from experiment, modelled by analogy, or invented as speculative ideas—and we make predictions and explanations; we build a sense of knowledge. But, to avoid the mistakes of the ancient philosophers, we must certainly test the predictions that emerge. We should also ask where the rules for the theory's starting point come from.

All through your study of science you should watch for the assumptions that are built into theories and check on their wisdom. Too many assumptions may lead to too much magic. "Words and magic were in the beginning one and the same thing." (Sigmund Freud, *First Lecture.*)

Look back on a simple question about demons.

Demons

Problem 1 in Chapter 7 asked, "How do you know that it is friction that brings a rolling ball to a stop and not demons?" Suppose you answer this, while a neighbor, Faustus, argues for demons. The discussion might run thus:

You. I don't believe in demons.

Faustus. I do.

You. Anyway, I don't see how demons can make friction.

Faustus. They just stand in front of things and push to stop them from moving.

You. I can't see any demons even on the roughest table.

Faustus. They are too small, also transparent.

Y. But there is more friction on rough surfaces.

F. More demons.

Y. Oil helps.

F. Oil drowns demons.

Y. If I polish the table, there is less friction and the ball rolls farther.

F. You are wiping the demons off; there are fewer to push.

Y. A heavier ball experiences more friction.

F. More demons push it; and it crushes their bones more.

Y. If I put a rough brick on the table I can push against friction with more and more force, up to a limit, and the block stays still, with friction just balancing my push.

F. Of course, the demons push just hard enough to stop you moving the brick; but there is a limit to their strength beyond which they collapse.

Y. But when I push hard enough and get the brick moving there is friction that drags the brick as it moves along.

F. Yes, once they have collapsed the demons are crushed by the brick. It is their crackling bones that oppose the sliding.[4]

Y. I cannot feel them.

F. Rub your finger along the table.

Y. Friction follows definite laws. For example, experiment shows that a brick sliding along the table is dragged by friction with a force independent of velocity.

F. Of course, same number of demons to crush, however fast you run over them.

Y. If I slide a brick along the table again and again, the friction is the same each time. Demons would be crushed in the first trip.

F. Yes, but they multiply incredibly fast.

Y. There are other laws of friction: for example, the drag is proportional to the pressure holding the surfaces together.

F. The demons live in the pores of the surface: more pressure makes more of them rush out to push and be crushed. Demons act in just the right way to push and drag with the forces you find in your experiments.

By this time, Faustus' game is clear. Whatever properties you ascribe to friction he will claim, in some form, for demons. At first his demons appear arbitrary and unreliable; but when you produce regular laws of friction he produces a regular sociology of demons. At that point there is a deadlock, with demons and friction serving as alternative names for a set of properties—and each debater is back to his first remark.

You realize that friction has only served you as a name: it has established no link with other properties of matter. Then, as a modern scientist, you start speculating on the molecular or atomic cause of friction, and experimenting to test your ideas. Solids are strong; they hang together. Their component atoms must attract with large forces at short distances. When solid surfaces slide or roll on each other, small humps on one get within the range of atomic attractions of local humps on the other and they drag each other when the motion tries to separate them. Friction, then, may be an atomic dragging, which is likely to make one surface drag small pieces off the other. That has been investigated experimentally. After a copper block has been dragged along a smooth steel table, microphotographs show tiny copper whiskers torn off on to the steel. Also chemical tests show that a little of each metal rubs off on to the other.[5]

At last you have a good case for *friction*: it is a scientific name for some well-ordered behavior that we can now link with other knowledge. It is atomic or molecular dragging, caused by the same forces that make wires strong and raindrops round. Its mechanism can be demonstrated by photographs and by chemical analysis. Its laws can even be predicted by applying our knowledge of elasticity to the small irregularities of surfaces. Friction has joined other phenomena in a general explanation.

[4] If Faustus has the equipment he should offer you a microphone attached to a glass table, with connections to an amplifier and loudspeaker. Then if you roll a steel ball along the table you will indeed hear noises like crushing demons.

[5] We can even show that when a *copper* block rubs on another block of *copper*, tiny pieces of copper—invisibly small—are exchanged between the two blocks. No chemical analysis could tell one block's copper from the other's; so this interchange seems impossible to detect. Yet it is now easy. By the end of the course you will know a good method.

And now we can state the full case against demons: they are arbitrary, unreasonable, multitudinous, and over-dressed. We need a special demon with peculiar behavior to explain each natural event in turn: therefore we need many kinds and vast numbers of them. And we have to clothe them with special behaviors to fit all the facts. We now prefer something more economical and comfortable: a consistent body of knowledge, with strong ties to experiment—and with cross checks and interlinkages to assure us of validity—all expressed in as few general laws as possible. Even where we meet new events that we cannot explain, we would rather speculate cautiously than invent a demon to calm some fear of mystery.

Good Theory

Now we can return to the contrast between plogglies and Universal Gravitation. The plogglies were specialized demons. The author of the fable, a psychologist who offers it for therapeutic purposes, discusses it as an example of prescientific or magical theories that explain the working of nature by unpredictable gods or demons. He states his overall objection to the theory: "The only thing wrong with it was that there aren't any plogglies." There many modern physicists would disagree. They would not mind the plogglies being a fiction (like any "model" in science) but they would call the plogglies bad theory because they are too expensive. The plogglies were invented and endowed with two special behaviors to explain two sets of events, and they do not explain anything else. They are an "*ad hoc*" theory, a theory concocted just "for this purpose." There is nothing wicked about *ad hoc* theories—they may even turn out to be true—but they are weak, usually little more than narrow hypotheses loaded with faith. Labelled merely "*ad hoc* assumptions" they may be useful signposts for honest speculation. And when they lead to explanations of *other* observed behavior we think better of them and may promote them to a respectable title.

Then, as theory grows from a single speculative guess to a general form of knowledge that fits many observed effects we trust it more and more. We are so pleased with its consistency and fruitfulness that we say, "It can't be wrong." Look at Newton's gravitational theory as an example of such a grand conceptual scheme. Newton started with a number of assumptions: vector properties of forces and motions; the behavior summed up in his Laws of Motion; gravitational pulls proportional to inertial masses; inverse-square law; Euclidean geometry. Some of these were extracted from experiment; others were little more than definitions (Law I defining "zero force") and rules of procedure (Law III). But whatever their origin they were stated as starting points for deductive theory. Then step by step with clear reasoning he drew his "explanation" of the solar system from them. We call this good theory because it is economical. Starting from general assumptions, Newton tied together in a single scheme many things that had seemed disconnected:

Moon's circular motion disturbances of Moon's simple motion planetary motions (Kepler's Laws I, II, III) planetary perturbations motion of comets tides bulge of the Earth differences of gravity precession of equinoxes	ALL RELATED by inverse-square-law gravitation and a spinning Earth

Deductive Theory and Scientific Knowledge.

The fanciful picture of Fig. 24-1 shows some of the construction of good theory. Inductive gathering of knowledge must come first. Then, when the time is ripe, the theory may be brewed from a complex mixture of ingredients. At early stages some of the assumptions should be drawn off into preliminary tests (as in Newton's test of inverse-square gravity with the Moon). At a later stage the predictions yielded by the theory should be put to experimental tests—as retrospective checks on the original assumptions. We judge a theory not by how "right" it is but by how helpful, how it suggests experiments or promotes thought. To many a scientist, however, the full value of a great theory is not just in fruitful predictions but in a deep sense of sureness of knowledge that it gives.

In a way the picture shows the making of scientific knowledge rather than just of theory. It is obviously a complex method that will take many forms.

"THE Scientific Method"

Now you can see why we say there is not one single scientific method but many. Francis Bacon ($\sim$ 1600) advocated a formal scientific procedure:

make observations and record the facts
perform many experiments and tabulate the results
extract rules and laws by induction

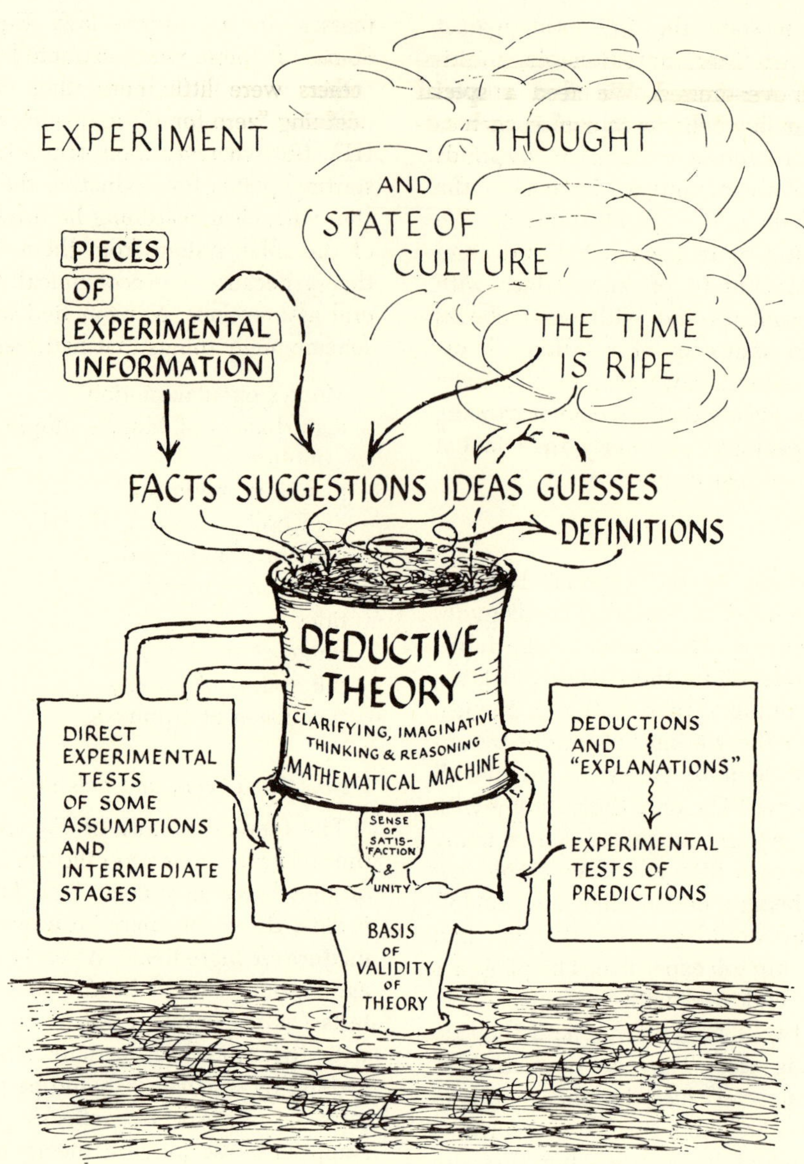

Fig. 24-1.

and this earnest beginning was elaborated into *THE Scientific Method*, advocated by logical enthusiasts to this day:[6]

make observations and extract rules or laws
formulate a tentative hypothesis (a guess, which may be purely speculative)
deduce the consequences of the hypothesis combined with known laws
devise experiments to test those consequences.

IF EXPERIMENT CONFIRMS	IF EXPERIMENT REFUTES
hypothesis, adopt the hypothesis as true law; and then proceed to frame and test more hypotheses.	hypothesis, look for an alternative hypothesis.

Real scientific enquiry is not so "scientifically logical" or so simple. (We do follow some such scheme unconsciously, as James B. Conant points out, when we look for a fault in our car lights, or when we deal with water dripping from the ceiling of our apartment—but we rightly say that our quick process of guessing and looking is plain common sense.)

Scientific Methods

In the real development of science we approach our problems and build our knowledge by many

[6] The same "scientific approach" is advocated by some experts in other fields—e.g. the social sciences. There it offers useful guidance and criticism, but if enforced with blind enthusiasm it will probably restrict progress just as it would in physics. Besides, how do we know that knowledge in other fields can be made to crystallize in the form that fits physics?

methods: sometimes we start by guessing freely; sometimes we build a model for mathematical investigation, and then make experimental tests; sometimes we just gather experimental information, with an eye open for the unexpected; sometimes we plan and perform one great experiment and obtain an important result directly or by statistical sorting of a wealth of measurements. Sometimes a progressive series of experiments carries us from stage to stage of knowledge—the results of each experiment guiding both our reasoning and our planning of the next experiment. Sometimes we carry out a grand analysis thinking from stage to stage with a gorgeous mixture of information, rules, guesses, and logic, with only an occasional experimental test. Yet experiment is the ultimate touchstone throughout good science, whether it comes at the beginning as a gathering of empirical facts or at the end in the final tests of a grand conceptual scheme.

How far scientists' theoretical thinking will develop at a given time depends on the state of knowledge and interest—on whether the time is ripe. When the general climate of opinion is ready for a change of outlook or a new idea, a scientific suggestion may take root where it might have starved a century before; and this control of the advance of understanding by the intellectual and social climate is still true today.

When the time is ripe, the same problem is often attacked by many scientists simultaneously and the same solution may be discovered by several. Yet one scientist may get the credit for reaping the harvest—quite rightly if he is the only man with enough insight or skill to carry the innovation through. In Newton's day new interest in motion, general thinking about the planets, Kepler's discoveries, new studies of magnetism and its forces, ... and new attitudes towards experimenting and scientific knowledge, all made contributions of facts and attitudes and interests—the time was ripe for the great development. Hooke, Wren, Halley, Huygens, and others were all jumping to reach a unified theory for celestial and earthly motions. Each succeeded in grasping some parts of the solution, but it was Newton who gave the complete solution in one grand scheme, making "not a leap but a flight."

Scientific Method: Sense of Certainty

Above all, most scientific knowledge—facts, concepts, conceptual schemes—is built up by a crisscross process of investigation and reasoning from several angles. We do not push straight ahead along one narrow path of brilliant discovery—like a romantic story of making better toothpaste—but we investigate nature first along one line, then along another; and then we make still another guess and test it; and so on. As time goes on, we gather our new concepts from several lines and check them from different viewpoints; and our strong sense of good knowledge is assured by the agreement from different lines of inquiry.

The modern physics of atoms and nuclei is a particularly good example. That region of study is like a great central room with, say, seven closed doors around it. Scientists looked in at one door, and got a glimpse of micro-nature and its mysterious ways. Then through another door: quite a different view. Then through another—and then a comparing of notes. (For example, radioactivity gave one view; electron streams quite another; and the photoelectric effect raised new problems. X-rays gave still another view; and presently some knowledge of X-rays linked up with radioactivity and some with the photoelectric effect, and in still another way X-rays confirmed earlier measurements of atomic sizes.) Finally by checks and comparisons between different views a consistent scheme emerged, a *picture* was formed, to describe micro-nature. We word that picture in everyday words that describe the large "macro" world around us (atoms are *round*, electrons are *small*, X-rays *travel like light*). So it is not a true description—whatever true may mean—but a "model" to enable us to describe our experience of the micro-world in everyday terms. This scheme, our model and its rules—our descriptive theory—is still being modified and extended. If we discover new experimental facts that fit, we enjoy the confirmation. If new facts conflict with predictions from our model, we modify the model—changing as little as possible, from natural conservatism. And if we discover new facts that go beyond the scope of our model we extend it. (When we found that atoms are easily pierced by fast alpha-rays, we stopped saying they are round balls and described them as hollow round balls.)

Our knowledge has grown already to be a comprehensive conceptual scheme that we trust because it fits with our views through many doors. Though we shall make many changes, though we may change our whole scheme of thinking about atomic physics, we already hold much knowledge with a confidence that comes from its consistency with every experimental aspect. To the outside critic seeing us looking through a single door, the evidence seems frail and the wealth of speculation too great. But those who are building the knowledge say, "We are sure we are on sound lines because if we were seriously wrong some inconsistency would show up somewhere, some clash with at least one of our experimental viewpoints—trouble will out."[7]

Building this sense of assurance is the essence of scientific methods. Professor Ernest Nagel has suggested that if there is a single scientific method it lies in the way in which scientists check and countercheck their knowledge by experiment and reasoning from several angles, so that they feel that their knowledge is warranted, that its validity is assured.

THE UNFINISHED PRODUCT—UNDERSTANDING

Models

That is why science seems complicated to learn, and even difficult to trust at first: we gain our knowledge by repeated attacks from different angles and we base our belief on the consistency of that knowledge. We do not necessarily believe that the picture of nature we thus form *is* the real world. Many scientists say it is simply a *model* that works.

It is easy to see that our picture of atomic structure is only a model—the invisible atom described in terms of large visible bullets and baseballs and large forces that we can feel like weights and the attraction of magnets. Yet it is uncomfortable to realize that we do not know what an atom is "really like," and can only say that it "behaves as if. . . ." Moreover, with the progress of invention, microscope . . . electron microscope . . . ion microscope . . . , you may decide that we can see real atoms and not just a model of them—there is a *photograph* of tungsten atoms at the end of this book. Yet all such "seeing" of the micro-world, however clear its results, is quite indirect: the images we obtain must be interpreted in terms of the models that guided our use of apparatus. In casual talk we gladly say, "Now we know what the atoms are really like, how they are arranged and how they move about"; but in serious discussion, most scientists say, "We have only shown that our model serves well, and we have obtained some measurements of parts of our model." In a way, we use models in almost all our scientific thinking: atoms, molecules, gravity, magnetic fields, perfect springs, . . . Our general system of models are what we use to replace plogglies.

Since theory is largely a reasoned model structure, based on some facts, we can always make changes in it. Popular writers describe scientists as gaily throwing away a theory when new discoveries conflict with it; but in fact most scientists cling to their old theory desperately. When scientists do change their theory to fit increasing knowledge, it is more often by progressive modification than by a revolution.

"Crucial Experiments"

Sometimes rival theories lead to different predictions so that a "crucial experiment"[8] can decide be-

---

[7] This is like our assurance of finances. It is easy to see whether the accounts of a small store are correct, but the financial statements of a big corporation are too complex for an amateur to analyze. Yet we are confident that any major financial wrongdoing, however well concealed, would show up in the course of time. After many years of watching a company's accounts without understanding them, most shareholders maintain they are sure the company is sound. We have long had that feeling of warranty for Newton's gravitational theory, and for some general principles such as the Conservation of Energy.

[8] A fine example of a crucial experiment occurred in the history of light. Two hundred years ago there were two views of the nature of light: (A) the bullet theory advocated by Newton, and (B) the wave theory developed by Hooke and Huygens. Both accounted for the general behavior of light-rays, such as reflection and refraction, but refraction also offered a crucial test.

When a slanting beam of light rays hits the surface of a pool of water, the rays are bent to a steeper slant as they enter the water. This bending of light at a boundary is called *refraction* and it has been well known as a property of light for thousands of years. Ptolemy gave an approximate law for the amount of bending, and Willebrod Snell discovered the exact law half a century before Newton wrote on optics. Both the theories of light accounted for refraction and both predicted the exact form of Snell's Law:

(A) The bullets of light must be attracted by water as they approach it (rather like a vapor molecule returning to the liquid surface). So their momentum is changed thus:
 (i) vertical component of momentum is increased (by the action of the attractive force);
 (ii) horizontal component remains unchanged (symmetry). The resultant momentum therefore runs steeper in the water, showing the refraction of the stream of bullets. The geometry predicts Snell's experimental law. With this change of momentum, the bullets must travel *faster* in water than in air.

(B) On wave theory, the advancing lines of crests must be delayed when they meet the water, so that they are slewed around and travel on through the water in a steeper direction. This requires light-waves to travel *slower* in water than in air.

A comparison of speeds of light in water and air would be a crucial experiment to decide between the two theories. It was not until 1850—a century and a half after Newton, Hooke, and Huygens—that Foucault made the crucial experiment. He showed that light moves *slower* in water than in air. That settled the case against bullets—but only against that particular model: bullets that have constant mass and move with increased velocity, momentum, and energy in water. Choose instead bullets that have the same energy in air and water but change their mass on entering water, and you can concoct a theory that predicts Snell's Law and makes the bullets move slower in water. In this case, the escape from failure is easy though the product proves to be unruly; but almost any theory can survive the condemnation of a "crucial" test, at a cost of complicated improvements.

tween them. Even then there is no absolute decision: the defeated theory can usually be pushed and twisted into a form that survives the test—just as demons could always be endowed with extra properties. For example, Newton's demonstration of a guinea and a feather falling freely in a vacuum decides between two theories of ordinary fall:

(A) "All bodies fall with the same acceleration, but for air resistance."
(B) "Bodies have natural downward motions proportional to their weights."

Yet (B) can be polished up into agreement with the experiment by blaming the vacuum pump instead of thanking it: (B) "Bodies . . . weights; but vacuum also exerts downward forces in inverse proportion." (When a barometer is demonstrated, the story will have to grow more fantastic still.)

Only in a few great cases does the decision seem final: as in the choice, for the form in which light carries energy, between smooth waves and compact bullets (quanta). There, the photo-electric effect decides resolutely in favor of quanta; or, again, when experiments with the speed of light support special Relativity, the decision seems certain. Yet even in those great cases it is the weight of several lines of evidence that decides rather than an inescapable proof by a single experiment.

Intellectual Satisfaction

Thus the test of good theory is not success *vs.* failure but remains simplicity and economy *vs.* increasing complexity or clumsiness. The best theory is the one that is most fruitful, economical, comprehensive, and intellectually satisfying.

We expect a theory—or grand conceptual scheme—to be fruitful in predictions and explanations, while keeping its assumptions as few and as general as possible.

Remember that a scientific "explanation" is neither an ultimate "reason why" from some inspired source nor a mere jumble of words describing observed behavior in technical terms. It is a linking of the observed behavior with some other well-known facts or with more general knowledge derived from observation. The greater the number and variety of the facts thus linked together the more satisfied we feel with our theory. As confidence in it increases we "explain" some facts by linking them to speculative guesses in our theory. Yet those guesses in turn are linked to experimental knowledge in our structure of theory; so it is much the same kind of "explaining," now vouched for by our belief in the validity of our theory.

As we build our theoretical treatment, we start with practical assumptions and simple concepts closely related to experiment; then we devise more general concepts to rule the simpler ones; and we aim finally at deducing our whole picture of nature from a few general concepts.

Above all, we value good theory for the sense of intellectual satisfaction it gives us—our feeling of confidence in our knowledge and pleasure in the compact way we can express it. There is an art in choosing theory so that it gives the strongest sense of consistent knowledge, and that is what we mean by understanding nature.

> "If God were to hold out enclosed in His right hand all Truth, and in His left hand just the ever-active search for Truth, though with the condition that I should forever err therein, and should say to me: 'Choose!' I should humbly take His left hand and say: 'Father! Give me this one; absolute Truth is for Thee alone.'"
>
> —G. E. Lessing, *Eine Duplik* (1778)

PART THREE

MOLECULES AND ENERGY

ASTRONOMY . . . MECHANICS . . . MOLECULES AND ENERGY . . . ELECTRICITY . . . ATOMS.

This Part and the next, both spanning a rich century of developing science, together set the stage for modern atomic physics. Two grand conceptual schemes, Kinetic Theory of Gases and Conservation of Energy, provide essential ideas and powerful tools of thought. And the new field of Electricity provides essential knowledge and instruments.

> "The things that any science discovers are beyond the reach of direct observation. We cannot see energy, nor the attraction of gravitation, nor the flying molecules of gases, nor the luminiferous ether, nor the forests of the carbonaceous era, nor the explosions in nerve cells. It is only the premisses of science, not its conclusions, which are directly observed."
>
> C. S. Peirce (1898)
> American Scientist-Philosopher

CHAPTER 25 · THE GREAT MOLECULAR THEORY OF GASES

"This is indeed a mystery," [remarked Watson] "What do you imagine that it means?"
"I have no data yet. It is a capital mistake to theorize before one has data. Insensibly one begins to twist facts to suit theories, instead of theories to suit facts."

—"Sherlock Holmes," A. CONAN DOYLE

Newton's theory of universal gravitation was a world-wide success. His book, the *Principia*, ran into three editions in his lifetime and popular studies of it were the fashion in the courts of Europe. Voltaire wrote an exposition of the *Principia* for the general reader; books were even published on "Newton's Theory expounded to Ladies." Newton's theory impressed educated people not only as a brilliant ordering of celestial Nature but as a model for other grand explanations yet to come. We consider Newton's theory a *good* one because it is simple and productive and links together many different phenomena, giving a general feeling of understanding. The theory is simple because its basic assumptions are a few clear statements. This simplicity is not spoiled by the fact that some of the deductions need difficult mathematics. The success of Newton's planetary theory led to attempts at more theories similarly based on the laws of motion. For example, gases seem simple in behavior. Could not some theory of gases be constructed, to account for Boyle's Law by "predicting" it, and to make other predictions and increase our general understanding?

Such attempts led to a great molecular theory of gases. As in most great inventions the essential discovery is a single idea which seems simple enough once it is thought of: the idea that gas pressure is due to bombardment by tiny moving particles, the "molecules" of gas. Gases have simple common properties. They always fill their container and exert a uniform pressure all over its top, bottom, and sides, unlike solids and liquids. At constant temperature, PRESSURE · VOLUME remains constant, however the gas is compressed or expanded. Heating a gas increases its pressure or volume or both—and the rate of increase with temperature is the same for all gases ("Charles' Law"). Gases move easily, diffuse among each other and seep through porous walls. Could these properties be "explained" in terms of some mechanical picture? Newton's contemporaries revived the Greek philosophers' idea of matter being made of "fiery atoms" in constant motion. Now, with a good system of mechanics they could treat such a picture realistically and ask what "atoms" would do. The most striking general property that a theory should explain was Boyle's Law.

Boyle's Law

In 1661 Boyle announced his discovery, "not without delight and satisfaction" that the pressures and volumes of air are "in reciprocal proportions." That was his way of saying: PRESSURE $\propto$ 1/VOLUME or PRESSURE · VOLUME remains constant, when air is compressed. It was well known that air expands when heated, so the restriction "at constant temperature" was obviously necessary for this simple law. This was Boyle's discovery of the "spring of the air"—a spring of variable strength compared with solid Hooke's Law springs.

In laboratory you should try a "Boyle's-Law experiment" with a sample of dry air, not to "discover" a law that you already know, but as a problem in precision, "your skill against nature." You

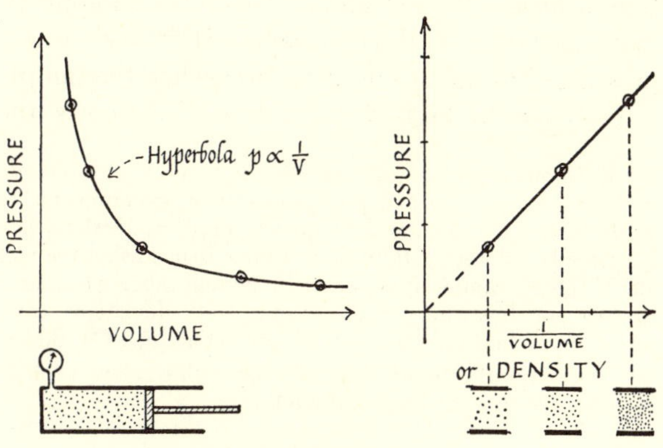

FIG. 25-1. BOYLE'S LAW

will be limited to a small range of pressures (say ½ atmosphere to 2 atm.) and your accuracy may be sabotaged by the room temperature changing or by a slight taper in the glass tube that contains the sample.[1] If you plot your measurements on a graph showing PRESSURE *vs.* VOLUME you will find they mark a hyperbola—but that is too difficult a curve to recognize for sure and claim as verification of Boyle's Law.[2] Then plot PRESSURE *vs.* 1/VOLUME and look for a straight line through the origin.

Boyle's measurements were fairly rough and extended only from a fraction of an atmosphere to about 4 atm. If you make precise measurements with air you will find that pV changes by only a few tenths of 1% at most, over that range. Your graph of p *vs.* $1/V$ will show your experimental points very close to a straight line through the origin. Since MASS/VOLUME is density and MASS is constant, values of $1/V$ represent DENSITY, and Boyle's Law says

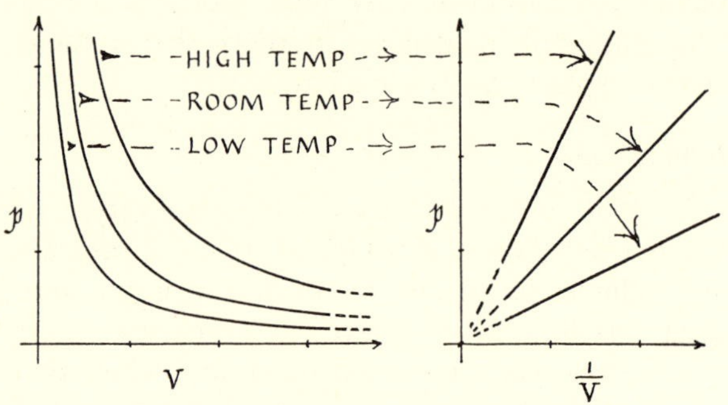

FIG. 25-2. BOYLE'S LAW ISOTHERMALS

PRESSURE ∝ DENSITY. This makes sense on many a simple theory of gas molecules: "put twice as many molecules in a box and you will double the pressure."

All the measurements on a Boyle's-Law graph line are made *at the same temperature*: it is an *isothermal* line. Of course we can draw several isothermals on one diagram, as in Fig. 25-2.

If the range of pressure is increased, larger deviations appear—Boyle's simple law is only an approximate account of real gas behavior. It fits well at low pressures but not at high pressures when the sample is crowded to high density. Fig. 25-3 shows the experimental facts for larger pressures, up to 3000 atmospheres. (For graphs of CO_2's behavior, including liquefaction, see Ch. 30.)

Theory

Boyle tried to guess at a mechanism underlying his experimental law. As a good chemist, he pictured tiny atomic particles as the responsible agents. He suggested that gas particles might be springy, like little balls of curly wool piled together, resisting compression. Newton placed gas particles farther apart, and calculated a law of repulsion-force to account for Boyle's Law. D. Bernoulli published a bombardment theory, without special force-laws, that predicted Boyle's Law. He pointed out that *moving* particles would produce pressure by bombarding the container; and he suggested that heating air must make its particles move faster. This was the real beginning of our present theory. He made a brave attempt, but his account was incomplete. A century later, in the 1840's, Joule and others set forth a successful "kinetic theory of gases," on this simple basic view:

> *A gas consists of small elastic particles in rapid motion: and the pressure on the walls is simply the effect of bombardment.*

Joule showed that this would "explain" Boyle's Law, and that it would yield important information about the gas particles themselves. This was soon polished by mathematicians and physicists into a large, powerful theory, capable of enriching our understanding.

In modern theories, we call the moving particles *molecules*, a name borrowed from chemistry, where it means the smallest particle of a substance that exists freely. Split a molecule and you have separate atoms, which may have quite different properties from the original substance. A molecule of water, H_2O, split into atoms yields two hydrogen atoms and one oxygen atom, quite different from the particles or molecules of water. Left alone, these separated atoms gang up in pairs, H_2, O_2—molecules of hydrogen and oxygen gas. In kinetic theory, we deal with the complete molecules, and assume they are not broken up by collisions. And we assume the molecules exert no forces on each other except during collisions; and then, when they are very close, they exert strong repulsive forces for a very short time: in fact that is all a collision *is*.

You yourself have the necessary tools for constructing a molecular theory of gases. Try it. Assume

[1] Even modern glass tubing is slightly tapered, unless made uniform by an expensive process; so when experiments "to verify Boyle's Law" show deviations from pV = constant they are usually exhibiting tube-taper rather than misbehavior of air. If the air sample is replaced by certain other gases such as CO_2, or by some organic vapor, real deviations from Boyle's Law become obvious and interesting. See Ch. 30.

[2] The only safe shapes of graphs for testing a law, or finding one, are straight lines and circles.

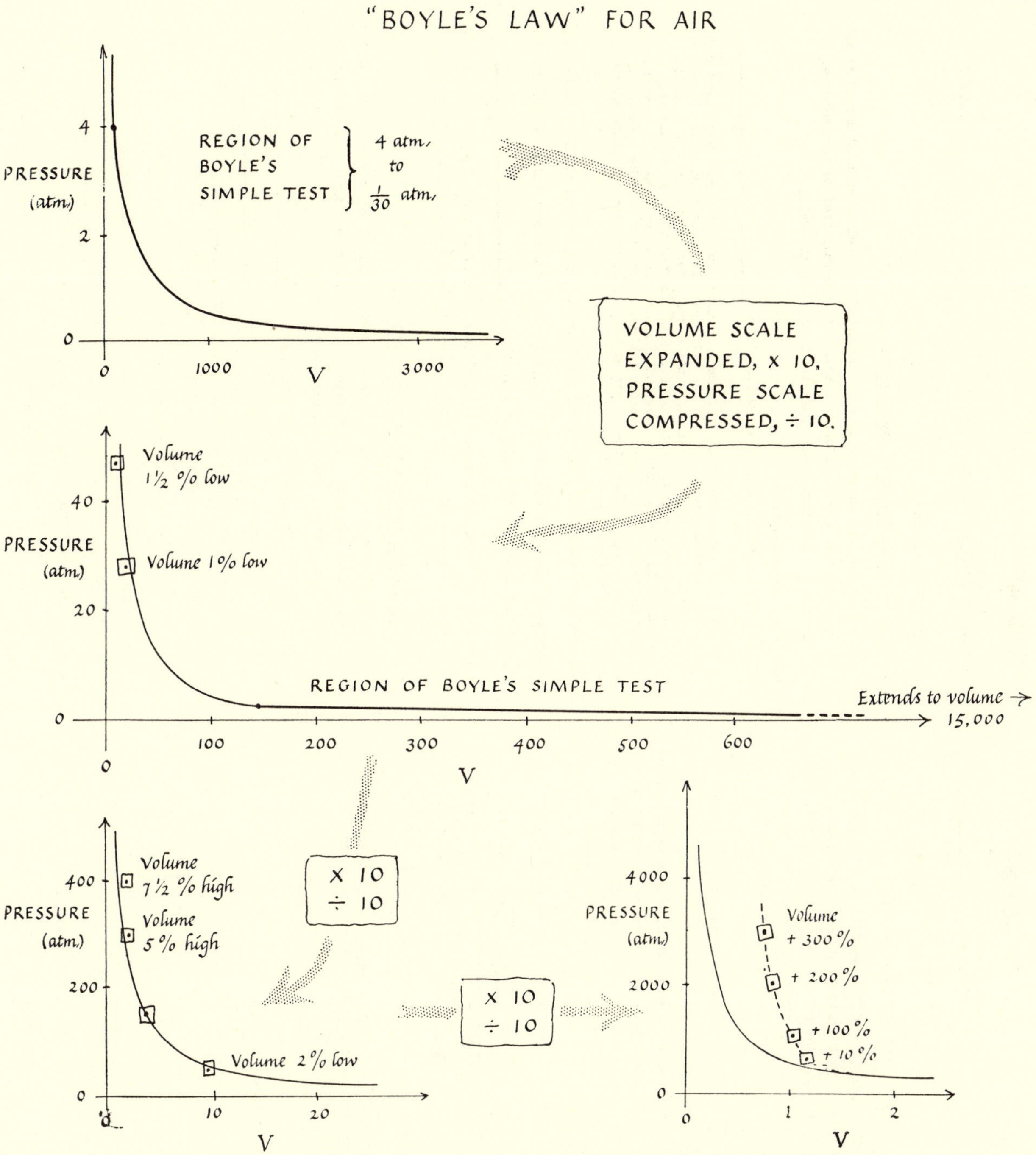

FIG. 25-3. DEVIATIONS FROM BOYLE'S LAW FOR AIR AT ROOM TEMPERATURE
The curve shows the PRESSURE:VOLUME relationship for an ideal gas obeying Boyle's Law.
The points show the behavior of air, indistinguishable from the curve at low pressures.

that gas pressure is due to molecules bouncing elastically on the containing walls. Carry out the first stages by working through Problems 1 and 2. They start with a bouncing ball and graduate to many bouncing molecules, to emerge with a prediction of the behavior of gases. After you have tried the problems, return to the discussion of details.

"PROBLEM SHEETS"

The problems here are intended to be answered on typewritten copies of these sheets. Work through the problems on the enlarged copies, filling in the blanks, (———), that are left for answers.

SPECIAL PROBLEMS ON MOLECULAR THEORY NAME ____________________

Introduction

These problems will help you to build a great molecular theory.

The success of Newton's planetary theory soon led to attempts at building more theories similarly based on Newton's Laws of Motion, again using a few clear assumptions. Theories of gases were tried with gas pictured as a cloud of many tiny molecules bouncing about very fast. Assuming a few simple properties for these molecules (including the basic assumption that they exist!) and assuming that Newton's Laws of Motion applied to molecules, scientists were able to deduce (predict) Boyle's Law and so many other properties of gases that this theory too has come to be regarded as a good one.

As in most theories, having made our assumptions we need to do some calculations to arrive at deductions. To make these calculations easier, a series of problems about bouncing balls is given you below. These lead on to a calculation about molecules which will in fact make valuable predictions. The main calculation may seem hard at first, perhaps just because it is about mysterious molecules, but if you will carry it through once and then leave it you will find it makes good sense the next time.

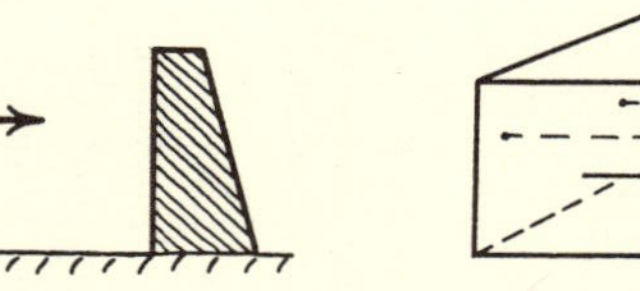

FIG. 25-4.

Box containing balls or molecules which move to-and-fro between front end and other end, making pressure by their impact.

PROBLEM 1.

I(a) EXCHANGE OF MOMENTUM

A ball of mass 2 kilograms moving 12 meters/sec hits a massive wall head-on and stops dead.

(i) The ball's momentum before impact is . . . ________ . ________ units

(ii) The ball's momentum after impact is. . . . ________ . ________ units

(iii) ∴ change of momentum suffered by ball is ________ . ________ units

(iv) If Newton's Law III, which summarizes universally observed behavior of colliding bodies, is correct and applies to this case, we can say that change of momentum suffered by the wall (and whatever it is attached to) must be ________ ________ units

SPECIAL PROBLEMS ON MOLECULAR THEORY SHEET 2

PROBLEM 1. (continued)

I(b) FORCE ON WALL DUE TO STREAM OF BALLS

Now suppose the wall is hit by a stream of such balls, each of mass 2 kg and speed 12 meters/sec. 1000 such balls hit the wall head-on in the course of 10 seconds and stop dead. What is the push on the wall?

The total momentum-change suffered by the wall (in the course of that 10-sec period) is ________ . ________ units

(NOTE: Actually the changes of momentum occur in bumps, one bump when each ball hits the wall, but you can still calculate the total change, and you can then use that to calculate the average force, averaging the sudden bump-forces over the whole 10 seconds. To find the size of the very short-lived bump forces, you would need to know, and use, the very short time taken by each ball to lose its momentum, that is, the duration of a single bump. This time is not given you, so you can calculate only the averaged-out value of the force.)

The average force on the wall, during the 10-sec time, due to all 1000 balls losing momentum, is given by applying $F \cdot t = \Delta(mv)$ to the whole collection of 1000 balls.

∴ Average force, F, on wall must be. . . . . ________ . ________ units

NOTE that since the relation $F \cdot t = \Delta(mv)$ is a form of Newton's Law II, all forces used in it must be in "absolute" units, as in $F = m \cdot a$, e.g., newtons.

II. FORCE ON WALL DUE TO STREAM OF ELASTIC BALLS

As in I(b) suppose that 1000 balls, each of mass 2 kg, hit a massive wall head-on in the course of 10 seconds; but this time they arrive with speed 12 meters/sec and bounce straight back with equal speed, 12 meters/sec.

(i) Each ball's momentum before impact is. . . ________ . ________ units

(ii) Each ball's momentum after impact is . . . ________ . ________ units

(Remember momentum is a vector. Use + and - signs.)

(iii) Change of momentum of one ball is. . . . . ________ . ________ units

(iv) ∴ change of momentum suffered by wall is ________ . ________ units

NOTE: ANSWER IS NOT ZERO

(v) If, during 10 seconds, 1000 balls strike the wall and rebound thus, total change of momentum suffered by wall is ________ . ________ units

(vi) ∴ Average force on wall during the 10-second period is . . ________ . ________ units

(vii) If the 1000 balls hit a patch of wall 2 meters high by 3 meters wide, average PRESSURE (= FORCE/AREA) on that patch is ________ . ________ units

PROBLEM 1. (continued)

III. MOTION INSIDE A BOX. Before changing from bouncing balls to bouncing molecules, we must put the moving things inside a closed box. Suppose we have an oblong box, 4 meters long from end to end, with only one ball in it moving to-and-fro from end to end with speed 12 meters/sec. The ball hits each end head-on and rebounds with speed 12 meters/sec to the other end. Now the same ball will hit the front end of the box many times in ten seconds. Instead of using the number of balls hitting the wall, we must calculate and use the number of hits made by this one ball. To find force on one end, we use the hits on that end only.

(i) Between successive hits on the front end of the box the ball travels one "round trip." It travels the whole length of the box from the front end to the other end and back to the front end. So it travels __________ meters

(ii) With its speed of 12 meters/sec, the total distance the ball travels in 10 seconds is . . . . __________ meters

(iii) ∴ the number of round trips the ball makes in 10 seconds is . . . __________ round trips

(iv) ∴ in 10 seconds the number of hits the ball makes on the front end is . . . __________

(So if we want to have 1000 hits on the front end wall in 10 sec, as in II above, we need more than one ball in the box. In fact we need about ______ balls all moving to-and-fro between the ends.)

(v) At each hit on the front end, the single ball suffers a change of momentum . . . . __________ · __________ units

(vi) ∴ in 10 secs the ball makes ______ hits on front end of box, suffering a momentum-change of __________ · __________ units at each hit.

∴ the total momentum-change suffered by front-end of box in 10 seconds is . . . . __________ · __________ units

∴ average force on front end during the 10-second period is. . . . . . __________ · __________ units

(PRESSURE is given by FORCE/AREA; and if you knew the area of the front end you could calculate the (average) pressure on it caused by the repeated impacts of the ball. In this case of a single large ball the pressure is not a sensible thing to calculate; but you can make the calculation for molecules and thus predict the pressure of a gas.)

PROBLEM 1. (continued)

We now apply a similar calculation to gas molecules in a box. Later we shall repeat it with algebra (Problem 2.)

IV. GAS MOLECULE IN A BOX. A metal box, 4 meters long, with ends 3 meters by 2 meters, contains one gas molecule which moves to-and-fro along the length with speed 500 meters/sec, bouncing back elastically at each end. The molecule approaches one end with speed 500 meters/sec, hits it, and bounces back with speed 500 meters/sec, travels to the other end, hits that and rebounds moving forward 500 meters/sec. The mass of the molecule is approximately 5×10^{-26} kilograms.*

(i) The molecule's change of momentum when it hits the first end and rebounds is. . . . __________ · __________ units

(ii) In 10 seconds the molecule travels total distance. . . __________ meters

(iii) Between successive impacts on the first end, molecule has to travel to other end and back, so travels a distance. . . . . . . __________ meters

(iv) ∴ in 10 seconds, molecule can make ________ trips to-and-fro, and so can make this number of impacts on the first end.

(v) ∴ in 10 seconds, the molecule makes ________ impacts on the first end of the box, suffering a change of momentum at each impact of. . . . . __________ · __________ units

(vi) ∴ total change of momentum suffered by first end-wall of box in 10 seconds is . . . __________ · __________ units

(vii) ∴ average FORCE, during 10 second period, on first end of box is . . . . __________ __________ units

(viii) PRESSURE is FORCE/AREA. The end wall has area 3 x 2 sq. meters.

∴ average PRESSURE on end wall is. . . . . __________ · __________ units

*NOTE: Simple chemical measurements suggest that the oxygen and nitrogen molecules (of air) are roughly 30 times as massive as a hydrogen atom. Difficult physical measurements tell us that a hydrogen atom has mass 1.67×10^{-27} kilograms. So the molecular mass suggested here, 5×10^{-26} kilograms, is a fair value for air.

PROBLEM 1. (continued)

V. MANY GAS MOLECULES IN A BOX

(i) Now suppose that this box contains 6×10^{26} molecules (= 600,000,000,000,000,000,000,000,000). That is roughly the actual number in such a box if filled with air at atmospheric pressure.* In reality these would be moving about in all directions at random; but to simplify the calculation pretend they are sorted out into three regimented groups, one lot moving up-and-down, one lot to-and-fro along the length, and one lot moving to-and-fro across the width. Symmetry considerations suggest we should have the molecules equally divided among the three groups (FIG. 25-5). The pressure on an END of the box will be solely due to impacts of molecules moving to-and-fro along the length. We now proceed to calculate that pressure, assuming there are only one-third of the molecules involved; that is, 2×10^{26} or 200,000,000,000,000,000,000,000,000 molecules, moving 500 meters/sec along the 4-meter length of the box, hitting the end, rebounding 500 meters/sec, hitting the other end, rebounding, and so on.

Using result of IV above, we predict that:

average pressure on end of box will be . . . . . . ____________ ____________ units

(The data are roughly right for ordinary air in a room. What value for atmospheric pressure, in the same units, is given by direct measurement with a barometer?) . . . . ____________ ____________ units

How does your result calculated above for the molecules compare with atmospheric pressure measured in lab? . . . . . . ________________________

(ii) Now suppose the box is gently squashed end-ways, so that its length is reduced to 2 meters (i.e., half the original length) without changing the number of molecules or their speeds, or the size of the end-wall.

Average pressure on end of box will be . . . ____________ ____________ units

NOTE: There is very little change in the arithmetic from (i) to (ii). Check through to get the new answer.

(iii) Comment on the answer to V(ii).

VI. OPTIONAL. On a separate sheet, repeat the calculation of average pressure, V(i), with algebra. Take a box of length a meters, width b meters, and height c meters, containing a TOTAL of N molecules moving with average speed v meters/sec. (i) Calculate pressure p. (ii) Calculate the product (PRESSURE) x (VOLUME), by multiplying p by abc.

*NOTE: Chemists often deal with a "mole" or "gram-molecule" of gas. A mole of any gas occupies 22.4 liters at atmospheric pressure and the temperature of melting ice. At room temperature (and one atmosphere) a mole occupies about 24 liters. Here we have chosen 1000 moles, a "kilo-mole" or "kilogram-molecule," which would occupy about 24,000 liters or 24 cu. meters at room temperature (about 20 °C.)

BOX CONTAINING 6×10^{26} MOLECULES IN RANDOM MOTION

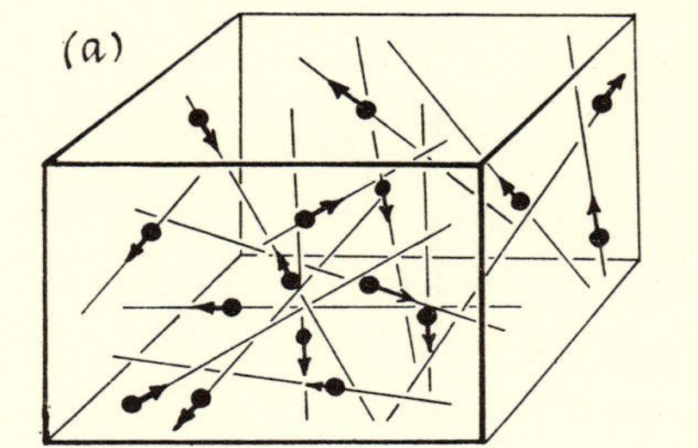

Instead of random directions of motion, as in (a), pretend there are three regimented groups as in (b), each group of 2×10^{26} having full velocity but moving parallel to one edge of the box.

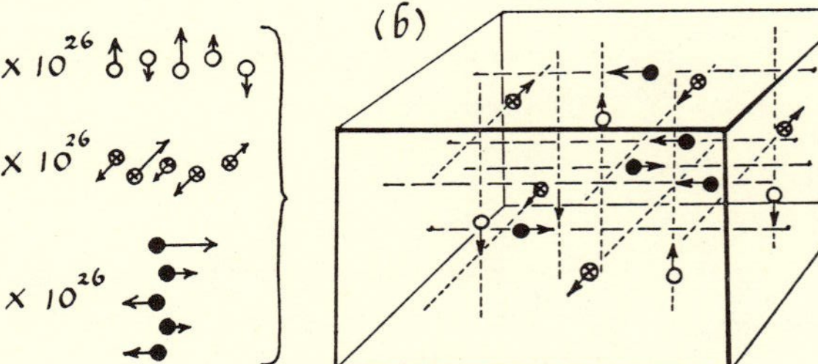

Then assume that the pressure on end face is due to impacts by one group, 2×10^{26} molecules, moving to-and-fro parallel to the length of the box, as in (c).

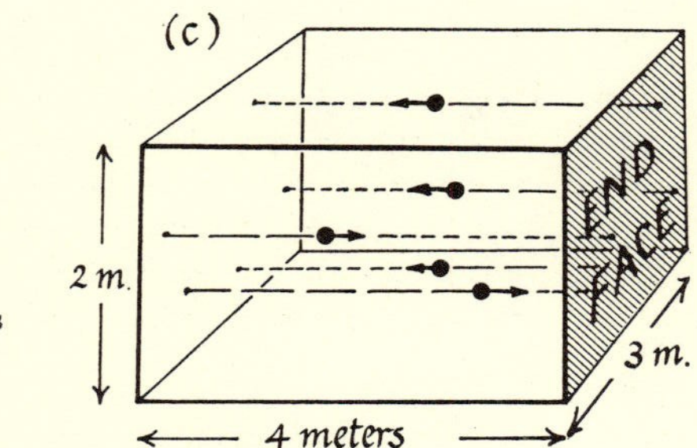

FIG.25-5.

SPECIAL PROBLEMS ON MOLECULAR THEORY NAME ____________

PROBLEM 2. KINETIC THEORY WITH ALGEBRA SHEET 6

(This treats many gas molecules in a box with algebra. It should be tried after Problem 1 has been answered and corrected.)

Suppose that the box contains N molecules (N in the whole box, not N molecules in each cubic meter as in some texts). Suppose that the box has length a meters and ends of dimensions b meters by c meters.
In the course of their random motion with many collisions the molecules will exchange momentum and will not all keep the same velocity. However, if the temperature is kept constant, we believe their velocities will range around a fixed average velocity, which we call v meters/sec. To calculate the pressure on one end of the box we deal only with molecular impacts on that end. So to simplify the problem we pretend that the N molecules are regimented in three equal groups, one lot moving up-and-down, one lot to-and-fro across the width, and one lot moving forwards-and-backwards along the length. For the pressure on one end we then consider the last lot only. Symmetry-considerations suggest we should imagine the molecules equally divided among the three groups. Making these assumptions, answer the questions below, using m kilograms for the mass of one molecule.

(i) When one molecule hits the front end head-on and rebounds, its change of momentum is ____________

(ii) Between successive impacts on the front end a molecule travels to the other end and back: a total distance ____________ meters

(iii) In a total time t seconds, a molecule moving with velocity v meters/sec travels a total distance ____________ meters

(iv) ∴ in t seconds, a molecule can make ____________ round trips and so can make this number of impacts on front end.

(v) ∴ in t seconds, a molecule makes ____________ impacts on front end of box, suffering at each impact a change of momentum ____________

(vi) ∴ total change of momentum, due to impacts of one molecule, suffered by front end in t seconds is ____________

(vii) But there are N molecules in the box, of which ____________ are in the group moving forward and backward between the ends.
∴ the total change of momentum, due to impacts of all molecules concerned, suffered by front end in t seconds is ____________

SPECIAL PROBLEMS ON MOLECULAR THEORY SHEET 7

PROBLEM 2. (continued)

(viii) But, F·t = Δ(MOMENTUM) ∴ F = Δ(MOMENTUM)/t,
and in this case the average FORCE, during this period of t seconds, on the front end of the box is* . . . . ____________

(ix) PRESSURE = FORCE/AREA, and the area of the end face is ____________
∴ average PRESSURE on end of box is . . . . ____________

(x) The volume of the box is . . . . ____________ cu. meters
∴ the product (PRESSURE)·(VOLUME) = ____________
But m is the mass of one molecule, and there are N molecules, so the total mass of gas in the box, M kilograms = ____________ kilograms
Substituting M into the algebra above, we have
(PRESSURE)·(VOLUME) = ____________

(xi) Providing we use a closed box or other apparatus allowing no leakage of gas, then M is constant. Suppose we keep temperature constant; then other experiments in physics suggest that the average velocity v remains constant. Then in this case when the volume is changed, the result of (x) above suggests that ____________

(xii) If we measure the volume of a sample of gas, say in a globe, and find its mass (by weighing the globe full of gas and then evacuated), and measure the pressure of the sample with a barometer, then the result of (x) above enables us to calculate a very important piece of information, the value of ____________, which is the ____________ of the molecules.

(xiii) UNITS TO BE USED. In making the calculation of (xii) above, if the volume is in cu. meters the mass should be in . . . . ____________
and the pressure should be in. . . . . ____________

(xiv) We have already derived two useful things from our molecular theory, a behavior-suggestion in (xi) and a very interesting measurement in (xii), and more results will emerge; but we must pay for them by the assumptions that go into the machine. List on a separate sheet as many assumptions as you can,
(a) of general physical laws assumed to apply to molecules,
(b) of special properties, of behavior, size, etc., assumed for molecules.

*Here, t is the time during which the average force would have to act to produce this momentum-change. Therefore t IS the time t seconds for which we have calculated the total momentum-change.

Difficulties of the Simple Theory

The relation you worked out in Problem 2 seems to predict a steady pressure and Boyle's-Law behavior, from molecular chaos. How can a rain of molecules hitting a wall make a steady pressure? Only if the collisions come in such rapid succession that their bumps seem to smooth out into a constant force. For that the

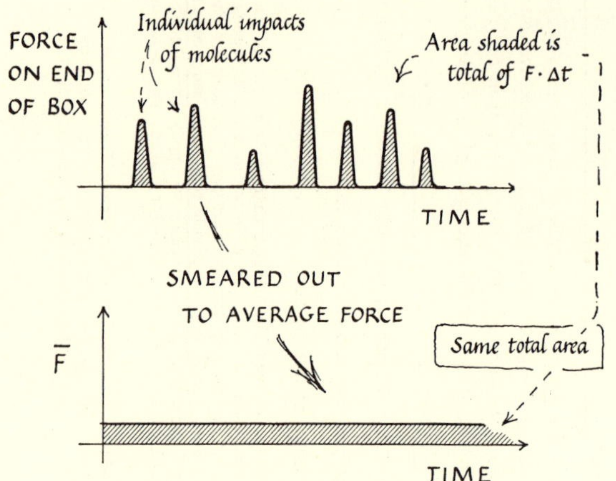

FIG. 25-6. SMOOTHING OUT IMPACTS

molecules of a gas must be exceedingly numerous, and very small. If they are small any solid pressure-gauge or container wall will be enormously massive compared with a single gas molecule, so that, as impacts bring it momentum, it will smooth them out to the steady pressure we observe. (What would you expect if the container wall were as light as a few molecules?)

The problem pretended that molecules travel straight from end to end and never collide with each other en route. They certainly do collide—though we cannot say how often without further information. How will that affect the prediction?

★ PROBLEM 3. COLLISIONS IN SIMPLE THEORY

(a) Show that it does not matter, in the simple derivation of Problems 1 and 2, whether molecules collide or not. (Consider two molecules moving to and fro from end to end, just missing each other as they cross. Then suppose they collide head-on and rebound. Why will their contribution to the pressure be unchanged? Explain with a diagram.)

(b) What special assumption about molecules is required for (a)?

(c) Suppose the molecules swelled up and became very bulky (but kept the same speed, mass, etc.), would the effect of mutual collisions be an increase of pressure (for the same volume etc.) or a decrease or what? (*Note:* "bulky" means large in size, not necessarily large in mass.)

(d) Give a clear reason for your answer to (c).

Molecular Chaos

Molecules hitting each other, and the walls, at random—some head on, some obliquely, some glancing—cannot all keep the same speed v. One will gain in a collision, and another lose, so that the gas is a chaos of molecules with random motions whose speeds (changing at every collision) cover a wide range. Yet they must preserve some constancy, because a gas exerts a steady pressure.

In the prediction $p \cdot V = (1/3)[N\, m\, \overline{v^2}]$, we do not have all N molecules moving with the same speed, each contributing $m\,\overline{v^2}$ inside the brackets. Instead we have molecule #1 with its speed v_1, molecule #2 with $v_2, \ldots,$ molecule N with speed v_N. Then

$$\begin{aligned} p \cdot V &= (1/3)\ [m\,v_1^2 + m\,v_2^2 + \ldots + mv_N^2] \\ &= (1/3)\ [m\ (v_1^2 + v_2^2 + \ldots + v_N^2)\ ] \\ &= (1/3)\ [m\ (N \cdot \text{AVERAGE}\ v^2)\ ] \quad \text{See note 3.} \end{aligned}$$

The v^2 in our prediction must therefore be an average v^2, so that we write a bar over it to show it is an average value. Our theoretical prediction now runs:

$$\text{PRESSURE} \cdot \text{VOLUME} = 1/3\, N \cdot m \cdot \overline{v^2}.$$

We know that if we keep a gas in a closed bottle its pressure does not jump up and down as time goes on; its pressure and volume stay constant. Therefore in spite of all the changes in collisions, the molecular $\overline{v^2}$ stays constant. Already our theory helps us to picture some order—constant $\overline{v^2}$—among molecular chaos.

A More Elegant Derivation

To most scientists the regimentation that leads to the factor 1/3 is too artificial a trick. Here is a more elegant method that treats the molecules' random velocities honestly with simple statistics. Suppose molecule #1 is moving in a slanting direction in the box, with velocity v_1. (See Fig. 25-7.) Resolve this vector v_1 into three

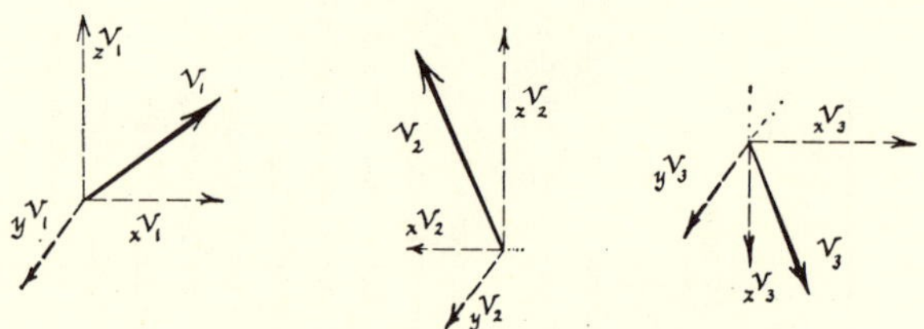

FIG. 25-7. ALTERNATIVE TREATMENT OF GAS MOLECULE MOTION
(More professional, less artificial.)
In this we keep the random velocities, avoiding regimentation, but split each velocity v into three components, $_xv$, $_yv$, $_zv$, parallel to the sides of the box. Then we deal with $_xv^2$ in calculating the pressure and arrive at the same result. Sketches show three molecules with velocities split into components.

components along directions x, y, z, parallel to the edges of the box. Then v_1 is the resultant of $_xv_1$ along x and $_yv_1$ along y and $_zv_1$ along z; and since these are mutually perpendicular, we have, by the three-dimensional form

[3] Because AVERAGE v^2 = (sum of all the v^2 values)/(number of v^2 values) = $(v_1^2 + v_2^2 + \ldots + v_N^2)/(N)$
$\therefore (v_1^2 + v_2^2 + \ldots + v_N^2) = N \cdot$ (AVERAGE v^2) or $N \cdot \overline{v^2}$
This $\overline{v_2}$ is called the "mean square velocity." To obtain it, take the speed of each molecule, at an instant, square it, add all the squares, and divide by the number of molecules. Or, choose one molecule and average its v^2 over a long time—say a billion collisions.

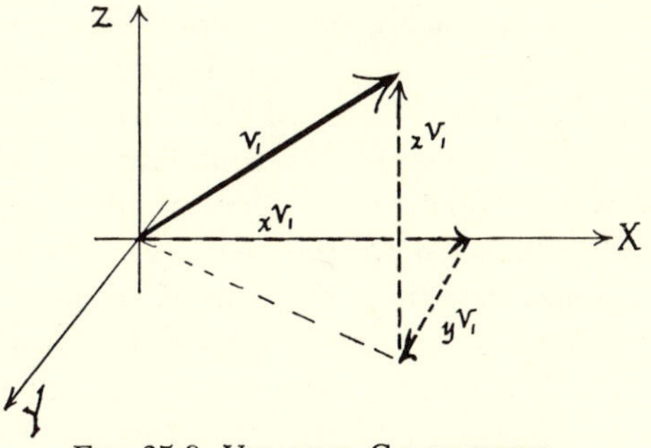

FIG. 25-8. VELOCITY COMPONENTS
PYTHAGORAS: $v_1^2 = {}_xv_1^2 + {}_yv_1^2 + {}_zv_1^2$

of Pythagoras' theorem: $v_1^2 = {}_xv_1^2 + {}_yv_1^2 + {}_zv_1^2$
And for molecule #2 $v_2^2 = {}_xv_2^2 + {}_yv_2^2 + {}_zv_2^2$
And for molecule #3 $v_3^2 = {}_xv_3^2 + {}_yv_3^2 + {}_zv_3^2$
and so on . . . .
And for molecule #N $v_N^2 = {}_xv_N^2 + {}_yv_N^2 + {}_zv_N^2$

Add all these equations:

$$\begin{aligned}(v_1^2 + v_2^2 + v_3^2 + \ldots + v_N^2) \\ = ({}_xv_1^2 + {}_xv_2^2 + {}_xv_3^2 + \ldots + {}_xv_N^2) \\ + ({}_yv_1^2 + {}_yv_2^2 + {}_yv_3^2 + \ldots + {}_yv_N^2) \\ + ({}_zv_1^2 + {}_zv_2^2 + {}_zv_3^2 + \ldots + {}_zv_N^2)\end{aligned}$$

Divide by the number of molecules, N, to get *average* values:

$$\therefore \quad \overline{v^2} = \overline{{}_xv^2} + \overline{{}_yv^2} + \overline{{}_zv^2}$$

Appealing to symmetry, and ignoring the small bias given by gravity, we claim that the three averages on the right are equal—the random motions of a statistically large number of molecules should have the same distribution of velocities in any direction.

$$\therefore \quad \overline{{}_xv^2} = \overline{{}_yv^2} = \overline{{}_zv^2}$$

$$\therefore \quad \overline{v^2} = 3\overline{{}_zv^2}$$

To predict the pressure on the end of the box we proceed as in Problem 2, but we use v_x for a molecule's velocity along the length of the box. (That is the velocity we need, because ${}_yv$ and ${}_zv$ do not help the motion from end to end and are not involved in the change of momentum at each end.) Then the contribution of molecule #1 to PRESSURE · VOLUME is $m \cdot {}_xv_1^2$ and the contribution of all N molecules is

$$m\,({}_xv_1^2 + {}_xv_2^2 + \ldots + {}_xv_N^2) \text{ or } m \cdot N \cdot \overline{{}_xv^2};$$

and by the argument above this is $m \cdot N \cdot (\overline{v^2}/3)$

$$\therefore \quad \text{PRESSURE} \cdot \text{VOLUME} = (1/3)\,N \cdot m \cdot \overline{v^2}$$

(If you adopt this derivation, you should carry through the algebra of number of hits in t secs, etc., as in Problem 2.)

Molecular Theory's Predictions

Thinking about molecular collisions and using Newton's Laws gave the $(1/3)\,N \cdot m \cdot \overline{v^2}$ prediction:

$$\text{PRESSURE} \cdot \text{VOLUME} = (1/3)\,N \cdot m \cdot \overline{v^2}$$

This looks like a prediction of Boyle's Law. The fraction (1/3) is a constant number; N, the number of molecules, is constant, unless they leak out or split up; m, the mass of a molecule, is constant. Then *if* the average speed remains unchanged, $(1/3)\,N \cdot m \cdot \overline{v^2}$ remains constant and therefore $p \cdot V$ should remain constant, as Boyle found it does. But *does* the speed of molecules remain fixed? At this stage, you have no guarantee. For the moment, anticipate later discussion and *assume* that molecular motion is connected with the heat-content of a gas, and that *at constant temperature gas molecules keep a constant average speed*, the same speed however much the gas is compressed or rarefied.[4] Later you will receive clear reasons for believing this. If you accept it now, you have predicted that:

The product $p \cdot V$ is constant for a gas at constant temperature.

You can see the prediction in simplest form by considering changes of DENSITY instead of VOLUME: just put twice as many molecules in the same box, and the pressure will be doubled.

A marvelous prediction of Boyle's Law? Hardly marvelous: we had to pour in many assumptions—with a careful eye on the desired result, we could scarcely help choosing wisely. A theory that gathers assumptions and predicts only one already-known law—and that under a further assumption regarding temperature—would not be worth keeping. But our new theory is just beginning: it is also helpful in "explaining" evaporation, diffusion, gas friction; it predicts effects of sudden compression; it makes vacuum-pumps easier to design and understand. And it leads to measurements that give validity to its own assumptions. Before discussing the development, we ask a basic question, "Are there really any such things as molecules?"

Are there really molecules?

> "That's the worst of circumstantial evidence. The prosecuting attorney has at his command all the facilities of organized investigation. He uncovers facts. He selects only those which, in his opinion, are significant. Once he's come to the conclusion the defendant is guilty, the only facts he considers significant are those which point to the guilt of the defendant. That's why circumstantial evidence is such a liar. Facts themselves are meaningless. It's only the interpretation we give those facts which counts."
>
> "Perry Mason"—Erle Stanley Gardner*

[4] Actually, compressing a gas warms it, but we believe that when it cools back to its original temperature its molecules, though still crowded close, return to the same average speed as before compression.

* *The Case of the Perjured Parrot*, Copyright 1939, by Erle Stanley Gardner.

A century ago, molecules seemed useful: a helpful concept that made the regularities of chemical combinations easy to understand and provided a good start for a simple theory of gases. But did they really exist? There was only circumstantial evidence that made the idea plausible. Many scientists were skeptical, and at least one great chemist maintained his right to disbelieve in molecules and atoms even until the beginning of this century. Yet one piece of experimental evidence appeared quite early, about 1827: the Brownian motion.

The Brownian Motion

The Scottish botanist Robert Brown (1773-1858) made an amazing discovery: he practically saw molecular motion. Looking through his microscope at small specks of solid suspended in water, he saw them dancing with an incessant jigging motion. The microscopic dance made the specks look alive, but it never stopped day after day. Heating made the dance more furious, but on cooling it returned to its original scale. We now know that any solid specks in any fluid will show such a dance, the smaller the speck the faster the dance, a random motion with no rhyme or reason. Brown was in fact watching the effects of water molecules jostling the solid specks. The specks were being pushed around like an elephant in the midst of a football game.

Watch this "Brownian motion" for yourself. Look at small specks of soot in water ("India ink") with a high-magnification microscope. More easily, look at smoke in air with a low-power microscope. Fill a small black box with smoke from a cigarette or a dying match, and illuminate it with strong white light from the side. The smoke scatters bluish-white light in all directions, some of it upward into the microscope. The microscope shows the smoke as a crowd of tiny specks of white ash which dance about with an entirely irregular motion.[5] (See FIG. 30-3 for an example)

Watching the ash specks, you can see why Brown at first thought he saw living things moving, but you can well imagine the motion to be due to chance bombardment by air molecules. Nowadays we not only think it may be that; we are sure it is, because we can calculate the effects of such bombardment and check them with observation. If air molecules were infinitely small and infinitely numerous, they would bombard a big speck symmetrically from all sides and there would be no Brownian motion to see. At the other extreme, if there were only a few very big molecules of surrounding air, the ash speck would make great violent jumps when it did get hit. From what we see, we infer something between these extremes; there must be many molecules in the box, hitting the ash speck from all sides, many times a second. In a short time, many hundreds of molecules hit the ash speck from every direction; and occasionally a few hundreds more hit one side of it than the other and drive it noticeably in one direction. A big jump is rare, but several tiny random motions in the same general direction may pile up into a visible shift.[6] Detailed watching and calculation from later knowledge tell us that what we see under the microscope are those gross resultant shifts; but, though the individual movements are too small to see, we can still estimate their speed by cataloguing the gross staggers and analysing them statistically.

You can see for yourself that smaller specks dance faster. Now carry out an imaginary extrapolation to smaller and smaller specks. Then what motion would you expect to see with specks as small as molecules if you could see them? But can we see molecules?

Seeing molecules?

Could we actually see a molecule? That would indeed be convincing—we feel sure that what we *see* is real, despite many an optical illusion. All through the last century's questioning of molecules, scientists agreed that seeing one is hopeless—not just unlikely but impossible, for a sound physical reason. Seeing uses light, which consists of waves of very short wavelength, only a few thousand Ångström Units[7] from crest to crest. We see by using these waves to form an image:

with the naked eye we can see the shape of a pin's head, a millimeter across, or 10,000,000 ÅU

with a magnifying glass we examine a fine hair, 1,000,000 ÅU thick

with a low-power microscope we see a speck of smoke ash, 100,000 ÅU

with a high-power microscope, we see bacteria, from 10,000 down to 1000 ÅU

but there the sequence stops. It must stop because the wavelength of visible light sets a limit there. Waves can make clear patterns of obstacles that are larger

[5] There may also be general drifting motions—convection currents—but these are easily distinguished. An ash speck in focus shows as a small sharp wisp of white, often oblong; but when it drifts or dances away out of focus the microscope shows it as a fuzzy round blob, just as camera pictures show distant street lights out of focus.

[6] Imagine an observer with poor sight tracing the motion of an active guest at a crowded party. He might fail to see the guest's detailed motion of small steps here and there, and yet after a while he would notice that the guest had wandered a considerable distance.

[7] 1 Ångström Unit, 1 ÅU, is 10^{-10} meter.

than their wavelength, or even about their wavelength in size. For example, ocean waves sweeping past an island show a clear "shadow" of calm beyond. But waves treat smaller obstacles quite differently. Ocean waves meeting a small wooden post show no calm behind. They just lollop around the post and join up beyond it as if there were no post there. A blind man paddling along a stormy seashore could infer the presence of an island nearby, but would never know about a small post just offshore from him.[8] Light waves range in wavelength from 7000 ÅU for red to 4000 for violet. An excursion into the short-wave ultraviolet, with photographic film instead of an eye, is brought to a stop by absorption before wavelength 1000 ÅU: lenses, specimen, even the air itself, are "black" for extreme ultraviolet light. X-rays, with shorter wavelength still, can pass through matter and show grey shadows, but they practically cannot be focused by lenses. So, although X-rays have the much shorter wavelength that could pry into much finer structures, they give us only unmagnified shadow pictures. Therefore the limit imposed by light's wavelength seemed impassable. Bacteria down to 1000 ÅU could be seen, but virus particles, ten times smaller, must remain invisible. And molecules, ten times smaller still, must be far beyond hope. Yet viruses, responsible for many diseases, are of intense medical interest—we now think they may mark the borderline between living organisms and plain chemical molecules. And many basic questions of chemistry might be answered by seeing molecules.

The invisibility of molecules was unwelcome, but seemed inescapable. Then, early in this century, X-rays offered indirect information. The well-ordered atoms and molecules of crystals can scatter X-rays into regular patterns, just as woven cloth can "diffract" light into regular patterns—look at a distant lamp at night through a fine handkerchief or an umbrella. X-ray patterns revealed both the arrangement of atoms in crystals and the spacing of their layers. Such measurements confirmed the oil-film estimates of molecular size. More recently, these X-ray diffraction-splash pictures have sketched the general *shape* of some big molecules—really only details of crystal structure, but still a good hint of molecular shape. Then when physicists still cried "no hope" the electron microscope was invented. Streams of electrons, instead of light-waves, pass through the tiny object under examination, and are focused by electric or magnetic fields to form a greatly magnified image on a photographic film. Electrons are incomparably smaller agents than light-waves,[9] so small that even "molecules" can be delineated. Then we can "see" virus particles and even big molecules in what seem to be reliable photographs with huge magnifications. These new glimpses of molecular structure agree well with the speculative pictures drawn by chemists arguing very cleverly from chemical behavior.

Recently, still sharper methods have been developed. At the end of this book you will see a picture of the individual atoms of metal in a needle point. Why not show that now? Because, like so much in atomic physics, the method needs a sophisticated knowledge of assumptions as well as techniques before you can decide in what sense the photograph tells the truth. Going still deeper, very-high-energy electrons are now being used to probe the structure of atomic nuclei, yielding indirect shadow pictures of them.

In the last 100 years, molecules have graduated from being tiny uncounted agents in a speculative theory to being so real that we even expect to "see" their shape. Most of the things we know about them—speed, number, mass, size—were obtained a century ago with the help of kinetic theory. *The theory promoted the measurements, then the measurements gave validity to the theory.* We shall now leave dreams of seeing molecules, and study what we can measure by simple experiments.

Measuring the Speed of Molecules

Returning to our prediction that:

$$\text{PRESSURE} \cdot \text{VOLUME} = (1/3)\, N \cdot m \cdot \overline{v^2}$$

We can use this if we trust it, to estimate the actual speed of the molecules. N is the number of molecules and m is the mass of one molecule so *Nm is the total mass M of all the molecules* in the box of gas. Then we can rewrite our prediction:

$$\text{PRESSURE} \cdot \text{VOLUME} = (1/3) \cdot M \cdot \overline{v^2}$$

where M is the total mass of gas. We can weigh a big sample of gas with measured volume at known pressure and substitute our measurements in the relation above to find the value of $\overline{v^2}$ and thus the value of the average speed.

Fig. 25-9 shows the necessary measurements. Using the ordinary air of the room, we measure its pressure by a mercury barometer. (Barometer height and the measured density of mercury and the measured value of the Earth's gravitational field strength, 9.8 newtons per kilogram, will give the pressure in absolute units, newtons per square meter.)[10] We weigh the air which fills a flask. For this, we weigh the flask first full of air at atmospheric pressure and second after a vacuum pump has taken out nearly all the air. Then we open the flask under water and let water enter to replace the air pumped

[8] Tiny obstacles do produce a small scattered ripple, but this tells nothing about their shape. Bluish light scattered by very fine smoke simply indicates there are very tiny specks there, but does not say whether they are round or sharp-pointed or oblong. The still more bluish light of the sky is sunlight scattered by air molecules.

[9] Electrons speeding through the electron microscope behave as if they too have a wavelength, but far shorter than the wavelength of light. So they offer new possibilities of "vision," whether you regard them as minute bullets smaller than atoms, or as ultra-short wave patterns. A technology of "electron optics" has developed, with "lenses" for electron microscopes and for television tubes (which are electron projection-lanterns).

[10] Since we made our kinetic theory prediction with the help of Newton's Law II, the predicted *force* must be in absolute units, newtons; and the predicted *pressure* must be in newtons per square meter.

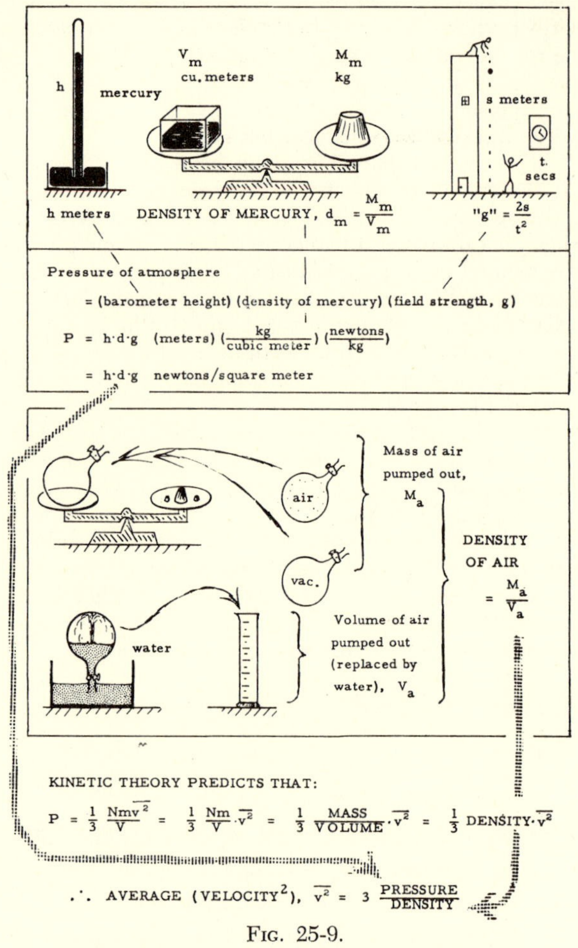

FIG. 25-9.
MEASURING MOLECULE VELOCITIES INDIRECTLY, BUT SIMPLY, ASSUMING KINETIC THEORY.

out. Measuring the volume of water that enters the flask tells us the volume of air which has a known mass. Inserting these measurements in the predicted relation we calculate $\overline{v^2}$ and thence its square root $\sqrt{(\overline{v^2})}$ which we may call the "average speed," $\bar{v}$ (or more strictly the "root mean square," or R.M.S. speed). You should see these measurements made and calculate the velocity, as in the following problem.

★ PROBLEM 4. SPEED OF OXYGEN MOLECULES

Experiment shows that 32 kg of oxygen occupy 24 cubic meters at atmospheric pressure, at room temperature.
(a) Calculate the density, *MASS/VOLUME*, of oxygen.
(b) Using the relation given by kinetic theory, calculate the *mean square velocity*, $\overline{v^2}$, of the molecules.
(c) Take the square root and find an *"average" velocity*, in meters/sec.
(d) Also express this very roughly in miles/hour. (Take 1 kilometer to be 5/8 mile)

Air molecules moving ¼ mile a second! Here is theory being fruitful and validating its own assumption, as theory should. We assumed that gases consist of molecules that are moving, probably moving fast; and our theory now tells us how fast, with the help of simple gross measurements. Yet theory cannot prove its own prediction is true—the result can only be true to the assumptions that went in. So we need experimental tests. If the theory passes one or two tests, we may trust its further predictions.

Speed of Molecules: experimental evidence

We have rough hints from the speed of sound and from the Brownian motion.

PROBLEM 5. SPEED OF SOUND

We believe that sound is carried by waves of compression and rarefaction, with the changes of crowding and motion handed on from molecule to molecule at collisions. If air does consist of moving molecules far apart, what can you say about molecular speed, given that the measured speed of sound in air is 340 meters/sec ($\approx$ 1100 ft/sec)?

PROBLEM 6. BROWNIAN MOTION

Looking at smoke under a microscope you will see large specks of ash jigging quite fast; small specks jig faster still.
(a) There may be specks too small to see. What motion would you expect them to have?
(b) Regarding a single air molecule as an even smaller "ash speck," what can you state about its motion?

The two problems above merely suggest general guesses. Here is a demonstration that shows that gas molecules move very fast. Liquid bromine is released at the bottom of a tall glass tube.* The

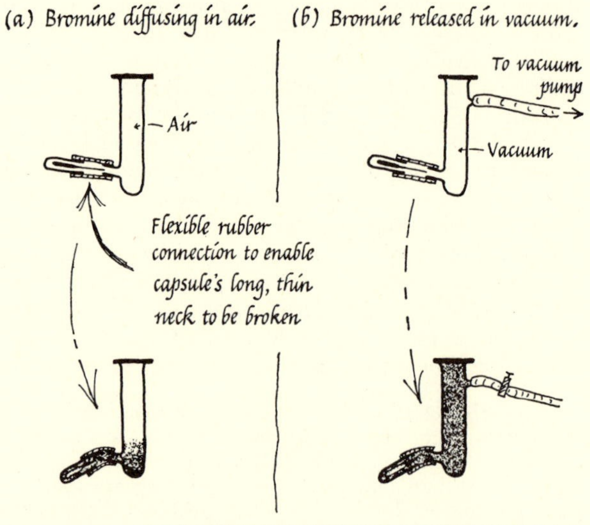

FIG. 25-10. MOTION OF BROMINE MOLECULES: DEMONSTRATION OF MOLECULAR SPEED.

* The bromine is inserted as liquid bromine in a small glass capsule with a long nose that can be broken easily.

liquid evaporates immediately to a brown vapor or "gas," which slowly spreads throughout the tube. The experiment is repeated in a tube from which all air has been pumped out. Now the brown gas moves very fast when released. (In air, its molecules still move fast, but their net progress is slow because of many collisions with air molecules.)

Direct Measurement

The real test must be a direct measurement. Molecular speeds have been measured by several experimenters. Here is a typical experiment, done by Zartman. He let a stream of molecules shoot through a slit in the side of a cylindrical drum that could be spun rapidly. The molecules were of bismuth metal, boiled off molten liquid in a tiny oven in a vacuum. A series of barriers with slits selected a narrow stream to hit the drum. Then each time the slit in the drum came around, it admitted a small flock of moving molecules. With the drum at rest, the molecules travelled across to the opposite wall inside the drum and made a mark on a receiving film opposite the slit. With the drum spinning, the film was carried around an appreciable distance while the molecules were travelling across to it, and the mark on it was shifted to a new position. The molecules' velocity could be calculated from the shift of the mark and the drum's diameter and spin-speed. When the recording film was taken out of the drum it showed a sharp central mark of deposited metal but the mark made while it spun was smeared out into a blur showing that the molecular velocities had not all been the same but were spread over a considerable range. Gas molecules have random motion with frequent collisions and we must expect to find a great variety of velocities at any instant. It is the *average* velocity, or rather the root-mean-square average, $\sqrt{(\overline{v^2})}$, that is involved in kinetic theory prediction. The probable distribution of velocities, clustering round that average, can be predicted by extending simple kinetic theory with the help of the mathematical statistics of chance. In Zartman's experiment, we expect the beam of hot vapor molecules to have the same chance distribution of velocities with its peak at an average value characteristic of the temperature. Measurements of the actual darkening of the recording film showed just such a distribution and gave an average that

ZARTMAN'S EXPERIMENT

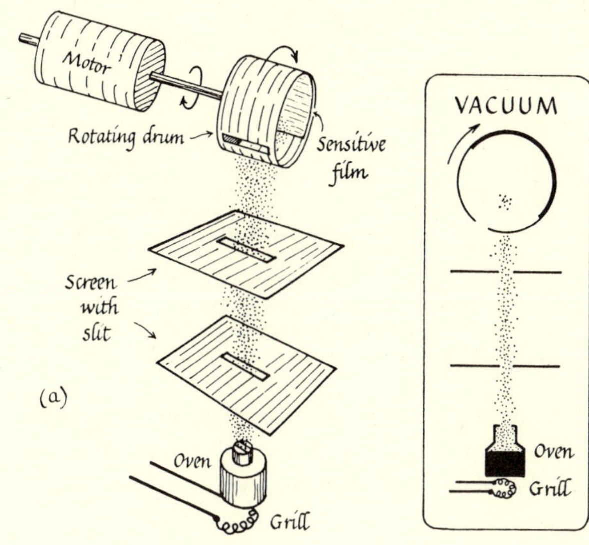

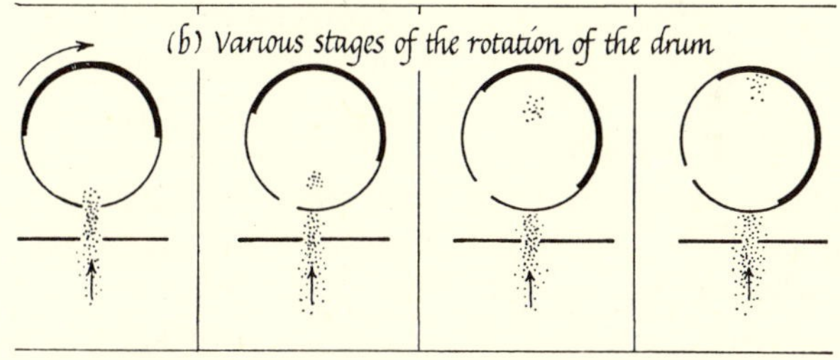

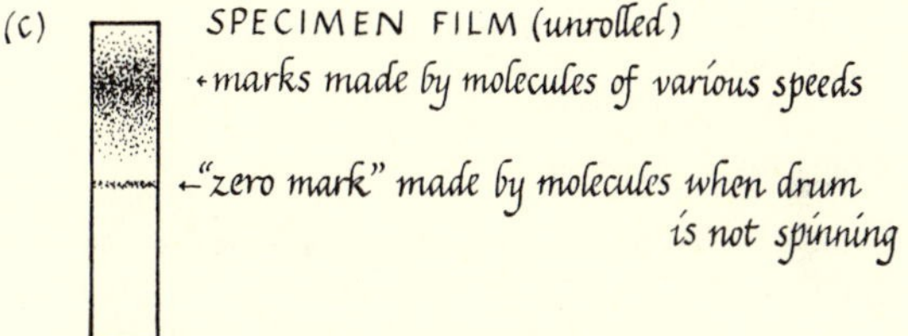

FIG. 25-11. MEASURING MOLECULE VELOCITIES DIRECTLY
(a) Sketch of Zartman's experiment.
(b) These sketches show various stages of the rotation of the drum.
(c) Specimen film (unrolled).

agreed well with the value predicted by simple theory (see sketch of graph in Fig. 25-12).[11]

Molecular Speeds in Other Gases. Diffusion

Weighing a bottle of hydrogen or helium at atmospheric pressure and room temperature shows these gases are much less dense than air; and carbon dioxide is much more dense. Then our predic-

[11] Zartman's method is not limited to this measurement. One method of separating uranium 235 used spinning slits, though the uranium atoms were electrically charged and were given high speeds by electric fields. And mechanical "chopper" systems are used to sort out moving neutrons. Such choppers operate like traffic lights set for some constant speed. The simplest prototype of Zartman's experiment is the scheme shown in Fig. 8-8 for measuring the speed of a rifle bullet.

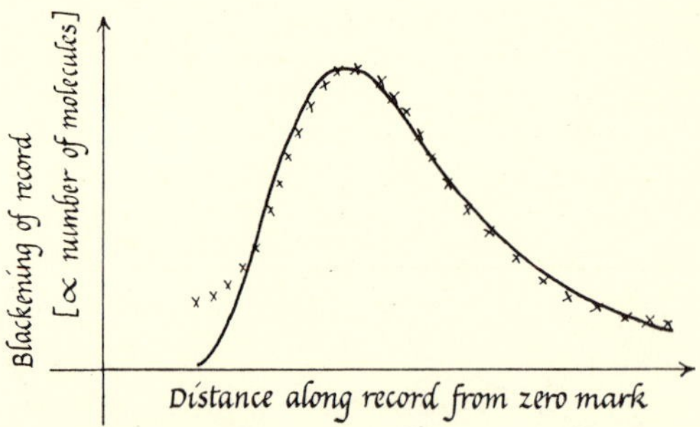

FIG. 25-12. RESULTS OF ZARTMAN'S EXPERIMENT
The curve, drawn by a grayness-measuring-machine, shows the experimental results. The crosses show values predicted by kinetic theory with simple statistics.

tion $pV = (1/3)\,M\,\overline{v^2}$ tells us that hydrogen and helium molecules move faster than air molecules (at the same temperature), and carbon dioxide molecules slower. Here are actual values:

Gas	Measurements at Room Temperature and Atmospheric Pressure	
	Volume	*Mass*
hydrogen	24 cu. meters	2.0 kilograms
helium	24 " "	4.0 kg
carbon dioxide	24 " "	44.0 kg
oxygen	24 " "	32.0 kg
nitrogen	24 " "	28.0 kg
air (1/5 oxygen 4/5 nitrogen)	24 " "	28.8 kg

★ PROBLEM 7. SPEEDS

(i) If oxygen molecules move about ¼ mile/sec at room temperature, how fast do hydrogen molecules move?
(ii) How does the average speed of helium molecules compare with that of hydrogen molecules at the same temperature? (Give the ratio of "average" speeds.)
(iii) How does the speed of carbon dioxide molecules compare with that of air molecules at the same temperature? (Give the ratio of "average" speeds.)

PROBLEM 8

Making a risky guess,* say whether you would expect the speed of sound in helium to be the same as in air, or bigger or smaller. Test your guess by blowing an organ pipe first with air, then with helium (or with carbon dioxide). Or breathe in helium and then talk, using your mouth and nose cavities as miniature echoing organ pipes. A change in the speed of sound changes the time taken by sound waves to bounce up and down the pipe, and thus changes the frequency at which sound pulses emerge from the mouth. And that changes the musical note of the vowel sounds, which rises to higher pitch at higher frequency.

* It is obviously risky, since we are not considering the mechanism of sound transmission in detail. In fact there is an unexpected factor, which is different for helium: the ease with which the gas heats up as sound-compressions pass through. This momentary rise of temperature makes sound compressions travel faster. The effect is more pronounced in helium than in air, making the speed of sound 8% bigger than simple comparison with air suggests. Kinetic theory can predict this effect of specific heat, telling us that helium must have a smaller heat capacity, for a good atomic-molecular reason.

PROBLEM 9

How would you expect the speed of sound in air to change when the pressure is changed without any change of temperature? (Try this question with the following data, for air at room temperature: 28.8 kg of air occupy 24 cubic meters at 1 atmosphere pressure; at 2 atmospheres they occupy 12 cubic meters.)

Diffusion

If molecules of different gases have such different speeds, one gas should outstrip another when they diffuse through long narrow pipes. The pipes must be very long and very narrow so that gas seeps through by the wandering of individual molecules and not in a wholesale rush. The pores of unglazed pottery make suitable "pipes" for this. See Fig. 25-13a, b. The white jar J has fine pores that run right through its walls. If it is filled with compressed gas and closed with a stopper S, the gas will slowly leak out through the pores into the atmosphere, as you would expect. But if the pressure is the same (atmospheric) inside and out you would not expect any leakage even if there are different gases inside and outside. Yet there are changes, showing the effects of different molecular speeds. The demonstrations sketched start with air inside the jar and another gas, also at atmospheric pressure, outside. You see the effects of hydrogen molecules whizzing into the jar faster than air can move out; or of air moving out faster than CO_2 molecules crawl in. These are just qualitative demonstrations of "diffusion," but they suggest a process for separating mixed gases. Put a mixture of hydrogen and CO_2 inside the jar; then, whether there is air or vacuum outside, the hydrogen will diffuse out faster than the CO_2, and by repeating the process in several stages

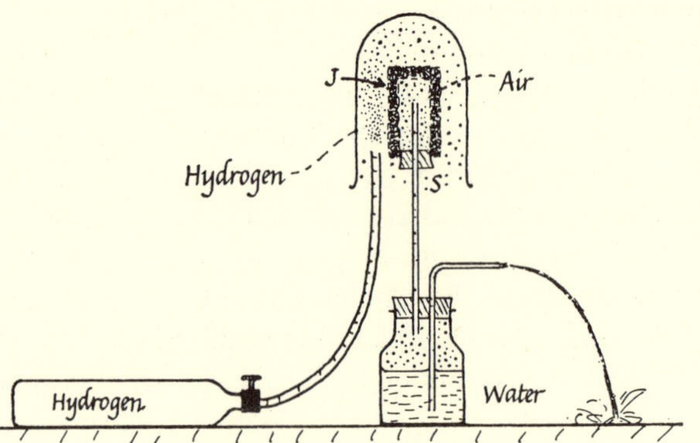

FIG. 25-13a. DIFFUSION OF GASES
Hydrogen diffuses in through the porous wall J faster than air diffuses out.

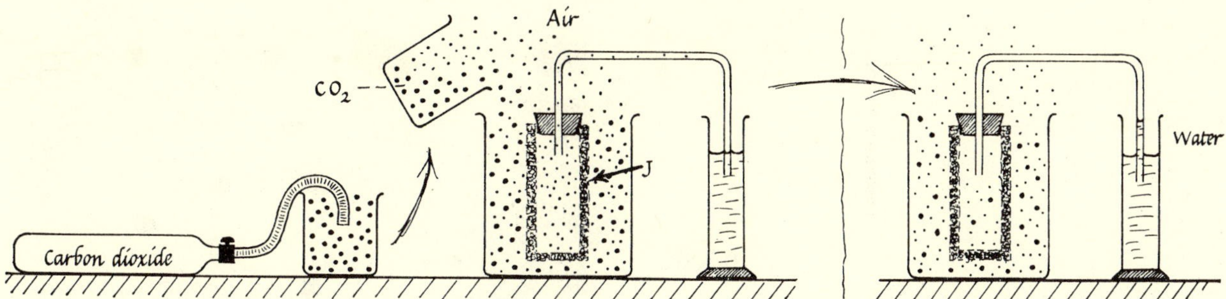

Fig. 25-13b. Diffusion of Gases
Carbon dioxide diffuses in through the porous wall, J, slower than air diffuses out.

you could obtain almost pure hydrogen. This is a *physical* method of separation depending on a difference of molecular speeds that goes with a difference of molecular masses (see Fig. 25-14). It does not require a difference of *chemical* properties; so it can be used to separate "isotopes," those twin-brothers that are chemically identical but differ slightly in atomic masses. When isotopes were first discovered, one neon gas 10% denser than the other, some atoms of lead heavier than the rest, they were interesting curiosities, worth trying to separate just to show. Diffusion of the natural neon mixture from the atmosphere proved the possibility. But now with two uranium isotopes hopelessly mixed as they come from the mines, one easily fissionable, the other not, the separation of the rare fissionable kind is a matter of prime importance. Gas diffusion is now used for this on an enormous scale. See Problem 11, and Figs. 25-15, 16 and 17. Also see Chs. 30 and 43.

Temperature

Heating a gas increases p or V or both. With a rise of temperature there is always an increase of pV, and therefore of $(\frac{1}{3})\, N\, m\, \overline{v^2}$. Therefore making a gas hotter increases v^2, makes its molecules move faster. This suggests some effects of temperature.

★ PROBLEM 10

(a) Would you expect the speed of sound to be greater, less, or the same in air at higher temperature? Explain.
(b) Would you expect diffusion of gases to proceed faster, slower, or at the same rate, at higher temperature? Explain.

Kinetic Theory To Be Continued

We cannot give more precise answers to such questions until we know more about heat and temperature and energy. Then we can extract more predictions concerning gas friction, heat conduction, specific heats; and we shall find a way of

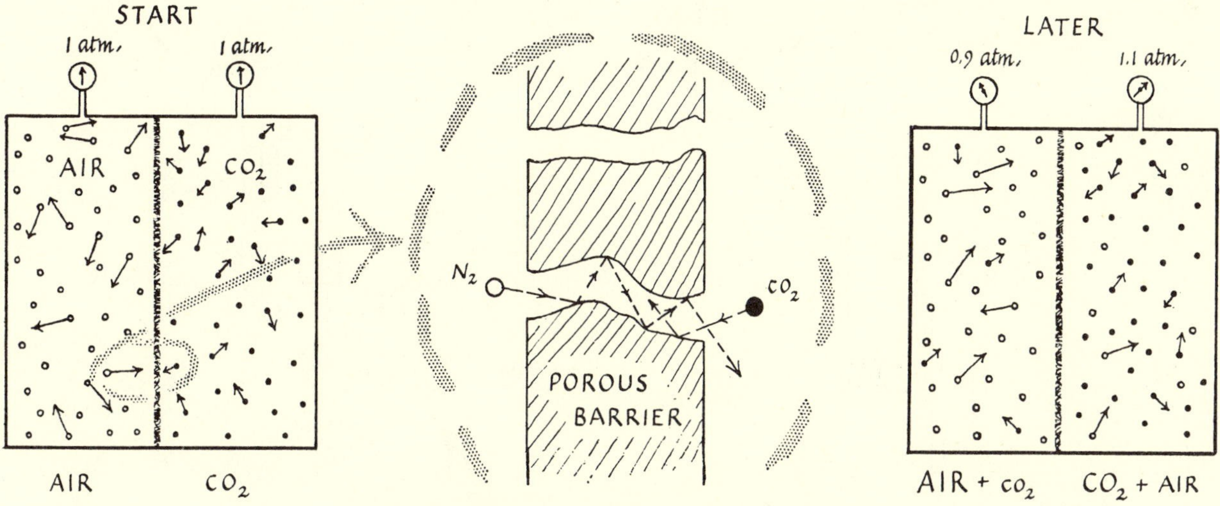

Fig. 25-14. Unequal Diffusion of Gases
Air and carbon dioxide, each originally at atmospheric pressure, are separated by a porous barrier.
At the start, with equal volumes at the same pressure, the two populations have equal numbers of molecules.
On the average, air molecules stagger through the pores faster than CO_2 molecules.
Then the populations are no longer equal so the pressures are unequal.

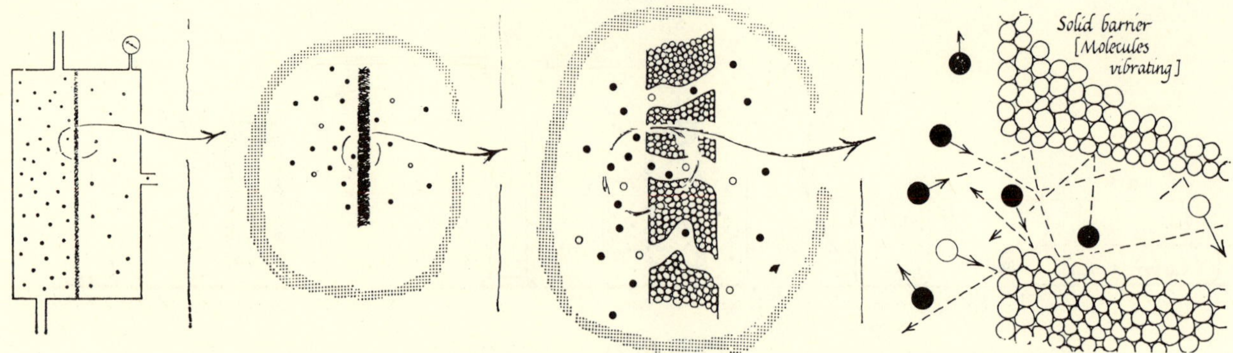

FIG. 25-15. SEPARATION OF URANIUM ISOTOPES BY DIFFUSION OF UF_6 THROUGH A POROUS BARRIER
Gas molecules hit the barrier, and the walls of its pores, many times—net result: a few get through.

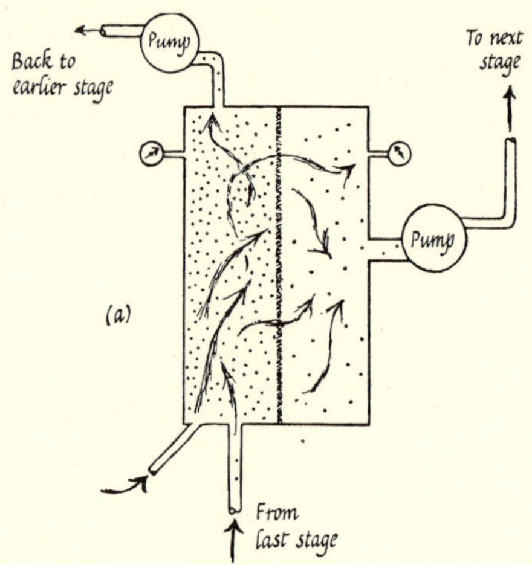

FIG. 25-16a. SEPARATION OF URANIUM ISOTOPES BY DIFFUSION OF UF_6 THROUGH A POROUS BARRIER.

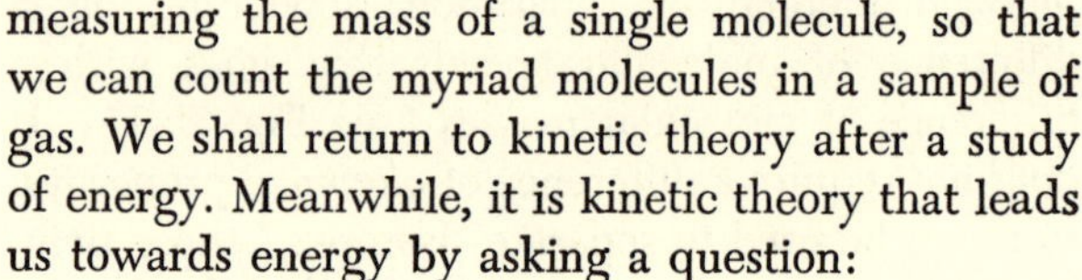

measuring the mass of a single molecule, so that we can count the myriad molecules in a sample of gas. We shall return to kinetic theory after a study of energy. Meanwhile, it is kinetic theory that leads us towards energy by asking a question:

What is mv^2?

The expression $(\frac{1}{3})\,N\,m\,v^2$ is very important in the study of all gases. Apart from the fraction $(\frac{1}{3})$ it is

THE NUMBER OF MOLECULES · (mv^2 for one molecule)
What is mv^2 for a moving molecule? It *is* just the mass multiplied by the square of the speed; but what kind of thing does it measure? What are its properties? Is it an important member of the series: $m \quad mv \quad mv^2 \quad \ldots\ldots$? We know m, mass, and treat

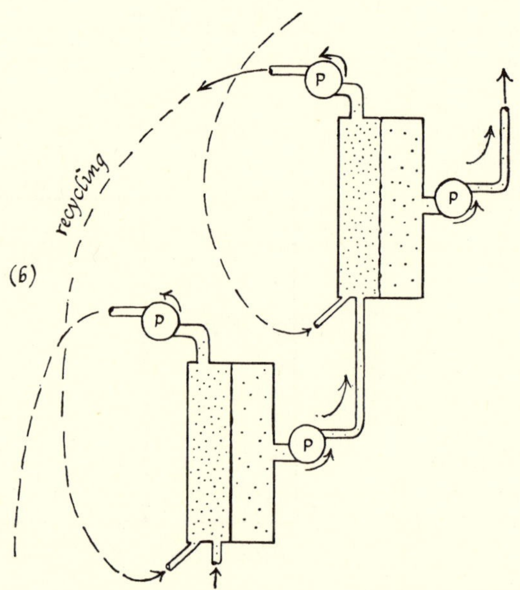

FIG. 25-16b. MULTI-STAGE DIFFUSION SEPARATION
Mixture diffusing through in one stage is pumped to the input of the next stage. Unused mixture from one stage is recycled, pumped back to the input of the preceding stage.

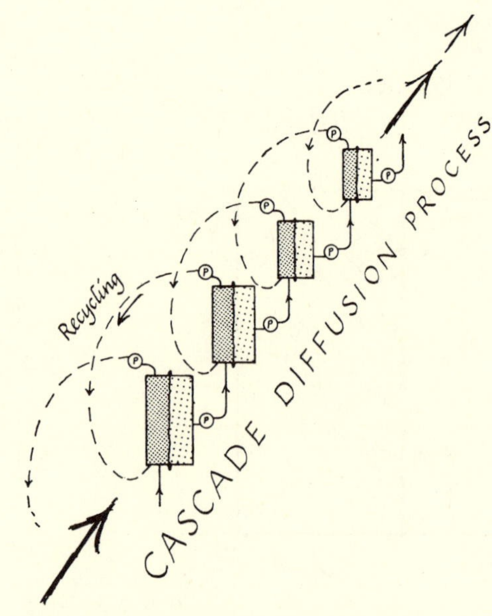

FIG. 25-17. SEPARATING URANIUM ISOTOPES BY DIFFUSION
To effect a fairly complete separation of $U^{235}\,F_6$, thousands of stages are needed.

it as a constant thing whose total is universally conserved. We know mv, momentum, and trust it as a vector that is universally conserved. Is mv^2 equally useful? Its structure is $mv \cdot v$ or $Ft \cdot v$ or

FORCE · TIME · DISTANCE/TIME.

Then mv^2 is of the form FORCE · DISTANCE. Is that product useful? To push with a force along some distance needs an engine that uses fuel. Fuel . . . money . . . energy. We shall find that mv^2 which appears in our theory of gases needs only a constant factor (½) to make it an expression of "energy."

PROBLEMS FOR CHAPTER 25

★ 1. DERIVING MOLECULAR PRESSURE

Work through the question sheets of Problem 1 shown earlier in this chapter. These lead up to the use of Newton's mechanics in a molecular picture of gases.

★ 2. KINETIC THEORY WITH ALGEBRA

Work through the question sheets of Problem 2.

Problems 3-10 are in the text of this chapter.

★ 11. URANIUM SEPARATION (For more professional version, see Problem 3 in Ch. 30)

Chemical experiments and arguments show that oxygen molecules contain two atoms so we write them O_2; hydrogen molecules have two atoms, written H_2; and the dense vapor of uranium flouride has structure UF_6.

Chemical experiments tell us that the *relative* masses of single atoms of O, H, F, and U are 16, 1, 19, 238. Chemical evidence and a brilliant guess (Avogadro's) led to the belief that a standard volume of any gas at one atmosphere and room temperature contains *the same number of molecules* whatever the gas (the same for O_2, H_2, or UF_6). Kinetic theory endorses this guess strongly (see Ch. 30).

(a) Looking back to your calculations in Problem 7 you will see that changing from O_2 to H_2 changes the mass of a molecule in the proportion 32 to 2. For the same temperature what change would you expect in the $\overline{v^2}$ and therefore what change in the average velocity? (That is, how fast are hydrogen molecules moving at room temperature compared with oxygen ones? Give a ratio showing the *proportion* of the new speed to the old. Note you do not have to repeat all the arithmetic, just consider the one factor that changes.)

(b) Repeat (a) for the change from oxygen to uranium fluoride vapor. Do *rough* arithmetic to find approximate numerical value.

(c) Actually there are several kinds of uranium atom. The common one has mass 238 (relative to oxygen 16) but a rare one (0.7% of the mixture got from rocks) which is in fact the one that undergoes fission, has mass 235. One of the (very slow) ways of separating this valuable rare uranium from the common one is by converting the mixture to fluoride and letting the fluoride vapor diffuse through a porous wall. Because the fluoride of U^{235} has a different molecular speed the mixture emerging after diffusing through has different proportions.
   (i) Does it become richer or poorer in U^{235}?
   (ii) Give reasons for your answer to (i).
   (iii) Estimate the percentage difference between average speeds of $[U^{235}_1F_6]$ and $[U^{238}_1F_6]$ molecules.

(*Note*: As discussed in Ch. 11, a change of x % in some measured quantity Q makes a change of about $\frac{1}{2}$ x % in $\sqrt{Q}$.)

12. Figs. 25-13a and 25-13b show two diffusion demonstrations. Describe what happens and interpret the experiments.

★ 13. MOLECULAR VIEW OF COMPRESSING GAS

(a) When an elastic ball hits a massive wall head-on it rebounds with much the same *speed* as its original speed. The same happens when a ball hits a massive bat which is held firmly. However, *if the bat is moving towards the ball*, the ball rebounds with a different speed. Does it move faster or slower?

(b) (Optional, hard: requires careful thought.) When the bat is moving towards the ball is the time of the elastic impact longer, shorter, or the same as when the bat is stationary? (*Hint*: If elastic . . . . S.H.M. . . .)

(c) When a gas in a cylinder is suddenly compressed by the pushing in of a piston, its temperature rises. Guess at an explanation of this in terms of the kinetic theory of gases, with the help of (a) above.

(d) Suppose a compressed gas, as in (c), is allowed to push a piston out, and expand. What would you expect to observe?

★ 14. MOLECULAR SIZE AND TRAVEL

A closed box contains a large number of gas molecules at fixed temperature. Suppose the molecules magically became more bulky by swelling up to greater volume, without any increase in number or speed, without any change of *mass*, and without any change in the volume of the box.

(a) How would this affect the *average distance apart* of the molecules, center to center (great increase, decrease, or little change)?

(b) Give a reason for your answer to (a).

(c) How would this affect the *average distance travelled by a molecule between one collision and the next* (the "mean free path")?

(d) Give a reason for your answer to (c).

CHAPTER 26 · ENERGY

"It's love that makes the world go round."
—Ancient ditty

"Energy makes the World go round. ENERGY explains EVERYTHING."
—Modern ditty to be found in general science books

Energy and Fuel

It is easy to say that Energy explains everything in physics, chemistry, . . . perhaps biology. It is also practically meaningless and in some cases downright misleading. To make good and honest use of the idea of energy, you need to know how that idea has been built up, and you need to understand what energy means. Then you can use it as a great concept in scientific thinking.

In this course, we have avoided discussing energy till we could view it from several aspects. We shall now start with a very earthy, and rather vague, description of energy as *something we pay for, the product of fuel.* As we proceed to fuller descriptions, the meaning and measurement and use of energy will become definite.

We live in a fuel-bound civilization. We cannot do without fuel in civilized life—in any life, for that matter, if we count food as a fuel. Coal for steam, oil for motors, hay for horses, and food for man all must be used and paid for. Fuels are used to do jobs; and we must buy an amount of fuel proportional to the amount of job to be done. The modern man who says, "I do not have to buy fuel, I use electricity from the wall," is not escaping the cost of fuel. He pays the power company, and they pay for coal to run their generators. True, we may use sunshine to heat our house, or a waterfall to run a generator, and then we do not pay money for the "free" fuel. Yet there is still a definite accounting. Heated roof or flowing river provides a definite stream of free fuel—like free pocket-money from a rich parent—and we can get a definite amount of useful jobs done by it; but, like pocket-money, it cannot be stretched at will to do unlimited jobs. In most cases, we are drawing indirectly on the Sun's nuclear fuel. The river draws on recent sunshine, via evaporation and winds; the coal has drawn on ancient sunshine. Cost in money is a poor measure of fuel for our present discussion, since fuels range in price from free sunshine through cheap firewood, to oil or coal that are expensive when far from their source. Instead, we clear up our general idea of ENERGY-provided-by-FUEL from the other end, from the point of view of what we get. There are certain kinds of job that require fuel; we study such jobs and find that they use amounts of fuel proportional to the amounts of job. For example, consider hauling loads up a hill or to the top of a building. Fuel must be used. Whatever fuel is used, two equal jobs of load-raising take twice as much fuel as one—twice as many lumps of coal or twice as many gallons of oil, or two waterwheels running for an hour instead of one, or twice as many hours of sunshine. Try what trick we will, we use fuel proportionally.

Energy, we say, is something involved in certain useful jobs and provided by fuel. We think of fuels as stores of usable energy, just as we think of our bank balance as a store of usable wealth. Not every useful job or important machine needs fuel. Which ones *do* need fuel? Do we have to use fuel to produce or maintain great forces or great motion, or what? Torturers with thumb-screws found that fuel is hardly needed; they could produce huge forces with practically no expenditure of fuel, certainly without requiring fuel in proportion to the force or to the time it lasts. A clamp or a vise needs only a slight turn of the screw and will then maintain a force indefinitely without any fuel supply. A weight sitting on a shelf provides a force free of charge, as long as we leave it there. A heavy piston, sitting on gas in a cylinder exerts a high pressure indefinitely at no continuing cost. Again, steady motion need not use fuel: it continues of its own accord (Newton, Law I). Planets and gas molecules remain in motion without drawing on fuel. What kinds of jobs, then, *do* use fuel *necessarily* and *proportionally*? Let us list some which we know from common experience:

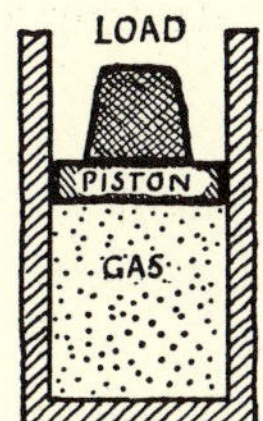

FIG. 26-1

(a) *Raising a load.* The earliest builders found that this needed fuel—food for slaves, corn for horses, waterfalls for wheels. A load will not raise itself—even if given a start it does not continue to rise on its own but slows to a stop.

(b) *Moving a cart along a rough road.* Fuel must be used to keep the cart moving. Of course if the cart stays at rest, pushing it need not cost any fuel. In that case, instead of hiring a man to push, we could make a lead statue and let it lean against the cart forever without fuel. But if the cart moves along, the leaden man falls down and is of no more use. A real man is needed to keep on shoving the cart along the rough road—or a steam engine burning coal, or an electric motor making a continuing demand on the power supply.

(c) *Winding up a clock-spring.* To wind a clock-spring we have to push the key around. Stretching or compressing any spiral spring needs a moving arm or some other engine using fuel. A wound-up spring can then raise loads or drag carts as it unwinds. (And a stretched or compressed coil can do similar jobs when released.) We seem to store up some job-doing-ability in the distorted spring—we can bank our fuel in it.

(d) *Making any object move faster.*

(e) *Heating up a bathtub of water.* This certainly takes fuel, not proportionally to temperature rise alone, but also in proportion to the mass heated. We could extend this example:

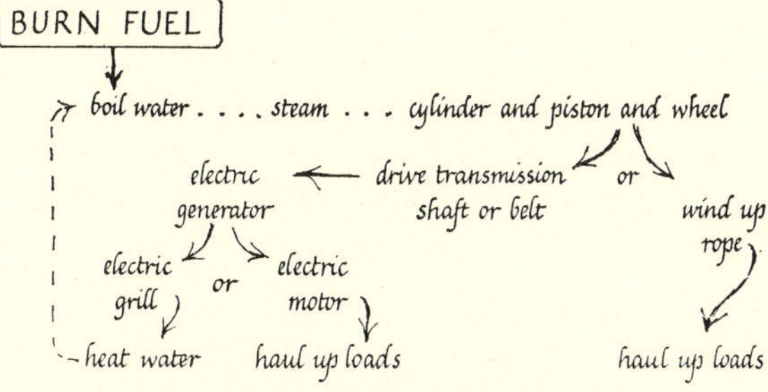

FIG. 26-2. ENERGY INTERCHANGES

Here is a hint for further thinking about *interchanges* between one form of energy and another.

"Work" and Energy Measurement

All schemes of raising loads, or stretching springs, or propelling carts on rough roads require a force that *moves along.* We haul a load upward, along the line of its motion; we pull the end of the spring outward along the line of its stretch. How can we make a general scheme of fuel-billing for such jobs, "energy-measurement" as we shall call it?

Consider the fuel-demand for raising loads in building and mining, the oldest mechanical examples of a clear need for fuel. Suppose we wish to raise a 2-pound load 3 ft vertically. Start with a 1-pound load and raise it just 1 ft. That will cost some fuel. Common sense tells us that if it costs so much fuel to raise the 1-pound load one foot it will cost the same for the next foot, and the same for the next, etc. Raising it 3 ft altogether will cost

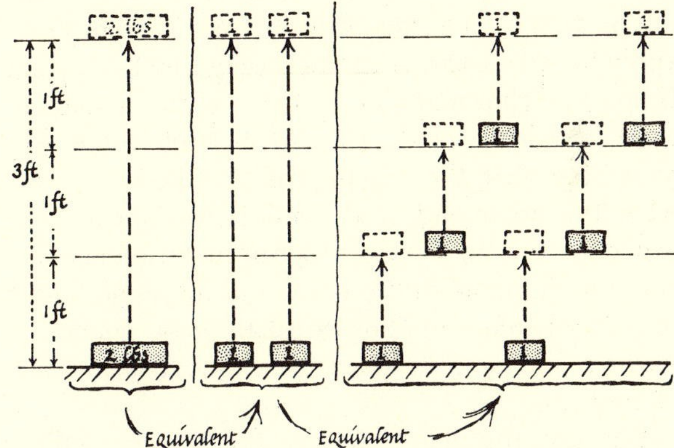

FIG. 26-3. SPLITTING WEIGHT-RAISING JOB UP INTO STAGES

3 lots of 1-foot-raising fuel. Now change to the larger load, 2 pounds. Raising a double load is like raising two single loads separately. Common sense and direct experiment tell us that the fuel requirements are additive: the raising job can be broken up into separate stages. Then, to find the cost of raising 2 pounds 3 ft, we split the job into 2 sub-jobs of raising 1 pound 3 ft and subdivide each sub-job into 3 unit-jobs of raising one pound one foot. Then we say, raising 2 pounds 3 feet will cost twice as much as raising *one pound,* and 3 times as much as for *one foot,* so it will cost 2×3 or 6 times as much as raising one pound one foot. It will cost 6 "foot · pounds." A foot · pound means:

1 foot DISTANCE · 1 pound-*weight* FORCE.

We might picture the total job being done by a crew of identical demons, each raising one pound to a shelf one foot higher—another demon taking over from there up to the next shelf—and each repaid by a standard grain of food. Common sense

FIG. 26-4. WORK
Units demons do weight-raising job.

suggests that this splitting-up into unit jobs is allowable: that fuel-consumptions are additive. Only experimental tests could justify this treatment completely for all varieties of load-raising. They do.

For load-raisings, then, the total fuel-demand could be measured by WEIGHT · HEIGHT RAISED. And this is FORCE · DISTANCE MOVED. Does that measure apply to other cases of engines using fuel to do jobs? If some engine hauls in a rope, we may make the rope raise loads or drag a cart or stretch a spring. Surmising that the engine will not know or care what the other end of its rope is working on, we expect to find the same amount of fuel used in every case for the same pull exerted and the same length of rope hauled in. So we shall now generalize WEIGHT · HEIGHT into FORCE · DISTANCE as a measure of fuel use.

We use the name *energy* for that something which fuel possesses in stored-up form and hands on through engines to raised loads, which again have something stored up which enables them to raise other loads if they fall themselves. In this way, energy is rather like money, which can exist in many different forms, can be stored in banks or put to use, and can be transferred to and fro in many ways. We do not call the useful product FORCE · DISTANCE MOVED energy, but name it "WORK" and regard it as a statement of "TRANSFER OF ENERGY," much as a bank check is a statement of transfer of money.[1] Note two things about this use of the word "work":

(i) It is much narrower than in ordinary talk, where we speak of working on a problem or of finding it hard work to hold a heavy load in our hand; though we also use "work" in the scientific sense, saying we do work when we *raise* a load. This is an unfortunate choice of a common word for a specialized use in science. We should have been wiser to invent a name for FORCE · DISTANCE, just as the name "gas" was coined by the early chemists. (Whether concentrating on mathematical problems does use extra food-fuel, and why holding a heavy load in our hands is fatiguing, are physiological questions we shall touch on later.)

(ii) In multiplying FORCE by DISTANCE we must use the *distance moved along the direction of the force* (or, what amounts to the same thing, we may multiply the DISTANCE MOVED by the COMPONENT OF FORCE *along the motion*).[2] A man keeping a cart moving along a rough road by pushing it must run along behind and push steadily forward, and that will cost him some food-fuel. But a *sideways* push neither helps the motion nor costs fuel. A leaden man could lean against the side of the cart and slide along beside it on roller skates keeping up with it at no cost except for friction losses.

If we wish to be careful and precise, we combine reminders of (i) and (ii) in the statements:

(I) "WORK" means, in science,
FORCE · DISTANCE MOVED BY THE POINT OF APPLICATION OF THE FORCE, IN THE DIRECTION OF THE FORCE

(II) "WORK," thus measured, shows the amount of ENERGY transferred *from* one place or form *to* another.

[1] Some texts treat work as a form of energy, and distinguish between work done *by* a spring and work done *on* it. It seems better to take work, FORCE · DISTANCE, as merely showing how much energy is transferred *from* one form or place *to* another. Then work has both a positive aspect and a negative one. It shows the energy lost by one agent, and the energy gained by another. A bank check is not itself money (nor is it wealth): it is just a statement of the amount of a transaction, of money lost from one account, money gained by another. Here we treat work as neither + nor —. It shows the energy lost by one party in some exchange, gained by the other.

[2] There is quite a different kind of FORCE · DISTANCE in which the FORCE and the DISTANCE are at *right angles to each other*. That measures the "moment" or "torque" or turning effect of a force around an axis, a useful concept in dealing with pulleys and levers.

FIG. 26-6.

We can exert a MOMENT (sometimes called a TORQUE) by pushing against the spoke of a wheel. If the wheel remains at rest, no work is involved. We measure moments in pound · feet, to distinguish them from foot · pounds of energy. In the MKS system, moments are measured in meter · newtons, energy in newton · meters. Moments (torques), and their applications to levers and to production of spin-acceleration, are not treated in this course.

FIG. 26-5. WORK
(a) WORK = F · s
(b) A man pushing forward against the end of the cart transfers energy F · s whether he accelerates the cart or merely keeps it moving steadily on a rough road.
(c) The leaden man pushing sideways against the cart as it moves along does not help the motion. He need not use energy from fuel. If on frictionless roller skates, he could just continue leaning on the cart.
The real man must push the cart to keep it moving steadily along rough road. The force F_M does transfer energy: the force F_L does not.

Now we are beginning to describe *energy* clearly. It is the thing whose amount of interchange is measured by work.

Economy Schemes ?

Fuel is terrifyingly important in our present civilization. In house-heating and cooking, rapid transport, easy communication, all moving machinery for manufacture, we transfer the energy stored in vast quantities of fuel to other forms. Deprived of adequate fuel, commercial and civil life would be brought to a standstill, and so would modern warfare. Food, the most necessary fuel of all, may yet be the limiting factor of all mankind's life in the very distant future—as it is now in many parts of the world.[3] Could we devise machines to save us using so much fuel? Levers and pulley-systems can produce big *forces* from small ones. Can such devices, "machines" as we call them in physics, also magnify our fuel resources and produce more *energy* from less fuel? We can insert a "machine" as intermediary between a fuel-using engine and the final job. Examine some typical schemes and see what help they give.

Machines

1. *Lever or see-saw.* The see-saw may be treated as a children's toy or a device for balancing forces, as in weighing machines; but it can also transfer energy when it turns: pushing one end down makes the other end raise a load. Suppose an engine (using fuel) applies a force F_1 at A, one end of the bar ABC, pivoted at B (see Fig. 26-7). Then the other end may support a much larger load at C. As the engine pushes A down, the big load is raised at C. But we calculate energy-changes by WORK which is not just FORCE: it is FORCE · DISTANCE. Suppose the end A moves down a distance s_1. Then the energy-transfer *from* the engine *into* the machine at A is $F_1 \cdot s_1$. The other end C pushes the load up with force F_2, rising a distance s_2. The energy transfer *from* the machine *to* the rising load at C is the work $F_2 \cdot s_2$. How do $F_1 \cdot s_1$ and $F_2 \cdot s_2$ compare? We shall find they are equal. If F_2 is many times

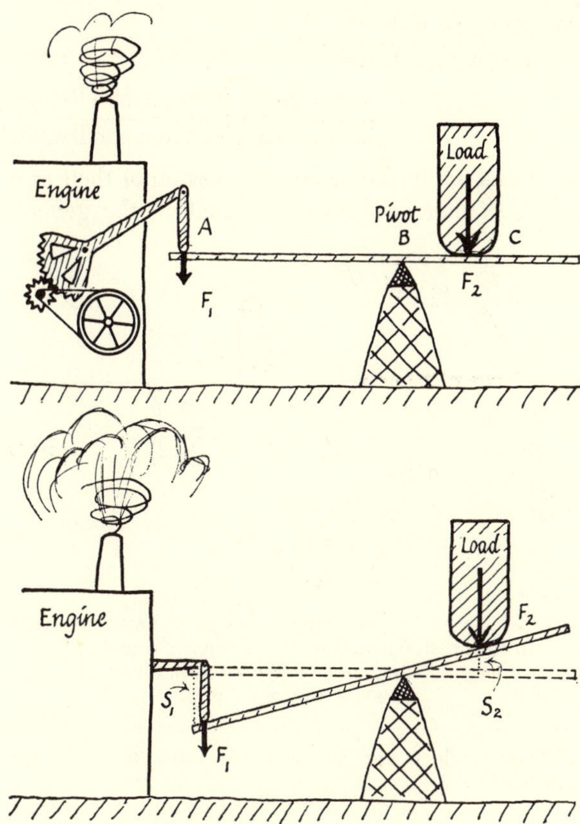

FIG. 26-7. ENGINE USING LEVER TO RAISE A LOAD

[3] See Sir Charles Darwin's fanciful but grave essay, *The Next Million Years* (New York, 1953).

bigger than F_1, then s_2 is just as many times smaller than s_1. Here is the argument. Simple experiments will show you the rule for balancing see-saws or levers if you do not already know it. Fig. 26-8 shows one case being tested. The bar ABC, pivoted at B,

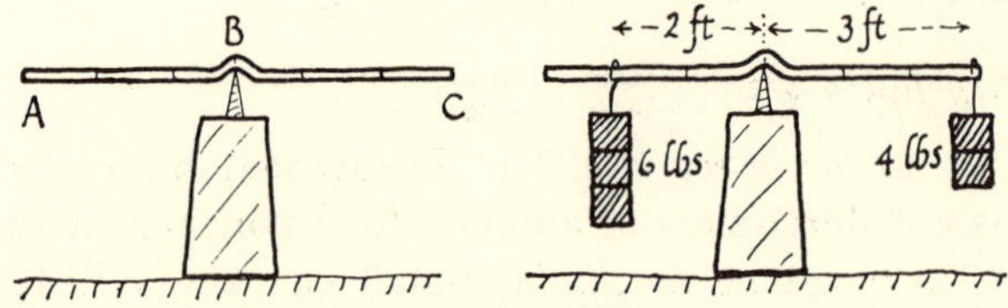

FIG. 26-8. SEE-SAW FOR TESTING THE BALANCING RULE "FORCE · ARM = FORCE · ARM"

remains balanced in equilibrium, if left unloaded. Loads are added as shown, 4 pounds at 3 ft from the pivot and another load 2 ft out on the other side, to balance. For balance, the second load must not be an equal one, 4 pounds. Experiment shows the right load is 6 pounds. In this example,

4 pounds-weight × 3 feet
= 6 pounds-weight × 2 feet. (See note 4)

FORCE · ARM-LENGTH = FORCE · ARM-LENGTH

each arm-length being the perpendicular distance from the axis, B, to the line of action of the force. For this balanced lever,

$$\frac{\text{FORCE of 6 pounds-wt.}}{\text{FORCE of 4 pounds-wt.}} = \frac{\text{ARM-LENGTH for force of 4 lbs-wt.}}{\text{ARM-LENGTH for force of 6 lbs-wt.}}$$

The loads are in the inverse proportion of their arm lengths. Other specimen tests confirm the general rule. Just for interest, Fig. 26-9 shows more complicated cases.

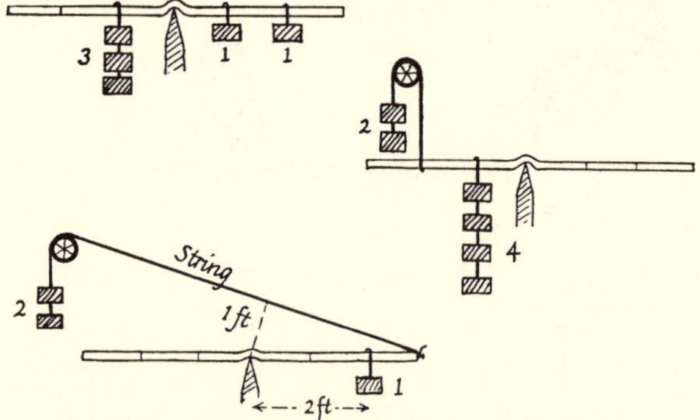

FIG. 26-9. MORE COMPLICATED CASES OF BALANCED LEVERS
In each case the *total* of the products (+ and −) FORCE · PERPENDICULAR ARM FROM PIVOT TO LINE OF FORCE is zero.

Returning to the lever connecting engine and load, we know from experiment that the forces F_1 and F_2 are in *inverse proportion* to the arms L_1 and L_2. That is, $F_2/F_1 = L_1/L_2$. But by geometry the distances moved, s_1 and s_2 have the same proportion as the arms, L_1 and L_2. (See the similar triangles shaded in Fig. 26-10.)

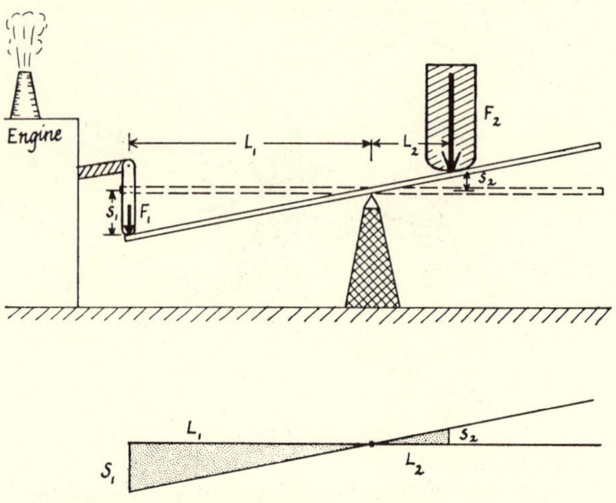

FIG. 26-10. SKETCH FOR DISCUSSION OF MACHINE
The shaded triangles are similar. Therefore, $L_1/L_2 = s_1/s_2$. The load and effort really move in arcs of circles; so s_1 and s_2 are slightly curved, and the "triangles" are really sectors. But the same reasoning applies.

$$\therefore \frac{L_1}{L_2} = \frac{s_1}{s_2};$$

$$\therefore \frac{F_2}{F_1} = \frac{L_1}{L_2} = \frac{s_1}{s_2};$$

$$\therefore F_1 \cdot s_1 = F_2 \cdot S_2 .$$

Therefore the two ENERGY-TRANSFERS, the WORK $F_1 \cdot s_1$ and the WORK $F_2 \cdot s_2$ are equal. The work, $F_1 \cdot s_1$ is the energy-transfer *to* the lever *from* the engine; and $F_2 \cdot s_2$ is the energy-transfer *from* the lever *to* the load. As $F_1 \cdot s_1 = F_2 \cdot s_2$ we say that the lever gains and loses equal amounts of energy.

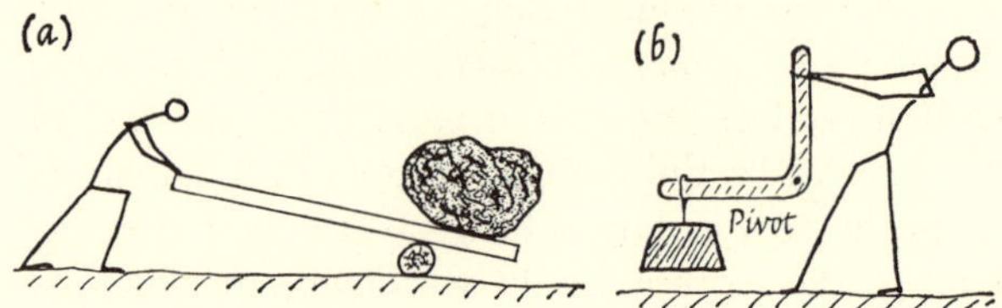

FIG. 26-11. LEVERS CAN BE USEFUL
(a) Raise big load easily. (b) Bent lever with equal arms.

[4] These products are "MOMENTS" measured in pound · ft as in footnote 2.

The lever's INPUT-ENERGY and its OUTPUT-ENERGY are equal. As a "machine" it merely hands energy on; it neither creates nor destroys any. This does not prevent a lever being very useful, to change the force to a more convenient size, or to alter its direction; but it does not solve the fuel problem. And, if there is friction at the pivot, there are tiny opposing forces and a little of the input-energy is wasted to some useless form.

2. *Pulleys.* A pulley wheel acts as an equal-armed lever.[5] It changes the *direction* of force, often greatly to our convenience, but if truly round and frictionless it does not change the *size* of force.

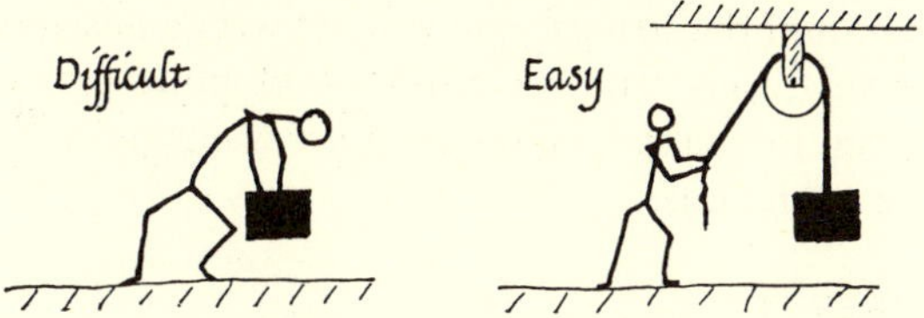

FIG. 26-12. PULLEYS ARE USEFUL
A pulley helps by changing the direction of force.

A single pulley, then, offers no hope of changing the value of FORCE · DISTANCE. But what about pulley systems, wonderful block-and-tackle arrangements which enable a man to raise huge loads far beyond his ordinary strength? Investigate the system sketched, using common sense (=, as usual, experimental knowledge accumulated in nursery and ordinary life). The pulley system of Fig. 26-14 is used by a man to raise a large load. What load can the machine raise if the man pulls down with a 10-pound pull? The pulley wheels act as equal-arm levers, changing the direction of the force but not its size, except for friction. The 10-pound tension continues through rope A, over the pulley to B, and so on; 10 pounds-wt. in ropes B, C, and D. Each of these is a tension, the force with which a rope pulls at each end.[6] The total force exerted by the ropes on the load is 10 pounds-wt. pull up of rope B + 10 pounds-wt. pull up of rope C + 10 pounds-wt. pull up of D. That makes 30 pounds-wt. in all. For equilibrium—rest or steady motion—this must just be balanced by the weight of the load. Therefore the man's 10-pound-wt. pull will

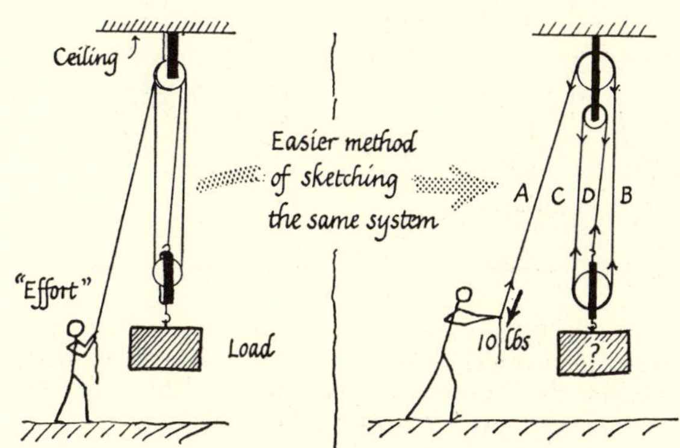

FIG. 26-14. PULLEY SYSTEM: FORCE RATIO
Finding the force ratio (mechanical advantage) of a pulley system. The same tension continues through all the parts of the rope. The arrows show the rope tension pulling inward on man's hand, upward on load, etc.

raise 30 pounds, except for friction. In real pulleys, friction takes a small tax and 10 pound-wt. pull will raise less than 30 pounds. In general this block and tackle arrangement gives a ratio of LOAD LIFTED : MAN'S PULL of 3 : 1, at best. This force ratio,

$$\frac{\text{"LOAD" (= THE WEIGHT OF THE LOAD RAISED)}}{\text{"EFFORT" (= THE PULL EXERTED BY THE MAN)}}$$

$$\text{or } \frac{\text{"OUTPUT FORCE"}}{\text{"INPUT FORCE"}}$$

is often called the *mechanical advantage* of the machine. This set of pulleys therefore has a mechanical advantage of 3/1. This is its "theoretical" mechanical advantage, as an ideal frictionless machine. Its "practical" mechanical advantage will be less than 3/1, because of pulley friction.

[5] To show that a pulley is like an equal armed lever, imagine the wheel to be of wood, with equal loads hung over it. Drive nails in at A and B to attach the rope to the wheel there. Then the rest of the wheel and the rest of the rope can be cut away, leaving an ordinary see-saw. To keep this balanced, the two loads must be equal; so we were correct in placing equal loads on the original wheel. If instead of one load, a man pulls on the rope, we still have a see-saw with equal arms, but one arm may now be bent up. In all cases the rope-tension is the same on both sides of the wheel. (If there is friction, the wheel will oppose turning, and the rope-tension on the "output" side will be slightly less than that on the "input" side.)

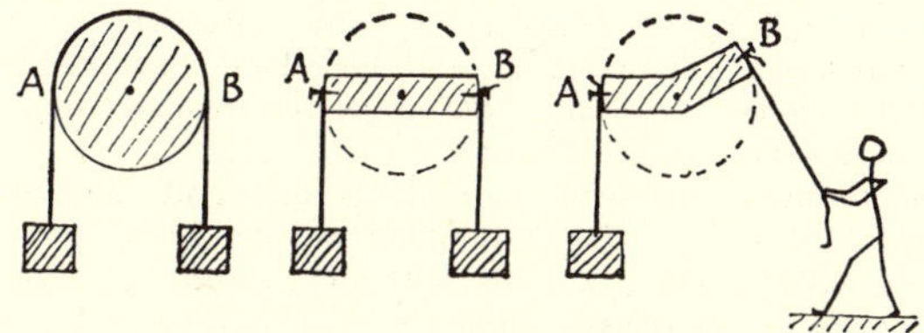

FIG. 26-13. A PULLEY IS LIKE AN EQUAL-ARMED LEVER

[6] Remember tension in a rope is the pull *at each end*, and it is *inward* along the rope, at each end—a rope could hardly push.

Now compare the DISTANCES MOVED by pulling hand and rising load—or, compare their SPEEDS OF MOTION. For this, it is easier to start the argument at the load. Suppose the load is raised 1 ft. Pretend it rises by magic, letting the ropes develop slack loops. Then remove the slack by cutting out a 1-ft length from each of the ropes B, C, D, and rejoining.

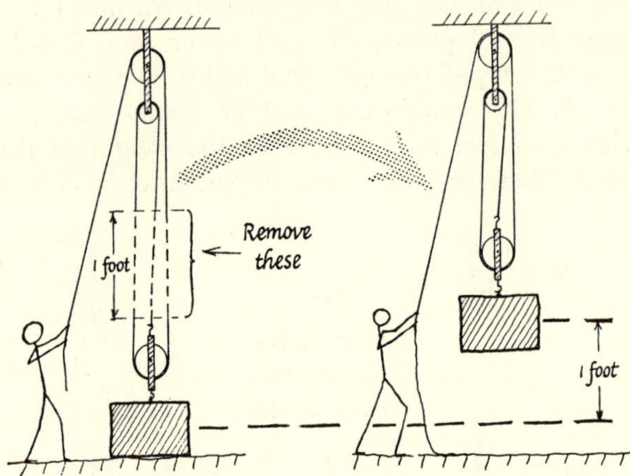

FIG. 26-15. PULLEY SYSTEM: DISTANCE RATIO
Finding the distance ratio (velocity ratio) of a pulley system. Imagine the load rises one foot and calculate how much slack the man pulls down.

If these three 1-ft lengths were not cut out, there would be 1 ft of slack in each of B, C, D, or 3 ft altogether. The load is really raised by pulling A down instead. Therefore the man must pull down 3 ft of slack. The distance ratio,

$$\frac{\text{DISTANCE PULLED BY MAN}}{\text{DISTANCE MOVED BY LOAD}},$$

is often called the *velocity ratio* of the machine, since it also gives

$$\frac{\text{SPEED OF MAN'S ROPE}}{\text{SPEED OF LOAD}}.$$

In this machine it is clearly 3/1. Once again "what is gained in force is lost in distance."

Now compare the energy-transfers in our example: the INPUT WORK is 10 pounds-wt. × 3 ft or 30 foot · pounds; the OUTPUT WORK is, at best, 30 pounds-wt. × 1 ft or 30 foot · pounds. Except for friction tax, the machine neither creates nor destroys energy, but only transmits it, transforming[7] the force of our arm pull to a bigger force for our convenience. With friction, some of the input energy seems to disappear—diverted into heat.

Make sure you understand this simple treatment of pulleys by working through Problems 1 and 2.

Science vs. Rules

Guess at the force ratio and distance ratio for the "Spanish Burton" pulleys of Fig. 26-17. It may be puzzling if you have studied pulleys before and trust the school-book rule of thumb, "count the strings." If you use the kind of argument given above (and asked for in the problem), you will find the force ratio and distance ratio correctly. If you "count the strings" you will get wrong answers—that rule, only true for one type of pulley system, is given to enable beginners to get the right answer in examinations.[8]

PROBLEM 3

If you want to drive yourself crazy in a short argument, work out the behavior of the "fool's tackle" sketched in Fig. 26-18. If necessary, try it.

FIG. 26-18. "FOOL'S TACKLE"

---

[7] The word "transforming" here matches its use in electrical engineering. A transformer hands electrical energy from one power line to another with little loss, but it changes the VOLTAGE very helpfully. It changes the CURRENT too, so that the POWER remains the same.

[8] Such blind parrot-rules are dangerous, and they give science a bad name among thinking people. They offer to answer examination questions without thought, and they provide the second-class engineer with a useful formula for his handbook. (A second-class engineer is one who trusts formulas blindly, hoping, for example, that a rule for strength of wooden beams will apply to concrete pillars. But he is a useful man for all that, and he is a good engineer if he knows his limitations. A first-class engineer wants to know what is happening in detail and why. He is interested in understanding and asks where formulas come from. In fact he is a scientist at heart, with an added practical turn—a rare man who should make a great career.)

"PROBLEM SHEETS"

The problems here are intended to be answered on typewritten copies of these sheets. Work through the problems on the enlarged copies, filling in the blanks, (———), that are left for answers.

PROBLEM ON MACHINES NAME________________________

Show that each of the "machines" below merely makes it easier to transfer some energy, and does not change the product FORCE · DISTANCE MOVED which measures the amount of energy transferred. (This is for an ideal case of no friction. With friction, some of the input is transferred to heat.)

PROBLEM 1. SIMPLE PULLEY SYSTEM

In the system sketched, one rope runs over several frictionless pulleys.

To calculate the change of (FORCE), complete the following argument:

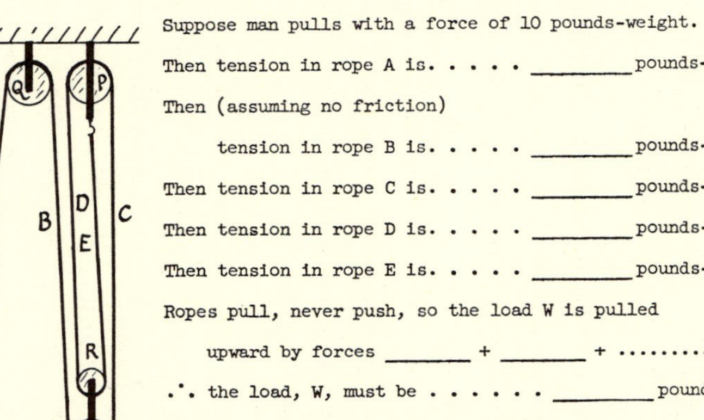

FIG. 26-16.

Suppose man pulls with a force of 10 pounds-weight.

Then tension in rope A is. . . . . ________pounds-weight

Then (assuming no friction)

tension in rope B is. . . . . ________pounds-weight

Then tension in rope C is. . . . . ________pounds-weight

Then tension in rope D is. . . . . ________pounds-weight

Then tension in rope E is. . . . . ________pounds-weight

Ropes pull, never push, so the load W is pulled

upward by forces ______ + ______ + ...?

∴ the load, W, must be . . . . . . ________pounds

∴ the force is increased by a factor________

To calculate the change of (DISTANCE MOVED), carry through the following argument. Suppose load W is raised 1 foot.

Then each of the ropes labeled _____, _____, __________

will be shortened by 1 foot.

∴ man pulls down ________ feet of slack.

∴ distance is changed in the ratio ________

If the extra pulley on the ceiling, Q, is removed and the man pulls the rope B upward instead of pulling A down, how does this affect the system's mechanical advantage and velocity ratio? How does this change affect the usefulness of the system?

PROBLEMS ON MACHINES SHEET 2

PROBLEM 2. SPANISH BURTON PULLEY SYSTEM

In the system sketched, there are three separate ropes, FGH, JK, LM. Each is anchored at one end to a fixed ceiling. The man pulls the other end of rope FGH. The other ends of ropes JK and LM are attached to the axles of pulleys Q and R as shown.

Suppose the man pulls 10 pounds-weight on rope F.

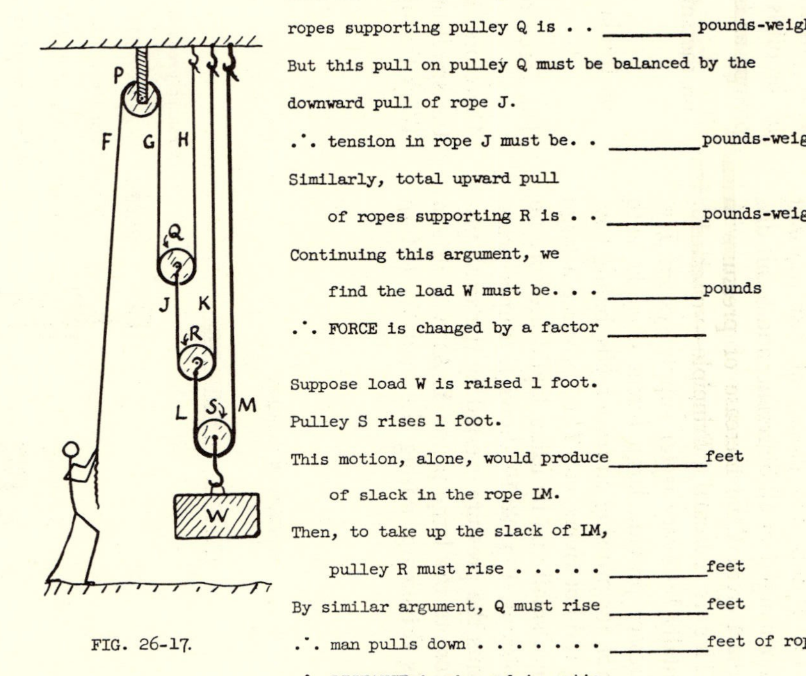

FIG. 26-17.

Then the total upward pull of the ropes supporting pulley Q is . . ________pounds-weight.

But this pull on pulley Q must be balanced by the downward pull of rope J.

∴ tension in rope J must be. . ________pounds-weight

Similarly, total upward pull

of ropes supporting R is . . ________pounds-weight

Continuing this argument, we

find the load W must be. . . ________pounds

∴ FORCE is changed by a factor ________

Suppose load W is raised 1 foot.

Pulley S rises 1 foot.

This motion, alone, would produce________feet of slack in the rope LM.

Then, to take up the slack of LM,

pulley R must rise . . . . . ________feet

By similar argument, Q must rise ________feet

∴ man pulls down . . . . ________feet of rope F.

∴ DISTANCE is changed in ratio ________

Apart from friction-taxes, does this pulley system change the amount of FORCE · DISTANCE?. . . . ________

Hydraulic Press

Fluids (= liquids and gases) at rest transmit pressure unchanged in all directions and to all distances (apart from the increase of pressure with depth). This is Pascal's Principle—vouched for experimentally by every water supply system. Hydraulic presses provide large forces cheaply with easy control, for pressing oil from seeds, baling hay, squeezing red-hot rivets, etc. They produce a large force from a small one by using liquid pressure on unequal pistons. Fig. 26-19 shows a simple press. Oil transmits the pressure exerted by the small piston X across to the large piston Y. As X is driven down Y is driven up, raising the load W. (More oil is taken in from a reservoir on the upstroke of X, to provide for repeated strokes.) If the pressure in the oil is p pounds-wt./sq. inch, it pushes on the pistons as follows:

on the small piston, AREA a sq. ins.,
with FORCE $F_1 = p \cdot a$ lbs-wt.

on the large piston, AREA A sq. ins.,
with FORCE $F_2 = p \cdot A$ lbs-wt.

The force ratio $= pA/pa = A/a$

Therefore, if the large piston is much larger than the small one, it can support a much larger load. Work through Problem 4.

PROBLEMS ON MACHINES SHEET 3

PROBLEM 4. HYDRAULIC PRESS

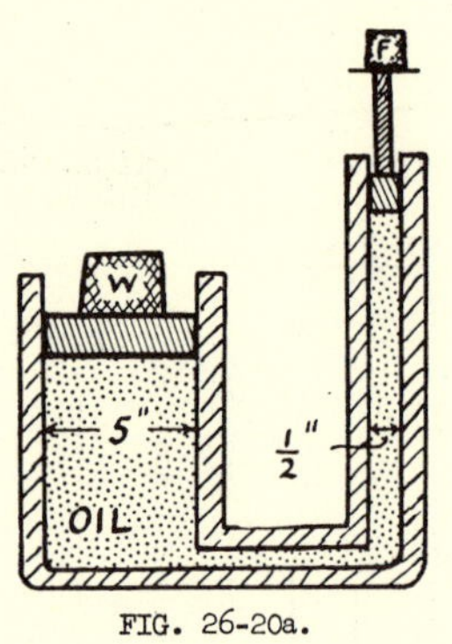

FIG. 26-20a.

In the hydraulic press shown, the small piston has diameter ½ inch, and the large piston has diameter 5 inches, ten times that of the small piston.

∴ area of big piston is . . . . _________ times as big as area of small piston.

The oil applies the same pressure to both pistons.

∴ force with which oil pushes on the big piston must be . . . . _________ times as big as the force with which it pushes on the small piston.

When the small piston is driven down, a certain volume of oil is driven across from the small cylinder to the large one, and drives the big piston up.

(NOTE: the volume of a cylinder = HEIGHT · AREA OF BASE)

If the oil does not change in volume during its flow,

volume lost by the small cylinder must = volume gained by the big cylinder;

and therefore, since the area of the big cylinder is . . . . _________ times the area of the small, the distance moved by the big piston is . . . _________ of the distance moved by the small piston.

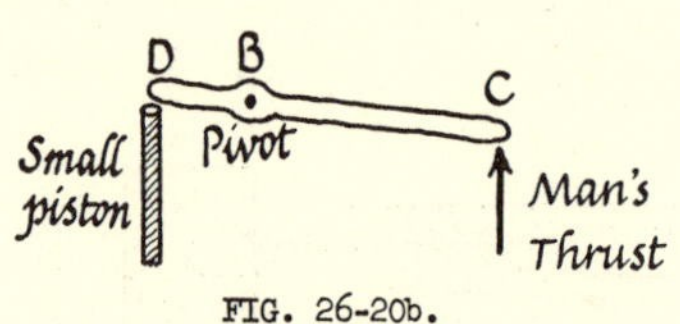

FIG. 26-20b.

Suppose that instead of pushing on the small piston directly, the man using the press has a lever, DBC, pivoted at B, that pushes the small piston down at D when he pulls it up at C. DB = 3 inches BC = 12 inches

The lever enables the man to raise _________ times as large a load as before.

The over-all force-ratio (mechanical advantage) is now _________.

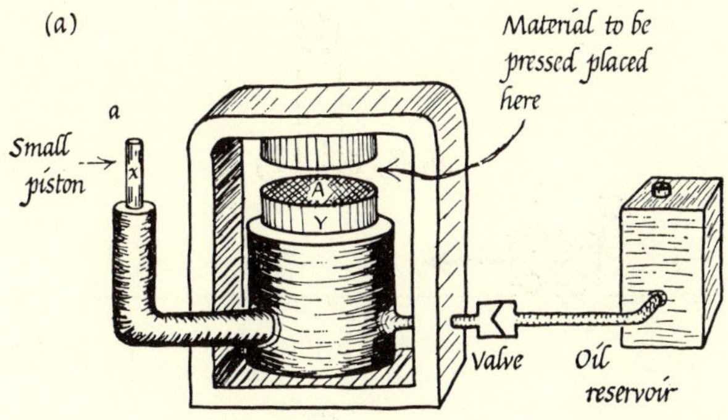

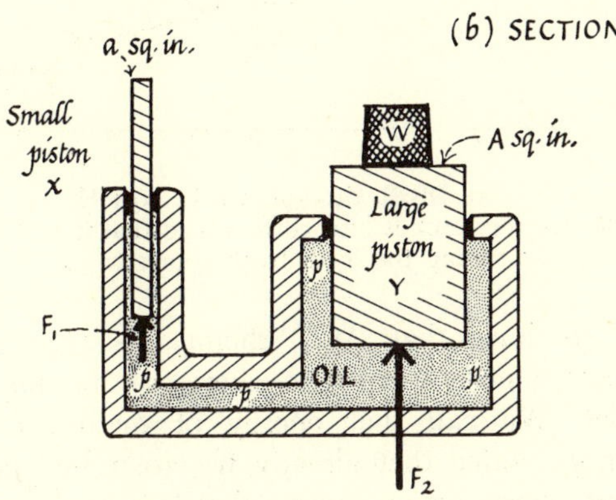

FIG. 26-19. HYDRAULIC PRESS

(a) The large piston and cylinder are contained in a frame which carries a roof above the large piston. The material to be compressed is squeezed between the large piston and the fixed roof.

(b) Section: the frame and roof are not shown. The important invention which makes the press work well is a leakproof piston design, rather like the piston arrangement in a bicycle pump.

However, the large piston moves far less *distance* than the small one. The VOLUME of oil pushed from the small cylinder by its piston is AREA · STROKE, s_1, or $a \cdot s_1$. The same volume moves across into the large cylinder, and pushes up the big piston. If the latter rises a DISTANCE s_2 the VOLUME of oil that fills the extra space is $A \cdot s_2$. As oil is almost incompressible, VOLUME $a \cdot s_1$ must $=$ VOLUME $A \cdot s_2$.

$\therefore$ the DISTANCE RATIO s_1/s_2 must $= A/a$.

But FORCE RATIO also $= A/a$

$$\therefore \quad \frac{F_2}{F_1} = \frac{A}{a} = \frac{s_1}{s_2}$$

Once again, "*what we gain in force, we lose in distance,*" and

INPUT WORK $F_1 s_1 =$ OUTPUT WORK $F_2 s_2$

Inclined Plane

We can use a hill to drag a load up in a cart instead of lifting the load vertically. But do we save any energy expenditure? When we raise a load of

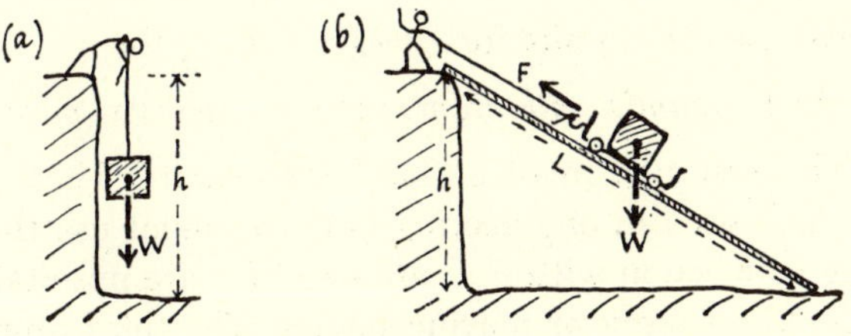

FIG. 26-21. RAISING A LOAD
(a) Direct raising of load, W, through height, *h*.
(b) Raising up along frictionless incline.

weight W a height h vertically the energy transfer is WORK $W \cdot h$. Hauling it up the slope involves WORK $F \cdot L$. Then the ENERGY TRANSFER by the vertical route is $W \cdot h$; by the slant route, $F \cdot L$. But we showed in Chapter 7 that, except for friction, $\frac{F}{W} = \frac{h}{L}$ on a hill.

$$\therefore \quad F \cdot L = W \cdot h$$

The energy transfer from man to load is the same by either method.

Summary

Each of these attempts to economize in energy fails.

What Is Energy?

We have not described energy clearly yet, and we may never be able to give a crisp definition. But you should now begin to feel familiar with energy from our discussion of jobs and fuel and work. In time, this acquaintance should mature into a clear understanding. As with your understanding of other general concepts such as Justice, Kindness, Love, we may never be able to agree on a definition, yet you will rightly claim, "I know perfectly well what it means; I *understand* it." For the present you should think of *energy* as *something supplied by fuels, something which can be interchanged between several forms,* and *something whose interchange-amount is measured by* FORCE *multiplied by* DISTANCE. Work measures interchange of energy. So the official definition of energy says, "*energy is capacity for doing work.*" Qualitatively, this is not a very helpful description; it merely says "a thing that has a lot of energy is able to give a lot of energy to something else." Quantitatively, this definition tells us to *measure energy in the same units as work.*

Units for Energy

Since we measure *transfers* of energy by FORCE · DISTANCE, which should have units

(*lbs-wt.*) (*feet*) or (*kg-wt.*) (*meters*)
or (*newtons*) (*meters*)

we use these units for energy, writing them

foot · pounds *kilogram · meters* *newton · meters*

The first two involve "bad" force units. They are much used in engineering but we cannot use them in connection with $F = Ma$—so they are unsuitable for the energy of moving bodies. The third unit, a *newton · meter*, is a "good" unit. It is used so much that we shorten it to a new name and call it a *joule.* Thus, saying that 1 newton · meter = 1 joule is no new science; it is mere dictionary-work. Since 1 newton is roughly ¼ pound-wt. and 1 meter is roughly 3 feet, 1 joule is roughly ¾ of a foot · pound; (more accurately 0.74).

Foot · pounds and kg · meters contain (bad) FORCE units. They mean foot · *pounds-weight* and *kilograms-weight* · meters, but the *kilogram* concealed in newton · meters is a MASS unit.

We still have no way to measure the total energy in some stock—whatever "total energy" means. We can only measure certain interchanges, using FORCE · DISTANCE OR WORK as a measure of the amount of energy transferred. Then energy-changes are measured in joules, ft · pounds, etc.

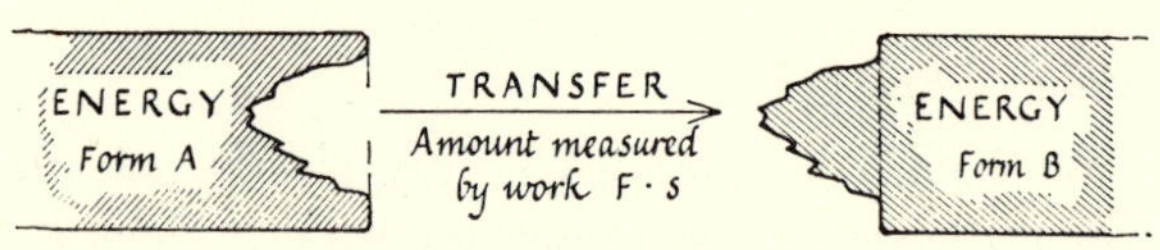

FIG. 26-22. WORK

Forms of Energy

Watching the energy interchange when jobs are done and measuring its amount by the work product FORCE · DISTANCE, we can describe several forms of energy. A wound-up spring has a store of *elastic energy* or *strain energy.* A load raised higher above the Earth has gained *gravitational energy.* Oil and gasoline have *chemical energy* that can be transferred to gravitational energy when they drive a machine that raises a load. Fireworks and explosives have *chemical energy* that can be released suddenly. Food has *chemical energy,* some of which our bodies turn into elastic energy or gravitational energy when we use muscles to wind springs or raise loads.

A moving truck has energy on account of its motion, because it could raise loads if attached to suitable ropes and pulleys, but it must lose its motion in doing that. And "chemical energy" in an explosive may end up in energy of motion in a bullet. We call that energy of motion *kinetic*[9] *energy.* Notice that already we are telling stories about energy which assume that it is never created or lost but only transformed.

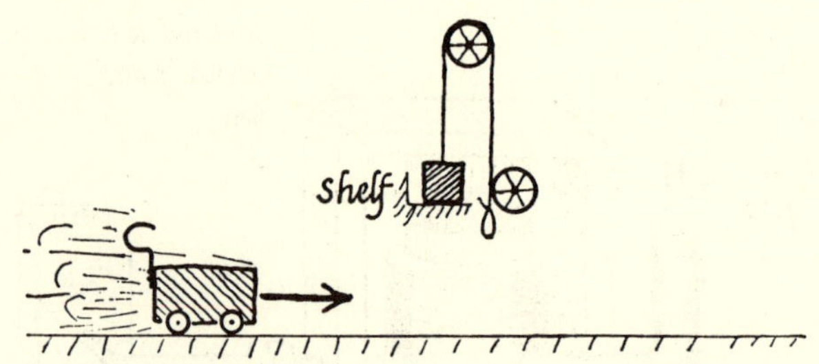

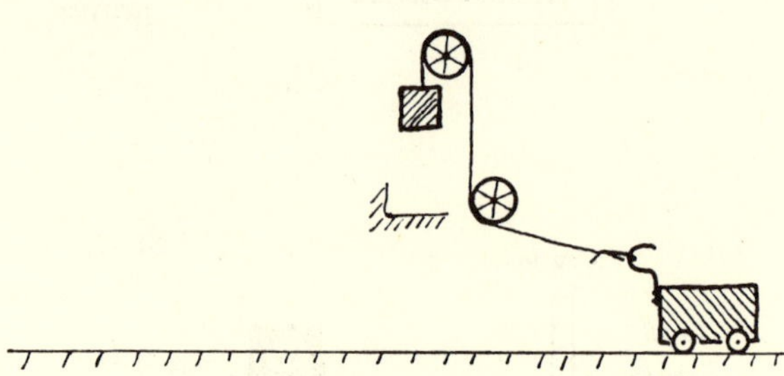

FIG. 26-23. CHANGING K.E. TO P.E.
The moving truck is hitched to a load that it can raise.
In raising the load, truck comes to rest.

If a truck is shifted from one place to another place *at the same height above ground,* there is no change of gravitational energy. If it starts and finishes at rest, we can imagine the move made "free," *with no net transfer of energy.* We pretend we give the truck a little kinetic energy (borrowed from some store of gravitational or chemical energy) and let it run along a frictionless level road to the new place. We bring it to a stop there, taking away the kinetic energy (and repaying the store). We might imagine a scheme like that in Fig. 26-23 arranged to do this.

We find that a truck moving with kinetic energy along a rough floor comes to rest without raising weights. Has its energy been destroyed or only transferred? Maintaining our belief, we look for the energy in some new form. The rubbing of rough surfaces has produced heat. Can we speak of *heat energy* too? Early studies of heat did not show it as a form of energy. Heat was regarded as a mysterious and indestructible fluid, like our present view of electricity. A great series of experiments in the last century showed that heat *is* a form of energy. And now the scheme has been extended to still more forms, a grand system of energies adding up

[9] From the Greek for motion; same origin as "cinema."

to a constant total, in any enclosed region. We shall discuss the "Law of Conservation of Energy" in Ch. 29. For the moment, consider the mechanical attempts to break that law with a perpetual motion machine.

Perpetual Motion

Simple machines such as levers and pulleys transform the size of force and the speed of pushing or pulling, often greatly to our comfort; but the ultimate fuel-demand, measured by FORCE · DISTANCE, is the same—or even greater on account of friction-tax. Useful ENERGY OUTPUT *from* machine = ENERGY INPUT *to* it, less some friction-tax. Yet for centuries men have wanted to save using fuel, and the temptation and hope are as strong as ever. Perhaps pulleys and levers are too straightforward. Could a more ingenious machine give out more energy than it takes in? This was the search for *perpetual motion*, now numbered among the "seven follies of science."

Such a machine would indeed make its inventor's fortune: it could change the course of civilized life. We speak of such machines as "perpetual motion machines"; and we are now convinced they can never succeed. The name "perpetual motion" is unfortunate because it misleads people into thinking that the essential problem is to remove friction and thus obtain what we might call, to distinguish it, *perpetual movement*. If that is all, we can nearly succeed with a wheel spinning on ball-bearings, and Nature does succeed with the spinning Earth and the moving planets, atoms, etc. These cases of *perpetual movement* would soon fail as *perpetual motion* schemes; if we tried to take energy from them they would slow down and stop.

A perpetual motion machine would, of course, run of its own accord, taking any little payment needed for friction-tax from its own extra output. But the essential characteristic of the hoped-for device is not just continual running, a steady spinning—that would be mere "perpetual movement," pretty but useless. To bring lasting fame and rich rewards it would have to do more than keep running steadily, with its output just feeding its input. Its output would have to exceed its input; so that it could supply its own input and still have some output left over to do useful jobs free of charge. It would have to yield a continual profit of energy without fuel. It should be able to run steadily *while lifting loads*; or if left unloaded it should run *faster and faster*. Such motion, providing a continual supply of energy from nowhere, is what we call perpetual motion. Could such a machine be made? For centuries this problem has offered great rewards for success, tempting the clever thinker, intriguing the ingenious designer, taunting the conqueror of nature. All efforts have failed. There is always a catch. Solutions have been offered by well-meaning scientific inventors, and by cheating charlatans. Business investors have been tempted by hopes of fabulous wealth; and some have risked their money, and lost it when the machine failed or its inventor absconded. In the course of time, suggestions grew more complicated; simple wheels with swinging spokes or sliding loads gave place to engines using complex gears, schemes for making an electric generator and motor run each other *with profit*,[10] and even an alluring use of liquid air. All that were actually tried failed, *failed to deliver a continual stream of energy without using further fuel.* Those that were examined critically without being tried in a working model have all shown some catch, so that our general knowledge predicts failure. Scientists now feel so surely convinced that perpetual motion (in its special sense) is impossible that they now state their belief as a general Principle, often called the First Law of Thermodynamics.

If we believe *mechanical* perpetual motion is impossible, we are half-way to assurance of conservation; machines cannot create energy, but they might destroy some. Only when we account for heat etc. as well do we gain complete assurance.

What view should you yourself take? Should you continue the search for perpetual motion yourself, or obediently accept scientists' dogmatic denial? In view of the evidence, the latter seems wiser. For once, we do not encourage you to start with doubts and do your own experimenting. However, in a way you are surveying the results of experiments when you pay your bills for gasoline or electric supply; machines to *replace* fuel have failed. Satisfy yourself, by arguing them through, that arrangements of levers and pulleys cannot be turned into a perpetual motion machine; and try to find the "catch" in diagrams of some of the more complicated devices suggested as perpetual motion machines. Then like professional scientists you may take refuge in a general statement of belief that perpetual motion seems utterly unlikely, unthinkable, impossible, "against Nature."

Note that this statement brings no proof. It is a mixture of a *summary of experimental tests* and a

[10] A motor and generator can almost keep each other running—*perpetual movement*—if no external demands are made. The motor drives the generator by a belt and the generator supplies current to run the motor, with a little help from a battery to pay for friction and heating of wires. This is used as an economical scheme for testing big machines.

Fig. 26-24. Perpetual Motion Machines

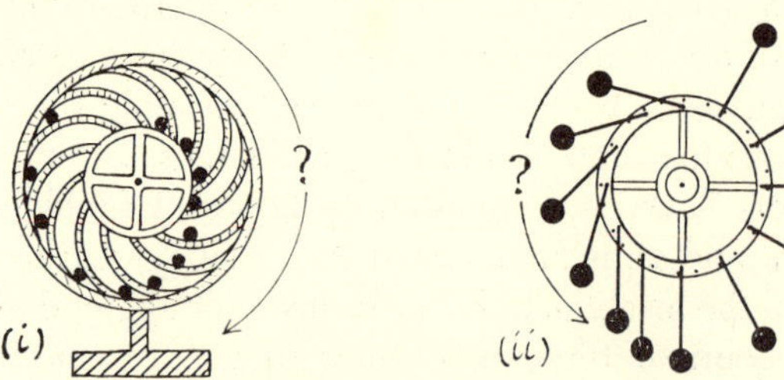

Fig. 26-24a. Two simple proposals
(i) A wheel carries balls in curved compartments.
(ii) A wheel carries weights on spokes pivoted between stops on its rim.

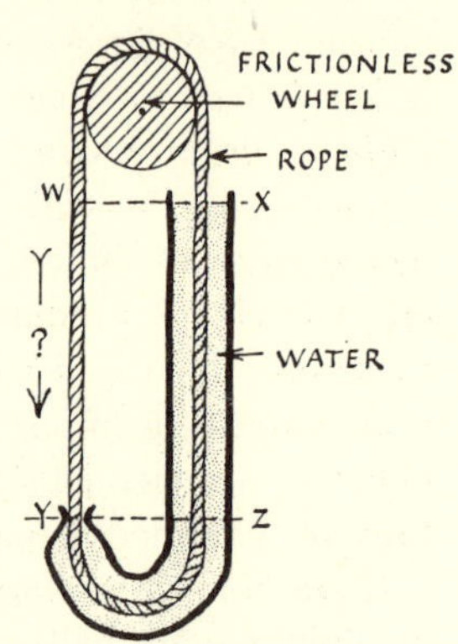

Fig. 26-24b.
An endless loop of rope partially immersed in water. (The joint at Y should be frictionless, yet leakproof; but this requirement is NOT the essential catch.)

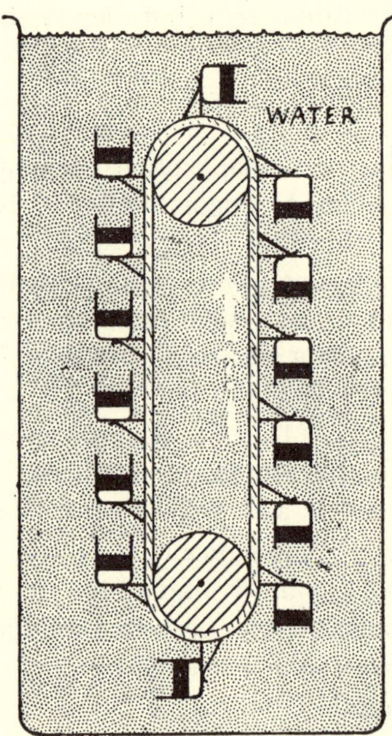

Fig. 26-24c.
An endless belt on frictionless pulleys under water carries cups with air enclosed by a heavy frictionless piston. Where the cups are upright (on left), the weight of the piston compresses the air more than where they are inverted. Therefore, the cups on the left displace less water than those on the right and should experience smaller buoyant force.
(The pulleys should be frictionless and so should the pistons, but this requirement is NOT the essential catch.)

statement of faith.[11] Scientists fully familiar with the interlacing threads of evidence hold this view confidently. Yet in this century two other strong scientific beliefs have been broken: we can now create and destroy matter, and we can now change one chemical element into another—we should stop calling the alchemists' dream of turning lead into gold another of the "seven follies of science." Why, then, are we so sure that energy is conserved, and perpetual motion impossible? That is partly because the evidence is so plentiful and varied; partly because the idea of energy is an artificial one, moulded to fit our impressions of the outside world and our thinking about it. So, unlike matter, which we might say just *is* there, energy is our own idea and we shall choose to preserve it even if we have to change the design from year to year. Yet each extension of our scheme of energies, however fantastic a guess at first, has in the course of time met test after test of validity. The Conservation of Energy *does* tell us something very valuable about the Universe.

FORMS OF ENERGY AND THEIR MEASUREMENT

When the earliest clear ideas of energy developed, only kinetic energy of motion and gravitational potential energy and elastic potential energy were considered. In distinction from kinetic energy, the other two were called potential energy.[12] The sum of kinetic energy (K.E.) and potential energy (P.E.) remains constant in many simple mechanical changes: rocks falling, springs unwinding, motion of systems of loads and pulleys. This conservation, [K.E. + P.E.] = constant, is not surprising, since, as you will see, K.E. is so defined, and the P.E.'s are so chosen, as to make this true. The interesting property of nature is that the resulting scheme of energies is a simple scheme, useful and easy to deal with. And the amazing and delightful thing is that the scheme can be expanded by adding other forms of energy to make a great universal

[11] "*You cannot get something for nothing*" is a very unwise attempt to support our belief. You can get a lot for practically nothing if you want force, happiness, blindness—anything except a few carefully defined quantities like momentum and energy. In fact, physics has been much concerned with separating out and defining those things of which you cannot get something for nothing. So, "can't get something for nothing" may *apply* to energy but it does not prove that perpetual motion is impossible. *You should avoid quoting this almost meaningless phrase in science.*

[12] Potential energy means energy that is capable of coming into action, but somewhat hidden. Kinetic energy is obviously there, whizzing along with the moving body.

conservation scheme, a very powerful part of physical theory.

Gravitational Potential Energy

By burning fuel or using some other energy-supply, we can raise a load vertically. The work is WEIGHT · Δ HEIGHT, which shows the energy transferred *from* the fuel *to* the gravitational field.

Δ (GRAVITATIONAL P.E.) = WEIGHT · Δ HEIGHT.

We find it hard to say where this gained P.E. resides, but its amount is definite enough, and the raised weight clearly "owns" it.

When P.E. is lost, Δ (P.E.) = WEIGHT · Δ HEIGHT, and both Δ (P.E.) and Δ HEIGHT are negative; or, LOSS OF P.E. = WEIGHT · VERTICAL DISTANCE FALLEN.

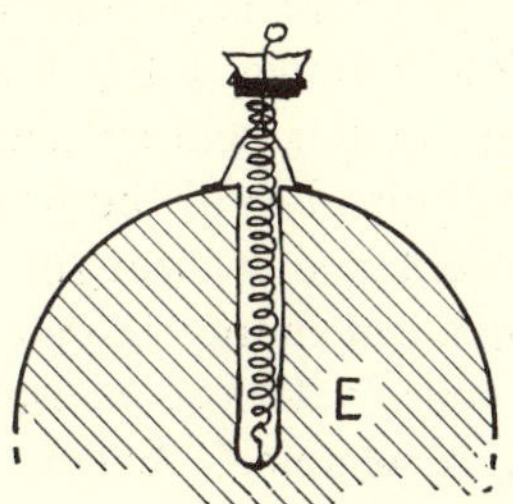

FIG. 26-25.

To give yourself a false but useful feeling of the reality of gravitational P.E., carry out this personal experiment: Stand with feet planted firmly on the floor, holding a heavy book with both hands. Shut your eyes and raise and lower the book several times slowly. As you feel the weight of the book, pretend that gravity is not really there. Pretend the book is pulled by a great spring running down to the center of the Earth instead. Try to picture the spring and feel it s-t-r-e-t-c-h when you raise the book. With a genuine attempt at imagining, you can make the spring feel real. Now, still with your eyes shut, think of the energy stored in the twisted coils of the spring when you stretch it.

Elastic or Strain Energy: Energy Stored in Stretched Springs, etc.

A spring stretched or compressed or twisted from its normal shape and size has potential energy stored in it. We say this not because we see any energy stored in the distorted metal but because we know we can get jobs done by letting the spring undo its strain, jobs measured by FORCE · DISTANCE. And we know that work, FORCE · DISTANCE, was involved in making the strain. We call this energy *strain energy*, or *elastic P.E.*, and may picture it as stored in force-fields between the atoms or molecules of the springy material.

STRAIN ENERGY

$$= \begin{bmatrix}\text{AVERAGE FORCE USED}\\ \text{IN STRETCHING SPRING}\end{bmatrix} \begin{bmatrix}\text{DISTANCE}\\ \text{STRETCHED}\end{bmatrix}$$

Gases are springy, and we might try to think of them as having strain energy too; but, when we picture their molecules buzzing about with no springy material packed in the space between, this idea seems untenable. As gas molecules are in constant motion, we suspect that energy stored in a gas is really kinetic energy of molecular motion, and not strain energy at all.[13]

Kinetic Energy: Energy of Motion

We shall now show that energy of motion, "kinetic energy," must be calculated, by using the rule: K.E. = ½ MASS · SPEED². This is derived from $F = Ma$. We let a force F accelerate a mass M driving it a distance s. If the mass starts from rest and reaches final velocity v, we shall find that the energy-transfer to it, $F \cdot s$, is $\frac{1}{2} Mv^2$.

If we push a moving body with a forward force F_1, the transfer of energy *from* pusher *to* body is $F_1 \cdot s$; and if some opposing force F_2 acts on the moving body, the body transfers energy $F_2 \cdot s$ *to* the opposing agent. The moving body gains the difference, $F_1 s - F_2 s$ or $(F_1 - F_2)s$.

But $(F_1 - F_2)$ is the resultant, F, of the two forces acting on the body. So the *net* TRANSFER OF ENERGY to the moving body is

$$(F_1 - F_2)s \quad \text{or} \quad \text{RESULTANT FORCE} \cdot s \quad \text{or} \quad F \cdot s$$

The resultant force, F, is wholly concerned in accelerating the body, making it move faster and in-

[13] The energy of a gas is the energy of molecular motion and we call this its *heat energy*. You may say "Heat is molecular motion"; but please avoid saying "Heat is due to collisions among molecules." That is entirely wrong. During a collision, one molecule may give some heat energy to another; but collisions cannot manufacture heat, and they certainly are not heat. If the moving molecules of a gas suddenly shrank, so that most hits became misses and they made far fewer collisions, they would still have the same heat energy at the same speed. The mistake probably arises from thinking of molecules as rough and rubbing each other with friction when they hit. When they collide they simply approach, repel, and move apart. Individual molecules are *not* rough. Roughness is made by great irregular piles of molecules on a surface. Real friction *is* just a dragging of many molecules by many others. Though it can reduce kinetic energy of a moving body to heat-motion of individual molecules, it cannot manufacture any new energy.

When gas molecules hit a *moving* piston they rebound with different speed—faster if the piston was approaching them. Such molecules do gain heat when they hit the piston, but they gain it from the the piston's energy, not from some mysterious spark called a collision.

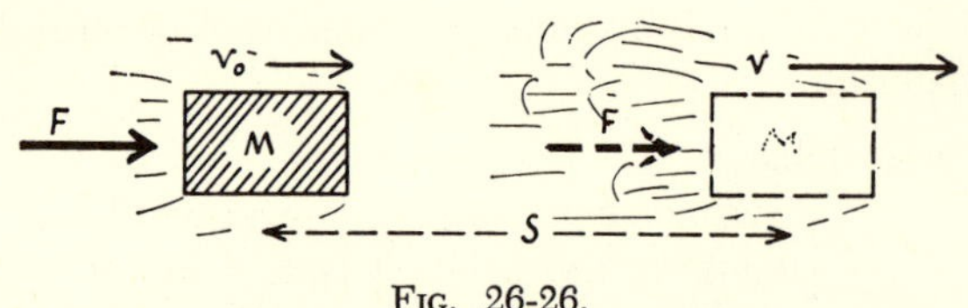

FIG. 26-26.

creasing its kinetic energy. So the work $F \cdot s$ shows the energy transferred to energy of motion—none of this RESULTANT FORCE · DISTANCE is wasted on friction, etc. Suppose then we give a mass M some kinetic energy by pushing it with a resultant force F for a distance s. The net transfer of energy to the moving body is $F \cdot s$; and, as F is the *resultant* force on the mass M we must have $F = Ma$. To deal with this accelerated motion we shall use the relation: $v^2 = v_0^2 + 2as$ which leads to: $as = \frac{1}{2} v^2 - \frac{1}{2} v_0^2$. (See Ch. 1, Appendix A for an "elegant" derivation of this important relation.)

$$F \cdot s = (Ma)s$$
$$= M(as), \text{ but } (as) = \tfrac{1}{2}v^2 - \tfrac{1}{2}v_0^2$$
$$\therefore F \cdot s = \tfrac{1}{2}Mv^2 - \tfrac{1}{2}Mv_0^2$$
$$= (\text{new } \tfrac{1}{2}Mv^2) - (\text{old } \tfrac{1}{2}Mv^2)$$
$$= \text{gain of } (\tfrac{1}{2}Mv^2)$$
$$= \Delta\, (\tfrac{1}{2}Mv^2).$$

But $F \cdot s$ is the interchange of energy, into energy of motion.

$\therefore$ gain of ENERGY-OF-MOTION = gain of ($\frac{1}{2}Mv^2$)

So we give $\frac{1}{2}Mv^2$ the name *energy of motion* or *kinetic energy*.

Therefore, we say, KINETIC ENERGY = $\frac{1}{2}Mv^2$. When the body was moving with speed v_0 it had kinetic energy $\frac{1}{2}Mv_0^2$. While it accelerated from v_0 to v it gained K.E. and ended with total K.E. = $\frac{1}{2}Mv^2$.

More simply, if M speeds up from rest to v, it acquires kinetic energy $\frac{1}{2}Mv^2$.

Units for Kinetic Energy

Since we used $F = Ma$ in deriving the expression $\frac{1}{2} Mv^2$ for kinetic energy, the F involved must be in absolute units, e.g., newtons; and the energy must emerge in absolute units, e.g., newton · meters.

But if M is in kilograms and v is in meters/sec, $\frac{1}{2} Mv^2$ is in kg · meters²/sec² or newton · meters. The change from obvious kg · (meter/sec)² for $\frac{1}{2} Mv^2$ to newton · meters makes sense, because: in

$$F = Ma$$
$$1 \text{ newton} = 1 \text{ kg} \cdot 1 \text{ meter/sec}^2$$

Then our energy unit

$$1 \text{ kg} \cdot \left(\frac{\text{meter}}{\text{sec}}\right)^2 = 1 \text{ kg} \cdot \frac{\text{meter}^2}{\text{sec}^2}$$
$$= 1 \text{ kg} \cdot \frac{\text{meter}}{\text{sec}^2} \cdot \text{meter}$$
$$= 1 \text{ newton} \cdot \text{meter}$$

K.E. emerges in newton · meters or joules, as it should. If we use pounds for M and ft/sec for v, the K.E. will emerge in foot · poundals, *not* foot · pounds. Although foot · pounds are useful in dealing with pulleys, etc., and are much used by engineers, use of them will lead to disagreements when K.E. is involved.

Heat

We now look upon heat as a major form of energy, ranking with gravitational P.E., K.E., and other mechanical forms. Only a century ago it was barely admitted as a full member of the brotherhood, but as it gained acceptance it brought with it other forms such as chemical energy and electrical energy; and the idea of conservation of energy grew rapidly from a narrow mechanical scheme to a general principle.

For calculations of heat energy, anticipate the discussion in Chapter 29: assume that heat is a form of energy, and that *1 Calorie = 4200 newton · meters or joules*. The Calorie[14] itself is a relic of the early history of heat measurement. Two centuries ago, a definite system of measuring heat was developed, based on the assumption that heat is conserved, never lost but only exchanged. We still use the following rules:

To measure heat, let the heat heat water and multiply mass of water by temperature-rise.

If the MASS is in kg and Δ (TEMPERATURE) in C°, the product is HEAT, in Calories or kilocalories.

If the heat is given to some other substance, first multiply mass by temperature rise, as for water, then multiply by "specific heat" of the substance.

To measure the heat supplied by a measured amount of some fuel, we need special apparatus to burn all the sample and give all the heat to water without having large unknown heat losses. Such measurements have been made for most standard

[14] Throughout this book we use the large Calorie, or kilocalorie, which is the heat that warms up 1 kilogram of water 1 Centigrade degree. This is 1000 times the small calorie, still in common use. Beware of confusion through this factor of 1000.

fuels, usually by enclosing a weighed sample with compressed oxygen in a strong metal "bomb." The "bomb" is placed in a jar of water, the sample ignited electrically, and the temperature-rise of the water measured. Since the "bomb" and its contents are heated as well as the water, allowance must be made for them. The tables show some results for fuels

SUBSTANCE	HEAT PROVIDED BY COMPLETE BURNING OF 1 KILOGRAM IN OXYGEN (*Calories per kg*)
Coal	7000-8000
Fuel oil	10,000
Gasoline	11,500
Hydrogen gas	34,000
Alcohol	6,500
Gunpowder	only 700

and, for the milder but thorough form of "burning" in our digestive system (though fast burning as a fuel gives the same amount of heat):

FOODSTUFF	(*Calories per kg*)
Cakes and cereals	3000-5000
Bread	2500
Sugar	4000
Butter	8000
Candy	3000-8000

Heat and Molecules

Any successful effort to give energy to a gas merely makes it hotter, increasing its PRESSURE · VOLUME. On kinetic theory, we see this as an increase of $(\frac{1}{3})\,N\,(m\overline{v^2})$ or of ⅔ of the kinetic energy of the N molecules' random motion. For a gas at any rate, heat is merely kinetic energy on a molecular scale. We guess that the same applies to liquids and solids, except that we must include kinetic energy of spin of molecules and their energy of vibration.

Imagine a bullet hurtling along with plenty of K.E. Now let friction bring it to rest. You may picture its K.E. being given to the molecules of surrounding air or rough wood, setting them going with extra motions. Bulk kinetic energy disappears; heat appears. Socialism! Heat, we might say, is socialized K.E., the wealth of a large batch of organized kinetic energy spread out among random motions of all the molecules, deserving or undeserving. When a lead bullet hits a wall, most of its rich load of K.E. is changed to energy of vibration of the individual atoms of lead, the energy of an advancing army of lead degraded into the jostling of a mob.

Chemical Energy

When an explosive is fired it produces hot gases which in turn can give a bullet kinetic energy. In this case, the bullet collects organized K.E. at the expense of the heat in the hot gases (their disorganized K.E.). How do the gases obtain the heat energy, which they give to the bullet? Before the explosive is fired it is a cool solid holding "chemical energy." The energy in our primary fuels—coal, wood, oil—is chemical energy. This is molecular energy, stored if you like to think of it so, in the fields of force between atoms. You might picture a chemical compound being made by shoving the atoms into place in molecules against springy interatomic forces and latching them there, with potential energy stored in compressed springs. Of course

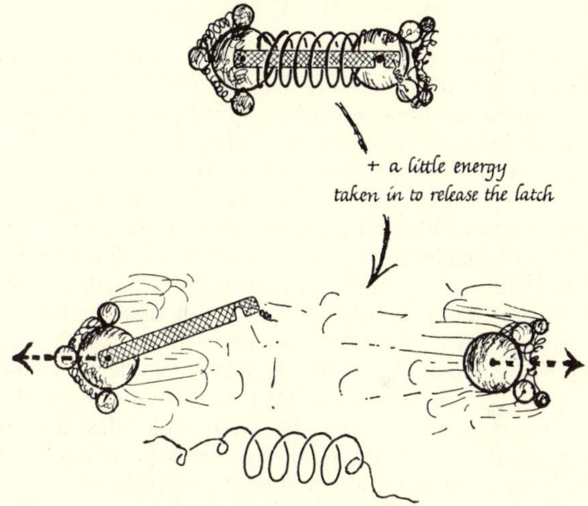

FIG. 26-27. FANCIFUL PICTURE OF A MOLECULE OF EXPLOSIVE, to illustrate description in the text.
(A more genuine model would have no latch but another spring in tension, in equilibrium with the compression spring but with a different law of force. Even with a scheme like that, models of fuel burning with oxygen are clumsy and unreal.)

chemical energy is really a much more complicated matter than this crude model, but the overall picture is clear: atoms and molecules do store energy which can be released in some chemical changes and taken in and stored in other changes. Most fuel releases its energy when it burns with oxygen; so fuel energy relates to the force-field between fuel molecules and oxygen. It is difficult to picture its location, though its amount is definite enough because we can measure the work, FORCE · DISTANCE, when it is transferred to other forms: so much transfer for every pound or kg of fuel burned completely. The chemical energy in a firework or explosive is easier to locate. It is all there, stored inside the molecules of material.

Food: a Supply of Chemical Energy

Food is fuel for man or beast; it provides chemical energy which is carried by the blood stream to muscles which need it. The muscles can transform some of the food-energy reaching them into mechanical energy for raising weights, etc.

Foodstuffs contain fuel molecules, mostly made of carbon, oxygen and hydrogen atoms. For example the simple sugar dextrose which runs muscle action is $C_6H_{12}O_6$ or

$$\begin{array}{ccccccccccccc} & & H & & H & & H & & H & & H & & H \\ & & | & & | & & | & & | & & | & & | \\ H & - & C & - & C & - & C & - & C & - & C & - & C{=}O \\ & & | & & | & & | & & | & & | & & \\ & & O & & O & & O & & O & & O & & \\ & & | & & | & & | & & | & & | & & \\ & & H & & H & & H & & H & & H & & \end{array}$$

In the process of muscle action and recovery, this fuel molecule is broken in half, then six molecules of H_2O are stripped off, and the carbon atoms join with oxygen from the lungs to make six molecules of CO_2.

Here is a short general account of living chemistry greatly over-simplified.[15] The chief food substances—starch, sugars, fats and proteins—have large molecules, built up of smaller molecular complexes which are themselves built up from atoms. In general, the smaller complexes are made by plants and linked in a particular pattern to form plant substances such as starch and cellulose. An animal, whether eating vegetable or animal food, breaks down these substances and rearranges the constituent parts to form the large molecules that it needs, but it does not synthesize such parts itself. It gets energy for movement and other activities by breaking down some of the molecular complexes yet further into carbon dioxide and water. This energy was originally obtained from sunlight by plants and stored in their preliminary syntheses. The linking and unlinking of the small complexes in the animal digestion is usually a simpler affair, not involving the transfer of much energy, and rendered swift by the action of enzymes or organic catalysts. The large molecules in our food are: starch and cellulose, both of them multiple gangs of simple sugar molecules like dextrose; fats with long CH_2 chains; and proteins, bigger still, very complicated, and essential for building and repairing tissue. The process by which chemical energy is released in warming the body or working its muscles is essentially the same as burning. When fuels burn in a fire they join with oxygen to form water and carbon dioxide. Simple body fuels such as dextrose sugar join with oxygen from the lungs to form water and carbon dioxide, but the process is gentler and less direct than in a fire: no fiery temperature is produced but the same energy is released. Plants take water and CO_2 from the air and join them together to make the sugars, starch, and cellulose that are the main energy-supply for animals.

The animal body gets its fuel for muscles like this: it extracts simple sugar molecules from food—much as a chemical factory extracts fuel alcohol from wood pulp—and keeps them stored in multiple gangs which are insoluble "animal starch" molecules. These stored starch molecules split up to maintain a supply of sugar. When muscles contract to do jobs, the sugar is converted, in two stages, into water and carbon dioxide. Animals also store fats from their vegetable food to be "burned" when needed for body heating.

Then, what man and animals have undone is re-made by plants, ready for animals to use again. How can plants do this? We cannot make a flame reverse its action; we cannot make a burnt fire "unburn" itself. How can plants do this vital synthesis for us, the latching-together of fuel molecules? Since the unlatching releases some chemical energy, the plants must put energy back into the combine. They need both a supply of energy and a mechanism for using it to re-latch H_2O and CO_2 into starch and cellulose. Sunlight provides the energy, packaged in minute chunks of light waves, and the plant provides clever molecules such as green chlorophyll to effect the manufacture. In sunlight, green leaves take in CO_2 and manufacture starch. Thus the plant and animal life of the world combine to provide a cycle, starting with *water and carbon dioxide and sunlight* and ending with *water and carbon dioxide and animal heat and mechanical energy.* As with all our

[15] Much of this paragraph is drawn from the excellent chapter on biochemistry in *Matter and Change* by Sir William Dampier (Cambridge, 1924).

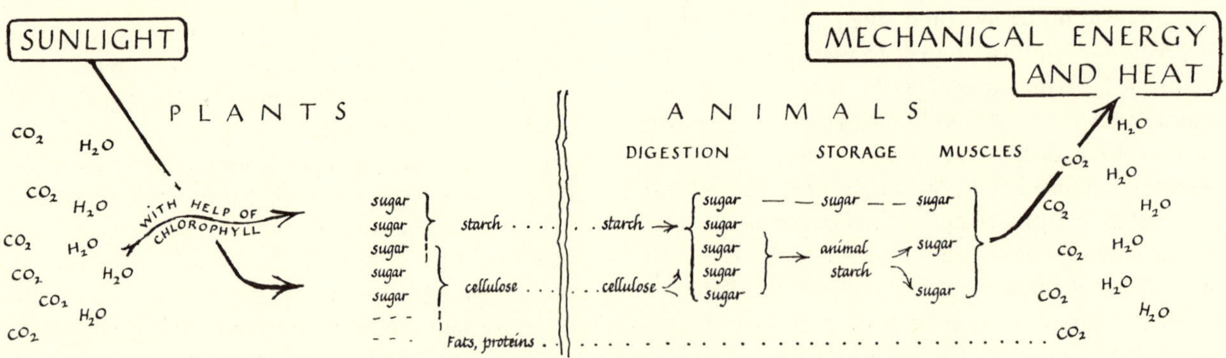

FIG. 26-28.

machinery that uses coal, oil, wind, or waterfall, animals get all their fuel, which they take in as food, originally from sunlight.

Fatigue

This brings us back to a problem: why do our bodies get tired when we merely hold a large load up in the air? That seems to take food energy though it does not provide any visible gain of gravitational P.E. Careful observation shows that we do not hold the load quite still. The load trembles up and down, our muscles carrying on an incessant twitching. If you clench your jaws and cover your ears you may hear a dull buzzing from muscular vibration. On each tiny raising of the load we use some chemical energy in our muscles. On the next tiny fall of the load the gravitational P.E. is returned but it does not go back into chemical energy—our fuel chemistry is not reversible. Instead it goes into heat which we waste.

Other Forms of Molecular Energy

Along with chemical energy, "molecular energy" should include the energy involved in melting solids and evaporating liquids or solids. Solids when warmed up to their melting point then require extra heat to unlink their molecules from an orderly crystalline arrangement into an irregular fluid crowd.

When a liquid evaporates, speedy molecules which manage to escape take away more than an average share of heat-energy; extra heat must be supplied to keep a liquid boiling or counteract the cooling of evaporation. These supplies of extra heat taken in for melting or evaporation disappear into the molecular force-fields as some kind of molecular potential energy. This extra heat is often called *latent heat*, meaning "heat that lies hidden." The latent heat taken in does not warm up the material—the temperature remains constant while melting or boiling proceeds. In the reverse process, solidification or condensation, the latent heat is released, *as heat*—a burn from live steam is far worse than one from boiling water.

Energy of Rotation: "Spin Energy"

Set a wheel spinning, hang a load on a cord attached to the wheel, and the wheel will wind up the cord and raise the load, slowing down as it does so. A spinning wheel evidently has energy to spend on raising a load. We call this spin energy or rotational energy.

The individual parts of a spinning body all move—those farthest from the axis moving fastest—

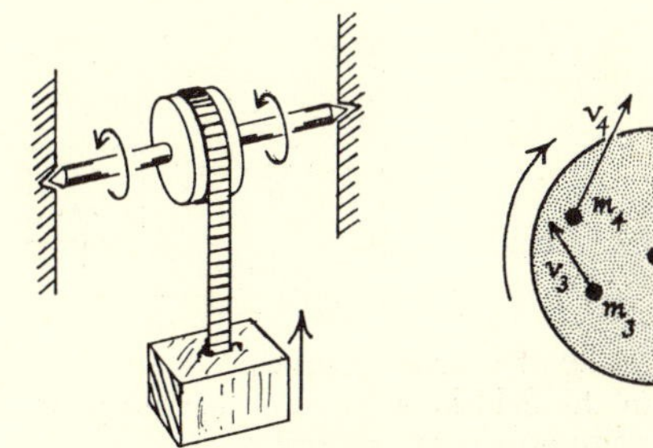

FIG. 26-29. SPINNING FLYWHEEL
(a) Spin energy of spinning flywheel can turn to P.E. of load lifted by tape wrapped around wheel.
(b) Velocities of various parts of spinning flywheel.

and they each have kinetic energy. Adding up the $\frac{1}{2}mv^2$ contributions from all the small pieces would tell us the total spin energy. Spin energy is really ordinary kinetic energy, part of the total K.E. if the body is spinning *and* moving along, but it is often convenient to give it a separate label.

Electric Energy

Electric batteries need chemicals to make them work, and while they supply current there are chemical changes inside. Chemical energy is released, but instead of just heating the battery it can be piped round an electric circuit by wires and delivered as heat in a grill, or as mechanical P.E. in a motor raising loads, or even as chemical energy in another battery being charged. We think of the energy being carried from the battery in the form of "electric energy."

Electric generators convert mechanical energy into electrical energy, and electric motors carry out the reverse process.

We can store electric energy. A battery can drive + and − electric charges on to the plates of a capacitor, storing electric energy in the electric field. When the capacitor is discharged with a wire or spark, heat is produced. This heat tells us there was energy stored in the capacitor. If the words are comforting, you may say the energy is stored "in the electric field" between and around the plates. A more direct demonstration is shown in Fig. 26-30: a small chip of paper or metal leaf, placed between a pair of plates with + and − charges, collects a little charge of its own and is hauled up by the electric field. At the top plate it trades its charge for an opposite one and is driven down. As it repeats the up and down motion, it gives energy to the air by gas-friction. We might imagine it geared to a tiny machine to use the energy taken from the charged plates and their field.

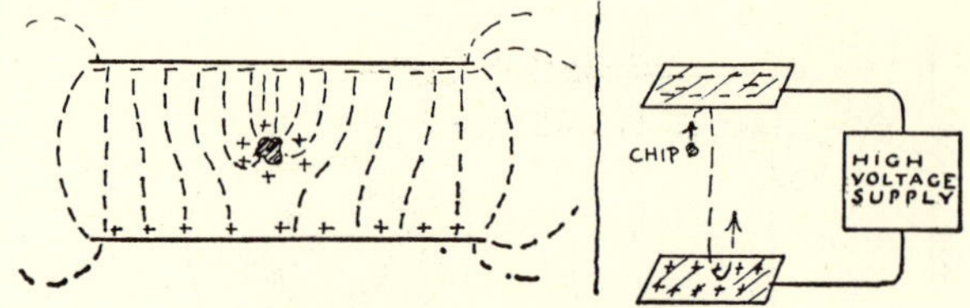

FIG. 26-30. ELECTRIC ENERGY
The energy stored in the field between two charged plates can raise up a chip of metal.

When electric currents flow around a circuit, we picture electrical energy being carried by moving electric fields from the battery or generator to various parts of the circuit where it turns into heat, mechanical energy, etc. When a generator supplies current, mechanical energy must be supplied to keep it running, or it will soon use up its spin K.E. and stop. The larger the current sent round the circuit, the harder the generator is to drive, the bigger its demand for mechanical energy. We picture the heat-energy in the steam-engine changing to mechanical energy (chiefly strain energy) in piston and moving belt; the generator changes that to electrical energy which travels to various parts of the circuit and changes to heat, etc. Electrical energy is a useful idea, but it is difficult to locate. Therefore, when we are asked just where it resides, we find it safest to think of it in the electromagnetic field all around the circuit, accompanying the current.

Magnetic Energy

Permanent magnets, if they have a special store of energy, do not easily exchange any of it, so we are not interested in a special label. But electric currents have strong magnetic fields, which extend far and wide around them. When the current is stopped the magnetic field disappears—"collapses"—and a considerable amount of energy is delivered to the circuit, not from the battery but, apparently, from the magnetic field. This happens during a short time while the current is dying down. You can feel a little of this "magnetic" energy transferring to chemical energy when you get an electric shock in the demonstration of Fig. 26-31.

Electromagnetic Energy

In many cases electric and magnetic energy are closely associated—each may almost be regarded as a one-sided view of the other. Alternating currents are driven by alternating electric fields and have alternating magnetic fields around them. When a radio station is broadcasting, rapidly alternating currents surge up and down its antenna, and the surrounding electric and magnetic fields have electric and magnetic energy; or, as we prefer to say, the field has electromagnetic energy. But these fields do not just pulsate in the neighborhood of the antenna. Their patterns break off in loops which travel away as electromagnetic waves[16] carrying electromagnetic energy with them (see Ch. 33 and Ch. 37). The waves move out with the speed of light.

Thus we think of radio waves as moving fields carrying energy. When they reach a receiving antenna, they try to produce tiny alternating currents with electric and magnetic fields, which ultimately turn most of their energy into a small heating of the wires of the receiving system.

Light produces heat when absorbed, so we think of light as carrying energy. We now know that light and infra-red radiation and ultraviolet light are all really streams of electromagnetic waves. We include them with radio waves under one name, *radiation.* Radiation carries energy, in the form of electromagnetic wave energy. When radiation is stopped (absorbed), its energy changes to another form, usually heat.

Wave Energy

Ocean waves can carry a large stream of energy. When a calm sea lapping the shore changes to a stormy ocean, the waves can tear ships apart, hurl rocks up on the shore, or carry water into a raised pool and store up P.E. The water in such waves has motion, yet an individual chunk of water does not move very far; it gains motion from neighbors when

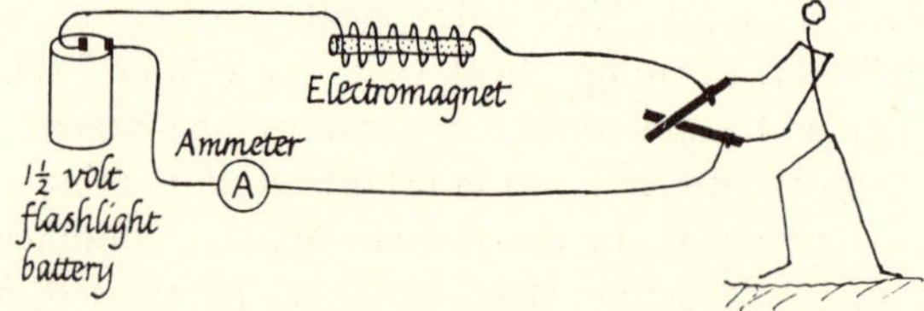

FIG. 26-31. MAGNETIC ENERGY
Experimenter completes the circuit by touching together two rods which he holds in his hands. He tries to break the circuit by separating the rods, but as he still has a rod in each hand the circuit continues to be complete, but with his high-resistance body included in it. When the experimenter separates the rods, the current drops to almost zero, the magnetic field of the electromagnet collapses, and in doing that, produces a very large extra voltage—which tries to maintain the current—and this drives a shocking current through the experimenter.

[16] For all their wave behavior, they carry their energy in compact quanta, "particles of energy." (See Ch. 44)

the wave arrives, then hands its motion on to the next chunk of water. Waves involve a complex pattern of K.E. and P.E., the one continually changing into the other, both being handed on with a characteristic wave-speed. The source of energy of most ocean waves is the wind that drives them. The wind gets its energy from warm land. Original source: sunshine.

Sound waves travelling in air carry energy. The alternating compressions and rarefactions of air which, following each other along, are the sound waves, involve small changes of molecular speed. The consequent changes in molecular kinetic energy are handed on as the wave travels; they add up to a tiny net increase, which is the energy of the sound wave. The details of energy-changes in a sound wave are difficult; you must take the account on trust. When someone shouts loudly in a room, he fills the air with slight extra outward motions, handed on rapidly with the speed of sound; and when the shout dies away the energy is left in the form of a minute heating of the room's walls which have absorbed the sound.

Electromagnetic waves, discussed above, carry energy in much the way that water waves and sound waves do. We call their energy "radiation energy" or just "radiation."

Water waves have K.E. of moving water and P.E. of crests raised above troughs. Sound waves have K.E. of tiny organized air motions forward and backward. They also have "P.E." that behaves like strain energy of a spring, but it is really just a little extra K.E. of *random* motion of molecules—a temporary store of extra heat. Electromagnetic waves seem to have P.E. in their electric fields and K.E. in their magnetic fields. It is not surprising to find magnetic fields associated with *kinetic* energy since they belong to electric charges in *motion*—currents in circuits, or whirling electrons in magnetic atoms, etc.

Nuclear Energy[17]

Radioactive atoms—natural ones found in uranium ores, and artificial ones manufactured from stable atoms by "big machines"—shoot out subatomic bullets, parts of their atoms, with incredible speed. The bullets carry enormous burdens of kinetic energy for their size. In our present picture of atomic structure there is no likely origin for these "bullets" in the outer regions of atoms, the regions where chemical changes swap electrons. They must come from the very small massive core (nucleus) of the radioactive parent atom. We therefore think of that core as having a great store of *nuclear energy*. Some of this energy is released willy-nilly in radioactive changes, but until recently it was thought to be inaccessible to human interference. Now we can precipitate a release by manufacturing unstable (radioactive) atoms.

In a few rare cases the nucleus, instead of shooting out a small chip, divides into almost equal pieces. This "fission" releases even more energy, and it provides spare neutrons to trigger fission in neighboring atomic nuclei, thus making an explosive chain reaction possible. Only certain very massive atoms are fissionable.

At the other extreme, nuclei of light atoms such as hydrogen release energy when several join to form a single more massive nucleus.[18] It is very difficult to push nuclei close enough to start this fusion process. So fusion seems quite impossible at ordinary temperatures. At *very* high temperatures, say in a star, molecular speeds may permit productive collisions, and the huge amount of energy then released in fusion might maintain the temperature for more fusions. If we can find some workable and controllable way to drive nuclei close enough for fusion, we shall have an almost inexhaustible supply of heat. We are already using controlled *fission*; and supplies of the "fuel," uranium, are unlikely to outlast oil and coal by many centuries. And it now seems likely that controlled *fusion* will succeed.

Uses of Energy

We have spread the forms of energy into a large list, anticipating experimental tests of our general belief in energy conservation. We are treating energy as something very useful, something that will do jobs for us, such as raising loads, something that can be measured by the job (in

[17] We call this energy "nuclear" rather than "atomic" since it seems to be located in the innermost core, nucleus, of atoms. However the older name "atomic energy" also remains in use, as in the "Atomic Energy Commission." Chemical energy (which is involved in changes of atoms' outer parts, their electrons) could well be called "atomic." And then we could make two suitable uses of "molecular energy"; for the energy of molecular motion (= heat) and for the potential energy involved in melting and evaporation.

[18] It may be misleading to draw a parallel between these nuclear changes that release "nuclear" energy and chemical changes that release "molecular" energy. Ordinary fuels and explosives that involve chemical changes make puny little energy releases compared with those in the nuclear events of fission and fusion. And the mechanism is essentially different. Yet we can illustrate the difference between fission and fusion by calling the explosion of a TNT molecule a case of "chemical fission"; and the explosion of oxygen and hydrogen to form water a case of "chemical fusion."

CLASSIFICATION OF ENERGY FORMS

NAME OF ENERGY	SHORT NAME	WHERE USED
Gravitational potential energy (often called simply "potential energy")	Grav. P.E. (P.E.)	Whenever a load is raised there is a gain of gravitational P.E. stored in the field.
Strain energy (often called simply "potential energy")	Strain E. (P.E.)	When a spring, or any other elastic material, is bent, stretched, twisted, or compressed, it stores up strain energy.
Kinetic energy	K.E.	The energy of motion of body. We can show that K.E. = (½) MASS · SPEED^2
Rotational kinetic energy or spin energy	Spin E.	Each part of a spinning object is moving, therefore has some K.E. Spin E. is the total of these K.E.'s.
Heat energy	Heat	We shall show that heat can be exchanged with K.E., P.E., etc. We now regard heat as the energy of molecular motions.
Chemical energy or Molecular energy (should be called atomic energy)	Chemical E.	Fireworks, explosives, fuels and food can release heat and other forms of energy in chemical changes. We picture them holding "chemical energy" stored "between atoms."
Molecular energy of melting & evaporation	Latent Heat	Extra heat is taken in, in melting or in evaporation. It does not make temperature rise but is stored in molecular fields.
Electric energy Magnetic energy	Electric E. Magnetic E.	Electric circuits, charged capacitors, electromagnets, involve electric energy and magnetic energy.
Electromagnetic energy	E-M E.	The energy may reside in the electric and magnetic fields, which are closely interconnected.
Electromagnetic wave energy	E-M Wave E.	Electromagnetic waves involve travelling electric and magnetic fields. They include visible light, infra-red, u.v., X-rays, and radio waves of all wavelengths.
Radiation energy (including light energy)	Radiation (Light E.)	Radiation includes light and all other electromagnetic waves.
Wave energy (including light and sound and ocean waves)	Wave E. (Sound E.) (P.E.+K.E.)	Most waves carry energy (e.g. light, sound, ocean waves).
Nuclear energy	Nuclear E.	Released in nuclear changes: radioactivity, nuclear fission, nuclear fusion.

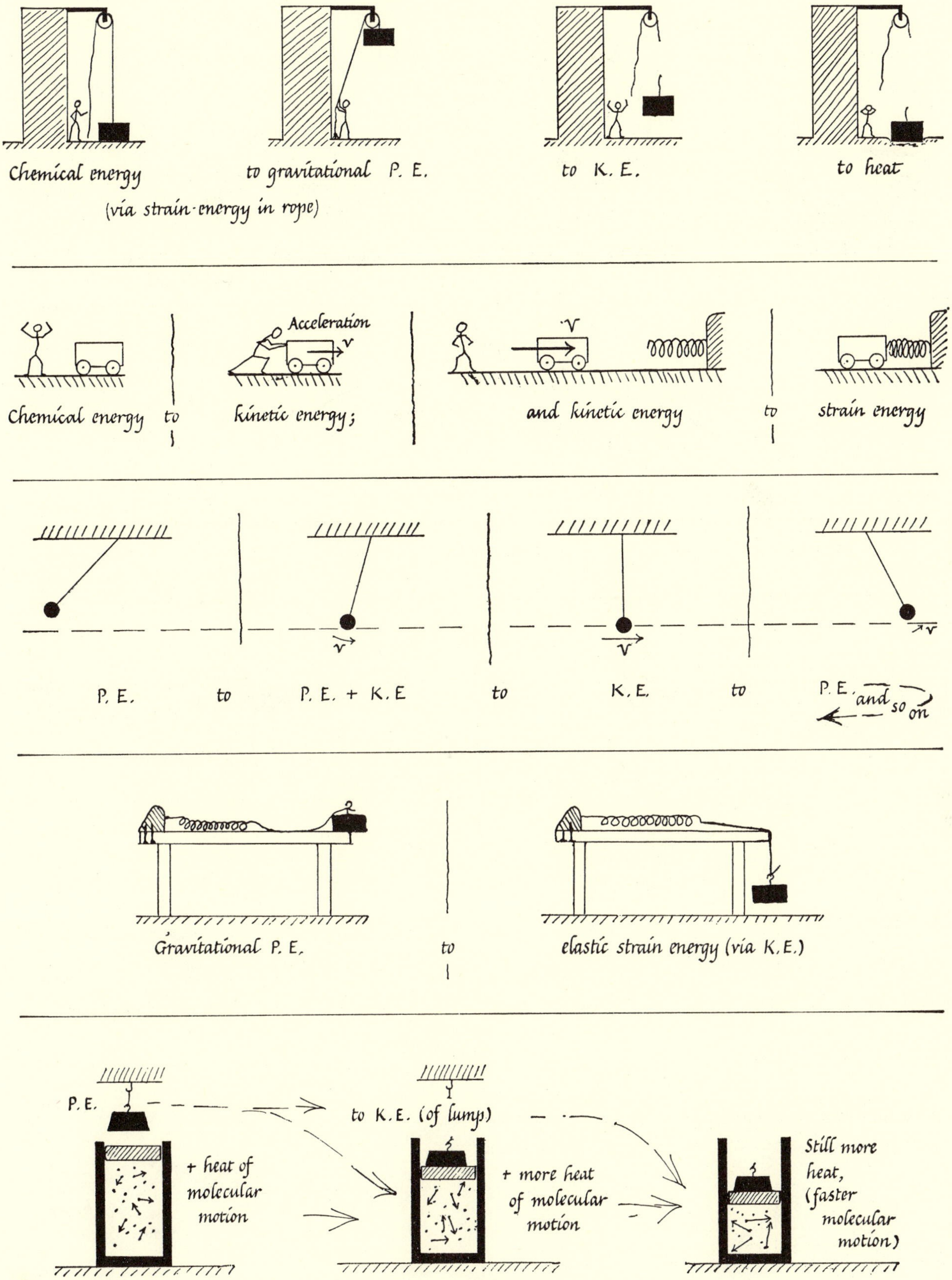

Chemical energy (via strain-energy in rope) to gravitational P. E. to K. E. to heat

Chemical energy to kinetic energy; and kinetic energy to strain energy

Acceleration

P. E. to P. E. + K. E to K. E to P. E. and so on

Gravitational P. E. to elastic strain energy (via K. E.)

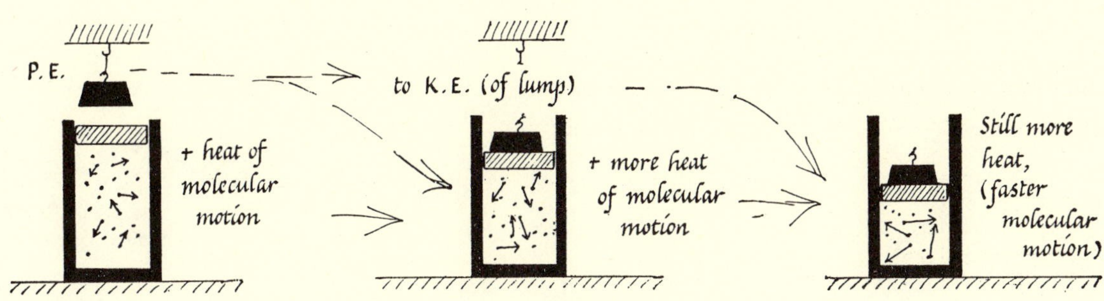

Fig. 26-32. Changes of Energy

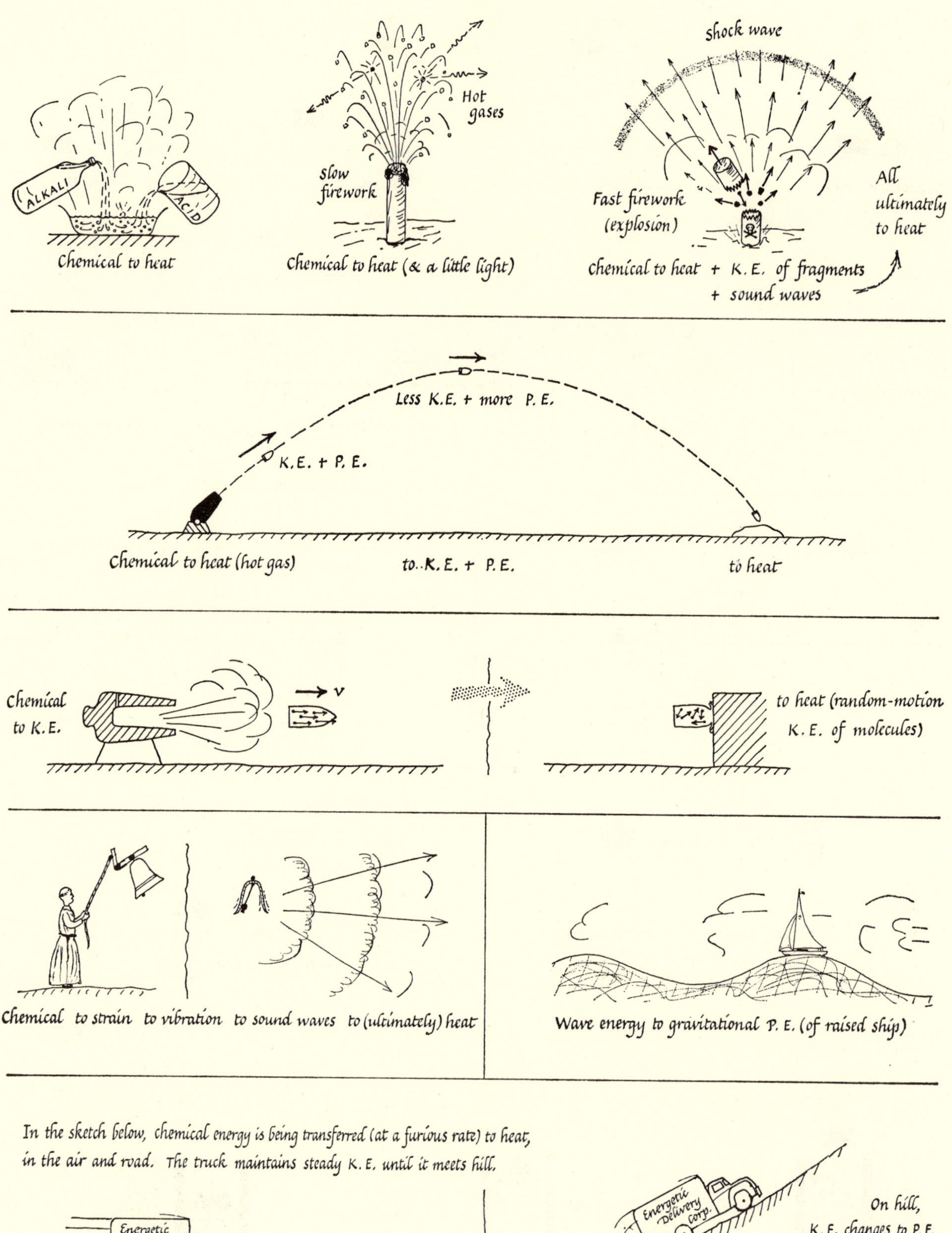

Fig. 26-32. Changes of Energy

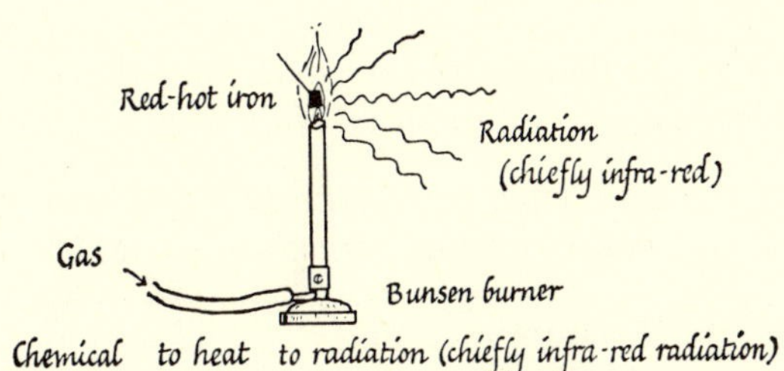

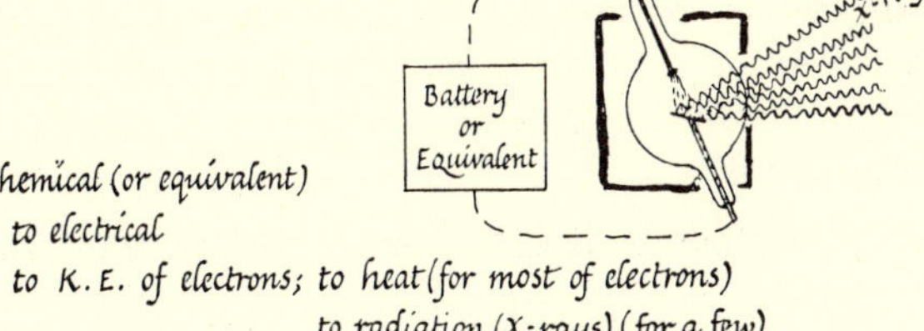

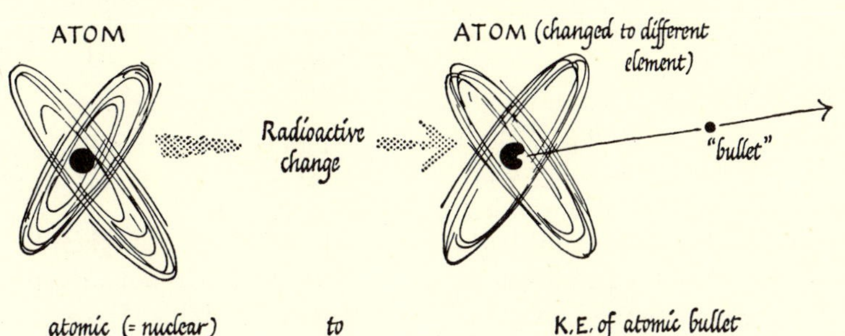

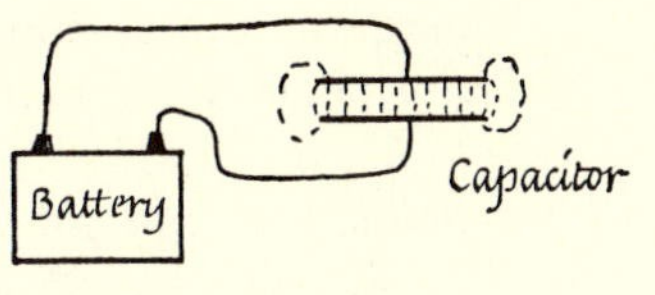

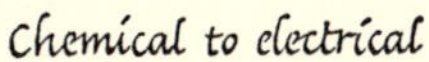

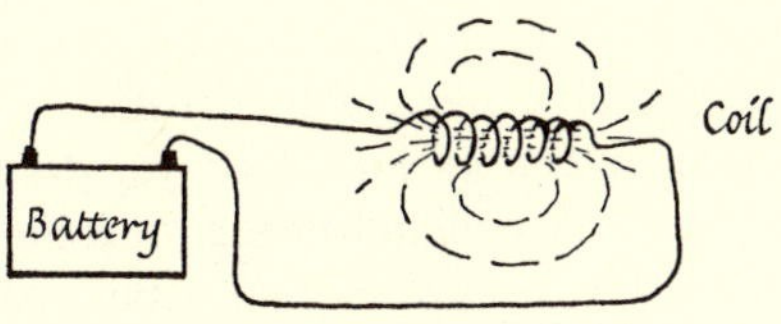

Chemical to magnetic

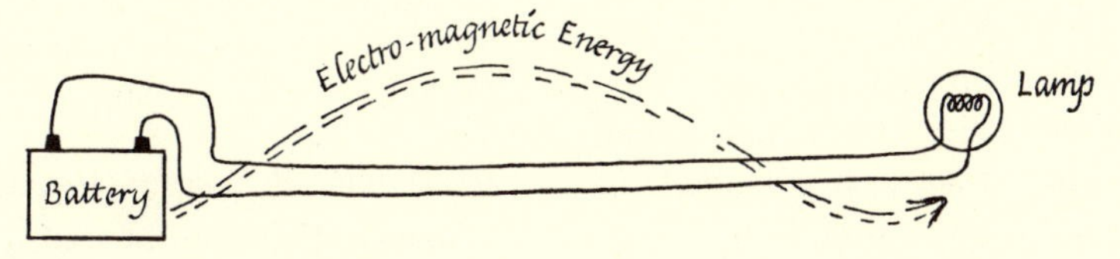

Chemical to electromagnetic

To heat and radiation

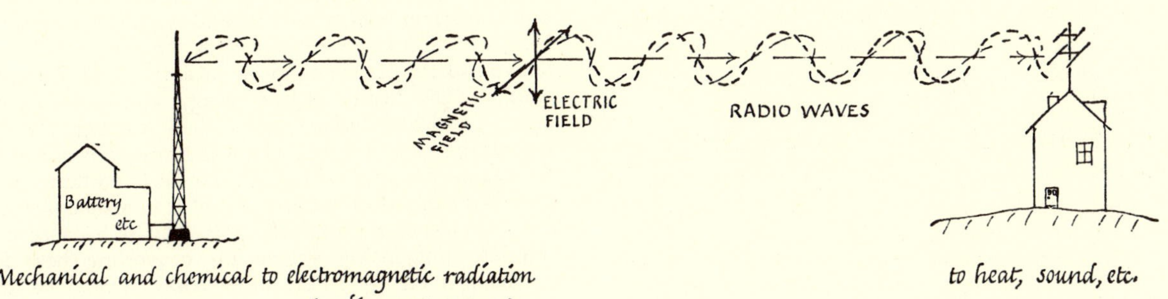

Fig. 26-32. Changes of Energy

units such as foot · pounds or newton · meters), using WORK, or FORCE · DISTANCE, to measure the energy transferred from one form or place to another. We need supplies of energy for a variety of useful jobs:

Heat for our bodies, houses and stoves, transportation.
Gains of P.E. for loads in elevators.
Gains of chemical energy in battery charging, or in fertilizer manufacture.
Supplies of electric energy for running machinery, lighting, cooking, &c. &c.

These uses fall into three general groups:

(a) *Chemical energy in food for living.* This is usually more expensive than energy from other fuels. Wheat costs more than coal, per Calorie yielded. Food provides necessary heat to maintain body temperature, necessary mechanical and chemical energy for body machinery; and sometimes external mechanical energy when men do jobs, with some waste heat.

(b) *Heat for house-warming and cooking.* This enables men to live in a wider variety of climates and to eat a greater variety of foodstuffs.

(c) *Mechanical energy and heat for transport, manufacture, etc.* All mankind needs food for living and more food for heavy manual labor. Modern civilizations use other fuels to replace man's labor, and increase the food supply, and extend his living area beyond the crude limits of climate. On a rough average over the whole world, each human being now uses eight times as much fuel, in energy value, as his food could supply. (The ratio ranges widely, from small values in countries like India which have little but animal-power as a supplement, to nearer 80 times in the U.S. In the U.S., the average mechanical power available for each worker using machines in a factory is 10 to 20 times his body's output.)

Although new fuel supplies are found, and the total world use of fuel increases yearly, the population increases almost as fast, and, on present prospects, will ultimately outstrip the fuel supply. How will this affect mankind?

A rough balance sheet of the whole world's energy uses runs something like the table below.[19] Each lot of energy there is shown by the amount of coal in millions of tons that would have to be burned to give the same energy (data refer to year 1937).

WORLD'S YEARLY FUEL CONSUMPTION

SOURCE	FORM OF FUEL PRODUCED	AMOUNT IN MT*
Current sunshine	Food for man, and Fodder for draft animals	625
	Firewood	150
	Hydroelectric power	35
Stored sunshine	Coal, oil, peat, natural gas	2200

* 1 MT means "1 million tons of equivalent coal" (about 8,000,000,000 kilowatt · hours).

Most of this energy taken by man is converted into useless heat without any profit, part of the waste being unavoidable. Here is what we get from it:

WORLD'S YEARLY USE

(a) *Food*: 250 MT go to feed humans; and, of that, about 25 MT emerge as useful human labor (replaceable by horses or engines and motors).

(b) *Useful household heat*: 100 MT.

(c) *Heat and mechanical energy for transport and manufacturing*: 350 MT (+ the 25 MT from men).

A total 700 MT out of a yearly supply of 3000 MT; the rest of it is "wasted" as heat. But we can hardly boast an efficiency of 700/3000, because we are omitting the sunshine wasted as heat, well over 10,000,000 MT a year.

Availability of Energy: Organized vs. Disorganized K.E.

In any discussion of usable energy, we must distinguish between heat, the energy of disorganized motion, and organized energy, known technically as free energy. The kinetic energy of a moving bullet is organized, all one way. We call it free energy because we can turn it all into potential energy—just fire the bullet straight up! The strain energy of a wound spring is well organized; we call it free energy because we can let the spring deliver all its strain energy in raising a load. Chemical energy is practically all free energy, and so are electric energy, and high-temperature radiation energy. We can put *all* the energy in any of these forms to use. But the disorganized energy of heat suffers an important restriction. *We can put only a fraction of it to mechanical use, however cleverly we go about it.* This is because even the best imaginable engine for converting heat into mechanical energy must throw away some heat to a cooling condenser—otherwise it could not repeat its cycle. We cannot organize all the random motion of the heat into free energy; there is always some disorganized motion left. A complex but convincing thought-experiment with an *ideal* heat-engine tells us that the maxi-

[19] Experts disagree on world energy supplies, partly because much of the information must be guessed at, partly because accounting policies differ—e.g., should we include farm wastes used as fuel, or is that counting fodder twice over? The guesses here are taken from two pamphlets issued by UNESCO on "Energy in the Service of Man"; No. 1 by Sir A. C. Egerton, No. 6 by Professor F. E. Simon (UNESCO, Paris, NS 74 and 79, 1951). (Data refer to the year 1937. Since then, world's yearly consumption of oil, gas, water-power energy has more than doubled.)

mum fraction of a heat supply that can possibly be used is $(\frac{T_1 - T_2}{T_1})$, where T_1 is the *absolute* temperature of the "furnace" or boiler and T_2 the *absolute* temperature of the local river or other cooling agent for the engine's condenser. (See Ch. 27 for meaning of absolute temperature.) For example, high-pressure steam at 500°K (227°C) and river water at 300°K (27°C) give an efficiency limit of (500 − 300)/500 or 40%. Such a steam engine *must* throw away 60% of its heat to the river, quite apart from practical losses.

The maximum fraction is $\frac{T_1 - T_2}{T_1}$ or $1 - \frac{T_2}{T_1}$. So, the larger T_1 is (and the smaller T_2), the nearer the fraction is to the desirable value 1, enjoyed by free energy. To lessen the waste, power-engineers aim at the highest possible T_1 for furnace or boiler. But they meet severe limits, set by oils that char and metals that melt; and they cannot lower T_2 below the surroundings with any permanent gain. We have practically no way of using the chemical (and nuclear) energy of fuels directly. We have to convert it to heat first; then we cannot avoid wasting a large fraction of that heat. Paradoxically, the same thought-experiment argument tells us that when our need runs the other way and we want heat from free energy—e.g., when we heat our house electrically—we can achieve an efficiency several times 100%. By using free energy to drive a small engine we can pump more heat from cool out-of-doors into our warm house. This "heat pump" is essentially an inside-out refrigerator with its freezer compartment out of doors. It is now coming into use and promises great saving of fuel.

Disorder, Entropy, Information

Use sunshine, or coal, or a water reservoir, to do some useful job, such as lighting lamps electrically, driving a lathe, pumping water uphill, . . . , and you will find that *heat* turns up again and again, as an almost unavoidable side-product (via friction) and as a very probable end-product. When the lamplight has been absorbed by walls, the lathe has shaped the metal, the water has flowed back to the lake, the energy provided by the original fuel has all ended up as heat. And if the original supply was in the form of heat, *the final heat resides at lower temperature*, less available for further use. Of course, we can contrive an ending in other forms: let the light escape into space, make the lathe wind up a spring, leave the water at the top of the hill; but heat is the usual product. (Reflect that all the energy released by burning gasoline in your car last year went ultimately into a slight heating of the air and earth.) This conversion to low-temperature heat means an increase of disorder among molecular motions. Even when heat is conserved, as when hot and cold air are mixed, there is an increase of disorder, from [a group of fast molecules in one region] + [a group of slow molecules in another region] to [a mixed group of molecules with random medium motions]. Considering both simple mixing of hot and cold and the general theory of heat-engines (thermodynamics), we conclude that the natural tendency among all molecular crowds is towards increased disorder, *as time goes on.* This gives time an important property, a forward direction *among statistical processes.* Simple mechanics, codified in Newton's Laws, lets time run forwards or backwards. A movie film of a collision between two molecules would look equally realistic run forwards or backwards. But a film of molecules of hot gas mixing with cold would look crazy if run backwards. Thus the jostling of myriads of molecules enforces "time's arrow" on our world. So we devise a physical measure of "disorder" and give it a name "entropy"; and we say "in the physical universe, entropy tends to increase." This is the "running down" of the universe[20] towards an ultimate "heat death" all at one low temperature of maximum disorder of matter and radiation.

Entropy may be defined as HEAT/ABSOLUTE TEMPERATURE, or by the probability of a molecular grouping. Details of its definition and use would take us too far beyond this course, but you should watch for this sophisticated concept in the science of the next half century. "The future belongs to those who can manipulate entropy; those who understand but energy will be only the accountants. . . . The early industrial revolution involved energy, but the automatic factory of the future is an entropy revolution."[21]

A group of gas molecules *might*, in the course of all their colliding motions, sort out, just once in a blue moon, into a fast (hot) group at one end of the box and a slow (cold) group at the other. That would be a *decrease* of disorder, the opposite of the likely thing that the entropy law predicts. But that pure chance event *is* unlikely—not impossible, but extremely unlikely. All the most likely groupings of molecular positions and speeds are the disordered ones; and a sorted-out, ordered, arrangement is likely to become disordered

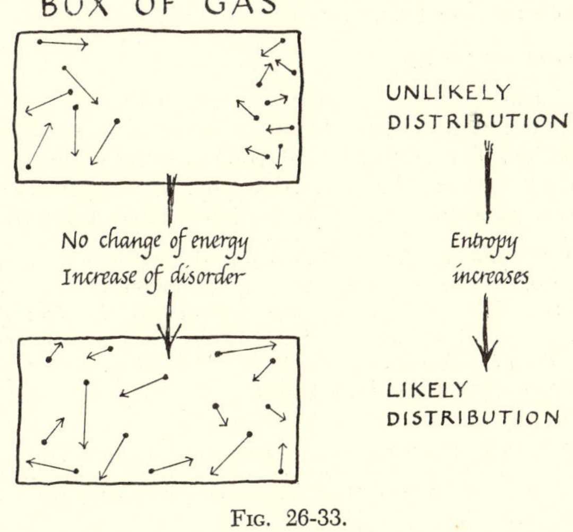

FIG. 26-33.

[20] How was this great machine wound up originally? Will it simply run down, or could some mysterious re-winding occur in long-time cycles? That would not violate the Law of Conservation of Energy, but it would violate our Entropy Law which seems equally true. Or is some continuous process of creation of matter replenishing the supply of available energy? These questions seem to be matters for philosophical or religious speculation rather than for science; yet astronomy and nuclear physics may be able to offer some answers.

[21] Frederic Keffer.

in the course of a few collisions, and is unlikely to occur again for a very long time. Sorted-out: unlikely . . . ; disordered: likely . . . : that is why we can define entropy in three ways, which are equivalent: (1) as a measure of disorder; (2) in terms of heat and temperature; (3) in terms of likeliness of a grouping, how statistically common it is likely to be. Law II of Thermodynamics says, essentially: "Entropy is likely to increase. Irreversible processes (such as heat-conduction, friction-rubbing, inelastic collisions, . . .) all make entropy increase. *The best we can hope for, in a perfect heat-engine, running continuously, is that entropy just stays constant.*"

Entropy is important in dealing with heat-engines, where we want to use all the available heat.[22] It seems very important in any discussion of Life, where time's arrow is paramount.

Recently, ideas of entropy have been transferred to "information theory," used in the communication industry and in designing machinery for automation. Suppose you could watch individual molecules of a gas and catalog the motion of each. With this detailed information you would no longer see the gas as a uniform community of maximum chaos, but would consider it highly uneven. *By gaining information you would decrease your estimate of the entropy of the gas.* Thus, information—pushed into a telephone channel as a message, or fed back from a thermometer to a thermostat—is like *negative entropy*. This analogy offers help in coding multiple telephone conversations efficiently, designing amplifiers, improving music/scratch ratios for phonographs, constructing automatic machinery; and in studying our own nervous system, vocabulary, memory, even reasoning, and perhaps "mind." Developed more loosely, it may be fruitful in social sciences.

Sources of Energy

When we look for energy sources, we seek stores of usable energy, not mysterious creators of it. Oil, coal, gasoline, illuminating gas, all contain stores of chemical energy that can be released to heat and other useful forms when they burn with oxygen.[23] These come from plants that flourished in sunshine a hundred million years or so ago. Wood to burn and crops to eat use sunshine to grow. They get their chemical energy from the Sun's radiation. Waterwheels and windmills provide mechanical energy that can be transferred to electrical and other useful forms and they too get their energy from sunshine. The winds are produced by unequal sunshine-heating; and the water that loses gravitational P.E. in a waterfall gains it in rising as vapor from lakes or oceans, under the action of wind and sunshine. All these energy supplies come from sunshine. The Sun's radiation brings us almost every energy "source" we use.

There are other sources independent of sunshine, such as volcanic heat, Moon's tidal action, and nuclear energy; but our present use of these is relatively small, and only nuclear energy promises to grow in use. Controlled fission, using uranium as "fuel," is already being used to propel submarines and even heat houses. We have ingenious "breeder" schemes to breed fissionable atoms from other isotopes that would not release their energy by fission. And we can foresee profitable extraction of uranium etc. from low grade ores, to provide a rich supply for centuries.

So we see the immediate future well supplied with three kinds of fuel: current sunshine, principally for food; stored sunshine in coal and oil; and nuclear energy. There is still a lot of coal. New oil fields are still being discovered faster than old ones are running dry—for several decades the experts have been saying "there's only enough left to last a dozen years." Prospectors are finding new deposits of uranium and thorium. However, none of these stores is likely to last more than a few centuries—perhaps ten at most—with a growing world population. Looking a thousand years and more into the future, when mankind has finished living on past capital, we see the Sun's current flow of radiation as the sole supply setting a limit to the feeding and comfort of all people—unless we realize our hope of almost unlimited energy from fusion of light elements.

We can speculate on the Sun's own source of energy for radiation. It is probably nuclear fusion: hydrogen atoms joining up in quartets to make helium atoms (perhaps by a roundabout process, with intermediate atoms) in the furnace[24] of the Sun's interior. This synthesis of helium yields heat on a vast scale, maintaining the Sun's huge outflow of radiation-energy at the expense of nuclear energy. It is fortunate that the Sun has a big supply of hydrogen to provide nuclear energy. "If the Sun had been a ball of ordinary fire, made of white-hot coal supplied with sufficient oxygen, it would have burned itself to ashes in a few thousand years." If, as we shall show later, radiation has mass, the Sun must be losing mass at a rate of some 300,000,000,000 tons or more per day! However, as the Sun's total mass is more than 2,000,000,000,000,000,000,000,000,000 tons, it will last a long time yet.

---

[22] Suppose we have a heat-engine that takes heat energy H_1 from the furnace and gives up H_2 to the cooling condenser. Then the engine gains ENTROPY H_1/T_1 from the furnace and loses ENTROPY H_2/T_2 to the condenser. At best, $H_1/T_1 - H_2/T_2$ is zero, or $H_2/H_1 = T_2/T_1$. If so, we can calculate the efficiency of the engine:

Efficiency

$$= \frac{\text{HEAT CONVERTED INTO MECHANICAL ENERGY, } H_1 - H_2}{\text{HEAT TAKEN IN FROM THE FURNACE, } H_1}$$

$$= \frac{H_1 - H_2}{H_1} = 1 - \frac{H_2}{H_1} = 1 - \frac{T_2}{T_1}, \text{ if entropy-change} = 0$$

$$= \frac{T_1 - T_2}{T_1}$$

That is for an ideal engine. Any real engine gains entropy in each cycle, by throwing away a larger H_2, so it must have even smaller efficiency.

[23] Strictly speaking, we should say that it is the fuel + oxygen together that have the store of chemical energy. Electron-shifts in combustion release it.

[24] ". . . a nuclear pot in which helium is being cooked up from pure hydrogen. . . ."—Atkinson and Houtermans. "There is one nuclear reactor which is enormously powerful, at a safe distance, and free for all: the Sun!"—Sir F. E. Simon (UNESCO, Paris, NS 79, 1951).

EXAMPLES OF ENERGY CALCULATIONS

Calculating Potential Energy: Δ (P.E.) $= F \cdot s$

We compute changes in P.E. easily, by the work involved, FORCE · DISTANCE. When a load is raised, its gain in P.E. is its WEIGHT, the pull of the Earth on it, multiplied by its VERTICAL RISE. If the body moves up along some slanting path, we still use WEIGHT · VERTICAL RISE. The Earth's pull exerts no sideways force on the load and so no work is involved in the sideways motion. If there is friction, of wheels on rough hill, or trudging foot on stairs, some work is certainly involved in horizontal motion as well; energy is transferred to useless heat in the road, tires, and shoes. But as this heat is not stored up so that it can be used on the return journey, we do not class it as P.E. Thus in calculating *useful P.E.*, stored energy that might be used to drive machinery, we do not count the horizontal motion. (See note on base-level for P.E. on page 398.)

EXAMPLE A: *Calculating Gain of* P.E.

(i) A 40-pound bag of corn is raised from the ground to the top of a barn 30 feet above the ground. Its gain of P.E. is WEIGHT · GAIN OF HEIGHT or WEIGHT · Δ HEIGHT.

Gain of P.E. $=$ (40 pounds-wt) (30 ft)
$=$ 1200 foot · pounds ["bad" units]

(ii) A 20-kg bag is raised from the ground to a height of 10 meters.

Gain of P.E. $=$ WEIGHT · Δ HEIGHT
$=$ 20 kg-wt. $\times$ 10 meters
$=$ 200 kilogram · meters.

If we want to express this gain in "good" units, which must always be used when there is motion involving K.E., we must express the weight in "good" force units, such as newtons, thus:

WEIGHT $=$ pull of Earth on 20 kg
$=$ (20 kg) (9.8 newtons/kg)
$=$ (20) (9.8) newtons $=$ 196 newtons

Gain of P.E. $=$ WEIGHT · Δ (HEIGHT)
$=$ (20 $\times$ 9.8 newtons) (10 meters)
$=$ 1960 newton · meters $=$ 1960 joules

(iii) Suppose the 20-kg bag is raised 10 meters by a curved route, guided by ropes and pulleys. It still gains *gravitational P.E.* (196 newtons) $\times$ (10 meters). That is the P.E. it would lose if it fell back to the ground vertically, and that is the P.E. it must gain for a 10-meter rise, however much it also moves

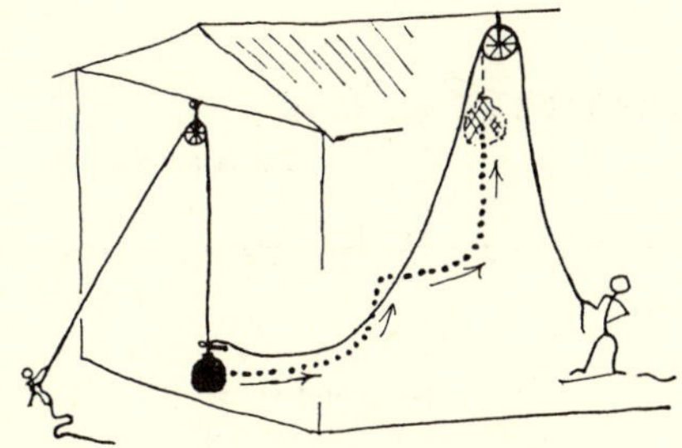

Fig. 26-34.

sideways. Test this in the following simple case. Suppose the load is hauled up a long incline, travelling 50 meters up the slope for the 10-meter vertical rise. In this case we know the uphill force needed to drag it up the slope without friction. That is given by

$$\frac{\text{FORCE } F}{\text{WEIGHT } W} = \frac{\text{VERTICAL HEIGHT}}{\text{SLANT LENGTH}}$$

$$\frac{F}{196 \text{ newtons}} = \frac{10}{50}$$

$$\therefore\ F = \frac{196}{5} = 39.2 \text{ newtons.}$$

But the man who pulls with only 196/5 newtons instead of the full 196 has to drag the bag 50 meters instead of 10 (or if he stands at the top he has to pull in 50 meters of rope, instead of 10). So his

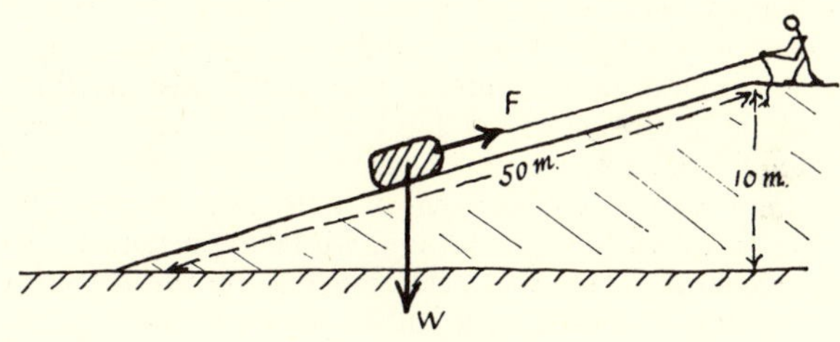

Fig. 26-35.

energy expenditure, measured by FORCE · DISTANCE is 196/5 newtons $\times$ 50 m., 1960 newton · meters as before.

If the slanting path suffers from friction, the man will have to use a bigger force than 39.2 newtons, but that is nothing to do with the gain in P.E. He uses extra force to drag the bag against friction, wasting some extra energy into heat, but not adding to the bag's gravitational P.E.

(iv) Stretching a spring. In this case the force changes, increasing steadily, and you must use a suitable *average* force. The best you can do is take the spring's pulls at the beginning and end of the

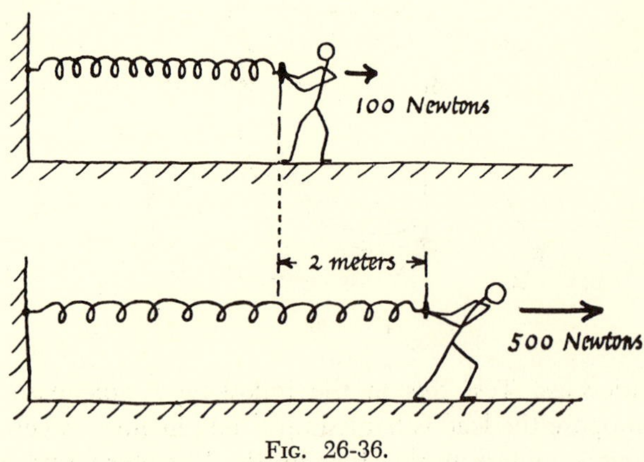

FIG. 26-36.

change and average them.[25] *Suppose a man holding a spring with tension 100 newtons pulls it out a further 2 meters, the tension rising to 500 newtons as he does this.*

Spring's gain of P.E. = (*average* FORCE) · DISTANCE

$$= \frac{(100 + 500 \text{ newtons})}{2} \cdot 2 \text{ meters}$$

$$= 300 \times 2 \text{ newton} \cdot \text{meters}$$

$$= 600 \text{ joules}$$

Note on Base-Level for Potential Energy

Gain of gravitational P.E. is given by: WEIGHT · HEIGHT RAISED. To calculate the *total* gravitational P.E. of some object, we must know its "total height," which seems meaningless. Where should we reckon the height from: ground, table, center of the Earth? Answer: there is no absolute base-level for P.E. in ordinary calculations. Fortunately we use *changes* of P.E.; and for those we may choose any convenient base-level. If we raise rocks from seashore to cliff-top, we take seashore as the base-level and give the rocks zero P.E. on the seashore. If we drop stones from a balloon down to the ground, we take the ground as base-level. If we drop stones down a well, we must *either* take the bottom of the well as base-level *or* take the ground as base-level and then use queer-looking negative potential energies for the stones when they are below ground. If we are doing experiments above a laboratory table, we may take the table top for base-level, or we may use the floor—in the latter case all heights will be greater by a few feet, but *differences* of height will of course be the same.

If we carry an object into the sky farther and farther from the Earth, we seem to give it more and more P.E. Just above the ground every kilogram gains 9.8 joules for each meter we raise it; but, when carrying it to great distances, we must allow for the changing pull of gravity, using the inverse-square law. A kilogram 4,000 miles above the ground gains only a quarter as much, 2.45 joules, for each meter we raise it. Because of this thinning of gravity, the P.E. of a body increases more and more slowly as it is carried away to greater and greater distances. Using the inverse-square law and calculus,[26] we find that at very great distances ("infinity" if you like) the P.E. creeps towards a limit, which turns out to be the P.E. the body would gain if raised *one Earth's radius against its full weight at the Earth's surface*. This means for one kilogram a gain of 9.8 newtons × 6,400,000 meters or 63,000,000 joules. If you take the absolute P.E. as zero at "infinity," then back at the Earth's surface each kilogram would have a negative P.E. = −63,000,000 joules. That would be clumsy and unhelpful for ordinary laboratory problems; but it tells us how much energy must be given to an object to get it away from the Earth. To hurl a rocket, or a gas molecule, out from the Earth so that it escapes permanently, we must give each kilogram more than 63,000,000 joules as kinetic energy to pay for potential energy. This sets a minimum "escape velocity." What speed must a mass M kg move with to have kinetic energy 63,000,000 M joules to spend on escaping?

A similar negative potential energy appears in simple atomic models like the Bohr atom. An electron has negative P.E. in the electric field of the attracting nucleus—reckoning from zero P.E. when the electron is far away, "at infinity" outside the atom. When the electron "falls" from some "outer" level to an "inner" one, the electric field loses some of its stored energy: there is less P.E., there is a *greater amount* of *negative* P.E. The energy released by the field is more than the electron needs for its increase of K.E. in a smaller "orbit" and the atom emits the balance as radiation.

EXAMPLE B. Calculations Involving Heat Energy

In calculating energy changes between heat and mechanical forms, etc., we must use the same units for both forms. Measurements discussed in a later chapter show that 1 Calorie ≈ 4200 joules.

A certain 4000-pound car drives one mile at 30 miles/hour on 1/20 of a gallon of gasoline, costing one or two cents.

---

[25] This averaging, to ½(initial force + final force) only holds for Hooke's Law springs. It can be justified by a geometrical argument, like Galileo's treatment of accelerated motion; or by calculus, as follows. Hooke's Law states: TENSION, $F = k$ (STRETCH, x). Then, in stretching from zero stretch to x_1, the spring gains P.E. given by:

$$\text{P.E.} = \int_0^{x_1} F\,dx = \int_0^{x_1} kx\,dx = \tfrac{1}{2}kx_1^2 = \tfrac{1}{2}(kx_1)x_1 = \tfrac{1}{2}F_1x_1$$

For stretch from x_1 to x_2,

$$\Delta(\text{P.E.}) = \tfrac{1}{2}kx_2^2 - \tfrac{1}{2}kx_1^2$$
$$= \tfrac{1}{2}(kx_2 + kx_1)(x_2 - x_1)$$
$$= \frac{(F_2 + F_1)}{2}(x_2 - x_1).$$

[26] $$\begin{matrix}\text{Gain of potential energy} \\ \text{Earth (R) to infinity}\end{matrix} = \int_{r=R}^{r=\infty} G\frac{Mm}{r^2}\,dr$$

$$= GMm\int_R^{\infty}\frac{dr}{r^2} = G\frac{Mm}{R} = G\frac{Mm}{R^2}R$$

$$= mgR = \text{WEIGHT} \cdot R$$

(a) *How much heat is released when this gasoline burns?*

(b) *What fraction of an average man's daily food would release as much energy when properly digested?* (Assume his diet is worth 3300 Cals/day.)

(c) *Make rough measurements on a decelerating car, and from them estimate the total "resistance" of road friction and wind-drag opposing the motion of the car at 30 miles/hr. Calculate the energy used to propel the car against that resistance for one mile.* (That is, the energy transferred from chemical energy to heat energy of road + air by means of the car's motion, excluding heat wasted directly from motor to air.)

(d) *What fraction of the heat released by the burning gasoline is taken to supply the energy needed for propulsion?*

Answers

(a) Gasoline is 0.74 times as dense as water. 1 gallon of water is 8 pounds of water.

$\therefore$ 1 gallon of gasoline is (0.74)(8) lbs. $\approx$ 6 lbs.

$\therefore$ 1/20 gallon of gasoline is 6/20 pounds
$\approx$ (6/20)/2.2 kg. $\approx$ 0.14 kg

$\therefore$ Heat released = 0.14 kg · 11,000 Cals/kg
$\approx$ 1500 Cals.

(b) This energy is 1500/3300 of the energy of an average day's food supply, or just under half. Therefore, *one mile in the car takes as much energy from the world's supply as one good lunch.*

(c) To calculate FORCE · DISTANCE for energy-transfer in propulsion, we need to estimate the force, *F*. Here is the record of an actual experiment with my car. The car was driven along a smooth, level road. When it was traveling just over 35 miles/hour, the gear was thrown into neutral and the car was allowed to coast along the road, with the resistance of air + road-friction decelerating it. It took 10 secs to decelerate from 35 to 25 miles/hour. (Thus the experiment was arranged to estimate the total resistance in the region of speed around 30 miles/hour.) The experiment was repeated with the car coasting in the opposite direction, so that effects of road-tilt or wind would be averaged out. Timings ranged from 10 secs to 14 secs, most of them between 12 and 13 secs.

$$\therefore \text{ The deceleration was } \frac{10 \text{ miles/hour}}{12.5 \text{ secs}}$$

$$\text{or } \frac{14.7 \text{ ft/sec}}{12.5 \text{ secs}} \text{ or about } 1.2 \text{ ft/sec}^2$$

$\therefore$ The decelerating force
$= 4000$ pounds $\times$ 1.2 ft/sec^2
$= 4800$ poundals $\approx$ 150 pounds-weight

This must be the force exerted by air resistance + road-friction.
In driving 1 mile against this resistance the energy-transfer (to heat, in air, road, tires, and the car's wheel bearings) is:

$$Fs, \text{ or } 150 \text{ pounds-weight} \times 5280 \text{ feet}$$
$$\approx \frac{150}{2.2} \text{ kg-wt.} \times \frac{5280}{3.3} \text{ meters}$$
$$= \frac{150}{2.2} \times \frac{5280}{3.3} \text{ kg} \cdot \text{meters} \times 9.8 \text{ newtons/kg}$$
$$\approx 1.1 \times 10^6 \text{ joules}$$

[Shorter method, using the value
1 joule $\approx$ ¾ ft · pound:

150 pounds-wt. $\times$ 5280 ft $\approx$ 800,000 ft · pounds
$\approx$ 800,000/(¾) joules
$\approx 1.1 \times 10^6$ joules]

(d) To compare this with the heat yielded by the gasoline, we must express both lots of energy in the same units.

Heat from gasoline = 1500 Cals
= 1500 · 4200 joules
= 6.3×10^6 joules

Therefore, out of 6.3 million joules provided by the gasoline, "useful" propulsion gets only 1.1 million. "Efficiency," for propulsion = 1.1/6.3 $\approx$ 17%, which is reasonable for a good gasoline motor.

EXAMPLE C. Calculation Using K.E. = ½ Mv^2

(Remember that since $F = Ma$ is used in deriving the expression ½ Mv^2 for K.E., kinetic energy will always emerge in absolute units such as joules or foot · poundals, and not in foot · pounds. Therefore, any force we use in *Fs* to calculate K.E. must be in absolute units.)

A rifle bullet emerges from a barrel 0.8 meter long with speed 400 meters/sec. The bullet has mass 0.002 kilograms. (a) *What is its energy?* (b) *What was the force*[27] *pushing it along the barrel, starting from rest?*

(a) Energy of bullet = ½ Mv^2
= ½ (0.002) $\times$ (400)^2 kilogram · meters^2/sec^2
= 0.001 $\times$ 160,000
= 160 newton · meters or joules

[27] This is the *average* force along the barrel. The explosive can be designed to burn at such a rate that the force is almost constant down the barrel.

(b) This energy was transferred from heat energy of the gases of the explosive. The WORK showing the transfer was: Fs or (F newtons) (0.80 meter)

$\therefore$ the K.E., 160 newton · meters, must = (F) (0.80). newton · meters, assuming that F was the resultant force, wholly concerned in accelerating the bullet, giving it K.E.

$\therefore$ FORCE, $F = 160/0.80 = 200$ newtons ≈ 45 pounds-weight, a large push on a small bullet.

EXAMPLE D. Illustration of conservation of energy.

(Try this for yourself, now. Then look at the answers if necessary.)

A man at the top of a cliff hauls a 2-kg rock up from the ground, 40 meters below.

(a) *How much P.E. does the rock gain?*

(b) *He lets it fall. When it has fallen 10 meters, how much of its recently gained P.E. does it still have? And how much K.E. does it have? Try adding P.E. and K.E.*

(c) *When it has fallen all the way, 40 meters, how much of its recently-gained P.E. does it still have? How much K.E. (before landing)?*

(d) *Describe the changes in P.E., K.E., and total [P.E. + K.E.] during the rock's fall.*

Answers to EXAMPLE D. Illustration of Conservation of Energy.

(a) Rock's gain of P.E.

= WEIGHT · HEIGHT RAISED
= (2 kg × 9.8 newtons/kg) · (40 meters)
= (19.6 newtons) (40 meters)
= 784 newtons · meters or 784 joules

(b) After falling 10 meters, the rock is only 30 meters above the ground, and its remaining P.E. is

(19.6 newtons) (30 meters) or 588 joules

To calculate its K.E., we need to know its speed, v. We first find t from $s = v_0t + \frac{1}{2}at^2$, thus:[28]

$$a = 9.8 \text{ m./sec}^2 \quad s = 10 \text{ meters} \quad v_0 = 0 \quad t = ?$$

$$s = v_0t + \tfrac{1}{2}at^2 \qquad 10 = 0 + \tfrac{1}{2}(9.8)\,t^2$$

$$t^2 = \frac{10}{4.9} = \frac{100}{49} \qquad t = 10/7 \text{ sec.}$$

$$\text{Then } v = v_0 + at = 0 + (9.8)\,(10/7) = 14 \text{ meters/sec}$$

$$\text{Then K.E.} = \frac{1}{2}Mv^2 = \frac{1}{2}(2 \text{ kg})\;(14^2 \text{ meters}^2/\text{sec}^2)$$
$$= 196 \text{ kg} \cdot \text{meters}^2/\text{sec}^2$$
$$= 196 \text{ newton} \cdot \text{meters}$$
$$= 196 \text{ joules}$$

$$\text{Total K.E.} + \text{P.E.} = 196 + 588 \text{ joules} = 784 \text{ joules}$$

(c) After falling 40 meters, the rock has lost all its P.E. It has fallen for 20/7 secs, and has speed 28 meters/sec.

$$\text{Its K.E. is } \frac{1}{2}(2 \text{ kg})\;(28^2 \text{ m}^2/\text{sec}^2) \text{ or } 784 \text{ joules}$$

$$\text{Total [K.E. + P.E.]} = 784 \text{ joules} + 0 = 784 \text{ joules}$$

(d) *General description of energy changes.* When the rock starts from rest, 40 meters above the ground, it starts with a store of P.E. 784 joules (supplied originally by the man, from chemical energy) and no kinetic energy. As it falls its P.E. decreases and its K.E. increases, but the total [K.E. + P.E.] remains the same, 784 joules, at all stages. Just before it lands it has no P.E. but maximum K.E. 784 joules. (Then when it hits the ground its K.E. disappears, and is probably changed into heat, warming rock, ground and air with nearly ⅙ Calorie.)

This problem is a simple, restricted example of the Conservation of Energy. Surprising? A delightful law of nature? Hardly: we have only just chosen the expression $\frac{1}{2}Mv^2$ for K.E. to make it equal to Fs and since we are using Fs to calculate changes of P.E. we must expect to find the total [K.E. + P.E.] constant, as a result of our choice.

If air friction took a large tax from the falling rock's motion, the total [K.E. + P.E.] would not stay fixed; it would grow less as friction took its energy-tax. Then energy would not seem to be conserved,

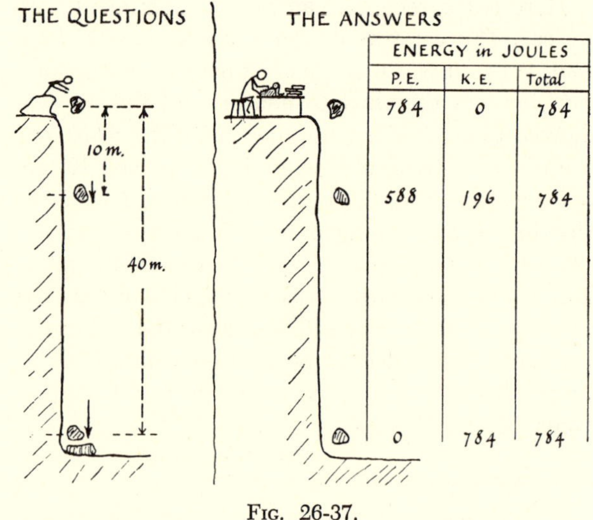

FIG. 26-37.

[28] For more direct calculation of v, see below.

until we took into account the heat energy (and K.E. of air currents) given to the air through friction.

In the calculation (b) above we used a clumsy way of calculating v, in steps, first the time by $s = v_0t + ½at^2$ then the velocity by $v = v_0 + at$. This was done to avoid the unfamiliar $v^2=v_0^2+2as$. If we use that, v emerges quickly:

$$v^2 = 0^2 + 2(9.8)\ (10) = 196\ \text{m.}^2/\text{sec}^2$$

$$v = \sqrt{196} = 14\ \text{meters/sec}$$

and, if we use that, it is all the clearer that we must expect a constant total, since that is the relation we actually used to derive K.E. $= ½\ Mv^2$.

In more general algebraic form: suppose the rock has mass m and is thrown down with initial velocity v_0 at height h_0 above the ground. When its height is h_1 above the ground, it has fallen a distance $(h_0 - h_1)$ from rest, with acceleration g downward; so its velocity v_1 is given by:

$$v_1{}^2 = v_0{}^2 + 2g(h_0 - h_1)$$

Then $[\text{K.E.} + \text{P.E.}]_1$

$$= ½\ mv_1{}^2 + mgh_1$$
$$= ½\ m\ \{v_0{}^2 + 2g(h_0 - h_1)\} + mgh_1$$
$$= ½\ mv_0{}^2 + mgh_0 - mgh_1 + mgh_1$$
$$= ½\ mv_0{}^2 + mgh_0$$
$$= \text{original } [\text{K.E.} + \text{P.E.}]_0.$$

So the total is the same at any h_1 as at the original h_0.

EXAMPLE E. Heat and K.E.

A lead bullet, mass 0.006 kg, moving 400 meters/sec hits a steel wall head-on and stops dead. Calculate its temperature rise. Specific heat of lead = 0.03 1 Calorie = 4200 joules. [*Note*: Specific heat 0.03 means that any piece of lead requires 0.03 times as much heat as the same mass of water for the same temperature-rise. See Chapter 27. For a rise of ΔT centigrade degrees, a mass M kg requires $M \cdot \Delta T$ Calories, if it is a mass of water. If it is lead it requires 0.03 times as much heat, or $M \cdot \Delta T \cdot (0.03)$ Calories.]

We assume that all the kinetic energy of the bullet turns into heat

$$\text{K.E.} = \frac{1}{2}Mv^2 = \frac{1}{2}(0.006)\ (400^2)$$
$$= (0.006)\ 80{,}000\ \text{joules.}$$

If the temperature rise is ΔT centigrade degrees, then heat acquired by lead

$$= (\text{MASS})\ (\text{TEMP.-RISE})\ (\text{SPECIFIC HEAT})$$
$$= (0.006)\ (\Delta T)\ (0.03)\ \text{Calories}$$
$$= (0.006)\ (\Delta T)\ (0.03)\ (4200)\ \text{joules}$$

If all the K.E. turns into heat, and *if* all that heat stays in the lead (0.006)(80,000) joules must be equal to (0.006)(ΔT)(0.03)(4200) joules. Cancelling the bullet's mass, 0.006—why should it cancel? —and solving for ΔT, we get

$$\Delta T = 80{,}000/(0.03)\ (4200) = 635\ \text{C}^\circ\ \text{temp.-rise}$$

Like many answers to textbook problems, this one is unrealistic, because such a rise would bring the lead above its melting point; and in a real collision some of the heat would appear in the wall.

Closed Systems

Any conservation law—for energy, momentum, water, cash, . . . —must refer to a "closed system." We draw an imaginary boundary around the region concerned, and make sure that none of the conserved item crosses the boundary. Then when we say something is conserved we mean it cannot be created or destroyed (except in equal + and − amounts) but only exchanged, inside the boundary. Pushed to the limit this requirement would make us take the whole universe as our closed system, but in most cases a small group serves as a practically closed system.

We could hardly maintain Conservation of Cash for a single person, or for a single town—in each case the system is not closed, but cash flows to and fro across its boundary. However, we might establish Conservation of Cash for a small island. This requirement seems obvious enough, yet it is easy to make paradoxes by forgetting it.

While a gun is firing a bullet, the bullet is *not* a closed system for either momentum or energy, but is gaining in both. But, if the gun is on wheels, gun & bullet & gas together form a practically closed system for *momentum*—they gain equal and opposite amounts, and the total momentum of the group remains constant. For energy we should have to take gun & explosive & bullet. Then we could claim conservation.

Conservation of Mechanical Energy: [P.E. + K.E.] = constant.

Suppose we have a closed system for energy—no forces to carry energy in or out across its boundary. The resultant force on the system from outside must be zero. All the forces inside must occur in pairs, F_1 and $-F_1$, F_2 and $-F_2$, &c. (Newton Law III). By splitting up each force into suitable components and multiplying by distance-moved we can calculate the energy-transfers for any change inside the system. This needs careful geometry and the knowledge that forces are vectors and act independently of each other. We shall not give details here, but the conclusion of the argument is that if all the forces are like the pulls of springs and gravitational fields, they will make equal and opposite energy-transfers between various kinds of P.E. and K.E. But the argument fails if there are forces like friction, which always opposes any sliding and does not oppose motion one way and help it the other as a spring would. If you drag a stone up a frictionless hill from A to B, the gain of gravitational P.E. is the same for a straight path from

A to B or a roundabout one. On a rough hill, the energy-transfer to heat is larger for a longer path. So the essential characteristics that enable us to say [P.E. + K.E.] is constant for some systems are:

P.E. depends on position only—the ends of the spring, the height in the gravitational field. Change of P.E. does not depend on the route taken.

K.E. depends on velocity only, not on the route or the time taken to acquire it.

A Powerful Tool

[P.E. + K.E.] = constant saves much calculation. For those "conservative systems" that have no friction, we can answer some problems quickly without calculating internal forces. For example: a pendulum 5 meters long with a 4-kg bob is pulled 4 meters horizontally to one side and released. What is the speed of the bob at the lowest point? This is like a body sliding down an in-

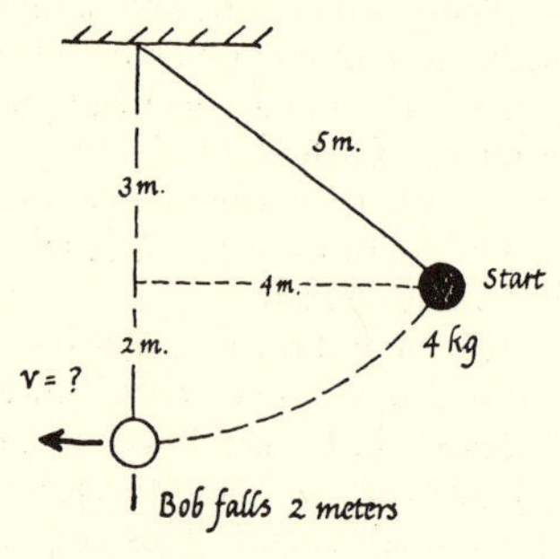

FIG. 26-38.

cline of *changing slope*. The accelerating force changes, and we may have some trouble in adding up all the gains of velocity to find the answer. But, trusting in conservation of energy, we find it quickly, thus:

$$\underset{\text{at starting point}}{\text{K.E.} + \text{P.E.}} = \underset{\text{at bottom of swing}}{\text{K.E.} + \text{P.E.}}$$

$$0 + (4\text{ kg})\ (9.8\text{ newtons/kg})\ (2\text{ meters}) = \frac{1}{2}\ (4\text{ kg})\ (\text{VELOCITY})^2 + 0$$

$$\therefore\ v^2 = 39.2 \qquad \therefore\ v = 6.26\text{ meters/sec}$$

This method applies to any path, straight, curved, even up and down, as long as there is negligible friction. So now we can regain from our thinking Galileo's general rule for motion down an incline (which went into our thinking at an earlier stage). A mass M starts from rest and slides down a frictionless incline of height h and slant length L. It loses P.E. Mgh and gains K.E. $\frac{1}{2}Mv^2$. Trusting in conservation, we say $\frac{1}{2}Mv^2 = Mgh$. Therefore $v^2 = 2gh$. This gives the final SPEED $= \sqrt{2gh}$. The mass has cancelled—all masses have the same motion. And there is only h, no sign of L: *the final speed is the same for all inclines of the same height.*

Human Health and Happiness

Food provides chemical energy for transfer to other forms, to keep us warm, maintain bodily machinery, and do mechanical jobs such as walking around and raising loads. All the energy we thus use must come from our food, or else we must draw it from our reserves by using some of our own fat. If you believe that energy is conserved you will agree that there is no escape from this choice: we must take in enough food (and digest it) or use our fat. A man can*not* do just one extra job without extra food; nor can he do normal jobs on subnormal diet, except by using his own fat, and that comes ultimately from his food. Our fuel-foods combine with oxygen to produce carbon dioxide and water. A given food releases the same amount of energy from chemical form, when "burned" to carbon dioxide and water, whatever the chain of chemical stages between—if not, we could disprove the conservation of energy by building up some food by one process and breaking it down by another process that released more energy! So we can measure energy-values of foods by burning samples in a laboratory. Then we can calculate the fuel *supply* in our diet and in the diets of people all over the world. We can calculate the fuel *used* in various activities from the CO_2 breathed out; in an hour's walking, in a night's sleeping, in a football game, in a day's shift in the machine shop, in a 3-hour mathematics examination. Then we can estimate the cost for a specimen day. The measurements are made by placing a mask on the victim's face and collecting the breath he breathes out in a short measured time. The volume collected is measured (with a gas meter) and a sample of it is analyzed to find how much CO_2 has been added to it in place of oxygen. This chemical change, multiplied up to the full volume of breath for the whole job, tells us the total amount of food fuel "burned" by the victim during the job. Basic living, keeping the machinery of heart, lungs, digestion just idling, demands a certain amount of energy each day. This minimum energy-turnover is called our *basal metabolism.* In cold weather, body heating demands some more. Walking and other mild activities add further demands, and violent exercise requires much more. For heavy physical work we must take in a great deal more energy in food than we need for the job itself, because our bodies are only about 25% efficient, wasting the other 75% as heat.

Basic living for a healthy man costs nearly 2000 Calories in food per day; swimming or football needs 500 Calories extra for every *hour*; 8 hours of heavy manual labor needs 2000 Calories extra for the day. Thus the heavy laborer uses twice as much fuel as the idle fellow of the same build; and therefore he must eat twice as much food. On the other hand, a student working hard at mathematics seems to use as little fuel, at the time, as if he were idle. Brain work makes little extra fuel-demand immediately—although it is so skillful, it is probably cheap in fuel; or perhaps it sends in its bill later on.

We all need some extra fuel beyond the "basic living" demand, unless we are utterly idle, in prison or sick in bed, and this must come from food. Where food is short because crops fail or other resources are low, people *must* work less. A starving people with insufficient diet cannot "make do." They must live a quiet life, or even stay in bed, or lose and lose their own body.

As populations increase it is food that sets the limit to growth. In the world as a whole there have always been large groups at the edge of starvation. Each time fuel-driven machinery makes it easier to grow food, or to do other jobs, the population rises to the new food + fuel level. At the present time, food and fuel are controlling factors in the peace and welfare of the world, though the supply of fresh water may hold even greater future threats. If we could develop a huge supply of cheap energy from atomic fusion, to be applied in agriculture as well as industry, we could make a much more comfortable world—or would the population simply increase to the edge of the new food supply?

MASS, MATTER, AND ENERGY. $E = mc^2$

Nowadays we hear strange statements such as "mass and energy are the same thing." And, "$E = mc^2$" is bandied about as if it explained an atomic bomb. When you are first gathering impressions of energy and its behavior for your scientific thinking it would indeed be misleading to throw the statement "mass = energy" into the discussion—and anyway that is a poor way of expressing a great discovery. It is perhaps the sharp wording of young reformers, the Galileos of the new age. What *has* been predicted by good theory and verified by many experiments is this: *energy has mass.*

Here, we shall first comment on this modern view, and then give some account of the way it developed.

When a material body gains energy (of any kind) its mass increases, and we consider the extra mass as belonging to the gained energy. For example, when radiation is absorbed the absorber grows hotter and gains a little mass—far too little to measure in ordinary experiments. Conversely, when a piece of matter emits radiation it loses a small bit of its mass, and the radiation itself seems to carry that mass away. The wider question arises: does *all* mass belong to some energy; that is, does all matter contain a huge store of energy locked in it? Long ago, radioactive changes suggested the answer yes. When a radioactive atom breaks up it releases a lot of energy as K.E., etc., and a small part of its mass disappears—measurements show that clearly. The energy seems to carry mass with it, leaving less matter-mass.

Thus, *some* of the mass of matter is interchangeable with mass of radiation, K.E., etc., and therefore we might say "matter and energy are partially interchangeable." Furthermore, we can now make pieces of matter, which *have* mass, change completely into radiation, which also has mass—and that radiation can transfer its energy to other forms, which also have mass. Conversely, radiation can turn into particles of matter. So, as well as "*energy has mass,*" we can now say "*particles of matter are interchangeable with radiation,* and thereby with other forms of energy." This is the creation and destruction of matter. Such weird events do not occur in the realm of everyday physics and chemistry and engineering; we must look for them in the small but violent actions studied in nuclear physics, or in the high-temperature cauldrons of atomic bombs or the Sun and stars. Even in these events mass is conserved: [MASS OF MATTER + MASS OF ENERGY] remains constant. (See Chapters 43 and 44.)

Yet it is still unwise to say "energy is mass." We say that energy *has* mass, just as matter *has* mass. And in the case of energy the mass of a chunk of energy is given by $m =$ ENERGY/(SPEED OF LIGHT)^2. With the standard symbol c for the velocity of light,[29] $m = E/c^2$, which rearranges to $E = mc^2$. In the case of matter, $E = mc^2$ suggests that a mass m of good solid matter has a huge internal store of energy given by MASS · SPEED OF LIGHT^2—but this store is locked and not to be released unless some mass of matter disappears.

How did this surprising idea arise, that energy has mass, and why was it not discovered earlier? It was suggested by both experiment and theory in several ways that we shall describe below; but it was not observed till this century, because energy-changes in ordinary experiments carry so little mass. Nowadays we are sure that a bullet in motion has the extra mass of its kinetic energy, but even at 4000 ft/sec a bullet that was exactly 1 ounce at rest would

[29] $c = 186{,}000$ miles/sec $= 3.0 \times 10^8$ meters/sec. Here we have changed from M to m for mass, to conform with custom.

have a total mass of only 1.00000000001 ounces. A kilogram of platinum heated white hot would gain 0.000000000004 kg—and no practical weighing could detect the change. Only when rich stores of energy are unlocked from atomic nuclei, or when atomic "bullets" are given speeds near that of light, does energy's mass become noticeable.[30] Plain kinetic energy contributes only too noticeably to the mass of high-speed protons that we accelerate with cyclotrons—and that increases the difficulties of working such machines.

Why Do We Believe That $E = mc^2$?

Where does $E = mc^2$ come from? Why do we believe that energy E has mass E/c^2? Nowadays we accept this as a direct deduction of Relativity theory, but the first hints came a century ago from the properties of radiation. It seemed very likely that radiation had mass. Since radiation carries energy, or *is* energy—on the wing at speed c—there was one case of mass belonging to something that is not matter. The behavior of such electromagnetic waves could be predicted from experimental laws of electromagnetism: they should have a "mass" given by their ENERGY/c^2. Only a wild speculative guess could extend $m = E/c^2$ to other forms of energy, until Relativity theory developed. (See Chapter 31.)

(1) *Hint of $E = mc^2$ for radiation*

All kinds of electromagnetic radiation (radio waves, infra-red, visible light, ultra-violet, . . .) have some simple common properties: they all move with the same speed, c, in vacuum; they all carry energy and momentum. We think of light and other radiation as wave disturbances travelling very fast, with a definite speed, $c = 3 \times 10^8$ meters/second. When light hits an absorbing wall, heat is produced, showing that the stream carries energy. That energy must whiz along in the stream at the speed of light—in fact that is how the speed is measured, by timing a chunk of light-energy over a huge distance.[31] When light is reflected from a mirror no heat is produced because the reflected beam carries the energy away again; but there is a pressure on the mirror, like the pressure of a stream of elastic balls or molecules. If the light hits a black, absorbing wall instead of a mirror, the pressure is only half as great. This shows that the stream carries momentum which is reversed by the mirror. Therefore light behaves *as if* it had mass. How else do you know a thing has mass? Does mass obviously exist in its own right, like length or green light or water? Or is it an artificial concept, described by behavior, like Justice or Humility? You really know mass by three descriptions:

(A) Vague statements about mass showing how much "stuff" is there. (On this view, mass belongs to matter—stuff that we can see, touch, push around.)

(B) Definite statements, $F = Ma$ and $F\,\Delta t = \Delta(Mv)$ and K.E. $= \frac{1}{2}Mv^2$.

(C) Mass is conserved.

If we forget the vague description, and shelve Conservation as a hope, we *define* mass in terms of momentum and energy. Then any moving thing with momentum and energy must have "mass." Its "MASS" should be MOMENTUM/VELOCITY or 2(K.E.)/VELOCITY^2.

We can find the "MASS" of radiation from its pressure, by a momentum calculation, as follows:

The actual pressure of a sunbeam on a mirror is very small, but it has been measured with a delicate "windmill" in a vacuum.[32] Then the energy-density was measured by letting the sunbeam warm up a small black absorber. The measurements showed that:

$$\left[\begin{array}{l}\text{PRESSURE OF}\\ \text{BEAM OF LIGHT}\\ \text{on mirror which}\\ \text{bounces}\\ \text{the light back}\end{array}\right] = \left[\begin{array}{c}\text{twice the ENERGY-DENSITY}\\ \text{which is } \dfrac{\text{ENERGY in beam}}{\text{VOLUME of beam}}\end{array}\right]$$

and this agrees with the prediction from laws of electromagnetism. Now try to predict that pressure by a kinetic theory calculation.

Suppose a beam of light of length a hits a patch of mirror of area $b \cdot d$ head-on and is reflected. Treat the light in the beam as equivalent to N elastic balls of mass m moving with speed c. Then the time taken for all the

[30] On the other hand, even the smallest noticeable difference of mass is a sign of a huge store of energy that *might* be released. For example: hydrogen and helium atoms have relative masses 1.008 and 4.004. If four hydrogen nuclei could be made to gang up into one helium nucleus, the mass-change would be 4.032 to 4.004, a small difference of 0.028 or 0.7%. But that would mean an enormous energy-release—primarily by radiation. 4.032 kg of hydrogen would then release 0.028 kg of radiation, which would have

$$0.028 \cdot (3 \times 10^8)^2 \text{ joules of energy,}$$
$$\text{or about } 600{,}000{,}000{,}000 \text{ Calories}$$

compared with a mere 140,000 Calories when the same amount of hydrogen combines with oxygen in a chemical explosion.

[31] Also when light falls on certain metal surfaces, it rips out electrons, and these come flying out as if they had been hit by compact bullets. The energy of light seems to travel in concentrated packets which we call "quanta." This is the "quantum" behavior of radiation—though some kind of wave seems to guide the paths of these bullets. For light of one color or wavelength, every bullet seems to carry the same energy, a definite "quantum." Such bullets of radiation whiz along with the speed of light—they *are* light—carrying energy and momentum. This makes it all the easier to ascribe mass to radiation, a definite mass to each bullet.

[32] The apparatus was like the "light-mill" now sold in toy stores. But the toy (a "radiometer") is turned by the much greater pressure of residual gas molecules hitting the vanes harder where they are heated by light. It turns the opposite way from the effect of radiation pressure.

Fig. 26-39. Radiation Pressure

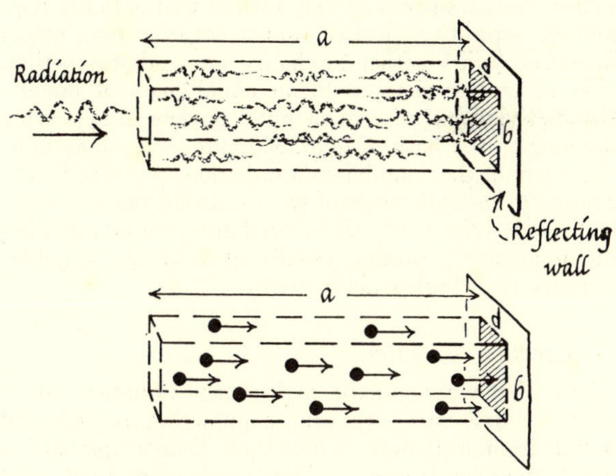

beam to reach the mirror is a/c. Total momentum change is $N \cdot 2mc$.

$$\therefore \text{ FORCE} = \frac{(\text{CHANGE OF MOMENTUM})}{\text{TIME}}$$

$$= \frac{N \cdot 2mc}{a/c} = \frac{2 \cdot N \cdot mc^2}{a}$$

$$\therefore \text{ PRESSURE} = \frac{\text{FORCE}}{bd} = \frac{2\,N\,mc^2}{abd} = \frac{2\,Mc^2}{\text{VOLUME}}$$

(See note 33.)

Compare this with experiment. Then:

$$\frac{2\text{ ENERGY}}{\text{VOLUME}} \text{ must} = \frac{2\,Mc^2}{\text{VOLUME}}$$

$\therefore$ ENERGY in beam must $= Mc^2$. $\quad\therefore\ M = E/c^2$

So, if radiation has a mass, like a stream of balls, its mass is E/c^2. This is not a rigorous proof, but a plausible suggestion based on some experimental knowledge. See the *end* of Ch. 31 for a much better derivation.

$E = mc^2$ resembles $E = \frac{1}{2}mv^2$. Since we are dealing with radiation, c has replaced v. But why has the ½ disappeared? The reason is that $\frac{1}{2}mv^2$ is not the correct expression for K.E. It is practically correct at all ordinary speeds, but at high speeds, approaching c, Relativity theory gives a new expression which will be discussed in the next section.[34]

Radiation has mass E/c^2. When radiation is emitted, the source feels a backward reaction and we guess it loses mass E/c^2. When radiation is absorbed the receiver should gain mass E/c^2. So $E = mc^2$ seems to apply to radiation, and its reactions with matter. Can we generalize and give *all* forms of energy a mass E/c^2? Relativity theory says yes.

(2) *Relativity*

Relativity theory was designed to straighten out a group of experimental paradoxes concerning absolute space and time. Two kinds of experiments with light gave conflicting results, and electrical experiments made the conflict seem worse. Einstein suggested that rather than put up with the conflict we should change our simple rules of geometry and vector addition. That change *is* his "special Relativity" theory, discussed in more detail in Chapter 31. We now adopt Einstein's scheme for dealing with lengths, times, velocities, and other vectors; then the conflicts disappear: experimental results and general laws fit into one consistent scheme.

For low speeds (from slowest crawl to the fastest rocket) the new scheme agrees with the old: K.E. is $\frac{1}{2}mv^2$ and matter keeps a constant mass, whether moving or not. At high speeds, comparable with the speed of light, our estimates of length or time are modified by motion of the apparatus relative to the observer; and in particular the mass of an object grows greater when it moves faster. Instead of the old rule for the mass of a piece of matter, $m =$ constant, Einstein showed we must use $m = (\text{constant})/\sqrt{1 - v^2/c^2}$ for an object moving past us at speed v. In this, c is the speed of light. We call the (constant) the "rest-mass" of the object, m_0. Then Einstein's mass formula becomes $m = m_0/\sqrt{1 - v^2/c^2}$. Electrical theory had already suggested this for the "mass" of a moving electric charge, and experiments confirmed it for high speed electrons. Then Relativity theory declared this increase of mass quite general. The change is trivial at ordinary speeds, and only rises to 1% at 100,000,000 miles/hour; but it becomes 10%, 100%, 1000% . . . for electrons and protons from radioactive atoms or modern accelerators. Experiments on these high-energy particles confirm the Relativity relation between mass and speed magnificently.

Any ordinary speed is so much smaller than c that v/c is a small fraction, v^2/c^2 still smaller and therefore $\sqrt{1 - v^2/c^2}$ is practically 1. At all small speeds mass seems constant, $m = m_0/1$.

---

[33] This is the gas-pressure prediction with the (⅓) left out, because the stream moves in one direction instead of at random, and with a 2 inserted, because we are counting the energy in the forward stream and not the forward and reflected streams.

[34] Before Relativity, K.E. was $\frac{1}{2}mv^2$ for any speed, and then the prediction ran

$$\text{PRESSURE} = 2\text{Nmc}^2/\text{VOLUME}$$
$$= 4\,(\text{K.E.})/\text{VOLUME}$$

for elastic balls; but experiment gave

$$\text{PRESSURE} = 2(\text{ENERGY})/\text{VOLUME}$$

for a beam of light. Here is the simple explanation of that disagreement that was given then—it is now out of date, and irritates most critics, but you may find it helpful. "A sunbeam is *not* a stream of moving bullets with kinetic energy: it is a stream of vibrating waves which, like all vibrating systems, have potential energy as well as kinetic energy, in equal shares.

"As pendulums oscillate, atoms vibrate, or waves travel, with S.H.M., energy swings to and fro from kinetic to potential to kinetic &c. Algebraic averaging of the K.E. and P.E. of such S.H.M.'s shows that their average values must be equal. For electromagnetic waves one might associate P.E. with the electric field and equal K.E. with the magnetic field. Therefore we must endow a mass M of radiation with 'kinetic' energy $\frac{1}{2}Mc^2$ and 'potential' energy $\frac{1}{2}Mc^2$ making total energy Mc^2. Now our prediction runs:

$$\text{PRESSURE} = 2Nmc^2/\text{VOLUME} = 2Mc^2/\text{VOLUME}$$
$$= 2(\text{ENERGY})/\text{VOLUME}$$

in agreement with experiment." That was thought to prove that light must be waves, with P.E. as well as K.E., and not bullets. Now we know better.

Then, maintaining Newton's definition of force, $F = \Delta(mv)/\Delta t$, and measuring work by $F \cdot \Delta s$, Einstein showed that the KINETIC ENERGY of any moving body $= (m - m_0)c^2$. Since $(m - m_0)$ is the gain of mass with motion, this says K.E. = (gain of mass) c^2. *Here is $E = mc^2$ for kinetic energy and its mass.* Adding a permanent store of locked-in energy m_0c^2, Einstein got TOTAL ENERGY $= (m - m_0)c^2 + m_0c^2 = mc^2$. (This is discussed in greater detail in Chapter 31, but even there the derivation is not given, because it needs calculus.)

K.E. $= (m - m_0)c^2$ looks quite different from K.E. $= \frac{1}{2}mv^2$. It *is* different, and needs to be, for high speeds. But look at its value for low speeds. Suppose v is small compared with c, so that v/c is small, and v^2/c^2 very small, compared to 1. Then we can use the binomial theorem (see footnote 2 in Chapter 22) thus:

$$\begin{aligned}
\text{K.E.} &= (m - m_0)c^2 \\
&= \frac{m_0}{\sqrt{1 - v^2/c^2}}c^2 - m_0c^2 \\
&= m_0c^2(1 - v^2/c^2)^{-\frac{1}{2}} - m_0c^2 \\
&= m_0c^2(1 - [-\tfrac{1}{2}]v^2/c^2 \\
&\qquad + \text{terms containing } v^4/c^4 \text{ \&c.}) - m_0c^2 \\
&= m_0c^2 + \tfrac{1}{2}m_0c^2\, v^2/c^2 + \text{negligible terms} - m_0c^2 \\
&\approx m_0c^2 + \tfrac{1}{2}m_0v^2 - m_0c^2 \\
&= \tfrac{1}{2}m_0v^2, \text{ the old expression for K.E.}
\end{aligned}$$

Here is good theory at work, producing the old result as a special case, and showing its limitation: low speeds.

We say that a moving body's extra mass *is* the mass[35] of its K.E. At *any* speed, a body of rest-mass m_0 has mass m_0 + (K.E./c^2), as you can see by using the Relativity form for K.E. At low speeds the binomial approximation gives $m = m_0 + (\frac{1}{2}mv^2/c^2) = m_0 + (\text{K.E.})/c^2$.

At the other extreme, radiation has no rest-mass, ($m_0 = 0$): it is not matter, and cannot be held still. It just has a mass m when it moves with its permanent speed c; so its energy is mc^2. We speak of the energy-bullets or quanta of light as *photons* whenever we want to emphasize the particle-behavior of light. Each photon has a definite mass m, definite energy $E = mc^2$ and momentum mc.

(3) *Nuclear transformations*

In some experiments with nuclei, the atomic masses fail to add up to the same total after a violent event. The energy released carries away some mass with it. What seemed a proper part of an atom's material mass disappears; but if we give the measured energy a mass E/c^2 we find *mass is conserved.*

(4) *Annihilation of matter*

Since we used to think of mass as the permanent mark of matter, this shifting of *mass* from matter to radiation—from the lamp to the moving light-beam—almost looks like *matter* being destroyed. It is just one jump further, and a surprising one, to find that actually happening: a positive electron and a negative one, pieces of matter, join together to change into radiation—their mass of matter changes to an equal mass of energy. This is literally a case of material objects disappearing like magic, in a flash of light. Measurements show that:

(ENERGY of the "annihilation radiation" produced)/c^2 is equal to the total MASS of two electrons, one + one −. And we find the newly-discovered anti-proton annihilating in joining a proton, usually in a splash of lighter particles with high kinetic energy.

(5) *Creation of matter*

Now that we command high-energy radiation—ultra short wave X-rays—we can manufacture particles of matter from radiation. When such X-rays bombard a target, they sometimes produce a pair of particles, e.g. + and − electrons. Again we find that, if we use $m = E/c^2$ for radiation and for kinetic energy, mass is conserved.

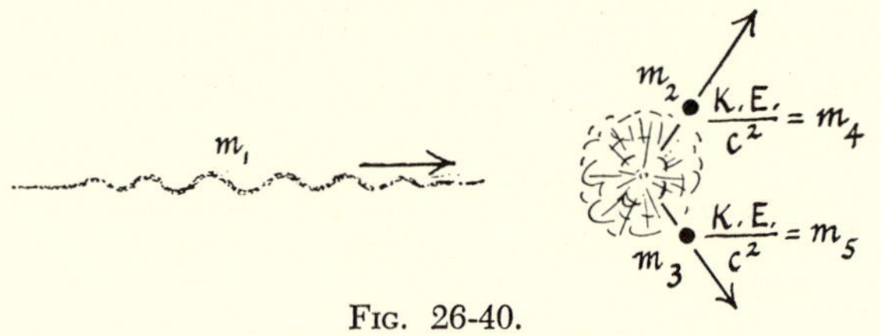

FIG. 26-40.

Conservation of Matter and Energy

Thus when matter gains or loses energy, or even appears or disappears completely, we expect to find mass preserved as the mass of radiation or some other form of energy. Such descriptions as "matter is frozen energy" seem childishly journalistic; but we do now believe that, come what may, mass is universally conserved, as a property of matter and energy combined. We have experimental confirmation for several varieties of energy: nuclear energy, radiation, ordinary kinetic energy, and we have no reason to doubt the general rule. We may if we like lump the Conservation of Mass and the Conservation of Energy into one grand Conservation of (MASS OF MATTER + MASS OF ENERGY).

Energy Locked up in Matter

Since there *are* exchanges between matter and energy, we may regard the mass of all matter as the mass of some "internal energy." However, none of that store is accessible in ordinary events. A little is released in some radioactive changes, more in nuclear fission and fusion; large fractions of it only, so far as we now know, in the annihilation of pairs of electrons, and other particle and anti-particle pairs.

---

[35] At higher speeds, with *extra* mass, the body gains *extra* K.E. on that account; therefore it has *more extra* mass, and makes its K.E. slightly greater still. This "series" converges to a definite, increased mass. But the true formula mounts to infinite mass (and infinite energy) for any piece of *matter* when v approaches c.

Still Other Forms of Energy?

What other forms of energy do we have? The scientist's usual answer is rather a surprising one. "Well, what other forms do we need?" Need? What for? When in the past we discovered new forms of energy we soon found them fitting into the general scheme and obeying the great accounting rule: "the total of all kinds of energy stays constant" or "*energy is never created or destroyed, it is only interchanged.*" In modern science this scheme has become so useful as a guiding principle that we should indeed be unhappy to see it fail—we should be scientists without a home, so to speak. If we do discover energy-changes that are unaccountable by our present listing of energy forms—if we find energy either vanishing or appearing from nowhere —we are tempted to imagine a new brand of energy to account for the difference. This looks dishonest—the banker inventing fictitious customers to square his accounts! It is not wicked, since we publish our assumption honestly and remember that we have made it; but it is risky. Looking back over the history of energy discussions, we can say that such risks have been taken before and have turned out to be magnificently justified. Only a century ago the idea of heat being a form of energy seemed queer. Many scientists were adopting it, but others spoke of it as a cult. The idea of nuclear energy locked in the core of atoms was suggested by radioactivity over half a century ago, but the full justification came only in recent years.

The neutron, now so important as the moving agent in atomic reactors, was discovered through a misfit in energy changes. Belief that energy and momentum are conserved (despite appearances) in certain atomic collisions, led to the conviction that an invisible particle, soon named the neutron, must be involved. Once guessed at, it was easily investigated experimentally. Then an energy-accounting problem appeared in another group of nuclear actions. Some radioactive nuclei emit electrons (beta rays); but, for the same final products, the electrons do not emerge with the same speed from every atom. Instead their kinetic energies range from zero to a characteristic maximum. Thus a large, variable, chunk of energy seems unaccounted for; and at the same time some momentum and some atomic spin-momentum seem to disappear. So physicists invented a tiny particle, perhaps tiniest of all the atomic hardware, the neutrino. It has no charge, is believed to have no rest-mass; it can whizz out from an atom unseen, almost undetectable, carrying just the right (!) amounts of K.E., momentum, and spin-momentum to keep the books balanced. A wicked invention? Hardly wicked, but certainly either risky or pigheaded. At worst it was like a treasury printing extra currency. At best it could lead to great experiments and growth of knowledge. Yet all this must seem a disturbing attitude, this readiness to invent bogus (?) forms of energy. In terms of banking this looks like a mysterious item marked "to value of goodwill" or an unexpected bogus customer on the books. Yet such things do happen, without putting banks out of business or breaking the good name of banking. In fact, the neutrino has turned out to be a genuine customer in our bank. For years, while it remained undetected, it continued to balance the books convincingly. Recently we have obtained clear experimental evidence that neutrinos exist.

With such doubts in mind, we return to the key question: "Is energy purely an experimental thing, and its universal conservation based on experiment all through? Or is it a scheme that we have dreamed up and agreed on, and will try to maintain?" In the limited region of mechanical energy, you should consider conservation assured by Newton's Laws of motion and the vector properties of forces. And you will certainly think energy and its conservation well vouched for and well founded on experiment when you study the last century's work on heat as energy (Chapter 29). You can be sure that if the general conservation of energy were a farrago of credulous thinking or wild imagining, the mistake would have been found out long ago.

But in its present fullest form you may consider it more than a generalization from experiment; it has expanded into a convention, an agreed scheme of energy now so defined that its total must, by definition, remain constant. If you feel disillusioned, read the following remark by Poincaré, one of the greatest mathematical physicists:

> "As we cannot give a general definition of energy, the principle of the conservation of energy simply signifies that there is *something* which remains constant. Well, whatever new notions of the world future experiments may give us, we know beforehand that there will be something which remains constant and which we shall be able to call *energy*."[36]

[36] *La Science et l'Hypothèse*, Ch. X (Flammarion, Paris, 1902).

PROBLEMS ON CHAPTER 26

Problems 1-4 are in the text.

★ 5. FORMULA

Derive, from the definition of acceleration, etc., the kinetic energy relation $F \cdot s = \Delta(mv^2/2)$.
(Show that $v^2 = v_0^2 + 2as$, and lead from that to the required result.)

★ 6. ENERGY CONSERVATION

A 100-kilogram bag of sand rests on a solid concrete sidewalk. A man on the top of a building hauls it up a vertical distance 20 meters ($\approx$ 66 ft) from the sidewalk. He then releases it and it falls freely to the sidewalk.

(a) Calculate the original energy transferred from man's chemical energy to gravitational potential energy in the raising of the bag. *State it in joules.*
(b) At the moment the man releases it, (i) how much K.E. has the bag? (ii) how much gained P.E. has it, reckoning sidewalk as zero level?
(c) When the bag has been falling freely for one second, from rest, (i) how far has it fallen? (ii) what is its velocity? (iii) what is its K.E.? (iv) what is its P.E. (above sidewalk level as zero)? (v) what is the total of its K.E. + P.E.?
(d) When the bag has been falling freely for 2 secs from rest, the answers to questions (i, ii, iii, iv, v) in (c) above may be different. Calculate each of them.
(e) Repeat (d) for the instant 3 secs from rest.

★ 7. ENERGY CHANGES

In each of the events listed below, energy is transferred from one form to another. For each of these, state the two forms, using names chosen from the following list. (Where there is an obvious, important, intermediate form, give that also, saying, "From . . . to . . . to . . .".)

List of forms and suggested abbreviations:

Chemical energy (Chem E.)	Electromagnetic radiation energy, in radio waves, infra-red, visible light, ultraviolet, X-rays, etc. (radiation)
Kinetic energy (K.E.)	
Gravitational potential energy (P.E.)	
Strain potential energy, as in compressed spring (strain E.)	Rotational-, or spin-, kinetic energy (spin E.)
Electrical energy (electric E.)	Nuclear, or "atomic" energy (nuclear E.)
Heat energy (heat)	

(*Note*: animals' energy, derived from food, is chemical energy.)

Example: Fast bullet hits wall, *stops* dead.
Answer: From K.E. to heat.

(a) Man *lifts* rock from floor to shelf.
(b) Rock has toppled off shelf and is *falling* (has not yet hit ground).
(c) Rock, falling fast, *lands* on ground.
(d) Man *throws* rock out horizontally.
(e) Rock, thrown along floor, *comes to rest*, sliding along floor.
(f) Explosive ammunition *explodes* producing highly compressed hot gas.
(g) Gas of (f) *pushes* bullet along barrel of gun.
(h) Child twists spinning-top with fingers, *makes it spin.*
(i) Spinning-top *comes to rest.*
(j) Car battery *runs* small cigarette lighter.
(k) Waterfall from high reservoir *charges* car battery (via generator).*
(l) Rapid river *charges* car battery (via turbine and generator).*
(m) Car battery *charges up* electrical capacitor (which can then give a spark).
(n) Hot furnace has a hole opened in its wall, so that *"red glow" escapes.*
(o) *Huge weights falling* run radio set which emits signals (via pulleys, gears, generator, tubes, etc.).
(p) Sunshine warms house.
(q) Sunshine helps plants grow.
(r) Radium atom emits fast helium nucleus (alpha-particle).

* Where machinery is spinning steadily, or moving steadily along, it has some K.E.; but this is not increasing or decreasing, and therefore does not enter into the balance sheet of energy-changes asked for in a question like this.

★ 8. A man pushes a 30-kilogram box along a rough floor. He exerts a horizontal push of 100 newtons. The box moves steadily at speed 3 meters/sec and does not change its speed.

(a) How big is the box's acceleration?
(b) What must be *resultant* force on it?
(c) How hard must the floor therefore drag it against the motion?
(d) Is the box *gaining* kinetic energy as it moves?
(e) How much energy does the man thus lose to the box in 10 seconds?
(f) Where does the energy lost by man go to?

★ 9. A man pushes a 30-kilogram box along a not-so-rough floor. He pushes with force of 100 newtons forward. Floor drags back 40 newtons. Box starts from rest. *Use absolute force units.*

(a) How big is the box's acceleration?
(b) How far will it move from rest in 3 seconds?
(c) How much energy does man lose to box in these 3 seconds?
(d) How much of this energy is delivered at the box/rough-floor surface, and in what form?
(e) How much energy is left over, as gained kinetic energy? (Get this by subtracting (d) from (c).)
(f) Calculate box's speed after the 3 seconds. Use the relation K.E. $= \frac{1}{2}mv^2$ to calculate its gained K.E.
(g) Does (f) agree with (e)? (If not . . . ?).

10. Suppose we have a frictionless incline which rises 3 m. vertically for every 5 m. up the slanting slope (or for every 4 m. along horizontal base). A 40-kilogram box rests on this incline.

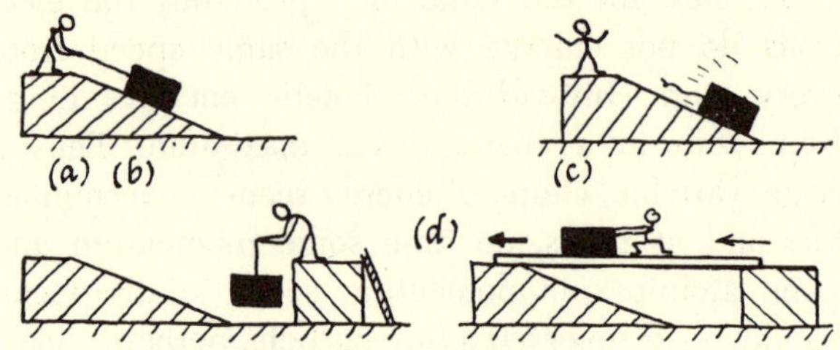

Fig. 26-41. Problem 10

(a) If a rope slanting up parallel to the incline holds the box at rest on it, how hard must the rope pull? (See Chapter 3.)

(b) Now suppose a man at top of hill pulls box up by means of rope, dragging it 10 meters up the slope without any gain of speed. How much energy does he lose?
(c) Now suppose he releases the box, and it slides the 10 meters down the slope again. Assuming the same facts, calculate the acceleration of box, and thence its final speed, and thence its *gained kinetic* energy.
(d) Suppose that instead of dragging box up slope, the man first raises it vertically to the same top level, then slides it along a frictionless horizontal plank placed there temporarily (shades of Galileo!). How much energy does he lose in thus raising the box?

★ 11. BULLET'S ENERGY

A 0.002-kg bullet moving 300 meters/sec is fired into a 1.998-kg block of wood at rest on a frictionless table. The bullet embeds itself in the block (which has such a providentially useful mass) and the combine proceeds slowly along the table. Calculate:

(a) The speed of the combine after impact (assume momentum conserved, as always).
(b) The kinetic energy of the bullet before impact.
(c) The kinetic energy of the combine after impact.
(d) The loss of kinetic energy in the course of impact.
(e) What fraction, as a %, of the *bullet's original K.E.* remains as *K.E. in the combine?*
(f) From (e), what fraction of original K.E. is lost in impact?
(g) Into what form of energy does the lost K.E. (probably) go?

★ 12. PERPETUAL MOTION

(a) Perpetual *movement* (in the sense of a thing going on moving without supplying or taking any extra energy) is regarded as possible.
 (i) Give one or two examples of almost perpetual movement.
 (ii) Why does perpetual movement not occur in most cases?
(b) Perpetual *motion* (in the sense of continual supply of extra energy) is regarded as impossible.
 (i) On what is this belief based? (e.g., belief in nature? teaching? *Encyclopedia Britannica*? experiment? hunch? U.S. Patent Office Rules?)
 (ii) On what is *your* belief in this based?

13. (A ski tow consists of a rope loop which is kept running by some engine to haul skiers up a hill.) A gasoline engine drives a ski tow which hauls a skier up a hill and he then slides down again on his skis on snow at 0°C. Trace the interchanges of energy involved in this story through as many stages as possible. Describe the forms the energy takes and, where it is not obvious, describe the mechanism of the change. *Note*: Where machinery is spinning steadily, or moving steadily along; it has some K.E., but this is not increasing or decreasing, and therefore does not enter into the balance sheet of energy changes asked for in a question like this. The ski tow can transmit strain E. along its rope, but it just keeps its K.E. constant. The engine's spin E. is constant, so *that* does not enter in.

14. Trace the forms of energy backwards from each of the following to the Sun: (i) coal; (ii) electric power lines supplied by power station driven by waterfall.

15. One gram of animal fat provides about 9.5 Calories when it is completely burnt. (1 Calorie is 1000 calories.) Suppose your present diet is 4000 Calories/day, and you cut your food (and candy, etc.) down to 3/4 of your present consumption in all respects, but keep up all your present physical activity. How many pounds of fat might you expect to lose in a month?

16. (a) A lead bullet of mass 0.010 kilograms moving 300 meters/sec hits a massive wall and stops dead. Calculate the temperature-rise of the bullet assuming that all the bullet's kinetic energy turns into heat and that all this heat resides in the bullet. Remember that the definition:

heat = *mass* · Δ (*temp.*) · *sp.ht.* gives heat in Cals if *m* is in kg and Δ *temp.* in C°.

Data: 1 Cal = 1 kilocalorie = 4200 joules (see footnote* below) specific heat of lead = 0.031.

(b) Explain why it is *not necessary* to know the mass of the bullet in calculating the temperature rise in (a).

* 1 kilocalorie, often called 1 Calorie, warms up 1 kg of water 1 C°. It is equal to 1000 "ordinary calories" used in the older system of units with grams and centimeters. "Calories" in diets are kilocalories.

17. To give people a feeling for the size of 1 joule, the demonstration shown in Fig. 26-42 is set up. A loop of string,

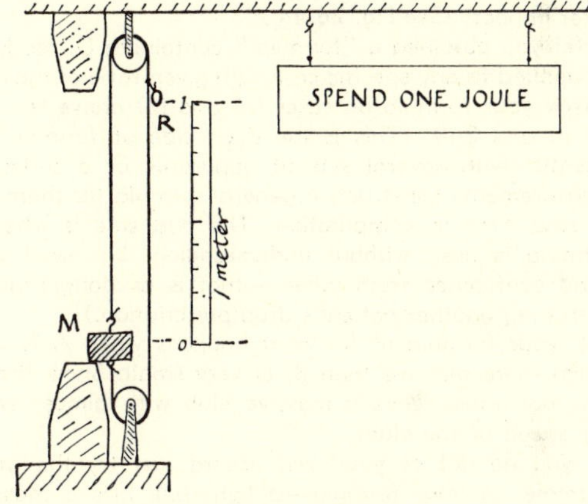

FIG. 26-42. PROBLEM 17

just taut, runs over two pulleys. A load *M* is tied to it at one point and a ring R is attached at another point so that pulling R down pulls *M* up. There are stops that limit the motion to 1 meter vertically. The apparatus is labelled:

"TO TRANSFER 1 JOULE PULL RING DOWN 1 METER."

What mass should *M* have?

18. A 1000-kilogram roller-coaster car starts from rest at A on the track whose vertical profile is sketched in Fig. 26-43 and coasts with negligible friction. Calculate its speed at B.

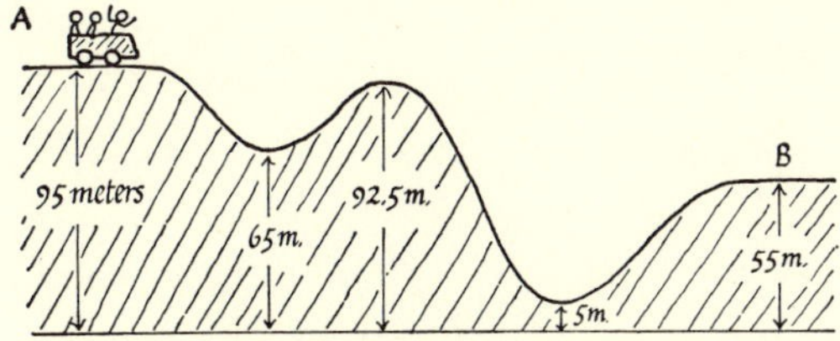

FIG. 26-43. PROBLEM 18

19. ELASTIC COLLISIONS: IMPORTANT FOR NUCLEAR PHYSICS

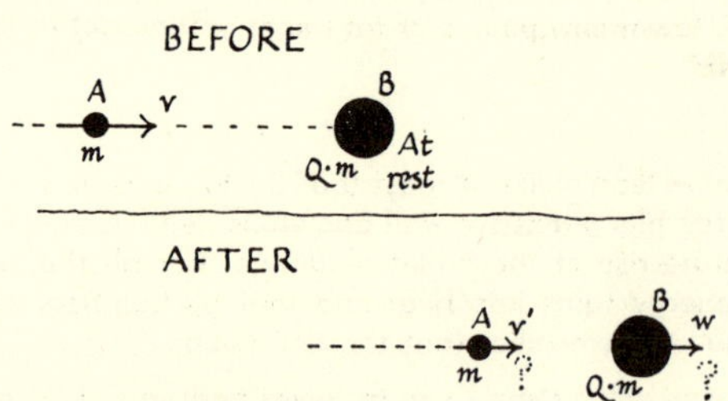

Fig. 26-44. Problem 19

(a) Suppose a particle A of mass m, moving with velocity v hits head-on a particle B, of equal mass m, at rest. The collision is elastic, so that momentum is conserved (as always) and K.E. is conserved. After collision, A moves with velocity v' and B with velocity w. Write two equations that show the collision is elastic, and solve them to find how A and B move after collision. (That is, solve the equations for v' and w in terms of v.)

(b) Suppose A hits B at rest, as in (a) above, but that B has mass $2m$, twice the mass of A. If it is a head-on elastic collision, how do A and B move afterwards?

(c) As in (b) except that B has Q times the mass of A; so that their masses are m and Qm. How do A and B move after impact? (See Fig. 26-44.)

(d) In (c) you obtained a "formula," containing Q, ready to be applied to any specific case with given ratio of masses. Check your formula on cases (a) and (b) above (m and m; m and $2m$). (This is the good side of formulas: a scientist with several sets of apparatus or a series of measurements constructs a general formula for them all, to save time in computation. The bad side is when a formula is used without understanding, borrowed with blind confidence from others—that is as dangerous as borrowing another patient's drug prescription.)

(e) Ask your formula of (c) what happens when A is very much more massive than B, Q very small. Show that a golf ball must leave a massive club with almost twice the speed of the club.
(If you do not of your own accord just try the other extreme on your formula—a light ball hits a massive wall—you are inhuman.)

20. ELASTIC COLLISIONS: APPLICATION TO NUCLEAR REACTORS

Use the results of Problem 19 to answer the following: A particle A moving with velocity v hits a particle B head-on in an elastic collision.

(a) What fraction of its initial K.E. does A lose, if:
 (i) A and B have equal masses, m and m
 (ii) B is twice as massive as A
 (iii) B has very small mass compared with A
 (iv) B has very large mass compared with A.

(b) In nuclear reactors, neutrons must be slowed down by (rare) elastic collisions with the nuclei of some material —the "moderator"—put there for the purpose. On the scale that gives a hydrogen nucleus mass 1, carbon nuclei have mass 12, aluminum 23, lead 208, and electrons about 0.0005. Which of these would be best for moderator, and which second best? The neutron itself has mass 1 on the scale. (In practice the "best" proves unsuitable because it absorbs neutrons, so we must move to the second best. Of course the neutron collisions are not all head-on but occur at all angles—yet this simple calculation indicates our choice.)

21. MATHEMATICS THE HONEST SERVANT

At one stage in finding your formula in Prob. 19 you probably cancelled out a common factor w. Before that you had a quadratic equation with two answers; and it became a simple equation with one answer when you cancelled w. What is the "other answer" and what does it mean?

22. GLANCING COLLISIONS: IMPORTANT FOR CLOUD CHAMBER ANALYSIS

Suppose a particle A moving with velocity v hits a particle B of equal mass at rest. The collision is not head-on: A glances off in one direction, B recoils in another (see Fig. 26-45). Show that if the collision is elastic, and if A and

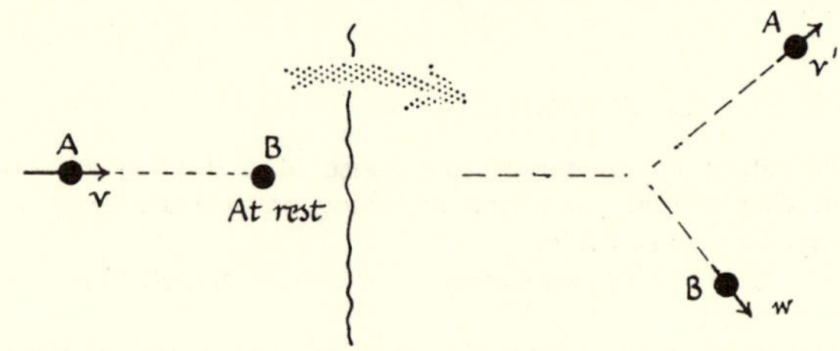

Fig. 26-45. Problem 22

B have equal masses, their paths after impact must make an angle of 90° with each other. (*Hint*: Since momentum is a vector, its conservation can be exhibited by a vector diagram of A's momentum before impact and the two lots of momentum after. If K.E. is conserved . . . Pythagoras. . . .)

Try this with elastic balls of steel or ivory on long pendulums. This provides an important test in nuclear physics. We can photograph the paths of alpha-particles (shot from radium) in a cloud chamber. When an alpha-particle makes a (rare) close collision with some other nucleus, the track shows a fork, like a Y. Alpha-particles are believed to be nuclei of helium; in which case the result above predicts a right angle between the arms of the fork when an alpha-particle collides with a helium nucleus. Tracks photographed with a cloud chamber containing helium gas (its atoms moving too slowly to spoil the assumption) show accurate 90° forks.

★23. DISCUSSION OF RADIATION: A GOOD EXAMPLE OF THEORETICAL PHYSICS

(This is a very long problem but it contains an important discussion. You will probably gain more by working through this problem than by reading the same discussion, with the answers given, in an advanced text.)

Some general properties of radiation can be predicted by an ingenious theoretical argument, appealing to certain common-sense ideas about heat and temperature. Complete the discussion outlined below, and you will have carried out that argument.

Imagine a large well-insulated box, maintained at constant temperature, perhaps a very high temperature so that the walls are white hot and the box is filled with radiation whizzing to and fro between the walls. There is a vacuum in the box, so there is no conduction or convection but only radiation to carry heat to any objects inside it.

Experiment shows that if several objects originally at different temperatures are placed in such a box, they all reach the *same* temperature eventually, the temperature of the walls of the box, even if they are of different sizes, shapes, materials, and surfaces. Consider the following cases, and guess at some properties of radiation:

(a) Suppose a small disk, B, is suspended in the box. Radiation, from the walls and other objects in the box, falls on the disk. The disk absorbs some of this radiation that hits it, and reflects the rest (or a transparent disk transmits it). The radiation that the disk absorbs produces heat that tends to make the disk hotter. But radiation that is reflected (or transmitted) contributes no heat to the disk—it just carries its energy on with it. At the same time, the disk itself is emitting radiation; and this *outflow* of radiation depends only on the disk's size, surface condition and temperature—it does not depend on the surroundings. This is an important idea, that a hot body emits radiation quite independently of the radiation it receives.

(i) When the disk has reached its ultimate constant temperature, getting neither hotter nor colder thereafter, how must the rates at which it *absorbs* radiation and *emits* radiation compare with each other? (*Note*: Just because it is no longer growing hotter, these two rates are not zero. It still receives radiation plentifully, from the walls, and it emits radiation, plentifully.)

(ii) Suppose it was cooler than the rest of the box when it was first put in. How did it become heated up, finally reaching a steady temperature?

(iii) Or if it was hotter than the rest of the box at first, say how it cooled down to lower temperature.

(iv) What (qualitative) inference can you make from (ii) and (iii) regarding RATE OF RADIATION and TEMPERATURE?

(b) Suppose now the same disk is suspended at several different places inside the box, long enough in each place for it to reach a steady temperature. *Experiment shows that it always reaches the same temperature*, that of the box walls, *wherever it is placed.*

(i) Since it is always at the same temperature as it is moved around, the amount of radiation that the disk *emits* remains . . . . ?

(ii) What can you then say about the intensity of radiation, travelling in any direction, in any part of the box, which falls on the disk wherever it is placed?

(c) Now suppose three disks A, B, and C of the same size but different materials are suspended in the box.

A has a brightly polished metal surface (almost perfect reflector)
B has a black surface (almost perfect absorber)
C is transparent (glass)

(i) Answers to (b) predict that the amounts of radiation *arriving at* the three different disks, A, B, and C, must be . . . ?

(ii) Of the radiation arriving at it, the transparent disk transmits most. What happens to *most* of the radiation reaching A and B?

(iii) What happens to the remainder of the radiation arriving at A, B and C?

(d) The three disks are not getting hotter, though they are absorbing radiation, but stay at a constant temperature because they are also emitting radiation. They are absorbing radiation at different rates, yet all reach the same temperature. From your earlier answers, what can you say about the relative amounts of radiation emitted by the three disks?

(e) What general property does this suggest regarding relative emitting and absorbing properties of different material or surfaces?

(f) Did any laboratory experiment agree with your conclusion for (e) above? If so, describe it briefly.

(g) Give a compact summary of the argument that led up to the important result (e). "Assuming that various bodies, black, bright, and transparent, all reach the same temperature in a hot oven full of radiation, wherever they are placed, we argue that . . ." (now continue).

(h) INVISIBILITY IN A FURNACE

Suppose you could look into the hot furnace through a peep-hole and peer at the glowing walls and the suspended objects A, B, C. In fact you would just see the bright glow of the walls everywhere inside, and you would be quite unable to distinguish A, B, and C. (This is what engineers find, when they look into a hot furnace or kiln; the objects inside are invisible. You can see something like this for yourself if you look into a brightly glowing fire; the outlines of glowing coals or logs disappear in a general glow.)

(i) Predict this indistinguishability from the discussion above.

(ii) Explain your argument.

CHAPTER 27 · MEASURING HEAT AND TEMPERATURE

A poet's view:

"Oh ye fire and heat, bless ye The Lord,
Praise Him and magnify Him for ever."

A scientist's view:

". . . when you can measure what you are speaking about, and express it in numbers, you know something about it; but when you cannot measure it, when you cannot express it in numbers, your knowledge is of a meagre and unsatisfactory kind: it may be the beginning of knowledge, but you have scarcely, in your thoughts, advanced to the stage of science."

—Lord Kelvin (1891)*

(This is a chapter to be studied on your own, without lectures. As a scientist studying energy, you must know how heat and temperature are measured. And if you are interested in the philosophy of science you will find that these simple measurements raise profound questions.

In this course we shall omit technical details of calorimetry and routine laboratory training, to save time for more important matters.

The first part of the chapter discusses crude measurement of heat, with experiments that you should try in the laboratory. The second part discusses temperature, starting with easy descriptions and leading to harder philosophy—read as far as you can.)

I. HEAT

Heat and Temperature

Place a pot of water over a flame and watch its temperature. The flame gives heat to the pot and contents and raises their temperature. A bigger pot with more water needs more flame-treatment, uses more fuel, for the same temperature rise. We say it needs more *heat*. In ordinary talk, the word "heat" is used as a rough synonym for temperature; but in science we use the two words for quite different things.[1] We use the name heat for the essential stuff that makes a thing hotter; temperature only shows how hot the thing is, how full of heat it is, or the "pressure of heat" in it.

Descriptions of Temperature and Heat

Temperature is hotness measured on some definite scale. Taking a hint from our own rough sense of hot and cold we develop thermometers with numbered scales. To make one thermometer agree with another, we assign standard numbers, 0 and 100, to two standard hotnesses, cold melting ice and scalding standard steam, and divide the thermometer scale into a hundred equal degrees between.[2] We shall discuss the philosophy of temperature-measurement later: for the moment we shall take thermometers for granted, much as we do stopwatches. They tell us how hot things are. They measure hotness on a definite scale, and we call their measurement "temperature."

Heat is something that makes things hotter, or melts solids, or boils away liquids. When we use a flame or an electric stove to heat a bar of iron or a tub of water, we think of heat pouring into the

* *Popular Lectures and Addresses*, Macmillan & Company Ltd.

[1] It is unfortunate that scientists have chosen to differ from ordinary talk and newspaper vocabulary over the common word "heat." We say "blood *heat*," "*heat* of bath-water" and read about "the *heat* of the day." In all these cases, scientists would say *temperature*. They reserve the word heat for the mysterious agent provided by fuels to make things rise in temperature. Between them, the scientists and the journalists have made a muddle of the word. It is not the journalists' fault. We too should accept the colloquial practice: use "heat" for the thing a thermometer measures and find some other word, perhaps the old name "caloric," for the thing that makes things hotter, the thing that we find is a form of energy. However, we cannot change the scientists' practice either, so we must learn to live with the quarrel. The best you can do is to use the words "temperature" and "heat" in the scientists' manner while you are dealing with scientific matters.

[2] This is the Centigrade or Celsius scale, used on scientific thermometers. See a later note on Fahrenheit scale. It is good scientific grammar to distinguish between 1°C and 1 C°. One degree Centigrade is a low temperature, just above the ice point. One centigrade degree is a temperature rise of one degree—anywhere on the scale. (C is practically an adverb in the first case, an adjective in the second.) A recent international agreement has declared Celsius the standard name instead of Centigrade, but the change will be slow.

victim, like a stream of an invisible weightless fluid, to make it hotter. The more iron or water to be cooked up, the more heat is needed. We get such heat from fuel and to warm up 2 pounds of water through some desired temperature-rise will cost twice as much as to warm up one pound; 5 pounds will cost 5 times as much as one pound. Or if we want a larger temperature-rise, we need more fuel. We guess that if so much fuel warms a victim from 10° to 20°, an equal quantity would then warm it the next ten degrees, from 20° to 30°. Experiment confirms this in most cases. If we gauge the mysterious fluid heat by fuel-use, we expect to find that, if HEAT $\propto$ FUEL USED, then

HEAT NEEDED $\propto$ MASS OF MATERIAL TO BE HEATED;

HEAT NEEDED $\propto$ TEMPERATURE-RISE OF MATERIAL.

Earlier experimenters some two hundred years ago were not so much concerned with fuel-use but arrived at the same result by thinking of heat as an invisible substance and experimenting on its interchange between hot and cold materials. We shall start with an experiment like theirs:

Finding a Satisfactory Scheme for Measuring Heat

EXPERIMENT A. Delivering a Small Dose of Heat to Water

(Try this in the laboratory. Failing that, see it demonstrated.) Give a definite dose of heat to water in a metal pot or glass beaker. A thimble-full of

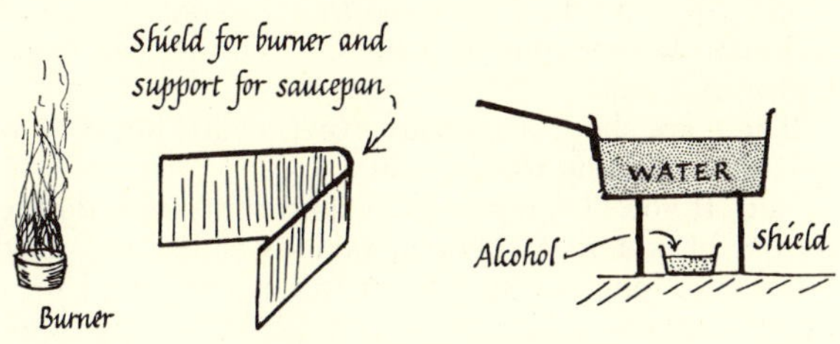

FIG. 27-1. EXPERIMENT A

alcohol burnt in an open metal cup provides a convenient dose which is much the same from one experiment to the next.[3]

(i) Give one "dose" of heat to 1 kilogram of water and measure the temperature-rise.

[3] A small dipper can be used to ladle 1 cubic centimeter (one millionth of a cubic meter) of denatured alcohol into a small dish pressed out of aluminum foil. The alcohol burns in less than a minute and gives a pot several Calories. This amount of alcohol costs about 0.1 cent, untaxed. So this "dose" of heat is given by 1 mill's worth of alcohol. Or a Bunsen burner can be kept burning steadily and placed under the pot for a minute's heating. Or we can run an electric heater in the water for one minute (using distilled water). But the use of a clock makes for confusion, because time is not really involved in heat measurement.

(ii) Give one "dose" to .5 kg of water. The flame can hardly mind whether it has half a kilogram of water over it or a kilogram. We are trying to regard heat as an invisible "substance" poured into the water and we are trying to find some rule or scheme for measuring the HEAT delivered by one "dose." If we find the correct way, it should therefore yield the same answer for experiment (ii) as for (i). Try some possible schemes, such as (a), (b), (c) below.

(a) Temperature-rise. Suppose we regard the TEMPERATURE-RISE as a sufficient measure of heat delivered. Does this meet our requirement: is it the same for (i) and (ii)? No.

(b) The amount of water heated is also involved, since with more water the temperature-rise is less. Try *adding* TEMPERATURE-RISE + MASS OF WATER. (With some juggling you could find an adding arrangement, such as

Δ TEMPERATURE · 43 + MASS OF WATER,

which would work; but would it work with other masses of water, such as 2 kg?)

(c) Try *multiplying* MASS OF WATER · TEMPERATURE-RISE.

You must be charitable in calculating these tests because the experiment is a rough one. Heat escapes from apparatus with exasperating ease. In most heat experiments the experimenter wages a desperate battle with heat-losses to the air, etc. So in our demonstration experiments, as in your own, you must be content with rough agreements.

Scheme (c) seems the most successful in a wide variety of heating experiments, and it agrees with our later treatment of heat as energy. So we adopt it and formulate a definite rule:

RULE: To measure heat, let the heat heat water, and multiply
MASS OF WATER *by* TEMPERATURE-RISE

Units for Heat

If the MASS OF WATER is measured in kilograms and the TEMPERATURE-RISE in degrees on the centigrade scale (= C°), the HEAT will be in (kilograms-of-water) · C° and we name these "kilocalories" or "Calories."[4]

[4] 1 Calorie = 1000 calories, the small calorie being the heat needed to warm up 1 *gram* of water 1 C°. Until recently the small calorie was always used, except in studies of food values. We use the kilocalorie to conform with our general use of kilograms. We shall spell it Calorie and abbreviate it to Cal. When a daily diet is said to provide 3000 Calories, that means 3000 kilocalories, whether the unit is spelled Calorie or calorie.

One kilocalorie is the heat needed to warm up 1 kilogram of water 1 centigrade degree. If we argue in terms of these unit-blocks of heat, Calories, the rule seems reasonable. For example: "How much heat is needed to warm up 3 kg of water through 5 C°?" Each kilogram, warmed 1 degree, needs 1 Calorie (definition of Calorie).

∴ One kilogram warmed 5 degrees needs 5 Calories.

∴ Three kilograms warmed 5 degrees need three lots of 5 Calories or 3×5 Calories, or 15 Calories.

Altogether, 3 kg warmed 5 degrees require 15 unit-warmings, each of one kg through 1 degree, or 15 Calories.

In general, M kg of water warmed through a temperature rise Δt centigrade degrees need $M \cdot \Delta t$ Calories. This reasoning tacitly assumes that quantities of heat, or fuel-use, are *additive*.

We choose water as a convenient standard substance to give the heat to for measurement, easily obtained, easily stirred. To see whether we need restrict the rule to water, repeat the experiment with 1 kg of some other safe substance such as aluminum or glycerin. Multiplying TEMPERATURE-RISE by MASS OF MATERIAL, as for water, does not give the same answer as before but a larger product (for aluminum, an answer roughly 5 times as big). To get the same value for heat with another substance we should first multiply MASS by TEMPERATURE-RISE, as for water, and then multiply the product by a special factor, a characteristic number (for aluminum, about 0.2), called the SPECIFIC HEAT of the substance. Specific heats are useful factors in some heating calculations, but we shall not use them seriously in this course.[5]

EXPERIMENT B. Demonstration Experiment: Mixing Hot and Cold Water

Let us try out our rule for measuring heat on a mixture experiment with hot and cold water—one of the early experiments which led to the idea of heat measurement or calorimetry.

Weigh out .3 kg of cold water in a glass beaker, and .4 kg of hot water in a big *thin*[6] beaker. Stir carefully and take their temperatures. Quickly pour the cold water into the hot water, stir, and take the final temperature. In the final mixture of warm water, the original hot water and cold water are inextricably mixed but we know their final temperature, the same for both, the temperature of the .7 kg of warm water. If heat is an indestructible fluid—or if we want to devise some such concept—we expect the cold and hot water to gain and lose equal amounts of heat. hopefully neglecting external heat losses. Calculate the TEMPERATURE-RISE of the cold water and TEMPERATURE-FALL of the hot water. Are *they* equal? No, they will not suffice as measures of this invisible HEAT. Now try using the products

MASS OF WATER · TEMPERATURE-CHANGE

Though these do not turn out to be equal and opposite, you will find this rule is the nearest simple one to anything satisfactory, and it is easy to think up excuses for its failure to work exactly!

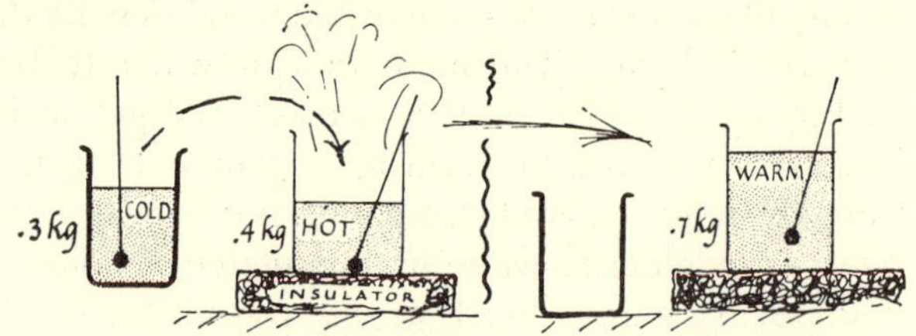

FIG. 27-2. DEMONSTRATION EXPERIMENT

EXPERIMENT C. Measuring Heat

(i) Measure the heat given to a cookpot of water by burning 1 mill's worth of alcohol.

(ii) Measure the heat given to a cookpot of water by a Bunsen burner running 1 minute (costing one or two-tenths of a mill).

These are short, inaccurate experiments, but they will give you a feeling for Calories.

(iii) If you like, repeat the experiments with different masses of water, or different periods of heating. In the latter case reduce your answer to heat delivered in one minute.

(iv) Burn 1 mill's worth of alcohol under a large block of aluminum, and measure the block's temperature-rise. *Assuming* the alcohol gives the same quantity of heat to the block as to a cookpot of water, estimate the specific heat of aluminum.

If you like, try various kinds of Bunsen flame: yellow smoky, clear quiet, and roaring. If you try these, you should also investigate the *temperature* of the flames by using a piece of iron wire or a dead match as a crude indicator.

In your calculation you may regard the water as the only victim heated *usefully*, or you may choose to include the heat given to the pot. In the latter case, ask for the specific heat of its material.

EXPERIMENT D (Optional). Rough Estimate of Temperature of Bunsen Flame by Calorimetry

In contrast with measurements of its heat supply, estimate the *temperature* of a Bunsen flame, thus:

[5] Measuring specific heats often bulks large in school laboratory work. This keeps alive a messy inaccurate business of heaving chunks of hot metal into cold liquid, that is over half a century out of date. For reliable measurements, a well-insulated sample is heated with a small electric heater.

[6] A thick beaker takes too much heat for itself. A thin plastic food-bag is better still.

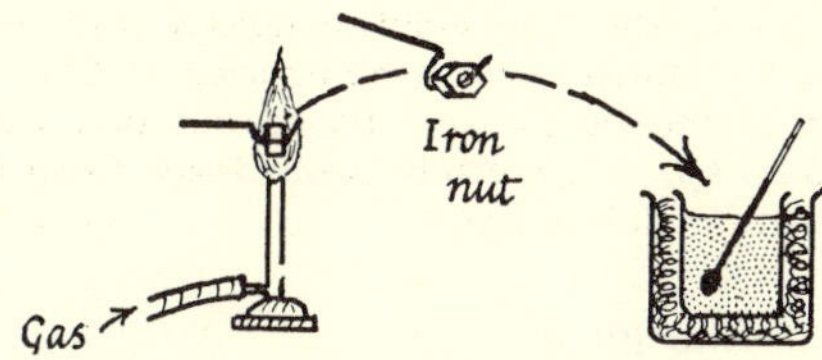

FIG. 27-3. EXPERIMENT D

Suspend a lump of iron, say a large iron nut, on an iron wire in the Bunsen flame. When the nut is as hot as you can get it, heave it into a small can of cold water (preferably a jacketed can called a "calorimeter"). Measure the initial and final temperatures of the water carefully. (The immersion of the nut is hazardous and exciting. Any sensible scientist would run through a trial first, with rough measurements, to see how much water is needed.)

To calculate the flame temperature from your measurements, follow the method of Problem 1 below. Use a value of specific heat of iron from a separate experiment.

PROBLEM 1 (Here is the flame temperature estimate of Experiment D on a larger scale.)

A 2-kilogram lump of iron is heated in a furnace and thrown into a bucket containing 30 kilograms of cold water at 15.0°C. After commotion and stirring the water temperature is 25.0°C. Assume the specific heat of iron is 0.159 for this high temperature range.

(a) Calculate the heat gained by the water.

(b) The heat lost by the iron is: MASS · TEMPERATURE FALL · SPECIFIC HEAT. Insert in this expression those data that you have.

(c) Assume that the heat lost by the iron = the heat gained by the water and calculate the temperature change of the iron.

(d) What was the temperature of the iron before immersion?

(e) Does this give an underestimate or an overestimate for the furnace temperature? Why?

(f) Is the experiment likely to be more accurate, or less, on a big scale like this than on a small scale? Give a clear reason for your answer. (*Hint*: Think carefully about heat losses.)

EXPERIMENT E. Heating Snow

(If snow is not available, shaved ice may do, but ordinary crushed ice is not fine enough.) Pack a small metal can with snow. Place a thermometer in it to take its temperature. Give the snow a "dose" of heat, by burning 1 mill's worth of alcohol under it.[7] Stir the slush till the thermometer reading does not change any more. Read the thermometer. Now give the can another dose of heat. Stir, and take the temperature. Continue this until the water is warm, or even until it has boiled away. It is essential to wait after each dose until efficient stirring has brought the temperature to a steady value. This may require a minute or more in early stages, but a fraction of a minute will suffice later, and no time at all later still. Plot a graph showing TEMPERATURE VS. NUMBER OF DOSES given to the can. What can you infer from the graph? (*Note*: Rough quantitative inferences can be made as well as qualitative ones.)

[7] A small Bunsen flame can be used instead. In that case a "dose" should be a quarter of a minute's heating with the flame. Then the flame must be removed while the stirring and temperature-taking are carried out. The flame must be kept burning steadily however, and this raises difficulties of organization—it is easy to let the flame change or forget the timing. A central clock, or an instructor issuing signals every ¼ minute, makes it easier to run this experiment.

Latent Heat

However rough and difficult, the experiment on heating snow (or a demonstration to replace it) shows that sometimes heat does not make things hotter. Sometimes HEAT is used for melting or for boiling away, in each case without change of temperature. Then we picture the heat energy being used to tear apart the molecules of a solid crystal, or to replace molecular kinetic energy stolen by extra-rich molecules evaporating. We call this disappearing heat *latent heat* which means "heat that lies hidden."

Experiment shows that 1 kg of ice needs 80 Calories to melt it, without change of temperature. To boil away into steam at 100°C, 1 kg of water takes 540 Calories. Starting with a kilogram of ice at say −10°C and supplying heat till it is all steam, we should have the following catalogue of requirements:

warming ice to melting point: (since sp. ht. of ice = 0.5)	5 Cals
melting the ice at 0°C: (turning solid to liquid without change of temperature)	80 Cals
warming melted ice to boiling point:	100 Cals
boiling away to steam at 100°C: (without any change of temp)	540 Cals

Notice how expensive it is to tear the molecules of liquid apart into vapor. From cold ice to boiling water costs 185 Calories, but boiling away costs almost three times that. In reverse, condensing steam delivers a lot more heat than cooling hot water: *a live steam burn is worse than a boiling water one.* The following problems describe experiments to measure these heat demands.

PROBLEM 2. HEAT NEEDED TO TURN WATER TO STEAM

An electric heater run for 2 minutes in a pail of water warmed up 10 kg of water from 20.0°C to 22.6°C. The same heater, running on the same electric supply, was immersed in a Thermos bottle of *boiling* water for the same time (2 minutes). The bottle + contents weighed 2.000 kg before and 1.950 kg after the heater was used.

(a) How much heat does the heater deliver in 2 minutes?*
(b) What mass of water was boiled away?

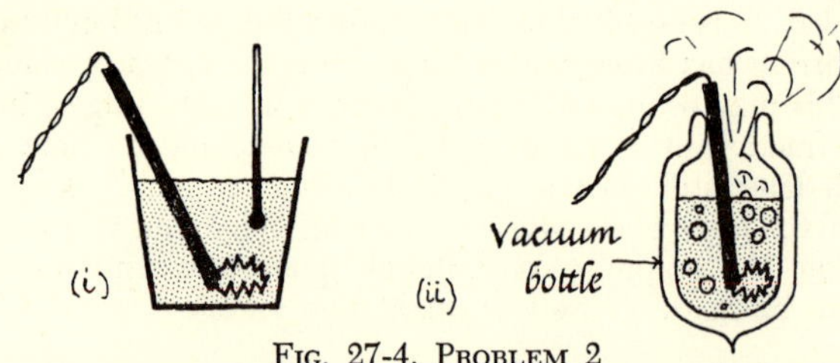

FIG. 27-4. PROBLEM 2

(c) Estimate the heat needed to boil away 1 kilogram of water.*
(d) Would you expect the answer to (c) to be an overestimate or underestimate? Why?

PROBLEM 3. HEAT DELIVERED BY CONDENSING STEAM

A bucket contained 5.00 kg of cold water at 18°C. A pipe from a large boiler blew steam into the water until its temperature was 30°C. The bucket then contained 5.10 kg of water.

(a) What mass of steam condensed?
(b) How much heat did the 5.00 kg of cold water gain?*
(c) Assume that the cold water received this heat from the steam which condensed and cooled down to 30°C. How much heat would 1 kg of steam give up in condensing and cooling thus?
(d) How much of that heat would 1 kg give up in just cooling from 100°C to 30°C?
(e) Estimate the heat delivered by 1 kg of steam in condensing without change of temperature.
(f) Is your estimate in (e) likely to be an overestimate or an underestimate? (Consider two likely troubles (i) heat losses, (ii) drops of water carried over with the steam, weighed as steam, but not effective as heat deliverers.)

PROBLEM 4. HEAT NEEDED TO MELT ICE

A hollow was scooped out of a big block of ice and its interior dried with a cloth. A beaker containing 2.0 kg of hot water at 50°C was then emptied quickly into the hollow. The water was stirred until it had cooled to 0°C. Then all the water in the hollow was tipped out and weighed. It weighed 3.25 kg. How much heat was taken to melt the ice? How much to melt one kg?

FIG. 27-5. PROBLEM 4

* In answering problems that ask "How much heat . . . ?" state the answers thus:

Heat needed = (15 kg) (35° − 5°) = (15) (30) = 450 Cal.
and not thus: Calories needed = 450.

The thing you are dealing with is heat, so it should be mentioned at the beginning. Calories are only the units, so they should go at the end. You would say, "My height is 6 feet" and not "my footage is 6," still less "my feet = 6." Putting the units first is an uneducated style: avoid it. There are a few exceptions where the practical men have got the better of good English and we bow to their jargon: "voltage," "tonnage," "acreage" may start a statement, but not yet "Calories."

Heat and Energy

Sudden compression makes a gas hotter—gives it more heat—yet when we ask "what does the moving piston do to the molecules," the reply is, "it only makes them move faster." So heat in a gas seems to be associated with molecular motion. Again, heat seems to appear when we hammer soft metal or rub rough surfaces. Heat may belong with motion of atoms and molecules in all cases. Careful measurements show that mechanical energy and heat are interchangeable at a fixed rate of exchange. Finally we decide that heat is a form of energy. These investigations of heat, and its incorporation into the general scheme of energy, will be discussed in a later chapter. For the present, to measure heat, let the heat heat water, and . . . .

II. TEMPERATURE

Thermometers and Temperature

"What is meant by a scale of temperature?" is a good question for a physicist at any level, from schoolboy to professor. A full reply might take a whole book, and it would illustrate the philosophical changes in physics over the last four centuries.

Temperature is hotness measured on some definite scale. Without thermometers, we can use our skin to make rough guesses, but our sense of hot and cold is limited and unreliable.[8] A thermometer tells us just how hot or cold a thing is; it can be used to compare different objects; and it can be used again and again, to link up observations made on different days. It provides *definite* numbers, which form a

[8] *Personal experiment on skin sense of hot and cold.* This experiment, which belongs in psychology as well as physics, gives a useful warning. Arrange three wash basins of water,

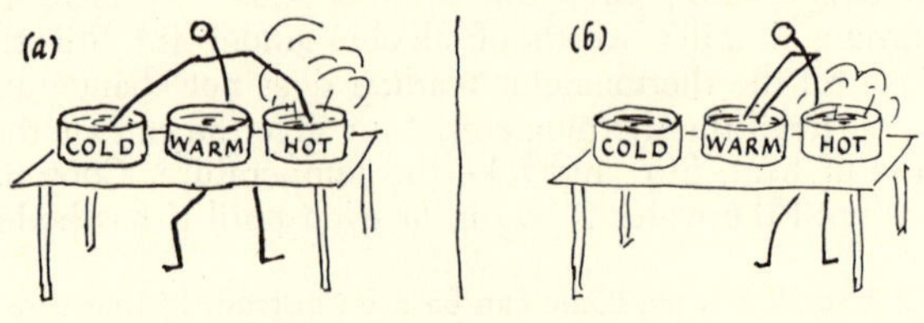

FIG. 27-6.

one very hot, one luke-warm, one very cold. Hold one of your hands in the hot water, and the other in the cold water, for three minutes or more. Then put both hands in the luke-warm water. Ask each hand what it tells you about the temperature of the luke-warm water.

permanent, reproducible scale—it has the characteristics of a good instrument. Yet when we devise a thermometer the instrument itself defines the scale and measurement-system we are to use. In passing from vague sensation to the definite scale of an instrument we are not just improving our incompetent skin's measuring system; we are inventing and putting into use a new concept: temperature.

Our rough sense of hot and cold suggests the concept of temperature. Investigations in physics and chemistry show that many important properties change when things get hotter, thus creating a demand for thermometers. Nowadays with thermometers in common use, philosophical questions about temperature are concealed by familiarity. We think a thermometer takes the temperature of our mouth or the air, or our bathwater; actually it only takes its own temperature. We assume that a temperature-rise from 60° to 70° is the same as a rise from 40° to 50°; but we seem to have no guarantee that they *are* the same, unless we declare them the same by definition. Yet thermometers are useful and trustworthy as minor servants. Is there behind their useful servile scales a master Temperature, a ruling scientific aristocrat?

Simple Thermometers and the Centigrade Scale

A bulb of liquid (mercury or colored alcohol) is attached to a graduated tube to show temperature by the expansion of liquid. To make the scales of thermometers agree with each other, we take two standard hotnesses, melting ice and standard steam, assign them standard numbers, 0 and 100, and divide the space between into 100 equal degrees.[9] With this arrangement, if one thermometer reads 30° in a bath of warm water, so will another, if both are correctly graduated, even if they have bulbs and tubes of different sizes. The mercury in the first thermometer has expanded 30/100 of its expansion from ice-point to steam-point. We can reasonably expect the mercury in the other thermometer to expand by the *same fraction*, and thus to read 30° on its scale. In this we are trusting to the Uniformity of Nature.[10]

Now suppose we use another liquid, say glycerin. With the same fixed points, would it give the same temperature scale? Our glycerin thermometer would read 0° in melting ice and 100° in standard steam, necessarily agreeing with the mercury one; but would they agree at intermediate temperatures? In fact, when the mercury thermometer reads 50.0°C, a glycerin thermometer in the same hot water would read 47.6°C—compared with mercury the glycerin seems to do a little less than half its expanding in the first half of its trip from the ice to the steam point. (Other thermometers can be made to show even greater disagreements; e.g., one using steam-pressure would read 12° when the mercury thermometer reads 50°!)

Which scale is the *right* one? Which thermometer reads the *true* temperature? Let us avoid this awkward question for the moment and ask which is the

Fig. 27-7.
Comparison of Expansion of Mercury and Glycerin when Heated
(The curvature is greatly exaggerated.)

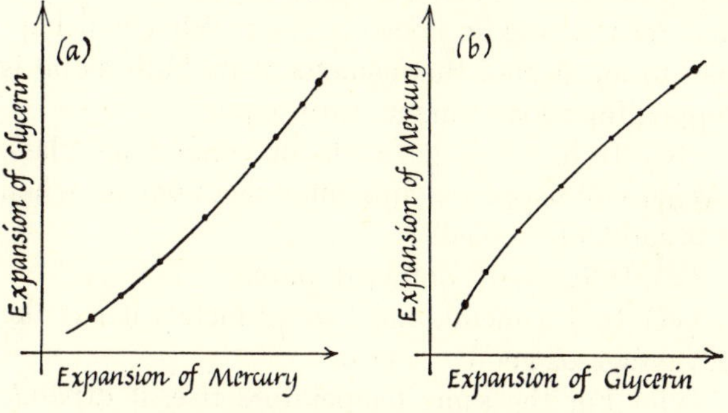

(a) Expansion of glycerin compared with expansion of mercury.
(b) Expansion of mercury compared with expansion of glycerin.
One liquid-expansion might be regarded as under investigation, while the other liquid-expansion is used to measure temperature; but which is which? Is there a *right* way around?

[9] This is the "centigrade" scale used in all scientific work except meteorology. The scale in domestic use is the Fahrenheit scale with the standard ice and steam points marked 32 and 212. Originally the Fahrenheit scale was constructed with different "fixed points": its zero in special freezing-mixture and its 96—a convenient number with many factors for subdivision—in the healthy human mouth. Modified to make whole numbers for the modern fixed points, it now places body temperature between 98 and 99. A room temperature of 68°F corresponds to 20°C. Changing between Fahrenheit and centigrade scales changes the numbers used for particular temperatures but indicates no change in our fundamental concept of temperature. A recent international agreement has decreed another change: instead of the standard ice-point and steam-point to define the scale, use "absolute zero" and the "triple point" of water—see Chapter 30. This *is* a fundamental change in temperature definition, but it makes no practical difference in ordinary scientific work. The number used for the triple point is chosen to make the new scale fit the old very closely.

[10] This is actually too trusting. The glass expands too. What does the glass-expansion do to the mercury-height in the tube? What, therefore, do the thermometer readings indicate, rather than the simple expansion of mercury? Now suppose the two thermometers under discussion contain pure mercury but have bulbs of different types of glass with different expansions. Would that affect the discussion?

most *convenient*, among, say, mercury, glycerin, and alcohol. The answer is: definitely mercury, on several scores. Below is a list of the "advantages" of mercury. The advantageous properties are given but the benefit of each is left as a problem for you to complete. The first ones are obvious. Numbers (vii) and (viii) are very hard, but when you have guessed the right answers you will know you are right. The answers to (viii) can be tested experimentally with a mercury thermometer.

"Advantages" of Mercury:

(i) It is opaque.

(ii) It remains liquid over a wide range of temperature.

(iii) It does not evaporate easily at ordinary temperatures. (Alcohol evaporates where it is warm and recondenses in cooler regions. What will happen to an alcohol thermometer if its bulb alone is dipped for a long time in warm liquid?)

(iv) It has a large angle of contact on glass. (What will happen in an alcohol thermometer when it is suddenly cooled?)

(v) It is easily obtained pure.

(vi) It is a metal, and like all metals it has the advantage of good . . . (?).

(vii) For the same temperature rise, it *expands less* than most liquids.

Mercury Has One "Disadvantage"

(viii) It is very dense.

PROBLEM 5. What are the benefits of (i)-(vii) above, and the trouble due to (viii)?

One more advantage is mentioned in some textbooks: *"mercury expands uniformly."* This is an unscientific statement. By itself it is not even wrong; it is nonsense. Can you see the catch? Consider a simple experiment that students used to do to examine expansion. A mercury-expansion apparatus is heated up in a bath of water, with a thermometer to show the temperature, as in Fig. 27-8(a).

A is the mercury-expansion apparatus: a glass bulb of mercury with a graduated tube to measure the expansion.

B is the thermometer: a glass bulb of mercury with graduated tube to show the temperature.

"Any resemblance between A and B is coincidental!" The obedient student who did this experiment heated the water bath, stirring carefully, took a series of readings, and plotted a graph of EXPANSION VS. TEMPERATURE. He was delighted to obtain a fine

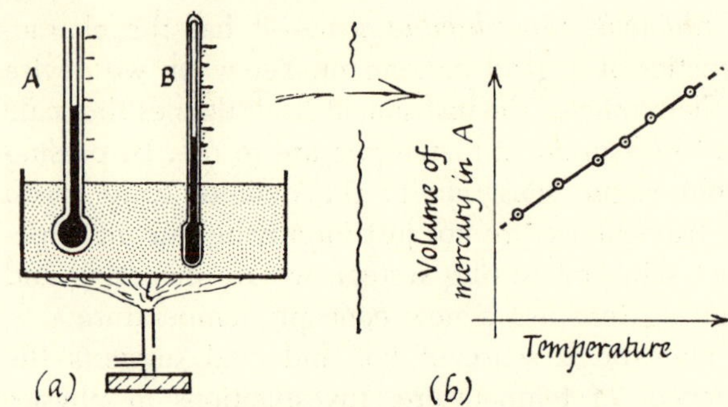

FIG. 27-8. ZANZIBAR EXPERIMENT

straight-line graph; but he should have realized that this gives no information about the expansion of mercury, beyond supporting a general belief in the Uniformity of Nature![11]

So we cannot say that one liquid gives the "right" temperature scale and others wrong scales. We can choose any one liquid and declare that one, by convention (a scientific lawyers' agreement), to be the standard scale. But we cannot then say we have the really true scale of temperature—any other liquid would be equally true.

If there is no "right" liquid, perhaps there is no right temperature. At first this seems a disappointing failure of science to reveal the truth. But reflecting on it brings two comforts. First, on the practical

[11] The fallacy is illustrated by the fable of Zanzibar, related by G.F.C. Searle. A ship's captain found his navigating clock, or chronometer, had stopped. This was a serious misfortune in the days before radio time-signals. He put into port at Zanzibar, hoping to re-set his chronometer with accurate local time. He inquired at the harbor and people told him, "Yes, there is a clockmaker down in the town, with a very accurate clock, his pride and joy." He carried his chronometer to the clockmaker, who assured him, "You can certainly trust my clock. It keeps wonderful time." The captain set his chronometer. To make sure, he asked, "How do you know your clock is reliable?" "I check it by the coastguard's noonday gun, and it is seldom a second wrong. When it is wrong, I adjust it." The captain was delighted. He carried his chronometer back to his ship. As he was going on board he met the coastguard, and, just to be sure, asked him about his gun.

"Good morning." " 'Morning sir."

"You have a gun you set off at noon?" "Yes sir, at noon, sir."

"Do you set it off just about noon time or exactly at noon?" "Exactly at noon, sir."

"Exactly?" "Yessir. Positive, sir."

"How do you know it's exactly at noon?" "Oh I make sure, sir; you see, there's a clockmaker down in the town with a very accurate clock, and . . ."

This trivial tale (which appeared more recently in another form in *The New Yorker* magazine) warns against something worse than circular argument; a spiral argument. You may think the warning unnecessary, but you will find similar fallacies in many fields of study, including science courses. In this course, we shall use the word "Zanzibar" as a name for such arguments or a warning against them. If you watch for it, you will be able to murmur a reproachful "Zanzibar" somewhere in almost any course where arguments are used.

side, we may decide on the mercury-in-glass thermometer. It is convenient and reproducible, and records temperatures on a definite system. On the theoretical side, we realize that temperature is an idea of our own manufacture, a concept whose definition is itself tied up with the instrument we choose for measuring it. Again and again in science we find that we must define a thing by the way we measure it—an "operational definition," this is called—and we try to avoid romantic ideas of an underlying really-truly piece of nature which our operational approach merely strives towards. Consider the differences between philosophers' long discussions of Justice, the ordinary man's strong belief that there is a single definite thing called Justice, and the lawmakers' operational approach in defining Justice in terms of practical law and its enforcement.

You may think that some measurements in science relate to really-true things which do not need to be defined operationally: but be careful. Consider surface area, for example. We know quite well what area is. Yet when we try to define a measured area such as 6 square feet we find ourselves drawing criss-cross lines and counting squares. Or, warned that that would be admitting an operational definition, we might talk about finding how much paint would be needed to cover the area; but *that* would be operational too—we can picture the paintbrush as the instrument.

Still we have a hankering to find a true temperature, or at least one that seems more basic, more fundamental, more general and less special than one based on the arbitrary choice of mercury. Gas thermometers bring us nearer that.

Gas thermometers measure temperature by the expansion of a sample of air or other gas; or they do what Boyle's Law assures us is equivalent: they use the PRESSURE-INCREASE of a sample kept enclosed at fixed volume. To lessen troubles due to glass, scientists have adopted gas thermometers as standards, because gas expansion, twenty times that of mercury, reduces glass misbehavior to an insignificant fraction. But gas thermometry brings philosophical comfort too, because we find that all gases have practically the same expansion. So gas thermometers with different gases all agree practically perfectly with each other. When we give up mercury in favor of air, nitrogen, hydrogen, helium . . . , the whole family of gases, we may feel we are a little nearer to the underlying true temperature, if we still believe in one.[12]

Gas Thermometers

In Fig. 27-9 (a) is a thermometer that measures the expansion of a gas sample. A small bead of

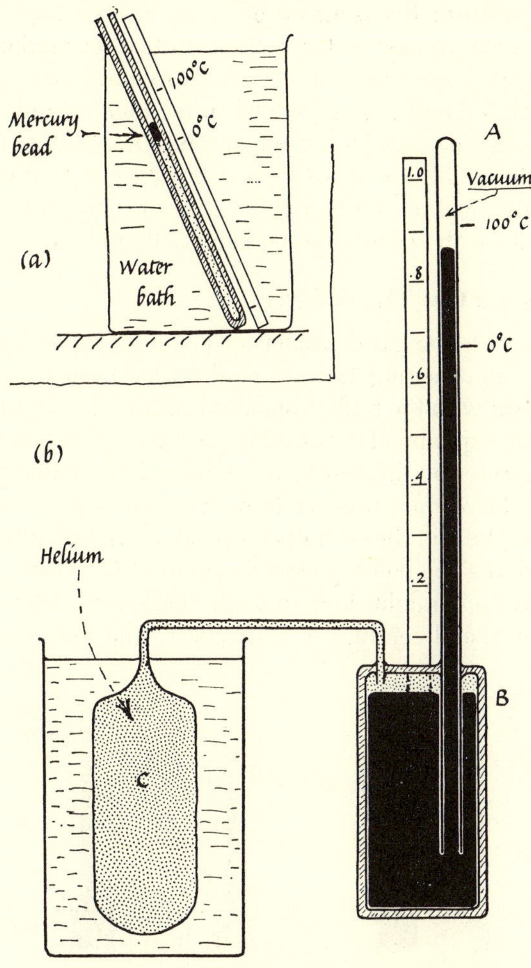

FIG. 27-9. GAS THERMOMETERS
Thermometer (a) uses the volume of a sample of gas, kept at atmospheric pressure, as a measure of temperature. A mercury bead acts as frictionless piston, enclosing the gas in a narrow uniform tube. Lengths from the closed end are used as measures of volume and thus of temperature. Thermometer (b) uses pressure of a sample of gas (helium) kept at constant volume as a measure of temperature.

---

[12] Many scientists say they don't, yet a ruthless adherence to operational science would stifle the more imaginative flights of theory—it would make science very right but very dull, and it might hinder its jumpy progress enormously. Max Born says, in his small book, *Experiment and Theory Physics* (Cambridge University Press: now reprinted by Dover Publications, Inc.): "This name [the 'operational method' of definition], has been given . . . to a procedure quite common among scientists. It consists in the demand that a physical quantity must not be defined by verbal reduction to other familiar conceptions but by prescribing the operations necessary to produce and to measure it. This is a sound rule, a reaction against verbalism and word fetishism. It is very useful in classical physics where one has to do with quantities accessible to direct measurements . . . but the operational definition is rather out of place if you wish to extend the idea of the field to atomic nuclei and electrons, and it comes to grief in quantum theory."

mercury traps the sample of, say, dry air in a narrow tube with one end closed. The whole thermometer must be immersed. Then the mercury marks the volume of the gas on a scale which is standardized by marking 0 and 100 in standard ice and steam, just as for a mercury thermometer. This thermometer is not used for accurate work—the mercury marker leaks or sticks—but it is mentioned because it shows the general idea clearly. (b) is a more usual form. The mercury pressure gauge AB is used to measure the pressure of a gas sample kept at constant volume in the bulb C. Instead of marking height or pressure units on the barometer, we can mark it 0 and 100 degrees when the bulb is in melting ice and standard steam, and fill in a centigrade scale of degrees. If we assume Boyle's Law and use some algebra, we can show that the temperature scale of (b) should be the same as that of (a).

Using a Gas Thermometer

To use the gas thermometer (b) immerse its bulb in clean melting ice and read its barometer; then repeat with the bulb in standard steam. Plot a graph with experimental pressures plotted against temperatures on the hereby-to-be-defined gas scale. We can leave the pressures in meters-of-mercury if we like. We plot the steam-point pressure against 100°C and the ice-point pressure against 0°C. Then we draw a straight line through these two defining points, and continue it if necessary. This graph line is straight by the definition of temperature on the gas scale. In fact it is the *temperature-defining line* arranged to give the standard values 0 and 100 at the ice and steam points. It is ready for us to read off the gas-thermometer-temperature of anything else, once we know the instrument's pressure reading at that temperature. In Fig. 27-10b the broken lines show how to find the temperature of a water bath in which the gas pressure is 0.6 meter-of-mercury.

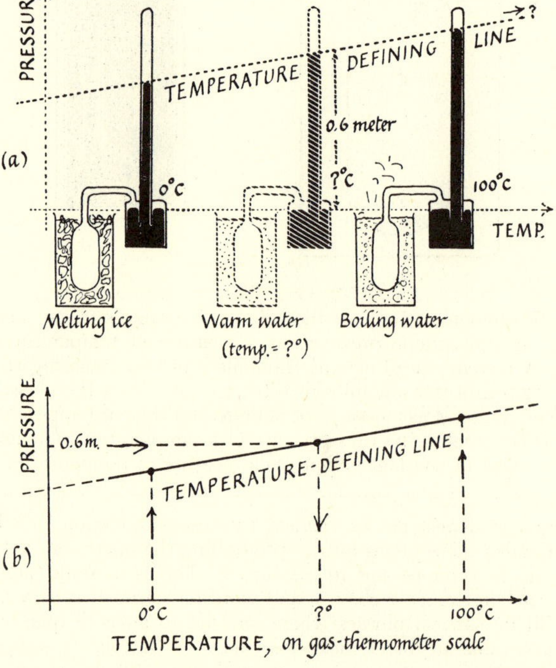

Fig. 27-10. Gas Thermometry

Now that we have switched to gas thermometers as standards, we can try mercury and glycerin against them. We find that most liquid expansions, plotted against gas-thermometer temperatures, give a slight curve—the two types of thermometer would therefore disagree somewhat, except at 0° and 100° where definition compels agreement. But surprisingly, mercury gives almost a straight line. *Now* we can rewrite mercury's "advantage" in a form that makes sense: "Mercury has uniform expansion *when compared with temperatures on the gas-thermometer scale.*" This genuine coincidence makes the early choice of mercury a happy one, because common thermometers can be used to give gas-scale temperatures directly in ordinary experiments. It was pure good luck.

Absolute Temperatures. ZERO

Another benefit of gas thermometers is their clear suggestion of an absolute zero. As we cool the thermometers in Fig. 27-9, the gas in (a) shrinks and the pressure in (b) grows smaller. Extrapolating this behavior to still lower temperatures we point to an *absolute zero* at which the gas would have collapsed to zero volume in (a) and to zero pressure too in (b). If the samples of gas really continued their behavior through the extrapolation, and remained properly behaved gases (which they do not), we could not expect to go below that absolute zero, or even to reach it. Real gases collapse into liquid and then solid before they have cooled down so far, but that does not prevent us dreaming of absolute zero as an intriguing idea. We can find out where it is on the ordinary centigrade scale by extrapolating our temperature-defining straight line on a gas thermometer's graph. Careful measurements with real gases agree, whatever the gas, on placing absolute zero at about −273°C. Practical attempts to reach this temperature by all kinds of cooling have brought us close to it but never quite there. It is probably an unattainable limit.

Scientists who have to calculate gas volumes at one temperature from measurements made at an-

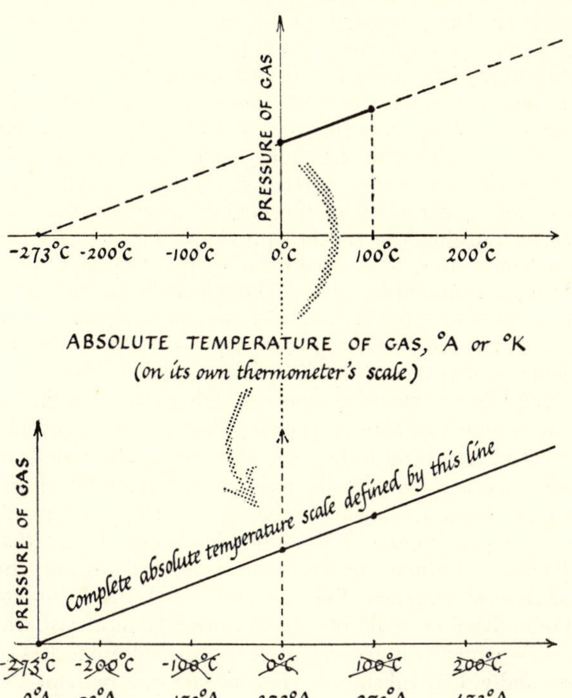

FIG. 27-11. THE GAS THERMOMETER'S TEMPERATURE SCALE

other temperature[13] make use of this straight-line-through-absolute zero to reduce gas-expansion problems to simple proportion sums, as follows. We take the temperature-defining graph and draw on it a new set of axes, with the origin at −273°C. We re-number the temperatures, beginning with 0 at the new origin (which is of course the "absolute zero," −273°C). We call the new temperature "absolute temperature" adding 273 to every ordinary temperature number. Thus we have moved the origin (but not the graph line) 273 degrees to the left. We have a straight-line graph running through the origin with gas-pressure upward and absolute temperature along. The GAS PRESSURE, p, varies directly as ABSOLUTE TEMPERATURE, T. For two different temperatures T_1 and T_2

$$p_1/p_2 = T_1/T_2$$

From using gas thermometer (a), or trusting Boyle's Law, we also find that for *volumes* V_1 and V_2, *at fixed pressure*

$$V_1/V_2 = T_1/T_2$$

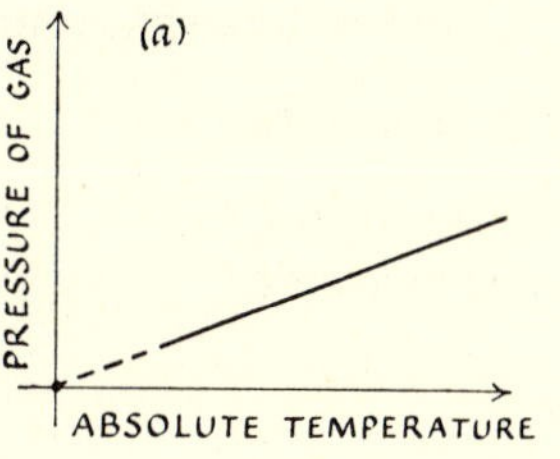

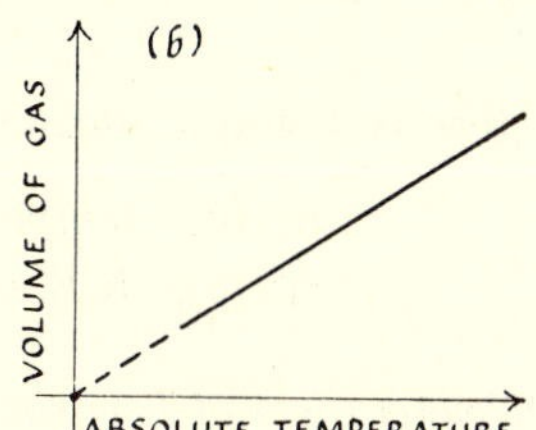

FIG. 27-12.
(a) PRESSURE OF GAS (fixed volume) *vs.* ABSOLUTE TEMPERATURE.
(b) VOLUME OF GAS (fixed pressure) *vs.* ABSOLUTE TEMPERATURE.

This behavior is true for gases in the ordinary temperature range, automatically true because the line is drawn straight to define gas temperature. At very low or high temperatures we could still insist it is true, but we should then find different real gases disagreeing about temperature scales. So we conceive of an ideal gas—"a favorite trick of theoretical thinking"—and use that to define a universal temperature scale, from absolute zero to anything we like. At ordinary temperatures our ideal gas agrees with most real ones, but it avoids the peculiarities of its weaker brethren like CO_2, and at extremes it pursues the simple gassy course where real gases would disagree or even liquefy.[14]

Kinetic Theory and Gas Temperature

Kinetic theory, which we regard as good theory from the success of its predictions, predicts that GAS PRESSURE should vary directly as the AVERAGE

---

[13] In chemistry, for example, some of the simplest and most important measurements are gas volumes. Gas is squashy stuff. The measured volume of gas manufactured in an experiment depends not only on the mass of ingredients but also on the pressure and temperature of the gas. To calculate the mass of gas in a sample (using a standard density), or to compare gas volumes from different experiments we must first cool the gas (in imagination) to a standard temperature such as 0°C and find the volume it would have then. Then we must do a Boyle's-Law calculation to reduce the volume to standard pressure.

[14] *Charles' Law*: Most texts quote a "Charles' Law" which states that all gases expand equally and uniformly, with volume varying directly as absolute temperature. This embodies an experimental fact, which we have already used: that *all gases agree in leading to the same absolute zero, −273°C*, when a straight line through their ice and steam point volumes is extrapolated; or in other words all gases expand by the same fraction from ice to steam points. But Charles' Law is often also made to say that the expansion is uniform. It should be clear from the discussion above that this might have two meanings:

(I) Gases expand uniformly when compared with mercury thermometer. This is an experimental fact arising from the chance choice of mercury!

(II) Gases expand uniformly when compared with the gas thermometer's temperature-scale. This is a necessary consequence of the definition of the gas thermometer's scale. To claim it as experimental fact of nature would look like "Zanzibar."

We may include either view (I) or view (II) in Charles' Law, but a vague mixture of the two would be unfortunate.

KINETIC ENERGY of the molecules. Now since defining our gas-thermometer temperature scale makes gas pressure vary as absolute temperature, we see that, on the gas-thermometer's scale,

GAS PRESSURE varies as ABSOLUTE TEMPERATURE.

Combining these two relations we have

MOLECULAR KINETIC ENERGY varies as ABSOLUTE TEMPERATURE

Here is a simple meaning for temperature:

Absolute temperature measures the average K.E. of a gas molecule.[15]

General Ideas of Temperature

Our idea of temperature now takes better shape. But lest it seem too satisfying, lest we say too dogmatically "*real* temperature is the average K.E. of a molecule," let us open our discussion into the widest general operational definition. To be honest, we do not know what temperature *is* and we are never likely to know more than is supplied by some operational definition we choose to adopt.[16] So we choose as a measure of temperature *any* physical quantity that changes reasonably[17] with heating and cooling. We measure this quantity at the ice point and steam point, plot the measurements on a graph against 0°C and 100°C (or, for an absolute scale, against 273°A and 373°A). Then we draw a temperature-scale-defining straight line through these two points. Thereafter we use the graph to read off temperatures on that scale.

Early thermometer-makers chose for their physical quantity the *volume* of a sample of water, then alcohol, then mercury; later, the pressure of a gas sample was chosen. Others that might be used are electrical resistance, thermo-electric voltage, differential expansion of two metals. In fact all these are used in certain secondary thermometers, which are not used to define temperature scales—instead the scales of such instruments are calibrated by comparison with mercury thermometers, which in turn agree well with the gas thermometer.

In making our choice of physical quantity, convenience of accurate thermometry is the criterion, and our hankering for a really true temperature scale must go unsatisfied, unless perhaps we can find a universal physical behavior that belongs to *all* substances, not just to a single substance like mercury or a class like gases. The miracle is that we can. A century ago, Kelvin, studying the properties of an ideal heat engine—the pluperfect of all steam engines, hot air engines, turbines, etc.—devised a temperature-scale using the heat intake of such an engine from a furnace as a measure of the temperature of that furnace. This hardly sounds like a very promising or satisfying scale, but it developed remarkable properties. First it is the same scale, whatever the construction of the engine, whatever the substance pushing the piston (provided the engine is an ideal one—that is, one with no friction, no heat losses through conduction, etc.). Though such an engine is impossible to make, it is easy to imagine, and the specifications for ideal working are simple and clear. Its efficiency in converting heat to useful mechanical energy—acting like any steam engine—is higher than that of any real engine that can be made. That is not surprising; but it is surprising to find the efficiency independent of the constructional details and working material of the engine. Lord Kelvin and others proved this independence by an ingenious thought-experiment. Using that efficiency (which involves the heat-intake from the furnace of course), Kelvin constructed his "absolute thermodynamic scale of temperature." Here at last was a scale which is independent of the special properties of any individual substance. Here at last was perhaps the really-true scale of temperature. Would scientists therefore keep it, and worship it? Not a bit. They would not have kept it for a moment on that account. They kept it, and now use it as the ultimate standard, for quite a different reason: it is of enormous use. (It also turned out to agree with the gas thermometer's scale, and that too made it easier to welcome!)

We can imagine all kinds of different substances put into an ideal engine and used in the cylinder. Then by letting the engine run, in imagination, and keeping track of heat-intake, etc., we can arrive at remarkable predictions for the substance we have chosen to put in the engine. The predicted relationships are surprising and useful. E.g., if a slush of ice and water is used in the cylinder the resulting equation tells us the change in the melting point of ice, per atmosphere of pressure, in terms of the density of ice, the density of water, and the heat needed to melt one kg of ice, and the temperature T of melting ice. The melting-point change is difficult to measure experimentally. We can now calculate it from four other, easy, measurements! Or we may fill the engine cylinder with radiation; and the resulting equation tells us that the RADIATION-FLOW from a hot surface varies as T^4, where T is the absolute temperature on the new scale. (Useful for measuring the Sun's temperature.) With a cloud of electrons in the cylinder we can find a relation that applies inside radio tubes! This game, of imagining some stuff in the cylinder, and obtaining a useful relation that applies to the stuff, is the science called Thermodynamics. It produces useful results for engineers, chemists, atomic physicists, astronomers. But *all its results are expressed in terms of its own temperature scale*, the Kelvin absolute thermodynamic scale. They would be impractical nonsense but for Kelvin's discovery that his ideal engine scale agrees

[15] See Ch. 30. Equipartition suggests this should be extended to other particles: the molecules and atoms of liquids and solids. But when we bring in other motions such as vibrations, as we must for atoms in solids, the quantum restriction appears and alters the picture gravely.

[16] We know that gas-thermometer temperature is a measure of molecular K.E. but we have not discovered that as a description of "true" temperature; we manufactured it when we settled on the gas scale for convenience.

[17] The volume of mercury is suitable. The volume of water is not. As water is warmed from 0°C to 100°C it contracts a little from 0° to 4°C and then expands a lot. A water thermometer would suffer from double-talk.

with the gas thermometer scale. So we now use gas thermometers as our standard, for their practical precision and for the use of their scale in all our thermodynamic predictions. In honor of Kelvin's work, we call the combined ideal-engine and gas scale the Kelvin scale of temperature, and instead of °A we write °K. To many of us, our exploring ship seeking the true temperature, after long wandering with little hope, has come to anchor on solid ground, solid ground of universal validity.

Interesting Temperatures

The chart (A) in Fig. 27-13 shows some interesting temperatures on the Kelvin absolute scale. Near absolute zero the very low temperatures seem crowded. This crowding, and the puzzle of the absolute ending of temperature at 0°K, are removed in (B) which is plotted on a logarithmic scale.

Absolute Zero seems a tantalizing limitation to most people when they first meet it, and an irritating one to some. The Kelvin scale sets the same absolute zero as the gas scale, and thermodynamic arguments make it clear that we could hardly hope to reach it. Temperatures below it neither exist nor have any ordinary meaning.[18] This limitation seems puzzling, but the puzzle begins to disappear when we try experimentally to reach very low temperatures. It requires considerable trouble and money to cool a piece of material down from 100°K to 10°K (from just above liquid air temperature to just below melting solid hydrogen). To cool it on down from 10°K to 1°K costs about as much. From 1°K to 0.1° again about as much, and again from 0.1° to 0.01°. . . . So, with mounting costs, absolute zero seems financially and physically unattainable.

This story of cost suggests a modification, which may remove the sense of difficulty. Instead of using these numbers, 100, 10, 1, 0.1 . . . for the temperatures, why not use some numbers more indicative of the equal steps of difficulty? We could get equal *steps* instead of equal *factors* of 1/10 by taking logarithms, thus: log 100, log 10, log 1, log 0.1, etc. These are 2, 1, 0, −1, . . . etc. As the temperatures on the old scale creep nearer to absolute zero by smaller and smaller steps, the logs stride on down undaunted, 2, 1, 0, −1, −2, −3, . . . to minus infinity. The old "absolute zero," now "minus infinity," looks quite unattainable. Such a logarithmic scale, a fairer one for showing very low temperatures, is shown on Chart B. What justification have we for using these logs to indicate temperatures? Well, what justification had we for using the plain gas-pressures originally for our temperature graph? We could have chosen to use PRESSURE^2 or $\sqrt{\text{PRESSURE}}$ or, as now, log (PRESSURE). It was just our prejudice for simplicity and preference for convenience that made us choose plain PRESSURE. And we still choose PRESSURE. The "log" scale is not in use; it was sketched here merely to encourage you to think about temperature. However, "logarithmic" scales are often used elsewhere in science—e.g. on graphs of radioactivity.

[18] Negative temperatures, and an extension of thermodynamics to go with them, have recently been dreamed up. They remain nonsense for ordinary thermometry but can have a real meaning in the strange boundary-region where the science of electronics meets theoretical physics.

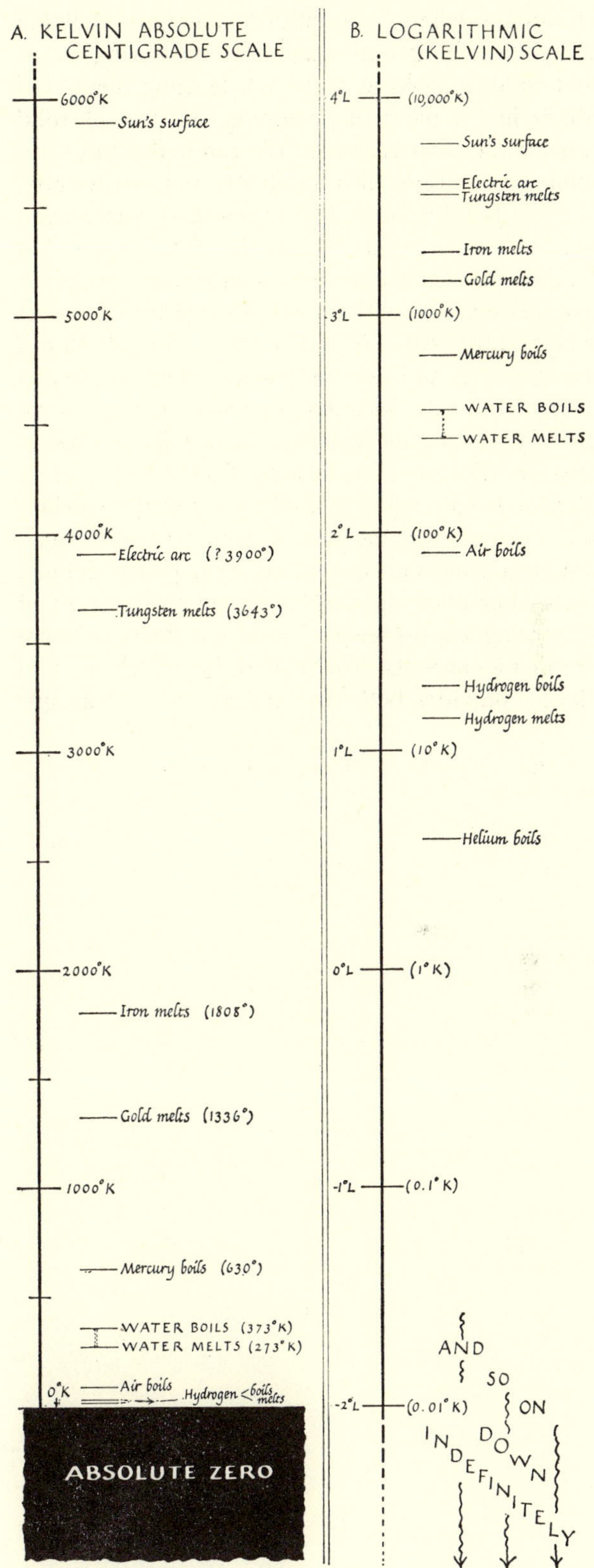

FIG. 27-13.

Temperature an Artificial Concept

Look back on this discussion of temperature. We started from a vague idea or sensation and manufactured a definite concept of temperature, defining it by describing our scheme for measuring it. At first sight, it looks as if the whole thing, scale and all, is just a piece of imagining, no part of solid science. Yet that is not so. We can make real thermometers and use them profitably and consistently. We can build a theoretical system that makes good predictions, in terms of a well-defined scale. Yet temperature itself remains a man-made concept, or man-chosen one, with an infinite choice of types of temperature scale still open to us. Not all things we measure and use in science seem artificially defined like this. Some seem obvious things, known already, things that common sense tells us how to measure. For example, LENGTH. People have a clear idea of length without needing a scientific definition. The size of the unit is arbitrary—and much practical confusion can come from poorly-defined units—but once the unit is settled the process of measuring lengths seems obvious and definite. There are many concepts, like LENGTH, for which we feel Nature provides both the concept and a clear system of measurement: for example, AREA; NUMBER of toes on a rabbit's foot; DENSITY, perhaps; possibly WEIGHT. But some philosophers condemn this view and claim that *all* measurements involve serious tacit assumptions and definitions. To them, all scientific measurements would be like temperature: concepts of our own making.

Thermometry for Rocket Travellers

A parting word on a practical question. If a rocket service is instituted for intercontinental travel, the ship will rise with great speed through the Earth's atmosphere, make most of its trip in very thin air indeed, then swoop down to its destination. *What temperature will a passenger observe in flight?* The temperature inside the rocket will depend on outside conditions and on the ship's air conditioning system. But what will he see if he looks through a window at a thermometer outside? A glimpse of a stationary thermometer hung on a floating balloon will probably tell him a very low temperature, especially if the instrument is shaded from sunshine. Perhaps about —50°C. But a thermometer just outside the ship, attached to it, and moving with it, would show a temperature of about 10,000° C. Why?

CHAPTER 28 · POWER. A CHAPTER FOR LABORATORY WORK

"In the public journals, and even in many books describing science for the general reader, the words, FORCE, ENERGY, and POWER were used interchangeably, with a preference for POWER in cases of doubt. The general impression of great technological wealth conveyed with this loose vocabulary was thought to outweigh the confusion maintained in the minds of readers (and writers) between three entirely distinct scientific concepts."

Preprint for "A History of Pidgin Science" ~ A.D. 2000

Power is *rate* of transfer of energy.

$$\text{POWER} = \frac{\text{ENERGY TRANSFERRED}}{\text{TIME}} = \frac{\Delta E}{\Delta t}.$$

Energy appears in many forms and its transfer from one place to another or from one form to another is of great importance to us. Energy being the prime thing we pay for, we often need to know *how much* energy is converted and we measure the transfer by FORCE · DISTANCE, in a variety of units: joules, foot · pounds, etc. We use engines to effect this conversion: gasoline engines, steam turbines, electric motors, fireworks, human bodies. Machines do little more than move energy from one place to another; e.g., pulleys and ropes, hydraulic press, levers.[1] We also want to know *how fast energy is being transferred,* or can be transferred, by some engine. A healthy man for example can transfer 50 foot · pounds of energy *every second,* from chemical form to mechanical form, drawing upon his food energy to raise heavy loads. He can keep this up for hours at a stretch; but if he tries to work ten times as fast, transferring 500 foot · pounds *in each second,* he cannot keep it up for many minutes; however well he is supplied with food to draw upon, he soon tires too much to continue. If he tries to work faster still, say at 5000 foot · pounds per second, he fails at once; his muscles refuse to do the job. Steam engines become more wasteful when required to work too fast. Electric motors are a little wasteful when running light, very efficient with medium loads; but they groan and stop when much overloaded. In general every engine has a limit to the rate at which it can transfer energy. Below this limit there is usually some optimum working rate which the user should know.

[1] A current of hot air is a "machine."

When any "engine" is running steadily—an electric lamp, for example, or electric motor, or a steam engine—the rate of transfer of energy tells us how fast we are running up the bill for fuel or electric supply. In particular, electric appliances, lamps, bulbs, heaters, etc., are labelled with their energy-transfer rate and we buy them according to that—not by *how much* electric energy they use, but by how much *energy per second.* We name[2] this useful rate-of-energy-transfer POWER.

Units of Power

We measure the ENERGY TRANSFERRED in newton · meters (= joules) and the TIME taken for the transfer in seconds; then the POWER or ENERGY-TRANSFER/TIME is in newton · meters/sec or joules/sec.

One joule/second is a useful size of unit, so we give it a name, one *watt.* This is mere dictionary-work, naming a useful unit. No experiment can prove that 1 watt is 1 joule/second: that is true automatically, by definition. For a larger unit we use 1 kilowatt, written 1 kw, which is 1000 watts. A strong man climbing a rope or running up stairs fast can transfer energy from chemical form to useful mechanical P.E. at a rate of 1 kilowatt for a few seconds. A large electric heater transfers energy from electrical form to heat at a rate of 1 kilowatt as long

[2] Power is often said to be the rate of *using* energy or sometimes rate of *supplying* energy. Strictly speaking, energy is never used up, never produced-from-nowhere, but only transferred from one form to another. Yet when we think of paying the bill for electric supply we naturally think of the exchange onesidedly and speak of *using* electric energy, or of the power station *supplying* it. Power is also said to be "rate of doing work." In this course, "work" is not treated as a form of energy, but only as a statement of energy interchange. On this view, the word "doing" is unsuitable, like the word "using" above; but it is excellent to say "*power is rate of working.*"

as it is kept turned on. Measuring energy in foot · pounds, we have 1 foot · pound/sec as a unit of power. Rough experiments with a strong horse raising loads suggested that a horse can convert food-energy to useful mechanical energy at a rate of 550 foot · pounds/sec for considerable stretches of time. This led to a unit for engineering use, 1 horsepower (1 HP) defined as 550 foot · pounds per second. Its value is kept fixed at 550 foot · pounds/sec, although real horses differ and most cannot work as fast as that.

The steam engine grew from earliest industrial use about 1700 to almost the modern form by 1800, thanks in great measure to the work of James Watt—hence the name watt for the power unit. He set up the unit 1 horsepower at a time when pumps, railroads, and other machinery were run by horses. "This is an awkward unit in more ways than one but a natural one to introduce at a time when every prospective customer of the engine-builder was asking the question, 'If I buy one of your engines, how many horses will it replace?' "[3]

PROBLEM 1. HP

The Cornish farm horse with which Watt experimented pulled on a rope which ran over a pulley to a 150-pound load in a mine shaft, and the horse walked steadily 220 feet in 1 minute.

(a) Show that these measurements lead to Watt's fixed value 550-ft · lbs/sec.
(b) Actually this is an overestimate, even for a farm horse, for continuous working. On that score, were Watt's claims of, say, 3 HP for one of his steam engines likely to turn out to be too large or too small?

Here is a list of energy-units and some corresponding units of power.

ENERGY	POWER
1 joule	1 joule/sec or 1 watt
1 ft · poundal	1 foot · poundal/sec
1 ft · pound	1 ft · pound/sec
	1 HP (= 550 ft · pounds/sec)
1 kilowatt · hour	1 kilowatt (= 1000 watts)

★ PROBLEM 2

(a) Show that 1 joule ≈ 3/4 foot · pound, using the following data:

1 pound = 0.454 kilogram
1 inch = 0.0254 meter

(b) Show that 1 HP is nearly 3/4 kilowatt. Taking 1 HP to be the same as 550 foot · pounds/sec, find its value in watts. (Express 550 pounds-wt. in kilograms-wt., then in newtons, and 1 foot in meters. Then express 550 foot · pounds in joules.)

[3] From "Physics, The Pioneer Science" by Lloyd William Taylor. (Reprinted by permission of Dover Publications, Inc., New York.)

PROBLEM 3

A 165-pound man climbs 20 feet vertically up a rope in 10 seconds. Calculate his "useful" power as follows:

(a) How much gravitational P.E. does he gain?
(b) What is his *useful* power, i.e., the rate at which he converts chemical energy into *useful* P.E.? Express this in (i) ft · lbs/sec; (ii) HP.

Note: Engineers often speak of the "horsepower" of an engine or even of a man, meaning the power expressed in HP. Though this is a vulgarism, like "rate of speed," it is a useful way of referring to the maximum power an engine can handle. However, it is wise to avoid saying "The HP of the climber is 0.5" and say instead "The power of the climber is 0.5 HP."

Efficiency

In Problem 3 the power calculated is the climber's rate of gaining P.E.—energy which he might then use, by means of ropes and wheels, to do some job such as raising a load of bricks. In climbing, he draws on his supply of chemical energy for this P.E., but the power at which he draws chemical energy is much greater because he is also producing some heat. It is possible to estimate the chemical energy used by catching his outgoing breath, measuring its volume, and analyzing its CO_2 content. From these data his fuel consumption can be calculated. And that in turn would tell us his total power use during climbing. For any engine the fraction USEFUL POWER OUTPUT/TOTAL POWER INPUT is called the efficiency. A rope climber is likely to waste a lot of food-energy as heat. Regarded as an engine for hauling up loads at the expense of food-fuel he has a poor efficiency.

PROBLEM 4

Physiological measurements suggest that the rope climber in Problem 3 would develop extra waste heat (beyond his normal, resting, use of food) at a rate of at least 1600 ft · lbs/sec. Taking this estimate, calculate his efficiency as an engine. (This is an unfair calculation, because we have also to provide food for his general living, to cover all 24 hours of the day. Since he can only climb ropes, with appropriate extra diet, for a small part of the 24 hours, his overall efficiency would be much lower still. If we compare a man doing considerable manual labor with a man doing little, we find the heavy worker needs more food, and the fuel value of his extra food is three to six times his extra output of mechanical energy. Men are at best 25% efficient in using extra fuel for extra jobs.)

Electric motors take more power from the electric supply than they yield to the machinery they drive, the difference appearing as an outflow of waste heat. A big motor may have an efficiency as high as 90%. An electric motor is a very flexible energy-converter. When it runs light, giving a small mechanical output, it demands little power from the electric supply.

When it is loaded heavily, it demands correspondingly more power, but runs nearly as fast. For a motor, the useful output of power can be measured by mechanical instruments, and the total electric input can be estimated from ammeter and voltmeter readings. We shall show later that with ammeter and voltmeter properly connected where the electric supply enters the motor, the POWER input is given, in watts,[4]

by AMMETER READING in amps · VOLTMETER READING in volts

For the present, in laboratory you should assume that this is true: but it will be discussed fully later.

Animals have great "overload capacity" and at the other extreme are economical when loaded lightly. A horse can be goaded into working at more than 1 HP for short stretches of time; or, if worked steadily each day at only a small fraction of a horsepower, will require correspondingly less feed.

Power in Human Activities

The table below shows estimates of rate of using chemical energy measured for various activities. These were made by measuring CO_2 breathed out so they represent overall use of fuel supplies, not just "useful power."

	RATE OF USING FUEL		
For a Healthy Student	*Calories/hour*	*Calories/minute*	*Watts*
Base rate (asleep)	60 to 80		80 to 95
Sitting still	95 to 115		110 to 135
Standing	115 to 125		270
Walking	230	3.8	
Running fast		8 to 12	700
Swimming		9	630
Playing soccer		11	770

By making up a schedule of a "typical day" we can use these data to estimate the total fuel used. The total expenditure in 24 hours varies greatly from one person to another. In one group of energetic healthy male students, totals ran from 3300 to 4900 Cals/day. Totals for females were 3000 to 3300, and their hourly rates were 10 to 20% smaller. The total expenditure, estimated from these data, should agree with the total fuel taken in as food during the same time, if the victim is healthy. This was tested by weighing the meals of the victims and measuring the fuel values of samples of these meals. Allowing for 5% wastage of food-fuel, the total intake of fuel was 1% to 8% greater than the total expenditure calculated from the CO_2 measurements. For such difficult measurements, this is remarkable agreement.

Steady work at heavy physical jobs makes a great demand of extra fuel. Here are data for factory workers.

	RATE OF USING FUEL	
Factory Workers (8-hour day)	*Calories/hour*	*Watts*
working at light manual work	190	220
medium	250	290
heavy	350	410

To see what a man can do, take 410 watts for heavy work, subtract 90 watts for base rate, leaving 320 watts as the cost of the actual job. Suppose the worker is 25% efficient, which is optimistic. Then he delivers, as *useful* power, 25% of 320 watts, or 80 watts. One man could just grind a generator running one 80-watt electric lamp (for an eight-hour working day), if he received enough food. Some men can work steadily with a useful output of 100 watts, others less than 80. In a civilization using slaves as prime movers for building, etc., one man-power might be reckoned as 90 watts, or ⅛ HP.

PROBLEM 5

Suppose an ancient civilization had modern electric lamps. What size of bulb could a capable slave keep alight from 4:00 p.m. till midnight daily by grinding a suitable generator which had an efficiency of 80 per cent in converting mechanical energy into electrical energy?

You can estimate your own *"useful"* power by timing yourself running upstairs. This will give a high estimate for a spurt which you could not maintain long. Or you can estimate your "useful" power for steady working by deciding the speed at which you are prepared to continue climbing an indefinitely high staircase for many hours. In raising your weight vertically, you are storing up useful P.E. In moving along horizontally you generate waste heat in your body and in the floor, but this is not likely to be useful, so we do not count the horizontal part of the climbing motion. Moreover, you do not know the forces involved in the horizontal motion, and even a rough estimate of its cost is difficult.

[4] *Watts.* Electrical devices are usually rated in watts, so most people think that watts are special electrical units. They are not. They are general units of power applicable to any engine, any interchanges.

PROBLEM 6

A 150-pound man runs up a staircase in 10 seconds. The staircase rises 30 feet vertically, extends 40 feet horizontally and measures 50 feet up the slope.

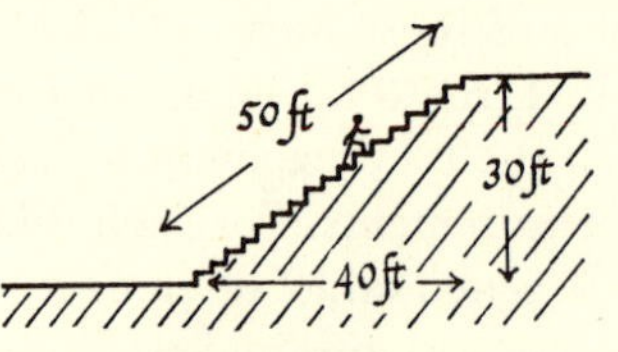

Fig. 28-1. Problem 6

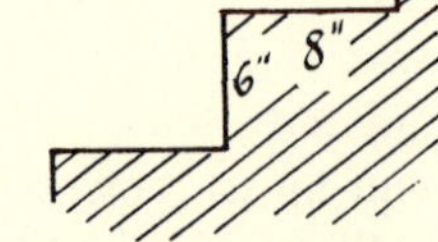

Fig. 28-2. Problem 7

(i) Calculate the man's useful power at which he transfers food energy to useful P.E. (a) in foot · pounds/sec; (b) in HP.
(ii) If his muscles are 25% efficient, he must waste three times this useful power as heat. What is the total power demanded for this vertical motion?

(In this calculation you should use the vertical height; not the slant height. Ignore the horizontal part of the motion.)

PROBLEM 7

The man in Problem 6 does waste some power in his horizontal motion. Use the guesses suggested below to make a very rough estimate of that loss. The man's foot skids a little, forward, as he puts it on each stair-step. The energy converted into heat by this shoe/floor friction is thus taken from him and wasted.

In 10 seconds the man climbs 30 ft up, moves 40 ft horizontally. Suppose this staircase has 60 steps (each rising 6 inches, which is reasonable). Each step is 8 inches long. The man's foot cannot skid more than 8 inches on each step; and it cannot fail to skid some distance, unless he takes special care. His foot probably skids 1 to 4 inches. (Try this yourself.) The friction-force that opposes the skid is likely to be only a few pounds-weight, since the man is still supported mainly by the other foot. Assume the friction force is 2 to 4 pounds-wt. (a) How much energy is wasted (as heat) in 10 secs, by these trudges? (b) What is the wasted power? (c) If three times this external waste is wasted in his body, what is the total power wasted by the horizontal motion?

PROBLEM 8

Guess at your own useful power for climbing stairs. Answer the following questions by making a sensible guess, or by rough trial.

(i) How much does one stair-step rise vertically?
(ii) How many stair-steps could you climb in each minute if you had to continue for many hours? (*Note*: You are likely to overestimate this, unless you make a rough trial or can draw on memories of mountain climbing.)
(iii) What would be your useful power climbing at this rate, in ft · lbs/sec, and in HP?

Sizes of Power Units

As rough indications of the size of some of these units, think what creatures could climb vertically with unit power. 1 watt is about ¾ foot · pound per second. So we want to find an animal weighing ¾ lb that can climb vertically 1 ft/sec. This would be a small rat, so we might call 1 watt a ratpower, and a kilowatt an apepower.

Small electric motors in domestic appliances run from 1/100 HP to ½ HP. Motors in machine shops run from ½ HP up to huge ratings. These ratings indicate the maximum power that it is wise to make the motor deliver—it can deliver more, at corresponding input, but may overheat.

Some motors are rated in horsepower, others in kilowatts. How do these two units compare? Make a rough calculation:

1 HP = 550 ft · pounds/sec.
1 kilogram ≈ 2.2 pounds
1 HP ≈ 550/2.2 ≈ 250 ft · kilogram-wt./sec
1 meter is about 10% more than 3 ft; take 10% off 250 and divide by 3
1 HP ≈ 225/3 ≈ 75 meter · kg-wt./sec
or "kilogram · meters"/sec
≈ 75 × 9.8 *newton* · meters/sec
≈ 735 newton · meters/sec
or joules/sec

Accurate arithmetic gives 746.

So 1 HP = 746 joules/sec

Then 1 HP = .746 kilowatt ≈ ¾ kilowatt, a useful conversion factor for engineers.

LABORATORY EXPERIMENTS ON POWER

The object of these experiments is to give you a feeling for power and its units, and to enable you to estimate your own useful power output in certain circumstances.

A. *Measure (roughly) your own useful power* in converting food-energy to *useful* mechanical energy in the following cases:

(i) Run up a flight of stairs as fast as you can.
(ii) Run up a very tall flight of stairs as fast as you can.
(iii) Climb a flight of stairs at a rate which you believe you could maintain for several hours (for fairness, carry out some preliminary conditioning exercises).
(iv) Climb a rope.
(v) Grind a wheel against a band-brake for several minutes. (See details of brake measurements below.)

The timing can be done by a partner—but each partner should do his own climbing. Since the measurements are necessarily rough, and will differ from one trial to the next, accurate timing is not justified —an ordinary watch with a seconds hand will suffice. In each case, calculate your useful power in each of the units below. (In most laboratory experiments, changing the result into a variety of units is just timewasting arithmetic, but here several units have important uses and you should know your power in each of them.)

(a) foot · pounds/second
(b) joules/second (watts)
(c) horsepower
(d) for (*iii*) and (*v*) Calories per working day of 8 hours.

B. *Measure the useful power of a bunsen burner* for heating water in a pan.

(i) Fill the pan with water and heat it for one minute; make the necessary measurements for calculating the heat gained according to the rule HEAT = MASS OF WATER · TEMPERATURE RISE.
If you consider that the heat gained by the metal is also useful, you can allow for that too: multiply MASS by TEMPERATURE RISE, as for water, and then multiply the product by the conversion factor for the metal, its SPECIFIC HEAT. (Take specific heat to be 0.2 for aluminum; 0.1 for copper and iron.) Express the useful power in (a) Calories/minute; (b) Calories/second; (c) watts; (d) HP (roughly).

(ii) (Optional) Repeat the experiment with various types of flame: smoky yellow flame, clear quiet flame, roaring flame (reputedly the hottest). Comment on their efficacy.

C. *If one is available, examine or use a "fuel calorimeter"* for measuring the *complete* heat output from a bunsen burner or the complete heat of combustion of a coal sample.

D. *Measure the power of an electric grill* (a) by using voltmeter and ammeter[5]; (b) by collecting its heat output in water.

(At this stage you should accept the heater and instruments already connected. Later, connect up such systems yourself.)

(a) and (b) can be done simultaneously. Meters can be read while the heater is immersed in a beaker of water for say 2 minutes. Express its power in watts and comment on the two values.

[5] At this stage, assume that the following relation is correct, (VOLTMETER READING, in volts) · (AMMETER READING, in amps) gives (POWER, in watts). It may seem strange that we can measure power with these two instruments alone, without a clock. Power is a rate and needs a clock for its measurement. But an ammeter is itself a rate-measurer: in a sense it incorporates a clock because it measures electric current in coulombs/sec.

E. *Measure the useful power of an electric motor. Also estimate its efficiency.* Instead of letting the motor do a useful job, which would make measurement difficult, make it drive a pulley against a band-brake (*see below*). Calculate its output power in kilogram · meters/sec, then in watts.

The ammeter and voltmeter (already connected for you at this stage) will tell you the total input power from the electric supply. Calculate the input power. Calculate the motor's efficiency. (*Note*: If you think about it, you will realize that a motor's efficiency cannot be a fixed value, belonging to that motor for all loads. What efficiency would you expect for a motor "running light" with *no* load, doing no useful work at all? What would you expect for a motor loaded so cruelly that it is unable to rotate?)

Using a band-brake to measure power

To measure a man's useful power directly we could make him haul up a load hung on a long rope. But even

FIG. 28-3.

a tall stairwell would not provide enough height for several minutes' hauling, unless we used a multiple pulley system and a huge load, which would be dangerous. So we use a circular arrangement, replacing the suspended load by the friction-drag of a band-brake on a wheel or drum. Fig. 28-4 shows one form; others use a bicycle frame and make the man pedal against a brake.

The band of rope or leather ABCDE is pulled upward by spring balances X_1 and X_2 so that it presses tightly on the circumference of the wheel all round the lower half-circle of it. The man turns the wheel steadily, by means of a crank handle. The belt is prevented from moving with the wheel by the spring balances, so it rubs against the wheel, opposing the wheel's motion with a friction-drag all along the half-circle BCD. The part DE, whose pull is measured by X_1, pulls hard *against* the motion of the wheel. The part AB, pulls with a smaller force, measured by X_2, trying to *help* the motion. If these two pulls are F_1 and F_2, the net drag is their difference, $F_1 - F_2$.

While the wheel makes one revolution, every part of its circumference, $2\pi R$, travels around under the belt, moving past the friction drag $(F_1 - F_2)$. To the man

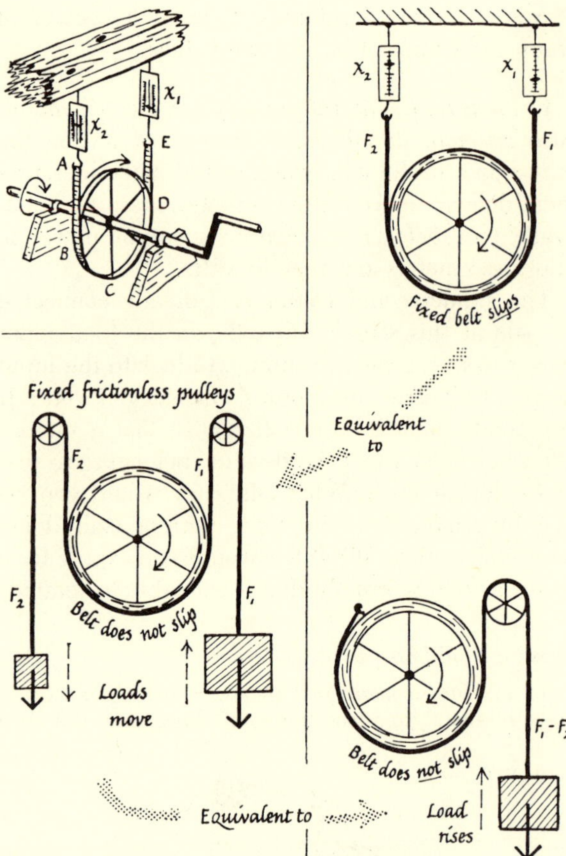

Fig. 28-4. Use of Band-Brake to Measure Power
When a wheel is ground around rubbing on a stationary belt (brake), the transfer of energy (from mechanical energy supplied to wheel, to heat in the brake) is given by: WORK = (FORCE) · (DISTANCE); and for one revolution, by: WORK = (CIRCUMFERENCE) (TENSION DIFFERENCE). These sketches are intended to show why this is so. The system with fixed spring-balances is equivalent to one with fixed loads hung over pulleys, and that is equivalent to one load $(F_1 - F_2)$ being raised by the wheel.

driving the wheel, this is the same as if the belt moved with the wheel, carrying loads pulling it with forces F_1 and F_2. This would be equivalent to having the wheel haul up a single rope with a load giving a pull $(F_1 - F_2)$. That would make the man give energy $2\pi R \cdot (F_1 - F_2)$ for each revolution. Return to the real arrangement with spring balances reading F_1 and F_2; for each revolution the man must supply energy (CIRCUMFERENCE, $2\pi R$) · (DIFFERENCE OF TENSIONS, $F_1 - F_2$).

Using this last statement, combined with a count of the number of revolutions and measurements of the wheel and balance readings for a timed grinding of the wheel, we can calculate the power being converted to heat under the brake. The brake turns this power into a stream of waste heat; but it could be useful power, if the belt stuck to the wheel and hauled up loads. So this scheme measures "useful" power.

PROBLEMS FOR CHAPTER 28

Problems 1-8 are in the text.

★ 9. Copy out and complete the following statements (*Model*: "1 knot is a unit of velocity. It is equal to 1 sea-mile/hour.")

(a) 1 joule is a unit of ———. It is equal to 1 ———.
(b) 1 watt is a unit of ———. It is equal to 1———.
(c) 1 kilowatt is a unit of ———. It is equal to ——— (give the number and units, using units that are based on seconds, meters, newtons, joules, etc.).
(d) 1 watt · second means 1 watt × 1 second, so it is the same as 1 ———. Therefore, 1 watt · second is a unit of ———.
(e) 1 kilowatt · hour is a unit of ———. It must be equal to ——— (give the number and units, using units that are based on seconds, meters, newtons, joules, etc.).

★ 10. If you buy energy from an electric power company you pay a few cents per kilowatt-hour. A common price is 4 cents per kilowatt-hour. If you could take your day's diet of 3300 Calories electrically how much would it cost at that rate?

"PROBLEM SHEET"

This problem is intended to be answered on a typewritten copy of this sheet. Work through the problem on the enlarged copy, filling in the blanks, (————), that are left for answers.

PROBLEM 11. DIET AND ENERGY

The average rations run below 2000 Calories per day in some parts of the world today. The average American diet provides 3300 Calories per day. These Calorie statements are estimates of the total energy obtainable from food by digesting it. There is no other supply of energy for keeping the body running, except for the using up of the body's own material of fat, etc. --- the wasting-away process of starvation.

To examine the problem of insufficient diet, work through the following estimates;

(Take 1 Calorie = 1 kilocalorie = 4200 joules)

FOR ANSWERS MARKED (*) IN THIS COLUMN, MAKE A ROUGH CALCULATION AND GIVE A ROUGH WORKED-OUT ANSWER. (See instructions for "Judging" in Ch. 11)

(a) UNITS: CONVERSION OF DIET-SUPPLY TO OTHER UNITS

Convert 2000 Calories per day to other units. (Note that this is a POWER).

2000 Cals/day = ____________ Cals/second (LEAVE THIS IN FACTORS, WITHOUT CANCELLING)

= ____________ joules/second (IN FACTORS) = roughly (*) ________ joules/sec or watts

A healthy man living an active life may dissipate 2000 Calories per day as heat wasted from his body; so it is difficult to estimate how much extra energy a 2000 Cal/day diet can provide for useful mechanical work. However, to examine the plight of a person living on such a diet, carry out the calculations asked for below.

(b) MINIMUM DIET

An average man asleep uses about 75 watts to keep alive and warm (heart, lungs, etc.); and while just sitting still very quietly he uses about 120 watts.

What food supply, in Calories/day, would such a man need if he sleeps 8 hours of each 24-hour day and just sits still for the remaining 16 hours? . . . . (*) ________ Calories/day

(c) STARVATION DIET

Suppose a certain rather small man has a 2000 Cal/day diet, and that, in his case, three quarters of this supply does just suffice to keep his body going and that the remaining quarter, 500 Cal, is available for extra activity.

(It is doubtful if the 3/4 would suffice. He would probably have to draw on his supply of fat, and then on even less safe stores.)

(i) With the quarter, how many hours per day could he walk slowly, using 100 watts for the extra activity involved in walking ? . . . . (*) ________ hours/day

(ii) Or, with the quarter, 500 Cals, how many hours per day could he produce 1/8 horsepower of useful power ? Take 1/8 HP ≈ 100 watts; and assume that he will also develop waste heat at a rate of 300 watts while he is providing that useful power. . . . . (*) ________ hours/day

(d) A GOOD DIET

Suppose you receive a 3300 Cal/day diet and use 2000 Cal/day just for ordinary living (eating, sleeping, walking, etc.). Suppose you use the remaining 1300 Cal/day for some activity such as ditch-digging or climbing stairs.

(i) Suppose your activity requires 200 watts useful output (≈ 1/4 HP) and in delivering that you produce waste heat at 600 watts. How many hours a day would the 1300 Cal/day allow you this activity ? . . . . (*) ________ hours

(ii) Find your own useful power output working against a band-brake in laboratory. Assume you are 25% efficient; so add three times the useful output to allow for extra heat generated and wasted. At that rate (including the waste heat) how many hours a day of grinding work would the 1300 Cals/day provide for ? . . . . (*) ________ hours

(iii) If you could use all the 1300 Cals/day for stair-climbing, converting it all to gravitational P.E. without heat wastage, how many 15-foot (or 5-meter) flights of stairs could you climb per day ? (Use your own weight in this estimate) . . . . (*) ________ flights

(iv) Assume you are only 25% efficient. How many flights like those in (iii) could you climb with a supply of 1300 Cals ? . . . . (*) ________ flights

CHAPTER 29 · THE PRINCIPLE OF CONSERVATION OF ENERGY – EXPERIMENTAL BASIS

> "We have examined the balance sheet . . . in accordance with generally accepted auditing standards, and accordingly included such tests of the accounting records and such other auditing procedures as we considered. . . . In our opinion. . . ."
>
> ——— & ———, Certified Public Accountants

From vague considerations of fuel, and a mechanical rule that WORK = FORCE · DISTANCE, we produced a definite concept, "energy," whose changes are measured by "work"; and we stated that energy is always conserved. We stated that principle—or, rather, quietly assumed it and gave examples which took it for granted. That was to give you a feeling for energy before discussing its prodigious history.

Early experimenters found it difficult to disentangle the various forms of energy and keep a balance-sheet. Soon after Newton had shown the importance of mv in mechanics, claims were made for mv^2 as a better measure of the "effect" of a force. While mv was named momentum, mv^2 was given the energetic name of *vis viva*, "living force." In the 1700's there were rival schools, one violently advocating mv, the other mv^2. Then it was realised that both have proper uses: gain of mv is FORCE · TIME and gain of mv^2 is twice FORCE · DISTANCE.

Machines

FORCE · DISTANCE had long been important in primitive engineering. Ancient machine designers used it unconsciously and Leonardo da Vinci ($\sim$ 1500) wrote about it clearly. For levers, wheels, pulleys, presses, etc., the input and output of FORCE · DISTANCE are equal, except for some friction tax. If we call INPUT-WORK and OUTPUT-WORK "energy changes" then frictionless machines conserve energy. Our assurance of this is based on experiment—either directly by measurements on real machines (with allowances for friction) or indirectly by deduction from experimental rules for levers, fluid pressure, etc.

There must be an experimental basis, somewhere. An arm-chair scientist could not be sure, from pure brooding, that for a balanced seesaw $F_1 \cdot \text{ARM}_1$ and $F_2 \cdot \text{ARM}_2$ must be equal (from which he could argue that input-work = output-work). Even if he claimed his thoughts made this seem very probable, he should be suspected of drawing on early laboratory work in his nursery years.[1]

Perpetual Motion Machines

Combining simple machines into complex schemes affords no better hope of getting more energy than we put in. Failures of perpetual-motion machines gave overall confirmation of energy conservation in the restricted mechanical sense. Huygens, contemporary of Newton, warned people clearly:

> "If any number of weights be set in motion by the force of gravity, the common center of gravity of the weights as a whole cannot possibly rise higher than the place which it occupied when the motion began. . . . If the devisors of new machines, who made such futile attempts to construct a perpetual motion, would acquaint themselves with this principle, they could easily be brought to see their errors and to understand that the thing is utterly impossible by mechanical means."[2]

P.E. + K.E.

The law of levers applies to a balanced seesaw at rest or moving steadily—while the fat boy sinks and the thin boy rises, each with constant speed, the lever law of arms and forces still holds, and therefore INPUT-WORK = OUTPUT-WORK.

It is quite easy to make this untrue; just move the fat boy farther out. But then the seesaw + boys accelerates, thin boy up and fat boy down till he hits the ground. Treating the boys' weights as input

[1] Archimedes with his "Grecian mania for geometrical proof" thought he had derived the lever rule by pure argument from simple self-evident axioms. Being a clever experimenter he probably tested his rule carefully in spite of his theoretical magic. He is said to have made excellent use of real levers and catapults in defending Alexandria against invaders.

[2] In "*The Pendulum Clock*" (1673).

and output forces, we no longer have INPUT WORK = OUTPUT WORK—the fat boy's weight puts in more than the thin boy's takes out. But we need not give up our new Principle of Conservation of Energy. We can invent another form, kinetic energy, and calculate it by the rule K.E. $= \frac{1}{2} mv^2$ got by combining $F = ma$ with the definition WORK $= Fs$. By the early 1800's conservation of energy meant:

Total P.E. (whose changes are calculated by FORCE · DISTANCE)	+ K.E. (whose value is $\frac{1}{2} mv^2$)	remains constant, in frictionless mechanical systems

This is a useful general principle for solving problems in physics and engineering. It really consists of Newton's Laws II and III, with the assumption that forces compound as vectors—so it is based on experiment to the same extent at $F = ma$. It shows up an important characteristic of such mechanical systems that was already there unrecognized: the energy-change in any move or alteration is independent of the route chosen. For example: suppose we shift a load from the barn door (A) to the far end of the barn loft (B). It gains the same P.E. whether we move it

up, then along
or along, then up
or up a slanting plane
or up in some fantastic curve, with pulleys
or even up above the roof, then down to the loft.

To see how energy-conservation requires this, consider the move from A to B by two routes, each frictionless, starting and ending at rest. Take the load from A to B by Route I, then back to A by Route II. Having returned to our starting-point, A, we have returned to the same potential energy. Therefore, Route I and Route II must cost the same energy, from A to B. If not, we could make a perpetual motion scheme, going up one route and down the other, gaining energy from nowhere on each cycle.

Believing in energy-conservation at this level, we find Galileo's inclined plane rule obvious: whatever the slope, a mass M sliding down a hill of vertical height h loses P.E. Mgh, and gains K.E. $\frac{1}{2} Mv^2$. If there is no friction-tax, these two changes must balance. $Mgh = \frac{1}{2} Mv^2$. Then $v^2 = 2gh$: the same v for all hills of height h, steep or shallow, straight or curved. *Galileo's pin-and-pendulum experiment was a foundation test of this conservation of energy.*

Given a Sun and one planet, started with some given motion, mathematicians can predict the planet's orbit. One of the easiest ways is to write an equation stating that the planet's total [K.E. + P.E.] (in Sun's variable gravitational field) remains constant all along the orbit. This, combined with an equation for some other conservation (e.g., rotational-momentum) will yield the equation of the orbit: an ellipse.[3]

So far the principle is useful, but hardly a great generalization. Its growth to full importance came with the inclusion of heat, chemical energy, etc., in one grand scheme.

HEAT AS A FORM OF ENERGY

Lucretius ($\sim$ 80 B.C.), describing the views of Greek philosophers several centuries earlier, wrote:

> "No rest is allowed the elemental-atoms moving in space. Driven by perpetual and diverse motion, some, when they collide, leap far apart; and others are thrust but a short way from the blow. Those which when driven closer together rebound only a short way and are caught by their own entangling shapes, these form the substances of strong rock and rigid iron. Others leap far apart, with great spaces between; these supply for us thin air. . . ."

The speculations of the Greek atomists seem to have remained forgotten or disliked for many centuries. Ideas of atoms were revived in Galileo's time. Descartes built a fanciful theory involving atoms, and Newton speculated on heat being atomic motion. Philosophers a century later had grandiose schemes for applying Newton's powerful mechanics to Descartes' universal atoms, so that, given the positions and motions of all atoms they then could predict the whole course of the future. But the atomic picture was still only intelligent speculation and the association between heat and "atomic" motion a vague one.

"Caloric"

The idea of heat itself was none too clear for a long time after Newton. About 1750 Joseph Black made the clear distinction between *quantity of heat* and *temperature.* He measured heat by heating water or by giving it to melting ice. In the latter case no thermometer was needed; the heat was measured by the mass of ice melted. He defined the useful quantity that we now call "specific heat" and

[3] For a sun with two "planets"—e.g., Sun, Earth, Moon—the problem grows much harder if full details of perturbations are to be worked out. Yet the first thing one would do in attacking this famous "three body problem" would be to write a statement that the total energy is conserved.

in general built up the concept of heat as a definite "fluid" that moves without loss from hot things to cold. Even if heat seemed to disappear when things melted or evaporated, it was hidden as "latent heat" which could be recovered on reversing the change.

This fluid was soon named "caloric." Heating a thing meant filling the spaces between atoms to a higher "pressure" of caloric. Specific heat was a measure of the amount of space between atoms to hold caloric. Water with its big specific heat had plenty of room between its "atoms." Lead with a small specific heat must have very small spaces for caloric—a little caloric would suffice to fill the spaces to a high temperature. There was much discussion of the weight of caloric. Some thought it had weight; others, finding hot bodies lighter, claimed a negative weight. Finally Count Rumford weighed some ice, supplied heat till it was warm water, re-weighed and found no change. This did not disprove the existence of caloric but merely gave it the interesting property of weightlessness. By 1800, the caloric theory was firmly installed in favor, with what seemed a good experimental basis. It made experiments on heating and cooling, and melting and boiling, comfortable to think about. It accounted for such things as expansion on heating—the caloric nudged the atoms farther apart, exercising force-fields similar to those now fashionable in atomic physics. And it easily explained heating of matter by friction. When a sailor slides down a rope he squeezes caloric out of the rope, the calorists said. One could almost picture the man's hands wringing the heat out from between the rope's atoms like water out of a wet sponge. Why did not the caloric rush back in after the sailor let go? "Well it doesn't" was probably the first reply. It certainly does not; because things which have been rubbed together remain hot afterwards and only slowly lose heat to the surroundings. "Well the process is this," the detailed explanation ran, "the friction squeezes the rope, making less room for caloric in it. So the caloric is pushed out and burns the man's hands. The change is permanent;[4] the rope is left with less room for caloric in it." Less room for caloric? But in that case the squeezed rope must be left with a smaller specific heat. This could provide a crucial test of the caloric view. Experiments showed no change; yet many calorists clung to their belief—as scientists will today. They probably excused any doubts about specific heat by saying that only a tiny fraction of the total caloric is squeezed out: the change in specific heat would be very small. Yet while Black and the calorists clarified and improved the measurement of heat others were suggesting, with increasing assurance, that heat is energy of molecular motion.

> "Heat is a very brisk agitation of the insensible parts of the object, . . . what in our sensation is heat, in the object is nothing but motion."
>
> —John Locke, 1796

> ". . . heat is the *vis viva* resulting from the insensible movements of the molecules of a body."
>
> —Lavoisier and Laplace, 1780[5]

By 1800, the growing use of steam engines and the new understanding of the chemistry of burning gave engineers and natural philosophers (chemists and physicists) a common interest in the nature of heat. Lavoisier and Laplace suggested that animals and men "burn" their food with oxygen to form water and carbon dioxide, gaining just as much heat as if the same food were burned in a small stove and used to heat water. They argued that measurements of either oxygen used up or carbon dioxide breathed out would show how much food had been "burned." They were suggesting the idea of chemical energy that could be released in burning. In 1779 Crawford kept a guinea pig in an insulated box and measured its heat output for a measured consumption of oxygen. He then replaced the guinea pig by a tiny coal fire, burning carbon. For the same consumption of oxygen, the carbon fire produced about the same amount of heat. Burning wax gave a similar result. His actual results, given as temperature-rise of a standard mass of water were:

Guinea pig 1.73 F° Carbon 1.93 F°
Wax 2.0 F°

These figures hardly proved the case, but they were suggestive. Such experiments are indeed difficult, but they have since been carried out with great precision, on animals and on men. Their results show that the heat developed by the animal agrees with the heat got by burning the same food within a fraction of 1%.

Evidence: Rumford

At the end of the 18th century, Count Rumford produced the first strong experimental evidence that heat is not a conserved fluid but something that can be manufactured in unlimited amounts at the expense of mechanical energy. Rumford, whose original name was Benjamin Thompson, was a

[4] As a model of this, crush a piece of "plastic foam," now used for Christmas tree snowballs.

[5] Quoted by Lloyd Taylor in *Physics, The Pioneer Science.*

remarkable man. Born in New England he made an unwise choice of sides in the Revolution and had to leave for England under some pressure. He proved to be a magnificent organizer with great ability and strong interest in scientific experimenting. He also showed skill in gaining popularity and honor. He was knighted for his services; and then he set out to travel across Europe. In Bavaria he made so favorable an impression as an organizer that he was given the post of war minister to reorganize the army. Having done that, he used the army to round up the many beggars who troubled Munich at that time, and put them in comfortable barracks and set them to work. He was made a Count by the grateful Bavarian government and chose the name of Rumford (near Concord, New Hampshire). He turned an inquiring scientific mind on many of his activities. He developed cheap wholesome diets and cooking equipment for his beggars; and he did so much research on economical stoves and good chimneys that he was called in as a consultant far and wide on his return to England.

In the Bavarian arsenal, Rumford investigated the heat developed when brass cannon were being bored. A blunt borer removed hardly any metal, but produced a huge supply of heat. Rumford boiled kettles on the cannon while horses drove a very blunt borer. He concluded that the supply of heat was inexhaustible, depending only on the horse continuing to work. He was feeling his way towards the idea of heat as a form of energy.[6] And he gave the calorists a severe blow by measuring the specific heat of the borings from the cannon. The chips had the *same* specific heat as the rest of the cannon, just as much room for "caloric."

1840-1860: Proof

By 1840 the caloric theory was under severe attack, though still a popular tradition generally held by scientists.[7] The time was ripe for the new belief that heat can be manufactured or destroyed in exchange with mechanical energy; but the idea was not clearly formulated—the name "energy" was new; there was still a confusing use of the word "force." Precise experiments, and a wide variety of them, were needed to establish heat as a form of energy. These came with a rush in the early 1840's.

To appreciate our firm belief in Conservation of Energy, you should treat these experiments as witnesses, testifying in court for the new young theory. To carry a case against a strong opponent, many witnesses are needed. They must all tell much the same story, and they are more convincing if they are different, not all from one family. If heat *is* a form of energy, interchangeable with P.E. and K.E., *every* experiment which makes such an interchange must show the same rate-of-exchange between heat and mechanical energy. Experiment after experiment was done to produce heat at the expense of some mechanical energy. Loss of mechanical energy was measured by FORCE · DISTANCE and gain of heat by MASS-OF-WATER · TEMPERATURE-RISE. Each time, the question was asked, "Does each unit of P.E. lost produce the same amount of heat?" Or, "Does the same number of newton · meters of P.E. disappear for every Calorie that appears, whatever the material or method?" If the same conversion-rate holds in all transactions—with heat and/or chemical energy and/or electrical energy—we can claim a general scheme of conservation.

Many of the experiments were done by J. P. Joule, the Manchester brewer, an amateur scientist who put his heart into proving his conviction that heat is a form of energy. Joule developed experimental proofs with great enthusiasm and incredible skill. He made his argument convincing both by the variety and by the precision of his measurements.

It is very difficult to measure heat accurately, because heat leaks away from any apparatus, unless it remains exactly at room temperature. Leakage can be reduced by using a very small temperature-rise, because the net rate of leakage varies (roughly) as the temperature-difference between apparatus and room. And leakage can be made less important by using large apparatus, because heat leakage is a

---

[6] Rumford even recorded some measurements: in $2\frac{1}{2}$ hours of boring the heat generated was enough to raise 12 kg of water from freezing point to boiling. And he stated that a single horse could easily have kept the borer running. If we want to extract a conversion-rate from this record we must make a guess about the horse. Using Watt's estimate: 1 horse provides 550 foot · pounds/sec. $\therefore$ in 2.5 hours, 1 horse provides $2.5 \times 550 \times 60 \times 60$ foot · pounds
$\approx 5{,}000{,}000$ foot · pounds $\approx 7{,}000{,}000$ joules.
Heat produced was 12 kg of water $\times$ 100 C° or 1200 Cal.
$\therefore$ cost of each Calorie was 7,000,000/1200 joules
or 6000 joules.

[7] Invisible fluids were the fashion in science: caloric, electricity, æther. If you despise this idea of mysterious fluids as archaic or savoring of witchcraft, remember that the imagined properties were only those which fitted observed nature—so any "operational" enthusiast should approve. These fluids were believed to be conserved, just as we believe energy and momentum are conserved. The mistakes the sponsors of caloric made were calling it "fluid" and choosing a quantity that turned out not to be conserved; they bet on the wrong horse, after giving it an unfortunate name. Benjamin Franklin and early electricians put their money on conservation of another fluid: electric charge, one of the surest winners in the history of physics.

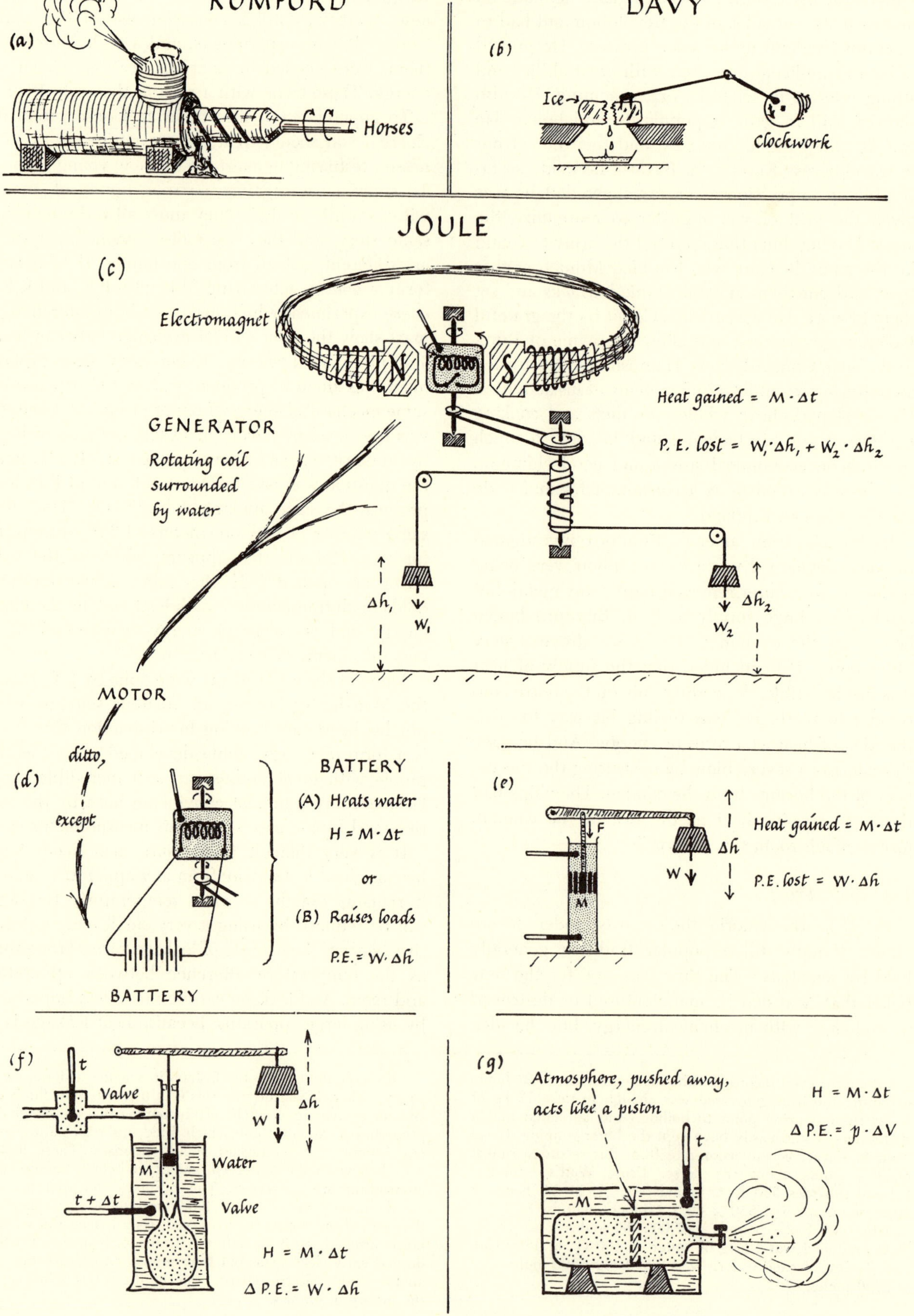

Fig. 29-1. Experiments on Energy Conversion

surface effect while the total stock of heat is a *volume* quantity, for a given temperature-rise. So if the apparatus is made larger leakage takes a smaller fraction of the total heat being measured. Joule used many pounds of water, and special thermometers graduated to 1/20 degree, so that he could estimate temperature to about 1/200 degree. He took great pains to lessen heat losses and to make estimates of them. In some cases, he tried to eliminate the effect of heat leakage by subtracting the results of two experiments, a "heavy" run and a "light" run, in which energy-interchanges were different but leakage much the same.

In one of his first experiments he developed heat in water by making it flow through very fine tubes. A perforated piston was driven through a cylinder of water by a weight falling a measured height. Then WEIGHT · Δ HEIGHT gave the P.E. lost (in, say, foot · pounds), and MASS-OF-WATER · TEMP.-RISE measured the heat produced. Joule found that for every unit of heat, 1 pound-of-water · 1 $F°$, produced 770 ft · lbs of P.E. disappeared.[8]

Joule continued his method of heating water by pushing it around. He churned water in an insulated container with a paddle wheel driven by falling weights. The container was shaped inside so that the paddles fitted closely and made a lot of water-friction and needed big weights to drive them. When the weights had fallen Joule disconnected them and raised them again for another fall. The water was heated by 20 falls (see Problem 2 at end of this chapter). The weights, descending slowly, lost P.E. in each fall, but finished with a little K.E. which they gave to the floor when they hit it. Joule made careful allowance for that K.E., which came from the P.E. lost but did not contribute to the heat he measured. He made careful measurements on the container cooling without churning so that he could allow for heat-leakage during the churning. He calculated the total P.E. lost and the total heat generated and found a conversion rate 780:1 in his units. Then, to prove this result was not a peculiarity of water, he churned mercury, whale oil, and even solid iron plates rubbing together.[9] Later still he returned to water churning with greater precision. His final experiment, 40 years after he began, was rivalled by Rowland's water-churning measurement at Johns Hopkins, in which a steam engine drove the paddle while a measured force held the container at rest.

In his earliest experiments, Joule used a very bold approach. The newly-discovered electric current was being studied and used. Great electromagnets were built by Joule in England, Henry in Princeton, and by others. Joule made one of the early "electromagnetic engines" that could be used as a generator or as a motor. It had a coil of copper wire that rotated in the field between the poles of an electromagnet. Joule drove it as a generator by falling weights. Running "light" with the coil supplying no current, it needed only small weights to provide for friction. Running "heavy" with the spinning coil producing a current, it needed much larger weights to drive it, but then the current supplied heat. Joule guessed that the extra P.E. used was delivered by the current as heat. He connected the coil's ends together to short-circuit it for maximum current and collected the heat developed in the coil itself by surrounding it with water. Then he calculated the proportion of P.E. lost by the falling weights to the heat gained by the water. He subtracted measurements of "light" run from "heavy" run, to eliminate friction that would otherwise go unaccounted. Here, with electrical energy as an intermediary, he obtained practically the same proportion 780:1.

Then he used his machine as an electric motor, with a battery to drive it. When the coil was clamped at rest, the current through it heated the water around it. When the coil was free to turn and raise loads, there was less heat, but the load gained P.E. Subtracting two runs made with the same chemical change in the battery gave him the ratio of alternative yields, P.E. or heat, again about 800:1. This time chemical energy was the common source, and Joule assumed that, for the same chemicals used, the same energy was released. (He assured himself by other chemical experiments that chemical, electrical, and heat energy balance their books truly when they interchange.)

Meanwhile, other experimenters were adding their testimony. Hirn, in France, made some comparisons like Joule's and added two which, though rough, were important because they were different. He hammered lead with a huge iron pendulum that swung down and crushed a lump of lead against a stone anvil. He calculated the K.E. of the pendulum before impact, allowed for residual K.E. afterward, and compared the loss with the heat developed in the inelastic lead. Hirn also made a measurement in reverse when heat *disappears* and mechani-

[8] Nowadays we measure heat in *kg-of-water* · $C°$ or *Calories*; and for P.E. we use the absolute unit *newton* · *meter*, later named in honor of Joule.

[9] He even allowed for the sound energy from the noisy grinding of the iron plates. He had a 'cello player match the noise in loudness, and measured the force and stroke of the 'cello bow. Resulting correction: 1%.

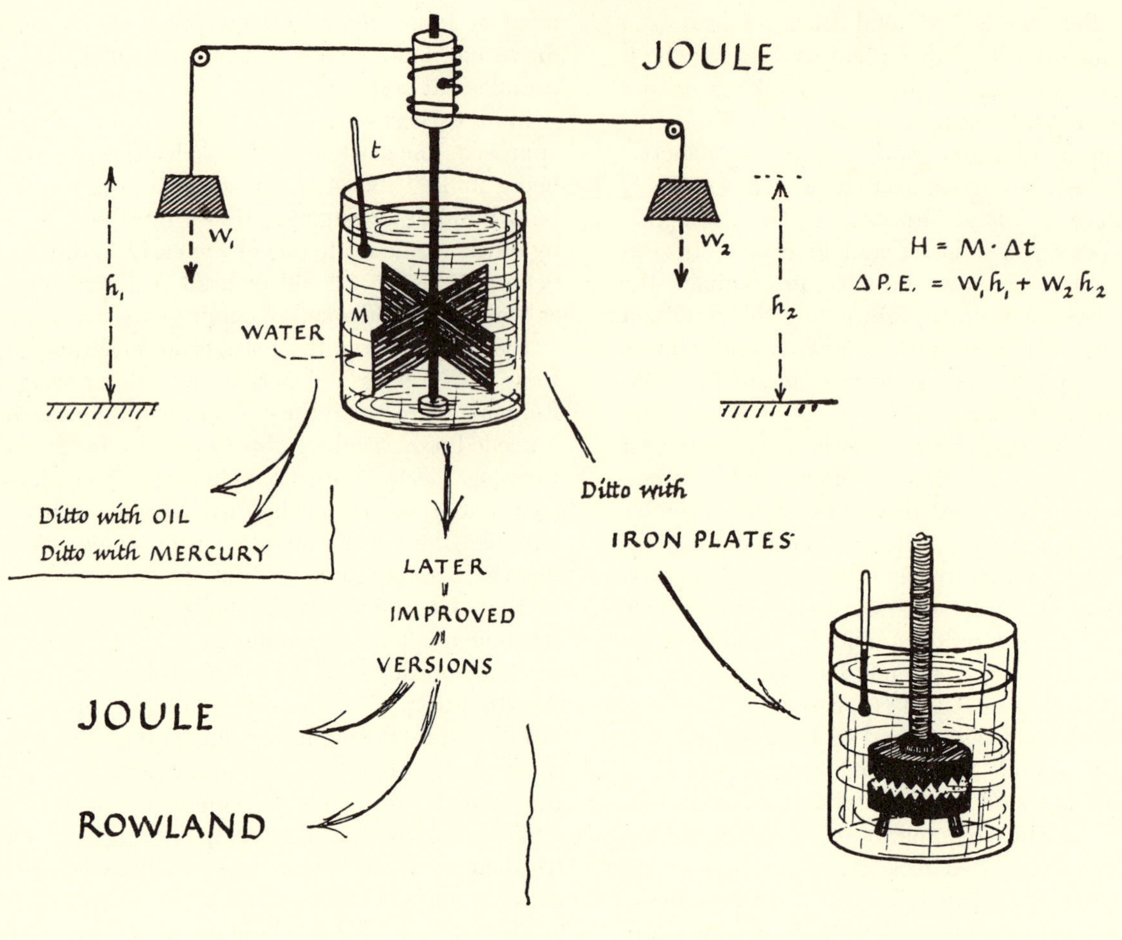

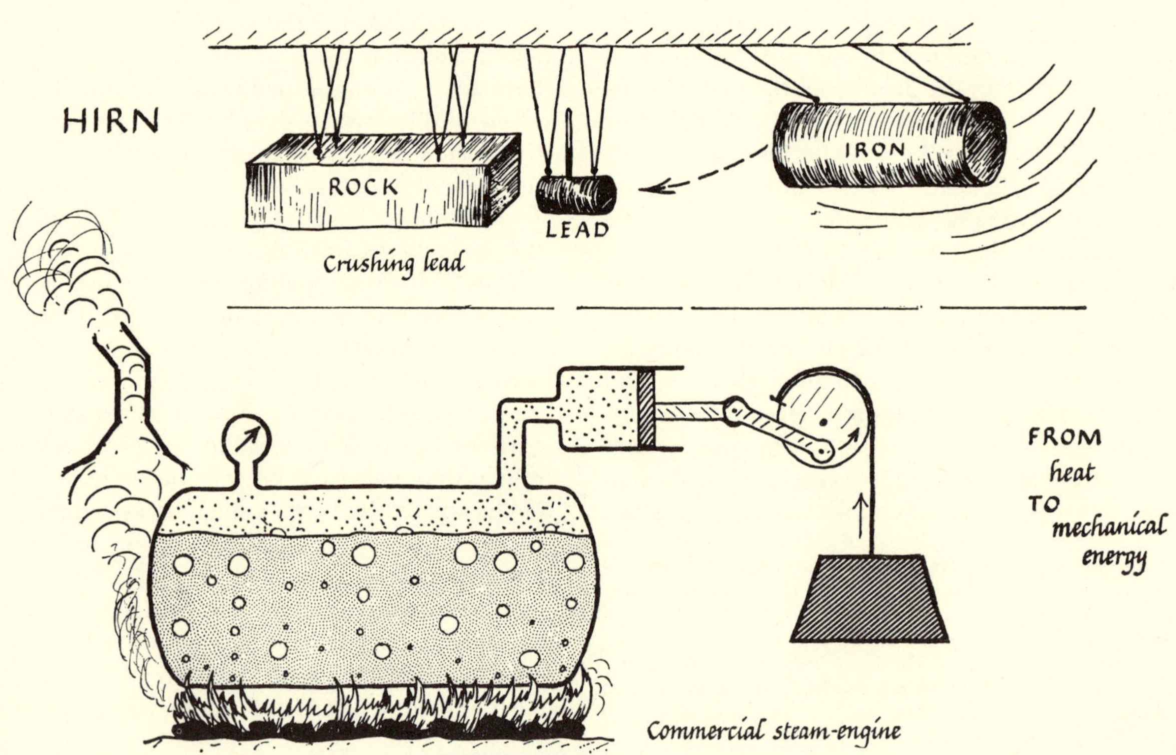

Fig. 29-1. Experiments on Energy Conversion

cal energy appears. He borrowed a commercial steam engine in a mill and measured the heat input and mechanical output. He measured the heat taken in with the hot steam, allowed for the heat wasted to the air and condenser, etc., and compared the balance with the mechanical energy yielded.

Be the jury and consider the evidence. It is spread in a long table of results, from early rough attempts to the latest precise measurements. The rate of exchange has been expressed in modern units, joules/Calorie. If you survey Joule's own work, you will see why we have named a unit of energy for him.

EXCHANGES BETWEEN MECHANICAL ENERGY AND HEAT
RECORD OF ACTUAL EXPERIMENTS

(This list gives a short description and the result, of some of the most famous experiments. Results are expressed in the form: value of *1 Calorie* in *thousands of joules.*)

Date	Experimenter	Method	Result *Value of 1 Calorie in thousands of joules*
1798	RUMFORD	Cannon boring with blunt tool. Horse driving boring machine produced "endless supply" of heat. [Rumford made no estimate of mechanical equivalent, but guesses based on his record of horse's work and water heating led, according to Joule, later, to a rough value:	5 or 6
1799	DAVY	Rubbed ice blocks together and believed he had melted ice by its own friction. Also melted wax by isolated clockwork in vacuum. [Experiments too insensitive, so gave no genuine test, but Davy's discussion and experiments had great effect on views of others.]	
1842	MAYER	Suggested the phrase, "MECHANICAL EQUIVALENT OF HEAT." Made an estimate from specific heats of gases, using rough data, making serious assumptions.	3.5
1839-1843	JOULE	Experimented with electric currents, and wrote reports that showed he was interpreting heating effects and chemical effects in terms of a growing belief in something like energy-conservation, with heat a mode of motion.	
1843	JOULE	Built simple electric machine which could be used as a generator or as a motor. Drove it as generator by falling weights, and measured heat produced when generator drove a current through coil immersed in water. (Coil was actually the rotating armature-coil of the machine.) Subtracted results of experiments with magnet turned off ("light run") from those with magnet on ("heavy"), to get rid of friction of bearings, etc.	4.76 5.38 5.60 4.90
1843	JOULE	Machine (above) used as a motor. (A) Battery drove motor which raised weights: *or* (B) The battery sent same current through a wire and heated it. [Actual arrangement was more indirect, but essentially like this.]	5.51 3.15
		ditto, improved apparatus	4.62, 4.62, 3.95
1843	JOULE	Water, driven through fine tubes, warmed up by fluid friction. Piston with very fine holes drilled through it was pushed by measured force through water in a cylinder.	4.22
1844	JOULE	Air, compressed by many successive strokes of piston-pump, warmed up. The compressed-air bottle was surrounded by large mass of water to remove and measure the heat developed. In calculating mechanical energy used, Joule allowed for changes of compressing force made by "Boyle's Law" changes of pressure.	4.42
1845	JOULE	ditto, greater compression	4.27
1845	JOULE	Compressed air, from bottle in water bath, expanded, pushing away atmosphere (as a piston) and thus cooled.	4.08, 4.37, 4.91

Date	Experimenter	Method	Result *Value of 1 Calorie in thousands of joules*
1845	JOULE	Paddle-wheel, driven by falling weights, stirred water and heated it by fluid friction. [The first form of Joule's great experiment.]	4.80
1847	JOULE	Improved paddle-wheel churned water. [Joule wound up the weights and let them fall again 20 times, to obtain enough temperature rise. He allowed for the heat lost meanwhile to the air, etc. He allowed for K.E. which the weights had when they hit the floor.]	4.21
		ditto, churning whale oil instead of water [used measured specific heat of oil]	4.22
		ditto, churning mercury	4.24
1848	JOULE	ditto, churning water. Forty more experiments with greater care. [Joule believed this result reliable to 0.5%]	4.15
1850	JOULE	ditto, churning mercury	4.16
1850	JOULE	Friction of iron plates rubbed together	4.21
1857	FAVRE	Battery produced (A) mechanical energy, or (B) heat, for same current and time.	4.17 to 4.54
1857	HIRN	Boring metal with blunt borer	4.16
1861	HIRN	Water-cooled metal brake	4.23
	HIRN	Liquid, driven through hole by high pressure, warmed up	4.16
	HIRN	Crushing lead [700-pound hammer moving 15 ft/sec smashed a 6-pound block of lead against a 1-ton anvil of stone. The lead warmed up about 5C°.]	4.17
	HIRN	Compressed air expanded against atmosphere, cooled	4.31
	HIRN	Steam engine (*From* HEAT *to* MECHANICAL ENERGY). Borrowed the use of a steam engine in a commercial mill; estimated total heat given to steam by furnace; allowed for heat wasted by radiation, condenser, etc.; estimated mechanical energy delivered.	4.12 to 4.23
1858	FAVRE	Friction of metals in mercury	4.05
1857 to 1859	QUINTUS ICILIUS WEBER FAVRE SILBERMAN JOULE BOSCHA LENZ & WEBER	Indirect electrical methods. Measured heat produced by current in a wire or battery, or heat produced in beaker of battery-chemicals. Estimated mechanical energy indirectly by electrical instruments: absolute ammeter, voltmeter, and/or ohm. The electrical units were still uncertain, so the results were not reliable.	3.9 4.2 4.2 4.2 4.1 4.1 3.9 to 4.7
1865	EDLUND	Expansion and contraction of metals	4.35, 4.21, 4.30
1867	JOULE	Heat produced by known electric current through known resistance	4.22
	WEBER	ditto	4.21
1870	VIOLLE	Disc rotated in magnetic field was heated by electrical "eddy currents." Measured mechanical drag and heat output—no electrical measurements.	4.26 4.26 4.27
1875	PULUJ	Friction of metals	4.167 to 4.180
1878	JOULE	Water churned by paddle: improved apparatus [weighted average of 34 experiments].	4.158(5)

By now, the case was proved, and the remaining question was the exact length of sentence. The value of the mechanical equivalent, "J" was now being measured so accurately that a careful measurement of "g" had to be used; and the value of 1 Calorie depended on whether the water was weighed against a brass kilogram in air or in vacuum without the buoyancy of air. And it had become clear that water does not take quite the same energy to heat it up from 10° to 11° as from 17° to 18°. If we make the specific heat of water 1, by definition, around 20° C (a comfortable room temperature) it is slightly bigger at lower temperatures. So, for statements accurate to 0.1% and better we must state the temperature-region used for the Calorie.

There have been many careful determinations of "J" in the last eighty years: A few are given below, with vacuum weighing, for 20° Calorie (1 kg water 19.5° to 20.5°C).

1878	JOULE (England): Water churned. Result of experiment above reduced to weighings in vacuum and corrected to gas-thermometer.	4.172
1879	ROWLAND (Johns Hopkins, U.S.A.): Water churned by paddle wheel driven by steam engine. Tremendous care over apparatus design and thermometer corrections.	4.179
1892	MICELESCU (France): Water churning	4.166
1899	CALLENDAR & BARNES (England): Continuous flow of water heated electrically. Temperature rise measured electrically.	4.183
1927	LABY & HERCUS (Australia): Water churned by paddle.	4.1802 ± .0001
1939	OSBORNE et al. (National Bureau of Standards, U.S.A.): Electrical heating of water.	4.1819

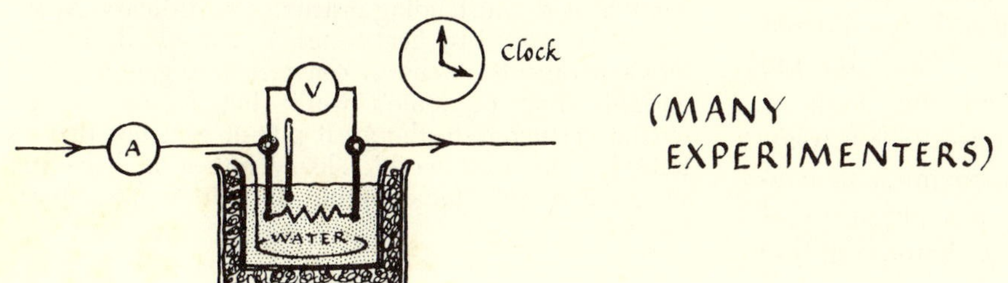

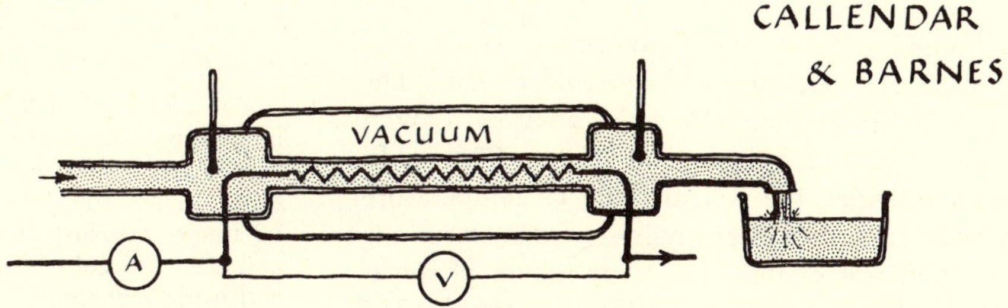

FIG. 29-2. EXPERIMENTS ON ENERGY CONVERSION
Indirect methods using electrical measurements. Ammeter standardized by comparison with apparatus that weighs the force between coils carrying currents. Voltmeter standardized by comparison with primitive generator that provides an EMF which can be calculated from geometry, a measured current, and spin-speed.

Finally, heat, chemical energy, electrical energy were established as interconvertible with P.E. and K.E. as forms of some universally conserved thing, energy. But energy was still being measured on two distinct systems: P.E. and K.E. in work units, such as newton · meters; and heat in kg-of-water · C° or Calories. Chemical energy was measured indirectly in heat units. Electrical energy could be measured in either. We have been using the ratio of those units (1 Calorie) : (1 newton · meter) as witness in our trial of caloric. If we now agree that Heat *is* Energy, that ratio must be universally the same; but we still need to know its value accurately. Taking a weighted mean of the most careful measurements, we may say,

the 20° Calorie = 4180 joules
the 15° Calorie = 4184 joules

Hence our rough value, 4200 joules/Cal for calculations of interchanges.

Thermodynamics

We now state a general law: "Heat and Mechanical Energy are interchangeable at a fixed rate-of-exchange." We call this the First Law of Thermodynamics. In its most general form, it includes such statements as "perpetual motion is impossible." We have extracted this law from a variety of experiments, but in doing so we took an overall view—"how much heat?," "how much P.E.?"; we did not inquire into detailed mechanism—"what did the chemicals do in the battery?," "are the atoms of hammered lead vibrating?" This overall-view treatment is characteristic of thermodynamics, in contrast with the approach of atomic physics that investigates detailed mechanisms before stating general results.

A similar overall survey of heat-engines yielded the Second Law of Thermodynamics: "Heat does not of its own accord flow from cold to hot." This simple platitude combines with the First Law to produce a powerful theoretical science. Thermodynamics provides the Kelvin scale of temperature, the basic theory of heat engines—from steam turbines to rocket motors—the basic theory of refrigerators and heat-pumps, and a great variety of useful predictions—such as a connection between battery-voltage and chemistry, or the relation RADIATION FLOW $\propto T^4$. Its foundation on overall views makes it all the more powerful; no change of detailed mechanism can upset its conclusions.

When molecular details are added, we develop a "Statistical Mechanics" that treats the probabilities of chaotic motions. That connects our thermodynamics with atomic physics. And recently, when applied to bits of information instead of molecules, offers to reform communication in theory and in practice.

Nineteenth-Century Physics

At the beginning of the last century, energy was an idea without a clear name. In the hands of Joule and many others the scheme of conservation was built up: mechanical energy into heat, and heat to mechanical energy: the books balanced; chemical energy to heat, or chemical energy to electrical energy and then to heat, or electrical energy to chemical energy and then to heat—all these were tried in a host of measurements that checked and cross checked: the books balanced.

It was a tremendous scientific century; at its beginning, chemistry growing to manhood, the electric current just discovered; in the middle, electrical science and engineering making huge strides, and at its end atomic physics just beginning to open up. The conservation of energy was perhaps the greatest development of all; it was the conceptual scheme that tied the others together.

"Joule Experiments" in the Laboratory

Joule's work was a miracle of careful experimenting, wrought with outstanding instruments. Ordinary experiments are spoiled by heat losses. You should do laboratory experiments on energy conversion, to gain a clearer understanding of Joule's work; but hardly to add reliable testimony to the great pile of evidence that he started—your own work is likely rather to make you sympathize with Joule in his difficulties and admire his skill.

A. ROUGH EXPERIMENT WITH FALLING SHOT.

Measure the conversion of gravitational P.E. into heat, by letting lead shot fall.

Put a handful of lead shot in a cardboard tube, close both ends, and starting with the tube vertical invert it quickly, so that the shot falls the whole height of the tube. Invert the tube sharply, again and again, say 50 times. Measure the temperature of the shot before and after this set of falls by pouring it into a paper cup and using a mercury thermometer. Each time the tube is inverted, the lead gains gravitational P.E. at the expense of your food energy. As the shot falls this P.E. is converted into K.E. which turns into heat when the lead arrives at the bottom with an inelastic squelch.

Calculate the total P.E. lost by the shot, and the heat gained by the shot. Assuming that all the lost P.E. is converted into heat and that no heat is lost, calculate the *mechanical equivalent*, J, that is, the P.E. in joules converted into 1 Calorie of heat.

Notes

(1) Weigh the shot if you like; or explain why this is unnecessary.

(2) Take the specific heat of lead to be 0.035; or look it up in tables.

(3) Sketch the tube, and show the handful of shot, (a) at the top, (b) at the bottom. *With the help of this sketch decide what height to use for the shot's fall.*

(4) *When the shot finishes its fall, the bottom of the tube should be resting on a firm table.* If you hold the tube in your hands and "cushion" the fall by letting your forearm move, the shot's K.E. is largely taken into your arm and changed there, thus giving heat to you rather than to the shot. At the other extreme, if you sweep the tube upward while inverting it, and then bring it down on the table with an angry bang, the shot will arrive with more than the calculated K.E.

(5) Why 50 falls? After 5 falls there is too small a temperature rise; after 5000 a steady temperature. (Why?) Which is best: 10, 20, 50, 100?

After trying the experiment, think about improvements. Would any of the following be better: more falls, more shot, stronger tube, longer tube, different thermometer? Some of these changes can be studied by imagination. Others deserve an experimental trial. A group of students once carried out a series of experiments that showed clearly how one of these changes could improve matters.

This is a *very* rough experiment with large unknown errors. You must not expect its result to agree with Joule's; and you cannot hope to remove the main errors by averaging many trials.

B. A GREAT EXPERIMENT.

If apparatus is available for a more serious measurement, you should try it.

PROBLEMS FOR CHAPTER 29

★ 1. JOULE AND THE WATERFALL

A waterfall provides a water-churning experiment on a huge scale. Joule (on his honeymoon in Switzerland) measured the temperature-difference between the top and bottom of a waterfall about 50 meters high.

(a) Assuming Joule's theory correct, estimate the temperature difference to be expected, thus:
  (i) Calculate the P.E. lost by 2 kg of water
  (ii) Calculate the temperature-rise if this P.E. is given, as heat, to 2 kg of water. (Assume you know that 1 Cal = 4200 joules.)

(b) Explain why the temperature-rise does not depend on the mass chosen for the calculation.

(c) Why must the measurements be made on a windless day? What would you expect on a windy day?

(d) Only some waterfalls should show the predicted temperature-difference, even on a calm day. Sketch or describe the type of waterfall that should show *no* temperature-difference.

★ 2. JOULE'S THERMOMETRY

In Joule's water-churning experiment, the paddle wheel was turned by *two* weights falling, each of them about 14 kg. Each fell nearly 2 meters. He wound them up and let them fall again, repeatedly, so that there were *twenty falls* in one experiment. The total effective mass of water in his calorimeter was about 7 kg. (This includes an allowance, in terms of kg of water, for the calorimeter and paddle, etc.)

Assume that heat and mechanical P.E. are exchangeable at a rate of 4200 joules for each Calorie, and *estimate* the temperature rise of the water. (This is, of course, a cruel reversal of the real order. Joule measured the temperature-rise and thence derived an early estimate of "J." However, it would be unfair to perform the calculation in Joule's way without taking into account his many allowances and corrections, and without using Joule's exact measurements. The calculation asked for here is intended to show the size of temperature-rises Joule had to measure.)

3. THERMODYNAMIC APPROACH vs. ATOMIC APPROACH

An investigator studying common clocks might arrive at two "laws of clocks":

I. The big hand goes around twelve times as fast as the little one.

II. The hands rotate with a right-handed screw motion, seen from the front.

Suppose these "laws" had never been formulated. Describe the way in which you think a "thermo-dynamic" investigator would discover them; and describe the way in which you think an "atomic" investigator would discover them.

CHAPTER 30 · KINETIC THEORY OF GASES: FRUITFUL EXPANSION

"*Barrier design.* . . . At atmospheric pressure, the mean free path of a molecule is of the order of a ten-thousandth of a millimeter. . . . To insure true "diffusive" flow of the gas, the diameter of the myriad holes of the barrier must be less than one tenth the mean free path. Therefore the barrier must have almost no holes which are appreciably larger than . . . 4 x 10^{-7} inch, but must have billions of holes of this size or smaller. . . . Even assuming full atmospheric pressure on one side and zero pressure on the other side, . . . calculations showed that many acres of barrier would be needed."

—HENRY D. SMYTH, *Atomic Energy for Military Purposes* (1945)

Molecular Speed and Temperature

Now we can expand our molecular theory in the light of energy.

$$\text{PRESSURE} \cdot \text{VOLUME} = \frac{1}{3} Nm\overline{v^2} = \frac{2}{3}\,\text{KINETIC ENERGY of molecules}$$

On the gas-thermometer scale, $P \cdot V$ is a measure of ABSOLUTE TEMPERATURE, T. Therefore,

$$\text{K.E. OF MOLECULES} \propto T;$$
$$\text{and AVERAGE SPEED} \propto \sqrt{T}.$$

PROBLEM 1. TEMPERATURE EFFECTS

(a) Now predict precisely how the speed of sound in air should be related to temperature.
(b) Predict how the rate at which a gas seeps through a porous barrier should be related to temperature. (This question is too vague for the answer to be safe for practical use. Assume the diffusion apparatus is heated up in an enclosure of fixed volume. Otherwise the gases will thin out to less density on heating, and you will not obtain the benefit you expect.)

How do different gases compare in their molecular motions, at the same temperature? We need a definite rule, to help in predicting properties, and to help in designing diffusion plants for separating uranium isotopes. The actual rule proves to be a simple one.

Speeds of Molecules

Even in a gas with molecules all alike, there is a great variety of speeds. Though the average speed of the molecules is definite for a given temperature, an individual molecule changes its motion at every collision, sometimes moving faster than average, sometimes slower. Suppose we could appoint a demon to watch a molecule and record its speed every millionth of a second. His statistical record would look like Graph I in Fig. 30-1, a bell-shaped probability curve showing random motion, with the speed near the average most of the time. Or, with a box of gas molecules, his snapshot census of the speeds of all of them at one instant would give the same distribution, Graph I. This is called the "Maxwell distribution," after James Clerk Maxwell who first described it for gases, nearly a century ago.

With a mixture of two gases, there would be two kinds of molecule to keep track of, exchanging some momentum and kinetic energy at every collision. Our demon, after a busy hour of watching and cataloging should be ready to report on the velocity, momentum, kinetic energy, of each kind of molecule. We have no such demon and we ourselves are too gross to observe individual molecules; but we can do the job in our imagination if we make some guiding assumptions. We assume that:

(i) Molecules *move at random* and are so *numerous* and *collide so often* that statistical treatment is justified.

(ii) At every collision, *momentum is conserved* as a vector; i.e., we assume molecules obey the rule that applies universally to large bodies colliding.

(iii) At every collision, *kinetic energy is conserved*; i.e., the collisions are elastic.[1] Otherwise the molecules would collapse on the floor in a fraction of a second.

[1] At present, we think of molecules as having no way to throw energy away or store extra P.E. permanently. So, K.E., after being stored for an instant in some potential form during collision-squashing, is released with the same total.

Then, picturing any two molecules, 1 and 2, approaching, colliding, recoiling, we write down simple algebraic equations:

vector total of **momentum** before collision = vector total of **momentum** after collision

$$m_1v_1 + m_2v_2 = m_1v'_1 + m_2v'_2$$

total K.E. before collision = total K.E. after collision

$$\frac{1}{2}m_1v_1^2 + \frac{1}{2}m_2v_2^2 = \frac{1}{2}m_1v'^2_1 + \frac{1}{2}m_2v'^2_2$$

Fig. 30-1. Distribution Graph I.

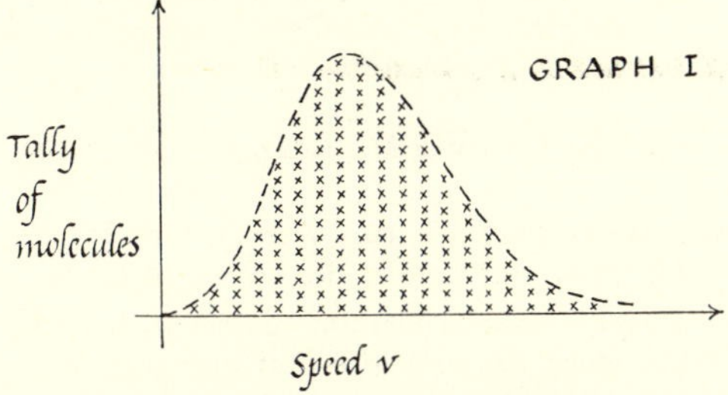

(a) Tally (or histogram) showing distribution of molecular speeds in a sample of gas. On this tally, each x shows a molecule whose speed is within a small standard range around the plotted speed v.

NOTE: Any x near the left shows a slow molecule: one near the right shows a molecule moving extra fast (at the moment). The hump shows the most popular speed; and the average speed is not far from that.

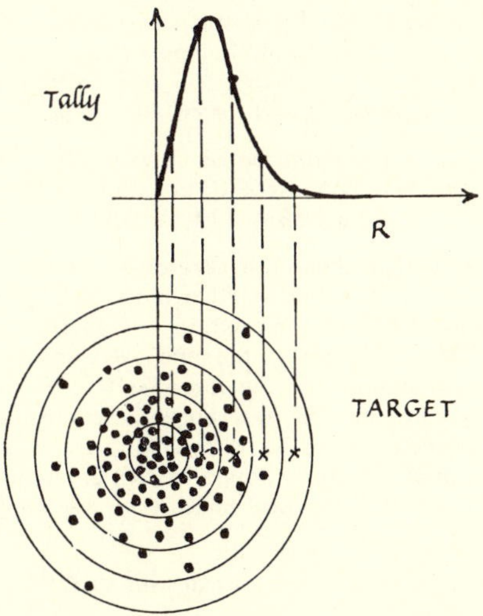

(b) A similar bell-shaped curve gives the tally to be expected for a marksman's errors when he aims at a bull's-eye target. (The same simple "law of chance" applies to his random wobbles of aim as the molecular velocities produced by random elastic collisions in a gas). The curve shown in (b) is drawn for an ordinary flat target with rings of equal width. The curve of (a) would give the tally successfully for a three-dimensional target of spherical shells.

One collision contributes little to the general picture. We have to write these equations for billions of collisions, and take the grand tally. The simple result: *in a mixture of gases, A and B, both types of molecule have the same AVERAGE kinetic energy*

$$\frac{1}{2}m_A\overline{v_A^2} = \frac{1}{2}m_B\overline{v_B^2}$$

To obtain this result, we use no mysterious advanced physics, beyond the assumptions above; but we do need advanced mathematical averaging. We must enlist the help of a statistician for that. He does a similar job in other fields. For insurance companies, he averages lifetimes of many people in different circumstances. One individual's lifetime may differ greatly from the average, but the averages themselves are amazingly reliable—insurance companies keep their million-dollar finances afloat by betting on those averages. A gas operates with even more customers and events than the biggest insurance company—a thimbleful of air contains some 50,000,000,000,000,000,000 molecules, each making several billion collisions a second. Although we expect many individual fluctuations, as in the Brownian motion, the averages should meet the statistician's prediction reliably.[2]

To see the statistical problem more clearly, consider an imaginary problem in sociology: we place a million giants and a million pygmies on an island, with money, fuel, food, etc. We ask a statistician, "What distributions shall we find, for giants and for pygmies, if we survey the population a few years later?" The statistician demands instructions and rules: "What do you want the distribution *of*, money? cloth? height? Am I to assume equality or exploitation?" We specify: free enterprise with equal opportunity; and we ask for distributions of (a) money, (b) cloth in personal clothing. Our statistician might return answers like Graphs II and III. Even if we require money to be conserved and cloth to be conserved—old garments re-sewn—we should expect the same averages for money, and different averages for cloth.

Returning to molecules: the statistician tells us that for any mixture of molecules, *kinetic energy* will show the same distribution, with the same

[2] A small insurance company might be ruined by chance deviations from the expected averages—or it might make a fortune—but a big company would find such fluctuations averaged to practically zero. Similarly, a bacillus floating in air suffers from Brownian motion; but a balloon feels a steady atmospheric pressure.

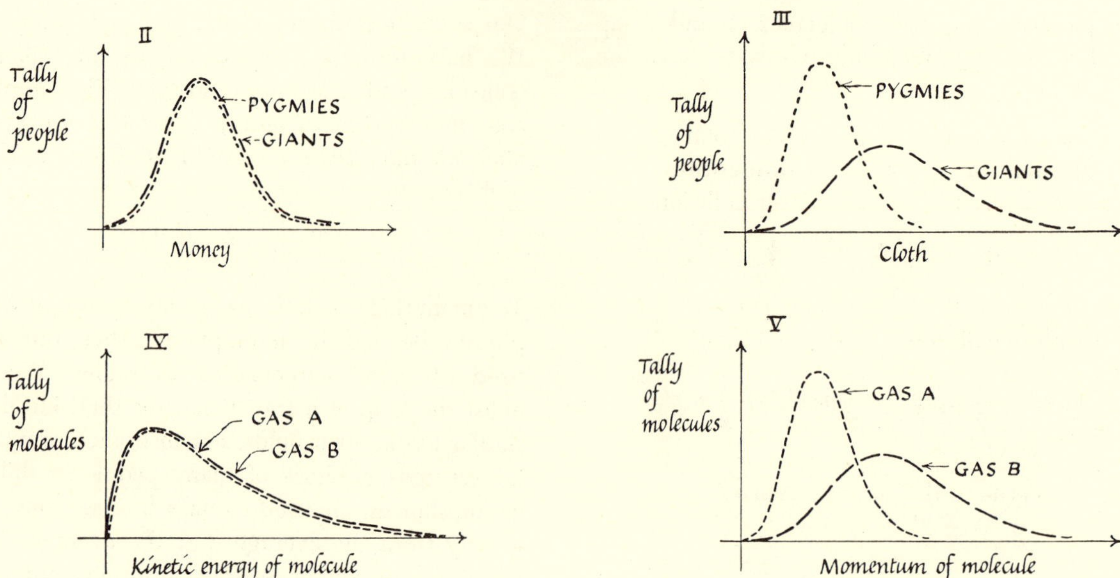

FIG. 30-2. DISTRIBUTION GRAPHS II, III, IV, V. (In Graph V, molecules are tallied by the *size* of Mv, irrespective of its direction.)

average for all types of molecules.[3] *Momentum will not.* (The *vector* average of momentum is zero—except in a wind. Here we refer to the *sizes* of Mv, irrespective of direction.)

Equipartition of energy

This statistical result is called the *equipartition of energy.* In any mixture of gases, the energy is so shared out that *the average K.E. of one type of molecule is the same as the average K.E. of any other type.* The same applies if we do not mix the gases but have them in separate containers at the same temperature—because we could then mix them without change. Therefore if two gases A and B are at the same temperature,

$$\text{average K.E. } \tfrac{1}{2} m_A \overline{v_A{}^2} = \text{average K.E. } \tfrac{1}{2} m_B \overline{v_B{}^2}$$

This equipartition (= "equal sharing") is very useful. See the problems that follow.

★ PROBLEM 2. DIFFUSION RATES

Assuming that equipartition applies, predict the ratio of diffusion rates, for two gases A and B, when their molecules stagger along thin pipes, from a high-pressure reservoir to a vacuum.

★ PROBLEM 3. URANIUM SEPARATION

For fission bombs and for enriched nuclear reactors, it is necessary to separate fissionable uranium^{235} from the plentiful isotope U^{238}, which sabotages fission. This is done by diffusion on a vast scale. Solid uranium is converted chemically to uranium fluoride, UF_6, which can be cooked up into a dense vapor that diffuses through fine pores of a special "barrier" (see Figs. 25-15, 16, 17 in Chapter 25). Estimate the possible gain, by following the discussion below.

(i) Chemical experiments and reasoning show that oxygen molecules contain two atoms, so we write them O_2; hydrogen molecules two atoms, H_2; and uranium fluoride vapor has the composition UF_6.

(ii) Chemical measurements tell us that the *relative* masses of single *atoms* of O, H, F, and common uranium are 16, 1, 19, and 238. These are on a scale that arbitrarily gives mass 1 to the lightest atom, H (or, more accurately, 16.0000 . . . to O^{16}).

(a) How would you expect the speed of oxygen molecules to compare with the speed of hydrogen molecules, at the same temperature? From the statements above, (mass of O_2)/(mass of H_2) = 32/2. Apply equipartition (without going back to $PV = 1/3$ . . .) and calculate the ratio of speeds (average speed for H_2)/(average speed for O_2).

(b) Repeat (a) for comparison of O_2 with UF_6. (The relative mass for UF_6 is *not* just 238, but 238 for U + 6 lots of 19 for F_6, making 238 + 114, or 352.)

(c) Now remember there are several varieties of uranium. The common one has relative mass 238, but the rare one (only 0.7% of the mixture refined from rocks) has mass 235, and that is the one we want to separate. Suppose a mixture of the fluorides ($U^{238}F_6$ and $U^{235}F_6$) diffuses through a porous barrier. Because the lighter UF_6 molecules have a different average speed, the mixture that diffuses through and emerges has different proportions. Will the new mixture be relatively richer or poorer in $U^{235}F_6$?

(d) Give a clear reason for your answer to (c).

(e) Estimate the percentage difference between average speeds of $U^{238}F_6$ and $U^{235}F_6$. (*Note:* As shown in Ch. 11, a change of x % in some measured quantity Q makes a change of about $\frac{1}{2}$ x % in the square root, $\sqrt{Q}$.)

(f) The difference found above shows the small step that diffusion can make from the raw mixture with 0.7 % $U^{235}F_6$ toward the desired product with, say 99 % $U^{235}F_6$. Therefore many stages of diffusion are needed,

[3] It is easy to say, "I could have predicted equal energy-distributions by common sense"; but the same common sense would suggest equal momentum-distributions, which is wrong. The statistical proof is neither easy nor obvious.

in a cascade system, with "recycling"—as sketched in Fig. 25-17. How many successive stages would you expect to need, dozens, hundreds, thousands, millions? (Make an intelligent guess.)

The Mass of a Molecule

Now with the help of equipartition we really can measure the mass of a molecule. Problem 4 shows the essential idea. The experimental treatment has to be less direct. What we see in the Brownian motion are the cumulative effects. We cannot see the frequent short staggers, so we underestimate the motion hopelessly. We *can* catalog the position of the speck at regular intervals of time, and obtain the *resultant* distance it wanders in each interval. These are statistical resultants, so again we need mathematical help. This time it was Einstein who, with others, showed how to extract the speck's real speed, v_1, from a catalog of wander-distances. Fig. 30-3 shows a map of the Brownian motion of a single speck, recorded by the great French physicist Perrin. He marked the position every 2 minutes.

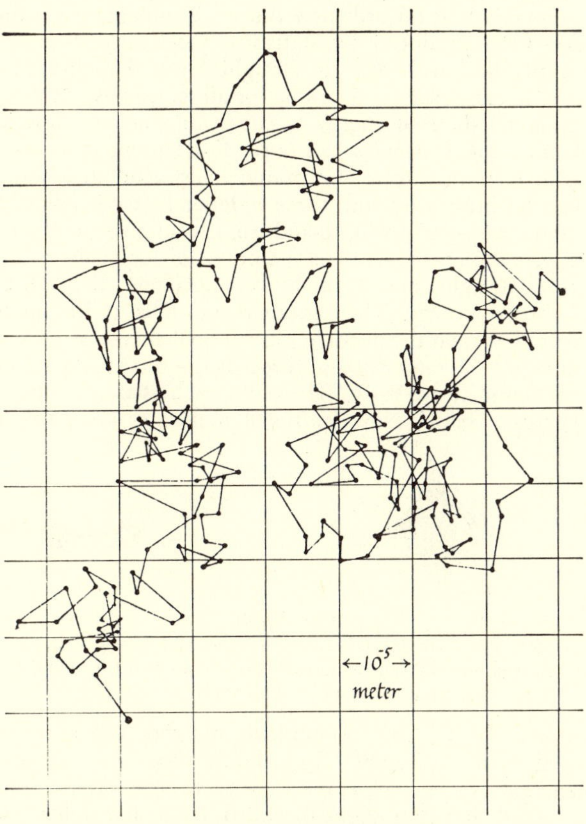

FIG. 30-3. PERRIN'S MAP
Perrin observed very small particles in water under a microscope, and recorded their positions every two minutes. In this map, the position of one particle is marked by a dot every 2 min. The grid lines have been drawn on Perrin's record to represent a spacing of about 10^{-5} meter (1/100 millimeter) in the water. Mass of particle $\sim 10^{-16}$ kg.

Then measurements of its fluctuations led to v_1 and thence to the mass of the surrounding molecules. The method is not very accurate, except in the hands of a great experimenter like Perrin, who devoted years of his life to it. The results agree with reliable estimates, made by combining CHARGE/MASS ratio for ions with the value of the electron CHARGE.

★ PROBLEM 4. MEASURING THE MASS OF A SINGLE MOLECULE

Equipartition is so general that we can apply it to a speck of ash dancing with Brownian motion among air molecules. Suppose you could measure the speck's average speed v_1, and its mass m_1.

(a) Explain how you could then calculate the mass of a single air molecule.
(b) Say what other experimental information you would need for (a).
(c) How is that other information obtained?

★ PROBLEM 5. MASS OF AN AIR MOLECULE

Here are data that a Brownian-motion experiment might yield. These are artificial, and not taken from a real experiment—which would have to be indirect—but they are typical of real Brownian motion. For an ash-speck of mass 10^{-14} kg (a hundred-millionth of a milligram) observations suggest average velocity of random-staggers is about 10^{-3} meters/sec (a millimeter/sec).

Estimate the mass of an individual air molecule. (Your result will be wrong by 10 to 30%; but even a rough estimate of such a fundamental quantity is welcome. This is a calculation of principle rather than precision.)

★ PROBLEM 6. AVOGADRO'S LAW

Over a century ago, the Italian scientist Avogadro rescued the progress of Chemistry from a serious block by a brilliant suggestion. He guessed that *equal volumes of different gases contain the same number of molecules* (at the same temperature and pressure). This gave a simple way of comparing masses of molecules, by weighing equal volumes of two gases.

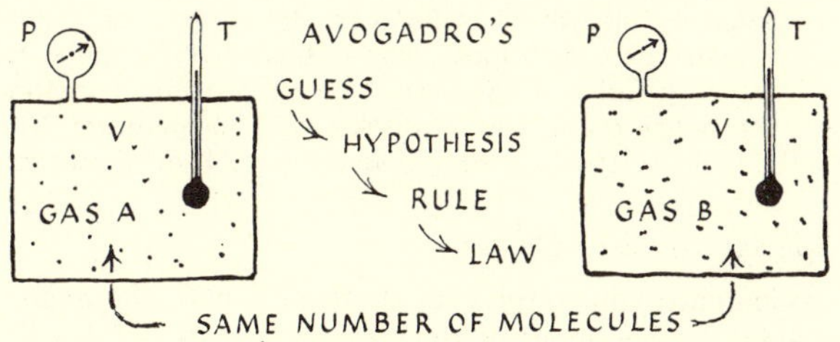

FIG. 30-4.

The results agreed with other evidence and chemists were tempted to trust Avogadro's "hypothesis" as a reliable "law." Now you can apply equipartition to prove it. Suppose gases A and B occupy equal volumes V and V at the same pressure P. Gas A has N_A molecules of mass m_A and gas B has N_B of mass m_B. Write down the kinetic theory prediction $PV = 1/3$ . . . for each gas and apply the equipartition relation to prove that $N_A = N_B$, which is Avogadro's Law.

PROBLEM 7. CHEMICAL LOGIC

Here is the use of Avogadro's Law in chemical theory. When the gases hydrogen and chlorine are mixed in equal proportions by *volume*, a flash of light starts an explosion in which they join to form a new gas, hydrochloric acid, a simple hydrogen + chlorine compound. (If the original volumes are not equal, the excess of one gas remains unused.) Then, 1 quart of hydrogen and 1 quart of chlorine make 2 quarts of mixture; and after the explosion there are 2 quarts of the compound, hydrochloric acid gas (when it has cooled back to the original temperature).

(a) If the quart of hydrogen contains N molecules, how many molecules does the quart of chlorine contain?
(b) How many molecules of compound are formed?
(c) Dividing by the number of compound molecules, find how many hydrogen molecules go into forming *one* molecule of hydrochloric acid.
(d) What can you infer concerning the number of hydrogen atoms in a molecule of hydrogen gas?
(e) Give a clear reason for your answer to (d).

★ PROBLEM 8. PREDICTING THE SPECIFIC HEAT OF HELIUM

Instead of measuring the specific heat of helium—difficult for a gas—we can pull it out of our theoretical hat, like magic. Assume that the heat energy of helium gas *is* the kinetic energy of its molecules, thus:

(i) From the prediction $PV = (1/3)$ . . . , the total kinetic energy of all N molecules, which is $N \cdot (\frac{1}{2}m\overline{v^2})$ must be equal to . . . ?
(ii) Assume that all this K.E. is the heat put into the gas to warm it up from absolute zero to whatever temperature T it is at. Measurements with a gas-thermometer show that if the gas is at the temperature of melting-ice, it is 273 centigrade degrees above absolute zero. The total mass is Nm. The rise of temperature is 273C°. The HEAT needed for this is MASS · Δ TEMPERATURE · SPECIFIC HEAT. Combine this with the result of (i) above to find an algebraic expression for the specific heat, in terms of pressure, P, volume V, mass Nm or M, etc.
(iii) Apply this to helium, with the following data: 4 kg of helium at melting-ice temperature and 1 atmosphere pressure (= approx. 100,000 newtons/square meter) occupy 22.4 cubic meters. Calculate the specific heat of helium. (Remember that the K.E. calculated by (i) will be in newton · meters or *joules*, but the heat calculated by (ii) will be in *Calories*. Before you can rightly equate these, you must express them both in the same units. 1 Cal = 4200 joules. (*Note*: Instead of melting ice temperature, you may take room temperature, 293 C° above absolute zero, and volume 24 cubic meters.)

Specific Heats of Gases

Compare your answer to Problem 8 with the experimental measurement of the specific heat, *0.74*.[4] Prediction and experiment agree well. Further, the measured value is the same at all temperatures, as the calculation suggests. Our theory is good.

Now try the same prediction for hydrogen. With 2 kg of hydrogen replacing 4 kg of helium in the same volume, the specific heat should be about 1.5. The experimental value is quite different, nearly 2.5. Our theory has soon broken down. But the failure yields a profitable hint towards new theory. Equipartition in the full form produced by statistical mechanics does not simply deal with K.E. of straight forward motion and say "average K.E. is the same for all molecules." It gives an equal share of energy to *every independent type of motion the molecules can have.* For helium molecules, which we picture as tiny round balls, random motions can be broken up into three independent components; motions up-and-down, to-and-fro, in-and-out, along, say, x, y, z directions. These are motions of molecules whizzing straight ahead, and we nickname their K.E. "whizzo" energy. Then equipartition tells us the molecules have, on the average, three equal shares of whizzo energy. The total of these three[5] is the total K.E., which we found to be $\frac{3}{2}PV$. Therefore each share of whizzo energy $= \frac{1}{2}PV$. Now *hydrogen* molecules are pairs of atoms, H—H, like a dumb-bell, and they can rotate as well as move bodily. They should also have "spinno" energy. We can see two independent axes of spin—we pretend that spin about the third axis is too difficult to generate by collisions. So we provide for two shares of spinno energy as well as three shares of whizzo, all at

Fig. 30-5. Separate Motions for Equipartition

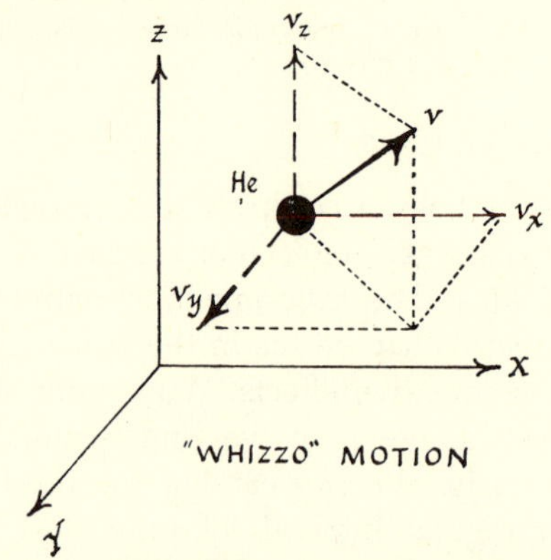

Fig. 30-5a. "Whizzo" Motion
A molecule, shown here as a helium atom, has K.E. of straight-ahead motion. Any such motion can be split into three perpendicular components along, say, x, y, z directions. Molecules are regarded as having *three independent* motions along, say, x, y, z.

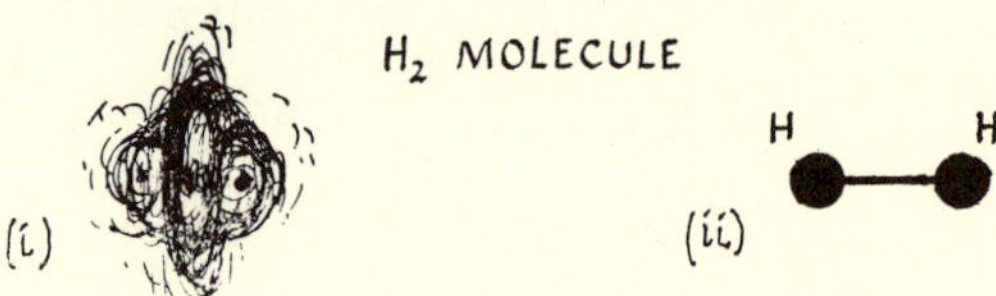

Fig. 30-5b.
A Molecule of Two Atoms (such as H_2) is really as fuzzy as is shown in (i), or more so. However, we represent it schematically by a dumb-bell, as in (ii).

[4] This is the specific heat "at constant volume." The measurement "at constant pressure" includes some piston work as the gas expands, unwanted here. The "constant volume" value is about 3/5 of the "constant pressure" value, for helium.

[5] A molecule's v is split into three component vectors, in x, y, z directions. Pythagoras tells us that the *squares* of these components add up to v^2. Therefore the *kinetic energies* contributed by these components add up to the K.E. of the molecule.

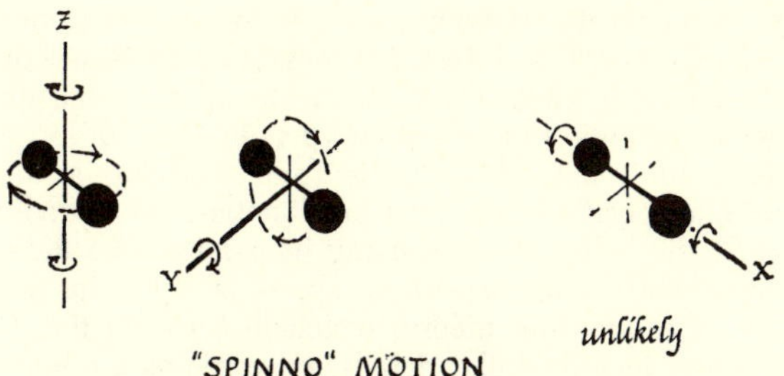

FIG. 30-5c. "SPINNO" MOTION
A molecule of two atoms, such as H_2, has two independent spinning motions, here shown with axes y and z. Spin around the axis of the molecule is difficult to generate.

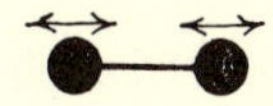

FIG. 30-5d. "VIBRO" MOTION, in and out along the axis of the molecule, would have P.E. and K.E.

½ *PV* a share. Returning to specific heat, that means hydrogen requires heat to increase spinno as well as whizzo, 5 shares instead of 3. Its specific heat should be 5/3 times our previous prediction. We have (5/3)(1.5) = 2.5 in good agreement with experimental 2.40. This should delight chemists who knew that hydrogen molecules must be H_2 *or* H_4 or H_6 . . . and now find proof that they are H_2. Unfortunately, we are still too simple: the pair of atoms can also vibrate in-and-out along the axis, so they need a share of "vibro" energy. They need a double share, because any vibration has P.E. as well as K.E. and equipartition promises a standard share to each. Then our prediction would be too great, 3.5. Looking at the experimental values, we find the specific heat of hydrogen changes with temperature: at very low temperatures, only 1.5; around room temperature, 2.4; and it climbs towards 3.5 at very high temperatures. These are values to be expected for heat energy if it is:

only whizzo (3 shares)	specific heat 1.5
whizzo (3) + spinno (2)	" " 2.5
whizzo (3) + spinno (2) + vibro (2)	" " 3.5

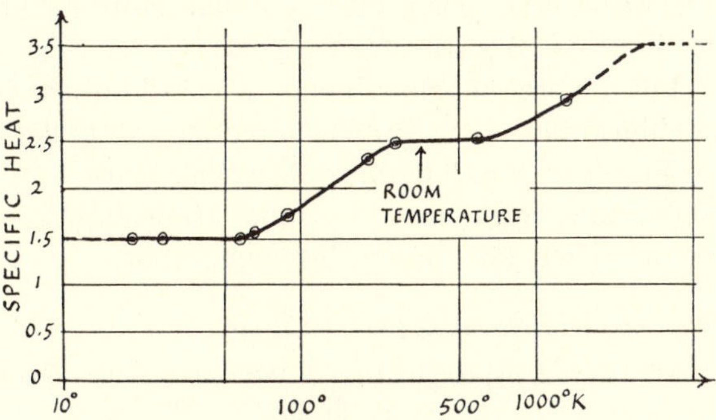

FIG. 30-6. THE SPECIFIC HEAT OF HYDROGEN
(After Richtmyer, Kennard, and Lauritsen, *Introduction to Modern Physics*, McGraw-Hill Book Company.)

Such changes, with a plateau at each expected value, were puzzling until it was realized that they imply restrictions on spinno and vibro energy—quantum restrictions. The quantum rule, introduced to deal with other unexpected behavior in radiation, insists that energy of any *repeating* motion, such as vibration or spin, must be taken up in standard "packages." The energy in each "package" or "quantum" is given by the rule[6]:

$$\left(\begin{matrix}\text{ENERGY IN A}\\ \text{QUANTUM}\end{matrix}\right) = \left(\begin{matrix}\text{Universal}\\ \text{Constant}\end{matrix}\right)\left(\begin{matrix}\text{REPEAT-FREQUENCY}\\ \text{OF MOTION}\end{matrix}\right)$$

Thus high-frequency spins or vibrations (or radiation-waves) must hold energy in 1, 2, 3 . . . large packets, perhaps too big for a molecule or atom to purchase any and maintain the average energy characteristic of the temperature. (If sugar could only be bought and eaten in 200-pound bags it would go out of the average person's diet. Giants might still gobble it.) This packaging restriction, imposed on top of equipartition, explains the peculiarity and predicts the experimental facts. Give a molecule one quantum of spinno. That will make it spin fast, because its rotational inertia is so small. (K.E. of spin = ½ ROTATIONAL INERTIA · SPIN SPEED².) However, *if* it spins fast, its quantum has to be a big one. Therefore molecules must take big quanta of spinno or remain at rest without spin. At low temperatures the share suggested by equipartition is much smaller than one quantum—so that few molecules take up any spinno. At room temperatures the average share is several quanta, so equipartition can be fulfilled.

Molecules vibrate with *very* high frequency: practically no molecules can afford to vibrate until the gas is at *very* high temperature.

Specific Heats of Solids

We can apply similar treatment to the vibrating atoms of a solid crystal. Imagine helium atoms condensed to a solid. Now each atom is tied to its place in the crystalline lattice, by elastic field forces. It has no whizzo or spinno motion, but can vibrate, in three independent dimensions; so it should have 6 shares of vibro energy instead of 3 of whizzo. The specific heat of solid helium should be twice 0.75 or 1.5. Helium does not behave so simply when frozen solid—quantum troubles again—but other solid elements at higher temperatures agree well with this prediction. Multiply the helium prediction 1.5 by its relative atomic mass 4 and we have 6.0. If you follow through the argument of Problem 8 you will find that changing to another element predicts the same value for SPECIFIC HEAT · RELATIVE ATOMIC MASS, a value about 6. This is the Law of Dulong and Petit, discovered a century ago and used in settling early squabbles over chemical atomic masses. It is fairly reliable except at low temperatures where the quantum rule makes itself felt.

At low enough temperatures, the quantum restriction brings the specific heat down towards zero, in a curve that can be predicted by combining the quantum rule with kinetic theory. "Low enough" differs from one solid to another, depending on the natural vibration-frequencies of atoms in the crystal. Therefore, to com-

[6] FREQUENCY = number of complete cycles of vibration, or rotation, per second.

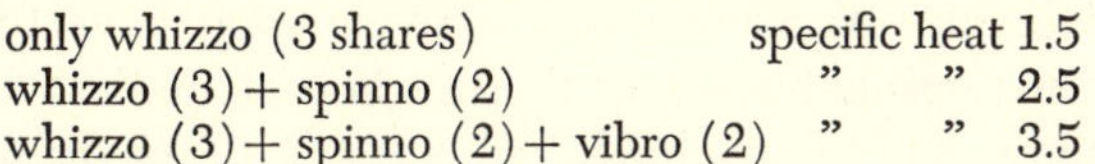

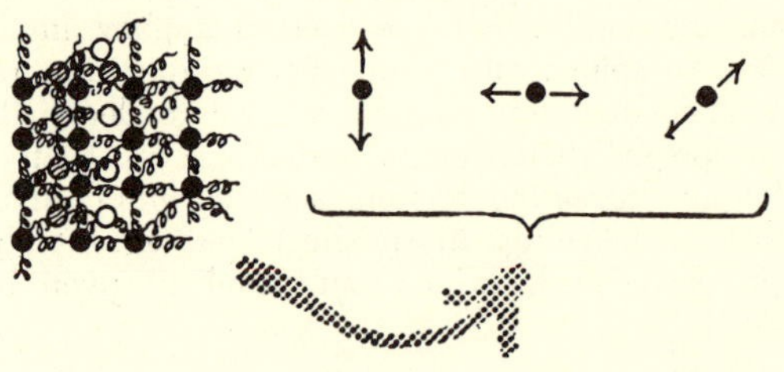

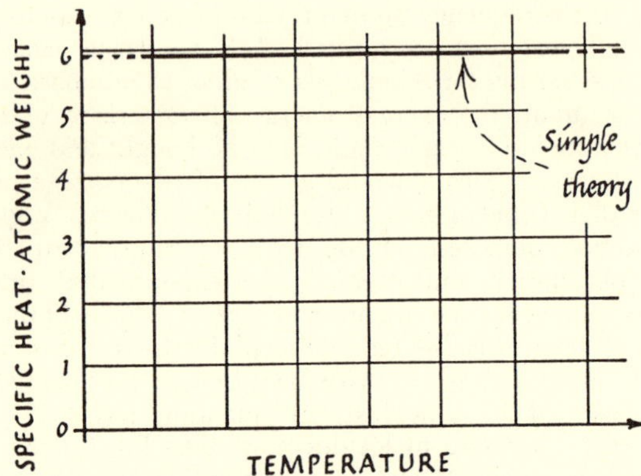

Fig. 30-7. Predicting the Specific Heat of a Solid
For a solid element, with atoms all alike in a crystal lattice, each atom can oscillate. Equipartition suggests a share of K.E. and an equal share of P.E. for each of the three independent directions of vibration that each atom has in three-dimensional space. From this we predict that: SPECIFIC HEAT · ATOMIC WEIGHT should have a constant value of nearly 6. However, experimental measurements give much smaller values at low temperatures, as shown by the marks on the graphs below.

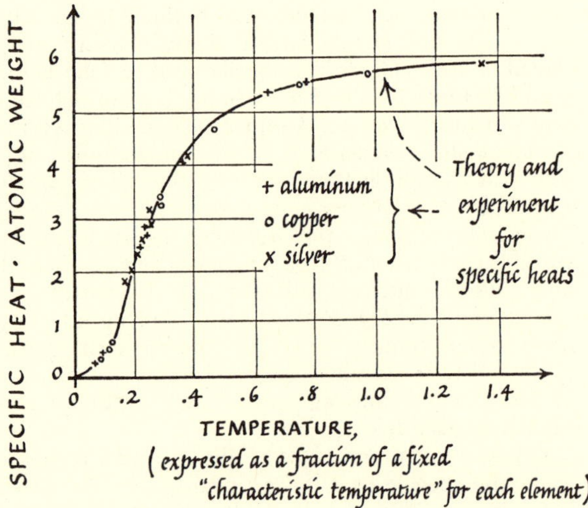

Fig. 30-8. Quantum Theory of Specific Heats
Imposing the quantum restriction for atomic vibrations on the equipartition prediction gives the theoretical prediction shown by the curve. In fitting the experimental measurements on the graph, a different temperature scale has been used for plotting points for each element.

pare experimental measurements with the theoretical curve we choose a different temperature-scale on the graph for each solid. Then we can fit all the measurements on a single theoretical curve as in Fig. 30-8.

Here are remarkable results from studying a dull detail like specific heats. First our kinetic-theory predictions fit the facts, thereby giving us a sense of validity for our theory. Then exceptions appear and ask for new theory, the quantum theory, which in turn fits the experimental facts in fuller detail, and explains a number of other puzzles.

MOLECULAR INFORMATION: SIZE, MASS, NUMBER, . . .

What is a Molecule's Diameter?

How large is an air molecule? This is a useless question until we say how violently we propose to treat the molecule in measuring it. A tailor *could* reduce his estimate of a man's waist toward zero by cutting him with a tight steel wire as measuring tape. We *can* reduce a molecule's "diameter" towards zero by investigating it with billion-volt electrons. Here we mean "diameter for ordinary collisions"—the distance between centers of colliding molecules at closest approach; or molecules packed close into liquid or solid. In terms of picture-models of atomic structure, this is the size of the outer electron-cloud around the atoms of the molecule.

We can derive a rough estimate from measurements of oil films (Ch. 6, Problems 5, 6, 7). Those give us a length about 24×10^{-10} meters or 24 Ångström Units for a chain-molecule 19 carbon atoms long. That suggests a "diameter" of 1 or 2 Å.U. for each carbon atom in a tightly linked chain. Molecules of oxygen and nitrogen each have two atoms (probably larger than carbon atoms); so we might round their dumb-bell into a "ball" of diameter 3 or 4 Å.U.[7] Then their "collision cross section" is about $\pi(3.5 \times 10^{-10}/2)^2$ sq. meters.

For a more reliable estimate, we shall not use the oil-film result or the Brownian-motion extrapolation of Problems 5 and 6 above. We shall make a fresh start now, with measurements using liquid air, common air, and brown bromine vapor.

Size of Air Molecules: Direct Estimate

We do two experiments: (i) We measure the change of volume from liquid air to air. (ii) We measure the diffusion of bromine in air, estimating its progress in a known time.

[7] In view of doubts concerning both definitions and measurements, a wise scientist might follow the suggestion of Professor J. R. Zacharias and adopt *few* as a number, saying, "diameter = few Å.U."

The first measurement enables us to compare the spacing-apart of molecules in gas with that in liquid. The second measurement yields an estimate of the "mean free path" of gas molecules—their average travel between successive collisions. Both results are connected with the target-size or diameter of molecules; and by combining them we can extract a good, though rough, estimate of molecular diameter. From that we can calculate the mass of a molecule and the number of molecules in any known volume.

This is "desperate physics," where a rough value that shows the size of a molecule within an "order of magnitude" is well worth having, whereas attempts at a more precise estimate would cloud the whole story with complex details. We shall cut corners in the experimental measurements and in the mathematical reasoning. We shall simplify and make approximations and use guesses, as good scientists often do when they first explore a strange field.[8] So our results can only be rough estimates. Yet in this matter of atomic measurements our rough estimates are very valuable, both to tell us the order of magnitude of sizes and numbers and to show how such measurements of the microphysical world can be made.

(i) Volume-change: Spacing of Molecules in Air. How far is a molecule in air from its near neighbors, on the average? To put this question in definite form, suppose we could momentarily discipline all the molecules in a sample of air into a completely regular spacing: put each molecule in a cubical prison cell of side D, in a cubical array of cells. Then we might say D is the "average distance apart" or "spacing" of molecules in air.

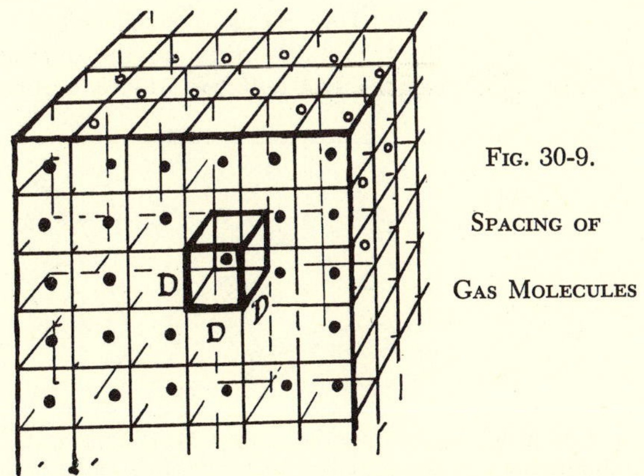

FIG. 30-9. SPACING OF GAS MOLECULES

In liquid air, we imagine molecules crowded till they "touch"—a molecule-diameter d from center to center. Liquids are dense and flow easily, but they are almost incompressible; so their molecules must be crowded close, yet neither packed solid-tight nor spread far out of touch as in a squashy gas. As a guess, we pretend the molecules in liquid air can be regimented in a cubical array of little cells each having side d and volume d^3. (That is *not* the closest possible packing—it just leaves room for fluidity.) Then, as liquid turns to gas, distance-

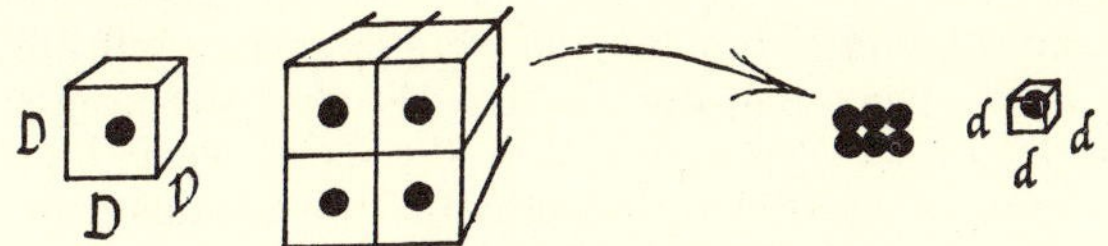

FIG. 30-10. GAS TO LIQUID

apart changes from d to D, the volume of each cell changes from d^3 to D^3. And that volume-change is the same for one molecule as for air in bulk. We can measure it quite easily. We fill a small measuring flask by dipping it in liquid air in a vacuum vessel: we take it out, full of boiling liquid air and quickly attach a flexible plastic tube to its neck. The tube runs to a large trough of water and ends under a huge inverted plastic box full of water. The liquid air boils away into gaseous air which bubbles up into the collecting box and is measured there.

EXAMPLE: 20 cubic centimeters of liquid air become 15000 cubic cm of common air at room temperature and atmospheric pressure.

$$\therefore\ D^3/d^3 = 15000/20 = 750$$

$$\therefore\ D/d = \sqrt[3]{750} \approx 9 \text{ (within 1\%)}$$

The average spacing between molecules in air = the side of the cell containing one molecule = $D \approx 9d$. At atmospheric pressure, air molecules are spaced 9 or 10 diameters apart. This gives some idea of the amount of empty space in a gas. It also gives a hint that molecular bulk will not interfere much with our simple kinetic-theory predictions.

PROBLEM 9

How many diameters apart are air molecules in a cylinder of compressed air at 125 atmospheres? (*Hint*: not sardine-packed at an impossible 10/125 of a diameter.)

---

[8] In the treatment that follows, we pretend that molecules of air, or bromine, are hard spheres with a definite diameter, d; though it would be more "realistic" to picture them as oblong and slightly fuzzy. We make an arbitrary guess for liquids, that molecules are crowded so that each "occupies" a cube of volume d^3. In deriving the "drunkard's walk" result, we assume all the strides are of the same length, L, whereas a molecule's path varies greatly from one step to the next, ranging around an average MFP L. We choose to take a root-mean-square average for the resultant walk: a plain arithmetical average would be about 20% smaller.

In guessing the half-brown distance for bromine, we are looking at random-walk components in only one dimension, upward, instead of three; furthermore, we do not know what type of average our eyes take. We could refine that estimate by comparing two rectangular tubes, one twice as wide as the other, but to make that attempt would be to miss the point of the experiment: "desperate physics." And precision in determining the half-brown distance would be overruled by uncertainty about the difference in size between bromine molecules and air molecules, and the absence of refining factors such as $\sqrt{\pi}$ and $\sqrt{2}$ in our simplified statistics.

Liquid bromine is over 3 times as dense as liquid air, and bromine molecules are 5½ times as massive as air molecules. From that we expect a bromine molecule's diameter to be 1.2 times that of an air molecule. The compromise diameter in a collision between a bromine molecule and an air molecule would then be about 1.1 times as great as for air and air; and the target area would be $(1.1)^2$ or 1.2 times as great. But here too we remember this is "desperate physics" and ignore the 20% difference.

(ii) Mean Free Path. How far does a molecule travel between one collision and the next, on the average? This distance, the *mean free path*, or MFP, is certainly

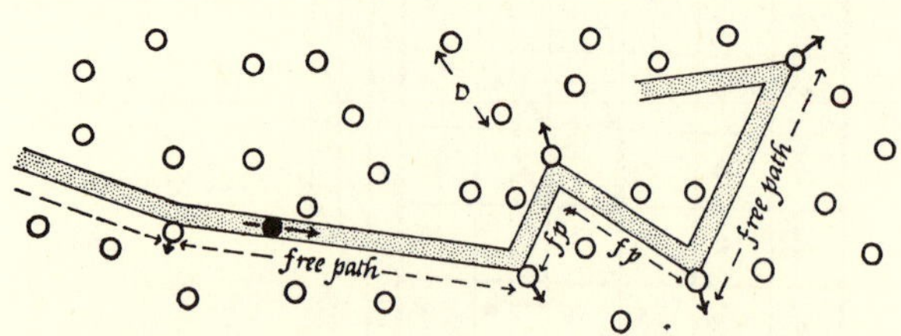

FIG. 30-11. MEAN FREE PATH OF A GAS MOLECULE
At atmospheric pressure, the mean (= average) free path is much longer than the spacing, D. (The shaded tube shows the volume swept out by one molecule moving through others.)

not the same as *D*, the average spacing apart; because, if molecules were perfect points of zero size, they would whizz past their neighbors and never collide at all. The fatter the molecules the bigger the target-area they offer to a moving neighbor, and the sooner a collision occurs.

We can estimate MFP by using the molecules of visible bromine vapor as markers. We repeat the demonstration of bromine diffusing in air in a tall tube. (Ch. 25, page 364, Fig. 25-10.) We time the progress of the brown vapor. We start the clock when liquid bromine is released at the bottom of the tall tube; some time later, say after 500 seconds, we guess the average *resultant* distance that bromine molecules have progressed up the tube. We do that by deciding where the mixture of bromine and air in the tube looks "half-brown," half as dark as the "full brown" just above the liquid bromine. We measure the height of that place above the bottom of the tube. That is obviously only a rough guess, a subjective estimate; but if each member of an audience watching the demonstration makes his own guess, the votes will seldom range more than 10% from some average distance.

Each bromine molecule reaches its final position as a result of a tremendous "random walk" of many steps.* To use our estimate of half-brown, we need the help of statistics. We need an expression for the average resultant of a large number of strides, *N*, each of length *L* taken in succession in random directions. This is the problem called "the random walk," nicknamed "the drunkard's walk." A statistical investigation shows that the expected average is $\sqrt{N}$ strides. See the calculation opposite, which shows how to predict this result for a walk in two dimensions. It holds equally well in three dimensions and is useful in a number of physical problems such as: the progress of photons of radiation staggering out from the core of the Sun; neutrons diffusing through the "moderator" in a nuclear reactor; the resultant sound wave from a singing chorus—and thence to the advantage of "coherent" light waves from lasers over the ordinary light from hot filaments or gases, where the atoms "sing like an incoherent chorus."

We imagine the random walk of a bromine molecule hurtling from collision to collision through a collection of air molecules. We pretend that each makes all its strides of equal length, the average free path, *L*. If each makes *N* strides in time *t*, the average resultant progress is $\sqrt{N} \cdot L$. We calculate the number of strides thus: in time *t* the total travel, *straightened-out path*, is *vt*, where *v* is the speed of a bromine molecule. And *N*, the number of strides in that travel, is *vt/L*.

$$\therefore \text{ average } \textit{resultant} \text{ travel, } S = \sqrt{N} \cdot L$$
$$= \sqrt{vt/L} \cdot L$$
$$= \sqrt{v \cdot t \cdot L}$$

The guess for half-brown gives us S, so if we know *v* we can calculate *L*. We obtain *v* from measurements of density and pressure for bromine vapor, as for other gases. For massive bromine molecules, these give $v = 210$ meters/sec at room temperature.**

Specimen Example: Suppose a group of observers estimates half-brown to be 9 centimeters above full brown, after 500 seconds. In that time, the straightened-out path of a bromine molecule would be (210 m./sec) × (500 secs); and the number of strides in its path would be (210 × 500)/(mean free path, *L*). Average progress S is given by

$$S = \sqrt{210 \cdot 500/L} \cdot L = \sqrt{210 \cdot 500 \cdot L}$$

$$\therefore \ (9/100) \text{ meter} = \sqrt{210 \cdot 500 \cdot L}$$

$$\therefore \ L \approx (81/10^4)/(210 \cdot 500) \approx 770 \times 10^{-10} \text{ meters}$$

or 770 Å.U.

We shall take 800 Å.U. as a rough measure of the MFP of a bromine molecule in air; and, without serious error, *we shall take the same value* for the MFP of an *air* molecule in air. (If the vote averaged 10 cm instead of 9, the MFP estimate would be 1000 Å.U.—we are indeed making rough judgments.)

Then we have as estimates for common air: (i) molecules are spaced about 9 diameters apart; (ii) mean free path is about 800 Å.U. (800×10^{-10} meters).

Mean Free Path and Pressure

Suppose we remove half the molecules from a box of gas. We halve the target chances, and so double the mean free path. In general, MFP must vary inversely as NUMBER OF MOLECULES PER UNIT VOLUME, or inversely as PRESSURE. In a high vacuum, of 1 billionth of an atmosphere, the MFP is a billion times greater, $10^9 \cdot 800 \times 10^{-10}$ m., or 80 meters. A molecule of

* See Fig. 30-3 on page 447 for a sketch of one random walk of a few hundred strides. In the time we allow, a bromine molecule makes an enormous number of collisions (hundreds of billions); and we are observing a vast number of molecules: so *N* is enormous and we are averaging a vast number of random walks—we need not fear poor statistics.

** Trusting Avogadro, we infer from density and pressure that bromine molecules are 5½ times as massive as air molecules; and we thus arrive at the chemical "molecular weight" 160 for Br_2. And simple kinetic theory tells us their average speed will be smaller by $\sqrt{5.5}$.

residual gas in a vacuum tube for radio sails bing-bang between the walls, seldom hitting another gas molecule on the way.

Deriving Molecular Diameter. Showing that $\pi d^2 L = D^3$

There is clearly a connection between MFP L and diameter d: the larger d is, the bigger the target-area for collisions and the smaller the free path. We can show that $\pi d^2 L =$ volume occupied by one molecule, D^3. The geometrical argument is given below. Then we shall use the result to calculate diameter d from L and the volume-ratio D^3/d^3.

When two spherical molecules are in collision the distance between their centers is RADIUS + RADIUS or one DIAMETER, d. To simplify the geometry, we pretend that one molecule in a collision has all the bulk, *radius* d, and we make the other molecule a point. The distance between centers at collision is still d. Then we imagine we fire a *point*-molecule straight at a prison-block of cells, each cell of width D, each containing one

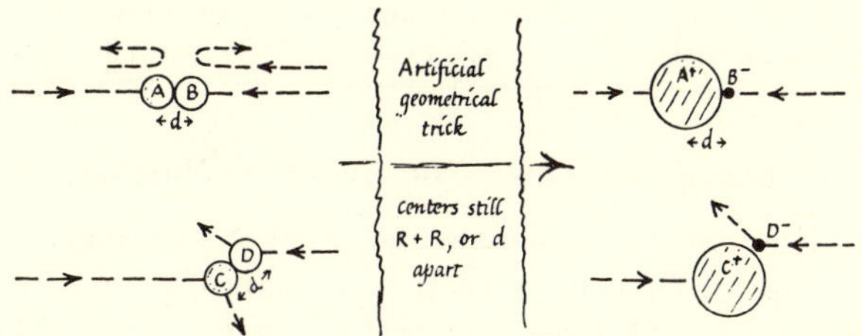

FIG. 30-12. SIMPLIFYING GEOMETRY FOR MEAN FREE PATH

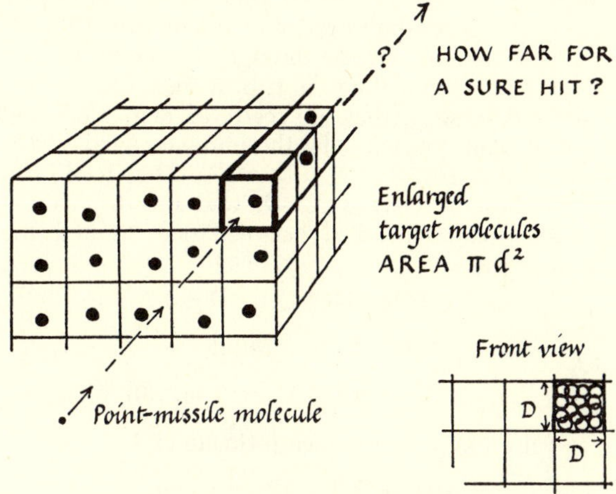

FIG. 30-13. MEAN FREE PATH
A missile point-molecule enters a cell face. How many cells must it traverse to be sure of a hit? Molecule entering face of area D^2 must see it, with cells behind it, just filled by bull's eyes, each of area πd^2.

The Random Walk ("Drunkard's Walk) Calculation

When a bromine molecule wanders among air molecules, making collision after collision and travelling in a new direction after each collision, how far does it succeed in travelling *"as the crow flies"*—on the average?

As a model for this, we picture a man returning from a party on a foggy night, in a completely vague or carefree frame of mind. He starts from a lamp post, takes one stride; then, forgetting his first stride he takes a second stride in any new direction; forgetting that, he takes a third stride; . . . and so on: altogether N strides in random directions. How far from his start at the lamp post will he end up? He might end up back at the lamp post, or *very* near it; and that is very likely. He might end up as far away as N strides (in the rare case when he happens to take all the strides in the same direction), but that is unlikely. His *resultant* progress (straight from start to finish) must lie somewhere between 0 and N strides. We want to predict its *average* value, averaged over many batches of N strides.

We imagine him repeating his walk again and again, starting afresh each time. For each walk we measure his resultant progress, S. After many walks, we find the average value of S for all those walks. For convenience, we find the average value of S^2 and take the square root, obtaining a root-mean-square (R.M.S.) average. We can show that this average should approach $\sqrt{N}$ strides. (For example, if he takes 100 strides we expect him to end up only 10 strides from his start—on the average.) Here is a two-dimensional proof. The 3-dimensional one is similar.

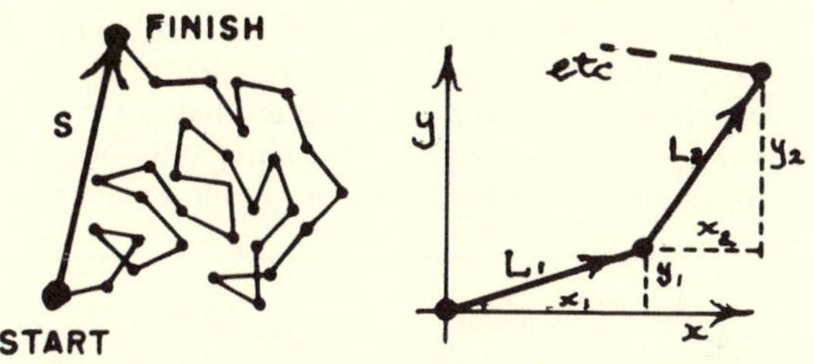

Sketch the first few strides of a random walk. Each stride is of length L, and there are N strides in the walk. Using x- and y-coordinates, resolve stride No. 1 into components x_1 and y_1; stride No. 2 into x_2 and y_2, and so on. For stride No. 1, $x_1^2 + y_1^2 = L^2$ and similarly for the rest. Then resultant, S, has
x-component $(x_1 + x_2 + \ldots + x_N)$ and
y-component $(y_1 + y_2 + \ldots + y_N)$; and

$$\begin{aligned} S^2 &= (x_1 + x_2 + \ldots + x_N)^2 + (y_1 + y_2 + \ldots y_N)^2 \\ &= x_1^2 + x_2^2 + \text{etc} \ldots + 2x_1x_2 + 2x_1x_3 + \text{etc} \\ &\quad + y_1^2 + y_2^2 + \text{etc} \ldots + 2y_1y_2 + 2y_1y_3 + \text{etc} \\ &= L^2 + L^2 + \text{etc} \ldots + \text{ZERO} \\ &= N L^2 \end{aligned}$$

The "cross terms," such as $2x_1x_2$, add up to zero *in averaging over many walks*, because those terms are as often negative as positive, and they range similarly from 0 to $2L^2$. Similarly for the y- "cross terms."
Then average value of $S = (\sqrt{N}) \cdot L$

The proof is better if you use trigonometry and resolve each stride, L, into horizontal and vertical components $L \cos \theta$ and $L \sin \theta$. Then pairs of "cross terms" such as $2L^2 \cos \theta_1 \cos \theta_2$ and $2L^2 \sin \theta_1 \sin \theta_2$ add to make $2L^2 \cos (\theta_1 - \theta_2)$, and those cosines are likely to be as often positive as negative and average to zero.

target molecule of *radius d*.[9] The missile enters one front cell but it will probably miss the target molecule which is at some random place in the cell. We ask how many cells the missile must continue through, to be sure of making one hit—that will tell us its mean free path. The cell face has area D^2, but the target molecule in each cell offers a bull's-eye of area only πd^2. Suppose the missile has to continue through X cells to be sure of a hit. In contemplating the trip, it sees X bull's-eyes ahead; and these, if they are scattered at random and do not overlap, must just fill the entrance window D^2. $\therefore\ X \cdot \pi d^2 = D^2 \quad \therefore\ X = D^2/\pi d^2$. The total trip through these X cells is $X \cdot D$; and this trip is the average distance travelled between one hit and the next. It is the MFP.

$$\therefore\ \text{MFP} = X \cdot D = (D^2/\pi d^2) \cdot D = D^3/\pi d^2$$

$$\therefore\ (\text{MFP}, L)(\pi d^2) = D^3 = 750\, d^3$$

$$\therefore\ d = L\,\pi/750 = 800 \text{ Å.U.} \times 3.14/750$$

(inserting our specimen estimate of L)

$$\therefore\ d \approx 3.4 \text{ Å.U. or } 3.4 \times 10^{-10} \text{ meter.}$$

Our specimen example yields 3.4 Å.U. for d. Such guesses could yield results anywhere between 2 and 7 Å.U. As we have described it, this is a token measurement, an "experiment of principle," to show how an estimate can be extracted: a microscopic (atom-sized) dimension from macroscopic (man-sized) measurements *and* theory.

Readers should see the experiment and make their own estimates. Professional measurements, with more refined theory, yield 3.72 Å.U. Even that is an artificial measure for an imagined hard sphere; but the ensuing estimates of molecule masses and numbers are confirmed by methods using Millikan's value for the electron charge (footnote 7, page 688). From now we shall accept the value 3.72, writing it 3¾ Å.U. as a reminder of doubts.

Then the spacing between molecules in air is about 9 × 3¾ or 35 Å.U., and for atmospheric air we have the following rough estimates:

SIZE *(nearest approach in collisions)*	SPACING *(av. distance between neighbors)*	MEAN FREE PATH *(between collisions)*	SPEED *(average)*
3¾ Å.U.	35 Å.U.	600 to 1000 Å.U.	500m/sec

Fuller statistical treatment reduces MFP to 650 Å.U.

PROBLEM 10

From the data above, calculate how many collisions one molecule makes in a second, in air at atmospheric pressure.

Number of Molecules. The Avogadro Number.

Now we can calculate the number of molecules in any given volume such as a roomful of air. We choose a small room 3 m. × 2 m. × 4 m., volume 24 cubic meters. We assign each air molecule a volume D^3 or 750 d^3. Then the number of molecules in the room is $\dfrac{24 \text{ cubic meters}}{750\,(3\tfrac{3}{4} \times 10^{-10}\text{ meters})^3}$ or about 6×10^{26} molecules. This small room (10 ft × 6½ ft × 13 ft) contains about 600000000000000000000000000 molecules of air, at room temperature and atmospheric

[9] *Alternative argument*: Instead of inflating the target-molecules, shoot one inflated missile molecule among the rest of the molecules frozen as points. Then the missile molecule has cross section area πd^2. As it flies along, it sweeps out a tube of that cross section, making an abrupt bend at each collision—like a bent stovepipe. If it meets any

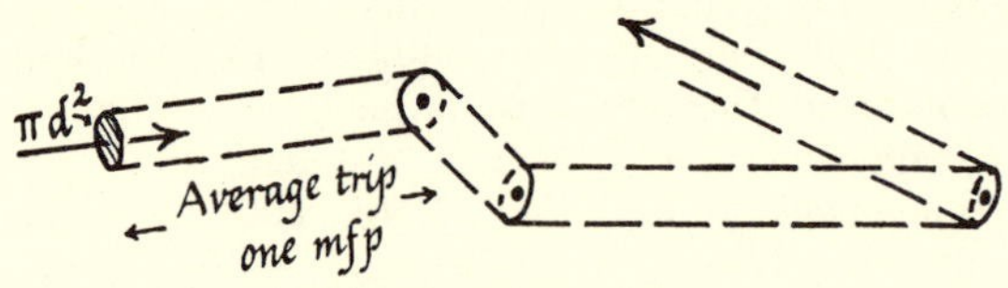

FIG. 30-14.

target-molecule (point) in that "tunnel of exclusion," it makes a collision: it misses any target points that are outside the pipe. Between one collision and the next, it travels on the average, one MFP and sweeps out a volume (cross section area) × (MFP) or $\pi d^2 L$. *It hits just one target molecule* in sweeping through that volume, so that volume is also D^3, the volume of one "prison cell" for a gas molecule.
$\therefore\ \pi d^2(\text{MFP}) = D^3 \qquad \therefore\ \text{MFP} = D^3/\pi d^2$, as above.

Still simpler method. At the expense of even more risky guessing, we can reach an estimate of d without the geometry of missile molecules and stovepipes. We argue thus: Place a sample of common air in a tall cylinder. Drive a piston in to compress the sample 750 to 1, so that its molecules are crowded as closely as in liquid air. (Also cool the sample till it *is* liquid air, if you like.) Crowding the molecules like that will bring all the targets for a moving molecule 750 times closer, so the mean free path is 750 times smaller. Now try to *guess what the mean free path will look like* among the molecules crowded to liquid density. Guess how much room one molecule has in traveling to hit another—the *surface-to-surface distance*, not the center-to-center distance. That is a difficult, doubtful guess.

You should make your own guess; but here is some guidance. If the molecules were so far apart that each had an average space of 1 diameter between its surface and the surface of the neighboring molecule it soon hits, the array would be so open that it would behave like a gas and not a liquid. (Remember, liquids are almost incompressible; a pressure of 20,000 atmospheres only compresses water to 75% of its normal volume.) If they were crowded so close that each had only ⅒ diameter to travel before hitting another, they would be practically locked in a solid. Sketch a lot of molecules on paper, or spread a collection of coins on the table, and play with the question: what distance, between 0.1 d and d, seems a fair guess for the MFP at liquid crowding?

Suppose, for example, we guess "³⁄₁₀ of d." Then we can say:

$$(\text{crowded MFP}, L/750) = \tfrac{3}{10}\, d$$

$$\therefore\ 800 \times 10^{-10}/750 = (3/10)\, d$$

$$\therefore\ d = 3.6 \times 10^{-10} \text{ m.} = 3.6 \text{ Å.U}$$

(It is easy to get the right answer by choosing the right fraction like that, but it is dishonest. The honest thing is to state extreme limits, or make a guess by picturing molecules and then remember that it *is* only a very rough guess.)

pressure. The number is the same for any other gas, in the same volume at the same temperature and pressure—Avogadro (see Problem 6 above)—and we call it the *Avogadro Number.*

We chose that sample volume[10] because the number of kilograms of air or other gas it holds is a very useful number: 2 kg of hydrogen, 4 kg of helium, 32 kg of oxygen, 44 kg of CO_2. These are the relative molecular masses, on the chemical scale (mass of H atom = 1). We name each such mass in kilograms a *kilomole.*[11] Then one kilomole of any substance is 6×10^{26} molecules. (Here we shall call it a mole for short.)

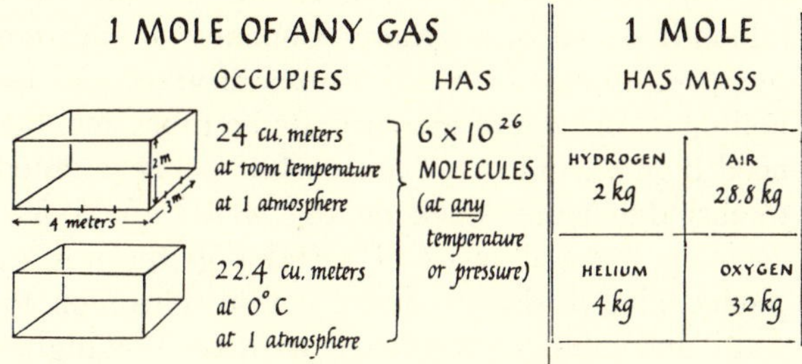

FIG. 30-15. A KILO-MOLE

Even if we remove the air with a very good pump, to make a "high vacuum," the pressure will still be a billionth of an atmosphere, and there will still be 600000000000000000 molecules in the room.

Mass of a Molecule

Now we can calculate the mass of a single molecule. Our sample roomful, 24 cubic meters, contains 6×10^{26} molecules, weighing altogether 28.8 kilograms. (That comes from weighing and measuring a sample of air, as on page 264.)

$$\therefore \text{ Mass of air molecule} = 28.8/(6 \times 10^{26}) = 4.8 \times 10^{-26} \text{ kg.}$$

If we had a roomful of hydrogen instead, 6×10^{26} molecules would weigh 2 kg.

$$\therefore \text{ mass of hydrogen molecule} = (2\text{ kg})/(6 \times 10^{26}) = 3.33 \times 10^{-27} \text{ kg;}$$

and knowing from chemical arguments (page 594) that a hydrogen molecule is a pair of atoms, H_2, we say:

$$\therefore \text{ mass of hydrogen } atom = 1.67 \times 10^{-27} \text{ kg.}$$

This is the mass of a "proton" (see footnote 4 on page 684), which we shall use for an essential calculation of energy (page 688).

[10] 22.4 cubic meters at atmospheric pressure and 0°C, which expands to 24 cubic meters—about 850 cubic feet—at room temperature of 20°C or 60°F.

[11] Note that the kilomole is 1000 times the usual chemical *mole* that uses grams. Our *Avogadro Number* is likewise 1000 times the number for the smaller mole.

Molecules and Temperature

On the universal gas-thermometer scale, absolute temperatures are measured by PV; so we write our definition: $PV = RT$, where R is a constant. If we use one mole of gas, R is the same for all gases. Kinetic theory gives:

$$PV = \tfrac{1}{3} N m\overline{v^2} = (\tfrac{2}{3})\, N(\tfrac{1}{2} m\overline{v^2}) = (\tfrac{2}{3}) \text{ K.E.}$$

$$\therefore \text{ K.E. of molecules} = (\tfrac{3}{2}) \cdot PV = (\tfrac{3}{2}) \cdot RT$$

Then,

$$\text{average K.E. of a } single\ molecule = (\tfrac{3}{2})\, RT/N$$

$$= (\tfrac{3}{2}) \cdot \frac{(R \text{ for one mole of gas})}{(\text{Avogadro Number for one mole})} \cdot T$$

$$= (\tfrac{3}{2}) \cdot k \cdot T,$$

where k is the "gas constant for a single molecule."

Equipartition gives the same K.E. to every kind of molecule at temperature T, so k is a general constant, the same for every kind of molecule.

Now we can give temperature a clear simple meaning: *"Absolute temperature is measured by the average K.E. of any molecule."* It is $2/(3k)$ times the K.E. We picture gas molecules sharing K.E. with molecules in the walls of the container or in a thermometer bulb, with something like equipartition for those exchanges too.

Uranium Separation

To separate U^{235} by diffusion of UF_6 we need a porous wall that will distinguish faster molecules from slower. A large hole in the wall will make no separation. The gas molecules will pour through, hitting each other as usual, gathering momentum from the difference of pressure in a common stream of molecules—the mixture of molecules will be carried through unchanged. Small holes, a fraction of one MFP across, offer individual treatment to different molecules. If the pores are long, each molecule that staggers through hits the walls of the pore many times and in that motion a faster molecule has the advantage over a slow one. Diffusion barriers should have holes much narrower than 1000 Å.U. but wider than the diffusing molecules, say 5 or 6 Å.U. for large UF_6. The holes must be between 100 Å.U. and 10 Å.U. in diameter—an extraordinary requirement of invention and manufacture. Such a barrier can be made—e.g. by pressing metal powder together to make a thin porous plate. It is in use on a vast scale, separating U^{235} by the kilogram, through a cascade of several thousand diffusion stages, with automatic pumps and controls to handle the dangerous poisonous vapor. The change of proportions is like the growth of a sum of money at compound interest over many years. There must be many stages, with recycling of the "also ran" molecules back to the input of earlier stages. (See Fig. 25-17 in Chapter 25.)

Vacuum Pumps and Gauges

Research in electronic, atomic, and nuclear physics needs a good vacuum in most apparatus, to avoid unwanted collisions with air molecules. Commercial manufacture of radio tubes, X-ray tubes, etc., requires a good vacuum on a production-line scale. How can we make suitable pumps and how can we measure the residual pressure of a millionth or even a billionth of an atmosphere?

Mechanical pumps, with a piston consisting of a rotating scraper sealed in oil, will pump from 1 atmosphere to 1/10 to 1/100, ... easily to 1/10,000 atmosphere. Beyond that, for high vacuum, we use faster pistons: *single moving molecules.* A stream of hot fast mercury vapor molecules sweeps down through a cylinder with water-cooled walls which slow them on collision to slow mercury molecules.

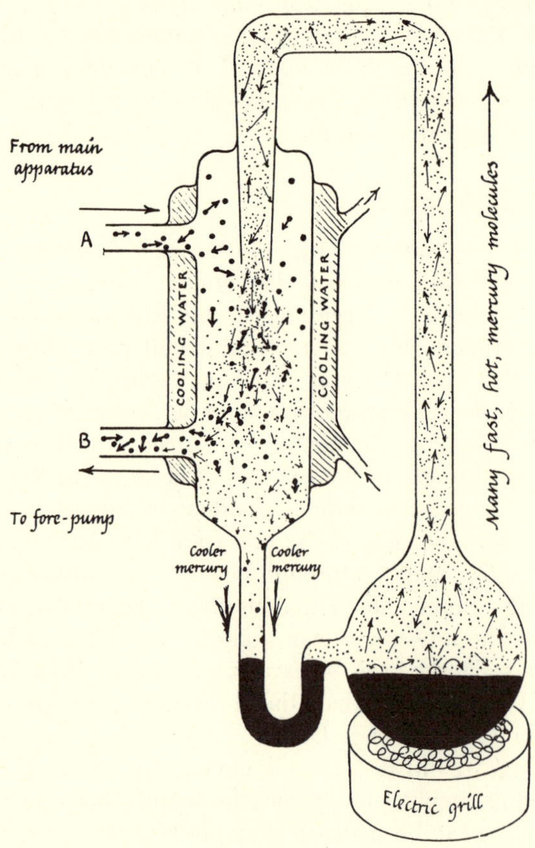

FIG. 30-19. MERCURY DIFFUSION PUMP
A "molecular" pump.

Thus in this region there are many speedy mercury molecules at the top, slow ones at the bottom. A stray air molecule wandering into this region of changing turmoil is hit harder and more often *downwards* by *hot* mercury molecules than *upwards* by *cold* ones—slam downwards and only flip upwards, so that its history runs: slam . . flip . . slam . . slam . . . flip . . slam . . . flip . . . slam . . slam . . . . . The wanderer is given downward momentum on the average. As air molecules wander in at A from the main apparatus they are slammed towards B to be removed by a mechanical pump. This does not sound a very promising system, but the mercury molecules are so numerous, and change their speed so much as they cool, that they are very effective. Backed by a good mechanical pump, such a molecular *diffusion pump* can reduce the pressure to a billionth of an atmosphere. Commercial diffusion pumps are made of metal instead of glass, and use boiling oil instead of mercury. Radio tubes, etc. that need high vacuum are baked while being pumped, to dislodge gases stuck on the walls. For a final clean-up a small piece of metal ("getter") is exploded electrically: it makes a thin mirror on the walls and clutches residual gas there. For highest vacuum of all, the residual gas molecules are ionized by electron bombardment and then gunned hard by an electric field into the walls, to stay there.

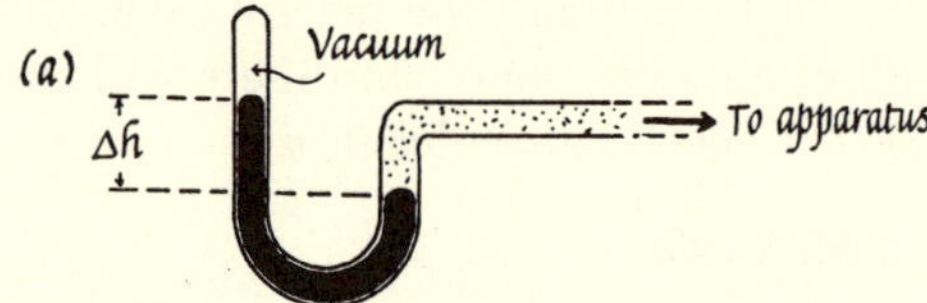

FIG. 30-20a. ROUGH PRESSURE-GAUGE FOR VACUUM

How can we measure the very low pressure in a "good" vacuum? For 1/100 atmosphere, a small mercury U-tube will suffice, with vacuum on one side and pressure-to-be-measured on the other. At 1/1,000,000 atm. or less, the level-difference is too small to see, and swamped by capillarity differences. For highest vacuum, we now use an intricate *ionization gauge*, in which we knock electrons off a few remaining molecules and measure the tiny current the ions thus formed can carry.

For an ordinary good vacuum, we use a M'Leod[12] gauge, devised by an ingenious Scotsman to make the measurement economically. A large sample of the residual gas is taken into a wide cylinder and compressed with a mercury piston in a measured ratio, say 10,000 to 1. Then pressure must have increased 1 : 10,000. The pressure in the squashed sample is measured by mercury level-difference. In the apparatus sketched in Fig. 30-20b the sample is squashed from a cylinder of 50 millimeters diameter to a capillary tube of 1 millimeter diameter, a change

[12] Pronounced "McLoud."

of area of $50^2 : 1$. The length of sample changes from cylinder length 10 cm to a length in the capillary 2.5 cm; compressed 4 : 1. Then the volume is compressed (2500) · (4) : 1 or 10,000 : 1. The magnified pressure is given by level-difference[13] between

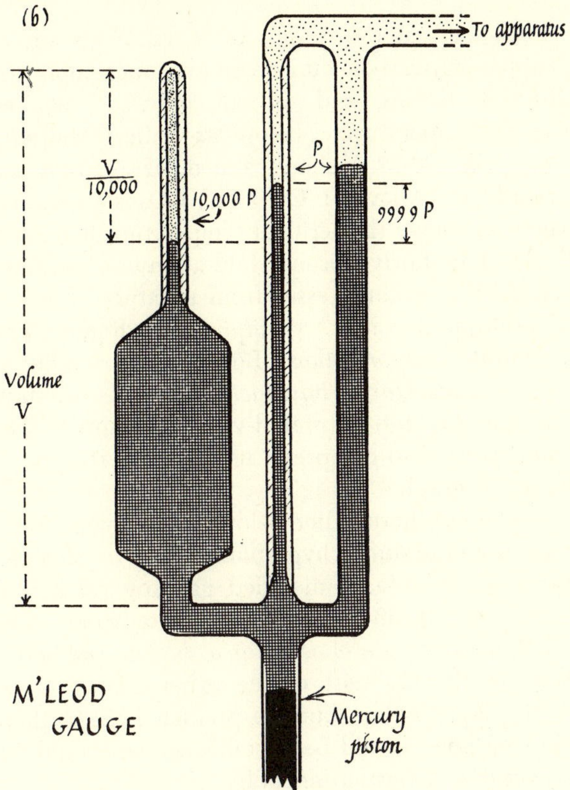

FIG. 30-20b. M'LEOD GAUGE
A sample of greatly rarefied gas in vacuum is compressed to, say, 1/10,000. Then its pressure, which is now 10,000 times as big, is measured. The shaded region is occupied by mercury after the trapping and compression of the sample of gas.

mercury against the squashed sample and mercury in a parallel tube connected to the main apparatus.

Realistic Theory

We can use our theory to polish up its own prediction into a world-wide rule for real gases. Molecules are elastic, free, small, ... , but not infinitely small. If they themselves take up some space, the space available for to-and-fro trips is not the observed volume, V, but some smaller volume $V - b$, where b is a fixed allowance for molecular bulk.[14] Again, molecules are not completely free from each other's influence. They attract when close—as we know from liquids. At high compression, their clutching-together should lessen the pressure on container walls. Plausible arguments suggest that the observed pressure, P, instead of being simply $[RT/V]$ should be $[RT/V - (\text{a constant, } a)/V^2]$.[15]

Our prediction changes from $PV = RT$ to

$$\left(P + \frac{a}{V^2}\right)(V - b) = RT.$$

This is Van der Waals' improved gas law. (See Fig. 30-22(ii) for graphs of its predictions.) The new "formula" fits the behavior of real gases reasonably well, predicting their deviations from Boyle's Law, etc., over a vast range—up to thousands of atmospheres, and even below critical temperature. It reduces to the old simple form when V is big: e.g., for air at atmospheric pressure or lower. (That is an example of Bohr's Correspondence Principle, that new theory must reduce to old at its boundary where the new conditions are unimportant.)

Here is good use of theory, adding realistic assumptions to produce an extended rule. We can fit experimental measurements to this new rule and find working values of a and b for each gas. Then we can use the rule to correct gas-thermometer measurements to the ideal gas scale. From b we extract good estimates of molecule diameters. And when we convert the gas into liquid, the attraction-correction, a/V^2, far outweighs ordinary pressure; it *is* the effect of surface-tension that holds liquids together.

A Clever Use of Theory

In using a M'Leod gauge, we trust Boyle's Law. How do we know that Boyle's Law holds at very low pressures, far beyond the range of ordinary tests of it? To make sure, we should have to measure P and V for a gas, ranging down to such pressures: but how could we measure P? Hardly with a M'Leod gauge! We are driven instead to ask kinetic theory whether we can trust Boyle's Law. It is usually risky to *extrapolate* theory; yet here our theory gives a clear answer: "If there is any region where molecular sizes and attractions are negligible, and the simple law should hold true, it is just there. You can be sure of Boyle's Law for very low pressures." This is an unusual case of theory guaranteeing its own

[13] Strictly speaking this difference shows $(10{,}000\ P - P)$ or $(9{,}999\ P)$ but it would be neurotic to call that anything but 10,000 P.

[14] Problem 14 in Ch. 25 asked you to predict this effect by imagining molecules swelled to greater bulk. That would make trip-times shorter, pressure greater.

[15] a/V^2 is a small correction that varies as the FREQUENCY OF COLLISIONS. We argue that mutual attractions weed out a few slow molecules in collisions just before they hit a wall. It takes two to make a collision, or a telephone call. Telephone traffic in urban areas varies roughly as $(\text{POPULATION-DENSITY})^2$. Similarly the chances, and therefore frequency, of molecular collisions vary as $(\text{DENSITY})^2$ or as $(1/V^2)$. So, (DECREASE OF MEASURED PRESSURE due to weeding-out) $\propto 1/V^2$. Therefore, we add a/V^2 to the *measured* pressure, P, to arrive at the simple predicted form.

extrapolation, with great assurance.

Compressing Gas to Liquid?

How is liquid air made? Not just by compression. Even if we compress a gas till it is as dense as a liquid, it still fills its container. Its molecules seem unable to gang up into liquid. However, if we cool the gas below a certain *critical temperature* it can form liquid when compressed. If cooled but not sufficiently compressed to liquefy, it still behaves like a gas, but it is called a *vapor.* A vapor can be liquefied by simple compression; but to liquefy a true *gas* we must first cool it below its critical temperature, then compress it (and continue to take heat away while it condenses). In reverse, any liquid evaporates to vapor, given space.

Thus each substance has a definite critical temperature, above which it is a *gas*—unliquefiable—and below which it is *vapor*, or *vapor + liquid*, or *liquid*, according to the pressure imposed. At room temperature, most of the gases you know are far above their critical temperature; and of course all the liquids are below theirs. Nitrogen is a gas, steam a vapor, mercury a liquid, lead a solid. On the Sun all these would be gases; on Neptune, all solids.

The critical temperature for air is $-140°C$; for helium, only a few degrees above absolute zero. For water, it is about $+365°C$. For carbon dioxide it is 31°C, (88°F). On ordinary days, CO_2 fire extinguishers are, say, ¾ full of liquid, with vapor above.[16] On a very hot day the surface of the liquid disappears—it is all gas. You should see this change

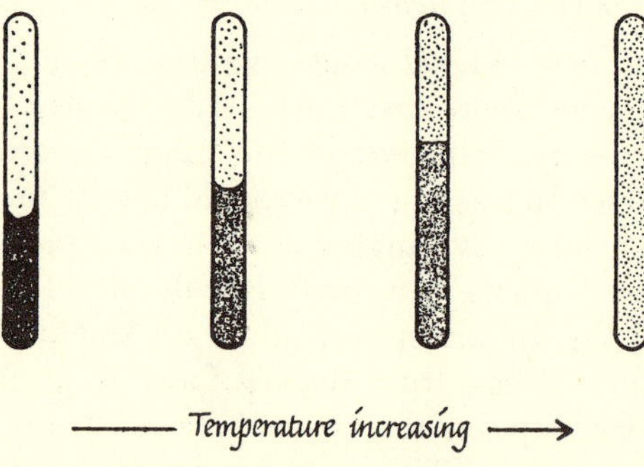

Fig. 30-21. Critical Temperature
A strong glass tube with liquid and vapor is heated, starting below the liquid's critical temperature and ending above it.

happen in a glass tube. As the temperature rises, the liquid expands greatly, growing less dense while the vapor grows denser; then the boundary disappears—and only reappears on cooling, in a sudden storm of mist drops. This is a dangerous experiment, but an amazing change to see.

We shall return to critical temperature after looking at the molecular view of evaporation.

[16] At atmospheric pressure, CO_2 cannot exist as a liquid. If we try to liquefy it by cooling, it solidifies. When CO_2 in a fire extinguisher (at some 60 atm.) is released to the open air, most of the liquid evaporates, taking away latent heat and freezing the rest to a cloud of snowy "dry ice."

"Boyle's Law" and CO_2

Look back at the graphs for P vs. V for air, in Chapter 25. Carbon dioxide shows much more exciting deviations, and so will *any* "gas" at low enough temperatures (below its critical temperature). Fig. 30-22(i) shows a comparison between air and CO_2. Above 31°C (88°F) CO_2 is a gas; and when far above that critical temperature it follows Boyle's Law fairly closely. Held at any temperature below 31°C it compresses from unsaturated vapor to (saturated vapor + liquid) to liquid. The *unsaturated vapor* follows Boyle's Law roughly at low pressures. *During liquefaction*, the pressure stays constant (at the saturated-vapor pressure). The *liquid* is hard to compress: its line rises steeply on the $P : V$ graph.

Thus an isothermal line below critical temperature is far from the simple hyperbola of $P \cdot V = \text{constant}$. Yet Van der Waals' modified gas law predicts it quite well. Fig. 30-22(ii) shows the prediction when suitable values are chosen for a, b. The prediction fails for the flat part of the experimental curves during liquefaction; but the predicted shape there (broken line) would be unstable, so we should not expect to see it experimentally.

Liquids and Vapors

Liquids have their molecules close together—note their incompressibility. Yet their molecules must be in motion, probably with the same share of K.E. that equipartition accords to gases. Liquid in an open dish slowly disappears, evaporates to invisible vapor. Put some liquid in a closed bottle and evaporation soon stops. Then there is vapor above (and there may be air too), liquid below, and glass walls around them, all at the same temperature. There is probably equipartition—same average K.E. for all: vapor molecules (air molecules too), liquid molecules in their short trips between collisions; and double shares of [K.E. + P.E.] for each vibrating molecule of the glass bottle. To a molecule of gas or vapor, the glass is not a quiet smooth wall, but a jangling array of vibrating atoms that "give as good as they get" when bombarded. That is why gas molecules rebound at the same speed from a solid wall; and rebound with increased energy from a hotter wall whose members average more K.E. To a gas molecule, a liquid surface is equally disturbed, a milling crowd, from which an occasional vapor molecule evaporates.

Fig. 30-22 (i). The Behavior of Carbon Dioxide Compared with That of Air
The graphs show experimental isothermals (constant-temperature curves of pressure plotted against volume).

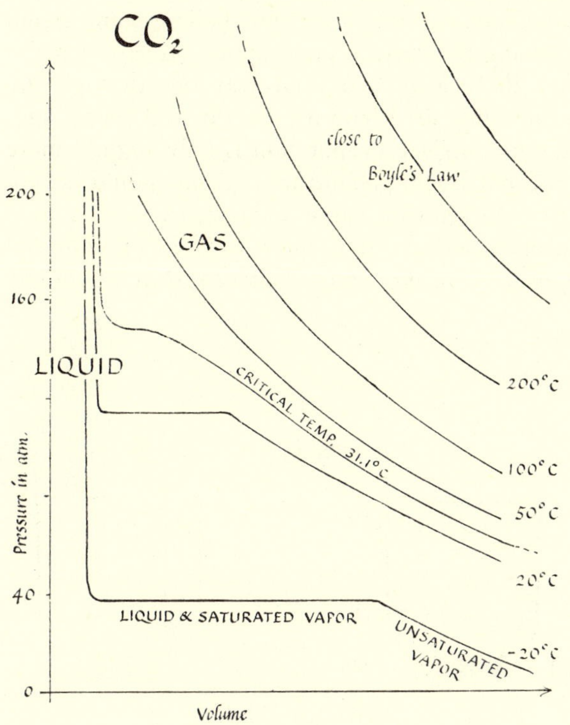

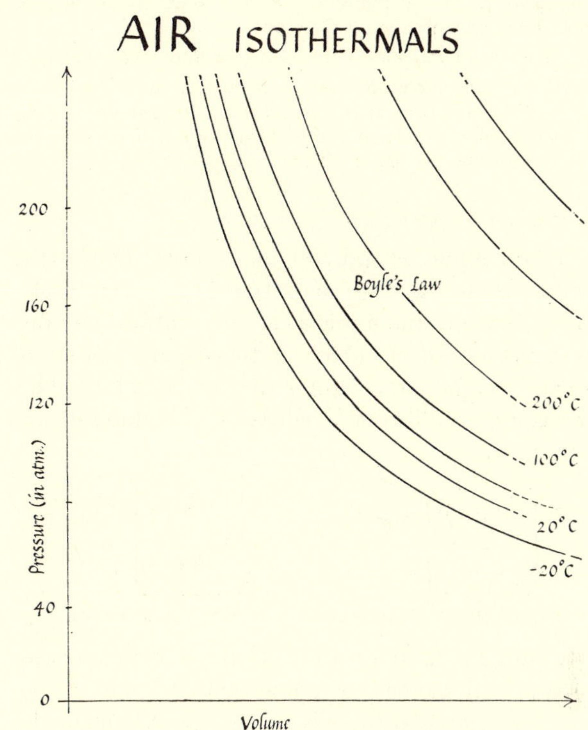

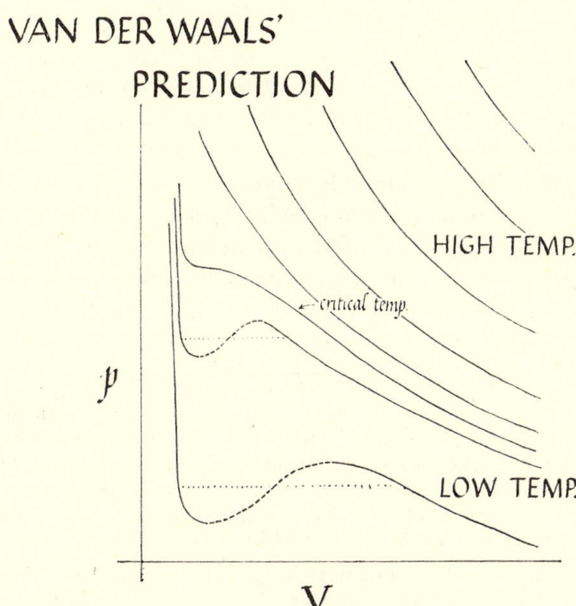

Fig. 30-22 (ii). Isothermals (constant temperature curves) predicted by Van der Waals' modified gas law.
The *full lines* and *broken lines* show predicted p:V graphs for various constant temperatures. The *dotted lines* show where the behavior of real substances differs seriously from the predictions.

Evaporation

Evaporation is the escape of molecules from their neighbors at a liquid surface. Problem 11 shows that evaporation must be a cooling process.

★ PROBLEM 11. EVAPORATION

(a) What experimental evidence suggests that molecules attract each other in a liquid?

(b) What experimental evidence suggests that molecules attract each other very little if at all in gases?

(c) (i) Give a sketch of a molecule leaving a liquid surface (evaporating), being attracted by all its neighbors within a certain very short range of attraction, only a few molecular diameters. (ii) Sketch the direction of the resultant attraction on an escaping molecule that has just left the surface. (iii) Most molecules that start to escape do not have enough K.E. to succeed. They are like a ball thrown into the air trying to escape from the Earth. What do such molecules do? Sketch the paths of several.

(d) If a molecule does escape, it does work against the resultant attraction, turning some of its supply of K.E. into potential energy (stored up in the intermolecular field of force). Many molecules have less K.E. than is needed for complete escape. In fact, the *average* K.E. is not enough. Only a few molecules, with more-than-average K.E., escape completely. What then happens to the *average K.E. of the remaining molecules* of the liquid?

(e) Interpreting molecular K.E. as heat-content, say what happens to the remaining liquid when some evaporates.

(f) Why are there some molecules which move extra fast? Are these special ones, a sort of atomic *aristocracy*?

★ PROBLEM 12. EVAPORATION AND TEMPERATURE

If we raise the temperature of liquid, we increase the average K.E. of the molecules, and a larger proportion of them have enough energy to escape.

(a) How would you expect this to affect rate of evaporation?

(b) If we could increase the temperature so much that even the *average* K.E. of a molecule is sufficient for escape, what would you expect to find happening? (This deserves a clear-headed guess. The answer is not "boiling.")

Saturated Vapor

When liquid evaporates in a closed bottle, the vapor reaches a stage of "saturation," where molecules return to the liquid (and are grabbed) as fast as molecules leave. (Here is the department-store-owner's nightmare: returns as fast as sales.) This "dynamic equilibrium" maintains a "saturated" vapor with a definite pressure. That pressure increases greatly with increase of temperature. Liquids evaporate very slowly if there is air outside. Air molecules cannot really discourage liquid molecules from evaporating, but they bounce many an emigrant back into the liquid. So air slows the process of making a saturated vapor, but does not affect the final vapor pressure.

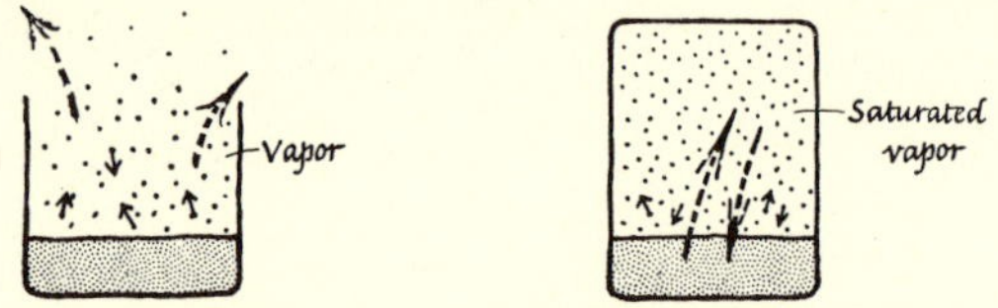

FIG. 30-23. EVAPORATION AND SATURATED VAPOR

Comfort

Cooling by evaporation is of great importance to us in maintaining a steady body temperature. When we are active, we burn food-fuel, and 75% of the energy released appears as waste heat. We must get rid of this heat; otherwise our body temperature will rise and make us uncomfortable. Air carries heat away slowly, but evaporation from moist skin absorbs heat fast. Sweat dripping from your brow does no good, but sweat evaporating into dry air cools you. (In refrigerators, too, the cooling is done by a liquid evaporating.) In a very crowded room, your face becomes blanketed with wet air. Then your temperature rises and you may have a headache. Experiments show that such discomfort is not due to carbon dioxide—even in a very stuffy room the concentration of CO_2 never reaches a quarter of that needed for a headache!—but it is due to humidity which prevents cooling. Fanning your face carries away some of the wet blanket and brings drier air that lets sweat evaporate successfully. Good air-conditioning provides air that is dry as well as cool. Some cheap schemes provide cool wet air!

Boiling

Vapor evaporates from a liquid at any temperature. When water is just below boiling-point, steam is leaving the surface in a copious stream of molecules that easily find their way out through the surrounding air. When it is raised to boiling point, the outer *surface* evaporation is only slightly more rapid, but boiling comes into action: liquid evaporates into growing vapor bubbles, as well as from the open surface. It is *these bubbles of saturated vapor that are the essential characteristic of boiling.*

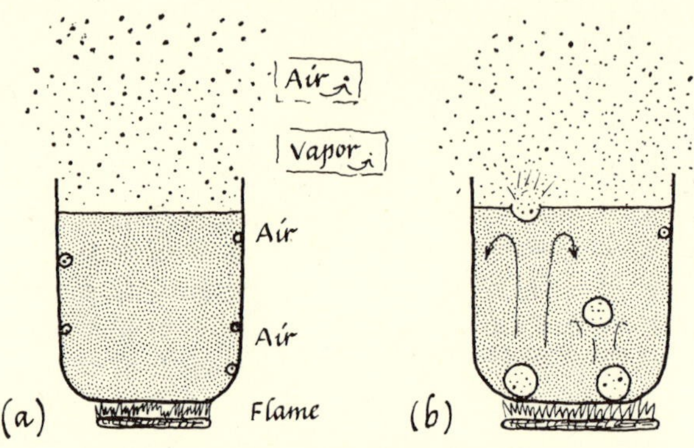

FIG. 30-24. BOILING
(a) Liquid just below boiling point evaporates from the surface only.
(b) Liquid boiling evaporates from surface as before, *and* evaporates into vapor bubbles.

The bubbles carry out great belches of vapor when they rise to the surface, so that boiling liquids evaporate much faster. However, to maintain boiling we must continually supply heat to make up for the evaporation-cooling. Why does a liquid show no boiling till it is heated to a definite boiling point, and why does its temperature then stay constant? Why is the boiling point lower on a mountain top?

★ PROBLEM 13. CONDITIONS FOR BOILING

Bubbles of vapor growing in the liquid are the only essential characteristic of boiling. They usually form at the bottom of a cookpot because the liquid is a little hotter there. All around the growing vapor bubble, there is liquid which presses on the bubble trying to make it collapse. Inside there is saturated vapor, trying to make it expand.

(a) If the bubble does not collapse but can just grow, how must the vapor pressure inside it compare with the liquid pressure outside it?

(b) What is the approximate value of the *liquid* pressure, if the liquid is not very deep?

(c) Therefore from (a) and (b), a liquid cannot begin to boil until it is so hot that its vapor pressure is equal to . . . ?

(d) During boiling, if the liquid ever gets hotter still, its vapor pressure will grow bigger. What will then happen to the bubbles? And what will then happen to the liquid's temperature? (Remember evaporation always takes away heat.) Draw a conclusion regarding boiling-temperature.

(e) (Hard) A surface tension demonstration showed that a small soap bubble has a bigger pressure inside than a large bubble; so we expect a small steam bubble in boiling to need extra pressure inside to push against surface tension. This extra pressure must be provided by the vapor pressure. Therefore, to start a very small steam bubble the liquid temperature must be . . . ?

(f) Boiling liquids often boil with an irregular bumpy motion. Explain this *in terms of (e).*

Watch a beaker of water heating over a gas flame. You will see how boiling starts, and you will see the part played by tiny air bubbles that act as starters.

By removing more and more of the outside pressure, we can make a liquid boil at lower and lower temperatures. On a high enough mountain, water boils when just warm to the touch.

We can even make water boil and freeze at the same time, at a temperature just above 0°C.[17] At *very* low pressure the boiling point is only just above 0°C, and (with atmospheric pressure removed) the freezing point rises a little above 0°C to meet the boiling point. As vapor is pumped away, fast evaporation cools the water to the common triple point and even freezes some to ice on the surface while boiling continues with bubbles rising through the ice.

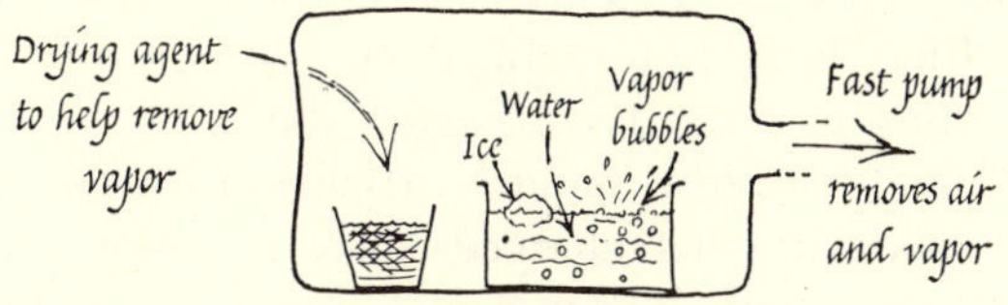

FIG. 30-25. TRIPLE POINT
At very low pressure, cold water boils, cooling the remaining water by evaporation until it freezes while still boiling, at the "triple point." The triple point is the combination of temperature and pressure at which solid, liquid, and vapor can exist in equilibrium. For water, the triple point temperature is just above 0° C.

Mirrors and Mean Free Path

The best modern mirrors are made by shooting metal atoms onto glass in a high vacuum. You should see this demonstrated. A small bead of aluminum is heated electrically by a tungsten wire till it melts and evaporates. Evaporating atoms fly away at high speed and cling to the surrounding walls. A sheet of glass placed nearby will catch them and acquire a fine mirror face—anything from a transparent metal film to a thick reflecting coating. If the vacuum is

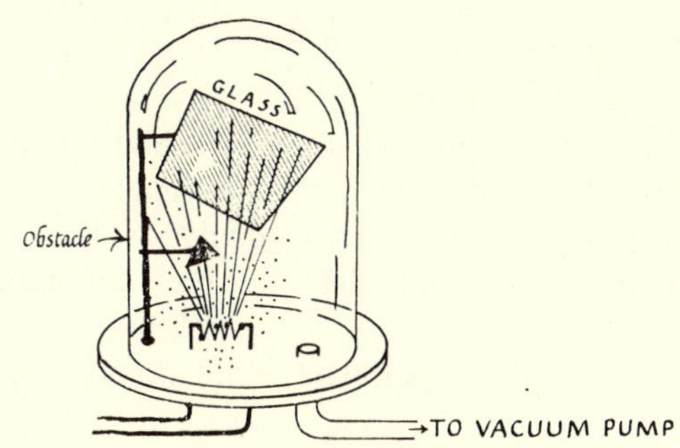

FIG. 30-26.
MAKING A MIRROR BY EVAPORATING ALUMINUM IN VACUUM
The obstacle casts a sharp "shadow" of clear glass on the glass plate.

very good, an obstacle will cast a sharp shadow of clear glass. This shows that the atoms of aluminum have shot straight out from the bead. If a little air remains to shorten the free path of the atoms, the shadow shows blurred edges.

Critical Temperature and Kinetic Theory

Now we can interpret critical temperature. For a given temperature, gas molecules keep the same average K.E. whatever the compression. At high temperatures, the molecules always have enough K.E. to escape from liquid bonding forces, however closely they are crowded. But at low enough temperature they can form a liquid—once we have pushed them together so that each molecule finds itself close to many neighbors. In a liquid, Van der Waals' "a/V^2" takes charge: the molecules are tied by attractive forces in a semi-organized, mobile, crowd; there must be some local order among neighbors, but not the permanent seating of a solid crystal. Thus the critical temperature marks the boundary where K.E. grows too large for ganging up. If we start with a liquid, from which a few "extra rich" molecules escape to form vapor, we can raise the temperature until even the "average" molecule has enough K.E. to escape: then the liquid will fall to pieces, as a gas!

PROBLEM 14. CRITICAL TEMPERATURE

(a) The surface tension of a liquid grows less as the temperature rises. What value should it have at critical temperature?

(b) How could you estimate critical temperature by surface tension measurements?

(c) At critical temperature, what value would you expect for the latent heat of vaporization (= heat to convert 1 kg to vapor, without change of temperature)?

(d) A liquid is usually much denser than its vapor. How should densities of vapor and liquid compare near critical temperature?

[17] This "triple point" has now been chosen as the defining "fixed point" for temperature scales. The new international agreement has thrown out the ice and steam points, and uses this single fixed point and absolute zero. For all ordinary uses, the temperature scale is the same as before.

Low Temperatures: Liquid Air

The critical temperature for air is −140°C. Air is cooled below this and then liquefied by further compression and cooling.[18] To lessen evaporation, it is kept in double-walled containers with vacuum between. Then, open to the air, it boils continuously at about −190°C.

Demonstration experiments with liquid air show that:

Liquid air (or oxygen or nitrogen): is very cold, though boiling. It freezes flesh, flowers; it makes rubber brittle, and lead more elastic. It finds an ordinary table very hot, so it runs over it with big angle of contact, cushioned by a layer of vapor (like water on a red-hot stove). Shows about −190°C (on a helium gas thermometer) when boiling into open air. Expands 800 : 1 when it evaporates and warms up to room temperature.

Liquid nitrogen: floats on water; has lower boiling point than oxygen, so it can condense liquid oxygen from the atmosphere.

Liquid oxygen: sinks in water; is blue; is slightly magnetic; provides a compact supply of oxygen for burning fuels. (Cigars, steel wool, etc., flare if lit after being soaked in liquid oxygen; and absorbent cotton makes a good explosive.)

High Temperatures: Ions in gases

As a solid is raised to high temperature, the radiation from it spreads from infra-red into the visible: red glow, orange, . . . , white hot. The light of a match flame comes from white hot soot. The way in which the electrons of a solid's crowded vibrating atoms emit light is complicated, and we shall not discuss it here. However, hot *gases* glow on a simpler scheme, which gives us more atomic information. At high enough temperatures, the atoms can be excited by common collisions.

Most collisions among molecules are elastic—K.E. is conserved—but a violent enough collision can use up K.E. in separating the atoms of a compound molecule; and a very violent collision may knock a negative electron off an atom, leaving the rest of the atom positively charged. The detached electron may then be picked up by another atom, which thus becomes negatively charged. This making of electrically charged particles in a gas is called *ionization*. It takes considerable energy to tear an electron

[18] The easiest scheme, used for hydrogen and helium, which are hardest to liquefy, is to let the gas push out a mechanical piston, thereby cooling.

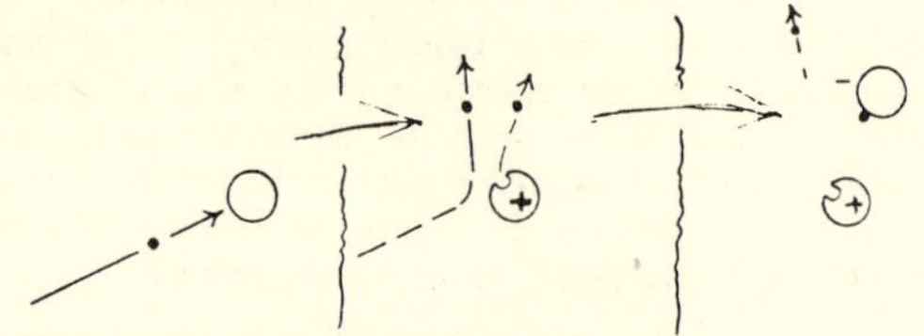

Fig. 30-27. Making Ions
Electron shot at an atom knocks an electron off it, leaving an ion⁺. The detached electron may join another atom, making an ion⁻.

off an atom—a transfer from K.E. to electrical P.E. At room temperature, collisions that ionize are too rare to count. At flame temperature, say 1200°C, some gas molecules have enough K.E. to make ions on collision—only a few in every billion, but these build up enough ions to show noticeable effects.

An atom can show intermediate damage: an electron is lifted out to a higher level, without being torn off entirely. That makes an "excited" atom. When the electron of an excited atom switches back to normal, it releases its extra energy as radiation. That is how atoms give out light, by an electron returning to a lower, more stable energy-level. The bright yellow light of a salted flame, the blue-green of a coppered one, the red glow of a neon sign, are emitted by excited atoms returning to normal—or by ions that have just grabbed an electron and are returning to normal via excited states.

Flame gases can conduct a trickle of electric current, showing that charged ions are there, ready to act as carriers. Ions in flames are chiefly made by releases of chemical energy that produce far speedier molecules locally than the average of the flame gases. At higher temperatures, such as the 6000°C of the Sun's surface, common collisions will excite enough atoms to make a visible glow. (In the case of the Sun and other stars, that glow is swamped by a great outpour of radiation from the far hotter furnace inside.)

However, gross heating is a clumsy way to manufacture ions. Electrical acceleration is easier. We apply a strong electric field to a gas. When an ion is made, both it and the detached electron are driven by the field so much that they gain enough energy to make *more ions* at the next collision with gas molecules. Thus one ion makes more, and those make still more, in a growing avalanche or chain reaction that we call a spark. This can be illustrated by the model sketched in Fig. 30-28. Rows of balls rest on small ridges on a sloping plank. These represent molecules, and the ridges are "one MFP" apart. A ball representing an ion placed at the top rolls down the plank till it meets a row of "molecules."

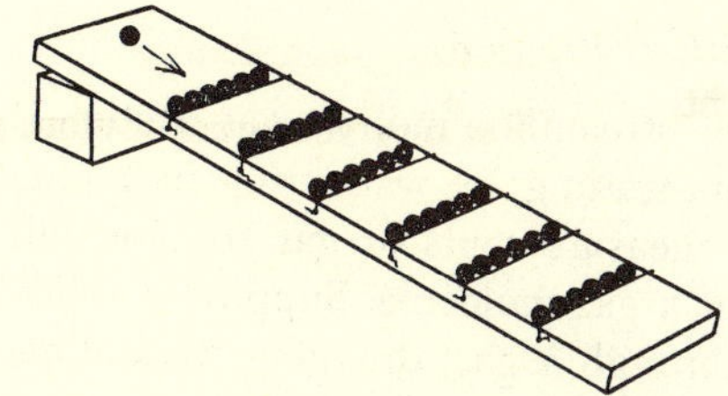

Fig. 30-28. Model of "Chain Reaction" Spark
Thin wires across plank make small ridges.

If the slope is very slight (small electric field), the "ion" stops. If the slope is small, the ball just starts another as it stops, and a tiny "leakage current" continues down the plank. If the slope is big ("big electric field") the "ion" starts several balls rolling, and each of those . . . . . . , in a growing chain reaction ("spark").

A good way to make a long thin row of ions is to fire a very high-energy charged particle, (nucleus, or electron), through gas (or liquid or solid). Using

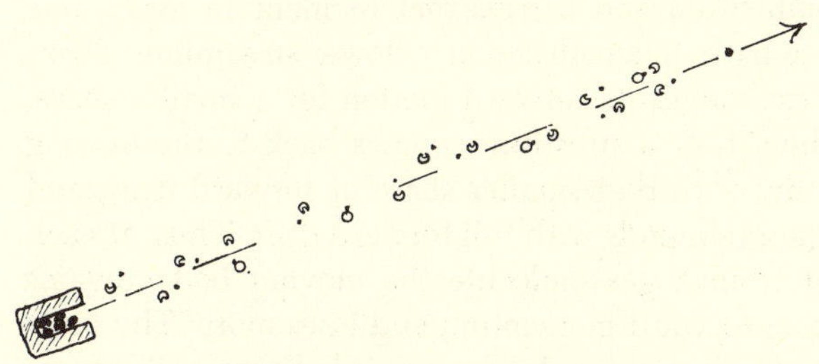

Fig. 30-29. High-Speed Charged Particle (nuclear bullet) leaves a trail of ions in its path.

its electric field, that bullet whips an electron off many an atom as it passes, leaving a narrow streak of ionized gas. That is what alpha rays and beta rays from radioactive atoms do. In a later chapter we shall show how the streaks of ions can be photographed, or put to use in a Geiger counter.

In a weak electric field an ion moves like a speck of dust falling in air. The gravitational field makes the speck accelerate, but air molecules take away a share of the gained kinetic energy at every collision. On the average the speck loses what it gains between collisions; so it seems to fall with constant speed, its weight balanced by air friction. Microscopically, its motion is a series of accelerated falls, retarded by collisions. If such a speck has an electric charge, it can also be pulled by an electric field. And, for a small speck, it is easy to make the pull of the electric field on its charge far outweigh the pull of gravity on its mass. Then the speck moves at a much greater steady speed—again a series of accelerated jumps retarded by air friction. (That is what happens to the extremely tiny charged oil-drop in Millikan's measurement of the electron charge.)

Ions, with their electric charge, do much the same. An ion in air accelerates in an electric field until it collides with a gas molecule, where it shares out the extra K.E. it has gained from the field. Then it accelerates again to the next collision, and so on (see Fig. 30-30). In *weak* fields, the ion's

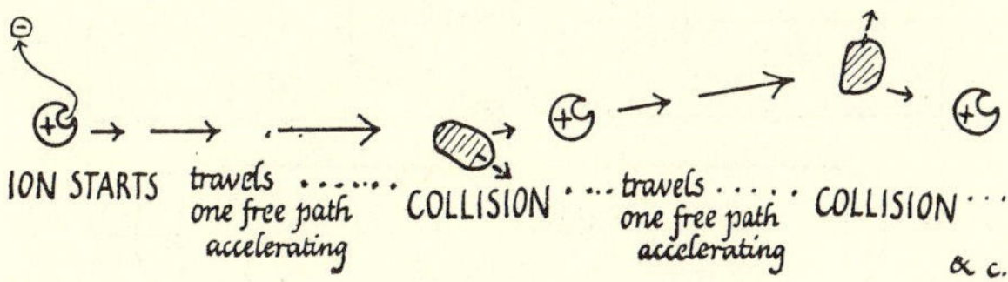

Fig. 30-30. Ion Travelling
(in a weak electric field in a gas).

collisions are elastic: it does not gain enough K.E. between collisions to make more ions; it just continues its staggering path, leaving the gas imperceptibly warmer. To make more ions, it must gain a lot of energy between one collision and the next; and for that we must either increase the electric field or lengthen the mean free path. A spark will start in ordinary air with a strong enough field—say 3 million newtons/coulomb. Pump away half the air, and the MFP is doubled. In this longer path the ion would gain twice as much energy from the field before a collision; so an electric field only half as strong will suffice for a spark. Pump away 99.9% of the air, till the pressure is 1/1000 atmosphere, and a far smaller field will make ions multiply. (Moral: compressed air is a better insulator than ordinary air.)

Apply 100,000 volts to metal electrodes in a long glass tube, and slowly pump the air out. At atmos-

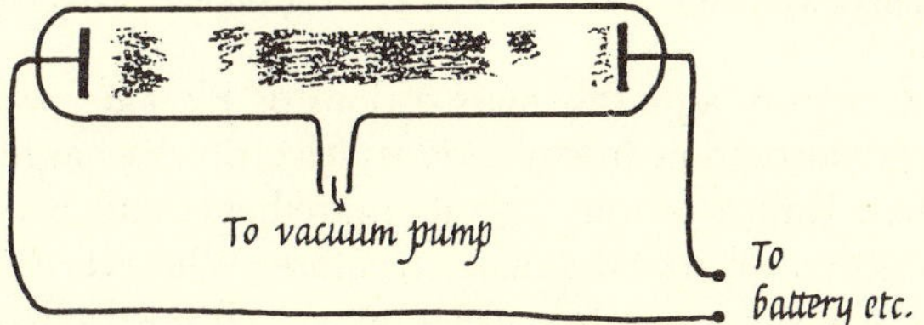

Fig. 30-31. Discharge Tube

pheric pressure, nothing happens in the tube; instead, sparks jump across a short alternative gap outside. At 1/100 atm., long streamer sparks choose the tube instead. At 1/1000 atm., the streamers have swelled to a glow that fills the tube. In the tube there is now a complicated mess of neutral molecules and atoms (some of them excited), positive ions moving one way, negative ions moving the other, whizzing electrons, X-rays, ultraviolet light, and visible light—the glow itself. Coat the tube inside with mineral that glows in ultraviolet light, and you have a modern fluorescent lamp. Pump it out to 1/1,000,000,000 atm. and there are too few molecules left to glow. Now insert 1/100 atm. of neon and you have a neon sign tube. Put in more gas and reduce the voltage till it is just insufficient for a spark, and you have a Geiger counter tube.

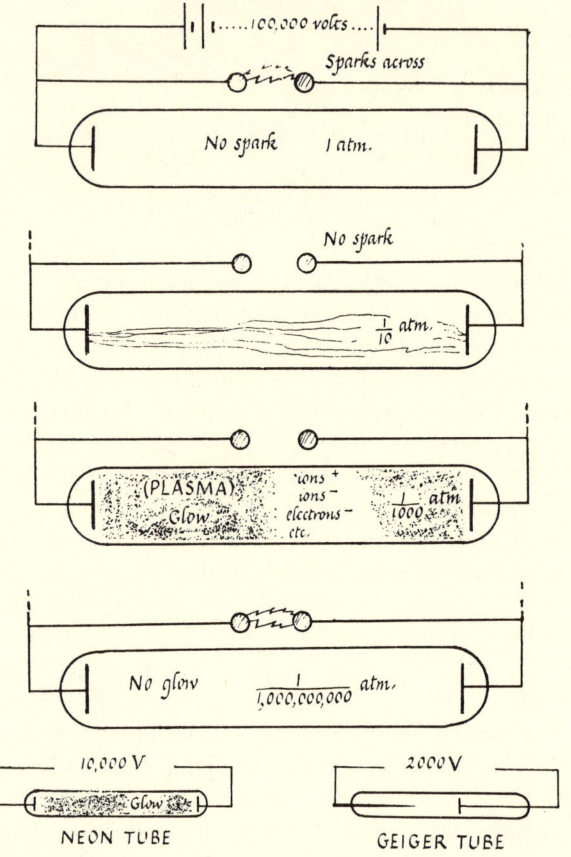

Fig. 30-32.
Discharge Tubes Containing Gas at Various Pressures

A century ago, this mess of ions or *plasma*[19] was mere decoration in mysterious "electric discharge" tubes. Half a century ago its ingredients had been sorted out to reveal atomic structure. Whatever the gas, the same negative electrons are chipped off in collisions and hurled backwards down the tube. But the atom-remainders, positive ions, differ from gas to gas, both in their mass and in their color of glow. To use electrons and ions in studying atomic physics, you need some knowledge of simple electric circuits (Chapter 32) and of electric fields (Chapter 33). Then we shall return to electrons, . . ions, . . atoms. After that, using still more violent collisions with atomic projectiles, we shall study atomic nuclei.

[19] Nowadays, the name *plasma* is usually reserved for stripped atoms, nuclei, and electrons: the torn-apart state of matter that exists in hot stars and in nuclear fusion experiments.

Viscosity (Gas Friction)

For slow streamline motion, gas friction is simply momentum-trading by wandering individual molecules. So measurements of gas friction will tell us the size of a gas molecule. Suppose a solid body is drifting through a gas, dragging streamlines of gas nearby. Each molecule that hits the body bounces

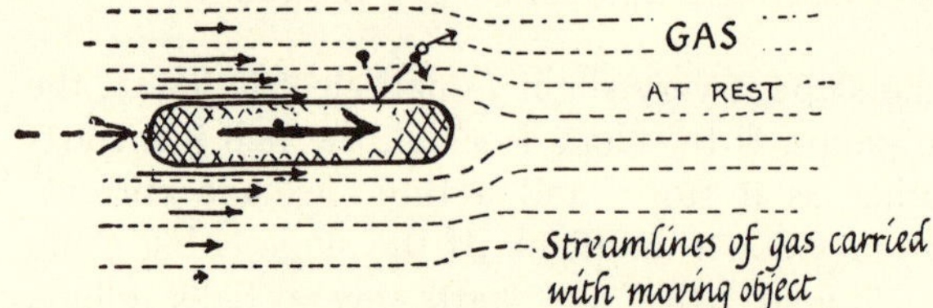

Fig. 30-15. Gas Friction
A solid body moving slowly through gas is dragged back by molecules, taking away momentum.

off with the body's forward drift added to its random motion. It rebounds the richer by a little forward momentum and carries that momentum away, one free path, to a collision in a slower streamline. There it exchanges its forward motion for a smaller share. Then it, or a substitute, comes back to the moving body, with that smaller share of forward drift, and again rebounds with full forward drift. Thus, at each hit from a gas molecule the moving body regains some forward momentum and loses more. The body thus losing forward momentum behaves as if it were being dragged back by a friction force. In fact this nibbling away of momentum *is* gas friction. Moving across the streamlines, molecules hand on this stolen momentum from collision to collision till it ultimately reaches the walls of the container.

Now suppose the gas is pumped out to half the pressure, only half as many molecules. How will this change the friction drag? There are only half as many molecules to hit the moving body and carry away momentum—on that score the drag should be halved. However, the MFP is twice as great, so the molecules bringing momentum back come from a streamline twice as far away, and therefore bring back much less. On that score, each molecule is much more effective in taking away forward momentum; so the friction should be greater. On detailed examination, the two factors cancel: the gain by having twice the MFP exactly compensates the loss by having half as many molecules. (Read the freight-train fable below that illustrates this.) In this momentum-robbery, we now have half as many thieves, but we have doubled the efficiency of each thief—twice as good a get-away. Hence a surprising prediction, *gas friction should be independent of pressure.* This seems absurd, that as we pump the air out of a box, dust will fall at the same speed, pendulum swings will die down at the same rate at 1 atmosphere, ½ atm., ¼, 1/10, 1/100, 1/1000 . . . . . ! In fact, for *small* objects, moving slowly, it is true over a vast range of pressures. Try it—a difficult demonstration.

Demonstration

We fix a small knob on a fine flexible stalk of quartz (pure melted sand) firmly anchored in a glass bottle. This tiny "wig-wag" continues to vibrate

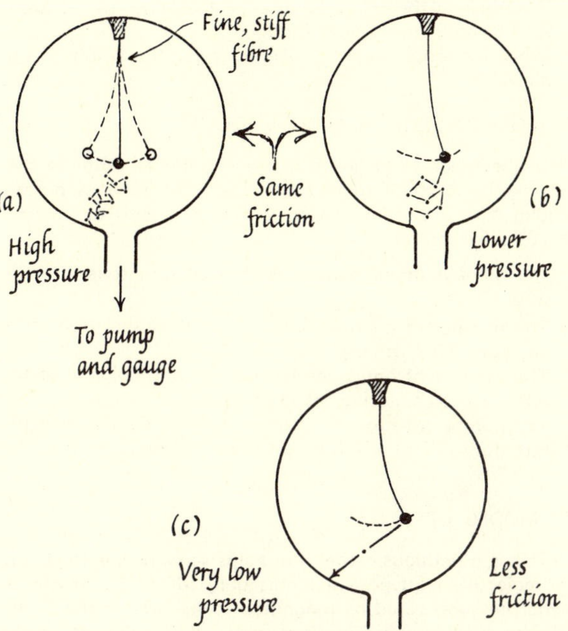

Fig. 30-16.
Test of Gas Friction Being Independent of Pressure
(a) High pressure. (b) Low pressure.
(c) At very low pressure, MFP is as large as the bottle.
After that, further removal of gas reduces the friction.

when given a start but the air friction damps the amplitude of vibration down as time goes on. With more and more of the air pumped out, the damping—due to air friction—stays almost the same, as kinetic theory predicted. This *cannot* be true to the bitter end, a perfect vacuum. Our theory must break down, and it foretells its own failure. Friction should remain unchanged as long as the MFP increases in the same proportion as the number of molecules decreases. But when so much gas has been removed that the MFP is as big as the bottle, then removing still more gas can not increase the MFP, but will only reduce the number of carriers. So we predict: constant friction, down to the pressure at which the MFP is about the size of the bottle; below that, friction will die towards zero. We can use this to estimate mean free path. Pump air from the bottle until the "wig-wag" changes its behavior and continues to vibrate. Measure the pressure of air left in the bottle, and measure the width of the bottle. Then argue back to what the MFP must be at atmospheric pressure.

Here is a specimen record: Bottle spherical, about 0.10 meter radius. Friction almost constant from 1 atm. to .1 to .00001 atm. Friction noticeably less when pressure is 1 millionth atmosphere: then, as a rough estimate, MFP at 1/1,000,000 atmosphere is 0.10 meter. At 1 atmosphere, pressure would be 1,000,000 times as great; and the density would be 1,000,000 times as great (Boyle's Law), and the mean free path 1,000,000 times smaller. The MFP would be (0.1 meter)/(1,000,000) or 10^{-7} meter, or 1000 Å.U. These are ideal data, chosen to give the right MFP, but they are quite possible. The actual experiment is a rough one, and its estimate of MFP may be several times too long or too short. With more geometrical reasoning, we can estimate the MFP from accurate measurements of gas flow through a capillary tube against this kind of friction. The result for atmospheric air: about 10^{-7} meter or 1000 A.U.

A Fable to Illustrate the Gas Friction Paradox

The mechanism of gas friction or viscosity can be illustrated by an analogy. Imagine a railroad of nine tracks, running South-North. A luxury train runs steadily 50 miles/hour Northward on the central track. Beside it, long freight trains of bare flatcars are running steadily Northward on the other tracks. They run 40 miles/hour on the tracks next to the central track, 30 miles/hour on the tracks next outside, then 20 miles/hour, and 10 miles/hour on the outermost. On every flatcar there are

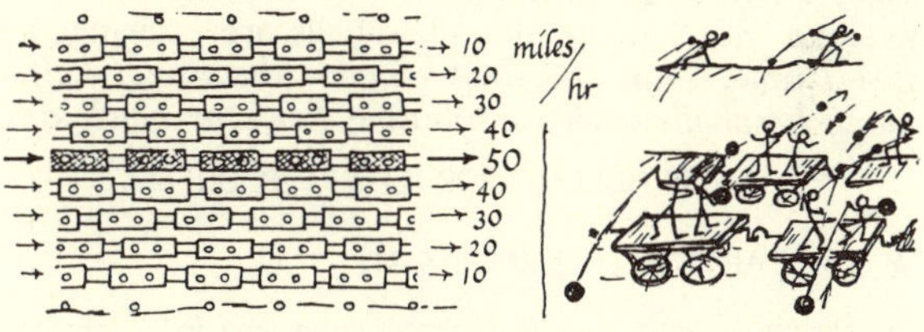

Fig. 30-17.

two lunatics throwing baseballs out sideways, East and West, at a regular rate. The lunatics also catch the baseballs thrown at them from the trains next to them—so the trains neither gain nor lose mass by the baseball-traffic. There are lunatics standing at each side of the railroad who catch the balls from the outermost freight trains, and throw them back. There are also lunatics on the central luxury train, who catch and throw baseballs sideways East and West. The lunatics on the trains are ignorant of relative-velocity physics, so they do not allow for their own motion but aim "straight at" the neighboring train; they throw sideways relative to their own train. Then a ball from a train carries the East-West momentum given by its thrower *and* the Northward momentum of the train's velocity.

On the average, the luxury train gains baseballs as fast as it loses them, so the baseball-traffic does not change its mass; but it does take away momentum. Every ball leaving the luxury train is moving *forward*

(North) 50 miles/hour, and carries away some forward momentum. Every ball joining the luxury train comes from a lunatic on the 40 mile/hour freight train beside it, and brings back a smaller amount of forward momentum. On the balance, the luxury train loses more forward momentum than it regains. This shows up as a dragging force on the train—and the train will slow down unless its locomotive pulls to compensate for the baseball-traffic. (The place where this backward force appears is at the lunatics' hands when they catch the balls returning to them—those balls are moving forward slower than the catchers, who must therefore tilt their hands to receive a backward blow. As they catch a ball their hands accelerate it forward to the full train-speed.) The intermediate trains gain and lose mass at equal rates. They also gain and lose forward momentum at equal rates, *in this model*, because they receive faster balls from one side, slower from the other. The lunatics on the ground at the sides receive forward momentum with every ball, and experience a drag forward.

This is a model of a solid body moving through a gas. The luxury train represents the moving body, the freight trains represent accompanying streamlines of gas, and the baseballs are sample molecules that happen to move sideways.

Now remove half the lunatics and half the baseballs, leaving one flatcar out of every two empty (so that a baseball will whizz across it uncaught, till it hits a car with a lunatic, on some remoter track). Before, a baseball was always caught by a waiting lunatic after whizzing one track across. On the average it will now whizz two tracks across before it is caught. The mean free path of baseballs for East-West motion has doubled! The luxury train loses baseballs moving 50 miles/hour Northward and regains them moving only 30 miles/hour, from two tracks away. The difference of forward motion is 50 − 30 instead of 50 − 40, so every exchange of balls takes away twice as much forward momentum on the average. But half as many baseballs are exchanged in given time. *Thus the drag remains the same: half as many baseballs each contributing twice as much drag.*

PROBLEMS FOR CHAPTER 30

★ 15. ZARTMAN'S EXPERIMENT

(a) Describe, with your own sketch, the way in which Zartman measured the speeds of molecules.
(b) In your sketch, show the recording film in the drum.
(c) Sketch the film unrolled, and mark with pencil the sort of record you would expect for a gas containing only one kind of molecule.
(d) Sketch the kind of record you would expect for a mixture of two gases one having molecules 4 times as massive as the other. Label your record marks clearly.
(e) Explain how you arrived at the spacing for the marks in (d).

16. A FORMULA FOR ZARTMAN

(a) Suppose in a Zartman experiment some molecules with speed v make a mark distance y from the "zero mark" on the film. How will y be changed if v is doubled? How will y be changed if each of following is doubled, without changing any of the others:
    the drum's spin-speed *R* revolutions/sec?
    the drum's radius *r*?
    the drum's length *L* along its axis?
(b) Obtain a formula for calculating v from measurements of y, *R*, etc.

17. VISCOSITY

(a) When a small object moves in air making a streamlined motion, the air drags it with "fluid friction" and the object is brought to rest. Describe in your own words the mechanism of gas friction from the point of view of molecules.
(b) If the air pressure is halved, how does this affect the rate at which its molecules bounce against the moving object, producing friction drag?
(c) If the air pressure is halved, how does this affect the mean free path of the molecules?
(d) Why would you expect a change of mean free path to affect the fluid friction on the moving object?
(e) What effect then would you expect halving the air pressure to have on its fluid friction?
(f) What limitation would you predict for the result of (e) above?

18. MOLECULES IN A HIGH VACUUM

Use the rough data given in this chapter to estimate the following for a small radio tube. Treat the tube as a rectangular box 2 centimeters × 2 cm × 5 cm, of volume $20/10^6$ cubic meter.

(a) The number of air molecules in it at atmospheric pressure.
(b) The number of air molecules left in it at the usual high vacuum, 10^{-9} atmosphere.
(c) The number of times per second any particular molecule hits another molecule, at high vacuum.
(d) The number of times per second any particular molecule hits the walls of the tube, at high vacuum.

19. MODEL OF A GAS

(This is a fictitious model that helps some people to picture gas molecules.) Suppose a tennis ball, full of air at atmospheric pressure, could be magnified to the size of the Earth, with the same number of "gas molecules" enlarged on the same scale. Take the original ball's internal radius as 0.032 meter ($\approx 1\frac{1}{4}$ inches), and the Earth's radius as 6.4×10^6 meters.

(a) Make rough estimates of the following for the "molecules" in the enlarged model: (i) diameter; (ii) average spacing apart; (iii) mean free path.
(b) Describe the estimates of (a) by common measures, such as "as big as a golf ball," "five yards apart."

20. BOUNCING AND ENERGY

(a) A thin rubber ball (with no leak) has atmospheric air inside. It is thrown fast against a massive vertical wall, and bounces back. Using your knowledge of kinetic theory, describe the changes in the air in the ball during the process of hitting the wall and rebounding.
(b) When a moving lump of lead hits a steel target and stops dead, it grows hotter. But when a rubber ball moving just as fast hits a target and bounces back (with almost the same speed), it grows hardly any hotter. Discuss the difference.

21. DIFFUSION PUMP

A mercury (or oil) vapor diffusion pump is remarkably good at pumping out residual hydrogen or helium. Suppose a massive mercury molecule and a light molecule approach with relative velocity v and hit head-on. Predict the motion after collision. (See a problem at the end of Ch. 26 for the essential hint.) Describe how such collisions produce rapid pumping.

INTERLUDE

MATHEMATICS AND RELATIVITY

"Mathematics is in a sense 'the language of science,' basic to advanced work in almost all the natural sciences and in some of the social sciences."

—Princeton University Catalogue (1947-1959)

This Interlude is not an essential part of the course, yet it touches the whole background of modern physical science.

"Ah, but a man's reach should exceed his grasp,
Or what's a heaven for?"

—Robert Browning

CHAPTER 31

MATHEMATICS: ACCURATE LANGUAGE, SHORTHAND MACHINE, AND BRILLIANT CHANCELLOR

RELATIVITY: NEW SCIENCE AND NEW PHILOSOPHY

"Mathematics is the queen of natural knowledge."

—C. F. Gauss ($\sim$ 1840)

"Mathematics deals exclusively with the relations of concepts to each other without consideration of their relation to experience. Physics too deals with mathematical concepts; however these concepts attain physical content only by the clear determination of their relation to the objects of experience. . . . The theory of relativity is that physical theory which is based on a consistent physical interpretation of . . . the concepts of motion, space, time."

—A. Einstein

Mathematics as Language

The scientist, collecting information, formulating schemes, building knowledge, needs to express himself in clear language; but ordinary languages are much more vague and unreliable than most people think. "I love vegetables" is so vague that it is almost a disgrace to a civilized language—a few savage cries could make as full a statement. "A thermometer told me the temperature of the bath water." Thermometers don't "tell." All you do is try to decide on its reading by staring at it—and you are almost certainly a little wrong. A thermometer does not show the temperature of the water; it shows its own temperature. Some of these quarrels relate to the physics of the matter, but they are certainly not helped by the wording. We can make our statements safer by being more careful; but our science still emerges with wording that needs a series of explanatory footnotes. In contrast, the language of mathematics says what it means with amazing brevity and honesty. When we write $2x^2 - 3x + 1 = 0$ we make a very definite, though very dull, statement about x. One advantage of using mathematics in science is that we can make it write what we want to say with accuracy, avoiding vagueness and unwanted extra meanings. The remark "$\Delta v/\Delta t = 32$" makes a clear statement without dragging in a long, wordy description of acceleration. $y = 16t^2$ tells us how a rock falls without adding any comments on mass or gravity.

Mathematics is of great use as a shorthand, both in stating relationships and in carrying out complicated arguments, as when we amalgamate several relationships. We can say, for uniformly accelerated motion, "the distance travelled is the sum of the product of initial velocity and time, and half the product of the acceleration and the square of the time," but it is shorter to say, "$s = v_0t + \frac{1}{2}at^2$." If we tried to operate with wordy statements instead of algebra, we should still be able to start with two accelerated-motion relations and extract a third one, as when we obtained $v^2 = v_0{}^2 + 2as$ in Chapter 1, Appendix A; but, without the compact shorthand of algebra, it would be a brain-twister argument. Going still further, into discussions where we use the razor-sharp algebra called calculus, arguing in words would be impossibly complex and cumbersome. In such cases mathematics is like a sausage-machine that operates with the rules of logical argument instead of wheels and pistons. It takes in the scientific information we provide—facts and relationships from experiment, and schemes from our minds, dreamed up as guesses to be tried—and rehashes them into new form. Like the real sausage-machine, it does not always deliver to the new sausage all the material fed in; but it never delivers anything that was not supplied to it originally. It cannot manufacture science of the real world from its own machinations.

Mathematics: the Good Servant

Yet in addition to routine services mathematics can indeed perform marvels for science. As a lesser marvel, it can present the new sausage in a form that suggests further uses. For example, suppose you had discovered that falling bodies have a constant acceleration of 32 ft/sec/sec, and that any downward motion they are given to start with is just added to the motion gained by acceleration. Then the mathematical machine could take your experimental discovery and measurement of "g" and predict the relationship $s = v_0t + \frac{1}{2}(32)t^2$. Now suppose you had never thought of including upward-thrown things in your study, had never seen a ball rise and fall in a parabola. The mathematical machine, not having been warned of any such restriction, would calmly offer its prediction as if unrestricted. Thus you might try putting in an upward start, giving v_0 a negative value in the formula. At once the formula tells a different-looking story. In that case, it says,

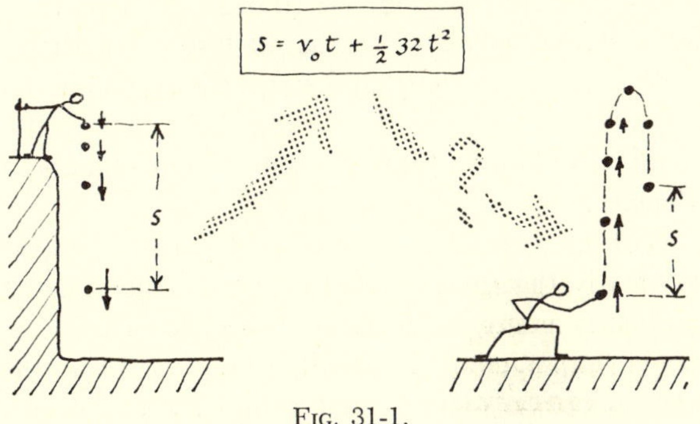

Fig. 31-1.

the stone would fly up slower and slower, reach a highest point, and then fall faster and faster. This is not a rash guess on the algebra's part. It is an unemotional routine statement. The algebra-machine's defense would be, "You never told me v_0 had to be downward. I do not know whether the new prediction is right. All I can say is that *IF an upward throw follows the rules I was told to use for downward throws, THEN an upward thrown ball will rise, stop, fall.*" It is we who make the rash guess that the basic rules may be general. It is we who welcome the machine's new hint; but we then go out and try it.[1] To take another example from projectile mathematics, the following problem, which you met earlier, has two answers.

Problem:

"A stone is thrown upward, with initial speed 64 ft/sec, at a bird in a tree. How long after its start will the stone hit the bird, which is 48 feet above the thrower?"

Fig. 31-2.

Answer:

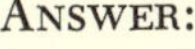

1 second *or* 3 seconds.

This shows algebra as a very honest, if rather dumb, servant. There are two answers and there should be, for the problem as presented to the machine. The stone may hit the bird as it goes up (1 sec from start), or as it falls down again (after 3 secs). The machine, if blamed for the second answer would complain, "But you never told me the stone had to *hit* the bird, still less that it must hit it on the way *up*. I only calculated *when the stone would be 48 feet above the thrower.* There are two such times." Looking back, we see we neither wrote anything in the mathematics to express contact between stone and bird nor said which way the stone was to be moving. It is *our* fault for giving incomplete instructions, and it is to the credit of the machine that it politely tells us all the answers which are possible within those instructions.

If the answer to some algebra problem on farming emerges as 3 cows *or* 2¼ cows, we rightly reject the second answer, but we blame ourselves for not telling the mathematical machine an important fact about cows. In physics problems where several answers emerge we are usually unwise to throw some of them away. They may all be quite true; or, if some are very queer, accepting them provisionally may lead to new knowledge. If you look back at the projectile problem, No. 7 in Chapter 1, Appendix B, you may now see what its second answer meant. Here is one like it:

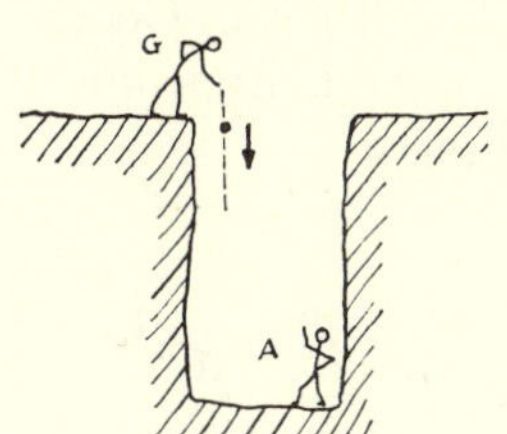

Problem:

A man throws a stone down a well which is 96 feet deep. It starts with *downward* velocity 16 ft/sec. When will it reach the bottom?

Fig. 31-4.

[1] This is a simple example, chosen to use physics you are familiar with—unfortunately so simple that you know the answer before you let the machine suggest it. There are many cases where the machine can produce suggestions that are quite unexpected and do indeed send us rushing to experiment. E.g., mathematical treatment of the wave theory of light suggested that when light casts a sharp shadow of a disc there will be a tiny bright spot of light in the middle of the shadow on a wall: "There is a hole in every coin."

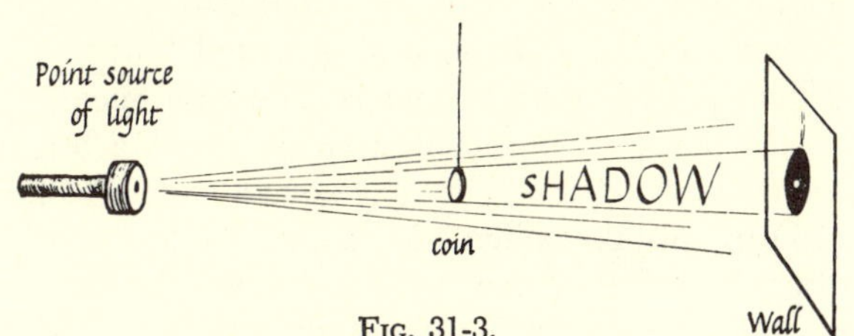

Fig. 31-3.

Assign suitable + and — signs to the data, substitute them in a suitable relation for free fall, and solve the equation. You will obtain two answers: One a sensible time with + sign (the "right" answer), the other a negative time. Is the negative answer necessarily meaningless and silly? A time such as "—3 seconds" simply means, "3 seconds before the clock was started." The algebra-machine is not told that *the stone was flung down by the man.* It is only told that when the clock started at zero *the stone was moving DOWN with speed 16 ft/sec*, and thereafter fell freely. For all the algebra knows, the stone may have just skimmed through the man's hand at time zero. It may have been started much earlier by an assistant at the bottom of the well who hurled it upward fast enough to have just the right velocity at time zero. So, while our story runs, "George, standing at the top of the well, hurled the stone down . . . ," an answer —3 seconds suggests an alternative story: "Alfred, at the bottom of the well, hurled the stone up with great speed. The stone rose up through the well and into the air above, with diminishing speed, reached a highest point, fell with increasing speed, moving down past George 3 seconds after Alfred threw it. George missed it (at $t = 0$), so it passed him at 16 ft/sec and fell on down the well again." According to the algebra, the stone will reach the bottom of the well one second after it leaves George, and it *might* have started from the bottom 3 seconds before it passes George.

Return to Problem 7 of Chapter 1, Appendix B and try to interpret its two answers.

PROBLEM 7:

A man standing on the top of a tower throws a stone up into the air with initial velocity 32 feet/sec upward. The man's hand is 48 feet above the ground. How long will the stone take to reach the ground?

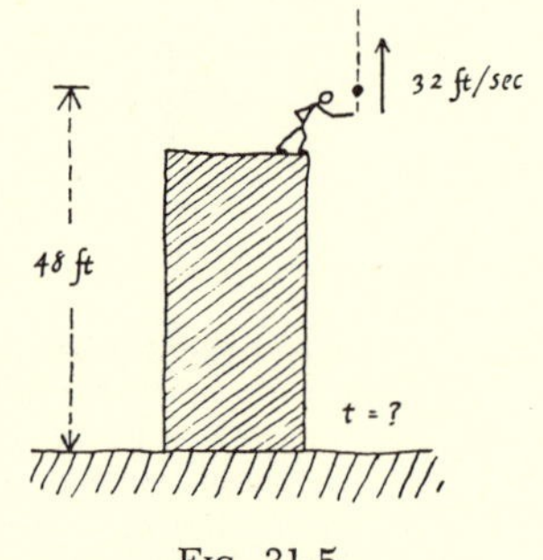

FIG. 31-5.

In these problems mathematics shows itself to be the completely honest servant—rather like the honest boy in one of G. K. Chesterton's "Father Brown" stories. (There, a slow-witted village lad delivered a telegram to a miser. The miser meant to tip the boy with the smallest English coin, a bright bronze farthing (⅓¢), but gave him a golden pound ($3) by mistake. What was the boy to do when he discovered the obvious mistake? Keep the pound, trading on the mistake dishonestly? Or bring it back with unctuous virtue and embarrass the miser into saying "Keep it, my boy"? He did neither. He simply brought the exact change, 19 shillings and 11¾ pence. The miser was delighted, saying, "At last I have found an honest man"; and he bequeathed to the boy all the gold he possessed. The boy, in wooden-headed honesty, interpreted the miser's will literally, even to the extent of taking gold fillings from his teeth.)

Mathematics: the Clever Servant

As a greater marvel, mathematics can present the new sausage in a form that suggests entirely new viewpoints. With vision of genius the scientist may see, in something new, a faint resemblance to something seen before—enough to suggest the next step in imaginative thinking and trial. If we tried to do without mathematics we should lose more than a clear language, a shorthand script for argument and a powerful tool for reshaping information. We should also lose an aid to scientific vision on a higher plane.

With mathematics, we can codify present science so clearly that it is easier to discover the essential simplicity many of us seek in science. That is no crude simplicity such as finding all planetary orbits circles, but a sophisticated simplicity to be read only in the language of mathematics itself. For example, imagine we make a hump in a taut rope by slapping it (Fig. 31-6). Using Newton's Law II, we

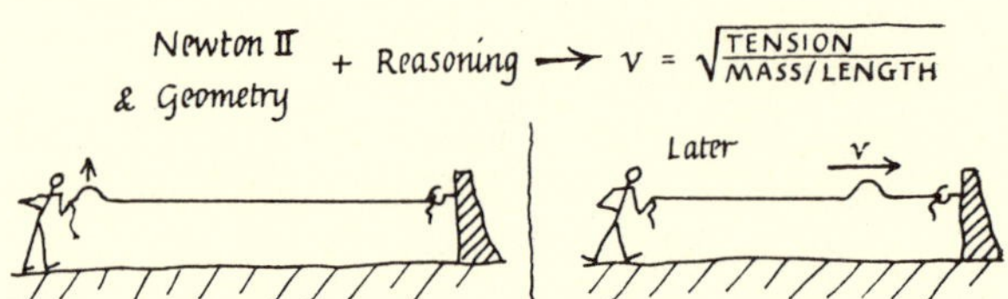

FIG. 31-6. WAVE TRAVELS ALONG A ROPE

can codify the behavior of the hump in compact mathematical form. There emerges, quite uninvited, the clear mathematical trademark of *wave* motion.* The mathematical form predicts that the hump will travel along as a wave, and tells us how to compute the wave's speed from the tension and mass of the

* The wave-equation reduces to the essential form:

$$\nabla^2 V = (1/c^2)\ d^2V/dt^2$$

For *any* wave of constant pattern that travels with speed *c*. (If you are familiar with calculus, ask a physicist to show you this remarkable piece of general mathematical physics.) This equation connects a spreading-in-space with a rate-of-change in time. $\nabla^2 V$ would be zero for an inverse-square field at rest in space: but here it has a value that looks like some acceleration. In the electromagnetic case, we may trace the d^2V/dt^2 back to an accelerating electron emitting the wave. (For the meaning of $\nabla^2 V = 0$, see p. 545. It is equivalent to relation I in Fig. 31-7 on opposite page, a statement of the inverse-square law).

rope. Another example: A century ago, Maxwell reduced the experimental laws of electromagnetism to especially simple forms by boiling them down mathematically. He removed the details of shape and size of apparatus, etc., much as we remove the shape and size of the sample when we calculate the density of a metal from some weighing and measuring. Having thus removed the "boundary conditions," he had electrical laws that are common to all apparatus and all circumstances, just as density is common to all samples of the same metal. His rules were boiled down by the calculus-process of differentiation to a final form called differential equations. You can inspect their form without understanding their terminology. Suppose that at time t there are fields due to electric charges and magnets, whether moving or not; an electric field of strength E, a vector with components E_x, E_y, E_z, and a magnetic field H with components H_x, H_y, H_z. Then, *in open space* (air or vacuum), the experimental laws known a century ago reduce to the relations shown in Fig. 31-7.

I	II
$\frac{dE_x}{dx} + \frac{dE_y}{dy} + \frac{dE_z}{dz} = 0$	$\frac{dH_x}{dx} + \frac{dH_y}{dy} + \frac{dH_z}{dz} = 0$
III	**IV**
$\left(\frac{dE_z}{dy} - \frac{dE_y}{dz}\right) = K_H \frac{dH_x}{dt}$	$-\left(\frac{dH_z}{dy} - \frac{dH_y}{dz}\right) = 0$
$\left(\frac{dE_x}{dz} - \frac{dE_z}{dx}\right) = K_H \frac{dH_y}{dt}$	$-\left(\frac{dH_x}{dz} - \frac{dH_z}{dx}\right) = 0$
$\left(\frac{dE_y}{dx} - \frac{dE_x}{dy}\right) = K_H \frac{dH_z}{dt}$	$-\left(\frac{dH_y}{dx} - \frac{dH_x}{dy}\right) = 0$

FIG. 31-7. MAXWELL'S EQUATIONS (INCOMPLETE)
The constant K_H relates to magnetic fields. It appears in the expression for the force exerted by a magnetic field on an electric current. (See the discussion in this chapter and in Ch. 37.) There is a corresponding electric constant, K_E, which appears in Coulomb's Law (See Ch. 33).

Look at IV and compare it with III. The equations of IV look incomplete, spoiling the general symmetry.[2] Maxwell saw the defect and filled it by inventing an extra electric current, a spooky one in space, quite unthought-of till then, but later observed experimentally. How would *you* change IV to match III if told that part of the algebra had been left out because it was then unknown? Try this.

The addition was neither a lucky guess nor a mysterious inspiration. To Maxwell, fully aware of the state of developing knowledge, it seemed *compulsory*, a necessary extension of symmetry—that is the difference between the scientific advance of the disciplined, educated expert and the free invention of the enthusiastic amateur.

Having made his addition, fantastic at the time, Maxwell could pour the whole bunch of equations into the mathematical sausage-machine and grind out a surprising equation which had a familiar look, the same trademark of wave-motion that appears for a hump on a rope. That new equation suggested strongly that changes of electric and magnetic fields would travel out as waves with speed $v = 1/\sqrt{K_H K_E}$. Here K_H is a constant involved in the magnetic effects of moving charges; and K_E is the corresponding electrostatic constant inserted by Maxwell in his improvement.[3] (K_E is involved in the inverse-square force between electric charges.)

An informal fanciful derivation is sketched near the end of Chapter 37.

To Maxwell's delight and the wonder of his contemporaries, the calculated v agreed with the speed of light, which was already known to consist of waves of some sort. This suggested that light might be one form of Maxwell's predicted electromagnetic waves.

It was many years before Maxwell's prediction was verified directly by generating electromagnetic waves with electric currents. As a brilliant intuitive guess, a piece of synthetic theory, Maxwell's work was one of the great developments of physics—its progeny, new guesses along equally fearless lines, are making the physics of today.

One of the great contributions of mathematics to physics is Relativity, which is both mathematics and physics: you need good knowledge of both mathematics and physics to understand it. We shall give an account of Einstein's "Special Relativity" and then return to comments on mathematics as a language.

[2] In completing IV, you will need to insert a constant K_E corresponding to K_H in III. The minus sign is obviously unnecessary in the present form of IV, and when IV is completed it will spoil the symmetry somewhat; but the experimental facts produce it, conservation of energy requires it, and without it there would be no radio waves.

[3] In this course we use different symbols. See Ch. 33 and Ch. 37. We write the force between electric charges $F = B\,(Q_1\,Q_2)/d^2$. Comparison with Maxwell's form shows our B is the same as $1/K_E$. Again, we write the force between two short pieces of current-carrying wire, due to magnetic-field effects, $F = B'\,(C_1\,L_1)\,(C_2\,L_2)/d^2$, and our B' is the same as K_H. Then Maxwell's prediction, $v = 1/\sqrt{K_E\,K_H}$, becomes, in our terminology, $v = 1/\sqrt{(1/B)\,(B')} = \sqrt{B/B'}$. So, if you measure B and B' you can predict the speed of electromagnetic waves. The arithmetic is easy. Try it and compare the result with the measured speed of light, 3.0×10^8 meters/sec. ($B = 9.00 \times 10^9$ and $B' = 10^{-7}$, in our units).

RELATIVITY

The theory of Relativity, which has modified our mechanics and clarified scientific thinking, arose from a simple question: "*How fast are we moving through space?*" Attempts to answer that by experiment led to a conflict that forced scientists to think out their system of knowledge afresh. Out of that reappraisal came Relativity, a brilliant application of mathematics and philosophy to our treatment of space, time, and motion. Since Relativity *is* a piece of mathematics, popular accounts that try to explain it without mathematics are almost certain to fail. To understand Relativity you should either follow its algebra through in standard texts, or, as here, examine the origins and final results, taking the mathematical machine-work on trust.

What can we find out about space? Where is its fixed framework and how fast are we moving through it? Nowadays we find the Copernican view comfortable, and picture the spinning Earth moving around the Sun with an orbital speed of about 70,000 miles/hour. The whole Solar system is moving towards the constellation Hercules at some 100,000 miles/hour, while our whole galaxy. . . .

We must be careering along a huge epicycloid through space without knowing it. Without knowing it, because, as Galileo pointed out, the mechanics of motion—projectiles, collisions, . . . , etc.—is the same in a steadily moving laboratory as in a stationary one.[4] Galileo quoted thought-experiments of men walking across the cabin of a sailing ship or dropping stones from the top of its mast. We illustrated this "Galilean relativity" in Chapter 2 by thought-experiments in moving trains. Suppose one train is passing another *at constant velocity* without bumps, and in a fog that conceals the countryside. Can the passengers really say which is moving? Can mechanical experiments in either train tell them? They can only observe their *relative* motion. In fact, we developed the rules of vectors and laws of motion in earthly labs that *are* moving; yet those statements show no effect of that motion.

We give the name *inertial frame* to any frame of reference or laboratory in which Newton's Laws seem to describe nature truly: objects left alone without force pursue straight lines with constant speed, or stay at rest; forces produce proportional accelerations. We find that any frame moving at constant velocity relative to an inertial frame is also an inertial frame—Newton's Laws hold there too. In all the following discussion that concerns Galilean relativity and Einstein's special Relativity, *we assume that every laboratory we discuss is an inertial frame*—as a laboratory at rest on Earth is, to a close approximation.[4] In our later discussion of General Relativity, we consider other laboratory frames, such as those which accelerate.

We are not supplied by nature with an obvious inertial frame. The spinning Earth is not a perfect inertial frame (because its spin imposes central accelerations), but if we could ever find one perfect one then our relativity view of nature assures us we could find any number of other inertial frames. Every frame moving with constant velocity relative to our first inertial frame proves to be an equally good inertial frame—Newton's laws of motion, which apply *by definition* in the original frame, apply in all the others. When we do experiments on force and motion and find that Newton's Laws seem to hold, we are, from the point of view of Relativity, simply showing that our earthly lab does provide a practically perfect inertial frame. Any experiments that demonstrate the Earth's rotation could be taken instead as showing the imperfection of our choice of frame. However, by saying "the Earth *is* rotating" and blaming that, we are able to imagine a perfect frame, in which Newton's Laws would hold exactly.

We incorporate Galilean Relativity in our formulas. When we write, $s = v_0 t + \frac{1}{2}at^2$ for a rocket accelerating horizontally, we are saying, "Start the rocket with v_0 and its effect will persist as a plain addition, $v_0 t$, to the distance travelled."

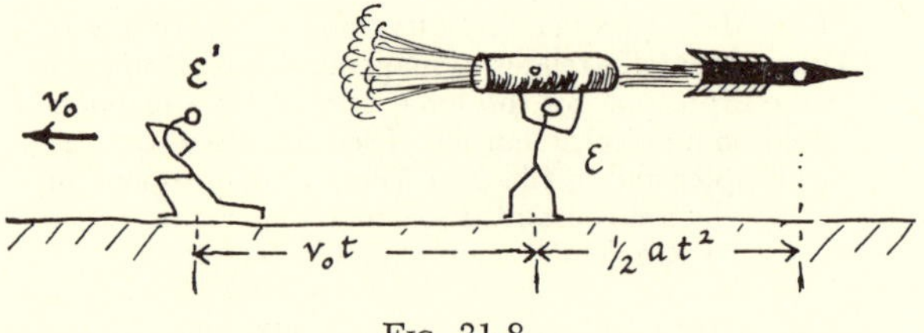

Fig. 31-8.

This can be reworded: "An experimenter ε, starts a rocket from rest and observes the motion: $s = \frac{1}{2}at^2$. Then another experimenter, ε', running away with speed v_0 will measure distances-travelled given by $s' = v_0 t + \frac{1}{2}at^2$. He will include $v_0 t$ due to his own motion."

We are saying that the effects of steady motion

[4] Though the Earth's velocity changes around its orbit, we think of it as steady enough during any short experiment. In fact, the steadiness is perfect, because any changes in the Earth's velocity exactly compensate the effect of the Sun's gravitation field that "causes" those changes. We see no effect on the Earth as a whole, at its center; but we do see *differential* effects on outlying parts—solar tides. The Earth's *rotation* does produce effects that can be seen and measured —Foucault's pendulum changes its line of swing, g shows differences between equator and poles, &c.—but we can make allowances for these where they matter.

and accelerated motion do not disturb each other; they just add.

ε and ε′ have the following statements for the distance the rocket travels in time t.

EXPERIMENTER ε	EXPERIMENTER ε′
$s = \frac{1}{2}at^2$	$s' = v_0t + \frac{1}{2}at^2$

Both statements say that the rocket travels with constant acceleration.[5]

Both statements say the rocket is at distance zero (the origin) at $t = 0$.

The *first statement* says ε sees the rocket start from rest. When the the clock starts at $t = 0$ the rocket has no velocity relative to him. At that instant, the rocket is moving with his motion, if any—so *he* sees it at rest—and he releases it to accelerate.

The difference between the two statements says the relative velocity between ε and ε′ is v_0. There is no information about absolute motion. ε may be at rest, in which case ε′ is running *backward* with speed v_0. Or ε′ may be at rest, and ε running *forward* v_0 (releasing the rocket as he runs, at $t = 0$). Or both ε and ε′ may be carried along in a moving train with terrific speed V, still with ε moving ahead with speed v_0 relative to ε′. In every case, v_0 is the relative velocity between the observers; and nothing in the analysis of their measurements can tell us (or them) who is "really" moving.

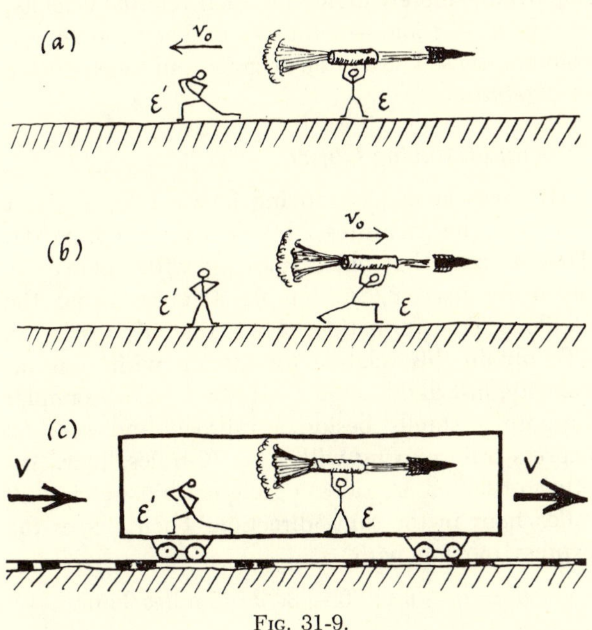

Fig. 31-9.

[5] The first statement is simpler because it belongs to the observer who releases the rocket from rest relative to him, at the instant the clock starts, $t = 0$.

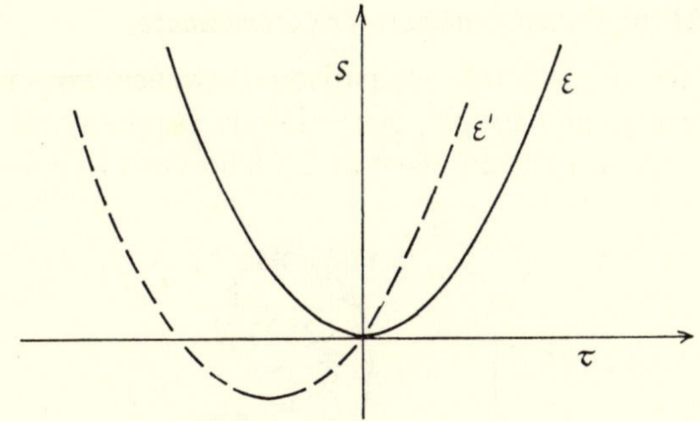

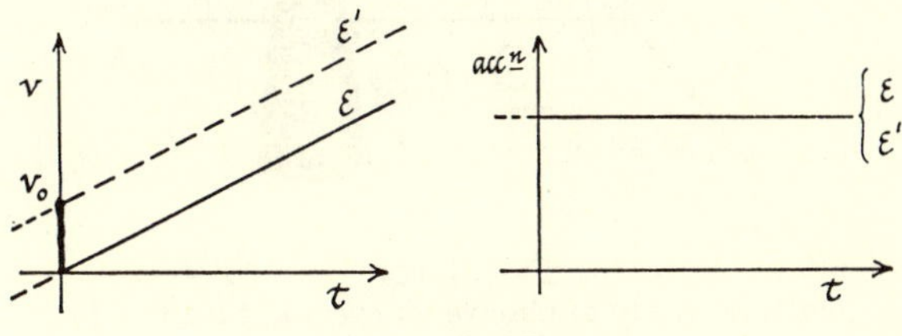

Fig. 31-10.

Adding v_0t only *shifts* the graph of s vs. t. It does not affect estimates of *acceleration, force,* etc. Then, to the question, "How fast are we moving through *space*?" simple mechanics replies, "No experiments with weights, springs, forces, . . . , can reveal our *velocity*. Accelerations could make themselves known, but uniform velocity would be unfelt." We could only measure our *relative velocity*—relative to some other object or material framework.

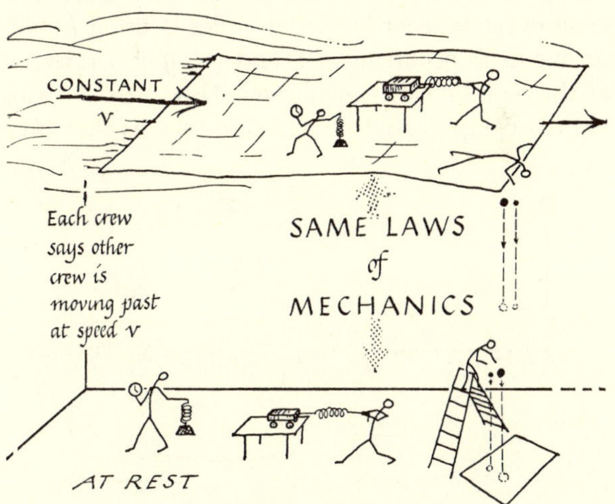

Fig. 31-11.
Observers in two laboratories, one moving with constant velocity v relative to the other, will find the same *mechanical* laws.

Yet we are still talking as if there *is* an absolute motion, past absolute landmarks in space, however hard to find. Before exploring that hope into greater disappointments, we shall codify rules of relative motion in simple algebraic form.

Galilean Transformation for Coordinates

We can put the comparison between two such observers in a simple, general way. Suppose an observer ε records an event in his laboratory. Another

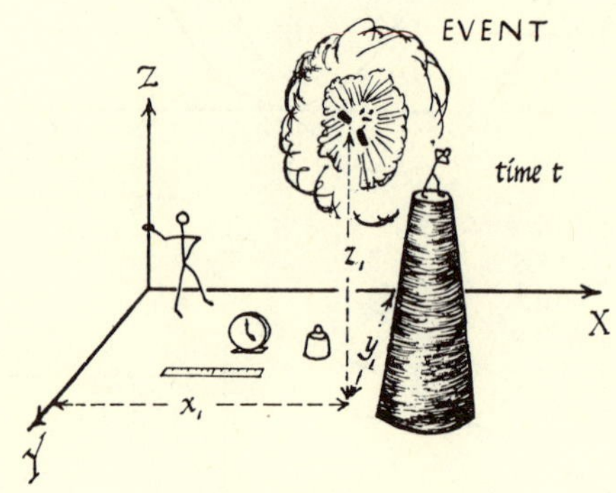

Fig. 31-12a
Observer ready to observe an event at time t and place x, y, z.

observer, ε′, flies through the laboratory with constant velocity and records the same event as he goes. As sensible scientists, ε and ε′ manufacture identical clocks and meter-sticks to measure with. Each carries a set of x-y-z-axes with him. For convenience, they start their clocks ($t = 0$ and $t' = 0$) at the instant they are together. At that instant their coordinate origins and axes coincide. Suppose ε records the event as happening at time t and place (x, y, z) referred to his axes-at-rest-with-him.[6] The same event is recorded by observer ε′ using *his* instruments as occurring at t' and (x', y', z') referred to the axes-he-carries-with-him. How will the two records compare? Common sense tells us that time

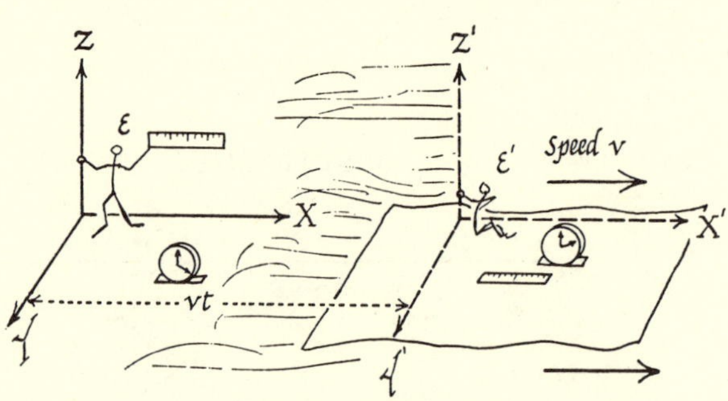

Fig. 31-12b.
Another observer, moving at constant velocity v relative to the first, also makes observations.

is the same for both, so $t' = t$. Suppose the relative velocity between the two observers is v meters/sec along OX. Measurements of y and z are the same for both: $y' = y$ and $z' = z$. But since ε′ and his coordinate framework travel ahead of ε by vt meters in t seconds, all his x'-measurements will be vt shorter. So every x' must $= x - vt$. Therefore:

$$x' = x - vt \qquad y' = y \qquad z' = z \qquad t' = t$$

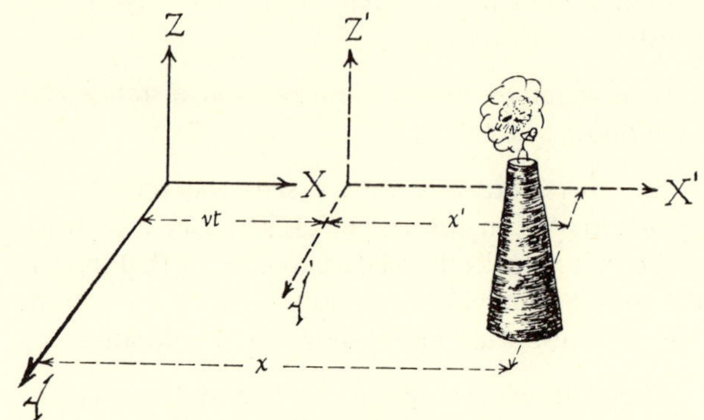

Fig. 31-12c.
For measurements along direction of relative motion v, the second observer measures x'; the first measures x. Then it seems obvious that $x' = x - vt$.

These relations, which connect the records made by ε′ and ε, are called the *Galilean Transformation.*

The reverse transformation, connecting the records of ε and ε′, is:

$$x = x' + vt \qquad y = y' \qquad z = z' \qquad t = t'$$

These two transformations treat the two observers impartially, merely indicating their *relative* velocity, $+v$ for ε′ — ε and $-v$ for ε — ε′. They contain our common-sense knowledge of space and time, written *in algebra.*

Velocity of Moving Object

If ε sees an object moving forward along the x direction, he measures its velocity, u, by $\Delta x/\Delta t$. Then ε′ sees that object moving with velocity u' given by his $\Delta x'/\Delta t'$. Simple algebra, using the Galilean Transformation, shows that $u' = u - v$. (To obtain this relation for motion with constant velocity, just divide $x' = x - vt$ by t.) For example: suppose ε stands beside a railroad and sees an express train moving with $u = 70$ miles/hour. Another observer, ε′, rides a freight train moving 30 miles/hour in the same direction. Then ε′ sees the express moving with

$$u' = u - v = 70 - 30 = 40 \text{ miles/hour.}$$

(If ε′ is moving the opposite way, as in a head-on collision, $v = -30$ miles/hour, and ε′ sees the express approaching with speed

$$u' = 70 - (-30) = 100 \text{ miles/hour.})$$

[6] For example: he fires a bullet along OX from the origin at $t = 0$ with speed 1000 m./sec. Then the event of the bullet reaching a target 3 meters away might be recorded as $x = 3$ meters, $y = 0$, $z = 0$, $t = 0.003$ sec.

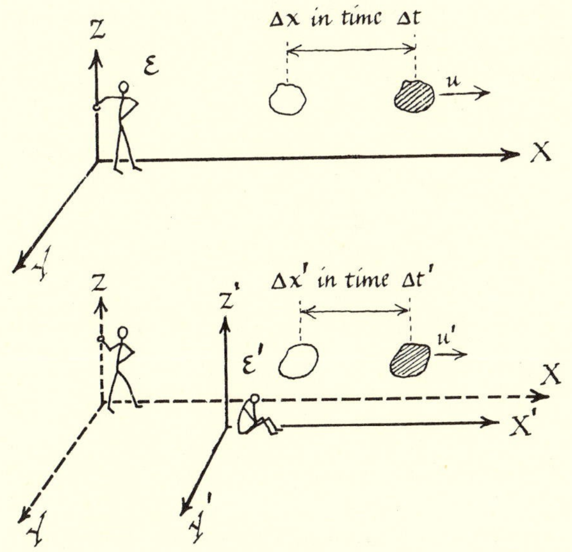

Fig. 31-13.
Each experimenter calculates the velocity of a moving object from his observations of time taken and distance travelled.

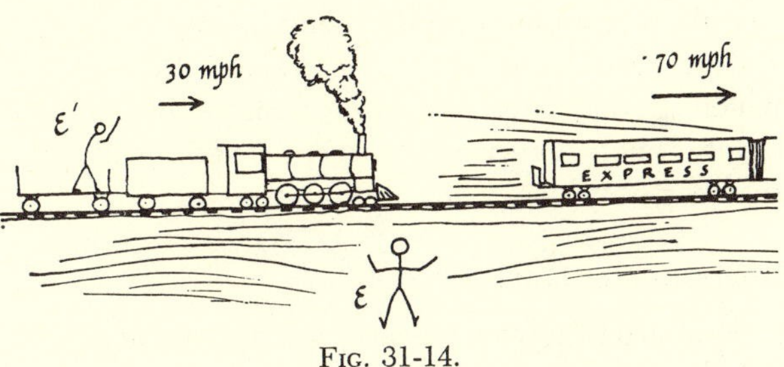

Fig. 31-14.

Stationary experimenter ε observes the velocities shown and calculates the relative velocity that moving experimenter ε′ should observe.

This is the "common sense" way of adding and subtracting velocities. It seems necessarily true, and we have taken it for granted in earlier chapters. Yet we shall find we must modify it for very high speeds.

?Absolute Motion?

If we discover our laboratory is in a moving train, we can add the train's velocity and refer our experiments to the solid ground. Finding the Earth moving, we can shift our "fixed" axes of space to the Sun, then to a star, then to the center of gravity of all the stars. If these changes do not affect our knowledge of mechanics, do they really matter? Is it honest to worry about finding an absolutely fixed framework? Curiosity makes us reply, "Yes. If we *are* moving through space it would be interesting to know how fast." Though mechanical experiments cannot tell us, could we not find out by electrical experiments? Electromagnetism is summed up in Maxwell's equations, for a stationary observer. Ask what a *moving* observer should find, by changing x to x', etc., with the Galilean Transformation: then Maxwell's equations take on a different, more complicated, form. An experimenter who trusted that transformation could decide which is really moving, himself or his apparatus: absolute motion would be revealed by the changed form of electrical laws. An easy way to look for such changes would be to use the travelling electric and magnetic fields of light waves—the electromagnetic waves predicted by Maxwell's equations. We might find our velocity through space by timing flashes of light. Seventy-five years ago such experiments were being tried. When the experiments yielded an unexpected result—failure to show any effect of motion—there were many attempts to produce an explanation. Fitzgerald in England suggested that whenever any piece of matter is set in motion through space it must contract, along the direction of motion, by a fraction that depended only on its speed. With the fraction properly chosen, the contraction of the apparatus used for timing light signals would prevent their revealing motion through space. This strange contraction, which would make even measuring rods such as meter-sticks shrink like everything else when in motion, was too surprising to be welcome; and it came with no suggestion of mechanism to produce it. Then the Dutch physicist Lorentz (also Larmor in England) worked out a successful electrical "explanation."

The Lorentz Transformation

Lorentz had been constructing an electrical theory of matter, with atoms containing small electric charges that could move and emit light waves. The experimental discovery of electron streams, soon after, had supported his speculations; so it was natural for Lorentz to try to explain the unexpected result with his electrical theory. He found that if Maxwell's equations are *not* to be changed in form by the motion of electrons and atoms of moving apparatus, then lengths along the motion must shrink, in changing from x to x', by the modifying factor:

$$\frac{1}{\sqrt{1-\left(\dfrac{\text{SPEED OF OBSERVER}}{\text{SPEED OF LIGHT}}\right)^2}}$$

He showed that this shrinkage (the same as Fitzgerald's) of the apparatus would just conceal any motion through absolute space and thus explain the experimental result. But he also gave a reason for the change: he showed how electrical forces—in the new form he took for Maxwell's equations—would compel the shrinkage to take place.

It was uncomfortable to have to picture matter in motion as invisibly shrunk—invisibly, because we should shrink too—but that was no worse than the previous discomfort that physicists with a sense of mathematical form got from the uncouth effect of the Galilean Transformation on Maxwell's equations. Lorentz's modifying factor has to be applied to t' as well as x', and a strange extra term must be added to t'. And then Maxwell's equations maintain their same simple symmetrical form for all observers moving with any constant velocity. You will see this "Lorentz Transformation" put to use in Relativity; but first see how the great experiments were made with light signals.

Measuring Our Speed through "Space"?

A century ago, it was clear that light consists of waves, which travel with very high speed through glass, water, air, even "empty space" between the stars and us. Scientists imagined space filled with "ether"[7] to carry light waves, much as air carries sound waves. Nowadays we think of light (and all other radio waves) as a travelling pattern of electric and magnetic fields and we need no "ether"; but before we reached that simple view a tremendous contradiction was discovered.

Experiments with light to find how fast *we* are moving through the "ether" gave a surprising result: "no comment." These attempts contrast with successful measurements with sound waves and air.

Sound travels as a wave in air. A trumpet-toot is handed on by air molecules at a definite speed *through the air*, the same speed whether the trumpet is moving or not. But a moving observer finds *his* motion added to the motion of sound waves. When he is running towards the trumpet, the toot passes by him faster. He can find how fast he is moving *through air* by timing sound signals passing him.

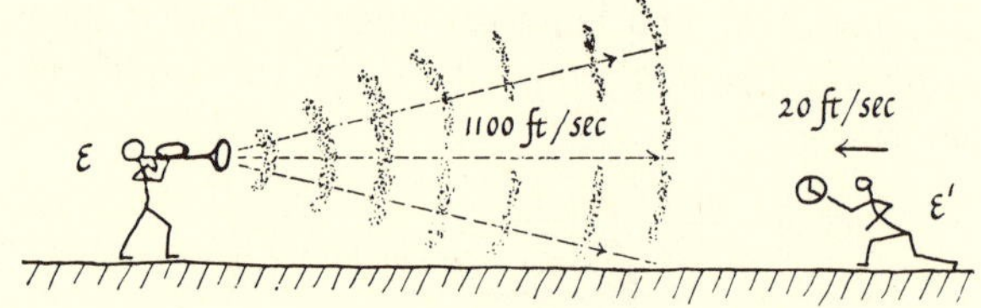

Fig. 31-15.
Experimenter running towards source of sound finds the speed of sound 1120 ft/sec, in excess of normal by his own speed.

[7] This ether or æther was named after the universal substance that Greek philosophers had pictured filling all space beyond the atmosphere.

A moving observer will notice another effect if he is out to one side, listening with a direction-finder. He will meet the sound slanting from a new

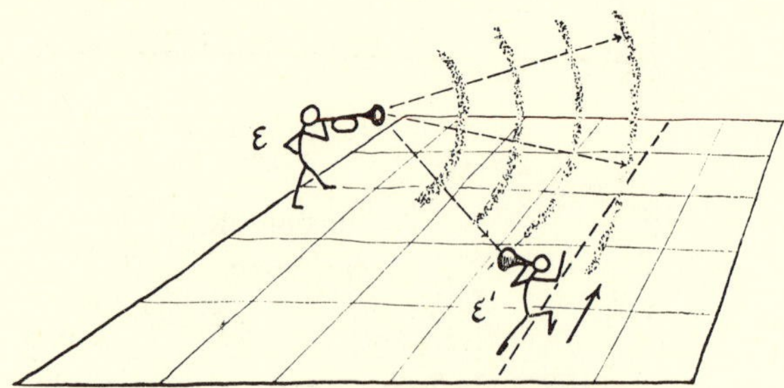

Fig. 31-16.
Observer running across the line-of-travel of sound notices a change of apparent direction of source.

direction if he runs. Again he can estimate his running speed if he knows the speed of sound.

In either case, his measurements would tell him his speed *relative to the air*. A steady wind blowing would produce the same effects and save him the trouble of running. Similar experiments with light should reveal our speed relative to the "ether," which is our only remaining symbol of absolute space. Such experiments were tried, with far-reaching results.

Aberration of Starlight

Soon after Newton's death, the astronomer Bradley discovered a tiny yearly to-and-fro motion of all stars that is clearly due to the Earth's motion around its orbit. Think of starlight as rain showering down (at great speed) from a star overhead. If you stand in vertical rain holding an umbrella upright, the rain will hit the umbrella top at right angles. Drops falling through a central gash will hit your head. Now run quite fast. To you the rain will seem slanting. To catch it squarely you must tilt the umbrella at the angle shown by the vectors in the sketch. Then drops falling through the gash will still hit your head. If you run around in a circular orbit, or to-and-fro along a line, you must wag the umbrella this way and that to fit your motion. This is what Bradley found when observing stars precisely with a telescope.* Stars near the ecliptic seemed to slide to-and-fro, their directions swinging through a small angle. Stars up near the pole of the ecliptic

* This *aberration* is quite distinct from *parallax*, the apparent motion of near stars against the background of remoter stars. Aberration makes a star seem to move in the same kind of pattern, but it applies to *all* stars; and it is dozens of times bigger than the parallax of even the nearest stars. (Also, a star's aberration, which goes with the Earth's *velocity*, is three months out of phase with its parallax.)

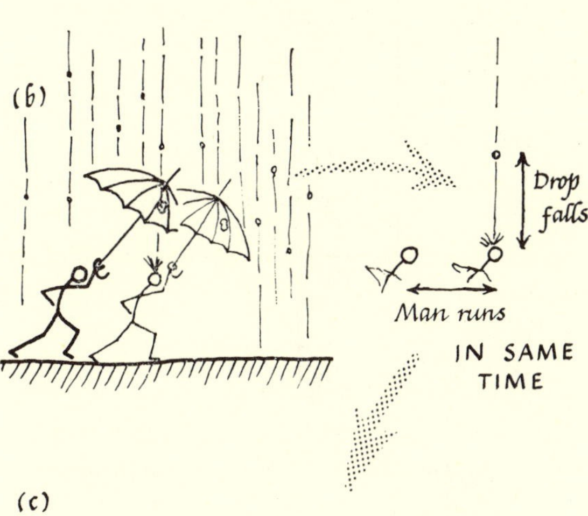

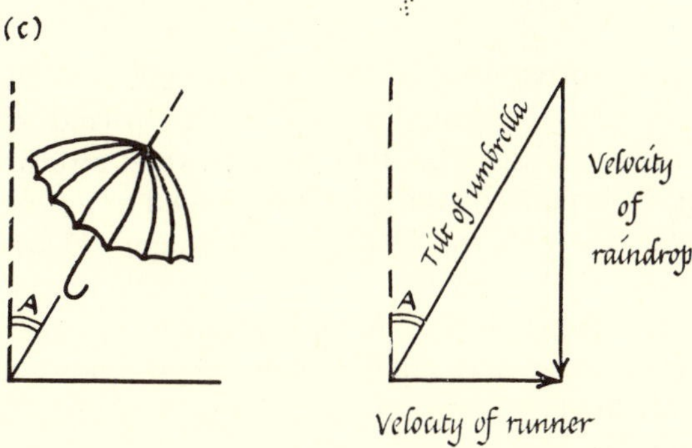

Fig. 31-17. "Aberration" of Rain

move in small circles in the course of a year. The telescope following the star is like the tilting umbrella. In six months, the Earth's velocity around the Sun changes from one direction to the reverse, so the telescope tilt must be reversed in that time. From the tiny measured change in 6 months, Bradley estimated the speed of light. It agreed with the only other estimate then available—based on the varying delays of seeing eclipses of Jupiter's moons, at varying distances across the Earth's orbit.[8]

Fig. 31-18. "Aberration" of Rain Falling in Wind
If you stand still but a steady wind carries the air past you, you should still tilt the umbrella.

To catch rain drops fair and square, you must tilt your umbrella if you are running *or* if there is a steady wind, but not if you are running *and* there is also a wind carrying the air and raindrops along *with you*—if you just stand in a shower inside a closed railroad coach speeding along, you do not tilt the umbrella. Therefore, Bradley's successful measurement of aberration showed that *as the Earth runs around its orbit it is moving through the "ether" in changing directions*, moving through space if you like, nearly 20 miles/sec.

An overall motion of the solar system towards some group of stars would remain concealed, since that would give a permanent slant to star directions,

[8] It was another century before terrestrial experiments succeeded.

(~ 1600): Galileo recorded an attempt with experimenters signalling by lantern flashes between two mountain tops. ε_1 sent a flash to ε_2 who immediately returned a flash to ε_1. At first ε_2 was clumsy and they obtained a medium speed for light. As they improved with practice, the estimated speed grew greater and greater, towards "infinity"—light travels too fast to clock by hand.

(~ 1700): Newton knew only Roemer's estimate from Jupiter's moons.

(1849): Fizeau succeeded, by using a distant mirror to return the light and a spinning toothed wheel as a chopper to make the flashes and catch them one tooth later on their return. His result confirmed the astronomical estimate. His and all later terrestrial methods use some form of chopper—as in some methods for the speeds of bullets, and electrons.

The result: speed of light is 300,000,000 meters/sec or 186,000 miles/sec.

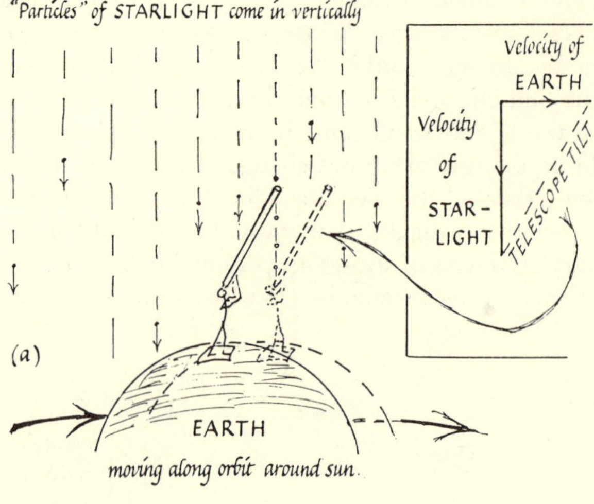

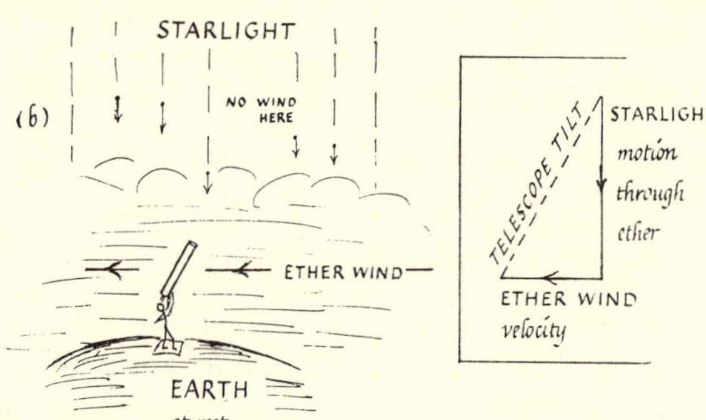

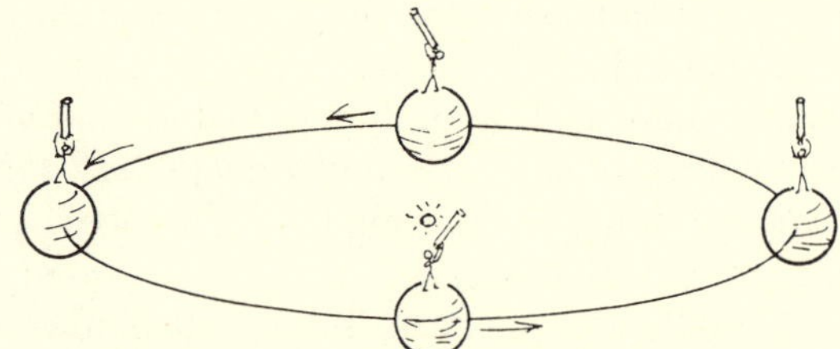

Fig. 31-19. Aberration of Starlight

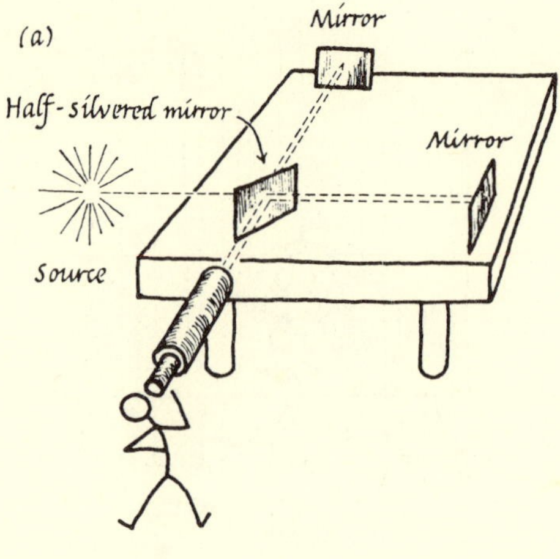

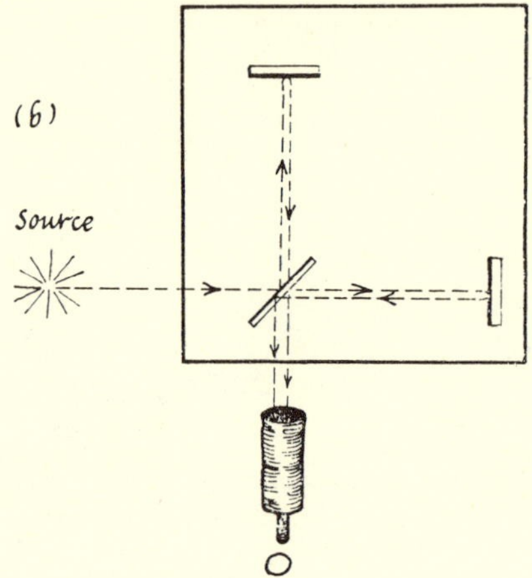

Fig. 31-20. The Michelson-Morley Experiment

whereas Bradley measured *changes* of slant from one season to another.

The Michelson-Morley Experiment

Then, seventy-five years ago, new experiments were devised to look for our absolute motion in space. One of the most famous and decisive was devised and carried out by A. A. Michelson and E. W. Morley in Cleveland; this was one of the first great scientific achievements in modern physics in the New World. In their experiment, two flashes of light travelling in different directions were made to pace each other. There was no longer a moving observer and fixed source, as with Bradley and a star. Both source and observer were carried in a laboratory, but the experimenters looked for motion of the intervening ether that carried the light waves. A semi-transparent mirror split the light into two beams, one travelling, say, North-South and the other East-West. The two beams were returned along their paths by mirrors and rejoined to form an interference pattern. The slightest change in trip-time for one beam compared with the other would shift the pattern. Now suppose at some season the whole apparatus is moving upward in space: an outside observer would see the light beams tilted up or down by the "ether-wind" the same tilt for both routes. At another season, suppose the whole Earth is moving due North horizontally in space, then the N-S light beam would take longer for its round trip than the E-W one. You will find the experiments described in standard texts, with the algebra to show that if the whole laboratory is sweeping through the ether, light must take longer on the trip along the stream and back than on the trip across and back.

You can see that this is so in the following example. Instead of light, consider a bird flying across a cage and back, when the cage is moving relative

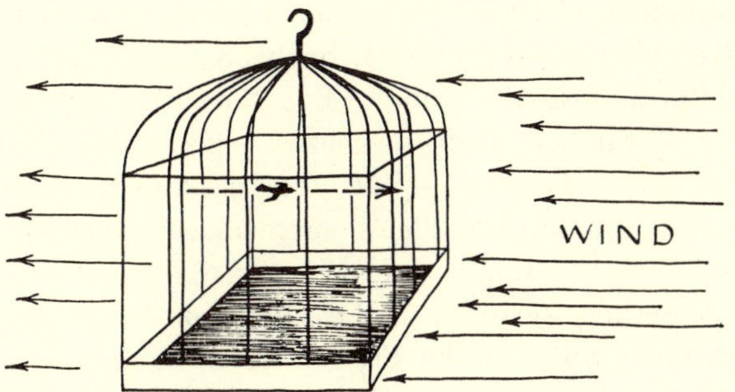

FIG. 31-21. GIANT BIRDCAGE IN WIND

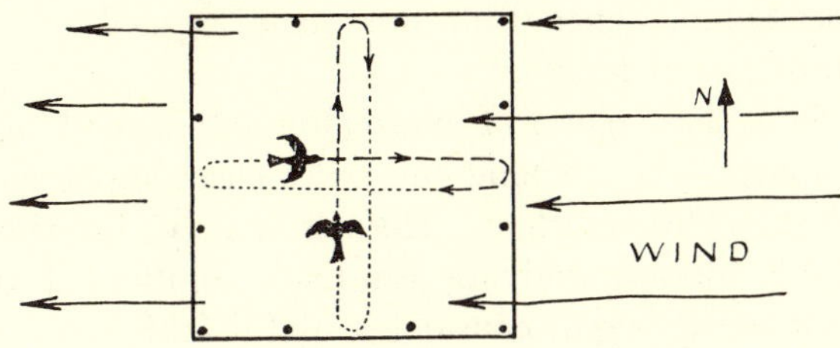

FIG. 31-22.
Bird flies either against the wind and back; or across the wind and back across the wind.

to the air. Either (a) drag the cage steadily along through still air, or (b) keep the cage still and have an equal wind blow through it the opposite way. We shall give the wind version; but you can re-tell the story for a moving cage, with the same results. Suppose the bird has air-speed 5 ft/sec, the cage is 40 ft square, and the wind blows through at 3 ft/sec. To fly across-stream from side to side and back takes the bird 10 sec + 10 sec,[9] or 20 sec for the round trip. To fly from end to end, upstream and back, takes

$$\frac{40\text{ ft}}{(5-3)\text{ft/sec}}+\frac{40\text{ ft}}{(5+3)\text{ft/sec}}$$

or 20 sec + 5 sec, a much longer time.[10] Put a bird in a cage like this and compare his round-trip times E-W and N-S, and you will be able to tell how fast the cage is moving through the air; or use twin birds and compare their returns. Twist the cage to different orientations, and returns of the twins will tell you which way the cage is travelling through air and how fast. A similar experiment with sound waves in an open laboratory moving through air would tell us the laboratory's velocity. Let a trumpeter stand in one corner and give a toot. The arrivals of returning echoes will reveal general motion, of lab or wind. (Of course, if the moving laboratory

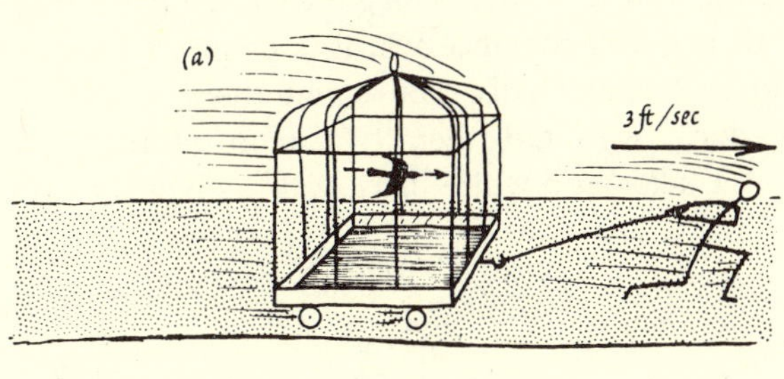

or

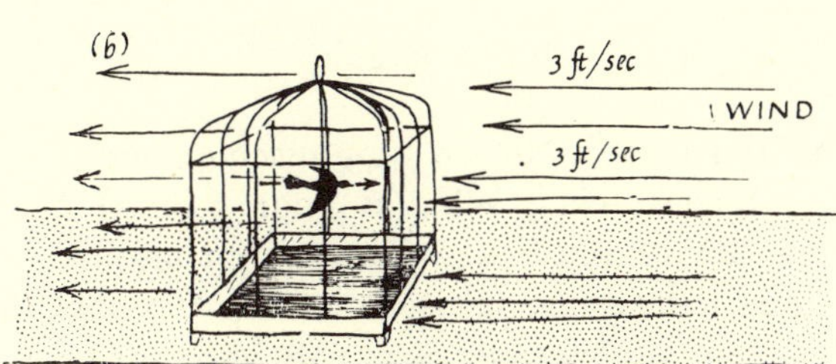

FIG. 31-23.
Cage moving 3 ft/sec through still air has same effect on bird's flight as wind blowing 3 ft/sec through stationary cage.

[9] This requires some geometrical thinking. The bird must fly a 50-ft hypotenuse to cross the 40-ft cage while the wind carries him 30 ft downstream. The simple answer 8 + 8 sec, which is incorrect, is even shorter.

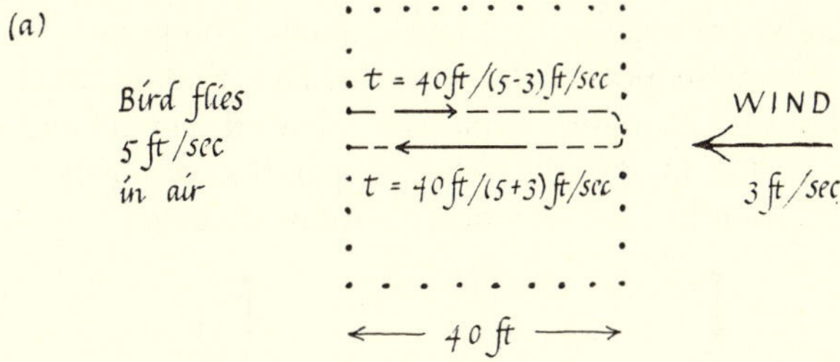

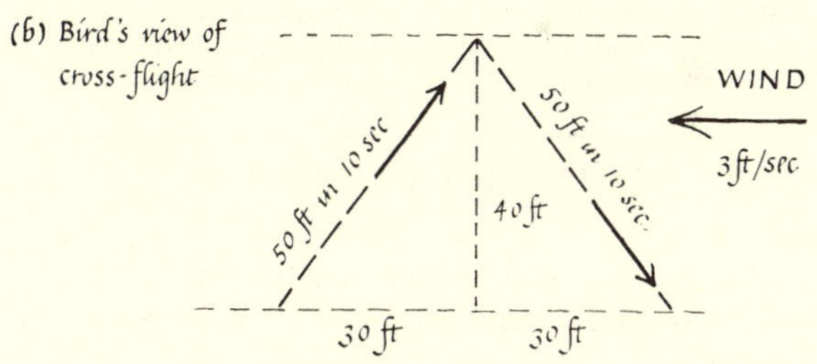

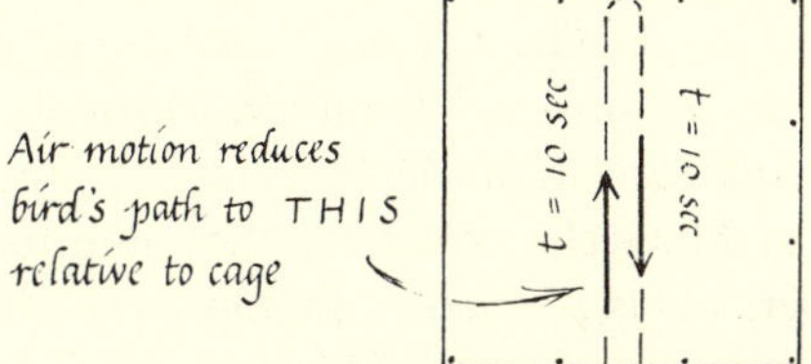

FIG. 31-24. DETAILS OF FLIGHTS
Bird flies 5 ft/sec. Steady wind 3 ft/sec.

[10] If you are still not convinced and feel sure the trips up and downstream should average out, try a thought-experiment with the wind blowing faster, say 6 ft/sec. Then the bird could never make the trip upstream—that time would be infinite!

is closed and carries its air with it, the echoes will show no motion.)

The corresponding test with light-signals is difficult, but the interference pattern affords a very delicate test of trip-timing. When it was tried by Michelson and Morley, and repeated by Miller, it gave a surprising answer: NO MOTION through the "ether." It was repeated in different orientations, at different seasons: always the same answer, NO MOTION. If you are a good scientist you will at once ask, "How big were the error-boxes? How sensitive was the experiment?" The answer: "It would have shown reliably ¼ of the Earth's orbital speed around the Sun, and in later[11] work, 1/10. Yet aberration shows us moving through the "ether" with 10/10 of that speed. Still more experiments added their testimony, some optical, some electrical. Again and again, the same "null result." Here then was a confusing contradiction:

"ABERRATION" OF STARLIGHT	MICHELSON, MORLEY, MILLER EXPERIMENTS
Light from star to telescope showed change of tilt in 6 months.	Light signals compared for perpendicular round trips: pattern showed no change when apparatus was rotated or as seasons changed.

EARTH, MOVING IN ORBIT AROUND SUN, IS MOVING FREELY THROUGH "ETHER"	EARTH IS NOT MOVING THROUGH "ETHER"; *or* EARTH IS CARRYING ETHER WITH IT

CONTRADICTION

Growing electrical theory added confusion, because Maxwell's equations seemed to refer to currents and fields in an absolute, fixed, space (= ether). Unlike Newton's Laws of Motion, they are changed by the Galilean Transformation to a different form in a moving laboratory. However, the modified transformation devised by Lorentz kept the form of Maxwell's equations the same for moving observers. This seemed to fit the facts—in "magnets and coils experiments" (Experiment C in Ch. 41), we get the same effects whether the magnet moves or the coil does. With the Lorentz Transformation, electrical experiments would show *relative* velocity (as they do), but would never reveal uniform *absolute* motion. But then the Lorentz Transformation made mechanics suffer; it twisted $F = Ma$ and $s = v_0 + \frac{1}{2}at^2$ into unfamiliar forms that contradicted Galileo's common-sense relativity and Newton's simple law of motion.

Some modifications of the Michelson-Morley experiment rule out the Fitzgerald contraction as a sufficient "explanation." For example, Kennedy and Thorndike repeated it with *unequal* lengths for the two perpendicular trips. Their null result requires the Lorentz change of time-scale as well as the shrinkage of length.

Pour these pieces of information into a good logic machine. The machine puts out a clear, strong conclusion: "Inconsistent." Here is a very disturbing result. Before studying Einstein's solution of the problem it posed, consider a useful fable.

A Fable

[This is an annoying, untrue, fable to warn you of the difficulty of accepting Relativity. Counting items is an absolute process that no change of viewpoint can alter, so this fable is very distressing to good mathematical physicists with a strong sense of nature—take it with a grain of tranquilizer. You will find, however, that what it alleges so impossibly for adding up balls does occur in relativistic adding of velocities.]

I ask you to watch a magic trick. I take a black cloth bag and convince you it is empty. I then put into it 2 white balls. You count them as they go in—one, two—and then two more—three, four. Now I take out 5 white balls, and the bag is empty.

FIG. 31-26.

Pour this record into the logic machine and it will say, "Inconsistent." What is your solution here? First, "It's an illusion." It is not. You are allowed to repeat the game yourself. (Miller repeated the Michelson-Morley experiment with great precision.) Next, "Let me re-examine the bag for concealed pockets." There are none. Now let us re-state the record. The bag is simple, the balls are solid, the

[11] The latest test (Townes, 1958) made by timing microwaves in a resonant cavity, gave a null result when it would have shown a velocity as small as 1/1000 of the Earth's orbital speed.

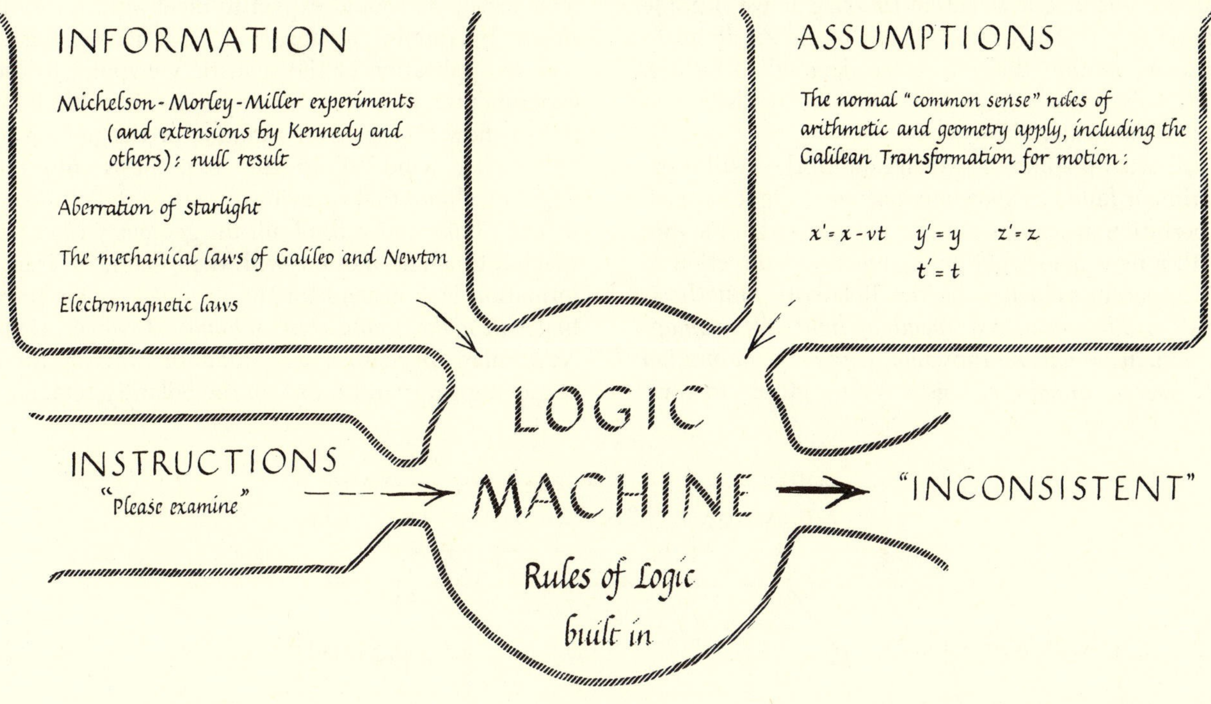

Fig. 31-25.

tally is true: 2 + 2 go in and 5 come out. What can you say now? If you cannot refute tried and true observations, you must either give up science—and go crazy—or attack the rules of logic, including the basic rules of arithmetic. Short of neurotic lunacy, you would have to say, *"In some cases,* 2 + 2 do not make 4." Rather than take neurotic refuge in a catch-phrase such as "It all adds up to anything," you might set yourself to cataloguing events in which 2 and 2 make 4—e.g. adding beans on a table, coins in a purse; and cataloguing events for which 2 + 2 make something else.[12]

[12] There *are* cases where 2 + 2 do not make 4. Vectors 2 + 2 may make anything between 0 and 4. Two quarts of alcohol + two quarts of water mix to make less than 4 quarts. In the circuit sketched, all the resistors, R, are identical but the heating effects do not add up. Two currents each delivering 2 joules/sec add to one delivering 8 joules/sec.

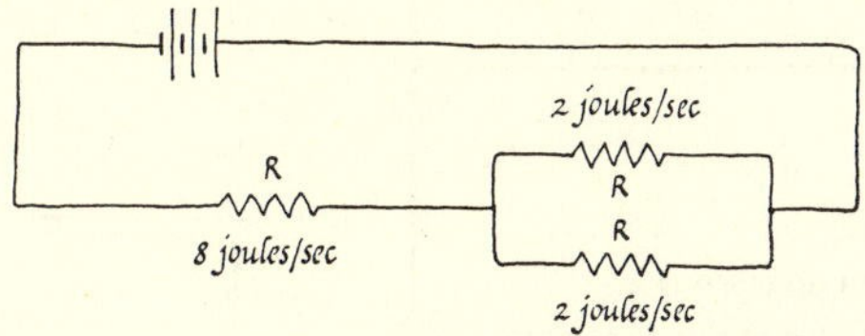

Fig. 31-27.

In studying Nature, scientists have been seeking and *selecting* quantities that do add simply, such as masses of liquids rather than volumes, copper-plating by currents rather than heating. The essence of the "exceptions" is that they are cases where the items to be added *interact*; they do not just act independently so that their effects can be superposed.

In this fable, you have three explanations to choose from:

(a) "It is witchcraft." That way madness lies.
(b) "There is a special invisible mechanism": hardly any better—it turns science into a horde of demons.
(c) "The rules of arithmetic must be modified."

However unpleasant (c) looks, you had better try it—desperate measures for desperate cases. Think carefully what you would do, in this plight.

You are not faced with that arithmetical paradox in real life, but now turn again to motion through space. Ruling out mistaken experimenting, there were similar choices: blame witchcraft, invent special mechanisms, or modify the physical rules of motion. At first, scientists invented mechanisms, such as electrons that squash into ellipsoids when moving; but even these led to more troubles. Poincaré and others prepared to change the rules for measuring time and space. Then Einstein made two brilliant suggestions: an *honest viewpoint,* and a *single hypothesis,* in his Theory of Relativity.

The Relativity *viewpoint* is this: scientific thinking should be built of things that can be observed in real experiments; details and pictures that cannot be observed must not be treated as real; questions about such details are not only unanswerable, they are improper and unscientific. On this view, fixed space (and the "ether" thought to fill it) must be thrown out of our scientific thinking if we become convinced that all experiments to detect it or to measure motion through it are doomed to failure. This viewpoint merely says, "let's be realistic," on a ruthless scale.

All attempts like the Michelson-Morley-Miller experiment failed to show any change of light's speed. Aberration measurements did not show light moving with a new speed, but only gave a new direction to its apparent velocity. So, the Relativity *hypothesis* is this: *The measured speed of light (electromagnetic waves) will be the same, whatever the motion of observer or source.* This is quite contrary to common sense; we should expect to meet light faster or slower by running against it or with it. Yet this is a clear application of the realistic viewpoint to the experimental fact that all experiments with light fail to show the observer's motion or the motion of any "ether wind." Pour this hypothesis into the logic machine that previously answered, "Inconsistent"; but remove the built-in "geometry rules" of space-&-time and motion, with their Galilean Transformation. Ask instead for the (*simplest*) *new rules that will make a consistent scheme*. However, since Newtonian mechanics has stood the test of time, in moving ships and trains, in the Solar System, etc.,

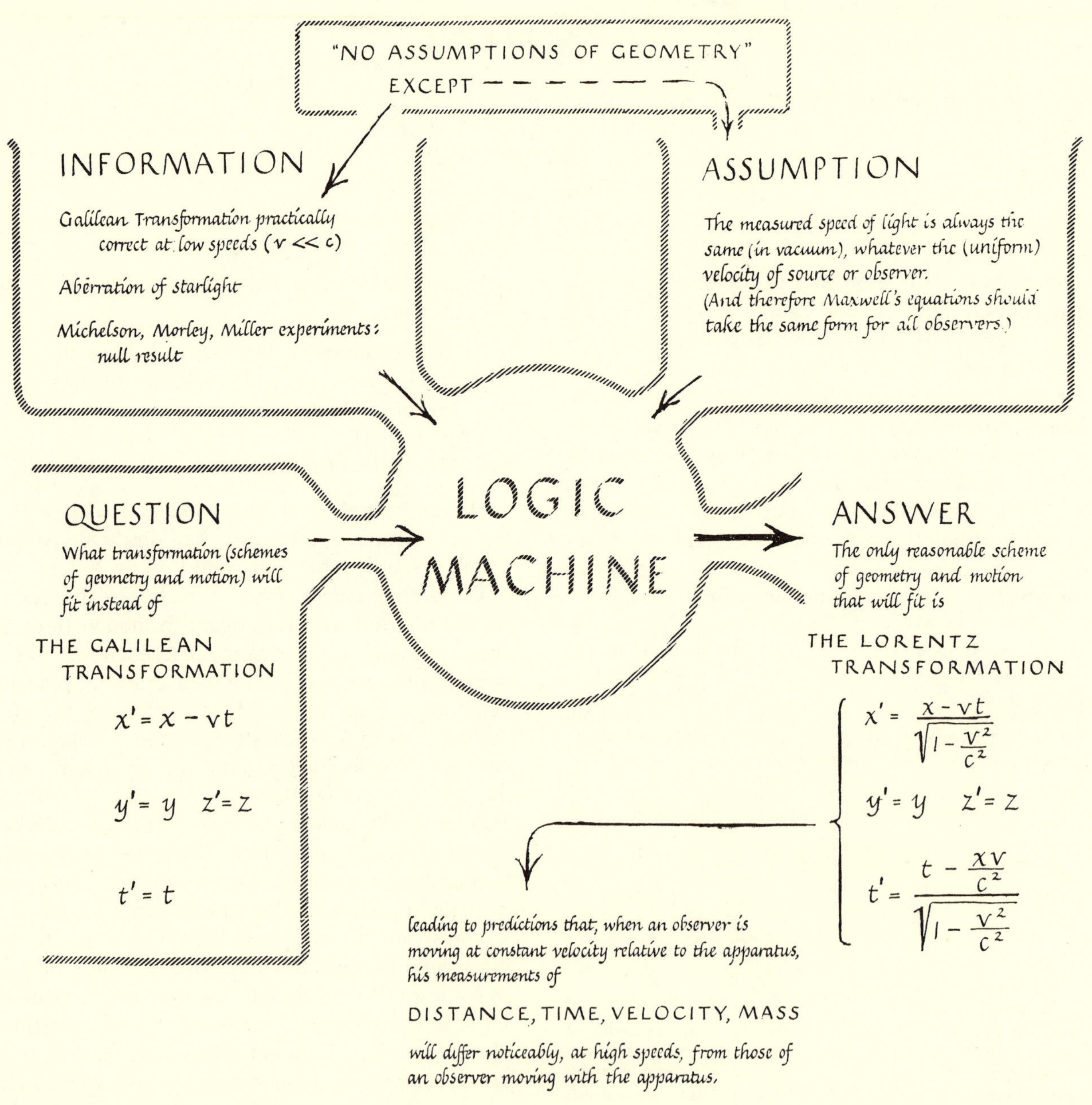

Fig. 31-28.

the new rules must reduce to the Galilean Transformation at low speeds.[13] The logic machine replies: "There is only one reasonable scheme: the transformation suggested by Lorentz and adopted by Einstein."

Instead of the GALILEAN TRANSFORMATION

$$x' = x - vt \qquad y' = y \qquad z' = z \qquad t' = t$$

the LORENTZ-EINSTEIN TRANSFORMATION runs

$$x' = \frac{x - vt}{\sqrt{1 - v^2/c^2}} \quad y' = y \quad z' = z \quad t' = \frac{t - xv/c^2}{\sqrt{1 - v^2/c^2}}$$

and these turn into the reverse transformation, with relative velocity v changing to $-v$

$$x = \frac{x' + vt'}{\sqrt{1 - v^2/c^2}} \quad y = y' \quad z = z' \quad t = \frac{t' + x'v/c^2}{\sqrt{1 - v^2/c^2}}$$

where c is the *speed of light in vacuum.* That speed is involved essentially in the new rules of measurement, because the new transformation was chosen to make all attempts to measure that speed yield the same answer. And the symmetrical form shows that absolute motion is never revealed by experiment. We can measure relative motion of one experimenter past another, but we can never say which is really moving.

Of course the new transformation accounts for the Michelson-Morley-Miller null result—it was chosen to do so. It also accounts for aberration, predicting the same aberration whether the star moves or we do. *But* it modifies Newtonian mechanics. In other words, we have a choice of troubles: the old transformation upsets the form of electromagnetic laws; the new transformation upsets the form of mechanical laws. Over the full range of experiment, high speeds as well as low, the old electromagnetic laws seem to remain good simple descriptions of nature; but the mechanical laws do fail, in their classical form, at high speeds. So we choose the new transformation, and let it modify mechanical laws, and are glad to find that the modified laws describe nature excellently when mechanical experiments are made with improved accuracy.

The new transformation looks unpleasant* because it is more complicated, and its implications are less pleasant. To maintain his Galilean relativity, Newton could assume that length, mass, and time are independent of the observer and of each other. He could assert that mechanical experiments will fail to reveal uniform motion through "space."[14] When Einstein extended the assertion of failure to experiments with light, he found it necessary to have measurements of length and time, and therefore mass, different for observers with different motions. We shall not show the steps of the logic machine grinding out the transformation and its implications, but you may trust them as routine algebra.[15] We shall follow custom and call it the Lorentz Transformation.

Implications of the Lorentz Transformation

Take the new modified geometry that will fit the experimental information, and argue from it how measurements by different observers will compare.

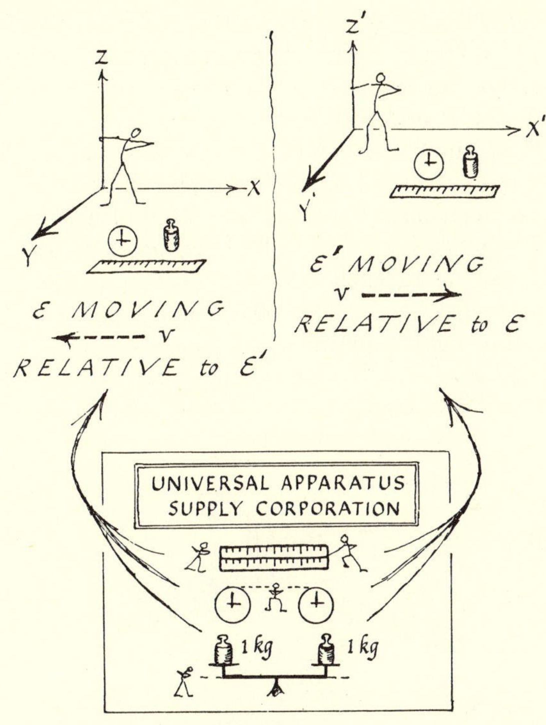

FIG. 31-29
One experimenter is moving with constant velocity relative to the other. They arrange to use standard measuring instruments of identical construction.

Return to our two observers ε and ε′, who operate with identical meter sticks, clocks, and standard kilograms. ε′ and his coordinate framework are moving with speed v relative to ε; and ε is moving backward with speed v relative to ε′. The trans-

[13] This is an application of Bohr's great "Principle of Correspondence": in any extreme case where the new requirement is trivial—here, at low speeds—the new theory must reduce to the old.

*This transformation may seem more reasonable if you see that it represents a rotation of axes in space-&-time. For that, see later in this chapter, page 495.

[14] When an experiment leads us to believe Newton's Laws I and II are valid, it is really just telling us that we are lucky enough to be in a laboratory that is (practically) an inertial frame. If we had always experimented in a tossing ship, we should not have formulated those simple laws.

[15] For details, see standard texts. There is a simple version in *Relativity . . . A Popular Exposition* by A. Einstein (published by Methuen, London, 15th edn., 1955).

formations $\varepsilon \to \varepsilon'$ and $\varepsilon' \to \varepsilon$ are completely symmetrical, and show only the relative velocity v—the same in both cases—with no indication of absolute motion, no hint as to which is "really moving."

The results of arguing from the transformation differ strangely from earlier common sense, but only at exceedingly high speeds. An observer flying past a laboratory in a plane, or rocket, would apply Galilean Transformations safely. He would agree to the ordinary rules of vectors and motion, the Newtonian laws of mechanics.
The speed of light, c, is huge:

$$c = 300{,}000{,}000 \text{ meters/sec} = 186{,}000 \text{ miles/sec}$$
$$\approx \text{a billion ft/sec} \approx 700 \text{ million miles/hour}$$
$$\approx 1 \text{ ft/nonasecond, in the latest terminology.}$$

For relative motion with any ordinary velocity v, the fraction v/c is tiny, v^2/c^2 still smaller. The factor $\sqrt{1 - v^2/c^2}$ is 1 for all practical purposes, and the time-lag xv/c^2 is negligible—so we have the Galilean Transformation.

Now suppose ε' moves at tremendous speed relative to ε. Each in his own local lab will observe the same mechanical laws; and any beam of light passing through both labs will show the same speed, universal c, to each observer. But at speeds like 20,000 miles/sec, 40,000, 60,000 and up towards the speed of light, experimenter ε would see surprising things as ε' and his lab whizz past. ε would say, "The silly fellow ε' is using inaccurate apparatus.

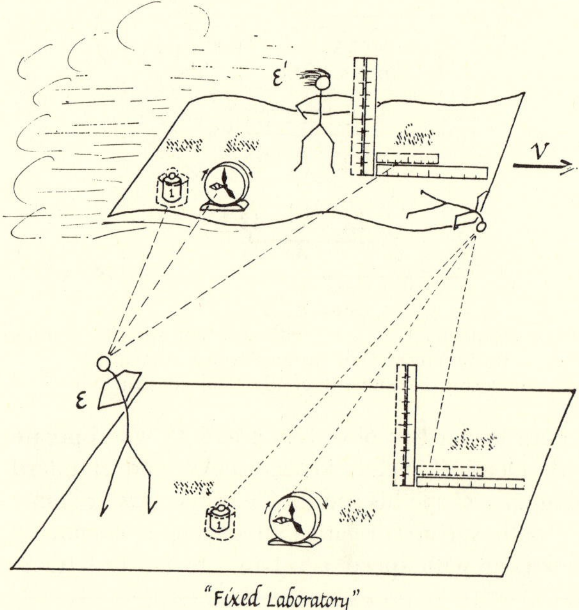

Fig. 31-30.
Each experimenter finds, by using his own standard instruments, that the other experimenter is using incorrect instruments: a shrunken meter stick, a clock that runs too slowly, and a standard mass that is too big.

His meter stick is shrunken—less than my true meter. His clock is running slow—taking more than one of my true seconds for each tick." Meanwhile ε' finds nothing wrong in his own laboratory, but sees ε and his lab moving away backwards, and says, "The silly fellow ε . . . his meter stick is shrunken . . . clock running slow."

Suppose ε measures and checks the apparatus used by ε' just as they are passing. ε finds the meter stick that ε' holds as standard shrunk to $\sqrt{1 - v^2/c^2}$ meter. ε finds the standard clock that ε' holds to tick seconds is ticking longer periods, of $1/\sqrt{1 - v^2/c^2}$ second. And ε finds the 1 kg standard mass that ε' holds is greater, $1/\sqrt{1 - v^2/c^2}$ kg. These are changes that a *"stationary"* observer sees in a *moving* laboratory; but, equally, a moving observer watching a "stationary" laboratory sees the *same* peculiarities: the stationary meter stick shorter, clock running slower, and masses increased. The Lorentz Transformations $\varepsilon' - \varepsilon$ and $\varepsilon - \varepsilon'$ are symmetrical. If ε' and ε compare notes they will quarrel hopelessly, since each imputes the *same* errors to the other! Along the direction of relative motion, each sees all the other's apparatus shrunk, even electrons. Each sees all the other's clocks running slowly, even the vibrations of atoms. (Across the motion, in y- and z-directions, ε and ε' agree.) In this symmetrical "relativity" we see the same thing in the other fellow's laboratory, *whether he is moving or we are.* Only the *relative* motion between us and apparatus matters—we are left without any hint of being able to distinguish *absolute* motion through space.

The shrinkage-factor and the slowing-factor are the same, $1/\sqrt{1 - v^2/c^2}$. This factor is practically 1 for all ordinary values of v, the relative speed between the two observers. Then the transformation reduces to Galilean form where geometry follows our old "common sense." Watch a supersonic 'plane flying away from you 1800 miles/hour ($= \frac{1}{2}$ mile/sec). For that speed, the factor is

$$\frac{1}{\sqrt{1 - \left(\dfrac{\frac{1}{2}\text{ mile/sec}}{186{,}000 \text{ miles/sec}}\right)^2}} \quad \text{or } 1.000000000004$$

The plane's length would seem shrunk, and its clock ticking slower, by less than half a billionth of 1%. At 7,000,000 miles/hour (nearly 1/100 of c) the factor rises to 1.00005. At 70,000,000 miles/hour it is 1.005, making a ½% change in length.

Until this century, scientists never experimented with speeds approaching the speed of light—except for light itself, where the difference is paramount.

c is speed of light, 186,000 miles/sec

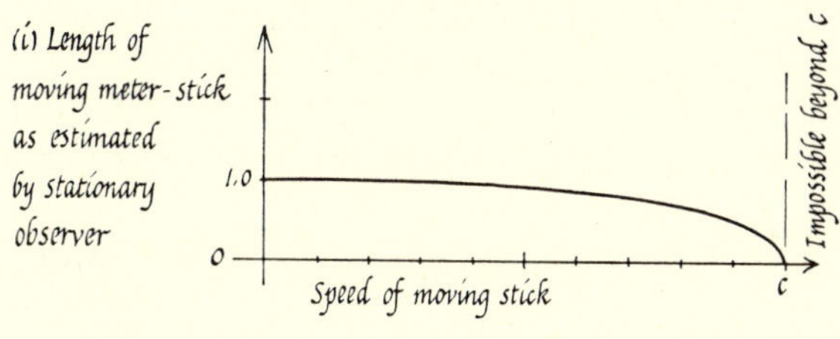

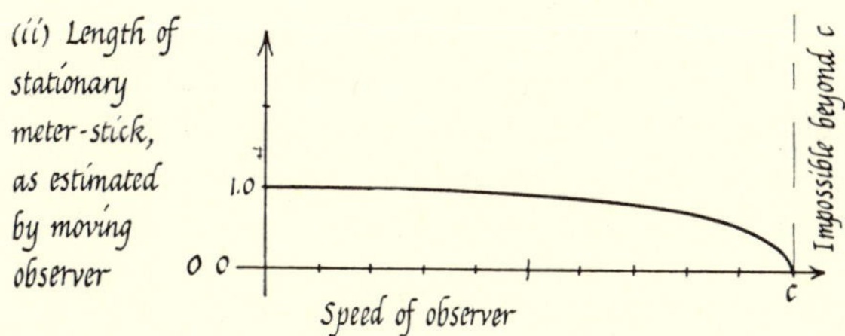

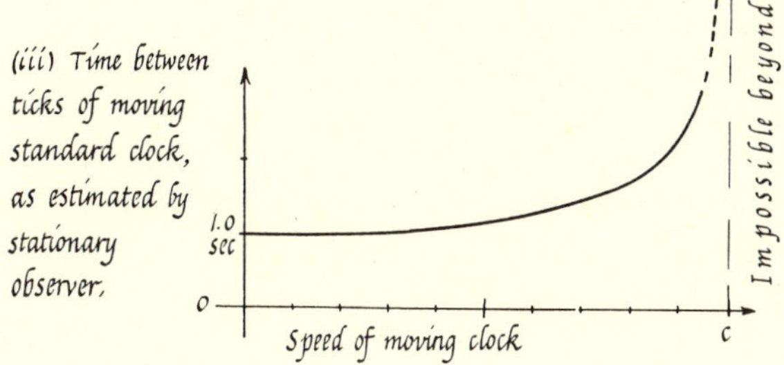

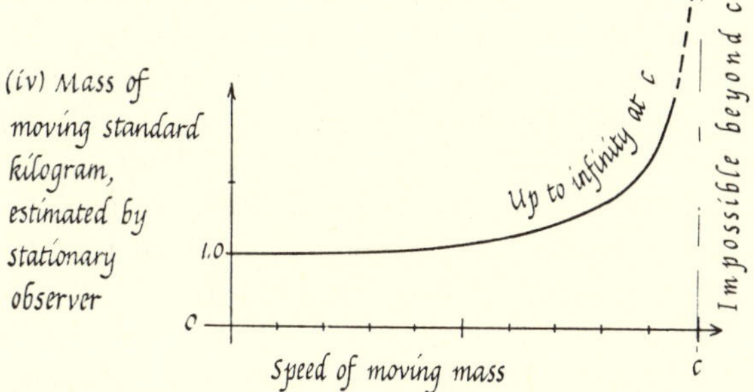

Fig. 31-31.
Changes of Measurement Predicted by Relativity

Nowadays we have protons hurled out from small cyclotrons at 2/10 of c, making the factor 1.02; electrons hitting an X-ray target at 6/10 of c, making the factor 1.2; beta-rays flung from radioactive atoms with 98/100 of c, making the factor 5; and billion-volt electrons from giant accelerators, with .99999988 c, factor 2000.

Among cosmic rays we find some very energetic particles, mu-mesons; some with energy about 1000 million electron · volts moving with 199/200 of the speed of light. For them

$1/\sqrt{1-v^2/c^2} = 1/\sqrt{1-199^2/200^2} = 1/\sqrt{\frac{1}{100}} = 10.$

Now these mesons are known to be unstable, with lifetime about 2×10^{-6} sec (2 microseconds). Yet they are manufactured by collisions high up in the atmosphere and take about 20×10^{-6} seconds on the trip down to us. It seemed puzzling that they could last so long and reach us. Relativity removes the puzzle: we are looking at the flying meson's internal life-time-clock. To us that is slowed by a factor of 10. So the flying meson's lifetime *should seem to us* 20×10^{-6} seconds. Or, from the meson's own point of view, its lifetime is a normal 2 microseconds, but the thickness of our atmosphere, which rushes past it, is foreshortened to 1/10 of *our* estimate—so it can make the shrunk trip in its short lifetime.

Measuring Rods and Clocks

We used to think of a measuring rod such as a meter stick as an unchanging standard, that could be moved about to step off lengths, or pointed in different directions, without any change of length. True, this was an idealized meter stick that would not warp with moisture or expand with some temperature change, but we felt no less confident of its properties. Its length was *invariant.* So was the time between the ticks of a good clock. (If we distrusted pendulum-regulated clocks, we could look forward to completely constant atomic clocks.) Now, Relativity warns us that measuring rods are *not* completely rigid with invariant length. The whole idea of a rigid body—a harmless and useful idealization to 19th-century physicists—now seems misleading. And so does the idea of an absolutely constant stream of time flowing independently of space. Instead, our measurements are affected by our motion, and only *the speed of light, c, is invariant.* A broader view treats c as merely a constant scale-factor for *our* choice of units in a compound space-&-time, which different observers slice differently.

Changes of Mass

If length- and time-measurements change, mass must change too. We shall now find out *how* mass must change, when a moving observer estimates it, by following a thought-experiment along lines suggested by Tolman. We shall assume that *the conservation of momentum holds true* in any (inertial frame) laboratory whatever its speed relative to the observer—we must cling to some of our working rules or we shall land in a confusion of unnecessary changes.

Consider ε and ε' in their labs, moving with relative velocity v in the x-direction. Suppose they make two platinum blocks, each a standard kilogram, that they know are *identical*—they can count the atoms if necessary. Each places a 1-kg block at rest in his lab on a frictionless table. Just as they are passing each other ε and ε′ stretch a long light spiral spring between their blocks, along the y-direction. They let the spring tug for a short while and then remove it, leaving each block with some y-momentum. Then each experimenter measures the y-velocity of his block and calculates its momentum.

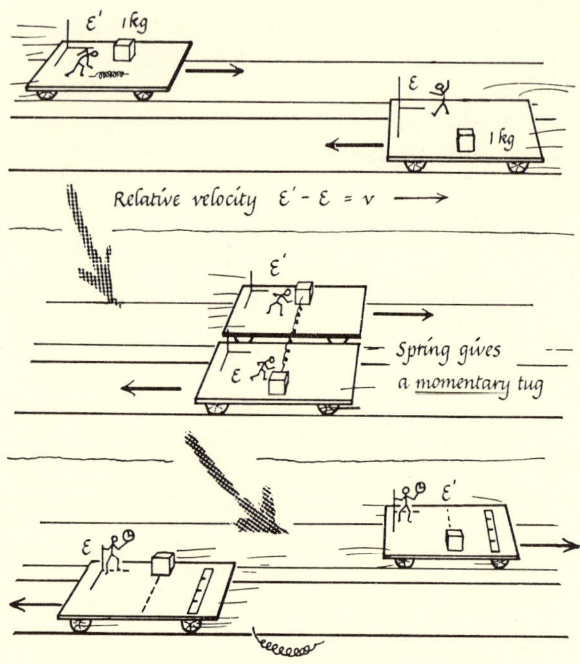

Fig. 31-32. Two Observers Measuring Masses (A thought-experiment to find how mass depends on speed of object relative to observer.) ε says: I have 1 kg, moving across my lab with velocity 3 meters/sec. I know ε′ has 1 kg, and I see that *he* records its velocity as 3 meters/sec; but I know his clock is ticking slowly, so that the velocity of his lump is *less* than 3 meters/sec. Therefore his lump has mass *more* than 1 kg.

They compare notes: each records 3 meters/sec for his block in his own framework. They conclude: equal and opposite velocities; equal and opposite momenta. They are pleased to adopt Newton's Law III as a workable rule. Then ε, watching ε′ at work, sees that ε′ uses a clock that runs slowly (but they agree on normal meter sticks in the y-directions). So ε sees that when ε′ said he measured 3 meters travel in 1 sec, it was "really" 3 meters in *more-than-1-second* as ε would measure it by *his* clock. Therefore ε computes that velocity as *smaller than 3 meters/second* by $\sqrt{1 - v^2/c^2}$. Still believing in Newton III and momentum-conservation, ε concludes that, since his own block acquired momentum 1 kg · 3 meters/sec, the other, which he calculates is moving slower must have greater mass[16]—increased by the factor $1/\sqrt{1 - v^2/c^2}$. While that block is drifting across the table after the spring's tug, ε also sees it whizzing along in the x-direction, table and all, with great speed v. Its owner, ε′, at rest with the table, calls his block 1 kg. But ε, who sees it whizzing past, estimates its mass as greater, by $1/\sqrt{1 - v^2/c^2}$.

This result applies to all moving masses: mass, as we commonly know it, has different values for different observers. Post an observer on a moving body and he will find a standard value, the "rest-mass," identical for every electron, the same for every proton, standard for every pint of water, etc. But an observer moving past the body, or seeing it move past him, will find it has greater mass $m = \dfrac{m_0}{\sqrt{1 - v^2/c^2}}$. Again, the factor $1/\sqrt{1 - v^2/c^2}$ makes practically no difference at ordinary speeds. However, in a cyclotron, accelerated ions increase their mass significantly. They take too long on their wider trips, and arrive late unless special measures are taken. Electrons from billion-volt accelerators are so massive that they practically masquerade as protons.

For example, an electron from a 2-million-volt gun emerges with speed about 294,000,000 meters/sec or 0.98 c. The factor $1/\sqrt{1 - (.98c)^2/c^2}$ is $1/\sqrt{1 - (98/100)^2} \approx 1/\sqrt{4/100} = 5$. To a stationary observer the electron has 5 times its rest-mass.* (Another way of putting this is: that electron's kinetic energy is 2 million electron · volts; the energy associated with an electron's rest-mass is half a million ev, and therefore this electron has K.E. that has mass 4 rest-masses; and that with the original rest-mass makes 5 rest-masses.)

This dependence on speed has been tested by deflecting very fast electrons (beta-rays) with electric and magnetic fields, and the results agree excellently with the prediction. Another test: in a cloud chamber a *very fast electron* hitting a *stationary electron* ("at rest" in some atom of the wet air) does not make the expected 90° fork. In the photograph of Fig. 31-34c, the measured angles

[16] Suppose ε and ε′ are passing each other with relative velocity 112,000 miles/sec. Then ε sees the clock used by ε′ running slow, ticking once every 1.2 seconds. So he knows the block belonging to ε′ has velocity 3 meters/1.2 secs or 2.5 meters/sec. His own block has momentum 1 kg · 3 m./sec. To preserve momentum conservation, he must say that the other block has momentum 1.2 kg. · 2.5 m./sec. So he estimates the mass of the other block as 1.2 kg, a 20% increase.

* To the moving electron, or to a neighbor flying along beside it, its mass is the normal rest-mass; and it is the experimenter rushing towards it who has 5 times *his* normal rest-mass and is squashed to 1/5 his normal thickness.

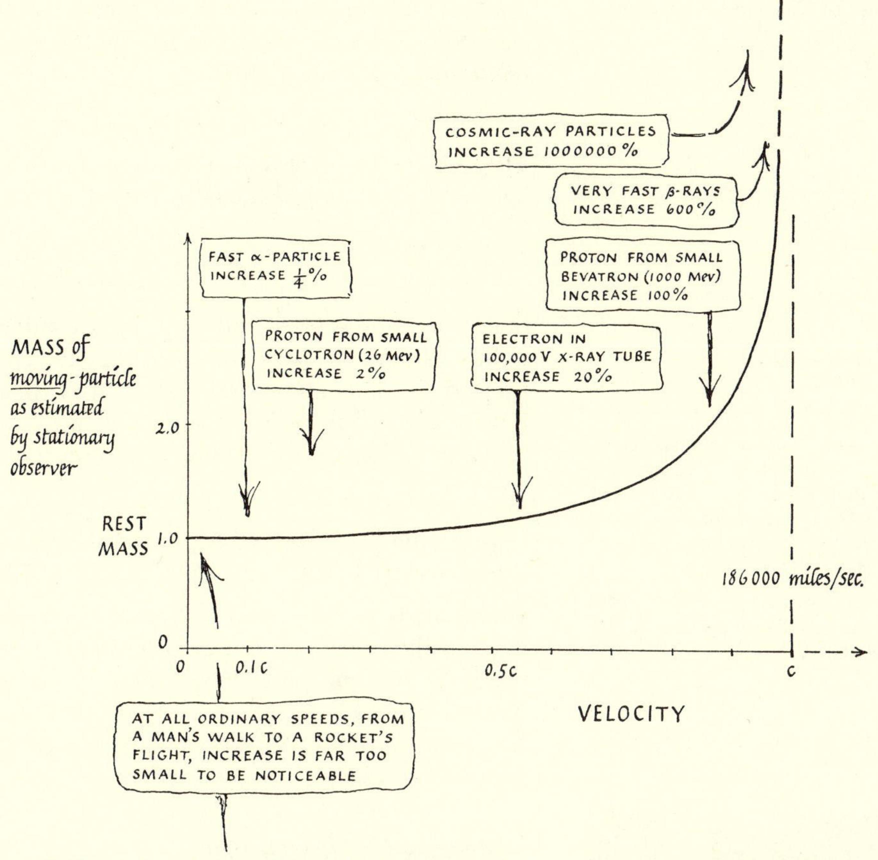

The graph below is the early part of the graph above magnified 1,000,000 times in horizontal scale.

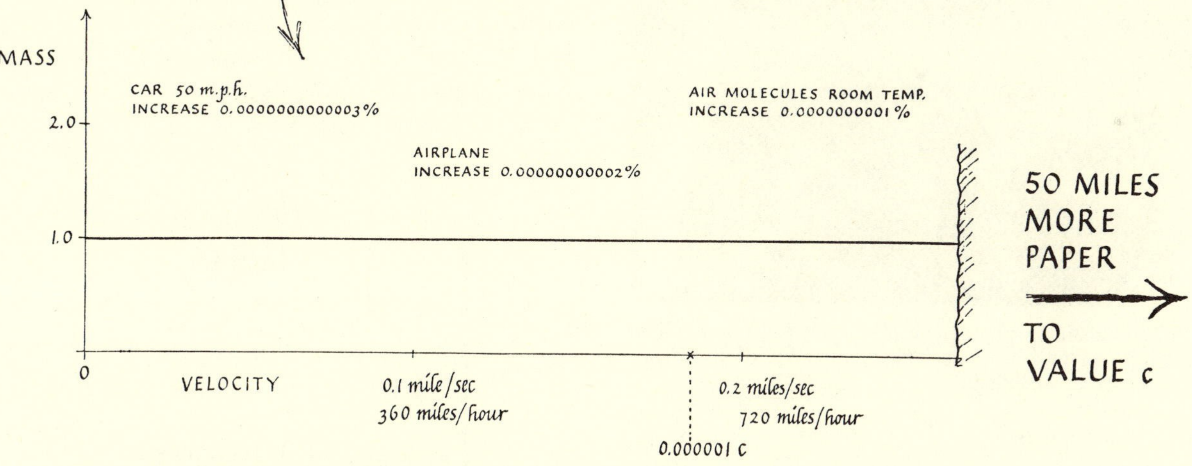

FIG. 31-33. CHANGES OF MASS OF OBJECTS MOVING RELATIVE TO OBSERVER
(The graphs of Fig. 31-31 cover the whole range of speeds from zero to the speed of light; and they may give a mistaken impression of noticeable increase of mass at ordinary speeds. This graph is a copy of the mass-graph there, with comments.)

agree well with those predicted by Relativity for a moving mass $12.7m$ hitting a stationary mass m, in an elastic collision. The tracks are curved because there was a strong magnetic field perpendicular to the picture. Measurements of the curvatures give the momentum of each electron after collision, and the momentum of the bombarding electron before collision. Measurements of the angles shown in the sketch confirm the proportions of these momenta. If non-relativistic mechanics [K.E. $= \frac{1}{2}mv^2$, etc.] is used to calculate the masses, assuming an elastic collision, the projectile's mass appears to be about four times the target particle's mass. Yet the tracks look like those of an electron-

FIG. 31-34. RELATIVISTIC MASS IN ELASTIC COLLISIONS

ELASTIC COLLISIONS

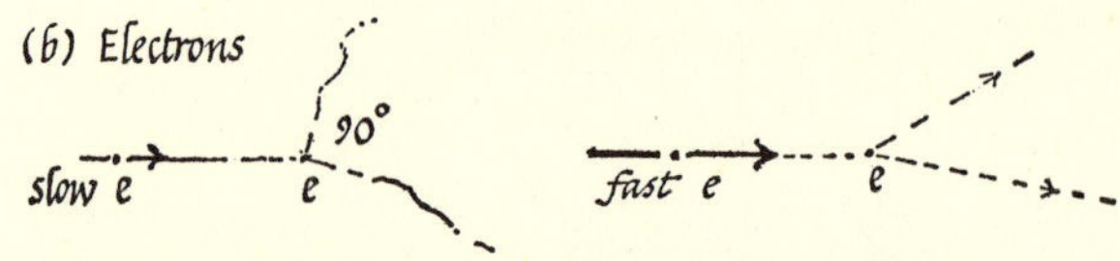

(a) Collision of alpha-particle with stationary atom. Even with its high energy, an alpha-particle from a radioactive atom has a speed that is less than 0.1 c, so its mass is not noticeably increased. It makes the expected 90° fork when it hits a stationary particle (He) of its own mass. With a hydrogen atom as target, it shows its greater mass.

(b) When a slow electron hits a stationary one, the fork shows the expected 90°. When a fast electron hits a stationary one, the angles show that the fast one has much greater mass.

ELECTRONS COLLIDE

(c) Cloud chamber photo

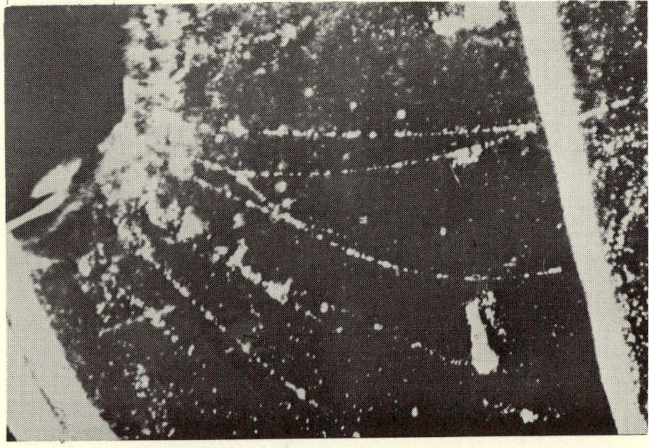

(c) Cloud-chamber photograph of very fast electron colliding with a stationary one. (Photograph by H. R. Crane, University of Michigan.)

(d) Measurements

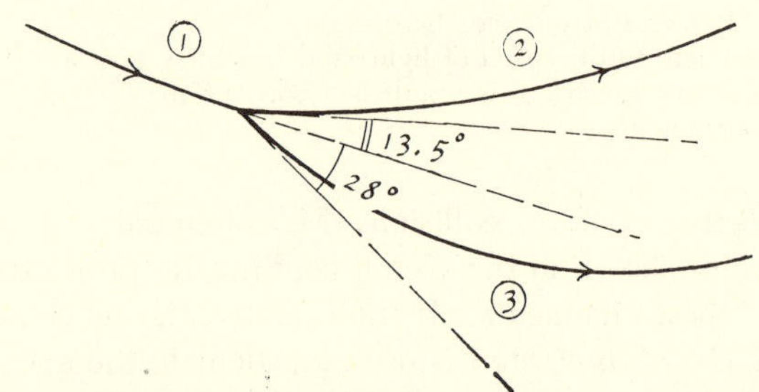

(d) Measurements of original photograph, (c), gave the following radii: (1) 0.15 ± 0.01 meter, (2) 0.105 m., (3) 0.050 m. Magnetic field strength was 1,425,000 (in our units for H in $F = 10^{-7}(Qv)(H)$, discussed in Ch. 37.)

electron collision; and we do not expect $4m$ and m *classically* for two electrons. So we try assuming *relativistic* mechanics [K.E. $= (m - m_0)c^2$, MOMENTUM $= mv$, with $m = m_0/\sqrt{1 - v^2/c^2}$]. Then we find a consistent story: from the magnetic field and our measurements of curvature we find:

BEFORE COLLISION:

projectile had mass $12.7m_0$, speed 0.9969 c;

Since the track is short and only slightly curved, its radius cannot be measured very precisely; so the projectile's momentum, and thence mass, is uncertain within about 6%. We should say

mass $= 12.7\, m_0 \pm 6\%$ or mass $= 12.7\, m_0 \pm 0.8\, m_0$

AFTER COLLISION:

projectile had mass $8.9\, m_0$, speed 0.9936 c;
target particle had mass $4.3\, m_0$, speed 0.9728 c,

where m_0 is the standard rest-mass of an electron and c is the speed of light. Before collision the total mass was $13.7\, m_0$ (including the target); after collision it was $13.2\, m_0$. Mass is conserved in this collision—within the 6% experimental uncertainty—and so is energy, now measured by mc^2.

A Meaning for Mass Change

There is an easy physical interpretation of the change of mass: the extra mass is the mass of the body's kinetic energy. Try some algebra, using the binomial theorem to express the $\sqrt{\ }$ as a series, for fairly low speeds:

$$m = \frac{m_0}{\sqrt{1 - v^2/c^2}}$$

$$= \mathrm{m}_0 \left[1 - \frac{v^2}{c^2}\right]^{-\frac{1}{2}}$$

$$= m_0 \left[1 - (-\tfrac{1}{2})\frac{v^2}{c^2} + \frac{(-\tfrac{1}{2})(-\tfrac{3}{2})}{2}\frac{v^4}{c^4} + \dots\right]$$

$$= \mathrm{m}_0 \left[\begin{array}{l} 1 + \tfrac{1}{2}\frac{v^2}{c^2} + \text{higher powers of } \frac{v^2}{c^2} \\ \text{which are very small at low speeds} \end{array}\right]$$

$= m_0 + \tfrac{1}{2}m_0v^2/c^2 +$ negligible terms at low speeds

$=$ REST-MASS $+$ K.E.$/c^2$

$=$ REST-MASS $+$ MASS OF K.E.

Maximum Speed: c

As a body's speed grows nearer to the speed of light, it becomes *increasingly* harder to accelerate—the mass sweeps up towards infinite mass at the speed of light. Experimenters using "linear accelerators" (which drive electrons straight ahead) find that at high energies their victims approach the speed of light but never exceed it. The electrons gain more energy at each successive push (and therefore more mass) but hardly move any faster (and therefore the accelerating "pushers" can be spaced evenly along the stream—a welcome simplification in design).

Mass growing towards infinity at the speed of light means unaccelerability growing to infinity. Our efforts at making an object move faster seem to run along the level of constant mass, till it reaches very high speeds; then they climb a steeper and steeper mountain towards an insurmountable wall at the speed of light itself. No wonder Relativity predicts that *no piece of matter can move faster than light*; since in attempting to accelerate it to that speed we should encounter more and more mass and thereby obtain less and less response to our accelerating force.

Adding Velocities, Relativistically

Faster than light? Surely that is possible: mount a gun on a rocket that travels with speed $\frac{3}{4}c$ and have the gun fire a bullet forward with muzzle velocity $\frac{1}{2}c$. The bullet's speed should be $\frac{1}{2}c + \frac{3}{4}c$ or $1\frac{1}{4}c$. No: that is a Galilean addition of velocities. We must find the relativistic rule.

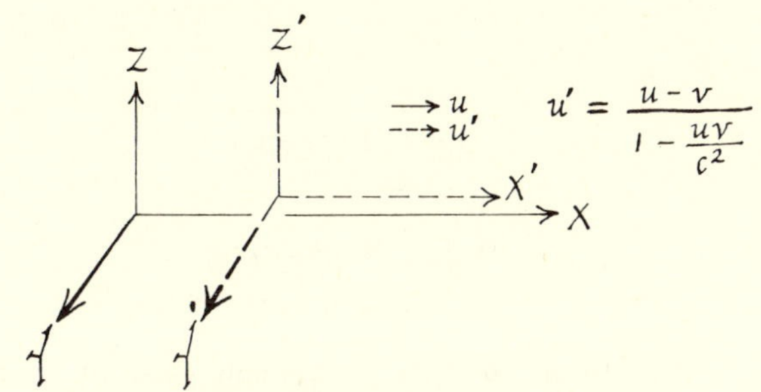

FIG. 31-35a. OBSERVERS MEASURE A VELOCITY
Two experimenters observe the same moving object. How do their estimates of its velocity compare? The Lorentz transformation leads to the relation shown, between u as measured by ε and u' as measured by ε'.

Suppose ε sees an object moving in his laboratory with velocity u, along the x-direction. What speed will ε' measure for the object? As measured by ε, $u = \Delta x/\Delta t$. As measured by ε', $u' = \Delta x'/\Delta t'$ and simple algebra leads from the Lorentz Transformation to

$$u' = \frac{(u - v)}{\left[1 - \frac{uv}{c^2}\right]}$$

instead of the Galilean $u' = (u - v)$. And the inverse relation runs:

$$u = \frac{(u' + v)}{\left[1 + \frac{u'v}{c^2}\right]}$$

The factor in [ ] is practically 1 for all ordinary speeds, and then the relations reduce to Galilean form. Try that on a bullet fired by an ordinary rifle inside an ordinary express train. ε', riding in the train, sees the rifle fire the bullet with speed u'. ε, sitting at the side of the track, sees the bullet move with speed u. He sees the train passing him with speed v. Then $u = (u' + v)/[1]$. The Galilean version fits closely:

SPEED OF BULLET RELATIVE TO GROUND

$= \frac{\text{SPEED OF BULLET}}{\text{RELATIVE TO TRAIN}} + \frac{\text{SPEED OF TRAIN}}{\text{RELATIVE TO GROUND}}$

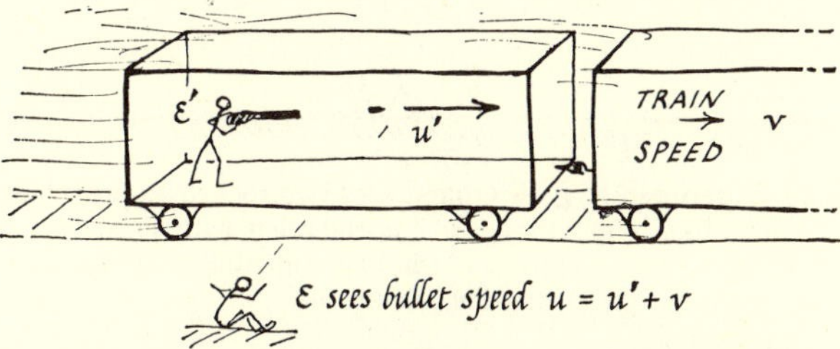

FIG. 31-35b. ADDING VELOCITIES AT ORDINARY SPEEDS
Two experimenters observe the same bullet, shot from a gun in a moving train. With such speeds, the Lorentz transformation leads to the simple Galilean relations: $u' = u - v$ and $u = u' + v$.

Now return to the gun on a $\frac{3}{4}c$ rocket firing a $\frac{1}{2}c$ bullet forward. ε' rides on the rocket and sees the bullet emerge with $u' = \frac{1}{2}c$. ε on the ground sees ε' and his rocket moving with speed $\frac{3}{4}c$; and ε learns from ε' how fast the gun fired the bullet. Then, using the relativity-formula above, ε predicts the bullet-speed that *he* will observe, thus:

FIG. 31-36. ADDING VELOCITIES AT VERY HIGH SPEEDS

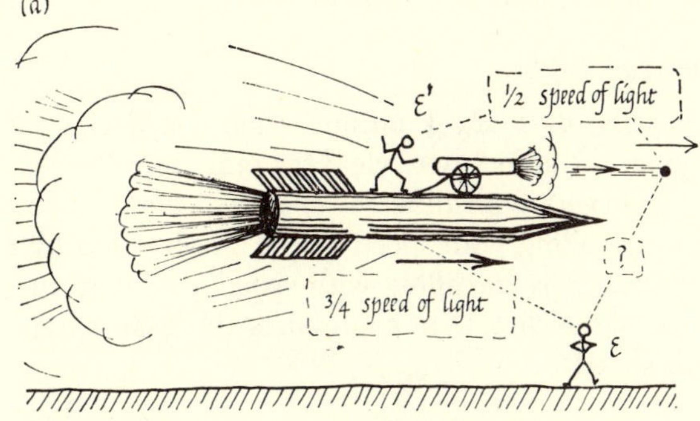

(a) Experimenter ε on ground observes a rocket moving at $\frac{3}{4}c$. Experimenter ε' riding on the rocket fires a bullet at $\frac{1}{2}$ c relative to the rocket. What will be the speed of the bullet, as measured by ε on the ground?

$$u = \frac{u' + v}{1 + u'v/c^2} = \frac{\tfrac{1}{2}c + \tfrac{3}{4}c}{1 + \tfrac{1}{2}c \cdot \tfrac{3}{4}c/c^2} = \frac{1\tfrac{1}{4}c}{1 + \tfrac{3}{8}}$$

$$= \frac{(\tfrac{5}{4})c}{(\tfrac{11}{8})} = \frac{10}{11}c, \text{ still just less than } c.$$

SPEED OF BULLET RELATIVE TO GROUND

$$= \frac{\text{SPEED OF GUN RELATIVE TO GROUND} + \text{SPEED OF BULLET RELATIVE TO GUN}}{1 + \dfrac{\text{SPEED OF BULLET}}{\text{SPEED OF LIGHT}} \cdot \dfrac{\text{SPEED OF GUN}}{\text{SPEED OF LIGHT}}}$$

Have another try at defeating the limit of velocity, c. Run two rockets head on at each other, with speeds $\tfrac{3}{4}c$ and $\tfrac{1}{2}c$. ε on the ground sees ε′ riding on

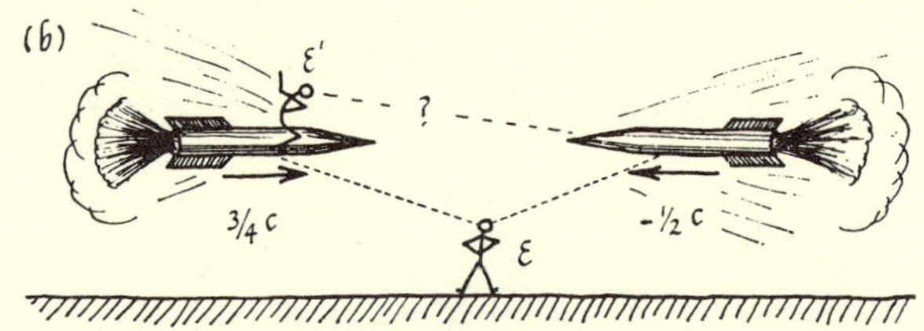

(b) Experimenter ε on ground sees two rockets approaching each other, one with speed $\tfrac{3}{4}c$, the other with speed $\tfrac{1}{2}c$. What speed of approach will experimenter ε′ riding on the first rocket see?

one rocket with velocity $v = \tfrac{3}{4}c$ and the other rocket travelling with $u = -\tfrac{1}{2}c$; and he thinks they must be approaching each other with relative velocity $1\tfrac{1}{4}c$. ε′, riding on the first rocket, sees the second rocket moving with predicted speed

$$u' = \frac{u - v}{1 - uv/c^2} = \frac{(-\tfrac{1}{2}c) - (\tfrac{3}{4}c)}{1 - (-\tfrac{1}{2}c)(\tfrac{3}{4}c)/c^2}$$

$$= \frac{-1\tfrac{1}{4}c}{1 + \tfrac{3}{8}} = -\frac{10}{11}c$$

Their rate of approach is less than c. Whatever we do, we cannot make a material object move faster than light—as seen by any observer.

Speed of Light

Finally, as a check on our velocity-addition formula, make sure it does yield the same speed of light for observers with different speeds. Take a flash of light travelling with speed $u = c$, as observed by ε. Observer ε′ is travelling with speed v relative to ε, in the same direction. ε′ observes the flash moving with speed

$$u' = \frac{u - v}{1 - uv/c^2} = \frac{c - v}{1 - cv/c^2} = \frac{c(1 - v/c)}{(1 - v/c)} = c$$

Every observer measures the same speed c for light.

Fig. 31-37.
Two experimenters measure the speed of the same sample of light. Experimenter ε sees that ε′ is running with velocity v in the direction the light is travelling.

(No wonder, since the Lorentz Transformation was chosen to produce this.) This certainly accounts for the Michelson-Morley-Miller null results.

Energy

We rebuild the Newtonian view of energy to fit Relativity as follows. Define MOMENTUM as mv, where m is the observed mass of the body in motion: $m = m_0/\sqrt{1 - v^2/c^2}$. Define force, F, as $\Delta(mv)/\Delta t$. Define change from potential energy to K.E. as WORK, $F \cdot \Delta s$. Combine these to calculate the K.E. of a mass m moving with speed v. We shall give the result, omitting the calculus derivation.

$$m = \frac{m_0}{\sqrt{1 - v^2/c^2}} \quad \text{[part of Lorentz Transformation]}$$

$$F = \frac{\Delta(mv)}{\Delta t} \quad \text{[Newton Law II Relativity form]}$$

$$\left.\begin{aligned} \Delta(\text{K.E.}) &= F \cdot \Delta s \\ &= F \cdot v \cdot \Delta t \\ \text{K.E.} &= 0 \text{ if } v = 0 \end{aligned}\right\} \quad \text{[Definition of K.E.]}$$

↓

CALCULUS

↓

$$\text{K.E.} = mc^2 - m_0c^2 = (m - m_0)c^2$$

We assign the body a permanent store of "rest-energy" m_0c^2—locked up in its atomic force-fields, perhaps. We add that to the K.E.; then the *total* energy, E, of the body is $m_0c^2 + (mc^2 - m_0c^2) = mc^2$. Therefore total $E = mc^2$. This applies whatever its speed—but remember that m itself changes with speed. At low speeds, mc^2 reduces[17] to

$$(\text{rest-energy } m_0c^2) + (\text{K.E. } \tfrac{1}{2}mv^2).$$

For a short, direct derivation of $E = mc^2$, see the note at the bottom of the next page.

This view that energy and mass go together according to $E = mc^2$ has been given many successful tests in nuclear physics. Again and again we find some mass of material particles disappears in a

[17] See the discussion above, with the binomial theorem.

nuclear break-up; but then we find a release of energy—radiation in some cases, K.E. of flying fragments in others—and that energy carries the missing mass.

The expression for mass, $m = m_0/\sqrt{1 - v^2/c^2}$ follows from the Lorentz Transformation and conservation of momentum. So $E = mc^2$ follows from Newton's Laws II and III combined with the Lorentz Transformation.

Then if an observer assigns to a moving body a mass m, momentum mv, and total energy mc^2 he finds that, in any closed system, *mass is conserved, momentum is conserved* (as a vector sum), and *energy is conserved.* In all this he must use the observed mass m, which is $m_0/\sqrt{1 - v^2/c^2}$ for any body moving with speed v relative to him. Then he is doubling up his claim of conservation because, if the sum of all the masses $(m_1 + m_2 + \ldots)$, is constant, the total energy $(m_1c^2 + m_2c^2 + \ldots)$ must also be constant. If energy is conserved, mass must also be conserved. One rule will cover both. That is why some scientists say rather carelessly, "mass and energy are the same, but for a factor c^2." In fact, since c^2 is universally constant, there is little harm in saying that mass and energy are the same thing, though commonly measured in different units. But there is also little harm if you prefer to think of them still with quite different flavors as physical concepts. And a very important distinction remains between *matter* and *radiation* (and other forms of energy). Matter comes in particles, *whose total number remains constant* if we count the production or destruction of a [particle + anti-particle] pair as no change. Radiation comes in photons; and the total number of photons *does* change when one is emitted or absorbed by matter.

Covariance

Finally, Einstein treated momentum as a vector with three components in space-&-time, and kinetic energy with them as a fourth, time-like, component of a "supervector." Thus, conservation rules for mass, momentum, and energy can be rolled into one great formula in relativistic mechanics. The Lorentz Transformation gives this formula *the same form* with respect to any (steadily moving) set of axes whatever their velocity. We say such a formula or relation is "covariant." We put great store by covariance: covariant laws have the most general form possible and we feel they are the most perfect mathematical statement of natural laws. "We lose a frame of reference, but we gain a universally valid symbolic form."[18]

"A Wrong Question"

The physical laws of mechanics and electromagnetism are covariant: they give no hope of telling how fast we move through absolute space. This brings us back to Einstein's basic principle of being realistic. Where the answer is "impossible," the question is a foolish one. We are unscientific to imply there *is* an absolute space, as we do when we ask "How fast . . . *through space*?" We are begging the question, inside our own question, by mentioning space. We are asking a wrong question,

[18] Frederic Keffer.

NOTE: Derivation of $E = mc^2$

This short derivation, due to Einstein, uses the experimental knowledge that when radiation with energy E joules is absorbed by matter, it delivers momentum E/c kg·m./sec. (Experiment shows that PRESSURE of radiation on an *absorbing* wall is ENERGY-PER-UNIT-VOLUME of radiation-beam. Suppose a beam of area A falls on an absorbing surface head-on. In time Δt, a length of beam $c \cdot \Delta t$ arrives. Then MOMENTUM delivered in Δt

$= \text{FORCE} \cdot \Delta t \qquad = \text{PRESSURE} \cdot \text{AREA} \cdot \Delta t$
$= (\text{ENERGY/VOLUME}) \cdot \text{AREA} \cdot \Delta t$
$= (\text{ENERGY}/A \cdot c \cdot \Delta t) \cdot A \cdot \Delta t$
$= \text{ENERGY}/\ c$

This also follows from Maxwell's equations).

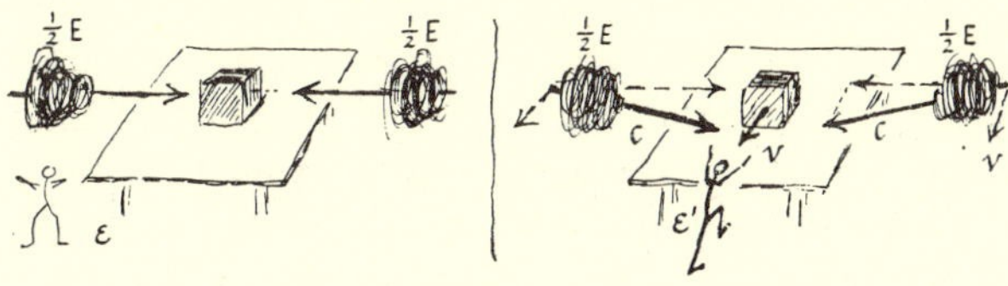

We take two views of the same thought-experiment: (A) Place a block of matter at rest on a frictionless table. Give it some energy E by firing two chunks of radiation at it, ½E from due East, ½E from due West. The block absorbs the radiation and gains energy E; but its net gain of momentum is zero: it stays at rest. (B) Now let a running observer watch the same event. He runs with speed v due North; but according to Relativity he can equally well think he is at rest and see the table, etc. moving towards him with speed v due South. Then he sees the block moving South with momentum Mv. He sees the two chunks of radiation moving towards the block, each with speed c but in directions, slanted southward with slope v/c. (This is like the *aberration of starlight.*) In his view, each chunk has momentum $(\frac{1}{2}E/c)$ with a southward component $(\frac{1}{2}E/c)(v/c)$. Thinking himself at rest, he sees total southward momentum $Mv + 2(\frac{1}{2}E/c)(v/c)$. After the block has absorbed the radiation, he still sees it moving South with the same speed v—since in version (A) we saw that the block gained no net momentum. However, the block may gain some mass, say m. Find out how big m is by trusting *conservation of momentum*:

$$Mv + 2(\tfrac{1}{2}E/c)(v/c) = (M + m)v$$
$$\therefore \quad m = E/c^2 \quad \text{or} \quad E = mc^2,$$

where m is the MASS gained when ENERGY E is gained.

like the lawyer who says, "Answer me 'yes' or 'no.' Have you stopped beating your wife?" The answer to *that* is, "A reasonable man does not answer unreasonable questions." And Einstein might suggest that a reasonable scientist does not *ask* unreasonable questions.

Simultaneity

The observers ε and ε′ do not merely see each other's clocks running slowly; worse still, clocks at different *distances* seem to disagree. Suppose each observer posts a series of clocks along the x-direction in his laboratory and sets them all going together. And when ε and ε′ pass each other at the origin, they set their central clocks in agreement. Then each will blame the other, saying: "*His* clocks are not even synchronized. He has set his distant clocks wrong by his own central clock—the greater the distance, the worse his mistake. The farther I look down his corridor, along the direction he is moving, the more he has set his clocks there back—they read early, behind my proper time. And looking back along his corridor, opposite to the direction of his motion, I see his clocks set more and more forward, to read later than my correct time." (That judgment, which each makes of the other's clocks, is not the result of forgetting the time-delay of seeing a clock that is far away. Each observer allows for such delays—or reads one of his own clocks that is close beside the other's—and then finds the disagreement. This disagreement about setting of remote clocks belongs with the view that each observer takes of clock *rates*. Each claims that all the other's clocks are running too slowly. Each says, "His center clock runs more slowly than mine. It agrees with mine *now*. But, *some while ago*, when it was out to one side, it must have read later than mine—so that mine could catch up by now. *Later on*, out to the other side, it will read earlier than mine—as its hands turn too slowly.")

ε observes his own row of clocks ticking simultaneously all in agreement. But ε′ does not find those ticks simultaneous. *Events that are simultaneous for* ε *are not simultaneous for* ε′. This is a serious change from our common-sense view of universal time; but it is a part of the Lorentz Transformation. In fact, the question of simultaneity played an essential role in the development of relativity by Poincaré and Einstein. Arguing with thought-experiments that keep "c" constant, you can show this change is necessary. The following example illustrates this.

Suppose ε and ε′ have their laboratories in two transparent railroad coaches on parallel tracks, one moving with speed v relative to the other. Just

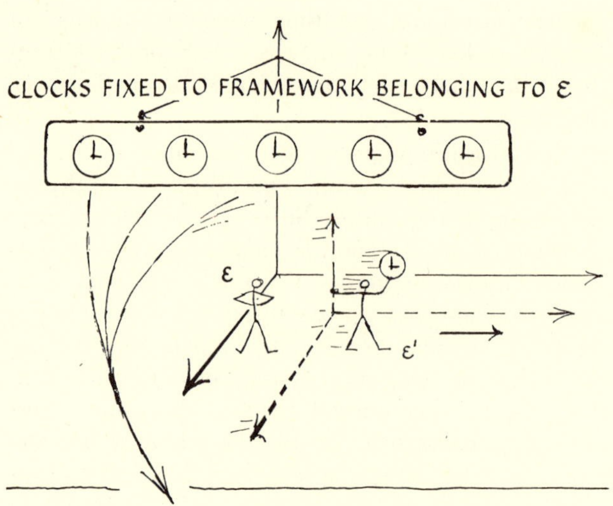

SAME CLOCKS AS REPORTED BY ε′

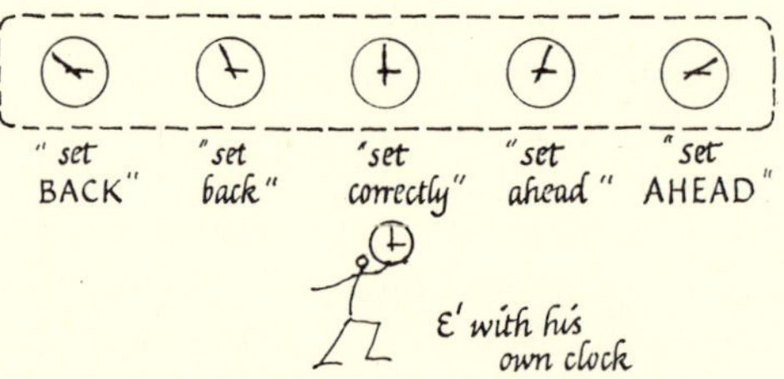

FIG. 31-38. "SIMULTANEOUS" CLOCK SETTINGS
Each experimenter sets his own clocks *all in agreement* (allowing carefully for the time taken by any light signals he uses in looking at them). Each experimenter finds that the other man's clocks *disagree among themselves*, progressively with distance. (That is, after he has allowed carefully for the time taken by the light signals he uses in checking the other man's clocks against his own.) The sketch shows a series of clocks all fixed in the framework belonging to ε. As adjusted and observed by ε, they all agree: they are synchronized. As investigated by ε′ those clocks disagree with each other. The lower sketch shows what ε′ finds by comparing those clocks simultaneously (as he, ε′, thinks) with his own clock. The two sketches of clocks disagree because each experimenter thinks *he* compares them all simultaneously but disagrees with the other man's idea of simultaneity.

as the coaches are passing, ε and ε′ lean out of their center windows and shake hands. They happen to be electrically charged, + and —, so there is a flash of light as they touch. Now consider the light from this flash. Some of it travels in each coach starting from the mid-point where the experimenter is standing. ε finds it reaches the front and hind ends of his coach simultaneously. And ε′ finds it reaches the ends of *his* coach simultaneously. Each considers he is in a stationary coach with light travelling out from the center with constant speed c. But ε can also observe the light flash reaching the ends of the other coach that carries ε′. He observes the events that ε′ observes; but he certainly does not find them simultaneous, as ε′ claims. By the time the flash has travelled a half-length of the ε′ coach, that coach has moved forward past ε. As ε sees it, the light travels farther to reach the front end of that moving coach, and less to the hind end. So ε sees the flash hit the hind end first, while ε′ claims the hits are simultaneous.[19] (Reciprocally, ε′ sees the light reach the ends of the coach carrying ε at different instants, while ε claims they are simultaneous.) You will meet no such confusion in ordinary life, because such disagreements over priority arise only when the events are very close in time, or very far apart in distance. Where events P and Q are closer in time than the travel-time for light between them, observers with different motions may take different views: one may find P and Q simultaneous, while another finds P occurs before Q, and still another finds P later than Q. To maintain Einstein's Relativity, we must regard time as interlocked with space in a compound space-time, whose slicing into separate time and space depends somewhat on the observer's motion. If we accept this compound space-time system, we must modify our philosophy of cause and effect.

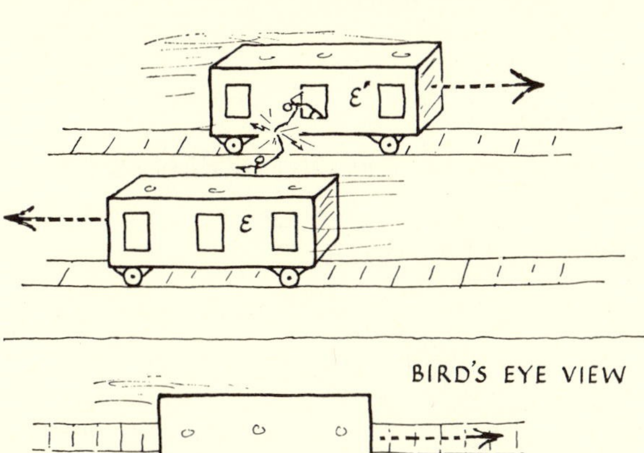

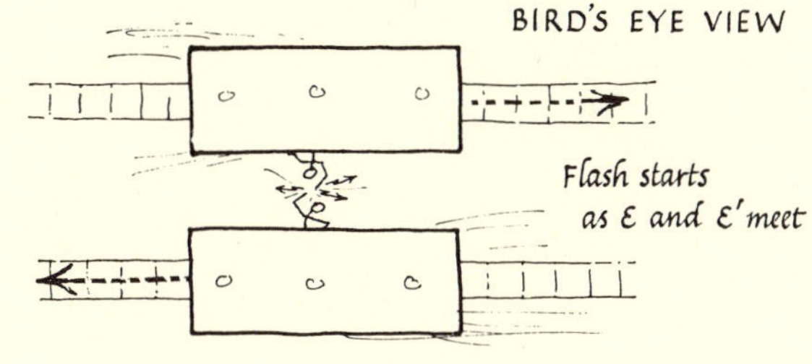

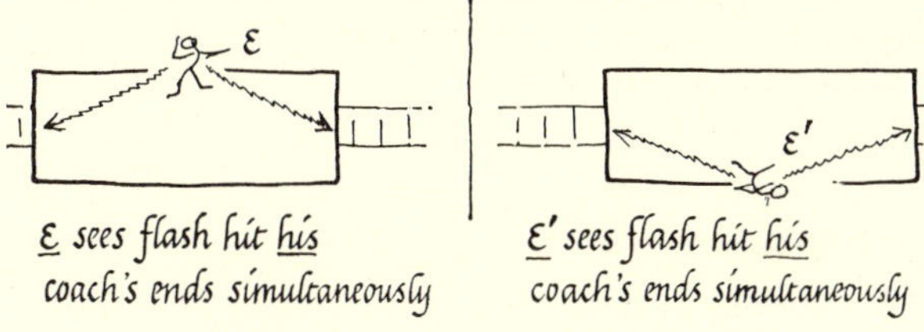

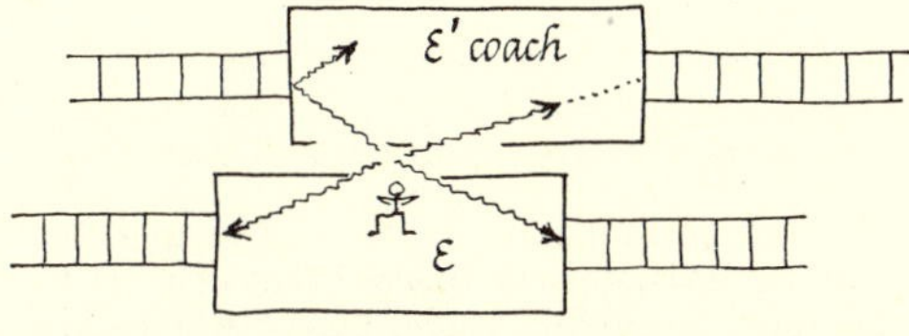

ε sees flash hit both ends of his coach simultaneously, but the ends of ε′ coach at different times. (Similarly for ε′)

FIG. 31-39. THOUGHT-EXPERIMENT
To show that events that are simultaneous for one observer are not simultaneous for an observer moving with a different velocity.

Cause and Effect

Earlier science was much concerned with causality. Greeks looked for "first causes"; later scientists looked for immediate causes—"the heating caused the rock to melt"; "the pressure caused the liquid to flow"; "the alpha-particle caused the ions to be formed." It is difficult to define cause and effect. "P causes Q": what does that mean? The best we can say is that cause is something that *precedes* the effect so consistently that we think there is a connection between them.

Even in common cases (like STRESS and STRAIN or P.D. and CURRENT), we prefer to say P and Q go together: we still look for *relationships* to codify our knowledge, but we treat P and Q as cousins rather than as parent and child.

And now Relativity tells us that *some* events can show a different order in time for different observers—and all observers are equally "right." The sketches of Fig. 31-40(e), below, show how various observers *at* an event P, *here-now*, must classify some other events (e.g., Q_1) as in the *absolute future*; some other events (e.g., Q_2) in the *absolute past*; and some events (e.g., Q_3) in the *absolute elsewhere* (as Eddington named it) where observers with different motions at P may disagree over the order of events P and Q.

[19] Note that the disagreement over simultaneity is not due to forgetting the time taken by light signals to bring the information to either observer. We treat the problem as if each observer had a whole gang of perfectly trained clockwatchers ranged along his coach to make observations without signal delays and then report at leisure. The observers compare notes (e.g. by radio). Then each has an obvious explanation of the other man's claim that he saw the light flash reach the ends of his own coach simultaneously: "Why, the silly fellow has set his clocks askew. He has a clock at each end of his coach, and when the light flash hit those end clocks they both showed the same instant of time—I saw that, too. But he is wrong in saying his end clocks are set in agreement: I can see that he has set his front-end clock back by my standard, and his hind-end clock ahead. *I* can see that the flash had to travel farther to reach his front end. And *my* clocks tell me it arrived there later, as *I* know it should. But since his clock is mis-set, early by mine, the lateness of arrival did not show on it. Those mistakes of his in setting his clocks just cover up the difference of transit-time for what I can see are different travel-distances to the ends of his coach." As in all such relativistic comparisons, *each observer blames the other for making exactly the same kind of mistake.*

PAIRS OF EVENTS

ON A TIME AND DISTANCE MAP

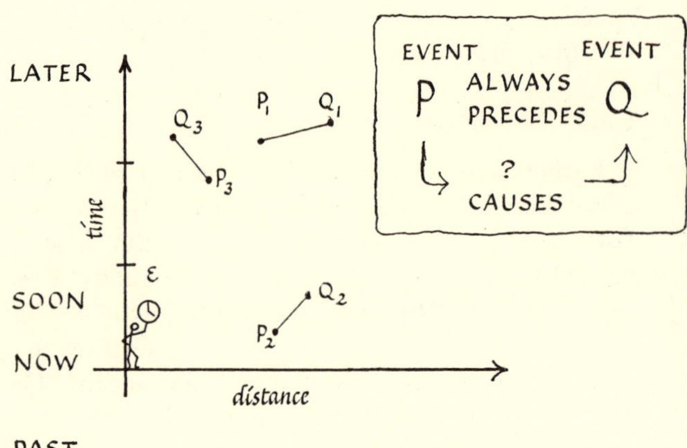

GALILEAN TIME AND DISTANCE MAP

for OBSERVER $\mathcal{E}$ *and* MOVING OBSERVER $\mathcal{E}'$

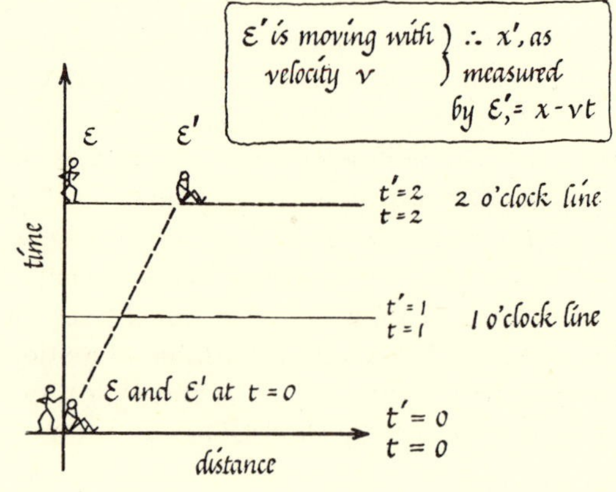

TWO OBSERVERS MOVING VERY FAST RELATIVE TO EACH OTHER RECORD EVENTS P & Q

IN GALILEAN WORLD

IN LORENTZ WORLD

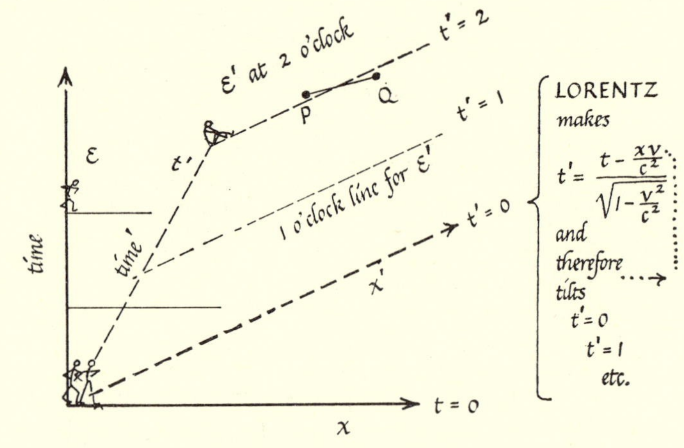

Fig. 31-40. Charts of Space (One Dimension) and Time

(These fanciful sketches are highly restricted: all the events shown occur in one straight line, in a one-dimensional space, along an x-axis.

In the Lorentz picture, a very high relative velocity between ε and ε' is assumed. The distortion of the x' and t' system in the Lorentz picture shows the view taken by ε. Of course, ε' himself would take an "undistorted" view of his own system, but ε' would find the x and t system "distorted."

It is not possible to show the essential symmetry here; so the Lorentz picture should only be taken as a suggestion: taken literally, it would be misleading.)

(a) An event that occurs on the straight line (x-axis) is shown by a point on this chart. Distance *along* shows *where* the event occurs on the line. Distance *up* shows *when* it occurs. Event P precedes event Q in time. It may be sensible to say that P causes Q, for some types of event.

(b) A moving experimenter carries his origin for distance with him. On the Galilean system he uses the same time-scale as a stationary experimenter.

(c) With a Galilean transformation between two experimenters, the lines for each hour by the clock are the same for both observers, and parallel to the axis, $t = 0$.

(d) The Lorentz transformation between two experimenters *tilts* one coordinate system of space-&-time relative to the other (through a negligible angle, except when speed of ε' relative to ε approaches c).

Then an event Q that follows event P in time for one experimenter may precede P for another—but only if the events are so far apart that a light signal from one event could not travel to the place of the other event and reach it before the other event occurred there.

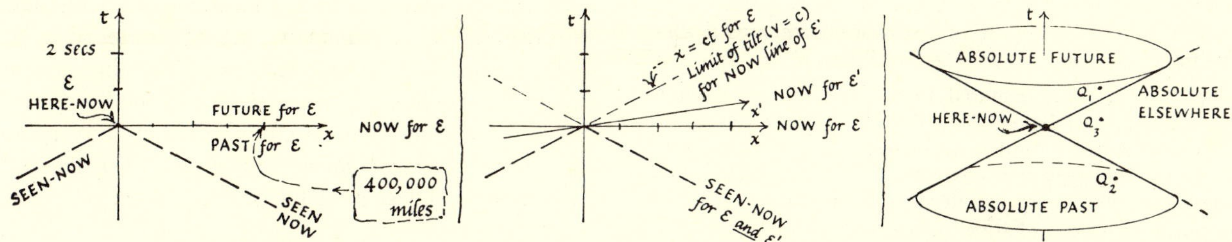

Fig. 31-40(e) [after Eddington]
Observer ε is at the origin; and so is ε′ who is moving fast along x-axis relative to ε. The line SEEN-NOW has equation $x = -ct$, and marks all events that ε (*or* ε′) sees at this instant NOW. ε, knowing the value of c, allows for travel-time and marks his axis of events that happen now along the x-axis. However, ε′ will make a different allowance from the same SEEN-NOW line and will mark a tilted "now" line as his x'-axis. The lines continuing SEEN-NOW in the forward direction of time mark the maximum tilt that ε′ could have for his NOW line—because ε′ can never have relative velocity greater than c; so his x'-axis can never tilt as much as those "light-lines" which have slope c. Rotate the picture around the t-axis and the light-lines make a double cone. Suppose an event P occurs at the origin, HERE-NOW, and another event at Q. If Q is within the upper light-cone (Q_1), it is definitely in the future of P for all observers. Similarly, all events in the lower light-cone (Q_2) are in the absolute past, earlier than P for all observers. But Q_3 in the space between the cones may be in the future for ε and yet be in the past for an observer ε′ whose x'-axis tilts above it. So we label that intermediate region ABSOLUTE ELSEWHERE. If Q falls there, *neither* P *nor* Q *can cause the other*—they simply occur at different places.

So now we must be more careful. We may keep cause and effect in simple cases such as apples and stomach-ache, or alpha-particles and ions; but we must be wary with events so close in time, for their distance apart, that they fall in each other's ABSOLUTE ELSEWHERE.

In atomic physics you will meet other doubts concerning cause and effect. Radioactive changes appear to be a matter of pure chance—the future lifetime of an individual atom being unpredictable. In the final chapter you will see that nature enforces partial unpredictability on all our knowledge, hedging individual atomic events with some unavoidable uncertainty, making it unwise to insist on exact "effects" from exact "causes."

The Lorentz Transformation as a Rotation

The sketches of Fig. 31-40 suggest we can throw light on the Lorentz transformation if we look at the effect of a simple rotation of the axes of a common x-, y-graph. Try the algebra, and find the "transformation" connecting the old coordinates of a point, x, y, with the new coordinates, x', y' of the *same* point, thus:

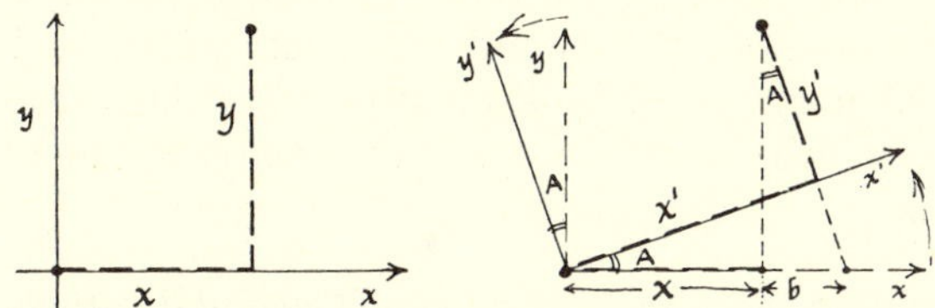

Refer a point in a plane to x-y-axes. Then rotate the axes through an angle A (around the z-axis). The point, remaining at its old position in space, has coordinates x', y' referred to the new axes. Use the symbol s for the *slope* of the new x-axis, so that s is tan A. Then, as the diagram shows

$$x' = (x + b)\cos A = (x + y\tan A)\cos A$$
$$= (x + sy)/\sec A = (x + sy)/\sqrt{(1 + \tan^2 A)}$$
$$\therefore\; x' = (x + sy)/\sqrt{(1 + s^2)}$$

Similarly, $y' = (y - sx)/\sqrt{(1 + s^2)}$

This transformation for a simple rotation of axes shows a square root playing much the same role as in the Lorentz transformation. In fact we obtain the Lorentz form if we replace y by a time coordinate, thus: instead of y, use t multiplied by constant c and by i the square root of (-1). And instead of slope s use $i(v/c)$. Then, with $y = ict$ and $y' = ict'$ and $s = iv/c$, the simple rotation-transformation *is* the Lorentz transformation. Try that. That shows how the Lorentz transformation can be regarded as a slicing of space-&-time with a different slant for different observers.

The Invariant "Interval" between Two Events

We can define the "interval" R between two events (x_1, t_1) and (x_2, t_2) by the Pythagorean form

$$R^2 = (x_1 - x_2)^2 + (ict_1 - ict_2)^2$$

Then we can also write the expression that gives R', the "interval" for another observer who records the same two events at (x_1', t_1') and (x_2', t_2') on his coordinates. If we then use the Lorentz transformation to express R' in terms of the first observer's coordinates, we find that R' is the same as R. The Lorentz transformation keeps that "interval" invariant. That states the Relativity assumption—measured c is always the same—in a different way.

John A. Wheeler suggests a fable to illustrate the role of c. Suppose the inhabitants of an island do their surveying with rectangular coordinates, but measure *North-South* distances in *miles* and *East-West* ones in *feet.* Then a sudden, permanent shift of magnetic North through an angle A makes them turn their system of axes to the new direction. They again measure in miles along the new N′-S′ direction, and in feet E′-W′. They try to compute the distance R between two points by Pythagoras: $R^2 = (\Delta x)^2 + (\Delta y)^2$; and they find that R takes a different value with the new coordinates.

Then they find that they obtain the same value for R (and a useful one) with both sets of coordinates if they define R by: $R^2 = (\Delta x)^2 + (5280\,\Delta y)^2$.

Their "mysterious essential factor," 5280, corresponds to c in the relativistic "interval" in the paragraph above. Moral: c is not so much a mysterious limiting velocity as a unit-changing factor, which suggests that time and space are not utterly different: they form one continuum, with both of them measurable in meters.

Is There a Framework of Fixed Space?

Thus we have devised, in special Relativity, a new geometry and physics of space-&-time with our clocks and measuring scales (basic instruments of physics), conspiring, by their changes when we change observers, to present us with a universally constant velocity of light, to limit all moving matter to lesser speeds, to reveal physical laws in the same form for all observers moving with constant velocities; and thus to conceal from us forever any absolute motion through a fixed framework of space; in fact, to render meaningless the question whether such a framework exists.

HIGHER VALUES OF MATHEMATICS AS A LANGUAGE

Mathematical Form and Beauty

As a language, algebra may be very truthful or accurate, and even fruitful, but is it not doomed to remain dull, uninteresting prose and never rise to poetry? Most mathematicians will deny that doubt and claim there is a great beauty in mathematics. One can learn to enjoy its form and elegance as much as those of poetry. As an example, watch a pair of simultaneous equations being polished up into elegance. Start with

$$2x + 3y = 9$$
$$4x - 2y = 10.$$

Then with some juggling we can get rid of y and find $x = 3$; and then $y = 1$. But these are lopsided, individual equations. Let us make them more general, replacing the coefficients, 2, 3, 9, etc., by letters a, b, c, etc., thus

$$ax + by = c \qquad\qquad dx + ey = f$$

After heavier juggling we find $x = \dfrac{ec - fb}{ae - db}$. Then more juggling is needed to find y. These solutions enable us to solve the earlier equations and others like them by substituting the number coefficients for a, b, c, etc. But unless we had many equations to solve that would hardly pay; and we seem no nearer to poetry. But now let us be more systematic. We are dealing with x and y as much the same things; so we might emphasize the similarity by calling them x_1 and x_2. To match that change, we use a_1, a_2, a_0 instead of a, b, c and write: $a_1x_1 + a_2x_2 = a_0$. But then we have the second equation's coefficients. We might call them a_1', etc., but even so the two equations do not look quite symmetrical. To be fairer still, we call the first lot a_1' etc. and the second lot a_1'' etc. Then:

$$a_1'x_1 + a_2'x_2 = a_0'$$
$$\text{and } a_1''x_1 + a_2''x_2 = a_0''$$

These look neat, but is their neatness much use? Solve for x. We obtain $x_1 = \dfrac{a_0'a_2'' - a_0''a_2'}{a_1'a_2'' - a_1''a_2'}$. Here is a gain: we need not solve for x_2 or y. Symmetry will show us the answer straight away. Note that x_1 and x_2 (the old x and y) and their coefficients are only distinguished by the subscripts $_1$ and $_2$. If we interchange the subscripts $_1$ and $_2$ throughout, we get the same equations again, and therefore we must have the same solutions. We make that interchange in the solution above and $x_1 = \dfrac{a_0'a_2'' - a_0''a_2'}{a_1'a_2'' - a_1''a_2'}$ becomes $x_2 = \dfrac{a_0'a_1'' - a_0''a_1'}{a_2'a_1'' - a_2''a_1'}$. Now we have the answer for x_2 (the old y), free of charge. The economy of working may seem small; but think of the increased complexity if we had, say, five unknowns and five simultaneous equations. With this symmetrical system of writing, we just solve for one unknown, and then write down the other four solutions by symmetry. Here is *form* playing a part that is useful for economy and pleasant in appearance to the mathematical eye. More than that, the new form of equations and answers is general and universal—in a sense this is a case of covariance. This is the kind of symmetrical form that appealed to Maxwell and Einstein.

This is only a little way towards finding poetry in the language of mathematics—about as far as well-metered verse. The next stage would be to use symmetrical *methods* rather than symmetrical *forms*, e.g. "determinants." As the professional mathematician develops the careful arguments which back up his methods, he builds a structure of logic and form which to his eye is as beautiful as the finest poem.

Geometry and Science: Truth and General Relativity

Thus, mathematics goes far beyond working arithmetic and sausage-grinding algebra. It even abandons pert definitions and some of the restrictions of logic, to encourage full flowering of its growth; but yet its whole scheme is based on its own starting points; the views its founders take of *numbers, points, parallel lines, vectors,* . . . . Pure mathematics is an ivory-tower science. The results, being derived by good logic, are automatically true to the original assumptions and definitions. Whether the real world fits the assumptions seems at first a matter for experiment. We certainly must not trust the assumptions just because they seem reasonable and obvious. However, they may be more like definitions of procedure, in which case mathematics, still true to those definitions, might interpret any world in terms of them.

We used to think that when the mathematician had developed his world of space and numbers, we then had to do experiments to find out whether the real world agrees with him. For example, Euclid made assumptions regarding points and lines, etc. and proved, or argued out, a consistent geometry. On the face of it, by rough comparison with real circles and triangles drawn on paper or surveyed on land, the results of his system seemed true to nature. But, one felt, more and more precise experiments were needed to test whether Euclid had chosen the right assumptions to imitate nature exactly; whether, for example, the three angles of a triangle do make just 180 degrees.[20] Relativity-mechanics and astronomical thinking about the universe have raised serious questions about the most fitting choice of geometry. Mathematicians have long known that Euclid's version is only one of several devisable geometries which agree on a small scale but differ radically on a large scale in their physical and philosophical nature.

Special Relativity deals with cases where an observer is moving with *constant velocity* relative to apparatus or to another observer. Einstein then developed *General Relativity* to deal with measurement in systems that are *accelerating.*

What is General Relativity, and how does it affect our views of physics—and of geometry?

[20] It probably seems obvious to you that they do. This may be because you have swallowed Euclid's proof whole—authoritarian deduction. Or you may have assured yourself inductively by making a paper triangle, tearing off the corners and assembling them. Suppose, however, we lived on a huge globe, *without knowing it.* Small triangles, confined to the schoolroom would have a 180° sum. But a huge triangle would have a bigger sum. For example, one with a 90° apex at the N-pole would have right angles at its base on the equator.

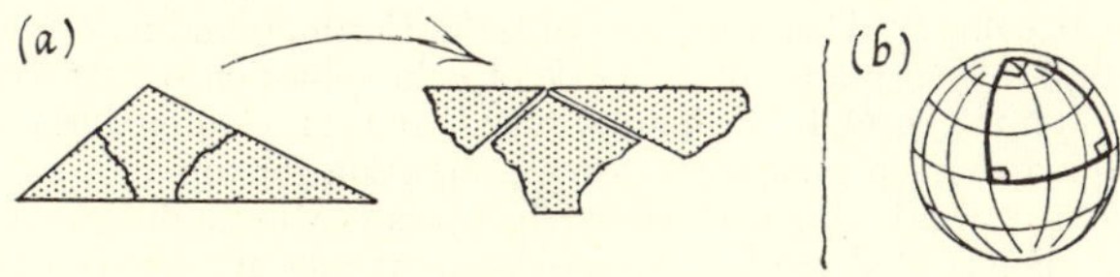

FIG. 31-41.
(a) Tearing a paper triangle. (b) Triangle on a sphere.

Einstein's Principle of Equivalence

Einstein was led to General Relativity by a single question: "Could an observer in a falling elevator or accelerating train really know he is accelerating?" Of course he would notice strange forces (as in the case of truck-and-track experiments to test $F = Ma$ in an accelerating railroad coach.* There strange forces act on the truck and make $F = Ma$ untrue). But could he decide by experiment between acceleration of his frame of reference and a new gravitational field? (If a carpenter builds a correctly tilted laboratory in the accelerating coach, the observer will again find $F = Ma$ holds, but he will find "g" different.)* Therefore, Einstein assumed that no local experiments—mechanical, electrical or optical—could decide: no experiments could tell an observer whether the forces he finds are due to his acceleration or to a local "gravitational" field. Then, Einstein said, *the laws of physics must take the same essential form for ALL observers,* even those who are accelerating. In other words, Einstein required all the laws of physics to be covariant for all transformations from one frame of reference (or laboratory) to another. That is the essential basis of General Relativity: all physical laws to keep the same form.

It was obvious long ago that for *mechanical behavior* a gravitational field and an accelerating frame of reference are equivalent. Einstein's great contribution was his assumption that they are *completely* equivalent, that even in optical and electrical experiments a gravitational field would have the same effect as an accelerated frame of reference. "This assertion supplied the long-sought-for link between gravitation and the rest of physics. . . ."[21]

Accelerating Local Observer ≡ "Gravitational Field"

The Principle of Equivalence influences our view of matter motion and geometry in several ways:

(1) *Local Physics for Accelerated Observers.* If the Principle of Equivalence is true, all the strange effects observed in an accelerating laboratory can be ascribed to an extra force-field. If the laboratory's acceleration is a meters/sec^2, we may treat the laboratory as at rest instead if we give every mass m kg an extra force $-ma$ newtons, presumably due to a force-field of strength $-a$ newtons/kg. Then, with this field included, the ordinary rules of mechanics should apply—or rather the Lorentz modification of Newtonian mechanics and Euclidean geometry, just as in Special Relativity.

* See Chapter 7, Problems 28 and 29, p. 133.

[21] Sir Edmund Whittaker, in *From Euclid to Eddington* (Cambridge University Press, 1949): now in Dover paperback edition.

Examples:

(i) Experimenters in a railroad coach that is accelerating—or in a rocket that is being driven by its fuel—will find Newton's Laws of motion applying at low speeds, provided they add to all visible forces on each mass m the extra (backward) force, $-ma$, due to the equivalent force-field.[22] Objects moving through the laboratory at very high speeds would seem to have increased mass, etc., just as we always expect from Special Relativity.

(ii) An experimenter weighing himself on a spring scale in an elevator moving with downward acceleration a would obtain the scale reading that he would expect in a gravitational field of strength $(g - a)$. (See Ch. 7, Problem 10.)

(iii) In a freely falling box the force exerted by the equivalent force-field on a mass m would be mg upward. Since this would exactly balance the weight of the body, mg downward, everything would appear to be weightless. The same applies to experiments inside a rocket when its fuel has stopped driving it, or to experiments on any satellite pursuing an orbit around the Earth: the pull of the Earth's controlling gravity is not felt, because the whole laboratory is accelerating too.

(iv) In a rotating laboratory, adding an outward force-field of strength v^2/R would reduce the local mechanical behavior to that of a stationary lab.

(2) *Interpreting Gravity.* All (real) gravitational fields can be reinterpreted as local modifications of space-&-time by changing to appropriate accelerating axes so that the field disappears. This change gives us no help in mechanical calculations, but it leads to a new meaning for gravity, to be discussed in the next section.

(3) *"Removing Gravity."* If a gravitational field is really equivalent to an accelerating frame, we can remove it by giving our laboratory an appropriate acceleration. Common gravity, the pull of the Earth, pulls vertically down. It is equivalent to an acceleration of our frame, g vertically up. If we then let our lab fall through our frame of reference with acceleration g vertically down, we observe no effects of gravity. Our lab has two accelerations, the "real" one of falling and the opposite one that replaces the gravitational field. The two just cancel and we have the equivalent of a *stationary lab* in *zero gravitational field.* That just means, "let the lab fall freely, and gravity is not felt in it." We do that physically when we travel in a space ship, or in a freely-falling elevator. Our accelerating framework removes all sign of the gravitational field of Earth or Sun[23] on a small local scale. Then we can leave a body to move with no forces and watch its path. We call its path in space-&-time a straight line and we expect to find simple mechanical laws obeyed. We have an inertial frame in our locality.

(4) *Artificial Gravity.* Conversely, by imposing a large real acceleration we can manufacture a strong force-field. If we trust the Principle of Equivalence we expect this force-field to treat matter in the same way as a very strong gravitational field. On this view, centrifuging increases available "g" many thousandfold.

(5) *Myth-and-Symbol Experiment.* To an observer with acceleration a every mass m^* seems to suffer an opposite force of size m^*a, in addition to the pushes and pulls exerted on it by known agents. In a gravitational field of strength g every mass $m^\dagger$ is pulled with a force $m^\dagger g$. Here, we are using m^* for inertial mass, the m in $F = ma$, and $m^\dagger$ for gravitational mass, the m in $F = GMm/d^2$. The Principle of Equivalence says that gravitational field of strength g can be replaced in effect by an opposite acceleration g of the observer.

$$\therefore\ m^\dagger g \text{ must be the same as } m^* g \quad \therefore\ m^\dagger \equiv m^*$$

The Principle of Equivalence requires gravitational mass and inertial mass to be the same; and the Myth-and-Symbol Experiment long ago told us that they are. As you will see in the discussion that follows, Einstein, in his development of General Relativity, gave a deeper meaning for this equality of the two kinds of mass.

General Relativity and Geometry

Over *small* regions of space-&-time, the Earth's gravity is practically uniform—and so is any other

[22] Over 200 years ago, the French philosopher and mathematician d'Alembert stated a general principle for solving problems that involve accelerated motion: *add* to all the known forces acting on an accelerating mass m an extra force $-ma$; then treat m as in equilibrium. By adding such "d'Alembert forces" to all the bodies of a complex system of masses in motion we can convert the *dynamical* problem of predicting forces or motion into a *statical* problem of forces in equilibrium. This is now common practice among professional physicists, but it is an artificial, sophisticated notion that is apt to be misleading; so we avoid it in elementary teaching. It is the basis of the "engineer's headache-cure" mentioned in Opinion III of centrifugal force, in Chapter 21.

[23] That is why the Sun's gravitational pull produces "no noticeable field" as we move with the Earth around its yearly orbit. (That phrase in the table of field values on p. 123 was a quibble!) Only if inertial mass and gravitational mass failed to keep *exactly* the same proportion for different substances would any noticeable effect occur. Minute differences of such a kind are being looked for—if any are discovered, they will have a profound effect on our theory.

gravitational field. So we can "remove" gravity for local experiments by having our lab accelerate freely; and it will behave like an inertial frame with no gravitational field: an object left alone will stay at rest or move in a straight line; and with forces applied we shall find $F = ma$. However, on a grander scale, say all around the Earth or the Sun, we should have to use many different accelerations for our local labs to remove gravity. In fitting a straight line defined in one lab by Newton's Law I to its continuation in a neighboring lab, also accelerating freely, we should find we have to "bend" our straight line to make it fit. The demands of bending would get worse as we proceeded from lab to lab around the gravitating mass. How can we explain that? Instead of saying "we have found there is gravity here after all" we might say "Euclidean geometry does not quite fit the real world near the massive Earth or Sun." The second choice is taken in developing General Relativity. As in devising Special Relativity, Einstein looked for the simplest geometry to fit the new assumption that the laws of physics should *always* take the same form. He arrived at a General-Relativity geometry in which gravity disappears as a strange force reaching out from matter; instead, it appears as a distortion of space-&-time around matter.

"From time immemorial the physicist and the pure mathematician had worked on a certain agreement as to the shares which they were respectively to take in the study of nature. The mathematician was to come first and analyse the properties of space and time building up the primary sciences of geometry and kinematics (pure motion); then, when the stage had thus been prepared, the physicist was to come along with the dramatis personae—material bodies, magnets, electric charges, light and so forth—and the play was to begin. But in Einstein's revolutionary conception the characters created the stage as they walked about on it: geometry was no longer antecedent to physics but indissolubly fused with it into a single discipline. The properties of space in General Relativity depend on the material bodies and the energy that are present. . . ."[24]

Is this new geometry right and the old wrong? Let us return to our view of mathematics as the obedient servant. Could we not use *any* system of geometry to carry out our description of the physical world, stretching the world picture to fit the geometry, so to speak? Then our search would not be to find the *right* geometry but to choose the *simplest or most convenient* one which would describe the world with least stretching.[25] If we do, we must realize that we *choose* our geometry but we *have* our universe; and if we ruthlessly make one fit the other by pushing and pulling and distorting, then we must take the consequences.

For example, if all the objects in our world consisted of some pieces of the elastic skin of an orange, the easiest geometrical model to fit them on would be a ball. But if we were brought up with an undying belief in plane geometry, we could press the peel down on a flat table and glue it to the surface, making it stretch where necessary to accommodate to the table. We might find the cells of the peel larger near the outer edge of our flattened piece, but we should announce that as a law of nature. We might find strange forces trying to make the middle of the patch bulge away from the table—again, a "law of nature." If we sought to simplify our view of nature, the peel's behavior would tempt us to use a spherical surface instead of a flat one, as our model of "surface-space." All this sounds fanciful, and it is; but just such a discussion on a three- or four-dimensional basis, instead of a two-dimensional one, has been used in General Relativity. The strange force of gravity may be a necessary result of trying to interpret nature with an unsuitable geometry—the system Euclid developed so beautifully. If we choose a different geometry, in which matter distorts the measurement system around it, then gravitation changes from a surprising set of forces to a mere matter of geometry. A cannon ball need no longer be regarded as being dragged by gravity in what the old geometry would call a "curve" in space. Instead, we may think of it as sailing serenely along what the new geometry considers a straight line in its space-&-time, as distorted by the neighboring Earth.

This would merely be a change of view (and as scientists we should hardly bother much about it), unless it could open our eyes to new knowledge or improve our comprehension of old knowledge. It can. On such a new geometrical view, the "curved" paths of freely moving bodies are inlaid in the new geometry of space-&-time and *all* projectiles, big and small, with given speed must follow the same path. Notice how the surprise of the Myth-and-Symbol fact disappears. The long-standing mystery of gravitational mass being equal to inertial mass is solved. Obviously a great property of nature, this equality was neglected for centuries until Einstein claimed it as a pattern property imposed on space-&-time by matter.

[24] Sir Edmund Whittaker, *From Euclid to Eddington*, *op.cit.*, p. 117.

[25] You can have your coffee served on any tray, but on some trays it wobbles less.

Even a light ray must follow a curve, just as much as a bullet moving at light speed. Near the Earth that curve would be imperceptible, but starlight streaming past the Sun should be deflected by an angle of about 0.0005 degrees, just measurable by modern instruments. Photographs taken during total eclipses show that stars very near the edge of the Sun seem shifted by about 0.0006°. On the traditional ("classical") view, the Sun has a gravitational field that appears to modify the straight-line law for light rays of the Euclidean geometrical scheme. On the General Relativity view, we replace the Sun's gravitational field by a crumpling of the local geometry from simple Euclidean form into a version where light seems to us to travel slower. Thus the light beam is curved slightly around as it passes the Sun—the reverse of the bending of light by hot air over a road, when it makes a mirage.

Finding this view of gravitation both simple and fruitful—when boiled down to simplest mathematical form—we would like to adopt it. In any ordinary laboratory experiments we find Euclid's geometry gives simple, accurate descriptions. But in astronomical cases with large gravitational fields we must either use a new geometry (in which the mesh of "straight lines" in space-&-time seems to us slightly crumpled) or else we must make some complicating changes in the laws of physics. As in Special Relativity, the modern fashion is to make the change in geometry. This enables us to polish up the laws of physics into simple forms which hold universally; and sometimes in doing that we can see the possibility of new knowledge.

In specifying gravitation on the new geometrical view, Einstein found that his simplest, most plausible form of law led to slightly different predictions from those produced by Newton's inverse-square law of gravitation. He did not "prove Newton's Law wrong" but offered a refining modification—though this involved a radical change in viewpoint. We must not think of either law as right because it is suggested by a great man or because it is enshrined in beautiful mathematics. We are offered it as a brilliant guess from a great mind unduly sensitive to the overtones of evidence from the real universe. We take it as a promising guess, even a likely one, but we then test it ruthlessly. The changes, from Newton's predictions to Einstein's, though fundamental in nature, are usually too small in effect to make any difference in laboratory experiments or even in most astronomical measurements. But there should be a noticeable effect in the rapid motion of the planet Mercury around its orbit. Newton predicted a simple ellipse, with other planets producing perturbations which could be calculated and observed. General Relativity theory predicts an extra motion, a very slow slewing around of the long axis of the ellipse by 0.0119 degree per century. When Einstein predicted it, this tiny motion was already known, discovered long before by Leverrier. The measured value (a tiny residual after allowing for perturbations) is close to 0.012°/century.

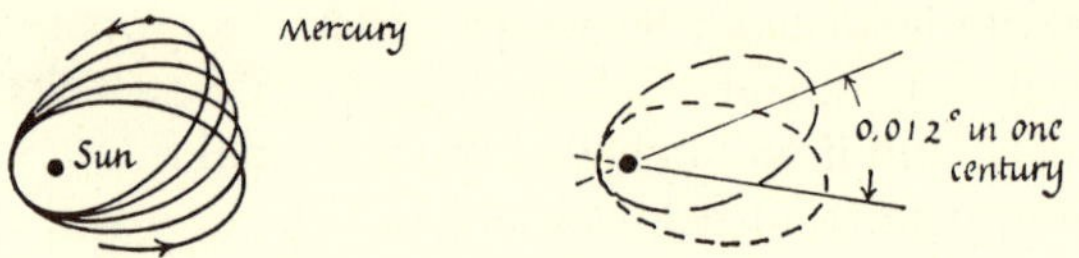

Fig. 31-42. Motion of Planet Mercury

Accepting this view of gravity, astronomers can speculate on the geometry of all space and ask whether the universe is infinite or bounded by its own geometric curvature (as a sphere is). We may yet be able to make some test of this question.

There are still difficulties and doubts about General Relativity. Even as we use it confidently to deal with Mercury's motion, or the light from a massive star, we *may* have to anchor our calculations to *some* frame of reference, perhaps the remotest regions of space far from gravitating matter, or perhaps the center of gravity of our universe. So space as we treat it, *may* have some kind of absolute milestones. This doubt, this threat to a powerful theory, does not irritate the wise scientist: he keeps it in mind with hopes of an interesting future for his thoughts.

New Mathematics for Nuclear Physics

In atomic and nuclear physics, mathematics now takes a strong hand. Instead of sketching a model with sharp bullet-like electrons whirling round an equally sharp nucleus, we express our knowledge of atoms in mathematical forms for which no picture can be drawn. These forms use unorthodox rules of algebra, dreamed up for the purpose; and some show the usual mathematical trademark of waves. Yet, although they remain mathematical forms, they yield fruitful predictions, ranging from the strength of metal wires and chemical energies to the behavior of radioactive nuclei.

We now see mathematics, pure thought and argument, again offering to present physics in clearer forms which help our thinking; but now far from a servant, it is rather a Lord Chancellor standing behind the throne of ruling Science to advise on law. Or, we might describe mathematics as a master architect designing the building in which science can grow to its best.

PART FOUR

ELECTRICITY AND MAGNETISM

As a householder, you should know more about electricity than how to change a fuse. You should understand the relationships among current, voltage, and power, the advantages and disadvantages of alternating current, . . . The first chapter of this Part offers "householders' electricity."

As a scientist interested in atomic physics, you need a knowledge of electricity and magnetism to understand how atomic knowledge is gathered. Several chapters here provide for that need.

As a philosopher interested in defining *theory*, you will find a useful example in the chapter on magnetism.

In reading this Part for the knowledge it offers, watch also for warnings of the *limits* of that knowledge. A wise scientist is fully aware of such limits. He must "know what he does not know"; because much of his work is at the frontier between known and unknown.

"Knowledge is proud that he has learned so much;
Wisdom is humble that he knows no more."

William Cowper ($\sim$ 1760)

CHAPTER 32
ELECTRIC CIRCUITS – A LABORATORY CHAPTER

". . . one must learn
By doing the thing; for though you think you know it
You have no certainty until you try."

Sophocles, translated by Sir George Young.
[Everyman Edition]

[This is a chapter of laboratory experiments with descriptions *to be read on your own, without lectures. Try the experiments suggested*—failing that, see them demonstrated—and combine them with your reading and general knowledge to obtain a good understanding of electric circuits.]

Early electrical knowledge, dating back many centuries, was concerned with electric "charges" obtained by rubbing things. Electric circuits, with currents to run lamps and motors, did not develop until batteries were invented about 1800. Then the growth was so rapid that it produced not just a part of physics but a new electrical civilization, in less than a century.

In this course, instead of following the history, we shall study electricity with modern apparatus in the laboratory; and we shall draw upon your general knowledge of circuits gained from living in a world of cars and electric lighting.

Here are some experimental facts about "electric circuits" used for house lighting, car lights, electric bells, etc. Some kind of supply—battery or generator or power lines from a power station—is necessary, or nothing happens. To make a lamp light or motor run a metal wire must go all the way from the supply to the lamp and on from the lamp back to the supply. Inside the lamp there is a thin metal filament; so there is metal wire of some kind all the way from the supply out through the lamp and back again. If this wire is cut, the lamp goes out. A switch is simply a device for making such a cut. A fuse does the same, when its wire melts. This continuous metal route is called an electric circuit. If we leave out the lamp and make the circuit of a long piece of thin wire, the whole wire gets warm—something is happening all the way along to warm it.[1] If some of the wire is thin and some thick, the thin wire gets warm, the thick wire less noticeably so; the lamp in the first circuit is an extreme case of this. If we shorten the whole length of wire, it gets hotter; still shorter, still hotter. A circuit made of a very short piece of wire all alone might get hot enough to melt or to set fire to other things. This is literally a short circuit, and the phrase "short-circuit" is used for any electric circuit which is made so short—or of such low resistance—that there is danger of damage.

[1] We have spread the lamp filament's heating all around the circuit.

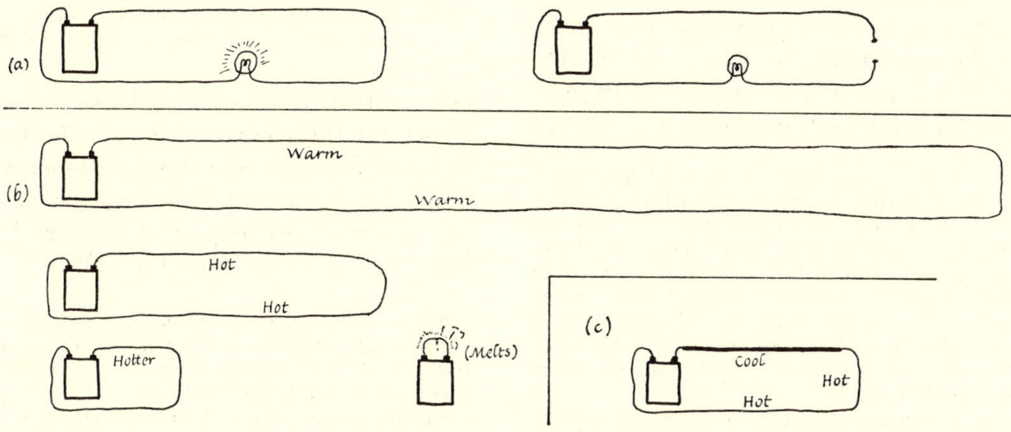

Fig. 32-1. Electric Circuits

To avoid some of the dangers of short circuits, wires are insulated, protected by coatings of non-metal such as rubber, waxed paper, cloth.[2]

★ PROBLEM 1. SHORT CIRCUIT

Suppose wires from some supply running to a lighted lamp and back as in Fig. 32-2 touch each other by accident, making good metallic contact at X.

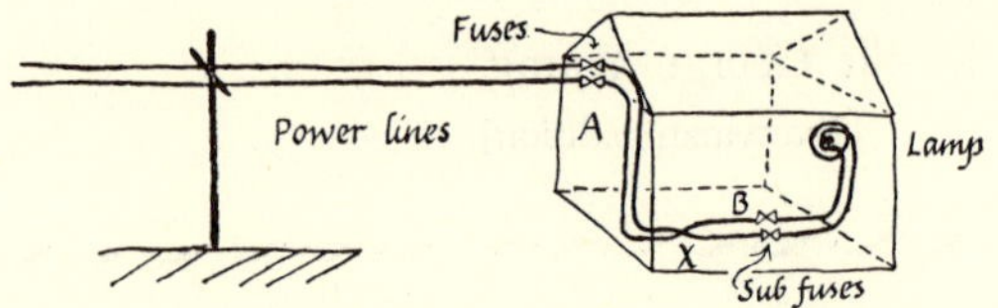

FIG. 32-2. SKETCH FOR PROBLEM 1

(a) Which parts of the circuit are likely to be hottest now?
(b) The fuses shown are wires of low-melting-point metal. Which one(s) will melt, if any, A or B?

Returning to the lamp circuit, we find that if the lamp is moved to some other position in the circuit it still glows. If several lamps are placed in the circuit, "in series," they will then all glow equally, but much less brightly than one lamp alone. Something

FIG. 32-3. ALL LAMPS LIGHT EQUALLY

seems to be available or happening all along the circuit, a ready-to-make-lamp-glow state of affairs. Separate experiments with lamp filaments show that this glowing is just a matter of supplying heat to the filament; if we could make it just as hot with a bunsen flame, it would glow as brightly. So the peculiar electrical property seems to be a supply-heat state of affairs, all around the circuit.[3]

Does such a circuit have other "electrical properties"? Without breaking the wire, wind it into a coil as Oersted and Ampère did a century ago. You will find the wire will magnetize an iron rod

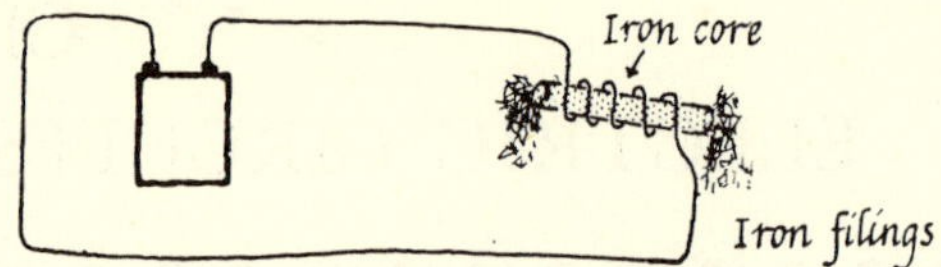

FIG. 32-4. ELECTROMAGNET

so that it will pick up chips of iron. With two such coils, each in a circuit, you can magnetize two rods and make them push or pull each other strongly. Without iron cores, the coils themselves push or pull each other weakly. Here is the working principle of electric motors, bells, telephone earpieces, and some ammeters: electromagnets pushing and pulling each other. Again, the coil can be anywhere in the circuit, so long as the circuit is complete. So the circuit has another "electrical property,"[4] a readiness-to-make-magnets, or magnetic effect.

Any more? Yes, but the third electrical effect seems more obscure, and it is surprising that it was discovered as early as the others, in the great rush of electrical discovery and invention 150 years ago. Cut the wire of a circuit anywhere and dip the cut ends into a glass of dirty water;[5] small bubbles of gas appear. Add salt or vinegar to the water and there is much more visible effect; bubbles rise from

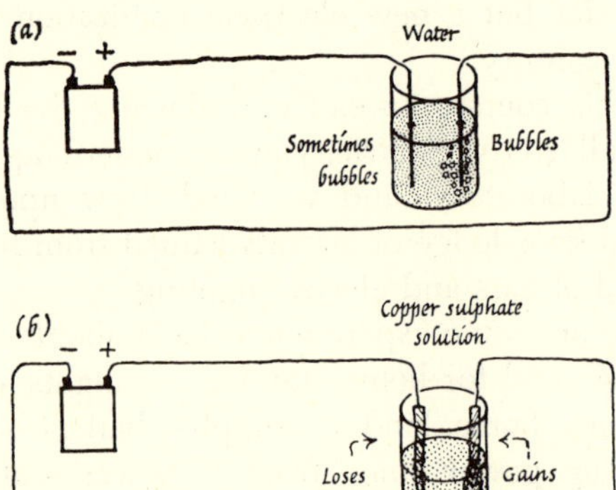

FIG. 32-5. CHEMICAL EFFECTS OF ELECTRIC CURRENT
(a) water (b) copper sulphate solution

one or both wire-ends, and there are chemical changes in the solution. Add crystals of copper sulphate to the water and dip the copper wires in the blue solution; one wire is eaten away and the other grows fatter with deposited copper: "electroplating." We call these "chemical effects."

[2] Nowadays these coatings are wound, braided, painted, or moulded onto the wires by automatic machinery; but in the early days of electric circuits insulated wire was not commercially available. Bare wire, drawn for use on farms and in manufacturing, had to be insulated by hand with string or strips of cloth. Professor Joseph Henry, experimenting at Princeton a century ago on the beginnings of radio, had to make large electromagnet windings. He made his insulation by tearing his wife's silk petticoats into strips to be wound round his wires.

[3] Notice we have already named this an *electrical* property. We know it is going to be important enough to need a name.

[4] We call this by the same name, electrical, since we find it always goes with the heating effect. We are implying that they are different aspects of the same thing—a risky guess which has proved to be a good one.

[5] Distilled water shows very little effect; it is almost an insulator.

Currents

All three effects can happen together in the same circuit. They are even found inside the battery or generator itself. They show that something is happening all around the circuit. The same-all-around-the-circuit behavior seemed to the early experimenters rather like the flow of liquid around a closed circuit of pipes. So they imagined some mysterious thing, electricity, flowing around the circuit. Their name for this flow, "electric current," turned out to be remarkably suitable, and we have kept it. There might have been nothing really flowing; then the words "current" and "flow" would have been prejudicial, making clear thinking harder. We now know there *is* a flow, usually of negative electrons; and we keep the borrowed terms in our electrical vocabulary. You have no evidence so far of such a real flow, yet we have been teaching you with those terms to indoctrinate you quickly!

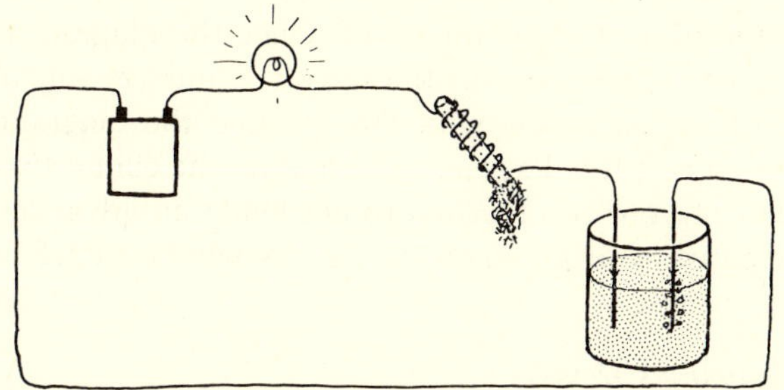

Fig. 32-6. Electric Circuit Can Show All Three Effects

To this day, in teaching elementary electricity we liken electric circuits to hydraulic circuits of water pipes full of water all the way around, with pumps, taps, flowmeters,[6] pressure gauges . . . to correspond to generators, switches, ammeters, voltmeters. . . . Like many uses of analogy in teaching, this does make things easier for the beginner to understand.

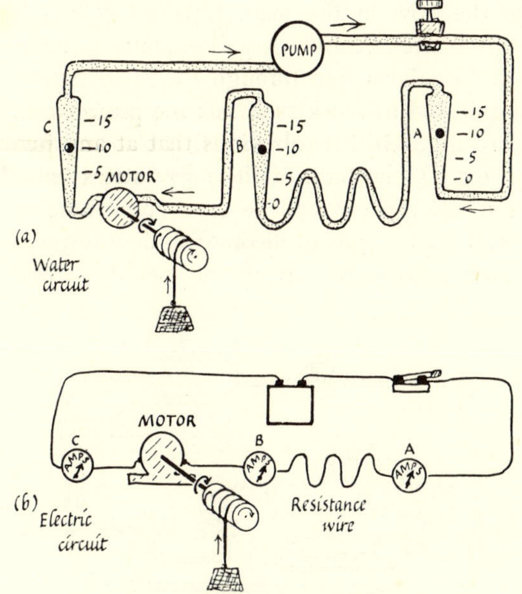

Fig. 32-8. Analogous Circuits

With pipes full of water, it is obvious that the flow (say 10 gallons/minute) will be the same all around the circuit; flowmeters at A, B, C, &c., will all read the same.[7] A reads 10 gallons/minute; B will read 10 gallons/minute and so will the rest. Even a flowmeter incorporated in the internal works of the pump would read 10 gallons/minute. If the pipe divides into several branches "in parallel" it is obvious that the flows in the separate branches will add

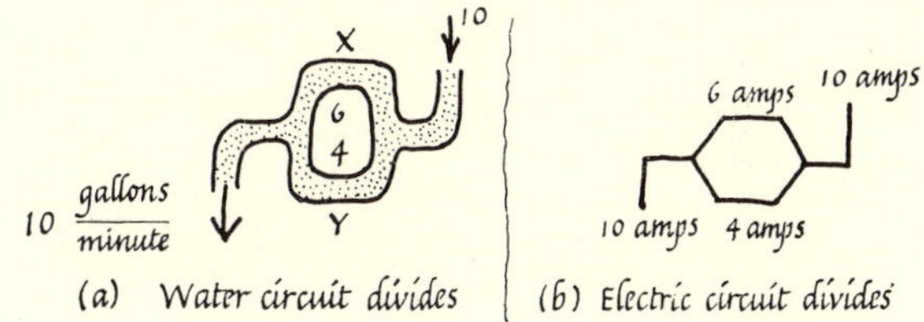

Fig. 32-9. Branching Circuits

[6] Fig. 32-7 shows a simple form of flowmeter to measure rate-of-flow of a fluid, in, say, gallons/minute. Water flows up a vertical tapered tube past a ball of glass or metal, B, which is a little smaller than the narrowest part of the tube. The ball is carried up the tube by the water-flow till it reaches a place where it hovers. This position indicates the flow-rate. The faster the flow, the higher the ball must rise before it reaches a place where there is enough room for the water to flow past it without pushing it still higher. This model is worth seeing. It makes the idea of an ammeter easier to grasp. Such meters are made commercially, under the name Rotammeters, with the ball replaced by a little metal peg-top with slanting cuts to make it spin as it hovers. The Rotammeter spins freely in the tube, while the simple glass ball is inclined to cling to one side of the tube. (Puzzle: Why does the ball cling thus? This relates to something much earlier in the course.)

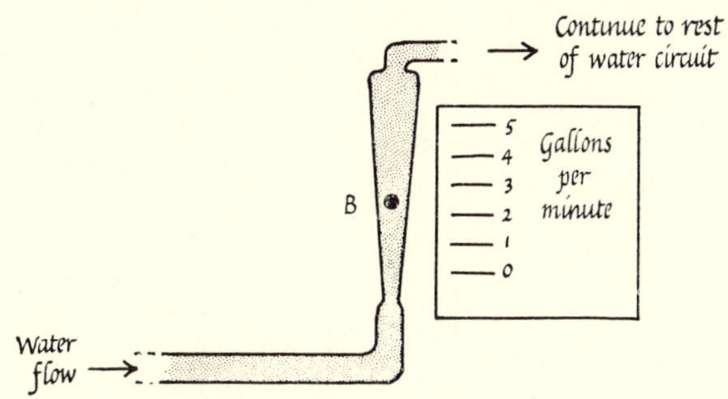

Fig. 32-7. Rotammeter—a flowmeter for fluids

[7] If not, water must be accumulating somewhere or leaking out somewhere. For flow of incompressible fluid in a closed pipe the flow rate must be the same at all places around the circuit.

up to the flow in the main pipe (6 gallons/min through branch X plus 4 gallons/min through Y add to 10 gallons/min through the main pipe). In a complicated network such as the plumber's paradise in Fig. 32-10 it is obvious that at any junction point, e.g. O, the total of the flow rates in all the pipes meeting there will be zero (reckoning with plus and minus signs to distinguish flow towards O and flow away). Yet saying the electric circuit is

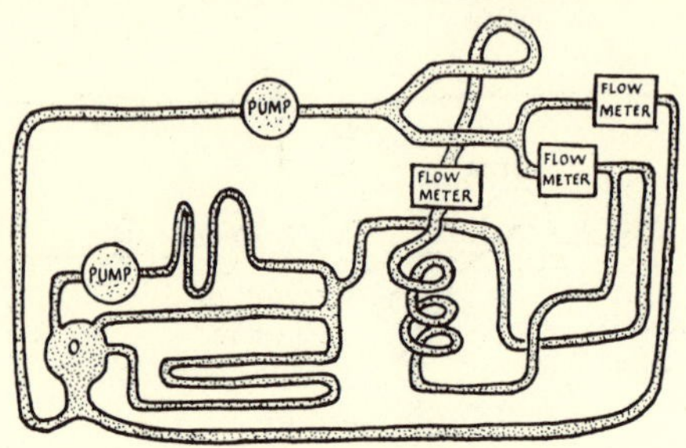

Fig. 32-10. Plumber's Paradise

"just like" the water circuit does not prove the electric circuit will have such behavior. "Current flow" is a hind-sight description, put in after we have found that the electric circuit does have experimental properties which resemble those of a water circuit. As such, it is good teaching, but if misused as an attempted proof, would be bad science.[8]

With the water-circuit analogy to give you useful prejudices, you should do your own experimenting at your own laboratory table if possible, not just watch demonstrations. If you work with partners, *each* should draw his own circuit diagram before connecting up any apparatus. Once the diagram is drawn, connecting up the circuit is nursery-school work if you follow your nose around the diagram. If you cast the diagram aside and do the connecting "out of your head," you are not being an advanced scientist: you are showing an unscientific belief in good luck.

Drawing Circuits

Physicists and electricians long ago systematized the drawing of electric circuits. Here are standard rules and symbols, which we shall use from now on.

[8] This seems a long complaint against a teacher's kindly illustration. Yet it was the great mistake of medieval science to argue "what must be" from some authoritarian statement; and present-day popularizers of science make the mistake of building hard-won knowledge into glib analogies from which the science is then apparently produced. Attempts to make physics clearer by analogies may mislead the reader unless he is warned.

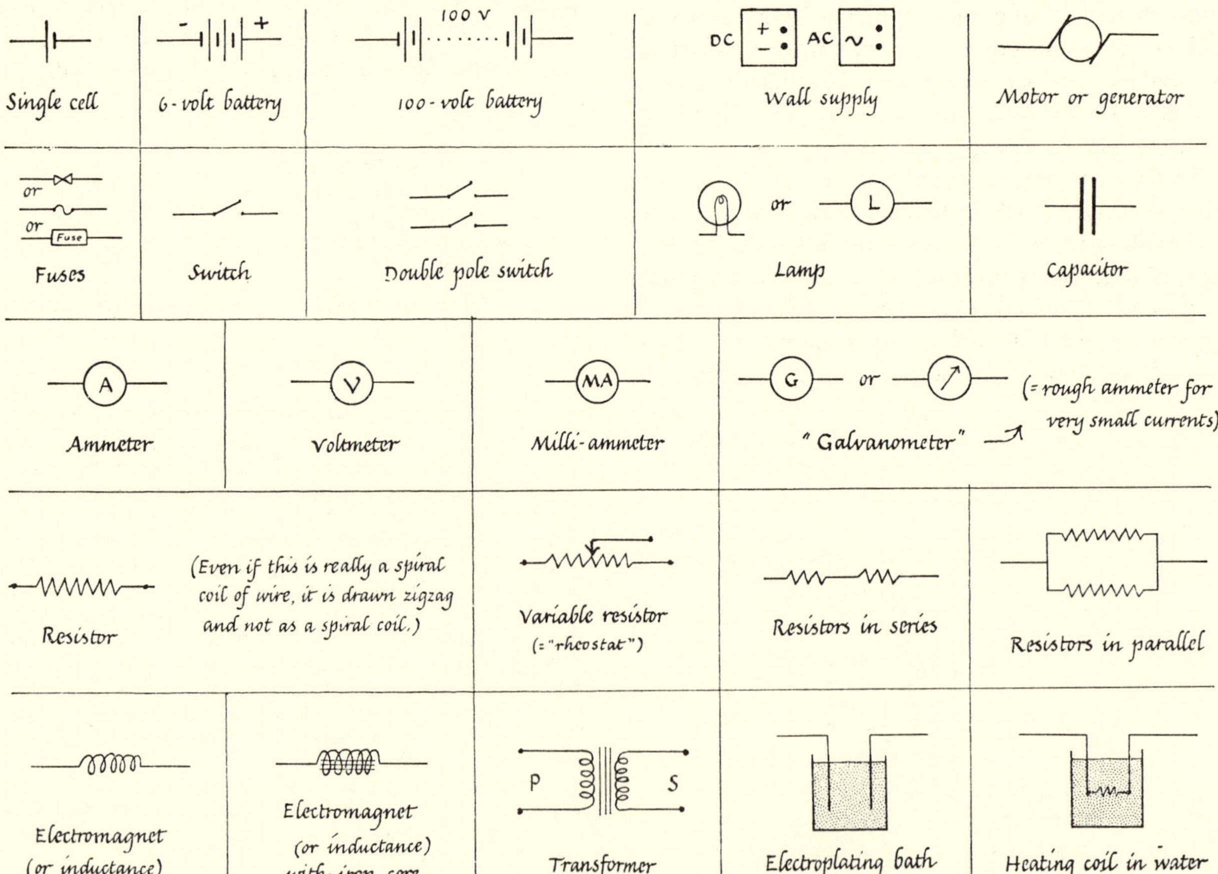

Fig. 32-11. Standard Symbols for Electric Circuits

All connecting wires, whether thick or thin, are shown as thin straight lines; vertical and horizontal lines as far as possible. Thus simple circuits are drawn as rectangles whether they look like that or not in the real experiment. Where a circuit branches, this scheme may be continued, or slanting lines used.[9]

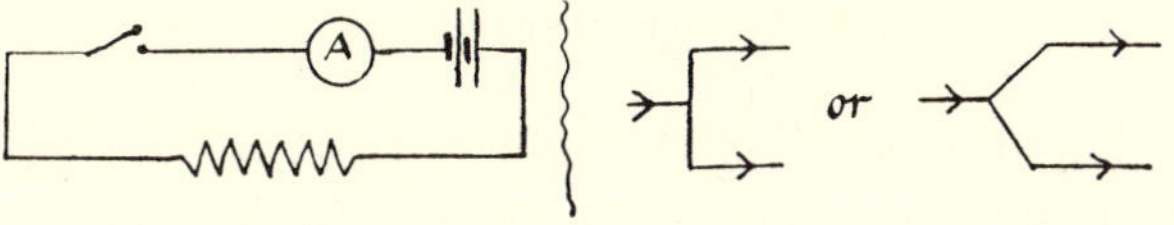

FIG. 32-12. FIG. 32-13.

Positive and Negative

In some electrical effects, there is evidence of a definite *direction* of flow. The two knobs of a battery are not identical. So they are distinguished by being painted red and black, or labelled + and —, and are called the positive and negative[10] binding posts. In drawing batteries, which are made up of cells in series, we use the following standard convention for positive and negative of each cell, so that it is unnecessary to add the + and — signs:

draw the negative binding post as a short thick vertical line (a minus sign, in fact)
draw the positive one as a long thin vertical line (enough material to make a + sign!)

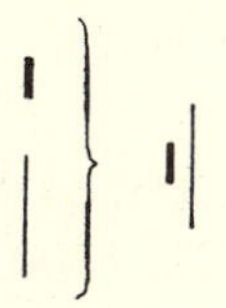

In drawing a battery of cells, joined in series, + of one to — of next, the joins between cell and cell are omitted.

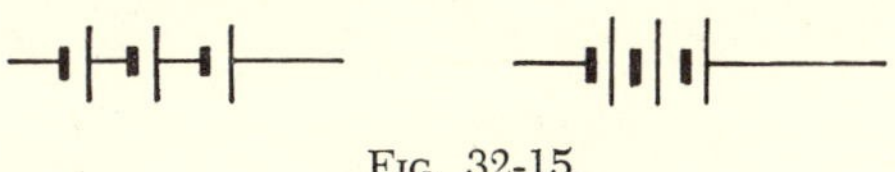

FIG. 32-15.

---

[9] Where one wire crosses another *without contact,* the skip-over (a) makes this obvious. Where the wires *are joined,* a blob of solder (b) makes that obvious. Avoid

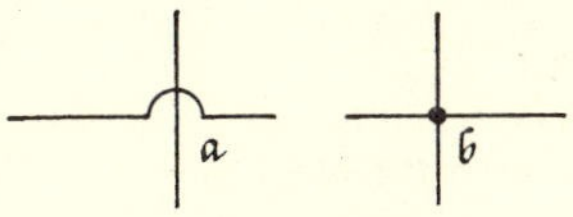

FIG. 32-14.

ambiguous junctions without either jump or blob. Some radio designers say snobbishly, "Oh, well, it's obvious to anyone that matters which is meant."

[10] The names positive and negative were introduced by Benjamin Franklin. They do not mean that one is superior. We could have called them A and B instead, but the arithmetical flavor of plus and minus carries a useful reminder that they are concerned with "opposite" electric charges which can neutralize each other just as positive and negative numbers, +6 and —6, can add up to zero.

EXPERIMENTS WITH CIRCUITS

(In *each* experiment, start by drawing your circuit.)

EXPERIMENT A. SIMPLE CIRCUIT

(i) Use an ammeter[11] to "measure the current" through a small lighted car lamp. To do this, connect in series battery, switch, ammeter, with wires to complete the circuit.

(ii) Use a "variable resistance" or "rheostat"[12] to change the current through the lamp. For that, add a variable resistance in the circuit.

(iii) Testing fuses. Given fuse wire labelled "1 amp fuse wire," investigate the makers' implied claim.

EXPERIMENT B. RESISTANCE

How does current coming *out* from the resistance compare with the current going *in*; i.e., how much does the resistance reduce the current in various parts of the circuit?

[11] *Use of ammeters in lab*: In most common forms of ammeter the pointer moves backwards if the connections from the battery are reversed. This is one good reason for saying the two battery terminals are different. To avoid mistakes and damage, ammeter manufacturers label one binding post plus. The ammeter does not have a real + and — like a battery, so this labelling is not a scientific description but just an arbitrary convention. The convention is: the binding post which should receive the wire from the + of the supply is labelled "+."

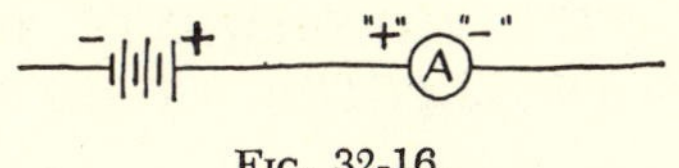

FIG. 32-16.

A good ammeter is easily ruined. Inside the meter, the pointer is attached to a delicately pivoted frame, running in jeweled bearings like those of the balance-wheel in a watch. The ends of the axle are not perfect points—too sharp a point would bend and blunt—but they have a very tiny area where they rest on the jewel. The weight of the moving system bearing on this tiny area makes a huge *pressure* which threatens to blunt the point. The pressure may be several tons/sq. inch in a good meter. If you put the meter down on the table suddenly, the abrupt negative acceleration, many times "g," brings pressures many times as great to act on the pivot. The pivot may be blunted and then the instrument will read stickily. Repairs are expensive. Meters should be treated very gently.

[12] "Rheostat" means "flow steadier" or "flow controller." Common laboratory rheostats are a coil of alloy wire connected to posts A and B at its ends, and a slider, S, making contact with the wire, connected to a post C. For a variable resistance, you must use *two* of these three binding posts.

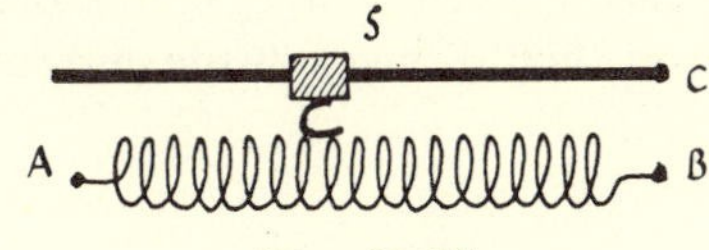

FIG. 32-17.

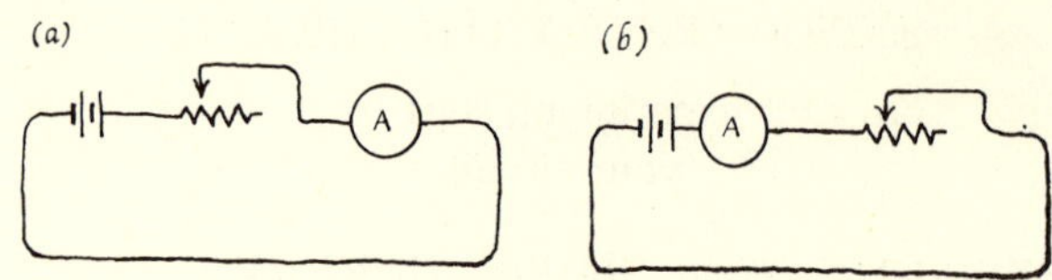

FIG. 32-18. EXPERIMENT B

Throughout this experiment *keep the variable resistance set at a fixed value.* Insert the ammeter (a) just before it, then (b) just after it. Compare the ammeter readings.

EXPERIMENT C. LAMPS AND CURRENT

Light each of the following lamp arrangements, using a car battery, without a rheostat. In each case, make a note of the current and the brightness of the lamps.

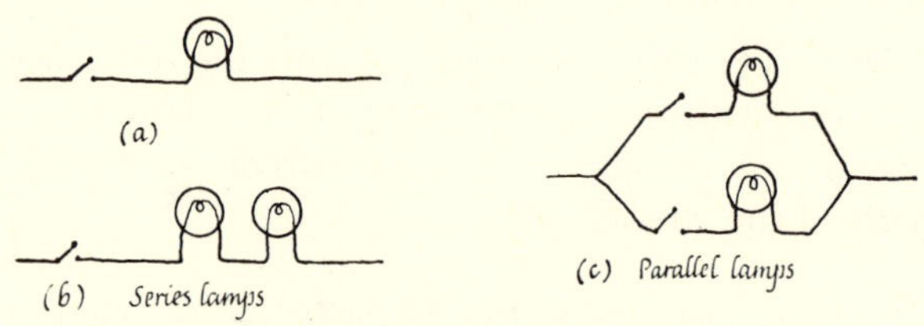

FIG. 32-19. EXPERIMENT C

(a) a single car lamp, with a switch
(b) two lamps in series, with a switch
(c) two lamps in parallel, with a switch for each.

EXPERIMENT D.

QUALITATIVE EXPERIMENTS TO SHOW EFFECTS OF ELECTRIC CURRENTS

Try some of these experiments with a simple circuit such as the one you used for testing fuse wires in Experiment A(iii). If you have no ammeter, you may use a circuit without one, but you must be careful not to overheat the rheostat: therefore, insert a suitable fuse for protection.

(i) *Heating effect.* You have already seen this, with lamps and fuses.

(ii) *Magnetic effect.* Insert a long piece of flexible insulated wire in your circuit and arrange to send a large current through it. Wind the wire around a bar of iron (e.g., a big carpenter's nail) to make a spiral coil with an iron core. Offer some small chips of iron to the coil when it is carrying a big current. Try turning the current on and off. (If you use a very large current you will obtain more striking effects but you should then run the current for only a few seconds, to avoid damage.) Try the coil without any iron core. Try the wire uncoiled. In drawing circuits for these tests use the official symbols for electromagnet coils with and without core.

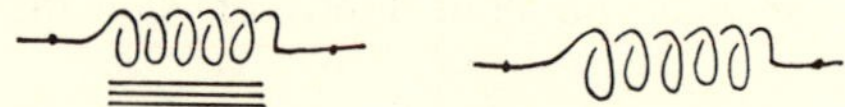

FIG. 32-20. EXPERIMENT D(ii)

(iii) *Chemical effects.* Make a break in your circuit and insert a beaker of water, with the connecting wires dipping in it. Try for any effects with

FIG. 32-21. EXPERIMENT D(iii)

(a) distilled water, (b) water + sulphuric acid, (c) copper sulphate solution. The copper wires may be involved in secondary chemical changes; so you should, if possible, repeat these tests with rods of inert carbon dipping in the beaker—pencil leads will do.

COMMUNAL OR DEMONSTRATION EXPERIMENT. AMMETERS AND LOOPS

Set up some complicated circuit such as the one in Fig. 32-22. Read the ammeters, and make whatever inferences you think justified.

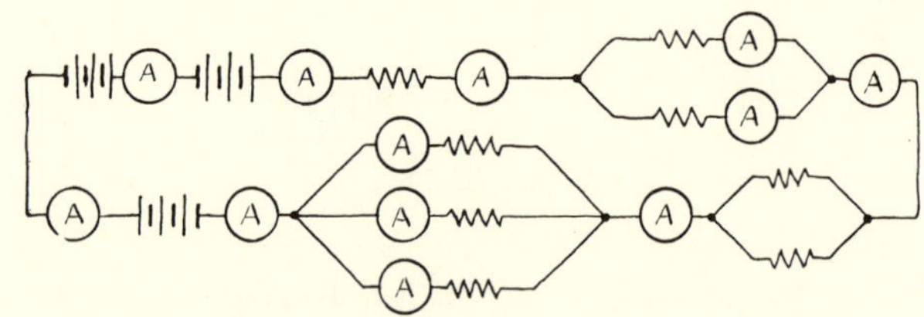

FIG. 32-22. COMMUNAL EXPERIMENT

★ PROBLEM 2. AMMETERS

If in the circuit sketched in Fig. 32-23 one ammeter reads considerably more than the other

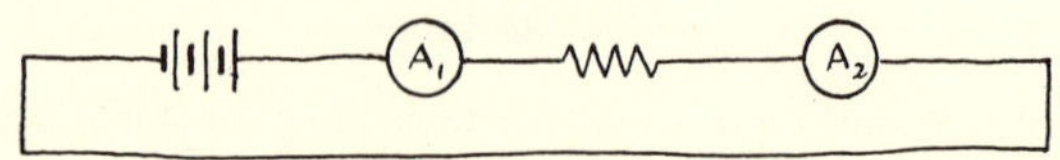

FIG. 32-23. PROBLEM 2

(a) Suggest an explanation.
(b) How would you test your explanation experimentally?

★ PROBLEM 3. POWER SUPPLY

Fig. 32-24 shows a power station supplying two villages through thick copper supply cables of *negligible resistance.*

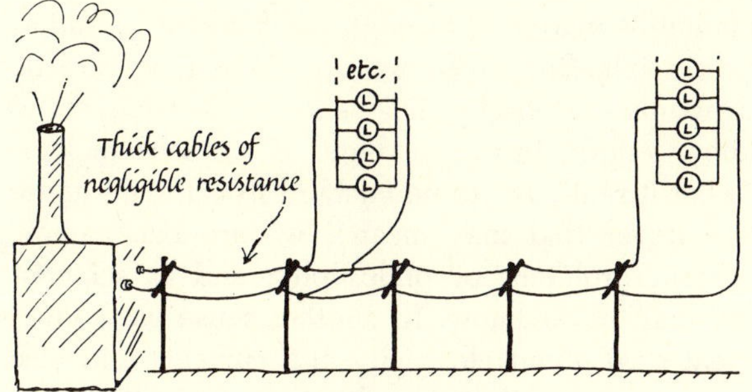

FIG. 32-24. PROBLEM 3

For obvious convenience in switching, as well as for important reasons involving current and resistance, the lighting lamps are arranged "in parallel" in each village. How would you expect the ammeter reading at power station to change
(a) When 100 lamps 1 mile away are switched off and at the same instant 100 lamps 2 miles away are switched on?
(b) When the total number of lamps lighted is doubled?

Notice that you have been using ammeters without either information or enquiry—but this is not much worse than using stop-watches without opening them. By learning how they behave in actual experiments you can accept and use them in terms of their behavior. To say they *measure the current* continues the suggestive analogy.

Measuring Current

Suppose we shorten the circuit and "increase the current," as shown by brighter lamps in the circuit. Experiments like D show that with the increased heating there are also increased magnetic and chemical effects: electromagnets in the circuit pull harder, and chemical baths in the circuit show faster chemical changes. Since these three effects are all we know of the electric current, we would be more realistic to say they *are* the current, rather than call them effects of some mysterious flow. And whatever view we take, if we want to devise a scheme of measuring currents we must use one or more of these effects. We describe CURRENT by all its effects; and we decide to measure it by one of them. This is like our treatment of temperature, where we invented the concept and defined its measurement by our choice of measuring effect (such as expansion) and measuring instrument (such as mercury thermometer). In this course, we choose to measure currents by their chemical effect. In particular, we choose a copper-plating bath and weigh the copper deposited on the receiving plate.

We use the RATE OF DEPOSITING COPPER in a copper-plating bath as our measure of the CURRENT. This is a statement *defining* what we mean by current. It is only an *experimental* statement to the following extent:

(i) Experiment shows that when we change a circuit to make the current bigger (as judged by heating or magnetic effects), the chemical effect is bigger (= faster) too.

(ii) Where a circuit branches, the sum of the copper-plating rates in the branches is the same as the plating rate in the main wire.

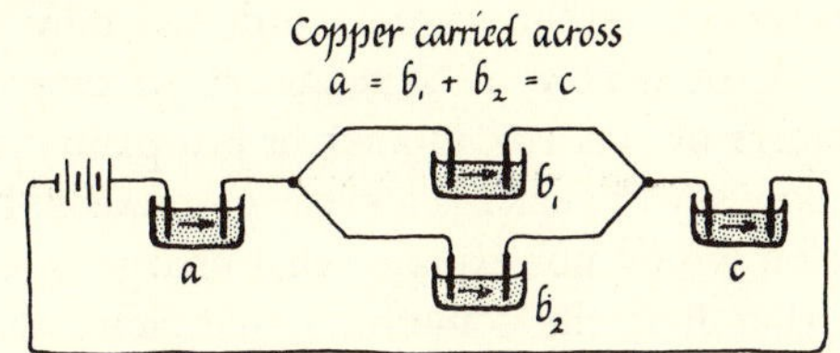

FIG. 32-25. ADDITIVE EFFECTS

These make us feel that we are making a reasonable choice. Yet since current, like temperature, is an artificial idea we may make what definition we choose (provided we do not use several conflicting ones together). The unit we choose comes from earlier history, so it seems an odd size. It is this:

> *1 ampere is that current which deposits copper in a copper-plating bath at a rate of 0.000000329 kg of copper per second.*

And we define current measurement in general thus: the current is measured by the rate of copper transfer in a bath of copper sulphate solution.
Then,

1 amp deposits 0.000000329 kg of copper per sec
2 amps deposit 0.000000658 kg of copper per sec
20 amps deposit 0.000006580 kg of copper per sec
C amps deposit (C) · (0.000000329) kg of copper per sec

This defines our unit of current, 1 amp (short for 1 ampere), and our general scheme of measurement; and it defines current[13] as something proportional to rate-of-copper-plating.

[13] With our definition, by copper-plating-rate, we can investigate magnetic effects of currents experimentally, as Ampère and Oersted did over a century ago. In recent years professional physicists have agreed to change the order of logic in standardizing the system of electrical concepts and units. The modern official agreement defines currents as proportional to their magnetic effects. And the amp is defined in terms of the force exerted by one current on another

PROBLEM 4. AMMETER CALIBRATION

(a) Suppose an unknown current is flowing in a circuit. We make a cut and insert a copper-plating bath. If we find 0.00658 kg carried across in 1000 secs (about ¼ hour), how big is the current?

(b) How could you make sure (without illogic) that the current when the bath is inserted is the same as before?

(c) A small copper refinery puts out 12,000 tons a year, purified by deposition in a plating bath. If the refinery runs day and night, what current does it use? (Make a rough estimate. Take 1 ton $\approx$ 900 kg.)

Quantity of Electricity: Charge

We called the thing that flows around a circuit "electricity," a name 300 years old. As electricity has become a rather vague word, we now speak instead of QUANTITY OF ELECTRICITY or CHARGE OF ELECTRICITY or ELECTRIC CHARGE or just plain CHARGE. So we now say CURRENT is a FLOW OF CHARGE. In this discussion we do not explain what charge is, except to say that it is the something that flows when a current occurs. We may suggest that a charge is a huge bunch of electrons. Later we shall meet charges in the way they were first studied—electricity at rest. We measure CHARGE in units ("chunks of electricity") called *coulombs*.[14] One coulomb is given by a flow of 1 amp for 1 second. Amps are the same as coulombs/sec.

CURRENT in amps
= (CHARGE in coulombs)/(TIME in secs)

With a current of 1 amp, we say that the flow carries one coulomb past each point in the circuit in every second. Post a demon observer at some place in the circuit, and tell him to count the coulombs as they pass. With 1 amp he will count 1 coulomb every second, 60 coulombs a minute, 3600 an hour. With a current of 2 amps, he will count 2 coulombs per second; and so on. Since the current is the same all around the circuit, the observer will count coulombs passing at the same rate whatever part of the circuit he chooses. In the course of time, any particular coulomb—supposing we could label it—will travel right around the circuit and start around again.[15]

A coulomb, then, is a chunk of electricity. When a current flows through a lamp, we picture the coulombs marching through the filament in endless procession. But, when we say all that, we are not explaining electricity. Do we *know* electricity as we know length, Justice, a dime? If we seek to know "what it really *is*," to understand its ultimate nature (whatever that may mean), we are back among ancient problems of philosophy; and as scientists we shall never know. In another sense we know a great deal about electricity and currents and electrons: we know them by what they *do*. In that sense we understand them well, and we expect to learn still more.

Flow of Coulombs

So, following early beliefs and modern knowledge of electrons, we say a current of, say, 5 amps is a *flow* of 5 *coulombs of electricity per second* past each point in the circuit. Flowmeters at different points in a water circuit will all read the same, unless a bulge is developing or there is a leak. An observer noting a flow of 5 gallons/second at one point would agree with an observer stationed to watch the flow anywhere else. We picture gallons of water whizzing past, at the rate of 5 a second. (The actual speed of the water itself involves the width of the pipe as well, and there is a similar aspect of electric flow.[15]) In an electric current of 5 amps we picture coulombs whizzing past any observer at the rate of 5 a second. *A current of C amps means a flow of C coulombs per second, past each point, all around the circuit.*

Then we define *one coulomb* as *that charge which will deposit a mass of 0.000000329 kg of copper*, or carry that mass across in a copper-plating bath. (Such a bath is unfortunately called a copper voltameter, not to be confused with a voltmeter.) Notice how much more careful we have to be in a

---

parallel one, via their magnetic field. (On that basis, magnetic effects are proportional to current by definition, but chemical effects must be investigated experimentally.) That new system makes it easier to maintain accurate standards. Our system here, using copper-plating, makes it easier to build up a general understanding; it makes amps, coulombs, etc., easier to think about—so we shall use it. The two systems are equally logical; a mixture of them would be horribly illogical.

[14] After the French experimenter, Coulomb, who established the inverse-square law of forces between such chunks of electricity some 170 years ago.

[15] Simple experiments tell us nothing about its actual speed, however, because we do not know the total number of coulombs engaged in the flow. We now believe, from indirect atomic evidence, that the current-carrying charge in a metal is enormous. A thin copper wire 1 ft long has some 10^{22} electrons to carry current (counting 2 "conduction electrons" per atom). Each electron has a charge 1.6×10^{-19} coulomb, so the total moving charge is about 1600 coulombs. When the wire carries 1 amp, those electrons must drift around the circuit at 1/1600 ft/sec—nearly half an inch a minute. (That drift, which *is* the current, is superimposed on their random motion through the crystal lattice of metal atoms. And *that* motion, quantum theory assures us, is unexpectedly fast, over 200 miles/sec.)

new strange science about clear consistent definitions—or at least descriptions—and honest statements of our ignorance, compared with our easy use of common-sense knowledge in dealing with, say, pressure measurements.

So far we have not made it clear what a coulomb is, but have simply said that it is something that . . .

PROBLEM 5

Continue the sentence just above.

Electrons

Other experiments, earlier in history, suggest that coulombs are "charges" of electricity, like the charges that we collect by rubbing fountain pens on sleeves or scuffing the floor, or flying a kite in a storm. Later experiments point to universal tiny particles, each carrying a negative charge of 1.6×10^{-19} coulombs, negative electrons, as common building blocks in atoms and the common carriers of currents in metals. Then we can say vulgarly "a coulomb is a bunch of electrons, a huge number of these tiny electric charges." This seems a comforting explanation to many people, and there is little harm in picturing coulombs as such crowds. In a wire a moving coulomb is indeed a moving army of electrons, an army some 6,000,000,000,000,000,000 electrons strong. As a crude picture—dangerous because it paints definite details that we cannot observe experimentally—we imagine some electrons buzzing about loose in a metal wire, like gas molecules.[16] They collide rarely with the massive remainders of the metal atoms (which are anchored in their crystalline lattice, but vibrate with some heat energy). When current flows, the "loose" electrons are driven along the wire by an electric field tugging them along the wire. They gain a little extra K.E. but on the average lose it again in collisions with metal atoms. So we imagine the electrons staggering down the wire, like a horde of excited children staggering down a wooded mountainside, gaining K.E., then losing it by collisions with trees. No wonder the metal atoms increase their vibration, showing a gain of heat-energy.

EXPERIMENT E. TESTING AMMETER

Accepting the copper-plating definition, test one mark on your ammeter. This may be done by the following demonstration experiment to avoid the difficulty of weighing small deposits of copper. Run as big a current as the ammeter will carry, through it and through a huge bath of copper sulphate solution, running the current in by a central wire and across the solution to a large sheet of copper. Wash, dry, and weigh the copper sheet before and after running the current for a measured time of say 1000 seconds. Record the weighings and the ammeter reading. Calculate the current and estimate the error and percentage error of the meter.[17]

PROBLEM 6. COMPARING AMMETERS

(a) Suppose several students want to calibrate their ammeters. Is it necessary to do the copper-plating experiment several times? Give a reason for your answer.

(b) Suppose some students miss the experiment, but find they need to calibrate their ammeters later. How can they safely avoid repeating the copper plating experiment?

(c) What assumptions are involved in your suggestion for (b)?

(d) If too small a copper plate is used, the electrochemical effects are the same but the new copper is crowded on in such a hurry that it makes copper trees and some twigs fall off. How would this affect the ammeter's reputation?

(e) A good way of drying the copper plate quickly is to wash it in tap water, then distilled water, then in alcohol, then set fire to the alcohol and burn it off. Even so, the copper tarnishes, picking up oxygen and getting heavier. How will this increase in weight affect the ammeter's reputation?

(f) Why is it wise to wash and dry the plate before the *first* weighing too?

(g) It is usual to set the apparatus up and run the current for a short time *before* the first weighing, then remove the plate, wash and dry and weigh it. Mention *two* advantages of this.

(h) In itself the copper-plating test of an ammeter is a dull formal experiment, a hairsplitting affair. What then is the point of doing it in this course?

In practice it would be a tedious job to calibrate an ammeter like this at many points on its scale. Trying it for one point shows you the method. In case of need we can obtain standard ammeters calibrated by the Bureau of Standards and compare ours with them.

What Are Voltmeters?

Engineers watching the distribution of electric power far and wide, and physicists making electrical

[16] Modern quantum theory insists that these "loose" electrons have unexpectedly high K.E., at all temperatures, 200 times the K.E. that "equipartition" would award them at room temperature. They also have an unexpectedly long MFP; this shortens with rise of temperature, making electrical resistance greater.

[17] This experiment should not be obscured by delicate weighing techniques. It is important as part of a consistent study of electricity. If ammeters are available with a high range, and a copper sheet of a square foot or more area is used, the gained copper can be weighed easily in public. As a communal or demonstration experiment the plating can be left to run for a long time while other things are discussed.

energy-measurements, need voltmeters as well as ammeters. What are voltmeters?

If we liken an electric circuit to a water circuit, we find a voltmeter corresponds to a pressure gauge. Soon you will find yourself using voltmeters to deal with energy transfers and that use, the real use of voltmeters, will be discussed in the next section. But a vague idea that a voltmeter is an electrical pressure-gauge will help you to start using one. Fig. 32-26 (a, b) shows an electric circuit and the analogous water circuit. The battery B corresponds

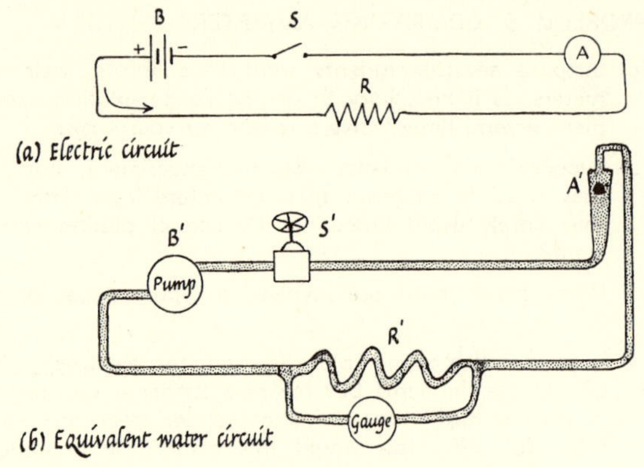

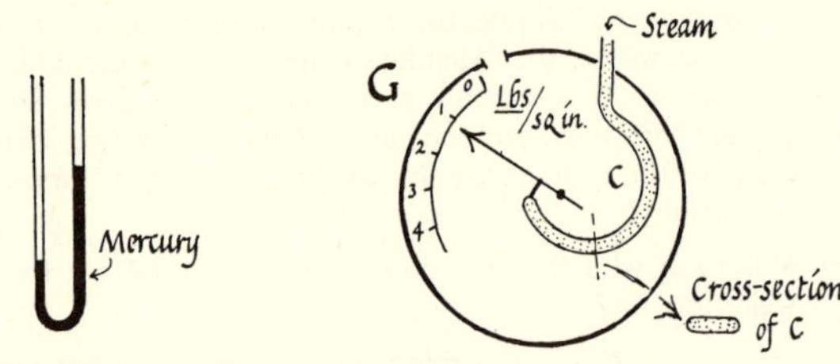

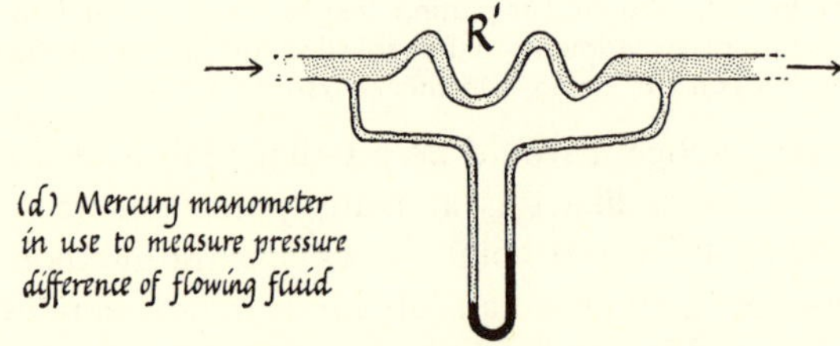

Fig. 32-26 Circuits

to the pump B′, the switch S to the tap S′. The ammeter A corresponds to the flowmeter A′. The resistance R is represented by the narrow pipe R′ full of obstructions.

As Joule found, such a pipe makes the moving fluid develop a little heat. If we wish to measure the pressure used to drive the flow through the pipe R′ we use some kind of pressure gauge. This might be a U-tube filled with mercury. In this, the level-difference measures the pressure-difference between the ends of R′. The U-tube really measures pressure-*difference*, even if there is vacuum on one side. The common steam pressure gauge, used on steam boilers has a flexible pipe C inside which is unbent by the steam pressure. The steam blows into one end of the pipe C and the other end is closed. The pipe itself is not made from round piping but is oval in cross-section, so the steam pressure makes it try to uncurl, like the paper toys which have tongues that uncurl. The uncurling of C measures pressure on a scale with pointer. The larger the pressure, the more C uncurls, against its own elastic forces.

PROBLEM 7. PRESSURE GAUGES

(a) In the pressure gauge G (Fig. 32-26c) when steam has pushed the pointer across, there are two things we could do to send the pointer back. One is remove the steam pressure. The other is maintain the steam pressure, and . . . ? (*Hint:* Such gauges have a small hole in their outer case.)

(b) In the light of (a) does the gauge really measure "absolute pressure" or pressure differences?

Most practical pressure-gauges are really pressure-difference gauges. In Fig. 32-26d such a gauge is shown connected to measure the pressure difference that maintains the flow through the resistance-pipe R′.

PROBLEM 8. CAUSE : EFFECT

Is it fair to say that pressure difference causes flow, or should we say flow causes pressure difference, or both? Discuss briefly

(a) from the point of view of a man driving a pump (Fig. 32-27a);

(b) from point of view of a pilot in a plane with a hole in its nose (Fig. 32-27b).

(c) In each case, could experiments determine which is which of cause and effect?

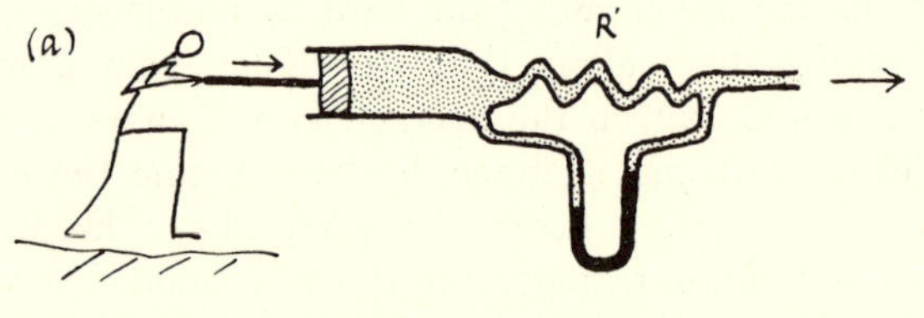

(b) Pilot in plane with hole in cockpit feels wind pressure

Fig. 32-27. Problem 8
Two views of pressure and flow.
Which is cause, which is effect?

EXPERIMENT F(i). VOLTMETER

Draw and set up circuit shown in Fig. 32-28 and *add* a voltmeter as an electrical pressure-gauge to measure the "pressure difference" used by the lamp L. (This is the time to explore with the voltmeter.

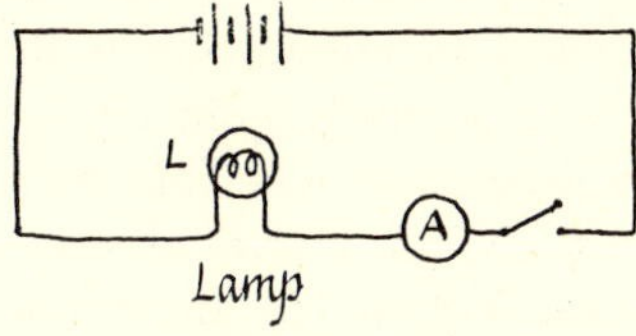

FIG. 32-28. EXPERIMENT F(i)

If you are not sure how it should be connected, try various arrangements for yourself. "Wrong" ones may be informative too.)

EXPERIMENT F (ii). VOLTMETER

Insert a variable resistance in the circuit to change the current. Note the effect of changing the current on the voltmeter reading. (This is a rough preliminary experiment; measurements will be asked for later.)

Electric Currents and Energy

The chief use of electric currents in modern civilization is to transfer energy quietly, flexibly, and conveniently. The cables from power station generators to factory motor act as an ideal transmission belt, with variable gear ratios included. Electric lamp filaments are heated with energy from coal burning, or water falling, miles away. Early in the last century, before the Conservation of Energy reached its full glory, it was found that electric batteries "use up" their chemicals when they run lamps or motors. When generators came into use, they proved harder to drive when they were supplying lamps or motors than when running unloaded. Joule's studies wove electrical energy, chemical energy and heat energy into a magnificent interchange system with mechanical energy.

Picture of an Electric Circuit

Picture electrical energy being carried to various parts of the circuit by the current. Coulombs, hordes of electrons, surge around the circuit in a steady stream. They come out of the battery well provided with potential energy (6 joules each coulomb, from a 6-volt car battery) which they receive via the electric field and spend as they stagger around the circuit. They skim gaily through thick wires, hardly paying out any energy for that part of their trip. In wires of poorer conductivity they push through the difficult terrain, stumbling, tripping, kicking—leaving the atoms of the wire violently vibrating with extra heat. In thinner wires the crowd of electrons must drift faster, for the same current, gaining and losing *more* K.E.; hence there is more heating. Passing through a motor they push uphill against counter pressures, as if against some imaginary piston, giving up energy to mechanical form. Finally they slide back into the battery, which draws on chemical supplies to give them a new stock of energy to make another trip.

Even inside the battery itself there is some resistance, leading to some waste of energy as heat. In good storage batteries used for cars, this resistance is small, so that heat wastage is small except when the current is very big. In running a car-starter the heating may be big. Or, if the battery's plates touch by mistake, making a short circuit inside, the heating may be very big, causing the plates to buckle—then the short circuit worsens and the battery is ruined.

Voltage

As coulombs trail through the resistance, we picture them delivering energy which they seem to have carried along the wire from the battery. The battery suffers chemical changes. Joule found that the same amount of chemical change can happen if the same chemical ingredients are just thrown into a beaker; but then heat is released.

PROBLEM 9

If you took battery and circuit and heaved the whole arrangement, chemicals, wire and all, into a beaker, heat would be developed in the beaker. When Joule put the chemical ingredients straight into a beaker and obtained heat, the hunks of metal—e.g., zinc and copper—that he used, touched. In what way could the latter "plain chemistry" experiment still be considered electrical?

With the circuit, the heat is not delivered among the battery chemicals; instead it appears in the wires (or, if a motor is substituted, mechanical P.E. may be provided instead). The ammeter tells us how fast coulombs travel through the wire, but we cannot tell how much heat (or other energy) is delivered until we also know *how much energy each coulomb delivers*, or releases in passing through the wire. This energy-delivered-by-each-coulomb, or energy-transfer per coulomb, is a very important and useful measurement for calculating energy transfers, and we shall call it *voltage*. Suppose we have some instrument to tell us that each coulomb delivers 4 joules of heat in passing through the wire; and an ammeter telling us that the current is 3 amps (or 3 coulombs per second) passing through the wire. Then *in every second 3 coulombs pass through the wire, each delivering 4 joules.* Therefore, the current is delivering energy at the rate of *12 joules per second,* or 12 watts.[18] The circuit is changing electric energy to heat at a rate of 12 watts.

The energy transfer per unit charge is a useful thing to know. Ideally, we could measure it by taking a few sample coulombs through an electric motor that is hauling up a load, and measuring the P.E. gained by the load. In practice this would be clumsy, and it would be inaccurate because real motors waste some of their electrical input-energy as heat. Instead we use later electrical knowledge to construct a measuring instrument, a *voltmeter.* It measures the energy (in joules) transferred by each coulomb from electrical form to heat, etc. It measures "joules/coulomb." We give the unit 1 joule/coulomb the name 1 *volt,* after the Italian scientist Alessandro Volta who made the first batteries.

There are voltmeters which make a genuine measurement of force and put it into a calculation of FORCE · DISTANCE/CHARGE, in coulombs. But ordinary voltmeters are really trickle-meters in disguise. They measure a tiny current that leaks through them and use their makers' superior knowledge to interpret this in volts. However, you should take your laboratory voltmeter as a closed box, not to be opened, and satisfy yourself *by direct tests* that it does read joules/coulomb. You may picture the voltmeter enticing an occasional specimen coulomb along the detour through it, and then—like a gangster on a side-road saying "shell out your money"—making the coulomb give up its energy. The meter is graduated to read the energy thus given up by each specimen coulomb of the trickle. Energy-transfer per unit charge is too long a name for the thing measured so we shall call it P.D. or VOLTAGE. "P.D." stands for electrical "POTENTIAL DIFFERENCE," an old name which was devised in connection with stationary electric charges. It is still the official name and we shall use it, contracted to P.D. You may think of it as standing vaguely for electrical *pressure difference* if you like. Voltage is an engineer's word, roughed out of the name volt for unit P.D. It has become quite respectable in common use, so we shall use voltage interchangeably with P.D.[19]

Voltmeters as Electrical Pressure Gauges

Voltmeters will still seem easier to use if you think of them as electrical pressure gauges. There is a real similarity. Here is the justification: we build voltmeters to read ENERGY DELIVERED BY EACH COULOMB, in some chosen portion of circuit. This is the ratio,

$$\frac{\text{ENERGY DELIVERED (in some portion of the circuit)}}{\text{NUMBER OF COULOMBS FLOWING THROUGH}}$$

or

$$\frac{\text{ENERGY DELIVERED (in joules)}}{\text{TOTAL CHARGE (in coulombs)}}$$

We can express pressure-difference in some part of a water circuit in a similar form:

$$\frac{\text{ENERGY DELIVERED (in joules)}}{\text{TOTAL VOLUME MOVING THROUGH (in cubic meters)}}$$

The following argument shows this. Suppose we push water through a long thin pipe, driving it forward with a piston P_1 at the beginning of the pipe, and letting it push a piston P_2 at the other end. The pipe has resistance, so that the water needs greater pressure p_1 at P_1 than p_2 at P_2. The pressure difference $p_1 - p_2$ drives the water through the pipe against resistance forces. As Joule found, the water emerges slightly warmer having gained heat at the expense of some of the energy supplied by piston P_1. Suppose piston P_1 has area A_1 square meters and is pushed a distance s_1 meters against water-pressure p_1 newtons/sq. meter. The force needed to push P_1 is

[18] This is the first of many very important electrical calculations. Do not be intimidated by the scientific names. Read "loaves of bread" for "joules of energy" and "trucks" for "coulombs" and the story runs: If 3 trucks a day pass through my yard, and each truck delivers 4 loaves of bread, how many loaves do I receive per day? Answer:

$$3\,\frac{\text{trucks}}{\text{day}} \cdot 4\,\frac{\text{loaves}}{\text{truck}} = 12\,\frac{\text{loaves}}{\text{day}}$$

This is arithmetic of earliest schooldays. Many important electrical calculations are just as easy. As a check, make sure the units cancel out sensibly as they did above. For example, if you lost your head and *divided* you would have had

$$\frac{4\,\dfrac{\text{loaves}}{\text{truck}}}{3\,\dfrac{\text{trucks}}{\text{day}}} \text{ or } \frac{4}{3}\,\frac{\text{loaves}}{\text{truck}} \cdot \frac{\text{days}}{\text{truck}}$$

or 1.33 loaf · days per square truck—completely wrong. Beware of the medical man who is afraid of mathematics and, dividing pills per patient by patients per day, prescribes pills · days per square patient.

[19] "Voltage" has crept in, like "footage" in the movie industry. But "amperage" is still far outside the bounds of good English. You would say, "My height is 68 inches," putting the thing first, then the number, and the unit last. You would not say, "My inchage = 68 inches." As an educated scientist, you should avoid "amperage."

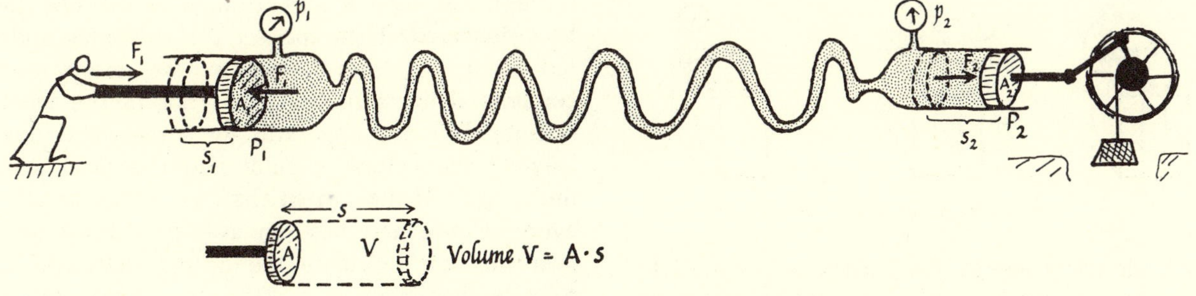

FIG. 32-29. HYDRAULIC ENERGY-TRANSFER

PRESSURE · AREA or $p_1 \cdot A_1$ newtons and the energy transferred *from* the agent pushing the piston *to* the water is: FORCE · DISTANCE MOVED or $p_1 \cdot A_1 \cdot s_1$ joules. Similarly, the energy transferred *from* the water to the victim pushed by piston P_2 is $p_2 \cdot A_2 \cdot s_2$. But the PISTON-AREA · DISTANCE MOVED is the VOLUME of water shoved along the pipe, V cubic meters.

$\therefore A_1 s_1 = V = A_2 s_2$ since water is not compressed on the way. Then the difference of the two energy-transfers, which is the mechanical energy lost to the water, is

$$p_1 A_1 s_1 - p_2 A_2 s_2 = p_1 V - p_2 V = (p_1 - p_2) \cdot V$$

$\therefore$ the energy delivered as heat in this region is $(p_1 - p_2) \cdot V$ joules. Then

$$\frac{\text{ENERGY DELIVERED (in joules)}}{\text{VOLUME MOVED (in cu. meters)}} = \frac{(p_1 - p_2) \cdot V}{V}$$

$$= p_1 - p_2, \text{ in } \frac{\text{newtons}}{\text{sq. meter}}$$

So, in a hydraulic system, PRESSURE DIFFERENCE, $p_1 - p_2$ newtons/sq. m., is given by ENERGY DELIVERED/VOLUME MOVED. This is analogous to the description:

electrical "PRESSURE DIFFERENCE"
= ENERGY DELIVERED/CHARGE MOVED.

In professional science we never speak of electrical pressure difference, but call it potential difference, or P.D.

PROBLEM 10

"PRESSURE DIFFERENCE = ENERGY DELIVERED PER UNIT VOLUME" is a perfectly good statement about pressure. What units does it give for pressure if

(a) energy is in foot · pounds and volume in cubic feet?

(b) energy is in joules and volume in cubic meters (justify your answer by showing how you obtain it).

(c) Is this view of pressure consistent as regards units with the kinetic theory statement for gases:

PRESSURE · VOLUME
$= \frac{1}{3}$ NUMBER OF MOLECULES · MASS · $\overline{v^2}$?

Connecting Voltmeters in Circuits

The thing that voltmeters measure is: electrical potential difference, loosely described as electrical pressure difference, accurately defined as energy-transfer per coulomb. The voltmeter deals only with the energy transfer in a specified region, such as a motor or a coil of wire. The voltmeter's binding posts must be connected to the ends of that region. Since it measures the *difference* in the coulomb's energy as it enters and leaves the region, it must be connected to both entering and leaving points, just as a pressure-difference gauge must be connected to two places—though often one place is the open atmosphere. So we speak of the voltage or P.D. *between* two points in the circuit, or *between* the ends of some apparatus. And we speak of connecting the voltmeter *across* the apparatus. This last wording is helpful in drawing or connecting up circuits containing voltmeters. Unless you are completely confident, *draw and set up the circuit first without any voltmeter*, and then *add the voltmeter* across the region where you want to measure the P.D.

Tests of Voltmeter's Behavior and Accuracy

Do the following experiments on voltmeters to test the word of the physicists who designed them, and the people who made them, that they do measure energy-delivered per coulomb.

EXPERIMENT G. TEST OF NATURE OF VOLTMETER SCALE

(Voltmeter range 6 volts or more.) Try connecting the voltmeter directly to a car battery of three cells; then to only two out of the three cells, then to one cell only. Record its readings.

Also make sure that all three cells are equal by a preliminary experimental test, recording the voltmeter readings. If the voltmeter does measure

ENERGY-DELIVERED PER COULOMB,

then when applied to a battery it must tell us how

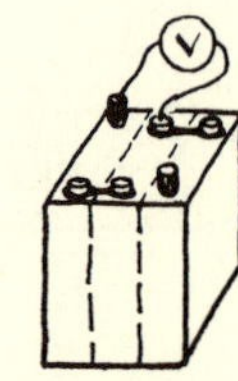
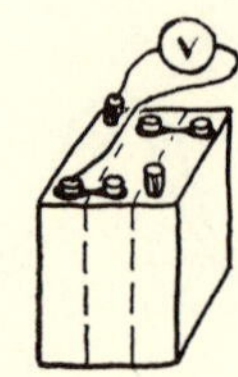

Fig. 32-30. Experiment G

much electrical energy the battery is prepared to give each coulomb, at the expense of chemical energy. The picture is this: the battery converts chemical energy to electrical potential energy which the coulomb then delivers in other forms (heat, etc.) on its way around the circuit. If the coulomb goes through only one cell it gains so much energy (one "boost"). If it goes through 2 cells in series, our energy ideas suggest it will gain twice as much potential energy—boost and boost, each cell providing its full share unaware of the other cell's help. With three equal cells in series, three times as much—boost, boost, and boost. This is justified by the experimental fact that the same amount of chemical change occurs in each cell, for one coulomb passing through, whether the cell is alone or there are two, three, . . . , cells in series. This experiment does not prove the meter measures energy/coulomb, but you can say whether your observations are consistent with such a view.

EXPERIMENT H. TESTING VOLTMETER SCALE AT ONE POINT

This is a precise experiment, giving an actual calibration. Make an arrangement in which the energy delivered per coulomb is measured directly, and let the voltmeter also "measure" it. The easiest arrangement is to have a resistance wire R deliver heat to water in a well-protected beaker ("calorimeter"). Draw and set up a circuit using a good battery, a compact resistance wire R immersed in water, and a variable resistance to keep the current at a fixed value during the experiment. Use ammeter and clock to measure the coulombs passing through the wire R and arrange to measure the heat delivered. Then connect the voltmeter under test so that it tries to measure the energy each coulomb delivers *in the wire* (not in the whole circuit). Run the apparatus for a measured time, keeping the voltmeter pointer at the fixed mark under test. (If the current changes, record its value every minute and take an average.) From your measurements calculate the energy delivered, in calories, then in joules, and the total charge delivering it, in coulombs. Thence calculate the *true voltage*, assuming your energy-measurements are accurate. Compare this with the *meter's* estimate and state the meter's error.

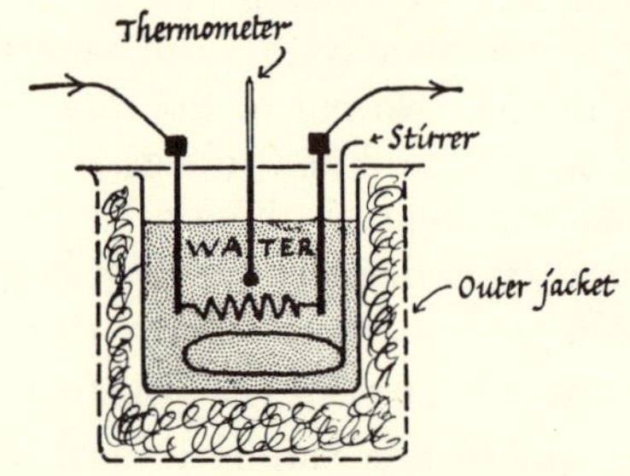

Fig. 32-31. Experiment H
"Calorimeter" with Resistor and Stirrer

This experiment is difficult to run. The apparatus needs adjusting beforehand, and these instructions will seem insufficient. How much water should be used, for instance? How long should the current run? How can you have the right voltmeter reading the moment the current is turned on? All these cry for either preliminary experimenting or much fuller instructions. Providing a set of detailed instructions for the apparatus of your laboratory would be cook-book treatment suitable for a class of slow children. It might test your obedience, but it would hardly increase your understanding of science. Here, as in most laboratory work, it is worth your while to experiment roughly and think about improvements of your own apparatus and procedure. Start by running the apparatus roughly, then discuss its best running for lessening errors. (In that discussion you may need considerable comment from those who know your laboratory's apparatus, but you can treat that as expert help rather than cook-book orders.)

In your experiment, certain errors are likely to be serious.

PROBLEM 11

What are the mysterious "errors" in the test above? Would they make the voltage calculated from your measurements higher than the true voltage or lower? How could you lessen them?

Much more reliable schemes for giving a voltmeter an absolute test can be constructed, but they are complicated and indirect.[20] They all involve

[20] One, an ultimate court of appeal in electrodynamics, consists of making a very primitive generator whose voltage can be calculated directly from its dimensions and its speed of running and a measured current which is used to produce a magnetic field. This will provide a standard voltage. (Usually this generator is made to provide a standard resistance without requiring any measurement of current. Its EMF is matched against the P.D. produced across the resistance by the current used for the magnetic field. That is how the standard ohm was arrived at).

some measures of force and distance, current and time to make a voltage known in joules per coulomb.

This experiment shows the underlying idea. It is not worth while repeating it for several points on the voltmeter scale—trust the makers instead, or ask to compare it with a standard voltmeter which in turn has been carefully calibrated.

PROBLEM 12. COMPARING VOLTMETERS

Suppose you had not carried out your voltmeter test but wanted to compare your instrument quickly with a standard voltmeter that had been tested. Draw the circuit you would use, for comparing them at several points on the scale.

NOTE: Some people like to use the voltmeter experiment with coil in water to "measure the mechanical equivalent of heat," as further testimony in favor of Joule's work and the great generalization of the conservation of energy. In that case they cannot *also* use this experiment to test the voltmeter. They must assume the voltmeter has already been tested, presumably against the absolute generator mentioned in footnote 20. The experimental work is the same, of course, but the subsequent argument differs. In this course it seems better to use the experiment to gain assurance over voltmeters. At any rate do not use it for a mixture of both purposes—that would indeed be unscientific.

If you agree with the view that voltmeters measure energy delivered per unit charge, each volt meaning a joule/coulomb, you can proceed with two important matters: power calculations, and an experimental investigation of "Ohm's Law."

Power

If you magnify your experience as an electrician into complete assurance that voltage *is* energy-transfer per unit charge, you can discuss electric power supply. Teach yourself by working through the following problems:

★ PROBLEM 13. CALCULATING POWER

Suppose an ammeter inserted in a lamp circuit shows the current is 3 amps and the lamp is running on a 120-volt supply.

Questions:

(a) How much energy is dissipated by the lamp, as heat and radiation, in 10 seconds?*

(b) What power is being "used" by the lamp; i.e., at what rate is the lamp transferring energy from electrical form to heat, etc.?*

* In answering these questions you should copy out, and complete, the specimen answers given here. Using these rigmaroles will make later problems easier.

Answers:

(a) (i) The current is 3 amps; *THIS MEANS THAT IN EVERY SECOND 3 COULOMBS FLOW THROUGH THE LAMP.*
∴ In 10 secs the charge passing through the lamp is ________ · ________.
units

(ii) The voltage across the lamp is 120 volts. *THIS MEANS THAT EVERY COULOMB PASSING THROUGH THE LAMP DELIVERS 120 JOULES* (in the lamp).
(or: *each coulomb transfers 120 joules from electrical energy to heat, etc., in the lamp.*)

Now combine the information of (i) and (ii).

∴ In 10 secs, ENERGY DELIVERED
= (______ · ______) · (______ · ______)
units
= ________ · ________
units

(b) ∴ POWER = RATE OF ENERGY-TRANSFER
= ENERGY-TRANSFER/TIME
= ______ · ______
units

But one watt is shorthand for one *joule/sec.*

∴ power . . . . = ______ · ______
new units

★ PROBLEM 14. POWER

An automobile lamp takes 12 amps from a 6-volt battery, through thick wires. Calculate the power being used, in watts. In your calculation, include rigmaroles like those which are in *CAPITALS* in the problem above, modifying them to fit this problem.

★ PROBLEM 15. POWER

An electric motor takes 20 amps from a 100-volt supply. Three quarters of the power is used to haul up a load. The remaining quarter goes into waste heat in the motor.

Questions:

(a) Calculate the useful power.*

(b) Motors are often rated in horsepower. Give a rough estimate of this motor's rating in HP, on the score of useful power.*

(c) If the load is 110 pounds, how fast does it rise vertically?*

(d) Suppose a 1000-volt supply is available for the same load-raising job, and a suitable motor is provided that runs normally on 1000 volts instead of 100. What current would the new motor take, and why?
(Assume: same 110-pound load is to be raised at same speed; new motor has same 75% efficiency.)

(e) What advantages and disadvantages do you think the 1000-volt arrangement would have compared with the 100-volt one?

(One HP is 746 watts, or about 3/4 kilowatt. Take 1 kilowatt as 4/3 HP; and 1 HP as 550 ft · pounds-wt./sec.)

Answers:

(a) The CURRENT is 20 amps. *THIS MEANS THAT* . . . . .

The VOLTAGE is 100 volts. *THIS MEANS THAT* . . . .

∴ In each second, energy delivered is . . . .

∴ POWER is ______ · ______
units

∴ USEFUL POWER is ______ · ______
units

* You should copy out, and complete, the specimen answers given here.

(b) On score of useful power, motor should be rated about ———— HP.

(c) USEFUL POWER, from (b), = ———— HP,

= ———— ft. · pounds-wt./sec.

FORCE exerted must be 110 pounds-weight; and, if the load rises with SPEED X ft/sec, power is, (using X),

———— ft. · pounds-wt./sec

∴ ———— ft · lbs-wt./sec must be the same as ———— ft · lbs-wt./sec

∴ SPEED OF LOAD, X, must be . . . . . ———— ft/sec.

(d) and (e) See question.

Algebra Formula for Power

Instead of repeating these rigmaroles concerning amps and volts for each problem, we can run through them once for all with symbols and arrive at an expression which will tell us the power in every case. If this seems to you sensible and a good economy, proceed; and use its result. If it seems puzzling mathematics, avoid it. (Do not learn it as a formula to get right answers quickly in examinations, because the examinations, if they are sensible ones, will ask you to show where the formula comes from!)

PROBLEM. A lamp (or motor, or something else) runs on a supply with voltage V volts, taking current C amps. Calculate the energy-transfer in t seconds, and the power.

ANSWER. The current is C amps. *THIS MEANS THAT, IN EACH SECOND, C COULOMBS FLOW THROUGH THE LAMP.*

∴ in t seconds charge flowing through lamp is $C \cdot t$ coulombs.

The voltage is V volts. *THIS MEANS THAT EACH COULOMB PASSING THROUGH LAMP DELIVERS V JOULES.*

∴ in t seconds, $C \cdot t$ coulombs pass, each delivering V joules.

∴ in t seconds, ENERGY delivered (transferred) is $Ct \cdot V$ or $V \cdot C \cdot t$ joules.

$$\text{POWER} = \text{RATE OF ENERGY-TRANSFER} = \frac{\text{ENERGY TRANSFERRED}}{\text{TIME}} = \frac{V \cdot C \cdot t}{t} = V \cdot C \text{ joules/sec or watts}$$

$$\therefore \text{POWER} = V \cdot C \text{ watts}$$

The power is VC watts: *THIS MEANS THAT IN EACH SECOND VC JOULES ARE DELIVERED TO THE LAMP AND DISSIPATED BY IT.*

PROBLEM 16. POWER LINE

Here is a simple model of a commercial power supply problem. We shall develop it into a more realistic form later. A farm with a power plant of its own supplies a cottage some distance away with electric power. The cottage takes 5 amps. This current runs from the power plant along a

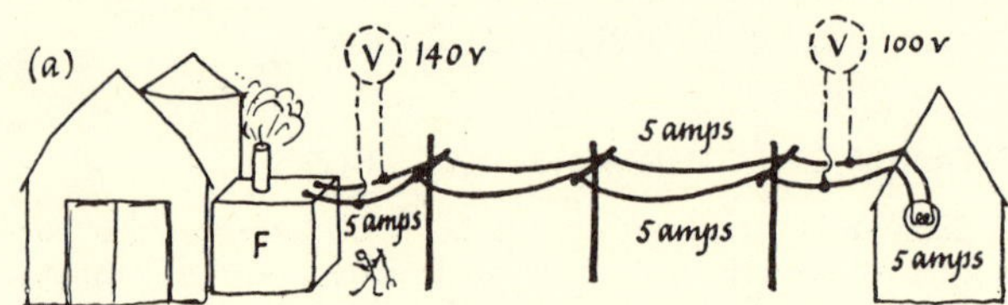

FIG. 32-32a. PROBLEM 16

copper wire on poles, through lamps in the cottage and back along an equal copper wire on poles. A voltmeter applied at the cottage shows the P.D. is 100 volts across each lamp (and therefore across the supply wires). A voltmeter across the power plant reads 140 volts.

(a) Calculate the power used by cottage.

(b) Calculate the power supplied by power plant.

(c) Where do you think the difference of power [(b) — (a)] goes? In what form and where?

(d) Fig. 32-32b shows the system, with two extra voltmeters installed. What would you expect V_1 and V_2 to read? (This requires careful calculation which should be explained.)

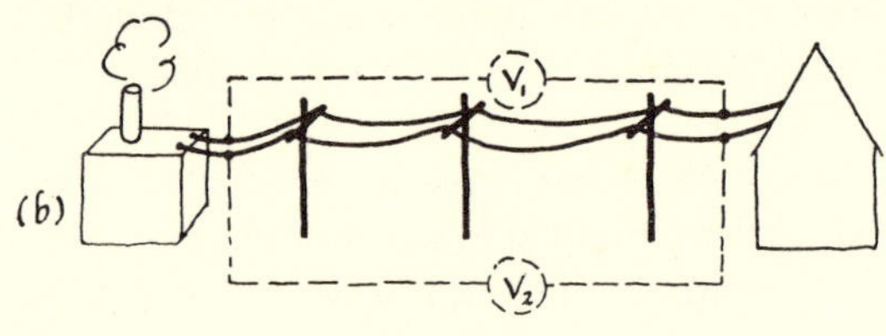

FIG. 32-32b.

PROBLEM 17. ECONOMY

An old theater wired for lighting stage plays with many 100-watt bulbs finds that modern progress demands brighter lighting, and proposes to replace every lamp by a 200-watt one.

(a) Assuming that the new lamps have the same efficiency (at best 2 candlepower per watt) say how this change should affect the total light. How would it affect the total current?

(b) Why would you expect this change to increase the heating of the *supply wires in the theater walls?* (A simple general reason is asked for. Do not give calculations. Even if you used engineers' formulas and calculated the heating, your answer would be wrong. With real copper wires the heating increases more than the formula says!)

(c) Suppose—as in the case of a real theater that met this problem—the increase of heating (b) is considered dangerous.
   (i) Where would the danger be, in lamps or in the walls?
   (ii) How could the danger be avoided by changing to a different supply-voltage (with an appropriate change of lamps)?

OHM'S LAW

Ohm's Work

Voltage and current are useful measures in electric circuits. One of the main uses of electricity is to carry energy quickly from place to place and deliver it in suitable form. The product P.D. · CURRENT gives the POWER, the rate of such energy-transfer. To keep track of power in a public supply system or in a private experiment, we seem to need two instruments, a voltmeter and an ammeter.[21] Could we do without one of the instruments and compute POWER from some characteristic of the circuit, such as its "resistance," together with the reading of the remaining meter? This proves to be a very profitable line of thought. What *is* the RESISTANCE of a wire, or a whole circuit, anyway? Do wires have a constant characteristic thing that could be called resistance—as water-pipes do, and pipes in a vacuum system? In such pipes, the ratio

$$\frac{\text{PRESSURE DIFFERENCE, "driving the flow"}}{(\text{FLOW})}$$

is usually a constant, characteristic of the pipe itself. Again, heat-flow in conduction along a bar follows a simple relationship involving temperature-difference, cross-section area, and length. The discovery of the equivalent relation for electric circuits is a rare example of a successful planned search based on a hopeful guess.

G. S. Ohm, a German schoolmaster, set out to look for such relationships in the 1820's. He longed for fame which would carry him into university life; and he chose a field of research where he had special advantages. He was the son of a locksmith, so he knew how to draw metal wires of different sizes for his experiment. At that time, wires were not available in stores in great variety as they are in our electrical civilization. Ohm made his own wires and experimented with them, trying different lengths, different thicknesses, different metals, and even different temperatures—varying each factor in turn, like any good scientist. Batteries were still incompetently weak and variable, so he used a thermocouple as his generator, the hot junction kept heated by a flame. He used a crude magnetic ammeter and he measured P.D.'s—which he called "tensions"—by changing the temperature or number of thermojunctions.

The science of electric circuits was still young. Started by the invention of batteries about 1800, it was growing like a tropical plant in a rich climate. Apparatus was made, often by hand, instruments designed, laws discovered, concepts and terminology framed, and general principles evolved—all leading to a greater understanding. This understanding of "electricity" took its place on the one hand as a new field of physical science and formed the basis, on the other hand, for a great development of engineering; batteries, generators, supply systems for lighting and power, furnaces, motors . . . the electrical civilization. Ohm's discoveries were of tremendous importance in both developments. They enabled circuit behavior to be predicted easily, first for steady direct currents, later (in equivalent form) for alternating currents. His book of some 250 pages, published in 1826 to set forth his theoretical discussion and experimental results, was received with scorn. Crude, directly planned experimenting was unwelcome in the philosophical fashion of the day; and the

> ". . . Minister of Education pronounced it as his opinion that 'a physicist who professed such heresies was unworthy to teach science.' There was nothing for Ohm to do but resign his secondary-school teaching position. Having failed to secure a university appointment because his work was not thought to be experimental, he now lost his job because in other quarters it *had* been considered experimental.
>
> "For six years Ohm lived from hand to mouth in obscurity and bitter disappointment. Gradually, however, his work became known, at first outside of Germany. This led to honors from abroad which forced unwilling recognition at home. At last, in 1849, twenty-two years after the publication of his book, he received a professorial appointment at the University of Munich, which he held, busily and happily, for the five years before his death in 1854."[22]

Ohm discovered a clear Law relating current and P.D. for a single wire (or part of a circuit, or a whole circuit). And he discovered rules for the effect of changing from one size of wire to another. He is justly famous; and his name is given to the Law. Try to remake his discovery for yourself, with modern apparatus, in the following laboratory experiment.

EXPERIMENT I. CURRENT AND P.D.

How does the P.D. *between the ends of a resistance wire depend on the* CURRENT *through it?* Using a

[21] A voltmeter and ammeter can be combined into a single instrument called a wattmeter but such instruments are expensive, rather clumsy, and not in general use.

[22] From *Physics, The Pioneer Science*, by Lloyd W. Taylor (Dover Publications). Reprinted by permission of Dover Publications, Inc., New York.

length of "resistance wire"[23] as victim, find how the P.D. between its ends depends on the current through it. Arrange a circuit to send a measured current through the victim, including some means for changing the current over a wide range. Arrange a voltmeter to measure the P.D. across the victim. Make a set of measurements over as wide a range of current as possible. In your record, examine your measurements looking for any simple relationship. Also plot a graph to exhibit your measurements. Draw what conclusions you can from your graph.

EXPERIMENT J. (Optional)

Repeat Experiment I using as victim a wire of some single metal, such as iron, or tungsten (in a vacuum).[24]

EXPERIMENT K. (Optional)

Repeat Experiment I with a carbon filament bulb.

EXPERIMENT L. (Optional)

Repeat Experiment I with a slab of "thyrite." (Thyrite is used as a safety device, to cut off excessive currents.)

EXPERIMENT M. RADIO TUBE

Repeat Experiment I with a simple radio tube ("diode"). This is a glass bulb, with a very good vacuum in it, with two pieces of metal sealed into it. One is a metal plate, the other a tungsten filament which can be heated electrically. Investigating the volt : amp characteristic of this tube will do far more than just tell you about one part of a radio set. It opens the study of electron streams, electron guns, oscilloscopes, television tubes, etc.

With the usual supply, such as a 120-volt battery or generator, varying the voltage for this victim would require a rheostat of huge resistance, which

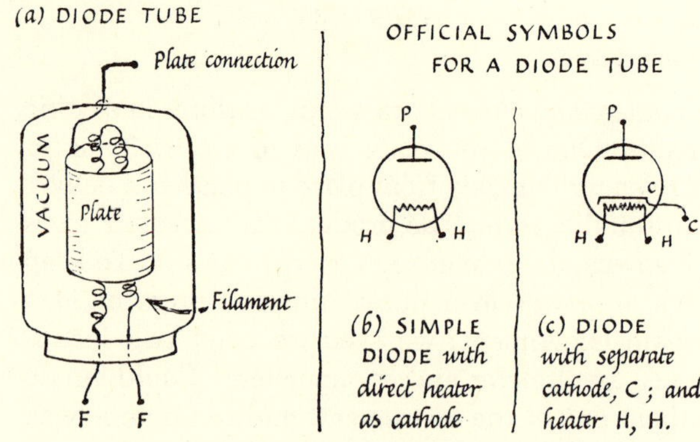

FIG. 32-33. A SIMPLE FORM OF RADIO TUBE (diode)

may not be available. Instead you should use an ingenious scheme called a "potential divider" described below.

Potential Divider

This is a scheme for providing a smoothly variable voltage from a fixed high-voltage supply. Choose a rheostat (variable resistance) that can safely have its complete resistance connected across the main supply. Connect its fixed ends, A, B, to the supply. Then run wires to the apparatus from the slider, S, and from one fixed end, B. When the slider is at the top, A, the apparatus is connected to A and B. What voltage is then applied to the apparatus? When the slider is at the bottom, B, the apparatus is connected to B and B. What voltage is then applied to the apparatus? When the slider is halfway down, say at C, the apparatus is connected to C and B. Guess the voltage applied to the apparatus. To check your guess, follow the argument below—and then go and try the arrangement, using a voltmeter as your apparatus.

Suppose in this scheme the full supply voltage is 120 volts. Then a coulomb traveling along the resistance from A to B delivers 120 joules. By half-way, C, the coulomb will have only 60 joules left to deliver. If it travels down the rest of the resistance, it will deliver 60 joules. If it travels, instead, from C around through

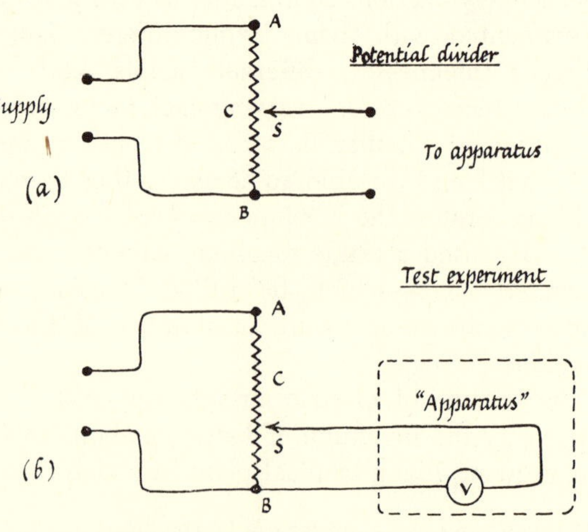

FIG. 32-34. POTENTIAL DIVIDER

[23] All wires have resistance. The term "resistance wire" is misleading. But thin wires, and wires made of special alloys, have so much greater resistance than the same lengths of thick copper wire that we use the phrase to emphasize the contrast. In this experiment you should be given a wire of special alloy that has relatively high resistance *and* changes its resistance very little when the temperature changes. Single metals, such as copper, iron, &c. increase their resistance about 4% for a 10 degree rise in temperature; but some alloys keep an almost constant resistance. "Constantan," an alloy of 60% copper and 40% nickel melted together, changes its resistance by 0.05% or less for a 10-degree rise. It is cheap; it should be used for this experiment.

[24] Tungsten wires, housed in vacuum or inert gas which will save them from oxidizing, can be obtained cheaply under the name of lamp bulbs. (The official name of this metal has recently been changed back to Wolfram.)

the apparatus and back to B, it still delivers the remaining 60 joules, so the apparatus finds a P.D. of 60 joules per coulomb, or 60 volts, applied to it. (This simple argument applies where the "apparatus" has high resistance so that relatively little current is diverted through it. If its resistance is comparable with that of AB, we must modify the argument—and we must be careful not to burn out the portion AS.)

★ PROBLEM 18. USE OF POTENTIAL DIVIDER

A uniform sliding resistance is used as a potential divider. It is connected to a 120-volt supply as in the sketch. When the slider is 3/4 of the way down from a to b,

(1) What is the voltage applied to the apparatus connected to b, s?

(2) Give an explanation or justification of your answer to (1).

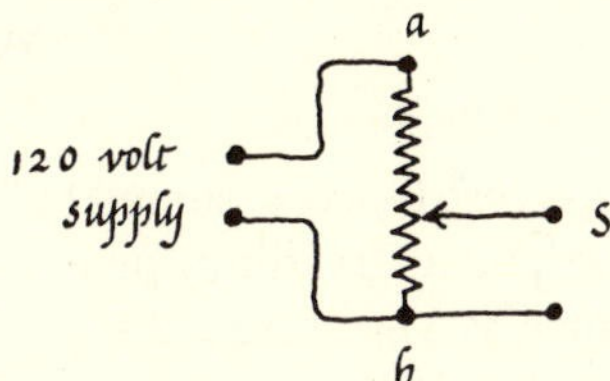

FIG. 32-35. PROBLEM 18

EXPERIMENT M (1). RADIO RESISTOR

Use a high-voltage supply, such as 120 volts D.C. with a potential divider to apply a variable voltage to the radio tube. Measure the P.D. across the victim and the current through it as before. To prevent the new circuit from obscuring the new victim's behavior, *try it first on a familiar victim, an ordinary wire resistance.* For comparison with the radio tube, this should have a very high resistance; a wire-wound radio resistor is suitable.

To see whether the victim behaves symmetrically, you should make measurements with the current running one way through it and then with the current running the opposite way through it. At first thought, it seems simple to arrange for reverse currents: just reverse the connections from battery or wall supply, as in Fig. 32-36. But then the voltmeter and ammeter will both give backward deflections. Quite right too: the P.D.'s and currents should now be reckoned negative. However, this is bad for the meters so you must also reverse the connections to each meter, but apply minus signs to their readings. Thus you should reverse connections *from* the supply and reverse connections *to* both meters. There is a much easier way to achieve the same result: use it! Then plot a graph to show the radio resistor's behavior. For most Ohm's Law investigations, we plot P.D. upward and current along. But radio engineers always plot current up and P.D. along in dealing with their tubes. So you should plot your graph with current up and P.D. along for comparison with your next graph of the radio tube's behavior. Arrange your graph with the origin in the center so that you can plot the negative values on it also. If your resistor "obeys" Ohm's Law, you would expect a graph like the sketch; but it may heat up with larger currents and change the pattern. (If you do some careful thinking, and quick experimenting, you can find out whether heating is responsible for any bad behavior that you observe.)

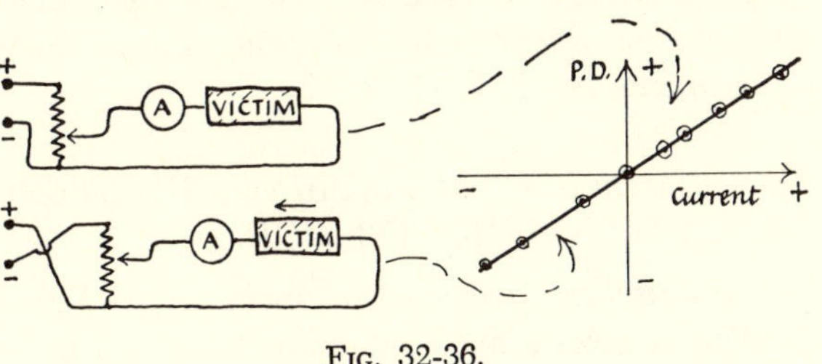

FIG. 32-36.

The Voltmeter Trickle Difficulty

With such victims of very high resistance, a new difficulty arises: the voltmeter itself lets a tiny current through it—in fact, common voltmeters are disguised trickle-meters. We must avoid mistaking the voltmeter trickle for the current through the victim. Two circuit arrangements, (a) and (b),

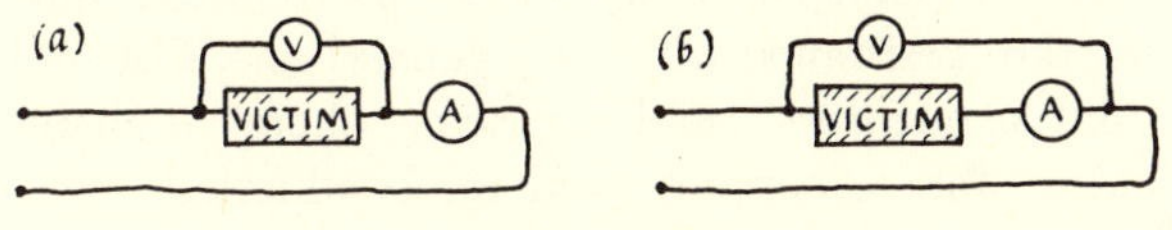

FIG. 32-37.

are shown. If there is a current of a few milliamps[25] through the victim and a current of a few milliamps through the voltmeter, which arrangement, (a) or (b), should you use? Be careful to use the right arrangement in this experiment and in later ones with small currents through the victim.[26]

EXPERIMENT M (2). RADIO TUBE (COLD)

When you have investigated the radio resistor, and plotted a quick graph for it, replace it with the diode radio tube. Connect your investigating circuit to the radio tube's plate and one end of its filament.[27]

25 A milliamp is a thousandth of an amp, 0.001 amp.

26 Ammeters have very low resistance; so the P.D. across the ammeter is likely to be very small.

27 Some tubes have a connection to the center of the filament, or a surrounding cover, labelled cathode. If your tube has such a connection, use that instead of one end of the filament.

Again try a full range of P.D.'s from 0 to +100 volts or more and from 0 to −100 volts. Plot a graph of your results, whatever they are.

EXPERIMENT M (3). RADIO TUBE (HOT)

Now try the tube again with its filament ("cathode") heated. The heating is done by an auxiliary current driven by a storage battery. Re-draw your circuit with a heating circuit consisting of battery, rheostat, ammeter, and switch. Run the heating current at the value suggested for your tube and repeat the previous measurements. Again plot a graph, with + and − values.

EXPERIMENT M (4). SIGN OF CHARGES

There is reason to think that the heated filament gives out some electrically charged particles which can carry current from filament to plate. Must such carriers have a plus charge, or a minus one? *Examine your apparatus carefully and argue out the answer to this important question from your own circuit.* You will make use of a radio tube like this in later experiments.

RESISTANCES

Resistance and Its Units

Experiments show that the simple relation found for the "resistance wire" is a general one. For most solid conductors (and for some chemical-change baths, and sometimes even for conducting gases) the ratio:

$$\frac{\text{P.D. BETWEEN THE ENDS OF THE CONDUCTOR}}{\text{CURRENT THROUGH CONDUCTOR}}$$

is constant, at constant temperature. We call the constant the "resistance" of the conductor. Therefore

$$\text{RESISTANCE} = \frac{\text{P.D. BETWEEN ENDS OF CONDUCTOR}}{\text{CURRENT THROUGH CONDUCTOR}},$$

and with the usual units this is $\frac{\text{P.D. in volts}}{\text{CURRENT in amps}}$.

Therefore RESISTANCE must be measured in $\frac{\text{volts}}{\text{amps}}$ or volts per amp. As we use resistances often in electrical calculations, we give a shorter name to these units and call one volt/amp one *ohm*. Then "ohms" are short for "volts/amp." We often shorten them still more to one Greek letter: ω or Ω (these are small omega and capital omega, the last letter of the Greek alphabet).

When we say a wire has resistance 5 ohms (or 5Ω) we mean that FOR EVERY AMP FLOWING THROUGH IT, THERE MUST BE A P.D. 5 VOLTS BETWEEN ITS ENDS.

A mile of copper telephone wire has a resistance of a few dozen ohms. A mile of power cable has a resistance of a fraction of an ohm, or a few ohms at most. An electric lamp has a resistance of a hundred ohms or so.

OHM'S LAW IN TABLOID FORM

$\frac{\text{P.D.}}{\text{C}} = \text{R}$	$\frac{\text{Volts}}{\text{Amps}} = \text{Ohms}$
RESPECTABLE	UNSPEAKABLY VULGAR

Measuring Resistances

To measure a resistance, assumed to "obey" Ohm's Law, take one pair of readings of P.D. and current. If the resistance is very large the ammeter should be a *milli*ammeter. If the resistance is very low the voltmeter should be a *milli*voltmeter.

If a volt:amp characteristic graph is plotted for the conductor, the resistance is P.D./current, or the slope of the graph.

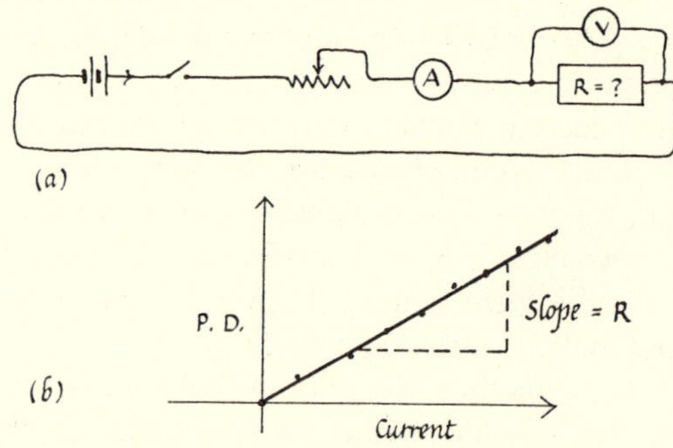

FIG. 32-38. MEASURING RESISTANCE

Electrical engineers often need to know the resistance of power lines, telephone wires, electric motor windings; and you will find resistances given to you as data in electrical problems. There are ingenious trick devices for measuring resistance without heating the victim with large currents, but the simple method you used in your investigation will often suffice. Use it in the following measurements, Experiments N, O, P, Q, R, S.

EXPERIMENT N. MEASURING RESISTANCE OF A GIVEN WIRE OR COIL

If you assume that a wire "obeys Ohm's Law"—i.e., that it gives a fixed value for the ratio P.D. over current—you need only take one pair of measurements of current and voltage. You might take another pair as an independent check, but unless you are going to open up the general investigation all over again for the new victim, there is no point in taking a whole series of measurements. Make a good pair of measurements and calculate the resistance. (Remember that measurements in which the meters give readings in the lower part of their scales, near their zeroes, are likely to give larger *percentage* errors than measurements with meter-readings near the top of their scale.)

EXPERIMENT O. Measure the resistance of a radio resistor (e.g., one *labelled "2,000 ohms."*) For a single measurement you do not need a potential divider; but you must be careful not to measure the trickle through the voltmeter.

EXPERIMENT P. Measure the resistance of your voltmeter.

ELECTROLYSIS AND OHM'S LAW

EXPERIMENT Q. Investigate the volt:amp characteristic of a copper-plating bath with copper electrodes.[28]

EXPERIMENT R. Investigate the volt:amp relationship for a bath of acid water in which oxygen and hydrogen are produced at platinum electrodes when the current flows.[29] This behaves in a confusing way unless you use the following procedure. Use a 6- or 8-volt battery to drive currents through this water-electrolysis apparatus, *without a rheostat.* You will be able to apply P.D.'s of 2 volts, 6 volts, and an intermediate value; but you may have to leave the region between 0 and 2 volts unexplored. Sketch a graph. Can you draw a straight line near its points? Can you interpret it? This will be discussed later.

EXPERIMENT S. LOCATING A FAULT IN A CABLE OR WIRE (Optional)

Assuming that Ohm's Law holds, locate the fault in a model telephone wire. Overhead wires and underground cables can suffer several kinds of disaster. The owners wish to locate the damage and send men to repair it without spending time and money on bicyclists to look for tangled wires, or diggers to unearth a whole cable. So they use electrical methods to locate the fault, to find out how far the trouble is from one end of the line. A clean break inside the cable, as in Fig. 32-39(a), is difficult to locate; but it is easy to locate a short-circuit

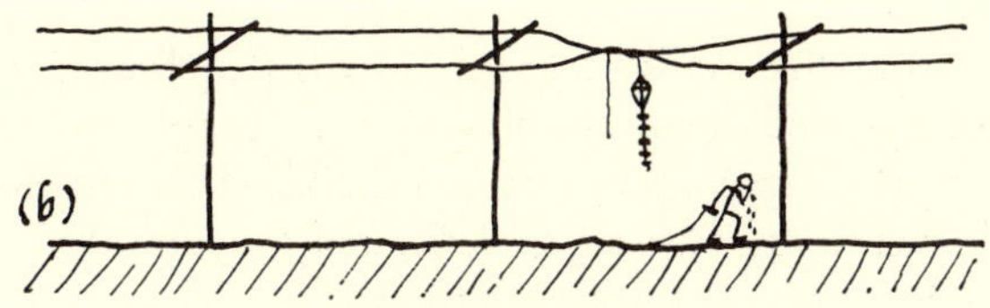

Fig. 32-39. Cable Faults

as in (b), due to two wires touching. (Cause: birds? gardeners with spades?)

You are provided with two wires, representing, say, a line from New York (A) to Boston (B) and return, twisted together at a concealed point (C) to make a short-circuit. At A make some measurement. Proceed with your apparatus to B and make a measurement. Measure the total length of wire from A to B and calculate the location of the fault, using common sense, arguing by simple proportions, and using arithmetic. Verify your prediction by examining the wires or by asking where the fault was placed. Professional fault-finders have ingenious devices for measuring resistances, but the principle is the same as here.

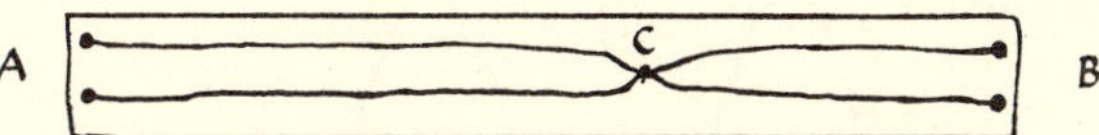

Fig. 32-40. Model Telephone Line with "Fault"

EXPERIMENTS WITH ALTERNATING CURRENT SUPPLY

For reasons to be discussed later, modern power systems do not maintain a steady voltage to drive a steady current through lamps, etc., but have an alternating voltage which swings from, say, +150 volts down to 100 volts, 50 volts, zero, —50 volts, —100 volts to —150 volts; then back —100, —50, 0 and up to +150; and so on, swinging to and fro with S.H.M. of amplitude 150 volts. (See Ch. 10) The graph of such P.D. vs. TIME is a "sine curve" (Fig. 32-41). Applied to an Ohm's Law resistor, the P.D. at each instant produces a corresponding

[28] Electrodes are wires or plates by which current enters and leaves the bath.

[29] Unfortunately called a water voltameter—nothing to do with voltmeter.

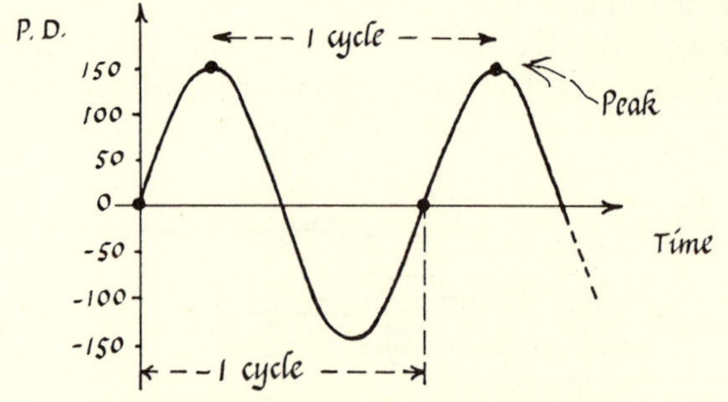

FIG. 32-41. ALTERNATING VOLTAGE

current—the electrons are skittery little fellows who can respond immediately—so the graph of the current plotted against time is similar. This is called an alternating current, or A.C., and the term A.C. is used in general for such supplies.

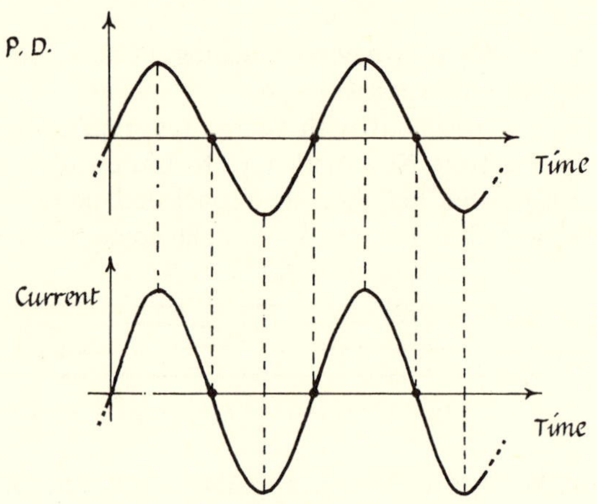

FIG. 32-42. P.D. AND CURRENT, when alternating voltage is applied to a resistor.

EXPERIMENT T. ALTERNATING CURRENT

What would you expect A.C. to do to a lamp? Try applying an A.C. supply to a suitable lamp (e.g. 5 or 6 volts to an automobile headlamp). What would you expect A.C. to do to an ordinary ammeter?[30] Try inserting an ammeter[31] in the circuit with the lamp. Compare its reading with the reading it gives when the same lamp is lit by a D.C. supply *of the same voltage.*

[30] Of course there are special ammeters which do read alternating currents. In one form, the current's magnetic field magnetizes two iron bars (temporarily). The bars repel each other, whichever way the current flows. One bar is fixed; the other is pivoted, to be pushed away against the increasing opposition of a hair-spring. The movable bar carries a pointer which reads on an uneven scale.

[31] *Beware of damaging the ammeter.* If it fails to indicate A.C. its pointer may hover near zero while the current is heating the internal coil just as it heats the lamp. To be safe, start with *D.C.* and then change to the same voltage *A.C.* but *do not shift to a lower range on the ammeter.*

EXPERIMENT U. ALTERNATING VOLTAGE AND A DIODE TUBE

What would you expect to find with an alternating voltage applied between plate and cathode (or filament) of a diode radio tube? (i) Set up the circuit shown below with the Radio Resistor (RR) as the victim. Note the behavior of the milliammeter.

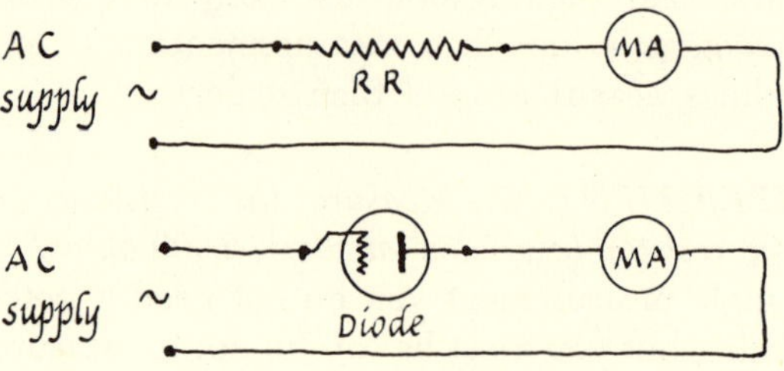

FIG. 32-43. EXPERIMENT U

(ii) Replace RR by the radio tube (with any auxiliary apparatus you consider necessary). Observe the milliammeter.

EXPERIMENT V. ELECTRONIC TIME-GRAPHS

Electron streams can be used to draw time-graphs, just as they paint television pictures. Later, we shall discuss the tubes which do this, but you can use one now to draw graphs for the A.C. supply. The device is called a cathode ray oscilloscope (CRO). The moving spot on its screen moves *steadily* across at constant speed, to give a time axis; and it is moved up or down proportionally to an applied P.D. so the spot traces the time-graph of that P.D. To see what the diode tube does, you must produce a small sample P.D. to be plotted. To draw the time-graph of a current, you must manufacture a P.D. which varies directly as the current; so you should use an Ohm's-Law "Sampling Resistor" (SR) as shown in Fig. 32-44. Set up your previous circuit,

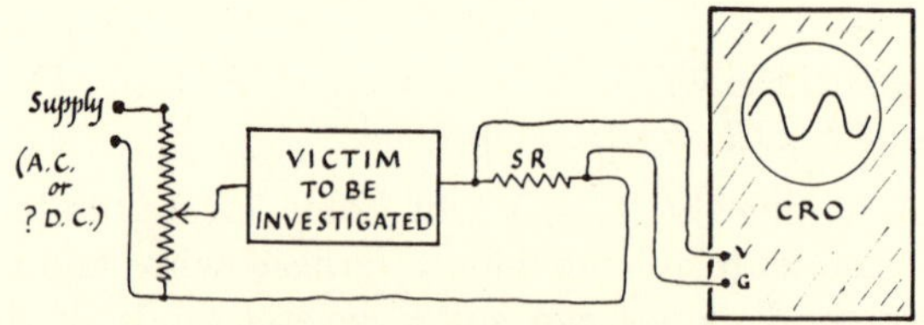

FIG. 32-44. EXPERIMENT V
STUDYING THE CURRENT THROUGH A "VICTIM" WITH A CRO.
The current produces, in the sampling resistor SR, a proportional P.D. which drives the writing spot of the CRO up or down.

with alternating P.D. applied to a radio resistor, RR, and insert SR as shown. Run leads from SR to the CRO and watch the pattern of the current. This shows the time-graph of the current through the circuit.[32] Now replace the Radio Resistor by the diode tube and again watch the pattern. (Think about the tube's action first, and see if you can predict the shape of the pattern before you actually try it.)

This action of the diode tube in producing bumps of one-way current is called "half-wave rectification."[33] It is very useful in cases where we want to produce a D.C. supply from an A.C. supply. For example, we charge batteries by driving a current "backwards" through them. We must have a direct current for this. An alternating current would do no good at all, in fact even some harm. Rectifying tubes can provide a bumpy direct current for battery charging from an alternating supply (Fig. 32-45).

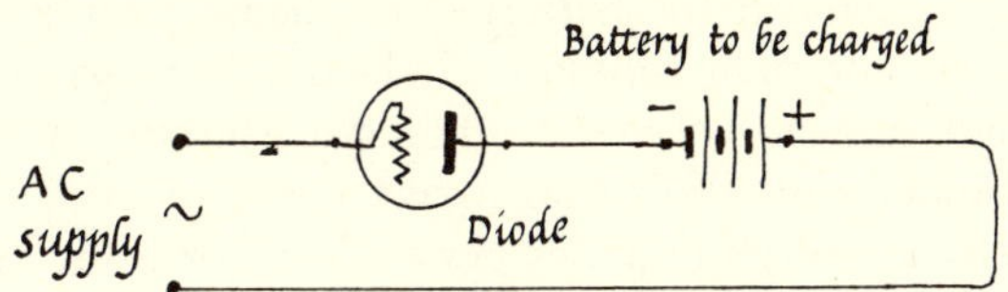

Fig. 32-45.
Diode Acting as a "Half-Wave Rectifier"
Charging a Battery

An ingenious arrangement of two diodes (sometimes both inside the same glass envelope) provides "full-wave rectification" which looks less bumpy. An electron stream is pulled across first in one tube then in the other, then in the first, and so on—like milking a cow with two hands, the milk always running the same way into the bucket. Thus the battery receives bumps of current always in the same direction, running to the tubes to provide the electron streams. Problem 34 discusses this. See a demonstration.

EXPERIMENT W. POWER LINE EXPERIMENT

We shall discuss power lines in later problems, and return to them with alternating supplies later still; but at this stage you should investigate a model power system in the laboratory with your own meters. In this experiment it is the *discussion* of the measurements that is fruitful, so we shall give you cookbook instructions. Arrange a pair of resistance wires on poles, to represent the power line between a power station (PS) and a village (V). Represent the village and its consumption by a lamp (L) and

Fig. 32-46. Power Line Laboratory Experiment

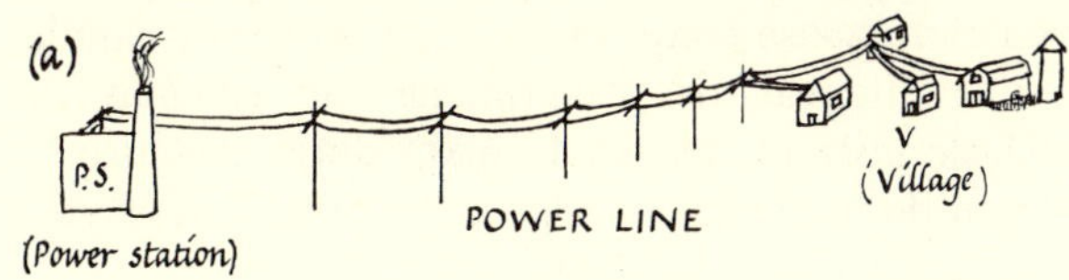

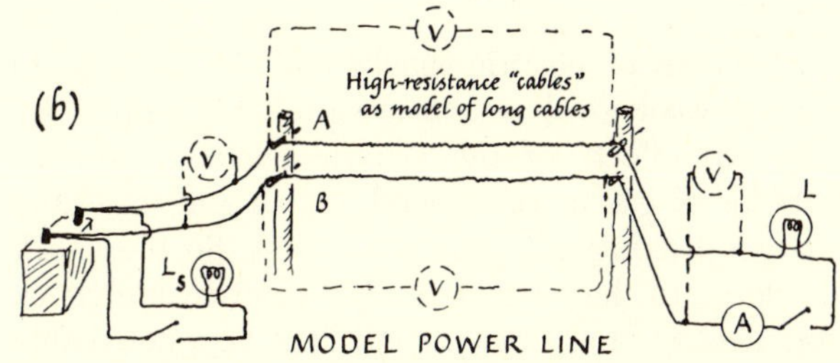

(a) Power line. (b) Model power line.
(In your laboratory record you should draw proper circuit-diagrams, not informal sketches like the one above. However, you may show the "cables" as resistors.)

Fig. 32-46c. Specimen Record Table
For use with power line laboratory experiment

PART OF CIRCUIT UNDER INVESTIGATION	CURRENT THROUGH PART amps	P.D. BETWEEN ENDS OF PART volts	POWER DELIVERED IN PART watts
village			
power wire A			
power wire B	SPECIMEN	Do not write on this table	
power system (village and power wires, by one direct measurement)			
check: total of village + A + B ——→			
(this table should then continue for the experiment with high voltage)			

the power station by a storage battery. In real power stations, the switchboard is lit by a lamp that does not use the same fuses or circuit-breakers as the village. You should install a switchboard lamp (L_s). This will serve for comparison with L (but you should not measure the current through L_s). Arrange to measure the current in the line from power station to village and back. Use a voltmeter with long leads (connecting wires), to make the measurements needed for the record shown in the specimen table.

Calculate the power delivered in each region. Make an arithmetical check of the obvious expectation. Calculate the efficiency of the system, defined as follows:

EFFICIENCY is

$$\frac{\text{USEFUL POWER used by village}}{\text{TOTAL POWER supplied to power line \& village}}$$

Now, keeping the same power line, change to a high voltage supply instead of the battery, say, a 120-volt

[32] The CRO has its own "time base" which swings the writing spot steadily across, to plot a time graph. The spot repeats this horizontal motion regularly, with such a frequency that the patterns of successive sweeps are superposed.

[33] "Rectification" means "straightening." Any one-way device for producing D.C. (bumpy or smooth) from A.C. is called a rectifier. A rectifier is essential in a radio set.

D.C. supply. This will require different lamps, but you can choose lamps of similar wattage or candle-power. Repeat the experiment and calculations. (These instructions should end, "draw conclusion." If you do not draw an important conclusion of your own accord, you are inhuman.)

Shocks

The pain of electric shocks and the damage to your physiological system are probably caused by chemical effects of the electric current—perhaps bubbles of gas in your blood stream and upsetting charges leaking along nerve systems. So the shock is related to the current. To get a noticeable shock you need a voltage big enough to drive a noticeable current through your resistance. The resistance of a human body varies from time to time and greatly from person to person. Most of the resistance is in the dry outer skin; inner fluids contain salts which make them good conductors. Try holding the binding posts of a 6-volt battery with your two dry hands. The current up one arm, across you, and down the other arm is too small to notice. But a milliammeter will show there is a current. A 100-volt battery will hurt dangerously. If you are healthy, and have a good heart, your instructor may provide you with a series of voltages, 6 volts, 10 volts, 20 volts, etc. (from a radio battery or from a potential divider) and allow you to make a record of voltage, current (with milliammeter), and pain. You can estimate your resistance. If you then get your hands wet your resistance will be much smaller. Then even small voltages may be dangerous or fatal. It is the *current* that matters. When you are wet, you will probably experience the same pain for the same current, but this may be driven by a *much* smaller voltage. A dozen milliamps can hurt. One amp would probably be fatal. One of the main effects of electric shock is to upset the nerve center that controls breathing. In rescuing a victim, the first thing to do is clear him from the electric supply; and then apply artificial respiration.

PROBLEM 19. VOLTMETER AND BATTERY

A voltmeter connected across a wire carrying a current tells us how many joules of energy each coulomb "delivers" in passing through the wire. The coulomb does not manufacture this energy. It brings it in the form of electrical energy (from the battery, probably transmitted by an electric field) and releases it as some other form of energy, often heat. So the voltmeter tells us how much energy each coulomb transfers from electrical form to some other form in passing through the wire. We may think of the voltmeter as diverting a few sample coulombs and having them pass through its mechanism and therein finding how much electrical energy each can release. When a voltmeter is connected across a battery or generator, it again reads volts or "joules/coulomb" but it no longer tells us how much energy each coulomb transfers from electrical form to heat, etc., of a battery or generator. What *does* the voltmeter tell us in this case?

The Coulomb's Energy-supply: EMF

However they work, supplies such as batteries and generators seem to be arrangements for giving energy (measured in joules) to electric charges (measured in coulombs) traveling around circuits. We have used a "6-volt battery" without explaining what the voltage means for a battery. And you have been using the information sensibly to mean the energy supplied to a coulomb for each trip. On its way round the circuit the coulomb transfers energy *from* electrical form to heat, etc. but in the supply it *receives* electrical energy at the expense of other forms, chemical for batteries, mechanical for generators. For such *supplies* of electrical energy, we do not call the voltmeter reading a P.D. but give it a grander name: ELECTROMOTIVE FORCE or EMF. An EMF is measured in joules per coulomb or volts; but it refers to energy transfers the opposite way, *from* chemical or mechanical energy *to* electrical energy.

You may liken the battery giving energy to coulombs to a roller-coaster engine that hauls each car up to the top of the first hill, giving it a store of P.E. for the trip. Imagine the engine, at the bottom of the first hill, saying to the car, "This is what you get for the trip. Use it as you like, but don't come back till you have spent it all."[34] When the car does return "empty handed," the engine replenishes its supply and pushes it out on another trip. We know from Joule's work that chemical changes and electrical heating, etc. play proper parts in the balance sheet of Conservation of Energy. And we know that for every coulomb passing through the battery the same amount of chemical change always occurs (like the depositing of 0.000000329 kg of copper, in fact, but in reverse direction). Therefore we expect the CHEMICAL ENERGY THAT IS LOST FOR EACH COULOMB to go into the ENERGY THE COULOMB DELIVERS, as heat, etc., on its way around + any ENERGY THE COULOMB KEEPS as "savings" after a trip. In normal circuits, we find no sign of an increasing store of energy as time goes on.[35] (But you will meet cases of that in

[34] In most roller-coasters, the car then draws on P.E. to develop more and more K.E. as it accelerates—and that corresponds to electrons drawing on a battery to accelerate in a vacuum. In some gentle roller-coasters for children there is so much friction that most of the downhill run is made at constant speed, P.E. disappearing to warm the rails and air. That corresponds to a battery driving a current along wires.

[35] In normal roller-coasters the car comes to rest as it returns to the start. It does not return with a stock of surplus K.E. and then coast faster and faster on subsequent trips.

the "big machines" that accelerate ions and electrons—see Chapter 42). So we are sure that each coulomb spends all its P.E. on the trip. Coulombs go around and around the circuit, spending one EMF's worth of joules on each trip, and finishing empty handed. Some of this energy is spent in the battery itself, producing unwelcome waste heat. In the same way a generator is warmed by its own current, making it less than 100% efficient.

Summary of Electric Circuit Behavior

We picture coulombs as great bunches of electrons flowing around the circuit, giving out the energy which they gained in each trip through the battery. The flow, which we call the current and measure in coulombs/second or amps, is the same at all points all around the circuit. If the circuit divides into several parallel branches the current divides into smaller currents which add up to the total current in the main circuit. Since we expect the coulomb to deliver all its stock of energy in a complete trip around the circuit, we expect the voltages across all parts of the external circuit to add up to the total voltage across the battery or generator. We find that this is so. Where a complicated circuit is made up of Ohm's-Law wires, we can apply these ideas of branching currents and adding voltages to predict the "resistance" of a group of wires in series or a group of wires in parallel. (The results are important in engineering but we do not need them here.)

Radio-tube experiments suggest that hot metal can boil off negatively charged carriers, which we call electrons, and we guess that these may be the carriers in metal wires. Several indirect lines of evidence support this guess. We picture a swarm of electrons whizzing at random through the metal's lattice of atoms and drifting under the influence of the applied voltage, and losing their gained K.E. at collisions, giving the atoms heat.

In electrolysis (Chapter 35), we believe the carriers are not electrons but positive and negative ions (charged atoms which have lost or gained electrons). These carriers too have to struggle against fluid friction—collisions with molecules—so they waste energy as heat; and they may also have to push their way against opposing electrical forces involved in chemical changes at the electrodes. So *some* of their energy is converted to chemical energy. In electric motors the carriers, presumably electrons, find magnetic fields sweeping across their paths and tugging with an EMF against their motion; so the outside voltage has to drive them *against* extra forces which make them transfer some energy to mechanical form.

Ohm's Law: Rules and Calculations

Ohm's experiments, later confirmed with high precision over a vast range of currents, showed that the ratio P.D./CURRENT is a constant for metals and some other conductors, at fixed temperature. This is likely to apply to each part of a circuit and to the complete circuit. In the latter case, we say:

$$\frac{\text{EMF}}{\text{CURRENT}} = \text{RESISTANCE OF COMPLETE CIRCUIT}$$

(including resistance of battery or generator)

The fact that P.D./CURRENT is *constant* is the important experimental result. In most investigations the name for a constant ratio is assigned *after* its constancy is discovered. Although Ohm did plan his work in terms of resistance to flow, it is unwise to say that Ohm proved that P.D./CURRENT is equal to an already-known RESISTANCE—as if the resistance were a definite characteristic provided (and labelled) by heaven long before Ohm and waiting to have P.D./CURRENT proved equal to it. It is better to say P.D./CURRENT = a constant, *called* RESISTANCE.

Ohm found that doubling the length of wire doubles the resistance: RESISTANCE varies directly as LENGTH. Doubling the diameter of wire gave a quarter of the resistance: RESISTANCE varies inversely as CROSS-SECTION AREA. Combining these, we say

$$\text{RESISTANCE} = (\text{constant, } s) \cdot \frac{\text{LENGTH of wire } L}{\text{CROSS-SECTION AREA, } A}$$

The constant, s, is called the SPECIFIC RESISTANCE or RESISTIVITY. It is characteristic of the material of the wire. It does not depend on the shape and size of sample chosen (though it may change with temperature). Its reciprocal, $1/s$, is called the CONDUCTIVITY; and it bears a remarkable similarity to HEAT-CONDUCTIVITY. To compare them, write

$$\frac{\text{P.D.}}{\text{CURRENT}} = \text{R} = s \cdot \frac{\text{LENGTH of wire}}{\text{AREA of section}}$$

$$\text{Then} \quad \text{CURRENT} = \frac{1}{s} \cdot \frac{\text{AREA of section} \cdot \text{P.D.}}{\text{LENGTH of wire}}$$

But CURRENT is flow of electricity or ELECTRIC CHARGE/TIME. So we re-write this:

$$\frac{\text{CHARGE FLOWING}}{\text{TIME}} = (1/s) \cdot \frac{\text{AREA of section} \cdot \text{POTENTIAL DIFFERENCE}}{\text{LENGTH of wire}}$$

But, for heat being conducted along a rod

$$\frac{\text{HEAT FLOWING}}{\text{TIME}} = (k) \cdot \frac{\text{AREA of section} \cdot \text{TEMPERATURE DIFFERENCE}}{\text{LENGTH of bar}}$$

So ELECTRICAL CONDUCTIVITY $(1/s)$ and HEAT CONDUCTIVITY (k) are analogous. In fact they show a surprising resemblance in their actual values. Metals, which are good electrical conductors, are also good heat conductors; and the best electrical ones, such as copper, silver, aluminum, are also the best thermal ones. The correspondence is so striking that we suspect the carriers of electric currents also carry heat.

Ohm's Law Not Universal

There are many materials and devices that do not obey Ohm's Law. Radio tubes, transistors, mineral crystals in loose contact, all give unsymmetrical graphs of P.D. *vs.* CURRENT; and their graphs are curved, often with marked "knees." Far from regarding these exceptions as misfortunes, we are dependent on them for rectifiers (in battery charging and in detecting radio signals) and other important uses in modern electronics. And power companies use a lightning arrestor, made of silicon carbide, that has a very high resistance for ordinary voltages but collapses to low resistance if lightning strikes a power line and produces very high voltage. With all these "non-linear" materials we can still calculate a RESISTANCE, but it does not have a constant value.

Series and Parallel

Experiments show that when several wires, of resistance R_1, R_2 . . . etc., are joined in series, the total resistance of the group, R, is $R_1 + R_2 + \ldots$ the sum of the separate resistances. When several

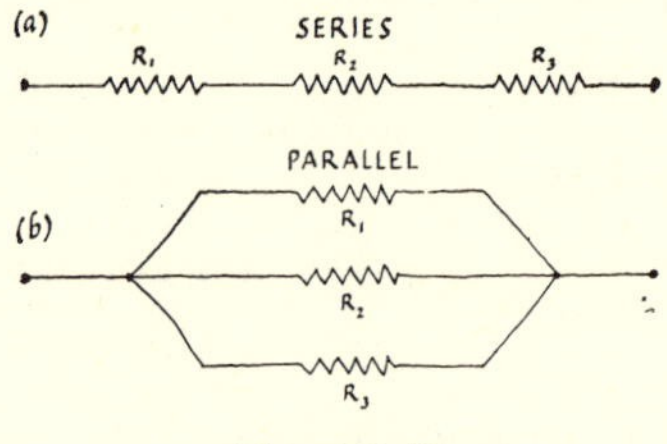

FIG. 32-47.

such wires are connected in parallel (all joining the same two points) the resistance R of the combine is given by $1/R = 1/R_1 + 1/R_2 + \ \ldots$ etc. The *conductance* of the group, $1/R$, is the sum of the conductances of the separate wires. These rules can be deduced from assumptions of conservation of energy and rules for adding currents.

Temperature

Metals change their resistance with temperature. The resistance of *elements* such as copper and tungsten increases with increasing temperature.[36] Roughly speaking, the resistance of most metal elements varies directly as absolute temperature over a wide range. Carbon's resistance decreases with rise of temperature. Alloys usually have greater specific resistance than their ingredient metals; they are useful for high-resistance coils in rheostats, etc. Some alloys show almost no change of resistance with temperature (e.g. "Constantan," 60% copper, 40% nickel, which you probably used in the laboratory for your first "Ohm's Law Experiment," to make it easier). These are useful for standard resistances.

Problems

Once we know the resistance of a power line, generator coil, or motor winding, we can calculate power losses, predict currents for a given voltage, and in general find out what will happen without having to do elaborate experiments. And telephone and power engineers measure and catalogue the resistances of their lines carefully, so that they can locate faults by simple resistance measurements from both ends. When you are asked to do arithmetical problems on circuits with Ohm's-Law behavior, the problems are meant to increase your understanding of electrical matters, not to waste your time on arithmetic or turn you into a formula-monger. Look at the *results* of your calculations carefully.

Unscientific Jargon

It is unscientific jargon—poor English for a scientist—to say "Calculate the watts" or "Calories = 63.4/4200" or "The amps are 6," instead of "Calculate the power," "Heat = 63.4/4200 Calories," "The current is 6 amps." Avoid using the name of the *unit* where you should use the name of the quantity measured. The first versions above have an uneducated flavor. They look earthy and clear, but they do not belong with clear thinking. (Engineers have dragged one unit-name into such common use that we now accept it as a quantity. Horsepower is a unit of power; it is also a cheap name for power. It is respectable to say, "The horsepower of the motor is 3000 watts.") Questions such as, "How many joules?" are respectable, but answers such as, "Number of joules = . . ." are not.

[36] If you observed carefully, you may have seen evidence of this in your first circuit experiment.

Unit Dictionary

Here is a dictionary of unit synonyms that will be useful. But in working problems these relations should not be used in unscientific jargon for the physical quantities they represent.

1 joule = 1 newton · meter
1 amp = 1 coulomb/sec
1 watt = 1 joule/sec
1 volt = 1 joule/coulomb
1 ohm = 1 volt/amp

PROBLEMS FOR CHAPTER 32

Problems 1-19 are in the text.

★ 20. MEANING OF "VOLTS," ETC.

A current of 5 amps in a circuit means that 5 *coulombs pass each point in the circuit every second.* Copy each of the following statements and complete them in similar manner.

(a) A current of 7 amps in a circuit means that . . .
(b) A power of 10 watts supplied to a small heater means that . . .
(c) A potential difference of 12 volts between the ends of a wire means that . . .

★ 21. LAMP FILAMENTS

Two lamps A and B are designed to run on different supply-voltages, but both give the same power when running properly. A runs at full brightness on 6 volts, and B runs at full brightness on 120 volts.

(i) Does B take a larger current or smaller than A?
(ii) Give a clear reason for your answer to (i).
(iii) Should B have a shorter filament or longer than A, for same thickness?
(iv) Should B have thicker filament or thinner than A, for same length?
(v) Give clear reasons for your answers to (iii) and (iv).

22. An arc lamp runs steadily when the P.D. across it is 70 volts. What resistance must be inserted in series with it, when it is used on a 100-volt supply?

(a) The supply has P.D. 100 volts. What does this mean?
(b) The P.D. across the arc is 70 volts. What does this mean?
(c) What, therefore, must be the P.D. across the extra resistance?
(d) If the arc takes 5 amps, what is the current through the resistance?
(e) What resistance must the resistor have?
(f) Draw the circuit for supply and arc, with the necessary resistor.

23. LAMP RESISTANCE

(a) A lamp uses 200 watts. What does this mean?
(b) It runs on a 100-volt supply. What does this mean?
(c) What current does it take?
(d) What is its resistance when running?

(Show how you arrive at your answers.)

★ 24. ELECTRON · VOLTS

When an electron moves along an electric field it gets energy from the field. When an electron moves from one point to another with P.D. of 1 volt between them it is said to have gained one "electron · volt" of energy (1 ev). Using the value of the electron charge and the definition of a volt, calculate the value of *one electron · volt* in *joules.*

[The charge of one electron is -1.60×10^{-19} coulomb. Although this is a negative charge (on the scheme of signs we happen to have chosen), an electron · volt is not a negative piece of energy. An electron with its *negative* charge *gains* energy when it falls through a *negative* P.D., such as -1 volt.]

25. One pound of uranium 235 contains over 10^{24} atoms. It is found by experiment that when one atom of U^{235} undergoes fission it releases about 200,000,000 electron · volts of energy.

(a) How much heat would be given by total fission of one pound of U^{235}?
(b) Coal, chiefly carbon, yields 7000 Cals when 1 kilogram burns to CO_2. How many times as much energy does 1 pound of U^{235} yield as 1 pound of coal? (Take 1 kg $\approx$ 2.2 pounds.)
(c) Carbon atoms are about 1/20 as massive as uranium atoms, so 1 pound contains over 20×10^{24} atoms. Estimate the energy-release in electron · volts per atom of coal. (The data are rough and your estimate may be 10 or 20% wrong.)

★ 26. (a) What is one kilowatt · hour a unit of?
(b) Express 1 kilowatt · hour in other common units (not using the word watt).
(c) In many towns 1 kilowatt · hour costs about 4 cents. Suppose, for convenience, that 1 kilowatt · hour costs 3.6 cents. Then 1 cent would buy . . . ? (State number and unit.)

27. A GENERAL PROPERTY OF P.D.

This problem discusses an important property of P.D. as an energy-reckoning. Assume there is no "source" with EMF in the paths taken from X to Y in the discussion below—no battery with chemical changes, no generator with moving magnetic field. Then we can say that each coulomb brings electrical potential energy (from elsewhere in the circuit) to spend, in obvious forms, between X and Y.

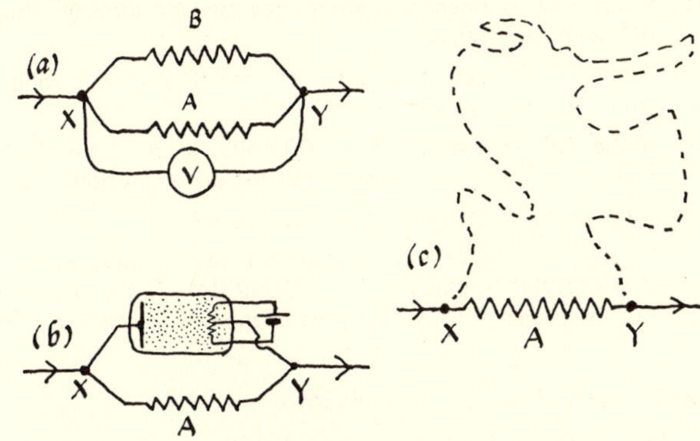

Fig. 32-48. Problem 27

(a) In a certain circuit there are two wires A and B in parallel as shown. A voltmeter is used to measure the P.D. between the ends of wire A. It reads 100 volts.
(i) What is the P.D. between the ends of wire B?
(ii) Suppose the P.D. between the ends of B were not the same as the P.D. between the ends of A. Suppose the P.D. across B were 101 volts. What would happen

if we let 1 coulomb flow down B, and then we paid (out of the energy thus delivered) enough energy to drag the coulomb backwards up A? How much energy should we gain (from nowhere) in such a round trip? What do you conclude, from this argument, concerning question (i) above?

(b) As in (a), there are two things in parallel. One of them is a wire A with its ends at X and Y. The other is a radio tube with electrons boiling off a heated filament driven across between X and Y in the tube. Following the argument in (a), if the P.D. between the ends of the wire A is 100 volts, the P.D. between the ends of the radio tube must be . . . ?

(c) As in (b), a wire A joins X and Y, but the other route from X to Y is not through a wire or through a radio tube, but through the air, along a wandering track traveled by a stray electron which manages to stagger from X to Y unscathed. If the P.D. between the ends of the wire is 100 volts, what is the P.D. for a charge moved along the wandering track?

(d) What general statement can you make about the electrical potential difference between any two points X and Y provided no "source" intervenes?

★ 28. An electric charge falling through a potential difference takes electrical energy and delivers it in some other form. In what form is the energy delivered

(a) in a resistance wire?

(b) when a P.D. is applied to a stream of electrons in a radio tube?

(c) when the victim is a storage battery with current being driven through it backwards to "charge" it?

29. POWER SYSTEM PROBLEM

A power station supplies a remote village through long thin copper cables which have resistance 0.1 ohm from power station to village and 0.1 ohm from village back to power station. The village takes 50 amps to light its lamps, etc.

(a) What is the CURRENT through the copper cable from power station to village?

(b) What P.D. is needed to drive this current through the 0.1 ohm resistance?

(c) What P.D. is used to drive the current through the cable from village to power station?

(d) If the P.D. in the village is 100 volts, what is the total P.D. provided by the generator at the power station?

(e) What is the POWER used in the village?

(f) What is the POWER wasted by the two copper cables? (For POWER multiply P.D. by CURRENT, but you must use the *proper P.D. for the part of the circuit concerned.*)

★ 30. THE FACTORY PROBLEM

This problem should be provided on a typewritten copy of the small version printed here. Write your answers on the sheet itself.

★31. POWER

The statement "volts times amps gives watts" is a poor, vulgar but useful way of stating the very important relation:

POWER in watts = (P.D. in volts) · (CURRENT in amps)

Give the argument to show that this relation is true. (Consider any device which has P.D. V volts between its ends when a current C amps flows through it. Explain what each of these pieces of information means, and show that the POWER must therefore be $V \cdot C$ watts.)

★ 32. POWER AND ENERGY DELIVERED IN RESISTOR

(Leave the working of this problem in *factors*.)

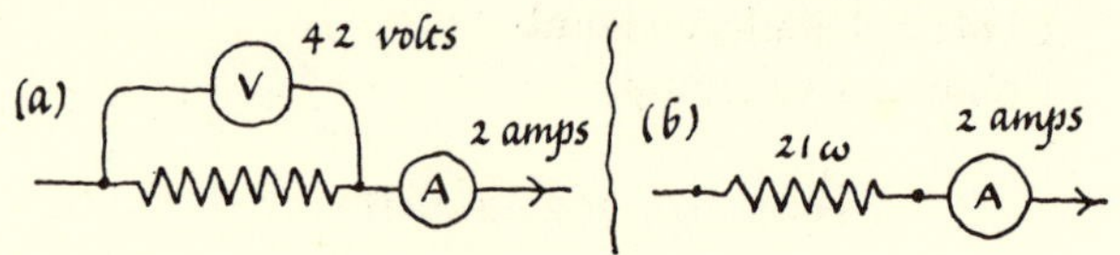

FIG. 32-49. PROBLEM 32

(a) A current of 2 amps flows through a wire, and the P.D. between its ends is 42 volts.
   (i) What is the POWER (rate of transfer of energy) in watts?
   (ii) What is the POWER in Calories per second?
   (iii) How much ENERGY, in joules, is dissipated from wire in 10 secs?
   (iv) How much ENERGY, in Calories, is dissipated from wire in 10 secs?

(b) A wire of resistance 21 ohms carries 2 amps.
   (o) Use Ohm's Law to find the P.D. between its ends.
   (i), (ii), (iii), (iv). Repeat (i)-(iv) of (a) above.

(c) Repeat (b) with 4 amps, twice the current, flowing through the 21-ohm wire.

(d) Comparing (c) and (b) how are POWER and ENERGY changed when CURRENT is doubled, resistance remaining constant?

★ 33. JOULE'S LAW OF ELECTRICAL HEATING (discovered by Joule: a great achievement in the early days of electric-circuit discoveries).

A current C amps in a wire with P.D. V volts between its ends transfers energy at a rate VC watts, and in t seconds it transfers VCt joules. In many cases we know the resistance of a wire and the current through it, but we do not know its P.D. directly. So we need to calculate power or energy from CURRENT and RESISTANCE. Suppose we have a wire of resistance R ohms carrying a current C amps.

(a) State Ohm's Law and express the P.D. in terms of R and C.

(b) Insert the result of (a) in the expression for power, P.D. · CURRENT, and give an expression (using R and C) for the rate at which energy is transferred, in watts.

(c) Express the result of (b) in Calories/second.

(d) Give an expression for the energy delivered in t seconds, (i) in joules, (ii) in Calories.

The answers to (b), (c), and (d) are expressions of Joule's Law (the "Joule heating"). These are the general forms of answers to Problem 32. They are the routine engineer's thumb-nail friends, the wireman's pride and joy. Creative physicists and engineers value them too, but only use such formulas when they understand how they are arrived at. Now, as regards yourself. . . .

SPECIAL PROBLEM ON ECONOMICS OF POWER NAME__________________

PROBLEM 30. EFFICIENCY OF POWER SUPPLY: "THE FACTORY PROBLEM"

200 kilowatts are supplied to a factory, through cables of total resistance 0.1 ohm, (a) at 100 volts P.D. at the factory, (b) at 1000 volts P.D. at the factory. In each case, (a) and (b), calculate:

(i) the CURRENT, (ii) the P.D. wasted in driving this current through the cables,
(iii) the POWER wasted in the cables,
(iv) the EFFICIENCY of the supply system, (defined as $\frac{\text{power used by factory}}{\text{total power supplied}}$).

To answer the problem, complete the following statements.

The factory uses electric energy at a rate of 200 kilowatts. This means

that ____

	SECOND CALCULATION, FOR P.D. AT FACTORY 1000 VOLTS
The voltmeter at the factory shows that the P.D. of the supply line is 100 volts at the factory. This means that __________________ ____________________________, in the factory.	
∴ the current through the factory, in coulombs/sec, must be $\frac{?}{?}$ or __________ coulombs/sec	___?___ amps
This current flows through supply cables of total resistance 0.1 ohm (for both cables together) from the remote power station. Ohm's Law states that for most conductors the ratio P.D./CURRENT is a constant, called RESISTANCE. In this case, then, the P.D. needed to drive the current through the supply cables must be	
______________ or ______________ volts. This means	___?___ volts
that ____________________________________ __________________________ <u>in the cables, as heat.</u>	
But, the current in the cables is ____________ coulombs/sec	___?___
∴ power wasted in the cables is (__________)·(__________) or ____________ joules/sec or ____________ kilowatts.	___?___ k.w.
∴ the efficiency of the supply system, or the fraction $\frac{\text{POWER USED BY FACTORY}}{\text{TOTAL POWER SUPPLIED BY POWER STATION}}$ is ____________________ or ___________ %	________ %

NOW REPEAT THE CALCULATION WITH THE SAME POWER USED IN FACTORY AND SAME CABLE RESISTANCE, BUT WITH 1000 VOLTS AT FACTORY. WRITE ANSWERS ABOVE.

34. FULL-WAVE RECTIFIER

Fig. 32-50 shows a pair of diode tubes which can be used to give "full-wave rectification." To use the system to charge a battery, X, from an alternating supply, we can use an arrangement like the one sketched (though really with a transformer instead of potential divider). An alternating voltage with peak value 200 volts is applied to the potential divider *ab*. Then when voltage between *a* and *b* is 200 volts with *a* positive, voltage between *a* and center c is 100 volts with *a* positive, and voltage between *b* and c is — 100 volts, *b* negative. A quarter of a cycle later these voltages are zero, and another quarter cycle later still the voltages are — 100, + 100.

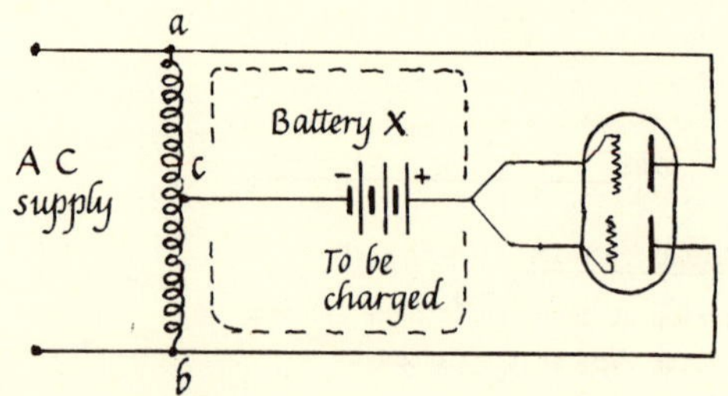

Fig. 32-50. Full-Wave Rectification System
For charging a battery

(a) Say what happens inside the tubes at each of the following stages:
 (i) When voltage *a* to c is + 100 volts and voltage *b* to c is — 100 volts.
 (ii) A quarter cycle later, when both voltages are zero.
 (iii) A quarter cycle later still when voltage *a* to c is — 100, *b* to c is + 100.

(b) Sketch a time graph (e.g., as shown by an oscilloscope) of electron stream through the combined tubes (or through the battery X).

(c) Sketch a time graph for the electron stream when one plate is removed (a simple diode).

35. POWER ECONOMICS

What factors should a power company consider in designing their system of power transmission for best use and economy? Answer this question by treating the following changes of design that they might make. In each case, (i) say whether the change would increase the quantity under consideration, decrease it, or not change it much; and (ii) give a reason for your choice in (i).

(a) With a given system of cables from a power station to a village, will the *efficiency* be changed by *increase of current* (turning on more lights)? Why?

(b) Will the efficiency be changed by an increase of supply voltage (say from 100 V to 1000 V), with suitable change of lamps? Why?

(c) Will a change to high voltage affect the costs of supporting the cables on towers (for SAME CABLES AS BEFORE); and of wiring houses? Why?

(d) Will a change to high voltage affect the safety of workers in power station, linemen, customers? Why?

(e) A change to high voltage in the village would necessitate a change in lamp filaments. How would the new filaments compare with the old ones: thicker? longer? Would lamps for much higher voltage be much harder to make or easier? Why?

(f) For a long power line, the copper wire is a major cost. What would you advise for economy? What other changes in the installation would your advice lead to?

(g) Why are alternating supplies so much more convenient?

(h) Mention some disadvantages of alternating supplies.

[★ 36. PROBLEMS ON ELECTRON STREAMS, needing only a knowledge of this Chapter: see Problems 1, 2, 3, 4 in Chapter 36.]

CHAPTER 33
ELECTRIC CHARGES AND FIELDS: ELECTROSTATICS

"... Coulomb's law. Like charges repel, and unlike charges attract each other, with a force that varies inversely as the square of the distance between them ... in *all* of atomic and molecular physics, in *all* solids, liquids, and gases, and in *all* things that involve our relationship with our environment, the *only* force law, besides gravity, is some manifestation of this simple law. Frictional forces, wind forces, chemical bonds, viscosity, magnetism, the forces that make the wheels of industry go round—all these are nothing but Coulomb's law ..."

—J. R. ZACHARIAS

In *Science*, March 8. 1957.

[THE STUDY of electricity at rest—"electrostatics"—used to bulk large in elementary physics. It was all that was known of electricity two centuries ago, and tradition dies hard. It makes a poor beginning for modern electric circuits, so we have avoided it. Now you need some knowledge of it for atomic physics. How much you see and learn of this part of physics will depend on apparatus, weather, and instructor. On the whole, the less the better.]

Electric Charges at Rest

When an electric circuit is incomplete, there is no steady current. But when a battery is first connected to an incomplete circuit, momentary currents flow, as shown in Fig. 33-1. If the wires at the break in the circuit are expanded into large sheets of metal (a "capacitor") the momentary "charging

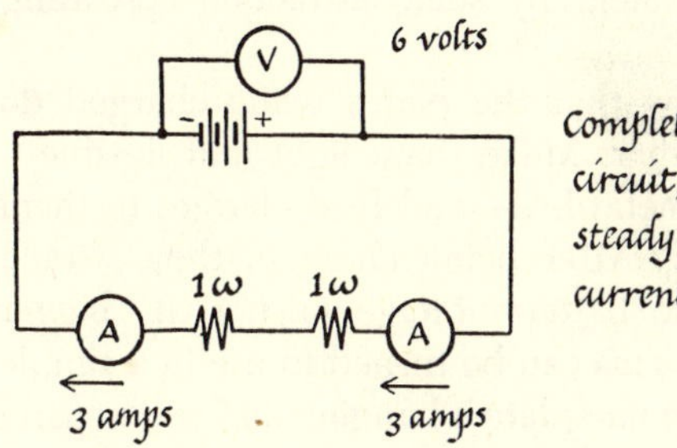

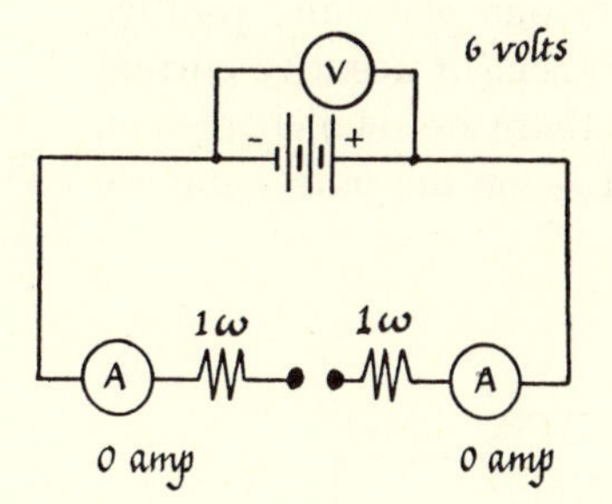

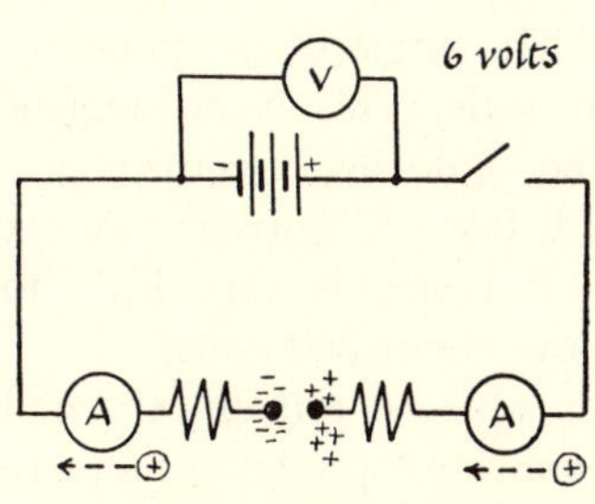

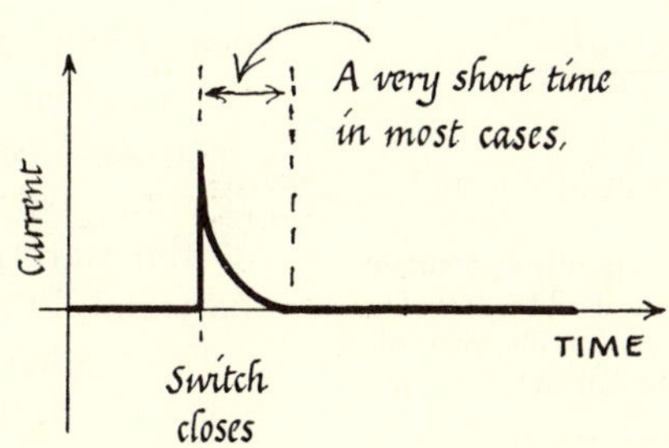

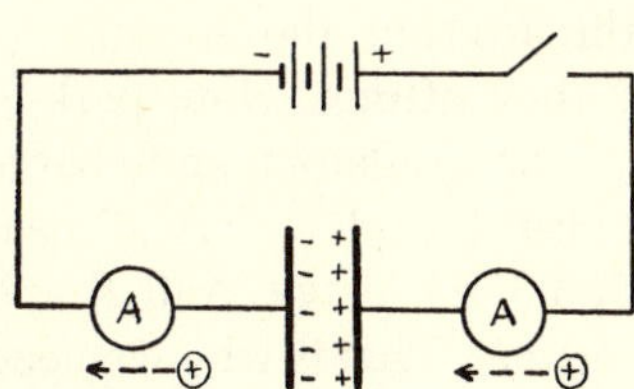

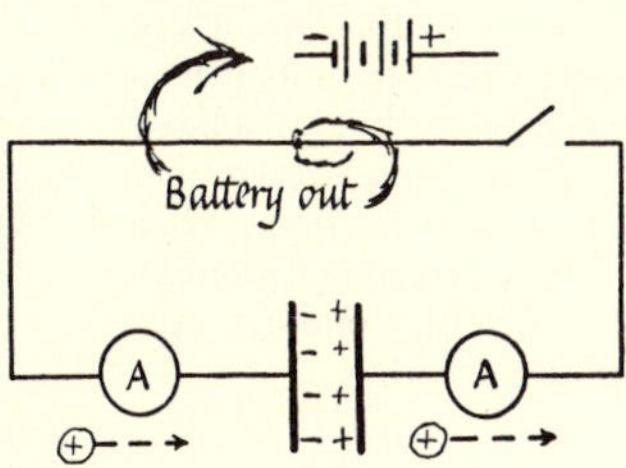

FIG. 33-1a. MOMENTARY CURRENTS

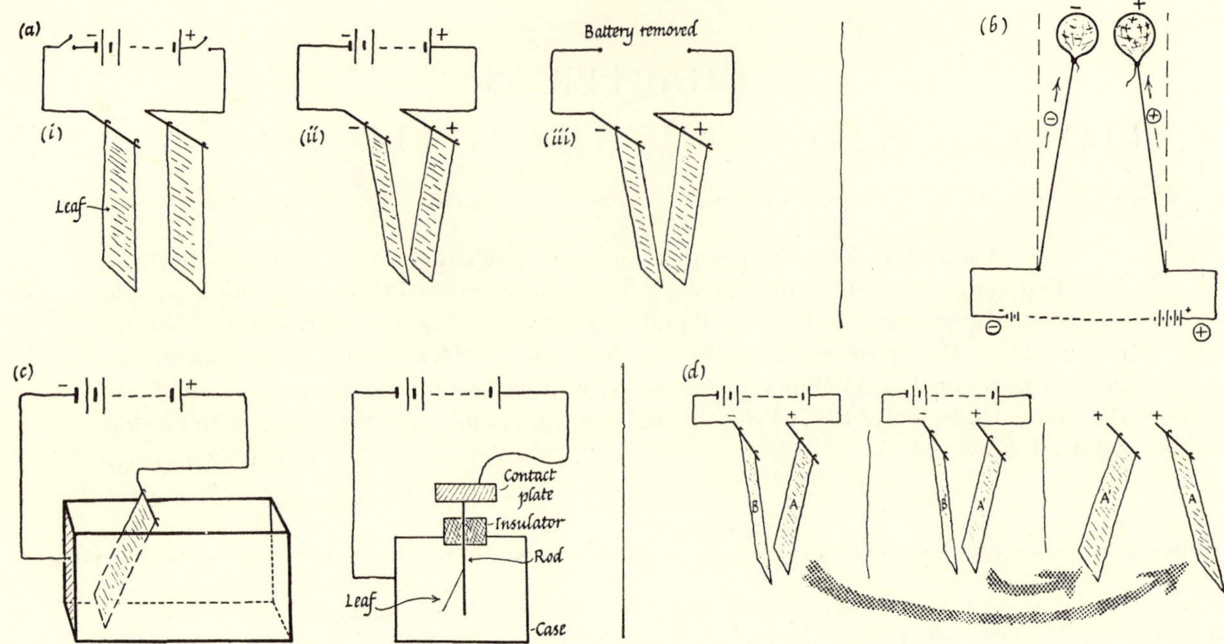

FIG. 33-3. ELECTROSCOPE

currents" are larger or last longer. A few "coulombs/sec" run for a fraction of a second. Therefore some "coulombs" must go from the battery to the plates and be left there when the flow stops. Meters show positive current flowing *to* one plate and positive current flowing *from* the other (or negative current *to* it). Therefore we think there are then charges on the plates, + on one and − on the other; and we say the plates are "charged."

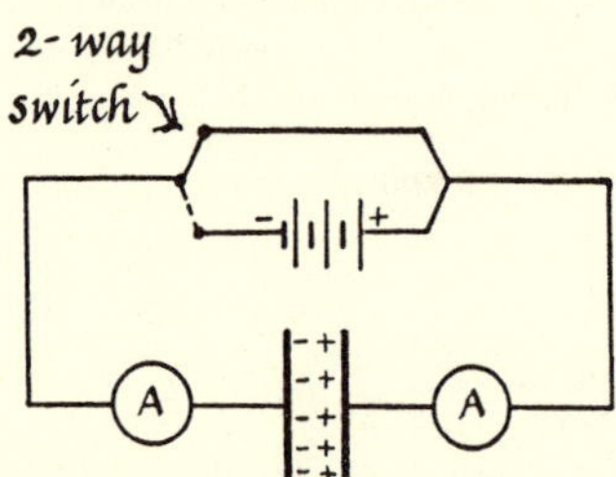

FIG. 33-1b. CIRCUIT FOR DEMONSTRATING MOMENTARY CURRENTS

The 2-way switch connects the battery in circuit, *or* removes battery and completes the circuit without it. The two ammeters, A and A, show similar momentary currents, *into* one plate, *out of* the other. (Experiment G in Ch. 41)

[2] The water-circuit analog of a break in the circuit (or a capacitor) is not just a blocked pipe, but an elastic diaphragm, like a rubber sheet, across the pipe. Then a pump can drive a momentary current, making the diaphragm bulge.

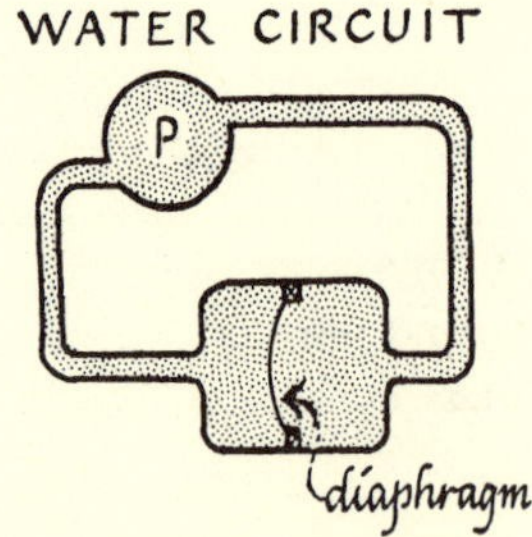

FIG. 33-2. HYDRAULIC ANALOG OF CAPACITOR

If the battery is disconnected and replaced by a wire, momentary currents flow in reverse, the plates "discharging" through the wire. But until that is done the charges remain on the plates, pushed by the battery or held by some attraction spreading across the gap.[2]

We can show that the plates while charged do attract each other. Make them light and flexible—strips of thin metal leaf—and feed charges to them from a battery. After being charged, they attract. The bigger the battery-voltage (EMF) the bigger the attraction. This can be turned to use in a simple voltmeter, keep one plate a hanging leaf and expand the other into a metal box. When a P.D. is applied between leaf and box, the leaf tilts out towards the nearest side of the box. The angle of tilt indicates the P.D.—on an uneven scale. This is an ancient instrument, still used, the "gold leaf electroscope." It is an ideal voltmeter: it takes no current (except at the start); however, it responds very little to potential differences below about 300 volts.

Once charged, metal objects remain charged when the battery is disconnected, provided they are kept on insulating supports and not connected by metal wires or damp human fingers. Connect two flexible plates A and B (or two metal coated balloons on conducting threads) to the + and − terminals of a battery: they attract. Now park A and B, charged + and −, on insulators; and charge another pair A′ and B′, also + and −. Try A′ near A: they repel. See Fig. 33-3(d) above. And B and B′ repel; while A and B attract. That is why we need the labels "plus" and "minus."

Two plates connected to a battery have opposite charges, and attract; and those charges disappear ("neutralize") when the battery is removed and the plates are connected by a wire.

There is no special electrical virtue in flexible leaves or floating balloons—those were chosen to balance small electrical forces with small gravity forces. Any metal objects will do (provided they are supported by insulators to prevent currents carrying charges away). Charges run freely over a conductor (metal, carbon, etc.). That is why we can charge a metal object completely with one touch of the battery wire. Insulators can gather charges too, but only where the wire touches them.

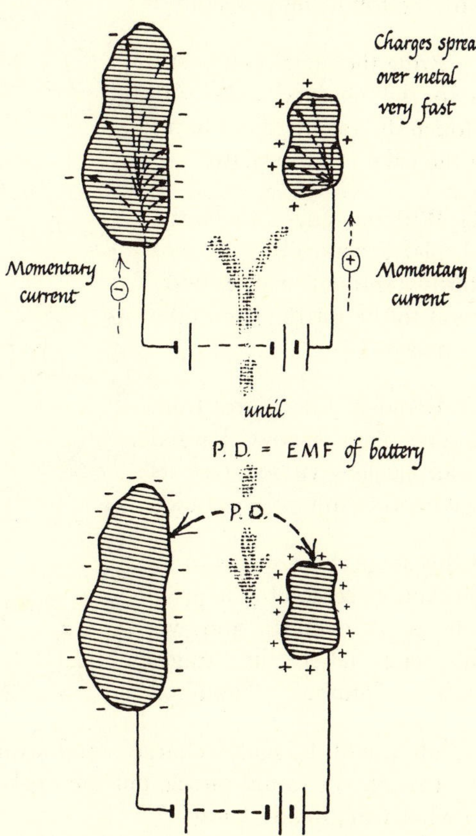

Fig. 33-4. Moving Charges

A battery connected to two insulated metal objects pushes charges quickly on to them, until the P.D. between them is the battery's EMF.

Sometimes we seem to have given charge to one object alone; and we wonder where the opposite charge has gone. We usually find it on the surroundings: the box of the electroscope, the walls of the room, the Earth itself. If we are careless, the other wire from the battery trails on a damp table or floor and drives the opposite charge to "ground." Or we may intentionally connect the other wire to the water pipe which runs to ground. With one

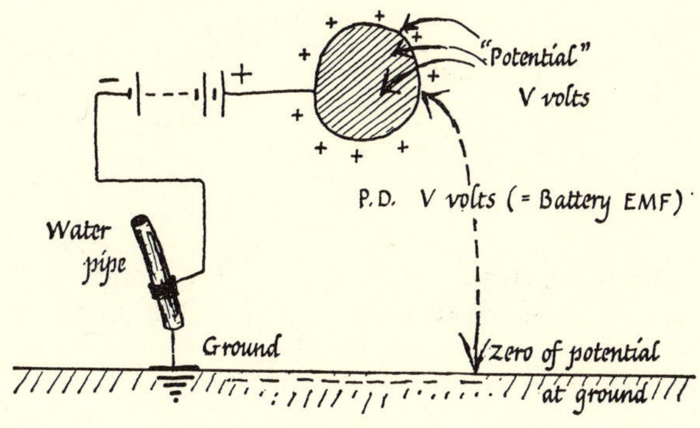

Fig. 33-5.

terminal of the battery "grounded," we can charge objects with the ungrounded wire. Then we speak of the *potential* of a charged object, meaning the P.D. between it and ground.[3]

However, there is another way of charging things, that works magnificently with all materials. It has been known for centuries as part of the science of electrostatics, electricity at rest. We shall now make a fresh start with those simple phenomena. Forget for the moment your knowledge of currents and charges and watch the ancient knowledge being built.

A Fresh Start: Charging by "Friction"

Rub a stick of plastic or hard rubber or dry glass with cloth or fur (or better still, "Saran Wrap"), and it will pick up dust and small pieces of paper. Two similar rubbed rods repel each other. We say they are *charged.* At this stage "charged" is merely a name for the properties "will pick up bits of paper" and "will repel similar things."[4] We imagine the rods have gathered something on their surface, something that we call electricity and we call this imagined collection of electricity a *charge of electricity,* or an *electric charge,* or just a *charge.* We can show by skinning the surface off a rod that the charge judged as above stays on the surface.

Conductors and Insulators

Wires can carry charges away from a surface and so can fingers or wet threads; but *insulators* such as glass or lucite do not do so. By putting samples on insulating handles, so that any charges they gain cannot run away, we find that *any two dissimilar materials become charged when touched together*—metals, non-metals, elements, compounds.

[3] In some cases—e.g., when dealing with atomic structure—we shift the zero of potential from "ground" to infinity.

[4] We use the word to recognize the special state of affairs, but we are only naming the state of affairs, not giving an explanation. An earlier age might have used "bewitched."

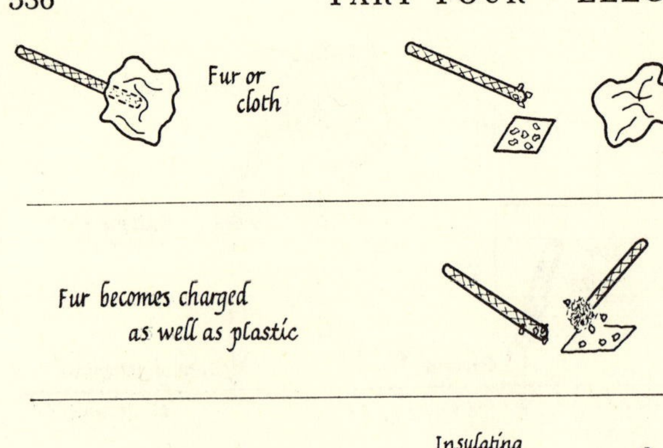

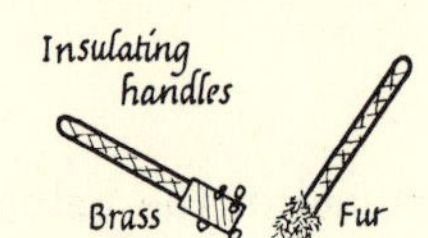

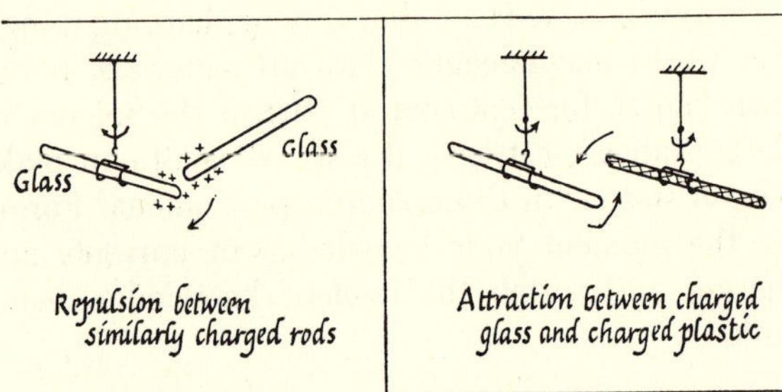

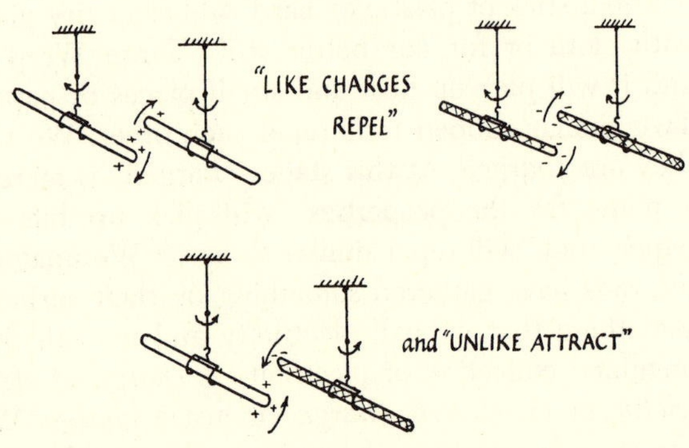

Fig. 33-6. Electric Charges

Forces: + and — Charges

We soon find there are two kinds of charge: hence the need for labels + and —. Pieces of charged lucite or hard rubber repel each other: pieces of charged glass or hard rubber repel each other: pieces of charged glass (rubbed on silk) also repel each other. But a charged lucite attracts charged glass. If we rub lucite and fur together, they both become charged and attract each other. We label *negative* the charges on hard rubber, lucite, sulfur; and *positive* the charge on glass and on fur that has rubbed lucite.[5] Charged things exert forces thus: + and + repel, — and — repel, + attracts —, and — attracts +. We sum this up in the ancient rubric:

LIKE CHARGES REPEL: UNLIKE CHARGES ATTRACT

Multiplying Charges: The Electrophorus

To obtain large charges easily we use an *electrophorus*—a cake of lucite (or hard rubber) charged by flicking it with fur. It then sits on the table with its top surface charged.

Fig. 33-7.

Take a metal plate with insulating handle and obtain a charge on it by the following procedure.

(1) Bring the metal plate very close to the charged cake. (It may touch the cake without harm since the cake is an insulator.)

(2) With one finger, touch the metal plate, connecting it momentarily (through arm & body & shoes) to the Earth. Then take finger away.

(3) Remove the plate from cake. The plate is now charged and can easily give some of its charge to other things by contact.

At this stage there seems to be little rhyme or reason to the process—it is witchcraft, and we might well label the stages "mumbo," "jumbo," "mumbo."

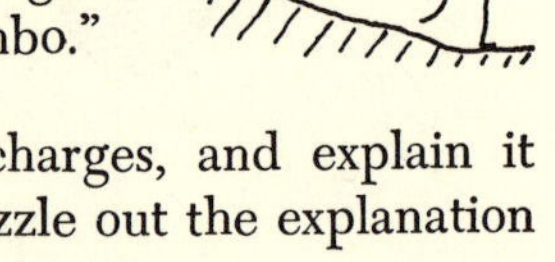

We shall use it to make charges, and explain it later—though you could puzzle out the explanation from what you already know.

The charge on the cake emerges unchanged; so the process can be repeated, giving an endless stream of charges on the plate. Where does the energy come from?

Experiments with Charged Plate, Electroscope, and Test Ball

Brought near a finger or nose, the charged electrophorus plate makes a small spark and loses its charge. The victim feels a tiny shock, but does not remain charged unless he stands on an insulating

[5] Benjamin Franklin made the choice of labels, long before batteries produced knowledge of electric currents.

Fig. 33-8. Electric Charges

(a, b) Shocks. (c) Balloons charged by repeated use of electrophorus repel visibly. The balloons should be coated with metal or carbon to give them conducting surfaces.

table. The plate can be recharged, and an insulated metal object can be fed with charge after charge till it gains little more from further applications.[6] Then the charged object can make a big spark and give a shock. Or it can share its charge with other insulated objects by contact, or by conduction along damp threads or metal wires. Offered a chance to share its charge with the Earth, it seems to lose all charge—if sharing is in proportion to size, there will indeed be little left on the original object.

[6] In terms of voltage: the plate is charged to a certain voltage (P.D. from ground). The victim's potential rises nearer and nearer to that of the plate, and the victim can remove a smaller and smaller share of the plate's charge.

Metal-coated balloons on silk threads can be given large charges and repel visibly. Or we can make a "gold leaf" electroscope[7] with one or two strips of metal leaf hung from an insulator. This is a sensitive instrument for testing charges.[8] It measures the force on its leaf due to the charge there by balancing it against gravity. The usual form has one leaf hung from a metal rod that passes through an insulator in the top of a metal case. A knob or plate on the top of the rod, outside the case, enables charges to be given to the leaf easily. Even bringing a charge near makes the leaf swing out—while it remains near.

We use a small test ball to bring *sample* charges from a big charged object to the leaf. Each dose of charge makes the leaf rise more. If after giving the leaf some positive charge, we bring it a small negative charge, the leaf falls slightly, suggesting that the new charge has neutralized some of the old one. So the labels + and − seem suitable. This provides an easy test for + and − charges.

[7] "Scope" means "look at." This name means "instrument for looking at effects of electric charge."

[8] The electrical force is due to repulsion by charges on the leaf's support and attraction by opposite charges on the walls of the electroscope's case. If the leaf is connected to some large charged object, it takes only a small fraction of the total charge; and the charge it takes is a measure of the voltage of the combined object & leaf. So the leaf's tilt indicates the P.D. between [object & leaf] and the surrounding walls or Earth.

ELECTROSCOPE

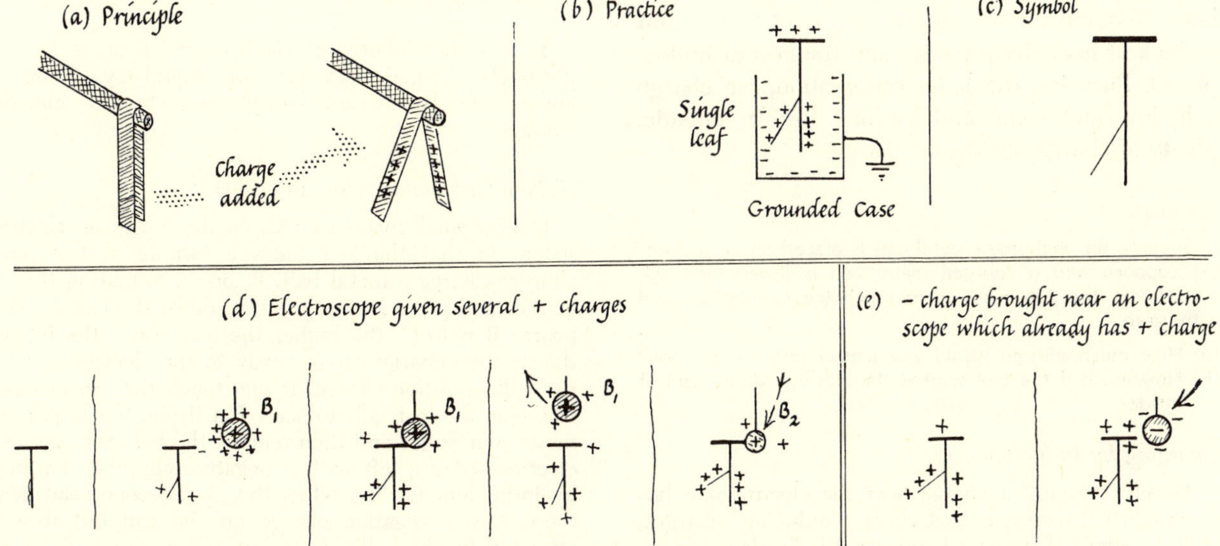

Fig. 33-9. The Electroscope

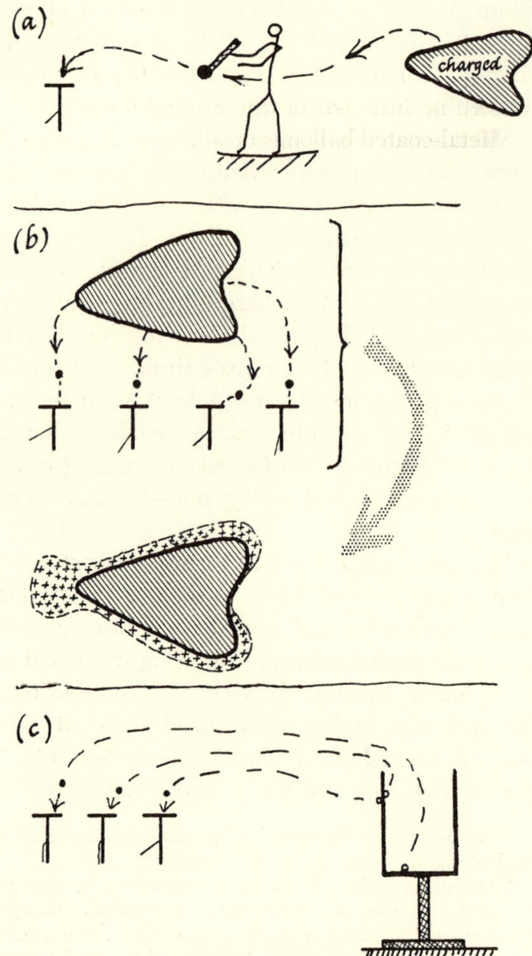

FIG. 33-10. CHARGE DISTRIBUTION
(a) Investigating charge distribution on conductor.
(b) Area with + signs indicates surface density of charge.
(c) Charge distribution *on* and *in* hollow, charged metal can.

We use a sampling ball and electroscope to explore the "charge density" all over some charged body. We find that charge is not spread evenly, but is thickest near sharp points, and thinnest at hollow places. Pursuing the latter observation, we charge a hollow metal can and explore. Result: outside, plenty of charge; inside, *no charge.*

PROBLEM 1

Suppose an uncharged metal can is placed on an insulating support, and a charged metal ball is lowered on silk thread into the can, allowed to touch the can *inside*, and withdrawn.

(a) How much charge would you expect left on the ball?
(b) How much if the ball touches the can outside instead of inside?

Charging by Induction

Merely bringing a charge *near* the electroscope has some effect. Investigate that effect ("inducing" charges, as it is called). Charge a large metal ball; place it near a long metal object. Use a small test ball and electroscope to look for charges on the "sausage." Suppose the large ball is charged +. Then − charges are found on the near end of the sausage; little or no charge on the middle, and + charge at the far end. We guess that these + and − charges were there originally in the uncharged sausage, and have been pulled apart by the charge on the big ball. Charges must travel easily on the sausage; so the big ball's + charge can pull negative charges towards it and push + charges away. Now join the sausage to the Earth by touching it with a finger. Remove the finger and re-test. The sausage still has the − charges near the ball; the + charges at the far end have disappeared. We say they have run away still farther, to ground through our finger. (Route: arm—body—legs—damp shoes—damp floor, etc.) As you know, this motion of charges can be shown with a microammeter. Now remove the original ball, with its + charge still on it. The sausage is left with a − charge distributed over it, extra thickly at the pointed ends. We have "manufactured" a − charge on the sausage (without losing the big ball's original + charge). Remove that − charge and put it to some use. The process can then be repeated any number of times, supplying a series of negative charges which could then be fed along a wire in a tiny stream of current. Notice the sequence, starting with the original charged ball: (1) bring sausage near; (2) touch sausage momentarily with finger; (3) remove sausage, and find that it has a charge available for use. This process is called *charging by induction.* Where have you met it before? Where does the energy gained with this charge come from?

PROBLEM 2

(a) Given a glass rod with a small + charge from rubbing, you can charge an electroscope by scraping some charge off the rod on to the electroscope. Would the electroscope be charged + or —?
(b) Or you can produce charges on the electroscope by holding the + charged rod near it. How would you then proceed to leave some charge permanently on the electroscope, without bringing the rod any nearer? Would that charge be + or —?

You can now charge an electroscope + or −, from a positively charged glass rod. You should try adding + and − charges to an electroscope which is already charged.

Hollow Conductors (See Fig. 33-12)

Place a small metal can, C, on the top of an electroscope, so that the leaf takes a sample of the can's charge. Charge a metal ball, B, on an insulating thread or rod, and place it near the outside of the can C. The nearer B is to C, the higher the leaf rises—the bigger the positive charge driven away to the electroscope by the ball's positive charge. If *you* touch the can or electroscope, the leaf falls to zero. But if you touch and remove your finger and then remove the ball, the can and electroscope are left with a negative charge—charging by induction. Earlier, when the electroscope read zero there was a negative charge on the can but it was attracted by the ball's + charge and none ran down to

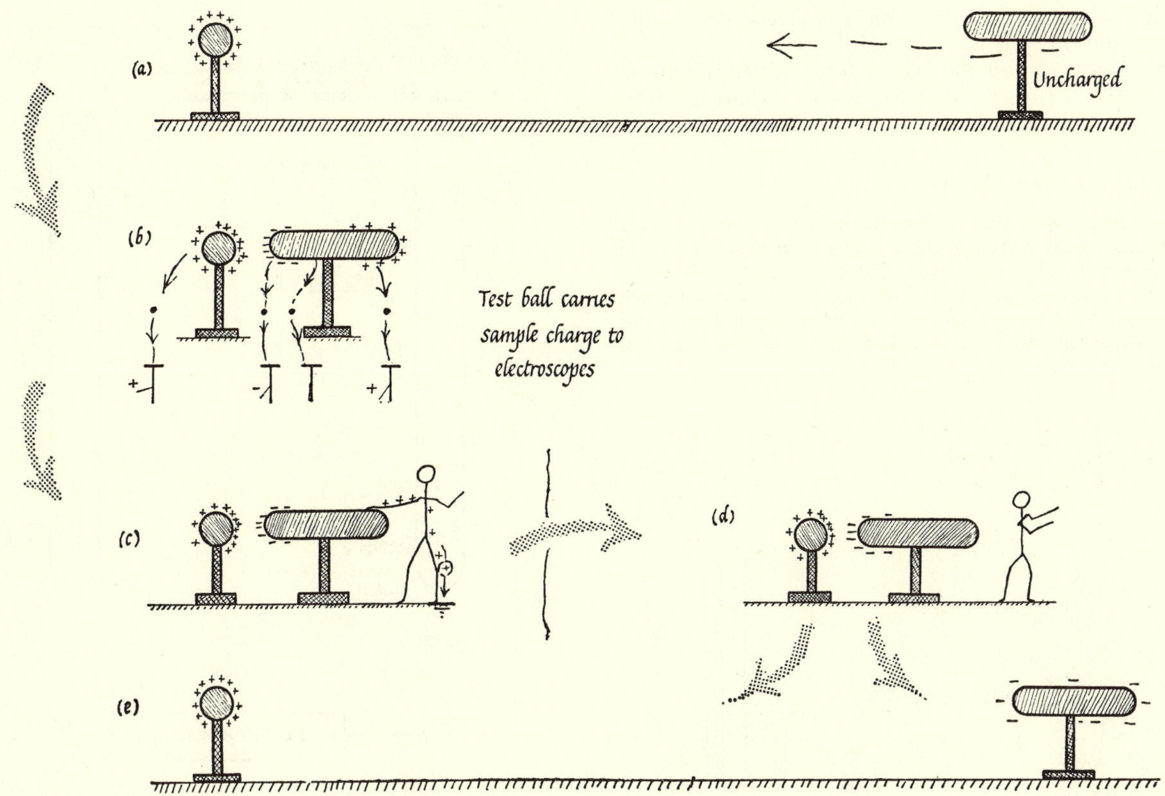

FIG. 33-11. STAGES OF CHARGING BY INDUCTION

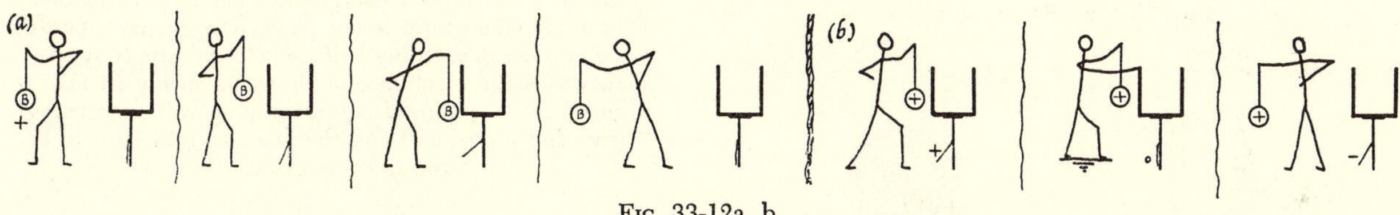

FIG. 33-12a, b.

the leaf. Such charges used to be called "bound charges" —an expressive description.

Now start again, with can and electroscope uncharged, and lower the + charged ball *inside* the can. The electroscope shows a large charge, which stays the same

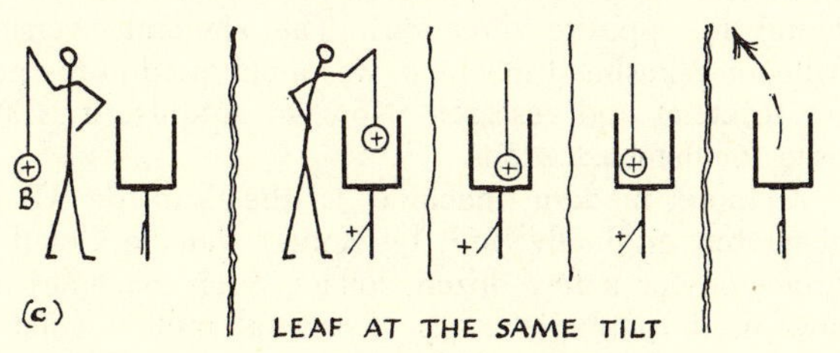

FIG. 33-12c.

wherever B is moved inside the can. If B is removed, the leaf falls. If B is put in again, the leaf returns to its full tilt. Now let B touch the can; the leaf stays at the same tilt. Remove B: the leaf stays up at full tilt. B must have given up all its charge. (Well, where *would* all B's charge go, if placed on the inside of the can?) We can "explain" this in detail by saying that when B is inside the can its + charge holds an equal and opposite charge "bound" on the inside of the can; and it drives an equal + charge (from the originally neutral can) to the outside of the can and the electroscope. When B touches the inside of the can the two inside charges have a chance to neutralize. The ball B loses all its charge, but an equal charge is left on the outside of the can.

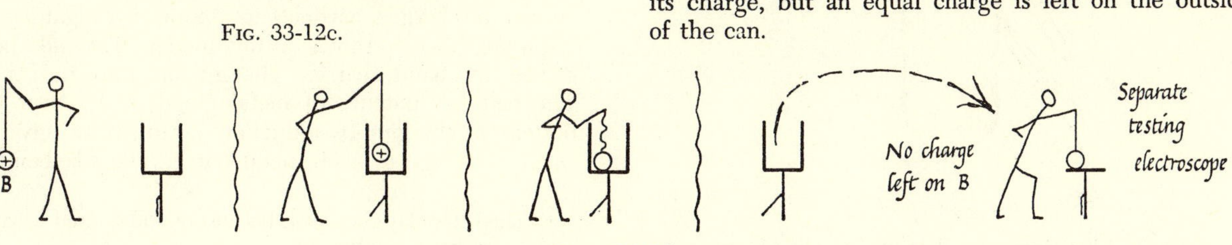

FIG. 33-12d

This behavior of a hollow can[9] has several uses. Problems 3 and 5 below show two of them.

We can even test the law of force between charges by looking for electrical effects inside a charged metal ball. This will be discussed later.

PROBLEM 3. (See Fig. 33-13a)

In a million-volt Van de Graaff machine, electric charges are built up on a huge copper ball as follows, for use in "atom-smashing" experiments. Charges (sometimes from a mumbo-jumbo-mumbo device) are placed on a moving silk belt which runs up and over a pulley wheel inside the copper ball. Inside, a trailing wire touches the belt. Why can the ball get the charges off the belt?

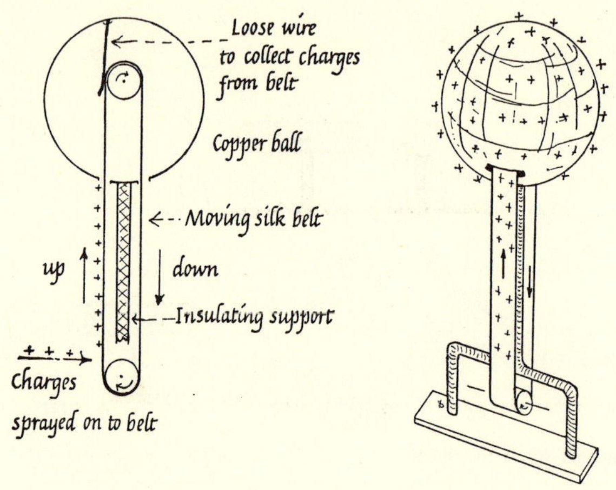

FIG. 33-13a (PROBLEM 3).

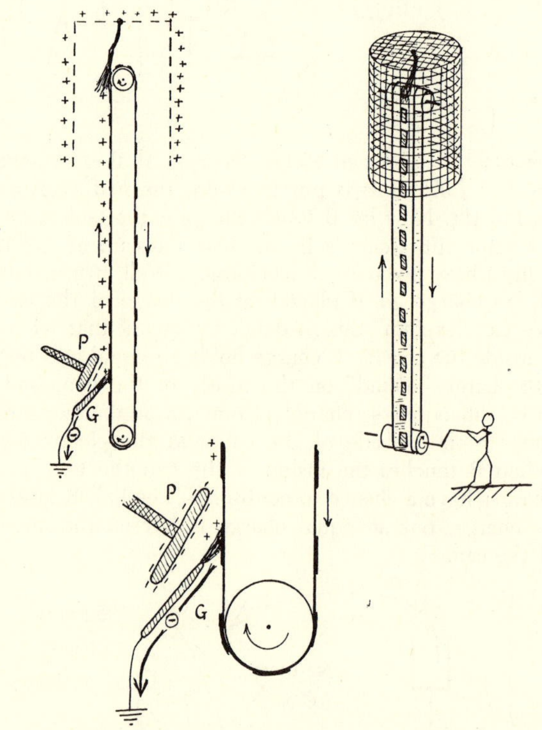

FIG. 33-13b. MODEL OF VAN DE GRAAFF MACHINE

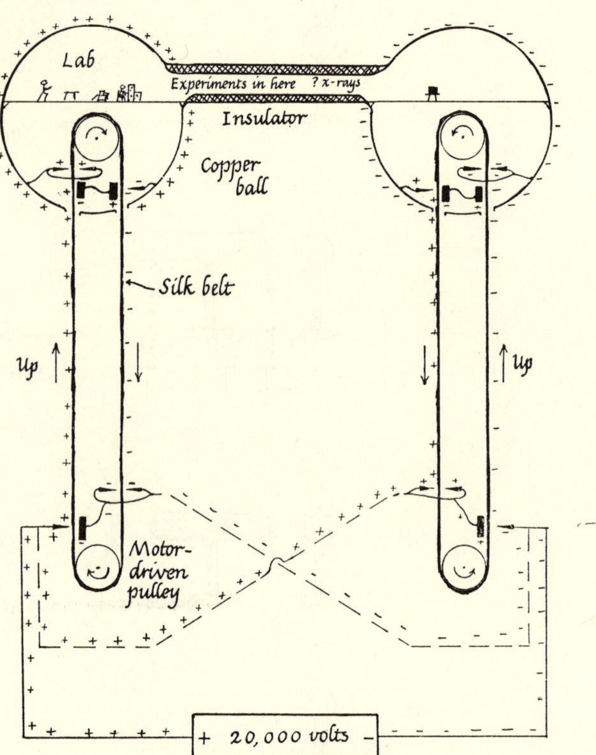

FIG. 33-14. TWIN VAN DE GRAAFF MACHINES
A 200,000-volt supply sprays + charges on one belt and — charges on the other, but inside the large sphere an ingenious charge-inducing scheme (shown in a simplified form here) places charges of opposite sign on the descending belt, to help the supply system. In very large machines, producing a steady trickle of current with a P.D. of several million volts, insulation problems are serious and the machine may have to be run in compressed gas.

The Van de Graaff Machine

The electrophorus, our mumbo-jumbo-mumbo machine, can be elaborated into a double-acting rotary version, somewhat as a hand-operated pair of cymbals could be elaborated into a musician's nightmare with two circles of players, each carrying a cymbal, marching round in opposite directions. The elaborate version, called a Wimshurst machine, was much used in the past for teaching and research. Now an antique, it is still used for demonstrations.

A more modern machine is the Van de Graaff generator, originally built by Robert Van de Graaff at Princeton for a few dozen dollars. Such machines are now used to obtain P.D.'s of several million volts to accelerate charged particles in atomic investigations.

Figure 33-13b shows a toy model. The silk belt, driven by hand, carries charges up into the cage. The belt has patches of metal glued on it. Near the bottom a charging-by-induction arrangement gives a small + charge to each metal patch. The silk being a

[9] Called Faraday's ice pail, because Faraday used a wine-cooling bucket a century ago.

good insulator prevents these charges running away, so the belt carries them up to the cage where a trailing wire touches them and collects their charge. A large metal leaf hung on the outside of the cage will show the charge building up.

PROBLEM 4.

Referring to Fig. 33-13b, explain briefly how the + charges are put on the metal patches. Note that P is an insulated metal plate fixed near the belt, carrying a negative charge; G is a metal brush, connected to *ground*, touching the metal patches near P.

In the big machine, the cage is a huge copper ball, big enough in some cases to house a laboratory for experimenters inside.[10] The silk belt has no metal patches; the charges rest on the silk itself. They are put on the silk at the bottom and removed at the top, without touching the silk, by sets of sharp points placed nearby.[11]

PROBLEM 5. TESTING FOR EQUAL AND OPPOSITE CHARGES (See Fig. 33-15)

Suppose a small metal can is attached to an electroscope. A small piece of lucite and a small piece of fur, on insulating handles, are held in the can uncharged. *If* the charges produced by rubbing are equal and opposite, what would you expect the electroscope to do at each of the following stages, *given that the leaf tilts at 60° in stage (b)*:

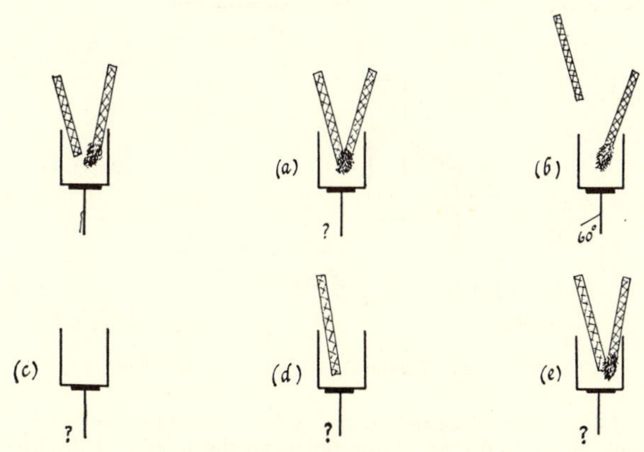

FIG. 33-15. PROBLEM 5

(a) fur and lucite rubbed together inside can (*not touching can*)
(b) lucite removed from can (leaf tilts at 60°)
(c) fur also removed without touching can
(d) lucite returned to can
(e) both returned to can

[10] The lab inside the ball proved uninhabitable: completely shielded from any electric field, but not from X-rays.

[11] See later for a discussion of the strange action of points.

Conservation of Charge

Ordinary things, left alone, seem to be uncharged: they do not attract or repel noticeably; gas molecules do not crowd over one way in a uniform electric field. If uncharged matter contains + and − charges that can be separated or transferred, they must be there in equal quantities. When we charge things by rubbing or by a battery, we expect to find equal and opposite quantities of charge. The experiment of Problem 5 shows a very delicate, important test. From this and Coulomb force experiments we conclude that *electric charge is conserved*: exchanged without gain or loss, never manufactured except in equal and opposite quantities. We maintain this as a basic principle, right through atomic and nuclear physics. We even find a high-energy photon of radiation (which is certainly neither matter nor electricity) turning into two electrons, but with equal and opposite charges. And we talk of "charge exchange" between some nuclear ingredients—a switch too rapid to be described in mechanical terms—but we still maintain charge conservation.

Forces Exerted by Charges on Each Other: Coulomb's Law

Electricity and *charges* are names for a state of affairs, of being able to push and pull. The pushes and pulls seem to radiate from the charged bodies. Two charges push or pull each other directly along the line joining them, and the forces get smaller when the distance increases. A century after Newton guessed at inverse-square law gravitation, the French physicist Coulomb experimented on the force between charges and showed that it obeys an inverse-square law. Fig. 33-16 shows the arrangement of his apparatus, essentially like the one Cavendish was using about the same time to measure the Gravitation constant, G. An insulating bar AB, hung on a fine twistable fibre CD, carries a

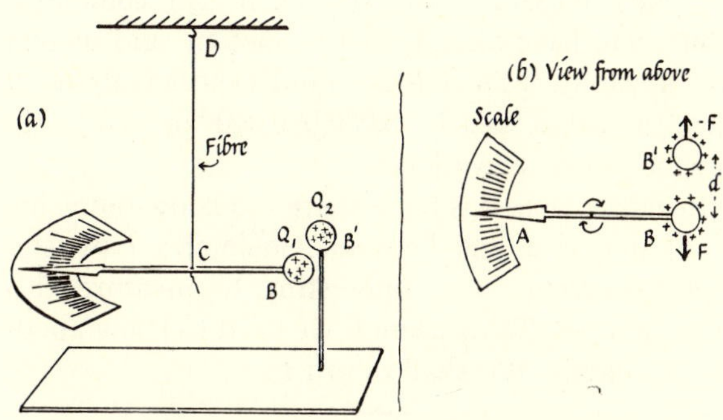

FIG. 33-16. COULOMB'S TORSION BALANCE
(a) Side view. (b) View from above.
The force was estimated by the angle through which the fibre twisted. The charged balls were carefully insulated. The whole apparatus was enclosed, to lessen effects of air currents.

metal ball B which can be given a charge. Another ball B′ is brought near B and the repulsion measured (proportionally) by the twist given to the fibre. Assuming Hooke's Law for the twisting fibre, Coulomb could compare the forces for different distances between B and B′. He found that F varies as $1/d^2$. He could change the charge on one ball, and then on the other, from charge Q to $\frac{1}{2}Q$, to $\frac{1}{4}Q$ and so on, measuring the force in each case. He found (or, rather, assumed and found no contradiction) that

F varies as charge on B, (Q_1) and *F varies as charge on B′, (Q_2)*. Putting these together he found that

$$F \propto \frac{Q_1 \cdot Q_2}{d^2} \text{ or } F = B \cdot \frac{Q_1 Q_2}{d^2}$$

This is Coulomb's Law. B is a universal constant playing a part similar to that of the gravitation constant, G.

★ PROBLEM 6.

Coulomb had no means of measuring the charges Q or Q′ which he started with. Even so, he could if he wished change from Q to $\frac{1}{2}$Q and thus investigate the part played by each Q in the law of force.

(a) (Guess) How could he reduce Q to $\frac{1}{2}$Q? (*Hint*: He had a spare metal ball of the same size.)

(b) What assumption concerning the nature of charge must be made in using the trick of (a)? (Actually this assumption was made almost axiomatically in building up the concept of electric charge. Nowadays we can vouch for it to some extent experimentally by counting electrons and observing positron and electron pairs.)

The value of B depends on the units chosen for measurement of charges; just as the value of G changes when we shift from kilograms to pounds for mass measurement. We shall use coulombs, which you have already met in motion, and meters and newtons.[12] Then B is found experimentally to have a value nearly 9,000,000,000 or 9.0×10^9 newton · meters²/coulomb².

In this course, it is not essential to know the value of B, but as a link between Coulomb's Law and electric-currents it is interesting to measure this huge number. To measure B we need to know about electric fields. We shall return to it.

[12] One scheme of units, much used in electrostatics in the last century, still appearing today, compelled B to have the simple value 1.000 by choosing the size of unit Q to make this so. But then this unit turned out to be inconveniently small for use in current-flow discussions. Choosing a unit may simplify calculations in one region of study, but we then meet more trouble somewhere else. Such choices cannot really affect the underlying science.

Test of Coulomb's Inverse-Square Law

DEMONSTRATION EXPERIMENT. We can make a rough direct test by measuring the force between two charges. We give large charges to two metal balls[13] and "weigh" one ball when the other is at various distances away, 0.1, 0.2, 0.3 meter, etc., center to center.[14]

You should see this done, even if only roughly, as witness to a great experimental law. In case you do not see it, there are specimen measurements in Problem 7.

★ PROBLEM 7. INVERSE-SQUARE LAW TEST (See Fig. 33-17)

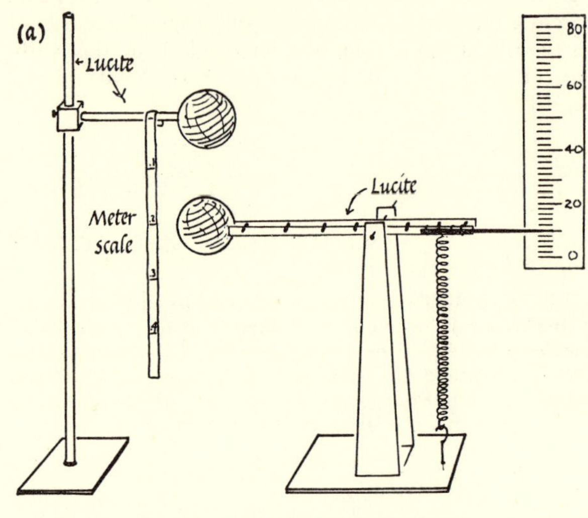

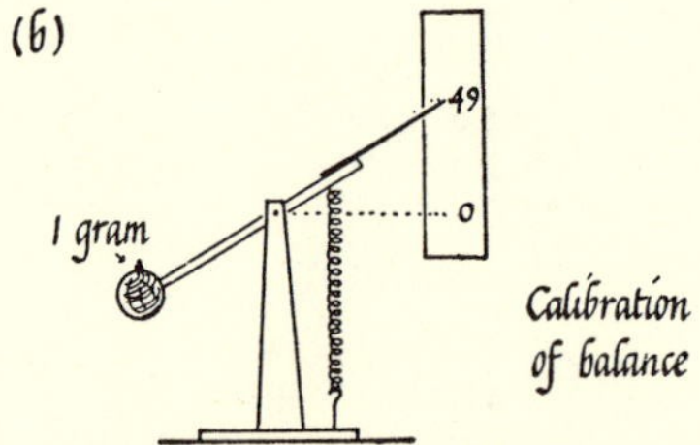

Fig. 33-17. Testing Inverse-Square Law

Ball A was attached to one end of a lucite see-saw. A weak (but good) steel spring balanced the weight of the ball and measured any extra force on it. A long pointer extended the other end of the see-saw to read on a vertical scale. Ball B was supported separately, vertically above A on a movable insulating rod. A lucite scale was hung from B's support to measure the distance between the balls. Each ball

[13] Ping-pong balls coated with metal will do. Thin glass Christmas tree ornaments are better.

[14] The charges on one ball repel those on the other and drive them away over the metal surface to the opposite side. Strictly speaking, we should not use the distance between centers of the balls, but some slightly greater distance. The experiment is also spoiled by charges leaking away, in any but very dry weather.

was charged by an electrophorus. With ball B moved far away, A's pointer was adjusted to read zero. Then ball B was placed at measured distances vertically above A, and the pointer reading was taken, to estimate the repulsion. (The scale was marked in arbitrary divisions, each about $\frac{1}{2}$ inch. To estimate the spring-strength—not needed here—a 1-gram load was placed on ball A; the pointer read 49.) The table shows specimen measurements. ("Infinity" means ball B far away.)

DISTANCE BETWEEN BALLS CENTER TO CENTER (*meters*)	*READING OF POINTER* (*scale divisions*)
"infinity"	0
0.10	26
0.20	7
0.30	3
0.40	? 2

(a) Copy the table adding a column for TEST OF LAW. Make calculations to test for an inverse-square law.

(b) *Assuming* that the value of *B* (which we shall measure later) is 9.0×10^9 and assuming the two charges are equal, estimate their size in coulombs.

(Note that the force must be expressed in *newtons*.)

Electric Fields

We think of every charge as having an electric field, rather like the gravitational field of a mass. *We define the strength of an electric field at any point as the force on a test coulomb placed there.* Since one coulomb is so huge, we word this more realistically

$$\text{ELECTRIC FIELD STRENGTH} = \frac{\text{FORCE ON A SMALL TEST CHARGE}}{\text{SIZE OF THE TEST CHARGE}}$$

which gives us the FORCE PER UNIT CHARGE, in newtons/coulomb.[15]

Fig. 33-18 shows demon experimenters, at work exploring gravitational and electric field strengths.

Around a small isolated charge, the field strength varies inversely as the square of the distance.

$$\text{FIELD STRENGTH} = \frac{\text{FORCE on test charge}}{\text{TEST CHARGE}}$$

$$= \frac{BQ \cdot (\text{test charge}, q)/d^2}{(\text{test charge}, q)}$$

$$= \frac{BQ \cdot q/d^2}{q}$$

$$= BQ/d^2$$

[15] With charges on metal objects, the insertion of a test charge in the space between would push the original charges around, and thus alter the field we are trying to measure. The smaller the test charge the less this alteration. So we pretend to use a smaller and smaller and smaller test charge, then jump, in mathematical imagination, to the limit and call the *LIMIT* of FORCE/CHARGE the FIELD-STRENGTH. Once again, a mathematical limit offers the ideal measure of a physical quantity, as mentioned in Ch.1 for acceleration, pressure, density; but here too the atomic nature of electricity would prevent us closing in to the limit.

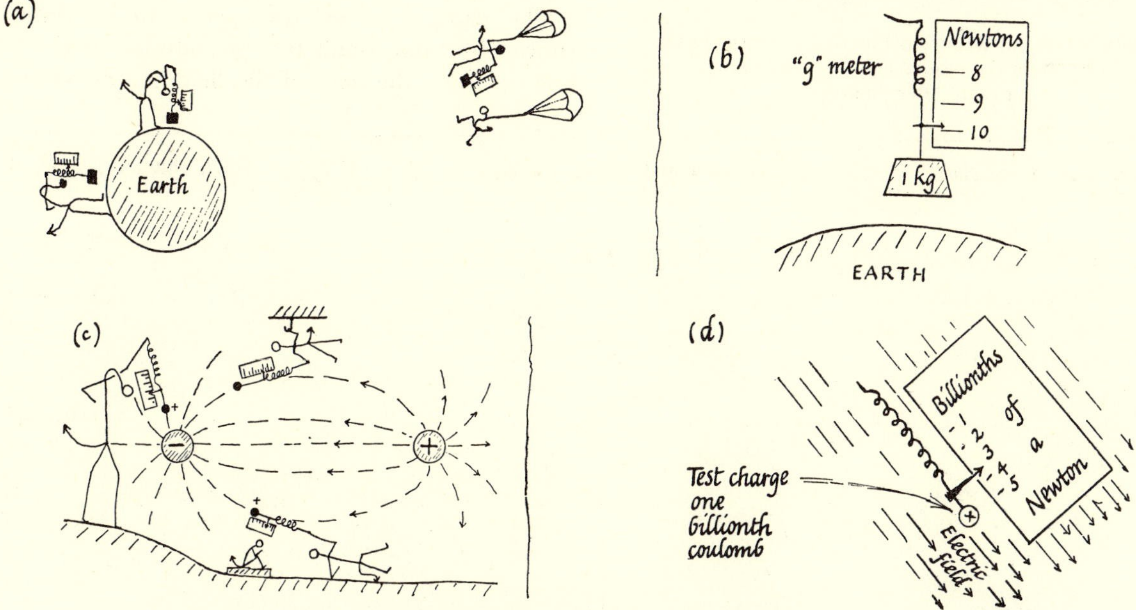

FIG. 33-18. MEASURING FIELD STRENGTH
(a, b) Measuring the Earth's gravitational field strength. The pull on a standard kilogram is measured, in newtons. (c) Measuring electric field strength. (d) "Electric field meter." To avoid disturbing the field that is to be measured, and to avoid having *huge* forces to measure, the test charge would have to be much smaller than 1 coulomb. Then we should measure the force on, say, 1 billionth of a coulomb, with a spring scale graduated in billionths of a newton. This would tell us the field strength in newtons/coulomb.

The same applies to an isolated charged sphere. Field strength is a vector. As well as size, the field has the direction of the force on a positive test charge. We can map the directions of electric fields by a trial balloon, real or imaginary, carrying a small test charge. Fig. 33-19 shows two huge metal balls charged + and —. A charged balloon will float across from one to the other along any of the paths shown by broken lines. These are called *lines of force*. They show the direction of the field; that is, the direction of the resultant force on a test charge. They are curved because the test charge is pushed by one main charge $+Q$ and pulled by the other $-Q$ with forces that change in direction *and size* from one place to the next. Using vector addition we can predict such patterns, though the process is tedious. Suppose the two main charges are equal and opposite. Then a test charge q placed at P experiences a repulsion F_1 due to Q_1 and a smaller attraction F_2 due to Q_2 (smaller because Q_2 is farther away). Adding these gives the resultant force, R, on q. At P the line of force of the field points along R. Now repeat the discussion at another point nearby, P′, and then at another, P″, and so on. We have chosen P′ so that it seems to be a little way along R (showing field at P), and we have chosen P″ on R'. We can then combine the pictures and predict part of a line of force. Quicker methods, using more difficult geometrical ideas, but the same basic inverse-square law, give the whole set of lines and can give other fields such as those shown in Fig. 33-20.

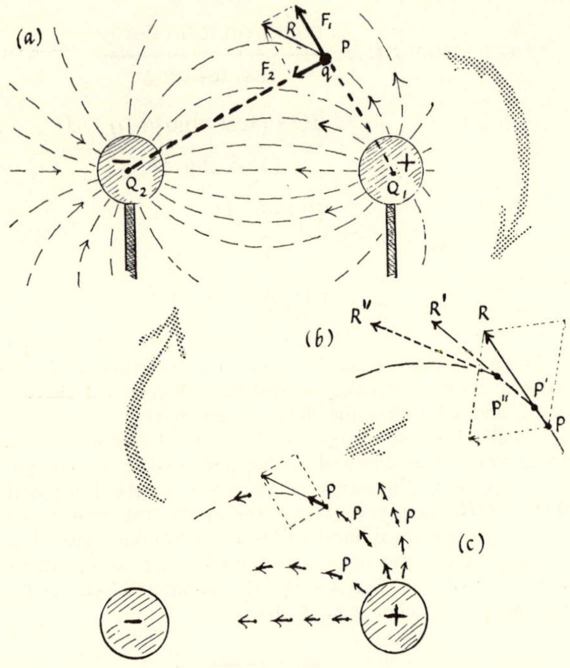

FIG. 33-19. MAPPING ELECTRIC FIELDS GEOMETRICALLY, by constructing the resultant pull on test charge at point after point.

PROBLEM 8.

(a) In Fig. 33-20(C) the charges are unequal. Which is larger?
(b) Give reason for your answer to (a).
(c) Answer (a) and (b) for Fig. 33-20(D).

Field Patterns

The patterns of electric fields can be demonstrated by using small bits of material which will line up along the lines of the field. No demonstra-

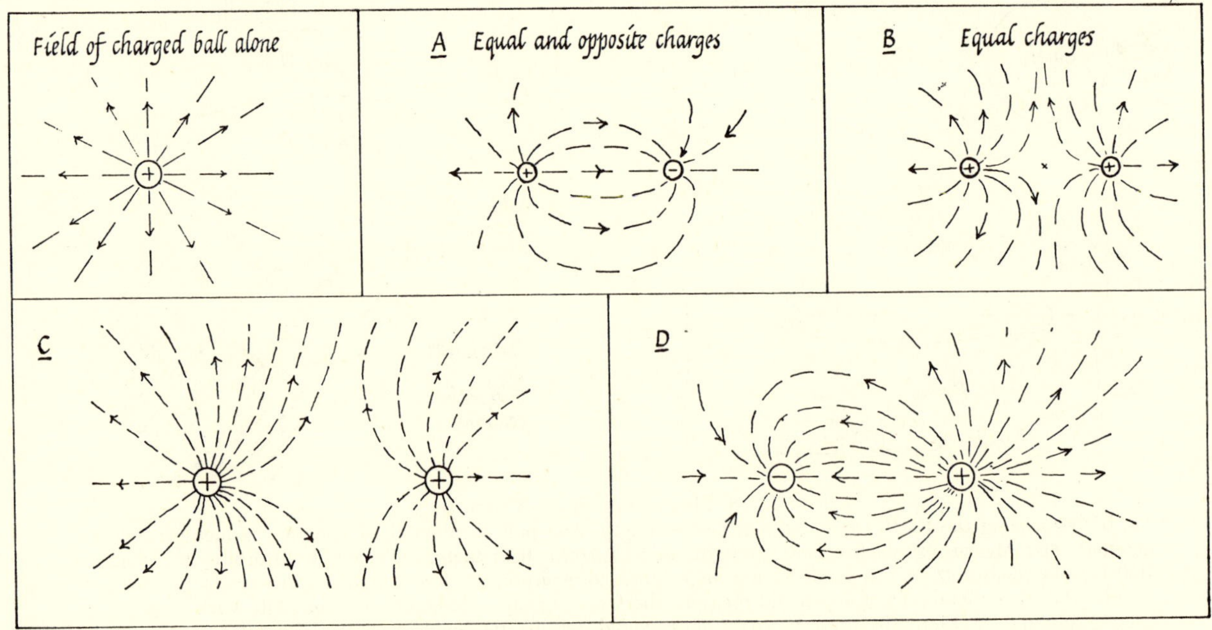

FIG. 33-20. ELECTRIC FIELDS

tion shows electric fields as clearly as iron filings can show magnetic fields. Metal objects representing Q_1 and Q_2 are charged by some machine, kept running to make up for leakage. A glass dish is placed around these metal electrodes, and filled with machine oil containing fine clipped hair. The hairs line up along lines of force of the electric field.

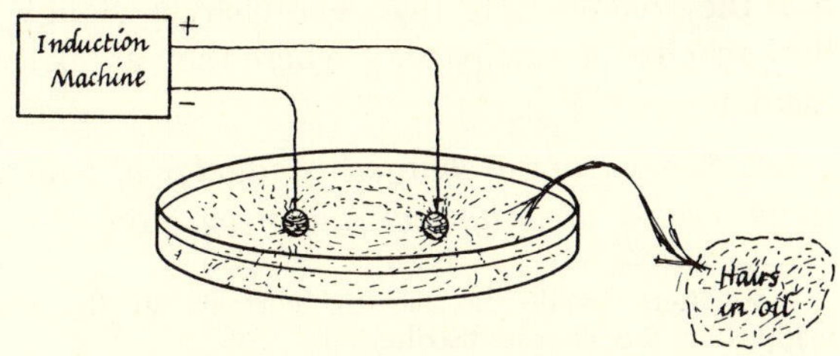

FIG. 33-21. MAPPING ELECTRIC FIELDS EXPERIMENTALLY, with short bits of hair in thick oil, using a strong electric field.

They develop pairs of charges and try to line up along the lines of the field. You should look at these field patterns and compare them with the equivalent magnetic fields (see Ch. 34). Here is the electric field due to equal and opposite charges +Q and −Q and the magnetic field round a bar magnet with "poles" +P and −P. These do not look exactly

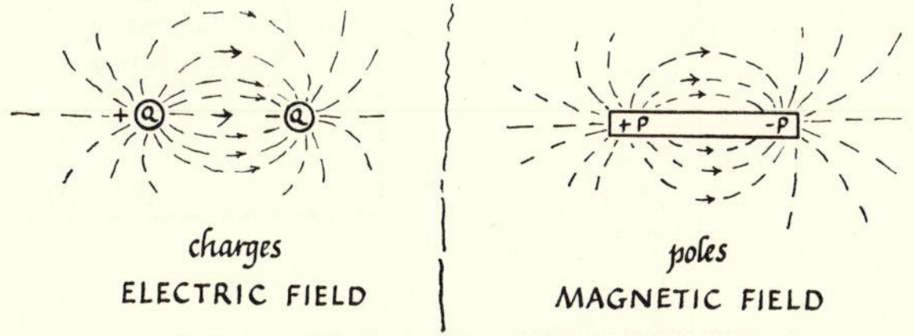

FIG. 33-22.

alike until we sketch the magnet's outline on top of the electric field. Both are constructed from inverse-square-law forces on a tiny imaginary test victim. The inverse-square law can be boiled down, mathematically, to a simple general form which can be used to predict the shape of any field in space, due to inverse-square-law forces. The general form looks simple to professional mathematicians, but difficult to amateurs. Just for fun, here it is

$$\frac{\partial^2 V}{\partial x^2} + \frac{\partial^2 V}{\partial y^2} + \frac{\partial^2 V}{\partial z^2} = 0$$

where V stands for the P.D. between that point in space and some standard zero-level, ground or infinity. This is so useful, so universal, that it is even given a condensed name $\nabla^2 V = 0$ (pronounced del-squared V equals zero). Whole books have been written to develop solutions of particular problems from it, for point charges, charged balls, cylinders, odd shapes, complicated arrangements. An inverse-square law applies also to gravitational fields, flow of heat by conduction, streamline problems of water, etc. and to magnetic fields (with important limitations). So solutions of $\nabla^2 V = 0$ which give the electric field of some arrangement of charges can also give a magnetic field, or the flow pattern of water with a corresponding arrangement of sources and sinks. Measurements of one kind of field can be translated to another kind of field. For example, a problem in heat conduction, important in some engine design, may be difficult to study experimentally, but it can easily be studied with electric fields instead (with the charged bodies shaped to imitate the heat supply). There seems little connection between water flow and electric fields—of course there *is* little real physical connection, if any, but only a formal connection of geometrical behavior—yet, turn back to the sketch of streamlines in a lake with a source and drain, in Chapter 9, Fig. 9-6, and you will see the pattern of the electric field of equal + and − charges.

This shows how mathematics can be used interchangeably. We never know when some piece of inverse-square-law behavior, studied and shelved in one region of physics, will suddenly become important and be dragged off the shelf for a new use in another region. For example, long ago physicists realized the inverse-square law must apply to the diffusion of dissolved salts in liquid—sugar spreads in coffee along streamlines like those of water in a lake—and speed of diffusion was related to salt concentration by "Fick's law" which is really a special case of the inverse-square law. Years later, the diffusion of neutrons in a nuclear reactor became a matter of great importance. Fick's law, and a host of ready made solutions were immediately available.

Field Patterns and Forces

Sometimes an artist's paintings give us uncanny insight into his character. The patterns of electric fields offer you just such hints and comments. Let them show you opposite charges attracting with clutching arms, similar charges pushing apart with buffers, or electroscope leaves swept up by a festoon of pulling cords. To Faraday, thinking about electric and magnetic fields a century ago these *lines of force* were very real indeed. He thought of them as elastic tubes, each starting on a + charge and ending on an equal − one, tugging with tension along their length and elbowing their neighbors with bulging expansion sideways. Such thinking helped him in his experimenting and enabled him to prepare the ground for Maxwell's electromagnetic theory of light. We still find them a useful concept. We even speak of a radio wave as a waggle travelling out along an electric line of force, like a loop flicked along a whip—though this is a misleading description. Most scientists do not think in terms of mysterious "action at a distance"—forces of gravity, or of electric attraction, skipping instantaneously across empty space to act on a victim. Instead we picture the effects being carried by a field (gravitational or electrical). Changes of force travel as field changes with definite speed.

Each line of force is a line along which a small test charge is urged by the electric field—the line of the resultant of attractions and repulsions of all other charges acting on the test charge. We can observe the patterns of such fields (with chopped hair in oil, or with a small pivoted pointer); or we

FIG. 33-23. EXPLORING LINES OF ELECTRIC FIELD, with a light pivoted pointer.

can construct their patterns geometrically (using tedious vector additions of inverse-square forces to find the direction of the resultant); or we can use a highbrow mixture of calculus and geometry to predict the patterns from $\nabla^2 V = 0$. All these methods agree in their results, of course. The sketches of Fig. 33-24 show some examples and the notes below comment on them in terms of Faraday's realistic thinking.

(a) *The lines run across from a + charge to a − charge.* They pull the unlike charges together. (Lines are in tension: they provide the attraction.)

(b) *Between the two + charges, the lines splay apart, like repelling buffers, with a "neutral point" where the repulsive fields of the two separate charges just cancel.* (Lines seem to bulge with pressure and push apart sideways; this provides or at least illustrates the repulsion.)

(c) A single charge alone gives lines that spread straight out as radii like the gravitational field of a small mass or a sphere. (As lines spread, their crowding decreases; this illustrates the weakening of field with increasing distance.)

(d) *From any charged piece of metal, the lines sprout out perpendicular to the surface.* They never sprout obliquely. (Lines at rest cannot have any component of pull *along* the metal surface—if they did, they would drag their end-charges along[16] till they reached a new pattern where they *are* perpendicular.)

(e) *Lines from a charged conductor of irregular shape crowd densely near any sharp protuberance*

[16] So there would be electric currents in the metal, readjusting the charge-distribution.

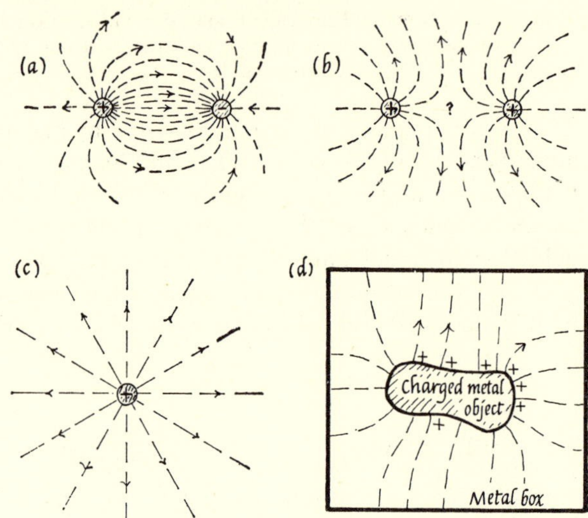

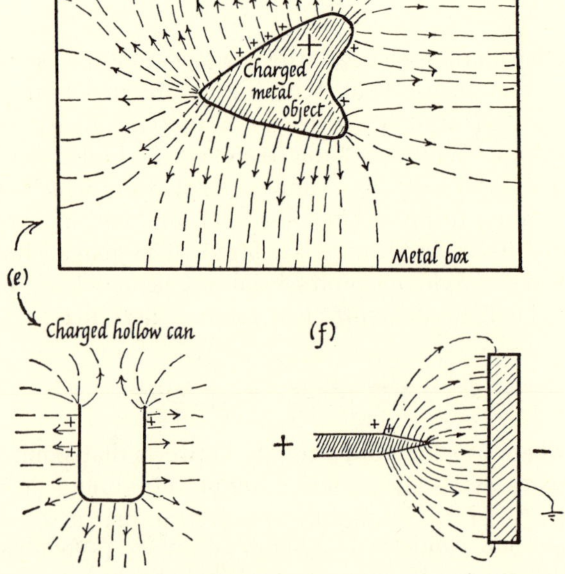

FIG. 33-24. ELECTRIC FIELDS

or point; they elbow away from any depression or hollow. (Near a point, lines can spread, radially, to less crowding; so the sideways elbowing of lines by their neighbors shoves them along from flat regions to pointed regions. This is not obvious, but the reverse effect, with hollows, is easy to see: if lines ran into a hollow region and anchored on charges there, they would elbow each other out of the hollow.)

(f) *When a point with a + charge is near a flat sheet with a — charge, the lines run from a very strong crowded field around the point to a large patch of field near the plane, where the field is perpendicular to the plane.* The field near the point is very strong. It may be too strong for the air there; it may drive a stray electron to bombard air molecules and release more charged particles—tiny electrified carriers. The main charge on the point then attracts those of the newly-made carriers that have charges of opposite sign. These are pulled to the point and neutralize some of the main charge there.

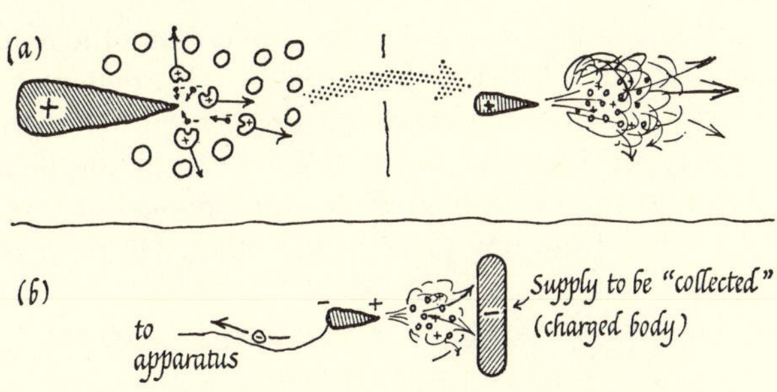

FIG. 33-25. ACTION OF CHARGED POINTS IN AIR

(a) A strong electric field near the point tears electrons off some atoms, leaving positively-charged atoms. Both the electrons and the charged atoms drift in the electric field. Carriers with opposite charge drift *to* the point and neutralize some of its charge. Carriers of same sign rush *away*, making "wind" which carries charge.

(b) When a point collects charge from some supply, wind from the point neutralizes some of the supply's charge and the point gains a corresponding charge. This is useful in collecting charge from moving things (e.g., a belt in Van de Graaff machine) where a rubbing collector would be harmful.

The carriers with charges like the point's are repelled and rush away, carrying more air with them by collision, making an "electric wind." This is the charged wind used to spray charge on the Van de Graaff belt. It can also be used to "collect" charge from a charged object by using the wind as a subtractor. At the end of this book you will find the strong field of a charged needle point used to photograph atoms!

On a more violent scale, the strong field near a point may start sparks in air, or electron avalanches in a Geiger tube, or even vast lighting flashes between charged clouds and a pointed tree or rooftop. The sharp pointed metal rods used to protect houses are just that, lighting conductors to start and carry small, invisible, harmless, reverse lightning flashes before the storm's electric field builds up to dangerous voltages. And even when there is a big flash, the point offers it an easy starting place and then the rod carries the discharge-current safely to the ground.

FIG. 33-26.
STORM CLOUDS OFTEN CARRY LARGE ELECTRIC CHARGES, possibly developed in the breaking-up of raindrops. These induce opposite charges on roofs and ground. If the field of the combined system grows strong enough to make charged carriers from air molecules, a lightning flash may start. Once started, it finds the air a fairly good conductor and a vast current may flow.

Every line of force of a field must start from a + charge and run, ultimately to a — charge. Remember that the line shows the direction of the resultant force on a small positive test charge; then you will see that every line must run thus, from a + to a —.

Then crowded lines near a charged surface must mean crowded charges on the surface. If so, we should expect to find charges unevenly distributed on an irregular metal body, most crowded near points, least crowded in hollows. We have already tested this expectation by carrying samples to an electroscope with a small test ball or disc (Fig. 33-10).

(g) *The field in the space between parallel charged plates is uniform.* This field is useful, be-

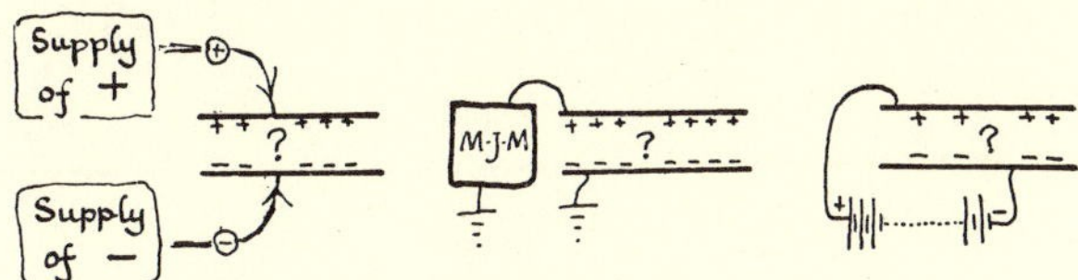

FIG. 33-27. CHARGING PARALLEL PLATES

cause we can calculate its strength from a voltmeter reading. We shall use it in measuring the Coulomb's Law constant, B, and later in Millikan's experiment to measure the charge of a single electron. Guess at the pattern of this field by looking at the field between two charged balls and imagining the balls growing bigger and bigger.

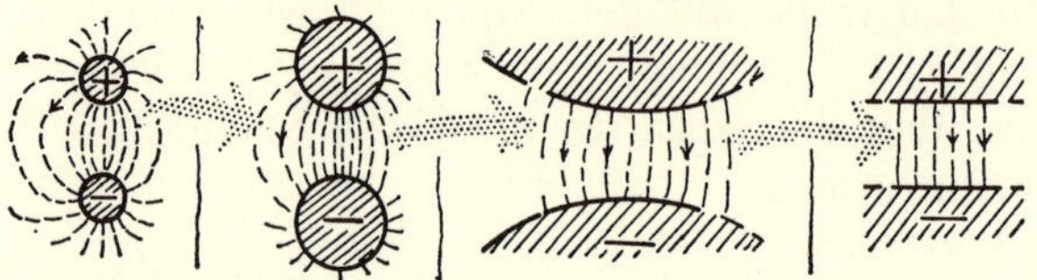

FIG. 33-28. EXTRAPOLATION

Extrapolate, in imagination, to the limit when the balls are infinitely big (so that their surfaces are flat) but still have a small air-gap between. This is only a guess. Getting help from algebra or from experimental mapping, we find the actual field is like Fig. 33-29.

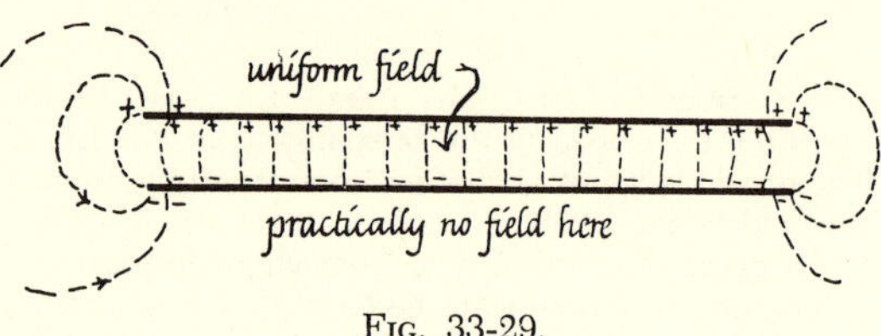

FIG. 33-29.

In the region between the plates, the lines are parallel and evenly spaced. (Curiously enough, algebra can prove, from $\nabla^2 V = 0$, that *if* the lines are parallel, they *must* be evenly spaced.) The field has the same direction and same strength everywhere[17]: how could a small test charge placed between the plates know or mind where it is among this forest of parallel palings? We call this a *uniform* electric field. Outside the plates there is practically no field except near the edges of the plates, where the lines of the inside field bulge out, indulging in "fringing."

A test charge of 1 coulomb would experience the same force anywhere in this uniform field. Suppose that force is X newtons. Then the field strength is X newtons per coulomb. Hire a demon to drag a

FIG. 33-31.
$X \cdot d$ newton · meters/coulomb $= V$ joules/coulomb

coulomb from one plate to the other, along the field, *against* the electrical force on it. The demon requires fuel for the work he does, to provide the energy he transfers to electrical form in the field. He must drag with force X newtons along a distance d meters, from one plate to the other. He must do work

$$\text{FORCE} \cdot \text{DISTANCE} = (X \text{ newtons})\ (d \text{ meters}) = Xd \text{ joules}$$

for one coulomb. But this *is* the energy-transfer for one coulomb passing from one plate to the other; it is the potential difference, in *joules per*

[17] Think of yourself as a + test charge of electricity. Climb to A, between a vast positive ceiling, and a vast negative floor, and look for charges to influence you. Upwards, you see a positive ceiling, an infinite sheet covered with an even layer of charges. When you shift sideways, you see the same, if the plates are infinitely wide. When you move higher, but still look upwards, you are nearer to some of the ceiling, but any cone of vision that you choose to use now takes in less ceiling; and these two changes, an inverse-square-law change for distance and a direct square change of ceiling area, just compensate. Therefore you would find the ceiling just as repulsive wherever you stood, at whatever height. The same with the floor, equally attractive. Therefore you would find yourself in a uniform field. (For a similar reason, with scattered light you will find a white ceiling above you looks just as bright but no brighter when you climb up nearer to it.)

If you stood outside the region between plates, say below them, you would look up towards two ceilings, a negative sheet above you, and a positive sheet above that, still higher but (if infinite in extent) just as repulsive as the lower one is attractive. You would therefore find yourself in zero resultant field.

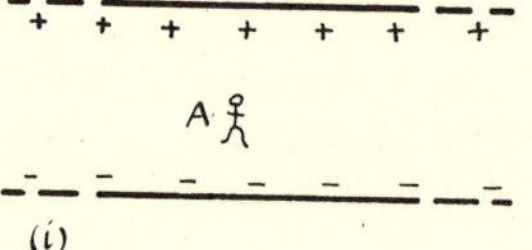

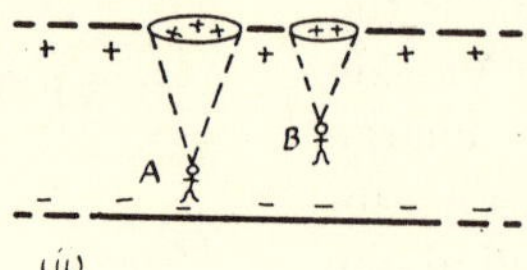

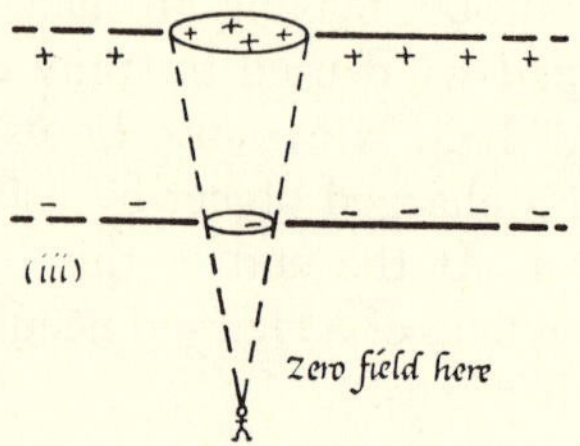

FIG. 33-30.

coulomb or *volts*. Therefore, if a voltmeter connected across the plates reads V volts,

$$Xd \text{ joules/coulomb} = V \text{ volts.}$$

$$\therefore \text{ FIELD STRENGTH, } X \text{ newtons/coulomb} = \frac{V}{d} \text{ volts/meter}$$

Therefore a voltmeter and a meter stick will enable us to measure the field strength, X; and this is useful in some important experiments.

Insulators and Conductors

Experiments show that charged bodies supported on dry lucite, or sulfur or hard rubber, remain charged. We call such substances insulators. Metal wires, carbon, wet string, etc. will let charges run along them. They "conduct" positive charges and negative charges towards each other, to neutralize; or they enable a charged metal body to share its charge with another conducting body. If the latter

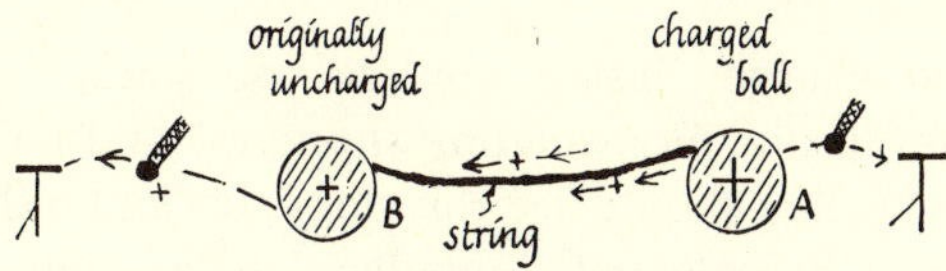

FIG. 33-32.
DEMONSTRATING THE SLOW MOVEMENT OF CHARGE THROUGH A WET STRING. A small ball is used to carry a sample of charge to each electroscope. Ball B gains charge and ball A loses some.

is a huge body such as the Earth, it takes so large a share that all the original charge seems to disappear. We call such carriers of electricity *conductors*. Water, though a very poor conductor compared with metals, carries charges away fast enough to spoil many an experiment. Some insulators, particularly glass, are fond of water and collect a skin of water molecules. Glass rods are almost useless as insulators for electrostatics experiments unless they are specially dried.

PROBLEM 9.

"Glass attracts water molecules strongly." This is a reminder from surface tension studies. What experimental test do you therefore suggest for selecting an insulator that will not suffer much moisture-trouble?

Current IS a Motion of Charges

While a conductor is carrying charges to or from a charged body, there is a current. If we look for the usual characteristics of current (heat, chemical effects, magnetic fields) we find them but they are small.[18] Rowland put this to a great test in 1876. He spun a wheel that carried charges fixed on its rim, and obtained the magnetic field that a current flowing around the rim would give. Nowadays, cyclotron engineers measure the current of their stream of charged atoms with a micro-ammeter! In lab, you should make a battery drive charges to the huge plates of a "capacitor" and watch the brief pulse of current with a meter (Experiment G in Chapter 41).

Now remember how we started by charging objects from a battery; and return to the meters and batteries of a modern current laboratory. Batteries supply static charges with just the same properties as charges from a rubbed rod. The coulombs of Q_1 and Q_2 in Coulomb's Law are the same as the coulombs in "C amps = C coulombs/sec." They are the same in nature, and we shall make them the same in size by our choice of B. (When we determine the value of B experimentally we shall use an ordinary voltmeter, calibrated in joules/coulomb, in measuring the electric field-strength.) From now on we shall use our full equipment, measuring P.D. equally with voltmeter or electroscope, taking charges equally from electrophorus, Van de Graaff machine, battery, or even electronic power-supply with transformer and rectifying diode.

Modernization

Why not modernize further and describe all electrostatics in terms of free electrons? Simply because no experiment in this chapter so far shows any behavior that necessitates electrons. Furthermore, there are *positive* charges that move in some substances.

We now know that the movable charges in metal conductors are negative electrons. They can run freely through metal, but positive charges remain anchored in atoms of the solid. So where we say "positive charges run down the wire to the ground," we should say "negative charges (electrons) run up the wire from the ground, to neutralize positive charges at the top." Where we say "a ball with positive charges on it" we should say, "a ball with negative electrons wiped off it." Where we say that a positively charged ball placed near a metal bar "pulls negative charges towards it and drives positive charges to the other end," we should say "pulls negative charges towards it, leaving positive charges (unbalanced) at the other end."

However, in our early experiments it is just as easy to think of both + and − charges moving. Then, *since it does not matter* whether both move or only negative electrons, it seems antiscientific to insist on the difference at this stage. Here you meet the modern problem: is good theory what is *useful, consistent, adequate,* for present use, or must it also be true? If you crave for true electrons at once, re-word all the early explanations in terms of them for yourself. If you adopt the tough attitude of many a modern theorist, use the old view until you find cases where it makes a difference that "only electrons move." You will find electrons necessary, and in full use, later in this book.

A hard rubber rod picks electrons off fur; a glass rod yields them to silk. We now know that any two substances in contact will exchange some electrons, one

[18] The charges involved in most electrostatic experiments are very small, less than a millionth of a coulomb. The P.D.'s are huge: 10,000 volts for a charged electrophorus plate, millions for big machines. As a very rough guide, a one-inch spark in air needs a P.D. of 50,000 volts.

becoming richer (thus acquiring a negative charge), the other poorer (positive). The exchange, which happens very quickly on contact, continues until a small P.D. is built up (by the + and − charges thus separated) which prevents more migration. If the two substances are moved apart, the mechanical dragging away increases the P.D. and the separated charges may even be driven back. To obtain big charges "by friction," contact is necessary; rubbing is no help, speed is useful: flick the glass rod with silk; don't rub it.

The Original Electrostatic Experiment: Picking up Chips

The ancient Greek electrical experiment, picking up light chips of wood, etc. with rubbed amber, is not so easily explained as one would expect. Why should a charged body pick up an uncharged one? Because it induces charges on the chip. Then the chip should be

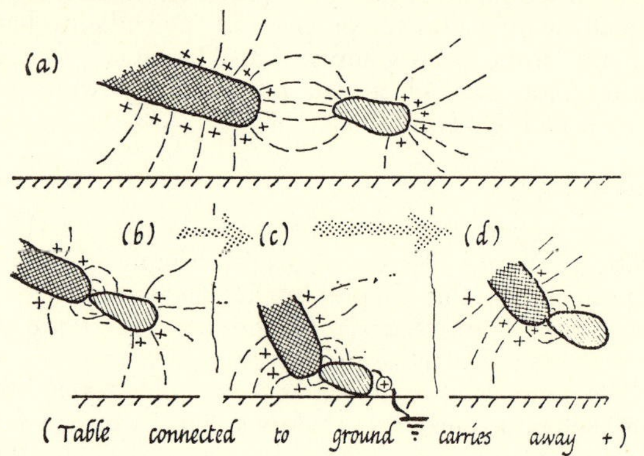

FIG. 33-33. CHARGED INSULATOR PICKS UP A CHIP OF METAL

made of metal, so that the induced charges can run apart. In fact, light metal chips—aluminum leaf, for example—are attracted very well by a charged rod: induced charges *are* the explanation. A chip of perfect insulator would show hardly any effect, but bits of wood and paper always have enough moisture to make them slightly conducting. If the chip rests on a table connected to ground, the "like" induced charge may run away leaving only the "opposite" charge; then the attraction is still greater. The sketches in Fig. 33-33 show the process.

In fact, however, even perfect insulators are attracted, though often not strongly. Their molecules seem to have charges within them that can be pulled apart slightly,

FIG. 33-34. POLARIZING A MOLECULE

making *electrically elongated*, or *polarized*, molecules, with end-charges for attraction. Even an atom can be polarized, its electron-cloud pulled one way and nucleus the other. That is often how atoms or molecules attract each other when close and cling together: the van der Waals forces of surface tension are electrical forces due to molecules polarizing in each other's field.

Experimental Proof of Inverse-Square Law

Instead of measuring the force between two small charges, we can test the inverse-square law indirectly but no less surely by an amazingly simple crucial experiment. *If the inverse-square law holds, there is no electric field inside an empty metal box however big a charge it carries*, and conversely. Make a closed metal box—cylindrical can, hollow ball, cubical box, any shape—and give it a large charge. Then investigate the state of affairs *inside* the box: look for electric fields there. Look for charges inside if you like, for where there are charges there will be fields. Experiment shows there

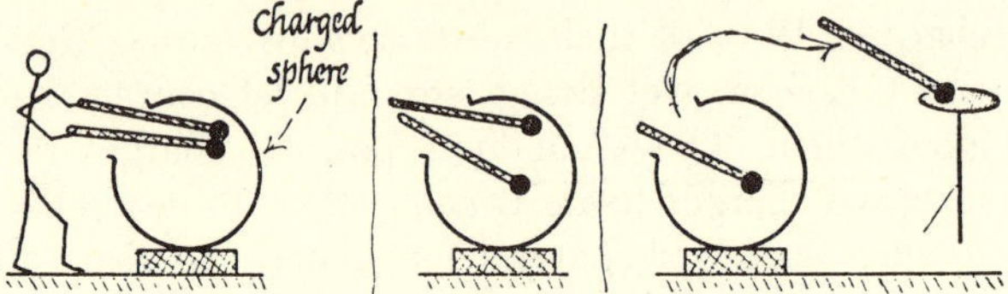

FIG. 33-35. TESTING INVERSE-SQUARE LAW

are no charges inside (unless you parked extra charges inside, *on insulating supports*) and no electric field. You should see this done. It can be shown on a small scale with a hollow metal globe and sampling balls. Or it can be done on the huge scale Faraday used when he climbed into an electrified wire cage: with sparks flying outside, he could find no effects inside. The experiment is clear and easy, but why does that show there is an inverse-square law?

We shall prove this for a round ball, though it can be extended to any shape of closed conducting box. We use a beautiful piece of geometry—which you will see Newton must have used long ago for a gravitational version of the problem. Suppose the ball, shown in section in Fig. 33-36, is positively charged. Then, from symmetry, the charge will spread evenly all over its surface. Suppose an observer is searching for an electric field at D, inside the ball. He looks out in a narrow cone towards a patch P_1 of the surface. This carries a charge Q_1, which would certainly repel a test + charge carried by the observer at D. So far as that charge Q_1 is concerned, there is certainly a field at D, a field pointing along P_1D. But, looking backwards, the observer sees an opposite patch P_2, whose charge Q_2 also contributes field at D, but pushing the opposite way. Now the observer defines the patches more carefully by drawing a cone from D to P_1, and extending the *same* cone backward to P_2. We can show that the effects of charges Q_1 and Q_2 will just

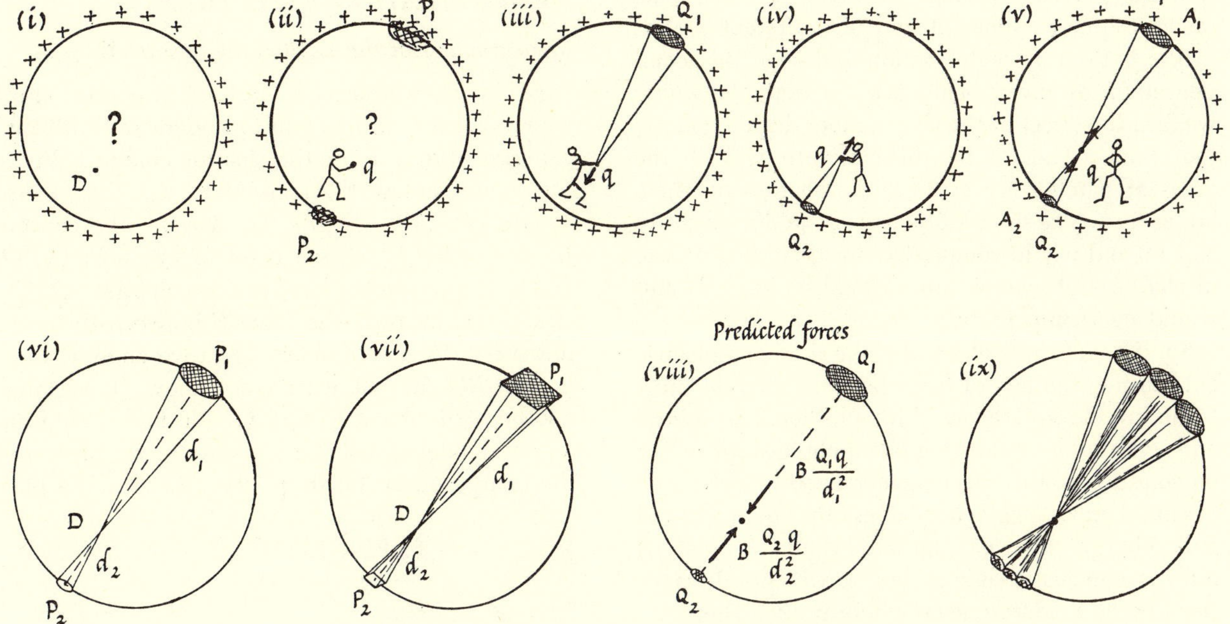

FIG. 33-36. ELECTRIC FIELD INSIDE A CHARGED METAL BALL

cancel. If the observer at D is nearer to P_2 than P_1 the area of patch P_2 will be smaller and will include less charge. On this score, Q_2 will be smaller than Q_1 and exert less force on the test charge at D. But on the score of inverse-square law, the charge Q_2 which is nearer will exert a larger force than Q_1. We want to show that the two factors compensate. For charge spread evenly all over the globe—assured by symmetry—the charge on one square inch will be the same everywhere; the charge on 2 square inches will be twice as much, &c. The charges on P_1 and P_2 will be proportional to their areas. Because these patches are marked out by the same cone, their areas are proportional to the squares of their distances from D.[19]

$$\frac{\text{Area of } P_1}{\text{Area of } P_2} = \frac{d_1^2}{d_2^2} \text{ by geometry.} \quad \therefore \frac{Q_1}{Q_2} = \frac{d_1^2}{d_2^2}$$

If Coulomb's Law holds, we should expect the forces that Q_1 and Q_2 exert on a tiny test charge, q, to be $B \cdot Q_1 \cdot q/d_1^2$, and $-B \cdot Q_2 \cdot q/d_2^2$ but we have shown that Q_1 and Q_2 are proportional to d_1^2 and d_2^2. So we call Q_1 and Q_2 Kd_1^2 and Kd_2^2.Then the forces on test q would be

$$\frac{B \cdot (Kd_1^2) \cdot q}{d_1^2} \quad \text{and} \quad -\frac{B \cdot (Kd_2^2) \cdot q}{d_2^2}$$

or BKq and $-BKq$ and these are equal and opposite and cancel each other. [See below for shorter algebra version.]

This is for only one little cone, marking out P_1 and P_2. We now imagine another cone next to the first one, also passing through D. The same argument applies to this and all the other cones with which we now imagine the whole globe filled.[20]

Test

Given the inverse-square law, we predict "no electric field inside a hollow charged ball." To test the inverse-square law we rely on the converse. Converses of true statements are not always true, but we can easily show that this one is. Geometry of the spreading cones provides factors d_1^2 and d_2^2 in the tops of the expressions above; the inverse-square law inserts the same factors in the bottom of each fraction. One charge is bigger than the other, but that effect is compensated by distance in just the

[19] For example, if P_2 is 3 times as far away from D as P_1, then the cone from D spreads till it is 3 times as wide at P_2 as at P_1. If it is a round cone, cutting off a circular patch, the patch at P_2 has 3 times the diameter of the patch at P_1. Therefore the diameters are as 3 to 1, radii as 3 to 1; areas are as 3^2 to 1 or 9 to 1.

The argument applies to cones of any shape. If the argument with round cones seems hard, think of small square cones, cutting rectangles on the globe's surface. These are drawn thus: start with a tiny rectangle, P_1. Join its corners to the test point D, and continue your lines straight on through D till they meet the globe again, at P_2. There they will mark out a rectangle of different size but of the same shape.

[20] Most of these tiny cones hit the surface of the globe on a slant, but that does not spoil the argument, because the slant is the same at both ends of the cone, P_1 and P_2. Any chord of a circle or sphere makes equal angles with the tangents at its ends.

same proportion. If some other law of force than inverse square applies, it will put different factors in the bottom of each fraction and spoil this compensation. An inverse-cube law, for example, would weaken the effect of the larger, more distant, charge too much. (Thus, if the distances are as 3 : 1, the areas of patches are as 9 : 1, the charges on them are as 9 : 1. The inverse squares of the distances are as 1 : 9 and would compensate for the different size of charge. But inverse cubes would be as 1 : 27 and would overcompensate.)

So, *if* a uniformly charged globe shows no electric field inside, the law of force between charges must be an inverse-square one. With additional geometry, and some cunning thinking about charges spreading on conductors, this can be generalized to any shape of closed metal box. With shapes other than a round ball, charge does not spread evenly; in fact, it crowds into just that distribution which will make the electric field zero everywhere inside. Therefore we have an easy, delicate test of the inverse-square law: give a large charge to a closed metal ball or box, and look for fields. Outside the box an electroscope shows large fields; and there may even be sparks. Inside the box there is no electric field from the external charges. For a clear search for *fields*, use two small metal balls on insulating handles. Place them, uncharged, in the region to be explored. Let them touch; separate them, and test each for charge (See Fig. 33-35).

A closed metal box makes a perfect shield for electric forces (and a cage of wire netting is almost as good). If it is large enough, experimenters can work quietly in it, completely shielded. You will find small metal shields like this around some parts of your radio.

The argument is given here fairly fully, not as something so important that you should remember it all your life, but as a sample of a scientific argument leading from a guessed-at law to a crucial test.

[*Algebra*

Here is a much shorter and neater discussion: Assume an inverse n^{th} power law of force, so that $F \propto 1/d^n$. The two ends of a narrow cone cut off patches of sphere with charges Q_1 and Q_2; and $Q_1/Q_2 = d_1^2/d_2^2$ by geometry and symmetry. These exert opposite forces F_1 and F_2 on test charge q, such that

$F_1/F_2 = (Q_1/d_1^n)/(Q_2/d_2^n) = (Q_1/Q_2)(d_2^n/d_1^n)$. If these forces are equal and opposite, $F_1/F_2 = 1$ and then $(Q_1/Q_2)\ (d_2^n/d_1^n) = 1$. Therefore $(d_1^n/d_2^n) = d_1^2/d_2^2$. So we must have $n = 2$ for the forces to cancel in detail.]

DEMONSTRATION EXPERIMENT.

Measurement of the Coulomb Constant B

(This demonstration is not itself important in this course but it will help you to understand Millikan's experiment to measure the electron charge.) We use the lucite seesaw that served to test the inverse-square law (Fig. 33-17). To measure the constant B in $F = BQ_1Q_2/d^2$ we need to know F, Q_1, Q_2, and d. We arrange to have the two charges (Q_1, Q_2) equal, and measure the force F between them at a measured distance d apart. Then we know F and d in $F = BQ^2/d^2$ but we do not know Q. We need another experiment to find Q. We find it by using a giant model of Millikan's apparatus for measuring the charge on an oil drop (Fig. 33-37). We place

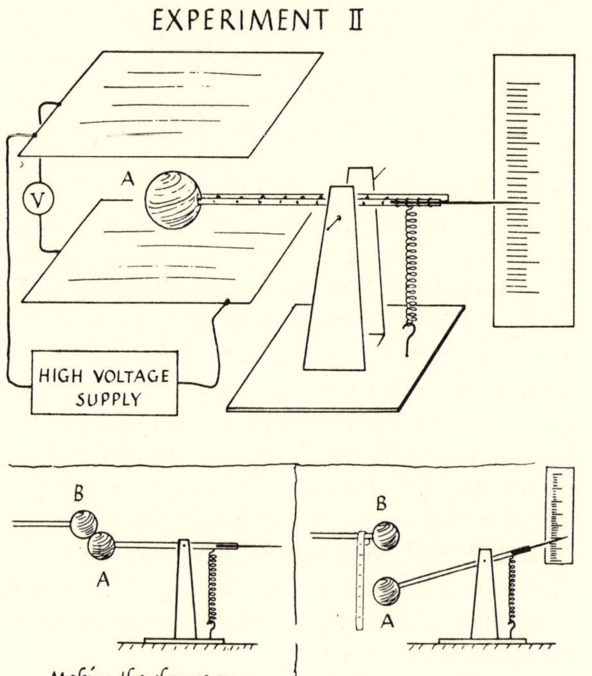

Fig. 33-37.
Measurement of Constant B in Coulomb's Law
$F = B \cdot Q_1Q_2/d^2$

ball A, still with the *same charge*, in a measured electric field and we measure the force on it. To do this we leave A on its seesaw, remove the other ball, and install a pair of metal plates above and below A. (The upper plate is split in half, for installation.) Then we connect a high-voltage supply to the plates, add a voltmeter to measure the P.D. We "weigh" the force exerted on A by the uniform electric field. Knowing the field strength, X newtons/coulomb, and the new force F newtons, we can calculate the charge Q in coulombs. But we arranged for the two balls to have equal charges, so we know *both Q*'s in the original relation and can now calculate *B*.

See this done if possible, as a rough illustration of an important measurement. If not, try the problem below.

★ PROBLEM 10. EXPERIMENT TO MEASURE *B*.

Using the specimen measurements given below, calculate:
(a) the FORCE on the ball in experiment II (in newtons)
(b) the FIELD-STRENGTH between the plates in experiment II
(c) and thence the CHARGE on ball (A) in experiment II
(d) the FORCE of repulsion in experiment I (in newtons)
(e) the value of *B* (by substituting the calculated charge and the measured force of experiment I in the inverse-square law)

Specimen measurements:

Calibrating seesaw. With no charges the seesaw pointer read zero. A load of 1-gram (= 0.001 kg) placed on ball A (uncharged) moved the pointer from 0 to 56.

Experiment I. Equal (unknown) charges Q and Q′ distance apart *d* = 0.10 meter. When second ball was brought into this position from infinity, pointer reading changed from 0 to 26.

Experiment II. Charge Q in field between plates 0.40 meter apart. Pointer reading changed from 0 to 16 when the field was applied, with P.D. = 16,000 volts between the plates.

(*Note*: The field had to be applied gradually and the plates moved to keep ball A midway between the plates, to avoid errors due to "mirror charges" induced on the plates by A's charge.)

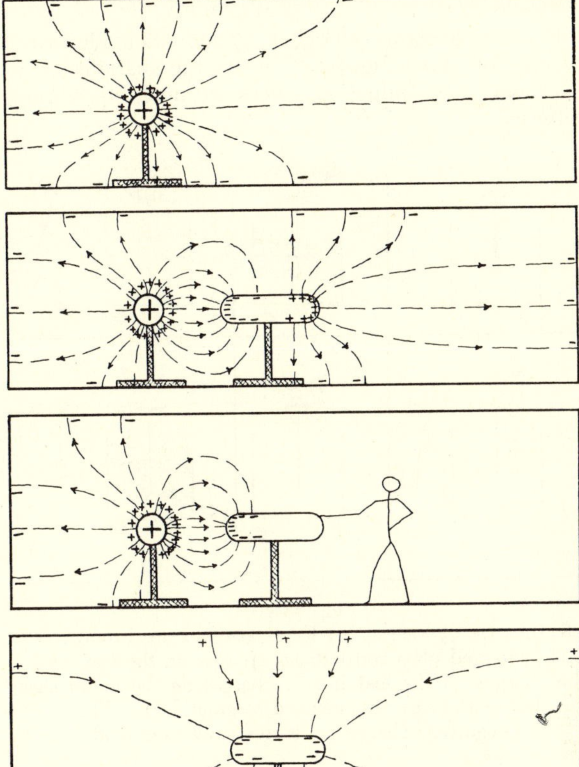

FIG. 33-38. CHARGING BY INDUCTION

Induced Charges and Potentials

Charges run freely along conductors. We do not expect to have an electric field in a metal wire or any other conductor, unless we provide a battery to maintain a continual motion of charge. When an uncharged metal "sausage" is moved into the electric field of a charged ball, charges immediately separate out in the sausage and run along its metal surface until the whole surface, and all the rest of the metal inside the surface, are free from electric field.[21] We then find that the whole sausage is at the same potential—there are no P.D.'s between different parts of it, and there is no electric field anywhere in its metal. If there were P.D.'s or field, readjusting currents would run in the metal until there was no P.D. or field.

The zero electric field, everywhere in the metal of the sausage, is the resultant of the outside field due to the charged ball nearby and the local field of the charges on the sausage.

Once the charges have settled down to rest, which they do very fast, the field lines tug on them *perpendicular to the metal surface* and cannot move them any more. If you look at the sketches of bodies with induced charges you will see that a metal body can have + charges in one region and − charges in another and be uncharged in some regions, and yet it must all be at the same potential (= same P.D. from ground). If this seems strange—no P.D.'s between parts of the body, despite both + and − charged regions—remember that these different patches of +, −, and some with zero charge, would not stay there but for the charges on other bodies nearby. The potential is due to the combined effects of neighboring charges and the charges on the victim.

[21] Except for field perpendicular to the surface, which begins *at* the surface and extends outward.

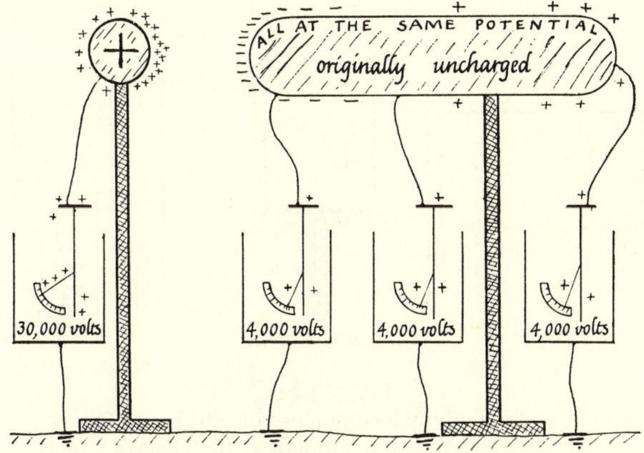

FIG. 33-39.
UNCHARGED SAUSAGE BROUGHT NEAR CHARGED BALL
Voltmeters show the same P.D. between all parts of the sausage and ground. Therefore all parts of sausage are at the same potential, having no P.D.'s between them. Note the use of electroscopes as non-leaking voltmeters.

Charging by Induction—Interpreted by Lines of Force

Fig. 33-38 shows charging by induction illustrated by lines of force. Fig. 33-40 shows the charging of an electroscope by induction. These are final, equilibrium patterns.

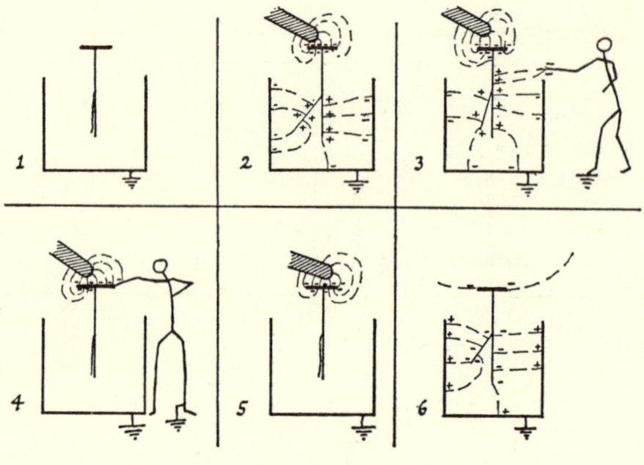

FIG. 33-40.
STAGES OF CHARGING AN ELECTROSCOPE BY INDUCTION
The charged glass rod induces charges on the leaf, etc. The charges on the leaf induce charges on the metal case; but as the case is connected to ground, the "like" (negative) charges on it run away to ground.

Faraday's Ice Pail

When we place a body with charge $+Q$ inside a closed metal box, all the lines of force from $+Q$ spread to the inside surface of the box and end there on charges that total $-Q$. (If they continued into the metal or through it there would be forces and fields making currents flow in the metal to readjust matters.) A tall metal can with open top behaves almost like a closed box. A charge $+Q$ held anywhere inside induces $-Q$ on the inner surface of the can and drives an equal and opposite $+Q$ away to the outside, or to ground if the can is not insulated.

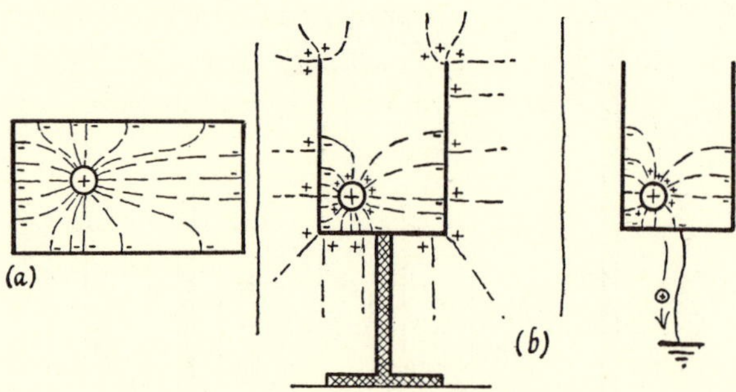

FIG. 33-41.
(a) Charged body placed in closed metal box. (The box is supposed connected to ground, so the charges which retire to the outer surface are not shown.)
(b) Charged ball placed in open metal can.
PROBLEM: In each of the three cases shown above, what will the pattern be like when the ball is moved sideways till it just touches the container?

Moving Lines: Momentary Currents

(See snapshots in Figs. 33-42 and 33-43.)

An isolated charged ball has a field of radial lines that spread out to opposite charges on the remote walls, ceiling, Earth. When the sausage is brought near, it, so to speak, makes a break in some lines of force; and the broken ends attach themselves to charges on the surface of the sausage and drag them along. Thus if the ball has a + charge, its broken lines snap on to − charges on the sausage and drag them towards the ball; and the other ends of the break snap on to + charges on the sausage and drag them away from the ball. As the sausage is brought closer to the ball, more

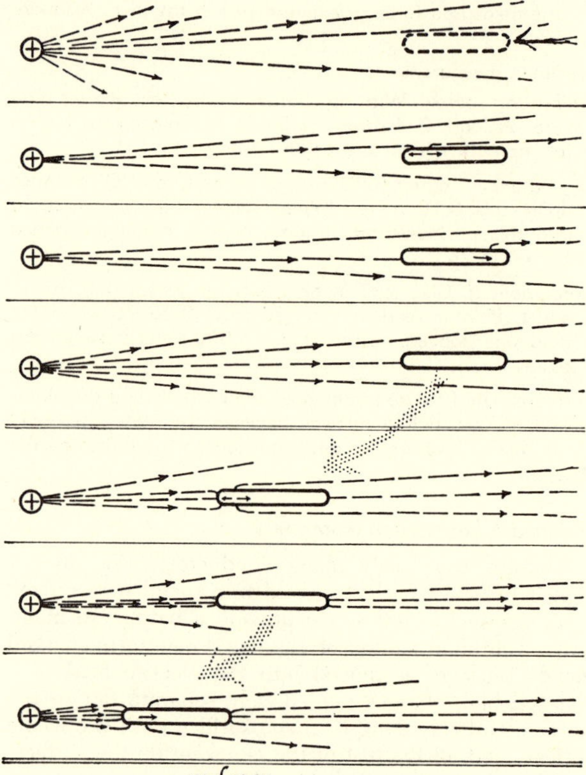

FIG. 33-42. MOVING LINES OF FORCE
When a metal sausage is brought near a charged ball, it upsets the ball's field. These are momentary snapshots showing the "broken" lines of force in motion. Currents flow as the moving ends of the lines drag charges to the new equilibrium positions. The lines need not be perpendicular to the sausage surface while they are moving—in fact, they *must* slant, with a component of pull along the surface. As the sausage is moved nearer, more lines break and drag their ends along the sausage.

The "momentary snapshots" show stages that last for only a very short time. The moving ends of the lines are moving charges, whose motion is a momentary current. These currents do flow, but they soon die out—as man and sausage all reach the same potential (ground potential). Then, with no potential difference, there are no currents.

FIG. 33-43.
LETTING INDUCED CHARGE "RUN TO GROUND" THROUGH FINGER
(a) As man approaches, more and more lines snap on to him.
(b) Momentary snapshot. Lines are dragging their ends along (and shortening) to get charges to his finger.
(c) Momentary snapshot. The last of the collapsing lines are just shown.

and more lines break and drag apart bigger induced charges.

A finger connected to ground offers the tense lines that run from sausage to walls a chance to grow shorter. True, they may have to lengthen during transit (if they already ran to − charges on a nearby table they may have to stretch while running up and over the experimenter's shoulder), but they will in the end be able to shorten to nothing and they do that. Before the finger was there they stretched from + on the sausage to − on the floors or walls. When the finger touches, they drag one or both ends (through finger, arm, body, floor) till the ends meet and neutralize. (In metals, it is the negative ends that move, dragging electrons.)

Battery

A battery all alone has a field between its binding posts. Give it a wire to join them and the field tries to collapse, its lines whipping across, dragging charges along the wire in a current which is a continuing attempt to get rid of the field.

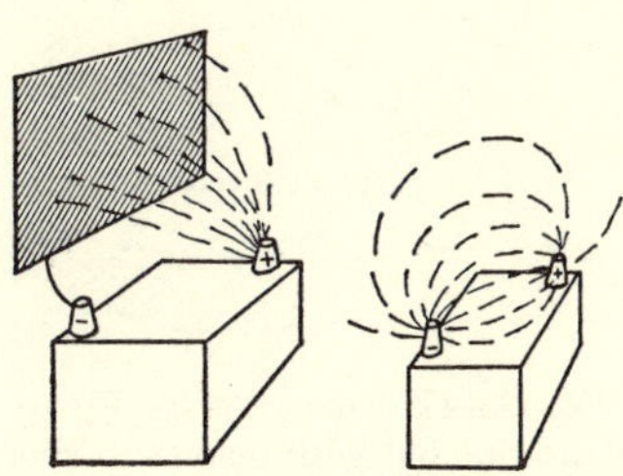

FIG. 33-44. A BATTERY HAS ITS OWN ELECTRIC FIELD

Lines of Force and Moving Charged Particles

The lines of a field show the direction of the force on a small specimen + charge at each point. Suppose we put a tiny charged particle in an electric field. Will it follow a line of force? Only at first.

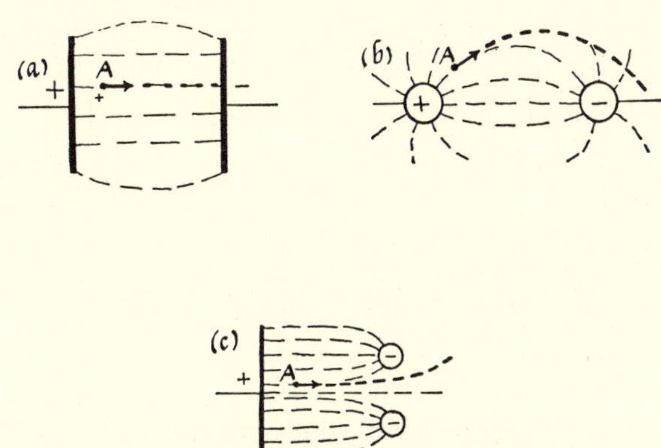

FIG. 33-45. SMALL POSITIVELY-CHARGED PARTICLE, POSSESSING MASS, MOVES IN ELECTRIC FIELD
In each case, the particle starts from rest at A and accelerates in the electric field, which pulls its positive charge. Its path is shown by the heavy line. (A negative particle, e.g. an electron, would need reversed field for this motion.) Sketch (c) shows a particle in the field between a positive plate and two negative rods, with a gap between them. The particle cannot make the curve and hurtles on through the gap.

The particle necessarily runs along the direction of its *resultant momentum.* At each place in its motion it is pulled along the line of force there and *gains momentum in that direction.* The gain of momentum is compounded with its previous momentum, which may have a different direction if the field-lines are curved. Electrons, for example, are accelerated along the lines of electric fields, but once moving they cannot follow curved field lines precisely—momentum carries them on.

So electric fields make electrons move: pull them across a radio diode, control their flow in an amplifying triode, accelerate a stream of them to the muzzle of an electron gun, sweep a stream sideways in an oscilloscope or television tube. Picture the fields pulling and tugging the electrons, controlling their motion, in the uses described below.

Electrons in an Electric Field

A charged particle placed in an electric field collects a few of the field's lines of force, which tug it along, as in Fig. 33-46(A). An electron, having

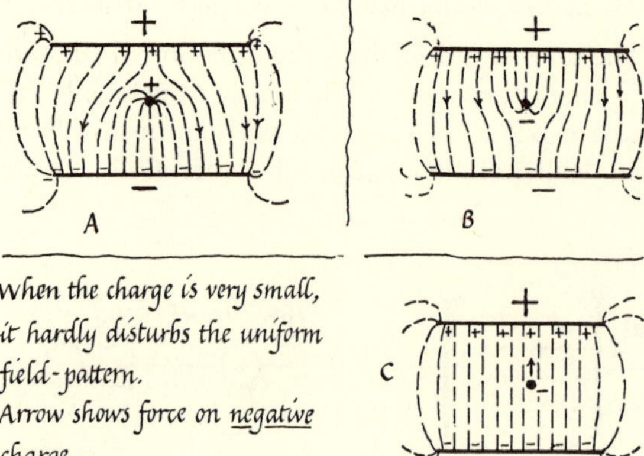

FIG. 33-46. SMALL CHARGED OBJECT IN ELECTRIC FIELD
In A and B the total field is sketched. In C the charge on the object is too small to disturb the uniform field. C also shows the external field *that acts on the object's charge* in all cases, whether the charge is large or small.

a *negative* charge, is pulled *against* the field,[22] as in (B). Of course, the electric field acting on the electron is the unmodified field, (C), without the electron's own "bootstrap contribution" to the overall pattern. The arrow in (C) shows which way the negative electron is urged.

Electrons from Hot Metal

Experiments with a simple radio tube show that a hot filament provides something that can carry current *one way*, across to the plate. This happens even though there is the best vacuum we can make in the tube, the ideal insulator between filament and plate. Since a current *is* a motion of charges, some current-carriers with electric charges must be appearing. A cold filament fails: only a hot one produces this effect. So we suspect the carriers come from the filament. A milliammeter and a voltmeter both tell us that if the carriers travel from filament to plate they must have *negative* charges. The current, when there is one, is *a positive current from plate to filament or a negative one from filament to plate.* And the P.D. that drives this current across makes the plate positive and the filament negative, so that the *electric field would pull negative charges from filament to plate.* If the P.D. is reversed, making the plate negative and filament positive, there is no current: the carriers are driven back and cannot travel. We call these carriers *negative electrons* or simply *electrons.*

The hot metal filament seems to have loose electrons in it, moving very fast, which can evaporate (rather like molecules evaporating from a liquid). The electrons can only escape when neighboring atoms can give them enough extra energy, when the metal is hot enough. The metal need not be heated *electrically* to make it emit electrons. A tungsten sheet with a little bunsen flame playing

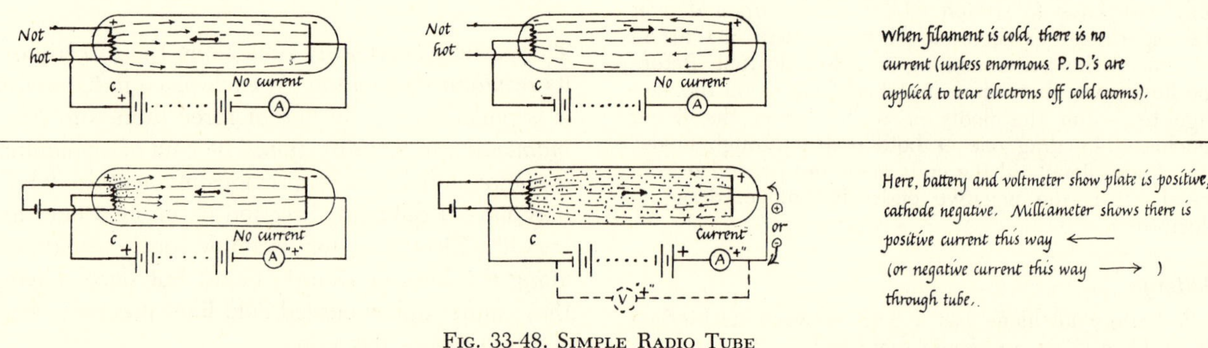

FIG. 33-48. SIMPLE RADIO TUBE

[22] This reversal is a consequence of the scientists' original choice for + and −. If only they had chosen the other way around, then arrows on electric fields (and current-arrows on circuits) would show the direction for common electrons. However it is too late to change now—electrical engineering has crystallized the original choice into a universal convention. And students who are clever enough to understand electric circuits and fields are clever enough to remember that electrons are pushed *against* the marked arrows that run from + to −. Anyway, it is not always negative electrons that move: in liquid solutions and in gases there are positive and negative carriers that both move, opposite ways of course; and we have positive electrons that follow the arrows in space (until they disappear in a violent encounter with a negative electron). And now we find "holes" in some crystalline semi-conductors, regions of too-few-electrons, that travel (by substitution) like a positive charge.

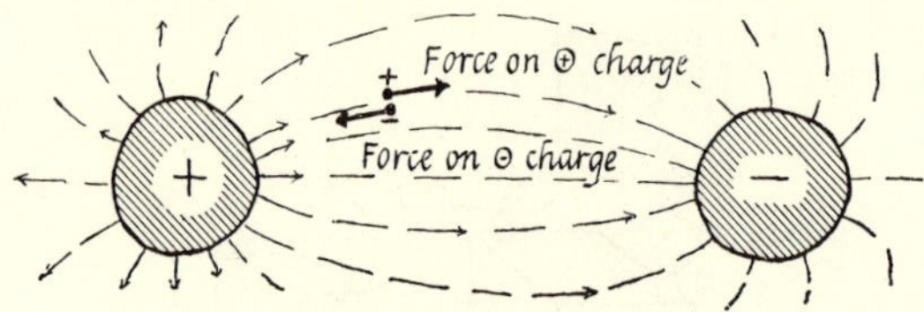

FIG. 33-47. ELECTRON IN FIELD
Electric fields are marked with arrows to show which way a small positive test charge would be pushed and pulled by the field. Electrons have *negative* charges, so they are pulled by the field in exactly the *opposite* direction.

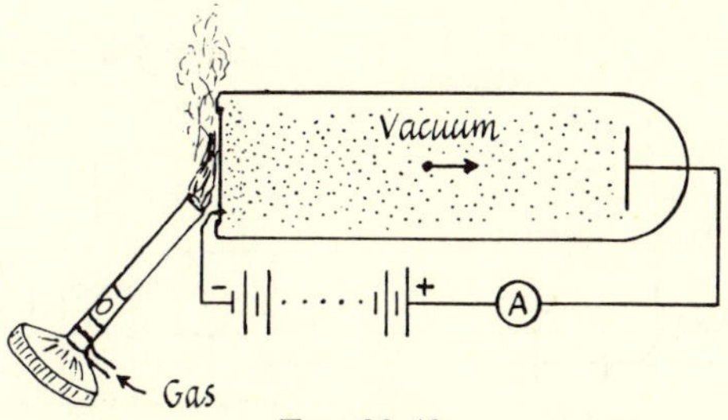

Fig. 33-49.
If a thin tungsten window, sealed in end of glass tube, could be made hot enough with a gas flame, an electron stream would flow across the tube if driven by suitable electric field.

on it from outside would emit just as well. In many modern radio tubes the heating is done separately. Then the heated surface that emits electrons is called the *cathode*;[23] and it is kept hot by a small electric grill very close to it. The cathode is usually a pipe with a grill wire running through it. Cathodes

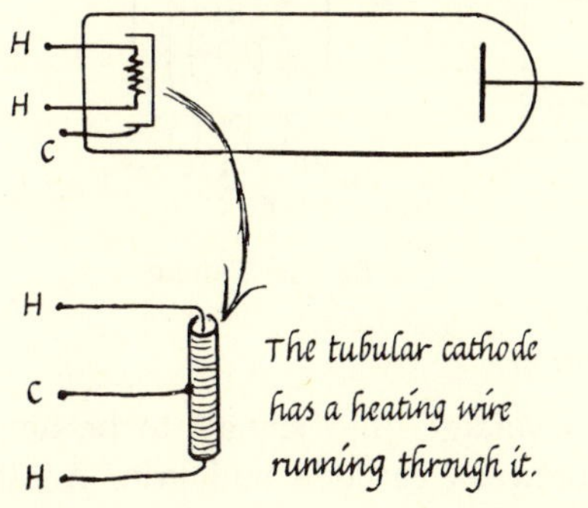

Fig. 33-50. Tube with Separate Cathode and Heater

are often coated with a special mineral which enables electrons to evaporate copiously at lower temperatures. From now on we shall sketch tubes with a cathode, C, and a separate heater, HH.

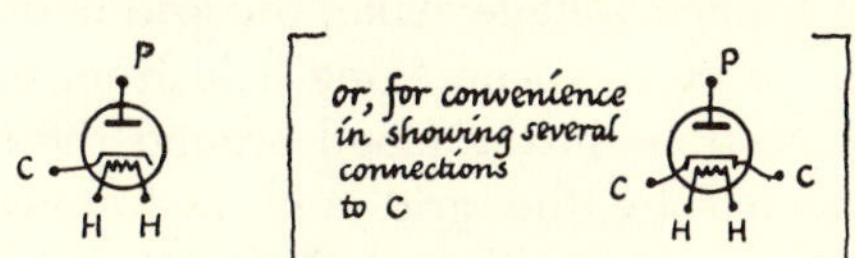

Fig. 33-51. Radio Tubes

RADIO TUBES

Diode Tube, Used for Rectifying

Electrons, boiling off the heated cathode, are pulled across by an electric field between cathode and plate. They accelerate and arrive at the plate with a bang, giving up their kinetic energy to the plate, adding to the random vibrations of its atoms, heating the plate.

The tube, originally with a good vacuum in it, becomes filled with a cloud of electrons, like molecules of a saturated vapor. When a suitable electric field is applied, the cloud drifts towards the plate, and there is a current. The cloud itself weakens the field somewhat; its negative charge drives some electrons back to the filament, so the current is only a trickle.

Fig. 33-52.
Simple Radio Tube (diode)
(The bulb has good vacuum.)

(1) When tungsten filament is white-hot, electrons boil off it copiously and form a cloud in the bulb. If they stay there, they make a discouraging electric field, opposing further evaporation.

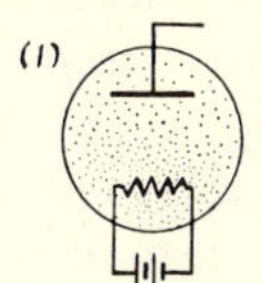

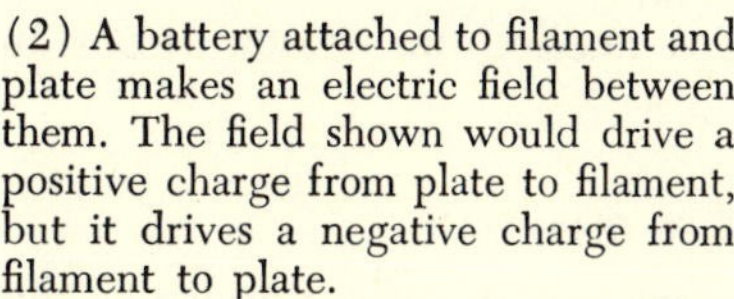
(2) A battery attached to filament and plate makes an electric field between them. The field shown would drive a positive charge from plate to filament, but it drives a negative charge from filament to plate.

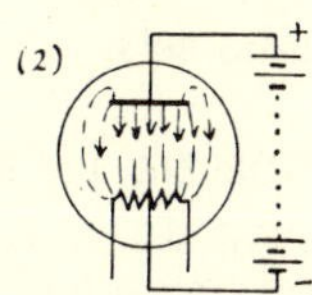

(3) With filament heated and field applied, electrons are swept across to the plate. If field is reversed the electrons are driven back and there is no flow. If the applied field is increased the flow reaches a maximum when electrons are all swept across as fast as they evaporate.

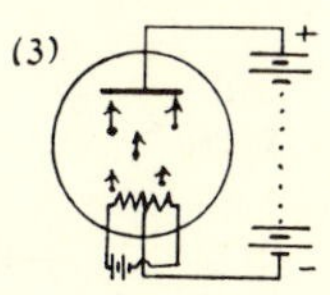

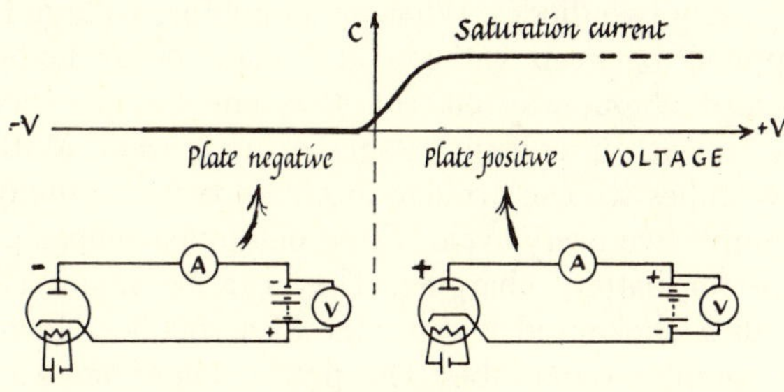

Fig. 33-53.
Diode Tube: "Characteristic" Graph of Behavior
Note: C is the *negative* current from cathode to plate.

Only when a very large voltage is applied do we get the maximum current, with electrons swept across to the plate just as fast as they evaporate; and then still larger voltages cannot increase the current. This constant maximum is called *saturation current*. (The word saturation here conflicts unfortunately with its use in "saturated vapor." Saturation *current* corresponds to vapor molecules being swept away by a strong wind as fast as they evaporate—the household laundress' dream.)

The electrons can only flow across one way, from cathode to plate. Thus the tube acts as a *valve*,

[23] Cathode was an early name for the negative electrode in an electroplating bath. The positive electrode, at which positive current was said to enter, was called the anode. A radio tube's plate, or an electron gun's muzzle is called its anode.

letting current through easily when the plate is positive and refusing when a reverse P.D. is applied. This is useful wherever we want to obtain a direct current from an alternating supply. Most public supplies give A.C., which is suitable for heating and lighting and properly designed motors; but it is hopeless for charging batteries or for the steady high-voltage supplies needed for triode tubes in radio sets.

PROBLEM 11. BATTERY CHARGING

The sketch shows a diode with its cathode heated by current from a 6-volt battery. The diode is being used to charge

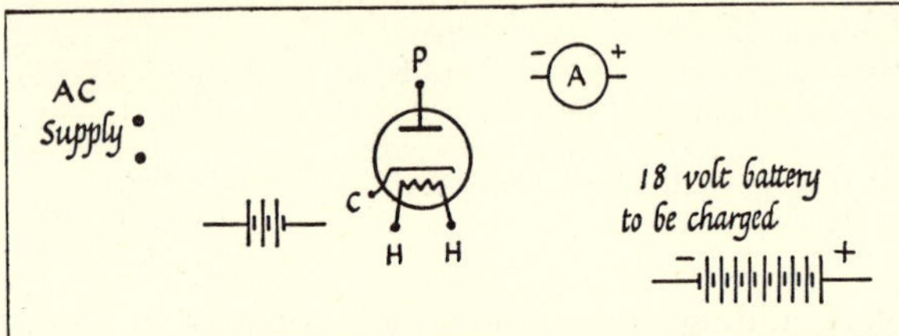

Fig. 33-54. Problem 11

an 18-volt battery (three car batteries in series) from an alternating supply. Copy the arrangement sketched and fill in the rest of the connections so that it would work properly. (*Note:* to *charge* a battery, positive current must go *in* at the battery's positive binding post, through the battery, and out at its negative post—or, of course, negative current the opposite way.)

The arrangement in Fig. 33-54 can be improved by using two diodes. When an alternating voltage is applied between cathode and plate of a diode, bumps of one-way current flow, one bump every cycle, with intervening stages of no current. With two tubes we can arrange to obtain twice as many bumps, two every cycle. These bumpy currents suffice for battery charging. They can be smoothed with a choke and a capacitor to a steady voltage for running radio tubes. This production of one-way current is called *rectification* (meaning "straightening out," because the reverse currents are stopped, leaving straight-ahead current only). You should see oscilloscope demonstrations of rectification.

Triode Tube, Used for Amplifying and Rectifying

We can turn the simple diode into a still more useful device, by inserting an extra plate with holes in it, the *grid,* near the cathode, to encourage or discourage the electron-flow to the plate. This is the triode tube, the main working device of radio sets. (There are still more complex tubes with such things as extra grids to catch stray electrons, but those are frills decorating a basic triode. Transistors do the same job, with less fuss.)

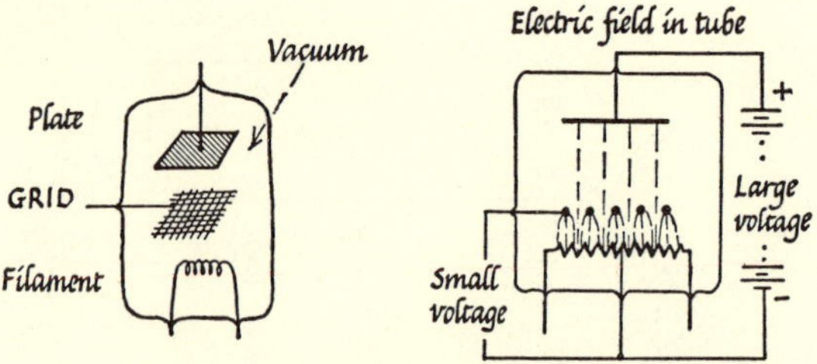

Fig. 33-55a. Triode
Skeleton scheme of triode radio tube, with filament, plate, and grid. The small P.D. between filament and grid, usually a discouraging one, has a strong controlling effect on the flow of electrons away from the filament. Many miss the grid and travel through its holes into the space beyond, where the encouraging field carries them across to the plate.

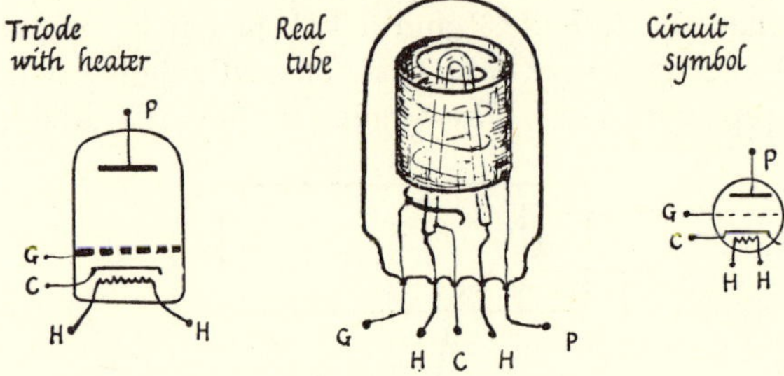

Fig. 33-35a. Triode

Amplification

A small voltage, the "signal" to be amplified, is applied between cathode and grid. Another large steady voltage between cathode and plate provides a strong field beyond the grid to sweep electrons across to the plate. With this arrangement, the small voltage between cathode and grid has a big effect on the electron stream to the plate: here is the basis of *amplification.*

If the applied voltage makes the grid positive, the electrons meet an encouraging field from the start. All the electrons in the cloud around the cathode are attracted by the grid and rush towards it, gaining momentum. They fail, for the most part, to make the curve along the field lines to the grid; so they whizz through the grid, are grabbed by the continuing field beyond, and hurled across to the plate. The tube becomes an overburdened diode. This violent scheme of positive grid is seldom used: the grid is usually negative.

If the grid is negative, the behavior is different, and more useful. A negative grid keeps the electron cloud in check. A few electrons poke their nose through the grid and are rushed to the plate by the strong field beyond. But if the grid is only slightly negative, some field-lines from the plate reach back through the grid and fasten on electrons. Then, all

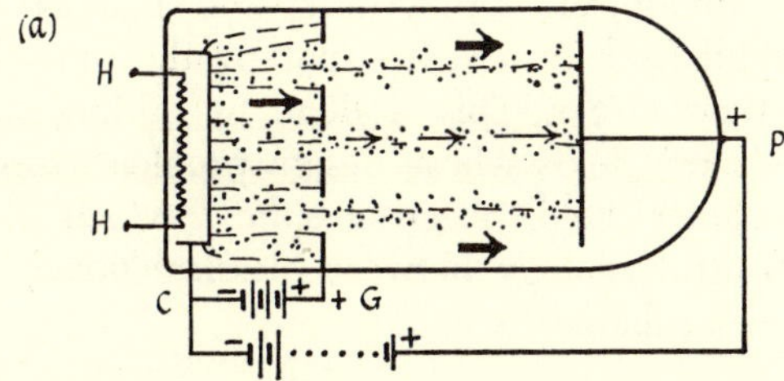

FIG. 33-56a. ELECTRIC FIELD IN TRIODE WITH GRID POSITIVE (unusual and unwise)
A roaring stream is carried through the grid and rushed on to plate—some electrons stopping off at G. The thin arrows show electron *velocities*. The heavy arrows show which way the electric field urges negative electrons.

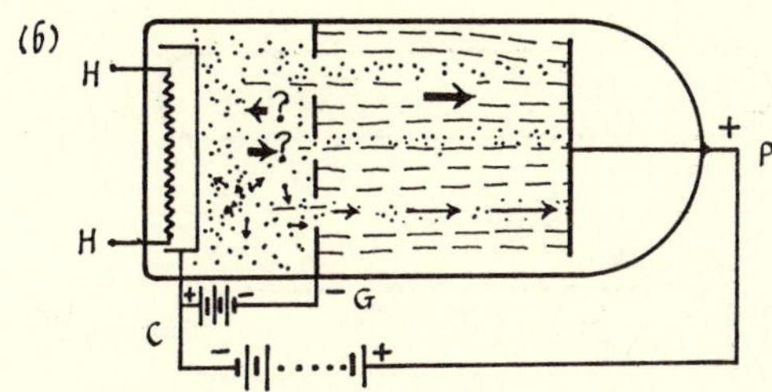

FIG. 33-56b. ELECTRIC FIELD IN TRIODE WITH GRID NEGATIVE (sensitive control)

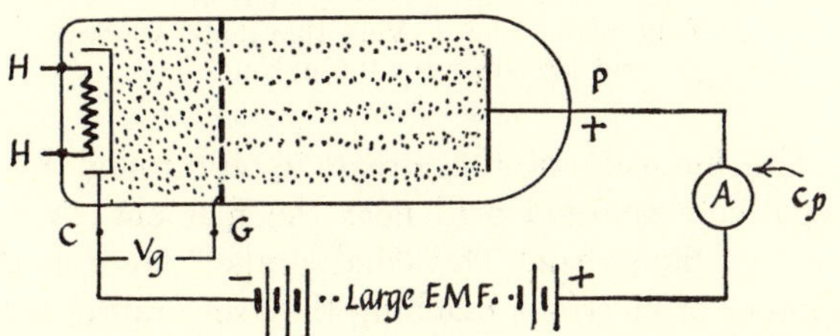

FIG. 33-56c. TRIODE IN ACTION
Electrons from a heated cathode form a cloud in the region between cathode and grid. Electrons that find their way through the grid holes are swept across to the plate by a strong field. The P.D. between cathode and grid controls the flow.

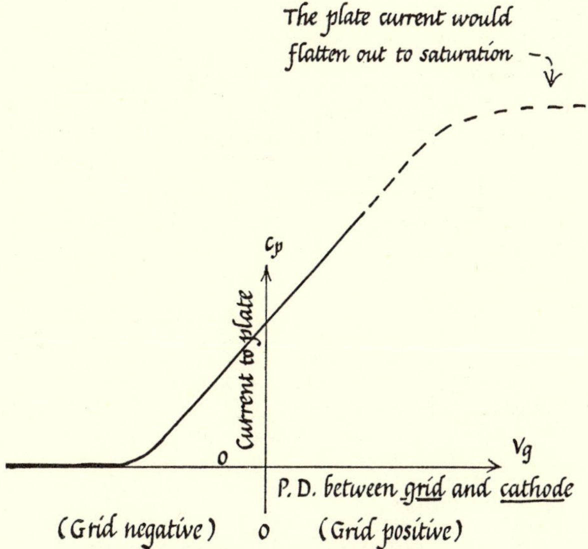

FIG. 33-56d. THE "CHARACTERISTIC GRAPH" OF A TRIODE
The plate voltage is kept constant during this experiment.

told, there is a slight encouragement, keeping electrons drifting from cathode to grid; and when they drift through the grid they are rushed to the plate. *The voltage applied between cathode and grid controls this drift, very sensitively. Small changes of voltage have large effects on the drift and thus have large effects on the current to the plate.* Here is a good way of making a small voltage control the stream. The same effects then would require a large change of the voltage between cathode and plate, if *that* voltage were to be changed instead. This tube can magnify (amplify) voltages.

Better still, it amplifies *power*. Only very small currents run to and from the grid itself; most of the electrons flow to the plate. Thus the grid receives a tiny current driven by small voltages; and the plate delivers a much bigger current that makes large voltage changes when it flows through a suitable resistor. So the triode delivers far more power and *changes of power* from its plate than are supplied to its grid. It is like modern push-button machinery where a small push is used to control vast amounts of energy. This is the triode's essential function as an amplifier. The extra energy comes from the constant high-voltage supply (battery or rectified A.C.) used in the plate circuit.

Triodes in Radio Sets

In the amplifier of a radio set, radio signals apply small voltages between cathode and grid of a triode,[24] and the consequent changes of electron-flow to the plate make large changes in the P.D. across a "sampling resistor" (technically a "load resistor") in the plate circuit. That P.D. can be applied to the grid and cathode of another triode for further amplification, and ultimately used to drive a loudspeaker.

If radio signals are to drive a loudspeaker successfully, they must not only be amplified but also be rectified—run through some one-way device. It is not obvious why rectification is necessary; the need for it will be explained in Chapter 41. A triode can

[24] Two auxiliary batteries, or other constant-voltage supplies must be inserted. To keep the electron flow sensitive to voltage changes, the grid must be slightly negative. For this a small battery is inserted in series with the changing signal—or an equivalent economy-scheme is used. To sweep electrons across to the plate, a high voltage must be included; and this is provided by a battery (called the "B" battery) or by a rectified A.C. supply.

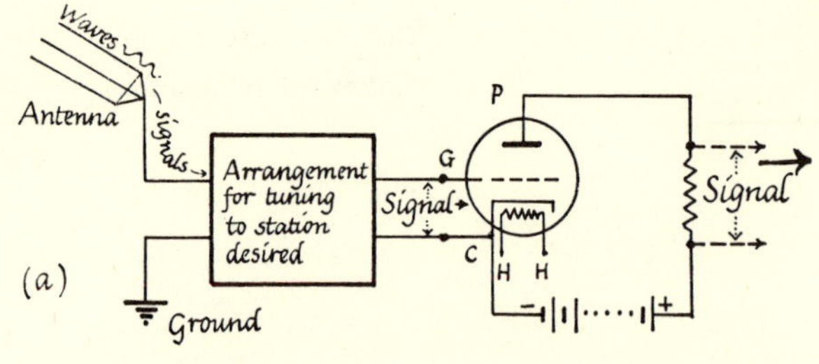

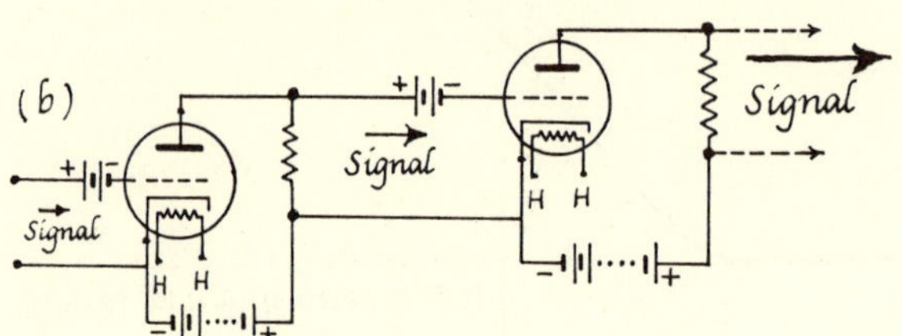

FIG. 33-57. AMPLIFYING A RADIO SIGNAL
(a) With one triode tube.
(b) With two stages of amplification.

be used for rectifying where the knee of its CURRENT : P.D. graph bends to zero current. Although it is possible to make the same tube do both rectifying and amplifying, it is better to separate the two jobs and use two tubes.

Triodes combine amplification with the virtues of diodes. Diodes are still used, in "power supplies" to provide D.C. from A.C., instead of batteries. Both kinds of tube are threatened by competition from the new small unheated "transistors," in which a lump of semi-conductor offers unsymmetrical, controllable opposition to electron motion across crystal boundaries.

Electron Guns

You are like the man who had been talking prose all his life without knowing it. You have been handling an electron gun without knowing it. The radio tube *is* an electron gun. Electrons evaporating from the hot cathode are accelerated by an electric field and bombard the plate. If there are holes in the plate a stream of electrons will shoot on through each hole. The electrons continue until they hit a solid wall—or, if there is residual gas in the tube, until they have lost their energy in collisions. If they have enough energy, the electrons can even travel through a thin wall of glass or metal and escape into the air where they are soon stopped.

An electron gun for projecting a narrow stream has some refinements. The plate has only one hole, and more grids are inserted to control the focusing and intensity of the beam. Focusing the beam means making sure the electrons emerge in a very narrow stream; or, if they do spread, making sure they bend back to a sharp spot by the time they reach their target. This is done by adding small extra electric fields whose design problems form a new technical field, "electron optics," which draws on a fruitful analogy between classical optics and electron mechanics.

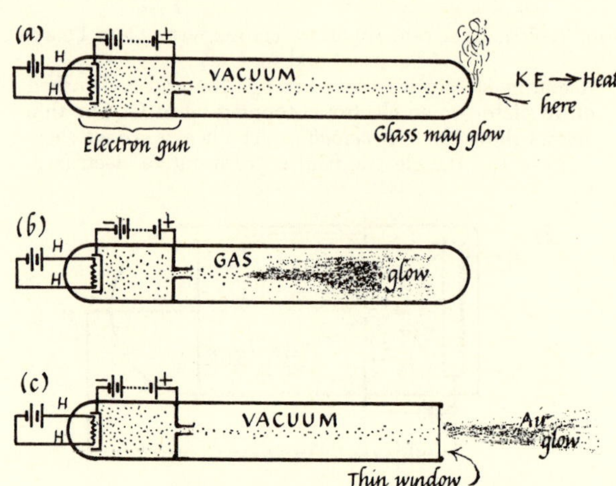

FIG. 33-58. ELECTRON GUNS
(a) Tube with vacuum.
(b) Tube with a little gas left in it.
(c) Tube with vacuum. Very thin metal window at end lets electrons through into air.

The intensity of the stream is controlled by a negatively charged grid near the filament, to discourage the stream. This "discourager" controls the number of electrons reaching the accelerating field. Made strongly negative, it pushes the whole bunch back into the filament. Made slightly negative it gives sensitive control. Made positive, it encourages such a deluge that the gun gets out of hand. In drawings of such guns we shall leave out these extra grids or anodes.

In the earliest studies of electrons, just before 1900, such copious sources of electron streams were not available. Instead, J. J. Thomson and others had to use electrons torn from metals or knocked off molecules of rarefied gas in discharge tubes and then dragged down the tube by a P.D. of several thousand volts. This gave a thin stream of electrons with a mixture of kinetic energies.

Oscilloscopes

An electron stream emerging from an electron gun will proceed in a straight line, with unchanging speed, if it travels in a vacuum and meets no more electric fields. When they reach the glass end of the confining tube, the electrons stop with a bang; their K.E. is turned to heat, except for rare events where the decelerating electron produces a photon (quantum) of X-ray radiation instead. If the glass is coated with a suitable mineral—like the coating inside fluorescent lamps—the electrons make the coating glow, not by heating it, but by displacing other electrons when they bombard it. This glowing spot can draw graphs and pictures for us. Electric fields can waggle the stream sideways or up and down to make it draw graphs or paint television pictures. The stream moves very fast and has very small mass so it responds incredibly quickly and sensitively to deflecting fields.

Fig. 33-59 shows a cathode ray oscilloscope tube (CRO), forerunner of television tubes.

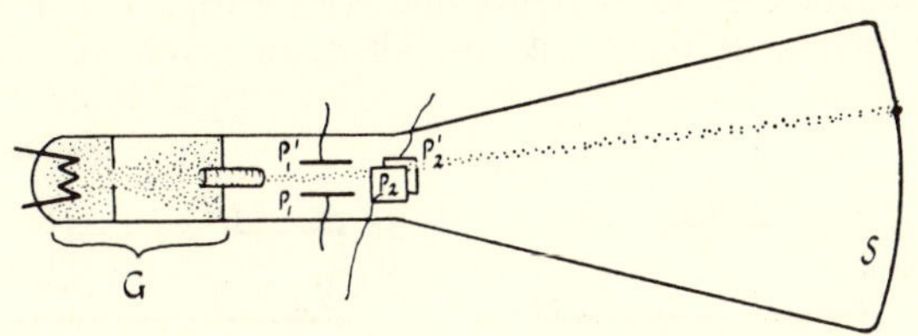

Fig. 33-59. Cathode Ray Oscilloscope Tube

Electrons are shot in a narrow stream by the gun G and hit a screen S at the other end. S is coated with a mineral which glows when bombarded with electrons. The stream is focused by the gun to make a small glowing spot on the screen. It passes through a deflecting electric field between plates P_1 and P'_1. This field bends the stream up or down, by giving the electrons some vertical momentum. There is another pair of plates P_2, P'_2 for a horizontal deflecting field. If P_1 and P'_1 are connected to a battery (say 45 volts EMF), the upper plate positive, the stream is deflected upward, and the spot hits a fixed place in the upper part of S as long as the battery is connected.[25] If an alternating voltage is connected to P_1 and P'_1, the spot moves up and down, following the changing voltage practically instantaneously. To draw a time graph of the vertical deflection, the horizontal electric field must swing the stream steadily *across* the tube. This "sweep" arrangement is discussed in Problem 11 in Chapter 36. You should experiment with it in the laboratory (Experiment J in Chapter 41).

In television tubes the spot must sweep up and down as well as to and fro, rapidly covering the whole picture area. As it does so, its *brightness* must be changed by the incoming radio waves to make the bright and dark parts of the picture.

[25] Incidentally, long bombardment is harmful; the stream "burns" the screen. In laboratory, avoid letting the spot remain at one position for long. Television camera tubes suffer similar damage, from which they recover in time, and their cameramen have many a trouble avoiding burns.

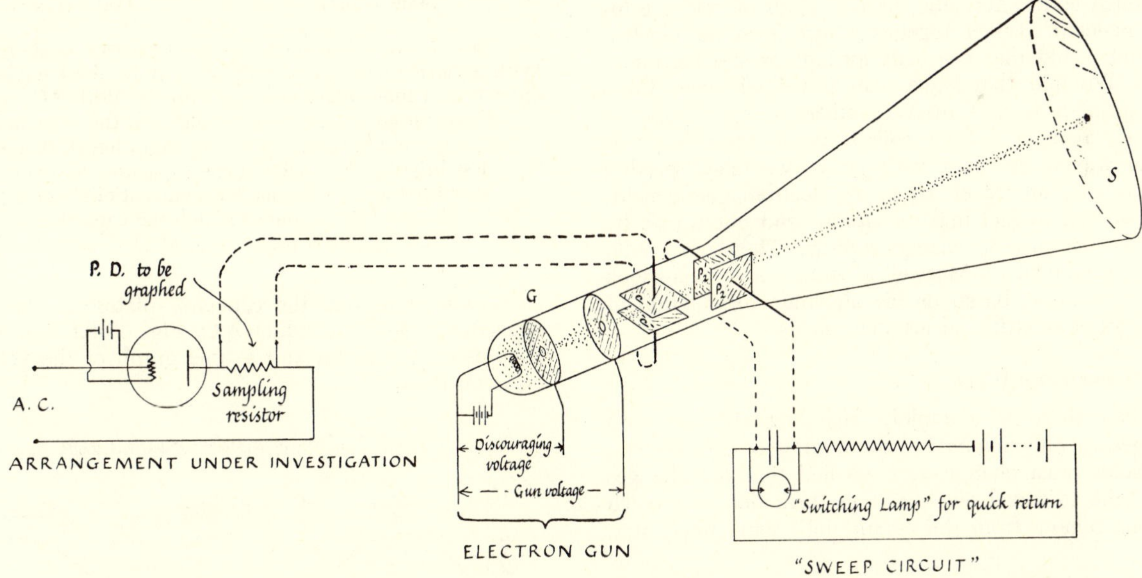

Fig. 33-60. A Simple Cathode Ray Oscilloscope Tube being used to sketch time-graphs of a changing current. The current passes through a "sampling resistor" producing proportional P.D.'s which are applied to the plates in the oscilloscope tube to produce a vertical deflecting electric field.
The "switching lamp" in the sweep circuit is a special radio tube with a little gas in it, like a neon glow lamp. It carries no current at low voltage. When the voltage across it reaches a certain value, it suddenly glows and carries current (with ions made by collision). The current through the high resistance charges up the capacitor until the switching lamp glows and lets the capacitor discharge quickly to zero. This provides a sawtooth P.D. for the sweep.

PROBLEM 12.

Given the basic design of an electron gun, and asked to design a television tube, how would you suggest changing the *brightness* of the spot? (There are several possible schemes, some of which would not be any use in a television tube.) *Hint:* Those electrons which do hit the screen must be moving at full speed, or they may fail to excite the screen at all.

In modern television tubes the deflection of the electron stream is not done by electric fields but by magnetic fields. You will see later that these could be used instead.

Energy of the Field; Waves

[The rest of this chapter gives a fanciful account, vouched for by more advanced studies, but too difficult to justify here. However it throws some light on electric energy, and gives a hint of electromagnetic waves.]

Moving charges carry their fields with them, their lines of force move with them like a man's whiskers with their owner. In all the space around a circuit where an electric current is flowing, lines of force must be moving.[26] It is these moving lines that carry the electrical energy from the battery to the various parts of the circuit across the intervening space. They tug and drag to keep the current going, and they transmit strain energy much as a moving belt does.

With this motion comes a new effect: magnetic fields, also with lines of force (lines of a quite different kind, though they follow similar patterns). The circuit's magnetic field is found to have a store of energy of its own, which varies as CURRENT^2. The likeness to kinetic energy, which also varies as SPEED^2, suggests that *magnetic-field* energy may be the KINETIC ENERGY of a circuit, perhaps related to the K.E. of the electrons' motion. This "kinetic energy" of a current's magnetic field suggests that it has something like mass. And the STRAIN-ENERGY of *electric* fields suggests that they behave rather like springs or stretched cords. Actually, the two kinds of field are interconnected. Taken together, they form an electromagnetic field that has both springiness and inertia or mass. We find that waves can travel wherever there are springiness and mass together, along a rope in tension, along a massive coiled spring, through air as sound waves. So, as a risky guess, we might predict that waves can travel out in an electromagnetic field; and we can in fact make a circuit send out waves by making its current change rapidly. These are radio waves predicted by Maxwell a century ago, in a much less risky guess based on his algebraic formulation of the laws of electric and magnetic fields.

Electromagnetic Waves

When there is a rapidly changing current, with charges not only *moving* but *accelerating* or *decelerating*, the fields must change, and we find that the changes take time to travel out—news of them does not reach regions remote from the circuit until some time later.

[26] If you picture the current as a swarm of negative electrons drifting along the wire in which an equal swarm of positively charged atoms remains at rest, you should still imagine the electrons carrying a moving sheaf of inward-pointing lines of force, while the positive atoms have a stationary sheaf of outward-pointing lines. The motion of field lines is still there.

To see how this happens, first look at a broken circuit, like the one at the beginning of this chapter (Fig. 33-1), with a battery applied; then with an alternating voltage applied. Fig. 33-61 shows this, with a capacitor, a pair of plates, at the break. An alternating voltage drives

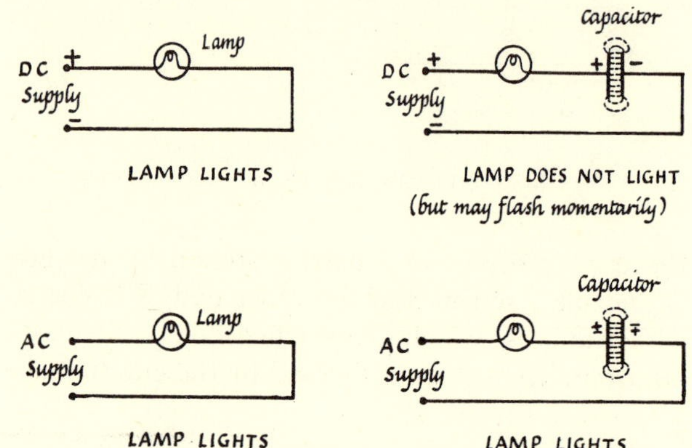

FIG. 33-61. A CAPACITOR IN AN ELECTRIC CIRCUIT
With a capacitor in series with DC supply, the lamp shows (at most) a momentary pulse of current. With AC supply, charges surge to and from the plates of the capacitor, flowing through the lamp. So the lamp lights, though less brightly than without the capacitor. Maxwell suggested we should imagine a current of changing electric field running through the capacitor and making a magnetic field of circles.

charges to and from the capacitor plates, so that the capacitor *seems* to conduct A.C. Fig. 33-62 shows the charges on the plates and a time graph of the voltage between them.

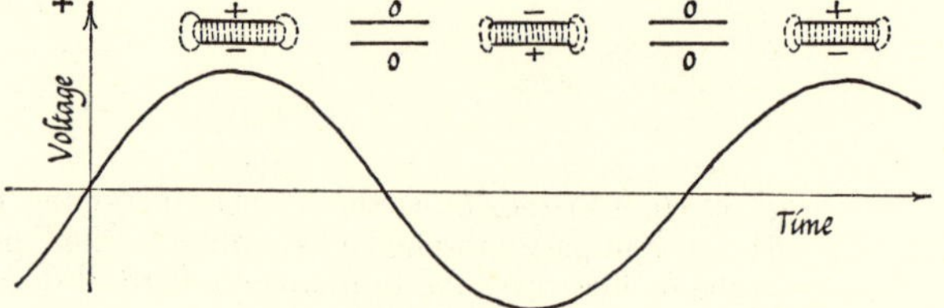

FIG. 33-62.
TIME-GRAPH OF ALTERNATING VOLTAGE ACROSS CAPACITOR

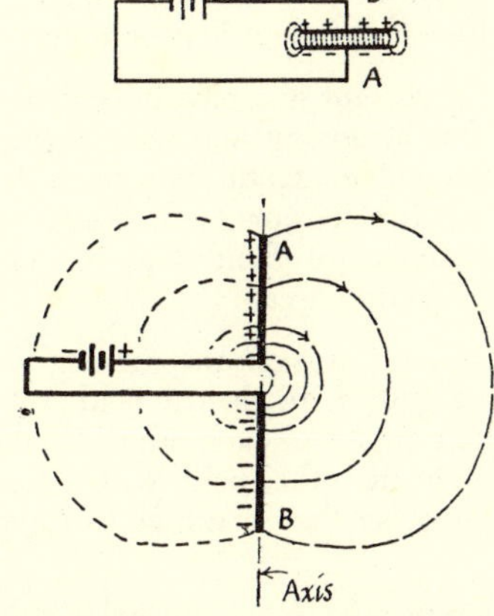

FIG. 33-63. CHARGING UP AN ANTENNA SYSTEM
A pair of plates, connected to battery, will charge up (+ and −) until the P.D. between them = EMF of battery. When the two plates are replaced by two rods, the rods become charged and have an electric field as shown. The field is symmetrical around the rods as axis. To picture the full field, imagine the sketch rotated around the axis. The rest of the circuit, wires and battery, has electric fields across it, but these are not shown here.

Now, to spread out the electric field between the plates, replace the plates by two antenna rods, A and B. Fig. 33-63 shows the electric field when the rods are charged by a battery. In Fig. 33-64 a *slowly* alternating voltage is applied to the rods. When the voltage is maximum the rods have + and − charges as shown in (a). In (b) the voltage has fallen to half, and in (c) to zero. From (a) to (b) to (c) the charges run back along the rods towards the supply, carrying their lines of force with them: the lines collapse. In the next quarter cycle, the voltage and charges and field grow to maximum with opposite signs. Then back to zero, and up to the first stage. Charges are pumped to and from the antenna rods. There is little if any radiation to notice far away.

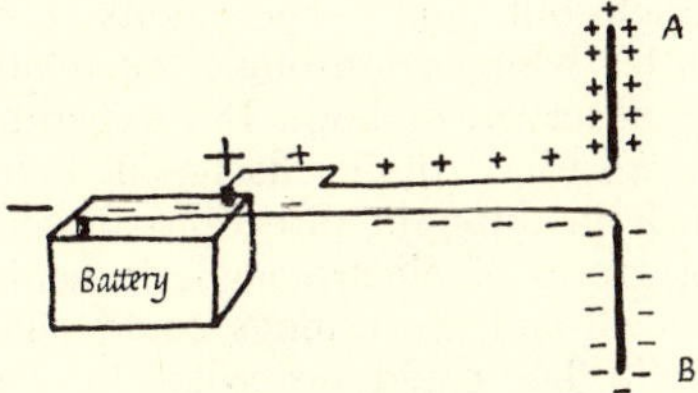

FIG. 33-64a.
BATTERY CHARGES ANTENNA RODS, + and −, making electric field.

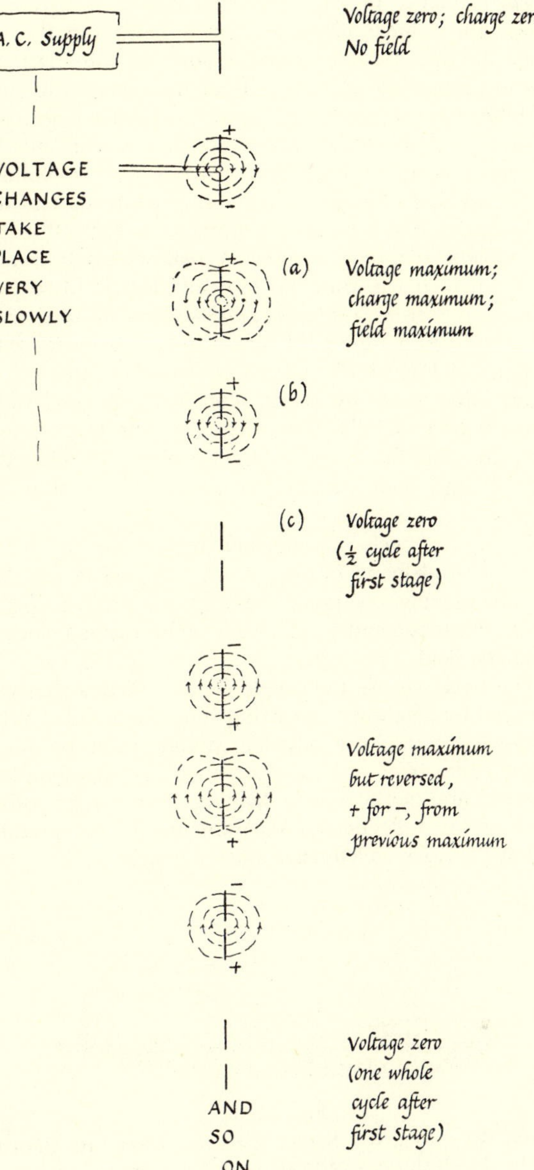

FIG. 33-64b. AC SUPPLY CONNECTED TO ANTENNA-RODS
These sketches show the electric field in the neighborhood of the two rods. (Of course there is also field across the rest of the circuit.) The frequency of the alternations is supposed to be very low, so that the changes of the field take place *very slowly*. As the voltage grows and dies, charges surge out along the rods and back: great umbrellas of lines of force swell out and shrink—a pulsating electric field, in phase with voltage. The lines sketched show a plane section: the real field is solid in three dimensions. To imagine it, picture the pattern spinning around the axis of the rods.

The sketches of Fig. 33-65 show the same stages when the voltage alternates much more rapidly (or when the rods are much taller). Now the outlying parts of the pattern do not learn about the changes in time to follow the nearer parts to oblivion. They get left behind as closed loops, which are then shoved out by the next group of lines that grow when the voltage builds up again. Imagine a long whiplash being cracked suddenly, so that it makes a loop which breaks off under the violent snapping motion. A real whiplash cannot detach a loop like this, but lines of electric force can. Calculations from a suitable modification of $\nabla^2 V = 0$ for charges in motion predicts just such a behavior. The sketches of Figs. 33-65, 67, 68 are based on such calculations, first made by Hertz, who first produced radio waves experimentally. Once detached, the loops career outward with the speed of light—in motion they *are* light though their wavelength may not be within the narrow visible range.

Cycle after cycle of voltage-changes drives out batch after batch of loops, in the pattern sketched in Fig. 33-67. The moving, changing, electric fields carry magnetic fields with them, and a radio wave *is* this moving electromagnetic field.

The two rods are the simple dipole antenna that you see used for sending or receiving short-wave radio. With long-wave radio, the antenna system must be on a larger scale; one "rod" is represented by the complete antenna and vertical wire leading from it, the other "rod" is the reflection of that antenna in the ground, which provides an effective mirror image.

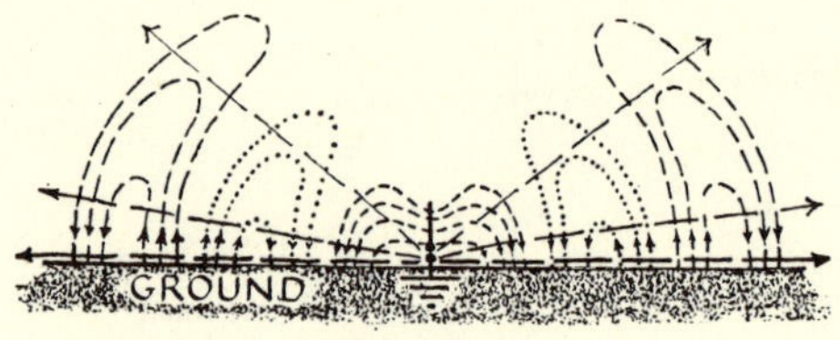

Fig. 33-68.
Radio Waves from a Small Antenna Near the Ground
The sketch shows a vertical antenna and ground (which provides the lower "rod" by reflection), radiating electromagnetic waves above the Earth's surface. The arrows show the way the patterns travel and grow.

Fields and Motion; Relativity

At first sight electric and magnetic fields seem quite separate as well as different; but connections appear.

(1) (a) Charges at rest have stationary electric fields.
(b) Magnets at rest have stationary magnetic fields.

(2) (a) A steady current in a circuit has a stationary magnetic field (and a moving electric field, though it may be disguised).
(b) A moving magnet has a moving magnetic field and provides a stationary electric field (the principle of the electric generator).

(3) A changing current in a circuit (e.g. an alternating current) has a moving and changing electric field and a changing magnetic one. Such a circuit will give out some electromagnetic waves, copiously if the alternations are rapid and the circuit has a suitable "radiating area."

(4) Electromagnetic waves are moving fields which travel out together, an electric field and a magnetic field at right angles to each other. Changes in each field maintain the other field, with a sort of mutual back-slapping, so they continue to travel.

How are electric and magnetic fields related? More advanced studies show that the motion of either field always produces the other. A moving electric field produces a magnetic field, and a moving magnetic field produces an electric field. But suppose the experimenter moves instead of the field. Will he observe the same magnetic field when *he* moves across a charge's electric field? If not, he can tell which is moving, he or the charge, and he has discovered a test of absolute motion; he can establish milestones in absolute space. Only experiment can answer this question for certain. However, we can *guess* by applying common-sense geometry of motion to Maxwell's laws—just as we might *guess* the path of a falling orange as seen by a running observer. Guessing thus, we find Maxwell's laws should change to a different, clumsier form for any observer except one "at rest," anchored to a milestone fixed in space. But actual experiments give us the laws of electrodynamics in the same simple form for all observers, whether at rest or moving across the laboratory. Which is at fault: the simple laws that experiment supports, or the simple geometry that we used for our guess?

Ignoring this strange conflict, 19th-century scientists tried to measure their motion through space by its effect on electromagnetic experiments. Any measured effect proportional to velocity v shows clearly the relative velocity between experimenter and apparatus, when motions in the laboratory were changed; but the effect of any overall motion of the laboratory through space simply subtracts out. Such experiments showed that it does not matter which contributes the relative motion, apparatus or observer, or both. But experiments of another type were tried: timing flashes of light forwards-and-backwards in different directions in space. A flash of light is a group of electromagnetic fields travelling through "space," and the timings could yield an effect involving v^2. So they could, according to common-sense geometry, reveal our motion through space. Yet, again and again they failed to show any effect. We are sure we are moving: yet we find no motion. What are we to conclude: wrong experiments, or wrong geometry, or wrong idea of milestones in space? This is the basic question that gave rise to the theory of Relativity.

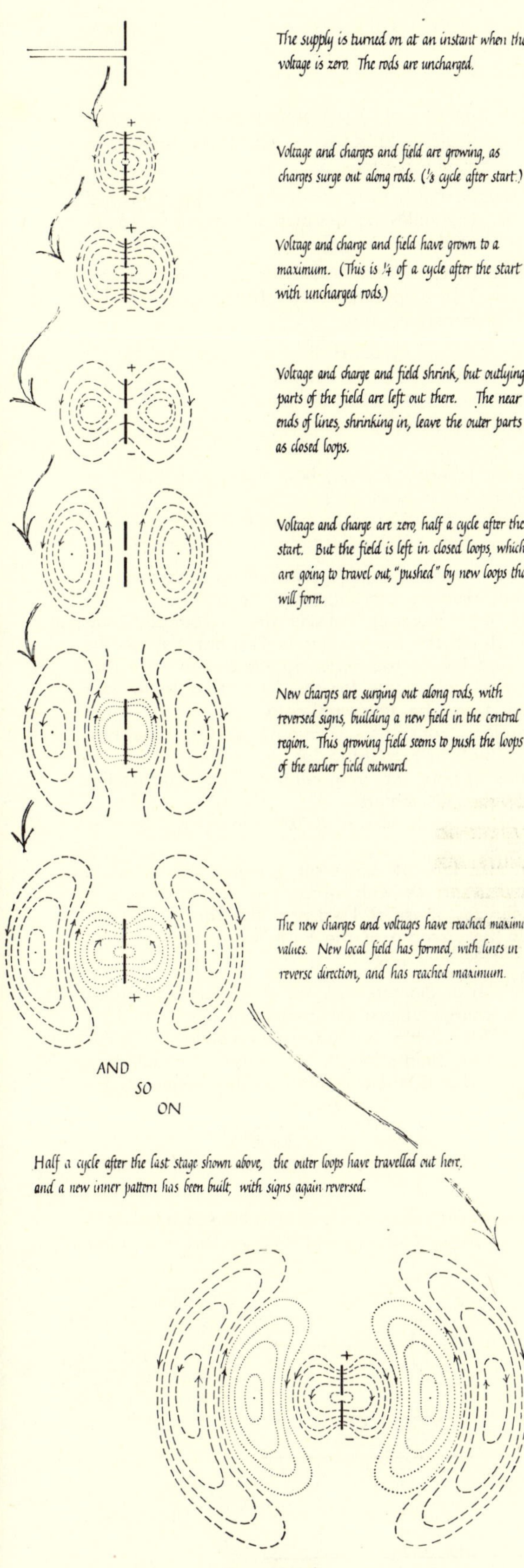

Fig. 33-65.
High-Frequency AC Supply Connected to Rods
Voltage changes taking place *very rapidly*.

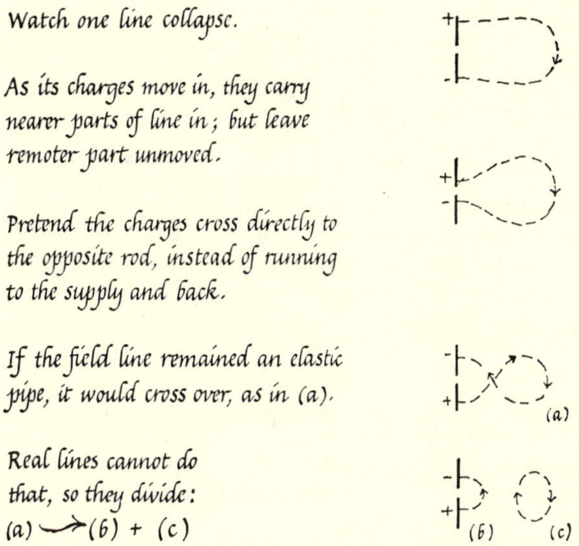

Watch one line collapse.

As its charges move in, they carry nearer parts of line in; but leave remoter part unmoved.

Pretend the charges cross directly to the opposite rod, instead of running to the supply and back.

If the field line remained an elastic pipe, it would cross over, as in (a).

Real lines cannot do that, so they divide:
(a) → (b) + (c)

Fig. 33-66. Rough Explanation of Loop Forming

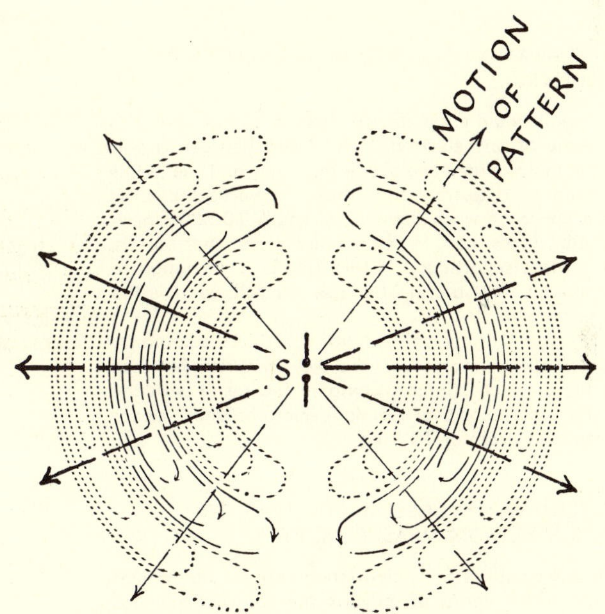

Fig. 33-67. Radio Waves
Instantaneous "snapshot" of the electric field radiated from two short rods (as in previous sketches) connected to a *rapidly* alternating voltage. These are the loops of lines of force that are pushed out, as new loops form at the rods, and travel away as radio waves. The pattern close to the rods, shown in previous sketches, is omitted here. The radiating arrows show the way the patterns travel and grow. You should picture this pattern traveling out from the source (the rods) with the speed of light, growing in size as it travels. The pattern travels out practically along radii from the source, once it is clear of the immediate neighborhood of the source. So we should expect an inverse-square law of radiation-intensity.

PROBLEMS FOR CHAPTER 33

Problems 1-12 are in the text.

13. ELECTROPHORUS

The electrophorus is an arrangement for obtaining an indefinite supply of electric charges which consists of:

(A) A lucite cake which is charged by rubbing it with fur. (This scrapes negative electrons off the fur onto the lucite where they remain trapped on or near the surface because lucite is so good an insulator.)

(B) A metal plate with insulating handle.

Charges are obtained by the following routine (once the lucite has been rubbed):

(1) Plate brought near lucite.
(2) Finger touched plate to connect it to ground (an enormous electrical conductor).
(3) Finger removed.
(4) Plate removed and found to have useful charge.

Even if the plate touches the lucite it removes very little of the lucite's charge since the lucite is such a bad conductor.

(a) Draw sketches of the stages of the process, marking them with + and — signs where you think there are charges. Add lines of force.
(b) Explain briefly what happens from one stage to the next.

★ 14. PREPARATION FOR MILLIKAN'S ELECTRON MEASUREMENT

A very small liquid drop (blown from a throat spray) is given a charge of one electron. It is placed between a pair of horizontal metal plates one above the other and the plates are connected to a battery that makes a vertical electric field in the space between them of strength 100,000 newtons/coulomb. (This can be calculated from the battery voltage and plate separation.) If this field is just enough to make the droplet float and neither rise nor fall, what is the mass of the droplet?

Given: electron charge $= -1.60 \times 10^{-19}$ coulomb.

(This calculation, *reversed*, is the way in which Millikan measured the electron charge.) Compare the result with the limit for the most sensitive chemical micro-balance, about one billionth of a gram or 10^{-12} kg.

★ 15. PRELIMINARY PROBLEM RELATING TO MILLIKAN'S ELECTRON MEASUREMENT

When a tiny raindrop falls, air-friction opposes the motion with a *force which varies directly as the speed.* (This has been checked carefully by experiment.) For drops of different sizes, friction drags are directly proportional to the radii. ∴ Friction drag $= K \cdot r \cdot v$, where K is a constant number of known value, obtained from experiments on air-flow in tubes. (Assume that the value of K is such that $K \cdot r \cdot v$ gives the force in *newtons.*) When a raindrop starts falling, it accelerates at first but soon reaches *constant speed.*

(a) What two forces act on a drop, *at any stage of its fall?*
(b) When it is *falling at constant speed,* what must be the *resultant* force?
(c) Write an equation arising from (a) and (b), using K, etc., mentioned above.
(d) Suppose another drop has *twice the radius.* What mass will it have? From your equation (c) show that it will fall *4 times as fast.*

★ 16. ELECTRIC FIELDS AND INVERSE-SQUARE LAW: PREPARATION FOR MILLIKAN'S ELECTRON MEASUREMENT

We use electric fields in a number of "atomic physics" measurements. The following question will help you to understand electric fields:

GRAVITATIONAL FIELD-STRENGTH is defined as the FORCE (newtons) on a specimen UNIT MASS (one kg). It is measured in newtons/kilogram.

ELECTRIC FIELD-STRENGTH is defined as the FORCE (newtons) on a specimen UNIT ELECTRIC CHARGE (one coulomb). It is measured in newtons/coulomb.

(a) What is the strength of the Earth's gravitational field, roughly
 (i) In your laboratory? (i.e., how much does the Earth pull one kg, in newtons?)
 (ii) At *any* point 4000 miles from Earth's center?
 (iii) At any point 8000 miles from Earth's center?

(b) We can calculate the Earth's gravitational field-strength by another method, starting with the gravitation constant, G. This is clumsier and silly when we already know the strength directly from falling bodies, but it shows the method used for electric fields. So you should show that you get the same value for the field strength in your laboratory from the following data:
 Gravitation constant, $G = 6.67 \times 10^{-11}$ (newton-meter^2)/(kg^2) in $F = GM_1M_2/d^2$
 Distance of your lab from center of E $\approx$ 4000 miles $\approx 6.34 \times 10^6$ meters.
 Mass of Earth $M_E \approx 6.6 \times 10^{21}$ tons $\approx 6.0 \times 10^{24}$ kilograms
 Using the law of universal gravitation, calculate *roughly* the force with which the Earth attracts a *specimen test mass of* 1 kg near the Earth's surface. This *is* the strength of the Earth's gravitational field there.

(c) A toy balloon of diameter 0.2 meter (radius 0.1 meter) is given a charge of one micro-coulomb, or 1×10^{-6} coulomb. (This is a big charge for such a balloon.) Calculate the electric field strength 0.5 meter from balloon's center. Instead of G, the constant B in the inverse-square law for charges is $9 \times 10^9 \frac{\text{(newton-m.}^2)}{\text{coulomb}^2}$.

(d) Calculate electric field strength 1.0 meter from balloon's center.

(e) When we deal with the electric field in the space between two parallel plates (such as Millikan used in measuring

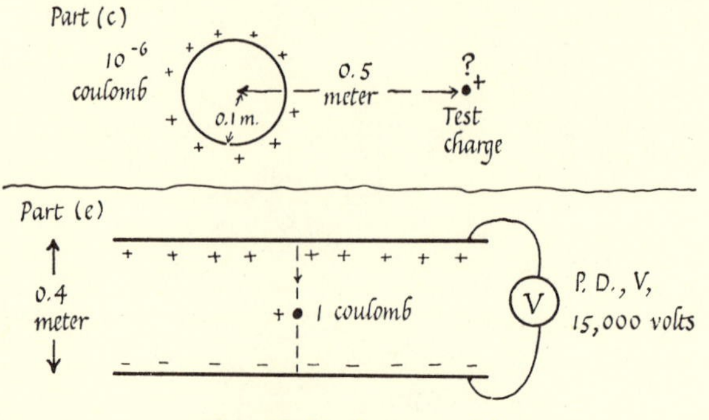

FIG. 33-69. PROBLEM 16

charge on electron; such as C.R.O. tubes use to deflect the electron stream), we use the experimental knowledge that such a field is *uniform*. This means the field has the same strength everywhere in the space between the plates; and it is directed straight across from one plate to the other, perpendicular to plates. If we know the voltage between the plates (as if a voltmeter were connected across them, in parallel with the battery which is attached to them to make the field) we can calculate the field, as in the following example.

Suppose the P.D. between two parallel plates 0.4 m. apart is 15,000 volts.

(i) P.D. 15,000 volts between plates means that . . . Copy this, complete it.

(ii) From (i), calculate the energy given by the electric field to 1 coulomb passing from one plate to the other, in joules.

(iii) Suppose 1 coulomb placed as a specimen in the field experiences a force X newtons. (i.e., suppose the field strength between the plates is X newtons/coulomb.) Imagine this force driving a coulomb across from one plate to another. How much energy will thus be given to the coulomb? (Note this is just FORCE · DISTANCE.)

(iv) Now write an equation stating that the answers to (ii) and (iii) are the same thing. Solve the equation and find the field strength X.

17. CALCULUS PROBLEMS (FOR ATOMIC THEORY)

To work out details of any picture of atomic structure, such as the simple Bohr model, we need to know the potential energy of a negative electron at distance r from a positive nucleus. (We should meet a similar need for gravitational P.E. in trying to predict a planet's orbit.) For that, we need to know the *P.D.* V between infinity and distance r from a nuclear charge Q. If a charge + 1 coulomb gains V joules of electrical P.E. when dragged in from infinity to r, an electron with charge e gains Ve joules. Since e is negative, -1.6×10^{-19} coulomb, an electron *loses* P.E. in moving in from infinity to an orbit of radius r. That is why it cannot leave the orbit unless given more energy by bombardment. So you will find an expression for V useful in studies of atoms. It is obtained by calculating the work (energy-transfer) in dragging + 1 coulomb against repulsion $BQ(1)/d^2$ from $d =$ infinity to $d = r$. That force changes with distance, d, so we cannot find V by simple multiplication of FORCE · DISTANCE. Instead we have to split the trip up into very short steps and calculate the work for each step and add up the answers. In the limit, that *is* the calculus method called integration. (There are tricks for doing this calculation without calculus —anyone who can follow the trick method could learn calculus, and enjoy it, in the same time.) Here is the calculus derivation, using x instead of d.

Suppose the specimen + 1 coulomb is part way in on the trip, at distance x, and we drag it in a further distance $-dx$. (The d is a limiting form of Δ, meaning an infinitely small step, and the minus sign is there because an inward step is a negative increase of x). Then the work is:

$$\text{FORCE} \cdot \text{DISTANCE or } [BQ(1)/x^2]\,[-dx].$$

Then, $V =$ total work

$=$ sum of all bits of work from infinity to $x = r$

$$= \int_{\infty}^{r} [B\,Q\,(1)/x^2]\,[-\,dx]$$

This is the energy-transfer, per unit charge, *FROM* the external dragging-agent *TO* potential energy of the electric field.

(a) Carry out the integration, remembering that B is a constant (9.0×10^9) and Q is constant (the central charge).

(b) The result of (a) gives V at distance r from a point-charge Q. It also gives V at the surface of a charged ball of radius r carrying charge Q. Why?

(c) From (b) estimate the potential of a metal ball the size of a baseball ($r \approx 0.05$ meter) carrying 1 micro-coulomb (10^{-6} coulomb). This will be in volts.

(d) According to simple atomic models, the "radius" of an atom of hydrogen, the distance at which the electron spends most of its time, is about 0.5 Ångstrom unit ($= 0.5 \times 10^{-10}$ meter). The electron has charge -1.6×10^{-19} coulomb, and looking towards the nucleus it sees a positive charge of the same size.

(i) Calculate V due to nuclear charge $+1.6 \times 10^{-19}$ coulomb at the atom's edge (for H atom).

(ii) Calculate the electron's P.E. there, in joules by multiplying V by the electron's charge, -1.6×10^{-19} coulomb. Then divide by electron's charge again, to express the P.E. in electron · volts. (Note that this P.E. is negative. In a Bohr model the electron also has K.E., just half as big as the P.E., and, of course, positive. So half your answer here shows the energy needed to knock the electron off the atom and make an ion. Bombardment experiments give 13.6 electron · volts for hydrogen atoms.)

(iii) Alpha-particles, carrying + 2e charge, are shot at atoms of gold. Very occasionally one bounces straight back. In such a case, we think it moved in against the repulsion of the big + charge on a gold atom's nucleus, until it had lost all its K.E. into electrostatic P.E.; then it backed out again. From the data below (all from experiments) estimate the alpha-particle's closest approach to the gold nucleus. Compare the answer with the traditional "size of an atom," radius about 0.5×10^{-10} meter.

DATA: Alpha-particle (from radium) has
speed $v \approx 1.6 \times 10^7$ meters/sec
mass $m \approx 6.6 \times 10^{-27}$ kg
charge $2e = +2 \times 1.6 \times 10^{-19}$ coulomb
Gold nucleus is much more massive and has
charge $79e = +79 \times 1.6 \times 10^{-19}$ coulomb

CHAPTER 34 • MAGNETISM FACTS AND THEORY

"The leading distinction of magnets is sex. . . . The kind that is found in Troas is black and of the female sex, and consequently destitute of attractive power."

—Pliny, *Natural History,* ~ A.D. 77

ELECTRIC FIELDS accelerate electron streams and deflect them, but they prove insufficient when we try to find out about the charge *and* mass *and* speed of moving electrons or charged atoms. We need another weapon too: *magnetic* fields. So we must still wait before attacking atoms, and study magnetism briefly. This chapter aims at showing what magnetic fields are like and how they can be used in the study of atoms. It shows how a simple theory can be developed for magnets—as an example of good theory.

Magnets

A magnet is a metal bar that will attract small pieces of iron such as iron filings. By the end of this discussion we shall want to modify that crude description; but with greater understanding we shall no longer require a formal definition of a magnet.[1] For the moment, we can list four common properties:

(i) Magnets attract and pick up small pieces of iron, etc.

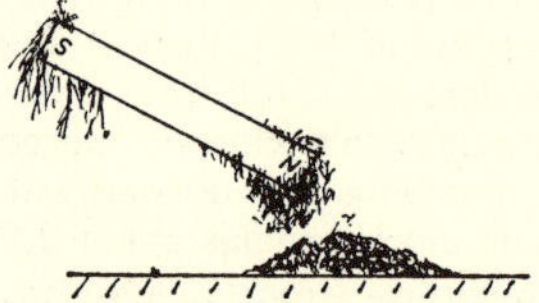

FIG. 34-1. A MAGNET COLLECTS IRON FILINGS

(ii) When suspended, a long magnet turns until it points in a direction which runs roughly North-South. Magnets of other shapes orient themselves, showing a definite "magnetic axis" which tends to take the N-S direction, however the magnet is suspended.

[1] Physics does not always need formal definitions, like those in a dictionary. An idea may be well understood and play a useful part, and yet elude simple definition. (Note the difficulty of defining JUSTICE in Religion or Philosophy.) On the other hand we take great care in framing clear definitions of *physical quantities* which we *measure* (e.g., temperature, electric field strength). "For of course the true meaning of a term is to be found by observing what a man does with it, not by what he says about it."—P. W. Bridgman, *The Logic of Modern Physics*

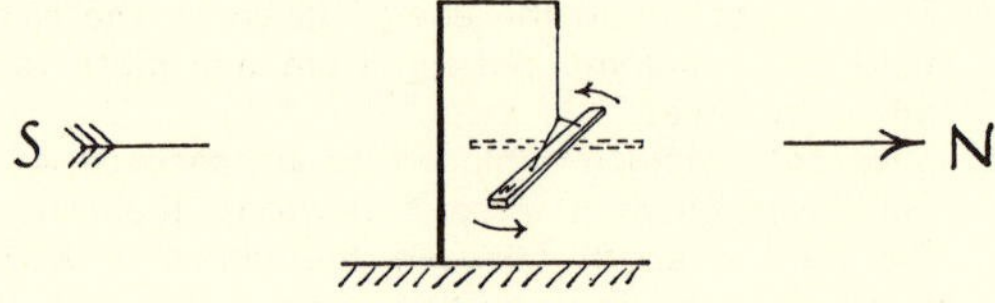

FIG. 34-2a.
A suspended magnet orients itself N-S. The end that points north is called the magnet's North-Seeking Pole.

(iii) A compass needle (or any other suspended magnet) brought near a large magnet tends to turn, pointing towards certain "poles" of the magnet.

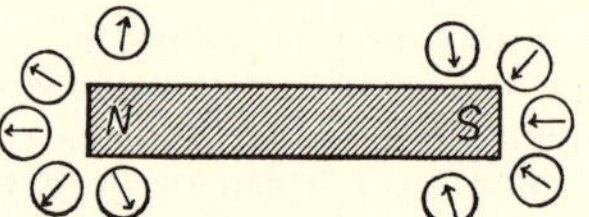

FIG. 34-3.
TESTING THE POLES OF A MAGNET WITH A COMPASS NEEDLE
The compass needle, itself a small suspended magnet, turns till its N pole is nearest to the big magnet's S pole, or its S pole nearest to the big magnet's N pole. The compass needle is really showing the direction of the big magnet's field. The field lines sprout out from the pole regions; but they do not leave the magnet at right angles—because, unlike metals for electric charges, magnetic materials are not perfect "conductors."

(iv) New magnets can be made by rubbing rods of suitable material with another magnet. This was the ancient way of manufacturing magnets—rubbing a steel rod with "lodestone," natural magnetic rock. It is now easier and better to place the rods inside a coil of wire carrying a current.

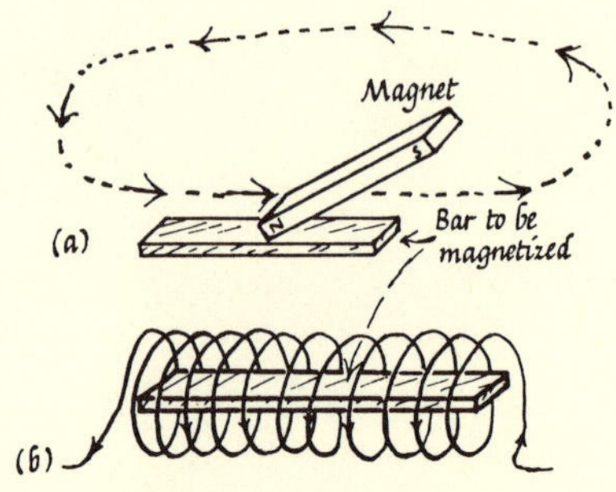

FIG. 34-4. MAGNETIZING A BAR OF STEEL
(a) The bar is stroked with a magnet.
(b) The bar is placed, momentarily, inside a coil carrying a current.

Poles[2]

The regions of a magnet where it picks up iron filings most copiously are called its *poles*. Long bar magnets usually have poles near their ends, though freak magnets with poles in other positions can be made. An exploring compass needle brought near a magnet's pole will point straight towards that pole. In general, the poles seem to be regions from which magnetic effects radiate. When a simple bar magnet is suspended free to turn, it twists until its poles lie on a line running roughly N-S; so its magnetic axis seems to be a line joining the two poles. We call the pole which settles farthest North the NORTH-SEEKING POLE of the magnet. We abbreviate this to NORTH POLE or N-POLE, but the meaning remains the same.[3]

EXPERIMENTS WITH MAGNETS

(See these demonstrated or try them yourself.)

EXPERIMENT A. MAKING MAGNETS

Place various metal rods inside a large hollow coil of wire carrying a current. Test the rods with iron filings or nails, with the current on and off. Also try using an alternating current. These experiments will show you that:

(i) Most materials cannot be "magnetized"; e.g., brass, glass, wood.

(ii) Iron, steel, certain alloys, can be magnetized—these are called magnetic materials—and retain some of their magnetization after the current is turned off, or after they are taken out of the coil.

(iii) A coil carrying alternating current can be used in a certain way to demagnetize a rod;

(iv) however, certain treatment with a coil carrying alternating current can leave a rod magnetized.

(v) Some hardened steels make good permanent magnets; and soft iron makes a strong temporary magnet while the surrounding electric current flows, but loses almost all its magnetization when the current is switched off.

(vi) A current in the coil alone, without any iron core, has some magnetic effect.

[2] Poles seem natural and real when you look at a magnet in action. Yet we now know they are not really there, because magnets derive their properties from electronic current-whirls which have no separable "poles." Though it is fashionable to condemn teaching in terms of poles as untrue, you can safely use them here, as a temporary idea, provided you remember they are artificial.

[3] Those who have studied .physics before may note how trivial these recitations of information now seem—yet they seemed an end in themselves in some elementary studies.

EXPERIMENT B. SUSPENDED MAGNETS

Hang a rod in a cradle on a silk thread. Observe its orientation, and bring up other rods near each end of the suspended rod. You will see that:

(vii) A magnet tries to settle pointing in a definite direction, roughly N-S.

(viii) A piece of soft iron (unmagnetized), or a piece of brass (nonmagnetic), does not try to point thus.[4]

(ix) A magnet's poles, where it picks up iron, will push and pull the poles of another magnet; magnets seem to have two kinds of pole, usually one at each end; and the pushes and pulls run thus:

N-pole repels N-pole
S-pole repels S-pole
N-pole and S-pole attract each other
} *"LIKE POLES REPEL AND UNLIKE ATTRACT"*

The force between two poles fits with an inverse-square law. (There is a difficulty in discovering or testing this law. We need two isolated poles free to move towards or away from each other. We cannot use two magnets because each magnet then imports its other pole as well. We have to resort to trick arrangements such as using very long magnets, so that the other poles are too remote to matter; or using short magnets and testing the forces that inverse-square law would predict for all four poles.) Careful experiments with forces measured by weighing (or by Coulomb torsion balance) show the forces fit adequately with an inverse-square law. There is no perfect test like the great test of no-field-inside-a-charged-metal-box in electrostatics; but there are satisfactory indirect tests.

EXPERIMENT C. PERMANENT AND TEMPORARY POLES

Suspend a soft iron bar.

(x) Soft iron is always attracted by both poles of a magnet.

Dip magnets in iron filings. Dip a soft iron rod.

[4] If *you* were suspended by a rope, you would point in a definite direction, and swing back to it if disturbed, but not for magnetic reasons. How would you prove, when investigating magnets, that the suspending thread was not spoiling your tests?

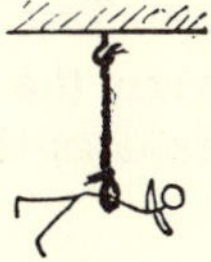

FIG. 34-2b.

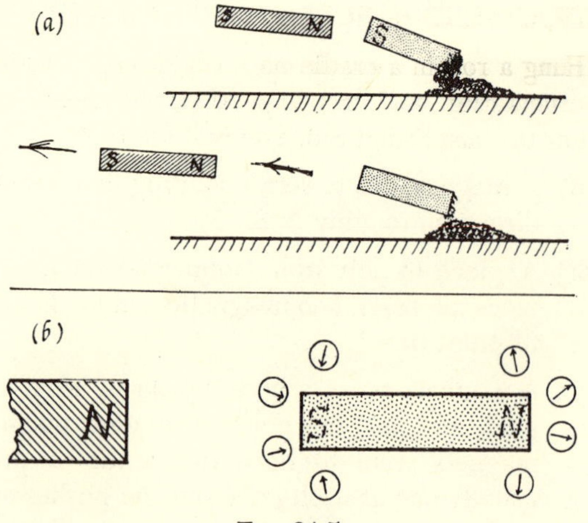

Fig. 34-5.
(a) Soft iron becomes a temporary magnet.
(b) The poles of a temporary soft iron magnet.

Dip one end of a soft iron rod while a magnet pole is held *near* the other end.

(xi) A magnet brought near a soft iron rod makes the rod able to pick up filings temporarily. (Fig. 34-5)

Further tests show that this is because the soft iron develops temporary magnetization in such a direction that it is attracted.

PROBLEM 1

What experimental evidence told you that soft iron develops this changeable magnetization before you ever tried a magnet on a suspended rod of iron?

EXPERIMENT D. USE OF COMPASS

Suspend a magnet on a good pivot so that it can turn in a horizontal plane. This is just a description of a compass needle! Some of the previous experiments can now be repeated by bringing a compass needle near a magnet (Fig. 34-3). We can then use the compass needle to label the poles of any magnet "N" and "S." We call the end of the compass which points (roughly) Northward, its N-pole and we call all similar magnet poles N-poles. (See later comment on Earth's magnetic poles.)

EXPERIMENT E. TEMPORARY MAGNETIZATION OF SOFT IRON

Use a compass to label the poles of a bar magnet N and S. Hold one end of the bar magnet near one end of a bar of soft iron. Test each end of the soft iron rod for poles. Reverse the magnet and test the iron again. What poles does the bar have in each case?

Magnetic Fields

We think of a magnet as having a magnetic field all around it, far and wide, much as electric charges have electric fields. We define the lines of magnetic force as *lines along which a small test N-pole would drift.* Or, if we have doubts about obtaining a test pole all by itself, we define them as lines along which a tiny test compass needle would point. Both definitions give the same patterns: the field that pulls the needle's N-pole forward along the line of force pushes its S-pole backward along the line, thus twisting the needle around to point *along the line.* (We could define the field strength as the resultant force on a unit test pole due to all the neighboring magnets, by analogy with electric field strength; but we shall not need it in this form.) The patterns of magnetic fields can be derived from the inverse-square law in the way used for electric fields; so most of the discussion of electric field patterns applies again here—except that there are no perfect conductors for magnetism, to correspond with metals for electricity. Though the patterns are often similar, magnetic fields are quite different from electric fields. They are different, separate, force-fields, one belonging to the special things we call magnets, the other to universal electric charges.

EXPERIMENT F. MAGNETIC FIELDS

In laboratory, trace out some magnetic fields with a small compass, to become familiar with their nature and patterns. The needle points along the field lines wherever it is put. Place a magnet on a big sheet of paper and put a small compass near it. Then move the compass, always steering "due North by the compass." This will make you move the compass along a line of force of the magnetic field. Mark the path of the compass on the paper. Best of all, make a pencil mark just ahead of the compass needle's point; move the compass forward till it points back at that mark; make a new mark, &c., as in Fig. 34-6.

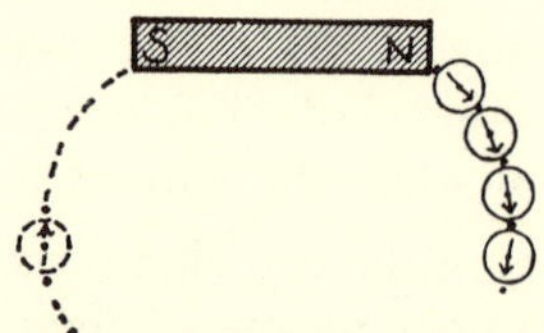

Fig. 34-6. Mapping with Compass
Using a small compass, make a dot ahead of the needle's N pole. Move compass until its S pole points back at that dot. Make a new dot ahead, . . . and so on.

Start again and mark another line from a different starting point and continue until you can see the full pattern of lines all over the paper. You may need to start mapping some lines from the edge of the sheet.

Or, instead of a compass, use a sprinkling of iron filings. These seem to act as small temporary compass needles and gang up in chains which mark lines of the field. The filings lack good pivots, so you must tap the paper to help them to arrange themselves. Make the sketches from life, the "size of the palm of your hand" or bigger. A few lines to show the general pattern far and wide (Fig. 34-8) are better than a crowded fuzz. Fig. 34-9 shows some arrangements for which you should sketch field lines.

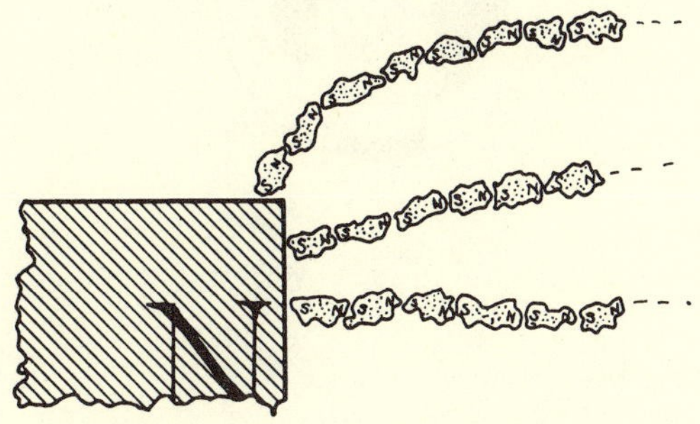

Fig. 34-7. Mechanism of Iron-Filing Maps

Fig. 34-8. Maps with Iron Filings

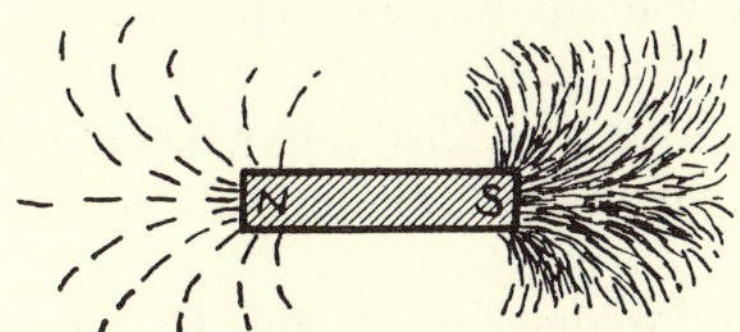

Fig. 34-8a.
Make a small sketch of each pattern, drawing the sketch from life. Sketches should be "size of palm of hand" or bigger. You are advised to use broken lines. A few main lines make a better map than a fuzz of many lines.

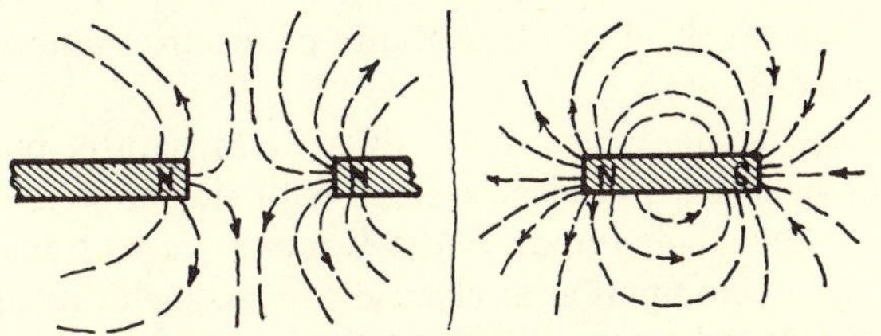

Fig. 34-8b. Samples of Magnetic-Field Sketches

Interpreting Magnetic Field Maps

When we make maps of a variety of fields, we find that the patterns can be interpreted to show something about the forces acting on the magnets that make the field. The field-lines seem to behave like elastic tubes pulling and trying to contract along their length, yet bulging and elbowing each other out sideways as if filled with fluid. Between a N-pole and a S-pole the lines run like clutching tentacles, hinting at the attractive force. Between N-pole and N-pole they spread and push against each other like buffers, hinting at repulsion. In more complicated arrangements we can see the lines tugging and twisting a magnet.

As they near a pole the lines crowd closer and closer. We know the magnetic field grows stronger nearer a pole (inverse-square law). So in this case the crowding of lines of force goes hand in hand with the field strength. If we investigate the patterns of lines in detail, we find that wherever they crowd

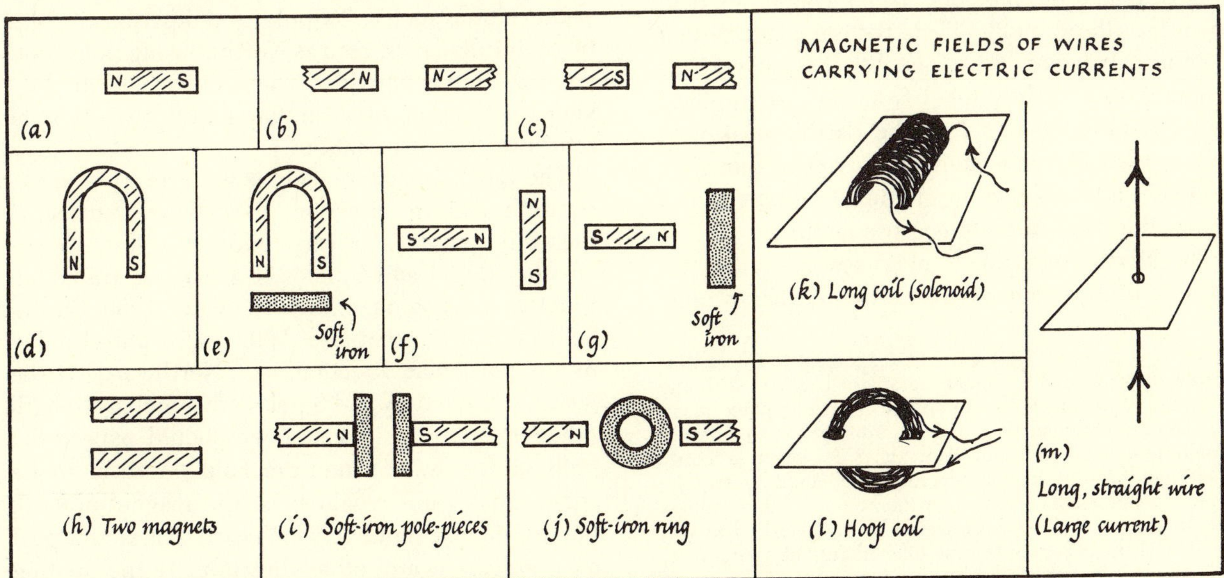

Fig. 34-9. Magnet Arrangements Suggested for Maps to be Sketched

more closely the field is stronger; so we can read general indications of field strength from a map of lines of force. (In more advanced studies of magnetism this idea is codified into a definite scheme for showing field-strength numerically by crowding of lines.)

It is useful to develop a feeling for these magnetic lines of force as agents by which magnets pull and push each other, because we shall find similar ideas apply to the magnetic-field forces that electric currents exert on each other and on magnets. Then magnetic field maps give us a way of picturing the actions involved in the working of electric motors, ammeters, etc.

Electric fields are quite different in nature, but they have lines of electric force with similar shapes that show their forces. Radio waves seem to travel along a combination of electric and magnetic fields—like waggling waves along taut ropes—so we may come to feel that the lines of these fields are very real. Neither set of lines is really there, of course, but the fields themselves are certainly there.

Earth's Magnetic Field

If we use a compass needle to map the magnetic field on a bare table, we find ourselves drawing a set of parallel lines running roughly North and South. A suspended bar magnet, free to turn, will set itself along the same direction—it *is* just a giant compass needle. This set of N-S lines is the magnetic field that is still there when we have removed all magnets. Exploring over the face of the Earth we find that the lines seem to converge towards a region in northern Canada and another region in Australia. In most places the lines are not horizontal but slant down into the ground.[5] Their pattern suggests that the Earth behaves like a vast bar magnet with its magnetic axis slightly tilted from the geographical spin-axis (Fig. 34-10). It is the Earth's weak field that is used for navigating by compass, though a steel ship adds its own magnetic field—part of it a changeable field, too—to confuse matters.

The North-seeking pole of a compass points towards northern Canada. Therefore the Earth must have a South-seeking type of pole there. (The Earth's pole there is sometimes called the "North magnetic pole." If you find this confusing, avoid the abbreviations such as "N-pole" and describe all poles by their full names, such as North-seeking pole," and you will find no paradox. When you have cleared up your thinking on this, you may find it saves time to return to the abbreviations.)

The Earth's magnetic field is uniform, constant in strength and in direction, over wide regions. It therefore provides a very important test, to see whether the N and S poles of a magnet are equal. Float a magnet on a cork in water. The Earth's field pulls it to point N-S. Will it also pull it along in any particular direction, e.g., Northward? If the floating magnet's N and S poles are equal in strength (though opposite in kind) we should expect the pulls of the Earth's uniform field on them to be equal. Such pulls would twist the magnet around, but would not move it Northward across the pool of water or in any other direction. If the floating magnet's poles are not equal, we should expect

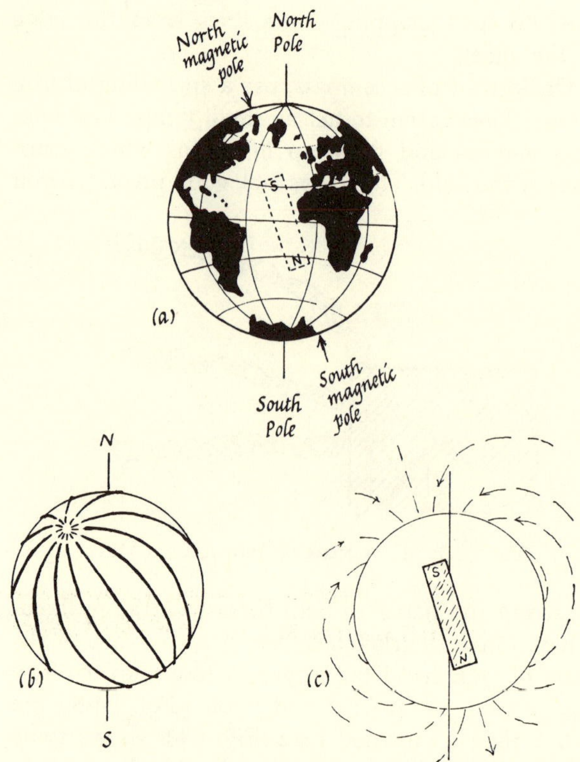

FIG. 34-10. THE EQUIVALENT MAGNET FOR THE EARTH'S EXTERNAL MAGNETIC FIELD
(a) Earth and equivalent magnet.
(b) Lines of magnetic field along the Earth's surface. (These show the direction, but not the strength, of the *horizontal* component of the Earth's magnetic field.)
(c) Lines of the Earth and its magnetic field, shown in section.

[5] In most parts of the United States the Earth's field dips steeply down into the ground, at angles like 70° with the horizontal. A magnet free to follow it should point along this slanting direction. Why does an ordinary compass needle remain horizontal and not slant thus? Its pivot certainly would not prevent it from slanting—we ought to see the N-pole of the needle scraping on the bottom of the compass box. If you guess the correct answer to this puzzle, you will see that there is a simple experimental test of your guess; and you should apply for the necessary apparatus and try the test.

unequal pulls to move it in some direction. Try this important test.

Though the Earth's magnetic field is weak, it can bend the path of electron streams appreciably. You will see in a later section how a magnetic field pushes an electric current sideways with a "catapult force." The streams of charged particles in cosmic rays come flying into the Earth's magnetic field from far away in space and are swung around by it. Experiments are being made now which use the Earth in this way as a giant control magnet for cosmic rays.

Making Magnets

In modern practice, electric currents are used to magnetize magnets. The current is not sent through the bar to be magnetized, but *around* it in a coil of wire around it. The magnetic field inside a long cylindrical coil (a "solenoid") is uniform; and it can be varied from very weak to very strong by increasing the current, so it provides a good field for experimenting on making magnets. We place a steel bar inside a solenoid, turn on a current through the coil and find that, while the current runs, the bar is a magnet. When the current is switched off, the bar remains a magnet, though not such a strong one. The current needs to run for only a very small fraction of a second to produce its full effect. There are several good materials for such "permanent" magnets. Most hardened steels are fairly good. Special steels containing tungsten or cobalt are better, while some new alloys containing aluminum, "Alnico," make stronger magnets still, though they need big fields to magnetize them. All these can be magnetized by being placed in a magnetic field for a short time. Reversing the magnetic field, by reversing the current in the coil, reverses the magnetization.

Demagnetizing Magnets

We can demagnetize a bar completely by placing it in such a coil carrying *alternating* current and then *slowly* taking it out and away from the coil. Or we may slowly reduce the alternating current to zero, with a rheostat.

Temporary Magnets, Soft Iron

When we attempt to magnetize a piece of soft iron (pure iron, or steel that has not been hardened), we find we are very successful as long as the current runs in the magnetizing coil. When the current is switched off the bar loses nearly all its magnetization. Soft iron makes an excellent temporary magnet, and it is therefore used for the cores of electromagnets in motors, etc.

We can magnetize a soft iron bar temporarily by bringing a magnet near it. If the N-pole of a magnet is placed near the end A of a soft iron bar AB, a compass needle shows that AB becomes a magnet, with a S-pole at A, near the magnet's N-pole. It develops a N-pole at the remote end B. When we remove the magnet from AB these poles disappear. Now you can see why unmagnetized iron filings are attracted to a magnet. In the magnet's field, near the magnet, these small chips of iron are magnetized temporarily, and their poles experience unequal pulls and pushes. A chip of iron near a magnet's N-pole will develop a temporary S-pole near the magnet's N-pole, and this S-pole will be attracted strongly by the magnet's N-pole. The chip also has a N-pole, but this is farther from the magnet, therefore in a weaker field. So the chip is more strongly attracted than repelled, and is pulled towards the magnet.[6]

On a larger scale, a magnet attracts any unmagnetized piece of iron, by producing temporary magnetization. Even a small compass needle will magnetize an iron bar temporarily. Being more easily movable than the large bar the needle will turn and point to it. Such turning merely tells us that both needle and bar can be magnetized and that at least one *is* magnetized. We cannot tell from such *attraction* whether both are magnets. We can tell, however, if we find *repulsion.*

Magnetic and Non-magnetic Materials

If we try to magnetize samples of brass, iron, glass, etc., in a current-carrying solenoid, only some materials show any magnetic effect. We call these magnetic materials. (Examples: iron and many iron alloys, nickel.) Some substances such as liquid oxygen and some iron compounds seem to be slightly magnetic, and many substances seem to be non-magnetic. On this basis, we say that a non-magnetic substance cannot be magnetized, while a magnetic substance can—and when it is magnetized we may call it a magnet. More delicate investigations upset this simple story. Many substances are slightly

[6] A similar reason applies to the action of electrically charged objects in attracting small bits of paper, etc. The field of the charged objects pulls and pushes "induced" charges across the bit of paper, leading to an attraction and a smaller repulsion. In many cases the induced charge which is repelled to the remote end of the bit of paper is able to leak away to ground, leaving the attraction unopposed to have even greater effect. Magnets keep both poles (or none) and never show this effect of a pole disappearing like a charge leaking to ground.

magnetic, showing temporary effects when placed in a field; and we can trace their magnetic properties back to their individual atoms. In fact we can even show that some individual atoms are magnets, and we can measure the strength of such atomic magnets in a way that will be described later. The few metals like iron which show huge effects and can make permanent magnets still owe their properties to atomic magnetism, but their atoms have a special property of ganging together with their atomic magnets aligned to make strong permanent groups. Atomic theory also makes a general prediction of another magnetic property for all kinds of atoms. Curiously enough, the predicted property is a sort of negative magnetism, quite unlike the kind we usually meet, and theory predicts it, in a very weak degree, for all substances. On what is this prediction based? Is it well-founded? Is this negative magnetism observed experimentally? If so, why not for all substances? We shall answer these questions briefly in Chapter 41.

Fields of Electric Currents

Experiments show that electric currents have magnetic fields. A long coil, often called a solenoil, has an outside magnetic field which resembles that of a bar magnet. Compare the two, and you will find that the solenoid has the same field, externally, as a bar magnet of the same shape and size. Inside the hollow coils you can see the lines running in a crowded parallel sheaf, showing a strong uniform magnetic field.

★ PROBLEM 2

Why would you expect to magnetize a steel rod better by putting it inside a solenoid carrying a current than by having it just outside?

★ PROBLEM 3

Sketch A in Fig. 34-11 shows the magnetic field of a long solenoid. If the solenoid is shortened by squashing it, concertina fashion, the field must be modified as in sketch B. Suppose the coil could be squashed completely, making just a ring-shaped coil C, like a hoop. Predict the shape of the magnetic field of C carrying a current, just by imagining the squashing of the field continued. Sketch your prediction. Does it agree with experiment?

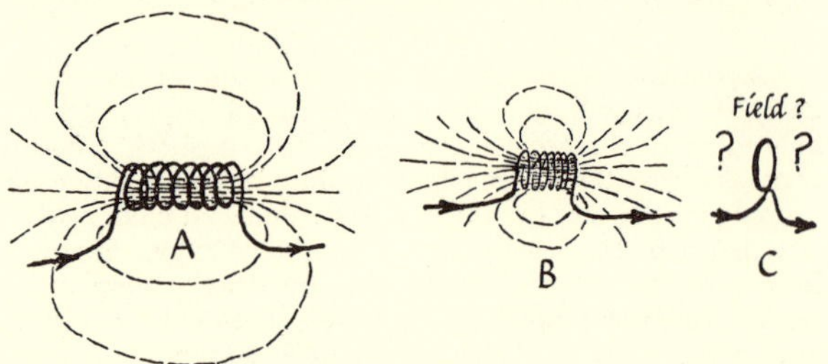

Fig. 34-11.

★ PROBLEM 4

The external magnetic field of a solenoid is the same as that of a bar magnet of the same shape and size. What shape of magnet should have the same field as a hoop coil, C, carrying a current? Sketch or describe that equivalent magnet.

Current Carrying Coils as Magnets: Demonstration Experiments

A coil carrying a current behaves like a magnet: suspended, it will turn and point its axis N-S; it seems to have "poles" near its ends which repel and attract other poles. A small coil placed in a magnetic field—the field of the Earth or a magnet or another coil—will try to twist round like any compass needle until its magnetic axis points along the field.

Magnetic Field of a Straight Wire Carrying a Current

There is one peculiar case, of great importance, where there is no "equivalent magnet" of the same shape and size. That is the field of a long straight wire carrying a current. Iron filings, or tiny exploring compasses, show that the lines of this field are circles around the wire—not just in one plane, of course, but in all the region around. The field is stronger near the wire, weaker at greater distances.

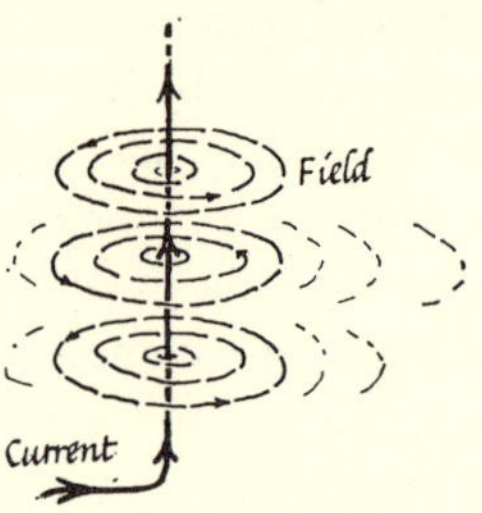

Fig. 34-12.

This is said to be the first magnetic effect of a current to be discovered. At the end of a lecture on electric currents, the Danish scientist Oersted placed a current-carrying wire near a compass needle and was amazed to see the needle turn. "The experiments which he began in April 1820 are among the most memorable in the whole history of science."[7]

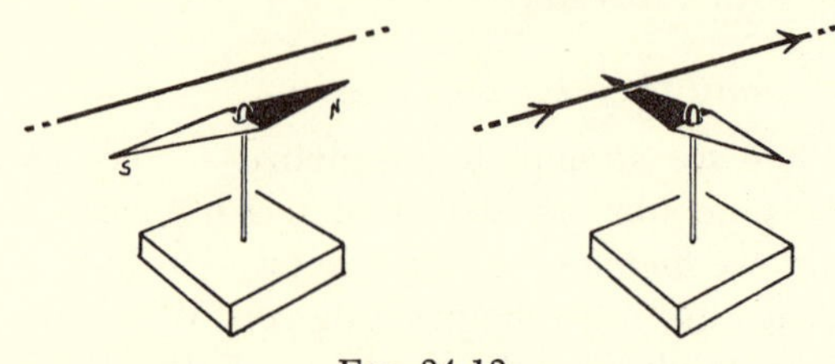

Fig. 34-13.

[7] G. Sarton.

When word of the discovery spread across Europe, it started an avalanche of experimenting. Ampère and others opened up the field of electromagnetism in a rapid growth of theory of experiment.

Oersted's original observation was a surprising one: a compass needle placed under the wire, parallel to it, was pulled around through 90°. The forces on its poles therefore seemed to be sideways forces, not in the straight line from wire to pole, but perpendicular to that line. Later experiments confirmed this and showed that the corresponding force exerted *by* a magnet *on* a current is also a sideways one; a current-carrying wire tries to move crabwise in a magnetic field. These new forces were quite

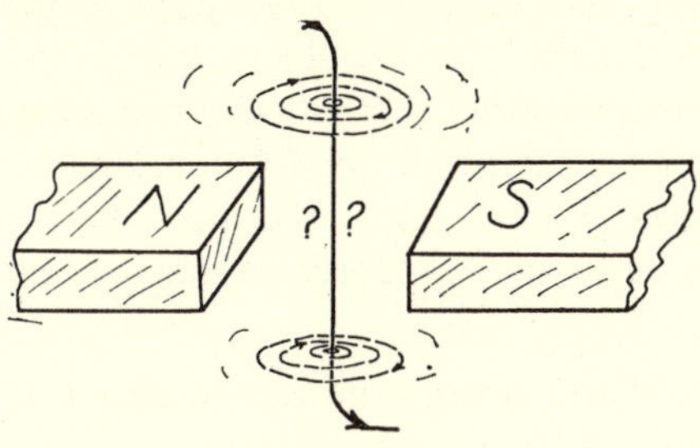

Fig. 34-14.

different from the direct forces already known, such as gravitational pulls (which run directly from one mass to the other) and forces between colliding balls or molecules (which we assume push directly apart). The same kind of direct forces, pushes and pulls, were found to act between electric charges, and between magnetic poles. Until this discovery, all known forces were direct. In fact a school of thinkers, including Voltaire and others, just before the French Revolution, had built up a mechanistic philosophy of a completely predictable universe, on the assumption of simple point-to-point forces. When the new electromagnetic forces were seen to depend on the *speed* of motion (current), they seemed queerer still. Here were forces that increased with speed and acted crab-fashion! These are in fact the forces that make an electric motor go.

We can illustrate these forces by a magnetic field map. Just by itself the circular field around a straight wire is peculiar but not very useful. Combined with a uniform magnetic field it shows the crabwise forces that run many machines: motors, meters, television tubes, and some of the "atom-smasher" accelerators. We shall now predict the pattern of that field, by vectors. Then you should see it demonstrated.

The "Catapult" Magnetic Field

We predict the pattern by adding vectors that represent the two separate fields. The pattern of the uniform field is a set of parallel lines, evenly spaced as in Fig. 34-15(a). And the straight wire's pattern (when it is carrying current) is a set of circles as in (b). We draw the circles more crowded near the wire, to show the field is stronger there. (Detailed investigation shows that

$$\text{FIELD STRENGTH} \propto \frac{1}{\text{DISTANCE OUT FROM WIRE}}$$

so we ought to draw the circles twice as crowded at half the distance.) Adding these two fields by vectors produces much the same pattern as the streamlines of air flowing past a spinning ball worked out in Chapter 9.[8] So we use the same scheme. Sketch both fields together, as in (c). At specimen point, A, sketch vectors to represent the two field-strengths, one along the uniform field, the other along the circle's circumference. Add these and *mark the resultant's direction by a short arrow* at A. At another point, B, the current's field is weaker, but the uniform field is the same. Again add their vectors and *mark a short arrow* at B to show the direction of the resultant (see sketch d). Mark many

[8] The streamline pattern of a spinning *cylinder* in uniform flow is exactly the same as the pattern of this magnetic field. (See Ch. 9 but note that the *force* is in opposite direction.)

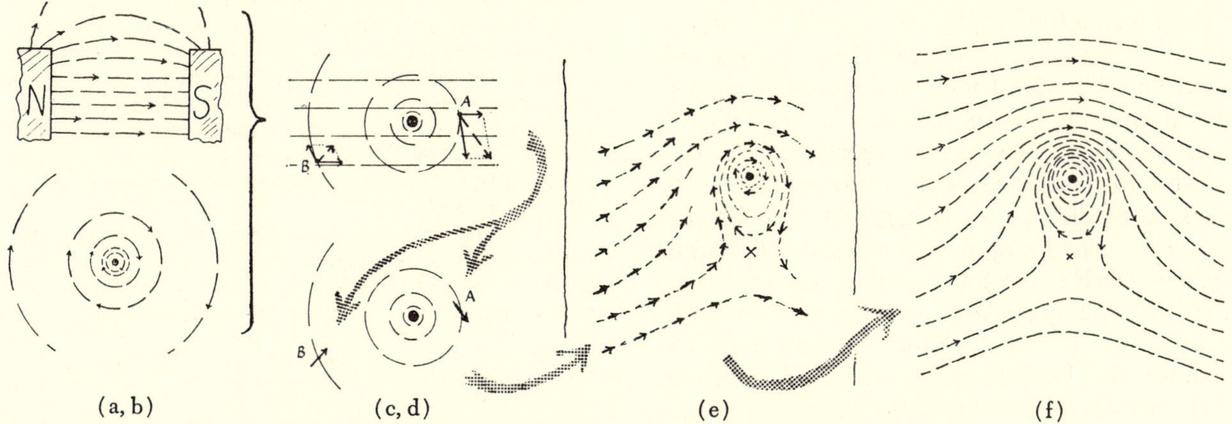

Fig. 34-15. Predicting the "Catapult Field" by Adding Vectors

such arrows, all over the diagram. These show the direction of the resultant field that we are looking for. Then sketch lines of the resultant field running through these arrows. We can appeal to common sense for some help:

(a) Near the wire, the current's field predominates and the resultant lines are practically circles round the wire.
(b) Far from the wire the current's field is negligible; and the field will have the straight lines of the uniform field.
(c) There will be a neutral point, of zero field, at some point X where the two fields just cancel.

To draw the combined field reliably needs much patience. Fortunately it can be sketched by indirect geometrical methods (based on $\nabla^2 V = 0$ as usual!) and then someone knowing the answer can make a plausible show of sketching the lines on our diagram. The picture is shown in Fig. 34-15(f).

If you follow Faraday in thinking of lines of force as showing graphically the forces that push and pull on magnets and circuits, you can see that this magnetic field will tug the wire *downwards* in the last sketch. This is the crabwise force, perpendicular to the wire and perpendicular to the uniform field. From the look of the pattern, we call this the "catapult" or slingshot" field. See it demonstrated (Fig. 34-16).

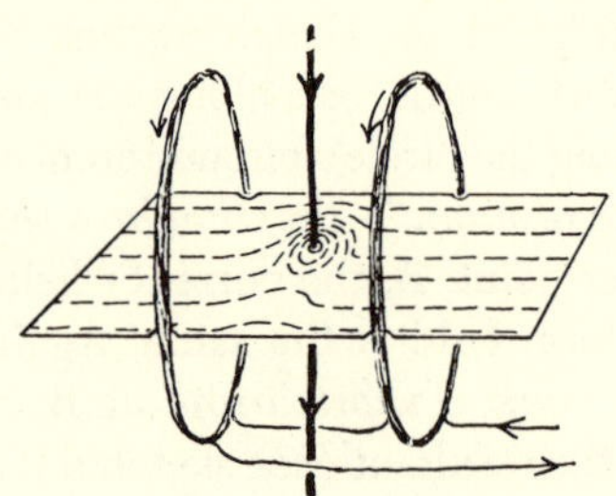

Fig. 34-16. Demonstration Experiment

The Catapult Force

Does such a force really act on a wire carrying a current across a magnetic field? Try it yourself with a small flexible wire, battery and horseshoe magnet. Switch on a big current through the wire when it is in various positions in the strong uniform field between the horseshoe's arms. There is a force, sideways, as predicted (see Experiment A in Chapter 41). The current may be just a stream of charged particles, such as electrons, without any wire. Such a stream is also pushed sideways by a magnetic field—a very useful effect that we shall discuss in Chapter 37. (For full effect, the outside magnetic field must be perpendicular to the stream or current—any component of magnetic field *along* the stream has no effect.) Try a bar magnet near the side of a cathode ray oscilloscope. The pattern looks like one studied in Ch. 9, but the *direction* of the catapult *force* is opposite to that of the Bernoulli force.

Attempts to Obtain a Single Pole; The Beginning of a Theory

Returning to steel magnets, try one more experiment. So far you have found pairs of poles in every magnet. Can you isolate a North pole, as you can isolate a positive electric charge?[9] Try cutting a magnet in half. Magnetize a piece of steel wire or clockspring. Use iron filings to show that it has poles at its ends, and label these poles N and S with a small compass. Then cut the magnet in the middle (with unmagnetized cutters) and test each half for poles. The faces of the break develop new poles, sabotaging your attempt. You have two complete magnets. This odd property of magnets raises two questions:

(i) Could the cutting up into smaller magnets be continued indefinitely?
(ii) Why are there always new poles at the cut faces?

Attempts to answer these have produced a very satisfying theory of magnetism, ranging from a crude picture of the structure of magnets to important comments on the structure of atoms.

SIMPLE THEORY OF MAGNETISM

Morris Cohen, commenting on the "popular myth that science consists in observing facts and ignoring theories," wrote:[10]

"If, however, we follow the progress of some actual scientific research, it becomes obvious that without some guiding principle, idea, or theory as hypothesis, we cannot even determine what facts to look for. Nor is it at all true that we can always find and recognize the truth or the facts by mere observation. If that were the case, the progress of science would be a much simpler affair than it is, and anyone could manage it. Actually, however, scientific research involves very difficult and elaborate methods for eliminating what seem to be facts to the ordinary observer. . . . Theories are points of view or perspectives for seeing things in their connections. They are, as Chauncey Wright pointed out, the true eyes of the scientist, wherewith to anticipate and discover things hitherto invisible.

"Of course, theories have a certain emotional grip which makes us sometimes cling to them despite evidence to the contrary. But do we not with equal obstinancy cling to our opinions on matters of fact? Two

[9] Make + and − induced charges on a conductor. Cut the conductor into two pieces, and you will have + on one, and − on the other.

[10] Morris R. Cohen, *The Faith of a Liberal.* (Henry Holt and Company, N.Y.)

points must be noted with regard to this. In the first place, in science as in other human enterprises we cannot well dispense with tenacity. For what seems to be evidence contrary to our theory, may, on careful re-examination, turn out to be rather a needed verification and extension of it. . . .

"The second and more important point, however, is that in actual scientific procedure an hypothesis is abandoned only when some other hypothesis is shown to be more in agreement with all of our previous knowledge as well as with new observations. The way to a real knowledge of the facts, therefore, is not by avoiding theories or anticipations of nature but by systematically multiplying the latter so that we may develop many points of view and thus overcome the tendency toward too much confidence in any one. That is why the logical or mathematical technique of physics, chemistry, general biology, and other theoretic sciences has been so fruitful in enabling us to discover hitherto unknown facts."

Making a Theory

What kind of a "theory of magnetism" can we construct? It is not worth while making one unless it guides our experimenting by fruitful predictions or increases our understanding by providing a helpful vocabulary of words and ideas. The previous sections gave some of the common properties of magnets that are found by experiment, most of them known for many centuries.[11] We could hardly have devised a profitable theory without some experimental information to start from. Of course we might have started at the very beginning with, "Magnets just *are* as they are. Whatever magnets have inside them, it is just what is needed to make them behave as they do. Steel has a 'magnetotropism,' a 'tendency towards magnetism.' That's my theory of magnets." Such a theory would necessarily be "true," but it would be no use and a wise scientist would not waste time over it.[12] So far we have been operating with a simple theory: that magnets have poles. A pole is not an experimental fact; it is a concept, an artificial idea that is useful when we discuss experiments. Further experiments lead us to believe there are no real poles. Poles are untrue, but that in itself would not kill our simple theory; we should keep it as long as it is useful and fruitful. The concept of poles helps our vocabulary, but it is not fruitful in suggesting new experiments or carrying our thinking much deeper. So, although we shall continue to talk in terms of magnet poles, we shall climb to a better theory. Now that we have some experimental information we can embark on some brave guesses, try to form some general scheme or picture, make deductions from that in turn, and then start new experiments to test the deductions.

"Molecular Theory"

Since we find that many chemical and electrical properties of substances can be referred back to the atoms inside them—and even some mechanical properties—it seems worth while to ask: "Do the properties of magnets come from some special behavior of the atoms or molecules inside them?" This leads to an experiment at once. We try to cut a magnet open to find what is inside. In particular we hope to obtain separate N and S poles by cutting a magnet in half. This experiment shows an unexpected result; cutting a magnet produces new pairs of poles at the cut faces, so that each piece is a new magnet. If we cut carefully, without jarring the bar much, we find the old poles remain as strong as before, and the new ones just match them. Thus we can cut a magnet into more and more pieces, and find each of the pieces is itself a magnet. If we

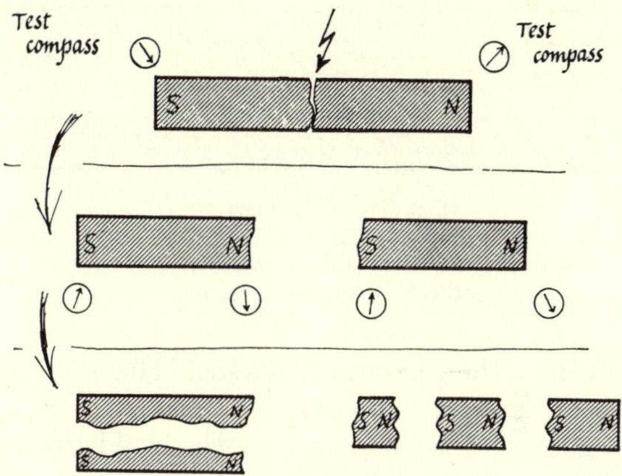

Fig. 34-17.
Cutting or Breaking a Magnet Produces New Pairs of Poles

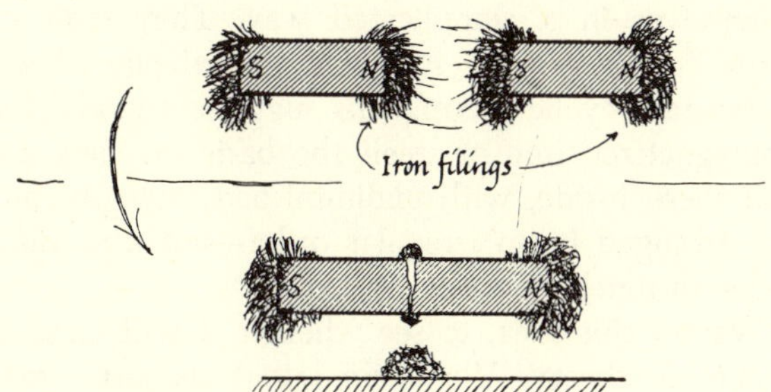

Fig. 34-18. New Poles Almost Disappear

[11] A generation before Galileo, the English physician Gilbert wrote a marvelous book on experimental magnetism and electricity. He recorded experiments and made sensible guesses. He likened the Earth to a ball of magnetized material.

[12] However, you will find such statements treated with respect in certain other fields of study.

try to put the pieces together again, the new N and S poles which appeared seem largely to disappear when their faces are placed in contact. Perhaps they are still there but have no external magnetic field, since their separate fields are opposite and practically neutralize each other.

Continuing in imagination the cutting up of a magnet, we think we should come to a stop at some stage when we had cut it into little "elementary" magnets. A century ago these were thought perhaps to be actual molecules or atoms of iron. We now prefer to think of them as groups of atoms—many millions in a group—to be called "domains" and looked for with a microscope. But at present we shall merely say they are very small, very numerous, basic magnets. So we think of a magnet being cut up to yield many tiny basic magnets. Putting them back together again to form one big magnet we should have them all aligned, with the N-pole of one close to the S-pole of the next, almost neutralizing each other's external effect, except at the ends

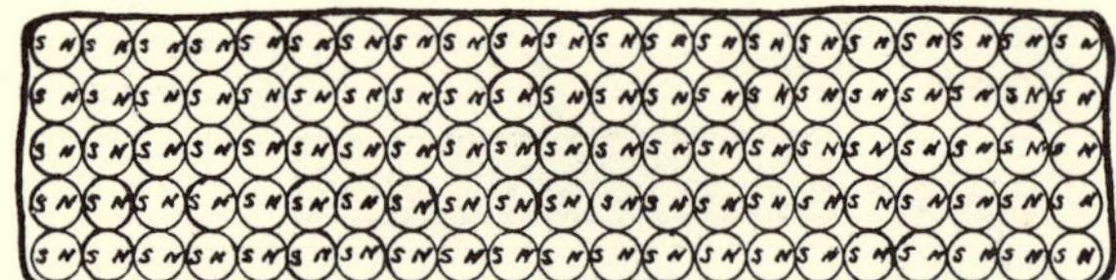

Fig. 34-19. Suggestion:
Magnets may be made up of tiny "internal magnets" aligned like this. Poles of neighbors would neutralize each other's effect, except at ends.

of the bar. There at one end we should have a whole face of exposed N-poles and at the other end a whole face of S-poles. In fact we should have an ordinary magnet. We may if we like imagine an ordinary magnet to be full of little magnets, aligned thus—though so far there is no profit in imagining such a complicated picture. We can even make a model of this arrangement with a large number of small compass needles which can be pulled into alignment by a magnetizing field. In such a model, the needles stay aligned as long as the magnetizing field is there, but when it is removed they rearrange themselves in a complicated way. They tend to form "family groups" in which several point head to tail in a cyclic group. This suggests a model for unmagnetized iron or steel: the basic magnets are still there inside, with undiminished strength, but are arranged in an irregular order—not a random order so much as cyclic groups.

We try this idea, to see whether it will make a profitable theory. We assume that magnetic materials consist of innumerable basic magnets which

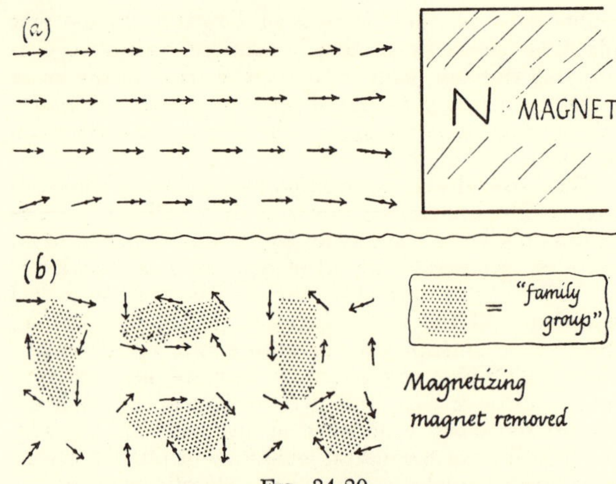

Fig. 34-20.
(a) Pivoted compass-needles aligned by large magnet.
(b) Same without magnet, jarred to get rearrangement.

are aligned in a magnetized bar, and disarranged in an unmagnetized bar. Experiments show that soft iron is easily magnetized and easily demagnetized while hardened steels need stronger fields to magnetize them and then retain some magnetization as permanent magnets. This suggests we should assume that the basic magnets in soft iron turn around easily, while in hard steel they stick, jammed against neighbors, perhaps, suffering from some forces like friction. Assuming this simple picture, what can we "explain" with it? First we see that it explains the appearance of new poles when we break a magnet. Unless we can cut through the basic magnets themselves, we are bound to produce new poles at the break. But this explaining is no praiseworthy achievement. Our theory has merely returned the

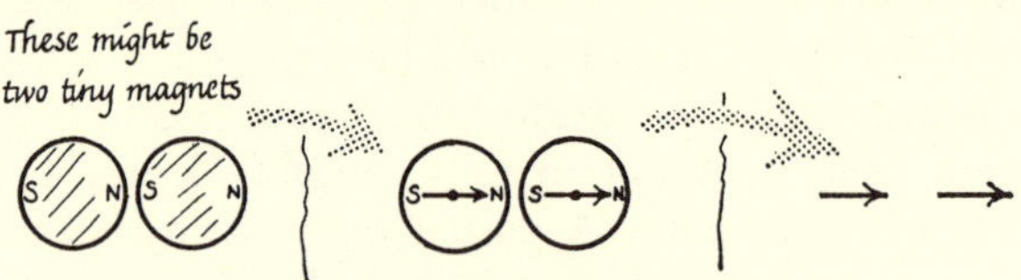

Fig. 34-21. Key,
showing how the drawing of the basic magnets is abbreviated in diagrams.

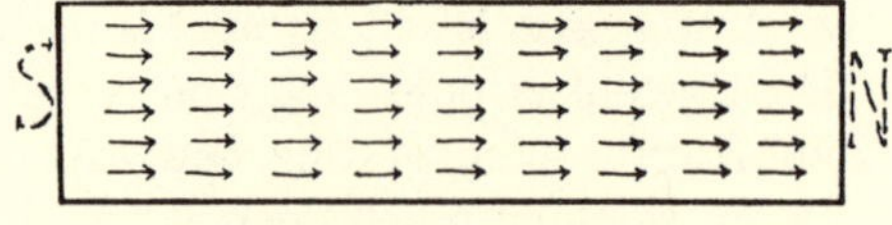

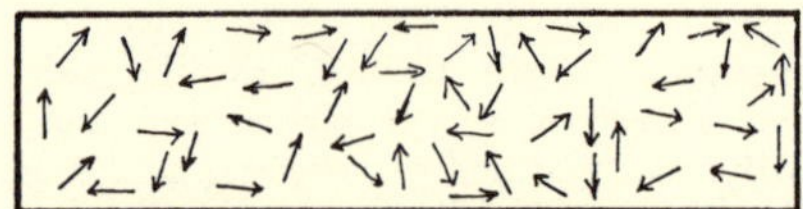

Fig. 34-22. "? Basic Magnets" in a Steel Bar

experimental fact that started it. It has shown that it contains the information it was built with. Furthermore, it has begged the question by stating, without any reason, that a basic magnet cannot itself be cut in half. Was this included in their definition? Defining them as having this virtue does not make them really have it.

Nowadays, however, we believe Ampère's suggestion of electric-current-whirls. We ascribe all magnetism to electrons etc. spinning and moving around orbits. Such current-whirls have their magnetic fields in closed loops and obviously cannot be cut to reveal "poles."

But without that special knowledge we should feel no satisfaction over this first success—if this is all the theory can do we should throw it away as cumbersome nonsense. Its value will hang on its success in giving good answers to new questions. Here are some questions.

(1) *A limit to magnetization?* We can produce huge electric currents—apart from overheating there is no limit to the current that can be driven through a wire. Is there a limit to the "magnetism that can be put into a bar"? Our theory at once answers, "Yes, when all the basic magnets are aligned, nothing better can be done." Here is a definite answer that can be tested. Fig. 34-23 shows the graph of such a test. The apparatus is sketched in Fig. 34-31. There is a limit.

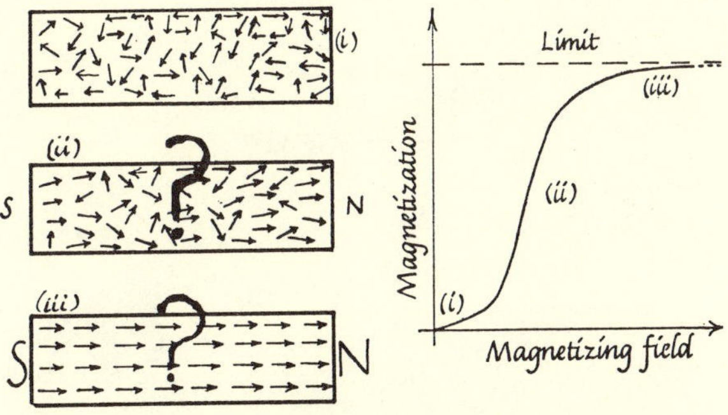

Fig. 34-23. Stages of Magnetization of Iron Bar
The graph shows the experimental record. The sketches of basic magnets show guesses of early theory. For a more modern view of "domains" see Fig. 34-33.

(2) *Location of poles?* If a steel bar is magnetized, and remains a magnet when the magnetizing field is removed, should the poles remain just at the ends? Theory suggests, "No. The basic magnets at the ends are likely to splay out, where they have no more magnets ahead of them to hold them in line. The N-poles at one end-face will repel each other and may succeed in spreading the N-pole region to the neighboring sides." Experiment shows

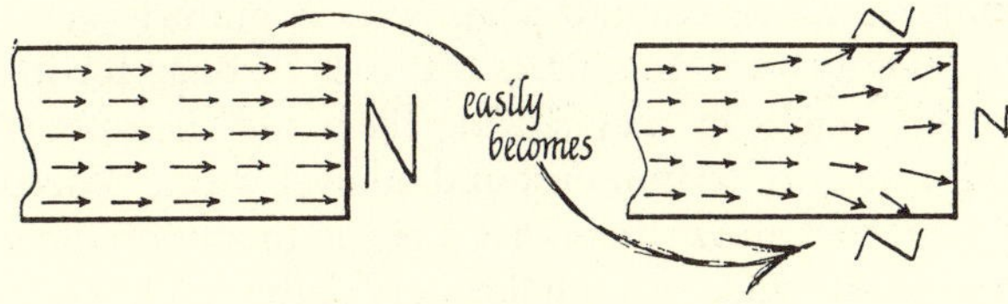

Fig. 34-24.
At Open Ends of Magnets the Poles May "Splay Apart."

that bar magnets do have spread poles. (Try a bar magnet with iron filings or a test compass.)

(3) *Keeping magnets.* Arising from (2) what could be done to prevent poles splaying thus, and even getting worse—magnets losing their magnetization? Theory does us a kindness, and suggests a cure. Another magnet placed ahead of the splaying poles would help to pull them into line, if placed the right way. This is found to be a good way of

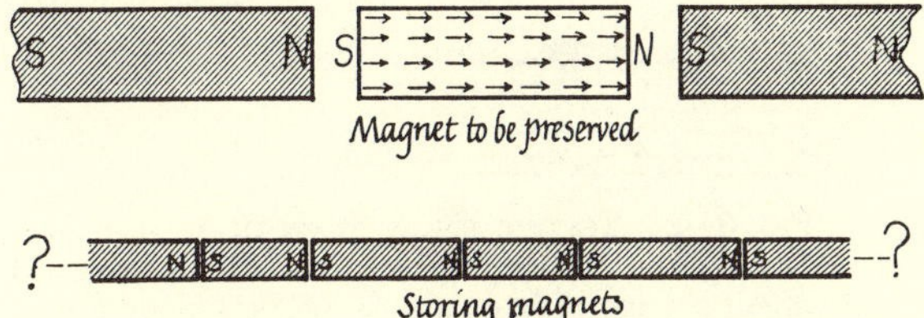

Fig. 34-25. Preserving the End-Face Poles of a Magnet

storing magnets, in a line tail to head; but it leaves an obvious problem: what should be done with the magnets at the end of such a line?

(4) *Effects of hammering?* Magnets can be harmed by hammering, heating, general rough treatment. (See a demonstration.) Can you explain this? Theory says "Yes," and draws a vivid picture of basic magnets being jarred out of alignment. "Once the magnetizing field is removed, the basic magnets tend to rearrange themselves in cyclic groups, but they may be jammed, leaving some permanent magnetization. Hammering may give them a chance to turn from alignment into disorder." This is all very well, but like many of theory's explanations it only gives a "reason" for what we already knew. Let us ask for more. *"Can hammering ever magnetize a bar, even though the hammer is non-magnetic?"* Without theory to help, we could hardly answer this by guessing, and experimental attempts would be unlikely. Theory suggests a clear answer: "Yes," *in certain circumstances*; and experiments verify this prediction. (What circumstances? If you guess correctly, you will know you are right. See a demonstration.)

(5) *Testing for flaws in iron castings.* Despite our contempt of theory's first prediction, that cutting a

magnet will reveal new poles, we can put it to good commercial use. Engineers test iron castings for invisible cracks by magnetizing them and then pouring a mixture of iron dust and oil over them. Where the magnetized material has a crack, theory predicts that poles will appear on the faces of the crack. Then iron dust should collect in a little ridge along the edge of the crack, a long semi-circular hump (like a bridge over a gulley, with the bridge extended to cover the whole length of the gulley). This works well, making an excellent method of detecting flaws that are otherwise invisible. (See Fig. 34-26.)

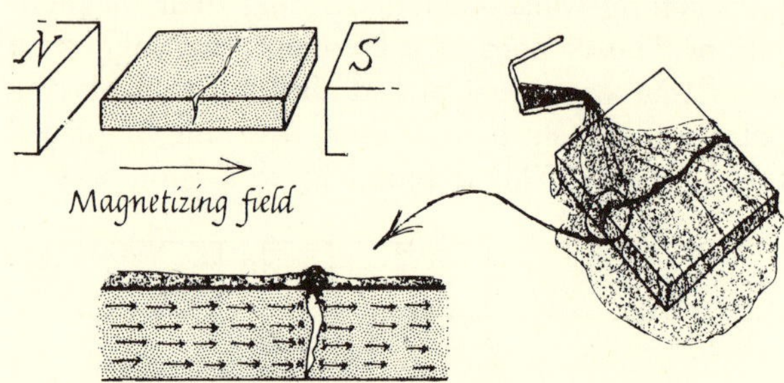

Fig. 34-26. Testing for a Flaw in Steel
Oil containing iron powder is poured over the magnetized specimen; a ridge of iron particles collects along the flaw where sets of opposite poles are exposed.

(6) *Alternating magnetization.* We can magnetize a bar forwards, then backwards, then forwards and so on, by placing it inside a solenoid carrying alternating current.[13] If we did this to comparable bars of soft iron and hard steel what difference might be expected? Theory suggests, "Since hard steel's basic magnets seem to suffer considerable friction-like opposition to turning, we might find a lot more heat developed in the steel bar." Experimentally, this predicted effect is often masked by others, but it certainly occurs. It is a matter of importance to engineers. The coils in electric motors and generators are wound on iron cores. If these coils carry alternating current it is most important to make their cores of soft iron. Otherwise the cores would heat up, endangering insulation and wasting energy. In D.C. machines, too, a *rotating* core is magnetized in varying directions, so it should be made of soft iron.

[13] The usual A.C. supply swings its voltage up to a maximum one way, down to zero, to maximum the other way, back to zero, up to the first maximum again, and so on, making 60 complete cycles per second. If we apply this alternating voltage to a coil, it drives through the coil an alternating current which swings through similar cycles, making a magnetic field which also alternates through similar cycles.

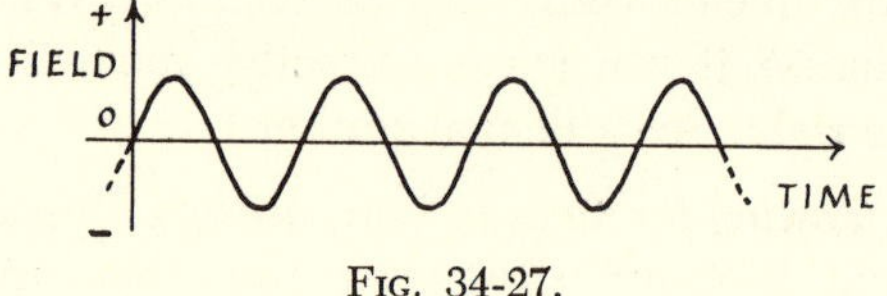

Fig. 34-27.

(7) *Theory at its best.* So far, theory has made helpful predictions, some of them merely fitting with what was already known: others soon verified. We now ask it a question which leads to one of the finest uses of theory. "*Suppose a man tries to magnetize a ring of steel, and thinks he has succeeded, yet he can find no poles, no external magnetic field. Is it possible that in any reasonable sense of the term the ring is "magnetized"?*

Without a theory, the answer seems clear: "*Nonsense. No poles, no magnet.*" With the theory in mind we have quite a different answer. "*Yes, the ring may be magnetized, cyclically, with its basic magnets arranged tail-to-head round the ring.*" Here is a remarkable achievement. Theory enables us to understand what would otherwise be incomprehensible.

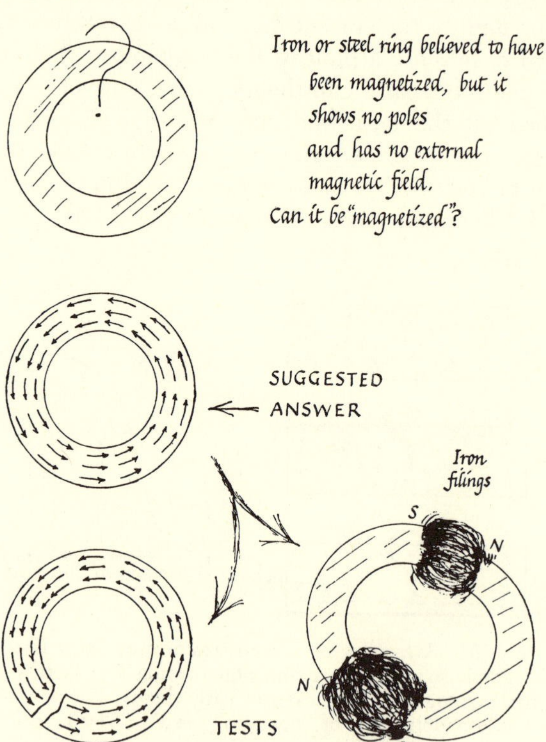

Fig. 34-28. A Question for Theory

Theory is at its best when it gives new meaning to a word or an idea—in this case "magnetization." Then it rises above its usual role of remembrancer and forecaster and becomes an artist to give us insight. Here, theory gives new knowledge and it deserves the praise given to the elephant's trunk: "You couldn't do *that* with a mere smere nose." Few theories rise to this height—or, rather, few show their advantage so clearly.[14]

Theory continues, "This explanation can be tested by cutting the ring. If it is magnetized in this way, poles will appear at the cut faces." The test can be carried out; and, if the ring has been prepared properly, strong poles do appear.

Such ring magnets are now common and important in engineering—they are not merely special magnets devised to put a theory to the test. The iron cores of transformers are often designed as complete rings to be magnetized in this way. Such magnetization is vital for good transformer action, and transformers are essential to our modern transmission of electric power. You may later guess a way of testing this ring magnetization without even cutting the ring.

(8) Now we can return to the problem of preserving magnets. Horseshoe magnets are often provided with keepers—a bar of soft iron which bridges across the poles. Similar bars are used in the storing of bar magnets. In both cases the soft iron becomes

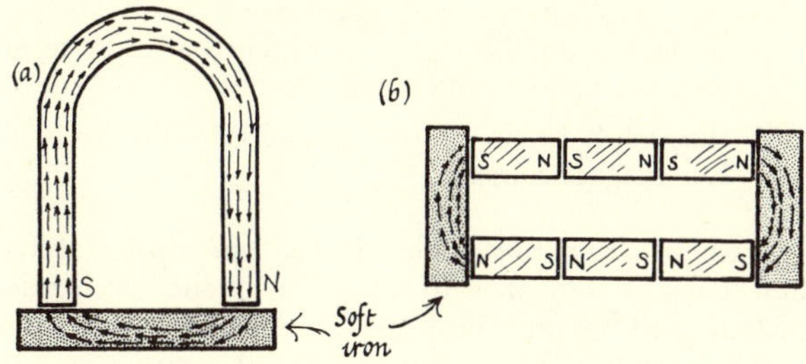

Fig. 34-29. Soft Iron "Keepers"

temporarily magnetized by the magnets in such a way that there is essentially a closed magnetized ring like the one discussed above. In the light of our theory, we expect the keeper to have a genuine beneficial effect.

In general, such sketches of the basic magnets aligned in various ways in various shapes of magnetized material help us to visualize the state of magnetization, though the picture of them is probably too crude and looks too real.

[14] In this, the crude theory of magnetism ranks with structural-formula theory in organic chemistry, which enables us to answer otherwise-meaningless questions about varieties of compounds having the same composition but different chemical properties. For example, benzene, C_6H_6, can form dichlor-benzene, $C_6H_4Cl_2$. If we ask, "How many *different* dichlor-benzenes should there be?" the question seems silly without the illumination of structural formula theory. But that theory says at once, "Three." And there are just three. Then theory continues, in its own terms, and tells us how to determine experimentally *which* of the three we have in a pure sample.

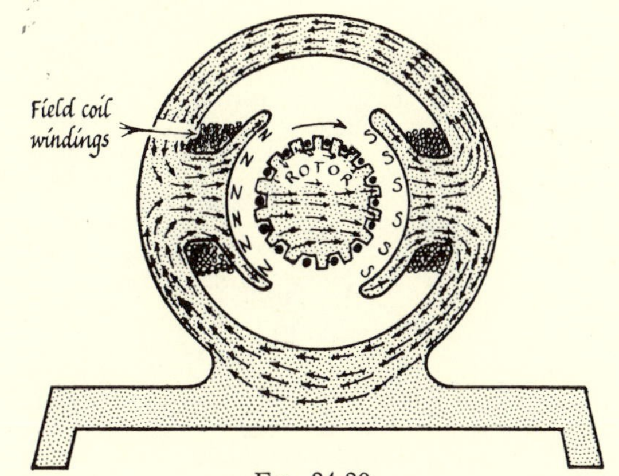

Fig. 34-30.
Magnetization of Soft Iron Frame of Electric Motor

5. QUESTIONS ON MAGNETIC THEORY

(a) Describe what happens when a magnetized bar of steel is cut into short pieces in an attempt to obtain isolated "poles." Show how this experiment suggests a "theory" of magnetism, using small arrows to indicate the magnets—or rather the domains that are now believed to be the basic units involved.

(b) Give sketches illustrating a piece of magnetized steel and a piece of unmagnetized steel; and show how this idea fits with the results of cutting a magnet.

(c) What difference of behavior between soft iron and hard steel must be assumed for the basic magnets to account for observed behavior?

(d) Use the theory to answer or comment on each of the following. Give sketches where possible.

(i) Is there a limit to the strength of magnetization of a steel bar?

(ii) In a bar magnet the "poles" are not quite at the ends, or spread over just the end surfaces. Explain why this is to be expected.

(iii) Heating a magnet tends to demagnetize it. Suggest an explanation.

(iv) Hammering a magnet tends to demagnetize it. Suggest an explanation.

(v) In certain circumstances, hammering a steel bar can magnetize it, even though the hammer used is not magnetic. What circumstances? Explain.

(vi) When bar magnets are stored in a box they are arranged with "keepers" across the ends. Can the keepers really help to preserve the magnetized state of the magnets? What should the keepers be made of?

(vii) Horseshoe magnets also have "keepers." Give a sketch showing their action.

(viii) A man thinks he has magnetized a ring of steel but he can find no poles, and the ring has no external magnetic field. Is it possible that in any reasonable sense of the term the ring is "magnetized"? Explain.

(ix) If your answer to (viii) is Yes, give a test of such magnetization.

(x) A bar of steel or iron is placed in a coil carrying alternating current. The bar is found to heat up. This heating comes from several effects, one being the alternating remagnetization of the bar. What difference in this respect would you expect between soft iron and hard steel?

(xi) A magnet is placed in a coil carrying alternating current and the current is slowly reduced to zero. Explain why this should demagnetize the magnet. Include a sketch of a graph.

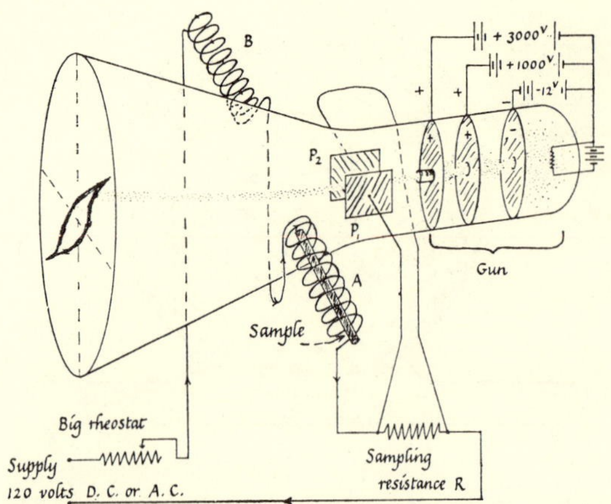

FIG. 34-31. DEMONSTRATION APPARATUS FOR INVESTIGATING MAGNETIZATION OF A SAMPLE OF IRON OR STEEL

The sample is placed in a magnetizing coil, A, which carries a current to make a magnetic field inside the coil. The sample when magnetized produces a magnetic field which deflects the electron beam up or down.

(The current-carrying coil, A, also has an external magnetic field; to avoid this field deflecting the electron beam, we place a "compensating coil," B, on the other side, carrying the same current, so that its field just neutralizes A's field in the region of the electron-beam.)

Thus the up-or-down motion of the electron beam tells us the magnetization of the sample. The electron beam is also moved sideways, by an electric field between plates P_1 and P_2, connected to a resistor, R, which carries the magnetizing current. R obeys Ohm's law, so the P.D. across it varies as the current, and so does the magnetizing field acting on sample. Therefore, the electric field between P_1 and P_2 varies as the magnetizing field acting on the sample. Therefore, the sideways deflection of the electron beam indicates the magnetizing field. Thus, *the electron beam draws a graph of magnetization (upward) vs. magnetizing field (along).*

If a D.C. supply is used and the current is steadily increased from small value by means of the "big rheostat," the increasing magnetization of the sample is shown by the shifting of the spot. If an A.C. supply is used, the supply itself varies the magnetizing current through repeated cycles; so the rheostat is left at a fixed value. The spot draws a curve for a complete cycle of magnetization so quickly and so often that it shows a fixed pattern on the screen.

Experimental Study of Stages of Magnetization

We can investigate the stages of magnetization of a metal bar by placing it in a solenoid and slowly increasing the current in the solenoid's wires. We assume the magnetizing field strength inside the solenoid varies directly as the current (this will be justified later); so we use the size of the current, or a sample from it, to represent the *magnetizing field.* We measure the *magnetization* of the bar by the bar's effect on a small compass needle or by its effect on an electron stream in a cathode-ray oscilloscope. We change the current slowly with a rheostat; or we use an alternating current, which will run the sample bar through a whole cycle of changes 60 times a second. With an oscilloscope we arrange for a sample *electric* field derived from the magnetizing current to swing the electron stream to and fro sideways 60 times a second; and we place the iron sample so that its *magnetic* field swings the electron beam up and down following the sample's magnetization. Thus the electron beam traces out a complete magnetization curve for the sample, plotting magnetization upwards and magnetizing field along.

Starting with an unmagnetized sample, with a magnetizing current increasing from zero, soft iron gives a curved graph showing three stages:

(i) for small magnetizing fields there is some (proportional ?) magnetization;

(ii) the graph bends and rises much more steeply through a middle region in which the response of magnetization seems much greater;

(iii) the graph levels off to "saturation"—the limit of full magnetization.

With an alternating field, soft iron runs up to saturation, back to zero, then to saturation in the reverse direction, and so on. Hardened steel gives a loop, the graph tracing a different path on the return trip—its magnetization lags behind. The sample retains some magnetization even when the magnetizing field is zero—a permanent magnet. This lagging of magnetization behind the magnetizing field is called "hysteresis." The bigger the loop the more "friction" there must be jamming the tiny magnets, and the more waste heat produced in each cycle of magnetization.

Now you can see why a bar magnet is demagnetized when it is placed in a solenoid carrying alternating current and slowly removed. The alternating magnetic field

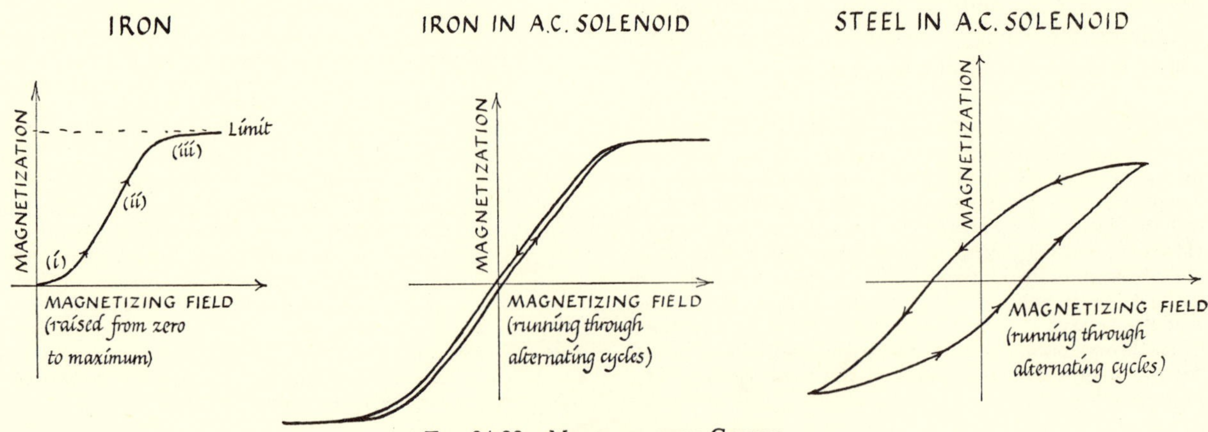

FIG. 34-32a. MAGNETIZATION CURVES

carries the magnet around and around a magnetization cycle 60 times a second. As the magnet is withdrawn from the solenoid it experiences a weaker and weaker magnetizing field, which carries it around a smaller and smaller cycle. Thus the magnetization pattern closes in, cycle after cycle (rather like the layers of an onion) until it reaches a point at the center showing no magnetization.

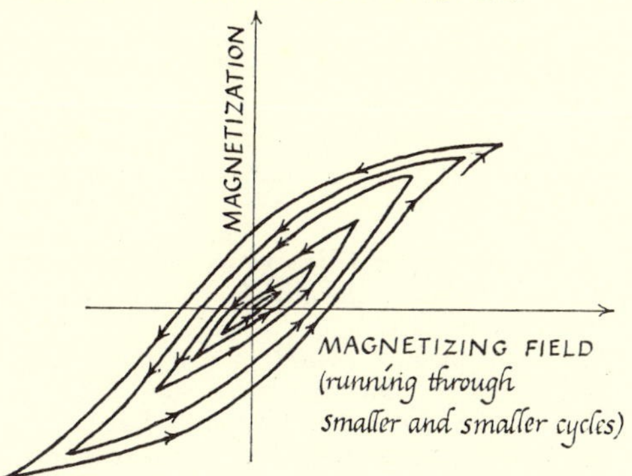

Fig. 34-32b. Steel Bar Being Demagnetized

More Modern Theory: Magnet Domains

So far we have made no clear assumptions about the size of the internal magnets. We have not said what we think they are. In recent years we have obtained clear evidence that the basic magnets are not separate molecules but quite large chunks of metal crystal—tiny sections as we see them under a microscope but huge crowds from an atomic point of view. We call those chunks "domains." Of course, we can cut up a domain into still smaller magnets and eventually reach atoms. So we still think of the ultimate elementary magnets as atoms.[15] We can see the domain boundaries under a microscope if we pour very fine iron powder on the surface—like testing a casting for cracks.

All the metal in a domain seems to be fully magnetized in one direction—usually along one of the main crystal axes of the metal. In unmagnetized metal, the domains have their directions of magnetization arranged indiscriminately up and down each of the crystal axis directions, probably making cyclic families of domains in three dimensions. When the metal is being magnetized two kinds of change occur:

(a) Some domains grow in size at the expense of others, adding atoms to their single-minded bloc, at the expense of other blocs. (Compare with changes in political boundaries.) The domains already magnetized in a direction near that of the magnetizing field are the ones that grow. In weak fields, such changes are small and are reversible—the pattern reverts if the field is removed [Stage (i) in Fig. 34-33].

Stronger fields make permanent changes of boundary. Many favorably oriented domains grow larger and we have a strong magnet [Stage (ii)].

(b) In strong fields, the magnetization of domains twists nearer to the direction of the field. It is the crystal axes of the metal crystals, and not the outside shape of the bar, that define the directions of easy magnetization. The atoms in a domain naturally lock their magnetization along an easy direction. However, the field we apply may not be parallel to any crystal axis. Then very big fields are needed to twist the magnetization of favored domains closer to the field's direction [Stage (iii)].

(c) In some cases, whole domains may switch their direction of magnetization abruptly from an unfavorable one to a favorable one—like a political revolution. This type of change is less common than we used to think.

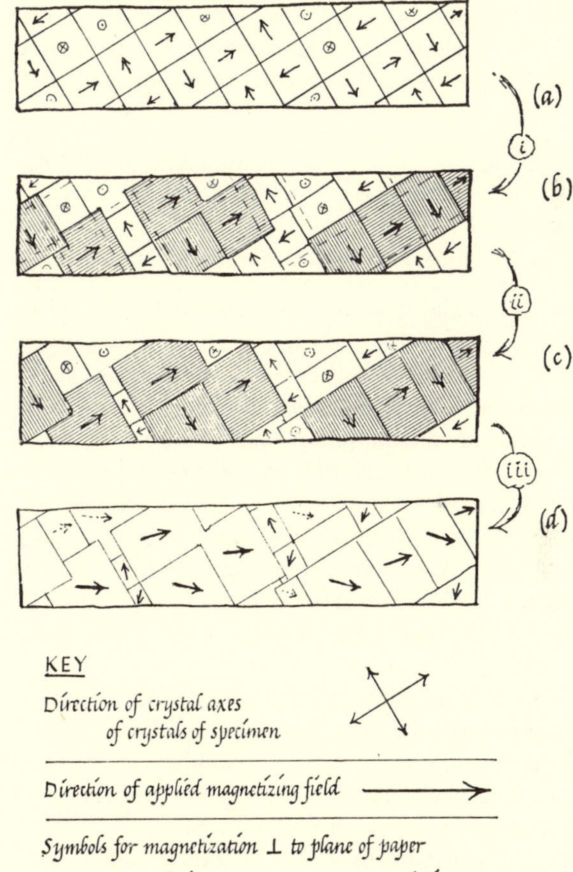

Fig. 34-33.
Magnetic Domains in a sample in stages of magnetization. (This is a simplified schematic picture to illustrate the mechanism of domain changes.)

[15] "Domains are just like nations: we always find people grouped in them, but the ultimate units of mankind are still individual men."—Frederic Keffer

Reality?

Again we seem to have stained our scientific honor by dragging in a new lot of assumptions (in the form of detailed stories about domains) merely to "explain things." We can defend our new form of details by pointing to the experimental evidence of iron-powder patterns which show the boundaries between domains. As we magnetize a bar we can watch the powder patterns shift, showing some domains growing at the expense of others. If you see this, you will indeed consider that it verifies our theory.

There is another remarkable demonstration of domain changes that are too small to see directly but are detectable electrically. We wind a small coil around a sample of iron and connect the coil to an amplifier to show tiny induced voltages when the bar *changes* in magnetization. We connect the amplifier to a loud speaker. As we magnetize the bar by bringing a magnet near we hear a curious hissing noise, like the noise of sand pouring on a drum. This noise is really a rapid series of tiny clicks, just what we should expect from a huge number of domain-jumps. With molecules jumping singly the clicks would be *far* too small and too numerous to make this noise. So the audible noise suggests that the domains are large groups of molecules. We have recently changed our interpretation of this stream of clicks. We used to think each click was a signal from a complete domain switching its magnetization. We now know the clicks are far more numerous than the actual domains in the sample. Each click probably shows a small change of a domain boundary, a reversible growth-&-shrinkage of two neighbors, as in stage (i).

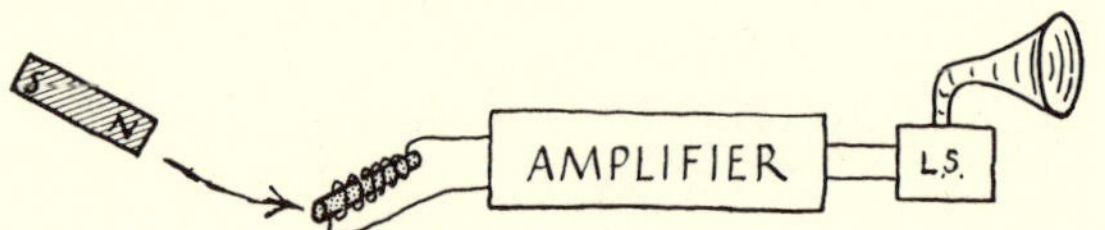

Fig. 34-34. An Experiment to listen for domain changes

In this small part of physics, a piece of theory has proved a useful guide to the experimenter, and a wise friend to the scientific thinker. If you ask, "Is it true?" the scientist first replies, "Well, it is useful." Then he claims, "It is at least partly true." Some of the theory's ideas are certainly true—as you can see for yourself in the laboratory. If some of the ideas seem more fanciful, such as details about magnetic atoms, we would rather ask whether they are helpful than whether they are true. Nevertheless, both our

Fig. 34-35. Analyzing a Stream of Magnetic Atoms

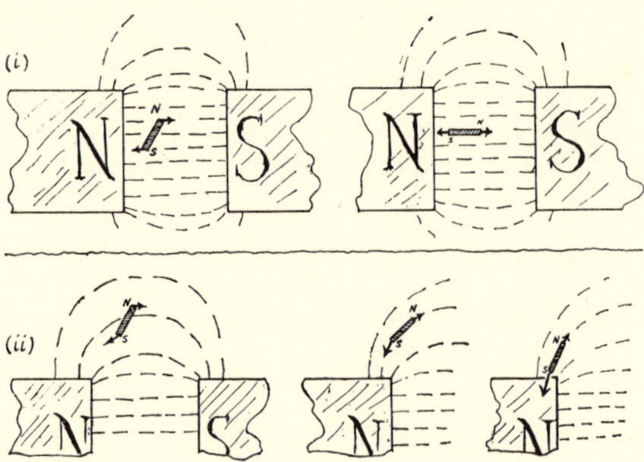

Fig. 34-35a. Some atoms behave as small magnets. To measure the magnetic strength of atoms we use a *non-uniform* magnetic field.

(i) Any small magnet placed in a *uniform* magnetic field twists but does not move along. It is pulled by a couple of equal and opposite forces.

(ii) Any small magnet placed in a *non-uniform* field twists and moves towards the strongest field.

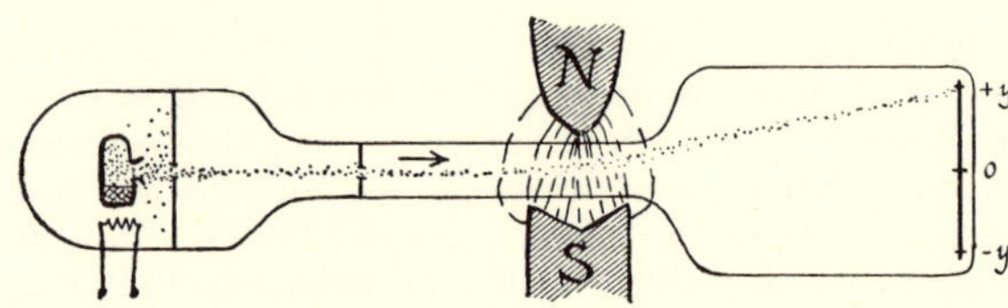

Fig. 34-35b. Experiment to measure the magnetic strength of single atoms. Uncharged atoms, shot from a hot oven in a vacuum, pass through a non-uniform magnetic field and are deflected towards the strongest field. The atoms continue to a recording film and make a mark there.
The surprising result of this experiment is that atoms make several clear marks (e.g., $+y$, 0, $-y$), showing that they are magnets but that they are mysteriously restricted to some fixed orientation—a "quantum" restriction.

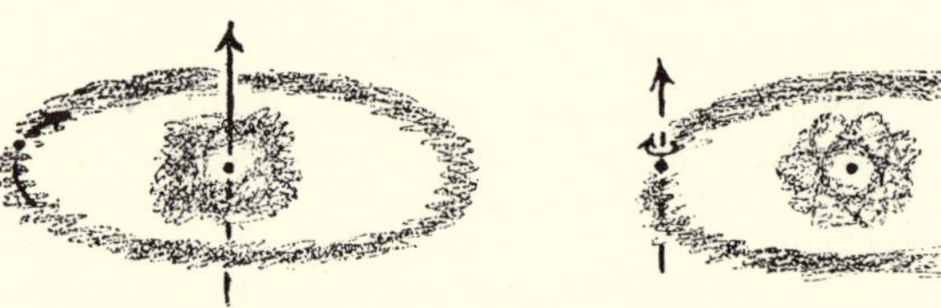

Fig. 34-35c. Magnetic Atoms. (Sketches very fanciful and far too definite.) An electron pursuing some large "orbit" would have a magnetic field like that of a current around a loop of wire, thus making the landlord atom a magnet. However, we also find the electron itself is spinning. That motion too makes a magnetic field; so it contributes to making the atom a magnet. Many atoms are non-magnetic, because the contributions from the atom's many electrons just cancel.

scientific curiosity and our romantic attraction towards atoms make us want to inquire further about the things inside the domains. And we can succeed in such enquiries. We fire streams of isolated atoms through *non-uniform* magnetic fields and find that some individual atoms are themselves magnets (Fig. 34-35).

Again, we can make atoms emit light when they are in strong magnetic fields, so that they tell us still more about their magnetic properties. In the end we infer that electrons, some nuclei, even the chargeless neutron, behave as minute magnets. Each of them has a definite, characteristic magnetic strength, which we associate with some mechanical "spin"—rotational momentum—belonging to the particle. These magnetic properties help us to disentangle nuclear structure.

Recently, streams of neutrons shot at magnetic material have shown a reflection property that depends on domain *boundaries*. We can count the actual number of domains by measuring the reflection of a neutron stream. Thus we now link together old magnetic theory and modern nuclear experimenting with a century of time between them.

In these last details of atoms we have merely announced results and have given no experimental or other basis for believing them; so you should take them merely as interesting anecdotes and as assurance that physicists have been able to carry their useful theory on to much finer detail.

CHAPTER 35 • CHEMISTRY AND ELECTROLYSIS

"I mean by elements . . . certain primitive and simple or perfectly unmingled bodies; which not being made of any other bodies, or of one another, are the ingredients of which all those called perfectly mixt bodies are immediately compounded, and into which they are ultimately resolved."

—Robert Boyle, 1661

"We cannot be sure that what we now regard as an element is in fact one. All we can say is that such a substance is the present limit of chemical analysis and cannot be split up further, so far as we know at present."

—Antoine Laurent Lavoisier, 1789

[This chapter is intended for readers who have not studied chemistry; it offers a background of chemical knowledge for their study of atomic physics in Part Five.

If you have not taken a chemistry course: read this chapter lightly to gather a general feeling for chemical changes, atomic weights, atomic numbers, ions.

If you have taken a chemistry course, your knowledge goes much farther and deeper than this chapter. If you read this chapter, take it with a grain of salt.]

CHEMICAL KNOWLEDGE

[Describing chemistry in one chapter is like describing a grocery store in one breath—"They-have-food-in-cans-arranged-on-shelves-. . . ." However, here we need only a summary of some of the facts and general ideas of chemistry and shall not set forth much of its great wealth of knowledge or its army of skillful techniques. Nor shall we follow the marvelous history of its emergence from crude alchemy, or the terrific forging of its knowledge in a century of logic and experiment. Here we shall announce[1] the results—the clear view with which chemists started the present century.]

[1] In this course, we have good reasons for teaching chemistry thus, by the "handout" method, to avoid delaying our study of physics. If we were using chemistry itself as our sample of science, such an approach to it would indeed be poor teaching. It would give you misleading ideas of science and the attitude taken by scientists.

Sadly enough there is a temptation to teach a beginning course of any science with a "handout" treatment that announces facts and conclusions. The beginner does need to acquire a considerable stock of the factual knowledge before he can discuss and enjoy the theory: so teachers are tempted to deliver information in the first round, to be learned as necessary equipment, and then everything will make grand sense in the second round. Then, for students who do not reach the second round, the first round gives a poor picture of science as a collection of information instead of a scheme of knowledge that makes sense. That is why we try to avoid such a treatment. Yet, when we offer a summary of another area of science—chemistry in this book, or nuclear physics in a chemistry book—we have to break our resolve and give information without explaining how it is obtained. This note is inserted as an apology and a warning.

Chemical Manufacture and Methods: Synthesis

Chemical artisans, and then chemical scientists, have learned how to manufacture many materials. In their manufacturing, chemists use "physical methods" such as dissolving in water, filtering out solids, evaporating to dryness, distilling, electrolysis. . . . And they use "chemical methods" such as heating two substances mixed together to produce a different substance, making interchanges occur in explosions or flames, producing a gas by mixing solutions of solids, ... The categories overlap, roasting a rock in air looks like physical treatment, but, when a gas is driven off, it may be produced by chemical changes.

The beginning of chemical knowledge dates from the beginning of civilization: the treasured records of craftsmen and alchemists were the early cookbooks of chemistry. One concern of chemistry today is super cookery, with analysis as well as recipes, reasons as well as rules. "To make an omelet," reads the ordinary cookbook "take four eggs, a dab of butter, a spoonful of salt." Apart from the vagueness of units, these are scientific instructions for putting together an omelet, for its synthesis. In chemistry many substances are built up by synthesis: for example, water by exploding a mixture of oxygen and hydrogen gases; carbon dioxide by burning coal in air; sulfuric acid by burning sulfur in air, adding some extra oxygen (with platinum to promote joining) and mixing the resulting sulfur trioxide with steam.

Analysis

We should find the reverse of omelet-making, picking the ingredients apart, or *analysis*, much harder, even before the omelet is cooked. But the chemist would want to carry the reverse process still farther, picking the omelet apart into its ingredients and then analyzing each of these into still simpler pure substances such as salt, water, ... These latter, chemical compounds, are the same however much subdivided. Egg yolk is not a single compound but a mixture to be separated into several compounds; but salt is already a single chemical compound. Salt keeps its *physical* properties right down to the smallest sample we can see. The smallest speck makes the same cubic crystals; has the same density, same melting point; bends a ray of light in the same way; and, as a fine test from physical chemistry, it has the same proportional effect in lowering the freezing point when dissolved in water—an effect which gives an estimate of molecular mass. Yet all these are gross tests: salt might still be a clever mixture like glass. It is *chemical* tests that lead us to assurance of a single compound, and on down to its unique molecule. Here are two examples among many: salt dissolves in water to make brine which tastes salty (a chemical test), which turns cloudy when mixed with solutions of silver salts (a test for chlorides). Every such test with salt runs the same way whether we use large samples or small, whether we recrystallize or otherwise refine the salt once or many times. As we go to smaller and smaller samples, and even more dilute solutions, we meet practical difficulties but there is no evidence of a change of nature. Viewing salt's universal chemical behavior over all the range we can handle, we extrapolate our belief down to the single molecule of salt. When we get to behavior that *is* different, we think we have split the molecules into atoms of constituent elements.

Although we have separated mixtures into simple uniform substances like salt, we need to go further, in building a science of chemistry, and tear the compounds apart—if they are compounds—into the ultimate elements of which they are made. These are the *elements* of which Robert Boyle wrote 300 years ago (see the heading of this chapter).

Returning to our specimen of a pure substance, salt, we ask: have we reached the limit; or can salt be taken apart into still simpler substances that *are* elements? Is salt a compound?

Electrolysis: a Tool for Separating Elements

Yes, we *can* tear salt apart by using more violent methods: for example, *electrolysis.* (This means passing an electric current through a solution or other liquid to release products at the electrodes where the current enters and leaves.) Melt the salt and drive an electric current through the molten liquid. Two new products appear, each with utterly different properties from salt. At one electrode chlorine bubbles out, a greenish, poisonous, choking gas; at the other electrode sodium is released, a silvery-grey metal. That is a complete break-up not a further separation into, say, some salt and some sand. Sodium and chlorine are the *elements* of which salt is composed. Elements are the simplest substances into which materials can be split up: they defy further separation. Once these elements are separated and their properties examined, we have basic building blocks for the synthesis of all kinds of materials. Common salt separates into the elements sodium and chlorine always in the same proportions: every 117 kg of salt will yield 46 kg of sodium and 71 kg of chlorine gas. The fact that the proportions of the constituents sodium and chlorine are always the same in all samples of salt, 46 : 71, is characteristic of a chemical compound; and here it is strong evidence that we are dealing with a simple compound. Such evidence also points to atoms as convenient fictions if no more: we may imagine sodium made up of myriads of equal atoms and chlorine made up of myriads of atoms all alike but 1½ times as massive as the sodium atoms. Then salt would consist of family groups (molecules) each having one sodium atom joined to one chlorine atom by some mysterious attraction. And if the masses of the two types of atom are in the ratio 46 : 71—as we nowadays know they are—it is easy to understand why salt always holds the two ingredients in these fixed proportions. Both the general idea of atoms and the mass ratio 46 : 71 would be unjustifiable assumptions if they just explained common salt. But when we find this fixed-proportion behavior holding true again and again, from one substance to the next, the atomic idea gains support and usefulness. We begin to look for the same characteristic "atomic masses" in every chemical change—and we find them. Then we are led to a framework of chemistry that can make predictions and interpretations—in fact, good science.

Mixtures . . . Compounds . . . Elements

Understanding the distinctions between mixtures, compounds, and elements forms a useful beginning for chemical knowledge, so we shall make a fresh start with a review of examples.

Shovel sand, salt, iron filings, and sawdust into a heap and mix them well. You can mix them in any proportions; and, given a magnet and a bucket of water, you can separate them again. They make a *mixture.* If the sand consists of big pebbles, it is a *coarse mixture* from which an ingredient can be picked by hand. If the sand and other particles are all very small, it is a *fine mixture* that can still be sorted out by hand under a microscope—or you can use more cunning methods.

Make still finer mixtures by dissolving salt in water or mixing water and alcohol, or melting zinc and copper together to make brass. These are *molecular mixtures* that cannot be sorted out by hand. Air is such a mixture of gases: oxygen, nitrogen, carbon dioxide, and a little helium, etc. The proportions in any of these mixtures are quite indefinite; you can mix as you like over a wide range. You can still separate the ingredients: evaporate the water and let the salt crystallize out; distill the alcohol; plate out the copper with an electric current; liquefy the air and boil off its gases separately. Now take each of these "ingredients" and ask whether it too is a mixture. Salt? alcohol? water? copper? . . . . All show constant properties, of density, melting point, crystal form, ... None shows any signs of being a mixture of variable proportions that could be sorted out by some molecule-cataloging demon. Each of these has the essential characteristic of a *compound* (or of an element): whatever ingredients *it* is made of are there in fixed proportions. Compounds are single substances that run true to form, from big chunks down to the last molecule—in fact, that is the definition of a molecule: the smallest individual particle of a substance that keeps the characteristic properties of that substance. A molecule is a definite group of atoms, held together by electrical forces—always the same group for the same chemical compound.

Try to analyze salt into simpler ingredients: gentle methods such as heating or freezing make no permanent change, but electrolysis of molten salt converts it to chlorine gas and sodium metal. Electrolysis of water also produces a complete change, to two gases: oxygen, in which burning things glow with a brighter flame; hydrogen, the lightest of all gases, which makes an explosive mixture with oxygen. Can these new "ingredients" be broken up into still more constituents? All attempts at bringing about permanent disruption fail: heating to high temperature, repeated electrolysis, even violent attacks by adding other chemical substances—all fail.[2] So we call these simplest substances, sodium, chlorine, oxygen, hydrogen, *elements.* Iron and copper are elements, too, and so are carbon, mercury, aluminum, iodine, but not brass and air, which are mixtures, not water and salt, which are compounds. Throughout chemistry, with all its linkings and unlinkings and transformations, we never change one element into another: elements are the ultimate ingredients of all chemical processes.[3]

"The formation of one substance from others by chemical change, then, is possible only if its elements are present in the other substances. Never expressed as a law, this statement is nevertheless the fundamental axiom which distinguishes chemistry from alchemy."[4]

Returning to our analysis of common salt, how do we know that sodium and chlorine are elements, the ultimate ingredients, and not just sub-groups? Though we now know for sure, we must admit that the evidence for elements came slowly, at first as negative evidence that nobody had ever succeeded in splitting them up into simpler substances still. Some substances emerged from the alchemists' furnaces long ago with the reputation of elements: gold, silver, lead, ... Much later when air was separated into several gases the nitrogen and oxygen proved unsplittable; they could show *gains* of weight by combining with other substances, but not subdivision into lesser weights of more elementary ingredients. What we now call carbon dioxide proved to be splitable into oxygen gas and black carbon (or even clear diamond); and it can easily be synthesized from these ingredients (by burning). Water was for a long time thought to be an element until Cavendish split it up into oxygen and hydrogen gases. Then it was labeled HO, meaning an atom of oxygen combined with an atom of hydrogen in each molecule of water. It was some time before

---

[2] Chemical attacks may produce new substances, but only by addition. For example, sodium is attacked by chlorine or oxygen, but in each case the product is a white crust whose mass is the sum of the mass of metal and mass of attacking gas used up. That is a putting-together to make compounds again, not a further splitting apart. Thus, careful weighing rescued early chemistry from confusion.

[3] In this century we have learned how to change the nuclei of an element's atoms, by bombarding them with high-energy atomic particles from accelerators (see Ch. 42 and Ch. 43). We can thus turn atoms into quite different atoms, reducing chemistry's simplicity to a new profusion. However in ordinary talk of atoms we prefer to maintain the chemical view, since atoms are not changed by ordinary chemical or physical treatment.

[4] K. B. Krauskopf, *Fundamentals of Physical Science,* 1st edn. (New York: McGraw-Hill, 1941).

chemical evidence and clever reasoning forced scientists to realize that it must be H_2O.

For a long time in the development of chemistry, repeated failure to split was the only guarantee chemists had that they had reached the ultimate simple elements. Then a century ago a systematic shelving system was devised for the elements, by Mendeleef and others. This scheme, now called the periodic table, will be described briefly later in this chapter. It provides a sort of family tree for the elements, arranging them by atomic mass and by chemical properties. It is a magnificent, illuminating guide, so every chemistry text gives it and uses it—rather as some mystery-stories give a family tree of people to guide the reader. All the substances believed to be elements now fall into remarkably clear order in this scheme, showing that earlier guesses of their nature were indeed correct.[5] Nowadays, the frequencies of spectrum lines—both X-rays and visible light—provide unique, identifying evidence of an element.

Chemical Changes or Reactions

As samples of chemical reactions—what a modern chemist might regard as borrowing and sharing of electrons between atoms—here are some changes, starting with carbon dioxide. Carbon dioxide, CO_2, is a dense colorless gas made by burning carbon with oxygen, or formed (together with water) when food fuels such as sugar are burnt or broken up in living bodies. In the burning, atoms of carbon pick up pairs of oxygen atoms to form heavy molecules, CO_2. This gas cannot normally release its oxygen for burning other things—its oxygen atoms are gripped too strongly—so it will extinguish a lighted match or even a large fire. It is only slightly poisonous; it harms living creatures rather by stifling, preventing the normal gas exchange in their lungs. It dissolves in water making a liquid with a faint sour taste ("carbonated water," popular in compounding drinks). There it makes a loosely-knit compound called carbonic acid ($CO_2 + H_2O = H_2CO_3$). This could also be called "hydrogen carbonate," since the group CO_3 (which does not normally exist by itself) is called carbonate. If we add common salt, sodium chloride, to the solution we have a gorgeous mixture of things in the solution. Electrical experiments on water solutions suggest that when in solution such compounds are largely broken up into electrically charged atoms or groups of atoms called ions. In this mixture we seem to have some sodium$^+$ ions, some chlorine$^-$ ions, some hydrogen$^+$ ions, and some CO_3^{--} ions. We cannot say whether we have sodium chloride and hydrogen carbonate or sodium carbonate and hydrogen chloride, or some mixture of all four, because each pair of molecules makes the same four ions when dissolved in water. There is probably a continual joining up of each combination and splitting again, giving all these molecules a transient existence.

The discussion is irrelevant, unless we can remove one of the compounds or ingredients and upset the swimming balance; and that is hard to do because all these ingredients remain dissolved. Now suppose we mix solutions of another carbonate and another chloride, sodium carbonate (washing soda) and calcium chloride (the white lumps that people put on sidewalks to melt ice). Again, the solution will contain a mixture of ions: sodium ions, carbonate ions, calcium ions, and chlorine ions. Again, we ask: have we got sodium carbonate and calcium chloride or sodium chloride and calcium carbonate, or some indeterminate compromise? In this case one of the possible groupings is insoluble. Chance collisions among ions will form some calcium carbonate molecules, and these are not soluble in water. They gather as specks of white powdery chalk which fall to the bottom of the solution as a solid "precipitate." This irreversible change continues, with more and more calcium carbonate being deposited, until the supply of ingredients runs out. The chalk can be filtered out, dried and sold to someone wanting pure chalk. Here is a piece of chemical manufacture; but it would be a silly and unsatisfactory one unless we want specially pure chalk, because chalk is available directly—we can dig it out of a hillside. Furthermore, such natural chalk was probably used in making the original materials for our "factory."

Starting with chalk, we can split it up by roasting it into carbon dioxide gas and lime. Lime is calcium oxide, a white "rust" of calcium metal that may form on it in air. 100 kg of chalk give 56 kg of lime and 44 kg of carbon dioxide gas. This is useful manufacture, not for the carbon dioxide but for the lime which is needed in agriculture and in chemical manufacture. Lime is slightly soluble in water, giving a dilute bitter solution containing calcium ions and, instead of the expected oxygen ions, "hydroxyl"

[5] For a long time there were still uneasy feelings. Even in this century, one element was "discovered," to fill a known gap, with what turned out later to be the wrong chemical properties. Then Bohr predicted the right properties for an element to fill the gap—with the help of his picture of atomic structure (Ch. 44). Then a careful search among suitable minerals produced the actual element, hafnium.

ions $(OH)^-$, made by some interchange with water on dissolving.

Having broken up the chalk by roasting—equivalent to giving its molecules smart blows with molecular hammers—we can put it together again. Lime + carbon dioxide rejoin to make chalk; but this is difficult to promote with solid lime in bulk—the inside is inaccessible. It is better to dissolve the lime in water and bubble carbon dioxide gas through the solution. White clouds of chalk are formed.[6]

Chemical Formulas and Equations

Already things are getting complicated, difficult to visualize or to remember. Chemists simplify things by using shorthand equations showing how atoms of elements combine or exchange places. They express their belief in the indestructibility of matter, and the unchangeable nature of elementary atoms, by making sure that for each element they write the same number of atoms on both sides of the equation. Atoms never get lost: the equations "balance." Each atom is represented by a letter (or two letters where necessary to avoid ambiguity). They use C for an atom of carbon, O for oxygen, Ca for calcium, Cl for chlorine, H, N, S, for hydrogen, nitrogen, sulfur. An atom of sodium is written Na, using the old Latin name for salt, *natrium*; and an atom of copper Cu for *cuprum*. A compound of one atom of calcium and one of oxygen, "calcium oxide" (lime), is written CaO. If a molecule of some compound contains two atoms of an element a 2 is written after that symbol,[7] as in CO_2 or H_2SO_4.

Here then are the changes already mentioned:

$$C + O_2 = CO_2$$

[carbon + oxygen make carbon dioxide]

[One atom of carbon combines with one molecule of oxygen (two atoms) to make one molecule of carbon dioxide.]

You will presently see good reason to think that in oxygen, hydrogen, chlorine, and many other gases, the atoms pair up into molecules with two atoms in each molecule, C_2, N_2, etc.: and it is considered good form to write them thus (good wisdom too: a single H instead of H_2 would mean an uncombined raw atom of hydrogen, freshly manufactured, a vicious little fellow with a dangerous chemical bite):

$$CO_2 + H_2O = H_2CO_3$$

[One molecule of carbon dioxide combines with one molecule of water, when it dissolves, to make a molecule of hydrogen carbonate.]

We find that H_2CO_3 sometimes splits up into hydrogen ions H^+, H^+, and CO_3 ions, CO_3^{--}.

$$H_2CO_3 \rightleftarrows H^+ + H^+ + CO_3^{--}$$

Mixing solutions of sodium chloride and hydrogen carbonate gives us an unexciting clear mixture. But mixing solutions of sodium carbonate, Na_2CO_3, and calcium chloride, $CaCl_2$, gives us a cloudy precipitate of chalk.

$$Na_2CO_3 + CaCl_2 = CaCO_3 + 2NaCl \text{ (see note 8)}$$

A careful chemist, knowing that in solution the substances are often split into ions would write his equation with electrically charged ions, and he would replace the = sign by arrows thus $\rightleftarrows$. That would show that the exchange might proceed in either direction. He would notice that one of the four groupings, $CaCO_3$, is insoluble, so he would expect the reaction to proceed in this $\rightarrow$ direction only. $Na_2CO_3 + CaCl_2 \rightarrow CaCO_3 + 2NaCl$

Now we can take our newly made chalk, $CaCO_3$, and roast it.

$$CaCO_3 \rightleftarrows CaO + CO_2$$

[chalk yields lime and carbon dioxide]

Here the arrows run both ways, since the reaction can run either way, according to the temperature.

Lime dissolves in water to make calcium hydroxide solution

$$CaO + H_2O \rightarrow Ca(OH)_2$$

This "lime water" will react with carbon dioxide bubbled into it thus:

$$Ca(OH)_2 + CO_2 \rightarrow CaCO_3 + H_2O$$

The OH, hydroxyl, group is characteristic of bitter caustic alkalies, though lime water itself is a very

[6] This provides a simple test for carbon dioxide. Try it. Make some lime water by shaking up lime and water, allowing the excess lime to settle and pouring off the clear solution. Breathe through a straw into this solution and watch the cloudiness grow. Air without carbon dioxide will not do this. Now continue blowing in air containing carbon dioxide for some time and watch for further changes. Chemistry is not so simple. . . .

[7] In chemical compounds, adjacent symbols are meant to be added, of course, not multiplied as in algebra. The law firm of Huggins, Huggins, Smith, Osborne, Osborne, Osborne, and Osborne would be written by a chemist as H_2SO_4.

[8] As usual, in chemical equations, we do not mention the water in which our reacting substances are usually dissolved. The same large number of water molecules is taken for granted on both sides of the equation. They do not affect the final product, though they facilitate the reaction by encouraging the formation of ions—and thus they play an essential part.

mild alkali. Put some sodium metal in water and it hurls a hydrogen atom out of that harmless liquid molecule and makes a strong alkali.

$$Na + H_2O \rightarrow NaOH + H$$

[sodium and water make sodium hydroxide and hydrogen]

Since bubbles of hydrogen emerge it is better form to double this equation and say

$$2Na + 2H_2O \rightarrow 2NaOH + H_2$$

Sodium hydroxide ("caustic soda") makes a slimy caustic alkaline solution that will neutralize acids

$$2NaOH + H_2SO_4 \rightarrow Na_2SO_4 + H_2O$$

[Sodium hydroxide (caustic soda) + sulfuric acid (hydrogen sulfate) form sodium sulfate + water]

Acids

All acids are compounds that contain hydrogen and non-metals (chlorine in hydrochloric acid, sulfur and oxygen in sulfuric acid). Their hydrogen becomes detached H^+ ions in solution: that is their essential characteristic; and that gives them their sour taste. (In contrast, alkalis contribute OH^- ions in solution.) Acids have sour taste—a dangerous test; they neutralize alkalis, producing water and heat; they react with some metals to produce hydrogen; and most of them drive carbon dioxide out in a stream of bubbles from a solution of a carbonate.

We can make a simple acid directly by mixing hydrogen gas and dense green chlorine gas in equal volumes, 2 : 71 by weight, and starting the explosive reaction with a flash of ultraviolet light.

$$H_2 + Cl_2 \rightarrow 2HCl$$

[hydrogen + chlorine explode to form hydrogen chloride or hydrochloric acid gas].

Hydrogen chloride is a gas that dissolves easily in water to make a strong acid solution with a sour taste. This is the ordinary form of "hydrochloric acid," really a solution. It is used in a number of chemical manufacturing processes, including human digestion.

The easy way to manufacture carbon dioxide in a chemistry laboratory is to pour hydrochloric acid on chalk.

$$CaCO_3 + 2HCl \rightarrow CaCl_2 + H_2O + CO_2$$

If the CO_2 is bubbled into sodium hydroxide solution, the hydroxide gathers it in to form carbonate. (If you prefer, regard the CO_2 as forming "carbonic acid," H_2CO_3, which then neutralizes the alkali NaOH.)

$$2NaOH + CO_2 \rightarrow Na_2CO_3 + H_2O$$

Sulfuric acid can be manufactured as follows, from raw sulfur dug out of the ground:

$$S + O_2 \rightarrow SO_2$$

[sulfur burns to form sulfur dioxide]

$$2SO_2 + O_2 \rightarrow 2SO_3$$

[sulfur dioxide takes on an additional oxygen atom]

In the last reaction we place the gases in contact with platinum. The platinum acts as a middle-man to expedite the reaction, and emerges unscathed at the end. We call it a *catalyst*. This reaction proceeds slowly to the right if the platinum is cold and at very high temperatures it proceeds from right to left. So we should write it

$$2SO_2 + O_2 \rightleftarrows 2SO_3$$

and we must choose a suitable temperature for the reaction to run fast in the forward direction if we wish to run an acid factory. Then:

$$SO_3 + H_2O \rightarrow H_2SO_4$$

[sulfur trioxide combines with water to form sulfuric acid]

Alkalis

Alkalis are substances which on dissolving in water yield $(OH)^-$ ions ("hydroxyl" ions). They are usually metal oxides or hydroxides. Most metals when exposed to air develop a coating that we call oxide: iron develops rust, copper tarnishes; calcium gathers a white crust of lime; sodium grabs oxygen and moisture so fast that it must be kept submerged in oil. Some of these oxides have no reaction with water, others dissolve or react with water to form alkaline hydroxides whose $(OH)^-$ ions are responsible for their characteristic bitter taste, burning attack on skin, and neutralizing action with acids. Examples: sodium hydroxide, NaOH; calcium hydroxide, $Ca(OH)_2$; and ammonium hydroxide, NH_4OH, in which the group NH_4 behaves like a metal.

When solutions of acid and alkali are mixed, H^+ ions and OH^- ions join to form water, with a considerable evolution of heat. If the right proportions are mixed, the product is neutral[9] and shows none of the strong properties of acid or alkali. (Of course if the wrong proportions are used an excess

[9] "Neutral," in chemistry, means neither acidic nor alkaline.

of one ingredient remains, with its characteristic properties.)

$$HCl + NaOH \rightarrow NaCl + H_2O + \{heat\}$$

[hydrogen chloride (hydrochloric acid) + sodium hydroxide (caustic soda) make sodium chloride (common salt) + water]

$$H_2SO_4 + 2NaOH \rightarrow Na_2SO_4 + 2H_2O + \{heat\}$$

[sulfuric acid (hydrogen sulfate) + sodium hydroxide make sodium sulfate + water]

Both these are examples of the general rule:

alkali + acid make salt + water + {heat}

Here "salt" includes many neutral compounds, with common table salt as an example. (Other examples: washing soda, copper sulfate, calcium chloride.) Salts are neutral but they have the *structure* of an acid, with a metal atom taking the place of the acid's essential hydrogen.

When sodium hydroxide reacts with the weak "fatty acids" that we meet in surface tension studies, the "salt" they form is soap. The actual process of soap manufacture is less direct.

ATOMS IN CHEMISTRY

The Evidence for Atoms

In our specimens of chemical changes we have been taking atoms for granted. In every equation the symbol for an element (such as H or Cl) means "one atom of" (hydrogen or chlorine). Of course it might be just a vague description (such as *some* hydrogen, a *little* chlorine), but in writing equations with each symbol appearing equally on both sides we have implied atoms and their conservation. We are sure that there *are* atoms of each element that can be exchanged between one molecule and another but never created or destroyed. That sure knowledge comes from accurate measurement of the ingredients in chemical changes. Nowadays, the easiest way to split compounds into elements for measurement is by electrolysis. Pass a current through water, and hydrogen and oxygen bubble out in proportions 1 : 8 by weight. And the water slowly decreases—very slowly because water is so much denser than the gases. (For every cubic meter of mixed gases only ½ kg of water is consumed: and these gases weigh just ½ kg; water : hydrogen : oxygen keep the proportions 9 : 1 : 8.) Water can be analyzed by other methods; e.g., when steam is passed over red hot iron, iron takes the oxygen and leaves hydrogen: 1 kg of hydrogen for every 9 kg of water used. Again, fluorine will grab hydrogen from water and leave oxygen. In every case we find the proportions run, by weight:

water, 9, yields hydrogen 1 + oxygen 8.

The same weight proportions hold in the reverse change, the synthesis of water. If we let 1 kg of hydrogen explode with 8 kg of oxygen, we obtain just 9 kg of water. (If we start with other proportions the excess of one ingredient is left over unused: e.g., 3 kg of hydrogen + 8 kg of oxygen → 9 kg of water + 2 kg of hydrogen.)

As we said earlier, electrolysis of molten sodium chloride produces chlorine gas and sodium in proportions 71 : 46 by weight. Burning carbon[10] with oxygen produces CO_2 with proportions 3 : 8 of carbon to oxygen. If the air supply to the burning carbon is reduced, we can make another gas, light, poisonous, colorless, explosive when mixed with air. We call this carbon monoxide, and find it contains a different proportion, 3 : 4 carbon : oxygen. Such constant proportions for a compound fit easily with the idea of uniform atoms for each ingredient element, but this evidence does not become compelling until we look for relations between the proportions in several different compounds. Look at the table.

Name of Compound	*Proportions by Weight*				
	H	O	Cl	Na	C
water	1	8			
hydrogen chloride	1		35.5		
sodium chloride			35.5	23	
sodium carbonate		48		46	12
carbon dioxide		32			12
sodium hydroxide	1	16		23	

When we look at the proportions of elements in many compounds, we see a pattern of proportions that suggests atoms very strongly. If all hydrogen is made of identical light atoms, oxygen of atoms 8 or perhaps 16 times as massive, chlorine of atoms 35.5 times as massive, sodium 23 times, etc., we can picture every compound molecule in turn built of such atoms. Clearly, in some compounds, an element must contribute several atoms to a molecule. (*If* sodium hydroxide is NaOH, water must be H_2O. If water is HO and not H_2O, sodium hydroxide must be NaO_2H.)

This use of a *quantitative, experimental* atomic theory to simplify chemical knowledge was a great

[10] This is difficult on a small scale, because the carbon must be made hot enough to start the burning, and then fed with air or oxygen at the rate it needs; but not with a blast that cools it too much. On a big scale this is easy: it is simply a coal fire.

advance in science. It was put forward with compelling clarity by John Dalton ($\sim$ 1808). Elaborations and improvements came fast, with careful measurements and discoveries of more elements by Berzelius, Davy, and many others. We can make the same relative mass serve for an element's atoms in every chemical activity. It is the force of that agreement among measurements that makes us really sure of atoms—and by that stage of reliable atomic masses the whole idea is too useful to give up even if the atoms prove to be an illusion!

If this idea is right, if there are simple, uniform atoms that gang together into molecules, we may guess at a further relation: if some elements combine together to make more than one compound, their proportions in those compounds should be very simply related. This is the Law of Multiple Proportions, one of the few laws that was first deduced theoretically instead of being extracted from experiment. Look at some examples:

Name of Compound	*Proportions by Weight*				
	H	O	Cl	Cu	C
water	1	8			
hydrogen peroxide	1	16			
there are two copper chlorides					
"cuprous" chloride			35.5	63.5	
"cupric" chloride			71	63.5	
carbon dioxide gas		32			12
carbon monoxide gas		16			12

(In this table we have used numbers that make the simple relationships show clearly. The actual measurements do not reveal the simple ratios straight away.)

If water is H_2O, cuprous chloride is CuCl, carbon dioxide is CO_2, do the others fit with simple formulas?

Not even this kind of evidence proves that there *must* be atoms—and at least one distinguished chemist maintained his right to disbelieve in them until practically the beginning of this century. Only modern measurements with atomic particles (Brownian motion, Millikan's *e*, *e/m* for electrons and ions, alpha-particle scattering) make atoms seem an absolutely necessary picture of matter.

We cannot assume that the weighings give the relative masses of *single* atoms. Is the mass of an oxygen atom 8 times that of H, or 16? A common compound of carbon with hydrogen is methane gas with carbon : hydrogen 3 : 1 by weight. Is methane, CH, with carbon only 3 times as massive as hydrogen? Chemists decided that it must be CH_4, in which case its carbon : hydrogen ratio should be written $(4 \times 3) : (4 \times 1)$, 12 : 4, with the carbon atom 12 times as massive as H. How was that important decision made? The case of methane can be settled by a clever chemical trick: substitute chlorine for the hydrogen. That can be done by treating methane with chlorine. If it is CH it must turn into CCl. If it is CH_4 it could swap one hydrogen for chlorine or 2 or 3 or 4. And, in fact, four distinct compounds with chlorine are found: the first is CH_3Cl, a dense gas; the next CH_2Cl_2, a denser gas; then $CHCl_3$, a liquid (chloroform); and the last is CCl_4, carbon tetrachloride, a poisonous fluid.

The proportions by weight run:

	Proportions			*Same Proportions Revised*		
	H	C	Cl	H	C	Cl
methane	1	3		4	12	
methyl chloride	1	4	11.8	3	12	35.5
dichloromethane	1	6	35.5	2	12	71
chloroform	1	12	106.5	1	12	106.5
carbon tetrachloride		3	35.5		12	142

Once you have guessed the pattern and put the proportions in the revised form it is clear that carbon should have 12 times the mass of H. (Unless, of course, methane is really CH_8, requiring 24 for carbon—but then we should expect still other chlorine substitutions, such as CH_7Cl, and we do not find them. We have other lines of evidence that confirm our choice of 12 completely.)

Now look back at water. If it is HO, hydrogen and oxygen atoms have masses 1 : 8. Why do we now say water must be H_2O and assign oxygen 16? When water was first found to be made of oxygen and hydrogen, the composition HO was adopted. Then came Avogadro's suggestion which helped so greatly to bring atomic chemistry to its present order.

Gas Volumes in Chemical Changes

Modern weighings and good arithmetic reduce the mass-proportions of elements to fairly simple numbers, but long ago the volumes of gases made a story that was more obvious and even simpler. If we explode hydrogen and oxygen at, say, 100° C and keep the product (water) in the form of steam, we find that:

> 2 quarts of hydrogen explode with 1 quart of oxygen to make
> 2 quarts of steam.

Again, 1 quart of chlorine and 1 quart of hydrogen explode to form
2 quarts of hydrochloric acid gas.

And, 2 quarts of carbon monoxide burn with 1 quart of oxygen to form 2 quarts of carbon dioxide.

There are dozens of simple volume stories like that. (Of course we must make all measurements at the same temperature and pressure, say atmospheric.)

Avogadro

Soon after Gay-Lussac pointed out the simple volume proportions for gas reactions, the Italian scientist Amadeo Avogadro made a brilliant guess (~ 1813) that (at any chosen temperature and pressure) *equal volumes of different gases contain the same number of molecules*: the same number of molecules in a box of gas, *whatever the gas*. He reasoned this guess out so well that if you read his original account you will think his conclusion inescapable. Nowadays physical arguments (see Ch. 30) assure us it is trustworthy and we call it *Avogadro's Law*.

Apply Avogadro's Law to a quart of chlorine (N molecules) combining with a quart of hydrogen. The argument is shown in Fig. 35-1.

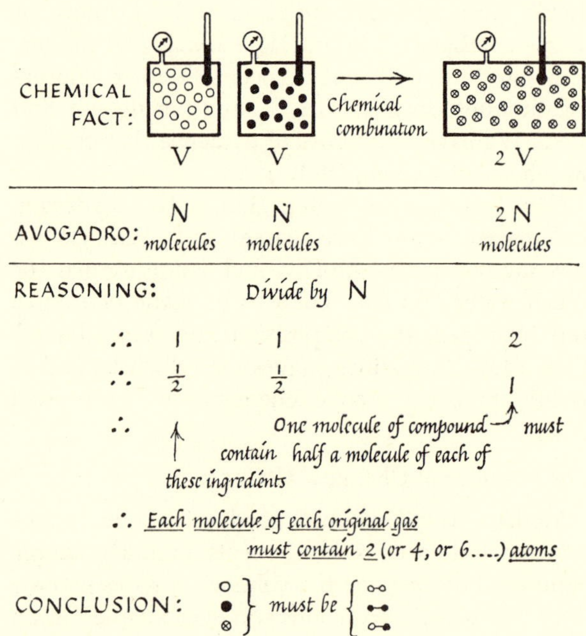

Fig. 35-1. Molecular Argument

Trusting the law, and the experimental facts, we reason that *one* molecule of hydrogen chloride contains *half* a molecule of hydrogen and *half* a molecule of chlorine. *Both those molecules must be halvable.* A hydrogen molecule must be H_2 or H_4 or H_6 and not H. Taking the simplest choice we say it is H_2—and that is supported by evidence of the type we used for methane; yet no purely chemical experiment can completely rule out H_4 or H_6 as the right answer. In the same way chlorine gas molecules must be Cl_2. *What can you conclude by a similar argument on the carbon dioxide reaction? What does the water reaction show?* You will find yourself compelled to write water H_2O.

Again and again, whenever a chemical reaction involves gases, in the initial ingredients or in the products, we find a simple ratio between the gas volumes 1 : 1, 2 : 1, 3 : 2, . . . (provided we measure them at the same temperature and pressure). Linking this experimental observation with Avogadro's rule, we feel sure we are seeing forming or reforming among gas molecules: 1 molecule makes 1 molecule, 2 make 1, 3 make 2, ...

Atomic Weights

Thus accurate weighing—with great help from gas volumes—brought the valuable idea of atoms to full use: the idea *that each element has atoms all alike, with a characteristic mass; and that molecules are standard family groups of atoms.* It was the quantitative simplicity provided by the atomic theory that made it so valuable in the development of chemistry. And for full quantitative use, chemists needed to know accurately the masses of different elementary atoms—or rather their relative masses on some arbitrary scale. They determined these masses by careful weighing of ingredients of compounds.

The relative masses, on a scale that takes the mass of a hydrogen atom as 1, are called *atomic weights.* Weighings gave 8 : 1 for O : H in water, the argument from gas volumes showed the water molecule must be H_2O; so the atomic weight of oxygen must be 16. From the proportions 23 : 16 : 1 in sodium hydroxide, the formula must be NaOH, and the atomic weight of sodium must be 23.

By 1810 Dalton had recognized 20 elements, and had rough values for their atomic weights. Since then great experimental skill and careful logic have gone into the measurement of atomic weights. A century ago, about three quarters of our present list of a hundred elements were known, and their atomic weights had been measured with considerable accuracy.

If we define a chemical compound as a substance having fixed proportions of ingredient elements, we must not be surprised to find that experiment shows that the proportions of elements in any compound we choose are fixed! What are surprising and important are the experimental facts that:

(1) There are an enormous number of such compounds, each with constant proportions of ingredients[11] and definite physical and chemical properties, whenever it is found, however it is made. (If there were only a few substances fulfilling this description, chemistry would be quite different, perhaps helped by the greater simplicity, perhaps overburdened by a wealth of mixtures.)

(2) There are very simple relationships among the proportions by weight[12] of ingredient elements in different compounds, suggesting that every atom of an element has the same characteristic mass, irrespective of its state of chemical combination. The miracle is that all chemical compounds fit into the same scheme, each element showing the same constant atomic weight in all its compounds.

How could the chemists, building their list of atomic weights, be sure they had not missed a factor of ½ or 2 or 3? Water might be HO, in which case oxygen would have atomic weight 8, or sodium hydroxide might be Na_2OH, making the atomic weight of Na 12.5 (after all, sodium sulfate *does* have Na_2 in its molecule, Na_2SO_4); or it might be $Na(OH)_2$, making Na 46. There were such doubts in some cases, but chemists developed good ways of deciding the proper factor. The argument with gas volumes settles oxygen and water. An empirical rule, the Law of Dulong and Petit, can help to decide the atomic weight of sodium. The rule says that, for most[13] solid elements

$$\text{ATOMIC WEIGHT} \cdot \text{SPECIFIC HEAT} \approx 6$$

The specific heat of sodium is about 0.3. Multiplying by possible atomic weights:

$$0.3 \times 12.5 = 3.7 \quad 0.3 \times 23 = 6.9 \quad 0.3 \times 46 = 14$$

When the chemical table was developed, a mistaken choice could be seen easily—there would be no vacancy on the shelf near the wrong atomic weight. And by the early part of this century, mass-spectrographs (Ch. 38) had been developed to measure actual atomic masses, atom by atom. With atoms of known masses—e.g. hydrogen and oxygen—mixed in with the sample as reference standards, the mass-spectrograph provides an irrefutable scale.

From known atomic weights we can calculate the *molecular* weight of any compound, if we know its chemical formula. The molecular weight is the mass of one molecule on the scale used for atomic weights. (Do not confuse this with a *mole*, which is a large gross sample of a compound which can be defined thus: "find the molecular weight of the compound, and take that number of grams." In this course, we use a kilo-mole: ". . . . that number of kilograms.")

If we do not know the chemical formula, but the substance is a gas, we can easily find its molecular weight (and thence its formula) experimentally. We have only to weigh equal volumes of that gas and of hydrogen, then apply Avogadro's Law and find at once the ratio of masses (compound molecule): (hydrogen molecule). Since we know the hydrogen molecule, H_2, has mass 2 on our scale (by definition), we then know the compound molecule's mass, the compound's molecular weight. That in turn may tell us the atomic weight of one of its ingredient elements.[14] For the inert gases such as helium, which make no chemical compounds, gas density gives us the only way of measuring their molecular weight, which for them is the same as their atomic weight.

Besides their great practical uses in quantitative analysis and chemical manufacture, atomic weights have been of great value in the development of chemistry. The periodic table was first set up by

[11] Thinking ahead in terms of isotopes (Chapters 38, 40, 43), we ask "is this entirely true?"

[12] There is nothing to tell us, in simple chemistry, just how small atoms are, though there are clear signs that atoms must be *very* small and numerous: microchemistry, handling billionths of a kilogram follows ordinary chemistry exactly and shows none of the bumpiness that would appear if one molecule less or more made a weighable difference. Evidently, the smallest pieces of matter we can weigh or handle directly contain huge numbers of atoms—as we now know (Ch. 30 and Ch. 36), millions of billions.

[13] Kinetic theory predicts the rule; quantum theory explains the exceptions. See Ch. 30 and Ch. 44.

[14] For example: weighings of samples give, at room temperature and atmospheric pressure,

density of hydrogen, H_2 = 0.0836 kg/cu. meter
density of hydrochloric acid, HCl (?) = 1.52 kg/cu. meter

$$\text{Then } \frac{\text{molecular weight of HCl}}{2(=\text{mol. wt. of } H_2)} = \frac{1.52}{0.0836}$$

$$\therefore \text{ molecular wt. of HCl} = 2 \times 1.52/0.0836 = 36.4$$

So, if the formula HCl is correctly chosen, the atomic weight of chlorine must be 36.4 − 1 or 35.4

Another example: we can predict gas densities correctly, without direct measurement. Suppose we wish to compare CO_2 with air. Air is ⅘ N_2 (molecular weight 28) and ⅕ O_2 (molecular weight 32); so air has an average molecular weight of about 28.8 and it must be about 14.4 times as dense as hydrogen (molecular weight 2). Carbon dioxide, CO_2, has molecular weight $(12 + 2 \times 16)$ or 44. So its density is 44/28.8 times that of air. It is 1½ times as dense—no wonder it settles into a suffocating pool on the floor in some caves.

CO, with molecular weight $(12 + 16)$ or 28 is a trifle lighter than air.

arranging elements in the order of increasing atomic weights. And the actual values of atomic weights (on the scale $H = 1$) showed a remarkable peculiarity: most of them were whole numbers. Less than ten years after Dalton set forth his atomic theory of chemistry, a British physician, William Prout, pointed out that peculiarity. He suggested it showed that every kind of chemical atom might be made up of basic units, each a hydrogen atom. The idea appeals not only to the liking for simplicity that is natural to scientists but also to the tradition of Greek philosophy. It is a tempting idea; look at a list

H	C	N	O	Na	S
(1)	12.0	14.0	16.0	23.0	32.0

There were enough examples, even in Prout's day, to make his hypothesis promising.[15] But soon exceptions appeared to spoil it: Cl 35.45 and Cu 63.5 were measured and re-measured. Prout's suggestion was doomed, a bright idea but. . . . It waited a century for an unexpected rescue (see Ch. 38 and Ch. 40). That rescue was no trivial affair. With it came new insight into the nature of atoms, and from it came predictions of "atomic energy."

Valency

Once we are sure of the atomic weights of the elements, analysis will tell us the proportions of different atoms in any compound molecule. For example, analysis of hydrogen chloride tells us it contains equal numbers of hydrogen and chlorine atoms; analysis of methane gas tells us it has four hydrogen atoms for every carbon atom; acetylene gas has equal numbers of hydrogen and carbon atoms; and aluminum chloride contains three chlorine atoms for every aluminum atom. Then we know the proportions, but we still do not know how many atoms of each kind the molecule contains. For example, hydrogen chloride might be HCl or H_2Cl_2 or H_3Cl_3, or . . . ; methane might be CH_4 or C_2H_8 or . . . ; acetylene might be CH or C_2H_2 or . . . ; etc. To settle that question, to find the actual molecular formula of a compound, we must measure its molecular weight.

If the compound is a gas, we just measure its density and trust Avogadro. For example, density measurements show that:

hydrochloric acid gas, hydrogen chloride, has molecular weight 36.4; so it is HCl
methane has molecular weight 16; so it is CH_4
acetylene has molecular weight 26; so it is C_2H_2

[15] We now reckon atomic weights with $O = 16$ as standard. That makes $H = 1.008$, but it brings most of the lighter elements' atomic weights even nearer to whole numbers.

When solids dissolve in a liquid, they produce some gas-like effects. For example they produce an extra pressure, on a suitable membrane, that can be measured ("osmotic pressure"); also they change the liquid's freezing point and boiling point slightly. Each of these effects is proportional to the number of molecules (or ions) of the dissolved solid. If we measure one of these effects we can obtain our needed estimate of molecular weight.

With a knowledge of molecular formulas, we can describe chemical-compound-forming in a new way, with the useful idea of *valency*. We see simple molecules of two atoms, HCl, NaCl, H_2, and we may fancy each atom is equipped with a single arm or hook to join with the corresponding arm or hook of the other atom. We might write the atoms H— Cl— Na— and the compounds H—Cl, etc. But we see H_2O, $CaCl_2$, CaO, . . . and these seem to require two arms on oxygen and calcium atoms, —O— (or O=), . . . We call the "number of arms" on each atom its *valency*. Hydrogen, chlorine, sodium, all have valency 1; they are *monovalent*, and so are some groups such as (OH). Oxygen and calcium have valency 2, and so does the sulfate group (SO_4). Aluminum has valency 3 (its chloride is $AlCl_3$). Carbon has valency 4 as shown by methane, CH_4, and by CO_2.

Carbon turns 4-valent behavior to an enormous variety of uses; the long chains of —CH_2— groups sketched in Chapter 6 are only one type of carbon compound out of a thousand, a few examples out of a million—so we shall return to "organic" compounds.

Some atoms seem to have a choice of valencies: copper has valency 2 in copper sulfate, $CuSO_4$, and in one of its oxides CuO, but it also forms Cu_2O, acting as a monovalent element. Again, nitrogen has valency 3 (e.g. in NH_3, ammonia gas) or 5, or sometimes 1 or 6.

The valency is a characteristic of an element's normal behavior that is useful in remembering and explaining chemical changes and in drawing structural pictures of complicated molecules. Here we have given an old-fashioned, crude account of the general idea of valency. Modern chemists distinguish several kinds of valency or rather types of *bonds* formed between atoms. They explain these bonds in terms of electrical forces: holding positive and negative ions together, or in other cases holding an electron shared between two atoms.

THE CHEMICAL TABLE

The Great Periodic Table

With the growth of chemical knowledge—properties of the elements and their atomic weights, properties of vast numbers of compounds, etc.—there was a great need for some overall scheme of arrangement. Think of a young stamp-collector faced with a profusion of stamps from a single country. He might devise an arrangement of rows and columns, with a row for each year of issue and a column for each denomination. Then he could not help noticing common properties, such as "all the 1-cent stamps are green," and he might be tempted to make rash predictions. His scheme would not be very useful—merely tidy. The chemical elements present a much greater and more promising problem. Lining them up *in order of atomic weights*, Mendeleef (~ 1869) and others saw some general chemical properties repeating at fairly regular intervals—like postage stamps beginning with 1-cent green in each new issue. Chemical properties are not as obvious as stamp colors; so if you have not studied chemistry you may not be impressed by the startling likenesses that recur. Here, instead of trying to convince you with chemical discussions, we shall describe some of the final arrangement and give examples of its virtues. See the table, two pages ahead.

Starting with the lightest elements, hydrogen (1), helium (4), lithium, beryllium, boron, carbon . . . in order of atomic weight, we have a series of entirely dissimilar light elements. So, leaving the primitive hydrogen and helium by themselves we string the next eight elements out along a line. After them we find the later elements repeat many of the properties of the beginning ones, so we write the next eight in a second line under the first, thus starting columns of "families" or element-groups having similar properties. We do this again and again as we proceed through the heavier elements, not always making the break after eight, but sometimes running out a longer string, guided by the chemical properties themselves.[16]

When all the arranging is done, we find one column contains all the "alkali metals": lithium, sodium, potassium, rubidium, cesium, and the recently discovered francium. These are soft, active metals which react violently with water, shoving out hydrogen and forming a caustic alkaline slimy solution. The atom of each is monovalent—has one chemical arm—so that it joins with one-armed chlorine to form simple chlorides such as sodium chloride. Physically all these metals form good surfaces for photoelectric cells (electric eyes)—light seems to find it easy to whip an electron off their atoms. We now regard these chemical and physical behaviors as closely connected; the electron that is easily whipped off by light is easily borrowed by another atom such as chlorine.

Lower down, the first column develops a second sub-family, of less active metals, that alternate with the first. We find copper, silver, gold, the "noble metals." The second column contains metals with valency 2, such as beryllium, calcium, barium, strontium, and radium, with insoluble sulfates (e.g. $CaSO_4$, plaster of Paris, blackboard-chalk; $BaSO_4$ used in medical X-raying to cast shadows of digestive tracts). Here, too, there is a second sub-family: zinc, cadmium, mercury (with soluble sulfates), also with valency 2.

To the right of the table we find another column with violent elements, this time *non-metals*; fluorine, chlorine, bromine, iodine. All combine easily with hydrogen to form an acid; all will scorch your skin off—although chlorine and iodine are helpful as germicides, if diluted enough. They show gradations of density and violence down the family column, with increasing atomic weight. Fluorine is a light, yellowish, corrosive gas, a wild spitfire that attacks almost everything, from platinum to human flesh. Chlorine is a heavy green poisonous gas, which can wreck your throat and lungs. Bromine, the redhead of the family, is a corrosive red-brown liquid, easily turning to brown vapor, ready to scorch your skin if you touch it. Iodine, more sedate, makes dark crystals at room temperature but can be warmed into a purple vapor. Each of these combines with potassium, sodium, calcium, . . . every metal, to form a salt.

The middle columns do not show such obvious common traits to a non-chemist; but to the expert judge they too are well ordered. The middle column of all begins well: carbon, silicon, both non-metals, both with valency 4, making CH_4, SiH_4, with hydrogen, and CO_2, SiO_2 (sand) with two-armed oxygen. But then the family seems to divide into two lines of half-brothers: titanium, zirconium, hafnium, . . . ; and germanium, tin, . . . lead; both branches grow more metallic in their later generations.

The inert gases, helium, neon, argon, . . . were unknown when the table was constructed. When they were discovered, some thirty years later, a new column had to be opened up for them. But once

[16] We might do better to write the elements around a cylinder, as if on a spiral tape. With such an arrangement the beauty of the scheme is even clearer.

given a home they make a consistent group, a whole family of chemical ne'er-do-wells, utterly inert, unable to make compounds with anything. We now call them the "noble gases." They even fail to join in pairs themselves. Unlike molecules of oxygen, hydrogen, etc., which are H_2, O_2, etc., their gas molecules buzz about as single atoms. This last comment seems an outrageous piece of romancing. If these gases make no compounds, if they indulge in no chemistry, how can we possibly settle their atomic weight and show that it is the same as their molecular weight? We have excellent assurance. In Chapter 30 you have seen how a physical measurement (specific heat) combined with reliable theory guarantees that their molecules are single atoms.

This chemical table is useful as a scheme for remembering properties and coordinating studies, but it also has some of the greater virtues of a good theory or conceptual scheme. It has been able to predict the properties of undiscovered elements where there were gaps. And it raises questions—from minor ones over mistakes in placing[17] to major questions of how to picture atomic structure in a way that explains the periodicities shown in the table itself.

Fig. 35-2 gives part of the table. Consult chemistry texts for complete versions and fuller discussions and for the marvellous stories of early research to fill in the gaps by using the table itself to predict the properties of missing elements.

ATOMIC NUMBERS

Atomic Numbers in the Chemical Table

The chemical table was first of all made by placing the elements in order of increasing atomic weights. Once a profitable arrangement emerged from this, it seemed natural to tag the elements with serial numbers in that order—just as a stamp collector might number all the varieties of stamps serially.[18]

Chemically, these are useful numbers, and we have added them above each element in our table. (Chemistry books will tell you these numbers have a much more exciting significance: numbers of electrons in atoms, etc., but that interpretation comes from measurements in atomic physics which we shall discuss in later chapters.) On this basis, we label hydrogen element #1, helium #2, . . . , and so on up to uranium #92.

For a long time there appeared to be neither any reason for stopping at 92 nor any hope of finding elements beyond. Today, we can make elements 93, 94 . . . by bombarding other atoms. There is a new "nuclear chemistry" of amazing prospects. We can even look farther and guess at the properties of atoms far beyond any we already know. Like the new ones we do have from 84 to 102, they will be unstable (radioactive)—so unstable that we are not surprised to find them absent in nature.

Atomic Numbers Today

Nowadays we regard the atomic number of an element as its most essential label. We know it is the amount of positive charge on the atom's nucleus, measured in electron charges. As such, it tells us how many electrons the neutral atom holds around its nucleus. The arrangement and binding energies of those electrons are determined by the nuclear charge that holds them; and since chemical reactions depend on changes and exchanges among the outermost electrons of atoms, chemical properties are determined by the nuclear charge, which is shown by the atomic number. An atom's innermost electrons, held strongly by the nuclear charge, are barely affected by chemical changes except in the very light atoms with few electrons—moving other atoms near to form a chemical compound cannot bring strong enough forces to play on the innermost electrons to affect their behavior noticeably. However, the innermost electrons are concerned when X-rays are emitted (or absorbed). Remove the innermost electron from an atom (e.g. by bombardment), and as a neighboring electron moves in to the innermost place the atom emits an X-ray quantum *whose wavelength is characteristic of the atom's nuclear charge*. Thus, by comparing the X-rays from targets made of various elements, we can determine the elements' atomic numbers. This method does not involve measuring atomic weights of many elements to arrange them in order and then hoping we have missed none; it is not affected by the element being tied up in a compound with others. It reveals atomic number, the nuclear charge, with absolute clarity, a unique number for each element.

Atoms of radioactive elements have unstable

[17] One discrepancy of placing occurs in the region shown in Fig. 35-2. The inert gas argon and the active alkali metal potassium have to be placed in the reverse order of their atomic weights, 40 and 39. Many a careful measurement was made, with the obvious hope, but failed to shift the order. Is this a trivial failure or a major blow to the whole scheme? It could not discredit the scheme completely because there is so much already to the scheme's credit. But it was a serious "exception" which, from chemical evidence, promised to remain a sore thumb. Yet we now have physical evidence to explain it away, or rather to make the whole problem disappear. See Ch. 38.

[18] There is a system of serial numbers for stamps called "Scott numbers." Atomic numbers are the "Scott numbers" of the elements.

THE PERIODIC TABLE OF CHEMICAL ELEMENTS

— METALS — — — — — — ·					- - — — — NON-METALS			
FAMILY I	FAMILY II	FAMILY III	FAMILY IV	FAMILY V	FAMILY VI	FAMILY VII	FAMILY VIII	FAMILY O
Subfamily A / Subfamily B; Alkali Metals / Noble Metals	Subfamily A / Subfamily B	A (B)	A (B)	(B) A	(B) A	(B) A Halogens,	Metal Triplets	Inert Gases
1 Hydrogen m = 1.008								2 Helium m = 4.00
3 Lithium m = 6.94	4 Beryllium m = 9.02	5 Boron m = 10.82	6 Carbon m = 12.01	7 Nitrogen m = 14.01	8 Oxygen m = 16	9 Fluorine m = 19.0		10 Neon m = 20.2
11 Sodium m = 23.0	12 Magnesium m = 24.3	13 Aluminum m = 27.0	14 Silicon m = 28.1	15 Phosphorus m = 31.0	16 Sulphur m = 32.1	17 Chlorine m = 35.46		18 Argon m = 39.94
19 Potassium m = 39.10	20 Calcium m = 40.1						26 Iron, 55.8 27 Cobalt, 58.9 28 Nickel, 58.7	
29 Copper m = 63.6	30 Zinc m = 65.4		In the middle region of the table the family likenesses are less obvious			35 Bromine m = 79.9		36 Krypton m = 83.7
37 Rubidium m = 85.5	38 Strontium m = 87.6						44 Ruthenium, 101.7 45 Rhodium, 102.9 46 Palladium, 106.7	
47 Silver m = 107.9	48 Cadmium m = 112.4		For the full table see any Chemistry text.			53 Iodine m = 126.9		54 Xenon m = 131.3
55 Cesium m = 132.9	56 Barium m = 137.4						76 Osmium, 190.2 77 Iridium, 193.1 78 Platinum, 195.2	
79 Gold m = 197.2	80 Mercury m = 200.6		82 Lead m = 207.2					86 Radon m = 222
87 Francium	88 Radium m = 226.0	89 Actinium m = 227	90 Thorium m = 232.1	91 Protoactinium	92 Uranium m = 238.1	93 Neptunium	94 Plutonium	

Fig. 35-2. The Periodic Table of the Elements

atoms. Although they have all properties of an element—definite chemistry, no sign of separating into still simpler ingredient elements, unique atomic number—they do not keep the same characteristic properties forever. Without warning, one atom of a stock-pile of a radioactive element changes to quite a different element . . . then another atom changes . . . then another. . . . After some time we have less of the original (parent) element and a corresponding amount of the different (daughter) element. The daughter (which may also be radioactive) is a perfectly proper element occupying a different place, in a different column, in the chemical table. As the change occurs, a minute bullet is hurled out of the atom, an alpha-particle or a beta-ray (and often a gamma-ray) carrying huge energy. These "radiations" were the first properties of such unstable atoms to be noticed, and they led to the name "radioactivity" for the breaking-up process. (See Chapters 39 and 43.)

An alpha- or beta-particle hurled out carries an electric charge from the atom's nucleus, thus changing its nuclear charge, and therefore its atomic number: hence the shift to a new place in the chemical table. (Of course, the chemical reason that makes the new atom deserve a place in the different column is that with a different nuclear charge it holds a different number and pattern of electrons and therefore has different chemistry.)

Radioactivity would make us revise the earliest definition of an element as never changing, but our labeling by atomic number saves us any trouble. Only an element has an atomic number; any compound will simply show the atomic numbers of its ingredient elements.

Until this century, atomic weights seemed equally good labels for elements. Chemical analysis and weighing carried out with great care gave completely consistent results—always the same atomic weight for an element. This seemed the natural thing to expect for atoms—something to be taken for granted. Then came two slightly different atomic weights for lead extracted from two kinds of rock. Could there be two kinds of lead atom, heavy and light? And then the mass-spectrograph and its forerunners showed that most elements come as a mixture of atoms with several distinct atomic weights. The atomic weight measured chemically is only an average of several atomic weights, mixed together in proportions which seem to be practically the same all the world over. What do the real atomic weights look like? How do new measurements of them affect Prout's discredited idea? See later chapters. Here is one more reason for trusting atomic numbers rather than atomic weights to characterize elements.

ELECTROLYSIS

Electrochemistry

Electric currents flow easily through water-solutions of acids, alkalis, and the salts that they make. Some electrically charged carriers seem to be available in such solutions. We name the carriers that we guess are there *ions*, from the Greek word for "travel." In many solutions, any small voltage drives a current without trouble or delay, so it seems likely that the ions are already swimming around in the water, ready to carry current when pushed by an electric field. In fact we believe that in some solid crystals + and − charges are separated on ions: e.g., a sodium chloride crystal is a cubical lattice of Na^+ ions and Cl^- ions. The sodium and chlorine atoms have already exchanged an electron and become ions with charges whose attractions hold the crystal together.[19]

Run an electric current through several tanks of solution in series, the current entering and leaving by suitable wires ("electrodes"). Try a solution of copper sulfate ($CuSO_4$) with copper electrodes; a tank of water (+ acid) with inert electrodes of platinum or carbon; a tank of brine with special electrodes to catch the products and prevent them reacting chemically with water. Then some material is released at each electrode as the current flows. If we run the current for twice the time, we collect twice the mass of products; if we double the current, we double the mass of products. The mass released at each electrode varies directly as CURRENT · TIME or as ELECTRIC CHARGE carried across the tank. But the mass of a product released is a measure of the number of atoms released;

$\therefore$ NUMBER OF ATOMS $\propto$ CHARGE CARRIED ACROSS
or CHARGE $\propto$ NUMBER OF ATOMS

This last version suggests that there are elementary carriers, atoms or groups of atoms, each carrying some standard charge—"atoms" of electricity riding on atoms of matter.

Furthermore, the masses of products released at different electrodes by the same charge passing through showed very significant proportions. If the current releases 1 kilogram of hydrogen, the same current for the same time releases 23 kg of sodium or 35.4 kg of chlorine or 108 kg of silver. These masses are proportional to the products' atomic weights. Therefore, all these masses released by the same electric charge contain the same number of

[19] This is not just a romantic guess but a good inference from X-ray exploration of atom-layers in salt crystals. X-rays reflected from slanting layers of a sodium chloride cube give a diffraction pattern that shows they have met a multi-deck sandwich of unequal layers, alternate layers being richer in electrons to scatter X-rays. These must be the layers of chlorine, atomic number 17, much richer than sodium #11. Repeating the experiment with *potassium* chloride, chemically similar with crystal structure almost identical, Sir William Bragg found a surprising difference. The slant layers gave a diffraction pattern that showed they are all *exactly* alike, all equally rich in electrons. Look at the atomic numbers of potassium and chlorine in the chemical table. They are the numbers of scattering electrons in neutral atoms. What must have happened, if all the atoms in this crystal have exactly the same number of electrons? And what kind of atoms would the resulting particles resemble in their grouping of electrons? (Look at the table again.) Would you expect them to try to change back or to remain as they are in the crystal?

atoms. Divide by that number of atoms, and we have a single atom carrying the same charge across in each case. This suggests that these ions are *single atoms* all carrying equal electric charges, the same charge for every atom of all these elements.[20] We exhibit that idea by labelling the ions with + and − signs to show one standard charge (actually, as we now know, one electron charge), e.g.:

H^+ Na^+ Cl^- Ag^+

But, while 1 kg of hydrogen is released in the electrolysis of water, only 8 kg of oxygen are released, not 16; and only 31.8 kg of copper (atomic weight 63.6) are deposited in the copper sulfate tank. This suggests *either* half-atoms of oxygen or copper carrying the same charge, *or* whole atoms with double charge. Whole atoms carrying double charge prove to be the more comfortable choice. Then we label the copper ion Cu^{++}. Half as many of these ions with double charge are needed to carry the same total charge across as, say, hydrogen ions. (We reserve judgment over O^{++} for the moment. In that case, the oxygen collected is a secondary product, and the actual carriers are not oxygen ions.) Notice that oxygen and copper have valency 2. Perhaps all ions from atoms with valency 2 carry double charge. Electrolysis measurements with calcium or zinc salts agree with Ca^{++} and Zn^{++} for the carriers. Try aluminum with valency 3 and atomic weight 27 and you will find that for the CURRENT · TIME that would release 1 kg of hydrogen we do not get 27 kg of aluminum but only 9 kg. The aluminum ion is Al^{+++}.

Some ions are groups of atoms. Water splits into ions H^+ and $(OH)^-$; caustic soda into Na^+ and OH^-; sulfuric acid into H^+ and H^+ and SO_4^{--}. Pure water is poor in ions: most of it remains molecules of H_2O, or groups thereof. So, to electrolyze water easily we provide more ions by adding H_2SO_4. This provides many more H^+ ions and many SO_4^{--} ions. When the SO_4^{--} ions arrive at the electrode, they give up their two negative charges and at the same time react with neighboring water to yield oxygen and H_2SO_4 (which breaks up into ions again).[21]

$$H_2SO_4 \rightarrow H^+ + H^+ + SO_4^{--}$$

$$SO_4^{--} \rightarrow SO_4 + 2\text{ charges}^- \qquad H^+ + H^+ \rightarrow H_2 + 2\text{ charges}^+$$

$$SO_4 + H_2O \rightarrow O + H_2SO_4$$

$$H_2SO_4 \rightarrow H^+ + H^+ + SO_4^{--}$$

and so on

Thus the acid is not used up: there is as much there at the end of a long electrolysis as at the start, but there is less water, and oxygen and hydrogen have appeared.

Electrolysis and Ions

If copper sulfate solution is electrolyzed, the Cu^{++} ions arriving give up their charges and become uncharged copper atoms which are deposited on the receiving electrode (whatever that is made of). At the other electrode, SO_4^{--} ions arrive, and if the electrode is copper they give up their charge and attack it to form copper sulfate. Then the overall effect is just a transfer of copper from one electrode to the other, delivering pure copper there, with no net loss of copper sulfate. The "products" are copper and removal-of-copper.

Thus we find solutions of acids, salts, etc. contain ions such as:

H^+	Na^+	$(NH_4)^+$	Cl^-	$(OH)^-$
Cu^{++}	Ca^{++}	Zn^{++}	$(SO_4)^{--}$	
Al^{+++}				

When a P.D. is applied to such a solution, it makes an electric field which drives the ions through the solution. Ions with a + charge move in the conventional current direction (from the battery + towards battery −). Ions with a − charge drive the opposite way. Both kinds of ion help to carry the current—the motion of their charges *is* the current if you like. Driven by the electric field, they drift with these opposite motions (usually at different speeds) while they also share the random motion of the water molecules. (Very near an electrode, ions of one sort are cleared away by the field and the ions of the other sort have to hurry up to carry the whole current.) As each sort of ion reaches its destination-electrode, it gives up its charge (loses or gains electrons in actual fact). The charge continues around the rest of the circuit, and the uncharged atoms that are released do whatever their own chemical nature makes them do when in contact with metal electrode and water. For example, copper ions (which are blue in solution) migrate to the electrode and pick up two electrons to become neutral atoms of pink copper which stick to the electrode. But sodium ions, which swim undisturbed

[20] The observations certainly do not *prove* such a simple story is true. The best we can say is that since the story offers the simplest interpretation we choose to take this view unless it is upset by further experimental findings. Such a choice is good science so long as we remember that we made it as a choice of convenience.

[21] If the electrodes are of lead, the SO_4 combines with lead to form lead sulfate—part of the action in charging a storage battery. If the electrodes are of copper, the SO_4 attacks the copper on arriving: copper sulfate is formed (and ionizes) instead of oxygen and sulphuric acid.

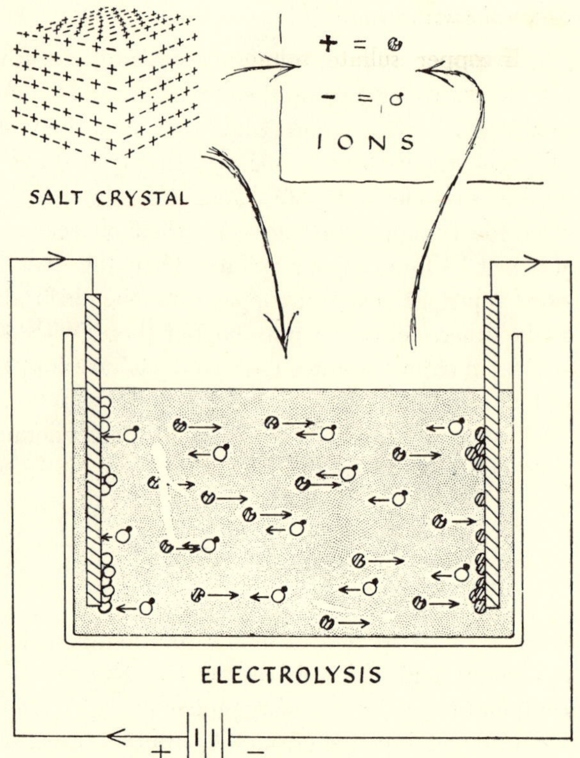

Fig. 35-3.
Salt, already ionized in the crystal, provides ions to carry a current in solution when an electric field is applied. (Sketch of salt crystal based on picture by P. R. Rowland, in *Science News* No. 15, March 1950)

in water, become atoms of sodium metal which react violently with water, throwing out hydrogen and producing caustic soda.

Faraday

Much of our understanding of electrolysis and ions was built up by Michael Faraday, from practically nothing, in the early days of experiments with electric currents. In a masterly series of experiments, guided by fine reasoning, he investigated electrolysis (which he named) and reduced the complicated observed behavior to two simple laws, which will always be known by his name. We have already discussed them informally. The laws he stated ($\sim$ 1833) can be put thus:

I. Whatever the nature of the solution, whatever the nature of the electrodes, the mass of substance liberated at an electrode varies directly as CURRENT·TIME or QUANTITY OF ELECTRICITY.

II. For the same QUANTITY OF ELECTRICITY the masses of substances liberated (at different electrodes) are proportional to the CHEMICAL EQUIVALENTS of the substances, that is, proportional to the values of ATOMIC WEIGHT/VALENCY.

Ions in Solution

There is good evidence that the ions involved in electrolysis are already there as soon as the solution is made. The dissolved molecules do not wait to be torn apart by the electric field that we apply to drive the current. Here are three lines of evidence:

(i) No extra P.D. is needed to start up electrolysis. (The only extra voltages observed are those of battery-effects due to the *products* setting up local batteries around the electrodes. Electrolysis of copper sulfate with copper electrodes shows no battery effects and its tank obeys Ohm's law; but with inert electrodes such as platinum there *is* a battery effect. Also there is one with the electrolysis of water. See Experiments *Q* and *R* in Ch. 32.)

(ii) When we estimate the molecular weight of salts in solution (by their "osmotic pressure" or by their effect on freezing point), we find there seem to be nearly twice as many dissolved particles as we would expect for whole molecules: the split into ions has already occurred. On the other hand, solutions that do not conduct electrically, sugar solution for example, show the normal molecular weight of the dissolved substance it has not broken up into ions.

(iii) The idea fits well with chemical behavior. When a solid is formed in a solution, or a gas is formed and bubbles out, we imagine ions of opposite sign meeting and neutralizing each other's charge and joining to make the product which leaves the solution. Ions are the active agents in many chemical actions occurring in solutions.

The "Faraday" of Charge, and e/M for an Ion

Measurements show that 96,500,000 coulombs will release 1.008 kg of hydrogen. The same charge releases 35.4 kg of chlorine or $^{16.0}\!/_2$ kg of oxygen or $^{63.6}\!/_2$ kg of copper. (We may call this charge that releases one chemical equivalent (in kg) one "kilo-faraday.") Then, *assuming all hydrogen ions are alike*, we calculate the ratio CHARGE/MASS for a single hydrogen ion (proton):

$$e/M = 96{,}000{,}000/1.008 = 95{,}700{,}000 \text{ coul/kg}$$

For a copper ion, CHARGE/MASS = 3,034,000 coul/kg. (The reciprocal of this was used in Ch. 32 to define the amp: 0.000000329 kg-of-copper/coulomb.)

Ions, Electrons, and the Chemical Table

Looking forward to atom-models, we can sketch a simple theory. We picture atoms with a cloud of electrons around a central core (nucleus), the innermost electrons held strongly, and the outer groups of electrons more loosely, all of them tied by electrical forces. We picture atoms in Column I (lithium, sodium, . . .) as having one "loose" electron that is easily borrowed by some more attractive neighboring atom. That electron spends most of its time outside an inner compact group of electrons around the nucleus. Such an inner compact group alone around a nucleus is our model of an atom in Column 0—no electrons that are easy to shift, no readiness to make ions, no chemical properties. An element in Column I loses one electron easily; its atom becomes an ion$^+$ to take part in the chemical operations. In the solid element these loose electrons are practically free to slide from atom to atom carrying an electric current when a P.D. is applied; we expect it to be a good conductor. All elements in Column I are metals, excellent conductors.

Atoms in Column II have two loose electrons around a stable inner array, and make ions^{++} by losing them. Again, these are metals, good conductors.

Now you can see a new interpretation of the columns of the chemical table. For Columns I, II, . . . the column-number gives the valency, the number of loose outer electrons. Look at aluminum in Column III with ions Al^{+++}. We expect to find three fairly loose electrons.

Meanwhile in Column VII instead of saying there are seven electrons that are fairly loose, we say there is *almost* a stable inert group enclosing the atom, like those in Column 0. In fact, the arrangement of electrons in VII *needs just one more* to make a stable group. No wonder fluorine, chlorine, etc. are ready to grab an electron (from sodium, from water, from almost anything) to make negative ions. How do we know that just one more electron would make a stable group? Look at Column 0, which is next door to Column VII.

On this basis, we see valency as a measure of an atom's greed for electrons; e.g., an electron from Na joins Cl in the ionic molecule Na^+—Cl^-. Those ions both look like an inert gas atom, except for their unbalanced charge. Then we have attraction—strong crystals, active ions, but no tendency to want further electron changes which would make for different chemistry. (PROBLEM: On this basis of electron-borrowing to make *stable* groups, what would you say happens when calcium and chlorine combine to make calcium chloride?)

Ion-making is not the only form of chemical bonding. Bonds in some compounds are due to atoms *sharing* one or more electrons mid-way between them. To explain that kind of valency and predict the behavior, we need to treat the electron-sharing by modern quantum theory or "wave mechanics."

Other cases of "physical" bonds between complete molecules are just due to a shifting of charges to make the molecules "oblong," [+ . . . −], when they are close, in each other's field. Then they twist around until they are side by side so that the charges of one molecule are nearest to opposite charges of the other, thus making an attraction that holds them together. Here too, quantum theory throws light on the mechanism.

Water Molecules

Ions seem to form easily when salts dissolve in water, but there is no sign of them in other solvents such as benzene. We think that water itself, which can form ions H^+ and OH^-, is an electrically oblong molecule, with one end + and the other −. We have evidence for that in the readiness of water vapor to condense on electrically charged particles—fog droplets form extra easily on gas ions. (See Ch. 39 for the use of this in cloud chambers.) Again, water placed between the plates of an electric capacitor makes it take 81 times more charge from a battery connected to it—the water molecules seem to orient in the electric field and add a sabotaging field of their own. When salts are dissolved, water molecules can help to make or preserve ions by clustering around them with ends of opposite charge pointing inward towards the ion. An ion will collect a cage of water molecules like that. This clutters up the ion with a big mass that moves with it. So when an electric field is applied the ions drift quite slowly along the field.

Seeing Ions Move

Like atoms, ions are far too small to see, but we can see their drifting motion by using a collection of colored ions. A patch of blue Cu^{++} ions or yellow chromate ions is placed in a solution of colorless ions. (Gelatin is added to stop convection, but the current runs, and the ions move, as easily as ever.) When an electric field is applied, the colored patch creeps slowly across—a millimeter a minute is fast, but you are watching the motion of real ions, a tidal wave of charged atoms.

ORGANIC CHEMISTRY

Carbon Chemistry: Organic Compounds

Carbon with valency 4 shows remarkable readiness to link up with elements on either side of its middle column in the table, and even with its own brother atoms of carbon. It joins with hydrogen to make methane, CH_4, with oxygen, nitrogen, sulfur, . . . with chlorine to make CCl_4. And long chains of carbon atoms form easily, with hydrogen at each side: methane expands to ethane, . . . , octane (a liquid, one of the constituents of gasoline), . . . still longer chains which make solid paraffin wax.

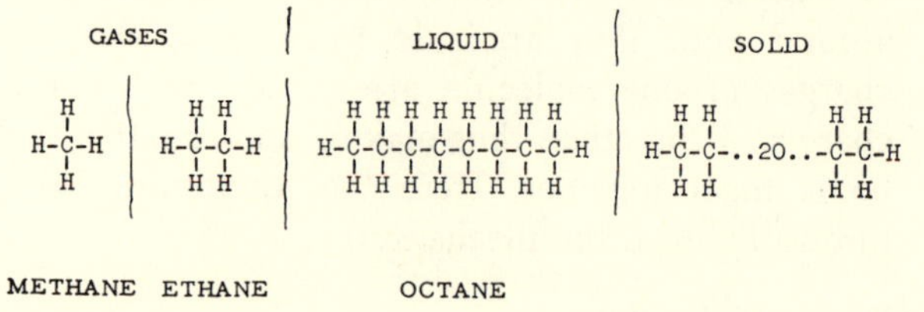

```
GASES                     LIQUID                      SOLID

  H        H H       H H H H H H H H        H H            H H
  |        | |       | | | | | | | |        | |            | |
H-C-H    H-C-C-H   H-C-C-C-C-C-C-C-C-H    H-C-C-..20..-C-C-H
  |        | |       | | | | | | | |        | |            | |
  H        H H       H H H H H H H H        H H            H H

METHANE  ETHANE          OCTANE
```

FIG. 35-4a.

These patterns are called structural formulas. They give us great help in understanding the chemical behavior of complicated organic compounds; and they enable us to understand why there are often several compounds with the same number of atoms of C, H, etc. but different chemical properties—and they even enable us to devise tests to find which pattern of these alternates we have in a given sample. In the carbon hydrogen chains other atoms such as chlorine may be substituted for H in a host of compounds; or a side chain can develop where a carbon atom of a new chain replaces one H. Again, adjacent carbon atoms can use more than one single arm each in joining and make double or triple bonds, as in acetylene H—C ≡ C—H.

When oxygen is added, its two valency arms take two carbon arms; or the oxygen can form a link with hydrogen beyond. Ethane takes on a spur of —O—H instead of one H, making alcohol, $CH_3 \cdot CH_2 \cdot OH$ (which looks like an alkali but is not alkaline).

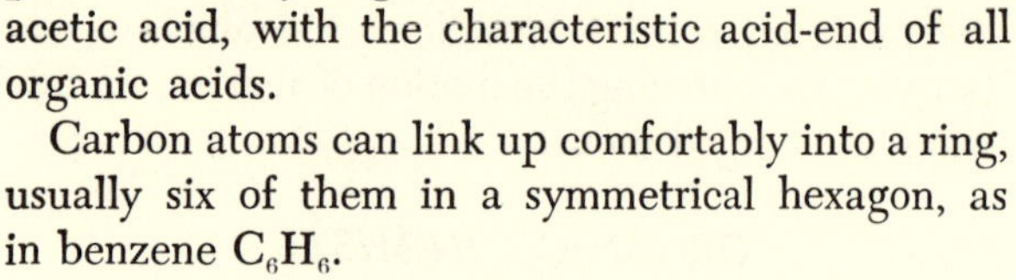

```
                                                          acid end
  H H            H H              H H                  H O
  | |            | |              | |                  | ‖
H-C—C-H        H-C—C═O          H-C—C—O-H            H-C—C—O-H
  | |            |                | |                  |
  H H            H                H H                  H

ETHANE         ALDEHYDE         ALCOHOL              ACETIC ACID
```

FIG. 35-4b.

If we add one more oxygen, held by two arms, in place of two hydrogens next to the —O—H we have acetic acid, with the characteristic acid-end of all organic acids.

Carbon atoms can link up comfortably into a ring, usually six of them in a symmetrical hexagon, as in benzene C_6H_6.

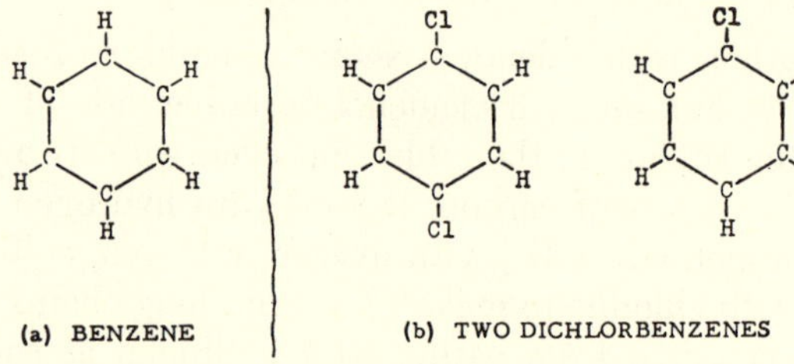

```
(a) BENZENE           (b) TWO DICHLORBENZENES
```

FIG. 35-5.

(In Fig. 35-5a each C has one arm to hold H and uses two to link in the ring. If the fourth arm seems unemployed, you must either use it in alternate double bonds around the ring or blame this as a defect of the flat model.) This pattern of a six-carbon ring is well vouched for by experiments in which other atoms are made to replace hydrogen in benzene. If you trust the picture of the benzene ring, you can use it as a piece of theory to answer questions such as the following:

PROBLEM

By treating benzene with chlorine it is possible to manufacture mono-chlor-benzene, C_6H_5Cl, di-chlor-benzene $C_6H_4Cl_2$, and tri-chlor-benzene. We can separate out these various products and choose the one whose composition is shown by analysis to be dichlorbenzene. In that two H atoms of benzene are replaced by Cl atoms; but the chemical and physical properties may well depend on just how these two Cl atoms are arranged in the ring. Fig. 35-5b shows two distinct patterns, a symmetrical one and a very unsymmetrical one.

In fact, several distinct dichlorbenzenes are known—paradichlorbenzene sold as a moth deterrent is one of them.

(a) *How many different dichlorbenzenes would you predict?*
(b) Chemists can verify the prediction of (a) and even find out *which* dichlorbenzene they have in a pure sample, in the following way. They treat the sample with chlorine in such a way as to convert it to trichlorbenzene. Then they find out how many different trichlorbenzenes they have in the product. *Sketch a structural formula for each of the dichlorbenzenes you guessed in (a). For each form sketched, say how many different trichlorbenzenes you would expect to find on addition of one chlorine.*

Organic and Inorganic Chemistry

Because many of these carbon compounds are found in living material, and the more complicated ones could at first be obtained only from plants and animals, the study of them was named *organic* chemistry. Now we find that carbon makes an enormous number of compounds, many of them very useful, and we keep the name for their study. Organic chemistry is as profuse and important as the chemistry of all other materials, which we lump together as *inorganic* chemistry.

From simple beginnings with alcohol, acetic acid, etc., organic chemistry goes rampaging away: analyzing, synthesizing, sketching compounds on paper and building them in the laboratory: solvents, dyes, soaps, plastics, drugs, . . . with molecular weights running to hundreds, even thousands. Scientists enjoy these successes not only for the wealth of resulting products but for the amazing skill of reasoning that lies behind this work. And now on to the proteins of food and living matter, with molecular weights like 35,000. Their composition is known, so we can guess at general formulas, but the architecture of their structural formulas is still being worked on.

PART FIVE

ATOMS AND NUCLEI

"Atomic Physics," the physics of this century, is a sophisticated science. Our knowledge is built up by several distinct lines of investigation and thought at once, each helping and criticizing the others. Therefore, no logical sequence of chapters, A, B, C, . . . will unfold its history simply or give an understanding that is immediately clear. Chapter B needs Chapter A as preparation, but also clears up some unexplained matters in A; and some of those matters may need further illumination from E and F. Moral: read this section twice—it makes richer sense on the second round.

To mention every important item, we should have to crowd this section with a stamp-collection of unexplained information. Instead of that insult to the good name of science, we choose some topics for careful study, so that you shall understand the basis of your own knowledge. In doing that, we shall cover many important areas, some reaching to the frontier of current science; and we shall leave you ready to explore other areas by your own reading in books: present books and future ones.

> Where is the wisdom we have lost in knowledge?
> Where is the knowledge we have lost in information?
> The cycles of Heaven in twenty centuries
> Bring us farther from GOD and nearer to the Dust.
>
> —T. S. Eliot, *The Rock*
> (Harcourt, Brace and Company)

PRELIMINARY PROBLEMS FOR CHAPTER 36

PROBLEM 1. SIMPLE ELECTRON GUN

An electron gun is a device for making a stream of (fast-moving) electrons. These electrons, which we believe are all alike, very small (mass 1/1840 of hydrogen atom mass), and negatively charged (-1.6×10^{-19} coulomb), can then be used for drawing television pictures, graphing waveforms, bombarding targets for atomic investigations, etc. This problem discusses the way in which such guns are arranged. You are asked to add the needed batteries to the diagram (see Chapters 32, 33).

An electron gun consists of a source of electrons, which is a hot cathode, C, heated by a heater HH; a gun muzzle M, through which the electrons emerge; and in many cases control grids or plates such as G.

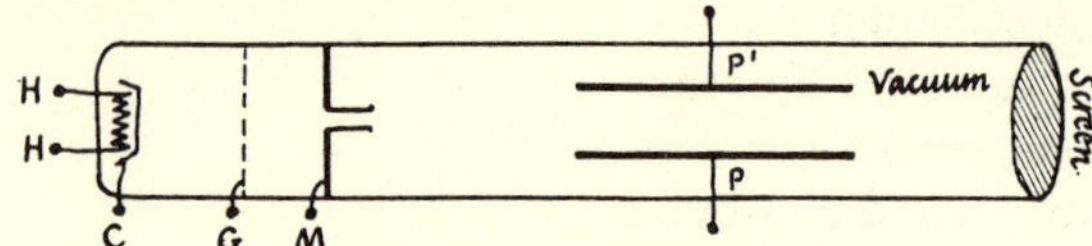

Fig. 36-1. Problem 1

(a) The cathode must be heated before it will boil off enough electrons.
(b) To emerge from the gun muzzle moving fast, the electrons must "fall through" a large P.D. This is provided by a battery (or equivalent) which makes an electric field, along the stream, to accelerate the electrons from cathode to gun-muzzle.
(c) If a terrific stream is thus rushed from cathode to gun muzzle, the gun is overheated by the K.E. of the many electrons that hit it and stop there (failing to get through the hole in it). To avoid this a small *discouraging* field is often applied between C and G. Then only a small stream proceeds beyond G, and thereafter is accelerated.
(d) To "deflect" the stream up or down, a vertical electric field is applied between plates P, P'.

All these electric fields can be applied by attaching suitable batteries to the appropriate points.

Re-draw the sketch and *add a battery* for each job that needs one, including these typical requirements:

(a) heating the filament (needs 5 amps, has 2 ohms resistance)
(b) discouraging the stream (use 12 volts)
(c) speeding up the stream to full speed (needs 2000 volts, say)
(d) deflecting stream (P and P' need 40 volts for their separation, 0.02 meter)

Make sure your batteries are suitably arranged for negative electrons. (*Mark + and − on them.*) Make deflection by PP' *upwards*.

PROBLEM 2. DEFLECTING A STREAM

A stream of electrons emerges from an electron gun and passes down a long tube. Plates P and P' connected to a battery apply an electric field, as shown.

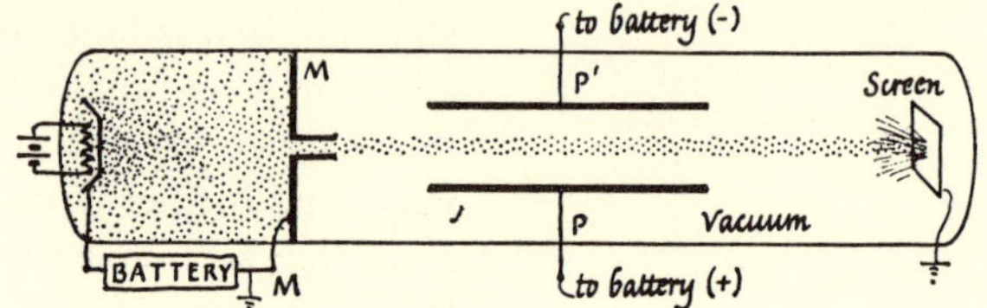

Fig. 36-2. Problem 2

(a) Once the electrons emerge from the gun at M, what path will they pursue if there is no electric field between the plates P and P'?
(b) If there is no field between P and P', how will the velocity of the electrons change as they continue down the tube from M?
(c) Name the general physical law (which sums up behavior observed in nature, and is used to predict similar behavior by inference) which you relied on in answering (a) and (b).
(d) Now suppose the field is turned on between plates P and P' and that there is then a strong uniform vertical field between P and P' but a negligible field before and beyond that region. What will the path of the stream be like in that field? (Give a definite *name* to the path. Also sketch it.)
(e) Where have you met a similar path for a body moving in a uniform field of force?
(f) When it emerges from the region between P and P' (with the field on) the stream is in a negligible field. Describe the stream's path there, and show it on your sketch.

CHAPTER 36 • ELECTRONS AND ELECTRIC FIELDS

". . . for his works on the elementary charge of electricity, and on photoelectric phenomena. . . ."

—from the citation for the award of the Nobel Prize in Physics (1923) to R. A. Millikan

". . . for his discoveries in physical science."

—from the citation for the award of the Copley Medal, the highest honor of the Royal Society, to J. J. Thomson

Investigating Atomic Particles

We know that electrons boiling off a hot filament carry a negative charge; but how can we measure the charge of single electrons, and their mass, to show they are all alike? How can we measure their speed when they emerge from a gun? Again, electrons chipped off molecules of residual gas in an electric discharge tube are said to be universal ingredients of matter. How can we prove they are the same? How can we show that they are far smaller than atoms? How can we dig into atomic structure by measuring positive ions, the remainders of atoms in discharge tubes? How can we measure the nuclear bullets (alpha- and bèta-particles) shot from radioactive atoms?

Electrons, ions, nuclei, . . . these are some of the tiny *particles* of modern physics; and we describe their character as individuals by measuring their charge, mass, speed, . . . We must catch these flying particles on the wing. Once stopped, they are usually lost[1] among the atoms of solid walls, wires, etc. So we gain our information by pulling them out of their path with fields—just as you can estimate the speed of a ball by the effect of a gravitational field on its path. Gravitational fields are too weak for measuring atomic particles. All such particles do fall under gravity as they whizz along; but, from any reasonable gun-voltage, they whizz too fast to fall measurably in any ordinary trip. (An alpha-particle from radium would fall 2 millionths of an inch in travelling a mile; and electrons in a television tube would fall a ten-millionth of an inch in a tube a mile long.) So we make electric and magnetic fields act on the electric charge carried by the particles.

In this chapter we shall show what can be done with *electric* fields. In the next chapter we shall return to the problem with magnetic fields and show how the effects of both kinds of field can be combined to yield the particle's speed and its essential identifying label, its CHARGE/MASS ratio, e/m.

[1] But not if they are radioactive—there lies a magnificent opportunity.

Fields and Streams

We have already applied electric fields to electrons. In an oscilloscope tube the gun voltage provides a field *along* the stream to speed it up, and the deflecting fields swing the stream up and down or sweep it across. In a "discharge tube" the electric field along the tube drives positive ions one way, electrons and negative ions the opposite way, producing a mess of excited atoms that emit a glow of light as their electrons settle back to lowest energy.

Stream from Electron Gun

The stream boiled off a hot cathode and accelerated by a gun-voltage carries a negative charge. You can see that by tracing the current across a diode tube through a milliammeter or by catching the stream in a small metal can connected to an electroscope. To show it is a stream of bullets, all alike

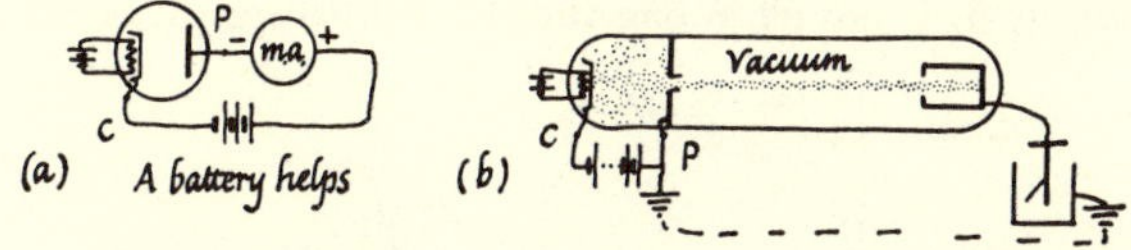

FIG. 36-3. STREAM CARRIES A NEGATIVE CHARGE

—"electrons"—we must make measurements with fields. The problems in this chapter have data typical of real experiments. Work through them to find how electrons can be investigated.

PROBLEM 3. DIRECT MEASUREMENT OF ELECTRON VELOCITY

This problem shows one method which has been used to measure the speed of electrons in a stream from an electron gun. We use a radio oscillator to provide deflecting electric fields to act on the stream.* With such an oscillator it is

* In actual experiment, an alternating **magnetic** field was used instead.

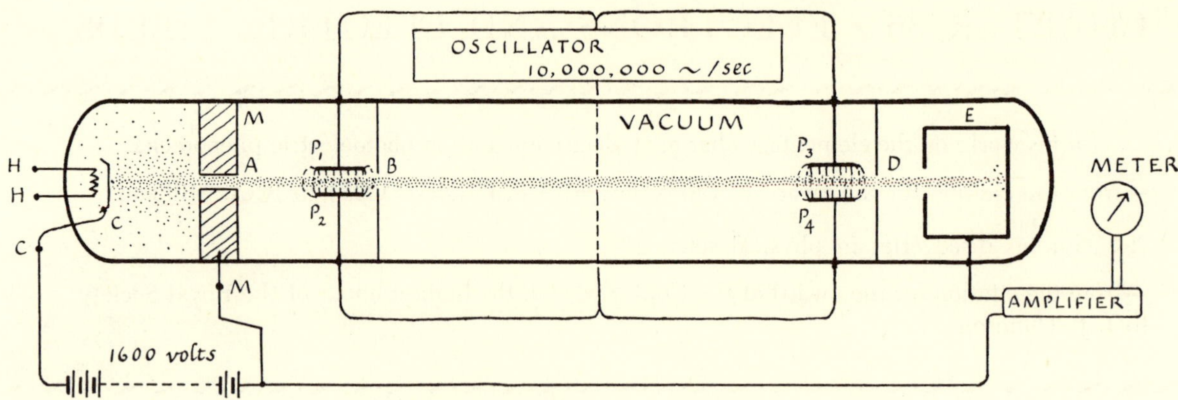

Fig. 36-4a. Problem 3: Measuring the Speed of Electrons

possible to charge a pair of metal plates + and —, then — and +, alternately, at a frequency of, say, 10,000,000 complete cycles per second. Then the electric field in the space between the plates changes direction from *down* to *up* to *down*, again and again, making 10,000,000 complete cycles in every second.

With modern pumps it is easy to make such a good vacuum in a long tube that an electron proceeds all the way without collisions. Fig. 36-4 shows such a tube with an electron gun near one end. A stream of electrons, boiled off the hot cathode C, accelerated by a gun-voltage of say 1600 volts between C and M, emerges through a small hole A in the gun muzzle M and passes on down the tube with no further change of speed. A pair of horizontal plates, P_1, P_2, is placed beyond the hole A, so that the electron stream passes through a vertical electric field between them. Just beyond these plates is a wall with a hole, B. If the alternating electric field described above is applied in the region between P_1 and P_2 it will swing the electron stream up and down so that electrons only get through the hole B in small spurts, once on the up swing, once on the down swing (*twice per cycle*), like a firehose being played up and down a fence with a knot hole. These spurts travel down the long tube with constant speed v. Far down the tube they pass between another pair of plates, P_3, P_4, with the same alternating electric field between them, synchronized with the field between P_1 and P_2. Just beyond is another wall with a hole D. Behind D is an electron collector, E, connected to an amplifier and a meter which shows whether electron spurts are arriving at E. The holes A, B, D are all in one straight line between C and E.

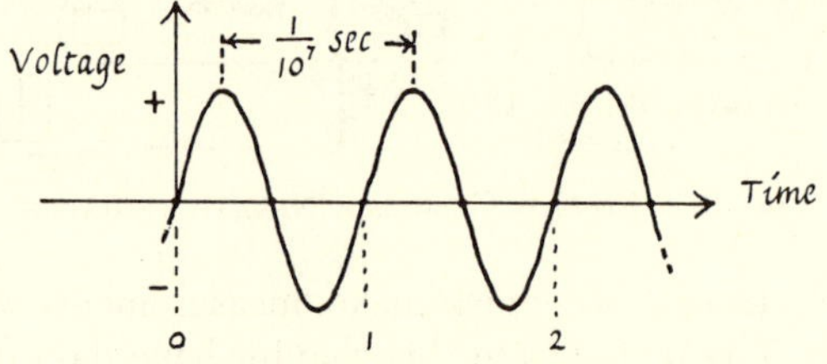

Fig. 36-4b. Voltage Supplied by Oscillator in Problem 3

DATA: The holes B and D are 1.20 meters apart along the length of the tube. The oscillator, connected to both pairs of plates, is known (from separate measurements) to have frequency of 10,000,000 cycles per second. If the gun voltage between cathode C and muzzle M is 1600 volts, it is found, in real experiments, that the collector D collects a lot of electrons. If the gun voltage is 1500 volts or 1700 volts, few electrons arrive.

NOTE that the deflecting field region between the plates is quite short, so that electrons get through it in a short time— a small fraction of one cycle of the oscillator's changes.

(a) Suggest a reason why no electrons reach E unless they have been made to move with certain definite speeds, by the gun (e.g., the speed imparted by 1600 volts on gun, not that from 1500 or 1700 volts).
(b) Estimate the speed of the electrons from the data above. (Make the simplest of any choice of assumptions. That will give you the *maximum* speed.)
(c) Explain why certain other speeds are consistent with the observations.
(d) Calculate one or more of the other speeds that fit with the data.
(e) How would you find out (by changing apparatus) which is the right choice among (b) and (d)? (This problem requires common sense + careful thinking.)

PROBLEM 4. ESTIMATING CHARGE/MASS RATIO FROM GUN-VOLTAGE AND VELOCITY

A stream of electrons is shot from an electron-gun near one end of a long tube. Suppose the speed, v, of the electrons has been measured, as in Problem 3. A battery provides gun-voltage V between C and M to speed up the electrons from rest to the final velocity v with which they emerge from the gun. Each electron, carrying a charge *e* coulombs "falls through" (or travels across) a P.D. of *V* volts in the gun, gaining K.E. In a good vacuum, all the energy given by the battery to an electron, through the action of the electric field in the gun, goes into the electron's K.E. Experiments show that:

when the P.D. applied to the gun is 100 volts
the electrons emerge with speed v which is
measured and found to be 6,000,000 meters/sec
(NO MISPRINT)

Use these data to estimate the *mass of a single electron* as follows:

(a) Call the mass of electron *m* kilograms. Write down its K.E. when it is moving along the tube, expressed in terms of *m*. (Give units.)
(b) Call the charge e coulombs. When it has "fallen through" 100 volts in the electron gun, how much energy has it gained, expressed in terms of *e*? (State units.)
(c) Make sure that the answers to (a) and (b) are in the same units.

Write an equation stating that the energy given to an electron by the 100-volt battery *is* its kinetic energy. Solve this to find the ratio *charge/mass* for the electron *e/m*.

(d) In a separate, different experiment, Millikan found that the *charge*, *e*, on an electron is 1.60×10^{-19} coulomb. If so, what is the value of its *mass* in kg?

Universal Electrons

Direct measurement of v is inconvenient, except for fairly slow electrons; but whenever they are made, measurements of Problems 3 and 4 combine to give the same value of e/m for electrons of different speeds—streams from guns with different gun-voltages. And the same value of e/m emerges whatever the materials used in the gun: all electrons have the same ratio of charge to mass. This conclusion that electrons are universal emerged from investigations at the beginning of this century. Uniform streams from known gun-voltage were not available then, so more complicated measurements had to be made, using an electric field to deflect the stream sideways, and then a magnetic field. Transverse deflections by electric and magnetic fields have been used ever since in fundamental atomic measurements. Problem 5 shows how a transverse electric field can be used.

The value of e/m can be combined with other measurements to yield fundamental atomic information; but, even by themselves measurements of e/m made two suggestions of tremendous importance:

(i) e/m has the same value for all electrons, from very slow to quite fast ones, whatever their source: boiled off a hot filament; flipped out of metal atoms by light waves (as in electric eye); ripped out of atoms by X-rays; fired out of radioactive nuclei as beta-rays; as well as the original method of knocking them off gas atoms by collisions in a discharge tube. This suggested that *electrons are all alike, a universal ingredient of matter.*

(ii) e/m has a smaller value than the standard one if measurements are made on very fast electrons. This suggested that, if e remains the same, *mass grows bigger at high speeds*—in just the way that fits with Relativity theory.

PROBLEM 5. DEFLECTION OF ELECTRON STREAM BY ELECTRIC FIELD

One of the ways in which cathode rays were first investigated with actual measurements was by sending a stream through an electric field as in preliminary Problem 2 above. The ratio *charge/mass* for the particles can be found from measurements of the deflection of the stream instead of using measured gun-voltage.

This question shows you one way of calculating e/m from measurements. Suppose a stream of electrons from a gun emerges with speed 2.4×10^7 m./sec, or 24,000,000 meters/sec, measured in the manner outlined in Problem 3. This stream is shot through a long region in which there is a transverse electric field, as in Fig. 36-5. Finally the stream hits a fluorescent screen, without emerging from the field-region.

The following data show the kind of measurements that might be obtained in real experiments. Pretend they are the

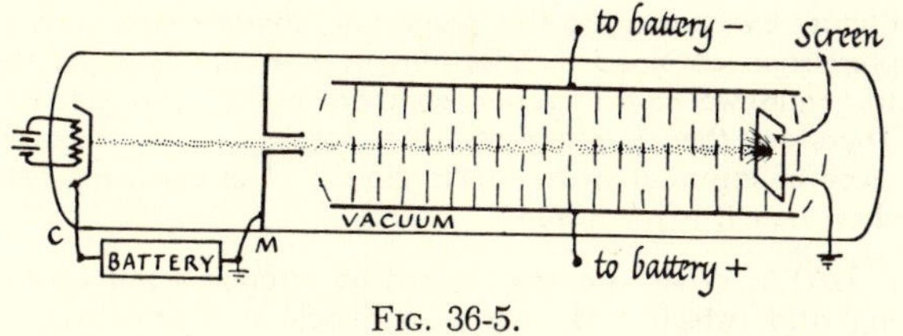

FIG. 36-5.

measurements in an attempt to measure the value of e/m for the electrons of the stream, and calculate e/m (in coulombs/kg).

Speed of particles in stream (see Problem 3) .... 2.4×10^7 m./sec
Deflection of stream when field is applied .... 0.015 meter vertically down
Distance between field plates .... 0.050 meter
Length of field region .... 0.20 meter
P.D. between plates .... 120 volts

(Note that the gun voltage is not given, and is not needed here.)

Calculate:

(a) The *electric field-strength* of the fields between plates: the force, in newtons, on one coulomb.
(b) The *force* exerted by field on a charge e coulombs, a single electron.
(c) The *acceleration* of the electron. Call its mass, m kg.
(d) The *time* the electron takes to travel through the field region, 0.20 meter. (*Note*: the distance 0.20 m. is the *horizontal* length of the field region, and the velocity given above is the *horizontal* velocity. Does it matter if the electrons are also gaining vertical velocity? Consult Galileo on independence of motion.)
(e) Hence calculate the distance, s, the electron moves vertically as a result of the field. Note that the electron enters the field with zero vertical velocity, moves with accelerated motion calculated in (b) (c) for time calculated in (d). (The answer to this will be a formula for s, containing the 0.20 meter, charge e, mass m, etc.)
(f) The value of e/m. The measured distance is, according to the data provided, 0.015 meter. Write an equation stating that the deflection, calculated in part (e), is 0.015 meter. Solve this equation for e/m.

In this problem, you were provided with the *velocity*, which could have been measured by the oscillating-electric-field method of Problem 3. Unless you know this velocity it remains in the bunch of unknowns, which will combine in the form e/mv^2 for the thing you can calculate from the deflection by an electric field. In most measurements of cathode rays, the velocity is not known until magnetic-field deflections are measured also.

PROBLEM 6. COMPARISON BETWEEN ATOMS AND ELECTRONS

The following discussion relates to the ions that enable solutions to conduct. Chemical evidence suggested long ago that such carriers are single atoms carrying charges, or sometimes groups of atoms. The charges were all equal. Each was a tiny universal "atomic" unit of electricity (named an "electron" long before *the* electron was discovered) or, in some cases the charge was 2 or 3 such units. Early this century, physical evidence (diffusion of ions) suggested that this basic unit of charge for chemical ions is the same size as the charge carried by the flying electrons measured by J. J. Thomson. So, when we measure the ratio *charge/mass* for products in electrolysis, we are probably making measurements that apply, *keeping the same proportion*, to single atoms. By comparing this proportion *charge/mass* with the proportion obtained for electrons in a stream from an electron gun we can therefore compare atoms with electrons. Therefore the *charge/mass* ratio for ions is an important piece of information in atomic physics. The questions below show how it is measured.

DATA: When a current is passed through water containing acid (which acts only as a middle-man providing ions that are replenished by the water) oxygen and hydrogen bubble off from the metal plates ("electrodes") by which the current enters and leaves. Experiment shows that a current of 10 amps flowing for 1000 seconds will release 0.001244 cubic meters of hydrogen gas, at room temperature at 1 atmosphere pressure.

In the same warm room a large glass globe weighed first full of hydrogen at atmospheric pressure, then empty, then full of water, might give weighings like the following:

Mass with hydrogen .... 1.60084 kg;
Mass empty .... 1.60000 kg;
Mass with water .... about 11.60 kg.

(a) Using these data, calculate the density of hydrogen, in kg/cubic meter.

(b) What *mass* of hydrogen is released in the "electrolysis" experiment described above?

(c) What *electric charge* (= quantity of electricity) passed through apparatus?

(d) At the plate where hydrogen bubbles appear, all the flow of charge to the plate is probably carried by hydrogen carriers. In that case, what is the *charge/mass* ratio for the *total charge* carried by the *total mass* of hydrogen in this experiment?

(e) *Assuming* that all the hydrogen carriers were exactly similar and all of them hydrogen atoms, what is the value of *charge/mass, e/M, for one hydrogen-atom ion?*

(f) Experiments with electron streams show that for electrons e/m is about 1.8×10^{11} coulomb/kg. (This means that 1 kg of electrons would have charge 180,000,000,000 coulombs.) Compare e/M for hydrogen-atom ion with e/m for electron, by giving the ratio of their values.

(g) Presuming e is the same for both, compare m and M.

Values of e/m

The measured deflection by an electric field enables us to calculate e/mv^2. The gun voltage *also* tells us e/mv^2. So if we know the gun voltage we gain no new information by deflecting the stream with an electric field.[2] To calculate e/m, the great characteristic ratio for charged atomic particles, we must also either measure v directly or measure the deflection by a magnetic field (Chapter 37) which yields the value of e/mv. For the present, assume that we *can* eliminate v and find a value of e/m.

[2] This disappointment is reasonable; because the gun voltage applies an electric field to the electrons, accelerating them along their path. We should not expect to get two different pieces of information from the same kind of agent, an *electric* field in both cases.

A hundred years ago, electrons as particles were unknown. The only sign of "atoms of electricity" was a strong hint from electrolysis that each atomic ion carries a standard electric charge—the same size for all chemical ions except for cases of double or triple charges. Fifty years ago electron streams had been produced and measured in a variety of ways. J. J. Thomson made the earliest consistent though rough measurements on charge and mass of electrons and of positive ions; he described them as the ingredients of atomic structure. The table shows some results.

TABLE OF VALUES OF *e/m*

(The earliest measurements were not very accurate. The table gives measurements made when good techniques had been developed; values usually calculated from deflections by electric and magnetic fields)

Particles	*Value of e/m in coulombs/kilogram*
Cathode rays in discharge tube: electrons knocked off gas atoms or metal electrodes by bombardment (Since this was the earliest method of producing electron streams, the method that gave them the name "cathode rays," the results from three experimenters are given.)	1.775×10^{11} 1.761×10^{11} 1.759×10^{11}
Electrons from white hot tungsten filament (as in diode)	1.76×10^{11}
Electrons from red hot oxide-coated cathode (as in modern radio tube)	1.78×10^{11}
Electrons pulled out of metal by ultraviolet light (the "photo-electric effect" of the electric eye)	1.756×10^{11}
Electrons inside atoms made to modify their "orbits" by external magnetic field (The "Zeeman effect")	1.761×10^{11}
Electrons in hydrogen and helium atoms: comparison of electron-mass with atom-mass by wavelengths of spectra, trusting Bohr's theory	1.761×10^{11}
Beta particles (slow) from radioactive atoms	1.763×10^{11}
Beta particles (slow . . . medium . . . fast) from radioactive atoms; a continuous range of *e/m* values from normal	1.76×10^{11} to 0.35×10^{11}

And, in later experiments where electrons from hot filaments are given huge energies by accelerators, values of *e/m* range on down to values thousands of times smaller—the change being accounted for by the relativistic increase of mass.

Positive rays: positive ions in discharge tubes. Value of e/M depends on residual gas in tube:	
hydrogen ion H^+	$\frac{1.76 \times 10^{11}}{1840}$
oxygen ion O^+	$\frac{1.76 \times 10^{11}}{\mathbf{16} \times 1840}$
oxygen ion O^{++}	$\frac{1.76 \times 10^{11} \times \mathbf{2}}{\mathbf{16} \times 1840}$
mercury ions Hg^+, Hg^{++} . . . to $Hg^{++++++++}$	$\frac{1.76 \times 10^{11} \times (\mathbf{1\ to\ 8})}{\mathbf{200} \times 1840}$
Positive ions in electrolysis: hydrogen ion H^+	$\frac{1.76 \times 10^{11}}{1840}$
copper ion Cu^{++}	$\frac{1.76 \times 10^{11} \times \mathbf{2}}{\mathbf{63.6} \times 1840}$
chlorine ion Cl^-	$\frac{1.76 \times 10^{11}}{\mathbf{35.5} \times 1840}$
Alpha particles from radioactive atoms	$\frac{1.76 \times 10^{11} \times \mathbf{2}}{\mathbf{4} \times 1840}$
And many recently discovered particles: [e.g. μ mesons, which have	$\frac{1.76 \times 10^{11}}{\sim \mathbf{200}}$]

Except for changes at very high speed, all electrons have the same e/m, 1.76×10^{11} coulombs/kilogram. Compare this with e/M for the lightest atomic ion, H^+, which carries across 96,500,000 coulombs for 1.008 kg of hydrogen released in electrolysis. So, for H^+, $e/M = 9.57 \times 10^7$ coulombs/kg. For electrons, e/m is nearly two thousand times larger. Then *if* e has the same size for both, *m for electrons must be nearly two thousand times smaller.* (Some time earlier, clever experimenting and reasoning had shown that $N \cdot e$ is the same for ions in gases and ions in electrolysis, where N is the Avogadro number. Neither N nor e was known, but $N \cdot e$ could be estimated: a gross measurement for electrolysis, simply the 95,700,000 coulombs carried by 1 kg of hydrogen ions; a tricky measurement for gas ions, involving diffusion. Then it was argued that e for gas ions *is* the charge of the electron that has been knocked off, and N is the same for both; therefore e is the same.) This is the primary evidence that electrons are tiny chips of atoms.[3]

[3] The time was ripe for this discovery at the end of the last century, but J. J. Thomson made the first clear measurements of e/m, and in that sense we say that he "discovered the electron."

Calculate the fractions more accurately:

$$\frac{m}{M} = \frac{(e/M)}{(e/m)} = \frac{9.57 \times 10^7}{1.76 \times 10^{11}} = \frac{1}{1840}$$

An electron and the atom that has lost it (the remaining positive ion) have equal and opposite charges, since matter is normally neutral. But they have vastly different masses. No wonder electrons move easily in electric fields: a speedy, deflectable stream in a television tube; a rapidly mobilized panic-crowd in a Geiger counter. With so much charge to their little mass, they accelerate in electric fields far faster than charged atoms. Only when electrons acquire huge kinetic energies, billions of ev, do they seem (to stationary observers) as massive as atoms.

THE ELECTRON CHARGE, e: THE ATOM OF ELECTRICITY

The use of knowing e

If we could measure e, we could divide it by measured e/m to find the mass of a single electron. It would tell us the mass of a single *atom* even more easily, since e/M for atomic ions is easily measured in electrolysis experiments. From the mass of an atom we could calculate the mass of any molecule, and thence the number of molecules in any sample of solid liquid or gas. Furthermore, theories of atomic structure involve calculations that use the actual size of e. An accurate knowledge of e is of paramount importance.

Measurements of e

By 1900, the electron was an established subatomic particle, with a definite e/m, but there was only a rough guess of e. The experimental facts of electrolysis had long before suggested that there are "atoms of electricity" all alike, with some ions carrying one such atom of electric charge, some ions doubly charged, &c. By 1910, the size of charge e was badly needed in developing atomic theories—the Bohr theory could never have reached its main test without a good knowledge of e as well as e/m. J. J. Thomson and others tried to measure e by forming a cloud of fine water drops, each drop on an ion with charge e, and collecting the cloud. This gave only a rough estimate and no assurance that all such charges are exactly equal.[4]

[4] It was also unproved that the *masses* of all atoms of a chemical element are exactly equal, but this seemed so likely that no one doubted it. We now know it is not true.

Then R. A. Millikan[5] performed his great experiment, using a tiny drop of oil that had collected a small charge from ions in the air. He measured the total charge on a drop again and again and found that it was always a small multiple (e.g., 1 or 2 or 10 lots) of a basic charge that was always the same. At the start, he did not know the size of that universal basic charge, "the electron," or how many lots of it his oil drop had. He had to measure many charged drops and carry out an arithmetical guessing game. The problem was "similar to the one of finding the weight of a single egg, given the weights of a large number of paper bags each containing a different and unknown number of eggs."[6]

PROBLEM 7

(a) Suppose paper bags of eggs weigh 12, 16, 28, 24 ounces. Guess the weight of an egg, and give the number of eggs in each bag.
(b) Suppose you are given one more bag, weighing 14 ounces. How does this affect your guesses?
(c) Suppose you are given one more bag weighing 12.1 ounces. What would you conclude?

The essential method used by Millikan and those before him was that shown in Chapter 33 for measuring the charge on a metal ball by weighing the force exerted on it by a uniform electric field. To measure *e*, a few electron charges were given to a tiny drop of liquid floating—or, rather, falling slowly—in air. The drop was placed in a vertical electric field which, acting on the drop's charge, hauled the drop upward. A single *e* is a very small charge, and a visible raindrop would be far too heavy; it would need a billion or so electron charges for a reasonable field to support its weight. So a very tiny drop was used, a minute droplet from a throat spray, so small that it was invisible except as a tiny bright star scattering light into an observer's microscope. Such a tiny droplet falls steadily in air with air friction balancing its weight. The heavier the drop, the faster the drift must be for air friction to balance its weight. The constant speed of downward drift can be measured and used to "weigh" the drop. When a vertical electric field is applied, it adds a new force, the pull of the field on the drop's electric charge. In early experiments, the electric field was adjusted to keep the drop from falling, so that it floated with its weight exactly supported. However, it proved more accurate to use a stronger field and haul the droplet up, then let it fall again with no field. Thus the measurements could be repeated, hauling the droplet up and letting it fall again and again, like a cat playing with a mouse. This was Millikan's measurement of the electron charge, a magnificent piece of experimenting which earned him lasting fame.

Work through Problem 8 below to see how Millikan used his measurements. His droplet (oil rather than water), usually started with an accidental charge given by friction at the spray nozzle—like a rubber rod being charged by fur. In the course of trips up and down, it would occasionally change its charge by meeting an ion or two in the air nearby. This change would at once give it a new upward drift-speed in the electric field. Sometimes Millikan made a quick change of charge by using a beam of X-rays to eject electrons from the drop itself. He tracked the same droplet through many changes of charge; then he had to solve an "egg-in-paper-bag" problem.

[5] Millikan has given an excellent popular account of his work—with convincing extracts from his own laboratory record—in his book, *Electrons* (+ *and* −), *&c.* (University of Chicago Press, 1947).

[6] F. A. Saunders in *A Survey of Physics* (New York, Henry Holt, 1930).

PROBLEM 8. MILLIKAN'S EXPERIMENT ON ELECTRON CHARGES

Millikan performed his experiment with a tiny oil drop which had collected a small charge from ions in the air. He could experiment for hours with the same droplet, hauling

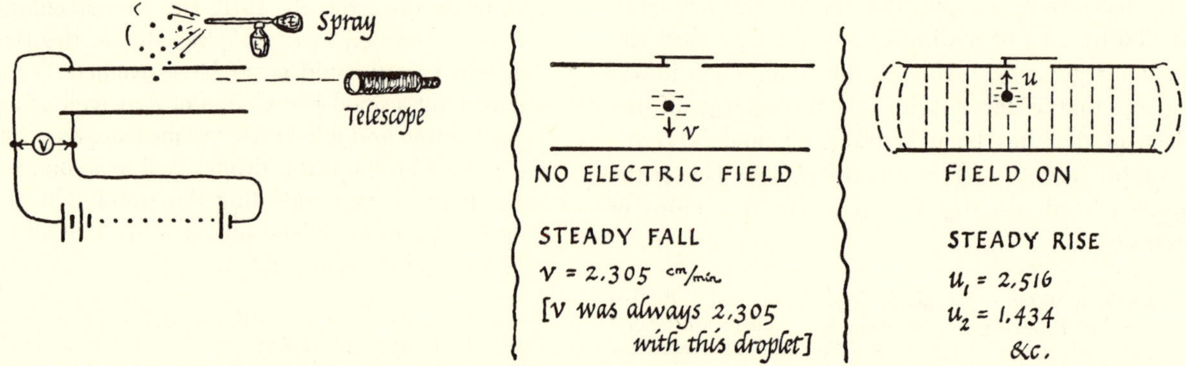

Fig. 36-6. Millikan's Experiment

it up and letting it fall again and again. Without any electric field, the droplet fell at a constant speed characteristic of the size of the drop.

(a) With a droplet of oil Millikan found the speed of fall remained the same, fall after fall, for many hours. However, with a water droplet, the time of fall increased as time went on. What can you infer regarding the oil droplet?

(b) With the electric field on, the droplet moved *up* with a (different) steady speed. This remained the same for many upward trips, then suddenly changed to a new speed. These abrupt changes happened more often just after he had run an X-ray tube nearby. Interpret these abrupt changes.

Here are some actual measurements for one oil droplet which fell again and again with speed $v = 2.305$ *centimeters per minute* (see note * at end of problem). With the electric field on, it rose with speed $u_1 = 2.516$ cm/min for several upward trips. Then it suddenly changed to new speed $u_2 = 1.434$ cm/min for one or more trips, then suddenly changed to $u_3 = 0.903$ cm/min, then to 0.369, then 0.903, then 1.958, 0.903, 1.434.

Both theory and experiment show that for very slow streamline motion of a sphere through viscous fluid (which includes air for small enough droplets) the drag force of fluid friction is given by:

$F = K \cdot$ (velocity), where K is a constant which involves the fluid's friction coefficient and the sphere's radius, which are constant all through one droplet experiment.

For falling with no field, the only forces on the droplet are its weight, $m(9.8)$ newtons, and the friction drag, $K \cdot v$. *The drop speeds up and then falls steadily without acceleration.*

(c) Write an equation, showing how these two forces are related for steady fall. (In this equation use the experimental value for v, 2.305 cm/min; see note *.)

(d) Suppose that the electric field, when it is on, has strength X newtons per coulomb and acts on the droplet's charge, Q coulombs. What force does the field exert on the droplet?

(e) With field on, droplet drifts upward with velocity u (such as 2.516 cm/min) and the friction force $K \cdot u$ acts downward, opposing this drift. The drop's weight, $m(9.8)$, also acts downward. The drop moves with constant velocity without acceleration, once started. Write an equation relating the *three* forces acting on the droplet.

(f) In the last equation get rid of the weight, $m(9.8)$, by substituting its value from the first equation, and re-write the result in the form $Q = \ldots$ This new equation should show that Q varies directly as $(v + u)$ if X is constant.

(g) Use the result of (f) to analyze Millikan's measurements given above. The value of v is 2.305 cm/min throughout, and different values of u are given above. If $(v + u)$ gives a measure of the total charge Q, the *change* of $(v + u)$ must give a measure of the *change of charge*, which is the charge the droplet gains from ions, etc.
$\therefore$ *Change* of charge, ΔQ, is measured by *change* of $(v + u)$ or change of v + change of u.
But v does not change, so change of v is zero, and ΔQ *is measured by change of u.*
$\therefore$ *Calculate the changes of upward speed u and use these as measures of change of charge*; i.e., charge picked up by droplet. *Calculate all the changes of u.* Find the basic change of u which will just fit into all these changes, and assume that this change corresponds to *one electron charge*. Then say how many electrons must have been involved in the change each time the charge changed.

(Millikan's result could not be "perfectly accurate." The last figure he gives is probably in doubt. So you should tolerate small differences. "How" small is a matter of guessing for you, of judgment for Millikan when he reviewed his own probable errors, and a matter of quarrel (eggs-in-a-bag) between Millikan and a rival who long claimed a "sub-electron." Doubting Millikan's last figure suggests that chance errors might make 1 or 2% variations in Δu.)

(h) Using that change in u which you have decided refers to a single electron charge, return to $(v + u)$, which measures the TOTAL charge, and calculate how many electron charges the drop started with when its rise speed, u_1, was 2.516 cm/min.

Calculations like those of (g) and (h) were Millikan's way of showing that all electrons had same charge.

* All the speeds are given in the odd units that Millikan used, centimeters/minute. There is no need to convert them to meters/sec because only comparative values are needed here. Assume that K is expressed in suitable units, newtons/(cm/min).

The Universal Atom of Electricity

Measurements on one droplet could give assurance that there is a basic atom of electric charge. Millikan had to do many experiments with droplets of different sizes, different liquids, and with different ionizing agents, to show that the basic atom of charge is a universal constant. If any experiment had yielded an odd fraction of his tentative atom, instead of a whole number of them, he would have been forced to change to a smaller atom of electricity—and a downward path to smaller and smaller atoms would have spoiled both his hopes of success and our present theories of atoms.

To determine the actual size of the charge in coulombs, he calculated the droplet's weight from its rate of fall and the general friction-coefficient for streamline flow of air. The algebra is roundabout but not difficult; and it is well explained in Millikan's book.

The result: *whatever the source of charges, whatever the material and size of droplet, the total charge on every droplet was always a whole multiple of the same basic electric charge*

$$1.60 \times 10^{-19} \text{ coulomb.}$$

This is the size of the negative charge of every electron (or of the positive charge left on an atom ionized by losing an electron, or the negative charge on an ion made by gaining an electron). It is the universal atom of electricity.

PROBLEMS FOR CHAPTER 36

Problems 1, 2: Preliminary Problems, at beginning of chapter
Problems 3-8: in text

9. ELECTRICITY vs. GRAVITATION

(a) A hydrogen atom with its one electron removed is called a hydrogen ion, or a "proton." It has a charge $+e$. Suppose a hydrogen ion and an electron are placed a distance *d* meters apart. Using the data below, write:
 (i) an expression for their electrical attraction
 (ii) an expression for their gravitational attraction
(b) Calculate the ratio of the electrical force in (a) to the gravitational force.
(c) Why do we usually neglect gravity in making calculations for atomic models?

DATA: Gravitation constant, $G = 6.6 \times 10^{-11}$ newton • meter²/kg²
Coulomb's Law constant, $B = 9.0 \times 10^9$ newton • meter²/coulomb²
Mass of electron = 1/1840 of mass of hydrogen ion (Call mass of hydrogen ion *M*)
Electron charge, $e = 1.6 \times 10^{-19}$ coulomb

10. ELECTRIC FIELDS vs. GRAVITY

A typical cathode ray oscilloscope applies electric fields of 10,000 volts/meter to swing the electron stream up to the top of the screen, or to sweep it across to one side. Compare the force that such a field exerts on an electron with the force that gravity exerts on it.

11. OSCILLOSCOPE

Inside the box of a cathode ray oscilloscope there are radio tubes (diodes with plate and heated cathode) for rectifying the A.C. supply to half-wave or full-wave "bumps" of D.C., with choke coils and condensers to smooth this to steady D.C. voltage. There are transformers for supplying A.C. to be rectified and for supplying the filament heaters at low voltage.* But the principal thing is the large cathode ray tube itself.

(a) Give a simple sketch of such a tube, in perspective or in section, labelling the main parts clearly.
(b) Explain what makes the green spot on the screen.
(c) There are at least two changes that might be made in the supplies to the tube to make the green spot *brighter*. Guess two, explaining why each change should have the desired effect. (In modern tubes, neither of the two most obvious changes is made when the intensity knob is turned to increase the brightness!)
(d) When a P.D. is supplied from the apparatus to be tested and applied to the two binding posts "V" and "G," the spot moves up or down (or for A.C. it moves up-and-down-and-up, &c.). Explain how this applied voltage produces this effect.
(e) When we wish to graph an up-and-down motion against *time*, we sweep the spot *steadily across* horizontally (then very quickly back, then across steadily again, and so on). The sweep is produced by a circuit (inside the box) essentially like the one in Fig. 36-7. *Look at the diagram and explain how the steady sweep across works.* (NOTE: The quick return motion requires a sudden switching by a radio tube. That need not be described here. It is the tube shown connected across the capacitor. It contains gas, whose resistance breaks down when the P.D. reaches a certain value.)

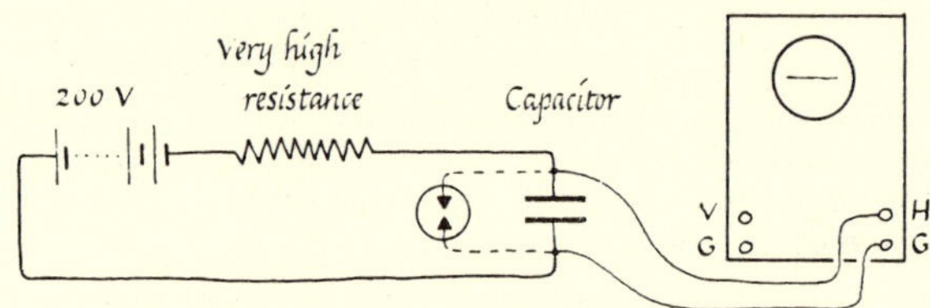

Fig. 36-7. Problem 11

* Most oscilloscopes also have amplifiers in the box, to increase the signal applied to the binding posts before it is applied to the deflecting plates. The controls for these amplifiers are usually marked "GAIN." Essentially, they use triode tubes much like that in Ch. 41, Experiment H.

CHAPTER 37 • MAGNETIC CATAPULTS: DRIVING MOTORS AND INVESTIGATING ATOMS

"The seashore crab is a politician. Threatened with danger from above, he looks straight ahead and runs away sideways."

—Anon.

Catapult Forces

A wire carrying a current across a magnetic field is pushed sideways with a force perpendicular to both wire and field. If free, it moves crab-fashion.[1] This is the "catapult force," mentioned in Chapter 34. If the current is reversed, or if the magnetic field is reversed, the force is reversed. The catapult force acts on a stream of electrons in a vacuum just as well as on a current in a wire.

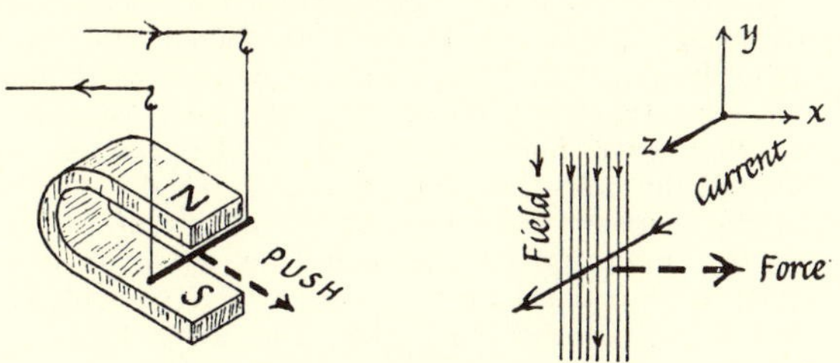

Fig. 37-1. Catapult Forces
Trapeze in magnetic field. Force, field, and current are all perpendicular to each other like x, y, z axes.

Demonstration Experiments

Bring a magnet near a cathode-ray oscilloscope—or a television tube—and see the spot move. Consider the direction of the stream and the direction of the magnet's field and decide whether the stream is pushed along the field (as it would be in an electric field), or across it, crab-fashion.

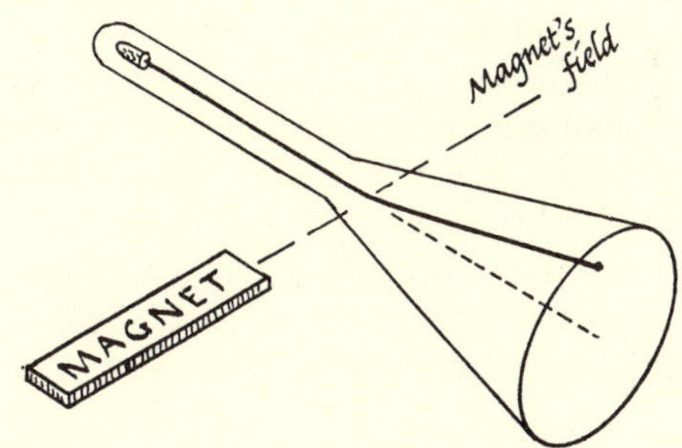

Fig. 37-4. Catapult Force
Electron stream in oscilloscope or television tube.

[1] Schoolbooks give memory-rules to tell which way the wire will move: little stories of a swimmer pointing this way or that, or a fist with fingers and thumb sprouting out all at right angles as in Fig. 37-2. These rules have gained foolish importance in elementary teaching, probably because they provide easy examination questions, while the underlying phenomenon is more embarrassing to examine. It is very important to know that the force is perpendicular to the magnetic field and perpendicular to the current—just as the thumb in Fig. 37-2 is perpendicular to both fingers. But

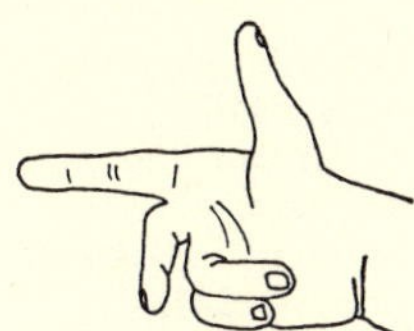

Fig. 37-2.
Three Mutually Perpendicular Directions Shown by Fingers

remembering whether the force is forwards or backwards along its strange crabwise direction is often a trivial detail. For example, if the magnetic field runs horizontally North and South and the current runs horizontally East and West, then the catapult force is vertical, pulling the wire up or

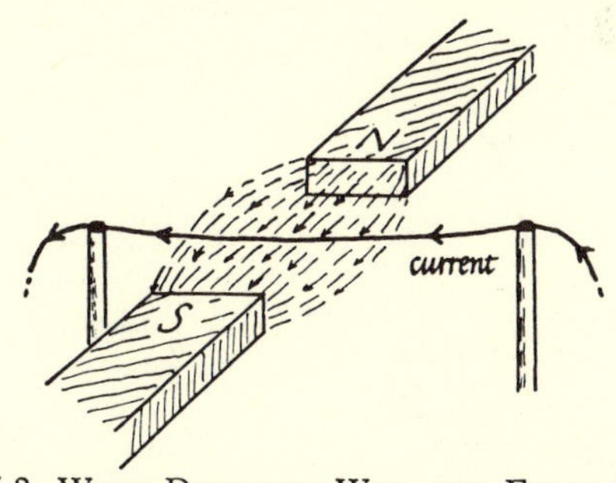

Fig. 37-3. What Direction Will the Force Have?

down (Fig. 37-3). It is important to know the force is *vertical*, but it is often trivial to bother whether it is *upward* or *downward*. In this course, the sensible answers to problems asking for directions of forces, current, etc. will be: *in-or-out*, *left-or-right*, *up-or-down*. In the example above the answer is "up-or-down."

When you need a rule, in more advanced physics—e.g. to demonstrate the essential minus sign in "Lenz' Law"—ask for the simple right-hand rule of fingers-curled-around-thumb for the circular magnetic field of current in a straight wire. That will enable you to predict the direction of catapult forces. For the single, simple right-hand rule, see Note in Problem 1.

An electron stream of "cathode rays" can be knocked off residual gas molecules (or the metal cathode) in a small discharge tube and hurled through a slit by a large gun voltage. The electrons hitting a slanting screen make it glow, showing the path of the stream by a splash of light. Try bringing a magnet near. Also try bringing a current-carrying wire near.

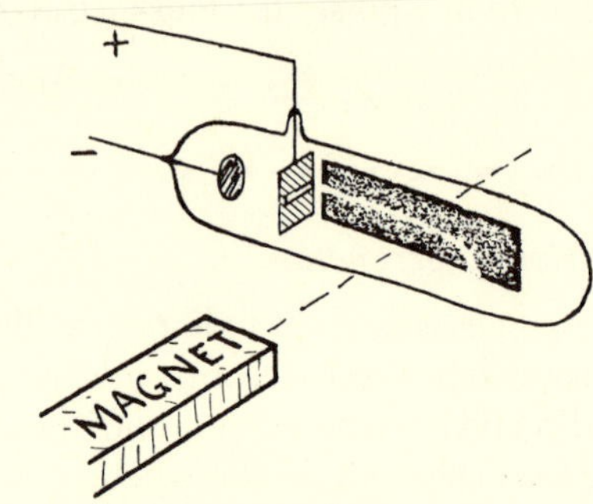

FIG. 37-5. CATAPULT FORCE
"Cathode rays" (electron stream) in discharge tube, with slanting screen.

The effect of the "catapult force" is even more clearly shown with a narrow stream of electrons from a small gun making a glow in mercury vapor or hydrogen (see Fig. 37-6). There a suitable magnetic field tugs the stream sideways, across the motion, like the Earth pulling the Moon. The streak makes a circle. For that, the magnetic field must be perpendicular to the motion. If the stream has any motion along the magnetic field that component continues without change, making the path of the stream a gorgeous spiral. This happens on a grand scale to streams of electrons ejected from the Sun when they meet the Earth's magnetic field.

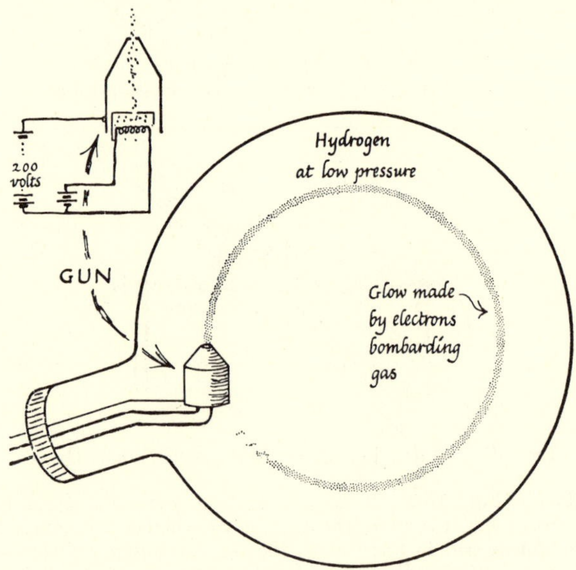

FIG. 37-6. MEASURING e/m FOR ELECTRONS
Current-carrying coils make a magnetic field perpendicular to plane of this sketch.

Uses of Catapult Forces

Catapult forces drive electric motors, work ammeters, oppose the driving of generators; they sort out isotopes of atoms; they hold the cyclotron beam in its path; and they provide measurements of e/m for atomic particles. We shall discuss "engineering" uses briefly, then "atomic" uses.

PROBLEM 1. PRELIMINARY PROBLEM ON CATAPULT FORCE

Note: In problems, use the following rules for the directions of magnetic fields (the arrows that mark the way a N-pole would move):

(a) The lines run *from* N-poles *to* S-poles.
(b) For the circles of magnetic field around a wire carrying current, use the following rule: use your RIGHT hand, curl your fingers around your thumb, point your thumb along the direction of the current; then your fingers point around the lines of force. (See Chapter 34.)
(c) For the magnetic field of a current running round a hoop coil, *either* argue from (b) above *or* use the following rule: use your RIGHT hand; stick its thumb out, as in "thumbing a ride," and curl your fingers, as if trying to wrap them around the thumb; hold your hand so that your fingers point along the current flow around the coil; then your thumb shows the direction of the magnetic field through the middle of the coil. (In fact, the roles of fingers and thumb are interchangeable: If thumb shows the current in a straight wire, fingers curl around the circles of magnetic field.)

Fig. 37-7 shows a wire, A, perpendicular to the plane of the paper, in the magnetic field of a horseshoe magnet NS.

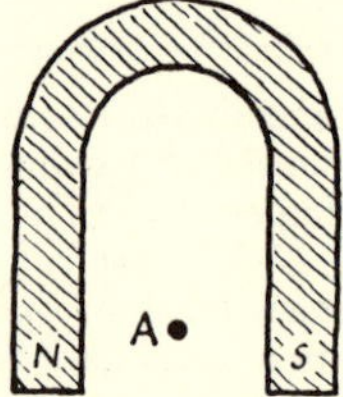

FIG. 37-7. PROBLEM 1

An electric current flows through the wire, down into the paper.

(i) Copy the sketch on a larger scale, *without the wire*, and sketch the magnet's field.
(ii) Copy the sketch *without the magnet* and sketch the current's magnetic field.
(iii) Copy the whole sketch and sketch the combined magnetic field (see Chapter 34).
(iv) Show the direction in which the wire will be pushed.

PROBLEM 2. AMMETER

The perspective sketch, Fig. 37-8, shows a coil of wire pivoted on an axle in the space between the poles of a horseshoe magnet. Current flows around the coil as shown by arrows.

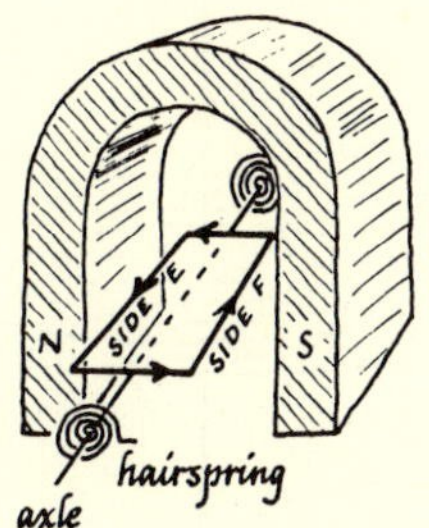

FIG. 37-8. PROBLEM 2

(i) Sketch a section, like Fig. 37-7 above but with two wires E, F (instead of wire A) to show the section of the coil.
(ii) On your sketch, show the combined magnetic field.
(iii) Which way is the side, E, of the coil pushed?
(iv) Which way is F pushed?
(v) What is the combined effect of these catapult forces?
(vi) The coil is restrained from turning by a hairspring which obeys Hooke's Law. Explain why the pointer, which is attached to the coil, indicates the strength of the current, on a uniform scale.

Motors

We can explain how a motor works by treating it as a modified ammeter.[2] The coil is no longer attached to hairsprings but is carried on an axle free to turn. The coil is embedded in soft iron, to add mass and magnetization. The permanent "field magnet" is replaced by an electromagnet, to provide a stronger field. Then catapult forces pull the coil around, as in the ammeter. When the rotor (coil + iron core) has been carried around beyond dead center by momentum, it would turn back but for an ingenious trick: the coil current is reversed. This happens every half turn, so that the coil is pulled around another half turn . . . and another . . . and so on. The reversing is done by an automatic switch, turning with the rotor itself. The switch, called a "commutator," consists of two half cylinders of copper mounted on insulator on the coil's axle. The current is carried in and out from the supply by "brushes" rubbing on these bits of copper. At one instant, the + brush feeds current into bit A and around the coil one way to B and out. Half a turn later the + brush feeds current into bit B and around the coil the opposite way—but in the coil's new position, that is just the way the current should run to maintain the motion. Practical motors have many coils with different orientations to make the motion smoother, and a more elaborate commutator reverses the current in each coil at the right moment. For details, see a real motor.

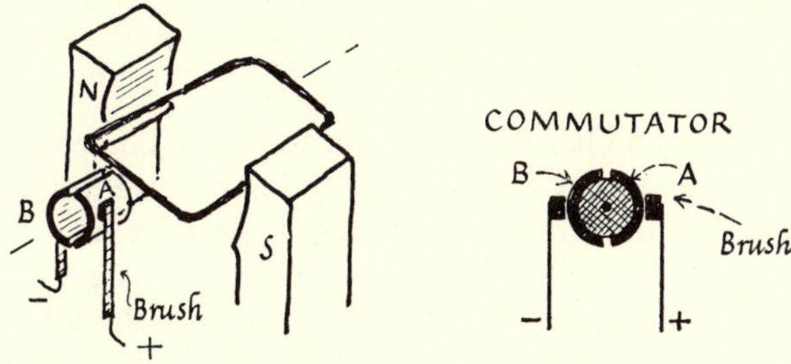

FIG. 37-9. PRIMITIVE MOTOR WITH COMMUTATOR

[2] Another view, the simplest explanation, runs: one electromagnet, the rotor, is pulled round by another electromagnet, the field magnet, with the commutator reversing the rotor current every half turn so that the motion continues.

The Law of Catapult Forces

For measurements and explanations in "atomic" uses of catapult forces, we need a definite rule expressing the force in terms of current, length of wire, etc. We shall produce a rule but its development will seem complicated—the most difficult "formula" in the course. However, the rule is essential for understanding atomic physics—without it, you would have to be given childish descriptions of apparatus without a genuine explanation. So you should study the rule's development below and learn to use it.[3]

In this course, we define current-measurement by copper plating,[4] so the magnetic effects of currents are matters for experimental investigation. Experiment shows that the FORCE on a wire carrying a current across a magnetic field varies directly as the CURRENT.

DEMONSTRATION EXPERIMENT. FORCE vs. CURRENT?

Hang a wire or a coil from a balance. Send measured currents through it. Apply a strong magnetic field across the wire, and weigh the catapult force.

[3] However you should *not* memorize the formula forms—that would not be learning science. As an assurance that such parrot-learning is not called for, all standard formulas, and especially these, ought to be printed on any examination paper that might need them.

[4] In advanced treatments of electrodynamics, current is now defined and measured by magnetic effects. On that system, a current's magnetic field varies directly as the current—automatically, *by definition*—and so does the force on a current in a magnetic field. And, on that system, Faraday's first law of electrolysis is an experimental law.

The basic *experimental* fact is that when we increase the "current," whatever that may mean, its chemical and magnetic effects both increase in the same proportion. If we choose one effect to measure "current," then the experimental fact shows that the other effect varies as that "current." Compare this with the position of Charles' Law in gas theory.

Fig. 37-10 shows suitable apparatus to show that FORCE ∝ CURRENT.

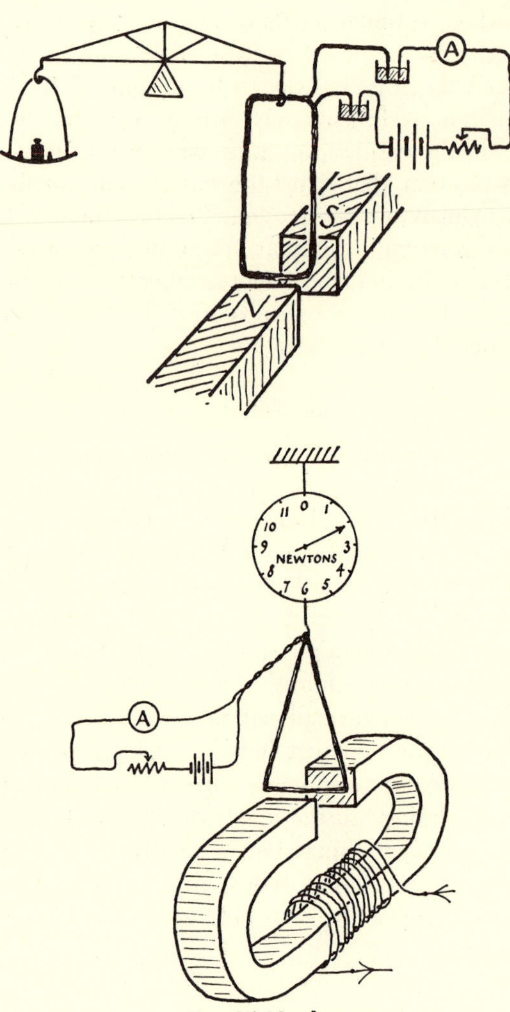

Fig. 37-10a, b.

To find the other factors determining catapult forces we shall survey some general information, then guess at a simple rule, then test the rule.

Solenoids carrying currents behave like bar-magnets of the same shape and size. Hoop coils behave like very short fat bar-magnets. Two coils attract, repel, and twist one another just like the equivalent bar magnets. Unlike bar magnets however, such coils have magnetic fields that run right through them inside, the lines of force making closed loops (see sketches asked for in Chapter 34). The magnetic field running through the center of a circular hoop is almost uniform over a considerable region, and we shall use this for measurements.

When currents run through two long parallel wires, A and B, each wire finds itself in the circles of the magnetic field of the other wire's current. So each exerts a catapult force on the other. The circles due to A's current cut B at right angles. The catapult force on B is perpendicular to those circles, and perpendicular to B. So it must be directed straight across to A. If you sketch the actual lines of the *combined* magnetic fields you will find that the wires attract if their currents run the same way, repel if opposite ways.

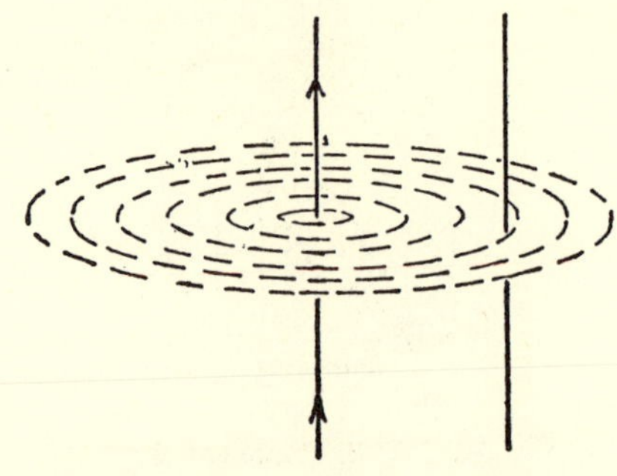

Fig. 37-11. Catapult Force Between Parallel Wires
Current in one wire makes field which cuts other wire at right angles. If that other wire also carries a current, it experiences a catapult force. And then the first wire experiences an equal and opposite force.

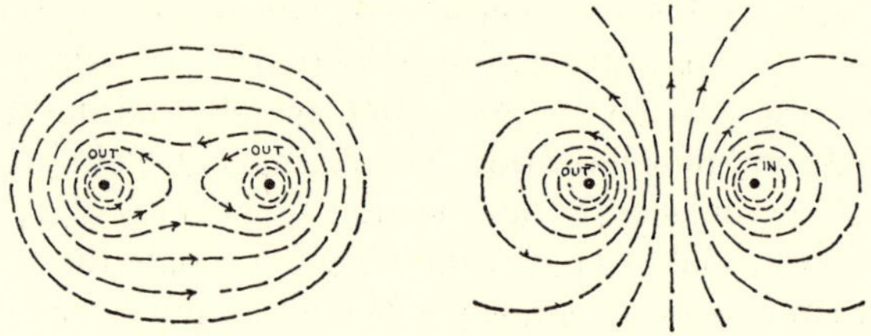

Fig. 37-12.
Combined Magnetic Fields of Parallel Currents
Suggest: attraction between like currents and repulsion between opposite currents.

For a simple rule of catapult forces we do not use long wires or complete coils. Instead we try to simplify the story by choosing a short bit of wire carrying current. Then we can treat any long wire, coil, or complete circuit as made up of short bits and find the force on the whole by adding up the forces on its bits. This is useful for calculating forces on the coils of motors, ammeters, etc. When we ex-

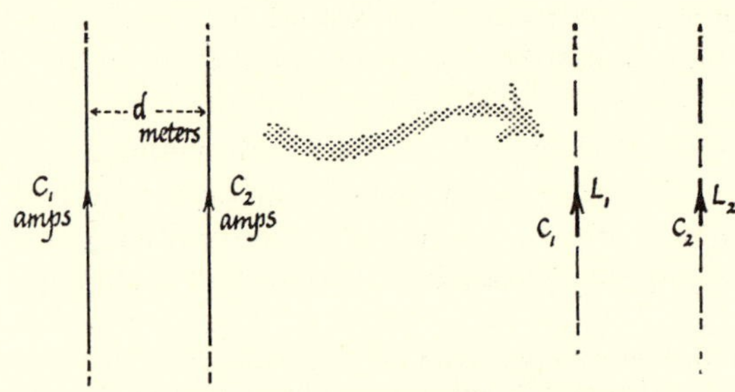

Fig. 37-13. "Current-Elements"
From two long wires, choose short bits L_1, L_2, opposite each other.

change a short bit of wire for a single moving electron, our simplified scheme comes into its own. Ampère and others made ingenious guesses at such a rule a century ago, but they had no way of testing their guess in detail, because they had only complete circuits. However, they did test their predictions successfully on circuits of various shapes.[5]

To guess our rule we start with two long parallel wires carrying currents C_1 amps and C_2 amps, d meters apart. These would exert sideways catapult forces on each other, as in Fig. 37-14. Then we

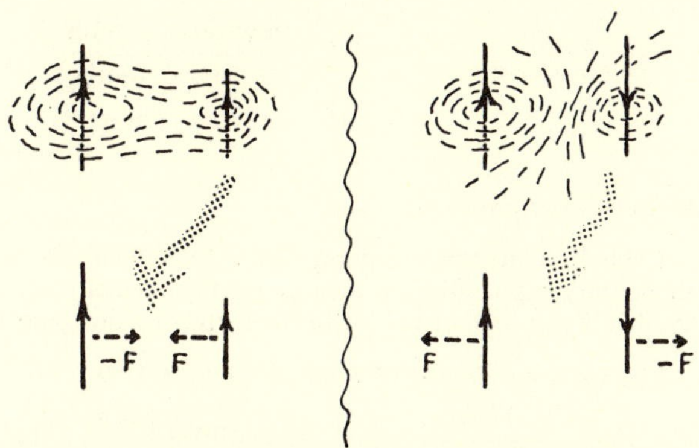

FIG. 37-14. FORCES BETWEEN CURRENT-ELEMENTS are suggested by their combined magnetic fields.

choose two very short bits of wire opposite each other, L_1 and L_2, and ignore the rest of each wire. We regard these as sections of long parallel wires and expect each to have a magnetic field of circles around it. With currents in the same direction, the catapult forces will pull these two "current-elements" towards each other. (Wire L_1, for example, carrying current C_1, finds itself cut by circles of magnetic field due to C_2 in wire L_2; so it experiences a catapult force F to the right.)

From the demonstration above we know that this FORCE varies directly as the CURRENT in the "victim" L_1.

$$\therefore\ F \propto C_1 \quad \text{(by experiment)}$$

If we made the "victim" twice as long by having two L_1's in series, we should have two F's, or double the force on double length. FORCE on "victim" should vary directly as LENGTH of victim.

$$\therefore\ F \propto L_1 \quad \text{(guess, by thought-experiment or common sense)}$$

$$\therefore\ F \propto C_1 \text{ and } F \propto L_1 \quad \text{or} \quad F \propto C_1 \cdot L_1$$

[5] Many texts simply announce "Ampère's formula" with little explanation or experimental justification. No wonder electrodynamics seems like authoritarian witchcraft. Follow the discussion through to satisfy yourself it is reasonable; but do not learn it.

But the scheme is symmetrical—who can say which is the "victim" and which is the "agent" providing a magnetic field?

$$\therefore\ F \propto C_2 \cdot L_2 \text{ as well as } F \propto C_1 \cdot L_1 \quad \text{or}$$
$$F \propto (C_1 \cdot L_1) \cdot (C_2 \cdot L_2)$$

The full catapult rule must involve the distance d between the bits of wire. Simple experiments show that F decreases as d increases. Knowing that, what would you guess? The likely guess of an inverse-square law proves successful when tested by experiment. Then:

$$F \propto \frac{(C_1L_1)(C_2L_2)}{d^2} \quad \text{or} \quad F = B\,\frac{(C_1L_1)(C_2L_2)}{d^2}$$

where B is a general constant.

The rule is of little use in this form. It needs a modifying factor for slanting directions—which we shall avoid by choosing simple geometry. In practice we use complete circuits; so we make L_1 the end of a long rectangular coil (but, later, a bit of an electron's path). For convenience we do not use one short L_2 but take many of them in series and wind them up into a hood coil with L_1 at its center. [For this, refer to Fig. 37-15 on next page.] Then each bit of the hoop, carrying C_2 amps, makes rings of magnetic field cutting across L_1 at the center, and each piece is at distance R, the radius of the hoop, from L_1. Then the force on L_1 is given by:

$$F = B\,\frac{(C_1L_1)(C_2 \cdot \text{first } L_2)}{R^2} + B\,\frac{(C_1L_1)(C_2 \cdot \text{second } L_2)}{R^2} + \text{etc. (for all } L\text{'s of the hoop)}$$

$$= B\,\frac{(C_1L_1)(C_2)}{R^2}\ (\text{first } L_2 + \text{second } L_2 + \ldots \text{ all the hoop})$$

$$= B\,\frac{(C_1L_1)(C_2 \cdot 2\pi R)}{R^2}$$

If the hoop coil has N turns of wire,

$$F = B\,\frac{(C_1L_1)(C_2 2\pi RN)}{R^2}$$

Having made a guess at a rule, we test it by measuring the force on the short end of a rectangular coil carrying current, at the center of a hoop coil carrying current. You may see tests such as those sketched in Fig. 37-15; otherwise you must take this "formula" on trust.

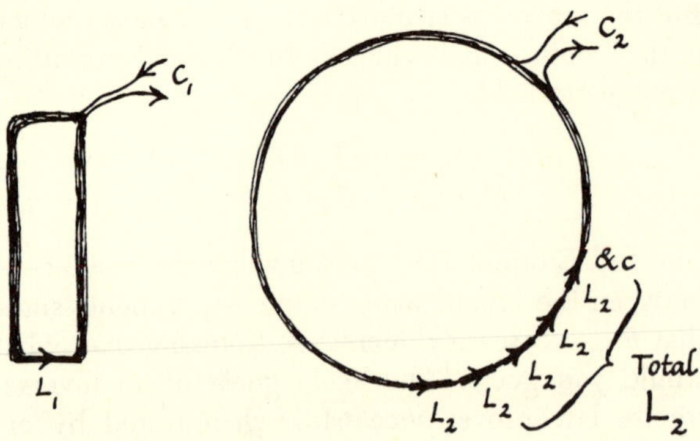

FIG. 37-15a.
CURRENT-ELEMENTS IN TESTS OF CATAPULT FORCE RULE
The force on a short end, L_1, is measured. Hoop-coil carrying current C_2 is treated as made up of many short elements, all at distance R from center.

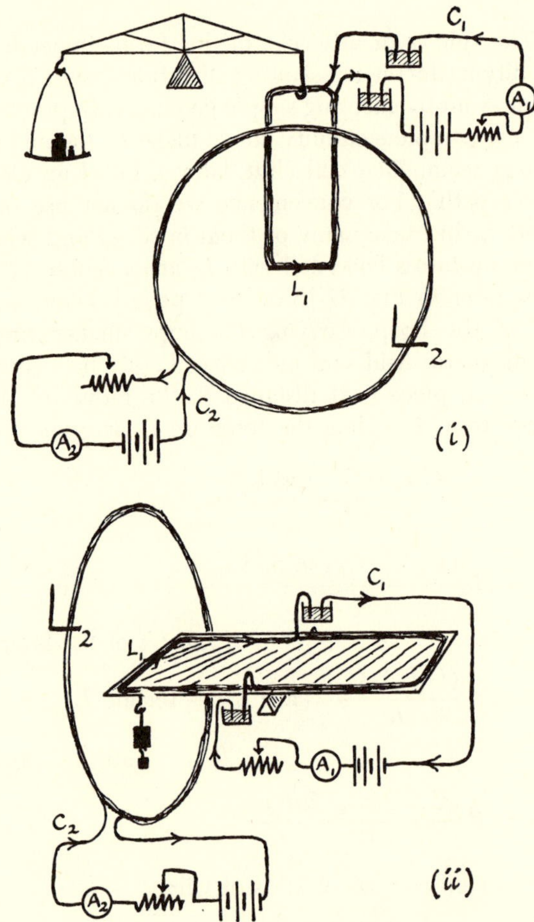

FIG. 37-15b.
WEIGHING A CATAPULT FORCE TO TEST PARTS OF THE RULE
(i) The long coil is hung on a weighing balance and fed via mercury cups, or (ii) the long coil is laid on a see-saw and fed via mercury cups.

Measurement of the Constant B

If we make measurements (dimensions, both currents, and force) in a demonstration such as that sketched in Fig. 37-15b we can estimate the constant B. Accurate measurements yield $B = 0.000000100$ or 10^{-7}. It is in fact a round number, 1/10,000,000 because the size of the ampere was chosen to make it so. In consequence, our definition in terms of copper plating has to use an unround number, 0.000000329 kg of copper per sec. Hereafter we shall write 10^{-7} instead of B, to avoid confusion with the other B used in Coulomb's law of force between charges. When using the catapult rule, remember that this 10^{-7} is not a pure number like 2π, but has units:

$$B_{\text{for catapult forces}} = 10^{-7}\,\frac{\text{newtons}\cdot(\text{meter})^2}{(\text{amp})^2\cdot(\text{meter})^2} = 10^{-7}\ \text{newtons}/(\text{amp})^2$$

Maxwell's extraordinary prediction

(The discussion of this paragraph is too difficult for an elementary explanation: it must remain a piece of magic. Omit it if you like, or read it for its surprising outcome.)

For catapult forces, of magnetic field on current:

$$F = B_M\,\frac{(C_1L_1)\,(C_2L_2)}{d^2}\ \text{where}\ B_M = 10^{-7}\,\frac{\text{newton}}{(\text{amp})^2}.\ \text{This}$$

B_M relates to *magnetic* fields.

For Coulomb's Law of force between electric charges:

$$F = B_E\,\frac{Q_1Q_2}{d^2}\ \text{where}\ B_E = 9.0\times10^9\,\frac{(\text{newton})\cdot(\text{meter})^2}{(\text{coulomb})^2}.$$

This B_E relates to *electric* fields. The two B's are quite different. We expect no relationship between them until we find that a changing magnetic field generates an electric field. (In magnets-and-coils experiments (C in Ch. 41) you find that a moving magnet generates an electric field that drives a current around a coil.) A century ago, Maxwell made a brilliant guess, that the reciprocal effect also occurs: a changing *electric* field generates a *magnetic* field. With a steady current—e.g. a stream of electrons with constant velocity—there is a moving electric field and a stationary magnetic field. But with a changing current—e.g. electrons *accelerating* up and down a radio antenna—there are changes in the moving electric field and changes in the accompanying magnetic field. Such *changes* in electric field and magnetic field ought to travel out together as an electromagnetic wave, changes in one field continually generating the other field. That was Maxwell's discovery of radio waves, from theory. The two B's, one for magnetic fields, one for electric, must be related. Look at the fraction:

$$\frac{B_E\ \text{for forces between electric charges}}{B_M\ \text{for catapult forces}}.$$

This has units $\dfrac{(\text{newton})\cdot(\text{meter})^2/(\text{coulomb})^2}{\text{newton}/(\text{amp})^2}$

or $\dfrac{(\text{meter})^2\cdot(\text{amp})^2}{(\text{coulomb})^2}$ or $\dfrac{(\text{meters})^2}{(\text{sec})^2}$ or $(\text{meters}/\text{sec})^2$.

These are units of (VELOCITY)^2

Take the value of this fraction $\frac{9.00 \times 10^9}{10^{-7}}$ or 9.00×10^{16} and take its square root, to find the value of that velocity: 3.0×10^8 meters/sec (= 186,000 miles/sec). This is a well-known velocity, the *speed of light.* Maxwell showed from his detailed theory that this fraction $\sqrt{B_E/B_M}$ not only has the units of velocity but must be the velocity of a wave of changing electric and magnetic fields. Light then appeared to be electromagnetic waves, with its measured speed agreeing with the speed calculated by pure theory from two constants measured in the electrical laboratory. Light, radio waves, X-rays, . . . all electromagnetic waves travel with this speed in space.

If you have seen the two B's measured, one by weighing the force between charged balls, the other by weighing the catapult force between currents, you can predict the speed of light from measurements you have seen.

Catapult Force on Moving Electron or Ion

We need to make one more change in our rule: replace the bit of wire L_1 carrying C_1 by a single moving charge, such as a flying electron. Catapult forces certainly act on moving charges; you can see a stream of electrons being bent in a cathode-ray tube. And you may imagine electrons streaming along inside the wire L_1 to make the current C_1.

Suppose C_1 is just a drifting of n charged particles, each with charge Q coulombs drifting with speed v along the wire L_1.

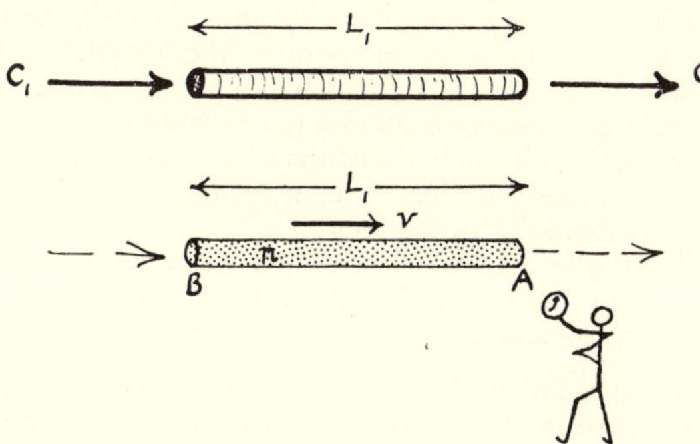

FIG. 37-17.

Post an observer at the outflow end of L_1 to count the charged particles and read a clock and calculate the current C_1. He starts his clock at the instant the first particle A emerges. He stops it when the last particle B emerges some time later because it has to travel along L_1 with speed v. That takes time L_1/v seconds. During that time the observer sees n particles emerge each with Q coulombs. He estimates the current $\frac{\text{CHARGE}}{\text{TIME}}$ as $\frac{nQ \text{ coulombs}}{L_1/v \text{ secs}}$ or $\frac{nQv}{L_1}$ amps. Then instead of C_1 amps we have nQv/L_1 amps. And instead of C_1L_1 in the catapult rule we should write $\frac{nQv}{L_1} \cdot L_1$ or nQv.

Then the catapult force on a batch of n charged particles is given by

$$F = 10^{-7}(nQv)(C_2 \cdot 2\pi RN)/R^2$$

The force on a *single* particle with charge Q, speed v is given by

$$F = 10^{-7}(Qv)(C_2 \cdot 2\pi RN)/R^2,$$

replacing C_1L_1 by Qv,

or $F = 10^{-7}(Qv)(H)$ where H is called the magnetic field; and at the center of a hoop coil, H has value $C_2 \cdot 2\pi RN/R^2$

This is the force that bends the path of a stream of electrons or any other charged particles. The force is always perpendicular to the motion, so it cannot change the speed of the particles. Their velocity changes in direction, not in size. This force, $10^{-7}QvH$, is called the "Lorentz force," in honor of the Dutch physicist H. A. Lorentz who first used it for electrons.

PROBLEM 3. PATH OF ELECTRON STREAM IN A MAGNETIC FIELD

A stream of electrons, each with *negative* charge is shot horizontally due North in a region where there is a vertical uniform magnetic field.

(a) What is the direction of the force exerted by the magnetic field on the moving electrons in the stream?

(b) When the direction of the stream has been changed by this force, the catapult force on the electrons must have a new direction. Describe the new direction.

(c) As the motion continues to change direction, the electrons keep the same speed (in a vacuum). Why?

(d) Is the catapult force larger, smaller, or same size in the new direction?

(e) As the motion continues to change direction, the stream's path is bent into a curve. What curve?

(f) Suppose you wished to make the electrons *slow down.* How could you do that? (This question is harder than it looks. It requires careful thought; you should be able to answer it successfully. There are several good methods. Include a sketch.)

Focussing

A stream of charged particles shot across a magnetic field pursues a circular-path. This has a peculiar extra benefit: focussing. Picture a small fan of electron streams emerging from a gun, all with the same speed. Now apply a magnetic field perpendicular to all the streams. Then each stream is bent into a circle. In Fig. 37-18, the middle stream of the fan, C, is bent into the circle shown. Then each of the streams is bent into a circle of the same

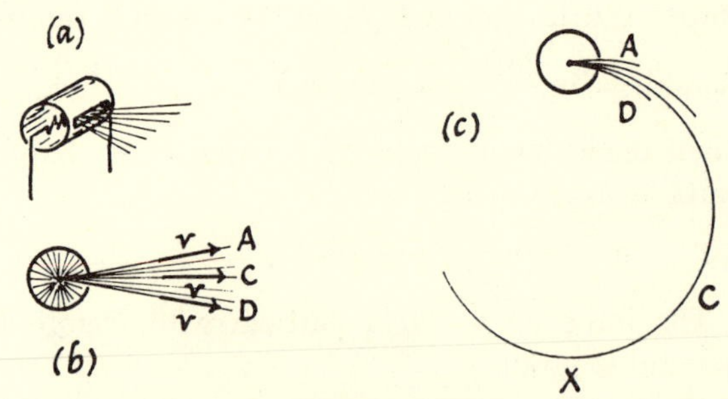

Fig. 37-18. Focussing
A fan of electron streams is "focussed" by a magnetic field. Try completing a sketch like this.

size, but tilted to start from the gun in its proper direction. Here, the streams A and D are shown starting on their circles. The complete circles practically intersect at X, the opposite end of a diameter from the gun. We use this focussing property in measurements and in machines.

PROBLEM 4. FOCUSSING

Draw a small fan of electron streams emerging from a small slit (all with the same speed). Draw their circular paths carefully with a compass (or run your pencil around a round can or bottle) and make a diagram to demonstrate focussing.

PROBLEM 5. STREAM OF CHARGED ATOMS IN MAGNETIC FIELD

A stream of charged hydrogen atoms (= H nuclei, protons) is shot at high speed across a uniform magnetic field.

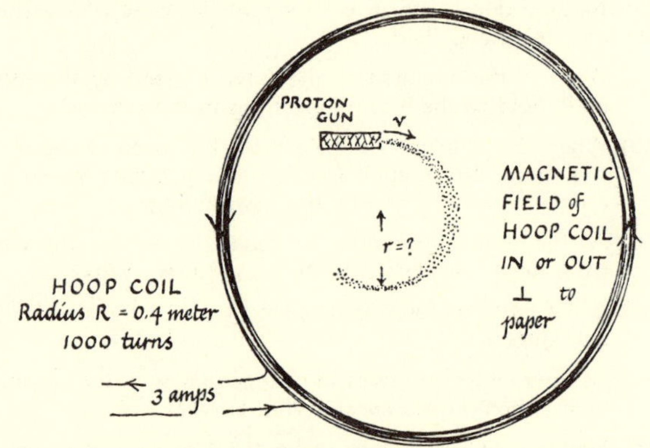

Fig. 37-19. Problem 5.

(a) What shape will the path of the stream take?
(b) Calculate the radius, r, of the path of the stream in the field, using the data given, by answering questions (i) and (ii) below.

DATA: Each atom has mass 1.66×10^{-27} kg, charge $+1.60 \times 10^{-19}$ coulomb, and speed 100,000 meters/sec.
The magnetic field is made by a current passing around a hoop coil. The coil has 1000 turns of wire, of radius 0.40 meter, and carries 3 amps.
The stream is shot across the central region of the coil, in the plane of the coil.

(i) Calculate the force on each moving atom, due to the magnetic field.
(ii) Remembering that this force is perpendicular to the path of the atoms, calculate the radius r of their path. (Remember that motion in a circle needs an inward force mv^2/r. The magnetic field, acting on the moving charge, provides this force.)
Warning: Do not confuse r for orbit with R for hoop windings.

PROBLEM 6. STREAM OF ELECTRONS IN MAGNETIC FIELD

Suppose the charged particles of Problem 5 are electrons instead of H nuclei, with the same speed, same charge (but negative) but mass only 1/1840 as big. Describe the path of this stream in the magnetic field.

Containing a Nuclear Fusion Furnace

If we succeed in running a controlled nuclear fusion "furnace" to supply vast quantities of energy, we cannot hold the reacting material in any solid container; because the reaction will proceed at a temperature of tens or hundreds of millions degrees. Problem 7 hints at a possible way of using a magnetic field to "contain" the charged particles in a nuclear fusion scheme.

PROBLEM 7. SLANTING STREAM OF ELECTRONS IN MAGNETIC FIELD

Suppose the electron stream of Problem 6 has the same velocity perpendicular to the field, but also some velocity *along the field*. (For example, it might be shot 1.41 times as fast as before, in a direction making 45° with the field. Then it would have two equal components of velocity: 100,000 m./sec across the field, and 100,000 m./sec along the field.) Describe the path of this stream.

Derivation of Speed of Electromagnetic Waves

[This is a crude attempt—at best, a plausible story—that neglects the restrictions that would be imposed by Relativity and modern electrodynamics. Ignore it or treat it lightly as a way of seeing how Maxwell's prediction could make sense in terms of the knowledge you already have. In this form it is not sound physics, but it is offered here as a reminder of two things:
(a) although such derivations may need advanced mathematics for a proper treatment, they are not essentially mysterious.
(b) the way in which a proper treatment is worked out often begins with a rough treatment like this—any means to the end, in the first round.]

We shall follow the lines of the argument we used to find the speed of waves along a taut rope. Before studying the derivation below, review that argument at the end of Chapter 10. There we derived the speed by considering a "knee" travelling along the rope. Here we shall calculate the speed of a "knee" along the lines of an electric field instead of a rope.

Suppose an experimenter E holds a charge $+Q$ at rest at the origin. Suddenly, E gives an upward tug and starts Q moving upwards with velocity u. Thereafter, E keeps Q moving up along the y-axis with constant velocity u. The message of the sudden starting of Q's motion, from rest to u, will travel out as a wave—along lines of Q's electric field, if you like.

Out along the x-axis there is originally a horizontal electric field, due to Q. But when Q has been moving for a time t, and has travelled up a distance ut, the line of electric field that was originally horizontal now slants down from Q to a "knee," K, and then continues along

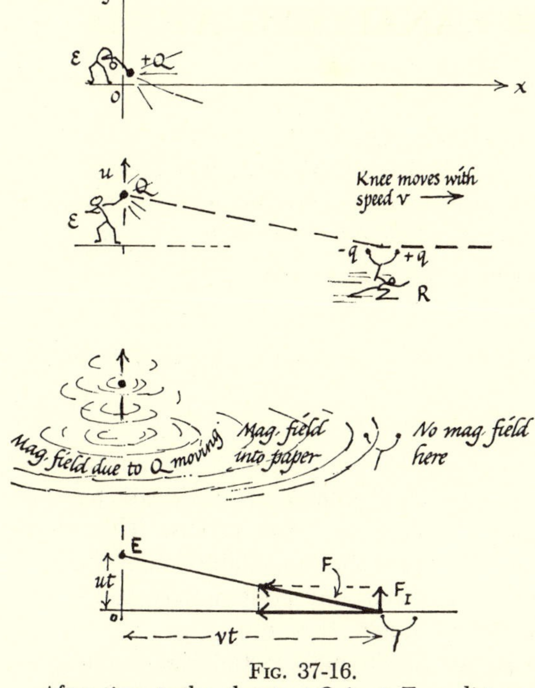

Fig. 37-16.
After time t, the charge $+Q$ is at E, a distance ut above the origin, O; and the knee, K, is at distance vt out from O.

the original horizontal line. The wave-message of Q's start has only just reached K. The bend of the knee is a piece of wave pattern, and it travels along with wave-speed, v.

Now suppose another experimenter, R, runs along beside the knee at wave-speed v, keeping level with it. For a rope-wave, the runner carried a box that enclosed the knee without exerting any force on it—so the rope tensions had to provide all the forces needed to maintain the knee and its motion. In this case, we imagine the runner carrying a small Y-shaped handle of insulator with test charges $+q$ and $-q$ on its prongs. We stand still and watch him running and we see him carrying *no net charge* on the Y, so we expect him to experience no net force on the Y due to Q or its motion. We say, "Look, the Y has no net charge; therefore it could be put inside a little black box which would coast along unaided, unaware of the knee in the electric field."

If the Y keeps completely ahead of the knee (or completely behind it), both $+q$ and $-q$ are in the same electromagnetic field; so we certainly find no net force on the Y, no evidence of the charge Q or its motion. But even if the Y just keeps level with the knee (with $+q$ ahead of the knee and $-q$ behind), the resultant force on the Y should still be zero—there, too, the Y should be able to coast along with speed v without help from the runner. A test-object coasting along with a wave like a surf-rider should feel no force from the wave-disturbance. (This is the essential assumption and it is *not* an obvious or convincing one.[6] If you accept it: read ahead to see how v is found. If you consider it unreasonable: you have seen the weak point in the argument.)

Now consider the forces we expect to see acting on $+q$ and $-q$ at the knee. The forward charge, $+q$, feels only a horizontal push outward from Q, since it is beyond the knee, in the original horizontal field. And the other charge, $-q$, is pulled along the slanting line, inward and upward towards Q. However, $-q$ feels a catapult force also, due to the motion of Q, since the news of Q's motion has reached it. Argue out the direction of that catapult force and you will find it is vertically *downward*. If Q's speed is small, so that its travel, ut, is small compared with the knee's travel, vt, ($u \ll v$), the horizontal forces on $+q$ and $-q$ practically cancel, because $+q$ and $-q$ are at practically the same distance from the charge Q that provides the field. There is no vertical force on $+q$; so the two vertical forces on $-q$ must just cancel if the resultant force on the Y is to be zero. Those two vertical forces are:

I. The vertical component of the electrostatic attraction between $+Q$ and $-q$.

This is the vertical component of $B_E \dfrac{Qq}{r^2}$

This is $B_E \dfrac{Qq}{r^2} \cdot \dfrac{OE}{EK}$ or $B_E \dfrac{Qq}{r^2} \cdot \dfrac{ut}{vt}$

since EK is practically[7] equal to the distance travelled by the wave, vt.

II. The catapult force on $+q$ moving with speed v due to the motion of Q with speed u. These are both *moving charges* instead of *short wires* carrying currents. For them, we must make the change described above and write Qu instead of C_1L_1 and qv instead of C_2L_2. Furthermore, these two currents" are not parallel but at right angles to each other in space. However, the magnetic field of (Qu), when Q is still near the origin, will be a set of circles that cut the "current" (qv) at right angles (Fig. 37-16d), so our expression for the catapult force should hold.

$$\text{Catapult force} = B_M \frac{(C_1L_1)(C_2L_2)}{r^2}$$

$$= B_M \frac{(Qu)(qv)}{r^2}$$

If there is no net vertical force on the Y, the forces I and II are equal and opposite:

$$B_E \frac{Qq}{r^2} \cdot \frac{ut}{vt} = B_M \frac{(Qu)(qv)}{r^2}$$

Cancel Q, q, t, u and r^2. All the details of our artifice disappear.

$$B_E/v = B_M v \quad \text{or} \quad v^2 = B_E/B_M.$$

Therefore, wave velocity, v, $= \sqrt{B_E/B_M}$.

This is the speed at which the knee must travel along. A wave of any other shape can be treated as made up of "knees," all travelling with this speed.
This is the universal speed of electromagnetic waves in free space, for any wave-form or frequency.

---

[6] If you like, reflect that Q will have no force exerted on it by the charges on the Y; because, although they may be big charges, they are infinitely close together. Therefore, E need not exert any force, need not provide any energy, to keep Q moving—Q will just coast upwards unaided. Therefore, we should hardly expect to find Q exerting a resultant force on the Y.

[7] Even this approximation is removed when the magnetic field is properly calculated for the *slanting* direction EK.

CHAPTER 38 • ANALYZING ATOMS

"For it is not knowing, but the love of learning, that characterizes the scientific man; while the "philosopher" is a man with a system which he thinks embodies all that is best worth knowing. If a man burns to learn and sets himself to comparing his ideas with experimental results in order that he may correct those ideas, every scientific man will recognize him as a brother, no matter how small his knowledge may be."

—C. S. Peirce

Sorting

EXTRACT FROM MANUFACTURER'S CATALOG SENT TO BANKS: "This motor driven machine sorts and counts batches of mixed coins with great speed . . . will count all coins from pennies to half dollars, inclusive, continuously into bags or in predetermined amounts into paper coin wrappers . . . two sets of counting dials: one records each denomination of coin in dollars and cents . . . second set records each denomination of coin numerically, as, 399 pennies, 204 nickels, etc., ..."

THE BANKER of earlier days took time and trouble to count his cash: a rough weighing of a bag of coins was followed by laborious counting. The modern banker buys a machine. The scientist of earlier days was in an even poorer position with atoms: he could separate them chemically into batches of each element, but then he could only weigh them by the bagful; he could not get his fingers on individual atoms. He had to assume that nature, like a good treasury, minted all atoms of one element alike—and that assumption was a mistake. Now we have a machine that sorts out atoms as efficiently as any coin-sorter. That machine, the *mass-spectrograph*, uses the tools discussed in the last two chapters (electric and magnetic fields) to sort any sample atom by atom, weighing the atoms as it does so. If it did no more than sort atoms by elements, that would be a masterpiece of chemical analysis; but it does far more than that for us. Early measurements with its ancestor gave a simple analysis of atoms into ingredients; the mass-spectrograph itself gave a simple rule for nuclear masses. Then more precise measurements showed small variations from the simple rule; and those variations, far from spoiling our faith in nature, have led to nuclear knowledge of great importance.

Before discussing modern mass-spectrographs, we shall look at the history of this "atomic analysis" that began early in this century.

Atomic Fragments

All negative electrons are alike, but positive ions —the remainders of atoms after an electron has been chipped off—come in great variety. This is the simplest splitting of the "uncuttable" units of matter, atoms, into electrons and positive ions, and it gave a first hint of atomic structure.

The easiest way to manufacture these remainders for analysis is to bombard gas molecules. Nowadays we fire a stream of electrons from a small gun at a sample of gas. Half a century ago when the analysis was begun, the bombardment was done by applying a large voltage to gas at low pressure so that electrons and ions made more ions by collision. Then in the glowing "electric discharge" between the electrodes there was a mess of moving things: positive ions, electrons, negative ions, neutral atoms and molecules, X-rays, and some visible light. The electrodes (metal plates) were moved in from the ends and a hole was drilled in each (Fig. 38-1). Then

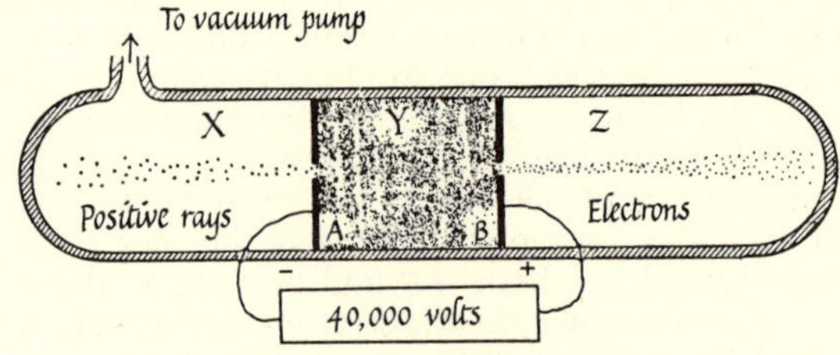

FIG. 38-1. DISCHARGE TUBE, with gas at low pressure, with holes in the electrodes to allow streams of electrons and positive ions to continue beyond the electric field.

a stream of charged particles continued through the hole in each plate into the space beyond. A stream of electrons shot through the hole in the positive plate. Quite a different stream shot out the opposite way through the other plate: a stream that proved to be made of much more massive particles carrying positive charges. Both streams were analyzed by the use of electric and magnetic fields.

The *electron stream* behaved best with an almost perfect vacuum so that there was little gas around to slow the electrons by collisions or to sabotage the electric field by a shield of ions, and then it showed clearly a single brand of particles—the same e/m—from every kind of material.

For a copious stream of *positive rays*, some gas had to be left in the tube to provide the ions which are the positive rays. That required continual pumping and a very thin hole in the electrode to maintain a good enough vacuum in the analysis region. Then measurements of deflections showed that the positive rays consist of particles of much greater mass than electrons; different masses according to the gases and vapors providing them:

hydrogen gave mass $1840m$, with charge $+e$, compared with the electron's m and $-e$;
oxygen gave mass $16 \times 1840m$ with charge $+e$ and sometimes $+2e$;
carbon (vaporized from a solid sample or broken off a gas such as CH_4) gave $12 \times 1840m$ and $+e$.

This was one of the great measurements of atomic physics, a weighing of the other ingredient of atoms beside electrons. The first guesses about the *internal structure of atoms* could be made from these measurements of the fragments of atoms: atoms must be made up of light electrons combined with positive remainders that carry most of the mass. Atoms, the "unsplittable" elements of chemistry, were being analyzed into components.

Problem 1 illustrates the general idea of this analysis, in a fictitious simplified form.

PROBLEM 1. ANALYZING ATOMS IN GAS DISCHARGE

Fig. 38-2 represents a discharge tube, with metal plates A and B as electrodes, with a large P.D. applied to make a strong electric field in the region Y between them. Assume there is no horizontal electric field in the outside regions, X and Z. There is a little gas in the region Y so that both electrons and positive ions are produced there. Most of the electrons are dragged across by the field to hit plate B, but some continue through the hole in a stream through the space Z. A vertical electric field between plates P_Z and P_Z' bends the stream downward. A magnetic field, perpendicular to the plane of the paper, acts on the stream all over the region Z and that too bends the stream downward. Now consider the effects of the same fields on positive ions traveling through the hole in A to the region X.

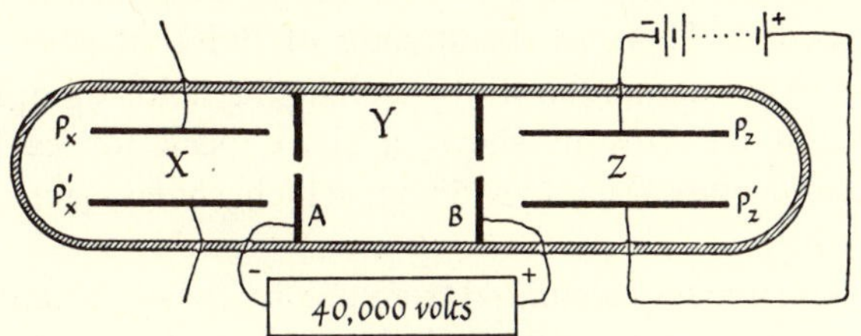

FIG. 38-2. DISCHARGE TUBE WITH DEFLECTING PLATES (See Problem 1.)

(a) Which way would the same electric field between P_X and P_X' bend the positive rays in X, up or down?
(b) Which way would the same magnetic field bend them, up or down? Why?
(c) Suppose some electrons whizzing through Z were originally released near Plate A and "fell" through the whole gun-voltage V in Y (as many do in such a tube). And suppose some positive ions, with an equal charge $+e$, started near B and emerged into X also having fallen through the full gun-voltage V. Applying equal *electric* fields between plates P and P′ to the streams in X and Z would give *equal deflections to both streams* and thus would fail to show the difference between particle masses. Explain why the deflections would be equal. (Do not give detailed calculations. Just devise a qualitative explanation by considering horizontal velocities; or argue with rough algebra.)
(d) (i) If the two types of particle have the same size of charge and fall through the same gun-voltage V, they all emerge with the same kinetic energy.
$$\tfrac{1}{2}Mv_i^2 = Ve \text{ for ions}$$
$$\tfrac{1}{2}mv_e^2 = (-V)(-e) \text{ for electrons}$$
Now suppose the same magnetic field, H, is applied to both streams (in regions X and Z), exerting a force on a particle with charge Q moving with speed v, given by
$$F = 10^{-7}(Qv)(H)$$
The path of each stream will be bent into an arc of a circle. Compare the bendings by expressing the ratio of the radii, R_i/R_e, in terms of M and m.
(ii) If M/m is about 26,000 (as it is for nitrogen ions) and a certain magnetic field gives the electron stream a radius 0.05 meters ($\approx$ 2 inches) what radius would it give the ion stream?
(e) If some of the particles start from the middle of Y region and fall through only $\tfrac{1}{2}V$, how will that affect the deflection by
(i) the electric field
(ii) the magnetic field?

As Problem 1 suggests, electric fields alone will not enable us to sort out masses of moving ions. This is because one electric field has already been applied along the stream by the accelerating gun-voltage: using the same tool again, in an electric field across the stream, will yield no new information. But by applying a magnetic field after the gun-voltage has produced the stream we can sort out the ions in the stream. If we apply the gun-voltage between A and B in Fig. 38-2 and a magnetic field across the stream in region X we get deflections proportional to values of e/Mv. Thus as well as raising problems of maintaining a good vacuum, positive rays need a much greater magnetic field for their analysis than electrons. As Problem 1 (e) suggests, ions that start in different parts of region Y will have different energies, so even for one e/M there will be a whole range of deflections. However we can devise schemes for producing a stream of ions which all have one velocity—or, by another scheme, all one kinetic energy—and then deflections give us direct measurements of e/M ion by ion.

Early Measurements

Rough measurements just before this century began showed that the particles in positive-rays have the masses of atoms and molecules (if their charge is $+e$ or a few $+e$'s). In 1910 J. J. Thomson fired a thin stream through electric and magnetic fields arranged to make a definite mark for each value of e/M, however varied the velocities. His measurements showed (as in the table in Ch. 36) that

> With hydrogen in the tube, e/M is about 10^8 coulombs/kg, the same as for hydrogen ions in electrolysis. His record showed no sign of H^{++} ions with a doubled value of CHARGE/MASS—no sign that a hydrogen atom has more than a single electron to lose. However, he did find a half value, which he correctly interpreted as belonging to ionized hydrogen molecules, H_2^+.
>
> With oxygen in the tube he recorded ions with 1/16 of the e/M of H^+ ions, suggesting O^+ ions. He also got double that value, which he could prove was due to O^{++}.

By inverting his e/M values, Thomson could compare the atomic masses of many elements, or, as chemists had long called them, the "atomic weights." Thus O ions have 16 times the mass of H ions.

> With methane gas, CH_4, in the tube, he got marks for masses 1, 12, 16, belonging to H^+, C^+, ionized methane molecules CH_4^+, and even unstable groups such as CH_2^+, then unknown to chemists as free groups.
>
> Mercury ions appeared to be able to carry up to eight + charges.

These measurements agreed with those of ions and atoms already known from electrolysis and chemistry.

Then came the delightful surprise: Neon gas in the tube gave a peculiar record. Its atomic weight was well known to be 20.2, yet its record showed *two* marks, one for ions of 20.0 times the mass of the hydrogen ion, and a fainter one for 22.0.

PROBLEM 2. A CRUCIAL TEST

Thomson was sure the "20" mark was due to neon ions, because its intensity was proportional to the amount of neon in any gas mixture he used. The fainter 22 line might show a heavier brand of neon atom or it might be due to some strange compound of neon with stray hydrogen in the tube, NeH_2. What simple experiment would make a crucial test between these two possibilities?

Isotopes

There are *two* neons—twin brother atoms of the same element—one 10% heavier than the other. They were named *isotopes*, a name already coined for such a case on the basis of evidence from radioactivity.

Mass-Spectrographs

Then the great field of atomic analysis opened up. One clever scheme after another was devised to "focus" streams of positive rays into clear, sharp marks that could yield precise measurements of atomic masses.

Here was a wonderful method of analyzing materials by actual atomic masses. There is no difficulty about separating out pure elements. Every type of atom makes its mark; and we can even estimate proportions of ingredient atoms by densities of marks. True, some molecular groups make marks too, but an experienced mass-spectrographer can recognize them as easily as a physician reads a blood sample.

Originally the ions were focussed on a mineral screen that glowed when ions smashed into it. Then a photographic plate or film was used, because a dark mark could be developed where ions hit it (see Fig. 38-3a, b). In all these experiments, the record has a mark for each atomic mass. The display of marks on a band of photographic film looked rather like the optical line-spectrum of light from a glowing gas. This suggested a suitable name, *mass-spectrograph*, for the apparatus that spreads out the record of atomic masses like that. We sometimes call it a *mass-spectrometer* when we are using it for very precise measurements. (Optical spectra themselves identify each element uniquely, but they barely distinguish between isotopes.)

Nowadays the ions of one mass are often focussed on a narrow slit so that when they pass through it they can be collected and their charge carried away in a tiny electric current that can be amplified and recorded. Then, as the streams of different masses are swept across the slit by a changing field, a graph of the ion current shows a sharp peak for each atomic mass. See Fig. 38-3c, which shows such a record.

The masses calculated from precise measurements are those of ions. To find the masses of a neutral atom we must add the mass of the missing electron(s), but that is easily done.

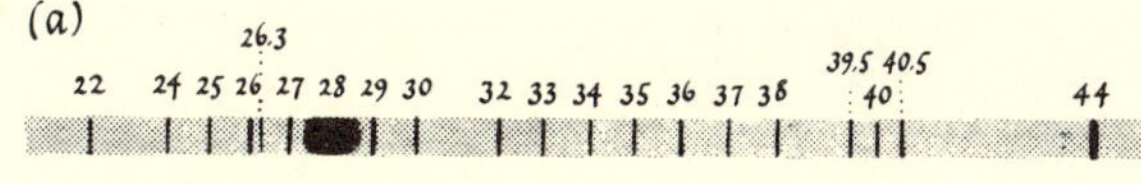

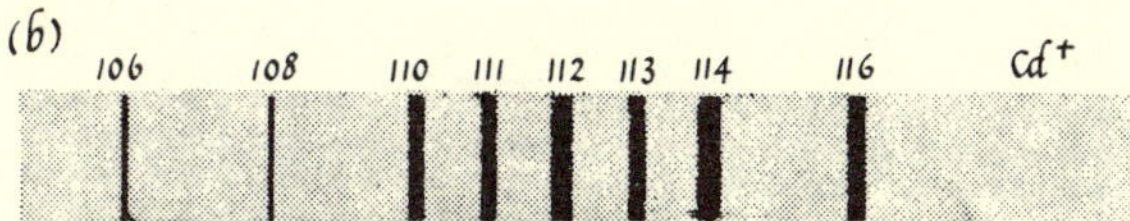

FIG. 38-3a, b. MASS-SPECTROGRAPH RECORDS
(Sketches from photographic records. The mark for each ion mass is gray or black on the original. Here it is shown by black marks.)
(a) Record by Aston, with the gases bromine (80 and 81) and carbon dioxide (44) fed into the tube. The CO_2 provided ions that made marks at 44 (CO_2^+), 22 (CO_2^{++}), and 28 (CO^+), and at 32 (O_2^+). The bromine provided ions that made marks at 39.5 and 40.5 (Br^{++}), and 26.3 and 27 (Br^{+++}). There were also traces of H, Cl, S, etc., in the tube; and these provided ions that made marks at 35 and 37 (Cl^+), and at 36 and 38 (HCl^+), etc.
From *Mass Spectra and Isotopes* by F. W. Aston (Edward Arnold Ltd., London.)
(b) Record by Dempster. Cadmium ions (Cd^+) were made by a "hot spark," between cadmium electrodes in a vacuum. The photograph shows all the isotopes and indicates their proportions.
From *Proceedings*, American Philosophical Society, Vol. 75 (1935).

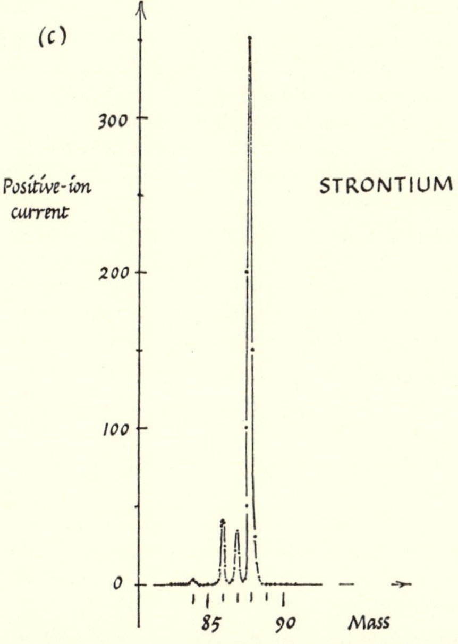

FIG. 38-3c. MASS-SPECTROGRAPH RECORD by M. B. Sampson, using an apparatus somewhat like that shown in FIG. 38-5. With strontium ions, the record showed the most prominent isotope at 88 (82% of total), other isotopes at 87, 86, 84. There was no sign of the radioactive isotope at 90, though the measurements would have shown the presence of 0.05%.

Prout

The discovery of isotopes offered a new answer to an ancient chemical puzzle concerning round-number atomic proportions among atomic masses. A century before, Prout had pointed out a clear simplicity among chemical "atomic weights." Taking the mass of a hydrogen atom as 1.01 we have (using modern values from precise chemical weighing):

lithium	6.94	carbon	12.01
nitrogen	14.01	oxygen	16.00
sodium	23.00	sulfur	32.06
tungsten	183.92	uranium	238.07

(The best collection of whole numbers seems to be obtained by using a scale with oxygen exactly 16, instead of hydrogen 1. That makes hydrogen 1.008. Chemical experts decided on that long ago, and we have used it here.)

This is an unfair list, a few lucky ones chosen to illustrate Prout's claim; but with his smaller list of rougher values the claim still made sense—there were more whole numbers than was likely by mere luck. He suggested that all atoms might be made of some hydrogen-like building block. But enough exceptions persisted to spoil the story: the sorest thumb of all was chlorine, for which careful chemical measurements gave 35½ again and again.

Then forty years ago mass-spectrographs came to the rescue and showed that there *is* no 35½ in chlorine, but a mixture of two isotopes, 35 and 37, in just the proportions that average out to 35.46.

PROBLEM 3

There were also marks on the record for chlorine at 36 and 38. These were correctly guessed to be due to another charged ion, not plain Cl^+, although they appeared when only chlorine was used.

(a) What ion would you suggest, knowing that other gases had been used in the apparatus for comparison?
(b) How could your suggestion be tested with the same apparatus?

Isotope Masses

Many elements proved to have two or more isotopes, all with masses nearly whole numbers relative to $H = 1$ or $O = 16$. Even oxygen, whose atomic weight chemists had decided to take as standard at 16, showed a fairly rare isotope of mass 17 (so we now use the common isotope O^{16} as the standard mass 16.0000).

More recently hydrogen showed a heavy isotope of double mass now called deuterium.[1] More recently still we have found a hydrogen isotope of triple mass, now called tritium. This is radioactive.

[1] This is not just a hydrogen molecule of two ordinary H atoms loosely joined. It is a single atom of double mass, that follows through all the chemistry of common hydrogen without ever splitting up. Only a violent attack with gamma rays —or a stupendous collision at temperatures undreamed of till récently—can make its nucleus break into a neutron and a proton (a common hydrogen nucleus).

TABLE

Examples of Mass-spectrograph Results

Here are a few examples, chosen to illustrate the mixtures of isotopes found in nature. The masses given are rough values, not the precise measurements discussed in the next section. The precise values seldom differ from these whole numbers by as much as 0.1. Scale: mass of O^{16} atom = 16.00000

ELEMENT	ISOTOPES: MASS	ISOTOPES: RELATIVE ABUNDANCE	CHEMICAL "ATOMIC WEIGHT" *for mixture found in nature*
hydrogen	1.008	99.98%	1.008
(deuterium)	2.015	0.02%	
(tritium*)	3.017	—	
helium	4	100%	4.00
	(3	very rare)	
lithium	6	8%	6.94
	7	92%	
chlorine	35	75%	35.46
	37	25%	
copper	63	68%	63.58
	65	32%	
iodine	127	100%	126.95
mercury	196	0.1%	200.6
	198	10.1%	
	199	17.0%	
	200	23.3%	
	201	13.2%	
	202	29.6%	
	204	6.7%	
uranium*	234	0.006%	238.1
	235	0.7%	
	238	99.3%	

* Radioactive.

Design of Mass-Spectrographs

Problems 4 and 5 discuss designs of real mass-spectrographs.

PROBLEM 4. BAINBRIDGE'S MASS-SPECTROGRAPH WITH VELOCITY SELECTOR

This apparatus, devised by J. H. Bainbridge, first selects all ions with one velocity and then analyzes them according to e/m by a magnetic field (see Fig 38-4). An electric field in region Y provides a mess of ions there. The strong accelerating field between electrodes A and B drives a stream of ions out through the slit S with a wide range of velocities. They pass through a narrow corridor between plates P and P′ which have a constant P.D. applied between them. Over the region of the corridor there is also a uniform magnetic field perpendicular to the plane of the diagram. The electric field between P and P′ pushes the ions sideways as they travel down the corridor, and the magnetic field is arranged to push them sideways in the opposite direction. Thus the ions

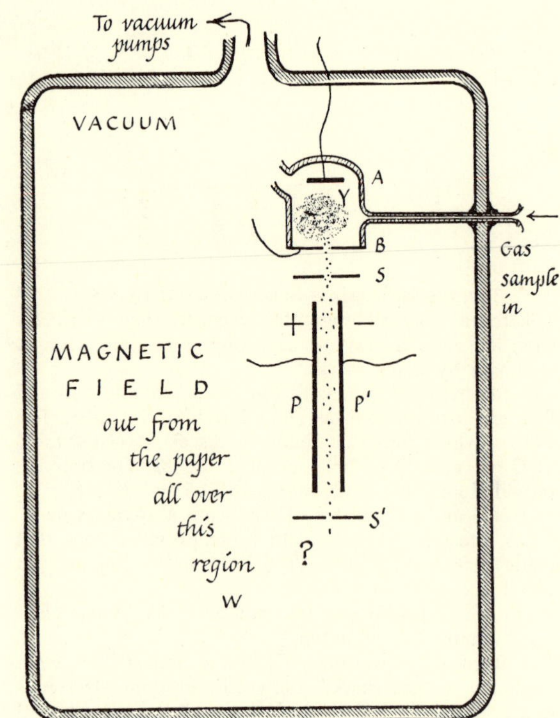

FIG. 38-4. MASS-SPECTROGRAPH, with velocity selector, Bainbridge's form (Problem 4).

in the stream have to run the gauntlet through the electric and magnetic fields. The only ions that can travel down the corridor and reach the far end, S′, successfully are those for which the pulls due to the electric field and magnetic field just cancel. To show that *all ions of one particular velocity can get through*, whatever their e and M, carry through the following calculation:

(a) If the P.D. between P and P′ is V volts, and the distance between the plates is d meters, the strength of the electric field between them is (V/d) volts/meter. What is the force exerted by this field on an ion with charge e coulombs moving v meters/sec? (See Ch. 33.)

(b) If the uniform magnetic field has strength H and the force on a moving charge is given by $F = 10^{-7}\,(Qv)\,(H)$ what is the force on the moving ion?

(c) Calculate the velocity of ions that can get through the corridor to S′.

Those ions which do get through to the second slit S′ emerge into the region W where there is no electric field. But *the same uniform magnetic field, H, acts (perpendicular to the diagram) all over the region* W.

(d) Predict the shape of the path in region W for a stream of ions all of one mass M_1 and the selected speed v worked out above.

(e) Where will such a stream, coming out from S′ in a narrow fan of directions, focus?

(f) Where will the stream focus if its ions have the same selected v as in (e) but twice the mass, $2M_1$?

(g) Sketch the diagram roughly and show a photographic film in the proper position to receive focussed streams of ions of mass M_1, $2M_1$, $3M_1$, etc., all moving with the selected speed v.

★ PROBLEM 5. MASS-SPECTROGRAPH WITH SINGLE-ENERGY ION SOURCE

In this form (see Fig. 38-5) all the ions, whatever their mass, are made to fall through the same gun-voltage. Then

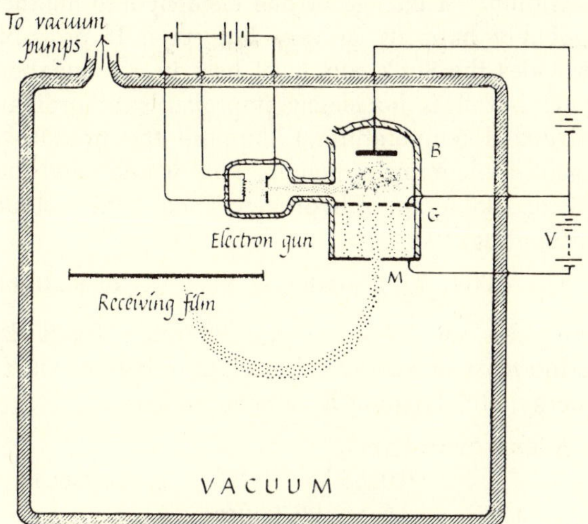

FIG. 38-5. MASS-SPECTROGRAPH, with ions all of same kinetic energy focussed by magnetic field (Problem 5).

all ions with single charge $+e$ have the same kinetic energy. (Ions with double charge $+2e$ will gain twice as much K.E., but their record can be diagnosed.) The ion gun has three parts: (i) A small electron gun to bombard the gas and make ions. The easy job of breaking the gas molecule into separate atoms is done by the same bombardment. (ii) A coaxing region between plate B and grid G. The ions are made here and driven gently towards G by a small P.D. between B and G. They drift through G and meet* a strong electric field provided by a large P.D., V, between G and gun-muzzle M. Thus, wherever an ion starts from in the region behind G, it gains the same K.E. by falling through the gun-voltage V. Then all ions emerge with practically the same K.E. Assume, then, that all the ions emerging have fallen through the same gun-voltage V volts.

Suppose that the stream from the muzzle contains ions of mass M kg and charge $+e$ coulombs. And suppose there is a uniform magnetic field of strength H perpendicular to their path. The path therefore becomes a circle of radius r.

(a) Explain why each ion's kinetic energy $\frac{1}{2}mv^2$ must equal eV joules.

(b) Show that the radius, r, of the circular path must be given by

$$r = \frac{\sqrt{2V \cdot M/e}}{10^{-7}H}$$

Note: The force on a moving charge Q moving with speed v across a magnetic field of strength H is given by $F = 10^{-7}(Qv)(H)$.

(c) Suppose that V and H are kept constant, and there are two kinds of ion with the same charge but different masses, one mass twice the other. How will the radii of their focussing circles compare?

(d) Suppose there are two kinds of ion, one with twice the mass of the other, as in (c), and the gun-voltage, V, is changed, while H is kept constant. By what factor must V be changed to make the heavier ion have the focussing radius that the lighter ion had?

* This scheme for making sure that all ions receive the full boost is somewhat like the technique a restaurant might use for getting rid of a difficult guest without disturbing the other customers. The manager engages him in gentle talk and ambles with him to the door. There the victim receives a full ejecting boost from the doorman.

In the form that Problem 5 (c) asks about, a photographic film placed in the focussing region receives ions of various masses and shows on development a clear mark for each mass. In the form of (d) streams of one ion-mass after another are brought to a fixed receiving slit by changing the gun-voltage. Then the current from the ions caught by the receiver behind the slit is amplified and indicates the number of ions of that particular mass. Thus the graph of ION CURRENT THROUGH SLIT against GUN-VOLTAGE shows the distribution of ion masses. Fig. 38-3c shows a typical graph.

Chemistry and Atomic Masses

Long before the measurements we have been describing, atomic masses had been compared accurately by chemical weighing; but these were gross weighings of vast bunches of atoms, assumed all uniform for each element. Through a century and more of the great development of chemistry, isotopes had never been revealed. All the isotopes of an element follow the same chemistry together—witness the constancy of the mixture of chlorines (35 and 37) which give the same "atomic weight" 35.46 in one chemical weighing after another, the same whatever the source of the chlorine, the same whatever chemical processes the chlorine has been pushed through before it is separated and weighed.

Again, half a century before gas ions were investigated there were good measurements for ions in salt solutions, the charged atoms that carry the current in electrolysis. It was easy to weigh the product released by electrolysis, measure the total charge transported, and thus obtain an accurate value of e/M for each kind of ion. When the first measurements of gas ions gave similar values for e/M it seemed a safe guess that the same atomic particles were carrying the same charges in both cases. But measurements of e/M by electrolysis are done by gross weighing, so they yield only average values for each element,

$$\frac{\text{charge } e}{(\text{average } M \text{ for ions of that element})}$$

That gave no worry as long as all atoms of an element were thought to be identical—an obvious assumption that seemed implicit in the whole use of atoms in 19th-century chemistry.

Even in the early 1900's, uniform atoms seemed a safe idea, except for some doubts from another quarter: radioactivity. The end-product of different series or family trees of radioactive elements is the metal lead; and it seemed likely that the several different series that had been discovered ended with lead atoms of slightly different masses. There was a further hint from some intermediate radioactive members of the series that behaved like lead but certainly had abnormal masses—we now know they are radioactive lead. The name *isotopes* was suggested for such atoms of the same element with (slightly) different masses. However this hint of unequal masses was regarded as a special idea peculiar to radioactivity; and uniform atoms remained the rule until the analysis of positive rays showed that there *are* isotopes.

The presence of such "uneven twins" among atoms is both unexpected and useful in chemistry. Now that we know there are isotopes and can separate them by physical means we can use them as special *tracers* in chemical exchanges. Deuterium, the heavy hydrogen isotope, is particularly useful in tracing the path of hydrogen atoms in the forming of organic molecules.

Isotopes can be separated by various physical means. See the description of diffusion separation in Chapters 25 and 30. If necessary, we can use a mass-spectrograph itself to separate isotopes and collect minute samples of each ion at its focussing point. That was used to separate small quantities of the light uranium isotope U^{235} for early experiments on fissionable material. It is now of great use in separating special isotopes for use as chemical tracers.

Minor Disagreements

Discoveries of isotopes brought Prout's idea back into favor. Yet very precise measurements with mass-spectrographs that bring each ion stream to a very sharp focus show that *the masses are not exactly whole numbers* (on any of the scales tried). The differences look trivial until we translate them into equivalent energy-differences, using $E = mc^2$. Then we find the mass-spectrograph providing the precise measurements needed for our energy-accounting in nuclear transformations.

Fission and Fusion Energy Releases: Examples of the Use of Precise Atomic Masses

The testing and use of $E = mc^2$ in nuclear physics will be discussed in Chapter 43. Meanwhile, if you like to take it on trust you can see it put to use with some precise mass-spectrograph measurements. We shall try two simplified nuclear examples, one of "fusion," one of "fission." They are neither correct in detail nor feasible in practice, but they give fair illustrations of our energy-accounting. We shall use the following masses[2] from modern mass-spectrograph measurements (which can give even greater precision, to one or two more decimal places):

TABLE OF ATOMIC MASSES

(on the scale that gives the oxygen atom 16.0000)

H^1 (common hydrogen)	1.0081
He^4 (common helium)	4.0039
Ag^{107} (the lighter isotope of silver)	106.939
Xe^{128} (one of the rarer isotopes of xenon)	127.944
U^{235} (the fissionable isotope of uranium)	235.116

[2] These are masses of complete atoms. In Ch. 43 we shall subtract the mass of all the atom's electrons and deal with nuclei. But leaving the electrons in, as we do here, does not alter the accounting materially.

Fusion

Suppose four hydrogen atoms could gang together to make one helium atom. (This "chemical impossibility," a change of one element into another, probably happens in very hot stars. It probably provides the Sun's supply of heat in a roundabout way; but it is hopelessly improbable at ordinary terrestrial temperatures.) Suppose the product is about 4 kilograms of helium. That would need four lots of 1 kilogram of hydrogen. Then, with accurate accounting:

$$4 \times 1.0081 \text{ kg of matter} \rightarrow 4.0039 \text{ kg of matter}$$

We guess, correctly, that the difference, 0.0285 kg, is the mass of energy released as radiation, kinetic energy, etc. Trusting $E = mc^2$, we expect:

$$\begin{aligned} \Delta\ \text{ENERGY} &= (\Delta m)c^2 \\ &= 0.0285 \text{ kg} \times (3.0 \times 10^8 \text{ m./sec})^2 \\ &\approx 2.6 \times 10^{15} \text{ joules} \\ &\approx 600{,}000{,}000{,}000 \text{ Calories} \end{aligned}$$

Compare this with the "molecular fusion" of hydrogen and oxygen atoms to form water:

CHEMICAL REACTION

$2H_2$	+	O_2	→	$2H_2O$
4 kg of hydrogen	+	32 kg of oxygen	→	36 kg of water

with a release of 140,000 Calories.

For the same amount of hydrogen used (with oxygen free from the air), the yield is four million times smaller.

Fission

Suppose an atom of U^{235} divides by fission into two lighter atoms and nothing else, an atom of silver, Ag^{107}, and an atom of xenon gas, Xe^{128}. (Most fission events release neutrons as well; so, although the xenon and silver are possible products, this is an artificial example. However it suffices for an estimate of energy-release.) Assume 235 kilograms of uranium make this change:

NUCLEAR CHANGE

U^{235}	→	Ag^{107}	+	Xe^{128}
235.116 kg	→	106.939 kg	+	127.944 kg

$$235.116 \text{ kg of matter} \rightarrow 234.883 \text{ kg of matter}$$

Trusting $E = mc^2$, we expect an energy release:

$$\begin{aligned} \Delta\ \text{ENERGY} &= (\Delta m)c^2 \\ &= 0.233 \text{ kg} \times (3.0 \times 10^8 \text{m./sec})^2 \\ &\approx 2 \times 10^{16} \text{ joules} \\ &\approx 5{,}000{,}000{,}000{,}000 \text{ Calories} \end{aligned}$$

Compare this with the chemical undoing of TNT molecules. When 235 kg of TNT explode, the energy-release is about 850,000 Cals. A TNT molecule is about 3% lighter than an atom of uranium. So one molecule of TNT exploding yields six million times less energy than the fission we have supposed for one uranium atom.

The Needle and the Haystack: Important Precision

If we want to know "the weight of a haystack" for some scientific purpose, it does not matter whether or not the weighing includes one needle embedded in the haystack. But if we want "the weight of a needle" *and can only weigh it in a haystack*, then we must make very precise weighings of [haystack + needle] and of [haystack] if we want to obtain the small difference with any precision.

Again and again in the history of science, small differences derived from precise measurements have yielded great new knowledge: early astronomical records enabled Hipparchus to discover precession of the equinoxes; Tycho's passion for accuracy gave Kepler his trusted 8′ difference, "on which . . ."; extremely precise measurements of optical spectra enabled the Bohr atom model to be extended; and small fractions from the mass-spectrograph's measurements of atomic masses foreshadowed the changes of nuclear energy that we now command.

PRELIMINARY PROBLEMS FOR CHAPTER 39

★ 1. QUESTION PREPARING FOR CLOUD CHAMBER USED IN NUCLEAR PHYSICS

A *very small* water drop seems to evaporate much more easily than a pool with a flat surface (or a large round drop). This is not just a matter of relative importance of surface and volume for different sizes of drop; because in fact a large drop or a pool left in *saturated* wet air (100% humidity) will just remain, while a *tiny* drop will still evaporate.

(a) Sketch a molecule evaporating from a VERY small round drop, and suggest why it can escape extra easily in this case. (*Hint*: Water molecules do attract each other—shown by surface tension—but this attraction extends only a short distance, a few molecule diameters. That limitation is shown by the fact that oil films, etc., have the same surface tension whether they are thick, thin or very thin—only if extremely thin, do they have smaller surface tension.)

(b) Thence suggest a reason why a cloud of water drops does not form easily, even in supersaturated air. (*Hint*: Each raindrop must begin as . . . ?)

(c) Ordinary dusty air has some dust specks in it that water can stick to easily. Though they look microscopic, these specks are large compared with the "very tiny water drop that does not form easily." Why does a fog form easily in supersaturated dusty air?

(d) A sample of wet dusty air is placed in a cylinder with a movable piston. The piston is suddenly drawn outward so that the air expands. Why does a cloud form? (Note that cold air needs less water vapor to saturate it than warm air.)

(e) Dust enables water drops to form; but, in dust-free air, drops can still form if there are electrically charged air molecules or atoms (= "ions"). (There is evidence that water-molecules are attracted by electrically charged objects. Possibly water molecules are oblong with + charge at one end and − charge at other. Or perhaps an outside charge can distort them easily into such a "shape"—in a way similar to the way in which a magnet makes soft iron a temporary magnet.) Explain why water drops form easily on ions that carry + or − charge.

(f) We believe that atoms have electrons which can be detached. Some atoms lose an electron very easily; other atoms grab an electron easily. (Then they are called "ions.") Normal air does not conduct a current, but air does conduct when a flame is lit in it, or when it is traversed by alpha-rays or beta-rays or gamma-rays or X-rays. Some of these rays are moving *charged particles* which make a trail of charged ions as they whizz through air. Such high speed bullets move too fast to collect any water themselves. Yet we expect to see their tracks in wet air that has been cooled suddenly. Explain how the tracks can become visible.

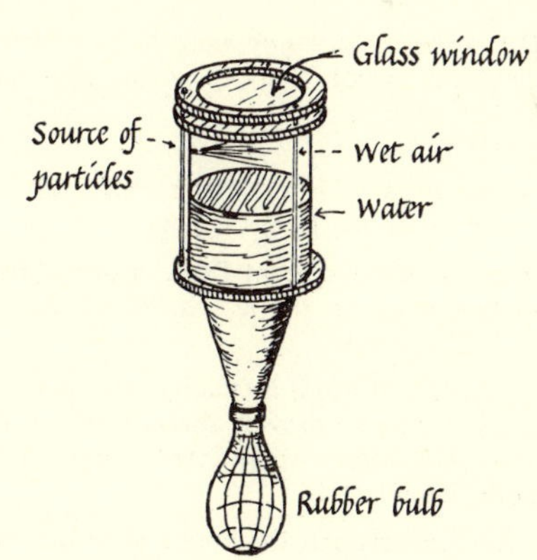

Fig. 39-1. Simple Cloud Chamber (Problem 1g)

(g) The apparatus that makes such tracks of "nuclear bullets" visible is called a "cloud chamber." In a simple form, a glass cylinder contains wet air over a piston of water that can be driven up by squeezing a rubber bulb. Explain why the following procedure makes the tracks visible: (1) squeeze the rubber bulb, (2) wait for some time, (3) release bulb suddenly, and the tracks appear as lines of droplets of water.

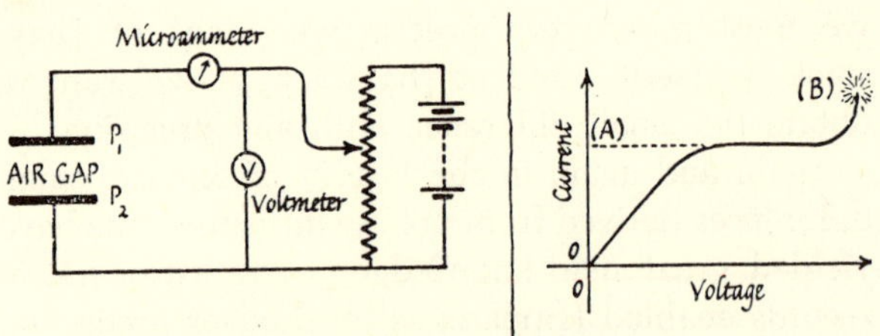

FIG. 39-2. PROBLEM 2.

★ 2. [This problem discusses ions (carriers) in gases. It is such ions that make the red glow in neon-sign tubes, that are the working agents in Geiger-counter tubes, that make electric sparks what they are. Try making guesses at the answers.]

If there is an air gap in a simple circuit with a battery, there is no steady current, because air is an insulator; it has "infinitely great" resistance. However, if a flame is placed in the gap, there is a tiny current around the circuit. (All currents in this probelm are VERY small, say 10^{-12} amp.) Or if a small stock of radium is placed in or near the air gap, there is a tiny current. This current does not seem to come *from* the flame or radium. The circuit just behaves as if the air could now conduct slightly.

(a) What does this suggest has happened to (some) air molecules?

If a stock of radium is kept permanently near the gap, the current is a steady one. If the voltage is now increased (by potential divider, or by adding a battery), the current increases in proportion (Ohm's Law), up to a certain stage. With still higher voltages, the current slopes off to a constant value (A), and remains constant for a wide range of voltages.

(b) What can you suggest is happening at this stage, (A)?

With very much higher voltages still, the current rises rapidly, (B), and soon there is a *spark*.

(c) What can you suggest is happening at this stage (B) when the battery is making very strong driving electric fields in the air?

(d) (HARD.) If the air pressure were halved (density halved, MFP doubled), what changes would you expect in the graph, and why?

3. If we make a very large electric field, say between two knobs of metal, a spark jumps across.

(a) During the spark a current flows, from whatever supply is used to charge up the knobs. What things carry this current in the spark?

(b) Once a spark is started, it usually continues if the charging supply is able to provide the current. In other words, once there has been a spark later sparks seem to find an easy path. Explain this.

(c) If the voltage between the knobs is so big that it nearly starts a spark but not quite, smoke or a match flame will often start one. Why?

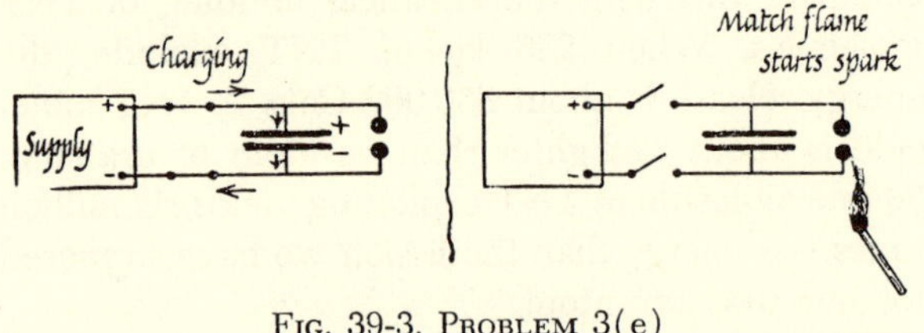

FIG. 39-3. PROBLEM 3(e)

(d) Instead of the match flame in (c), a small piece of radium brought near the gap may start a spark. Why?

(e) Suppose a large capacitor is connected to the 2 knobs, but the supply charging it is removed before the spark starts, and then a spark is started with a lighted match. There is a noisy spark but it is soon over. Why does the spark not continue in this case? (The answer is not, "The spark runs out of ions.")

★ 4. When an electric charge is given to an irregularly shaped conducting body, the charge does not spread evenly all over the surface (see Chapter 33).

(a) Where is the charge density thickest on the surface (and the field just outside consequently strongest)?

(b) In a Geiger-counter tube, one of the electrodes is a very thin wire suspended in the center of the tube. Suggest a reason for this.

★ 5. Alpha-particles are known to be charged helium atoms.

(a) Electric- and magnetic-field deflections show that they have e/m half as great as hydrogen ions. "Chemical" evidence tells us that helium atoms have 4 times the mass of hydrogen atoms. So, instead of saying an alpha-particle has $\frac{1}{2}(e/m)$ compared with hydrogen, we say it has $(2e)/(4m)$, and guess that its charge is 2e, twice the hydrogen ion charge, or twice the electron charge.

(b) By using a Geiger counter we can count the number of alpha-particles shot out by a small sample of radium in a measured time.

(c) Alternatively, we can fire the same stream of alpha-particles (in a vacuum) into a small metal can, collect the charge for the same period of time and measure it (or even measure the tiny current that trickles away from can to ground).

(i) *What important piece of atomic information can be obtained by combining the measurements of (b) and (c)? (Note that (b) and (c) refer to the same stream of α-particles.)*

(ii) *What further atomic information can be obtained by combining the observation and reasoning of (a) with the result of question (i) above?*

CHAPTER 39 • RADIOACTIVITY AND THE TOOLS OF NUCLEAR PHYSICS

1892 While it is never safe to affirm that the future of Physical Science has no marvels in store even more astonishing than those of the past, it seems probable that most of the grand underlying principles have been firmly established, and that further advances are to be sought chiefly in the rigorous application of these principles to all the phenomena which come under our notice. . . . An eminent physicist has remarked that the future truths of Physical Science are to be looked for in the sixth place of decimals.

—A. A. Michelson
(Professor of Physics, Case Institute, Clark University, University of Chicago)

1909 The new discoveries made in physics in the last few years, and the ideas and potentialities suggested by them, have had an effect upon the workers in that subject akin to that produced in literature by the Renaissance. . . . In the distance tower still higher peaks, which will yield to those who ascend them still wider prospects, and deepen the feeling, whose truth is emphasized by every advance in science, that "Great are the Works of the Lord."

—J. J. Thomson
(Cavendish Professor of Experimental Physics, Cambridge University)

SOMETIMES a householder's peaceful assurance is broken by several independent clues ganging up to reveal a crime. The discovery of radioactivity joined those of X-rays, cathode rays, photoelectrons and other ions in opening up suspicions, then knowledge, of the internal structure of atoms.

Radioactivity shows us fragments of atoms shot from their innermost region; it shows that some atoms are unstable and do not remain unalterable. It shows how one chemical element is related to the next, even revealing *transmutation* from one element to another. And it provides bullets to probe the structure of all atoms. Furthermore, in experimenting on radioactive materials we have developed tools of great use in modern physics: the cloud chamber, Geiger counter, and other machines that deal with individual atomic particles.

Ionizing Rays

When radioactivity was discovered, shortly before the beginning of this century, the exciting thing was the effect of the radiations on charged electroscopes and on photographic plates. Electroscopes lost their charge when radioactive substances were held near them; and photographic plates developed dark patches as if they had been exposed to light.[1] When the responsible chemical elements (uranium, radium, etc.) were isolated, these effects were still stronger, and in some cases the surrounding air even glowed. Clearly, *these substances emitted something that made ions* in the air, or in the photographic emulsion. In fact radioactivity, as first known, *was* this aura of ionization. The "amount of radioactivity" was measured by the ionization pro-

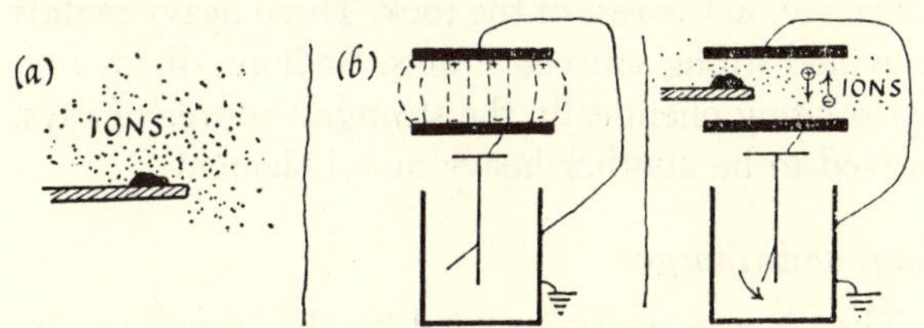

FIG. 39-4. RADIOACTIVITY AND IONIZATION
(a) Radioactive material makes ions in the air around it.
(b) A charged electroscope is discharged when radioactive material is placed near it.

[1] That was how radioactivity was discovered, soon after X-rays, in a search for other "penetrating radiations" that would fog a photographic plate. Uranium ore placed near a plate for some hours made a mark, even through thick black paper. The parts of the rock richest in uranium developed the blackest spots, and regions protected by a metal shield remained clear. Such "auto-radiographs" are made today by holding photographic film against biological samples containing a "tracer" of some radioactive element. For example: radioactive phosphorus, P_{32}, added to an animal's food, has been traced into a growing tooth, by taking a slice from the tooth, some days later, and holding it against photographic film.

duced.[2] The ions seemed to be manufactured by some things shot straight out from the substances; so these ionizing agents were called "rays." (We now know that the rays make ions by knocking electrons off atoms in their path.)

Experiments with slabs of absorbing material suggested three kinds of rays, according to penetrating power:

(α) intensely ionizing rays that travel *only an inch or two in air*, along straight paths. These were called *alpha-rays* (later, *alpha-particles*). A thick sheet of paper stops them, but they can get through cigarette paper or a few sheets of gold leaf.

(β) rays that travel farther, through *a foot of air*, many sheets of paper, a few millimeters of light metal. These spread their ionizing effect over ten times the distance in air, and their paths are not so straight. These were named *beta-rays.*

(γ) rays that travel much farther, straight out through *many feet of air* (with only an inverse-square law "weakening," due to spreading with distance), through inches of lead (with an *exponential* absorption, each inch of lead cutting out the same fraction). These rays, soon recognized to be electromagnetic radiation like X-rays, were called *gamma-rays.*

A narrow tube with a little radium at the closed end provided a "gun" that shot out a stream of rays. In a vacuum, all the rays continued straight indefinitely.

At first these "rays" were mysterious in origin as well as nature. When the heavy metals uranium and thorium were refined out of the ores, it was found that the rays came from them and not from oxygen, silica, etc. in the rest of the rock. These heavy metals provided strong sources of rays. Radium, discovered by isolating chemically the strongest source of rays, proved to be another heavy metal element.

Rays and Charges

The electric charge carried by the "rays" was investigated by collecting them in a metal can in a vacuum: alpha-particles carry a positive charge; beta-rays a negative charge; and gamma-rays carry no charge.

Individual alpha-particles could be counted by a trained observer watching the tiny flashes of light they make on hitting a mineral screen. Such counts provided very important atomic information. (See Problem 5; also Chapter 40.)

We now know these rays are missiles from the innermost cores, or nuclei, of atoms. You shall see how they were identified, measured, and used. But first look at actual pictures of their flight. These amazing photographs give some of the most valuable evidence in all atomic physics. They are like photographs laid before a jury in court, clearer than a dozen talking witnesses.

Cloud Chamber Pictures

How can we photograph the flight of a single atomic particle and a very fast one at that? A direct photograph is hopeless—the particle is far too small and far too fast. But we can obtain a picture from the damage it leaves behind along its path. Here is an illustration due to Professor Andrade: Shoot a cannon ball through standing wheat and try to photograph its path from an airplane. The ball is past before the picture can be taken; the broken stalks are invisible against the rest; but a little later you can photograph a line of blackbirds settled on the fallen ears. Rays from a radioactive source leave a litter of ions in the air they pass through. The trails of ions can be made visible by letting small drops of water condense on them—a thin line of droplets, that can be seen or photographed. These are the atomic versions of the giant cloud-streaks left in the sky by jet planes.

Our device for making the tracks of atomic particles visible was invented and developed by C. T. R. Wilson. The *Wilson cloud-chamber* enables us to see and photograph the paths of single particles flying through air—particles that are fractions of an atom—electrons, α-particles, or larger nuclei. We can thereby "see" individual nuclear events: collisions, transmutations, explosions. . . .

The Working of a Cloud Chamber

When water condenses into drops of ordinary fog, each drop forms around a speck of dust which acts as a starting core.[3] A fog droplet that started by itself would necessarily begin as an extremely small drop, a chance gathering of a few molecules. Such a gathering would evaporate extra easily; in fact, it would never form without excessive provocation. It is the sharp curvature of a tiny drop that makes evaporation extra easy. A small drop will not form in saturated air, but a large drop will re-

[2] We still use electroscopes to estimate a big radioactive stockpile by measuring its rate-of-production-of-ions. Small ones, working like fountain pens, are used to check health hazards. Geiger counters have largely supplanted them for small quantities or accurate measurements.

[3] Not every kind of dust will serve. Should the speck be of material on which water has small angle of contact or big? This question is an important one for cities that wish to avoid "smog."

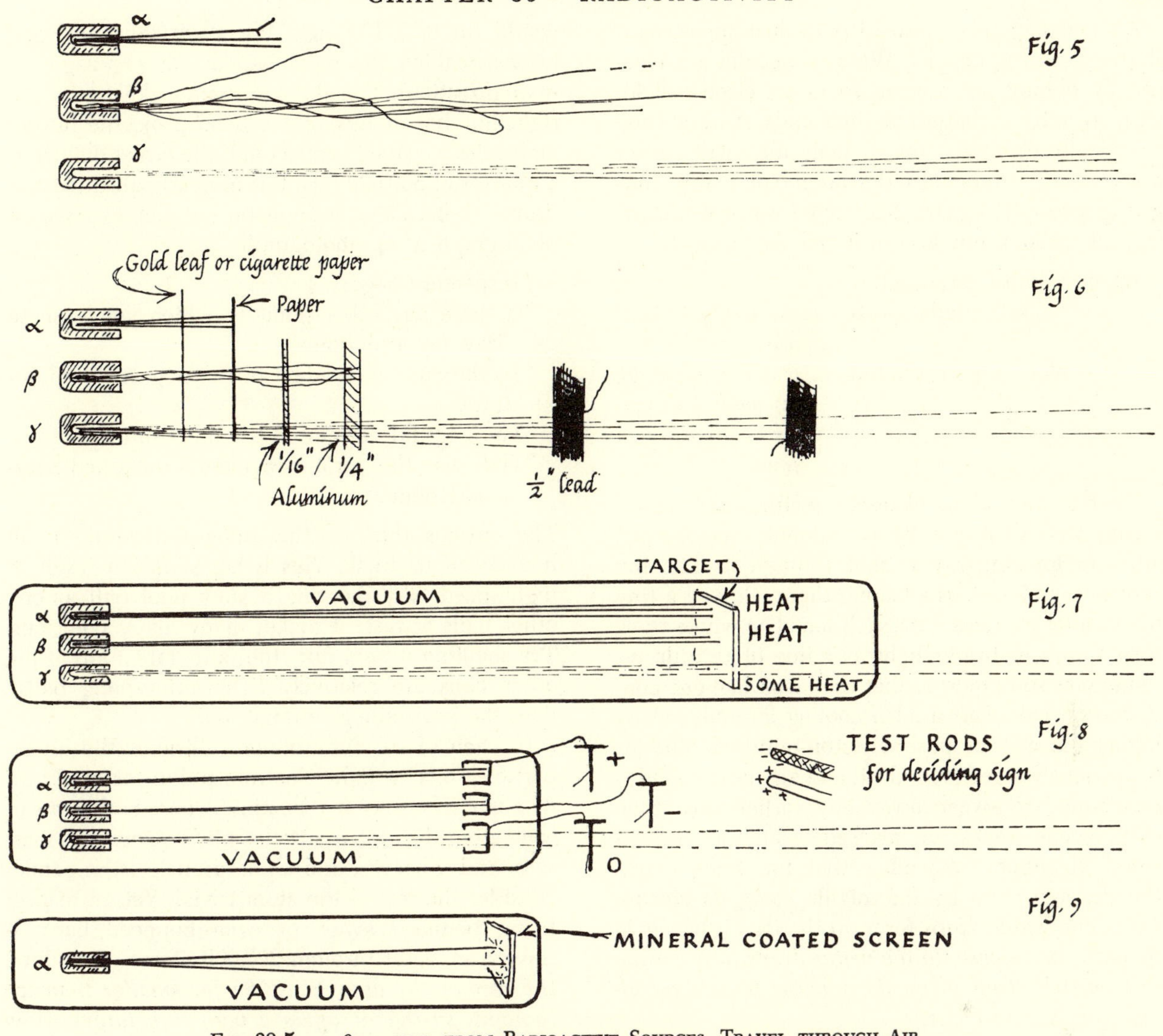

Fig. 39-5. α-, β-, γ-rays, from Radioactive Sources, Travel through Air
Fig. 39-6. α-, β-, γ-rays are Stopped by Sheets of Various Solids
Fig. 39-7. α-, β-, γ-rays Travel Straight in Vacuum; Produce Heat When Stopped
Fig. 39-8. α- and β-rays Carry Charges; γ-rays Carry No Charge
Fig. 39-9. α-rays Make Tiny Flashes of Light When They Hit a Suitable Screen

main, and so will a large water-covered speck of dust. (On the scale of this molecular discussion, a drop or a dust speck that is barely big enough to be seen is a *large* one.)

Air that contains a little water-vapor is only slightly "wet"—low relative humidity. If more and more water-vapor is added, it approaches saturation—relative humidity 100%—beyond which adding more water-vapor only causes condensation. The saturation limit depends on temperature. On a hot day, a cubic meter of *saturated* air (1.2 kg) contains about 23 grams of water, but on a cold day only 10 grams. Take some warm, wet, dusty air, and cool it. Then there is more than enough water-vapor for saturation and it starts to condense in a fog on dust specks, as well as on the walls of the container. Take some warm, wet air that is *dust-free* and cool it quickly. Then the changes run thus:

warm............almost saturated→ *(no fog, dry walls)*
cooler............saturated ——→ *water ready to condense*
cooler............supersaturated → *water condensing on walls; ready to make fog, but lacking starters*
still cooler....supersaturated → *water condensing on walls starting to make fog, somehow (?)*

An ion may serve instead of a dust speck, as a starter for a fog droplet. Water molecules are electrically oblong, or become so in an electric field, with + and − charges at their ends. An ion (any electrically charged atom or molecule) will collect a crowd of water molecules around it easily, thus giving a droplet a start. Take some warm wet dust-free air, make a few ions in it and then cool it:

warm....almost saturated
cooler....saturated ⟶ *water ready to condense*
cooler....supersaturated → *water condensing on walls and water condensing on ions*

That is how cloud-chamber pictures are made: warm, wet, dust-free air is suddenly cooled, just after an ionizing ray is shot through it. The ray leaves + and − ions all along the track, and a tiny fog droplet condenses on each ion. The whole track then shows in bright light as a line of tiny drops, sometimes so numerous that they look like one continuous streak. The sudden cooling is produced by letting the wet air push a piston outward. The air is previously freed from dust, and an electric field is maintained to sweep away any earlier ions: then only recent tracks are marked. Watch a working cloud chamber. Remember that the track marks the damage done by the missile *using its electric field.* The water drops form on the ions it leaves in its path. *In no case do the water drops form on the fast particle itself or on the nuclear targets recoiling from severe collisions.*

PROBLEM 6. CLOUD CHAMBER
(Alternative to Problem 1)

In a simple form of cloud chamber, a glass cylinder contains wet air over a piston of water that can be driven up by squeezing a rubber bulb. (Fig. 39-1). Explain why the following procedure makes the tracks visible: (1) squeeze the rubber bulb; (2) wait for some time; (3) release the bulb suddenly, and the tracks appear. What happens in each stage, (1), (2), (3)?

Alpha-particle Tracks in Cloud Chamber

Alpha-particles make straight tracks a few inches long, so crowded with water drops that they look like a miniature firehose jet. We can count the ions by the total number of water drops, or by wholesale electrical methods: two hundred thousand pairs of ions in a 2-inch track. The alpha-particle tugs an electron off 200,000 "air" atoms in travelling 2 inches or 0.05 meter. Here is extraordinary behavior, no major collisions, but 200,000 minor ones. What would happen to an ordinary air molecule in a 2-inch path? With mean free path 10^{-7} meter, it would hit $0.05/10^{-7}$ neighbors, 500,000 major collisions, making the path wander zig/zag/zog . . . in all directions. The alpha-particle makes about the right number of hits—for a *point*-projectile hitting molecules we should expect half the target diameter, a quarter of 500,000 hits. But in nearly all these collisions it blunders straight on, as you can see in photograph after photograph.

(Inspector Gregory:)
"Is there any other point to which you wish to draw my attention?"
"To the curious incident of the dog in the night-time."
"The dog did nothing in the night-time."
"That was the curious incident," remarked Sherlock Holmes.[4]

The curious thing is the alpha-particle does not bounce off its track. This is not simply a result of high speed. Try shooting a slow pool ball among other balls at rest: it makes many major collisions. Try shooting it *very fast*: the same fate. Only if the other balls are relatively light (ping-pong balls) does the projectile go straight on.

In each of its many minor collisions, the alpha-particle "hits" a light electron, pulling it easily off its atom. The mystery remains: where is the *rest* of each atom the alpha-particle "hits"? Atoms are massive, and whatever the alpha-particle is, it cannot consider the rest of the atom trivial. Yet, instead of being bounced away, or even bounced back, it travels on, right through 200,000 atoms. Therefore, *the rest of the atom must be far smaller than we thought, so that it offers a very rare target.* How small? Are there never any major collisions? In the photographs you will occasionally see a bend showing a major collision with something massive. Then the recoiling victim also leaves a track. Thus every alpha-particle makes a track crowded with minor collisions, but there are only rare cases of a bent track that shows a major collision. In air, the alpha-particle may even bounce back, while the victim makes a thick forward track. In helium, the tracks of a fork look alike and show a characteristic angle between them. In hydrogen, the alpha-particle always continues forward, and the target (H) also slants forward making a poorer track.

Atomic Structure

Here, in these photographs of atomic events, we have clear evidence that atoms are mostly hollow, with light, removable, electrons in their outer regions. Atoms must have a very small massive nucleus containing most of the mass, on which speedy alpha-

[4] "Silver Blaze," Sir Arthur Conan Doyle.

Fig. 39-10. Cloud Chamber Photographs

[In (a) and (c) the radioactive source is at the bottom of the picture. In all others it is out to the left of the picture.]

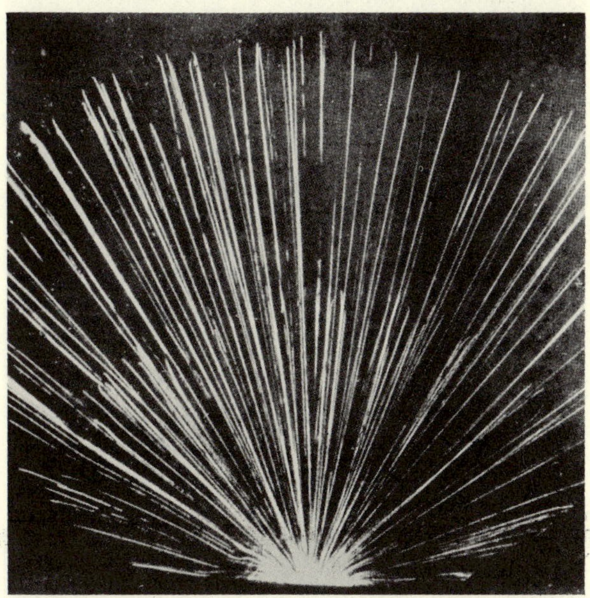

(a) Alpha-rays from a small source: a mixture of thorium C and C′. Note there are two lots of alpha-rays, each with a definite range in air. (J. Chadwick.) (From *Radioactive Substances and Their Radiations*, by Rutherford, Chadwick, and Ellis; Cambridge University Press.)

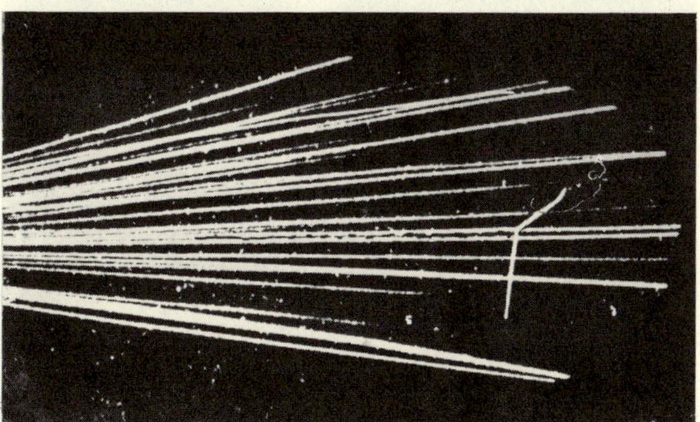

(b) Alpha-rays in wet nitrogen. One suffered severe collision with nitrogen nucleus and moved downward. The recoiling nitrogen nucleus made the short thick track. (P. M. S. Blackett.)*

(c) Stereoscopic photos of alpha-rays in wet helium. The pictures show on analysis that the two tracks after collision make 90° with each other. (P. M. S. Blackett.)*

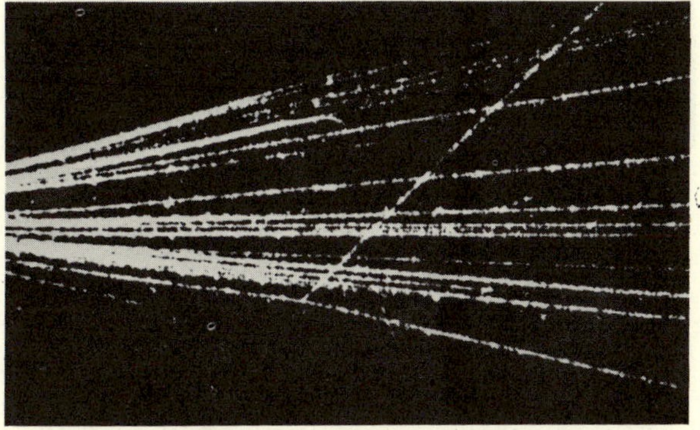

(d) Alpha-rays in wet hydrogen. One suffered severe collision with hydrogen nucleus, which recoiled forward and upward, making a thinner track. (P. M. S. Blackett.)*

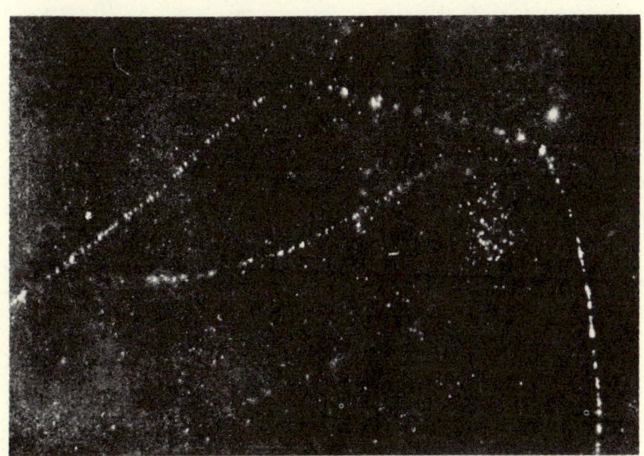

(e) Track of beta-ray in wet air. (C. T. R. Wilson.)*

* From *Proceedings of the Royal Society of London.*

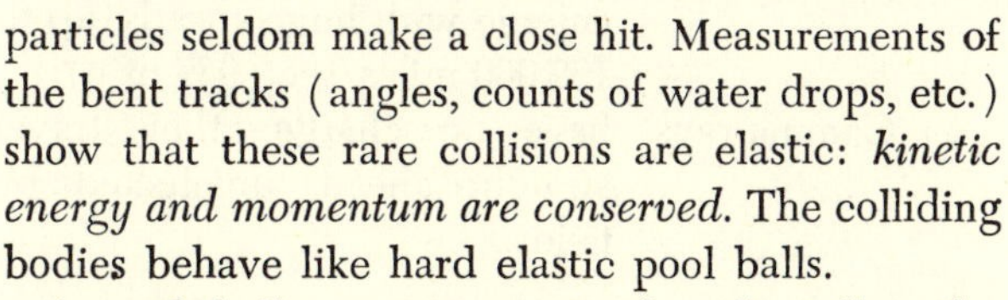

particles seldom make a close hit. Measurements of the bent tracks (angles, counts of water drops, etc.) show that these rare collisions are elastic: *kinetic energy and momentum are conserved.* The colliding bodies behave like hard elastic pool balls.

As with balls, measurements of angles tell us the relative masses. In air the alpha-particle hits something several times its mass. In helium, the fork always has an angle 90°, and from this we can infer the bullet hit something of its own mass at rest. (See Ch. 26, Problem 22) In hydrogen, the angles show the bullet hit something only a quarter as massive. Remember the relative atomic masses in chemistry:

hydrogen 1   helium 4   nitrogen 14   oxygen 16
and electrons (on the same scale) 1/1840

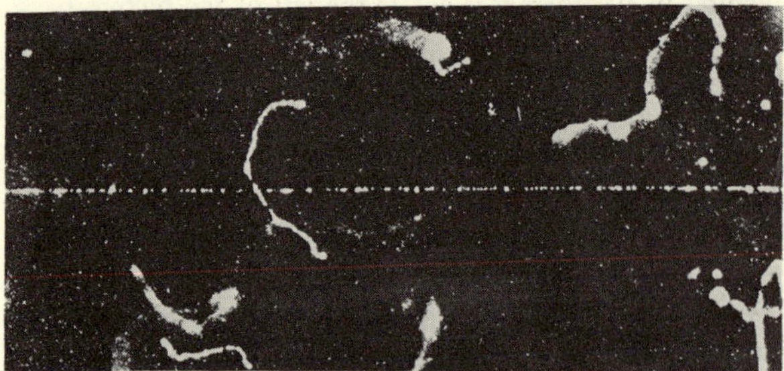

(f) Track of beta-rays. One *fast* beta-ray crossed the field: other tracks show slow ones. (C. T. Wilson.)*

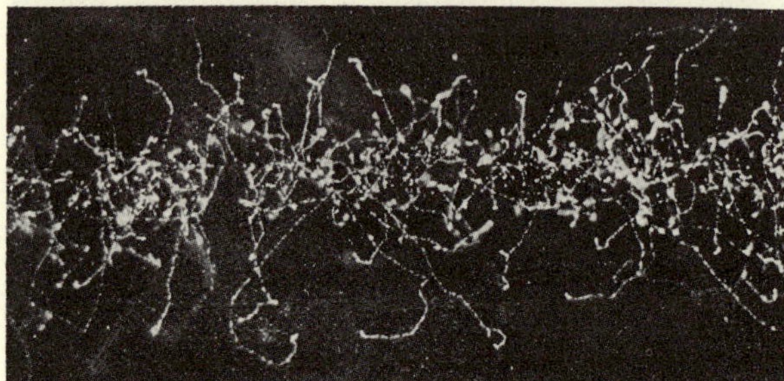

(g) Electrons ejected by a beam of X-rays passing through wet air (from left to right). A beam of gamma-rays produce a similar effect, with fewer, but longer, tracks. (C. T. R. Wilson.)*

* From *Proceedings of the Royal Society of London.*

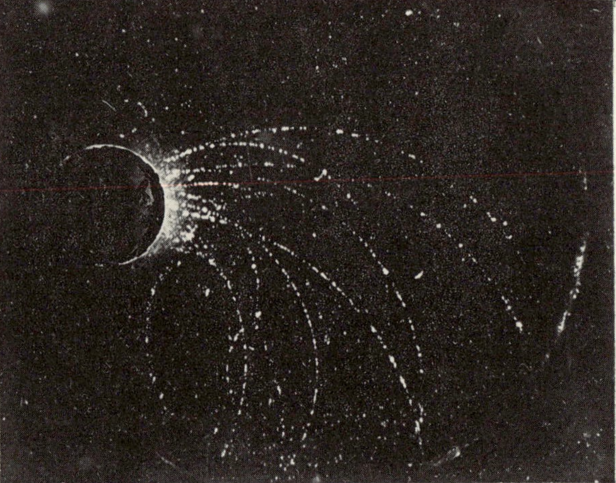

(h) Beta-rays in magnetic field. The field is not very strong. The radioactive source is on a cylinder at the left. (E. C. Crittenden, Jr.)

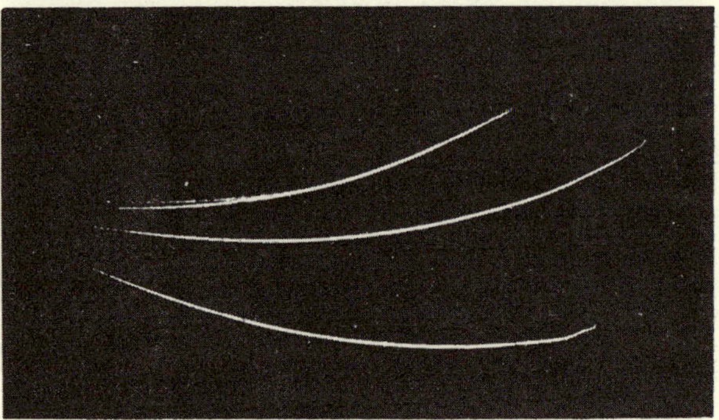

(i) Alpha-rays in magnetic field. The field is *very* strong. (Note the increasing curvature and pronounced bends near the end, where the particles have been slowed by many collisions. (P. L. Kapitza) (From *Radioactive Substances and their Radiations*, by Rutherford, Chadwick, and Ellis; Cambridge University Press.)

The fork measurements point to a mass 4 for alpha-particles, suggesting a helium atom. No wonder they make such short straight tracks—a sturdy helium atom, electrically charged, blundering through air and tugging out electrons 7000 times lighter.

Beta-ray Tracks

Look at a photo of beta-ray paths in wet air: long straggly tracks, with an ion here and an ion there, and many small bends. The picture almost suffices to identify the criminal: a fast electron skittering along among other electrons of similar mass, at the mercy of every local electric field.

Gamma-ray Tracks?

A stream of gamma-rays makes no visible track of its own path. A gamma-ray usually passes straight on, like light through glass, leaving no effect on matter. Sometimes it bounces off an electron which is left with a little recoil energy. Finally the gamma-ray picks on some unsuspecting electron in an atom and hurls it out with all its energy. Such electrons, ejected more or less sideways from the beam, make straggly tracks like beta-rays.

Analysis by Electric and Magnetic Fields

Before cloud-chamber pictures made things so clear, streams of "rays" were analyzed by passing them (in a vacuum) through electric and magnetic fields. Electric field deflections vary as e/mv^2; magnetic field deflections vary as e/mv; and the combined measurements yield v and e/m

(α) Alpha-particles have positive charge;
have e/M just half e/M for hydrogen ions, H^+;
are emitted with various speeds, up to 10,000 miles/second.

(β) Beta-particles have negative charge;
have same e/m as electrons from hot filaments,[5] etc. They *are* electrons;
emerge with huge speeds, up to 182,000 miles/sec (98% of c).

(γ) Gamma-rays have no charge; they travel straight ahead, unaffected by fields.

Thus α- and β-rays are speeding particles.

[5] That is, when they have slowed down. At very high speeds, e/m is abnormally small—an example of relativistic mass-change.

Gamma-rays

γ-rays behave like X-rays of very short wavelength. They are diffracted by a crystal—the regularly spaced atom-layers of the crystal act like a microscopic diffraction grating. They travel with the same speed as light. And they can hurl electrons out of any kind of matter—a gigantic photo-electric effect.

Identifying Alpha-particles

Several lines of evidence suggested that an alpha-particle is a doubly charged helium atom, He^{++}—collision angles, counting and collecting charges, e/M measurements. Rutherford and Royds clinched this guess by collecting spent alpha-particles and proving that helium gas appeared. A sample of radon gas (daughter of radium), which emits alpha-particles, was sealed in a glass tube with extremely thin walls. Some alpha-particles from the radon traveled through the thin wall into an outer tube, where a spark showed an increasingly strong yellow glow, characteristic of helium gas. They made sure, by a common-sense test, that the helium had not merely leaked in from the air.

Origin

Here then were violent events, electrons from some atoms with speeds nearing that of light; helium with ++ charge shot from others with huge speed. These particles could not be products of ordinary chemical or physical actions, like CO_2 being roasted out of chalk or bubbling out of soda water. Wherever the radium or uranium went in chemical manipulations, the radioactivity went with it. Whereabouts in these parent atoms did the particles come from? To answer that, some picture of atomic structure was needed; and alpha-particles themselves helped, as investigating bullets, to make that picture. We shall discuss it in Chapter 40.

Radioactivity and Chemistry

The changes in the radioactive atoms themselves opened up a great field. In the early 1900's physicists and chemists joined in examining the chemical nature of radioactive elements. Not only did uranium, radium, etc. emit ionizing rays but *at the same time they changed their chemistry—completely*. They appeared to change from being one chemical element to being a different chemical element. Radium, itself a dense metal, with chemistry rather like that of the metals barium and calcium, emits alpha-particles and slowly disappears. A new element appears instead, a very dense, completely inert, gas, now called radon. This gas belongs to the helium, neon, . . . family: all so inert that they have no chemistry. We now know that when a single

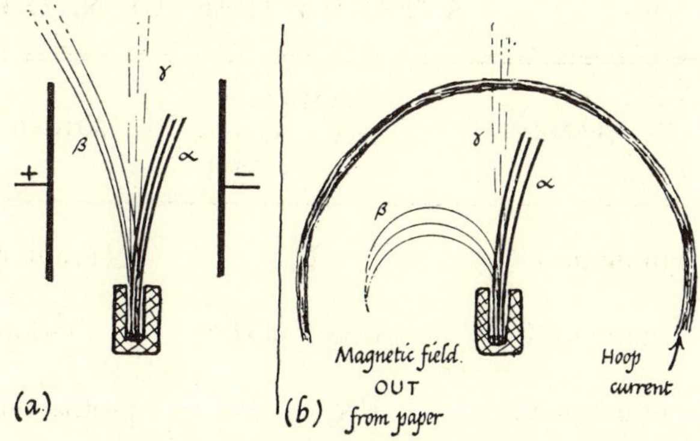

FIG. 39-11a, b.
CONVENTIONAL DIAGRAMS SHOWING PATHS OF α-, β-, and γ-rays
(a) in electric field and (b) in magnetic field.
In magnetic field, α-rays are bent much less than β-rays (about 1/100).

atom of radium makes its explosive "radioactive" break-up it emits one alpha-particle and becomes an atom of radon.

Radium atoms, therefore, are somewhat unstable. Buy a stock of radium and try to keep it. Only half of it will remain, as radium, 1600 years later. The betting on a particular atom of radium breaking up[6] in the next 1600 years is 50:50 for and against. Its daughter, radon, is much more unstable. The betting is 50:50 for and against a break-up in the next 4 days. That 4-day period is its *half-life*. Try to preserve a sample of pure radon gas, and four days later you will find half of it has disappeared[6] into a deposit of solid elements—and its gas pressure will have dropped to half. Keep it four days more and half that remainder will be gone, and so on.

As long as a radium atom is still a radium atom—that is, until the instant when it blows up—it has the fixed chemical properties of a proper element, occupying a proper place in the chemical table. When the atom has ejected an alpha-particle, the new radon atom has different properties and fits into another place in the chemical table—and it remains so until it, in turn, disintegrates. The new atom is a proper element—albeit unstable. Radon emits an alpha-particle and becomes "radium-A," with a 3-minute half-life—still more unstable. The family tree continues through a series of unstable elements, some emitting an alpha-particle, some a beta-ray, until it ends at "radium-G," which is one form of common, stable lead. Thus radioactivity not only supplies interesting projectiles; it also provides evidence of one chemical element changing spon-

[6] How should the change be described? Should we say the radium atoms *disappear*, or *change*, or *decay*, or *break up*, or *blow up*, or *disintegrate*? "Disappear" is rather misleading, "change" is vague. The other four expressions are all used. "Disintegrate" means "come to pieces."

A "FAMILY TREE" OF NATURALLY RADIOACTIVE ELEMENTS

ELEMENT	SYMBOL EARLY FORM	SYMBOL MODERN FORM	ISOTOPE OF	PARTICLE EJECTED	HALF LIFE
uranium I	${}_{92}U_1{}^{238}$		(uranium)	α	4.5×10^9 years
uranium X_1	${}_{90}UX_1{}^{234}$	${}_{90}Th^{234}$	thorium	β, γ	24 days
uranium X_2	${}_{91}UX_2{}^{234}$	${}_{91}Pa^{234}$	protoactinium	β, γ	1.2 min
uranium II	${}_{92}U_{II}{}^{234}$	${}_{92}U^{234}$	(uranium)	α	2.5×10^5 years
ionium	${}_{90}Io^{230}$	${}_{90}Th^{230}$	thorium	α, γ	8×10^4 years
radium	${}_{88}Ra^{226}$		(radium)	α, γ	1620 years
radon	${}_{86}Rn^{222}$		(radon)	α	3.82 days
radium A	${}_{84}Ra\text{–}A^{218}$	${}_{84}Po^{218}$	polonium	α	3 min
radium B	${}_{82}Ra\text{–}B^{214}$	${}_{82}Pb^{214}$	lead	β, γ	27 min
radium C	${}_{83}Ra\text{–}C^{214}$	${}_{83}Bi^{214}$	bismuth	β, γ	19 min
radium C^1	${}_{84}Ra\text{–}C^{1\,214}$	${}_{84}Po^{214}$	polonium	α	0.00016 sec
radium D	${}_{82}Ra\text{–}D^{210}$	${}_{82}Pb^{210}$	lead	β, γ	22 years
radium E	${}_{83}Ra\text{–}E^{210}$	${}_{83}Bi^{210}$	bismuth	β	5 days
radium F	${}_{84}Ra\text{–}F^{210}$	${}_{84}Po^{210}$	(polonium)	α, γ	138 days
radium G (lead)	${}_{82}Ra\text{–}G^{206}$	${}_{82}Pb^{206}$	(lead)	—	stable

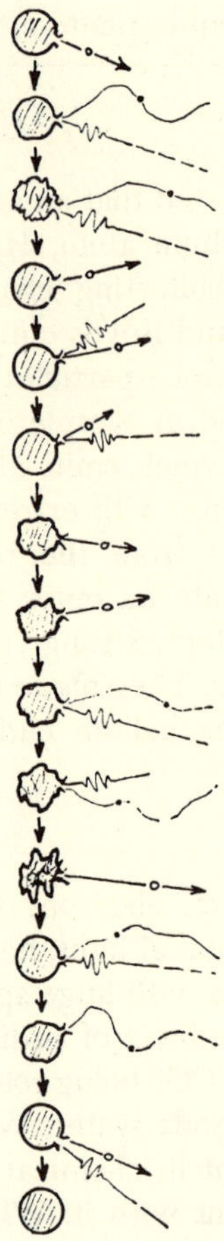

taneously to another—natural *transmutation*, unaided by man (and it seemed at first uncontrollable by man). The "family tree" shows the radium series from uranium to lead. (The actual tree is more complicated: there is a side branch which rejoins the main pattern. And there are several such families, running more or less parallel, from beginnings like uranium to endings in lead.)

Nowadays, by bombarding stable atoms of almost any element with high-energy particles from a "big machine," we can make new radioactive elements, each with a family tree—though in many cases there are only one or two stages before the family ends at a stable atom. Some of the new radioactive elements have places beyond uranium in the chemical series —e.g., plutonium. These disintegrate (by radioactive changes like those listed above) to other atoms along family trees that run down through uranium, like the radium family. So the family tree of radium is typical both of the old radioactive series found in nature and of many new series started with parents we make by our bombardments.

In the tree's list, symbols like Ra-E are old-fashioned ones dating from early days of sorting out radioactive substances chemically—they all seemed to be daughters, granddaughters, etc., of radium; and, although they are quite different chemically, they were given family names with Ra in them. Now, with tremendous profusion of radioactive atoms, natural and artificial, we prefer to label them by their standard chemical-element symbol. For example radium-B is chemically identical with lead, so we label it Pb (lead = *plumbum* in Latin). In statements of radioactive changes, we usually add more information to the label. We call radium-B $_{82}Pb^{214}$. The prefix 82 is the element's "atomic number," its serial number in the chemical series. *ALL varieties of lead have number 82.* The suffix, 214, is the element's "mass number," its atom's mass *in round numbers,* on the scale that gives hydrogen 1. Different isotopes of lead have different atomic weights. 214 is abnormally high for lead and would make us expect instability due to overcrowding in the nuclear household.

There is nothing mysterious about the chemistry of radioactive atoms—except the extreme smallness of the amounts that can be detected and measured precisely by their radioactivity. Mix some radium-B with common lead as "carrier," and you will never separate them chemically—as long as the radium-B remains itself. Melt the mixture: the radioactivity due to the Ra-B remains. Dissolve it in nitric acid and crystallize out lead nitrate: the lead nitrate has all the radioactivity with it. Mix the crystals with other salts (of potassium, aluminum, tungsten, . . .), then analyze for lead: all the radioactivity separates out with the lead. However, you would have to hurry these chemical experiments, because Ra-B decays with half-life 27 minutes into a radioactive bismuth. Running your chemistry at a leisurely rate, you would find less and less radioactivity accompanying the lead and more and more following the chemistry of bismuth. After half an hour, half of the original *lead* radioactivity would have disappeared.

Radium-B and radium-G are both isotopes of lead. Chemically and in ordinary physical properties, they *are* lead; but one is unstable with a 27-minute half-life and the other is stable. We call the second one "natural lead" only because mankind found it first and still finds much more of it than the unstable one. As we now survey the elements, each of them has several isotopes (same atomic number, different atomic mass), some of them stable (separated out by mass-spectrograph) and some unstable (manufactured with the help of cyclotrons etc.). And we find a few unstable ones occurring in nature —the original radioactive elements.

Changes in Atomic Number and Atomic Weight

An alpha-particle carries away mass 4, so we should expect the atomic weight of a daughter atom to be 4 less than that of a parent atom that emits an alpha-particle. Direct chemical weighing gave radium atomic weight 226. Then its daughter, radon, should have atomic weight 226 — 4 or 222. This prediction was tested by measuring the density of radon gas.[7] Result: atomic weight very close to 222. Where a beta-ray is emitted, the atomic weight should not change. The nucleus emits an electron, but the daughter atom needs one more *outer* electron, and collects one from neighbors. In any case, one electron's mass is trivial, 1/7000 of that of an alpha-particle.

When an atom emits an alpha-particle (α^{++} or He^{++} or $_2He^4$), it shifts to a new box in the chemical

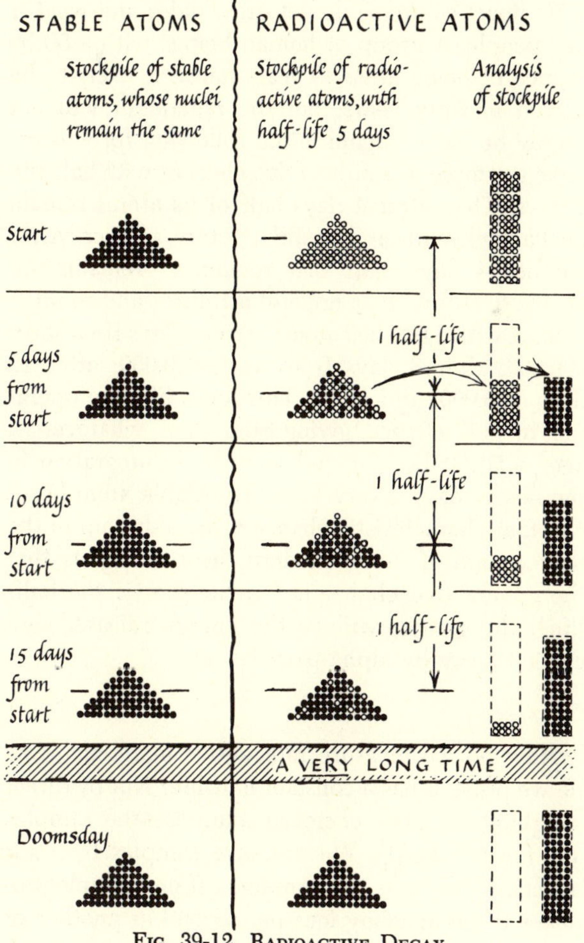

FIG. 39-12. RADIOACTIVE DECAY

[7] Avogadro's Law—vouched for by equipartition—assures us that a cubic meter of one gas contains the same number of molecules as a cubic meter of any other gas (at the same temperature and pressure). Therefore, comparing two gases, the ratio of densities = ratio of masses of equal numbers of molecules = ratio of masses of single molecules. For radon with monatomic molecules of mass 222 we should expect density 111 times that of hydrogen H_2 with molecules of mass 2. Experiment gave values around 111.5—surprisingly close for a difficult measurement with a minute sample.

FIG. 39-13. PRELIMINARY STUDY OF COUNTERS
(a) Connect a battery to a lamp through a bath of clean water. Add a handful of salt to provide ions in the water: The lamp lights. This will serve as a COUNTER for handfuls of salt, if a quenching and replenishing arrangement is provided.

table, always two columns to the left, a change of two down in atomic number. When we discuss atomic theory you will see an explanation of this.

When an atom emits a beta ray (β^-, or e^-), it shifts one up in atomic number. And the many new unstable atoms made by proton bombardment that emit positrons (β^+) must make a shift one down in atomic number.

Half-life: Pure Chance Instability

Radioactive atoms do not grow older and weaker like people. A group of human beings with a 50:50 chance of being alive 40 years hence is likely to be halved in forty years; but the remainder will not simply be halved again in the following forty years. Take a sample of a radioactive element with half-life 4 days. Then after 4 days half of its atoms remain unchanged, with as hopeful a future as ever. After another 4 days, half that remainder remains unchanged, still with as hopeful a future: and so on.

Start with 1,000,000 atoms. After 4 days from start, 500,000; after 8 days from start, 250,000; after 12 days, 125,000; and so on. This looks like the operation of pure chance, giving each atom, whatever its age, a 50:50 bet for and against disintegration in the next 4 days. Every kind of unstable atom has a constant characteristic chance of breaking up in the next second of time.[8] Modern theory suggests that the greater that chance is (= the shorter the half-life), the greater will be the energy released and carried away by alpha-particles, etc.

Counters

How do we measure radioactive material? How do we prove it has a constant half-life? Not by direct weighing or other chemical analysis—the samples are far too small. We measure samples by their effects, by the ions they produce. If one sample produces twice as many ions per second as another of the same material, we infer it is twice the size, twice as many atoms. We argue thus because we believe

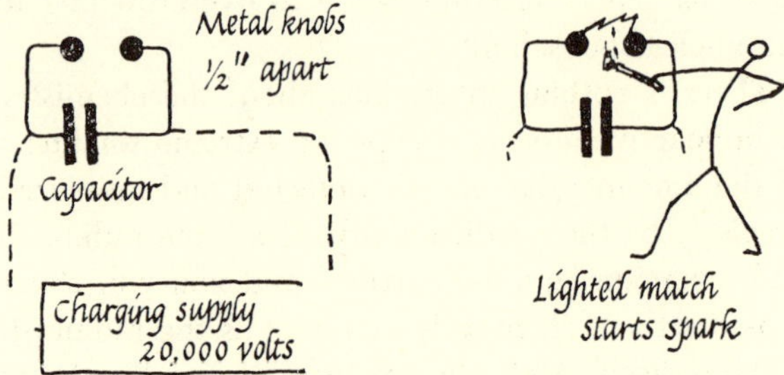

(b) Store big charges in a capacitor from a high-voltage supply until the field across a gap is just too small to start a spark. Hold a lighted match under the gap: a noisy spark occurs. This will serve as a COUNTER for lighted matches. It has rather a long "recovery time," while the supply recharges the capacitor.

all the atoms have the same probability of breaking up, so that twice the death-rate (measured by ions made by rays shot out) means twice the living population.[9]

Then SPEED-OF-MOVEMENT-OF-ELECTROSCOPE-LEAF
$\propto$ RATE AT WHICH IONS CARRY ITS CHARGE AWAY
$\propto$ NUMBER OF IONS PRODUCED PER SECOND IN ELECTROSCOPE'S BOX
$\propto$ NUMBER OF ALPHA- (OR β-) PARTICLES SHOOTING THROUGH BOX PER SECOND
$\propto$ NUMBER OF RADIOACTIVE ATOMS DISINTEGRATING PER SECOND
$\propto$ NUMBER OF ATOMS NOW IN STOCKPILE

Thus SPEED OF MOVEMENT of electroscope-leaf measures our STOCKPILE. By measuring that speed again and again as time goes on, we can see how a stockpile decays, and we can estimate its half-life.

---

[8] Some scientific philosophers, questioning how we know that time flows evenly, how we are sure that successive identical cycles of a pendulum take equal times, suggest we should assume constant instability of some radioactive material as an axiom on which to base time-measurement.

[9] Suppose there are two healthy villages, A and B. A has population 6000, but B's population is not known. In A, 80 people die per year; and in B, 40 people per year. Estimate the population of B.

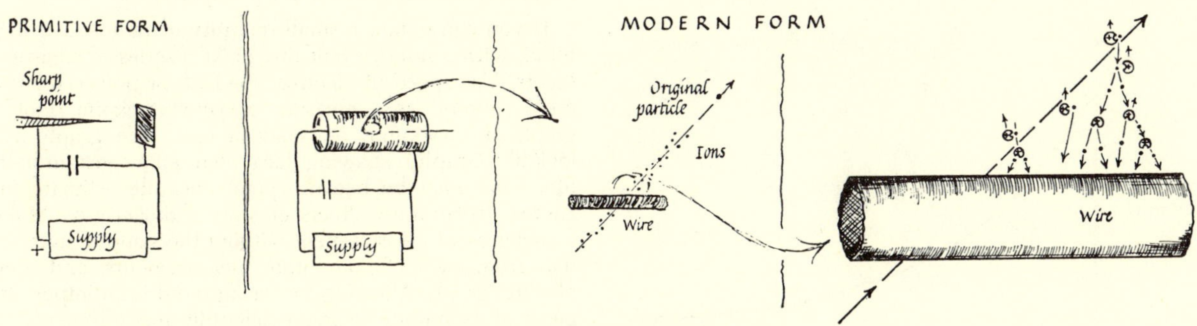

FIG. 39-14. GEIGER COUNTER TUBE

Geiger Counters

For accurate measurements we now use a Geiger counter. This multiplies up the ions made by each particle to a big pulse of standard size that can be counted. Alpha-particles (or β- or γ-rays) from the stockpile shoot through a small box or tube containing a suitable gas. They make ions in that gas. The tube has an insulated wire inside, with a large P.D. between wire and walls of tube. This makes a strong electric field in the gas, especially strong near the curved surface of the thin wire. The field is not quite big enough to start a spark by itself. If some ions are made in the gas, they are driven so hard by the electric field that they make more ions by collision, and more, and more, in a chain reaction. The tubeful of ions is swept across to wire and walls by the electric field and delivers a pulse of charge that can be amplified and made to work a mechanical counter or a loudspeaker. The simplest form, the Geiger counter proper, uses the electrons knocked off gas atoms before they have time to join other atoms and make heavy negative ions. These electrons, driven by the intense field near the central wire, knock electrons off other atoms and those . . . . . and an avalanche of electrons is collected by the central wire.

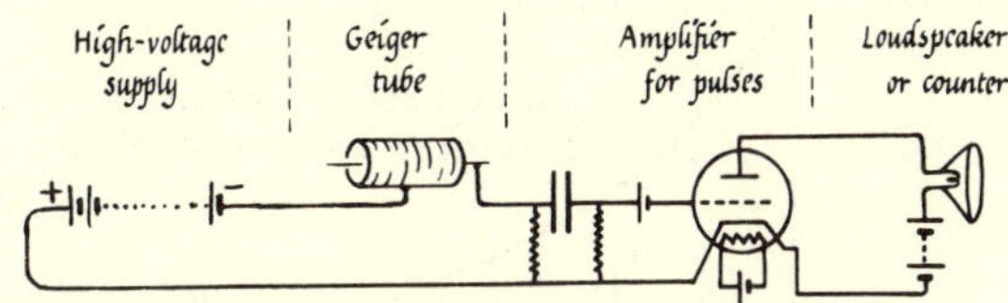

FIG. 39-15. GEIGER COUNTER ARRANGEMENT

Each ionizing "ray" entering the tube acts as a trigger. When you see or hear a counter in action you are watching the *effect* of single radioactive atoms blowing up. With a Geiger counter, the number of pulses per second is used as a measure of the stockpile. The tube is connected through an amplifier to a "scaler" which records the pulses, often by lighting small neon lamps.[10]

Other counters use a smaller electric field—below any threat of avalanche—so that the pulse can measure the total energy of the ionizing particle. It shows how many ions the particle made and thus indicates its original energy. This is the electroscope method, on a microscopic scale that can record single events.

---

[10] The simplest arrangement of lamps uses a "scale of two" in which

the first lamp flips on/off/on/ . . . at successive pulses;
the second lamp flips on/off/ . . . after every 2 pulses;
the third lamp flips on/off/ . . . after every 4 pulses;
. . . . . . and so on, counting pulses in batches of 1, 2, 4, 8, 16, and 32, instead of on the decimal scheme in batches of 1, 10, 100, ...

It is a more economical scale, easily read with a little training. It is also used by communication engineers and automation designers. You will find it in most "electronic brains," computers that use *flip-flop* circuits in which a radio tube is either *on* or *off*—a scheme that obviously lends itself to "scale-of-two" counting. The lights on the scaler sketched read: one 1, no 2's, one 4, no 8's, no 16's, one 32, totaling $1 + 4 + 32 = 37$. The mechanical counter on the right scores "1" each time all the lamps have lit together and gone out, for [1 + 2 + 4 + 8 + 16 + 32 + one more] or [64] pulses. In the sketch it reads 152 lots of 64, and the total score is $37 + (152 \times 64)$ or 9765 pulses.

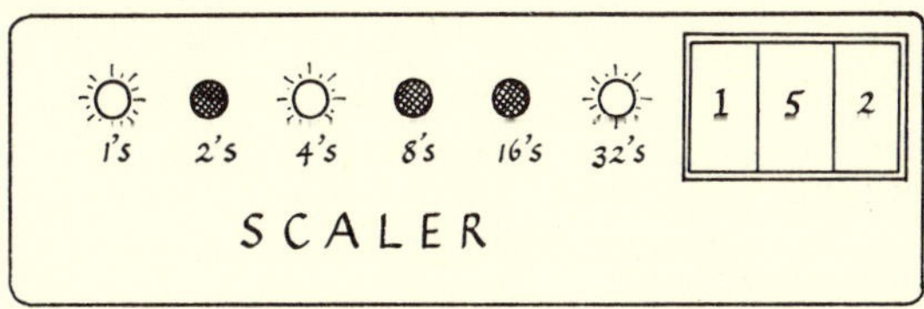

FIG. 39-16.

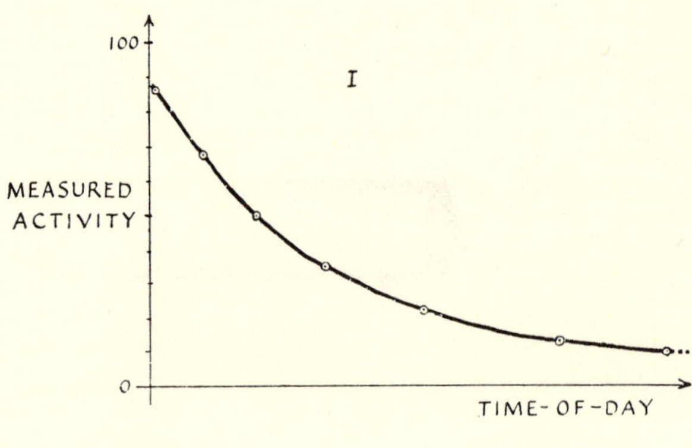

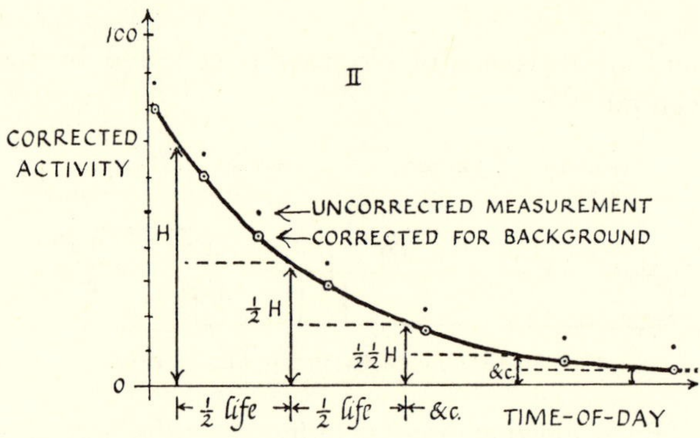

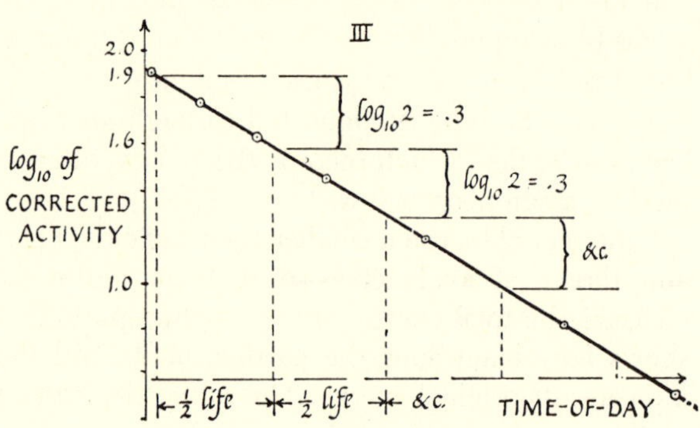

FIG. 39-17. RADIOACTIVE DECAY

Radioactive Decay: Laboratory Experiment

If you can obtain a small quantity of radioactive material, with a suitable half-life, make a series of measurements. Use speed of electroscope leaf, or pulses/minute on a counter, as a measure of your stockpile. Plot a graph of STOCKPILE *vs.* TIME-OF-DAY. The graph will look like Graph I, showing decay, but not a constant half-life. That may be because your stockpile estimate includes *background* effects of stray radioactivity. Make a background measurement without the sample, subtract that from each of your other measurements, and then plot the graph. Allowing for background is an important piece of technique in many scientific measurements: a "blank test" with the reagents in chemical analysis, a "control group" in biology, psychology, or sociology.

With a pure sample, you will be able to estimate the half-life from the graph, as in Graph II, and you will find it remains constant as time goes on.

Now take the logarithm of each stockpile-estimate. In each half-life the size of stockpile is *divided* by 2. When you use logarithms, division is done by subtracting logs. In each half-life the log of the stockpile should have log 2 subtracted from it. With a pure sample, the graph of log (STOCKPILE) *vs.* TIME-OF-DAY should drop by log 2 in one half-life, by log 2 in the next half-life, by log 2 in the next, . . . and so on. It will be a straight line. This gives the best test of pure-chance (*exponential*) decay of a radioactive substance.[11] (See Graph III.)

---

[11] Using calculus, we can express radioactive changes neatly. If we have N atoms, all unstable, all alike, at time t, then dN/dt is the rate of *growth* of stockpile, the increase of number of atoms per second. The rate of *breaking up* is $-dN/dt$. For pure-chance distintegration, the

RATE OF BREAKING UP
is a constant fraction of the
POPULATION-AT-THE-MOMENT

$\therefore -dN/dt = kN$ where k is a constant fraction

$\therefore \int dN/N = \int(-k)dt$

$\therefore \log_e N = -kt +$ constant (and log N, to *any* base, plotted against t will give a slanting straight line)

$\therefore \log_e N = -kt + \log_e N_0$ where N_0 is the original stockpile at $t = 0$

$\therefore N = N_0 e^{-kt}$
This exponental decay law appears often in science.

Atomic Structure

Electrons are easily chipped off atoms by bombardment. That suggests a crude picture of atoms as made up of several electrons loosely tied to a massive remainder.

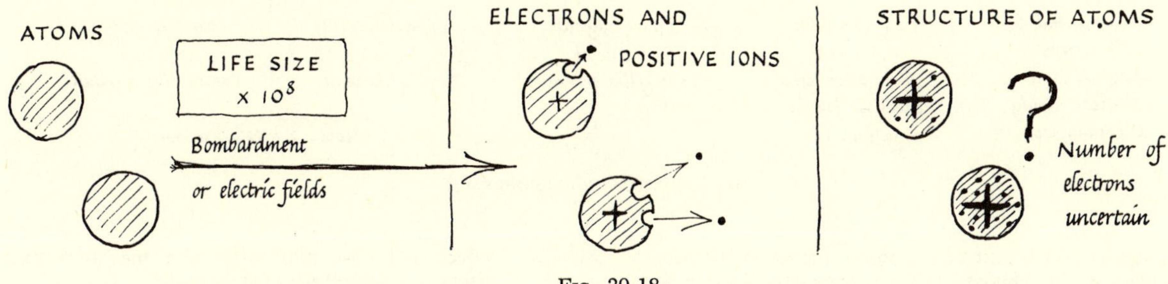

Fig. 39-18.

Using alpha-particles as probes, we find evidence that all atoms are mostly "hollow," with their light, loosely tied electrons far outside a compact massive core.

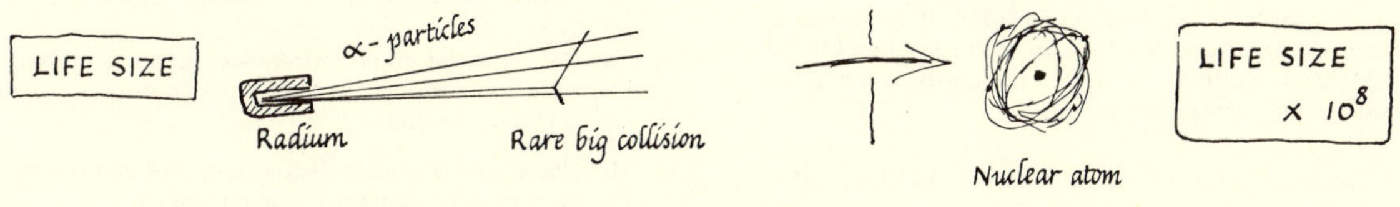

Fig. 39-19.

The radioactive atoms show that they themselves are unstable and able to release missiles with huge energy. We suspect that the missiles—and their energy—come from the central core.

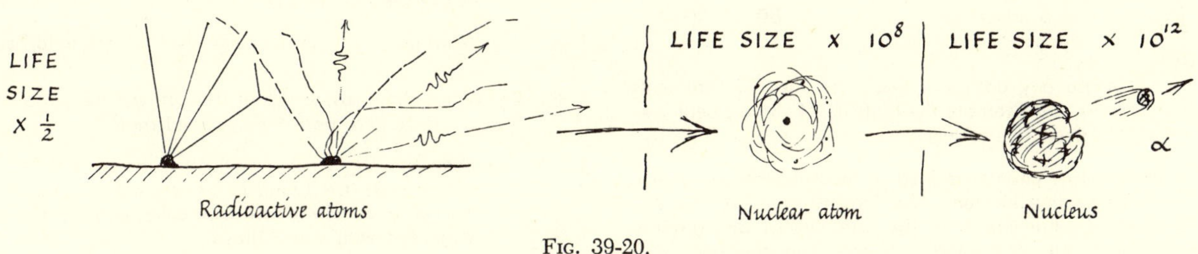

Fig. 39-20.

We even surmise that all atoms, stable as well as unstable, may have sub-atomic particles (α-particles? electrons? . . . . ? . .), locked in their cores, with vast stores of energy.

Half a century ago there were only vague guesses. Further experiments and reasoning yielded a wealth of detail and enabled us to build a useful *atomic model*, the "nuclear atom."

PROBLEMS FOR CHAPTER 39

Problems 1-5. Preliminary problems, at beginning of chapter.

Problem 6. In text.

7. Suppose you are supplied with radioactive copper with half-life about 10 minutes.
   (a) For this copper, how many half-lives are there in one hour?
   (b) What fraction of your original supply would you have after one hour?
   (c) What fraction of your original supply would you have after two hours from start?

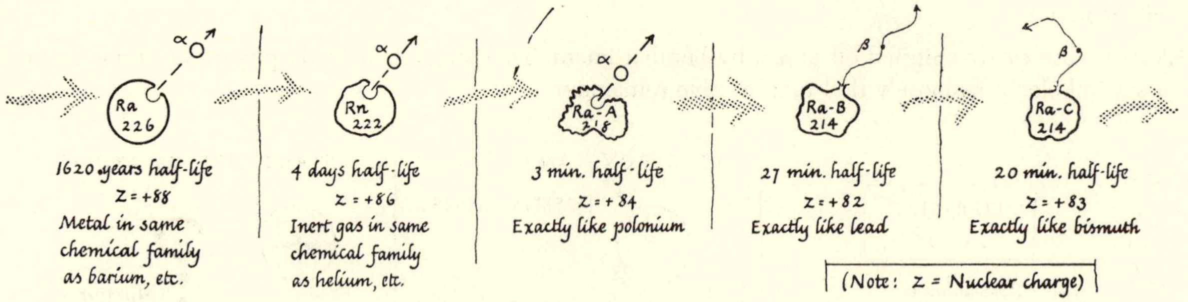

Fig. 39-21. Sketch for Problem 8.

8. Fig. 39-21 shows part of a diagram based on the *table* supplied in this chapter. Make a complete diagram for the whole table, drawing on the data there. You may show the instability of an atom (which is indicated by the shortness of its half-life) by the wobbliness of the boundary you draw. Note that radium belongs to the same group of chemical elements as barium and so has chemical properties rather like those of barium, but it is not the same chemical element as barium. It can be separated chemically from barium. (Similarly, radon is *somewhat* like helium. Both are inert gases.) However, radium-B is *exactly* like lead, chemically. It is the same chemical element, and the two are quite inseparable by any chemical method. However, radium-B is unstable (= radioactive) while lead is not.

9. (a) A Geiger counter is used to measure the decay of a radioactive element. The background makes 20 counts (pulses) per minute. With the sample in position the measurements are:

TIME OF DAY	"COUNTING RATE"
	in counts (pulses) per minute
0 (start)	120
6 hours from start	70
8 hours	60
10.5 hours	50
20 hours	30

Choose any data you like from the measurements above and estimate the half-life of the element . . . (No need to use logs)

(b) A Geiger counter is used to measure the decay of a radioactive element. Use logs to estimate the element's half-life from the data below. Background: negligible, compared with the high counting rates with sample.

Counting rate at start .... 1000 counts/minute
Counting rate 2 hours later ...... 100 counts/minute
log 2 = 0.301 log 3 = 0.477 log 5 = 0.700

10. Suppose you have a mixture of equal numbers of atoms of two radioactive elements, A with half-life 6 minutes, B with half-life 60 minutes; and both emit β-rays. You are making measurements of the β-rays from the mixture with an electroscope.

(a) Which of the two, A or B, will affect your electroscope most during the first few minutes?

(b) Say, roughly, what fraction of the total effect is due to the one you have named in (a), at first.

(c) Which will have most effect (of the small total effect remaining) after two hours?

(d) Roughly, what fraction of the total effect is accounted for by your answer to (c)?

(e) (HARD. MAKE AN INTELLIGENT GUESS.) Suppose you measured the activity of such a mixture for 2 hours or more. Sketch roughly the graph you would expect to get if you plotted log of the *rate-of-movement* of electroscope (corrected for background) upward and *time-of-day* along.

11. (a) Describe the main differences of appearance between alpha-particle tracks and beta-ray tracks in a cloud chamber.

(b) Use your knowledge of the nature of alpha-rays and beta-rays to explain these differences.

★ 12. In watching alpha-ray tracks in a cloud chamber you will see all the tracks are straight. You will seldom, if ever, see a bent track. When a great number of tracks are examined in photographs, a few are seen to show abrupt bends, with another track also running away from the bend, making a forked track.

(a) What can you infer from the fact that so many tracks are straight?

(b) What do you think happens when a track is forked?

★ 13. When alpha-rays are shot through wet helium a few tracks are forked; and it is observed that:

(i) when the angle between the two arms of the fork is measured it is found to be very close to 90°.
(ii) the two arms of the fork look alike, equally densely populated with water drops.

(a) What can you infer from observation (i) alone? What assumptions do you have to make?

(b) Granted the inference (a), what can you then infer from observation (ii)?

14. (a) When a fast electron is shot through a cloud chamber, its thin, fairly straight track sometimes shows a fork with arms making 90°. Both arms look about equally thin. Interpret this collision.

(b) When a very fast electron makes a collision like that in (a) the arms make less than 90°, and one arm points nearly forward along the original track. Supposing that other evidence suggests that this is the same kind of collision as in (a), interpret the change in geometry.

★ 15. Cosmic rays are streams of very high energy particles (and γ-rays) which travel to us, and through us, from outer space. There are atomic nuclei, electrons, and other particles in this mixed stream. *Positive electrons (positrons)* were discovered in cloud-chamber tracks made by cosmic rays (in 1932). Very high energy electrons pass right through a cloud chamber (wall and all) in a straight track which can be photographed by having water droplets condense on the ions formed by the minor collisions between the speedy electron and electrons of atoms in the chamber. The ions are rather sparse, when made by so light and so speedy a projectile, so the track looks rather "thin," but it is easily photographed. If a strong magnetic field is applied (perpendicular to the track) the track is slightly curved. These very high energy electrons can pass easily through thick metal plates, though they will lose some of their energy as they do so.

(a) If we know from the look of a track that it is made by an electron, what can we find by measuring the *curvature* of the track?

(b) If the track curves the opposite way from most tracks, then we guess that *either* it is made by a positive electron instead of by a negative one like the rest, or that it was made by a negative electron which was . . . ?

(c) To decide between the two guesses in (b), we put a metal plate across the middle of the chamber and hope that such a track will pass through the plate. This has succeeded. (See Fig. 39-22 and photograph in Fig. 43-3.) *How does this enable us to decide?*

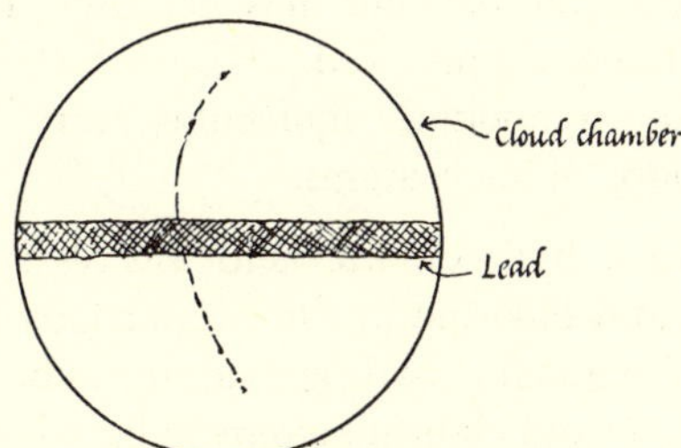

Fig. 39-22. Problem 15.

16. In Chapter 43 (Fig. 43-3) there is a cloud-chamber picture of a collision involving a "neutron" coming in a known direction from a source. The track of the "neutron" itself is invisible. What do you infer from that? Give a clear reason for your inference.

17. The damage caused to human tissue by radioactive materials (which make sores like burns on the surface and other irritations in deeper layers) is mainly due to ionization of atoms in the tissue.

(a) Which, therefore, do the damage to flesh: those rays which are stopped by the flesh, or those which pass through successfully?

(b) Comparing an alpha-particle, a beta-ray, and a gamma-ray of the same total energy, we find that an alpha-particle loses all its energy in making many ions in a few centimeters of air, a beta-ray makes far fewer ions per centimeter and struggles on for much greater distances in air, while gamma-rays whiz through—making ions very rarely. A gamma-ray has no effect—except for trivial recoil-electrons in occasional collisions—until it picks on an electron in some atom and gives up all its energy. The electron is whipped out with high speed just like a beta-ray; it makes many ions in coming to rest. (Though gamma-rays thus make "beta"-rays when they give up all their energy, they take the "none" option in so many collisions before taking the "all" option that they penetrate huge distances.) The disturbance caused by such rays is sometimes useful in discouraging malignant growths.

(i) Which would you advise, *and why*, for treatment of the skin of a patient, alpha, beta, or gamma radiation?

(ii) Which for treating the skin and underlying flesh?

(iii) Which for treating internal organs?

18. (a) Which is more dangerous to carry around with you, one gram of an unstable element with short half-life or one gram of an element with long half-life?

(b) The different isotopes of plutonium have different half-lives. The one first manufactured by cyclotron bombardment has a relatively short half-life. Does this make it safer for the preliminary experiment that had to be done with it to find its chemical properties while the first uranium reactors were being built? (See Ch. 43.)

19. A pipe from a boiler squirts out a small jet of steam. For the first few inches beyond the nozzle, the jet is practically invisible, but a small cloud forms farther out. An inventor trying to "eliminate the cloud electrically" passes sparks through the steam just outside the nozzle. Predict and explain his disconcerting failure. (See a demonstration of this.)

20. POSSIBLE USES OF TRACERS

Suppose your uncle, vice president of a small corporation, asks you how a radioactive tracer could be used to solve a problem in his business. He studied some physics long ago, but mostly of the older kind, with little atomic physics, and nothing about radioactive tracers. Explain to him how you would use a tracer in one of the examples below, and answer his two objections: (1) "But that would leave things radioactive forever"; (2) "Why would I have to know the half-life—what do you mean by 'allowing for decay'?"

(a) Your uncle runs a chicken farm and finds that too many eggs have soft shells. He says, "It's no good putting more oyster shells in the food. The hens only use the calcium they are hatched with!" Tell him how you would use a tracer to find whether chalk (calcium carbonate) fed to the hens now will go into egg shells soon. (Assume that you can obtain small quantities of Ca^{47} with half-life 4.8 days. There is such an isotope, now available commercially.)

(b) Your uncle runs a steel company that rolls out thin steel strips one foot wide by a few thousandths of an inch thick. How can he make sure the strip is being made of constant thickness, while it is still hot and moving? Assume you have supplies of Fe^{59} with half-life 46 days. Explain why the tracer need not make any danger for the ultimate users of the steel.

CHAPTER 40 • ATOMS: EXPERIMENT AND THEORY

"Theory, glamorous mother of the drudge experiment"
—Harlan Mayes, Jr., Princeton '52

(From a final examination in Elementary Physics, 1949)

[THIS CHAPTER makes a fresh start. It is a survey of atomic theory from a Greek idea to the early model of the nuclear atom.]

Early Ideas of Atoms and Molecules

Long before there was any real evidence pointing towards atoms and molecules, Greek thinkers pictured matter as made up of fiery moving particles, called "atoms," meaning indivisible—infinite subdivision seemed unthinkable. Perhaps this picture arose from a real interest in scientific speculation—thinking about what goes on in nature, or perhaps it was favored by a childish craving for definite rules and simplifying pictures which would make the hard external world seem less complex and thus easier to face. We should not despise this belief that nature is simple, even if we suspect that its roots lie in some sense of insecurity. Modern science is built on a more adult form, that nature is *reasonable*. We constantly look for simple laws running through our experimental data, and in building a theory we seek the simplest assumptions. Ultimately, our whole science may consist of simple, unproved statements used as starting points for a great deductive scheme whose results when tested are found to express the behavior of the natural world.

Greek ideas of atoms were only lucky guesses; but they provided a prejudiced background for later scientists, making it easier to set up atomic theories when the experimental evidence was gathered and the time was ripe for them as scientific theories.[1]

In reducing nature to four elements, Earth, Air, Fire, and Water, the Greek simplifying urge went further. The same urge remains with us and makes us ready to welcome Prout's hypothesis and later its brilliant justification by the mass-spectrograph. Yet as scientists we are not content with the simple story that emerges, but look for its slight failings and are thereby led to new knowledge of atomic energy.

[1] Some liquids mix freely with others; many solids dissolve in liquids; colored dyes spread uniformly in liquid; gases diffuse and compress. These common observations suggested intermingling of tiny particles, such as atoms or molecules. Dancing dust motes in a sunbeam suggested moving atoms.

Evidence of Atoms and Molecules, 1700-1900

As chemical knowledge developed out of alchemy, it encouraged the idea of atoms as basic building blocks from which molecules of compounds are built. Chemical compounds can be split up into their constituent elements whose proportions can be measured by direct weighing. Experiments show that a given compound always has the same proportions of ingredients—a result that is easily "explained" by the assumptions:

(a) that each element has its own brand of identical atoms, and
(b) that compound molecules are identical groups of such atoms.

For example, hydrochloric acid always splits into hydrogen and chlorine in the proportions 1 to 35.5 by mass. We imagine hydrogen atoms to be all alike, with mass M, and chlorine atoms to be all alike with mass $(35.5M)$, and hydrochloric acid molecules to be all alike, made up of an atom of hydrogen and an atom of chlorine joined together, with mass $(M + 35.5M)$ or $(36.5M)$. With this picture, we should then *expect* every analysis of hydrochloric acid to give the same proportions 1 to 35.5, and this is so. However, it would not be worth while making up this picture ("theory"?) just to "explain" such observations alone. In fact, it comes *from* such observations. Chemical analyses tell us two other things:

(i) Some sets of elements make several different compounds; and in such cases the proportions in one compound are very simply related to the proportions in another. For example: hydrogen and oxygen combine to form water in the proportion 2 to 16 by weight. They also combine to form another compound in the proportion 2 to 32. These two compounds are called, in chemical shorthand, H_2O and H_2O_2.

(ii) The *same* relative mass can be used for an element's atoms in all the compounds it forms, provided we assume that in some of the compound molecules it contributes more than one atom, say two or three. (For example, we can call the masses of hydrogen atom and oxygen atom and chlorine atom 1, 16, 35.5. These will fit the facts for *all* compounds involving H and/or O and/or Cl, such as HCl, H_2O, H_2O_2, $HClO_4$.) It is as if the castles in a nursery school showed by their simple structure that they could all be built from just a few varieties of identical blocks. Such chemical evidence supported the early ideas of atoms and molecules; it was easily understood in terms of atoms but it did not *prove* the existence of atoms. The kinetic theory of gases, thought about by Newton, developed by Joule and others, showed that a theory of elastic molecules in motion would "explain" Boyle's Law and make other predictions which agreed well with experiment. Again, this supported the idea of molecules but did not prove their existence. Yet, even then, it made thinking about gases easier.

The Brownian motion did indeed make the idea of moving molecules in gases and liquids seem real. Observers felt they could almost see the molecular bombardments. By assuming equipartition of energy in kinetic theory, they could use measurements of Brownian motion to estimate the mass of a gas molecule. The masses obtained were incredibly small, $\frac{1}{3000000000000000000000000000}$ of a kilogram for a hydrogen *molecule* (and larger for other molecules by factors easily obtained from chemical weighings).

Meanwhile, the experimental facts of electrolysis suggested the idea of charged ions in solutions, ready to carry current across when an electric field was applied. If an element's ions are all alike, identical atoms (or groups of atoms), the facts of electrolysis show that their electric charges must be all alike. Every ion of the same kind must have the same charge. Some ions have + charge, some —, and some kinds have double or triple charge; except for that, all must have identical charges—the basic charge must be universal. Thus an idea of "atoms" of electric charge joined the idea of atoms of matter a century ago.

Atomic Weights and Atomic Numbers

In the last century, chemists weighed the atoms of the many elements and catalogued their properties. The absolute masses of atoms were only vaguely guessed at, but their *relative* masses were measured accurately by chemical separation and weighing. These, on a scale that took the hydrogen atom's mass as 1, were called *atomic weights* (A).[2]

Chemists, professionally systematic people, shelved their elements neatly in order of atomic weights; hydrogen 1.0 [helium, discovered later, 4.0], lithium 6.9, beryllium 9.0, boron 10.8, carbon 12.0, nitrogen 14.0, oxygen 16.0, fluorine 19.0, and so on. Then they numbered their elements serially: hydrogen No. 1, helium No. 2, lithium No. 3, and so on. These serial numbers are called atomic numbers (Z). They seemed to be about half the atomic weight.

Prout's Hypothesis

A further suggestion arose: that even atoms, the basic building-blocks of matter, the atoms of the elements, are themselves composed of bunches of one primitive block, the hydrogen atom. The many whole-number values of atomic masses made Prout suggest this. It happens too often for mere chance. Examples: hydrogen 1, carbon 12.0, oxygen 16.0. But there were unwelcome exceptions, for example chlorine, whose mass was carefully investigated and emerged clearly fractional: 35.5; and copper 63.6. So Prout's hypothesis was dropped, to be revived in this century when isotopes were discovered. We now know that chlorine has two kinds of atom, of relative masses 35.0 and 37.0. Common chlorine gas is a mixture of these. Wherever chlorine is found, this mixture occurs in the same proportions of the two twin chlorines. As the twins are chemically inseparable, chemists were convinced they had a single chlorine of atomic weight 35.5.

Sizes of Atoms

The size of atoms, imagined as round solid lumps, was known roughly in the last century: diameter a few ÅU (10^{-10} meter). This could be estimated in several ways:

(i) For gases, an estimate of the mean free path at measured pressure leads to the "diameter" of the colliding molecules or their closest approach. The MFP could be estimated from gas-friction measurements. (Late 19th century)

(ii) Surface tension experiments with spreading oil films gave an estimate of the *length* of long organic molecules. (Late 19th century)

[2] Nowadays, A, read to the nearest whole number, is called the "mass number."

(iii) In this century we obtained more definite estimates. Knowing the masses of single atoms (e.g., from Brownian motion), we could calculate the *number* of them in a block of solid of known size. From this number we could calculate the average distance between individual atoms. For a solid we could call this distance the "diameter" of one atom. More reliable data for this calculation were obtained by combining measurements of e/M for ions with a value of e from Millikan's experiment.

In all these estimates there was considerable uncertainty; and some of them referred to atoms while others referred to molecules, which are groups of atoms. But the evidence pointed fairly consistently to atoms about 10^{-10} meter in diameter and molecules somewhat larger. In this century these estimates were confirmed by X-ray measurements of the spacing between atom-layers in a crystal.

Structure of Atoms, 1890-1910

Towards the end of the last century, cathode rays and positive rays were discovered and investigated. In a discharge tube containing gas at low pressure a large electric field produces two streams of particles:

(i) *Cathode rays.* These seem to start near the electrode connected to the negative of the battery. They travel down the tube and can pass through the holes in the positive electrode as a stream of charged particles. Electric and magnetic field deflections show they are negatively charged, moving very fast, with an e/m which is the same for all of them, whatever the gas.

(ii) *Positive rays.* These travel the opposite way, and can pass through holes in the negative electrode. Deflections by fields show they are positively charged and have high speeds, and various values of e/M, many times smaller than e/m for cathode rays.

We call the negative particles of cathode rays "electrons," and we obtain the same electrons, with the same e/m, by many processes. They can be boiled off hot filaments, whipped out of metal by light, ripped out of atoms by X-rays. They are also shot out by some radioactive atoms as beta-rays. Their e/m is some 1840 times larger than e/M for hydrogen ions in electrolysis. We guess (on the strength of strong indirect evidence) that the charge, e, has the same size for both electrons and hydrogen ions in gas or solution; and therefore we conclude that electrons have about 1/1840 of the mass of a hydrogen atom. Electrons appear to be universal identical constituents of atoms, fairly easily detached by bombardment, etc.

Positive-ray particles seem to be the remainder of the atom after it has lost one or more electrons. Their e/M is the same as e/M for corresponding ions in electrolysis. In fact positive rays *are* just fast-moving ions. They may be groups of atoms or single atoms, and may have several + charges (e.g., H^+, O^+, O^{++}, H_2O^+, CH_3^+).

The early picture of atomic structure used these two ingredients. To preserve the stability of the system, the electrons were imagined to be embedded in a large ball of positively charged material which clutched them there like raisins in a pudding. This was the model suggested by J. J. Thomson, and generally held in the early years of this century: a massive positive pudding, 10^{-10} meter or more in diameter, with enough tiny, light, negative electrons embedded in it to make it just neutral. Such a picture explained the effects observed in discharge tubes, and made it easy to understand why very fast particles such as beta-rays and alpha-rays could travel straight through matter. They went right through the pudding, nowhere encountering a large enough mass with a big enough charge close enough to deflect them much.

Like archeologists reconstructing from fragments, Thomson and others had made their atom model from the pieces of shattered atoms found in electric discharge tubes. Yet this simple reassembling of ingredients raised a serious problem. The negative electrons could not be left lying loose outside the positive remainder. They would be pulled in, with huge forces at such tiny distances; they would crash into the positive core. Nor could they be arranged in same fancy pattern of + and — particles to keep each other in equilibrium with inverse-square law forces. Electrical attractions and repulsions could hold them in an equilibrium pattern; but the arrangement would be *unstable*—a small disturbance would grow worse and lead to collapse. (The schoolboy experimenting with magnets can make one of them float in mid-air momentarily, but it soon topples sideways unless he applies non-inverse-square forces with his fingers or with wooden pillars.) Earnshaw had shown that this instability is unavoidable. Any arrangement of bodies at rest *exerting only inverse-square-law forces*—electric charges, magnets, gravitating masses—is in unstable equilibrium. He developed a theorem showing this from $\nabla^2 V = 0$, the mathematical statement of the inverse-square law. Earnshaw's objection does not apply to a system with *acceleration*, such as one with the electrons circulating like planets in orbits; but that motion raises another serious objection. An electron running around an orbit has an acceleration, v^2/r, and it was known that accelerating charges would emit electromagnetic waves. Therefore, an electron in an orbit would radiate, lose energy, and the orbit would collapse towards the center in a tiny fraction of a second. Early radio experiments had shown that when charges *do* accelerate (in alternating currents surging up and down a radio antenna) waves *are* emitted. Light was known to be very short wavelength radio waves, presumably emitted by electrons accelerating somewhere in an atom. Atoms do radiate light *sometimes*; but we could not picture them radiating all the time—they would soon collapse. To avoid this trouble, Thomson imagined the electrons embedded in a positive pudding which was held together by mysterious non-inverse-square forces that provided stability.

By 1910, however, this picture was proving unsatisfactory. Alpha-particles, used as investigating projectiles to pry into the structure of atoms, gave results that could not be explained by a pudding atom. Rutherford suggested a new *atom* model, chiefly hollow instead of a solid pudding, a tiny nucleus surrounded by electrons moving in orbits—with nothing said about the unfortunate question of emitting radiation.

Alpha-Particle Scattering and The Rutherford Atom, 1910-1915

A stream of α-particles can shoot right through a thin layer of atoms, such as a sheet of gold leaf. But some of the alpha-particles are deflected from their path by a small angle, say, 5°, 10°, while a few are deflected through big angles of 60° or 80° and a very few are deflected through very big angles, such as 150°. You can see such collisions as rare forks in cloud-chamber tracks. Rutherford, examining the experimental counts of scattering, saw that the Thomson atom model could not fit with the rare large deflections. If the pudding were a hard lump, *all* alpha-particles would bounce back. If it were a soft lump *none* would bounce back. Bouncing back required a collision with something massive (larger mass than alpha-particle, which was known to be a charged helium atom), and it required a strong force. Rutherford guessed the force might be provided by the ordinary inverse-square-law repulsion between the alpha-particle's charge and the charge on the positive part of the gold atom. If so, the alpha-particle must approach far closer to that positive charge than 1 or 2 AU (the "size of an atom") to experience the forces to slow it down and bounce it back. Further, it would need the repulsion of the gold atom's + charge, unshielded by its negative electrons. So Rutherford suggested a new model of the atom: a very tiny core, positively charged, with almost all the mass of the atom concentrated in it; and a set of electrons, far out from the core, circulating in orbits like the planets around the Sun. He assumed that:

a hydrogen atom has a core with a single charge, $+e$, and one outer electron with charge $-e$;
a helium atom a core with double charge, $+2e$, and two outer electrons each with $-e$;
a lithium atom, core charge $+3e$ with three outer electrons;
and so on up the chemical series.

He suggested that the zth atom, with "atomic number" Z to show its place in the series, would have core charge $+Ze$ and Z outer electrons.

Rutherford then asked the following mathematical questions:

"(i) If an alpha-particle with ++ charge is shot towards such an atom what shape will its orbit take, when it is inside the region of electrons, and near the nucleus?

(ii) If a thin gold sheet contains such atoms, how will the scattered alpha-particles be distributed in different directions when a stream of them hits the sheet?

Assume an inverse-square-law repulsion between the alpha-particle and the atom core. Assume the gold atom core carries charge $+Ze$."

Mathematical machinery then produced clear predictions:

(i) The orbits will be hyperbolas[3] (for repulsion, just as Kepler's are ellipses for attraction).

(ii) The distribution of scattered alpha-particles will follow a definite relationship with direction and velocity, which can be tested experimentally.

Around 1910, alpha-particles were not counted with Geiger counters, and there were not enough photographs of collisions in cloud chambers to give statistical information, so Rutherford had to rely on the tiny flashes of light ("scintillations"), made when alpha-particles strike a mineral-coated screen. The observer must wait for 20 minutes in the dark

[3] Calculus predicts:
For attraction, orbit $x^2/a^2 + y^2/b^2 = 1$, an ellipse or, (for speeds above escape-velocity),
$x^2/a^2 - y^2/b^2 = 1$, a hyperbola.
For repulsion, always a hyperbola.

until his eyes are sensitive; then, watching the screen through a microscope, he sees a faint flash for each alpha-particle that hits the screen.[4]

So the mathematical machine was made to predict the number of scintillations to be expected on a small movable screen placed in various positions, to receive alpha-particles whose path had been bent

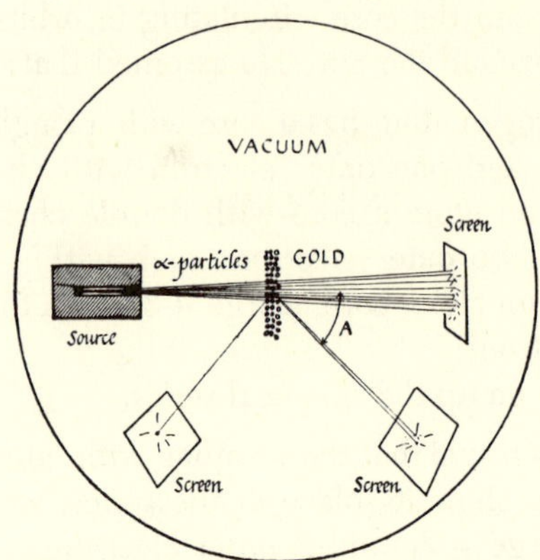

FIG. 40-1. SCATTERING OF ALPHA PARTICLES

through various angles. This prediction was then tested carefully. The number predicted was the following fraction of the total number fired at the gold leaf (most of them going almost straight through):

$$\frac{K \cdot (\text{AREA OF SCREEN})(2e)^2(Ze)^2}{(\text{VELOCITY OF ALPHA-RAYS})^4} \cdot \frac{1}{(\sin \frac{1}{2}A)^4}$$

where K is a constant which can be calculated from the geometry (distance of screen, and the thickness of gold); $+Ze$ is the charge on the gold nucleus, that is, Z positive electron charges); A is the angle of deflection. This prediction was based on the inverse-square law, without which neither the $\frac{1}{v^4}$ nor the $\frac{1}{(\sin \frac{1}{2}A)^4}$ factor would be the right one.

Rutherford and his colleagues tested both these factors of the prediction.

Using fast, medium and slow α-particles of known speeds, they tested the prediction

$$\text{NUMBER HITTING SCREEN} \propto (1/v^4)$$

by multiplying the observed numbers hitting the screen by v^4. You have already met their measurements in Problem 3 of Chapter 18. The measurements agreed quite well with the prediction. This in itself gives a clear indication that the inverse-square law is the right one. For the same scattering angle, faster particles must be ones that passed nearer the nucleus—thus experiencing bigger forces in their shorter time-of-encounter—and we should expect a smaller number to be aimed (by chance) at that smaller target-ring.

But the grand test of the whole state of affairs deep inside the gold atom was to see whether (NUMBER HITTING SCREEN) does vary as $\frac{1}{(\sin \frac{1}{2}A)^4}$, as predicted.

The α-particles fired at the sheet of gold acted as electric-field-investigators, showing by their deflections what forces they had experienced. In a thin sheet, most of them fail to pass very near any nucleus, so they are only deflected through small angles; some pass fairly near and are deflected noticeably; and a few are deflected through large angles because they happen to be aimed very close to some gold nucleus. The table shows the test. Here was a result to delight Kepler. The agreement of the ratios in the final column gives clear testimony in favor of the inverse-square law over a huge region inside the gold atom.

SCATTERING OF ALPHA-PARTICLES BY GOLD
(Experimental Test by Geiger and Marsden)

EXPERIMENTAL MEASUREMENTS		TEST OF THEORETICAL PREDICTION	
*Angle of Deflection** A°	*Experimental Count*† *N*	*Value of* $\frac{1}{(\sin \frac{1}{2}A)^4}$ from tables	*Test* $\frac{N}{1/(\sin \frac{1}{2}A)^4}$
150	33	1.15	29
135	43	1.38	31
120	52	1.79	29
105	69.5	2.53	28
75	211	7.25	29
60	477	16.0	30
45	1,435	46.6	31
30	7,800	233.3	33
15	120,570	3,445	35
10	502,570	17,330	29
5	8,289,000	276,300	30

* Of path of alpha-particles.

† Number of scintillations seen, for deflection A°, in a standard time.

Note: In the actual experiments Geiger and Marsden made one set of measurements for the larger angles of deflection, and another set, with a much smaller radioactive source, for the smaller angles. To make one complete set in the table above, the numbers for smaller angles have been multiplied up to fit the set for larger angles.

The original account may be found in *Philosophical Magazine*, Vol. 25 (1913), p. 610, Table II.

[4] Nowadays we are using scintillations again, but we use an electronic eye to see them and increase their effects by electron multiplier tubes.

Rutherford could even estimate the charge on the nucleus. His first calculations pointed to the atomic number, the serial number of the scattering element, in the chemical list. This serial number, which was about half the atomic weight of the lighter elements, was "in the air" as a useful idea in treating atoms. It seemed possible that the number of electrons in an atom was about half the atomic weight. This was not so for hydrogen which showed every sign of having only one electron to lose. However, helium (mass 4 times hydrogen) could clearly lose two electrons and showed no sign of ever losing more. An attempt was made to count the number of electrons in a carbon atom by letting them scatter X-rays. X-rays, probably emitted by "? vibrations" of atomic electrons, could also be scattered by solid material, and it seemed probable that the responsive "? vibrators" were electrons. A rough, difficult estimate gave about six electrons to a carbon atom. The number of electrons circulating around the nucleus in Rutherford's model must be equal to the total number of positive electronic charges, Z, on the nucleus. Rutherford, therefore, made this guess, that his Z was the same as the chemical serial number, the atomic number.[5] This could be tested by α-particle scattering because the constant K in the prediction could be calculated, so that everything in the formula was known except Z. Thus from observed numbers of alpha-particles scattered the value of Z could be calculated. This was done for scattering by thin sheets of copper, silver, platinum. The chemical serial numbers or "atomic numbers" for these elements are 29, 47, 78. Alpha-ray scattering by these metals gave values of Z 29.3, 46.3, 77.4, with a probable error of 1%.

Further, we can calculate how close the α-particle passes to the nucleus, now that we are assured of the inverse-square law and know the value of the nuclear charge. We find the closest approach is about 10^{-14} meter or .0001 ÅU. This is 10,000 times smaller than the estimates of atomic size, 1 or 2 AU. It seems as if 9999/10000 of the atom is hollow. See Problem 17 in Chapter 33.

So we have a clear picture of an atom, a tiny massive nucleus carrying a positive charge +Z electron charges, with Z electrons circulating around very far out from the nucleus. Hydrogen atoms have $Z = 1$, a single positive nucleus with one electron; helium atoms, $Z = 2$, a nucleus with charge ++ and two electrons; and so on. Stripped of its electron, the hydrogen atom becomes a hydrogen ion, H^+, which we now call a *proton*. Stripped of both its electrons, the helium atom is an alpha-particle, He^{++}. (No wonder alpha-particles whizzing along without electrons have just two + charges.) Ions of other atoms have usually lost only one or two of their many electrons.

But perhaps the picture of an atom as a miniature solar system is *too clear*: Further investigations show that electrons do *not* revolve in planetary ellipses, and are not arranged with a housekeeper's neatness on shelf-like orbits. That early model contained too much imaginative picturing. Yet α-particle scattering has given some clear information: that atoms are mostly hollow with small, massive, charged nuclei exerting inverse-square law electric fields over a great range of distances inside the old "atomic" region. The Rutherford atom picture was clearly incomplete and over-described even when it was first made. More thinking was needed and more inquiring experiments. The thinking, started by Bohr, led to new theory, to which we shall return.

[5] When Bohr was first devising his atomic model, Moseley tried to use it for characteristic X-rays emitted by electrons nearest the nucleus. He made a bold prediction and carried out a marvelous set of measurements with X-ray targets drawn along in a miniature freight train in a vacuum. He emerged with the earliest clear estimates of nuclear charges for many elements.

PROBLEMS FOR CHAPTER 40

★ 1. (a) What experimental evidence assures us that the sun's gravitational field obeys an *inverse-square law*, over a large region ranging from about 36,000,000 miles from the sun to 2,800,000,000 miles?

(b) What observations can be made (occasionally) to extend this investigation of the sun's gravitational field *nearer the sun and farther away*, outside the range mentioned in (a)?

(c) What experiments show that atomic nuclei (e.g., the core of a gold atom) exert *inverse-square law* forces on other electric charges?

(d) What *other* information about atomic nuclei is provided by the experiments referred to in (c)?

2. ALPHA-PARTICLE SCATTERING AND SPEED

If you did not solve Problem 3 in Ch. 18, you should now analyze its data in the light of further knowledge.

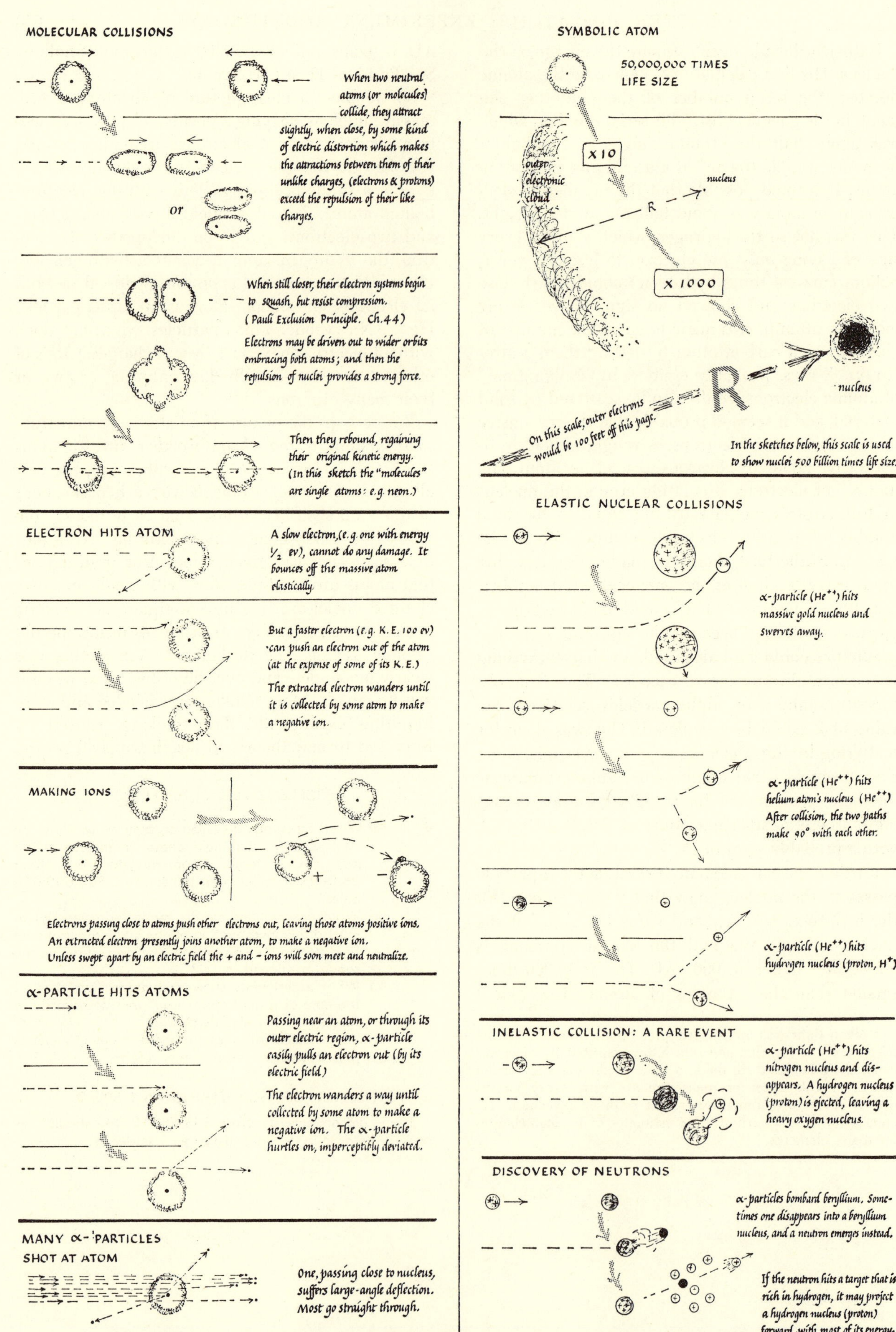

Fig. 40-2. Pictures of Atomic Collisions and Nuclear Collisions.
(These are fanciful sketches that contain unwarranted details of atomic structure; but they illustrate genuine atomic events.)

CHAPTER 41 • LABORATORY WORK WITH ELECTRONS: FROM GENERATORS TO OSCILLOSCOPES

"Sine experientia nihil sufficienter sciri potest"
"Without experience, nothing can be certainly known"

—Roger Bacon (~ 1250), on experience, both spiritual and practical.

". . . encourage the love of observing and investigating natural phenomena, which is the mainspring of scientific life. Given that love, the mastery of scientific method becomes a natural incident of . . . progress; without it, scientific method, however scrupulously "cultivated," is sterile. . . . The student must learn . . . what it is to face problems in the position of a real investigator, left largely to his own wits to wrest from nature the answers to the questions he puts to her."

—Sir Percy Nunn (in *The New Teaching*, ~ 1918)

[How do generators "produce current"? Why is power transmitted more cheaply by alternating current? What are the essential components inside radio sets, amplifiers, television tubes, . . . and how do they work? And how do we measure the properties of the basic electrons in this "machinery"? These questions of the present electrical age go beyond the "householders' electricity" of Chapter 32, and need more information before they can be answered. A good physics course should give you clear answers to some of these; but that cannot be done satisfactorily by simple pictures or glib stories. Instead, you should extract your own understanding from some of the *experiments* in this chapter. *Do them yourself*, or see demonstrations, or leave the topics untouched.]

EXPERIMENT A. Catapult Force

Arrange a circuit to send a large steady current through a flexible wire that runs across a magnetic field. Or use a metal trapeze with its crossbar across

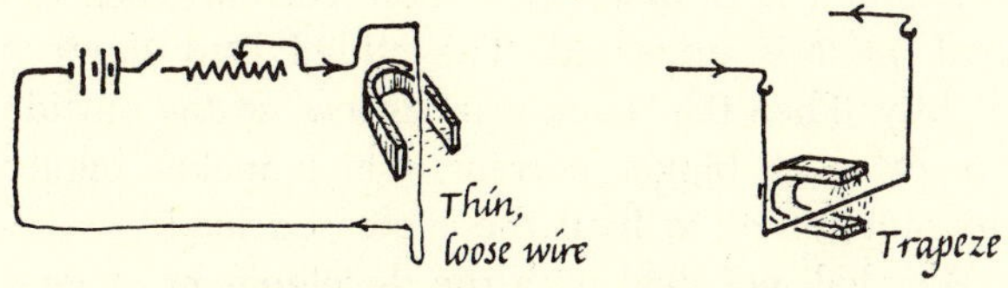

Fig. 41-1. Catapult Force

the field. Watch the effect of the catapult force when the current is switched on. (This catapult force will seem small and gentle. To feel it on a big scale, hold the shaft of an electric motor and try to prevent it from spinning.)

EXPERIMENT B. Catapult Force on Electrons

Make a stream of electrons cross a magnetic field. To do this, bring a magnet near a cathode ray oscilloscope tube. Watch and record the effect on the stream. Try the same magnet on a flexible wire carrying current in a known direction (Experiment A) and compare the effect with the deflection of the cathode ray stream. Does this prove that the cathode rays carry negative charge?

C. EXPERIMENTS WITH MAGNETS AND COILS

(These form a series of experiments with some explanations. Work through them yourself, studying the explanations.)

EXPERIMENT C (1). Preliminary Experiments with Magnets and Coils

Connect a small coil of insulated wire (say 40 turns) to a sensitive microammeter.[1] There is no battery in this simple circuit, so you would expect no current. However, if voltage *is* somehow generated in the coil, it will drive a current through the meter. Try the following:

[1] The early name for a primitive ammeter was "galvanometer." We now use this name for any sensitive instrument to indicate or measure small currents. Microammeter is a modern name meaning much the same.

(a) Move N-pole of a bar magnet into the coil. Record the direction and amplitude of the microammeter deflection. Remove N-pole, and record the deflection.
(b) Repeat (1) with the S-pole.
(c) Repeat (1) with double the number of turns. (Double up your coil as in Fig. 41-2.)
(d) Run the coil over one leg of horseshoe magnet. Remove coil.

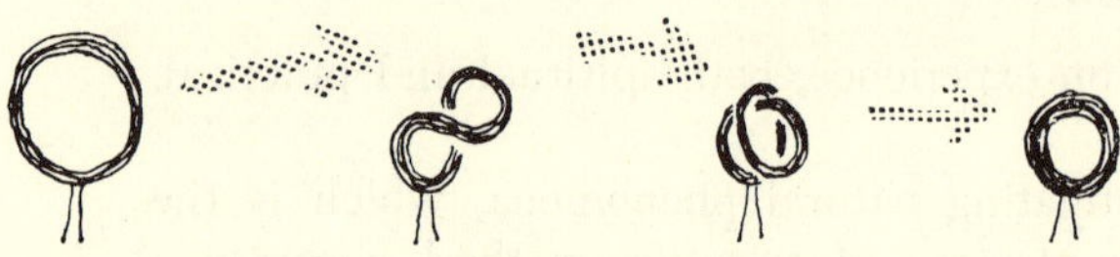

Fig. 41-2.
Preliminary Experiments with Magnets and Coils
(Convert coil to one with twice as many turns, by twisting it into a figure of 8 and doubling it up like this.)

Magnets and Coils: Theoretical Interlude

In the preliminary experiments, the meter showed a small current whenever a magnet was *moving* near the coil. This was driven by a *voltage "induced" by the motion* of the magnet. The essential requirement seemed to be either motion or some other changing of a magnetic field. The coil itself cannot see the magnet or smell it; so the coil only knows of the magnet's motion by experiencing a changing magnetic field. Suppose we bring a coil near a magnet as in the move from A to B in Fig. 41-3. During

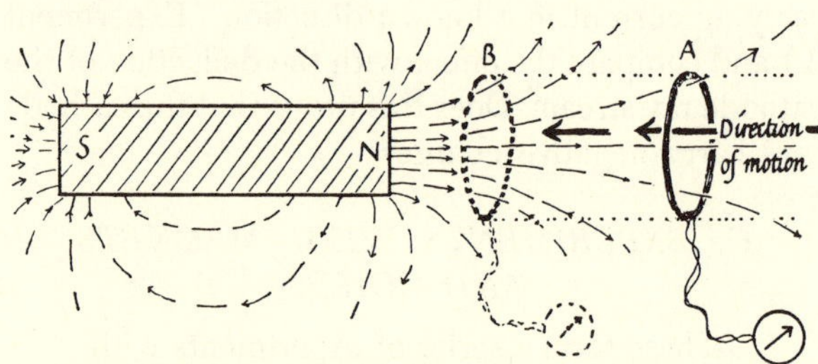

Fig. 41-3. Coil and Magnet's Lines of Force
When coil moves relative to magnet, it cuts lines of magnetic field and total number of magnetic field lines threading the coil is changed.

the motion, there is an induced voltage driving current around the coil; and from the coil's point of view the only change is either:

(I) the wires of the coil are cutting across the lines of force of the magnetic field, as shown by the breaks marked on the lines where the coil cuts through them; or,
(II) the total number of lines threading through the coil's loop is changing.

If you think about it, you will see that (I) and (II) tell much the same story—the coil cannot cut across lines without changing the number of lines that thread through it.

Faraday's Discovery

Experiments investigating these induced effects were started by Michael Faraday and Joseph Henry over a century ago.

They arrived at the general result *that whenever a wire is cutting the lines of a magnetic field, or whenever the total number of lines threading a wire circuit is changing, there is an induced voltage in the wire, trying to drive a current through it.* If the wire circuit is complete, there will be a current—and it is this current that made your meter's pointer move. If the circuit is broken there is no current, but we can show that the induced voltage is still there during the change. Ordinary microvoltmeters are microammeters in disguise, letting a current pass, so you cannot test this last statement on a small scale. But there is a device available for testing it on a large scale, in which many coils of wire are moved fast in a strong magnetic field—this is simply a big generator! By connecting a running generator to a true voltmeter we can show that the induced voltage is really there. This voltage *is* the driving EMF of the generator.

In the 1820's, the time was ripe for this discovery. Ampère and Oersted were looking for it—but did not see the need for a moving magnet—and Faraday in England and Henry in America were trying to "convert magnetism into electricity." In 1832 both announced the discovery of what we now call *electromagnetic induction.* Like your work with "magnets and coils," their simple experiments seemed a long way from our vast electric power systems, yet they had discovered the principle on which all electric generators are based today. It is also an essential principle for electric motors: in the spinning coils of motors, the field magnets induce a "back EMF" which limits the current and gives motors their great flexibility for varying loads.

Watch a D.C. motor take more current when the load on it is increased. The added load slows it slightly. Then the "back EMF" is less, so the outside P.D. drives a bigger current, which makes bigger catapult forces to haul the increased load.

We shall not deal with the development of generators, but you should see a demonstration of a simple D.C. generator: a spinning coil with a "commutator" to reverse its output every half cycle.

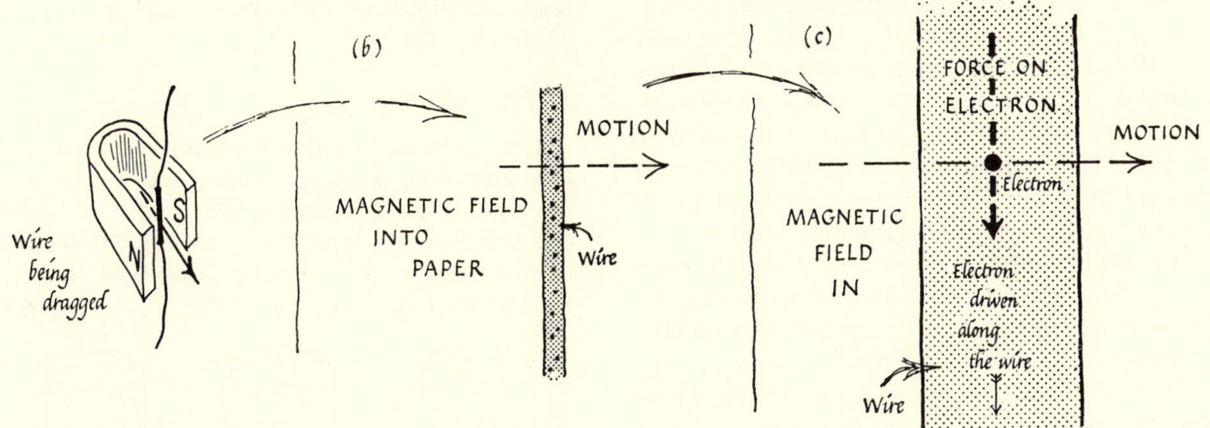

FIG. 41-4.

Electron Theory Explanation

Simple electron theory gives a clear interpretation of induced voltages. We picture metal wires as containing a cloud of loose electrons ready to carry a current. When we drag a wire across a magnetic field, we drag its loose electrons with it, *across* the field. Each moving electron constitutes an electric current, *across* the field. We expect each electron, therefore, to experience a catapult force, crabwise, perpendicular to the motion and perpendicular to the field. The force is therefore *along the wire.* This force acts on the electrons, pushing them along the wire, thus making an EMF that tries to drive a current, just as if there were a battery. So we regard the induced EMF as due to catapult forces exerted on the loose electrons as they are dragged across the magnetic field.[2] (Positive charges are catapulted the opposite way: the same EMF acts on them—and moves them, *if* they are free.)

Lenz's Law

Which way does the induced current flow? To find out, you need to experiment with a moving magnet and a coil and compare the meter's movement with the effect of a *known* current through it. You will then find that in every case the induced current runs through the coil (or plain wire) in such a direction that the current's own magnetic field opposes the change being made: that is, if the magnet is approaching the coil, the current through the coil makes a magnetic field to repel the magnet; if the magnet is receding from the coil, the current makes the coil attract the magnet; if the coil is twisting and thus changing the number of magnetic field lines threading it, the current makes a field to oppose the twist. Induced effects always oppose the changes that cause them. This is general "inertia" behavior. It is called Lenz's Law after Emil Lenz who formulated it. It can clearly be inferred from experiment; or, if you believe in the conservation of energy, you can deduce it from that. When the circuit is incomplete, there is no induced current, but still an induced EMF in a direction such that, *if* it drove a current. . . .

Negative Magnetism: Universal Diamagnetism

Now we can throw some light on "negative magnetism" of all substances, mentioned in Ch. 34. Each electron pursuing some kind of "orbit" around the nucleus in an atom is equivalent to a tiny electric

---

[2] We can predict the actual value of the induced voltage—in full agreement with experiment—from the catapult force. Here is an outline of the derivation: If the wire moves (sideways) with velocity v meters/sec, an electron with charge e coulombs experiences a catapult force $F = 10^{-7}\, evH$, where H is the magnetic field ($= C_2 \cdot 2\pi N/R$ at the center of a hoop-coil). Therefore, the strength of *electric field* produced is:

$$X = \text{FORCE/CHARGE} = 10^{-7}\, evH/e = 10^{-7}\, vH.$$

Suppose the moving wire has length L, and the EMF induced between its ends is E volts. Then (as shown in Ch. 33)

$$\text{FIELD STRENGTH, } X = \text{P.D./DISTANCE} = E/L$$
$$\therefore\ E = XL = 10^{-7}\, vHL.$$

But vL is the area swept across by the wire per second.

$$\therefore\ E = 10^{-7}\, H \cdot (\text{AREA SWEPT ACROSS PER SECOND})$$

If we show the size of the magnetic field by drawing H lines through each unit area, $(H)(\text{AREA})$ is the number of lines cut across.

$$\therefore\ E = 10^{-7}\ (\text{RATE OF CUTTING ACROSS LINES OF MAGNETIC FIELD})$$

This gives the induced EMF. A full discussion shows it should have a minus sign (Lenz's Law): $E = -10^{-7}\, dn/dt$, where dn/dt means either (RATE OF CUTTING LINES OF H) or (RATE OF CHANGE OF NUMBER OF LINES, n, THREADING CIRCUIT).

circuit. As we bring in a magnetic field from outside, its lines of force threading the electron orbit make an induced EMF that hurries or slows the electron *in such a way as to oppose the increase of magnetic field in the locality.* Then, as long as the imposed magnetic field remains, the orbit should retain its changed motion. (The magnetic contribution of electron *spins,* however, does not change.)

We should expect all electron orbits of all atoms to contribute to this opposition or weakening effect—all substances should be repelled, very slightly, by a magnet. This "diamagnetism" is masked by the positive effect of atoms like iron and oxygen, whose electrons provide a resultant magnetic field that can be aligned with the outside field and thereby add to it. Iron and oxygen are attracted by a magnet. But in substances with non-magnetic atoms (where both spins and orbits balance out magnetically), diamagnetism does show up as a simple general property of matter.

Lines-of-force Picture

You may, if you like, think of the moving wire dragging magnetic lines of force with it stretching them and failing to cut them until later, when it has dragged them some way. This fanciful picture actually gives a hint predicting both the existence and the direction of induced currents. For example, if a wire perpendicular to the paper is moved upwards in the field of a horseshoe magnet, as indicated in Fig. 41-5, we might imagine the wire carrying some of the field's lines up with it in a festoon as in sketch (ii). If we add a few details, making sketch (iii),

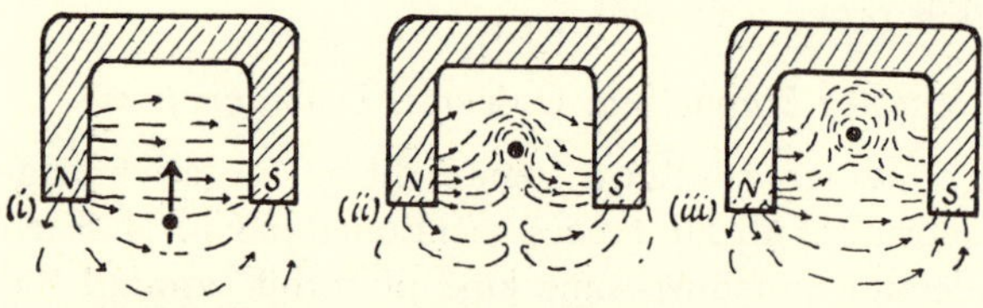

Fig. 41-5. Wire Moving Across Lines of Magnetic Field
A fanciful picture of the mechanism of induction.

we have the field that we should expect if the wire itself carried a current. That suggested current *is* the induced current, when there is one. The combined field suggests the wire would be catapulted *downward.* The induced current predicted by sketch (iii) would oppose the motion.

EXPERIMENT C (2)

Continuing your early experiments, connect a coil to microammeter and run one leg of the horseshoe magnet in and out of the coil. Can you see why, in terms of the discussion above, this produces an alternating current?

EXPERIMENT C (3)

You could also change the number of magnetic lines threading a coil by placing the coil between the arms of the horseshoe magnet and then moving the coil. Try this. Squash the coil till it is small or narrow enough to fit in the space and then try twisting it (Fig. 41-6). This makes a primitive A.C.

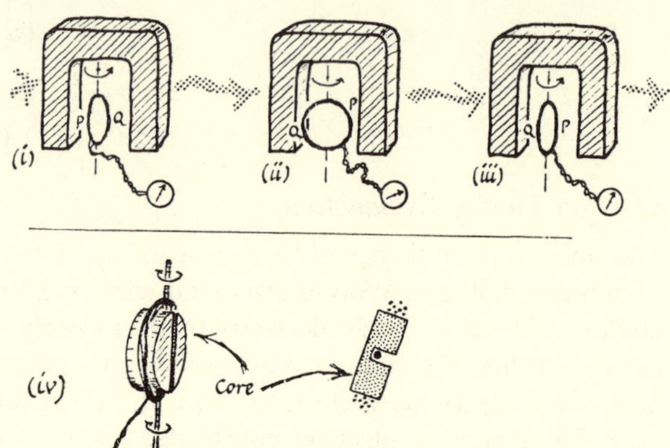

Fig. 41-6. Experiments $C(3)$ and $C(5)$

generator. [Note when the coil is in position (i) there are, say, +100 lines of magnetic field through it; in position (ii) there are zero lines through it; in position (iii) there are −100 lines through it. The change from (i) to (iii) is −200 lines; and the *rate-of-change* is fastest at stage (ii).]

The primitive generator in Experiment B has the obvious disadvantage that the wires leading from the coil get more and more twisted together. In real generators this is avoided by connecting the coil to two "slip rings" which spin with the coil and have fixed metal "brushes" rubbing on them to connect them to the outside world. See a demonstration of this.

EXPERIMENT C (4)

You can avoid the twisting of the wires by twisting the horseshoe magnet instead. Try this. This arrangement is used in many large modern A.C. generators.

EXPERIMENT C (5). Generator with Iron Core in Coil

Obtain a small iron core to fit inside the coil. Push a brass axle into a slot in the core, and fix it there with a piece of wax. Repeat Experiment C (3) or C (4), comparing the effects with and without core.

EXPERIMENT C (6). Alternative Form of Experiment C (5)

Instead of a special core, use the bar of soft iron which acts as "keeper" for your horseshoe magnet. Push the keeper through the coil and hold it *near* the magnet. Try twisting the keeper around and around. Try rotating the magnet instead.

EXPERIMENT C (7)

Place the coil on the horseshoe magnet's keeper as in C (6). Hold the keeper close to the magnet's poles and try moving it nearer and farther. Repeat, with the coil on one leg of the magnet. This shows the principle of a telephone *earpiece* when used as a microphone, as it was in early days.

Transformers

Instead of bringing a magnet near a coil, bring another coil near. When that coil carries a current it behaves like a magnet, and you will find it produces similar effects in the original coil. We can make this bringing near of a current-carrying coil happen electrically by the simple trick of switching the current on suddenly. And we can increase the magnetic linkage between the current-carrying coil and the coil in which we are inducing currents by running a soft iron bar through both.

EXPERIMENT C (8)

Connect a coil to the meter as before. To allow changes from 40 turns to 80 turns without changes of resistance, connect two such coils in series (Fig. 41-7). These two constitute the "secondary" coil in

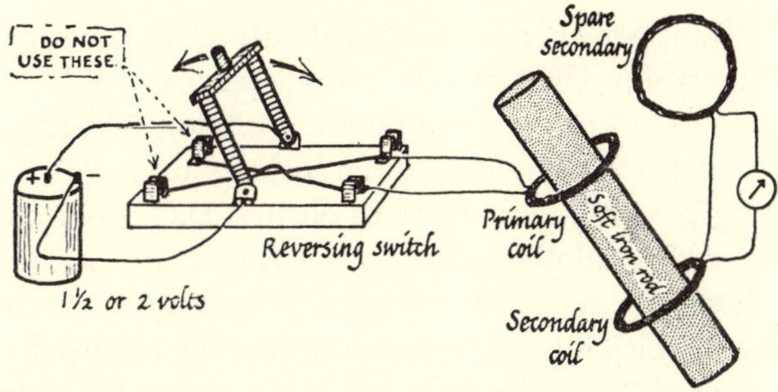

Fig. 41-7. Experiment *C*(8): Primitive Transformer (One form of reversing switch is shown. Whatever form you meet, examine it, use common sense, and try to decide for yourself the proper way to use it. This is an intelligence test.)

which you hope to find induced currents. Now instead of the magnet use a "primary" coil connected to a battery with a special reversing switch to enable you to switch the primary current on, off, on in the opposite direction, off, on in the first direction, &c.[3]

With the switch, run an "alternating" current through the primary coil. Put the primary coil close to the secondary, and increase the linkage with a soft iron "core." Watch the microammeter in the secondary when you are using 40 turns of secondary, then 80 turns. By keeping both secondary coils in series all the time you keep the resistance constant. Note that there are two ways of having 80 turns, one with both 40-turn coils wound the same way, one with their windings running opposite ways.

This arrangement is a primitive transformer.

Transformers. Further Discussion

A transformer consists of two coils insulated from each other wound on the same iron core. One good form of core to make a magnetic linkage between the coils is a simple ring of soft iron. As the current

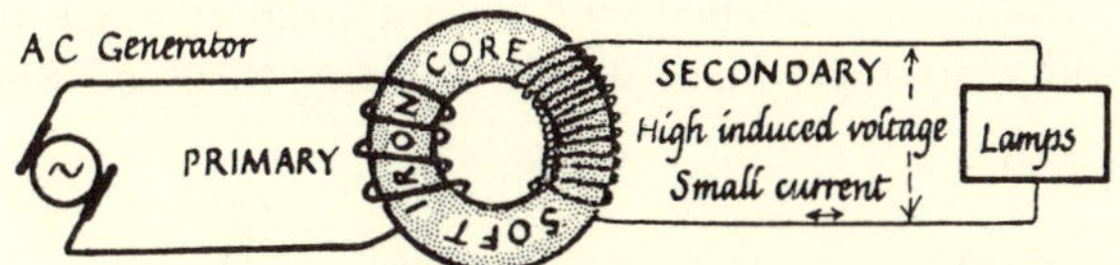

Fig. 41-8. Use of Transformer
The primary has few turns, fat wire.
The secondary has many turns, thin wire.

in one coil alternates, the ring is magnetized cyclically, first clockwise then counter-clockwise, then clockwise, &c. The other coil then has an alternating voltage induced in it.

By winding many turns in the other coil (secondary) we can obtain a large voltage, volt after volt, in series in turn after turn of wire. A step-up transformer has a few turns of fat wire in the primary and many turns of thin wire in the secondary. A small alternating voltage applied to the primary drives a large alternating current through it, and a large alternating voltage is induced in the secondary. From considerations of energy-conservation we expect to get as much power (or less) out of the secondary as we put into the primary. So we should expect the secondary with its high voltage to drive a small current. Many shapes of core are used, often with a double ring as in sketch (ii), with both coils wound on the central leg. The arrows show the

[3] The reversing switch may seem complicated. Examine it in detail if you like or take our word for it that it does do this if you connect

the two "input wires," from battery to the *center* binding posts;

the two "output wires," from *one pair* of *end* binding posts.

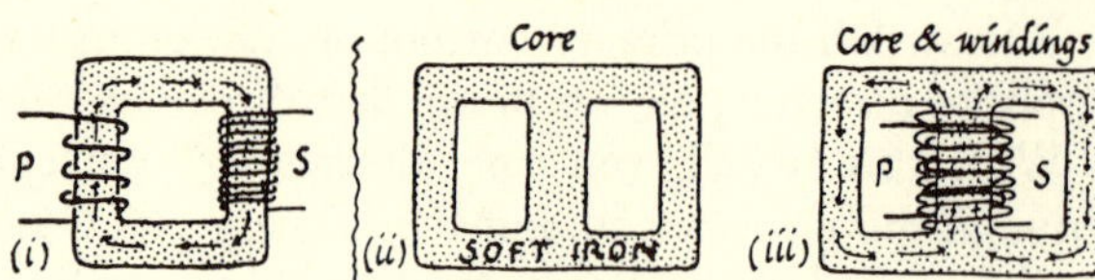

FIG. 41-9. TRANSFORMER DESIGNS

state of magnetization of the core at some particular instant. In circuit diagrams, transformers are shown by the symbols of Fig. 41-10, with the core shown as a bundle of iron wires—an ancient form for cores.

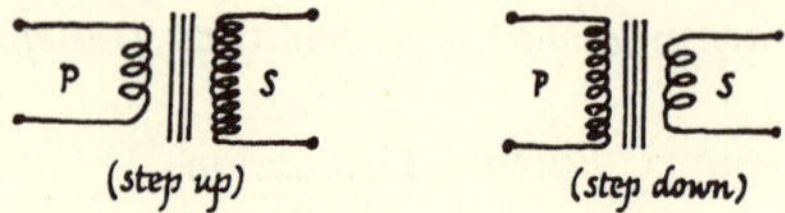

FIG. 41-10. CIRCUIT SYMBOLS FOR TRANSFORMERS

EXPERIMENT D. Transformer. (Try this or see demonstration)

Obtain a U-shaped core with a straight block to bridge its top.[4] Place a coil of many turns on one

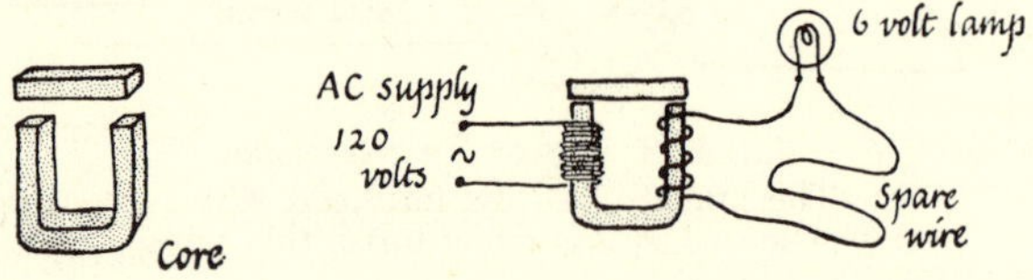

FIG. 41-11. EXPERIMENT *D*. "HOME-MADE TRANSFORMER."

leg to serve as primary and wind a few turns of insulated wire on the other leg, as secondary. Add the top bridge, to make the core a complete ring. Connect the primary to a 120-volt A.C. supply. Try the secondary with a 6-volt lamp. If it glows faintly, add more turns to the secondary.

EXPERIMENT E. Model Power Line with Alternating Current

Use the model power lines that you used with D.C. (6 volts and then 120 volts) in Experiment W of Chapter 32. Start with a 6-volt A.C. supply as your "power station." First use it for a low-voltage system without transformers. Run one 6-volt lamp at the power station, and another 6-volt lamp at the "village" at the far end of the power lines. You will find that this low-voltage A. C. system does as poorly as the low-voltage D.C. system.

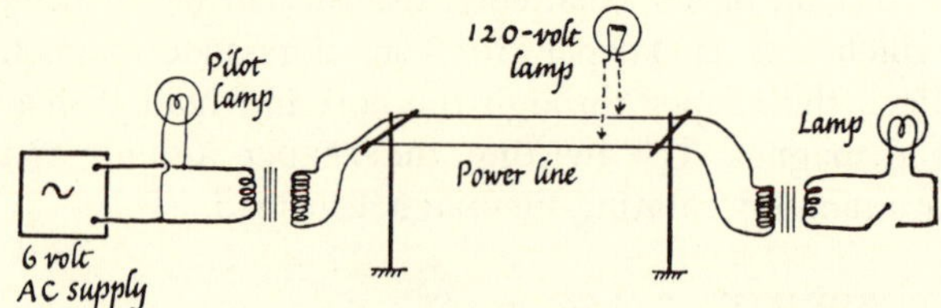

FIG. 41-12.
EXPERIMENT *E*. MODEL POWER LINE WITH ALTERNATING CURRENT

Then install small transformers to step up the voltage from low at the power station to high between the lines, and step down to low at the village. (Use transformers intended for running bells or heating radio tubes. In these, one coil has about 20 times as many turns in its windings as the other. The coil with few turns is made of fat wire.)

Record your observations. You cannot make measurements unless you have special A.C. meters. However, you can test for high voltage by using a 120-volt lamp as a voltage indicator.

Alternating Current and Power Distribution

For efficient power lines high voltages are needed. Yet at either end low voltages are essential for safety, and for good machines and motors. Low-voltage power lines are hopelessly wasteful, unless they are made of very thick wire to have very low resistance. Such heavy wires would be far too expensive for long lines, both in cost of their metal and in cost of supporting their weight. The efficient solution is *A.C. with transformers.*

Transformers have two great virtues: (1) they are remarkably efficient—the output power may be as high as 95% of the input power, with only 5% wasted as heat; (2) they require no looking after—a spinning motor + generator to change D.C. from one voltage to another requires much care and service. Therefore, wherever changes of voltage must be made efficiently, A.C. and transformers are used.

For distribution in towns, A.C. power lines run with several thousand volts between them, and for distribution between big power stations and remote towns, the alternating P.D.'s may run up to a million volts.

EXPERIMENT F. "Electromagnetic Inertia"

When the current through a coil of wire changes, the coil's magnetic field changes its strength. That changing field induces voltages *in the coil itself* opposing the

[4] These are made up of "U and I" core stampings of thin iron. All transformer cores are built up of thin sheets to avoid the wasteful "eddy currents" that would be induced in a solid core. The core itself is an iron "secondary coil" all in one block instead of in many turns. Unless it is broken up into many thin leaves, the small voltage induced in it will make huge alternating currents circulate and generate much waste heat.

change. Thus any coil can show "self induction," opposing changes of current. This behavior resembles the inertia or mass of any piece of matter opposing changes of velocity. If the coil has an iron core the effect is much stronger.

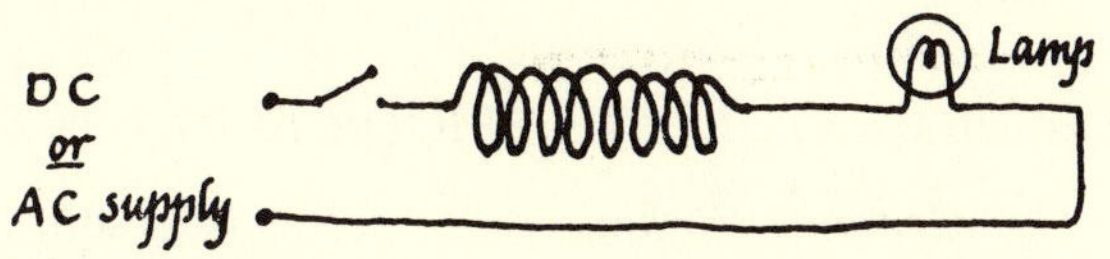

FIG. 41-13. EXPERIMENT *F*. "ELECTROMAGNETIC INERTIA." A coil of wire opposes any change of current. The opposition (a momentary reverse voltage) is much larger if there is an iron core. With D.C., the coil has no effect on the lamp, except for its added resistance. With A.C., the lamp shows a noticeable effect.

Connect a lamp to a suitable D.C. supply. Try adding a big coil of wire in the circuit. Except for the coil's resistance, the coil does not affect the lighting of the lamp. Now change to an A.C. supply of similar voltage and try the lamp with and without the coil in series. Try inserting a soft iron core in the coil.

The primary of a transformer shows great self-induction *when the secondary is not supplying power*. Then the primary current is much smaller than it would be with a D.C. supply—that is part of the reason why a transformer wastes practically no power.

A coil (usually with an iron core) used for the sake of its self-induction, is called an inductance, or a "choke." Chokes are used, with capacitors, to smooth out the bumps in D.C. from a rectifier.

EXPERIMENT G (1). Capacitor

(This has no direct connection with "magnets and coils," but it is given here because capacitors are used with A.C. and in radio sets.)

A capacitor (old name "condenser") consists of two sheets of metal separated by a slab of insulator. It is often made by placing thin sheets of metal foil on each side of a sheet of wax paper. Then this sandwich is rolled up and put in a box to protect it. You should use a small capacitor of this kind. The binding posts on the box are connected to the metal sheets, one to each. Such capacitors are of great use in radio. When connected to a battery, the plates become charged + and −, and then there is an electric field between them. In this way the plates store charges. A capacitor can hardly carry a current, since the plates are separated by insulator. Yet the charges must travel to the plates somehow. Investigate the matter by using the circuit shown in Fig. 41-14.

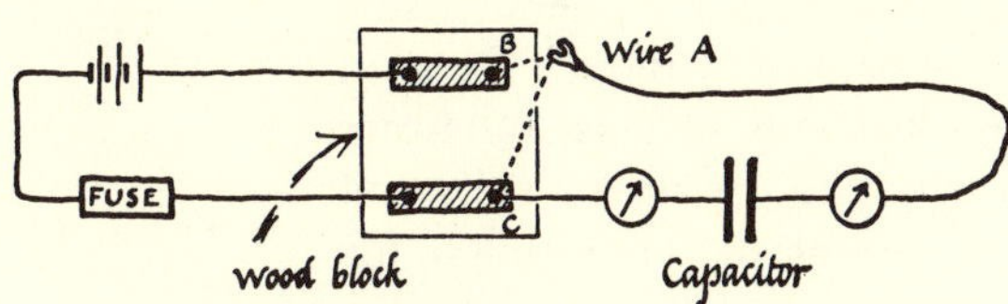

FIG. 41-14. CHARGING A CAPACITOR
To charge capacitor, connect wire-end A to post B. To discharge it, connect A to C.

Connect a battery (through a fuse) to a set of binding posts, B and C, on a wooden block. Arrange two microammeters, one each side of the capacitor. Complete the circuit as shown in the sketch.

The microammeters should tell you whether there is any *momentary* current when you charge the capacitor by connecting A to binding post B. Then you can remove the battery and "discharge" the capacitor by completing the circuit without the battery. Do this by connecting wire A to binding post C. (The easy way to make these changes: hold A in your hand and make it touch B, then C, then B, ...) Repeat the experiment with four volts instead of six, then with two volts.

Electrical Springiness. Oscillations

A capacitor in an electric system is like a spring in a mechanical system (See Fig. 33-2). A capacitor connected to a coil with self-induction is like a spring with a mass hung on it. The loaded spring can oscillate

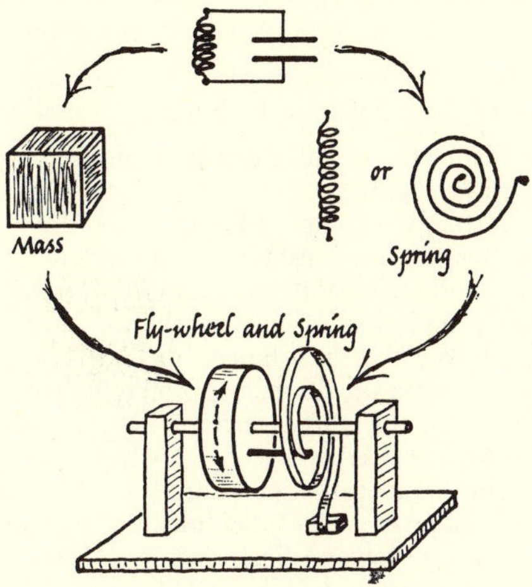

FIG. 41-15a.
FLYWHEEL CIRCUIT and ANALOGOUS MASSIVE FLYWHEEL AND SPRING
Note: The electrical inductance coil corresponds to a mass, not to a coiled spring.

with simple harmonic motion. (Ch. 10). So can the combine [capacitor + coil]. Simple harmonic currents can oscillate to and fro through the coil, charging the capacitor plates to a voltage which oscillates with S.H.M. Such "oscillatory circuits" are essential in radio, for sending out radio waves, and for tuning receivers to respond to waves. We shall not study them in this course.

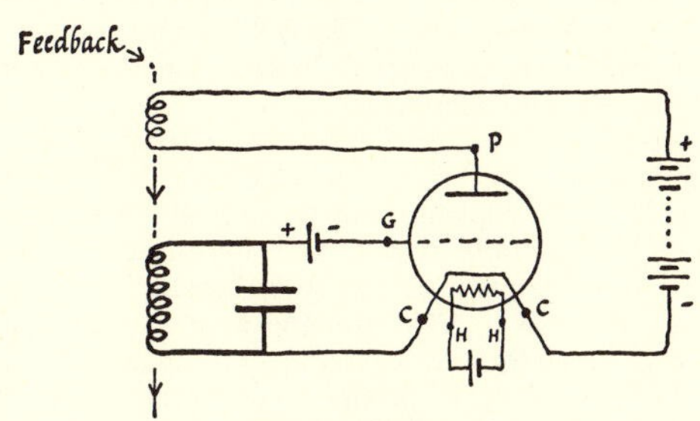

FIG. 41-15b.
FLYWHEEL CIRCUIT DRIVING AMPLIFYING TUBE, WITH FEEDBACK TO MAINTAIN OSCILLATIONS

EXPERIMENT G (2). Capacitor as A.C./D.C. Filter

Try inserting a capacitor in series in a circuit with a lamp lighted by D.C.; then repeat with A.C. The D.C. supply drives charges on to the plates of the capacitor with momentary currents as in G (1). If the capacitor is big, the lamp may show a momentary flash as the charging pulse of current runs through it. After that there is no current and the lamp does not light.

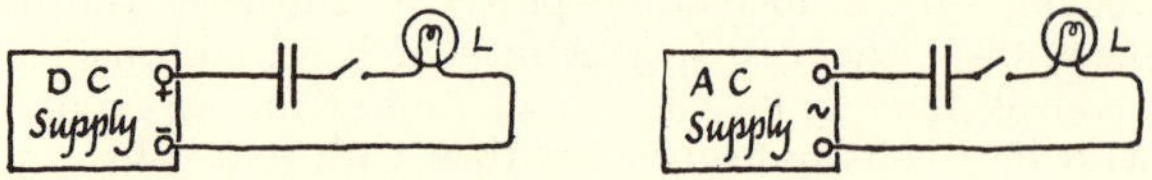

FIG. 41-16. EXPERIMENT G(2)

With alternating supply, the lamp continues to carry current, which may keep it brightly lit. Current does not really go *through*[5] the capacitor, but runs to-and-from one plate, and from-and-to the other as the voltage alternates. Inside the capacitor there are charges on the plates and an electric field between them: these alternate

from ZERO to $^{+}$MAXIMUM^{-} to ZERO to $^{-}$MAXIMUM^{+} to ZERO to $^{+}$MAXIMUM^{-} to ZERO . . . and so on.

In the rest of the circuit there is an alternating current carrying these charges.

With a steady voltage, which would drive D.C., a capacitor has infinite resistance—no current through its insulating sandwich. With an alternating voltage, it behaves like a medium resistance[6]—the bigger the capacitance the lower the "resistance" or "impedance." It is the rapid changes of alternating voltage that make the capacitor seem to carry current. With more rapid changes—higher frequency A.C.—the "resistance" is lower. (The same pulse of charge would run to the plates in a shorter cycle-time, making current bigger, thus "resistance" lower.) Therefore a capacitor can act as a filter to separate A.C. from D.C. And it can act as a preferential filter for a mixture of alternating frequencies, letting high frequency A.C. through easily, offering much more "resistance" to low frequency A.C. In this it is the opposite of a choke which opposes changes of current and, therefore, lets D.C. through easily, offers opposition to low frequency A.C. and great opposition to high frequency A.C.

[5] Maxwell in his electromagnetic theory suggested that the changing electric field which pumps forward and back in the insulator of the capacitor *is* a form of electric current. On that basis, current does run all the way around the circuit. Since a capacitor may have a vacuum between its plates, this new form of current is not leakage through a poor insulator, but a current in space. How do we know such a current is there? By its magnetic field. Maxwell's fantastic idea enabled him to predict electromagnetic waves; but we do not use it in elementary discussions.

[6] There are important differences: current generates heat in a resistance, whether it is D.C. or A.C.; but an alternating P.D. applied to a capacitor makes no heat—it merely pumps charges in and out, storing and unstoring electric strain energy. And there is a difference of "phase." In a resistance, current alternates *in step* with the applied alternating voltage; in a capacitor, the charging current is a quarter of a cycle early, reaching its maximum (charg*ing*) a quarter of a cycle before the capacitor's voltage reaches its maximum (charg*ed*). The phase difference is connected by simple algebra with the lack of heating.

In radio circuits you will often see a capacitor used for such filtering. You will see combinations of capacitor and choke used to smooth bumpy D.C. from a rectifier into steady D.C.; or to separate A.C. from D.C., or high-frequency A.C. ("radio") from low-frequency A.C. ("audio").

EXPERIMENT H. THE TRIODE RADIO TUBE

The General Working of a Triode is described and explained in Chapter 33. (You should re-read that section.)

The standard radio tube, which plays an essential part in radio receiving sets, and in transmitters, amplifiers, etc., is the triode, with heated cathode, grid and plate. Other tubes with fancier names and more electrodes (e.g., "pentodes") are really triodes with extra frills. The essential action is the same.

Students learning to be radio engineers do long experiments with triodes to plot graphs of PLATE CURRENT *vs.* GRID VOLTAGE; and the results can be made to yield interesting information such as the tube's "resistance," amplification factor, capacitances, etc. *In this course you should do a more direct experiment: make the tube go, and hear it amplifying.* (Use a simple triode, such as 6J5.)

Construction of the Tube. The innermost part is a heating wire, or filament (connected to the binding posts H, H). The filament heats a surrounding tube, the cathode, which gives out electrons when hot (connected to the binding post C). Use this cathode, C, for all connections involving the stream of electrons emitted. Around the cathode is a spiral of wire, the grid (connected to binding post G), insulated from the cathode. Outside that is the plate, a box of dark metal. Unfortunately, the plate-box is practically all you can see through the glass bulb. There is a good vacuum in the bulb. Ask to see a broken tube and look at the grid and filament—the whole arrangement is a marvel of mechanical assembly.

EXPERIMENT H (1). Making the Tube Go (see Fig. 41-17)

(a) Arrange to heat the filament, and include a switch and an ammeter in the circuit. Use a 6-volt battery, without a rheostat. The filaments of most tubes are designed to run on 6.3 volts, but the current driven through the heater by 6 volts will make the cathode hot enough to release all the electrons you need.

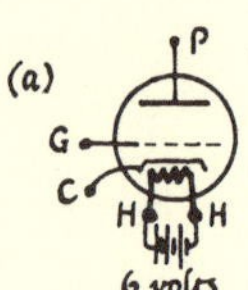

FIG. 41-17a. HEATING THE CATHODE

(b) There must be a suitable P.D. between grid and cathode to control the flow of electrons. If the grid is positive, it encourages a damaging torrent of electrons—so grids are practically never made positive in electronic practice. For sensitive control, the grid should be a few volts negative relative to the cathode.

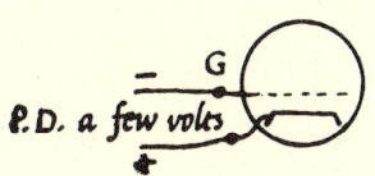

FIG. 41-17b. GRID VOLTAGE

(c) Use a potential divider and a battery (say with EMF 10 volts), to provide a suitable variable P.D. between grid and cathode.

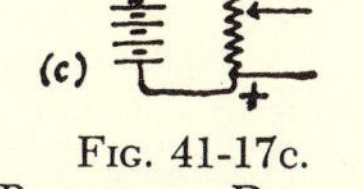

FIG. 41-17c. POTENTIAL DIVIDER FOR GRID

(d) To pull across to the plate those electrons that wander through the grid, there should be a large P.D. between plate and cathode. Should the plate be + or —? Use a 120-volt D.C. supply, and insert a milliammeter. You need not add a voltmeter yet. Draw your circuit, set it up, and try it, to make sure there is some plate current. Then put your tube to use, as in H (2) and H (3).

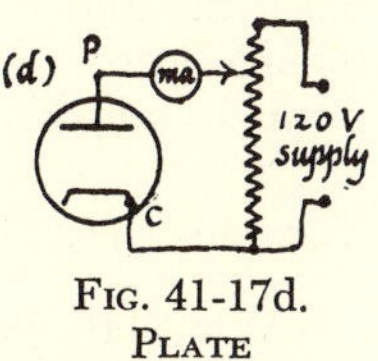

FIG. 41-17d. PLATE CONNECTIONS

Amplifying

When a triode amplifies, it magnifies any voltage that is applied between the grid and the cathode into a bigger voltage across a resistor in the plate circuit. The essential mechanism (described in Chapter 33) is this: changes of grid-voltage make great changes in the electron stream through the grid. When it gets through the grid, this stream is sped on across to the plate by the field due to the large plate-voltage. Thus changes in grid-voltage make large changes in the "plate current," the electron stream from cathode, through grid, to plate, on through plate-battery (or equivalent) and back to the cathode. That plate current also goes through any other apparatus inserted in series in the plate circuit and it is there that amplified voltage-changes may appear. The plate circuit shows more than amplification of VOLTAGE. It also shows a much larger CURRENT. Therefore, the available OUTPUT POWER in the plate circuit is much greater than the INPUT POWER in the grid circuit. Unlike a transformer, the tube amplifies *power*, the extra energy coming from the battery in the plate circuit.

EXPERIMENT H (2). Triode Amplifying

Make the following changes in your circuit:

(i) Insert a sampling resistor in the plate circuit to make a P.D. proportional to the electron stream.

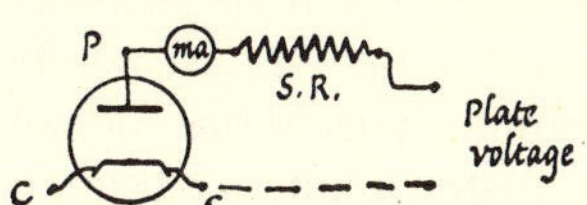

FIG. 41-18. EXPERIMENT *H*(2)

This is the "output" P.D., which might drive the grid of another tube in a several-stage amplifier.

(ii) Connect a voltmeter V_1 across the variable voltage added between grid and cathode in H(i). This will show the "input changes" of voltage.

(iii) Connect another voltmeter V_2 across the sampling resistor, S.R. in the plate circuit. This will show "output changes."

Slide the 10-volt potential divider in the grid input up and down and watch the two voltmeters. You should see "amplification": several volts *change* on V_2 for a 1-volt *change* on V_1. With a small triode such as 6J5, the amplification factor is between 5 and 10. With some tubes it can be 10 or 20 or even higher. Make a measurement to estimate the amplification, by comparing the changes on V_1 and V_2. Choose a region for your specimen change where there is a plate current—avoid the "cut-off" region where plate current falls to zero. Record your estimate of "voltage amplification factor" for your tube thus used. If you want to make a more professional measurement of the tube's amplifying factor for voltage, try optional experiment H (4).

EXPERIMENT H (3). Amplifying a Musical Sound

Use the same circuit to make the tube amplify a rapidly alternating current that can make a musical sound in a loudspeaker or telephone. An alternating P.D. inserted in the grid circuit should be amplified to a larger alternating P.D. across the sampling resistor in the plate circuit. An oscillator will give you a small alternating voltage with frequency several hundred cycles per second. When this voltage drives an alternating current through a telephone earpiece, it produces a singing musical note—the higher the frequency of oscillations, the higher the musical pitch. (Instead of the oscillator, which makes a steady note, you could use the output of a small radio or victrola that is playing music.)

Make a break in the connection from grid to cathode and insert the leads from the oscillator.[7] (Retain the "grid bias" provided by battery and potential divider. It is not essential, but it will make sure the grid remains negative. Try varying this bias voltage.) Then hear and see the tube amplifying, as follows.

(a) Test with telephone. Connect a telephone earpiece directly across the supply-leads from the oscillator, and listen to the "input" sound. Then move the earpiece connections to the ends of the sampling resistor in the plate circuit and listen to the "output" sound. (The earpiece should be a high-resistance one, so that its sampling does not disturb the circuit.)

(b) Test with cathode-ray oscilloscope (CRO). Connect the oscilloscope's vertical deflection plates (binding posts marked V and G) to the supply leads from the oscillator, and adjust the CRO's vertical gain to make the spot move up and down about half an inch. At the same time run the CRO's horizontal sweep (which carries the spot steadily across again and again) so that it plots a time-graph of the oscillating voltage.[8] Then move the CRO's V and G connections to the ends of the sampling resistor, and see the time-graph of the output voltage. Estimate the voltage amplification by comparing the *heights* of the two graphs. Record your estimate and compare it with your estimate from voltmeter readings.

EXPERIMENT H (4). (OPTIONAL) Amplifying Factor. Modification of H (2)

This is a more professional estimate. Install a potential divider in the plate-voltage supply and move the voltmeter in the plate circuit from its position across the sampling resistor to the position shown in Fig. 41-19.

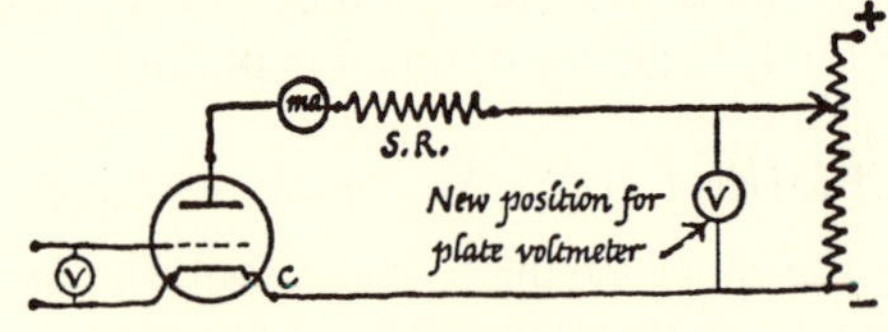

Fig. 41-19. Experiment $H(4)$

Now try to make *the same change of plate current* first by changing the grid voltage, then by changing the plate voltage instead. That will show how changes of voltage in the plate circuit compare with equivalent changes of voltage in the grid circuit. Proceed thus:

(a) Find a region of your range of grid voltage in which its changes seem to control changes of plate current fairly smoothly; avoid the "cut-off" region where the plate current falls to zero. Then change the grid voltage by a small measured amount, e.g. from -3 volts to -1 volt. Observe the change of plate current. (*Record* the actual meter readings, not just differences.)

(b) Then, keeping your grid voltage fixed at one of the two values you just used, make the same change of plate current occur by changing the plate voltage. (Again record meter readings.) Then calculate the amplifying factor, as in the following fictitious example.

Example: Suppose a change of grid voltage from -3 volts to -1 increases the plate current from 3 milliamps to 7 milliamps. Suppose that with grid voltage kept at -1 volt the plate current can be brought back from 7 milliamps to 3 milliamps by reducing the plate voltage from 120 volts to 80 volts. Then the same change in current is produced (in this example) by a 40-volt change in plate voltage as by a 2-volt change in grid voltage: 40 volts needed instead of a mere 2 volts. Then the grid is twenty times as effective as the plate voltage in controlling plate current. Conclusion for this example: amplifying factor 40/2 or 20.

The Need for Detection (= Rectification) in Radio

(We shall not discuss radio in detail, but you should now be ready to understand the function of the chief components of a radio set. However, you need the following explanation.)

The vibrations of audible sound and music have frequencies ranging from a few dozen cycles per second to several thousand. There are two objections to making radio waves travel out with just these frequencies: (i) it is very difficult to make a station radiate much power at these low frequencies, unless it has a stupendously big antenna system; (ii) the customers would hear all neighboring stations at once, in a hopeless jangle.

It is difficult to radiate much power in radio waves at 1000 cycles/sec, but easy at 1,000,000 cycles/sec. Therefore, radio stations produce waves of high frequency (radiofrequency, or "r.f."), and make the *amplitude* of these waves swell and shrink according to the oscillations of the speech or music (audiofrequency, or "a.f.").

The basic wave ("carrier") looks like Fig. 41-20 when it is not carrying any musical message. The audio waves whose pattern must be impressed on the radio wave look like Fig. 41-21. The amplitude of the basic radio wave is made to follow the audio pattern: it is "modulated" as in Fig. 41-22.

[7] If several groups of experimenters share the oscillator and use the same plate voltage supply, there is a danger of reversed connections short-circuiting the supply. As a safeguard, the oscillator supply should be fed through a small isolating transformer to each group.

[8] A special control marked "sync signal" enables you to make the pattern repeat from one sweep to the next, so that you see a constant picture. Do not change this, but ask for help if it needs adjusting. Its action will be explained later.

RADIO FREQUENCY
CARRIER WAVE

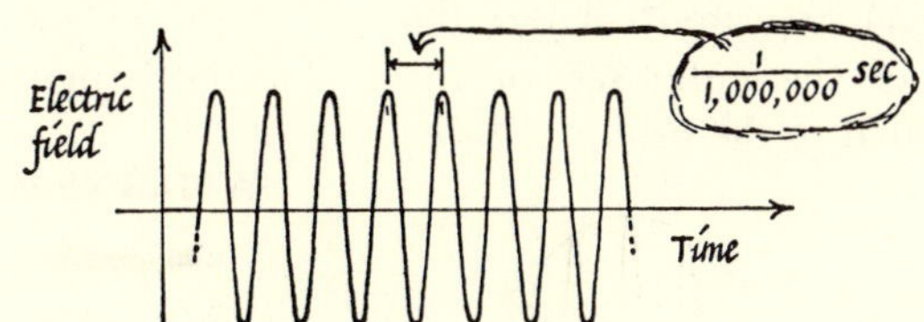

Fig. 41-20. Time-Graph of Radio Wave, frequency 1,000,000 cycles/sec, constant amplitude.

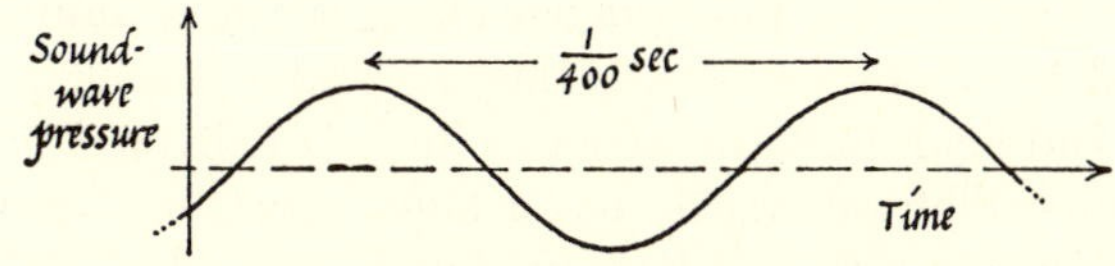

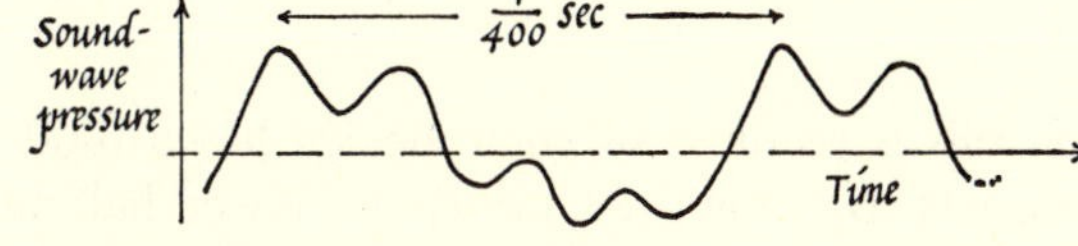

Fig. 41-21. Time-Graphs of Audio Waves, frequency a few hundred cycles/sec.

(a) Pure musical note: a sine graph repeating, say, 400 times/sec.

(b) A vowel being sung, or a note from some musical instrument. The pattern is more complex, again repeating, say, 400 times/sec.

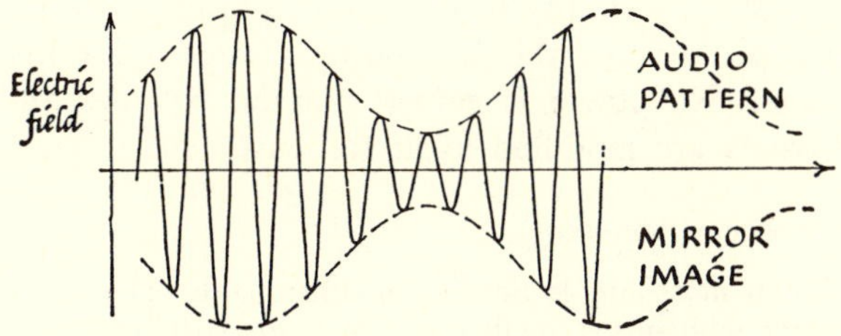

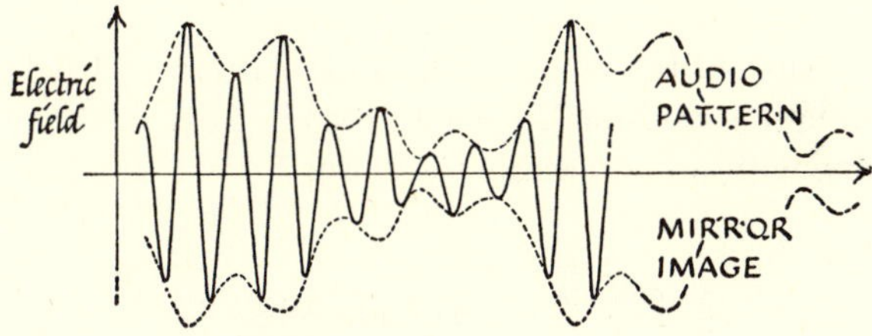

Fig. 41-22. Radio Wave "modulated" by Audio Pattern.
Radio wave "modulated" by audio pattern. (The frequency of the radio wave is really a million or more cycles/sec; and therefore one cycle of the audio pattern should contain thousands of cycles of the radio waves. In these sketches, that true proportion is not shown.)

This is the pattern of the radio wave broadcast by the station. When such a wave reaches a customer, it sweeps across his antenna, producing radiofrequency voltages between antenna and ground. There is little responsive r.f. current between antenna and ground unless that system is arranged so that it can oscillate naturally with just that frequency. Then there is "resonance,"[9] and the incoming radio waves build up a large responsive oscillation. The customer tunes his antenna-ground system to the frequency of the waves coming in from the station he wants. He does this by turning a knob to change a capacitor in the oscillatory circuit which his set inserts between antenna and ground.

[9] Antique name, "sympathetic vibrations." An adult pushing a child in a swing builds up a large oscillation by resonance, using repeated small pushes of the right frequency.

When the modulated radiofrequency pattern (Fig. 41-22) is thus received, it is applied to the grid of a triode and appears amplified as changes in the tube's plate current—a gain of *power*, at the expense of the plate battery. From there you might imagine stage after stage of amplification until the final plate current is fed through a loudspeaker. This would be a complete failure. The massive coil or cone of the loudspeaker cannot follow the rapid demands of r.f. oscillations; and, even if it could, it would give out no sound waves but merely some inaudibly rapid disturbances that swell and shrink with the audio pattern. It is essential to change the r.f. pattern to something that is just the audio pattern before the loudspeaker can use it. This is done by rectifying the r.f. pattern (called "detecting," when thus used in radio).

Fig. 41-23 [Fig. 41-22(a) repeated] shows the current in an early part of the radio set, or the corresponding voltage applied to the grid of a tube.

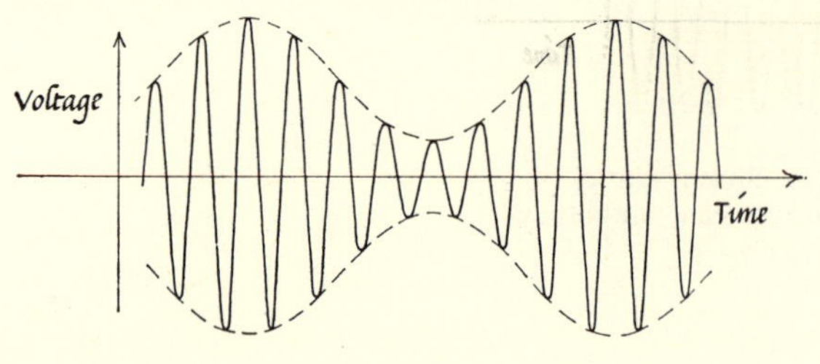

Fig. 41-23.

When this is rectified by some device like a diode which acts as a one-way valve, the lower half of the pattern is cut out and the upper half remains.

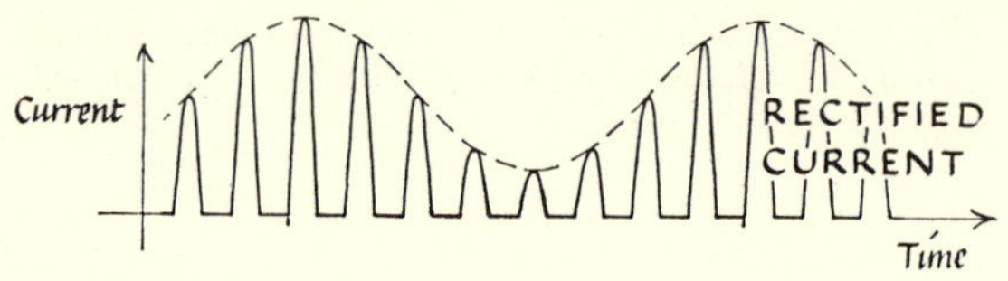

A current following the original pattern if driven through a loudspeaker would merely treat the cone thus, with no noticeable response:

PUSH PULL PUSH PULL PUSH PULL PUSH PULL PUSH PULL PUSH PULL PUSH PULL PUSH PULL PUSH PULL PUSH PULL PUSH PULL PUSH PULL &c

However, a current following the rectified pattern would treat the cone thus:

PUSH PUSH --PUSH PUSH PUSH PUSH PUSH PUSH PUSH -PUSH PUSH PUSH &c

The massive cone would add up these million-a-second pushes and respond to their general average like this:

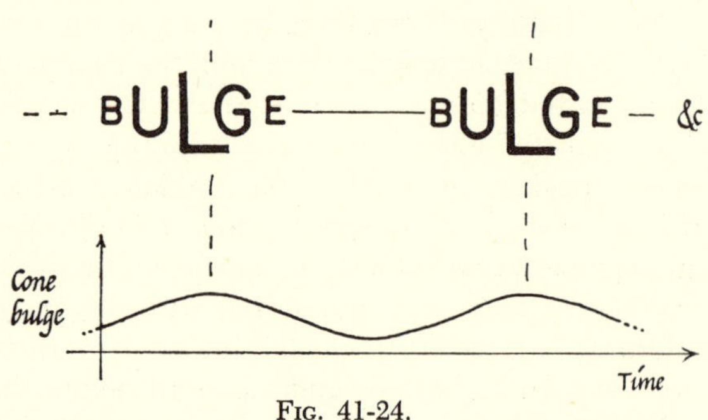

Fig. 41-24.

This smoothed average of the rectified r.f. makes the loudspeaker cone follow a diminished version of the audio wave pattern. Therefore the loudspeaker emits sound waves that are a good copy of the original audio waves used to modulate the radiofrequency wave.

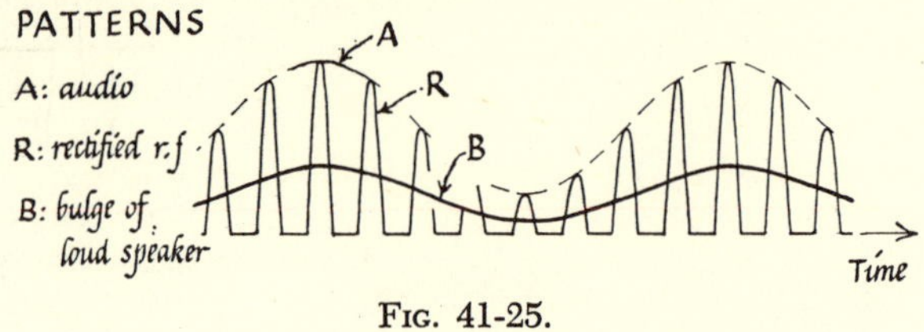

Fig. 41-25.

These slow audiofrequency changes can be amplified by another triode, acting as "audio amplifier."

The final audiofrequency voltage is delivered by a tube that necessarily has a high "resistance," but loudspeakers[10] have low resistance. Instead of "resistance" we should rather say "impedance" for the more general opposition offered to *changing* currents.[11] Loudspeakers need a large current driven by a small voltage instead of what the tube offers, a small current driven by a large voltage. We must "match" the impedance of the amplifier to the loudspeaker, and this is done by a stepdown transformer. No power is gained but the voltages and currents are modified to fit the customer.

Impedance Matching

Such *matching* is needed in other fields. The prizefighter who hits a small boy (on skates) delivers little damage—the boy bounces away. If the prizefighter hits an elephant, he bounces back. When he hits a man of his own mass he can deliver a fully effective smash. In much the same way, to transfer maximum energy or power from an electrical supply we must match the consumer to the supply, give them equal impedances. Take a simple example: a battery feeding power to an electric grill. How must the grill be designed to take maximum power from the battery? The current in this simple circuit [battery + grill] delivers heat in the grill. However, any real battery has some resistance in its own materials, so the current delivers heat in the battery too—quite useless heat. Try constructing the grill with very high resistance: then the current is small and delivers very little heat altogether, most of it in the grill. Try constructing the grill with very low resistance: then the current is huge and delivers a lot of heat, but most of it in the battery. At either extreme, the grill gets little heat; but, with some intermediate resistance, the grill gets much more heat. Calculus, or trial-and-error examples, will show that the grill gets most heat if it has the same resistance as the battery. Then the same amount of heat is developed in each. *The consumer gets maximum power when its resistance*

[10] The cone of an ordinary loudspeaker is moved by a small coil in the field of a magnet—an in-and-out-vibrating form of electric motor.

[11] Impedance includes ordinary Ohm's-Law resistance and the opposition of voltages induced by the magnetic fields of *changing* currents.

(impedance) matches that of the supply—and then it gets just 50% of the total power supplied. (This is not the suitable condition for a power station feeding lamps in a town—then we want *constant voltage* rather than maximum power so we make the lamp resistance much greater than that of the supply system.) In transferring power from stage to stage of a radio, or from car battery to starting motor, we want maximum power-transfer and must arrange to match impedances. We do this by building them equal (man hits man) or by interposing a matching mechanism.

When there is a mismatch, a mechanism can provide matching. Suppose we want to transfer energy from a moving ball to a stationary one in an elastic collision. If the masses are matched (equal) there is good transfer of energy—100% in a head-on collision, an average of 50% in assorted impacts. However, if the masses are badly mismatched, the moving ball retains a large fraction of its energy. Then, to transfer much of it, we must use a mechanism, such as a lever (like a see-saw with a vertical axle).

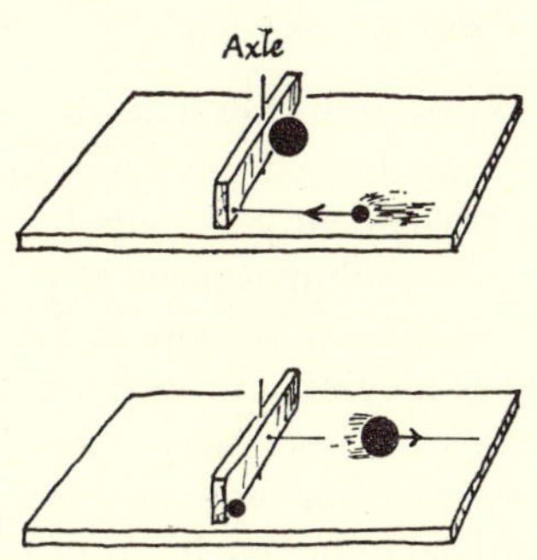

FIG. 41-26. IMPEDANCE MATCHING

The stationary ball (Fig. 41-26) is placed touching the lever, which the moving one hits at a different distance from the axle. The distances are chosen so that [ball + lever] matches the other ball in effective mass. That is what a transformer does for electrical devices: it is chosen so that [device + transformer] matches the other device in effective impedance.

Impedance-matching occurs in many fields. The prize-fighter uses a loose wrist as a lever when he slaps a child. Builders of nuclear reactors choose hydrogen or carbon atoms as "moderators" to slow the neutrons. The neutron that hits a nucleus of gold, or an electron, loses little energy. Only when it hits something of mass like its own can it lose much. A long, flared horn matches a small loudspeaker to the *open* air. The surgeon uses his stethoscope to match chest to air to ear. The human ear has three little rocking bones to match air-driven eardrum to inner liquid mechanism. The human heart is well matched to the artery+vein system until aging arteries harden. Even in business management, executives are beginning to speak of matching the impedance of the Vice-President to the impedance of the plant-supervisor—perhaps by the mechanism of a labor-relations expert.

Radio Sets

Real radio circuits are more complicated but their essential components are those described above, ingeniously arranged in elaborate schemes to give greater economy, sensitivity, selectivity, and amplification. An ordinary radio set must carry out the following processes:

Reception: Electromagnetic fields produce currents in antenna.

Tuning: Oscillatory circuits favor desired station by resonance.

Detection: Radiofrequency oscillations are rectified.

Amplification: At r.f. or a.f. stage or both, oscillations are increased in amplitude *with gain of power.*

Matching: Impedance of one stage is matched to fit with next.

Radiation: Loudspeaker emits sound waves.

The sketches of Fig. 41-27 show a series of radio circuits. None of these circuits would work well. They are intentionally simplified to show the essential action.

EXPERIMENT I. Measuring Velocity and e/m of Electrons

(Details depend on apparatus available. You should either make these measurements or watch a demonstration.)

Electrons were first recognized as tiny chips of atoms by their extraordinary value of CHARGE/MASS, e/m. The same measurements showed that even when shot by a low-voltage gun they move with stupendous speed.

Make these basic measurements on a stream of electrons from a small gun. To find their e/m and their speed, v, you must make two distinct measurements:

(i) The effect of an electric field, measured by the gun voltage, used to accelerate the electrons in the gun.

(ii) The effect of a magnetic field, which bends the stream into a circle.

Apparatus and Measurements

A narrow beam of electrons is fired from a simple gun, like that sketched in Fig. 41-28. The electrons evaporate from the hot filament and are accelerated by the electric field from filament to outer cylinder. A beam of these electrons shoots on out through a slit in the cylinder. All emerge with the same K.E., and thereafter move with constant speed. Measure the gun voltage.

The beam is pulled into a circular orbit by the magnetic field of a large hoop coil. Measure the radius of this orbit. For that, the beam must be made visible. This is done by making it cause a glow. It is splashed against a flat mineral-coated screen, or fired through gas—mercury vapor or hy-

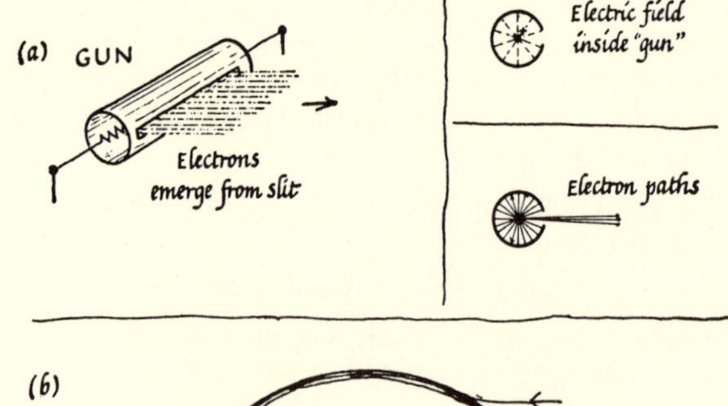

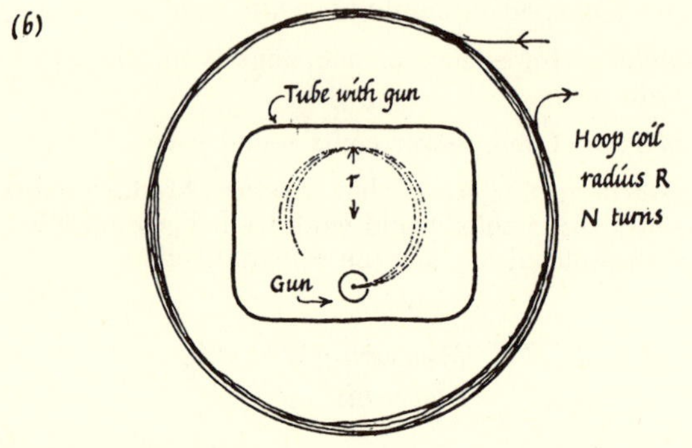

FIG. 41-28. MEASURING v AND e/m OF ELECTRON

drogen, at very low pressure. Measure the orbit diameter, by comparing it with a meter scale held near it. Also make the measurements you need to calculate the hoop-coil's magnetic field.

Calculations

(1) *From measured gun voltage*

(a) Start with algebra. Suppose the electron has charge e coulombs, mass m kg, and "falls" through gun voltage V volts. Write an equation stating that the energy given to the electron by the gun provides its K.E. in the emerging beam.

(b) Insert the measured value of gun voltage and calculate e/mv^2.

(2) *From measured orbit in magnetic field*

Remember that a charge e coulombs moving with speed v across a magnetic field at the center of a hoop coil experiences a catapult force given by

$F = 10^{-7}\,(ev)(C_2 \cdot 2\pi R \cdot N)/R^2$ where hoop coil has N turns of radius R, and carries C_2 amps

$= 10^{-7}\,(ev)(H)$, where H, the magnetic field, is $(C_2 \cdot 2\pi N)/R$

(a) Calculate the value of H, in ampere · turns per meter, from your measurements.

(b) Write an equation stating that the force $10^{-7}\,(ev)(H)$ is the real force that pulls the electron's path into a circle of radius r. (Note that r is orbit radius, while R is hoop radius.)

(c) Insert your value of H and measured value of r in this equation and calculate the value of e/mv.

(3) From the two results, values for e/mv^2 and e/mv, calculate v.

(4) Calculate e/m.

(5) (a) Express your e/m as a multiple of e/M for hydrogen ions, from electrolysis, 9.57×10^7 coulombs/kg.

(b) Express your v as a fraction of the speed of light, 3.0×10^8 meters/sec.

EXPERIMENT J. ADJUSTING AND USING A CATHODE-RAY OSCILLOSCOPE

Warning. The big tube, like all television tubes, is expensive. *Do not "burn" the screen: keep the spot moving.*

If the electron beam makes a small bright spot at the same place on the screen for some time, the screen may be permanently damaged. This is rather like the danger of spoiling a piece of paper by casting the Sun's image on it with a burning glass. When the spot is moving on the screen, there is no danger of damage. Do not keep a bright spot motionless on the screen for more than a few seconds. Keep the spot moving or put it out of focus or turn the intensity knob till the spot disappears.

Use of INTENSITY knob. Do not make the spot brighter than necessary. A very bright spot increases the danger to the screen, but does not hurt the rest of the tube.

EXPERIMENT J (1). Use of Oscilloscope Controls

Experimenting. An oscilloscope has various control knobs on its front, most of them attached to voltage-dividers inside. Find out by trial what each does, and teach yourself to adjust the pattern quickly. (Get help from the labels near the knobs and from the notes below.)

Record. In your report, sketch the face of the box with the control knobs and binding posts. Add clear labels to show the function of each of these.

Quiz. When you can make the 'scope show a good time-graph of an alternating voltage, and are sure of the controls, ask for a short practical test. You will receive the 'scope with all the controls askew—the pattern either missing or in a mess—and you will be asked to adjust it quickly to some simple specified pattern. You could probably succeed by haphazard juggling, but that would not pass the test. Think out a logical attack.

Notes on Oscilloscopes

(a) *Deflection plates.* The tube contains two pairs of "deflection plates" to apply electric fields to deflect the beam. Each pair is connected to a pair of binding posts on the box—V and G for vertical deflections, H and G for horizontal. Binding post G of *each* pair is connected to the metal case of the box, and is intended to be connected to ground. So the two pairs have this connection in common. In most oscilloscopes the connections do not run straight from binding posts to deflecting plates, but an amplifier is interposed inside the box. Such an amplifier will take a small voltage applied to the binding posts and increase it to a larger voltage to be applied to the plates.

The amplifier is usually one designed for alternating voltages. Then this arrangement is not suitable for a constant voltage; e.g., from a battery. That should be connected straight to the plates.

(b) *"Test signal" binding post.* A small alternating voltage is provided for tests and adjustments. This runs

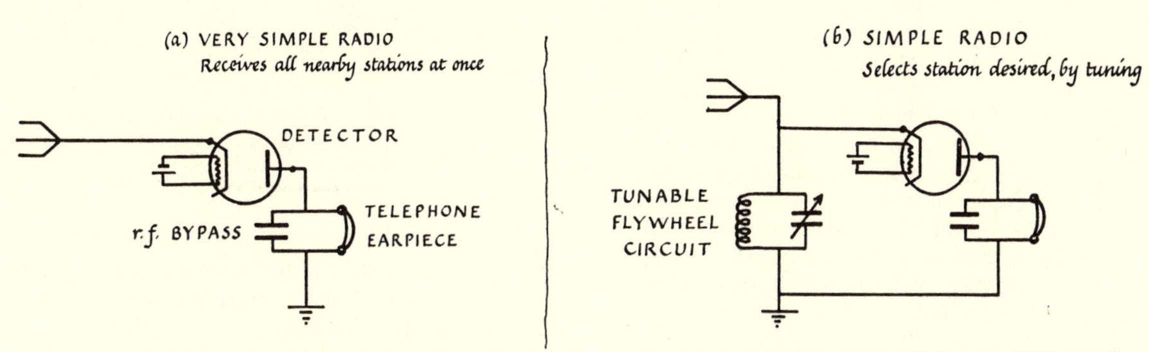

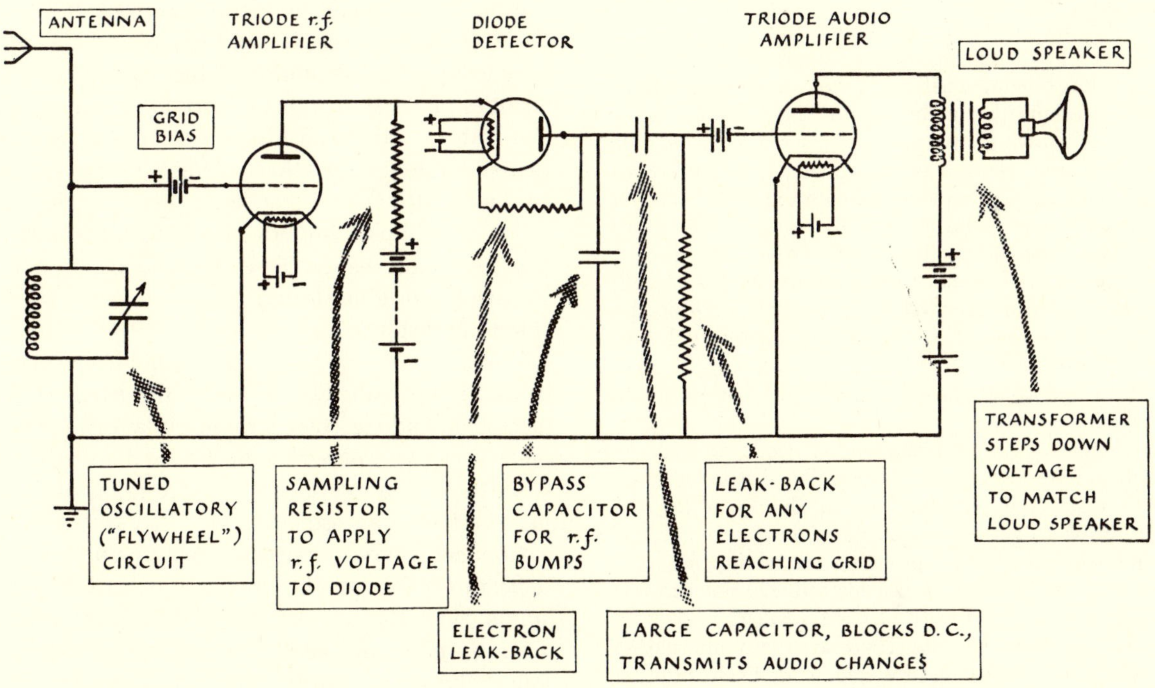

FIG. 41-27. SIMPLE RADIO SETS.

Set (c) works well in principle, and would barely work in practice. (The first triode would have to be safeguarded against starting oscillations of its own. The plate batteries would normally be replaced by a "power supply" [= transformer + diode + smoothing choke + capacitor]. The cathode heaters would be run on A.C. from a transformer. The grid-bias voltages would be provided by an ingenious trick. For reasonable selectivity, the single tuning device would be expanded to a more complex gang.)

from a transformer inside to the binding posts marked TEST and G (= Ground). You can apply this test signal to the deflection system by a single wire from TEST to V (or H), because the other connection is already made—G is common to one side of the test voltage, one side of vertical deflection system, one side of horizontal deflection system and the metal case of the whole instrument. Use this test signal (about 6 volts) to provide alternating deflections and examine the wave-form of this sample.

(c) *Synchronizing signal.* The sweep motion is repeated regularly, but even so you would not see a clear pattern unless each sweep drew its pattern exactly on top of the pattern made by the previous sweep. Thus we want to compel the spot to make its quick switch back at exactly the same stage in the pattern each time. This is done by an ingenious internal device. The spot is swept across by an increasing electric field, caused by an increasing voltage in a device inside the box. When this voltage reaches a large value, the spot has moved right across the screen and should be switched back. A special tube containing ionizable gas does the switching, by removing the sweeping voltage at a standard large value. The increasing sweep voltage is applied to the gas tube, and when it reaches a certain critical value the gas flashes into a glow, and "short-circuits" the sweep voltage. The voltage is removed, so the spot flies back and starts a new sweep. How can we make sure that this switch is made at exactly the same stage in the pattern each sweep? That is done by adding a "synchronizing signal" to the increasing sweep voltage before it is fed to the gas switching tube.

The synchronizing signal is a small sample of the actual voltage being analyzed (the voltage applied to the vertical-deflection plates). Thus the gas tube, waiting for the growing sweep voltage to reach to its tripping value, receives the sweep voltage plus the added sample. The added sample varies with the pattern on the screen, and when it has an upward bump it adds a lot, thus threatening to trip the switch. The sample adds a series of "straws," some positive, some negative and, towards the end of the sweep, an upward bump on the sample will be the last straw and the switch will trip, starting the sweep afresh. Then the sweep will be restarted by the same upward bump in the pattern again and again.

For this, the synchronizing signal needs to be large enough to determine the switching, but if it is too large it may upset the pattern. So you should *adjust the "sync." knob to get a steady pattern*, but no more.

Arrangements are also provided for taking the sync. signal from an external device instead. Do not use this but keep the switch at "*INTERNAL SYNC.*"

(d) *Mysterious patterns.* There are many alternating electric fields around, mostly due to 60-cycle lighting supplies, and your body may act as an antenna. If you touch one wire from the deflection plates you may get a strange pattern. This is due to stray fields, with some filtering action by your body, which may act as a capacitor. These patterns look interesting but are of little diagnostic value. You are advised to ignore them.

EXPERIMENT J (2). Wave Forms of Speech and Music

Use a telephone microphone to show the "wave-form" of your voice on the oscilloscope.

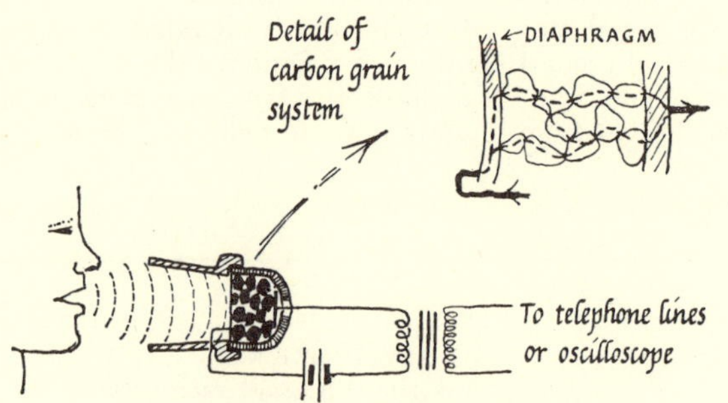

FIG. 41-29. CARBON GRAIN MICROPHONE
Sound waves make the pack of carbon granules behave as a variable resistance.

When sound waves from your mouth enter a telephone mouthpiece (microphone), they push and pull a thin metal sheet, the diaphragm. The sheet follows the wave motion of your voice, more or less. Behind it is a box of small pieces of carbon, loosely packed. The motion of the diaphragm alternately increases the packing of the carbon and releases it. Carbon will conduct an electric current, but where there are poor contacts between pieces the resistance is high. So the motion of the diaphragm makes changes of resistance in the carbon system.

A battery drives an electric current through the carbon system, and that current changes in accordance with the resistance changes, which in turn follow the diaphragm movements which in turn follow the sound waves. Thus, the *changes* in current imitate the original sound waves, though with considerable distortion.

The changing current is led through the primary of a step-up transformer where the small changes in current induce considerable voltages in the secondary. These voltages may be used to drive currents through a telephone earpiece or, as here, they may be fed to an oscilloscope.

Connect the output from your transformer to the deflecting plates of the 'scope, and try talking and singing into the telephone mouthpiece. If you sing a steady note, and *keep to a single vowel*, you can adjust the sweep and its synchronizing signal to get a steady pattern, showing the wave-form of your voice. Sketch a few patterns.

(If you speak French, try the difficult light "u" sound, as in "*tu.*" The CRO will show you it has a characteristic high frequency component. Then whisper it, and you will hear that component as a whistling musical sound. Practice with this can help you to pronounce that difficult vowel.)

If tuning forks, organ pipes, etc., are available, try them too.

Try making "beats" by sounding two different high frequencies together. Two forks, or two pipes, or two students whistling out of tune will do. Pitches about two octaves above middle C, and one or two whole tones apart, show well.

MORE MODERN EXPERIMENTS

Other experiments that explore or demonstrate the physics of this century are being re-designed in forms that may make them available for your laboratory work. (The original forms are too expensive or too complicated to do in a limited period.) Some of the following are already available, and there is promise of the others.

K. *Millikan's Experiment: Measurement of Electron's Charge*

L. *Radioactive Decay: Measurement of Half-life (See Ch. 39)*

M. *Alpha-particles: Range, and Scattering Experiments*

N. *Properties of Beta-rays and Gamma-rays*

O. *Experiments with Neutrons*

P. *Cloud-chambers (Expansion type and Diffusion type)*

Q. *Study of Atomic Particle Tracks in Photographic Emulsions*

R. *Home-made Radio Tube (with its high vacuum made by vaporizing a metal wire to trap the residual gas on the walls)*

S. *Photoelectric effect and the Quantum Constant: Measurement of h/e*

T. *Size of Molecules from Mean Free Path, by shooting Atoms through Gas at Low Pressures*

U. *Energy Levels: Exciting Atoms by Bombardment with Electrons (Franck and Hertz)*

V. *Indoor Measurement of Speed of Light*

W. *Relativistic Change of Mass with Speed: e/m for Electrons with Various Energies*

X. *? Mass-spectograph*

Y. *? Diffraction of Electrons*

Z. *? ?*

PRELIMINARY PROBLEMS FOR CHAPTER 42

★ PROBLEM 1. CYCLOTRON ALGEBRA

[A cyclotron is a device designed to give high energy to an ion (charged atom) by repeated accelerations in each of which the ion falls through a P.D. To keep the ion available for many impulses, a magnetic field is applied to hold the ion in a circular orbit. Here you are asked to investigate the *timing* of the accelerating impulses—an essential problem in design.]

Suppose an ion of mass m kg with charge Q coulombs moves with speed v meters/sec across a uniform magnetic field H ampere · turns/meter. The magnetic field exerts a catapult force on the moving charge. (The magnetic field must be perpendicular to the motion. Any component of field along the velocity will make no contribution to the force.) The force is given by $F = 10^{-7}(Qv)(H)$. (When H is the field at the center of a hoop coil its value is $2\pi NC_2/R$, so it must be measured in amp · turns/meter. Here we produce the magnetic field by an iron core and two coils, so we just call it H.)

(*Note*: in other books, you may find 10^{-7} replaced by $1/c$; and a 4π may appear. These differences merely relate to other units for Q instead of coulombs.)

The force is perpendicular to both the field and the velocity of the ion, so it bends the path into a circle without changing the ion's speed.

(a) Write an equation stating that the catapult force given above (in terms of H) just provides the centripetal force to make the ion move in a circle of radius r.

(b) Solve the equation in (a) for the speed v, obtaining $v = \dots$

(c) How much time is taken by anything moving with speed v to go around a semicircle of radius r?

(d) Combine the results of (b) and (c) to obtain the time for a semicircle trip in terms of Q, H, etc., without r in the expression.

(e) Suppose an electrical impulse accelerates the ion and increases v after each semicircle of travel. How will the time between one impulse and the next have to be changed for successive semicircles, increased or decreased or kept the same? Give a clear reason for your answer.

★ 2. CYCLOTRON AND RELATIVITY

In large cyclotrons, the ions gain much energy, in the course of many trips around, and that energy possesses mass, so the mass of an ion increases slightly. Will the ions get around on time, in the later stages, or should the electric impulses be slowed or speeded up?

CHAPTER 42 • ATOM-ACCELERATORS– THE BIG MACHINES

"The physicist . . . accumulates experiences, and fits and strings them together by artificial experiments . . . but we must meet the bold claim that this is nature with . . . a good-humored smile and some measure of doubt."

—Goethe, *Contemplations of Nature*

"The brain is continually searching for fresh information about the rhythm and regularity of what goes on around us. This is the process I call doubting, seeking for significant new resemblances. Once they are found, they provide us with our system of law, of certainty. We decide that this is what the world is like and proceed to talk about it in those terms. Then sooner or later someone comes along who doubts, someone who tries to make a new comparison; when he is successful, mankind learns to communicate better and to see more."

—J. Z. Young, *Doubt and Certainty in Science*

Accelerators Manufacture Tools

This is an age of big machines, "atom smashers" to the general public. These are atom-accelerators (or more strictly ion-accelerators) and electron-accelerators. Charged atoms and electrons moving at high speed with terrific energy are tools of modern physics. We use them to investigate other atoms by bombardment. By mapping their rebounds from stationary atoms we can discern the inner structure of those atoms, even their nuclear structure.

Sometimes our bombarding atomic projectiles bounce off the target atoms elastically; kinetic energy is conserved, and both projectile and target remain unchanged. Sometimes the collision is inelastic, leaving the target atom "excited" so that it later emits a gamma-ray, and perhaps even then remains unstable. And sometimes there are violent nuclear changes: the projectile goes in and does not come out; but some other particle may emerge from the target instead—nuclear billiards. Then we have a new atom, often an unstable one. Thus, by firing atomic projectiles from machines into suitable targets, we can manufacture new radioactive atoms, to be used for atomic research and for tagging ordinary elements in chemical reactions, in biological processes, in medical treatment, in metallurgical manufacture. . . .

Nature provides some natural atomic bullets, radiations from radioactive atoms. These are alpha-particles, which are helium atoms shot out with energies of several million electron•volts, and beta-rays, electrons produced and ejected with speeds up to 98% of the speed of light. But with these ready-made tools we must take what we are offered. The choice of energies is limited, and the number of bullets in the stream depends on the quantity of radioactive material obtainable—usually small. Until 1917 the choice of radioactive sources seemed fixed: atoms were either stable—permanent and quite unsplittable—or radioactive with an instability that we could not alter. Roasting, hammering, violent chemical treatment, all had no effect; the radioactive changes just continued. Then Rutherford succeeded in making ordinary nitrogen atoms unstable by firing alpha-particles into them. The victim absorbed the bullet, emitted a hydrogen nucleus (*proton*) and changed into an oxygen atom, O^{17}. This first artificial transmutation of one element into another gave hopes of making other nuclear changes by our own design. For a dozen years such hopes hovered over the world of atomic physics. Experimenters hunted for unusual effects of bombardment, with occasional success. Thinkers dreamed of strange atomic particles, even uncharged "neutrons." Designers, turning from the natural guns of radioactive nuclei to artificial accelerators, sketched machines to speed up streams of charged atoms to incredible energies. Shooting extra protons into other target nuclei promised to produce important changes. So the idea of speeding up a stream of hydrogen ions for use as bombarding bullets was specially tempting. A man-made accelerator which could fire such atomic bullets, with control over their energy, was indeed a tempting scheme, especially if it would provide copious

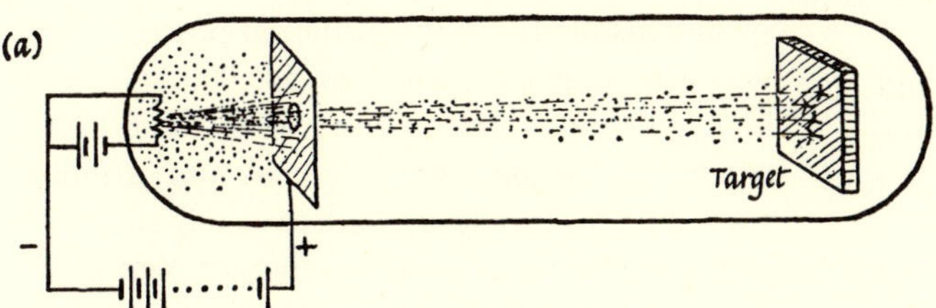

(a) An electron accelerator is simply an electron gun with some arrangement to provide a huge accelerating voltage or its equivalent.

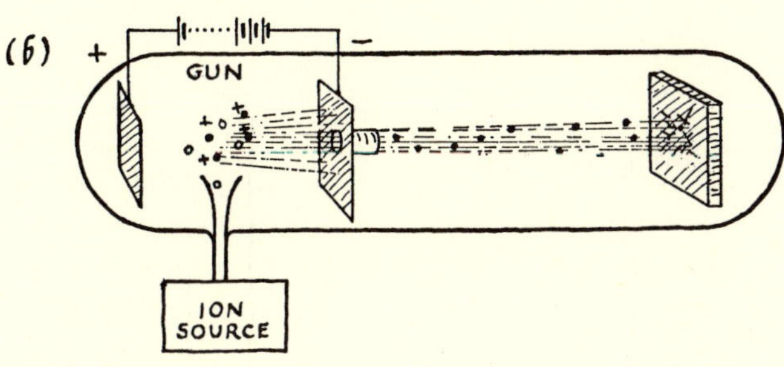

(b) An ion accelerator (Van de Graaff, cyclotron, linear accelerator, bevatron, etc.) is simply a huge ion-gun, accelerating positive ions in a stream to bombard a target. The ions are atoms of hydrogen, helium, . . . , stripped of electrons, so they are really atomic nuclei. The machine has some arrangement to provide a huge accelerating voltage or its equivalent.

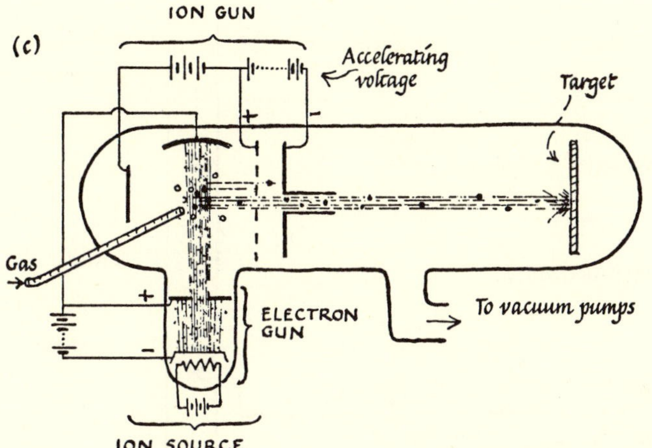

(c) The complete ion accelerator consists of ion-source, ion-gun, and target. The ions are made by bombarding gas atoms with electrons from an electron gun. A thin stream of gas is fed in to provide ions, and large pumps maintain a good vacuum.

Fig. 42-1. Accelerators: Basic Scheme

streams of bullets, and run them up to even higher energies than the alpha-particles already in use.

Alpha-particles from radioactive atoms must have been speeded up in the electric field of their parent nucleus. If we want to manufacture our own atomic bullets, we must give them an electric charge and then use an electric field in the gun to accelerate them. There is little hope of accelerating *uncharged* atoms directly—and, still less, chargeless neutrons—because we cannot pull or push them with a big force. Gravity grips any matter, charged or uncharged, but its pull is very small.[1] We need the big pull of *electric* fields on atomic charges. So the problems were: to make charged atoms for bullets—that is easy—and to manufacture huge potential differences to drive the bullets—and that promised to be difficult and expensive.

[1] For example, a hydrogen atom falling freely from rest in a vacuum gains a speed of only 180 meters/sec in a mile of fall, thereby gaining less than 1/1000 electron · volt of K.E.!

Nuclear Vocabulary: Protons et al.

Hydrogen nuclei, H^+ (or ${}_1H^1$) are so important, both in discussing nuclear structure and as bombarding bullets, that they are given a special name, which we have already been using, *protons.* A proton is a hydrogen atom stripped of its only electron. When we picture heavier nuclei as made of protons and neutrons, we lump both kinds of particle together under the name *nucleons.* (We have special names for the heavier isotopes of hydrogen. The nucleus of "heavy hydrogen" or "deuterium," ${}_1H^2$, is often called a *deuteron*; and the radioactive nucleus of still heavier "tritium," ${}_1H^3$, is called a *triton.*)

Accelerators

In the early 1930's, with the newly-discovered neutron proving useful for sneak raids on atomic nuclei, the first big machines were built and started pouring out rich streams of investigating bullets. Van de Graaff, in Princeton, devised a continuous high-voltage supply in which charges are carried by a silk belt into a large copper ball—essentially a continuous charging-by-induction process. A pair of such machines gave a P.D. of 500,000 volts between their collecting balls. These, equivalent to a giant battery, could be used to accelerate protons and other ions to an energy of half a million electron · volts, or 0.5 Mev. Nowadays bigger Van de Graaff machines, requiring a whole building, can run up to several million volts and supply quite large ion-

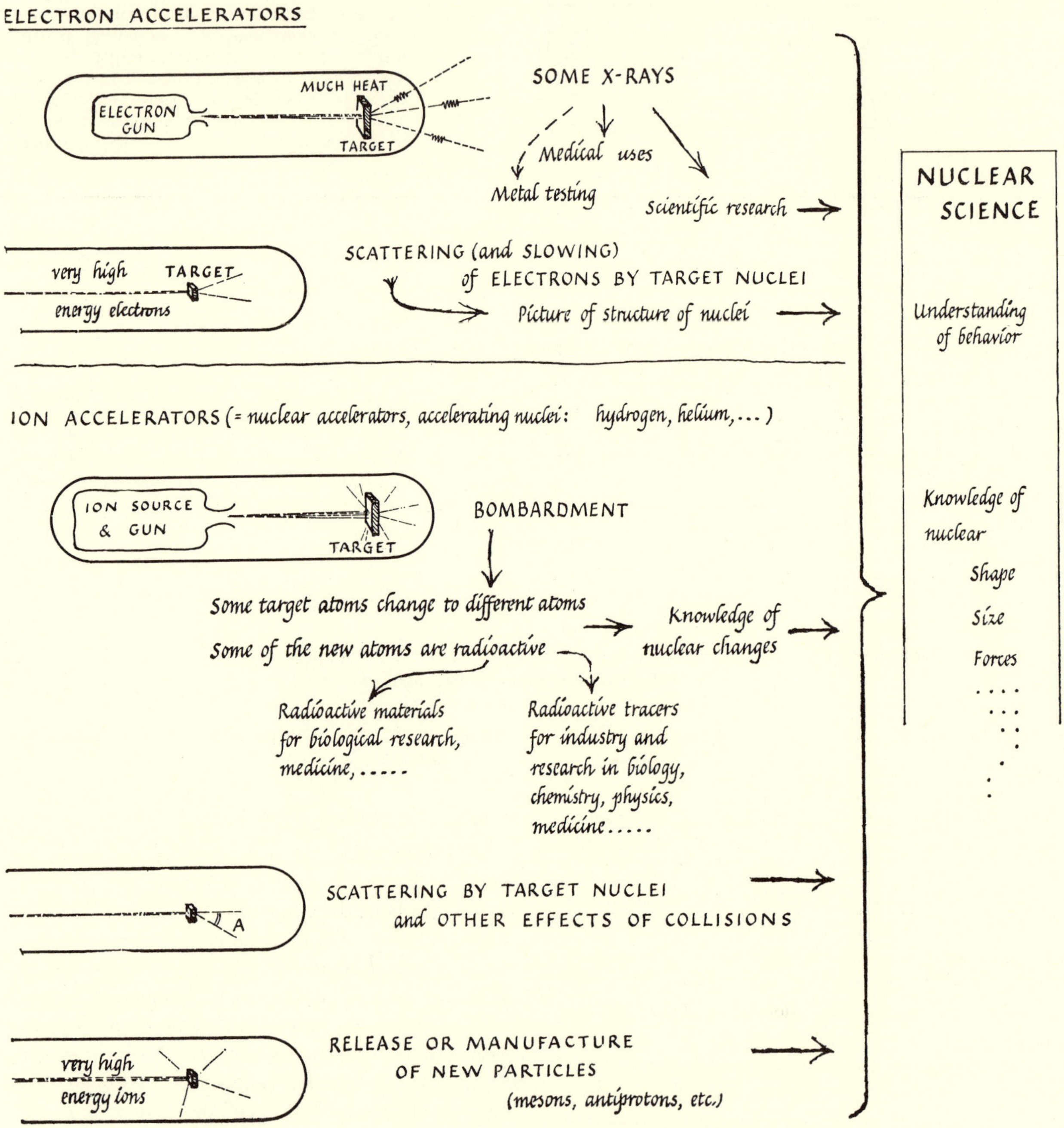

FIG. 42-2. USES OF ACCELERATORS
The "big machines" are not experimental investigators themselves. They provide "tools" for experiments on atomic nuclei. These tools are high-speed charged particles (electrons or ions). The machines provide streams of these high-energy particles to bombard targets. Here are some of the effects produced.

currents. Still greater voltages were hoped for; but costs of construction and insulation increase sharply with the voltage required. Therefore, instead of trying to produce a huge voltage directly experimenters tried to add up the effects of many smaller voltages. In California the first large cyclotron was being built. In the Cavendish laboratory at Cambridge, Rutherford's young colleagues were building an array which would double and redouble the voltage from a rectified A.C. supply. They were adding up a staircase of steps of voltage. Special radio tubes ingeniously switched the charging voltage from one capacitor to the next in a tall tower of capacitors, thus building a large voltage up the whole tower. Rutherford, eager as ever to try new bombardments, and impatient with delays, said, essentially, "Well, try what you have got now, anyway." Joyful success: they found that protons driven by the machine's mere 150,000 volts could upset the nuclei of lithium atoms and make them break up into pairs of helium atoms. This was a man-made nuclear smash-up, on a much more copious scale than Rutherford's earlier occasional upset of nitrogen atoms (see Fig. 42-3).

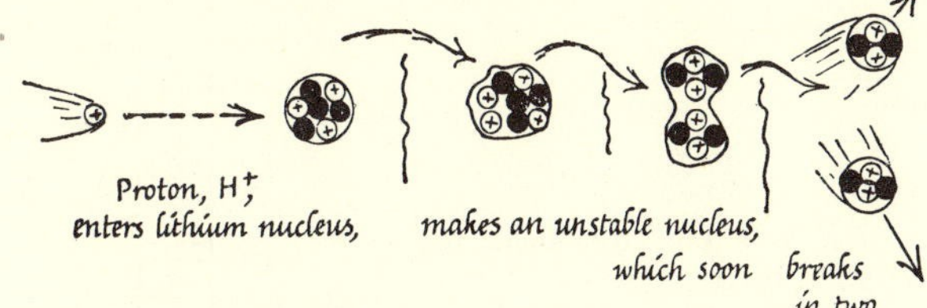

Fig. 42-3. Breaking Up Lithium Atom's Nucleus

Mankind could now make the small lithium atom nucleus undergo "fission."[2]

Cyclotrons

Meanwhile, the *cyclotron* was being developed to produce streams of high energy ions by an entirely different method. Instead of building up a P.D. of a million volts or more and applying it to a great ion-gun, with severe troubles of insulation, could we apply a much smaller voltage, say 30,000 volts *again and again*, building up big ion energy without using a big voltage anywhere in the machine? In terms of hurling a ball very fast, instead of using a giant to give it one tremendous throw, could we speed it up in stages by hitting it moderately many times? For a ball one such scheme would be to tie it by a string to a post, as for children's tennis practice, and hit it again and again every time it comes around. This is the basic principle of

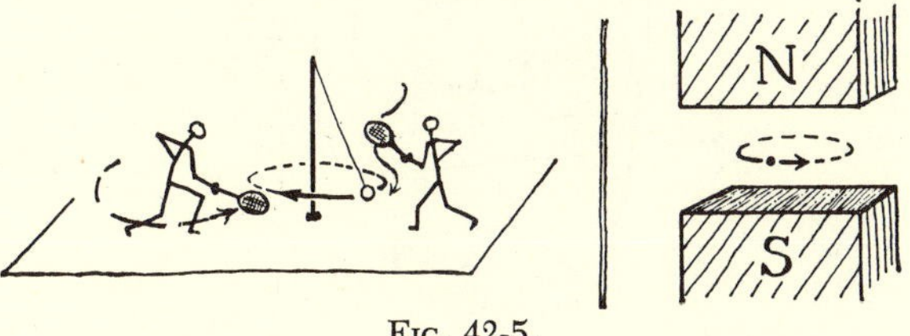

Fig. 42-5.

the cyclotron. We cannot tie a charged atom to a post, but we can make it move around a circular path by using a magnetic field to pull it sideways as it travels. Then if it travels around and around such a path we might speed it up by applying an electric field again and again. To pull the moving ion sideways we need a magnetic field perpendicular to its path. So a cyclotron has an enormous magnet whose sole job is to keep the moving ions swinging around again and again so that they can be given an accelerating push many times by an applied electric field. To use the same moderate voltage over and over again, we cannot just apply it steadily at some region in the ions' path or they would cross it backwards as well as forwards and find it slowing them down as often as speeding them up. Some trick must be used to turn the voltage off and then on at a suitable instant when the ions have come around again; and the ions must be protected from unwanted deceleration at other stages. To see how this is done, study the following simplified version.

In the central space between the magnet poles there is a huge evacuated box, B, in which the ions are accelerated. We shall call this box the can. There is some device inside the can for manufacturing protons, H^+, to be accelerated. There are two accelerating electrodes, D and D′, stretching across the box. For the moment, pretend these are metal sheets facing each other and connected to a battery, with D positive and D′ negative. (This is not the real arrangement, but a bogus provisional story to introduce the real one.) Suppose that D and D′ are connected to a 20,000-volt battery. Then they are parallel charged plates, with a strong uniform electric field across the space between them, but practically no field outside them. Suppose a positive ion is formed somewhere in the central region between D and D′. It is pulled across towards D′.

[2] Fission is a term used by biologists to mean "dividing in two." Certain very simple animals, e.g., the amoeba, reproduce by fission. A parent animal moves into an oblong shape, develops a narrow waist, and divides into two new animals.

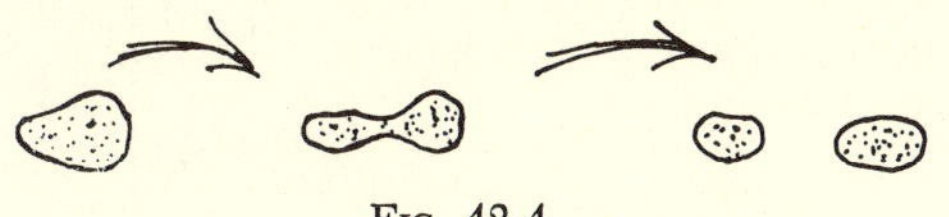

Fig. 42-4.

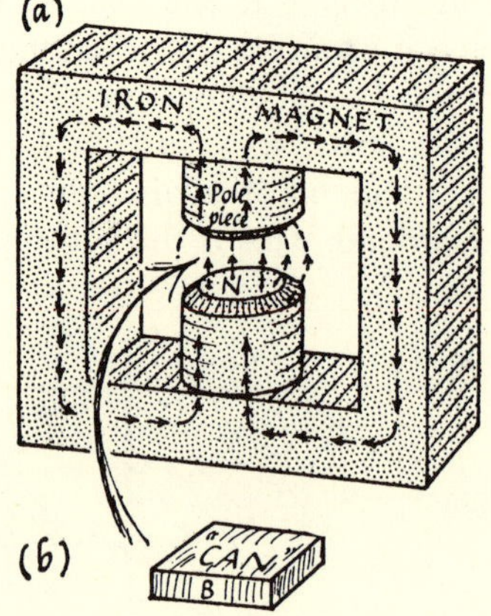

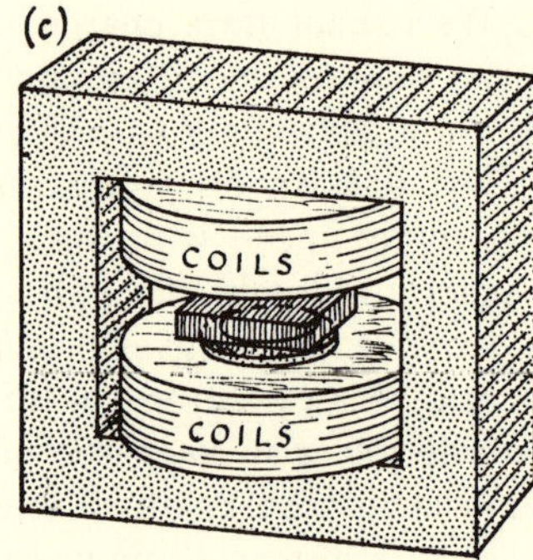

Fig. 42-6. Cyclotron Magnet
(a) The soft iron magnet is magnetized by currents in a pair of coils, so that the magnetic field in the gap between its pole pieces is a strong vertical field, practically uniform.
(b) The box B (the can) is a rectangular box containing ion source, dees to provide the accelerating electric fields, and probes to catch the accelerated ions. A good vacuum is maintained in the box.
(c) Currents in the coils keep the iron magnetized and maintain the magnetic field. The coils are usually water cooled.

Just before it hits D′ we imagine a hole drilled in D′ to let it through. It will whiz on through (and the electric field it meets beyond will be negligible). The *magnetic* field is perpendicular to its path, so the ion will swing around in a circle behind D′. After a half circle trip the ion will hit D′ again, from behind. Imagine a hole there. Whizzing through the hole in D′, the ion would find itself in the electric field between D and D′ moving *against* the field and being *de*celerated. That would undo the earlier gain. To give the ion more energy, the electric field must be reversed by now. Somewhere during the ion's semicircle trip behind D′ the battery must be disconnected and reconnected the other way. Then the ion will receive a useful accelerating "spank" when it returns to the D-D′ region. Whizzing faster than before through a hole in D, the ion will swing around a larger semicircle than before,

Fig. 42-7. Progress of an Ion in a Simplified Cyclotron
Pretend the ion is accelerated in an electric field between two plates D and D′ inside the can.

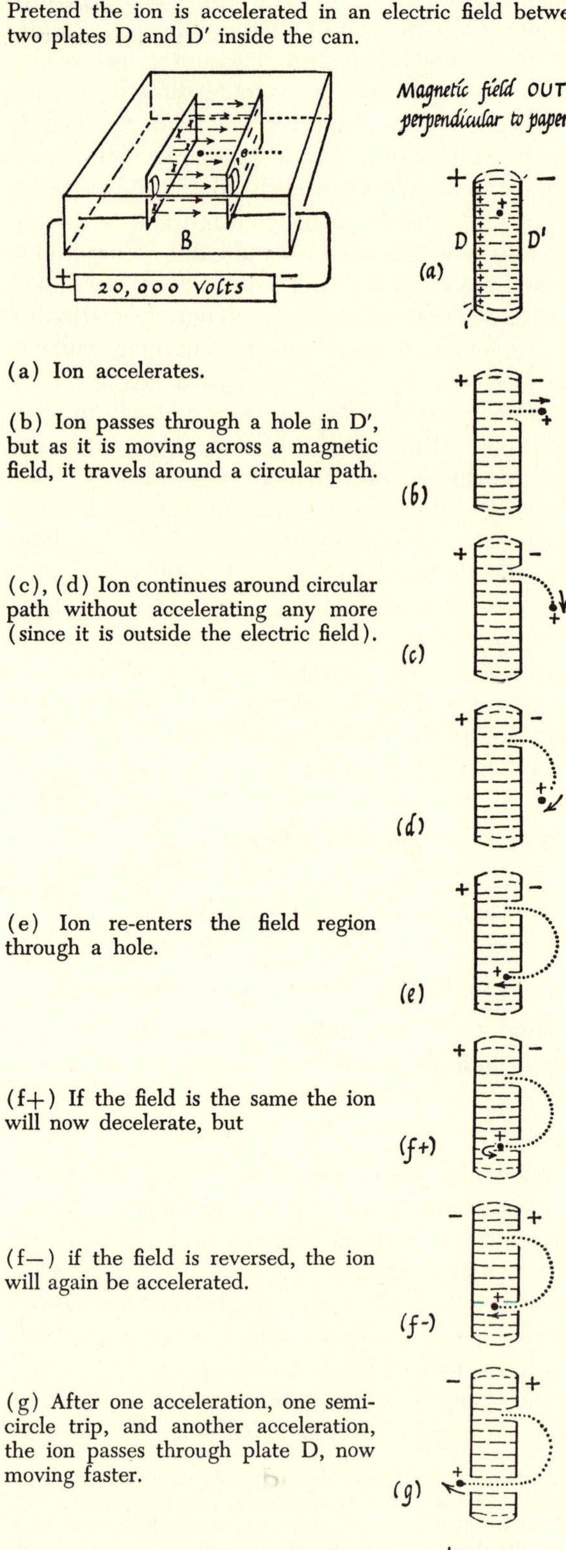

(a) Ion accelerates.

(b) Ion passes through a hole in D′, but as it is moving across a magnetic field, it travels around a circular path.

(c), (d) Ion continues around circular path without accelerating any more (since it is outside the electric field).

(e) Ion re-enters the field region through a hole.

(f+) If the field is the same the ion will now decelerate, but

(f−) if the field is reversed, the ion will again be accelerated.

(g) After one acceleration, one semi-circle trip, and another acceleration, the ion passes through plate D, now moving faster.

(h) The magnetic field bends the ion's path into another semi-circle—larger radius because the ion is moving faster, and

so

on.

because of its greater speed. When it reaches D again and passes through a hole into the electric field, it would find that field pointing the wrong way; so the battery connections should again be reversed sometime during the ion's trip behind D.

Now you can see the story. The ion goes around and around in a growing "spiral" of semicircles, moving faster after every spank by the electric field; but that electric field must be reversed during every semicircle trip of the ions. If the ion has one electron charge a 20,000 volt battery will give it 20,000 electron · volts every time it passes through the field between D and D′. If the ion starts from rest near D it will gain 20,000 electron · volts in moving across to D′ then another 20,000 ev after one semicircle trip, another 20,000 ev after the next semicircle trip, &c., gaining 20,000 ev every half circle or 40,000 ev every circle. After 1000 semicircular trips it should have gained 1000 × 20,000 or 20 million electron · volts. It would have as much energy as if driven by a single huge spank from a 20-million volt battery. Such a battery would be terribly costly to insulate and maintain. Here we use only 20,000 volts, which can be provided fairly easily; yet the ion emerges with the energy and speed for bombarding other atoms that a 20-million volt P.D. would give it.

This is wonderful—"Now we can smash atoms"—but there are serious difficulties. How can the battery be reversed, quickly and at the right moment? How can the timing of battery reversals be adjusted to the ion's larger and larger semicircular trips? How can the ion be protected from electric fields when it is on its way around, *not* in the accelerating region between D and D′? And how can we avoid the wasting of ions that miss the holes in the plates D-D′ and are stopped by the metal? These are not minor questions: they are essential problems of design which must be solved if the machine is to work. We now proceed to the real design, which does solve them. We want the field voltage reversed, forwards-backwards-forwards, etc., every half circle of the ion's motion. Instead of a huge battery with a clumsy reversing switch, this calls for a supply with alternating voltage. A radio-tube oscillator is used, not unlike the high-frequency supply for radio stations. The oscillator swings the field voltage to and fro regularly. But this makes the second question seem more severe. How can the ion get around the next half circle in time?

To receive a full accelerating spank it must arrive in the space between D and D′ at the right instant, when the voltage is at a maximum. If it arrives

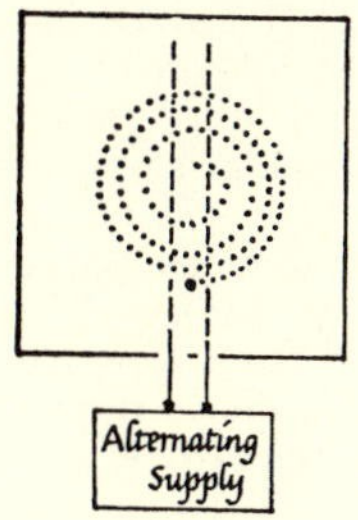

Replace the plates D, D′ by screens, D, D′, so that an ion can get through them in a spiral of semi-circular trips. Instead of switching a battery to and fro, connect an oscillator to D and D′ to apply an alternating electric field in the region between them. The alternating electric field must have its maxima, + and −, at the right instants when the ion needs them.

FIG. 42-8. IMPROVING THE SIMPLIFIED CYCLOTRON

early or late the changing field will give it a smaller spank. If as much as ¼ cycle late it receives no spank at all and if half a cycle late it will receive a reverse spank decelerating it. Obviously, the rate of oscillation must be adjusted so that while the ion travels its first semicircle the voltage runs through half a cycle from maximum one way to maximum the other way. But how about the next trip, and the next, larger and larger semicircles? Can the ion travel around them and arrive on time, in step with the oscillator's change of voltage? Must the oscillator change its timing in some remarkable way or does every semicircle take the same time? The latter would be wonderful luck; it would solve the problem. Does it happen? In other words, as the ion moves faster around the bigger semicircles is the distance enough longer to take the *same time*? To examine this we must do the "cyclotron algebra" of Problem 1. Carry through the algebra if you have not already done so, and you will find that if the magnetic field is constant all over the region of the ion's path, every semicircle does take the same time!

Here then is the basic reason why the cyclotron works easily. An oscillator running steadily, at the right frequency for the magnetic field and the ion's e/M, will deliver spank after spank at just the right instants to a bunch of ions as they run around a spiral of semicircles. Return to the other questions: How can we shield the ion from electric fields on its semicircular trips outside the region D-D′? How can we avoid wasting ions that hit the plates D and D′?

The same trick solves both problems. We replace the plates D and D′ by hollow boxes of a much more suitable shape. Inside any empty closed metal box, the electric field is zero; any charges given to the box run to the outside and arrange themselves on the surface to make the electric field zero both in the metal walls of the box and in the space inside. Inside an *almost-closed* box the field is almost zero, except near its mouth. So instead of plates D and D′ we use copper boxes shaped to accommodate the ion's largest semicircles, closed

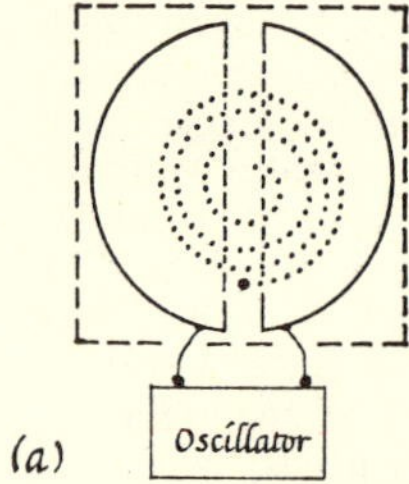

Instead of the screens D and D′, install a pair of dees (hollow, semi-circular boxes with open mouths) facing each other where the original plates were placed. Then there will be an alternating electric field in the space between the mouths of the dees. The hollow dees will shield the moving ion from unwanted electric fields during its trip around each semi-circle.

Fig. 42-9a.
Improving the Simplified Cyclotron to the Real Design

except for open mouths facing each other at D and D′. To picture these boxes think of a squat round tin can cut in half by a vertical slice. These two boxes, shaped like ᗡ and D, are called "dees" on account of their shape. Where there was a metal plate with holes in our bogus picture, the real machine has the open mouth of a dee. Ions are not lost by hitting metal because they hit an open mouth; and they are protected from stray electric fields during their trips by the enclosing shield of the copper dees.

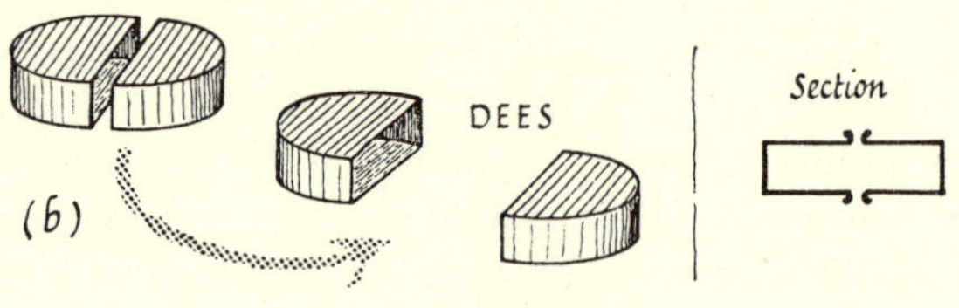

Fig. 42-9b.

There are other problems of design: how to manufacture the ions, how to avoid wasteful collisions with unwanted gas molecules, and how to "get the beam out"—that is, how to divert the ions to the experimental target when they have been accelerated. The ions are made by bombarding hydrogen gas with electrons from a small gun near the center of the can. A tiny stream of hydrogen is fed in to provide ions; but huge pumps are used to remove excess hydrogen and keep the can as free as possible from lounging gas molecules.

Here then is the complete picture: ions of hydrogen atoms stripped of their electron by bombardment start in the center of the can; they are given a little energy by the electric field, whirl around a semicircle under the protection of the dee, back into the electric field which has been reversed just in time, are accelerated again, whirl around a bigger semicircle inside the other dee, back into the electric field (again reversed), whirl around a still bigger semicircle with still bigger speed; and so on, until their semicircles are almost as big as the dees. In the outermost orbit the ions have reached enormous energies, and specimens to be bombarded

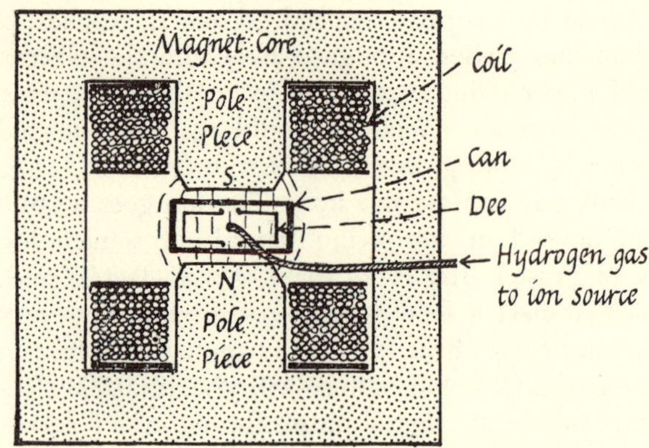

Fig. 42-10. Section Through a Cyclotron
The ion source at the center of the can consists of an electron gun which bombards hydrogen molecules (fed in as a tiny stream through a pipe) knocks them apart and ionizes them to make H^+ ions. This ion-making process is conducted on such a rich scale that the ion source can be seen as a glowing arc.

can be placed there to intercept them inside the can. However, this requires vacuum traps (like an escape hatch on a submarine) for inserting and removing the target; and for many experiments it is better to bend the stream of ions away out of its largest semicircle so that they shoot out through a window of thin metal in the side of the can and on down a pipe to apparatus where they can be used. The bending is done by a "deflector plate" which supplies a suitable electric field. Then the high energy ions, usually protons, can be used to bombard any target we choose, producing an amazing variety of nuclear changes.

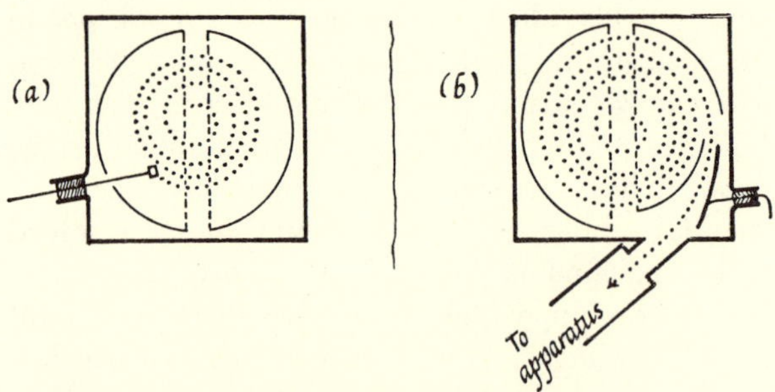

Fig. 42-11. Using the Cyclotron Beam
(a) A "probe" is inserted so that its end catches the beam inside a dee when it is making a large semi-circle (therefore, with large ion energy).
(b) A "deflector plate" with a strong negative voltage bends the beam so that it comes out through a window in the can and passes down a long evacuated tube to other apparatus where it is used.

Cyclotrons for Higher Energies

The simple cyclotron will provide a stream of ions each with energy a few million electron · volts.

If we build a bigger machine aiming at higher energies we find that ions with many Mev energy fail to make their trips on time. In the outer semicircles where we hope they will gain still more energy, they arrive later and later. The simple cyclotron algebra seems to have misled us. What has changed? The ions still have the same charge, the magnetic field has the same strength (supposing we have built a bigger magnet to extend far out), and Newtonian mechanics suggests each semicircle trip should still take the same time. But Relativity suggests that the ions will have a bigger mass, because their kinetic energy itself has some mass to be added to their original mass. As a result of increased mass, the semicircles are a little too big and the ion arrives late. We can now cure this by modifying the oscillator. A flock of ions starts from the source and is accelerated regularly in the early stages. Then the oscillator slows down slightly, delivering its spanks less often, spanking the extra-massive high-energy ions just when they need it. Then the ions emerge after many successful accelerations with huge energies—20 million electron · volts from a small cyclotron with pole-pieces 3 ft across. The oscillator then speeds up to normal rate to operate on another flock of ions. The changing of the oscillator's rate is done by a changing capacitor in the oscillating circuit. A big *frequency modulated* cyclotron can produce ions with energies of several hundred Mev. But for that it must have an enormous magnet—thousands of tons.

Relativity

According to Relativity, when we push a particle to higher and higher kinetic energy, its speed should trail up towards the speed of light, *c*, but never go beyond—the particle's observed mass should increase instead. And that is just what *does* happen in the big accelerators. It costs money and affects the design of machines. At high speeds the mass of a proton *is* greater; and we must pay for the device to slow the oscillations in a "frequency modulated" cyclotron. In big racetrack machines the ions have huge energy in the later stages of "acceleration," and they *do* make trip after trip around in almost the same time, as their speed nears *c*; so the oscillator's timing must be arranged for that.[3] When the ions emerge, they *do* show their increased mass by the collisions they make.

In *electron* accelerators, the speed limit even simplifies the design: after the early stages, the electrons' speed is so near *c* that every trip around a constant-radius orbit *does* take the same time, $2\pi R/c$. Yet, in each trip the electrons are given more energy—and mass—and the magnetic field must increase to hold them in orbit.

Relativity mechanics does fit the facts.

The Need for Still Higher Energies

Projectiles with still higher energy would be very valuable investigating tools. They might make new changes inside the nuclei they hit. Physicists wondered whether, given high enough energy, they could manufacture new matter. Pairs of positive and negative electrons can be manufactured from gamma-ray photons of energy one Mev or more. What would it cost to manufacture a complete hydrogen atom, nearly two thousand times as massive? Presumably, at least two thousand times the energy, or 2,000 Mev. Simple theory of collisions tells us even this would not suffice. We need more, nearer 5,000 Mev, to manufacture a "proton pair," a proton together with an "anti-proton"—a mysterious negative counterpart much dreamed of, only recently observed. When such pairs are manufactured they are new atomic particles, "matter from nowhere." There are two new particles in addition to the original bombarding particle, and their mass comes from the mass of the kinetic energy of the bombarding particle.

There are other new atomic particles that we can manufacture: *mesons*, with various masses intermediate between electron and proton, with charges $+e$, $-e$ and zero. These strange short-lived particles seem to play an important part in binding atomic nuclei together, so nuclear physicists are anxious to study their properties. Instead of waiting for a few mesons to arrive in cosmic ray streams, we can now manufacture them by bombarding targets with high-energy ions from accelerators. But, for a copious stream of mesons, we need ions of energy 500 to 1000 Mev and even more. Thus we now have Bev (billion electron · volts) as a unit[4] for new and important energy goals.

The "Racetrack" Machines

A cyclotron to deliver ions with several Bev energy would demand far too much metal for its magnet. But we now turn the magnet inside out, so

[3] For example: as a proton's kinetic energy is increased from 1 billion ev to 2, 3, 4 billion, its speed rises from 87/100 of *c* to 94/100, 97/100, 98/100. It reaches 98% of *c* instead of the speed 2.8*c* predicted by the classical formula K.E. $= \frac{1}{2}mv^2$.

[4] Although Mev was first used to mean "million electron · volts," strict scientific grammarians read it "mega-electron · volts." And there is a new agreement to call 1000 Mev "1 Gev," with G for "giga-" as in "gigantic."

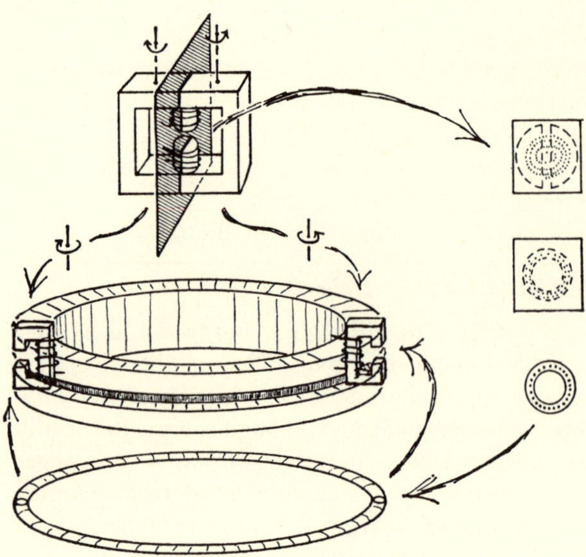

FIG. 42-12. RACETRACK ACCELERATOR
The giant ring-magnet is built up of many thin C-shaped sections. The doughnut tube is placed between the jaws of the ring-magnet.

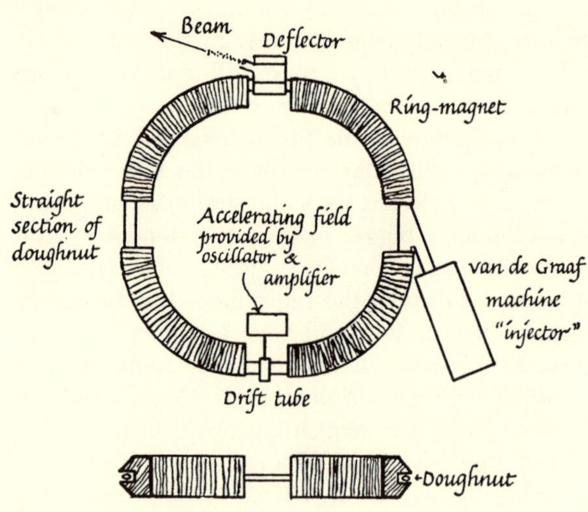

FIG. 42-13. RACETRACK MACHINE
The ions, injected at high speed from a Van de Graaff machine, make a million trips or so before they are deflected to become an external beam.

that the central gap becomes a ring-gap, dozens of feet in diameter; and the can becomes a ring-tube like a doughnut. This is the "racetrack" machine. Some have already been built and are in use, under the names "cosmotron," "bevatron." The giant ring-magnet is built up of many thin C-shaped sections. The doughnut tube is placed between the jaws of the ring-magnet. One dee shrinks into a short shielding "drift tube," and the other dee becomes the whole of the rest of the doughnut. The ions are accelerated by an oscillating electric field in the gaps at the ends of the drift tube. The particles to be accelerated are ionized and given a preliminary acceleration to an energy of a few Mev with a small machine such as a Van de Graaff. Then they are injected into the doughnut tube and circulate around it, being given a further dose of energy once a trip by an electric field derived from an oscillator. Now, with only a narrow doughnut pipe for their orbit, the ions cannot move to larger orbits at higher energies as in the cyclotron. Instead, the magnetic field that holds them in their orbit must be changed to fit their growing energy. It must start small, to hold the slow ions in the doughnut orbit as they enter from the injector. It must be increased as the ions move faster, always holding them in the same orbit. When the ions reach maximum energy, after a million trips or so around the doughnut, they are deflected to hit a target. After that, the magnetic field must be reduced to a small value, ready for the next bunch of injected ions.

Where does the energy of the huge magnet go to when the field is reduced? We cannot afford to waste it and then pay for a new supply when the field is increased for the next bunch of ions. Therefore, we store the magnet's energy in a huge mechanical flywheel or in a huge electric capacitor and choke.

Since the ions continue around the same orbit, the oscillator's frequency must change as the ions gain energy and move faster and faster. Special schemes are needed to make the oscillator oscillate more and more rapidly *and* the magnetic field grow stronger and stronger "hand-in-hand." These schemes succeed and the machines already in use are producing bunches of ions with energies as high as 5 Bev. Protons, emerging with so much kinetic energy that their mass is six times normal, smash into targets so violently that they produce flocks of mesons and even anti-protons—on which investigations are now proceeding.

PROBLEMS FOR CHAPTER 42

1, 2. Preliminary problems (on the last page of previous chapter).

3. CYCLOTRON AND ALPHA-PARTICLES

Suppose a cyclotron that has been used to accelerate protons is used to accelerate He^{++}, helium nuclei with 4 times the mass and twice the charge of a proton (making a stream of "alpha-particles"). The magnetic field is kept the same.

(a) What change, if any, must be made in the oscillator frequency?

(b) How will the K.E. of each "alpha-particle" produced compare with the K.E. of the protons produced previously? (Ignore relativistic corrections.)

4. LINEAR ACCELERATOR

Imagine the "spiral" path of the ions in a cyclotron unwound into a straight line. Then, with no tethering magnet, the machine would be a "linear accelerator." Such machines have been designed and work well, but they are expensive. They need a very long pipe with a good vacuum in it. Ions are made at one end of the pipe and accelerated down the pipe to hit a target at the other end. The acceleration is done by electric fields applied at intervals down the pipe. In one form, a series of metal tubes, A, B, C, D, E, . . . shield the ion while it is travelling along inside but apply an accelerating electric field in the short gap between one tube and the next. To apply the field, a high-frequency radio-oscillator is connected to the tubes, one terminal of the oscillator to every second tube, A, C, E, . . . and the other to the alternate tubes, B, D, ...

(a) If the ion meets a full accelerating field in each gap ($A_2 - B_1$, $B_2 - C_1$ . . .) how must the spacing of those gaps, the ion's speed (constant inside any one tube), and the oscillator timing be related?

(b) For an oscillator running at constant frequency, how must the gaps be spaced (or the tube lengths be chosen) for slow ions starting from rest?

(c) For an oscillator running at constant frequency, how must the gaps be spaced for particles with very high energy (e.g., electrons that have already gained many Mev)?

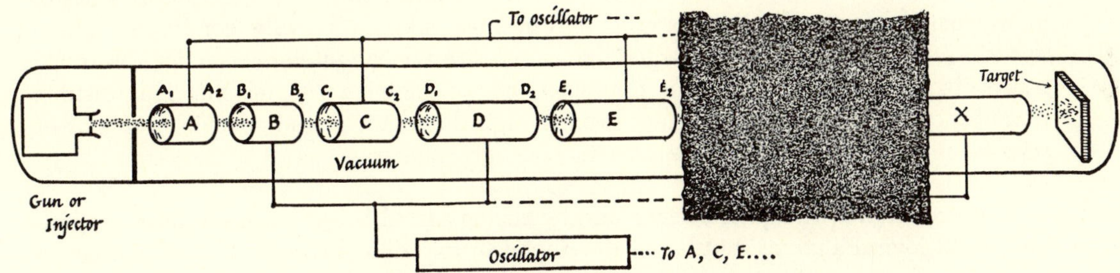

Fig. 42-14. Problem 4. Linear Accelerator

CHAPTER 43 • NUCLEAR PHYSICS

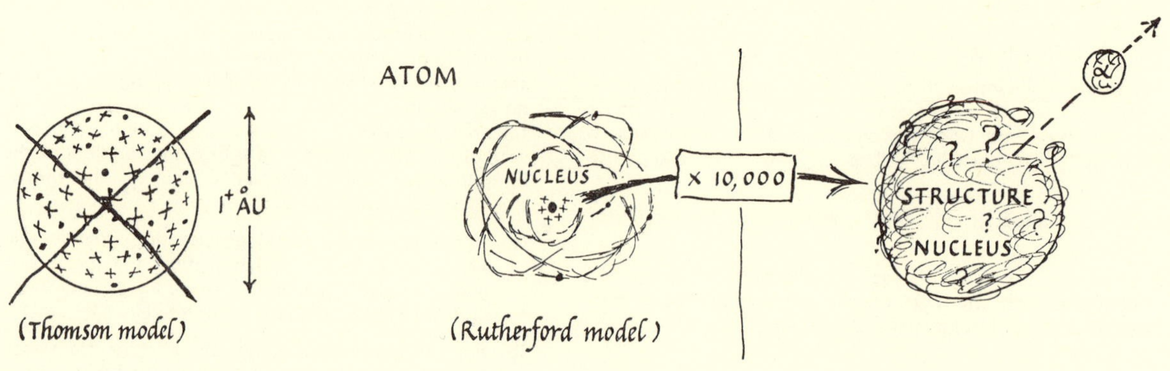

"The object of scientific and engineering research is to augment and clarify our knowledge of the natural world. Our eagerness to pursue such knowledge arises from two sources, either an intellectual interest in understanding the laws of nature or a desire to use natural forces for the material improvement of man's living conditions. These two motives can be crudely characterized as a desire for enlightenment and a desire for utility. In the scientist the first is dominant, in the engineer the second. This distinction of motivation remains important even though the methods and specific problems of science and engineering have become increasingly similar.

"Our present knowledge of science and technology is all based on observation, experiment, and logical analysis, but it has developed in two different ways, corresponding to the two different motivations I have cited. Technology has been much more empirical. An engineer or inventor is primarily interested in some practical problem. If he can develop a machine or a process that will give the result he wants, it may not be important to him how it works. In contrast it is just exactly the 'how of things' that does interest the scientist. Often to make a step towards understanding, the scientist must simplify the conditions he is studying. He must reduce the number of variables. In doing so he often finds himself investigating something that has no immediate bearing on the practicalities of life. Historically, science has progressed by avoiding problems too difficult for it to understand. Technology has progressed by utilizing the success of process or machines regardless of whether they were fully understood. For an engineer the utility of a machine or process is the first consideration. For a scientist, understanding comes first. . . .

"The whole relation between the separate sciences and between science and technology has changed over the past fifty years. As the body of scientific knowledge has increased and the methods of science have become more powerful it has become possible and fruitful to apply the methods and knowledge of one science to another. . . . At the same time, technology has become so intricate that empirical methods have become inadequate. In short, it has become both possible and necessary to apply science to engineering. The difference of motivation still remains and often has importance but the methods and problems have become more and more alike."

—Henry D. Smyth

"Atomic Energy"? "Transmutation"?

The discovery of radioactivity raised exciting questions. An α- β- or γ-ray carries an enormous amount of energy out of an atom. Could mankind tap that rich supply of energy which seems to be locked up in radioactive atoms?

Again, a radioactive parent element turns into an entirely different daughter element. Could man make such changes happen profitably, say from lead to gold? To succeed, he would need to influence radioactive changes; he would have to speed up some known ones or to start some unknown ones. Early experimenters tried to modify radioactive changes by violent treatment. They soon decided that radioactive instability is fixed and unchangeable and the stability of non-radioactive elements equally unassailable. Yet they continued to try, and now we know how to succeed: *use particles with huge bombarding energies.* When we consider the energies involved, we can understand the earlier failures.

Energies of Radioactive Changes

Alpha- beta- and gamma-rays, essential symptoms of radioactive changes, emerge with millions of electron · volts. An energetic α-particle makes 200,000 pairs of ions in travelling 2 inches through air, at a cost of about 30 electron · volts a pair: 6 million electron · volts.[1]

These rays cannot have started from the outer regions of the parent atom, in the electron cloud far out from the nucleus. There is too little matter there to provide an α-particle's mass. The electric field runs the wrong way to eject a beta-ray; and it is too weak to give any of the rays several million electron · volts. They must come from the nucleus. So to affect their production we might well need comparable energies, say a few million electron · volts. That rules out hopes of creating radioactivity by ordinary heating: gas molecules have an average K.E. of about 1/30 electron · volt at room temperature, and only 1/3 ev at white-hot furnace temperature. The innermost furnace of a star might be another matter.

[1] As α-particles slow down and stop in matter, they spend their K.E. in making ions; and those, in returning to neutral atoms, deliver most of the energy ultimately as heat. Thus, radioactive material kept in a bottle generates heat. Radioactive rocks help to keep the Earth warm.

Try to keep 2 kilograms of a radioactive element that emits an α-particle. After one half-life, 1 kg will have disintegrated, yielding, say, 500,000,000 Calories. Compare this with the yield of ordinary fuels such as oil or coal; 10,000 Cals/kg at most. A pound of radioactive material yields as much heat as tons of coal, but it takes its own time.

Chemical Energies

Nor are chemical changes energetic enough; a molecule loses or gains a few electron · volts in chemical changes. For example: when coal burns, heat is released.

C	+	O_2	→	CO_2	+	heat
1 atom of carbon	+	1 molecule of oxygen	YIELD	1 molecule of carbon dioxide	+	4 ev of heat

Other fuel-burnings are similar. Even an explosion is merely a very quick burning in a confined space, a sudden release of heat that makes a compression wave in the air. The energy released by one molecule is still quite small. Gasoline exploding with oxygen releases 40 to 50 electron · volts of heat *per molecule.*[2] And TNT releases only 30 ev per molecule when its self-sufficient molecule decomposes. The loud bang when hydrogen and oxygen explode to form water is financed by a heat-release of less than 2 ev per molecule.[3] These statements come from measurements of heat absorbed or released in chemical reactions, but they agree with direct atomic measurements by electron bombardment. We can gun electrons gently at neutral atoms and molecules. Electrons with energy a few electron · volts bounce off the target atoms elastically; but if we shoot them with more energy, a few dozen electron · volts, they knock an electron off the victim and make a positive ion. This making of ions costs only 4 or 5 ev for sodium or potassium atoms which lose an electron easily; 13.6 ev for a hydrogen

[2] More energy from one gasoline molecule than from one carbon atom in burning coal, because the molecule is bigger. Pound for pound, gasoline and coal release comparable amounts of heat when burned, but it is difficult to burn coal fast enough to make its fire sound like an explosion.

[3] Therefore, we might expect the splitting of a water molecule into hydrogen and oxygen to require an energy-supply of one or two ev per molecule. That is why in the electrolysis of water (Ch. 32, *EXPERIMENT R*) the "Ohm's-law-line" does not pass through the origin, but seems to require a starting P.D., between 1 and 2 volts. That is the energy each coulomb transfers *from* electrical *to* chemical energy—the reverse of a simple "water battery" EMF.

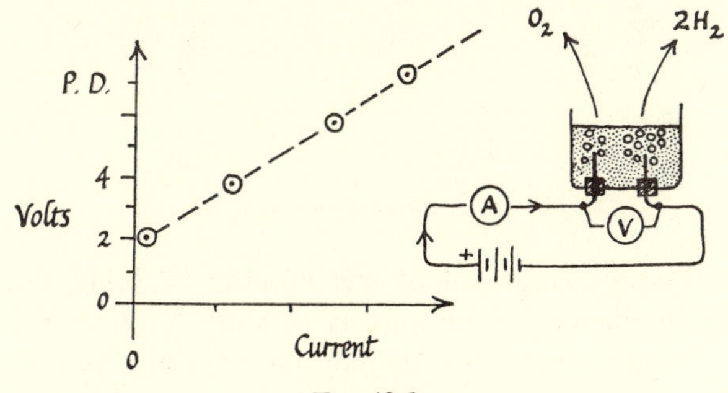

FIG. 43-1.

atom; 15 ev for a hydrogen *molecule*; 25 for a helium atom, and nearly 80 to strip *both* electrons from a helium atom in one shot and make an α-particle. We believe that chemical reactions consist of mild borrowings and exchanges of outermost electrons; so we are glad to find the two lines of evidence agree: outer-electron-changes absorb or release a few ev to a few dozen ev.

We can explore deeper into the electron-cloud of the heavier atoms, which have many electrons: we can shoot in faster electrons and excite X-rays, or we can shoot in X-rays and rip out inner electrons. But even these events involve only ten thousand or so electron · volts (until we try the heaviest atoms of all and run to 100,000 ev).

Chemistry and Radioactivity

So we recognize α-particles and β-rays as missiles from the nucleus. As they leave the nucleus its MASS changes—an α-particle takes away four hydrogen-atom units. A β-particle takes away a trivial amount, which is made good when the new atom captures the extra outer electron it needs. And the nuclear CHARGE changes. An α-particle carries away charge $+2e$, leaving the nuclear charge smaller (the atomic number down) by two units. A β-particle carries away charge $-e$, and the nuclear charge rises by $+e$; its atomic number goes up one.

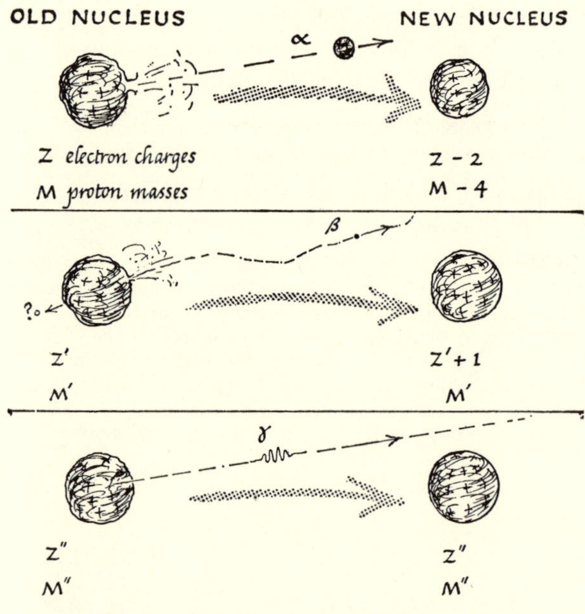

FIG. 43-2. RADIOACTIVE CHANGES

These changes of atomic number Z, bring the same changes in the number of outer electrons for a neutral atom, and thereby changes in chemical properties. Chemical properties are controlled by the outer electrons, but the number and arrangement of those electrons is controlled by the nuclear charge $+Ze$, and we cannot hope to change one chemical element into another unless we can change the nuclear charge. The alchemists' dream of transmuting lead ($Z = 82$) into gold ($Z = 79$) might be realized, if we could take three $+e$ charges from each lead nucleus. Radioactive elements *do* change nuclear charge. Could we *produce* or *influence* those changes? Early experiments said no; and we can see that there was no hope unless bombarding bullets with huge energies were available. Electrons have too small a mass: so they easily swoop around a nucleus like a comet. Alpha-particles carry a ++ charge, so they are repelled. They come *out* from a radioactive nucleus with K.E. of several million electron · volts: they should need as much energy to get back *in*. (A complete, neutral, atom offers a vain hope as a missile: quite early in its approach to a nucleus its electrons would be torn off and its nucleus would be repelled like an α-particle.) However, there were some hopes for fast α-particles bombarding light atoms with small atomic number, low nuclear charge; and these led to the first success in man-made transmutation of elements.

Artificial Disintegration. Man-Made Transmutation

A quarter of a century after the discovery of radioactivity, Rutherford succeeded in "smashing" the nuclei of a few nitrogen atoms by bombarding them with fast α-particles. Alpha-particles from a radioactive source were hurled by their parent nuclei through nitrogen gas. Occasionally some lighter particle was knocked forward, beyond the range of the α-particles. A magnetic field bent the paths of those ejected particles and showed that they were protons,[4] H^+. Though this event was rare, it was photographed. A quarter of a million cloud-chamber tracks were photographed on movie film and showed seven examples of it (see Fig. 43-3a). They showed that a light particle bounced out, almost certainly a proton, and a heavy atom recoiled a short distance; but *the original α-particle was seen no more.* Measurements of angles and track-lengths suggested that momentum was conserved in the collision; but kinetic energy did not tally well.

[4] From now on, we shall always call a hydrogen nucleus a *proton*, the modern name for that fundamental atomic particle. Proton, hydrogen ion, H^+, hydrogen nucleus, ${}_1H^1$, all mean the same thing: a hydrogen atom with its only electron removed.

FIG. 43-3. CLOUD CHAMBER PHOTOGRAPHS

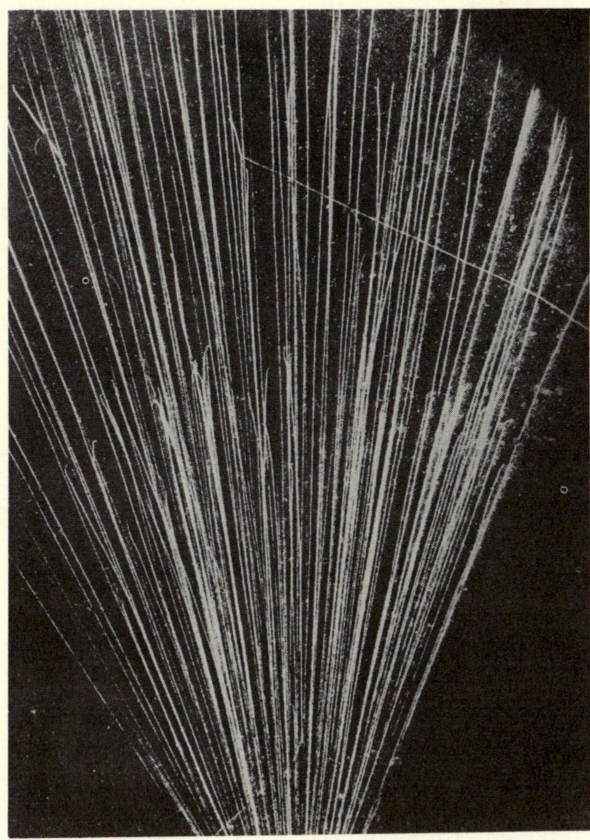

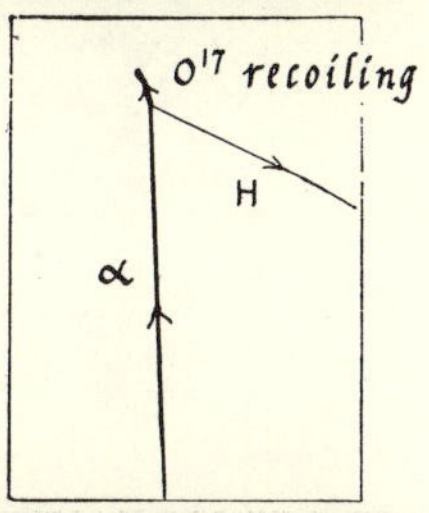

(a) NUCLEAR CHANGE BY BOMBARDMENT:

Blackett's photograph of Rutherford's discovery. Alpha-particle hits nitrogen nucleus and disappears. A proton (hydrogen nucleus) is emitted and the resulting oxygen nucleus recoils. (P. M. S. Blackett, *Proceedings of the Royal Society of London.*)

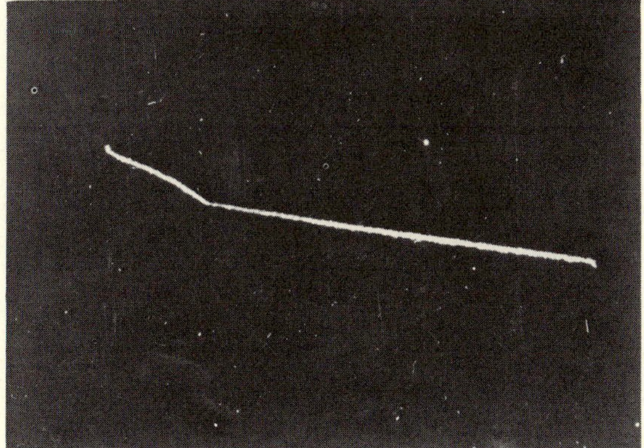

(d) A NEUTRON CAUSES A NUCLEAR DISINTEGRATION. A neutron hit a nitrogen nucleus and was absorbed: the resulting nucleus ejected an alpha-particle, and the remainder recoiled. The original neutron, whose path is invisible, came from a source below the picture—along arrow. (N. Feather)*

* From *Proceedings of the Royal Society of London.*

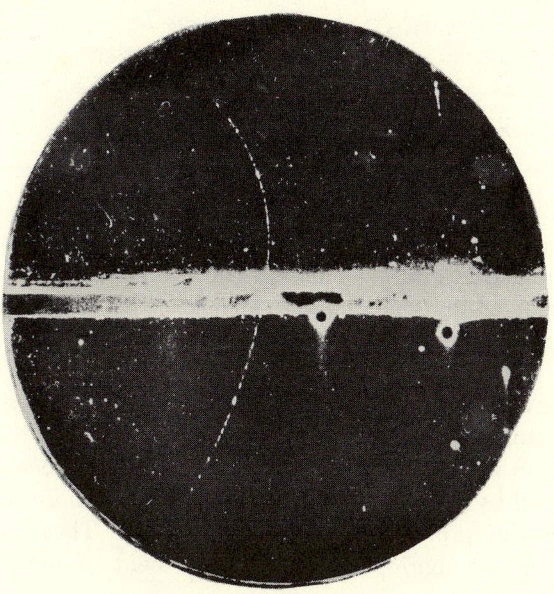

(b) THE POSITRON

Anderson's discovery of the positive electron. The track passes through a lead plate. There is a strong magnetic field perpendicular to the picture, inward. (Photograph by Carl D. Anderson: print from Science Museum, London.)

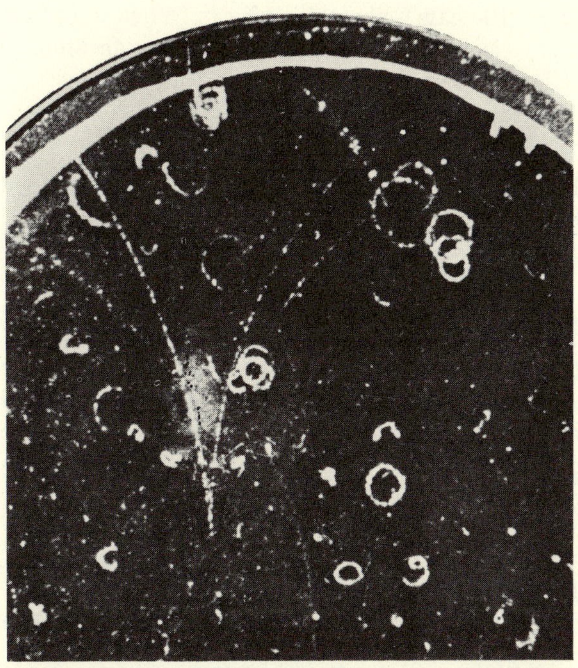

(c) PAIR PRODUCTION. THE CREATION OF MATTER

A negative electron and a positive electron appear, at the expense of a gamma-ray photon, coming from below the picture. There is a strong magnetic field perpendicular to the picture. (W. A. Fowler, E. R. Gaerttner and C. C. Lauritsen.)

We now record this event thus:

α-particle	HITS	EJECTS	LEAVES
(He nucleus)	nucleus of	proton	new nucleus
(charge = +2e)	nitrogen atom	(H nucleus)	of ??????
(mass = 4 times proton mass)	(charge = +7e)	(charge = +e)	(charge = 7e + 2e − 1e)
	(mass = 14)	(mass = 1)	(mass = 14 + 4 − 1)

Then the new nucleus must have charge $+8e$, characteristic of oxygen, and mass 17, unusual but not unheard-of for oxygen. (The mass spectrograph had showed that common oxygen^{16} is accompanied by a few heavier atoms of oxygen^{17}.)

Since Rutherford's announcement in 1919, we have made many such "nuclear reactions," first with natural bombarding projectiles (α-particles) then with still more energetic projectiles, protons accelerated by big machines; and now with still more successful projectiles, chargeless neutrons. These nuclear changes provide a prolific field of nuclear "chemistry."

Nuclear "Chemistry"

We now adapt chemical practice to nuclear "equations." For example, we write a radium nucleus with its chemical symbol Ra, and state its atomic number or nuclear charge +88 electron charges, and its atomic mass, 226 (on the $H \approx 1$, $O = 16$ scale) like this: ${}_{88}Ra^{226}$. When a radium atom ejects an α-particle and becomes an atom of radon gas, we describe the disintegration thus:

$${}_{88}Ra^{226} = {}_{86}Rn^{222} + {}_{2}He^{4}$$

Rutherford's transmutation of nitrogen is described thus:

$${}_{2}He^{4} + {}_{7}N^{14} = {}_{8}O^{17} + {}_{1}H^{1}$$

The first "big machine" was not very big: it accelerated protons till they had energy 150,000 electron · volts, but even these could surge into the atoms of a lithium target and make their nuclei break up. Cloud-chamber pictures confirmed the description: "proton enters lithium atom, two α-particles produced, with huge energy."

$${}_{1}H^{1} + {}_{3}Li^{7} = {}_{2}He^{4} + {}_{2}He^{4}$$

The protons went in with energy about 150,000 electron · volts. Each α-particle emerged with energy 8,500,000 electron · volts, or 17 Mev for the two. So we can say:

$${}_{1}H^{1} + {}_{3}Li^{7} + (0.15)\ \text{Mev} = {}_{2}He^{4} + {}_{2}He^{4} + (17)\ \text{Mev}$$

where Mev means million electron · volts.

The α-particles emerge with far more K.E. than the

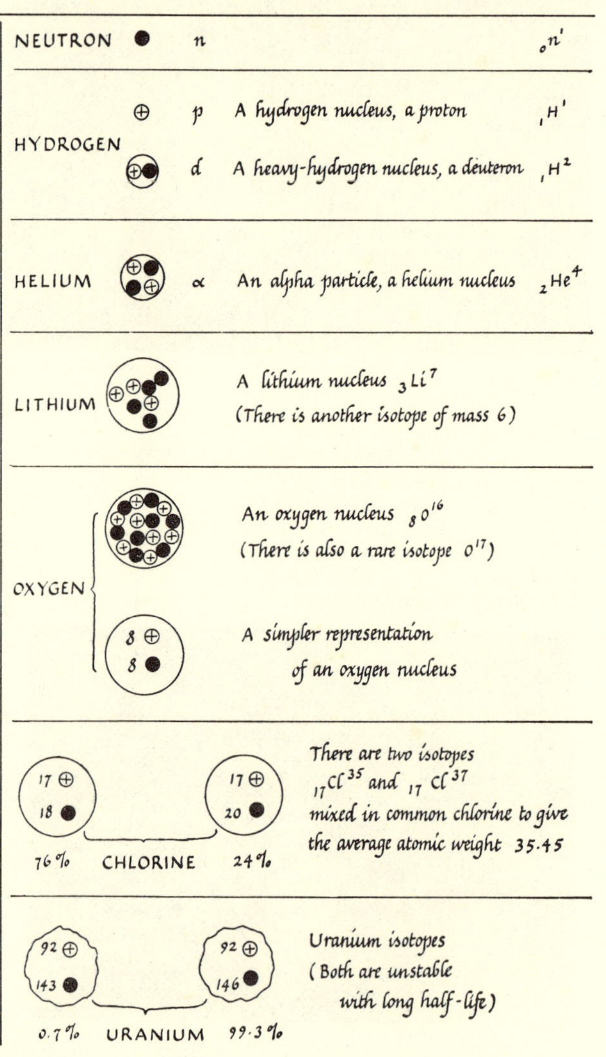

Fig. 43-4a. Schemes for Representing Atomic Nuclei

(Note: the rest of the atom—its electronic cloud—extends for many hundreds of feet from this paper—on this scale). The sketches that follow, using this style for representing nuclear events, are adapted from illustrations in *Classical and Modern Physics* by Harvey E. White, published by D. Van Nostrand Company, Inc., Princeton, 1940.

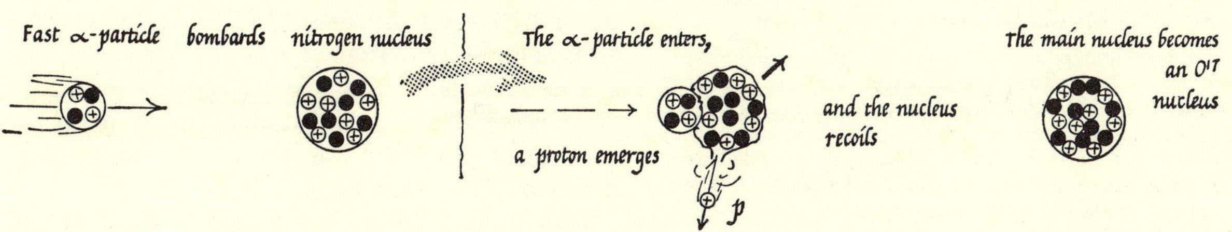

FIG. 43-4d. THE FIRST ARTIFICIAL DISINTEGRATION

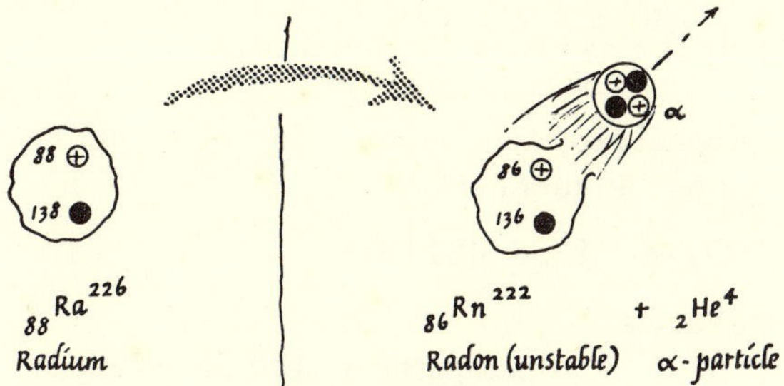

FIG. 43-4b. α-PARTICLE EMISSION

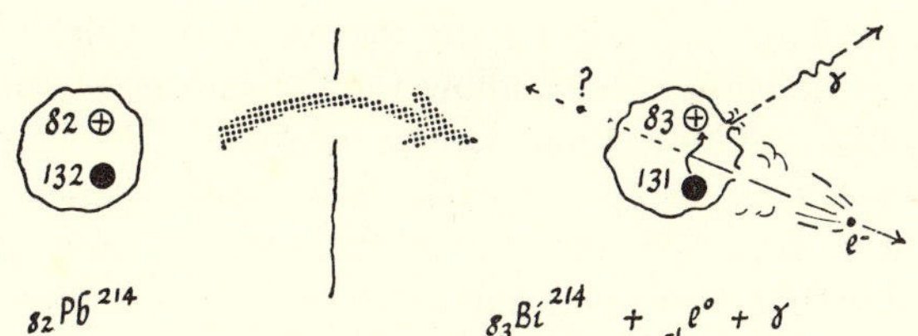

FIG. 43-4c. β-RAY EMISSION

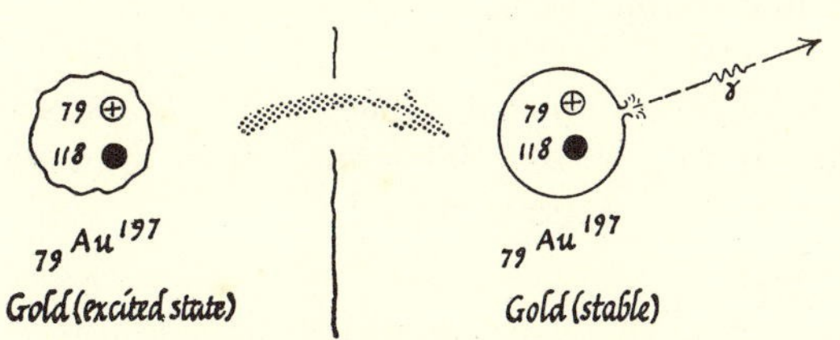

FIG. 43-4f. γ-RAY EMISSION

proton carried in. As they collide with surrounding air they lose this energy in making ions, and occasionally by nuclear collisions; and it all appears ultimately as heat. Compare that heat output, 17,000,000 ev from one lithium atom, with that of burning coal, 4 ev per carbon atom.

Nuclear Energy: No Hope of Practical Use

Here is a huge release of "nuclear energy." Unlike the huge releases from naturally radioactive atoms, it is under our control: we can trigger it by firing protons. Could we use it to run a power plant? No; because, although this is a huge energy-release, we can only obtain one atom's worth at a time. We have to use an accelerator, which is expensive to set up and run. Only one proton in tens of thousands succeeds in splitting a lithium atom; so each successful proton costs far more money than the value of its kinetic energy. The yield of energy is much too expensive, and on far too small a scale, to be profitable and useful. If only the release were *self-maintaining*, with one exploding lithium triggering the fission of neighbors—like the burning of ordinary fuel, where one piece sets light to the next—then we should have a wonderful supply of power. Then a piece of lithium triggered by a proton would flare up in a growing chain-reaction[5] of fissions, releasing energy

[5] As a simple demonstration of a chain reaction, apply a flame to the head of one match at the end of a book of matches. For a *growing* chain reaction, light one match in a whole array or box full of matches.

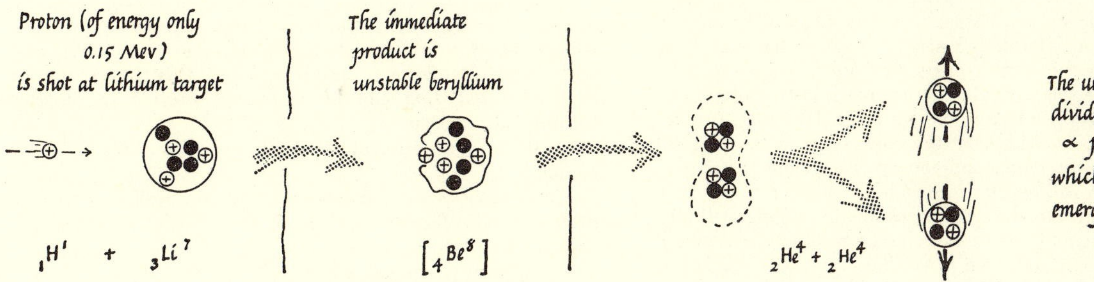

FIG. 43-4e. FISSION OF LITHIUM ATOM BOMBARDED BY PROTON

FIG. 43-3(e). CLOUD CHAMBER PHOTOGRAPH

"Fission" of lithium nuclei into pairs of alpha-particles by bombardment with hydrogen nuclei from an accelerator. In this picture, the bombarding particles were *heavy*-hydrogen nuclei (deuterons, H^2), and the lithium nuclei involved were *light* lithium, Li^6. Several pairs of alpha-particle tracks can be seen going out to the edge of the cloud chamber. (The alpha-particles come out with energy 11. Mev each, in this case.) There are also shorter tracks of alpha-particles from another nuclear reaction in which a neutron is released as well. (P. I. Dee and E. T. S. Walton, *Proceedings of The Royal Society of London.*)

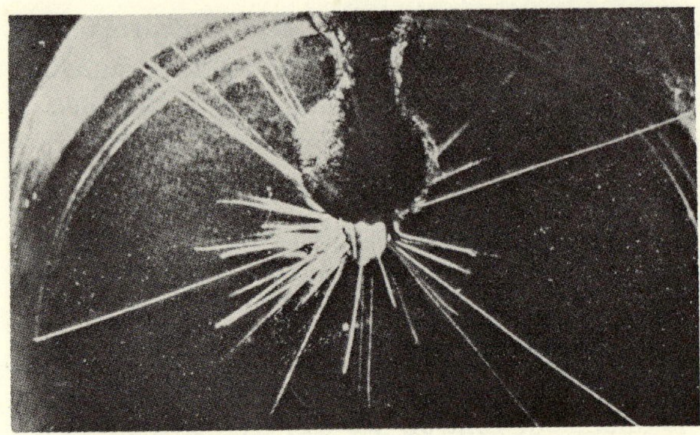

on a scale undreamed of by any fuel merchant. But one atom of lithium cannot trigger a neighbor: the splitting does not provide a new proton to do that.

Nor could the many other nuclear changes produced by the bigger accelerators offer any hope of practical power, for the same reasons. However, they did produce extremely valuable *information*—the internal structure of nuclei began to show.

Measurements of these reactions showed that mass of matter is not conserved exactly. If we reckon the masses of *matter*, adding first the masses of the target nuclei and bombarding particle then the emerging particles, the total mass afterwards is not quite equal to the total mass before. But, if we assign mass to *energy*, as well as matter, the accounting is perfect: the total [MASS OF MATTER + MASS OF ENERGY] is the same before and after every nuclear event.

Mass and Energy. $E = mc^2$

All through this century there has been a suspicion that energy has mass.[6] In Chapter 26 we gave a plausible story to suggest that any ENERGY, E, has a MASS of amount E/c^2. We are now convinced by many experimental tests that all forms of energy *do* have mass E/c^2 and therefore that the relation $E = mc^2$ is universally true.

The fission of lithium atoms gave an excellent test of this view that energy has mass. The masses of the particles involved in that event had been measured very carefully with a mass spectrograph. On a standard scale that gives the oxygen^{16} *atom* a mass 16.0000, the particles had the following masses:

original projectile: proton	$_1H^1$ $M = 1.0076$	
target: lithium nucleus	$_3Li^7$ $M = 7.0165$	
TOTAL MASS BEFORE		8.0241
products:		
alpha-particle	$_2He^4$ $M = 4.0028$	
alpha-particle	$_2He^4$ $M = 4.0028$	
TOTAL MASS AFTER		8.0056

The total mass after the event is less than the mass before if we just add up the masses of particles as pieces of matter. Now allow for the masses of the ingoing and outcoming kinetic energies. We shall say MASS = K.E.$/c^2$ for each such mass. Then we must reduce the answers to the special scale of masses that we use for chemistry and nuclear physics, which calls the mass of an O^{16} atom 16.0000 or a hydrogen atom 1.0081. To make the change of units, calculate the amount of energy that has the same mass as one proton and express it in joules, then in electron · volts.

Finding the Conversion Factor. The mass of one proton is 1.67×10^{-27} kilogram.[7] Then the ENERGY which has that MASS is given by:

$$\begin{aligned} E &= Mc^2 \\ &= (1.67 \times 10^{-27}\ \text{kg})(3.00 \times 10^8\ \text{meters/sec})^2 \\ &= 1.50 \times 10^{-10}\ \text{kg} \cdot \text{m.}^2/\text{sec}^2 \end{aligned}$$

or (kg · meter/sec²) · meters
or newton · meters or joules.

[6] Some physicists say "Mass is Energy" and "Energy is Mass," and even "Matter is Energy." Such statements seem to confuse properties with owners, and are not likely to help you to be a wise scientist. Energy has mass, just as aluminum has mass, and copper has an orange color. However, professional scientists, knowing that the conversion factor, c^2 (between mass-measurements and energy-measurements in usual units) is a universal constant, consider it sensible to say that mass and energy *are* different measures of the same thing. Yet, matter appears to us as particles (e.g., atoms, electrons) with definite mass; and even if we like to imagine this is the mass of some energy locked up inside the particle, we have not abolished the reality of the particle and reduced its matter to energy by taking that view. We sometimes find matter (two particles) disappearing, and something else with mass (radiation) appearing—and this conversion of matter to *radiation* can be carelessly worded as "matter turning into *energy*."

[7] Millikan's experiment gives charge $e = 1.60 \times 10^{-19}$ coulombs.
Electrolysis of water (or measurements on positive ions in a discharge tube) give for e/M 9.57×10^7 coulombs/kg.*
Arithmetic gives

$$M = \frac{1.60 \times 10^{-19}\ \text{coulomb}}{9.57 \times 10^7\ \text{coulomb/kg}} = 1.67 \times 10^{-27}\ \text{kg for a proton.}$$

Any form of energy, K.E., P.E., etc., of amount 1.50×10^{-10} joules will have a mass 1.67×10^{-27} kilograms, the mass of a proton. But we usually express the energies of a nuclear event in ev or Mev. Remember that one electron · volt is the energy gained by one electron charge, 1.60×10^{-19} coulomb, falling through a P.D. of one volt.

$\therefore$ 1 ev $= (1.60 \times 10^{-19}$ coulomb) (1 joule/coulomb)
$= 1.60 \times 10^{-19}$ joules.

$\therefore$ ENERGY which has same mass as one proton, 1.67×10^{-27} kg, is:

$$\frac{1.50 \times 10^{-10} \text{ joules}}{1.60 \times 10^{-19} \text{ joules/electron} \cdot \text{volt}}$$

or $(1.50/1.60)\ 10^{9}$ electron · volts,
or 0.94×10^{9} ev or 940 Mev.

Computation from precise data gives the value: *938 Mev for the energy that has the same mass as one proton.*

[*Better Method.* The method we used above is the simplest to understand but not the neatest. (We used the electron charge twice over, and concealed the fact that it cancels out.) The better method runs thus:

The energy that has mass M kilograms is Mc^2 or $M\ (3.0 \times 10^8)^2$ joules.

1 electron · volt is
(electron charge, e coulombs) (1 joule/coulomb) or e joules

$\therefore$ ENERGY that has MASS M kilograms

$$= \frac{M\ (3.0 \times 10^8)^2}{e} \text{ electron} \cdot \text{volts}$$

$$\text{or} \quad \frac{(3.0 \times 10^8)^2}{e/M} \text{ ev}$$

If M is the mass of a proton and e the electron charge, e/M is the CHARGE/MASS ratio for hydrogen ions; and this is 95,700,000 coulombs/kilogram,* from simple measurements of electrolysis of water,

$\therefore$ ENERGY that has same mass as proton

$$= \frac{(3.0 \times 10^8)^2}{95{,}700{,}000} \text{ electron} \cdot \text{volts} = \frac{9.0 \times 10^{16}}{9.57 \times 10^7} \text{ ev}$$

$= 0.94 \times 10^9$ ev, or 940 Mev.]

Atomic Mass Units and Energy. Measurements of (relative) masses of molecules, atoms and nuclei are now expressed on a scale of "Atomic Mass Units," (a.m.u.), that gives an atom of O^{16} a mass 16.0000. On this scale, a hydrogen *atom* has mass 1.0081; and a proton or hydrogen nucleus has mass $1.0081 - 1/1840$ or 1.0076. Precise measurements with mass spectrographs give the following masses, on this scale:

	Mass of Atom	*Mass of Nucleus*
hydrogen	1.0081	1.0076
helium	4.0039	4.0028
lithium^7	7.0182	7.0165
oxygen^{16}	16.0000 by definition	15.9957

On this scale, the ENERGY that has MASS 1 unit is a little less than the 938 Mev for one-proton-mass. It is $938 \times (1.0000/1.0076)$ or 931 Mev.

This is a very important conversion factor in dealing with release of atomic energy:

ENERGY 931 Mev HAS MASS
1 ATOMIC MASS UNIT,
(on scale that gives an O^{16} atom a mass 16.0000)

Test of $E = mc^2$ with Lithium Fission. Now we can write down the masses of kinetic energies in a nuclear event, taking 931 million electron · volts of energy to have the mass of one atomic mass unit. Try this on the bombardment of lithium:

EVENT:

$${}_1H^1 + {}_3Li^7 + \underset{\text{K.E. of proton}}{0.15 \text{ Mev}} \;=?=\; {}_2He^4 + {}_2He^4 + \underset{\text{K.E. of } \alpha + \alpha}{17 \text{ Mev}}$$

$$\text{MASSES: } 1.0076 + 7.0165 + \frac{0.15}{931} \begin{pmatrix}\text{mass of} \\ \text{K.E. in} \\ \text{same units}\end{pmatrix}$$

$$=?= 4.0028 + 4.0028 + \frac{17.0}{931} \begin{pmatrix}\text{mass of} \\ \text{K.E. in} \\ \text{same units}\end{pmatrix}$$

$$1.0076 + 7.0165 + 0.0002$$

$$=?= 4.0028 + 4.0028 + 0.0183$$

TOTAL MASSES: 8.0243 =?= 8.0239

This time the totals are much nearer. The total mass of *matter* changed from 8.0241 to 8.0056, *a loss of 0.0185.* The mass of *kinetic energy* changed from 0.0002 to 0.0183, *a gain of 0.0181.* This gain accounts for 98% of the loss. The 2% difference is well within the uncertainty of the difficult measurements of K.E. For the balance to be perfect the α-particles' K.E. should have mass 0.0187: their measured energy should be 17.4 instead of 17 Mev.

Notice how we needed very precise measurements of atomic masses for this test of $E = mc^2$. The mass-spectrograph has gone far beyond its early successes—which were: its proof that there are isotopes

* The usual number 96,500,000 is the charge carried across by a *kilomole*:
96,500,000 coulombs/1.008 kg. = 95,700,000 coulomb/kg.

with whole-number masses and its skill as a chemical analyst. Now it shows with precise measurements that atomic masses are not exactly whole-number multiples of a hydrogen atom's mass, or of any other basic unit. There are small differences, which have great meaning if we trust $E = mc^2$, and we do trust it. We have many detailed checks like the early one on lithium, and all support it. So do the strange events that involve "positrons"—electrons with a positive charge—discussed in a later section. With the backing of Relativity theory and experimental checks, we trust $E = mc^2$ to predict energy changes in other nuclear events. We can even predict the huge release of energy when a massive nucleus divides into two smaller ones—fission—or when light nuclei join together to make one big nucleus—fusion.

Structure of the Nucleus. The Neutron

Before studying the behavior of neutrons, you should either review Problems 20 and 21 in Ch. 8 or try the easier Problem 1 below.

PROBLEM 1. LOSS OF KINETIC ENERGY BY ELASTIC COLLISIONS

Draw upon your common-sense knowledge to answer the following:

(a) A ping-pong ball moving 20 meters/sec due North hits a stationary elephant head-on. Suppose the collision is completely elastic—no kinetic energy is converted into heat—and the elephant has frictionless feet.
 (i) After impact, will the ball move North or South?
 (ii) Will the elephant move N or S?

b) Suppose the ball is a 2-gram ball (0.002 kg) and the elephant is a 2000 kg elephant. Make *rough* estimates (within, say, 1 %) of the following:
 (i) The ball's change of momentum.
 (ii) The elephant's change of momentum. (Assume conservation of momentum.)
 (iii) The elephant's speed after impact.
 (iv) The elephant's K.E. after impact.
 (v) The ball's K.E. after impact.

(c) Does the ball keep most of its K.E., give most of it away, or share equally on impact with the elephant?

(d) Now reverse the proportions of masses: suppose the elephant is sliding very fast due North and hits a ping-pong ball suspended in the air, waiting to be hit.
 (i) How much would you expect the elephant to change its motion on impact?
 (ii) What velocity would you expect the ball to have after impact? (To solve this without algebra, imagine you are riding on the elephant, *in a fog*, unaware of the elephant's motion. You think you are at rest, and watch the ball's motion relative to you and the elephant. What motion would you think the ball had before impact? What motion would you see the ball move with after impact? Now remember the elephant is moving, and say what motion a stationary observer on the ground would see the ball have after impact.)
 (iii) Does the elephant keep most of its K.E., give most of it away, or share it equally, on impact?

(e) Now suppose the two masses are equal: a perfectly elastic ball moving fast due North hits an equal ball, at rest, head-on.
 (i) What motion would you expect after impact? (If in doubt try a rough experiment and extrapolate to the ideal case.)
 (ii) Does the moving ball keep most of its K.E. or give most away or share equally, on impact?

(f) MORAL: If we wish to slow down a moving body by elastic collisions with a stationary body, the masses should be in the proportion ? : ?

In 1930, when the first big accelerators were being built, atomic speculations were waiting on experiment. The Rutherford atom picture had been modified—the outer electrons were weaving a wavy cloud, instead of pursuing sharp planetary orbits; but it was still a hollow atom with a tiny positive nucleus. Since electrons, protons, and α-particles come out of some nuclei, every nucleus was thought to be made up of protons and electrons (with an α-particle regarded as a compact subassembly). But this picture of the nucleus was an uneasy one. Such a nucleus seemed more likely to collapse than to remain stable; the electrons would not fit inside—they are so light that their wavelength is too big—and there were more serious worries about conserving rotational momentum. Then neutrons were discovered—and changed the picture.

Certain bombardment experiments were producing curious results. Alpha-particles could knock protons (occasionally) out of the nucleus of nitrogen and some other light elements. Beryllium, a very light metal with $Z = 4$, was a promising target. But when beryllium was bombarded with α-particles it produced some strange rays that made no ions, no track in a cloud chamber—gamma-rays, perhaps. However, thick sheets of lead did not stop the new rays much—perhaps they were very energetic, penetrating gamma-rays; but that raised a serious question: "where does the ray's energy come from?" In contrast, a block of paraffin wax or a tank of water stopped many of the mysterious rays—in conflict with the evidence from lead. And the rays drove protons forward from the absorbing wax or water—a serious conflict with energy and momentum considerations for gamma-rays. Then Rutherford's colleague, James Chadwick, wrote a short letter to the journal *Nature*, soon to be a very famous letter. He suggested that these conflicts would be resolved if the new "rays" were not gamma-rays but neutral, uncharged, particles with mass almost the same as a proton's. These would not be neutral atoms (positive nuclei accompanied by outer electrons) but, so to speak, "neutral nuclei." Such *neutrons* with no

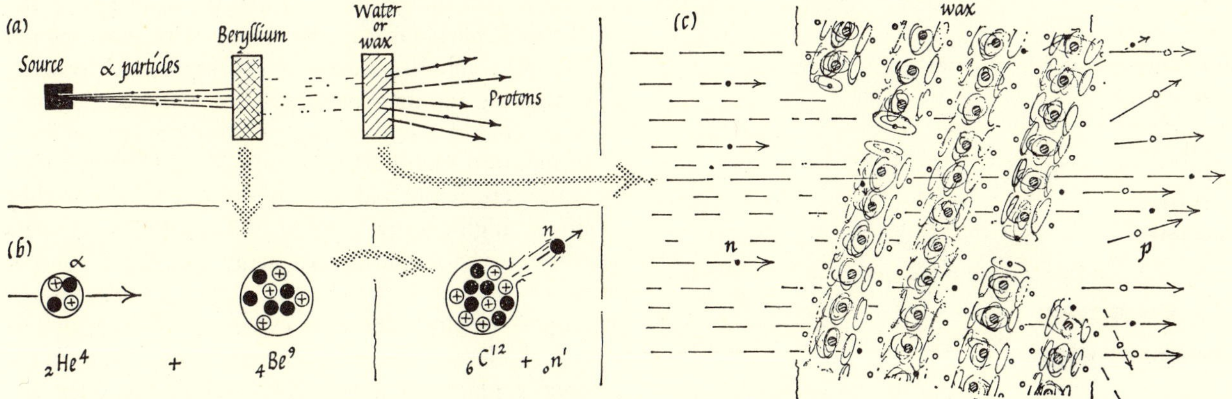

Fig. 43-5. Neutron Production

electric field would make no ions, suffer no violent electrical repulsions from nuclei; so they would pass through matter easily, losing no energy, making no tracks. If one did hit a heavy nucleus in some strange "head-on" collision, it would bounce off with little loss of energy. But if one ever hit a light nucleus, best of all a proton, it would knock that nucleus forward, giving it a considerable share of energy. Neutrons *do* bounce off lead, *do* knock protons forward.

Soon the neutron's mass was measured, by simple momentum-calculations from collisions, 1.0089 as against 1.0076 for a proton. Its bombarding uses were explored far and wide and it became a very important atomic particle. When energetic neutrons bombard a target they are not repelled by the target nuclei, so they usually go straight on, even when they pass very close to those nuclei. Occasionally, one enters a target nucleus—having sneaked up to it unseen, unfelt, it is suddenly clutched by some strange nuclear force. Then we have a new nucleus. Adding a neutron does not change the nuclear charge, so the new nucleus is an isotope of the old one, but one unit heavier. It may be an unknown isotope, and often an unstable radioactive one at that. Here, then, is an easy way of making new radioactive atoms. We can now produce isotopes of almost any element we like by bombardment with neutrons. At first, the neutrons were obtained from a radium-beryllium source. Then cyclotrons began to knock neutrons out of many a nucleus by means of fast protons; and these neutrons proved useful servants. Nowadays the huge nuclear reactors contain a swarm of neutrons, fast, medium, and slow; so neutron bombardment is easily done by hanging a specimen inside a reactor. For example, radioactive phosphorus, P^{32}, is made by neutron bombardment of ordinary phosphorus, P^{31}, in a reactor.

Nuclear Composition

Now return to the structure of nuclei. With neutrons available as ingredients for nuclei, a much more satisfactory structure was imagined: protons and neutrons in every nucleus, and no electrons. The atomic number, Z, gives the number of protons, and the rest of the nuclear mass ("atomic weight" — Z) is neutrons. For example, a helium nucleus (α-particle, with charge $+2e$ and mass 4) was no longer pictured as an uneasy association of 4 protons with 2 electrons to make the charge right; now it became a tight clump of 2 protons and 2 neutrons. Special forces have to be imagined to hold these together, but the alliance seems more thinkable and it conserves spin momentum. So we now picture every nucleus as made of protons, to provide its charge and some of its mass, combined with enough neutrons to provide the rest of its mass. The nuclei have about equal numbers of neutrons and protons in light elements, up to 50% excess neutrons in the heaviest elements. (See next diagram for examples.)

This picture is confirmed by experiments with very-high-energy protons from big accelerators. Whatever target atoms they hit, one of them often knocks out a neutron and about equally often one knocks out a proton. (Project a pool ball across a table covered with an even mixture of black balls and white balls, well separated; it may make no hit. But if it *does* hit, it is about equally likely to knock a black ball forward as a white one.) That seems to show protons and neutrons as the standard ingredients of nuclei.[8] How these ingredients are

[8] This is also described by a more ingenious picture: the moving particle (proton or neutron) goes straight through the target-nucleus without faltering, except that it may exchange a $+e$ charge with a particle in the target nucleus (so that it emerges as a neutron or a proton). Such "charge exchange" still makes us picture nuclei as made up of protons and neutrons.

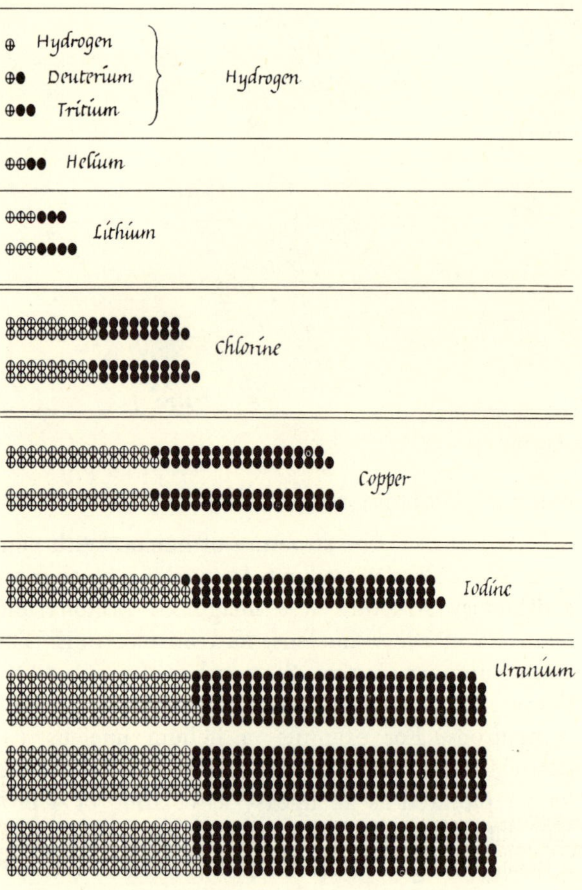

Fig. 43-6. Ingredients of Nuclei

arranged and how they interact, inside the nucleus, are questions that we cannot yet answer fully. In other words, we have not arrived at a completely satisfactory "model" for nuclear structure—although several rival ones are promising—and we cannot describe nuclear forces completely.

Positive Electrons, "Positrons"

In 1932 a remarkable cloud chamber picture showed that a positive electron can exist. Carl Anderson was recording the tracks of very-high-energy particles that come flying in from outer space (and straight on through us). This all-pervading stream of *cosmic rays* reaches us as a mixture of protons, electrons, neutrons, and other particles, charged and uncharged. The primary particles are protons and other atomic nuclei, with energies of millions of Mev, and even billions of Mev. Though many of these hurtle unscathed through our atmosphere, some are stopped in a confusion of collisions which give rise to "secondary cosmic rays": gamma-rays, showers of electrons, protons, neutrons, and mesons. The Earth's magnetic field bends the tracks of the charged particles so that those with low energy cannot reach the Earth at the equator. Physicists examining the mixture of primaries and secondaries with a cloud chamber apply a strong magnetic field, to measure each particle's momentum. Heavy metal walls screen out particles of local origin, but they allow high-energy cosmic ray particles to pass through the chamber. An array of Geiger counter tubes can be arranged as a "sighting telescope" to trigger the cloud chamber expansion when a particle passes through in a chosen direction. Very-high-energy electrons pass right through the chamber, walls and all, making a track that can be photographed by having water droplets condense on the ions in the chamber. The ions are rather sparse when made by so speedy a particle and the track looks "thin," but it is easily photographed. Anderson photographed a track that looked like the track of a fast electron. (Bending by a magnetic field that ran through the chamber gave the particle's momentum. The long track then ruled out a more massive particle such as a proton.) Anderson decided he had found a *positive* electron.

PROBLEM 2. PROVING A POSITIVE ELECTRON (See Anderson's photograph, Fig. 43-3b)

(a) If Anderson knew from the look of the track that it was made by an electron, what could he find by measuring the *curvature* of the track at various places?

(b) If the track curved the opposite way from most such tracks in similar pictures, then *either* it was made by a positive electron instead of a negative one like the rest, or it was made by a negative electron which was . . . ?

(c) To decide between the two guesses in (b), Anderson had a metal plate across the middle of the chamber, hoping to photograph a track that passed through the plate. See his successful photograph. How can this picture prove that the track belongs to a positive electron?

Positive electrons, now called *positrons*, had almost been predicted a few years earlier by P.A.M. Dirac, in a flight of speculative theory. So when Anderson made his discovery, theory was ready to interpret it. For the sake of mathematical completeness and symmetry Dirac imagined there are electrons of *negative kinetic energy* swimming about in great profusion, unreal if you like, and certainly undetectable. (How can ½ · MASS · VELOCITY^2 be negative?) However, if enough energy ($2mc^2$) could be given to one of these mythical "negative-energy" electrons, $^-(e^-)$, to raise it to positive energy it would appear as an ordinary electron e^-, leaving a "hole" in the ocean of mythical $^-(e^-)$ electrons. Then that "hole" would appear as a real positive electron, e^+. Thus, theory predicted a positive electron as a sort of mirror image of an ordinary electron. In our new vocabulary of "anti-particles" a positron is an anti-electron.

Once a real e^+ had been seen in a cloud chamber, the predicted event of a pair, e^+ and e^-, appearing "from nowhere" was looked for, and seen. A high-energy γ-ray hitting a massive atom can yield its energy to produce a pair of electrons, e^+ and e^-. This is the *creation of matter*, material particles from radiation (Fig. 43-8b).

PROBLEM 3. ENERGY FOR PAIR CREATION

A positron and an electron have the same mass (at rest), each about 1/1840 of a proton's mass (which is the mass of energy 940 Mev).

(a) How much energy must a single gamma-ray have (at least), to make a pair, e^+ and e^-?
(b) How much K.E. must an electron have, in ev, to make its *total* mass, to a stationary observer, twice its rest-mass?
(c) How massive would a beta-ray with kinetic energy 2 Mev seem to be, compared with an electron at low speed?
(d) How much energy, in ev, must a gamma-ray have to be able to manufacture a pair, e^+ and e^-, and hurl them forward *each* with K.E. 2 Mev?

In fact, a γ-ray with just the minimum energy would give the pair no motion to show in a cloud chamber. A γ-ray with more energy (bigger quantum, shorter wavelength) can give its spare energy to K.E., and hurl them forward. Then the pair can be seen as a V in a cloud-chamber with a magnetic field. (See Fig. 43-3c.) The γ-ray needs a massive atomic nucleus as an "anvil" on which to forge out the pair—an impedance-matching requirement for energy- and momentum-conservation. In a cloud chamber, pair production happens at the nucleus of some atom of gas, or of a metal sheet placed there to offer opportunities.

Positrons, e^+ (or β^+), became common commodities when it was found that many man-made radioactive atoms emit them. An ordinary atom that has had an extra proton pushed into its nucleus by a cyclotron is often unstable. The new nucleus feels over-rich in protons (= over-poor in neutrons) for a nucleus of that size. Therefore it has a good chance of making the change:

[A] a proton → neutron + electron; and the electron must carry away a + charge: it must be a β^+, to maintain the universal conservation of electric charge.[9] In other radioactive atoms the nucleus may be over-rich in neutrons, and then the instability leads to:

[B] a neutron → proton + electron with the electron carrying away a negative charge, β^- (see footnote 10).

Thus, it seems unwise to say that a neutron is "composed of" a proton and an electron in close alliance. Instead, we talk more generally, thus:

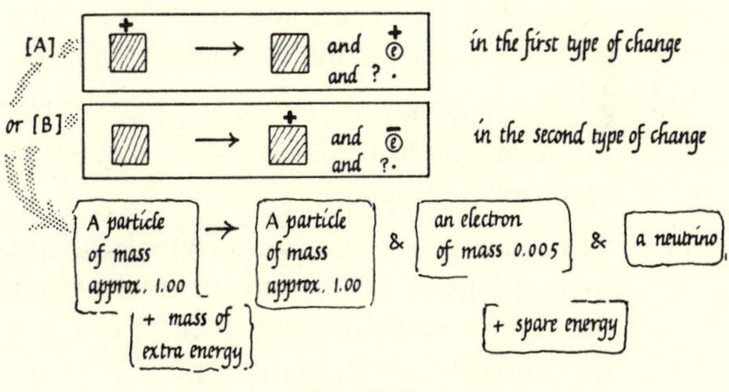

Fig. 43-7.

The statements [A] and [B] above are mere shadows of the more complex events that we now believe occur among particles and charges. Change [B] also happens spontaneously for neutrons *free in space*: they decay radioactively with a half-life about 12 minutes, into proton and electron (H^+ and e^-), *and a neutrino*. The neutrino must emerge also to carry away spin and spare energy. The change [A] cannot occur without a supply of extra energy. And we are now sure that a neutrino is involved as well. However, in *all* changes, two rules hold:

(i) Charge is conserved: when new charges appear, they are equal and opposite.
(ii) Particles are conserved, provided we count a particle and its antiparticle (e.g., e^+ and e^-) as "equal and opposite," and cancel them in the bookkeeping.[11]

The general moral is "use the new conversion rules that we are now discovering for nuclear events, but not the old idea of one particle being a package

[9] Example: copper ^{62}, made by knocking a neutron out of common copper ^{63} with a fast proton from a cyclotron.

$${}_{29}Cu^{62} \rightarrow {}_{28}Ni^{62} + {}_{+1}e^0$$

[10] Examples: radium B (isotope of lead) emits a β-ray and becomes an isotope of bismuth

$${}_{82}Pb^{214} \rightarrow {}_{83}Bi^{214} + {}_{-1}e^0$$

Common phosphorus, P^{31}, can collect an extra neutron from the swarm inside a nuclear reactor and become radioactive P^{32}, which emits β^-

$${}_{15}P^{32} \rightarrow {}_{16}S^{32} + {}_{-1}e^0$$

with a half-life of 14 days.

[11] A further rule seems necessary: the number of "heavy" particles (protons, etc.) is conserved, separately from light ones (electrons, etc.). In a very-high-energy collision, a nucleus may be shattered so that a spray of ingredient particles splashes out, some heavy and some light, but the score of heavy particles still tallies; we never see one of them turning completely into light particles.

composed of some others." We extend the kitchen idea of *ingredients* of a cake to molecules in chemistry—we say, for example, that a water molecule is *composed* of two H atoms and one O—but we may falsify our thinking if we try to extend it down to the sub-atomic scale and say, "neutron is *composed* of a proton and an electron." There, habit has carried us too far beyond our kitchen analog. We should treat such statements as warily as the child who has bitten a worm in an apple and is told, "The worm is just made of apple."

Yet we do have some experimental knowledge of neutron structure:

(i) Its mass is 0.001 a.m.u. more than the mass of a proton; suggesting an extra store of about 1 Mev of energy—so the release of energy in the neutron→proton change is not unexpected. However, that change is not a simple coming-to-pieces of an unstable group, because

(ii) although a free neutron is unstable, a neutron in an atomic nucleus lasts indefinitely.

(iii) Although it has no net charge, a neutron has a magnetic field around it, suggesting some moving charge(s) inside.

(iv) Electrons have been used to bombard neutrons (captive inside atomic nuclei), and the results seem to show that there is a magnetic field inside it, and yet no sign of any charges.

The suggestion is that a neutron *does* have a structure, possibly of a negative meson revolving around a proton—but that is risky picturing which, if taken literally, conflicts with some of our experimental knowledge.

Annihilation of Matter

The inverse of the pair production event can occur easily. A positron meets an ordinary negative electron and they disappear into γ-rays.

$$e^+ + e^- \rightarrow \gamma + \gamma$$

To conserve energy and momentum, the product must be *two* γ-rays, moving out in opposite directions. These are observed, when a radioactive sample emitting β^+ is placed between two suitable counter tubes. The counters then show *simultaneous pairs* of γ-rays of just the expected energy, 0.5 Mev each. (The energy can be measured by the ions made when the γ-rays eject electrons in ionization chambers.)

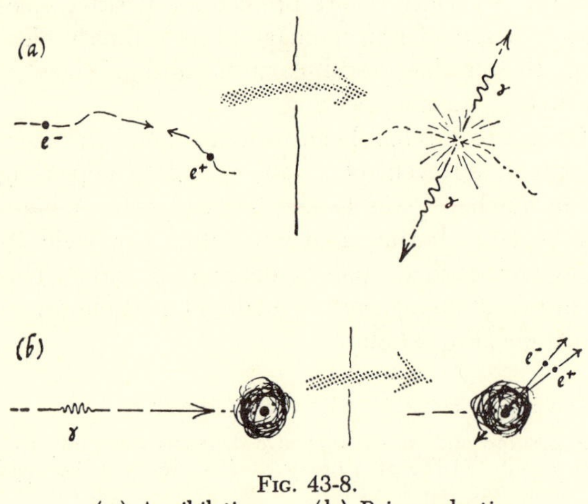

FIG. 43-8.
(a) Annihilation. (b) Pair production.

PROBLEM 4. ANNIHILATION OF ELECTRONS

Show that a pair of electrons, converted into a pair of γ-rays, without loss of mass, would give the γ-rays an energy of about 0.5 Mev each.

Romantic Thinking

Thus $E = mc^2$ is not only verified for speedy particles but even applies to the creation and annihilation of matter. If critics denied that, by saying "electrons are not solid matter," they had to wait from 1930 to 1955 to see the manufacture of hydrogen nuclei—protons and antiprotons. The biggest accelerators today can provide particles with energy several billion electron · volts to forge out such pairs from their kinetic energy.

Now we know a series of "mirror image" pairs: electrons and positrons, protons and antiprotons, and neutrons and antineutrons. The last pair cannot have opposite charges, since they are uncharged, but they differ in magnetic and spin-momentum properties. On meeting, particle and antiparticle annihilate violently. Fanciful thinkers can speculate on "anti matter" and "anti worlds." Scientists may laugh at such fictions; yet they should not sneer. Dirac's vision of positive electrons seemed preposterous at first.

New Radioactive Nuclei

Nowadays we can easily manufacture unstable nuclei—artificially radioactive ones—by pushing in an extra neutron. The nucleus finds itself with the same charge (*same element, same chemistry*), but *too much mass*, too many neutrons for its protons. This discomfort suggests the change mentioned above: change one neutron in the nucleus to a proton. That happens when the unstable nucleus disintegrates. Equal + and — charges are manufactured, and the — charge is carried away by a small particle: an electron is born and hurled out as a beta-ray. (A neutrino must emerge, unseen, as well.)

So, from 1930 to 1940 a rich stream of nuclear physics came from bombardment experiments. The chemical list of elements splurged: the square hutch containing each element grew into a whole row of kennels housing different isotopes, some stable, some unstable, all of them atoms of the same element, but with different masses. New information was posted for each unstable isotope: its half-life, the particle it radiates and the energy released when it disintegrates.

An example of bombardment is shown in Fig. 43-10. When protons with energy 20 Mev from a small cyclotron bombard a copper target, they are brought to rest in a millimeter of travel in copper by a multitude of minor collisions; and their K.E. is converted to heat in nearly every case. However, just a few in passing close to a copper nucleus are clutched by strong nuclear forces and enter the nucleus. Then, in some cases, a neutron emerges instead, leaving a radioactive *zinc* nucleus; in other cases the entering proton knocks out both a neutron and a proton, leaving a *copper* nucleus, but an abnormally light one (Cu^{62}) which is radioactive.

Another example is shown in Fig. 43-11. Heavy hydrogen is fed into a cyclotron and its nuclei (deuterons) accelerated and used to bombard a target of ordinary sodium. In some of the target atoms a deuteron goes into the nucleus and knocks

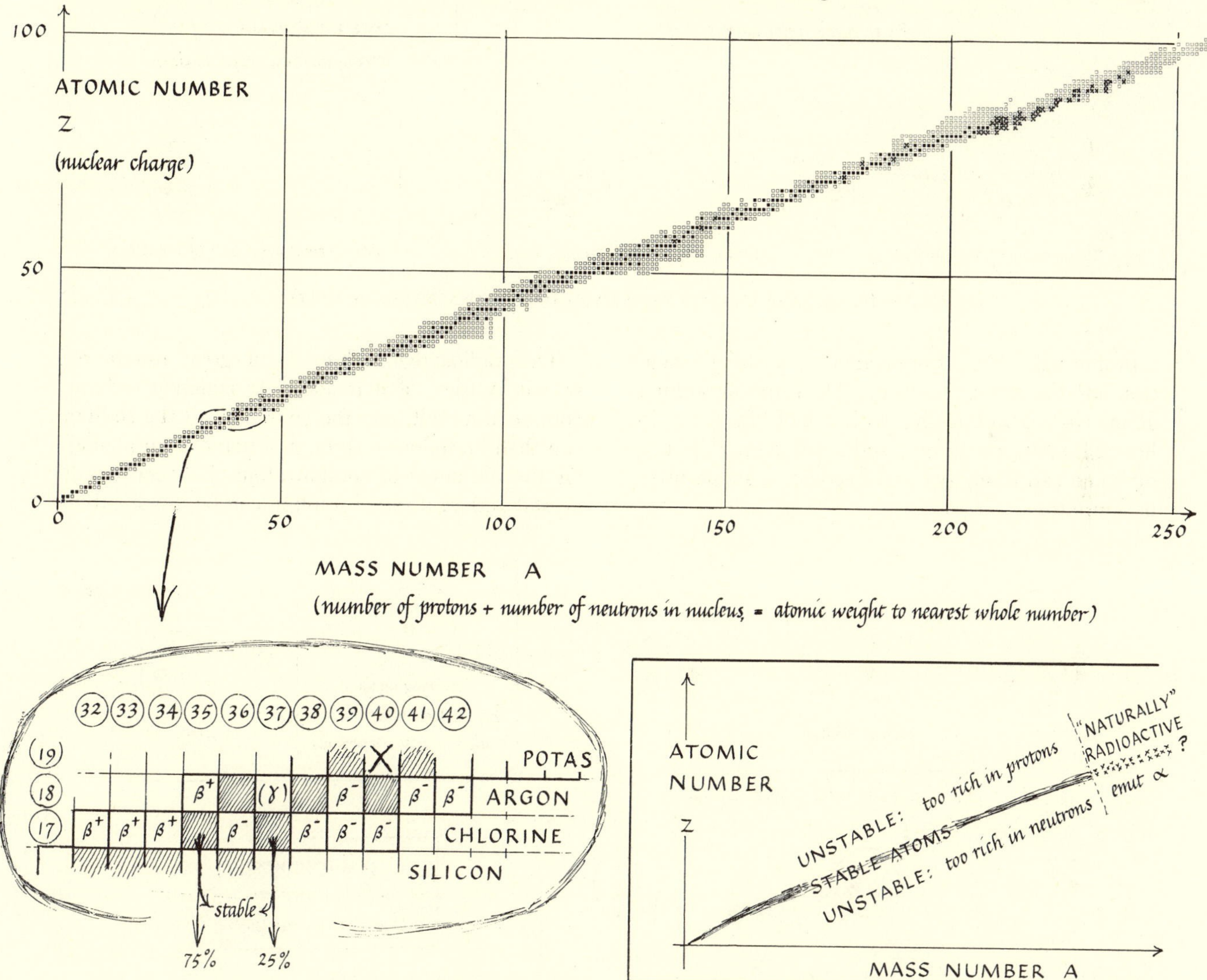

Fig. 43-9. A Listing of Isotopes
Solid black squares show stable isotopes. Hollow squares show "artificial," unstable, radioactive isotopes, manufactured by bombardment, etc. Squares with X in them show "natural" radioactive isotopes, found in nature (most of them in the highest atomic numbers—but common potassium includes one radioactive isotope).

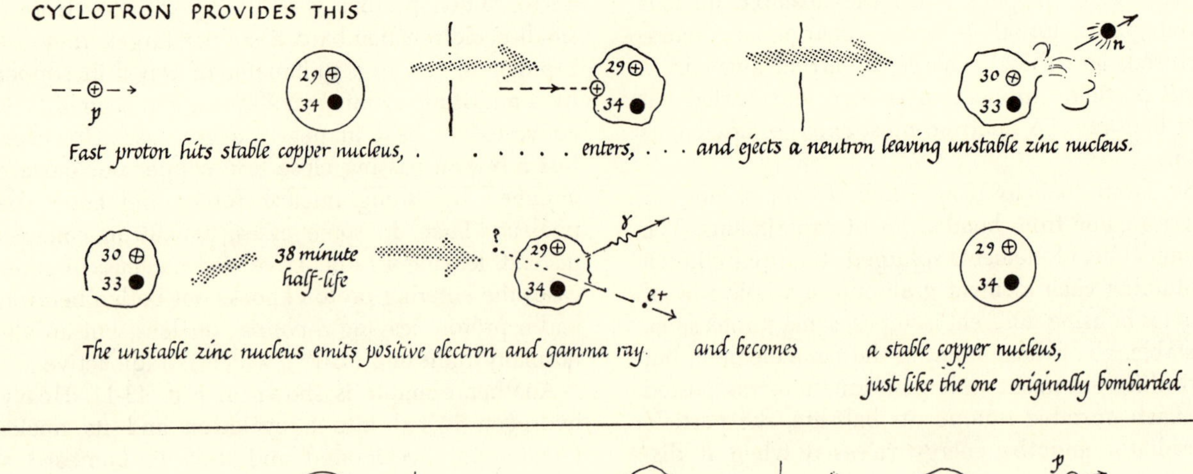

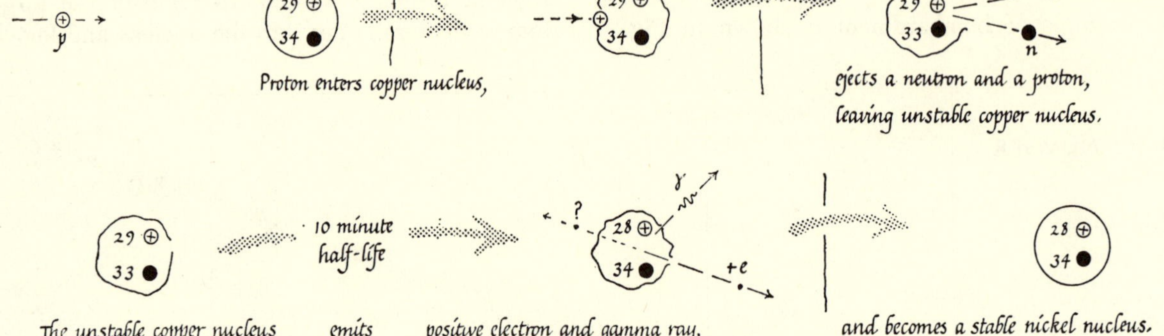

Fig. 43-10. Nuclear Changes from Cyclotron Bombardment of Copper

a proton out. (This is equivalent to pushing a neutron into the sodium nucleus.) The resulting sodium atom, Na^{24}, is radioactive with a half-life about 15 hours. It emits a beta-ray (and a gamma-ray to carry off some extra energy) and becomes a stable magnesium atom.

This radioactive sodium is of great use in research. A trace of it is added to common sodium, combined as salt, and the course of all the sodium can then be followed through a plant or an animal. Or the exchange of sodium atoms between a salt crystal and a saturated solution can be investigated.

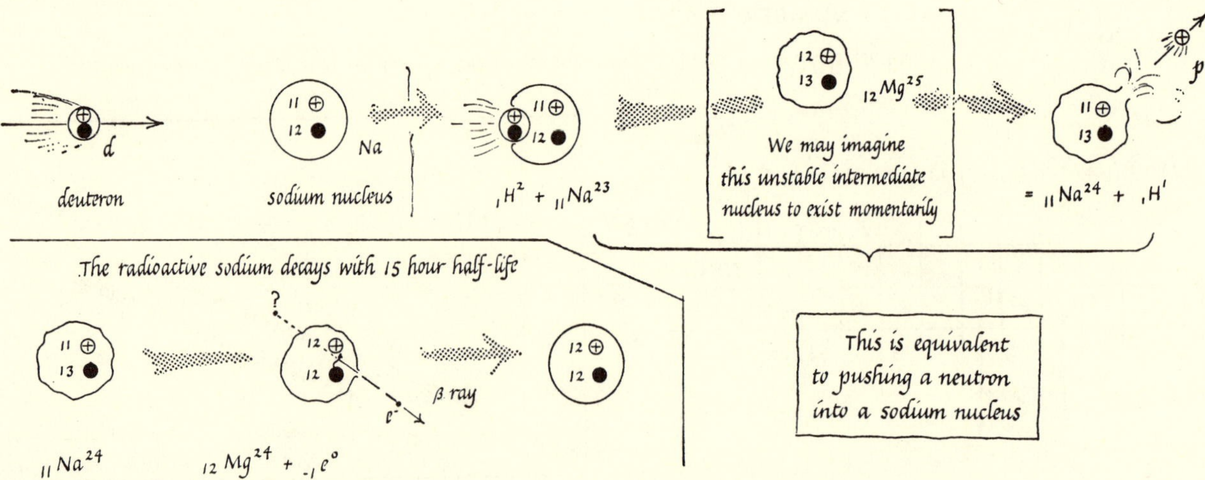

Fig. 43-11. Making Radioactive Sodium with Deuterons from a Cyclotron
Heavy-hydrogen (deuterium) is fed into a cyclotron and its nuclei accelerated and used to bombard common sodium. In some of the atoms of the target a deuteron goes into the nucleus and knocks a proton out.

Or the replacement of other atoms by sodium in an organic molecule can be measured. A Geiger counter can show the presence of radioactive sodium, by catching and recording the beta-ray that signals the decay of one sodium atom. A good counter can detect 0.00000000000000000001 kg, 10^{-20} kilogram, of radioactive sodium against normal background.

Nuclear Forces

We learn from the scattering of alpha-particles (and of protons, etc. from accelerators) that the force exerted by any target nucleus is a simple inverse-square electric force over a vast range of distances. Outside an atom, the electric force due to the nuclear charge is practically shielded by the atom's cloud of electrons. Once within that outer cloud, a charged projectile finds only the inverse-square electric force described by Coulomb's Law. For a light target atom, "radius" ~ 1 Å.U., the only noticeable field is a Coulomb field from just inside the atom's electron cloud, $r \sim 0.1$ Å.U. to thousands of times closer, a few 0.00001 Å.U. In the electron cloud the field is weaker because the electrons provide opposite forces, but it is still made of Coulomb forces. So the whole atom provides Coulomb fields from its outside, $r \sim 1$ Å.U. up to closer than 1/10,000 Å.U. With heavy atoms such as gold, the same kind of picture holds, but the cloud of many electrons extends farther in.

Looking back on the hill models for force-fields

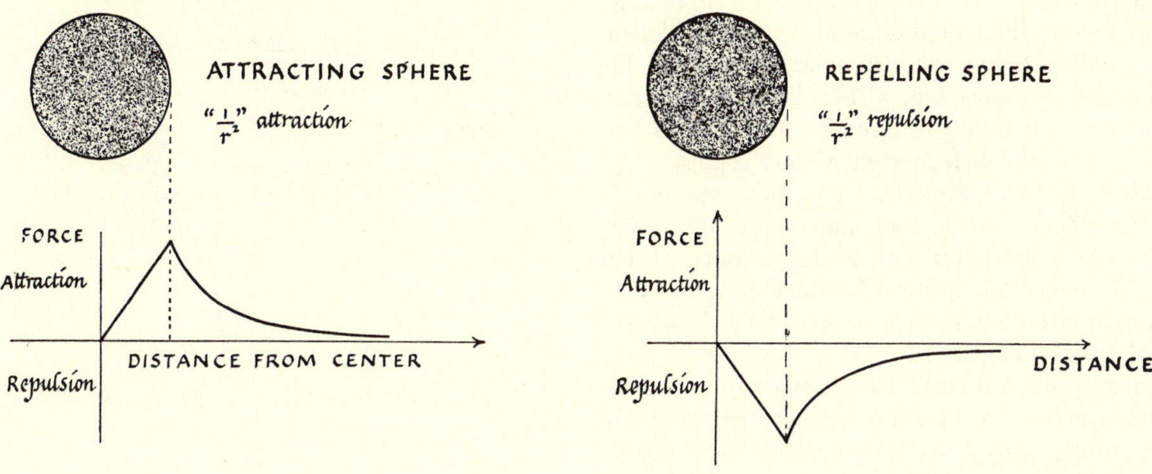

FORCES ON EXPLORING ALPHA PARTICLE

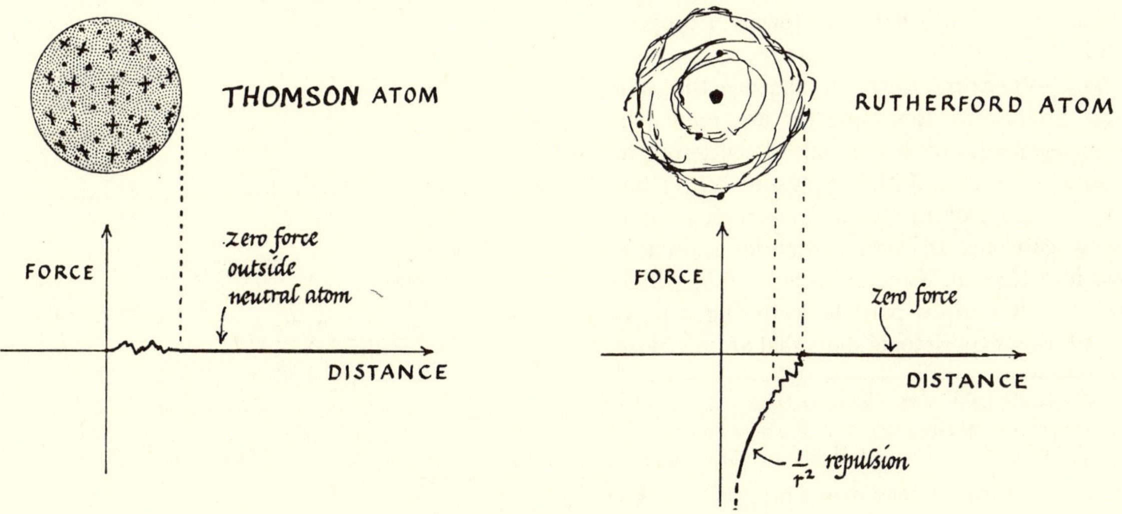

Fig. 43-12. Atomic Forces

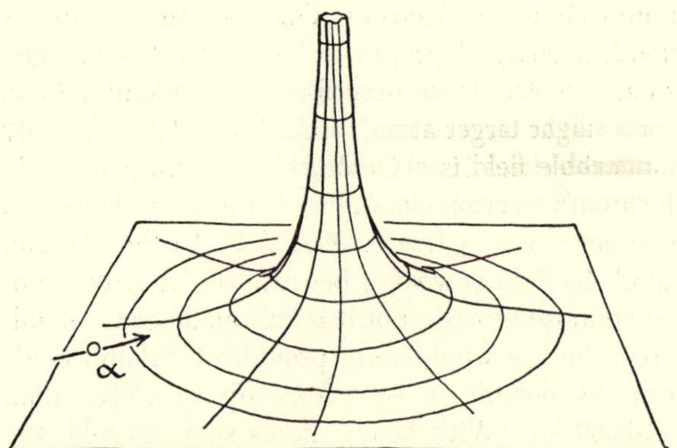

Fig. 43-13. Hill Diagram: Inverse Square Law Repulsion in Rutherford atom.

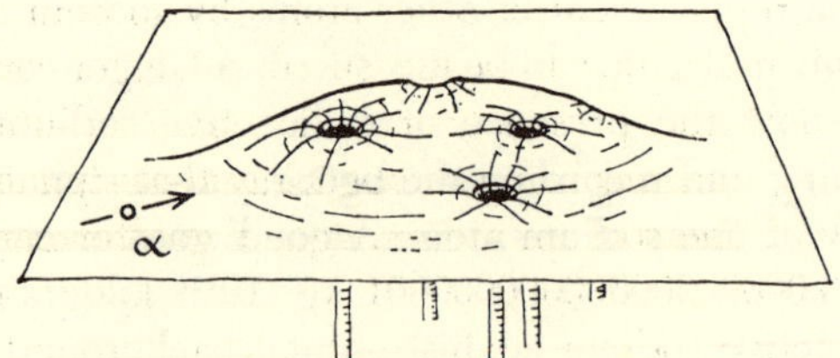

Fig. 43-14. Hill Diagram: Thomson Atom

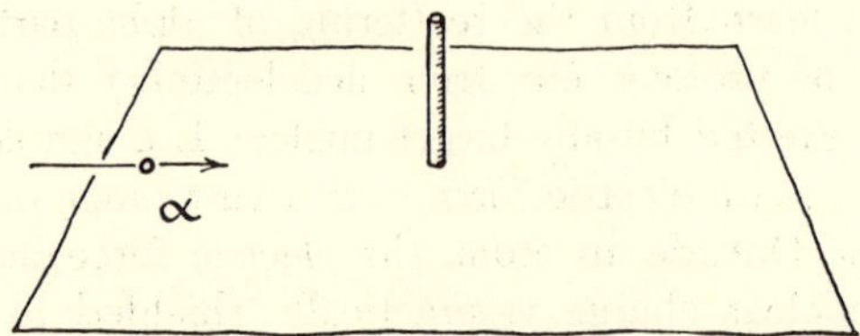

Fig. 43-15. Hill Diagram
Small, hard, unyielding core as target.

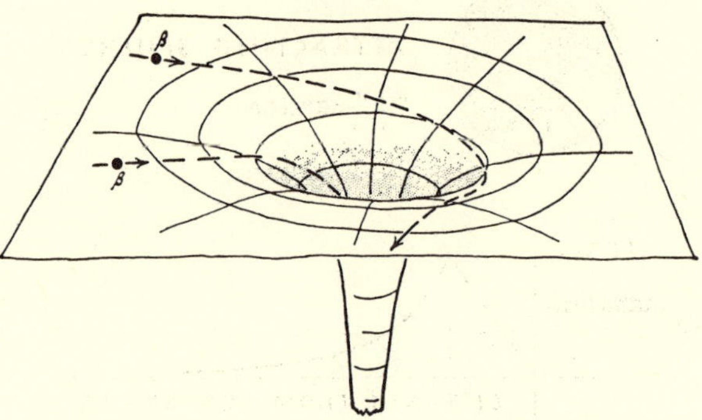

Fig. 43-16. Hill Diagram: Attracting Force

sketched in Chapter 8, we can make a model of the Rutherford atom with a curved pillar,[12] as in Fig. 43-13 (while the Thomson atom would be shown by a shallow hump, with irregular dimples for the embedded electrons Fig. 43-14). Suppose, instead, there were no Coulomb force but the nucleus behaved as a solid ball exerting violent repulsion on "contact" (a silly word for "very close approach" on the atomic scale). Then the energy hill model would be a flat plain with a narrow pillar rising sharply from it to represent "contact" (Fig. 43-15). A strong *attractive* force field would be shown by a steep well (Fig. 43-16).

Imagine we roll small balls along various paths towards several such models, each constructed with many force pillars or wells to represent many atoms in a scattering target. Then models with different pillar shapes will scatter the projectiles in various directions in different proportions. The Coulomb-force model will scatter some out at right angles (Fig. 43-17) and very few backwards. But the sharp-pillar model will scatter far fewer in proportion—and so would the steep negative version (a narrow well)—because nearly all the projectiles will travel on undeflected. Reversing the argument, we can use experimental measurements of scattering to choose among models. With bigger and bigger accelerators to hurl protons etc. as projectiles, we can investigate scattering at closer and closer approach. Then we find that the Coulomb model hill, which does well for slow alpha-particle scattering, has its limits. Scattering experiments show that at very close

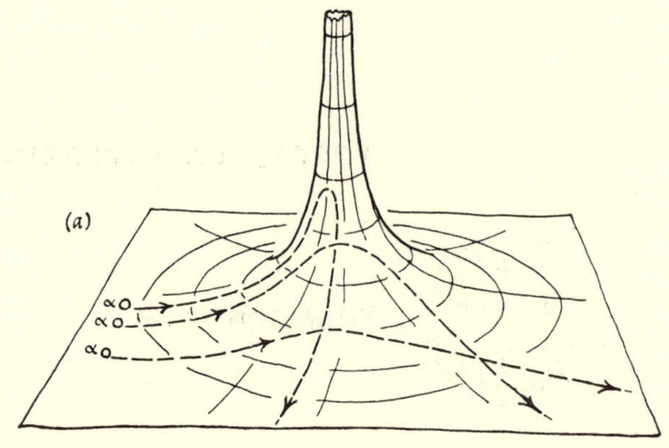

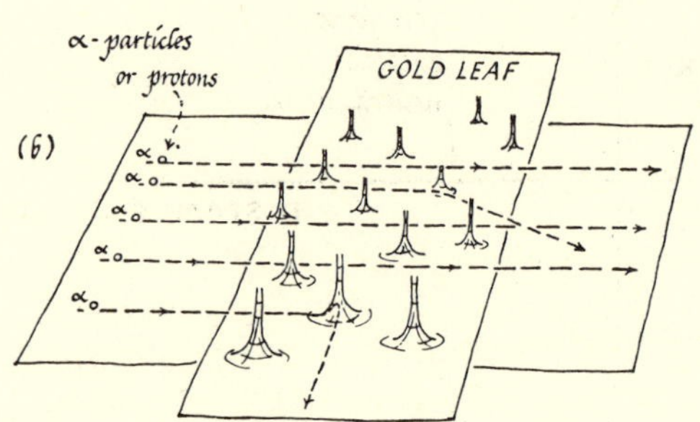

Fig. 43-17. Scattering of α-Particles or Protons

[12] In a Coulomb field with forces on charged particles that vary as $1/r^2$, the potential energy of a test charge q is given by $\int_{\infty}^{r}(BQq/r^2)(-dr)$ and therefore varies as $1/r$. To construct a hill model we must draw a hyperbola $y = k/x$ and rotate it around the y-axis to make a hill whose height, y, varies as $1/r$.

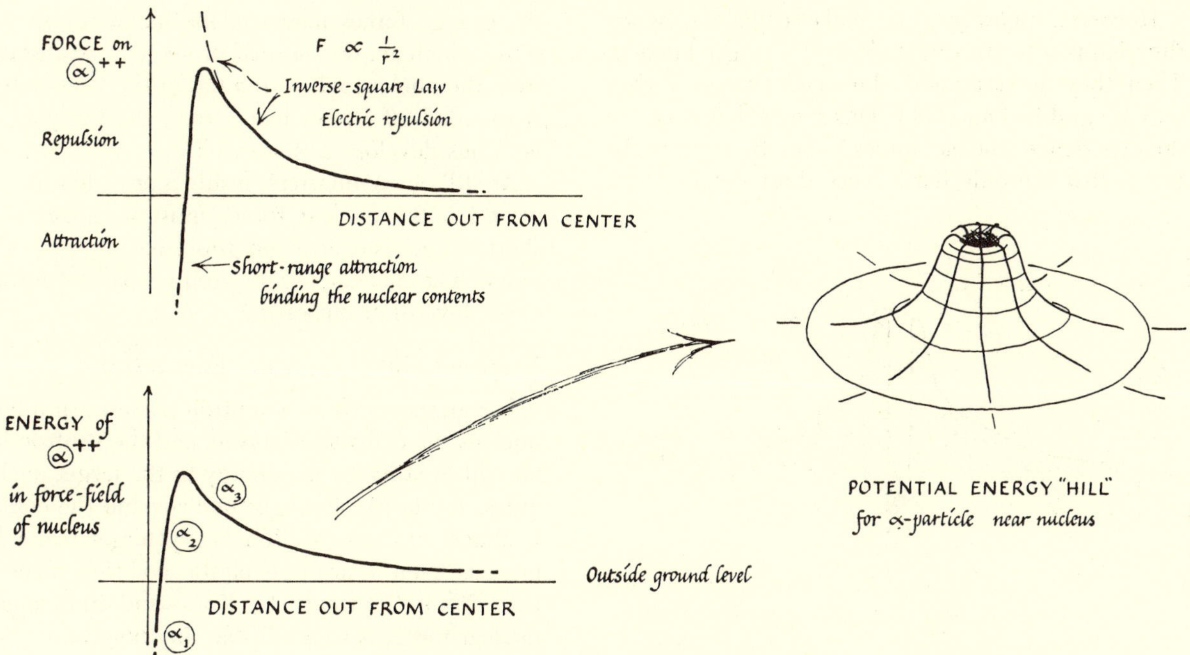

FIG. 43-18a. NUCLEAR FORCES AND ENERGY
α_1: α-particle group in nucleus with this energy, below "ground-level," will stay in—the nucleus is stable.
α_2: α-particle with this energy might escape. α_3: this α-particle is outside.

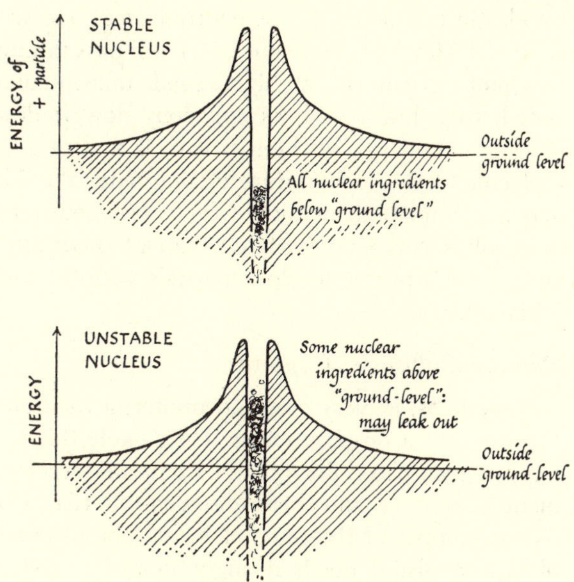

FIG. 43-18b. NUCLEAR HILL MODELS

approaches a positively charged projectile experiences less force than Coulomb's Law alone would provide. The particles that are bounced backwards are the ones that have made closest approaches; and there are fewer of these than expected.

New forces begin to make themselves felt strongly by 0.00001 Å.U. (10^{-15} meter, or 1 fermi as it is now called by nuclear physicists). There must be very short-range attractive forces that make the hill level off to a rim and then drop into a well. That central well is the nuclear well in which the nuclear ingredients live, presumably in considerable turmoil. Inside stable nuclei we picture the ingredients far down in the well, with no chance of escape. In radioactive nuclei they fill the well to a level not far from the rim—anyway, above ground level—so that there are chances of escape. Note that the hill and well are not a material container like a coffee cup, but only a shape in an energy-diagram—yet they do show how the ingredients stay in.

Neutron Bombardment

When we use neutrons, instead of alpha-particles or protons, as investigating projectiles the picture is quite different. A neutron with no electric charge, and therefore no electric field around it, does not whip out electrons as it passes by an atom. Nor does it feel a deflecting Coulomb force when it passes near a nucleus. In most cases, neutrons travel straight on, near an atom, through an atom, even close to a nucleus. For neutron projectiles, the "hill model" of an atom is just a level plain.

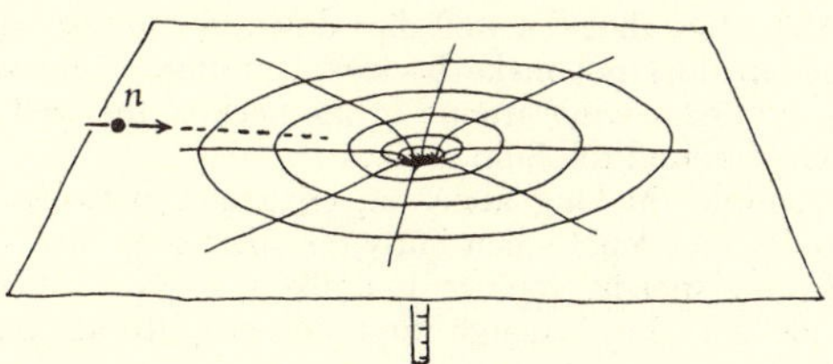

FIG. 43-19. "HILL" DIAGRAM FOR NEUTRON APPROACHING NUCLEUS

However, neutrons can make collisions, when they happen to pass very close to a target nucleus. Then they are scattered—bounced away—or they may be grabbed and held. This shows that neutrons do experience nuclear forces,[13] but these must be forces that are only felt at very short range. These short-range forces may well be much the same as those which act on *charged* projectiles. At distances from the nucleus where the top of the Coulomb hill is rounded off by nuclear forces, the flat plain for neutrons develops a steep well.[14]

At still closer quarters, inside a crowded nucleus, we think the nuclear forces must be attractive at short range and become repulsive at very short range. The nuclear family group must not collapse either inward or outward.

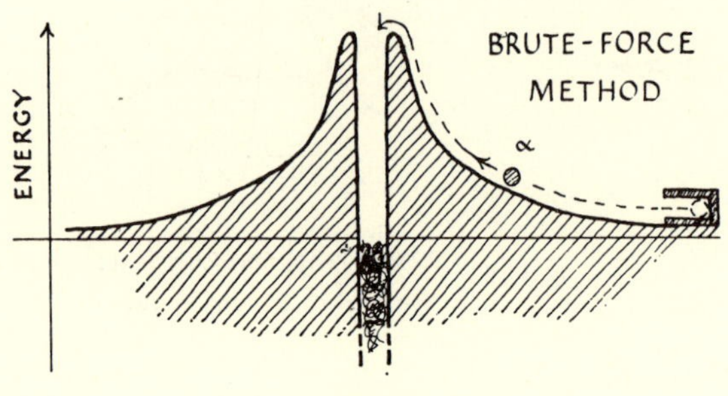

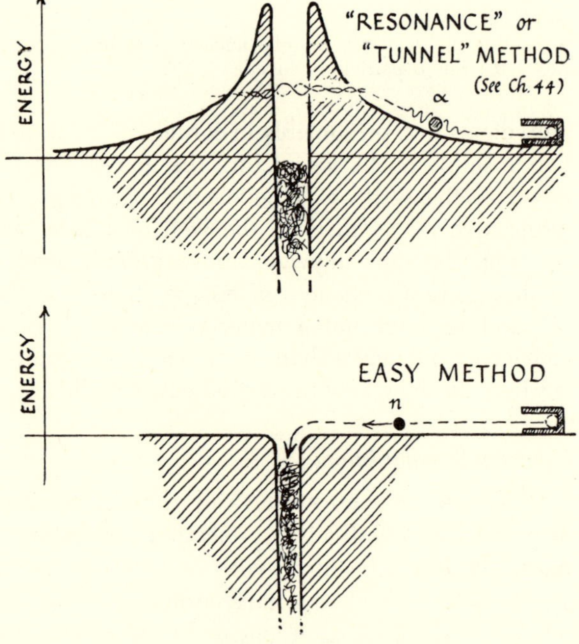

Fig. 43-20. Bombarding Nuclei

Neutron Collisions: Elastic Encounters

In most cases when a neutron passes near a target nucleus the collision is elastic, and the neutron loses very little of its kinetic energy to the target nucleus, unless the target is a light nucleus and the collision is almost head-on. With a heavy target like a lead nucleus even a head-on elastic collision takes less than 2% of the neutron's K.E.—and the range of nuclear forces is so small that anything like a head-on collision is very rare. Yet atoms are numerous, and a chunk of matter that seems small to us contains such vast numbers of atoms that a neutron passing through is soon slowed down appreciably by elastic collisions. A fast neutron (e.g. one with K.E. $\sim$ 1 Mev from fission of U^{235}) will pass through neighboring material at high speed, then medium speed, then low speed, as collisions slow it down to "thermal energy" (when its K.E. is that of a gas molecule at the temperature of the material). This makes a total path of inches for a fast neutron in most solids, compared with less than a thousandth of an inch for a proton or alpha-particle with the same initial energy.

Neutron Collisions: Capture

Sometimes, at very close encounter, a neutron is captured by a target nucleus. How likely this is to happen seems to differ greatly from one target element to another, and there are even differences between isotopes of the same element. The likelihood of capture also depends strongly, in an odd way, on the speed of the neutron.[15]

---

[13] These are probably partly electrical, due to some internal electrical structure of neutrons.

[14] The fact that the well dips down in contrast with a hill rising up does not make backward scattering impossible. A projectile can swing around in the neck of the well just like a comet around the Sun.

[15] Simple thinking about an encounter between *charged* particles that repel when fairly far apart suggests:

At low speeds, capture is unlikely because the particles cannot get close enough to experience strong attractions.

At high speeds, capture is unlikely because the particles do not spend a long enough time close together for strong forces to have enough effect. ($F \cdot \Delta t = \Delta(mv)$ and, if Δt is very small, . . .)

With neutrons, the low-speed ban does not apply and we might expect a simple increase of capture with decreasing speed. The actual behavior is much more complicated: there are clear signs of easy capture at specially favorable speeds which we should interpret as "resonance" if we thought neutrons were wave vibrations.

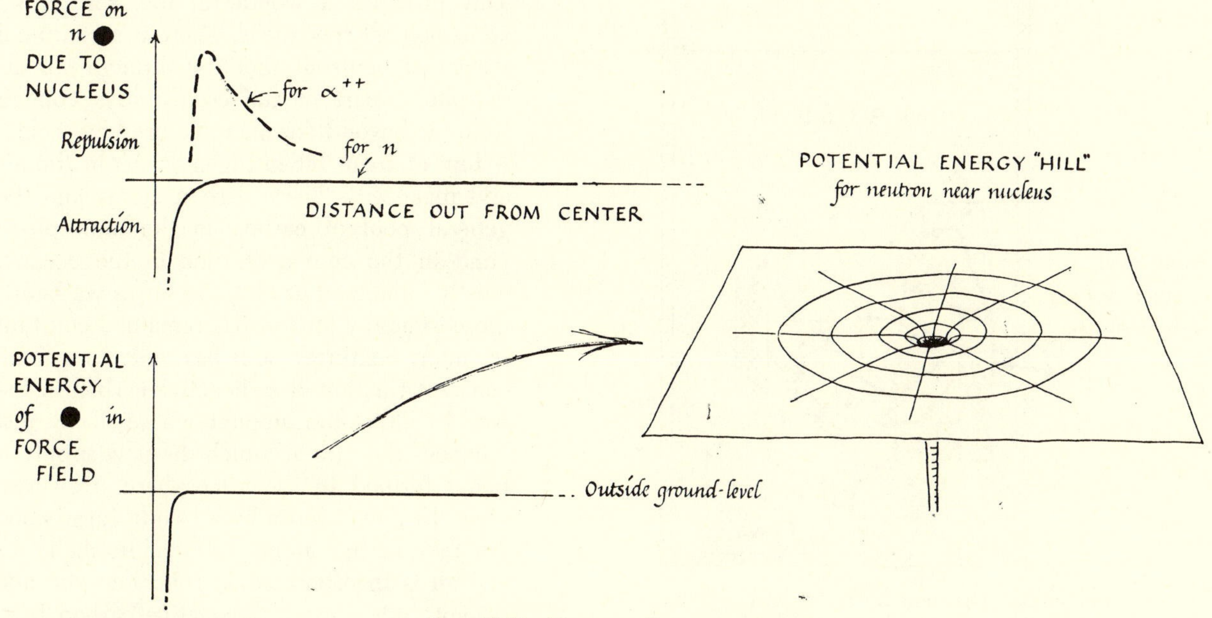

Fig. 43-21. Neutron Bombardment

The new nucleus after capture is often unstable, radioactive. Experiments on neutron capture not only explore nuclear structure; they also provide new unstable atoms. Here are a few examples of such events, out of the many hundreds now known.

(i) A hydrogen nucleus may absorb a neutron and become a "heavy hydrogen" nucleus consisting of a proton and a neutron closely bound (deuterium).

$$_0n^1 + {}_1H^1 \rightarrow {}_1H^2$$

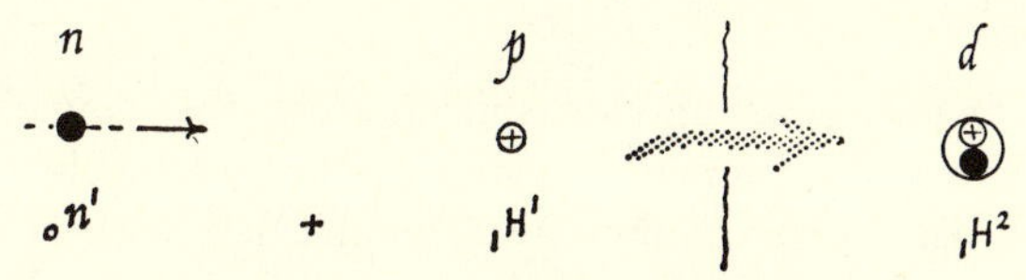

Fig. 43-22a. Making Deuterons

(ii) A silver nucleus may absorb a neutron and become a radioactive silver nucleus. This occurs most easily with *slow* neutrons, and makes a good demonstration: moderate fast neutrons with a bucket of water and make a silver coin radioactive. (See Figs. 43-22b below and c on next page.)

(iii) An aluminum nucleus may absorb a neutron, lose an α-particle and become radioactive sodium—the same useful isotope as that made by deuteron bombardment of a sodium target.

$$_0n^1 + {}_{13}Al^{27} \rightarrow {}_{11}Na^{24} + {}_2He^4$$

(iv) A boron[10] nucleus may absorb a slow neutron and break up into a lithium nucleus and an alpha-particle, which fly apart with a total K.E. of 2.8 Mev. (So a gas containing boron is placed in ionization chambers for counting slow neutrons.)

$$_0n^1 + {}_5B^{10} \rightarrow {}_3Li^7 + {}_2He^4$$

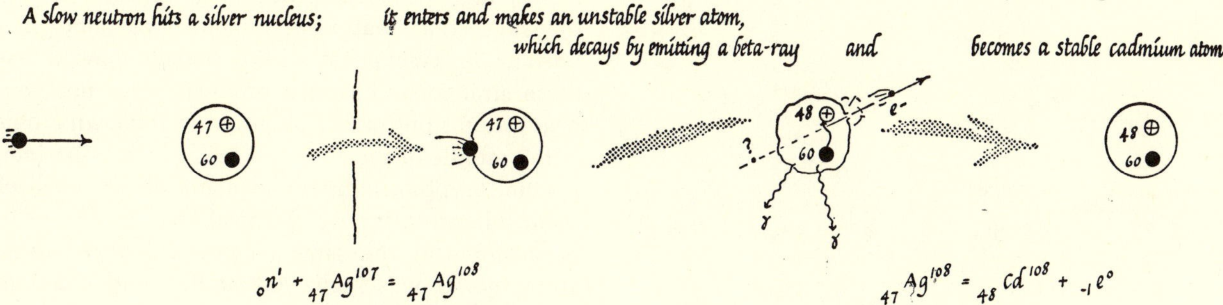

Fig. 43-22b. Manufacturing Radioactive Silver by Neutron Bombardment

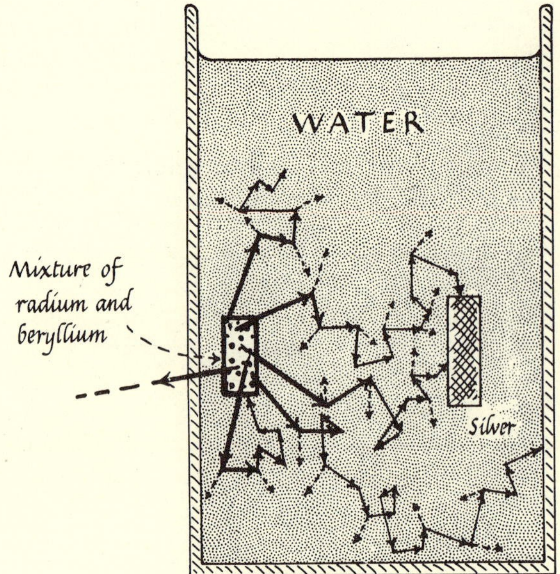

FIG. 43-22c. NEUTRON BOMBARDMENT
A mixture of radium and beryllium provides fast neutrons. The neutrons colliding with hydrogen nuclei in the water lose energy at each collision and are slowed down, in a dozen collisions or so, to thermal speeds. Then they have a good chance of entering a silver atom's nucleus if they pass close.
The neutrons also hit oxygen nuclei but bounce off with much less loss of K.E. Sometimes they are captured by an H nucleus to form a heavy-hydrogen nucleus.

(v) Cadmium shows a surprisingly large "capture cross section" for *slow* neutrons, making it a very valuable absorber to control a nuclear reactor.

(vi) *Radioactive carbon and "carbon dating."* When a neutron hits a nitrogen nucleus, it sometimes knocks out a proton, leaving a radioactive carbon nucleus:

$$_0n^1 + {}_7N^{14} \rightarrow {}_6C^{14} + {}_1H^1$$

Radioactive C^{14} decays with a half-life 5600 years, emitting a β-ray and returning to the original nitrogen.

$$_6C^{14} \rightarrow {}_7N^{14} + {}_{-1}e^0$$

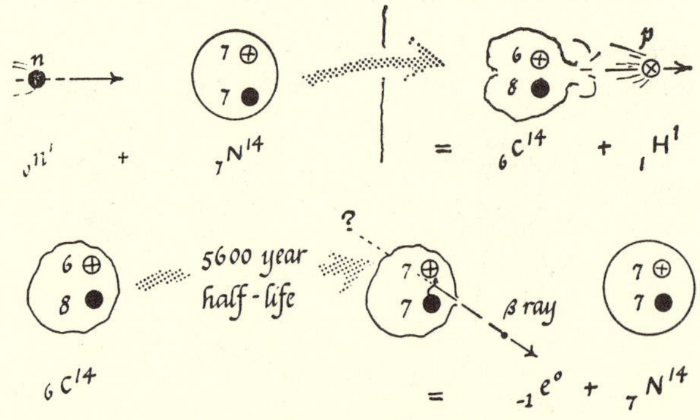

FIG. 43-23. RADIOACTIVE CARBON
(a) formation: neutrons bombard nitrogen;
(b) decay

This provides a wonderful tool for dating archeological specimens. There is a continual stream of neutrons shooting through our atmosphere, part of the cosmic rays coming from far beyond (or manufactured by them). A few of them hit nitrogen atoms in the air and make C^{14}. These carbon atoms join the general pool of carbon in the atmosphere (and in the course of time in the ocean), mostly combined as CO_2. So far as we know, the cosmic-ray stream has remained constant for many centuries; so it has built up a small constant fraction of radioactive carbon in the world's CO_2—the amount whose decay just balances the rate at which the new supply is being formed in the atmosphere. However, once the CO_2 is used by a tree to form wood (or by a marine animal to make its shell), its carbon is incorporated in solid material and cannot collect more radioactive carbon from nitrogen. Then that carbon's radioactive fraction decreases with half-life 5600 years. Thus by measuring the radioactive *fraction* of the carbon in a specimen of wood (or sea shell) we can tell how old it is—how much time has elapsed since its carbon was mixed in the atmosphere's pool. Sensitive counters have been developed to measure this β-ray activity and it is now possible to date samples of wood and cloth from the earliest civilizations.[16]

(vii) *Fission.* Uranium sometimes shows a strange new behavior on absorbing a neutron. We shall now discuss that.

Fission

Twenty years ago a suspicion arose, that new elements had been made, beyond the last on the chemical list when uranium was bombarded with slow neutrons. It was clear that some other elements had appeared in the bombarded uranium; and chemical tests showed that they were not neighboring elements, like the usual products of capture. Hence the suggestion that new, unknown elements had been made. Then chemical experiments showed that these strange, radioactive products were not very heavy atoms but were isotopes of well known atoms in the middle of the chemical table, such as barium, cesium, krypton, iodine, and many others, each of them following, in any chemical analysis, the ordinary atoms of the same element wherever those atoms go. It was realized that the huge uranium

[16] For a very good account by W. F. Libby, who discovered the method, see *Amer. Journal of Physics*, Vol. 26, No. 8, Nov. 1958.

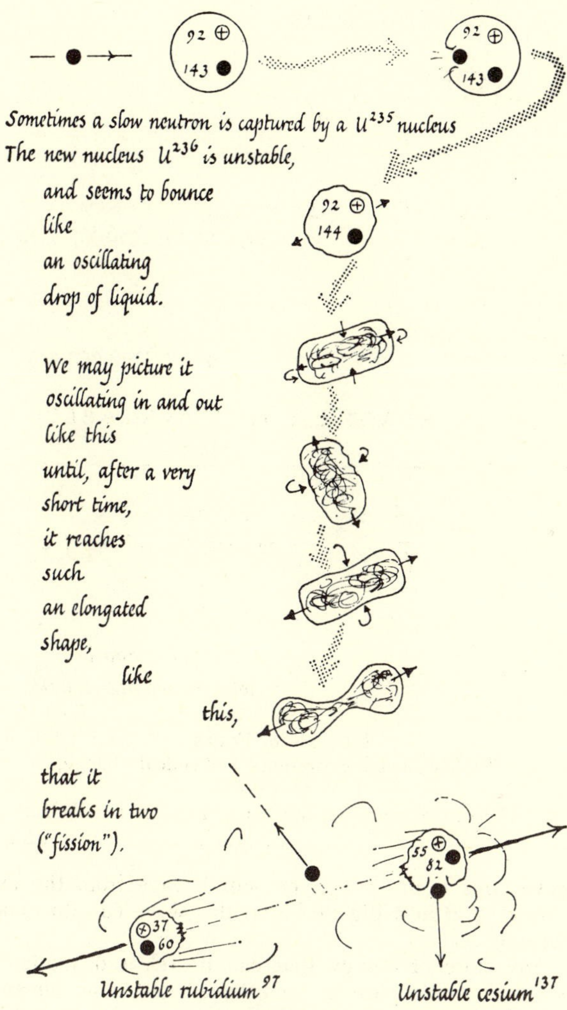

Fig. 43-24. Uranium and Neutrons
This picture of an oscillating drop may be an entirely incorrect picture of the mechanism; or it may even be unjustifiable to try to draw a mechanical picture of the microscopic process inside a nucleus—though the use of this analogy, with mathematical treatment, has produced helpful suggestions. Anyway, the observed fact is that the unstable nucleus U^{236} divides into two large unequal fragments in a *very* short time, releasing neutrons too.

nucleus had divided into two large fragments, not equal halves but still of comparable size. This was called "fission," the name for a biological cell dividing. The neutron that had entered a nucleus of uranium had started some instability and had unlatched the two "halves" so that their electric charges could drive them apart with enormous force.

The massive *fission fragments* were hurled apart, appearing with huge kinetic energy, a total of nearly 200 million electron·volts. This event was photographed in cloud chambers—a terrifying mess of huge tracks—and measured with ionization counters.

Most of the energy is released by plain electrostatic repulsion between the two "half-size" nuclei, with their positive charges (each a few dozen), pushing apart from appallingly close quarters. Not only is the energy-release terrifyingly big, it also offers a possibility of use in a chain reaction, *because a few neutrons are released in addition to the fission fragments.* Could these neutrons be used to trigger the fission of more uranium atoms, so that a block of uranium would blow up explosively? Not with ordinary uranium, a mixture of isotopes, because a neutron causes fission easily only in the rare U^{235}.

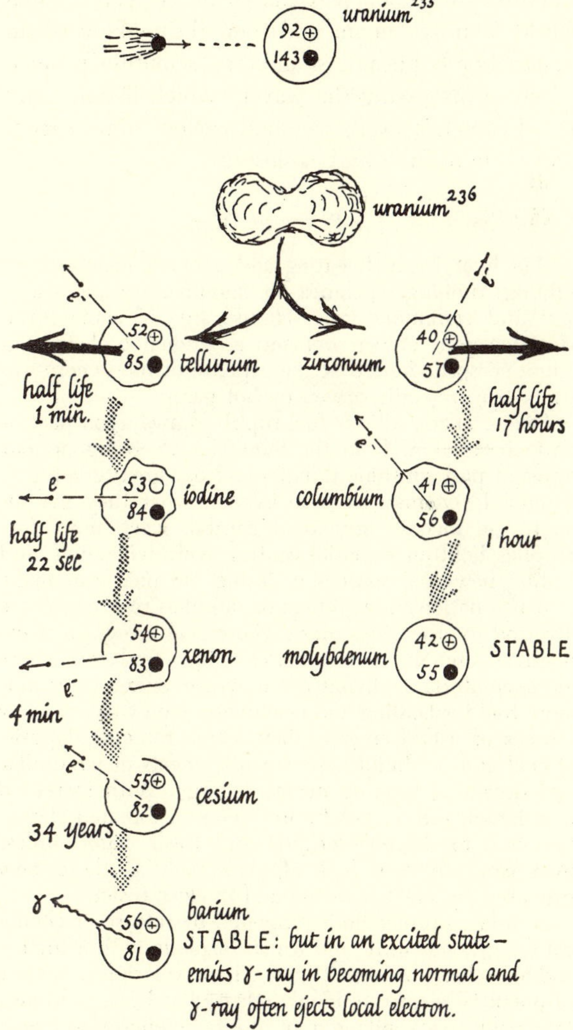

Fig. 43-25. Uranium Fission. A Specimen "Family Tree" for a Pair of Possible Fission Products
A U^{236} atom divides by fission in a variety of ways. The sketches here show one of these ways and the subsequent changes of the two radioactive "fission fragments."

A neutron is usually absorbed by the common isotope, U^{238}. The mass spectrograph had shown that uranium consists of two isotopes of masses U^{235} and U^{238} (and a very rare U^{234}). Minute samples of the isotopes were separated by a mass spectrograph; and, when they were tested for fission, only U^{235} showed it. And fission occurred more frequently with slow neutrons than with fast ones. The U^{238} just absorbed neutrons quite easily, particularly fairly fast ones. Common uranium is a mixture of 99.3% of U^{238} and only 0.7% of U^{235} which shows fission easily. If one fission occurred in a lump of common uranium, the fast neutrons it releases would almost certainly be grabbed by U^{238} atoms and there would be no chain reaction. Only a lump of pure U^{235} could give a chain reaction and act as a bomb. This much was known or suspected when World War II began; and then the making of an atomic bomb became a military scientific project.

Before discussing the way in which fission (and now fusion) is used, we shall review some earlier discussion of fuels and explosives.

A Note on Fuels for Peace or War

Man lives by fuel: eating and cooking, heating and lighting, running machines for manufacture and transport and communication—all use fuel in some form. Most weapons of war demand additional fuel, from a primitive spear hurled at the expense of food-energy to a modern projectile driven by hot gases.

Today, nearly all our fuel supply draws upon past or present radiation from the Sun. We are using up our stores of past sunshine rapidly, and our great-great- . . . grandchildren may have to live on what they get directly, on income instead of capital. Even now, fuel supplies hold an essential control over civilization, and finding new fuel sources or losing old ones can make or mar a nation's life. Where oil supplies run out, or are diverted to other customers; where coal grows more expensive as miners require better living standards; where more comfortable living for a whole people demands more fuel for heating and machinery, men welcome new sources of useful energy: they search for new deposits of coal and oil, build new dams, try fleets of windmills, and dream of tapping nuclear energy. With increased population, the distant future seems to threaten starvation due to shortage of fuel and fresh water, unless man can use new fuel supplies such as those now promised by nuclear fission and nuclear fusion.

A modern projectile is propelled by release of chemical energy into heat, and it may explode with a further sudden release of heat. That is all an explosion is: a store of potential energy suddenly released as heat; suddenly, in a small space, giving a lot of kinetic energy to everything around such as gas molecules, bomb fragments, . . . , automobile pistons. In an explosion, gas molecules are pushed to high temperatures, their high speeds making high pressure. This bunch of hot compressed gas pushes the air around it by molecular collisions and hands its energy on outward in a sharp compression wave: this *is* the bang, the sound wave from the explosion; and in a big explosion this wave can do great damage.

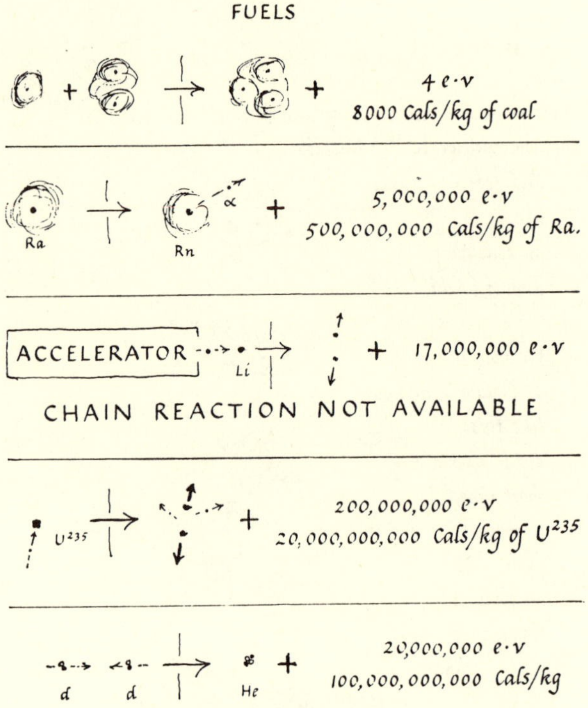

Fig. 43-26. Fuels
The first sketch represents a chemical change,
$C + O_2 \rightarrow CO_2$.
The others represent nuclear changes as labelled.

Any store of energy that can be released suddenly can make an explosion: compressed gas in an air-gun or a bottle of soda-pop (K.E. of molecular motion); some air sandwiched between hands that are being clapped (K.E.); gasoline and oxygen, or gunpowder or dynamite (chemical energy). Explosives do not always make an explosion—carbon dioxide can leak around the cork of a pop bottle; a loose pile of gunpowder burns quietly. To make a bang, the energy must be released quickly so that it cannot creep away in a slow stream. And, to make a big bang, a lot of energy must be released quickly—a long series of tiny explosions will only make a firework-flare. And the energy must be released in a small space, so that it produces a sharp, pushing explosion wave instead of a slow belch.

Therefore a bomb must be made of compact material, and when it explodes the burning must spread very quickly through all the material; one piece of burning material must ignite the next, and that piece the next, and so on. But such a *steady chain reaction*[17] of ignition would make a poor explosion. The burning should increase rapidly, as it will if one piece of burning material ignites several neighbors, and each of them ignites several others, in an *increasing chain reaction*. As an illustration, take a single page from a book of matches, and light a match at one end. A steady chain reaction will travel along. Take a complete book or a large box of

[17] A steadily running chain reaction is what we use in nuclear engines—reactors.

matches, and light one match in it: an increasing chain reaction will develop. Of course that chain reaction will not grow indefinitely, since the total material is limited. It will grow to a maximum rate, then dwindle to nothing for lack of material.

Fission and Neutron Capture

Fission releases very fast neutrons, with K.E. $\sim$ 1 Mev. These will travel several inches in the surrounding uranium before they are slowed right down to "thermal" speeds, and have the average K.E. of a neighboring atom. To understand the making of both fission bombs and nuclear reactors you need to know how a neutron's chances of being caught by uranium nuclei depend on its speed.

A ROUGH SUMMARY OF NEUTRON CAPTURE BY URANIUM NUCLEI

NEUTRON	TARGET NUCLEUS U^{238}	TARGET NUCLEUS U^{235}
high speed (K.E. $\sim$ 1 Mev)	capture unlikely (if capture occurs, it may lead to fission)	capture unlikely (if capture occurs, it usually leads to fission)
medium speed (K.E. $\sim$ 5 ev)	CAPTURE VERY LIKELY leading to U^{239}, radioactive emitting β^-	capture unlikely (if capture occurs, it usually leads to fission)
low speed "thermal neutrons" (K.E. $\sim$ 0.03 ev)	capture unlikely (if capture occurs, it leads to U^{239})	CAPTURE VERY LIKELY leading to fission

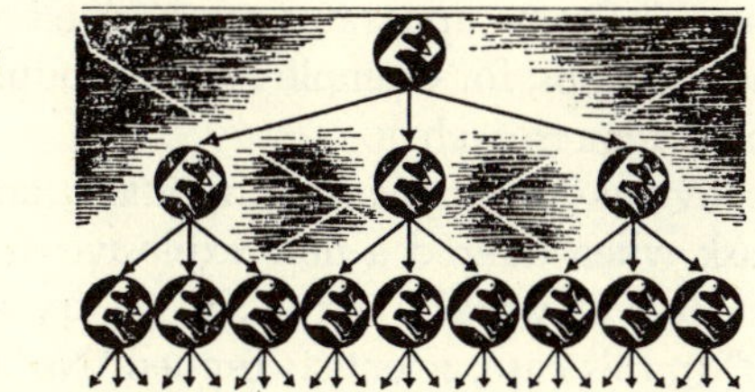

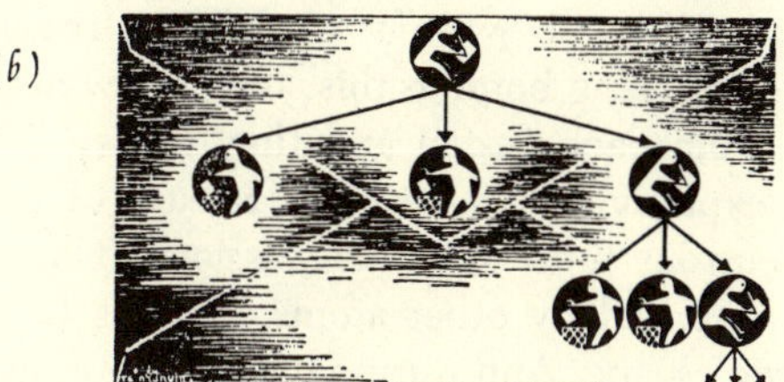

FIG. 43-27. CHAIN REACTIONS
ILLUSTRATED BY THE WRITING OF "CHAIN LETTERS"
In this model, each writer sends out three letters, asking the person receiving a letter to send out three more.
(a) INCREASING CHAIN REACTION.
The rate of the reaction depends on the number of letters written at each stage. Here, the rate increases rapidly, and the reaction is "explosive." (This is a model of a bomb.)
(b) STEADY CHAIN REACTION
Here, the rate of reaction remains the same from one stage to the next. (This is a model of a reactor running steadily.)
(Sketches from *What is Atomic Energy?*, by K. Mendelssohn, published by Martin, Secker, and Warburg, Ltd., London; Drawings by Victor Reinganum and the author.)

Atomic Bombs. The Separation of U^{235}

To make an atomic bomb, it was necessary to separate the rare U^{235} from the common U^{238} which would sabotage a chain reaction by absorbing neutrons. Chemical separation was hopeless, since both are uranium, the same chemical element with identical chemical properties. Physical separation of the isotopes (e.g. by gas diffusion) was known to be possible; but could it be done on the vast scale needed to produce material for a bomb? (For details of the methods tried, difficulties, successes, consult books on "Atomic Energy.")[18] One successful method is to let uranium hexafluoride vapor diffuse through a barrier with very fine pores. Molecules containing U^{235} have higher average speed than those containing U^{238}, so they stagger through the pores faster. A single stage produces a little preferential sorting. Thousands of stages—cycle after cycle, with useful recycling of rejected fractions—succeeded in producing a stream of sufficiently pure U^{235}. (See diagrams in Ch. 25, Problem 11 in Ch. 25, and Problem 3 in Ch. 30.)

Another method uses a mass spectrograph scheme on a huge scale. Uranium ions are manufactured, accelerated to a suitable energy by an electric field, then held in circular orbits by a large uniform magnetic field. The orbits focus a semicircle across from the gun muzzle, and a small can placed at the focus collects the ions. The lighter U^{235} ions make a smaller circle than the rest and are collected in a separate can. This scheme is expensive for the amount of its yield; but it probably produced the pure samples that were badly needed for early ex-

[18] Primarily: *Atomic Energy for Military Purposes* by Henry D. Smyth (Princeton, 1945).

periments. The same apparatus can be used to separate other isotopes, for example fission products that are wanted for research.

The obvious questions any military authority would ask when offered a new explosive are, "Are you sure it will work?" and "Can we try a small sample?" In this case, scientists replied "No" to both questions. They added, "We believe the bomb *will* work, and our belief is based on sound theory. We know a small bomb will *not* work." The reason for insistence on a big bomb is this: only a few neutrons emerge from each fission and they are easily lost. For an explosive chain-reaction release of energy, these neutrons must start new fissions. They must not be absorbed by other atoms, such as U^{238}, nor must they escape. And escape is only too easy for a neutron—it whizzes so freely through matter. In a small bomb most of the neutrons released by a fission would escape and any attempted explosion would fizzle out. In a very large bomb, with a huge block of U^{235}, a neutron released by fission would travel on and on through uranium atoms until it did make a hit and could start another fission. A small bomb *cannot* explode; a big bomb *must* explode. Somewhere between the extremes, there is a definite "critical size" such that a lump of U^{235} below critical size will not explode, and one above critical size will.

The critical size is a few pounds—a few cubic inches, since uranium is so dense that a cubic inch weighs about a pound. The critical size has been regarded as a military secret; but, given the "size" of nuclei,[19] one can make a good guess. Now that we have good measurements of the target area of the U^{235} nucleus for neutrons (its "cross-section for neutron capture"), the exact critical size can be

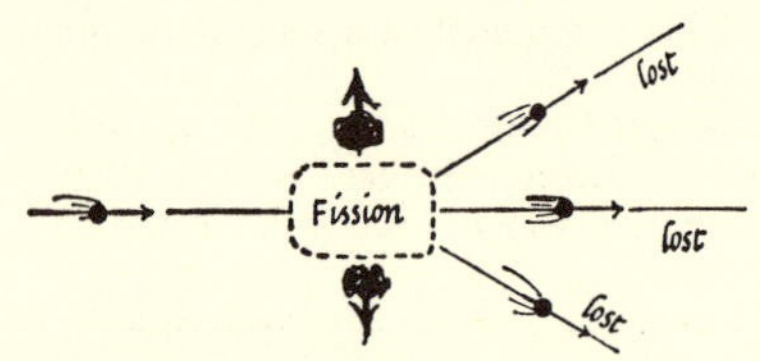

(a) Single stage: one fission in a small piece of U^{235}

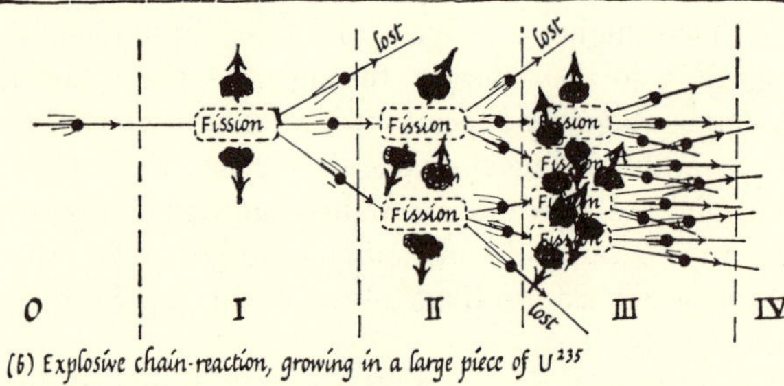

(b) Explosive chain-reaction, growing in a large piece of U^{235}

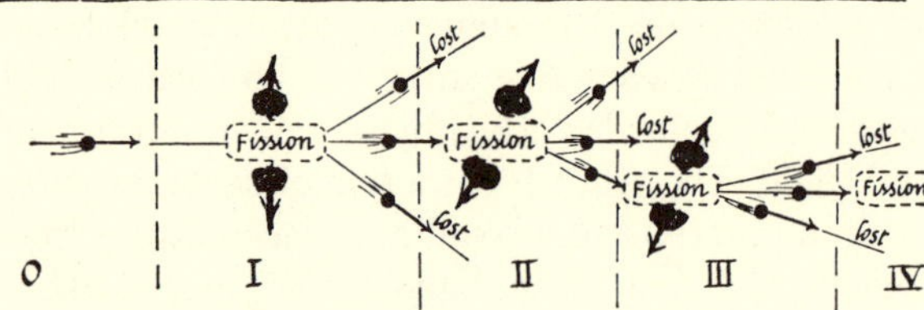

(c) Steady chain-reaction, as in a reactor

FIG. 43-28. URANIUM FISSION: CHAIN REACTIONS
These sketches are purely schematic. The box marked "fission" is just a label for that event. The neutrons released in fission fly out in all directions. They are shown shooting forward here, to enable the sketches to show successive stages.

FIG. 43-29. SKETCHES OF ARRANGEMENT FOR ATOMIC BOMB
(This is a purely schematic arrangement.)

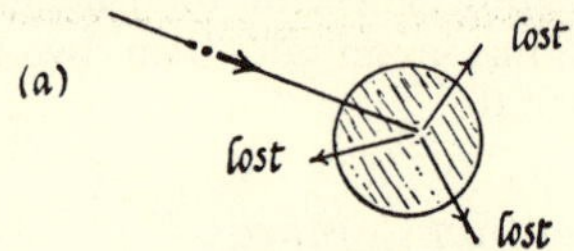

FIG. 43-29a. In a piece of fissionable material, e.g. U^{235}, smaller than critical size a chain reaction *cannot* grow: neutrons are lost.

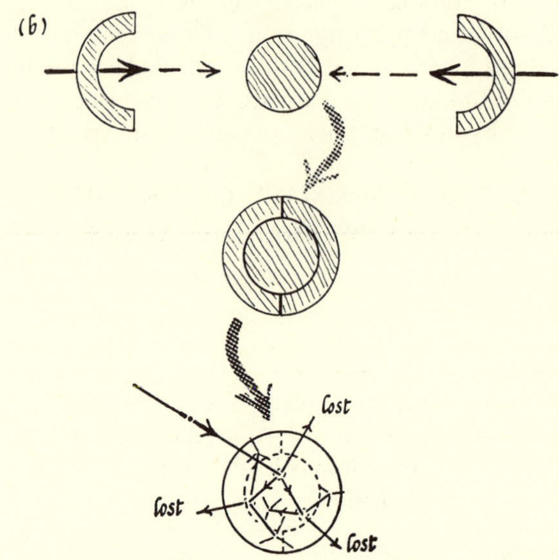

FIG. 43-29b. A piece of fissionable material larger than critical size can be made by assembling smaller pieces. In the large assembly a chain reaction, once started, *will* grow explosively. In the growing chain reaction one fission provides neutrons that trigger more than one new fission. However, many a neutron is lost.

19 Scattering experiments with high-energy protons or alpha-particles show that the simple Coulomb inverse-square force is all that matters down to about 10^{-14} meter from the center of an atom. From then on in, "nuclear binding forces" compete with it, and swamp it at shorter ranges. So we say that, for a light nucleus, $r \sim 10^{-15}$ meter, and a few times larger for heavy nuclei. Scattering experiments with *neutrons* give radii of this kind of size for many nuclei as targets, but much bigger radii for good absorbers such as cadmium.

calculated by any good theoretical physicist in any country that wants it.

Ordinary bombs need a fuse or detonator to start them. An atomic bomb presents a different problem. A big bomb will explode immediately, as soon as a stray neutron, always available from cosmic rays (or a spontaneous fission) starts one fission. So the bomb must be prepared in pieces smaller than critical size—pieces that will not explode spontaneously. Then the pieces must be put together quickly—so fast that the bomb does not explode before it is completely assembled. For example: take two blocks of U^{235} each ¾ critical size and slam them together, suddenly, into a block 1½ times critical size, which will then explode. The slamming together can be done just quickly enough, possibly by firing one block at the other with a small cannon. How this is really done is presumably a military secret; but the prime "secret,"[20] the fact that this can be done successfully, was revealed by the explosion of atomic bombs.

[20] Many non-scientists underestimate the difficulty (and cost) of scientific secrecy—and they even mistake the nature of the problem—and many scientists regret it because they consider it confuses the advance of man's knowledge. Here is a fable to illustrate the difficulty: Suppose that a new fatal disease were to sweep across the world, with no known cure. Suddenly, in one country, many people are cured, but the cure is held secret except for a vague rumor: "it is connected with cocoanut palms." How long would it then take for the medical schools of any major country to solve the secret? Given the hint about cocoanut palms, they could narrow their field of research to reasonable size; and given the fact that a cure is possible, their government would provide, all the more willingly and quickly, all the money and facilities they needed.

When a fission bomb of a few pounds of uranium explodes, it scatters a lot of its material without fission; so one cannot estimate its energy-release by reckoning 200 Mev for every U^{235} atom. However, even after allowing for wasted material and lost neutrons, we arrive at an enormous energy-release from a single bomb. The release (by K.E. of fission fragments) is so big that a scorching stream of radiation flies out; and a violent compression wave drives out from the heated gas. And fission products and spare neutrons[21] and gamma-rays all produce ionization damage far and wide.

Reactors: Making Plutonium

When fission was discovered, the fate of a U^{238} atom that absorbs a neutron was investigated, with a surprising result.[22] The new nucleus, U^{239}, is unstable and soon emits a beta-ray, becoming a new atom of an unknown element, beyond uranium. The nucleus ${}_{92}U^{239}$ emits β^- and becomes ${}_{93}?^{239}$. That new element is unstable and emits a β-ray to become another unknown element. Following an astronomical precedent, these new elements beyond uranium were named neptunium and plutonium.

$$ {}_0n^1 + {}_{92}U^{238} \rightarrow {}_{92}U^{239} \rightarrow {}_{-1}e^0 + {}_{93}Np^{239} \rightarrow {}_{-1}e^0 + {}_{94}Pu^{239} $$

[21] Neutrons make no ions directly, but they knock protons forward and the protons make ions. That is how neutrons damage living cells.

[22] When a very fast neutron is absorbed by U^{238}, fission sometimes occurs, but not often enough to maintain a chain reaction in a block of U^{238} or of mixed uranium. Occasionally a U^{238} atom fissions spontaneously, without any triggering neutron. A spontaneous fission may provide the initial neutron in starting up a nuclear reactor.

Fig. 43-30. Production of Plutonium

Fig. 43-30a. Uranium and neutrons. Sometimes a fast neutron is captured by a U^{238} nucleus. It enters, making a more massive uranium nucleus, which is unstable (= radioactive) and emits a beta ray.
Since the new uranium nucleus is extra rich in neutrons, it succeeds in changing one neutron to a proton, the extra + charge being balanced by the − charge of the beta ray emitted. Then the new nucleus has charge +93e, so it is the nucleus of a new element, one place beyond uranium. It is named neptunium.

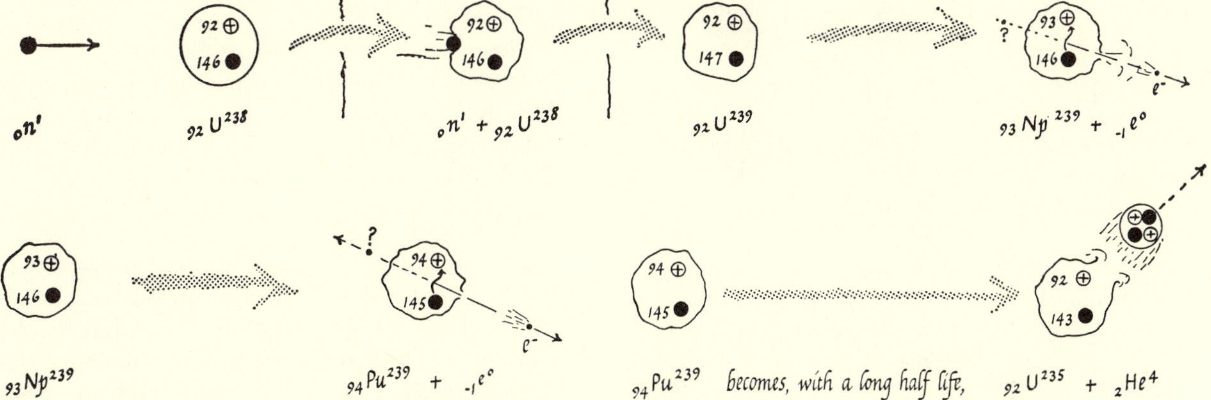

Fig. 43-30b. The neptunium nucleus is unstable and emits a beta ray, becoming plutonium, which can be separated chemically from unused uranium.

Fig. 43-30c. Plutonium is itself fissionable, like U^{235}, but if left alone it is unstable like uranium or radium and decays into a U^{235} nucleus, by emitting an alpha-particle.

Good speculative theory of nuclear behavior suggested that plutonium should be fissionable like U^{235}. Experiments with a small sample made by bombardment with a cyclotron confirmed this. This offered another bomb material, and one that could be separated much more easily. Plutonium is a different chemical element from uranium, so it can be separated from all uranium by *chemical* methods.

Here was a new element, previously unknown, manufactured in minute quantities by a cyclotron, a few atoms at a time. Could man manufacture large quantities for a bomb? If neutrons are shot through a big block of ordinary uranium, most of them will be absorbed by U^{238} atoms, leading ultimately to plutonium, which could be separated; but where can the neutrons be obtained in the enormous numbers needed? From fission of U^{235}.

Fission provides very fast neutrons, which are soon slowed down by collisions with neighboring nuclei. Unfortunately, U^{238} nuclei absorb neutrons with *medium* speeds so easily that there would be no neutron left to start another U^{235} fission and keep up the supply of neutrons. *Very slow* neutrons, however, are much more easily absorbed by U^{235}, leading to fission. So the problem was to slow neutrons down without having them all captured by U^{238} at intermediate speeds. The following scheme was proposed and tried, and it ultimately succeeded. Chunks of uranium (U^{238} and U^{235} in the natural mixture) are embedded in a huge pile of a suitable light element which acts as a *moderator*. The moderator slows down neutrons that pass through it, without absorbing them; neutrons hitting its nuclei in (rare) elastic collisions lose a little energy each time. Hydrogen, in water, would be an ideal moderator but for its tendency to grab a neutron and become "heavy hydrogen," deuterium. Deuterium itself in "heavy water" does well, but heavy water is expensive to separate from water. Pure carbon, with nuclei only 12 times as massive as a neutron, does quite well. (A *head-on* collision takes away 15% of the neutron's K.E.) When a fission of U^{235} occurs, the neutrons released career about in the uranium and moderator until they move as slowly as gas molecules at room temperature and have "thermal energy" of about 1/30 ev. Then they are more likely to be trapped by U^{235} (and start another fission) than by U^{238}. The U^{238} atoms are much more numerous, but they trap medium-speed neutrons and are poor at catching slow ones.

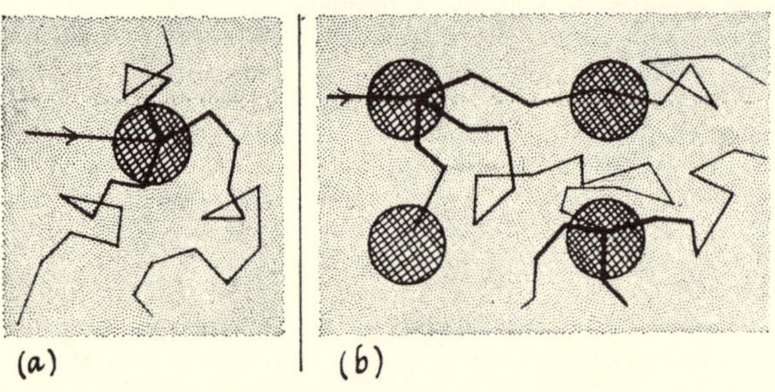

Fig. 43-31. Reactor: Neutrons in a "Pile" of Uranium Slugs Surrounded by Moderator

(The straight lines show a neutron's path between collisions. Their thickness is intended to indicate the neutron's speed.)

(a) One neutron causes a fission and the 3 neutrons released escape.

(b) One neutron causes a fission, and one of the 3 neutrons released causes another fission. One is absorbed by U^{238}.

With the proper proportion of uranium blocks to surrounding moderator in this huge *reactor*, the neutron housekeeping averages out as follows: one neutron from each U^{235} fission is slowed down and finally starts another U^{235} fission. The other neutrons from the original fission get caught, fast or slow, by a U^{238} atom, and lead to plutonium. Like a bomb, such a self-sustaining reactor must be larger than a certain critical size, or too many neutrons will escape to the outside world.[23] For a carbon-moderated reactor with common uranium, the critical size is about the size of a cottage, even with a reflecting shield of heavy metal to bounce neutrons back in. The continual "burning" of U^{235} to provide neutrons releases a tremendous stream of heat from recoiling fission-fragments. Great air fans, or a river of water, must be used to cool the reactor. The heat can be harnessed for useful power production on a huge scale, but the problems of shielding from radioactivity are serious.

To obtain fissionable plutonium, we must remove the uranium chunks from the reactor, dissolve them, and use chemical treatment to separate plutonium from unused uranium and a host of fission products. Since this mixture is strongly radioactive, the separation must be done by remote control. Plutonium then serves as material for atomic bombs or for a compact reactor to supply useful heat.

The making of plutonium is an extraordinary achievement. Man is making an unknown *element* out of old known ones, not just by the atom or two, but by the pound. This goes beyond reasonable expectations: it is more like a tale from the First Book of Genesis.

Neutron Economics

The possibility of successful reactors depends on the fact that one U^{235} fission releases *several* neutrons. There are not enough to waste, but there are enough to run a big reactor. Some fissions release 1 spare neutron, some 3 or more, and the average

[23] The way in which the neutrons that are released in the body of a reactor diffuse out to the surface is a case of "Fick's Law" for the diffusion of salt dissolving in water and that law is one more example of $\nabla^2 V = 0$.

Fig. 43-32. Uranium Reactor.

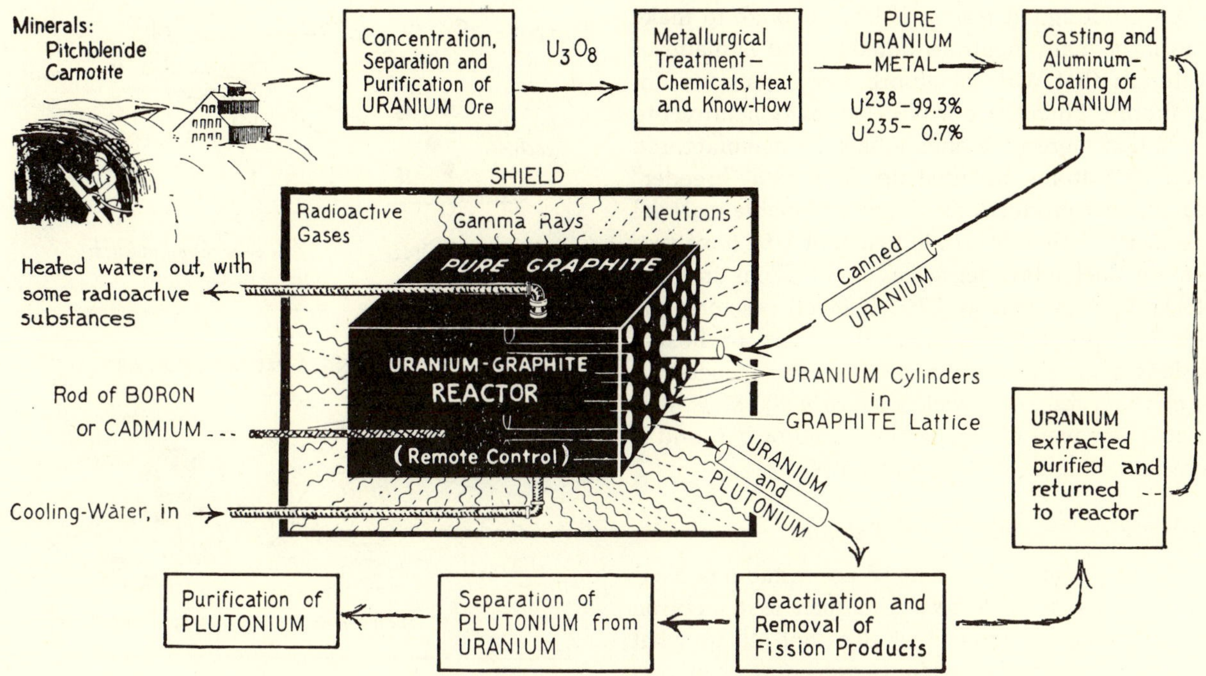

Fig. 43-32a. This fanciful sketch, intended to give a general idea of the processes, is based on sketches published by the Westinghouse Corporation.

Fig. 43-32b, c. Reactor. (Fanciful sketches based in part on sketches issued by the Atomic Energy Commission). (b) The "pile" of uranium slugs embedded in graphite moderator, surrounded by concrete shield, delivers an enormous stream of heat to the cooling water. Fission of U^{235} produces fission products, heat, and neutrons. One neutron per fission causes another fission, maintaining a steady chain reaction. The other neutrons are either lost or absorbed by U^{238} which becomes, after two changes, plutonium. The slugs are removed and treated chemically, by remote control, to separate plutonium and fission products from the remaining uranium.
(c) Reactor in use for making radioactive atoms. Inside the concrete shield there is a swarm of neutrons moving through the graphite moderator and uranium slugs, like molecules of a gas. A sample hung in this "neutron gas" may be bombarded by so many neutrons that some of its atoms change to other atoms, which may be radioactive. The stream of neutrons emerging through a hole in the shield can also be used to bombard samples.

number for U^{235} fission is about 2½. This is an important number for neutron economics. (If this number were 1.00 or less a chain reaction could not occur. If it were large, say 10, chain reactions would be very easy and critical sizes very small.)

At an average of 2.5, each fission yields:
one neutron to trigger another fission,
and one or two neutrons that could:

(a) escape and be lost, or be absorbed without further effect;

or (b) trigger another fission, thus contributing to an *explosive* chain reaction;

or (c) manufacture a new fissionable nucleus; e.g., by entering a U^{238} nucleus, which ultimately becomes a plutonium nucleus.

Breeder Reactors

A well-designed reactor is large enough to make escape of many neutrons unlikely, and is made of materials that will not absorb many neutrons unprofitably. Then there may be so many neutrons to spare that more plutonium atoms are manufactured than U^{235} atoms are used up. That is a "breeder" reactor that produces more new fissionable material (from U^{238}) than it uses in fission of U^{235} or of plutonium. Such a breeder makes use of all our uranium stock, U^{238} as well as U^{235}. Like all reactors, the breeders produce enormous quantities of heat—fissions must continue, generation after generation *at a steady rate,* each yielding nearly 200 Mev. This heat is removed by circulating liquid or gas, and it may be made to yield useful power.

Forecasting Nuclear Energy-Releases

We may know how to make an unstable nucleus by shooting a small projectile such as a neutron into a stable atom; but that does not tell us what energy-release to hope for. However, we can predict the energy-release in *nuclear* events from accurate *atomic* mass measurements with a mass spectrograph. We use $E = mc^2$. If we know how a lithium nucleus splits up when bombarded, we can calculate the energy-release from mass measurements. Similarly, if we know how U^{235} splits in fission, we can predict the release of energy. We can even go further and say clearly which elements *could* release energy if they did split by fission.

We treat a heavy nucleus as made up of *protons* and *neutrons,* which we lump together under a common name, *nucleons.* Imagine the process of making a heavy nucleus from separate nucleons. In assembling to form that nucleus, the nucleons must attract, when close enough, and as they move still closer *they must give out some energy.* Then they are tightly bound together, and energy from outside would be needed to tear them apart again. If the nucleons give out energy in assembling, they should end up in the nucleus with less mass. However, if the nucleons could be piled together loosely in the nucleus without meeting strong forces, the mass of the nucleus would just be the sum of the masses of the ingredients. Try an example, say lithium: the nucleus $_3Li^7$ consists of 3 protons and 4 neutrons. Its measured mass is 7.0165 Atomic Mass Units (from mass-spectrograph measurements). When alone, a proton has mass 1.0076 and a neutron 1.0089. Adding the masses of ingredients gives

$$3 \times 1.0076 + 4 \times 1.0089 = 7.0588.$$

This sum is greater than the actual mass, 7.0165; so in assembling to form a nucleus of Li^7 the nu-

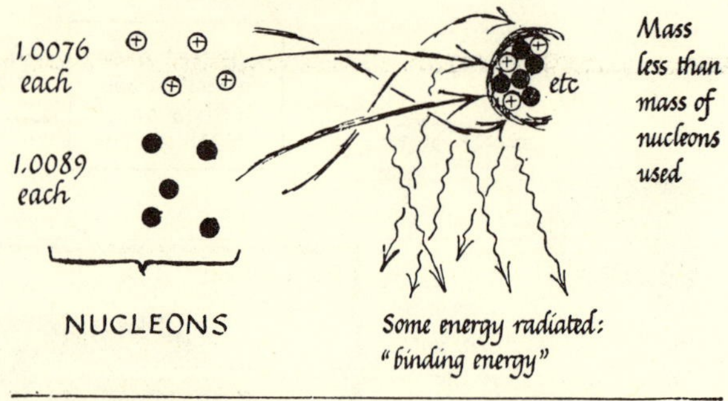

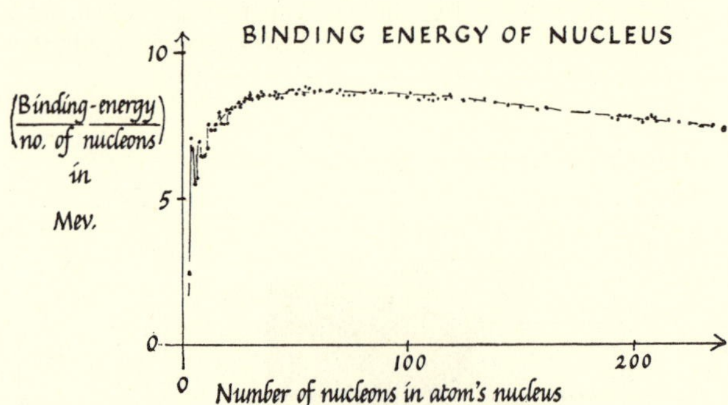

Fig. 43-33. Binding Energy
(a) Binding energy is energy that would be released if protons and neutrons could be assembled to form a compound nucleus.
(b) It is (therefore) the energy needed to pull the nucleus to pieces.
(c) Binding energy *per nucleon* is highest for the most stable nuclei—the middle elements.

cleons have "lost" some mass, or, rather some energy has been released, carrying mass away. The loss of mass looks small, but it represents an enormous amount of energy (45 million electron · volts per Li^7 nucleus formed).

Every nucleus in the whole list of atoms shows this (except H^1): its mass is less than the sum of the masses of its ingredient nucleons. So the assembling of *any* nucleus from protons and neutrons—if it could be done—would release an enormous amount of energy. We call this the *binding energy* of the nucleus. In reverse, the *binding energy* is also the *energy needed from outside to tear the nucleus apart into nucleons.*

In assembling to form a nucleus, the ingredient nucleons must lose some mass, the mass of the binding energy. From atomic mass measurements we calculate binding energies, and from binding energies we can predict the energy-release in any nuclear event, whether a minor change in bombardment or fission or fusion.

However, since we are dealing with one atom breaking up into others, the energy-accounting is easier if, instead of using the mass of the whole

FIG. 43-3 (f, g). CLOUD CHAMBER PHOTOGRAPHS OF TRACKS OF FISSION FRAGMENTS. In these, fission fragments made thick tracks, showing they carried large charges. Some of the thinner tracks were made by protons hit by neutrons of the original bombarding stream, others by α-particles from the uranium.
(J. K. Bøggild, K. J. Brostrøm, and T. Lauritsen, Royal Danish Academy of Arts and Sciences)

(f) Track of fission fragment from uranium (at left) bombarded with neutrons. The fragment, moving through a mixture of hydrogen and water vapor, knocked many protons forward and sideways (small spurs); and it made one severe collision with an oxygen nucleus (large spur).

(g) A pair of fission fragments from uranium in a thin sheet across the middle of the chamber.

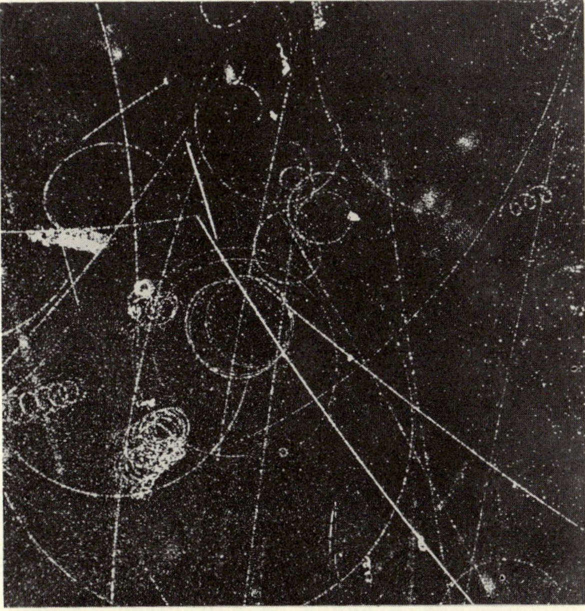

FIG. 43-3 (h). CLOUD CHAMBER PHOTOGRAPH: COSMIC RAYS. Cosmic-ray particles, streaming through a slab of dense material above the cloud-chamber, produced electrons and positrons (some fast, some slow), which made tracks in the chamber, in a strong magnetic field. The picture also shows an unusual event: the decay of an unstable particle (which made the two heavier tracks that form a shallow V). (E. W. Cowan)

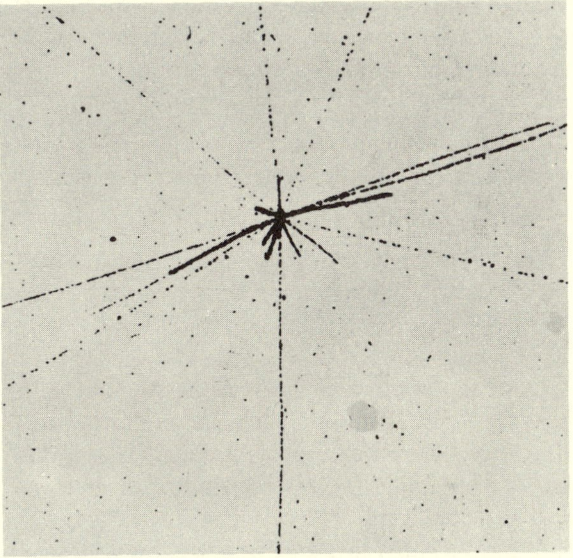

FIG. 43-3 (i). EMULSION TRACK PICTURE. Here the tracks are not shown by drops of water in a cloud-chamber, but by black specks of silver developed in photographic emulsion. This picture, greatly magnified, shows a "star" or nuclear "explosion": cosmic-ray particle (about 1000 Mev) hit a nucleus, probably silver, in the emulsion and shattered it into 7 protons, 5 alpha-particles, and several heavier fragments. The track of the primary bombarding particle is invisible. (J. Hornbostel, Brookhaven National Laboratory)

nucleus, we divide by the number of nucleons in it and keep track of the MASS PER NUCLEON. Then when the MASS PER NUCLEON decreases in some event, we know that the nucleons altogether have lost some mass and therefore some energy has been released. So we plot a very important graph of:

MASS PER NUCLEON *vs.* MASS NUMBER,

for all the elements.

Then we can see at once for any atom how much mass (and thence energy) each nucleon has lost in assembling: we compare the value of the MASS PER NUCLEON for that atom with the "average mass" of the separate nucleons—somewhere between 1.0076 for a proton and 1.0089 for neutrons, say 1.0083. The lower an atom's place on this graph, the greater its binding energy.[24]

We calculate MASS PER NUCLEON thus:

$$\text{MASS PER NUCLEON} = \frac{\text{MASS OF NUCLEUS}}{\text{NUMBER OF NUCLEONS}}, \text{ where}$$

$$\text{MASS OF NUCLEUS} = \text{MASS OF NEUTRAL ATOM} - \text{MASS OF ITS ELECTRONS}$$

MASS OF ATOM (or rather its ION$^+$) is measured with great precision by mass spectrograph, in Atomic Mass Units (which give O^{16} mass 16.0000)

NUMBER OF NUCLEONS (= [PROTONS + NEUTRONS] in atom's nucleus) *is* the atom's MASS NUMBER; that is, the MASS OF ATOM, in a.m.u., ("atomic weight"), rounded off to the nearest whole number.

In the whole list of elements with all their isotopes the NUCLEAR MASS (in Atomic Mass Units) never gets far from a whole number. For example:

hydrogen is above 1, at 1.0076
lithium^7 has 7.0165
iron^{56} falls below 56 to 55.938
uranium ^{235} has 235.068

That whole number (1, . . . 7, . . . 56, . . . 235, . . .), tells us the number of nucleons in the nucleus, the MASS NUMBER. The differences between ATOMIC (or NUCLEAR) MASSES and whole numbers reveal differences of binding energy—differences of energy released as nucleons are packed in.

When we divide NUCLEAR MASS by NUMBER OF NUCLEONS or MASS NUMBER we obtain values that start at 1.009 and 1.008 for neutron and proton, and drop down to a minimum 0.9993 for "middle elements" such as iron, copper, bromine, krypton, and then rise slowly to about 1.0003 for uranium. Therefore *if* a heavy nucleus could be persuaded to undergo fission into two middle elements, its nucleons would lose some mass, in the course of a large energy release. From uranium to the middle elements the MASS PER NUCLEON drops by nearly 0.001, judging from the graph. For the 235 nucleons in a U^{235} nucleus, the mass of energy released should be 235 × 0.001, or 0.235 Atomic Mass Units. The energy which has that mass is 0.235 × 931 Mev or about 200 Mev. From the graph you can see that an energy-release by fission could only occur in the heaviest nuclei. The nuclei of middle elements are the most stable: their nucleons could not lose mass in a move to left or right on the graph, but have the greatest tear-apart energy of all.

Nuclear Energy from Fusion

Instead of fission, there might be energy release by *fusion*, joining together of light nuclei. The graph runs downhill from lightest elements to medium, showing that fusion would release energy. Fusion needs no disturbing neutron to start it like fission. Instead, the problem is to shove the light nuclei, against electric repulsion, close enough together to get them within reach of attractive nuclear binding forces. If we could compel two protons and two neutrons to gang up into a helium nucleus—or four protons to join up with appropriate changes—we could obtain an enormous release of energy. Such a pushing together could be done by heating them to stupendous temperatures until ordinary collisions bring them close enough together for nuclear forces to produce fusion. Once started, the fusion process would perhaps provide enough heat to maintain the high temperature for more fusions. We should have a vast firework, limited in size only by the amount of material available. This probably happens in hot stars. A several-stage process of "burning" hydrogen into helium by nuclear fusion is probably the source of the Sun's continuing stream of radiation.

For our own engineering uses, the pushing together of plain hydrogen nuclei requires far too high a temperature—or else too long a time for chance collisions of extra violence—to be of use to us. Heavy hydrogen nuclei, deuterons, are easier to push into fusion, but still extremely hard. Tritons, the still heavier hydrogen nuclei, $_1H^3$, offer much better chances of joining with hydrogen or deuterium nuclei. However, tritium (extra-heavy hydrogen) has to be manufactured in a reactor at the expense of neutrons. Could still heavier atoms be used? The next in the list are two lithium isotopes and these might offer material for a compact fusion bomb. Presumably any such bomb would have to be started by a *fission* bomb.

[24] Many authorities plot BINDING ENERGY PER NUCLEON instead of MASS PER NUCLEON against MASS NUMBER. The graph is like this one, but inverted, and the zero-level is different.

To obtain useful energy from fusion for peaceful uses—electric power production—raises very difficult problems of *containing* the reaction. The gases must be heated to, say, 50,000,000° C; and any solid container close to them will vaporize. If there *is* a useful release of heat by fusion that will make the container problem worse. However, there is hope of containing the reacting material by using electromagnetic fields. A magnet may be kept floating by other magnets—though its equilibrium is unstable. An electric current—a stream of electrons in a wire or a stream of ions alone—is "pinched" by its own magnetic field: neighboring streams of flowing charges attract (like currents in parallel wires) and squeeze the whole flow into a narrower stream. This pinch-effect can be combined with other catapult forces exerted by outside magnetic fields to confine the *plasma*, a mess of high-speed nuclei and electrons, in a "magnetic bottle," where fusion may succeed.

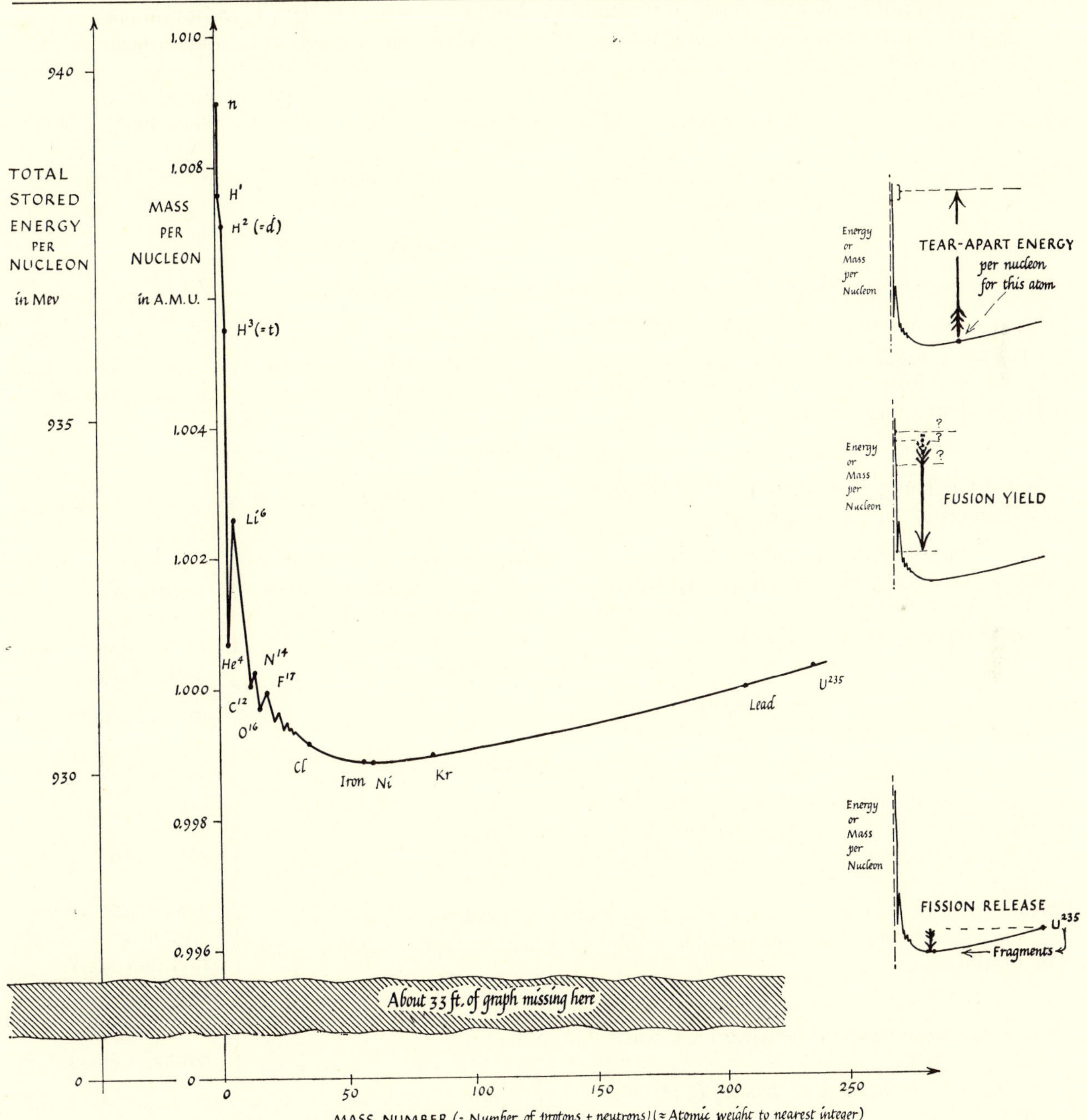

Fig. 43-34. Graph of Mass-per-Nucleon of nuclei *vs.* Mass Number

$$\text{The mass-per-nucleon is } \frac{\text{mass of nucleus (by mass spectrograph)}}{\text{total number of protons+neutrons}}$$

CHAPTER 44 • MORE THEORY AND EXPERIMENT: PHYSICS TODAY

"Every new theory . . . believes that at last it is the fortunate theory to achieve the 'right' answer. . . . When will we learn that logic, mathematics, physical theory, are all only our inventions for formulating in compact and manageable form what we already know, and like all inventions do not achieve complete success in what they were designed to do, much less complete success beyond the scope of the original design, and that our only justification for hoping to penetrate at all into the unknown with these inventions is our past experience that sometimes we have been fortunate enough to be able to push on a short distance by acquired momentum?"

—P. W. Bridgman, 1936
The Nature of Physical Theory (Princeton University Press, 1936)

"At the midnight in the silence of the sleep time
When you set your fancies free. . . ."

—Robert Browning

[This final chapter does not so much end the course as link it to future inquiries and reading. It is a chapter to read for yourself at leisure. It cannot bring you completely up to date with right knowledge at last. Instead it will leave you with doubts and ragged ends of unfinished knowledge—and that is characteristic of the frontiers of any growing science.]

CLASSICAL PHYSICS

By the beginning of this century mankind had built up, in the course of a hundred or so generations, a great framework of physical science:

Statics of ropes and pulleys, pillars, bridges, . . . ; equilibrium rules *Dynamics* of moving bodies, force and mass, Laws of Motion; Momentum and Energy *Hydrostatics* of pumps, air pressure; rules for floating ships and their stability . . . laws of fluid pressure; Boyle's Law, . . . *Hydrodynamics* of fluid flow, in streamlines and in vortex motion	ALL TIED TOGETHER BY NEWTONIAN LAWS
Electromagnetism of charges, currents, magnets, . . . fields . . . *Optics* of light rays traveling straight and bent by lenses to form images; of wave-behavior in diffraction and interference; electromagnetic theory of light	TIED TOGETHER BY MAXWELL'S LAWS
Acoustics: the physics of musical instruments and sound waves *Heat*: thermometry and calorimetry; heat as a form of energy *Properties of Matter*: Elasticity; Friction of solids and fluids; Surface Tension; etc.	TIED TO MECHANICS
Kinetic Theory of Gases, and heat treated as molecular motion *The behavior of Atoms and Molecules* in crystal structure, in surface tension, elasticity, diffusion	MECHANICS APPLIED TO COLLECTIONS OF INVISIBLY-SMALL PARTICLES
Thermodynamics: relations between heat, work and matter	TREATED BY OVERALL RULES

Side by side with all this Physics, *CHEMISTRY* grew into a great science of molecular structure and behavior:

Inorganic Chemistry: of chemical reactions and properties, interpreted in terms of elements, compounds, atoms and molecules.

Organic Chemistry: the study of the carbon compounds involved in living substances: an almost infinite family of molecules—investigated, catalogued, displayed by "structural formulas," and even synthesized from elements—ranging from simple CO_2 to complex and vast protein molecules.

Physical Chemistry: the study of the physical effects of chemical operations: heats of reaction, heats of solution; molecular mass estimates by vapor pressure, by change of osmotic pressure, by changes of freezing point of solutions; the mechanics and statistics of chemical reactions; etc.

Yet a big gap remained between Physics and Chemistry. Chemists made increasing use of physical instruments, but physicists remained curiously aloof in many cases, and missed good chances to link chemical changes with the developments of physical knowledge.

Other sciences, *ASTRONOMY, MINERALOGY*, &c., had branched out, maintaining strong links with Physics and Chemistry.

In all this PHYSICAL SCIENCE certain overall rules or Principles emerged as reliable summaries and guides: Vector addition of Velocities, Forces, etc., and Galilean Relativity; Newton's Laws of Motion; the Constancy of Mass; Conservation of Momentum: Conservation of Energy; Inverse-Square-Law Gravitation; Coulomb's Law and Maxwell's Equations which contain it; Light, etc., treated as Electromagnetic Waves; the Indestructibility of Atoms; the Identical Nature of each Element's Atoms.

All the physics of these developments is now called *Classical Physics*. It seemed well understood, complete except for minor details, and capable, and satisfying. Some of it was extended downward to atoms and molecules and upwards to the solar system on the assumption that the same general rules and principles apply. The physics of falling rocks, bouncing balls, etc., was confidently extrapolated[1] to the planets and to gas molecules.

THE NEW PHYSICS

In this century, the assurance and completeness of classical physics were upset by five great developments:

(1) *Atomic structure* was revealed by the discovery of electrons and radioactivity. Atoms can be broken and they can even change to other atoms. *The nuclear atom model emerged.*

(2) *Relativity* cleared up some paradoxes and modified our treatment of space, time, mass, and fields.

(3) Light (and all other radiation) was found to have its energy packaged in bullets although traveling by waves. *Quantum theory emerged.* This led to Bohr's atom-model, with the guidance of his Correspondence Principle.

(4) *The particles of atomic physics (electrons, nuclei, . . .) were found to behave both as waves and as bullets.* The wave:particle double-behavior now covers both radiation and material particles. This led to new theory, "Quantum Mechanics," with the important philosophical ideas of the Uncertainty Principle and Complementarity.

(5) *More and more sub-atomic particles were discovered*: electrons, nuclei, neutrons, mesons, neutrinos, and recently many others.

Of these developments, (1) has been described in recent chapters, and (2) in Chapter 31. The developments of (3) and (4) are discussed in this chapter. We shall not deal with the newest particles of (5)—experimental and theoretical attacks are proceeding, and we still look forward to solving present puzzles of nuclear forces and structure.

Atomic Physics 1890-1915

At the beginning of this century, "atomic physics" was a young science, rapidly growing from new electrical experiments. The older science of electricity and magnetism had been well built in the last century, providing consistent knowledge of charges, currents and fields. On the practical side, scientists and engineers developed—by interpolation—its industrial uses, electric motors, meters, lamps, power systems, and communications. On the theoretical side, the experimental laws, codified in Maxwell's equations, led logically to a prediction of radio waves. At the turn of the century, radio waves had been produced electrically but not yet put to use; and light was recognized to be very-short-wave radio. Then, when the picture seemed nearly complete, new knowledge, of atoms and electrons, came

[1] "Extrapolation" means trading on known information to make a guess beyond the known range. In contrast, "interpolation" means making a guess within the known range. If we know the times when a train leaves Boston and arrives at New York, we can guess by interpolation the time it passes through New Haven; but we extrapolate when we guess its time of reaching Washington. Interpolation is safer—with sufficient data we can plot a reliable graph and read off intermediate information quite reliably. Extrapolation is risky—the train might end at Philadelphia! Yet the most fruitful speculations in science are made by wise extrapolation—by what John A. Wheeler calls "daring conservatism."

pouring in from several different directions: the discoveries of X-rays, radioactivity, the photo-electric effect, and the emission of electrons from hot metals; and measurements of ions and electrons in discharge tubes. Atoms could be dissected into positive ions and universal electrons, with measurable properties. Pictures of the internal structure of atoms were being devised and tried out as the century began.

In the first quarter of this century, the new atomic knowledge grew; but some serious paradoxes appeared. Rutherford suggested a good theoretical model of the atom: a small massive nucleus surrounded by moving electrons, like a minute solar system. The electrons are all the same, 1/1840 of the mass of a hydrogen atom, with a universal charge $-e = -1.6 \times 10^{-19}$ coulomb. The nucleus is incredibly small, 1/10,000 of an atom's 1-Ångström-Unit-diameter. The nucleus carries a positive charge, ranging from $+e$ for a hydrogen nucleus (proton) to $+92e$ for uranium, with an inverse-square electric field pointing out from it. The number of $+e$ charges, Z, gives the "atomic number" of the element, and determines its place in the chemical table. That is because it also gives the number of electrons surrounding the nucleus in a neutral atom. These electrons, grouped in some scheme of layers or shells, are responsible for chemical properties. Atoms that lose or gain electrons become ions, the active agents in many chemical processes. Electrical forces between + and − ions bind some molecules together—e.g., the salt molecule Na^+——Cl^-. In other chemical compounds, electrons must be shared between atoms rather than handed over outright. *Electrical forces hold all atoms and molecules together.*

FIG. 44-1. ATOMIC PICTURES

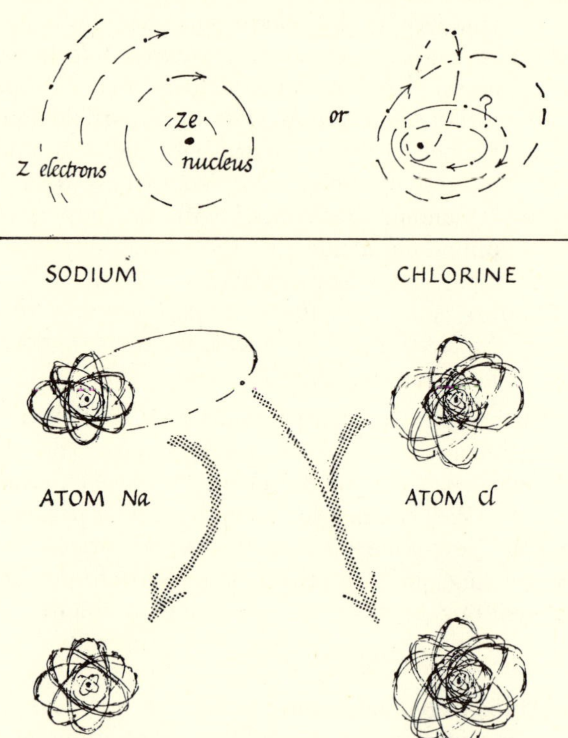

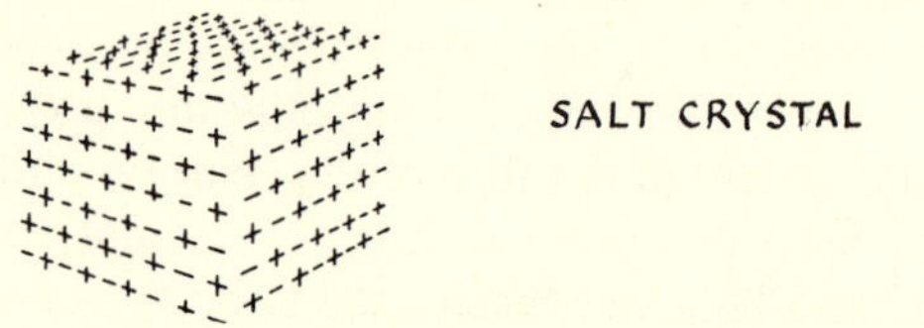

(a) Rutherford atom. (b) Bohr atom pictures. (c) Ions in a salt crystal (see Ch. 31).

The lightest atoms in the list (hydrogen, helium, lithium, . . .) have their few electrons far out from the nucleus, with a vast intervening region of empty space[2] and inverse-square-law field. The heaviest (gold, uranium, etc.), with nearly a hundred electrons, hold them in several layers. The *innermost* group (still far out from the nucleus) are strongly bound by the electric field of the big nuclear charge and are not disturbed in ordinary chemical actions. However, they can be upset by bombardment with energetic electrons, X-ray photons, etc.

Only the *outermost* group of electrons are involved in chemical changes. They find most of the nuclear attraction neutralized by the repulsion from electrons farther in—the inner electrons exert a "shielding" action. So the outermost electrons are loosely bound, and easily exchanged or shared, making the forces that bond atoms in chemical compounds and the fields that store "chemical energy." Chemical evidence suggests that only a few electrons belong to that outermost group, and later reasoning by Bohr confirmed the picture in detail: one in hydrogen, sodium, potassium, and other metals that make + ions by losing one electron; two in copper, etc., that make ++ ions; three in aluminum. . . . In chlorine, the outermost electrons form a group of 7 which grabs one more to make a compact stable group of 8 in the ion Cl^-, a structure electronically like neutral argon. (That is why sodium will combine so easily with chlorine to make

[2] There is no matter there, but the space is crammed full of energy—the strain-energy of the electric field of the nucleus.

a salt in which the atoms remain ionized even in the solid crystal.)

Beyond these suggestions from chemistry, there was no clear architectural scheme for the array of electrons. And the Rutherford model left an awkward paradox unsolved: the circulating electrons should radiate electromagnetic waves[3] and spiral in faster and faster—infra-red light, then red, green, . . . , ultraviolet, . . . X-rays, . . . in a rapid collapse of the atom. Obviously, atoms do *not* collapse like that. We do not see the glow, and we do find atoms stable. Why?

Something was known about the structure of the nucleus itself. Radioactive elements emit α, β-, and γ-rays with such energy that they must come from the nucleus. Mass spectrographs showed nuclear masses to be almost whole multiples of the proton mass. So theories of *nuclear* structure started by picturing a compact group of protons and electrons held by special forces. We now consider that model uncomfortable: electrons are so light that their wavelength is too big to fit into the measured limit of nuclear size. Besides there were problems of spin-conservation. The discovery of the neutron offered a better ingredient. Nuclei are now regarded as made up of protons and neutrons,[4] somehow bound very closely.

When a radioactive atom disintegrates there is a huge release of energy that can be measured precisely. But, given a single radioactive atom, we cannot predict its exact lifetime before disintegration. We can only give a probability, such as a half-life for a collection of atoms. Watch a Geiger counter recording α-particles: you will see them arrive with the haphazard timing of pure chance, like raindrops on a tin roof. In the early 1900's, physics was taking a further turn towards a statistical viewpoint. That was already well established in kinetic theory, where the regularities such as constant gas-pressure, steady flow of gases, . . . , Boyle's Law . . . , were seen to be statistical averages covering a vast crowd of molecules. Now chance seemed to determine the break-up of radioactive nuclei, the escape of α-particles from nuclear turmoil.

How do most nuclei hold together permanently, while others blow up on pure chance? This question asked for a new viewpoint, a new theory.

Relativity 1905

Meanwhile, Relativity was developed and accepted. On the philosophical side, it preached a reforming discipline: do not endow pictures of nature with untestable details. *Do not even ask questions that assume such details.* (For example, do not decorate electrons with unobservable properties by asking what color they are or by sketching sharp orbits for them.) On the hardheaded physics side, Relativity predicted different measurements for experiments on moving objects. Here are some of them:

(i) A moving object will show, to an observer at rest (or moving past it), an increased mass m, greater than its "rest-mass," m_0. Its mass m will increase with speed, rising to infinity at the speed of light, c. Therefore no *material object* can be accelerated to move faster than light, because an infinite force would be needed.

(ii) Energy of any kind, has mass, of amount ENERGY/(SPEED OF LIGHT)^2 : $m = E/c^2$.

(iii) Any body of mass m should be regarded as having TOTAL ENERGY mc^2. (This includes the body's K.E. and its "rest energy," m_0c^2, locked up in its internal structure.)

(iv) Past, present, future are not always divided absolutely. For some events (far apart in space or very close in time) observers with different motions will make different decisions. One observer may find that events P and Q occur simultaneously, another observer, moving with different velocity may see Q happen before P; and still another knows that P happens before Q. Therefore, Relativity warned us not to treat *cause and effect* always with such confidence.

(v) All observers, whatever their motion, will measure the same speed for light—running towards the source or away will have no effect. This was the original assumption from which Relativity rules were derived. We now develop it into a more general requirement of all measurements, that ALL LAWS OF PHYSICS TAKE THE SAME *FORM* FOR ALL OBSERVERS, WHATEVER THEIR MOTION RELATIVE TO THE APPARATUS.

[3] In modern "race-track" accelerators, electrons pursuing great circular orbits not only "should" radiate but do—at enormous cost of wasted power.

[4] The neutron itself *may* be regarded as made up of a proton, an electron and a neutrino—but those are the products when a *free* neutron disintegrates. A neutron *inside a nucleus* proves elusive when we investigate its structure. It behaves like a single particle, or perhaps like a proton in close association with a meson—which again might be imagined to yield an electron and neutrino. On that last view, all nuclei, in fact all matter, can be said to be "made of" protons, electrons, and neutrinos, with the nuclear electrons bound in particles called neutrons; and mesons acting as nuclear mortar. However, by the time we get into the interior of nuclei, haggling about whether electrons are "really" there carries us too far from gross observables. It should remain a matter of choosing the most convenient theoretical view.

Models

Early in this century, it had become clear that in building "models" of nature, whether of tiny atoms or of great galaxies of stars, we were making huge assumptions. We were using the rules gathered from man-sized experiments[5] and taking a risk that they hold for atoms. We were applying macroscopic (= large size) physics to microscopic nature. We interpreted microscopic nature in terms of macroscopic machinery—accelerating trucks, flying baseballs, ... So long as the models were *fruitful*—suggested experiments, helped interpretation, encouraged speculation—well and good. But, with the warnings of Relativity ringing in their ears, scientists grew much more careful about calling their models true. Their students were ready to believe the model tells why, but the best thinkers were once again stepping down from trying to say WHY ("we know from our model that this happens *because* . . ."), to saying WHAT ("in terms of our model, the behavior looks like . . . and perhaps we should experiment to see whether . . ."). That is an old lesson, never fully learned—if it were, scientists would lapse from credulous glorying into over-cautious denial—but a lesson to be applied again and again. As a good scientist, you should be suspicious of models, particularly if they are treated as true descriptions. Yet you should not despise models as childish. They play an essential part in the human mind's method of perceiving and learning. When our senses tell us about something quite new, our first mental move is to find something familiar that it reminds us of. We attach the old familiar label very strongly to the new thing, and only slowly change to a new view.[6] Even the able modern scientists who cry loudest for operational methods—"describe everything in terms of methods of observation"—set their fancy free with models when they speculate on new developments.

THE DEVELOPMENT OF THE QUANTUM IDEA, 1900-1915

In several fields of physics, traditional models were upset by discoveries pointing towards a strange property of light and other radiation: *bullets of energy.* Just when light seemed established as electromagnetic waves, experiments began to show that it also consists of small definite packets of energy, like particles. This *quantum*[7] view arose from several paradoxical conflicts between experiments and classical theory.[8] It resolved the conflicts by a single rule modifying classical physics:

any interchange of energy between matter and radiation occurs in definite packets of energy, "quanta." For each packet or quantum:

ENERGY = (universal constant, h) · (FREQUENCY OF RADIATION).

Thus, not only are matter and electricity discontinuous, parcelled in atoms and electron-charges, but energy too is discontinuous in certain important circumstances. There is no atomic unit of energy; but (for certain forms of energy) the quantity ENERGY/FREQUENCY has a universal "atom" or unit, h. That quantum restriction looks harmless—particularly to readers who have heard about it before—but it conflicts with Newtonian mechanics when applied to molecules, atoms, electrons, . . . :

[5] Philip Morrison points out that the things men make and work with fall in a narrow range. The biggest skyscraper is only 300 men high. If we tried to build one 3000 men high our materials would fail—unless we made it solid like a mountain. The ball from a ball-point pen is about 1/3000 man wide. We are rash to assume our direct experimental rules extend up to stars 3000000000000000000 men apart, or down to electrons $\frac{1}{300000000000000}$ of a man wide.

[6] Psychologists let an observer peep into a distorted room that is built askew, with slanting walls and floor and decorated in perspective to conceal the distortion. The normal observer interprets what he sees in terms of a familiar model, the usual undistorted room. Then when ordinary people enter at different corners, he sticks to his model and insists that the people are giants and dwarfs (see Fig. 44-2).

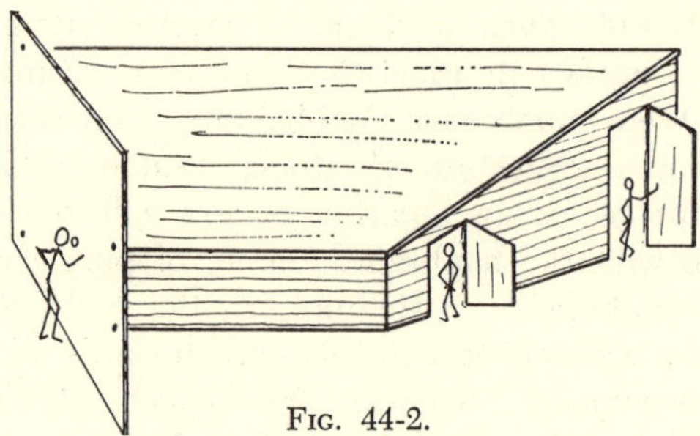

FIG. 44-2.

[7] From Latin *quantum* meaning "how much" or "as much as." This word generally means a share or allowed portion. In physics it means a very definite chunk, "just so much by the formula."

[8] This idea made a revolution, first in atomic physics, then in all basic physics and chemistry, and finally in the philosophy of science. Yet it is too long a story to open up properly here. It would require much discussion of experimental evidence, with considerable mathematical analysis. It would require detailed treatment of waves; it would involve mathematics of chance distributions; and then the argument would force you to accept a new view of nature on the atomic scale: on one hand, police rules of quantum restrictions; and, on the other hand, a free-enterprise system of chance behavior in the atomic world that would make the Newtonian determinism look like slavery. Here we shall only mention some results. We cannot even refer you confidently to other books. The elementary books hand out information and state rules. The advanced ones are weighed down by the mathematical form that the new explanations seem to require.

(1) Heat a block of blackened metal white hot so that it emits a copious stream of radiation: ultra-violet + visible light + infra-red + radio waves. General arguments with Newtonian mechanics predict that nearly all the radiation-energy will flow out in the ultra-violet (shortest wavelength, highest frequency). Not true. An absorbing thermopile exploring the spectrum shows a hump of maximum energy-flow in the middle. That conflict was known in 1900, and led to the first suggestion of the quantum restriction. With that restriction imposed, mechanical theory predicts the observed spectrum.

(2) Warm up a sample of solid or gas and measure its specific heat around various temperatures. Newtonian physics predicts, through equipartition, a constant specific heat, independent of temperature. Not true. Measured values of specific heats rise from very small values at very low temperatures towards the classical predicted value at very high temperatures. The quantum restriction predicts that (See Ch. 30).

(3) Light can eject electrons from certain metal surfaces. Classical mechanics would picture the incoming light waves building up a bigger and bigger vibration of an electron tethered to a metal atom, until the electron breaks loose. On that view, very weak light should always show a long delay in building up enough vibration; also, very strong light might hurl electrons out with greater energy. Not true. Electrons are ejected with the same full energy however bright or faint the light. And they emerge with random timing, as often as not just after the light is first turned on. As Einstein soon suggested, light packaged in bullets of energy would easily account for this "photoelectric effect."

(4) Other experimental properties of spectra seem strange to classical physics. The spacing of the bright lines in the spectrum of glowing gases had been measured in the last century and reduced to simple formulas that could not be "explained" classically. X-rays exhibited similar spectrum regularities in their far-shorter wavelengths. Bohr showed how a quantum theory could provide good explanations and a rich field for further interpretations.

All these, which formed the basis for quantum theory, are discussed in greater detail in the following sections.

The Spectrum of White Light

Heat a block of blackened metal white hot and analyze the radiation from it. Better still, heat up a furnace, and let radiation pour out through a hole in its wall. Remember that a good absorber must

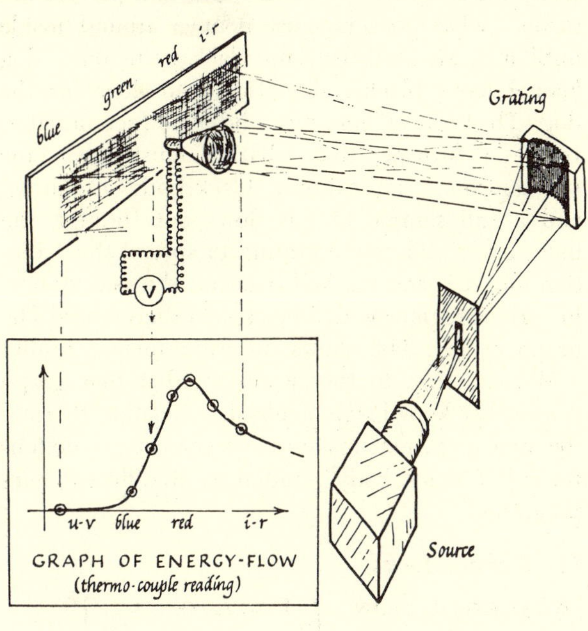

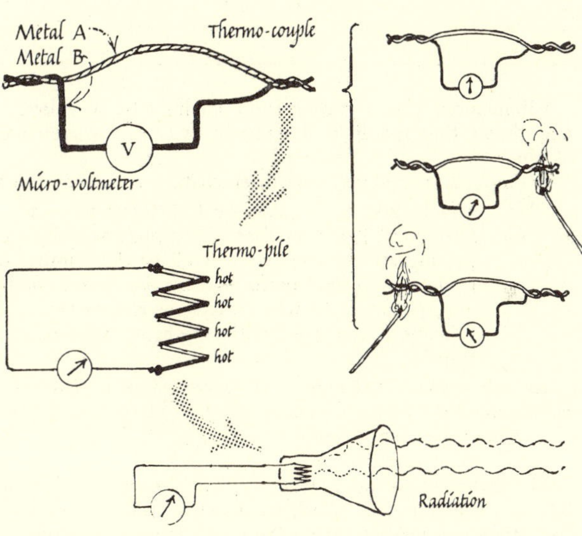

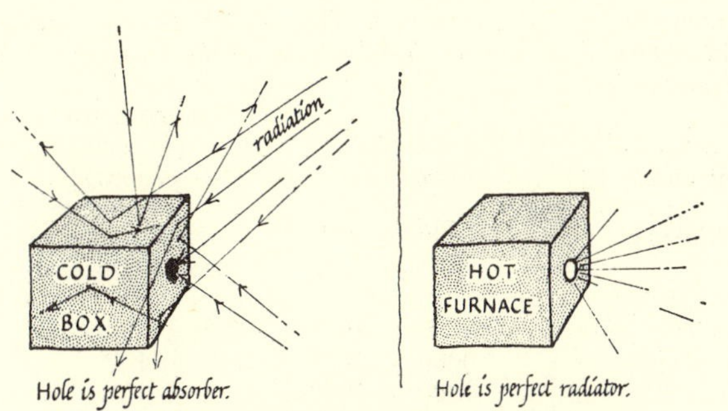

Fig. 44-3. Energy-Spectrum of Radiation
(a) Apparatus. (b) Details of thermopile.
(c) Hole is perfect "black body" radiator.

be a good emitter (Ch. 26, Problem 23 and Ch. 4, Experiments F 7 and 8). The best emitter is a perfectly black body. A hole in a box is a perfect absorber: what goes in must bounce around inside until it is all absorbed—no black paint on a dog kennel looks blacker than the open door for the dog. Therefore a hole must be a perfect radiator. Inside a furnace, the radiation must reach the full assortment typical of a "black body" radiating, and a full sample of this flows out through the hole. Use a diffraction grating to spread the radiation into a spectrum, and measure the energy-flow in various regions with a blackened thermopile. The graph of Fig. 44-3 shows the experimental result.[9]

We can turn to theory and predict that graph from other knowledge of physics. Reliable thermodynamic argument makes some correct predictions for full ("black body") radiators at different temperatures:

(i) Stefan's Law:

$$\begin{bmatrix}\text{TOTAL ENERGY FLOW}\\ \text{in the whole spectrum}\end{bmatrix} \propto \begin{bmatrix}\text{ABSOLUTE TEMP.}\\ \text{of radiator}\end{bmatrix}^4$$

$$\text{or } E \propto T^4$$

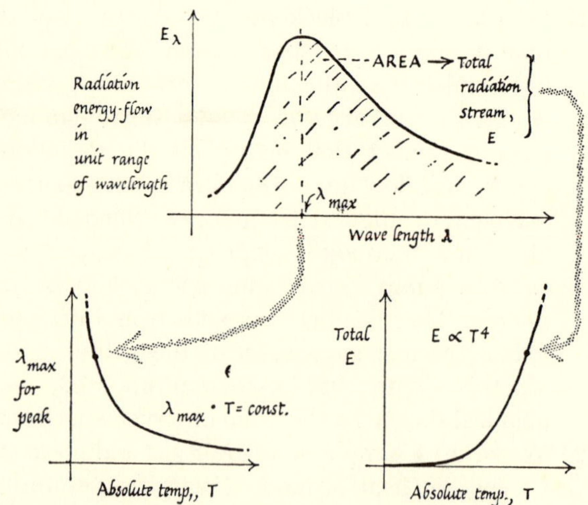

FIG. 44-4a. RADIATION: EXPERIMENTAL FACTS

(ii) Wien's Law:

The peak of the graph of ENERGY *vs.* WAVELENGTH occurs at a wavelength, λ_{max} that varies as 1/ABSOLUTE TEMPERATURE, SO

$$\lambda_{max} \cdot T = \text{constant}$$

Experiments support (i) and (ii) up to the highest temperatures for which we have measurements with gas thermometers. We trust both (i) and (ii), and use them to estimate temperatures of furnaces, and Sun and stars, on the Kelvin scale. Both laws yield the same temperature for the Sun's *surface*, $\sim 6000°$ K.

But a further discussion assuming traditional mechanics[10] predicts the detailed shape of the graph: an entirely wrong shape. The prediction, in glaring contradiction of fact, says: Exchanges inside the furnace must transfer energy from any wavelength to a shorter wavelength to still shorter, until *practically all the energy is in the ultraviolet or beyond.* Thus the predicted graph rises towards infinity in the ultraviolet. This "ultraviolet catastrophe" is certainly not observed with real radiators, from red-hot iron to white-hot Sun. They radiate an orange glow and do not cool in a rapid ultraviolet flash. Physicists made repeated attempts to derive the experimental energy-distribution from standard mechanics of waves, and Maxwell's equations, and electron theory. All failed: again and again, prediction agreed well with experiment at

---

[9] Remember that a grating sorts out light by wavelength, and shows that red light has nearly twice the wavelength of blue.

A simple harmonic wave has three characteristic quantities:

(i) wavelength, λ, the distance from crest to crest

(ii) frequency, f, the number of complete wavelengths passing an observer per second; or the number of cycles per second performed by the source, or by any specimen the wave excites as it goes by.

(iii) velocity, v, the speed with which the wave pattern travels.

In one second, a length v of wave pattern passes by, containing f wavelengths. Therefore, $v = f\lambda$

VELOCITY = FREQUENCY · WAVELENGTH,

for any periodic wave.

For *light* in air or vacuum, v has the universal value 3.0×10^8 meters/sec, which we label c. And we label the frequency ν instead of f. (ν is the Greek letter n, pronounced "new.")

Then $c = \nu\lambda$ and frequency $\nu = c/\lambda$. Since c is constant, FREQUENCY $\propto$ 1/WAVELENGTH. The smaller the wavelength, the higher the frequency. The table below shows some rough values.

COLOR OF LIGHT	WAVELENGTH (*in meters*)	FREQUENCY (*in cycles per sec*)
ultraviolet	3000×10^{-10} and down . . .	10×10^{14} and up to $30{,}000 \times 10^{14}$ for x-rays & γ-rays
violet	4000×10^{-10}	7.5×10^{14}
green	5000×10^{-10}	6×10^{14}
red	7000×10^{-10}	4×10^{14}
infra-red	$10{,}000 \times 10^{-10}$ and up to thousands	3×10^{14} and down to 10^6 for long radio waves

[10] Pour in Newton's Laws, and thence equipartition, and geometry of waves. Then calculus + algebra + geometry will grind out the result with assurance. The line of argument runs thus: Imagine a wave setting up a standing-wave pattern as it zig-zags around inside the furnace. Only some wavelengths can fit into a stationary pattern in the furnace box, *many* more in short wavelength ranges than in long; still more in very-short wave ranges, and so on. Equipartition awards *each* standing-wave pattern an equal share of radiation energy and therefore. . . .

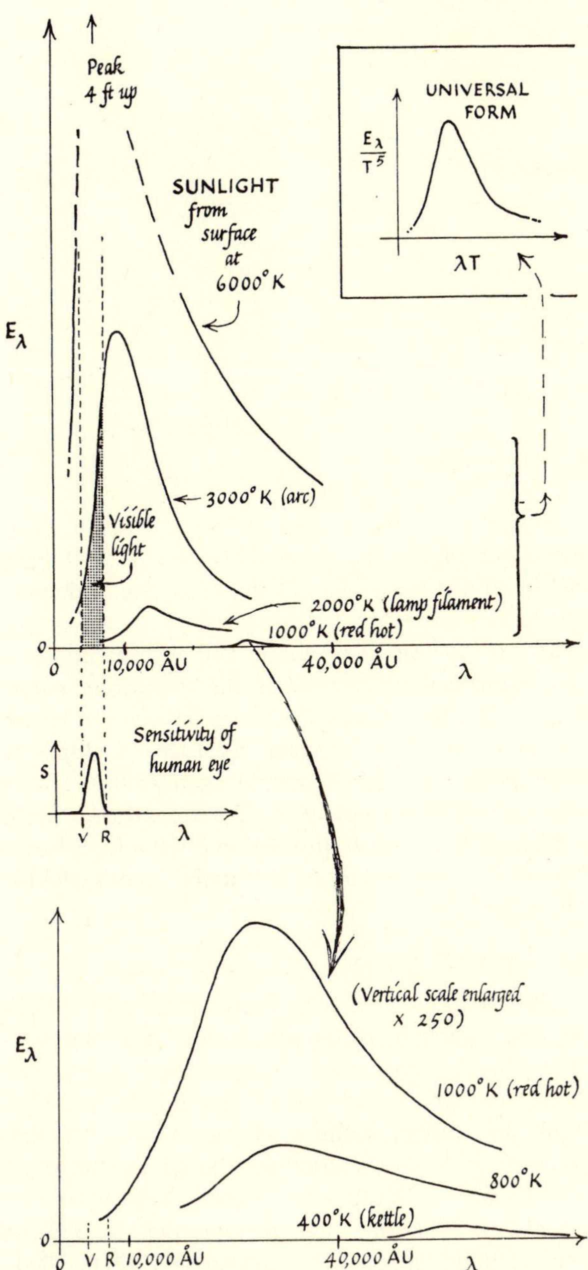

Fig. 44-4b. Radiation: Experimental Facts
Energy-flow, in various regions of spectrum, from perfect "black body" radiator at various temperatures.

the red end of the spectrum; but at the other end the ultraviolet catastrophe emerged. Then, about 1900, the German physicist Max Planck attacked the problem in reverse and asked—as Einstein did in Relativity—what is the minimum modification that will make theory fit the facts? He had only the experimental curve, with no algebraic formula for it, so he could not make the logic-machine grind out the exact modification. Instead, he had to guess and try, like Kepler. In a brilliant speculative analysis, he found a successful rule.

Planck saw that some rule was needed that would leave red light practically alone and discriminate against violet and ultraviolet. Consider an analogous problem in a big grocery store. How could one discriminate against some particular item without changing its price? That could be done by restricting its sale to huge packages. For example: it matters little to the distribution of money in family shopping if rice, sugar, and salt are not loose but must be bought in 1-pound packages. But if sugar is put up in unbreakable bags of 200 lbs. only a few families with strong arms, large cars, and other resources will buy sugar. The traffic in sugar could be ruined by restricting it to big enough bags.

Planck guessed, essentially, that radiation *energy* is packaged in minute, definite atomic chunks, to be called "quanta." The quanta are not the same size for all colors, but are tiny for infra-red, small for green, big for ultraviolet. How does such a packaging affect the predicted radiation-spectrum? Suppose the radiation comes from a hole in a furnace; and consider the traffic in energy inside, between radiation and furnace walls. The quantum-restriction will make itself felt at the ultraviolet end of the spectrum where the quanta are big. Infra-red light will continue to pour out in a copius stream of tiny quanta, too tiny and too numerous to modify the traffic. But ultraviolet light must be emitted in big quanta or not at all. Blue, violet, and above all ultraviolet will be seriously limited and the ultraviolet catastrophe averted. Here is Planck's rule in detail:

Radiation is packaged in chunks (= *"quanta"*)—a natural guess to try in this age of atoms of matter, atoms of electric charge, ...

Each quantum consists of radiation of a single frequency (and thus of single wavelength, light "all of one color"—*monochromatic radiation*).

The rule defining sizes of quanta is:

ENERGY IN QUANTUM $\propto$ FREQUENCY OF RADIATION IN QUANTUM

or, ENERGY $= h \cdot$ FREQUENCY or $E = h\nu$

(and therefore $E \propto 1/\lambda$) where h is a universal constant (now called Planck's constant) and ν is the traditional symbol for radiation frequency.

From this, Planck predicted the energy-distribution in black-body radiation. His prediction fits the experimental graph at the ultraviolet end as well as everywhere else. For the infra-red end his formula boils down to the traditional prediction that already fitted there. This was a remarkable guess, in an unlimited field. Of course, it leads to agreement with experiment: otherwise, Planck, like Kepler,

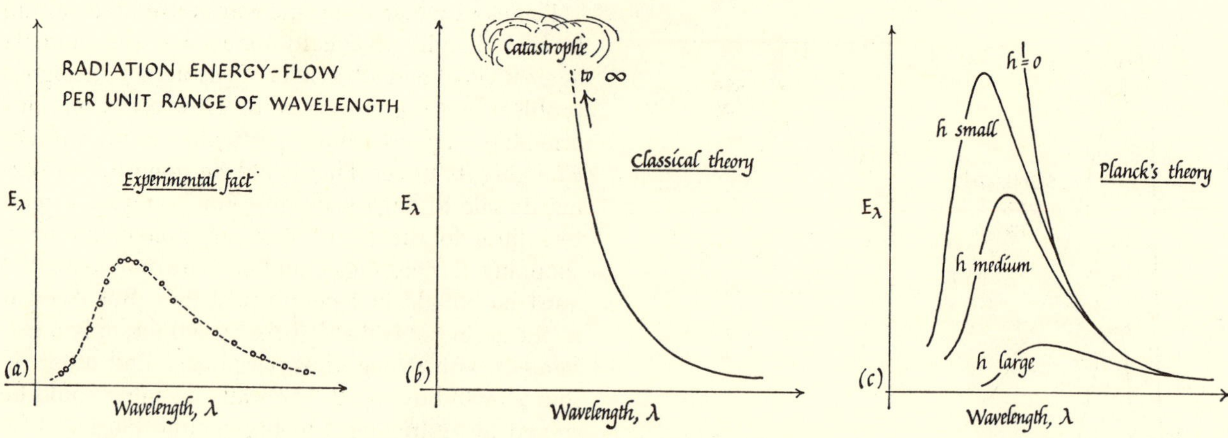

Fig. 44-5a. Radiation: Facts *vs.* Theory

would have moved on to different guesses. The amazing thing is that the same rule resolves other paradoxes that seemed quite separate.

The Size of h, the Quantum Constant

The value of the universal constant, h, is

$$6.62 \times 10^{-34} \text{ in MKS units.}$$

As h = energy of quantum / frequency, it must be measured in $\frac{\text{joules}}{\text{cycles/second}}$ or $\frac{\text{joule} \cdot \text{sec}}{\text{cycle}}$ or in plain joule · sec, since cycles have no measured dimensions. So,

$$h = 6.62 \times 10^{-34} \text{ joule} \cdot \text{sec.}$$

Planck did not predict this value by speculation: it emerged from comparison with experiment. Planck changed the picture of radiation from a smooth stream like a wind, to a grainy stream like a sandblast. The bigger the value chosen for h, the bigger *all* the grains must be, and the more the graininess would make itself felt. If h were zero, all the grains would be too small to show, and we should have the traditional prediction of the ultraviolet catastrophe. If h were huge, the short wavelength grains would be too big for ordinary furnace atoms to manufacture, and there would be no ultraviolet at all—except at extremely high temperatures. With a medium value for h, prediction fits the facts beautifully. The value 6.6×10^{-34} joule · sec is yielded by trial.

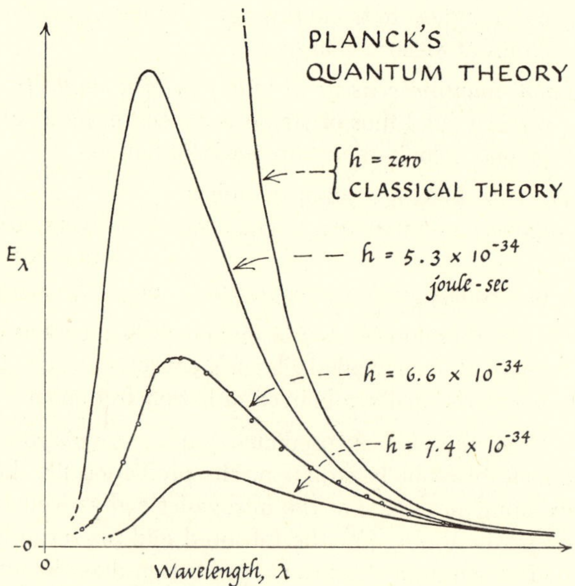

Fig. 44-5b. Radiation Theory: fitting Planck's theory to experimental facts by choice of "*h*."

The Sizes of Quanta

The quantum rule, $E = h\nu$, puts green light in small packets of energy, about 2.5 electron · volts. Red light with longer wavelength has lower frequency, and smaller quanta, 1.8 ev. Blue light has bigger quanta, 3 ev. These are tiny packets: look at a lighted candle across the room and your eye is taking in *visible* light at a rate of about 10,000,000,000 quanta per second. A camera taking a snapshot uses some 1,000,000,000,000 quanta. Yet a dozen blue quanta can develop a mark on a photographic film; and the human eye is so sensitive that a nerve in its retina can almost respond to a single quantum.

Continuing outside the visible spectrum, we find infra-red radiation in very small packets of energy, and radio waves in such small ones that we could hardly expect to notice the bumps of individual quanta directly—yet they are clearly detected by indirect experiments where they tip over the spin of atom-gyroscopes.

On the other hand, ultraviolet light comes in large quanta, a dozen or so electron · volts. X-rays come in huge quanta, such as 50,000 ev; and γ-rays in

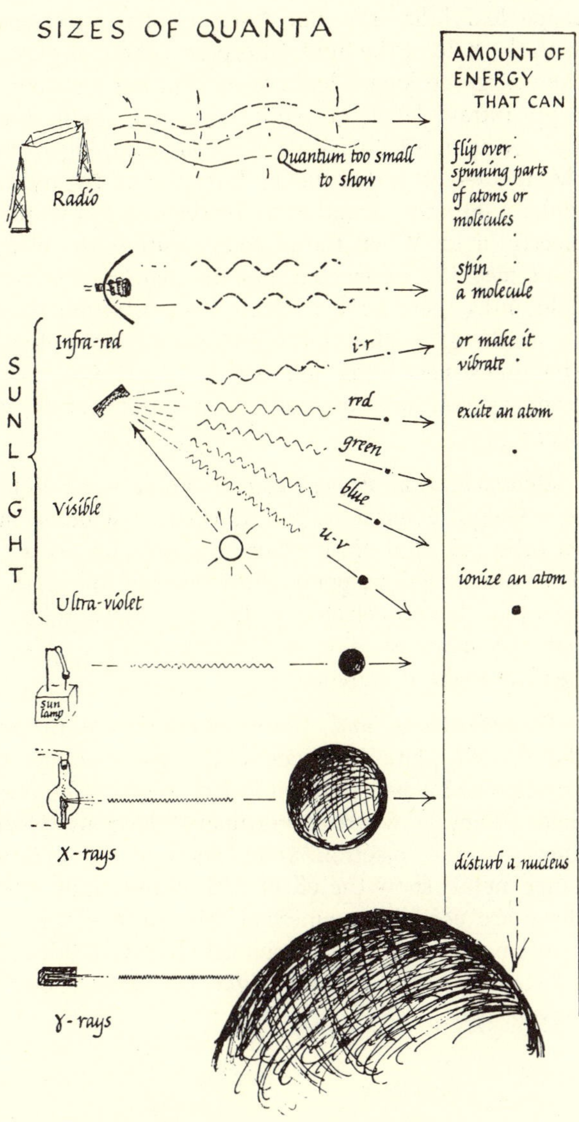

Fig. 44-6. Energy-sizes of Quanta
The *volume* of the ball sketched indicates the energy of one quantum.

gigantic quanta, up into millions of ev and beyond. The absorption of a single large quantum can change an inheritance-gene in a living cell, or even kill the cell.

The Quantum Revolution

This may seem a lot of worry about a single misfit between experiment and theory; but it was a vital misfit, concerning the radiation-spectrum of every furnace, from steam boiler to star, showing that something was badly wrong in the combined theory of mechanics, light-waves and electrons. Though that defect was revealed by large-scale experiments, Planck traced its origin to the atomic scale: radiation must be emitted and absorbed by atoms (and presumably must travel) in atomic packages of energy.

Scientists are conservative: they do not at once welcome revolutionary changes in theory, but hang on to their old views as lifelines of belief. Many were shocked by the idea of bullets of radiation, and doubted whether Planck's rule, $E = h\nu$, gave a proper or necessary description of nature. Then Einstein rescued it from opposition by showing that it explained other puzzles too: specific heat changes, and the photoelectric effect. And Poincaré gave a general mathematical proof that radiation *must* have some atomic packaging if it is to fit the experimental facts. Nowadays we have so many experimental evidences of energy packets in radiation that we take the quantum view for granted. Yet for many years the basic puzzle remained: *how can radiation be both a smooth stream of waves and a shower of bullets?*

Specific Heats[11]

Measured specific heats of elementary solids and gases do not agree with the predictions of traditional kinetic theory. Einstein and others tried applying the quantum restriction, $E = h\nu$, to that heat-energy which resides in the *vibrations* of atoms and *spinning* of molecules. Result: excellent agreement with experiment. This extended the quantum rule to other things besides radiation: to any *repeating* motion such as vibration or rotation, that has a definite frequency. (Kinetic energy of plain motion straight ahead, "whizzo," was left unquantized.) (See Ch. 30.)

Photoelectric Effect

One paradoxical phenomenon had been waiting for quantum theory to clear it up; its experimental facts almost shouted "quanta." That was the *photoelectric effect*, now put to use in electric eyes, and probably always used, in complex form, in our own eyes. When light falls on clean metal, it may eject electrons. Faint light does not eject electrons with less kinetic energy than bright light: it merely ejects fewer electrons. Even with light so faint that it only

[11] See Ch. 27 and Ch. 30. Specific heat is a comparison-number. It shows how much heat a sample of substance requires to warm it, compared with the same mass of water for the same temperature-rise. Numerically, it is the heat, in Calories, needed to raise the temperature of 1 kg of substance by 1 C°. At very low temperatures, a quantum restriction would make itself felt: the average heat-content of an atom for each equipartition-share is very small. One atom having a quantum of some repeating motion would have far more than average, so very few atoms would have such a quantum. *That type of motion would be practically ruled out*, making the specific heat unexpectedly small.

ejects an electron every minute or so, those electrons still emerge with their full speed. And with very faint light an electron does not always wait as long as it would need to accumulate enough energy from a steady stream so weak: sometimes it emerges *at once*, full speed when the light is turned on; another

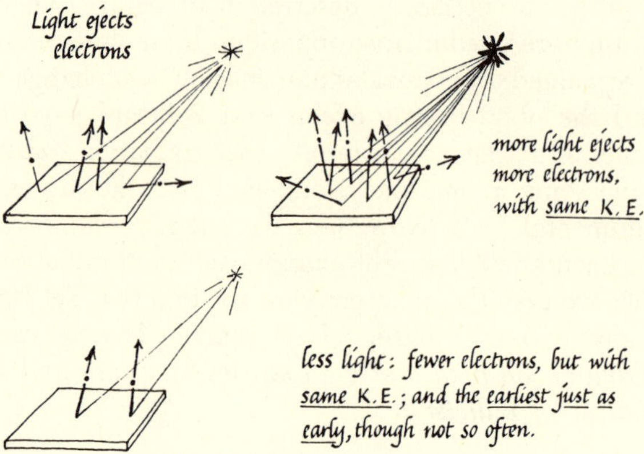

FIG. 44-7. PHOTO-ELECTRIC EFFECT

may wait longer than average—a matter of pure chance. This is not the behavior we should expect from a steady stream of waves pushing an electron to and fro until it breaks loose—like a child in a bathtub building up damaging oscillations of water. This is more the behavior of a detaching agent that arrives in a lump, like a stick of dynamite. Here is the most direct evidence of quantum packets—the graininess of light.

Demonstration Experiment I. Connect a sheet of *clean* zinc to an electroscope, charge it positively and shine white light on it. Nothing happens. Repeat with the zinc negatively charged. The electroscope leaf falls—the zinc loses negative electrons with the help of the light. Interpose a sheet of glass: the charge no longer leaks away. Further tests confirm: ultraviolet light is the essential agent, the leakage is due to negative electrons escaping from the metal—or rather being flipped out by ultraviolet light, and carried away by the charged plate's electric field. When the plate is positive, the electrons may still be ejected, but the plate's attractive field pulls them back in. This demonstration only shows the gross effect—the graininess is not noticeable unless very weak light is used. Then, with electrons ejected one by one, great amplification is needed.

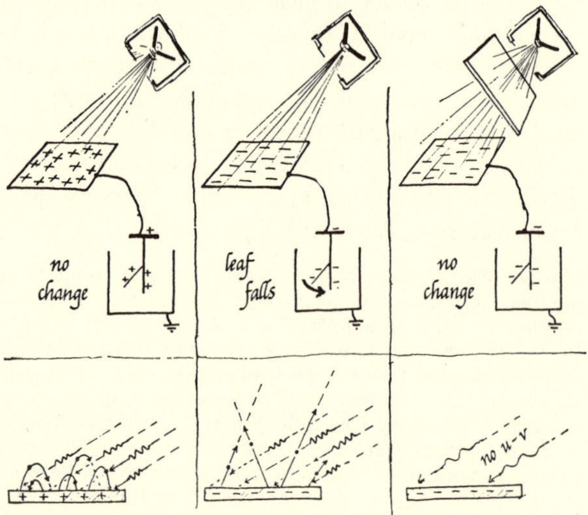

FIG. 44-8a. PHOTO-ELECTRIC DEMONSTRATION

Demonstration Experiment II. Offer weak light to a Geiger counter with a transparent window in its tube (or offer X-rays from a source far away). The counter will respond to the burst of ions made by each electron ejected in the gas in the tube—but this demonstration needs considerable refinement to make it convincing.

Photoelectrons and Color of Light. With an illuminated plate in vacuum, the particles flying out can be investigated with electric and magnetic fields. They prove to be ordinary electrons, with energy a few electron · volts. Sodium and a few other metals show the effect with visible light, and these are used in commercial "electric eye" tubes. Most metals fail until ultraviolet light is used. *All* substances show it with light of short enough wavelength.

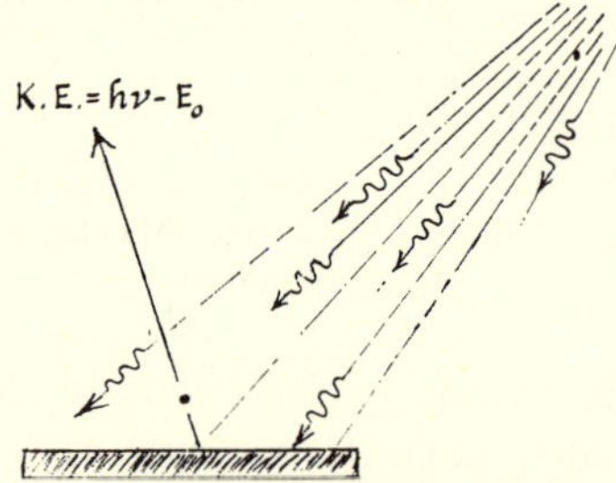

FIG. 44-8b.
PHOTO-ELECTRIC EFFECT: EINSTEIN'S DESCRIPTION OF MECHANISM

With any particular metal and one color of light, all the electrons are released with the same energy[12] (a standard energy characteristic of that color, less a standard tax paid to detach an electron from that metal). Inference: light carries its energy in standard packets (see Fig. 44-8b).

[12] In actual experiments the electrons range in speed from a definite maximum down to zero; but the lower speeds are clearly due to electrons from deeper layers having to struggle out past other atoms. To test this, illuminate a *gas*.

For other colors the size of the energy-packet must be different. Photoelectrons emerge with more energy when blue light is used than with green light: and in most cases red light is too poor to pay the exit-tax. Ultraviolet light ejects electrons with more energy than any visible color. X-rays and γ-rays show the effect still more violently: they will hurl electrons out of any substance, though they often travel a long way through matter before they pick on an electron for this treatment. Cloud-chamber pictures show the tracks of such "photoelectrons" ejected from air molecules by X-rays (see Fig. 39-10g). Each track shows an electron with the same energy—X-rays are grainy too.

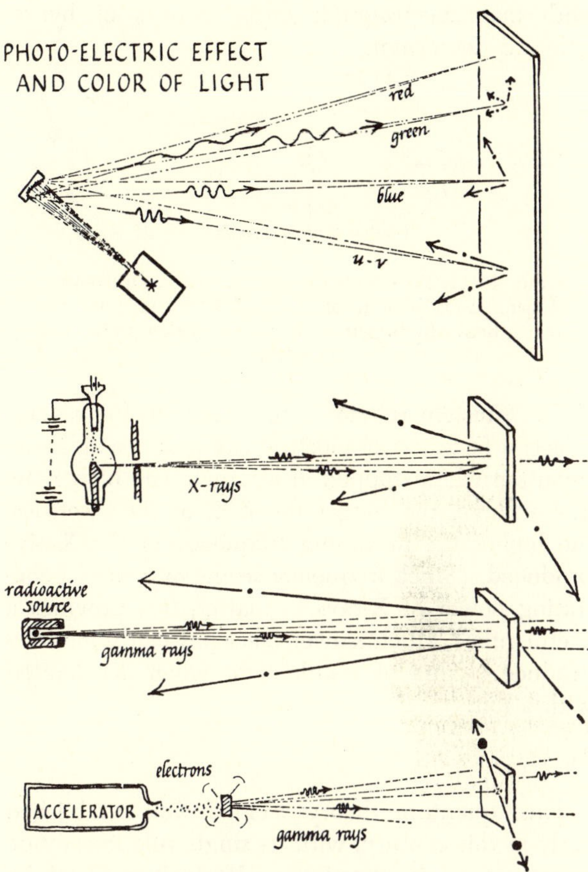

FIG. 44-9a. PHOTO-ELECTRIC EFFECT AND COLOR OF LIGHT
Red light is unable to eject any electrons from most surfaces; blue light ejects electrons with a little K.E., from some metals; ultra-violet light, electrons with more K.E.; X-rays and gamma-rays, electrons with still more K.E., from *any* material. And gamma-rays of high energy can even tear the nucleons of nuclei apart (photo-disintegration).

This behavior seemed very strange to classical physicists. It was as if ocean waves entering a harbor failed to rock the ships anchored there: but instead nothing happened for a while, then one ship would be bounced up 100 ft in the air; again a calm pause, then another ship . . . 100 ft. Soon it became clear that the photoelectric effect could not be a case of trigger-action, with the light

> "releasing an electron already loaded up with energy for its journey; [because] it is the [color] of the light which settles the amount of load. *The light calls the tune, therefore the light must pay the piper.* Only, traditional theory does not provide light with a pocket to pay from."[13]

The photoelectric effect remained a puzzle until Einstein applied the quantum theory to it. Suppose the arriving light delivers its energy in quanta. A single electron takes in a whole quantum, pays some as tax to escape from its own atom and neighbors, and flies away with the rest as K.E. Suppose the energy taken to free the electron is E_0—an exit-tax like the tax for a molecule evaporating from liquid. Then a light-quantum can flip an electron out if its energy E exceeds E_0. The electron will emerge with

$$\text{KINETIC ENERGY} = E - E_0 = h\nu - E_0.$$

On this view, all photoelectrons *should* emerge with the same energy, for light of one color. Most materials have E_0 bigger than a blue-light quantum; and then only ultraviolet or X-rays can succeed.

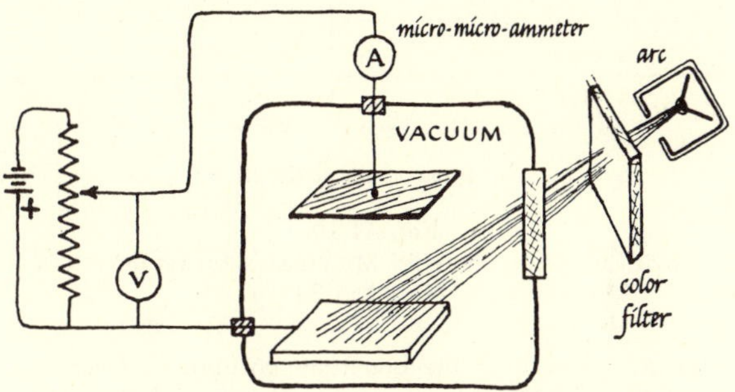

FIG. 44-9b. PHOTO-ELECTRIC EFFECT AND COLOR OF LIGHT: APPARATUS FOR MEASUREMENTS

To test Einstein's suggested relation, we apply a repulsive electric field sufficient to stop electrons reaching a collecting plate nearby. If P.D. V volts just suffices, the electrons must have emerged with kinetic energy (V volts)(e coulombs), or Ve joules. Therefore, Ve must $= h\nu - E_0$. The graph of stopping-voltage V against frequency, ν, obtained by using several different colors of light, should be a *straight line*. Its slope should be

$$h/e, \text{ PLANCK'S CONSTANT/ELECTRON CHARGE.}$$

[13] A. S. Eddington, The Nature of the Physical World (Cambridge University Press, 1928) where there is an excellent discussion with helpful analogies—the "collection-box" theory *vs.* the "sweepstake" theory.

PROBLEM 1

Earlier in the course, you probably found the use of "should" condemned ("Scientists say what *does* happen, not what should . . ."). In what sense is the use of "should" (twice), just above, good science?

PROBLEM 2

Give the algebra to show that the graph-slope should be h/e.

Millikan made careful measurements with photo-electrons from sodium, etc. To lessen sabotage by variable tarnish, he cut clean metal surfaces with a little turret lathe operated *in a vacuum* by magnets from outside! He not only verified Einstein's prediction, but measured h/e and obtained a value of h that agreed with the value Planck had obtained in a different way.

PROBLEM 3. PHOTOELECTRIC EFFECT AND QUANTUM CONSTANT

The graph of Fig 44-10 shows Millikan's measurements (*Physical Review*, VII, 1916, page 362). From the graph,

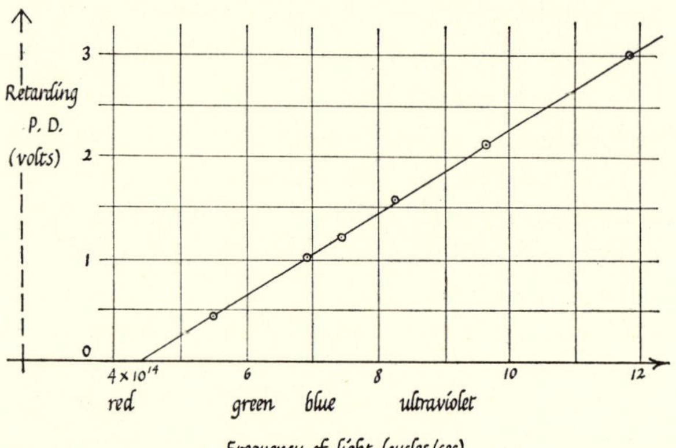

Fig. 44-10.
Photo-electric Effect: Millikan's Measurements (Problem 3)

make an estimate of the quantum-constant, h, given that $e = 1.6 \times 10^{-19}$ coulomb. The frequencies of the various colors of light he used are given in cycles/sec, calculated from the velocity of light and wavelengths measured with a diffraction grating. The P.D. plotted is the retarding voltage that was just sufficient to prevent the electrons ejected by the light from reaching a collector. (To find these P.D.'s from his measured values, Millikan had to allow for the constant "battery EMF" produced by the two different metals he used for photoelectric surface and collector; but that does not affect the slope of the graph.)

The Einstein relation holds for ultraviolet light, X-rays and γ-rays; and for all substances. Thus the photoelectric effect shows the quantum behavior of radiation clearly. Visible light gives electrons a little flip at best, ultraviolet light flings them, X-rays hurl them out, gamma-rays smash them out with the crack of a bull-whip. And very short wavelength gamma-rays have such a huge quantum that it can break up a nucleus. For example, *photo-disintegration of deuterium*:

gamma ray	hits	deuterium nucleus	splits it into	neutron and proton which fly apart
γ	$+$	${}_1H^2$	$\rightarrow$	${}_0n^1 + {}_1H^1 +$ KINETIC ENERGY

In fact this provides a good way of measuring the mass of a neutron. We find the minimum energy that a γ-ray must have for this, and combine that with mass-spectrograph measurements of hydrogen and deuterium.

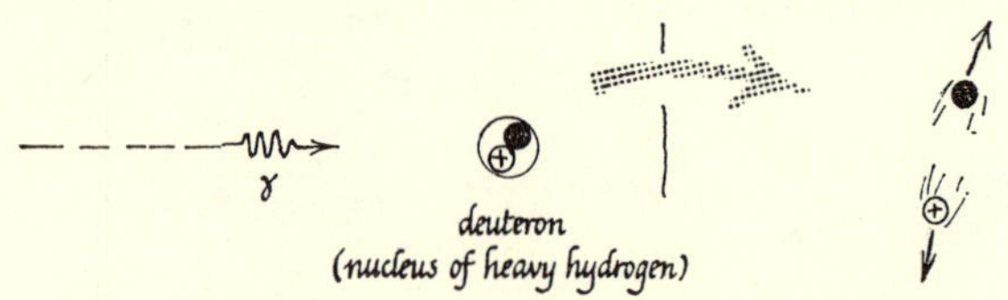

Fig. 44-11. Photo-disintegration of a Deuteron
A gamma-ray photon, of energy 2.2 Mev or more, can tear a heavy-hydrogen nucleus apart into a proton and a neutron.

The Einstein relation also applies to the reverse effect: a fast-moving electron can produce an X-ray quantum when stopped at a target. The higher the gun voltage, the bigger the K.E. of the electrons, the higher the maximum frequency of the X-rays produced. (Since frequency seems to control penetrating power of X-rays in matter, this provides a useful rule.) Much slower electrons can sometimes produce quanta of visible light when decelerated with a jolt.

Photons

Thus quantum theory seemed well established early in this century, with its single rule ENERGY OF QUANTUM $= h \cdot$ FREQUENCY. It dealt successfully with the radiation spectrum, specific heats, photoelectric effect, X-ray production. Planck had pictured quantum-packaging of energy in the radiating atoms. Then Einstein made a further great advance by showing that radiation itself must be packaged in quanta. That was in 1905, the year in which he also published his theory of the Brownian motion and set forth his system of special Relativity! Thus, with Einstein's help, quantum theory changed from a packaging rule to a clear treatment of radiation as small particles. To emphasize the particle picture, we speak of *photons* (to match *electrons, nucleons,* etc.) *whenever we are thinking of the particle aspect of radiation behavior.* All photons travel (in air or vacuum) with the speed of light, c. From the relativistic mass-formula they must therefore have zero rest-mass. That does not matter, since we never find them at rest. In flight they have a mass m such that

$$\text{ENERGY, } mc^2 = \text{ENERGY } h\nu$$
$$\text{and MOMENTUM} = mc = \text{ENERGY}/c = h\nu/c.$$

Thus we think of radiation as a stream of *photons*, each carrying MASS, MOMENTUM, ENERGY ($h\nu$), guided by waves that travel with the speed of light, c.

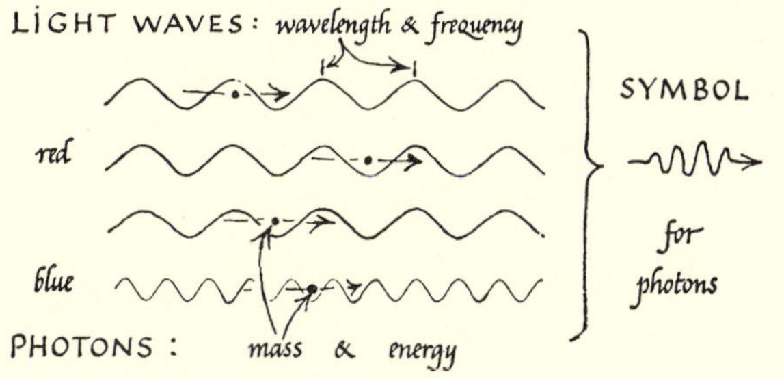

Fig. 44-12a. The Nature of Light

Photon Collisions: the Compton Effect

If photons carry momentum, they should exert pressure on an absorbing wall, and double pressure on a reflecting one just as waves would. They do. The minute pressure of a beam of light has been measured and confirms the momentum expression E/c. If photons are compact particles carrying momentum, they should exchange some of it in a collision with another particle, say an electron. They do. X-ray photons bounce off an electron loosely held by some atom and travel away in a new direction, with less energy, and therefore longer wavelength. The electron bounces away with the rest of the photon's original energy, and its short recoil track can be seen in a cloud chamber. This recoil effect was discovered by A. H. Compton, and his measurements agreed with the predictions for such a collision between "particles." (Relativity mechanics must be used, of course, since photons move with the highest speed.) This is one of the best proofs that a photon is a particle obeying the conservation laws of mechanics, able to deliver momentum and energy at once in a collision with a material particle.

h, an atom of "ACTION"

Planck's constant h is a universal "atomic" constant. Like the speed of light, its value is unchanged by the relativity transformation—it is the same to all observers. It is not an atom of energy. It is an atom of [ENERGY/FREQUENCY] or ENERGY · PERIOD or ENERGY · TIME, which is called "action." If you think it out you will find that ACTION is of the form FORCE · DISTANCE · TIME in Newtonian mechanics. From the look of it, such a quantity might well have useful properties. It does. Newton's laws, etc., can be reworded to say, "Projectiles, planets, electrons, light waves—*all* choose for their motion that path which keeps ACTION either a minimum or a maximum—a "stationary value," in terms of calculus. Thus nature behaves as if it were unwilling to let ACTION slide downhill or up.

RUTHERFORD-BOHR ATOM, 1915-1925

The Rutherford atom was a success—it promoted thinking and experimenting—but a paradox remained. Atoms do have an inverse-square-law field around the nucleus (vouched for by alpha-particle scattering); and negative electrons do reside far out in that field (also vouched for by alpha-particle scattering, and later by X-ray spectra). Therefore, (I): the electrons cannot be at rest in stable equilibrium (Earnshaw). Atoms do not collapse by radiating electromagnetic waves; therefore, (II): the electrons cannot be in motion, pursuing elliptical

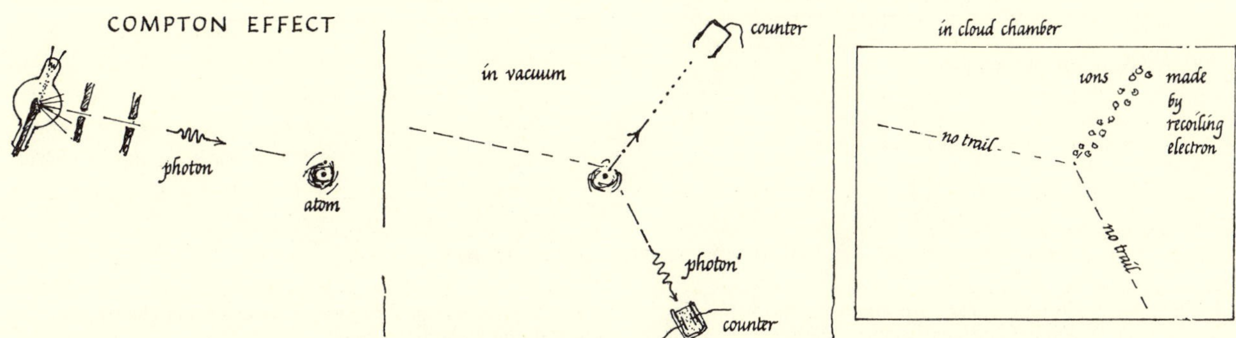

Fig. 44-13. The Compton Effect.
An X-ray photon colliding with an electron loosely bound to an atom moves away with less energy (longer wavelength); the electron recoils.

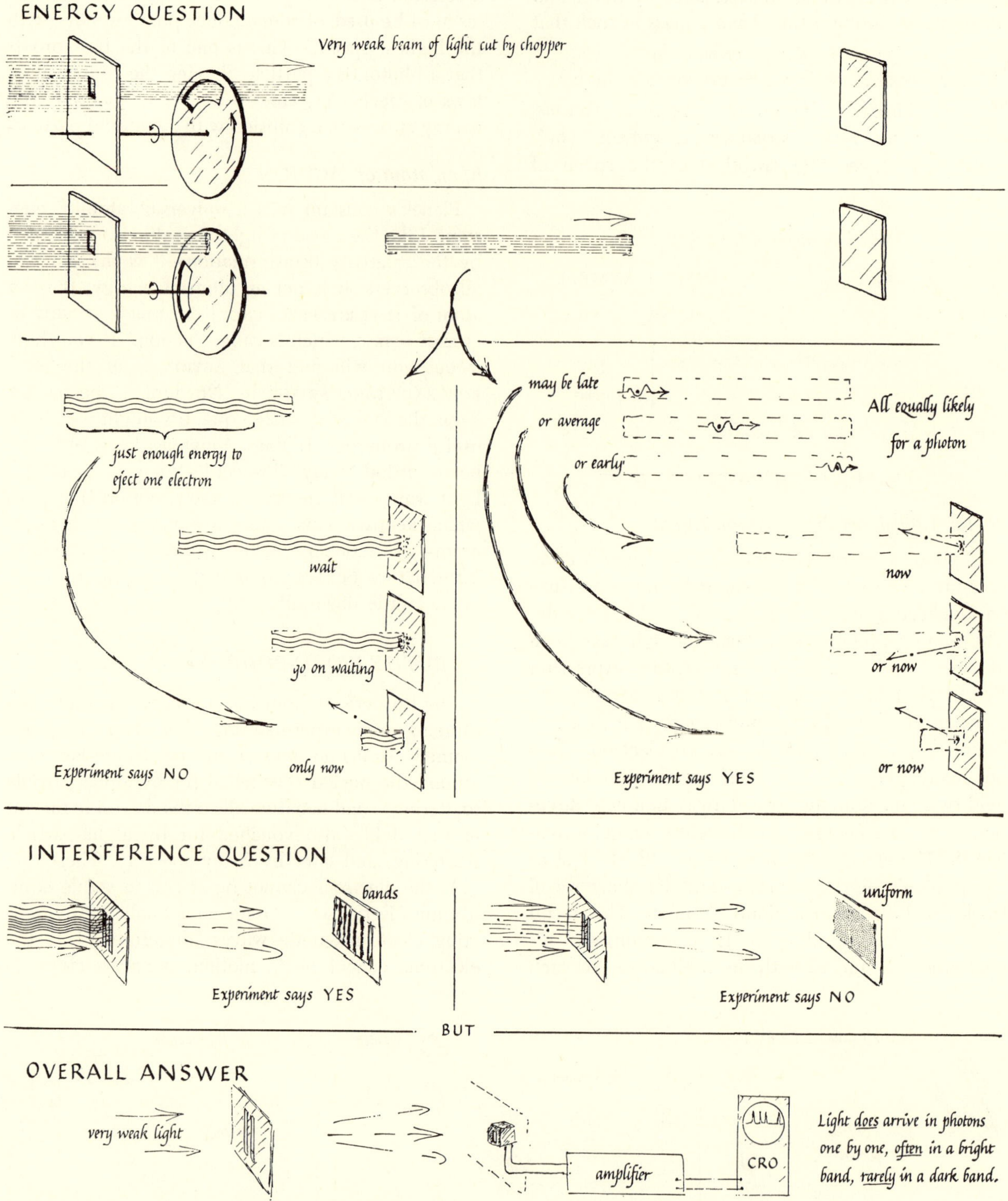

ENERGY QUESTION

Very weak beam of light cut by chopper

just enough energy to eject one electron

wait

go on waiting

only now

Experiment says NO

may be late

or average

or early

All equally likely for a photon

now

or now

or now

Experiment says YES

INTERFERENCE QUESTION

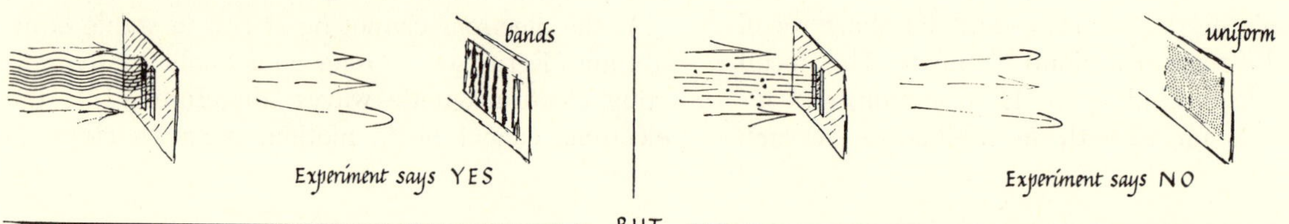

BUT

OVERALL ANSWER

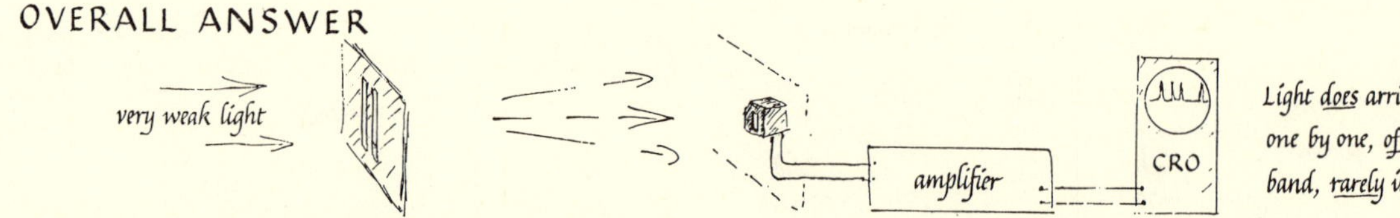

FIG. 44-12b. THE NATURE OF LIGHT*

* Experiments on light reveal two characteristic behaviors, which appear to conflict but can be put together in a single consistent story.

We pass a beam of light through a "chopper" (that selects a short sample of it) and on to a metal surface from which the light can eject electrons. We use so weak a beam of light that *if the light is a steady stream of waves*, we should always have to wait until almost the end of the sample before enough wave-energy has arrived to eject one electron. However, *if the light is a stream of particles*, we expect to find electrons ejected at arbitrary stages: sometimes at the end of the sample, sometimes in the middle, and sometimes when the very beginning of the chopped sample arrives. This experiment has been done. Each electron that is ejected is accelerated to a target where it knocks out more electrons, in an "electron multiplier tube" that acts as amplifier for the

Kepler orbits. I and II conflict. Further, atoms do radiate, *sometimes*; they emit light. The colored light from glowing *gases* splits into very sharply defined colors, spectrum "lines," with definite wavelengths and frequencies of vibration. These frequencies from excited atoms are obviously arranged in groups, several characteristic series for each element's atom. By 1905 the general formulas for spectral series were known, and the measured frequencies of some series had been decoded into a simple law—for which theory offered no satisfactory explanation. That simple law probably involved the quantum restriction in some way, since it dealt with photons. (Each spectrum line is light of a single color, a single frequency, so it must be made by a stream of photons that all have the same energy.) The simple law contained a constant that seemed to belong to many spectra. If one could manufacture this universal spectrum-constant out of other general constants, such as e the electron charge, c the speed of light, h the quantum constant, etc. (with π or 2 or $\sqrt{2}$ thrown in to help), one would have made a delightful discovery. If one could also give a clear theoretical reason for the choice of ingredients, it would be a great discovery. There were attempts and claims of success—from Pythagoras to Kepler to the present day, scientists have found the fitting together of important numbers a fascinating game the results of which range from nonsense to famous discoveries. Bohr not only found the combination for the spectrum-constant, but did so in a reasoned development that brought lasting fame.

The Bohr Atom: Rules

In 1913, the young Danish physicist, Niels Bohr, fearless and mystical, wrote the minimum modifications of traditional physics needed to fit the facts,

single ejected electron. The pulse of charge from that amplifier is viewed on an oscilloscope whose sweep is synchronized with the chopper. The pulses are seen to arrive arbitrarily, at all stages in the sample. See a remarkable movie film of this demonstration made by P.S.S.C.—"Photons," by John King. (In that film you will see this question illustrated by the problem of a man who wants to obtain a quart of milk from a supply-stream that comes out to him through a chopper. There are two forms of supply: (i) a steady stream along a spillway, which makes him take time to fill his quart bottle; (ii) a conveyor belt carrying quart bottles full of milk to him—with the bottles spaced at random along the belt.)

We let light from a single source pass through a pair of slits and look at the pattern on a remote screen. *If the light is a stream of particles*, we expect a uniform bright patch where the light from both slits reaches the screen. *If the light is a steady stream of waves*, we expect interference-bands on the screen. In fact, we do see bands: Young's fringes.

But, if we look for the energy arriving at the interference-bands, we find it arrives in particles (quanta or photons). Many photons arrive at a bright band while very few arrive at a dark one. So we may say that light consists of particles which are guided to an interference pattern by waves. (See another very fine P.S.S.C. film: "Interference of Photons" by John King).

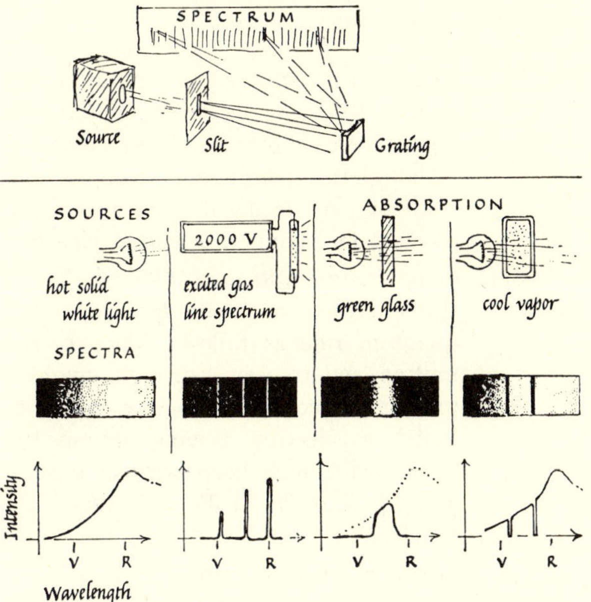

FIG. 44-18. MAKING AND ANALYZING SPECTRA

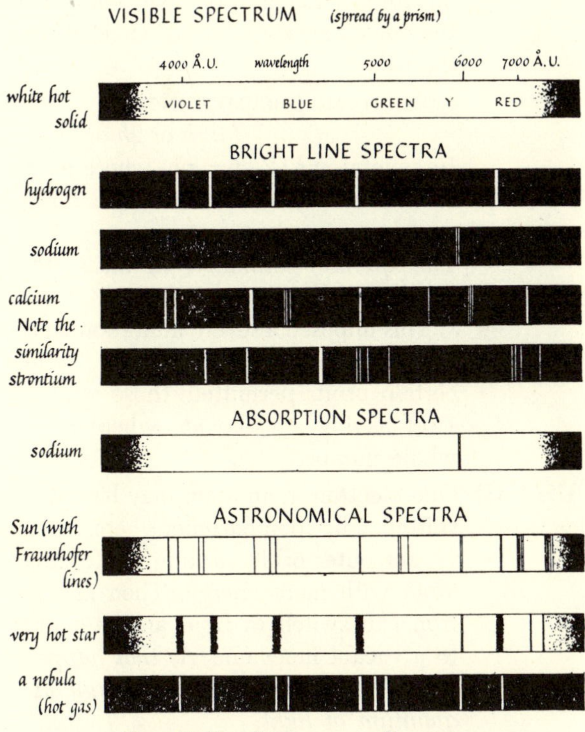

FIG. 44-19a. SKETCHES OF SPECTRA

These black-and-white sketches are dull imitations. See the real thing: look at a neon sign through a prism or a diffraction grating: hold a bright sewing-needle in sunshine and view it through a prism held close to your eye and you will see the Fraunhofer absorption lines. (Real spectra usually show many more lines than the few sketched here. The relative brightness of lines depends on conditions of excitation.)

and combined them to make magnificent predictions. Faced with the paradox of stable atoms that should quickly collapse, he announced, essentially, the following new rules:

RULE I. Atoms are constructed on the Rutherford model, BUT *the electrons continue to move in stable orbits without radiating.* (Though this merely sanctioned the contradiction by stating it, there was much comfort in having it admitted clearly.)

RULE II. *Only some orbits are allowed.* Those stable orbits are to be specified by a quantum rule as follows. Since h, the unit of "action," seems strongly involved in atomic-electronic behavior, we state that[14] an electron moving around the stable orbit must keep ACTION constant at value h, or $2h$, or $3h$, . . . , nh, . . . We have:

ACTION = ENERGY · TIME
= [FORCE · DISTANCE] · TIME
= [FORCE · TIME] · DISTANCE
= MOMENTUM · DISTANCE

So, for a circular orbit, we boldly take the CIRCUMFERENCE for DISTANCE, and try the rule:

MOMENTUM · CIRCUMFERENCE
= h or $2h$ or $3h$, . . .

In general, $mv \cdot 2\pi R = nh$, where $n = 1$ for the smallest allowed orbit, 2 for the next larger, and so on.

The quantum number, n, must be a whole number.

This imposed a severe limitation on the "solar-system" model for atoms: only certain orbits permitted, those with ACTION having a value nh, where n is a whole number.

RULE III. One electron in an atom may have been shifted (by bombardment, etc.) to a vacant outer orbit, making an "excited" atom with more energy. Then an electron can switch in from an outer orbit to a vacant inner one. *As that happens, the atom emits the spare energy as a quantum of light.*

h · FREQUENCY OF LIGHT EMITTED
= ELECTRON'S SPARE ENERGY
= ELECTRON'S ENERGY in "outer" orbit
— ELECTRON'S ENERGY in "inner" orbit

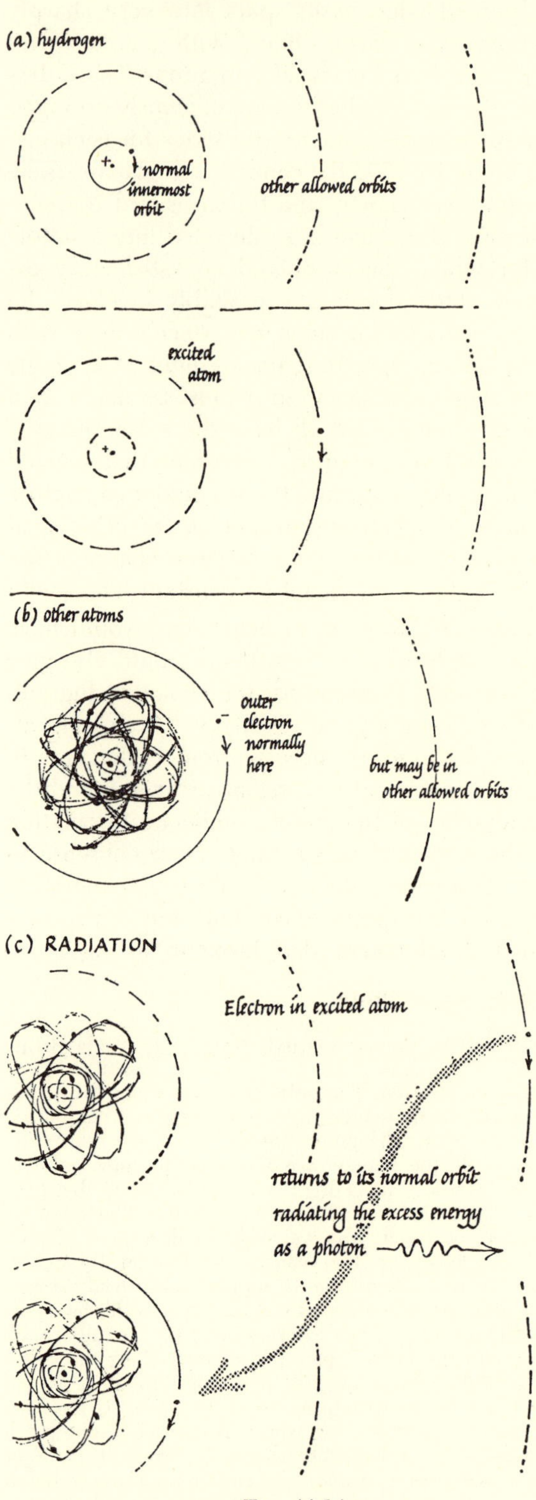

FIG. 44-14.

[14] The choice is not obvious or unique. Like Kepler, Bohr looked for a *simple* rule that would *fit the facts.* Kepler had circular orbits to start from. Bohr had classical physics to start from.

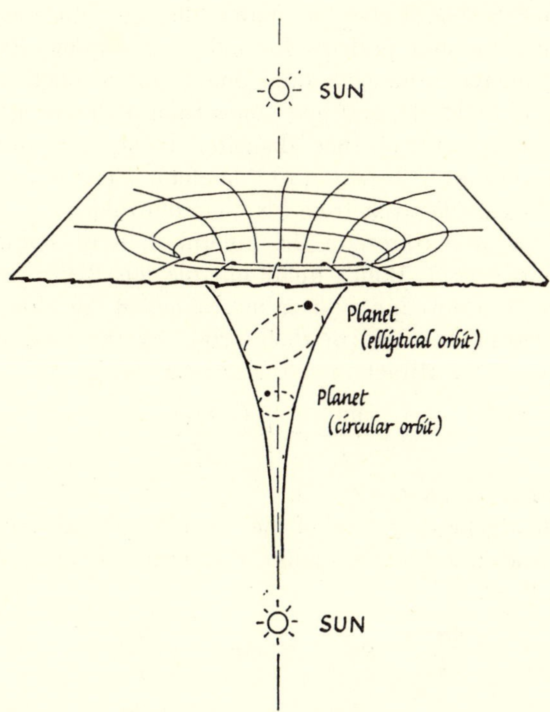

FIG. 44-16. HILL (WELL) DIAGRAM FOR SOLAR SYSTEM*

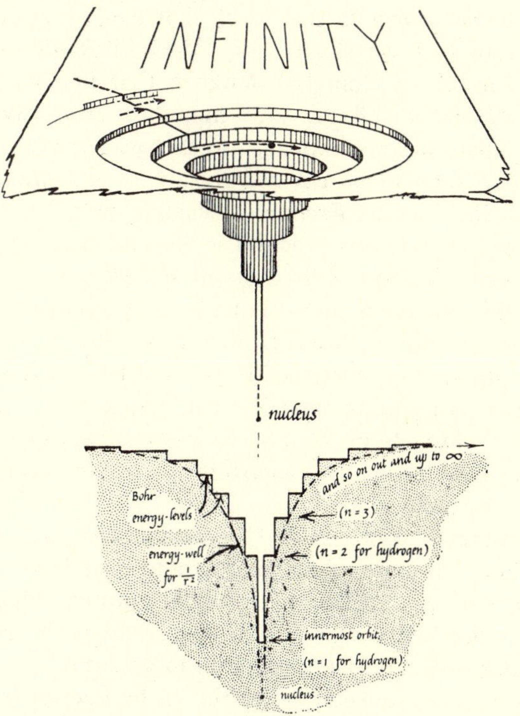

FIG. 44-17. HILL (WELL) DIAGRAM FOR BOHR ATOM-MODEL*

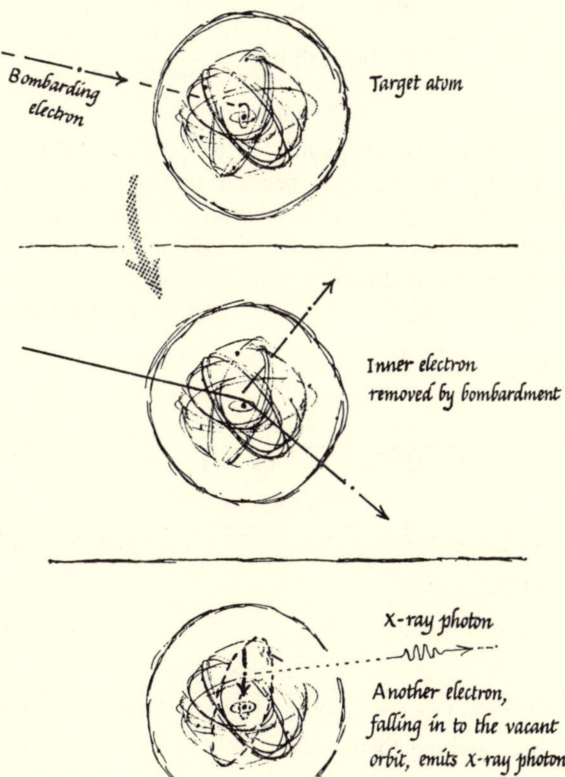

FIG. 44-15.

The Bohr Atom: a Fruitful Theory

Bohr kept the Rutherford picture of electrons pursuing orbits around the nucleus which pulled them with an inverse-square force. So a hill-diagram could still be drawn for their energy. As for the real Solar system, the diagram for an *attractive* force is a well instead of a hill. But Bohr's quantum-restriction rule (II) carves shelves around the well's wall and limits the orbits to those shelves. Fig. 44-16 shows a sketch of a hill-diagram for the Sun with one planet pursuing a circular orbit and another pursuing an ellipse. The "hill" has to be a well for the Sun's *attractive* force. A "Rutherford atom" without restrictions would have a similar hill-diagram for nucleus and electrons. Fig. 44-17 shows a simple atom's "hill"-diagram with a set of quantum-rule shelves for circular orbits. Other shelves (energy levels), for ellipses, were soon added to the Bohr scheme. Every orbit was defined, in size, shape, etc., by several quantum-numbers.

The lowest shelf, defined by $n = 1$ in Bohr's quantum-rule, gives the smallest allowed orbit. There an electron has the least energy; and that is the most stable orbit—we should expect the elec-

(* After K. Mendelssohn, in *What Is Atomic Energy*, Sigma Books, Ltd.)

tron to fall down to that level, if it can, and stay there. In fact we should ask, "Why don't *all* the electrons of an atom fall down to that first shelf and all atoms collapse to a minimum size?" We know that does not happen, and that was one more mystery, soon to be covered by a rule announced by W. Pauli as his *Exclusion Principle*: which said essentially "only one electron is allowed per shelf: only one electron with a given set of quantum-numbers."An atom never allows several electrons to occupy exactly the same orbit. But we now double each shelf, to accommodate two electrons distinguished by opposite spins. We can give a reason for Pauli's rule in terms of modern knowledge, but for a long time it remained an arbitrary but very useful guide for atom models.

Essentially, Bohr said: "Electrons don't radiate continually: atoms don't obey traditional laws of physics. If you will accept my bold guesses, things will make better sense." And Pauli added the restricting rule, "only one electron to an orbit."

These rules looked like legislation by decree, but they calmed the irritation over atom-model paradoxes, by producing new knowledge. They enabled Bohr to calculate the general spectrum-constant from the fundamental ingredients e, h, c, e/m. They produced a working explanation of all spectra that soon went beyond explaining solved problems and illuminated unsolved ones. Even while Bohr was developing his rules, Moseley put them to use in a crude theory of X-ray spectra that enabled him to measure atomic numbers—almost before it was agreed what atomic numbers mean in terms of the nuclear atom.

As an example, watch Bohr *predict* the diameter of an ordinary hydrogen atom, already known experimentally from gross measurements.

Size of Hydrogen Atom

The single electron circulating around the nucleus of a hydrogen atom (in Bohr's model) should mark the general size of the atom by its orbit-diameter, $2r$. Two atoms more than $2r$ apart would be neutral to each other and not exert much force. Two atoms closer than $2r$ would find their electron orbits interpenetrating and there would be disturbances leading to attractions, and, when still closer, to repulsion.[15]

[15] The repulsion is not an obvious consequence. A classical physicist might guess that the light electrons are shoved out of the way and the positive nuclei repel. A modern physicist would point to Pauli's exclusion principle; in a collision the electrons of one atom *have* to steer clear of the electrons of the other, leaving nuclei to repel. In any case, real atoms *do* repel when pushed close enough.

Thus $2r$ should give the atom's "diameter" for mild collisions, and perhaps for molecule-forming. Experimental estimates (rough ones from the length of an oil molecule, and good ones from hydrogen gas friction) agree on that "diameter" being for hydrogen just over 1 Å.U. (10^{-10} meter). Here is how Bohr calculated $2r$ from his quantum rule:

For an electron in the innermost, most stable, orbit, $n = 1$. Bohr's quantum rule $mv \cdot 2\pi r = nh$ becomes $mv \cdot 2\pi r = h$ or $mv = h/2\pi r$. An atom's electrons are held in their orbits by the inverse-square law attraction of the nuclear charge, Ze.

$$\therefore \frac{mv^2}{r} = B\frac{(Ze)(e)}{r^2}$$

and for hydrogen $Z = 1$.
Use algebra to get rid of the electron's orbital speed v, and find the orbit-radius, r, in terms of constants h, B, etc.

$$\frac{mv^2}{r} = B\frac{e^2}{r^2} \quad \therefore mv^2 = \frac{Be^2}{r}$$

$$mv = \frac{h}{2\pi r} \quad \therefore m^2v^2 = \frac{h^2}{4\pi^2 r^2}$$

Divide: therefore,

$$\frac{mv^2}{m^2v^2} = \frac{1}{m} = \frac{Be^2 4\pi^2 r}{h^2}$$

$$\therefore \text{ orbit-radius } r = \frac{h^2}{4\pi^2 Be^2 m}$$

Use the values from experiment:

h (Planck or Millikan) 6.62×10^{-34} joule · sec
B (by weighing forces on charges in electric fields) 9.00×10^{9} newton · meters²/coulomb²
e (Millikan) 1.60×10^{-19} coulomb
m (from Millikan's e and e/m by deflecting fields) 9.11×10^{-31} kilogram

$$\text{Then } r = \frac{h^2}{4\pi^2 Be^2 m}$$

$$= \frac{(6.62 \times 10^{-34})^2}{4(3.14)^2(9.00 \times 10^9)(1.60 \times 10^{-19})^2(9.11 \times 10^{-31})}$$

$$= \frac{43.8 \times 10^{-68}}{4 \times 9.86 \times 9.00 \times 10^9 \times 2.56 \times 10^{-38} \times 9.11 \times 10^{-31}}$$

$$= 0.53 \times 10^{-10}$$

and the units for r are

$$\frac{(\text{joule} \cdot \text{sec})^2}{(\text{newton} \cdot \text{meters}^2/\text{coulomb}^2)(\text{coulomb}^2)(\text{kilogram})}$$

which reduce to *meters*.

Here was Bohr's prediction of the size of a hydrogen atom, pulled out of his mathematical hat. Look at the ingredients: h, Planck's constant for radiation quanta; *B*, the Coulomb's-law-constant; the charge and mass of an electron, *e* and *m*. The outcome: *diameter of the electron* orbit in an unexcited H atom, 1.06×10^{-10} meter or 1 Å.U., an "atom-size" in good agreement with direct measurements.

Bohr and Spectra—Electronic Harmonics?

The greatest achievement of Bohr's theory was its prediction of the bright spectrum lines from glowing hydrogen. Such "line spectra" were another paradox that had been crying for an explanation—experimental formulas that demanded simple theory. While white-hot solids emit a general spectrum of "black body" radiation, glowing gases (heated or excited electrically) give quite a different spectrum. Their glow when analyzed shows no light at all of most colors, only some very narrow, very bright patches of particular colors—a spectrum of bright "lines" each of practically a single wavelength (see Ch. 10). Each gas has its characteristic array of lines: red (and many others) for neon; yellow (and several ultraviolet lines) for sodium vapor in a salted flame; etc. The wavelengths of bright spectrum lines have unique values for the atoms of each element. They provide a first-class tool for analyzing materials—whether engineering samples or gases of distant stars. Each element's lines are arranged in regular series with graded steps across the spectrum. Many such series had been sorted out from very careful measurements of wavelengths in the last century. There is obviously some systematic scheme in the spacing of each series. In the case of glowing hydrogen, and a few others, a very simple underlying rule had been extracted from measurements.

When white light from a hot solid passes through cooler gas we see an "absorption spectrum": the full white-light spectrum except for black lines, light absent, at the wavelengths the gas would emit if glowing alone. These absorption lines can be used to analyze the atmosphere around solid stars, including our own Sun, whose white-light spectrum shows dark lines characteristic of hydrogen, sodium, calcium vapor, etc. The lines are obviously due to the cooler gas absorbing light that fits their own "natural vibrations."[16] They must re-emit the same kind of light, but in all directions, so the line looks dark in contrast with other colors that come straight through from the hotter core. This again hints at some vibratory mechanism able to respond to waves of the right frequency.

It was clear, then, that atoms stimulated by bombardment or electric fields could vibrate and emit waves of light with several frequencies. Are these the atoms' own "natural frequencies," like those of a vibrating string? The string of a harp or violin can vibrate with any of a series of frequencies, according as it makes a standing-wave pattern of one loop,

[16] Here again, a good emitter is a good absorber, for the same kind of radiation.

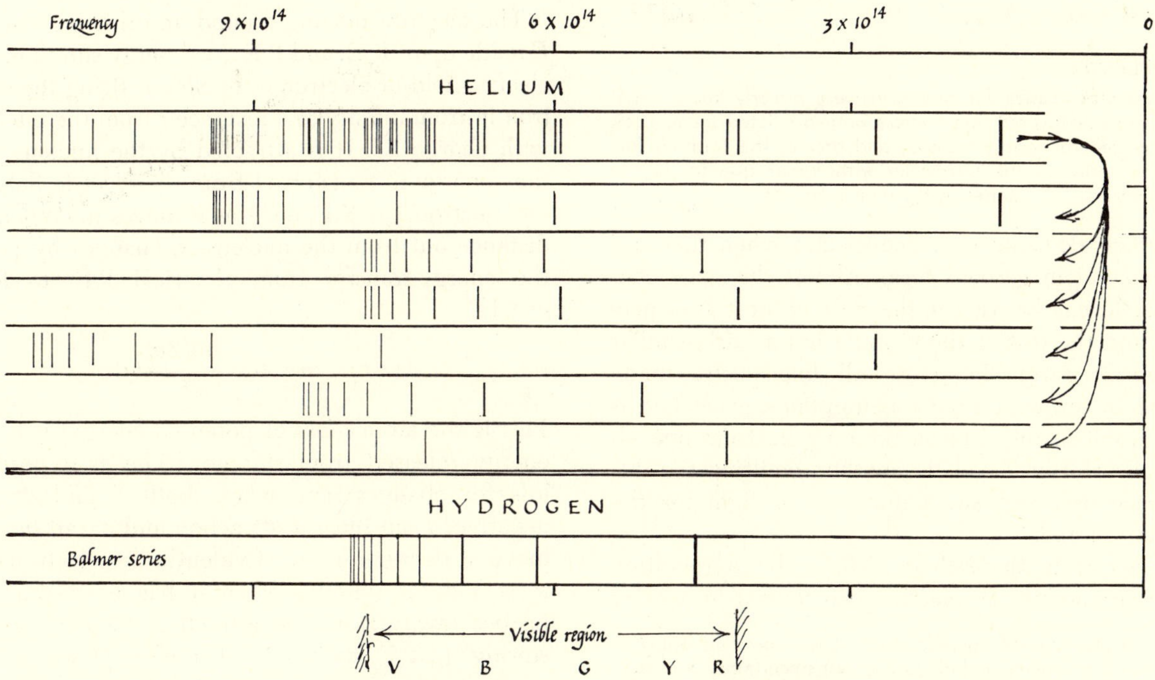

FIG. 44-19b. LINE-SPECTRA MAPPED BY FREQUENCY
(a) The spectrum of helium analyzed into series. (b) The visible series of atomic hydrogen.

or 2,3,4, . . . loops. The frequencies then run in the same proportions, 1:2:3:4: . . . and the musical notes emitted have those frequencies. Experiment and theory agree for strings. Newton's Law II predicts the series: FREQUENCY $=$ (constant, k)(n), with $n = 1,2,3, \ldots$. The frequencies of the colors in the hydrogen spectrum made almost as simple a series, whose formula (from experiment) runs:

$$\text{FREQUENCY} = (\text{constant, } K)\left(\frac{1}{2^2} - \frac{1}{n^2}\right), \text{ with}$$

$n = 3,4,5, \ldots$ for successive lines of the series. Precise measurements[17] with fine diffraction gratings had yielded the value

$$K = 3{,}290{,}000{,}000{,}000{,}000$$

Bohr *predicted* the formula, with the constant, K, given by

$$2\pi^2 mB^2(e)^2(Ze)^2/h^3$$

(see the next section)

This, with the measured values of h, e, m, and $Z = 1$, and $B = 9.00 \times 10^9$, predicted

$$K = 3{,}286{,}000{,}000{,}000{,}000.$$

In fuller form Bohr's prediction was

$$\nu = K\left(\frac{1}{n_f^2} - \frac{1}{n_i^2}\right)$$

where n_f is the "final" quantum number, for the electron's "home-orbit," the same for all lines of a series, and n_i stands for the "initial" quantum number of another orbit, that the electron falls in from. The visible series for hydrogen has $n_f = 2$, and $n_i = 3, 4, 5, \ldots$ &c.

PROBLEM 4

Two other series for hydrogen were already known, and two more discovered later, where Bohr predicted them. Look at the general formula above, and predict the formula for: (i) the series in the ultraviolet with larger quanta, higher frequencies; (ii) a series in the near infra red.

Other elements give line-spectra when they are stimulated in gaseous form. Almost the same constant, K, can be used in the code of most line-spectra, and the rest of the formula has a fairly similar algebraic form. Therefore all line-spectra made sense in terms of Bohr's assumptions: each line is the result of an electron skipping in, from one allowed orbit or "level" to another, emitting the difference of energy as a quantum of light for the spectrum line.

The orbits themselves, defined by a quantum rule, are stable: an electron can remain in one for a long time without radiating. That is "why" a gas does not glow until it has been bombarded. Radiation is emitted only during a change of orbits. Therefore, to emit light, an electron must first be moved to an outer orbit, a higher energy-level, so that it can fall back to a lower one. And that is "why" atoms in a gas emit sharp spectrum lines, each of a single wavelength because each line comes from a switch between definite orbits. Thus, spectra come from excited atoms.

Ions (atoms that have lost an electron) can emit one line, or several in stages, when they have recovered that lost electron, as it "falls" in from one energy level to another.

Nowadays stable "orbits" sound too definite, but we still use their essential property, a definite energy for each; so we speak of "*energy levels.*"

Thus Bohr changed spectroscopy from an empirical scheme for recognizing gases to an essential tool for probing atomic structure.

Bohr's Spectrum-Constant: On Trust, or by Calculus

The prediction of K requires calculus to compute the potential energy of the electron in the field of the nucleus. Take it on trust, as a piece of mathematical machine-shop work (Fig. 44-20), or follow the calculation below.

We need to know the energy of the electron for $n = n_i$ and $n = n_f$. Then we can predict the frequency, ν, of the photon radiated, because

$$h \cdot \nu_{\text{emitted}} = \Delta E = E_i - E_f$$

The electron moving around an orbit has energy E made up of K.E. and P.E. ($=$ energy stored in the electric field of electron $+$ nucleus). Bring the electron in from infinity to a distance r from the nucleus with charge Ze. It is attracted by the nucleus and thus energy is transferred from the electric field to the electron (as K.E. etc.) as it moves in. When its distance out from the nucleus, x, changes by (dx), the energy-transfer from electrical P.E. to K.E. etc. is

$$F(dx) \quad \text{or} \quad B\frac{(-e)(Ze)}{x^2}(dx)$$

This is the atom's *loss* of potential energy; it is the energy released from storage. (This is a positive loss for changes here where both F and dx are negative: F an *inward* attraction and dx an *inward* move, a decrease of x.) Evidently the electron has most P.E. at infinity, when it has been dragged farthest away from the attractive nuclear charge, storing up P.E. as it goes. Therefore if we reckon

[17] Spectroscopists' measurements gave the value of K/c (where $c =$ speed of light), with an uncertainty less than 1 in a million. Bohr's calculated value agreed within 0.1%.

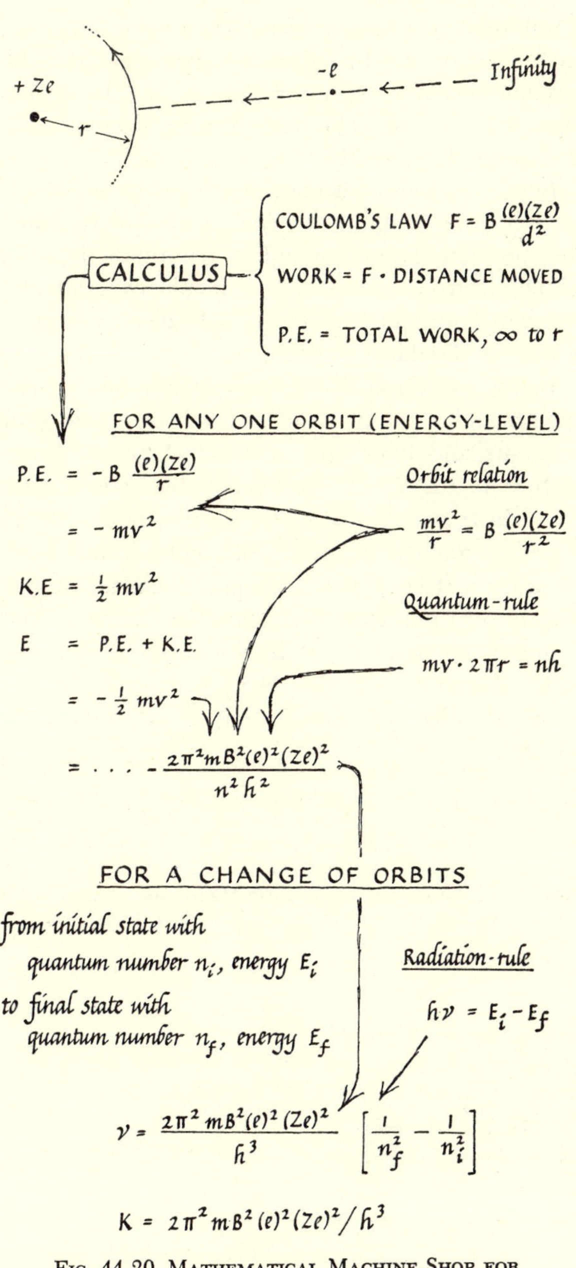

FIG. 44-20. MATHEMATICAL MACHINE SHOP FOR BOHR ATOM

the electron's P.E. from zero at infinity, it must be negative near the nucleus.

The electron's P.E. when it has come in from infinity to orbit-radius r is:

$$0 - \int_{x=\infty}^{x=r} B\frac{(-e)(Ze)}{x^2}(dx)$$

and the value of this, by integration, is:

$$0 - B\frac{(e)(Ze)}{r}$$

$$\therefore \text{ P.E. of electron} = -B\frac{(e)(Ze)}{r}$$

Since $-\dfrac{mv^2}{r} = B\dfrac{(-e)(Ze)}{r^2}$ for orbital motion (the — sign showing inward force), we have

$$\text{P.E.} = -B\frac{(e)(Ze)}{r} = -mv^2$$
$$= -(\text{twice K.E.})$$

$$\therefore\ E = \text{K.E.} + \text{P.E.}$$
$$= \tfrac{1}{2}mv^2 + (-mv^2) = -\tfrac{1}{2}mv^2$$

We need to know energy E in terms of quantum-number n, etc., without v or r. We use:

the orbit-relation, $-\dfrac{mv^2}{r} = B\dfrac{(-e)(Ze)}{r^2}$

and eliminate r with the help of:

the quantum-rule, $mv \cdot 2\pi r = nh$

$$mv^2 = B\frac{(e)(Ze)}{r}$$

$$mv = \frac{nh}{2\pi r}$$

Therefore, dividing,

$$v = \frac{B(e)(Ze)(2\pi)}{nh}$$

$$\text{Then } E = -\tfrac{1}{2}mv^2 = -\tfrac{1}{2}m\left[\frac{B(e)(Ze)(2\pi)}{nh}\right]^2$$

$$= -\frac{2\pi^2 mB^2(e)^2(Ze)^2}{n^2h^2}$$

Put $n = n_i$ for the start and $n = n_f$ for the finish of the electron's switch of levels; and use

$$h \cdot \nu_{\text{emitted}} = E_i - E_f$$

$$\therefore\ \nu_{\text{emitted}} = \frac{E_i - E_f}{h}$$

$$= \frac{2\pi^2 mB^2(e)^2(Ze)^2}{h^3}\left[\frac{1}{n_f^2} - \frac{1}{n_i^2}\right]$$

$$\therefore \text{ the spectrum-constant } K = \frac{2\pi^2 mB^2(e)^2(Ze)^2}{h^3}$$

With measured values of e, e/m, h, B, and $Z = 1$ for hydrogen, this gives magnificent agreement the value obtained experimentally from measurements of spectra, 3286 trillion compared with 3290.

More Predictions

Bohr's predictions went further. Suppose a helium atom has lost one of its two electrons completely, and the remaining electron emits light when switching orbits. Then, with its one remaining electron, the ionized helium atom is like a hydrogen atom with a double controlling charge, $+2e$, on the nucleus. With $Z = 2$, the predicted frequencies in the glow are just 4 times the hydrogen frequencies (see Bohr's formula for K). Such lines had already been discovered in the spectrum of sparks in helium and mistakenly ascribed to hydrogen in some strange state. Bohr not only identified them as coming from half-stripped helium but, using a minor correction from nuclear mass that entered the prediction, he estimated the electron/proton mass-ratio—result, 1/1830.

Again, for switches of the *innermost* electrons of heavy atoms, Bohr's hydrogen formula predicted X-ray frequencies, far greater than visible light frequencies, because K contains Z^2 as a factor.[18] That is how Moseley could compare atomic numbers (Z), by taking square roots of characteristic X-ray frequencies from targets of different elements. This was the earliest easy measurement of atomic numbers.

Atoms of an element can emit several series, whose codes are connected by simple arithmetical differences. Bohr at once explained these connections by referring all spectrum lines to *changes* among a set of energy levels. Each *series* belongs to a different *final* energy level. Then subtracting the frequency of one line from another gives the difference between two upper levels, and that difference can be obtained from a variety of subtractions—see Fig. 44-21.

Bohr could also calculate the "ionization energy" of a hydrogen atom, the energy needed to knock the electron away completely. His prediction was 13.54 electron · volts.[19] A difficult experiment—bombarding atomic hydrogen with electrons—gave a measured value 13.6 ev.

Using measured X-ray frequencies and visible spectra, Bohr mapped energy-levels of atoms of all elements and built up a scheme of atomic architecture to justify the chemical table. His scheme had electrons circulating in neat circles (or ellipses), arranged in groups or shells. It was fairly successful. It "explained" general chemical properties. It even predicted the properties of one missing element and enabled it to be discovered (hafnium).

The fruitfulness of the theory justified Bohr's announced rules; yet they remained arbitrary, rules without reasons.

[18] Tungsten, a common target in X-ray tubes has $Z = 74$; so Z^2 makes frequencies 74^2 or 5500 times bigger, wavelengths 5500 times shorter than for hydrogen with $Z = 1$. That shifts the line-spectrum from the visible ($\sim$ 5000 AU for green light) to the region of penetrating X-rays ($\sim$ 1 AU).

[19] Simply put $n = 1$ and $n = \infty$ in the expression for E, the electron's energy in the orbit with quantum-number n. Subtract, to find the ionization energy. Try this, with measured values of e, e/m, etc. and $Z = 1$ for hydrogen. The answer will be in joules if you use MKS units. Remember to divide by 1.60×10^{-19} joules/ev to reduce it to electron · volts.

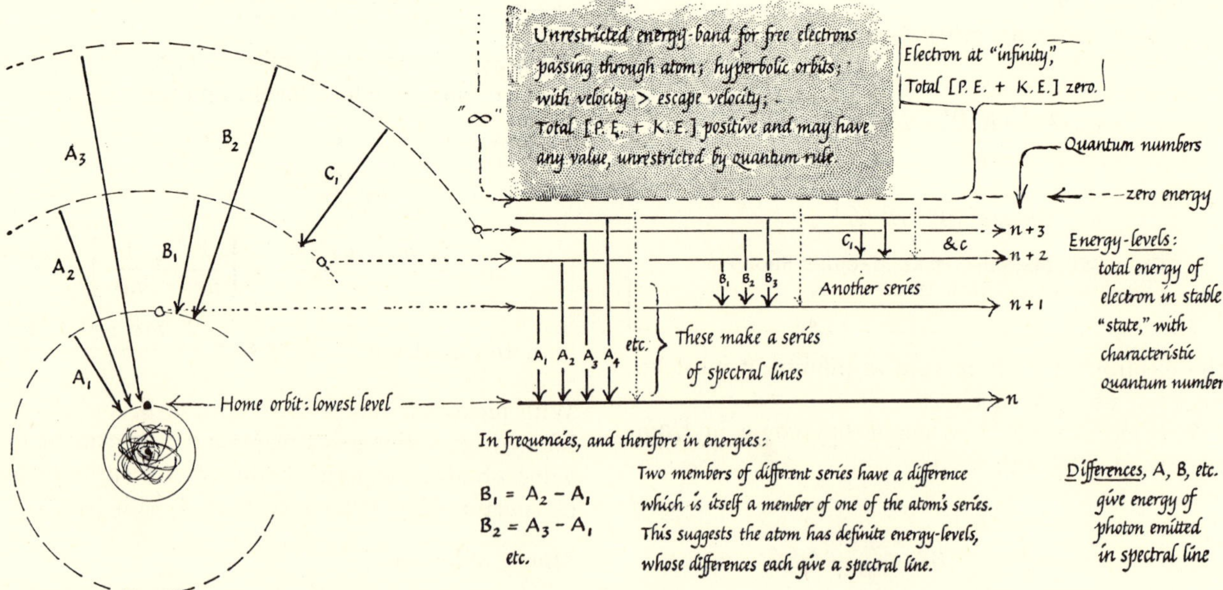

FIG. 44-21. ENERGY-LEVELS IN BOHR ATOM MODEL (simplified view)

Bohr's Assurance : The Correspondence Principle

Not entirely arbitrary, because from the first Bohr maintained one safe anchor with earlier knowledge. Consider some *extreme* circumstance where the quantum-restriction becomes unimportant. There the prediction made by new theory should agree with the prediction by old, classical theory. Where an old theoretical fabric fails to fit the facts it has to be sheared away; but the new fabric that is used to repair the gap or extend the whole design must weave into the old texture at the edges.

This *Correspondence Principle* says: *New theory and old must overlap and agree in the region where the difference between their assumptions does not matter.* It seems obvious when stated thus, but it is a powerful guide in new speculation. It is a general rule for good science, for all good theory. You have met examples:

Relativity rules boil down to ordinary geometry and mechanics for small velocities (in the extreme case where the factor $\sqrt{1 - v^2/c^2} \approx 1$).

Planck's formula for black-body radiation agrees with the classical prediction in the long wavelength region (in the extreme case where quanta are small and numerous).

Newton's Law II, $F = Ma$ reduces to Newton's Law I for a body moving steadily (in the extreme case of $a = 0$).

The theory of Kepler's laws for elliptical orbits agrees with simpler theory of circular orbits (in that extreme case).

The wise thinker will carry the Correspondence Principle farther afield: in devising, *or criticizing*, new theories of geometry, population growth, business cycles, government, ethics, ...

Thus, Bohr saw that his quantum rule for orbits would impose less peculiarity with larger quantum numbers. For an orbit with, say, $n = 1000$, the switch to the next orbit would make only a small *percentage* change in the electron's "action" (MOMENTUM · CIRCUMFERENCE). The graininess of action imposed by quantum theory would be unimportant. The electron would have almost the same frequency of revolution before and after the switch; and the only frequency that the emitted light could have is that roughly-constant frequency of the circulating electron—as traditional theory would require. In such extreme cases of very large n, Bohr's rules *do* lead to this agreement with old theory. In fact, by requiring correspondence, Bohr could find the needed new rule $mv \cdot 2\pi r = nh$; but although the correspondence pointed to the right rule to choose, it gave no reason for it. Ever since, in brilliant suggestions for new theory, Bohr and others have continued to require such "correspondence" or fitting, where new meets old.

Difficulties

This fruitful theory suffered from three kinds of difficulty: (a) It failed to predict the spectra of other elements in detail; it failed to predict chemical forces and arrangements in molecules. (b) In the hands of less cautious philosophers than Bohr, the *model* took on an air of *reality*. The orbits became sharply engraved in space—as they are still embossed on some textbook covers! Scientists were forgetting Einstein's general admonition: not to decorate theory with details that cannot be checked. We can never see the orbits. We can only construct them by *assuming* that macroscopic (= large scale) planetary rules apply. All we really know is that atoms have definite energy-levels, and emit or absorb a photon when switching levels. (c) Furthermore, there was still no plausible explanation of Bohr's quantum-rule for choosing orbits or levels and no reason for electrons to remain in them without radiating.

As with many a good theory, further use had revealed weaknesses—a modified version was needed, and soon made, to save the phenomena. The modification was a sweeping one, started by an extraordinary suggestion by the French physicist Prince Louis de Broglie.[20]

PARTICLES AND WAVES:

New Theory

"Treat electrons like light waves," was de Broglie's suggestion. *"Though they are bullets, they are also waves."* This meant treating electrons like light.

Light has two sets of properties that seem utterly in conflict: wave properties and bullet (particle) properties. Of course, many things that light does are common to both. Light travels in straight lines, with definite speed; it carries energy and momentum, and exerts pressure; its rays are reflected and refracted (bent) at boundaries of different materials. A stream of bullets can do all these: and so can a train of waves. But there are essential differences:

[20] Pronounced "dĕ Broy" or "dĕ Bro-ee."

BULLETS	*WAVES*
CARRY THEIR ENERGY (K.E.) AND MOMENTUM IN A COMPACT BUNDLE	CARRY THEIR ENERGY SPREAD OVER THE WHOLE "WAVE-FRONT"
ADD THEIR CONTRIBUTIONS WHEN TWO STREAMS OVERLAP: BULLETS + BULLETS = MORE BULLETS	"*INTERFERE*" WHEN TWO STREAMS (FROM THE SAME SOURCE) OVERLAP WAVES + WAVES = { *MORE WAVES* IN SOME PLACES BUT *NO WAVES* IN OTHER PLACES
CAST SHARP SHADOWS	BEND AROUND CORNERS
EITHER PASS THROUGH A HOLE IN A WALL OR DO NOT—A BULLET CANNOT PASS PARTLY THROUGH ONE HOLE AND PARTLY THROUGH ANOTHER HOLE IN THE SAME WALL.	CAN PASS THROUGH ANY NUMBER OF HOLES SIDE BY SIDE IN A WALL
	TRANSVERSE WAVES CAN SHOW POLARIZATION

Light and all other electromagnetic radiations show properties from both lists. Light from a faint star ejects photo-electrons with full energy, here and there over any area it falls on. Yet the same light can be gathered by a great telescope lens to form an interference pattern that .shows the whole wide wave-front cooperating.

X-rays, too, eject photo-electrons with full energy; and when X-rays are scattered by a bunch of atoms, each quantum of them bounces off some electron, giving it a little recoil momentum, just as a bullet would (Compton effect, Fig. 44-13). Yet X-rays show smooth, cooperative interference effects when scattered by a grating of atom-layers in a crystal; and they show polarization like light waves. So we now describe light, X-rays, etc., as "photons," *bullets of energy and momentum*, with a *wave to guide their paths*, somewhat like Newton's guess long ago.

Then de Broglie made his outrageous suggestion: *credit real bullets with the same wave quality.* Endow every particle of matter with a wave to guide it as it travels, a secret phase-wave of definite wavelength and strange speed, able to guide an electron, an atom, *any* moving thing, into an interference pattern! In some of its behavior, then, we should say *an electron IS a wave.* In other behavior, *it IS a particle.* When de Broglie wrote the suggestion in a short startling letter to public scientific journals, it seemed almost crazy. In due course, it won him a Nobel Prize. Here is his prescription: for light, we already take both particle view and wave view. Use the quantum rule to relate these views, by expressing wavelength in particle terms:

LINKING TOGETHER VIEWS OF LIGHT

Particle View	*Quantum Rule*	*Wave-View*
ENERGY $= mc^2$ MOMENTUM $= mc$	ENERGY $= h\nu$	WAVELENGTH $= c/\nu$

$$\therefore\ mc^2 = h\nu$$

$$\therefore\ \text{MOMENTUM} = mc = \frac{mc^2}{c} = \frac{h\nu}{c} = \frac{h}{\lambda}$$

$$\therefore\ \text{WAVELENGTH, } \lambda = \frac{h}{\text{MOMENTUM}}$$

Now assign a wavelength λ to a *particle of matter* in the same way:

$$\text{WAVELENGTH} = h/\text{MOMENTUM} = h/mv.$$

Then a particle of big mass and ordinary speed would have such a small λ that interference and diffraction effects would be negligible—rifle bullets should fly straight, as we know they do, and not pepper targets far and wide with interference patches. But small particles, such as electrons should show wave effects. Electrons from a 50- to 100-volt gun should have a wavelength like that of X-rays, comparable with the spacing of atom-layers in crystals. Davisson and Germer, in the Bell Telephone Laboratories, observed interference of "electron waves" by firing a stream of electrons at a small crystal of nickel.[21] When de Broglie made his sug-

[21] See an excellent popular account of their work, by Karl K. Darrow, in *Scientific American*, May 1948, Vol. 178, No. 5.

gestion, they had already discovered that the electrons, instead of bouncing off in a general blur of directions, showed definite preferences—a mysterious behavior for bullets. Then, taking the wave suggestion, they made careful measurements: electrons from a 54-volt gun fired straight at the nickel crystal bounced away copiously at an angle of 50°—but very few at other angles, except those that bounced straight back.

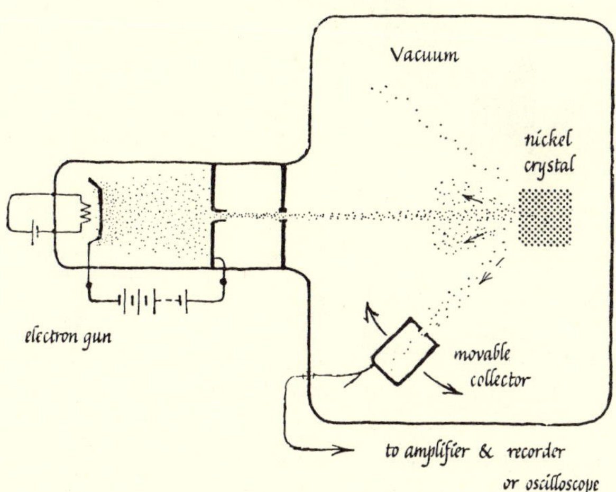

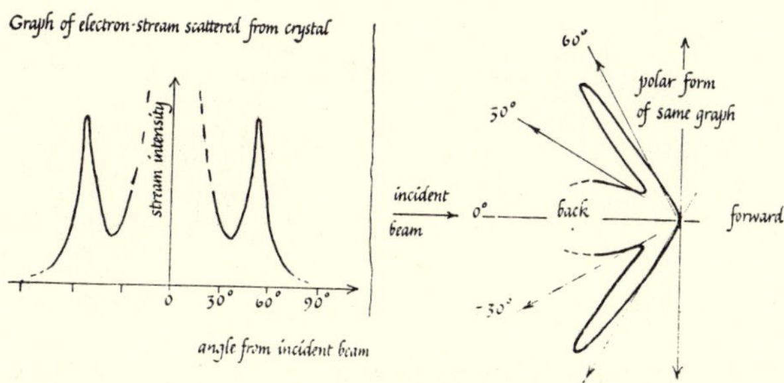

FIG. 44-22.
ELECTRONS AS WAVES: DAVISSON AND GERMER'S EXPERIMENT

PROBLEM 5. ELECTRON WAVES?

(a) Calculate the de Broglie wavelength of a "54-volt" electron as follows:
The electron's K.E. is the energy gained by its charge e falling through 54 volts.
Calculate:

(i) its velocity, v (Use ordinary mechanics, since it is traveling slowly compared with light. Take $e/m = 1.8 \times 10^{11}$ coulombs/kilogram).
(ii) its mass, m (from e/m and e. Take $e = 1.6 \times 10^{-19}$ coulomb),
(iii) its momentum, mv,
(iv) its wavelength, λ. (Take $h = 6.6 \times 10^{-34}$ joule · sec or (meters) (kg · meters/sec.)

The answer to (iv) will be in meters. Express it also in Å.U. (10^{-10} meter).

(b) Calculate out Davisson and Germer's test. They knew from X-ray measurements that the spacing between layers of atoms in their nickel crystal was 2.15 Å.U. For waves of wavelength λ hitting a grating of spacing d, wave theory predicts a diffracted beam at an angle A given by $d \sin A = \lambda$.

(As with X-rays, a strong diffracted beam of electrons is due to *several layers* of atoms cooperating to reflect waves all in phase. However, we can use the plane-grating formula above, that was derived in Ch. 10. There, d was the distance between furrows; here it is the distance between rows of atoms on the exposed face of the crystal. This calculation for a surface grating requires the same wavelength for a diffracted beam as the full three-dimensional treatment.)

(i) Calculate λ for waves that are diffracted at 50° by a crystal with spacing 2.15×10^{-10} meter. ($\sin 50° = 0.766$)
(ii) Compare this with the "wavelength" of 54-volt electrons, you predicted in (a).

The interference pattern was measured and agreed with $\lambda = h/mv$. G. P. Thomson in England made a similar test at the same time. Since then, many similar patterns have been measured for streams of electrons, protons, atoms, even neutrons. The patterns are there: moving particles *do* follow wave directions. The waves tell us where we shall find the particles.

(We expect no exceptions. Having seen wave behavior for atoms, we believe it applies to baseballs and rifle bullets, but a rough estimate of their wavelength, even when moving slowly, shows it would be hopelessly too small for us to measure or for any wave-effects to be noticeable.)

The most startling demonstration of all is one devised by G. Möllenstedt: an extremely thin wire is strung across a stream of electrons from a gun. The stream casts a shadow of the wire on a distant photographic film. Then the wire is given a small positive charge. The electric field of that charge bends the electron stream on either side of the wire; so the shadow disappears and is replaced by a "bright" patch where the streams passing the two sides of the wire overlap. That patch shows Young's fringes —places where

"electrons + electrons = more electrons"

and other places where

"electrons + electrons = no electrons."

In the dark fringes, electrons do not annihilate each other—they just fail to arrive there because their wave patterns combine to guide them elsewhere.

The techniques are difficult—the picture must be enlarged by electron microscope lenses—but the result is magnificent. See Fig. 44-23, and compare the pattern made by electrons with that made by light-waves.

Awkward Conflict

The idea of wave:particle duality is hard to swallow for light, but harder still for atoms and electrons and all chunks of matter. Light passing

Fig. 44-23.
Möllenstedt's Experiment: Young's Fringes by Electrons

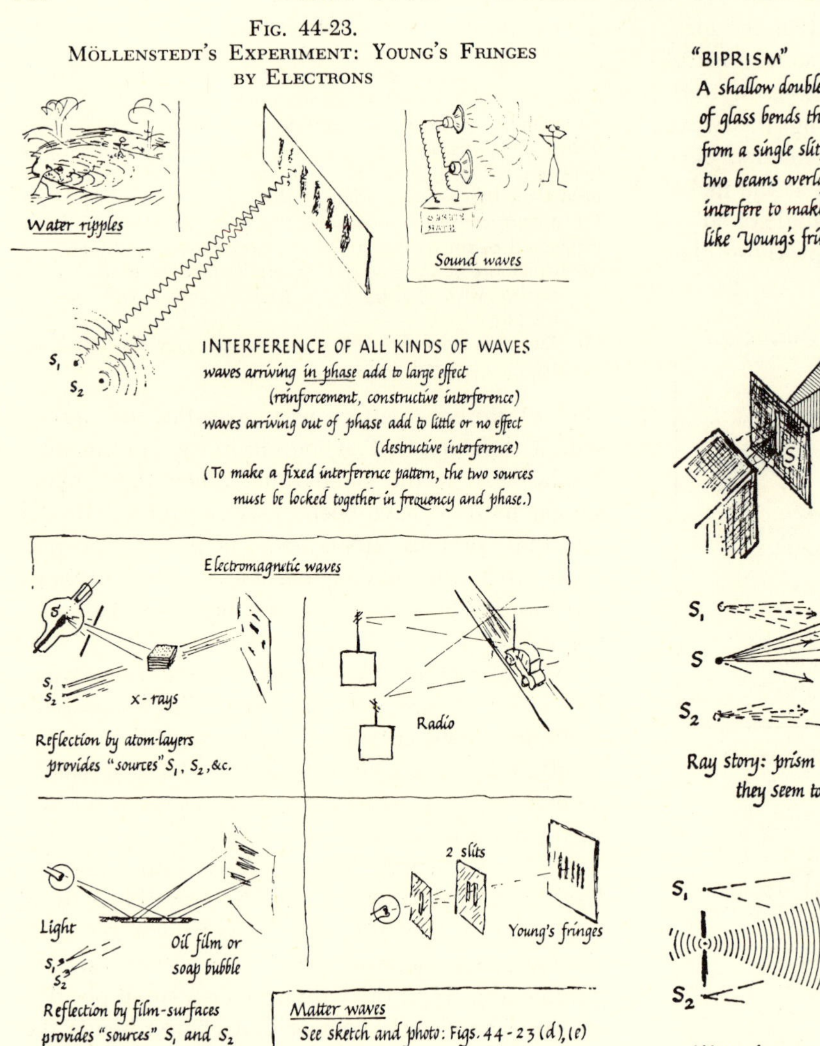

(a) Waves from two "coherent" sources make interference bands.

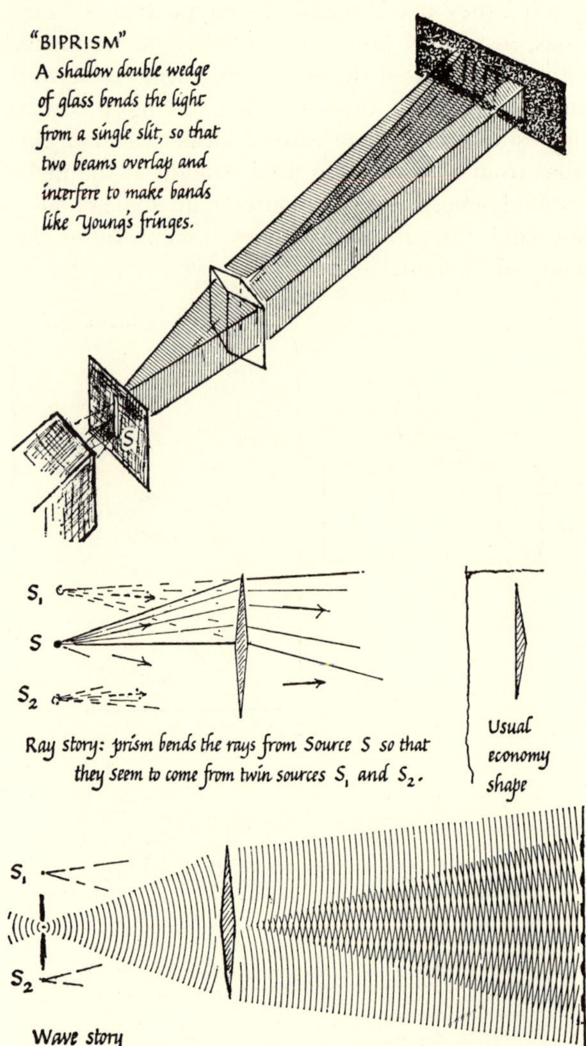

(b) For light, a "biprism" of glass converts a source into a pair of sources, so that the light makes interference bands.

through a pair of slits in a wall makes Young's fringes on a distant screen. Yet clearly its energy is carried in bullet-like quanta, most of them arriving in the bright fringes, few in the dark ones. If an electron stream passes through two slits,[22] there is again an interference pattern. So each individual electron must somehow pass through *both* slits.[23] Its wave must go through both slits, or how could it make the pattern? Yet how can a bullet do that?

Wave Packets

De Broglie gave some comfort towards this paradox—but the argument is difficult and requires more mathematics than we shall give. His idea was this: both for photons and for particles of matter, think of the wave as a compact group of ripples—like the group that you make with a single splash in a bathtub or pond. Watch such a group of ripples travel out on water; you will see it spread as a ring of disturbance with definite speed v. Now watch the details in the group as it travels. It consists of small ripples traveling *slower* than the group itself. Ripples are constantly dying at the back of the advancing group while new ones are being born at the front. This idea of a group with different speed from that of its component waves is essential to de Broglie's scheme. Electrons and light quanta (pho-

[22] We cannot make a real pair of slits for electrons, but two layers of atoms in a crystal fulfill the same purpose, or we can use the charged-wire scheme described above.

[23] As in the cartoon of a skier's tracks downhill that separate around a pine tree and rejoin below it.

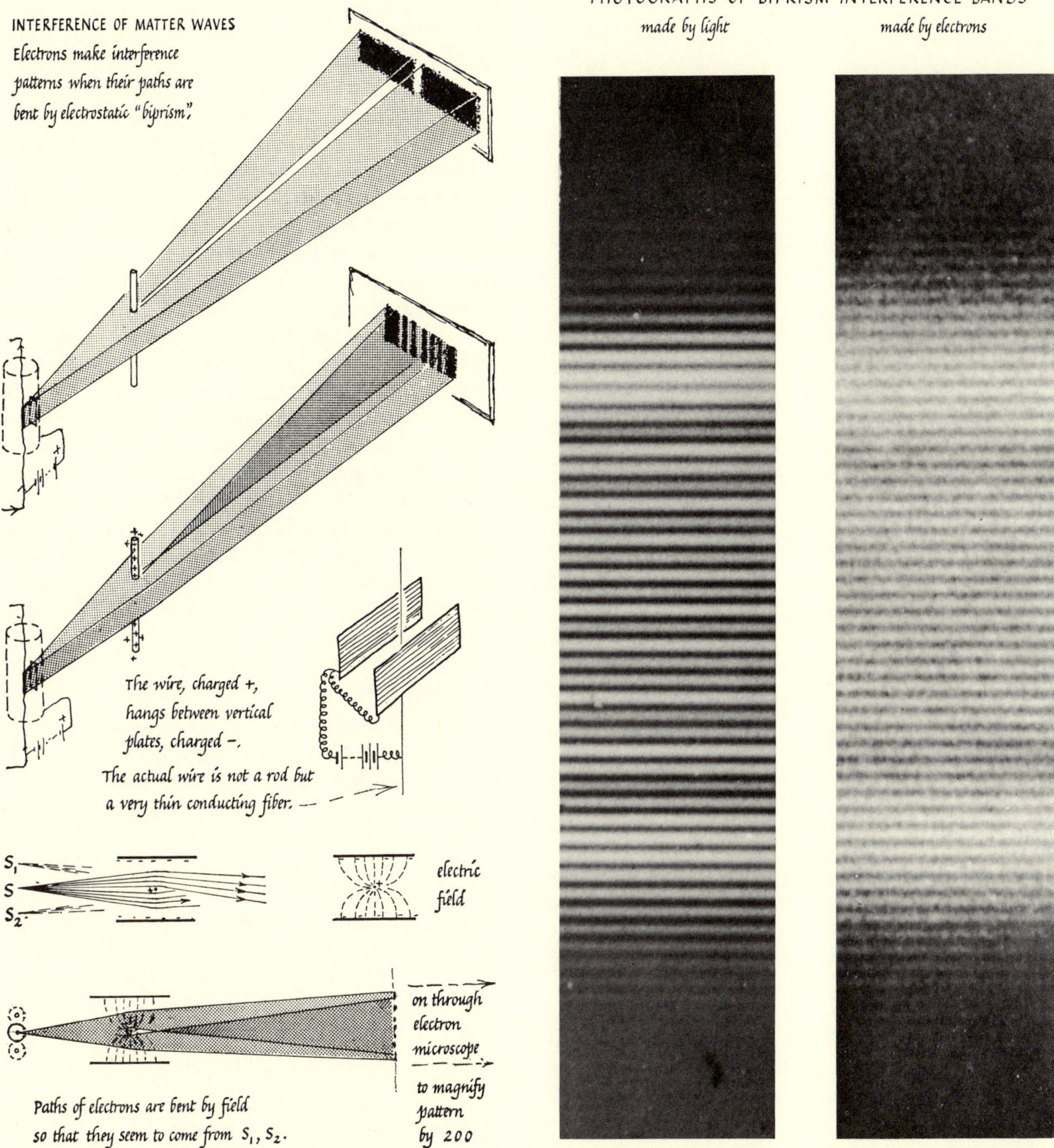

(d) Möllenstedt"s experiment. For electrons, the electric field near a positively charged wire acts as a "biprism." As with photons, there are "bright" bands, where many electrons arrive at the screen, and "dark" bands where few arrive—the guiding "matter waves" form an interference pattern. The pattern is so small that it must be magnified by electron-microscope lenses and then enlarged optically.

(c) Interference pattern made by visible light (about 30 times life size). Photograph by Henry A. Hill, Princeton University.

(e) Interference pattern made by electrons (about 5,000 times life size). Photograph by G. Möllenstedt and H. Düker, University of Tübingen. Experiment published in Zeitschrift für Physik, Vol. 145, 1956.

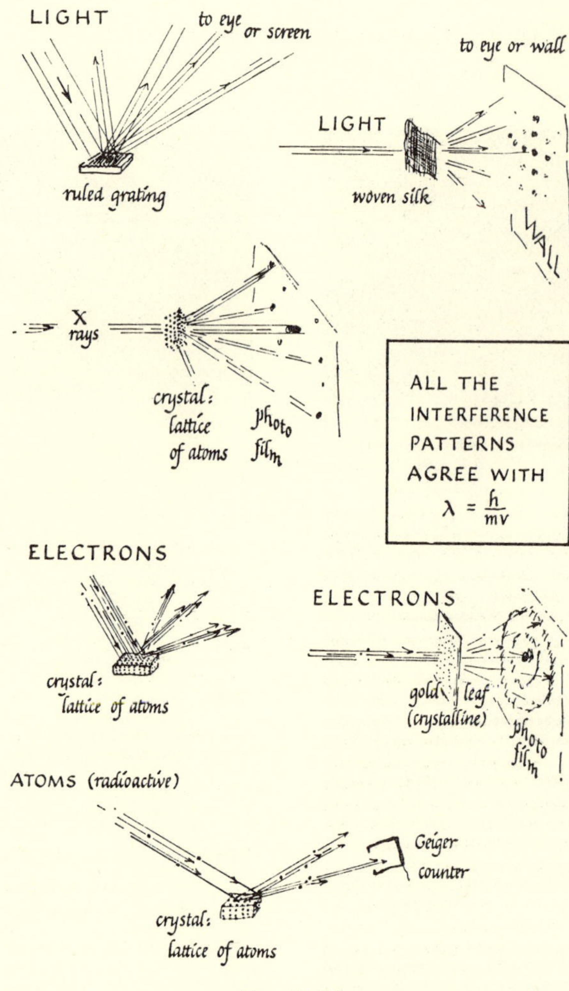

Fig. 44-24.

tons) are wave-groups carrying energy and momentum compactly. The "wave-packet" is what we normally observe as the "particle" (electron, proton, or photon . . .). (See Fig. 10-14.) However, the guiding ripples inside it are made up of many neighboring wavelengths, which gang together to make the resultant pattern. These sets of ripples are in phase near the center of the group, but elsewhere they get out of step and cancel. Yet each individual component-wave may be regarded as a guide that extends far ahead and behind. It is the wave that guides the group-particle to a bright fringe in an interference pattern. The guide-ripples of a moving particle *travel* faster than the group—the reverse of the case for water ripples. A particle of mass m and velocity v has momentum mv; its *wave-packet* travels with *group-speed* v; its ripples themselves have wavelength h/mv. And ripples travel at speed V, greater than v. In fact $Vv = c^2$ so V is greater than c itself. This breaks no Relativity rule, since the guide ripples are only phase-waves, an all-pervading pattern that carries no energy with that speed. You might fancy the ripples being there ahead of the particle to mark out the interference pattern and tell the particle where to go. This still leaves the wave:particle duality a mystery but it makes it easier to fit the wave idea to particles by saying "what you observe as a particle is a compact wave-packet." (Look at Fig. 10-14.)

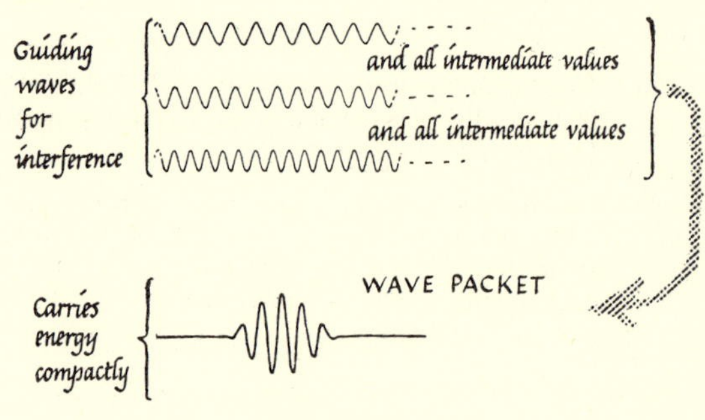

Fig. 44-25.

The de Broglie idea not only led to experiments that showed the wave behavior of electrons, etc., but also threw wonderful light on the Bohr model. And then it led to great developments of modern atomic theory.

Waves and the Bohr Atom

De Broglie offered a delightful reason for Bohr's mysterious rule of allowed orbits, $mv \cdot 2\pi r = nh$. An electron pursuing such an orbit has wavelength $\lambda = h/mv$. So $mv = h/\lambda$. Then Bohr's rule becomes $(h/\lambda) \cdot 2\pi r = nh$ or $2\pi r = n\lambda$.

According to this view, the only orbits allowed as stable are those in which the electron wavelength just fits into the circumference n times: $2\pi r_1 = \lambda_1$, $2\pi r_2 = 2\lambda_2$, $2\pi r_3 = 3\lambda_3$, etc. The electron must weave its wavy way around the orbit (like a snake eating its own tail) and form a *standing wave* (see Ch. 10), with a whole number of wavelengths in the circumference[24] (Fig. 44-26). We no longer

[24] To illustrate a circular standing wave, fill a large round glass bottle partly full of water. Give the bottle a small rapid rocking motion, and find the frequency that will build up a standing wave around the rim of the water surface. For details of standing waves, see Figs. 10-32, 10-33.

Fig. 44-27.

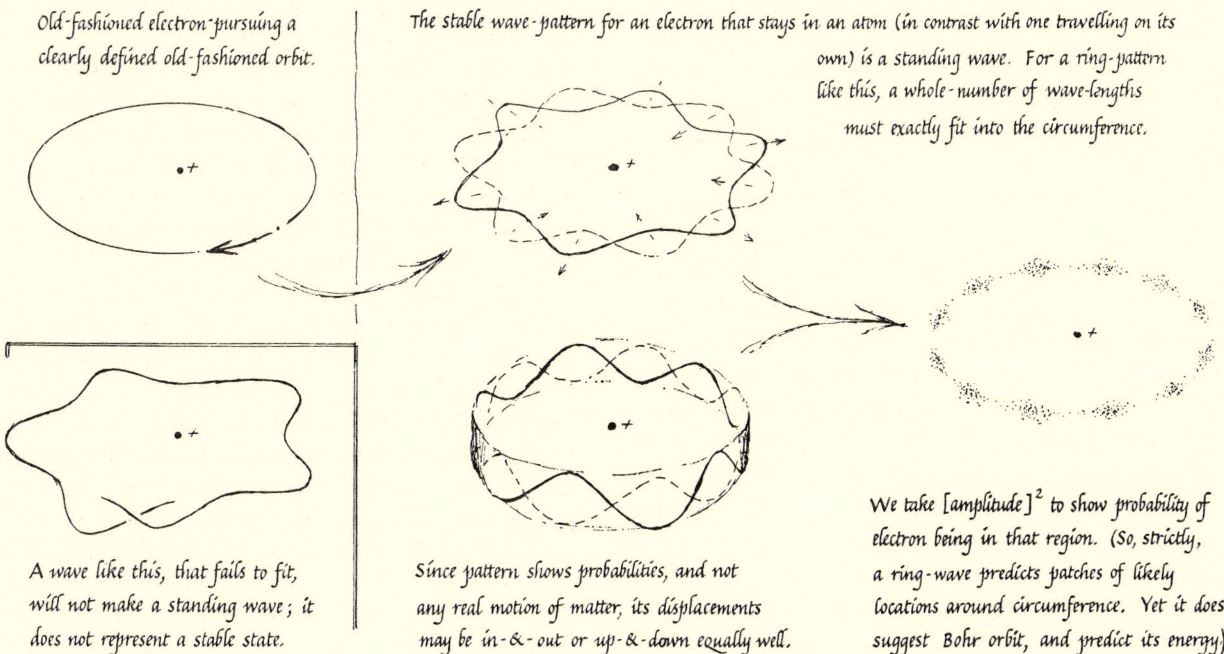

FIG. 44-26. BOHR ATOM-MODEL WITH ELECTRON REPLACED BY DE BROGLIE WAVE

see a solid particle with central acceleration v^2/r that would make it radiate, but a vibrating pattern that occupies the orbit. To many a physicist here was at last a comforting reason for Bohr's arbitrary rule: the "allowed orbits" are the possible standing waves.

We now regard these de Broglie waves as a scheme to tell us the electron's probable location: the stronger[25] the wave in any region the more *likely* we are to find the electron there when we look for it. These waves—a running wave for an electron moving alone in space, a standing wave for an electron bound in an atom—are not waves of moving matter or waves of changing fields: they are "probability" waves. The ring-waves that were first suggested to represent Bohr orbits might place the electron's probable locations in patches around the ring. Or they might be equivalent to travelling-waves running opposite ways around the ring. Then we could no longer profitably ask at what point in the ring the electron is located. But we now have other wave-patterns to specify location-probabilities in various states: *radial* standing waves as well as those *circumferential* ones. Bohr's innermost ring for hydrogen has changed to a radial line that goes right in to the nucleus. Yet the electron spends most of its time some distance out on that line, at an average distance that fits Bohr's early prediction.

[25] Here, "strength" is measured by (AMPLITUDE)^2 of these waves.

And when that line is swung in all directions, for symmetry, it paints for the electron a ball of probability fuzz around the nucleus. Some of the wave patterns for more complex atoms, and for higher states, give a more complicated shape of fuzz. Each pattern shows only a probability fuzz for an electron's *location*; but the frequency of the wave-motion in the pattern is definite enough, and that predicts a definite energy-level for the electron.

In particular, this view shows why an atom cannot collapse with electrons moving into smaller and smaller orbits without limit. If each electron's location *is* described by a standing wave, the circumference of the smallest orbit must be covered by just one wavelength—no fraction of a wavelength is thinkable in a ring standing wave—and that must be the smallest pattern the atom can be squeezed into. (With circular "orbits" replaced by a looser choice of patterns nowadays, a corresponding limitation for the simplest standing wave still holds.)

And Pauli's Exclusion Principle makes some sense: put several identical electrons in the same "orbit" and their standing wave patterns would add to a single pattern, and we might expect to find one electron there after all!

New Atomic Theory

A powerful treatment was developed by Schrödinger. He formed a general wave-equation (see page 470) for electrons, starting with the de Broglie quantum-wavelength rule. Then he looked for solutions in the form of standing waves that would fit the condition of an inverse-square coulomb force-field inside an atom. This corresponds to: finding the speed of waves along a taut string, (see Ch. 10: SPEED^2 = TENSION/MASS PER UNIT LENGTH); putting that into a general wave equation ($\nabla^2 V = (1/c^2)\ d^2V/dt^2$); imposing "boundary conditions" for, say, a harp string of length L (fixed ends that prevent motion at $x = 0$ and $x = L$); and then finding the frequencies of the possible standing waves (as in Ch. 10, where frequencies for 1 loop, 2 loops, etc. are calculated). In Schrödinger's case, the frequencies give the energies, by the quantum rule. The wave equations and the boundary conditions are more complex than for a string, but the results are rich.

This new interpretation was soon elaborated mathematically into a wave treatment of electrons in atoms that far outdoes Bohr in successful predictions. It succeeds with hydrogen as Bohr did, but also with finer details and with complex atoms. It yields good predictions in treating electron-sharing in chemical changes with waves. It extends successfully into nuclear physics and interprets radioactivity in terms of particle-waves seeping through a nuclear barrier. All this is at the expense of any definite model or picture. The Bohr atom took on unnecessary realistic details—uncheckable ones, and therefore scientifically immoral ones for permanent theory. Now the clear orbits disappear and are replaced by mathematical statements of wave-patterns that yield definite energy levels, just like the energies of electrons in the old Bohr orbits. But we have no picture of the waves involved. They are not like water waves or light waves at all—only their mathematical form is wavy. Our new treatment uses waves whose strength[25] shows where the electrons are most likely to be. The strength of the wave in a chosen region gives the betting of finding an electron there. We say for light waves interfering that the mathematical description tells us the betting of a photon arriving in a particular part of the pattern—low betting in a dark band, high in a bright one. Now for electrons bound in *STABLE STATES* in atoms we have statements of *STANDING WAVES* which place the electrons only in a fuzz of betting; *most likely* here, *likely* there, *less likely* elsewhere.

Inside a radioactive nucleus there is a similar fuzz for, say, an α-particle group of nucleons: a strong probability that the α-particle is inside, but, if the waves run on out, there is some chance the α-particle is outside. One day, the α-particle takes up that bet; it *is* outside, and thereafter is hurled away by the electric field. That is a rough hint of the "explanation" of radioactivity.

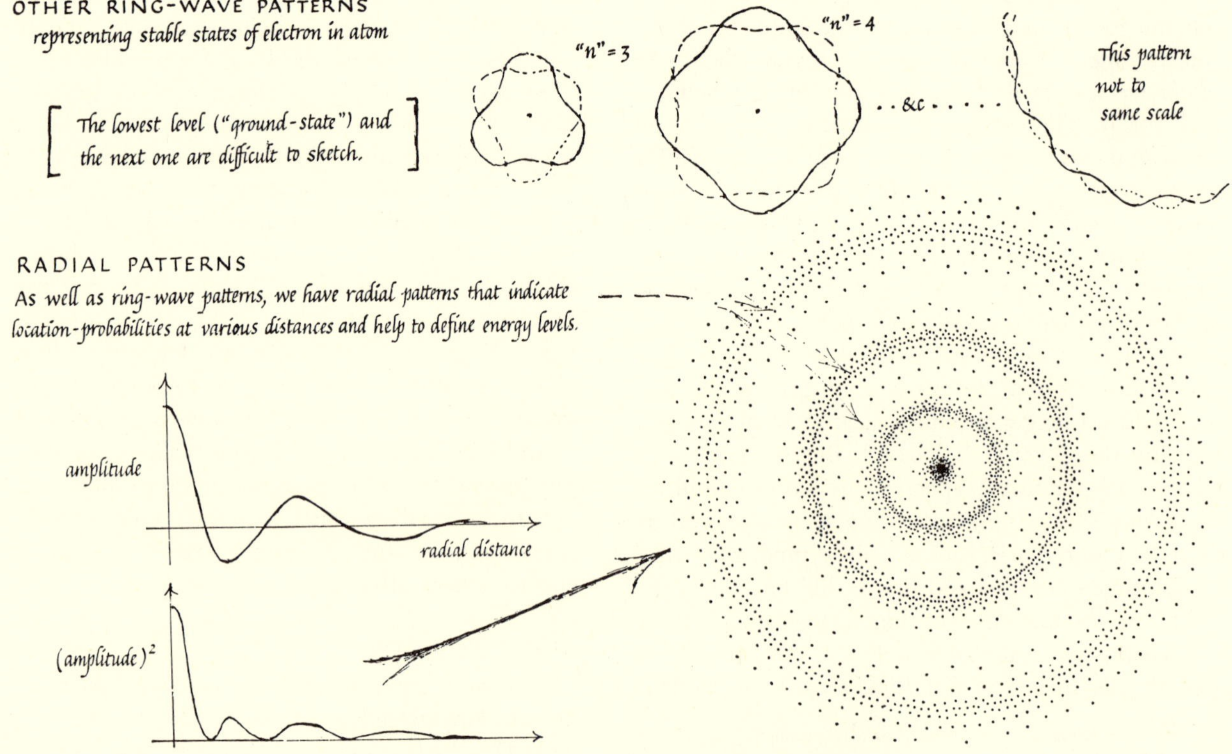

FIG. 44-28. OTHER WAVE-PATTERNS FOR ATOM-MODELS

The new treatment of atoms is neat but not simple. In more detailed development, the waves become more useful but no more real. We have good mathematical machinery to handle them, but no model to interpret the results. In fact, models seem doomed to failure, by being too misleading. Our picture of atoms and their behavior remains a *mathematical* pattern—a complex interlocking pattern that yields fruitful results. So we entrust atomic theory to mathematicians. They have developed new tools—a complex algebra, for example, with unorthodox rules.[26] However strange the methods, the results are excellent, tied down to experimental tests at one check-point after another:

> The wave-forms for a complex atom yield energy levels *and* the probability of an electron being at each level; and these lead to successful predictions of frequencies of spectrum lines *and* their brightness.
>
> Combined wave-forms for atoms in a molecule predict chemical energies and surface-tension forces. They even predict distances and angles between atoms in long chain molecules—checked by X-ray measurements on crystals.
>
> Calculated probabilities of finding an electron in some region of a particular atom agree with measured scattering of an electron stream by the atomic electrons in a target of those atoms.
>
> The K.E. of α-particles tells us their wavelength. That gives their probability of leakage through the nuclear barrier and thus predicts the half-life of the parent nucleus with some success.

Yet we have to leave the machinery of all this fruitful theory in mathematical form and can offer no good model to make it seem "reasonable." We are back to the childhood reason "BECAUSE IT DOES," and the Greek view, "It is NATURAL." Like the Greeks, our atomic physicists have clear rules—rules for

[26] To follow the treatment, you need mathematical tools that we cannot develop here. Hence this vague description, which does the new theory no justice.

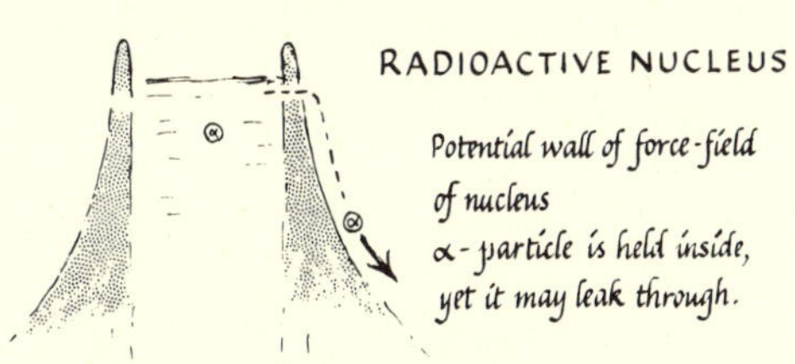

Nuclear ingredients are so energetic [excited, "at high temperature"], that they reach to upper level where wall is thin.

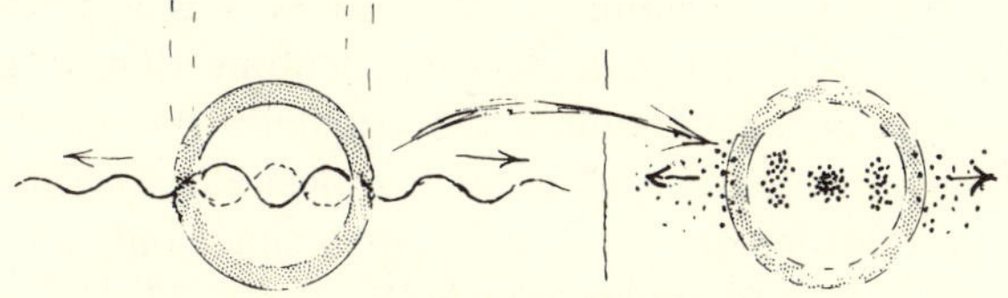

Snapshot of location-wave for α-particle group in nucleus. Wave merely specifies likely "locations" of particle:
MOST LIKELY INSIDE *but* POSSIBLY OUTSIDE

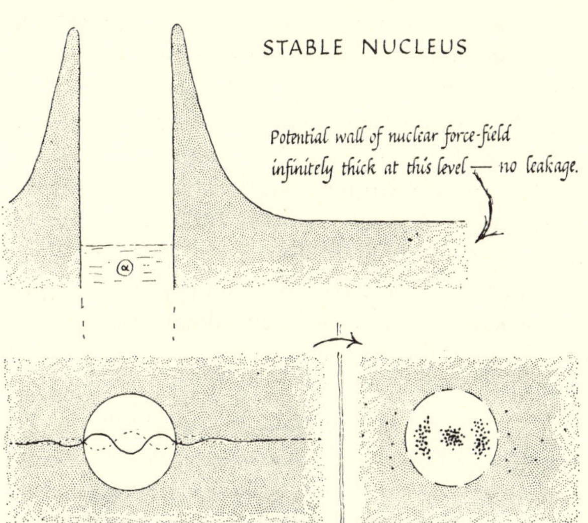

Snapshot of location-wave for α-particle group in nucleus. Wave merely shows likely locations.

FIG. 44-29. WAVE PATTERNS IN NUCLEI

quantizing and for symmetry, rules that work—but no first causes for them. This is still good science, saying "Here is how nature behaves" in a few rules rather than blaming demons of a hundred colors.

From this point on, theory and experiment seem to divide the people who work in physics today. The experimental man continues to search, to test predictions, and to look for new behavior, often with large expensive machinery for his indirect attack on the submicroscopic world of atoms and nuclei. Theoretical physicists use tough mathematical tools and try to avoid the thinking in models that more practical minds would like. Here and there we find a greater speculator and wiser scientist who can weld theory and experiment and advance our understanding.

Uncertainty Principle

If a moving electron is guided by its own waves, we shall have trouble when we try to pin it down precisely—the trouble the pinhole-camera maker meets if he makes the hole too small. Fire a stream of electrons through a narrow hole in a wall: a narrow stream passes through. Make the hole still smaller, till its diameter is as small as an electron wavelength: the electrons passing through will spread in all directions. They *must* do so. This is not just an avoidable deflection by neighboring atoms in the wall: it is an unavoidable part of electron wave-geometry. Try to predict what a particular electron that gets through will do beyond the wall, and you are strangely powerless. You know accurately where it will cross the wall—through the tiny hole—but you cannot tell how much sideways momentum it will acquire. Or, if you want to be sure of its momentum, and say confidently "it will emerge with momentum mv straight ahead; that's all," you must enlarge the hole so that the electron-wave passes straight on, with little spreading by diffraction. But then you do not know exactly where the electron-particle passed through the wall—just somewhere in the wide hole. What you gain in knowledge of momentum you lose in knowledge of location.

From $\lambda = h/mv$ we can calculate limits to our certainty. The general geometry of waves shows that if they pass through a hole of diameter λ or less they spread completely; waves passing through a hole a few λ wide spread noticeably; and through holes a dozen λ's or more wide they pass practically straight on (see Fig. 10-22). If we pin down a moving particle's position at the wall to a few λ we must risk adding sideways momentum that is a large *unknown* fraction of the original forward mv. If we are content to know its position roughly, within dozens of λ's, we risk only a small unknown fraction in subsequent knowledge of mv. The table below is rigged to look more definite than is natural, but it illustrates the results of a full investigation with detailed wave-geometry.

"UNCERTAINTY PRINCIPLE"

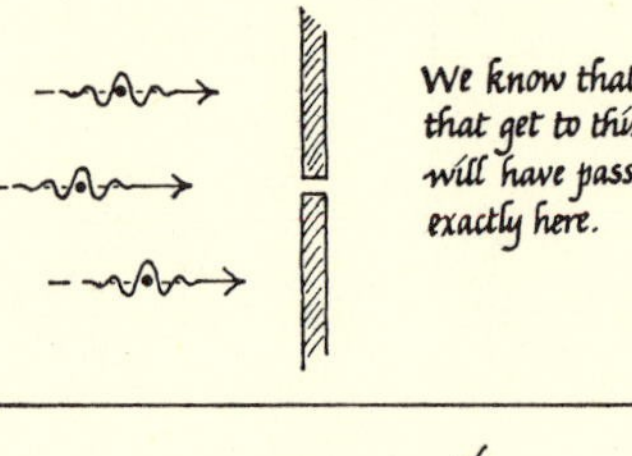

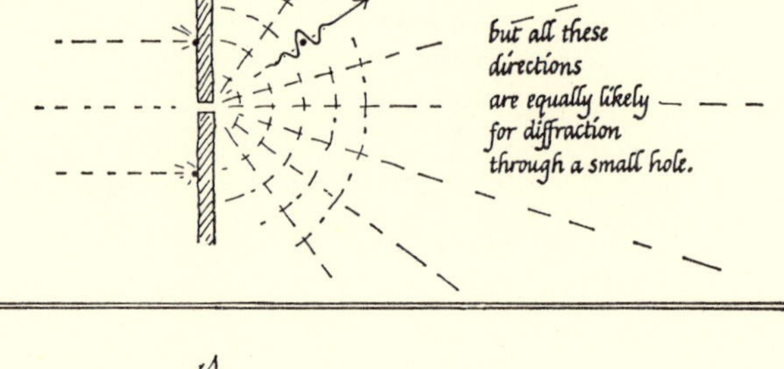

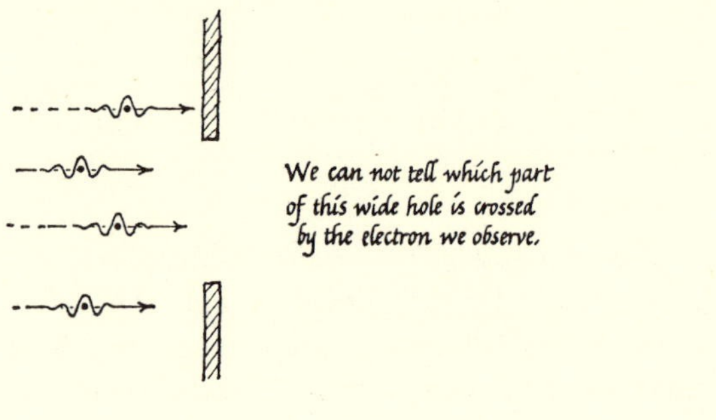

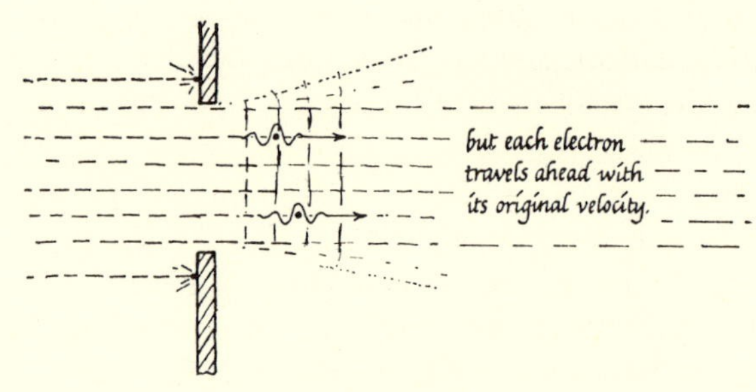

Fig. 44-30.

UNCERTAINTY ABOUT POSITION	UNCERTAINTY ABOUT MOMENTUM	PRODUCT OF UNCERTAINTIES	
("*Where* does the electron pass through the screen—how big is the defining hole?")	("How fast will it move afterwards, and in what direction? What sideways *momentum* may it have?")	("Can we find a simple rule?")	
Δx	$\Delta(mv)$	$(\Delta x) \cdot (\Delta mv)$	
many λ's say $N\lambda$	a small fraction of mv say $(1/N) \cdot mv$	$(N\lambda) \cdot (1/N\ mv)$	$= \lambda \cdot mv$
a few λ's say 3λ	a fraction of mv say $(⅓)mv$	$(3\lambda) \cdot (⅓ mv)$	$= (h/mv) \cdot mv$
about λ	about all mv	$(\lambda) \cdot (mv)$	$= h$ in each case

These are rough statements about our roughness of information. However, detailed examination leads to the same conclusion: that in every case the two uncertainties multiplied together yield $\lambda \cdot mv$, which is $(h/mv)(mv)$, which is h.

UNCERTAINTY ABOUT POSITION, Δx	$\cdot$	UNCERTAINTY ABOUT MOMENTUM, $\Delta(mv)$	$\approx$	QUANTUM CONSTANT h

These are uncertainties which doing one measurement imposes on the other. The more precisely we make one measurement, the less precisely we can predict the outcome of the other. The uncertainty is not produced by careless techniques; it is built into nature. The action of making one measurement precisely destroys the necessary facilities for the other measurement. Since each uncertainty is a statement about our roughness of information, it is necessarily rather rough itself.[27] Therefore we should not say that $(\Delta x) \cdot (\Delta mv)$ is exactly h, but that it is about the size of h, "of the order of h."

This is Heisenberg's *Uncertainty Principle.* It has been of great use in building mathematical machinery for treating waves and particles in atoms. Its comment on electrons in experiments is stern: like light waves, they will defeat attempts to measure with complete precision. It also changes the Bohr atom picture. We may state the electron's momentum (and therefore its energy-level) in an orbit precisely; but then we must be quite vague about its location—we cannot say where it is. So we certainly should not draw a sharp orbit and mark the electron as a blob in it.

The change in philosophical outlook goes even further. We *cannot* push knowledge to the utmost precision we like: it is not just a question of skill, patience, machines, money. . . . In imagination build a super microscope to watch an electron. Surely *then* you could see both its position and its momentum? No, because any microscope must use some "light" for seeing. If you are to "see" the electron at all, at least one quantum must bounce off it into the super-microscope. That collision would upset the electron's motion, giving it an unpredictable change of momentum (by the Compton effect). And, to locate the electron precisely, the "light" must itself have very short wavelength or *its* diffraction pattern will blur the image. Therefore that light-quantum must have very small λ, very high frequency: it must be a huge quantum, an X-ray photon. In that case the collision will be a violent one, and the electron's unknown recoil-momentum makes a great uncertainty. A detailed calculation, using the proper expressions for Compton recoil and diffraction of light in a microscope, leads to $\Delta x \cdot \Delta(mv) \approx h$.

This uncertainty-rule of nature operates similarly with ENERGY and TIME. We cannot measure a particle's K.E. with complete precision in an infinitely short region of time. The uncertainty of our knowledge of energy, ΔE and the spread of time taken by

[27] Remember the admonition in Ch. 11: it is anti-scientific to be *over-precise* in stating a % difference which expresses some *imprecision.*

the measurement of energy, Δt, are related thus $\Delta E \cdot \Delta t \sim h$. (See note 28.)

In all these uncertainty statements, even the sign $\approx$ is too definite. We ought to say $\Delta x \cdot \Delta mv \sim h$ (is about the size of h). Even that is talking about the *best* we can hope for, so we should say "about the size of h, or greater."[29]

The tiny atom of *action*, h, seems to be the area of the holes in the finest net with which we can fish for information in Nature. The error-boxes around any patch of information that we can catch must be bigger than h in area, or we shall fail to catch it. We can strain the net to pull its holes narrow one way, to catch a fine piece of detail, but the holes lengthen the other way. The detail that is fine in "width" must be, so to speak, coarse in "length," or it will slip through. We can measure both momentum and position *roughly*; but only one of them precisely, at the expense of poorer knowledge of the other.

We see that the experimenter's apparatus and line-of-inquiry are strongly interlocked with the item he is investigating. Once he decides to pinpoint one measurement, he loses hope of knowing another measurement accurately. This is not the beginner's catch-all excuse, "experimental error," or the professional scientist's "probable error" of $\pm$ so many % for his particular apparatus; it is an essential interaction between apparatus and observer.

The corresponding restriction in biology was pointed out by H. G. Wells long ago: the zoologist experimenting on a living animal never has a completely normal healthy specimen, because his very act of experimenting modifies the animal under test. Modern biologists and poll-takers, and all psychologists meet similar effects.

This limitation of accuracy in physical experiments remained undiscovered as long as experiments were done with large bodies containing many atoms. With them, statistical averages (which we gross experimenters usually observe) smooth out the fluctuations and present us with measurements that make easy constant laws.[30] You have already seen such smoothing when the impacts of gas molecules smooth out to a steady pressure. But remember that if a single bacillus could conduct physical experiments it would give quite a different description of Nature. Surface forces would outweigh gravity; the Brownian motion would be paramount; photons would operate by ones and twos. If we could watch, we could still hope to see and predict these irregularities; but on the still smaller scale of a single atom the *essential* uncertainty-linkage of nature would operate with full force to restrict our prying.

Uncertainty and Wave:Particle Nature

When we mark the position of an electron accurately at some instant, that act wipes out our chances of measuring the velocity it then had. This is repeating, in a new form, the wave:particle duality of matter. If we decide to measure an electron's velocity precisely, that tells us its momentum and wavelength, its essential wave property. We are finding its wave size by measuring a long train and that cuts us off from treating it as a compact particle finding its position. If, instead, we mark its position,

[28] Derive this for the case of a photon, thus: a quantum of light has energy $E = h\nu = mc^2$ and momentum mc or E/c, where m is its mass.

$\therefore$ uncertainty of momentum $= \Delta(mc) = \Delta(E/c)$.
The photon moves with speed c, so, if we are uncertain of its travel-distance by an amount $\Delta(x)$ we are uncertain of time by an amount $\Delta(t) = \Delta(x)/c$

$\therefore$ uncertainty $\Delta(x) = c\Delta(t)$

$\therefore$ $\Delta(mc) \cdot \Delta(x) = \Delta(E/c) \cdot c\Delta(t) = \Delta(E) \cdot \Delta(t)$

If $\Delta(mv) \cdot \Delta(x) \sim h$, then $\Delta(E) \cdot \Delta(t) \sim h$

If we measure the photon's energy (and thence its frequency and wavelength) fairly precisely, say to 1%, then $\Delta(E) = E/100$, and $\Delta(t)$ must be bigger than $h/(E/100)$ or $100\ h/E$ or $100\ h/h\nu$ or $100/\nu$, or 100 complete periods of the light-wave. We must waste that much time to measure E (or ν) to 1%.

If we use a chopper to select a short sample of a beam of light, we *must* lose precision in our knowledge of its energy, frequency or wavelength. See Fig. 10-14.

[29] A re-examination of best hopes leads to $\Delta x \cdot \Delta mv \approx h/2\pi$ in some cases, a gain of a factor of 6 at best.

[30] With man-sized apparatus we can make uncertainty Δx a tiny fraction of the object's height and uncertainty $\Delta(mv)$ a tiny fraction of its momentum—remember its mass is huge—and yet find their product is many times h. For example, try to observe a baseball, mass 0.2 kg, moving 3 meters/sec. Suppose we pinpoint its position to one wavelength of green light (which is all we could hope for with an optical microscope). The uncertainty Δx is 5000 Å.U. or 5×10^{-7} meter. And suppose we time its flight over 1 meter to the nearest millionth of a second (all we can hope to do with such a big object). Then our uncertainty in velocity is $1/10^6$ sec in $1/3$ sec or 3 in a million. Then our uncertainty in mv is also 3 in a million, or $3/10^6$ of the measured value, 0.2×3 kg · meters/sec. So $\Delta(mv) \approx 2 \times 10^{-6}$. Then $\Delta x \cdot \Delta(mv) \approx (5 \times 10^{-7})(2 \times 10^{-6}) \approx 10^{-12}$. In any attack on nature, we cannot make this product less than h, 6.6×10^{-34}. Here, our product is more than a thousand billion billion times that minimum limit—we shall meet no trace of that restriction, but shall merely encounter experimental difficulties, which we may succeed in reducing.

On the other hand, choose an electron emerging with speed 6 million meters/sec from a 100-volt gun. Try to pinpoint its path to the nearest atom-diameter, $\Delta x \approx 10^{-10}$ meter; and measure its velocity within 10%, giving an uncertainty of 0.6 million meters/sec. Then:

$$\Delta(mv) \approx (\text{mass } 9 \times 10^{-31} \text{ kg})(\Delta v,\ 0.6 \times 10^6 \text{ m./sec}) \approx 5 \times 10^{-25}$$

$\Delta x \cdot (\Delta mv) \approx 10^{-10} \cdot 5 \times 10^{-25}$ or less than one tenth of h We have already set our hopes too high. We cannot track the electron to the nearest atom without making its velocity uncertain by more than 10%. And that is an unavoidable uncertainty.

we thereby state it is a particle and cloud over any wave view that would tell us its momentum. (See Fig. 10-14 for pictures of this.)

This is not a careless dishonesty between Nature and us. Far from that; *this is the result of our attempts to force Nature into an unnatural mold!* At the microscopic level of atoms, and electrons and quanta, *Nature's behavior is not that of a particle, or a wave.* If we insist on looking for particle behavior because something reminds us of it, we measure some properties that the electron or atom would have if it were a particle. Man has always tried to press the gods into his own likeness, and now we try to press atoms into the likeness of our own playthings: baseballs, drops of water, ocean waves, . . .

The Principle of Complementarity

From early experimenting and simplifying assumptions, we built up Newton's Laws, Maxwell's equations, the theory of Relativity, which describe the behavior of large moving masses, charges, etc. quite well. Then we assume these ideas are suitable for electrons, nuclei, quanta, . . . Or, rather, since we have no direct contact with that micro-world we *choose* to apply our large-scale laws whether they are suitable or not, and we must agree to take the consequences. Now we see that it is silly to ask whether these laws are true. We may assume they are, and then we get a view of Nature revealed in terms of that assumption. (If we ask, "How angry is the storm," the reply for a thunderstorm is, "Very angry." Yet we should be unwise to let that answer prove that storms have tempers.) Whatever the micro-world really is—and that "really" may be itself a macro-man's mistake—it is not a world of waves or particles. Forced into a wave description by a wave question, it gives a wave answer, and a complaint of particle ignorance. Or, asked a particle question, it gives a particle answer with wave ignorance. It is not Nature that pulls down a curtain over particle facts when we ask a wave question. It is our questioning that forces a non-wave-non-particle electron into an uncomfortable wave form, or an equally artificial particle shape. In fact, one uses a wave picture "to describe accurately, not the *electron*, but the *state of one's knowledge about the electron*."[31]

We are wiser not to push our assumption-laden questions so harshly, but to realize that in such cases where two different conflicting descriptions can be made to fit partially:

[31] John A. Wheeler.

(i) Each description complements the other: each is "true" when we are dealing with *its* aspect of the matter.

(ii) Each description excludes the other. While we are using one description, we must not try to use the other as well.

This is Bohr's great Principle of Complementarity, stated by him as follows (using the word "classical" to refer to traditional rules of old established physics): ". . . any given application of classical concepts precludes the simultaneous use of other classical concepts, which in a different connection are equally necessary for the elucidation of the phenomena."[32]

Consider four short tales, invented to show the flavor of the Complementarity view.

FIRST TALE. Suppose you receive a toy "atom" made of a handful of modelling clay. Asked the length of an "atom," you could squeeze it into a worm a foot long. Asked the diameter of an "atom," you might think of a ball, shape it into a sphere, and measure two inches. Treating it like that, you cannot answer both questions at once. This tale shows the way in which our choice of question controls the method of answer and inhibits alternative experiments. However, this illustration is misleading because the clay remains with you and you are not prevented from trying both experiments in turn *with the same clay.*

SECOND TALE. Suppose you have a newly-hatched beetle, the last of his species. If you want to know how long a newly-hatched beetle of that species can live without food, you can keep him in a box and record how long he lives. If you want to find out how long he takes to double his weight on a diet of sugar, you can feed him and record his progress. But you cannot make both measurements on the same beetle. The two experiments are mutually exclusive. This tale shows how completely the choice of one question can cut us off from answering the other. However, this illustration too is misleading because it is the situation that is frustrating and not the behavior of Nature—both questions are normal, but we have manufactured an artificial difficulty.

THIRD TALE. A nervous depositor who has put his savings in a bank is anxious about his money and wants to make sure it is really there, safe in the bank. He tries two methods of reassuring himself:

(a) He watches for the regular payments of interest, and infers from them that the bank has his money.

(b) He revives a childhood memory of peering into a piggy-bank with a flashlight, and he goes to the bank and asks to have his money out to look at it, from time to time. The bank agrees, but warns him that then there will be no interest. (The bank further tells him that if all customers made such demands frequently, banks could not operate at all.)

The depositor can use either method successfully, but each prevents the other being a fair test. He must revise

[32] Niels Bohr, *Atomic Theory and the Description of Nature* (Cambridge, 1934).

his picture of banking and learn that although his money is safe it is not really there as cash in a vault.

FOURTH TALE. Suppose we have a small box, like a match box, that emits a thin chirping sound. Two people are allowed to examine it, to discover what is inside, without opening it. B, an amateur biologist, listens and decides there is an insect like a cricket inside. Having made that decision, B can ask scientific questions and obtain answers. He listens to the musical note of the squeaks and their duration and, knowing that the cricket makes the music by rubbing its legs, he estimates (we may imagine) the number of ridges on the cricket's legs, and their spacing. That may even help him to guess what species of cricket is inside. That is good scientific guessing *on the basis of his hypothesis*—but in his enthusiasm for peering into the box indirectly, B may not notice the restriction he imposed on himself by making his first guess.

Meanwhile, E, a radio engineer, looks and listens and decides that there is a little transistor radio oscillator inside the box. From the size of the box and the musical pitch of the squeaks, he estimates (we may imagine) the size of the electrical capacitor that controls the squeaking.

Then B and E get together. Though they disagree completely on the explanation, they agree to try a further experiment. They are still forbidden to open the box, but they receive permission to pour mineral oil in through a crack in the edge. The squeak drops to a lower musical pitch. The biologist, B, says: "Just what I expected: he finds it harder to move his legs in viscous oil." The engineer, E, says: "That settles it; oil poured between the plates of a capacitor always increases its capacitance and that should lower the musical note." Each has confirmed his decision, and trusts the measurement he has made—and each forgets that his measurement only has meaning in terms of his main choice of explanation.

General Complementarity

We meet complementarity at many major points in our intellectual world: in science, in philosophy, in ethics—in life itself. We meet many a matter that has different, exclusive, aspects that complement each other. Our atomic experience suggests we should not say, "These are conflicting views. It is annoying that each seems right, and it is mysterious to find that using one view seems to shut us off from the other." Instead, we should blame our Greek inheritance of arguing to an absolute yes or no—discussed in Chapter 1. We should see that our discomfort comes from forcing a behavior that is neither A nor B, but something unknown, into one form, say A—in which case we must lose sight of B. Instead we should learn to live with both viewpoints, giving each its best use.

Fig. 44-31 shows a model constructed to illustrate Complementarity.[33]

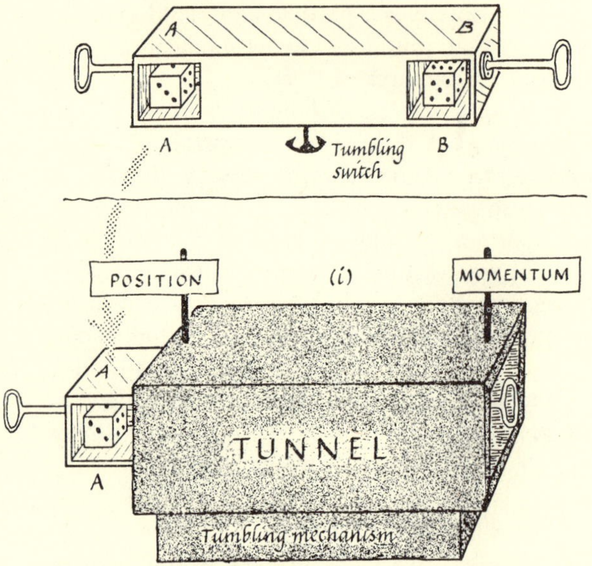

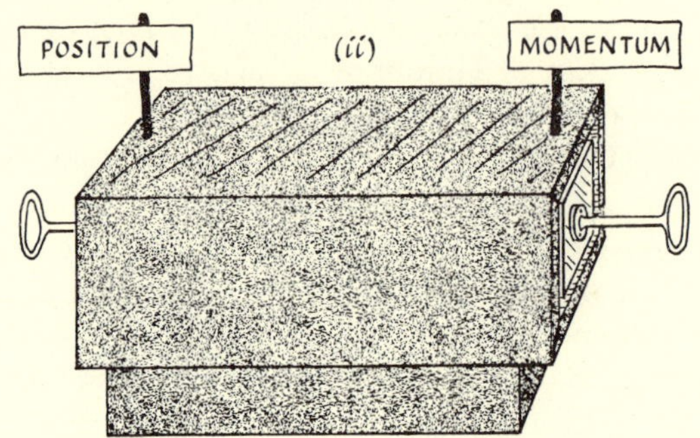

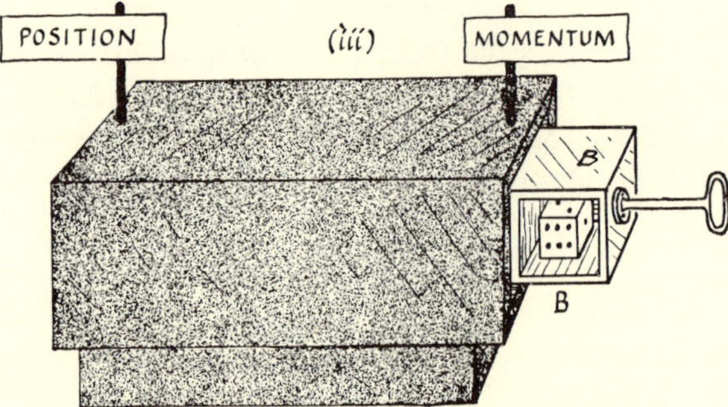

Fig. 44-31. "Die-Box" to illustrate Complementarity

Two dice A and B are placed in a long drawer *AB*, one at each end, in compartments with glass windows. The drawer is housed in a tunnel, so that you can pull it out at either end to see the die there. To see die A, you pull the drawer till end *A* is out; A is visible and B is hidden. Then to see die B, you

[33] Constructed, probably at the suggestion of Bohr, for the World's Fair, 1939. The model described is a large version, designed by H. M. Waage and J. A. Wheeler. See *The American Scientist*, Vol. 44, No. 4, October 1956, "A Septet of Sybils" by J. A. Wheeler.

must push the drawer through the tunnel until end *B* is out and B is visible. Pulling or pushing the drawer right through operates a tumbling mechanism in the tunnel so that *as one die comes out into view the other die is thrown.* (This is done by an arm that hits an elastic rubber floor under the die.) The tumbler does not act until the drawer is more than half way through. So if you shove the drawer just into the tunnel (both dice hidden) and pull the same end out again, you will see the same score—you can pull end *A* out and look at die A repeatedly and see the same face. But when you shove the drawer right through until you see B, you can hear A being tumbled before B emerges; and then you cannot predict what A will show when you next pull it out. Looking at one die prevents you knowing what the other die will show.

This is not a "model" in the scientific sense, but a mechanical toy that illustrates the choices of Complementarity. If we label end *A* "POSITION OF ELECTRON" (or "WHERE?"), end *B* should be labeled "MOMENTUM" (or "HOW FAST?"). But if we label end *A*, "K.E. OF PARTICLE," we must label *B* "TIME." There are many other pairs of aspects that are thus complementary—important, but mutually exclusive, so that dealing with one cuts us off from the other. Here are suggestions listed by John A. Wheeler.[34]

LABEL AT END *A* FOR MEANING OF DIE A	LABEL AT END *B* FOR MEANING OF DIE B
Position (e.g. of an electron) or *where*?	Momentum (of electron) or *how fast*?
Energy (e.g. of an electron)	Time (at which energy measured)
EXPERIMENT TO OBSERVE wave-aspect of matter	EXPERIMENT TO OBSERVE particle-aspect of matter
Use of word to communicate *information*	Analysis of the *meaning* of the word
Love of fellow man	Justice to all men
Free will	Determinism

Complementarity does not contrast opposites, such as love and hate, but only joins exclusive aspects of the same thing, such as love and justice. We must learn to live with complementarity.

Free Will and Determinism

(If you read this section, take it with grains of salt. It offers comments of physicists thinking as amateurs in a difficult field of philosophy.)

[34] In "A Septet of Sybils," *loc.cit.*

At last we can add a useful comment on the puzzle of free will *vs.* determinism. After the great development of Newtonian mechanics it seemed clear that given the position and velocity, *now*, of Sun, planets, etc., one could *in theory* predict all future motions in the Solar system. The future of human actions then seemed equally predictable, in theory, however complicated the problem might be in detailed structure. One could imagine cataloguing the position and motion, *now*, of every electron in a man's brain and elsewhere in his body, and everywhere in his surroundings, too. Then, with complete knowledge of those starting conditions, we could use the laws of classical physics to predict every change in the man's brain, every decision, every muscle movement, etc.—the whole future in due course. Though we may lack the knowledge of details (or even knowledge of some necessary laws), the facts are there; the laws are true, as many a scientist thought; and therefore the future is already settled, completely determined by nature. A man could not alter the future by changing his mind. Or rather, if he did change his mind, that change too would be part of an inevitable predetermined chain. Free will—our ability to choose rightly or wrongly, to decide on our actions—seemed to be an illusion, mere wishful thinking.

Then quantum-mechanics appeared to release individual electrons from a complete determinism, and that in turn might break the rigid chain in man's determinate future. Complete prediction would require a precise knowledge of the present position *and* momentum of every particle, and the uncertainty principle tells us that if we know one of these measurements with great precision we can hardly know the other at all. Thus the greatest collection of information we could hope to gather is only half what classical physics suggested as both possible and necessary. However, human decisions involve hordes of electrons, not just one or two, and their statistical averages would bring us back towards practical certainty—determinism. Now, however, we see in living processes and mental decisions a complementarity-pair of free will:determinate fate. If we look for determinism, we shall find fixed fate, *in the area of discussion defined by our search.* But if we look for free will, we shall consider our choice still open, in the scheme we have then defined.

If we opened up a man's brain we might in theory be able to catalogue all the atomic states now, and average out the uncertainties and then predict future actions. But that very process of investigation (an experiment) will stop the man from going on living and making his decisions with free will. Free will and determinism are a complementarity pair; each may well be true in its own way, but whenever we adopt one view we cut ourselves off from discussing the other in that particular case.

ONE MORE TASTE OF THEORY

Foreseeing a New Particle: Yukawa's Meson

As a final sample of theoretical physics, watch the wave:particle theory predict an entirely unexpected sub-atomic particle: a *meson*, intermediate between electrons and atoms. (The following is a crude sketch of a great piece of speculative theory. The version here is far from complete, and it is not even sound physics; but it is offered as a hint of the line of argument. Remember that a proper treatment is often discovered by trying a rough treatment like this—any means to the end, in the first round.)

Thirty years ago, when the structure of atoms was being worked out with great success by the new wave:particle physics, the structure of nuclei remained a puzzle, partly illuminated, partly a mystery demanding some unknown mechanism of forces to hold the nucleus together. Inverse-square-law electric forces link electrons and nuclei in atoms; and de Broglie wave patterns describe the probabilities of their arrangement. But there must be other forces inside a nucleus to outweigh the electric repulsion between protons and to hold protons and neutrons together. In fact, scattering experiments show signs of such forces just outside nuclei. When very energetic alpha-particles or protons bombard nuclei, the few that are scattered back at big angles after close encounters arrive in unexpected proportions. There are even fewer of them than we predict for an inverse-square-law repulsion alone; so new attractive forces seem to be acting: forces that are only felt very close to the nucleus. In the hill-model for energy in such collisions, these "short-range" forces make the hill level off to a rim and drop down into a crater. At the rim, the short-range attractions just balance the electric repulsion that acts on a bombarding charged particle. Outside the rim, they soon become trivial: within the crater, they are very important for any nuclear particles. From bombardment experiments we find that the rim is about 1.4×10^{-15} meter from the center, for an average nucleus. We shall call that distance the nuclear radius, r. Thus, something was known about the *range* of nuclear forces, and measurements of binding-energy could give an estimate of their *size* at closer quarters; but nothing was known about their essential nature or mechanism—there was no picture to link them with the macro-world.

Then the Japanese physicist Yukawa made the startling suggestion: sub-atomic particles are continually being created inside the nucleus, tossed across from one nucleon to another, and there absorbed. Such an "exchange process," with particles that had never been imagined before, could provide nuclear binding forces if the particles were endowed with a suitable mass and in some cases, carried a charge. (This is somewhat like the binding effect of a tennis ball on two players who dislike each other. As long as the ball flies to-and-fro in the game, they are held on the court.) This picture of *exchange forces* followed a scheme that theoretical physicists had developed for "explaining" ordinary electric forces such as the repulsion between electrons. We shall look at that scheme first.

How does one electron repel another? We can say that each is acted on by the other's electric field; but in explaining how an electric field and an electron connect with each other we now treat photons as essential agents of mechanism. When an electron that is accelerating emits (or absorbs) a photon, there is an exchange of energy and momentum between the electron and radiation; and the electron feels a force. When an electron is not accelerating but feels the force of an electric field, theoretical physicists could still say it is transferring energy and momentum to and from photons. But these are "virtual photons," emitted-and-grabbed-in so rapidly that we could never have a chance to observe them. Such photons, "out on a leash" from an electron (or some other charged particle) would carry momentum and exert forces as they come and go; and these would be the observed forces that we say the electric field exerts on a charge. Occasionally, when a charge suffers a violent acceleration, one such photon breaks loose from the leash and flies away as an observable quantum of light. Except for that, we do not observe the virtual photons, but we imagine them trading energy with electrons—so that each electron must spend a small fraction of its time (1%, we think) in a state of higher energy, when it has just absorbed one of these photons. (If this idea seems wild and even irritatingly unreal to you, remember it is a theoretical device to extend our methods of speculating about atoms. We use it because it is successful: it helps us to make predictions and it increases our vocabulary of understanding.)

Yukawa looked for a corresponding mechanism to

hold protons and neutrons together in nuclei. There the forces have a different distance law: they fall off far more quickly than a $1/r^2$ force, dropping from huge values inside a nucleus to negligible values just outside the "rim." Yukawa found that for such forces a particle like a photon with zero rest-mass would not work as the binding agent. The particle must have some rest-mass to fit with the experimental knowledge of forces. Then it could provide the forces as it is manufactured, tossed across to another nucleon, and absorbed before even the sharpest experimenter has time to notice anything wrong in the energy accounting. Here is the prediction of the particle's mass. (Anticipating present-day knowledge that there are such particles, we shall use the name we now give them: "mesons.")

Suppose a meson is created and spends its short life before absorption circulating around the nuclear rim, like an electron in a Bohr orbit much farther out. (See below for alternative suppositions.) To form such an "orbit" its de Broglie wave would have to form a ring standing wave, of radius r, and the simplest wave would be one with a single wavelength, λ, in the circumference $2\pi r$. Then:

$$\lambda = 2\pi r$$

But, for any particle, $\lambda = h/mv$

$$\therefore\ mv = h/2\pi r$$

We are only speculating about the particle (and speculating in a careless way, too), and we do not know its m or its v. We know that if we say v is c, that is an overestimate (unless the particle is a photon with zero rest-mass, in which case v is exactly c). And if we say m is m_0, the rest-mass of the meson, that is an underestimate. For a rough hint, we make both changes and hope they will compensate somewhat: we write m_0c instead of mv, and have

$$m_0c = h/2\pi r$$

Then $m_0 = h/2\pi rc$, and we can actually predict the size of m_0 from known values of h, c, and the estimate of nuclear radius $r \approx 1.4 \times 10^{-15}$ meter.

$$m_0 \approx \frac{6.62 \times 10^{-34}\ \text{joule} \cdot \text{sec}}{2 \times 3.14 \times (3.0 \times 10^8\ \text{m/sec}) \times (1.4 \times 10^{-15}\ \text{m})}$$

$$\approx 250 \times 10^{-30}\ \text{kilogram.}$$

Compare this with the rest-mass of an electron, about 0.9×10^{-30} kg. To live (briefly) inside a nucleus and provide the right kind of binding force at the right kind of distance, the exchange-particle must have a rest-mass of several hundred electron masses. Yet that makes it five or ten times lighter than the lightest atom.

Not long after this prediction, which seemed so strange and unreal, intermediate particles were observed in cloud-chamber tracks made by cosmic rays. At first this seemed a wonderful confirmation of the prediction—even the mass was roughly right. Then the new particles, obviously unstable, proved to have the wrong half-life and other characteristics that failed to fit the theory. However, further experimental searches revealed still more types of particles; and some of them do fit Yukawa's prediction and are, as we now think, the essential binding agents in nuclei. We can now produce these varieties of mesons with accelerators for further study, free from the nuclear leash.

Mesons are real enough as particles, now commonplace in sub-atomic physics; and meson theory plays an important part in nuclear physics. Their measured mass ($\sim$270 electron masses*) fulfils the prediction that Yukawa made when such particles had never been observed or even thought of.

[Note on alternative suppositions:

(i) The ring standing wave is too artificial. It would be better physics to picture the meson travelling straight across the nucleus, to and fro inside the bounding wall of its crater. Then the de Broglie standing wave pattern would be more like that of a vibrating string. The simplest standing wave would have one loop, half a wavelength, along a diameter, nodes at the rim. That would make $\frac{1}{2}\lambda =$ diameter of nuclear crater, $2r$, or $\lambda = 4r$ instead of $\lambda = 2\pi r$.

(ii) A really sound treatment avoids this realistic talk of waves and walls, and appeals to the Uncertainty Principle in its form for ENERGY and TIME:

$$\Delta E \cdot \Delta t = \text{at least}\ h/2\pi$$

Suppose one nucleon in a nucleus manufactures a meson and tosses it across to another somewhere between 0 and $2r$ away. The change in energy, the momentary uncertainty in accounting, ΔE, is the cost of manufacturing the meson, m_0c^2. Assume its trip across to the receiving nucleon is a distance r on the average, made with speed c. Then the trip takes time $\Delta t = r/c$. If the meson lasts just long enough to make the trip without any chance of being caught by an experimenter, $\Delta E \cdot \Delta t$ must be just $h/2\pi$ and therefore

$$(m_0c^2) \cdot (r/c) = h/2\pi$$

and this leads to the same estimate, $m_0 = h/2\pi rc$.]

* The lighter mesons (210 electron masses) first discovered behave differently, more like heavy unstable electrons.

ONE MORE EXPERIMENT

Seeing Atoms

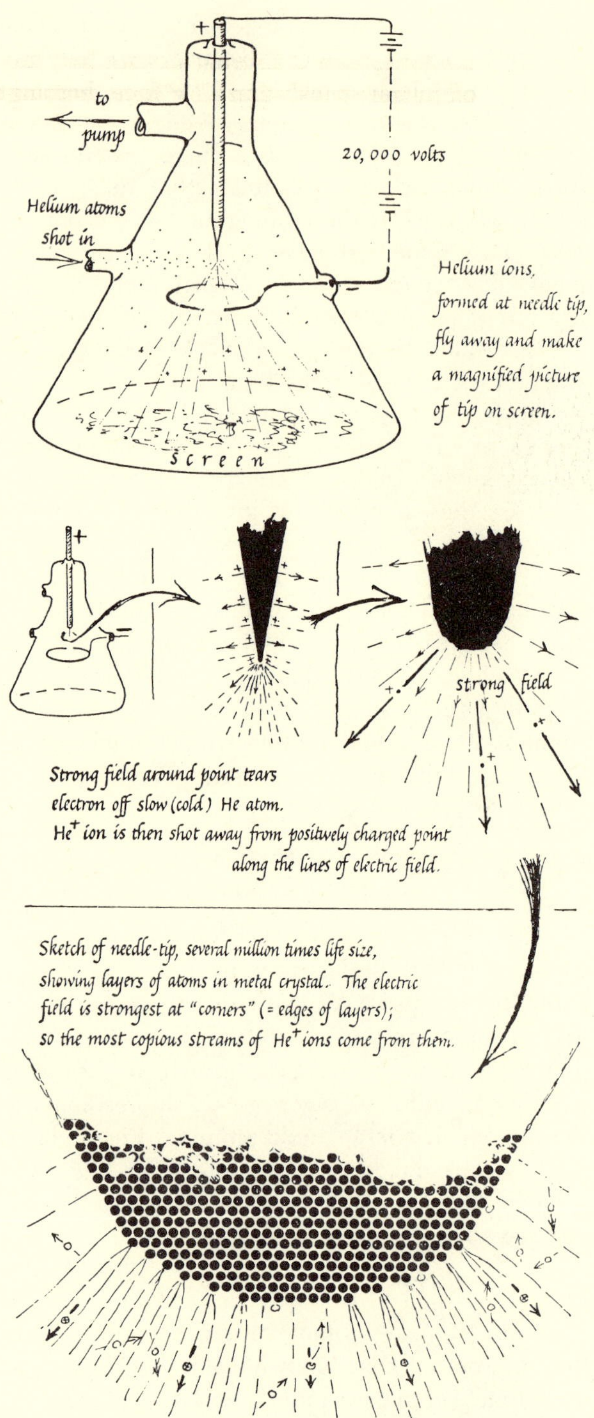

Fig. 44-32. Müller's Ion Microscope for Seeing Atoms
(This sketch gives a simplified view of the apparatus and its method of working. For a much fuller account, see *Scientific American*, Vol. 196, June 1957.)

As a final sample of experimental physics, come back to Earth and look at a photograph of individual atoms, now that we can discuss in technical terms the sophisticated method of obtaining it. Fig. 44-35 shows a photograph of atoms that form the end of a needle of tungsten. The pattern is made on a fluorescent screen, by helium ions that have bounced off the needle point. Fig. 44-32 shows the apparatus devised by Erwin Müller.[35] A tungsten needle, reduced to a very fine point, is placed at the center of a glass bottle, with a good vacuum except for a little helium gas. A large P.D. is applied between needle and neighboring metal ring, R. Because the needle is so sharp, the electric *field* near its point is extremely strong. It is strong enough to tear an electron off a helium atom that strays near the point.[36]

The ion thus made is then driven away violently by the field. It travels straight out from the needle point's surface in a straight line and hits the fluorescent screen painted on the inner surface of the globe and makes the screen glow where it hits. (The bombarding ion displaces electrons in the mineral coating; as the mineral's atoms recover they emit a glow.) The point's electric field is very strong close to it, and its lines of force sprout out perpendicular to its surface (so far as it can be called a surface on the atomic scale); so the ions are accelerated straight out, perpendicular to the surface, at first. In the weaker field beyond, they travel on with little further change of speed or direction so the marks where they hit the screen give a faithful

[35] For a fine description of the apparatus, by Professor Müller himself, see *Scientific American*, Vol. 196, p. 113, June 19, 1957, where there are additional pictures and diagrams.

[36] We can make a rough estimate, using "judging": We first guess the "radius" of the point. By thinking about layers of atoms (with a glance at the picture itself) we guess that the end layers are at most a dozen atoms wide and rest on a layer two atoms wider, and that on a layer two more atoms wider, etc. Rough sketching then tells us the radius of the point—which looks rough and rounded on an atomic scale—is about 30 atom diameters, $\approx 30 \times 3$ Å.U. $\approx 100 \times 10^{-10}$ meter. A helium atom is small, less than 1 Å.U. in diameter. Take its radius as ¼ Å.U. We know that the energy to remove one electron from it is about 25 ev. Thus, 25 volts is the P.D. needed to drag an electron off a helium atom to infinity—most of this P.D. being needed close to the atom, where the field due to its nucleus is strong. The end of the tungsten needle is also "round" but it has 400 times the radius, so to make the same electric field near it the P.D. between it and infinity must be 400 times as big, 400×25 volts or 10,000 volts. In fact the P.D. used in the apparatus is a few 10,000 volts between needle and ring. (Most of this P.D. occurs near the needle, say within one radius. Then the FIELD STRENGTH there is $= 10{,}000$ volts/100×10^{-10} meter or 10^{12} volts/meter. Across one helium atom's radius this field gives [10^{12} volts/meter][$\frac{1}{4} \times 10^{-10}$ meter] or 25 volts, just the P.D. needed to make an ion. This last reckoning is of course a disguised version of the earlier one, in reverse.)

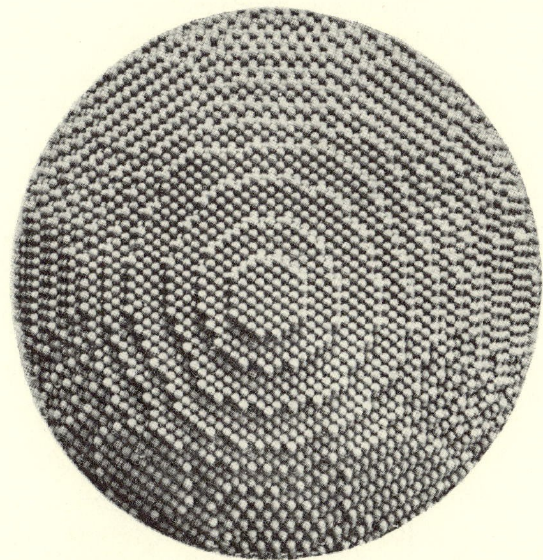

FIG. 44-34.
INTERPRETING ION MICROSCOPE PICTURES OF NEEDLE-POINT. (From *Scientific American*, Vol. 196, June 1957.)
(a) A model made of cork balls arranged in layers to represent the layers of atoms at the point of a very sharp tungsten needle.

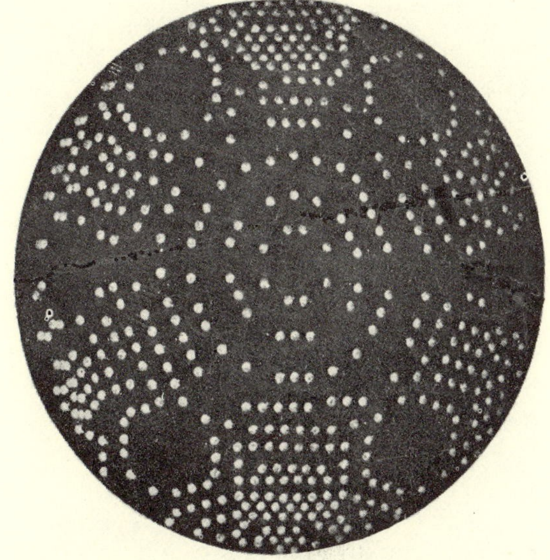

(b) The balls at the *edges* of the layers in the model, (a), were painted with fluorescent material so that they would glow under ultraviolet light. This is a photograph of the same model, with ultraviolet light.

enlarged picture of their starting points on the needle. Where the needle's surface is "sharpest," the electric field sprouting from it is strongest and will make and drive most helium ions. Thus a bright spot on the screen implies a sharp curve or edge at the corresponding starting place on the needle.

Now think of the needle on an atomic scale. Its "end" is just the last patch of tungsten atoms, sitting on a slightly bigger patch behind, which sits on a bigger patch still, . . . &c. The point can be made so small that the final patch at its tip is just a dozen atoms or so wide. However, even that does not look very sharp to the helium atoms. What does look sharp is the *edge* of such a layer of atoms, where the patch ends and the "surface" drops back to the next layer behind.

X-ray diffraction pictures tell us that tungsten metal can crystallize as a regular array of layers of atoms like a neat pile of oranges or ancient cannon balls. Fig. 44-34(a) shows a model of such regular piling, made with small balls of cork. As well as the end face of the needle point with sharp edges around each layer, the same regular piling reveals slanting faces at the sides, where the outermost patch is again a few atoms wide. By labelling the balls around the edges of layers, we can predict the pattern to be expected from a real needle point of tungsten atoms. Such edge-balls in the model were painted with fluorescent paint that glows in ultraviolet light. Fig. 44-34(b) shows a photograph of the model taken in u.v. light.

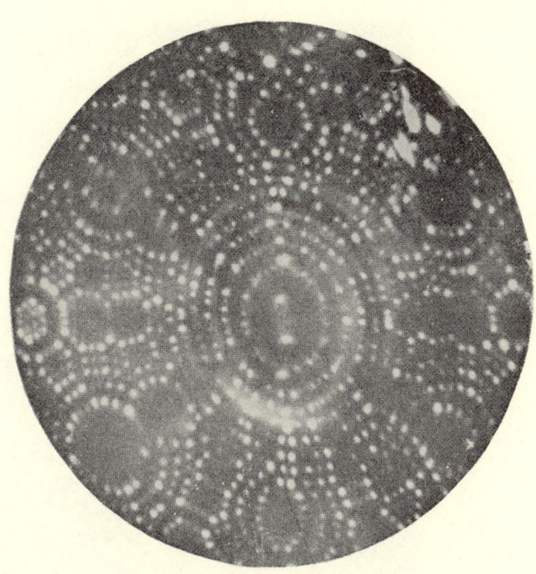

(c) Ion Microscope photograph: the pattern made by ions from tungsten needle point.

Comparing that with the photo, (c), from the original apparatus, we are ready to accept the picture painted by the stream of helium ions as a genuine picture of tungsten atoms arranged in layers.

The experiment has to be done at very low tem-

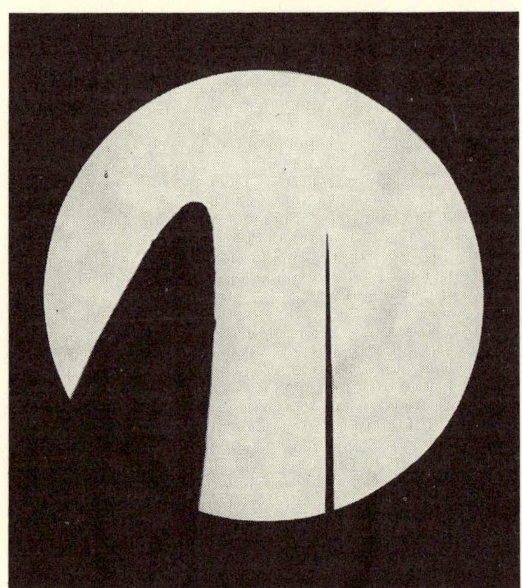

Fig. 44-33.
Needle-point of tungsten (used in Ion Microscope) and the point of an ordinary, sharp pin (left) seen under ordinary optical microscope. Photograph by Erwin W. Müller.

peratures, or the helium atoms contribute too much motion of their own—we want the motion of the ions to be solely that due to the electric field. And at ordinary temperatures the helium atoms seem to bounce off the tungsten without becoming ionized—perhaps they need more time in the electric field, and they get that when moving slowly. So the apparatus is cooled with liquid hydrogen (boiling quietly in a jacket of boiling nitrogen). Then we see pictures such as Fig. 44-35.

If the tungsten is allowed to warm up a little the pattern can still be seen but some of its atoms wander *visibly*: we can see them migrating and evaporating.

Can you truly say, "In this photograph I am seeing individual atoms directly"? That depends on your own choice of meaning. If you did not understand electric fields, charge and field distribution at sharp points, ionization, mean free path, glowing screens . . . molecules and temperature . . . , you would have to say, "Not even indirectly: the picture is just a set of blobs with magic talk about atoms." With a scientific background, you can say with

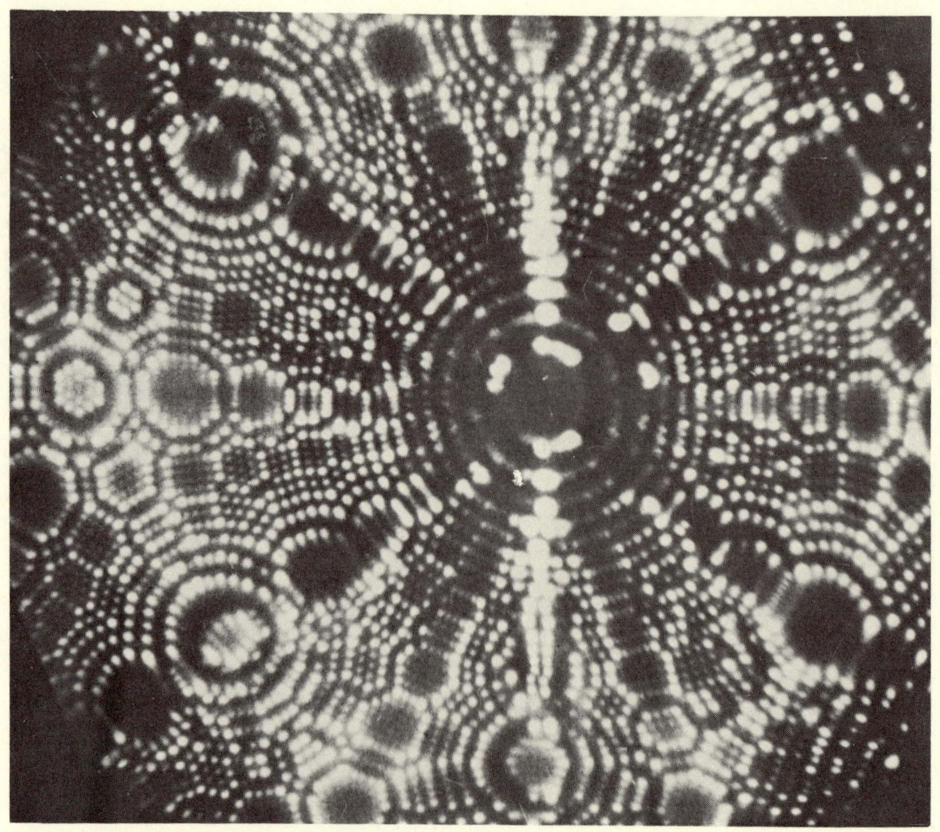

Fig. 44-35. Ion Microscope Photograph of Tungsten Needle Point
The picture shows layers of atoms at the point, and slanting layers of atoms exposed at the sides on the needle, with a magnification about 3,000,000 (Photograph by Erwin W. Müller, Pennsylvania State University)

delight as well as conviction, "I have seen atoms, directly"—and there is a wealth of sophisticated understanding behind that statement.

SCIENTIFIC PEOPLE

Scientists

Now at the end of the course, I hope you will think about scientists and their delight in discovery and understanding. Pythagoras . . . Ptolemy . . . Copernicus . . . Tycho . . . Kepler . . . Galileo . . . Newton . . . Joule . . . Maxwell . . . Rutherford . . . Einstein . . . de Broglie . . . Bohr—all had great delight, all contributed more to our intellectual heritage than to practical comfort or success, and what they gave us will outlast material gains by countless generations.

In looking to the future, we see a wide field of facts, laws, theory, speculation—scientific knowledge—that will grow in extent and, we hope, be reduced in complexity. This study of nature that is a part of man's intellectual life is a delight to all scientists. In their hands, science is not just a business of collecting facts or stating laws or directing experiments. It is above all an *art* of sensing the best choice of view or the most fruitful line of investigation for a growing understanding of nature.

Pure science, the pursuit of knowledge and understanding of nature, will always be a delight. Then, are scientists in their ivory tower of experiment and theory just selfish fellows of no value to the rest of mankind—like silent poets in seclusion? The thinker's answer is: "No, they add to man's intellectual resources."

The practical man's answer is: "No, the scientists are useful because in the course of time their discoveries are put to use in engineering." But nowadays we have a new group, *technologists,* who combine the activities of scientist and engineer; and our question becomes more general: How are scientists, and engineers, and technologists related in their importance to mankind in general?

Scientists, Engineers, and Technologists

Science enables engineering to develop and flourish. The modern engineer[37] does a magnificent job, using scientific knowledge, and techniques at every turn. For centuries, good engineers have relied chiefly on a great store of experience inherited from generations of empirical experimenting and learning by trial. But now, as needs and methods grow more complex and change more rapidly, good engineers draw directly on scientific knowledge. This change is a lessening of the "degree of empiricism," as James B. Conant calls it,[38] in modern technology. It is to the advantage of engineers, but it makes them more dependent on the help of scientists. They now draw on so much scientific material—condensed into compact formulas or tables in handbooks—that they have little time to study its origins; so they must either take it on trust or consult scientific advisers.

The technologist[37]—the creative engineer or applied scientist—combines engineering with a thorough knowledge of science. He is clever, creative, and equipped with scientific knowledge right up to the frontiers of research. He is a scientist by training and activity, but his interest is in new *uses* for science rather than entirely new science. Thus, the technologist *interpolates* in the field of existing science, while the pure scientist at his happiest is pushing on towards new knowledge and deeper understanding and thus tries to *extrapolate.* With that distinction, civilization should value the pure scientists with the poets and artists. All scientists contribute somehow to man's intellectual and emotional growth.

Yet, the practical man still questions: "The technologist takes all science as grist to his mill, and he works hard to advance science itself. Why, then, with many of the best brains put to technology, is the pure scientist necessary?" I think the fair reply to that is: "The scientist in his ivory tower makes a practical contribution that no nation, no civilization, can long do without: he provides the intellectual nursery for the next generation of scientists *and* technologists. Technology, like the mule, is strong and clever, but cannot breed its own next generation. This is because the next generation of first-class technologists needs an endowment of fresh outlook and knowledge, and new wisdom, if they are to work as creative scientists rather than as routine engineering directors. And that fresh outlook comes from the philosophic interest that the scientist maintains." The pure scientist is necessarily somewhat of a philosopher; he thinks about his own thoughts, and criticizes his own experiments, and develops a sense of good taste for his own art. He knows something about *how he knows*: he knows a lot about *what he knows*; and he is ready to say clearly *what he does not know.* And in his growing knowledge he finds intense delight.

[37] Of course, engineer, technologist, and scientist are sometimes combined in one man—then the comments here refer to him in each role separately.

[38] See his excellent book, *On Understanding Science* (Yale University Press, New Haven, 1947).

The Expert Scientist

What makes scientists *experts* in contrast with laymen? We are all of us laymen, amateur thinkers and doers, in every field except our own. We can learn the rules and even use them without realizing that we are amateurs, but we make grave mistakes with them. The non-scientist can buy a small book that tells him "all the laws of physics"—he can even number the important ones on the fingers of two hands. The professional physicist does not know more laws—he may remember even fewer, since he trusts the books to remind him—but he carries in his mind two vast commentaries on the facts and laws and principles, . . . , commentaries that are the product of his training, experience, and thinking.

(I) He knows the limitations of each fact or law, its range of application, the meaning of its terms; he has a general sense of its ties with real materials. Such knowledge forms what we have called "the little black notebook in the scientist's pocket"—a symbol for the scientist's wealth of experience that distinguishes him from the confident layman who can quote the laws but cannot safely use them. (He knows, for example, that, in some formula for the breaking of a beam, *P* is *stress* and not *force*, that the formula is intended for concrete with a low limit for tension, much higher limit for compression. He will neither mistake *P* for force nor apply the formula to wood, which creaks and splinters in quite a different way. Again, he knows *why* Boyle's law is safe for gases at very low pressures; and so he will not expect it to hold if the temperature is also very low.) This pocket book might be titled DETAILED KNOWLEDGE AND UNDERSTANDING.

(II) We now see that every good scientist has another "notebook" that tells him how his factual knowledge is connected together. This is a vast unfinished volume. It might be titled THEORY.

It is possession of these two imaginary notebooks, acquired in a lifetime of training, that makes the scientist an expert. As an educated layman you can go far towards joining the scientist if you understand his use of those "notebooks."

The Genius and the Crank

All of us—technologists, scientists, engineers, laymen—all meet enthusiasts offering new scientific views, not a good invention or a clever gadget but a revolutionary theory. How are we to distinguish between a brilliant advance and phony nonsense? Either may be dressed up to seem attractive and reasonable to the laymen. In fact, either may look bold and possible, even to experts. We *all* make mistakes in spotting cranks: we laugh at unrecognized genius, or fall for well-worded nonsense; but it is the scientist, with the sensitive instincts of his art, who has the best chance of picking the prophet from the crank. The bones of the scientist's wisdom are built of past knowledge, but its nerves extend into the future. The technologist, looking mostly to the present and the past, may find it harder to discriminate.

A crank is often a sincere scientist in his intentions, and sometimes even by training, with an urge towards a new view that carries him away from a consistent knowledge of Nature. There is no harm in his bold thinking—that belongs with good science—but the danger lies in his uncritical assurance that he is right. If we say the scientist *extrapolates* and the technologist *interpolates*, then the crank *guesses and believes emotionally with his eyes shut*. The curious thing is that the crank's enthusiasm convinces many a layman with the same blind belief. This course, we hope, will make *you* an open-eyed layman.

Wonder and Delight: Intellectual Progress

Instead of an intense emotional conviction that his view of Nature is the right one at last, the scientist's chief feeling is one of enjoying finding out and enjoying gaining a wider understanding of Nature. The *growth* of knowledge is his main concern, not its storage. He shares with our earliest ancestors a sense of curiosity and feelings of wonder and delight; and he extends these into a great sense of intellectual progress. You can share that with him.

Yourself and Science

Thus our sampling of physics in the course has come full circle, back to the first questions about *reasons why, explanations, experiment and theory.* You will have reached no crisp conclusions to be labelled as the "right" science; in fact, you may be aware of more unfinished discussions than when you started and more knowledge still to be pursued. But a compact memory of settled facts must now look less important than a sense that *you understand scientists and their work*, or a belief that *science makes sense as part of the wisdom of mankind.*

If you have enjoyed watching scientific inquiry at work, and have taken some questions to your own heart for future thought; if you enjoy meeting and arguing and working with scientists; if you like experimenting; if you can distinguish the "daring conservatism" of the speculative scientist from the loose enthusiasm of the crank; and above all if you will some day read more science for yourself as an educated person in a scientific age, then this course has done its work.

The future of science, both as practical knowledge and as an intellectual heritage, depends greatly on the attitude of laymen: parents, teachers, administrators, governors, . . . all educated people; so it is in your hands, as a member of a scientific civilization, to maintain the good name of science.

"GENERAL PROBLEMS"

This is a collection of problems and questions that draw on material from several chapters, or even the whole book. The preliminary problems before Chapter 1, and most of the problems in the chapters, belong in this collection too—there is no essential difference. The problems in this collection provide for important teaching, often in the later stages of the course.

This collection contains: some lists of small items for review; some general problems (or specimen "examination questions"), which ask for descriptions, calculations, reasoning, and courageous guessing; some questions that need considerable time for thought, and ask for opinions rather than a single "right answer"; and some longer essays.

Where the reason for an answer is asked for, that is usually much more important than the answer itself.

Large simple sketches will often save time, make the answer clearer, and show that you appreciate scientists' use of diagrams.

In some questions where a short answer will suffice, its suggested length is indicated thus: (~ 3 lines).

NAMES AND UNITS

1. For each of the following, (a) give a short definition or description saying clearly what it is; (in 2 lines. Use words rather than symbols). (b) Give its usual units in the MKS system (based on meters, kilograms, seconds).

(a) velocity
(b) speed
(c) acceleration
(d) density
(e) specific gravity
(f) force
(g) momentum
(h) weight
(i) mass
(j) stress
(k) strain
(l) pressure
(m) wavelength
(n) frequency
(o) current
(p) resistance
(q) charge
(r) kinetic energy
(s) P.D. (potential difference)
(t) E.M.F. (electromotive force)
(u) gravitational field strength
(v) electric field strength
(w) angle of contact (include a sketch)
(x) mean free path of gas molecule
(y) "e/m" for an atomic particle
(z) the "quantum constant," h

2. Write a short note to identify or describe or explain each of the following (~ 3 lines each):

(a) equinox (b) planet (c) comet (d) pole star
(e) quadrant (instrument) (f) parallax (g) ecliptic

3. In each of the cases below, show clearly that the two *units* given are exactly the same, or at least equivalent.

(a) Electric field strength can be measured in *volts/meter* or *newtons/coulomb*.
(b) Gravitational field strength measured in *newtons/kg* is given by an acceleration in *meters/sec*^{2}.
(c) Momentum-change in *kg* · *meters/sec* gives impulse in *newton* · *secs*.
(d) Surface tension in *newtons/meter* (the pull on unit edge of meniscus) is also mechanical energy needed to make unit area of new surface, in *joules/sq. meter*.
(e) The rate of energy dissipation in a resistor is usually expressed in *watts*, but can be expressed in *amp*^{2} · *ohms* (using $P = C^2R$).

4. State the physical quantity that each of the following units is used to measure, and give its value in terms of the common units of the MKS (meter-kilogram-second) system. [Example: 1 Å.U. is a unit of *length*. It is 10^{-10} meter.]

(a) 1 electron · volt (b) 1 kilowatt · hour (c) 1 light-year.

EXPERIMENTS AND KNOWLEDGE

5. Describe experiments that you have seen (either demonstration experiments or ones that you have done in laboratory) that show:

(a) Momentum is conserved in a collision (even an inelastic one).
(b) A constant force produces constant acceleration.

6. Give a short account of an early example of any three of the following, explaining clearly how the result was reached (~ 1 page for each of the three):

(i) Measurement of the radius of the Earth.
(ii) Use of the Moon's motion in a test of inverse-square gravitation.
(iii) Measurement of Moon's distance from Earth.
(iv) Measurement of distance of Venus from Sun, in terms of Earth's distance.
(v) A measurement of the value of the gravitation constant, G, in $F = G \cdot M_1M_2/d^2$.

7. KNOWLEDGE OF MOLECULES

What qualitative or quantitative information about molecules is contributed by each of the following? (Do not answer with a single word, but give a short explanatory statement.)

(a) Brownian motion.
(b) Surface tension.
(c) Diffusion (spreading) of bromine vapor.
(d) Area of oil film on water.
(e) Measurements of pressure, volume, and mass of a sample of air.
(f) Measurements of specific heat of gas at various temperatures.

8. KNOWLEDGE OF ATOMS

Write a short note (a few lines for each) describing the contribution of each of the following to our knowledge or picture of atoms and/or radiation.

(a) Alpha-particle scattering by gold.
(b) Cloud-chamber pictures of alpha-particle tracks in wet helium.
(c) Measurement of e/m for electrons.
(d) Experiments on photo-electric effect.
(e) "Young's fringes" made by light.

(f) Diffraction of X-rays by crystal.
(g) Diffraction of electrons by nickel crystal.
(h) Application of a magnetic field to cloud-chamber in which tracks are photographed.

9. Write a short note saying how each of the following experimental discoveries affected or contributed to astronomical beliefs (~ 3 lines for each).

(a) Galileo's observations of Venus.
(b) Galileo's observation of sunspots.
(c) Galileo's discovery of Jupiter's moons.
(d) The discovery of Uranus (about 1780).
(e) The discovery of Neptune (about 1840).
(f) Tycho's very accurate measurements of the positions of Mars.
(g) Measurements by Halley and others which showed that comets move in elongated ellipses.

10. When streams of electrons are accelerated or deflected by electric fields and magnetic fields, measurements can be made which yield information concerning electrons. They do not yield the *charge* of an electron, but they do yield another quantity which is the *same for all electrons.*

(a) What is that quantity, which can be calculated from the measurements?
(b) Why do the measurements only tell us that quantity and not the charge? (~ 4 lines)

11. Physicists often present a detailed picture of atoms such as the following:

> Atoms are small (few, 10^{-10} meter in linear dimensions). They have a number of easily detached electrons which have small mass and carry negative charges. They are mostly hollow, having nearly all their mass concentrated in a very small nucleus which carries a positive charge (+ Z electron charges, where Z is the serial number of the element in the chemical periodic table, where elements are arranged, more or less, by atomic weights). Some elements have several kinds of nuclei (isotopes) with different masses but the same charge. Some atoms are unstable and disintegrate, in a chance or random manner, ejecting alpha-particles, beta-rays, etc. Alpha-particles are helium nuclei with charge +2e. The electrons of atoms are much less massive (mass about $\frac{1}{1800}$ of mass of hydrogen atom), have negative charges, are all alike. Outer, loosely bound electrons are easily removed from atoms by ultraviolet light.

The details of this picture are not wild embellishments drawn from imagination to "explain" nature. They are based on experimental investigations.
Choose several details of the picture above, and describe the experiments on which they are based. (Indicate the kind of apparatus used. Indicate the general way in which the picture-information was obtained from the measurements or observations. You need not give mathematical details. The word "several" gives considerable latitude of choice. You should give at least three distinct kinds of experiment.)

12. PREDICTIONS FOR A SPRING

The data below show the TOTAL LENGTH of a certain steel spring, when carrying various loads.

LOAD (in pounds)	TOTAL LENGTH (in inches)
0	20
2	23
8	32

(a) Predict total length with load 4 pounds.
(b) Predict total length with load 12 pounds.
(c) On what general physical knowledge did you base your predictions (a) and (b)?
(d) Do you consider your two predictions (a) and (b) equally reliable? Explain.
(e) A 2-pound load is hung on the spring, pulled down below its equilibrium position and released to move up and down vertically. Why would you expect its motion thereafter to be S.H.M.?
(f) Predict the maximum amplitude (distance of load from rest-position at each end of its cycle) for which the motion in (e) would be perfect S.H.M. Give a clear reason for your prediction.
(g) Predict the maximum amplitude for perfect S.H.M. with a total load of 4 pounds on the spring.
(h) A theoretical physicist notices that the same line of thought is used in answering (g) as in answering (f), so he uses it to predict the maximum amplitude for perfect S.H.M. with a total load of 6 pounds. Experiment shows his prediction is wrong, for a particular spring with the behavior listed. Suggest an explanation.

13. MOTION

Five identical rectangular bricks of aluminum are treated as follows:

one is used as a single block, A;
two are joined by a thin horizontal rod, to make a dumb-bell, B;
two are melted together and re-cast as a single brick, C, of similar shape and density, but larger. (See Fig. P-1.)

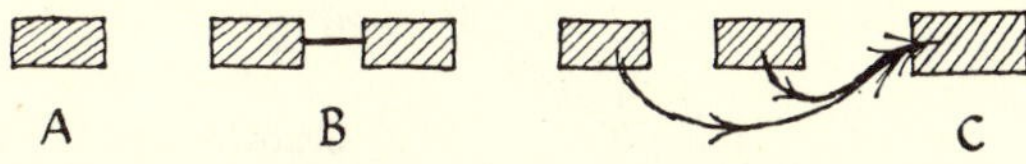

Fig. P-1. Problem 13.

(a) When A, B, C are released together in air, they all fall with the same motion. What do you infer from this observation? (~ 3 lines)
(b) When A, B, C are released in water, each accelerates at first but then reaches a constant speed, which it maintains thereafter. Why is there no acceleration in this later stage? (~ 3 lines)
(c) Falling in water, A and B reach the same final speed, but the larger block, C, reaches a different speed.
  (i) Will C's speed be greater or less than A's?
  (ii) Explain carefully why C's speed is different. (~ 5 lines)

14. BREAKING FORCE OF WIRES

(a) What relation does "common sense" suggest between *breaking force*, F_B, for a round wire and its *diameter*, *d*?
(b) Suppose an experimenter who has measured F_B for several steel wires of different diameters, but all of steel with the same composition, wishes to plot his measurements on a graph in which the plotted points are *likely to lie in a straight line.* If he plots F_B upwards, what should he plot along, to make a straight line likely (d, or $1/d$, or $\sqrt{d}$, or what)?
(c) Now suppose the graph sketched in Fig. P-2. shows his measurements plotted on such a graph. Note that the points in the upper right corner lie far from the line.
  (i) Are those points probably mistakes? (Give a clear reason for your answer) (~ 2 lines)
  (ii) What sizes of wire do those points refer to: thickest, medium, or thinnest?

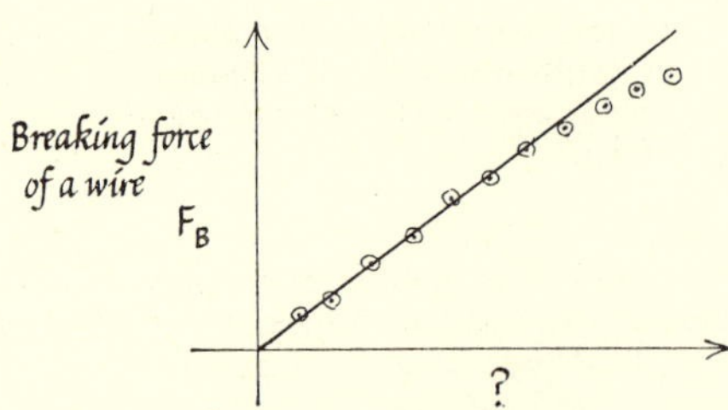

Fig. P-2. Problem 14.

(d) Suppose you are now told that all the wires had been "hardened" before the tests plotted on the graph, by being heated and then plunged immediately in cold oil. Suggest—making a courageous guess—an explanation for the points in the upper right corner lying off the straight line. (~ 2 lines)
(e) Give reasoning to defend your guess for (d). (~ 3 lines)
(f) What experiments could be tried to test your guess for (d)? (~ 5 lines)

15. SCALE MODEL

An engineer designs a crane, to be used to lift huge rectangular blocks of concrete. The crane lifts the block by a single steel rope. To test his design, he constructs a scale model in which everything (including the rope and the concrete block) is made with the same shape and of the same materials as in his proposed crane, but on a scale of 1 inch to one foot. (i.e. reduced to $\frac{1}{12}$ linear size.)

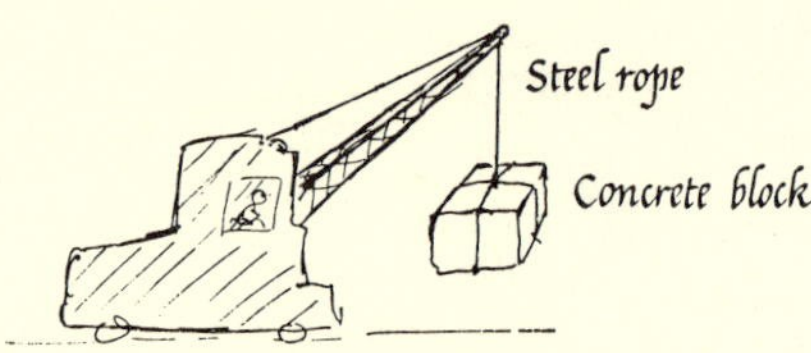

Fig. P-3. Problem 15.

The model works well, and is able to lift five concrete blocks without breaking the rope. When the real crane is made, the rope snaps when a single concrete block is used. Comment on and explain the failure of the crane. (~ 5 lines) (Assume the materials are just as good, and thoroughly tempered, in both model and crane.)

16. CAPABLE ESTIMATE

Making some reasonable assumptions about mass, speed, time-of-impact, etc., estimate the force exerted by an average carpenter's hammer when driving a nail.

(You are not given actual measurements but are asked to invent numbers that seem reasonable, and then calculate the answer. Since you must make your own guesses, you must state these guesses clearly before you start using them. Since they are only guesses, a reader will accept any reasonable answer, provided you explain how you obtain it.)

17. ACCELEROMETER

A steel spring is hung from the roof of an elevator which is at rest. When unloaded, the spring is 0.18m long. When loaded with 1 kilogram hung on its lower end, its total length is 0.24m. If another 1 kilogram is added, the length becomes 0.30m. Now suppose the elevator accelerates steadily. When the spring is loaded with just 1 kilogram the spring's length is observed to be 0.25m, and it remains constant at this value.

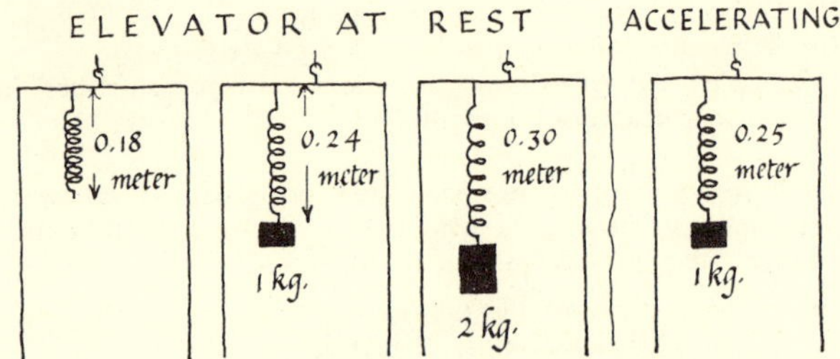

Fig. P-4. Problem 17.

(a) Is the elevator's acceleration upward or downward? up? down? cannot tell?
(b) Is the elevator *moving* (up? down? at rest? cannot tell?)?
(c) Calculate the *acceleration* of the elevator. Give its units.

18. A sports car of well-known make can accelerate from rest to 60 miles/hour (88 ft/sec) in 10 seconds. Estimate:

(a) its acceleration.
(b) the distance it requires to reach 60 miles/hour from rest.
(c) To test the brakes, the car (running at 60 miles/hour) is put in neutral and brought to a stop on a level road. The best the brakes can do is to stop the car in a travel-distance of 300 feet. The car is a 2000-pound car. Estimate the (average) decelerating force acting on the car during stopping.
(d) Which problem in Ch. 7 is the same as (c) just above in essential structure? (Note: this question does not ask you to find another problem on a braking car or starting runner. It asks you to use a flexible mind and find a problem that looks quite different, yet needs the same type of solution.)

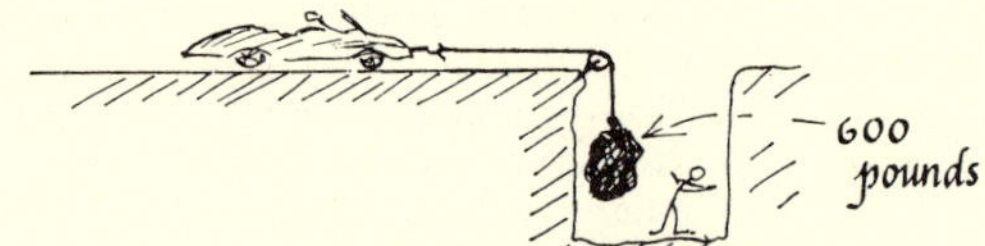

Fig. P-5. Problem 18.

(e) The same brakes are tested in a garage as follows: The car is put in neutral, the brakes are applied fully, and the car is pulled forward by a rope that runs over a (frictionless) pulley and carries a load of iron. With 600 pounds of iron hung on the rope, the car moves forward at a constant velocity with a creaking noise. Estimate the braking force.
(f) If your answers to (c) and (e) disagree, suggest a good reason, other than incompetent measurements.

19. MASS AND WEIGHT

Write a short note contrasting *mass* and *weight*. Describe the properties of each. (~ 1 page)

20. A scientific crank, intent on encouraging the confusion between mass and weight, devises a scheme in which forces are expressed in pounds and masses are expressed in pounds, and yet $F = Ma$ holds true. He achieves this horrible scheme by manufacturing a special unit of length, the "link," which makes $F = Ma$ hold true. Then his accelerations are measured in links/sec/sec.

(a) If the crank operates in New York, 1 link = . . . ? . . . meters.
(b) Give a clear reason for your answer to (a).

21. FIREHOSE PROBLEM

A fireman plays a horizontal stream of water on the plain, vertical back of a 4000-pound truck. The water hitting the truck splashes out along the back of the truck without rebounding and runs down it to the ground. The hose delivers 150 gallons (= 1200 pounds of water) in 10 secs, in a horizontal stream moving with speed 30 ft/sec.

Fig. P-6. Problem 21.

(a) The force on the truck (which has its brakes on) is . . . ? . . .
(b) If the brakes are taken off and the truck is free to move with *no friction* on a level road, its acceleration at first will be . . . ? . . .
(c) As the truck moves ahead (with all the water still hitting its back), its acceleration will (increase? decrease? remain same?) because . . . ? . . .
(d) Suppose, instead, that the fireman directs the horizontal stream through an open door in the back of the truck so that all the water arrives inside the truck and collects there. How will the accelerated motion (with brakes off) compare with the motion in (b) and (c) above? Give clear reason(s). (~ 4 lines)

22. A man standing still on a frictionless frozen lake is given a long horizontal box which contains an animal that runs to and fro, fast, from end to end of the box. When the man receives the box, the animal is running, about midway between ends. See sketch (i) of Fig. P-7. Suppose I could observe the man as he continues to hold the box (but could not see inside the box). (Box and man are at rest when he first gets box.)

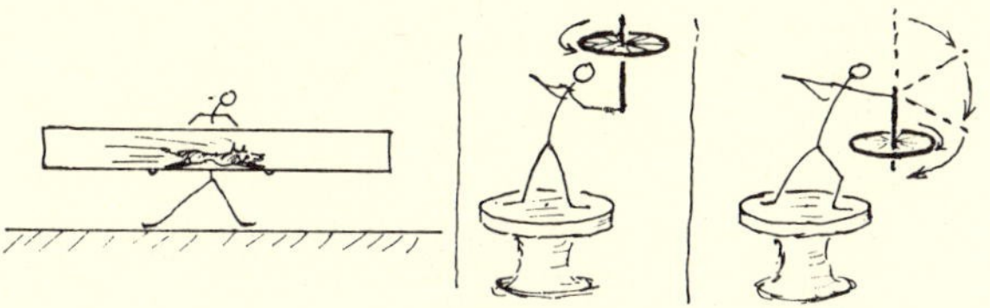

Fig. P-7. Problem 22.

(a) Shall I see the man move? If so, with what motion? Give reasons for your answer.
(b) Presently the animal gets tired and lies down in the middle of the box. What motion will the man then have?
(c) Does your answer to (b) depend on whether the animal was moving forward (original direction) or backward when it decided to stop? Why?
(d) A man standing at rest on a turntable with frictionless bearings is given (by a neighbor who does not touch him) a spinning bicycle wheel, with vertical axis, to hold like an umbrella. See sketch (ii). When he has received the wheel he is still at rest. Then he puts his other hand on the rim and stops the wheel. Predict what will happen.
(e) Instead of stopping the wheel, the man tilts the axle down from *vertically up*, to slanting, to horizontal, to slanting down, until it is pointing *vertically down*, still with its spin. What would you expect to see happening when he holds it in the new position, still spinning?
(f) Give your reasoning for your answer to (e).

23. INFERENCES

(a) The details of Jupiter's surface are obscured by clouds; yet we know that Jupiter is spinning. What observation tells us that? Explain how we can estimate the rate of spin.
(b) We cannot see gas molecules; yet we know they are in rapid motion. What observation tells us that?
(c) We cannot see gas molecules; yet we know that helium molecules are single atoms and hydrogen molecules are pairs of atoms. What observations or measurements in *physics* (as distinct from *chemical* experiments and arguments) tell us that helium molecules are single atoms?

24. EVAPORATION

A liquid is evaporating from an open dish, and a fan blows away the vapor.

(a) Why does the liquid seem to disappear faster than it does when the fan is not running? (~ 1 line)
(b) Why does the liquid disappear faster if the temperature of the liquid is raised? (Note: "The molecules move away faster" is not an adequate explanation.) (~ 3 lines)

25. AIR MOLECULES AND GRAVITY

(a) When a barometer is carried to the top of a tall building, it reads less; and still greater changes are noticed in climbing mountains, etc. Interpret this change in terms of the kinetic theory of gases. (~ 2 lines)
(b) A stone falls to the ground if released above the ground. From the viewpoint of the kinetic theory of gases, why do not all the air molecules fall to the ground?

26. MEAN FREE PATH OF MOLECULES

Molecules of air at atmospheric pressure have a mean free path of about 10^{-7} meter.

(a) What does that mean?
(b) If the air pressure is reduced to half an atmosphere, at the same temperature:
  (i) What change is made in the mean free path?
  (ii) Would the average distance of a molecule from its neighbors be about the same, or considerably more, or considerably less?
  (iii) What change would you expect in air-friction (for slow, streamline motion)?
  (iv) Give a clear reason for your answer to (iii).
(c) Suppose we could change to molecules that have twice the diameter (but are otherwise the same, with unchanged mass, etc.), at atmospheric pressure.
  (i) How would the mean free path be changed?
  (ii) How would the average distance apart be changed?
  (iii) Suppose you could liquefy the new gas and compare it with liquid air. How would you expect the two liquids' densities to compare?

27. HEAT CONDUCTION BY A GAS (Difficult speculation)

When a gas carries heat away from a hot body, most of the transfer is by convection—large portions of the gas moving as a whole. However, if convection is prevented,* gases do show some conduction, like solids, though they turn out to be very poor conductors. (*Convection can be prevented by entangling the gas in wool or cork dust, or by having the hot body in the form of a hot horizontal plate above a cold horizontal plate, with the gas between. In this latter arrangement, used for measuring conductivity of gases, there is little or no convection, since such motion would have to consist of currents of warmed gas moving downward.)

(a) In the light of the Kinetic Theory, suggest a mechanism of heat *conduction* by gases. (Make a guess.) (~ 2 lines)
(b) Suppose we have two gases, A and B, whose molecules are the same size and have the same general characteristics, except that B's molecules have four times as great a mass as A's. How would you expect the *heat conductivities* of the two gases to compare? (Which would be the better conductor? If you can, say how much better.)
(c) Give a reason for your answer to (b). (~ 2 lines)
[(d) As a rough experimental test, make a small piece of wire glow red-hot in air by heating it electrically; then plunge it, glowing, into a beaker of CO_2; then into a beaker of helium.]

28. KEPLER'S LAWS

Suppose the universe were constructed quite differently, with an *inverse cube law* of gravitation instead of an inverse square one, ($F = G \cdot M_1 \cdot M_2/d^3$). Kepler's Law I for planets would not hold true: orbits would not be ellipses.

(a) Would Kepler's Law II ("equal areas") hold? Why?
(b) What form would Kepler's Law III now take for circular orbits? (Give algebra)

29. EXPERIMENT WITH ORBITS

Suppose an experimenter tries to demonstrate Kepler's Laws I, II, III with the apparatus shown in Fig. 21-12c in Ch. 21. A "planet" of mass *M*, with a frictionless base of dry ice, coasts around an orbit on a horizontal table of aluminum. This "planet" is pulled by a thread that runs to the center of the table, over a small frictionless pulley and down through a hole to a load of mass *m* that hangs on it.

The experimenter can start the planet off in orbits of various radii, circular or otherwise. He finds that those orbits which are not circles have a strange shape. They are not ellipses: Kepler's Law I *does not apply*.

(a) Will Kepler's Law II (equal areas) apply to some of the orbits with this arrangement, or all, or none? Why?
(b) For *circular* orbits of various sizes, the experimenter finds Kepler's Law III does not hold. R^3/T^2 is *not* the same for all. Work out the rule that will hold for this arrangement, if *M* and *m* remain unchanged.

30. MOON'S MASS

(a) Explain why we cannot estimate the mass of the Moon (in terms of mass of Earth, Sun, etc.) nearly so easily as the mass of Jupiter. (~ 2 lines)
(b) Suppose an artificial satellite is projected to the Moon and left to travel around the Moon in a small circular orbit. From the data below for this supposed satellite, estimate the Moon's mass, as a fraction of the Earth's mass.

DATA. Satellite travels around Moon once every 6 hours. Satellite's distance from Moon's center subtends angle $\frac{1}{100}$ radian at Earth, so the radius of its orbit is $\frac{1}{100}$ of Moon's distance from Earth. Moon's period of revolution (relative to stars) is 27.3 days.

31. "GRAVITY" ON JUPITER

(a) Make a rough estimate of the strength of Jupiter's gravitational field, at the surface of Jupiter, using the data below, and state the units of your answer.
Mass of Jupiter = about 300 times mass of Earth.
Radius of Jupiter = about 10 times radius of Earth.
(b) As your calculation above will have shown, the gravitational field is considerably greater than the gravitational field at the Earth's surface.
Suppose Jupiter had an atmosphere suitable for life, and that animals like terrestrial ones developed on Jupiter. Would you expect the largest land animals to be larger than our largest elephants, or smaller, or same size?
(c) Give reason for answer to (b). (~ 4 lines)

32. A one-man flying platform consists of an inverted tub, with a huge fan inside to drive a blast of air downward. The pilot stands on top of the tub. The fan inside takes in air through holes in the sides and drives it out through the open bottom with a velocity 50 ft/sec downward. Suppose the platform is made of 150 pounds of metal and fuel, and the pilot is a 200-pound man, making 350 pounds altogether.

Fig. P-8. Problem 32. Fig. P-9. Problem 33.

(a) Calculate how much air must be handled by the fan *in 10 seconds* while the platform+flyer hovers motionless for 10 seconds. (Explain your calculation fully.)
(b) Suppose the fan *speed* could be *doubled*. (Make guesses at the following.)
  (i) How would this probably affect the exit-speed of the air?
  (ii) How would you expect this to affect the amount of air handled per sec?
  (iii) How would this, therefore, affect the thrust?
  (iv) What motion would you expect platform+pilot to have just after this change? (Give a detailed statement if you can.) (~ 2 lines)

33. PHYSICS IN A SPACE SHIP

Suppose a huge space ship (the form of a doughnut ring-tube of 100 ft radius and 10 ft internal diameter) is projected far into space (where no gravitational effects are to be noticed). There is no atmosphere outside. For the convenience of the passengers, the ship is spinning rapidly about its central axis perpendicular to the ring. Passengers inside the ring note that a plumb-line (pendulum) takes a position with its string along a diameter such as AB (See Fig. P-9).

(a) Sketch a small passenger standing "the right way up" as he feels it—on his feet—inside the ring, near A; and another passenger near B and another at C.
(b) One of these passengers holds an apple in front of him at shoulder height and releases it (without pushing it in any direction). Describe the motion (if any) of the apple, as the passenger sees it. (Use words or sketch)
(c) Suppose the ship is transparent and that there is an outside observer nearby, moving with the ship's center, but not revolving. Describe the apple's motion as seen by the outside observer. (Use a sketch if you like)

(d) Now suppose the passenger opens a window or trapdoor in the ship and releases an apple and watches it for some time, say while the ship makes ¼ revolution. What path will the outside observer see the apple take? What path will the passenger see the apple take? (Give sketch)
(e) Now suppose the passenger releases a luminous apple outside the ship as in (d), and waits for several revolutions, then observes it for some time with a telescope. Describe the path he will observe.

34. FUEL MEASUREMENT

If a man could live on alcohol alone, how much would he need per day, for a 3000-Cal per day diet? Use the information from the experiment described below to make your estimate, and *leave your answer in factors without cancelling.*

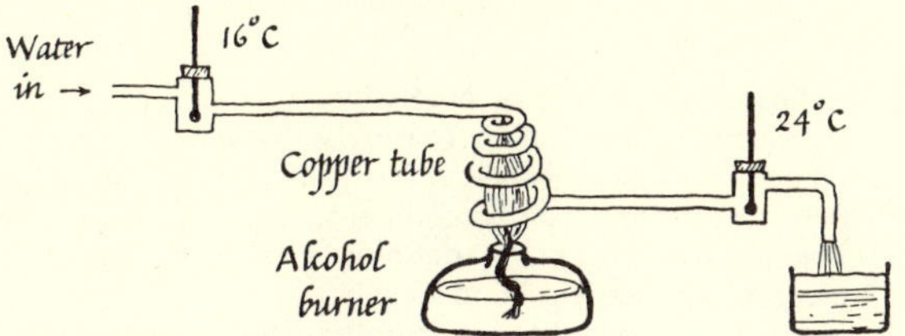

Fig. P-10. Problem 34.

A small lamp burns alcohol steadily. It uses 2 kilograms of alcohol in a 24-hour day. Its flame is surrounded by a coiled copper pipe through which water flows steadily, entering at 16.0°C and leaving at 24°C. In 10 minutes, 11.0 kilograms of water flow through the pipe. (The pipe is closely coiled around the flame, with just enough space for ventilation; and the whole arrangement is clothed with asbestos; so you may assume that heat loss to the air is negligible.) (See Fig. P-10)

35. LABORATORY PROBLEM

Suppose in an investigation you need to know the specific heat of some oil, and decide to measure it with the help of a small electric heating coil. Describe briefly the measurements you would make, and say how you would calculate the specific heat. Assume you have any apparatus you want, including a suitable heater coil and battery, but no voltmeter. (You may have an ammeter if you like.) (~ ½ page)

36. LABORATORY QUESTION

In the circuit shown in Fig. P-11, R_1, R_2, R_3 are three equal lengths of the same kind of resistance wire, each with resistance 1.00 ohm. The battery is a 6-volt car battery. The other connecting wires and battery have negligible resistance. P and Q are common electrical measuring instruments (meters), inserted for sensible use.

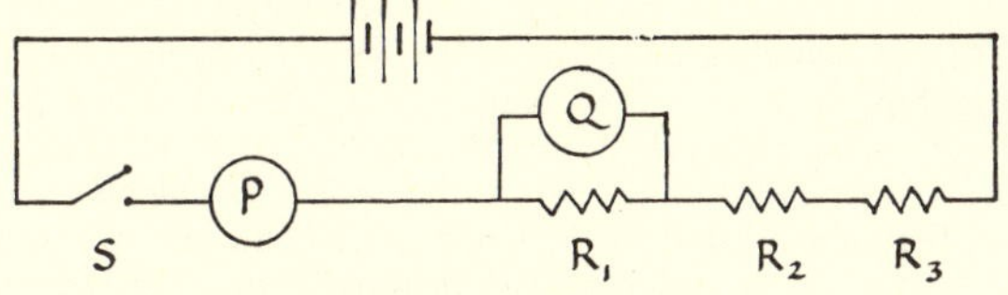

Fig. P-11. Problem 36.

(a) What kind of meter is P, most probably?
(b) What kind of meter is Q, most probably?
(c) When the switch S is off ("open"), what will the meters P and Q read?
(d) When switch S is on, what will the meters P and Q read?
(e) When switch S is on, but wire R_1 is removed or cut, what will the meters read?

SCIENTIFIC METHODS AND IDEAS

37. "In building a piece of science, theory and experiment often play complementary roles." Discuss this statement carefully, illustrating your discussion with examples from the course.

38. What is meant by a *law* in science? (~ 2 pages)

(Obviously there is no one "right" answer to this question. You are invited to give your own opinion and comments to, say, an intelligent non-scientist who asks questions such as: what *are* laws; why does apparatus [or Nature] obey them; how are they used; are they true? Give examples from this course if possible.)

39. GOOD EXPERIMENTS

(a) What would you mean, as a scientist, if you described an experiment as a *good* experiment, or a *successful* one? (~ 1 page)

(Obviously, there is no single right answer to this. You are invited to give your own opinions and comments to, say, an intelligent non-scientist who says, "How do you know whether you are experimenting or just playing around with apparatus; and what is the difference between research and routine technical measurements?" Give examples from your work in this course if you like.)

(b) Some years ago a State Motor Vehicle authority marked several unusual systems of white lines on stretches of road containing bad corners, and they put up signs saying "Motor Vehicle Authority. Experimental Road Markings." Discuss the meaning of "success" for such an experiment.

40. Write an essay on one of the following pairs of activities. In your discussion, compare such things as: methods used, the value of what is achieved, subject matter, and purpose in carrying out the activity.

(i) Physics and Poetry.
(ii) Physics and History.
(iii) Physics and Art Criticism.

41. What, in your opinion, are the general goals and functions of *theoretical* physics? In answering this question, discuss some of the following as examples:

The Rutherford atom picture.
Kinetic theory of gases.
Relativity.
Early work on Radioactivity or on X-Rays (from outside reading).

42. Consult "*A History of Science*" by Sir William Dampier
"*Out of My Later Years*" by Albert Einstein
"*Experiment and Theory in Physics*" by Max Born
"*The Edge of Objectivity*" by Charles Gillispie
and other books on the Philosophy of Science. Then write a short essay answering the question below *in your own words*: "What are the goals of *theoretical* physics?"

43. SCIENTIFIC KNOWLEDGE

(a) How is scientific *knowledge* arrived at in physics? Discuss the parts played by observation, experiment, and mathematics. Give examples.

(b) Can mathematics *alone* produce new knowledge of the real world? If so, give an example and discuss it critically. If not, explain why mathematics is important, if it only produces what experiment has already told us.
(c) Consult people who are studying biology, geology, or psychology, and find how knowledge is arrived at in one of those sciences. Compare the methods in that science with the methods in physics.

44. "Newton's theory of universal gravitation made a grand conceptual scheme for the Solar System and the Earth." What was the theory, and what were its achievements, and why was it so grand?
(In discussing achievements, you should not simply name them but should indicate their arguments. However, you should not spend time giving full mathematical derivations; except, if you like, for any one achievement you choose.)

45. "The Kinetic Theory of Gases produces a false sense of familiarity with molecules. It cannot be called good science, because the picture it presents is only a framework of imaginative assumptions."

Suppose you are asked to make a speech or write a long letter *opposing* the view stated above, or refuting it. Give an outline or notes, showing the important points of your speech or letter. (~ 1 page)

46. (a) Suppose you made a very careful measurement of the density of a sample of chemically pure copper, with evidence that your experimental error was about 0.001% of your result. Would you then be willing to publish a statement about the density of pure copper?
(b) Suppose your measurement had been made on the density of a local restaurant's vanilla ice cream. Would you publish a statement about the density of ice cream and vouch for it?
(c) On the basis of your answers to (a) and (b) as examples, comment on the Principle of Uniformity of Nature (i.e. the principle that things in Nature repeat themselves exactly in exactly similar circumstances). Discuss the question whether the Principle is based on experience or philosophical assumption.

47. OCCAM'S RAZOR

Consult books on Philosophy (especially Philosophy of Science) or on History of Science, and find out what "Occam's Razor" means.

(a) Write a short description of it.
(b) Find, and state, Newton's version of it.
(c) Is it useful today?

48. Why do we consider astrology* unscientific?

(Write a short essay. This asks for reasoned argument, with discussion of definitions, not just sweeping condemnation or unsupported claims.)

* Astrology is an organized system, applying detailed, traditional rules to good astronomical information, that claims to foretell or reveal human fate or character from star and planet positions.

49. SUPERSTITION

What do you consider "superstition" means? Discuss the relationship between superstition and science in matters of material comfort and intellectual happiness.

50. (a) We say that we can derive Boyle's Law, $P \cdot V =$ constant, from Kinetic Theory. Suppose a distinguished scientist derived from another theory (not kinetic theory) a relation $P/V =$ constant. That result is in conflict with experimental tests on gases. Still, suppose the mathematical deduction in each case seems completely sound. Would you believe the new theory? If so, why? If not, what would you suspect?

(b) Suppose two theoretical "derivations" of Boyle's Law are shown you. The two are entirely different; they make different assumptions and involve entirely different ideas of what a gas is, but they both lead to the relation $P \cdot V =$ constant. Might you be able to form a preference for one view and its derivation over the other? If not, why not? If so, why, and on what grounds?

(c) What comments can you make concerning the validity of theoretical proofs?

51. The following items give some characteristics of physical science. Give an example, in a sentence, of each characteristic, by referring to something studied in the material of the course. (~ 2 lines each)

(a) Experimental testing of prediction.
(b) Mathematical deduction from assumptions.
(c) Emphasis on experience common to most people.
(d) Extracting a law from experiment.
(e) Inventing a concept, or defining a term, to make some piece of science easier to develop.
(f) Inventing a quantity, calculated from measurements, which is independent of the shape and size of the sample.
(g) Measuring a physical constant by many varieties of apparatus to show that the law embodying it is true.
(h) Suppressing a personal wish.

52. What can you infer from each of the following facts?

(a) Liquid air exists.
(b) Some of the "naturally" radioactive elements emit alpha-particles, but none of them emit protons or deuterons.
(c) Air at room temperature "obeys" Boyle's law very closely over a range of pressures from several atmospheres to a small fraction of an atmosphere.
(d) Very-high-energy electron accelerators, that make the electrons travel around the same orbit many times, work properly when the accelerating impulses are spaced at equal time intervals (after the early stages of acceleration).
(e) In the "guinea and feather experiment" the feather falls as fast as the guinea at any low pressure, such as $1/100$ atmosphere—a high vacuum is not necessary for this demonstration, but it is necessary in the "wig-wag" measurement of mean free path.

FORMULA-MAKING

53. Good scientists often use formulas when they work out results of experiments. They assign a letter-symbol to each thing they measure; they express the rules they are going to assume in algebraic form; then they use algebra to produce a formula for the result they wish to calculate; and *then* they use arithmetic. This has two advantages over direct arithmetic: (i) it reduces the arithmetic to more compact form—and some factors may even cancel—so that it saves time and lessens the risk of errors; (ii) the final formula is ready for use with many different sets of measurements. Such a formula is not something to be treasured in a handbook or memorized and trusted blindly for answering problems: it is simply part of a scientist's way of handling his work.

Try the following examples of formula-making.

(a) A ballistics expert fires a rifle bullet of mass m kilograms and speed v meters/sec horizontally into a truck of mass M kg at rest on a frictionless track. The bullet stays in the truck. The expert measures the time, t secs, taken by the recoiling truck to travel s meters. Work out a formula, beginning $v = \ldots$, to express v in terms of the measurements m, M, s, t.

(b) Suppose in Zartman's experiment, (Ch. 25), the drum has diameter d meters, and makes n revolutions per second. Molecules with speed v meters/sec make a mark on the recording film a distance y meters from the zero-mark. Work out a formula for calculating v from the measurements.

(c) The friction-drag on a small sphere moving slowly in air is given by $F = Krv$, where K is a constant for air, r and v are the radius and speed of the sphere. Suppose a very small drop of liquid of density d falls with a constant "terminal velocity" v_t in air. A measurement of that velocity can be used to "weigh" the drop. (See Millikan's experiment in Ch. 36.) Work out a formula for calculating the drop's mass m in terms of the measured values of d, v_t and K (which must be measured in a separate experiment).

(d) In (c) above, r may be in meters, v in meters/sec, and F in newtons. Then K must be expressed in newton · sec/meters^2. Suppose you have continued to use such units in (c). Then your answer, the formula for m, must have units that reduce to kilograms. Work through the units of your answer and find out whether that is so.

(e) An experimenter measures "g" by timing a pendulum consisting of a small spherical bob of diameter d hung on a thread of length x (from support to top of bob). For small amplitudes, n complete swings to-and-fro take t seconds. Assuming that period $T = 2\pi\sqrt{(L/g)}$ applies, work out a formula beginning $g = \ldots$ for calculating g from his measurements n, t, d, etc.

(f) If the experimenter in (e) uses MKS (meter-kilogram-second) units for his measurements, will your formula yield g in suitable units? Show your checking-through of units.

VOCABULARY

[See also Problems 1 and 5 preceding Chapter 1, and Problems 12, 13, 16 at end of Chapter 1]

54. INTERPRETING

The passage printed below is extracted from SCIENTIFIC AMERICAN, May 1950, Vol. 182, No. 5, pages 27 and 28. That magazine publishes excellent popular scientific articles, taking great care to make them reliable and honest.

Suppose that you wish to explain the news in the extract to someone who knows much less physics than you. Write short notes on each of the underlined words or phrases to explain their meaning as clearly as possible. (Your notes should help your reader to understand, but they should also impress him with your knowledge.)

"Element 98

A few days after SCIENTIFIC AMERICAN had gone to press with the article in the April issue entitled 'The Synthetic Elements,' which stated that 'at the moment the list of identified *elements* stands at 97,' the number was raised to 98. The creation of *element 98,* the sixth *synthetic* addition to the *periodic table,* was announced at the University of California. . . .

The new element was produced in Berkeley's *60-inch* Crocker *cyclotron* by bombarding a few millionths of a gram of curium *242* with *alpha-particles* with an energy of 35 *million electron · volts.* The 60-inch cyclotron, in which five other synthetic elements were also first created, is used for this work in preference to the giant 184-inch because the 400-*Mev* alpha-particles produced in the latter would *shatter* the *nuclei* of the *target* material instead of being *captured* by them.

Element 98 has a *half-life* of only 45 minutes, and the tiny amount produced was difficult to detect. It was identified only because the California workers had been able to predict its half-life and chemical properties. Although the element will probably never be made in visible amounts, its properties are of considerable *theoretical importance.* . . . It is a homologue of dysprosium, element 66, which is part of the rare-earth series that begins with lanthanum. The successful *identification* of element 98 on the basis of the predictions bears out the *theory* that the synthetic elements beyond uranium are part of a second rare-earth series whose prototype is element 89, actinium. If this theory is correct, five more new elements may be expected to complete the series."

ENERGY PROPOSALS

55. HUMAN ENERGY

A man receives a diet worth 3000 Calories per day. Keeping his breathing, etc. going costs him 1500 Cals/day, and walking around in his spare time costs a further 1000 Cals/day. He does a job which involves *carrying heavy loads up a tall flight of stairs.* Each day's work costs him 200 Cals extra for the work of climbing and lifting, and he wastes a further 800 Cals as extra heat.

(a) By how much does his food fail to provide for his total needs?

(b) A number of suggestions are given below of ways in which he might be helped to do the job. For each of these, say whether you think it *would* help him (yes or no), and give a brief reason (~ 2 lines) for your answer. ASSUME THAT ONLY ONE SUGGESTION OPERATES AT A TIME.

- (i) He should eat extra candy.
- (ii) He can succeed by sheer grim determination not to be beaten by life (with no other measures such as those listed here).
- (iii) He can use up some of his fat and possibly some of his other body tissues.
- (iv) He should drink water whenever he feels tired.
- (v) He should hold on to the stair rails when he climbs, and pull himself up with his hands to help his legs.
- (vi) He should ride up on an escalator driven by electric motor.
- (vii) He should stay at the bottom and use a pulley system to raise the loads he is asked to carry to the top.
- (viii) He should stand still or stay in bed in his spare time instead of walking around.
- (ix) He should arrange a counterpoise for himself by running a rope from his shoulders up to the top of the stairway, over a (frictionless) pulley, and putting a weight a little less than his own weight (unloaded) on the other end of the rope.

56. THE SUNSHINE UTILIZATION RESEARCH AND DEVELOPMENT COMMITTEE.

(This is an entirely fictitious situation, though some of its propositions might well arise, and one of the schemes has been used.)

In this discussion, assume that:

(I) You are invited to advise an important Administrative Committee which is examining schemes to utilize the energy of sunshine;

and (II) owing to some serious necessity, it is advisable to develop such schemes, if workable, even if they are inefficient; and vast sums of money may be available for Research and Development of these schemes;

and (III) engineering systems *can* be produced to utilize the heat from water, etc., warmed by sunshine;

and (IV) the Committee has received suggestions and comments from a variety of people. You have been asked to criticize or comment on these, using your knowledge of science.

(Of course, in real life your first comment might well be, "Let me see the experimental record," or "I must do some experimenting myself," or "I suggest consulting specialists in this field." However in answering this problem you should merely find specific fault, or give approval, or state some general reason for doubt.)

To save trouble you should assume, during this discussion, that those statements which are underlined are true and need not be criticized.

(i) For each suggestion write a comment in code, saying whether you consider the suggestion good, doubtful, or bad. Use the following code:

OK = probably good, or worth trying.
? = doubtful.
X = wrong, misleading, probably nonsense.

(ii) Give a brief supporting reason (say, 1 to 5 lines) for your comment.

(Your answers should be judged by your reason rather than by your comment.)

SCHEME A. A HUGE SHALLOW TANK FULL OF WATER, WITH ALUMINUM LID, SPREAD OVER AN OPEN FIELD, WILL COLLECT HEAT WHICH CAN BE USED.

Suggestion A. 1. "Paint the roof of the tank black on top."

Suggestion A. 2. "It is true that black paint is beneficial, but the same mass of paint will attract as much heat if left in a paintpot. Place an open can of black paint on the lid of the tank instead."

Suggestion A. 3. "The XYZ Corporation offers their special paint No. 477B, which is ten times as good as black paint at drawing heat to a surface. Paint it with that."

Suggestion A. 4. "Replace the water by the same mass of solid aluminum, so that five times as much heat will be collected." (Aluminum has specific heat 0.2; so the temperature-rise may be expected to be five times as big as for water.)

Suggestion A. 5. "Aluminum generates heat when placed in contact with sulphuric acid, by dissolving to form sulphate. Paint the roof regularly with acid."

Suggestion A. 6. "Some of the heat collected by the tank should be used to run electric generators to charge batteries to run lamps which will shine on the tank on cloudy days."

SCHEME B. GROW TREES. "Instead of having the tank grow trees all over the area, cut them, dry them, and burn them. This will yield ten times as much heat."

SCHEME C. SUNLIGHT CAN RIP ELECTRONS OUT OF CERTAIN METAL SURFACES. THE ELECTRON STREAM FROM SUCH SURFACES COULD BE COLLECTED AND MIGHT BE USED TO RUN MACHINERY. HOWEVER THE ELECTRONS EMERGE WITH VERY SMALL VELOCITY.

Suggestion C. 1. "The stream can be collected and fed to triode radio tube amplifiers which will deliver a larger stream at higher voltage. Use this."

Suggestion C. 2. "Magnets placed on top of the surface will act on the electrons and speed them up."

Suggestion C. 3. "Let the released electrons operate across junctions in semi-conductors. Connect many patches of semi-conductor in series to gain more voltage, and use that 'battery' to run a motor."

SCHEME D. MIRRORS CAN BE USED TO REFLECT SUNLIGHT ONTO A BOILER TO MAKE STEAM.

Suggestion D. 1. "Paint the boiler black."

Suggestion D. 2. "Paint the mirror black."

Suggestion D. 3. "Instead of water in the boiler use a liquid with large molecules. When the liquid has boiled and become a vapor, the friction between the big molecules will generate a lot of extra heat."

SCHEME E. The Committee has been approached by a Corporation which has acquired from an inventor a secret process that makes a sheet of iron glow red-hot in sunshine. A demonstration was made of a pilot model, and the iron sheet, placed on top of the inventor's "processor box," glowed red-hot and boiled a big can of water.

The inventor explained that a full-sized model would work just as well, because he has a mathematical formula which applies to all sizes of machine. He also explained that, while there is no hint of his device in the scientific journals, it is 20 years old and has merely been kept secret; so it has stood the test of time.

The Committee is satisfied that the inventor is reliable, because his recent book on some branches of physics is selling in vast numbers and has received prominent reviews (albeit somewhat adverse) in newspapers of wide reputation.

The Committee, impressed by this scheme, proposes to authorize buying the secret and building a full-scale plant. (Give a longer comment on this if you like.)

ROUGH ESTIMATES

57. "JUDGING"

Make a very rough estimate in some of the following. Obviously, your method of estimating is just as interesting as the result; so you should explain how you make your guesses.

(a) What fraction of the total surface area of the United States is the area of paved roads of all kinds? (Is it 10%, which would make living barely possible, or 0.0001%, which would be negligible, or where between the two?)
(b) Supposing that pouring oil on a stormy sea does have a calming effect, estimate the area that might be calmed by 10 gallons of suitable oil.
(c) How many drug stores are there in the United States?
(d) What is the maximum temperature you could give an iron nail by hitting it with an ordinary hammer?
(e) Farmyard grindstones have been known to "burst" when rotated too fast. Estimate the speed likely to cause bursting.
(f) Suppose as a wild assumption that a car tire running on smooth road loses one molecule-width of rubber in each revolution. Estimate the molecule width. (Note that rubber forms enormous "molecules," yet we may picture it with component molecule chains not unlike those of oil molecules.)
(g) Estimate the mass of a mountain.
(h) Estimate the energy in a hurricane. Compare it with man-made explosions.

"READING PAPERS"—LONG ESSAYS

A "reading paper" asks you to do scientific reading and writing in the study of new knowledge. If, after reading this book, or studying its course, you have gained an understanding of physical science as well as a knowledge of facts, you may want to prove your gains—to yourself or to others—by showing how you can read farther in some unfamiliar field of physics and expound it. For that, you should choose some topic of physics that has not been treated in this book—but yet one that can be attached to the work of the course and draw on material already treated. Spend some weeks studying your topic in books and scientific journals; and then write your own account of it. That will enable you to fill in a gap in the course, and, far more important, it will give you an opportunity to show how well you can now learn more science by your own reading. In that way, a "reading paper" is a test of the course.

Some topics are suggested below. Lists of suggested books are not given, because the best choice depends on library facilities and may change as new books appear. Furthermore, there is great value in looking for books, in consulting research papers, and in comparing the views of one book with those of several others. That makes the difference between a dutiful report on a single book and an essay that gives a sense of thorough knowledge.

In writing your paper, you should keep in mind a picture of the reader for whom it is intended. Your "target reader" might be a neighbor in the course who has chosen another topic for his own paper but wishes to learn about your topic by reading your paper. Or he might be a student who has studied physics elsewhere in a straightforward course of factual material and wishes to learn from your treatment. Or you might choose to write for a professional physics teacher who proposes to read your paper and grade it—but that choice is likely to lead to less ultimate benefit to you.

Even when you have chosen your topic, you have a wide choice between treating a narrow piece of it in detail or covering the whole topic superficially. That should be a matter of taste for you. A wise reader will accept your choice of width—and will expect a corresponding adjustment of depth of treatment. His main basis of judgment should be the sense you give of having mastered the material and made it your own so that you can expound it to others. You are invited to set forth thorough knowledge and careful thought rather than make a collection of information. Thus, quality, rather than length of writing, is the important thing.

(i) A history of Radioactivity (1890-1915 and/or 1915-now).

A rich field with good books. This is an interesting study, ranging in difficulty from quite easy to as hard as you like to make it. It is difficult to write an unsatisfactory paper on this, because once you start reading you will see the scope of the topic. And a short acceptable paper can easily be expanded to one of greater and greater scope. So this is a good choice if you are uncertain what to choose.

(ii) Millikan's experiment to measure *e*.

Millikan wrote an excellent account of his own work. Consult his book first. It may suffice.

(iii) Cloud-chambers: working, pictures, and interpretation of pictures.

The construction and working of the instrument are easily described. The main work of making this paper would consist of collecting interesting results, and showing how detailed information about atomic or nuclear changes can be derived from them, and describing the results thus derived. This paper could easily be a disappointingly thin, descriptive list; but, with care and emphasis on inferences, it could be a very good paper.

(iv) Particle tracks in photographic emulsions, and/or in bubble chambers.

These are the successors of cloud chambers. They yield new rich stores of nuclear information. This paper would require more detailed description of instruments and techniques; but, as in (iii) above, its success will depend largely on good descriptions of results and explanations of inferences.

(v) Discovery and properties of X-rays.

X-rays have a romantic history—and good accounts of Roentgen's discovery are available—and they have very important applications in medicine and in atomic physics. This paper should pay considerable attention to the use of X-rays in physics, for example in the sorting out of crystal-structure. This paper will be easier and more productive if you have studied optics.

(vi) Cosmic Rays.

High-energy particles come shooting in from remote space, in a complicated mixture of electrons, nuclei, mesons, etc. They can be used to produce violent atomic changes, though the experimenter is at the mercy of the meagre supply. Experimenters use cloud chambers, ionization chambers, counters . . . up in balloons, down in mines, away on ships—so that the effects of the atmosphere and of the Earth's magnetic field can be studied or used. In the study of cosmic rays, the reading is fairly hard; but this is a very rich field. It would be wise to choose part of the field for a paper. An excellent choice if you have courage and an inquiring mind.

(vii) High-energy particle physics.

Electrons, nuclei, mesons, etc., moving with very high energy (and speed almost c), are provided by cosmic rays, and by the biggest accelerators. A paper on experiments, results, and interpretations would be difficult but rewarding.

(viii) Mass-spectrographs: design, working, results.

Details of simple designs are given in Chapter 38, but modern working designs are much more complicated and ingenious. This is a rather technical topic, but you have the necessary background for its study, and there are several good accounts of modern instruments.

(ix) Accelerators: design and working.

There are now many types of accelerator that differ from the Van de Graaff machine and cyclotron described in the text. The study of these needs a good grasp of electromagnetism. It would be hard but interesting. (Examples: linear accelerator, "betatron," race-track "cosmotron.")

(x) Properties of atomic particles.

New varieties of particle are still being discovered: alpha-, beta-particles, electrons, positrons, neutrons, neutrinos, mesons; and the list continues to still more strange particles. This topic has some ground in common with (i), but it extends to the frontier of present research. Choose several particles or survey them all. How are these particles produced, detected, investigated? What do we thus find out about atoms?

(xi) Properties of the electron.

Discovery, measurement of charge, and other properties; wave properties, photoelectric effect. This would go a considerable way outside the content of the course. With considerable reading, this would be a difficult but very interesting topic.

(xii) Wave : particle behavior.

The great revolution in physics a quarter of a century ago grew from the discovery that both photons and material particles have a "dual nature": they behave as particles and have wave properties. Explore the experimental evidence for this. If you are a good mathematician, then see how the new ideas affected theoretical physics.

(xiii) Experiments to show that "classical physics is wrong."

An account of the experiments that led to the development of Quantum Theory—e.g. experiments on specific heats, black-body radiation, photoelectric effect, discovery of "matter waves," etc. A difficult topic, needing hard study, that could be very rewarding. It would probably be best to restrict the paper to two of the topics mentioned and treat them in detail.

(xiv) Very low temperatures: production, and experiments that use them.

This is the region below the temperatures of common freezing mixtures or dry ice. It includes "liquid air" and extends on down practically to absolute zero. A variety of methods and apparatus have been used for making liquid air, liquid hydrogen, liquid helium; and finally, ingenious demagnetization schemes have produced (and measured) still lower temperatures very near absolute zero. At very low temperatures most substances freeze into brittle solids, but there are other more unexpected effects which make a romantic study. A difficult topic.

(xv) The physics of stars.

How have the masses, temperatures, sizes, . . . of stars been estimated and stellar evolution investigated? A very interesting subject of almost infinite range. There are short descriptive accounts of "astrophysics" in some textbooks; but you would need to read much further to find how our knowledge of stars is argued out from observations, how observations are interpreted by theory.

(xvi) Spectroscopy in Astronomy.

Compositions, speeds, temperatures, . . . of planets, stars, nebulae, are investigated by attaching a spectroscope to a telescope. Stellar spectrocopy has also extended our knowledge of atomic spectra beyond the stages of excitation and ionization available in ordinary terrestrial furnaces. In a way we might say that such a spectroscope is THE astronomical instrument of this age. This paper requires some descriptions of the methods of using spectroscopes, then discussion of the physical interpretation of the results. Romantic, modern, astronomical topic; but quite difficult to cover well.

(xvii) Age of the Earth: physical estimates.

This is not just a matter of geological knowledge or speculation. Much physics has been used in dating—some of the recent methods are radioactive ones. To do justice to this topic as a physics essay, you would need to set forth the physical methods and show how their results and interpretations are connected with geological evidence.

(xviii) Biography: The life and work *in physics* of one of the scientists listed below:

(a) Ernest Rutherford, (b) J. J. Thomson, (c) A. A. Michelson.

Since this is an account of the man and his work, it should contain at least a short biography. Since it is intended to be a study of important developments in physical science, this paper should give a full and clear account of the man's work in physics—his physical discoveries should be explained as well as described. It would be easy to underestimate the serious possibilities of this topic—or to underestimate the quantity of physics expected. Treated thoroughly, one of these would be a very interesting study.

(xix) Newton's Mechanics and Philosophy.

How was Newton's physics related to the philosophical views of his day? How did his work affect the philosophy of later generations? This is a paper that should be read by a physicist and a philosopher together. To write it, you should be a keen amateur philosopher, but you do not need a technical knowledge of philosophy.

(xx) The Philosophy of physical science—from an amateur's point of view.

If you have studied philosophy of science in a formal course, you are not likely to find this essay very rewarding—it would seem more like a course review than new thinking, reading, and writing. But if you are an amateur with no training but with an interest in philosophy, you may enjoy reading and writing on this subject—and you should expect readers to be sympathetic to amateur views. Look at the books listed below and others that you can find on Philosophy of Science. You should base your essay on at least two different books.
"Philosophy of Science," by Stephen Toulmin;
"What is Science?," by N. R. Campbell;
"Experiment and Theory in Physics," by Max Born;
"The Metaphysical Foundations of Modern Physical Science," by E. A. Burtt.

(xxi) The Physics of Sound and Music.

There are many good books describing and expounding the physics of sound. In writing on this topic you should make sure that you are treating physics thoroughly and not just describing results or stating rules. As a test, your paper should give both pleasure and profitable knowledge to a musician who reads it.

(xxii) If you choose a topic of your own, the following conditions are important:

(a) It should be a topic sufficiently rich in physics so that when you have written your paper, you will consider it solid science.
(b) It should be a topic for which there are several modern books available.

INDEX